ICHEP 2002

# ICHEP 2002

Proceedings of the 31st International Conference
on High Energy Physics

Amsterdam, The Netherlands
25-31 July 2002

Edited by

S. Bentvelsen
NIKHEF, Amsterdam
The Netherlands

P. de Jong
NIKHEF, Amsterdam
The Netherlands

J. Koch
NIKHEF, Amsterdam and University of Amsterdam
The Netherlands

and

E. Laenen
NIKHEF, Amsterdam and Utrecht University
The Netherlands

2003

ELSEVIER

Amsterdam · Boston · London · New York · Oxford · Paris
San Diego · San Francisco · Singapore · Sydney · Tokyo

ELSEVIER SCIENCE B.V.
Sara Burgerhartstraat 25
P.O. Box 211, 1000 AE Amsterdam, The Netherlands



First edition 2003

**Library of Congress Cataloging in Publication Data**
A catalog record from the Library of Congress has been applied for.

**British Library Cataloguing in Publication Data**
A catalogue record from the British Library has been applied for.

ISBN: 0 444 51343 4

⊗ The paper used in this publication meets the requirements of ANSI/NISO Z39.48-1992 (Permanence of Paper), where ⊗ = the mathematical symbol denoting infinity (like the number 8 on its side) set inside a circle. For more information re "Permanence of Paper" and an example of the symbol used to show compliance with this standard you are referred to http://www.niso.org/standards/resources/Z39-48.pdf.

Printed in the Netherlands.

## FOREWORD

The thirty-first International Conference on High Energy Physics was held in Amsterdam from July 25 through July 31, 2002. It was the first time that this biennial conference took place in The Netherlands.

The 933 participants from 47 countries met for three days of parallel sessions and, after the traditional break on Sunday, for another three days with plenary talks. All sessions were held in the RAI Congress Centre, south of the city centre. The conference was opened by Professor Jos J. Engelen, director of the Dutch National Institute for Particle Physics NIKHEF, the host of the meeting.

We were pleased to see the first precision measurements on CP violation in the B system. Both the BELLE and the BaBar collaboration presented among others results for $\sin 2\beta$ with much improved accuracy. Results from the Sudbury Neutrino Observatory, SNO, also deserve to be mentioned. Although the convincing evidence of solar neutrino oscillations had been presented by SNO prior to the conference, its full presentation attracted a large audience. It was also good news for the Solar Model. The participants were further treated to an incredibly precise measurement of the anomalous magnetic moment of the muon, a fresh result from the Brookhaven National Laboratory. Apart from these distinct physics highlights, there were also the first results from the new Tevatron run and from the relativistic heavy ion collider RHIC. Theorists spoke of our ever better understanding of the Standard Model and of what might lie beyond. Risky as it is to highlight only a couple of exciting subjects, it is merely meant to whet your appetite for further reading! Apart from these proceedings we also maintain our website www.ichep02.nl with the presented material.

The conference was held under the auspices of the International Union of Pure and Applied Physics, IUPAP. We thank the C11 Commission for their support and advice. We also gratefully acknowledge the generous support of NIKHEF, including its four partner universities, as well as of the Physics Funding Agency FOM, the Netherlands Organization for Scientific Research NWO, and the Ministry of Education, Culture and Science. Grants were also obtained from the Foundation Physica, and the Foundations for High Energy Physics and for Nuclear Physics conferences. On top of that several industrial companies and publishers lent their support.

It is a good tradition of IUPAP to require a public lecture. More than 600 people, among which many participants of the conference, were entertained by Nobel Laureate, Professor Gerard 't Hooft, in a lecture entitled “The universe inside the atom”. The line-up at the microphone during the discussion session marked the enthusiasm among the public. The social programme included a concert in the Westerkerk church and the dinner in the “Muziektheater” opera.

We hope that your memories of the Amsterdam Conference are as good as ours; we greatly enjoyed being your hosts!

Ger van Middelkoop

## NOTE FROM THE EDITORS

This book contains the contributed papers of the 31st International Conference on High Energy Physics ICHEP 2002, as presented during the parallel sessions.

The invited papers, presented during the plenary sessions, are published separately as vol. 117 of Nuclear Physics B (Proc. Suppl.). Both invited and contributed papers are accessible on the world wide web, through the Science Direct service of Elsevier.

# COMMITTEES

## Local Organizing Committee

Ger van Middelkoop (chair)
Stan Bentvelsen
Paul de Jong
Justus Koch
Els Koffeman
Eric Laenen
Frank Linde
Gerard van der Steenhoven
Anja van Dulmen

## Program Committee

R. Cashmore (CERN)
S. de Jong (Nijmegen)
R.H. Dijkgraaf (Amsterdam)
J. Ellis (CERN)
J.J. Engelen (NIKHEF)
A. Goshaw (Durham N.C.)
E. Iarocci (INFN)
R. Klanner (DESY)
P.J. Mulders (Amsterdam)
M. Peskin (SLAC)
M. Pohl (Genève, Nijmegen)
M. Spiro (DAPNIA)
G. 't Hooft (Utrecht)
N.G. Turok (Cambridge)
P.J. van Baal (Leiden)
F. Verbeure (Antwerpen)
F. Wilczek (MIT)

## International Advisory Committee

A. Astbury (TRIUMPH)
J.J. Aubert (IN2P3)
A. Bettini (Gran Sasso)
H. Chen (Beijing)
P. Darriulat (Hanoi)
J. Dorfan (SLAC)
L. Fo (Pisa)
V.G. Kadyshevsky (JINR)
T. Kirk (Brookhaven)
L. Maiani (CERN)
G. Mikenberg (Weizmann Institute)
Y. Nagashima (Osaka)
K.J. Peach (RAL)
M. Roos (Helsinki)
A.N. Shrinsky (Budker)
H. Sugawara (KEK)
S.N. Tovey (Melbourne)
M. Veltman (Ann Arbor, NIKHEF)
A. Wagner (DESY)
M.S. Witherell (Fermilab)
A.K. Wroblewski (Warszawa)

## Acknowledgements

The Local Organizing Committee gratefully acknowledges the support of the following companies and organizations:

NIKHEF
Physics Funding Agency FOM
Foundation Physica
Free University of Amsterdam, VU
Utrecht University, UU
Leiden University
High Energy Physics Foundation
IUPAP
Ministry of Education, Culture and Science, OC&W
University of Nijmegen, KUN
University of Amsterdam, UvA
University of Technology Twente
Nuclear Physics Institute KVI, Groningen
Foundation for Nuclear Physics Conferences
Netherlands Organization for Scientific Research, NWO

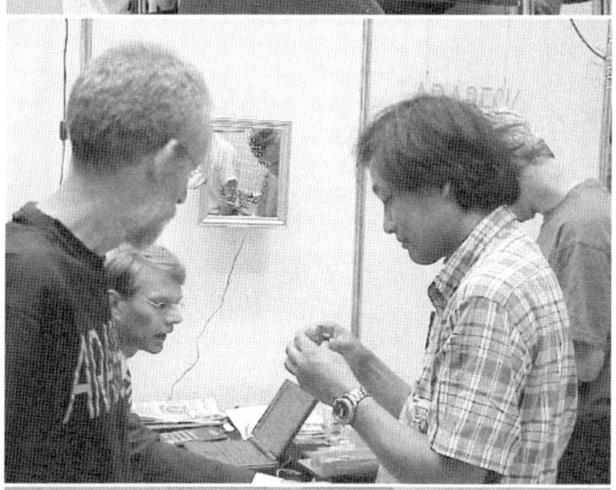

UNITY in DIVERSITY

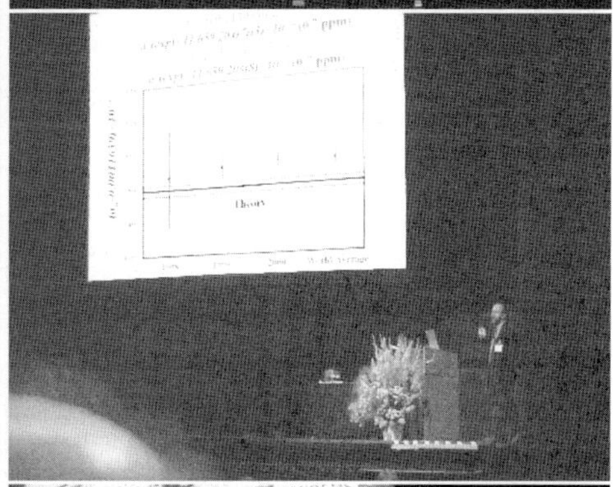

TANDEMS
ICHEP 2002, Amsterda

Gerard 't Hooft

BANG !!

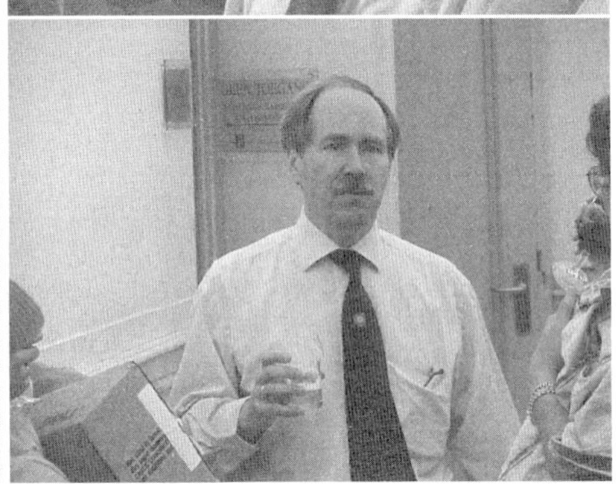

SPEAKERS
CENTRE
SPEAKERS
CENTRE

## CONTENTS



*Manuscript not received.

**Session 2: Quark Matter; Heavy Ion Collisions**
*Session conveners: P. Giubellino, University of Torino/CERN; R. Snellings, NIKHEF*

**Session 3: Particle Astrophysics and Cosmology**
*Session conveners: A. Achucarro, University of Leiden; E. do Couto e Silva, Stanford University/SLAC*

**Session 4: Electroweak Physics**

*Session conveners: J. Hobbs, SUNY at Stony Brook; D. Bardin, JINR, Dubna*

**Session 5: QCD: Hard Interactions**

*Session conveners: C. Niebuhr, DESY Hamburg; L. Reina, Florida State University, Tallahassee; J. Huston, Michigan State University*

**Session 6: QCD: Soft Interactions**

*Session conveners: W. Kittel, University of Nijmegen; M. Diehl, RWTH Aachen; A. Kaidalov, ITEP, Moscow*

**Session 11: Computational Quantum Field Theory (methods)**
*Session conveners: J. Vermaseren, NIKHEF; E. Laermann, University of Bielefeld*



**Session 12: String Theory and Mathematical Quantum Field Theory**
*Session conveners: S. Vandoren, Utrecht University; J. Louis, University of Halle*



**Session 13: R&D for Future Accelerators and Detectors**
*Session conveners: Y. Kuno, Osaka University; T. Behnke, DESY Hamburg*

# Session 1

# Neutrino Masses and Mixings

*Session conveners:*
*Y. Grossman, Technion–Israel Institute of Technology, Haifa*
*Y. Suzuki, ICRR, University of Tokyo*

ELSEVIER

Proceedings of ICHEP 2002, pp. 3–7
*S. Bentvelsen, P. de Jong, J. Koch and E. Laenen (Editors)*

31st INTERNATIONAL CONFERENCE ON HIGH ENERGY PHYSICS AMSTERDAM

# Neutrino Oscillation Results from the Sudbury Neutrino Observatory

Scott M. Oser[a], for the SNO collaboration

[a]Department of Physics and Astronomy, University of Pennsylvania, Philadelphia, PA 19104-6396, USA

The Sudbury Neutrino Observatory (SNO) has determined the flavor content of the $^8$B solar neutrino flux by measuring the rates of charged current and neutral current neutrino interactions on deuterium. These results directly demonstrate neutrino flavor transformation at greater than $5\sigma$ significance. The total flux of $^8$B neutrinos is found to be in good agreement with solar model predictions. Measurements of the day and night neutrino energy spectra probe models of neutrino oscillation. A global fit of SNO data and results from other solar neutrino experiments to neutrino oscillation models strongly favors the Large Mixing Angle (LMA) MSW solution.

## 1. Introduction

The Sudbury Neutrino Observatory (SNO) detects $^8$B solar neutrinos through the reactions:

$$\begin{array}{lcll} \nu_e + \mathrm{d} & \rightarrow & \mathrm{p}+\mathrm{p}+\mathrm{e}^- & \text{(CC)}, \\ \nu_x + \mathrm{d} & \rightarrow & \mathrm{p}+\mathrm{n}+\nu_x & \text{(NC)}, \\ \nu_x + \mathrm{e}^- & \rightarrow & \nu_x + e^- & \text{(ES)}. \end{array}$$

The charged current reaction (CC) is sensitive exclusively to electron-type neutrinos, while the neutral current reaction (NC) is equally sensitive to all active neutrino flavors ($x = e, \mu, \tau$). The elastic scattering reaction (ES) is sensitive to all flavors as well, but with reduced sensitivity to $\nu_\mu$ and $\nu_\tau$. Sensitivity to these three reactions allows SNO to determine the electron and non-electron active neutrino components of the solar flux [1].

SNO [2] is a water Cherenkov detector located at a depth of 6010 m of water equivalent in the INCO, Ltd. Creighton mine near Sudbury, Ontario, Canada. The detector uses ultra-pure heavy water contained in a transparent acrylic spherical shell 12 m in diameter to detect solar neutrinos. Cherenkov photons generated in the heavy water are detected by 9456 photomultiplier tubes (PMTs) mounted on a stainless steel geodesic sphere 17.8 m in diameter. The geodesic sphere is immersed in ultra-pure light water to provide shielding from radioactivity in both the PMT array and the cavity rock.

The data reported here represent a total of 306.4 live days, spanning the entire first phase of the experiment, in which only $D_2O$ was present in the sensitive volume. These analyses are described in more detail in [3], [4], and [5]. PMT times and hit patterns were used to reconstruct event vertices and directions and to assign to each event a most probable kinetic energy, $T_{\mathrm{eff}}$. The total flux of active $^8$B solar neutrinos with energies greater than 2.2 MeV (the NC reaction threshold) was measured with the NC signal (Cherenkov photons resulting from the 6.25 MeV $\gamma$ ray from neutron capture on deuterium.) The analysis threshold was $T_{\mathrm{eff}} \geq 5$ MeV, providing sensitivity to neutrons from the NC reaction. Above this energy threshold, there were contributions from CC events in the $D_2O$, ES events in the $D_2O$ and $H_2O$, capture of neutrons (both from the NC reaction and backgrounds), and low energy Cherenkov background events.

A fiducial volume was defined to only accept events which had reconstructed vertices within 550 cm from the detector center to reduce external backgrounds and systematic uncertainties associated with optics and event reconstruction near the acrylic vessel. The neutron response and systematic uncertainty was calibrated with a $^{252}$Cf source. The deduced efficiency for neutron captures on deuterium is 29.9±1.1% for a uniform source of neutrons in the $D_2O$. The neutron detection efficiency within the fiducial volume and above the energy threshold is 14.4%. The energy scale uncertainty is 1.2%.


doi:10.1016/S0920-5632(02)02061-3

## 2. Backgrounds

The primary backgrounds to the NC signal are due to low levels of uranium and thorium decay chain daughters ($^{214}$Bi and $^{208}$Tl) in the detector materials. These activities generate free neutrons in the $D_2O$, from deuteron photodisintegration (pd), and low energy Cherenkov events. *Ex-situ* assays and *in-situ* analysis of the low energy (4 – 4.5 MeV) Cherenkov signal region provide independent uranium and thorium photodisintegration background measurements.

Two *ex situ* assay techniques were employed to determine average levels of uranium and thorium in water. Radium ions were directly extracted from the water onto either $MnO_x$ or hydrous Ti oxide (HTiO) ion exchange media. Radon daughters in the U and Th chains were subsequently released, identified by $\alpha$ spectroscopy, or the radium was concentrated and the number of decay daughter $\beta$-$\alpha$ coincidences determined. These techniques provide isotopic identification of the decay daughters and contamination levels in the assayed water volumes, presented in Fig. 1 (a). Secular equilibrium in the U decay chain was broken by the ingress of long-lived (3.8 day half-life) $^{222}$Rn in the experiment. Measurements of this background were made by periodically extracting and cryogenically concentrating $^{222}$Rn from water degassers. Radon from several tonne assays was subsequently counted in ZnS(Ag) scintillation cells [6]. The Radon results are presented (as mass fractions in g(U)/g($D_2O$)) in Fig. 1(b).

Independent measurements of U and Th decay chains were made by analyzing Cherenkov light produced by the radioactive decays. The $\beta$ and $\beta$-$\gamma$ decays from the U and Th chains dominate the low energy monitoring window. Events in this window monitor $\gamma$ rays that produce photodisintegration in these chains ($E_\gamma > 2.2$ MeV). Cherenkov events fitted within 450 cm from the detector center and extracted from the neutrino data set provide a time-integrated measure of these backgrounds over the same time period and within the fiducial volume of the neutrino analysis. Statistical separation of *in situ* Tl and Bi events was obtained by analyzing the Cherenkov signal isotropy. Tl decays always result in a $\beta$ and a 2.614 MeV $\gamma$, while in this energy window Bi decays are dominated by decays with only a $\beta$, and produce, on average, more anisotropic hit patterns.

Results from the *ex situ* and *in situ* methods are consistent with each other as shown on the right hand side of Figs. 1(a) and 1(b). For the $^{232}$Th chain, the weighted mean (including additional sampling systematic uncertainty) of the two determinations was used for the analysis. The $^{238}$U chain activity is dominated by Rn ingress which is highly time dependent. Therefore the *in-situ* determination was used for this activity as it provides the appropriate time weighting. The average rate of background neutron production from activities in the $D_2O$ region is $1.0 \pm 0.2$ neutrons per day, leading to $44^{+8}_{-9}$ detected background events. The production rate from external activities is $1.3^{+0.4}_{-0.5}$ neutrons per day, which leads to $27 \pm 8$ background events since the neutron capture efficiency is reduced for neutrons born near the heavy water boundary. The total photodisintegration background corresponds to approximately 12% of the number of NC neutrons predicted by the standard solar model from $^8$B neutrinos.

Low energy backgrounds from Cherenkov events in the signal region were evaluated by using acrylic encapsulated sources of U and Th deployed throughout the detector volume and by Monte Carlo calculations. Probability density functions (pdfs) in reconstructed vertex radius derived from U and Th calibration data were used to determine the number of background Cherenkov events from external regions which either entered or mis-reconstructed into the fiducial volume. Cherenkov event backgrounds from activities in the $D_2O$ were evaluated with Monte Carlo calculations. Together the low energy background from Cherenkov events totaled $45^{+18}_{-12}$ events.

Other sources of free neutrons in the $D_2O$ region are cosmic ray events and atmospheric neutrinos. To reduce these backgrounds, a neutron background cut imposed a 250-ms deadtime (in software) following every event in which the total number of PMTs which registered a hit was greater than 60. In addition, a "muon follower" cut removed every event following within 20 s of a throughgoing muon. The number of remaining NC atmospheric neutrino events and background events generated by sub-Cherenkov threshold muons, reactor anti-neutrinos, and $\alpha$-induced disassociation of deuterons is estimated to be $7^{+3}_{-2}$.

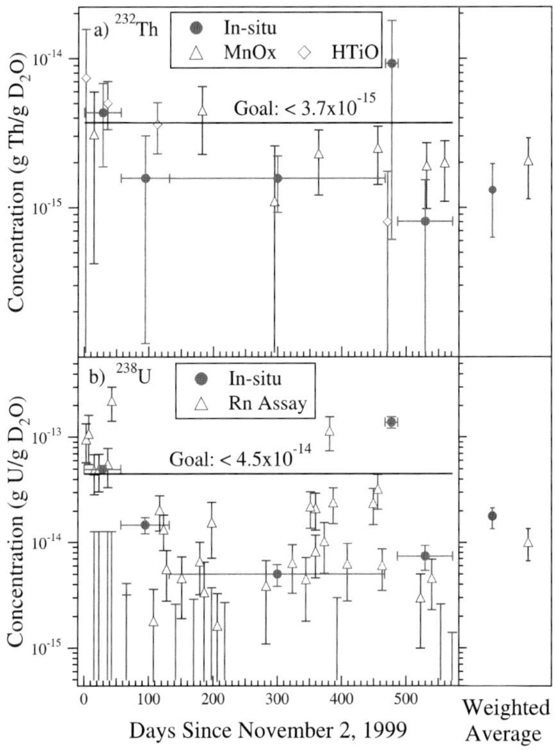


Figure 1. Thorium (a) and uranium (b) backgrounds (equivalent equilibrium concentrations) in the $D_2O$ deduced by *in situ* and *ex situ* techniques.

## 3. Integral Flux Analysis

The data recorded during the pure $D_2O$ detector phase are shown in Figure 2. There are 2928 events in the energy region selected for analysis, 5 to 20 MeV. Fig. 2(a) shows the distribution of selected events in the cosine of the angle between the Cherenkov event direction and the direction from the sun ($\cos\theta_{\odot}$) for the analysis threshold of $T_{\rm eff} \geq 5$ MeV and fiducial volume selection of $R \leq 550$ cm, where $R$ is the reconstructed event radius. Fig. 2(b) shows the distribution of events in the volume-weighted radial variable $(R/R_{\rm AV})^3$, where $R_{\rm AV} = 600$ cm is the radius of the acrylic vessel. Figure 2(c) shows the kinetic energy spectrum of the selected events.

In order to test the null hypothesis, the assumption that there are only electron neutrinos in the solar neutrino flux, the data are resolved into contributions from CC, ES, and NC events above threshold using pdfs in $T_{\rm eff}$, $\cos\theta_{\odot}$, and $(R/R_{\rm AV})^3$, derived from Monte Carlo calculations generated assuming no flavor transformation and the standard $^8$B spectral shape [7]. Background event pdfs are included in the analysis with fixed amplitudes determined by the background calibration. The extended maximum likelihood method used in the signal decomposition yields $1967.7^{+61.9}_{-60.9}$ CC events, $263.6^{+26.4}_{-25.6}$ ES events, and $576.5^{+49.5}_{-48.9}$ NC events, where only statistical uncertainties are given.

Normalized to the integrated rates above the kinetic energy threshold of $T_{\rm eff} \geq 5$ MeV, the flux of $^8$B neutrinos measured with each reaction in SNO, assuming the standard spectrum shape [7] is (all fluxes are presented in units of $10^6$ cm$^{-2}$s$^{-1}$):

$$\phi^{\rm SNO}_{\rm CC} = 1.76^{+0.06}_{-0.05}({\rm stat.})^{+0.09}_{-0.09}\ ({\rm syst.})$$
$$\phi^{\rm SNO}_{\rm ES} = 2.39^{+0.24}_{-0.23}({\rm stat.})^{+0.12}_{-0.12}\ ({\rm syst.})$$
$$\phi^{\rm SNO}_{\rm NC} = 5.09^{+0.44}_{-0.43}({\rm stat.})^{+0.46}_{-0.43}\ ({\rm syst.}).$$

Electron neutrino cross sections are used to calculate all fluxes. The excess of the NC flux over the CC and ES fluxes implies neutrino flavor transformations. The NC flux is in excellent agreement with the predicted solar model flux of $5.05^{+1.01}_{-0.81} \times 10^6$ cm$^{-2}$s$^{-1}$ [8].

A simple change of variables resolves the data directly into electron ($\phi_e$) and non-electron ($\phi_{\mu\tau}$) components [1],

$$\phi_e = 1.76^{+0.05}_{-0.05}({\rm stat.})^{+0.09}_{-0.09}\ ({\rm syst.})$$

[1] This change of variables allows a direct test of the null

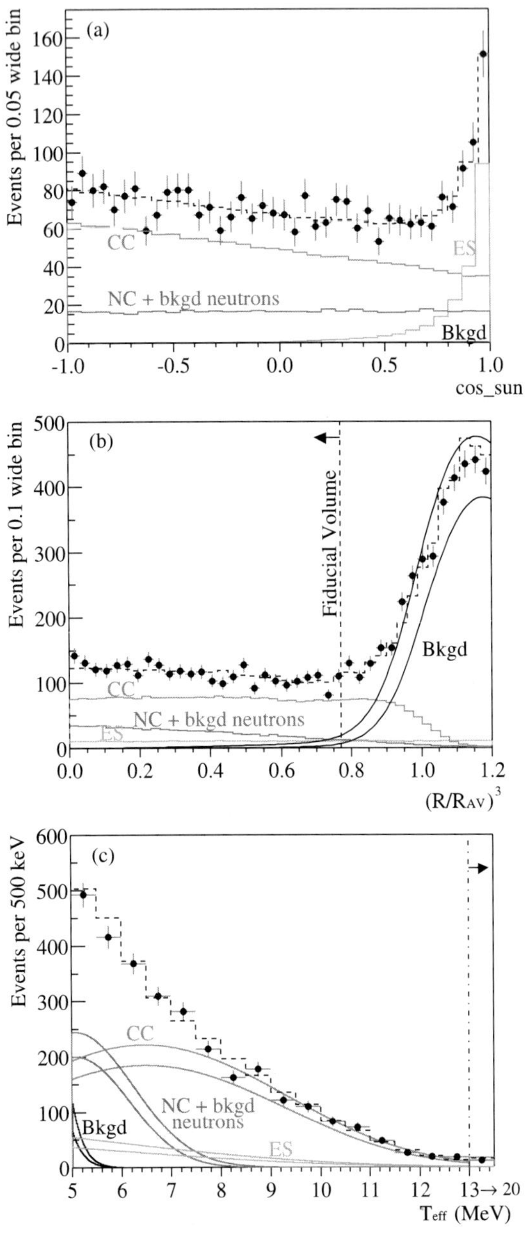


Figure 2. (a) Distribution of $\cos\theta_{\odot}$ for $R \leq 550$ cm. (b) Distribution of the volume weighted radial variable $(R/R_{\rm AV})^3$. (c) Kinetic energy for $R \leq 550$ cm. Also shown are the Monte Carlo predictions for CC, ES and NC + bkgd neutron events scaled to the fit results, and the distribution of Cherenkov background (Bkgd) events.

$$\phi_{\mu\tau} = 3.41^{+0.45}_{-0.45}(\text{stat.})^{+0.48}_{-0.45}\ (\text{syst.})$$

assuming the standard $^8$B shape. Combining the statistical and systematic uncertainties in quadrature, $\phi_{\mu\tau}$ is $3.41^{+0.66}_{-0.64}$, which is $5.3\sigma$ above zero, providing strong evidence for flavor transformation consistent with neutrino oscillations [9,10].

## 4. Day-night Results and MSW Contours

To search for day-night asymmetries in the neutrino flavor content, the data set was divided into separate day and night portions according to the elevation of the Sun relative to the horizon. The electron neutrino flux and total neutrino flux were determined for the day and night data, and their day-night asymmetries $A$, defined as the night rate minus the day rate divided by their average, were calculated. Most systematic uncertainties cancel in the day-night asymmetry ratio, and the result is statistics-limited. The derived asymmetries are:

$$\begin{array}{lcrcrl} A_e & = & 12.8\% & \pm & 6.2\% & ^{+1.5}_{-1.4}\% \\ A_{tot} & = & -24.2\% & \pm & 16.1\% & ^{+2.4}_{-2.5}\% \end{array}$$

The statistical correlation between the two asymmetries is -0.602. Both asymmetries are consistent with zero. If the total flux asymmetry is constrained to equal zero, as predicted for neutrino oscillation models with only active neutrino flavors, then a value for $A_e$ of $7.0\% \pm 4.9\%^{+1.3}_{-1.2}\%$ is obtained (see Figure 3).

Solar neutrino data from the chlorine, gallium, Super-Kamiokande, and SNO experiments has been fit to a two-flavor model of MSW-enhanced neutrino oscillations in order to determine the allowed mixing parameters (see Figure 4) [5]. The Large Mixing Angle (LMA) solution is strongly favored, and is the only surviving solution at 99% CL. Values of $\tan^2\theta > 1$ are ruled out at $\sim 3\sigma$ significance.

## 5. Conclusion

In summary, the results presented here are the first direct measurement of the total flux of ac-

hypothesis of no flavor transformation ($\phi_{\mu\tau} = 0$) without requiring calculation of the CC, ES, and NC signal correlations.

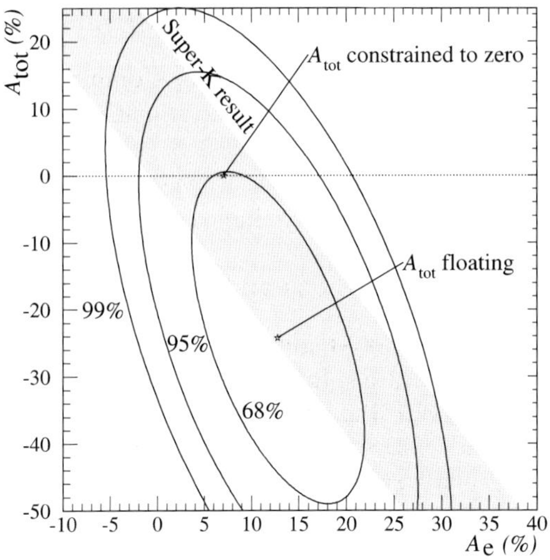


Figure 3. Joint probability contours for $A_{tot}$ and $A_e$. The points indicate the results when $A_{tot}$ is allowed to float and when it is constrained to zero. The diagonal band indicates the 68% joint contour for the Super-K $A_{ES}$ measurement [11].

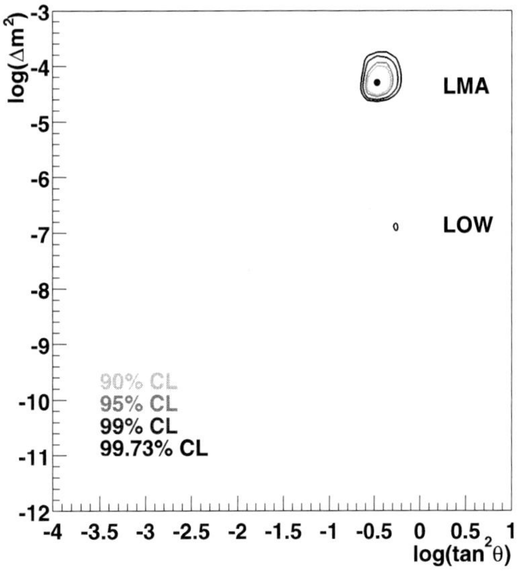


Figure 4. Allowed regions of the MSW plane determined by a $\chi^2$ fit of all solar neutrino data to a two-flavor MSW model.

tive $^8B$ neutrinos arriving from the Sun and provide strong evidence for neutrino flavor transformation. The total flux of $^8B$ neutrinos measured with the NC reaction is in agreement with the SSM prediction. Measurements of day and night rates show no significant asymmetries, and a global fit of solar neutrino data to MSW models strongly favors the Large Mixing Angle region of parameter space.

## 6. Acknowledgements

This research was supported by: Canada: NSERC, Industry Canada, NRC, Northern Ontario Heritage Fund Corporation, Inco, AECL, Ontario Power Generation; US: Dept. of Energy; UK: PPARC. We thank the SNO technical staff for their strong contributions.

## REFERENCES

1. H. Chen, Phys. Rev. Lett. **55**, 1534 (1985).
2. The SNO collaboration, Nucl. Instr. and Meth. **A449**, 172 (2000).
3. Q.R. Ahmad *et al.*, Phys. Rev. Lett. **87**, 071301 (2001).
4. Q.R. Ahmad *et al.*, Phys. Rev. Lett. **89**, 011301 (2002).
5. Q.R. Ahmad *et al.*, Phys. Rev. Lett. **89**, 011302 (2002).
6. M.-Q. Liu, H.W. Lee, and A.B. McDonald, Nucl. Inst. Meth. **A329**, 291 (1993).
7. C. E. Ortiz *et al.*, Phys. Rev. Lett. **85**, 2909 (2000).
8. John N. Bahcall, M.H. Pinsonneault, and Sarbani Basu, Astrophys. J. **555**, 990 (2001).
9. Z. Maki, N. Nakagawa, and S. Sakata, Prog. Theor. Phys. **28**, 870 (1962).
10. V. Gribov and B. Pontecorvo, Phys. Lett. **B28**, 493 (1969).
11. S. Fukuda *et al.*, Phys. Rev. Lett. **86**, 5651 (2001).

ELSEVIER

Proceedings of ICHEP 2002, pp. 8–12
*S. Bentvelsen, P. de Jong, J. Koch and E. Laenen (Editors)*

31st INTERNATIONAL CONFERENCE ON HIGH ENERGY PHYSICS AMSTERDAM

# Solar Neutrinos in Super–Kamiokande-I

M. R. Vagins[a] for the Super–Kamiokande Collaboration

[a]Department of Physics and Astronomy, University of California, Irvine
4129 Reines Hall, Irvine, CA 92697 USA

Results from the entire 1496 live days of solar neutrino data from Super–Kamiokande-I's 1996-2001 run are presented. These results include our measurements of the Sun's absolute Boron-8 neutrino flux and energy spectrum, as well as studies of their variation with respect to the solar zenith angle (day/night asymmetry) and Earth's distance to the Sun. The possibility of neutrino oscillations is discussed in light of these results and those of other experiments, and the MSW Large Mixing Angle solution is found to be strongly favored.

## 1. Introduction

Super–Kamiokande is the product of the collaborative effort of approximately 120 Japanese and American astrophysicists, many of whom previously worked on either the Kamiokande or IMB water Cherenkov experiments. The Super–Kamiokande site is about 300 kilometers west–northwest of Tokyo near the small town of Mozumi in the Japanese Alps. Located under 1 kilometer of rock (2,700 meters water equivalent) in the same ancient zinc mine as the old Kamiokande detector and the new KamLAND experiment, Super–K shares the same basic design as its former neighbor and namesake: a cylinder of ultra–pure water surrounded with inward–facing photomultiplier tubes [PMT's], a light barrier, a layer of outward–facing PMT's, and a veto region of water, all contained within a stainless steel tank.

Roughly an order of magnitude larger than its predecessors, Super–K has been designed to be a premier facility for studying solar neutrinos, atmospheric neutrinos, nucleon decay, and neutrinos from galactic supernovae. Weighing in at 50,000 tons of water, and holding over 11,000 inward-facing fifty–centimeter diameter PMT's and 1,850 outward-facing twenty–centimeter PMT's, Super–K is the world's largest underground water Cherenkov detector. Figure 1 shows a cutaway view of the detector.

On November 12th, 2001, while the detector

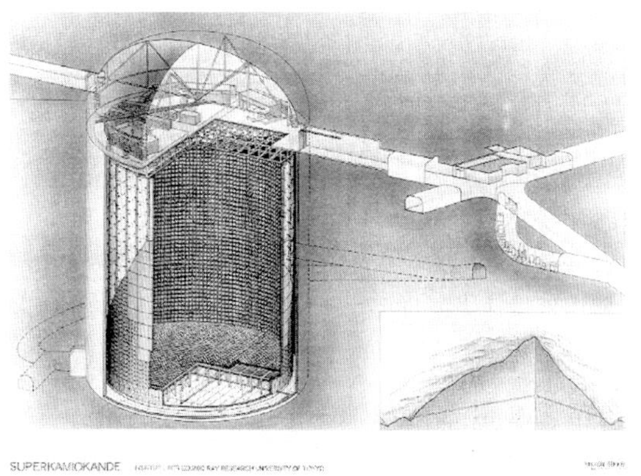


Figure 1. Cutaway view of the Super–Kamiokande detector. The detector itself stands some 42 meters tall — to get a sense of the scale, consider that the tunnels shown in the drawing are large enough for full-sized trucks to drive through. The curving tunnel on the right side of the sketch contains our water purification plant; just above that is our main control room.


doi:10.1016/S0920-5632(02)02065-0

was being refilled with water after a summer of refurbishment work, a chain reaction of imploding PMT's destroyed 6777 of the inner tubes and about 1100 of the outer tubes. An intensive investigation into the cause(s) and future prevention of such a calamity was immediately begun. As a result of this inquiry, additional pressure housings were designed and tested for the large inner PMT's, and rebuilding began in early 2002. Super–Kamiokande-II is expected to be on line with its original complement of outer tubes and about 50% of its difficult-to-manufacture inner tubes (rearranged to provide even coverage) by the end of 2002. Its physics reach, especially at higher energies, will be only modestly affected. Nevertheless, before the new JHF accelerator turns on and sends an intense long-baseline neutrino beam to Mozumi in 2007, the original inner PMT coverage will be restored and Super–Kamiokande-III will begin operations.

## 2. Solar Neutrino Data in Super–Kamiokande

With an endpoint energy of about 15.0 MeV, Super–Kamiokande's primary source of solar neutrinos is the following nuclear reaction in the Sun:

$$^8B \rightarrow ^8 Be^* + e^+ + \nu_e \quad (1)$$

These $^8$B neutrinos are seen in Super–K via elastic scattering:

$$\nu_e + e^- \rightarrow \nu_e + e^- \quad (2)$$

We have obtained what is by far the largest single sample of solar neutrino events in the world. Our most recent results, representing 1496 live days of analyzed low-energy data in the range 5.0 MeV to 20.0 MeV, spanning the period of May 31, 1996, to July 15, 2001, are presented in this paper.

Our standard method of displaying the solar neutrino signal is through the use of $\cos\theta_{\rm sun}$ plots, where $\theta_{\rm sun}$ is the angle between a reconstructed low-energy event's direction and the direction defined by a line drawn between the Sun's current position and the vertex position.

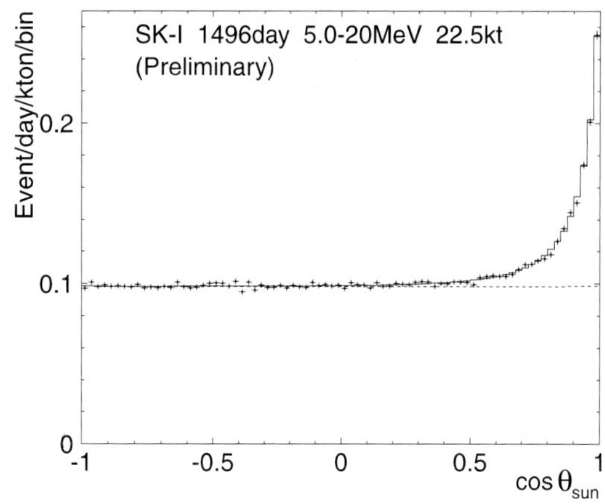


Figure 2. Solar neutrino signal between 5.0 MeV and 20.0 MeV. The line is a fit to 46.5% of the BP2000 SSM. This plot is the result of 1496 live days of data and a 22.5 kton fiducial volume.

The solar neutrino signal is shown in Figure 2, where the peak above the background in the direction of $\cos\theta_{\rm sun} = 1$ (i.e., originating from the direction of the Sun) are our solar neutrinos. There are some 22,400 events under the peak and above the background. Note that, unlike atmospheric neutrinos, one can only identify solar neutrinos in a statistical fashion. No one has yet devised a way to prove that any given event in our detector actually originated from the Sun. For this reason, reducing the sea of background events under the solar peak is of central importance in all low-energy investigations.

The best fit to the data points is given by the flux predicted by the BP2000 version of the Standard Solar Model [SSM] multiplied by a factor of 46.5%. More specifically, we measure a flux of:

$$2.35 \pm 0.02(st.) \pm 0.08(sys.) X 10^6 cm^{-2} s^{-1} \quad (3)$$

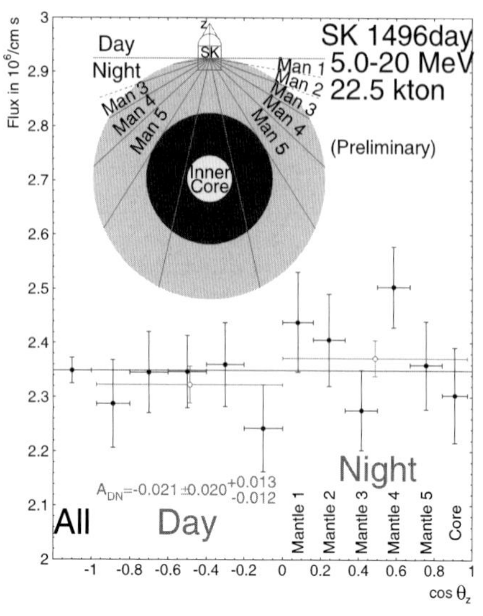


Figure 3. Variation of the solar neutrino flux as a function of zenith angle.

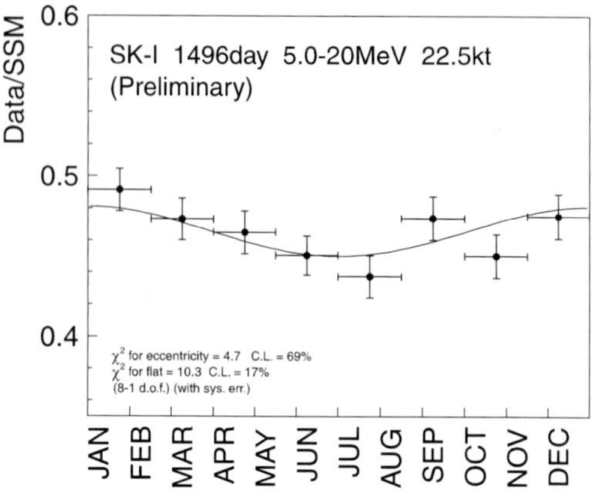


Figure 4. Seasonal variation in solar neutrino signal between 5.0 MeV and 20.0 MeV. The wavy line represents the expected $\frac{1}{r^2}$ variation in the flux due to the eccentricity of Earth's orbit around the Sun.

and find that

$$\frac{\text{Data}}{\text{SSM}_{BP2000}} = 0.465 \pm 0.005(st.) \pm 0.016(sys.) \quad (4)$$

Figure 2 contains all the low-energy data – if the data are broken down into bins based on where the Sun was in relation to the horizon at the time the signal was received we get Figure 3. The bins on the right side of the plot are defined within the figure. At present our value for the overall day/night difference is as follows:

$$\frac{\text{D - N}}{(\text{D + N})/2} = -0.021 \pm 0.020(st.) \pm 0.013(sys.) \quad (5)$$

Another interesting study which can be performed by breaking up the data is the search for seasonal variations in the flux. Such variations would be due to vacuum oscillations as the Earth moves around the Sun. Our results are shown in Figure 4. The wavy line represents the expected $\frac{1}{r^2}$ variation in the flux due to eccentricity of the Earth's orbit. It can be seen that the fit to the expected no-vacuum-oscillation line is rather good, much better than the fit to flat. As such, this represents the first ever demonstration of the eccentricity of the Earth's orbit via inverse square variations in the solar neutrino flux.

Perhaps the most powerful test of oscillations, however, is made by looking at the energy spectrum of the recoil electrons from the $^8$B solar neutrinos. Assuming that neutrinos are massive, neutrinos of a given energy will have an opportunity to execute a given number (or fractional number) of oscillations before reaching Super–Kamiokande. Therefore, deviations from the predicted spectral shape would constitute rather strong evidence of oscillations, since neutrinos of certain energies would then be more (or less) likely to be seen in our detector than neutrinos of other energies.

The results of our energy spectrum analysis can be seen in Figure 5, where the data points have been divided by the non-oscillating SSM prediction for each bin. If these points fell in a straight,

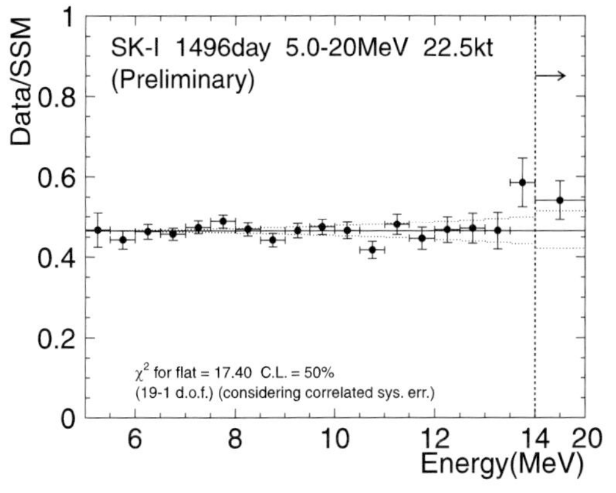


Figure 5. Energy spectrum of solar neutrino recoil electrons, divided by theoretical predictions.

flat line then they would be consistent with an unoscillated spectrum. In fact, the present shape seen in Figure 5 has a very good fit to flat. This lack of deviations will allow us to rule out certain oscillation hypotheses in the next section.

## 3. Solar Neutrino Oscillation Analysis

The probability of flavor oscillation (in the simplest, two-component case) is given by the well-known expression

$$P(\nu_a \to \nu_b) = \sin^2 2\Theta \sin^2(1.27\Delta m^2 L/E) \qquad (6)$$

where $\Delta m^2$ is in $eV^2$, $L$ is in kilometers, and $E$ is in GeV. Because of the $\Delta m^2$ term, proof of oscillations would provide evidence of at least one non-zero neutrino mass. Indeed, these two phenomena, oscillations and massive neutrinos, are inextricably linked.

Using Super–Kamiokande's measured flux and zenith angle spectral information (essentially, how the energy spectrum varies with zenith angle) along with a constraint on the $^8$B flux provided by the SSM we arrive at Super–K's allowed regions in $\tan^2\Theta$ and $\Delta m^2$ phase space for oscillations into active neutrino species. Figure 6 shows that, using experimental data from Super–K alone, large solar neutrino mixing is selected. Although not shown, oscillations into purely sterile neutrinos are completely ruled out by Super–K's data at the 95% level everywhere in phase space.

By combining Super–Kamiokande's elastic scattering results in a global fit with the rate data of all other solar neutrino experiments (SNO's neutral current and charged current, Homestake, Gallex/GNO and SAGE), we are left with Figure 7. To provide the most general possible result, here the $^8$B and *hep* fluxes are free parameters and are not constrained to any particular solar model. At the 95% confidence level, only the LMA solution survives.

## REFERENCES

1. "Measurements of the Solar Neutrino Flux from Super–Kamiokande's First 300 Days," The S–K Collab., Physical Review Letters **81**, 1158 (1998)
2. "Constraints on Neutrino Oscillation Parameters from the Measurement of Day Night Solar Fluxes at Super–Kamiokande," The S–K Collab., Physical Review Letters **82**, 1810 (1999)
3. "Measurement of the Solar Neutrino Energy Spectrum Using Neutrino Electron Scattering," The S–K Collab., Physical Review Letters **82**, 2430 (1999)
4. "Solar B-8 and hep Neutrino Measurements from 1258 Days of Super–Kamiokande Data," The S–K Collab., Physical Review Letters **86**, 5651 (2001)
5. "Constraints on Neutrino Oscillations Using 1258 Days of Super–Kamiokande Solar Neutrino Data," The S–K Collab., Physical Review Letters **86**, 5656 (2001)
6. "Determination of Solar Neutrino Oscillation Parameters Using 1496 Days of Super–Kamiokande-I Data," The S–K Collab., Physics Letters **B539**, 179 (2002)

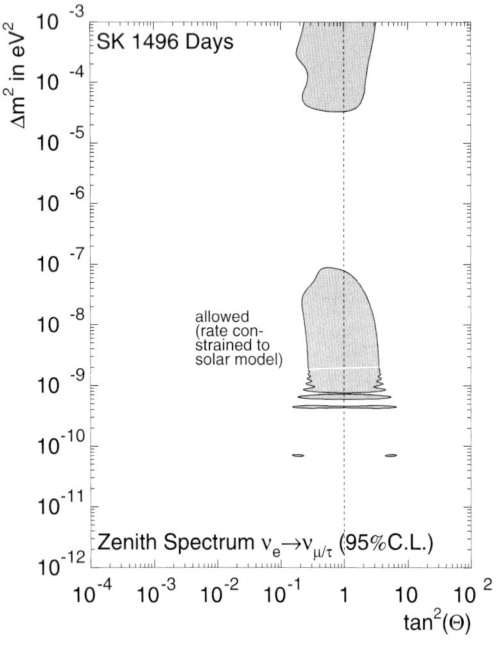


Figure 6. The allowed oscillation regions in ($\tan^2 \Theta$,$\Delta m^2$) phase space based on Super–K's flux and zenith angle information, as well as SSM $^8$B flux predictions. Shaded areas are allowed at the 95% confidence level for oscillations into active neutrino species. Large mixing is uniquely selected without inputs from any other experiments.

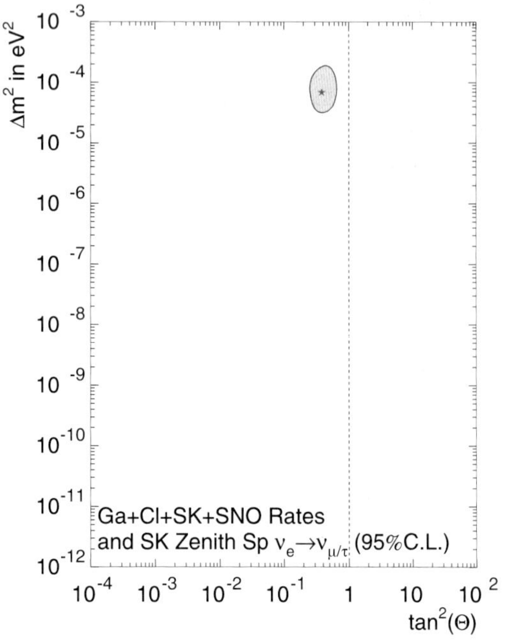


Figure 7. The remaining phase space for oscillations after Super–K's flux and zenith angle information are combined with all other solar neutrino data. The shaded area is allowed at the 95% confidence level for oscillations into active neutrino species, while the star marks the best fit point. Only a portion of the LMA region survives.

ELSEVIER

Proceedings of ICHEP 2002, pp. 13–17
*S. Bentvelsen, P. de Jong, J. Koch and E. Laenen (Editors)*

31st INTERNATIONAL CONFERENCE ON HIGH ENERGY PHYSICS AMSTERDAM

# KamLAND experiment

Tadao Mitsui [a] for the KamLAND collaboration

[a]Research Center for Neutrino Science,
Faculty of Science, Tohoku University, Sendai 980-8578, Japan

A low-energy and low-background neutrino experiment, "KamLAND" started data acquisition in January 2002, to search for a long base-line neutrino oscillation using the nuclear reactors as sources. With good detector performance very close to the design values, the data analysis is progressing rapidly, in order to test the LMA solution of the solar neutrino problem, by an experiment with artificial sources for the first time.

## 1. Introduction

The electron neutrino ($\nu_e$) oscillation is, at present, the key to understanding the lepton mixing matrix and its origin. For studies of all the exciting physics, such as three-generation mixing involving $U_{e3}$, or further possibility of CP violation, one has to, first of all, determine the $\nu_e$ oscillation parameters. Although $\nu_e$ experiments have depended on the solar model for a long time, recent analyses are making allowed solutions narrower [1,2] in a way independent from the solar model, i.e., observing the "day-night" effect, or measuring neutral and charged currents individually. The experiment with artificial sources, i.e., nuclear reactors[3], is another independent method, although no positive result has been obtained so far.

A high sensitive detector with 1-kton liquid scintillator, "KamLAND" has been constructed to perform a new reactor neutrino experiment, in which the sensitivity for $\Delta m^2$ will be improved by 2 orders of magnitude, thus reaching the "large mixing angle (LMA) solution" for the solar neutrino problem. By precisely examining this most favored solution at present [1,2], we can expect a large progress in studies of neutrino masses and mixings. KamLAND just started data acquisition in January 2002, after 4 years of construction period. Here the current status of data analysis is presented after overview of detector and physics targets.

## 2. Detector Overview

Figure 1 shows a schematic of the KamLAND detector. All components are contained in a 6,200-m$^3$ water pool (diameter, 20 m; and height, 20 m), whose inner surface is surrounding rock with polyurethane lining. The rock-overburden is 2,700 m.w.e. on average and the cosmic-ray muon rate is about 0.3 Hz inside the active region of the detector. In the pool, a spherical tank with an 18-m diameter made of stainless steel is placed to divide the "inner detector" and the "outer detector." The outer detector is filled with 3,000-m$^3$ pure water, whose Čerenkov lights are directed to 225 photomultiplier tubes (PMTs) attached inside the pool, producing a veto signal against penetrating cosmic-ray muons.

The inner detector consists of 2 layers, i.e., a 1,200-m$^3$ liquid scintillator and a 1,900-m$^3$ "buffer oil," which are divided by a 13-m diameter plastic balloon. The balloon is suspended with 44 braided ropes. The buffer oil layer with a 2.5-m thickness, not being an active detector, effectively reduces the background radiation from PMTs and stainless steel tank. Radon (Rn) gas also emanates from the glass of PMTs, stainless steel, welding, and residual dusts. To prevent such Rn from diffusing toward the scintillator balloon, buffer oil is further divided into 2 layers by an acrylic sphere (radius, 8.25 m; and thickness, 3 mm). Light pulses from scintillator are directed to 1879 PMTs attached inside the stainless steel tank.


doi:10.1016/S0920-5632(02)02073-X

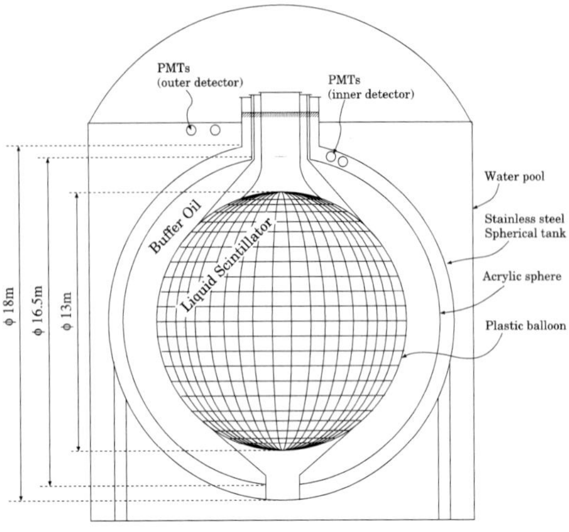


Figure 1. KamLAND detector

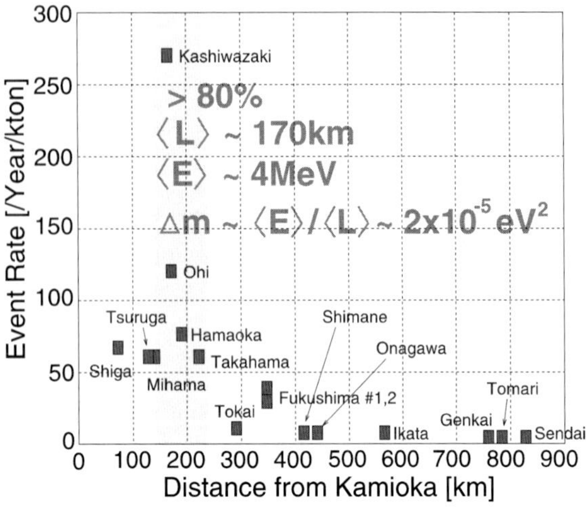


Figure 2. Expected rates of reactor neutrino events observed by KamLAND; contributions of each reactors as a function of the distance

In the front-end electronics, the PMT signal is discriminated by a threshold of 1/3 p.e. to provide the total number of hit tubes for event triggering. The threshold of the event trigger corresponds to ~0.8 MeV for the "global trigger," while the "delayed trigger" with ~0.5 MeV is issued only in a period of 1 ms after each global trigger. Each PMT pulse is then amplified in three different gains depending on the amplitude, and the pulse shapes are digitized by "Analogue Transient Waveform Digitizers (ATWD)" with 1.5-ns step of 128 6-ns width gates. Two identical ATWDs are prepared for each PMT channel to make the system practically deadtime-less for succeeding events coming within a shot period.

## 3. Long Baseline Neutrino Oscillation using Reactors

Nuclear reactor is well understood neutrino source, whose flux can be predicted within a few percent error. By detecting disappearance of reactor-$\bar{\nu}_e$, therefore, long baseline $\bar{\nu}_e \to \nu_x$ oscillation can be searched for.

In the previous reactor neutrino experiments [3], the baseline was limited within about 1 km. In order to realize a longer baseline, one needs a detector with much larger mass, because of uniformly emitted reactor $\bar{\nu}_e$ whose flux decays as $r^{-2}$. Here we will perform an experiment with baseline larger than 100 km by employing 1-kton scintillator, with which 2 orders improvement on the $\Delta m^2$ sensitivity is expected. There are powerful commercial reactors around Kamioka, such as Kashiwazaki (24.6 GW). In addition, as shown in Figure 2, reactors between 140 and 210 km are expected to contribute to 85% of the total reactor $\bar{\nu}_e$ events observed in KamLAND. This condition is equivalent to that there is a 67-GW large reactor at the distance (with an error) of 175±35 km, then we can obtain clear result for neutrino oscillation without much interference between reactors with different distances.

Reactor-$\bar{\nu}_e$ with energies above 1.8 MeV can be detected by inverse $\beta$ decay. Incoming $\bar{\nu}_e$ interacts with a proton in the scintillator as $\bar{\nu}_e + p \to e^+ + n$, the emitted neutron is then thermalized and captured by another proton, i.e., $n + p \to d + \gamma$, where the energy of $\gamma$ is 2.2-MeV. It takes about 200 $\mu$s for the neutron to be captured by a proton, so the delayed coincidence between $e^+$ and 2.2-MeV $\gamma$ signals can be required to identify this event, by which backgrounds will be strongly suppressed. Utilizing this mode also

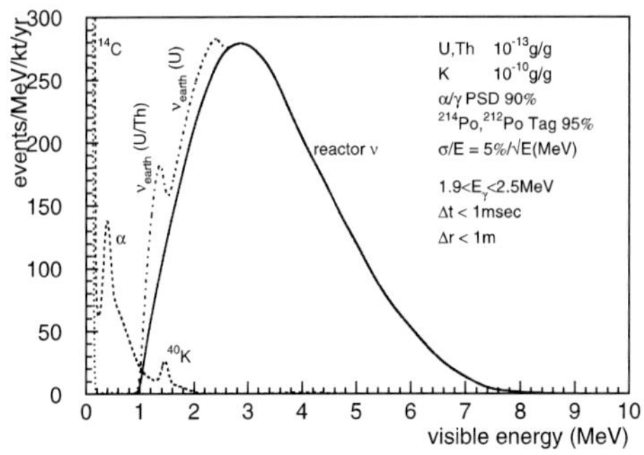


Figure 3. Expected energy spectra of $e^+$ emitted in the inverse $\beta$ decay events.

enables us to unambiguously identify $\bar{\nu}_e$, because inverse $\beta$ decays do not occur for other neutrinos.

Figure 3 shows an expected energy spectra of positrons (including 2 511-keV annihilation $\gamma$'s) from the reactor-$\bar{\nu}_e$ events (no oscillation assumed) together with backgrounds. As shown, the reactor-$\bar{\nu}_e$ events will dominate on backgrounds from scintillator (assumed impurity is shown in the figure) and rocks. The low energy background of "geo-$\bar{\nu}_e$" will be discussed in section 5. With 3-years data-taking in KamLAND, more than 1000 reactor-$\bar{\nu}_e$ events are expected assuming no oscillation. By comparing the event rate with the expectation, a missing less than 10% can be detected considering all errors of both observation and prediction. Then a highly sensitive search for neutrino oscillation can be performed as shown in Figure 4.

As another analysis independent from the absolute flux, we can use the energy spectrum distortion due to the neutrino oscillation. As shown in Figure 4, some of the solution point in currently allowed LMA region, can be confirmed by this analysis with high precision, provided that a few percent error of the energy scale is attained.

## 4. Detector Performance and Data Analysis

Since our observation started, we have been successfully accumulating data, as well as cali-

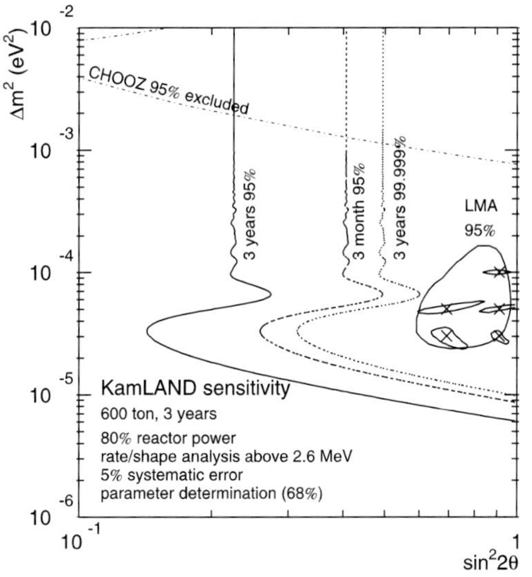


Figure 4. Expected sensitivity of KamLAND

bration data. Gain and timing calibration of each PMT was performed using a "laser diffuser ball" located at the center of the balloon, After that, calibration of vertex position and energy was performed using two sources, i.e., $^{65}$Zn (1.116 MeV) and $^{60}$Co (1.173 + 1.333 MeV) located at various positions along z-axis (vertical direction).

In off-line analysis, charge and timing of each PMT are obtained from waveform analysis, then timing is used for vertex reconstruction, while charge is for energy estimation. Only 1325 17-inch PMTs (out of total 1879 PMTs) are used now, and photon yield is $\sim$250 p.e./MeV.

In our preliminary analysis tools, we obtained position resolution of 60 cm at 1 MeV with systematic error of 10 cm, energy resolution of $7.5\%/\sqrt{E/[\mathrm{MeV}]}$, and systematic error of energy scale 2%. Those values are being improved by further calibration and improvement of tools.

Figure 5 shows the single energy spectra with various fiducial volumes. By comparing those spectra, we can see most of the background comes from outside of the balloon or from the balloon itself (film and ropes). Especially, 1.4- and 2.6-MeV peaks are seen in the spectra with no

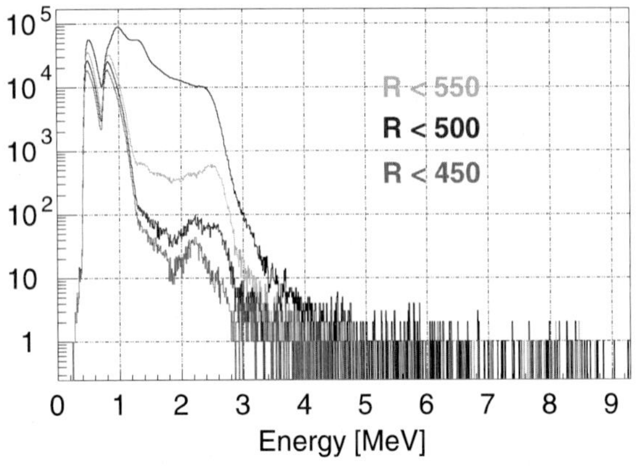


Figure 5. Energy spectra with various fiducial volumes. The curves correspond, from above, no fiducial cut (R < 6.5 m; the balloon radius), R < 5.5 m, R < 5.0 m, and R < 4.5 m. Edges at ~0.5 and ~0.8 MeV are due to threshold effect from global and delayed trigger, respectively.

fiducial-cut and R < 5.5 m cut respectively, which are not seen in spectra with more stringent cuts. These peaks correspond to $\gamma$'s from $^{40}$K and $^{208}$Tl respectively, and these spectra are consistent with the expectation that $^{40}$K $\gamma$ should come from the balloon ropes, and $^{208}$Tl $\gamma$ should come from the rocks surrounding the detecctor.

In the spectrum with R < 4.5 m cut, still there is a peak at 2.2 MeV, corresponding to the gamma ray energy in the neutron capture events, n + p $\rightarrow$ d + $\gamma$. This neutron should come from $^{12}$C nuclear spallation caused by cosmic-ray muons, $^{12}C \rightarrow {}^{11}C + n$, because the 2.2-MeV peak is not removed with any fiducial cuts. To confirm this, the spatial distribution of the 2.2-MeV events is examined and it was found that most (80%) of the neutron capture events are correlated with the previous muon track within the distance of 2 m and time difference of 2 ms.

The "muon veto" is, then, necessary to reject not only those neutrons but also more serious spallation products such as $^{8}$He and $^{9}$Li [6]. The timing and spatial cuts of the muon veto should be tuned based on the analyses of the spallation products, as well as an existing precise study of the muon spallation [6]. Our spallation analysis is progressing at the moment.

The radio-impurity in the scintillator is estimated by analyzing the single spectrum with stringent fiducial cut and muon veto, also with subtraction of the contributions from identified spallation products. At the moment we don't find any significant contributions from $^{214}$Bi, $^{208}$Tl, and $^{40}$K, thus obtaining the upper limit on the impurities, $^{238}\text{U} < 3.0 \times 10^{-16}$ g/g, $^{232}\text{Th} < 1.8\times 10^{-16}$ g/g, and $^{40}\text{K} < 2.3\times 10^{-16}$ g/g, which are sufficiently better than the requirement of the reactor neutrino experiment (Figure 3).

With this very low background level, there is no serious problem in searching for $\bar{\nu}_e$ events. The main selections are (i) fiducial cut for prompt events (positron) (ii) spatial distance between prompt and delayed events (2.2-MeV gamma) (iii) time difference of them, (iv) energy of the delayed events, and (v) muon veto. We are trying analysis with more than one criteria for above selections. For each of them, efficiency, backgrounds, and systematic errors are estimated, by which the stability of the analysis is checked.

## 5. Background from Geo-neutrino

Since the concentration of radio-impurity is extremely low as described above, only inevitable background will be "geo-neutrino" (geo-$\bar{\nu}_e$), which comes from the decay of $^{238}$U and $^{232}$Th in the crust and mantle of the Earth [4]. Although observation of geo-$\bar{\nu}_e$ is one of the purposes of our experiment [5,4], here we treat it as a background because, to know the uncertainty of geo-$\bar{\nu}_e$ flux and what the flux depends on is important both for geo- and reactor-$\bar{\nu}_e$ analysis.

In Figure 3, the expected geo-$\bar{\nu}_e$ flux is shown together with reactor-$\bar{\nu}_e$. The two components are comparable below ~2.6 MeV. This flux is calculated in the U and Th distribution of "Model I," shown in Figure 6, in which 3 other models are also shown [5]. In all models, the total amounts of U and Th are conserved respectively, while only distribution in the Earth is different. This approach is based on the fact that, the decay of U and Th contributes to ~90% of radiogenic heat in the Earth [4,5], and radiogenic heat is expected to contribute ~40% of total heat source in

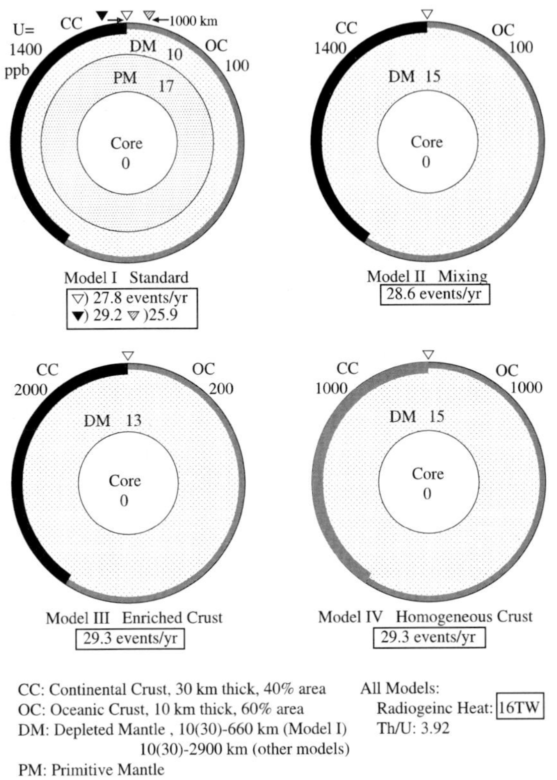


Figure 6. Models of U and Th distribution in the Earth, shown with the expected number of $\bar{\nu}_e$ events observed in KamLAND (0.5 kt fiducial). In Model I, three observation points are tested, to estimate the uncertainty from CC/OC structure around Japanese Island Arc. The effect of changing Th/U ratio is discussed elswhere [5,4], although it doesn't change the conclusion here.

the Earth. The rest of the heat (~60%) is considered to be derived from the secular cooling of the Earth after a high temperature event in the early Earth. Then, the total amount of U and Th is a good parameter which is closely related to dynamics and heat balance in the Earth. On the other hand, the distribution of U and Th in mantle and crust depends on models of chemical process of element concentration, and also on mantle convection, which are more uncertain.

As shown in Figure 6, geo-$\bar{\nu}_e$ flux doesn't much depend on the distribution of the elements (only ±10%). So, the normalization of geo-$\bar{\nu}_e$ flux can essentially be expressed by one parameter of radiogenic element abundances. Figure 3 and 6 correpond to radiogenic heat of 16 TW, while its upper limit should be ~40 TW, which is the measured heat flow from inside of the Earth. Expressing geo-$\bar{\nu}_e$ flux by one parameter in this way helps us in an inclusive analysis for geo- and reactor-$\bar{\nu}_e$ fluxes, because there are already 2 parameters concerning neutrino oscillation, and too many parameters can be an obstacle to a transparent analysis. It is an advantage also in geophysics that our observation of geo-$\bar{\nu}_e$ is almost directly connected to the abundance of the radiogenic elements, which is related to the heat flux and thermal state of the present Earth.

Sincere gratitude is to the organizing comittee and staffs of ICHEP 2002 for kind hospitality. The author also thanks all KamLAND collaborators from Japan and U.S.A. for helpful discussions, and Professor E. Ohtani for valuable suggestion on geophysics. KamLAND project is supported by the Japanese ministry of Education, Culture, Sports, Science and Technology and the U.S. Department of Energy.

## REFERENCES

1. M.R. Vagins (Super-Kamiokande Collaboration), this proceedings, and references therein.
2. S. Oser (SNO Collaboration), this proceedings, and references therein.
3. M. Apollonio et al. (CHOOZ Collaboration), Phys. Lett. B420 (1998) 397, F.Boehm et al. (Palo Verde Collaboration), Phys. Rev. D64 (2001) 112001.
4. R.S. Raghavan, S. Schoenert, S. Enomoto, J. Shirai, F. Suekane and A. Suzuki, Phys. Rev. Lett. 80 (1998) 635., C.G. Rothschild, M.C. Chen and F.P. Calaprice, Geophys. Res. Lett. 25 (1998) 1083., and references therein.
5. E. Ohtani, private communication, Proc. 1st Sendai Intern. Conf. on Neutrino Science (2002).
6. H. Hagner et al., Astr. Part. Phys. 14 (2000) 33.

ELSEVIER

Proceedings of ICHEP 2002, pp. 18–23
*S. Bentvelsen, P. de Jong, J. Koch and E. Laenen (Editors)*

31st INTERNATIONAL CONFERENCE ON HIGH ENERGY PHYSICS AMSTERDAM

# Recent results from the K2K experiment

Y.Hayato for the K2K collaboration

Institute of Nuclear and Particle Studies, High Energy Accelerator Research Organization, Oho 1-1, Tsukuba, Ibaraki, Japan

Since 1999, the KEK to Kamioka long-baseline neutrino experiment (K2K) has begun its investigation of neutrino oscillations suggested by atmospheric neutrino observations. Until July 2001, we have accumulated $4.8 \times 10^{19}$ protons on target and observed 56 in-fiducial events at Super-Kamiokande detector, which is located 250km from KEK. The expected number of events is estimated to be $80.1^{+6.2}_{-5.4}$. We fit the neutrino oscillation parameters by using the observed number of events and measured energy spectrum of neutrinos. The probability to observe only 56 events and such an energy spectrum due to statistical fluctuation is less than 1%. Also, the favored oscillation parameter region agrees well with the one suggested by the atmospheric neutrino experiments.

## 1. Introduction

Since the late 1980s, some experiments on atmospheric neutrinos reported deficits of $\nu_\mu$ with energies around 1GeV. In 1998, the Super-Kamiokande(SK) collaboration not only confirmed a deficit of atmospheric $\nu_\mu$ with high statistics, but they also reported the clear zenith angle dependence of the deficit.[1] This means that the disappearance probability of $\nu_\mu$ depends on the distance between points where the neutrino was generated and the detector. These experimental results are well explained by neutrino oscillation and imply that neutrinos should have non-zero masses. Therefore, various experiments, which use accelerators as the neutrino source, were proposed to investigate the oscillation and various characteristics of neutrino. Among of them, the K2K experiment became the first accelerator-based long baseline neutrino oscillation experiment which covers the parameter region suggested by the atmospheric neutrino experiments. The K2K experiment started in April 1999 and took data until July 2001. This time, we used the data taken from June 1999 to July 2001, corresponding to $4.8 \times 10^{19}$ protons on target(POT).

## 2. Overview of the K2K experiment

The K2K neutrino beam is produced by 12 GeV protons from the KEK proton synchrotron.[2][3] The extracted protons hit the aluminum target and generate the particles. Among of the generated particles, positive charged ones, mainly $\pi^+$, are focused by a pair of pulsed magnetic horns.[4] The neutrino beam is produced by the decay of these particles and the generated beam is nearly pure $\nu_\mu$ beam ( 98.2% $\nu_\mu$, 1.3% $\nu_e$ and 0.5% $\bar{\nu}_\mu$) The mean energy of the neutrino is about 1.3GeV.

From time to time, we measured the momentum and directional distributions of pions just after the horn. The energy spectrums of the neutrino beam at near(KEK) and far(Kamioka) sites and the relative flux ratios between the two sites can be obtained by using these distributions. For this measurement, we have developed the gas-Cherenkov type detector called pion monitor(PIMON).[5] The PIMON can measure the pions, whose momentum is above 2GeV/c. This corresponds to ~1GeV for the neutrinos.

The direction of the neutrino beam should be stable and well under control. Therefore, we placed the muon monitor(MUMON) just after the beamdump.[5] This detector is able to measure the profile of the muons, which are generated from the decay of pions, spill by spill basis. The center of the beam is found to be very stable and controlled within 1mrad, which satisfies our re-


doi:10.1016/S0920-5632(02)02074-1

quirement.

We also placed the near neutrino detectors in KEK to measure the neutrino beam itself. There are two different types of the detectors, 1kt water Cherenkov detector(1KT) and fine grained detectors(FGD). 1KT is the miniature of the SK detector. This detector consists of the cylindrical water tank, which can held 1kt of pure water, equipped with 680 20inch Photo multiplier tubes(PMTs). Since 1KT is the same type of the detector used in SK, most of the systematic uncertainties are canceled. FGD consists of a scintillating fiber tracking detector[6], a lead-grass calorimeter(LG) and a muon range detector.[7] SCIFI consists of the 20 layers of scintillating fiber tracking modules and 19 layers of 6cm thick water containers. The diameter of the scintillating fiber is 0.7mm and this detector has very fine position resolution, less than 1mm. MRD consists of the drift chambers and 12 iron plates. In order to get better energy resolution, the upstream 4 iron plates are 10cm thick and the downstream 8 plates are 20cm thick. The total thickness of iron is 2.0m and it is possible to measure muons up to ~3GeV/c. Because this detector has large coverage and volume, the stability of the neutrino beam direction has been confirmed on daily basis and the stability of the energy spectrum has been confirmed on monthly basis. 1KT has high efficiency for muons below 1GeV/c and full $4\pi$ coverage in solid angle. On the other hand, the FGD has high efficiency for measuring muons above 1GeV/c and these two complimentary detectors cover the relevant energy range.

The neutrinos from KEK travel through the Earth for 250km and are detected by the SK detector in the Kamioka-mine, Gifu. At the K2K experiment, baseline is fixed to 250km and thus, the oscillation probability of $\nu_\mu$ depends on the energy of neutrino. This means that, if the $\nu_\mu$ oscillation occurs, not only the observed number of events at SK is decreased but also the observed energy spectrum of neutrino is distorted compared to the expectation with null-oscillation assumption.

The flux and spectrum of neutrino at SK is obtained by multiplying the so-called Far to Near (F/N) ratio by the measured energy spectrum at the near detectors. This F/N ratio is an energy dependent function, which is derived from our neutrino beam Monte-Carlo simulation. The reliability of our beam simulation has been checked by using the results from the PIMON and the near neutrino detectors.

Finally, we compare the data from SK and the expectations to search for the effect of the neutrino oscillations. At the same time, we also fit the data with the oscillated neutrino flux and obtain the allowed regions of the oscillation parameters.

## 3. Measurements of the neutrino flux and energy spectrum

We use 1KT to obtain the flux normalization factor and to measure the spectrum. In the 1KT analysis, we use the events whose interaction vertices are in the fiducial volume. The fiducial volume is defined to be a sideways cylindrical volume of 25tons in the upstream part of the detector. For the flux normalization, we use the same criteria used in the previous report[8]:(a) there is no detector activity in the 1.2 $\mu$s preceding the beam spill. (b)Only a single event is observed in the spill. (c) The reconstructed vertex is in the fiducial volume. The measurement has a 5% systematic uncertainty, of which the largest contribution comes from the vertex reconstruction. For the spectrum measurements with 1KT, we want to increase the fraction of the events from charged current quasi-elastic scattering(CCQE). Because this interaction is rather simple compared to the other interactions and it allows us to reduce the uncertainties coming from the neutrino interaction models. Also, the CCQE events are important because it is possible to reconstruct the energy of neutrino by using the measured momentum and direction of the muon as follows:

$$E_\nu^{rec} = \frac{m_N E_l - m_l^2/2}{m_N - E_l + P_l \cos\theta_l}, \tag{1}$$

where $m_N, m_l, E_l, P_l$ and $\theta_l$ are the nucleon mass, the lepton mass, the lepton energy, the lepton momentum, and the scattering angle relative to the neutrino beam direction, respectively. In order to increase the fraction of the CCQE events, we select the 1 ring $\mu$-like events. Also we require the muon in an event to be stopped in the detector to measure the momentum of muon. The largest uncertainty for the spectrum measurement is an uncertainty on the energy scale of the detector. The energy scale has been calibrated within $^{+2}_{-3}\%$ by using the cosmic ray muons.

For the spectrum analysis with FGD data, we use the events occurred in the fiducial volume of SCIFI (5.9 ton). Because the target material of SCIFI is mainly water, which is same as the water Cherenkov detectors. The actual event selection criteria are as follows: (a)one or two tracks are identified, (b)at least one track has to be reached at the MRD. The criterion (b) is used to select charged current $\nu_\mu$ interactions. The momentum of a track is measured by its range with an error of 3%. If the momentum of proton is larger than 500MeV/c, the proton track can be also identified. The efficiencies for reconstructing CCQE events are 93% for one track events and 86% for two-track events. For the two-track events, we apply additional cut to make CCQE enriched sample and non-CCQE enriched sample. Assuming the interaction to be CCQE, direction of the proton can be predicted by using direction and momentum of the muon. If the direction of the second track in an event agrees with the predicted direction within 25 degree, the event is identified to be CCQE-like. Whereas, the observed direction and the predicted one differs by more than 30 degree, the event is identified as non-CCQE-like. In the CCQE enriched sample, 60% of the events are estimated to be CCQE by the Monte-Carlo simulation. As for the non-CCQE enriched sample, 85.7% of events are generated by the interactions other than CCQE. In total, we have three event samples, one track, two-track CCQE enriched, and two-track non-CCQE enriched for the spectrum measurement.

Here, we have four event categories, one from 1KT and three from FGD, which can be used to obtain the energy spectrum of neutrino. In order to fit the energy spectrum, we use two dimensional distributions of the muon momentum versus angle respect to the beam direction for each categories. A $\chi^2$ fitting method is used to obtain the best set of parameters to match the Monte-Carlo distributions to the data. The neutrino

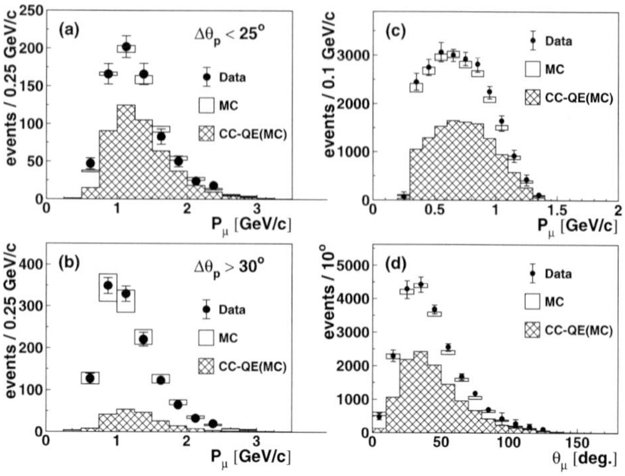


Figure 1. The muon momentum distributions of (a) SCIFI CCQE enhanced sample, (b) SCIFI non-CCQE enhanced sample, (c) 1KT 1 ring $\mu$-like sample, respectively. (d) The angular distribution of the 1KT 1 ring $\mu$-like sample. The crosses are data and the boxes correspond MC simulation with the best fit parameters. The hatched histograms show the CCQE events estimated by the MC simulation.

spectrum is divided into 8 energy bins as defined in Table 1. The neutrino fluxes in each energy bin, the ratio between the CCQE and non-CCQE cross-sections($R$) are the parameters of the fit. The systematic uncertainties of the near detectors, such as energy scale, track finding efficiencies, and energy thresholds, are incorporated into the fitting parameters. In the fitting, the results from PIMON are used as a constraint. The value of $\chi^2$ is 227.2/197 d.o.f.. The fit results agree well with the data as shown in Fig.1.

The errors of the measurements are provided in the form of an error matrix. The diagonal elements in the error matrix on the determination of the spectrum at SK are shown in Table 1. As a result, all the parameters are found to be in the range of the known uncertainties.

We have also studied the uncertainties on the spectrum measurements due to neutrino interac-

| $E_\nu$(GeV) | $\Phi_{\rm ND}$(%) | F/N ratio(%) | $\epsilon_{\rm SK}$(%) |
|---|---|---|---|
| 0 - 0.5 | 37 | 2.6 | 8.7 |
| 0.5 - 0.75 | 12 | 4.3 | 4.3 |
| 0.75 - 1.0 | 6.7 | 4.3 | 4.3 |
| 1.0 - 1.5 | 3.1 | 6.5 | 8.9 |
| 1.5 - 2.0 | 4.7 | 10 | 10 |
| 2.0 - 2.5 | 8.2 | 11 | 9.8 |
| 2.5 - 3.0 | 15 | 12 | 9.9 |
| 3.0 - | 17 | 12 | 9.9 |

Table 1
The size of the energy dependent systematic errors on the predicted neutrino spectrum at SK.

tion models. For each interaction mode considered in the Monte-Carlo simulation, we changed the parameters or models to estimate the systematic errors. These changes do not affect to the spectrum beyond the size of the fitting errors. However the change of the value $R$ was larger compare to the others and thus, we assigned additional error of $R$ by $\pm$ 20% in the error matrix.

As described, the F/N ratio is used to obtain the expected neutrino spectrum at SK. The errors on the F/N ratio is estimated based on the PIMON measurements for $E_\nu$ above 1GeV. The errors on the ratio for $E_\nu$ below 1GeV are estimated based on the uncertainties in the hadron production models used in the neutrino beam Monte-Carlo.[5] Thus, the errors of F/N ratio between above and below 1GeV are treated to be non-correlated.

## 4. Results from the Super-Kamiokande detector

In order to select the events at SK, which were generated by the neutrinos from KEK, we use the timing information provided by the GPS system.[9] By using the absolute timing information provided by GPS system, we calculate the time differences between the beam spill timing and the trigger timing of the events in SK to identify the events which were generated by the neutrino from KEK. Here, we use the same criteria used in the previous report.[8] The actual criteria are defined as follows: (a)the timing of the event is consistent with the beam, (b)there should be no events before 30 $\mu$ s, (c)electron equivalent energy of a event should be larger than 30MeV, (d)there should be no activities in the outer detector, (e)the reconstructed vertex must be in the fiducial volume (22.5kt). The detection efficiency of charged current interactions is estimated to be 93%. After applying these cuts to the data, 56 events passed the selection($N_{obs}$) and the expected number of background is about $10^{-3}$ events.

For the oscillation analysis, which utilize the energy spectrum, we use the 1 ring $\mu$-like events to increase the fraction of CCQE interactions. At the same time, we excluded the data taken in June 1999. Because the diameter of the target and the horn current were different at that time and thus, the energy spectrum is also different. The data in June 1999 correspond to $3.1 \times 10^{18}$ POT and the fraction to the total POT is 6.5%. The number of 1 ring $\mu$-like events excluding the data taken in June 1999($N_{1R\mu}$) is 29.

## 5. Oscillation analysis

The expected number of events at SK without oscillation is estimated to be $80.1^{+6.2}_{-5.4}$, which should be compared to the $N_{obs} = 56$. This value is obtained by extrapolating the observed number of events at 1KT as described before. The correlations between the energy bins from the energy spectrum measurement at the near detectors and the F/N ratio are taken into account to estimate the systematic errors. The major contributions to the errors on the expected number of events are the uncertainties of the F/N ratio($^{+4.9}_{-5.0}$%) and the normalization error($\pm$5%) dominated by the uncertainties of the fiducial volumes at 1KT and SK. The search for the oscillation parameters were performed by the maximum-likelihood method. The likelihood function $L$ is defined as $L = L_{norm} \times L_{shape}$. The normalization term ($L_{norm}(N_{obs}, N_{exp})$) is the Poisson probability function to observe $N_{obs}$ events when the expected number of events is $N_{exp}$. Here, $N_{exp}(\sin^2 2\theta, \Delta m^2, f)$ is the function of oscillation parameters ($\sin^2 2\theta, \Delta m^2$) and the set of parameters($f$) to consider the systematic errors in the fitting. The parameters $f$ consist of the measured energy spectrum of neutrino($\Phi_{ND}$), the F/N ratio, the reconstruction efficiency of 1 ring $\mu$-like events at SK, the CCQE to non-CCQE ratio($R$), the energy scale of SK, and the overall normalization. The errors of the first three items are energy dependent and the correlations are taken into account. The diagonal part of the error matrices are summarized in Table 1. As described, the error on $R$ is set to 20% from the results at the near detectors. The error on the energy scale of SK is 3%. The errors on the normalizations for the data taken in June 1999 and the rest period are $^{+1.00}_{-0.85}$% and 5.3%, respectively. This normalization error only affect to the normalization term of the likelihood function. The shape term ($L_{shape}$) is the product of the probabilities of each 1 ring $\mu$-like event to be observed at $E_\nu^{rec}$ ($L_{shape} \equiv \prod_{i=1}^{N^{1R\mu}} P(E_i; \sin^2 2\theta, \Delta m^2, f)$), where $P$ is the normalized $E_\nu^{rec}$ distribution estimated by the Monte-Carlo simulation and $N^{1R\mu}$ is the number of 1 ring $\mu$-like events.

In the actual fitting, two different approaches are taken for the treatment of systematic errors in the likelihood function. One of the method treats the parameter $f$ as fitting parameters with an additional constraint term in the likelihood functions(method 1).[1][10] In the other method, we take the average of $L(f)$ sampled over many random trials weighted according to the probability density distribution of the systematic parameters $f$(method 2).[11] In order to search for the best fit point in the oscillation parameter space($\sin^2 2\theta, \Delta m^2$), the likelihood function is calculated at each point in the parameter space. The best fit points in the physical region are found to be at $(\sin^2 2\theta, \Delta m^2) = (1.0, 2.8 \times 10^{-3}\text{eV}^2)$ in method 1 and $(1.0, 2.7 \times 10^{-3}\text{eV}^2)$ in method 2, respectively. At the best fit point, the total number of event is expected to be 54.2, which agrees with the observation of 56 within statistical error. The observed $E_\nu^{rec}$ distribution of 1 ring $\mu$-like events together with the expected distributions with and without oscillation are shown in Fig.2. The consistency is checked by the KS test and KS probability is 79% for the best fit oscillation parameters. The allowed regions of oscillation parameters are shown in Fig.3. The two methods give consistent results.

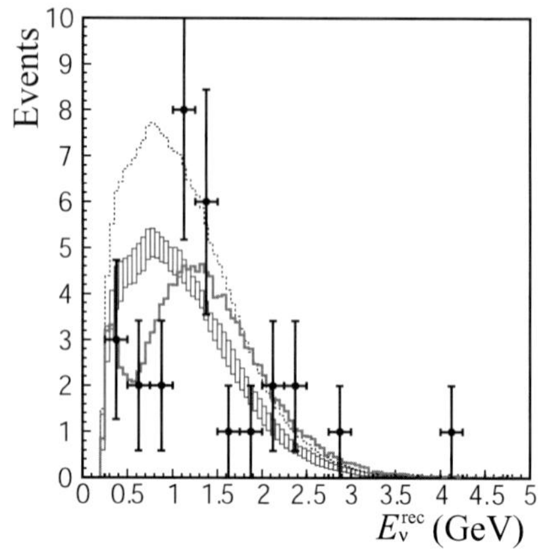


Figure 2. The reconstructed $E_\nu$ distribution from the method 1. Points with error bars are data, the boxes are the expected spectrum without oscillations, where the height of the box is the size of systematic error. The open histogram is the best fit spectrum. Both histograms are normalized by area to the observation. The dashed line shows the expectation without any oscillations.

We have also obtained the null-oscillation probability by computing the likelihood ratio of the best fit point to the no-oscillation case. As a result, the probabilities are found to be 0.7% and 0.4% for method 1 and 2, respectively. If we use only the normalization(shape) information alone for the fitting, we obtained 1.3%(16%) and 0.7%(14.3%) for method 1 and 2, respectively. We have also checked the effect of the uncertainties in neutrino interaction models. As a result, the effect is found to be negligible. Because the same model are used for the near detectors and SK and thus, the uncertainties are almost canceled.

## 6. summary

The K2K experiment was successfully started in March 1999. We have accumulated $4.8 \times 10^{19}$

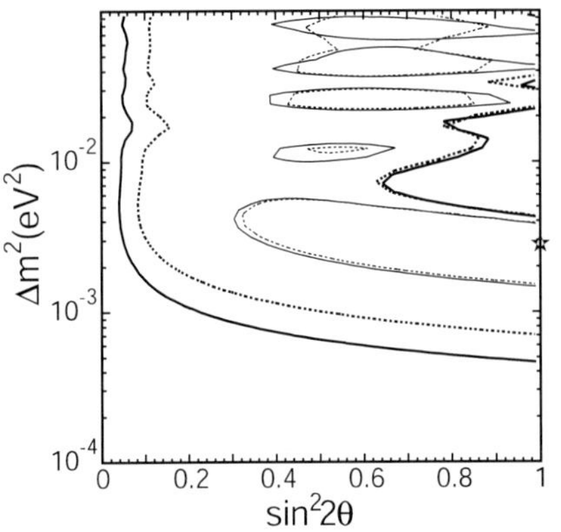


Figure 3. Allowed regions of oscillation parameters. Thick (thin) lines are 99% (90%)C.L. Solid lines are from the method 1 and dashed ones are from method 2. The best fit point is shown by a star.

POT from June 1999 to July 2001 and observed 56 in-fiducial events at SK detector. We have measured the neutrino energy spectrum by the near detectors in KEK. In order to obtain the energy spectrum at SK, we extrapolated the results from near detectors. In the extrapolation, we used the beam Monte-Carlo, which has been confirmed by the PIMON. Without considering neutrino oscillations, the expected number of events at SK is $80.1^{+6.2}_{-5.4}$. The probability to observe only 56 events and the observed energy spectrum due to statistical fluctuation is less than 1%. Also, the obtained oscillation parameter regions from this analysis are consistent with the ones suggested by the atmospheric neutrino experiments. The K2K experiment expects to accumulate $10^{20}$ POT and will obtain sufficient data for the further studies of neutrino oscillations.

We thank the KEK and ICRR Directorates for their strong support. We also thank the KEK PS machine and beam channel group. Without their inventiveness and efforts, it is not possible to run this experiment. We gratefully acknowledge the cooperation of the Kamioka Mining and Smelting Company. This work has been supported by the Ministry of Education, Culture, Sports, Science and Technology, Government of Japan and its grants for Scientific Research, the Japan Society for Promotion of Science, the U.S. Department of Energy, the Korea Research Foundation, and the Korea Science and Engineering Foundation.

## REFERENCES

1. Y.Fukuda *et al.*, Phys.Rev.Lett. **81**, 1562 (1998).
   S.Fukuda et al., Phys. Lett. **B539**, 179 (2002)
2. M.Ieiri *et al.*, Proc. 1st Asian Pacific Accelerator Conference, 579 (1998).
3. H. Sato, Proc. Particle Accelerator Conference (1999).
   K. Takayama, ICFA Beam Dynamics Newsletter No.20, (1999).
4. Y.Yamanoi *et al.*, Proc. 15th International Conference on Magnet Technology (1997).
   Y.Yamanoi *et al.*, IEEE Transactions on Applied Superconductivity **10**, 252 (2000).
   Y.Suzuki *et al.*, Proc. International Conference on Accelerator and Large Experimental Physics Control Systems (ICALEPCS) (1997).
   Y.Suzuki *et al.*, Proc. of ICALEPCS (1999).
5. T.Maruyama, Ph.D Thesis, Tohoku University (2000).
6. A.Suzuki *et al.*, NIM **A453**, 165 (2000).
7. T.Ishii et al., NIM **A482**, 244 (2002)
8. S.H.Ahn et al., Phys. Lett. **B511**, 178 (2001)
9. H.G.Berns and R.J.Wilkes, IEEE NS **47**, 340 (2000).
10. T.Affolder et al., Phys. Rev. **D61**, 072005 (2000).
11. J.Swain and L.Taylor, NIM **A411**, 153 (1998).
    W.T.Giele and S.Keller, Phys. Rev. **D58**, 094023 (1998).

ELSEVIER

Proceedings of ICHEP 2002, pp. 24–28
*S. Bentvelsen, P. de Jong, J. Koch and E. Laenen (Editors)*

31[st] INTERNATIONAL CONFERENCE ON HIGH ENERGY PHYSICS AMSTERDAM

# Current Status of Neutrino Masses and Mixings

C. Giunti[a]

[a]INFN, Sezione di Torino, and Dipartimento di Fisica Teorica, Università di Torino, Via P. Giuria 1, I–10125 Torino, Italy

The evidences in favor of solar and atmospheric neutrino oscillations are briefly reviewed and shown to be gracefully accommodated in the framework of three-neutrino mixing with bilarge mixing. Talk presented at the 31$^{st}$ International Conference on High Energy Physics "ICHEP02", 24–31 July 2002, Amsterdam.

## 1. Introduction

Neutrino physicists are living a very interesting time, in which several fundamental facts,

(F1) there are three active light neutrinos ($\nu_e$, $\nu_\mu$, $\nu_\tau$),

(F2) the mass of light neutrinos is smaller than a few eV,

(F3) there are solar $\nu_e \to \nu_\mu, \nu_\tau$ transitions,

(F4) there are atmospheric $\nu_\mu \to \nu_\tau$ transitions,

(F5) atmospheric and long-baseline $\nu_e$ oscillations are suppressed,

(F6) short-baseline oscillations of $\nu_e$ and $\nu_\mu$ are suppressed,

have been experimentally established with a reasonable degree of reliability and several fundamental mysteries,

(M1) which is the absolute scale of neutrino masses?

(M2) are neutrinos Dirac or Majorana particles?

(M3) which is the number of massive neutrinos?

(M4) are there transitions of active neutrinos into sterile states?

(M5) is there CP violation in the lepton sector?

are under investigation. The known facts about neutrino masses and mixings are important for the formulation of theoretical frameworks and of experimental strategies for the future clarification of the mysteries. For an extensive and updated list of references see Ref. [1].

In this review, I will briefly summarize the experimental evidences in favor of the above-mentioned facts ((F2))–((F5)) and I will show that they can be accommodated elegantly in the framework of three-neutrino mixing. I will not discuss the already well known experimental proof that there are only three active light neutrinos (fact (F1) above, see Ref. [2]). For lack of space I will not review the abundant experimental evidences of the fact (F6) above (see Refs. [1–3]) and I will not discuss the status of the four-neutrino mixing hypothesis (see Ref. [1,3]), which was proposed in order to accommodate also the claim of short-baseline $\bar{\nu}_\mu \to \bar{\nu}_e$ transitions in the LSND experiment (see Ref. [4]). This hypothesis is now disfavored by data (see Ref. [5]) and my opinion is that it should be put under hibernation waiting for its final elimination or resurrection by new data (especially crucial will be the check of the LSND result in the MiniBooNE experiment [6]).

## 2. Solar Neutrinos

All solar neutrino experiments (Homestake [7], SAGE [8], GALLEX [9], GNO [10], Super-Kamiokande [11], SNO [12]) measured a suppression of the observed rate with respect to the one predicted by the Standard Solar Model (SSM) [13]. The only exception is the neutral-current reaction $\nu + d \to p + n + \nu$ in the SNO experiment [12], which is equally sensitive to all active


doi:10.1016/S0920-5632(02)02076-5

neutrino flavors, whose rate is in good agreement with the SSM. This measurement has marked the triumph of the SNO experiment, which measured also a suppression of a factor $0.35 \pm 0.02$ of the flux of solar $\nu_e$'s with respect to the SSM through the charged-current reaction $\nu_e + d \to p + p + e^-$ [12]. The discrepancy between the neutral-current and charged-current suppressions is a clear and impressive model-independent evidence of transitions of solar $\nu_e$'s into active $\nu_\mu$ or $\nu_\tau$ (fact (F3) above). The existence of solar $\nu_e \to \nu_\mu, \nu_\tau$ transitions is also supported by the comparison of the different suppressions of the rate of the SNO charged-current reaction [12] and the rate of the elastic scattering reaction $\nu + e^- \to \nu + e^-$, which is mainly sensitive to $\nu_e$ but has also a small sensitivity to $\nu_\mu$ and $\nu_\tau$ ($\sigma^{\mathrm{ES}}_{\nu_\mu,\nu_\tau} \simeq \sigma^{\mathrm{ES}}_{\nu_e}/6$), measured in Super-Kamiokande [11] and SNO [12].

Let me emphasize also the importance of the Homestake experiment [7], in which the solar neutrino problem was discovered, and the Gallium SAGE [8], GALLEX [9] and GNO [10] experiments, which are sensitive to the fundamental flux of *pp* neutrinos (see Ref. [13]). Moreover, the results of all neutrino experiments are necessary in order to get information on the neutrino mixing parameters.

The global analysis of all solar neutrino data in terms of $\nu_e \to \nu_\mu, \nu_\tau$ performed in Ref. [14] yielded

$$\begin{aligned} &0.24 < \tan^2\vartheta_{\mathrm{S}} < 0.89\,, \\ &2.3 \times 10^{-5} < \Delta m^2_{\mathrm{S}}/\mathrm{eV}^2 < 3.7 \times 10^{-4}\,, \end{aligned} \tag{1}$$

at 99.73% C.L. ($3\sigma$), where $\Delta m^2_{\mathrm{S}}$ is the relevant neutrino squared-mass difference and $\vartheta_{\mathrm{S}}$ is the effective mixing angle in two-generation analyses of solar neutrino data. In Eq. (1) I reported only the boundaries of the so-called LMA region (see Ref. [3]), which is currently favored, because it is much larger than other regions (a LOW region and three VAC regions appear at 99% C.L.) and it contains the minimum of the $\chi^2$ (best fit) at

$$\tan^2\vartheta_{\mathrm{S}} \simeq 0.42\,, \; \Delta m^2_{\mathrm{S}} \simeq 5.0 \times 10^{-5}\,\mathrm{eV}^2\,. \tag{2}$$

The limits in Eq. (1) show that the mixing relevant for solar neutrino oscillations is large. However, maximal mixing seems strongly disfavored from the analysis of solar neutrino data in Ref. [14]. This conclusion is supported by the results of some other authors [15,16,18,11], whereas the authors of Refs. [17,19–21] found slightly larger allowed regions, with maximal mixing marginally allowed. Therefore, it is not clear at present if maximal mixing in solar neutrino oscillations is excluded or not.

In particular, the authors of Refs. [20,21] found a larger LMA region and several allowed LOW+QVO and VAC regions, where QVO indicates the region at $\Delta m^2_{\mathrm{S}} \sim 10^{-8}\,\mathrm{eV}^2$ where both matter effect and vacuum oscillations are important.

The authors of Ref. [19] found that taking into account only the most recent Gallium data (taken in the period 1998-2001, with improved systematic uncertainty with respect to the earlier data) the LOW region is less disfavored with respect to the LMA region.

In any case, there are good reasons to believe that the SNO experiment has solved the solar neutrino problem proving its neutrino physics origin and the presence of $\nu_e \to \nu_\mu, \nu_\tau$ transitions, and that the current data indicate a large mixing. The final confirmation of the true allowed region will hopefully come from the results of the KamLAND [22] or BOREXINO [23] experiments.

## 3. Atmospheric Neutrinos

In 1998 the Super-Kamiokande experiment found model independent evidence of disappearance of atmospheric $\nu_\mu$'s [24], which is supported by the results of the K2K long-baseline experiment [25], of the Soudan 2 [26] and MACRO [27] atmospheric neutrino experiment. The atmospheric neutrino data of the Super-Kamiokande experiment are well fitted by $\nu_\mu \to \nu_\tau$ transitions with large mixing (fact (F4) above):

$$\begin{aligned} &1.2 \times 10^{-3} < \Delta m^2_{\mathrm{A}}/\mathrm{eV}^2 < 5.0 \times 10^{-3}\,, \\ &\sin^2 2\vartheta_{\mathrm{A}} > 0.84\,, \end{aligned} \tag{3}$$

at 99% C.L. [28], where $\Delta m^2_{\mathrm{A}}$ is the relevant neutrino squared-mass difference and $\vartheta_{\mathrm{A}}$ is the effective mixing angle in two-generation analyses of atmospheric neutrino data. The best fit is at

$$\sin^2 2\vartheta_{\mathrm{A}} = 1.0\,, \; \Delta m^2_{\mathrm{A}} \simeq 2.5 \times 10^{-3}\,\mathrm{eV}^2\,. \tag{4}$$

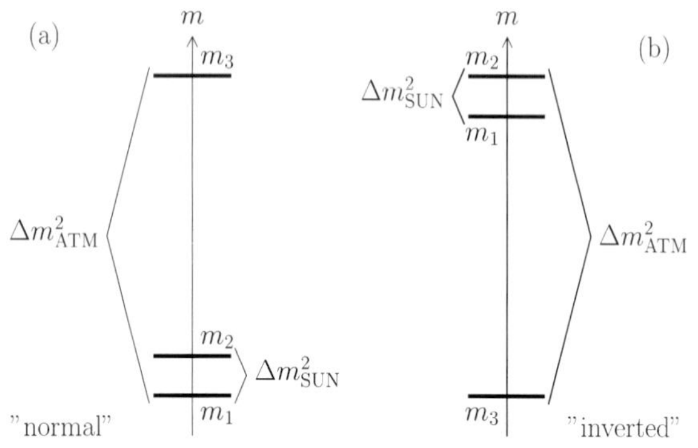


Figure 1. The two possible types of three-neutrino schemes.

Furthermore, other oscillation channels ($\nu_\mu \leftrightarrows \nu_e$ or $\nu_\mu \to \nu_s$, where $\nu_s$ is a sterile neutrino) as well as other mechanisms (as $\nu$ decay) are disfavored.

In the future, the K2K experiment will improve its results taking more data and the MINOS [29] long-baseline experiment will measure with better accuracy the mixing parameters in Eq. (3). The long-baseline experiment ICARUS [30] and OPERA [31] will be aimed at direct observation of $\nu_\mu \to \nu_\tau$ transitions.

## 4. Three-Neutrino Mixing

Solar and atmospheric neutrino data can be well fitted in the framework of three-neutrino mixing (see Ref. [3]) with the hierarchy of squared-mass differences

$$\Delta m^2_{\mathrm{S}} = \Delta m^2_{21} \ll |\Delta m^2_{31}| \simeq \Delta m^2_{\mathrm{A}} \,, \tag{5}$$

where $\Delta m^2_{kj} \equiv m^2_k - m^2_j$ for $k, j = 1, 2, 3$, and $m_1, m_2, m_3$ are the three neutrino masses.

Figure 1 shows the two possible types of three-neutrino schemes: a "normal" type (a) and an "inverted" type (b). The absolute scale of these schemes (mystery (M1) above) is not fixed by neutrino oscillation experiments, which can measure only squared-mass differences. The normal three-neutrino scheme allows the natural mass hierarchy

$$m_1 \ll m_2 \ll m_3 \,, \tag{6}$$

whereas the inverted scheme allows the so-called "inverted hierarchy" $m_3 \ll m_1 \simeq m_2$.

The absolute scale of neutrino masses is bounded from above by the results of Tritium $\beta$-decay experiments, which found [32]

$$m_{\nu_e} < 2.2\,\mathrm{eV} \quad (95\%\ \mathrm{C.L.}) \,. \tag{7}$$

Since in the case of neutrino mixing $\nu_e$ does not have a definite mass, but is a superposition of massive neutrinos, the limit in Eq. (7) is a bound on the masses of the main massive neutrino components of $\nu_e$. In the three-neutrino schemes in Fig. 1 all the neutrino masses are constrained by the limit in Eq. (7). An independent confirmation of this limit comes from the study of the formation of large-scale structures in the Universe [33,34]. In the future the KATRIN experiment [35] will probe the effective electron neutrino mass in Tritium decay down to about 0.3 eV. Unfortunately, in the case of the natural scheme in Fig. 1a with the neutrino mass hierarchy (6), Eqs. (3) and (5) imply that $m_3 \lesssim 7 \times 10^{-2}\,\mathrm{eV}$, which is beyond the reach of the KATRIN experiment.

Let us discuss now the current information on the neutrino mixing matrix $U$. Since the solar neutrino transitions of $\nu_e$ in $\nu_\mu$ or $\nu_\tau$ cannot be experimentally distinguished because the energy is well below the threshold for charged-current reactions with production of $\mu^-$ or $\tau^-$, solar neutrino oscillations depend only on the elements $U_{e1}$, $U_{e2}$, $U_{e3}$ of the mixing matrix. On the other hand, the hierarchy (5) of squared-mass differences in the two possible three-neutrino schemes in Fig. 1 implies that atmospheric (and long-baseline) neutrino oscillations depend only on the elements $U_{e3}$, $U_{\mu 3}$, $U_{\tau 3}$ (see Ref. [3]). Since $\Delta m^2_{\mathrm{S}} = \Delta m^2_{21}$ and $\Delta m^2_{\mathrm{A}} \simeq |\Delta m^2_{31}|$ are independent, the only quantity that connects solar and atmospheric neutrino oscillations is $U_{e3}$. Therefore, any information on the value of $U_{e3}$ is of crucial importance.

From the results of the CHOOZ long-baseline reactor neutrino experiment [36], it is known that the element $U_{e3}$ of the three-generation neutrino mixing matrix is small [21]:

$$|U_{e3}|^2 < 5 \times 10^{-2} \quad (99.73\%\ \mathrm{C.L.}) \,. \tag{8}$$

The results of the CHOOZ experiment have been confirmed by the Palo Verde experiment [37], and by the absence of $\nu_e$ transitions in the Super-Kamiokande atmospheric neutrino data [28].

An important consequence of the smallness of $U_{e3}$ is the practical decoupling of solar and atmospheric neutrino oscillations [38], which can be analyzed in terms of two-neutrino oscillations with the effective mixing angles $\vartheta_S$ and $\vartheta_A$ given by

$$\sin^2\vartheta_S = \frac{|U_{e2}|^2}{1-|U_{e3}|^2}\,, \quad \sin^2\vartheta_A = \frac{|U_{\mu 3}|^2}{1-|U_{e3}|^2}\,. \tag{9}$$

Neglecting a possible small value of $|U_{e3}|$, the mixing matrix can be written as

$$U \simeq \begin{pmatrix} c_{\vartheta_S} & s_{\vartheta_S} & 0 \\ -s_{\vartheta_S}c_{\vartheta_A} & c_{\vartheta_S}c_{\vartheta_A} & s_{\vartheta_A} \\ s_{\vartheta_S}s_{\vartheta_A} & -c_{\vartheta_S}s_{\vartheta_A} & c_{\vartheta_A} \end{pmatrix}, \tag{10}$$

with $c_\vartheta \equiv \cos\vartheta$ and $s_\vartheta \equiv \sin\vartheta$. In this case, solar neutrino transitions occur from $\nu_e = c_{\vartheta_S}\nu_1 + s_{\vartheta_S}\nu_2$ to the orthogonal state $-s_{\vartheta_S}\nu_1 + c_{\vartheta_S}\nu_2 = c_{\vartheta_A}\nu_\mu - s_{\vartheta_A}\nu_\tau$. Hence, the fractions of $\nu_e \to \nu_\mu$ and $\nu_e \to \nu_\tau$ transitions are determined by the value of the atmospheric effective mixing angle $\vartheta_A$.

From the experimental best-fit value $\vartheta_A = \pi/4$ in Eq. (4), we see that solar $\nu_e$ transform approximately into $(\nu_\mu - \nu_\tau)/\sqrt{2}$, which is an equal superposition of $\nu_\mu$ and $\nu_\tau$. Since the suppression factor of solar $\nu_e$'s measured in the charged-current SNO reaction is about 1/3 [12], the flux of $\nu_e$, $\nu_\mu$ and $\nu_\tau$ on Earth is approximately equal. The mixing matrix is approximately given by

$$U \approx \begin{pmatrix} c_{\vartheta_S} & s_{\vartheta_S} & 0 \\ -s_{\vartheta_S}/\sqrt{2} & c_{\vartheta_S}/\sqrt{2} & 1/\sqrt{2} \\ s_{\vartheta_S}/\sqrt{2} & -c_{\vartheta_S}/\sqrt{2} & 1/\sqrt{2} \end{pmatrix}. \tag{11}$$

Let us now consider the best fit value in Eq. (2) of the effective solar mixing angle, which implies that $\vartheta_S \simeq \pi/6$, leading to the approximate bilarge mixing matrix

$$U \approx \begin{pmatrix} \sqrt{3}/2 & 1/2 & 0 \\ -1/2\sqrt{2} & \sqrt{3}/2\sqrt{2} & 1/\sqrt{2} \\ 1/2\sqrt{2} & -\sqrt{3}/2\sqrt{2} & 1/\sqrt{2} \end{pmatrix}. \tag{12}$$

However, it is widely hoped that $|U_{e3}|^2$ is not much smaller than the current upper limit in Eq. (8), because a non-vanishing value of $U_{e3}$ is essential for the possibility of measuring CP violation in neutrino oscillations (mystery (M5) above, see Ref. [39]) and for the measurement of the sign of $\Delta m^2_{31}$, which distinguishes the normal and inverted schemes in Fig. 1. The sign of $\Delta m^2_{31}$ can be measured in long-baseline neutrino experiments through matter effects in the Earth, which need the participation to the oscillations of $\nu_e$ through $U_{e3}$ ($\nu_\mu$ and $\nu_\tau$ have the same neutral-current interaction with matter, whereas $\nu_e$ has also charged-current interactions; neutrino oscillations are affected only by the difference of interactions of different flavors).

The matrix in Eq. (12) can be considered as the approximate current best-fit mixing matrix. The limits on $\vartheta_S$, $\vartheta_A$ and $U_{e3}$ in Eqs. (1), (3) and (8), respectively, allow to derive the following allowed intervals for the absolute values of the elements of the mixing matrix (the intervals are correlated, because of unitarity):

$$|U| \approx \begin{pmatrix} 0.71-0.90 & 0.43-0.69 & 0.00-0.22 \\ 0.24-0.66 & 0.40-0.81 & 0.53-0.84 \\ 0.24-0.66 & 0.40-0.81 & 0.53-0.84 \end{pmatrix}. \tag{13}$$

The best known way to investigate the Majorana nature of neutrinos (mystery (M2) above) is the search for neutrinoless double-$\beta$ decay, whose amplitude is proportional to the effective Majorana mass

$$|\langle m\rangle| = |\sum_k U^2_{ek}\, m_k|\,, \tag{14}$$

for which the current experimental upper limit is between about 0.3 and 1 eV, taking into account a theoretical uncertainty of about a factor 3 in the calculation of the nuclear matrix element (see Ref. [40]). In general, since the elements $U_{ek}$ of the mixing matrix are complex, cancellations among the contributions of the three massive neutrinos are possible. The current allowed interval for $|\langle m\rangle|$ in the case of the two three-neutrino mixing schemes in Fig. 1 can be found in Ref. [41].

Here let me consider the case of the natural scheme in Fig. 1a with the neutrino mass hierarchy (6). In this case, the limits in Eqs. (1) and

(3) imply that the contribution of $m_1$ to $|\langle m\rangle|$ is negligible and the absolute values of the contributions of $m_2$ and $m_3$ are limited by

$$1.5 \times 10^{-3}\,\text{eV} \lesssim |U_{e2}^2\, m_2| \lesssim 9.5 \times 10^{-3}\,\text{eV}\,, \quad (15)$$

$$|U_{e3}^2\, m_3| \lesssim 3.5 \times 10^{-3}\,\text{eV}\,. \quad (16)$$

Therefore, it is possible that the dominant contribution to $|\langle m\rangle|$ comes from $m_2$, and not from the largest mass $m_3$. If $|U_{e3}^2\, m_3| \ll |U_{e2}^2\, m_2|$, there are no cancellations in Eq. (14) [42] and the effective Majorana mass is expected in the interval in Eq. (15). Unfortunately, such values of $|\langle m\rangle|$ are beyond the reach of proposed experiments (see Ref. [40]).

## 5. Conclusions

The results of all neutrino experiments, except LSND, are in agreement with the hypothesis of three-neutrino mixing. If future data will confirm this scenario the most important phenomenological task will be to refine the measurement of the neutrino mixing parameters, especially the measurement of $U_{e3}$, whose value is related to the possibility of distinguishing the two schemes in Fig. 1 and of measuring CP violation in neutrino oscillations. The measurements of the absolute scale of neutrino masses and of the effective Majorana mass in neutrinoless double-$\beta$ decay seem to be more difficult tasks (beyond the reach of existing projects if the neutrino masses satisfy the natural hierarchy (6)).

## REFERENCES


1. C. Giunti and M. Laveder, Neutrino Unbound: http://www.to.infn.it/~giunti/NU.
2. D. Groom et al., Eur. Phys. J. C15 (2000) 1.
3. S.M. Bilenky et al., Prog. Part. Nucl. Phys. 43 (1999) 1.
4. E. Church et al., Phys. Rev. D66 (2002), 013001.
5. M. Maltoni et al., hep-ph/0207157.
6. A. Bazarko, these proceedings.
7. B.T. Cleveland et al., Astrophys. J. 496 (1998) 505.
8. J.N. Abdurashitov et al., astro-ph/0204245.
9. W. Hampel et al., Phys. Lett. B447 (1999) 127.
10. M. Altmann et al., Phys. Lett. B490 (2000) 16.
11. S. Fukuda et al., Phys. Lett. B539 (2002) 179.
12. Q.R. Ahmad et al., Phys. Rev. Lett. 89 (2002) 011301.
13. J. Bahcall et al., Astrophys. J. 555 (2001) 990.
14. J. Bahcall et al., JHEP 0207 (2002) 054.
15. Q.R. Ahmad et al., Phys. Rev. Lett. 89 (2002) 011302.
16. V. Barger et al., Phys. Lett. B537 (2002) 179.
17. A. Bandyopadhyay et al., Phys. Lett. B540 (2002) 14.
18. P.C. de Holanda and A.Y. Smirnov, hep-ph/0205241.
19. A. Strumia et al., Phys. Lett. B541 (2002) 327.
20. G.L. Fogli et al., hep-ph/0206162.
21. G.L. Fogli et al., hep-ph/0208026.
22. T. Mitsui, these proceedings.
23. BOREXINO, `http://almime.mi.infn.it`.
24. Y. Fukuda et al., Phys. Rev. Lett. 81 (1998) 1562.
25. K. Nishikawa, Talk at Neutrino 2002.
26. W. Allison et al., Phys. Lett. B449 (1999) 137.
27. M. Ambrosio et al., Phys. Lett. B478 (2000) 5.
28. M. Shiozawa, Talk at Neutrino 2002.
29. MINOS, `http://www-numi.fnal.gov`.
30. ICARUS, `http://pcnometh4.cern.ch`.
31. OPERA, `http://operaweb.web.cern.ch`.
32. C. Weinheimer, Talk at Neutrino 2002.
33. O. Elgaroy et al., Phys. Rev. Lett. 89 (2002) 061301.
34. S. Hannestad, astro-ph/0205223.
35. A. Osipowicz et al., hep-ex/0109033.
36. M. Apollonio et al., Phys. Lett. B466 (1999) 415.
37. F. Boehm et al., Phys. Rev. D64 (2001) 112001.
38. S.M. Bilenky and C. Giunti, Phys. Lett. B444 (1998) 379.
39. C. Albright et al., hep-ex/0008064.
40. S.R. Elliott and P. Vogel, hep-ph/0202264.
41. S. Pascoli and S.T. Petcov, hep-ph/0205022.
42. C. Giunti, Phys. Rev. D61 (2000) 036002.

ELSEVIER

Proceedings of ICHEP 2002, pp. 29–32
*S. Bentvelsen, P. de Jong, J. Koch and E. Laenen (Editors)*

31st INTERNATIONAL CONFERENCE ON HIGH ENERGY PHYSICS AMSTERDAM

# JHF-Kamioka neutrino oscillation experiment

Yoshitaka Itow [a]

[a]Institute for Cosmic Ray Research, Universtiy of Tokyo,
Kamioka, Yoshiki-gun, Gifu, 506-1205, Japan

A new high intensity proton facility, Japan Hadron Facility (JHF), started its construction from 2001. The long baseline neutrino oscillation experiment using Off-Axis $\nu_\mu$-beam from 50 GeV PS at JHF to Super-Kamiokande will open the new field of high precision study of neutrino mixing parameters. The overview of the experiment is described.

## 1. Introduction

The mass and mixing of neutrinos is very fundamental problem in particle physics. It is also very interesting for the contrast of their mixing structure to the one for quark sector. Neutrino oscillations is an unique probe to measure neutrino masses and their mixing. For example, in the case of neutrino oscillation between two flavors, the probability of oscillation, $\nu_\alpha \to \nu_\beta$ is given by;

$$P(\nu_\alpha \to \nu_\beta) = \sin^2 2\theta \sin^2(1.27 \frac{L(km)}{E_\nu(GeV)} \Delta m^2(eV^2)),$$

where $\theta$ is the mixing angle between two neutrinos, $L$ is the flight length of neutrino, $E_\nu$ is the energy of neutrino, $\Delta m^2$ is the difference of mass squares in unit of $\mathrm{eV}^2$.

In 1998, the first evidence of neutrino oscillations was found in atmospheric neutrinos at Super-Kamiokande [1–4]. This is the discovery of $\nu_\mu$-$\nu_\tau$ oscillations with $\Delta m^2$=1.2∼3.8×$10^{-3}$ $\mathrm{eV}^2$ and with $\sin^2 2\theta > 0.93$ [5]. This results was also confirmed by the K2K experiment, the first long baseline neutrino oscillation experiment with 100km scale [6,7]. In 2001, solar neutrino deficit was proven as a result of $\nu_e$-$\nu_\mu$,$\nu_\tau$ oscillations by combining Super-Kamiokande [8] and SNO results [9]. The recent best fit of all the existing solar neutrino measurements suggests $\nu_e$-$\nu_\mu$,$\nu_\tau$ oscillations with $\Delta m^2$=0.2 $\sim$ 2.5 $\times$ $10^{-5}$ $\mathrm{eV}^2$ and with $\sin^2 2\theta$ = 0.71∼0.93 [8].

Next step of neutrino oscillation experiments will be high precision measurements of oscillation parameters in 3-flavor mixing scheme. For this purpose, large statistics provided by a high intensity neutrino beam from a strong proton driver is essentially needed. A new high intensity proton facility, Japan Hadron Facility (JHF), gives a nice opportunity to do it. The construction of main facility was approved in Dec,2000 and started from 2001. We have proposed a high precision neutrino oscillation experiment from JHF to Super-Kamiokande, namely the JHF-Kamioka neutrino experiment [10]. This experiment is expected to start in 2007 at the same time of completion of the JHF facility. Here overview of the experiment is briefly described.

## 2. Overview of the experimental facility

The site for JHF is at Japan Atomic Energy Research Institute (JAERI) in Tokai about 60km north-east of KEK. The accelerator complex of JHF consists of 400 MeV LINAC, 3 GeV booster and 50 GeV PS. Total power is 0.75 MW, corresponding to 3.3×$10^{14}$ ppp in 3.94 sec cycle. Here typical 1 year run corresponds 130 days run providing 1×$10^{21}$ POT. Proton beam bunches (8 bunches in ∼5.2 $\mu$sec spill) are extracted by a kicker magnet and bend by almost 90-degree toward Kamioka direction then transported to the target station. This 90-degree bending section will use large aperture super-conducting magnets which are newly developed. Finally produced $\pi$'s are focused by conventional double horn magnets then decay in the decay volume of 130m length.


doi:10.1016/S0920-5632(02)02079-0

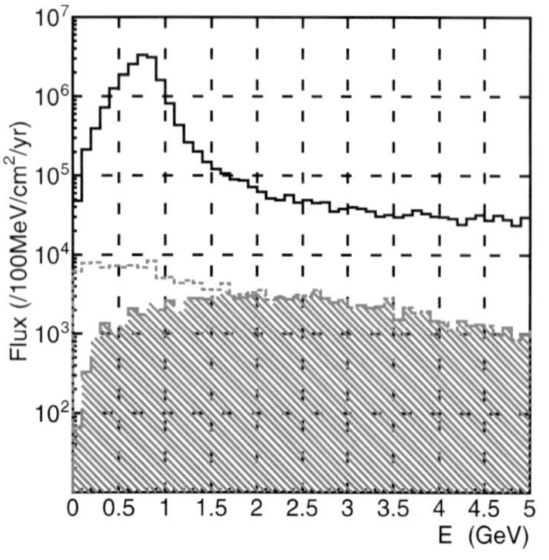


Figure 1. *Beam energy spectrum for 2 degree Off Axis Beam in JHF-Kamioka experiment. The solid (black) histogram shows* $\nu_\mu$ *component. The dashed (red) histogram shows* $\nu_e$ *component in the beam. Hatched region corresponds to* $\nu_e$*'s from K-decays.*

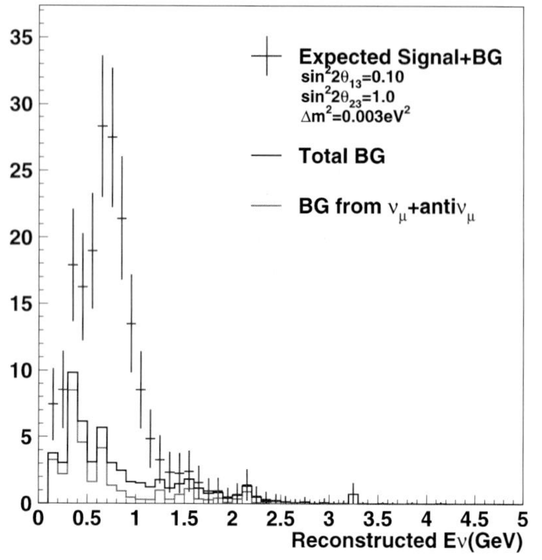


Figure 2. *The reconstructed* $E_\nu$ *of 1-R e-like events as* $\nu_e$ *candidates at Super-Kamiokande with JHF 2 degree OAB beam in 5 years run. Crosses shows expected signal+background with* $\sin^2 2\theta_{13}=0.10$, $\sin^2 2\theta_{23}=1.0$ *and* $\Delta m^2=0.003$ *eV*$^2$. *The solid (black) histogram shows total background, mainly beam* $\nu_e$ *component and miss-identified* $\pi^0$. *The lower (red) lines show background from* $\pi^0$ *only.*

Contrast to the K2K experiment, center of neutrino beam direction is intentionally set to off-direction to Super-Kamiokande. This technique is known as off-axis neutrino beam (OAB) [11] to produce intense low energy neutrino beam with narrow energy spread. It is crucial for clear measurement of neutrino oscillations to tune neutrino beam energy at oscillation maximum which is about 0.7 GeV for 290 km distance between Tokai and Kamioka, Since OAB utilizes neutrinos from off-axis decay of secondary $\pi$'s, the mean energy of neutrinos can be controlled by the off-axis angle. For example, the off-axis of 2 degree gives 0.7 GeV in mean energy as shown in Fig 1. Contamination of $\nu_e$ in the beam is about 0.2 % around peak region of energy spectrum. About 4,500 $\nu_\mu$ interactions and 3,000 $\nu_\mu$ CC interactions are expected in 22.5kt fiducial volume of Super-Kamiokande for 1 year run. Compare to K2K, we can expect almost 100 times larger statistics. Higher energy neutrinos which is irrelevant to neutrino oscillations are less in OAB so that possible contamination or smearing of oscillation signal can be suppressed. Two near detector systems at 280 m and 2 km are considered. The detector at 280 m will be mainly used to monitor the profile and spectrum of neutrino beam at on-axis and off-axis. The detector at 2 km will be a complex of a water Cherenkov detector and a fine-grained detector which cover the off-axis direction. They will be used for absolute flux normalization, detail energy spectrum measurements and background estimations. Since energy spectrum shape at 2 km and at Super-Kamiokande is almost identical, near site measurement at 2 km can be used with small systematic error for oscillation analysis. The far detector is Super-Kamiokande, the 50 kt gigantic underground water Cherenkov detector. After recovery from the PMT explosion accident at Nov 2001,

Super-Kamiokande has started its operation with half of designed PMT coverage after Oct 2002. It is expected that the original PMT coverage will be restored by the start of the JHF-Kamioka experiment.

## 3. Discovery of $\theta_{13}$ by $\nu_e$ appearance

Major physics motivation of the JHF-Kamioka experiment should be the discovery of $\theta_{13}$ by measuring appearance of $\nu_e$. This will be the first evidence of 3-flavor neutrino mixing. For $\Delta m^2_{23} \gg \Delta m^2_{12}$, probability of 3-flavor neutrino oscillation can be approximately written as;

$$P(\nu_\mu \to \nu_e) = \sin^2 2\theta_{13} \cdot \sin^2 \theta_{23} \cdot \xi \quad (1)$$

$$P(\nu_\mu \to \nu_\tau) = \cos^4 \theta_{13} \cdot \sin^2 2\theta_{23} \cdot \xi \quad (2)$$

where $\xi = \sin^2(1.27 \cdot \Delta m^2_{23} \cdot L/E)$. In Super-Kamiokande, $\nu_e$'s are detected by charged current quasi elastic interactions ( CC QE ). They are observed as events with single fuzzy Cherenkov ring identified as EM-showering, namely 1-ring $e$-like ring events. By using Cherenkov ring pattern and the angle of Cherenkov cone, Super-Kamiokande has achieved excellent particle-ID capability for the 1 GeV energy region. With 5 years run of 2 degree OAB to Super-Kamiokande, total 115 $\nu_e$ events are expected from $\nu_\mu$ oscillations with $\sin^2 2\theta_{13}$=0.1 as the current limit given by CHOOZ [12]. Figure 2 shows reconstructed $E_\nu$ distribution for $\nu_e$ candidates assuming 2-body reactions. Expected background are 25 events, about half of which come from original $\nu_e$ component in a beam. Rest of background come from $\pi^0$ events identified as single ring in the case of $\pi^0$ asymmetric decay or 2 $\gamma$'s emitted in very forward directions. Addition to standard Super-Kamiokande analysis, more tight $\pi^0$ ring-finding method is applied to improve $\pi^0$ rejection efficiency. Also events followed by decay electrons which are produced by invisible charged $\mu$ or $\pi$ decays are rejected in order to reduce background events due to $\pi$ productions. About 90 % of background $\pi^0$ events come from neutral current (NC) resonance production of $\nu_\mu$. Background from NC deep inelastic $\pi^0$ productions is small because of less high energy tail of OAB beam. These background events should be measured by near detectors and subtracted with good accuracy. Compare to the two background category above, background events due to miss-identification of $\mu$ Cherenkov ring are small thanks to excellent performance of particle-ID method in Super-Kamiokande. Figure 3 shows sensitivity at 90 % C.L. and $3\sigma$ discovery of $\theta_{13}$ as $\sin^2 2\theta_{\mu e} \sim \frac{1}{2}\sin^2 2\theta_{13}$ with 5 years run of 2 degree OAB. Assuming 10% systematic uncertainty for the background subtraction, the sensitivity of $\sin^2 2\theta_{13}$ will be >0.006 at 90 % C.L. The JHF-Kamioka experiment will improve the results of existing CHOOZ experiment by more than one order of magnitude.

## 4. Precision study of $\nu_\mu$ disappearance

High precision measurement of $\nu_\mu$ disappearance will be also available by the OAB beam. In the analysis of disappearance, energy spectrum of $\nu_\mu$ at Super-Kamiokande is reconstructed for single ring $\mu$-like events via $\nu_\mu$ CC QE interactions by assuming 2-body kinematics. There is contamination to single ring $\mu$-like events from CC non-QE interactions associated with invisible pion productions below Cherenkov threshold. For such a event, neutrino energy is reconstructed lower than true value. Careful subtraction of CC non-QE events is needed for correct determination of oscillation parameters. Since CC QE interactions is dominant fraction of CC interactions at the mean energy of OAB beam, systematic error due to the subtraction is small. Considering 10 % systematic uncertainty for flux normalization, 4 % systematic error for energy scale and 20 % error for the subtraction of CC non-QE interaction events, the total error is estimated to be less than 1 % for $\sin^2 2\theta_{23}$ and less than $1\times10^{-4}\text{eV}^2$ for $\Delta m^2_{23}$ for 5 years run of OAB 2 degree.

## 5. Conclusion

In the JHF-Kamioka neutrino experiment, with the intense low energy $\nu_\mu$-beam from JHF 50GeV PS and with the gigantic underground water Cherenkov detector Super-Kamiokande, neutrino oscillation physics will get into a new era of precision study. The OAB $\nu_\mu$ beam will be tuned

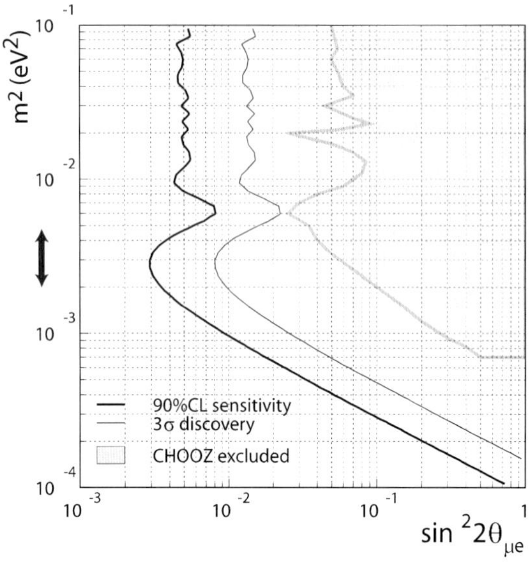


Figure 3. *Sensitivity at 90% C.L. and 3$\sigma$ discovery for* $\sin^2 2\theta_{\mu e}$ ($\sim \frac{1}{2}\sin^2 2\theta_{13}$) *by* $\nu_e$ *appearance measurement. Also shown 90 % C.L. excluded region by CHOOZ. The Up-down arrow along the vertical axis shows 90% C.L. alowed region of* $\Delta m^2$ *from atmospheric neutrinos.*

at oscillation maximum, $\sim$0.7GeV/c with narrow energy spread. This configuration enhances oscillation signals and suppress any systematic uncertainty. Also this neutrino energy is good for water Cherenkov technique because of large fraction of CC QE interactions. The first evidence of 3-flavor mixing of neutrinos can be provided by observation of non-zero $\theta_{13}$ vis $\nu_e$ appearance measurement. If non-zero $\theta_{13}$ is observed in the JHF-Kamioka neutrino project, it will be the first evidence of 3-flavor mixing. Moreover, it will be the first step of CP-violation experiments, since CP-violation phase only appears in appearance channel. The presented experiment has future potential of CP-violation measurements with possible future super $\nu$-beam from 4MW-upgraded JHF and with more gigantic water Cherenkov detector in mega-ton size, Hyper-Kamiokande. Also precision measurement of $\theta_{23}$ or $\Delta m^2$ would provide a good test of theoretical consideration why quark and lepton mixing are so different. These data will be indispensable for understanding of mixing nature.

## REFERENCES

1. Super-Kamiokande Collaboration, Y.Fukuda et al., Phys. Lett. **B433**, 9(1998).
2. Super-Kamiokande Collaboration, Y.Fukuda et al., Phys. Lett. **B436**, 33 (1998).
3. Super-Kamiokande Collaboration, Y.Fukuda et al., Phys. Rev. Lett. **82**, 2644(1998).
4. Super-Kamiokande Collaboration, Y.Fukuda et al., Phys. Rev. Lett. **81**, 1562(1998).
5. M.Shiozawa, Talk in XXth International Conference on Neutrino Physics and Astrophysics, May 25-30,2002,Munich,Germany.
6. K2K Collaboration, S.H.Ahn et al., Phys. Lett. **B511**, 178(2001).
7. K.Nishikawa, Talk in XXth International Conference on Neutrino Physics and Astrophysics, May 25-30,2002,Munich,Germany.
8. Super-Kamiokande Collaboration, S.Fukuda et al., Phys. Lett. **B539**, 179(2002).
9. SNO Collaboration, Q.R.Ahmad et al., Phys. Rev. Lett. **87** 071301(2001).
10. JHF-Kamioka neutrino working group, hep-ex/0106019.
11. D.Beavis, A.Carroll, I.Chiang, et al., Proposal of BNL AGS E-889(1995).
12. CHOOZ Collaboration, M.Apollonio et al., Phys. Lett. **B466**, 415(1999).

ELSEVIER

Proceedings of ICHEP 2002, pp. 33–36
*S. Bentvelsen, P. de Jong, J. Koch and E. Laenen (Editors)*

31st INTERNATIONAL CONFERENCE ON HIGH ENERGY PHYSICS AMSTERDAM

# Status of MiniBooNE

Andrew O. Bazarko[a] for the BooNE Collaboration

[a]Department of Physics, Princeton University, Princeton, New Jersey 08544-0708, USA

MiniBooNE is a neutrino oscillation experiment now running at Fermilab. The experiment will search for $\nu_\mu \rightarrow \nu_e$ oscillations in order to make a conclusive statement about the yet-unconfirmed evidence for oscillations presented by the LSND experiment. Preparations for the start of running were completed over the summer of 2002, and MiniBooNE observed its first neutrino events in late August.

## 1. LSND EVIDENCE

The Liquid Scintillator Neutrino Detector ran at the Los Alamos Neutron Science Center over the years 1993 to 1998. Using the accelerator's 800 MeV proton beam, the experiment studied neutrinos coming from the decay $\mu^+ \rightarrow e^+\nu_e\bar{\nu}_\mu$ of muons at rest and from the decay $\pi^+ \rightarrow \mu^+\nu_\mu$ of pions in flight. The muon decay at rest (DAR) process allowed a search for $\bar{\nu}_\mu \rightarrow \bar{\nu}_e$ appearance via the reaction $\bar{\nu}_e p \rightarrow e^+ n$; neutrinos from pion decay in flight (DIF) provided a search for $\nu_\mu \rightarrow \nu_e$ appearance via the reaction $\nu_e C \rightarrow e^- N$.

The LSND detector was a cylindrical tank of 167 tons of mineral oil doped with a small amount of scintillator viewed by 1220 photomultiplier tubes. It was positioned approximately 30 m from the beam target area. For the 1993-95 running period the primary proton target was a water target, and in 1996-98 the target was made mostly of tungsten. The neutrino yield from the water target was higher, so that about 59% for the total DAR flux and 62% of the total DIF flux came from the 1993-95 period.

The DAR process yields neutrinos with energies up to the 52.8 MeV decay endpoint, and the DIF process produced $\nu_\mu$ up to 300 MeV. The detector uses Cherenkov light for electron particle identification. The reaction $\bar{\nu}_e p \rightarrow e^+ n$ is observed by correlating the electron track with the 2.2 MeV gamma ray that follows from neutron capture on free protons in the liquid, $np \rightarrow d\gamma$.

LSND has presented results periodically. Analysis of 1993-95 DAR data produced an excess above backgrounds of $51.0^{+20.2}_{-19.5} \pm 8.0$ events, corresponding to an oscillation probability of $(0.31 \pm 0.13)\%$ (combined statistical and systematic errors) [1]. The DIF data from the same period yielded an excess of $18.1 \pm 6.6 \pm 4.0$ events, and an oscillation probability of $(0.26 \pm 0.11)\%$ [2].

In 2001, LSND presented final results using all of its data, combining the $\bar{\nu}_\mu \rightarrow \bar{\nu}_e$ and $\nu_\mu \rightarrow \nu_e$ searches into a single analysis, and employing new event reconstruction with better spatial resolution [3]. A total excess of $87.9 \pm 22.4 \pm 6.0$ events was observed in the $\bar{\nu}_\mu \rightarrow \bar{\nu}_e$ search, corresponding to an oscillation probability of $(0.264 \pm 0.081)\%$. The common event selection was optimized for the DAR region below 60 MeV, and was therefore less effective than the previous DIF analysis in removing background events above 60 MeV. This analysis found no significant signal in the DIF energy region, where the observed excess was $8.1 \pm 12.2 \pm 1.7$ events.

The Karlsruhe Rutherford Medium Energy Neutrino experiment at the ISIS facility of the Rutherford Lab also searched for $\bar{\nu}_\mu \rightarrow \bar{\nu}_e$ oscillations using $\bar{\nu}_\mu$ from $\mu^+$ decay at rest [4]. The experiment employed a 56 ton segmented liquid scintillator detector located about 18 m from the beam target. The 1997-2001 run, known at KARMEN 2, did not observe an excess of events, finding 15 events while expecting a background of $15.8 \pm 0.5$ events. However, KARMEN 2 only had the sensitivity to rule out a portion of the LSND allowed parameter space. A comparison of the favored LSND oscillation parameters with the KARMEN 2 limit is shown in Fig. 1. A com-


doi:10.1016/S0920-5632(02)02081-9

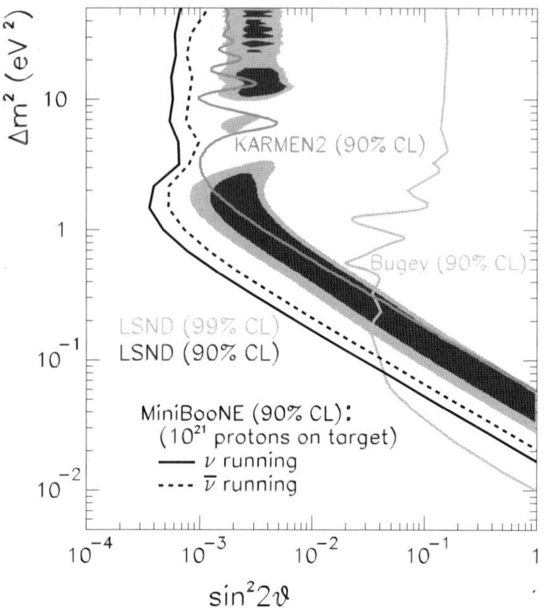


Figure 1. The LSND 90% and 99% C.L. allowed regions compared with the 90% C.L. exclusion contours from KARMEN 2 ($\bar{\nu}_\mu \to \bar{\nu}_e$), Bugey ($\bar{\nu}_e \to \bar{\nu}_\mu$), and the expected MiniBooNE contours using a $\nu_\mu$ or a $\bar{\nu}_\mu$ beam.

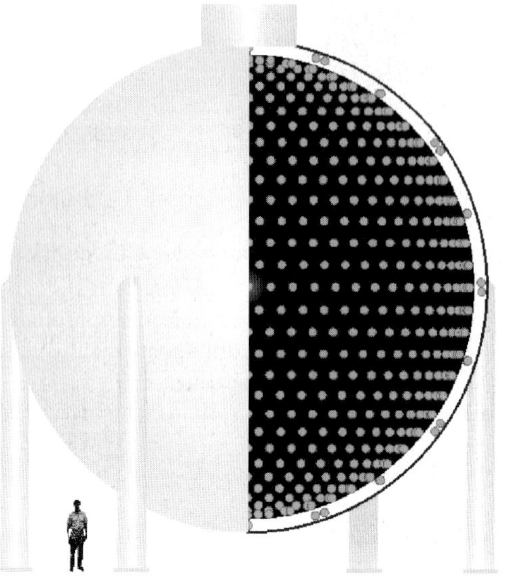

Figure 2. A schematic of the MiniBooNE detector. The cutaway shows the layout of 20-cm phototubes in the black main region and in the white veto region.

bined statistical analysis of LSND and KARMEN 2 finds regions of oscillation parameters compatible with both experiments [5].

The LSND neutrino mass difference, $\Delta m^2_{LSND} > 0.1$ eV$^2$, is distinct from those indicated by oscillations of solar and atmospheric neutrinos [6]. This poses the problem that three neutrino masses cannot explain the three $\Delta m^2$'s. Proposed solutions include consideration of a sterile neutrino, supersymmetry, $CPT$ violation, or lepton number violating muon decay [7]. In addition, the relatively high neutrino mass would contribute to the question of dark matter. It remains important to confirm or refute the LSND evidence, which MiniBooNE sets out to do.

## 2. MINIBOONE OVERVIEW

At MiniBooNE we have begun collecting data to search for $\nu_\mu \to \nu_e$ oscillations. The $\nu_\mu$ beam energy is in the range 0.5 – 1.5 GeV with a small intrinsic $\nu_e$ component and we will search for an excess of electron neutrino events in a detector located approximately 500 m from the neutrino source. The baseline to neutrino energy ratio is thereby similar to that of LSND, $L/E \sim 1$ m/MeV, while neutrino energies are more than an order of magnitude higher.

The MiniBooNE neutrino beam is initiated by a primary beam of 8 GeV protons from the Fermilab Booster. The Booster is a reliable, high intensity machine, expected to run at least $2 \times 10^7$ s per year, while delivering $5 \times 10^{12}$ protons per 1.6 $\mu$s pulse at a rate of 5 Hz to MiniBooNE. By this proton accounting, one nominal year for the experiment is expected to supply $5 \times 10^{20}$ protons on target. The Booster has the capacity to provide protons for several Fermilab efforts, and the Booster continues to deliver protons to the TeVatron collider program.

A secondary beam is produced when the 8 GeV protons strike a beryllium target positioned inside a magnetic horn. At present, positively charged

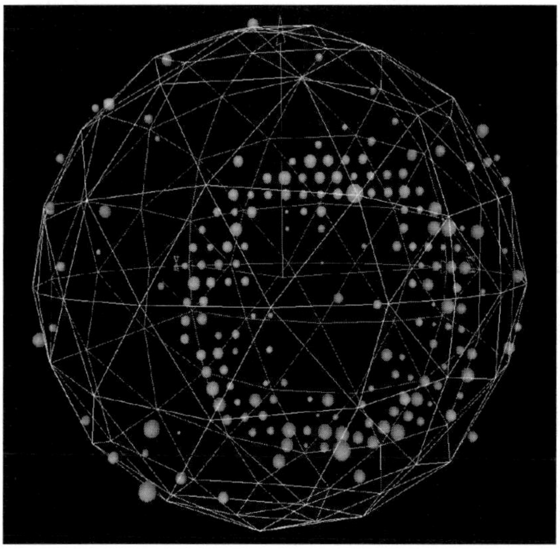

Figure 3. Event display of the muon signature from a candidate $\nu_\mu$ interaction. Each dot represents a hit photomultiplier tube with the size proportional to the charge collected. This event comes from the set of the first few beam events observed by MiniBooNE in late August 2002.

particles (mostly $\pi^+$'s) from the target are focused forward by the single horn into a 50 m decay tunnel. The $\nu_\mu$ beam is produced from the decay of these secondary particles. At a later time, the polarity of the horn can be changed in order to focus $\pi^-$'s to generate a $\bar{\nu}_\mu$ beam. Decay lengths of 50 and 25 m are possible through the use of two steel and concrete beam absorbers. One absorber is permanently positioned at the end of the decay tunnel. The intermediate absorber can either be lowered into the decay tunnel or raised out of the way. This ability to vary the decay length will provide a check of experimental systematics associated with $\nu_e$ contamination.

The MiniBooNE neutrino detector consists of 800 tons of pure mineral oil contained in a 40-foot (12.2 m) diameter spherical tank. A structure in the tank supports phototubes, and optically isolates the most outer 35 cm of oil from the rest, turning the outer oil into a veto region that should stay quiet while a neutrino produces light only in the inner, main region. The main region is viewed by 1280 20-cm phototubes, providing 10% photocathode coverage of the 445 ton fiducial volume. The veto region contains 240 20-cm phototubes mounted in pairs on the tank wall. A schematic of the detector is shown in Fig. 2. The center of the detector is positioned about 490 m from the end of the decay tunnel, and about 6 m below ground, corresponding to the level of the neutrino beam. An overburden of about 3 m of soil shields the detector enclosure.

The detector records the rise time and total charge for each hit phototube. From this information, the track position and direction are determined. Electrons from $\nu_e$ interactions are identified via their characteristic Cherenkov and scintillation light signatures (the undoped oil scintillates modestly). Backgrounds to the oscillation search will be due to $\nu_e$ contamination in the beam and to the misidentification as electrons of muons and $\pi^0$'s produced in the detector. A display of one of MiniBooNE's first candidate neutrino events is shown in Fig. 3.

## 3. BEAMLINE STATUS

Intense activity over summer 2002 focused on preparing the new 8 GeV proton beamline coming off of the Fermilab Booster to the MiniBooNE production target and secondary beam focusing horn. MiniBooNE started receiving proton beam on target in late August.

The focusing horn was designed to run at 5 Hz and for 200 Mpulses — a higher pulse rate and more pulses than any previous horn. In fall 2001 the horn was successfully bench-tested with more than 10 Mpulses, a record already making it the world's most-pulsed horn.

In summer 2002 the proton beam was commissioned in a few stages. In June protons were sent through the new beamline to a stop ahead of the MiniBooNE target area. Next the target and horn were installed and then removed in a dry run of procedures to be used for removing a damaged, radioactive horn. In July, without the target and horn in place, protons were sent through the target location, where a temporary multiwire beam monitor recorded the beam profile. The multiwire measurements were used to commission the final focus and to begin calibration of permanent beam position monitors (BPM's). In August the target and horn were reinstalled. The final configuration was modified slightly to include a multiwire just upstream of the target, which was used to complete the calibration of the target BPM's.

High intensity Booster operation is restricted by radiation limits, and beam losses will have to be reduced before MiniBooNE can reach its full beam intensity. In September the Booster was typically supplying $4 \times 10^{12}$ protons per pulse. (Losses were observed to grow disproportionately when trying to run with $5 \times 10^{12}$ protons per pulse.) By mid-September the pulses were arriving on target at 1 Hz. The experimental priority is to increase the repetition rate to 5 Hz (whereas $4 \times 10^{12}$ protons per pulse is adequate). Operation above 1 Hz will become possible with the planned installation of a new extraction magnet in the winter. However, under current run conditions 5 Hz operation would exceed the radiation limits for beamline elements.

The Booster now supplies protons to both the TeVatron collider and to MiniBooNE. The start of MiniBooNE running has had negligible effect on collider operation.

## 4. DETECTOR STATUS

Installation of phototubes and of all related in-tank calibration and monitoring hardware was completed in September 2001. Filling the tank with oil started in December and completed in May 2002. The data acquisition electronics were commissioned over the same period, and phototube signals produced by cosmic rays and by a laser calibration system indicated the rising oil level.

The laser calibration system provides pulses to four light-diffusing flasks at various locations in the detector. The system is being used to determine phototube time offsets and time slewing corrections. Energy calibrations are being performed using electrons from the decay of stopped cosmic ray muons. In addition, a cosmic ray calibration system is being commissioned, which consists of a hodoscope muon tracker above the detector with seven scintillator cubes (7.6 and 10 cm on a side) located under the tracker inside the detector. The system will provide the entering position and direction of cosmic ray muons. Cosmic rays that stop in a cube are a sample of muons with well-known path length in the detector, and will be used to calibrate the position, energy, and direction determination of the reconstruction algorithms.

## 5. PROSPECTS

As the first neutrinos were delivered to MiniBooNE the detector was ready to record them. It now remains to be seen how reliable and intense the beam will be. The beam intensity came up quickly, and within about three weeks of the start of the run, $4 \times 10^{12}$ protons per pulse at a rate of 1 Hz were being delivered. The 1 Hz operation is currently limited by beamline hardware, which will be replaced this winter to allow 5 Hz operation. Work is ongoing to control beam losses so that 5 Hz beam operation is possible within radiation limits.

The experiment is currently running with a $\nu_\mu$ beam. If oscillations occur as indicated by LSND, with $10^{21}$ protons on target MiniBooNE will observe an excess of several hundred electron events that will be significantly above background. The option will exist during the run to reverse the horn polarity to provide a $\bar{\nu}_\mu$ beam. The $\bar{\nu}_\mu$ beam will have a reduced flux, and the antineutrino cross section is smaller, but the backgrounds are also expected to be smaller, so that the experiment would retain significant sensitivity. MiniBooNE's expected 90% C.L. exclusion contours using a $\nu_\mu$ and a $\bar{\nu}_\mu$ beam are shown in Fig. 1.

## REFERENCES

1. LSND, Phys. Rev. C **54** (1996) 2685.
2. LSND, Phys. Rev. C **58** (1998) 2489.
3. LSND, Phys. Rev. D **64** (2001) 112007.
4. KARMEN, Phys. Rev. D **65** (2002) 112001.
5. E.D. Church, et al. Phys. Rev. D **66** (2002) 013001.
6. C. Giunti, these proceedings.
7. K.S. Babu, these proceedings.

ELSEVIER

Proceedings of ICHEP 2002, pp. 37–42
*S. Bentvelsen, P. de Jong, J. Koch and E. Laenen (Editors)*

31st INTERNATIONAL CONFERENCE ON HIGH ENERGY PHYSICS AMSTERDAM

# Status of the MINOS Experiment and Review of Long-Baseline Neutrino Oscillation Experiments in Europe and North America

Jon Urheim, representing the MINOS Collaboration [a]

[a]School of Physics and Astronomy, University of Minnesota,
116 Church St. SE, Minneapols, Minnesota 55455 USA

Together with K2K in Japan, MINOS (Main Injector Neutrino Oscillation Search) and the CNGS (CERN Neutrinos to Gran Sasso) program represents the first generation of accelerator-based long-baseline neutrino oscillation experiments. I report on the status of MINOS and CNGS – their goals, construction status, and prospects. In particular, construction and commissioning of the MINOS Far Detector is far along, and detector performance is being studied with cosmic ray data. Finally, I summarize ideas for the next generation of experiments currently being considered in Europe and North America.

## 1. Introduction

At this conference, we have heard results from non-accelerator experiments on the atmospheric (Super-Kamiokande [1]) and solar neutrino (SNO [2] and Super-K [3]) anomalies, as well as results from K2K [4], an accelerator-based 'long-baseline' experiment bearing on the former. We have also heard status reports on the KamLAND [5] and MiniBoone [6] experiments which are in early stages of collecting data with reactor and accelerator neutrinos, probing oscillations in the 'solar' and 'LSND' regions of squared mass difference $\Delta m^2$, respectively. Together, these experiments are beginning to provide a coherent, though incomplete, picture [7–9] of neutrino masses and mixings.

This contribution focuses on long-baseline experiments, which, within the next several years, will begin to collect data with intense high-energy $\nu_\mu$ beams generated by accelerators at Fermilab (NuMI/MINOS) and CERN (CNGS/OPERA and potentially CNGS/ICARUS). These experiments aim to study oscillations of $\nu_\mu$'s with frequencies dictated by the 'atmospheric' $\Delta m^2$, ($0.0016 < \Delta m^2_{atm} < 0.0039\ \mathrm{eV}^2$, 90% CL [1]).

These experiments will be able to confront, with precision data, critical aspects of the emerging picture. The oscillation interpretation of atmospheric neutrino data from Super-K [1] is very strong, leading to the allowed $\Delta m^2_{atm}$ region given above, and is supported by data from K2K [4]. Super-K data also suggests that the primary oscillation mode is $\nu_\mu \to \nu_\tau$, and that the mixing angle $\theta_{23}$ is consistent with maximal ($\sin^2 2\theta_{23} > 0.92$ at 90% CL [1]). Direct confirmation of the $\nu_\mu \to \nu_\tau$ hypothesis will come from OPERA and ICARUS which are capable of identifying $\nu_\tau$ interactions in their detectors. MINOS will measure $\Delta m^2$ and $\sin^2 2\theta_{23}$ for this mode precisely, and will test CPT symmetry in the neutrino sector by measuring $\nu_\mu$ and $\overline{\nu}_\mu$ oscillation parameters separately.

A key element missing from the current picture of neutrino masses and mixings is the third mixing angle, $\theta_{13}$. A non-zero value is a necessary condition for CP non-conservation via neutrino mixing. Since $\theta_{23}$ is nearly maximal and since $\theta_{12}$ appears to be large ($\tan^2\theta_{12} \sim 0.34$) [2,7,8], not only might the CP-violating effects be observable (provided $\theta_{13}$ is not too much smaller than present experimental limits), but also they might have played a significant role in leptogenesis in the early universe [10]. Experimentally, $\theta_{13}$ would give rise to a subdominant $\nu_\mu \to \nu_e$ oscillation at $\Delta m^2_{atm}$. Clearly, observing such oscillations is an important goal for the long-baseline experiments: the three experiments discussed here are well-positioned to do this.


doi:10.1016/S0920-5632(02)02088-1

Furthermore, new experiments to measure $\theta_{13}$ and study CP-violating effects are being conceived. The JHF-Kamioka project in Japan was discussed in a separate talk [11]; at the end of this contribution, I will describe some of the ideas for new experiments in Europe and North America.

## 2. The NuMI/MINOS Project

The NuMI project consists of an intense $\nu_\mu$ beam produced by 120-GeV protons extracted from the Fermilab Main Injector (MI), directed northwest and downward by 57 mrad, and re-emerging from the earth in northern Minnesota. The MINOS 'Far' Detector is a 5.4-kton magnetized steel and plastic scintillator tracker and calorimenter under construction in the Soudan iron mine 2341 ft below ground. With a baseline of 735 km, the first maximum in the oscillation probability occurs at $E_\nu \sim 1.5$ GeV for $\Delta m^2 = 0.0025$ eV$^2$. The signature for neutrino oscillations is a dip in the energy spectrum of $\nu_\mu$'s detected at Soudan relative to extrapolations from measurements by a 1-kton 'Near' Detector located on the Fermilab site. The location of the dip depends on the value of $\Delta m^2$, while its depth gives a measure of $\sin^2 2\theta$. Below, I summarize the status of the beam and detector construction, as well as the physics capabilities and outlook.

### 2.1. The NuMI Beam Line

The MI will provide $4 \times 10^{13}$ 120-GeV protons in an $8\,\mu$s single-turn-extracted pulse onto a graphite target once every 1.9 seconds, for an average beam power of 0.4 MW. Just beyond the target are two parabolic aluminum magnetic horns, focusing charged hadrons of energies that depend on the relative positions of target and horns, as shown in Fig. 1. Downstream of the 2nd horn is a 675 m vacuum decay pipe, followed by a steel hadron absorber and rock for muon absorption. The Near Detector is located 317 m beyond the end of the decay pipe.

The excavation of the beam line and detector hall areas is essentially complete, and the decay pipe has been installed and encased in concrete. Work will begin on the outfitting of these areas and the construction of surface buildings above

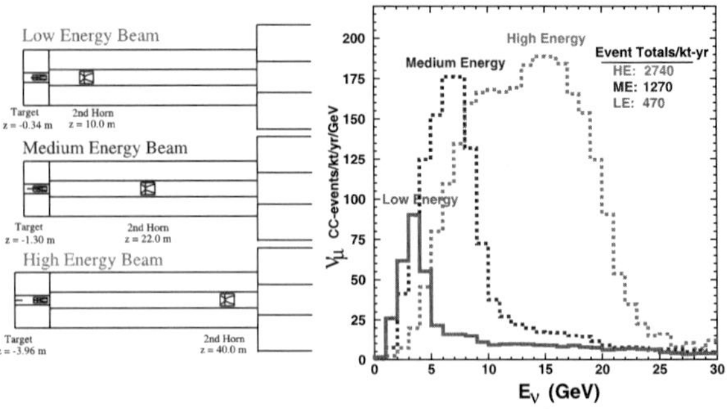


Figure 1. Neutrino energy spectra of the NuMI beam for three horn/target configurations, in units of charged-current (CC) events per kton-yr observed in the MINOS Far Detector, assuming no oscillations.

the target area and the Near Detector hall in the coming months. Engineering is complete for most of the technical components, and many components are being fabricated. Overall, the project is on schedule to be completed in late 2004, with beam on target expected by the end of the 2004.

### 2.2. The MINOS Far Detector

The MINOS Far Detector is organized as two supermodules, each with 243 planes of steel 2.54 cm thick and 8m × 8m wide in an octagonal shape. Each plane is formed from eight 1.27-cm thick steel plates, cut so as to fit in the mine shaft elevator; these sheets are plug-welded underground. Onto each plane of steel is attached eight *scintillator modules* as shown in Fig. 2. The resulting steel plus scintillator module assemblies are then hung from their 'ears' on a steel support structure, much like file folders in a filing cabinet. In successive planes, the scintilator modules alternate between the orientation shown in the figure and one in which they are rotated by 90°. Through the center of each supermodule is a hole for its magnet coil: this is energized to 15,000 Amp-turns, generating a uniform toroidal magnetic field in the steel.

A scintillator module consists of 20 or 28 polystyrene strips 4.1 cm × 1 cm in cross-section (co-extruded with a reflecting layer of $TiO_2$) and up to 8m long, encased in a light-tight aluminum

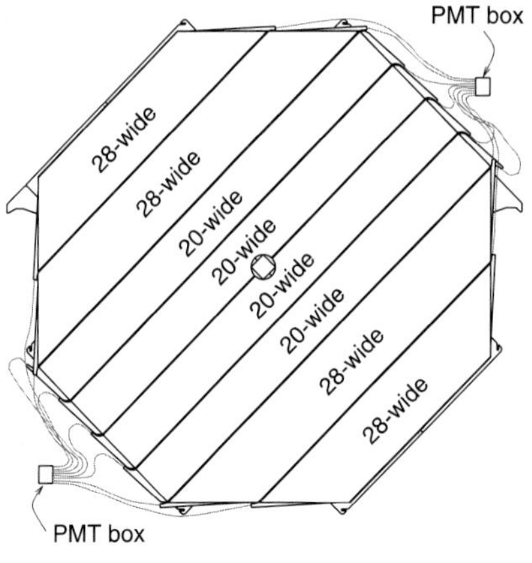


Figure 2. Layout of scintillator modules for a plane in the MINOS Far Detector

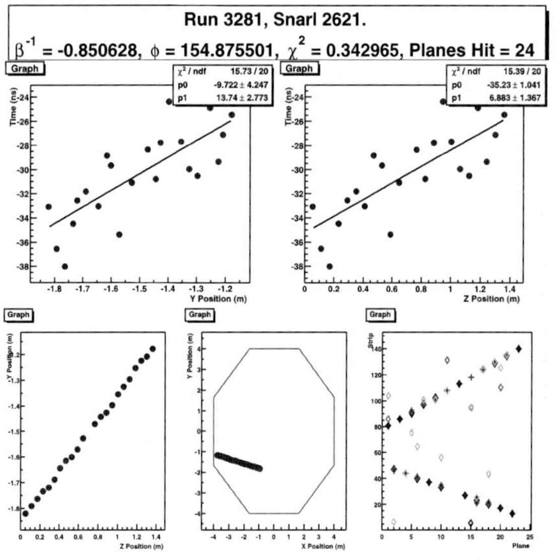


Figure 3. First upward-going muon candidate in the MINOS Far Detector. Top: plots of time (ns) vs. vertical ($y$) and longitudinal ($z$) hit positions. Bottom: $y$ vs. $z$, $y$ vs. $x$, and strip vs. plane.

sheet metal enclosure. Readout is achieved by wavelength-shifting (WLS) fibers glued into a groove in one face of each scintillator strip. The WLS fibers are routed through plastic manifolds to connectors at each end of a module, from which clear optical fiber cables carry the wavelength-shifted scintillation light to Hamamatsu M-16 multi-channel photomultiplier tubes housed in racks next to the detector. For normally-incident minimum-ionizing particles, the signals from the two ends of the fiber in a given strip typically sum to 10 – 15 photoelectrons.

We record pulseheight and timing information. For beam running, detector readout is triggered by a timing system GPS-synched to the MI beam-spill. For cosmics the detector is self-triggering, utilizing dynode signals from the PMT's.

### 2.2.1. Status and commissioning

A technical support crew of thirty, divided into two shifts, carries out the welding, assembly and rigging of the detector planes. Typically, seven planes are erected per week, with electronics and high voltage connected and tested as they go up. Testing is performed with electronic and optical pulsing systems as well as with cosmic ray muons.

Construction of the first supermodule was completed in July, and its coil energized for the first time the day before my presentation at ICHEP. Completion of the second supermodule is scheduled for April 2003.

Cosmic ray (CR) muons have provided pulse-height and timing calibrations for the first supermodule. Over periods of months, the detector response is stable: to within 1% in energy and to within $\sim$ 1 ns in timing. The single-strip timing resolution has been measured to be $\sim$ 2.6 ns from residuals to fits of time versus position for CR muons. We are currently evaluating resolutions in muon energy (expected to be better than 12%) by cross-calibrating curvature and range measurements for stopping muons, and also evaluating the angular resolution using multiple-muon events.

The excellent timing resolution is critical for the study of atmospheric neutrinos with MINOS. Indeed, we have observed a handful of upward-going muon candidates, using timing to determine the muon flight direction. In Fig. 3, distributions for the first upward-going muon candidate are plotted. This muon entered the upstream face of the detector (plane 0) and exited from one side after traversing 23 planes.

Assembly of MINOS Near Detector planes is also under way, being done on the surface at Fermilab. These will be lowered into the detector hall once outfitting is complete and the support structure has been installed there. A challenging aspect of the Near Detector is the large instantaneous rate from neutrino interactions, which could be as large as 50 events within the 8$\mu$s beam spill. Multi-hit front-end electronics are necessary so as to avoid potential biases in the comparison between the busy Near Detector and the low-occupancy Far Detector event environments.

Finally, the MINOS detector will be used as a calorimeter for (1) identifying CC $\nu_e$ interactions and measuring the electron energy (resol'n$\sim$ $23\%/\sqrt{E}$), (2) identifying neutral-current (NC) interactions, and (3) measuring the energy of hadronic showers (resol'n$\sim 55\%/\sqrt{E}$) in both CC and NC events to help in the determination of the energy of the interacting neutrino. Stopping CR muons provide an excellent calibration of energy deposited through ionization, however there is no natural source for calibrating the response to electromagnetic and hadronic showers. To provide this, a 60-plane 1m$\times$1m version of the MINOS detector has been running in test beams at CERN. Already, the energy resolution on EM showers has been demonstrated with this data.

### 2.3. Physics Capabilities and Outlook

As mentioned, the primary goal for MINOS is to precisely measure the parameters for $\nu_\mu \to \nu_\tau$ oscillations (assuming this is the mechanism behind the observed atmospheric $\nu_\mu$ deficits). Initial running will be with the *low energy* beam configuration (see Fig. 1), which, despite yielding a smaller flux, is best matched to the values of $\Delta m^2$ favored by Super-K. Fig. 4 gives an idea of what our reconstructed neutrino energy spectrum might look like, relative to expectations assuming no oscillations, after a two-year run. The oscillatory behavior is clear in all three cases, as exemplified by the bottom row of plots. The sensitivity to oscillation parameters, accounting for expected systematic uncertainties, is shown in Fig. 5.

Overall the outlook for MINOS is quite good: by 2007, oscillation parameters for $\nu_\mu \to \nu_\tau$ will be measured with good precision. In subsequent running, horn currents can be reversed to take data with anti-neutrinos, allowing searches for possible violations of CPT symmetry. Also by 2007, the search for $\nu_e$ appearance will be able to extend the reach in $\sin^2 2\theta_{13}$ relative to the sensitivity of CHOOZ [12] by a factor of 2 to 4, depending on $\Delta m^2$. In addition, we will by then have had $\sim$ 20 kton-yrs of exposure to atmospheric $\nu$'s. This will provide additional constraints on oscillation parameter measurements, and an early look at CPT asymmetries will be possible thanks to the magnetic field. Finally, unique opportunities exist for conventional neutrino physics with the MINOS Near Detector and cosmic ray physics with the Far Detector.

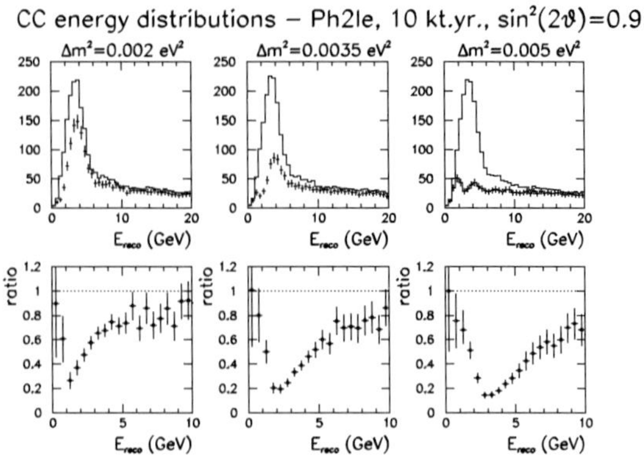


Figure 4. Top: expected CC $\nu_\mu$ interaction yields after two years of data-taking with MINOS in the case of no oscillations (solid histograms), and in the case of oscillations (points) with $\sin^2 2\theta = 0.9$ for three different values of $\Delta m^2$, as a function of reconstructed neutrino energy. Bottom: ratios of 'measured' (oscillated) to 'expected' (not oscillated) yields, as given by the top plots.

## 3. The CNGS Project

The CERN to Gran Sasso project also aims to produce an intense source of neutrinos, in this case using protons from the 400-GeV CERN SPS. The beam is directed toward the Gran Sasso Laboratory (LNGS), 732 km to the southeast in Italy. Two experiments will use this beam to search for $\tau$-lepton signatures from $\nu_\mu \to \nu_\tau$ oscillations and

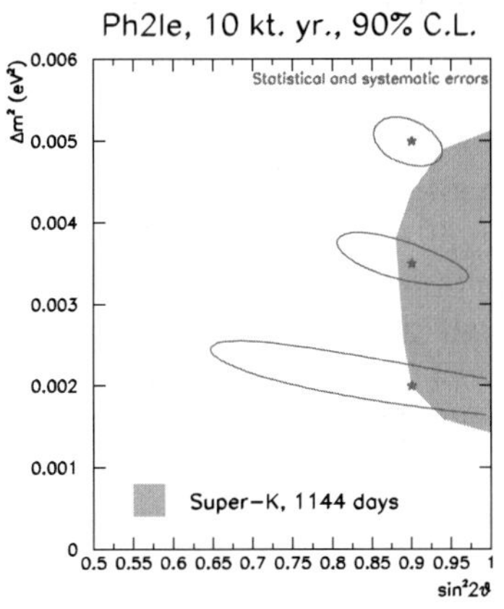


Figure 5. MINOS 90% CL intervals for oscillation parameters after two years of data-taking at full beam intensity for three parameter sets. Overlaid is the 90% CL interval from the 1144-day Super-K atmospheric $\nu$ measurements. (Note that the constraints reported by Super-K at this conference [1] are more restrictive than shown here.)

for $\nu_e$ appearance from the subdominant oscillation mode. The OPERA experiment is an approved CNGS experiment based on photographic emulsions, to be located in Hall C. The ICARUS collaboration has been approved by LNGS to deploy detector modules in Hall B.

Excellent progress has been made on the CNGS beam line. The target chamber and decay tunnel have been excavated, and the civil construction is expected to be complete by May 2003. The present schedule has the SPS delivering $4.5 \times 10^{19}$ protons per year to the CNGS target beginning in 2006. Ideas exist for increasing the proton flux by $30 - 50\%$ with minimal investments.

The mean neutrino energy of the CNGS beam is 17 GeV, well above the location of the first oscillation peak for even the most favorable values of $\Delta m^2$. However, the $\nu_\tau$ CC interaction cross-section rises dramatically in the region just above $\tau$ production threshold: the cross-section times oscillation probability peaks around $10 - 12$ GeV. Thus the CNGS beam is well-suited for observing $\nu_\tau$'s: for $\Delta m^2 = 0.0024\,\mathrm{eV}^2$ and maximal mixing, there will be 13.4 $\nu_\tau$ CC interactions per kton-yr of detector mass-exposure at nominal intensity, along with 3660 NC and $\nu_\mu$ CC interactions.

Space constraints allow only brief descriptions of the OPERA and ICARUS experiments, given below. For more information on these experiments, as well as on the CNGS project, the reader is referred to the presentation by S. Katsanevas at the Neutrino 2002 conference [13].

The OPERA detector consists of Pb/emulsion 'bricks', stacked into 62 'walls', with $52 \times 64$ bricks per wall, for a total target mass of 1.8 kton. Each brick contains 56 emulsion films deposited on thin plastic bases, separated by Pb sheets 1 mm thick. MINOS-like plastic scintillator planes are positioned between the walls to help locate neutrino interaction vertices in the emulsion.

The ICARUS detector is a liq.-Argon TPC, providing bubble-chamber-like imaging of particle trajectories with sub-mm resolutions in three dimensions. The ICARUS collaboration has recently tested a 300-ton prototype. A 600-ton version is being installed in LNGS (completion expected in summer 2003). The full proposal calls for a 3-kton detector to be installed by 2006.

With the currently favored oscillation parameters, the two detectors should see of order 25 $\tau$ events in five years of running, with negligible backgrounds. The prospects for $\nu_e$ appearance are especially interesting: sensitivities to $\theta_{13}^2$ around $5 \times 10^{-3}$ are expected.

## 4. Ideas for Future Efforts

With the recent spate of exciting results on solar and atmospheric $\nu$'s, the topic of neutrino mass and mixing has become an area where experiments have a chance to really see something new. MINOS and the CNGS experiments will contribute significantly to this field, but key questions will remain – particularly on the possibility of CP violation due to neutrino flavor mixing. It is not too early to prepare for the next generation of experiments to explore these questions.

### 4.1. Use of Existing Facilities

One possible step is to extend the capabilities of existing neutrino beams and/or detectors. The CNGS and NuMI beams are obvious resources, and ideas to use them for dedicated $\nu_e$ appearance experiments are being advanced. In both cases, new detectors, optimized for electron ID and rejection of NC backgrounds, would be sited between 0.5° and 2° off-axis relative to the beam centerline (this is also the plan for JHF-Kamioka [11]). By going off-axis, one gains an advantage due to kinematics: the resultant $\nu$ energy spectrum is dominated by a Jacobean peak, effectively giving a narrow-band beam. Furthermore, $\nu$'s with energies above the peak are naturally suppressed at these angles, thereby reducing otherwise severe NC backgrounds coming from higher energy $\nu$'s.

A 20-kton off-axis NuMI detector [14] could be located on the surface either in Minnesota near the Soudan mine, or further north along the trans-Canada highway (where the beam centerline is overhead at $\sim$ 15 km). The sites in Minnesota are attractive because of existing infrastructure at Soudan, while the Canada sites benefit from the longer baseline ($\sim$ 950 km), allowing matter effects to become more prominent. Detector technologies being considered include RPC's, liquid scintillator with fiber readout, water Cherenkov, and ICARUS-like liquid-argon TPC's. The projected sensitivity to $\theta_{13}^2$ is at the level of $1.5 \times 10^{-3}$.

The idea being floated for a new CNGS experiment involves an off-shore water Cherenkov detector in the Gulf of Taranto [15], with a baseline of 1200 km. The CNGS beam would be tuned to a lower energy, so that for a detector 2° off axis the mean neutrino energy would be 0.8 GeV – at the peak of the *second* oscillation maximum.

### 4.2. New Facilities

The construction of new neutrino beams for long-baseline experiments is also under consideration. Most of these ideas make use of existing accelerators or components to capitalize on infrastructure and keep costs low.

One possibility [16] being explored is that of upgrading the BNL AGS to a beam power of 0.5 – 1.3 MW, and building one or more neutrino beam lines directed toward sites, such as Soudan, Homestake, or the Waste Isolation Pilot Plant (WIPP) in New Mexico, where multi-function underground detectors could be built.

In Europe, several interesting ideas involve beams directed from CERN to the Frejus tunnel. One involves a new 2-GeV superconducting proton linac, based on LEP RF cavities, which would be used to generate a neutrino beam of 300 MeV, appropriate for the 130 km baseline. Another novel idea is that of 'beta beams' in which $\nu_e$ and $\overline{\nu}_e$ beams are generated from accelerated $^{18}$Ne and $^6$He ions which decay in flight.

Finally I have not mentioned projects that might be realized on a longer timescale, such as neutrino factories from muon storage rings [17].

## 5. Summary

The summary is that we're at a very exciting stage in the study of neutrino mass and mixing. Bring on the neutrino beams!

## 6. Acknowledgements

I thank S. Katsanevas for providing me with information and figures on the CNGS program.

## REFERENCES

1. C. Mauger, these proceedings.
2. S. Oser, these proceedings.
3. M. Vagins, these proceedings.
4. Y. Hayato, these proceedings.
5. T. Mitsui, these proceedings.
6. A. Bazarko, these proceedings.
7. C. Giunti, these proceedings.
8. D. Wark, these proceedings.
9. C. Gonzalez-Garcia, these proceedings.
10. Z. Xing, these proceedings.
11. Y. Itow, these proceedings.
12. M. Apollonio *et al.* (CHOOZ Collab.), Phys. Lett. B **466** (1999) 415.
13. http://neutrino2002.ph.tum.de
14. A letter of intent was submitted recently to the Fermilab PAC: http://www-numi.fnal.gov/fnal_minos/new_initiatives/
15. See the expression of interest posted at: http://home.cern.ch/dydak/oscexp.ps
16. D. Beavis *et al.* (BNL Neutrino Working Group), hep-ex/0205040, (2002).
17. G. Hanson, these proceedings.

ELSEVIER

Proceedings of ICHEP 2002, pp. 43–46
*S. Bentvelsen, P. de Jong, J. Koch and E. Laenen (Editors)*

31st INTERNATIONAL CONFERENCE ON HIGH ENERGY PHYSICS AMSTERDAM

# Leptogenesis and CP Violation in Neutrino Oscillations

Zhi-zhong Xing [a]

[a]Institute of High Energy Physics, P.O. Box 918 (4), Beijing 100039, China

Assuming the seesaw and leptogenesis mechanisms, I give some general remarks on possible connection or disconnection between CP violation in heavy Majorana neutrino decays and that in neutrino oscillations. A simple but predictive ansatz is proposed to simultaneously interpret the observed baryon asymmetry of the universe and oscillations of solar and atmospheric neutrinos.

The density of baryons compared to that of photons in the universe is extremely small: $\eta \equiv n_{\rm B}/n_\gamma = (2.6 - 6.3) \times 10^{-10}$, extracted from the Big-Bang nucleosynthesis [1]. This tiny quantity is related to the observed matter-antimatter or baryon-antibaryon asymmetry of the universe,

$$Y_{\rm B} \equiv \frac{n_{\rm B}}{\rm s} \approx \frac{\eta}{7.04} = (3.7 - 8.9) \times 10^{-11}, \qquad (1)$$

where $\rm s$ denotes the entropy density. In order to produce a net baryon asymmetry in the standard Big-Bang model, three Sakharov necessary conditions must be satisfied [2]: (a) baryon number nonconservation, (b) C and CP violation, and (c) a departure from thermal equilibrium. Among several interesting and viable baryogenesis scenarios proposed in the literature, Fukugita and Yanagida's leptogenesis mechanism [3] has attracted a lot of attention – due partly to the fact that neutrino physics is entering a flourishing era.

Indeed the Super-K [4] and SNO [5] data have provided very convincing evidence that neutrinos are massive and lepton flavors are mixed – a kind of new physics which can only be understood beyond the standard electroweak model. A simple extension of the standard model is to include one right-handed neutrino in each of three lepton families, while the Lagrangian of electroweak interactions keeps invariant under the $\rm SU(2)_L \times U(1)_Y$ gauge transformation. In this case, the Yukawa interactions are described by

$$-\mathcal{L}_{\rm Y} = \bar{l}_{\rm L}\tilde{\phi}Y_l e_{\rm R} + \bar{l}_{\rm L}\phi Y_\nu \nu_{\rm R} + \frac{1}{2}\overline{\nu^{\rm c}_{\rm R}} M_{\rm R}\nu_{\rm R} + {\rm h.c}, \qquad (2)$$

where $l_{\rm L}$ denotes the left-handed lepton doublet, $e_{\rm R}$ and $\nu_{\rm R}$ stand respectively for the right-handed charged lepton and Majorana neutrino singlets, and $\phi$ is the standard-model Higgs doublet. The lepton number violation induced by the third term of $\mathcal{L}_{\rm Y}$ allows decays of the heavy right-handed Majorana neutrinos $N_i$ (for $i = 1, 2, 3$) to occur:

$$N_i \rightarrow l + \phi^\dagger \quad {\rm vs} \quad N_i \rightarrow l^{\rm c} + \phi \, . \qquad (3)$$

As each decay can happen both at the tree level and at the one-loop level (via the self-energy and vertex corrections), the interference between tree-level and one-loop decay amplitudes may give rise to a CP-violating asymmetry $\varepsilon_i$ between the two CP-conjugated processes in Eq. (3) [3]. If the masses of three heavy Majorana neutrinos $N_i$ are hierarchical ($M_1 \ll M_2 \ll M_3$) and the interactions of $N_1$ are in thermal equilibrium when $N_2$ and $N_3$ decay, the asymmetries produced by $N_2$ and $N_3$ can be erased before $N_1$ decays. The CP asymmetry $\varepsilon_1$ produced by the out-of-equilibrium decay of $N_1$ survives, and it results in a lepton-antilepton asymmetry $Y_{\rm L} \equiv n_{\rm L}/{\rm s} = \varepsilon_1 d/g_*$, where $g_* \sim 100$ is an effective number characterizing the relativistic degrees of freedom which contribute to the entropy $\rm s$, and $d$ accounts for the dilution effects induced by the lepton-number-violating wash-out processes. Finally the lepton asymmetry $Y_{\rm L}$ is converted into a net baryon asymmetry $Y_{\rm B}$ through the $\rm (B + L)$-violating sphaleron processes [6]:

$$Y_{\rm B} = \frac{c}{c-1} Y_{\rm L} = \frac{c}{c-1} \cdot \frac{d}{g_*}\varepsilon_1 \, , \qquad (4)$$

where $c = (8{\rm N}_f + 4{\rm N}_\phi)/(22{\rm N}_f + 13{\rm N}_\phi)$ with $\rm N_\mathit{f}$ being the number of fermion families and $\rm N_\phi$ be-


doi:10.1016/S0920-5632(02)02108-4

ing the number of Higgs doublets. Taking $N_f = 3$ and $N_\phi = 1$ for example, we obtain $c \approx 1/3$.

After spontaneous symmetry breaking, we get the charged lepton mass matrix $M_l = Y_l \langle \phi \rangle$ and the Dirac neutrino mass matrix $M_{\rm D} = Y_\nu \langle \phi \rangle$ from $\mathcal{L}_{\rm Y}$, in addition to the right-handed Majorana neutrino mass matrix $M_{\rm R}$. The scale of $M_l$ and $M_{\rm D}$ is characterized by the gauge symmetry breaking scale $v \equiv \langle \phi \rangle \approx 175$ GeV, but that of $M_{\rm R}$ may be much higher than $v$, because right-handed neutrinos are $SU(2)_{\rm L}$ singlets and their mass term is not subject to the electroweak symmetry breaking. As a consequence, the light (and essentially left-handed) neutrino mass matrix $M_\nu$ can be obtained via the seesaw mechanism [7]:

$$M_\nu \approx -M_{\rm D} M_{\rm R}^{-1} M_{\rm D}^{\rm T} \; . \tag{5}$$

Note that lepton flavor mixing at low energy scales stems from a nontrivial mismatch between the diagonalizations of $M_\nu$ and $M_l$, while the baryon asymmetry at high energy scales depends on complex $M_{\rm D}$ and $M_{\rm R}$ in the leptogenesis scenario. To see these points more clearly, let us diagonalize three of the four lepton mass matrices: $U_l^\dagger M_l \tilde{U}_l = {\rm Diag}\{m_e, m_\mu, m_\tau\}$ and $U_\nu^\dagger M_\nu \tilde{U}_\nu^* = {\rm Diag}\{m_1, m_2, m_3\}$ as well as $U_{\rm R}^\dagger M_{\rm R} \tilde{U}_{\rm R}^* = {\rm Diag}\{M_1, M_2, M_3\}$. Then the lepton flavor mixing matrix at low energy scales is given by $V = U_l^\dagger U_\nu$, and the leptonic CP violation in neutrino oscillations is measured by the Jarlskog parameter $\mathcal{J}$ defined through

$${\rm Im}\left(V_{\alpha i} V_{\beta j} V_{\alpha j}^* V_{\beta i}^*\right) = \mathcal{J} \sum_{\gamma, k} \left(\epsilon_{\alpha\beta\gamma} \cdot \epsilon_{ijk}\right) , \tag{6}$$

where $(\alpha, \beta)$ run over $(e, \mu, \tau)$ and $(i, j)$ run over $(1, 2, 3)$. On the other hand, the CP asymmetry $\varepsilon_1$ between $N_1 \rightarrow l + \phi^\dagger$ and $N_1 \rightarrow l^{\rm c} + \phi$ decays at high energy scales can be given as [8]

$$\varepsilon_1 = -\frac{3}{16\pi v^2} \cdot \frac{M_1}{\left[U_{\rm R}^{\rm T} M_{\rm D}^\dagger M_{\rm D} U_{\rm R}^*\right]_{11}} \cdot \sum_{j=2}^{3} \frac{{\rm Im}\left(\left[U_{\rm R}^{\rm T} M_{\rm D}^\dagger M_{\rm D} U_{\rm R}^*\right]_{1j}\right)^2}{M_j} , \tag{7}$$

where the mass hierarchy of three heavy Majorana neutrinos ($M_1 \ll M_2 \ll M_3$) has been assumed. We see that $\mathcal{J}$ depends on $U_l$ and $U_\nu$ or equivalently on $M_l$ and $M_\nu$, while $\varepsilon_1$ depends on $M_{\rm D}$ and $M_{\rm R}$ (or $U_{\rm R}$ and $M_i$). The only possible relationship between $\mathcal{J}$ and $\varepsilon_1$ is due to the seesaw mechanism in Eq. (5), which links $M_\nu$ to $M_{\rm D}$ and $M_{\rm R}$. Therefore one can conclude that there is no direct connection between CP violation in heavy Majorana neutrino decays ($\varepsilon_1$) and that in neutrino oscillations ($\mathcal{J}$). Such a general conclusion was first drawn by Buchmüller and Plümacher [9]. Recently a few other authors have carried out some more delicate analyses and reached the same conclusion [10].

Depending on the specific flavor basis that we choose in model building, $\mathcal{J}$ and $\varepsilon_1$ can either be completely disconnected or maximally connected. To illustrate, let us consider two extreme cases:

(1) In the basis where $U_\nu = \mathbf{1}$ holds (i.e., $V = U_l^\dagger$ – lepton flavor mixing and CP violation in neutrino oscillations arise solely from the charged lepton sector [11]), we find that $\varepsilon_1$ has nothing to do with $V$ or $\mathcal{J}$. In this special case, less fine-tuning is expected in building a phenomenological model which can simultaneously interpret the baryon asymmetry of the universe and lepton flavor mixing at low energy scales.

(2) In the basis where both $U_{\rm R} = \mathbf{1}$ and $U_l = \mathbf{1}$ hold (i.e., $V = U_\nu$ – lepton flavor mixing and CP violation in neutrino oscillations arise solely from the neutrino sector), we find that $\varepsilon_1$ can indirectly be linked to $V$ and $\mathcal{J}$ through the seesaw relation in Eq. (5). It is worth remarking that the correlation between high- and low-energy observable quantities requires quite nontrivial attempts in model building. In particular, the textures of $M_{\rm R}$ and $M_{\rm D}$ have to be carefully chosen or fine-tuned to guarantee acceptable agreement between the model predictions and the observational or experimental data.

In the second part of this talk, I propose a phenomenological ansatz to simultaneously interpret the observed baryon asymmetry of the universe and neutrino oscillations. Our simple ansatz [8] is essentially a non-SO(10) modification of the Buchmüller-Wyler ansatz [12], but it has more powerful predictability and its predictions are in better agreement with current data.

First of all, we assume $M_l$ and $M_{\rm D}$ to be symmetric matrices, just like $M_{\rm R}$. Second, we assume that the (1,1), (1,3) and (3,1) elements of $M_l$, $M_{\rm D}$ and $M_{\rm R}$ are all vanishing in a specific flavor basis, like a phenomenologically-favored texture of quark mass matrices $M_{\rm u}$ and $M_{\rm d}$ [13]. Third, we assume that the non-zero elements of $M_{\rm D}$ and $M_l$ can be expanded in terms of the Wolfenstein parameter $\lambda \approx 0.22$. To be explicit, we conjecture that $M_{\rm D}$ and $M_l$ have the following patterns:

$$\frac{M_{\rm D}}{m_0} = \begin{bmatrix} 0 & \hat{\lambda}^3 & 0 \\ \hat{\lambda}^3 & x\hat{\lambda}^2 & \hat{\lambda}^2 \\ 0 & \hat{\lambda}^2 & e^{i\zeta} \end{bmatrix}, \frac{M_l}{m_\tau} = \begin{bmatrix} 0 & \lambda^4 & 0 \\ \lambda^4 & y\lambda^2 & \lambda^3 \\ 0 & \lambda^3 & 1 \end{bmatrix}$$

with $m_0 \approx v$, $\hat{\lambda} \equiv \lambda e^{i\omega}$ and $(x, y)$ being real and positive coefficients of $\mathcal{O}(1)$. It is easy to check that three mass eigenvalues of $M_{\rm D}$ have the hierarchy $\lambda^4 : \lambda^2 : 1$, and those of $M_l$ have the hierarchy compatible with our experimental data on $m_e$, $m_\mu$ and $m_\tau$. As the (1,1), (1,3) and (3,1) elements of both $M_{\rm R}$ and $M_{\rm D}$ are vanishing, $M_\nu$ must have the same texture zeros via the seesaw relation in Eq. (5) [14]. To generate a sufficiently large mixing angle in the $\nu_\mu$-$\nu_\tau$ sector, (2,3), (3,2) and (3,3) elements of $M_\nu$ should be comparable in magnitude. This requirement is actually strong enough to constrain the texture of $M_{\rm R}$ in a quite unique way [12]. For our purpose, we obtain

$$\frac{M_{\rm R}}{M_0} = \begin{bmatrix} 0 & \lambda^5 & 0 \\ \lambda^5 & z\lambda^4 & \lambda^4 \\ 0 & \lambda^4 & 1 \end{bmatrix}, \frac{M_\nu}{m'_0} = \begin{bmatrix} 0 & \hat{\lambda} & 0 \\ \hat{\lambda} & z' & 1 \\ 0 & 1 & e^{i2\varphi} \end{bmatrix},$$

where $M_0 \gg v$, $m'_0 = v^2/M_0$, $z$ is a real and positive coefficient of $\mathcal{O}(1)$, $z' \equiv 2x - ze^{i\omega}$ with $|z'| \sim \mathcal{O}(1)$, and $2\varphi \equiv 2\zeta - 5\omega$. Note that an overall phase factor $e^{i(5\omega-\pi)}$ has been omitted from the right-hand side of $M_\nu$, since it has no contribution to lepton flavor mixing and CP violation. To generate a large mixing angle in the $\nu_e$-$\nu_\mu$ sector, the condition $|z'e^{i2\varphi} - 1| \equiv \delta \sim \mathcal{O}(\lambda)$ must be satisfied. There exists an interesting parameter space, in which [8] $x = 1/\sqrt{2}$, $z = 1 + \sqrt{2}\lambda$ and $\zeta = -\omega = \pi/4$. One may check that $\delta = \sqrt{2}\lambda$ holds in this parameter space.

As mentioned above, complex $M_\nu$ can be diagonalized by a unitary matrix $U_\nu$, On the other hand, the strong hierarchy of real $M_l$ under consideration implies that its contribution to lepton flavor mixing is small and negligible. Then we arrive at $V \approx U_\nu$, which links the neutrino mass eigenstates $(\nu_1, \nu_2, \nu_3)$ to the neutrino flavor eigenstates $(\nu_e, \nu_\mu, \nu_\tau)$. Current data on solar and atmospheric neutrino oscillations suggest $|V_{e3}| \ll 1$, $|V_{e1}| \sim |V_{e2}|$ and $|V_{\mu 3}| \sim |V_{\tau 3}|$. Hence a parametrization of $V$ needs two big mixing angles ($\theta_x$ and $\theta_y$) and one small mixing angle ($\theta_z$), in addition to a few complex phases. Following Ref. [8], we obtain

$$\tan 2\theta_x \approx 2\sqrt{2}\frac{\lambda}{\delta}\ , \quad \tan 2\theta_y \approx \frac{2}{\delta}\ , \quad \tan 2\theta_z \approx \frac{\lambda}{\sqrt{2}}\ .$$

Taking $\delta = \sqrt{2}\lambda$, we explicitly obtain $\theta_x \approx 31.7°$, $\theta_y \approx 40.6°$ and $\theta_z \approx 4.4°$, favored by current experimental data. In addition, the masses of three light neutrinos are given by

$$\frac{m_1}{m_3} \approx \frac{\lambda \tan\theta_x}{2\sqrt{2}}\ , \quad \frac{m_2}{m_3} \approx \frac{\lambda \cot\theta_x}{2\sqrt{2}}\ , \quad m_3 \approx \frac{2m_0^2}{M_0}\ .$$

We see that a normal neutrino mass hierarchy $m_1 : m_2 : m_3 \sim \lambda : \lambda : 1$ shows up. Then the absolute value of $m_3$ can be determined from the observed mass-squared difference of atmospheric neutrino oscillations $\Delta m^2_{\rm atm} \equiv |m_3^2 - m_2^2| \approx m_3^2$. Using $\Delta m^2_{\rm atm} = (1.6 - 3.9) \times 10^{-3}$ eV$^2$ [15], we obtain $m_3 \approx \sqrt{\Delta m^2_{\rm atm}} \approx (4.0 - 6.2) \times 10^{-2}$ eV. Given $m_0 \approx v$ for the Dirac neutrino mass matrix $M_{\rm D}$, the mass scale of three heavy Majorana neutrinos turns out to be $M_0 \approx 2v^2/m_3 \approx (4.9 - 7.6) \times 10^{14}$ GeV, which is quite close to the scale of grand unified theories $\Lambda_{\rm GUT} \sim 10^{16}$ GeV.

Note that our ansatz predicts a very small value for the effective mass term of the neutrinoless double-$\beta$ decay: $\langle m \rangle_{ee} \approx \lambda^2 m_3/8 \approx (2.4 - 3.8) \times 10^{-4}$ eV, which is much lower than the present experimental upper bound ($\langle m \rangle_{ee} < 0.35$ eV at the 90% C.L.[16]) and seems hopeless to be detected in practice. We also obtain the Jarlskog parameter, which measures CP and T violation in neutrino oscillations, as $\mathcal{J} \approx \lambda/(4\sqrt{10}) \approx 2\%$. Leptonic CP violation at the percent level could be measured in the future at neutrino factories.

The symmetric mass matrix $M_{\rm R}$ can be diagonalized by a unitary matrix $U_{\rm R}$, as pointed out above. To leading order, we find $M_1 \approx \lambda^6 M_0/z$, $M_2 \approx z\lambda^4 M_0$ and $M_3 \approx M_0$ as well as $U_{\rm R11} \approx i$, $U_{\rm R22} \approx U_{\rm R33} \approx 1$, $U_{\rm R12} \approx iU_{\rm R21} \approx \lambda/z$, $U_{\rm R13} \approx$

0, $U_{R31} \approx i\lambda^5/z$, and $U_{R23} \approx -U_{R32} \approx \lambda^4$. One can see that the masses of three heavy Majorana neutrinos perform a strong hierarchy. Note that $\{M_1, M_2, M_3\} \approx \{5.2\times10^{10}, 1.8\times10^{12}, 6.0\times10^{14}\}$ GeV, if we typically take $M_0 = 6.0 \times 10^{14}$ GeV and $z = 1+\sqrt{2}\lambda$.

A CP-violating asymmetry ($\varepsilon_1$) may result from the interference between tree-level and one-loop amplitudes of the decay of the *lightest* heavy Majorana neutrino $N_1$, as already presented in Eq. (7). In our ansatz, we can explicitly obtain

$$\varepsilon_1 \approx -\frac{3\lambda^6}{16\pi}\left[\sin 2(2\omega-\zeta) - 2x(1+x^2)\sin\omega \right. \\ \left. + x^2 z\sin 2\omega\right]\left[z\left(1+x^2+z^2-2xz\cos\omega\right)\right]^{-1} .$$

For illustration, we adopt the specific parameter space chosen above to evaluate the size of $\varepsilon_1$. The result is $\varepsilon_1 \approx -5.2\times10^{-6}$. To translate this CP asymmetry into the lepton asymmetry $Y_L$ and the baryon asymmetry $Y_B$, it is necessary to calculate the dilution factor $d$ appearing in Eq. (4). Note that $d$ depends closely on the following quantity: $K_R \equiv [U^T M_D^\dagger M_D U^*]_{11} M_{Pl}/(8\pi v^2 \cdot 1.66\sqrt{g_*}M_1)$ with $M_{Pl} \approx 1.22\times10^{19}$ GeV, which characterizes the out-of-equilibrium decay rate of $N_1$. Making use of the typical inputs taken above, we arrive at $K_R \approx 73$. The dilution factor $d$ can then be calculated with the help of an approximate parametrization [17] obtained from integrating the Boltzmann equations (for the range $10 \le K_R \le 10^6$): $d \approx 0.3/[K_R(\ln K_R)^{0.6}] \approx 1.7\times10^{-3}$. Finally we get a very instructive prediction for the baryon asymmetry of the universe from Eq. (4): $Y_B \approx 4.7\times10^{-11}$. One can see that this result is consistent quite well with the observational value of $Y_B$ quoted in Eq. (1).

One may go beyond the typical parameter space taken in this talk to make a delicate analysis of all measurables or observables. It is remarkable that we can quantitatively interpret both the baryon asymmetry of the universe and the small mass-squared differences and large mixing factors of solar and atmospheric neutrino oscillations. In this sense, our ansatz is a *complete* phenomenological ansatz favored by current experimental and observational data, although it has not been incorporated into a convincing theoretical model.

This work was supported in part by the National Natural Science Foundation of China.

## REFERENCES


1. Particle Data Group, K. Hagiwara *et al.*, Phys. Rev. D **66** (2002) 010001; K.A. Olive, G. Steigman, and T.P. Walker, Phys. Rep. **333** (2000) 389.
2. A.D. Sakharov, JETP Lett. **5** (1967) 24.
3. M. Fukugita and T. Yanagida, Phys. Lett. B **174** (1986) 45.
4. S. Fukuda *et al.*, Phys. Rev. Lett. **85** (2000) 3999; **86** (2001) 5656.
5. Q.R. Ahmad *et al.*, Phys. Rev. Lett. **87** (2001) 071301; nucl-ex/0204008.
6. V.A. Kuzmin, V.A. Rubakov, and M.E. Shaposhnikov, Phys. Lett. B **155** (1985) 36.
7. T. Yanagida, in *Proceedings of the Workshop on Unified Theory and Baryon Number in the Universe*, KEK report 79-18 (1979), p. 95; M. Gell-Mann, P. Ramond, and R. Slansky, in *Supergravity*, edited by P. van Nieuwenhuizen and D. Freedman (North Holland, 1979), p. 315; R.N. Mohapatra and G. Senjanovic, Phys. Rev. Lett. **44** (1980) 912.
8. Z.Z. Xing, hep-ph/0206245; to appear in Phys. Lett. B.
9. W. Buchmüller and M. Plümacher, Phys. Lett. B **389** (1996) 73; Int. J. Mod. Phys. A **15** (2000) 5047.
10. See, e.g., G.C. Branco *et al.*, hep-ph/0202030; J. Ellis and M. Raidal, hep-ph/0206174; S. Davidson and A. Ibarra, hep-ph/0206304.
11. See, e.g., H. Fritzsch and Z.Z. Xing, Phys. Lett. B **372** (1996) 265; Phys. Lett. B **440** (1998) 313; Phys. Rev. D **61** (2000) 073016.
12. W. Buchmüller and D. Wyler, Phys. Lett. B **521** (2001) 291.
13. See, e.g., H. Fritzsch and Z.Z. Xing, Phys. Lett. B **353** (1995) 114; D. Du and Z.Z. Xing, Phys. Rev. D **48** (1993) 2349.
14. H. Fritzsch and Z.Z. Xing, Prog. Part. Nucl. Phys. **45** (2000) 1.
15. M. Shiozawa, talk at *Neutrino 2002*, Munich.
16. H.V. Klapdor, hep-ph/0103074.
17. E.W. Kolb and M.S. Turner, in *The early universe*, Addison-Wesley (1990).

ELSEVIER

Proceedings of ICHEP 2002, pp. 47–52
*S. Bentvelsen, P. de Jong, J. Koch and E. Laenen (Editors)*

31st INTERNATIONAL CONFERENCE ON HIGH ENERGY PHYSICS AMSTERDAM

# Neutrino oscillations and warped geometry

Stephan J. Huber[a] and Qaisar Shafi[b]

[a]Deutsches Elektronen-Synchrotron DESY, Hamburg, Germany

[b]Bartol Research Institute, University of Delaware, Newark, USA

It is shown how small Dirac and Majorana masses for neutrinos arise in five dimensional models with warped geometry. The atmospheric and solar neutrino anomalies can be satisfactorily resolved. Rare processes, such as proton decay, neutrinoless double beta decay and flavor violation, may provide valuable information on the higher dimensional theory.

## 1. Introduction

The subject of extra dimensions has created much recent excitement for a variety of reasons. Before discussing them, let us recall some earlier motivations:

• Extra dimensions may lead to 'deeper' unification than grand unification.
For instance, through inclusion of gravity, and unification of 'matter', gauge
and 'Higgs' sectors .

• Superstring theories for their consistency require 10 space-time dimensions.

• A particularly promising approach is based on the presumed existence of
M-theory , whose low energy limit , it is speculated, is 11D supergravity.

Alas, the predictive power of these higher dimensional theories is quite limited. Indeed, it is far from clear how this top-down approach leads to the SM (or MSSM) at low energies. It may greatly help matters if a bridge can be developed between the low energy world of the SM and MSSM and physics at the highest energies, such that the presence of the extra dimensions is somehow clearly exploited.

A number of recent attempts try to show, for instance, how the presence of one or more extra dimension can be exploited to yield possible resolution of the gauge hierarchy problem, the doublet-triplet problem, the apparent stability of the proton, and so on.

The idea that one or more of the extra dimensions may be warped can be motivated as follows: – It provides possible resolution of the gauge hierarchy problem, namely how $M_W \ll M_P$ [1]. Note that this can be achieved without invoking SUSY; – The wave functions of bulk fields are far more constrained compared to the case of 'flat' geometry. In principle, this could lead to more predictive power; – The observed atmospheric and solar neutrino oscillations can be accommodated within the SM framework without invoking additional fields such as singlet (right handed) neutrinos; – The approach may be consistent with grand unification, which is not the case say when dealing with large extra dimensions; – There exists the intriguing possibility that 'warping' could be exploited to understand the 'tiny' vacuum energy density $\rho_V \sim 10^{-120} M_P^4$ [2]; – There are experimental predictions which can be experimentally tested. For instance, the KK excitations of bulk SM fields are expected to have masses in the several TeV range which, hopefully, can be tested at the LHC. Since there are no large extra dimensions, it is warping which makes these excitations experimentally accessible (without warping the masses would be of order $M_P$ and therefore inaccessible for ever!).


doi:10.1016/S0920-5632(02)02112-6

## 2. Fermions in a warped background

By the Kaluza-Klein (KK) procedure the 5D fields are decomposed into an infinite tower of 4D fields. The wave functions encode information on where the KK states are localized in the extra dimension. Together with the KK masses they are obtained by solving the 5D equations of motion. In five dimensions fermions are vector-like, and we can associate with them a 5D Dirac mass term, parameterized by $m_\Psi = c \cdot k\ \mathrm{sgn}(y)$, where $y$ is the 5th coordinate. Depending on the $Z_2$ orbifold transformation property of the fermion, $\Psi(-y)_\pm = \pm\gamma_5\Psi(y)_\pm$, the left-handed (right-handed) zero mode, $f_0 \sim e^{(2-c)k|y|}$, of the KK decomposition is projected out by the boundary conditions [3–5]. The KK excited states come in left- and right-handed pairs, which are degenerate in mass. Note that the 5D Dirac mass parameter regulates whether the zero mode is localized towards the Planck-brane ($c > 1/2$) or the TeV-brane ($c < 1/2$).

## 3. The Dirac case

It was suggested in ref. [3] that in warped models small Dirac neutrino masses can arise from a coupling to sterile (right-handed) bulk neutrinos. In order to generate masses in the sub-eV range, the sterile neutrinos have to be localized close to the Planck-brane, while the SM model neutrinos were confined to the TeV-brane. We generalized this scenario to incorporate bulk SM neutrinos [6].

After electroweak symmetry breaking, the Dirac neutrino masses are generated from the Yukawa-type coupling between the SM neutrinos, contained in the SU(2) doublets $L_i$, and the three sterile neutrinos $\psi_i$

$$\mathcal{L} = h_{ij}^{(5)} \bar{L}_i \psi_j H + \text{h.c.} + \ \ldots \tag{1}$$

where $h_{ij}^{(5)}$ are the 5D Yukawa couplings. The 4D Dirac masses are given by

$$m_{ij} = \int_{-\pi R}^{\pi R} \frac{dy}{2\pi R} h_{ij}^{(5)} e^{-4\sigma} H(y) f_{0iL}(y) f_{0j\psi}(y), \tag{2}$$

where the Higgs field is confined to the TeV-brane, i.e. $H(y) = H_0\delta(y - \pi R)$, to maintain the solution of the hierarchy problem.

The fermion wave functions and consequently the induced 4D Dirac masses crucially depend on the 5D mass parameters of the left- and right-handed fermions, $c_L$ and $c_\psi$ respectively, which enter eq. (2). As the 5D Dirac mass, i.e. $c$ parameter of the fermion increases, it gets localized closer towards the Planck-brane. Its overlap with the Higgs profile at the TeV-brane is consequently reduced, which is reflected in a smaller 4D fermion mass from electroweak symmetry breaking. In ref. [5] we have shown that this geometrical picture nicely generates the charged lepton and quark mass hierarchies, as well as quark mixings, by employing $c$-parameters and 5D Yukawa couplings of order unity.

From the KK reduction of the left-handed neutrino field $\nu_L$ we obtain a left-handed zero mode $\nu_L^{(0)}$, corresponding to the SM neutrino, and an infinite tower of left- and right-handed KK excited states $\nu_L^{(i)}$ and $\nu_R^{(i)}$, where we omit flavor indices. The sterile neutrinos decompose into the right-handed zero mode $\psi_R^{(0)}$ and the KK excited states $\psi_L^{(i)}$ and $\psi_R^{(i)}$. After electroweak symmetry breaking the mass matrix takes the form

$$\begin{pmatrix} \nu_L^{(0)} \\ \nu_L^{(i)} \\ \psi_L^{(i)} \end{pmatrix}^T \begin{pmatrix} m^{(0,0)} & 0 & m^{(0,j)} \\ m^{(i,0)} & m_{L,i} & m^{(i,j)} \\ 0 & 0 & m_{\psi,i} \end{pmatrix} \begin{pmatrix} \psi_R^{(0)} \\ \nu_R^{(j)} \\ \psi_R^{(j)} \end{pmatrix} \tag{3}$$

where we again suppress flavor indices, i.e. every entry represents a $3 \times 3$ matrix in flavor space. The masses $m^{(i,j)}$ are obtained by inserting the relevant wave functions into eq. (2), and $m_{L,i}$ and $m_{\psi,i}$ denote the KK masses of the excited neutrino states. The zeros in (3) follow from the $Z_2$ orbifold properties of the wave functions.

The squares of the physical neutrino masses are the eigenvalues of the hermitian mass matrix $M_\nu M_\nu^\dagger$. The unitary matrix $U_\nu$, such that $U_\nu^\dagger M_\nu M_\nu^\dagger U_\nu$ is diagonal, relates the left-handed mass eigenstates $N_L^{\text{phys}}$ to the interaction eigenstates $N_L = (\nu_L^{(0)}, \nu_L^{(i)}, \psi_L^{(i)})$ via $N_L = U_\nu N_L^{\text{phys}}$. The observed neutrinos $\nu_L^{\text{phys}}$ correspond to the three lightest states in $N_L^{\text{phys}}$. The right-handed mass eigenstates are obtained from a unitary matrix $V_\psi$ which diagonalizes $M_\nu^\dagger M_\nu$. The physical neutrino mixings

$$U = U_l^\dagger U_\nu \tag{4}$$

also depend on the rotations of the left-handed charged leptons $U_l$. If the Yukawa couplings of the charged leptons are diagonal, the mixing angles of the physical neutrinos can be directly read off from $U$. Like in the CKM matrix, there are three mixing angles, and a single complex phase $\delta$ which induces CP violation in the lepton sector.

The atmospheric neutrino data imply [8]

$$1 \cdot 10^{-3}\mathrm{eV}^2 < \Delta m^2_{\mathrm{atm}} < 5 \cdot 10^{-3}\mathrm{eV}^2, \quad \sin^2 2\theta_{\mathrm{atm}} > 0.85. \tag{5}$$

There are several solutions to the solar neutrino anomaly [7]. Most favored is the LMA solution

$$2.5 \cdot 10^{-5}\mathrm{eV}^2 < \Delta m^2_{\mathrm{sol}} < 4 \cdot 10^{-4}\mathrm{eV}^2, \quad 0.3 < \sin^2 2\theta_{\mathrm{sol}} < 0.93. \tag{6}$$

The SMA solution gives only a poor fit to the data. The CHOOZ reactor experiment constrains $|U_{e3}|^2 \equiv s_2^2$ to be at most a few percent [9]. Nothing is known about the CP violating phase $\delta$.

Various constraints on the scenario with bulk gauge and fermion fields have been discussed in the literature [10–12]. With bulk gauge fields for instance, the SM relationship between the gauge couplings and masses of the Z and W bosons gets modified. The electroweak precision data then requires the lowest KK excitation of the gauge bosons to be heavier than about 10 TeV [10,11]. This bound becomes especially important if the fermions are localized towards the Planck-brane ($c > 1/2$). As discussed in ref. [5], this applies to all SM fermions, with the exception of the top-quark, if the fermion mass hierarchy arises from their different locations in the extra dimension. In this case the bounds induced by the contribution of KK excitations of the SM gauge bosons to the electroweak precision observables are weak. They only require the KK masses to be above about 1 TeV [4]. In the following we will therefore assume that the mass of the first KK gauge boson is $m_1^{(G)} = 10$ TeV. The corresponding masses of the KK fermions are then in the range 10 to 16 TeV, for $0 < c < 1.5$. The mass of the lightest KK graviton is 16 TeV.

In order to induce large mixings, the neutrinos have to be located at similar positions in the extra dimension. Therefore, we take $c_L^{(1)} = c_L^{(2)} = c_L^{(3)} \equiv c_L$. The charged lepton masses are, for instance, reproduced by $c_L = 0.567$, $c_E^{(1)} = 0.787$, $c_E^{(2)} = 0.614$ and $c_E^{(3)} = 0.50$, where for simplicity we have assumed diagonal Yukawa couplings for the charged leptons. The neutrino mixings are then solely governed by the neutrino mass matrix $M_\nu$. We take $k = M_5 = M_{\mathrm{Pl}}$. From $m_1^{(G)} = 10$ TeV we determine the brane separation $kR = 10.83$. A parameter set which implements the large angle MSW solution is

$$c_\psi^{(1)} = 1.43,\ c_\psi^{(2)} = 1.36,\ c_\psi^{(3)} = 1.30,$$

$$\frac{h_{ij}^{(5)}}{g_2^{(5)}} = \begin{pmatrix} -2.0 & 1.5 & -0.5 \\ -1.8 & -1.1 & 1.9 \\ 0.5 & 1.9 & 1.7 \end{pmatrix}. \tag{7}$$

We obtain the light neutrino masses $m_{\nu 1} = 1.0 \cdot 10^{-3}$ eV, $m_{\nu 2} = 1.0 \cdot 10^{-2}$ eV and $m_{\nu 3} = 7.1 \cdot 10^{-2}$ eV. For the neutrino oscillation parameters we find $\Delta m^2_{\mathrm{atm}} = 4.9 \cdot 10^{-3}$ eV$^2$, $\sin^2 2\theta_{\mathrm{atm}} = 0.99$, $\Delta m^2_{\mathrm{sol}} = 1.0 \cdot 10^{-4}$ eV$^2$ and $\sin^2 2\theta_{\mathrm{sol}} = 0.90$. Also, $|U_{e3}|^2 = 0.036$ is close to the experimental bound which is typical for the solutions we find.

The SM neutrinos mix not only with each other but also with the left-handed KK states of the sterile neutrinos $\psi_L^{(i)}$ [3]. This effect diminishes the effective weak charge of the light neutrinos. Thus, the effective number of neutrinos contributing to the width of the Z boson is reduced to $n_{\mathrm{eff}} = 3 - \delta n$, where $\delta n$ is obtained from summing the relevant squared entries of $U_\nu$. Measurements of the Z width induce the constraint $\delta n \lesssim 0.005$. For the parameter set (7) we find the result $\delta n = 2 \cdot 10^{-5}$, well below the experimental sensitivity.

## 4. The Majorana case

In the warped SM the suppression scale for non-renormalizable operators can be anywhere between a few TeV and the Planck scale, depending on the localization of the fermions [4,5]. Majorana masses for the left-handed neutrinos are generated by the dimension-five operator [4,5,13],

$$\int d^4x \int dy \sqrt{-g} \frac{l_{ij}}{M_5^2} H^2 \Psi_{iL} C \Psi_{jL} \equiv$$
$$\equiv \int d^4x \, M_{ij}^{(\nu)} \Psi_{iL}^{(0)} C \Psi_{jL}^{(0)}, \quad (8)$$

where $l_{ij}$ are dimensionless couplings constants and $C$ is the charge conjugation operator. The neutrino mass matrix reads

$$M_{ij}^{(\nu)} = \int_{-\pi R}^{\pi R} \frac{dy}{2\pi R} \frac{l_{ij}}{M_5^2} e^{-4\sigma} H^2 f_{0iL} f_{0jL}. \quad (9)$$

By localizing $f_{0L}$ close to the Planck-brane ($c_L \gg 1/2$), $M^{(\nu)}$ can be made very small. However, at the same time, the masses of the charged leptons, residing in SU(2) multiplets with the neutrinos, can become very small as well [4,5]. Since the tau is the heaviest lepton, it has to reside closest to the TeV-brane, and induces the most stringent constraint on $l$. In ref. [13] we considered the somewhat favorable case $\lambda_\tau^{(5)} = 10 g_2^{(5)}$ and $k/M_5 = 0.01$, obtaining $m_{\nu,\tau} = l \cdot 25$ eV. A neutrino mass of 50 meV can then be generated by $l = 0.002$, which is only a moderately small number.

Following the spirit of "anarchy", we parametrize the coefficients in (9) by

$$l_{ij} \equiv \Lambda \cdot \tilde{l}_{ij}. \quad (10)$$

The supposedly fundamental theory responsible for the effective interaction (8) is represented through the order unity coefficients $\tilde{l}_{ij}$. In order to generate large mixings we locate the left-handed doublets at the same position in the extra dimension,

$$c_{e,L} = c_{\mu,L} = c_{\tau,L} = 0.72. \quad (11)$$

To accommodate the tau, muon and electron masses we are led to $c_{\tau,R} = 1/2$, $c_{\mu,R} = 0.64$ and $c_{e,R} = 0.85$. Taking the order unity coefficients to be homogeneously distributed $1/2 < |\tilde{l}_{ij}| < 2$ with random phases from 0 to $2\pi$, we find the most favorable value for the overall scale to be $\Lambda = 1.9 \cdot 10^{-3}$. We randomly generate parameter sets for $\tilde{l}_{ij}$, calculate the neutrino mass matrix from (9) and compute the neutrino masses and mixings. The left-handed charged lepton rotations are similar to the neutrino rotations because both fields sit together in the extra dimension. We include them by generating random Yukawa couplings for the charged leptons as well.

| | $\Delta m^2_{\rm atm,sol}$ | $+\sin^2 2\theta_{\rm atm,sol}$ | $+\|U_{e3}\|^2$ |
|---|---|---|---|
| LMA | 28.1 | 5.9 | 0.7 |
| SMA | 0.04 | < 0.001 | < 0.001 |
| LOW | < 0.001 | < 0.001 | < 0.001 |

Table 1
Probability in percent that a randomly generated set of coefficients $1/2 < |\tilde{l}_{ij}| < 2$ satisfies the constraints from $\Delta m^2_{\rm atm,sol}$ (first column) and $\sin^2 2\theta_{\rm atm,sol}$ (second column) and $|U_{e3}|^2 < 0.05$ (third column). The results are given for the case $c_{e,L} = c_{\mu,L} = c_{\tau,L}$.

Focusing on the LMA solution of the solar neutrino anomaly, which turns out to be clearly favored, we find the following picture [13]. From our neutrino parameter sets about 70 % reproduce $\Delta m^2_{\rm atm}$ (5). Imposing in addition the constraint from $\Delta m^2_{\rm sol}$ (6), we are left with about 28 % of the parameter sets. The solar and atmospheric mixings angles bring this number down to about 6 %, which is still a considerable fraction given the number of constraints. The most stringent constraint turns out to come from the CHOOZ experiment. Adding the requirement $|U_{e3}|^2 < 0.05$, the fraction of viable parameter sets shrinks to about 0.7 %. This result is closely related to an average $\langle |U_{e3}|^2 \rangle = 0.22$, which is considerably above the experimental bound. Averaging only over the sets which pass all experimental constraints, we find $\langle |U_{e3}|^2 \rangle = 0.022$. This means that $|U_{e3}|^2$ is probably close to the experimental bound and can likely be tested at future neutrino experiments, such as MINOS, ICARUS and OPERA which are sensitive down to $|U_{e3}|^2 \sim 0.01$ [14]. Since we typically start with large phases in the neutrino and charged lepton mass matrices, the complex phase $\delta$ in the mixing matrix $U$ is found to be most likely of order unity.

The SMA, LOW and VAC solutions to the solar neutrino anomaly can be realized only with severe fine-tuning. Our results are summarized in table 1.

So far the constraint from the CHOOZ experiment is the most stringent one. Its inclusion reduces the probability that a parameter set realizes the LMA solution from about 6 % to less than one percent. With all entries in the neutrino mass matrix being of similar magnitude, the ensuing mixing angles are typically large. The fit to the neutrino data improves considerably if the electron neutrino is somewhat separated from the other two neutrino species. Shifting the electron neutrino closer to the Planck-brane induces small elements in the neutrino mass matrix. As a result, small values of $\Delta m^2_{\rm sol}$ and $|U_{e3}|^2$ become more probable. The neutrino mass matrix (9) acquires the following structure

$$M^{(\nu)} \sim \begin{pmatrix} \epsilon^2 & \epsilon & \epsilon \\ \epsilon & 1 & 1 \\ \epsilon & 1 & 1 \end{pmatrix}, \tag{12}$$

where $\epsilon \approx f^{(\nu)}_{0,e}(\pi R)/f^{(\nu)}_{0,\mu}(\pi R) \approx \exp(-(c_{e,L} - c_{\mu,L})\pi k R)$.

We keep the muon and tau neutrinos at the previous locations $c_{\mu,L} = c_{\tau,L} = 0.72$. A separation of these two would make the fit to the atmospheric neutrino data more difficult since the corresponding mixing is reduced. For the electron neutrino the most favorable choice turns out to be $c_{e,L} = 0.79$ leading to $\epsilon = 0.15$. The positions of the right-handed leptons are found to be $c_{e,R} = 0.79$, $c_{\mu,R} = 0.62$ and $c_{\tau,R} = 0.49$.

Let us again focus on the LMA solution. Assuming $1/2 < |\tilde{l}_{ij}| < 2$ for the order unity coefficients, we find the best value $\Lambda = 2.6 \cdot 10^{-3}$. The correct values of the $\Delta m^2$'s are fitted by 44 % of the parameter sets. Taking into account also the constraints from the solar and atmospheric mixing angles, this fraction shrinks to 5.8 %. This reduction is mostly due to the solar mixing angle which is suppressed if $\epsilon$ is small. Most important, however, the CHOOZ constraint $|U_{e3}|^2 < 0.05$ is now satisfied almost automatically, and we are finally left with 5.0 % of the parameter sets. Compared to the case of $\epsilon = 1$, the probability for a parameter set to satisfy all observational constraints is enhanced by almost an order of magnitude. The distribution of $|U_{e3}|^2$ is now peaked at small values. Averaging only over parameter sets which satisfy all constraints, we find a moderately small value $\langle |U_{e3}|^2 \rangle = 0.010$. As a consequence, the next generation neutrino experiments still has a chance to detect a non-vanishing $|U_{e3}|^2$ [14].

Taking even smaller values of $\epsilon$, it is possible to implement the SMA solution to the solar neutrino problem. The LOW and VAC solutions are very difficult to accommodate because of their small solar $\Delta m^2$. Thus, the LMA solution is by far the most favored scenario. A collection of our results is given table 2.

| | $\Delta m^2_{\rm atm,sol}$ | $+\sin^2 2\theta_{\rm atm,sol}$ | $+\lvert U_{e3}\rvert^2$ |
|---|---|---|---|
| LMA | 44.4 | 5.8 | 5.0 |
| SMA | 1.3 | 0.3 | 0.3 |
| LOW | 0.008 | 0.002 | 0.002 |

Table 2
As in table 1, but for $c_{e,L} > c_{\mu,L} = c_{\tau,L}$.

## 5. Rare processes

The most direct evidence for a Majorana mass of neutrinos would be the discovery of neutrinoless double beta decay ($0\nu\beta\beta$). The non-observation of this process in the Heidelberg-Moscow experiment implies an upper bound on $|M^{(\nu)}_{11}| \equiv m_{ee} < 0.35$ eV [15]. There are plans to bring this limit down to about 0.01 eV. If all neutrinos are at the same position in the extra dimension, we expect $m_{ee} \sim 0.02$ eV, which is far below the current experimental sensitivity but within reach of future experiments. Once the electron neutrino is localized closer towards the Planck-brane $0\nu\beta\beta$ becomes drastically suppressed proportional to $\epsilon^2$. For the clearly favored LMA scenario we find $m_{ee} \sim 0.001$ eV, which is too small to be detected in the near future. For the SMA, LOW and VAC solutions $m_{ee}$ is even smaller. It is therefore questionable if our model induces $0\nu\beta\beta$ at a detectable level.

The low cut-off scale dramatically amplifies the impact of non-renormalizable operators on rare processes, such as proton decay and flavor violation. With bulk fields localized towards the Planck-brane the corresponding suppression scales can be significantly enhanced [4, 5]. However, there are limits because the SM fermions need to have sufficient overlap with the Higgs field at the TeV-brane to acquire their observed masses. In the case of proton decay, typical dimension-six operators still have to be suppressed by small couplings of order $10^{-8}$ to be compatible with observations. One may be tempted to completely eliminate these small numbers by imposing additional symmetries. In ref. [6] we used lepton number conservation to make the proton stable and to forbid the potentially large Majorana neutrino masses of eq. (8). In the case of Majorana neutrinos proton stability can be ensured by lepton parity. In both cases baryon number violation may still be present, and processes like neutron anti-neutron oscillations and double nucleon decay could be close to their experimental bounds.

In the scenario of Dirac neutrino masses the rate of $\mu \to e\gamma$ transitions is considerably enhanced by the presence of heavy sterile neutrino states. If the SM neutrinos are confined to the TeV-brane, its large branching ratio pushes the KK scale up to 25 TeV and thus imposes the most stringent constraint on the model [16]. However, the rate for $\mu \to e\gamma$ is very sensitive to the mixing between light and heavy neutrino states. With bulk neutrinos the mixing with heavy states is considerably reduced. In the case of the large angle MSW solution (7) we obtain $\mathrm{Br}(\mu \to e\gamma) \approx 10^{-15}$ [6].

## 6. Conclusions

We have discussed how the atmospheric and solar neutrino oscillations, especially the bimaximal mixing scenario, are incorporated in models with a small but warped extra dimension. In one scenario Dirac neutrino masses arise from a coupling to right-handed neutrinos in the bulk. Alternatively, dimension five interactions may induce a Majorana mass for the neutrinos. In both cases the mixing angle $U_{e3}$ is expected be close to its present bound. While the direct detection of KK states is still waiting, rare processes, such as proton decay, neutrinoless double beta decay and flavor violation can provide important hints to the higher dimensional theory.

## REFERENCES

1. L. Randall and R. Sundrum, *Phys. Rev. Lett.* **83** (1999) 3370; M. Gogberashvili, hep-ph/9812296.
2. S J. Huber and Q. Shafi, hep-ph/0207232.
3. Y. Grossman and M. Neubert, *Phys. Lett.* **B474** (2000) 361.
4. T. Gherghetta and A. Pomarol, *Nucl. Phys.* **B586** (2000) 141.
5. S.J. Huber and Q. Shafi, *Phys. Lett.* **B498** (2001) 256.
6. S.J. Huber and Q. Shafi, *Phys. Lett.* **B512** (2001) 365.
7. V. Barger, D. Marfatia, K. Whisnant and B.P. Wood, hep-ph/0204253; J.N. Bahcall, M.C. Gonzalez-Garcia, C. Pena-Garay, hep-ph/0204314.
8. T. Toshito [SuperKamiokande Collaboration], hep-ex/0105023.
9. M. Apollonio *et al.* [CHOOZ Collaboration], *Phys. Lett.* **B466** (1999) 415.
10. S.J. Huber and Q. Shafi, *Phys. Rev.* **D63** (2001) 045010.
11. S.J. Huber, C.-A. Lee and Q. Shafi, *Phys. Lett.* **B531** (2002) 112.
12. J.L. Hewett, F.J. Petriello and T.G. Rizzo, *JHEP* **0209** (2002) 030.
13. S.J. Huber and Qaisar Shafi *Phys. Lett.* **B544** (2002) 295.
14. M. Komatsu, P. Migliozzi and F. Terranova, hep-ph/0210043.
15. The Heidelberg-Moscow collaboration, H.V. Klapdor-Kleingrothaus *el al.* in the proceedings of *DARK 2000*, Heidelberg, Germany, 10-16 July 2000, 520-533 [hep-ph/0103062].
16. R. Kitano, *Phys. Lett.* **B481** (2000) 39 [hep-ph/0002279].

ELSEVIER

Proceedings of ICHEP 2002, pp. 53–56
*S. Bentvelsen, P. de Jong, J. Koch and E. Laenen (Editors)*

31st INTERNATIONAL CONFERENCE ON HIGH ENERGY PHYSICS AMSTERDAM

# Testing CPT Invariance with Neutrinos

Tommy Ohlsson[a*†]

[a]Division of Mathematical Physics, Department of Physics, Royal Institute of Technology (KTH) – Stockholm Center for Physics, Astronomy, and Biotechnology (SCFAB), Roslagstullsbacken 11, 106 91 Stockholm, Sweden

We investigate possible tests of CPT invariance on the level of event rates at neutrino factories. We do not assume any specific model, but phenomenological differences in the neutrino-antineutrino masses and mixing angles in a Lorentz invariance preserving context, which could be induced by physics beyond the Standard Model. We especially focus on the muon neutrino and antineutrino disappearance channels in order to obtain constraints on the neutrino-antineutrino mass and mixing angle differences. In a typical neutrino factory setup simulation, we find, for example, that $|m_3 - \overline{m}_3| \lesssim 1.9 \cdot 10^{-4}$ eV and $|\theta_{23} - \bar{\theta}_{23}| \lesssim 2^\circ$.

## 1. INTRODUCTION

The theme of the current presentation is to discuss the plausibility of testing possible CPT invariance violation in the neutrino sector using future so-called neutrino factories.

The CPT theorem is one of the milestones of local quantum field theory (QFT). Moreover, the standard model (SM) of elementary particle physics is in very good agreement with all (at least most) existing experimental data. The CPT theorem is valid for the SM. Thus, CPT violation is connected with the search for physics beyond the SM. So far, no CPT violation has been found. However, neutrinos have been suggested as a source of CPT violation.

Prior papers on CPT violation with neutrinos include Refs. [1–4]. Recently, neutrinos as a source of CPT violation have been investigated by several authors [5–14]. Some of the models of CPT violation with neutrinos (beyond the SM) are the following:

- S.R. Coleman and S.L. Glashow [1,2]: A Lorentz and CPT-violating model is introduced in which the most general CPT-violating interaction $u^\dagger b u$ is allowed, where $b$ is a Hermitian matrix [15,16]. This interaction implies that the energies of the ultra-relativistic neutrinos with definite momentum $p$ are the eigenvalues of the matrix:

$$cp + \frac{m^2}{2p} + b,$$

where $m^2 \equiv mm^\dagger$ is the Hermitian mass squared matrix ($m$ is a complex symmetric mass matrix) and $c$ is also a Hermitian matrix, which describes velocity-mixing. Limiting the discussion to neutrino oscillations with two flavors, the neutrino oscillation transition probability formula becomes

$$\mathrm{P}(\nu_\alpha \to \nu_{\alpha'}) = 1 - \sin^2 2\theta \times \sin^2\left[\left(\frac{\Delta m^2}{4E} + \frac{\Delta b}{2} + \frac{\Delta c E}{2}\right) L\right],$$

where $\Delta m^2$, $\Delta b$, and $\Delta c$ are the differences between the eigenvalues of the matrices $m^2$, $b$, and $c$, respectively. Note that in order for the above formula to hold the mixing angles that diagonalize the matrices $m^2$, $b$, and $c$ need all to be equal to each other.

*E-mail: tommy@theophys.kth.se

†In collaboration with: S.M. Bilenky, M. Freund, M. Lindner, and W. Winter.


doi:10.1016/S0920-5632(02)02116-3

- V.D. Barger *et al.* [3]: The effective Lorentz and CPT-violating interaction for neutrinos is: $\bar{\nu}_L^{\alpha} b^{\mu}_{\alpha\alpha'} \gamma_\mu \nu_L^{\alpha'}$, which implies that

$$\begin{aligned} \Delta \mathrm{P}^{\mathrm{CPT}}_{\alpha\alpha} &\equiv \mathrm{P}(\nu_\alpha \to \nu_\alpha) - \mathrm{P}(\bar{\nu}_\alpha \to \bar{\nu}_\alpha) \\ &= -2\sin^2 2\theta \sin\left(\frac{\Delta m^2 L}{2E}\right) \\ &\times \sin(\Delta b L) \end{aligned}$$

when the matrices $m^2$ and $b$ are diagonalized by the same mixing angle $\theta$.

- G. Barenboim *et al.* [6]: CPT violation (but no Lorentz violation) is suggested to be so strong that the mass spectra of neutrinos and antineutrinos are completely different. This means that it would be possible to describe solar, atmospheric, and LSND neutrino data at the same time. Such a model is accomplished by changing the Hamiltonian from

$$H_0 = \int \frac{d^3p}{(2\pi)^3}(\mathbf{p}^2+m^2)\sum_s \left[a_{\mathbf{p}}^{s\,\dagger} a_{\mathbf{p}}^s + b_{\mathbf{p}}^{s\,\dagger} b_{\mathbf{p}}^s\right]$$

to

$$\begin{aligned} H_0 &= \int \frac{d^3p}{(2\pi)^3}\sum_s \Big[(\mathbf{p}^2+m^2) a_{\mathbf{p}}^{s\,\dagger} a_{\mathbf{p}}^s \\ &+ (\mathbf{p}^2+\overline{m}^2) b_{\mathbf{p}}^{s\,\dagger} b_{\mathbf{p}}^s\Big], \end{aligned}$$

where $m \neq \overline{m}$, *i.e.*, the mass of a neutrino is not equal to the mass of the corresponding antineutrino. For $m \neq \overline{m}$ the new Hamiltonian violates CPT invariance and locality, since there is no possibility to derive it from any local QFT.

## 2. CPT TESTS AT A NEUTRINO FACTORY

As we have seen, CPT violation implies physics beyond local QFT, which means that we have to consider Planck scale physics, large extra dimensions, or string theory. In Ref. [6], CPT violation was used to accommodate the LSND result [17–19]. Here we will study precision measurements at a future neutrino factory, which will lead to interesting phenomenological limits on the differences in the neutrino-antineutrino masses and mixing angles.

Now, CPT invariance implies that

$$\mathrm{P}(\nu_\alpha \to \nu_{\alpha'}) = \mathrm{P}(\bar{\nu}_{\alpha'} \to \bar{\nu}_\alpha), \tag{1}$$

where $\mathrm{P}(\nu_\alpha \to \nu_{\alpha'})$ is the neutrino oscillation transition probability that $\nu_\alpha \to \nu_{\alpha'}$ will occur. In the present discussion, we will violate CPT invariance and locality (but not necessarily Lorentz invariance) by assuming that masses and mixings differ for neutrinos and antineutrinos. For neutrinos we have $m_a$ and $U$, whereas for antineutrinos we have $\overline{m}_a$ and $\bar{U}$. Thus, as a consequence,

$$\mathrm{P}(\nu_\alpha \to \nu_{\alpha'}) \neq \mathrm{P}(\bar{\nu}_{\alpha'} \to \bar{\nu}_\alpha), \tag{2}$$

where for two neutrino flavors

$$\begin{aligned} \mathrm{P}(\nu_\alpha \to \nu_{\alpha'}) &\equiv \delta_{\alpha\alpha'} \\ &- (2\delta_{\alpha\alpha'} - 1)\sin^2 2\theta \sin^2 \frac{\Delta m^2 L}{4E}, \\ \mathrm{P}(\bar{\nu}_\alpha \to \bar{\nu}_{\alpha'}) &\equiv \delta_{\alpha\alpha'} \\ &- (2\delta_{\alpha\alpha'} - 1)\sin^2 2\bar{\theta} \sin^2 \frac{\Delta \overline{m}^2 L}{4E}, \end{aligned}$$

which means that

$$\begin{aligned} \Delta \mathrm{P}^{\mathrm{CPT}}_{\alpha\alpha} &\equiv \mathrm{P}(\nu_\alpha \to \nu_\alpha) - \mathrm{P}(\bar{\nu}_\alpha \to \bar{\nu}_\alpha) \\ &= -\sin^2 2\theta \sin^2 \frac{\Delta m^2 L}{4E} \\ &- \sin^2 2\bar{\theta} \sin^2 \frac{\Delta \overline{m}^2 L}{4E}. \end{aligned}$$

For three neutrino flavors we have more complicated formulas. However, we will carry out our numerical calculations using three neutrino flavors.

At a neutrino factory, neutrinos would be produced in muon decays $\mu^+ \to e^+ \nu_e \bar{\nu}_\mu$ (or $\mu^- \to e^- \bar{\nu}_e \nu_\mu$). The straightforward test of CPT violation would be to check the appearance relation $\mathrm{P}(\nu_e \to \nu_\mu) = \mathrm{P}(\bar{\nu}_\mu \to \bar{\nu}_e)$ [or $\mathrm{P}(\bar{\nu}_e \to \bar{\nu}_\mu) = \mathrm{P}(\nu_\mu \to \nu_e)$]. However, this would require to measure the sign of the charge of the produced lepton, which could be hard. We propose instead to check the equality

$$\mathrm{P}(\nu_\mu \to \nu_\mu) = \mathrm{P}(\bar{\nu}_\mu \to \bar{\nu}_\mu), \tag{3}$$

*i.e.*, the $\nu_\mu$ and $\bar{\nu}_\mu$ disappearance channels. The advantages of checking these disappearance channels are, for example:

- high event rates,
- no beam contamination, *i.e.*, no relevant background is present,
- small matter effects,
- large neutrino oscillation effects.

Thus, we could obtain exclusion limits for tiny CPT-violating effects.

If we consider the $\nu_\mu$ and $\bar{\nu}_\mu$ channels as independent experiments, then CPT violation in neutrino oscillations can be quantified by the following asymmetries:

$$\delta \equiv |\Delta m_{32}^2 - \Delta \overline{m}_{32}^2|, \quad (4)$$

$$\epsilon \equiv |\sin^2 2\theta_{23} - \sin^2 2\bar{\theta}_{23}|. \quad (5)$$

For $m_1 \ll \sqrt{\Delta m_\odot^2}$ (hierarchical) and $|m_3 - \overline{m}_3| \ll (m_3)_{\text{average}}$, we have [9]

$$\delta \simeq 2\, a_{\text{CPT}}\, \Delta m_{32}^2, \quad (6)$$

$$\epsilon \simeq 2\, b_{\text{CPT}} \sqrt{\sin^2 2\theta_{23}} \sqrt{1 - \sin^2 2\theta_{23}} \times \arcsin \sqrt{\sin^2 2\theta_{23}}, \quad (7)$$

where the asymmetry parameters are

$$a_{\text{CPT}} \equiv \frac{|m_3 - \overline{m}_3|}{(m_3)_{\text{average}}}$$

and

$$b_{\text{CPT}} \equiv \frac{|\theta_{23} - \bar{\theta}_{23}|}{(\theta_{23})_{\text{average}}}.$$

Next, we want to estimate the sensitivities $\delta a_{\text{CPT}}$ and $\delta b_{\text{CPT}}$ of the asymmetry parameters $a_{\text{CPT}}$ and $b_{\text{CPT}}$. The sensitivities to possible CPT violation are given by the accuracies with which $a_{\text{CPT}}$ and $b_{\text{CPT}}$ can be measured. Comparing $a_{\text{CPT}}$ and $b_{\text{CPT}}$ with the corresponding relative statistical errors $\delta\Delta m_{32}^2$ and $\delta\theta_{23}$ of the measurements of $\Delta m_{32}^2$ and $\theta_{23}$, we obtain the sensitivities of the asymmetry parameters [20]. Thus, the sensitivities $\delta a_{\text{CPT}}$ and $\delta b_{\text{CPT}}$ for the asymmetry parameters $a_{\text{CPT}}$ and $b_{\text{CPT}}$ are given by:

$$\delta a_{\text{CPT}} \sim \frac{\delta\Delta m_{32}^2}{2}, \quad (8)$$

$$\delta b_{\text{CPT}} \sim \delta\theta_{23}. \quad (9)$$

The result of a possible future neutrino factory setup simulation is presented in Fig. 1. For our

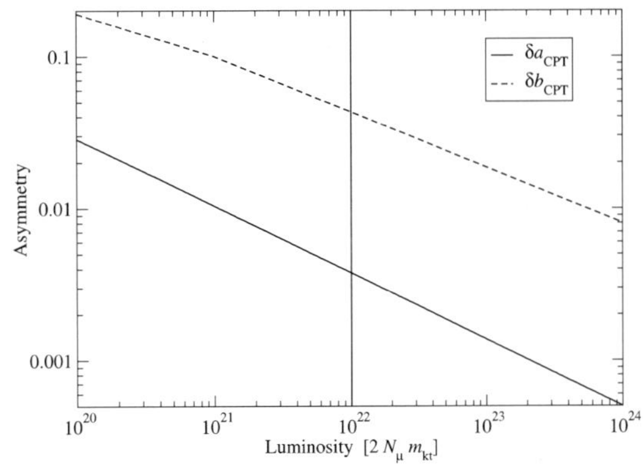


Figure 1. The sensitivities $\delta a_{\text{CPT}}$ (solid curve) and $\delta b_{\text{CPT}}$ (dashed curve) of an estimate of the asymmetries $a_{\text{CPT}}$ and $b_{\text{CPT}}$ as a function of the luminosity $\mathcal{L} \equiv 2N_\mu m_{\text{kt}}$. Parameter values: $E_\nu = 50\,\text{GeV}$ (muon energy), $L = 3000\,\text{km}$ ($\Delta m_{23}^2$, solid curve) / $L = 7000\,\text{km}$ ($\theta_{23}$, dashed curve), $m_{\text{kt}} = 10\,\text{kt}$ (mass of detector), $10^{20}$ muons/year, and 5 years (running time). The figure has been adopted from Ref. [9].

numerical calculations we assumed a 10 kt detector and $10^{20}$ stored muons per year during 5 years. Using the obtained upper bounds in Fig. 1 ($a_{CPT} \lesssim 3.8 \cdot 10^{-3}$ and $b_{CPT} \lesssim 4.3 \cdot 10^{-2}$) as well as $(m_3)_{\text{average}} \simeq \sqrt{\Delta m_{\text{atm}}^2} \leq 5 \cdot 10^{-2}$ eV and $(\theta_{23})_{\text{average}} \simeq \theta_{\text{atm}} = 45°$, we find that

$$|m_3 - \overline{m}_3| \lesssim 1.9 \cdot 10^{-4}\ \text{eV}, \quad (10)$$

$$|\theta_{23} - \bar{\theta}_{23}| \lesssim 2°, \quad (11)$$

which correspond to

$$a_{\text{CPT}} \lesssim 0.38\%, \quad (12)$$

$$b_{\text{CPT}} \lesssim 4.3\%. \quad (13)$$

## 3. SUMMARY & CONCLUSIONS

In summary, CPT violation is not allowed in local QFT. Thus, fundamental CPT violation would mean physics beyond local QFT and the SM, such as Planck scale physics, large extra dimensions, or string theory. Finally, we have shown in our neutrino factory setup simulation that CPT violation is detectable if $|m_3 - \overline{m}_3| \lesssim 1.9 \cdot 10^{-4}$ eV and $|\theta_{23} - \bar{\theta}_{23}| \lesssim 2°$.

### Acknowledgments

I would like to thank my co-workers Samoil M. Bilenky, Martin Freund, Manfred Lindner, and Walter Winter for fruitful collaboration and Håkan Snellman for proof-reading this proceeding.

This work was supported by the Swedish Foundation for International Cooperation in Research and Higher Education (STINT), the Wenner-Gren Foundations, the Swedish Research Council (Vetenskapsrådet), Contract No. 621-2001-1611, and the Magnus Bergvall Foundation (Magn. Bergvalls Stiftelse).

## REFERENCES

1. S.R. Coleman and S.L. Glashow, Phys. Lett. B405 (1997) 249, hep-ph/9703240.
2. S.R. Coleman and S.L. Glashow, Phys. Rev. D59 (1999) 116008, hep-ph/9812418.
3. V.D. Barger et al., Phys. Rev. Lett. 85 (2000) 5055, hep-ph/0005197.
4. H. Murayama and T. Yanagida, Phys. Lett. B520 (2001) 263, hep-ph/0010178.
5. M.C. Bañuls, G. Barenboim and J. Bernabeu, Phys. Lett. B513 (2001) 391, hep-ph/0102184.
6. G. Barenboim et al., hep-ph/0108199.
7. Z.z. Xing, J. Phys. G28 (2002) B7, hep-ph/0112120.
8. S. Skadhauge, Nucl. Phys. B639 (2002) 281, hep-ph/0112189.
9. S.M. Bilenky et al., Phys. Rev. D65 (2002) 073024, hep-ph/0112226.
10. G. Barenboim, L. Borissov and J. Lykken, Phys. Lett. B534 (2002) 106, hep-ph/0201080.
11. J.N. Bahcall, V. Barger and D. Marfatia, Phys. Lett. B534 (2002) 120, hep-ph/0201211.
12. A. Strumia, Phys. Lett. B539 (2002) 91, hep-ph/0201134.
13. I. Mocioiu and M. Pospelov, Phys. Lett. B534 (2002) 114, hep-ph/0202160.
14. G. Barenboim et al., Phys. Lett. B537 (2002) 227, hep-ph/0203261.
15. D. Colladay and V.A. Kostelecký, Phys. Rev. D55 (1997) 6760, hep-ph/9703464.
16. D. Colladay and V.A. Kostelecký, Phys. Rev. D58 (1998) 116002, hep-ph/9809521.
17. LSND Collaboration, C. Athanassopoulos et al., Phys. Rev. Lett. 77 (1996) 3082, nucl-ex/9605003.
18. LSND Collaboration, C. Athanassopoulos et al., Phys. Rev. Lett. 81 (1998) 1774, nucl-ex/9709006.
19. LSND Collaboration, A. Aguilar et al., Phys. Rev. D64 (2001) 112007, hep-ex/0104049.
20. M. Freund, P. Huber and M. Lindner, Nucl. Phys. B615 (2001) 331, hep-ph/0105071.

## Session 2

# Quark Matter; Heavy Ion Collisions

*Session conveners:*
*P. Giubellino, University of Torino/CERN*
*R. Snellings, NIKHEF*

ELSEVIER

Proceedings of ICHEP 2002, pp. 59–61
*S. Bentvelsen, P. de Jong, J. Koch and E. Laenen (Editors)*

31st INTERNATIONAL CONFERENCE ON HIGH ENERGY PHYSICS AMSTERDAM

# Heavy Ion Physics with the CMS experiment at the LHC

B. Wysłouch for the CMS Collaboration[a]

[a]Massachusetts Institute of Technology, Cambridge, MA 02139, USA

The LHC will collide protons at $\sqrt{s} = 14.0$ TeV and lead beams at $\sqrt{s_{NN}} = 5.5$ TeV as well as other AA and pA combinations. The physics program of the Compact Muon Solenoid (CMS) includes the study of heavy ion collisions. The collision energy, much higher than is available at RHIC, will allow the study the dense partonic system with very hard probes: heavy quarks and quarkonia with emphasis on the $b$ and $\Upsilon$, $W^{\pm}$ and $Z^0$ bosons, high $p_T$ jets, photons. The production of these particles is likely to be modified, relative to pp collisions, by the presence of the hot hadronic medium. The CMS detector consists of a 13 m long, 6 m diameter superconducting solenoid which provides a uniform 4T magnetic field. The silicon tracker and electromagnetic and hadronic calorimeters are located inside the magnet while the muon detector is outside. The high granularity, high resolution calorimeters will provide hermetic coverage for $|\eta| \leq 5.5$, making CMS an ideal detector to study high $p_T$ nuclear phenomena.

## 1. CMS Detector

The Large Hadron Collider (LHC) is scheduled to operate as a heavy ion collider for one month each year starting in 2008. The LHC will be capable of accelerating a variety of ions as heavy as Pb up to beam energies per nucleon of $7(Z/A)$ TeV[1].

The Compact Muon Solenoid (CMS) [2] experiment is a general-purpose facility for the study of hadronic collisions at the LHC. The detector consists of a tracking system, electromagnetic and hadronic calorimeters, and muon detectors. A solenoidal magnet provides a 4 Tesla magnetic field surrounding the tracking and calorimetric systems. The tracker which covers the rapidity region $|\eta| < 2.5$, is based on silicon technology. The electromagnetic calorimeter consists of about 83000 lead-tungstenate crystals arranged in a central barrel covering $|\eta| <1.48$ along with endcaps which extend the range to $|\eta| <3$. In the central barrel, the granularity is as high as $\Delta\eta \times \Delta\phi = 0.0175 \times 0.0175$. The hadronic calorimeter consists of barrel and endcap sections, each made of sandwiches of copper plates and plastic scintillator. In the central region, $|\eta| < 2$ the $\Delta\eta \times \Delta\phi$ segmentation is $0.087 \times 0.087$. The combination of electromagnetic and hadronic calorimeters thus provides coverage of the central rapidity region with excellent energy resolution. In addition, coverage at large rapidity, $3 < |\eta| < 5$, is achieved by two very forward calorimeters. This large coverage and good energy resolution makes CMS an optimal detector for jet studies.

The muon system covers the region $|\eta| < 2.5$. In the barrel, $|\eta| < 1.5$, muons must have a $p_T$ larger than 3.5 GeV/$c$ to be efficiently detected.

A set of forward calorimeters including two Zero Degree Calorimeters (ZDC) [3] and CASTOR[4] are proposed to complement the existing detector for heavy ion studies. The ZDC will improve collision centrality determination and CASTOR, covering $5 < |\eta| < 7$, will study energy fluctuations of charged and neutral particles in the very forward region.

The CMS Data Acquistion and Triggering system is designed to handle high luminosity pp collisions, resulting in a 40 MHz event rate at the lowest trigger level. The large event multiplicities present in heavy ion collisions can be handled by the exisiting DAQ due to a much lower collision rate of 7.6 kHz for Pb+Pb collisions. In fact the computer processing power available at the High Level Trigger online farm should allow execution of sophisticated triggering algorithms that will efficiently reject backgrounds. Studies of how to take full advantage of the DAQ and HLT capabilities are presently in progress.


doi:10.1016/S0920-5632(02)02119-9

## 2. Quarkonia Production

The suppression of charmonium production has provided strong evidence for the formation of hot dense hadronic matter at the SPS energies [5]. At RHIC, $J/\Psi$ production is an important part of the PHENIX [6] physics program and STAR is also expected to study charmonium [7]. At the LHC, CMS is well equipped to study quarkonium production. The study of $\Upsilon$ yields, which has a negligible cross section at lower energies, will allow further understanding of the characteristics of dense hadronic matter.

A detailed study of the feasibility of quarkonia detection in CMS has been carried out. Production cross sections were estimated assuming $\sigma_{AA} = A^{2\alpha}\sigma_{pp}$ with $\alpha$=0.9 (0.95) for $J/\Psi(\Upsilon)$. The pp cross sections, $\sigma_{pp}$ at 5.5 TeV were taken from [8]. The assumed luminosities and quarkonia yields are shown in Table 1. The yields are calculated for a one month LHC run assuming 50% accelerator and detector efficiency. The yields are corrected for the detector and selection algorithm acceptance but they are not corrected for the trigger efficiency.

| | Reconstructed kEvents in 1 Month Run | | | |
|---|---|---|---|---|
| | Pb+Pb | Sn+Sn | Kr+Kr | Ar+Ar |
| $< L >$ | $10^{27}$ | $1.7 \times 10^{28}$ | $6.6 \times 10^{28}$ | $10^{30}$ |
| $J/\Psi$ | 28.7 | 210 | 470 | 2200 |
| $\Psi'$ | 0.8 | 5.5 | 12 | 57 |
| $\Upsilon$ | 22.6 | 120 | 320 | 1400 |
| $\Upsilon'$ | 12.4 | 80 | 180 | 770 |
| $\Upsilon''$ | 7 | 45 | 100 | 440 |

Table 1
Expected machine luminosity[1] and projected number of reconstructed quarkonia events in CMS in the dimuon channel for different ion species.

The dimuon acceptance for quarkonia decay to dimuons was studied using the barrel and the endcap detectors. In the barrel, only $J/\Psi$ with $p_T > 5$ GeV/$c$ are detected, with an acceptance of only a few percent. When the endcaps are included, $J/\Psi$ with $p_T = 0$ have a 2% acceptance. For $\Upsilon$, the acceptance is 30(15)% when the endcaps are (are not) modeled, with little dependence on $p_T$. Quarkonia reconstruction was studied with detailed parametrized simulations and full reconstruction programs. Muon backgrounds from $\pi$, $K$ and heavy quark leptonic decays were included. The $p_T$ spectra for pions and kaons were obtained using SHAKER [9] where $\langle p_T \rangle_\pi = 0.48$ GeV/$c$ and $\langle p_T \rangle_K = 0.67$ GeV/$c$ are higher than in other event generators. The multiplicity in central Pb+Pb events was assumed to be $dN/dy = 5000$. Given that our estimates are probably pessimistic, the backgrounds are most likely overestimated. Figure 1 depicts the expected dimuon invariant mass spectrum in the $J/\psi$ and $\Upsilon$ regions for a one month Pb+Pb run.

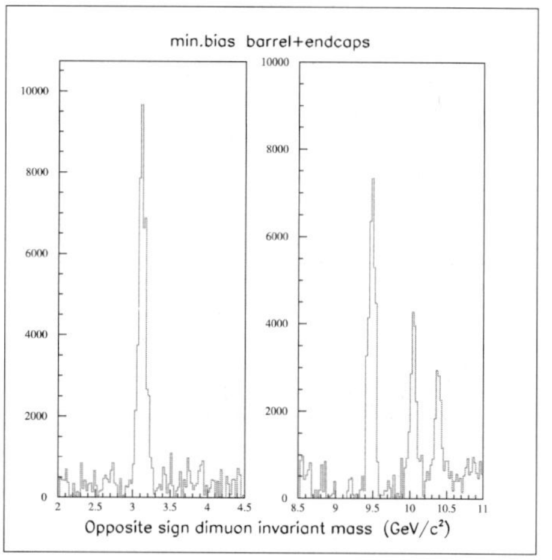


Figure 1. Invariant mass spectrum for dimuons (for a one month Pb+Pb run)in the regions of $J/\psi$ (left) and $\Upsilon$ (right) families reconstructed in the barrel and endcaps.

## 3. Jet Studies

Recent results from RHIC [6,10] indicate significant softening of the transverse momentum spectra of particles above 3 GeV/$c$ for central AA collisions relative to pp. Also, the study of production of back-to-back high $p_T$ particles [11] as well as the scaling of the high $p_T$ particle production with the number of participants [12] indicates that the strongest quenching effect occurs in the central region of the collision. This may be attributed to jet quenching [13] in the hot hadronic media produced after the collision. At the LHC, $10^7$ dijets with $E_T^{\text{jet}}$ >100 GeV are produced in the region $|\eta|$ <2.6 in a one month Pb+Pb run. The number is reduced by about a factor of 2 if only the barrel is considered, $|\eta|$ <1.5. At the LHC, jet reconstruction in Pb+Pb is possible, even for central collisions. A modified version of the window-type sliding jet-finding algorithm was used to reconstruct jets in central Pb+Pb collisions using the information provided by the electromagnetic and hadronic calorimeters. For this study the multiplicity in central Pb+Pb events was assumed to be $dN/dy = 8000$. Table 2 shows

| $E_T^{\text{min}}$ (GeV) | Efficiency (%) | Contamination (%) | $\frac{\sigma(E_T)}{E_T}$ (%) |
|---|---|---|---|
| 75 | 88±3 | 8.3±0.9 | 17.8 |
| 100 | 97±3 | 1.1±0.3 | 18.4 |
| 125 | 99±3 | 0.4±0.2 | 16.8 |
| 200 | 99±3 | 0.1±0.1 | 12.7 |

Table 2
Efficiency, contamination, and energy resolution for jets above different transverse momentum thresholds in central Pb+Pb collisions.

the efficiency, contamination and energy resolution for finding jets above certain thresholds. As can be seen, even the lowest threshold jets can be reconstructed with good efficiency and relatively low background. This good performance will allow CMS to study jet quenching in dijet production [14]. Jet energy loss in hadronic matter can also be measured by tagging jets opposite electroweakly interacting particles such as $Z^0$ or photons, produced in the reactions $qg \to qV$ and $q\bar{q} \to gV$ where $V = Z^0$ or $\gamma$.

## 4. Summary

CMS provides a unique tool to study heavy ion collisions at the LHC. It will probe hot hadronic matter by studying J/$\Psi$ and $\Upsilon$ production. Its excellent calorimetry will provide large coverage and good energy resolution for jet quenching studies. The capabilities of CMS for the physics of quarkonia and jets has been extensively simulated and evaluated. In addition the studies of the capabilities for measuring event-by-event multiplicity and particle flow, $b$-quark production and jet fragmentation are presently in progress.

## REFERENCES


1. D. Brandt, LHC Project Report 450, 2000.
2. The Compact Muon Solenoid, Technical Proposal, CERN/LHCC 94-38, LHCC/P1, Dec 15, 1994.
3. C. Adler *et al*, Nucl. Inst. and Meth. **A470** 488-499 (2001)
4. A. Angelis and A.D. Panagiotou, J. Phys. G **23** (1997)18
5. M. C. Abreu *et al.* [NA50 Collaboration], Phys. Lett. B **477**, 28 (2000).
6. W. Zajc, Recent Results from PHENIX, Nucl. Phys. **A698** 39c (2002).
7. T. J. LeCompte, Proceedings of the RHIC/INT 1998 Winter Workshop on *Quarkonium Production in Relativistic Nuclear Collisions* Eds X-N Wang, B. Jacak, World Scientific, 1999, p. 160.
8. G. Baur *et al* CMS/2000-60, 2000.
9. F. Antinori, CERN/ALICE 93-01 (MC).
10. J. Harris, Nucl. Phys. **A698** 64c (2002).
11. D. Hardtke, proceedings of Quark Matter '02.
12. C. Roland, proceedings of Quark Matter '02.
13. M. Gyulassy and M. Plumer, Phys. Lett. B **243**, 432 (1990).
14. I. Lokhtin, e-print hep-ph/0210010 and references therein.

ELSEVIER

Proceedings of ICHEP 2002, pp. 62–64
*S. Bentvelsen, P. de Jong, J. Koch and E. Laenen (Editors)*

31st INTERNATIONAL CONFERENCE ON HIGH ENERGY PHYSICS AMSTERDAM

# The Alice Experiment at the CERN LHC

P. Kuijer[a]
for The Alice collaboration

[a] NIKHEF,
P.O. Box 41882, 1009 DB Amsterdam, The Netherlands

ALICE is a general-purpose heavy-ion detector designed to study the physics of strongly interacting matter and the quark-gluon plasma in nucleus-nucleus collisions at the LHC at a center of mass energy of 5.5 TeV/nucleon. It is currently under construction and will be ready for data taking at LHC turn-on in 2007. The collaboration currently includes more than 900 physicists, both from nuclear physics and high-energy physics, from over 70 institutions in 27 countries.

## 1. INTRODUCTION

The Alice detector is designed to be a general purpose heavy ion detector, with high granularity tracking, good particle identification in the central region and a very low $p_T$ threshold. The LHC will run heavy ion beams approximately $10^6$ seconds per year, providing i n excess of $10^7$ central events per year. The low event rate (10 kHz with Pb beams) allows the use of relatively slow detectors like the TPC. The detector is designed to cope with the highest predicted particle multiplicity, 8000 particle per unit of rapidity for central Pb-Pb events.

Most of the Alice system [1] is placed in a moderate (0.5 T) magnetic field, provided by the L3 magnet. The central detectors measure event-by-event hadrons, electrons and photons. They cover polar angles from 45 to 135 degrees over the full azimuth. The tracking system consists of an inner tracking system (ITS) of high-resolution silicon tracking detectors, a cylindrical TPC, three particle identification arrays of time-of-flight (TOF), transition radiation (TRD) and Cerenkov counters (HMPID) and a single arm electromagnetic calorimeter (PHOS).

In the forward region an arrangement of absorbers, a large dipole magnet and triggering chambers constitutes a muon spectrometer. Several smaller trigger detectors (ZDC, PMD, FMD, T0, V0) are located at small angles.

## 2. CENTRAL DETECTORS

The design of the tracking system in the central part is primarily driven by the requirement for robust track finding in the central rapidity region ($|\eta| \leq 0.9$). It consists of a six-layer silicon tracker and a TPC. The performance of the ALICE tracking system was simulated in detail [2].

### 2.1. Time Projection Chamber

The central time projection chamber [2] provides track finding, momentum measurement and particle identification through $\mathrm{d}E/\mathrm{d}x$. It has an inner radius of 90 cm and an outer radius of 250 cm. The maximum hit density, for events with 8000 particles per unit of rapidity, is 0.1 $\mathrm{cm}^{-2}$. The $\mathrm{d}E/\mathrm{d}x$ resolution is better than 10%. The use of $Ne/CO_2$ minimizes the electron diffusion and reduces space charge, optimizing the double track resolution. The pad-readout chambers feature 570,000 channels. Construction of the chambers is near completion.

The field cage is assembled close to the ALICE pit. With a volume of 88 $m^3$ it is the largest field cage ever built. Nevertheless, the accuracy in planarity between the central electrode and the read-out chambers is 250 $\mu$m.

The read-out of the TPC performs digitization, tail cancellation, baseline restoration, data compression and multi-event buffering on-line. The track finding efficiency in the TPC is better than 90% down to transverse momenta of 100 MeV/c.


doi:10.1016/S0920-5632(02)02056-X

### 2.2. ITS

The inner tracking system [3] (ITS) consists of six cylindrical layers of silicon detectors located at 4, 7, 15, 24, 39 and 44 cm radius. The ITS covers $|\eta| \leq 0.9$ for all vertexes located within the interaction diamond, i.e. 10.6 cm along the beam direction. Three different technologies were used to optimize the performance and cost for the different regimes of particle densities.

The inner two layers of the ITS consist of pixel detectors, with a more extended coverage, $|\eta| \leq 1.98$ which, in combination with the forward multiplicity detectors, provides a continuous coverage for the measurement of charged particle multiplicities. The middle two layers of the ITS consist of silicon drift detectors. The outer two layers consist of double sided silicon strip detectors.

### 2.3. Time of flight detector

The time of flight detector is placed at 4 m from the interaction point. Using a large (140 $m^2$) array of multi-gap resistive plate chambers (MRPC), excellent timing characteristics are obtained. The MRPC design has been improved during 2001 and has now reached its final form, consisting of a double stack of gaps between glass plates separated by spacers. The multiple gaps are used in avalanche mode, where breakdown is prevented because of the resistive electrodes. The gaps are read-out in parallel by a single set of read-out strips. The hight gain and the parallel read-out of the gaps ensure a high (better than 99%) efficiency. Test beam measurements have shown that an overall time resolution better than the required 100 ps can be achieved.

### 2.4. High Momentum Particle ID

High momentum particles are identified using a RHICH of the proximity focusing type. The distance from the interaction point is approximately 4.5 m. About 15 $m^2$ of liquid $C_6F_{14}$ radiator is backed with multi-wire proportional chambers with pad read-out as UV detectors. The pads of the MWPC are covered with CsI serving as a photo-cathode. The RICH provides $\pi/K$ separation (at $3\sigma$) up to 3 GeV/c and K/p separation up to 5 GeV/c.A large prototype is producing physics results in the STAR experiment, showing the viability of this approach [5].

### 2.5. Transition Radiation Detector

The transition radiation detector (TRD), completes the electron identification capability of the ITS, TPC and TOF system. The TRD provides an $e/\pi$ rejection power of 100 using six layers of radiator foil stacks followed by time expansion chambers. In addition the TRD provides an essential trigger for high $p_t$ electrons. In its full configuration the TRD will consist of 800 $m^2$ of high granularity chambers with over one million channels, covering the central barrel region. The electronics provides analog and time information. A fast local reconstruction produces a level-1 trigger. Tests have shown that, for momenta larger than 1 GeV/c, the TRD has an efficiency of less than 1% at 90% electron detection efficiency.

### 2.6. PHOS

The photon spectrometer (PHOS) will be located 5 m below the interaction region. It uses $PbWO_4$, which has a small Moliere radius and a large light output. The cell size is less than the Moliere radius. The occupancy will be less than 3%. The electronics will provide both energy and time information. Thus, antineutron rejection and a trigger for high $p_t$ photons will be provided. A prototype detector was tested with high energy (0.6 - 30 GeV/c) electrons and the crystal production has started.

## 3. MUON SPECTROMETER

The muon spectrometer [6] consists of a 10 interaction length absorber placed in a magnetic field and five stations of multi-wire proportional chambers placed in front, inside and behind the absorber. The dipole magnet provides a 3 Tm field integral. The acceptance of the system extends from 2 degrees up to 9 degrees. A hole through the center of the absorber allows the LHC beams to pass through, wile additional shielding protects the read-out chambers from particles emitted in the 0 to 2 degree region. The mass resolution is better than 100 MeV in the Y region and better than 70 MeV in the $J/\Psi$ region at an efficiency better than 90%. The momentum resolution is approximately 1%. The main purpose of the muon spectrometer is the study vector resonances through their muon pair decays superimposed on the continuum produced by B and D meson decays and Drell-Yan processes.

## 4. SMALL ANGLE DETECTORS

The impact parameter of the events is measured at small angles using small, but dense, calorimeters (ZDC). Neutron and proton spectators and and beam particles can be separated because these calorimeters are placed behind the first LHC dipole. The photon multiplicity detector (PMD) measures the multiplicity of charged and neutral particles and photons at large rapidity. The forward multiplicity detector (FMD) measures the pseudo rapidity distribution of charged particles over the full azimuth in the regions $-5.1 \leq \eta \leq -1.7$ and $1.7 \leq \eta \leq 3.4$. In combination with the ITS the full rapidity range from -5.1 up to +3.4 is covered for measurements of the charged particle rapidity distribution. The event time is determined better than 50 ps by 24 Cherenkov counters (T0). Scintillator petals (V0) are used to provide the main event trigger and to locate the event vertex. Beam-gas events are rejected efficiently by requiring coincidence of both sides of the V0 counters.

## 5. TRIGGER, DATA ACQUISITION AND OFFLINE

On-line selection of events is achieved through three levels of hard-ware triggers, followed by a software generated high level trigger (HLT). In addition a pre-trigger, derived from the small angle counters alert the TRD. Then first two trigger levels (L0, L1) serve to gate the fast detectors. The final trigger level (L2), which includes past-future protection is used to start read-out of the relatively slow TPC. The HLT uses a dedicated computer farm and provides flexible trigger algorithms.

The existing Alice Data Acquisition Test Environment (DATE) is the basis for prototyping the final data acquisition (DAQ) system. The concepts of the DAQ system are continuously tested because this software is used for test-beam measurements. All read-out hardware in the Alice experiment will interface to the DAQ through a specially developed optical detector data link (DDL). The Alice experiment will produce up to 1.25 GB/s of data and the DAQ system is designed to allow storage of all this data. A series of "'data challenges"' will be used to develop and verify the huge system required to achieve this goal.

The Alice off-line software is entirely based on ROOT [7]. The system will be extended to work on parallel systems and it will be integrated in the GRID environment. A first test using the complete system, from simulation to reconstruction and analysis produced 5 TB of data and showed the viability of this concept.

## 6. CONCLUSION

The Alice detector is an all-round detector designed to study nucleus-nucleus collisions at the LHC. The R&D phase of the Alice project is near completion. The design has been verified extensively by simulations and prototype (beam) tests. Several larger components are already in production.

## REFERENCES


1. CERN/LHCC 1995-071 ALICE Technical Proposal.
2. CERN/LHCC 2000-001 ALICE TDR 7, 7 January 2000.
3. CERN/LHCC 1999-012 ALICE TDR 4, 18 June 1999.
4. CERN/LHCC 1999-005 ALICE TDR 3, 5 March 1999.
5. CERN/LHCC 1999-013, ALICE TP addendum.
6. CERN/LHCC 1996-032, ALICE TP addendum.
7. http://root.cern.ch/

Proceedings of ICHEP 2002, pp. 65–67
*S. Bentvelsen, P. de Jong, J. Koch and E. Laenen (Editors)*

ELSEVIER

31st INTERNATIONAL CONFERENCE ON HIGH ENERGY PHYSICS AMSTERDAM

# Color deconfinement transition in full QCD

J.M. Carmona[a], M. D'Elia[b], L. Del Debbio[c], A. Di Giacomo[c] *, B. Lucini[d], G. Paffuti[c], C. Pica[c]

[a]Departamento de Física Teórica, Universidad de Zaragoza, 50007 Zaragoza, Spain

[b]Dipartimento di Fisica dell'Università di Genova and INFN, I-16146, Genova, Italy

[c]Dipartimento di Fisica dell'Università and INFN, Via Buonarroti 2 Ed. C, I-56127 Pisa, Italy

[d]Theoretical Physics, University of Oxford, 1 Keble Road, OX1 3NP Oxford, UK

We report on evidence from lattice simulations that confinement is produced by monopole condensation in full QCD, as in quenched QCD. The vacuum is a dual superconductor in the confining phase, whilst the U(1) magnetic symmetry is realised à la Wigner in the deconfined phase.

Solid evidence exists that confinement in quenched QCD (pure gauge SU(3) theory) is produced by condensation of magnetic monopoles [1,2]. The tool used to produce that evidence is a disorder parameter $\langle\mu\rangle$, which is the vacuum expectation value of an operator carrying a non zero magnetic charge, as defined by some abelian projection. Numerical results show that $\langle\mu\rangle$ is independent on the choice of the abelian projection [3], a property that has been recently proven analytically [4]. The result is that $\langle\mu\rangle \neq 0$ in the confined phase, indicating Higgs breaking of a magnetic U(1) symmetry; $\langle\mu\rangle = 0$ in the deconfined phase, where the magnetic charge is superselected. $\mu$ is an $SU(3)_c$ gauge invariant, singlet operator. It is also invariant under the magnetic U(1) gauge symmetry, but has a global magnetic charge, it is in fact a Dirac operator [5]. All this confirms the original idea [6,7] that the vacuum is a dual superconductor in the confined phase, in which the electric field between a $q\bar{q}$ pair is channeled into an Abrikosov flux tube, so that the energy is proportional to the distance, $E = \sigma R$. The disorder parameter $\mu$ can be equally well defined in full QCD, i.e. in the presence of dynamical quarks [8].

In the pure gauge theory there exists a traditional order parameter, the Polyakov line $\langle L\rangle$, and a disorder parameter, the 't Hooft line $\langle\tilde{L}\rangle$. The comparison between the 't Hooft line and the magnetic disorder parameter shows that, within numerical errors, $\langle\mu\rangle = \langle\tilde{L}\rangle$, a property which confirms the validity of the mechanism described above [9,10].

In full QCD the traditional description is rather confusing. The $Z_N$ (or $\tilde{Z}_N$) symmetry, whose breaking is monitored by $\langle L\rangle$ (or $\langle\tilde{L}\rangle$), is broken by the coupling to the quarks. At $m_q = 0$ a symmetry exists, chiral symmetry, of which $\langle\bar{\psi}\psi\rangle$ is the order parameter. However this symmetry is also broken in the realistic case, where $m_q \neq 0$. The situation is depicted in Fig. 1, which refers to a model QCD with two degenerate flavours. There the transition line is defined by the maxima of a number of susceptibilities [11]:

$$\chi_{LL}(T, m_q) = \int dx \langle L^\dagger(x) L(0)\rangle$$

$$\chi_{\bar{\psi}\psi,\bar{\psi}\psi}(T, m_q) = \int dx \langle\bar{\psi}\psi(x)\bar{\psi}\psi(0)\rangle$$

as functions of $T$ for fixed $m_q$. All the $\chi$'s have a maximum at the same value $T_c(m_q)$, which defines the line in Fig. 1. The maximum of $\chi_{LL}$ diverges proportionally to the spatial volume in the infinite volume limit, for $m_q \geq 3GeV$, indicating a first order transition. It has been argued in Ref. [12] that the transition is second order at $m_q = 0$. For intermediate values of $m_q$, the susceptibilities do not diverge as $V \to$

*Speaker at the Conference (Email: digiaco@df.unipi.it). Partially supported by MIUR and by EC, FMRX-CT97-0122


doi:10.1016/S0920-5632(02)02057-1

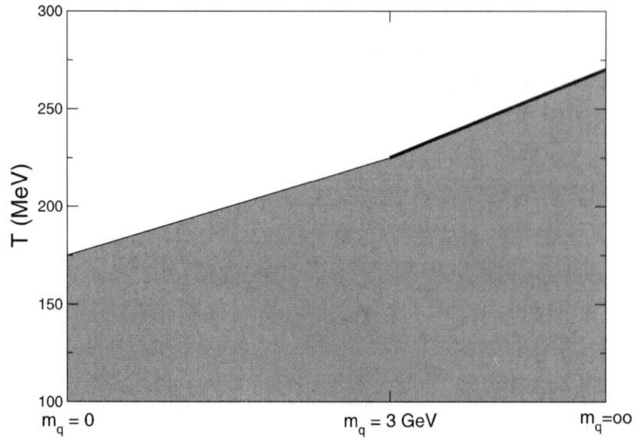


Figure 1. Phase diagram for two degenerate flavours, $m_u = m_d = m_q$.

$\infty$, and this has been interpreted as signalling a crossover. This would be a safe statement if $\langle L \rangle$ and $\langle \bar{\psi}\psi \rangle$ were order parameters. We have checked this statement using our disorder parameter $\langle \mu \rangle$. Firstly, we have shown that indeed $\langle \mu \rangle \neq 0$ in the confined phase, and that it vanishes in the deconfined phase [8]. We have then investigated the the bahaviour as $V \to \infty$ of the susceptibility $\rho = \frac{d}{d\beta} \log\langle \mu \rangle$, which is legitimate to indicate the order of the transition, $\langle \mu \rangle$ being a true order parameter. For $T < T_c$ we find that, as the spatial volume $V$ is increased, $\rho$ reaches a finite limit $\bar{\rho}$. This is shown in Fig. 2, where data from $4 \times 12^3, 16^3, 32^3$ are displayed, and it implies that $\lim_{V\to\infty} \langle \mu \rangle \neq 0$. For $T > T_c$, we find that, whenever the magnetic charge created by the operator $\mu$ is non zero, $\rho$ diverges in the thermodynamic limit as $\rho = -kV^{1/3} + C, (k > 0)$, while it stays constant when the magnetic charge vanishes, as displayed in Fig. 3. Hence, for $T > T_c$, the Hilbert space is superselected [8]. At the critical point $T_c$, we expect by dimensional analysis:

$$\langle \mu \rangle = \tau^\delta \phi \left( \frac{a}{\xi}, \frac{N_s}{\xi}, m_q V^{\gamma/3} \right) \tag{1}$$

where $a$ is the lattice spacing, $\tau = 1 - T/T_c$, and $\xi \sim \tau^{-\nu}$ is the correlation length with the corresponding critical index. If $\xi$ becomes large, the dependence on $a/\xi$ drops out of Eq. (1). In the quenched case, this implies

$$\langle \mu \rangle = \tau^\delta \phi \left( 0, \tau V^{1/3\nu} \right) \tag{2}$$

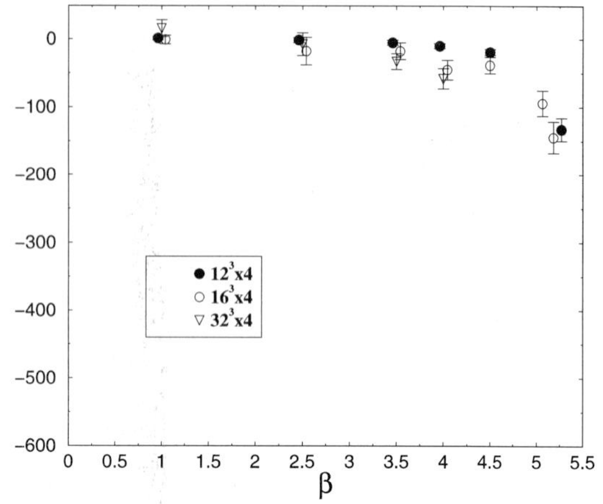


Figure 2. $\rho$ as a function of $N_s$ for $T < T_c$.

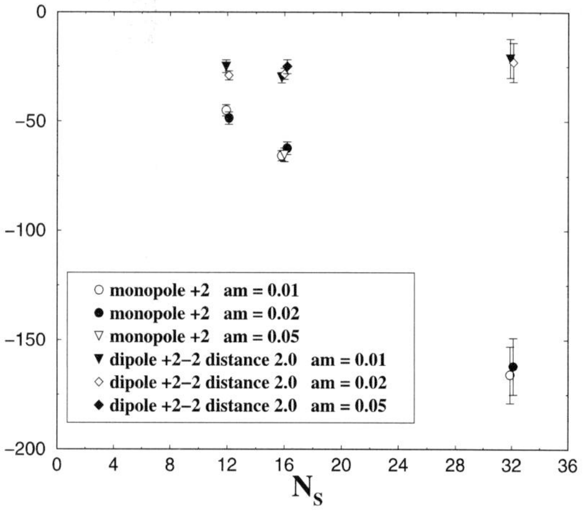


Figure 3. $\rho$ as a function of $N_s$ for $T > T_c$. As $N_s \to \infty$, $\rho \to -\infty$ if the magnetic charge is non-zero and stays constant otherwise.

or, equivalently, the scaling law

$$\rho / V^{1/3\nu} = f \left( \tau V^{1/3\nu} \right) \tag{3}$$

Hence $T_c$, $\nu$, and $\delta$ can be determined by fitting Eq. (3) to the numerical data. In full QCD the additional dependence on the quark mass, with the index $\gamma$ related to the anomalous dimension of $\bar{\psi}\psi$, makes the scaling law more complicated. However $\gamma$ is known, and choosing values of $m_q$ and $V$ such that $m_q V^{\gamma/3}$ is constant, we are left again with a scaling law of the form in Eq. (3), which allows a determination of $\nu$. The dependence of $\bar{\rho}$ on $V$, shown in Fig. 4, clearly gives

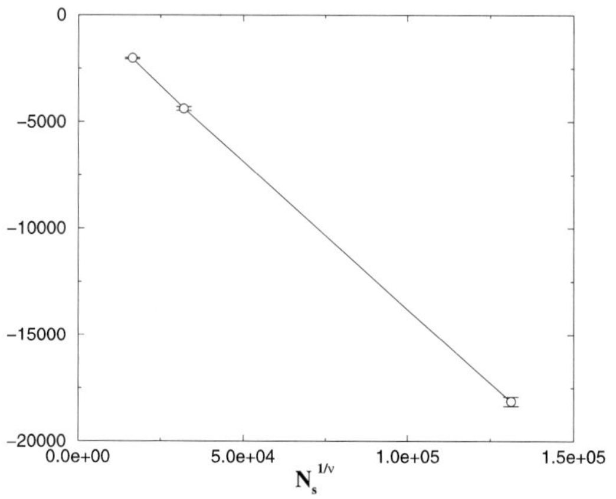


Figure 4. Values of $\rho$ at the peak as a function of $N_s^{1/\nu}$, with $\nu = 1/3$. The height of the peak clearly scales as $N_s^{1/\nu}$.

$\nu = 1/3$ in agreement with a first order transition. In conclusion we found that, also in full QCD, the vacuum is a dual superconductor in the confined phase, and makes a transition to normal at $T_c$. The deconfining phase transition for $N_f = 2$ and $m_u = m_d$ is first order, with a second order end-point at $m_q = 0$. We still consider the latter result as preliminary. A more systematic analysis of finite size scaling, including more data and the use of improved actions is on the way.

## REFERENCES

1. A. Di Giacomo, B. Lucini, L. Montesi, G. Paffuti, Phys. Rev. D **61**, 034503 (2000).
2. A. Di Giacomo, B. Lucini, L. Montesi, G. Paffuti, Phys. Rev. D **61**, 034504 (2000).
3. J.M. Carmona, M. D'Elia, A. Di Giacomo, B. Lucini, G. Paffuti, D **64**, 114507 (2002).
4. A. Di Giacomo, hep-lat/0206018.
5. A. Di Giacomo, G. Paffuti, Nucl. Phys. Proc. Suppl. **106** 604 (2002).
6. G. 't Hooft in High Energy Physics, ed. A. Zichichi, EPS Internatinal Conference, Palermo 1975.
7. S. Mandelstam, Phys. Rep. **23 C** 245 (1976).
8. J.M. Carmona, M. D'Elia, L. Del Debbio, A. Di Giacomo, B. Lucini, G. Paffuti, Phys. Rev. D **66**, 011503 (2002).
9. L. Del Debbio, A. Di Giacomo, B. Lucini, Phys. Lett. **B 500** 236 (2001).
10. L. Del Debbio, A. Di Giacomo, B. Lucini, Nucl. Phys. **B 594** 287 (2001).
11. F. Karsch and E. Laermann, Phys. Rev. D **50**, 6954 (1994).
12. R. Pisarski and R. Wilczek, Phys. Rev. D **29**, 338 (1984).

Proceedings of ICHEP 2002, pp. 68–70
*S. Bentvelsen, P. de Jong, J. Koch and E. Laenen (Editors)*

ELSEVIER

31st INTERNATIONAL CONFERENCE ON HIGH ENERGY PHYSICS AMSTERDAM

# New results on J/$\psi$ from the NA50 experiment

presented by C. Quintans[7)] for the NA50 Collaboration:
B. Alessandro[11)], C. Alexa[4)], R. Arnaldi[11)], M. Atayan[13)], C. Baglin[2)], A. Baldit[3)], M. Bedjidian[12)], S. Beolè[11)], V. Boldea[4)], P. Bordalo[7,a)], S.R. Borenstein[10,b)], G. Borges[7)], A. Bussière[2)], L. Capelli[12)], C. Castanier[3)], J. Castor[3)], B. Chaurand[10)], B. Cheynis[12)], E. Chiavassa[11)], C. Cicalò[5)], T. Claudino[7)], M.P. Comets[9)], S. Constantinescu[4)], P. Cortese[1)], J. Cruz[7)], A. De Falco[5)], N. De Marco[11)], G. Dellacasa[1)], A. Devaux[3)], S. Dita[4)], O. Drapier[10)], B. Espagnon[3)], J. Fargeix[3)], P. Force[3)], M. Gallio[11)], Y.K. Gavrilov[8)], C. Gerschel[9)], P. Giubellino[11,c)], M.B. Golubeva[8)], M. Gonin[10)], A.A. Grigorian[13)], S. Grigorian[13)], J.Y. Grossiord[12)], F.F. Guber[8)], A. Guichard[12)], H. Gulkanyan[13)], R. Hakobyan[13)], R. Haroutunian[12)], M. Idzik[11,d)], D. Jouan[9)], T.L. Karavitcheva[8)], L. Kluberg[10)], A.B. Kurepin[8)], Y. Le Bornec[9)], C. Lourenço[6)], P. Macciotta[5)], M. Mac Cormick[9)], A. Marzari-Chiesa[11)], M. Masera[11)], A. Masoni[5)], M. Monteno[11)], A. Musso[11)], P. Petiau[10)], A. Piccotti[11)], J.R. Pizzi[12)], W.L. Prado da Silva[11,e)], F. Prino[11)], G. Puddu[5)], C. Quintans[7)], L. Ramello[1)], S. Ramos[7,a)], P. Rato Mendes[7)], L. Riccati[11)], A. Romana[10)], H. Santos[7)], P. Saturnini[3)], E. Scalas[1)] E. Scomparin[11)], S. Serci[5)], R. Shahoyan[7,f)], F. Sigaudo[11)], M. Sitta[1)], P. Sonderegger[6,a)], X. Tarrago[9)], N.S. Topilskaya[8)], G.L. Usai[5)], E. Vercellin[11)], L. Villatte[9)], N. Willis[9)], T. Wu[9)].

1) Università del Piemonte Orientale, Alessandria and INFN-Torino, Italy. 2) LAPP, CNRS-IN2P3, Annecy-le-Vieux, France. 3) LPC, Univ. Blaise Pascal and CNRS-IN2P3, Aubière, France. 4) IFA, Bucharest, Romania. 5) Università di Cagliari/INFN, Cagliari, Italy. 6) CERN, Geneva, Switzerland. 7) LIP, Lisbon, Portugal. 8) INR, Moscow, Russia. 9) IPN, Univ. de Paris-Sud and CNRS-IN2P3, Orsay, France. 10) LPNHE, Ecole Polytechnique and CNRS-IN2P3, Palaiseau, France. 11) Università di Torino/INFN, Torino, Italy. 12) IPN, Univ. Claude Bernard Lyon-I and CNRS-IN2P3, Villeurbanne, France. 13) YerPhI, Yerevan, Armenia.

[a)] also at IST, Universidade Técnica de Lisboa, Lisbon, Portugal [b)] on leave of absence from York College, CUNY, New York, USA [c)] also at CERN, Geneva, Switzerland [d)] also at Faculty of Physics and Nuclear Techniques, University of Mining and Metallurgy, Cracow, Poland [e)] now at UERJ, Rio de Janeiro, Brazil [f)] on leave of absence from YerPhI, Yerevan, Armenia

The NA50 experiment studies the production of J/$\psi$ in Pb+Pb collisions. The anomalous J/$\psi$ suppression in relativistic heavy ion collisions, considered an unambiguous signature for the formation of a quark gluon plasma, is studied in the dimuon channel, using as reference the Drell-Yan muon pairs. Preliminary results from the analysis of the 2000 data, with the target region under vacuum, are presented and compared to an update of those obtained from previous data samples.

## 1. Introduction

The formation of a deconfined state of quarks and gluons in the early stages of heavy ion collisions can be probed through charmonia production (see e.g. Ref. [1]), since in an extremely dense medium a Debye colour screening occurs that prevents the binding of $c\bar{c}$ pairs. The study of J/$\psi$ and $\psi'$ vector-mesons via their decay into a pair of muons has the advantage that these survive the hadronization stage, carrying their original message. The NA50 experiment studies dimuons produced in Pb+Pb collisions at 158 GeV/nucleon laboratory energy. The details on the experimental apparatus can be found in [2]. In this paper we present the measurement of the J/$\psi$ yield as a function of the collision centrality, from data collected in year 2000, with the target region under vacuum.


doi:10.1016/S0920-5632(02)02058-3

## 2. Published results on J/$\psi$ suppression

The analysis of the J/$\psi$ production in Pb+Pb collisions as a function of centrality was already published by NA50 for data collected from 1995 to 1998 [2], with the observation of an anomalous J/$\psi$ suppression, i.e. a sharp decrease of the J/$\psi$ / Drell-Yan cross-section ratio at $E_T \simeq$ 40 GeV, without saturation for highly central collisions. This ratio shows a clear departure from the normal nuclear absorption curve for charmonia, established from proton and sulphur induced collisions. Nevertheless, a deeper understanding of Pb+Pb collisions revealed that there could be some contamination from Pb+Air interactions, in particular for peripheral collisions. In order to clarify this point, in 2000 the target region was placed under vacuum up to and including the pre-absorber of the experiment.

## 3. New results from the year 2000 run

To further investigate whether peripheral Pb+Pb collisions (i.e. the $E_T$ region before the onset of the anomalous suppression) are compatible with the results obtained from lighter systems, and since the available statistics in proton induced collisions was poor, more data was collected on several p+A systems, to add to the previous statistics from the NA51 [3] and the NA38 [4] experiments. The J/$\psi$ / Drell-Yan cross-section ratios measured by NA51 in p+p and p+d collisions and by NA50 in p+A collisions, at 450 GeV, in the kinematical domain $-0.5 < \cos\theta_{CS} < 0.5$ and $-0.4 < y_{cm} < 0.6$, together with NA38 data for p+W, p+U and S+U systems at 200 GeV per nucleon, in the domain $-0.5 < \cos\theta_{CS} < 0.5$ and $0 < y_{cm} < 1$ are used to extract the charmonia normal nuclear absorption. A Glauber fit to these data plotted as a function of the average path length of nuclear matter crossed by the $c\bar{c}$ after its formation, $L$, is performed. A common value for the J/$\psi$ absorption cross-section in the two kinematical conditions is extracted from this fit to be $\sigma_{abs} = 4.4 \pm 0.5$ mb [5].

The standard NA50 analysis method consists of a fit to the $\mu^+\mu^+$ and $\mu^-\mu^-$ mass spectra to de-

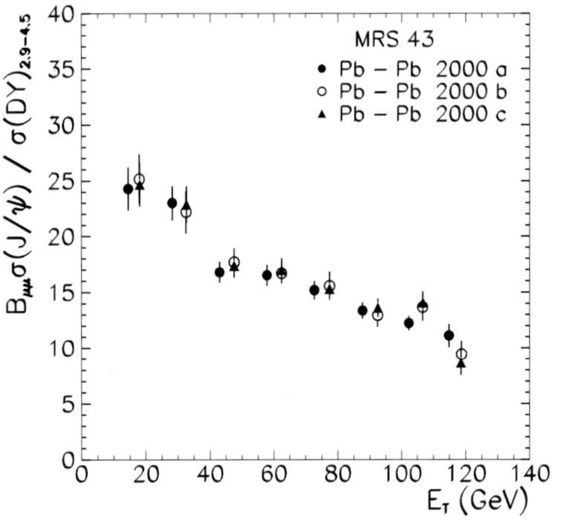


Figure 1. J/$\psi$ / Drell-Yan cross-section ratio vs. $E_{\rm T}$ from three independent 2000 data analyses.

termine the combinatorial background from $\pi$ and $K$ decays followed by a fit to the $\mu^+\mu^-$ mass spectra to extract J/$\psi$ $\psi'$, Drell-Yan and open charm yields. Finally, after correcting for acceptance, we obtain $B_{\mu\mu}\sigma(J/\psi)/\sigma(DY)$. This method has the advantage that most of the experimental efficiencies and the absolute normalizations cancel in the ratio. The Drell-Yan process is used as a reference to study the J/$\psi$ production, since its cross section scales linearly with the product $A \times B$ of the projectile and target mass numbers, as confirmed by NA51, NA38 and NA50 experiments.

The preliminary results from the Pb+Pb 2000 data are presented in Fig. 1, using the MRS 43 set of parton distribution functions (PDFs) for the extraction of the Drell-Yan cross-section in the mass range 2.9–4.5 GeV/$c^2$. Results from three independent analyses, using different data selections and fit methods are shown to be compatible within a few percent. As can be seen in Fig. 2, a good agreement is found between these preliminary 2000 data results vs. $E_{\rm T}$ and 1996 and 1998 results. The 1996 published data has a 3.5% correction to put it in the same analysis conditions as the 2000 data; and the 1998 data was re-analyzed with the 2000 data criteria; while the $E_{\rm T}$ scales for 1998 and 2000 data are renormalized to the 1996 one. The departure from the normal nuclear absorption curves, either the previous or the newly measured one, occurs at $\sim 40$ GeV, and the non-saturation at high $E_T$ are seen in all data sets.

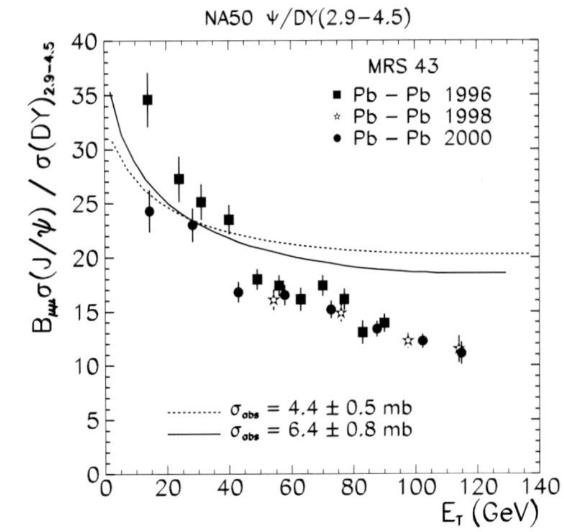


Figure 2. The J/$\psi$ / DY ratio vs. $E_T$ for the 2000, 1996 and 1998 year runs.

The numerical values for the J/$\psi$ / DY ratio we have published in the past are now understood to be sensitive to the parametrisation used for the PDFs (GRV 92 LO PDFs for the published analysis of the S+U data and 1995 Pb+Pb data; and MRS A (LQ) PDFs, i.e. MRS 43, in later Pb+Pb analyses). The reason for the discrepancy between the GRV LO and the MRS 43 analyses comes from the fact that the cross section $\sigma_{DY}(2.9 - 4.5)$, which we use as a denominator in the cross-section ratio, depends on the extrapolation of $dN_{DY}/dM$ from the high mass region, where Drell-Yan is directly measurable, to the region under the J/$\psi$ . To overcome this problem we decided to redefine the Drell-Yan reference, using now only the Drell-Yan directly measurable, $4.2 < M_{\mu\mu} < 7.0$ GeV/$c^2$, and where the two PDFs coincide. The result is shown in Fig. 3, where we can see that the two analyses for the 2000 year, using MRS 43 or GRV LO as PDFs, almost coincide. All the analyses are in agreement, except for the most peripheral points collected in year 1996, maybe due to Pb+Air contamination.

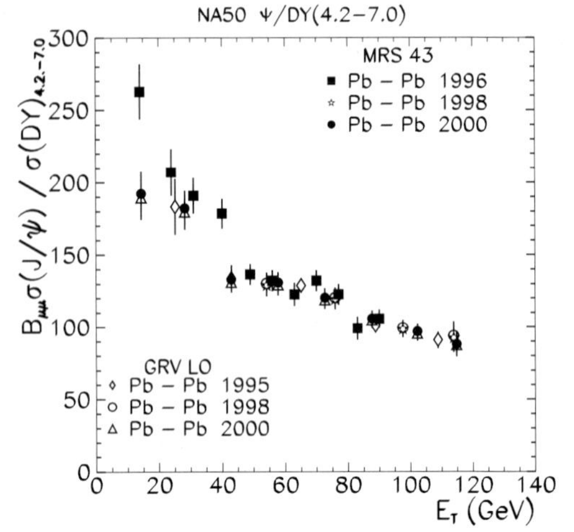


Figure 3. The J/$\psi$ / DY(4.2-7.0) ratio for several data sets.

## 4. Conclusions

The analysis of the most recent Pb+Pb data from NA50, with the target region under vacuum, allows to measure the J/$\psi$ / Drell-Yan cross section ratio in a centrality range extended down to the very peripheral collisions region. The preliminary results obtained confirm a clear departure from the normal nuclear absorption curve, now calculated with greater precision from new together with old light systems data. The anomalous J/$\psi$ suppression for central collisions is clear from all data sets, with a threshold effect around 40 GeV and with the peripheral data points (impact parameter $b > 8.5$ fm) following the normal nuclear absorption pattern.

## REFERENCES

1. T. Matsui and H. Satz, Phys. Lett. B178 (1986) 416.
2. M.C. Abreu et al. (NA50 Coll.), Phys. Lett. B410 (1997) 327; Phys. Lett. B410 (1997) 337; Phys. Lett. B450 (1999) 456; Phys. Lett. B477 (2000) 28; Phys. Lett. B521 (2001) 195.
3. M.C. Abreu et al. (NA51 Coll.), Phys. Lett. B438 (1998) 35.
4. M.C. Abreu et al. (NA38 Coll.), Phys. Lett. B449 (1999) 128; Phys. Lett. B466 (1999) 408.
5. P. Cortese for the NA50 Coll., proceedings of Quark Matter 2002; M.C. Abreu et al., submitted to Phys. Lett. B.

Proceedings of ICHEP 2002, pp. 71–73
*S. Bentvelsen, P. de Jong, J. Koch and E. Laenen (Editors)*

ELSEVIER

31st INTERNATIONAL CONFERENCE ON HIGH ENERGY PHYSICS AMSTERDAM

# Baryon and Antibaryon Production in Hadron-Hadron and Hadron-Nucleus Collisions at 158 GeV/$c$

Helena Bialkowska for NA49[a]

[a]Institute for Nuclear Studies, Warsaw, Poland

Baryon and antibaryon yields in the strangeness range from $-3$ to 3 have been measured in elementary and hadron-nucleus collisions at 158 GeV/$c$.The enhancement factors and their relation to baryon stopping and net baryon density evaluation are discussed.

## 1. Data

The NA49 experiment [1] measures hadrons produced in elementary and nuclear collisions at 158 GeV/$c$ incident momentum. These studies are performed with identical experimental apparatus, thus largely eliminating possible systematic biases in comparing different reactions.
The acceptance covers most of the forward hemisphere in the nucleon-nucleon CM system. The momentum determination is performed via track reconstruction in the time projection chambers, TPC's, placed in a magnetic field. Charged particle identification uses multiple energy loss sampling in the TPCs, accompanied by the time of flight measurements. Neutral strange particle are registered via charged decay modes.
For the hadron-nucleus interactions, centrality selection is performed, based on the measurement of the number of slow target protons. This allows for a selection of interactions characterized by a well defined average number of intranuclear collisions $\nu$.

## 2. Physics motivation

The value of net baryon density in high energy nuclear collisions is a very important quantity, as it may determine the conditions for eventual phase transition. The evaluation of such density is intrinsically connected with baryon stopping, as the maxima in baryon rapidity or Feynman x distributions evolve from peripheral to central interactions.
Strangeness production is obviously related to the mechanism of particle production in nuclear collisions. Thus it is of paramount importance to study baryon and antibaryon production, for strange and non-strange baryons, in elementary, hadron-nucleus and nucleus-nucleus collisions.
Of particular interest is the study of the $\Omega$ and $\bar{\Omega}$ production. A recent work of M. Bleicher et al., [2] suggests that the $\bar{\Omega}/\Omega$ ratio in pp collisions may serve as a discerning tool, discriminating between statistical and string model descriptions of hadronization.

## 3. Results

### 3.1. Net protons

Fig. 1a shows the net proton density as a function of Feynman x, for p-p and p-Pb collisions with controlled centrality ($\nu$ from 3.1 to 6.3). In order to compare it with corresponding distribution from central Pb-Pb collisions ($\nu$ from 3.2 to 4.6), we evaluate separately the projectile and target part of such distributions, assuming factorization, and using data on pion-proton and pion-nucleus collisions [7], [3]. In Fig. 1b the evolution of the projectile component of net proton density in p-p and pPb collisions is shown, and one may compare it with the net proton distribution for Pb-Pb interactions with different $\nu$ values (the centrality of nucleus-nucleus collisions is determined via the measurement of spectator energy deposited in the forward calorimeter).
Clearly the longitudinal spectra of net protons follow a continuous evolution from peripheral to central nuclear collisions, with the number of collisions $\nu$ determining the stopping.


doi:10.1016/S0920-5632(02)02059-5

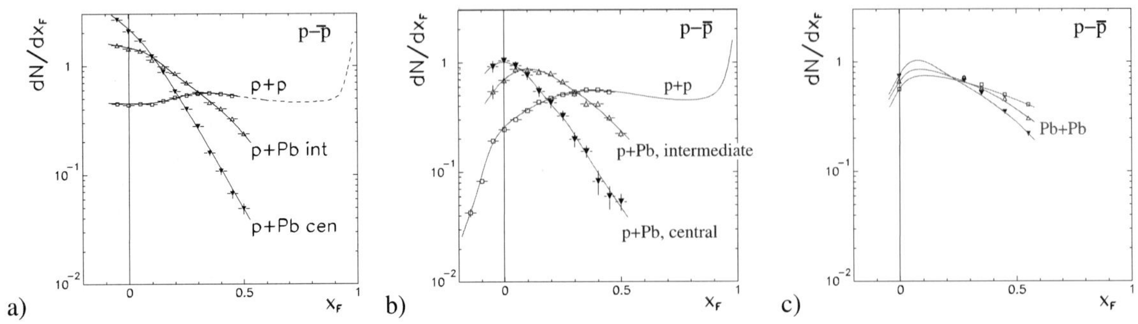


Figure 1. a) Net proton spectra from pp and pPb interactions. b) Projectile component of net proton spectra from pp and pPb interactions. c) Projectile component of net proton spectra from Pb-Pb interactions of different centrality.

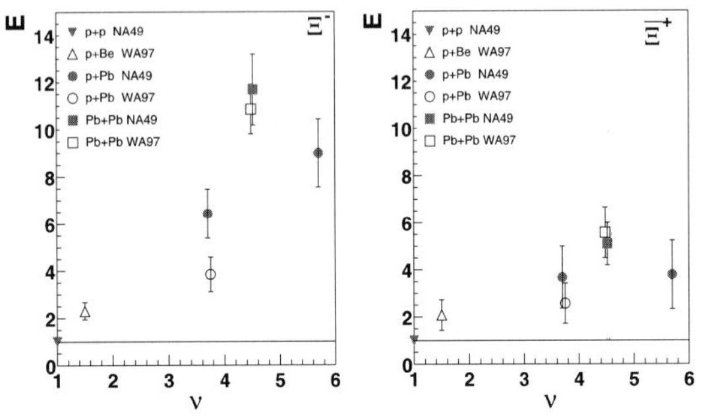


Figure 2. Enhancement factor of $\Xi$ (a) and $\bar{\Xi}$ (b) for pp, pBe, pPb and PbPb collisions as a function of the number of collisions.

### 3.2. Hyperons

There is ample experimental evidence for strange hyperon enhancement in nucleus–nucleus collisions (with respect to elementary collisions) [5], [6]. A meaningful analysis of strangeness enhancement in hadron-nucleus collisions requires proper treatment of both the target and projectile contribution (important for an asymmetric system), and of the isospin effects (important for heavy target with 60% of neutrons).

Figs. 2a and 2b show the enhancement factor $E$ for the $\Xi$ and $\bar{\Xi}$ (with respect to pp data; this is the first measurement of cascade production in elementary collisions at $158\,\mathrm{GeV}/c$), for pBe, p-Pb and Pb-Pb collisions, measured at midrapidity, as a function of the number of collisions $\nu$. Here the enhancement factor $E$ for hadron-nucleus collisions is assumed to come from the projectile contribution, according to the formula:

$$\frac{dN}{dy_{(pA)}} = \left[\frac{\nu}{2} + 1 \cdot \frac{E}{2}\right] \frac{dN}{dy_{(pp)}} \tag{1}$$

As seen in Fig. 2, the enhancement factors in hadron–nucleus and nucleus–nucleus collisions seem to be determined by the number of collisions, for both kinds of the reaction.

A question of isospin effects is related to the eventual asymmetry of baryon-antibaryon production. One can expect that in proton fragmentation, the antiproton yield will underestimate the number of pair produced protons, and in neutron fragmentation it will overestimate it. This is confirmed by first measurements of antiprotons in pp and np reactions at $158\,\mathrm{GeV}/c$ [4]. In absence of measurements for $\Xi$ from neutron projectile, we can assume that similar asymmetry appears for $\Xi$ and $\bar{\Xi}$. This leads to isospin-corrected enhancement factors shown in Fig. 3. Notice that now the enhancement factors for both $\Xi$ and $\bar{\Xi}$ are comparable [8].

A study of net strange baryon density has to take into account both the (properly corrected) enhancements — and the effect of stopping proper, that is, the evolution of net baryon longitudinal distributions.

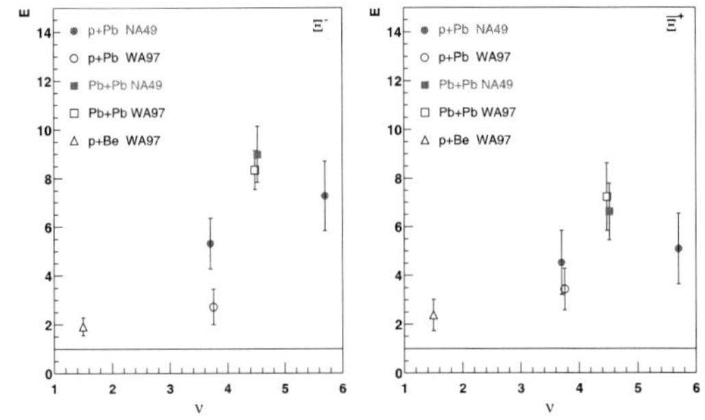


Figure 3. Isospin corrected ehancement at $y^* = 0$.

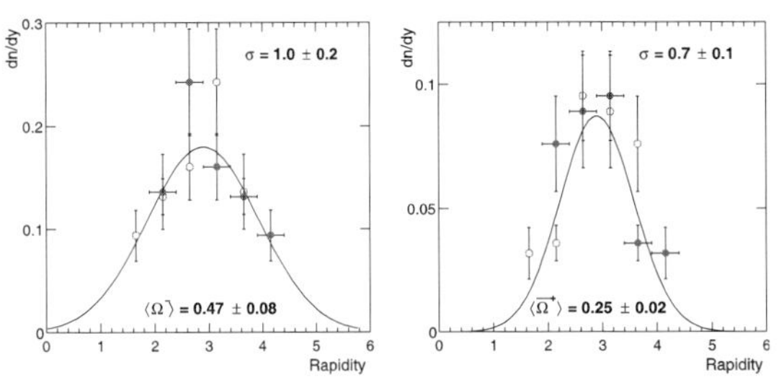


Figure 4. $\Omega$ and $\bar{\Omega}$ rapidity distributions from 20% central Pb-Pb collisions at $158\,\mathrm{GeV}/c$.

### 3.3. Omega production

The NA49 experiment has measured 2.5 million proton-proton interactions, searching for $\Omega$ and $\bar{\Omega}$ production. At present, an upper limit of 0.5 for the ratio $\bar{\Omega}/\Omega$ (at 95% confidence level)can be set. Only a substantial modification of string type models could account for this value. New data taken this year will double the statistics.

A study of $\Omega$ and $\bar{\Omega}$ in 20% most central Pb-Pb interactions has yielded the ratio of about 0.32, with the rapidity distributions as shown in Fig. 4, [9]. New data from NA49 on hyperon production at $40\,\mathrm{GeV}/c$ will soon contribute towards the study of hadronization mechanism in nuclear collisions.

## 4. Summary

A comprehensive study of elementary hadron-hadron and hadron-nucleus collisions is mandatory for proper analysis of nucleus-nucleus data, in particular when searching for new phenomena, exclusive for such reactions. We have shown that net baryon distributions evolve smoothly from elementary to central nuclear collisions. Strange hyperon enhancement in nuclear collisions seems to depend predominantly on the centrality of collision, as measured by the number of nucleon-nucleon collisions, and is observed already in proton-nucleus collisions. An interplay of baryon stopping and (properly evaluated) strangeness enhancement determines the effective antibaryon/baryon ratios.

## 5. Acknowledgements

This work was supported by the Director, Office of Energy Research, Division of Nuclear Physics of the Office of High Energy and Nuclear Physics of the US Department of Energy (DE-ACO3-76SFOOO98 and DE-FG02-91ER40609), the US National Science Foundation, the Bundesministerium fur Bildung und Forschung, Germany, the Alexander von Humboldt Foundation, the UK Engineering and Physical Sciences Research Council, the Polish State Committee for Scientific Research (2 P03B 130 23 and 2 P03B 02418), the Hungarian Scientific Research Foundation (T14920 and T32293), Hungarian National Science Foundation, OTKA (F034707), the EC Marie Curie Foundation, and the Polish-German Foundation.

## REFERENCES

1. S. Afanasiev et al., Nucl. Instr. Meth. A430 (1999), 210.
2. M. Bleicher et al., J. Phys. G28 (2002), 1965.
3. A. Rybicki, PhD Thesis, Cracow Institute for Nuclear Physics, 2002.
4. H.G. Fischer, Acta Phys. Pol. B33 (2002), 1473.
5. F. Antinori et al., Nucl. Phys. A661 (1999), 130c.
6. T. Susa etal., Nucl. Phys. A 698(2002), 491c.
7. H.G. Fischer, QM02,arXiv:hep-ex/0209043.
8. M. Kreps, QM02,arXiv:nucl-ex/0209003,
9. M. van Leeuwen et al., QM02,arXiv:nucl-ex0208014.

ELSEVIER

Proceedings of ICHEP 2002, pp. 74–76
*S. Bentvelsen, P. de Jong, J. Koch and E. Laenen (Editors)*

31st INTERNATIONAL CONFERENCE ON HIGH ENERGY PHYSICS AMSTERDAM

# Hyperon yields in Pb-Pb collisions from the NA57 experiment

F. Antinori[a], A. Badalà[b], R. Barbera[b], A. Belogianni[c], A. Bhasin[d], I.J. Bloodworth[d], G.E. Bruno[e], S.A. Bull[d], R. Caliandro[e], M. Campbell[f], W. Carena[f], N. Carrer[f], R.F. Clarke[d], A. Dainese[a], A.P. de Haas[g], P.C. de Rijke[g], D. Di Bari[e], S. Di Liberto[h], R. Divia[f], D. Elia[e], D. Evans[d], K. Fanebust[i], F. Fayazzadeh[j], J. Fedorišin[k], G.A. Feofilov[l], R.A. Fini[e], P. Ganoti[c], B. Ghidini[e], G. Grella[m], H. Helstrup[i], M. Henriquez[j], A.K. Holme[j], A. Jacholkowski[e], G.T. Jones[d], P. Jovanović[d], A. Jusko[n], R. Kammermans[g], J.B. Kinson[d], K. Knudson[f], A.A. Kolozhvari[l], V. Kondratiev[l], I. Králik[n], A. Kravčáková[k], P. Kuijer[g], V. Lenti[e], R. Lietava[d], G. Løvhøiden[j], V. Manzari[e], G. Martinská[k], M.A. Mazzoni[h], F. Meddi[h], A. Michalon[o], M. Morando[a], E. Nappi[e], F. Navach[e], P.I. Norman[d], A. Palmeri[b], G.S. Pappalardo[b], B. Pastirčák[n], J. Pišút[p], N. Pišútová[p], F. Posa[e], E. Quercigh[a], F. Riggi[b], D. Röhrich[i], G. Romano[m], K. Šafařík[f], L. Šándor[f], E. Schillings[g], G. Segato[a], R. Senè[q], M. Senè[q], W. Snoeyes[f], F. Soramel[a], M. Spyropoulou-Stassinaki[c], P. Staroba[r], T.A. Toulina[l], R. Turrisi[a], T.S. Tveter[j], J. Urbán[k], F.F. Valiev[l], A. van den Brink[g], P. van den Ven[g], P. Vande Vyvre[f], N. van Eijndhoven[g], J. van Hunen[f], A. Vascotto[f], T. Vik[j], O. Villalobos-Baillie[d], L. Vinogradov[l], T. Virgili[m], M.F. Votruba[d], J. Vrláková[k], P. Závada[r]

[a]University of Padua and INFN, Padua, Italy

[b]University of Catania and INFN, Catania, Italy

[c]Physics Department, University of Athens, Athens, Greece

[d]University of Birmingham, Birmingham, UK

[e]Dipartimento IA di Fisica dell'Università e del Politecnico di Bari and INFN, Bari,Italy

[f]CERN, European Laboratory for Particle Physics, Geneva, Switzerland

[g]Utrecht University and NIKHEF, Utrecht, The Netherlands

[h]University "La Sapienza" and INFN, Rome, Italy

[i]Høgskolen i Bergen, Bergen, Norway

[j]Fysisk Institutt, Universitetet i Oslo, Oslo, Norway

[k]P.J. Šafárik University, Košice, Slovakia

[l]State University of St. Petersburg, St.Petersburg, Russia

[m]Dip. di Scienze Fisiche, "E.R. Caianiello" dell'Università and INFN, Salerno, Italy

[n]Institute of Experimental Physics, Slovak Academy of Science, Košice, Slovakia

[o]Institut de Recherches Subatomiques, IN2P3/ULP, Strasbourg, France

[p]Comenius University, Bratislava, Slovakia

[q]Collège de France, Paris, France

[r]Institute of Physics, Academy of Sciences of the Czech Republic, Prague, Czech Republic

The new results on hyperon production in Pb-Pb collisions at 158 A GeV/c and 40 A GeV/c are presented.


doi:10.1016/S0920-5632(02)02060-1

## 1. Introduction

The study of strange particle production in ultra-relativistic nuclear collisions plays an important role in the search for deconfined quark-gluon matter. The strangeness enhancement, i.e. the enhanced abundance of strange particles in central nucleus-nucleus with respect to proton–proton and proton–nucleus interactions, has been suggested as a sensitive probe of the transition to quark-gluon plasma (QGP) phase [1–3]. In the QGP scenario, such an enhancement could be created as a result of rapidly developing flavour equilibration. The equilibration time is estimated to be comparable with the interaction time of the high energy heavy-ion collision (a few fm/c). For a purely hadronic system the characteristic time scale for equilibration is expected to be an order of magnitude longer. It is predicted that the enhancement should be more pronounced for multi-strange than for singly-strange particles [4].

The aim of NA57 is to study the dependence of the strangeness enhancement on the interaction volume (measured by the number of wounded nucleons, i.e. nucleons which undergo at least one primary inelastic collision with another nucleon [5]) and on the collision energy per incoming nucleon. The centrality range of the NA57 experiment extends down to a limit of about 50 wounded nucleons. The experiment has collected data using 158 and 40 *A* GeV/c beams at the CERN SPS.

## 2. The NA57 experiment

A detailed description of the NA57 set-up can be found in [6,7].

Strange and multi-strange particles are identified by reconstructing their weak decays into final states containing only charged particles (e.g. $\Xi^- \rightarrow \Lambda + \pi^-, \Lambda \rightarrow p + \pi^-$) in the silicon telescope. The acceptance windows cover about one unit of rapidity, centered at mid-rapidity for transverse momenta down to a few hundred MeV/*c* [6]. The centrality trigger for Pb-Pb, based on a system of scintillator detectors, selects about the 60% most central collisions at both 158 A GeV/c and 40 A GeV/c. The centrality of the collision is determined from the charged particle multiplicity, sampled at central rapidity by silicon strip detectors. Reference p–Be data at 40 GeV/c were also collected. The analysis is in progress. As a reference data at 158 *A* GeV/c we use the WA97 p–Be and p–Pb data samples [8].

In the following, we are describing the results on the $\Xi$ and $\Omega$ production at both 158 A GeV/c and 40 A GeV/c.

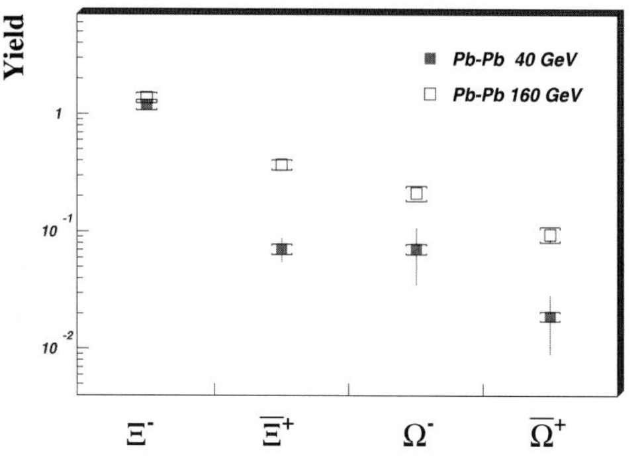


Figure 1. The $\Xi^-$, $\overline{\Xi}^+$, $\Omega^-$, $\overline{\Omega}^+$ yields in 25% most central Pb-Pb collisions at 40 A GeV/c and 158 A GeV/c. Square bars represents systematic errors.

## 3. Results

The hyperon signals in Pb-Pb collisions at 158 A GeV/c and at 40 A GeV/c are shown in [9]. Data were corrected for acceptance and efficiency losses by calculating a weight for each particle with a GEANT based Monte-Carlo simulation.

The fully corrected yields extrapolated to $p_T > 0$ and $|y - y_{cm}| < 0.5$ for 25% most central Pb-Pb collisions, both at 158 A GeV/c and 40 A GeV/c, are shown in Fig. 1. While the $\Xi^-$ production rates are similar at both energies, the yields for $\overline{\Xi}^+$, $\Omega^-$, $\overline{\Omega}^+$ are smaller by a factor 3 to 5 at 40 A GeV/c relative to 160 A GeV/c. Such a behaviour can be expected due to the increase of the baryon density as the beam momentum decreases.

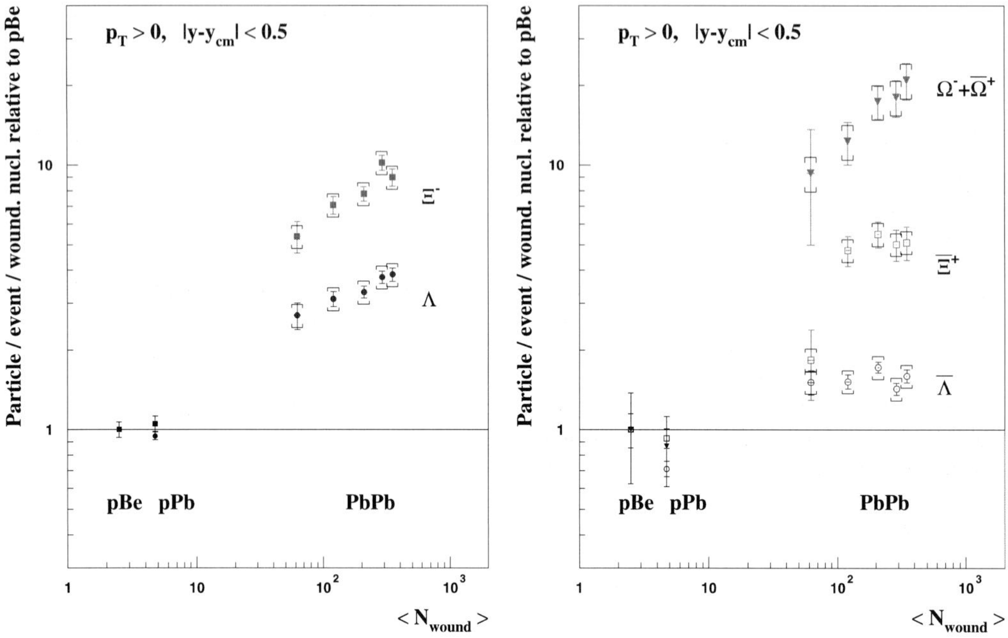


Figure 2. Yields per wounded nucleon in Pb-Pb relative to the p-Be at 158 A GeV/c as a function of number of wounded nucleons. Square bars represent systematic error.

The Pb-Pb collisions data at 158 A GeV/c were divided into five centrality bins. For the details of the multiplicity analysis see [9,10]. The dependence of the yields per wounded nucleon in Pb-Pb relative to p-Be on the centrality of Pb-Pb collisions is shown in Fig. 2.

## 4. Conclusions

The NA57 results at 158 A GeV/c confirm the hyperon enhancements measured by the WA97 experiment [8]. As already observed by the WA97 experiment, the enhancements increase with a strangeness content up to a factor 20 for Ω. There is indication of an increase of the enhancements with centrality of the collisions.

Ξ and Ω yields indicate larger baryon density at 40 A GeV/c with respect to 158 A GeV/c beam energy.

## REFERENCES


1. J. Rafelski and B. Müller, Phys. Rev. Lett. **48**(1982) 1066; Phys. Rev. Lett. **56**(1986) 2234.
2. P. Koch, B. Müller and J. Rafelski, Phys. Rep. **142**(1986) 167.
3. P. Koch, B. Müller, H. Stöcker and W. Greiner, Mod. Phys. Lett.**A 3** (1988) 737.
4. J. Rafelski, Phys. Lett. **B 262** (1991) 333.
5. A. Białas, M.Bleszyǹski and W. Czyz, Nucl. Phys **111** (1975) 411
6. V. Manzari *et al.*, Nucl. Phys. **A 661** (1999) 716c.
7. V. Manzari *et al.*, J. Phys. **G 27** (2001) 383.
8. E. Andersen *et al.*, Phys. Lett. **B 449** (1999) 401;
9. V. Manzari *et al.*, Proceedings of the Quark Matter 2002 Conference, Nantes, France July 18-24 2002, to be published in Nucl. Phys. **A**
10. N. Carrer *et al.*, J. Phys. **G 27** (2001) 391.

ELSEVIER

Proceedings of ICHEP 2002, pp. 77–79
*S. Bentvelsen, P. de Jong, J. Koch and E. Laenen (Editors)*

31st INTERNATIONAL CONFERENCE ON HIGH ENERGY PHYSICS AMSTERDAM

# Charged hadron transverse momentum distributions in Au+Au collisions at $\sqrt{s_{NN}} = 200$ GeV

Gerrit van Nieuwenhuizen for the PHOBOS Collaboration

B.B.Back[1], M.D.Baker[2], D.S.Barton[2], R.R.Betts[6], M.Ballintijn[4], A.A.Bickley[7], R.Bindel[7], A.Budzanowski[3], W.Busza[4], A.Carroll[2], M.P.Decowski[4], E.García[6], N.George[1,2], K.Gulbrandsen[4], S.Gushue[2], C.Halliwell[6], J.Hamblen[8], G.A.Heintzelman[2], C.Henderson[4], D.J.Hofman[6], R.S.Hollis[6], R.Hołyński[3], B.Holzman[2], A.Iordanova[6], E.Johnson[8], J.L.Kane[4], J.Katzy[4,6], N.Khan[8], W.Kucewicz[6], P.Kulinich[4], C.M.Kuo[5], W.T.Lin[5], S.Manly[8], D.McLeod[6], J.Michałowski[3], A.C.Mignerey[7], R.Nouicer[6], A.Olszewski[3], R.Pak[2], I.C.Park[8], H.Pernegger[4], C.Reed[4], L.P.Remsberg[2], M.Reuter[6], C.Roland[4], G.Roland[4], L.Rosenberg[4], J.Sagerer[6], P.Sarin[4], P.Sawicki[3], W.Skulski[8], S.G.Steadman[4], P.Steinberg[2], G.S.F.Stephans[4], M.Stodulski[3], A.Sukhanov[2], J.-L.Tang[5], R.Teng[8], A.Trzupek[3], C.Vale[4], G.J.van Nieuwenhuizen[4], R.Verdier[4], B.Wadsworth[4], F.L.H.Wolfs[8], B.Wosiek[3], K.Woźniak[3], A.H.Wuosmaa[1], B.Wysłouch[4]

[1] Argonne National Laboratory, Argonne, IL 60439-4843, USA
[2] Brookhaven National Laboratory, Upton, NY 11973-5000, USA
[3] Institute of Nuclear Physics, Kraków, Poland
[4] Massachusetts Institute of Technology, Cambridge, MA 02139-4307, USA
[5] National Central University, Chung-Li, Taiwan
[6] University of Illinois at Chicago, Chicago, IL 60607-7059, USA
[7] University of Maryland, College Park, MD 20742, USA
[8] University of Rochester, Rochester, NY 14627, USA

We present transverse momentum distributions of charged hadrons produced in Au+Au collisions at $\sqrt{s_{NN}} = 200$ GeV. The evolution of the spectra for transverse momenta $p_T$ from 0.25 to 5 GeV/C is studied as a function of collision centrality. We find a significant change of the spectral shape between proton-antiproton and peripheral Au+Au collisions. When comparing peripheral to central Au+Au collisions, we find that the yields at the highest $p_T$ exhibit approximate scaling with the number of participating nucleons, rather than scaling with the number of binary collisions.

In this paper, the yield of charged hadrons produced in collisions of gold nuclei at an energy of $\sqrt{s_{NN}} = 200$ GeV is presented as a function of collision centrality and transverse momentum $p_T$. The data were taken with the PHOBOS detector [1,2] during the second run of the Relativistic Heavy-Ion Collider (RHIC) at the Brookhaven National Laboratory. In the theoretical analysis of particle production in hadronic and nuclear collisions, a distinction is often made between the relative contributions of "hard" parton-parton scattering processes and "soft" processes. Hard processes are expected to contribute an increasingly larger fraction of particle production with increasing collision energy and at higher $p_T$ of the produced particles.

Collisions of heavy nuclei offer ideal conditions to test our understanding of this picture, as "hard" processes are expected to scale with the number of binary nucleon-nucleon collisions $N_{coll}$, whereas "soft" particle production is expected to exhibit scaling with the number of participating nucleons $N_{part}$. In the Glauber picture of nuclear collisions, $N_{coll}$ approximately scales as $(N_{part})^{4/3}$. For central collisions of Au nuclei, one therefore obtains an increase in the ratio of $N_{coll}/(N_{part}/2)$ by a factor of six, relative to proton-proton collisions.

In Fig. 1 we present the invariant cross-section of charged hadrons as a function of transverse momentum for particles with a rapidity of $0.2 < y_\pi < 1.4$ assuming the pion mass for calculat-


doi:10.1016/S0920-5632(02)02064-9

ing $y_\pi$. The cross-sections were corrected for the geometrical acceptance of the detector, the reconstruction efficiency, the distortion due to binning and momentum resolution, "ghost" tracks, secondaries and weak decays [3,4].

Data are shown for 6 centrality bins, with $65 < \langle N_{part} \rangle < 344$ corresponding, in the Glauber picture, to $107 < \langle N_{coll} \rangle < 1050$. The increase with centrality of the integrated yields, over the studied centrality range, is consistent with the centrality evolution of the mid-rapidity particle density presented in [5].

The top left panel in Fig. 2 shows the ratio of the most peripheral bin ($\langle N_{part} \rangle = 65 \pm 4$) divided by $\langle N_{part}/2 \rangle \times f(p_T)$, where $f(p_T)$ is a fit of the invariant cross-section in proton-antiproton collisions at the same energy [6]. This comparison shows that already in Au+Au collisions with an impact parameter of about 10 fm, the spectral shape is dramatically modified relative to that in $p + \bar{p}$ collisions. It is worth noting that the ratio $\langle N_{coll} \rangle / \langle N_{part}/2 \rangle$ increases by a factor of about three from $p + \bar{p}$ to peripheral Au+Au collisions. The other panels in Fig. 2 show the evolution of same ratio upto the most central bin ($\langle N_{part} \rangle = 344 \pm 12$). We observe only a moderate change in spectral shape, relative to $p + \bar{p}$, when going from peripheral to central events.

Fig. 3 shows the evolution of the spectra with centrality. The spectra, normalized by $\langle N_{part}/2 \rangle$, in the six centrality bins have been divided by a fit to the normalized spectrum measured in the most peripheral bin. The centrality range in the Au+Au collisions covers a change in $\langle N_{coll} \rangle / \langle N_{part}/2 \rangle$ by about a factor of two. It is remarkable that the change in spectral shape over this range of centralities is small compared to that between peripheral Au+Au collisions and $p + \bar{p}$ collisions. In particular at high $p_T$, above 3 GeV/c, the yields scale to a good approximation as $\langle N_{part}/2 \rangle$ as a function of centrality.

Fig. 4 shows the scaling behaviour of the charged hadron relative yields in Au+Au and $p+\bar{p}$ collisions for different values of $p_T$. The yields were determined by interpolating the data shown in Fig. 1 using power law fits.

The observed trends can be contrasted with the expectation that particle production should be characterized by a change from $\langle N_{part} \rangle$ scaling at low $p_T$ to $\langle N_{coll} \rangle$ scaling at high $p_T$. No corresponding increase in particle production per participant at $p_T = 3$ GeV/c and above is observed. Rather, the yields in this region scale approximately with the number of participating nucleons. Upcoming studies of d+Au collisions at RHIC will provide further insight into the modification of particle spectra in a nuclear environment.

This work was partially supported by US DoE grants DE-AC02-98CH10886, DE-FG02-93ER-404802, DE-FC02-94ER40818, DE-FG02-94ER40865, DE-FG02-99ER41099, W-31-109-ENG-38. NSF grants 9603486, 9722606 and 0072204. The Polish groups were partially supported by KBN grant 2-P03B-103203. The NCU group was partially supported by NSC of Taiwan under contract NSC 89-2112-M-008-024.

## REFERENCES

1. B. Back et al., Nucl. Phys. **A698** (2002) 416.
2. H. Pernegger et al., Nucl. Instrum. Methods. **A419** (1998) 549.
3. PHOBOS Collaboration (In preparation).
4. B. Back et al.,nucl-ex/0206012, submited to Phys. Rev. **C**
5. B. B. Back *et al.*, Phys. Rev. **C65**, (2002) 061901.
6. C. Albajar *et al.*, Nucl. Phys. **B335**, (1990) 261.

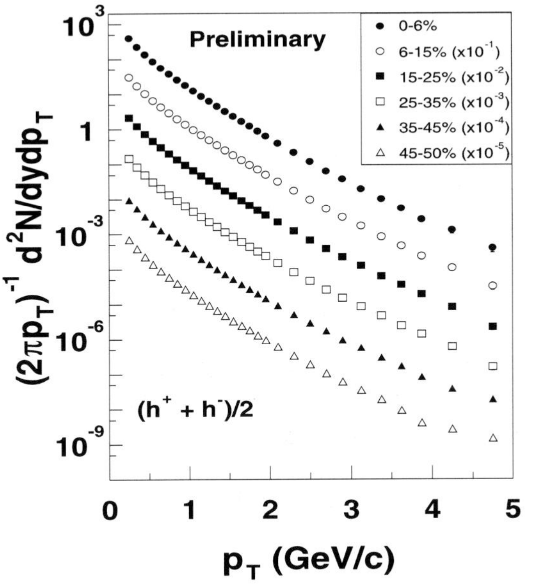


Figure 1. Invariant yields for charged hadrons as a function of $p_T$ for 6 centrality bins. For clarity, consecutive bins are scaled by factors of 10.

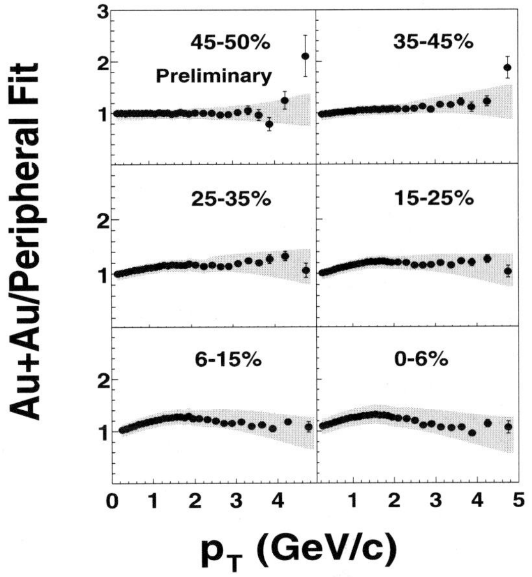


Figure 3. Charged hadron yield in Au+Au in six centrality bins, divided by a fit to the most peripheral bin and scaled by $\langle N_{part}/2\rangle$.

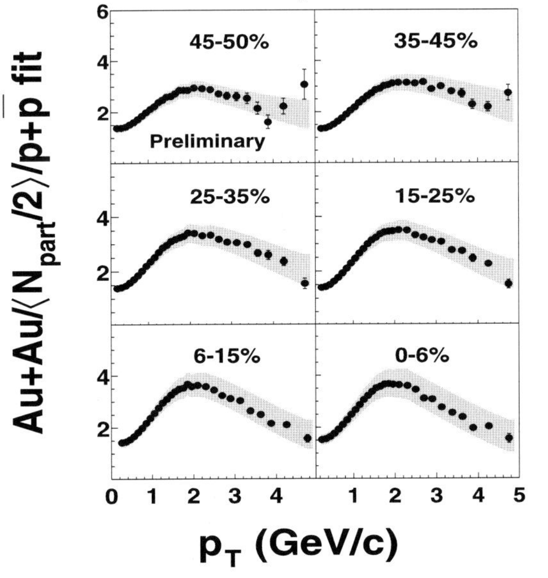


Figure 2. Ratio of the yield of charged hadrons in Au+Au collisions to a fit of proton-antiproton data scaled by $\langle N_{part}/2\rangle$ as a function of $p_T$.

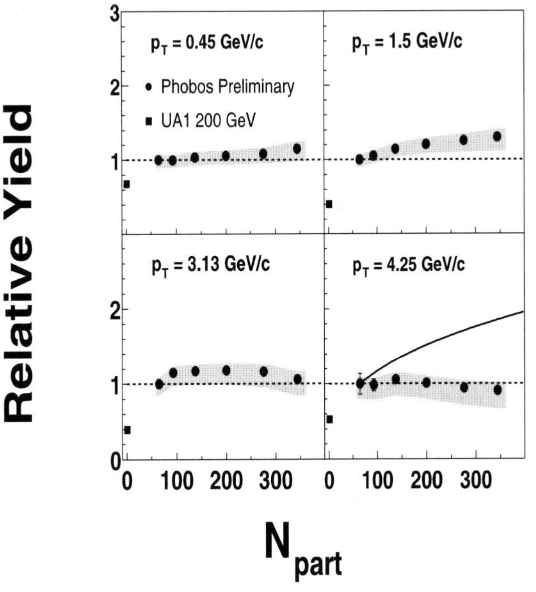


Figure 4. Charged hadron yield in Au+Au, normalized to the average yield in the 6 centrality bins, as a function of $N_{part}$ at different $p_T$. The solid line in the right bottom panel shows the expectation from $\langle N_{coll}\rangle$ scaling.

ELSEVIER

Proceedings of ICHEP 2002, pp. 80–82
*S. Bentvelsen, P. de Jong, J. Koch and E. Laenen (Editors)*

31st INTERNATIONAL CONFERENCE ON HIGH ENERGY PHYSICS AMSTERDAM

# Results from the BRAHMS experiment at RHIC

Dieter Röhrich
for the BRAHMS collaboration:
I. G. Bearden[8], D. Beavis[1], C. Besliu[11], Y. Blyakhman[7], J. Brzychczyk[4], B. Budick[7], H. Bøggild[8], C. Chasman[1], C. H. Christensen[8], P. Christiansen[8], J. Cibor[5], R. Debbe[1], E. Enger[2], J. J. Gaardhøje[8], M. Germinario[8], K. Grotowski[4], K. Hagel[9], O. Hansen[8], A. K. Holme[2], H. Ito[1], E. Jakobsen[8], A. Jipa[11], J. I. Jordre[10], F. Jundt[3], C. E. Jørgensen[8], R. Karabowicz[4], T. Keutgen[9], E. J. Kim[6], T. Kozik[4], T. M. Larsen[2], J. H. Lee[1], Y. K. Lee[6], G. Løvhøiden[2], Z. Majka[4], A. Makeev[9], B. McBreen[1], M. Mikkelsen[2], M. Murray[9], J. Natowitz[9], B. S. Nielsen[8], J. Norris[12], K. Olchanski[1], D. Ouerdane[8], R. Płaneta[4], F. Rami[3], D. Roehrich[10], B. Samset[2], D. Sandberg[8], S. Sanders[12], R. A. Sheetz[1], Z. Sosin[4], P. Staszel[8], T. S. Tveter[2], F. Videbæk[1], R. Wada[9], A. Wieloch[4], and Z. Yin[10], I. S. Zgura[11]

[1] Brookhaven National Laboratory, Upton, USA
[2] Fysisk Institutt, University of Oslo, Blindern, Norway
[3] IReS, University Louis Pasteur, Strasbourg, France
[4] Jagiellonian University, Krakow, Poland
[5] Institute of Nuclear Physics, Krakow, Poland
[6] Johns Hopkins University, Baltimore, USA
[7] New York University, New York, USA
[8] Niels Bohr Institute, University of Copenhagen, Denmark
[9] Texas A&M University, College Station, USA
[10] University of Bergen, Norway
[11] University of Bucharest, Romania
[12] University of Kansas, USA

Results from the BRAHMS collaboration for Au+Au collisions at $\sqrt{s_{NN}} = 200$ GeV are presented, focussing on the measurement of the net-proton rapidity distribution, the determination of the pseudorapidity distribution of charged particles and on the rapidity distributions of pions and kaons.

## 1. The BRAHMS experiment

The BRAHMS detector system [1] consists of two independent magnetic spectrometers that can be positioned over an angular range from 2.3° to 30° (forward spectrometer, FS) and from 30° to 90° (midrapidity spectrometer, MRS) with respect to the beam line. The spectrometers are used to fully identify charged particle over a large range in rapidity $y$, and transverse momentum $p_t$. Due to the small solid angle of each spectrometer (6 msr and 0.5 msr, respectively) the acceptance must be explored through numerous combinations of angular and field settings.

The BRAHMS experiment has four different global detector systems. The Beam-Beam Counters (BB) measure charged hadrons in the pseudorapidity range of $2.1 < |\eta| < 4.7$. The Zero Degree Calorimeters (ZDC) measure non-interacting neutrons from the collisions. The time difference between the two ZDC signals allows a determination of the collision vertex with a precision of $\approx 2.5$ cm. A scintillator tile multiplicity array (TMA) measures charged particle emission in the central pseudorapidity region ($-2.0 < \eta < +2.0$) and is used for centrality selection and determination of the multiplicity of charged particles as a function of pseudorapidity. A silicon multiplicity array (SiMA) ($-2.2 < \eta < +2.2$) provides additional information on charged particle multiplicity in the central region. Due to


doi:10.1016/S0920-5632(02)02067-4

the spread in collision vertices these two arrays provide charged particle multiplicity information for $-3 < \eta < +3$; combined with the beam-beam information the three arrays cover the range $-4.7 < \eta < +4.7$.

## 2. Stopping

Stopping, the slowing down of incoming nucleons, is a measure of the conversion of the incoming longitudinal energy of a projectile and target into transverse degrees of freedom. BRAHMS is well suited to study the mechanism underlying the stopping of the baryons of the colliding nuclei due to its ability to identify particles over a broad range in rapidity. Fig. 1 shows the measurement [8] of the net-proton distribution at RHIC energies. The rapidity density is about 8 at midrapidity, stays constant for approximately $\pm$ 2 units of rapidity and increases to about 17 at $y = 3$, a value close to midrapidity densities at the SPS. The data point at $y = 3$ is derived from an extrapolation in $m_T$; a $p_T$ extrapolation would give a 25% higher yield.

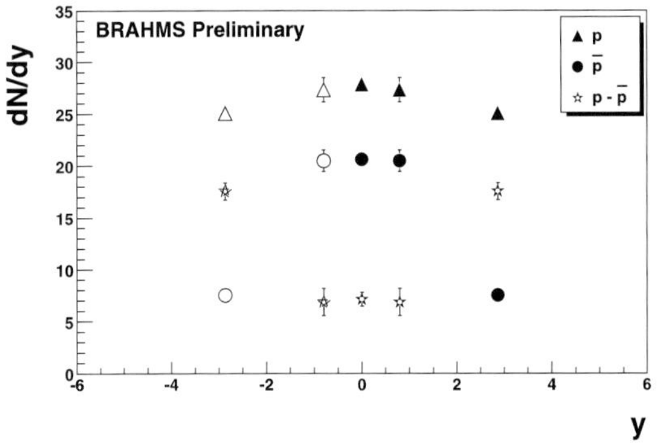


Figure 1. Rapidity distribution of net protons ($p$ - $\bar{p}$) for central (10%) Au+Au collisions at RHIC.

## 3. Pseudorapidity densities of charged particles

Fig. 2 shows the pseudorapidity density distributions for different collision centralities for $100AGeV + 100AGeV$ Au+Au collisions. In fig. 3

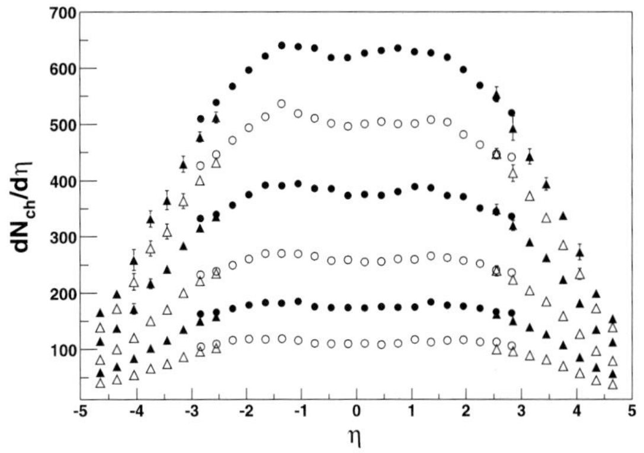


Figure 2. Distributions of $dN_{ch}/d\eta$ vs $\eta$ for centrality ranges of, top to bottom, 0-5%, 5-10%, 10-20%, 20-30%, 30-40%, and 40-50% at $\sqrt{s_{NN}}$ = 200 GeV. Statistical uncertainties are shown where larger than the symbol size.

one can see the charged particle densities normalized to the number of participant pairs $N_{part}/2$ for the two RHIC energies. Data are plotted as a function of pseudorapidity and shifted by the beam rapidity. Data from 9.4% central Pb+Pb reactions at CERN-SPS are also shown [4]. The figure illustrates that the particle production is practically identical at rapidities closer to the beam rapidity. This supports the notion of a limiting behavior of particle production in the fragmentation region [2,3].

## 4. Rapidity dependence of ratios and yields of identified particles

Fig. 4 shows the rapidity dependence of various antiparticle/particle ratios [6]. The $N(\bar{p})/N(p)$ ratio drops when going to forward rapidities, the $N(K^-)/N(K^+)$ ratio follows this trend while the $N(\pi^-)/N(\pi^+)$ ratio stays constant. Fig. 5 shows the rapidity spectra of charged pions and kaons [7,9]. All peak at midrapidity, and the $K^-$ distribution is narrower than that of $K^+$.

## 5. Acknowledgements

This work was supported by the Division of Nuclear Physics of the Office of Science of the U.S. Department of Energy, the Danish Natural

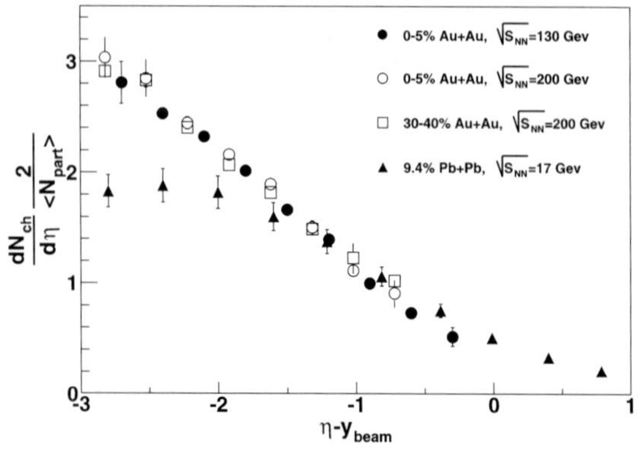


Figure 3. Distribution of $dN_{ch}/d\eta$ divided by the number of pairs of participant baryons plotted as a function of pseudorapidity minus beam rapidity for BRAHMS data at $\sqrt{s_{NN}} = 130$, 200 GeV and SPS data.

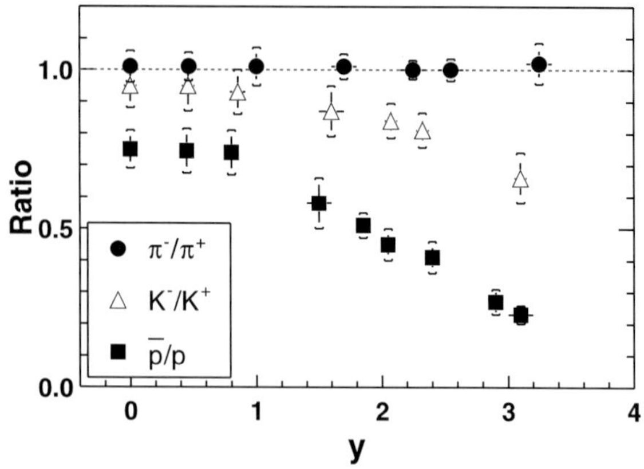


Figure 4. Rapidity dependence of the $N(\bar{p})/N(p)$, $N(K^-)/N(K^+)$ and $N(\pi^-)/N(\pi^+)$ ratios for central (20%) Au+Au collisions.

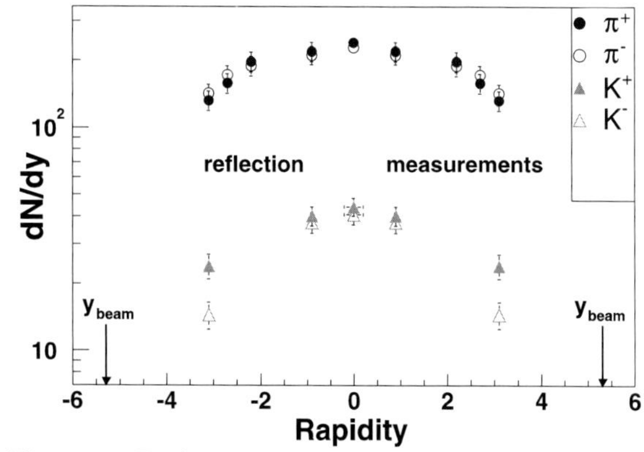


Figure 5. Preliminiary rapidity spectra of charged pions and kaons for central (10%) Au+Au collisions.

Science Research Council, the Norwegian Natural Science Research Council, the Polish Ministry of Research and the Korea Research Foundation Grant.

## REFERENCES

1. I.G. Bearden *et al.* (BRAHMS collaboration), Nucl. Instrum. Meth., to be published.
2. I. G. Bearden *et al.* (BRAHMS collaboration), Phys. Lett. **B253** (2001) 227.
3. I. G. Bearden *et al.* (BRAHMS collaboration), Phys. Rev. Lett. **88** (2002) 202301.
4. P. Deines-Jones *et al.*, Phys. Rev. **C62** (2000) 014903.
5. I.G. Bearden *et al.* (BRAHMS collaboration), Phys. Rev. Lett. **87** (2001) 112305.
6. I.G. Bearden *et al.* (BRAHMS collaboration), submitted to Phys. Rev. Lett. (2002), nucl-ex/0207006.
7. I.G. Bearden *et al.* (BRAHMS collaboration), *Recent results from BRAHMS*, Proceedings Quark Matter 2002, Nantes, France, July 2002, to be published.
8. J.H. Lee *et al.* (BRAHMS collaboration), *Where do the nucleons go at RHIC?*, Proceedings Quark Matter 2002, Nantes, France, July 2002, to be published.
9. D. Ouerdane *et al.* (BRAHMS collaboration), *Rapidity and transverse mass spectra in Au+Au collisions at 200 GeV*, Proceedings Quark Matter 2002, Nantes, France, July 2002, to be published.

Proceedings of ICHEP 2002, pp. 83–85
*S. Bentvelsen, P. de Jong, J. Koch and E. Laenen (Editors)*

ELSEVIER

31st INTERNATIONAL CONFERENCE ON HIGH ENERGY PHYSICS AMSTERDAM

# Universal Behavior of Charged Particle Production in Heavy Ion Collisions at RHIC Energies

Peter A. Steinberg[2] * for the PHOBOS collaboration

B.B.Back[1], M.D.Baker[2], D.S.Barton[2], R.R.Betts[6], M.Ballintijn[4], A.A.Bickley[7], R.Bindel[7], A.Budzanowski[3], W.Busza[4], A.Carroll[2], M.P.Decowski[4], E.García[6], N.George[1,2], K.Gulbrandsen[4], S.Gushue[2], C.Halliwell[6], J.Hamblen[8], G.A.Heintzelman[2], C.Henderson[4], D.J.Hofman[6], R.S.Hollis[6], R.Hołyński[3], B.Holzman[2], A.Iordanova[6], E.Johnson[8], J.L.Kane[4], J.Katzy[4,6], N.Khan[8], W.Kucewicz[6], P.Kulinich[4], C.M.Kuo[5], W.T.Lin[5], S.Manly[8], D.McLeod[6], J.Michałowski[3], A.C.Mignerey[7], R.Nouicer[6], A.Olszewski[3], R.Pak[2], I.C.Park[8], H.Pernegger[4], C.Reed[4], L.P.Remsberg[2], M.Reuter[6], C.Roland[4], G.Roland[4], L.Rosenberg[4], J.Sagerer[6], P.Sarin[4], P.Sawicki[3], W.Skulski[8], S.G.Steadman[4], P.Steinberg[2], G.S.F.Stephans[4], M.Stodulski[3], A.Sukhanov[2], J.-L.Tang[5], R.Teng[8], A.Trzupek[3], C.Vale[4], G.J.van Nieuwenhuizen[4], R.Verdier[4], B.Wadsworth[4], F.L.H.Wolfs[8], B.Wosiek[3], K.Woźniak[3], A.H.Wuosmaa[1], B.Wysłouch[4]

[1] Argonne National Laboratory, Argonne, IL 60439-4843, USA [2] Brookhaven National Laboratory, Upton, NY 11973-5000, USA [3] Institute of Nuclear Physics, Kraków, Poland [4] Massachusetts Institute of Technology, Cambridge, MA 02139-4307, USA [5] National Central University, Chung-Li, Taiwan [6] University of Illinois at Chicago, Chicago, IL 60607-7059, USA [7] University of Maryland, College Park, MD 20742, USA [8] University of Rochester, Rochester, NY 14627, USA

The PHOBOS experiment at RHIC has measured the multiplicity of primary charged particles as a function of centrality and pseudorapidity in Au+Au collisions at $\sqrt{s_{NN}}$ = 19.6, 130 and 200 GeV. Two kinds of universal behavior are observed in charged particle production in heavy ion collisions. The first is that forward particle production, over a range of energies, follows a universal limiting curve with a non-trivial centrality dependence. The second arises from comparisons with $pp/\bar{p}p$ and $e^+e^-$ data. $\langle N_{ch}\rangle/\langle N_{part}/2\rangle$ in nuclear collisions at high energy scales with $\sqrt{s}$ in a similar way as $N_{ch}$ in $e^+e^-$ collisions and has a very weak centrality dependence. This feature may be related to a reduction in the leading particle effect due to the multiple collisions suffered per participant in heavy ion collisions.

## 1. Introduction

The PHOBOS experiment has measured $dN_{ch}/d\eta$ and the average multiplicity of charged particles $\langle N_{ch}\rangle$ produced in heavy ion collisions for center of mass energies in the nucleon-nucleon center of mass system, $\sqrt{s_{NN}}$, of 19.6, 130 and 200 GeV. The data is also binned as a function of event centrality (impact parameter) characterized by the number of participating nucleons, $N_{part}$, allowing comparisons to elementary systems, like $pp/\bar{p}p$ and $e^+e^- \to$ hadrons.

The PHOBOS multiplicity detector consists of several arrays of silicon detectors which cover nearly the full solid angle for collision events. The event centrality is characterized by the multiplicity of charged particles measured by two sets of 16 paddle counters covering $3 < |\eta| < 4.5$. The methods used for measuring the multiplicity of charged particles as well as for extracting $N_{part}$ has been described in more detail in Ref. [1].

## 2. Limiting Behavior in Pseudorapidity Distributions

Fig. 1 shows $dN_{ch}/d\eta'/\langle N_{part}/2\rangle$ ($\eta' = \eta - y_{\text{beam}}$) measured at three different RHIC energies for peripheral ($\langle N_{part}\rangle \sim 100$) and central events ($\langle N_{part}\rangle \sim 355$), in the left and right panels, respectively. These show a clear "limiting behavior" in the fragmentation region. That is, the distributions are independent of beam energy in a substantial range in $\eta'$. As the beam energy increases, $dN/d\eta'$ follows the universal trend until it reaches 85-90% of it's maximum value at midrapidity, at which point it stops following the trend. Similar behavior has been observed in el-

*Current address: Physics Department, University of Cape Town, South Africa


doi:10.1016/S0920-5632(02)02069-8

ementary collisions as well, both in $\overline{p}p$ collisions [2] and in $e^+e^-$ collisions over a large range of energies [3].

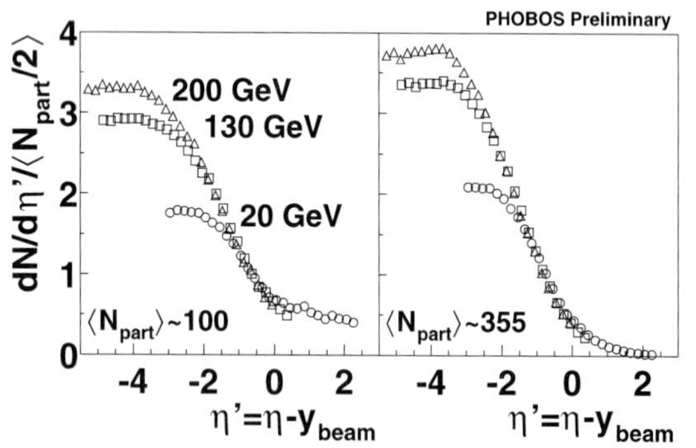


Figure 1. $dN/d\eta'/\langle N_{part}/2\rangle$ for peripheral ($\langle N_{part}\rangle \sim 100$) and central ($\langle N_{part}\rangle \sim 355$) events at three RHIC energies.

The limiting curve constrains the energy dependence of the charged particle multiplicity. It also varies with centrality in such a way that the increases seen at low $\eta'$ (which is midrapidity in $\eta$) as $N_{part}$ increases, are accompanied by decreases near $\eta' \sim 0$ (forward rapidities), as seen in Fig. 1. It is not clear why this behavior occurs, e.g. whether it is from energy conservation or a true long-range correlation.

## 3. Comparison with Elementary Systems

The angular distributions of charged particles for different strongly-interacting systems are shown in Fig. 2, where central Au+Au (divided by $\langle N_{part}/2\rangle$), $\overline{p}p$ [2] and $e^+e^-$ [4] data are compared, all at $\sqrt{s} = 200$ GeV. In this figure, the Au+Au and $\overline{p}p$ data are shown as $dN_{ch}/d\eta$, while the $e^+e^-$ data (with cuts applied to reject events with large initial-state photon radiation) is shown as $dN/dy_T$, the rapidity distribution along the event thrust axis, calculated assuming the pion mass. The shapes of Au+Au and $e^+e^-$ are similar (within 10%) in shape and magnitude, especially within $|\eta| < 4$. In the lower panel, we also

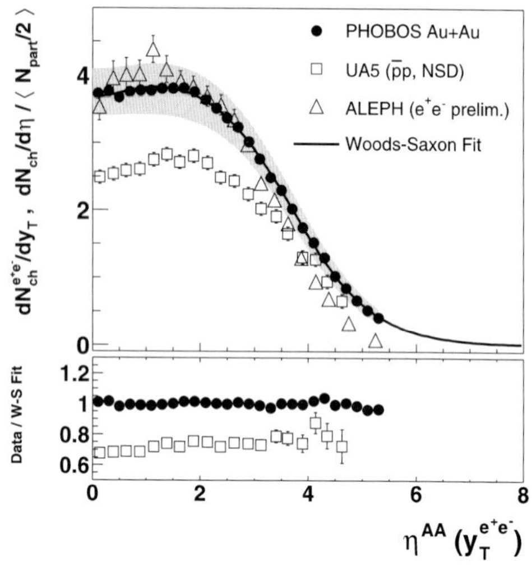


Figure 2. Top: $dN_{ch}/d\eta/\langle N_{part}/2\rangle$ for central Au+Au collisions at $\sqrt{s_{NN}} = 200$ GeV compared with $\overline{p}p$ and $e^+e^-$ data. Bottom: Au+Au and $\overline{p}p$ data divided by a fit to the Au+Au data.

observe that the shapes of the Au+Au and $\overline{p}p$ distributions are also very similar over a large range in $\eta$, but the integrals differ by $\sim 40\%$.

In Fig. 3, we compare $\langle N_{ch}\rangle/\langle N_{part}/2\rangle$ in heavy ion collisions [5] to $e^+e^-$ and $pp/\overline{p}p$ data over a large range in $\sqrt{s}$ [6]. It is observed that $\langle N_{ch}\rangle/\langle N_{part}/2\rangle$ lies below $pp$ at low energies, passes through the $pp$ data around $\sqrt{s} \sim 10$ GeV, and then gradually joins with the $e^+e^-$ trend above CERN SPS energies. These comparisons can be seen more clearly by dividing all of the data by a fit to the $e^+e^-$ data [7].

The $pp/\overline{p}p$ data follows the same trend as $e^+e^-$, but it can be shown that it matches very well if the "effective energy" $\sqrt{s}_{\text{eff}} = \sqrt{s}/2$ is used, which accounts for the leading particle effect seen in $pp$ collisions [8]. Ref. [8] finds that bulk particle production in $pp$ and $e^+e^-$ data does not depend in detail on the collision system but rather the energy available for particle production. In this scenario, the Au+Au data suggests a substantially reduced leading particle effect in central collisions of heavy nuclei at high energy.

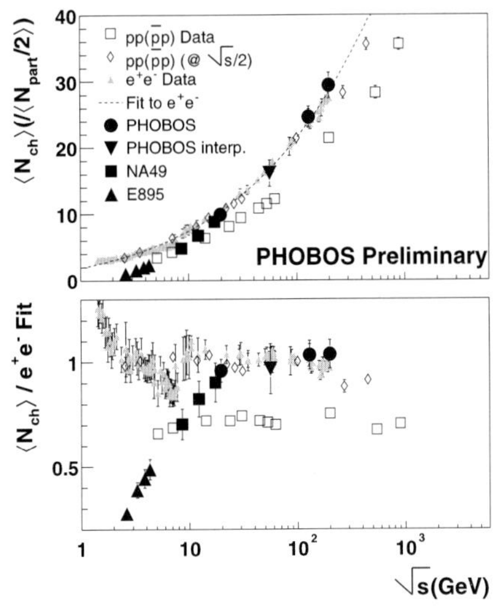


Figure 3. Comparison of $\langle N_{ch}\rangle/\langle N_{part}/2\rangle$ for A+A, $pp/\overline{p}p$, and $e^+e^-$ data compared with a fit to the $e^+e^-$ data.

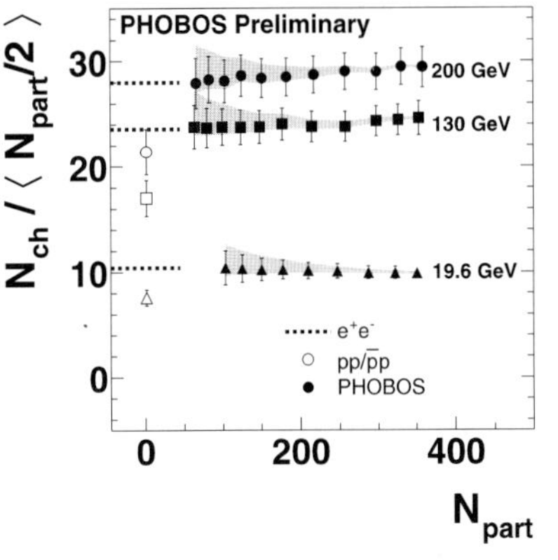


Figure 4. $\langle N_{ch}\rangle/\langle N_{part}/2\rangle$ vs. $N_{part}$ for $\sqrt{s_{NN}} =$ 19.6, 130, and 200 GeV.

The alleviation of the leading particle effect might not be so surprising in nuclear collisions. Each participating nucleon is typically struck $\overline{\nu} = \frac{N_{coll}}{N_{part}/2} > 3$ times on average as it passes through the oncoming gold nucleus for $N_{part} > 65$. One could speculate that the multiple collisions transfer much more of the initial longitudinal energy into particle production. This naturally leads to the scaling of total particle production in heavy ion collisions with $N_{part}$, as seen in Fig. 4, reminiscent of the "wounded nucleon model" [9] but with the scaling factor determined by $e^+e^-$ rather than $pp$.

In conclusion, PHOBOS has observed two kinds of universal behavior. The first is an energy-independent, but centrality-dependent, universal limiting distribution of charged particle production away from midrapidity. The second is that the total charged particle multiplicity per participant pair in heavy ion collisions above CERN SPS energies scales with sqrt(s) in a similar way as $e^+e^-$ collisions. These features may offer a new perspective on particle production in heavy ion collisions.

Special thanks to the ICHEP organizers and Heavy Ion convenors for the invitation to speak at the conference. This work was partially supported by U.S. DOE grants DE-AC02-98CH10886, DE-FG02-93ER40802, DE-FC02-94ER40818, DE-FG02-94ER40865, DE-FG02-99ER41099, and W-31-109-ENG-38 as well as NSF grants 9603486, 9722606 and 0072204. The Polish groups were partially supported by KBN grant 2 PO3B 10323. The NCU group was partially supported by NSC of Taiwan under contract NSC 89-2112-M-008-024.

## REFERENCES

1. B. B. Back *et al.*, to be submitted to PRL.
2. G. J. Alner *et al.*, Z. Phys. C **33**, 1 (1986).
3. P. Abreu *et al.*, Phys. Lett. B **459**, 397 (1999).
4. H. Stenzel, ALEPH Collaboration, Contributed paper to ICHEP2000 (2000).
5. J. Klay, U.C. Davis PhD. Thesis (2001). S. V. Afanasiev *et al.*, arXiv:nucl-ex/0205002 (2002).
6. D. E. Groom *et al.*, Eur. Phys. J. C **15**, 1 (2000).
7. A. H. Mueller, Nucl. Phys. B **213**, 85 (1983).
8. M. Basile *et al.*, Phys. Lett. B **92**, 367 (1980). M. Basile *et al.*, Phys. Lett. B **95**, 311 (1980).
9. J. E. Elias *et al.* Phys. Rev. Lett. **41**, 285 (1978). A. Białas, B. Bleszyński and W. Czyż, Nucl. Phys. **B111**, 461 (1976).

ELSEVIER

Proceedings of ICHEP 2002, pp. 86–88
*S. Bentvelsen, P. de Jong, J. Koch and E. Laenen (Editors)*

31st INTERNATIONAL CONFERENCE ON HIGH ENERGY PHYSICS AMSTERDAM

# High momentum particle suppression in Au-Au collisions at RHIC.

Federica Messer[a]* for the PHENIX Collaboration.

[a]Departement of Physics and Astronomy,
Stony Brook University, 11794-3800 Stony Brook, NY, USA

Phenix has measured at mid-rapidity tranverse momentum spectra of charged particles in Au-Au collisions at $\sqrt{s_{NN}}$ = 130 [1] and 200 GeV/c. At both energies, in central collisions Phenix observes a suppression of high momentum particles relative to a binary collision scaling of nucleon-nucleon interactions. The evolution with the event centrality has been investigated [2].

## 1. Introduction

At $\sqrt{s_{NN}}$ = 130 and 200 GeV/c jet fragmentation is expected to contribute substantially to particle production. According to several calculations, jets, produced early in the collisions, will suffer significant energy loss when traversing a hot and dense medium. As a consequence, the particle yield at high $p_T$ should be suppressed. The study of high $p_T$ particle production as function of centrality may help to define the characteristic of the medium formed in the very energetic heavy ion collisions and to contrain models and calculations.

## 2. Experimental method

A large data set of 27.6 (1.5) million minbias events in Au-Au collisions at 200 (130) GeV/c has been analyzed. The triggering detectors are represented by two Zero-Degree Calorimeters and two Beam-Beam Counters used also to determine the centrality on a event-by-event basis. A Glauber model calculation, combined with the simulated response of the trigger detectors, is used to estimate the average number of partecipant $N_{part}$ and binary collisions $N_{coll}$ corresponding to a selected centrality interval. Charged particles are reconstructed in the two central arms of PHENIX by a Drift Chamber (DC) and three multi-wires proprotional chambers (PC1,PC2,PC3). Particle momenta are measured at high momentum with resolution equal to 1% p(GeV/c) at 200 GeV/c ( and 3.5% p at 130 GeV/c).
Decay and conversion products generated near the DC are only partially deflected in the magnetic field and may be falsely reconstructed as high $p_T$ tracks. At the highest $p_T$ (at each energies) this kind of background introduces a (conservatively estimated) systematic error of 30%.

To correct the raw momentum distribution for acceptance, decay in flight, reconstruction efficiency and limited momentum resolution, events with single particles are simulated through the GEANT implementation of the PHENIX detector. The effect of increasing occupancy in more central collisions is studied by embedding single simulated tracks into real events. This corrections accounts for a factor of 1.4 (independent of $p_T$) in the 5% most central collisions.

## 3. Results and Discussion

Invariant charged hadron spectra at mid-rapidity are presented in Fig. 1 for two different centrality selections (0 − 10% and 60 − 92%) for two energies. The spectra obtained at 200 GeV develop a clear power law shape at high $p_T$ and are higher in yield relative to 130 GeV[1]. The increase in yield is consistent with the increase of total multiplicity in PHENIX [3–5]. The ratio between central and peripheral collisions, both normalized by the mean number of binary collisions,

*RIKEN-BNL, Brookhaven National Laboratory, Upton 11973 NY, USA.


doi:10.1016/S0920-5632(02)02070-4

is plotted in Fig. 2. The ratio at 200 GeV is somewhat lower than at 130 GeV, possibly reflecting a stronger suppression or a change in particle contribution from 130 to 200 GeV [6,7]. In general, the ratios for two energies have similar trends, namely an increase up to 2 GeV/c followed by a decrease as a function of $p_T$. Within errors, the ratio is approximately constant above 4 GeV/c.

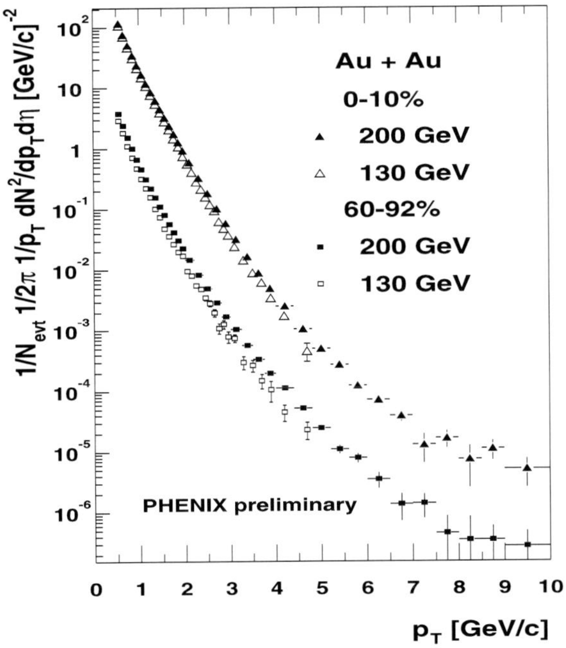


Figure 1. Charged hadron spectra for central and peripheral collisions for both $\sqrt{s_{nn}} = 200$ GeV and $\sqrt{s_{nn}} = 130$ GeV. The systematic errors (which have not been shown for clarity) are roughly 16% below 8 GeV/c and 33% above 8GeV/c.

The suppression effect can be quantified by $R_{AA}$, the *nuclear modification factor*, which is defined as

$$R_{AA}(p_\perp) = \frac{dN_{AA}/dyd^2p_\perp}{\langle N_{binary}\rangle dN_{pp}/dyd^2p_\perp} \quad (1)$$

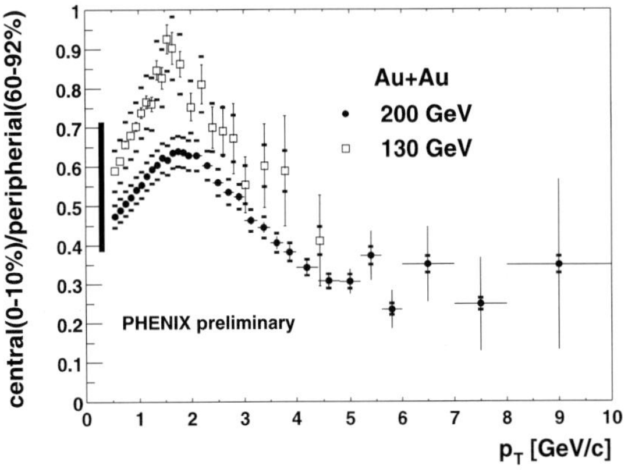


Figure 2. Ratio of central to peripheral charged particle spectra for two energies. The solid error bar on each point is the statistical error, the brackets around each point indicate the systematic error, the vertical thick line at the left edge indicates the overall scale uncertainty on the ratio due to the uncertainty on the number of binary collisions.

The p+p reference spectrum was obtained from a simultaneous fit to 200 GeV UA1 data [8] and PHENIX $\pi^0$ data scaled by a factor of 1.6 [9]. The $p_T$ dependence of $R_{AA}$ is shown for central collisions in Fig. 3. The $R_{AA}$ at 200 GeV is significantly lower than at 130 GeV over the entire $p_T$ range.

Finally, the centrality dependence of the suppression is shown by Fig. 4 where the charged particle yield integrated above 4 GeV/c and normalized to the number of binary collisions is plotted as a function of centrality. We observe a continuous decrease of the yield per-collision with increasing $N_{part}$; the per-collision yield in the most central collisions is a factor of 3.5 smaller than in the most peripheral collisions.

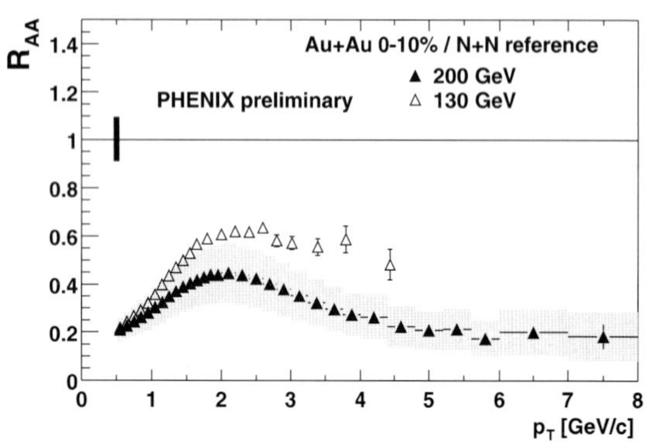


Figure 3. Ratio of Au+Au collisions to the p+p reference [9] in central collisions for both $\sqrt{s_{nn}} = 200$ GeV and $\sqrt{s_{nn}} = 130$ GeV. The shaded band indicates the uncertainty on the p+p reference and systematic error of the spectra. For $\sqrt{s_{nn}} = 130$ GeV, this uncertainty is somewhat smaller. The vertical bar at the left edge indicates the $p_T$ independent scale error due to uncertainty on the number of collisions.

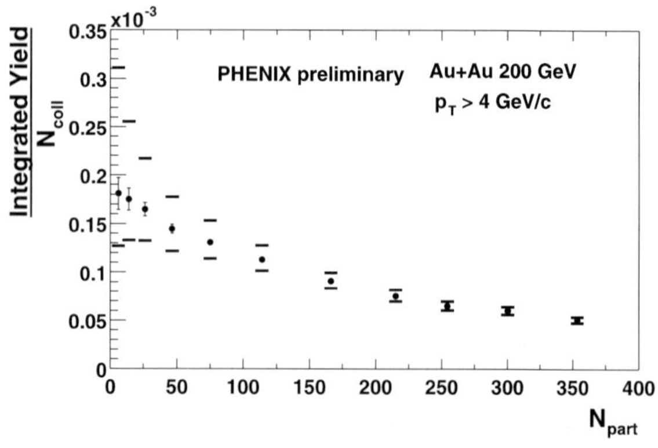


Figure 4. Integrated charged hadron yield above 4 GeV/c nomalized by $\langle N_{coll} \rangle$. Brackets around data points indicates the systematic errors due to the uncertainty of $\langle N_{coll} \rangle$ and the occupancy correction.

## 4. Conclusions

We have observed that in the most central collision the particle production at high transverse momentum is suppressed in reference to a binary collision scaling of nucleon-nucleon reaction as well as to peripheral interaction measured by the same experiment. We observe a continuous and gradual decrease in the behaviour of the high momentum particle yield (normalized by the average number of binary collisions) as function of centrality.

## REFERENCES

1. K. Adcox *et al.*, PHENIX collaboration Phys. Rev. Lett. 88, 22301 (2002).
2. K. Adcox *et al.*, PHENIX collaboration nucl-ex/0207009.
3. B.B. Back *et al.*, Phys. Rev. Lett. 88,3100 (2000); B.B. Back *et al.*, Phys. Rev. **C**65,061901 (2002).
4. K. Adcox *et al.*, PHENIX collaboration Phys. Rev. Lett. 86,3500 (2001).
5. A. Bazilevsky for the PHENIX collaboration, these proceedings.
6. T. Chujo for the PHENIX Collaboration, Nuclear Physics A, in print.
7. S. Mioduszewski for the PHENIX Collaboration, Nuclear Physics A, in print
8. C. Albajar *et al.*, Nucl. Phys. **B335**,261 (1990).
9. H. Torii for the PHENIX Collaboration, Nuclear Physics A, in print.

ELSEVIER

Proceedings of ICHEP 2002, pp. 89–94
*S. Bentvelsen, P. de Jong, J. Koch and E. Laenen (Editors)*

31st INTERNATIONAL CONFERENCE ON HIGH ENERGY PHYSICS AMSTERDAM

# Results on Au+Au collisions at $\sqrt{s_{NN}} = 130$ GeV and $200$ GeV from the PHENIX experiment

T. Peitzmann[a] for the PHENIX collaboration*

[a]Institut für Kernphysik, University of Münster, 48149 Münster, Germany

Recent results from the PHENIX experiment at RHIC on Au+Au and p+p collisions are presented, including identified hadron spectra, ratios and particle correlations. The production of hadrons at high transverse momenta is investigated in detail. A strong suppression of high $p_T$ hadrons in central heavy ion collisions compared to extrapolations from p+p collisions is observed. Evidence of jet structures in heavy ion collisions and the status of di-electron, di-muon and photon measurements is discussed.

## 1. INTRODUCTION

The PHENIX detector [1] at RHIC, the Relativistic Heavy Ion Collider at Brookhaven National Laboratory, is designed to measure the properties of nuclear matter at the highest temperatures and energy densities with a variety of experimental probes, and in particular to search for the predicted transition into a phase of deconfined quarks and gluons, the "quark-gluon-plasma" (QGP). The particular emphasis of PHENIX is the investigation of penetrating probes, like photons, electrons and muons. In addition, it has excellent hadron identification capabilities out to high transverse momenta.

PHENIX has measured Au+Au collisions at $\sqrt{s_{NN}} = 130$ GeV and at $\sqrt{s_{NN}} = 200$ GeV and p+p collisions at $\sqrt{s} = 200$ GeV. The PHENIX experiment consists of two central spectrometer arms that were completely instrumented in the most recent beam time (Run II) at RHIC. The central arms each cover pseudorapidity ($|\eta| < 0.35$), transverse momentum ($p_T > 0.2$ GeV/c), and 90 degrees in azimuthal angle $\phi$. They are comprised from the inner radius outward of a Multiplicity and Vertex Detector (MVD), Drift Chambers (DC), Pixel Pad Chambers (PC), Ring Imaging Cherenkov Counter (RICH), Time Expansion Chamber (TEC), Time-of-Flight Scintillator Wall (TOF), and two types of Electromagnetic Calorimeters (EMCal). PHENIX has two forward muon spectrometers consisting of hadronic absorbers, a cathode strip chamber muon tracking system, and interleaved Iarocci tubes with steel plates for muon identification and triggering. Each spectrometer covers approximately $1.2 < |\eta| < 2.2$, and $p_{tot} > 2$ GeV/c. The south muon spectrometer was commissioned and operational in Run II, while the north arm is currently being completed.

Au-Au and proton-proton collisions are characterized with minimum bias triggers based on a set of zero degree calorimeters (ZDC), beam-beam counters (BBC), and a scintillator multiplicity counter (NTC) for larger coverage in proton-proton collisions.

The present paper reports on recent results regarding identified hadron spectra and ratios and correlations. Particular emphasis is laid on the investigation of hadron suppression at high transverse momentum and on evidence for jets in heavy ion collisions. The status of the measurements of penetrating probes like J/$\psi$, open charm and direct photons is discussed. The PHENIX experiment has obtained a large number of other interesting results, which can not be covered in these proceedings, recent reports can be found in [2].

## 2. IDENTIFIED CHARGED HADRONS

Using the central arm tracking detectors and the high resolution TOF PHENIX can identify charged hadrons over a large momentum range. Figure 1 shows $p_T$ distributions in the 5% most central Au+Au collisions at 200 GeV. The spectra extend up to 1.8

*For full author and institution list see the Collaborations Appendix to the proceedings of the XVI International Conference on Ultrarelativistic Nucleus-Nucleus Collisions, Quark Matter 2002, Nantes, to be published in Nucl. Phys. A.


doi:10.1016/S0920-5632(02)02075-3

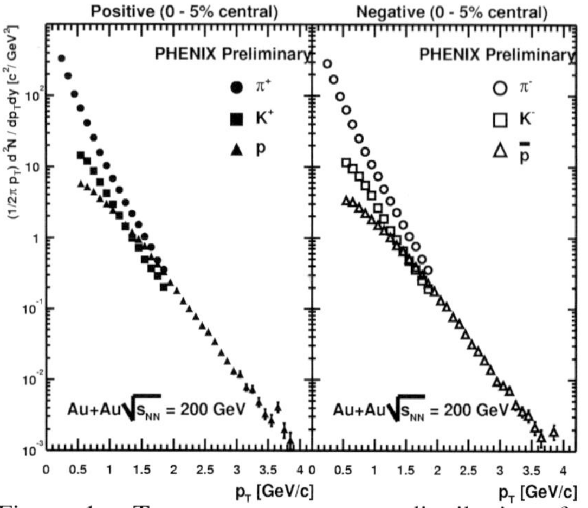


Figure 1. Transverse momentum distributions for identified charged hadrons in the 5% most central Au+Au collisions at $\sqrt{s_{NN}}$ = 200 GeV. The error bars are statistical only.

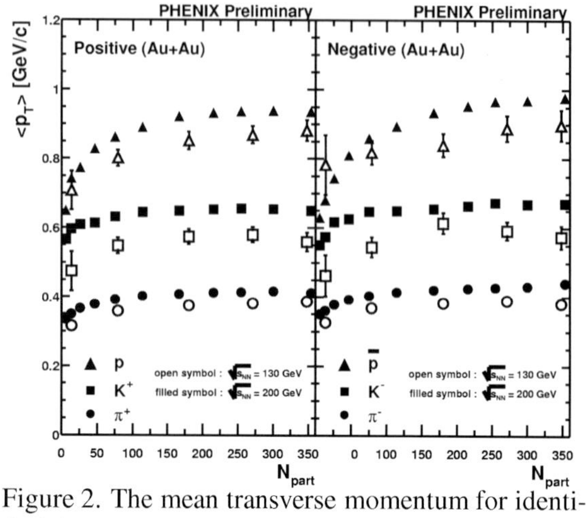


Figure 2. The mean transverse momentum for identified charged hadrons as a function of $N_{part}$ in Au+Au collisions at 130 GeV [3] (open symbols) and 200 GeV (filled symbols).

GeV/$c$ for charged pions and kaons and 3.8 GeV/$c$ for (anti-)protons. Around 2.0 GeV/$c$ in $p_T$, the proton yield becomes comparable to the pion yield in central collisions, and a similar observation is made for negative particles. This behavior is very different from the one observed in p+p or $e^+$+$e^-$ collisions, where the (anti)proton to pion ratio is much lower. It might be explained by a collectively expanding source, as will be discussed below.

This observation is of course related to the different shape of the spectra for different species, which can e.g. be quantified by the average transverse momentum. In figure 2, the centrality dependence of $\langle p_T \rangle$ for identified charged hadrons is shown, together with data taken at 130 GeV [3]. In both the 200 GeV and 130 GeV data, the $\langle p_T \rangle$ of all particles increases from the peripheral to the central events. The $\langle p_T \rangle$ also increases with increasing particle mass. This dependence would also be naturally explained by assuming a collective hydro-dynamical expansion. In such a scenario, the dependence on the number of participant nucleons ($N_{part}$) may be due to an increasing radial expansion from peripheral to central events.

## 3. HIGH $P_T$ HADRON PRODUCTION

The experiments at RHIC have truely opened up the domain of hard scattering in heavy ion collisions. This is of particular interest as jets originating from hard scattered partons and their possible modification in nuclear collisions compared to p+p via energy loss is considered as a new signal for the quark-gluon-plasma [4]. PHENIX has the great advantage to be able to identify neutral pions with the EDMCal out to extremely high $p_T$, the range being essentially limited by statistics, which has been greatly enhanced in Run II. Fig. 3 shows the neutral pion spectrum for p+p collisions, which reaches out to 13 GeV/$c$. The data agree with pQCD calculations [5] (dotted lines) and also with the data from UA1 [6] (not shown), which were measured only for $p_T < 6\,\mathrm{GeV}/c$. Besides the important information this measurement provides by itself, the p+p measurement is also of great interest as a baseline for comparisons to the heavy ion results.

Fig. 4 shows the transverse momentum spectra of neutral pions for Au+Au collisions at $\sqrt{s_{NN}}$ = 200 GeV and 130 GeV, the one for the 10% most central collisions at 200 GeV reaches out to $p_T$ = $8\,\mathrm{GeV}/c$.

Already in earlier measurements at 130 GeV [7,8] a suppression of the high $p_T$ hadron yield relative to the expectation from naive estimates from perturbative QCD has been observed, which is qualitatively similar to the predicted jet-quenching from the plasma [4]. To quantify such effects a nuclear modification factor is usually introduced:

$$R_{AA}(p_T) = \frac{\text{Yield per A + A collision}}{\langle N_{coll} \rangle (\text{Yield per p + p collision})}$$

Hard processes have a small cross section and are gen-

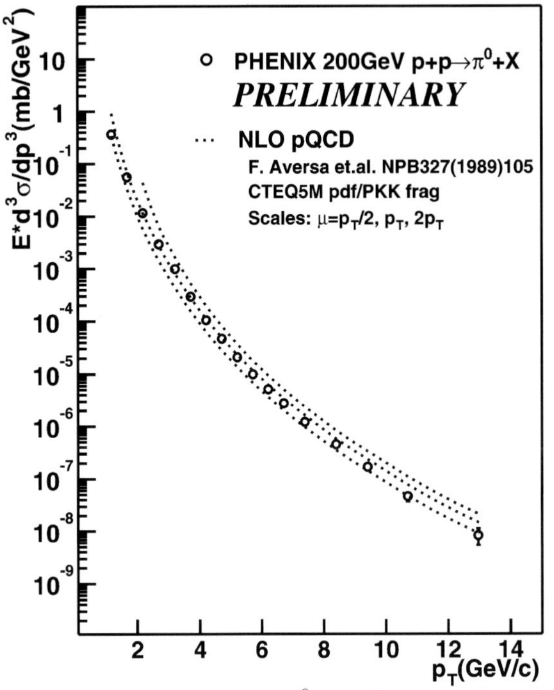


Figure 3. Cross section of $\pi^0$ as a function of $p_T$ in p+p collisions at $\sqrt{s_{NN}} = 200$ GeV. The dotted lines show NLO pQCD calculations using $p_T/2$ (top line), $p_T$ (middle line), and $2p_T$ (bottom line) renormalization and factorization scales.

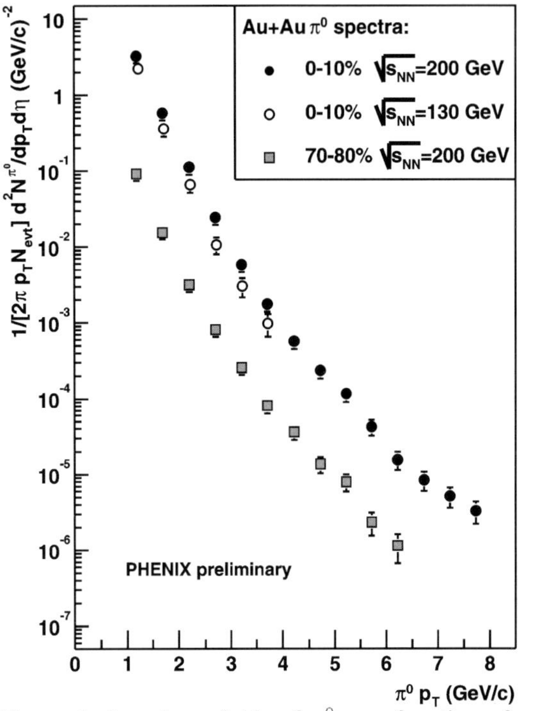


Figure 4. Invariant yields of $\pi^0$ as a function of $p_T$ in central (0-10%) Au+Au collisions at $\sqrt{s_{NN}} = 200$ GeV and 130 GeV and peripheral (70-80%) collisions at $\sqrt{s_{NN}} = 200$ GeV.

erally believed to be incoherent. Their probability in Au+Au collisions should thus be proportional to the average number of binary collisions $N_{coll}$. In the absence of any nuclear modification the ratio $R_{AA}$ is therefore expected to be one for particles produced from such processes (i.e. at high $p_T$).

Fig. 5 shows $R_{AA}(p_T)$ for the 10% most central Au+Au collisions at $\sqrt{s_{NN}} = 200\,\mathrm{GeV}$. A strong suppression relative to a scaling with $N_{coll}$ is observed, which is qualitatively similar to the result at 130 GeV. In this measurement the suppression persists up to $p_T \approx 8\,\mathrm{GeV}/c$ reaching values of $1/5 - 1/6$. As the p+p reference spectrum has now been measured with the same detector, the systematic errors have been significantly reduced. In Fig. 5 this is compared to $R_{AA}$ from central Pb+Pb collisions at $\sqrt{s_{NN}} = 17\,\mathrm{GeV}$ from WA98 [9], which shows a quite different behavior, namely a strong enhancement with values much larger than 1 increasing with increasing $p_T$. This enhancement at lower energies is similar to the "Cronin effect" [10] observed in p+A collisions, which has been attributed to $p_T$ broadening due to initial state scatterings. One should however note that, if heavy-ion collisions do exhibit collective flow increasing for central collisions, this would also lead to $p_T$ broadening and be similarly reflected in $R_{AA}$.

The spectrum for peripheral collisions (70-80%) is consistent with p+p collisions scaled by $N_{coll}$. It is of interest to investigate how the transition from these peripheral collisions to central collisions, where the suppression has been observed, takes place, and whether one can locate a centrality, where the suppression sets in. Fig. 6 shows $R_{AA}$ as a function of the mean number of participating nucleons $N_{part}$ for neutral pions with $p_T > 4\,\mathrm{GeV}/c$. The suppression smoothly increases from peripheral to central events. Results have also been obtained with an analog analysis using non-identified charged hadrons, where a qualitatively similar suppression in central reactions and the same trend with centrality is observed [11].

From the identified charged hadron spectra (see above) one has made the interesting observation that

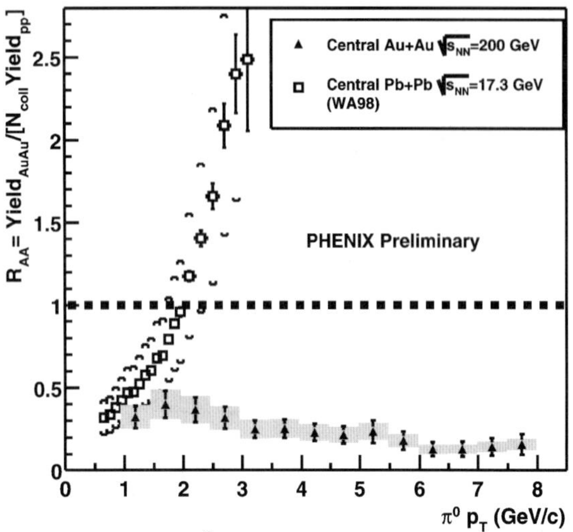


Figure 5. $R_{AA}$ ($\pi^0$) for central Au+Au collisions at $\sqrt{s_{NN}} = 130$ GeV and central Pb+Pb collisions at $\sqrt{s_{NN}} = 17$ GeV. The error bars are the quadratic sum of statistical and $p_T$-dependent systematic errors. The brackets/boxes are the errors on the normalization of this ratio.

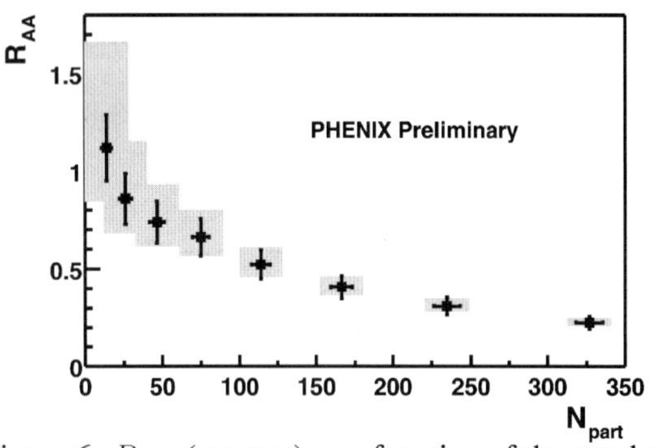


Figure 6. $R_{AA}$ (see text) as a function of the number of participants for neutral pions with $p_T > 4$ GeV/c.

the proton and antiproton yield is comparable to the pion yield for $p_T = 3 - 4\,\mathrm{GeV}/c$. While at these transverse momenta one might imagine that hydrodynamics could be invoked as a possible explanation, it would be expected that this should break down at still higher $p_T$, where eventually perturbative QCD should start to dominate. The neutral pions allow to study the chemical composition of the spectra in this high $p_T$ region via the ratio of $\pi^0$ to nonidentified charged hadrons $(\mathrm{h}^+ + \mathrm{h}^-)/2$. This ratio is shown in Fig. 7 for minimum bias events. From the knowledge of particle production in jets (as measured e.g. in $e^+e^-$ collisions [12]) one would expect this ratio to approach

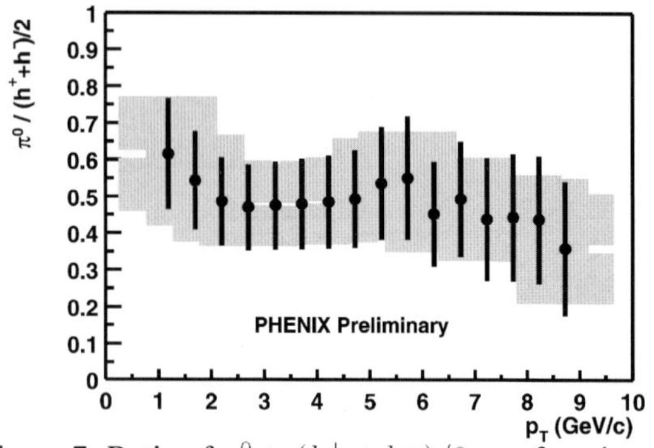


Figure 7. Ratio of $\pi^0$ to $(h^+ + h^-)/2$ as a function of $p_T$ for minimum bias Au+Au collisions.

one at high $p_T$ (assuming $\pi^0 = (\pi^+ + \pi^-)/2$). Even for $p_T > 4\,\mathrm{GeV}/c$ this is not observed. The ratio remains nearly constant at a value around 0.5, indicating that approximately half of the charged hadrons at high $p_T$ are protons and/or kaons. This may call for some other production mechanism for protons and/or kaons in Au+Au collisions at these large transverse momenta. It remains puzzling especially in view of the fact, that correlation studies (as discussed below) indicate, that jet fragmentation may still be the dominant production mechanism at these $p_T$. Alternatively, the large proton and/or kaon content at high $p_T$ could be due to a difference of the modifications of jet fragmentation into pions relative to protons/kaons.

To understand particle production at high $p_T$ it is of utmost importance to get more direct information on the hard scattering component. The high multiplicity in Au+Au collisions does not allow to identify individual jets based on event topology as in p+p. Therefore one has to search for hints of jet structures in a angular correlations. PHENIX has selected events with at least one neutral particle with $E > 2.5\,\mathrm{GeV}$ measured in the EMCal (mostly photons from neutral pion decays). In these events correlation functions of charged particles relative to the trigger photon are calculated. Measurements of such correlation functions in p+p exhibit peaks due to jets, which agree with those from the PYTHIA event generator [13]. One can then study the same correlation in Au+Au and again compare to PYTHIA to look for possible modifications. Figure 8 shows the background-subtracted correlation as a function of the relative azimuthal angle $\Delta\phi$ for charged particles with $p_T = 2 - 4\,\mathrm{GeV}/c$. Also shown is a correlation function combining effects of jets as produced by PYTHIA

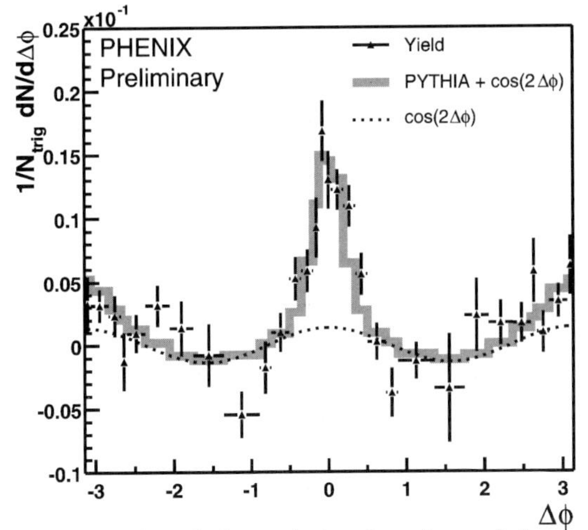


Figure 8. Azimuthal correlation functions of charged particles relative to a high $p_T$ photon in triggered Au+Au collisions (acceptance corrected). The effect of elliptic flow (dotted line) and a fit of PYTHIA + elliptic flow (grey line) is also shown.

with elliptic flow characterized by a $\cos(2\phi)$ term, which has been fit to the data. Apparently, the strong correlation at $\Delta\phi = 0$ is dominated by particles from jet fragmentation, which in shape looks very similar to the PYTHIA calculation. Thus there is high probability that a high $p_T$ $\pi^0$ as represented by the trigger photon originates from a jet.

## 4. ELECTROMAGNETIC PROBES

PHENIX has excellent electron identification capabilities and has thus measured inclusive electron spectra. Electrons resulting from semi-leptonic decays of charm D mesons ($D \longrightarrow e + K + \nu$) and beauty B mesons ($B \longrightarrow e + D + \nu$) have significant contributions to these inclusive electrons above 1.0 and 2.5 GeV, respectively. The dominant sources of "background" electrons are from neutral pion Dalitz decay ($\pi^0 \longrightarrow \gamma e^+ e^-$) and photon conversions in material in the PHENIX aperture. One can account for these sources using the measured $\pi^0$ distributions. In addition, although their contributions are relatively small, additional electrons from $\eta, \eta', \omega, \phi, \rho$ must be estimated. Results from Au-Au collisions at $\sqrt{s_{NN}} = 130$ GeV have been published [14] and one finds a significant excess of electrons over "background" above $p_T \approx 0.7$ GeV/c that increases with $p_T$.

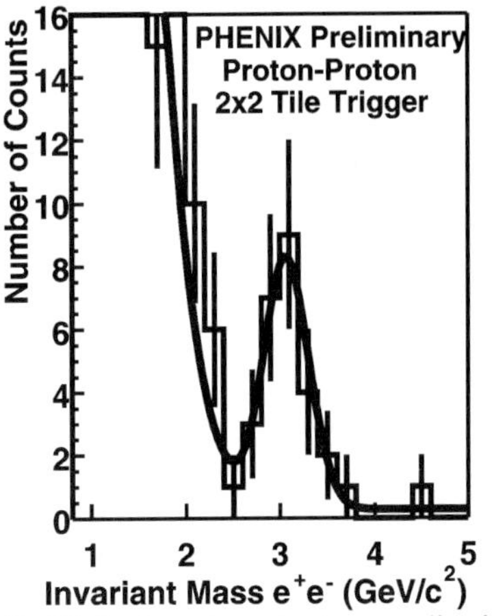


Figure 9. Dielectron invariant mass distribution from proton-proton events.

In Run II, PHENIX has a more powerful method of extracting the single electron charm contribution. Using a brass photon converter wrapped around the beam pipe near the interaction point for a fraction of the beam time, a good estimate of the photonic contributions to the background can be obtained. Since there is a fixed relation between the $\gamma$ from $\pi^0$ decay (which is the dominant source of conversion photons), and the $\pi^0$ Dalitz contribution, we can subtract out both contributions.

After this subtraction the data appear reasonably described by a PYTHIA calculation of the expected charm contribution assuming scaling with $N_{coll}$, similar to the 130 GeV data analysis. The electron data in four centrality bins appears to qualitatively follow a scaling with $N_{coll}$. Significant work remains to reduce the systematic errors in order to exactly quantify this scaling relation. PHENIX is currently analyzing the proton-proton data set so that we can calculate the charm scaling without relying on the PYTHIA baseline.

Phenix has also performed the first measurements of $J/\psi$ production in proton-proton collisions at $\sqrt{s} = 200$ GeV. As an example Figure 9 shows the dielectron invariant mass distribution for proton-proton collisions with at least one 2x2 calorimeter tower tile above our Level-1 trigger threshold. This represents about half of our proton-proton statistics. Similarly, measurements in the dimuon channel have been performed for proton-proton collisions.

We have corrected for the central and muon arm acceptances, tracking and particle identification efficiencies, and Level-1 trigger efficiencies and have thus measured the cross section $d\sigma/dy$ as a function of rapidity. The data are fit to a simple Gaussian form and also to the predicted rapidity shape from PYTHIA using parton distribution function GRV94LO. Integrating over rapidity, we find a preliminary result of:

$$\sigma_{pp}{}^{J/\psi} = 3.8 \pm 0.6(\mathrm{stat}) \pm 1.3(\mathrm{sys})\ \ \mu\mathrm{b}.$$

Preliminary results on $J/\psi$ production from the di-electron channel have also been obtained in Au+Au collisions. The limited statistics does presently not allow to draw any firm conclusions from this measurement. Increased statistics and the di-muon measurements are currently being analyzed. Future beam times with increased luminosity and a fully equipped PHENIX detector should significantly enhance this measurement.

The measurement of direct photons is another important goal of PHENIX. Present preliminary results show no significant excess of photons over those from hadron decays, which is mainly due to the large systematic error. This difficult analysis is under way and should allow to significantly reduce the systematic errors in the future.

## 5. CONCLUSIONS

PHENIX has obtained a wealth of interesting new results from the last RHIC beam time. Identified hadron spectra out to high $p_T$ have been presented. High $p_T$ neutral pion production in central Au+Au collisions at 200 GeV is suppressed by a factor of $1/5 - 1/6$ compared to p+p collisions scaled by the number of binary collisions. The suppression increases gradually with number of participants, or decreasing impact parameter, no threshold behavior is observed. High $p_T$ neutral particles, which are predominantly photons from $\pi^0$ decays are accompanied by correlated charged particles in a jet-like pattern, which indicates that neutral pions at high $p_T$ should originate to a large extent from jet fragmentation. However, the ratio of neutral pions to charged hadrons implies a large non-pionic contribution to hadrons out to $p_T = 8\,\mathrm{GeV}/c$, which is not in line with expectations from unmodified hard scattering production. Single electron spectra have been measured, which are consistent with predictions of charm production by perturbative QCD. A measurement of the cross section of $J/\psi$ in p+p has been presented, and the status of the measurement of $J/\psi$ and direct photons in Au+Au collisions has been discussed.

## REFERENCES


1. PHENIX Collaboration, accepted for publication in NIM-A Special Issue.
2. PHENIX collaboration, Talks by T. Chujo *et al.*, S. Mioduszewski *et al.*, J.L. Nagle *et al.*, N.N. Ajitanand *et al.*, R. Averbeck *et al.*, A. Bazilevsky *et al.*, J. Burward-Hoy *et al.*, M. Chiu *et al.*, D. d'Enterria *et al.*, A. Enokizono *et al.*, S. Esumi *et al.*, A.D. Frawley *et al.*, J. Jia *et al.*, D. Mukhopadhyay *et al.*, K. Reygers *et al.*, T. Sakaguchi *et al.*, H. Torii *et al.*, Proceedings of Quark Matter 2002, Nantes, to be published in Nucl. Phys. A.
3. K. Adcox *et al.*, Phys. Rev. Lett. **88**, 242301 (2002).
4. M. Gyulassy and M. Plümer, Phys. Lett. B **243**, 432 (1990); R. Baier *et al.*, Phys. Lett. B **345**, 277 (1995); X.N. Wang, Phys. Rev. C **58**, 2321 (1998).
5. Private communication with W. Vogelsang.
6. C. Albajar *et al.*, Nucl. Phys. B **335**, 261 (1990).
7. K. Adcox *et al.* (PHENIX Collaboration), Phys. Rev. Lett. **88**, 022301 (2002).
8. C. Adler *et al.* (STAR Collboration), nucl-ex/0206011.
9. M.M. Aggarwal *et al.* (WA98 Collaboration), Eur. Phys. J. C **23**, 225-236 (2002).
10. D. Antreasyan *et al.*, Phys. rev D **19**, 764 (1979).
11. F. Ceretto *et al.* (PHENIX Collaboration), these proceedings.
12. P. Abreu *et al.* (DELPHI Collaboration), Eur. Phys. J. C **17**, 207 (2000).
13. T. Sjostrand, Comput. Phys. Commun. **82**, 74 (1994).
14. K. Adcox *et al.* (PHENIX Collaboration), Phys. Rev. Lett. **88**, 192303 (2002).

Proceedings of ICHEP 2002, pp. 95–100
*S. Bentvelsen, P. de Jong, J. Koch and E. Laenen (Editors)*

ELSEVIER

31st INTERNATIONAL CONFERENCE ON HIGH ENERGY PHYSICS AMSTERDAM

# Recent Results from STAR

J.H. Thomas [a] for the STAR Collaboration

[a]Lawrence Berkeley National Laboratory
B510A, 1 Cyclotron Rd, Berkeley, CA 94720

Recent results from the STAR experiment are presented. The data indicate that nuclear matter produced in ultra-relativistic heavy ion collisions is hot, dense, and possibly opaque.

STAR is a wide aperature detector with full acceptance for hadronic observables out to $\approx 45°$ with respect to the beamline[1]. It is one of two large detectors at the Relativistic Heavy Ion Collider (RHIC) at Brookhaven National Laboratory. At the heart of STAR is the world's largest time projection chamber[2]; see figure 1. A silicon drift detector (SVT) surrounds the interaction vertex, and in the forward and backward directions there are radial drift TPCs (FTPCs) to cover the high rapidity region. Outside of the main TPC is a RICH detector and in the future STAR will be surrounded by a full barrel of electromagnetic calorimeters.

The data acquisition system is triggered by two independent signals: neutral energy measured in a pair of calorimeters at zero degrees with respect to the beam (Zcal) and by the multiplicity of particles that hit the central trigger barrel (CTB). The combination of the Zcal and CTB signals allow us to select events by centrality (or impact parameter).

STAR detected the first collisions at RHIC in June of 2000. Since then, we have made measurements during two very successful physics runs; Au on Au at $\sqrt{s}$ = 130 GeV in 2000 and Au on Au at $\sqrt{s}$ = 19.6 and 200 GeV in 2001-2002. We also ran polarized protons at 200 GeV in 2001-2002 but the proton data are beyond the scope of this brief report.

Typical spectra from the Au-Au runs cover $\pm 1$ unit of rapidity. The measured momenta range from 150 MeV up to 1.0 GeV for identified particles in the TPC, up to 3 GeV for identified particles in the RICH, and out to 12 GeV for all charged particles.

STAR has collected a large sample of spectra for identified protons, pions, kaons (charged and neutral), and their anti-particles. And in the higher statistic runs we have spectra for lambdas, cascades, and omegas. A recently prepared set of anti-proton spectra at mid-rapidity are shown in figure 2.

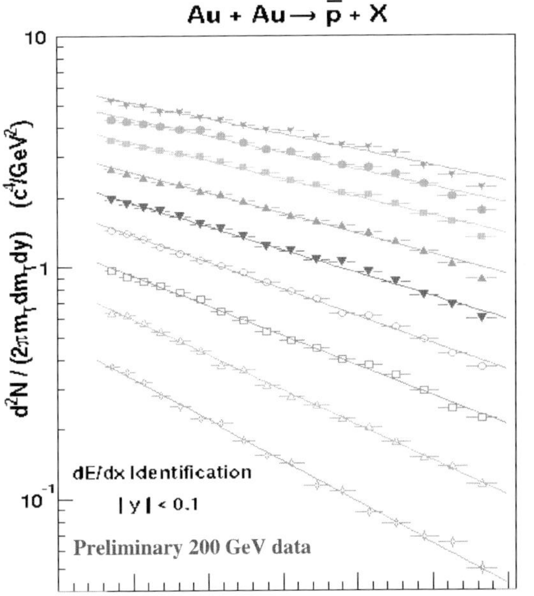


Figure 2. The multiplicity of anti-protons at $\sqrt{s}$ = 200 GeV are shown as a function of centrality. The lines are Gaussian fits to the spectra with respect to the transverse mass of the anti-protons. Each line represents a different centrality class.

The data are plotted as a function of centrality


doi:10.1016/S0920-5632(02)02078-9

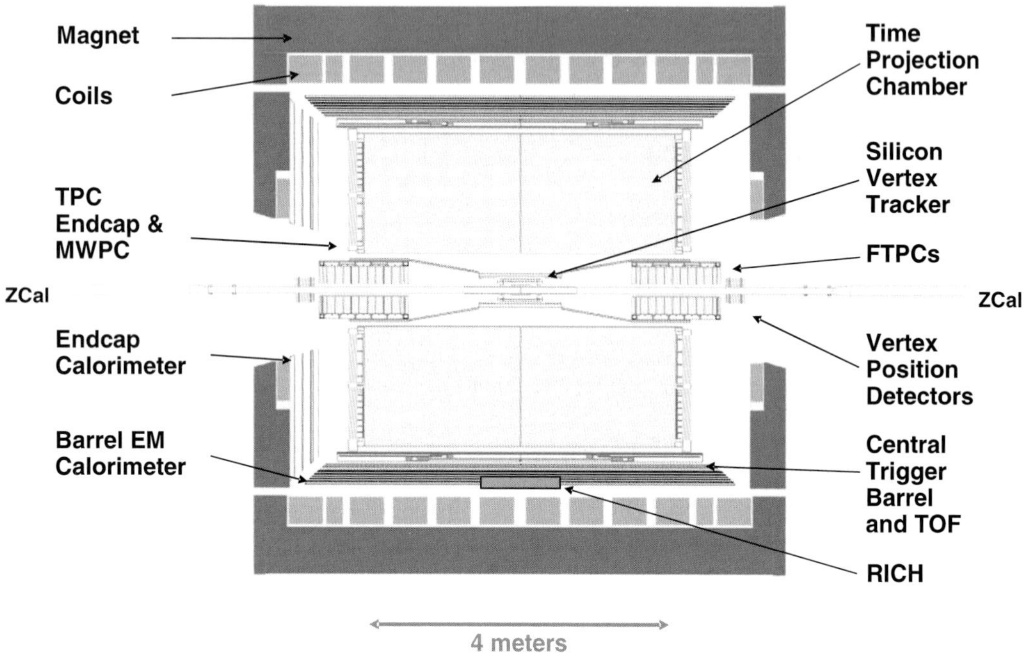


Figure 1. The STAR detector at RHIC. The full suite of detectors makes it possible to measure hadronic and leptonic observables over a wide rapidity interval and into $2\pi$ azimuth.

with the most peripheral collisions represented by the bottom curve on the plot and the most central collisions by the top-most curve. The top curve represents the top 5% of the cross-section for Au-Au interactions at 200 GeV, the next line represents the bin between 5 and 10% of the cross-section, and the bins are 10% wide after that. STAR has measured p and $\overline{\text{p}}$ spectra at $\sqrt{s} =$ 19.6, 130, and 200 GeV. The 19.6 and 200 GeV data are new, while the 130 GeV data are already in the literature[3].

The yield of anti-protons as a function of collision energy is interesting because there aren't any anti-baryons in the initial state at RHIC. Examination of the $\overline{\text{p}}$/p ratio is one way to summarize the collision data. We have measured ratios of 0.11±0.01, 0.71±0.05, and 0.80±0.05 at 19.6, 130, and 200 GeV, respectively[4]. In the early Universe, we believe that the anti-baryon to baryon ratio was unity and so the data show a clear trend towards the early universe values for the $\overline{\text{p}}$/p ratio as a function of $\sqrt{s}$. Another way to appreciate this trend is to calculate the fraction of net baryons that are created by pair production. The observed $\overline{\text{p}}$/p ratio suggests that 80% of the protons come from pair production and the remaining 20% are transported into the collision zone by the beam. So for the first time at a heavy ion collider, more baryons are pair produced than are present in the initial state.

The $\overline{\text{p}}$/p ratio, and other particle ratios measured by STAR at 130 GeV, can be used as input to a simple model[5] to estimate the temperature of the system at the time of hadro-chemical freeze-out. The result is that T = 175 ± 7 MeV and $\mu_B = 51 \pm 6$ MeV. This temperature is not so different than what was seen at the SPS, however the baryo-chemical potential, $\mu_B$, is much lower. ($\mu_B \cong 270$ MeV at the SPS). Hadron resonance ideal gas models yield very similar results[6] but, perhaps, slightly lower chemical potentials $\mu_B \cong 40$ MeV. It is interesting to note that these models assume thermal and chemical equilibrium at hadro-chemical freeze-out. The models also have predictive power and can accurately predict the particle ratios for species not included in the input data.

Overall, the particle spectra at RHIC tell us where we are on the phase diagram of T .vs. $\mu_B$. The chemical freeze-out temperature is high, 175 MeV, while the chemical potential is low, 40 MeV. The temperature is very close to the phase transition temperature predicted by lattice QCD, 173 MeV in a two flavor calculation[7], and it is unlikely that any hadronic system will achieve a higher temperature than this. So we know where we are on the phase diagram, but we don't know what other features are on the diagram ... such as the location of the tri-critical point and whether the phase transition to the QGP is first order or second order and in a region accessible to RHIC. These are mysteries to be resolved in the future.

There is another freeze-out temperature in relativistic heavy ion collisions and it is the temperature of kinetic freeze-out. In elementary particle collisions, chemical and kinetic freeze-out occur at the same temperature; but in heavy ion collisions they occur at different temperatures. So if chemical freeze-out marks the end of inelastic collisions and the freezing of the number of chemical species, then kinetic freeze-out marks the end of elastic collisions and the freezing of particle momenta.

We can estimate the kinetic freeze-out temperature by studying the slopes and shapes of the particle spectra. The spectra in figure 2, for example, were plotted as a function of the transverse mass of the anti-proton and then fit with a Gaussian curve. You might expect the fundamental spectrum shape would be a Boltzman distribution or perhaps a Bose Einstein distribution (for pions or kaons). But a thermal distribution is not enough. These spectra are boosted by the hydrodyamical flow of the participants because heavy ion collisions generate pressure gradients inside the collision zone and these gradients drive a radial transverse expansion of the source. The symmetric expansion term is called radial flow.

Radial flow adds kinetic energy to the particle distribution and therefore modifies the spectrum shape. In figure 3, the effective temperature for the pions is 215 MeV, for the kaons it is 310 MeV, and for the anti-protons it is 575 MeV. The $< p_T >$ and the effective temperatures increase with mass. Thus, the classical expression for total energy

$$E_{tot} \quad = \quad \frac{3}{2}T \; + \; \frac{1}{2}mv^2 \tag{1}$$

suggests a linear relationship[8]

$$T_{obs} = \quad T_{KFO} \; + \; mass \; \times \bar{\beta}^2 \tag{2}$$

where the observed temperature is the sum of a universal freeze-out temperature plus the addition of a kinetic energy term due to the high velocity of the expanding shock front.

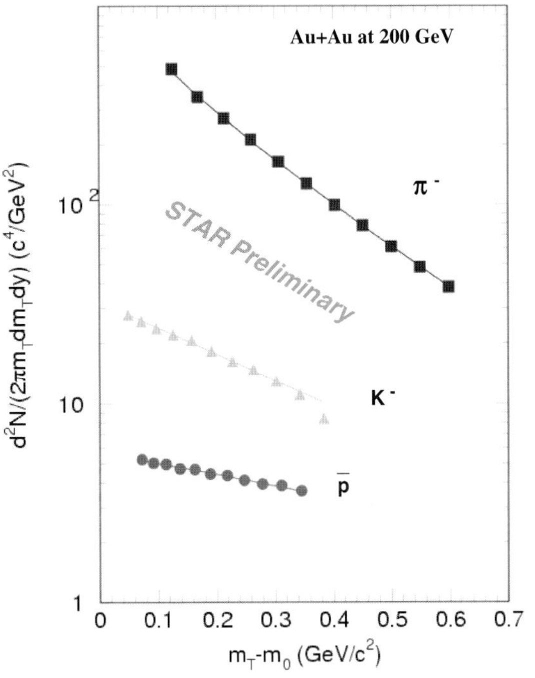


Figure 3. The slopes of the particle spectra decrease with mass, however the $< p_T >$ and the effective temperature increase with mass.

The freeze-out temperature for kinetic processes at RHIC is $\approx$ 100 MeV and this is approximately the same as the freeze-out temperature measured at the SPS; as a function of $\sqrt{s}$, freeze-out is constant from 10 GeV to 200 GeV. The mean expansion velocity is not constant, however. It increases with $\sqrt{s}$ and, at RHIC Energies, the mean radial expansion velocity is 60% of the speed of light.

There are other hydrodynamic effects in the data. Elliptic flow measures the anisotropic shape of the transverse flow field. It is the $2^{nd}$ harmonic Fourier coefficient of the momentum distribution of particles with respect to the reaction plane[9]. The ellipticity in spatial coordinates comes about because of the asymmetric overlap of two spheres in a non-central collision. However, a finite cross-section for re-interactions is required for this initial state spatial asymmetry to translate itself into a final state momentum space asymmetry; and the cross-sections must be large at early times or the details of the initial collision geometry could not manifest themselves in the final state momentum distributions.

We see elliptic flow in pion, kaon, proton, and lambda spectra. The amount of elliptic flow is large and in good agreement with hydrodynamic models for ultra-relativistic collisions. Even the mass dependence is well described by hydrodynamics[10]. This is interesting because hydro assumes local thermal equilibrium and large amounts of elliptic flow can only be generated if equilibrium is established very early in the time history of the collisions.

The surprising thing about our data is that elliptic flow persists to high momentum and then it saturates. The flow pattern shown in figure 4 follows the prediction of hydrodynamics up to about 2 GeV/c but hydro predicts that the flow curve should continue to rise above 2 GeV/c.

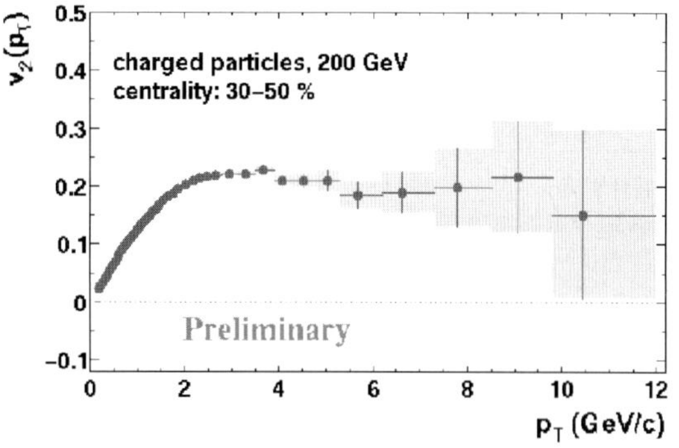


Figure 4. Elliptic flow is plotted as a function of $p_T$. The flow parameter, $v_2$, saturates at high $p_T$.

We observe that elliptic flow saturates at 2 GeV/c. Thus, there is some mechanism that builds pressure gradients, even at high $p_T$, and allows the asymmetry in the emission pattern to persist to the highest measured transverse momenta. This mechanism may be partonic energy loss or another exotic mechanism.

High transverse momentum spectra are interesting in other ways. For example, it is tempting to hypothesize that Au-Au collisions are a simple superposition of p-p collisions. We can test this hypothesis by comparing the normalized yield of charged particles for Au-Au collisions to those for p-p collisions as shown in figure 5. The normalization factor is taken to be the number of binary collisions in a Au-Au collision. Previous experience at the SPS has shown that soft collision processes scale as the number of participants while hard scattering processes scale as the number of binary collisions. Thus, neglecting shadowing and the Cronin effect, we expect the spectra to start from participant scaling at low $p_T$ and evolve towards binary scaling at high $p_T$.

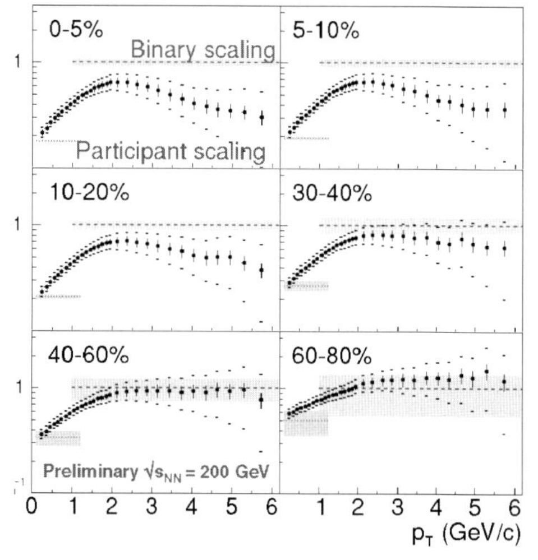


Figure 5. The normalized ratio of the Au-Au to the p-p yields for collisions at 200 GeV. Charged hadrons were measured in the range from 0 to 6 GeV/c. Systematic errors are indicated by the bars and the caps show the quadrature sum of the error bars with the systematic errors on the reference data.

The different panels in the figure show the ratio of cross-sections at different collision centralities.

We expect this ratio to be 1.0 at *high* $p_T$ if a Au-Au collision is a trivial scaling of a p-p collision. So, for example, the *peripheral* collision data, in the lower right panel, show a smooth evolution from participant scaling at low $p_T$ to binary scaling at high $p_T$. This is what we expect in a simple scaling hypothesis. But further examination of the data shows a clear violation of this hypothesis as the collisions become more and more central. The top left hand panel in figure 5 indicates that the ratio is suppressed by a factor of 3 in the most central collisions.

It isn't necessary to have a p-p reference spectrum in order to see the suppression of the high $p_T$ spectra. The suppression can also be seen when the spectra are compared by centrality (ie. impact parameter). Figure 6 shows the ratio of central collisions to peripheral collisions at 130 and 200 GeV. The yields are normalized by the number of binary collisions expected in each centrality class.

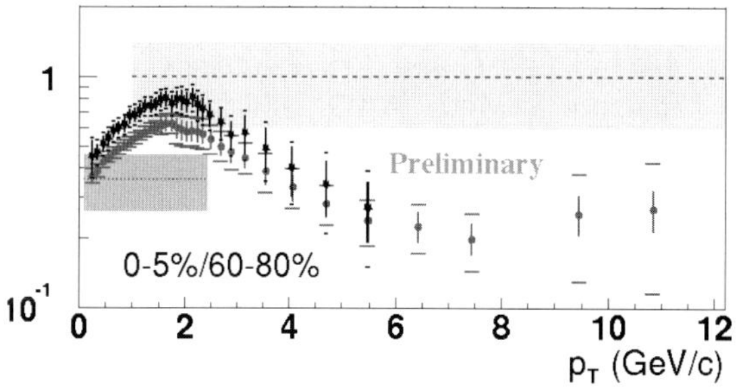


Figure 6. The normalized ratio of yields for central to peripheral Au-Au collisions at 130 and 200 GeV. The most central collisions ( 0-5%) are compared to the most peripheral collisions (60-80%). The 130 GeV data extend to 6 GeV and the 200 GeV data extend to 12 GeV.

The suppression of this ratio is especially interesting in view of the fact that the Cronin effect enhances the heavy ion cross-sections at high $p_T$ and therefore the ratio should rise above one. We see the suppression effect in our 200 GeV data and in our 130 GeV data[11].

One possible explanation for the suppressed yield in central collisions is that there are new energy loss mechanisms that come into play at high $p_T$ which distort the spectrum. A candidate mechanism is gluon bremmstrahlung. If a sufficiently dense medium exists in a heavy ion collision, then gluon bremsstrahlung and other dE/dx processes will shift particles out of the high $p_T$ part of the spectrum and add them to the low $p_T$ portion of the spectrum. Due to the exponential decrease in the cross-sections, a small shift out of the high $p_T$ tail will look like a significant decrease in the ratio of yields in figures 5 and 6. This mechanism may also explain the persistence of elliptic flow at high $p_T$ because the flow data require that primary particles interact with the expanding medium, even at high $p_T$.

The comparison of Au-Au collisions to p-p collisions is interesting in another way. How do jets in Au-Au collisions compare to jets in p-p collisions?

Jets are hard to find in heavy ion collisions because the high multiplicity of particles hides the jets, so we have done it using a statistical correlation technique. The correlation function is built by identifying a particle with a transverse momentum that exceeds a trigger threshold and then looking for associated high $p_T$ particles at similar angles and rapidity intervals.

$$C_2(\Delta\phi, \Delta\eta) = \frac{1}{N_{trigger}} \frac{N(\Delta\phi, \Delta\eta)}{Efficiency} \quad (3)$$

The correlation data show that jets exist in central collisions at RHIC and we have previously reported on them at 130 GeV[12]. However, other correlations exist in heavy ion collisions which may not exist in p-p collisions. Elliptic flow is an example. So, we need a technique to distinguish one correlation from the other.

We have done this by looking in the non-jet region ($45° < \Delta\phi < 125°$) and building up the correlation function in and out of the jet cone. The new correlation function is shown below:

$$C_2(Au + Au) = C_2(p + p) + A^*(1 + 2v_2^2 \cos(2\Delta\phi)) \quad (4)$$

It includes the p-p correlation and the effects of elliptic flow. The $v_2$ parameter was determined independently by a reaction plane analysis and

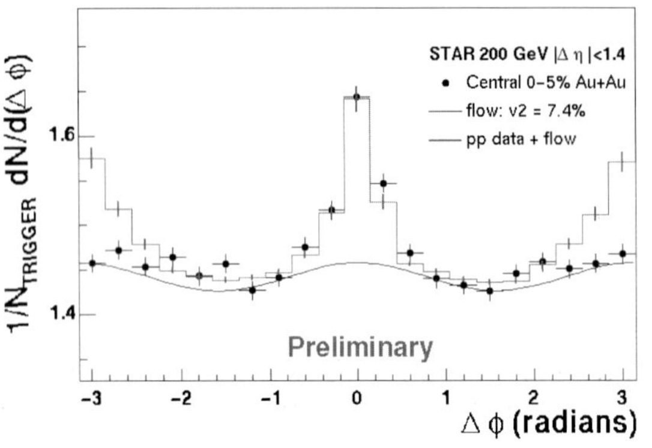


Figure 7. The observed jet like correlation function in heavy ion collisions is compared to the correlations seen in p-p collisions. The effects of elliptic flow are added to the p-p reference function. Note that the correlation is suppressed at 180 degrees in Au-Au.

the magnitude, $A^*$, of the flow term was fit in the non-jet cone region. The fit is shown in figure 7.

The correlation functions for Au-Au and $p-p(+\text{flow})$ collisions are very similar in the forward jet cone region ($|\Delta\phi| < 1$ radian ). But the correlation functions are different in the backward cone region and, in particular, the backward going jets in Au-Au are suppressed. What does this mean? It may mean that a jet can only be seen in a heavy ion collision when it originates near the surface of the collision zone. The backward going jet is suppressed; either because the fireball is opaque to high $p_T$ particles, or, perhaps, the angular correlation with the away side jet is destroyed by multiple gluon exchange in the gluon saturated core of the fireball.

## Conclusions

Nuclear matter created by collisions of heavy ions is not a trivial convolution of nuclear matter created by collisions of protons. Our analyses at 130 and 200 GeV show that the fireball created in RHIC collisions is very dense and possibly opaque. Novel and new energy loss mechanisms will be required to explain how high $p_T$ particles interact with this hot and dense matter.

Our data show that nuclear matter produced in RHIC collisions is accurately described by hydrodynamic models which assume local thermodyamic equlibrium at very early times in the collision sequence. And we have achieved the highest temperature possible for chemical freeze-out. A higher freeze-out temperatures cannot be achieved in heavy ion collisions.

The properties of the material produced in these collisions lie in an entirely new region of the phase diagram for nuclear matter and it lies at the boundary between hadronic matter and the quark gluon plasma.

In summary, nuclear matter at RHIC is hot, it is dense, it's expanding rapidly, and its properties are not inconsistent with local thermal equilibrium.

## REFERENCES


1. The STAR Collaboration homepage is located at http://www.star.bnl.gov. Our journal contributions can be found by clicking on "Publications".
2. H.H. Wieman *et al.*, IEEE Trans. Nuc. Sci. **44**, 671 (1997) and J.H. Thomas *et al.*, Nucl. Instrum & Meth. **A478**, 166 (2002).
3. C. Adler *et al.*, Phys. Rev. Lett. **87**, 262302-1 (2001).
4. We previously reported the $\overline{p}/p$ ratio to be $0.60 \pm 0.06$ at 130 GeV. This analysis was based on a limited data sample and a more extensive dataset is now available.
5. Braun-Munzinger *et al.*, Phys. Lett. **B518**, 41-46 (2001) and hep-ph/105229.
6. N. Xu and M. Kaneta, Nucl. Phys. **A698** , 306-313 (2002) and nucl-ex/0104021.
7. F. Karsch, Nucl. Phys. **A698**, 199 (2002) and hep-lat/0106019, hep-ph/0103314.
8. N. Xu, Nucl. Phys. **A610**, 175c (1996).
9. K.H. Ackermann *et al.*, Phys. Rev. Lett. **86,** 402-407 (2001).
10. C. Adler *et al.*, Phys. Rev. Lett. **87**, 182301 (2001).
11. C. Adler *et al.*, nucl-ex/0206011.
12. C. Adler *et al.*, submitted to Phys. Rev. Lett. and nucl-ex/0206006.

## Session 3

# Particle Astrophysics and Cosmology

*Session conveners:*
*A. Achucarro, University of Leiden*
*E. do Couto e Silva, Stanford University/SLAC*

ELSEVIER

Proceedings of ICHEP 2002, pp. 103–105
*S. Bentvelsen, P. de Jong, J. Koch and E. Laenen (Editors)*

31st INTERNATIONAL CONFERENCE ON HIGH ENERGY PHYSICS AMSTERDAM

# The Auger Observatory for high-energy cosmic rays

G. Matthiae
University of Roma II and Sezione INFN, Roma, Italy (*for the Pierre Auger Collaboration*)

The Southern Observatory of the Pierre Auger project for the study of cosmic rays of very high energy, above $10^{18}$ eV, is now under construction in Argentina. Some examples of the first data are shown.

## 1. INTRODUCTION

The Auger Observatory [1] is a *large acceptance "hybrid" system*. Large acceptance is needed because in the energy region close to $10^{20}$ eV the flux of the primaries is extremely low, of the order of one particle/km$^2$/sr/century. The hybrid concept, combination of a surface array for detecting the shower front at ground level and telescopes to observe the longitudinal profile of the shower from the fluorescence light produced in the atmosphere, is a unique feature of the Auger project. It allows control of the systematic uncertainties and cross-calibration of the measurement of the energy.

The energy region above $10^{20}$ eV is of special interest because of the GZK cut-off caused by the interaction of the primary particles with the microwave background. Evidence was reported [2] for the presence of events in the region which is expected to be depleted if a uniform distribution of sources is assumed. This evidence is controversial [3], because of problems in the determination of the energy scale. If confirmed, a natural implication is that the very high-energy particles come from distances which are relatively small on astrophysical scale, less than about 100 Mpc. Being very energetic, these particles, if protons, are magnetically rigid enough to point back to their sources which could then be identified.

In any case the accurate study of the energy, direction and mass composition of the primary particles in the region above $10^{18}$ eV with statistics at least one order of magnitude larger than present experiments, will provide important information for the understanding of the origin and acceleration mechanism of these particles.

## 2. THE OBSERVATORY

In the Auger project two Observatories are foreseen, in the Northern and Southern hemispheres. At present the Southern Observatory is being constructed in Argentina, in the province of Mendoza, near the small town of Malargüe (Fig.1). The site, Pampa Amarilla, covers an

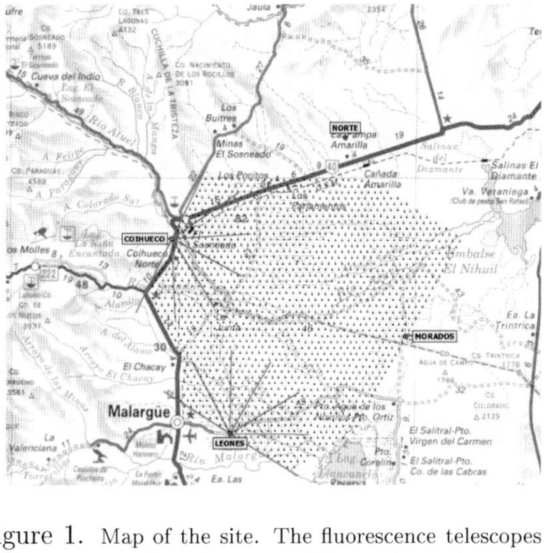


Figure 1. Map of the site. The fluorescence telescopes overlook the site from four small hills on the perimeter.

area of 3000 km$^2$ at the altitude of 1400 m (atmospheric depth of 875 g/cm$^2$). It is almost flat with very low population density and favourable atmospheric conditions in terms of man-produced light and aerosol pollution. The detector data is transmitted by radio technology in a network and collected at the central data acquisition system in the Observatory central building in Malargüe.


doi:10.1016/S0920-5632(02)02080-7

### 2.1. The Surface Detector (SD)

The Surface Detector is a grid of 1600 water Cherenkov counters spaced at a distance of 1.5 km. Each counter (Fig.2) is a polyethylene tank of cylindrical shape and size of 10 $m^2$ x 1.2 m, filled with purified water. The internal wall is lined with diffuse reflector. The Cherenkov light produced by the shower particles in the water is detected by three 9" photomultipliers looking downward. Each unit is autonomous with a battery and a solar panel.

The time structure of the photomultiplier signals is analyzed by Flash ADCs and these data are transmitted together with the timing provided by a GPS receiver. Cosmic ray muons provide the calibration which is expressed in terms of "vertical equivalent muons" (VEM).

Figure 2. Picture of a water Cherenkov tank installed on the site. The solar panel and the antenna are visible

### 2.2. The Fluorescence Detector (FD)

The Fluorescence Detector (FD) consists of 24 telescopes located in four stations ("eyes"). Each station is on top of a small elevation on the perimeter of the site and contains 6 telescopes.

A sketch of a telescope is shown in Fig.3 where from left to right we see the shutter, the circular diaphragm with a UV filter transmitting in the spectral region of the fluorescence light from nitrogen, the camera and the large spherical mirror with curvature radius of 3.4 m.

The sensitive element of the camera is a matrix of 440 hexagonal photomultipliers placed on the focal surface. The pixel size is 45 mm, corresponding to the angular size of $1.5^o$. The field of view is $30^o$ in azimuth and $28.6^o$ in elevation. The PMT signals are sampled by Flash ADCs.

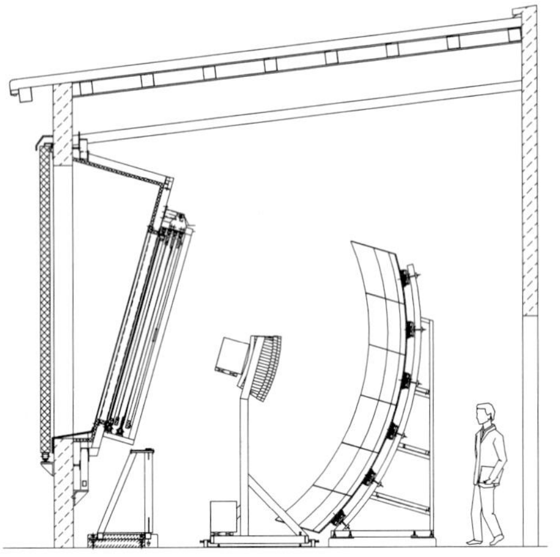

Figure 3. Sketch of a fluorescence telescope.

### 2.3. General features

Various trigger schemes have been devised using the SD and FD either independently or combined together.

The efficiency of a trigger requiring a coincidence of at least 3 adjacent tanks reaches 100% at $3x10^{18}$ eV. Requiring a coincidence of at least 5 tanks, the efficiency is 40% at $3x10^{18}$ eV and becomes 100% at $10^{19}$ eV.

The FD may provide a trigger with threshold at the level of a few $10^{17}$ eV.

The expected number of events for one year operation of the Observatory is shown in the Table.

| Energy | Number of events |
|---|---|
| E > $3x10^{18}$ eV | ~ 35,000 |
| E > $10^{19}$ eV | ~ 5,000 |
| E > $5x10^{19}$ eV | ~ 500 |
| E > $10^{20}$ eV | ~ 50-100 ? |

The FD duty cycle is 12-15% due to the requirement of operating in moonless nights.

For hybrid operation the error on the shower energy and direction is less than 5% and $1^o$, respectively, while the position of the shower axis at the ground level is determined with accuracy of about 50 m. The depth of the shower maximum is measured by the FD with error of 15-20 g/$cm^2$.

## 3. THE ENGINEERING ARRAY

The first element of the Observatory, the "Engineering Array",was completed in 2001 and has taken data for several months. It consists of 40 tanks overlooked by 2 telescopes installed on the top of the hill of Los Leones on the southern boundary of the site. The Engineering Array is only a small fraction of the final system but it contains all basic elements of the SD, FD, communication network and data acquisition.
About 80 hybrid events were collected in 3 months using the hybrid trigger combining the information from both SD and FD.
In Fig.4 the observed signals for an event with 11 active tanks are shown. A preliminary analysis [4] of this event indicates that the energy of the shower is (2.5±0.4)x$10^{19}$ eV and the zenith angle about $54^o$.

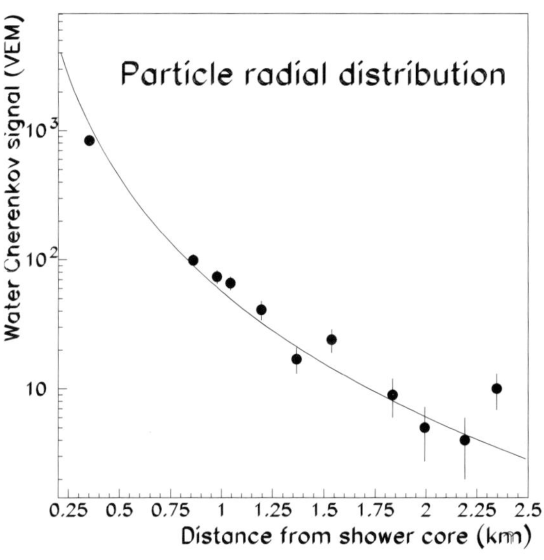


Figure 4. Radial distribution of particles observed by the SD for a 11-tank event with E≃2.5x$10^{19}$ eV. The data are expressed in terms of vertical equivalent muons (VEM). The solid line is the result of a fit using a standard form.

Two examples of longitudinal profiles measured by the FD are shown in Fig.5 and 6. The observed number of charged particles is presented as a function of the atmospheric depth together with the result of fitting using the Gaisser-Hillas form. A preliminary estimate of the energy is also given together with the depth $X_{max}$ where the development of the shower reaches its maximum.

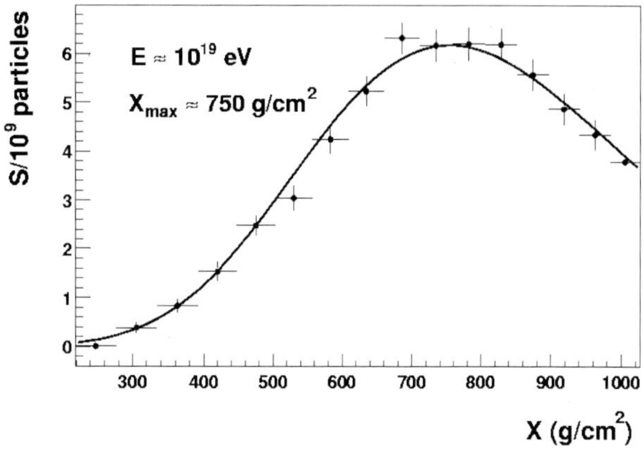


Figure 5. Profile of a shower with E∼$10^{19}$ eV.

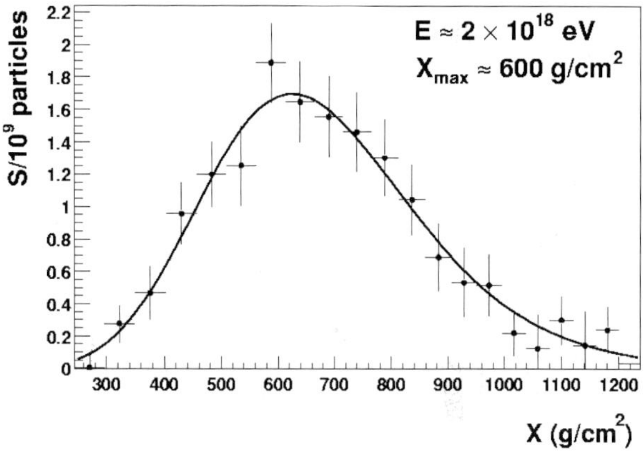


Figure 6. The profile of a shower of relatively low energy which appears to be fully contained in the atmosphere.

## 4. CONCLUSIONS

The successful operation of the Engineering Array has been extremely useful to assess the properties and reliability of all components and the feasibility of operating the hybrid system. The construction is going on with the aim of completing the Observatory by the year 2005.

## REFERENCES

1. J.W.Cronin, Proc. of ICRC 2001, Hamburg.
2. N.Sakaki et al.(AGASA), Proc. of ICRC 2001.
3. D.R.Bergman (HiRes), Proc. of this Conference.
4. M.Ave et al., Auger technical note GAP-2002-020.

Proceedings of ICHEP 2002, pp. 106–108
S. Bentvelsen, P. de Jong, J. Koch and E. Laenen (Editors)

31st INTERNATIONAL CONFERENCE ON HIGH ENERGY PHYSICS AMSTERDAM

# Monocular UHECR Spectra as Measured by HiRes

D. R. Bergman [a] presented on behalf of the High Resolution Fly's Eye Collaboration

[a]Rutgers, The State Univeristy of New Jersey
Department of Physics and Astronomy
Piscataway, New Jersey, USA 08854

We have measured the spectrum of UHE cosmic rays in monocular mode using separately both detectors the High Resolution Fly's Eye experiment. We describe the two detectors and the basic methods of analysis, and we present our measured spectra. We compare these spectra with that produced by an astrophysical source model with galactic and uniformly distributed extra-galactic sources. We also compare our spectra to the spectra produced by the AGASA experiment.

## 1. Detectors

The HiRes experiment consists of two air-fluorescence detectors separated by 12.6 km and located at the U.S. Army Dugway Proving Ground in Utah. One detector, HiRes-I, consists of 22 mirrors covering 3–17° in elevation and nearly 360° in azimuth[1]. It uses a sample-and-hold data acquisition system to save the photomultiplier tubes' pulse height and time information. HiRes-I has been running since June of 1997. The second detector, HiRes-II, located 12.6 km away, consists of 42 mirrors, covers 3–31° in elevation, and uses a flash ADC (FADC) system to save pulse height and time information from its phototubes [2]. The sampling period of the FADC electronics is 100 ns. HiRes-II has been operating since October of 1999. The mirror units at each site are identical, with a 5 m$^2$ mirror collecting light onto an array of 256 phototubes. Each phototube covers about one square degree of the sky.

## 2. Monocular Analysis

The two detectors are designed to observe cosmic ray showers steroscopically. However, HiRes-I has been running for much longer than HiRes-II and at energies below about $10^{18}$ eV, events can only be observed by one detector. Thus we are presenting here spectra from monocular analyses.

Determination of the shower geometry in monocular mode is done by fitting the trigger times, $t_i$, of hits to the following function of the angles, $\chi_i$, of hits above the horizon,within the event plane:

$$t_i = t_0 + \frac{R_p}{c} \tan\left(\frac{\pi - \psi - \chi_i}{2}\right) \quad (1)$$

Here $R_p$ is the impact parameter, $\psi$ the in-plane angle between the shower and the horizon, and $t_0$ the time of closest approach.

With limited elevation coverage, HiRes-I monocular events are too short in angular spread for reliable determination of $\psi$ and $R_p$ by timing alone. For this analysis, the expected form of shower development itself was used to constrain the time fit to yield realistic geometries. The shower profile was assumed to be described by the Gaisser-Hillas parameterization

$$N(x) = N_m \left(\frac{x - x_0}{x_m - x_0}\right)^{(x_m - x_0)/\lambda} e^{(x_m - x)/\lambda} \quad (2)$$

where $N(x)$, $N_m$ are the number of particles at depth $x$ and at shower maximum depth $x_m$, respectively. The first-interaction depth and shower elongation constant are denoted by $x_0$, and $\lambda$ [3].This technique is called the profile-constrained fit (PCF). Equation 2 is in agreement with previous HiRes measurements [4], and with CORSIKA/QGSJET simulations [5–7]. Based on these, we fixed $x_0$ and $\lambda$ at 40 and 70 g/cm$^2$, respectively. We allowed $x_m$ to vary in 35 g/cm$^2$ steps between 680 and 900 g/cm$^2$, matching the


doi:10.1016/S0920-5632(02)02082-0

expected range for proton to iron primaries in this energy range. This procedure breaks down for events with energies less than $3 \times 10^{18}$ eV, providing the lower limit to the HiRes-I spectrum.

The greater elevation coverage of HiRes-II allows for the reconstruction of the shower geometry from timing alone. Equation 1 is linear in $R_p$ and $t_0$, we find the best value for these variables analytically for each $\psi$; the best $\psi$ is then found by $\chi^2$ minimization. With the geometry of the shower known, we fit the observed light signal to the Gaisser-Hillas parameterization of Eqn. 2. We collected photo-electrons from all tubes into a sequence of time bins. This exploited the FADC data acquisition system and lessened our sensitivity to PMT acceptance.

Two calibration issues effect both sites: the absolute calibration of the phototube gains and the determination of the transparency of the atmosphere. The phototube gains are determined through a nightly illumination of the phototube clusters with a YAG laser and a monthly illumination by a Xenon flash bulb. The atmospheric transparency is determined by probing the atmosphere hourly with a set of laser shots from each of the sites which is observed and reconstructed by the other site.

## 3. Data-MC Comparisons

The aperture as a function of energy is a necessary component of the spectrum calculation. It is determined by Monte Carlo calculation. To verify our calculated aperture and analysis procedures, we undertook an extensive program of Monte Carlo simulation development and comparison of simulated data with real data. Two figures exemplify this; both are from the HiRes-II analysis. In figure 1 we compare data and MC distributions of two geometric variables: $r_p$ and $\psi$. In figure 2 we compare the distributions of reconstructed energies. Note that any large discrepancy between the atmospheric parameters used in MC and the conditions under which the data were actually taken would heavily influence the distribution of impact parameters.

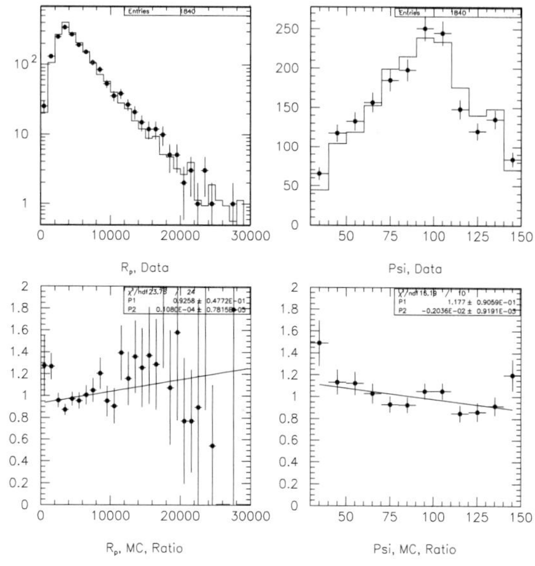


Figure 1. Data/MC comparisons of $r_p$ and $\psi$.

## 4. The HiRes Spectra

The two HiRes monocular spectra are shown in figure 3 [8,9] along with what one would expect from a model combining galactic sources with uniformly distributed extra-galactic sources[10]. The GZK cutoff is evident, and there is no need from this data to invoke extraordinary means in order to evade it.

In figure 4 we compare our spectra with that of AGASA[11]. The shape of the two spectra agree quite well for energies below 100 EeV, and reducing the AGASA energies by 20%, which is within AGASA's stated systematic uncertainty, makes this comparison explicit. The famed discrepancy between HiRes and AGASA has thus been reduced to a few points above 100 EeV, where AGASA has five events (with they're energy scale reduced by the 20%) and HiRes has only one.

## REFERENCES


1. T. Abu-Zayyad *et al.*, Proc. 26th Int. Cosmic Ray Conf. (Salt Lake City), **5**, 349 (1999).

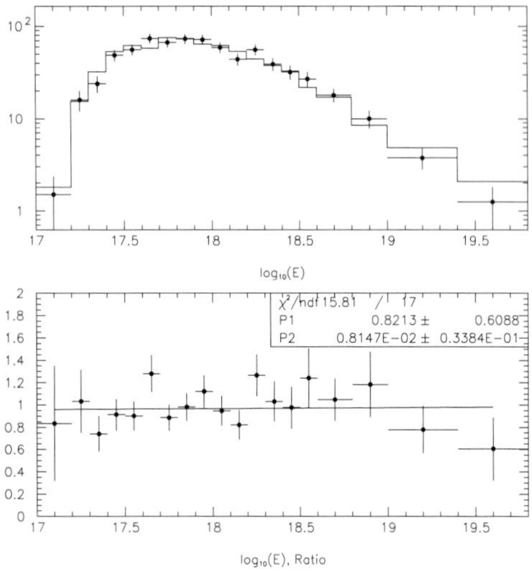


Figure 2. Data/MC comparison of reconstructed energies.

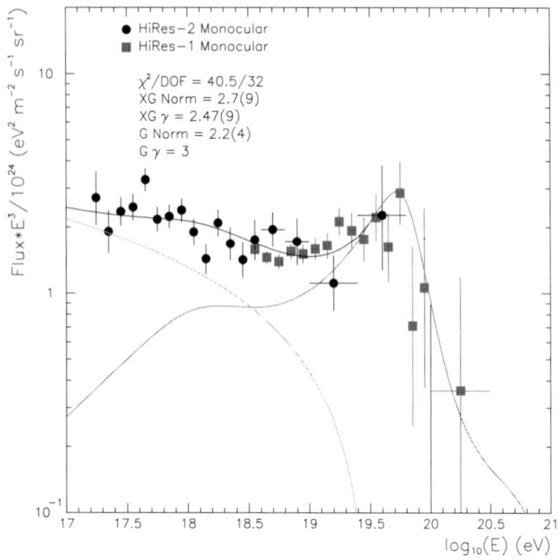


Figure 3. HiRes-I and HiRes-II spectra fit to galactic and uniform extragalactic source model.


2. J. Boyer, B. Knapp, E. Mannel, and M. Seman, Nucl. Instr. Meth. **A482**, 457 (2002).
3. T. Gaisser and A.M. Hillas, Proc. 15th Int. Cosmic Ray Conf. (Plovdiv), **8**, 353, (1977).
4. T. Abu-Zayyad *et al.*, Astropart. Phys. **16**, 1, (2001).
5. C. Song, Z. Cao *et al.*, Astropart. Phys. **14**, 7, (2000).
6. D. Heck, J. Knapp, J.N. Capdevielle, G. Schatz and T. Thouw "CORSIKA : A Monte Carlo Code to Simulate Extensive Air Showers", Report FZKA 6019 (1998), Forschungszentrum Karlsruhe.
7. N.N. Kalmykov, S.S. Ostapchenko and A.I. Pavlov, Nucl. Phys. B (Proc. Suppl.) **52B**, 17, (1997).
8. T. Abu-Zayyad *et al.*, submitted to Phys. Rev. Lett., astro-ph/0208243.
9. T. Abu-Zayyad *et al.*, submitted to Astropart. Phys., astro-ph/0208301.
10. V. Berezinsky, A.Z. Gazizov, S.I Grigorieva, hep-ph/0204357.
11. M. Teshima, private communication.


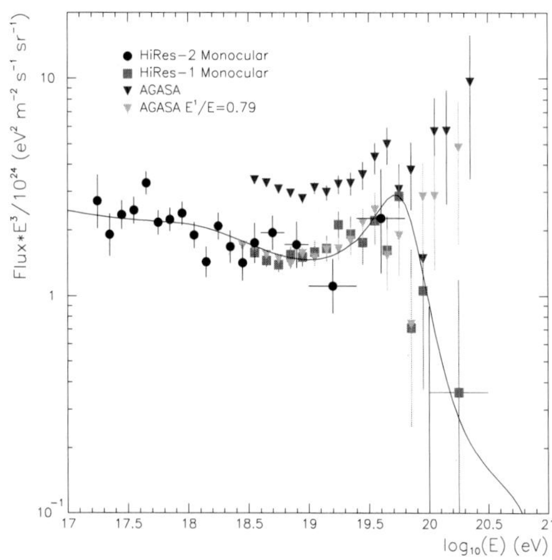


Figure 4. HiRes-I and HiRes-II spectra compared with AGASA.

ELSEVIER

Proceedings of ICHEP 2002, pp. 109–112
*S. Bentvelsen, P. de Jong, J. Koch and E. Laenen (Editors)*

31st INTERNATIONAL CONFERENCE ON HIGH ENERGY PHYSICS AMSTERDAM

# Cosmic Rays with the LEP detectors

C. Timmermans[a]

[a]University of Nijmegen,
Toernooiveld 1, 6525 ED, Nijmegen, the Netherlands

The LEP detectors are well suited for different studies on cosmic ray induced muons. In here, several aspects of the muon spectrum are reported. The vertical as well as the zenith-angle dependence of both the absolute muon spectrum and the muon charge ratio have been measured. High and low muon multiplicities have been measured, the latter in combination with the electromagnetic shower size. All of this has been compared to simulations.

## 1. Introduction

The study of the cosmic ray induced muons at ground level contributes to the knowledge on a wide variety of physics aspects. A measurement of the absolute flux provides input to the calculation of the expected neutrino flux. The measurement of the muon multiplicity distribution gives information with respect to the average composition of the primary cosmic rays. When this information is accompanied by the electromagnetic shower size, the sensitivity to the primary composition increases.

Two collaborations (L3+C and Delphi) presented results to this conference regarding the study of high energy muons ($E_\mu > 20\,\mathrm{GeV}$ or $E_\mu > 50\,\mathrm{GeV}$), where the energy cut depends on the amount of overburden above the detector.

### 1.1. Detector Characteristics

Both L3 and DELPHI are located at the LEP accelerator near Geneva, Switzerland. However, the amount of overburden above the detectors is very different, as well as the type of detector used for this studies. The main characteristics of importance for the analyses reported here are shown in table 1. The detector differences are reflected in the different analyses performed. A thorough description of these detectors can be found in [1] and [2].

## 2. The muon spectrum

### 2.1. Simulation

A detailed CORSIKA [3] simulation has been parametrized into a muon flux description at sea level (L3CGEN) by the L3+C collaboration. This description is used to generate muons which pass through the overburden above the L3 detector, and finally through the detector itself. The simulation of the overburden and detector is based upon the GEANT [4] package.

### 2.2. The vertical flux and charge ratio

L3+C is able to use the unique features of the L3 muon spectrometer to measure the differential muon spectrum and charge ratio as function of the muon momentum. The vertical spectrum (zenith angle less than 10 degrees) and charge ratio are shown in figures 1 and 2. In figure 1 the uncertainty shown at each point shows the point-to-point uncertainty. The box shows the uncertainty in the overall normalization. It should be noted that the vertical spectrum is only shown for muon momenta above 80 GeV. This is due to uncertainties in the effects of the overburden, which are still under study. Above 200 GeV, the slope of the sharp drop in the flux depends heavily on the knowledge of the momentum resolution. Furthermore, possible momentum dependence of reconstruction or detector response are still under study. When comparing the flux measurement with modern calculations made by Bugaev et al. [5] and Naumov [6,7], a good agreement


doi:10.1016/S0920-5632(02)02084-4

Table 1
Detector Characteristics

| | L3+C | DELPHI |
|---|---|---|
| Momentum cutoff | 15 GeV | 50 GeV |
| Detector used | High precision drift chamber in magnetic field | fine granularity calorimeter |
| Readout | Dedicated DAQ | Cosmic trigger in standard readout |
| Data sample | $12 \cdot 10^9$ events | $90 \cdot 10^3$ multi-muon events |

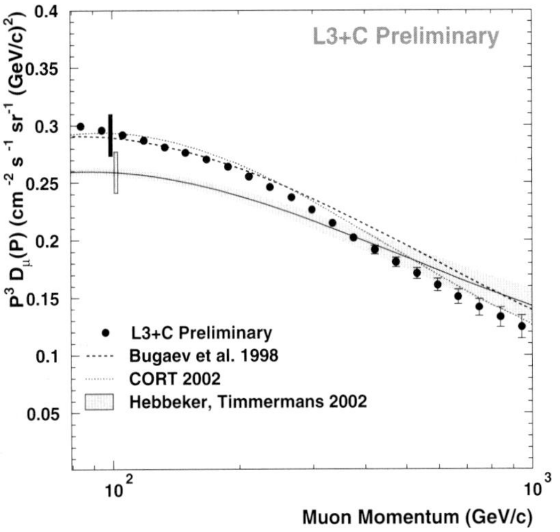


Figure 1. The vertical muon spectrum. The box represents the uncertainty of the normalization.

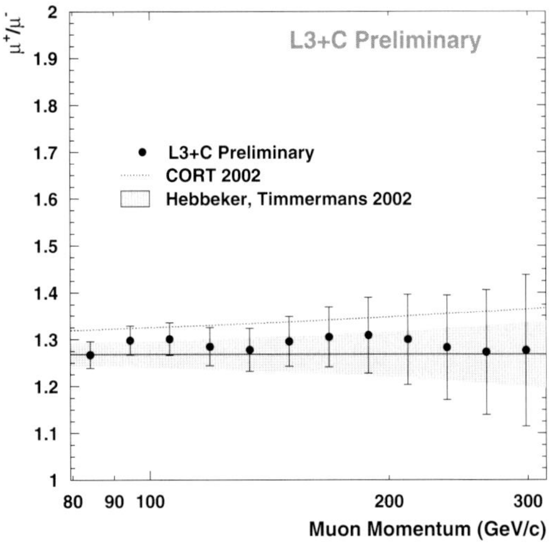


Figure 2. The Vertical charge ratio compared to theory and previous measurements.

can be seen. However, a comparison with a compilation of available data above 10 GeV [8] shows only marginal agreement in both the normalization and the slope. It should be noted that in the compilation of Hebbeker and Timmermans different datasets with incompatible normalizations were averaged. Furthermore, the compilation of data above 200 GeV is dominated by a single measurement.

The knowledge of the charge ratio is bound by the exact knowledge on the position of the different detector elements. The current understanding limits a precise measurement above 300 GeV. The charge ratio is in agreement with previous measurements. The CORT-calculation is also in agreement with the measurements, but systematically high.

### 2.3. Zenith Angle dependence

L3+C is the first detector to provide an absolute measurement of the vertical spectrum, and a continuous measurement of the zenith dependence down to $\cos(\Theta) = 0.6$. Examples of the zenith angle dependence of the charge ratio and

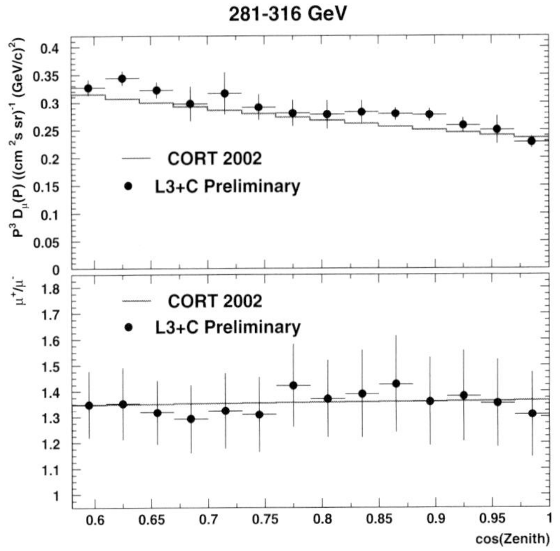


Figure 3. Zenith dependence of spectrum and charge ratio for a muon momentum around 300 GeV.

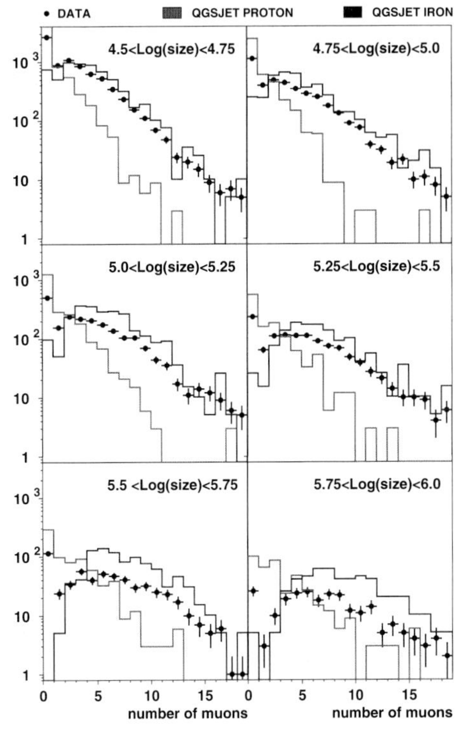


Figure 4. Muon multiplicity as function of shower size. Preliminary measurement from L3+C.

the flux can be seen in figure 3. This measurement is performed for all momentum bins (from 80 GeV onward), and a good agreement with theory can be seen in every instance.

## 3. Muon Multiplicity Studies

### 3.1. Simulations

The measurements discussed below are compared to air-shower simulations performed with CORSIKA. This package allows the usage of several hadronic interaction models. The one used here is QGSJET [9]. Furthermore, the overburden above L3 or DELPHI was simulated using GEANT, and the detector response itself was simulated using a dedicated detector simulation package. This package has been used and tested extensively in the analysis of the $e^+e^-$ collisions at LEP.

### 3.2. Low multiplicities

The study of low multiplicity events (thus low muon densities) can only add to the discussion of the primary composition if the primary energy, or alternatively the air shower size, is known. During the 2000 run, L3+C had a small air shower array to measure this quantity. Offline, the data from this array and the muon detector were combined using a common GPS time-stamp as well as a flag in each data-stream indicating a common event in the other detector. The resulting muon multiplicity distribution as function of shower size is compared to a CORSIKA simulation. Even though the simulation shows the expected difference between iron and proton primaries, the simulations do not agree with the measurements. Figure 4 shows that the measured muon multiplicity resembles simulations of iron primaries, even at small shower sizes (low energies), where we know the chemical composition to consist mostly of protons and helium nuclei.

### 3.3. High Multiplicities

A multiplicity distribution extending to larger multiplicities, or higher muon densities, has also been compared to predictions of air shower simulations. The input energy of the simulations extend over several orders of magnitude, where the highest multiplicities depend on the composition of the cosmic radiation around the knee-region. The results of DELPHI are compared to a pure proton simulation and a pure iron simulation in figure 5. From this figure it is clear that the highest multiplicities cannot be explained even when a pure iron flux is assumed. These data are in qualitative agreement with the results reported earlier by ALEPH [10].

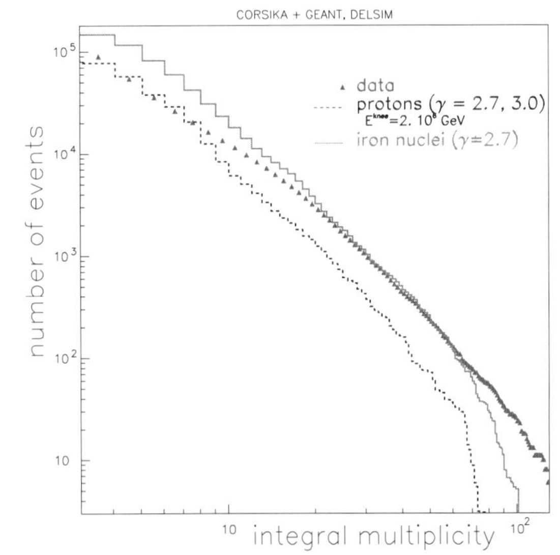


Figure 5. Muon multiplicity distribution compared to simulation of iron and proton primaries. The spectral indices used in the simulations are shown in the plot. Preliminary measurement from DELPHI.

## 4. Conclusion

Two aspects of cosmic ray induced sea-level muons have been discussed. The total flux of muons and charge ratio are presented for the vertical direction, and as function of the zenith angle. All aspects are well described by current calculations.

The muon density (or detected multiplicity) has been discussed for high energy muons (above 20 or 50 GeV). The interpretation of this measurement is problematic due to the fact that the simulations fail to describe the measurement already at a very global level.

## REFERENCES

1. Adriani O. et al,. Nucl. Instr. Meth. A **488** (2002) 209.
2. Abreu P. et al., Nucl. Instr. Meth. A **378** (1996) 57.
3. Heck D. et al., Corsika: A Monte Carlo code to simulate extensive air showers, Technical report, Forschungszentrum Karlsruhe (1998).
4. CERN Application Software Group, GEANT Detector description and simulation tool, CERN internal note (1994).
5. Bugaev E. et al., Phys. Rev. **D58** (1998) 054001.
6. Naumov V. A., hep-ph/0201310 to be published in the Proceedings of the 2nd Workshop on Methodical Aspects of Underwater/Underice Neutrino Telescopes, Hamburg, August 15-16 (2001).
7. Fiorentini G. et al., Phys. Lett. **B510** (2001) 173.
8. Hebbeker T. and Timmermans C., Astropart. Phys. 18 (2002).
9. N. N. Kalmykov, S. S. Ostapchenko and A. I. Pavlov, Nucl. Phys. Proc. Suppl. **52B** (1997) 17.
10. V. Avati *et al.*, "Cosmic multi-muon events observed in the underground CERN-LEP tunnel with the ALEPH experiment," CERN-EP-2000-152.

ELSEVIER

Proceedings of ICHEP 2002, pp. 113–115
*S. Bentvelsen, P. de Jong, J. Koch and E. Laenen (Editors)*

31st INTERNATIONAL CONFERENCE ON HIGH ENERGY PHYSICS AMSTERDAM

# Perturbative QCD Predictions and Giant Air Showers

Ralph Engel[a*]

[a]Bartol Research Institute, University of Delaware, Newark, DE 19716, U.S.A.

We discuss QCD-inspired hadronic interaction models used for simulating extensive air showers. The role of perturbative QCD predictions in the extrapolation of modern interaction models to ultra-high energy is summarized, including corresponding uncertainties. Two possible extrapolations of the SIBYLL model are presented and predictions for extensive air showers are discussed, emphasizing the relation between the minijet cross section and observable shower features at ultra-high energy.

## 1. Introduction

Extensive air showers (EAS) are currently the only method to study cosmic rays at energies above $10^{15}$eV. Using the atmosphere as target, air shower arrays are restricted to measure secondary particles arriving at the detector level. Fluorescence detectors employ the atmosphere as calorimeter by mapping the energy deposit along the shower core. Both methods do not allow the direct measurement of primary particle properties and hence rely on our understanding of cosmic ray interactions in the atmosphere. On the other hand, cosmic ray interactions in the atmosphere cover an energy range exceeding by far currently accessible accelerator energies. They are also sensitive of a phase space region of final state particles complementary to that measured in collider experiments.

## 2. QCD-inspired models and minijet cross section

The evolution of air showers is governed by the most energetic secondary particles of cosmic ray interactions. These are typically hadrons produced with small transverse momentum. However, soft hadron production cannot be calculated from first principles and models have to be used. Therefore modern hadronic interaction models such as DPMJET, NEXUS, QGSJET, and SIBYLL combine perturbative QCD predictions with phenomenological assumptions to generate complete hadronic final states [1]. In the construction of such models the QCD minijet cross section plays an important role. Though it cannot be used to predict leading particle distributions it is the most important theoretical input for extrapolating to ultra-high energies.

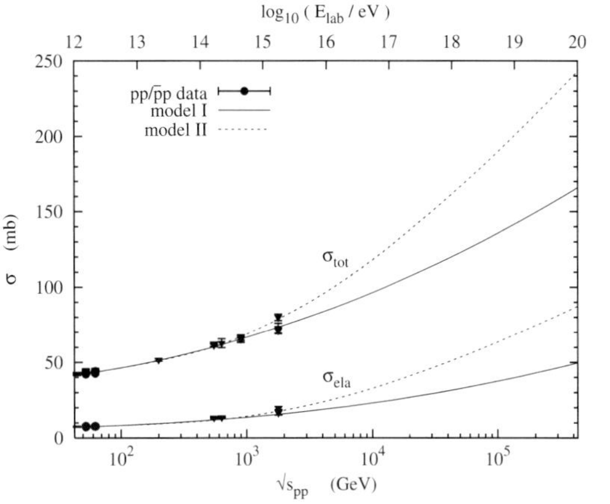


Figure 1. Model predictions for $p\bar{p}$ total and elastic cross sections (see text).

Measurements at the HERA collider revealed a steep rise of the gluon density at low $x$ in the proton, $xg(x) \sim x^{-\Delta}$, $\Delta = 0.3\ldots0.4$ [2]. Previously $\Delta \approx 0$ was assumed in all had. interaction models. Whereas a model with $\Delta \approx 0$ could be extrapolated to high energy in a straight-

*Now at Institut für Kernphysik, Forschungszentrum Karlsruhe, Postfach 3640, 76021 Karlsruhe


doi:10.1016/S0920-5632(02)02086-8

forward manner, the steep rise of the parton densities does not allow such a simple extrapolation [3]. The linear parton density evolution equations are expected to break down and non-linear equations, including parton-parton screening, have to be used for the description of low-$x$ processes [4]. Furthermore recent RHIC measurements underline the importance of modifications of nucleon parton densities in nuclei and increase the uncertainty in the extrapolation of parton densities. In addition to theoretical and experimental difficulties related to the low-$x$ region there is the problem of the poorly known range of applicability of pert. QCD calculations to processes with a hard scale of only a few $\mathrm{GeV}^2$.

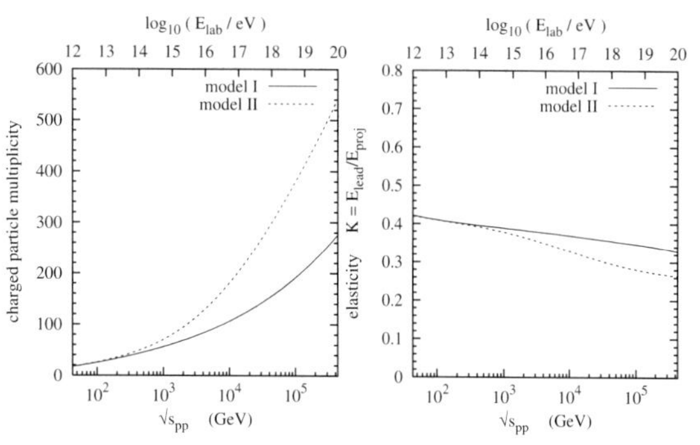


Figure 2. Model predictions for $p$-air interactions. The left panel shows the mean charged particle multiplicity and the right panel the elasticity.

To study the impact of such uncertainties we consider in the following two variants of the SIBYLL 2.1 interaction model. In model I a minijet cross section is implemented which rises moderately with energy. This is achieved by increasing with energy the transverse momentum cutoff applied in the pert. QCD calculation. In model II a rapidly increasing minijet cross section is employed. By construction both models describe all collider data including Tevatron measurements reasonably well. In particular, the current total cross section measurements at $\sqrt{s} = 1800$ GeV do not allow to discriminate between a high- or low-cross section extrapolation (see Fig. 1).

All QCD-inspired models use $pp/p\bar{p}$ interactions as building blocks which are generalized for the description of hadron-air interactions. The uncertainties in the description of $pp/p\bar{p}$ interactions directly enter the modeling of hadron-air and nucleus-air interactions. Fig. 2 shows the predicted mean multiplicity of charged particles produced in $p$-air collisions and the corresponding elasticity. The elasticity is defined as the energy fraction carried by the most-energetic hadronic particle in the final state. Although the production of minijets does influence the energy fraction carried by fast secondaries only through energy-momentum conservation, the impact is clearly visible. A large minijet cross section increases not only the multiplicity of secondaries but also reduces significantly the elasticity.

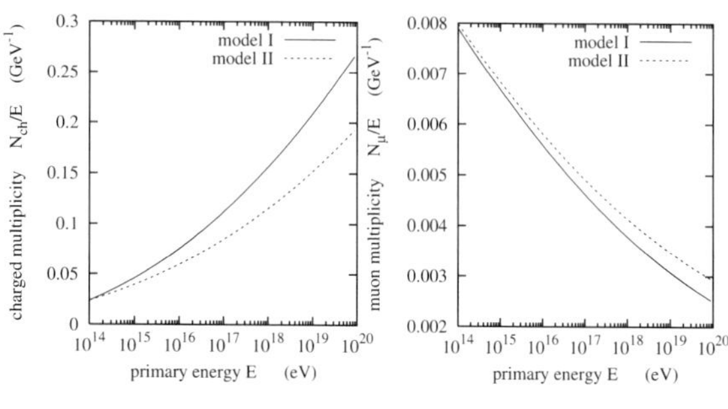


Figure 3. Model predictions for proton-induced EAS at 45° zenith angle. The left panel shows the mean charged particle multiplicity (mostly electrons). The right panel gives the multiplicity of muons with $E_\mu > 300\mathrm{MeV}$.

## 3. Extensive air shower predictions

To estimate the importance of the minijet cross section for EAS we implemented the two new variants of SIBYLL in CORSIKA [5] and simulated showers for the altitude of the Pierre Auger Observatory in Argentina.

Fig. 3 shows the expected mean number of shower particles reaching observation level. Model II with a large minijet cross section predicts an earlier development of the showers (i.e. at higher altitude) which leads to a stronger absorption of the electron component and a higher muon multiplicity. The differences between model I and II are small at low energy ($10^{14} - 10^{15}$eV). At the highest energies the differences grow to $\sim 25\%$ resulting in considerable uncertainty in a mass and composition analysis of showers.

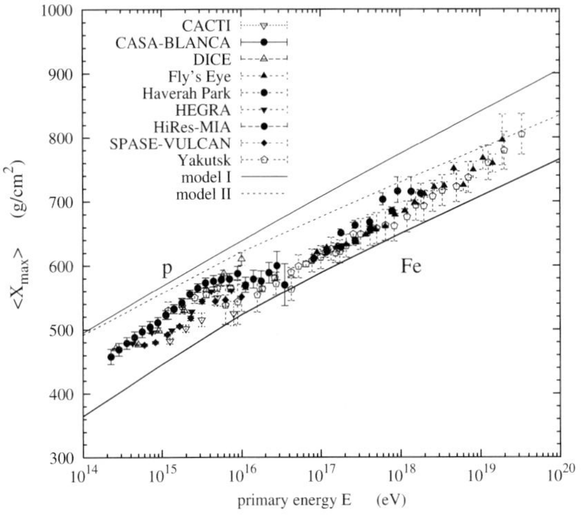


Figure 4. World data on $X_{\max}$ measurements [6] and model predictions.

Fluorescence detectors such as HiRes and Auger can measure the longitudinal shower profile. A quantity particularly well-suited for primary mass discrimination is the position of the shower maximum, $X_{\max}$. In Fig. 4 a compilation of $X_{\max}$ measurements is compared to predictions of the models I and II for proton and iron primaries. Only one representative result for iron primaries is shown. In the case of heavy primaries nuclear effects become very important which are beyond the scope of this work. The model-dependence of the interpretation of the measurements in terms of primary cosmic ray mass is striking. Model II predicts a rather light (i.e. proton-like) composition whereas model I would correspond to a mean mass number close to that of carbon.

## 4. Conclusions

The extrapolation of QCD-inspired models depends on the minijet cross section at high energy. This cross section is subject to large uncertainties for the following reasons

(i) parton densities at very low $x$ are not measured,
(ii) the extrapolation of parton densities is difficult because of expected non-linear effects such as saturation,
(iii) the nuclear environment in heavy nuclei leads to an early onset of nonlinear effects, and
(iv) the range of applicability of pert. QCD calculations is not well defined and minijet cross sections depend very sensitively on the applied transverse momentum cutoff.

The large uncertainty of the model extrapolation is most pronounced in quantities related to the production of slow particles, for example, the charged particle multiplicity. Although reduced, it also affects predictions for extensive air showers. Systematic studies of model predictions and their theoretical uncertainties are needed. For the foreseeable future measurements of leading/forward particle production at high-energy accelerators will be a decisive, urgently needed input for tuning hadronic interaction models and understanding EAS.

**Acknowledgments** The author thanks J. Alvarez-Muniz, T.K. Gaisser, D. Heck, J. Ortiz, S. Ostapchenko, J. Ranft, S. Roesler and T. Stanev for many fruitful discussions on air shower models. This work is supported by the US Department of Energy contract DE-FG02 91ER 40626.

## REFERENCES

1. J. Ranft: Phys. Rev. D51 (1995) 64; H. J. Drescher *et al.* Phys. Rept. 350 (2001) 93; N. N. Kalmykov, S. Ostapchenko and A. I. Pavlov, Nucl. Phys. B (Proc. Suppl.) 52B (1997) 17; R. S. Fletcher *et al.* Phys. Rev. D50 (1994) 5710 and R. Engel *et al.* Proc. of 26th ICRC (Salt Lake City) 1, (1999) p. 415.
2. H. Abramowicz and A. Caldwell, Rev. Mod. Phys. 71 (1999) 1275.
3. R. Engel, T. K. Gaisser and T. Stanev, Proc. of 27th ICRC (Hamburg, 2001) 431.
4. A. H. Mueller, hep-ph/0111244.
5. D. Heck *et al.* Wissenschaftliche Berichte FZKA 6019, 1998.
6. T. K. Gaisser, AIP Conf. Proc. 558 (2001) 27.

Proceedings of ICHEP 2002, pp. 116–118
*S. Bentvelsen, P. de Jong, J. Koch and E. Laenen (Editors)*

ELSEVIER

31st INTERNATIONAL CONFERENCE ON HIGH ENERGY PHYSICS AMSTERDAM

# Positron Fraction from Dark Matter Annihilation in the CMSSM

W. de Boer[a], C. Sander[a], M.Horn[a] and D. Kazakov[b]

[a]Institut für Experimentelle Kernphysik, Universität Karlsruhe (TH), P.O. Box 6980, 76128 Karlsruhe, Germany

[b]Bogoliubov Laboratory of Theoretical Physics, JINR, 141980 Dubna, Moscow Region, Russian Federation

We investigate, if the cosmic ray positron fraction, as reported by the HEAT and AMS collaborations, is compatible with the annihilation of neutralinos in the supergravity inspired Constrained Minimal Supersymmetric Model (CMSSM), thus complementing previous investigations, which did not consider constraints from unification, electroweak symmetry breaking and the present Higgs limits at LEP. We consider in a global analysis the complete SUSY parameter space and find that in the acceptable regions the neutralino annihilate dominantly into $b\bar{b}$ quark pairs. The positrons from $b\bar{b}$ improve the fit to the experimental positron fraction data considerably compared to a fit with background only.

## 1. Introduction

The cosmic ray positron fraction at momenta above 7 GeV, as reported by the HEAT collaboration, is difficult to describe by the background only hypothesis[1]. The data at lower momenta agree with previous data from the AMS experiment[2]. A contribution from the annihilation of neutralinos, which are the leading candidates to explain the cold dark matter in the universe, can improve the fits considerably[3,4]. The neutralinos are the Lightest Supersymmetric Particles (LSP) in supersymmetric extensions of the Standard Model, which are stable, if R-parity is conserved. This new multiplicative quantum number for the supersymmetric partners of the Standard Model (SM) particles is needed to prevent proton decay and simultaneously prevents the LSP a) to decay into the lighter SM particles and b) can only interact with normal matter by producing additional supersymmetric particles. The cross sections for the latter are typically of the order of the weak cross sections, so the LSP is "neutrino-like", i.e. it would form halos around the galaxies and consequently, it is an excellent candidate for dark matter.

In addition to being of interest for cosmology, supersymmetry solves also many outstanding problems in particle physics, among them[5]: 1) it provides a unification of the strong and electroweak forces, thus being a prototype theory for a Grand Unified Theory (GUT) 2) it predicts spontaneous electroweak symmetry breaking (EWSB) by radiative corrections through the heavy top quark, thus providing a relation between the GUT scale, the electroweak scale and the top mass, which is perfectly fullfilled 3) it includes gravity 4) it cancels the quadratic divergencies in the Higgs mass in the SM 5) the lightest Higgs mass can be calculated to be below 125 GeV in perfect agreement with electroweak precision data, which prefer indeed a light Higgs mass. Therefore it is interesting to study the positron fraction from neutralino annihilation in the reduced region of SUSY parameter space, where these constraints are satisfied and compare it with the AMS and HEAT data.

## 2. Neutralino Annihilation in the CMSSM

In the Constrained Minimal Supersymmetric Model (CMSSM) with supergravity mediated breaking terms all sparticle masses are related by the usual GUT scale boundary conditions of a common mass $m_0$ for the squarks and sleptons and a common mass $m_{1/2}$ for the gauginos. In


doi:10.1016/S0920-5632(02)02089-3

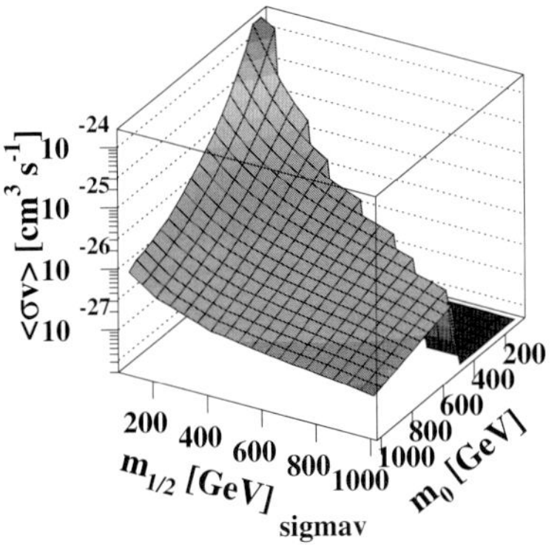


Figure 1. *The thermally averaged cross section times velocity for neutralino annihilation as function of $m_0$ and $m_{1/2}$ for* $\tan\beta = 50$ *The neutralino mass equals* $\approx 0.4m_{1/2}$ *in the CMSSM.*

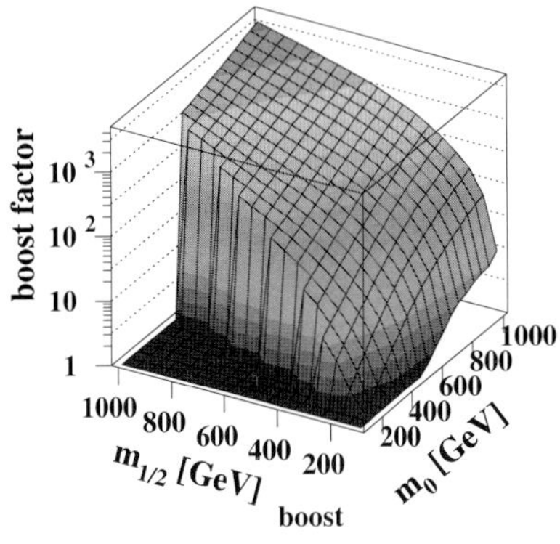


Figure 2. *The boost factor from the best fit to the combined AMS and HEAT data as function of $m_0$ and $m_{1/2}$ for* $\tan\beta = 50$. *The neutralino mass equals* $\approx 0.4m_{1/2}$ *in the CMSSM.*

addition, we assume a common trilinear coupling $A_0$ at the GUT scale, but we leave the Yukawa couplings free and neglect the Yukawa and trilinear coupling from the first two generations of quarks and leptons. The parameter $\tan\beta$ is varied between 1 and 50 to study the different annihilation channels, which strongly depend on $\tan\beta$ . In addition we scan over the remaining parameters $m_0$, $m_{1/2}$. For details we refer to a previous publication[6] and references therein. Neutralino annihilation can occur through Z- and Higgs exchanges in the s-channel and sfermion exchange in the t-channel. This annihilation in the halo of the galaxies will produce antimatter at high momenta, thus anomalies in the spectra of positrons and antiprotons provide an excellent signal for dark matter annihilation. The neutralino is a spin 1/2 Majorano particle, so it obeys the Pauli principle, which implies an asymmetric wave function or spins antiparallel for annihilation at rest. This results in a p-wave amplitude for the fermion-antifermion final states, which is proportional to the mass of the fermion in the final state. Therefore heavy final states are enhanced at low momenta. The total annihilation cross section is a strong function of the SUSY mass scale, since the t-channel contribution is proportional to one over the sfermion mass squared, as demonstrated in the $m_0, m_{1/2}$ plane in Fig. 1. For values of $\tan\beta > 5$ the $b\bar{b}$ final state dominates and the annihilation cross sections quickly increases with $\tan\beta$ [6]. The annihilation rate is proportional to the cross section and the neutralino density squared. The latter is unknown, since the clumpiness of the dark matter is unkown. To take this uncertainty into account, a boost factor for the cross section is introduced, which is used as a free parameter in the fit to the data. The boost factors needed to explain the possible excess in the HEAT data[1] are plotted in Fig. 2. For large $\tan\beta$ it can become in a reasonable range between 10 and 200, in which case the annihilation cross section is also sufficiently large to obtain a relic density of dark matter in the cosmologically interesting range of $\Omega$ between 0.1 and 0.3[7]. Fig. 3 shows a typical fit to the data. Clearly a contribution from neutralino annihilation is preferred: it reduces the $\chi^2$ from 48 for background only to around 26 for 18 data points. The fits were done with DarkSusy[8] and follow

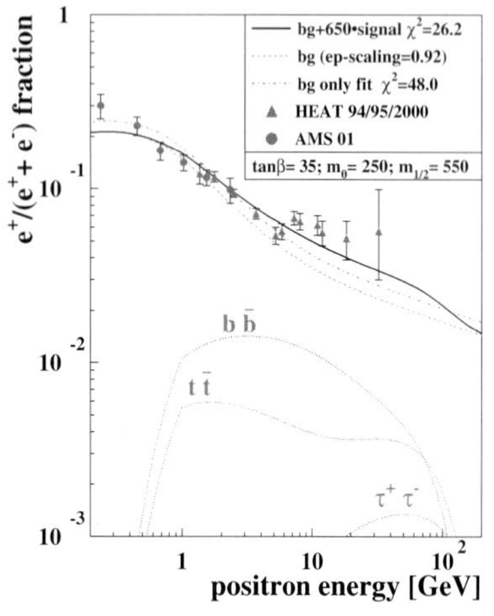


Figure 3. *The best fit to the combined AMS and HEAT data for* $\tan\beta = 35$. *The* $\chi^2=48$ *for background only and around 26 if neutralino annihilation is included.*

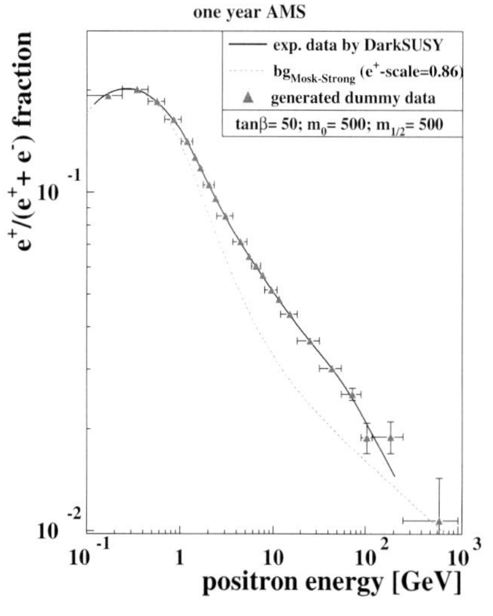


Figure 4. *Expected statistics after one year of data taking with AMS-02 for a neutralino of 200 GeV and assuming a boost factor of 150 given by the best fit to the present data.*

the same principle as the ones from Ref. [3], i.e. the background is left free within a normalization error of about 20% and the signal can be enhanced by a boost factor. Unfortunately, the errors in the present balloon experiments are quite large. For future space experiments[9,10], which will take data over several years instead of several days, the significance will improve dramatically, as shown in Fig. 4. Clearly, if the background estimates will be confirmed by future measurements of the electron spectrum, then the positron spectra can give a clear indication of neutralino annihilation with a rather precise determination of the neutralino mass. We like to thank Drs. L. Bergström, J. Edsjö and P. Ullio for helpful discussions.

## REFERENCES


1. S. W. Barwick *et al.* [HEAT Collaboration], Astrophys. J. **482** (1997) L191 [arXiv:astro-ph/9703192]. M. A. DuVernois *et al.*, Astrophys. J. **559** (2001) 296.
2. J. Alcaraz *et al.* [AMS Collaboration], Phys. Lett. B **484** (2000) 10 [Erratum-ibid. B **495** (2000) 440].
3. E. A. Baltz *et al.*, Phys. Rev. D **65** (2002) 063511 [arXiv:astro-ph/0109318].
4. G. L. Kane, L. T. Wang and T. T. Wang, Phys. Lett. B **536** (2002) 263 [arXiv:hep-ph/0202156].
   J. R. Ellis, J. L. Feng, A. Ferstl, K. T. Matchev and K. A. Olive, arXiv:astro-ph/0110225.
   M. Kamionkowski and M. S. Turner, Phys. Rev. D **43** (1991) 1774.
5. W. de Boer, Prog. Part. Nucl. Phys. **33** (1994) 201 [arXiv:hep-ph/9402266] and references therein.
6. W. de Boer et al., arXiv:astro-ph/0207557
7. A. H. Jaffe *et al.* [Boomerang Collaboration], Phys. Rev. Lett. **86** (2001) 3475
8. DarkSUSY, P. Gondolo,*et al.*, arXiv:astro-ph/0012234.
9. V. Bonvicini *et al.* [PAMELA Collaboration], Nucl. Instrum. Meth. A **461** (2001) 262.
10. J. Alcaraz *et al.* [AMS Collaboration], Nucl. Instrum. Meth. A **478** (2002) 119.

Proceedings of ICHEP 2002, pp. 119–121
*S. Bentvelsen, P. de Jong, J. Koch and E. Laenen (Editors)*

ELSEVIER

31st INTERNATIONAL CONFERENCE ON HIGH ENERGY PHYSICS AMSTERDAM

# AMS-high energy gamma-ray physics potential

G. Lamanna[a]

[a]CERN, European Organization for Nuclear Research,
1211 Geneva 23, Switzerland

The Alpha Magnetic Spectrometer (AMS) is a large acceptance, superconducting magnetic spectrometer designed to search for anti-matter and to study, with high accuracy, the origin of dark-matter and the composition of cosmic rays. AMS will be able to provide accurate measurements of high-energy $\gamma$-ray spectra over a broad energy range (up to hundreds of GeV). The $\gamma$-ray sky survey sensitivity of AMS and a few preliminary examples of physics objectives are discussed.

## 1. Introduction

The dynamic universe emerging from a decade of operation of the Energetic Gamma Ray Experiment Telescope (EGRET) [1], has generated much interest in high energy $\gamma$-ray detection: a new generation of space-born $\gamma$-ray telescopes AGILE [2] and GLAST [3] are currently under construction. In this context, it is interesting to analyze the $\gamma$-ray detection potential of the Alpha Magnetic Spectrometer, a new generation space-born cosmic-ray detector. AMS is a large acceptance particle detector which will be operating on the International Space Station for not less than three years, starting in 2005. The AMS experiment has been proposed to search for antimatter in the Universe on the level of less than $10^{-9}$, to search for dark matter and to make high statistic measurements of cosmic-ray composition over a broad energy range (up to the $TeV$ region).

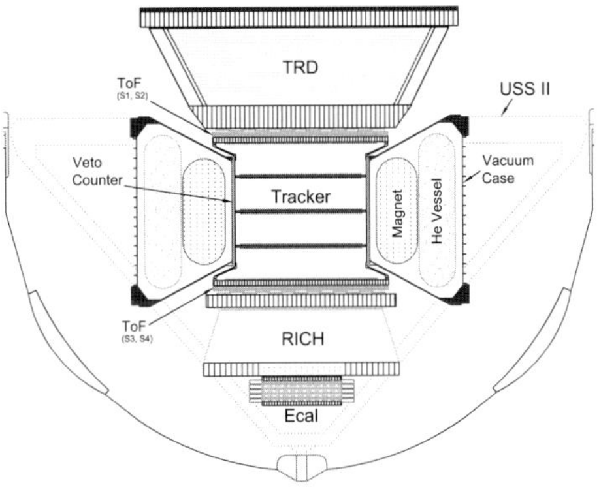


Figure 1. Schematic view of the AMS-02 experiment which will operate on the International Space Station.

## 2. The AMS-02 detector

The major elements of AMS-02[1] detector are shown in Fig.1 and consist of a superconducting magnet with a cylindrical shell shape with inner diameter 1.2 m and length 0.8 m and provides a central dipole field of 0.8 Tesla., a gaseous transition radiation detector (TRD), a silicon tracker (Tracker), time of flight hodoscopes (ToF), a ring imaging Cerenkov detector (RICH), an electromagnetic calorimeter (ECAL) and anticoincidence Veto counters. More detailed description of the detector can be found elsewhere [5].

[1]A reduced version of the detector, AMS-01, was flown and operated successfully during a 10-day shuttle flight in June 1998, NASA mission STS-91 [4].

## 3. AMS-02-$\gamma$ detection capabilities

To detect photons in the AMS experiment two complementary methods were explored through a Monte Carlo simulation study (see [6] for details). The first method is the *conversion mode*: it consists in the identification and reconstruction of $e^+e^-$ pairs from $\gamma$ conversions happened somewhere in the material upstream of the first


doi:10.1016/S0920-5632(02)02104-7

silicon tracker layer[2]. The event signature for this mode was two reconstructed tracks in the Tracker coming from a common vertex. Primary $\gamma$-ray energy and incidence direction were determined by adding the fitted momenta vectors of $e^{\pm}$ pair, evaluated at the entrance of the AMS detector. The second method is the *single photon mode*: it is based on the detection of photons in the electromagnetic calorimeter. The event signature for this mode was the presence of electromagnetic-type energy deposition in the ECAL, while almost nothing was found in other AMS sub-detectors.

### 3.1. Simulated performance

The dominant background to the (tiny) $\gamma$ signals is due to the most abundant cosmic rays ($p$, $He$, $C$, $e^{-}$s). An O($10^4$) to O($10^6$) background rejection level was obtained, which would allow a background-to-signal ratio of the order of a few per cent. The average detector acceptance as a function of the $\gamma$-ray energy (Fig.2), after all selection cuts have been applied, is comparable for the two methods (0.05-0.06 $m^2$ sr), with a main difference in the respective working energy range. Because of the different working principles and the geometry factors (the maximal opening angle being $\sim 42°$ for the *conversion mode* and 22° for the *single photon mode*) of the two sub-detectors (Si-Tracker and ECAL), the resulting *effective areas* are less comparable (Fig.2.middle). The $\gamma$ energy resolution is dominated by multiple scattering and tracker measurement errors in the *conversion mode* and it is parametrized as $\sigma(\mathrm{E})/\mathrm{E} = 0.03 \oplus 0.5\mathrm{E(TeV)}$. While in the *single photon mode* the relative energy resolution was parametrized as $0.03 \oplus 0.13/\sqrt{\mathrm{E}(GeV)}$ up to 2 TeV. An important figure of merit for the gamma sources detection is the angular resolution (Fig. 2.bottom). The ECAL detector would allow a maximal 0.9° for a 68% containement ($\sigma_{68}(E)$=$0.9° \oplus 8.5°/\sqrt{\mathrm{E}(GeV)}$), while more promizing resolution as a function of energy can be obtained by the *conversion mode*: $\sigma_{68}(E)$=$0.018° \oplus 0.85°/E(GeV)$.

[2]The material in front of the first silicon tracker plane, consisting of the TRD, the first two layers of ToF scintillators, and mechanical supports, represents $\simeq 0.23X_0$ [5].

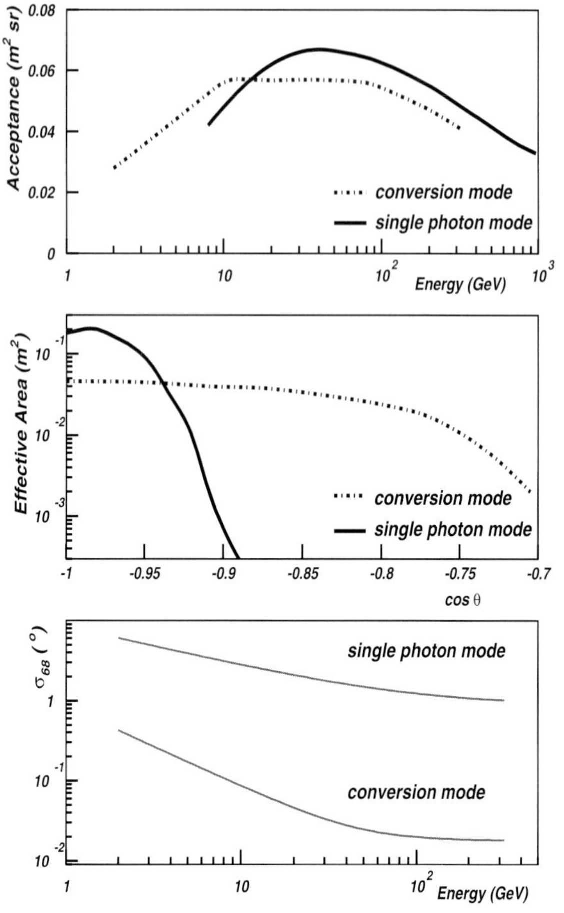


Figure 2. Acceptance as a function of energy (top panel). Effective area $A(E,\theta)$ versus zenith angle at 50 GeV (middle panel). Angular resolution as a function of energy (bottom panel).

## 4. Sensitivity

One year ISS orbits were simulated and a preliminary estimation of a $5\sigma$ point source sensitivity was computed [5]. The point source sensitivity for any given position in the sky was defined as the computed number of photons detected by AMS, $5\sigma$s over the diffuse gamma background starting from a chosen threshold energy $E_t$. Averaging over all longitudes for two latitude ranges, galactic ($|b|<10^o$) and extra galactic ($|b|>10^o$), the AMS-*conversion mode* one year ($E_t$=1 $GeV$) sensitivities are about $1.5\times10^{-8}$ ph. cm$^{-2}$ s$^{-1}$ and $10^{-9}$ ph. cm$^{-2}$ s$^{-1}$ respectively. While the AMS-*single photon mode* one year ($E_t$=10 $GeV$) sensitivities are $10^{-8}$ ph. cm$^{-2}$ s$^{-1}$ and $2.5\times10^{-9}$ ph. cm$^{-2}$ s$^{-1}$ respectively. Such sensitivity allows to pursue some fashinating physics objectives.

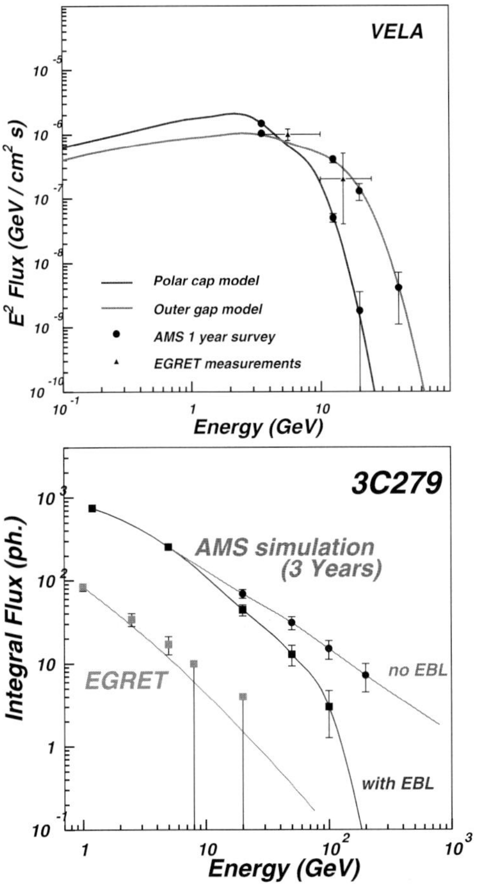


Figure 3. Top: Vela spectrum expected from AMS in one year . Bottom: 3C279 spectrum expected from AMS in three years $\gamma$-ray sky survey. EGRET measurements are also shown

EGRET has detected more than 60 blazar AGNs with a mean redshift near $z$=1 and more than 10 unidentified likely blazar $\gamma$-ray sources. AMS will be able to improve the identification of the EGRET sources and increase up to about a thousand of new classes of such objects. The AMS photons detection would be essential for answering questions about the mechanism of $\gamma$s ejection from the highest energy particle accelerators, the Pulsars. In Fig.3 we compare the statistical sensitivity of AMS to EGRET observation of higher energy part of the Vela spectrum. In one year sky survey it will be possible to distinguish between the two ejection models: *Polar Cap* and *Outer Gap* [7]. In three years the most brilliant blazars would be studied by AMS in detail, as it is the case for the 3C279 AGN. Through the measurements of its energy spectrum and a higher energy cutoff, it will be possible to disantangle the expected extintion of $\gamma$ rays off the diffuse soft (IR-optical) photons background (EBL effect)(Fig.3).

## 5. Conclusion

The expected performance and sensitivity of AMS as a high-energy $\gamma$-ray detector were presented. The AMS effective area together with its angular resolution and long exposure time, represent an useful combination to contribute to the investigation of the $\gamma$-ray atrophysics: a few preliminary examples were also addressed.

## 6. Acknowledgements

I wish to thank the AMS collaboration for having supported my participation at this conference.

## REFERENCES

1. R.C. Hartman et al., Ap.J.S. 123,79,1999
2. M. Tavani, Int.J.Mod.Ph. A17,1829, 2002
3. L. Latronico, this Conference Proc.
4. J. Alcaraz et al., Ph. Rep. 366/6, 331, 2002
5. V. Choutko, G. Lamanna, A. Malinin, Int.J.Mod.Ph. A17,1817,2002.
6. G.Lamanna, Proc. SPACEPART Int. Conf., Elba *in press on Nucl. Phys. B*
7. GLAST Science (http://glast.gsfc.nasa.gov)

ELSEVIER

Proceedings of ICHEP 2002, pp. 122–125
*S. Bentvelsen, P. de Jong, J. Koch and E. Laenen (Editors)*

31st INTERNATIONAL CONFERENCE ON HIGH ENERGY PHYSICS AMSTERDAM

# The Gamma Ray Large Area Space Telescope - an astro-particle mission to explore the high energy gamma ray sky

L. Latronico[a] *

[a]Istituto Nazionale di Fisica Nucleare, Sez. Pisa,
Via Livornse 1291, 56010 S. Piero a Grado (Pisa), Italy

The *Large Area Telescope (LAT)*, the main instrument on the *Gamma-ray Large Area Space Telescope (GLAST)* mission under development for launch by NASA in 2006, will survey the sky in the rich yet poorly explored high-energy band between 20 MeV and 1 TeV with unprecedented resolution and sensitivity. The LAT, which benefits from the application of particle physics detection technologies to astrophysics instrumentation, has a pair-conversion silicon-strips tracker and an imaging CsI calorimeter to measure the direction and energies of gamma rays. The relatively very intense background of charged particles will be rejected primarily by an anticoincidence shield of segmented plastic scintillators. In this paper we give an overview of the many physics goals and potential reach of the LAT, describe the instrument design and performance, and report on the status of the tracker construction.

## 1. Motivation and instrument design

The first complete survey of the sky in high-energy (30 MeV-10 GeV) $\gamma$-rays was performed by the Energetic Gamma Ray Experiment Telescope (EGRET, [1]), which was launched in 1991 on board of the Compton Gamma Ray Observatory. It showed the high energy $\gamma$-ray sky to be surprisingly dynamic and was able to detect a few hundred $\gamma$-ray point sources, most of which still remain unidentified. The GLAST mission ([2]) has been conceived to answer many of the important questions in high energy astrophysics raised by the EGRET results. Like EGRET, the Large Area Telescope, the main instrument on-board GLAST, will be a pair conversion telescope with a tracker, calorimeter, and an anticoincidence shield. But the latest-generation technology applied to the design of the LAT results in a remarkable advance in performance. EGRET had a gas-filled spark chamber tracker, and monolithic calorimeter and anticoincidence shield. The LAT will have a very large (83 $m^2$ total area in 36 layers) silicon strip tracker with fine detector pitch (228 $\mu m$), a hodoscopic, 8-layer calorimeter that will permit shower profiling, and an anticoincidence shield that is segmented to minimize self-veto from high energy electron backsplash from the calorimeter.

The performance of the LAT relative to EGRET is compared in Table 1. The result is that the LAT will have unprecedented resolution and an advance in sensitivity of a factor 20 or more with respect to EGRET.

## 2. GLAST Science Program

### 2.1. Sky map and unidentified sources

A full-sky survey with the LAT is planned for the first year of the GLAST mission. The excellent angular resolution and sensitivity of the LAT telescope will provide high-quality data with large statistics for detailed sky mapping, with source localization to the level of arc-minutes. The LAT will also offer dramatic improvements in ability to detect $\gamma$-ray flares and perform direct pulsation searches and critical spectral measurements.

Most of the 271 $\gamma$-ray sources in the EGRET catalog are unidentified; the large error boxes in the localization do not allow unambiguous association of these sources to known objects in the optical, radio, or X-ray bands. Moreover the distribution of the sources on the sky and their variable temporal properties suggest that more than one source class is required.

*on behalf of the GLAST-LAT collaboration


doi:10.1016/S0920-5632(02)02106-0

Table 1
GLAST LAT performances compared to EGRET

| Quantity | EGRET | LAT(minimum spec.) |
|---|---|---|
| Energy range | 20MeV - 30 GeV | 20MeV - 300 GeV |
| Peak effective area[1] | 1500 $cm^2$ | 8000 $cm^2$ |
| Field of view | 0.5 sr | $> 2sr$ |
| Angular resolution[2] | $5.8^o$ (100MeV) | $< 3.5^o$ (100MeV)<br>$< 0.15^o$ ($>$ 10GeV) |
| Energy resolution[3] | 10% | 10% |
| Dead time per event | 100ms | $< 100\mu s$ |
| Source location determination[4] | $15^{'}$ | $< 0.5^{'}$ |
| Point source sensitivity[5] | $\simeq 10^{-7}cm^{-2}s^{-1}$ | $< 6x10^{-9}cm^{-2}s^{-1}$ |

[1]After background rejection. [2]Single photon, 68 % containment, on-axis. [3]1-$\sigma$, on-axis. [4]1-$\sigma$ radius, flux $10^5 cm^2 s^{-1} (> 100MeV)$, high $|b|$. [5] $> 100MeV$, at high $|b|$, for exposure of one-year all sky survey

While there are indications that many of these sources belong to well established classes like blazars, radio-quiet pulsars, or binary systems, some of them may very well be examples of previously undetected $\gamma$-ray emitters, like forming clusters of galaxies or the recently discovered Galactic microquasars.

### 2.2. Active galactic nuclei

A large fraction of the known high-energy sources is represented by AGNs, most of them belonging to the blazar class and characterized by jet-like emission and broadband spectra peaking at gamma rays and between radio and X-ray.

The high sensitivity and large effective area of the LAT will be crucial for testing different models proposed for the central engine of blazars. Detailed spectral studies of gammas emitted should allow distinction between pure leptonic and hadronic particle acceleration mechanism.

The two models can be identified by tracking time variation of AGN spectra and studying the decay characteristics of the synchrotron and $\gamma$-ray components during flares, which usually occur on time scales of the order of minutes. Flaring states, when the most energetic phenomena occur inside the cores of AGNs, are in fact closely related to the nature of AGNs central engine, and more information can be obtained after correlation with observations at other wavelength, which will be done exploiting a network of other satellite and ground-based experiments alerted by the LAT.

Extrapolation of the log $N$-log $S$ relation ([3]) for AGNs suggests that thousands will be detected by the LAT. With such a large population, the redshift dependence of the $\gamma$-$\gamma$ attenuation of AGN spectra by the infrared-UV Extragalactic Background Light may be useful for measuring this cosmic background.

### 2.3. Pulsars

It is widely projected that GLAST will discover $\simeq$ 250 new pulsars. The high time resolution and detection efficiency of the LAT will in fact allow direct pulsation searches in the $\gamma$-ray band, with a possible identification of many Galactic, bright, unidentified sources with X or radio-quiet pulsars like Geminga ([4]). The very high photon statistics provided by the LAT large effective area will bring accurate spectra, opening the way to a clear discrimination between the *outer gap* and *polar cap* models of $\gamma$-ray production for rotation-powered pulsars.

### 2.4. Diffuse background and dark matter

An apparently isotropic background of high-energy gamma rays was confirmed by EGRET data. However, the relatively poor angular resolution and photon statistics has left open the possibility that the isotropic background is unresolved emission from AGNs. A truly diffuse background could instead have a more exotic interpretation, such as relic radiation from some very high energy primordial process in the early universe. The LAT will resolve a large fraction of the AGN component of the high-energy, diffuse background, and spectral and fluctuation analyses should permit sensitive limits to be determined for any cosmic diffuse background.

A similar excess was found in the emission from the galactic centre ([5]), suggesting a possible connection with the existence of dark matter at the centre of the galaxy. The most promising candidate process is neutralino ($\chi$) decay, as foreseen by supersymmetric extensions of the elementary particles standard model ([6]). Besides theoretical reasons, the hypothesis of neutralinos as the main dark matter constituents in galaxies is particularly attractive as searches for supersymmetric particles can be correlated to current limits from accelerator and underground experiments as well as from other space missions.

Figure 1 shows a simulation of the galactic centre spectrum obtained with standard propagation models and a contribution from neutralino decay. GLAST will have enough sensitivity to disentangle the two contributions.

The supersymmetric phase space that GLAST will be able to probe is model-dependent. In the context of the Minimal SuperGravity Model, recent preliminary calculations for the $\chi\chi \rightarrow \gamma\gamma$ or $\chi\chi \rightarrow \gamma Z$ cross sections, which would produce a spectacular monochromatic line as distinctive signature, indicate that a large region in the $m_0 - M_{1/2}$ space will be accessible to GLAST ([7]), providing an alternative test field for the supersymmetric hypothesis.

### 2.5. Gamma-ray bursts

In recent years, Gamma-ray bursts (GRBs) have been shown to be immensely energetic ($10^{54}$ erg) outbursts that occur at cosmological distances. Bursts have highly variable temporal structures (~ms) and typical durations of seconds. The large effective area and field of view of the LAT, together with the unprecedently short instrumental deadtime, will greatly advance knowledge of GRBs at high energies. The highest-energy emissions constrain the acceleration processes and the bulk motions within the fireballs. LAT data will permit distinguishing between internal and external shock models for the production of gamma rays, and evaluation of the baryon fraction in the fireballs.

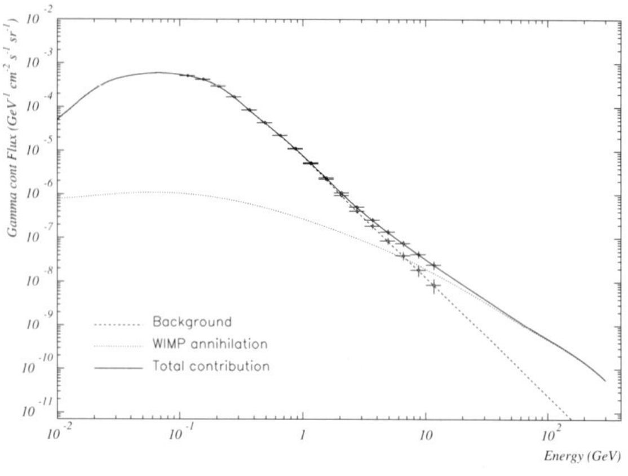


Figure 1. Simulation of the GLAST sensitivity to neutralino decay contributions to the total photon spectrum from the galactic centre

### 2.6. Cosmic rays

Cosmic rays (CR) have been known and measured for a long time. A fundamental outstanding question is the origin of the hadronic CRs, which is widely believed, but not yet proven to be supernova remnants (SNR). The SNR origin for nucleons would have a signature in a broad peak at 68 MeV from by the decay of $\pi^0$ produced by the interaction of freshly-accelerated CR nucleons with interstellar gas in the vicinity of an SNR. The LAT should be able to spatially resolve several SNR with degree-scale angular extents and spectrally resolve the $\pi^0$ decay peak from other $\gamma$-ray components.

## 3. Status of the LAT tracker construction

In order to maximize production efficiency with constrained cost, the LAT tracker is organized in a highly modular structure of $4x4$ identical towers of $37x37$ cm$^2$ area, each composed of a stack of 19 composite panels (*trays*) equipped with silicon-strip layers on both sides and a thin W converter, for a total of 18 bidimensional detection planes and an overall number of channels approaching $10^6$ (see fig.2). When a high energy gamma crosses the tracker, it produces an $e^+e^-$ pair after interacting with one of the converters, and the direction of the incoming photon can be evaluated by reconstructing the leptons tracks. Construction

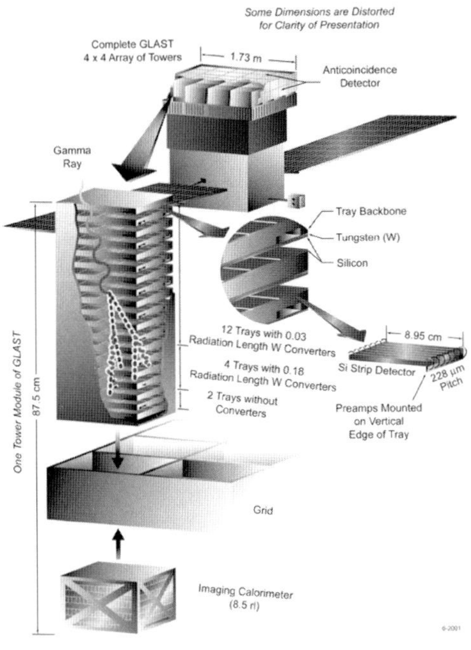


Figure 2. The GLAST-LAT detector

of the LAT tracker has begun with procurement and test of more than 3000 6" silicon sensors, representing more than 25% of the whole production. The electrical tests performed on the sensors show a high uniformity across the production (see table 2).

Construction of the baseline detector, obtained by wire bonding 4 silicon sensors together, has finished the prototyping phase with excellent results, while the engineering of the tray structure was confirmed after thorough qualification tests([8]).

## 4. Conclusions

The LAT is the product of a partnership of HEP and Astrophysics communities that will ex-

Table 2
Test results from the first 3000 silicon sensors

| Measurement | average | RMS |
|---|---|---|
| Leakage current[1] (nA) | 115 | 50 |
| Depletion voltage (V) | 70 | 20 |
| Bulk capacitance[1] (pF) | 1800 | 10 |
| Wafer cut alignment ($\mu$m) | 0 | 2 |
| Broken strips (overall rate) | $10^{-4}$ | |

[1]measured at a bias of 150V

plore the high-energy $\gamma$-ray sky. Diverse astrophysical objects like AGN, GRBs, pulsars, and SNR, radiate in this energy range, and observations in this regime are key to understanding the nonthermal processes involved. Observation of the diffuse $\gamma$-ray halo of the Milky Way will also allow testing the hypothesis of a supersymmetric origin of dark matter, probing a large part of the supersymmetric phase space. Today only a little is known about the $\gamma$-ray sky, mainly because of technological limitations of previous missions. GLAST will be equipped with state of the art particle detectors, resulting in greater than an order of magnitude advance in sensitivity and resolution.

**Acknowledgements**

I want to thank the GLAST-LAT collaboration for the material provided. Special aknowledgements to Seth Digel for his precious suggestions.

## REFERENCES

1. P. Nolan et al. IEEE Trans., Nucl. Sci.,39, 993-996
2. E. Bloom et al.,*Proposal for the Gamma-ray Large Area Telescope*, SLAC-R-22.February 1998
3. Stecker F.W., Salamon M.H. 1996, ApJ 464, 600
4. McLaughlin M.A. et al., 2000, ApJ 538, 818
5. H. Mayer-Hasselwander et al., Astron.Astrophys. 335, 161 (1998)
6. Bergstrom L. et al. Astropart. Phys. 1998, 9, 137
7. Morselli A. et al., *Search for Dark Matter with GLAST*, submitted to Nucl. Phys. B (*SpacePart* Conference Proceedings, Elba Island, 14/19 May 2002)
8. Bellazzini R. et al., *The GLAST Tracker design and construction*, submitted to Nucl. Phys. B (*SpacePart* Conference Proceedings)

ELSEVIER

Proceedings of ICHEP 2002, pp. 126–128
*S. Bentvelsen, P. de Jong, J. Koch and E. Laenen (Editors)*

31st INTERNATIONAL CONFERENCE ON HIGH ENERGY PHYSICS AMSTERDAM

# Recent Results from AMANDA II

K. Hanson for the AMANDA Collaboration:
J. Ahrens[a], X. Bai[b], S.W. Barwick[c], T. Becka[a], K.-H. Becker[d], E. Bernardini[e], D. Bertrand[f], F. Binon[f], A. Biron[e], S. Böser[e], O. Botner[g], O. Bouhali[f], T. Burgess[h], S. Carius[i], T. Castermans[j], A. Chen[k], D. Chirkin[l], J. Conrad[g], J. Cooley[k], D.F. Cowen[m], A. Davour[g], C. De Clercq[n], T. DeYoung[k*], P. Desiati[k], J.-P. Dewulf[f], P. Doksus[k], P. Ekström[h], T. Feser[a], T.K. Gaisser[b], M. Gaug[e], L. Gerhardt[c], A. Goldschmidt[o], A. Hallgren[g], F. Halzen[k], K. Hanson[k], R. Hardtke[k], T. Hauschildt[e], M. Hellwig[a], P. Herquet[j], G.C. Hill[k], P.O. Hulth[h], S. Hundertmark[h], J. Jacobsen[o], A. Karle[k], B. Koci[k], L. Köpke[a], K. Kuehn[c], M. Kowalski[e], J.I. Lamoureux[o], H. Leich[e], M. Leuthold[e], P. Lindahl[i], I. Liubarsky[k], J. Madsen[p], P. Marciniewski[g], H.S. Matis[o], C.P. McParland[o], Y. Minaeva[h], P. Miočinović[l], R. Morse[k], R. Nahnhauer[e], T. Neunhöffer[a], P. Niessen[n], D.R. Nygren[o], H. Ogelman[k], Ph. Olbrechts[n], C. Pérez de los Heros[g], A.C. Pohl[i], P.B. Price[l], G.T. Przybylski[o], K. Rawlins[k], E. Resconi[e], W. Rhode[d], M. Ribordy[e], S. Richter[k], J. Rodríguez Martino[h], D. Ross[c], H.-G. Sander[a], T. Schmidt[e], D. Schneider[k], R. Schwarz[k], A. Silvestri[c], M. Solarz[l], G.M. Spiczak[p], C. Spiering[e], D. Steele[k], P. Steffen[e], R.G. Stokstad[o], P. Sudhoff[e], K.-H. Sulanke[e], I. Taboada[q], L. Thollander[h], S. Tilav[b], C. Walck[h], C. Weinheimer[a], C.H. Wiebusch[e†], C. Wiedemann[h], R. Wischnewski[e], H. Wissing[e], K. Woschnagg[l], G. Yodh[c], S. Young[c]

[a]Institute of Physics, University of Mainz, Staudinger Weg 7, D-55099 Mainz, Germany
[b]Bartol Research Institute, University of Delaware, Newark, DE 19716, USA
[c]Dept. of Physics and Astronomy, University of California, Irvine, CA 92697, USA
[d]Fachbereich 8 Physik, BUGH Wuppertal, D-42097 Wuppertal, Germany
[e]DESY-Zeuthen, D-15735 Zeuthen, Germany
[f]Université Libre de Bruxelles, Science Faculty CP230, B-1050 Brussels, Belgium
[g]Division of High Energy Physics, Uppsala University, S-75121 Uppsala, Sweden
[h]Dept. of Physics, Stockholm University, SCFAB, SE-10691 Stockholm, Sweden
[i]Dept. of Technology, Kalmar University, S-39182 Kalmar, Sweden
[j]University of Mons-Hainaut, 7000 Mons, Belgium
[k]Dept. of Physics, University of Wisconsin, Madison, WI 53706, USA
[l]Dept. of Physics, University of California, Berkeley, CA 94720, USA
[m]Dept. of Physics, Pennsylvania State University, University Park, PA 16802, USA
[n]Vrije Universiteit Brussel, Dienst ELEM, B-1050 Brussels, Belgium
[o]Lawrence Berkeley National Laboratory, Berkeley, CA 94720, USA
[p]Physics Dept., University of Wisconsin, River Falls, WI 54022, USA
[q]Departamento de Física, Universidad Simón Bolívar, Apdo. Postal 89000, Caracas, Venezuela

We present new data taken with the AMANDA-II neutrino telescope array. The AMANDA-II upgrade was completed at the beginning of 2000. It significantly extends the sensitivity of the 10-string AMANDA-B10 detector to high- and ultrahigh-energy neutrino fluxes into regions of interest for probing current astrophysical models which remain unexplored by other experiments.

## 1. INTRODUCTION

The AMANDA II neutrino telescope is an array of 677 photomultipliers deposited approximately 2 km deep in the glacial ice of the South Pole. These sensors detect the passage of high-

*Present address: Santa Cruz Inst. for Particle Physics, University of California, Santa Cruz, CA 95064, USA.
†Present address: CERN, CH-1211, Genève 23, Switzerland.


doi:10.1016/S0920-5632(02)02111-4

energy charged particles as the particles radiate Cherenkov light in this natural medium. While the greatest portion of these charged particles are cosmic ray muons from hadronic primaries, striking AMANDA more than 100 times per second, a detectable fraction are muons from atmospheric neutrinos interacting in the ice or rock surrounding and beneath the detector. Even more rare, and, as yet undetected, are the muons and electrons and taus induced by neutrinos of extraterrestrial origin or neutrinos from annihilations of hypothesized dark matter particles gravitationally trapped at the centers of the Earth and Sun.

## 2. ATMOSPHERIC $\nu_\mu$

In the energy regime that AMANDA is sensitive to throughgoing, atmospheric $\nu$-induced muons, $\mathcal{O}(E_\nu) = 100\,\text{GeV}$, neutrino oscillations produce only negligible modulation of the atmospheric $\nu$

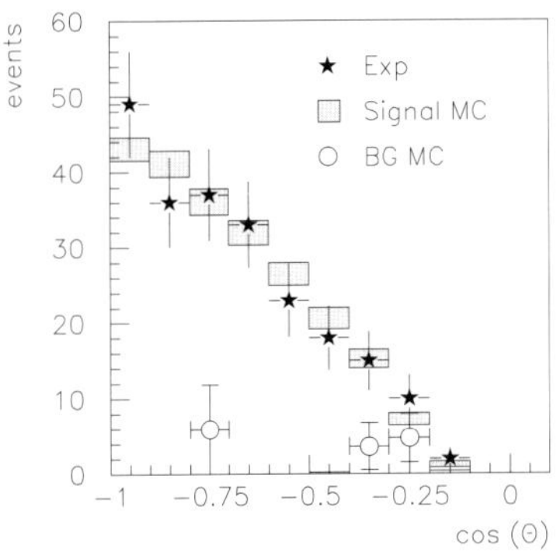


Figure 1. AMANDA B10 $\nu_{\rm ATM}$ events versus $\cos\theta$. While the $\nu$-induced muon flux steepens at the horizon, the AMANDA B10 acceptance vanishes there, thus most events are vertical. A total of 223 neutrino events were measured in 130 days of detector livetime.

flux. Nevertheless, atmospheric neutrinos provide a test beam of known angular shape, albeit of less-well-known absolute normalization at these energies. These particles have been measured by AMANDA-B4 [1], AMANDA-B10 [2], and most recently, AMANDA-II. The zenith angle spectra measured by B10 and AMANDA-II are shown in Figs. 1 and 2 In the analysis of the AMANDA-II

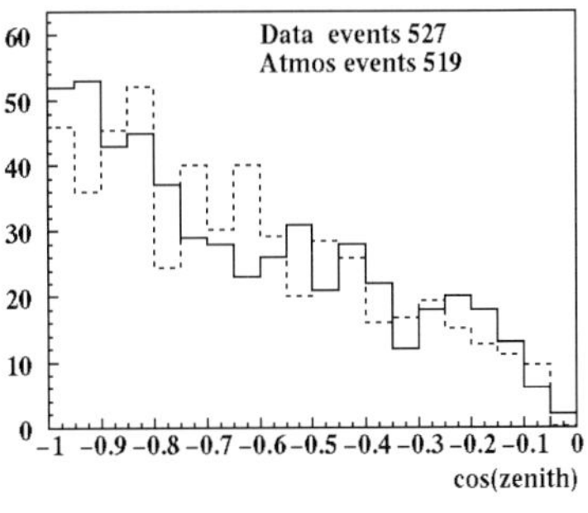


Figure 2. AMANDA II $\nu_{\rm ATM}$ events versus $\cos\theta$, measured during the 2000 run season (204 d livetime). Note that the AMANDA II acceptance is greatly enhanced with respect to B10 at the horizon. At this cut level, background contamination is estimated to be negligible.

data, the Monte Carlo neutrino flux of Lipari [3] has been scaled to agree with data at a very tight cut level containing just 56 events. The reported flux of Fig. 2 shows continued agreement in shape and normalization between data and MC at a much looser cut level, implying that background remains at a negligible level.

## 3. UHE DIFFUSE $\nu_\mu$

Unresolved, extragalactic sources are expected to produce neutrinos with a softer energy spectrum than the atmospheric neutrinos which are described by the spectal index $\gamma \approx -3.7$. These neutrinos therefore would appear as high energy tails in the atmospheric neutrino data. A search for this signal is underway with AMANDA-II data; currently we have analyzed 20% of the data and preserved 80% for 'unblinding' in the near future. Using just this fractional sample, we observe the 'N-Channel' spectrum of Fig. 3. The dif-

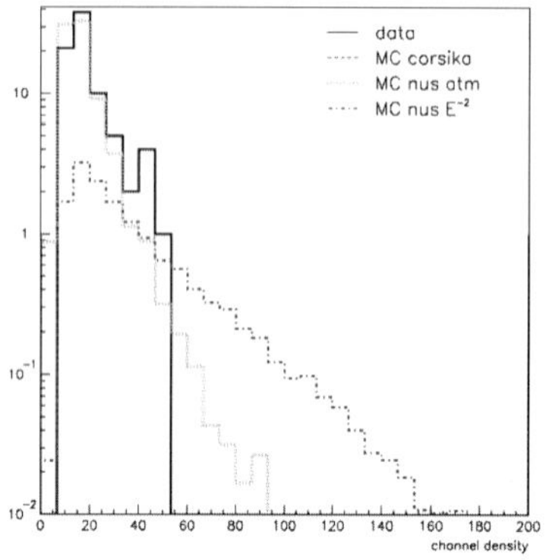


Figure 3. N Channel plot of events selected in the diffuse $\nu$ search. The MC line is for a model of "injected" AGN: $E_\nu^2\Phi_\nu = 10^{-5}\,\mathrm{GeV}\cdot\mathrm{cm}^{-2}\mathrm{s}^{-1}\mathrm{sr}^{-1}$.

fuse flux signal is separated from the atmospheric neutrino background by a cut on the number of hit channels at 50. Here, for the purpose of estimating the sensitivity of AMANDA-II (*i.e.*, under the assumption of no signal observed), the data is truncated to $N_{\mathrm{CH}} < 50$. The resulting sensitivity from the 20% is

$$E_\nu^2\Phi_\nu < 2.0 \times 10^{-6}\,\mathrm{GeV}\cdot\mathrm{cm}^{-2}\mathrm{s}^{-1}\mathrm{sr}^{-1} \quad (1)$$

which remains, still, some two orders of magnitude above the Waxman-Bahcall limit [4].

## 4. SEARCH FOR $\nu_\mu$ POINT SOURCES

Whereas diffuse signals of UHE neutrino sources are limited in energy range by the background of atmospheric neutrinos, point sources are identifiable at all energies by the signature of excess events clustering in angular bins on the sky. Table 1 lists the preliminary results of a search for these excess events in the upward muon sample around six preselected, candidate neutrino sources.

## 5. CONCLUSIONS AND OUTLOOK

The results from the atmospheric, diffuse, and point source analyses of the first of the AMANDA-II data sets have been presented.

Table 1
Selected point source 90% C.L. flux limits from analysis of AMANDA-II 2000 data. The limits are in units of $10^{-15}\,\mathrm{cm}^{-2}\mathrm{s}^{-1}\mathrm{sr}{-1}$ for $\Phi_\mu$ and $10^{-8}\,\mathrm{cm}^{-2}\mathrm{s}^{-1}\mathrm{sr}{-1}$ for $\Phi_\nu$ for integrated spectra $\propto E^{-2}$, $E > 10\,\mathrm{GeV}$. The muon angular resolution of AMANDA-II for this analysis $\approx 2.5°$.

| Source | $\delta(°)$ | $\alpha$(hr) | $\Phi_\mu$ | $\Phi_\nu$ |
|---|---|---|---|---|
| Mrk421 | 38.2 | 11.1 | 2.05 | 1.28 |
| Mrk501 | 39.8 | 16.9 | 1.71 | 1.06 |
| Crab | 22.0 | 5.6 | 6.76 | 3.57 |
| CasA | 58.8 | 23.4 | 2.09 | 1.57 |
| SS433 | 5.0 | 19.2 | 10.5 | 3.49 |
| CygX3 | 41.0 | 20.5 | 2.67 | 1.68 |

AMANDA-II analyses, still not as mature as the B10 analyses, nevertheless compare very favorably due to the superior performance of a bigger detector array. Filtering of the 2001 season data is almost completed and online, realtime filtering of the 2002 is returning analysis quality data from the South Pole daily.

## 6. ACKNOWLEDGMENTS

This research was supported by the following agencies: U.S. National Science Foundation, Office of Polar Programs; U.S. National Science Foundation, Physics Division; University of Wisconsin Alumni Research Foundation; U.S. Department of Energy; Swedish Research Council; Swedish Polar Research Secretariat; Knut and Alice Wallenberg Foundation, Sweden; German Ministry for Education and Research; U.S. National Energy Research Scientific Computing Center (supported by the Office of Energy Research of the U.S. Department of Energy); UC-Irvine AENEAS Supercomputer Facility; Deutsche Forschungsgemeinschaft (DFG)

## REFERENCES

1. E. Andrés *et al.*, Astrop. Phys. **13** 1 (2000)
2. J. Ahrens *et al.*, Phys. Rev. D. **66** 012005 (2002)
3. P. Lipari, Astrop. Phys. **1** 195 (1993)
4. E. Waxman and J. Bahcall, Phys. Rev. D. **59** 023002 (1999)

ELSEVIER

Proceedings of ICHEP 2002, pp. 129–131
*S. Bentvelsen, P. de Jong, J. Koch and E. Laenen (Editors)*

31st INTERNATIONAL CONFERENCE ON HIGH ENERGY PHYSICS AMSTERDAM

# Search for GUT Magnetic Monopoles with the MACRO Experiment at the Gran Sasso Lab.

Giorgio Giacomelli, for the MACRO Collaboration * [a]

[a]Dipartimento di Fisica dell'Università di Bologna and INFN, I-40127 Bologna, Italy
e-mail: giacomelli@bo.infn.it

The final results of the MACRO experiment on the search for GUT supermassive magnetic monopoles in the penetrating cosmic radiation are presented and discussed. The 90% CL upper limits are at the level of 1.5 - $2 \times 10^{-16}$ $cm^{-2}$ $s^{-1}$ $sr^{-1}$ for $0.0004 < \beta < 1$. Similar limits were obtained for nuclearites. Limits for the MM catalysis of proton decay are at 3 - 4 $10^{-16}$ $cm^{-2}$ $s^{-1}$ $sr^{-2}$ for $10^{-4} < \beta < 6 \cdot 10^{3}$.

One of the primary aims of the MACRO detector at the Gran Sasso lab. (at an average depth of 3700 mwe) was the search in the penetrating cosmic radiation for the supermassive magnetic monopoles (MMs) predicted by Grand Unified Theories of the electroweak and strong interactions; the search was planned for a sensitivity level well below the Parker bound ($\sim 10^{-15}$ $cm^{-2}$ $s^{-1}$ $sr^{-1}$) and for a large range of velocities, $4 \cdot 10^{-5} < \beta < 1$, $\beta$= v/c.

MACRO used three different types of detectors: liquid scintillators, limited steamer tubes and nuclear track detectors (NTDs; CR39 and Lexan) arranged in a closed-box modular structure, divided into a lower and an upper ("Attico") part. The overall dimensions of the apparatus were $76.5 \times 12 \times 9.3$ $m^3$. MACRO took data from early 1989 till december 2000 [1].

The response of the three types of detectors to slow and fast particles was experimentally studied [2 - 4]. The use of three types of subdetectors with redundant electronics, in stand alone and combined modes, ensured redundancy of information, cross-checks and indipendent signatures for possible MM candidates; it also allowed to cover the complete $\beta - range$, $4 \times 10^{-5} - 1$.

The searches performed with the liquid scintillator subdetector used different specialized triggers to cover the $10^{-4} < \beta < 4 \times 10^{-2}$ range [2]; a custom made 200 MHz wave form digitizer was used for the lower $\beta - range$, while the PHRASE system was used in the intermediate $\beta - range$.

The searches performed in the low $\beta - region$ with the limited streamer tubes were based on the search for single tracks and on the measurement of the velocity by means of a "time track" [3].

The searches with the CR39 NTDs required chemical etching of each individual sheet ($24.5 \times 24.5 \times 0.14$ $cm^3$), double scannings, measurements and eventual cross checks with other sheets of the same module. We analized 845.5 $m^2$ of sheets, exposed for an average time of 9.5 y [4].

Combined analyses with 2 or 3 types of detectors were performed as described in ref. [5].

No MM candidate was found; the obtained limits were presented and discussed at many conferences and compared with other results [6] [7].

Fig. 1 shows the 90% C.L. flux upper limits for $g = g_D$ MMs (one unit Dirac magnetic charge) obtained by indipendent methods. The final limit, computed by combining the indipendent limits obtained by the different analyses, is shown in Figs. 1 and 2; the MACRO limit is compared with previous limits in Fig. 2 [8].

Fig. 3 shows the flux upper limits obtained with the CR39 nuclear track detector for MMs with different magnetic charges, $g = g_D$, $2g_D$, $3g_D$ and for M + p composites.

The interaction of the GUT monopole core with a nucleon can lead to a reaction in which

*See ref. 1 for the list of MACRO collaborators and Institutions


doi:10.1016/S0920-5632(02)02114-X

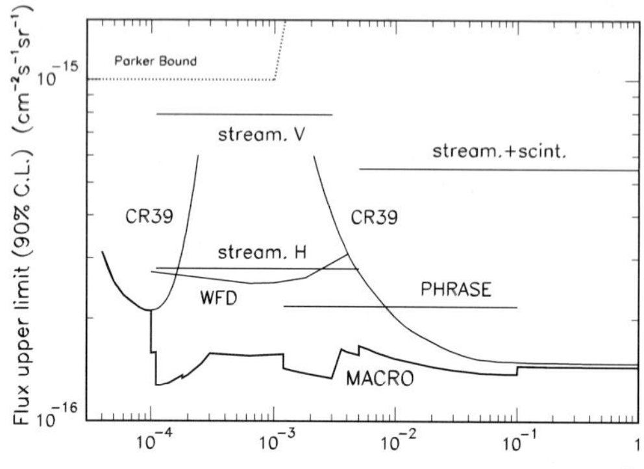


Figure 1. 90% C.L. upper limits vs $\beta$ for an isotropic flux of MMs obtained from different searches with the 3 subdetectors (both in stand alone and combined ways), and the MACRO global limit.

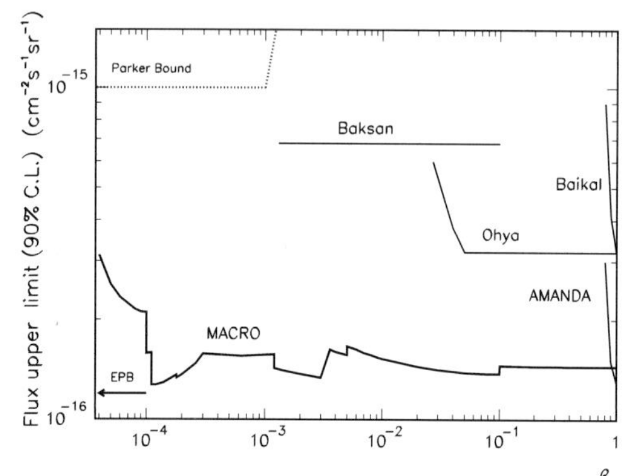


Figure 2. The global MACRO MM upper limit compared with limits from other experiments [8].

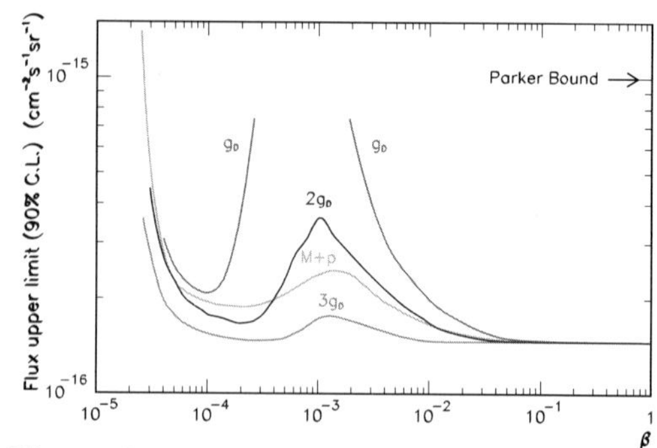


Figure 3. Upper limits (90% C.L.) for an isotropic flux of MMs obtained with the CR39 subdetector of MACRO, for poles with magnetic charge $g = g_D$, $2g_D$, $3g_D$ and for M+p composites.

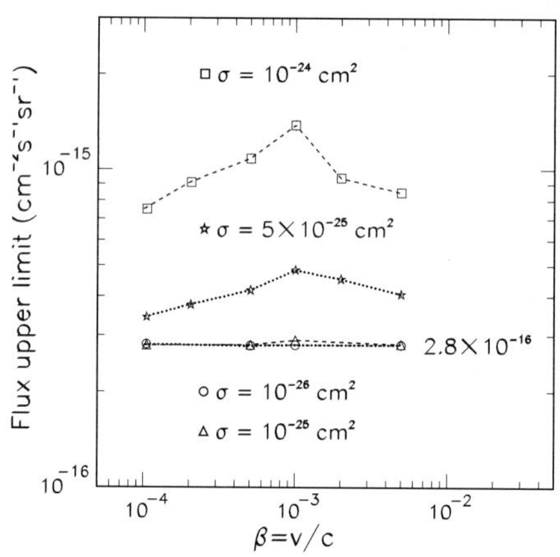


Figure 4. Upper limits for a MM flux as a function of the MM velocity for various catalysis cross sections.

the nucleon decays (monopole catalysis of nucleon decay), f. e. $M + p \rightarrow M + e^+ + \pi^0$ [9]. The cross section for this process is of the order of the core size, $\sigma \sim 10^{-56} cm^2$, practically negligible. But the catalysis process could proceed via the Rubakov-Callan mechanism with a cross section of the order of the strong interaction cross section [10]. We developed a dedicated analysis procedure aiming to detect nucleon decays induced by the passage of a GUT MM in our streamer tube system. The flux upper limit results of this search as a function of the MM velocity and of the catalysis cross section is shown in Fig. 4 [11].

MMs with intermediate masses, $10^5 < m_M < 10^{12}$ GeV, could be present in the cosmic radiation; underground and underwater experiments have a limited sensitivity to such MMs.

As a byproduct of our scintillator and nuclear track detector searches, we obtained upper limits for a flux of nuclearites and of Q-balls in the penetrating cosmic radiation.

Nuclearites (Strangelets, Strange Quark Matter) should consist of aggregates of u, d and s quarks in almost equal proportion. They could have been produced shortly after the Big Bang or in violent astrophysical processes, and may have survived as remnants; they could be part of the cold dark matter and would have typical galactic velocities, $\beta \sim 10^{-3}$. They should yield a large amount of light in scintillators and large Restricted Energy Losses in the nuclear track detectors. Fig. 5 shows our upper limits for an isotropic flux of nuclearites obtained with the scintillator and the CR39 subdetectors; also the

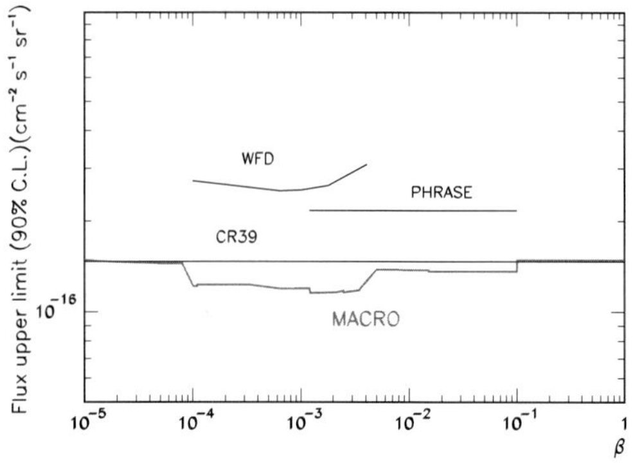


Figure 5. The 90% C.L. upper limits for an isotropic flux of nuclearites obtained with the liquid scintillators (WFD and PHRASE) and the CR39 subdetectors. The bottom line is the MACRO global limit.

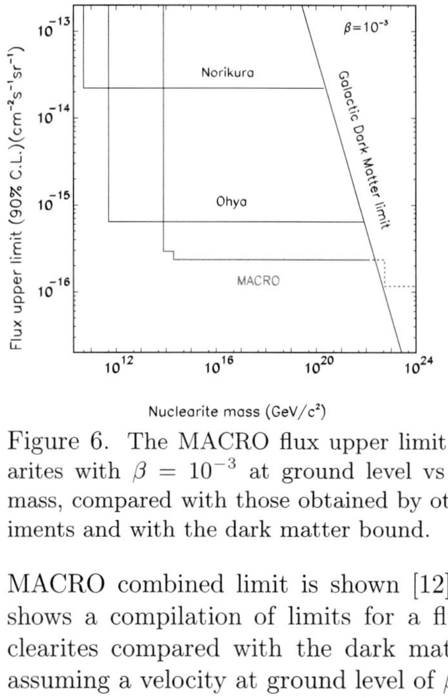


Figure 6. The MACRO flux upper limit for nuclearites with $\beta = 10^{-3}$ at ground level vs nuclearite mass, compared with those obtained by other experiments and with the dark matter bound.

MACRO combined limit is shown [12]. Fig. 6 shows a compilation of limits for a flux of nuclearites compared with the dark matter limit, assuming a velocity at ground level of $\beta = 10^{-3}$.

Q-balls should be aggregates of squarks $\tilde{q}$, sleptons $\tilde{l}$ and Higgs fields [13]. The scalar condensate inside a Q-ball core should have a global baryon number Q. There could exist neutral and charged Q-balls and they could be possible cold dark matter candidates. Charged Q-balls could have been detected, with reduced sensitivity, by our scintillators and nuclear track detectors.

In conclusion, we have presented the final results of the MACRO searches for superheavy MMs in the penetrating cosmic radiation. The flux limits are at the level of $1-2 \times 10^{-16}\ cm^{-2}\ s^{-1}\ sr^{-1}$ for $\beta > 4 \times 10^{-5}$. The obtained limits are the best existing and cover the widest $\beta - range$. It will be difficult to do much better since one would require detectors of much larger areas.

Similar considerations apply to massive nuclearites with $\beta \sim 10^{-3}$ and to charged Q-balls.

I thank the colleagues of the MACRO collaboration, in particular those from Bologna. I thank ms. Anastasia Casoni for her cooperation.

## REFERENCES

1. S.Ahlen et al., Nucl. Instr. Meth. Phys. Res. A324(1993)337; M.Ambrosio et al., Nucl. Instr. Meth. Phys. Res. A486(2000)663.
2. S.Ahlen et al., Phys. Rev. Lett. 72(1994)608; M.Ambrosio et al., Astroparticle Phys. 6(1997)113.
3. M.Ambrosio et al., Astroparticle Phys. 4(1995)33.
4. S.Cecchini et al, Il Nuovo Cimento A109(1996)1119; Radiat. Meas. 34(2001)54; G.Giacomelli et al., Nucl. Instr. Meth. A411(1998)41.
5. M.Ambrosio et al., hep-ex/0110083.
6. M.Ambrosio et al., Phys. Lett. B406(1997)249; hep-ex/0206027.
7. G.Giacomelli et al., hep-ex/0002032; hep-ex/0005041; hep-ex/0112009.
8. E.N.Alexeyev et al. ("Baksan"), $21^{st}$ ICRC, 10(1990)83. S.Orito et al., ("Ohya"), Phys. Rev. Lett. 66(1991)1951. V.A.Balkanov et al., ("Baikal", "Amanda" ), Nucl. Phys. B91(2001)438.
9. G.Giacomelli et al., hep-ex/0004019.
10. V.A.Rubakov, JETP Lett. B219(1981)644. G.G.Callan, Phys. Rev. D26(1982)2058.
11. M.Ambrosio et al., hep-ex/0207024.
12. S.Ahlen et al., Phys. Rev. Lett. 69(1992)1860; M.Ambrosio et al., Eur. Phys. J. C13(2000)453.
13. S.Coleman, Nucl. Phys. B262(1985)293. A.Kusenko, Phys. Lett. B405(1997)108.

Proceedings of ICHEP 2002, pp. 132–134
*S. Bentvelsen, P. de Jong, J. Koch and E. Laenen (Editors)*

31st INTERNATIONAL CONFERENCE ON HIGH ENERGY PHYSICS AMSTERDAM

# Signatures of primordial helicity in the CMBR

L. Pogosian[a], T. Vachaspati[b] and S. Winitzki[c]

[a]Theoretical Physics Group, Imperial College, London SW7 2BZ, United Kingdom

[b]Department of Physics, Case Western Reserve University, Cleveland, OH 44106-7079, USA

[c]Ludwig Maximilians Universität, Sektion Physik, 80333 München, Germany

We have studied ways in which helical primordial magnetic fields could be constrained by measurements of the cosmic microwave background radiation (CMBR). If there were helical flows in the primordial plasma at the time of recombination, they would produce parity violating temperature-polarization correlations ($C_l^{TB}$ and $C_l^{EB}$). However, the magnitude of helical flows induced by helical magnetic fields is unobservably small. We discuss an alternate scheme for extracting the helicity of a stochastically homogeneous and isotropic primordial magnetic field using Faraday rotation measure maps of the CMBR and the power spectrum of B-type polarization ($C_l^{BB}$).

## 1. Introduction

The improving quality of the CMBR measurements has made it possible to test different models of Early Universe physics. As more refined observations are made, a larger array of theoretical ideas will be put to the test and more details of the history of the universe will emerge.

A possibility that has already received some attention is that large-scale Parity (P) violation may be observed via the CMBR [1,2]. In Ref. [1] it was shown that a coherent magnetic field would induce non-zero P-violating correlations in the CMBR through Faraday rotation, while in Ref. [2] the P violation was due to the dynamics of a pseudoscalar field. In Ref. [3] we have examined the consequence of yet another possible source of P-violation (and also CP-violation), namely large-scale primordial helical magnetic fields[1]. Helical fields could be expected from cosmic events such as electroweak (EW) baryogenesis [4–7]. The reason for these expectations is that within the EW model production of baryons is accompanied by a change in the Chern-Simons number (CS), which can be interpreted as the net helicity in the $SU(2)_L \times U(1)_Y$ gauge fields. Changes in the CS are achieved via the production and decay of non-perturbative field configurations, such as sphalerons and linked loops of electroweak strings [8]. Because these configurations produce magnetic fields, it is possible that the helicity in the non-Abelian fields associated with the CS could be inherited by magnetic fields present immediately after the EW phase transition. It has also been argued in Ref. [6] that such helical magnetic fields could remain frozen in the primordial plasma and, provided sufficiently effective inverse cascading, develop cosmologically interesting strengths at the time of recombination.

Our approach in this work was to assume existence of a stochastic homogeneous and isotropic helical magnetic field at recombination and try to answer the following three questions, treated as independent from each other: **1)** Could helical magnetic fields produce helical flows (kinetic helicity) in the primordial plasma? **2)** Can one constrain primordial kinetic helicity using the CMBR? **3)** Can one detect helical magnetic fields using the CMBR? We provide answers in the following sections.

## 2. Kinetic helicity from magnetic helicity?

We are interested in the effects of a statistically homogeneous and isotropic magnetic field, with possibly non-vanishing helicity. If we denote the Fourier amplitudes of the magnetic field by $\mathbf{b}(\mathbf{k})$,

[1]A vector field $\mathbf{v}$ is helical if $\langle \mathbf{v} \cdot (\nabla \times \mathbf{v}) \rangle \neq 0$


doi:10.1016/S0920-5632(02)02062-5

then

$$\langle b_i(\mathbf{k}) b_j(\mathbf{k}')\rangle = (2\pi)^3 \delta^{(3)}(\mathbf{k}+\mathbf{k}') \times [(\delta_{ij} - \hat{k}_i \hat{k}_j) S(k) + i\varepsilon_{ijl} \hat{k}_l A(k)]. \quad (1)$$

Here $S(k)$ denotes the symmetric part and $A(k)$ the antisymmetric part of the correlator. One can check that $\langle \mathbf{B}(\mathbf{x}) \cdot [\nabla \times \mathbf{B}(\mathbf{x})]\rangle$ only depends on $A(k)$ and not on $S(k)$. Therefore $A(k)$ represents the helical component of the magnetic field and $S(k)$ the non-helical component.

In the tight-coupling approximation it is the Lorentz force that drives flows in neutral plasma. An evaluation shows that the Lorentz force depends on $S(k)$ but has no dependence on $A(k)$. Therefore the velocity flow at last scattering is unaffected by the helical component of the magnetic field. In reality the coupling of photons to electrons is much stronger than that to protons and so the plasma at recombination is better treated as composed of two fluids: the electron-photon fluid and the proton fluid. A calculation, presented in full in Ref. [3] and closely following that of Harrison[10], shows that the electron-photon fluid will gain an angular velocity $\omega_e = (en_e)^{-1}\nabla^2\mathbf{B}$, where $e$ and $n_e$ are the electron charge and the number density. If we estimate $|\nabla^2\mathbf{B}| \sim B/L^2$, where $L$ is the coherence scale of the field, we find $|\mathbf{v}| \sim |L\omega_e| \sim 10^{-18}\left(B_0/10^{-9}\text{G}\right)(1\text{kpc}/L_0)$ where $B_0$ and $L_0$ are the magnetic field strength and coherence scale at the present epoch. Compared to the velocities induced by gravitational perturbations ($\sim 10^{-5}$) the velocities induced by helical fields are insignificant.

## 3. Signatures of kinetic helicity

The CMBR anisotropies sourced by velocity flows are predominantly due to the Doppler effect. Observations of CMBR are usually presented in the form of spectral functions $C_l^{X,Y}$, where $X$ and $Y$ stand for $T$ (brightness temperature), $E$ or $B$ (so-called E- or B-type polarization) [9]. The correlators $C_l^{TB}$ and $C_l^{EB}$ are parity-odd, while all other correlators $C_l^{XY}$ are parity-even. The presence of parity-violating (helical) flows will produce nonzero $C_l^{TB}$ and $C_l^{EB}$.

In Ref. [3] we have shown that $C_l^{TB}$ and $C_l^{EB}$ can be written as integral expressions depending on the Fourier transform of the average kinetic helicity $\langle \mathbf{v} \cdot (\nabla \times \mathbf{v})\rangle$. We have assumed a power law $k$-dependence for the relevant power spectra with a spectral index $n$, and introduced a characteristic scale $k_*$ and a characteristic strength $v_0$ of the helical flow. We then evaluated $C_l^{TB}$ and $C_l^{EB}$ for several different values of $n$ and $k_*$ to see if observations could, in principle, constrain $v_0$. We found that for $n > -3$ the bound is set by the cut-off provided by $k_*$. Only for smaller $n$ it is possible to constrain $v_0$.

Causality does, in general, constrain the value of the spectral index: $n \geq 2$. Thus, our analysis suggests that the CMBR will not be able to constrain primordial kinetic helicity unless helical flows were correlated on superhorizon scales.

## 4. A strategy to detect magnetic helicity

From previous sections one concludes that only the non-helical component of the magnetic field can have a signature in the Doppler contribution to the CMBR. If we could find another observable that is sensitive to both the non-helical and the helical components, we could combine observations and extract the helical component of the magnetic field. Such an observable is the Faraday rotation of linearly polarized sources due to light propagation through a magnetized plasma. The CMBR is expected to be linearly polarized and so any intervening magnetic fields will rotate the polarization vector by an angle

$$\theta = \frac{3}{2\pi e}\lambda_0^2 \int \dot{\tau}(\mathbf{x})\, \tilde{\mathbf{B}} \cdot d\mathbf{l} \quad (2)$$

where $\dot{\tau}(\mathbf{x}) \equiv n_e \sigma_T a$ is the differential optical depth along the line of sight, $\lambda_0$ is the observed wavelength of the radiation and $\tilde{\mathbf{B}} \equiv \mathbf{B}a^2$ is the "comoving" magnetic field.

Faraday rotation depends on the free electron density, which becomes negligible towards the end of recombination. Therefore, the bulk of the rotation is produced during a relatively brief period of time when the electron density is sufficiently low for polarization to be produced and yet sufficiently high for the Faraday rotation to occur. The average Faraday rotation (in radians) between Thomson scatterings due to a tangled magnetic field was calculated in Ref. [11] to be $\approx 0.08\left(\frac{B_0}{10^{-9}\mathrm{G}}\right)\left(\frac{30\mathrm{GHz}}{\nu_0}\right)^2$, where $B_0$ is the current amplitude of the field and $\nu_0$ is the radiation frequency observed today. The amplitude of the CMB polarization fluctuations is expected to be an order of magnitude lower than that of the temperature fluctuations. As discussed in Ref. [12], detecting a Faraday rotation of order $1^{\circ}$ will require a measurement which is superior in sensitivity by another factor of $10^2$. Such accuracy is at the limit of current experimental proposals but there is a hope that it will eventually be accomplished.

A polarization map of the CMBR at several wavelengths will (in principle) make it possible to obtain a wavelength independent "rotation measure" (RM). The expression for RM is that of eq. (2) divided by $\lambda_0^2$. One could then use a "rotation measure map" to find correlations of the RM: $RR' \equiv \langle \mathrm{RM}(\hat{\mathbf{n}})\mathrm{RM}(\hat{\mathbf{n}}')\rangle$, where $\hat{\mathbf{n}}$ and $\hat{\mathbf{n}}'$ are two directions on the sky. Using Eq. (1) we find

$$RR' = \left(\frac{3}{2\pi e}\right)^2 \int \frac{d^3k}{(2\pi)^3}\left[\alpha S(k) + \beta A(k)\right], \quad (3)$$

where $\alpha$ and $\beta$ are calculable functions given explicitly in Ref. [3]. A crucial feature of $RR'$ is that it depends on both the helical and non-helical spectral functions $S(k)$ and $A(k)$.

The polarization spectra due to Doppler effect from plasma flows induced by tangled magnetic fields have already been calculated by Seshadri and Subramanian [13]. They computed the correlator $C_l^{BB}$, which only depends on the non-helical spectral function $S(k)$. This is because, as discussed in the previous section, $A(k)$ is the force-free component of the magnetic field and does not induce any velocity in the last scattering surface. Hence, if we could use $C_l^{BB}$ to obtain $S(k)$ – which would only be possible assuming some functional form (such as a power law) for $S(k)$ since $S(k)$ occurs within some integrals – we could insert the result in the expression for $RR'$ given in Eq. (3). This will isolate $A(k)$ in Eq. (3) and, with some assumptions about the functional form of $A(k)$, the cosmic magnetic helicity can be evaluated.

## 5. Summary

If there is kinetic helicity at last scattering, it would imprint a signature in the cross-correlators $C_l^{TB}$ and $C_l^{EB}$. Kinetic helicity can be induced by helical magnetic fields but the effect is too small to be significant since the helical component of magnetic fields is force-free. Instead we have proposed another strategy for detecting the helicity of primordial magnetic fields using polarization and rotation measure maps of the CMBR.

## REFERENCES

1. E. S. Scannapieco and P. G. Ferreira, Phys. Rev. **D56**, 7493 (1997).
2. A. Lue, L. Wang, and M. Kamionkowski, Phys. Rev. Lett. **83**, 1506 (1999).
3. L. Pogosian, T. Vachaspati and S. Winitzki, Phys. Rev. **D65**, 083502 (2002).
4. G. B. Field and S. M. Carroll, Phys. Rev. **D62**, 103008 (2000).
5. J. M. Cornwall, Phys. Rev. **D56**, 6146 (1997).
6. T. Vachaspati, Phys. Rev. Lett. **87** 251302 (2001).
7. T. Vachaspati, astro-ph/0111124, proceedings of the Cosmo-01 workshop.
8. A. Achúcarro and T. Vachaspati, Phys. Rep. **327**, 347 (2000).
9. W. Hu and M. White, Phys. Rev. **D56**, 596 (1997).
10. E. R. Harrison, Mon. Not. R. Astr. Soc. **147**, 279 (1970).
11. D. D. Harari, J. D. Hayward, and M. Zaldarriaga, Phys. Rev. **D55** (1997).
12. A. Kosowsky and A. Loeb, Ap. J. **469**, 1 (1996).
13. T. R. Seshadri and K. Subramanian, Phys. Rev. Lett. **87**, 101301 (2001).

Proceedings of ICHEP 2002, pp. 135–138
*S. Bentvelsen, P. de Jong, J. Koch and E. Laenen (Editors)*

ELSEVIER

31st INTERNATIONAL CONFERENCE ON HIGH ENERGY PHYSICS AMSTERDAM

# Inflationary cosmology - a dissipative quantum field theory process

Arjun Berera[a]*

[a]School of Physics, University of Edinburgh, Edinburgh, EH9 3JZ, United Kingdom

The current status of early universe cosmology can be summarized as precision, but inherently limited, data requiring explanation from precision and robust theory. Specifically inflation is presently the single concrete hope for solving the early universe problem, since it is generically realizable in quantum field theory, but treatment of interactions needs considerable attention. This talk reviews the importance and progress in this direction.

## 1. Introduction

Early universe cosmology is a subject with limited access to observation, since there is only so much information that can be acquired about this time period from our one local region in the universe. As such, a convincing theory about this time period must rely substantially on the soundness and unambiguity of its mathematical foundation. At present, inflation is the most hopeful idea for developing into a theory, since it is based on quantum field theory, and quantum fields so far have been exclusively successful in explaining the high energy world, up to at least the TeV scale. However, consistent solutions of inflation founded on quantum field theory have been allusive. Inflationary dynamics inherently is a multifield problem, since the vacuum energy that drives inflation eventually must convert to radiation, which generally is comprised of a variety of particle species. In the earliest conception of inflation [1], it was pictured that inflation would result in a isentropic expansion that would rapidly put the universe in a supercooled thermodynamic phase. Subsequently it was observed that supercooled inflation is not mandatory and that nonisentropic inflationary expansion, warm inflation, also is possible [2]. Moreover, since the main condition for inflation is $\rho_v > \rho_r$, where $\rho_v$,$\rho_r$ are the vacuum and radiation energy densities respectively, supercooled inflation appears as a limiting case within the general regime of warm inflation.

The most nontrivial aspect of the inflaton model is understanding the energy transfer dynamics from potential energy to radiation. In the most commonly followed inflation picture, dissipative effects of the inflaton field are ignored throughout the inflation period, leading to a supercooled inflationary regime. However, from a thermodynamic perspective, this picture appears very restrictive. The point being, even if the inflaton were to allow a minuscule fraction of the energy to be released, say one part in $10^{20}$, it still would constitute a significant radiation energy density component in the universe. For example, for inflation with vacuum (i.e. potential) energy at the GUT scale $\sim 10^{15-16}$GeV, leaking one part in $10^{20}$ of this energy density into radiation corresponds to a temperature of $10^{11}$GeV, which is nonnegligible. In fact, the most relevant lower bound that cosmology places on the temperature after inflation comes from the success of hot Big-Bang nucleosynthesis, which requires the universe to be within the radiation dominated regime by $T \gtrsim 1$GeV. This limit can be met in the above example by dissipating as little as one part in $10^{60}$ of the vacuum energy density into radiation. Thus, from the perspective of both interacting field theory and basic notions of equipartition, it appears to be a highly tuned requirement of supercooled inflation to prohibit the inflaton from such tiny amounts of dissipation.

These considerations have led to examining the possibility of warm inflation, an inflationary regime in which radiation also is present. Warm inflation is comprised of non-isentropic expansion in the background cosmology [2] and ther-

*PPARC Advanced Fellow


doi:10.1016/S0920-5632(02)02063-7

mal seeds of density perturbations [3,4]. During warm inflation, interactions between the inflaton and other fields cause the radiation energy density to remain substantial due to its constant production from conversion of vacuum energy. This expansion regime is intrinsically different from the supercooled inflation regime, since warm inflation smoothly terminates into a subsequent radiation dominated regime, without a reheating period.

The warm inflation picture has one immediate conceptual advantage, that the dynamics is free of questions about quantum-to-classical transition. The scalar inflaton field is in a classical state, thus justifying the application of a classical evolution equation, and the inflaton fluctuations, which induce the metric perturbations, are classical. Another problem alleviated by dissipative effects is the initial condition problem of inflation [5].

Warm inflation continues proving to be a promising solution to the early universe problem, with now several variants and perspectives on the basic idea [4,6]. Development of it has had two directions, which will be addressed in this talk. First is deriving a first principle quantum field theory realization of its dynamics. Second is determining observational tests for warm inflation.

## 2. First Principles Origin

The earliest works looked for high temperature warm inflation solutions, under rigid adiabatic, equilibrium conditions [7]. Within this limited framework, one type of warm inflation solution was obtained [7]. The high-T regime was examined first, since considerable methodology was already available for treating it. However, intrinsically, the statistical state relevant for warm inflation is not required to be an equilibrium state. The slowly varying nature of the macroscopic variables in warm inflation cosmology suggest that the statistical state may not be far from equilibrium, although this is something that should be proven from the dynamics. Much work remains in order to develop the mathematical formalism necessary to address this problem. As one step in this direction to fill the missing gaps, recently we studied the zero temperature dissipative dynamics of interacting scalar field systems in Minkowski spacetime [8] (for another interesting direction see [9]). This is useful to understand, since the zero temperature limit constitutes a baseline effect that will be prevalent in any general statistical state. The key result presented in this talk is that for a broad range of cases, involving interaction with as few as one or two fields, we find dissipative regimes for the scalar field system. This is important for inflationary cosmology, since it suggests that dissipation may be the norm not exception for an interacting scalar field system, thus warm inflation could be a natural dynamics once proper treatment of interactions is done.

We study the Lagrangian, where

$$\mathcal{L}_I = -\frac{\lambda}{4!}\Phi^4 - \sum_{j=1}^{N_\chi}\left\{\frac{f_j}{4!}\chi_j^4 + \frac{g_j^2}{2}\Phi^2\chi_j^2\right\} - \sum_{k=1}^{N_\psi}\bar{\psi}_k\left[h_{k\phi}\Phi + \sum_{j=1}^{N_\chi}h_{kj\chi}\chi_j\right]\psi_k \quad (1)$$

and $\Phi \equiv \varphi + \phi$ such that $\langle\Phi\rangle = \varphi$. Our aim is to obtain the effective equation of motion for $\varphi(t)$ and from that determine the energy dissipated from the $\varphi(t)$ system into radiation.

The tadpole method, which requires $\langle\phi\rangle = 0$, gives effective equation of motion for $\varphi(t)$,

$$\ddot{\varphi}(t) + m_\phi^2\varphi(t) + \frac{\lambda}{6}\varphi^3(t) + \frac{\lambda}{2}\varphi(t)\langle\phi^2\rangle + \frac{\lambda}{6}\langle\phi^3\rangle + \sum_{j=1}^{N_\chi} g_j^2\left[\varphi(t)\langle\chi_j^2\rangle + \langle\phi\chi_j^2\rangle\right] + \sum_{k=1}^{N_\psi} h_{k\phi}\langle\bar{\psi}_k\psi_k\rangle = 0 \,. \quad (2)$$

The field expectation values in this equation are obtained by solving the coupled set of field equations. In our calculation, we have evaluated them in a perturbative expansion using dressed Green's functions. One general feature of these expectation values is they will depend of the causal history of $\varphi(t)$, so that Eq. (2) is a temporally nonlocal equation of motion for $\varphi(t)$.

The general expression for the effective equation of motion is given in [8] and is very complicated. Formally, we can examine Eq. (2) within a Markovian-adiabatic approximation, in which the equation of motion is local in time and the motion of $\varphi(t)$ is slow. At $T = 0$, such an approximation is not rigorously valid. Nevertheless, this approximation allows understanding the magnitude of dissipative effects. Furthermore, we have shown in [8] that the nonlocal effects tend to filter only increasingly higher frequency components of $\varphi(t)$ from nonlocal effects increasingly further back in time. Thus for low frequency components of $\varphi(t)$, memory only is retained to some short interval in the past. Since within the adiabatic approximation, $\varphi(t)$ only has low frequency components, we believe the Markovian-adiabatic approximation is legitimate at least for order of magnitude estimates. Within this approximation, the effective equation of motion for $\varphi(t)$ has the general form

$$\ddot{\varphi}(t) + m_\phi^2\, \varphi(t) + \frac{\lambda}{6}\varphi^3(t) + \eta(\varphi)\dot{\varphi}(t) = 0\ , \qquad (3)$$

where

$$\eta(\varphi) \approx \varphi^2(t)\frac{N_\psi}{256\pi^2}\left(\frac{\lambda^2 h_{\phi\psi}^2}{4m_\phi}. + \frac{N_\chi g^4 h_{\chi\psi}^2}{m_\chi}\right).$$

An alternative to the above Lagrangian based derivation, is the canonical derivation following formalism developed in the mid 80's [10]. Although the canonical and Lagrangian approaches should agree, the former is far less developed in dissipative quantum field theory, in particular for treating interactions. Nevertheless, the canonical approach provides useful insight, especially for understanding the origin of particle creation. For example, consider in the canonical approach the expectation value of $\langle\chi_i^2\rangle$,

$$\int \frac{d^3q}{(2\pi)^3 2\mathrm{Re}\omega_{\mathbf{q},\chi_j}}\left[2x_{\mathbf{q},\chi_j}(t) + 2\mathrm{Re}y_{\mathbf{q},\chi_j}(t) + 1\right],$$

where $x_{\mathbf{q},\chi_j}(t) = \langle a^\dagger_{\mathbf{q},\chi_j}(t)a_{\mathbf{q},\chi_j}(t)\rangle$ is the particle number density and $y_{\mathbf{q},\chi_j}(t) = \langle a_{\mathbf{q},\chi_j}(t)a_{-\mathbf{q},\chi_j}(t)\rangle$ is the off-diagonal correlation. The evolution equations for $x_{\mathbf{q},\chi_j}(t)$ and $y_{\mathbf{q},\chi_j}(t)$ can be obtained from the field equation for $\chi_j$ to give

$$\begin{aligned}\dot{x}_{\mathbf{q},\chi_j} &= \frac{\dot{\omega}_{\mathbf{q},\chi_j}}{\omega_{\mathbf{q},\chi_j}}\mathrm{Re}\,y_{\mathbf{q},\chi_j}\ ,\\ \dot{y}_{\mathbf{q},\chi_j} &= \frac{\dot{\omega}_{\mathbf{q},\chi_j}}{\omega_{\mathbf{q},\chi_j}}\left[x_{\mathbf{q},\chi_j} + \frac{1}{2}\right] - 2i\omega_{\mathbf{q},\chi_j}y_{\mathbf{q},\chi_j}\ . \qquad (4)\end{aligned}$$

To yield dissipation, it is noted in [10] that the correlation amongst produced particles needs to be destroyed sufficiently rapidly and $\dot{x}_{\mathbf{q},\chi_j}(t)$ and $\langle\chi_j(t)\rangle$ should become local functions of time involving $\varphi(t)$ and $\dot{\varphi}(t)$. Based on this requirement, [10] asserts that $\omega_{\mathbf{q},\chi_j}$ in the above equation for $\dot{y}_{\mathbf{q},\chi_j}(t)$ must have an imaginary part, which in fact should be the $\chi_j$-particle decay width, $\mathrm{Im}\omega_{\mathbf{q},\chi_j} \propto \Gamma_{\mathbf{q},\chi_j}$. Applying these assumptions, the contribution to $\eta(\varphi)$ in Eq. (3) from the $\chi_j$ field, say $\eta_{\chi_j}(\varphi)$, once again is obtained (up to O(1) factors). Since this approach has ad-hoc assumptions, it still is incomplete and requires development. Nevertheless, the approach is interesting and for now accepting the assumptions, the origin of particle creation and energy conservation are clearly seen. In particular, the particle production rate is given by $\int(d^3k/(2\pi)^3)\dot{x}_{\mathbf{q},\chi_j}(t)\omega_{\mathbf{q},\chi_j}$ and similar to [10], it can be shown this is equal to the vacuum energy loss rate from the $\chi_j$ field contribution, $\eta_{\chi_j}(\varphi)\dot{\varphi}^2$.

Returning to Eq. (3), estimated magnitudes of energy production will be obtained in the overdamped regime of warm inflation, $m^2(\phi) = m_\phi^2 + \lambda\varphi^2/2 < \eta^2$. In this regime, the energy dissipated by the scalar field goes into radiation energy density $\rho_r$ at the rate $\dot{\rho}_r = -dE_\phi/dt = \eta(\varphi)\dot{\varphi}^2$.

In [8] we have determined radiation production for two cases (a). $m(\varphi) > m_\chi > 2m_\psi$, (b). $m_\chi > 2m_\psi > m(\varphi)$. To focus on a case typical for inflation, suppose the potential energy is at the GUT scale $V(\varphi)^{1/4} \sim 10^{15}$GeV and we consider the other parameters in a regime consistent with the e-fold and density fluctuation requirements of inflation. Note, although this is a flat nonexpanding spacetime analysis, since the dissipative effects will be at subhorizon scale, one expects these estimates to give a reasonable idea of what to expect from a similar calculation done in expanding spacetime. Expressing the radiation in terms of a temperature scale as $T \sim \rho_r^{1/4}$, we find for case (a) $1\mathrm{GeV} < T < 10^7\mathrm{GeV} < H$ and for case (b) $T \gtrsim 10^{14}\mathrm{GeV} > H$, where $H = \sqrt{8\pi V/(3m_p^2)}$.

## 3. Observational Tests

Supercooled inflation has three parameters, related to the potential energy magnitude $V_0$, slope $\epsilon$, and curvature $\eta$, whereas there are four observable constraints ($\delta_H$, $A_g$, $n_s$, $n_g$). This implies a redundancy in the observations and allows for a consistency relation [11]. This is usually expressed as a relationship between the tensor-to-scalar ratio and the slope of the tensor spectrum. Warm inflation has an extra parameter, the dissipation factor, which implies four constraints for four parameters. Hence we do not expect the consistency relation of standard inflation to hold in warm inflation [12]. Thus, to discriminate between warm and standard inflation, it requires a measure of all four observables. The upcoming MAP and Planck satellite missions should provide strong constraints on the scalar spectrum and being equipped with polarization detectors, it is hoped the tensor spectrum also will be measured. Recently nongaussian effects from warm inflation models were computed and found to be too small to measure [13].

## 4. Conclusion

The key point emphasized in this talk has been that any typical inflaton field theory model has interactions and so dissipation can be an unexceptional consequence. These dissipative effects have always been ignored, unjustifiably, in supercooled inflation, but as shown here their effects can be nontrivial. The most significant effect of dissipation is to completely alter the inflationary regime from supercooled to warm. However, irrespective of the inflation regime, as discussed above and in [5], dissipative effects before the inflation period may be important in alleviating the initial condition problem. Also, dissipative effects imply new realizations of thermal inflation [14]. Moreover, dissipative effects can alter density perturbations. In this talk we discussed the case where dissipation leads to an ideal thermalized state, thus with thermal density perturbations. However the opposite limit can be considered from ideal thermalization to that of negligible influence from the radiation field on the density perturbations, so that only the quantum fluctuations contribute. Then simple calculations show warm inflation models would *not* require fine tuning, so that scalar potentials of reasonably large curvature would be acceptable. The nature of density perturbations during warm inflation still requires much investigation in particular to understand the interplay between radiation and the scalar field. In conclusion, this talk discussed how dissipation during inflation could be robust and possibly solve inflation's most challenging dilemma, the scalar field fine tuning problem.

## REFERENCES

1. A. H. Guth, Phys. Rev **D23** (1981) 347; K. Sato, Phys. Lett. B **99** (1981) 66; A. Linde, Phys. Lett. B**108** (1982) 389; A. Albrecht and P. J. Steinhardt, Phys. Rev. Lett. **48** (1982) 1220.
2. A. Berera, Phys. Rev. Lett. **75** (1995) 3218; Phys. Rev. D **54** (1996) 2519; Phys. Rev. D **55** (1997) 3346.
3. A. Berera and L. Z. Fang, Phys. Rev. Lett. **74** (1995) 1912; A. Berera, Nucl. Phys. B **585** (2002) 666.
4. H. P. De Oliveira and S. E. Joras, Phys. Rev. D **64** (2001) 063513; H. P. De Oliveira, Phys. Lett. B**526** (2002) 1; L. P. Chimento *et. al.*, Phys. Rev. D **65** (2002) 083510.
5. A. Berera and C. Gordon, Phys. Rev. D **63** (2001) 063505.
6. H. P. de Oliveira and R. O. Ramos, Phys. Rev. **D57** (!998) 741; J. M. F. Maia and J. A. S. Lima, Phys. Rev. **D60** (1999) 101301; W. Lee and L.-Z. Fang, Phys. Rev. **D59** (1999) 083503.
7. A. Berera, M. Gleiser and R. O. Ramos, Phys. Rev. D **58** (1998) 123508; Phys. Rev. Lett. **83** (1999) 264.
8. A. Berera and R. O. Ramos, Phys. Rev. D **63** (2001) 103509.
9. I. D. Lawrie, Phys. Rev. D **66** (2002) 041702.
10. M. Morikawa and S. Sasaki, Prog. Theor. Phys. **72** (1984) 782.
11. A. R. Liddle and D. H. Lyth, Phys. Rept. **231** (1993) 1.
12. A. N. Taylor and A. Berera, Phys. Rev. D **62** (2000) 083517.
13. S. Gupta *et. al.*, Phys. Rev. D**66** (2002) 043510.
14. A. Berera and T. W. Kephart, hep-ph/0207208.

ELSEVIER

Proceedings of ICHEP 2002, pp. 139–141
*S. Bentvelsen, P. de Jong, J. Koch and E. Laenen (Editors)*

31st INTERNATIONAL CONFERENCE ON HIGH ENERGY PHYSICS AMSTERDAM

# Cosmic Solutions in the Einstein-Weinberg-Salam Theory and the Generation of Large Electric and Magnetic Fields

Yutaka Hosotani[a], Hiroki Emoto[b], and Takahiro Kubota[a]

[a]Department of Physics, Osaka University, Toyonaka, Osaka 560-0043, Japan

[b]High Energy Accelerator Research Organization (KEK), Tsukuba, Ibaraki 305-0801, Japan

In the $SU(2)_L \times U(1)_Y$ standard electroweak theory coupled with the Einstein gravity, new topological configurations naturally emerge, if the spatial section of the universe is globally a three-sphere ($S^3$) with a small radius. The $SU(2)_L$ gauge fields wrap the space nontrivially, producing homogeneous but anisotropic space. As the universe expands, large electric and magnetic fields are produced. The electromagnetic field configuration is characterized by the Hopf map.

## 1. STANDARD MODEL IN GRAVITY

It has been firmly established that the standard model of electroweak interactions based on the $SU(2)_L \times U(1)_Y$ gauge symmetry describes the Nature at low energies, and the evolution of the universe is well described by Einstein's theory of gravity. Given these two facts, it is natural to ask if there was strong interplay between the Einstein gravity and electroweak interactions in the early universe. We report that indeed such strong interplay may have existed, provided the universe is spatially closed.[1].

## 2. TOPOLOGICAL CONFIGURATIONS

The Lagrangian density is given by

$$\mathcal{L} = \frac{1}{16\pi G}(R - 2\Lambda) - \frac{1}{4}F^a_{\mu\nu}F^{a\mu\nu} - \frac{1}{4}G_{\mu\nu}G^{\mu\nu} - (D_\mu\Phi)^\dagger(D^\mu\Phi) - \lambda\Big(\Phi^\dagger\Phi - \frac{v_0^2}{2}\Big)^2$$

where $D_\mu\Phi = (\partial_\mu - i\frac{g}{2}\tau^a A^a_\mu - i\frac{g'}{2}B_\mu)\Phi$. $\Lambda$ is the effective cosmological constant at the time the universe was wrapped with nontrivial field configurations discussed below.

Recall that the metric of an unit three-sphere $S^3$ can be written as $d\Omega_3{}^2 = \sum_{j=1}^3 \sigma^j \otimes \sigma^j$ where the $SU(2)$ Maurer-Cartan 1-forms $\sigma^j = \sigma^j{}_\mu dx^\mu$ satisfy $d\sigma^j = \epsilon^{jkl}\sigma^k \wedge \sigma^l$. The metric of the spacetime we consider is

$$ds^2 = -e^0 \otimes e^0 + \sum_{j=1}^3 e^j \otimes e^j$$

$$e^0 = dt \ , \ e^j = a_j(t)\sigma^j \ . \tag{1}$$

It gives a homogeneous but anisotropic space. The $SU(2)$ and $U(1)$ gauge fields and the Higgs fields take

$$A = \frac{1}{2g}\sum_{j=1}^3 f_j(t)\sigma^j\tau^j \quad , \quad \Phi = \frac{1}{\sqrt{2}}\begin{pmatrix} 0 \\ v(t) \end{pmatrix}$$

$$B = h(t)\sigma^3 \tag{2}$$

where $\tau^j$'s are Pauli matrices. The Ricci tensors and the energy-momentum tensors in these tetrads become diagonal and independent of the spatial coordinates. Both the Einstein equations and field equations are consistently solved.

The spatially isotropic universe where all scale factors $a_j(t)$'s are equal to each other is not permitted in the presence of the $U(1)$ gauge interactions. It still is legitimate to set $a_1 = a_2$ and $f_1 = f_2$ when the Higgs field and the $U(1)$ gauge field are aligned as in (2). The ansatz of this type has been investigated before in the pure Einstein-Yang-Mills theory in [2].


doi:10.1016/S0920-5632(02)02066-2

## 3. LOCAL MINIMUM OF THE POTENTIAL

The configuration (2) is topological in the sense that the $SU(2)$ gauge fields wrap the space non-trivially if $a_j(t)$'s are sufficiently small. However it is not absolutely stable as it can be continuously deformed to the trivial configuration $A = B = 0$. To see how it may become important we insert the ansatz (2) into the potential:

$$V = \frac{\lambda}{4}(v^2 - v_0^2)^2 + \frac{v^2}{8}\left\{\frac{2f_1^2}{a_1^2} + \frac{(f_3 - g'h)^2}{a_3^2}\right\}$$
$$+\frac{(2f_3 - f_1^2)^2}{2g^2a_1^4} + \frac{2f_1^2(f_3-2)^2}{2g^2a_1^2a_3^2} + \frac{2h^2}{a_1^4} \quad . \qquad (3)$$

The global minimum of the potential is always located at $v = v_0$, $f_1 = f_3 = h = 0$. When the scale factors $a_1$ and $a_3$ are sufficiently small, there appears a new local minimum located at $1.5 < f_1, f_3 \le 2$ and $0 \le v < v_0$.

It could well be that the very early universe was settled near the local minimum of the potential. The local minimum has positive energy density so that the Einstein equations drive the universe to expand. As $a_j$'s become larger, the terms of $O(a^{-2})$ and $O(a^{-4})$ in $V$ become less important. The barrier separating the local minimum from the global minimum quickly disappears.

## 4. ELECTROMAGNETIC FIELDS

After the barrier disappears, the field configuration starts to roll down the hill of the potential. It is in this period of the evolution of the universe that some of interesting physical consequences emerge. One salient feature of the electroweak theory is that there remains the electromagnetic $U(1)_{\rm EM}$ gauge invariance after the Higgs field develops a nonvanishing vacuum expectation value, i.e. $v \neq 0$. The electromagnetic fields are given by

$$F_{\rm EM} = \frac{\dot{h}_{\rm EM}}{a_3}e^0\wedge e^3 + \frac{2h_{\rm EM}}{a_1^2}e^1\wedge e^2 \quad ,$$
$$h_{\rm EM}(t) = \frac{1}{\sqrt{g^2+g'^2}}\left\{gh + g'\frac{f_3}{g}\right\} \quad . \qquad (4)$$

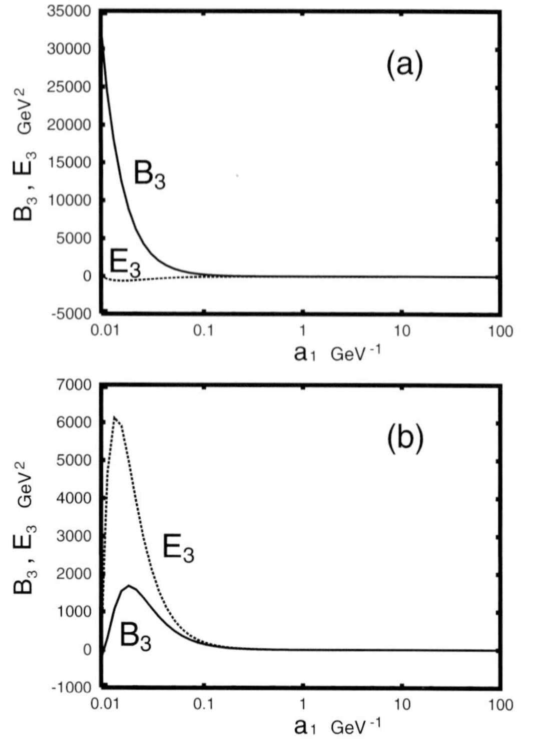


Figure 1. Generated electromagnetic fields as functions of $a_1(t)$. $\Lambda = 1.0\times10^5{\rm GeV}^2$ and $a_j(t_0) = 9\times10^{-3}{\rm GeV}^{-1}$. (a) The fields started from the local minimum of the potential at $t_0$. (b) Initially $B_3 = E_3 = 0$.

In the potential (3), the terms of $O(a^{-4})$ quickly become irrelevant as $a_j$'s increase. The terms of $O(a^{-2})$ are independent of $h_{\rm EM}$, reflecting the $U(1)_{\rm EM}$ invariance. In other words, the potential has approximately flat direction along $f_1 = f_3 - g'h = 0$. During the expansion the field configuration heads for this shallow valley of the potential, but not necessarily for the absolute minimum. $h_{\rm EM}$ approaches a nonvanishing value, producing large electromagnetic fields. Typical behavior is depicted in fig. 1.

A typical value for the generated magnetic fields is about $(50{\rm GeV})^2$. It dies away as $a(t)^{-2}$ as the universe further expands. The final value of $h_{\rm EM}$ depends on the initial condition, but the fact

$h_{\rm EM}(t=\infty) \neq 0$ does not. In fig. 2 the final values of $h_{\rm EM}$ and $f_3$ are plotted as functions of the initial value of $h_{\rm EM}$. The figure shows how generic it is to generate large electromagnetic fields.

## 5. FIELDS AS THE HOPF MAP

As is evident in eq. (4), the magnitude of the generated electromagnetic fields is independent of the spatial position. Both the electric and magnetic fields are in the $e^3 = a_3\sigma^3$ direction where $\sigma^3$ varies as the spatial position. They define smooth, regular vector fields without any singularity.

The space is topologically isomorphic to $S^3$. At each instant $t$, the vector $\vec{E}(x)$ or $\vec{B}(x)$ is on $S^2$ specified in the field space by $|\vec{E}|$ or $|\vec{B}|$ =constant. The vector field defines a map from $S^3$ onto $S^2$. It is nothing but the Hopf map.

This is how it becomes possible to have spatially homogeneous universe with nonvanishing $U(1)$ gauge field strengths. If we were living in two-dimensional space, the gauge field strengths with constant magnitude would necessarily induce singularities, either in the form of sources/sinks or vortices.

## 6. A COLD ERA OF THE EARLY UNIVERSE

When does the field configuration (2) become important in the history of the universe? In the standard scenario of the early universe, the temperature effect plays an important role. It modifies the effective potential and the time-evolution. What we have in mind here is an era preceding the hot universe which continuously evolves to the current universe. Suppose that at one instant the universe was very small and cold, and the gauge and Higgs fields assumed the nontrivial configuration under discussions. Driven by an effective cosmological constant the universe underwent inflation, and large electromagnetic fields were generated. Eventually the inflation stopped and the universe was reheated to the temperature about $(\Lambda/8\pi G)^{1/4}$. The universe continued to expand by the radiation dominance since then.

When large electromagnetic fields were generated, quarks and leptons must have been copiously created.[3] As the field configuration was not invariant under $CP$, there may have been asymmetry in the baryon and lepton numbers as well. Such asymmetry could be washed out in the later stage of the evolution of the universe, but a part of it may have survived.

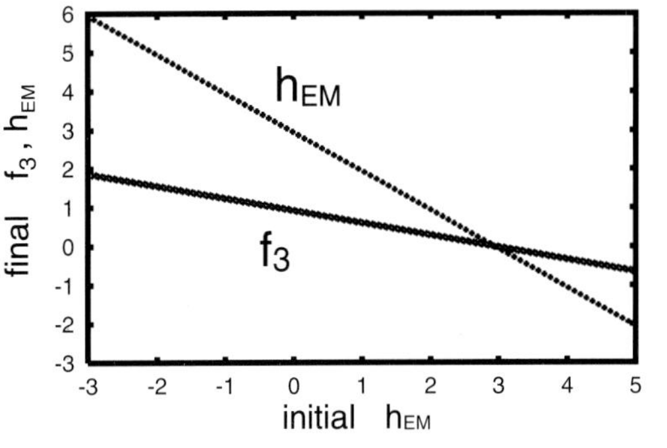


Figure 2. The final $h_{\rm EM}$ and $f_3$ v.s. the initial $h_{\rm EM}$. $\Lambda = 1\times 10^6$ GeV$^2$ and $a_j(t_0) = \sqrt{3/\Lambda}$.

In this report we have assumed the homogeneity of the space. In reality inhomogeneity has to be generated either classically or quantum mechanically. The generated magnetic flux, which is taken to be homogeneous in our ansatz (2), may be squeezed in space. Could such magnetic flux have remnant in the present universe?[4] We do not have an answer at the moment.

## REFERENCES


1. H. Emoto, Y. Hosotani and T. Kubota, Prog. Theoret. Phys. 108 (2002) 157.
2. Y. Hosotani, Phys. Lett.B147 (1984) 44; A. Hosoya and W. Ogura, Phys. Lett. B225 (1989) 117; S.J. Rey, Nucl. Phys. B336 (1990) 146; M.S. Volkov and D.V. Gal'tsov, Phys. Rep. 319 (1999) 141.
3. G.W. Gibbons and A.R. Steif, Phys. Lett. B320 (1994) 245.
4. J.D. Barrow, Phys. Rev. D55 (1997) 7451; A.D Dolgov, hep-ph/0110293.

## Session 4

# Electroweak Physics

*Session conveners:*
*J. Hobbs, SUNY at Stony Brook*
*D. Bardin, JINR, Dubna*

ELSEVIER

Proceedings of ICHEP 2002, pp. 145–148
*S. Bentvelsen, P. de Jong, J. Koch and E. Laenen (Editors)*

31st INTERNATIONAL CONFERENCE ON HIGH ENERGY PHYSICS AMSTERDAM

# Measurement of the W Boson Mass and Width at LEP2

Jeremy Nowell[a]*

[a]Blackett Laboratory, Imperial College, London SW7 2BZ
email: Jeremy.Nowell@cern.ch

The measurement of the mass of the W boson at the LEP2 collider is described, in particular the technique of direct reconstruction. Preliminary results are presented using data collected at centre of mass energies up to 207 GeV.

## 1. Introduction

### 1.1. Motivation

The mass of the W boson is a central component of the Standard Model of particle physics. It can be determined indirectly by a global fit to electroweak data using the following relationship:

$$M_{\mathrm{W}}^2\left(1-\frac{M_{\mathrm{W}}^2}{M_{\mathrm{Z}}^2}\right)=\frac{\pi\alpha}{G_F\sqrt{2}}\left(1+\Delta r\right)\ , \qquad (1)$$

where $\Delta r$ includes the effect of higher order corrections depending logarithmically on the mass of the Higgs boson and quadratically on the mass of the top quark. The current indirect measurement of the W mass has an uncertainty of 23 MeV/$c^2$ [1]. A major goal of the LEP2 physics programme is to make a direct measurement of the W mass to a similar level of precision as the indirect measurement, thus providing a test of the Standard Model and allowing the Higgs mass to be constrained through the effect of the radiative corrections.

### 1.2. W Pair Production at LEP2

At LEP2 pairs of W bosons were produced by electron positron annihilation, proceeding at tree level through the CC03 diagrams, as shown in Figure 1. The first two of these diagrams are the *annihilation* diagrams, with either a $Z^0$ boson or a photon produced in the $s$-channel, while the third diagram is the *conversion* diagram with the exchange of a neutrino in the $t$-channel. Each W then decays to either a quark and antiquark or lepton and associated neutrino. There are

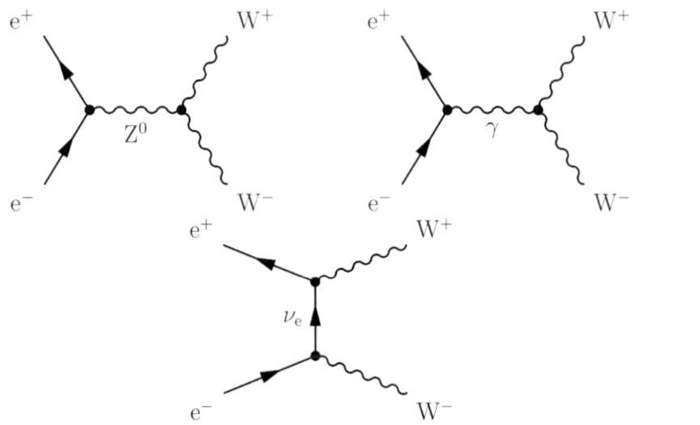


Figure 1. The CC03 diagrams of W pair production.

thus three distinct decay channels; fully-hadronic, semi-leptonic and fully-leptonic, with approximate branching ratios of 46%, 44% and 10% respectively.

### 1.3. LEP2 Programme

The LEP2 physics programme started in 1996 when the centre of mass energy of the collider was increased from the $Z^0$ resonance past the threshold for W pair production, 161 GeV. Although only a small amount of data was collected at this luminosity, about 10 $\mathrm{pb}^{-1}$ per experiment, a measurement of the W mass was made using the sensitivity of the WW cross section at this energy to the W mass [2]. In subsequent years the energy of the collider was steadily increased until a value of 207 GeV was reached at the end of LEP data taking in the year 2000. Each of the four LEP ex-

*On behalf of the LEP collaborations.


doi:10.1016/S0920-5632(02)02071-6

periments, ALEPH, DELPHI, L3 and OPAL has collected approximately 700 $pb^{-1}$ of data above the WW threshold, corresponding to over 10000 WW events per experiment.

## 2. W Mass Measurement

The four LEP collaborations have used, to one degree or another, differing techniques to measure the W mass at LEP2. In the short space available here only a brief overview can be given. More details can be found in References [3–6]. Above the threshold the main technique used to measure the W mass is that of direct reconstruction. The general procedure is as follows:

- Select WW events.
- Reconstruct W's from decay products.
- Form invariant mass distribution.
- Extract W mass from this distribution.

Each of these steps will now be described in more detail.

### 2.1. Event Selection

The event selection takes into account the characteristics of each individual decay channel, for example semi-leptonic events are typified by two hadronic jets, one isolated, energetic lepton and missing energy due to the presence of the neutrino. Fully-hadronic events have at least four hadronic jets; separation from the QCD $e^+e^- \to q\bar{q}(\gamma)$ background is typically achieved using multivariate algorithms such as neural networks.

### 2.2. Invariant Mass Reconstruction

Once W pair candidate events have been selected the next step is to reconstruct the individual W bosons from their decay products and then calculate their invariant mass.

After identification of the lepton in semi-leptonic events, the next step is jet clustering, in which the remaining particles are clustered to form two hadronic jets. In fully hadronic events the particles are clustered into four jets, although some analyses allow the possibility of five jets to allow for the chance of hard gluon radiation taking place.

The most important part of the reconstruction is the use of a kinematic fit which significantly improves the invariant mass resolution. Since LEP was an electron-positron collider the centre of mass energy and momentum are known with a high degree of precision, The kinematic fit exploits this knowledge by imposing the energy and momentum constraints on the final state particles. The additional constraint of equal W boson masses may also be applied in the kinematic fit.

In the fully hadronic channel the jets have to be paired to form the two W bosons due to the three-fold combinatorial ambiguity. Several different algorithms are used to do this.

After the kinematic fit and jet pairing the invariant mass may be formed:

$$M_{\mathrm{W}}^2 = \left(\sum_{i=1}^{n} E_i\right)^2 - \left(\sum_{i=1}^{n} \mathbf{p}_i\right)^2 , \qquad (2)$$

where the sum is carried out over all particles assigned to the W being considered. It is from the distribution of the invariant mass, together with other mass related variables that the W mass is extracted.

### 2.3. W Mass Extraction

Two main methods are used for the extraction of the W mass from the invariant mass distribution. The first is the technique of Monte Carlo reweighting, where Monte Carlo simulated events generated at a fixed W mass are weighted using a ratio of CC03 matrix elements until their invariant mass distribution best fits that of the data. The second method is to fit the invariant mass distribution using a function formed by the convolution of a Breit-Wigner type function with the detector resolution. This technique requires calibration with Monte Carlo. The W width can be extracted from the invariant mass distribution in the same way.

## 3. Systematic Uncertainties

The stage has now been reached where the total systematic uncertainty on the LEP combined measurement of the W mass is of the same order as the statistical uncertainty, therefore much study of the major systematic uncertainties is ongoing. The systematic uncertainties may be divided into two categories, those that are taken as correlated between experiments and those that are uncorrelated.

The uncertainties which are uncorrelated between experiments include background contamination of the data sample and effects related to the experimental detectors, such as uncertainties in the calorimeter energy scale. The uncertainty due to background contamination is typically small and that due to detector related effects is $\sim 10$ MeV$/c^2$.

Uncertainties that are taken to be fully correlated between all decay channels and experiments include those due to the modelling of the initial state radiation and fragmentation, and the uncertainty to the measurement of the LEP beam energy.

The uncertainty due to the modelling of the fragmentation is the largest arising from the analysis. The current estimate of 18 MeV$/c^2$ is obtained by comparing different fragmentation models, such as those in the JETSET [7], HERWIG [8] and ARIADNE [9] Monte Carlos. It is expected that this will be reduced by a careful comparison of data and Monte Carlo simulation, e.g. in the rate of Baryon production.

The value of the LEP beam energy, $E_{beam}$, is used in the kinematic fit, so the uncertainty in its measurement translates directly into an uncertainty on the measured W mass:

$$\Delta M_{\rm W} = M_{\rm W} \frac{\Delta E_{beam}}{E_{beam}} \,. \qquad (3)$$

The current estimate of 21 MeV$/c^2$ [10] gives an uncertainty on the W mass of 17 MeV$/c^2$.

There are two sources of possible systematic uncertainty which affect the fully-hadronic decay channel only. These are the final state interactions (FSI); colour reconnection and Bose-Einstein correlations. The decay of the two W bosons takes place over a much shorter space time distance, $\sim 0.1$ fm, than the hadronisation scale, $\sim 1$ fm. Therefore it is expected that the hadronisation of each W is not independent, with possible colour strings being formed between the partons from different W's, leading to a possible shift in the reconstructed mass compared to the Monte Carlo simulation.

It is predicted that Bose-Einstein effects exist between identical bosons (pions and kaons) from different W boson decays, possibly leading to a shift in the reconstructed mass. The uncertainty due to the full BE effect is estimated to be $\sim 35$ MeV$/c^2$, although dedicated studies [11] suggest that the observed BE effect is small, which would lead to a reduction in the uncertainty on the W mass.

Current dedicated studies of colour reconnection [12] place limits on the SK-I model [13], leading to an upper limit on the uncertainty on the W mass of 90 MeV$/c^2$. Since colour reconnection is expected to affect low momentum particles in the interjet region studies are ongoing to try and reduce the uncertainty on the W mass by removing these particles from the reconstruction by, e.g., applying a momentum cut or using a cone jet clustering algorithm [14]. These studies promise to reduce the uncertainty significantly.

## 4. Results

The results presented here are preliminary, and are produced by the LEP W Working Group [15]. The combination of the W mass measurements of the four LEP experiments gives a value of

$$M_{\rm W} = 80.447 \pm 0.030(stat) \pm 0.029(syst)\ {\rm GeV}/c^2, \qquad (4)$$

as illustrated in Figure 2. The difference in the semi-leptonic and fully hadronic masses can be obtained in a fit neglecting the systematic uncertainties due to final state interactions, giving a value of

$$\Delta M_{\rm W} = 9 \pm 44\ {\rm MeV}/c^2 \,, \qquad (5)$$

indicating that the effects of final state interactions are small. The measured W width is

$$\Gamma_{\rm W} = 2.150 \pm 0.068(stat) \pm 0.060(syst)\ {\rm GeV}/c^2 \,. \qquad (6)$$

## 5. Summary

An excellent performance of the LEP collider has allowed the four experiments to produce a preliminary measurement of the W boson mass of

$$M_{\rm W} = 80.447 \pm 0.042\ {\rm GeV}/c^2 \,. \qquad (7)$$

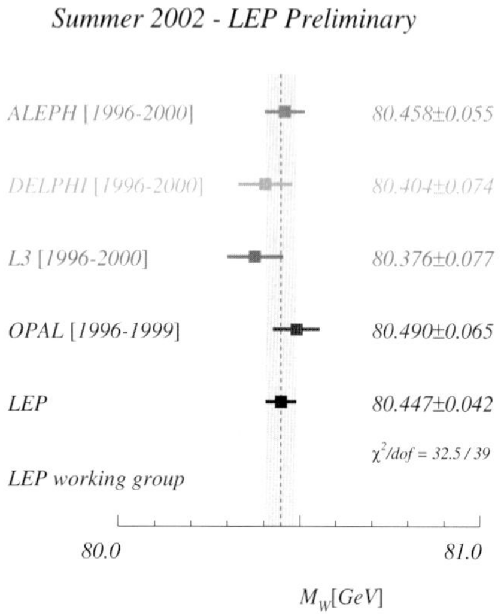


Figure 2. The preliminary LEP combined W mass measurement. The combined value takes into account correlations between experiments and years and so, in general, is not the same as a simple average.

The statistical uncertainty will be reduced by the inclusion of the final data and fine tuning of analyses, whilst the systematic uncertainty is expected to be reduced by further detailed study.

I would like to thank the organisers for inviting me to take part in an such an enjoyable conference. I would like to thank the LEP collaborations for providing results for the talk.

## REFERENCES

1. The LEP Collaborations ALEPH, DELPHI, L3, OPAL, the LEP Electroweak Working Group and the SLD Heavy Flavour and Electroweak Groups, *A Combination of Preliminary Electroweak Measurements and Constraints on the Standard Model*, CERN-EP-2001-098.
2. ALEPH Collaboration, *Phys. Lett.* **B401** (1997) 347;
   DELPHI Collaboration, *Phys. Lett.* **B397** (1997) 158;
   L3 Collaboration, *Phys. Lett.* **B398** (1997) 223;
   OPAL Collaboration, *Phys. Lett.* **B389** (1996) 416.
3. ALEPH Collaboration, *Phys. Lett.* **B453** (1999) 121; *Eur. Phys. J.* **C17** (2000) 241; ALEPH note CONF 2001-017.
4. DELPHI Collaboration, *Eur. Phys. J.* **C2** (1998) 581; *Phys. Lett.* **B462** (1999) 410; *Phys. Lett.* **B511** (2001) 159.
5. L3 Collaboration, *Phys. Lett.* **B407** (1997) 419; *Phys. Lett.* **B454** (1999) 386; L3 Note 2377; L3 Note 2575; L3 Note 2637.
6. OPAL Collaboration, *Eur. Phys. J.* **C1** (1998) 395; *Phys. Lett.* **B453** (1999) 138; *Phys. Lett.* **B507** (1999) 29; OPAL PN422; OPAL PN480.
7. Torbjorn Sjöstrand, *Comput. Phys. Commun.* **82** (1994) 74.
8. G. Marchesini *et. al.*, *Comput. Phys. Commun.* **67** (1992) 465 and hep-ph/9607393.
9. L. Lönnblad, *Comput. Phys. Commun.* **71** (1992) 15 and *Z. Phys.* **C70** (1996) 107.
10. The LEP Energy Working Group, LEPEWG 01/01, March 2001
11. S. Todorova-Nova, *Bose-Einstein Correlations at LEP and HERA*, these proceedings.
12. E. Bouhova-Thacker, *Colour Reconnection Effects in WW Production at LEP*, these proceedings.
13. T. Sjöstrand and V.A. Khoze, *Z. Phys.* **C62** (1994) 281 and *Phys. Rev. Lett.* **72** (1994) 28.
14. DELPHI Collaboration, *Measurement of Colour Reconnection model parameters using $M_W$ analyses*, DELPHI 2002-048-CONF-582, July 2002
15. The LEP Collaborations ALEPH, DELPHI, L3, OPAL, and the LEP W Working Group, *Combined Preliminary Results on the Mass and Width of the W Boson Measured by the LEP Experiments*, LEPEWWG/MASS/2002-02, September 2002

ELSEVIER

Proceedings of ICHEP 2002, pp. 149–152
*S. Bentvelsen, P. de Jong, J. Koch and E. Laenen (Editors)*

31st INTERNATIONAL CONFERENCE ON HIGH ENERGY PHYSICS AMSTERDAM

# 4-fermion production at LEP2

E. Migliore[a]

[a]Dipartimento di Fisica Sperimentale dell'Università di Torino and INFN Sezione di Torino, via P. Giuria 1, 10125 Turin, Italy

The running of LEP at centre-of-mass energies $\sqrt{s} \approx 200$ GeV offered the first opportunity to study four-fermion (4f) processes in $e^+e^-$ collisions. In this paper the non resonant 4f production, the main source of weak bosons at the future Linear Collider, is discussed.

## 1. Introduction

The running of the LEP collider at centre-of-mass energies above the Z mass (LEP2) offered a unique opportunity to study the production of four-fermion final states (Table 1).

Table 1
The running of LEP at high energy.

| $\sqrt{s}$ (GeV) | 183 | 189 | 192-202 | 205 | >207 |
|---|---|---|---|---|---|
| $\mathcal{L}_{\rm int}$/exp (pb$^{-1}$) | 55 | 170 | 230 | 75 | 130 |

The $e^+e^- \rightarrow 4f$ reactions are usually classified according to the subset of diagrams which are more relevant for a particular final state. In the LEP2 energy range, depending on the behavior of the cross-section as a function of $\sqrt{s}$, one can identify doubly resonant processes as WW or ZZ production, proceeding through conversion type diagrams (cf. Figure 1.b) and non resonant processes like single-W, single-Z and $Z\gamma^*$ production. In this contribution the non resonant processes are discussed. In the LEP2 energy range their cross-section has a weak dependence on $\sqrt{s}$ not exhibiting a threshold behavior. The expected cross-sections for these processes are typically below 1 pb. Therefore common definitions of the cross-sections were agreed among the LEP experiments to increase the significance of the comparison with the theoretical predictions [1].

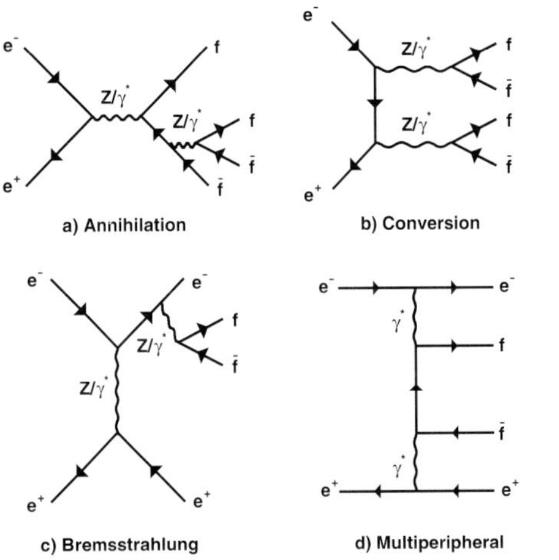


Figure 1. Diagrams contributing to $e^+e^- \rightarrow 4f$ at the lowest order.

## 2. Single-W

The study of single-W production is motivated mainly by testing the non-abelian sector of the electroweak Standard Model (SM). The final states investigated [1], $e^-\bar{\nu}_e q\bar{q}'$, $e^-\bar{\nu}_e \ell^+\nu_\ell$ ($\ell = \mu, \tau$) and $e^-\bar{\nu}_e e^+\nu_e$, are the same of the WW physics but here the analysis focuses on topologies with an electron lost along the beam-line. To enhance this contribution, the signal was defined considering only $t$-channel graphs with additional

[1]Charge conjugation is implied throughout the text.


doi:10.1016/S0920-5632(02)02072-8

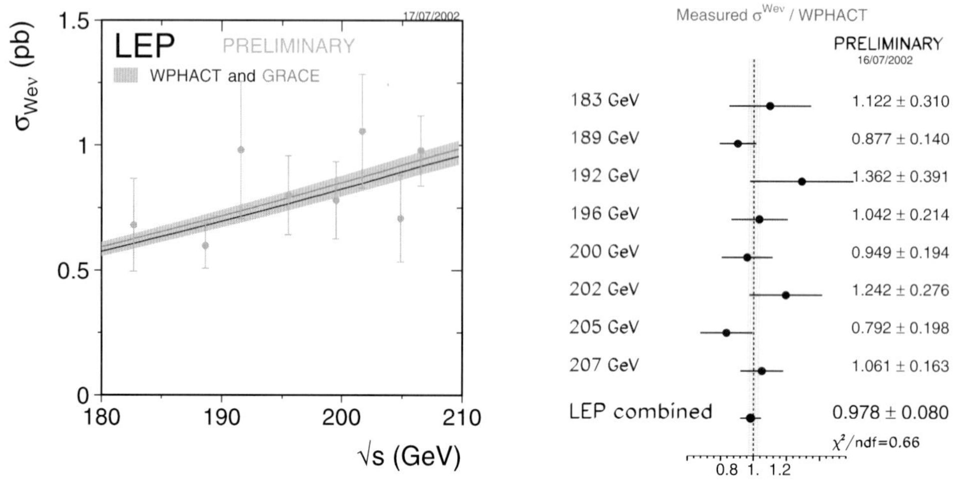


Figure 2. Measured single-W cross-sections as a function of $\sqrt{s}$ together with the SM predictions computed with GRC4f [3] and WPHACT [4] (left). Ratio of the measured to predicted cross-sections (right).

cuts on the phase space to reject the multiperipheral contributions:

- $m_{q\bar{q}'} > 45$ GeV/$c^2$ for $e^-\bar{\nu}_e q\bar{q}'$,
- $E_{\ell^+} > 20$ GeV for $e^-\bar{\nu}_e \ell^+ \nu_\ell$,
- $|\cos\theta_{e^+}| < 0.95$, $E_{e^+} > 20$ GeV and $|\cos\theta_{e^-}| > 0.95$ (veto on $e^-$) for $e^-\bar{\nu}_e e^+ \nu_e$.

From the technical point of view, the accurate prediction of the cross-section is challenging because having a collinear singularity, it requires a fully massive calculation of the amplitudes. From the physics point of view, the process is interesting because of the different energy scales involved:

- the scale of the electromagnetic coupling which could range from $\alpha_{QED}(t \sim 0) = 1/137$ to $\alpha_{QED}(s \sim M_W^2) = 1/128$, leading to a change of the cross-section

$$\left.\frac{\delta\sigma}{\sigma}\right|_{\alpha_{QED}(M_W^2)\to\alpha_{QED}(0)} \approx -5\%;$$

- the scale of the QED Initial State Radiation (ISR)

$$\left.\frac{\delta\sigma}{\sigma}\right|_{ISR(q^2=s)\to ISR(q^2=0)} \approx +8\%.$$

The accuracy of the codes used by the experiments is determined by the comparison with higher order calculations [2]. In the case of the electromagnetic coupling, approximate methods exist which are in agreement with higher calculations better than 2%. In the case of ISR no higher order prediction is available. An overall uncertainty of 5% is assumed on the expected cross-section.

Experimentally the measurement is dominated by the $e^-\bar{\nu}_e q\bar{q}'$ channel ($\approx$ 400-600 fb) which represents about 80% of the overall cross-section. The largest experimental error is the limited data statistics. The measured cross-sections as a function of $\sqrt{s}$ are shown in Figure 2 together with the theoretical predictions. The agreement with the SM expectations is expressed by the energy averaged ratio of the measured to the predicted cross-sections: $R = 0.978 \pm 0.080$ with about 80% of the full LEP2 sample analyzed by the experiments so far.

## 3. Single-Z

The process of single-Z production is dominated by the scattering of a quasi-real photon

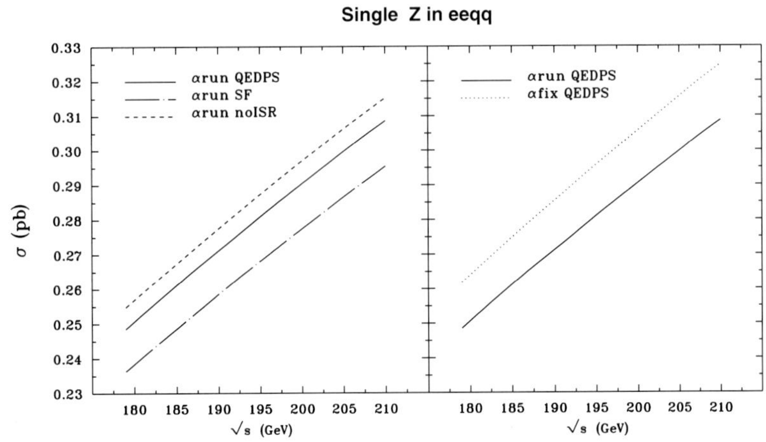


Figure 3. Predicted single-Z cross-section, $(e^-)e^+q\bar{q}$ channel positron tag only, as a function of $\sqrt{s}$: effect of different ISR implementations (left) and of running $\alpha_{QED}$ (right) [4].

emitted by an electron on a positron of the other beam. The process is forward peaked, characterized by a backscattered positron, an electron lost along the beam-line and the decay products of the Z in the barrel. A common definition of the cross-section was established for $(e^-)e^+q\bar{q}$ and $(e^-)e^+\mu^+\mu^-$ channels. It includes the entire set of 48 diagrams with additional cuts on the phase space to suppress the multiperipheral and $s$-channel contributions [2]:

- $m_{f\bar{f}} > 60$ GeV/$c^2$ $(f = \mu, q)$,
- $\theta_{e^-} < 12°$,
- $12° < \theta_{e^+} < 120°$, $E_{e^+} > 3$ GeV,

corresponding to a cross-section of about 500 fb for $(e^-)e^+q\bar{q}$ and 50 fb for $(e^-)e^+\mu^+\mu^-$. within this kinematic limits At $\sqrt{s} = 200$ GeV the bremsstrahlung contribution, Figure 1.c, amounts to about 97% for $(e^-)e^+q\bar{q}$ and to 67% for $(e^-)e^+\mu^+\mu^-$ final states. The sensitivity of the process to the scale of the electromagnetic coupling and of the ISR is similar to that of the single-W (Figure 3).

[2]LEP experiments assume the $z$-axis oriented along the direction of the incoming electron beam.

Experimentally single-Z events are selected requiring 2 jets or 2 identified muons accompanied with an isolated electron and the missing momentum of the event pointing in the beam-line direction. Signal efficiency is about 30-40% in the $(e^-)e^+q\bar{q}$ channel and about 25% in the $(e^-)e^+\mu^+\mu^-$ channel. Background level is about 20% in the $(e^-)e^+q\bar{q}$ channel, mainly from $q\bar{q}(\gamma)$ and $e^-\bar{\nu}_e q\bar{q}'$ events, while it is below 2% for $(e^-)e^+\mu^+\mu^-$. The error is dominated by the data statistics, the largest experimental systematics being related to the electron selection. The measured cross-sections as a function of $\sqrt{s}$ are shown in Figure 4 together with the theoretical predictions. The energy averaged ratio of the measured to the predicted cross-sections is $R = 0.951 \pm 0.083$ for $(e^-)e^+q\bar{q}$ and $R = 1.42 \pm 0.26$ for $(e^-)e^+\mu^+\mu^-$ with about 50% of the full LEP2 sample analyzed by the experiments so far.

## 4. $Z\gamma^*$

Four-fermion final states from $Z\gamma^*$ production were the first non resonant 4f processes observed already in the LEP pilot run at 130-136 GeV in November 1995. They are characterized by all the four fermions detected with one pair, typi-

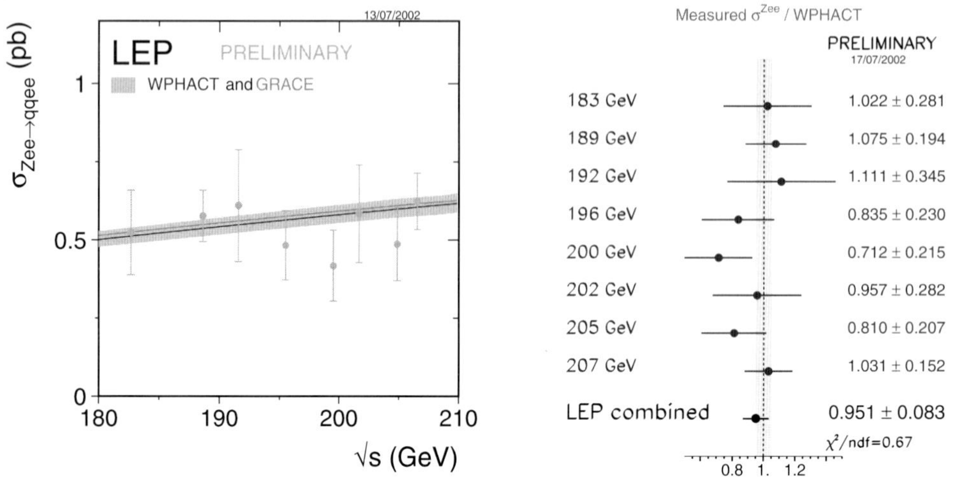


Figure 4. Measured single-Z cross-sections as a function of $\sqrt{s}$ together with SM predictions computed with GRC4f [3] and WPHACT [4] (left). Ratio of the measured to predicted cross-sections (right).

cally 2 jets, forming an invariant mass closed to the Z mass and the other, usually 2 leptons, at lower invariant masses. Currently only the OPAL collaboration has completed the analysis of the full LEP2 sample [5]. The analysis is done on $\mu^+\mu^-\mathrm{q\bar{q}}$ and $\mathrm{e^-e^+q\bar{q}}$ final states. The signal is defined topologically with the following cuts:

- $|\cos\theta_\ell| < 0.95 \quad (\ell = \mathrm{e}, \mu)$,
- $m_{\mathrm{q\bar{q}}} > 5\ \mathrm{GeV}/c^2$,
- $m_{\mathrm{ee}} > 2\ \mathrm{GeV}/c^2$, no cut on $m_{\mu\mu}$.

For $\mu^+\mu^-\mathrm{q\bar{q}}$ all the 8 contributing diagrams are included in the signal definition (Figure 1.a and 1.b). For the $\mathrm{e^-e^+q\bar{q}}$ final state two additional classes of diagrams contribute (Figure 1.c and 1.d): the bremsstrahlung is retained in the signal while the multiperipheral contribution is subtracted. The energy averaged cross-sections are:

$$\sigma_{\mathrm{eeqq}} = 199\pm27(\mathrm{stat.})\pm30(\mathrm{syst.})\ \mathrm{fb}\ (\mathrm{SM{:}}\ 180\ \mathrm{fb}),$$
$$\sigma_{\mu\mu\mathrm{qq}} = 160\pm26(\mathrm{stat.})\pm13(\mathrm{syst.})\ \mathrm{fb}\ (\mathrm{SM{:}}\ 165\ \mathrm{fb}),$$

where the SM predictions are computed with GRC4f and $\alpha_{\mathrm{QED}} = 1/128$.

## 5. Conclusions

The LEP2 run offered a unique opportunity to study non resonant 4f production which will be the main source of weak bosons at the future Linear Collider. The expected final accuracy on these processes is $\sim 7\%$ for single-W and $\sim 6.5\%$ on single-Z.

## REFERENCES


1. The LEP collaborations and the LEP WW working group, note in preparation.
2. M.W. Grünewald et al., Report on *Four-fermion production in electron-positron collisions* in *Reports of the working groups on precision calculations for LEP2 physics*, S. Jadach, G. Passarino and R. Pittau eds., (2000) CERN 2000-009.
3. J. Fujimoto et al., Comp. Phys. Comm. 100 (1997) 128.
4. E. Accomando et al., Comp. Phys. Comm. 99 (1997) 270;
   E. Accomando et al., hep-ph/0204052 (2002).
5. OPAL Collaboration, G. Abbiendi et al., CERN-EP/2002-052 (2002).

Proceedings of ICHEP 2002, pp. 153–157
*S. Bentvelsen, P. de Jong, J. Koch and E. Laenen (Editors)*

31st INTERNATIONAL CONFERENCE ON HIGH ENERGY PHYSICS AMSTERDAM

# Boson Pair Cross Sections at LEP

Luca Malgeri[a]

[a]University of Geneva - DPNC
24, Quai Ernest-Ansermet - 1211 Genève 4 - Switzerland

In five years of running at energies above WW and ZZ production thresholds, LEP–II provided each experiment an integrated luminosity of $\sim$ 700 pb$^{-1}$. The collected statistics allows to measure with high precision the total $e^+e^- \to WW, ZZ$, and $\gamma\gamma$ cross sections. Derived quantities as leptonic and hadronic branching fractions of the W boson, differential distributions of the polar scattering angle and constraints on new physics are also herein reported. All results, unless differently stated, have to be considered preliminary.

## 1. Introduction

After an incredibly successful running at energies around the $Z^0$ resonance, LEP started to investigate a new range of energies just above the $W$–pair production threshold in 1996. In the following years it continued on overtaking the threshold for $Z$–pair production up to a center-of-mass energy of 209 GeV.
The integrated luminosities collected by the four LEP experiments, ALEPH [1], DELPHI [2], L3 [3] and OPAL [4] in the five years of running is in the height of 700pb$^{-1}$/exp.

The Standard Model [5] processes $e^+e^- \to WW, ZZ$, and $\gamma\gamma$ are studied at all center-of-mass energies. The comparison of the measured cross sections with their expectations represents a powerful test of the Standard Model theory. The number of events collected by each experiment amounts to 10000 W-pairs, 500 Z-pairs and 5000 $\gamma$-pairs. This statistics pushes tests at the level of a percent allowing to probe the first order correction beyond the born level expectations.

For references, this report cover the following abstracts sent to the ICHEP02 conference: 234, 450, 236, 422, 532, 523, 443, 379.

## 2. W-pair final state

$W$–pairs at $e^+e^-$ colliders are produced via three main diagrams, defined as CC03 (Charged Current 03), involving the $Z/\gamma$ s–channel exchange and the neutrino t–channel exchange (Fig. 1). The current accuracy on the theoretical calculation for $\sigma_{WW}$, which include radiative corrections up to $\mathcal{O}(\alpha_{em})$, is about 0.5 % [6,7].

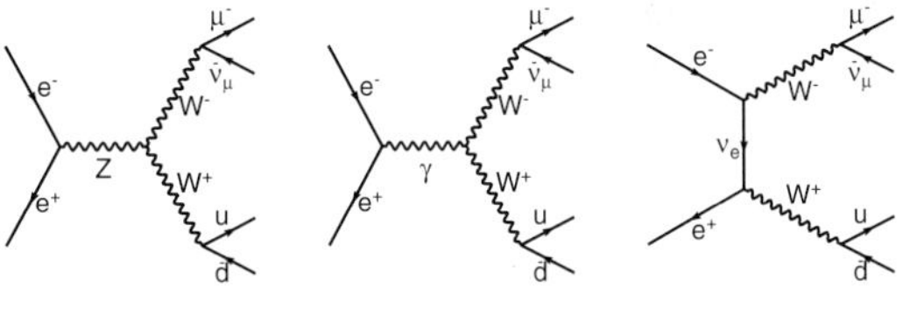


Figure 1. CC03 diagrams for the production of $W$–pair events.

In order to compare directly the measurements between different channels and experiments it has been conventionally chosen to quote CC03–level cross sections. It means that the actually measured cross sections have to be corrected. The correction factors may be as high as $\sim$ 10 %.

The selection of $e^+e^- \to WW$ events is conventionally divided in five main classes: the fully hadronic decays $e^+e^- \to qqqq$, the three semileptonic decays $WW \to qq\ell\nu$ with $\ell$ being electron, muon or tau and the fully leptonic decays $e^+e^- \to \ell\nu\ell\nu$. The Standard Model branching fractions for each of the five classes are shown in Tab. 1 together with a summary of the typical performances of selections.


doi:10.1016/S0920-5632(02)02077-7

Table 1
Typical efficiency and accepted non–$WW$ background for each of the 5 classes. The last column shows the Standard Model branching fractions for each class.

| Selection | Eff. (%) | Bkg (pb) | SM BR |
|---|---|---|---|
| $e^+e^- \to qqqq$ | 85 | 1.5 | 45.6 % |
| $e^+e^- \to qqe\nu_e$ | 80 | 0.15 | 14.6 % |
| $e^+e^- \to qq\mu\nu_\mu$ | 80 | 0.15 | 14.6 % |
| $e^+e^- \to qq\tau\nu_\tau$ | 50–60 | 0.2–0.3 | 14.6 % |
| $e^+e^- \to \ell\nu\ell\nu$ | 50–60 | 0.1–0.2 | 10.6 % |

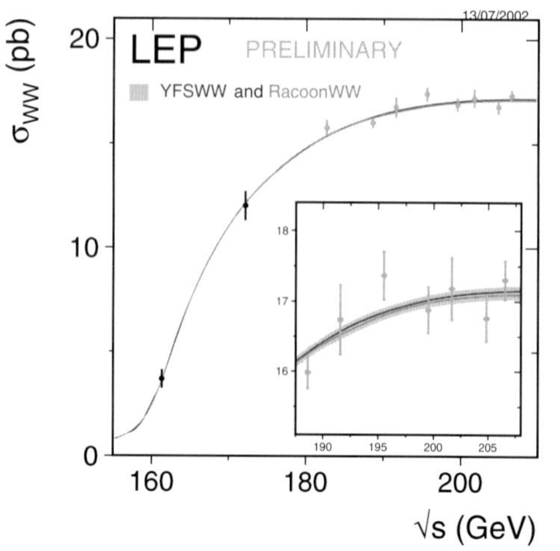


Figure 2. Total $e^+e^- \to WW$ cross section from LEP at $\sqrt{s} = 161 - 208$ GeV compared with theoretical predictions.

### 2.1. Cross sections and branching fractions determination

Once the selections have been applied and their performances are known, each experiment performs a fit, properly taking into account the correlations among channels, in order to get the single channel cross section. The same approach is used to get the total $WW$ cross sections which are then combined among the four LEP collaborations.

The detailed description of the combination procedure can be find in [8]. It takes into account correlated sistematic errors, among which the hadronization modeling plays a leading role.

In Fig. 2 the combined cross section measurements are shown together with the "status of the art" for theoretical predictions. The ratio R between the measured values and the predictions as calculated with YFSWW [6] is

$$R = 0.997 \pm 0.011$$

showing a very good agreement with the Standard Model at the 1 % level. The uncertainty on the ratio R is dominated by the common systematic effect amounting to 0.009.

The leptonic and hadronic branching fractions, $\mathrm{BR}(W \to \mathrm{qq})$ and $\mathrm{BR}(W \to \ell\nu)$ are directly derived from the single channel cross sections. The LEP combined values, using the whole statistics and assuming unitary sum and lepton universality, are:

$$\mathrm{BR}(W \to \mathrm{qq}) = 67.92 \pm 0.27\%$$

$$\mathrm{BR}(W \to \ell\nu) = 10.69 \pm 0.09\%$$

In the framework of the Standard Model the branching fractions of the $W$ boson are related to the six Cabibbo–Kobayashi–Maskawa matrix elements [9] not involving the top quark. Neglecting higher order corrections one can write:

$$\frac{\mathrm{BR}(W \to qq)}{1 - \mathrm{BR}(W \to qq)} = \left(1 + \frac{\alpha_s(M_W^2)}{\pi}\right) \sum |\mathrm{V}_{ij}^{CKM}|^2$$

Fixing the precisely measured $\mathrm{V}_{ud}$ and the other matrix elements to their current world average [10], a value for $|\mathrm{V_{cs}}|$ can be derived, without the assumption of unitarity:

$$|\mathrm{V_{cs}}| = 0.996 \pm 0.013 \,.$$

Inside the LEP community there are also plans to combine differential cross section measurements in function of the polar scattering angle of the $W^-$ boson: $d\sigma/d\cos\theta_W$. This distribution contains much more information than the total cross section and hence represents a powerful test of the SM expectation. In Fig. 3, as an example, the differential cross section as obtained by the L3 experiment is shown averaged on all energy points. Both the Standard Model prediction and

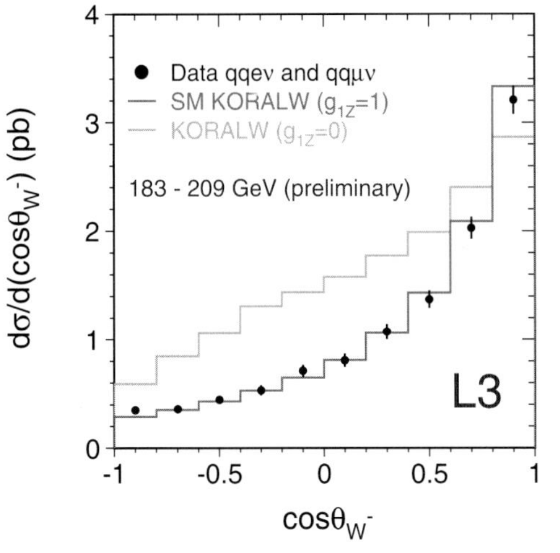


Figure 3. Differential cross section as a function of the W boson polar scattering angle as measured by L3. The Standard Model prediction and the prediction in the presence of an anomalous coupling are shown as solid lines.

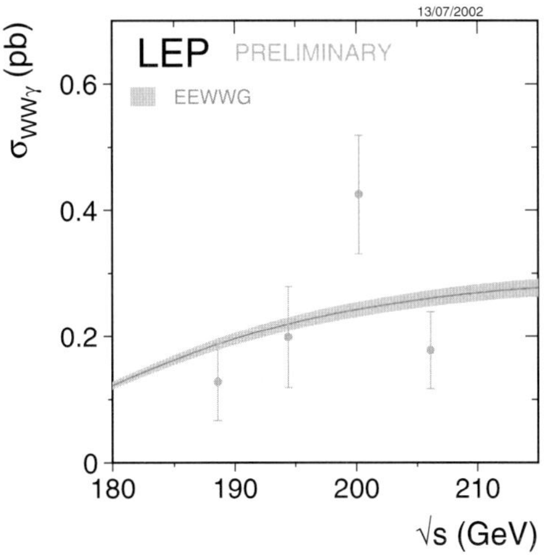


Figure 4. Combined $WW\gamma$ cross section as a function of the center-of-mass energy. Only L3 and DELPHI results are used.

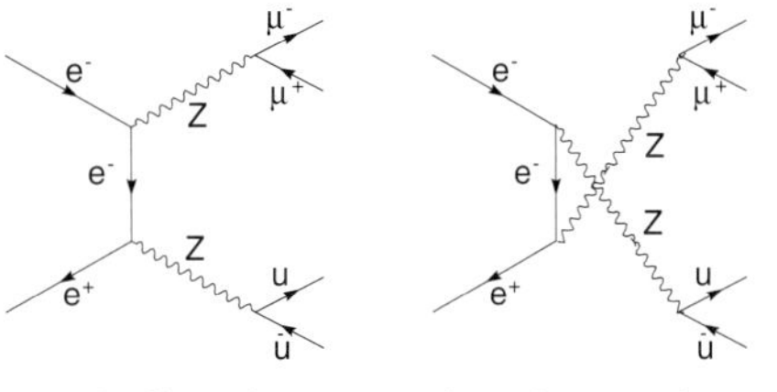

Figure 5. NC02 diagrams for the production of $Z$–pair events.

the prediction in the presence of an anomalous coupling are superimposed.

The presence of a detectable photon in the final state is also looked at in the LEP collaborations, due to its distinctive signature in case of the presence of an Anomalous Quartic Gauge Coupling in the $WW\gamma\gamma$ or $WWZ\gamma$ vertices. Although predicted in the Standard Model, the quartic vertices have a negligible contribution at LEP2 energies. A number of events in excess of the expected yield from radiation off the initial or final state particles, would directly point to new physics.

The results obtained by the DELPHI and L3 collaborations have been combined and they are shown in Fig. 4. No statistical significant deviation is observed.

## 3. Z-pair final state

At LEP, the Z-pair final state is produced by means of two main born level diagrams referred to as NC02 (Neutral Current) and shown in Fig. 5

The signal definition for the LEP combination is based on the NC02 set of diagrams, including ISR corrections.

### 3.1. Selections and results

The experimental investigations of Z-pair production is made difficult by its rather low cross section, compared with competing processes that consitute large ans sometimes irreducible backgrounds. All 19 visible Z-pair decay combinations are analysed. In all cases multi-variate techniques are used, to enhance the separation power against the most important backgrounds, namely W-pair production.

The combination of the Z-pair cross sections as a function of the center-of-mass energy is shown in Fig. 6. The ratio R, described in the previous section, obtained using YFSZZ [11] as theoretical

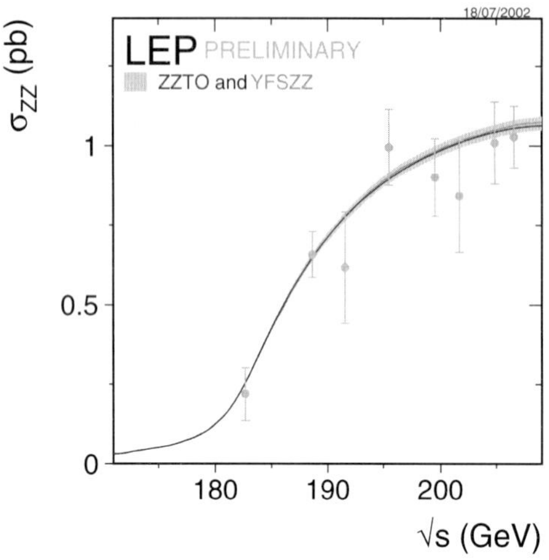


Figure 6. Combined Z-pair cross sections compared to predicted values using the YFSZZ Montecarlo program.

prediction, is $R = 0.962 \pm 0.055$. Good agreement with the Standard Model is observed.

## 4. $\gamma$-pair final state

The process $e^+e^- \to \gamma\gamma$ receives its main contribution from QED by means of the exchange of an electron via $t$-channel.

The lowest order contribution to the cross section is:

$$\frac{d\sigma}{d\Omega} = \frac{\alpha^2}{s} \frac{(1 + \cos^2\theta)}{(1 - \cos^2\theta)}$$

where $\theta$ is the polar angle of the photons.

The experimental signature of the final state is clean, allowing the analysis of event samples with negligible background. The sensitivity of this process to deviations with respect to the QED predictions grows with $\sqrt{s}$ and, in addition, non-QED contributions are small. Any deviation can be therefore interpreted as a sign of new physics: QED cut-off, low scale gravity, spin 3/2 leptons, excited electrons, etc.

The total cross sections are combined among the LEP collaborations [12] and the ratio between the measured values and the prediction from QED is found to be $\sigma_{meas}/\sigma_{th} = 0.982 \pm 0.010$.

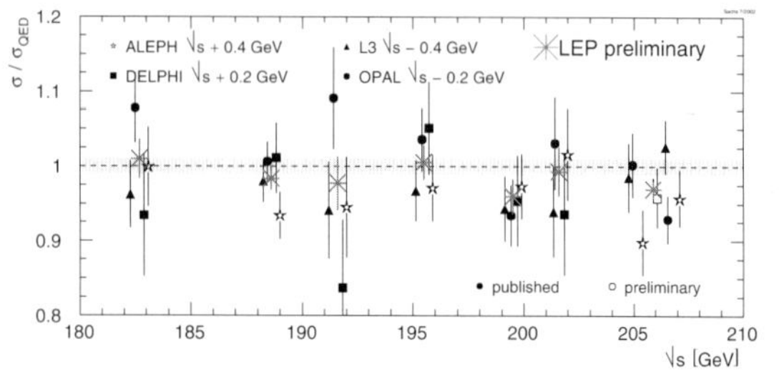


Figure 7. Ratios $\sigma_{meas}/\sigma_{th}$ for the $e^+e^- \to \gamma\gamma$ process at LEP.

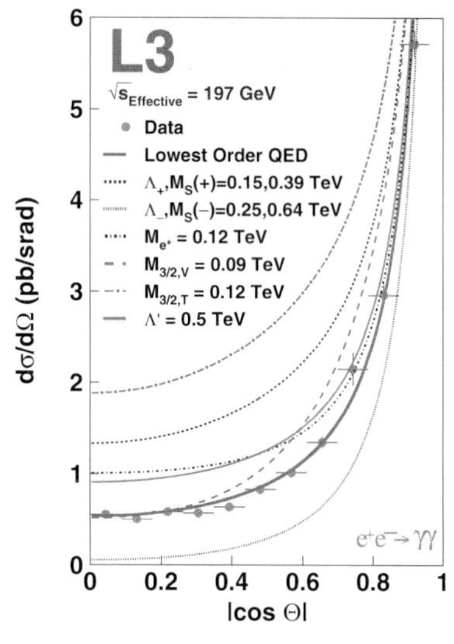


Figure 8. Differential $e^+e^- \to \gamma\gamma$ cross section as measured by the L3 experiment compared to alternative theoretical models.

The good agreement among the LEP esperiments and with theory as a function of the center-of mass energy is visible in Fig. 7.

For the search of new physics effects the differential cross section as a function of the polar angle, which is mostly sensitive to such effects, is used. The sensitivity of the differential cross section to new physics is shown in Fig. 8, where L3 data are compared to several alternative theoretical models.

Typical limits obtained by the LEP collaborations are of the order of 1 TeV for the graviton mass, and 300 GeV for the excited electron mass. See [13] for a comprehensive list of limits.

## 5. Conclusions

The amount of data collected by the LEP experiments in the high energy running period is beeing extensively analysed and provide a powerful tool the test the Standard Model. In many cases, as for the W-pair cross section, the level of accuracy has reached the sensitivity threshold for raditive corrections. Although not yet final, the results herein presented are based on the full available statistics and present no sign of deviation from the Standard Model.


## Acknowledgements

I would like to thank the organisers of ICHEP02 for their warm hospitality.

I would like to thank also my colleagues from the four LEP experiments and the members of the LEP Electroweak Working Group for providing the data which were presented in this report.


## REFERENCES


1. The ALEPH Collaboration, Nucl. Inst. Meth. **A**294 (1990) 121
2. The DELPHI Collaboration, Nucl. Inst. Meth. **A**303 (1991) 233
3. The L3 Collaboration, Nucl. Inst. Meth. **A**289 (1990) 35
4. The OPAL Collaboration, Nucl. Inst. Meth. **A**305 (1991) 271
5. S. L. Glashow, Nucl. Phys.**22** (1961) 579; S. Weinberg, Phys. Rev. Lett.**19** (1967) 1264; A. Salam, in *Elementary Particle Theory*, ed. N. Svartholm, Stockholm, Almquist and Wiksell (1968), 367
6. S. Jadach *et al.*, Phys. Lett. **B 417** (1998) 326; Phys. Rev. **D61** (2000) 113010; UTHEP-00-0101, CERN-TH/2000-337
7. A. Denner *et al.*, Nucl. Phys. **B 587** (2000) 67, *hep-ph/0006307*
8. The LEP Electroweak Working Group, note in preparation.
9. N. Cabibbo, Phys. Rev. Lett. **10** (1963) 531; M. Kobayashi and T. Maskawa, Prog. Theor. Phys. **49** (1973) 652
10. D.E. Groom *et al.*, *The 2000 Review of Particle Physics*, Euro. Phys. Jour. **C 15** (2000) 1
11. S. Jadach *et al.*, Phys. Rev. **D 56** (1997) 6939
12. LEP-II Diphoton group, LEP2FF note 02-02
13. R. McPherson, Beyond the Standard Model, these proceedings

Proceedings of ICHEP 2002, pp. 158–160
*S. Bentvelsen, P. de Jong, J. Koch and E. Laenen (Editors)*

ELSEVIER

31st INTERNATIONAL CONFERENCE ON HIGH ENERGY PHYSICS AMSTERDAM

# Final $R$-value results from 2-5 GeV from BES and QCD test with $R$ scan data

W.B. Yan, [a] W.G. Li, [a] Z.G. Zhao[a] [b] , Representing BES Collaboration

[a]Institute of High Energy Physics, Chinese Academy of Sciences, Beijing 100039, P.R.C.

[b]University of Michigan, Ann Arbor, MI48109, USA

Final results of the measurement of $R = \sigma(e^+e^- \to hadrons)/\sigma(e^+e^- \to \mu^+\mu^-)$ in the energy region from 2 to 5 GeV with the upgraded Beijing Spectrometer (BESII) at the Beijing Electron Positron Collider (BEPC) are presented. Preliminary results of the inclusive momentum spectra, multiplicity distribution, and second binomial moment measured with the $R$ scan data at 2.2, 2.6, 3.0, 3.2, 4.6 and 4.8 GeV are reported.

## 1. Introduction

The lowest order cross section for $e^+e^- \to \gamma^* \to$ hadrons in units of the lowest-order QED cross section for $e^+e^- \to \mu^+\mu^-$ is defined as $R$, namely $R = \sigma(e^+e^- \to \text{hadrons})/\sigma(e^+e^- \to \mu^+\mu^-)$, where $\sigma(e^+e^- \to \mu^+\mu^-) = \sigma^0_{\mu\mu} = 4\pi\alpha^2(0)/3s$. Presently the uncertainty in $R$ in the energy region below 5 GeV dominates the uncertainties in both $\alpha(M_Z^2)$, the QED running coupling constant evaluated at the Z pole, and $a_\mu^{SM}$, the value of $(g-2)_\mu$ based on the Standard Model calculation [1].

Hadron production from $e^+e^-$ annihilation is one of the most valuable testing grounds for Quantum Chromodynamics (QCD). Particularly, it is interesting and important to analyze low energy $e^+e^-$ collision data, for example, the inclusive momentum spectrum, defined as $\xi = -\ln(2p/\sqrt{s})$, where $p$ and $\sqrt{s}$ are the momentum of the charged particles and center-of-mass (c.m.) energy respectively, with today's knowledge. A purely analytical approach giving quantitative predictions for $\xi$ is the QCD calculation using the so-called Modified Leading Logarithmic Approximation (MLLA) [2] under the assumption of Local Parton Hadron Duality (LPHD) [3], which expresses the limiting spectrum for hadrons as

$$\frac{1}{\sigma_{had}}\frac{d\sigma}{d\xi} = K_{LPHD} \times f_{MLLA}(\xi, \Lambda_{eff}) \qquad (1)$$

where $K_{LPHD}$ is an overall normalization factor describing hadronization and $f$ is a complex function of $\xi$ and effective scale parameter $\Lambda_{eff}$ [2]. Eq. 1. is valid in the range of $0 \leq \xi \leq \ln(0.5\sqrt{s}/\Lambda_{eff})$.

Another example is the second binomial moments, which is a measure of the strength of hadron-hadron correlations and a sensitive probe for higher order QCD or non-perturbative effect. It is defined as $R_2 = \langle n_{ch}(n_{ch}-1)\rangle/\langle n_{ch}\rangle^2$, where $n_{ch}$ is the charged particle multiplicity. According to the next leading order QCD calculation (NLO), $R_2$ is given by

$$R_2 = \frac{11}{8}(1 - c\sqrt{\alpha_s(\sqrt{s})}) \qquad (2)$$

with $c = 0.55(0.56)$ for five (three) active flavors. There has been a long standing discrepancy between the value of $R_2$ calculated by NLO and that measured with $e^+e^-$, $\mu^+p$ and $\nu_\mu p$ experiments. In addition, there is relatively little data in the energy region below 5 GeV to compare with QCD calculations.

This paper presents the final results of the $R$ values measured with BESII [4] at BEPC in the energy region from 2-5 GeV. We also report preliminary results on the inclusive momentum spectra, the momentum distribution of charged particles, and second binomial moments $R_2$ obtained from the analysis of $R$ scan data.


doi:10.1016/S0920-5632(02)02092-3

## 2. $R$ values in 2-5 GeV

Refs. [5,6] describe in detail the $R$ scan performed by BES at 91 energy points between 2-5 GeV, and the experimental study of the background, particularly the beam associated background. The triggers and the determination of the trigger efficiency; the measurement of luminosity; the hadronic event selection and background subtraction; the determination of the detection efficiency for hadronic events; and the initial state radiative correction can be also found in Refs. [5,6].

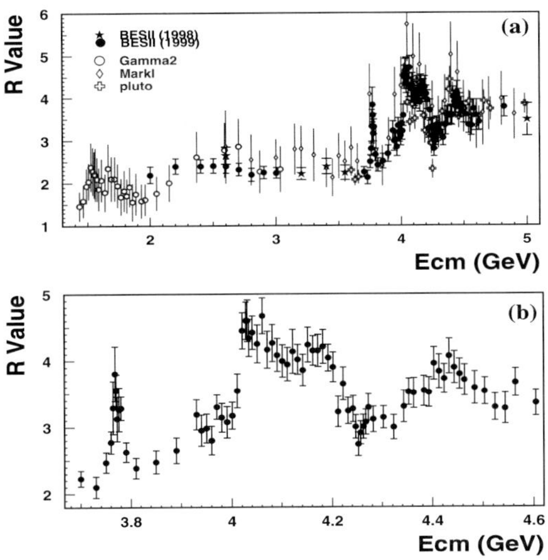


Figure 1. (a) A compilation of measurements of $R$ in the cm energy range from 1.4 to 5 GeV. (b) $R$ values from this experiment in the resonance region between 3.75 and 4.6 GeV.

The final $R$ values measured by BES in this experiment are displayed in Fig. 1, together with those measured by MarkI [7], $\gamma\gamma2$ [8] and Pluto [9]. The $R$ values from BESII have an average uncertainty of about 6.6%, which represents a factor of two to three improvement in precision in the 2 to 5 GeV energy region. These improved measurements have a significant impact on the global fit to the electroweak data and the determination of the SM prediction for the mass of the Higgs particle [10]. In addition, they are expected to provide an improvement in the precision of the calculated value of $a_\mu^{SM}$, and test the QCD sum rules down to 2 GeV [11–13].

## 3. Test of QCD models with $R$ scan data

### 3.1. $\xi$ spectrum

The measured $\xi$ spectrum at five different energies between 2.6 and 4.8 GeV are shown in Figure 2. The errors in the spectrum include errors from hadronic event selection and uncertainties from the event generators.

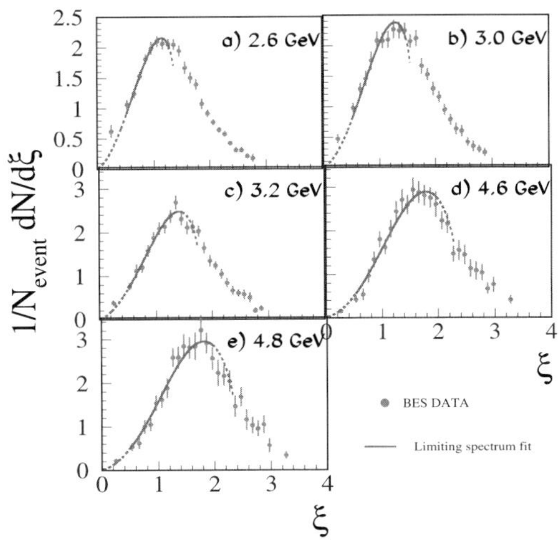


Figure 2. Measured $\xi$ spectrum (solid dot) at 2.6, 3.0, 3.2, 4.6 and 4.8 GeV. Solid curves are the fitting of the data for limiting spectrum with a Gaussian. The dotted line is an extrapolation of the fitted result.

### 3.2. Momentum spectrum

The momentum spectra at the five energy points are shown in Figure 3, together with those measured at higher energy up to 130 GeV in other experiments. The momentum spectra show that hadron production at very small momentum $p \leq 0.1$ GeV is approximately energy independent. This behavior has been explained in Ref. [14] to be due to the coherent emission of low energetic (i. e. long wavelength) gluons by the total color current. Correspondingly the number of produced hadrons at small momentum is approximately constant.

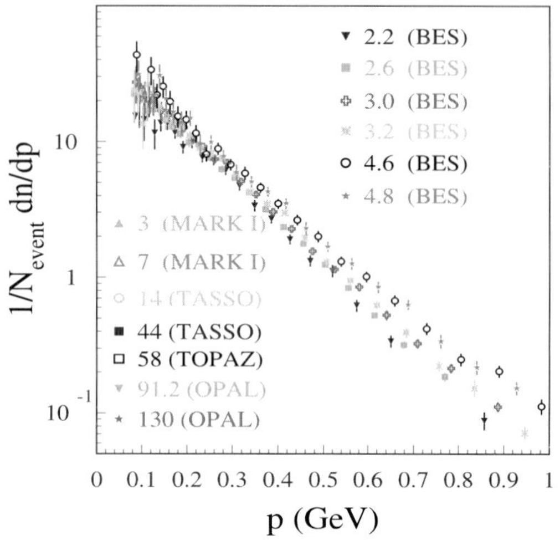


Figure 3. Charged particle momentum spectrum

### 3.3. Second binomial moment

Based on the multiplicity measured, we can obtain the second binomial moment $R_2$. The results are displayed in Figure 4, together with both NLO calculations and published data at higher energies up to 100 GeV from $e^+e^-$, $\mu^+p$ and $\nu_\mu p$ experiments [15,16]. Our measured $R_2$, although with large errors, are consistent with those of other measurements done at higher energies.

It's interesting to see that $R_2$ predicted by LO QCD are significantly higher than the measured data, while the NLO calculation comes closer to the data, although disagreement ($\sim 0.07$ in $R_2$) remains sizeable.

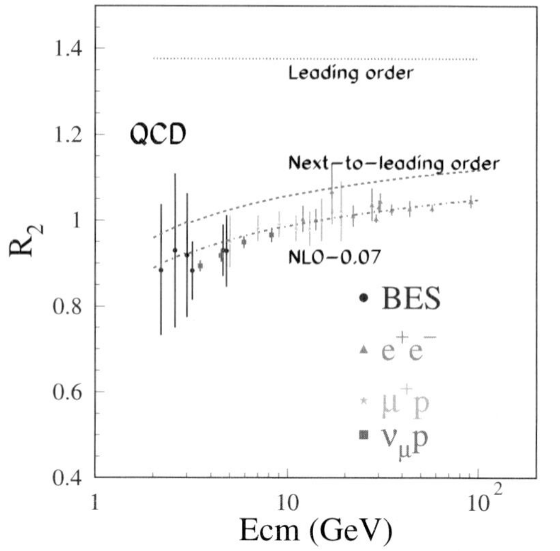


Figure 4. Energy dependence of second binomial moments $R_2$

## REFERENCES


1. Z.G. Zhao, International Journal of Modern Physics A15 (2000)3739.
2. V. A. Khoze and W. Ochs, Int. J. Mod. Phys. A12 (1997) 2949
3. Ya. I. Azimov, Yu. L. Dokshitzer, V. A. Khoze and S. I. Troyan, Phys. Lett. B165 (1985) 147; Z. Phys. C27 (1985) 65
4. J.Z. Bai *et al.*, (BES Collab.), *Nucl. Instrum. Methods* **A458**, 627 (2001).
5. BES Collab., J. Z. Bai et al., Phys. Rew. lett. 84 (2000) 594
6. BES Collab., J. Z. Bai et al., Phys. Rew. lett. 88 (2002) 101802
7. J. L. Siegrist *et al.*, (Mark I Collab.), *Phys. Lett.* B **26**, 969 (1982).
8. C. Bacci *et al.*, ($\gamma\gamma$2 Collab.), *Phys. Lett.* B **86**, 234 (1979).
9. L. Criegee and G. Knies, (Pluto Collab.), *Phys. Rep.* **83**, 151 (1982);
   Ch. Berger *et al.*, *Phys. Lett.* B **81**, 410 (1979).
10. H. Burkhardt and B. Pietrzyk, *Phys. Lett.* B **513**, 46 (2001).
11. M. Davier and A. Hoecker, *Phys. Lett.* B **419**, 419 (1998).
12. A. Martin *et al.*, *Phys. Lett.* B **492**, 69 (2000).
13. J.H. Kuehn and M. Steinhauser, *Phys. Lett.* B **437**, 425 (1998).
14. V. A. Khoze, S. Lupia and W. Ochs, Eur. Phys. J. C25 (1988) 77
15. EMC Collaboration, M. Arneodo et al., Nucl. Phys. B258 (1985) 249
16. WA21 Collaboration, G. T. Jones et al., Z. Phys. C54 (1992) 45

ELSEVIER

Proceedings of ICHEP 2002, pp. 161–163
*S. Bentvelsen, P. de Jong, J. Koch and E. Laenen (Editors)*

31st INTERNATIONAL CONFERENCE ON HIGH ENERGY PHYSICS AMSTERDAM

# CEEX Exponentiation in QED*

S. Jadach[a]

[a]Henryk Niewodniczanski Institute of Nuclear Physics,
ul. Radzikowskiego 152, 31-342 Cracow, Poland

The aim is to summarize briefly on: (a) Main features of YFS Coherent Exclusive EXponentiation in QED, (b) Examples of recent results relevant for LEP/LC physics program.

This presentation is based mainly on three recent papers [1–3]. Standard Model calculations for LEP with YFS exponentiation are implemented in many Monte Carlo event generators: for $e^+e^- \to f\bar{f} + n\gamma$, $f = \tau, \mu, d, u, s, c$ process: (1a) YFS1 (1987-1989) $\mathcal{O}(\alpha^1)_{exp}$ ISR [4], (1b) YFS2∈KORALZ (1989-1990) $\mathcal{O}(\alpha^1 + h.o.LL)_{exp}$ ISR [5,6], (1c) YFS3∈KORALZ (1990-1998) $\mathcal{O}(\alpha^1 + h.o.LL)_{exp}$ ISR+FSR [7,8], (1d) KKMC (1998-02) $\mathcal{O}(\alpha^2 + h.o.LL)_{exp}$ ISR+FSR+Interf. ($d\sigma/\sigma = 0.2\%$) [9]; for the low angle Bhabha process $e^+e^- \to e^+e^- + n\gamma$ for $\theta < 6°$ the following ones: (2a) BHLUMI 1.x, (1987-1990) $\mathcal{O}(\alpha^1)_{exp}$ [10], (2b) BHLUMI 2.x, (1990-1996) $\mathcal{O}(\alpha^1 + h.o.LL)_{exp}$ ($d\sigma/\sigma = 0.07\%$), [11,12]; for the large angle Bhabha $e^+e^- \to e^+e^- + n\gamma$ for $\theta > 6°$ (3) BHWIDE (1994-1998) $\mathcal{O}(\alpha^1 + h.o.LL)_{exp}$ [13]; and finally for $WW$ boson production processes $e^+e^- \to W^+W^- + n\gamma$, $W^\pm \to f\bar{f}$ (4a) KORALW (1994-2001) ISR YFS LL ($d\sigma/\sigma = 2\%$) [14,15] and (4b) YFS3WW (1995-2001) YFS expon. + Leading Pole Approx. ($d\sigma/\sigma = 0.4\%$) [16,17].

The detailed description of the Coherent Exclusive Exponentiation (CEEX) can be found in ref. [1], along with the description of the older Exclusive Exponentiation (EEX), closer to the original Yennie-Frautschi-Suura (YFS) scheme [18]. What is in a nutshell YFS exponentiation? The main steps in YFS exponentiation are:

- Reorganization of the perturbative complete $\mathcal{O}(\alpha^\infty)$ series such that IR-finite $\bar{\beta}$ components are isolated (factorization theorem).
- Truncation of the IR-finite $\bar{\beta}$s to finite $\mathcal{O}(\alpha^n)$ and calculation of them from Feynman diagrams recursively.

*Work partly supported by the European Community's Human Potential Programme under contract HPRN-CT-2000-00149 "Physics at Colliders", by Polish Government grants KBN 5P03B09320 and 2P03B00122, and by NATO grant PST.CLG.977751.


doi:10.1016/S0920-5632(02)02300-9

Apart from disentangling properly ultraviolet divergences from the infrared (IR) ones YFS deals consistently with the overlapping and sub-leading IR singularities. To illustrate this point let is show the "map of IR singularities" for $2\gamma$ distribution $D_2(k_1, k_2)$ in the plane of the energies of two real photons:

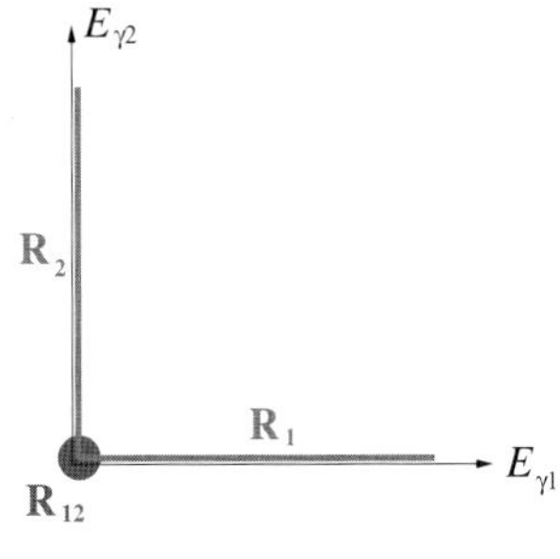


The point $R_{12}$ is the location of the leading double IR singularity and the lines $R_1$ and $R_2$ host non-leading single IR singularities. The above topological picture translates into a decomposition of the differential distributions:
$D_0(p_{f_1}, p_{f_2}, p_{f_3}, p_{f_4}) = \bar{\beta}_0(p_{f_1}, p_{f_2}, p_{f_3}, p_{f_4})$,
$D_1(p_f; \mathbf{k_1}) = \bar{\beta}_0(p_f)\tilde{S}(\mathbf{k_1}) + \bar{\beta}_1(p_f; k_1)$,and
$$D_2(\mathbf{k_1}, \mathbf{k_2}) = \bar{\beta}_0 \tilde{S}(\mathbf{k_1})\tilde{S}(\mathbf{k_2}) + \bar{\beta}_1(k_1)\tilde{S}(\mathbf{k_2}) + \bar{\beta}_1(k_2)\tilde{S}(\mathbf{k_1}) + \bar{\beta}_2(k_1, k_2)$$
where typicaly $p_{f_1} + p_{f_2} = p_{f_3} + p_{f_4}$ for $\bar{\beta}_0$ in $D_0$

and $p_{f_1}+p_{f_2} \neq p_{f_3}+p_{f_4}$ for $\bar{\beta}_0$ in $D_1$ and $D_2$. As we see the IR-finite part $\bar{\beta}_1$ in the decomposition of the 1-photon $D_1$ distribution into IR-leading and subleading parts is present as an element in the subleading 2-photon $D_2$ distribution. This nontrivial fact allows one to find out the IR-finite $\bar{\beta}_i$ functions order by order from standard perturbative calculations. In this particular case the recursive, order-by-order definition of IR-finite $\bar{\beta}$s looks as follows:

$\bar{\beta}_0(p_{f_1}, p_{f_2}, p_{f_3}, p_{f_4}) = D_0(p_{f_1}, p_{f_2}, p_{f_3}, p_{f_4}),$
$\bar{\beta}_1(p_f; k_1) = D_1(p_f; \mathbf{k_1}) - \bar{\beta}_0(p_f)\tilde{S}(\mathbf{k_1})$
$\bar{\beta}_2(k_1, k_2) = D_2(\mathbf{k_1}, \mathbf{k_2}) - \bar{\beta}_0\tilde{S}(\mathbf{k_1})\tilde{S}(\mathbf{k_2}) - \bar{\beta}_1(k_1)\tilde{S}(\mathbf{k_2}) + \bar{\beta}_1(k_2)\tilde{S}(\mathbf{k_1})$

Having at hand the IR-free $\bar{\beta}_i$ the YFS scheme provides immediately for the multi-photon differential distributions all over the phase space and the phase space is integraded without any approximations using MC method.

Below we show an example of the ISR $\mathcal{O}(\alpha^1)$ exponentiated multi-photon fully differential distributions for the CEEX variant of YFS exponentiation which is implemented in terms of the IR-free $\beta_i$ objects constructed at the amplitude level, rather than using the spin-summed squared $\bar{\beta}_i$s. The CEEX we present here very schematically for the process $e^-(p_1, \lambda_1) + e^+(p_2, \lambda_2) \rightarrow f(q_1, \lambda_1') + \bar{f}(q_2, \lambda_2') + \gamma(k_1, \sigma_1) + ...\gamma(k_n, \sigma_n)$:

$$\sigma = \sum_{n=0}^{\infty} \int_{m_\gamma} d\Phi_{n+2} \sum_{\lambda, \sigma_1, \ldots, \sigma_n} |e^{\alpha B(m_\gamma)} \mathfrak{M}^{\lambda}_{n, \sigma_1, \ldots, \sigma_n}(k_1, ..., k_n)|^2,$$

$\mathfrak{M}^{\lambda}_0 = \hat{\beta}^{\lambda}_0, \quad \lambda$=fermion helicities,
$\mathfrak{M}^{\lambda}_{1,\sigma_1}(k_1) = \hat{\beta}^{\lambda}_0 \mathfrak{s}_{\sigma_1}(k_1) + \hat{\beta}^{\lambda}_{1,\sigma_1}(k_1)$
$\mathfrak{M}^{\lambda}_{2,\sigma_1,\sigma_2}(k_1, k_2) = \hat{\beta}^{\lambda}_0 \mathfrak{s}_{\sigma_1}(k_1)\mathfrak{s}_{\sigma_2}(k_2)$
$\quad + \hat{\beta}^{\lambda}_{1,\sigma_1}(k_1)\mathfrak{s}_{\sigma_2}(k_2) + \hat{\beta}^{\lambda}_{1,\sigma_2}(k_2)\mathfrak{s}_{\sigma_1}(k_1)$
$\mathfrak{M}^{\lambda}_{n,\sigma_1,\ldots\sigma_n}(k_1, k_2, ...k_n) = \hat{\beta}^{\lambda}_0 \mathfrak{s}_{\sigma_1}(k_1)\mathfrak{s}_{\sigma_2}(k_2)...\mathfrak{s}_{\sigma_n}(k_n)$
$\quad + \hat{\beta}^{\lambda}_{1,\sigma_1}(k_1)\mathfrak{s}_{\sigma_2}(k_2)\ldots\mathfrak{s}_{\sigma_n}(k_n)$
$\quad + \mathfrak{s}_{\sigma_1}(k_1)\hat{\beta}^{\lambda}_{1,\sigma_2}(k_2)\ldots\mathfrak{s}_{\sigma_n}(k_n)\ldots$
$\quad \cdots + \mathfrak{s}_{\sigma_1}(k_1)\mathfrak{s}_{\sigma_2}(k_2)...\hat{\beta}^{\lambda}_{1,\sigma_2}(k_2)$ The $\mathcal{O}(\alpha^1)$ IR-finite building blocks are:
$\hat{\beta}^{\lambda}_0 = \left(e^{-\alpha B_4}\mathfrak{M}^{\mathrm{Born+Virt.}}_{\lambda}\right)\big|_{\mathcal{O}(\alpha^1)},$
$\hat{\beta}^{\lambda}_{1,\sigma}(k) = \mathfrak{M}^{\lambda}_{1,\sigma}(k) - \hat{\beta}^{\lambda}_0 \mathfrak{s}_{\sigma}(k)$

It is to be stressed that everything here was done in terms of the spin $\mathfrak{M}$-amplitudes! Multiphoton differential distributions are positive by construction (which is not the case for EEX). In KKMC the above CEEX is implemented up to $\mathcal{O}(\alpha^2)$ for ISR and FSR[2].

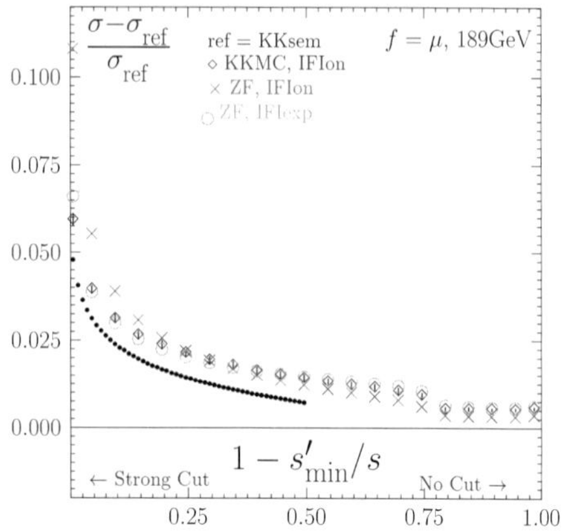


The example of the CEEX aplication is shown in the figure above from ref. [19] where we present the "state of art" comparison of KK MC vs. Zfitter for ISR+FSR, including IFI=ISR⊗FSR. We conclude that the precision $d\sigma/\sigma = 0.2\%$ was reached, which is enough for LEP2 data analysis. Similar high precision was obtained for the forward-backward asymmetry [19].

Recently, the CEEX matrix element was extended in KKMC to the neutrino channel. The contributions from following diagrams are now included:

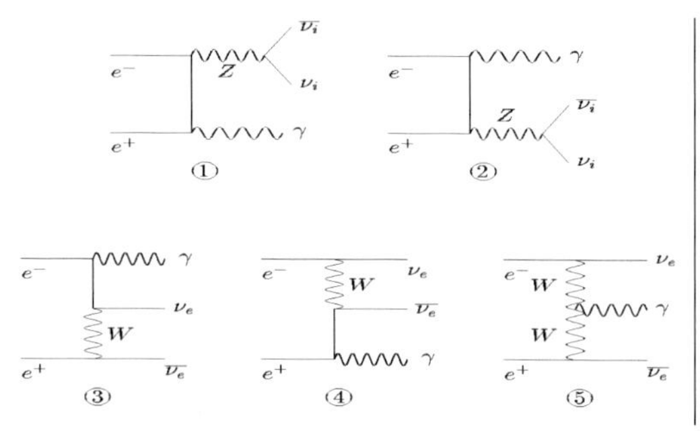


and also all double photon emission contributions feature the exact matrix element. We summarize below on the improvements in KKMC for the neutrino channel.

- The systematic error is estimated to be 1.3% for $\nu_e\bar{\nu}_e\gamma$ and 0.8% for $\nu_\mu\bar{\nu}_\mu\gamma$ and $\nu_\tau\bar{\nu}_\tau\gamma$.
- For observables with two observed photons we estimate the uncertainty to be about 5%.
- These new improved results were obtained thanks to the inclusion of non-photonic electroweak corrections of the ZFITTER package

[2]Multiperipheral contributions and non-IR ISR⊗FSR interferences are still incomplete at the $\mathcal{O}(\alpha^2)$ level. They are unimportant for most of LEP observables.

and due to newly constructed, exact, single and double emission photon amplitudes in the KKMC for the contribution with the $t$-channel $W$ exchange.

- The virtual corrections for the $W$ exchange are at present introduced in the approximated form. The exponentiation scheme CEEX is the same as in the original KKMC program

In the first version of KKMC of ref. [9] the $\mathcal{O}(\alpha^2)$ next-to-leading contributions were incomplete. In ref. [3] the $\mathcal{O}(\alpha^2)$ contribution from the following diagrams were evaluated

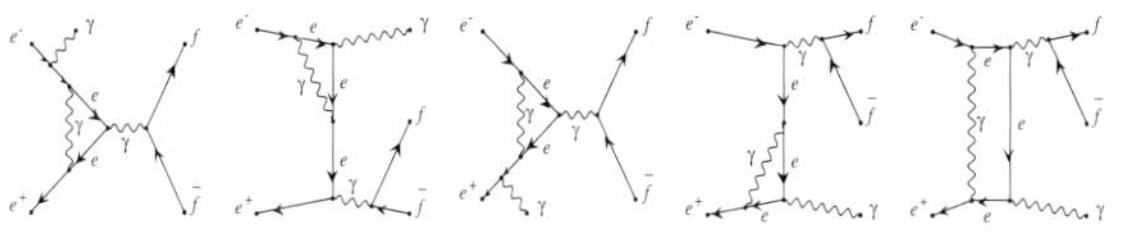

and partly included in the development of KKMC – and the remaining contributions were found to be negligible. The contributions in questions [3] are due to one real and one virtual photon emission and the new numerical results are shown in the following figure

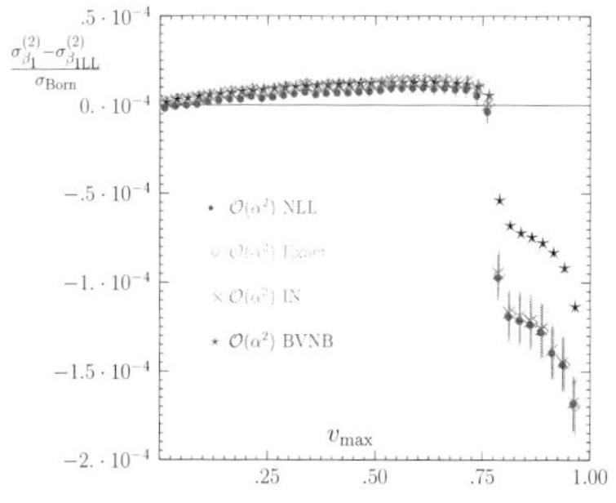


where the new calculation is shown to agree with the older calculations of the refs. [20,21]. NB. These contributions were also recently (re)calculated in ref. [22] for the purpose of the low energy electron collider experiments.

*Conclusions* The YFS inspired EEX and CEEX schemes are successful examples of the Monte Carlos based directly on the factorization theorem (albeit for IR soft case for Abelian QED only). They Work well in practice: KORALZ, BHLUMI, YWSWW3, BHWIDE, KK MC. Extension (as far as possible) to all collinear singularities would be very desirable and practically important! The KKMC program is extended to the neutrino channel. Missing fully differential $2f + 1\gamma_{virt} + 1\gamma_{real}$ distributions for $\mathcal{O}(\alpha^2)$ CEEX are now available.

## REFERENCES

1. S. Jadach, B.F.L. Ward and Z. Was, Phys. Rev. D63 (2001) 113009, hep-ph/0006359.
2. D. Bardin et al., Eur. Phys. J. C24 (2002) 373, hep-ph/0110371.
3. S. Jadach et al., Phys. Rev. D65 (2002) 073030, hep-ph/0109279.
4. S. Jadach and B.F.L. Ward, Phys. Rev. D38 (1988) 2897.
5. S. Jadach, B.F.L. Ward and Z. Was, The monte carlo program koralz, for the lepton or quark pair production at lep/slc energies: From version 4.0 to version 4.04, 1999, hep-ph/9905205, Computer. Phys. Commun. in print.
6. S. Jadach and B.F.L. Ward, Comput. Phys. Commun. 56 (1990) 351.
7. S. Jadach, B.F.L. Ward and Z. Wąs, Comput. Phys. Commun. 79 (1994) 503.
8. S. Jadach and B.F.L. Ward, Phys. Lett. B274 (1992) 470.
9. S. Jadach, B.F.L. Ward and Z. Was, Comput. Phys. Commun. 130 (2000) 260, hep-ph/9912214.
10. S. Jadach and B.F.L. Ward, Phys. Rev. D40 (1989) 3582.
11. S. Jadach et al., Comput. Phys. Commun. 70 (1992) 305.
12. S. Jadach et al., Comput. Phys. Commun. 102 (1997) 229.
13. S. Jadach, W. Płaczek and B.F.L. Ward, Phys. Lett. B390 (1997) 298, also hep-ph/9608412; The Monte Carlo program BHWIDE is available from `http://hephp01.phys.utk.edu/pub/BHWIDE`.
14. S. Jadach et al., Comput. Phys. Commun. 119 (1999) 272, hep-ph/9906277.
15. S. Jadach et al., Comput. Phys. Commun. 140 (2001) 475, hep-ph/0104049.
16. S. Jadach et al., Phys. Lett. B417 (1998) 326.
17. S. Jadach et al., Comput. Phys. Commun. 140 (2001) 432, hep-ph/0103163.
18. D.R. Yennie, S. Frautschi and H. Suura, Ann. Phys. (NY) 13 (1961) 379.
19. Two Fermion Working Group, M. Kobel et al., (2000), hep-ph/0007180.
20. M. Igarashi and N. Nakazawa, Nucl. Phys. B288 (1987) 301.
21. F. Berends, G. Burgers and W.V. Neerven, Phys. Lett. 177 (1986) 1191.
22. G. Rodrigo et al., Eur. Phys. J. C24 (2002) 71, hep-ph/0112184.

ELSEVIER

Proceedings of ICHEP 2002, pp. 164–167
*S. Bentvelsen, P. de Jong, J. Koch and E. Laenen (Editors)*

31st INTERNATIONAL CONFERENCE ON HIGH ENERGY PHYSICS AMSTERDAM

# LEP II beam energy measurement using radiative return events

C.G. Ainsley[a]*

[a]Cavendish Laboratory, University of Cambridge,
Madingley Road, Cambridge CB3 0HE, UK
E-mail: ainsley@hep.phy.cam.ac.uk

Using fermion-pair events radiative to the Z pole, recorded between the years 1997 and 2000, the four LEP experiments have been able to perform preliminary measurements of the LEP beam energy [1–4]. The results are found to be consistent with the standard calibration undertaken by the LEP Energy Working Group [5].

## 1. INTRODUCTION

One of the principal objectives of the LEP II program is the precise determination of the mass of the W boson, $M_{\mathrm{W}}$. Studies have shown that the resolution on the measured $M_{\mathrm{W}}$ is greatly improved by employing kinematic fits, in which the constraints of energy and momentum conservation are imposed [6]. This technique therefore offers a powerful means of extracting $M_{\mathrm{W}}$, provided the LEP II beam energy, $E_{\mathrm{b}}$, is known accurately. An uncertainty $\Delta E_{\mathrm{b}}$ on the beam energy would translate into an error $\Delta M_{\mathrm{W}}$ on the W mass measurement through

$$\frac{\Delta M_{\mathrm{W}}}{M_{\mathrm{W}}} = \frac{\Delta E_{\mathrm{b}}}{E_{\mathrm{b}}}. \quad (1)$$

The standard approach used to determine $E_{\mathrm{b}}$ at LEP II involves precise measurements based on resonant depolarisation of the beams at energies of 44–61 GeV (at higher energies the technique fails), combined with magnetic extrapolation to LEP II energies using NMR probe measurements [5]. This leads to a systematic uncertainty on $E_{\mathrm{b}}$ of approximately 20 MeV, dominated by the precision to which the magnetic extrapolation can be carried out.

To verify the reliability of the standard procedure, it is clearly desirable to have a direct measure of $E_{\mathrm{b}}$ at LEP II energies. The method presented here describes the use of radiative return events of the type

$$\mathrm{e^+e^-} \rightarrow \mathrm{Z}\gamma \quad ; \quad \mathrm{Z} \rightarrow \mathrm{f\bar{f}} \quad , \quad (2)$$

(Fig. 1) where the fermion f is a quark, muon or $\tau$-lepton, to perform such a check.

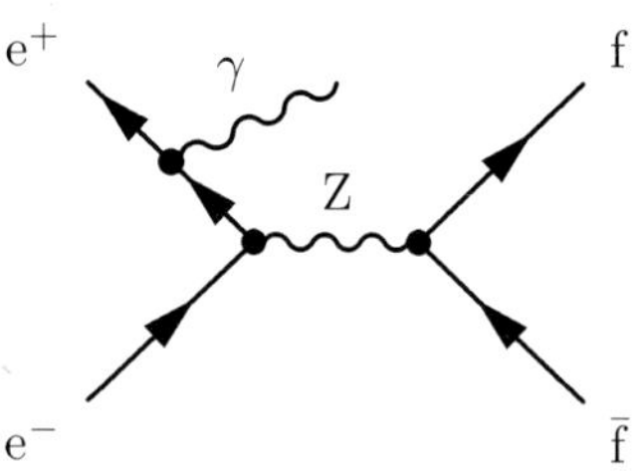


Figure 1. Radiative return fermion-pair production.

This technique was first applied by ALEPH to hadronic data collected in 1997 [1], where no evidence was found to refute the standard calibration; the results reported in this paper represent the outcome of work which has since been undertaken by DELPHI (1997–2000 data) [2], L3 (1998–2000 data) [3] and OPAL (1997–2000 data) [4] in order to substantiate this claim.

## 2. RADIATIVE RETURN APPROACH

At LEP II centre-of-mass energies, $\sqrt{s}(= 2E_{\mathrm{b}})$, above the Z peak, emission of hard initial state radiation (ISR) is common such that the *effective*

*Supported by a research studentship from the Particle Physics and Astronomy Research Council, UK.


doi:10.1016/S0920-5632(02)02097-2

*centre-of-mass energy after ISR* for the electron-positron system, denoted analogously by $\sqrt{s'}$, is close to the mass of the Z boson, $M_Z$. Utilising the precise knowledge of $M_Z$ obtained from LEP I [7], the spectrum of $\sqrt{s'}$ can be reconstructed both in fully simulated Monte Carlo (where $E_b$ is known *exactly*) and in data (where $E_b$ is *assumed* to be that computed by the standard technique). An example of the resulting $\sqrt{s'}$ distributions is given in Fig. 2: the contribution from backgrounds can be seen to be small, especially under the peak. Any systematic difference in the position of the radiative return peak in Monte Carlo and in data can be used to infer the *true* $E_b$ of the data.

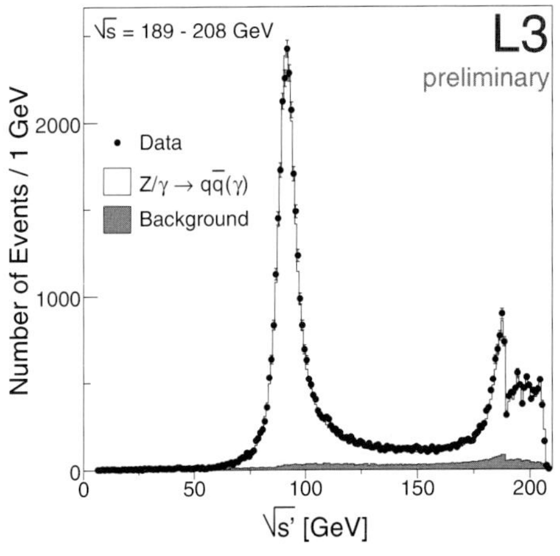


Figure 2. The $\sqrt{s'}$ distribution summed over all energy bins from 189 GeV to 208 GeV for the hadronic final state in L3 data. The return to the Z peak is clearly visible.

## 2.1. Reconstruction of $\sqrt{s'}$

Two alternative procedures have been adopted by the experiments to estimate $\sqrt{s'}$. For quark-pair events (detected as jets of hadrons), OPAL uses a kinematic fitting technique, constraining the total energy of all the particles in a given event (allowing for possible undetected photons escaping along the beam pipe) to equal $\sqrt{s}$. Combined with the constraint of three-momentum conservation, this improves the estimates of the jet four-momenta. The value of $\sqrt{s'}$, sensitive to $E_b$ as a result of this, is then computed as the invariant mass of the jets resulting from the fit. By contrast, DELPHI, L3 and OPAL in the leptonic channels calculate $\sqrt{s'}$ from the three-body event kinematics, using angles alone:

$$\sqrt{\frac{s'}{s}} = \sqrt{\frac{\sin\theta_{\ell+\gamma} + \sin\theta_{\ell-\gamma} - |\sin(\theta_{\ell+\gamma} + \theta_{\ell-\gamma})|}{\sin\theta_{\ell+\gamma} + \sin\theta_{\ell-\gamma} + |\sin(\theta_{\ell+\gamma} + \theta_{\ell-\gamma})|}} \tag{3}$$

where $\theta_{\ell+\gamma}$ and $\theta_{\ell-\gamma}$ are the angles between the leptons and the photon. Again, the dependence on $E_b$ is manifest.

## 2.2. Extraction of $E_b$

The problem of extracting $E_b$ from the data can be tackled in one of two ways. Either one can assume $M_Z$ from LEP I and measure $E_b$ directly, else one can assume $E_b$ in the data and measure $M_Z$, attributing any discrepancy, $\Delta M_Z$, from the nominal value to a mismeasurement, $\Delta E_b$, of the beam energy:

$$\Delta E_b = -E_b \left( \frac{\Delta M_Z}{M_Z} \right). \tag{4}$$

The first strategy is used, for example, in the hadronic channel of OPAL. A pair of matched relativistic Breit-Wigner functions:

$$S(s') = A \left( \frac{2\sqrt{s}}{s - s'} \right) \frac{s'\Gamma_\pm^2}{(s' - M^{*2})^2 + s'\Gamma_\pm^2} \tag{5}$$

with different widths, $\Gamma_\pm$, above and below the peak, $M^*$, were found to describe the distributions of $\sqrt{s'}$ well in the radiative return region. The application of (5) to data collected in the year 2000, together with the associated Monte Carlo, is illustrated in Fig. 3 for the analysis region $87 < \sqrt{s'}/\text{GeV} < 100$.

The fitting functions are scanned across the data as a function of $E_b$, and the optimal value is extracted at the equality between $M^*$ in data and $M^*$ in Monte Carlo.

By way of contrast, the second strategy is exemplified in the analysis of the hadronic channel by both DELPHI and L3. The value for the mass of the Z boson, $M_Z^{\text{new}}$, is varied in the Monte Carlo simulation to determine the point

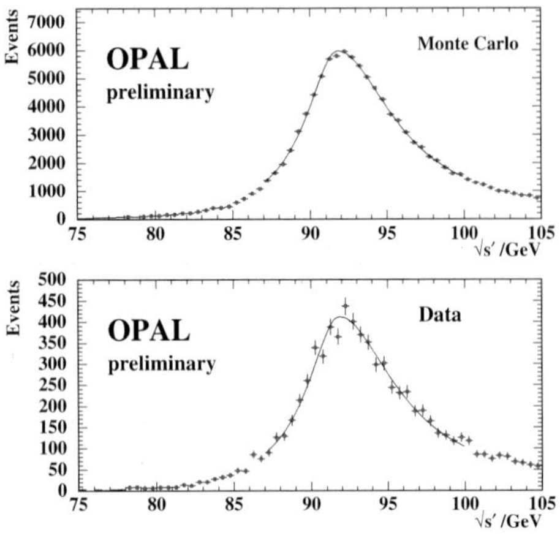


Figure 3. Fitting functions, used to estimate $M^*$, in simulation and year 2000 OPAL hadronic data.

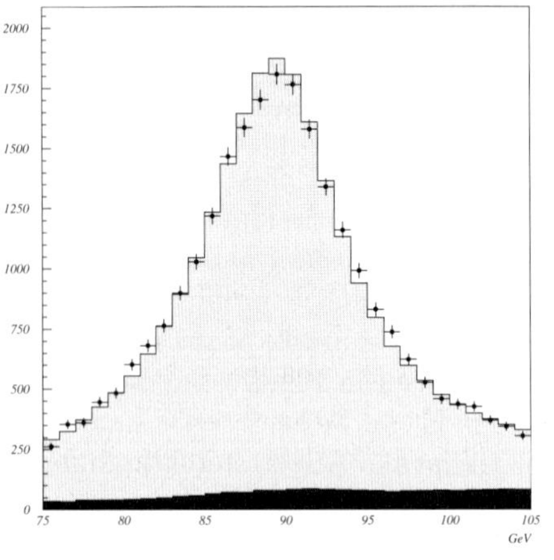


Figure 4. Reconstructed $\sqrt{s'}$ distributions for combined 1997–2000 DELPHI hadronic data (points) and simulation, with signal in light grey (yellow), background in dark grey (blue).

which yields the best agreement of the $\sqrt{s'}$ distribution between data and simulation. The variation is achieved by re-weighting the simulated $\sqrt{s'}$ spectrum using a Breit-Wigner-like function $f(s', M_Z)$, which is found to describe the shape of the radiative return peak well. For example, DELPHI employ

$$f(s', M_Z) = \frac{(1 + P(\sqrt{s'} - M_Z))s'}{(s' - M_Z^2)^2 + (s'\Gamma_Z/M_Z)^2}, \tag{6}$$

(where $P$ is a skewness parameter). The corresponding event weights are then $w(s', M_Z^{\rm new})$:

$$w(s', M_Z^{\rm new}) = \frac{f(s', M_Z^{\rm new})}{f(s', M_Z^{\rm LEP})}. \tag{7}$$

The distributions of $\sqrt{s'}$ for data and Monte Carlo (before re-weighting) are illustrated by Fig. 4.

After the re-weighting procedure has been carried out, the extracted value of $M_Z$ is compared to the nominal value, $M_Z^{\rm LEP}$, and $E_{\rm b}$ in the data is ascertained through (4).

Analogous techniques are employed in the analysis of the leptonic channels.

## 3. Systematic uncertainties

As is clear from section 2.2, the extraction of $E_{\rm b}$ relies on the direct comparison of data with Monte Carlo prediction. Thus the dominant systematic uncertainties are due to the uncertainties in the simulation.

- Detector modelling: Each year, calibration data at the Z peak are used to measure the energy scale and the energy and angular resolution for jets and leptons, together with their uncertainties. In addition, the uncertainty in the aspect ratio of each detector (the ratio of its length to width), which affects the angular reconstruction (and hence $\sqrt{s'}$), is accounted for. These amount to uncertainties of ~31 MeV in the hadronic channel and ~16 MeV in the leptonic channel, averaged over experiments.

- Hadronisation: Alternative phenomenological models, incorporated into the PYTHIA [8], HERWIG [9] and ARIADNE [10] Monte Carlo generators, are used to estimate the uncertainty in this process. By definition, this only affects the hadronic channel; the size of the effect, averaged over experiments, is ~29 MeV.

- ISR modelling: The difference between the ISR treatment in $\mathcal{KK}2f$ [11] and PYTHIA or between different orders in $\mathcal{KK}2f$ are used to estimate this uncertainty. In the hadronic channel this amounts to ~12 MeV and in the leptonic channel ~3 MeV, averaged over experiments.

## 4. Results

Fig. 5 shows the measurements of $\Delta E_{\rm b}$ completed by DELPHI, L3 and OPAL in both the hadronic and leptonic channels, together with their combinations, averaged over all years of data-taking. Fully taking into account the correlated systematic uncertainties, the combined results per channel and the overall combined measurement are also illustrated. These are summarised in Table 1.

| Experiment | f | $\Delta E_{\rm b}$ /MeV |
|---|---|---|
| DELPHI | q | −55±53±65 |
| DELPHI | $\mu$ | +113±75±27 |
| L3 | q | −83±37±75 |
| OPAL | q | −65±34±70 |
| OPAL | e, $\mu$, $\tau$ | −1±68±26 |
| Combined | q | −65±24±45 |
| Combined | e, $\mu$, $\tau$ | +51±50±19 |
| Combined | q, e, $\mu$, $\tau$ | −10±27±26 |

Table 1
Values of $\Delta E_{\rm b}$ derived from radiative return events: $e^+e^- \rightarrow Z\gamma \rightarrow f\bar{f}\gamma$. The first error is statistical, and the second systematic.

## 5. Conclusion

This paper has presented new preliminary measurements of the LEP II beam energy using fermion-pair events radiative to the Z pole. The findings are in good agreement with the standard calibration—a result which bodes well for many precision measurements at LEP, and, in particular, for the determination of $M_{\rm W}$. Furthermore, this method has been able to achieve, in the leptonic channel at least, a level of systematic uncertainty on a par with the ~20 MeV associated with the standard technique. Moreover, the latter relies on a *circulating* beam, rendering it unusable at a future linear collider. In a high statistics regime, the method of using radiative return events, by contrast, offers a viable option that might well be further investigated.

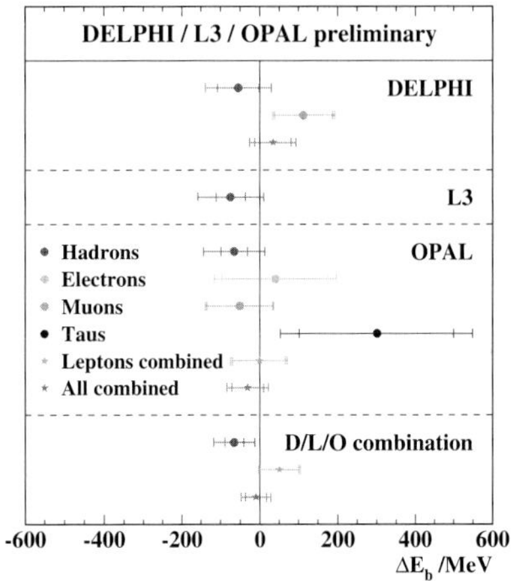


Figure 5. Measurements of $E_{\rm b}$ made by the LEP experiments relative to that made by the LEP Energy Working Group.

## REFERENCES


1. ALEPH Collab.; R. Barate et al., Phys. Lett. B464 (1999) 339.
2. DELPHI Collab.; *Preliminary determination of $E_{beam}$ at LEP2 using radiative 2-fermion events in DELPHI*, DELPHI 2002-084 CONF 618.
3. L3 Collab.; *Measurement of the Z boson mass using Z decays in $e^+e^- \rightarrow Z\gamma$ events at LEP 2 energies*, L3 Note 2733.
4. OPAL Collab.; *Determination of the LEP Beam Energy using Radiative Fermion-pair Events*, OPAL Physics Note PN493.
5. LEP Energy Working Group notes 98-02, 99-01, 00-01, 01-01, `http://lepecal.web.cern.ch/LEPECAL/reports/reports.html`
6. A. Ballestrero et al., "Physics at LEP2" p.141; CERN 96-01 vol 1.
7. Particle Data Group, D. E. Groom et al., Eur. Phys. J. C15 (2000) 1.
8. T. Sjöstrand et al., Comput. Phys. Commun. 135 (2001) 238.
9. G. Corcella et al., J. High Energy Phys. 0101 (2001) 010.
10. L. Lönnblad, Comput. Phys. Commun. 71 (1992) 15.
11. S. Jadach, B.F.L. Ward and Z. Wąs, Phys. Lett. B449 (1999) 97.

ELSEVIER

Proceedings of ICHEP 2002, pp. 168–171
*S. Bentvelsen, P. de Jong, J. Koch and E. Laenen (Editors)*

31st INTERNATIONAL CONFERENCE ON HIGH ENERGY PHYSICS AMSTERDAM

# W Polarisation and Spin Density Matrix Measurements at LEP

H. Voss[a] *

[a]Institut de Physique des Hautes Energies (IPHE), Lausanne University, Switzerland

The helicity structure of the W-pair production is studied by the LEP experiments using either measurements of spin density matrix elements or a fit of the expected angular distribution of the different helicity states to the observed data. The measurements are in good agreement with the Standard Model expectation, clearly confirming the longitudinal component of the W-gauge-boson. The W-helicity is either observed for each W-boson individually or in correlation of the two bosons. The mesurements also provide a test for genuine CP violation, with the measurements confirming CP invariance.

## 1. Introduction

At LEP2 W-bosons are produced in pairs via the s-channel three gauge boson vertex or the t-channel neutrino exchange. The three gauge boson vertex is a direct consequence of the non-Abelian structure of the $SU(2)_L \otimes U(1)_Y$ gauge group describing the electroweak interaction in the Standard Model. Deviations from the Standard Model in the W-pair production are often modeled using various coupling parameters introducing a model dependence. The measurement of the helicity structure of the W-pairs, in terms of their polarisation or spin density matrix, offers a model independent approach to the triple gauge boson vertex, as a test of the Standard Model and a search window to physics beyond it. It also provides indirect access to the electroweak symmetry breaking mechanism, responsible for the W-mass and hence the longitudinal polarisation of the W-boson.

The LEP experiments have each collected about 700 $pb^{-1}$ of data above the WW-production threshold during the years 1997-2000. The preliminary W-polarisation analysis from L3 [1] is already based on the whole available data set, while the spin density matrix measurements reported here from DELPHI [2], L3 [3] and OPAL [4] so far only use about 180 $pb^{-1}$ of data taken at a centre-of-mass energy of 189 GeV during 1998.

*CERN EP, 1211 Geneva 23, Switzerland


doi:10.1016/S0920-5632(02)02101-1

The W-bosons are produced in one of the three possible helicity states $\tau = -1, 0, +1$. The W-spin is reflected in the angular distribution of its decay products, typically measured in the W-rest frame. Following the well known V-A coupling of W-boson to fermions, the decay distribution of the fermion from a $W^-(W^+)$ with helicity $\pm1$ $(\mp1)$is $\propto (1 \mp \cos\theta^*)^2$, with $\cos\theta^*$ being the polar angle of the decay fermion in the W rest frame. The decay distribution for the longitudinal W-bosons with helicity $\tau = 0$ is $\propto \sin^2\theta^*$. In order to unambigiously measure the decay angles, it is necessary to distinguish between fermion and anti-fermion in the decay. With each of the W-bosons decaying either hadronically or leptonically, the WW event is either fully hadronic, fully leptonic or "semi-leptonic". As it is not feasible to identify the quark flavour for the hadronically decaying W-boson, all experiments use the "semi-leptonic" events for their measurements. Here the event topology can be fully reconstructed in contrast to the fully leptonic events, and the decay angles can be unambiguosly reconstructed at least for the leptonically decaying W-boson.

The spin density matrix (SDM) elements $\rho_{\tau_+,\tau'_+,\tau_-,\tau'_-}$ are the sums of the normalised products of the respective helicity amplitudes of the $W^+$ and $W^-$. While the full-SDM contains all the interferences and correlations between the two W-bosons and each of its three possible helicity states, with the limited size of the data sample, it is also usefull to study the single particle spin density matrix $\rho_{\tau,\tau'}$ only. This is obtained by looking at one of the W-bosons, averaging over the helicity states of the second W. This reduces the number of elements in the hermitian SDM from 81 to 9. The density matrix elements are extracted from the data using projection operators [5], exploiting the symmetry properties of the decay distribution for the various W-helicity states. The diagonal elements $\rho_{\tau\tau}$ of the spin density matrix are real and can be interpreted as the probability to produce a W boson with helicity $\tau$.

## 2. Polarisation Measurement

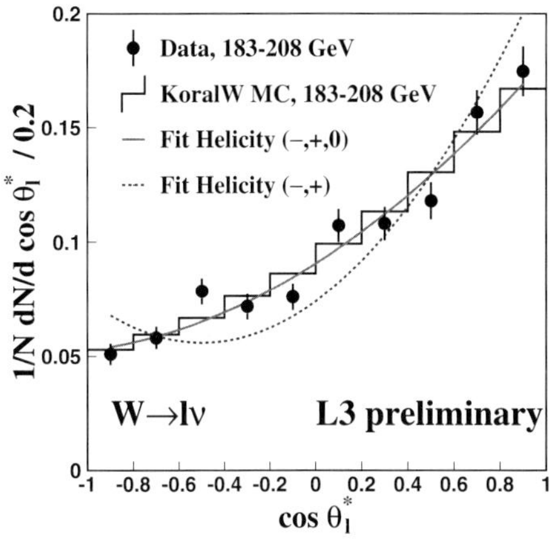


Figure 1. Corrected $\cos\theta_\ell^*$ distribution for the leptonic W-decays at $\sqrt{s}$=183-208 GeV, showing also the fit results for the different W-helicities.

The helicity fractions in L3 are obtained from a fit of the theoretical $\cos\theta^*$ distributions for the different W-helicity states to the efficiency and background corrected data. The measurements are performed separately for the leptonically and the hadronically decaying W. The latter allows only for a determination of the longitudinal and the sum of the two transverse helicity fractions, because only these give decay distributions symmetric under exchange of the indistinguishable quark and anti-quark. The two results are later combined for the longitudinal polarisation state exploiting the fact that all the helicity fractions need to sum up to one. The results clearly establish the existence of the longitudinal helicity state, as can be seen in Figure 1 where the failing fit result without allowing for longitudinally polarised W-bosons is shown. The results are also given in Table 1. The polarisation of the W-bosons has also been measured by OPAL using the SDM analysis. For this measurement, the SDM elements are efficiency and background corrected using Monte Carlo estimates. The polarisation of the W-bosons depends on the W-production angle $\cos\theta_W$, hence do the SDM elements, which are determined in 8 bins of $\cos\theta_W$. The polarised differential cross sections are obtained by multiplying the measured differential cross section with the corresponding diagonal elements of the single particle SDM element $\rho_{\tau,\tau}(\cos\theta_W)$. The overall polarisation given in Table 1 is obtained by integrating over $\cos\theta_W$. The polarisation dependence on $\cos\theta_W$ has also been measured by L3, dividing their data sample into 4 bins of $\cos\theta_W$. Both measurements by L3 and OPAL are in good agreement with the Standard Model expectations.

## 3. Spin Correlations

While the previously mentioned measurements focused on one of the W-bosons in the event only, both W-bosons can also be treated correlated. This is done by OPAL in the SDM approach by looking at combinations of full-SDM elements that are symmetric under the exchange of fermion and anti-fermion in one W-decay, because the quark flavour is not identified in the hadronically decaying W. The differential polarised cross sections for W-pairs with both W-bosons being transversely polarised, both being longitudinally polarised or a mixed combination is given in figure 2. In L3 the correlated spin states of the W-bosons are observed in regions of $\cos\theta_W$ where it is more likely to find a transverse polarised W-boson than a longitudial one. As shown in Figure 3, the enrichement or depletion of the transverse w.r.t. the longitudinal polarisation for the hadronically decaying W-boson, here leads to a difference in the measured fraction of transverse polarisation in the leptonically decaying W-boson of up to -0.378±0.134±0.060.

Table1: The measured overall helicity fractions of the W-bosons in L3 and OPAL. The expectation given for L3 corresponds to a luminosity-weighted average over the different beam energies.

| L3 $\sqrt{s}$ = (183-208) GeV | | | |
|---|---|---|---|
| helicity fraction | $\tau(W^-) = -1$ ; $\tau(W^+) = +1$ | $\tau(W^-) = +1$ ; $\tau(W^+) = -1$ | $\tau = 0$ |
| DATA $W^-$ | 0.547±-0.046±0.012 | 0.198±-0.027±0.013 | 0.255 ±0.038 ±0.006 |
| DATA $W^+$ | 0.631±-0.048±0.023 | 0.180±-0.024±0.014 | 0.189 ±0.039 ±0.015 |
| DATA $W^\pm$ | 0.585±-0.033±0.016 | 0.188±-0.018±0.015 | 0.228 ±0.027 ±0.012 |
| expected | 0.590 | 0.169 | 0.241 |

| OPAL $\sqrt{s}$ = 189 GeV | | |
|---|---|---|
| helicity fraction | $\tau = \pm 1$ | $\tau = 0$ |
| DATA | 0.790±0.033 ±0.016 | 0.210 ±0.033 ±0.016 |
| expected | 0.746±0.006 | 0.257 ±0.004 |

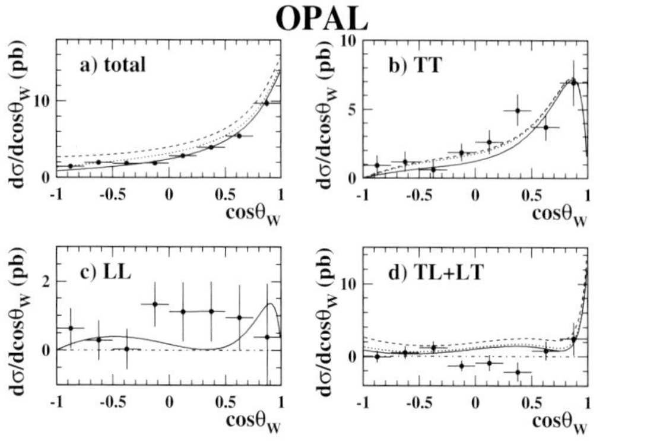


Figure 2. Differential production cross section for W-pairs, with both W-bosons having transverse($\sigma_{TT}$), longitudinal ($\sigma_{LL}$) or mixed polarisations ($\sigma_{TL}$). The points are OPAL data including statistical and systematic uncertainties. The solid line corresponds to the Standard Model and dotted line to anomalous couplings $\tilde{\kappa}_Z = \pm 0.5$

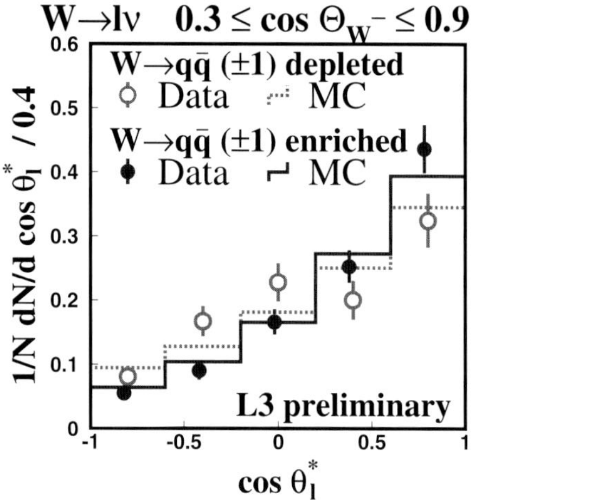


Figure 3. Corrected $\cos\theta^*_\ell$ distribution for the leptonic W-decays at $\sqrt{s}$=183-208 GeV. The distributions are shown for subsamples with enriched and depleted events with transverse helicity in the hadronically decaying W.

## 4. Spin Density Matrix and CP test

In Figure 4 an example from DELPHI is given for the whole single particle SDM elements without detector correction, in comparison with the Standard Model Monte Carlo expectation. All three experiments have results for this SDM matrix, all showing good agreement with the Standard Model expecation. The current SDM analysis by L3 (DELPHI) is restricted to the clean and almost background free $q\bar{q}e\bar{\nu}_e/q\bar{q}\mu\bar{\nu}_\mu$ ($q\bar{q}\mu\bar{\nu}_\mu$) events. The dependence of the various helicity contributions on the W-production angle $\cos\theta_W$ is clearly visible. In this measurement, CP invariance is assumed relating the elements for the $W^-$ to the $W^+$. With CP invariance, the fractions helicities (+) and (−) for the $W^-$ are equal to the fractions of opposite helicities (−) and (+) of the $W^+$. The fractions are separately listed in Table 1, showing that they agree with the assumption of CP invariance. Certain combinations of the imaginary parts of the off-diagonal SDM elements, shown in Figure 5, have to be fulfilled

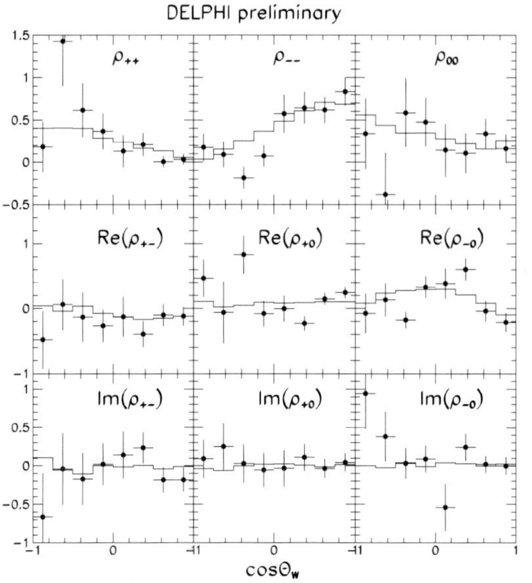


Figure 4. The single particle SDM measured in $q\bar{q}\mu\bar{\nu}_\mu$ events from DELPHI. The points show the data with the statistical errors only and the solid line corresponds to the Monte Carlo expectation.

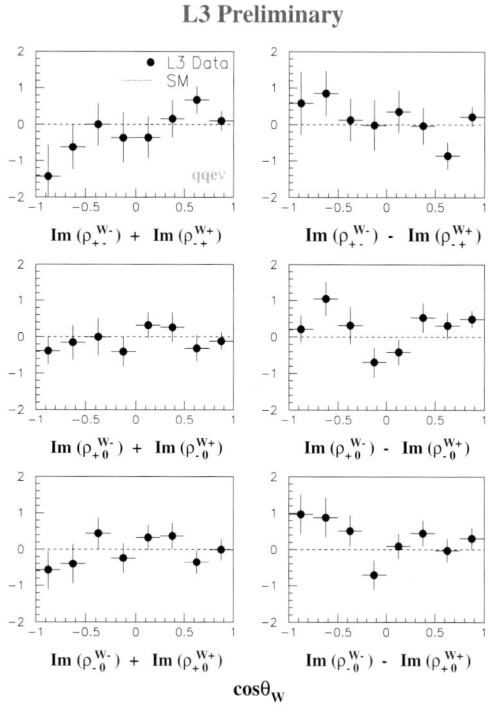


Figure 5. The difference (sum) of the imaginary parts of the off-diagonal SDM elements, sensitive to the CP-violation at tree (loop) level. The points are the L3 data with statistical errors only from $q\bar{q}e\bar{\nu}_e$ events at 189 GeV. The Standard Model prediction is zero.

for CP-invariance at tree level. The sum of the corresponding elements needs to be zero for more general CPT invariance. A deviation from zero here would indicate possible CP violation at the loop level. None of the tests showed signs for CP violation in any of the experiments.

## 5. Summary

The polarisation and SDM analyses offer a model independent way of studying the W-pair production properties. The measurements are consistent with the helicity distributions expected in the Standard Model. Moreover, the measurements passed the test of CP invariance of the three gauge boson vertex.

## REFERENCES

1. The L3 Collaboration, *Measurement of the* W-*Polarisation and the* WW *Spin Correlations in* $e^+e^-\to W^+W^-$ *events at LEP*, L3 Note 2755 (June 2002), submitted to Summer Conferences 2002; The L3 Collaboration, M. Acciarri *et al.*, Phys. Lett. **B474** (2000) 194.
2. The DELPHI Collaboration, *Spin Density Matrix analysis of the reaction* $e^+e^-\to W^+W^-$ *at 189* GeV, DELPHI Note 2001-098, submitted to Summer Converences 2001.
3. The L3 Collaboration, *Spin Density Matrix Analysis in* W-*Pair Production at* $\sqrt{s}$*=189* GeV, L3 Note 2752 (June 2002), submitted to Summer Conferences 2002.
4. The OPAL Collaboration, G. Abbiendi et al., Eur. Phys. J. C19 (2001) 229.
5. G. Gounaris, J. Layssac, G. Moultaka and F.M. Renard, Int. J. Mod. Phys. **A19** (1993) 3285.

ELSEVIER

Proceedings of ICHEP 2002, pp. 172–176
*S. Bentvelsen, P. de Jong, J. Koch and E. Laenen (Editors)*

31st INTERNATIONAL CONFERENCE ON HIGH ENERGY PHYSICS AMSTERDAM

# Precision predictions for $W$-pair production at LEP2

W. Płaczek[a b *]

[a]Institute of Computer Science, Jagellonian University, ul. Nawojki 11, 30-072 Cracow, Poland.

[b]CERN, TH Division, CH-1211 Geneva 23, Switzerland.

Theoretical calculations for the $W$-pair production process at LEP2 in terms of Monte Carlo event generators `RacoonWW` and `KorlaW&YFSWW3` are reviewed. The discussion concentrates on precision predictions for the main LEP2 $WW$ observables. The theoretical precision of the above programs is estimated to be $\sim 0.5\%$ for the total $WW$ cross section $\sigma_{WW}$, $\sim 5\,\mathrm{MeV}$ for the $W$-boson mass $M_W$, and $\sim 0.005$ for the triple-gauge-boson coupling $\lambda = \lambda_\gamma = \lambda_Z$, which is sufficient for the final LEP2 data analyses.

## 1. Introduction

The process of $W$-pair production in electron–positron colliders is very important for testing the Standard Model (SM) and searching for signals of possible "new physics"; see e.g. Refs. [1,2]. One of the main goals when investigating this process at present and future $e^+e^-$ experiments is to measure precisely the basic properties of the $W$ boson, such as its mass $M_W$ and width $\Gamma_W$. This process also allows for a study of the triple and quartic gauge-boson couplings at the tree level, where small deviations from the subtle SM gauge cancellations can lead to significant effects on physical observables – these can be signals of "new physics". On the other hand, the process $W$-pair production and decay constitutes a significant background to other processes, in particular to Higgs-boson production [3].

During the 1996–2000 LEP run, the four LEP2 experiments collected about 10,000 $W$-pair events each. This allows a test of the $W$-boson physics at the precision level of $\sim 1\%$ [4]. The main $WW$ observables investigated at LEP2 are (1) the total $WW$ cross section $\sigma_{WW}$, (2) a distribution of the $W$ invariant mass, which is used to extract the $W$-boson mass $M_W$, and (3) a distribution of the $W$ production polar angle $\cos\Theta_W$, which is the most sensitive observable to the triple-gauge-boson couplings (TGCs). The expected experimental accuracies of the final LEP2 data analyses for $\sigma_{WW}$, $M_W$ and the C- and P-conserving TGC $\lambda = \lambda_\gamma = \lambda_Z$ are given in Table 1. In this table we also present requirements for the theoretical precision of these quantities. For the theoretical errors ($\delta_{th}$) it is commonly required that they should be smaller than the experimental ones ($\delta_{ex}$) by a factor of at least 2.

Table 1:

| Quantity | Final LEP2 $\delta_{ex}$ | Required $\delta_{th}$ |
|---|---|---|
| $\sigma_{WW}$ | $\sim 1\%$ | $\leq 0.5\%$ |
| $M_W$ | $\sim 30\,\mathrm{MeV}$ | $\leq 15\,\mathrm{MeV}$ |
| $\lambda$ | $\sim 0.01$ | $\leq 0.005$ |

The final LEP2 data analyses need theoretical predictions in terms of Monte Carlo event generator(s) (MCEG) that meet these precision requirements.

## 2. Theoretical description of $W$-pair production

The basic lowest-order process in which $W$ pairs were produced at LEP2 is:

$$e^+ + e^- \to W^+ + W^- \to f_1 + \bar{f}_2 + f_3 + \bar{f}_4. \quad (1)$$

This can be described by three Feynman diagrams, called CC03: two annihilation diagrams with $\gamma$ and $Z$ in the $s$-channel (this is where the TGCs appear), and a conversion diagram with an electron neutrino in the $t$-channel (this contribution dominates at LEP2 energies). And here the

*Work partly supported by the European Commission 5th framework contract HPRN-CT-2000-00149.


doi:10.1016/S0920-5632(02)02103-5

first theoretical problem appears. Namely, these three diagrams alone are not gauge-invariant. In order to obtain a gauge-invariant scattering amplitude, one has to include a certain class of singly $W$-resonant diagrams. Therefore, the minimal gauge-invariant subset of Feynman diagrams containing $W$ pairs is the so-called CC11-class – for more details see e.g. [5]. This leads in practice to the necessity of considering the full $4f$-process, i.e. $e^+e^- \to 4f$, which considerably complicates the theoretical description.

The next problem that appears in the theoretical description of this process is the inclusion of a finite $W$-boson width $\Gamma_W$. This inclusion is necessary to avoid singularities in the scattering amplitude coming from $W$-boson propagators. Up to now there is no fully satisfactory approach to deal with this in the general $4f$-process. Some simple schemes, such as the "fixed-width scheme" and the "running-width scheme" are known to violate gauge invariance. The schemes that do not violate gauge invariance, such as the "fermion-loop scheme" or the "complex-mass scheme" possess other drawbacks, see e.g. [5,6]. In practice one commonly uses the "fixed-width scheme", for which gauge-violating effects are numerically negligible at LEP2 energies [5].

All the above problems increase dramatically when one goes to $\mathcal{O}(\alpha)$ calculations. At the one-loop level, the number of Feynman diagrams increases to several thousands per $4f$-channel (final state) [5] and the inclusion of the finite $W$-boson width gets much more difficult.

One may ask the question: Do we need to go beyond the Born level? So let us briefly discuss the various radiative corrections and their effects on the main LEP2 $WW$ measurements.

**Pure QED correction:**

QED corrections can be divided into a few classes: (1) initial-state radiation (ISR), i.e. photon radiation from incoming beams; (2) Coulomb correction – electromagnetic interaction of slowly moving $W^+W^-$; (3) final-state radiation (FSR), i.e. photon radiation in $W$ decays; and (4) non-factorizable (NF) corrections corresponding to interconnections of various stages of the process. They affect the $WW$ observables in different ways. The most sizeable numerical effects come from the ISR. For $\sigma_{WW}$ they amount to between about $-20\%$ near the $WW$ threshold and about $-5\%$ at $E_{CM} = 200\,\mathrm{GeV}$ (they depend stronlgy on $E_{CM}$). $M_W$ is affected at $\sim 10\,\mathrm{MeV}$, but this may be enhanced by experimental reconstruction effects (kinematic fits, etc.). The TGC $\lambda$ is shifted by the ISR by $\sim 0.07$. The Coulomb correction modifies mainly $\sigma_{WW}$ at the level of $\sim 6\%$ near the $WW$ threshold. The FSR influences considerably the $M_W$ measurement, at the level of $\sim$ 10–80 MeV, which strongly depends on experimental acceptances [7]. The NF corrections affect only $M_W$ at the $\sim$ 1–5 MeV level, when treated inclusively [8] (which is a good approximation for LEP2 experimental acceptances).

**Electroweak (EW) corrections:**

The leading EW corrections connected with an effective scale of the $W$-pair production and decay process can be taken into account by using the so-called $G_\mu$ scheme. This corresponds to parametrizing the cross section by the Fermi constant $G_\mu$ instead of the fine-structure constant $\alpha$. Numerically, this changes the overall normalization, i.e. $\sigma_{WW}$ by $\sim 15\%$. Then the $\mathcal{O}(\alpha)$ non-leading (NL) EW corrections amount to 1–2% for $\sigma_{WW}$ at LEP2 energies. They also affect the TGCs, e.g. $\lambda$ at the level of $\sim$ 0.01–0.02.

**QCD corrections:**

QCD corrections affect normalizations as well as events shapes of the hadronic $WW$ channels. They are usually accounted for by including the so-called naive QCD correction, i.e. the normalization factor $\alpha_s(M_W^2)/\pi \simeq 3.8\%$ for each final-state quark pair and $2\alpha_s(M_W^2)/3\pi$ for $\Gamma_W$, in the parton-level calculation, while leaving the exclusive QCD effects to be modelled by dedicated MC packages, such as PYTHIA, etc.

All the above effects are necessary in a MCEG for $W$-pair production at LEP2. The most difficult are $\mathcal{O}(\alpha)$ NL EW corrections. To date, there exist no complete one-loop calculations, even for the simplest CC11-type channels. Even if such calculations existed, they would probably be very complex and slow in numerical computation. Therefore they would not be useful for MC event generation. These are the reasons why some efficient approximations for the $WW$ process are necessary, at least for LEP2.

In the case of $W$-pair production, one is interested in the doubly $W$-resonant process. Therefore one can make use of another expansion parameter, which is $\Gamma_W/M_W \simeq 2.5\%$, and apply the so-called pole expansion, i.e. an expansion about the complex pole corresponding to an unstable $W$. This is the expansion in increasing powers of $\Gamma_W/M_W$, which corresponds to decreasing powers of the resonance. Then, one usually applies the so-called leading-pole approximation (LPA) in which only the highest-pole (resonant) contributions are retained, i.e. terms $\sim (\Gamma_W/M_W)^0$. For the $W$-pair production process this means retaining only double-pole contributions, i.e. applying a double-pole approximation (DPA). In this context LPA means just DPA; however, LPA itself has a more general meaning. The resulting matrix element is gauge-invariant, and the imaginary part of the pole position corresponds to the finite $W$-boson width, see e.g. [9,5,6]. In general, the LPA is however not sufficient at the Born level, because the error introduced by this approximation can be here $\sim \Gamma_W/M_W$, i.e. $\sim 2.5\%$. It may also not be sufficient for the leading corrections because of big-log enhancements, i.e. $\sim (\Gamma_W/M_W)(\alpha/\pi)\log(Q^2/m^2)$, where $Q$ is some large momentum transfer and $m$ is a small fermion mass. Here lower-pole terms might be necessary for the desired experimental precision. In practice, it is usually simpler to take the whole $e^+e^- \to 4f$, including the leading corrections. Then, for the genuine $\mathcal{O}(\alpha)$ NL EW corrections the uncertainty introduced by the LPA is $\sim (\Gamma_W/M_W)(\alpha/\pi) < 10^{-4}$, therefore negligible for the LEP2 precision.

The above solutions have been implemented in two MC event generators: `RacoonWW` [10–13] and `KoralW&YFSWW3` [14–20,7,21,22]. These programs are briefly described in the next two sections.

## 3. The MC program `RacoonWW`

The main features of the program `RacoonWW` are:

- Matrix elements for all $e^+e^- \to 4f$ and $e^+e^- \to 4f\gamma$ processes in massless-fermion approximation.
- The ISR in the $\mathcal{O}(\alpha^2)$ leading-log (LL) approximation with soft-photon exponentiation through QED structure functions.
- Coulomb correction for off-shell $W$'s.
- The NF virtual corrections in the DPA.
- The virtual $\mathcal{O}(\alpha)$ NL EW corrections in the DPA (for which one-loop calculations of on-shell $WW$ production and decay are used).
- Two QCD-inspired methods of treating soft and collinear photon singularities: dipole subtraction and phase-space slicing (to get a proper matching between virtual and real-photon corrections).
- Anomalous Triple and Quartic gauge-boson couplings.
- Multichannel MC algorithm for integration and event generation.

More details can be found in Refs. [10–13].

## 4. The MC programs `KoralW` and `YFSWW3`

`KoralW` and `YFSWW3` are actually two independent MCEGs, with some specific and some common features, which are listed below.

**`KoralW` specific features:**

- The fully massive matrix element for all $e^+e^- \to 4f$ processes (generated by the GRACE System [23]).
- Two independent efficient multichannel MC algorithms for the full $4f$ phase space.

**`YFSWW3` specific features:**

- Multiphoton radiation in the $WW$ production stage in the Yennie–Frautschi–Suura (YFS) exponentiation scheme.
- The $\mathcal{O}(\alpha)$ NL EW corrections in the LPA (based on the $\mathcal{O}(\alpha)$ calculations for on-shell $WW$ production of Refs. [24,25]).

**Common features:**

- The ISR in the YFS exponentiation scheme up to $\mathcal{O}(\alpha^3)$ LL with non-zero $p_T$ multiphoton radiation.
- Coulomb correction for off-shell $W$'s.
- Non-factorizable QED corrections in an inclusive approximation of the "screened-Coulomb" ansatz [8].
- Anomalous TGCs (three parametrizations).

- The FSR at $\mathcal{O}(\alpha^2)$ LL generated by `PHOTOS` [26] (with non-zero $p_T$ photons).
- $\tau$ decays done by `TAUOLA` [27], quark fragmentation and hadronization managed by `JETSET` [28].
- The semi-analytical program `KorWan` for the $WW$ process including leading corrections (for test, fits, etc.).

As can be seen from the above description, `KoralW` is devoted to the full $e^+e^- \to 4f$ process and it includes the leading corrections, i.e. all except $\mathcal{O}(\alpha)$ NL. `YFSWW3`, on the other hand, is dedicated to a precision description of the signal $WW$ process, i.e. it includes all numerically important radiative corrections to this process treated in the LPA, but does not include the $4f$-background contributions. However, thanks to the common features of the two programs, it is easy to correct the predictions of one of them, so that the final results include all the necessary contributions. In fact, the two programs implement special reweighting routines for this purpose. Moreover, we managed to combine the results of the two programs in a real-time execution (event-by-event) using the Unix/Linux named (FIFO) pipe mechanism. This resulted in the so-called Concurrent Monte Carlo (CMC) `KoralW&YFSWW3`. This CMC works effectively as a single MCEG and provides the description of the $W$-pair production process with all contributions/corrections needed for the LEP2 precision – for more details see Ref. [18].

## 5. Theoretical precision of the main LEP2 $WW$ observables

The precision requirements of the theoretical predictions for some of the main LEP2 observables has been shown in Table 1. Any MCEG to be used for the final LEP2 data analysis should satisfy these requirements. Estimating the precision of theoretical calculations is a difficult task, because it requires assessing the size of missing higher-order contributions. In practice, this is done by comparing different calculations (programs), switching on and off various contributions/corrections and looking at the effects for a given observable. In some cases this can be supplemented with a simplified scale-parameter analysis. All such investigations can give a hint on the size of the missing effects in the theoretical description.

The programs `RacoonWW` and `YFSWW3` have been extensively cross-checked and compared for many observables, see e.g. [6]. It has been found that for $\sigma_{WW}$ the two programs agree within 0.3% over the LEP2 energy range and they describe the LEP2 data very well [29]. From the comparisons between these two programs, the comparisons with other calculations and investigations of various effects, the theoretical precision for $\sigma_{WW}$ at LEP2 has been estimated at $\sim 0.5\%$ [6]. Similar studies, although more involved, were also performed for the $W$ mass $M_W$ and the TGC $\lambda$. In the case of the $W$ mass, we performed $M_W$ fits using `KorWan` to the invariant-mass distributions from `KoralW&YFSWW3` and `RacoonWW`, where various effects have been switched on and off. We found, for example, that results of the two programs differ in terms of the fitted $M_W$ by $\leq$ 3 MeV. We also performed a scale-parameter estimation of the missing effects. From all these studies we estimated the theoretical precision of $M_W$ for the above programs at $\sim 5\,\mathrm{MeV}$. All this is described in detail in Ref. [30]. A similar analysis for the TGC $\lambda$ was done in Ref. [31]. In this case the fits of $\lambda$ to the $\cos\Theta_W$ distributions were performed using not only `KorWan` but also dedicated polynomial fitting functions. These fitting functions were constructed using MC-genarated data from `KoralW&YFSWW3` and `RacoonWW`, not only at the parton level but also including the ALEPH detector-simulation effects. Based on these studies, we estimated the theoretical precision for $\lambda$ at $\sim 0.005$.

## 6. Conclusions

There are two independent Monte Carlo programs for the precision predictions of $W$-pair production at LEP2: `RacoonWW` [10–13] and `KoralW&YFSWW3` [14–20,7,21,22]. They include the $4f$-background contributions as well as all the necessary radiative corrections at the precision level required by LEP2. The agreement between these programs for the main observables is within the required accuracy of the LEP2 experiments. From comparisons between these programs, comparisons with other calculations, and investigations of various effects, we have estimated the theoretical precision ($\delta_{th}$) for three of the main LEP2 $WW$ observables: $\delta_{th}\sigma_{WW} \sim 0.5\%$, $\delta_{th}M_W \sim 5\,\mathrm{MeV}$ and $\delta_{th}\lambda \sim 0.005$. Comparing this with Table 1, it is clear that this satisfies the LEP2 precision requirements. However, for the future linear colliders (LC) this is not sufficient and further improvements in the theoretical calculations are necessary.

## REFERENCES


1. G. Altarelli, T. Sjöstrand, and F. Zwirner, editors, *Physics at LEP2* (CERN 96-01, Geneva, 1996), 2 vols.
2. S. Jadach, G. Passarino, and R. Pittau, editors, *Reports of the Working Groups on Precision Calculations for LEP2 Physics* (CERN 2000-009, Geneva, 2000).
3. C. Mariotti, Search for the SM Higgs Boson at LEP, these proceedings.
4. M. Grünewald, Electroweak Physics, these proceedings.
5. W. Beenakker *et al.*, in Ref. [1], Vol. 1, p. 79.
6. M. Grünewald *et al.*, Four-fermion production in electron–positron collisions, in Ref. [2], p. 1, hep-ph/0005309.
7. S. Jadach, W. Płaczek, M. Skrzypek, B. F. L. Ward, and Z. Wąs, Phys. Rev. **D61**, 113010 (2000).
8. A. P. Chapovsky and V. A. Khoze, Eur. Phys. J. **C9**, 449 (1999), hep-ph/9902343.
9. R. Stuart, Nucl. Phys. **B498**, 28 (1997), and references therein.
10. A. Denner, S. Dittmaier, M. Roth, and D. Wackeroth, Nucl. Phys. **B560**, 33 (1999), hep-ph/9904472.
11. A. Denner, S. Dittmaier, M. Roth, and D. Wackeroth, Phys. Lett. **B475**, 127 (2000), hep-ph/9912261.
12. A. Denner, S. Dittmaier, M. Roth, and D. Wackeroth, Nucl. Phys. **B587**, 67 (2000), hep-ph/0006307.
13. A. Denner, S. Dittmaier, M. Roth, and D. Wackeroth, Eur. Phys. J. **C20**, 201 (2001), hep-ph/0104057.
14. M. Skrzypek, S. Jadach, W. Płaczek, and Z. Wąs, Comput. Phys. Commun. **94**, 216 (1996).
15. M. Skrzypek, S. Jadach, M. Martinez, W. Płaczek, and Z. Wąs, Phys. Lett. **B372**, 289 (1996).
16. T. Ishikawa, Y. Kurichara, M. Skrzypek, and Z. Wąs, Eur. Phys. J. **C4**, 75 (1998).
17. S. Jadach, W. Płaczek, M. Skrzypek, Z. Wąs, and B. F. L. Ward, Comput. Phys. Commun. **119**, 272 (1999).
18. S. Jadach, W. Płaczek, M. Skrzypek, Z. Wąs, and B. F. L. Ward, Comput. Phys. Commun. **140**, 475 (2001), hep-ph/0104049.
19. S. Jadach, W. Płaczek, M. Skrzypek, and B. F. L. Ward, Phys. Rev. **D54**, 5434 (1996).
20. S. Jadach, W. Płaczek, M. Skrzypek, B. F. L. Ward, and Z. Wąs, Phys. Lett. **B417**, 326 (1998).
21. S. Jadach, W. Płaczek, M. Skrzypek, B. F. L. Ward, and Z. Wąs, Phys. Rev. **D65**, 093010 (2002), hep-ph/0007012.
22. S. Jadach, W. Płaczek, M. Skrzypek, B. F. L. Ward, and Z. Wąs, Comput. Phys. Commun. **140**, 432 (2001), hep-ph/0103163.
23. J. Fujimoto *et al.*, GRACE User's manual, version 2.0, MINAMI-TATEYA collaboration.
24. J. Fleischer, F. Jegerlehner, and M. Zrałek, Z. Phys. **C42**, 409 (1989).
25. K. Kołodziej and M. Zrałek, Phys. Rev. **D43**, 3619 (1991).
26. E. Barberio, B. van Eijk, and Z. Wąs, Comput. Phys. Commun. **66**, 115 (1991), *ibid.* **79**, 291 (1994).
27. R. Decker, S. Jadach, J. H. Kühn, and Z. Wąs, Comput. Phys. Commun. **76**, 361 (1993).
28. T. Sjöstrand and M. Bengtsson, Comput. Phys. Commun. **43**, 367 (1987).
29. L. Malgeri, Measurement of cross sections $\sigma_{WW}$, $\sigma_{ZZ}$, $\sigma_{\gamma\gamma}$ at LEP, these proceedings.
30. S. Jadach, W. Płaczek, M. Skrzypek, B. F. L. Ward, and Z. Wąs, Phys. Lett. **B523**, 117 (2001), hep-ph/0109072.
31. R. Brunelière *et al.*, Phys. Lett. **B533**, 75 (2002), hep-ph/0201304.

ELSEVIER

Proceedings of ICHEP 2002, pp. 177–181
*S. Bentvelsen, P. de Jong, J. Koch and E. Laenen (Editors)*

31st INTERNATIONAL CONFERENCE ON HIGH ENERGY PHYSICS AMSTERDAM

# Prospects for Higgs search at the LHC

F. Piccinini[a*] [†]

[a]CERN, Theoretical Physics Division, CH 1211 Geneva 23, Switzerland

The present status of theoretical calculations for signal and background processes relevant to the Higgs boson search at the LHC is reviewed, with special emphasis on recent developments. The issue of Higgs properties determination at the LHC is addressed.

## 1. Introduction

The search for the Higgs boson will be one of the primary goals of the LHC experiments. The main Higgs production channels at the LHC are gluon fusion, weak boson fusion (WBF) and $t\bar{t}H$ and $VH(V = W, Z)$ associated productions. Complete analyses with full detector simulation within the experimental collaborations have shown that, with only 10 $\text{fb}^{-1}$ of integrated luminosity per experiment, corresponding to one year of running at low luminosity ($\mathcal{L} = 10^{33}$ $\text{cm}^{-2}\,\text{s}^{-1}$), there is a signal significance above the $5\sigma$ level over a wide range of Higgs masses, the most difficult region being the window between the lower bound given by LEP and 140 GeV [1]. Recent detailed simulations, with realistic experimental conditions of the WBF processes, have proved the importance of these channels to improve significantly the sensitivity in that difficult region [2,3]. Crucial to this aim, on the experimental side, are the forward jet reconstruction and central jet veto efficiencies. On the theoretical side, a large amount of work has been done to reduce the uncertainties on theoretical predictions, both for the signals and the backgrounds. Further studies have been performed to improve on the strategy originally proposed in ref. [4] for the determination of Higgs boson properties, such as couplings to fermions and gauge bosons, and total width. Moreover, very recently, the first analyses on the LHC potential for the Higgs self-coupling determination have been carried out. In the following the present status of theoretical calculations and of the prospects for Higgs properties determination is reviewed. The main focus will be on the mass window 115-200 GeV, which is the preferred one by electroweak precision data and also partially by supersymmetry. During the last year a lot of effort has been concentrated on the Standard Model (SM) Higgs boson, but the strategies and results can be translated to the lightest scalar Susy Higgs. For these reasons this brief review deals only with the case of the SM Higgs boson.

## 2. Theoretical calculations

In order to disentangle a signal from backgrounds, a good understanding of uncertainties in theoretical predictions is necessary. Predictions based on leading order (LO) calculations are plagued by considerable uncertainties due to the strong dependence on the renormalization and factorization scales, introduced by the QCD coupling and the parton densities. At present the QCD corrections, at least at next-to-leading order (NLO), are known for all production channels, the most recent calculations being the NNLO calculation for the gluon fusion process in the limit of heavy top-quark mass [5,6] and the NLO corrections for the process $pp/p\bar{p} \to t\bar{t}H + X$ [7]. The higher-order corrections reduce the renormalization and factorization scale dependence of the theoretical predictions. In the case of $pp/p\bar{p} \to$

*On leave of absence from INFN Sezione di Pavia, Italy.
†Work supported in part by the EU Fourth Framework Programme "Training and Mobility of Researchers", Network "Quantum Chromodynamics and the Deep Structure of Elementary Particles", contract FMRX–CT98–0194 (DG 12 – MIHT).


doi:10.1016/S0920-5632(02)02107-2

$t\bar{t}H+X$, the NLO corrections increase the rate by roughly 20% with respect to the LO predictions over the entire intermediate Higgs mass range at the LHC. In the case of Higgs production through gluon fusion, the NLO corrections give a $K$ factor of the order of 2, so that NNLO corrections are needed [5,6]. At present, the uncertainties arising from QCD uncertainties (combining the residual scale dependence with the error from parton distribution functions) can be estimated to be of the order of $\pm 20\%$ for gluon fusion, $\pm 5\%$ for WBF, $\pm 10\%$ for associated production.

Concerning the backgrounds, several processes with low final state parton multiplicity (corresponding to important irreducible backgrounds) are available at NLO, namely $q\bar{q} \to \gamma\gamma$ [8], $gg \to \gamma\gamma$ [9], $pp(\bar{p}) \to Wb\bar{b}$, $pp(\bar{p}) \to Zb\bar{b}$ [10], $pp(\bar{p}) \to Wjj$, $pp(\bar{p}) \to Zjj$ [11], $pp(\bar{p}) \to VV$ [12] and QCD $H + jj$ production via gluon fusion [13]. Some of these calculations are already implemented in NLO Monte Carlo programs. In the case of multiparton final states, the methods developed up to now for NLO calculations cannot be applied, because of the complexity of the calculations for processes with many external legs. Recently some effort has been devoted to the realization of LO Monte Carlo event generators based on exact matrix element calculations, suitably interfaced to the shower evolution Monte Carlo programs producing the real final state hadrons [14–18].

## 3. Higgs couplings to fermions and gauge bosons

The LHC will allow not only the discovery of the Higgs boson, but also the study of its properties, such as mass, width and couplings to fermions and gauge bosons. While the decay channels $H \to \gamma\gamma$ and $H \to ZZ^{(*)} \to 4l$ will allow a direct mass measurement at the 0.1% level over a wide range of masses, the total width can only be determined with about 10% accuracy by direct measurement with the decay $H \to ZZ^{(*)} \to 4l$ for $m_H > 200$ GeV, the Higgs width for lower Higgs masses being too small with respect to the detector resolution. As will be shown below, by exploiting the available production and decay mechanisms at the LHC, an indirect measurement of the total width can be performed also in the low mass region. In principle, the Higgs coupling, for

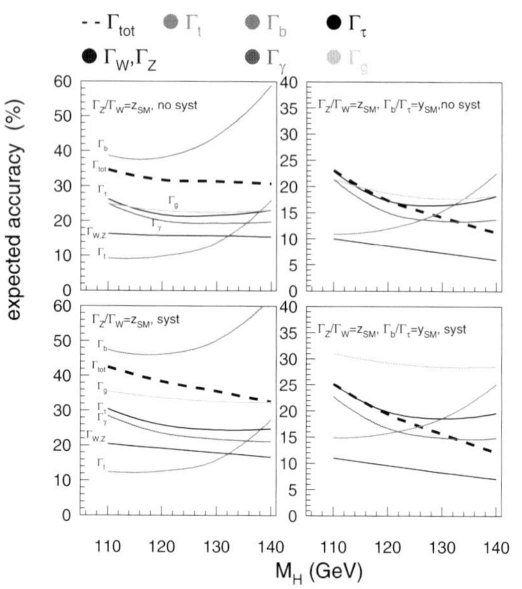


Figure 1. Relative accuracy (%) on the individual rates $\Gamma_i$ expected at the LHC (from ref. [30]). See the text for a detailed description of the panels.

instance to a given fermion family $f$, could be obtained from the following relation:

$$R(H \to f\bar{f}) = \int Ldt \cdot \sigma(pp \to H) \cdot \frac{\Gamma_f}{\Gamma},$$

where $R(H \to f\bar{f})$ is the Higgs production rate in a given final state, which can be measured experimentally, $\int Ldt$ is the integrated luminosity, $\sigma(pp \to H)$ is the Higgs production cross section, and $\Gamma$ and $\Gamma_f$ are the total and partial Higgs widths respectively. Hence, a measurement of the Higgs production rate in a given channel allows the extraction of the partial width for that channel, and therefore of the Higgs coupling $g_f$ to the involved decay particles ($\Gamma_f \sim g_f^2$), provided that the Higgs production cross-section and the total Higgs width are known from theory. Aiming at model-independent coupling determinations, one needs to consider ratios of couplings, which are experimentally accessible through the measurements of ratios of rates for different final states, because in the ratio the total Higgs cross-section and width cancel (as well as luminosity and all QCD uncertainties related to the initial state). In spite of the fact that the gluon fusion mechanism is the leading scalar Higgs production mode at the LHC, other subleading production modes, such as weak boson fusion and associated production, are extremely important to provide complementary information and allow unique determinations of ratios of Higgs boson couplings. Up to now detailed studies on signal and backgrounds for several channels have been performed, namely $gg \to H, (H \to \gamma\gamma, ZZ, WW)$ [1, 19–21], $qq \to qqH, (H \to \gamma\gamma, \tau\tau, WW)$ [22–26], $pp \to t\bar{t}H, (H \to b\bar{b}, WW, \tau\tau)$ [27–30] and $pp \to WH, H \to b\bar{b}$ [31]. Each process depends on two Higgs couplings, one from the Higgs boson production and one from the Higgs boson decay, with the exception of the weak boson fusion channels, for which it is experimentally impossible to distinguish between $WW \to H$ and $ZZ \to H$ production mechanisms. However, since the couplings of a scalar Higgs boson to the $Z$ and $W$ gauge bosons are closely related to the electroweak $SU(2)$ gauge symmetry, which has been very successfully tested by the LEP experiments, and since in a large class of models the ratio of $HWW$ and $HZZ$ couplings is identical to the one in the SM, including the MSSM, it is reasonable to assume $\Gamma_Z/\Gamma_W = z_{SM}$. Under this hypothesis, every production and decay channel provides a measurement of the ratio $Z_j^{(i)} = \Gamma_i\Gamma_j/\Gamma$, where $i = g, W, t$ indicates the production process and $j = b, \tau, W, Z, g, \gamma$ indicates the decay process. For the case $m_H < 140$ GeV, the above mentioned channels allow us to express the individual rates $\Gamma_t$, $\Gamma_b$, $\Gamma_\tau$, $\Gamma_W$, $\Gamma_g$ and $\Gamma_\gamma$ as functions of the observables $Z_j^{(i)}$ and the total Higgs width $\Gamma$ [30]. With the additional assumption that the total width is saturated by the known channels $\Gamma = \Gamma_b + \Gamma_\tau + \Gamma_W + \Gamma_Z + \Gamma_g + \Gamma_\gamma$ (otherwise new processes would be observed independently of any precision study), an expression for $\Gamma$ can be obtained in terms of measured quantities, namely [30]

$$\sqrt{\Gamma} = \frac{1}{\sqrt{Z_W^{(W)}}}\left[Z_\tau^{(W)}\left(1+\frac{Z_b^{(t)}}{Z_\tau^{(t)}}\right) + Z_W^{(W)}(1+z_{SM}) + \frac{Z_W^{(W)}Z_\gamma^{(g)}}{Z_\gamma^{(W)}} + Z_\gamma^{(W)}\right].$$

Figure 1 [30] summarizes the relative accuracy on the individual rates $\Gamma_i$ expected in the model-independent scenario as well as in a scenario with $\Gamma_b/\Gamma_\tau$ fixed to its SM value, assuming a total integrated luminosity of 200 fb$^{-1}$. The upper plots show the accuracies obtained without including any theoretical systematic error, while the lower plots show the same accuracies when a systematic theoretical error of 20% for the $gg \to H$ channel, of 5% for the $qq \to qqH$, and of 10% for the $pp \to t\bar{t}H$ channel are included. As can be seen, the total Higgs width can be indirectly determined in the low mass region with a precision of the order of 30% in a model-independent way, and the Higgs couplings can be determined with accuracies between 7% and 25%. In the case of $140 < m_H < 200$ GeV, the gluon fusion, weak boson fusion and $t\bar{t}H$ associated production processes can be exploited, with the Higgs boson decaying only to gauge bosons, allowing an indirect determination of $\Gamma_W$ and $\Gamma$ with a precision of the order of 10% [4,32]. In this Higgs mass range, however, there is no handle to study the Higgs Yukawa coulings to $b$ quarks and $\tau$ leptons. The assumption $\Gamma_Z/\Gamma_W = z_{SM}$ can be tested at the 20–30% level for $m_H > 130$ GeV by measuring the ratio $Z_Z^{(g)}/Z_W^{(g)}$ [32], and it can even be tested with the same level of accuracy for lower Higgs boson masses by comparing the two ratios $Z_b^{(WH)}/Z_b^{(t)}$ and $Z_\tau^{(W)}/Z_\tau^{(t)}$ [30]. For $m_H > 140$ GeV, with luminosities of the order of 300 fb$^{-1}$, the ratio $\Gamma_t/\Gamma_g$ can be tested in a model-independent way through a measurement of $Z_W^{(t)}/Z_W^{(g)}$ [29].

## 4. Higgs self-couplings

A complete determination of the parameters of the SM would require the measurement of the Higgs self-couplings. These include trilinear and

Table 1
Expected numbers of signal and background events after all cuts for the $gg \to HH \to 4W \to l^+l'^+4j\nu\nu$ final state, for $\int \mathcal{L} = 6000$ fb$^{-1}$ [34].

| $m_H$ | Signal | $t\bar{t}$ | $W^{\pm}Z$ | $W^{\pm}W^+W^-$ | $t\bar{t}W^{\pm}$ | $t\bar{t}t\bar{t}$ | $S/\sqrt{B}$ |
|---|---|---|---|---|---|---|---|
| 170 GeV | 350 | 90 | 60 | 2400 | 1600 | 30 | 5.4 |
| 200 GeV | 220 | 90 | 60 | 1500 | 1600 | 30 | 3.8 |

quadrilinear interactions. In the SM the corresponding couplings are fixed at LO in terms of the Higgs mass and vacuum expectation value $v$, namely $\lambda^{SM}_{HHH} = 3m_H^2/v$, $\lambda^{SM}_{HHHH} = 3m_H^2/v^2$. A direct measurement of $\lambda_{HHH}$ could be obtained via the detection of Higgs pair production, where a contribution is expected from the production of a single off-shell Higgs which decays into a pair of Higgses. This contribution is always accompanied by diagrams where the two Higgs bosons are radiated independently, with couplings proportional to the Yukawa couplings or the gauge couplings. As a result, different production mechanisms will lead to different sensitivities of the $HH$ rate to the value of $\lambda_{HHH}$. In the literature the following SM channels have been considered [33]: inclusive $HH$ production, dominated by the partonic process $gg \to HH$; vector boson fusion $qq \to qqV^*V^* \to qqHH$, associated production with $W$ or $Z$ bosons $q\bar{q} \to VHH$; associated production with top-quark pairs $gg/q\bar{q} \to t\bar{t}HH$. With the exception of the gluon fusion process, which has a total cross section at the level of few tens of fb, the cross section for all other channels is of the order of 1 fb over the intermediate Higgs mass range [33]. Given these low production rates, and the potentially large backgrounds associated to the $HH$ final states, a quantitative study of the Higgs self-coupling is very hard at the LHC. Recently a study of signal and backgrounds has been performed for the $gg \to HH$ channel [34], both for a standard LHC luminosity of $10^{34}\,\mathrm{cm}^{-2}\mathrm{s}^{-1}$ [35] and for a possible future upgrade of the luminosity to $10^{35}\,\mathrm{cm}^{-2}\mathrm{s}^{-1}$ [34]. Among all possible decay channels, the most interesting one turned out to be $gg \to HH \to W^+W^-W^+W^- \to l^{\pm}\nu jjl^{\pm}\nu jj$, which has a good branching ratio for $m_H \geq 170$ GeV. The like-sign lepton requirement is essential to reduce the high-rate opposite-sign lepton final states from Drell–Yan and $t\bar{t}$ production. Potential backgrounds to the considered signature are given by $t\bar{t}$+jets, $WZ$+jets, $t\bar{t}W$, $WWWjj$ including the resonant channel $W(H \to WW)jj$ and $t\bar{t}t\bar{t}$. By applying the cuts described in ref. [34], the number of events for signal and backgrounds are summarized in Table 1 for an integrated luminosity of 6000 fb$^{-1}$, where a signal significance of 5.3 (3.8) $\sigma$ for $m_H = 170\,(200)$ GeV can be reached, optimistically assuming that the main parameters of the detector performance will remain the same as those expected at $10^{34}\,\mathrm{cm}^{-2}\mathrm{s}^{-1}$. This would lead to a determination of the total production cross-section with a statistical uncertainty of $\pm 20\%$ ($\pm 26\%$) for $m_H = 170$ GeV (200 GeV), allowing a determination of $\lambda_{HHH}$ with statistical errors of 19% (25%) [34]. In the case of 300 fb$^{-1}$ only the non-vanishing of the Higgs self-coupling could be established at 95% C.L. for 150 GeV $< m_H <$ 200 GeV [35].

## 5. Summary

During the last few years there has been a dramatic improvement in both theoretical and experimental studies of several Higgs boson production and decay channels at the LHC. On the theoretical side, the corrections at NLO (in one case even at NNLO) have been calculated for the main Higgs production processes and for many irreducible backgrounds. Several LO Monte Carlo event generators based on exact matrix elements have been developed very recently to give prediction for multiparton final states, which represent important backgrounds for several Higgs signatures. On the experimental side, complete simulations, including full detector simulation, have been carried out for all production processes, pointing out the relevance of the weak boson fusion processes as discovery channels in the low Higgs mass region. Considering all channels, a signal significance above $5\sigma$ over the entire mass spectrum is well established already with only 10 $fb^{-1}$ of integrated luminosity per experiment. A strategy has been designed to study, in a model-independent way, the Higgs couplings to fermions and bosons, which allows also, with little theoretical assumption, an indirect determination of the total Higgs width. Recently the potential of the LHC in the determination of the Higgs self-coupling has been investigated, but only with an integrated luminosity of 6000 $fb^{-1}$, and in the mass range $170 \leq m_H \leq 200$ GeV a quantitative study could be performed.

The author wishes to thank S. Catani, F. Gianotti, R. Harlander, K. Jakobs, M.L. Mangano and M.T. Zeyrek for many useful discussions. M.L. Mangano is also acknowledged for a careful reading of the manuscript.

## REFERENCES

1. ATLAS Collaboration, Tech. Rep. CERN/LHCC/99-15, CERN, 1999.
2. K. Jakobs, talk given at ATLAS week, Clermont-Ferrand, June 2002.
3. H.D. Yildiz, R. Kinnunen and M.T. Zeyrek, CMS Note 2001/050; N. Akchurin *et al.*, CMS Note 2002/016.
4. R. Kinnunen *et al.*, Phys. Rev. D62 (2000) 013009.
5. M. Grazzini, hep-ph/0209302; S. Catani, D. de Florian and M. Grazzini, JHEP 0105 (2001) 025; and 0201 (2002) 015.
6. W.B. Kilgore, these proceedings; R.V. Harlander and W.B. Kilgore, Phys. Rev. D64 (2001) 025; Phys. Rev. Lett. 88 (2002) 201801; C. Anastasiou and K. Melnikov, hep-ph/0207004.
7. W. Beenakker *et al.*, Phys. Rev. Lett. 87 (2001) 201805; S. Dawson and L. Reina, Phys. Rev. Lett. 87 (2001) 201804; S. Dawson, L. Reina and D. Wackeroth, Phys. Rev. D65 (2002) 053017.
8. T. Binoth *et al.*, Eur. Phys. J. C16 (2000) 311; hep-ph/0203064.
9. Z. Bern, L. Dixon and C. Schmidt, hep-ph/0206194.
10. R.K. Ellis and S. Veseli, Phys. Rev. D60 (1999) 011501; J. Campbell, hep-ph/0105226.
11. J. Campbell and R.K. Ellis, hep-ph/0202176.
12. S. Frixione and B.R. Webber, hep-ph/0204244.
13. V. Del Duca *et al.*, N. P. B616 (2001) 367.
14. M.L. Mangano *et al.*, hep-ph/0206293.
15. A. Pukhov *et al.*, hep-ph/9908288.
16. T. Ishikawa *et al.*, MINAMI-TATEYA Group Coll., KEK-92-19; S. Tsuno *et al.*, hep-ph/0204222.
17. T. Stelzer and W.F. Long, Comput. Phys. Commun. 81 (1994) 357; F. Maltoni and T. Stelzer, hep-ph/0208156.
18. B.P. Kersevan and E. Richter-Was, hep-ph/0201302.
19. CMS Collaboration, Tech. Rep. CERN/LHCC/94-38, CERN, 1994.
20. M. Dittmar, and H.K. Dreiner, Phys. Rev. D55 (1997) 167; hep-ph/9703401.
21. D. Denegri *et al.*, hep-ph/0112045.
22. D. Rainwater and D. Zeppenfeld, JHEP 12 (1997) 005.
23. D. Rainwater, D. Zeppenfeld and K. Hagiwara, Phys. Rev. D59 (1999) 014037.
24. T. Plehn, D. Rainwater and D. Zeppenfeld, Phys. Rev. D61 (2000) 093005.
25. D. Rainwater and D. Zeppenfeld, Phys. Rev. D60 (1999) 113004.
26. N. Kauer *et al.*, Phys. Lett. B503 (2001) 113.
27. E. Richter-Was and M. Sapinski, Acta Phys. Polon. B30 (1999) 1001.
28. V. Drollinger, T. Müller and D. Denegri, hep-ph/0111312.
29. F. Maltoni, D. Rainwater and S. Willenbrock, hep-ph/0202205.
30. A. Belyaev and L. Reina, hep-ph/0205270.
31. V. Drollinger, T. Müller and D. Denegri, hep-ph/0201249.
32. D. Zeppenfeld, hep-ph/0203123.
33. A. Djouadi *et al.*, E. P. J. C10 (1999) 45.
34. F. Gianotti, M.L. Mangano, T. Virdee (conveners), hep-ph/0204087; A. Clark, A. Blondel and F. Mazzucato, ATL-PHYS-2002-005.
35. U. Baur, T. Plehn and D. Rainwater, hep-ph/0206024.

Proceedings of ICHEP 2002, pp. 182–185
*S. Bentvelsen, P. de Jong, J. Koch and E. Laenen (Editors)*

31st INTERNATIONAL CONFERENCE ON HIGH ENERGY PHYSICS AMSTERDAM

# Fermion pair production at LEP2

I. De Bonis[a]

[a]Laboratoire d'Annecy-le-Vieux de Physique des Particules,
74941 Annecy-Le-Vieux Cedex, France

Preliminary results on combined cross-sections and forward-backward asymmetries from $e^+e^- \to f\bar{f}$ processes at LEP2 for center-of-mass energy $\sqrt{s}$ from 130 to 207 GeV, are summarized. These results provide a test of the Standard Model to $\mathcal{O}(1\%)$ and allow to constraint physics beyond the Standard Model, such as four-fermion contact interactions, gravitational interaction in extra dimensions and leptoquarks models.

## 1. INTRODUCTION

Since the start of LEP2, each experiment has collected data for a total integrated luminosity of about 700 $pb^{-1}$, at center-of-mass energy $\sqrt{s}$ from 130 to 207 GeV [1]. In this paper, combined di-fermion results from the four LEP experiments up to 207 GeV are presented. Then, constraints on physics beyond the Standard Model such as four-fermion contact interactions, gravitational interaction in extra dimensions and leptoquarks are reported.

## 2. FERMION PAIR PRODUCTION AT LEP2

The process $e^+e^- \to f\bar{f}$ at energies above the $Z^0$ resonance is dominated by the $\gamma$ exchange. On the other hand the $\gamma - Z^0$ interference is responsible for the large forward-backward asymmetries in all channels. At LEP2 energies, compared to LEP1, the importance of radiative processes in which an initial state photon is emitted, is much increased. This initial-state radiation (ISR) reduces the available center-of-mass energy $\sqrt{s'}$ down to $Z^0$ resonance and QED radiative corrections lead to a factor five enhancement of the di-fermion Born cross-section. The reduced center-of-mass energy $\sqrt{s'}$ is defined as the invariant mass of the outgoing $l\bar{l}$ pair or the mass of the s-channel propagator for $q\bar{q}$ pair.
The di-fermion measurements presented below are provided for the *exclusive* samples of events with $\sqrt{s'/s} > 0.85$, so excluding the radiative events.

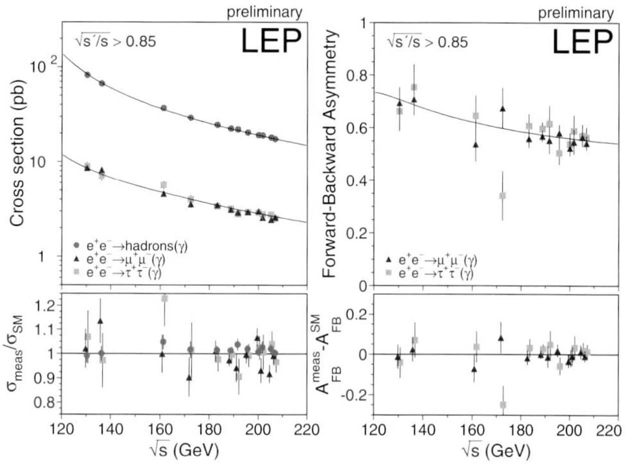


Figure 1. Preliminary LEP combined cross-sections and asymmetries results

## 3. DI-FERMION CROSS-SECTIONS AND LEPTONIC ASYMMETRIES WITH $\sqrt{s'/s} > 0.85$

Each LEP experiment provided its measurements of the $q\bar{q}$, $\mu^+\mu^-$ and $\tau^+\tau^-$ cross-sections and of $\mu^+\mu^-$ and $\tau^+\tau^-$ asymmetries. Before combination, those measurements are corrected


doi:10.1016/S0920-5632(02)02110-2

to have a common signal definition. The combination includes data from 183 to 207 GeV. All cross-sections and asymmetries are treated together in a single fit using a covariance matrix (detectors effects and ISR modeling being the main errors). The LEP preliminary combined results are shown in Figure 1. The difference between the measured hadronic cross section and the Standard Model expectation (ZFITTER v6.36 [2]), averaged over all energies, is the largest one and is about 1.6 standard deviation. Theoretical precisions are respectively 0.3% and 0.4% for the hadronic and leptonic cross sections. They are obtained from the difference between ZFITTER and KK2f [3] predictions.

Differential cross-sections $d\sigma/dcos\theta$ for $e^+e^- \to l\bar{l}$ processes are measured by the four LEP experiments. For $\mu^+\mu^-$ and $\tau^+\tau^-$ final states, preliminary results combination gives a good agreement with the Standard Model expectation (ZFITTER). A preliminary combination (ALEPH-L3-OPAL) of the $e^+e^-$ final states results have been made for the first time, averaged data and predictions from Monte Carlo generator BHWIDE [4] are in good agreement.

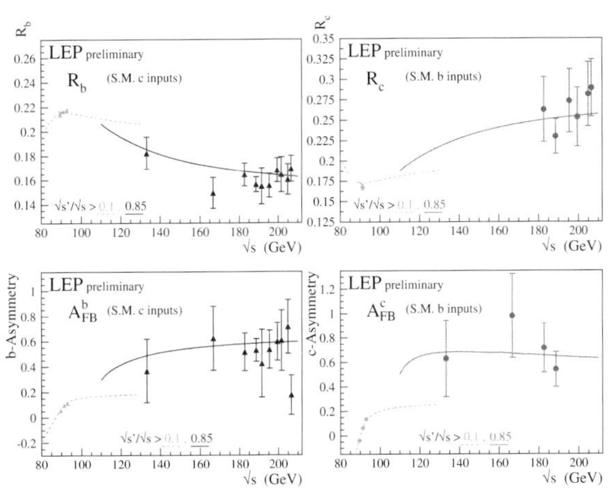


Figure 2. Preliminary LEP combined $R_q$ and $A_{FB}^{q\bar{q}}$ results for $b\bar{b}$ and $c\bar{c}$ production

## 4. HEAVY FLAVOURS CROSS-SECTIONS AND ASYMMETRIES

The measured observables are the cross section ratios $R_b = \sigma(b\bar{b})/\sigma(\text{had})$ and $R_c = \sigma(c\bar{c})/\sigma(\text{had})$, where $\sigma(\text{had})$ is the total $q\bar{q}$ cross section, and the forward-backward asymmetries $A_{FB}^{b\bar{b}}$ and $A_{FB}^{c\bar{c}}$ (Figure 2). The available heavy flavours results are combined, $R_c$ measurements coming only from ALEPH experiment. A simultaneous fit of $R_b$ and $A_{FB}^{b\bar{b}}$ is performed, $R_c$ and $A_{FB}^{c\bar{c}}$ being fixed to the Standard model values. A similar procedure is used for the $c\bar{c}$ heavy flavour measurements. The results are consistent with the Standard Model predictions of ZFITTER.

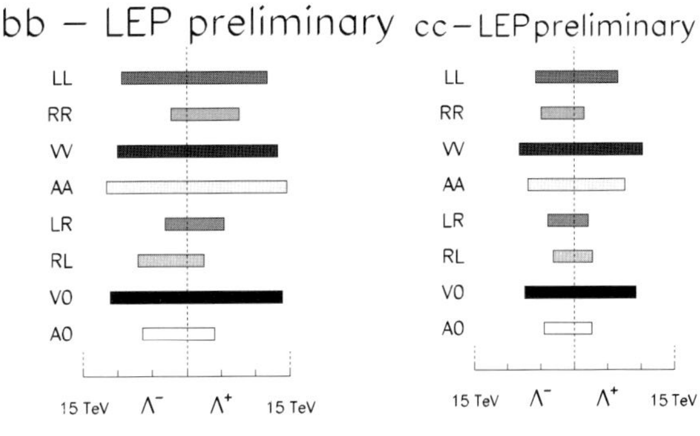


Figure 3. Preliminary combined 95% confidence limits on $\Lambda$ for $b\bar{b}$ and $c\bar{c}$. The $\pm$ sign on $\Lambda$ denotes positive and negative interference with Standard Model amplitudes.

## 5. INTERPRETATIONS

Comparaison of the measured di-fermion cross-sections and asymmetries with Standard Model predictions allows to put limits on many possible extensions of the Standard Model. The sensitivity to these new models comes from their interference with the Standard Model processes through virtual exchange of new particles.

### 5.1. Four-fermions contact interactions

Four-fermions contact interactions are characterized by an energy scale $\Lambda$, and a coupling $g$. $\Lambda$ is interpreted as the mass of a new heavy particle exchanged between the incoming and the outgoing fermion pairs. Different models correspond to the different helicity states involved in the interaction. LEP2 combined di-fermion measurements allow to constrain the scale $\Lambda$ for $e^+e^- \to l\bar{l}, b\bar{b}, c\bar{c}$ final states only reachable at LEP. Contact interactions would affect the total cross section and the asymmetries for $l\bar{l}$ $(\mu^+\mu^-, \tau^+\tau^-)$ and heavy flavours final states. Preliminary combined 95% confidence lower limits on $\Lambda$ are derived, assuming a coupling $g^2 = 4\pi$ (Figure 3). For the first time at LEP2, $u\bar{u}$ and $d\bar{d}$ final states are studied using the hadronic cross section and assuming that only one quark flavour of up-type or down-type is affected by the contact interactions (Figure 4).

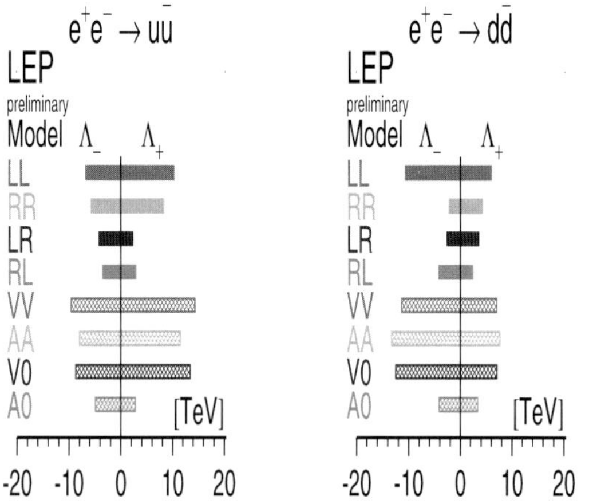


Figure 4. Preliminary combined 95% confidence limits on $\Lambda$ for up-type and down-type quarks

### 5.2. Gravitational interaction in extra dimensions

The gravitational interaction in extra dimensions are a new approach to understand the Hierarchy problem. These models, in particular the "Arkani-Ahmed, Dimopoulos, Dvali" model [5] assume $\delta$ extra space dimensions compactified at the size R and a fundamental scale $M_D$ close to the electroweak scale. Results are for $\delta = 2$ and in term of string mass scale $M_S \sim M_D$ [6]. The $e^+e^- \to f\bar{f}$ Standard Model differential cross section is then modified due to additional s-channel virtual graviton exchange amplitudes. Especially the $e^+e^-$ final state where the interference between t-channel $\gamma$ exchange and graviton exchange is expected to be large. The combined (ALEPH-L3-OPAL) preliminary 95% confidence lower limits on $M_S$ from the $e^+e^- \to e^+e^-$ measurements are the following:

$$M_S = 1.20\ TeV \quad for\ \lambda = +1$$
$$M_S = 1.09\ TeV \quad for\ \lambda = -1$$

where $\lambda$ is the sign of the interference with the SM amplitudes. The Figure 3 shows the measured $e^+e^- \to e^+e^-$ differential cross-section normalized to the Standard Model expectation (Figure 5).

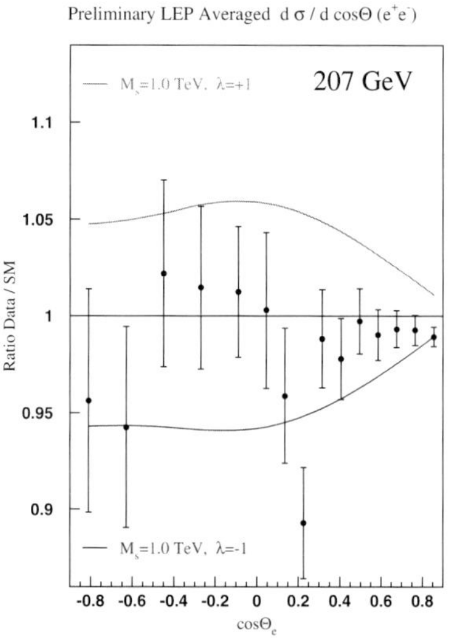


Figure 5. Measured $e^+e^- \to e^+e^-$ differential cross-section normalized to the Standard Model expectation at $\sqrt{s} = 207\ GeV$. The predictions for graviton interactions with $M_S = 1$ GeV are indicated.

### 5.3. Leptoquarks

A leptoquark (LQ) is a high-mass resonant state carrying baryon and lepton numbers (F=3B+L, where F is the LQ's fermion number)

Table 1
Preliminary indirect limits on the first generation leptoquark mass, assuming $\lambda = \sqrt{4\pi\alpha}$.

| LQ Type | $m_{LQ}$ $(GeV/c^2)$ | LQ Type | $m_{LQ}$ $(GeV/c^2)$ |
|---|---|---|---|
| $S_0(L) \rightarrow eu$ | 655 | $V_{1/2}(L) \rightarrow ed$ | 303 |
| $S_0(R) \rightarrow eu$ | 520 | $V_{1/2}(R) \rightarrow eu, ed$ | 227 |
| $\tilde{S}_0(R) \rightarrow ed$ | 202 | $\tilde{V}_{1/2}(L) \rightarrow eu$ | 176 |
| $S_1(L) \rightarrow eu, ed$ | 361 | $V_0(L) \rightarrow e\bar{d}$ | 917 |
| $S_{1/2}(L) \rightarrow e\bar{u}$ | 178 | $V_0(R) \rightarrow e\bar{d}$ | 165 |
| $S_{1/2}(R) \rightarrow e\bar{u}, e\bar{d}$ | 232 | $\tilde{V}_0(R) \rightarrow e\bar{u}$ | 489 |
| $\tilde{S}_{1/2}(L) \rightarrow e\bar{d}$ | – | $V_1(L) \rightarrow e\bar{u}, e\bar{d}$ | 659 |

and coupling to a lepton and a quark. They are predicted by theories such as the one proposed by W. Buchmüller, R. Rückl and D. Wyler [7]. Ten leptoquarks are predicted, five scalars ($S_0$, $S_{1/2}$,...), five vectors ($V_0$, $V_{1/2}$,...). There are three generations of LQ couplings, $\lambda_l q$. The

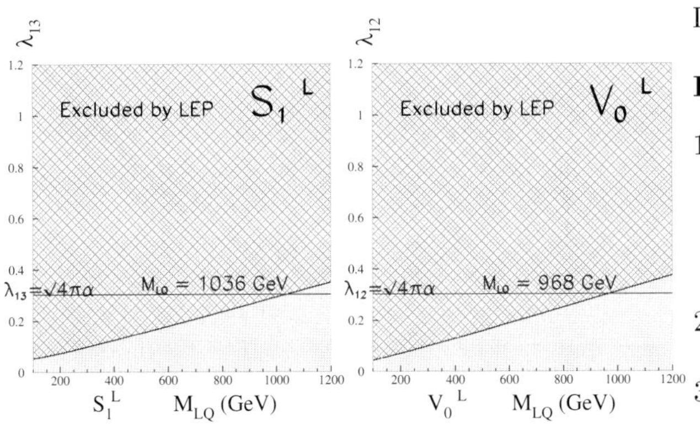


Figure 6. Preliminary limits in the plan $\lambda_{lq}^2$ versus LQ mass for $V_0(L)$ ($2^{nd}$ generation coupling) and $S_1(L)$ ($3^{rd}$ generation coupling)

$e^+e^- \rightarrow q\bar{q}$, leptoquarks can be exchanged in the t-channel or u-channel. The hadronic cross-section is modified by the first generation leptoquark exchange. $R_c$ and $A_{FB}^{c\bar{c}}$ measurements are modified by the second generation leptoquark exchange. $R_b$ and $A_{FB}^{b\bar{b}}$ measurements are sensitive to the third generation leptoquarks exchange. Assuming $\lambda = \sqrt{4\pi\alpha}$, the indirect limits on the first generation leptoquark mass, are given in Table 1. For the second and third generation leptoquark exchange, Figure 6 shows the best LEP2 limits in the plan $\lambda_{lq}^2$ versus LQ mass.

## 6. ACKNOWLEDGMENTS

I would like to thank the members of the LEP2 Di-Fermion Electroweak Working Group.

## REFERENCES

1. ALEPH collab. EPS-HEP2002 388, ALEPH CONF 2002-021, DELPHI 2001-168 PHYS 904, L3 collab. EPS-HEP2002, L3 Note 2763, OPAL TN704, LEPEWWG $f\bar{f}$ Subgroup, C. Geweniger *et al.*, LEP2FF/01/02
2. D. Bardin *et al*, CERN-TH 6443/92; http://www.ifh.de/~riemann/Zfitter/zf.html.
3. S.Jadach, B.F. Ward and Z. Was, "The precision Monte Carlo event generator KK for two-fermion final states in e+ e- collisions," Comput. Phys. Commun. **130** (2000) 260
4. S. Jadach, W. Placzek and B. Ward, Phys. Lett. **B390** (1997) 298
5. N. Arkani-Hamed, S. Dimopoulos and G.Dvali, Phys. Lett. **B429** (1998) 263
6. Phys. Rev. Lett., **82** (1999) 4765
7. W. Buchmüller, R. Rückl and D. Wyler, Phys. Lett **B191** (1987) 442

ELSEVIER

Proceedings of ICHEP 2002, pp. 186–189
*S. Bentvelsen, P. de Jong, J. Koch and E. Laenen (Editors)*

31st INTERNATIONAL CONFERENCE ON HIGH ENERGY PHYSICS AMSTERDAM

# Properties of tau lepton

Vadym Zhuravlov[a]

[a]Joint Institute for Nuclear Research

This note cover the new results obtained by LEP experiments in tau-physics.

## 1. Introduction

At LEP tau lepton is produced either in $Z^0$ boson decay or in two photon collision. DELPHI and L3 collaborations explored two photon mode of tau pair production and measured the cross-section of this process. Both experiments derived the limits on magnetic and electric dipole moments of tau lepton.

Weak dipole moments of tau lepton were studied by ALEPH and the most stringent limits were derived on these quantities.

DELPHI collaboration finished the study of tau hadronic branching ratios. In this analysis there was no attempt to separate charged kaons. Separate paper on charged kaon branching ratios is planed.

Tau lepton lifetime has been measured by DELPHI using three different methods. Two of them are based on one-prong decays and one is based on three-prong decays. Full data set of LEP1 was reanalysed using an improved three-prong method and new measurement of one-prong decays was combined with the previously published DELPHI result.

Tau decays provide the direct limit on the tau neutrino mass. The limit is derived from two-dimensional distribution of the invariant mass and energy of tau decay products. New limit on the tau neutrino mass is obtained by DELPHI from five-prong decay channel.

## 2. Hadronic branching ratios

The definition of the exclusive hadronic channel is the tau decay with determined charged multiplicity and counted number of $\pi^0$. Several effects complicate the charged multiplicity determination: photon conversion, Dalitz decay of $\pi^0$, nuclear re-interaction, $\delta$-rays, two-track resolution. All these effects were carefully studied by DELPHI [1] in data and Monte-Carlo and all necessary correction were applied to Monte Carlo to achieve a reasonable description of the data.

New DELPHI results are compared to other measurements in Fig. 1 and 2. Standard Model (SM) prediction [2] calculated from decay rate of charged pions and kaons is also shown on Fig. 1. $h\pi^0\nu_\tau$ experimental results are compared to Conserved Vector Current (CVC) prediction [3] in Fig. 2. New world average values were calculated for both decay channels. Preliminary L3 result [4] on $Br(\tau \to h\nu_\tau)$ is included into averaging in spite of low $\chi^2$ probability. CVC prediction is $2\sigma$ away from averaged experimental value of $\mathrm{Br}(\tau \to h\pi^0\nu_\tau))$ and dominating uncertainty comes from the error of CVC prediction.

## 3. Tau lepton lifetime

Tau lepton lifetime is measured by DELPHI detector with three methods [5] . One of them uses 3-prong tau decays and the other two explore 1-prong topology. 3-prong decay is a classical method which uses the position of the reconstructed vertex in $r - \phi$ plane. The measured transverse decay length is converted to the three-dimensional decay length using the polar angle of the event thrust. Then the fit is performed to the obtained decay time distribution with the likelihood function built as a convolution of physical function of the vertices distribution with the resolution function. The result of the method obtained on the full data set of LEP1


doi:10.1016/S0920-5632(02)02113-8

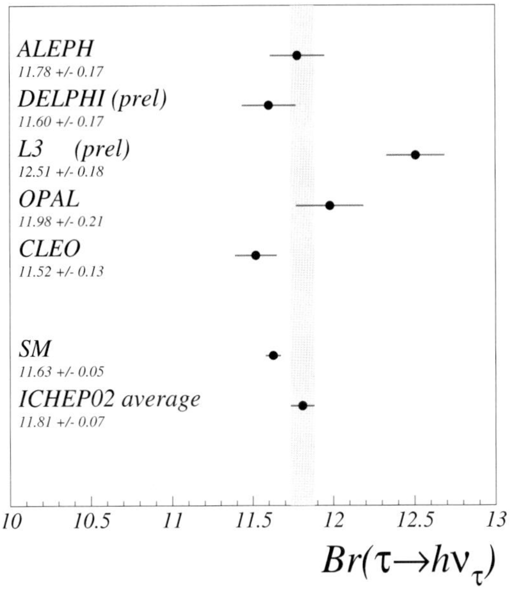


Figure 1. $\tau \to h\nu_\tau$ branching ratios

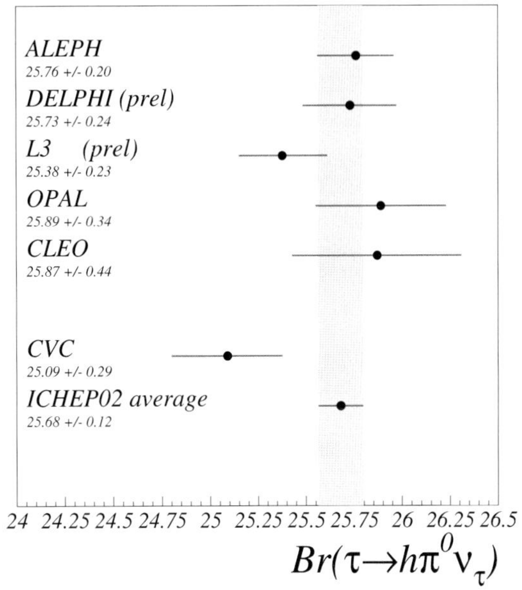


Figure 2. $\tau \to h\nu_\tau$ branching ratios

is $\tau_{3-prong} = 288.6 \pm 2.4_{stat} \pm 1.4_{syst}\ fs$.

Impact parameter difference method explores the the 1 versus 1 track topology. Impact parameter is defined as a distance of the closest approach of the track to the interaction point and can be calculated by the formula

$$d = L \sin\theta_\tau \sin(\phi - \phi_T)$$

where $\theta_\tau$ is a polar angle of the tau estimated as the polar angle of the thrust axis and $\phi - \phi_T$ is the difference in azimuthal angles of the tau and decay product. Taking the difference of the $\tau^+$ and $\tau^-$ impact parameters and averaging one can see that that the impact parameter difference is proportional to the projected accoplanarity:

$$< d_+ - d_- >=< L > \delta\phi \sin\theta_\tau$$

Lifetime is extracted from the straight line fit to the impact parameter difference as a function of the projected accoplanarity. The fit to the data of years '94 and '95 results to the value $\tau_{ipd} = 291.4 \pm 3.6_{stat} \pm 1.5_{syst}$ fs. The disadvantage of the method is the "double" sensitivity to the interaction point spread. To overcome this the miss distance method is used.

Miss distance is defined as a sum of impact parameters of positive and negative taus. The lifetime is determined from the fit to the observed distribution with the likelihood function build as the convolution of physical probability density function and the detector resolution function. The correct description of the resolution function is vitally important in this analysis because the information of the tau lifetime is contained in the width of the miss distance distribution. The resolution function is parametrised as the sum of three gaussians with the parameters depending on the kinematic variables of the event and on identified tau decay modes. This method gives the most precise result. For the data of years '94 and '95 it yielded to $\tau_{miss} = 292.0 \pm 2.3_{stat} \pm 2.1_{syst}$.

Averaging of one-prong methods is done taking into account 36 % statistical correlation between them. The correlation between 1-prong and 3-prong methods is found to be 5% dominated by vertex detector alignment. The combined result

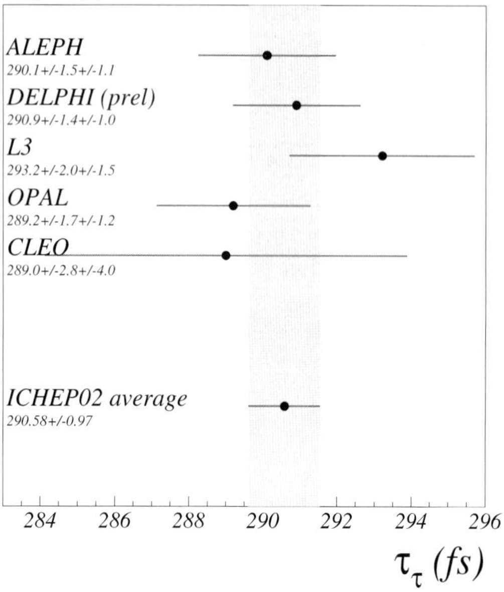


Figure 3. Tau lepton lifetime

$\tau_\tau = 290.9 \pm 1.4_{stat} \pm 1.0_{syst}$ is the most precise measurement up to date. Fig. 3 compares this measurement to the other measurements.

## 4. Tau neutrino mass limit

Study of tau multiprong hadronic decays allows to set a limit on tau neutrino mass from the endpoint of the hadronic system invariant mass spectrum The tau lepton direction can not be reconstructed but the lack of the information can by partially compensated by measuring the energy of hadronic system. Thus the mass of the tau neutrino is derived from the two-dimensional distribution of invariant mass and energy of hadronic system.

DELPHI has analysed the decays of the tau into 5 charged pions [6]. 47 5-pions decays were found in full set LEP1 data, the background was estimated to be 24%. The fit region was constrained by $m_{inv} > 1.6\ GeV/c^2$ and $0.85 < E/E_{beam} < 1.10$. The correct choice of the spectral function of $\tau \rightarrow 5\pi$ decay is an important issue in this analysis as imperfect knowledge of it can shift the expected distribution in the endpoint region. Several models were considered in the analysis and final result was obtained with pure phase space modes which gives the most conservative limit of 48.0 MeV at 95% C.L. The expected limit was of the order of 30 MeV. The best limit from the spectrum endpoint study so far was obtained by ALEPH [7]. The limit was set to 18.2 MeV at 95% C.L. and was derived from the combination of $\tau \rightarrow 3\pi$ and $\tau \rightarrow 5\pi$ decay channels.

## 5. Tau pair production in two-photon collisions

Besides $Z^0$ decay tau pairs are produced at LEP in collisions of two virtual photons emitted by electron and positron. DELPHI and L3 have analysed LEP2 data in order to extract the cross-section of this process. To suppress the severe background arising from the other two-photon final states one has to choose different decay modes for $\tau^+$ and $\tau^-$. Both experiments selected one-prong tau decays, DELPHI selected electron mode versus inclusive non-electron mode and L3 searched for electron mode versus $\rho$ mode. The electron identification in DELPHI analysis relies on dE/dx measurement while in L3 analysis electrons were identified by the response in electromagnetic calorimeter. The $\rho$ meson decay mode was identified by the requirement of a track and two photons with invariant mass consistent with $\pi^0$ mass. In addition the mass of $\pi^\pm\pi^0$ was required to be consistent with the $\rho$-meson mass.

Fig. 4 shows the measured cross-section as a function of c.m. energy. Comparison the measured cross-sections to the SM prediction the two experiments have derived the following limits on the dipole electromagnetic moments of the tau lepton: $-0.017 < a_\tau < 0.019$, $|d_\tau| < 3.8 \cdot 10^{-16}$ $e \cdot cm$ at 95% C.L. (DELPHI [8]) and $-0.062 < a_\tau < 0.044$, $|d_\tau| < 6.7 \cdot 10^{-16}$ $e \cdot cm$ at 95% C.L. (L3 [9]). Here $a_\tau$ denotes the anomalous magnetic moment and $d_\tau$ stands for electric dipole moment of the tau lepton. The regions allowed for anomalous magnetic and electric dipole moments are shown in Fig.4.

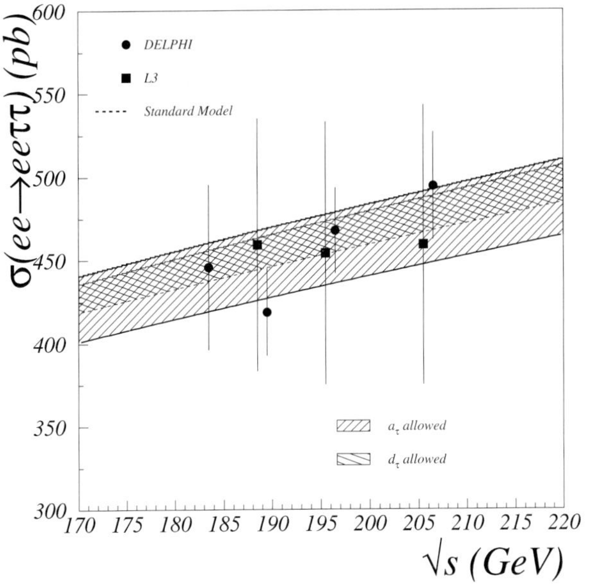


Figure 4. $e^+e^- \to e^+e^-\tau^+\tau^-$ cross-section as a function of c.m. energy. Regions allowed for variation due to anomalous magnetic and electrical dipole moment are also shown.

## 6. Weak dipole moments

Similar to the electromagnetic moments one can introduce the weak dipole moments by means of additional terms in neutral current and in charged current.

Neutral current moments were measured by OPAL [10], L3 [11] and recently by ALEPH [12]. ALEPH has extracted both real and imaginary parts of weak dipole moments from the analysis of spin-dependent differential cross-sections. Obtained results are the most precise measurement of these parameters up to date.

Weak magnetic moment of the charged current was measured by DELPHI [13]. Weak dipole moment of the charged current $d_\tau^W$ is not measured at all so far.

## 7. Conclusion

Tau lepton electromagnetic moments are measured by DELPHI and L3 from the study of the tau pair production in two-photon collision. The obtained limits are competitive with the world best limits obtained from measurement of $e^+e^- \to \tau\tau\gamma$ process [14]. ALEPH has measured tau lepton weak dipole moments of neutral current. The branching ratios of the hadronic tau decays and tau lifetime have been measured by DELPHI. New limit has been set on the tau neutrino mass from the study the endpoint of the $\tau \to 5\pi$ decay spectrum by DELPHI.

## REFERENCES

1. DELPHI coll., Paper 190 submitted to this conference (DELPHI note CONF-576)
2. R.Decker, M.Finkemeier Phys. Lett. B 334
3. S.I. Eidelman, "Status of CVC Tests from $e^+e^- \to$ *Hadrons* and $\tau$-lepton Decays", *Proceedings of "$e^+e^-$ Physics at Intermediate Energies" conference*, SLAC, Stanford, California, 30 Apr - 2 May 2001
4. G.Chen, Nuclear Phys. B (Proc. Suppl.) 98 (2001) 201-209
5. DELPHI coll., Paper 221 submitted to this conference (DELPHI note CONF-577)
6. DELPHI coll., Paper 221 submitted to this conference (DELPHI note CONF-578)
7. ALEPH Collab., Phys. Lett. B 414 (1997) 362
8. DELPHI coll., Paper 261 submitted to this conference (DELPHI note CONF-595)
9. L3 Collab., Paper 480 submitted to this conference (L3 note 2747)
10. OPAL Collab., K. Ackerstaff *et al.*, Zeit. fur Phys. C74 (1997) 403 - 412
11. L3 Collab., M. Acciarri *et al.*, Phys. Lett. B 426 (1998) 207-216
12. ALEPH Collab., "Search for anomalous weak dipole moments of the $\tau$ lepton" *To be submitted to Eur. Phys. J. C*
13. DELPHI Collab., P.Abreu *et al.*, E. Phys.J.C16 (2000) 229
14. L3 Collab., M. Acciarri *et al.*, Phys. Lett. B 434 (1998) 169-179; OPAL Collab., K. Ackerstaff *et al.*, Phys. Lett. B431 (1998) 188-198

Proceedings of ICHEP 2002, pp. 190–193
S. Bentvelsen, P. de Jong, J. Koch and E. Laenen (Editors)

ELSEVIER

31st INTERNATIONAL CONFERENCE ON HIGH ENERGY PHYSICS AMSTERDAM

# Heavy flavour electroweak physics at LEP1 and SLD

R. Hawkings[a]

[a]CERN, EP division, CH 1211 Geneva 23, Switzerland

Heavy flavour electroweak measurements from LEP1 and SLD are discussed, with an emphasis on the latest b quark asymmetry results from OPAL and DELPHI. The heavy flavour electroweak fit results are interpreted in terms of coupling parameters and the effective electroweak mixing angle $\sin^2\theta_W^{\rm eff}$.

## 1. Introduction

Measurements of the couplings of the $Z^0$ to heavy quarks (b and c) provide stringent tests of the Standard Model and allow parameters of the theory (for example the top and Higgs masses) to be constrained via their influence on radiative corrections. The fraction of hadronic $Z^0$ decays to b quarks ($\Gamma(b\bar{b})/\Gamma(\text{hadrons}) \approx R_b$) is sensitive to vertex corrections involving top quarks and W bosons, whilst the parity violating couplings of the $Z^0$ to fermions results in an asymmetry $A_{FB}^b$ in the differential cross-section for b quark production:

$$\frac{d\sigma}{d\cos\theta} \propto 1+\cos^2\theta+\frac{8}{3}A_{FB}^b\cos\theta\,, \qquad (1)$$

where $\theta$ is the angle between the directions of the incoming electron and outgoing b quark, and where initial and final state radiation, quark mass and higher order terms have been neglected. At the $Z^0$ pole, the asymmetry $A_{FB}^{b,0}$ is given by $A_{FB}^{b,0} = \frac{3}{4}A_eA_b$ where the coupling parameters of the $Z^0$ to electrons and b quarks, $A_e$ and $A_b$ depend on the effective vector and axial vector couplings of the $Z^0$. Within the context of the Standard Model, these depend on the effective electroweak mixing angle $\sin^2\theta_W^{\rm eff}$ and are sensitive to the unknown Higgs mass. At SLD, the high electron beam polarisation of the SLC linear collider allows the measurement of asymmetries separately for left- and right-polarised electrons. The resulting 'left-right-forward-backward asymmetry' is directly sensitive to the quark coupling $A_b$ or $A_c$, rather than the product of quark and electron couplings measured at LEP.

The measurements of $R_b$ and $R_c (\approx \Gamma(c\bar{c})/\Gamma(\text{hadrons})$ have been stable for some time, and no new results were presented. However, two new asymmetry measurements were submitted to this conference, namely a measurement of the b quark asymmetry using an inclusive tag from OPAL [1], and a simultaneous measurement of the b and c quark asymmetries using semileptonic decays from DELPHI [2]. These measurements are discussed in more detail below, before a presentation and interpretation of the fit to all available heavy flavour electroweak data.

## 2. New analyses

The new OPAL b quark asymmetry analysis improves on a previous analysis by using a more efficient and pure b tag developed for the $R_b$ measurement, b quark charge tagging developed for $B^0$ oscillation and CP violation measurements, an increased $\cos\theta$ acceptance and a more sophisticated fit procedure. Approximately 0.5M hadronic $Z^0$ decays recorded primarily for calibration purposes during the LEP2 era were also added to the 4 M events from LEP1.

Like other inclusive b asymmetry measurements, the OPAL analysis exploits the structure of $Z^0 \to b\bar{b}$ events with two back-to-back jets to measure almost all the required event properties from the data themselves. Hemispheres in $Z^0 \to b\bar{b}$ events are tagged using a combination of secondary vertex and lepton information, and the tagging efficiency for b and c quarks determined from ratios of single and double tagged events. The b quark charge in each hemisphere is deter-


doi:10.1016/S0920-5632(02)02118-7

mined using a combination of jet charges (a sum of all track charges in each jet, weighted according to the momentum of each track), secondary vertex charges and charged kaon information, exploiting the decay chain $b \to c \to s \to K^-$. The fraction of events in which the charge determinations in both hemispheres agree is used to measure the reliability of the charge determination (and hence the dilution of the true b quark asymmetry) directly from the data. The b quark asymmetry is then extracted in 70 analysis bins of differing b purity and $\cos\theta$, and the results combined. The final result [1] has an uncertainty of just over half that of the previous OPAL inclusive result [3].

The new DELPHI analysis [2] measures both the b and c quark forward-backward asymmetries using the charge of leptons produced in semileptonic $b \to \ell^-$ and $c \to \ell^+$ decays. The two sources of leptons are separated from 'cascade' decays ($b \to c \to \ell^+$, which gives the 'wrong' lepton charge) and background using a neural network exploiting lepton momentum, jet shape, lifetime and opposite hemisphere charge information. The latter is used to control the dilution due to $B-\bar{B}$ mixing, which can also be measured using the fraction of events with two leptons of the same charge. This analysis updates previous DELPHI results using a more sophisticated fitting procedure, and is now close to finalisation.

## 3. Heavy flavour electroweak fit

The experimental measurements of the heavy flavour partial widths and asymmetries are correlated, both due to direct dependencies of the measurement of one parameter on the assumed value of another, and because of common external inputs (including gluon splitting rates, bottom and charm hadron fragmentation functions, lifetimes and decay branching ratios). The values of the partial widths and asymmetries are therefore extracted using a simultaneous fit to all electroweak data, taking into account correlations and common systematic uncertainties. Several auxiliary parameters are also determined in this fit, including charm hadron production fractions, and b semileptonic branching ratios [4]. The asymmetries measured at LEP ($A^{\rm b}_{\rm FB}$ and $A^{\rm c}_{\rm FB}$) and SLD ($A_{\rm b}$ and $A_{\rm c}$) are treated as independent parameters.

In a first fit, the LEP asymmetries measured both on and off ($\pm 2\,\rm GeV$) the $Z^0$ peak are treated separately. Since the resulting variations with centre-of-mass energy are consistent with the Standard Model expectation, the measurements are all translated to the $Z^0$ peak, corrected for photon radiation and quark mass effects and combined. The resulting measurements of the pole asymmetries are:

$$A^{\rm b,0}_{\rm FB} = 0.0995 \pm 0.0015 \pm 0.0007$$
$$A^{\rm c,0}_{\rm FB} = 0.0713 \pm 0.0031 \pm 0.0018$$

where the first error is statistical and the second systematic in each case. The individual measurements of $A^{\rm b,0}_{\rm FB}$ [1,2,5,6] are shown in Figure 1, and the combined result is dominated by the inclusive jet charge analyses from ALEPH, DELPHI and OPAL. The final uncertainty is dominated by data statistics, and the common experimental systematic uncertainty is only 0.0004, resulting mainly from the understanding of corrections due to gluon radiation.

The individual measurements of $A^{\rm c,0}_{\rm FB}$ [2,6,7] are shown in Figure 2. The measurements using leptons and $D^*$ mesons are consistent, providing confidence that the flavour separation between semileptonic b and c decays is well understood. Again, the uncertainty is statistics dominated, and the common systematic uncertainty is 0.0009.

The combined results for the partial widths are

$$R_{\rm b} = 0.21644 \pm 0.00065$$
$$R_{\rm c} = 0.1718 \pm 0.0031$$

based entirely on final published analyses [8]. These results are compatible with the expectation from the Standard Model and the direct top mass measurement from the Tevatron.

## 4. Interpretation of asymmetry results

The measurements of $A^{\rm b,0}_{\rm FB}$ from LEP, and $A_{\rm b}$ from SLD are shown in Figure 3, together with the measurement of $A_{\rm e}$ from the inclusive left-

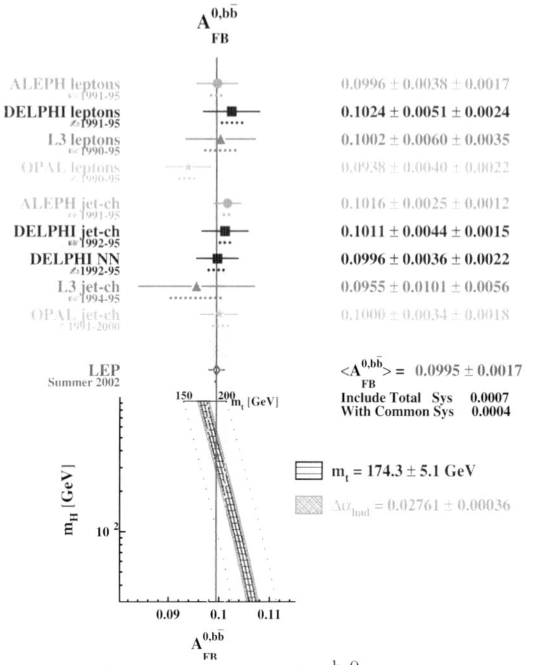


Figure 1. Measurements of $A_{\mathrm{FB}}^{\mathrm{b},0}$, the b quark asymmetry at the $\mathrm{Z}^0$ pole.

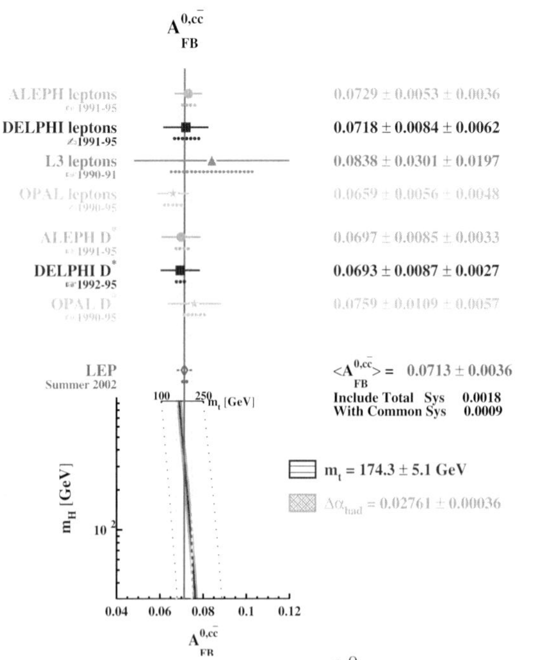


Figure 2. Measurements of $A_{\mathrm{FB}}^{\mathrm{c},0}$, the c quark asymmetry at the $\mathrm{Z}^0$ pole.

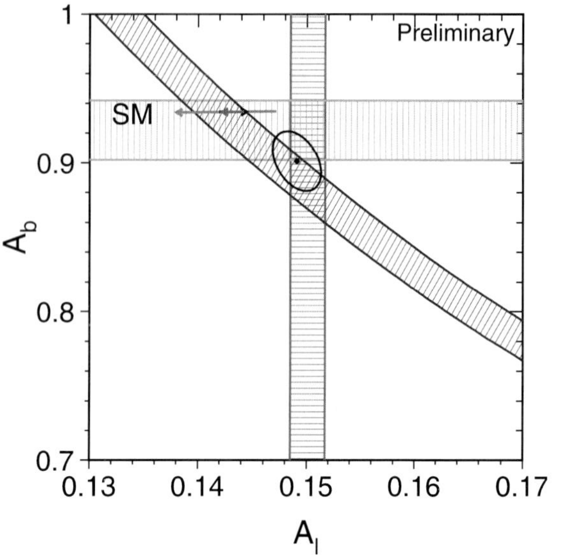


Figure 3. Combined results for the coupling parameters $A_{\mathrm{e}}$ ($\equiv A_{\mathrm{l}}$) and $A_{\mathrm{b}}$, and the asymmetry $A_{\mathrm{FB}}^{\mathrm{b},0}$. The bands show the $\pm 1$ standard deviation uncertainties, and the Standard Model expectation is shown by the arrow, with $m_{\mathrm{t}} = 174.3 \pm 5.1\,\mathrm{GeV}$ and $m_H = 300^{+700}_{-186}\,\mathrm{GeV}$.

right asymmetry ($A_{\mathrm{LR}}$) at SLD and leptonic observables at LEP. It can be seen that the LEP $A_{\mathrm{FB}}^{\mathrm{b},0}$ (the diagonal band) is consistent with the Standard Model expectation, as is the SLD $A_{\mathrm{b}}$ measurement of $0.922 \pm 0.020$ [9]. However, using the measured $A_{\mathrm{e}}$ (which is somewhat above the Standard Model expectation) to extract $A_{\mathrm{b}}$ from $A_{\mathrm{FB}}^{\mathrm{b},0}$ results in a value of $A_{\mathrm{b}}$ 2.6 standard deviations below the expected value. In contrast, the corresponding measurements for $A_{\mathrm{FB}}^{\mathrm{c},0}$ and $A_{\mathrm{c}}$ show good agreement with the Standard Model expectation.

A similar effect is visible on Figure 4, which shows the various asymmetry measurements interpreted within the Standard Model as determinations of the effective electroweak mixing angle for leptons $\sin^2\theta_{\mathrm{W}}^{\mathrm{eff}}$. The overall average from all measurements is

$$\sin^2\theta_{\mathrm{W}}^{\mathrm{eff}} = 0.23148 \pm 0.00017$$

but the fit $\chi^2$ is poor, at 10.2 for 5 degrees of freedom. This is entirely driven by the 2.9 stan-

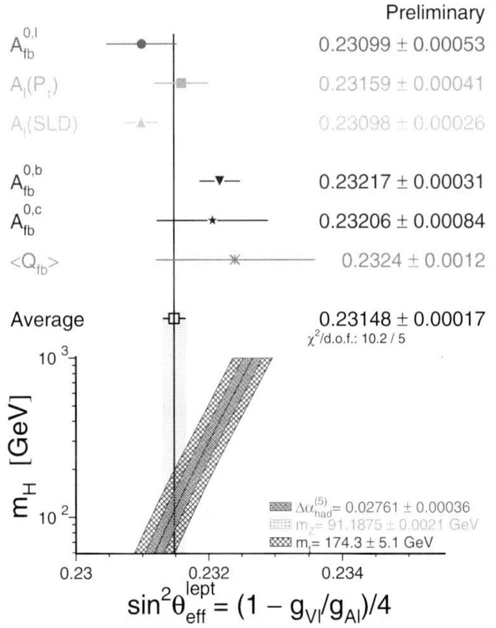


Figure 4. Determinations of the effective electroweak mixing angle $\sin^2\theta_W^{eff}$.

dard deviation discrepancy between $\sin^2\theta_W^{eff}$ measured from $A_{LR}$ and $A_{FB}^{b,0}$. The effect on the fitted Higgs mass is also shown—$A_{FB}^{b,0}$ prefers a mass of around 400 GeV, whilst $A_{LR}$ prefers a mass below 100 GeV.

## 5. Discussion

The LEP and SLD heavy flavour electroweak measurements have continued to improve over the last few years, with new improved analyses getting the maximum possible information from the data. However, there is now little scope for further reduction in the uncertainties, and almost all analyses have been finalised. Motivated by the three standard deviation discrepancy between $A_{LR}$ and $A_{FB}^{b,0}$, much effort has been invested in improving and checking the b quark asymmetry analyses in particular, where the uncertainty is still dominated by statistics. This discrepancy will most likely be left as a legacy from the LEP/SLD era—a statistical fluctuation, or perhaps a hint of new physics.

## Acknowledgements

The author thanks the members of the LEP/SLD heavy flavour electroweak working group for the averages presented here and many useful discussions.

## REFERENCES

1. G. Abbiendi et al., Phys. Lett. B546 (2002) 29.
2. DELPHI collab., DELPHI 2002-028 CONF 562.
3. K. Ackerstaff et al., Z. Phys. C75 (1997) 385.
4. The LEP collaborations, ALEPH, DELPHI, L3 and OPAL, Nucl. Instrum. Methods 378 (1996) 101.
5. A. Heister et al., Eur. Phys. J. C22 (2001) 201; DELPHI collaboration, DELPHI 2001-048 CONF 476; M. Acciarri et al., Phys. Lett. B439 (1998) 225.
6. A. Heister et al., Eur. Phys. J. C24 (2002) 177; M. Acciarri et al., Phys. Lett. B448 (1999) 152; OPAL collab., OPAL Physics Note 226.
7. R. Barate et al., Phys. Lett. B434 (1998) 415; P. Abreu et al., Eur. Phys. J. C10 (1999) 219; G. Alexander et al., Z. Phys. C73 (1996) 379.
8. R. Barate et al., Phys. Lett. B401 (1997) 163; P. Abreu et al., Eur. Phys. J. C10 (1999) 415; M. Acciarri et al., Eur. Phys. J. C13 (2000) 47; G. Abbiendi et al., Eur. Phys. J. C8 (1999) 217; K. Abe et al., Phys. Rev. Lett. 80 (1998) 660; R. Barate et al., Eur. Phys. J. C4 (1998) 557; R. Barate et al., Eur. Phys. J. C16 (2000) 597; P. Abreu et al., Eur. Phys. J. C12 (2000) 209; P. Abreu et al., Eur. Phys. J. C12 (2000) 225; K. Ackerstaff et al., Eur. Phys. J. C1 (1998) 439; G. Alexander et al., Z. Phys. C72 (1996) 1; SLD collab., contributed paper 174 to ICHEP98 Vancouver, Canada.
9. K. Abe et al., Phys. Rev. Lett. 88 (2002) 151801.

ELSEVIER

Proceedings of ICHEP 2002, pp. 194–197
*S. Bentvelsen, P. de Jong, J. Koch and E. Laenen (Editors)*

31st INTERNATIONAL CONFERENCE ON HIGH ENERGY PHYSICS AMSTERDAM

# Measurement of charged triple gauge-boson couplings(cTGC) at LEP2

S. Jézéquel[a]

[a]LAPP-IN2P3 Chemin de Bellevue 74941 Annecy-Le-Vieux Cedex, France
E-mail: jezequel@lapp.in2p3.fr

The measurement of charged triple gauge-boson couplings (WWZ and WW$\gamma$ vertices) at LEP2 is presented, in this note, based on the whole data collected between 1997 and 2000. It corresponds to $\simeq$700 $pb^{-1}$ per experiment with energies ranging from 183 to 209 GeV. These couplings are measured through W-pair and single W events and are compared to the Standard Model expectation. Deviations would indicate new physics at higher energies than for direct searches. This note will show that these measurements achieve a precision of a few percent and are limited by systematics.

## 1. Introduction

Due to the non-abelian nature of $SU(2)_L \times U(1)_Y$, WWV (V=Z or $\gamma$) vertices exist at tree level. In the Standard Model, the effective coupling is affected by vertex corrections including known particles ($\simeq 10^{-3}$[1]). It can also have contributions from new particles existing at higher scale than LEP2 threshold. Deviations from the Standard Model expectations would indicate new physics at higher scale. For the Minimal SuperSymetric Model, it was estimated to be of the order of $10^{-3}$[1].

The most general Lagrangian being Lorentz invariant for the WWV interaction is described in these papers[1],[2],[3],[4]. It contains seven terms:

- ($g_1^V$, $\kappa_V$ and $\lambda_V$) conserve C and P
- $g_5^V$ violates C and P but conserves CP
- ($g_4^V$, $\tilde{\kappa}_V$ and $\tilde{\lambda}_V$) violate CP

For ICHEP2002, the only new measurement of individual coupling provided by LEP experiments comes from OPAL on $g_5^Z$[5]. Because there is no new LEP combined result on individual parameters, no number is presented in this note.

Taking into account usual constraints (C and P conservation, $U(1)_{em}$ gauge symmetry) and LEP1 indirect measurements, the base line analysis[1] at LEP2 measures 3 parameters ($\Delta g_1^Z$ , $\Delta\kappa_\gamma$, $\lambda_\gamma$) which are the most likely to deviate from the Standard Model. They are equal to zero at the Standard Model. The $SU(2)_L \times U(1)_Y$ gauge symmetry implies the following constraint:

$$\Delta\kappa_Z = \Delta g_1^Z + \Delta\kappa_\gamma \tan^2\theta_w; \lambda_Z = \lambda_\gamma$$

Results are computed for the following configurations :

- three measurements of a single parameter (implying that the two others are fixed at their Standard Model value)
- simultaneous measurement of the three parameters

These couplings are evaluated through the single W and W-pair productions. A new result is presented by OPAL using single-$\gamma$ channel[6]. It is not used for the final LEP combination because of its low sensitivity for cTGC compared to the two other channels with equivalent luminosity.

## 2. WW analysis

W pair production is sensitive to $\Delta g_1^Z$, $\Delta\kappa_\gamma$ and $\lambda_\gamma$ through the $s$-channel contrary to the $t$-channel. Since the $s$-channel contribution increases with energy, the sensitivity to cTGC improves also with energy.

The selection of WW events is the same as for the cross section[7] but restricted to well reconstructed events[8]. Anomalous couplings affect quadratically the differential cross-section as summarized in the following equation.

$$d\sigma(\Omega,\alpha) = S^0(\Omega)+\Sigma_i\alpha_i.S_i^1(\Omega)+\Sigma_{ij}\alpha_i.\alpha_j.S_{ij}^2(\Omega) \quad (1)$$


doi:10.1016/S0920-5632(02)02122-9

with $\alpha_i = g_1^z$, $\kappa_\gamma$ and $\lambda_\gamma$ and $\Omega$ describes phase space. This effect is looked for through the total WW cross section and the kinematical distributions of events. Cross section are extracted from adjustment of the expected cross-section which is a function of couplings to the observed one. The kinematic information is described by 5 angles ($\cos\theta_W$,$\cos\theta^*$ and $\phi^*$ of the fermion in both Ws). To have the most precise measurement of one/many couplings from the five angles, an unbinned maximum likelihood is performed by ALEPH (analytical Particle Density Function convoluted with a detector resolution function), DELPHI and L3 (PDF mapped from simulated events). A second solution is the Optimal Observable[9] method where the five kinematic variables are projected onto 1 or 2 parameters per TGC coupling. OPAL and ALEPH fits the $\mathcal{O}_i^1$ and $\mathcal{O}_{ij}^2$ averages.

The main sensitive channel is the semileptonic one because the lepton tags the W charge and the jet pair gives the W direction. In the fully hadronic channel, the mispairing of jets to reconstruct Ws (80% efficiency) and the mischarge assignment of Ws (80 % efficiency) decrease the sensitivity. The fully leptonic channel is handicapped by its low branching ratio compared to the two other ones. The figure 1 gives the relative sensitivity for $\lambda_\gamma$ for each channel.

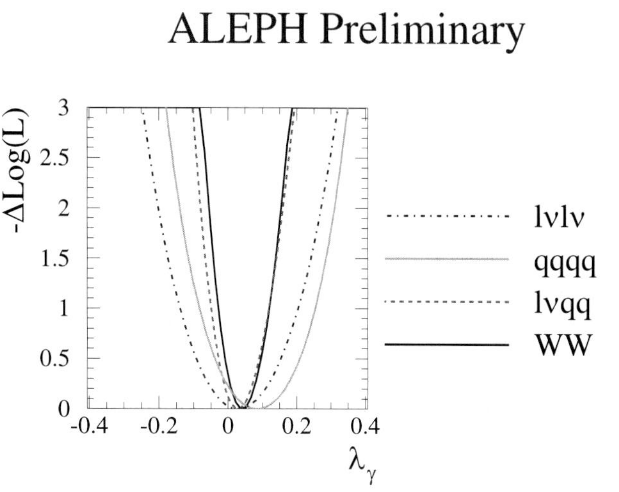


Figure 1. Loglikelihood distributions for fully hadronic, semileptonic, fully leptonic channels to $\lambda_\gamma$ and the combined one (WW)

## 3. Single W analysis

The single W channel was defined by the LEP TGC Working Group to deal with the many $t$-channel diagram contributions. It is asked that:

- the electron is undetected ($|\cos\theta| > 0.95$)
- in case of $W \to l\nu$ the lepton is within the detector acceptance and with an energy greater than 20 GeV
- in case of $W \to q\bar{q}'$ , the invariant mass should be greater than 45 GeV

The selection for the hadronic and leptonic decays of the W is identical to the cross section one [10].

This channel is sensitive to couplings mainly through the cross-section measurement. The kinematic information improves slightly the measurement [11] and is given by:

- $Pt_W$,$|\cos\theta_{jet1} - \cos\theta_{jet2}|$, NN output for W→2 jets
- $E_l$,$\cos\theta_l$ for W→l$\nu$

## 4. Systematics

The main systematic affecting the single-W channel is the theoretical uncertainty on the total cross section (± 5%). It is smaller than the statistical uncertainty since one measures experimentally the cross section with a precision of 8%.

For the TGC measurement using W-pair events, there are five systematics:

- the fragmentation model of jets (comparison JETSET/HERWIG),
- the empirical models on final state interaction (Bose-Einstein and Color Reconnection effect) affecting the fully hadronic channel
- the theoretical uncertainty on the total W

pair cross-section ($\pm 0.5\%$) after $\mathcal{O}(\alpha)$ correction

- the theoretical uncertainty on the $\mathcal{O}(\alpha)$ correction to the angle distributions after $\mathcal{O}(\alpha)$ correction

Although there is an estimation of the last systematics[12][13], the LEP-TGC Working Group decided to be conservative and quotes as systematics the full effect. The numerical values of the systematics are listed in table 1.

Table 1
List of systematics for cTGC

| Systematics | $\Delta g_1^Z$ | $\lambda_\gamma$ | $\Delta\kappa_\gamma$ |
|---|---|---|---|
| $\sigma$ (We$\nu$) | - | - | 0.011 |
| Fragmentation | 0.004 | 0.002 | 0.004 |
| Color-Reconnection | 0.005 | 0.004 | 0.010 |
| Bose-Einstein | 0.005 | 0.004 | 0.009 |
| $\sigma$ ($W^+W^-$) | 0.003 | 0.005 | 0.014 |
| Full $\mathcal{O}(\alpha)$ corr. on angular distribution | 0.015 | 0.015 | 0.039 |

The $\mathcal{O}(\alpha)$ uncertainty on the angular distribution of $\cos\theta_W$ is the main contributor to systematics.

## 5. Global results

For the W-pair analysis, the presented results[14] include all LEP2 data up to 209 GeV ($\simeq$700 $pb^{-1}$ per experiment). Most of the experiments analyzed the single W channel up to 202 GeV. For the LEP combination, the likelihood (statistical+uncorrelated systematics) are provided by each experiment for the W pair and single W channels. Due to the non gaussian behavior of the curves especially in multidimensional fits, a special minimization method[15] has been performed to improve the treatment of correlated systematics which are not small compared to the statistical errors. The results are presented for 1D or 3D fit in figures 2 and 3. In the 3D fit, ALEPH data are not included.

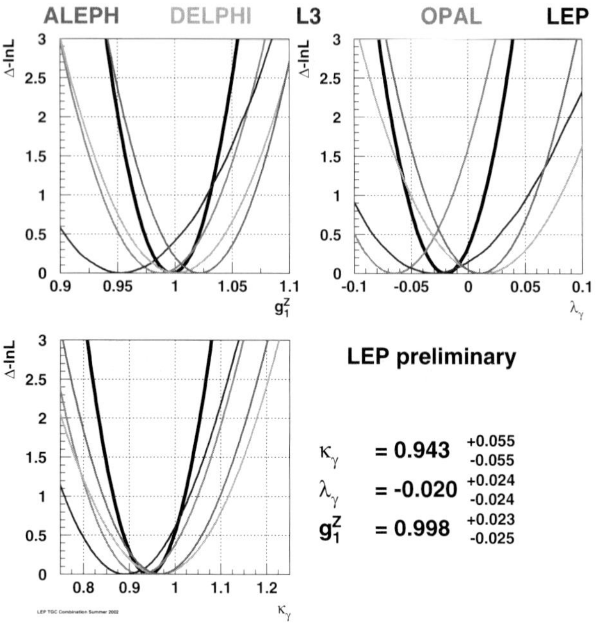


Figure 2. Results for 1 parameter fit for the 4 LEP experiments and the combined one

## Conclusion

For the baseline analysis, supposing that only one parameter deviates from Standard Model, the limits at 95% confidence level, are:

$$0.951 < g_1^Z < 1.043$$

$$0.835 < \kappa_\gamma < 1.052$$

$$-0.067 < \lambda_\gamma < +0.028$$

Since almost all LEP2 data have been analyzed (only single W has not been analyzed at all energies), no major improvement on the statistical error is foreseen at LEP2. But, the measurement is still limited by the full $O(\alpha)$ correction systematics. Experiments are cross checking the results based on YFSWW/RACOONWW comparisons to reduce it. In the mean time, experiments are finalizing their results. The LEP2 data analysis for charged triple gauge-boson couplings rejects models changing the effective coupling by more than a few percent.

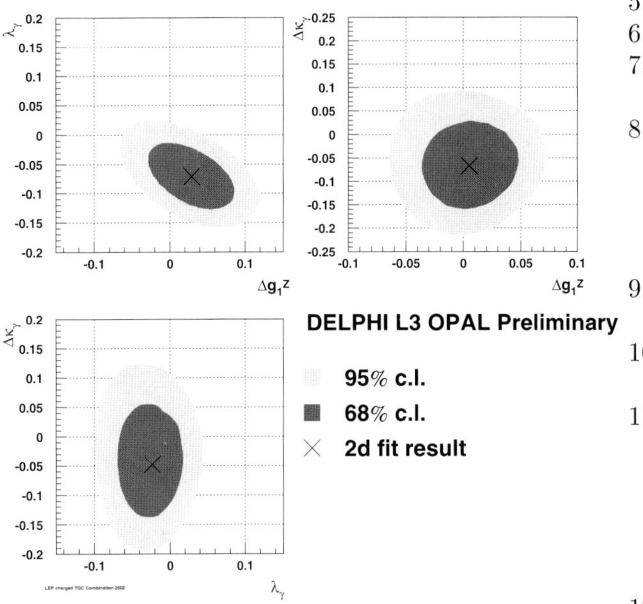


Figure 3. Result for 3 parameter fit. Figures are plotted with the third parameter fixed at the smallest $\chi^2$. ALEPH results are not included

## Acknowledgments

I want to thank my ALEPH, DELPHI, L3 and OPAL colleagues for making available their preliminary results, and the TGC LEP Working group for the combination results.

## REFERENCES

1. Physics at LEP2, edited by G. Altarelli, T. Sjostrand and F. Zwirner, CERN 96-01 Vol.1, 525.
2. K. Hagiwara, R.D. Peccei, D. Zeppenfeld and H. Hikasa, *Nucl Phys.* B **282** (1987) 253.
3. M. Bilenky, J.L. Kneur, F.M. Renard and D. Schldknecht, *Nucl. Phys.* B **409** (1993) 22; *Nucl. Phys.* B **419** (1994) 240.
4. K. Gaemers and G. Gounaris, *Z. Phys.* C **1** (1979) 259.
5. OPAL PN501.
6. OPAL PN508.
7. L. Maglieri "Measurement of cross-sections $\sigma_W W$, $\sigma_Z Z$ and $\sigma_{\gamma\gamma}$", these proceedings.
8. ALEPH Collaboration, ALEPH 2000-055 CONF 2000-037.
   DELPHI 2002-034 CONF 568
   L3 Note 2734
   OPAL PN501.
9. M. Diehl and O. Nachtmann, *Z. Phys.* C **62** (1994) 397.
10. E. Migliore "4-fermion production at LEP2", these proceedings.
11. ALEPH Collaboration, ALEPH 2000-054 CONF 2000-036.
    DELPHI Collaboration DELPHI 2000-143 CONF 442.
    L3 Collaboration L3 Note 2518.
    OPAL Collaboration PN427.
12. W. Placzek "Precision predictions for W-pair production at LEP2", these proceedings.
13. R. Brunelière et al "On Theoretical Uncertainties of the W Angular Distribution in W-Pair Production at LEP2 Energies" , *Phys. Lett.* B **533** (2002) 75; Phys.Lett.B533:75-84,2002
14. The LEP Collaborations and the LEP TGC Working Group LEPEWWG/TGC/2000-02.
15. J. Alcaraz, *A proposal for the combination of TGC measurements*, L3 Internal Note 2718.
    R. Barlow and C. Beeston, *Fitting using finite Monte Carlo samples*, Comp. Phys. Comm. 77 (1993) 219-228.
    R. Brunelière, *Test of LEP Triple Gauge-Boson Couplings combination procedures*, ALEPH 2002-007 PHYSICS 2002-006

Proceedings of ICHEP 2002, pp. 198–201
S. Bentvelsen, P. de Jong, J. Koch and E. Laenen (Editors)

ELSEVIER

31st INTERNATIONAL CONFERENCE ON HIGH ENERGY PHYSICS AMSTERDAM

# Anomalous Triple Neutral and Quartic Gauge Boson Couplings

St. Wynhoff[a]

[a]Princeton University - Mail: CERN-EP, 1211 Geneva 23, Switzerland, Stephan.Wynhoff@cern.ch

At LEP anomalous neutral triple gauge boson couplings are studied in Z$\gamma$ and ZZ production. Quartic gauge boson couplings are analysed in WW$\gamma$, Z$\gamma\gamma$ and $\nu\nu\gamma\gamma$ events. No significant anomalies are found.

## 1. Introduction

The non-abelian structure of the Standard Model (SM) leads to couplings between the gauge bosons. Two categories of vertices exist: Triple couplings, i.e. charged couplings $\gamma$WW and ZWW that exist in the SM and are measured in $e^+e^-$reactions at $\sqrt{s} > 161$ GeV , see Reference [1], and neutral couplings ZZZ, ZZ$\gamma$ and Z$\gamma\gamma$ that don't exist at tree level. SM contributions from quartic couplings, i.e. WW$\gamma\gamma$ and WWZ$\gamma$ that are negligible at LEP; the vertex ZZ$\gamma\gamma$ that doesn't exist at tree level, and WWZZ and WWWW are not accessible at LEP.

The SM does not predict any detectable effects (at LEP) for any of the triple neutral and quartic couplings. However, New Physics effects may enhance existing couplings or introduce new, additional couplings. Possible New Physics effects on physical observables are parametrised with an effective Lagrangian introducing anomalous couplings.

In this article the searches for anomalous triple neutral and quartic couplings performed by the experiments ALEPH [4], DELPHI [5], L3 [6–8] and OPAL [9,10] at the LEP $e^+e^-$collider are presented. Each of the four experiments analysed about 700 pb$^{-1}$ of data taken at centre-of-mass energies from 161 GeV to 208 GeV.

## 2. Neutral Triple Gauge Boson Couplings

While neutral triple gauge boson couplings don't exist in the SM at tree level, anomalous couplings may leed to vertices ZZZ, ZZ$\gamma$ and Z$\gamma\gamma$ that can be described by 12 parameters using the most general Lagrangian and assuming Lorentz invariance, $U(1)_{em}$ invariance and Bose symmetry [2]. The parameters $h_1^\gamma$, $h_2^\gamma$, $h_1^Z$, $h_2^Z$describe the strength of the CP-violating contribution in the reaction $e^+e^- \rightarrow \gamma/Z \rightarrow Z\gamma$; $f_4^\gamma$and $f_4^Z$the CP-violating contributions in the reaction $e^+e^- \rightarrow \gamma/Z \rightarrow ZZ$. The corresponding parameters for the CP-conserving contributions are $h_3^\gamma$, $h_4^\gamma$, $h_3^Z$, $h_4^Z$, $f_5^\gamma$and $f_5^Z$. In the SM all these parameters are zero at tree level.

To measure the size of eventual anomalous couplings events with two Z-bosons or one Z-boson and a hard photon are studied. All five visible decay channels of the ZZ events are investigated using artificial-neutral-net (NN) and optimal-observable (OO) techniques. In particular the event rates and the Z production angle $|\cos\vartheta_Z|$ are sensitive to anomalous couplings. For Z$\gamma$ events similar studies of $|\cos\vartheta_Z|$ and $E_\gamma$ with OO are used for the $q\bar{q}\gamma$ and $\nu\bar{\nu}\gamma$ channel.

The results of the four LEP experiments are combined [12]. Table 1 shows the 1-dimensional limits on the 12 parameters. Also 2-dimensional

| | | | |
|---|---|---|---|
| $h_1^\gamma$ | [-0.06;0.05] | $h_3^\gamma$ | [-0.05;0.00] |
| $h_2^\gamma$ | [-0.05;0.02] | $h_4^\gamma$ | [-0.00;0.04] |
| $h_1^Z$ | [-0.17;0.15] | $h_3^Z$ | [-0.23;0.11] |
| $h_2^Z$ | [-0.10;0.10] | $h_4^Z$ | [-0.07;0.15] |
| $f_4^\gamma$ | [-0.17;0.19] | $f_5^\gamma$ | [-0.34;0.38] |
| $f_4^Z$ | [-0.30;0.28] | $f_5^Z$ | [-0.36;0.38] |

Table 1
Limits on neutral triple gauge boson couplings at 95% confidence level.


doi:10.1016/S0920-5632(02)02127-8

fits are performed. The resulting limits for ($f_4^\gamma$,$f_4^Z$) and ($f_5^\gamma$,$f_5^Z$) are shown in Figure 1. All values are consistent with the SM values of zero (no anomalous couplings).

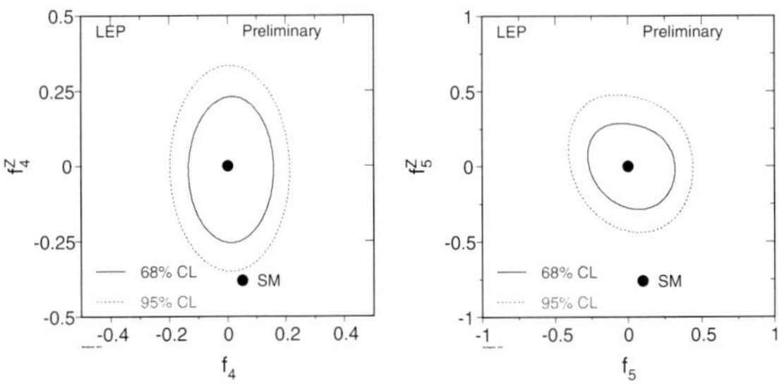


Figure 1. Contour curves of 68% and 95% confidence level in the $f_4^\gamma f_4^Z$ and $f_5^\gamma f_5^Z$ planes showing the combined LEP result.

## 3. Quartic Gauge Boson Couplings

At LEP only those vertices of four gauge bosons that involve at least one photon are accessible. Potential New Physics effects leading to anomalous couplings are parametrised with $a_0^W$ and $a_c^W$ for the WW$\gamma\gamma$ vertex, $a_0^Z$ and $a_c^Z$ for the ZZ$\gamma\gamma$ vertex, and $a_n$ for the WWZ$\gamma$ vertex. The corresponding effective Lagrangians [3] are proportional to the ratio of the coupling parameters to the energy scale for New Physics, $\Lambda$: $\mathcal{L} \propto a/\Lambda^2$. As for anomalous triple neutral gauge couplings event rates would be changed by anomalous quartic couplings. Figure 2 shows the LEP combined measurements of the cross section for the production of a W pair in association with a hard photon[1]. The measurements, determined selecting semi-leptonic and hadronic events, are compared to the SM prediction and predictions for different values of $a_n$. More information can be obtained by analysing the energy spectrum of the photon in these events. An anomalous coupling modifies not only the normalisation, but also the shape of the spectrum. As an example Figure 3 shows the energy distribution of the photon in comparison with the SM expectation and in case of anomalous couplings.

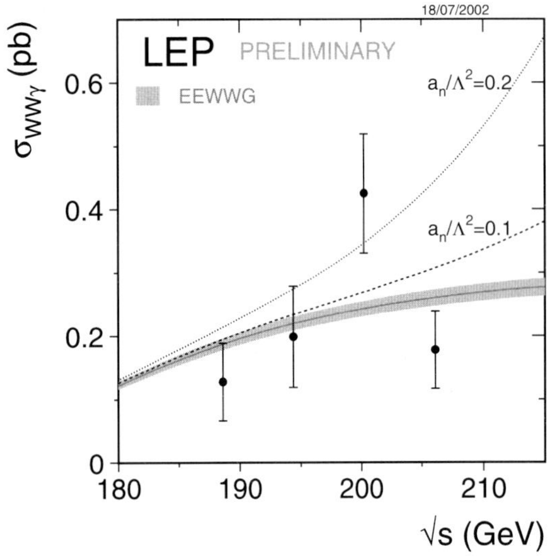


Figure 2. Measured cross section for the production of W-pairs and a hard, isolated photon. The results from DELPHI and L3 have been combined [13].

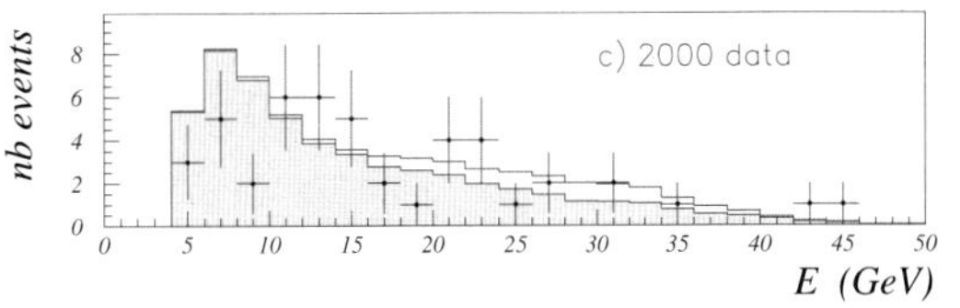


Figure 3. Photon energy spectrum for selected WW$\gamma$ events measured by DELPHI [5] at $\sqrt{s}$ = 206 GeV. Data (dots) are compared to SM predictions (dark histogram) and expected results for $a_n/\Lambda^2 = -0.16$ GeV$^{-2}$ (light histogram).

Another channel to search for quartic gauge couplings is $e^+e^- \to \nu\bar{\nu}\gamma\gamma$. The experimental signature are two acoplanar photons and missing energy. Figure 4 shows the missing mass distribution as measured by ALEPH. In the SM these events are produced with double initial state bremsstrahlung and the production of two neutrinos via s-channel Z- or t-channel W-exchange. The Feynman diagrams with the corresponding quartic gauge coupling vertices are shown in Figure 5. The existence of anomalous couplings

[1] Signal definition: $E_\gamma > 5$ GeV, $|\cos\vartheta_\gamma| < 0.95$, $\cos\alpha < 0.90$ ($\alpha$ = angle between $\gamma$ and closest charged fermion), $|m(f,f') - m_W| < 2\Gamma_W$

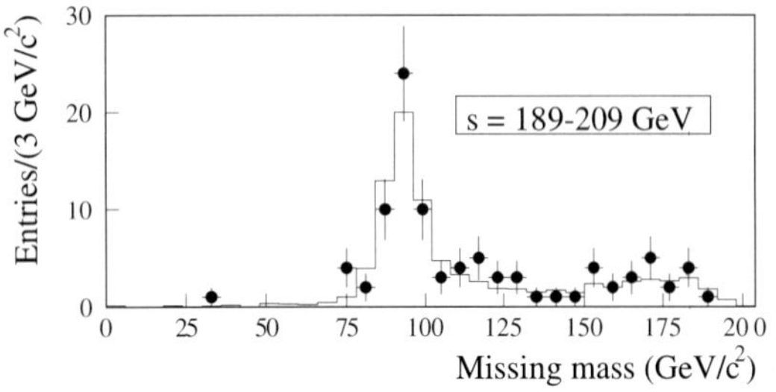


Figure 4. Missing mass spectrum in events with 2 acoplanar photons as measured by ALEPH [4]. The peak at 91 GeV is in the SM produced by s-channel Z exchange, the events at high missing masses by t-channel W exchange.

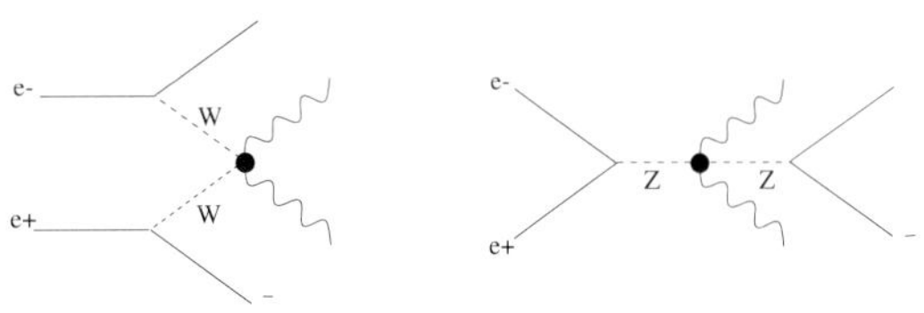


Figure 5. The Feynman diagrams for the production of $\nu\overline{\nu}\gamma\gamma$ final state via the $WW\gamma\gamma$ and $ZZ\gamma\gamma$ vertices.

would modify the missing mass spectrum and the energy spectra of the two photons. These distributions are used to fit the anomalous couplings. The changes of the spectra are different for variations of $a_0^W$, $a_c^W$ and $a_0^Z$, $a_c^Z$. This allows to fit W and Z related couplings separately. The results for the analysis of the $\nu\overline{\nu}\gamma\gamma$ and $WW\gamma$ final states can be combined. Figure 6 shows the resulting likelihood curves as measured by OPAL [9]. Both analyses show a small excess of events over the SM expectations. After combination this lead to a disvafouring of the SM at the level of $\Delta \log \mathcal{L}$ =4.0 (3.7) for $a_0^W$ ($a_c^W$) corresponding to 2.8 (2.7) standard deviations in the gaussian approximation. OPAL determines limits on $a_0^W$ and $a_c^W$ that are given in Table 2. The extension of the $WW\gamma$ analysis to the full LEP2 dataset is expected to clarify the situation.

In Table 2 the limits obtained by DELPHI, L3 and OPAL for the W-related anomalous couplings are given as well as the previous LEP combination.

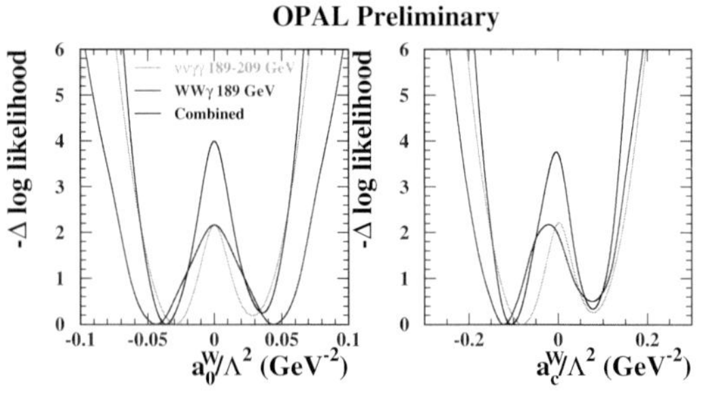


Figure 6. Likelihood curves for $a_0^W$ and $a_c^W$ obtained from $WW\gamma$ events and $\nu\overline{\nu}\gamma\gamma$ events and their combination from OPAL [9].

| | $a_0^W/\Lambda^2$ | $a_c^W/\Lambda^2$ | $a_n/\Lambda^2$ |
|---|---|---|---|
| DELPHI | [-0.018;0.018] | [-0.057;0.030] | [-0.16;0.12] |
| L3 | [-0.015;0.015] | [-0.048;0.026] | [-0.14;0.13] |
| OPAL | [-0.054;0.052] | [ -0.15;0.14 ] | [-0.61;0.57] |
| LEP '01 | [-0.031;0.030] | [-0.069;0.070] | [-0.45;0.41] |

Table 2
Limits in GeV$^{-2}$ on quartic gauge boson couplings at 95% confidence level. The following event samples have been used: L3 ($WW\gamma \to$ 207 GeV, $\nu\nu\gamma\gamma \to$ 207 GeV), OPAL ($WW\gamma \to$ 189 GeV, $\nu\nu\gamma\gamma \to$ 209 GeV), DELPHI ($WW\gamma \to$ 209 GeV)

L3 and OPAL use also hadronic events with 2 identified photons to investigate Z-related anomalous couplings. In Figure 7 the energy spectrum of the lower-energetic photon as measured by L3 is compared to the SM expectation and to modifications expected by quartic gauge couplings. This spectrum as well as angular distributions are used to set limits on $a_0^Z$ and $a_c^Z$. For the first time a LEP combined result is obtained combining the likelihood curves from L3 ($q\overline{q}\gamma\gamma$) and OPAL ($q\overline{q}\gamma\gamma$, $\nu\overline{\nu}\gamma\gamma$). The result – well compatible with the SM expectation of zero – is shown in Figure 8. The limits at 95% confidence level that are determined from these curves are listed in Table 3.

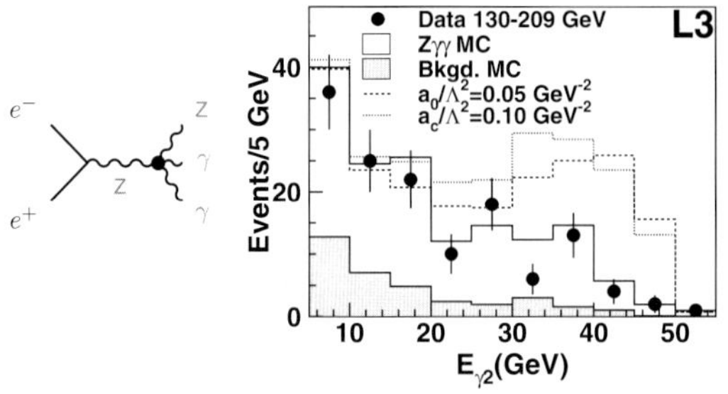


Figure 7. Left: The Feynman diagram for the production of $Z\gamma\gamma$ final state via the $ZZ\gamma\gamma$ vertex. Right: Energy spectrum of the second photon in $q\bar{q}\gamma\gamma$ events as measured by L3 [7]. The data are compared to the SM expectation and predictions assuming anomalous quartic gauge couplings.

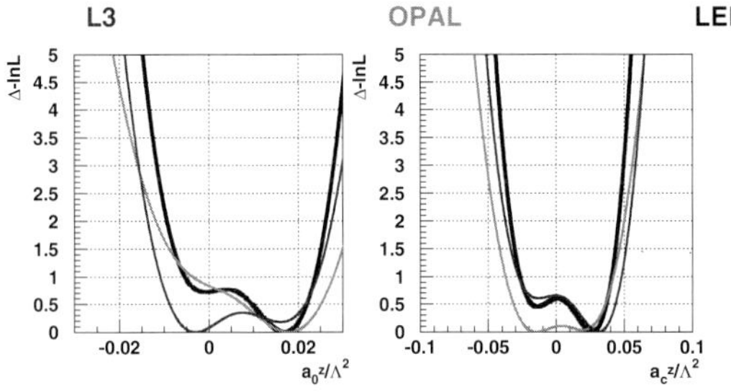


Figure 8. Likelihood curves for $a_0^Z$ and $a_c^Z$ obtained from $q\bar{q}\gamma\gamma$ and $\nu\bar{\nu}\gamma\gamma$ events by L3 and OPAL and their combination [13].

## 4. Conclusions

LEP data taken at centre-of-mass energies from 161 GeV to 208 GeV have been used to search for anomalous neutral triple and quartic gauge boson couplings. Events of the reactions $e^+e^- \to Z\gamma$, ZZ, $WW\gamma$, $Z\gamma\gamma$ and $\nu\nu\gamma\gamma$ have been used. The measurements are found to be in good agreement with Standard Model predictions and limits on the anomalous couplings are set. Further improvement of these limits can be expected when the full LEP data sample will be analysed by all four experiments in all channels.

| | |
|---|---|
| $a_0^Z/\Lambda^2$ | [-0.009;0.026] |
| $a_c^Z/\Lambda^2$ | [-0.034;0.046] |

Table 3
LEP combined (L3, OPAL) limits in $GeV^{-2}$ on quartic gauge boson couplings at 95% confidence level.


## Acknowledgments

I would like to thank my collegues from L3 and the convenors of the LEP four-fermion and gauge-coupling working groups for their support and fruitful discussions.


## REFERENCES


1. S.Jézéquel, these proceedings.
2. G.J.Gounaris, J.Layssac, F.M.Renard, Phys.Rev. D62(2000) 073013.
3. G.Montagna et al., Phys.Lett. B515 (2001),
4. ALEPH Collab., CERN-EP-2002-033.
5. DELPHI Collab., DELPHI 2002-059 CONF 593 (2002), Contributed paper for ICHEP 2002, Amsterdam, Netherlands.
6. L3 Collab., Phys.Lett. B527 (2001), 29.
7. L3 Collab., Phys.Lett. B540 (2002), 43.
8. L3 Collab., L3 Note 2672, Contributed paper to EPS 2001, Budapest, Hungary, 12-18 July 2001.
9. OPAL Collab., OPAL PN-510 (2002), Contributed paper to ICHEP 2002, Amsterdam, Netherlands.
10. OPAL Collab., OPAL PN-452 (2000), Contributed paper to ICHEP 2000, Osaka, Japan.
11. OPAL Collab., Phys.Lett. B471 (1999), 293.
12. The LEP electroweak working group, LEPEWWG/TGC/2002-01.
13. U.Parzefall, CERN-Seminar 22 July 2002.

ELSEVIER

Proceedings of ICHEP 2002, pp. 202–205
*S. Bentvelsen, P. de Jong, J. Koch and E. Laenen (Editors)*

31st INTERNATIONAL CONFERENCE ON HIGH ENERGY PHYSICS AMSTERDAM

# Search of the Standard Model Higgs at LEP

C. Mariotti [a]

[a] CERN, Geneva and INFN Torino

The final result on the direct search of the Standard Model Higgs at LEP are reviewed. A lower bound on its mass of 114.4 GeV/c$^2$ has been established, at 95% confidence level.

## 1. THE STANDARD MODEL HIGGS

The experimental data test the Standard Model (SM) at the per mill level and data on radiative corrections support strongly the existence of the Higgs boson. Yet the Higgs boson has not been discovered.

The value of its mass is not predicted by the theory, and the precision electroweak measurements suggest its mass to be light, i.e. less than 193 GeV at 95% CL, with a $\chi^2$ minimum at M(H)=81 $^{+53}_{-30}$ GeV/c$^2$.

In the following the direct search of the Higgs boson at LEP will be presented.

## 2. LEP: THE IDEAL PLACE

At LEP the simplicity of the initial state propagates through the final state. The dominant Higgs production mechanisms at LEP is the Higgstrahlung diagram, where the Higgs is produced together with a Z. The cross section of this process rapidly decreases at a value of the Higgs mass $M(H) \leq \sqrt{s} - M_Z$. The WW-fusion production diagram positively interferes with the Higgstrahlung one, resulting in a total cross section of the order of 40 fb for $\sqrt{s}$ = 206 GeV and M(H)=115 GeV.

Since the dominant decay mode of the Higgs is into a couple of beauty quarks, the final states are clean topologies characterised by the decay mode of the $Z$: 2jet+2b-jets, missing energy ($2\nu$)+2b-jets or 2 leptons+2b-jets.

From the LEP run at high energies, i.e. $\sqrt{s} = 130 - 209$ GeV and with a total integrated luminosity per experiment of more than 2.5 fb$^{-1}$, the sensitivity of the four LEP experiments together, reached 115 GeV on the Higgs mass.

For this very high value of masses the cross section is very small, and very few events are then expected to be produced: as an example only $\approx$ 8 events per experiment for M(H)=115 GeV during the year 2000. The cross sections for the other 4-fermions, 2-fermions and gamma-gamma processes are order of magnitude larger than the Higgs one. The gamma-gamma and the $qq(\gamma)$ processes are easy to reduce provided an hermetic detector. Within the 4-fermions processes, the single boson ones are an important background for the missing energy channel and the energy flow plays the most important role in the event reconstruction; the $ZZ$ and $WW$ are dominant at relative large signal to noise values and are irreducible for some values of the Higgs masses.

## 3. THE CHALLENGE OF THE SEARCH

Since only very few events are expected for a very high Higgs mass, the background should be reduced as much as possible and the tail of the distributions well under control, i.e. the resolution of the detector and the Monte Carlo modelling should be understood at the per cent level.

One of the fundamental tool of the search is the b-tagging. In the search of the SM Higgs all what was learned in 13 years of LEP has been used: the understanding of the silicon microvertex resolution to reach the per cent level in the data/Monte Carlo comparison, the sophisticated software techniques to reject by large factor (> 500) the WW background, the good modelling of the b physics (gluon splitting, Vcb, fragmenta-


doi:10.1016/S0920-5632(02)02083-2

tion) to have the high b-purity tails of the distributions under control, e.g. the semileptonic WW.

### 3.1. The four jets channel

One of the biggest difficulty in this channel is the pairing of the jets. In a genuine 4-jet event there are 3 possible di-jet combinations, while in a 5-jet event there are 10 combinations. Different solutions using more than one criteria are pursued by the different experiments: forcing one of the di-jet to the Z boson; pairing jets with the highest b-tag probability as the ones coming from the Higgs; making all the possible combinations with the WW, ZZ and HZ hypotheses and looking at the minimal $\chi^2$ of the fit; checking the spin of the boson with the jet decay angles. The irreducible background in this channel comes from ZZ and WW production having done the wrong pairing and from the so-called QCD events, i.e. $q\bar{q}gg$ final state. In these events, one of the combination has a high mass di-jet recoiling to a Z and the other correspond to the production of ZZ or WW bosons.

### 3.2. The lepton channel

The final state with 2 leptons and 2 jets it is almost background-free. It was the "golden candidate" that every experiment dreamt of. Indeed it is possible to fake a high mass candidate in this channel if a photon from initial state radiation or emitted from one of the lepton is wrongly associated to one of the jets, artificially increasing the di-jet Higgs mass.

### 3.3. The missing energy channel

It is probably the most difficult one because of the high background contamination and of the poor di-jet mass resolution. In order to increase the mass resolution, the missing mass is forced to be the one of the Z. By doing so part of the background is also pushed to the kinematic limit: M(jj)=$\sqrt{s}$-M(Z). For this reason collinear events $q\bar{q}(\gamma)$ are the background most difficult to control. The high mass Higgs signal is never collinear since the convolution of the rapidly falling cross section and the finite width of the Z results always in a boost to the Higgs. At acollinearity $\leq 5°$, it is expected to have only 5% of the Higgs signal (with $\sigma$=0.015 pb) and 30% of $q\bar{q}(\gamma)$ (with $\sigma$=80 pb). Of these a non negligible fraction has more than 60 GeV of energy lost in neutrinos: $6\times10^{-4}$ of $e^+e^- \rightarrow q\bar{q}$ of which $5\times 10^{-4}$ of $e^+e^- \rightarrow b\bar{b}$ at $\sqrt{s}$=206 GeV. In the year 2000, every experiment had $\approx$10 events $e^+e^- \rightarrow b\bar{b}$ (for L=220 pb−1) that lost more than 60 GeV in neutrinos [5].

## 4. THE FINAL LEP RESULTS

Since the end of the data in 2000 all the experiments implemented various modifications in order to improve the sensitivity and/or better control the backgrounds [1–4]: the full data set was reprocessed with final detector calibration, alignment, b-tagging; new Monte Carlo generators were used in DELPHI; more Monte Carlo statistics was produced; the precise knowledge of the LEP centre of mass energy was used. In OPAL new analyses were developed resulting in a better sensitivity: a new jet-pairing algorithm in the 4-jet channel and a new Neural Network in the missing energy channel. ALEPH implemented a better rejection of the beam related background. DELPHI extended the analysis down to low masses.

L3 had its final results already in 2001, showing few events compatible with the Higgs hypothesis.

ALEPH presents an excess of events compared to what is expected from the SM background, suggesting a Higgs boson of mass $\approx$ 114 GeV/c$^2$.

DELPHI has no evidence for any Higgs signal and sets the limit to $m_H$ >114.1 GeV/c$^2$.

OPAL has no evidence for any Higgs signal and sets the limit to $m_H$ >112.7 GeV/c$^2$.

### 4.1. The Higgs working group combination

The data from all the experiments, all the channels and at all the centre-of-mass-energies are combined in a 2-dimensional space: the reconstructed mass and the discriminating variables [6]. In each bin of this 2-dimensional space 3 quantities are given: the number of background events, the number of signal events and the observed data. By constructing a likelihood with these quantities, for any value of the Higgs masses it is possible to test if the data are more compatible with the background only or with the background+signal hypothesis. By integrating the probability density function of the signal+background or of the background only hypothesis the corresponding confidence level are computed.

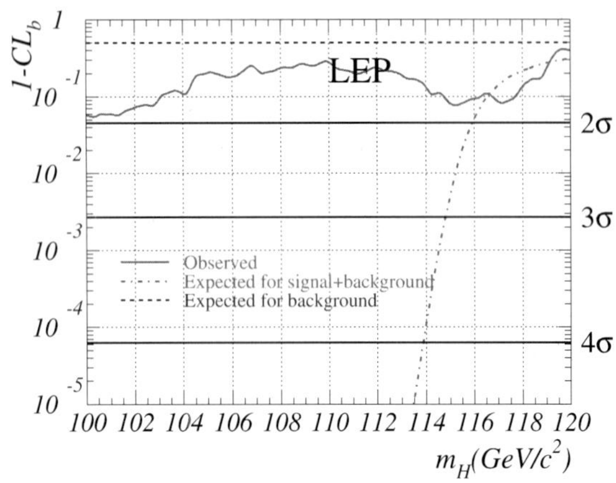


Figure 1. Probability 1-$CL_b$ as a function of the test-mass $m_H$. Solid line: observation; dashed/dash-dotted lines: expected probability for the background/signal+background hypotheses. The horizontal solid lines indicate the levels for $2\sigma$, $3\sigma$ and $4\sigma$ deviations from the background hypothesis.

The compatibility with the background only hypothesis for the LEP combined result is 1-$CL_b$=0.09, for a Higgs mass of 116 GeV/$c^2$, as shown in figure 1. The compatibility with the signal+background hypothesis (for $m_H$=116 GeV/$c^2$) is $CL_{s+b}$ = 0.37. The values for each experiments and for each channels separately are shown in table 1. The largest deviation comes from the 4-jet channel, that presents an excess of Higgs candidates with respect to the background hypothesis of less than $3\sigma$. The other channels do not show any deviation. It should be noted that the 4-jet channel and the lepton+missing energy channel have a similar potential for both the discovery and the exclusion as shown in table 2. In table 2 the expected and observed limit for the 4 experiments, the different channels and the combined result are shown. Figure 2 show the $CL_s$ as a function of the Higgs mass. The intersection of the horizontal line at $CL_s$ = 0.05 with the observed curve defines the 95% confidence level lower bound for the mass of the SM Higgs boson.

Table 1
The background confidence level 1-$CL_b$ and the signal+background confidence level $CL_{s+b}$ at $m_H$ =116 GeV.

| | 1-$CL_b$ | $CL_{s+b}$ |
|---|---|---|
| LEP | 0.099 | 0.369 |
| ALEPH | $2.41\times10^{-3}$ | 0.956 |
| DELPHI | 0.874 | 0.033 |
| L3 | 0.348 | 0.408 |
| OPAL | 0.543 | 0.208 |
| Four-jet | $5.70\times10^{-2}$ | 0.676 |
| All but four-jet | 0.368 | 0.217 |

Table 2
Expected (median) and observed 95% confidence level lower bounds on the SM Higgs boson mass.

| | Exp. limit (GeV/$c^2$) | Obs. limit (GeV/$c^2$) |
|---|---|---|
| LEP | 115.3 | 114.4 |
| ALEPH | 113.5 | 111.5 |
| DELPHI | 113.3 | 114.3 |
| L3 | 112.4 | 112.0 |
| OPAL | 112.7 | 112.8 |
| Four-jet channel | 114.5 | 113.3 |
| All but four-jet | 114.2 | 114.2 |

The LEP combined result is then compatible with the background only expectation, showing a maximal deviation of less than $2\sigma$ for Higgs masses larger than 114 GeV/$c^2$. The candidate events that contribute mostly to this deviations are the three 4-jets events from ALEPH and one missing energy event from L3.

## 5. FROM NOVEMBER 2000 TO THE FINAL RESULT

The first exciting LEP experiments combination presented in November 2000 at CERN [7] showed a very low probability, 1-$CL_b$ = $4.2\times10^{-3}$, for the background only hypothesis. For

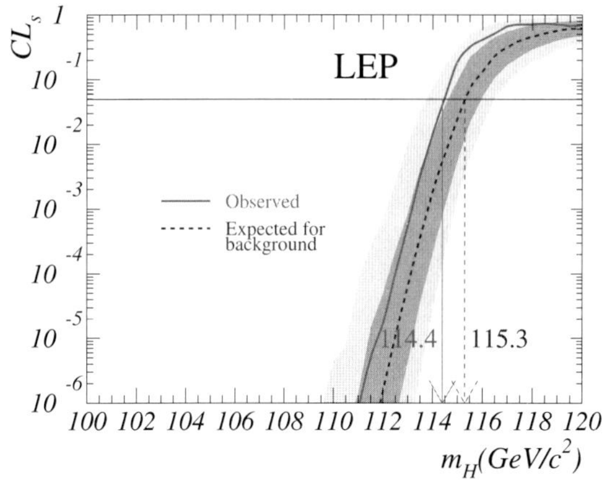


Figure 2. Confidence level $CL_s$ for the signal+background hypothesis. Solid/red line: observation; dashed line: median background expectation. The dark/green and light/yellow shaded bands around the median expected line correspond to the 68% and 95% probability bands computed with a large number of simulated background experiments.

the July 2001 combination, all the available luminosity was added into the analyses (about 10% more for each experiment), ALEPH implemented, for the LEP combination ntuples, the correction for the correlation between the mass and the discriminant variable, DELPHI optimised the Hee analysis, L3 included more simulated events and published the final result, OPAL improved the background estimation. The result of this second LEP combination was an increase of the background only hypothesis probability with respect to the November 2000 result, i.e. 1-$CL_b$= 0.03 for M(H)=115 GeV/$c^2$ [8].

For this summer the final LEP combination has been performed, with the individual experiment final analyses presented in the previous chapter and resulted in a even higher probability for the background only hypothesis of 1-$CL_b$= 0.09 for M(H)=115 GeV/$c^2$. While DELPHI has always shown a background like behaviour, and a $\approx 1\sigma$ deficit with respect to expectations, the significance of the small excess observed by the other experiments gradually decreased and the probability of the background only hypothesis increased [1–4,6]. The total LEP effect is of less than $2\sigma$.

## 6. THE STANDARD MODEL HIGGS AS OF TODAY

The experimental limits coming from the direct search, i.e. m(H) $\geq$ 114.4 GeV/$c^2$ at 95%CL, and from the Electroweak precision measurements, i.e. m(H) $\leq$ 193 GeV/$c^2$ at 95%CL, are consistent with a theoretical picture where the Standard Model stays valid up to very high energy. On the other hand the Electroweak precision measurements suggest a very light Higgs, M(H)=81 $^{+53}_{-30}$ GeV/$c^2$, that LEP maybe missed for very few GeV. Up to now there is no evidence for a discrepancy between the Electroweak precision measurements and the direct search, but it will be only in $\approx$5 year from now that the data from the colliders at FNAL (only for very low Higgs masses) and CERN will tell if the Higgs exists, and if it does, the value of its mass.

## REFERENCES


1. ALEPH Collab., R. Barate et al., Phys. Lett. **B526** (2002) 191.
2. DELPHI Collab., *Final result on SM and MSSM neutral Higgs bosons*, DELPHI 2002-041-CONF-575.
3. L3 Collab., M. Acciarri et al., Phys. Lett. **B517** (2001) 319.
4. OPAL Collab., *" Search for the Standard Model Higgs Boson with the OPAL Detector at LEP"*, OPAL Physics Report PR367 (July 2002), to be submitted to Eur. Phys. Journal **C**.
5. DELPHI Collab., A study of collinear events in the framework of the Higgs analysis , DELPHI 2002-085-CONF-619
6. ALEPH, DELPHI, L3, OPAL and the LEP Higgs Working Group, July 19, 2002 , LHWG Note 2002/01
7. P. Igo-Kemenes for the LEP Higgs working group, LEP Committee Open Session, 3 November, 2000.
8. ALEPH, DELPHI, L3, OPAL and the LEP Higgs Working Group, July 11, 2001 , CERN-EP/2001-055

ELSEVIER

Proceedings of ICHEP 2002, pp. 206–210
S. Bentvelsen, P. de Jong, J. Koch and E. Laenen (Editors)

31st INTERNATIONAL CONFERENCE ON HIGH ENERGY PHYSICS AMSTERDAM

# The Frontier of Multi-Loop Feynman Diagrams: (Semi-) Numerical Techniques

A. Ferroglia[a] *, G. Passarino[b] †, M. Passera[c], S. Uccirati[b]

[a]Department of Physics, New York University, U.S.A

[b]Dipartimento di Fisica Teorica, Università di Torino and INFN, Italy

[c]Dipartimento di Fisica G. Galilei, Università di Padova and INFN, Italy

Fast and efficient ways to compute multi-loop and multi-leg Feynman diagrams are discussed.

Why do we need fast and efficient ways to deal with multi-loop, multi-leg Feynman diagrams? Because computing higher order corrections is complex and the complexity is reflected by the fact that already in the Standard Model (SM) there are 11 lines and 57 vertices (modulo fermion multiplicity) but a full two-loop electroweak calculation is needed for GigaZ ($10^9$ $Z$ events). There is more, even for LEP physics the so-called theoretical uncertainty is not yet at a satisfactory level and also requires a full two-loop analysis of the SM (or Beyond).

The main message is that we have been moving from a mediaeval architecture of radiative corrections to the renaissance of LEP period and now one should not indulge in deciphering the intimate meaning of the last digit in the last fit but prepare for the new era (hopefully not baroque), although many auxiliary features are far from satisfactory, e.g. diagrammatica versus numerica, new algorithms versus standardization and the growth in complexity versus various software problems .

Analytically (quasi) solvable cases are: QCD, all masses $\equiv 0$; QED, only $m_e \neq 0$ and 0 or mass shell for external scales; SM $\oplus$ simplifying conditions, e.g. $m/m_{\rm top} \ll 1$. For the general case (arbitray int/ext masses) the analytical way (AW) is not feasible and, although a lot can be done, what is so special about it? Of course the analytical properties are explicit and expansions are available, cf.

$$\mathrm{Li}_2(z) \;=\; \sum_{n=0}^{\infty}(-1)^{n+1}\,\frac{B_n}{(n+1)\,!}\,\ln^{n+1}(1-z),$$

where the expansion parameter has the same cut of the special function itself. However, beyond one-loop and for the general mass case polylogarithms do not suffice. The AW is, in general, possible only for small number of scales; for instance the sunset topology is a sum of generalized hypergeometric series i.e. is accessible for small or large $p^2$ but not in the intermediate region of thresholds.

An alternative strategy is represented by the numerical way (NW). Here we summarize the main contributions available in the literature.

In ref. [1] the use of dispersion relations produces one-dimensional integral representation for two-loop self-energies. In re. [2] all diagrams are reduced to a limited number of master integrals to be solved via differential equations in $p_i^2$. Results are available for the sunset, the difficulty is in determining the initial conditions that are to be provided by other methods. Ref. [3] uses fully numerical integration below thresholds. Above thresholds, numerical integration is done after symmetric methods and/or hybrid methods, i.e. integrals for a few variables analytically until an

*From October 1 2002: Fakultät für Physik, Albert-Ludwigs Universität, Freiburg, Germany.

†Work supported by the European Union under contract HPRN-CT-2000-00149 and by MIUR under contract 2001023713006. Talk presented by G. Passarino.


doi:10.1016/S0920-5632(02)02085-6

integrable function is obtained. In the approach of ref. [4] calculations (in QCD) are performed using numerical integration over FS particles and analytical integration over virtual particles. It requires to deform some of the integration contours into the complex plane. In ref. [5] all diagrams are reduced to generalized sunsets and a one-dimensional integral representation is produced for the 10 generalized sunset master integrals needed in a renormalizable theory, with numerical integration over the external Feynman parameters.

Our conclusion is that, because of the size of the problem, full numerical calculability can hardly be avoided.

A generic (scalar) $N_L$ - leg, $n_l$ - loop diagram is

$$
\begin{aligned}
G &= \left(i\,\pi^{n/2}\right)^{n_l}\,\Gamma\left(N_L - \frac{n}{2}\,n_l\right) \\
&\times \int \frac{dx_G\,\delta\,(1-x_G)}{U^{n/2}\,(V - i\,\delta)^{N_L - n n_l/2}}, \\
&\equiv \int_S \prod_{i=1}^{N_l - 1} dx_i\, I_G(\{x_i\}), \\
dx_G &= \prod_{i=1}^{N_l} dx_i, \qquad x_G = \sum_{i=1}^{N_l} x_i. \qquad (1)
\end{aligned}
$$

Note that the parametric function $I_G$ has not a definite sign, that is poles inside the integration region that hinder the numerical integration. Example:

$$
\begin{aligned}
I(\alpha) &= \int_0^1 dx\,\Big[\chi_B(x) - i\,\delta\Big]^{-\alpha} \\
\chi_B(x) &= a\,(x + \frac{b}{2\,a})^2 - \frac{b^2 - 4\,a\,c}{4a}. \qquad (2)
\end{aligned}
$$

When the roots of $\chi_B - i\,\delta = 0$ are real and $\in [0,1]$ the integral can be computed by distorting the integration contour unless a pinch will occur; in this case the integral is singular.

A novel numerical approach has been introduced by the TOPSIDE Collaboration [6]: the method makes a search for algorithms of smoothness that avoid spurious singularites and can be summarized in the following transformation:

$$
G = \frac{1}{B_G}\int_S dx\,\mathcal{F}\Big[I_G\Big], \qquad (3)
$$

where $\mathcal{F}$ is a functional of $I_G$ that makes the whole integrand smooth in parametric space (from poles to cuts) and $B_G$ is a function of masses and momenta that contains the true singularities of $G$ (if any).

The prototype of all algorithms of smothness is represented by the Bernstein-Tkachov (BT) theorem [7]: for any finite set of polynomials $V_i(x)$, $x = (x_1, \dots, x_N)$ there exists an identity of the following form:

$$
\mathcal{P}\,(x,\partial)\prod_i V_i^{\mu_i+1}(x) = B\prod_i V_i^{\mu_i}(x). \qquad (4)
$$

where $\mathcal{P}$ is a polynomial of $x$, $\partial_i = \partial/\partial x_i$; $B$ and all coefficients of $\mathcal{P}$ are polynomials of $\mu_i$ and of the coefficients of $V_i(x)$. Iterative applicationsof the functional relations, followed by integration by parts, allow us to express the integrand in parametric space as a combination of $P^N(x) \times \ln P(x)$. Let us illustrate the strategy in a simple example: fermionic self-energy in QED. First we introduce a parametric representation

$$
\begin{aligned}
\Sigma &= -\,ie^2\,\pi^{2-\epsilon/2}\,\mu^\epsilon\,\Gamma\left(\frac{\epsilon}{2}\right) \\
&\times \int_0^1 dx\,\Big[i\,(2-\epsilon)\,x\not{p} + (4-\epsilon)\,m_e\Big]\,\chi^{-\epsilon/2}, \\
\chi(x) &= (1-x)\,\left(x\,p^2 + m_e^2\right). \qquad (5)
\end{aligned}
$$

Then we perform a reduction to form - factors

$$
\Sigma = -\,i\pi^2 e^2\,\Big[\Sigma_p\, i\not{p} + \Sigma_m\, m_e\Big], \qquad (6)
$$

followed by BT - iteration and integration - by - parts

$$
\begin{aligned}
\Sigma_p &= \frac{1}{\bar{\varepsilon}} + \frac{p^2}{P_e^2}\left[\frac{\gamma}{2}\,\frac{P_e^2}{p^2} - \frac{1}{3}\,m_e^2 - \frac{1}{6}\,p^2\right. \\
&+ \left.\int_0^1 dx\,\left(\frac{1}{2} - \frac{1}{2}\,\frac{m_e^2}{p^2} - 4\,x\right)\,\chi\,\ln\frac{\chi}{\mu^2}\right], \\
\Sigma_m &= \frac{4}{\bar{\varepsilon}} + \frac{p^2}{P_e^2}\,\left[\left(2\,\gamma - \ln\frac{m_e^2}{\mu^2} + \frac{1}{2}\right)\,\frac{P_e^2}{p^2}\right. \\
&+ p^2\,\ln\frac{m_e^2}{\mu^2} - \frac{2}{3}\,P_e\,\left(3\,\ln\frac{m_e^2}{\mu^2} - 2\right) \\
&- \left. 6\int_0^1 dx\,\chi\,\ln\frac{\chi}{\mu^2}\right], \qquad (7)
\end{aligned}
$$

where $P_e = p^2 + m_e^2$.

The solution for the BT - polynomial $\mathcal{P}$ is known $\forall$ all quadrics, i.e. for all $1l$ & $nL$ diagrams. However, moving to multi-loop, we see that already the sunset topology (the easiest one) requires degree $= 4$. Furthermore, around - threshold behavior remains unsolved in BT approach. We propose a minimal BT - approach where

$$G = \int \frac{dx_S}{B(x_S)} \int dx_L\, \mathcal{F}\Big[I_G(x_S, x_L)\Big]. \quad (8)$$

The analytical structure of $G_S$ (the sub-diagram with the smallest number of internal lines) is simpler, the $x_S$ integration can be performed with contour - distortion which is always possible unless a pinch (or an end-point) singularity occur but this is a real Landau singularity. Our analysis has shown that, $\forall$ two-loop two-point functions 2l&2L and $\forall$ two-loop three-point functions 2l&3L (but one) contour - distortion is not needed because a parametrization can be found that allows for a BT - procedure with $B \equiv B(\text{parameters})$. These $2l$ diagrams are no more involved that a $1l$ diagram and require normal BT $\oplus$ integration by parts.

Clearly, numerical evaluation requires a precise knowledge of the analytical structure of the diagram, i.e. a solution of the corresponding et of Landau equations. For instance, for the scalar one-loop vertex ($C_0$) the leading singularity $\equiv B_3 = 0 \equiv$ anomalous threshold (AT).

We now addres the question of what to do around $B_G = 0$. Consider the simple $C_0$ example: if $B_3 \neq 0$, then we use $V = (x-X)^t\, H\,(x-X) + B_3$ to obtain

$$\begin{aligned} C_0 &= \frac{1}{B_3}\Big\{\frac{1}{2} + \int_0^1 dx_1 \Big[\Big[\int_0^{x_1} dx_2 \ln V(x_1, x_2) \\ &+ \frac{1}{2}(X_2 - X_1)\ln V(x_1, x_1) \\ &- \frac{1}{2} X_2 \ln V(x_1, 0) \\ &- \frac{1}{2}(1 - X_1)\ln V(1, x_1)\Big]\Big\}. \end{aligned} \quad (9)$$

If $B_3 \approx 0$ and $0 < X_2 < X_1 < 1$ not fulfilled: $C_0$ regular, standard BT not applicable. Solution: Taylor expansion plus special BT - procedure.

If $B_3 \approx 0$ and $0 < X_2 < X_1 < 1$ fulfilled: $C_0$ singular, standard BT not applicable. Solution: Mellin - Barnes splitting plus sector decomposition plus BT - algorithm. For more general cases there is a tree of $B$ coefficients: for the scalar box we obtain

$$D_0(n=4) = \frac{1}{2\,B_4}\Big\{D_0^{n=6} \oplus \sum C_0\Big\}. \quad (10)$$

At the next iteration each reduced $C_0$ brings a sub-leading $B_3$ and the full chain $B_{\text{leading}} = B_3, B_{i,\text{subl}} = B_{3,i}$ must be treated.

To summarize the Mellin - Barnes technique we consider a simple example:

$$I(\alpha) = \int_0^1 dx\,\Big[\chi_B(x) - i\,\delta\Big]^{-\alpha}. \quad (11)$$

$$\begin{aligned} \chi_B(x) &= -P^2\,(x-a)^2 + \frac{\lambda}{4\,P^2} - i\,\delta, \\ a &= \frac{1}{2}\,\frac{P^2 + m_2^2 - m_1^2}{P^2}. \end{aligned} \quad (12)$$

A pinch will occur for $\lambda(-P^2, m_1^2, m_2^2) = 0$. Assume $P^2 = -s$, $s \geq 0$ and $0 \leq a \leq 1$ and obtain

$$\begin{aligned} I(\alpha) &= \Big(\frac{\rho}{s}\Big)^{\alpha} \sum_{X=a}^{1-a} \int_0^X dx\,(1 + \rho\,x^2)^{-\alpha} \\ &= \Big(\frac{\rho}{s}\Big)^{\alpha} \sum_{X=a}^{1-a} J_X(\alpha, \rho), \end{aligned} \quad (13)$$

where $\rho = -4\,s^2/\lambda$. We are now interested in the behavior for $|\rho| \to \infty$. Consider the case $\lambda < 0$ :

$$\begin{aligned} j_X(\alpha, s) &= \int_0^\infty d\rho\,\rho^{s-1}\,J_X(\alpha, \rho) \\ &= \frac{B(s, \alpha - s)}{1 - 2\,s}\,X^{1-2\,s}, \end{aligned} \quad (14)$$

which is defined for $0 < \text{Re}\,s < \min\{\frac{1}{2}, \text{Re}\,\alpha\}$ and $|\arg x^2| < \pi$. The inverse Mellin transform will be

$$J_X(\alpha, \rho) = \frac{1}{2\,\pi\,i}\int_{-i\,\infty}^{+i\,\infty} ds\,\rho^{-s}\,j_X(\alpha, s). \quad (15)$$

The integration contour will be closed over the right-hand complex half-plane at infinity and leading behavior comes from the pole at $s = 1/2$,

$$J_X(\alpha, \rho) = \frac{1}{2}\,B(\frac{1}{2}, \alpha - \frac{1}{2})\,\rho^{-1/2} + \dots, \quad (16)$$

with $1 < \rho < \infty$.

A complete project must be more ambitious: any realistic numerical strategy must be able to deal with infrared (IR) configurations, collinear configurations and, possibly, Sudakov configurations. Our numerical way to IR configurations is as follows. A necessary condition for the presence of infrared divergences is that the Landau equations are fulfilled. We regularize in $n$-dimensions and isolate the infrared poles by using the technique of iterated sector decomposition [8] (or defoliation). A $k/n$ defoliation is defined by the following relation

$$\begin{aligned} & \int_0^1 \prod_{i=1}^{k} dx_i \prod_{j=k+1}^{n} dy_j\, F(\{x\}\,;\,\{y\}) \\ = & \sum_{l=1}^{k} \int_0^1 \prod_{i=1}^{k} dx_i\, x_l^{k-1} \prod_{j=k+1}^{n} dy_j \\ \times & \;\; F(x_l\, x_1, \dots, x_l, \dots x_l\, x_k\,;\,\{y\}). \end{aligned} \tag{17}$$

A chain of defoliations will stop when all the singularites of $F$ which are located at the vertices of the simplex $[0,1]^k$ have been factorized,

$$\begin{aligned} & F(x_l\, x_1, \dots, x_l, \dots x_l\, x_k\,;\,\{y\}) \\ = & \;\; x_l^{\mu_l}\, F(x_1, \dots, 1, \dots x_k\,;\,\{y\}). \end{aligned} \tag{18}$$

To find the sectors, although far from unique, is not the difficult part of the evaluation. The resulting expression will contain infrared poles as well as an infrared finite part. The latter and the residues are given by integrals of inverse powers of polynomials that are not (usually) positive defined. BT - iterations or Mellin - Barnes splittings are still required.

Finally, the standard procedure in dealing with loop calculations is based on the introduction of scalar functions and of tensor integrals that are reduced to linear combinations of scalar ones. The new procedure is different since the BT functional relation can be written in a more general form starting from

$$\int_S dx\, Q(x) \prod_i V_i^{\mu_i}(x),$$
$$Q(x) \;\; = \;\; \text{an arbitrary polynomial of}\, x. \tag{19}$$

Hence we are led to consider all integrals in parametric space with no pre-reduction in momentum space. The final strategy will be to map as many integrals as possible in one integral,

$$\text{Process} \;\; \equiv \;\; \int_{\text{momenta}} \sum \text{diagrams} \equiv I \otimes FF$$

$$I \;\; = \;\; \int_{S_{\text{mx}}} \sum_{ij} Q_{ij}(X)\, V_i^{\mu_i}(X), \tag{20}$$

where $X = x_{\text{sub}_i} \in \{x_{\text{mx}}\}$ and where $I$ evaluation is performed in one stroke.

In conclusion we stress that, at the moment, our results allow us to deal with one-loop (up to six legs) and two-loop (up to three legs) Feynman diagrams with up to 9 different scales, 6 normal thresholds (plus pseudo and anomalous ones); the kinematics is unconstrained and IR - configurations are accessible.

## REFERENCES

1. S. Bauberger, F. A. Berends, M. Bohm and M. Buza, Nucl. Phys. B **434** (1995) 383 [hep-ph/9409388].
2. M. Caffo, H. Czyz, S. Laporta and E. Remiddi, Nuovo Cim. A **111** (1998) 365 [hep-th/9805118].
3. J. Fujimoto, Y. Shimizu, K. Kato and T. Kaneko, Int. J. Mod. Phys. C **6** (1995) 525 [arXiv:hep-ph/9505270].
4. D. E. Soper, Nucl. Phys. Proc. Suppl. **79** (1999) 444.
5. A. Ghinculov and Y. P. Yao, Phys. Rev. D **63** (2001) 054510 [arXiv:hep-ph/0006314].
6. G. Passarino, Nucl. Phys. B **619** (2001) 257 [arXiv:hep-ph/0108252];
G. Passarino and S. Uccirati, Nucl. Phys. B **629** (2002) 97 [arXiv:hep-ph/0112004];
A. Ferroglia, G. Passarino, M. Passera and S. Uccirati, [arXiv:hep-ph/0209219].
7. F. V. Tkachov, Nucl. Instrum. Meth. A **389** (1997) 309 [hep-ph/9609429];
L. N. Bertstein, Functional Analysis and its Applications, **6**(1972)66.
8. T. Binoth and G. Heinrich, Nucl. Phys. B **585** (2000) 741 [hep-ph/0004013].

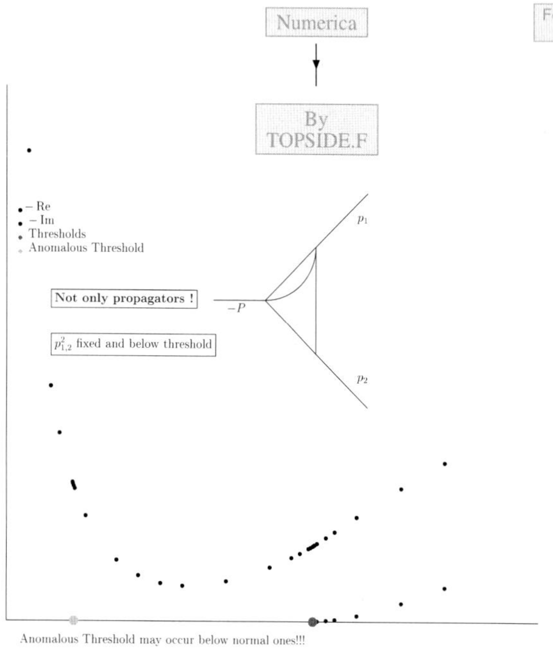


Figure 1: Evaluation of a two-loop vertex function.

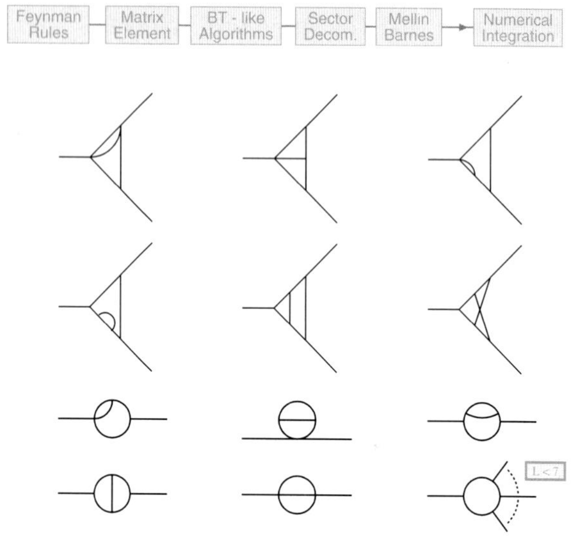


Figure 2: Availability in TOPSIDE.

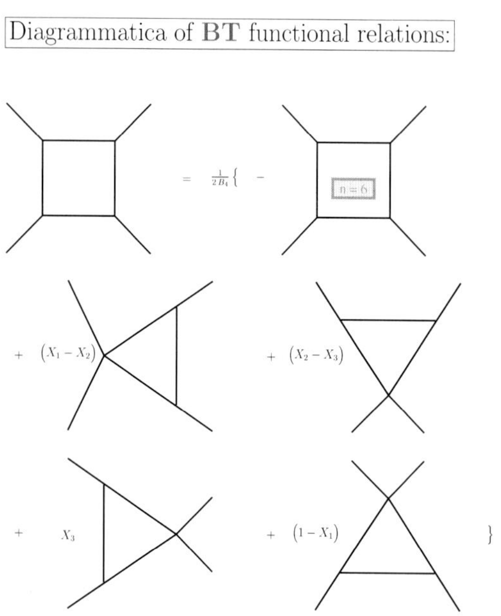


Figure 3: *Diagrammatica* of BT relations.

Proceedings of ICHEP 2002, pp. 211–215
*S. Bentvelsen, P. de Jong, J. Koch and E. Laenen (Editors)*

31st INTERNATIONAL CONFERENCE ON HIGH ENERGY PHYSICS AMSTERDAM

# Precision Electroweak Measurements: a Theorist point of view

Paolo Gambino[a] *

[a] Theory Division - CERN
CH 1211, Geneve 23, Switzerland

Present global fits to electroweak data are characterized by two results that differ from SM expectations by about $3\sigma$, the NuTeV result and the $b$ quark asymmetries measured at LEP. I review possible SM and new physics explanations of these anomalies and the implications for the indirect determination of the Higgs mass.

## 1. The global Standard Model fit

There is not so much going on in electroweak physics nowadays, apart from the muon $g-2$, it is tempting to say. Not quite so: the latest Standard Model (SM) fit performed by the LEP Electroweak Working Group [1] looks remarkably different from the one from last year. The main new result comes from the NuTeV Collaboration [2]: their measurement of the electroweak (e.w.) mixing angle in $\nu$-N Deep Inelastic Scattering (DIS) differs about 3 sigmas from theoretical expectations. The $\chi^2$/d.o.f. of the global fit is 29.7/15, corresponding to 1.3% probability. NuTeV result shares the responsibility for the degradation of the fit with another deviant measurement, that of the bottom quark Forward-Backward asymmetry, $A^b_{FB}$, at LEP. The best fit [1] points to a fairly light Higgs boson, with mass $M_H = 81$ GeV, while the 95% CL upper bound on $M_H$, including an estimate of theoretical uncertainty, is about 190 GeV. Interestingly, the information on the Higgs mass is almost insensitive to the NuTeV result: a fit performed excluding this new result gives practically the same constraints on $M_H$, but of course the quality of the fit improves sensibly, with $\chi^2$/d.o.f.=20.5/14, corresponding to a probability of 11.4%. Onw would conclude that the SM fit is quite satisfatory, if not for NuTeV. Let us therefore start this (incomplete) status review with a look at NuTeV intriguing *anomaly*.

## 2. The NuTeV electroweak result

NuTeV measures ratios of Neutral (NC) to Charged Current (CC) cross sections in $\nu N$ DIS. Ideally, in the parton model with only one generation of quarks and isoscalar target

$$R_\nu \equiv \frac{\sigma(\nu N \to \nu X)}{\sigma(\nu N \to \mu X)} = g_L^2 + r g_R^2$$

$$R_{\bar\nu} \equiv \frac{\sigma(\bar\nu N \to \bar\nu X)}{\sigma(\bar\nu N \to \bar\mu X)} = g_L^2 + \frac{1}{r} g_R^2,$$

where $r \equiv \frac{\sigma(\bar\nu N \to \bar\mu X)}{\sigma(\nu N \to \mu X)}$ and $g^2_{L,R}$ are average effective left and right-handed $\nu$-quark couplings. The actual experimental ratios $R^{exp}_{\nu,\bar\nu}$ differ from $R_{\nu,\bar\nu}$ because of $\nu_e$ contamination, experimental cuts, NC/CC misidentification, the presence of second generation quarks, the non-isoscalarity of steel target, QCD and e.w. corrections *etc.* In the NuTeV analysis, a MonteCarlo including most of these effects relates $R^{exp}_{\nu,\bar\nu}$ to $R_{\nu,\bar\nu}$. It is useful to note that most uncertainties and $O(\alpha_s)$ effects drop in the Paschos-Wolfenstein (PW) ratio [3]

$$R_{PW} \equiv \frac{R_\nu - r R_{\bar\nu}}{1-r} = \frac{\sigma(\nu N \to \nu X) - \sigma(\bar\nu N \to \bar\nu X)}{\sigma(\nu N \to \ell X) - \sigma(\bar\nu N \to \bar\ell X)}$$

which equals $g_L^2 - g_R^2 = \frac{1}{2} - \sin^2\theta_W$ and therefore could provide a clean measurement of $\sin^2\theta_W$, if experimentally accessible. NuTeV do not measure $R_{PW}$ directly, but, using the fact that $R_{\bar\nu}$ is almost insensitive to $\sin^2\theta_W$, they extract from it the main hadronic uncertainty, an effective charm mass. The weak mixing angle is then obtained from $R_\nu$. In practice, NuTeV fits for $m_c^{eff}$ and

*Marie Curie Fellow


doi:10.1016/S0920-5632(02)02087-X

$\sin^2\theta_W$. At first approximation, NuTeV procedure corresponds to a measurement of $R_{PW}$. The result is expressed as a test on the on-shell $s_W^2 \equiv 1 - M_W^2/M_Z^2$ definition of $\sin^2\theta_W$:

$$s_W^2(\text{NuTeV}) = 0.2277 \pm 0.0013 \pm 0.0006 \pm 0.0006,$$

where the three errors are statistical, systematic, and theoretical, respectively. Because of accidental cancellations, the choice of the on-shell scheme implies very small top and Higgs mass dependence in the above equation. The above value must be compared to the one obtained using the results of the global fit, $s_W^2 = 0.2226 \pm 0.0004$, which is about $3\sigma$ away.

QED corrections are important and their implementation in NuTeV could be improved, but they seem at the moment an unlikely explanation. Electroweak corrections, on the other hand, are small and under control.

A potentially very important source of uncertainty are the parton distribution functions (PDFs) employed in the analysis. NuTeV work at Leading Order (LO) in QCD in the context of a *cross section model* which effectively introduces some Next to Leading Order (NLO) improvement. They use LO PDFs self-consistently fitted in the experiment, with little external input. Is NuTeV estimate of the related uncertainty reliable? As we have seen, $R_{PW}$ is strictly independent of the details of first generation PDFs. As long as NuTeV result is equivalent to a measurement of $R_{PW}$, even with cuts and second generation quarks, the small uncertainty attributed by NuTeV might be realistic [4]. The problem is that NuTeV does *not* really measure $R_{PW}$ and there are indications [4] that this might be relevant at the required level of accuracy.

We have seen that NuTeV does not employ NLO QCD corrections. Are they necessary? The answer is very similar to the previous one: no, if you are measuring $R_{PW}$, which is not corrected at $O(\alpha_s)$. But any CC/NC or $\nu/\bar{\nu}$ asymmetry (introduced by cuts, differences in the energy spectra and in the sensitivity, *etc.*) spoils delicate cancellations (ordinary NLO corrections are 5-10%, while here a better than 0.5% precision is required). As NuTeV measurement seems to differ enough from that of $R_{PW}$, the analysis needs to be consistently upgraded to NLO. As we will see in a moment, this would also simplify the discussion of other issues.

### 2.1. Asymmetric sea

In the previous section I have implicitly used the assumptions, generally made in the extraction of PDFs from the data, of isospin symmetry and of a symmetric strange and charm sea ($s = \bar{s}$, $c = \bar{c}$). If we drop these assumptions, the PW relation is explicitly violated by new terms [4]

$$R_{PW} = \frac{1}{2} - s_W^2 + \frac{\tilde{g}^2}{Q^-}(u^- - d^- + c^- - s^-), \quad (1)$$

where $q^-$ is the asymmetry in the momentum carried by the quark species $q$ in an isoscalar target, $q^- = \int_0^1 x\,[q(x) - \bar{q}(x)]\,dx$, $\tilde{g}^2 \approx .23$ a coupling factor, and $Q^- = (u^- + d^-)/2 \approx 0.18$. There is no reason in QCD to expect $s^- = 0$, or $u^- = d^-$. In fact, eq.(1) tells us that unless they are very small, these two asymmetries could change significantly the value of $s_W^2$ measured by NuTeV.

What do we know about the strange quark asymmetry? An asymmetry $s^-$ of the sign needed to explain NuTeV can be induced non-perturbatively [5]. Unfortunately, the strange quark sea is mainly constrained by (mostly old) $\nu N$ DIS data, which are usually not included in standard PDFs fits. In fact, MRST and CTEQ use an *ansatz* $s = \bar{s} = \frac{\bar{u}+\bar{d}}{4}$. Barone et al. (BPZ) [6] have reanalysed at NLO all $\nu N$ DIS together with $\ell N$ and Drell-Yan data. They have a much higher sensitivity to strange sea than the standard fits and find a strange $s(x)$ larger than usual at high-$x$. This feature contrasts with NuTeV dimuon results, not included in their fit, but agrees well with positivity constraints from polarized DIS. Allowing for a strange asymmetry improves BPZ best fit drastically and could explain a large fraction of the discrepancy. The result, $s^- \approx 0.002$, agrees well with theory estimates [5].

I have already mentioned that BPZ do not include NuTeV data, especially those on dimuon events (tagged charm production), a rather sensitive probe of the strange sea. NuTeV has analyzed them, finding $s^- = -0.0027 \pm 0.0013$ and claiming that this result would *increase* the anomaly [7]. In fact, the above estimate *can-*

*not* be interpreted as a measurement of $s^-$ and should not be compared to that of BPZ [4]. Its use in the $s_W^2$ extraction is also highly questionable, even in the context of NuTeV improved LO model, because it assumes that the restricted set of dimuon events available be representative of the whole kinematical range employed in the $s_W^2$ analysis. At the very least, a generous theory error should be attached to this procedure. A preliminary NLO analysis of NuTeV dimuons [8] is in better agreement with BPZ, both on the total size and on the asymmetry of $s(x)$.

The bottom line is that we presently know very little on the strange sea. Before any conclusion can be drawn on its asymmetry and the effect on NuTeV $s_W^2$ result, a global NLO fit including all dimuons and $\nu$-N DIS data is needed. A precise $s(x), \bar{s}(x)$ determination will be possible at a neutrino factory [9].

A violation of isospin of the form $u_p(x) \neq d_n(x)$ would also affect the PW relation according to eq. (1). A rough estimate for its size is $(m_u - m_d)/\Lambda_{QCD} \approx 1\%$. So small a violation of charge symmetry would give no visible effect in any present experiment, apart from NuTeV measurement of $s_W^2$, where it could explain a fraction of the anomaly. Explicit model calculations [10] vary widely in their results for a shift in $s_W^2$. Small estimates are generally due to subtle cancellations of much larger contributions and should be taken with care.

The relevant momentum asymmetries in the quark sea are therefore only weakly constrained and could have a significant impact on $s_W^2$ extracted by NuTeV. Although it has been shown [7] that these effects are somewhat diluted in the actual NuTeV analysis, precisely because it differs from a measurement of $R_{PW}$, they introduce an unwelcome uncertainty very hard to estimate.

I should also mention that several attempts at explaining the NuTeV anomaly with nuclear effects like nuclear shadowing have been made [11], but no convincing case has so far been presented.

## 3. New Physics vs NuTeV

A New Physics explanation of the NuTeV anomaly requires a $\sim$ 2% effect, and naturally calls for tree level physics. It is very difficult to build realistic models that satisfy all present experimental constraints and explain a large fraction of the anomaly (see [4] for an overview).

In particular, supersymmetry, with or without R parity, cannot help, because it is strongly contrained by other precision measurements (often at the permille level) and by direct searches. The same is generally true of models inducing only oblique corrections or only anomalous $Z$ couplings. Realistic and well-motivated examples of the latter are models with $\nu_R$ mixing [4,12]. Models with $\nu_R$ mixing *and* oblique corrections are considered in [13]. On the other hand, the required new physics can be parameterized by a contact interaction of the form $G_F\,[\bar{L}_2\gamma_\mu L_2][\bar{Q}_1\gamma_\mu Q_1]$. This operator might be induced by different kinds of short-distance physics. Leptoquarks generally also induce another operator, which overcontributes to $\pi \to \mu\bar{\nu}_\mu$, or lead to opposite effects, but SU(2) triplet leptoquarks with non-degenerate masses could fit NuTeV, albeit not very naturally. Another possible new physics behind the above contact interactions is an unmixed $Z'$ boson. It could be either light ($2 \lesssim M_{Z'} \lesssim 10$ GeV) or heavy ($M_{Z'} \gtrsim 600$ GeV). Viable possibilities that could alleviate but not wholly explain the NuTeV anomaly and at the same time explain part of the $(g-2)_\mu$ anomaly, are based on abelian gauge symmetries $B - 3L_\mu$ [4] and $L_\mu - L_\tau$ [14].

## 4. The SM fit to $M_H$ is not satisfactory

The global fit without NuTeV has an 11% probability. This gives us an idea of the overall consistency of the data, but if we are interested in extracting information on the Higgs mass, it is clear that we should concentrate only on the subset of observables that are really sensitive to $M_H$ and, because of a strong correlation, to the top mass, $M_t$. Using only $M_W$,$M_t$,$\Gamma_\ell$, the $Z$-pole asymmetries, and $R_b$, one obtains $M_H^{fit} = 90$ GeV, $M_H < 195$ GeV at 95% C.L., and $\chi^2$/dof=13/4, corresponding to a 0.9% probability. In other words, the restricted fit gives the same constraints on $M_H$ of the global fit. However, it is now obvious that the SM fit to the Higgs mass is *not*

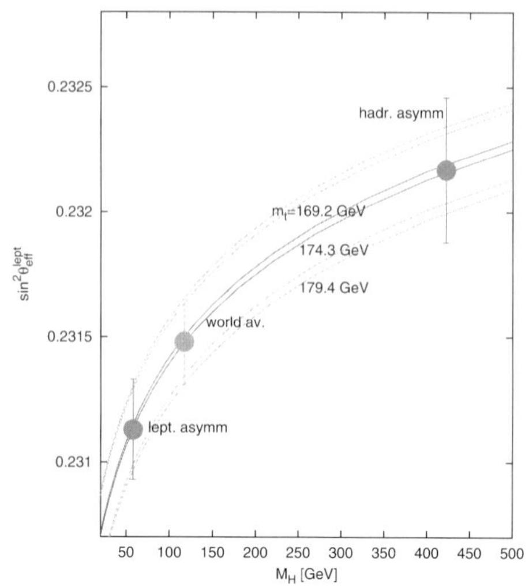


Figure 1. Higgs mass dependence of $\sin^2\theta_{eff}^{lept}$ extracted from leptonic and hadronic asymmetries, for three $M_t$ values.

satisfactory, even without NuTeV.

## 5. Another unwelcome anomaly

The root of the problem is an old $3\sigma$ discrepancy between the Left-Right asymmetry, $A_{LR}$, measured by SLD and $A_{FB}^b$ measured by the LEP experiments. In the SM these asymmetries measure the *same* quantity, $\sin^2\theta_{eff}^{lept}$, related to the lepton couplings. It now happens that all leptonic asymmetries, measured both at LEP and SLD, are mutually consistent and prefer a very *light Higgs* mass. In this sense, they are also consistent with $M_W$ measured at LEP and Tevatron. Only the asymmetries into hadronic final states prefer a *heavy Higgs* (see fig. 1).

Since the hadronic asymmetries are dominated by $A_{FB}^b$, and the third generation is naturally singled out in many extensions of the SM, could this be a signal of new physics in the $b$ couplings? It seems unlikely for several reasons: (i) fixing $\sin^2\theta_{eff}^{lept}$ at the value measured by the leptonic asymmetries, $A_{FB}^b$ corresponds to a measurement of a combination of $b$ couplings, $\mathcal{A}_b$, which is also tested by $A_{LR}^{FB}$ at SLD. The latter however agrees perfectly with the SM [1], and is compatible only at 1.5 $\sigma$ with $A_{FB}^b$; (ii) the value of $\mathcal{A}_b$ extracted from $A_{FB}^b$ would require a $\sim$ 30% correction to the $b$ vertex, i.e. tree level physics; (iii) $R_b$ agrees well with the SM and tests an orthogonal combination of $b$ couplings; it follows that new physics should predominantly affect the right-handed $b$ coupling, $|\delta g_R^b| \gg |\delta g_L^b|$. All this places strong restrictions on the extensions of the SM that can explain $A_{FB}^b$. Exotic scenarios that shift only the $b_R$ coupling include mirror vector-like fermions mixing with $b$ quark [15], and LR models that single out the third generation [16], but even these *ad-hoc* models have problems in passing all experimental tests. Difficult to explain in the most popular new physics models, both NuTeV and $A_{FB}^b$ are in a sense *unwelcome* anomalies.

## 6. Too light a Higgs

It is instructive to see what happens if one excludes the hadronic asymmetries from the above restricted Higgs mass fit. A consistent picture emerges: a very light Higgs with $M_H^{fit} = 40$ GeV fits perfectly all data and one obtains an upper bound $M_H < 109$ GeV at 95% CL. If really $M_W$, $\Gamma_\ell$, $M_t$, and the leptonic asymmetries are consistent data and the SM is correct, why hasn't the Higgs been found at LEP, which set a lower bound $M_H > 114$ GeV [17,18]?

The inconsistency with the direct lower bound marginally depends on the value of the hadronic contributions to $\alpha(M_Z)$ used in the fit, but even in the most unfavorable case the 95% upper bound is no more than 120 GeV. Similarly, present day estimate of the theoretical error agree that it cannot shift up $M_H^{95\%}$ more than $\sim$ 20 GeV [19]. The inconsistency would be alleviated if the top mass turned out to be heavier than the current central value, a possibility soon to be tested at Tevatron, but the fit does not suggest this possibility at all. An interesting way to quantify the inconsistency is to compute the combined probability of the fit *and* of having $M_H > 114$ GeV: it is the same with or without $A_{FB}^b$ [17].

We have seen that excluding $A_{FB}^b$ and NuTeV from the global fit the quality of the fit improves considerably, but $M_H^{fit}$ becomes very small. Finding New Physics that simulates a very light Higgs is much easier than fixing the two anomalies. An example are oblique corrections: in general it just requires $S < 0 (T > 0)$ or $\epsilon_{2,3} < 0$ [17,18]. A non-degenerate unmixed 4th generation with a heavy neutrino with $m_N \approx 50$ GeV would easily work [20]. More interestingly, the MSSM offers rapid decoupling (small corrections), $M_W$ always higher than in SM, and $\sin^2\theta_{eff}^{lept}$ lower than in SM. A plausible MSSM scenario involves light sneutrinos and sleptons, heavy squarks, and $\tan\beta \gtrsim 5$ [18]. The required mass spectrum cannot be obtained in minimal SUGRA models with universal soft masses, though alternatives exist, and could be discovered at Tevatron. Other susy scenarios have also been presented [12].

## 7. CONCLUSIONS

NuTeV aims at high precision in a complex hadronic environment. Theoretical systematics not fully under control or untested include a small strange/antistrange asymmetry and isospin violation. The analysis should be upgraded to NLO.

Even without NuTeV, the SM fit to $M_H$ is not satisfactory. What we know on $M_H$ depends heavily on the $b$ quark FB asymmetries, an even more puzzling experimental anomaly. Removing the two deviant results from the SM fit leads however to inconsistency with the direct lower bound on $M_H$.

Both the NuTeV result and $A_{FB}^b$ require new tree level new effects which are difficult to accomodate in reference new physics scenarios. For instance, supersymmetry with or without R parity cannot explain them. Proposed explanations for both NuTeV and $A_{FB}^b$ rely on *ad-hoc* exotic models and it is always problematic to reconcile them with other precision data. Keeping also in mind the discrepancy of the measured $(g-2)_\mu$ with the SM prediction, the SM looks definitely under strain, although a clear-cut, compelling case for new physics has yet to be made.

## REFERENCES

1. M. W. Grünewald, these proceedings, arXiv:hep-ex/0210003.
2. G. P. Zeller *et al.* [NuTeV Coll.], Phys. Rev. Lett. **88** (2002) 091802 [hep-ex/0110059].
3. E. A. Paschos and L. Wolfenstein, Phys. Rev. D **7** (1973) 91.
4. S. Davidson, *et al.* JHEP **0202** (2002) 037 [hep-ph/0112302].
5. S.J. Brodsky, B.Q. Ma, Phys. Lett. B **381** (1996) 317; A. I. Signal and A. W. Thomas, Phys. Lett. B **191** (1987) 205.
6. V. Barone, C. Pascaud and F. Zomer, Eur. Phys. J. C **12** (2000) 243 [hep-ph/9907512].
7. G. P. Zeller *et al.* [NuTeV Coll.], Phys. Rev. D **65** (2002) 111103 [hep-ex/0203004]; M. Goncharov *et al.* [NuTeV Coll.], Phys. Rev. D **64** (2001) 112006 [hep-ex/0102049].
8. D. Mason, for the NuTeV Coll., these proceedings.
9. M. L. Mangano *et al.*, hep-ph/0105155.
10. E. Sather, Phys. Lett. B **274** (1992) 433; F. G. Cao and A. I. Signal, *Phys. Rev. C* **62** (2000) 015203; E. N. Rodionov, A. W. Thomas and J. T. Londergan, Mod. Phys. Lett. A **9** (1994) 1799.
11. G. A. Miller and A. W. Thomas, hep-ex/0204007; S. Kovalenko, I. Schmidt and J. J. Yang, Phys. Lett. B **546** (2002) 68 [hep-ph/0207158]; S. Kumano, hep-ph/0209200.
12. K. S. Babu and J. C. Pati, hep-ph/0203029.
13. W. Loinaz, *et al.*, hep-ph/0210193.
14. E. Ma and D. P. Roy, Phys. Rev. D **65** (2002) 075021.
15. D. Choudhury, T. M. Tait and C. E. Wagner, Phys. Rev. D **65** (2002) 053002 [hep-ph/0109097].
16. X. G. He and G. Valencia, Phys. Rev. D **66** (2002) 013004 [Erratum-ibid. D **66** (2002) 079901].
17. M. S. Chanowitz, Phys. Rev. D **66** (2002) 073002 [hep-ph/0207123]; Phys. Rev. Lett. **87** (2001) 231802 [hep-ph/0104024].
18. G. Altarelli, *et al.* JHEP **0106** (2001) 018 [hep-ph/0106029].
19. Freitas *et al.* hep-ph/0202131, P. Gambino, hep-ph/9812332.
20. V. A. Novikov, *et al.* Phys. Lett. B **529** (2002) 111 [hep-ph/0111028] and hep-ph/0203132.

Proceedings of ICHEP 2002, pp. 216–219
*S. Bentvelsen, P. de Jong, J. Koch and E. Laenen (Editors)*

ELSEVIER

31st INTERNATIONAL CONFERENCE ON HIGH ENERGY PHYSICS AMSTERDAM

# The SM prediction of $g-2$ of the muon

K. Hagiwara[a], A.D. Martin[b], Daisuke Nomura[b] and T. Teubner[c]

[a]Theory Group, KEK, Tsukuba, Ibaraki 305-0801, Japan

[b]Department of Physics and Institute for Particle Physics Phenomenology, University of Durham, Durham DH1 3LE, U.K.

[c]Theory Division, CERN, CH-1211 Geneva 23, Switzerland

We calculate $(g-2)/2$ of the muon, paying particular attention to the hadronic vacuum polarisation contribution and its uncertainties. The different data sets for each $e^+e^-$ exclusive channel (as well as for the inclusive $e^+e^- \to$ *hadrons* channel) have been combined in order to produce the optimum estimate of the cross sections and their uncertainties. QCD sum rules are evaluated in order to resolve an apparent discrepancy between the inclusive data and the sum of the exclusive channels. We conclude $a_\mu^{\rm had,LO} = (683.1 \pm 5.9_{\rm exp} \pm 2.0_{\rm rad}) \times 10^{-10}$ which, when combined with the other contributions to $(g-2)/2$, is about $3\sigma$ below the present world average measurement as reported at this conference.

## 1. Introduction

The muon anomalous magnetic moment, $a_\mu \equiv (g_\mu - 2)/2$, is one of the most precisely measured quantities in particle physics. The world average of the existing measurements is

$$a_\mu^{\rm exp} = 11659203(8) \times 10^{-10}, \qquad (1)$$

which is dominated by the recent value obtained by the E821 collaboration at BNL[1]. To test the Standard Model (SM) at the quantum level and to constrain possible New Physics it is therefore important to evaluate the SM prediction of $a_\mu$ as accurately as possible. It may be written as the sum of three terms, $a_\mu^{\rm SM} = a_\mu^{\rm QED} + a_\mu^{\rm EW} + a_\mu^{\rm had}$. The QED contribution has been calculated up to and including estimates of the 5-loop contribution, see reviews [2,3], and reads $a_\mu^{\rm QED} = 116\,584\,705.6(2.9) \times 10^{-11}$. In comparison with the experimental error in eq. (1), and with the hadronic contribution error discussed later, the uncertainty in $a_\mu^{\rm QED}$ is much less important than other sources of uncertainty. The electroweak contribution is calculated through second order to be [4–6]: $a_\mu^{\rm EW} = 152(1) \times 10^{-11}$. Here again the error is negligibly small. Less accurately known is the hadronic contribution. It can be divided into three pieces, $a_\mu^{\rm had} = a_\mu^{\rm had,LO} + a_\mu^{\rm had,NLO} + a_\mu^{\rm had,l\text{-}b\text{-}l}$. The lowest-order (vacuum polarisation) hadronic contribution, $a_\mu^{\rm had,LO}$, has been calculated by a number of groups. The value $a_\mu^{\rm had,LO} = 6\,924(62) \times 10^{-11}$ taken from Ref. [7] has been frequently used in making comparisons with the data. The next-to-leading order hadronic contribution, $a_\mu^{\rm had,NLO}$, is evaluated to be [8,9] $a_\mu^{\rm had,NLO} = -100(6) \times 10^{-11}$. The hadronic light-by-light scattering contribution $a_\mu^{\rm had,l\text{-}b\text{-}l}$ has been recently reevaluated [10]–[15], and it is found to be $a_\mu^{\rm had,l\text{-}b\text{-}l} = 80(40) \times 10^{-11}$, where we quote the estimate of the full hadronic light-by-light contributions given in Ref. [16]. We can see that $a_\mu^{\rm had,LO}$ has the largest uncertainty, although the uncertainty in the light-by-light contribution $a_\mu^{\rm had,l\text{-}b\text{-}l}$ is also large.

In the following we discuss an update of $a_\mu^{\rm had,LO}$, which is given by the dispersion relation

$$a_\mu^{\rm had,LO} = \frac{1}{4\pi^3} \int_{s_{\rm th}}^{\infty} ds\, \sigma_{\rm had}(s)\, \frac{m_\mu^2}{3s}\, K(s)\,, \qquad (2)$$

where $\sigma_{\rm had}(s)$ is the total cross section for $e^+e^- \to$ hadrons$\,(+\gamma)$ at centre-of-mass energy $\sqrt{s}$, and the well known kernel function $K$ (see e.g. [17]) increases monotonically from 0.63 to 1 in the range $4m_\pi^2 < s < \infty$. To evaluate $\sigma_{\rm had}(s)$, we use experimental data up to 11.09 GeV and


doi:10.1016/S0920-5632(02)02090-X

perturbative QCD thereafter. The most important contribution comes from the $e^+e^- \to \pi^+\pi^-$ channel; the channels $\pi^+\pi^-\pi^0$, $K^+K^-$, $K_S^0K_L^0$, $\pi^+\pi^-\pi^+\pi^-$, $\pi^+\pi^-\pi^0\pi^0$, etc. give subleading contributions.

The special features of our analysis are (i) that it is data-driven, based on all available data, including the new data on exclusive channels from Novosibirsk, particularly $\pi^+\pi^-$ [18], and the BES data on the $R$ ratio [19], (ii) the careful application of a clustering method, so that data of differing precision can be combined consistently, and (iii) the use of QCD sum rules to resolve an apparent discrepancy between the inclusive and exclusive determination of $\sigma_{\rm had}(s)$ for $1.4 \lesssim \sqrt{s} \lesssim 2$ GeV.

We chose not to use data on hadronic $\tau$ decays to further constrain the $e^+e^- \to 2\pi, 4\pi$ channels for $\sqrt{s} \lesssim m_\tau$, because of the possible uncertainties connected with isospin-breaking effects. A careful study of the effects of including $\tau$ data has been made recently in Ref. [20].

## 2. Processing the hadronic data

We applied the hadronic vacuum polarisation corrections as given by Swartz [21] and calculated the final state radiative effects for all $e^+e^- \to \pi^+\pi^-$ data, except for the new (already corrected) CMD–2 data [18], based on eq. (45) of Ref. [22]. These two corrections increase the $\pi^+\pi^-$ contribution by about $1.1 \times 10^{-10}$, to which we assign a 50% error. For data where only insufficient information is available to make reliable radiative corrections we assigned an additional $\pm 1\%$ uncertainty to their contribution to $a_\mu^{\rm had,LO}$. The net effect is an error of about $\pm 2 \times 10^{-10}$ due to radiative corrections; a further discussion will be given in [23].

We now come to the important problem of 'clustering' data from different experiments (for the same hadronic channel). To combine all data points for the same channel which fall in suitably chosen (narrow) energy bins, we determine the mean $R$ values and their errors for all clusters by minimising the non-linear $\chi^2$ function $\chi^2(R_m, f_k) = \sum_{k=1}^{N_{\rm exp}} \left[(1-f_k)/{\rm d}f_k\right]^2 + \sum_{m=1}^{N_{\rm clust}} \sum_{i=1}^{N_{\{k,m\}}} \left[\left(R_i^{\{k,m\}} - f_k R_m\right)/{\rm d}R_i^{\{k,m\}}\right]^2$. Here $R_m$ and $f_k$ are the fit parameters for the mean $R$ value of the $m^{\rm th}$ cluster and the overall normalization factor of the $k^{\rm th}$ experiment, respectively. $R_i^{\{k,m\}}$ and ${\rm d}R_i^{\{k,m\}}$ are the $R$ values and errors from experiment $k$ contributing to cluster $m$. For ${\rm d}R_i^{\{k,m\}}$ the statistical and point-to-point systematic errors are added in quadrature, whereas ${\rm d}f_k$ is the overall systematic error of the $k^{\rm th}$ experiment. Minimization of $\chi^2$ with respect to the $(N_{\rm exp} + N_{\rm clust})$ parameters, $f_k$ and $R_m$, gives our best estimates for these parameters together with their error correlations.

Our definition of $\chi^2$ implies piecewise constant $R$ values but imposes no further constraints on the form of the hadronic cross section ('minimum bias'). Due to use of the overall normalization factors it also results in an adjustment of the different sets within their systematic uncertainties. This means, for example, that sparse but precise data will dominate the normalization, but that the information on the shape of $R$ from sets with larger systematic uncertainties is preserved.

The error estimate for each hadronic channel is then done using the complete covariance matrix returned by our $\chi^2$ minimization. Therefore statistical and systematic (point-to-point as well as overall) errors from the different sets are taken into account including correlations between different energies (clusters). The dispersion integral (2) is performed integrating (using the trapezoidal rule) over the clustered data directly for all hadronic channels, including the $\omega$ and $\phi$ resonances. Thus we avoid possible problems due to missing or double-counting of non-resonant backgrounds, and interference effects are taken into account automatically.

In the region between 1.43 and $\sim 2$ GeV we have the choice between summing up the exclusive channels or relying on the inclusive measurements. Surprisingly, the sum of the exclusive channels overshoots the inclusive data, even after having corrected the latter for missing two-body and (some) purely neutral modes. The discrepancy is shown in Fig. 1, where we display data points with errors after application of our clustering algorithm.

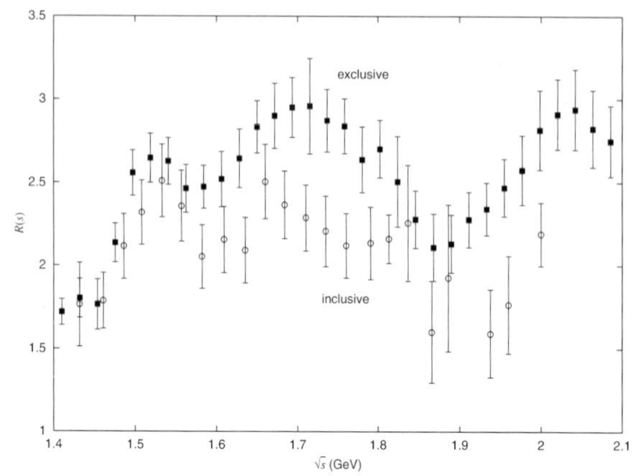


Figure 1. The inclusive and the sum of exclusive channel values of $R$, *after* the data have been clustered.

## 3. Results

In Table 1 we list the contributions to $a_\mu^{\mathrm{had,LO}}$ from different energy regimes: From the two-pion threshold up to 0.32 GeV, chiral perturbation theory is applied and for the high energy tail above 11.09 GeV, $R$ is calculated using perturbative QCD. The $J/\psi$, $\psi(2S)$ and $\Upsilon(1S-6S)$ resonance contributions are evaluated in narrow-width-approximation. Apart from those contributions we use the direct integration of clustered data as described above. For the controversial region from 1.43 to 2 GeV we present two results: if we use the lower lying inclusive data the corresponding contribution is considerably smaller than the one resulting from the use of the sum over the exclusive channels.

## 4. Resolution of the ambiguity: QCD sum rules

To resolve the ambiguity between the exclusive and inclusive data, we evaluate QCD sum rules of the form $\int_{s_{\mathrm{th}}}^{s_0} \mathrm{d}s\, R(s)f(s) = \int_C \mathrm{d}s\, D(s)g(s)$. $\sqrt{s_0} = 3.7$ GeV is chosen to be just below the open charm threshold and $C$ is a circular contour of radius $s_0$. $D(s)$ is the Adler $D$ function, $D(s) \equiv -12\pi^2 s\, \mathrm{d}\left(\Pi(s)/s\right)/\mathrm{d}s$, where $R(s) = 12\pi\, \mathrm{Im}\, \Pi(s)/s$. Experimental data for $R(s)$ are used to evaluate the left-hand-side, while QCD is used to determine $D(s)$. We take $f(s)$ to be of the form $(1-s/s_0)^n(s/s_0)^m$, with $n+m=0,1$ or 2. Once $f(s)$ is chosen, the functional form of $g(s)$ is readily evaluated. The sum rules with $n=1$ or 2 and $m=0$ are found to maximize the fractional contribution of the left-hand-side coming from the relevant $1.43 < \sqrt{s} < 2$ GeV interval. The evaluations of these two sum rules are shown in Table 2. Consistency clearly selects the inclusive, as opposed to the exclusive, determination of $R(s)$.

Table 1
Contributions to $a_\mu^{\mathrm{had,LO}}$ from different energy intervals. The second line and the total also include a small $0.13 \times 10^{-10}$ contribution from the $\pi^0\gamma$ channel near its threshold.

| energy (GeV) | comments | $a_\mu^{\mathrm{had,LO}} \times 10^{10}$ |
|---|---|---|
| $2m_\pi \ldots 0.32$ | chiral PT | $2.30 \pm 0.05$ |
| $0.32 \ldots 1.43$ | excl. only | $596.73 \pm 5.18$ |
| $1.43 \ldots 2.00$ | excl. only | $38.14 \pm 1.72$ |
| | incl. only | $32.43 \pm 2.46$ |
| $2.00 \ldots 11.09$ | incl. only | $42.09 \pm 1.25$ |
| $J/\psi$ and $\psi(2S)$ | nar. width | $7.31 \pm 0.43$ |
| $\Upsilon(1S-6S)$ | nar. width | $0.10 \pm 0.00$ |
| $11.09 \ldots \infty$ | pQCD | $2.14 \pm 0.01$ |
| $\sum$ of all | ex-ex-in | $688.81 \pm 6.17$ |
| | ex-in-in | $683.11 \pm 5.89$ |

Table 2
Sum rule results as discussed in the text. The main QCD error comes from $\alpha_S(M_Z^2) = 0.117 \pm 0.002$.

| sum rule | l.h.s. (data) | r.h.s. (QCD) |
|---|---|---|
| $n=1$ | $15.34 \pm 0.39$ (incl) | $15.34 \pm 0.08$ |
| | $15.99 \pm 0.35$ (excl) | |
| $n=2$ | $10.40 \pm 0.25$ (incl) | $10.30 \pm 0.06$ |
| | $10.90 \pm 0.22$ (excl) | |

## 5. Conclusions

We have undertaken a data-driven determination of the hadronic vacuum polarisation contribution to $a_\mu^{\mathrm{had,LO}}$, using all available $e^+e^-$ data and a non-linear $\chi^2$ approach to cluster data for the same channel in narrow bins. We found that there was a discrepancy between the inclusive value for $\sigma(e^+e^- \to \text{hadrons})$ and the sum of the exclusive channels in the region $1.4 \lesssim \sqrt{s} \lesssim 2$ GeV, which gave an uncertainty of about $6 \times 10^{-10}$ in $a_\mu^{\text{had,LO}}$. We used a QCD sum rule analysis to resolve the discrepancy in favour of the inclusive data. Thus finally we find that the SM predicts

$$a_\mu^{\text{had,LO}} = (683.1 \pm 5.9_{\text{exp}} \pm 2.0_{\text{rad}}) \times 10^{-10}. \quad (3)$$

Summing up all SM contributions to $a_\mu^{\text{SM}}$ as given in section 1, but with $a_\mu^{\text{had,LO}}$ from (3), we conclude that

$$a_\mu^{\text{SM}} = (11659166.9 \pm 7.4) \times 10^{-10}, \quad (4)$$

which is $36.1 \times 10^{-10}$ ($3.3\sigma$) below the world average experimental measurement. If, on the other hand, we were to take the value of $a_\mu^{\text{had,LO}}$ obtained using the sum of the exclusive data in the interval $1.43 < \sqrt{s} < 2$ GeV then we would find $a_\mu^{\text{SM}} = (11659172.6 \pm 7.7) \times 10^{-10}$, which is $30.4 \times 10^{-10}$ ($2.7\sigma$) below $a_\mu^{\text{exp}}$.

Our result (3) agrees also fairly well with a recent reevaluation of the leading hadronic contribution from F. Jegerlehner, who also used the recent CMD-2 data [18] and obtained $(688.9 \pm 5.8) \times 10^{-10}$, see [24]. Another independent SM prediction has very recently been made [20]. Their total final $e^+e^-$-based result, $(684.7 \pm 6.0_{\text{exp}} \pm 3.6_{\text{rad}}) \times 10^{-10}$, is very similar to ours. A detailed comparison with these two predictions will be presented elsewhere [23].

For the future, we can expect further improvement in the accuracy of the experimental $(g-2)/2$ measurement. As far as the SM prediction is concerned, we anticipate new low energy data for a variety of $e^+e^-$ channels, partly from *radiative return* at the $\phi$- and $B$-factories.

**Acknowledgements:** We thank Simon Eidelman for numerous helpful discussions concerning the data.

## REFERENCES


1. E821 Collaboration, G.W. Bennett et al., Phys. Rev. Lett. **89** (2002) 101804; see also: Y. Semertzidis, *in these proceedings*.
2. V.W. Hughes and T. Kinoshita, Rev. Mod. Phys. **71** (1999) S133.
3. A. Czarnecki and W.J. Marciano, Nucl. Phys. Proc. Suppl. **76** (1999) 245.
4. A. Czarnecki, B. Krause and W.J. Marciano, Phys. Rev. **D52** (1995) 2619; Phys. Rev. Lett. **76** (1996) 3267.
5. S. Peris, M. Perrottet and E. de Rafael, Phys. Lett. **B355** (1995) 523.
6. M. Knecht, S. Peris, M. Perrottet and E. de Rafael, hep-ph/0205102.
7. M. Davier and A. Höcker, Phys. Lett. **B435** (1998) 427.
8. B. Krause, Phys. Lett. **B390** (1997) 392.
9. R. Alemany, M. Davier and A. Höcker, Eur. Phys. J. **C2** (1998) 123.
10. M. Knecht and A. Nyffeler, Phys. Rev. **D65** (2002) 073034.
11. M. Knecht, A. Nyffeler, M. Perrottet and E. de Rafael, Phys. Rev. Lett. **88** (2002) 071802.
12. M. Hayakawa and T. Kinoshita, hep-ph/0112102 and erratum Phys. Rev. **D66** (2002) 019902 (E).
13. J. Bijnens, E. Pallante and J. Prades, Nucl. Phys. **B626** (2002) 410.
14. I. Blokland, A. Czarnecki and K. Melnikov, Phys. Rev. Lett. **88** (2002) 071803.
15. M. Ramsey-Musolf and M.B. Wise, Phys. Rev. Lett. **89** (2002) 041601.
16. A. Nyffeler, hep-ph/0203243.
17. K. Hagiwara, A.D. Martin, Daisuke Nomura and T. Teubner, hep-ph/0209187.
18. R.R. Akhmetshin et al., Phys. Lett. **B527** (2002) 161.
19. BES Collaboration, J.Z. Bai et al., Phys. Rev. Lett. **88** (2002) 101802.
20. M. Davier, S. Eidelman, A. Höcker and Z. Zhang, hep-ph/0208177.
21. M.L. Swartz, Phys. Rev. **D53** (1996) 5268.
22. A. Hoefer, J. Gluza and F. Jegerlehner, Eur. Phys. J. **C24** (2002) 51.
23. K. Hagiwara, A.D. Martin, D. Nomura and T. Teubner, in preparation.
24. F. Jegerlehner, talk at the workshop "Hadronic Contributions to the Anomalous Magnetic Moment of the Muon", Centre de Physique Théorique, Marseille, France, 14 – 16 March 2002.

ELSEVIER

Proceedings of ICHEP 2002, pp. 220–223
*S. Bentvelsen, P. de Jong, J. Koch and E. Laenen (Editors)*

31st INTERNATIONAL CONFERENCE ON HIGH ENERGY PHYSICS AMSTERDAM

# Recent Results on Sudakov Logarithms in Electroweak Processes

J.H. Kühn[a]

[a]Institut für Theoretische Teilchenphysik, Universität Karlsruhe, 76128 Karlsruhe, Germany

Recent results for the asymptotic behavior of fermion scattering amplitudes in the Sudakov limit are discussed including next-to-leading logarithmic (NLL) and NNLL corrections. These are used for the analysis of the dominant electroweak corrections to the fermion-antifermion pair production in $e^+e^-$ annihilation at high energy. The importance of angular dependent as well as subleading terms is emphasized.

## 1. Introduction

Four fermion processes are generally considered as benchmark processes at high energy colliders, with electron positron annihilation into muon or quark pairs at LEP and the Drell Yan process at hadron colliders as characteristic examples. Within the presently accessible energy region, typically up to 200 GeV, radiative corrections are dominated by the shift in the $W$ and $Z$ masses as parametrized by the $\rho$ parameter and by the running of the coupling constant. Vertex corrections and box diagrams involving gauge bosons are generally of minor importance. In the TeV region, accessible at future colliders like the LHC or TESLA, this picture changes drastically. A new class of effects starts to become relevant and rapidly dominant which are generally denoted as double logarithmic corrections and which were first observed by Sudakov in the context of quantum electrodynamics for reactions with a tight cut on the radiated energy of the photons. For electroweak interactions large negative corrections arise from the exchange of gauge bosons which remain uncompensated if one restricts the analysis to exclusive final states, consisting e.g. of a fermion antifermion pair only. The discussion of double logarithmic corrections is fairly straightforward for a theory with massive gauge bosons only. An important complication arises from the presence of massless photons in the final state. Events with soft and hard photon radiation are normally included in the sample – whence a "semi-inclusive" definition of the cross section is closest to the the experimental analysis.

One loop corrections to the four fermion process are available since long (see e.g. [1]). In this case the separation of photonic and weak corrections is still possible and, employing axial gauge, the leading logarithms can be trivially attributed to fermion self energy diagrams with virtual $W$ or $Z$ boson emission [2].

Results for the leading logarithmic corrections in higher orders have been presented in [2,3]. In [4], which includes NLL corrections and angular dependent terms, it was demonstrated that the approach developed in [2], where axial gauge was adopted and both real and virtual photon emission neglected, is numerically nearly identical to the results of [3] where virtual photon corrections and real photon emmission are included – a consequence of the smallness of the weak mixing angle, $\sin^2\theta_W = 0.23$. The analysis of subleading logarithms for the form factor, for the four fermion scattering in a spontaneously broken $SU(2)$ gauge theory and, last not least, for the Standard Model with $W$, $Z$ and the massless photon was performed in [4].

Before presenting a brief review (an extended review of this field can also be found in [5]) of the techniques and some recent results let us emphasize the importance of subleading corrections. In the energy region relevant at LHC or TESLA, i.e. around 1 to 2 TeV, this is already evident from the one-loop vertex correction

$$|F_B+\delta F|^2 \approx F_B\left(1+2\frac{\alpha_W}{\pi}C_F\rho(s/M^2)\right), \qquad (1)$$

$$\rho(u) = -\frac{1}{4}\ln^2 u+\frac{3}{4}\ln u+\frac{\pi^2}{12}-\frac{7}{8}.$$


doi:10.1016/S0920-5632(02)02093-5

For the timelike region and for characteristic values of the coupling, energy and mass of $\alpha_W/\pi = 10^{-2}$, $C_F = 3/4$ and $s/M^2 = 10^2$ one finds

$$2\frac{\alpha_W}{\pi}C_F\rho = 1.5\times 10^{-2}(-5.30+3.45-0.05) \quad (2)$$

Large compensations between leading and subleading terms are observed and, in fact, this pattern will reappear for the Standard Model as discussed below.

The calculation of the leading logarithmic terms in [3] is based on evolution equations that govern the dynamics of the amplitudes in the Sudakov limit. In [4] and [6] this approach was applied to the next-to-leading logarithmic and the NNLL analysis respectively of the Abelian form factor and the four fermion amplitude in the $SU(N)$ gauge theory. Functions that enter the evolution equations in the next-to-leading logarithmic approximation were evaluated by using, as an input, asymptotic expansions of one-loop diagrams. The solution of these equations lead to a summation of the leading and subleading Sudakov logarithms. The expansion of one-loop diagrams through the so-called generalized strategy of regions [7,8] identifies in a systematic way the nature of various contributions and the origin of logarithms.

## 2. The Abelian form factor in the Sudakov limit

Let us first analyze the (vector) form factor which determines the amplitude of the fermion scattering in the external Abelian field. In Born approximation

$$\mathcal{F}_B = \bar\psi(p_2)\gamma_\mu\psi(p_1). \quad (3)$$

We consider the limit $s = (p_1-p_2)^2 \to -\infty$ with on-shell massless fermions, $p_1^2 = p_2^2 = 0$, and gauge bosons with a small non-zero mass $M^2 \ll -s$. For convenience $p_{1,2} = (Q/2, 0, 0, \mp Q/2)$ so that $2p_1p_2 = Q^2 = -s$. The asymptotic behavior of the form factor is obtained from

$$\mathcal{F} = F_0(\alpha(M^2))\exp\Big\{\int_{M^2}^{Q^2}\frac{dx}{x} \times\Big[\int_{M^2}^{x}\frac{dx'}{x'}\gamma(\alpha(x')) + \zeta(\alpha(x)) + \xi(\alpha(M^2))\Big]\Big\}\mathcal{F}_B. \quad (4)$$

The NLL approximation includes all the terms of the form $\alpha^n \log^{2n-m}(Q^2/M^2)$ with $m = 0,\ 1$. This requires the one-loop values of $\gamma(\alpha)$ and $\zeta(\alpha)+\xi(\alpha)$ and using the one-loop running of $\alpha$ in $\gamma(\alpha)$. The NNLL approximation includes all the terms of the form $\alpha^n \log^{2n-m}(Q^2/M^2)$ with $m = 0,\ 1,\ 2$. In this case $\gamma(\alpha)$ is required up to $\mathcal{O}(\alpha^2)$, $\zeta(\alpha)$, $\xi(\alpha)$ and $F_0(\alpha)$ up to $\mathcal{O}(\alpha)$ together with the two-loop running of $\alpha$ in $\gamma(\alpha)$ and one-loop running of $\alpha$ in $\zeta(\alpha)$.

The expansion by regions consists of the following steps: (i) consider various regions of a loop momentum $k$ and expand, in every region, the integrand in Taylor series with respect to the parameters that are there considered small, (ii) integrate the expanded integrand over the whole integration domain of the loop momenta, (iii) put to zero any scaleless integral. In step (ii) dimensional regularization with $d = 4-2\epsilon$ space-time dimensions is used to handle the divergences. The following regions are of relevance in this context:

$$\begin{aligned} &\textit{hard (h):} && k\sim Q,\\ &\textit{1-collinear (1c):} && k_+\sim Q,\ k_-\sim M^2/Q,\ \underline{k}\sim M,\\ &\textit{2-collinear (2c):} && k_-\sim Q,\ k_+\sim M^2/Q,\ \underline{k}\sim M,\\ &\textit{soft (s):} && k\sim M. \end{aligned} \quad (5)$$

with $k_\pm = k_0 \pm k_3$, $\underline{k} = (k_1, k_2)$.

From the one-loop result and the hard part in two-loop approximation one finds

$$\begin{aligned} \gamma^{(1)} &= -2C_F\,, \quad \zeta^{(1)} = 3C_F\,, \quad \xi^{(1)} = 0\,,\\ F_0^{(1)} &= -C_F\left(\frac{7}{2}+\frac{2\pi^2}{3}\right), \\ \gamma^{(2)} &= -2C_F\left[\left(\frac{67}{9}-\frac{\pi^2}{3}\right)C_A - \frac{20}{9}T_F n_f\right], \end{aligned} \quad (6)$$

with $\gamma = \left(\frac{\alpha}{4\pi}\right)\gamma^{(1)} + \left(\frac{\alpha}{4\pi}\right)^2\gamma^{(2)} + \ldots$ and similarly for $\zeta$, $\xi$ and $F_0$.

Let us also consider the two-loop corrections. The LL, NLL and NNLL approximations are given respectively by

$$\mathcal{F}^{(2)}_{LL} = \frac{1}{8}(\gamma^{(1)})^2\ln^4\left(\frac{Q^2}{M^2}\right)\mathcal{F}_B\,, \quad (7)$$

$$\mathcal{F}^{(2)}_{NLL} = \frac{1}{2}\left(\zeta^{(1)} - \frac{1}{3}\beta_0\right)\gamma^{(1)}\ln^3\left(\frac{Q^2}{M^2}\right)\mathcal{F}_B\,, \quad (8)$$

$$\mathcal{F}^{(2)}_{NNLL} = \frac{1}{2}\Big(\gamma^{(2)} + \Big(\zeta^{(1)} - \beta_0\Big)\zeta^{(1)} \\ +F_0^{(1)}\gamma^{(1)}\Big)\ln^2\left(\frac{Q^2}{M^2}\right)\mathcal{F}_B\,. \quad (9)$$

The relatively small coefficient of the LL terms and the large coefficient of the NNLL terms in the form factor are clearly indicative of the importance of the NNLL corrections and, as we will see, reflect the general structure of the logarithmically enhanced electroweak corrections.

## 3. The four fermion amplitude

Let us now investigate the four fermion scattering at fixed angles in the limit when all the invariant energy and momentum transfers of the process are far larger than the gauge boson mass, $|s| \sim |t| \sim |u| \gg M^2$. The analysis of the four fermion amplitude is complicated by the extra kinematical variable and the presence of different "color" and Lorentz structures. The complete basis consists of four independent chiral amplitudes, each of them of two possible color structures.

The collinear divergences in the hard part of the virtual corrections and the corresponding "collinear" logarithms are known to factorize. They are responsible, in particular, for the double logarithmic contribution, depend only on the properties of the external on-shell particles but not on a specific process and are the same as for the form factor $\mathcal{F}$ discussed in the previous section. Let us denote by $\tilde{\mathcal{A}}$ the amplitude with the collinear logarithms factored out. It contains the logarithms of "soft" nature corresponding to the soft divergences of the hard region contribution and the renormalization group logarithms. It can be represented as a vector in the color/chiral basis and satisfies the evolution equation

$$\frac{\partial}{\partial \ln Q^2}\tilde{\mathcal{A}} = \chi(\alpha(Q^2))\tilde{\mathcal{A}}\,, \quad (10)$$

where $\chi(\alpha)$ is the matrix of the soft anomalous dimensions. It can be extracted from the coefficient in front of the single pole of the hard contribution to $\tilde{\mathcal{A}}$. The matrix elements of $\chi(\alpha)$ depend on the scattering angle. These angular dependent terms as well as those in the starting values of $\tilde{\mathcal{A}}$ discussed below are important for the calculation of NLL (see also [9]) and NNLL terms in cross sections and asymmetries. The solution of eq. (10) reads

$$\tilde{\mathcal{A}} = \sum_i \tilde{\mathcal{A}}_{0i}(\alpha(M^2))\exp\left[\int_{M^2}^{Q^2}\frac{\mathrm{d}x}{x}\chi_i(\alpha(x))\right], \quad (11)$$

where $\chi_i(\alpha)$ are eigenvalues of $\chi(\alpha)$ and $\tilde{\mathcal{A}}_{0i}(\alpha)$ are $Q$-independent eigenvectors of $\chi(\alpha)$ which determine the initial conditions for the evolution equation at $Q = M$.

To illustrate the significance of the subleading contributions let us again discuss the standard model inspired example considered in the previous section. For the annihilation process one has to make the analytical continuation to the Minkowskian region of negative $Q^2 = -s$. After integrating the differential cross section over all angles one obtains the result with logarithmic accuracy, e.g. for the case of $SU(2)_L$ group with $n_f = 6$ and one scalar boson

$$\sigma^{(1)} = \left[-3\ln^2\left(\frac{s}{M^2}\right) + \frac{80}{3}\ln\left(\frac{s}{M^2}\right) - \left(\frac{25}{9} + 3\pi^2\right)\right]\sigma_B\,,$$

$$\sigma^{(2)} = \left[\frac{9}{2}\ln^4\left(\frac{s}{M^2}\right) - \frac{449}{6}\ln^3\left(\frac{s}{M^2}\right) + \left(\frac{4855}{9} + \frac{37\pi^2}{3}\right)\ln^2\left(\frac{s}{M^2}\right)\right]\sigma_B\,, \quad (12)$$

if initial and final state fermions have the same isospin. Again we observe a relatively small coefficient of the LL and a large coefficient of the NNLL terms.

## 4. Electroweak processes

To analyze the electroweak correction to the process $f'\bar{f}' \to f\bar{f}$ the approximation with the $W$ and $Z$ bosons of the same mass $M$, the Higgs boson of the mass $M_H \sim M$ and massless quarks and leptons is employed. A fictitious photon mass $\lambda$ has to be introduced to regularize the infrared divergences. The NNLL approximation is not sensitive to the fine details of the mass generation because we need only the hard part of the potentially dangerous self-energy insertion to the gauge boson propagator.

The photon is, however, massless and the corresponding infrared divergent contributions should be accompanied by the real soft photon radiation integrated to some resolution energy $\omega_{res}$ to get an infrared safe cross section independent on an auxiliary photon mass. To study the virtual corrections in the limit of the vanishing photon mass the general approach of the infrared evolution equations developed in [3] (see also references therein) is particularly useful.

Explicit expressions for the NNLL results are given in [6]. Numerically, for the reaction $e^+e^- \to \mu^+\mu^-$ as a typical example, one obtains a correction factor

$$R^{\mu^+\mu^-} = 1 - 1.39\,L(s) + 10.12\,\ell(s) - 31.33\,a \\ +1.42\,L^2(s) - 18.43\,L(s)\ell(s) + 99.89\,\ell^2(s) \quad (13)$$

with $a = g^2/16\pi^2 = 2.69{\cdot}10^{-3}$, $\ell(s) = a \ln s/M^2$, $L(s) = a \ln^2 s/M^2$. To get the infrared safe result for the semi-inclusive cross sections one has to include the standard QED corrections due to soft photon emission and the pure QED virtual corrections.

In view of the large subleading terms the evaluation of two-loop terms linear in the logarithm or even of the constant terms seems desirable. As a first step the fermionic corrections to the form factor, i.e. terms resulting from the inclusion of massless fermions in loop insertions in the boson propagator, have been calculated for the Abelian form factor [10]:

$$F_{2,n_f} = n_f \frac{g^2}{16\pi^2}\left\{-\frac{4}{9}\ln^3\left(\frac{-s}{M^2}\right) + \frac{38}{9}\ln^2\left(\frac{-s}{M^2}\right) \right. \\ \left. -\frac{34}{9}\ln\left(\frac{-s}{M^2}\right) + \frac{16}{27}\pi^2 + \frac{115}{9}\right\} \quad (14)$$

The oscillating pattern of the signs in subleading terms continues, and the linear logarithm as well as the constant terms are sizeable. These results indicate the need for the evaluation of two-loop corrections for realistic reactions, including the linear logarithm and the constant term.

In addition to four fermion processes, which are clearly the benchmark reactions at high energies, also other reactions have been investigated in the high energy limit. In the case of gauge boson pair production LL and NLL terms have been evaluated by several authors [11]. Exclusive production of transversly and longitudinally polarized $W$ is suppressed by $\exp(-C_A \ln^2 s/M^2)$ and $\exp(-C_F \ln^2 s/M^2)$ respectively, as expected from the equivalence theorem. The implications of these results for gauge boson pair production at hadron colliders were studied in [12]. As noted in [13], enhanced electroweak corrections are also present in inclusive boson fusion processes, a consequence of the lack of compensation between virtual and real emission due to the non-Abelian charges of the incoming particles.

*Acknowledgement:* Pleasant collaboration with B. Feucht, S. Moch, A.A Penin and V.A. Smirnov is gratefully acknowledged.

## REFERENCES

1. W. Beenakker, W. Hollik and Van der Mark, *Nucl. Phys.* B365 (1991) 24.
2. J.H. Kühn and A.A. Penin, Preprint TTP/99–28, hep-ph/9906545.
3. V.S. Fadin, L.N. Lipatov, A.D. Martin and M. Melles, *Phys. Rev.* D61 (2000) 094002.
4. J.H. Kühn, A.A. Penin and V.A. Smirnov, *Eur. Phys. J.* C17 (2000) 97.
5. M. Melles, hep-ph/0104232.
6. J.H. Kühn, S. Moch, A.A. Penin and V.A. Smirnov, *Nucl. Phys.* B616 (2001) 286.
7. M. Beneke and V.A. Smirnov, *Nucl. Phys.* B522 (1998) 321.
8. V.A. Smirnov and E.R. Rakhmetov, *Teor. Mat. Fiz.* 120 (1999) 64; V.A. Smirnov, *Phys. Lett.* B465 (1999) 226.
9. M. Melles, *Eur. Phys. J.* C24 (2002) 193; M. Beccaria, M. Melles, F.M. Renard and C. Verzegnassi, *Phys. Rev.* D65 (2002) 093007.
10. B. Feucht, J.H. Kühn, S. Moch, A.A. Penin and V.A. Smirnov, in preparation.
11. M. Melles, *Phys. Rev.* D63 (2001) 034003; A. Denner and S. Pozzorini, *Eur. Phys. J.* C18 (2001) 461; M. Melles, *Phys. Rev.* D64 (2001) 014011.
12. E. Accomando, A. Denner and S. Pozzorini, *Phys. Rev.* D65 (2002) 073003.
13. M. Ciafaloni, P. Ciafaloni and D. Comelli, *Nucl. Phys.* B613 (2001) 382.

Proceedings of ICHEP 2002, pp. 224–227
*S. Bentvelsen, P. de Jong, J. Koch and E. Laenen (Editors)*

# Multi-lepton Production at High Transverse Momentum in $ep$ collisions at HERA

Cristinel Diaconu[a]
on behalf of H1 and ZEUS Collaborations

[a]Centre de Physique des Particules de Marseille,
163 Avenue de Luminy, 13288 Marseille cedex 09, France

Multi-electron and multi-muon production have been measured at high transverse momentum in electron-proton collisions at HERA. Good overall agreement is found with the Standard Model predictions, dominated by photon-photon interactions. Events are observed with a di-electron mass above 100 GeV, a domain where the Standard Model prediction is low.

## 1. Introduction

The measurement of rare processes provides a unique method to search for new physics. At HERA, two experiments (H1 and ZEUS) study electron-proton collisions with a center of mass energy of 300 to 320 GeV. In this paper we shortly describe the first measurement of multi-electron and multi-muon production at high transverse momentum ($P_T$) in $ep$ collisions at HERA [1–3].

## 2. Multi-lepton Processes

Within the Standard Model (SM) the production of multilepton events in $ep$ collisions is possible mainly through photon-photon interactions, where quasi-real photons radiated from the incident electron and proton interact to produce a pair of leptons $\gamma\gamma \rightarrow \ell^+\ell^-$ [4]. The GRAPE Monte Carlo generator [5] interfaced to full detector simulation has been used to simulate multi-lepton production. This generator is based on the full electro-weak matrix element calculation at tree level. The scattered proton or its remnants is treated in three phase space regions: elastic, quasielastic and inelastic.

The main experimental backgrounds to multi-electron production are processes where, in addition to a true electron, one or more fake electrons may be reconstructed from the final state particles. The dominant contribution is expected from Neutral Current Deep Inelastic Scattering (NC-DIS) where a fake electron from the hadrons or a radiated photon is selected together with the scattered electron. Elastic Compton scattering can also contribute if the high $P_T$ photon is misidentified as an electron. Background for multi-muon events are processes where fake muons are reconstructed from isolated hadrons. This contribution is negligible.

## 3. Measurement of multi-electron events

Both the H1 and ZEUS experiments are complex detectors with very good lepton identification capabilities. The electron identification is based on calorimetric information together with tracking conditions for efficient background rejection.

Electrons are measured by both H1 and ZEUS in a large acceptance range $5^\circ < \theta_e < 175^\circ$, where $\theta_e$ is the electron polar angle measured with respect to the outgoing proton direction. The electron energy measured from calorimetric information has to be above 5 GeV. This energy threshold is increased in H1 analysis to 10 GeV for electrons candidates with $\theta_e < 20^\circ$. The ZEUS analysis requires the electron energy above 10 GeV for $\theta_e < 164^\circ$. The electron candidates have to be isolated from other calorimetric deposits. In the central region defined by $20^\circ < \theta_e < 150^\circ$ for the H1 analysis and $17^\circ < \theta_e < 164^\circ$ for the ZEUS


doi:10.1016/S0920-5632(02)02096-0

| Selection | DATA | SM | GRAPE | NC-DIS + Compton |
|---|---|---|---|---|
| **H1 115 pb$^{-1}$** | $20° < \theta^{e1,2} < 150°$, $P_T^{e1} > 10$ GeV, $P_T^{e2} > 5$ GeV | | | |
| Visible 2e | 105 | $118.2 \pm 12.8$ | $93.3 \pm 11.5$ | $25.0 \pm 5.5$ |
| Visible 3e | 16 | $21.6 \pm 3.0$ | $21.5 \pm 3.0$ | $0.1 \pm 0.1$ |
| Visible 2e $M_{12} > 100$ GeV | 3 | $0.25 \pm 0.05$ | $0.21 \pm 0.04$ | $0.04 \pm 0.03$ |
| Visible 3e $M_{12} > 100$ GeV | 3 | $0.23 \pm 0.04$ | $0.23 \pm 0.04$ | $0.00 \pm 0.00$ |
| **ZEUS 130 pb$^{-1}$** | $17° < \theta^{e1,2} < 167°$, $P_T^{e1} > 10$ GeV, $E^{e2} > 10$ GeV | | | |
| Visible 2e | 191 | $213.9 \pm 3.9$ | $182.2 \pm 1.2$ | $31.7 \pm 3.7$ |
| Visible 3e | 26 | $34.7 \pm 0.5$ | $34.7 \pm 0.5$ | – |
| Visible 2e $M_{12} > 100$ GeV | 2 | $0.77 \pm 0.08$ | $0.47 \pm 0.05$ | $0.30 \pm 0.07$ |
| Visible 3e $M_{12} > 100$ GeV | 0 | $0.37 \pm 0.04$ | $0.37 \pm 0.04$ | – |

Table 1
Observed and predicted multi-electron event rates for all selected events and for events with masses $M_{12} > 100$ GeV as function of the number of identified electrons. The prediction errors for the H1 analysis include model uncertainties and experimental systematical errors added in quadrature. For ZEUS analysis, the predicted rates are shown with the statistical errors of the Monte Carlo only.

analysis, an isolated charged track measured in the central tracking system has to be associated to the calorimetric deposit. The identified electrons are indexed in decreasing transverse momentum $P_T$: $P_T^{e_i} > P_T^{e_{i+1}}$.

The selection of multi-electron events is based on the requirement of two central electrons with large energy or transverse momentum. Both H1 and ZEUS analysis require the first central electron to have transverse momentum above 10 GeV. The second central electron is required to have $P_T^{e2} > 5$ GeV ($E^{e2} > 10$ GeV) in H1 (ZEUS) analysis. Any other electron identified is also counted and the selected events are classified by the number of identified electrons in the event.

The results of H1 and ZEUS analyses are presented in the table 1. The H1 analysis on an event sample corresponding to 115 pb$^{-1}$ measured 121 multi-electron events, while ZEUS, with an integrated luminosity of 130 pb$^{-1}$ detected 217 such events. The di-electron sample is dominated by the signal with a 15-20% contribution from the background. In the tri-electron sample, the background contribution is negligible. Both H1 and ZEUS observations are in good agreement with the predicted rates. The main difference between H1 and ZEUS acceptances for the signal is due to different angular range for the central electrons.

The distributions of the invariant mass of the two highest $P_T$ electrons are shown in figures 1 and 2. Data are in good overall agreement with the Standard Model prediction. A few events with masses $M_{12} > 100$ GeV are observed in a region where the standard model prediction is low. H1 measured 3 di-electron events for 0.25 expected. ZEUS observed 2 di-electron events for 0.77 expected. In the tri-electron sample, H1 observed 3 events with $M_{12} > 100$ GeV for an expectation of 0.23 while ZEUS do not observe events in that mass region where the expectation is 0.37. For the high mass di-electron events, the transverse momenta of the two electrons is also large (above 50 GeV). The topology of the observed tri-electron high-mass events is different: the transverse momenta of the two highest $P_T$ electrons is lower (around 30 GeV) and the high mass value is associated with a larger polar opening angle between the two electrons ("forward-backward" topology).

## 4. Measurement of multi-muon events

Muons are identified using central tracker reconstructed tracks, calorimetric deposits and muon chamber signals. A search for multi-muon events has been performed by the H1 and ZEUS

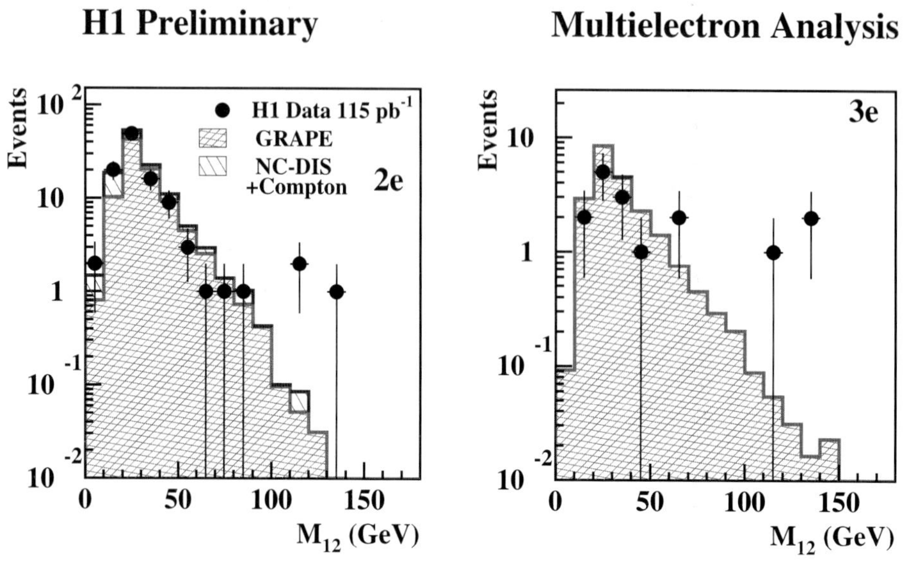


Figure 1. Distribution of the invariant mass $M_{12}$ of the two highest $P_T$ electrons for the H1 analysis. Events classified as di-electrons (left) and tri-electrons (right) are shown.

experiments in the angular range $20° < \theta < 160°$. At least two muons are required with $P_T^{\mu 1,2} > 2$ GeV ($P_T^{\mu 1,2} > 5$ GeV) in H1 (ZEUS) analyses. With an analyzed data sample corresponding to an integrated luminosity of 105 $pb^{-1}$, the ZEUS analysis detects 200 events for an expectation of $213 \pm 11_{stat}$. The H1 analysis of an integrated luminosity of 70 $pb^{-1}$ found 1242 multi-muon events in good agreement with the prediction of $1253 \pm 125_{stat. \oplus syst.}$.

No event with two muons at high mass $M_{\mu\mu} > 100$ GeV is observed by either experiment. Starting from the 3 high mass di-electron events observed by H1 the expected observation should be of the order of one di-muon event with $M_{12} > 100$ GeV, due to the lower efficiency and luminosity in the multi-muon channel. The expected increase in luminosity at HERA II will improve the knowledge of the high mass region and allow a better comparison between electron and muon channels.

Figure 3 presents the visible cross section measured by H1 as a function of the invariant mass of the muon pair compared to the Standard Model prediction. Backgrounds and also other sources of muon pair production like heavy hadron decays are negligible. Very good agreement with the Standard Model prediction is observed up to 80 GeV over a four order of magnitude decrease in the cross section. The integrated cross section in the visible phase space has been measured to be $46.5 \pm 4.7$ pb in good agreement with the prediction of 46.2 pb.

The H1 collaboration has measured separate cross-section for inelastic muon pair production. Inelastic events contain hadrons attributed to the proton dissociation detected in the main calorimeter or in the forward components of the detector. The inelastic cross section has been measured by H1 to be $20.8 \pm 3.3$ pb in agreement with the expected cross section of 21.5 pb.

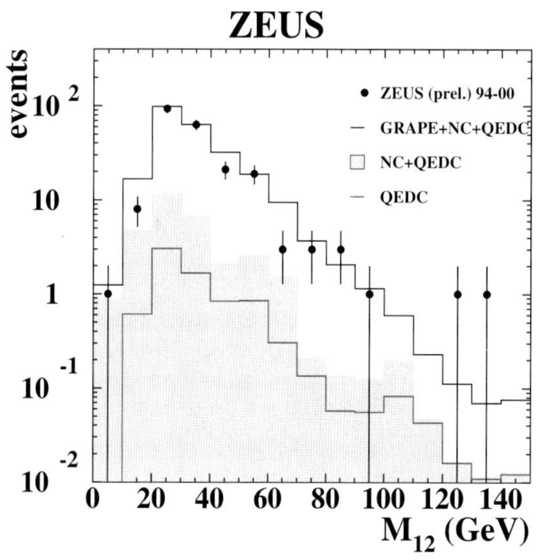


Figure 2. Distribution of the invariant mass $M_{12}$ of the two highest $P_T$ electrons for the ZEUS analysis. All observed events with more than two electrons are shown.

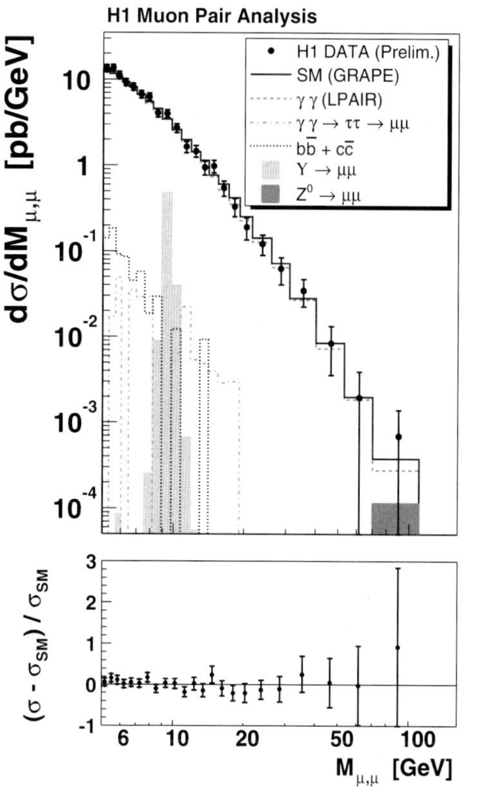


Figure 3. Differential muon pair production cross-section as a function of di-muon mass. The bottom plot shows the relative difference between the data and the Standard Model prediction.

## 5. Conclusions

Events with two or three visible leptons(electrons or muons) have been measured for the first time in electron-proton collisions by H1 and ZEUS experiments. Good overall agreement with the Standard Model prediction is observed. In the multi-electron analysis, several events with invariant mass of the two highest $P_T$ electrons $M_{12} > 100$ GeV have been observed. Multi-muon events have also been detected and the differential cross section has been measured and found in very good agreement with the Standard Model. No event with two muons at high mass $M_{\mu\mu} > 100$ GeV is observed by either experiment. The increase in luminosity expected at HERA II will improve the knowledge of multi-lepton production at high transverse momenta and high mass.

## REFERENCES

1. H1 Collaboration, Contributed paper **1019** to XXXIst ICHEP, July 2002, Amsterdam.
2. H1 Collaboration, Contributed paper **1021** to XXXIst ICHEP, July 2002, Amsterdam.
3. ZEUS Collaboration,Contributed paper **910** to XXXIst ICHEP, July 2002, Amsterdam.
4. J.A.M. Vermaseren, Nucl. Phys. **B229** (1983) 347; N. Artega-Romero, C. Carimalo, P. Kessler, Zeit.f.Phys **C52** (1991) 289.
5. T. Abe, Comp. Phys. Comm. **136** (2001) 126, http://www.awa.tohoku.ac.jp/~tabe/grape/.

ELSEVIER

Proceedings of ICHEP 2002, pp. 228–231
*S. Bentvelsen, P. de Jong, J. Koch and E. Laenen (Editors)*

31st INTERNATIONAL CONFERENCE ON HIGH ENERGY PHYSICS AMSTERDAM

# Top quark physics at the Tevatron

I. Iashvili[a] for the CDF and DØ Collaborations

[a] University of California, Riverside, CA 92521, USA

After a brief overview of top quark physics at the Tevatron collider in Run I, we discuss upgrades of the CDF and DØ detectors for Tevatron Run II which started in Spring of 2001. We present the first preliminary Run II results on the production of $W/Z$ bosons + jets, which constitute the dominant background to top quark production. Preliminary estimates of the detector performances relevant to $b$-tagging are presented.

## 1. INTRODUCTION

The top quark, heaviest among the six quarks, was discovered at the Tevatron by the CDF and DØ Collaborations in 1995 [1,2]. In $p\bar{p}$ collisions top quarks can be produced in pairs through the strong interaction processes $q\bar{q} \to t\bar{t}$ and $gg \to t\bar{t}$. For a top quark with mass $M_t = 174.3$ GeV, the theoretical cross section is $\simeq$5 pb at Run I energies of $\sqrt{s} = 1.8$ TeV and about 38% higher at Run II with $\sqrt{s} = 2.0$ TeV. Depending on the decay mode of the $W$-boson that comes from the $t \to Wb$ decay, one expects three distinctive event topologies from the $t\bar{t}$ signal: a dilepton final state with two high-$p_T$ isolated leptons, two jets, and missing $E_T$ ($\not\!\!E_T$) from escaping neutrinos; a lepton+jets final state with a high-$p_T$ lepton, four jets and $\not\!\!E_T$; and an all-jets final state consisting of six jets. Two jets in these final states are $b$-jets. While the dilepton final state gives the cleanest signal, it amounts to only 5% of the $t\bar{t}$ sample (considering only e and $\mu$ leptons), whereas the all-jets state constitutes 44% of the branching fraction, but suffers from a huge multijet background. The lepton+jets final state, forming 30% of $t\bar{t}$ events, is the best compromise between purity of the sample and the statistics.

The second top quark production mechanism at the Tevatron involves the electroweak production of a single top quark through a $Wtb$ vertex. There are three ways of producing a single top quark: an s-channel process $q'\bar{q} \to t\bar{b}$, a t-channel mode $q'g \to tq\bar{b}$, and a final state generated via both s- and t-channels, $bg \to tW$. The predicted single top production cross section is about 2.2 pb at Run I energies and by $\sim$30% higher at Run II. Searches for the single top quark production at the Tevatron are so far restricted to $W \to e\nu, \mu\nu$ decays. Thus, the final states have a high-$p_T$ lepton, $\not\!\!E_T$, and two or three jets.

## 2. TOP QUARK PHYSICS AT TEVATRON RUN I

At the Tevatron in Run I, $t\bar{t}$ production has been observed in all three classes of events, dilepton, lepton+jets and all-jets. Furthermore, a wealth of information has been extracted from only about 100 $t\bar{t}$ events per experiment based on $\sim$110 pb$^{-1}$ of data. The $t\bar{t}$ cross section measurements by CDF, $\sigma(t\bar{t}) = 6.5^{+1.7}_{-1.4}$ pb [3], and DØ $\sigma(t\bar{t}) = 5.7 \pm 1.6$ pb [4], together with the combined CDF and DØ mass measurement of $M_t = 174.3 \pm 5.1$ GeV [5], are in agreement with the Standard Model (SM) expectations.

According to the SM, the top quark decays before hadronization and spin information at $t\bar{t}$ production is therefore preserved. The spin correlation of the $t\bar{t}$ system has been studied at DØ using 6 dilepton events. A 68% CL limit on the spin correlation factor of $\kappa > -0.25$ [6] has been set.

The measurement of the ratio $R = B(t \to Wb)/B(t \to Wq)$ can be used to extract the CKM element $|V_{tb}|$. Using events with 0, 1, and 2 $b$-tagged jets in the dilepton and lepton+jets samples, CDF obtained $R = 0.94^{+0.31}_{-0.24}$ [7]. Assuming existence of three generations of fermions


doi:10.1016/S0920-5632(02)02098-4

Table 1
Measurement precisions of top quark properties at Run I, and expected accuracies at Run II with 2 fb$^{-1}$ and 15 fb$^{-1}$ integrated luminosities.

| | Run I precision | 2 fb$^{-1}$ | 15 fb$^{-1}$ |
|---|---|---|---|
| $M_t$ | 2.9% | 1.5% | 0.8% |
| $\sigma(t\bar{t})$ | 25% | 10% | 5% |
| $W$ helicity, $F_0$, $F_+$ | 0.5, 0.15 | 0.09, 0.03 | 0.04, 0.01 |
| $R = B(t \to Wb)/B(t \to Wq)$ | 30% | 4.5% | 0.8% |
| $\|V_{tb}\|$ inferred from $R$, 90% CL limit | >0.05 | >0.25 | >0.50 |
| $\sigma$(single-top) | — | 20% | 8% |
| $\Gamma(t \to Wb)$ inferred from $\sigma$(single-top) | — | 25% | 10% |
| $\|V_{tb}\|$ inferred from $\sigma$(single-top) | — | 12% | 5% |
| $B(t \to \gamma q)$ | 0.03 | $2\times10^{-3}$ | $2\times10^{-4}$ |
| $B(t \to Zq)$ | 0.30 | 0.02 | $2\times10^{-3}$ |

this gives $|V_{tb}| = 0.97^{+0.16}_{-0.12}$ [7].

The zero helicity mode of the W boson couples to fermion mass, therefore top quark decays provide a good tool to extract the fraction of longitudinal $W$s, which in the SM is expected to be $F_0 = M_t^2/(M_t^2 + 2M_W^2) \simeq 0.7$. Analyzing lepton $p_T$ distributions in dilepton and lepton+jets samples, CDF has obtained $F_0 = 0.91 \pm 0.37 \pm 0.13$ and $F_+ = 0.11 \pm 0.15 \pm 0.06$ [8], where $F_+$, the fraction of right handed $W$s, is zero in the SM.

CDF and DØ both have studied the top quark $p_T$ distribution. This has been found to be in agreement with the theory [9,10].

While the top quark discovery at the Tevatron relied on $t\bar{t}$ production, single top quark production has not yet been observed. Single top processes are more difficult to observe than $t\bar{t}$ events since they contain fewer jets and suffer from a larger background. Searches for single top by DØ and CDF have both led to negative outcomes, and 95% CL limits have been set on the production cross sections: $\sigma$(s-channel) < 17 pb and $\sigma$(t-channel) < 22 pb by DØ [12], and $\sigma$(s-channel) < 18 pb and $\sigma$(t-channel) < 13 pb by CDF [11].

Models with dynamically broken Electroweak symmetry predict the existence of a top quark condensate, topcolor, X, that can decay to $t\bar{t}$. With no indication for the topcolor signal in Run I data, the experiments have set 95% CL lower limit on its mass: $M_X > 480$ GeV by CDF [14] and $M_X > 560$ GeV by DØ . These limits are valid within a particular model of leptophobic topcolor [13] assuming $\Gamma_X = 0.012 M_X$ and $M_t = 175$ GeV.

## 3. TOP QUARK PHYSICS AT TEVATRON RUN II

During Run II, the Tevatron collider will deliver more than two orders of magnitude more integrated luminosity than in Run I, thanks to the new Main Injector and Antiproton Recycler. Thus, with Run II, the era of precision top quark physics begins. Table 1 summarizes the expected accuracies on top quark property measurements at Run II with 2 fb$^{-1}$ and 15 fb$^{-1}$ luminosities, together with the achieved precisions in Run I. Measurements of the single top quark cross section and top quark mass are viewed as the two most important top physics measurements at Run II. The $\sigma$(single-top) can be used to infer $|V_{tb}|$ with fewer assumptions than when using branching ratio measurements. Precise measurement of $M_t$ together with $M_W$ will considerably tighten the constraint on Higgs boson mass through radiative corrections. The expected precisions of $\Delta M_t \simeq$1.4 GeV and $\Delta M_W \simeq$16 MeV by the end of Run II will provide indirect constraints of $\Delta M_H/M_H \sim 25\%$ [15] and thereby will allow a consistency check of the SM.

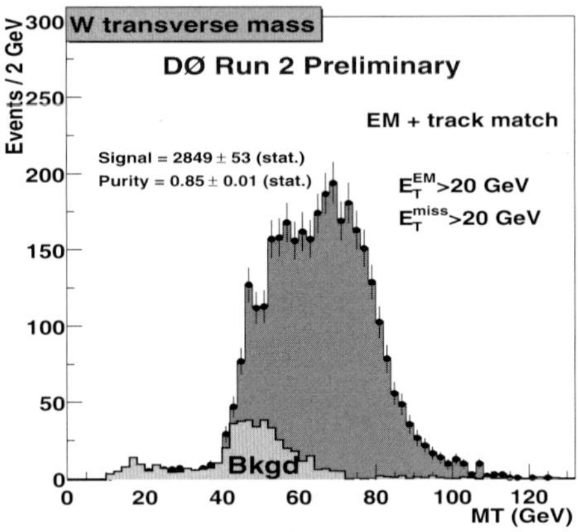


Figure 1. Distribution of $M_T(e,\not{E}_T)$ in events with $p_T^e > 20$ GeV and $\not{E}_T > 20$ GeV at Run II in DØ .

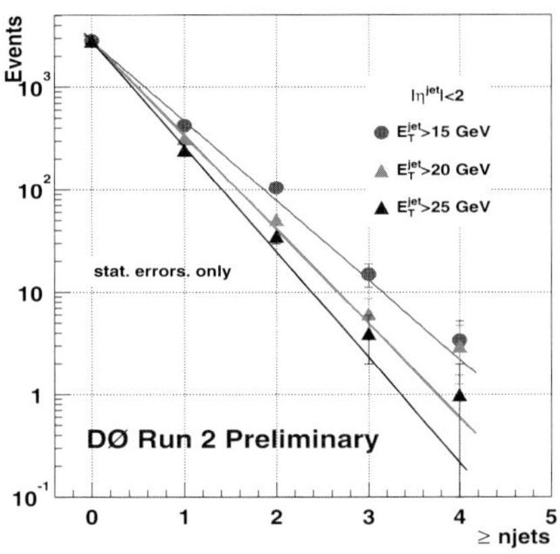


Figure 2. Inclusive jet multiplicity in mutijet-background-subtracted W→ $e\nu$ events at Run II in DØ .

The CDF and DØ detectors have both undergone extensive upgrades for Run II. For CDF this consists of replacing the entire tracking system with a large drift chamber and comprehensive silicon detectors, extending the muon system coverage, and adding a time-of-flight system. DØ also has a new tracking system with a silicon microvertex detector surrounded by a scintillating fiber tracker inside a new superconducting solenoid. The DØ upgrade also includes an improved muon detection system, a new forward shielding and new preshower detectors.

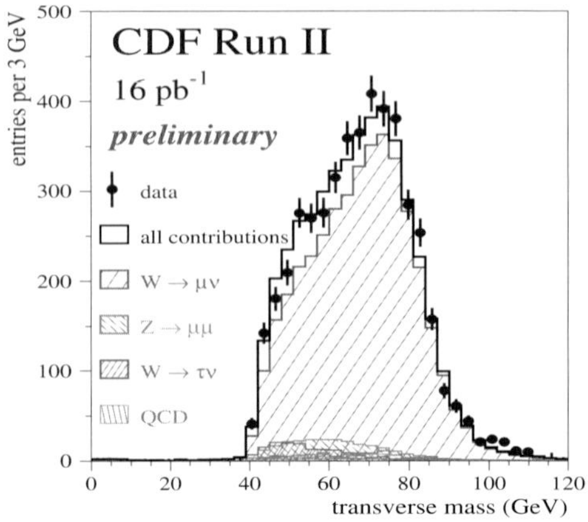


Figure 3. Distribution of $M_T(\mu,\not{E}_T)$ in events with $p_T^\mu > 20$ GeV and $\not{E}_T > 20$ GeV at Run II in CDF.

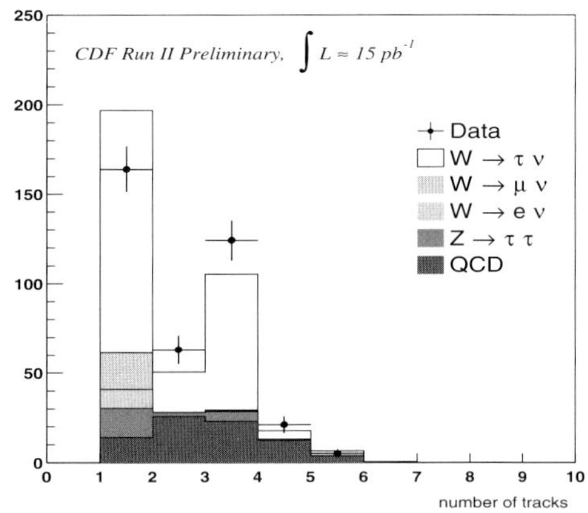


Figure 4. Multiplicity of tracks associated with tau candidates from $W \to \tau\nu$ events in Run II at CDF.

Run II started in Spring 2001. While most of the initial data-taking has been dedicated to detector commissioning, by Spring 2002 several pb$^{-1}$ of physics quality data has been accumulated by the experiments. Although the current amount of data is not yet sufficient to observe the top quark, experiments have studied production of $W$ and $Z$ bosons, as these, when produced together with jets, present the dominant background to $t\bar{t}$ events. Figure 1 shows the distribution of the transverse mass $M_T(e,\not{E}_T)$ in the events with $p_T^e > 20$ GeV and $\not{E}_T > 20$ GeV at

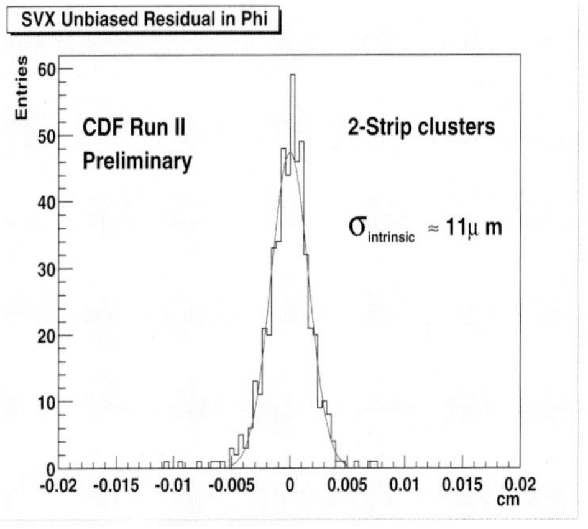


Figure 5. Resolution of 2-strip clusters for $p_T >$ 15 GeV tracks in CDF silicon detector.

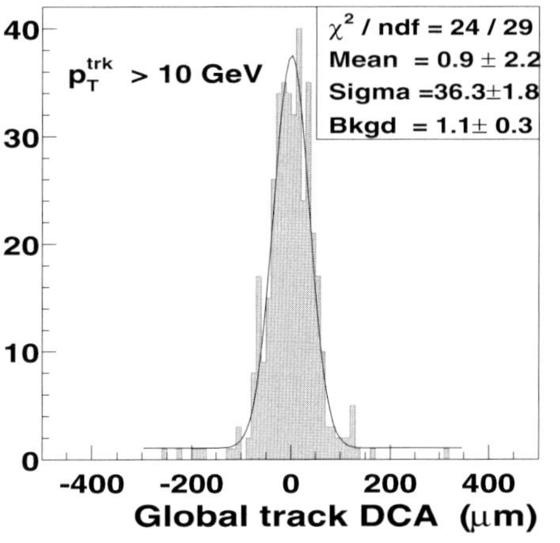


Figure 6. Distance-of-Closets-Approach distribution for tracks with $p_T > 10$ GeV in DØ .

DØ . The background from jets misidentified as electrons in multijet events has been estimated from the data and is shown in the figure. Distributions of inclusive central jet ($|\eta^{\rm jet}| < 2$) multiplicities in these $W$ events are shown in Figure 2 for different $E_T^{\rm jet}$ thresholds. With higher statistics and a jet $b$-tagging requirement, an enhancement of $\geq 4$ jets events would indicate a $t\bar{t}$ contribution. Figure 3 shows $M_T(\mu,\not{E}_T)$ distribution in events with $p_T^\mu > 20$ GeV and $\not{E}_T > 20$ GeV at CDF. Contributions from different processes are also indicated. Figure 4 shows the number of tracks associated with tau objects, identified as narrow calorimetric jets, in $W \rightarrow \tau\nu$ candidate events at CDF.

The Run II detectors both have $b$-tagging capability via reconstructed displaced vertices, which offer very important promise for efficient identification of high purity top quark samples. The performance of the silicon microvertex detectors in CDF and DØ are under detailed investigation. Figure 5 indicates that the intrinsic resolution of 2-strip clusters for $p_T > 15$ GeV tracks in the CD-F silicon detector is in excellent agreement with the expected value of 11 $\mu$m. Figure 6 shows the distance of closest approach (DCA) for tracks with $p_T > 10$ GeV in DØ after initial first-pass alignment of the silicon detector. Taking into account the beam width of ~30 $\mu$m, a track impact parameter resolution of ~20 $\mu$m is obtained.

To summarize, the CDF and DØ collaborations have already demonstrated that their detectors are approaching the performance needed for precision top quarks physics at Run II.

## REFERENCES

1. CDF Coll., Phys. Rev. Lett. 74, 2626, 1995.
2. DØ Coll., Phys. Rev. Lett. 74, 2632, 1995.
3. CDF Coll., Phys. Rev. D64 032022, 2001.
4. DØ Coll., hep-ex/0205019, 2002.
5. CDF and DØ Coll., Fermilab-TM-2084, 1999.
6. DØ Coll., Phys. Rev. Lett. 85, 256, 2000.
7. CDF Coll., Phys. Rev. Lett. 86, 3233, 2001.
8. CDF Coll., Phys. Rev. Lett. 84, 216, 2000.
9. CDF Coll., Phys. Rev. Lett. 87, 102001, 2001.
10. DØ Coll., Phys. Rev. D58, 052001, 1998.
11. CDF Coll., Phys. Rev. D65, 091102, 2002.
12. DØ Coll., Phys. Lett. B517, 282, 2001.
13. C. T. Hill, and S. Parke, hep-ph/9911288.
14. CDF Coll., Phys. Rev. Lett. 85, 2062, 2000.
15. Snowmass Working Group on Precision EW measurements, hep-ph/0202001.

ELSEVIER

Proceedings of ICHEP 2002, pp. 232–232
*S. Bentvelsen, P. de Jong, J. Koch and E. Laenen (Editors)*

31st INTERNATIONAL CONFERENCE ON HIGH ENERGY PHYSICS AMSTERDAM

# Updated measurements of the W mass and width from the Tevatron

Sarah Eno[a] for the DØ Collaboration

[a]Department of Physics
University of Maryland
College Park, MD 20742 USA

The DØ collaboration has recently published updates to their measurement of the mass and width of the W boson. This paper presents these results.

## 1. W mass

The DØ collaboration has published an update of their measurement of the W mass [1]. The new measurement, based on the 85 $pb^{-1}$ of $p\bar{p}$ data taken at a center-of-mass energy of 1.8 TeV during the 1994-1995 run of the Tevatron at Fermi National Accelerator Laboratory, includes electrons from a region of the calorimeter which was not used in previous analyses of these data[2]. The DØ central calorimeter, which covers the pseudorapidity region $|\eta|$ <1.1, consists of 32 modules in $\phi$. The new events have electrons which are close to the module boundaries. The new result for the W mass is 80.482 $\pm$ 0.084 $GeV/c^2$.

## 2. Direct Measurement of the W Width

The DØ collaboration has published an update of their measurement of the W width [3]. While DØ has previously published indirect determinations of the width based on the ratio of the W and Z production cross sections[4], this is the first direct determination from DØ . This measurement uses the shape of the transverse mass distribution to determine the width. The result is $2.23^{+0.15}_{0.14} \pm 0.01\ GeV/c^2$.

## 3. Combined World Averages

The Tevatron Electroweak Working group has updated the world averages for the W mass and W width to include these new results [5]. The result for the mass is $80.450 \pm 0.034\ GeV/c^2$. The result for the width is $2.158 \pm 0.042\ GeV/c^2$.

## REFERENCES

1. V. Abazov *et al.* (the DØ collaboration), Phys. Rev. D **66**, 012001 (2002).
2. B. Abbot *et al.*, Phys. Rev. D **62**, 09003 (2000).
3. V. Abazov *et al.* (the DØ collaboration), Phys. Rev. D **66**, 032008 (2002).
4. B. Abbott *et al.* (the DØ collaboration), Phys. Rev. D bf 61, 072001 (2000).
5. Tevatron Electroweak Working Group (for the CDF and DØ collaborations), FERMILAB-FN-716.


doi:10.1016/S0920-5632(02)02099-6

ELSEVIER

Proceedings of ICHEP 2002, pp. 233–236
*S. Bentvelsen, P. de Jong, J. Koch and E. Laenen (Editors)*

31st INTERNATIONAL CONFERENCE ON HIGH ENERGY PHYSICS AMSTERDAM

# Electroweak Prospects for Tevatron RunII

D. Glenzinski (on behalf of the CDF and D0 Collaborations) [a]

[a]Fermi National Accelerator Laboratory, P.O. Box 500, Batavia, Il. 60501-0500 U.S.A

The prospects of precision electroweak measurements from CDF and D0 using RunII data is reviewed.

## 1. Introduction

In RunI each experiment collected about 100 $\mathrm{pb}^{-1}$ of data. During RunIIa, each experiment is expected to collect about 2 $\mathrm{fb}^{-1}$ of data. The center-of-mass energy for RunII, $\sqrt{s} = 2.0$ TeV, is a bit larger than the 1.8 TeV of RunI and results in an increase of about 10% (35%) in the production cross-sections for $W$ and $Z$ ($t\bar{t}$) events. Additional gains in the event yield are expected due to improvements in the detector acceptance and performance. Taken together, the RunIIa upgrades are expected to yield 2300k (800) $W$ ($t\bar{t}$) events per experiement, including the effects of event selection and triggering, which can be compared to the RunI yields of 77k (20) events. With the RunI data-set, CDF and D0 produced a breadth of electroweak results and obtained the world's only sample of top quarks. While the RunII electroweak physics program is very similar, the RunII upgrade improvements should yield many precision results.

The Tevatron began delivering steady data in about June, 2001. The first six months of data taking was "commissioning dominated" for CDF and D0. Starting around January, 2002, the experiments were largely commissioned and began taking "analysis quality" data. The physics results reported at this conference are based on about $10-20pb^{-1}$ (depending on the data-set) per experiment. Thus, the presently available event samples are smaller than those available in RunI. At this early stage of RunII, it is interesting to compare the present detector performance to that assumed when making the RunII physics projections.

In the following sections I discuss some RunII projections for a few electroweak measurements of particular importance, namely the precision determinations of the W-boson mass, $M_W$, and the top-quark mass, $M_t$.

## 2. Precision Measurement of the W-Boson Mass

In RunI, CDF and D0 each measured the W-Boson mass with an uncertainty of about 80 MeV, which together yield a combined Tevatron result of $80.456 \pm 0.059$ GeV [1]. For comparison, the (preliminary) LEP combined result is $80.447 \pm 0.042$ GeV [2].

At the Tevatron the W-bosons are produced by hard collisions between the constituent (anti-)quarks of the proton and anti-proton. Thus, the center-of-mass energy for this hard collision cannot be known *a priori* event-by-event. As a consequence of this, only the constraints in the transverse plane remain. Due to overwhelming QCD backgrounds, only leptonically decaying Ws are used. The W-boson mass can be extracted from a fit to the transverse mass, $M_T = \sqrt{(E_T^\ell + E_T^\nu)^2 - (\overline{P}_T^\ell + \overline{P}_T^\nu)^2}$. It is obvious that a detailed understanding of the energy and momentum scales and resolutions is fundamental to this measurement. In order to accurately estimate the neutrino momentum, it is additionally important to have a detailed understanding of the underlying event and recoil distributions. Finally, since the shape of the $M_T$ distribution is affected by the parton-distribution-functions (PDFs), there is a model-dependent systematic uncertainty associated with this method. For a detailed discussion of these measurements, see references [1] and [3].

The sources of uncertainty in the combined


doi:10.1016/S0920-5632(02)02100-X

Table 1
Uncertainties in the CDF/D0 combined $M_W$ measurement. The contribution of each systematic source is approximate.

| Source | $\Delta M_W$ (MeV) |
|---|---|
| Statistical: | $\pm 40$ |
| Systematic | |
| scale: | $\pm 40^{\dagger}$ |
| recoil: | $\pm 20^{\dagger}$ |
| modeling: | $\pm 15^{*}$ |
| other: | $\pm 15^{\dagger}$ |
| total: | $\pm 43$ |

† dominated by statistics of control sample.
$*$ correlated between experiments.

Tevatron $M_W$ determination are given in Table 1. The "other" category includes uncertainties from background shape and normalization and residual fit biases. As is evidenced in the Table, even the Tevatron combined measurement is dominated by uncertainties which are expected to scale with the statistics of the relevant data-sets. The only exceptions to this are the "modeling" uncertainties, dominated by contributions from PDF uncertainties, but also including contributions from higher-order radiative corrections. As these are the only sources of correlated uncertainties between the experiments, they may well come to limit the ultimate precision with which $M_W$ can be determined at the Tevatron. Assuming no improvement in these modeling uncertainties (*i.e.* a conservative assumption), it is rather straight- forward to appropriately scale the other uncertainties to arrive at a projection of $\Delta M_W(2\,\text{fb}^{-1}) = \pm 30\,\text{MeV}$ per experiment and a Tevatron combined uncertainty of about 25 MeV. The only important caveat in this projection is the assumed resolution on $P_T^{\nu}$, which degrades with the mean number of additional interactions per event. This functional dependence was studied using RunI data. Although the RunIIa instantaneous luminosity increases, the mean number of additional interactions per event is comparable to that of RunI because the Tevatron is running with more bunches. Thus, for the RunIIa projection, this assumption is on fairly solid footing. Finally, depending on the evolution of the combined LEP uncertainty, and assuming the LEP and Tevatron measurements are completely uncorrelated, this gives an expected world average of $\Delta M_W(\text{LEPII} + \text{TeVIIa}) = \pm 15 - 20\,\text{MeV}$.

## 3. Precision Measurement of the Top-Quark Mass

In RunI, CDF and D0 each measured the top-quark mass with an uncertainty of about 7 GeV, which together yield a combined Tevatron result of $174.3 \pm 5.1$ GeV [4].

At the Tevatron, top quarks are predominantly pair-produced, with each top quark predominantly decaying to a $W$ and a b quark. The final-state topology is determined by the decay of the two $W$s, with the "di-lepton", "lepton plus jets" and "fully hadronic" final states corresponding to both, one, or neither of the $W$s decaying leptonically, respectively. While the di-lepton final state is the most pure, it's branching ratio is smallest (owing to the $BR(W \to \ell\nu)^2$ factor) and it's kinematics are under constrained (owing to the two neutrinos). On the other hand, the fully hadronic final state suffers from a large QCD background. Consequently, the most significant channel for the determination of $M_t$ is the lepton-plus-jets channel. The dominant background contributions to this channel are from $W$+$\geq$4 jet events, which can be suppressed by requiring that $\geq 1$ jet in the event is identified as a b-quark jet ("B-tagged").

The top-quark mass for each candidate event is determined from a kinematic fit which employs momentum constraints and requires that the two $W$ candidates have a mass consistent with the world average $M_W$, and that the two top quarks have the same mass. In order to perform the fit, jet-parton assignments must be made, thus giving rise to a combinatoric background, which greatly degrades the resolution of the kinematic fit. For lepton-plus-jet events with 0, 1 or 2 B-tagged jets, there are 12, 6 or 2 possible jet-parton combinations. Thus, the single most important factor in improving the $M_t$ determination for RunII, is the expected improvement in the B-tagging performance, which should yield a more efficient and pure event selection, and should reduce the combinatoric background, effectively enhancing the per event $M_t$ sensitivity. For a detailed discussion of these measurements, see references [4] and [5].

The sources of uncertainty in the "typical" RunI $M_t$ determination are given in Table 2. The "other" category includes uncertainties from background shape and normalization and residual fit biases. The "modeling" uncertainties are dominated by contributions from hadronization and fragmentation modeling, and modeling of final state gluon radiation and are the only source of correlated uncertainty between the experiments. The dominant systematic uncertainty is due to uncertainties in the jet energy scale and associated corrections, which, in RunI, were determined from low statistics control samples. For 2 $\text{fb}^{-1}$ of RunII data, the statistical uncertainty is expected to be $< 1$ GeV per experiment and the measurements are expected to be systematic limited. The RunIIa projections assume that the total systematic uncertainties can be reduced to the $2-3\,\text{GeV}$ per experiment. Reducing the systematic uncertainty to that level will require the use of special control samples ($Z+$ jets, $Z \to b\bar{b}$, and $W \to q\bar{q}$) which, in general, were too small to be of use in RunI. Thus, the RunII projections have been based on detailed Monte Carlo simulations of these data-sets. The efficiency and purity with which many of these controls samples are collected are contingent upon the performance of the silicon vertex detectors. There have not been, to the author's knowledge, any detailed study to estimate the projected Tevatron combined $M_t$ uncertainty after 2 $\text{fb}^{-1}$ of data. Obviously the projection is strongly dependent upon the assumed evolution of the modeling uncertainty. The most conservative projection would assume no improvement, so that each experiment would have a measurement uncertainty dominated by these modeling uncertainties. In that scenario, the combination would yield very little improvement over the $2-3\,\text{GeV}$ uncertainty per experiment.

## 4. Initial Detector Performance

The CDF and D0 detector upgrades have been described many times and the details are available in references [6] [7]. It is principally important to note that the RunII electroweak projections assume i) the energy and momentum resolutions are no worse than those of RunI, ii) B-jet and lepton[1] identification are extended to the $|\eta| > 1$ forward regions [2], and iii) the trigger performance allows efficient collection of the relevant samples up to instantaneous luminosities of about $2\times10^{32}\text{cm}^{-2}\text{sec}^{-1}$. The performance of the relevant detector components is briefly discussed here.

Both CDF and D0 upgrades include silicon micro-strip detectors at inner radii, surrounded by large volume tracking chambers, all inside a magnetic field. Both experiements have finished initial alignments of their silicon detectors and are measuring high signal-to-noise ratios ($\geq 12$) with the expected intrinsic resolution and high hit efficiency ($\geq 98\%$). This bodes well for the B-tag performance of the two experiments. Both experiments have also collected large statistics samples of $J/\Psi \to \mu\mu$ events, which are used to perform a variety of systematic studies to limit residual mis-

[1] Through the whole of this note, it should be understood that "lepton" means an electron or a muon, unless otherwise stated.

[2] The co-ordinate system has the z-axis parallel to the beam-axis and pointing in the proton flight direction, the x-axis orthogonal to the z-axis and pointing to the center of the Tevatron, and the y-axis defined to yield a right-handed co-ordinate system; the angles $\theta$ and $\phi$ are the traditionally defined polar and azimuth cylindrical co-ordinates, respectively, and the pseudo-rapidity, $\eta$, is defined as $\eta = -\ln(\tan(\theta/2))$.

Table 2
"Typical" per experiment uncertainties in the RunI $M_t$ measurement.

| Source | $\Delta M_t$ (GeV) |
|---|---|
| Statistical: | $\pm5$ |
| Systematic | |
| scale: | $\pm4$ |
| modeling: | $\pm2^*$ |
| other: | $\pm2$ |
| total: | $\pm5$ |

$*$ correlated between experiments.

alignments, and determine energy-loss and B-field corrections. Although still at an early stage, the present $P_T$ resolution for CDF's COT is better than $\sigma_{P_T}/P_T^2 < 0.13\%\,\mathrm{GeV}^{-1}$, comparable to the 0.10% design goal [8]; this is expected to improve as the alignment matures. Similarly, D0 expects the CFT to meet design goals once alignment and calibrations are finalized.

Since both experiments left their calorimetry largely unchanged relative to RunI (CDF replaced their forward calorimetry), the resolution should be well understood. The observed width of the invariant mass spectrum in $Z \to e^-e^+$ events can be used to estimate this resolution. Including all major corrections for both the forward and central calorimeters, CDF observes a width within 5% of that expected [9]. At the time of the conference, D0 had not yet included the full set of corrections and, consequently, were observing a width 30% larger than expected; this is roughly consistent with the contribution expected from the excluded corrections.

Despite some initial problems, both experiments have demonstrated their ability to efficiently trigger on events of interest. CDF has measured a Level 1 tracking efficiency $> 95\%$ (important for triggering on high-momentum leptons) and D0 showed similarly high effiencies for high-momentum electrons. CDF has also collected large samples enriched in heavy-flavor (c- and b-quark jets) by use of a displaced track trigger at Level 2 (SVT) - the first such trigger at a hadron collider [10]. This is an important first step in accumulating the $Z \to b\bar{b}$ control sample mentioned above.

Although it is still a bit early to draw any definitive conclusions, both the CDF and D0 detectors look to be on track to meet their design goals and fulfill the RunIIa Electroweak physics projections.

## REFERENCES


1. *Combination of CDF and D0 Results on W Boson Mass and Width*, Fermilab-FN-0716 (2002), and references therein.
2. *A Combination of Preliminary Electroweak Measurements and Constraints on the the Standard Model*, LEPEWWG/2002-01 (2002), and references therein.
3. Glenzinski DA, Heintz U. *Annu. Rev. Nucl. Part. Sci.* 50:207 (2000).
4. CDF Collaboration, *Phys. Rev. Lett.* 82:271 (1999). D0 Collaboration *Phys. Rev. Lett.* 84:222 (2000).
5. Tollefson K, Varnes E. *Annu. Rev. Nucl. Part. Sci.* 49:435 (1999).
6. *The CDFII Detector Technical Design Report*, Fermilab-Pub-96/390-E, (1996) 234p.
7. L. Babukhadia, *these procedings. The D0 Upgrade: The Detector and its Physics*, Fermilab-Pub-96/357-E, (1996) 87p.
8. S. Nahn, *these procedings.*
9. R.D. Erbacher, *these procedings.*
10. A. Cerri, *these procedings.*

ELSEVIER

Proceedings of ICHEP 2002, pp. 237–240
S. Bentvelsen, P. de Jong, J. Koch and E. Laenen (Editors)

31st INTERNATIONAL CONFERENCE ON HIGH ENERGY PHYSICS AMSTERDAM

# W Boson Cross Section and Decay Properties at the Tevatron

Kenneth Bloom[a]

[a]University of Michigan, for the CDF and D0 Collaborations

We present the first measurements of $\sigma(p\bar{p} \to W \to \ell\nu)$ and $\sigma(p\bar{p} \to Z \to \ell\ell)$ at $\sqrt{s} = 1.96$ TeV, along with new measurements of $W$ angular-decay distributions in $p\bar{p}$ collisions at $\sqrt{s} = 1.8$ TeV.

## 1. $W/Z$ production cross sections

$W$ and $Z$ boson production cross-section measurements in $p\bar{p}$ collisions are a test of the consistency of Standard-Model couplings, constrain proton parton distribution functions, and provide information on higher-order QCD corrections. They also test the mettle of an experiment, as the measurements require good understanding of detection efficiencies, backgrounds, and luminosity. If experimental uncertainties are small, and the cross sections can be well-estimated from theory, the boson production rates can be interpreted as a measure of luminosity, and can also be used to normalize measurements of other production cross sections. Finally, $W$ and $Z$ bosons provide a path to the physics of Run II at the Tevatron, where many signatures of top-quark and Higgs-boson production can include these bosons.

At the Tevatron, protons and antiprotons collide at a center-of-mass energy of 1.96 TeV. A $W$ boson appears in the detector as a high-momentum lepton and large missing energy due to the undetected neutrino. As the $z$ component of $p^\nu$ is unmeasured, all quantites are measured in the transverse plane. A $Z$ boson appears as two high-momentum opposite-signed leptons with an invariant mass around 90 GeV. The cross section can be expressed as

$$\sigma \cdot B = \frac{N_{obs} - N_{bg}}{A\epsilon \int \mathcal{L}dt}, \quad (1)$$

where $N_{obs}$ is the number of observed boson events, $N_{bg}$ is the estimated number of background events, $A$ is the kinematic and geometric acceptance, $\epsilon$ is the total efficiency, and $\int \mathcal{L}dt$ is the integrated luminosity.

The lepton plus missing transverse-energy sample can be accounted for by $W \to \ell\nu$, $Z \to \ell\ell$, $W \to \tau\nu$, QCD jets (fake leptons), and (for muons) cosmic rays, as illustrated in Figure 1. CDF measures the $W$ cross section in the electron and muon decay channels, and D0 does so in the electron channel. (CDF also looks at mono-jet events with large missing energy, and sees an enhancement in the number of jets with one and three charged tracks, evidence for $W \to \tau\nu$ decays.) The inputs and results for the cross-section measurements are given in Table 1. The measurements are systematics-limited, with the overall uncertainty typically dominated by the uncertainty in the integrated-luminosity measurement. The quoted uncertainty on the luminosity should be considered an upper limit.

The dilepton samples are quite clean and allow for easy identification of the $Z$ resonance, as seen in Figure 2. D0 measures $\sigma \cdot B(Z \to ee)$, with the result shown in Table 2. From this and the $W$ cross-section measurement, D0 calculates the ratio of cross sections $R_\ell = \sigma^W/\sigma^Z$, also given in Table 2; CDF measures the same ratio in the muon channel. This ratio can be expressed as

$$R_\ell = \frac{\sigma(p\bar{p} \to W)}{\sigma(p\bar{p} \to Z)} \frac{\Gamma(W \to \ell\nu)}{\Gamma(Z \to \ell\ell)} \frac{\Gamma(Z)}{\Gamma(W)}. \quad (2)$$

Taking the total cross-section values and $\Gamma(W \to \ell\nu)$ from theory, and the $Z$ branching fraction from measurements at the $Z$ pole, $\Gamma(W)$ can be extracted. The D0 $R$ measurement gives $\Gamma(W) = 2.26 \pm 0.18_{\text{stat}} \pm 0.29_{\text{sys}} \pm 0.04_{\text{external}}$ GeV, and the CDF measurement of $R$ implies $\Gamma(W) = 1.67 \pm 0.24^{+0.14}_{-0.13} \pm 0.01$ GeV. The current world average


doi:10.1016/S0920-5632(02)02102-3

Table 1
Measurement of the $W$ cross section at CDF and D0.

| | CDF $W \to e\nu$ | CDF $W \to \mu\nu$ | D0 $W \to e\nu$ |
|---|---|---|---|
| $N_{obs}$ | 5547 | 4561 | 9205 |
| $N_{bg}$ | $409 \pm 85$ | $569 \pm 63$ | $5782 \pm 357$ |
| $A$ (%) | $23.4 \pm 0.9$ | $14.2 \pm 0.4$ | $19.6 \pm 0.9$ |
| $\epsilon$ (%) | $81.1 \pm 1.8$ | $63.2 \pm 3.8$ | $86.5 \pm 3.6$ |
| $\int \mathcal{L}dt$ ($pb^{-1}$) | $10.4 \pm 1.0$ | $16.5 \pm 1.6$ | $7.5 \pm 0.8$ |
| $\sigma \cdot B$ (nb) | $2.60 \pm 0.03_{\text{stat}} \pm 0.13_{\text{sys}} \pm 0.26_{\text{lum}}$ | $2.70 \pm 0.04_{\text{stat}} \pm 0.19_{\text{sys}} \pm 0.27_{\text{lum}}$ | $2.67 \pm 0.06_{\text{stat}} \pm 0.33_{\text{sys}} \pm 0.27_{\text{lum}}$ |

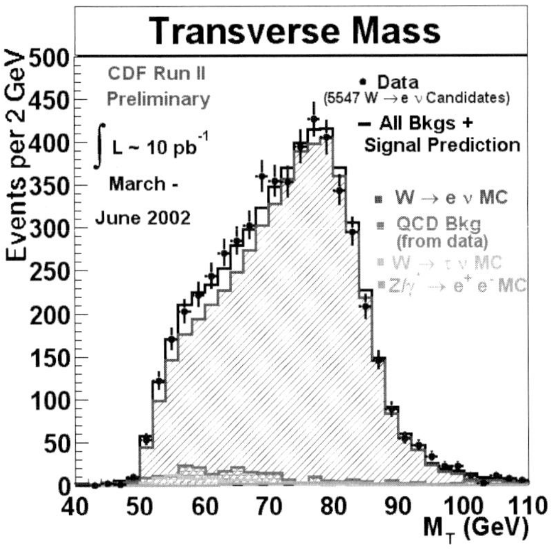


Figure 1. Transverse-mass distribution for the CDF electron plus missing energy sample.

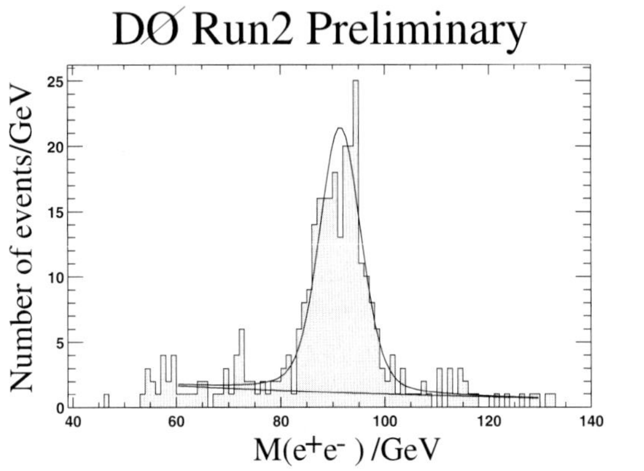


Figure 2. Invariant-mass distribution for the D0 dielectron sample.

is $\Gamma(W) = 2.114 \pm 0.043$ GeV [1].

All of the CDF and D0 cross-section measurements are summarized in Table 2. For comparison, the best measurements of these quantities in Run I data are $\sigma^W = 2.49 \pm 0.12$ nb [2], $\sigma^Z = 249 \pm 11$ pb [3], and $R = 10.90 \pm 0.43$ [4]. The Run II measurements have not reached that level of precision, but they should with additional data. In addition, the $W$ and $Z$ production cross sections at the Run II energy of 1.96 TeV are expected to be about 10% larger than those at the Run I energy of 1.8 TeV; theory predicts values of $\sigma^W = 2.73$ (2.50) nb and $\sigma^Z = 250$ (230) pb at $\sqrt{s} = 1.96$ (1.8) TeV [5]. The measured values are consistent with this predicted rise, as illustrated in Figure 3.

## 2. $W$-decay angular distributions

Understanding QCD effects in $W$ production helps reduce uncertainties in $M_W$ and other electroweak measurements. Without QCD, when the $W$ has no transverse momentum ($p_T^W$), the differential cross section for the charged lepton in the $W$ decay is $\propto (1 + q_\ell \cos\theta)^2$, as predicted by the $V - A$ interaction. But in NLO QCD, the differential cross section in the Collins-Soper rest frame is given as

$$\frac{d^4\sigma}{dp_T^W dy d(\cos\theta) d\phi} \quad \propto \quad 1 + \cos^2\theta +$$

Table 2
Summary of cross-section measurements; uncertainties are from statistics, systematics, and luminosity normalization.

| Quantity | Source | Value |
|---|---|---|
| $\sigma_e^W$ (nb) | CDF | $2.60 \pm 0.03 \pm 0.13 \pm 0.26$ |
| $\sigma_\mu^W$ (nb) | CDF | $2.70 \pm 0.04 \pm 0.19 \pm 0.27$ |
| $\sigma_e^W$ (nb) | D0 | $2.67 \pm 0.06 \pm 0.33 \pm 0.27$ |
| $\sigma_e^Z$ (pb) | D0 | $266 \pm 20 \pm 20 \pm 27$ |
| $R_e$ | D0 | $10.0 \pm 0.8 \pm 1.3$ |
| $R_\mu$ | CDF | $13.7 \pm 1.9^{+1.1}_{-1.2}$ |

$$\begin{aligned} & A_0(1-3\cos^2\theta)/2 + \\ & A_1 \sin 2\theta \cos\phi + \\ & A_2(\sin^2\theta\cos 2\phi)/2 + \\ & A_3 \sin\theta\cos\phi + \\ & A_4 q_\ell \cos\theta + \\ & A_5 \sin^2\theta \cos 2\phi + \\ & A_6 \sin 2\theta \sin\phi + \\ & A_7 \sin\theta\sin\phi, \end{aligned} \tag{3}$$

where the $A_i$ depend on $p_T^W$ and the boson rapidity; the latter is typically integrated out. (The $A_1$, $A_5$, $A_6$ and $A_7$ terms can be safely neglected.) Thus, angular distributions when $p_T^W \neq 0$ probe the values of the $A_i$, which can then be compared to predictions from NLO QCD. The D0 experiment has made a measurement of $A_0$ in Run I data[6], and CDF now has measurements of $A_0$, $A_2$ and $A_3$ from Run I data.

The $\theta$ distribution of the decay leptons is obtained by integrating Equation 3 over $\phi$:

$$\frac{d\sigma}{d(\cos\theta)} \propto 1 + q_\ell \alpha_1 \cos\theta + \alpha_2 \cos^2\theta, \tag{4}$$

with $\alpha_1 = A_4/(2+A_0)$, $\alpha_2 = (2-3A_0)/(2+A_0)$. There is no sensitivity to $\alpha_1$ due to the unknown $p_z^\nu$, but there is sensitivity to $\alpha_2$ through the shape of the $W$ transverse-mass spectrum, which changes with $p_T^W$. The Run I CDF $W$ sample is separated into four bins of $p_T^W$, and in each one a maximum-likelihood fit to the $M_T$ spectrum is performed to extract $\alpha_2$, accounting for background contamination and detector acceptance. The results are shown in Figure 4, along with the

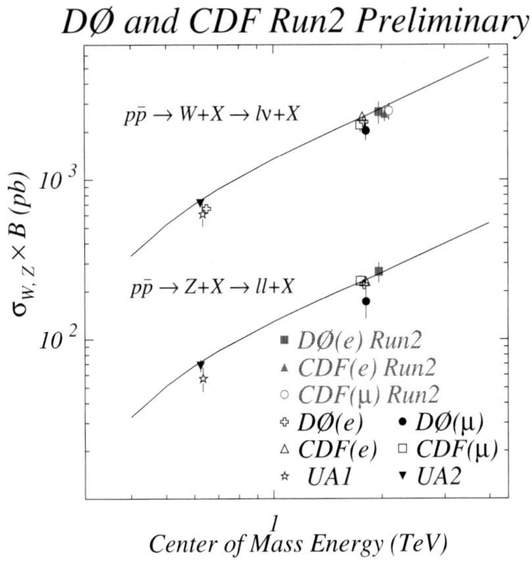


Figure 3. $W$ and $Z$ cross-section measurements at different center-of-mass energies, with theoretical prediction.

earlier D0 results and the prediction from QCD. There is good agreement among all three; the decrease in $\alpha_2$ with $p_T^W$ indicates increasing longitudinal polarization, as is expected. The CDF measurement is statistics-limited, with systematic uncertainties dominated by uncertainty in $M_W$, $p_z^W$, and the model of the $W$ recoil.

Similarly, one can integrate over $\theta$ in Equation 3 to obtain an expression that depends on $\phi$. Measuring the $\phi$ distribution of the lepton from $W$ decay gives information on the $A_2$ and $A_3$ coefficients. Again, events from the Run I CDF $W$ sample are separated into four bins of $p_T^W$; the events are required to have at least one hadronic jet to ensure that $p_T^W > 0$. A maximum-likelihood fit to the $\phi$ distribution is performed to extract $A_2$ and $A_3$. The results are shown in Figure 5, along with the prediction from QCD. As before, the experimental results are in agreement with the theory prediction. The $A_2$ measurement is systematics-limited, with the dominant uncertainties coming from the knowledge of $A_0$ and $A_4$, and the renormalization and factor-

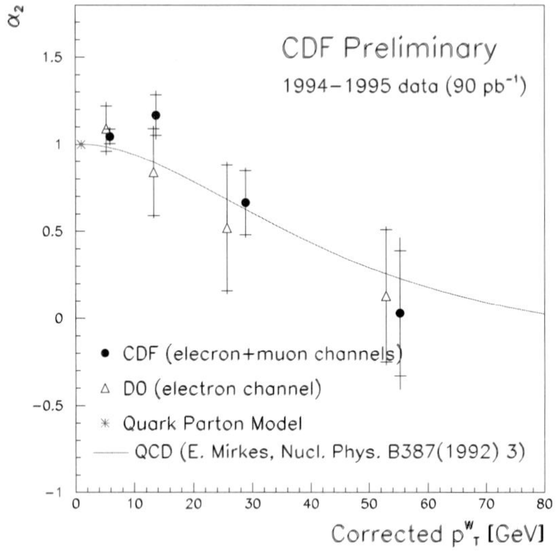


Figure 4. CDF measurement of $\alpha_2$ as a function of $p_T^W$, along with earlier D0 measurement and QCD prediction.

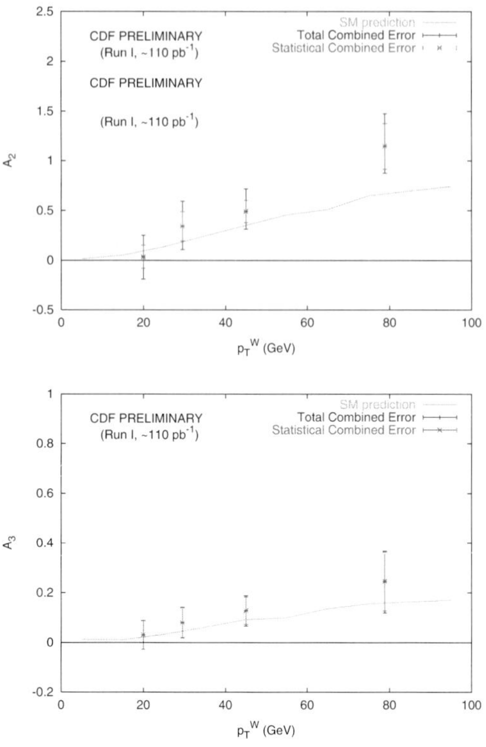


Figure 5. CDF measurement of $A_2$ (top) and $A_3$ (bottom) as a function of $p_T^W$, along with the QCD prediction.

ization scale. By contrast, the $A_3$ measurement is statistics-limited.

## 3. Summary

Run I at the Tevatron has produced a trove of knowledge about the $W$ and $Z$ bosons, as evidenced by the new CDF measurement of the $W$ angular-decay distributions. These sophisticated analyses will be completed soon. The CDF and D0 collaborations are now turning their attention to the Run II data, and their $W$ and $Z$-based analyses are underway. Both experiments have established techniques to identify the bosons with high efficiency. They have a good understanding of the acceptances and efficiencies of their detector, and this understanding will improve as more data are collected. The $W$ cross-section measurements are consistent with the theory predictions and expectations from earlier measurements, and are already systematics-limited. Measurements of the $Z$ cross section and $R$ tend to be statistics-limited, but that too will change as the datasets grow. This work provides a solid foundation for the studies of high-$p_T$ physics that are among the goals of Run II.

## REFERENCES

1. K. Hagiwara *et al.* (PDG), Phys. Rev. **D66**, 010001 (2002).
2. F. Abe *et al.* (CDF), Phys. Rev. Lett **76**, 3070 (1996).
3. T. Affolder *et al.* (CDF), Phys. Rev. Lett **84**, 245 (2000).
4. F. Abe *et al.* (CDF), Phys. Rev. **D52**, 2624 (1995).
5. W.J. Stirling, private communication.
6. B. Abbott *et al.* (D0), Phys. Lett. **B513**, 292 (2001).

ELSEVIER

Proceedings of ICHEP 2002, pp. 241–244
*S. Bentvelsen, P. de Jong, J. Koch and E. Laenen (Editors)*

31st INTERNATIONAL CONFERENCE ON HIGH ENERGY PHYSICS AMSTERDAM

# Higgs physics at a high luminosity $e^+e^-$ linear collider

André Sopczak*, Lancaster University, UK

Since the TESLA Technical Design Report (TDR) was published in 2001, the physics programme of an electron-positron linear collider (LC) has been further developed and a wide consensus has been reached on the physics case and the need for a high luminosity LC with center-of-mass energy up to about 1 TeV as the next worldwide high-energy physics project. The study of the Higgs boson properties represents a significant part of this physics programme. New studies demonstrate that a LC providing 1000 $\text{fb}^{-1}$ of data at center-of-mass energies of at least 500 GeV, is an excellent Higgs boson analyzer for a wide range of masses. A summary of preliminary results of these studies and their implications for identifying the nature of the Higgs sector and for constraining the parameter space of extended models is given. The focus is on recent developments and the relation to the LHC is addressed.

## 1. INTRODUCTION

A linear collider of at least 500 GeV and a total luminosity of at least 1000 $\text{fb}^{-1}$ has much potential for studying Higgs bosons and understanding the electroweak symmetry breaking and mass generation. Already at the CERN $e^+e^-$ collider, LEP, enormous progress for Higgs boson searches has been made. At LEP-1 [1] many search channels were almost background free and also at LEP-2 [2] the sensitivity exceeded expectations, leading to a Standard Model (SM) Higgs boson mass limit of 114 GeV at 95% CL. A small indication at LEP of a 115 GeV SM Higgs boson could also be interpreted in extended models [3].

During the last 10 years, LC studies have evolved from discovery studies to studies of precision measurements. Recent milestones were set with the TESLA TDR [4], the Snowmass Study [5] and the international workshop on LC in Korea [6].

The figures for this report are included in Ref. [7]. The importance of beam polarization and the option of a $\gamma\gamma$ collider are reviewed elsewhere [8,9]. Future LC Higgs studies will concentrate on more detailed detector simulations reflecting the progress in detector technologies, the second phase of a LC with center-of-mass energies up to 1 TeV for TESLA, and new theoretical developments, such as extra dimensions.

*On behalf of the ECFA/DESY Higgs boson study group with contributions from US and Asian Higgs study groups.


doi:10.1016/S0920-5632(02)02105-9

## 2. STANDARD MODEL PHYSICS

### 2.1. Higgs boson production mechanism

The expected SM Higgs boson production rate has a very large significance over the background ($\sigma \equiv N_{\rm sig}/\sqrt{N_{\rm bg}}$) [10]. More than $100\sigma$ is obtainable in the $\rm H \to b\bar{b}$ decay channel. For heavier Higgs bosons, the WW decay mode takes over. In relation to the LHC [11], a LC has a much larger sensitivity in the lower mass range, while the LHC can probe heavier Higgs bosons.

The sensitivity to the general production cross section has been studied [10]. The expected SM production cross section is much larger than the $5\sigma$ sensitivity to the cross section over a wide mass range. Even for models beyond the SM, the sensitivity to the cross section is much better than the expected minimal production cross section.

The Higgs boson will be produced via Higgs-strahlung and WW fusion by the two processes

$$\rm e^+e^- \to HZ \to H\nu\bar{\nu} \to b\bar{b}\nu\bar{\nu} \quad and$$
$$\rm e^+e^- \to \nu\bar{\nu}WW \to \nu\bar{\nu}H \to \nu\bar{\nu}b\bar{b}.$$

These production mechanisms can be distinguished by fitting the missing mass distribution [12,13].

### 2.2. Indirect and direct branching ratio measurements

A LC will perform very high precision measurements of the Higgs boson branching fractions. The underlying production process is the Higgs-strahlung,

$$\rm e^+e^- \to HZ \to H\ell^+\ell^-,$$

Table 1
Expected precision on branching ratios (in %) for a 120 GeV Higgs boson from the direct method and a preliminary combination with the indirect method.

| Decay | SM $BR$ | $\Delta BR_{\rm d}/BR_{\rm d}$ | $\Delta BR_{\rm c}/BR_{\rm c}$ |
|---|---|---|---|
| bb | 68 | 1.9 | 1.5 |
| $\tau\tau$ | 6.9 | 7.1 | 4.1 |
| cc | 3.1 | 8.1 | 5.8 |
| gluons | 7.0 | 4.8 | 3.6 |
| $\gamma\gamma$ | 0.22 | 35 | 21 |
| WW$^\star$ | 13 | 3.6 | 2.7 |

where the associated Z boson decays into a lepton pair.

In the indirect method, the inclusive production cross section

$$\sigma_{\rm inc} = \sigma_{\rm HZ} BR({\rm Z} \to \ell^+\ell^-)$$

is determined. This measurement is independent of the Higgs boson decay mode. The mass recoiling to the lepton pair corresponds to the Higgs boson mass. An individual Higgs boson decay cross section is measured

$$\sigma({\rm X}) = \sigma_{\rm HZ} BR({\rm Z} \to {\rm Y}) BR({\rm H} \to {\rm X}).$$

By taking the ratio of both cross sections the Higgs branching ratio can be determined, since the LEP experiments measured the Z decay branching fractions with high precision.

The Higgs boson branching ratios can also be determined directly from the HZ(Z $\to \ell^+\ell^-$) event sample [14]. The resulting precision on the Higgs boson decay branching ratios as well as a preliminary combination of both methods [14] are listed in Table 1.

In relation to the LHC, a LC will achieve a much higher precision and will cover all decay modes. This allows precision testing of the fundamental relation between the Yukawa coupling and the Higgs boson mass:

$$g_{\rm Hff} \propto m_{\rm f}.$$

### 2.3. Mass and width determination

Further properties of the Higgs boson, like its mass and decay width, can be determined with high precision. A study of a 240 GeV SM Higgs boson in the reactions

$$e^+e^- \to {\rm HZ} \to {\rm ZWW} \quad \text{and}$$

$$e^+e^- \to {\rm HZ} \to {\rm ZZZ}$$

has been performed [15]. A very clear signal with only a few background events is expected. The fit of the reconstructed mass peak leads to a mass resolution of 0.08% and an 11% error on the total decay width.

This result gives slightly higher sensitivity compared to results from other reactions on the determination of the Higgs boson mass [16].

### 2.4. Characterization of the Higgs boson potential

A LC will be able to measure fundamental properties of the Higgs boson potential. The Higgs boson can decay into a pair of Higgs bosons:

$$e^+e^- \to {\rm HZ} \to {\rm HHZ}.$$

The self coupling interaction can probe the shape of the Higgs boson potential through the relation

$$g_{\rm HHH} = 3m_{\rm H}^2/2v,$$

where $v = 246$ GeV. A sensitivity of $\Delta g/g = 29\%$ was obtained for $m_{\rm H} = 120$ GeV, $\sqrt{s} = 800$ GeV and $\mathcal{L} = 1000$ fb$^{-1}$ [17,18]. Higher sensitivity of $\Delta g/g = 7\%$ could be reached for a LC with $\sqrt{s} = 3$ TeV and $\mathcal{L} = 5000$ fb$^{-1}$.

### 2.5. Higgs-strahlung off top quarks

Owing to the strong coupling of Higgs bosons to top quarks, the radiation of Higgs boson off top quarks is a possible production mechanism. The decay modes involving a pair of b quarks and W bosons were studied [19] in the reactions

$$e^+e^- \to {\rm t\bar{t}H} \to {\rm t\bar{t}b\bar{b}} \quad \text{and}$$

$$e^+e^- \to {\rm t\bar{t}H} \to {\rm t\bar{t}WW}.$$

A challenge is the precision determination of the background to a level of 5% uncertainty, leading to

$$\Delta g_{\rm ttH}/g_{\rm ttH} = 7.5\% \text{ to } 20\%.$$

A previous study at 120 GeV achieved slightly higher sensitivity [20].

## 3. BEYOND THE STANDARD MODEL

### 3.1. Charged Higgs bosons

The discovery of charged Higgs bosons would immediately prove that physics beyond the SM exists. The reaction

$$e^+e^- \to Z \to H^+H^- \to t\bar{b}\bar{t}b$$

can be observed at a LC [21] and recent high-luminosity simulations [22] for $m_{H^\pm} = 300$ GeV, $\sqrt{s} = 800$ GeV and $\mathcal{L} = 1000$ fb$^{-1}$ show that the production cross section times branching ratio can be measured very precisely:

$$\Delta(\sigma BR(H^+ \to t\bar{b}))/\sigma BR(H^+ \to t\bar{b}) = 8.8\%.$$

### 3.2. Determination of tan $\beta$

The ratio of the vacuum expectation values $\tan\beta$ can be measured with several methods. The pseudoscalar Higgs boson, A, could be produced via radiation off a pair of b-quarks:

$$e^+e^- \to b\bar{b} \to b\bar{b}A \to b\bar{b}b\bar{b}.$$

The $b\bar{b}A$ coupling is proportional to $\tan\beta$ and thus the expected production rate is proportional to $\tan^2\beta$. A precision better than 10% can be achieved for a Higgs boson mass of 100 GeV and large $\tan\beta$ values [23]. The sensitivity decreases with increasing Higgs boson masses and decreasing $\tan\beta$ values. This study assumes a luminosity of 2000 fb$^{-1}$, corresponding to several years of data taking.

There are further methods to determine $\tan\beta$:

- The $b\bar{b}\bar{b}b$ rate from HA production

  $$e^+e^- \to HA \to b\bar{b}b\bar{b}$$

  can be exploited. While the HA production rate is almost independent of $\tan\beta$ the sensitivity is achieved owing to the variation of the decay branching ratios with $\tan\beta$.

- The value of $\tan\beta$ can also be determined from the H and A decay widths, which can be obtained from the previous reaction.

- The pair-production rate and total decay width of charged Higgs bosons can contribute to the determination of $\tan\beta$. Charged Higgs boson production can also be used at the LHC to measure $\tan\beta$ [24].

### 3.3. Invisible Higgs boson decays

Among other possibilities, invisible Higgs boson decays could occur from the reaction involving neutralinos:

$$e^+e^- \to HZ \to \tilde{\chi}^0\tilde{\chi}^0 Z.$$

In this case the Higgs boson can be reconstructed from the recoiling mass of the visible decay products

$$m_H = m_Z^{recoil}.$$

At LEP all Z decay modes involving charged fermions contributed to the search, while in a recent LC study [25] so far only the hadronic decay mode $Z \to q\bar{q}$ has been investigated. For Higgs boson branching ratios $BR_i$ into invisible decay products larger than 20% and a SM production rate,

$$\Delta BR_i/BR_i < 4\%$$

can be achieved.

### 3.4. Higgs boson parity

After the discovery of one or several Higgs bosons, it is very important to determine the parity of the Higgs bosons and distinguish a CP-even H boson from a CP-odd A boson. This could be achieved by investigating the Higgs boson decay properties into $\tau$-leptons. The subsequent decay of the $\tau$'s into $\rho$'s and pions through the reaction

$$H/A \to \tau^+\tau^- \to \rho^+\bar{\nu}_\tau\rho^-\nu_\tau \to \pi^+\pi^0\bar{\nu}_\tau\pi^-\pi^0\nu_\tau$$

clearly distinguishes the two parity states [26].

### 3.5. Distinction of Higgs boson models

The distinction of Higgs boson models is very important and could be based on precision branching ratio measurements. The ratio of the b-quark and $\tau$-lepton decay modes

$$R \equiv BR(H/A \to b\bar{b})/BR(H/A \to \tau^+\tau^-),$$

normalized to the SM expectation, can be used to distinguish a general two Higgs doublet model (2HDM) from the Minimal Supersymmetric extension of the SM (MSSM). Large deviations from $R = 1$ are expected in the MSSM [27]. In relation to the LHC, where the models can only be distinguished for $\tan\beta > 25$, a LC covers the entire $\tan\beta$ range.

Important predictions on the pseudoscalar Higgs boson mass can be made from the precision measurements of the scalar Higgs boson branching ratios:

$$BR(\mathrm{H} \to \mathrm{b\bar{b}})/BR(\mathrm{H} \to \mathrm{WW}).$$

A LC obtains a precision normalized to the SM expectation of less than 3.5%, while the expected precision at the LHC is less than 20%. For a 400 GeV pseudoscalar Higgs boson A, its mass can be predicted at a LC with an error of about 50 GeV, while at the LHC only a lower mass limit can be set [28].

Beyond the MSSM, the Higgs boson particle spectrum is enriched for example in the framework of the Non-Minimal Supersymmetric extension of the SM (NMSSM) [29] where an extra Higgs boson singlet $\lambda N H_1 H_2$ is present. Such a model could be distinguished from the MSSM by precision measurements of the Higgs boson masses and comparison with the predictions. Moreover, additional light neutral Higgs bosons may be observed and the mass sum rules are modified, leading, for example, to a reduced mass of the charged Higgs boson.

## 4. CONCLUSIONS

After a first discovery and initial precision measurements at the Tevatron or the LHC in some decay modes, already in the first phase of a LC, many Higgs boson decay modes will be measured with very high precision. The precise LC data will allow the determination of the nature of the Higgs sector. Models like the SM, the general 2HDM, the MSSM and the NMSSM will be distinguished for a wide range of parameters. The underlying mechanism of symmetry breaking and mass generation will be tested. Like for the top quark (LEP mass prediction, Tevatron observation), important consistencies of the model can be probed with combined LC and LHC physics. After 10 years of preparatory studies the LC has a solid case and the high-energy physics community is prepared to answer fundamental questions over the coming decades.

## ACKNOWLEDGEMENTS

I would like to thank M. Battaglia, J.-C. Brient, K. Desch, F. Gianotti, A. Gay, E. Gross, J. Guasch, A. Kiiskinen, R. Van Kooten, N. Meyer, D. Miller, S. Peñaranda, J. Schreiber, M. Schumacher, S. Yamashita, and P. Zerwas for help with the preparation of this presentation.

## REFERENCES

1. A. Sopczak, Phys. Rep. **359** (2002) 169.
2. LEP Higgs Working Group, July 2002.
3. A. Sopczak, ABS684, ICHEP02, Amsterdam, 2002.
4. TESLA TDR, DESY (2001).
5. American Physical Society study on the future of particle physics, Snowmass, USA, July 2001.
6. Int. Workshop on LC, Jeju, Korea, Aug. 2002.
7. A. Sopczak, hep-ph/0209372.
8. G. Moortgat-Pick, ABS23, ICHEP02, Amsterdam, 2002.
9. S. Soldner-Rembold, ABS770, ICHEP02, Amsterdam, 2002.
10. S. Yamashita et. al., hep-ph/0109166.
11. F. Gianotti et. al., LHCC, July 2000.
12. K. Desch, N. Meyer, LC-PHSM-2001-025.
13. R. Van Kooten, LC workshop, Baltimore, March 2001.
14. J.-C. Brient, LC-PHSM-2002-003.
15. N. Meyer, K. Desch, LC workshop, St. Malo, March 2002; www-dapnia.cea.fr/ecfadesy-stmalo.
16. V. Drollinger, A. Sopczak, EPJdirect **C-N 1** (2001) 1.
17. C. Castanier, P. Gay, P. Lutz, J. Orloff, hep-ex/0101028.
18. M. Battaglia, E. Boos, W. Yao, hep-ph/0111276.
19. A. Gay, LC workshop, St. Malo, March 2002; www-dapnia.cea.fr/ecfadesy-stmalo.
20. A. Juste, G. Merino, hep-ph/9910301.
21. A. Sopczak, Z. Phys. **C 65** (1995) 449.
22. M. Battaglia, A. Ferrari, A. Kiiskinen, T. Mäki, hep-ex/0112015.
23. J. Gunion, T. Han, J. Jiang, S. Mrenna, A. Sopczak, hep-ph/0112334.
24. K.A. Assamagan, Y. Coadou, A. Deandrea, EPJdirect **C 9** (2002) 1.
25. M. Schumacher, LC workshop, Cracow, Sep. 2001.
26. G.R. Bower, T. Pierzchal, Z. Wąs, M. Worek, hep-ph/0204292.
27. J. Guasch, W. Hollik, S. Peñaranda, Phys. Lett. **B 515** (2001) 367.
28. E.Gross, S.Heinemeyer, G.Weiglein (in preparation).
29. D. Miller, R.B. Nevzorov, P. Zerwas, SUSY02, DESY, June 2002; www.desy.de/susy02.

# Session 5

# QCD: Hard Interactions

*Session conveners:*
*C. Niebuhr, DESY Hamburg*
*L. Reina, Florida State University, Tallahassee*
*J. Huston, Michigan State University*

ELSEVIER

Proceedings of ICHEP 2002, pp. 247–250
S. Bentvelsen, P. de Jong, J. Koch and E. Laenen (Editors)

31st INTERNATIONAL CONFERENCE ON HIGH ENERGY PHYSICS AMSTERDAM

# Structure Function Results from ZEUS

A. Kappes[a] for the ZEUS Collaboration

[a]Physikalisches Institut der Universität Bonn, Nußallee 12, D-53115 Bonn

This contribution presents recent ZEUS results on proton structure functions at HERA. The inclusive $\phi(1020)$-meson cross section was measured, and it was used to determine the $s$-quark content of the proton. The structure function $F_2$ was extracted using initial-state radiative events. Neutral and charged current cross sections were used to extract the structure function $xF_3$ and measure the mass of the $W$ boson, respectively. A NLO QCD fit to ZEUS data and fixed target cross sections was employed to determine the parton density functions of the quarks and of the gluon inside the proton.

## 1. Deep inelastic $ep$ scattering (DIS)

The Hadron-Elektron-Ringanlage HERA at DESY in Hamburg accelerates electrons/positrons and protons to energies of $E_e = 27.5\,\text{GeV}$ and $E_p = 920\,\text{GeV}$ (900 GeV before 1998), yielding a center-of-mass energy, $\sqrt{s}$, of 318 GeV (300 GeV before 1998). In DIS events the interaction between the two particles can either occur via a neutral (NC, $\gamma$ or $Z$) or a charged current (CC, $W$) process. The reaction can be described by two of the three variables $Q^2$, $x$-Bjorken and $y$. Here, $Q^2$ is the negative of the squared 4-momentum transfer between lepton and proton and $y$ is the inelasticity.

The NC cross section can be written as

$$\frac{d^2\sigma(e^{\pm}p)}{dx\,dQ^2} = \frac{2\pi\alpha^2}{x\,Q^4}\left[Y_+ F_2{}^{\rm NC} \mp Y_- xF_3{}^{\rm NC} - y^2 F_L^{\rm NC}\right]$$

$$(F_2^{\rm NC}, xF_3^{\rm NC}) = x \sum_{q=u\ldots b} (A_f, B_f)\,[(q+\bar{q}, q-\bar{q})]\,,$$

where $q$ and $\bar{q}$ are the parton density functions (PDFs) of quarks and antiquarks and $Y_\pm = 1 \pm (1-y)^2$. $A_f$ and $B_f$ depend on $Q^2$ only and contain the vector and axial-vector couplings as well as the propagator terms. The reduced cross section is defined as $\tilde{\sigma}^{\rm NC} = \frac{d^2\sigma^{\rm NC}}{dx\,dQ^2} \times \frac{xQ^4}{2\pi\alpha^2}\frac{1}{Y_+}$.

The CC cross section is similar to that of NC

$$\frac{d^2\sigma(e^-p)}{dx\,dQ^2} = \frac{G_F^2}{4\pi x}\frac{M_W^4}{(Q^2+M_W^2)^2} \times \left[Y_+ F_2{}^{\rm CC} + Y_- xF_3{}^{\rm CC} - y^2 F_L^{\rm CC}\right]$$

$$(F_2^{\rm CC}, xF_3^{\rm CC}) = x\left[(u+c+\bar{d}+\bar{s}, u+c-\bar{d}-\bar{s})\right],$$

however, for electrons the $W$ couples only to positively charged quarks. The reduced cross section is defined as $\tilde{\sigma}^{\rm CC} = \frac{d^2\sigma^{\rm CC}}{dx\,dQ^2} \times \frac{4\pi x}{G_F^2}\frac{(Q^2+M_W^2)^2}{M_W^4}$.

## 2. Inclusive $\phi(1020)$-meson production

$\phi$ mesons in DIS can be either formed directly from $s$ quarks inside the proton or indirectly from $s$ quarks produced via BGF or in the hadronization process. Generally, the cross section for indirect production is much larger than that for the direct one, rendering the measurement of the $s$-quark content of the proton rather difficult. However, for direct $\phi$ production the variable $x_p(\phi) = 2p(\phi)/Q$ clusters around 1, where $p(\phi)$ is the $\phi$ momentum in the Breit frame. Therefore, at high $x_p(\phi)$ inclusive $\phi$ production is sensitive to the $s$-quark content of the proton.

The $e^+p$ data set used was recorded during the run periods 1995–97 at a center of mass energy of 300 GeV and corresponds to an integrated luminosity of $\mathcal{L} = 45.0\,\text{pb}^{-1}$. The cross section was measured in the range $10 < Q^2 < 100\,\text{GeV}^2$ and $2\times10^{-4} < x < 10^{-2}$. The preliminary cross section is $\sigma(e^+p \to e^+\phi p) = 0.506 \pm 0.022\,\text{nb}$ [1]. In Fig. 1 for $x_p \lesssim 0.8$, where the indirect contribution to $\phi$ production is expected to dominate and theoretical predictions are somewhat uncertain, large differences between the various models can be observed. However, for $x_p > 0.8$ the differ-


doi:10.1016/S0920-5632(02)02109-6

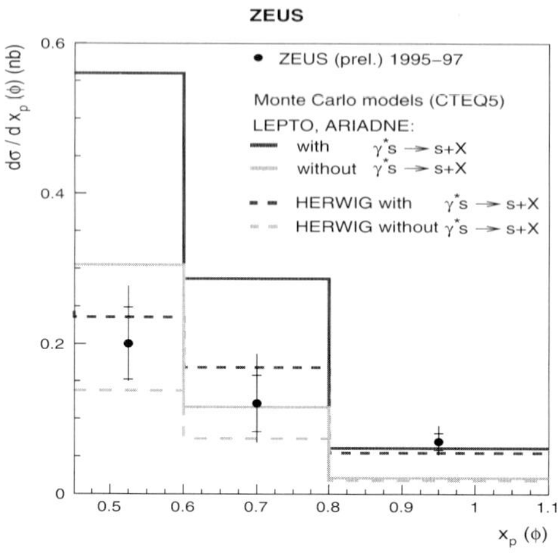


Figure 1. $d\sigma/dx_p$ as a function of $x_p$. The dark (light) lines correspond to predictions using LEPTO/ARIADNE and HERWIG with (without) inclusion of an $s$-quark content of the proton.

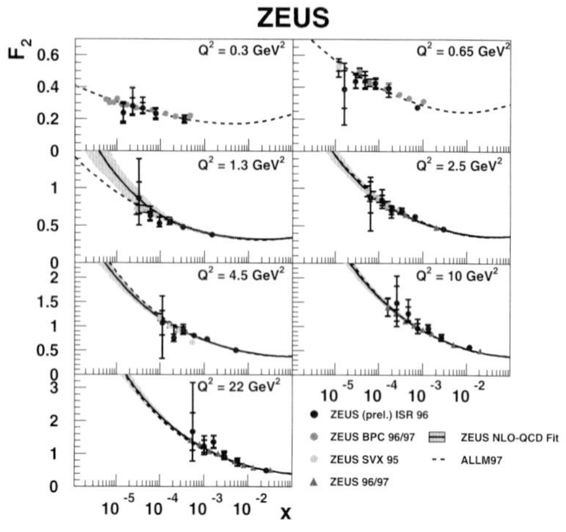


Figure 2. $F_2$ as a function of $x$ in different $Q^2$ bins. In addition to the ISR data also former measurements are shown together with theoretical predictions.

ences among models with $s$-quark content on the one hand and among those without $s$-quark content on the other hand are small compared to the measurement error. Here, models featuring an $s$-quark content of the proton are clearly required by the data.

## 3. $F_2$ from initial-state radiative events

The acceptance of the ZEUS main detector for scattered electrons is limited towards small scattering angles to $\theta^\star \gtrsim 4^\circ$ ($Q^2 \gtrsim 2\,\mathrm{GeV}^2$). One possibility to measure $F_2$ at even lower $Q^2$ values is to use initial-state radiative (ISR) events where a photon was radiated from the incoming lepton. Due to the energy carried by the photon, $E_\gamma$, $\sqrt{s}$ of the reaction is reduced by a factor $(E_e - E_\gamma)/E_e$. As a consequence, electrons with low $Q^2$ are scattered into the main detector.

The $F_2$ measurement [2] was performed on a small fraction of the 1996 $e^+p$ data set ($\mathcal{L} = 3.8\,\mathrm{pb}^{-1}$). It covers the range $0.3 < Q^2 < 22\,\mathrm{GeV}^2$, $1\times10^{-5} < x < 3\times10^{-2}$ and fills the gap in the $F_2$ coverage of the kinematic plane around $1\,\mathrm{GeV}^2$ and $x > 3\times10^{-4}$. Fig. 2 shows that the ISR $F_2$ values agree well with other measurements both in the lower and upper $Q^2$ region. Theoretical predictions based on the ZEUS NLO-QCD fit are in good agreement with the ISR data for $Q^2 > 2.5\,\mathrm{GeV}^2$. For lower $Q^2$ a Regge inspired fit (ALLM97) yields a good description of the data. ISR events will allow a direct measurements of $F_L$ in the future.

## 4. High-$Q^2$ NC cross sections from $e^-p$ DIS

Electromagnetic interactions are invariant under $P$ and $C$ transformations and hence, $\sigma^{\mathrm{NC}}(e^-p) \approx \sigma^{\mathrm{NC}}(e^+p)$ for $Q^2 \ll M_Z^2$ ($\gamma$-only exchange), where $M_Z$ is the mass of the $Z$ boson. However, weak interactions do not preserve $P$ and $C$ but only approximately $CP$. As a consequence $\sigma^{\mathrm{NC}}(e^-p) > \sigma^{\mathrm{NC}}(e^+p)$ for $Q^2 \gtrsim M_Z^2$, where in addition to the $\gamma$ exchange also the $Z$ exchange gives a considerable contribution to the cross section. The parity violating terms of $\sigma^{\mathrm{NC}}(e^\pm p)$ are combined in the structure function $xF_3$, whereas those in $F_2$ are invariant under $P$. A comparison of $e^-p$ and $e^+p$ cross sections yields a direct test of the electroweak sector of the Standard Model.

The $e^-p$ data set used was recorded in 1998–99 ($\mathcal{L} = 15.9\,\mathrm{pb}^{-1}$). The NC cross sections were measured in the range $185 < Q^2 < 50\,000\,\mathrm{GeV}^2$ and $0.0037 < x < 0.75$ [3]. Figure 3 shows good agreement between the theory computed with the ZEUS fit to the 1996–97 data (ZEUS-S) and the

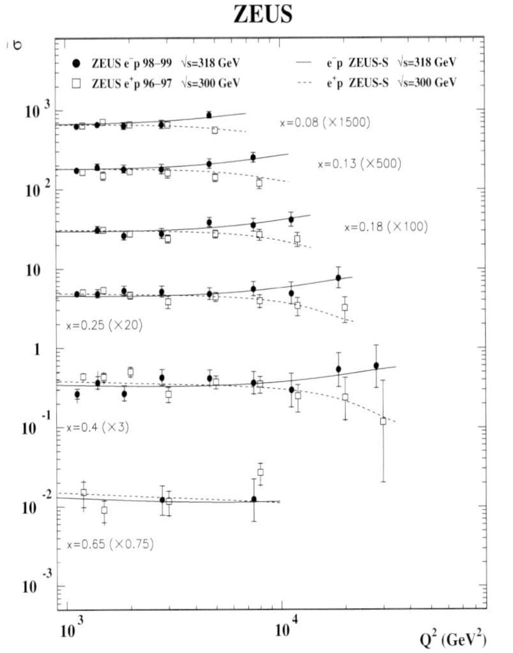


Figure 3. The reduced cross section $\tilde{\sigma}$ as a function of $Q^2$ in different $x$ bins.

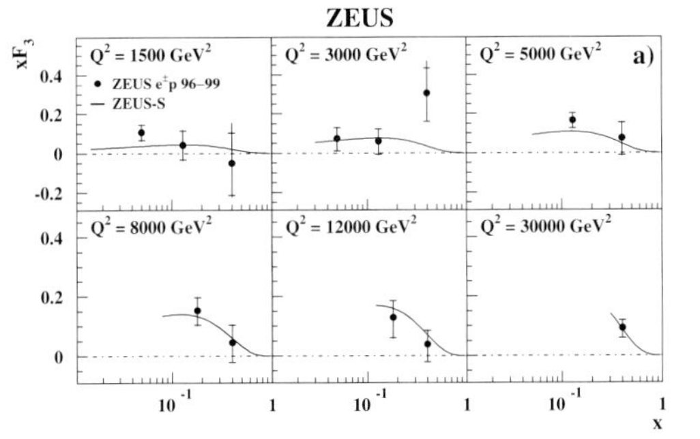


Figure 4. $xF_3$ as a function of $x$ in $Q^2$ bins.

data, where only points with $Q^2 > 1\,000\,\text{GeV}^2$ are shown. Also plotted is the $e^+p$ measurement from 1996–97 data based on $30\,\text{pb}^{-1}$. A comparison of $e^-p$ and $e^+p$ cross sections shows an increase of the former and a decrease of the latter as expected from the $Z$ exchange. The structure function $xF_3$ extracted from these data is shown in Fig. 4 and is compared to the prediction from the ZEUS-S NLO fit. The measurement yields $xF_3 \neq 0$ and the data is well described by the theory using ZEUS-S. The large errors of the measurement are dominated by the low $e^-p$ statistics.

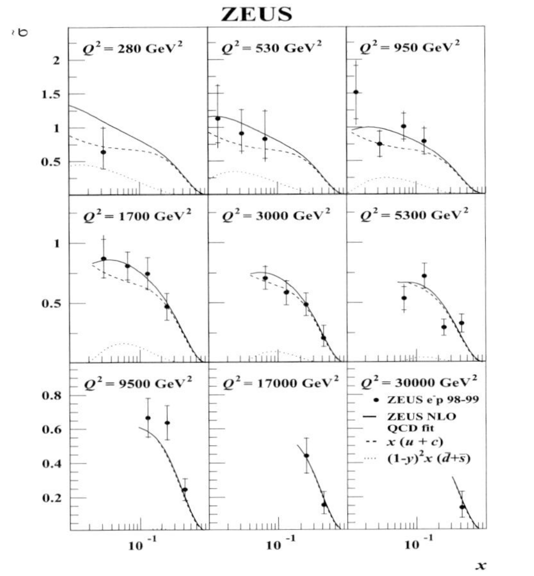


Figure 5. Reduced charged current cross section as a function of $x$ in different $Q^2$ bins. Shown are the measured and expected cross sections and the contributions from $u + c$ and $\bar{d} + \bar{s}$ quarks.

## 5. High-$Q^2$ CC cross sections from $e^-p$ DIS

In contrast to NC reactions CC reactions are purely weak and exhibit therefore a more direct dependence on the weak parameters. Also, due to the charge-selective coupling of the $W$ boson, measuring CC cross sections yields information about the flavor content of the proton.

The $e^-p$ data set used for the measurement [4] was recorded in 1998–99 ($\mathcal{L} = 16.4\,\text{pb}^{-1}$). The measurement was performed for $280 < Q^2 < 30\,000\,\text{GeV}^2$ and $0.015 < x < 0.42$. Figure 5 shows good agreement between data and predictions using ZEUS-S. At high $x$ the cross section is dominated by the $u$ (valence) quarks and a measurement in this region allows a direct determination of the $u$ quark PDF of the proton. At low $x$ the contribution from the $(\bar{d} + \bar{s})$ sea is also small due to the helicity structure of the reaction. As the CC cross section directly depends on the mass of the $W$ boson, $M_W$, a fit of $M_W$ to $d\sigma/dQ^2 \propto M_W^4/\left(Q^2 + M_W^2\right)^2$ was per-

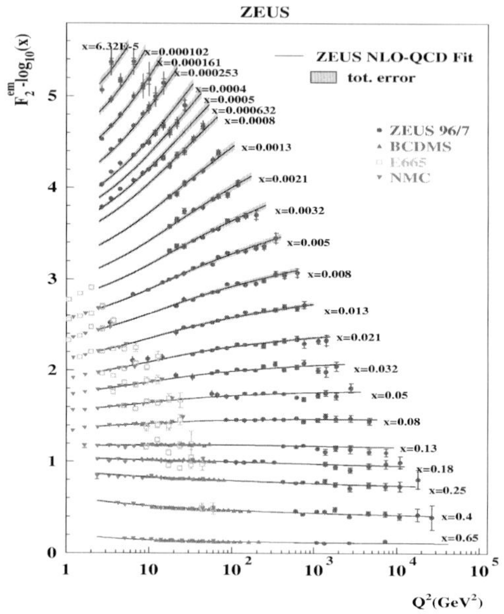


Figure 6. $F_2$ as a function of $Q^2$ in bins of $x$.

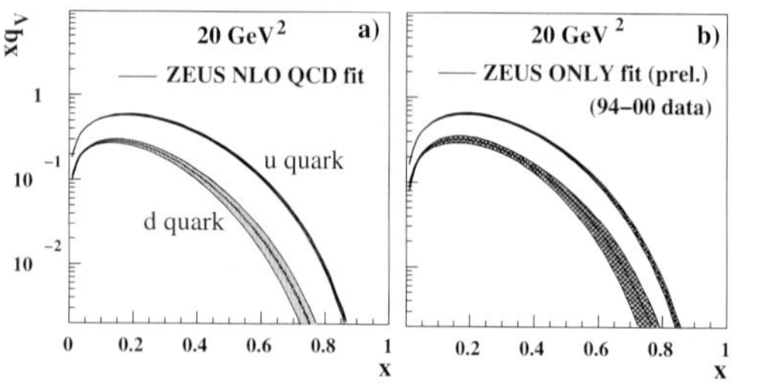


Figure 7. Valence quark distributions from a) the standard ZEUS fit and b) the fit to ZEUS data only, including the 1998–2000 data.

formed. The fit yielded $M_W = 80.3 \pm 2.1\,(\text{stat}) \pm 1.2\,(\text{sys}) \pm 1\,(\text{PDF})$ GeV. A comparison with the fit to the $e^+p$ data ($M_W = 81.4^{+2.7+2.0+3.3}_{-2.6-2.0-3.0}$ GeV) shows good agreement within errors. The systematic and PDF uncertainties are much smaller for the more recent measurement. In case of the systematic uncertainty, this is due to a much improved understanding of the energy scale uncertainty. On the other hand, the $u$-quark PDF, dominating the $e^-p$ cross section, is much better known than the $d$-quark PDF which mainly determines the $e^+p$ cross section. The $M_W$ values, measured in the space-like region, are in good agreement with and complementary to the time-like measurements from LEP and Tevatron.

## 6. NLO QCD analysis of data on DIS

The global NLO QCD fit to ZEUS data [5], including measurements until 1997, and to fixed target data from BCDMS, NMC, E665 and CCFR uses the full information on point-to-point correlations of systematic uncertainties. The data used cover the kinematic range $2.5 < Q^2 < 30\,000\,\text{GeV}^2$ and $6.3 \times 10^{-5} < x < 0.65$. The fit, displayed in Fig. 6, gives an excellent description of the data down to $Q^2 \approx 0.8\,\text{GeV}^2$. The new precise ZEUS data yield information on the gluon distribution and the quark density at low $x$. A fit to the data with $\alpha_s$ free gives $\alpha_s = 0.1166 \pm 0.0053$. The uncertainty of the PDFs is dominated by the correlated systematics. The ZEUS fit is compatible with both the MRST2001 and CTEQ6M parameterizations.

A fit to ZEUS only data including the newest data from 1998–2000, displayed in Fig. 7 b) yields similar uncertainties on the valence quark PDFs as the standard fit, shown in Fig. 7 a), which includes the fixed target data.

## 7. Summary

Recent precision DIS data from ZEUS complete the coverage of a large area of the kinematic plane spanning 6 orders of magnitude in $x$ and $Q^2$. All ZEUS results are fitted consistently by NLO QCD and yield precise PDFs.

## REFERENCES

1. ZEUS Coll., Abstract 850, 31st International Conference on High Energy Physics, Amsterdam, Netherlands, 2002.
2. ZEUS Coll., Abstract 771, 31st International Conference on High Energy Physics, Amsterdam, Netherlands, 2002.
3. ZEUS Coll., S. Chekanov et al., Preprint DESY-02-113, 2002, ICHEP abstract 766.
4. ZEUS Coll., S. Chekanov et al., Phys. Lett. **B 539**, 197 (2002), ICHEP abstract 763.
5. ZEUS Coll., S. Chekanov et al., Preprint DESY-02-105, 2002, ICHEP abstract 765.

ELSEVIER

Proceedings of ICHEP 2002, pp. 251–254
*S. Bentvelsen, P. de Jong, J. Koch and E. Laenen (Editors)*

31st INTERNATIONAL CONFERENCE ON HIGH ENERGY PHYSICS AMSTERDAM

# Structure function results from H1

Z. Zhang[a]*

[a]Laboratoire de l'Accélérateur Linéaire, BP 34, IN2P3-CNRS et Université de Paris-Sud, 91898 Orsay Cedex, France

New structure function results from H1 are presented. The measurements cover a huge kinematical range for $Q^2$, the four momentum transfer squared, from $0.35\,\mathrm{GeV}^2$ to $30\,000\,\mathrm{GeV}^2$, and for Bjorken $x$ between $\sim 5 \cdot 10^{-6}$ and 0.65. At $Q^2 > 100\,\mathrm{GeV}^2$, full HERA I data have been analyzed. The data are compared with a new QCD analysis. The impact of the HERA I data on the parton density functions is discussed.

## 1. INTRODUCTION

Deep inelastic scattering (DIS) experiments have played an important role in the understanding of the partonic structure of matter and in establishing the strong interaction sector of the Standard Model (SM): QCD.

In the past decade, the HERA experiments H1 and ZEUS have been providing more and more precise data for modern global fits allowing parton density functions (PDFs) be determined with steadily increasing precision. The precision of the PDFs is of importance as it is needed to have reliable predictions for precision measurements and for new physics searches at future hadron colliders such as LHC.

In this talk, I shall first present the latest new structure function measurements from H1, then try to answer two following questions: i) Can HERA I cross section or structure function (SF) data measured so far be described by QCD? and ii) What is the impact of the HERA I data?

## 2. NEW MEASUREMENTS AND IMPACT OF HERA I DATA

Four new preliminary measurements [1] from H1 have been submitted to this conference. Thanks to the increased proton beam energy from 820 GeV to 920 GeV since 1998 (the center-of-mass energy $\sqrt{s}$ is accordingly increased from 300 GeV to 320 GeV), the fully installed backward silicon tracker, the new capability of triggering low energy scattered electrons[2] down to 3 GeV, and the special runs where data were taken by shifting the nominal interaction vertex in the proton beam direction by 70 cm, the kinematical domain has been substantially extended with respect to previous measurements both to lower $Q^2$ and to smaller $x$. On the other extreme at high $Q^2$ and large $x$, the new measurements of both neutral current (NC) and charged current (CC) processes are obtained with improved precision using the highest data sample taken before the HERA machine was upgraded for the second phase.

Part of the new measurements at low $Q^2$, the so-called reduced cross sections $\tilde{\sigma}$ (or $\sigma_{\rm r}$), are shown in Fig. 1. The reduced cross section for the NC interaction may be expressed in terms of SFs as

$$\tilde{\sigma}_{\rm NC}(e^{\pm}p) = F_2 - \frac{y^2}{Y_+}F_L \mp \frac{Y_-}{Y_+}xF_3 \qquad (1)$$

where $Y_{\pm} = 1 \pm (1-y)^2$ with $y = Q^2/(xs)$, $F_2$ is the dominant contribution, the contribution of the longitudinal SF $F_L$ is suppressed at low $y$ by $y^2/Y_+$, and the SF $xF_3$, which arises mainly from the photon-$Z^0$ interference and has been previously determined [5] at $Q^2 \geq 1500\,\mathrm{GeV}^2$, becomes non-negligible only at such large $Q^2$ values. The reduced cross section differs from the experimentally measured double differential cross

*On behalf of the H1 collaboration.

[2]It refers generically to both electrons and positrons unless stated otherwise.


doi:10.1016/S0920-5632(02)02115-1

section $d^2\sigma/dxdQ^2$ by a kinematical factor of $2\pi\alpha^2 Y_+/xQ^4$, where $\alpha$ is the fine structure constant.

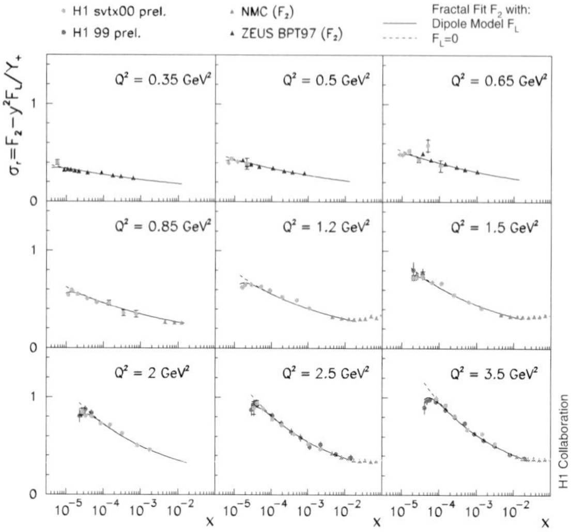


Figure 1. Preliminary reduced cross section (see text) measurements from H1 shown as a function of $x$ for $Q^2 \leq 3.5\,\mathrm{GeV}^2$. Also shown are measurements from NMC [2] and ZEUS [3] and expectations based on a phenomenological model [4].

From Fig. 1, one observes that the new measurements extend now down to $x \sim 5 \cdot 10^{-6}$ at $Q^2 = 0.35\,\mathrm{GeV}^2$. The rise of $F_2$ as $x$ decreases, the first important observation at HERA and being related to increasing gluon density $xg$ at small $x$, persists even at $Q^2 < 1\,\mathrm{GeV}^2$ although with a reduced magnitude. The same observation is also obtained from the measurement of $\mathrm{dln}\,F_2/\mathrm{dln}\,x$ (not shown here).

As $Q^2$ increases, the measured $\sigma_\mathrm{r}$ is observed to deviate from the dashed curves at high $y$ in which the $F_L$ is set to zero, but agree with the full curves where a non-zero $F_L$ according to a dipole model is included. Such a sensitivity to $F_L$ has been exploited by H1 to determine experimentally $F_L$, which is shown in Fig. 2. The new determination (shown in triangle) thus extends the previous determination [6] to smaller $x$ and higher $Q^2$. It should be pointed out that such a determination, though not a direct measurement which needs different beam energies, does provide a non-trivial consistency test as the determination gives a direct measure of $xg$ while the expectation from a QCD fit was based on an indirect $xg$ derived from the scaling violation of the $F_2$.

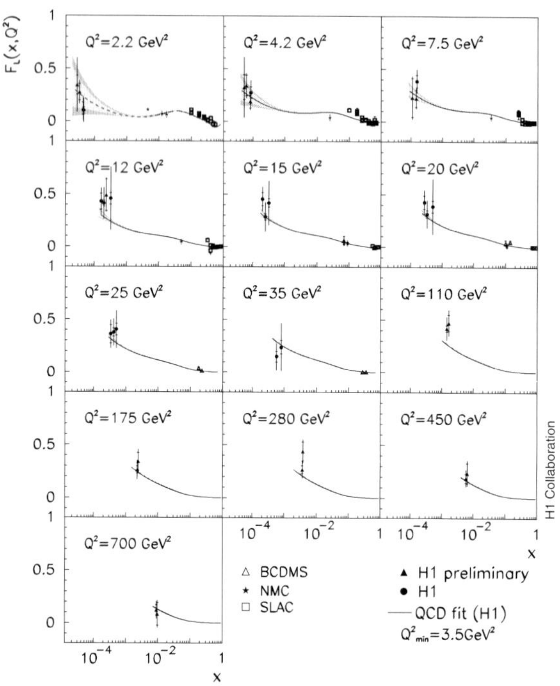


Figure 2. Preliminary determination of $F_L$ shown together with those by H1 [6] and BCDMS, NMC, and SLAC [7]. Also shown is the SM expectation - QCD fit (H1) [6].

Moving to even higher $Q^2$ ($100$–$30\,000\,\mathrm{GeV}^2$) region, new preliminary cross section measurements based on the $e^+p$ 1999-2000 at $\sqrt{s} = 320\,\mathrm{GeV}$ [1] are combined with those measured with the $e^+p$ 1994-1997 at 320 GeV [8] taking into account a small correction due to the increased $\sqrt{s}$. The combined cross sections in $e^+p$ scattering can be compared with those in $e^-p$ scattering [5]. In Fig. 3, the single differential NC and CC cross sections from H1 [1,5] and ZEUS [10,5] are shown and compared with the SM expectation from a global QCD fit (CTEQ5 [9]). While the NC $e^+p$ and $e^-p$ cross sections are undistinguishable at low $Q^2$ as expected from the photon

($\gamma$) exchange, they are different at high $Q^2$ due to the positive (negative) $\gamma Z^0$ exchange in $e^-p(e^+p)$ interactions. The CC $e^+p$ and $e^-p$ cross sections are different over the entire $Q^2$ region measured originating from different quarks probed by negatively and positively charged W bosons exchanged in the interactions and from different helicity structures involved. At $Q^2 \sim M_{Z,W}^2$, all four cross sections (NC, CC and $e^+p$, $e^-p$) are comparable demonstrating the universal coupling strength of electroweak interactions. It is also remarkable that the cross sections which vary over many orders of magnitude are described by the global fit which to some extent is independent of the data.

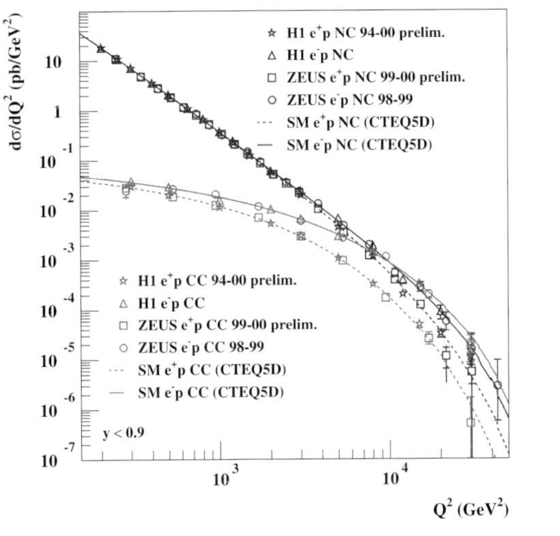


Figure 3. NC and CC cross sections $d\sigma/dQ^2$ measured with $e^+p$ and $e^-p$ data by H1 [1,5] and ZEUS [10,5] in comparison with the SM expectations from CTEQ5 [9].

HERA data have been important for constraining the gluon density $xg$ at small $x$. Indeed, in an earlier QCD analysis performed by H1 using the H1 low $Q^2$ precision data and part of the H1 high $Q^2$ $e^+p$ data available then, the gluon density $xg$ was determined with an experimental precision of about 3% [6]. In addition, the strong coupling constant $\alpha_s$ was determined with an experimental uncertainty of 1.5% [6] using the same data together with the proton data from BCDMS [11] for constraining $xg$ at large $x$. This determination at next-to-leading order (NLO) is unfortunately limited theoretically by the large scale uncertainty showing the necessity of having next-to-NLO calculations for improvements.

Presented at this conference is a further NLO QCD analysis performed by H1 using the low $Q^2$ precision data and all high $Q^2$ cross section data. The analysis aims at a determination of the quark momentum distributions, besides $xg$. A novel feature of the analysis is the way in which the various structure functions are decomposed in terms of different parton densities. The quark densities used by H1, the upper and down types of quarks ($xU$, $xD$) and their anti-quarks ($x\overline{U}$, $x\overline{D}$), directly enter the NC and CC cross sections contrary to an effective sea distribution or valence quark distributions. The latter follow from the differences $u_v = U - \overline{U}$ and $d_v = D - \overline{D}$. The initial distributions are parametrized in an MRST-like functional form for $Q_0^2$ at $4\,\mathrm{GeV}^2$. The results of the fit are shown in Fig. 4 as a function of $x$. In the fit with the full HERA I data plus the BCDMS proton and deuterium data [11], the experimental uncertainty shown in dark error band reaches a precision of a few percent. Shown in light error band is the model uncertainty, which includes variation of $Q_0^2$ and other parameters in the fits. The results of the fit using the H1 data only is shown with the solid curves. It is remarkable that the fit using HERA data only is able to give consistent results on the parton density functions with the combined fit which uses additional data at low $Q^2$ and large $x$.

The impact of the HERA I data at large $x$ is best seen on the u and d quark densities at valence quark region, determined by two complementary methods (Fig. 5). In one method, the densities are determined from the fit, where the precision is optimal, while in the second method, the densities ($xq_{\mathrm{exp}}$) are extracted locally from the measured cross sections $\tilde{\sigma}_{\mathrm{meas}}(x, Q^2)$, $xq_{\mathrm{exp}} = \tilde{\sigma}_{\mathrm{meas}}[xq/\tilde{\sigma}]_{\mathrm{fit}}$. The latter method has the advantage that the resulting $u$ and $d$ densities do not depend on the $Q^2$ evolution and are thus free from any nuclear corrections.

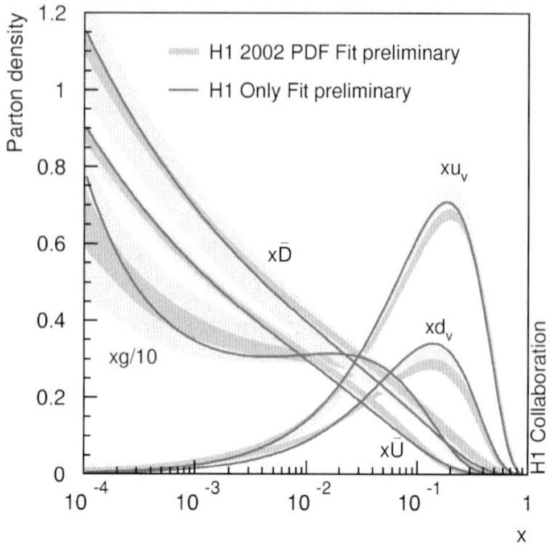


Figure 4. The valence quark densities $xu_v, xd_v$, the anti-up and down type densities $x\overline{U}, x\overline{D}$ and the gluon density $xg$ (reduced by a factor of 10), from the combined fit (error bands) using the H1 and BCDMS data [11] are compared with those from the fit (solid curves) using the H1 data only.

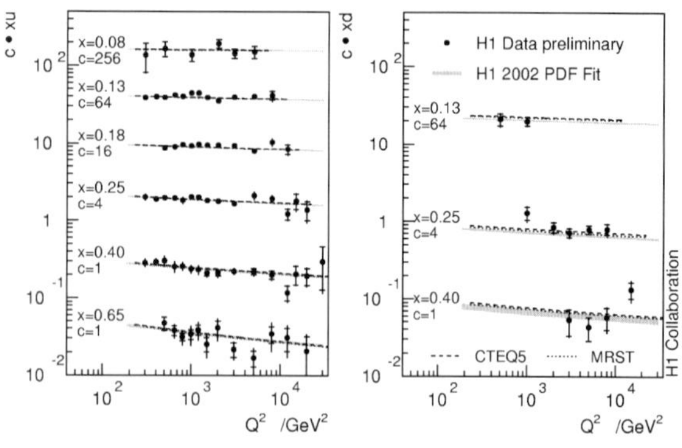


Figure 5. The $u$ and $d$ quark densities determined from the local extraction method (full points) and from the fit are compared with those from CTEQ5 [9] and MRST [12]. For displaying purpose, the densities are scaled by different factors $c$ for fixed $x$ values.

## 3. CONCLUSION

To conclude, the structure function data from almost a decade's operation in HERA phase I have enabled the theory of QCD be tested to a new level of precision in deep inelastic scattering and revealed its impact on the parton densities both for the gluon at small $x$ and for the $u$ and $d$ at large $x$. The full HERA I data at large $x$ and high $Q^2$ are still statistically limited. New data to be taken during the next years in HERA phase II are therefore important not only to improve the precision of the structure functions but also to fully explore the very high $Q^2$ region up to the kinematical limit $s \sim 10^5\,\mathrm{GeV}^2$.

## REFERENCES

1. H1 Collab., "A new measurement of the DIS cross section and of $F_L$ at low $Q^2$ and Bjorken $x$ at HERA", contri. paper # 799; "Measurement of the inclusive DIS cross section at $Q^2 \sim 1\,\mathrm{GeV}^2$ with the H1 experiment", contri. paper # 975; "Measurement of the proton structure function using radiative events at HERA", contri. paper # 976; "Measurement of inclusive DIS at high $Q^2$ and large $x$", contri. paper # 978.
2. NMC Collab., A. Arneodo et al., Phys. Lett. **B364** (1997) 107, Nucl. Phys. **B483** (1997) 3.
3. ZEUS Collab., J. Breitweg et al., Phys. Lett. **B487** (2000) 53.
4. T. Lastovicka, hep-ph/0203260.
5. H1 Collab., C. Adloff et al., Eur. Phys. J. **C19** (2001) 269; ZEUS Collab., DESY-02-113.
6. H1 Collab., C. Adloff et al., Eur. Phys. J. **C21** (2001) 33.
7. BCDMS Collab.,A. C. Benvenuti et al., Phys. Lett. **B223** (1989) 485; NMC Collab., 2nd ref. of [2]; L. W. Whitlow et al., Phys. Lett. **B250** (1990) 193.
8. H1 Collab., C. Adloff et al., Eur. Phys. J. **C13** (2000) 609.
9. CTEQ Collab., http://www.phys.psu.edu/~cteq.
10. ZEUS Collab., J. Breitweg et al., Eur. Phys. J. **C11** (1999) 427, **C12** (2000) 411.
11. BCDMS Collab., 1st ref. of [7].
12. A. D. Martin et al., Eur. Phys. J. **C4** (1998) 463.

ELSEVIER

Proceedings of ICHEP 2002, pp. 255–257
*S. Bentvelsen, P. de Jong, J. Koch and E. Laenen (Editors)*

31st INTERNATIONAL CONFERENCE ON HIGH ENERGY PHYSICS AMSTERDAM

# First Results from the NLO Dimuon Analysis at NuTeV

D. Mason[a] for the NuTeV Collaboration

[a]University of Oregon, Eugene OR 97403

First results from next to leading order (NLO) fits of the DIS charged current dimuon cross section data from the NuTeV experiment at FNAL are presented. Neutrino production of dimuon events provides a direct means of studying charm production and the strange sea. Here we present an NLO analysis of these data which utilizes a new calculation including all NLO diagrams and that is differential in all variables required to model detector acceptance. NuTeV used a sign selected beam, which allows independent measurement of the strange and antistrange sea, for which an asymmetry is potentially important in interpreting neutrino electroweak coupling results.

## 1. Introduction

Dimuons from charged current neutrino deep inelastic scattering provide a unique set of events with which to probe QCD and study nucleon structure. These events occur predominantly in charm production from charged current interactions with strange or down quarks. Charm production from down quarks however is Cabibbo suppressed, allowing dimuon events to be a direct means of extracting the strange parton distribution function (PDF). The charmed hadron produced in this interaction decays semileptonically with a muon ≈10% of the time, leading to a final state with two oppositely charged muons. These events provide a clearly distinguishable signal in the NuTeV detector[1],[2].

Of particular topical interest is whether the strange sea is different from the antistrange sea. It has been proposed[3] that the proton wavefunction has a virtual $K^+\Lambda$ pair component which would lead to an intrinsic strange sea. A strange-antistrange asymmetry has also been invoked[4], [5], [6] to explain the NuTeV electroweak result [7].

The NuTeV experiment's neutrino beam was "sign selected". A dipole in the secondary beam bent right sign pions and kaons towards the detector while opposite sign and neutral charges were dumped, providing high purity $\nu$ or $\overline{\nu}$ beams. This a priori knowledge of whether the event was the result of a neutrino or antineutrino interaction allows separate analysis of dimuon events for each polarity, making it possible to measure the strange sea independently of the antistrange sea.

Measuring the strange sea is complicated somewhat due to the large gluon content of the nucleon. Its magnitude requires study of the strange sea and charm production be done at next to leading order in QCD to have meaning beyond the context of neutrino-nucleon scattering[8]. In this analysis we employ a NLO calculation incorporating all relevant diagrams [9],[10].

## 2. NLO fit to the Dimuon Cross Section

The NuTeV collaboration recently published its measurement of the dimuon cross section in the form of a cross section table[11]. The cross section of dimuons with charm decay muons of energy greater than 5 GeV was extracted in bins in Bjorken $x$, inelasticity $y$, and neutrino energy $E_\nu$. When fitting a charm production cross section model to this cross section, one must be able to simulate the effects of fragmentation and the subsequent semileptonic decay. This requires a cross section model dependent on the fragmentation momentum ratio $z$. In addition, at NLO, the charmed quark has transverse momentum due to the interaction with the gluon which must also be modeled. A new calculation, implemented into a FORTRAN routine, DISCO[12], which calculates the charm production cross section differentially in $x$, $y$, $z$, and charm rapidity $\eta_c$ was em-


doi:10.1016/S0920-5632(02)02117-5

ployed for this analysis. At $z \sim 1$ and $\eta_c$ near the collinear limit, negative Sudakov logarithms need to be regularized, which DISCO achieves by integrating the cross section within $z$ and $\eta_c$ bins.

The NLO fit to the dimuon cross section is made up of the following components:

$$\frac{d\sigma_{2\mu}}{dxdy} = \frac{d\sigma_{charm}}{dxdy} \cdot EMC \cdot B_c \cdot \mathcal{A} \quad (1)$$

where $\frac{d\sigma_{2\mu}}{dxdy}$ is the measured NuTeV dimuon cross section, $\frac{d\sigma_{charm}}{dxdy}$ is the NLO cross section model[9],[10], $EMC$ is an $x$ dependent correction for the EMC effect and nuclear shadowing[13], $B_c$ is the semileptonic charm branching ratio (here 0.093 [14]), and $\mathcal{A}$ is an acceptance correction resulting from the 5 GeV cut on the energy of the muon from the semileptonic charm decay. The cross sections and $\mathcal{A}$ are functions of the neutrino energy $E_\nu$.

The key component in this fitting procedure is the $E_{decay} > 5\ GeV$ acceptance correction ($\mathcal{A}$). In addition to the kinematic variables $E_\nu$, $x$ and $y$, $\mathcal{A}$ depends both on the fragmentation parameter $\epsilon$ and on the charm mass. It is calculated using a Monte Carlo simulation that brings the charm quark through fragmentation to the observable semileptonic decay muon. For both neutrino and antineutrino modes, in each of the 3 $E_\nu$, 3 $y$ and 5 $x$ bins in the cross section table, $z$ and $\eta_c$ are thrown for large samples of events, and the charmed particle's kinematics for each are calculated. Each event is given a weight proportional to $\frac{d\sigma_{charm}}{dxdydzd\eta_c}$, and is then allowed to decay semileptonically following reference [15]. The correction is calculated by dividing the number of events that pass a 5 GeV cut on the decay muon energy by the total number of events in the bin.

The DISCO cross section calculation used for this analysis contains a two dimensional numerical integration, and generating corrections at any $\epsilon$ and $m_c$ point requires significant CPU time. A grid of fixed points in $\epsilon$ and $m_c$ were calculated with the Monte Carlo, and acceptance corrections were interpolated as the fit moved between different values of $\epsilon$ and $m_c$.

## 3. Preliminary Fit Results

A preliminary fit to the NuTeV dimuon cross section table using the NLO cross section model described in [10],[12] has been performed. Charm mass, the Collins Spiller fragmentation parameter $\epsilon$, and the strange and antistrange seas were fit. The GRV94HO [16] PDF's were used and the strange sea was parameterized with separate $\kappa$ and $\alpha$'s for neutrino and antineutrino modes:

$$s(x) = \kappa \frac{\overline{u}(x) + \overline{d}(x)}{2}(1-x)^{\alpha} \quad (2)$$

$$\overline{s}(x) = \overline{\kappa} \frac{\overline{u}(x) + \overline{d}(x)}{2}(1-x)^{\overline{\alpha}} \quad (3)$$

Table 1
Results from the preliminary 6 parameter fit to the NuTeV dimuon cross section tables

| parameter | result | ± stat. | ± syst. |
|---|---|---|---|
| $m_c$ | 1.46 | ± 0.24 | ± 0.07 |
| $\kappa$ | 0.516 | ± 0.033 | ± 0.031 |
| $\overline{\kappa}$ | 0.511 | ± 0.038 | ± 0.040 |
| $\alpha$ | 0.73 | ± 0.47 | ± 0.52 |
| $\overline{\alpha}$ | 0.92 | ± 0.43 | ± 0.06 |
| $\epsilon$ | 0.203 | ± 0.13 | ± 0.04 |
| $\chi^2/NDF$ | | 36.2/38 | |

As can be seen from the parameterization, fit values for $\kappa$, $\alpha$, $\overline{\kappa}$ and $\overline{\alpha}$ will depend on the PDF used, and therefore only have meaning within that context. Preliminary results from the full 6 parameter fit are as shown in Table 1. Figure 1 shows the fitted cross section with these parameters (curve) on the dimuon table cross section (points) in both neutrino and antineutrino modes. Figure 2 shows the parameterized $s$ and $\overline{s}$ seas. They are consistent with each other within errors.

## 4. Conclusion

We have presented a first NLO fit to the NuTeV dimuon cross section table, resulting in a look at

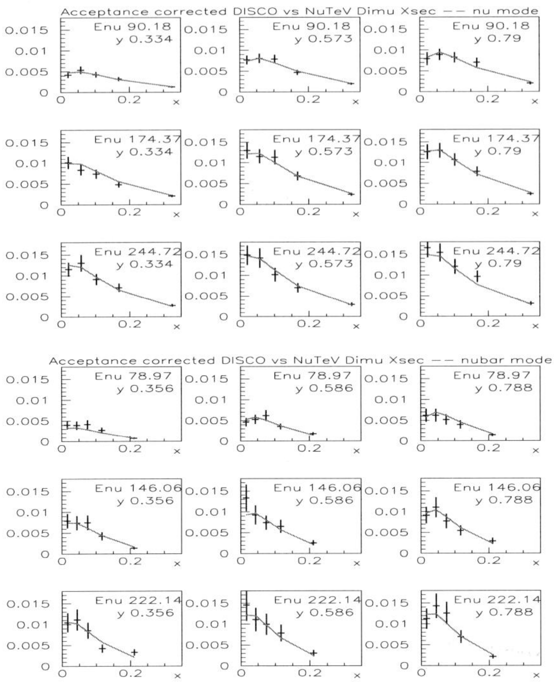


Figure 1. Top: Fitted cross section curve on $\nu$ mode cross section data points. Bottom: Same for $\overline{\nu}$ mode.

the strange and antistrange seas, and a measurement of the charm mass. Within errors, there is no evidence of any $s$-$\overline{s}$ asymmetry.

This is a preliminary analysis for which several improvements are already underway, including using other PDF's and studying different fragmentation and strange sea parameterizations. Although any differences are expected to be small, the dimuon cross section will also be re-extracted with full NLO Monte Carlo as a final consistency check.

## REFERENCES


1. J.M. Conrad, M.H. Shaevitz, T.A. Bolton, Rev.Mod.Phys. 70 (1998) 1341-1392
2. D. Harris, J. Yu, et al., Nucl. Instrum. Methods A447 377 (2000)


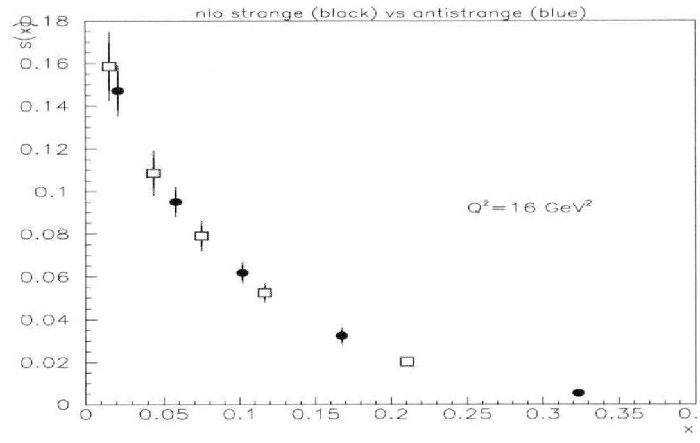


Figure 2. Fit strange and antistrange parameterizations evaluated at dimuon cross section $x$ bin centers for $Q^2 = 16$ GeV$^2$. Solid points are for $s(x)$, hollow squares are for $\overline{s}(x)$.


3. S. Brodsky, B.Q. Ma, Phys.Lett. B381 (1996) pp. 317-324
4. S. Davidson, S. Forte, P. Gambino, N. Rius, A. Strumia, JHEP 0202 (2002) 037
5. V. Barone, C. Pascaud, F. Zomer, Eur. Phys. J. C12 (2000) p. 243
6. G.P. Zeller *et. al* [NuTeV Collaboration], Phys. Rev. D65, 111103 (2002)
7. G.P. Zeller *et. al* [NuTeV Collaboration], Phys. Rev. Lett. 88, 091802 (2002)
8. M. Aivazis, J.C. Collins, F. Olness, W.K. Tung, Phys. Rev. D50, 3102 (1994)
9. M. Gluck, S. Kretzer, E. Reya, Phys. Lett. B380, 171-176,1996, Erratum-ibid.B405:391,1997
10. S. Kretzer, I. Schienbein, Phys. Rev. D58, 094035 (1998)
11. M. Goncharov *et al.* [NuTeV Collaboration], Phys. Rev. D62, 112006 (2001)
12. S. Kretzer, D. Mason, F. Olness, Phys. Rev. D65, 074010 (2002)
13. B. Seligman PHD thesis, Columbia University, page 250
14. T. Bolton, hep-ex/9708014
15. V. Barger, R.J.N. Phillips, Phys. Rev. Lett. D14, 80 (1976)
16. M. Gluck, E. Reya, A. Vogt, Z.Phys C67 433-448 (1995)

ELSEVIER

Proceedings of ICHEP 2002, pp. 258–261
*S. Bentvelsen, P. de Jong, J. Koch and E. Laenen (Editors)*

31st INTERNATIONAL CONFERENCE ON HIGH ENERGY PHYSICS AMSTERDAM

# Photon and electron structure from $e^+e^-$ interactions

R. Nisius[a]

[a]Max-Planck-Institut für Physik (Werner-Heisenberg-Institut), Föhringer Ring 6, D-80805 München, Germany, E-mail:nisius@mppmu.mpg.de.

The status of the measurements and the theoretical developments concerning the hadronic structure of the photon are briefly summarised.

## 1. Introduction

For more than 20 years measurements of photon structure functions give deep insight into the rich structure of a fundamental gauge boson, the photon. A recent review on this subject can be found in [1].

The main idea is that by measuring the differential cross-section

$$\frac{d^2\sigma_{e\gamma\to eX}}{dxdQ^2} = \frac{2\pi\alpha^2}{x\,Q^4}\left[\left(1+(1-y)^2\right)F_2^\gamma(x,Q^2) - y^2 F_L^\gamma(x,Q^2)\right]$$

one obtains the photon structure function $F_2^\gamma$, see Figure 1 for an illustration. Here $Q^2$ and $P^2$ are the absolute values of the four momentum squared of the virtual and quasi-real photons, with $P^2 \ll Q^2$. The symbols $x$ and $y$ denote the usual dimensionless variables of deep-inelastic scattering, and $\alpha$ is the fine structure constant. The flux of the incoming photons, $f_\gamma(z,P^2)$, where $z$ is the fraction of the electron energy carried by the photon, is usually taken from the equivalent photon approximation, EPA. In leading order, the structure function $F_2^\gamma$ is proportional to the parton content of the photon and therefore reveals the structure of the photon. In the region of small $y$ studied ($y \ll 1$), the contribution of the term proportional to $F_L^\gamma(x,Q^2)$ is small, and is usually neglected. Because the energy of the quasi-real photon is not known, $x$ has to be derived by measuring the invariant mass of the hadronic final state $X$, which is a source of significant uncertainties, and makes measurements of $F_2^\gamma$ mainly limited by the systematic error, except for large values of $Q^2$.

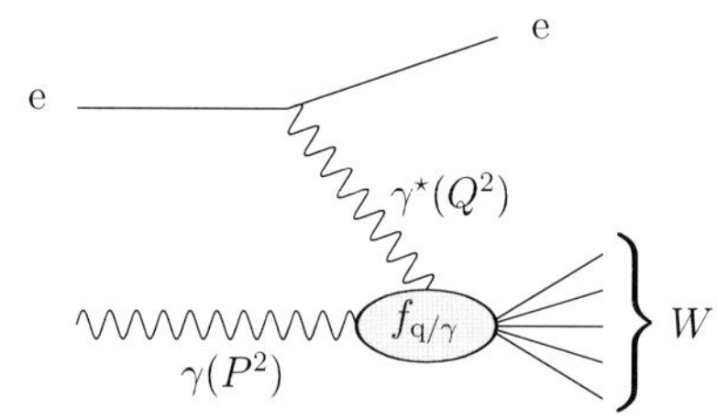


Figure 1. *Deep-inelastic electron photon scattering.*

At this conference new measurements of the hadronic structure function $F_2^\gamma$ and its charm component $F_{2,c}^\gamma$ have been presented. They are discussed, together with the most recent fits to the $F_2^\gamma$ data and the prospects for measurements of $F_2^\gamma$ at a future Linear Collider. In addition, an attempt by DELPHI is presented to investigate the photon structure by measuring the electron structure function $F_2^e$.

## 2. Photon structure function

The improvement in the measurement of $F_2^\gamma$ since the first result by PLUTO in 1981 is quite impressive, see Figure 2. The analysis of the LEP data has extended the kinematic coverage by about two orders in magnitude, both to larger $Q^2$ and to smaller $x$. In addition, due to continuous improvements of the analyses and a LEP combined effort to obtain a better description of the data by the Monte Carlo models, the precision of the measurements has been improved considerably.

For this conference the final OPAL result for


doi:10.1016/S0920-5632(02)02120-5

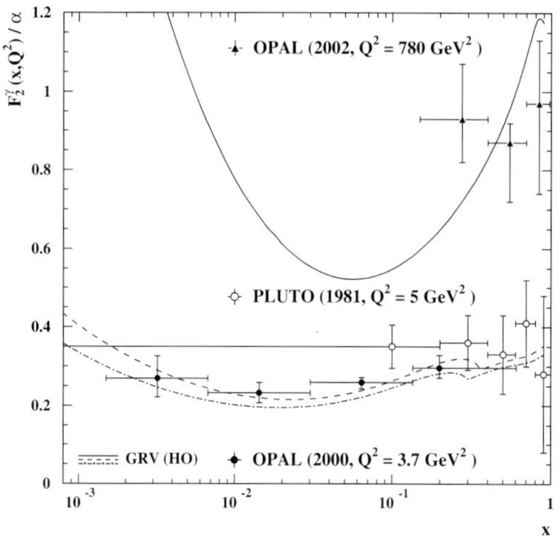


Figure 2. *The improvements in $F_2^\gamma$ at LEP.*

the measurement of the hadronic structure function $F_2^\gamma$ at high $Q^2$ has been available. This measurement is based on the complete LEP2 data and extends the measurement of $F_2^\gamma$ to $\langle Q^2 \rangle = 780\ \mathrm{GeV}^2$, the largest scale ever probed. As can be seen from Figure 3 the measured $F_2^\gamma$ is rather flat, and, within errors, the parameterisations from GRSc [2], SaS1D [3] and WHIT [4] are in agreement with the data. The already available

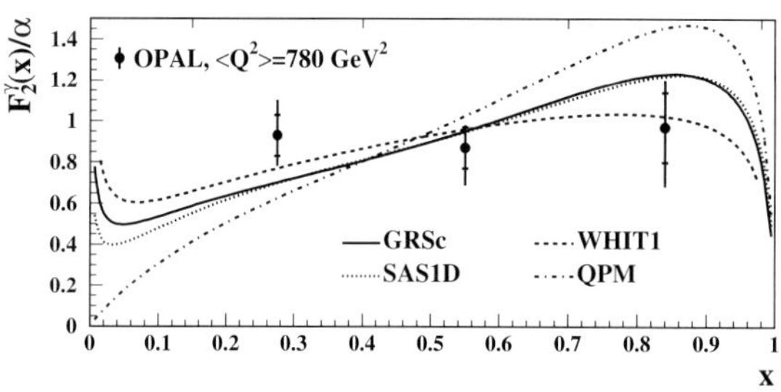


Figure 3. *The measurement of $F_2^\gamma$ at high $Q^2$.*

preliminary result from DELPHI [5] at a slightly lower $\langle Q^2 \rangle$ basically shows the same trend. Since the measurement at high $Q^2$ is mainly limited by the statistical error, it is very desirable to combine the OPAL result with the measurements to come from the other LEP experiments. To facilitate the combination, the analyses should be performed for the same bins in $x$ and $Q^2$.

Also the investigation of the evolution of $F_2^\gamma$ with $Q^2$ in ranges of $x$ has been continued using the LEP2 data. For medium values of $x$, the pre-

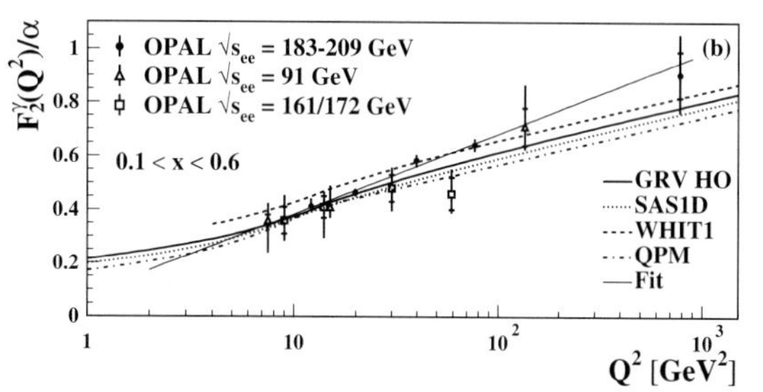


Figure 4. *The $Q^2$ evolution of $F_2^\gamma$ from OPAL.*

cision of the OPAL results based on LEP1 data has been improved considerably by using the large luminosity available at LEP2 energies. With the present level of precision the data start to challenge the existing parameterisations of $F_2^\gamma$. Given this, several theoretical as well as experimental issues have to be addressed in more detail when interpreting the data. Examples are the suppression of $F_2^\gamma$ with $P^2$ and radiative corrections to the deep-inelastic scattering process. A summary of the present status of all measurements of the $Q^2$ evolution of $F_2^\gamma$ is shown in Figure 5.

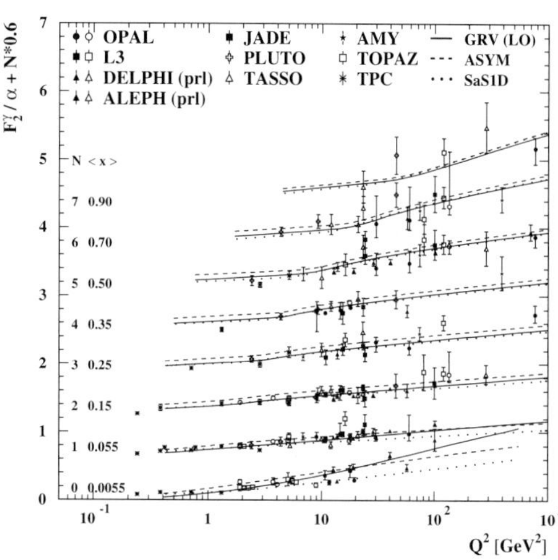


Figure 5. *The world data on $F_2^\gamma$ as a function of $Q^2$ in bins of $x$.*

## 3. Fits to $F_2^\gamma$ data

There have been recent fits [6] to $F_2^\gamma$ based on all available data, except for the TPC/$2\gamma$ results. To facilitate the analysis, some simplifications are made in the treatment of the experimental results. In the analysis the correlation matrix of the various points is used if it is provided by the experiment. However, the systematic errors are treated as uncorrelated, the $P^2$ effect and the radiative corrections are neglected, and finally, in case of asymmetric errors, the data points are moved to the central value and symmetric error are assumed. Given the precision of the data mentioned above, this procedure should most likely be improved for future analyses. The fits are done in a fixed flavour scheme with $uds$ as active flavours and charm treated as a Bethe-Heitler contribution with $m_c = 1.5 \pm 0.1$ GeV. Two different types of fits are performed in leading and next-to-leading order, NLO, using the DIS$_\gamma$ and $\overline{\text{MS}}$ schemes. In the first fit the hadron-like part of the photon structure is neglected, and only the point-like part, which is evolved from a starting scale of $Q_0^2 = \Lambda^2$, is taken into account. Consequently the only free parameter of the fit is $\alpha_s(M_Z)$. Since the hadron-like part dominates at low $x$ and $Q^2$, only the data in the region $x > 0.45$ and $Q^2 > 59$ GeV$^2$ are used in this fit. The second fit uses all data, takes into account both components and fits for ($N$, $\alpha$, $\beta$, $\alpha_s$, $Q_0^2$) using the assumptions, $uds(Q_0^2) = Nx^\alpha(1-x)^\beta$ and $g(Q_0^2) = 0$. Both types of fits give a reasonable description of the data. Within the assumptions made, the quoted theoretical precision on $\alpha_s(M_Z)$ is about 3% and predicted to shrink to less than 2% for $Q^2$ values larger than 300 GeV$^2$.

## 4. Prospects for $F_2^\gamma$ measurements

The prospects of future investigations of the photon structure in the context of the planned Linear Collider programme are very promising. The $e^+e^-$ Linear Collider will extend the available phase space, as shown in Figure 6, for the measurement of the $Q^2$ evolution of $F_2^\gamma$ at medium $x$, see [1]. The higher beam energy and luminosity compared to LEP also allows for the

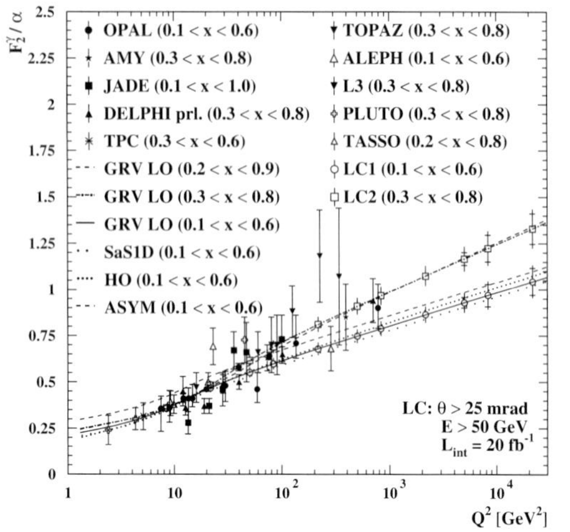


Figure 6. *The expected measurement of $F_2^\gamma$ at a future Linear Collider.*

investigations of novel features like the measurement of the flavour decomposition of $F_2^\gamma$ by exploring the exchange of $W$ and $Z$ bosons [7].

## 5. The charm contribution to $F_2^\gamma$

The final OPAL result [8] has been presented of the measurement of the charm component $F_{2,c}^\gamma$ using $D^*$ mesons to identify charm quarks. Compared to the first OPAL result on $F_{2,c}^\gamma$, this analysis is based on improved Monte Carlo models and higher statistics, leading to a more precise measurement presented in Figure 7. In a simi-

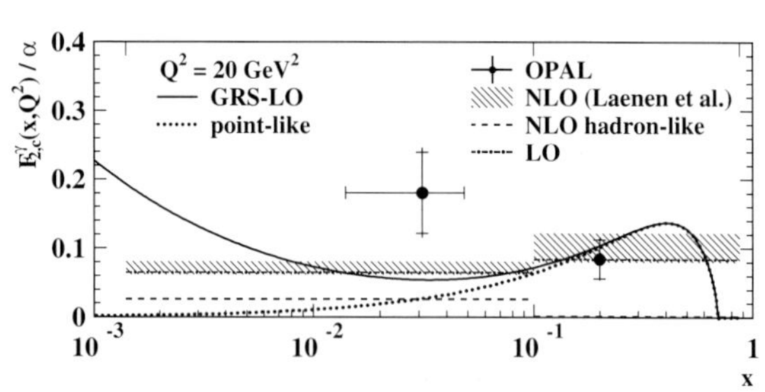


Figure 7. *The measurement of $F_{2,c}^\gamma$ from OPAL.*

lar way to the structure function for light quarks, $F_{2,c}^\gamma$ receives contributions from the point-like and the hadron-like components of the photon structure. These two contributions are predicted [9] to have different dependences on $x$, with the hadron-like component dominating at very low values of

$x$ and the point-like part accounting for most of $F_{2,c}^{\gamma}$ at $x > 0.1$. For $x > 0.1$ the OPAL measurement is described by perturbative QCD at next-to-leading order. For $x < 0.1$ the measurement is poorly described by the NLO prediction using the point-like component alone, and therefore the measurement suggests a non-zero hadron-like component of $F_{2,c}^{\gamma}$. Increased statistics and a better understanding of the dynamics for $x < 0.1$ are needed to get a more precise result in this region. Also here, to increase the statistics, it would be advantageous to combine the data from all four LEP experiments taken in the same phase space.

## 6. Electron structure function

For this conference there has been an attempt by DELPHI to access the photon structure by measuring the electron structure function. In this analysis, instead of measuring the electron-photon scattering cross-section by utilising the EPA to account for the flux of the quasi-real photons, the cross-section of electron-electron scattering is studied as functions of $Q^2$ and $x_e$, the fractional momentum of the parton with respect to the electron. This quantity is related to $x$ by $x_e = zx$. The main advantage of this approach is experimental, i.e. the incoming particle probed by the virtual photon is the electron and not the photon. This means that its energy is known, and therefore $x_e$ can be obtained without measuring $W$. However, there is also a disadvantage in this measurement. The photon structure is obscured because, e.g. the region of low values of $x_e$ receives contributions from both, the region of large momentum fraction $x$ and low scaled photon energy $z$, and the region of small $x$ and large $z$. The measurement of $F_2^e$ is performed with a precisions of about 3-20% both for the statistical and the systematic error. So far, no radiative corrections and no bin-centre corrections are applied. The preliminary DELPHI result is consistent with several existing parameterisations of $F_2^e$, obtained from $F_2^{\gamma}$ and $f_{\gamma}(z, P^2)$ by convolution. The usual exception is the LAC1 parameterisation, which is disfavoured. This investigation serves as a valuable cross-check of the $F_2^{\gamma}$ measurements, but does not give more insight into the photon structure.

## 7. Conclusion

Given the large statistics available at LEP2 energies, the region of phase space covered in the investigations of the structure of the photon is constantly increasing. Despite these large luminosities, for some of the measurements the results are still limited by the statistical error and a combination of the results from several experiments is desirable. This is particularly true for the measurement of $F_2^{\gamma}$ at large $Q^2$ and the determination of $F_{2,c}^{\gamma}$. For the first time, the high precision data from LEP have been used in NLO fits to $F_2^{\gamma}$ results. With an improved treatment of the experimental data, and the inclusion of the jet data from HERA to better constrain the gluon distribution in the photon, there are good prospects to achieve the first parametrisation of the parton distributions of the photon based on LEP and HERA data in the near future.

## REFERENCES

1. R. Nisius, Phys. Rep. **332**, 165–317 (2000), updated figures available at: *http://www.mppmu.mpg.de/~nisius/welcomeastruc.html.*
2. M. Glück, E. Reya, and I. Schienbein, Phys. Rev. **D60**, 054019 (1999).
3. G.A. Schuler and T. Sjöstrand, Z. Phys. **C68**, 607–624 (1995).
4. K. Hagiwara et al., Phys. Rev. **D51**, 3197–3219 (1995).
5. I. Tyapkin, DELPHI Collab., in proceedings of PHOTON 2001, Ascona.
6. S. Albino, M. Klasen, and S. Söldner-Rembold, Phys. Rev. Lett. **89**, 122004 1–4 (2002).
7. A. Gehrmann-De Ridder, H. Spiesberger, and P.M. Zerwas, Phys. Lett. **B469**, 259–262 (1999).
8. G. Abbiendi et al., OPAL Collab., Phys. Lett. **B539**, 13–24 (2002).
9. E. Laenen et al., Phys. Rev. **D49**, 5753–5768 (1994); E. Laenen and S. Riemersma, Phys. Lett. **B376**, 169–176 (1996).

Proceedings of ICHEP 2002, pp. 262–264
S. Bentvelsen, P. de Jong, J. Koch and E. Laenen (Editors)

ELSEVIER

31st INTERNATIONAL CONFERENCE ON HIGH ENERGY PHYSICS AMSTERDAM

# Structure Functions are not Parton Probabilities

Paul Hoyer[a]*

[a]Department of Physical Sciences and Helsinki Institute of Physics
P.O. Box 64, FIN-00014 Helsinki University, Finland

Parton distributions given by deep inelastic lepton scattering (DIS) are not equal to the probabilities of finding those partons in the parent wave function. Soft rescattering of the struck parton within the coherence length of the hard process influences the DIS cross section and gives dynamical phases to the scattering amplitudes. This gives rise to diffractive DIS, shadowing in nuclear targets and transverse spin asymmetry.

## 1. DIS DYNAMICS

Following the pioneering work of Drell, Levy and Yan [1] in 1969, the deep inelastic scattering ($\ell N \to \ell + X$, DIS) cross section has been commonly thought to be given by the probability of finding a parton in the target $N$ wave function, *i.e.*, by the parton density. In the Bjorken limit of photon energy $\nu$ and virtuality $Q^2$ tending to infinity at fixed $x_B = Q^2/2m_N\nu$, final state interactions do not affect the DIS cross section – in a scalar theory such as considered in Ref. [1]. In QCD, gluon exchange modifies the relation between the cross section and the parton density [2], and causes observed effects such as diffraction, shadowing and transverse spin asymmetry. In this talk I give a brief account of this.

The standard DIS 'handbag' shown in Fig. 1a may be viewed as a Light-Cone (LC) time-ordered diagram where the ordering of the vertices from left to right corresponds to an ordering in LC time $y^+ = y^0 + y^3$. In a frame where the virtual photon moves along the negative $z$-axis ($q^- \simeq 2\nu$, $q^+ \simeq -m_N x_B$) there are no $\gamma^* \to q\bar{q}$ splittings of the kind shown in Fig. 1b, since the quark lines must have $p^+ > 0$. Due to the large value of $q^-$ the scattering takes place almost instantaneously in LC time, $y^+ \sim 1/\nu$.

The dashed lines in Fig. 1a indicate that the

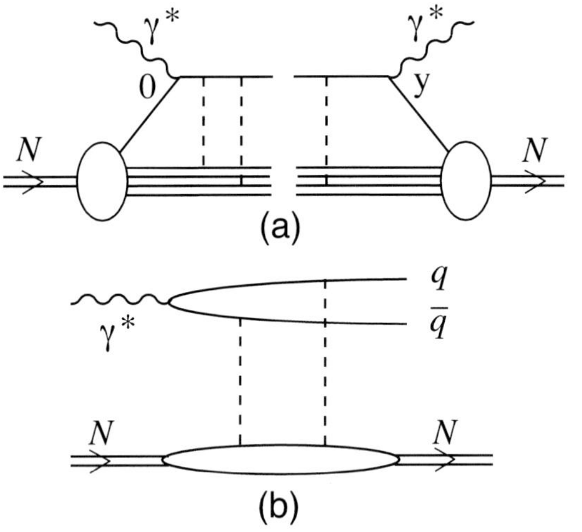


Figure 1. (a) The DIS cross section and (b) a diffractive DIS amplitude. The photon moves along the negative $z$-axis in (a) and in the opposite direction in (b).

struck quark can (and typically does) rescatter elastically on the color field of the target. Since the exchanged gluon has spin 1 this soft (instantaneous, coulombic) scattering does not decrease with the energy $\nu$ of the quark – whereas scalar exchange would bring a factor $\propto 1/\nu$. Thus rescattering affects the DIS cross section at leading twist if it occurs within the coherence length $\nu/Q^2 = 1/2m_N x_B$ of the virtual photon. Even at moderately small values of $x_B$ the coherence length is comparable to the size (several fm) of a

*Address until 1 August 2002: Nordita, Blegdamsvej 17, DK-2100 Copenhagen, Denmark. Research supported in part by the European Commission under contract HPRN-CT-2000-00130.


doi:10.1016/S0920-5632(02)02126-6

nuclear target. Note that the rescattering occurs in vanishing LC time $y^+$ due to the relativistic motion of the struck quark. The DIS amplitudes nevertheless acquire complex dynamical phases from contributions of on-shell intermediate states between the rescatterings. The interference between amplitudes with and without rescattering gives rise to diffraction and shadowing [3].

## 2. SUBTLETIES OF LC GAUGE

QCD factorization in DIS determines the expression for the parton distribution in a general gauge to be [4]

$$f_{q/N}(x_B, Q^2) = \frac{1}{8\pi}\int dy^- \exp(-ix_B p^+ y^-/2) \times \langle N(p)|\bar{q}(y^-)\gamma^+ [y^-;0] q(0)|N(p)\rangle \quad (1)$$

where all fields are evaluated at $y^+ = y_\perp = 0$. The path-ordered exponential

$$[y^-;0] \equiv \mathrm{P}\exp\left[ig\int_0^{y^-} dw^- A^+(w^-)\right] \quad (2)$$

ensures the gauge invariance of the matrix element and contains the rescattering effects discussed above.

The exponential (2) reduces to unity in LC gauge ($A^+ = 0$). The matrix element (1) can then be expressed as an overlap of the target Fock state amplitudes $\psi_n$ at $y^+ = 0$ [5],

$$f_{q/N} \stackrel{A^+\to 0}{\longrightarrow} \mathcal{P}_{q/N} = \sum_n \int \prod_i dx_i\, d^2k_{\perp i} \times |\psi_n(x_i, k_{\perp i})|^2 \sum_{j=q} \delta(x_B - x_j) \quad (3)$$

This appears like the probability of finding a quark in the initial state and is the origin of the misconception that the DIS cross section is an incoherent sum over the target quark density. However, the rescattering and interference effects discussed above have not vanished, they are just hidden by the peculiarities of LC gauge.

In Ref. [2] the properties of a perturbative, abelian model of DIS was studied in Feynman and LC gauge. In the $x_B \to 0$ limit the dominant amplitudes were found to have *on-shell* intermediate states between the Coulomb gluon exchanges (dashed lines in Fig. 1). Such states can kinematically occur only *after* the $\gamma^*$ has been absorbed, and thus belong to the final state. Each on-shell rescattering amplitude is gauge invariant - hence the on-shell contributions cannot be eliminated by a gauge choice.

The model furthermore showed that in Feynman gauge the DIS cross section is affected only by diagrams where the struck quark rescatters. Indeed, final state interactions (FSI) between spectators cannot affect the DIS cross section at leading twist in this gauge (see, *e.g.*, [2] for a proof).

In LC gauge the situation was found to be reversed. Rescattering of the *struck quark* does not affect the DIS cross section, which is consistent with the fact that the path-ordered exponential (2) reduces to unity for $A^+ = 0$. Hence the reinteractions producing on-shell intermediate states in LC gauge *occur between spectators*. The singularity of the gluon propagator $\mathcal{D}^{\mu\nu}_{LC}(k)$ at $k^+ = 0$ in LC gauge enhances the $k^+ \simeq 0$ exchanges between spectators so that they modify the DIS cross section at leading twist [6].

The spectator interactions occur within a vanishing LC time $y^+$ of $\mathcal{O}(1/\nu)$ after the $\gamma^*$ interacts, even though they are spread over a longitudinal distance of $\mathcal{O}(\mathrm{fm})$ in the target. Hence they can *formally* be included in the target Fock amplitudes $\psi_n(y^+ = 0)$. The $k^+ = 0$ gluon exchanges are then 'zero modes' which propagate along the light-like ($y^+ = 0$) quantization surface of the Fock states. The Fock state wave functions thus defined are complex, with the rescattering and interference effects contained within their norm. However, the inclusion of the rescattering effects as zero modes of the target LC wave function appears artificial from a physical point of view.

## 3. REMARKS

The insight that DIS dynamics involves rescattering and interference effects has led to interesting further developments. Brodsky, Hwang and Schmidt [8] showed that the complex phase arising from rescattering causes a transverse spin asymmetry at leading twist in semi-inclusive DIS.

Thus the azimuthal distribution of the current jet around the virtual photon momentum depends on the transverse polarization of the target nucleon in $\gamma^* + N_\uparrow \to jet + X$. This "Sivers effect" [9] had previously been thought to be of higher twist [10]. The effect is unrelated to the transverse spin carried by the target quark and thus complicates the measurement of the quark transversity distribution.

In semi-inclusive DIS the quark fields of the parton distribution (1) are at non-zero transverse separation, $y_\perp \neq 0$. The integral in the path-ordered exponential (2) must then cover also this transverse distance (at asymptotic LC time) [11]. The transverse path involves $A^\perp$ and does not vanish in LC gauge – in fact it is the origin of the transverse spin asymmetry.

Diffractive DIS is described by two-gluon (Pomeron) exchange amplitudes like the one in Fig. 1b. The scattering is most naturally viewed in a frame where the virtual photon moves along the positive $z$-axis, *i.e.*, with $q^+ \simeq 2\nu$. In this frame there is a $\gamma^* \to q\bar{q}$ transition which at low $x_B$ typically occurs long before the target. Crossing symmetry requires the amplitude to be dominantly imaginary at high energy.

The frame of Fig. 1b is related to that of Fig. 1a (where $q^- \simeq 2\nu$) by a rotation of 180° around the $y$-axis. Due to the nonperturbative nature of the target this rotation cannot in general be performed. The amplitude of Fig. 1b is often interpreted as corresponding, in the frame of Fig. 1a, to scattering off a 'Pomeron component' in the target wave function. The dynamics in the two frames can, however, be unambiguously related within the model of Ref. [2]. There are no miracles – the amplitude remains imaginary after the rotation, and this phase arises from a rescattering in the final state. Thus views of the Pomeron as residing in the nucleon wave function before the $\gamma^*$ interacts are incorrect.

The above developments motivate comparisons of DIS with other hard processes and renewed checks of the universality of parton distributions. It has become clear that the transverse spin asymmetry has opposite sign in DIS and the Drell-Yan process [10,12]. Furthermore, the dynamics of the rescattering which gives rise to shadowing is quite different in these two processes [13].

**Acknowledgments.** The work I have described was done in collaboration with S. J. Brodsky, N. Marchal, S. Peigné and F. Sannino. I am also grateful for helpful discussions with M. Diehl and X. Ji.

## REFERENCES

1. S. D. Drell and T. M. Yan, Phys. Rev. Lett. **24** (1970) 181;
   S. D. Drell, D. J. Levy and T. M. Yan, Phys. Rev. D **1** (1970) 1035.
2. S. J. Brodsky, P. Hoyer, N. Marchal, S. Peigne and F. Sannino, Phys. Rev. D **65** (2002) 114025 [arXiv:hep-ph/0104291].
3. G. Piller and W. Weise, Phys. Rept. **330**, (2000) 1 [arXiv:hep-ph/9908230].
4. J. C. Collins and D. E. Soper, Nucl. Phys. B **194** (1982) 445.
5. S. J. Brodsky, M. Diehl and D. S. Hwang, Nucl. Phys. B **596** (2001) 99 [arXiv:hep-ph/0009254];
   M. Diehl, T. Feldmann, R. Jakob and P. Kroll, Nucl. Phys. B **596** (2001) 33 [arXiv:hep-ph/0009255].
6. A. Metz and S. Peigné, Private communication.
7. S. J. Brodsky, Private communication.
8. S. J. Brodsky, D. S. Hwang and I. Schmidt, Phys. Lett. B **530** (2002) 99 [arXiv:hep-ph/0201296].
9. D. W. Sivers, Phys. Rev. D **41** (1990) 83 and Phys. Rev. D **43** (1991) 261.
10. J. C. Collins, Nucl. Phys. B **396** (1993) 161 [arXiv:9208213] and Phys. Lett. B **536** (2002) 43 [arXiv:hep-ph/0204004].
11. X. Ji and F. Yuan, Phys. Lett. B **543** (2002) 66 [arXiv:hep-ph/0206057];
    A. V. Belitsky, X. Ji and F. Yuan, arXiv:hep-ph/0208038.
12. S. J. Brodsky, D. S. Hwang and I. Schmidt, arXiv:hep-ph/0206259.
13. S. Peigne, arXiv:hep-ph/0206138.

Proceedings of ICHEP 2002, pp. 265–267
*S. Bentvelsen, P. de Jong, J. Koch and E. Laenen (Editors)*

31st INTERNATIONAL CONFERENCE ON HIGH ENERGY PHYSICS AMSTERDAM

# A New Generation of CTEQ Parton Distribution Functions with Uncertainty Analysis

Daniel R. Stump [a]

[a]Department of Physics and Astronomy
Michigan State University
East Lansing, MI 48824, USA

The CTEQ6 parton distribution functions are described. The PDF's are constructed by global analysis of data from many experimental collaborations and theoretical calculations based on QCD.

## 1. Introduction

For any short-distance scattering process involving an initial nucleon, the factorization theorem of QCD relates the experimental cross section to products of theoretical parton factors, calculable in perturbation theory, and parton distribution functions $\{f_i(x,Q)\}$ of the nucleon. The momentum fraction of the parton is $x$, the momentum scale of the process is $Q$, and parton types are labeled by $i$. The goal of the CTEQ global analysis is to use data from many scattering processes to determine the universal parton distributions (PDF's).

A new generation of PDF's has recently been developed, named CTEQ6.[1] Several advances have occured since the previous generation (CTEQ5). Significant new data are available: (i) for deep-inelastic $ep$ and $\bar{e}p$ scattering from the H1 and ZEUS collaborations at HERA; and (ii) for inclusive jet production as a function of rapidity and transverse energy from the DØ collaboration at the Tevatron. These new data have also been used in the recent MRST2001 parton distributions.[2] Further, new methods of analysis are used in CTEQ6. The published information on correlated systematic errors is used in the fitting procedure; and uncertainties of the PDF's due to propagation of experimental errors are analyzed.

Data sets for which detailed systematic errors have been published and used in the fit are listed in Table 1, from the BCDMS, H1, ZEUS, NMC, DØ and CDF collaborations. Other data sets that are fit without systematics are from the CCFR, E605, E866, and CDF (for $W$ production) collaborations.

Table 1
Data sets that include systematics.

| process | data set | $\chi_e^2/N_e$ |
|---|---|---|
| DIS $\mu p$ | BCDMS p | 378/339 |
| DIS $\mu d$ | BCDMS d | 280/251 |
| DIS $ep$ | H1a | 99/104 |
| DIS $ep$ | H1b | 129/126 |
| DIS $ep$ | ZEUS | 263/229 |
| DIS $\mu p$ | NMC F2p | 305/201 |
| DIS $\mu d$ | NMC F2d/p | 112/123 |
| $p\bar{p} \to$ jet | DØ jet | 69/90 |
| $p\bar{p} \to$ jet | CDF jet | 49/33 |

## 2. $\chi^2$ minimization with systematic errors

For each experiment in Table 1, we have a set of data values $\{D_i\}$ for $i = 1,\ldots,N$; and also standard deviations of uncorrelated errors $\{\alpha_i\}$ and of $K$ sources of correlated systematic errors $\{\beta_{ki}\}$ where $k = 1,\ldots,K$. The theoretical value $T_i(a)$ depends on a set of PDF model parameters $\{a_\lambda\}$ for $\lambda = 1,\ldots,d$. The data value $D_i$ differs from $T_i(a)$ (for the optimal PDF parameters $\{a_\lambda\}$) because of experimental errors, statistical


doi:10.1016/S0920-5632(02)02128-X

and systematic,

$$D_i = T_i(a) + \alpha_i r_{\text{stat},i} + \sum_{k=1}^{K} r_k \beta_{ki} \ ; \qquad (1)$$

here $r_{\text{stat},i}$ and $r_k$ are random variables, with standard deviation 1, representing the statistical and systematic measurement errors.

To fit the PDF model to the data, we define a $\chi^2$ function for each data set by

$$\chi'^2(a,r) = \sum_{i=1}^{N} \frac{(D_i - \sum_k r_k \beta_{ki} - T_i)^2}{\alpha_i^2} + \sum_k r_k^2 . (2)$$

$\chi'^2(a,r)$ depends on the PDF model parameters $a \equiv \{a_\lambda\}$ and the set of *systematic shifts* $r \equiv \{r_k\}$. We then minimize the global $\chi^2$ function—the sum of $\chi'^2$'s over all data sets—with respect to the shifts and model parameters.[1] Because we use a quadratic penalty term $r_k^2$ for each systematic shift, the minimization with respect to $\{r_k\}$ can be done analytically for arbitrary $a$. Then the minimization in the $d$-dimensional parameter space of $a$ is done numerically. For CTEQ6 the total number of fitting parameters for the PDF's is $d = 20$.

Table 1 lists the final $\chi^2$ values for the data sets for which systematic errors are included in the fitting procedure. All $\chi^2$'s are near 1 per data point, so the PDF model fits the data within the measurement errors. The global $\chi^2$ (summed over all experiments) is 1954 for 1811 data points.

More detailed comparisons between theory and data are made by comparing theory to *data shifted by the optimized systematic errors*. Figure 1 is the case of the CDF inclusive jet cross section, showing the fractional difference between data and theory; the optimized systematic errors $\sum_k r_k \beta_{ki}$ have been subtracted from the data values, and the error bars are the statistical errors $\{\alpha_i\}$ only. Similarly, for the other data sets in Table 1, the PDF model consistently fits the data with the systematic shifts subtracted, within statistical errors.

The CTEQ6 PDF's differ from the previous generation (CTEQ5), slightly for the quark distributions in the intermediate range of $x$, and more significantly for the gluon distribution $g(x,Q)$ especially at large $x$. Figure 2 shows $xg(x,Q)$ versus $x$ for $Q = 2$ and 100 GeV, focusing on large $x$. The dashed and dot-dashed curves are CTEQ6 and CTEQ5, respectively. (All comparisons in this paper are for parton distributions in the $\overline{MS}$ renormalization and factorization scheme, e.g., CTEQ6M and CTEQ5M1.) CTEQ6 has a hard gluon distribution, i.e., larger gluon density at large $x$ than the previous standard PDF set. The DØ data on jet production in $p\bar{p}$ collisions, as a function of rapidity and transverse energy, imply this hard gluon distribution.

Figure 1. Fractional difference between data and theory for the CDF inclusive jet cross section.

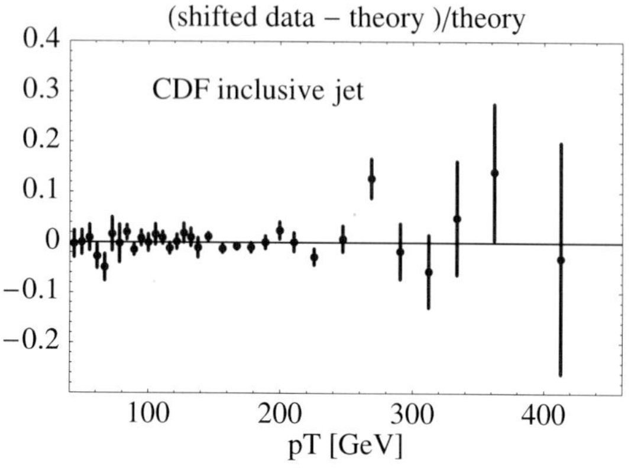


## 3. Quantitative uncertainties of PDF's

We have developed two methods for analyzing PDF uncertainties, Hessian [3] and Lagrange multiplier [4]. The Hessian method constructs alternative PDF sets by varying the position in the $d$-dimensional parameter space along the +/− directions of the $d$ eigenvectors of the Hessian, for a specified increase in $\chi^2_{\text{global}}$. From detailed comparisons we conclude that a rather large $\Delta\chi^2_{\text{global}}$, of order 100, is a reasonable estimate of the PDF uncertainty.[1]

As one example of uncertainty analysis, we may consider the accuracy of the determination

Figure 2. Comparison of CTEQ6 and CTEQ5 gluon distributions.

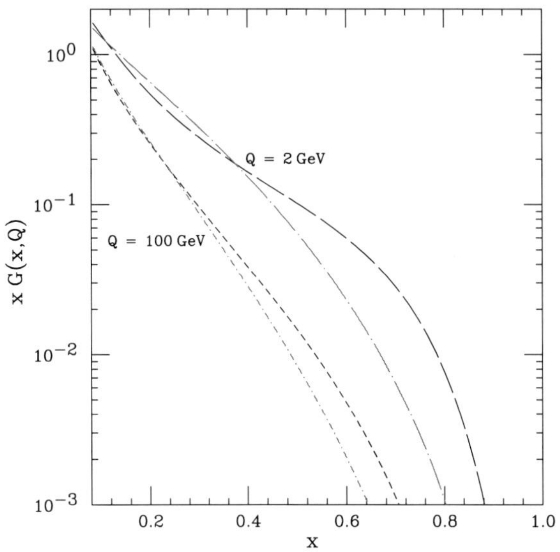


Figure 3. $\alpha_S(M_Z)$ from global analysis.

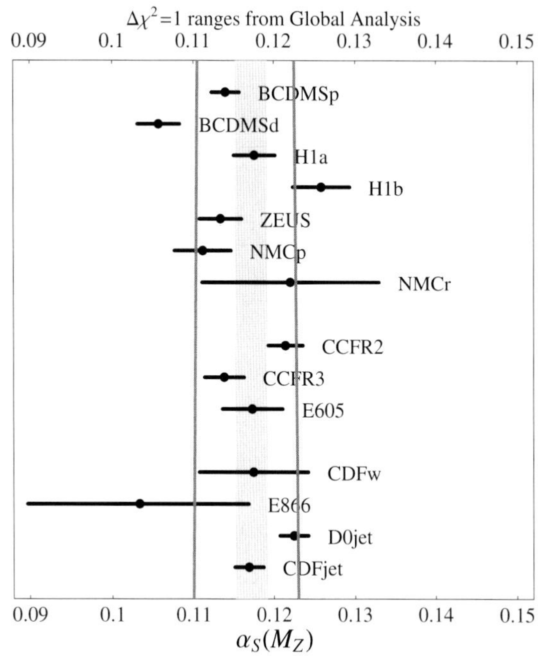


of $\alpha_S(M_Z)$ from the global analysis. For the standard CTEQ6 fit the value of $\alpha_S(M_Z)$ is fixed at 0.118, consistent with the Particle Data Group value. For the uncertainty analysis we vary $\alpha_S(M_Z)$, refit the data, and inspect the quality of the fit as a function of $\alpha_S$. Each data set implies a best value of $\alpha_S$, from the minimum of $\chi^2$ for that data; and an allowed range of $\alpha_S$, from the criterion $\chi^2 - \chi_0^2 \leq 1$ where $\chi_0^2$ is the minimum $\chi^2$ for that data.

The $\alpha_S$ values and ranges are plotted in Fig. 3 for 14 data sets used in the CTEQ6 fitting. The narrow band is the Particle Data Group value $0.117 \pm 0.002$.[5] The fluctuations in $\alpha_S$ are larger than expected for normal statistics, but this is not surprising because systematic errors exist. The vertical lines in Fig. 3 have $\Delta\chi^2_{\text{global}} = 100$, corresponding to our limits on $\alpha_S$ from the global analysis, $\alpha_S(M_Z) = 0.1165 \pm 0.0065$.

The CTEQ6 parton distributions are available at www.cteq.org in various forms: $\overline{MS}$ scheme, DIS scheme, for leading-order hard cross sections, and 40 +/− sets in the eigenvector basis. They are also available in the LHAPDF Interface.

## REFERENCES

1. J. Pumplin, D. R. Stump, J. Huston, H.L. Lai, P. Nadolsky and W.K. Tung, J. High Energy Phys. JHEP **07** (2002) 012.
2. A. D. Martin, R. G. Roberts, W. J. Stirling and R. S. Thorne, Eur. Phys. J. **C23** (2002) 73.
3. J. Pumplin, D. Stump, R. Brock, D. Casey, J. Huston, J. Kalk, H. L. Lai, and W. K. Tung, Phys. Rev. **D65** (2002) 014013.
4. D. Stump, J. Pumplin, R. Brock, D. Casey, J. Huston, J. Kalk, H. L. Lai, and W. K. Tung, Phys. Rev. **D65** (2002) 014012.
5. K. Hagiwara et al., Phys. Rev. **D66** (2002) 010001.

ELSEVIER

Proceedings of ICHEP 2002, pp. 268–271
*S. Bentvelsen, P. de Jong, J. Koch and E. Laenen (Editors)*

31st INTERNATIONAL CONFERENCE ON HIGH ENERGY PHYSICS AMSTERDAM

# The Les Houches accords: new tools for high energy physics

J. Huston [a*]

[a]Physics and Astronomy Dept., Michigan State University
East Lansing, Michigan 48824 USA
huston@msu.edu

Two new tools developed at the 2001 Les Houches *Physics at TeV Colliders Workshop* are described. First, generic Fortran common blocks are presented for use by high energy physics event generators for the transfer of event configurations from parton level generators to showering and hadronization event generators, and second, a universal interface for the inclusion of parton distribution functions into any matrix element or Monte Carlo program is described.

## 1. Introduction

There are a variety of theoretical tools that have been developed to enable experimentalists and phenomenologists to make predictions for final states at hadron-hadron, lepton-hadron and lepton-lepton colliders. These tools include leading order tree-level matrix element generators (MEG), parton showering and hadronization Monte Carlo programs (SHG), higher order (NLO and above) calculations, and programs in which the effects of soft gluon radiation are resummed [1]. It is important to understand the strengths and weaknesses of each approach and, in particular, the venues where two different approaches can be combined.

For example, matrix element generators (MEG) describe multi-body final states more accurately than do parton showering programs (SHG). But the SHG programs allow for the production of additional jets due to the effects of gluon radiation, as well as allow for the inclusion of hadronization effects. The latter is crucial, in particular, if the output is then to be processed through a detector simulation program.

The strengths of both MEG and SHG programs can be utilized if an interface is adopted for the transfer of parton level event configurations from matrix element generators to showering and hadronization generators. Specific interfaces, such as from the matrix element program Wbbgen [2] to the Herwig [3] parton showering Monte Carlo, or from the program CompHEP [4] to the Monte Carlo program Pythia [5] have already been in use.

## 2. The Les Houches accord

At the 2001 Les Houches workshop [6], authors of popular MEG and SHG programs, as well as interested users, agreed on a generic format (*the Les Houches accord)* for the transfer of parton level event configurations from matrix element generators to parton showering and hadronization event generators. Such an interface is intended to be universal and allows the transferal of the output of any MEG program to the input of any SHG program. The interface provides information on the parton 4-vectors, mother-daughter relationships, spins/helicities and color flows to the SHG programs, as well as pointing to intermediate particles whose mass should be preserved in the parton showering process.

This interface is not intended as a replacement for `HEPEVT` [7], which is the standard Fortran common block for interfacing between generators and and analysis/detector simulation. Instead, the Les Houches interface addresses communication between MEG and SHG programs only and not the communication of event generators with the

*for the Les Houches Monte Carlo and Parton Distribution Working Groups


doi:10.1016/S0920-5632(02)02129-1

outside world.

There are two stages for information transfer: (1) at initialization, to establish the basic parameters of the run as a whole and (2), for each event that is to be transferred from the MEG to the SHG. Each of these two stages has its own Fortran common block. A complete description of the Les Houches accord is given in Ref. [[8]].

With this interface the SHG can unweight events through the MEG, mix different subprocesses from the MEG or just read events straight from a file. If unweighting/mixing is required, then the SHG needs information about the subprocess cross sections and/or maximum weights. Note that negative weights are allowed as well. Such an option may be crucial for the interface to a SHG of a NLO calculation. [2]

**Example: hadronic $t\bar{t}$ production**

Consider the specific example of $t\bar{t}$ production.

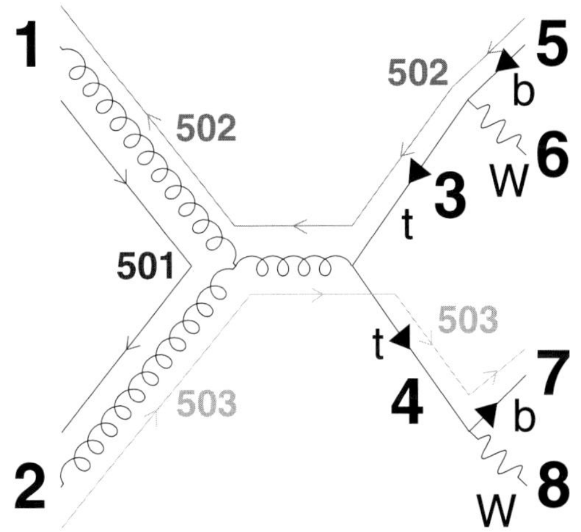


[b]

| I | ISTUP(I) | IDUP(I) | MOTHUP(1,I) | MOTHUP(2,I) | ICOLUP(1,I) | ICOLUP(2,I) |
|---|---|---|---|---|---|---|
| 1 | −1 | 21 ($g$) | 0 | 0 | 501 | 502 |
| 2 | −1 | 21 ($g$) | 0 | 0 | 503 | 501 |
| 3 | +2 | −6 ($\bar{t}$) | 1 | 2 | 0 | 502 |
| 4 | +2 | 6 ($t$) | 1 | 2 | 503 | 0 |
| 5 | +1 | −5 ($\bar{b}$) | 3 | 3 | 0 | 502 |
| 6 | +1 | −24 ($W^-$) | 3 | 3 | 0 | 0 |
| 7 | +1 | 5 ($b$) | 4 | 4 | 503 | 0 |
| 8 | +1 | 24 ($W^+$) | 4 | 4 | 0 | 0 |

The figure and the table above display some of the conventions adopted in the Les Houches accord. In the first column, I refers to the parton number in the Feynman diagram above. The second column in the table (ISTUP(I)) refers to the status code of the particles in the event: -1 refers to the initial state, +1 to the final state and +2 to those intermediate particles whose mass must be preserved when showering and hadronizing the event. Thus, in this example, the mass of both the $t$ and $\bar{t}$ are preserved. IDUP(I) is the particle ID, using the Particle Data Group convention [9] and MOTHUP(1,I) and MOTHUP(2,I) refer to the indices of the first and last mothers for each particle.

A specific choice of color flow for a particular event is often unphysical due to interference effects, but SHG's require a specific color state from which to begin the shower. The color flow for this example is given in the last two columns and is also indicated on the figure. The definition of a line as 'color' or 'anti-color' depends on the orientation of the graph. This ambiguity is resolved by defining color and anti-color according to the physical time order. A quark will always have its color tag ICOLUP(1,I) filled, but never its anti-color tag ICOLUP(2,I). The reverse is true for an anti-quark, and a gluon will always have information in both ICOLUP(1,I) and ICOLUP(2,I) tags.

As noted earlier, to be most useful, the Les Houches accord needs to be universal. Thus far, the accord has been incorporated into the MEG programs ALPGEN [10], CompHEP [4], GR@PPA [11], MADEVENT [12], MCFM [13] and the SHG programs HERWIG [14] and PYTHIA [5].

[2]Don't try this at home!

## 3. LHAPDF

Information on up-to-date parton distribution functions (pdf's) is also essential for predictions at hadron-hadron and lepton-hadron colliders. Such pdf's have been provided by a number of groups and are updated when new experimental data and/or theoretical treatments become available. Libraries such as PDFLIB [15] have been established that maintain a large collection of available pdf's.

However, PDFLIB is no longer supported making it more difficult for easy access to the most up-to-date pdf's. In addition, the determination of the pdf uncertainty of any cross section typically involves the use of a large number of pdf's (on the order of 40-100) and the manner in which the pdf's are stored in PDFLIB (grids in $x$ and $Q$) make storage of such ensembles very unwieldy.

At Les Houches in 2001, representatives from a number of pdf groups were present and an interface (Les Houches accord 2, or LHAPDF) [6] that allows the compact storage of the information needed to define a pdf. Each pdf is determined by only a few lines of information (basically the starting values of the parameters at $Q = Q_o$) and the interface carries out the evolution to any $x$ and $Q$ value, at either LO or NLO as appropriate for each pdf.

The interface is as easy to use as PDFLIB [15] and consists essentially of 3 subroutine calls:

- call InitPDFset(*name*): called once at the beginning of the code; *name* is the file name of the external pdf file that defines the pdf set (for example, CTEQ [16], GKK [17] or MRST [18])

- call InitPDF(*mem*): *mem* specifies the individual member of the pdf set

- call evolvePDF($x$,$Q$,$f$): returns the pdf momentum densities for flavor $f$ at a momentum fraction $x$ and scale $Q$

The interface can be downloaded at pdf.fnal.gov. It is currently included in the matrix element program MCFM and will be included in future versions of the other MEG and SHG programs. More details can also be found in the talk by W. Giele at this conference. [19]

## 4. Conclusions

A reminder of the big idea: the Les Houches accords will be implemented in all MEG/SHW programs used by experimentalists and phenomenologists. The accords will make is easier to generate the multi-particle final states crucial to understand Tevatron Run 2/HERA/LHC physics, as well as to generate the central value predictions and pdf uncertainties for any cross section.

## REFERENCES


1. See for example, the talks and archived video for the *Workshop on Monte Carlo Physics for Run 2 at the Tevatron*, http://thpc20.fnal.gov/runiimc/ .
2. F. Caravaglios, M. L. Mangano, M. Moretti and R. Pittau, Nucl. Phys. B **539**, 215 (1999), [hep-ph/9807570].
3. G. Corcella *et al.*, JHEP **0101**, 010 (2001), [hep-ph/0011363].
4. A. Pukhov *et al.*, [hep-ph/9908288]; A. S. Belyaev *et al.*, [hep-ph/0101232].
5. T. Sjöstrand, L. Lönnblad and S. Mrenna, LU TP 01-21, [hep-ph/0108264].
6. See the website for the 2001 Workshop on Physics at TeV Colliders, http://wwwlapp.in2p3.fr/conferences/Leshouches/Houches2001/.
7. T. Sjöstrand et al., in "Z physics at LEP 1", eds. G. Altarelli, R. Kleiss and C. Verzegnassi, CERN 89-08 (Geneva, 1989), Vol. 3, p. 327.
8. Generic User Process Interface for Event Generators, E. Boos, M. Dobbs, W. Giele, I. Hinchliffe, J. Huston, V. Ilyin, J. Kanzaki, K. Kato, Y. Kurihara, L. Lonnblad, M. Mangano, S. Mrenna, F. Paige, E. Richter-Was, M. Seymour, T. Sjostrand, B. Webber, D. Zeppenfeld, [hep-ph/0109068].
9. L. Garren, I. G. Knowles, T. Sjostrand and T. Trippe, Eur. Phys. J. C **15**, 205 (2000).
10. M.L. Mangano, M. Moretti, F. Piccinini, R. Pittau, A.D. Polosa, [hep-ph/0206293].
11. S. Tsuno et al., submitted to Comput. Phys.

Commun., [hep-ph/0204222; K. Sato et al., [hep-ph/0104237].
12. T. Stelzer and W. F. Long, Comput. Phys. Commun. **81**, 357 (1994), [hep-ph/9401258]; F. Maltoni, T. Stelzer, [hep-ph/0208156].
13. J. Campbell, K. Ellis, Phys.Rev.D**65**: 113007,2002, [hep-ph/0202176].
14. The Les Houches accord is already included in the version of Herwig bundled with Alpgen and will appear in the next regular release, Herwig 6.5; P. Richardson, private communication.
15. H. Plothow-Besch, Comput. Phys. Commun. **75**, 396 (1993); CERN Program Library Long Writeup W5051 (2000); refer to `http://consult.cern.ch/writeup/pdflib/`.
16. J. Pumplin, D.R. Stump, J. Huston, H.L. Lai, P. Nadolsky, W.K. Tung, JHEP **0207**, 012 (2002), [hep-ph/0201195]..
17. W.T. Giele, S.A. Keller, D.A. Kosower, [hep-ph/0104052].
18. A.D. Martin, R.G. Roberts, W.J. Stirling, R.S. Thorne, Eur. Phys. J **C23**, 73 (2002), [hep-ph/0110215].
19. W. Giele, these proceedings.

ELSEVIER

Proceedings of ICHEP 2002, pp. 272–274
*S. Bentvelsen, P. de Jong, J. Koch and E. Laenen (Editors)*

31st INTERNATIONAL CONFERENCE ON HIGH ENERGY PHYSICS AMSTERDAM

# Production of gauge bosons plus jets in hadronic collisions *

R. Pittau[a]

[a]Dipartimento di Fisica Teorica, Università di Torino and INFN, sezione di Torino, Italy

## 1. Introduction

Multijet final states are characteristic of a large class of important phenomena present in high-energy hadronic collisions. QCD interactions generate multijet final states via radiative processes at high orders of perturbation theory (pure QCD processes). Heavy particles in the Standard Model (SM), such as $W$ and $Z$ bosons or the top quark, $t$, decay to multiquark configurations (eventually leading to jets) via electroweak (EW) interactions (mixed EW-QCD processes).

In addition to the above SM sources, particles possibly present in theories beyond the SM are expected to decay to multiparton final states, and therefore to lead to multijets. Typical examples are the cascade decays to quarks and gluons of supersymmetric strongly interacting particles, such as squarks and gluinos (SUSY-QCD processes).

In addition to fully hadronic multijet final states, a special interest exists in final states where the jets are accompanied by gauge bosons. For example

$$p \overset{(-)}{p} \to Z + \ jets \to \nu\bar{\nu} + jets \tag{1}$$

is an important background to SUSY searches. Likewise,

$$p \overset{(-)}{p} \to W + \ 4 \ jets \to \ell \ \bar{\nu}_\ell + 4 \ jets \tag{2}$$

provides the leading source of backgrounds to the identification and study of top quark pairs in hadronic collisions.

Several parton-level Monte Carlo (MC) event generators exist in the literature (such as VECBOS [1], MADGRAPH [2], CompHEP [3], GRACE [4], PAPAGENO [5]) and studies can be performed by assuming that hard partons can be identified with jets, and that the jets' momenta are equal to those of the parent partons. This simplification gives rough estimates, but cannot be used in the context of realistic detector simulations, for which a representation of the full structure of the final state (in terms of hadrons) is required.

This full description can be obtained by merging the partonic final states with shower MC programs where partons are perturbatively evolved through emission of gluons, and subsequently hadronized.

However, this merging is not always possible, since common parton-level MC's sum and average over flavours and colours, and do not usually provide sufficient information on the flavour and colour content of the events.

In this contribution we review a strategy for the construction of event generators for multijet final states, based on the exact leading-order evaluation of the matrix elements for assigned flavour and colour configurations, and the subsequent shower development and transition into a fully hadronized final state.

We also present ALPGEN [6], a collection of codes realized within the framework of the presented approach.

As a case study, we then focus on the production of $WQ\bar{Q}+n$ jet final states [7] (with $Q$ being a massive quark, and $n \leq 4$) and present results for several production rates and distributions of interest at the LHC and at the TEVATRON.

## 2. The computational strategy

There are two main sources of problems: the matrix-element evaluation and the reconstruction of the colour flow. To illustrate our solutions we

*Work supported by the European Union under contract HPRN-CT-2000-00149.


doi:10.1016/S0920-5632(02)02130-8

take, for simplicity, the case of multigluon processes, as the extensions to cases with quarks and EW particles follow the same pattern.

The process $gg \to 10\, g$ has $5\times 10^9$ contributing Feynman diagrams. To efficiently carry out the computation, an algorithm is therefore needed which does not make explicit use of Feynman diagrams. Fortunately, such an algorithm exists [8,9]. The key ingredient is the use of a recursive procedure to determine the matrix element directly from the effective action.

To better illustrate the problem with the color flow, we take again the case of multigluon production. The scattering amplitude for $n$ gluons with momenta $p_i^\mu$, helicities $\epsilon_i^\mu$ and colours $a_i$ (with $i = 1, \ldots, n$), can be written as

$$M(\{p_i\},\{\epsilon_i\},\{a_i\}) = \sum_{P(2,3,\ldots,n)} \mathrm{tr}(\lambda^{a_{i_1}}\lambda^{a_{i_2}}\ldots\lambda^{a_{i_n}}) \quad A(\{p_{i_1}\},\{\epsilon_{i_1}\};\ldots\{p_{i_n}\},\{\epsilon_{i_n}\}), \qquad (3)$$

where the sum extends over all permutations $P_i$ of indices $(2, 3, \ldots, n)$, and the functions $A(\{P_i\})$ (known as *dual* or *colour-ordered* amplitudes) are gauge invariant, cyclically-symmetric functions of the gluons' momenta and helicities.

Each dual amplitude $A(\{P_i\})$ corresponds to colour flows from one gluon to the next, according to the ordering specified by the permutation of indices. Furthermore, each different colour ordering corresponds to a different shower evolution of the event, once the hadronization programs are turned on. Therefore, any realistic event generator should keep track, event by event, of the colour flow (and flavour content) to perform a reliable simulation of the subsequent evolution into fully hadronized final states.

More details on our algorithm for the colour reconstruction can be found in ref. [10].

## 3. The ALPGEN project

ALPGEN [6] is a collection of codes for the generation of multi-parton processes in hadronic collisions implementing the computational strategy described in the previous section.

The processes currently available in the package are:

1) $p\overset{(-)}{p} \to WQ\bar{Q} + n$ jets $(n \le 4)$
2) $p\overset{(-)}{p} \to Z/\gamma^* Q\bar{Q} + n$ jets $(n \le 4)$
3) $p\overset{(-)}{p} \to W + n$ jets $(n \le 6)$
4) $p\overset{(-)}{p} \to Z/\gamma^* + n$ jets $(n \le 6)$
5) $p\overset{(-)}{p} \to n_z\, Z + n_w\, W + n_h\, H + n$ jets $(n_z + n_w + n_h + n \le 8,\, n \le 3)$
6) $p\overset{(-)}{p} \to Q\bar{Q} + n$ jets $(n \le 6)$
7) $p\overset{(-)}{p} \to Q\bar{Q}Q'\bar{Q}' + n$ jets $(n \le 4)$
8) $p\overset{(-)}{p} \to HQ\bar{Q} + n$ jets $(n \le 4)$

$Q$ and $Q'$ being massive quarks. Anywhere, except for processes 5), $W$ and $Z/\gamma^*$ are considered off-shell, meaning that the corresponding produced final state is a pair of leptons.

The documentation and the codes can be found at `http://mlm.home.cern.ch/mlm/alpgen/`.

## 4. Results

We show a few illustrative examples obtained with ALPGEN for the process $p\overset{(-)}{p} \to Wb\bar{b} + n$ jets. At the LHC one gets the partonic rates of Table 1. The numbers in the first column refer to the subprocesses given in ref. [7]. The qualitatively new processes appearing when $N_J$ increases are large and are responsible for the growth of the partonic rates when going from $N_J = 2$ to $N_J = 3$.

In Figure 1 we show a comparison between the $p_T$ distributions of jets at the TEVATRON before and after the perturbative phase of the shower evolution performed by HERWIG [11]. A good matching of the jet spectra is achieved when using generation (parton level) cuts loser than jet-defining cuts. Finally, we address the issue of the ability of the shower MC to correctly predict the rate for hard radiation leading to extra final-state jets. In Figure 2 we compare the jet rates evaluated with exact matrix elements with those obtained from hard radiation during the shower evolution of lower-order parton-level processes. We find a good agreement in the range $p_T \lesssim 45$ GeV, for the radiation of one extra jet by HERWIG, while not enough hard radiation is emitted by HERWIG to correctly predict the emission of 2 extra jets.

Table 1
Partonic rates in pb at the LHC, as a function of the total number of jets $N_J$ (including $b$ and $\bar{b}$).

| Process | $N_J = 2$ | $N_J = 3$ | $N_J = 4$ | $N_J = 5$ | $N_J = 6$ |
|---|---|---|---|---|---|
| 1 | 2.60(1) | 0.63(1) | 0.144(3) | 0.036(2) | 0.008(1) |
| 2 | – | 2.97(1) | 2.11 (2) | 1.08(2) | 0.47(2) |
| 3+4 | – | – | 0.288(1) | 0.24(1) | 0.13(2) |
| 5 | – | – | – | 0.030(1) | 0.031(4) |
| 6 | – | – | – | – | 0.0010(3) |
| Total | 2.60(1) | 3.60(1) | 2.54(2) | 1.38(2) | 0.64(3) |

The jet defining cuts are $p_T^i > 20$ GeV, $|\eta_i| < 2.5$ and $\Delta R_{ij} > 0.4$.

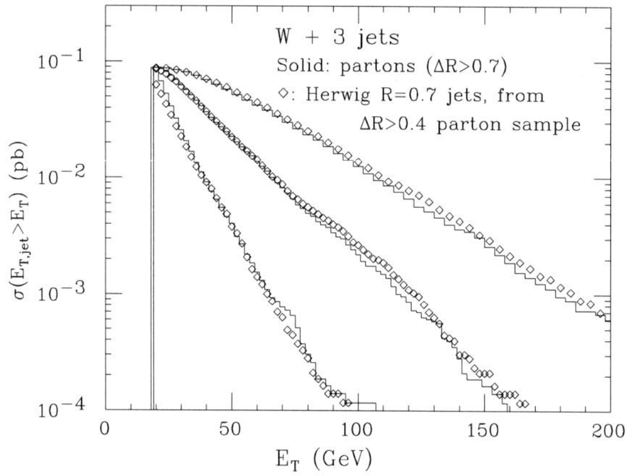


Figure 1. Inclusive $p_T$ distributions of jets at the parton level, with separation cut $\Delta R > 0.7$ (solid curves), and of fully-showered $R_{jet} = 0.7$ jets (dashed curves). These last are obtained starting from a full sample of $\Delta R > 0.4$ partonic events.

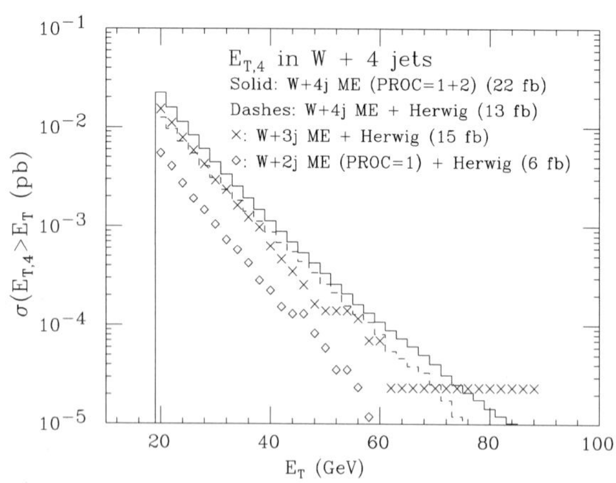


Figure 2. $p_T$ distributions of jets at the TEVATRON: Matrix Element versus Parton Shower.

## 5. Summary

The formalism we have outlined allows us to compute several complex multiparton processes.

ALPGEN is a collection of codes realized within the framework of the presented approach. It can be used to perform realistic studies of multiparticle processes at TEVATRON and LHC.

ALPGEN is ready-to-use and available at `http//mlm.home.cern.ch/mlm/alpgen/`.

## REFERENCES


1. F. A. Berends, H. Kuijf, B. Tausk and W. T. Giele, Nucl. Phys. B **357**, (1991) 32;
2. T. Stelzer and W. F. Long, Comput. Phys. Commun. **81**, (1994) 357;
3. A. Pukhof *et al.*, hep-ph/9908288;
4. T. Ishikawa *et al.*, [MINAMI-TATEYA group Coll.], KEK-92-19;
5. I. Hinchliffe, LBL-34372, *Submitted to Workshop on Physics at Current Accelerators and the Supercollider*, Argonne, IL, 2-5 June 1993;
6. M.L. Mangano, M. Moretti, F. Piccinini, R. Pittau, A. Polosa, hep-ph/0206293;
7. M.L. Mangano, M. Moretti, R. Pittau, Nucl. Phys. B **632**, (2002) 343;
8. F. Caravaglios and M. Moretti, Phys. Lett. B **358**, (1995) 332;
9. P. D. Draggiotis, R. H. Kleiss and C. G. Papadopoulos, Eur. Phys. J. C **24**, (2002) 447;
10. F. Caravaglios, M. L. Mangano, M. Moretti, R. Pittau, Nucl. Phys. B **539**, (1999) 215;
11. G. Marchesini *et al.*, Comput. Phys. Commun. **67**, (1992) 465.

Proceedings of ICHEP 2002, pp. 275–278
*S. Bentvelsen, P. de Jong, J. Koch and E. Laenen (Editors)*

# QCD soft gluon exponentiation: YFS MC Approach*

B.F.L. Ward[a†] and S. Jadach[b]

[a]Werner-Heisenberg-Institut, Max-Planck-Institut feur Physik,
Foehringer Ring 6, Muenchen, Germany

[b]Henryk Niewodniczanski Institute of Nuclear Physics,
ul. Radzikowskiego 152, 31-342 Cracow, Poland

We develop and prove the theory of the QCD extension of the YFS Monte Carlo approach to higher order QED radiative corrections. As a corollary, a new approach to quantum gravity by one of us (B.F.L.W.) is illustrated. Semi-analytical results and preliminary explicit Monte Carlo data are presented for the processes $p\bar{p} \rightarrow t\bar{t} + X$ at FNAL energies. We comment briefly on the implications of our results on the CDF/D0 observations and on RHIC/LHC physics.

## 1. Introduction

The problem of soft gluon resummation is well known [1,2] and some of its many phenomenological applications are also: the FNAL $t\bar{t}$ production cross section higher order corrections ( the current situation [3,4] has the experimental cross section $6.2^{+1.2}_{-1.1}$pb to be compared with a theoretical prediction [5] of 5.1±0.5pb) and the attendant soft gluon uncertainty in the extracted value of $m_t = 0.1743 \pm 0.0051$TeV, where [6] $\sim 2-3$GeV of the latter error could be due to soft gluon uncertainties; RHIC hard scattering polarized pp scattering processes, etc. For the LHC/TESLA/LC, the requirements on the corresponding theoretical precisions will be even more demanding and the QCD soft $n(G)$ MC exponentiation which we discuss in the following will be an important part of the necessary theory – YFS [7] exponentiated $\mathcal{O}(\alpha_s^2)L$ corrections realized on an event-by-event basis.

The results which we present also will allow us to investigate from a different perspective some of the outstanding theoretical issues in perturbative QCD, such as the treatment of phase space, no-go theorems for the soft regime, etc. – see ref. [8].

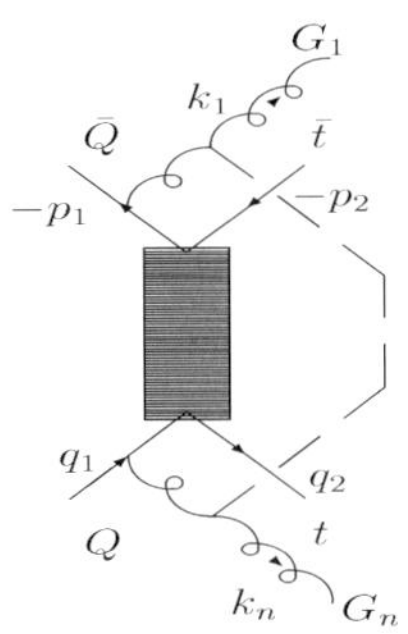


Figure 1. The process $\bar{Q}Q \rightarrow \bar{t} + t + n(G)$. The four–momenta are indicated in the standard manner: $q_1$ is the four–momentum of the incoming $Q$, $q_2$ is the four–momentum of the outgoing $t$, etc., and $Q = u, d, s, c, b, G$.

For definiteness, we will use the process in Fig. 1 , $\bar{Q}(p_1)Q(q_1) \rightarrow \bar{t}(p_2)t(p_1) + G_1(k_1)\cdots G_n(k_n)$, as our proto-typical process. Extension of the methods we develop to other related processes will be immediate – one of us (B.F.L.W.) has realized a new approach to quan-

*Work partly supported by the US Department of Energy Contract DE-FG05-91ER40627 , by NATO Grant PST.CLG.977751, and by Polish Government grant 5P03B09320.
†Permanent Address: Department of Physics, University of Tennessee, Knoxville, TN 37996-1200, USA.


doi:10.1016/S0920-5632(02)02131-X

tum gravity as a by-product. Although we shall use the older EEX formulation of YFS MC exponentiation as defined in ref. [9], the realization of our results via the the newer CEEX formulation of YFS exponentiation in ref. [10] is also possible and is in progress [11].

After a brief review of the QED case, we prove our result for the QCD case and conclude with some illstrative results.

## 2. Review of YFS Theory: An Abelian Gauge Theory Example

The Abelian gauge theory example of QED has been worked-out and realized in many applications for SLC/LEP1 and LEP2 physics in the MC's YFS2, YFS3, BHLUMI, BHWIDE, KORALZ, $\mathcal{KK}$ MC, YFSWW3, YFSZZ and KoralW by S. Jadach *et al.* in refs. [9,10,12]. For example, for the process $e^+(p_1)e^-(q_1) \to \bar{f}(p_2)f(q_2) + n(\gamma)(k_1, \cdot, k_n)$, renormalization group improved YFS theory [13] gives

$$d\sigma_{exp} = e^{2\alpha\ Re\ B + 2\alpha\ \tilde{B}} \sum_{n=0}^{\infty} \frac{1}{n!} \int \prod_{j=1}^{n} \frac{d^3k_j}{k_j^0} \int \frac{d^4y}{(2\pi)^4} e^{iy(p_1+q_1-p_2-q_2-\sum_j k_j)+D} \bar{\beta}_n(k_1,\ldots,k_n) \frac{d^3p_2 d^3q_2}{p_2^0 q_2^0}$$

where the YFS infrared functions $\tilde{B}$, $B$ and the IR finite function $D$ are known [7]. For example, the YFS hard photon residuals $\bar{\beta}_i$ in (1), $i = 0, 1, 2$, are given in refs. [12] for BHLUMI 4.04 realizing the YFS exponentiated exact $\mathcal{O}(\alpha)$ and LL $\mathcal{O}(\alpha^2)$ cross section for Bhabha scattering via a corresponding Monte Carlo implementation of (1). Here, we will develop and apply the analogous theoretical paradigm to the prototypical QCD higher order radiative corrections problem for $Q\bar{Q} \to t\bar{t} + n(G)$.

## 3. Extension to non-Abelian Gauge Theories: Proof

Viewing the YFS theory as a general rearrangement of renormalized perturbation theory based on its IR behavior just as the renormalization group as a general property of renormalized perturbation theory based on its UV(ultra-violet) behavior, for the process $q_i + \bar{q}_i \to t\bar{t} + n(G)$, let the amplitude be $\mathcal{M}^{(n)\alpha\bar{\alpha}}_{\gamma\bar{\gamma}} = \sum_\ell M^{(n)\alpha\bar{\alpha}}_{\gamma\bar{\gamma}\ell}$, where $M^{(n)}_\ell$ is the contribution to $\mathcal{M}^{(n)}$ from Feynman diagrams with $\ell$ virtual loops. If $S_{QCD}(k)$, as defined in ref. [8] is the virtual gluon IR emission factor, we get the "YFS representation", by analogy with the YFS analysis,

$$\mathcal{M}^{(n)} = exp(\alpha_s B_{QCD}) \sum_{j=0}^{\infty} \mathsf{m}_j^{(n)}, \tag{1}$$

where $\alpha_s(Q)B_{QCD}$ is the usual YFS integral [7,8] of $S_{QCD}(k)$ and $\mathsf{m}_j^{(n)}$ do not contain the virtual IR singularities in the product $S_{QCD}(k_1)\cdots S_{QCD}(k_i)$.

For the respective real IR singularities, we proceed as in ref. [8] so that, for the real emission factor $\tilde{S}_{QCD}(k)$ defined in ref. [8], upon effecting the analogous YFS expansion in $\tilde{S}_{QCD}(k)$ and summing on $n$, we arrive at the "YFS-like" result

$$d\hat{\sigma}_{\rm exp} = \sum_n d\hat{\sigma}^n = e^{\rm SUM_{IR}(QCD)} \sum_{n=0}^{\infty} \int \prod_{j=1}^{n} \frac{d^3k_j}{k^0{}_j} \int \frac{d^4y}{(2\pi)^4} e^{iy\cdot(p_1+q_1-p_2-q_2-\sum k_j)+D_{\rm QCD}} * \bar{\beta}_n(k_1,\ldots,k_n) \frac{d^3p_2}{p_2^0}\frac{d^3q_2}{q_2^0}, \tag{2}$$

where the IR finite functions $SUM_{IR}(QCD)$, $D_{\rm QCD}$ are defined in ref. [8] and the $\bar{\beta}_n$ are the QCD hard gluon residuals [8].

Each order in $\alpha_s$ must make an infrared finite contribution to $d\bar{\hat{\sigma}}_{\rm exp} \equiv \exp[-{\rm SUM_{IR}(QCD)}]d\hat{\sigma}_{\rm exp}$ since both $SUM_{IR}(QCD)$ and $d\hat{\sigma}_{\rm exp}$ are IR finite. If $\bar{\beta}^{(\ell)}_n = \tilde{\bar{\beta}}^{(\ell)}_n + D\bar{\beta}^{(\ell)}_n$, where $\tilde{\bar{\beta}}^{(\ell)}_n$ is completely free of IR divergences, then the IR finiteness just noted allows [8] us to conclude that the contributions from the $\{D\bar{\beta}^{(\ell)}_n\}$ cancel in $d\hat{\sigma}_{\rm exp}$ so that we arrive at the new result

$$d\hat{\sigma}_{\rm exp} = \sum_n d\hat{\sigma}^n = e^{\rm SUM_{IR}(QCD)} \sum_{n=0}^{\infty} \int \prod_{j=1}^{n} \frac{d^3k_j}{k_j} \int \frac{d^4y}{(2\pi)^4} e^{iy\cdot(p_1+q_1-p_2-q_2-\sum k_j)+D_{\rm QCD}} \tilde{\bar{\beta}}_n(k_1,\ldots,k_n)\frac{d^3p_2}{p_2^0}\frac{d^3q_2}{q_2^0}, \quad (3)$$

where the new hard gluon residuals $\tilde{\bar{\beta}}_n(k_1,\ldots,k_n) = \sum_{\ell=0}^{\infty} \tilde{\bar{\beta}}_n^{(\ell)}(k_1,\ldots,k_n)$ are completely free of IR divergences. Earlier arguments in ref. [14] were insufficient to derive the analog of (3) [8].

There are many consequences of (3) and the theory which underlies it [11]. We illustrate some of them in the next section. In addition, one of us (B.F.L.W.) recently pointed-out [15] that the YFS resummed propagators in quantum gravity, via the formula

$$i\Delta'_F(k)|_{YFS-resummed} = \frac{ie^{B''_g(k)}}{(k^2-m^2-\Sigma'_s+i\epsilon)}, \quad (4)$$

with the graviton exponent, in the deep UV, ( here, $\kappa^2 = 8\pi G_N$ )

$$B''_g(k) = \frac{\kappa^2|k^2|}{8\pi^2}\ln\left(\frac{m^2}{m^2+|k^2|}\right), \quad (5)$$

leads to exponential damping of all propagators in the deep UV and thereby by to UV finite loop corrections for quantum gravity. This is then a new approach [15] to quantum gravity.

## 4. YFS Exponentiated QCD Corrections to $t\bar{t}$ Production at High Energies

We have realized the result (3) via semi-analytical methods and via MC methods [8]. Here we illustrate these realizations.

For the process $p\bar{p} \to t\bar{t} + X$ at FNAL energies, we use a semi-analytical realization of (3) together with the standard formula structure function formula [8] for $\sigma(t\bar{t}+X)$, where the DGLAP synthesization procedure presented in ref. [16] is applied to avoid over-counting resummation effects already included in the structure function DGLAP evolution. In the MC realization, we employ the MC methods of ref. [9] to get the MC ttp1.0 as the respective event-by-event simulation. For the semi-analytical analysis, we have found in ref. [17] the normalization $n(G)$ effect ($r^{nls}_{exp}$ is the ratio of the exponentiated and Born cross sections) $r^{nls}_{exp} = 1.086$, $1.103$, $1.110$ for $\alpha_s$ evaluated at the scales $\sqrt{s}$, $2m_t$, $m_t$, respectively, which implies [8] that the corrections for $\mathcal{O}(\alpha_s^n, n \geq 2)$ give $0.006 - 0.008$ of the NLO cross section, in agreement with ref. [18]. This is an important cross check of our methods.

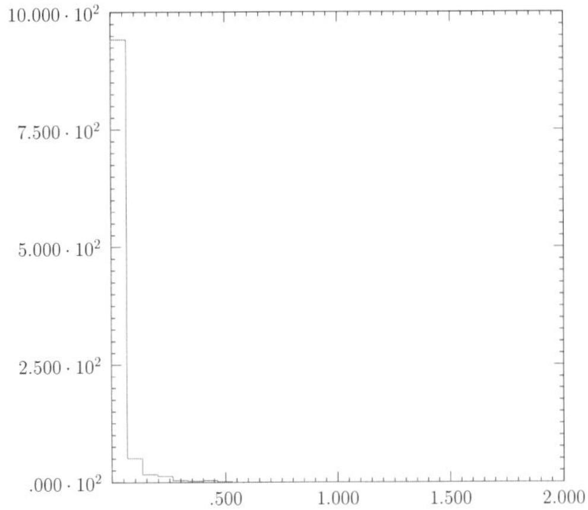


Figure 2. Gluon transverse momentum (TeV)

For the MC data from ttp1.0, which are preliminary, show in Fig. 2 where we see in the gluon transverse momentum distribution that $< p_\perp >$ is indeed large at FNAL and its effect must be taken into account in precision studies of the top quark production and decay systematics.

One of us (B.F.L.W.) thanks Prof. S. Bethke for the support and kind hospitality of the MPI, Munich, during the final stages of this work. The authors also thank Profs. G. Altarelli and Prof. Wolf-Dieter Schlatter for the support and kind hospitality of CERN while a part of this work was completed. These same two authors also thank Profs. F. Gilman and W. Bardeen of the former SSCL for their kind hospitality while this work was in its development stages.

## REFERENCES

1. G. Sterman, Nucl.Phys. **B281** (1987) 310, and references therein.
2. S. Catani and L. Trentadue, Nucl.Phys. **B327**

(1989) 323; *ibid.***B353** (1991) 183.
3. S. R. Blusk, in *Proc. ICHEP2000*, ed. C.S. Lim and T. Yamanaka, ( World Scientific, Singapore, 2001) p. 811.
4. D. Chakraborty, in *Proc. ICHEP2000*, ed. C.S. Lim and T. Yamanaka, ( World Scientific, Singapore, 2001) p. 814.
5. S. R. Willenbrock, in *Proc. RADCOR2000, Carmel, 2000*, eds. H. Haber and S. Brodsky, eConf C000911, (2001) 379; Rev. Mod. Phys. **72** (2000) 1141.
6. P. Grannis, private communication.
7. D. R. Yennie, S. C. Frautschi, and H. Suura, Ann. Phys. **13** (1961) 379; see also K. T. Mahanthappa, Phys. Rev. **126** (1962) 329, for a related analysis.
8. B.F.L. Ward and S. Jadach, Acta Phys. Pol. **B33** (2002) 1543, and references therein.
9. S.Jadach and B.F.L. Ward, Phys.Rev. **D38** (1988) 2897; *ibid.* **D39** (1989) 1471;*ibid.* **D40** (1989) 3582; Comput. Phys. Commun. **56** (1990) 351.
10. S. Jadach, B.F.L. Ward and Z. Was, Phys. Rev. D**63** (2001) 113009.
11. S. Jadach *et al.*, to appear.
12. See, for example, S. Jadach *et al.*, Phys. Lett. **B417** (1998) 326; Comput. Phys. Commun. **119** (1999) 272; Phys. Rev. **D61** (2000) 113010; Phys. Rev. **D65** (2002) 093010; Comput. Phys. Commun. **140** (2001) 432, 475; and references therein.
13. B.F.L. Ward, Phys. Rev. **D36** (1987) 939.
14. D. DeLaney *et al.*,Phys. Rev. D**52** (1995) 108, Phys. Lett. **B342** (1995) 239.
15. B.F.L. Ward, hep-ph/0204102.
16. B.F.L. ward and S. Jadach, Mod. Phys. Lett. **A14** (1999) 491.
17. S. Jadach *et al.*, Mod. Phys. Lett. **A12** (1997) 242.
18. S. Catani *et al.*, Phys. Lett. **B378** (1996) 329.

Proceedings of ICHEP 2002, pp. 279–281
*S. Bentvelsen, P. de Jong, J. Koch and E. Laenen (Editors)*

ELSEVIER

31st INTERNATIONAL CONFERENCE ON HIGH ENERGY PHYSICS AMSTERDAM

# All-orders infrared freezing of $R_{e^+e^-}$ in perturbative QCD

C.J. Maxwell[a]

[a]Centre for Particle Theory,University of Durham,
Durham DH1 3LE UK

Writing $R_{e^+e^-}(s) = 3\sum_f Q_f^2(1+\mathcal{R}(s))$, with $Q_f$ denoting the electric charge of quark flavour $f$, we show that for $N_f < 9$ flavours of massless quark, the perturbative correction $\mathcal{R}(s)$ to the parton model result smoothly approaches from below the infra-red limit $\mathcal{R}(0) = 2/b$, as $s{\to}0$. Here $b = (33-2N_f)/6$ is the first QCD beta-function coefficient. This freezing holds to all-orders in perturbation theory. The $s$-dependence can be written in closed form in terms of the Lambert $W$ function.

In the ultraviolet $s{\to}\infty$ limit of QCD the renormalized coupling vanishes. This property of Asymptotic Freedom means that perturbation theory can be used to make useful prodictions in this high energy regime. In the low energy $s{\to}0$ infrared limit, however, one may expect that perturbation theory will break down, with a Landau pole singularity in the coupling when $s{\sim}\Lambda^2_{QCD}$, and that non-perturbative effects will be important. In this talk, based on recent work with David Howe [1], I hope to convince you that in fact all-orders perturbation theory for the $R_{e^+e^-}(s)$ ratio is well-defined at all energies, and that there is a smooth freezing, with the perturbative corrections approaching from below the infra-red limit $\mathcal{R}(0) = 2/b$, with $b = (33-2N_f)/6$ the first beta-function coefficient for QCD with $N_f$ massless quark flavours. I will return later to discuss possible phenomenological applications of this seemingly rather abstract result.

We begin by defining the ratio $R_{e^+e^-}$ at c.m. energy $\sqrt{s}$,

$$R_{e^+e^-}(s) \equiv \frac{\sigma(e^+e^-{\to}\text{hadrons})}{\sigma(e^+e^-{\to}\mu^+\mu^-)} = 3\sum_f Q_f^2(1+\mathcal{R}(s))\,. \quad (1)$$

With $Q_f$ denoting the charge of quark flavour $f$. The perturbative correction to the parton model result, $\mathcal{R}(s)$, has the perturbation series,

$$\mathcal{R}(s) = a + \sum_{n>0} r_n a^{n+1}\,, \quad (2)$$

where $a{\equiv}\alpha_s(\mu^2)/\pi$. $r_1$ and $r_2$ have been computed in the $\overline{MS}$ scheme with renormalization scale $\mu^2 = s$ [2,3]. We can consider the $s$-dependence of $\mathcal{R}(s)$ at next-next-to-leading order (NNLO),

$$s\frac{d\mathcal{R}(s)}{ds} = -\frac{b}{2}\rho(\mathcal{R}) = -\frac{b}{2}\mathcal{R}^2(1+c\mathcal{R}+\rho_2\mathcal{R}^2)\,. \quad (3)$$

Here $c = (153-19N_f)/12b$ is the second universal QCD beta-function coefficient, and $\rho_2$ is the NNLO effective charge beta-function coefficient [4], an RS-invariant combination of $r_1, r_2$ and beta-function coefficients. The condition for $\mathcal{R}(s)$ to approach the infra-red limit $\mathcal{R}^*$ as $s \to 0$ is for the Effective Charge beta-function to have a non-trivial zero, $\rho(\mathcal{R}^*) = 0$. At NNLO the condition for such a zero is $\rho_2 < 0$. Putting $N_f = 2$ active flavours we find for the NNLO RS-invariant $\rho_2 = -9.72$. So that $\mathcal{R}(s)$ apparently freezes in the infra-red to $\mathcal{R}^* = 0.43$. The freezing behaviour was first investigated in a pioneering paper by Mattingly and Stevenson [5] in the context of the Principle of Minimal Sensitivity (PMS) approach. However, it is not obvious that we should believe this apparent NNLO freezing. In fact $\rho_2$ is dominated by a large $b^2\pi^2$ term arising from analytical continuation (AC) of the Euclidean Adler $D(-s)$ function to the Minkowskian $R(s)$, with $\rho_2 = 9.40 - \pi^2 b^2/12$. Similarly $\rho_3$ will contain the large AC term $-5c\pi^2 b^2/12$. This suggests that in order to check freezing we need to


doi:10.1016/S0920-5632(02)02132-1

resum the AC terms to *all-orders*.

The Adler $D$-function is the logarithmic energy derivative of the correlator of two vector currents $\Pi(s)$,

$$D(s) = -s\frac{d}{ds}\Pi(s). \quad (4)$$

The Minkowskian $R(s)$ is obtained by analytical continuation of the perturbative corections to $D(-s)$,

$$\begin{aligned}\mathcal{R}(s) &= \frac{1}{2\pi i}\int_{-s-i\epsilon}^{-s+i\epsilon} dt\frac{\mathcal{D}(t)}{t} \\ &= \frac{1}{2\pi}\int_{-\pi}^{\pi} d\theta\mathcal{D}(se^{i\theta})\,. \end{aligned} \quad (5)$$

Where $\mathcal{D}$ has the perturbation series,

$$\mathcal{D}(s) = a + \sum_{n>0} d_n a^{n+1}. \quad (6)$$

Expanding $\mathcal{D}(se^{i\theta})$ in powers of $a(se^{i\theta})$ and integrating term-by-term we then obtain the "contour-improved" perturbation series,

$$\mathcal{R}(s) = A_1(s) + \sum_{n+1}^{\infty} d_n A_{n+1}(s)\,. \quad (7)$$

The functions $A_n(s)$ resum at each order an infinite subset of AC terms present in the conventional perturbation series for $\mathcal{R}(s)$,

$$A_n(s) \equiv \frac{1}{2\pi}\int_{-\pi}^{\pi} d\theta a^n(se^{i\theta})\,. \quad (8)$$

For the simplified case of a one-loop coupling one has,

$$a(s) = \frac{2}{b\ln(s/\tilde{\Lambda}^2_{\overline{MS}})}\,. \quad (9)$$

The integrals are straightforward and one obtains,

$$\begin{aligned}A_1(s) &= \frac{2}{\pi b}\arctan\left(\frac{\pi b a(s)}{2}\right) \\ A_n(s) &= \frac{2a^{n-1}(s)}{b\pi(1-n)} Im\left[\left(1+\frac{ib\pi a(s)}{2}\right)^{1-n}\right] \\ &(n>1)\,. \end{aligned} \quad (10)$$

We then obtain the one-loop "contour-improved" perturbation series for $\mathcal{R}(s)$,

$$\begin{aligned}\mathcal{R}(s) &= \frac{2}{\pi b}\arctan\left(\frac{\pi b a(s)}{2}\right) \\ &+ d_1\left[\frac{a^2(s)}{(1+b^2\pi^2a^2(s)/4)}\right] \\ &+ d_2\left[\frac{a^3(s)}{(1+b^2\pi^2a^2(s)/4)^2}\right] + \ldots\,. \end{aligned} \quad (11)$$

As $s\to\infty$ the $A_n(s)$ vanish as required by Asymptotic Freedom. However, although the one-loop coupling has a "Landau pole" at $s = \tilde{\Lambda}^2_{\overline{MS}}$, the $A_n(s)$ are well-defined for all real $s$. As $s\to 0$, $A_1(s)$ smoothly approaches from below the infra-red value $2/b$, whilst for $n > 1$ the $A_n(s)$ vanish. Thus, as claimed, in the infra-red limit $\mathcal{R}(s)$ is asymptotic to $\mathcal{R}(0) = 2/b$ to all-orders in perturbation theory. A subtlety is that the contour integral definition of $A_n(s)$ in Eq.(8) is only valid for $s$ above the Landau pole. To define it for all $s$ one should use the dispersion relation relating $\mathcal{R}$ to $\mathcal{D}$, the first equality in Eq.(5). Correspondingly the second equality in Eq.(5) only holds above the Landau pole. Defining $A_n(s)$ using the dispersion relation for the Adler $D$-function results in an expansion equivalent to that in the Analytic Perturbation Theory approach of Shirkov and collaborators [6].

We now consider realistic QCD, beyond the one-loop approximation. It is convenient to work in a so-called 't Hooft scheme, where the non-universal beta-function coefficients are all zero, and the beta function has its two-loop form,

$$\frac{\partial a(\mu^2)}{\partial \ln\mu^2} = -\frac{b}{2}a^2(\mu^2)(1+ca(\mu^2))\,. \quad (12)$$

Crucially in such an RS we can express $a(\mu^2)$ analytically in closed form as [7],

$$\begin{aligned}a(\mu^2) &= -\frac{1}{c[1+W_{-1}(A(\mu^2))]} \\ A(\mu^2) &\equiv -\frac{1}{e}\left(\frac{\mu^2}{\tilde{\Lambda}^2_{\overline{MS}}}\right)^{-b/2c}\,. \end{aligned} \quad (13)$$

Here $W$ denotes the Lambert $W$ function, defined implicitly by $W(z)\exp(W(z)) = z$. The "$-1$"

subscript denotes the branch of the $W$ function required for Asymptotic Freedom, the nomenclature being that of [8]. Using the analytic form for the coupling in Eq.(8), and making the change of variable $w = W(A(se^{i\theta}))$, one arrives at the elementary integral,

$$A_n(s) = \frac{(-1)^{n+1}}{ic^{n-1}b}\int_{W_1(A(se^{-i\pi}))}^{W_{-1}(A(se^{i\pi}))}\frac{dw}{w(1+w)^{n-1}} . \quad (14)$$

We can then obtain the explicit results,

$$A_1(s) = \frac{2}{b} + \frac{2}{\pi b}Im[\ln(W_{-1}(A(se^{i\pi}))] , \quad (15)$$

and for $n > 1$,

$$A_n(s) = \frac{2(-1)^n}{bc^{n-1}\pi}Im[\ln\left(\frac{W_{-1}(A(se^{i\pi}))}{1+W_{-1}(A(se^{i\pi}))}\right) + \sum_{k=1}^{n-2}\frac{1}{k(1+W_{-1}(A(se^{i\pi})))^k}] . \quad (16)$$

Using properties of the Lambert $W$-function it is easy to check that the $A_n(s)$ vanish as $s \to \infty$, consistent with Asymptotic Freedom. As at one-loop, as $s \to 0$ the $A_n(s)$ for $n > 1$ vanish, whilst $A_1(s)$ approaches the infra-red limit $2/b$. So the 't Hooft scheme result approaches the infra-red limit $\mathcal{R}(0) = 2/b$ to all-orders in perturbation theory. Because it is an all-orders result the freezing behaviour holds with any choice of renormalization scheme.

Of course, the freezing result as it stands is of abstract significance since $s = 4m_\pi^2$ is the threshold in full QCD with massive quarks. For phenomenological studies one needs, as in Ref.[5], to include quark mass effects, and to employ a smearing procedure such as that introduced by Poggio, Quinn and Weinberg (PQW) [9]. One can then compare PQW smeared data for $R_{e^+e^-}$ with resonances averaged out, with the similarly smeared contour-improved perturbative prediction. One can also attempt to extract the Adler $D$-function in the Euclidean region from data and the contour-improved perturbation theory, and directly compare them. We are undertaking such a study at present [10]. A further interesting question concerns the meaning of the all-orders freezing result, given that all-orders perturbation theory is inherently ambiguous due to infrared renormalon effects. We hope to clarify this aspect of the problem further in forthcoming work [10].

## REFERENCES


1. D.M. Howe and C.J. Maxwell, Phys. Lett. **B541** (2002) 129.
2. S.L. Gorishny, A.L. Kataev, S.A. Larin, Phys. Lett. **B212** (1988) 238.
3. L.R. Surguladze, M.A. Samuel, Phys. Rev. Lett. **66** (1990) 560; S.G. Gorishny, A.L. Kataev, S.A. Larin, Phys. Let. **B259** (1991) 144.
4. G. Grunberg, Phys. Lett. **B95** (1980) 70; Phys. Rev. **D29** (1984) 2315.
5. A.C. Mattingly, P.M. Stevenson, Phys. Rev. **D49** (1994) 437.
6. D.V. Shirkov, Eur. Phys. J. 22 (2001) 331.
7. Einan Gardi, Georges Grunberg and Marek Karliner, JHEP 07 (1998) 007; M.A. Magradze, Int. J. Mod. Phys. A15 (2000) 2715.
8. R.M. Corless, G.H. Gonnett, D.E.G. Hare, D.J. Jeffrey, D.E. Knuth, Adv. Comput. Math. 5 (1996) 329.
9. E.C. Poggio, H.R. Quinn, S. Weinberg, Phys. Rev **D13** (1976) 1958.
10. D.M. Howe and C.J. Maxwell in preparation.

ELSEVIER

Proceedings of ICHEP 2002, pp. 282–284
*S. Bentvelsen, P. de Jong, J. Koch and E. Laenen (Editors)*

31st INTERNATIONAL CONFERENCE ON HIGH ENERGY PHYSICS AMSTERDAM

# Higgs Boson Production at Hadron Colliders

W.B. Kilgore[a]*

[a]Physics Department, Building 510A,
Brookhaven National Laboratory, Upton, New York 11973, USA

I report on a calculation of the inclusive Higgs boson production cross section at hadron colliders at next-to-next-to-leading order in QCD. The result is computed as an expansion about the threshold region. By continuing the expansion to very high order, we map the result onto basis functions and obtain the result in closed analytic form.

## 1. Introduction

At the LHC, gluon fusion will be the most important mechanism for Higgs boson production and discovery for masses below $\sim 700$ GeV. The discovery of the Higgs boson and the subsequent study of its properties therefore relies on a solid theoretical understanding of the gluon fusion production mechanism. Unfortunately, next-to-leading order (NLO) studies [1] of inclusive Higgs production do not provide this solid understanding. The NLO corrections are so large (of order $70 - 100\%$) that one cannot assume that they provide a reliable estimate of the total cross section. The unsettled nature of such an important signal clearly calls for a renewed effort to bring this process under control.

Earlier this year, we computed the full NNLO corrections to the hadronic cross section for Higgs boson production [2] as an expansion about the threshold region. The expansion was carried out to very high (18th!) order so that any uncertainty due to uncalculated higher terms would be very small. Very recently, this calculation has been confirmed by an exact NNLO calculation of the partonic cross section [3]. Shortly before this conference began, we extended our calculation of the expansion to sufficiently high order as to allow us to invert the series and obtain the exact result for the hadronic cross section. In addition, we have used the same methods to compute the NNLO corrections to pseudo-scalar Higgs production, for which we present preliminary results [4].

## 2. The Calculation

In the limit that all quark masses except that of the top quark vanish, gluons couple to Higgs only via top quark loops. This coupling can be approximated by an effective Lagrangian [5] corresponding to the limit $m_t \to \infty$, which is valid for a large range of $M_H$, including the currently favored region between 100 and 200 GeV. The effective Lagrangian is

$$\mathcal{L}_{\text{eff}} = -\frac{H}{4v} C_1(\alpha_s)\, G^a_{\mu\nu} G^{a\,\mu\nu}\,, \tag{1}$$

where $G^a_{\mu\nu}$ is the gluon field strength tensor, $H$ is the Higgs field, $v \approx 246\,\text{GeV}$ is the vacuum expectation value of the Higgs field and $C_1(\alpha_s)$ is the Wilson coefficient, which for this calculation we need to order $(\alpha_s^3)$ [6].

For the pseudo-scalar Higgs, we again assume that gluon fusion through top-quark loops is the dominant production mechanism. This assumption is not valid in all models, especially when the ratio of vacuum expectation values is large. The effective Lagrangian for pseudo-scalar production is [7]

$$\begin{aligned} \mathcal{L}_{Agg} &= -g_t\,\frac{A}{v}\left[\tilde{C}_1\,\tilde{\mathcal{O}}_1 + \tilde{C}_2\,\tilde{\mathcal{O}}_2\right]\,, \\ \tilde{\mathcal{O}}_1 &= G^a_{\mu\nu}\tilde{G}^{a,\mu\nu}\,, \quad \tilde{\mathcal{O}}_2 = \partial_\mu\left(\sum_q \bar{q}\gamma^\mu\,\gamma_5 q\right)\,. \end{aligned} \tag{2}$$

*Supported by the U. S. Department of Energy under Contract No. DE-AC02-98CH10886.


doi:10.1016/S0920-5632(02)02133-3

where $A$ is the pseudo-scalar Higgs field, $g_t$ is a model-dependent coupling constant and $\tilde{G}^a_{\mu\nu}$ is the dual of the field strength tensor:

$$\tilde{G}^a_{\mu\nu} = \epsilon_{\mu\nu\alpha\beta}\, G^{a,\alpha\beta}\,. \tag{3}$$

The calculation breaks down into four contributions: virtual corrections to two loops, single-real-emission to one loop, double-real-emission at tree-level and mass factorization. The virtual, single-real and mass factorization terms are computed exactly in closed analytic form. The double-real contribution is by far the hardest part of the calculation because of the complicated phase space integrals. We have computed this contribution by expanding the phase space integration about the threshold limit, where the partonic center-of-mass energy is close to the Higgs mass ($M_H^2/\hat{s} \equiv x \to 1$). Because Higgs production is dominated by threshold corrections, the series expansion converges quite rapidly. So, we compute double-real-emission, and thus the partonic cross section, as an expansion in $(1-x)$ and $\ln(1-x)$. Note that if all coefficients are computed, this is an exact expression. In fact, one can obtain the exact result for the partonic cross section from a finite number of terms.

## 3. Inverting the series

If one were to know the basis functions that make up the exact result and one could expand the series out to enough terms, one could invert the series, mapping it onto the basis functions. One can obtain a reasonable ansatz for the basis functions by examining the result for Drell-Yan production [8]. Using polylogagrithm identities [9] this result can be expressed in terms of functions which are analytic in $(1-x)$ and powers of $\ln(1-x)$ times functions which are analytic in $(1-x)$. Each of these basis functions can be multiplied by a pre-factor. Again using the Drell-Yan result as a guide and knowledge of the gluon splitting function, one can make an ansatz of the possible pre-factors. The ansätze for pre-factors and basis functions are shown in Table 1.

One sees that there are 7 pre-factors and 14 functions meaning that if one can expand the series result out to 98 terms, one can map the result onto these 98 functions. The mapping can be verified by computing still higher terms and comparing to the expansion of the mapped functions. To carry out this program, we have computed the double-real radiation terms out to order $(1-x)^{100}$. Since the series starts at order $(1-x)^{-1}$, this gives us 102 terms.

Table 1
Ansatz pre-factors and functions for exact result

| | | |
|---|---|---|
| $1$ | $1$ | $\ln(x)$ |
| $\frac{1}{x}$ | $\ln^2(x)$ | $\ln^3(x)$ |
| $\frac{1}{1-x}$ | $\mathrm{Li}_2(1-x)$ | $\mathrm{Li}_2(1-x)\ln(x)$ |
| $\frac{1}{1+x}$ | $\mathrm{Li}_2(1-x^2)$ | $\mathrm{Li}_2(1-x^2)\ln(x)$ |
| $1-x$ | $\mathrm{Li}_3(1-x)$ | $\mathrm{Li}_3\left(-\frac{1-x}{x}\right)$ |
| $(1-x)^2$ | $\mathrm{Li}_3(1-x^2)$ | $\mathrm{Li}_3\left(-\frac{1-x^2}{x^2}\right)$ |
| $(1-x)^3$ | $\mathrm{Li}_3\left(\frac{1-x}{1+x}\right)$ | $\mathrm{Li}_3\left(-\frac{1-x}{1+x}\right)$ |

It turns out that the $1/x$ pre-factor never occurs in the cross section (in Ref. [8] it appears in the Drell-Yan correction term, not the Drell-Yan cross section) and the last two basis functions, $\mathrm{Li}_3((1-x)/(1+x))$ and $\mathrm{Li}_3(-(1-x)/(1+x))$ always occur together as the difference. Thus, after the fact, we see that 78 terms would have sufficed to determine the functional form. The additional 24 terms significantly over-determine the system and provide a strong verification of the result. In addition, we have compared our result with that recently reported in Ref. [3] using a completely independent method, and find exact agreement.

## 4. Hadronic Results

In Ref. [2], the partonic cross section was computed using a series expansion out to order $(1-x)^{16}$. The difference between using that series expansion and the exact calculation is quite small (less than $\sim 1\%$) as one would expect. In light of the intrinsic uncertainty due to scale dependence of order ±10%, this difference is completely negligible. The convergence of the series for scalar Higgs production is shown in Fig. 1. Preliminary results for pseudo-scalar Higgs production is shown in Fig. 2

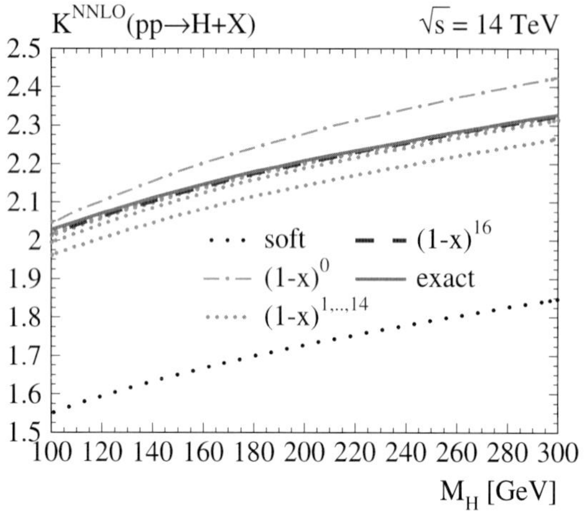


Figure 1. NNLO cross section for Higgs production using the exact partonic cross section and series expansions truncated at the indicated values.

We obtain our hadronic results by folding the partonic cross section with appropriate parton distribution functions (PDFs). We use approximate NNLO parton distribution functions [10] based on an approximation of the evolution equation [11,12].

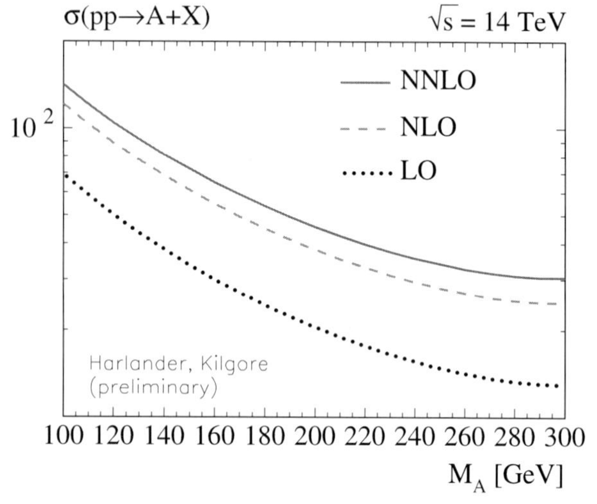


Figure 2. Cross section for pseudo-scalar Higgs production at leading order (LO), NLO and NNLO (preliminary) [4]. The model dependent couling $g_t$ has been set to unity.

## REFERENCES


1. S. Dawson, Nucl. Phys. **B359**, 283 (1991); A. Djouadi et al., Phys. Lett. **B264**, 440 (1991); D. Graudenz et al., Phys. Rev. Lett. **70**, 1372 (1993); D. Graudenz et al., Nucl. Phys. **B453**, 17 (1995).
2. R. V. Harlander and W. B. Kilgore, Phys. Rev. Lett. **88**, 201801 (2002).
3. C. Anastasiou et al. (2002), hep-ph/0207004.
4. R. V. Harlander and W. B. Kilgore (2002), hep-ph/0208096.
5. J. Ellis et al., Nucl. Phys. **B106**, 292 (1976); A. Vainshtein et al., Yad. Fiz. **30**, 1368 (1979); Usp. Fiz. Nauk **131**, 537 (1980); M. Voloshin, Yad. Fiz. **44**, 738 (1986).
6. K. Chetyrkin et al., Phys. Rev. Lett. **79**, 353 (1997); Nucl. Phys. **B510**, 61 (1998); M. Krämer et al., Nucl. Phys. **B511**, 523 (1998).
7. K. Chetyrkin et al., Nucl. Phys. **B535**, 3 (1998).
8. R. Hamberg et al., Nucl. Phys. **B359**, 343 (1991).
9. L. Lewin, *Polylogarithms and Associated Functions* (Elsevier North Holland, Inc., New York, 1981).
10. A. Martin et al., Phys. Lett. **B531**, 216 (2002).
11. S. Larin et al., Nucl. Phys. **B427**, 41 (1994); Nucl. Phys. **B492**, 338 (1997); A. Retey et al., Nucl. Phys. **B604**, 281 (2001).
12. W. L. van Neerven et al., Phys. Lett. **B490**, 111 (2000).

Proceedings of ICHEP 2002, pp. 285–287
*S. Bentvelsen, P. de Jong, J. Koch and E. Laenen (Editors)*

31[st] INTERNATIONAL CONFERENCE ON HIGH ENERGY PHYSICS AMSTERDAM

# Direct Higgs production at hadron colliders *

M. Grazzini[a]

[a]Dip. di Fisica, Univ. di Firenze and INFN, Sezione di Firenze, I-50019 Sesto Fiorentino, Florence, Italy

We consider QCD corrections to Higgs boson production through gluon-gluon fusion in hadron collisions. We compute the effects of soft-gluon emission to all orders. We present numerical results at the Tevatron and the LHC.

The most important mechanism for SM Higgs boson production at hadron colliders is gluon–gluon fusion through a heavy (top) quark loop. NLO QCD corrections to this process are known to be large [1–3]: their effect increases the LO inclusive cross section by about 80-100%. Recently, the calculation of the NNLO corrections has been completed in the large-$M_t$ approximation [4–8]. This approximation has been shown to work well at NLO provided the exact dependence on the mass $M_t$ of the top quark is retained in the LO result.

In Fig. 1 we show the NNLO effect with respect to NLO at the Tevatron and the LHC, as a function of the mass $M_H$ of the SM Higgs boson. We use the MRST2001 set [9], which includes approximated NNLO parton distributions, with $\alpha_S$ consistently evaluated at one-loop, two-loop, three-loop order for the LO, NLO, NNLO results, respectively. The factorization and renormalization scales $\mu_F$ and $\mu_R$ are fixed to $M_H$. The solid line is the full NNLO result [8] while the dashed line is the result including only soft and virtual plus leading collinear contributions (SVC) [4]. We see that the NNLO effect is moderate at the LHC: in the case of a light Higgs, the $K$-factor is about 2.1–2.2, corresponding to an increase of about 20% with respect to NLO. The NNLO effect is more sizeable at the Tevatron where $K \sim 3$, the increase being of about 40% with respect to NLO. Fig. 1 shows that the SVC approximation works remarkably well, the differences with respect to the full result being only about 2% at the LHC and 4% at the Tevatron. Thus the bulk of the NNLO contributions is due to soft and collinear radiation [4–6], which factorizes from the heavy-quark loop, whereas the hard radiation [7] gives only a very small effect. The dominance of soft and collinear radiation has two important consequences. First, it can be considered as a justification of the use of the large $M_t$-approximation at NNLO. Second, it suggests that multiple soft-gluon emission beyond NNLO can give a relevant effect. Here we discuss the effects of resummation of soft (Sudakov) emission to all orders [10].

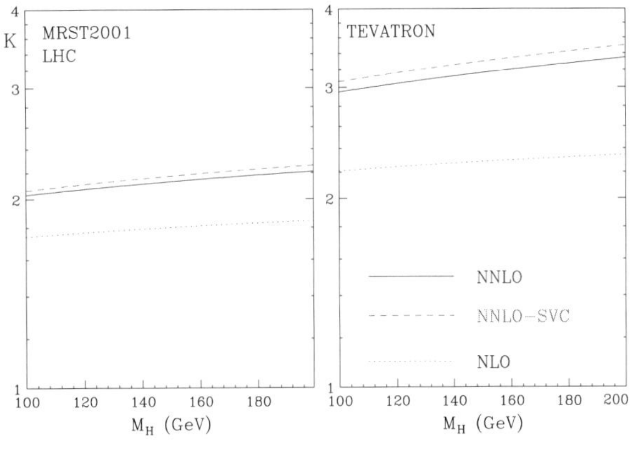


Figure 1. *K factors at the Tevatron and the LHC.*

The cross section $\hat{\sigma}_{gg}$ for the partonic subprocess $gg \to H + X$ at the centre–of–mass energy

*Work done in collaboration with S. Catani, D. de Florian and P. Nason


doi:10.1016/S0920-5632(02)02134-5

$\hat{s} = M_H^2/z$ can be written as

$$\hat{\sigma}_{gg}(\hat{s}, M_H^2) = \sigma_0 \, z \, G_{gg}(z) \,, \tag{1}$$

where $M_H$ is the Higgs mass, $\sigma_0$ is the Born-level cross section and $G_{gg}$ is the perturbatively computable coefficient function. Soft-gluon resummation is performed [11] in Mellin (or $N$-moment) space ($N$ is the moment variable conjugate to $z$). The *all-order* resummation formula for the coefficient function $G_{gg}$ is [4,12]:

$$G^{(\mathrm{res})}_{gg,\,N} = \alpha_S^2(\mu_R^2)\, C_{gg}(\alpha_S(\mu_R^2), M_H^2/\mu_R^2; M_H^2/\mu_F^2) \\ \Delta_N^H(\alpha_S(\mu_R^2), M_H^2/\mu_R^2; M_H^2/\mu_F^2) \,. \tag{2}$$

The function $C_{gg}(\alpha_S)$ contains all the terms that are constant in the large-$N$ limit, produced by hard virtual contributions and non-logarithmic soft corrections. It can be computed as a power series expansion in $\alpha_S$ as

$$C_{gg}(\alpha_S(\mu_R^2), M_H^2/\mu_R^2; M_H^2/\mu_F^2) = \\ 1 + \sum_{n=1}^{+\infty} \left(\frac{\alpha_S(\mu_R^2)}{\pi}\right)^n C_{gg}^{(n)}(M_H^2/\mu_R^2; M_H^2/\mu_F^2) \,, \tag{3}$$

where the perturbative coefficients $C_{gg}^{(n)}$ are closely related to those of the $\delta(1-z)$ contribution to $G_{gg}(z)$. The radiative factor $\Delta_N^H$ embodies the large logarithmic terms due to soft-gluon radiation. To implement resummation, the radiative factor is expanded to a given logarithmic accuracy as

$$\Delta_N^H = \exp\Big\{ \ln N \, g^{(1)}(\lambda) + g^{(2)}\left(\lambda, \frac{M_H^2}{\mu_R^2}; \frac{M_H^2}{\mu_F^2}\right) \\ + \alpha_S(\mu_R^2) g^{(3)}\left(\lambda, \frac{M_H^2}{\mu_R^2}; \frac{M_H^2}{\mu_F^2}\right) + \mathcal{O}(\alpha_S^2(\alpha_S \ln N)^k)\Big\}, \tag{4}$$

such that the functions $g^{(1)}, g^{(2)}$ and $g^{(3)}$ respectively collect the leading logarithmic (LL), next-to-leading logarithmic (NLL) and next-to-next-to-leading logarithmic (NNLL) terms with respect to the expansion parameter $\lambda = \alpha_S(\mu_R^2)\ln N$.

NLL resummation [4] is controlled by three perturbative coefficients, $A_g^{(1)}, A_g^{(2)}$ and $C_{gg}^{(1)}$. The coefficients $A_g^{(1)}$ and $A_g^{(2)}$, which appear in the functions $g^{(1)}$ and $g^{(2)}$, are well known [11]. The coefficient $C_{gg}^{(1)}$ in Eq. (3) is extracted from the NLO result.

At NNLL accuracy three new coefficients are needed [4]: the coefficient $C_{gg}^{(2)}$ in Eq. (3) and two coefficients, $D^{(2)}$ and $A_g^{(3)}$, which appear in the NNLL function $g^{(3)}$. The functional form of $g^{(3)}$ was computed in Ref. [13]. The coefficients $D^{(2)}$ and $C_{gg}^{(2)}$ are obtained [4] from the NNLO result. The coefficient $A_g^{(3)}$ is not yet fully known: we use its exact $N_f^2$-dependence [14] and the approximate numerical estimate of Ref. [15].

Finally, the dominant collinear logarithmic terms can be accounted for by modifying the coefficient $C_{gg}^{(1)}$ in the resummation formula as [4]

$$C_{gg}^{(1)} \to C_{gg}^{(1)} + 2A_g^{(1)} \frac{\ln N}{N} \,. \tag{5}$$

In the following we present a preliminary study of the resummation effect at the Tevatron and the LHC. The hadron-level cross section is obtained by convoluting the partonic cross section in Eq. (1) with the parton distributions of the colliding hadrons. As in Fig. 1 we use the MRST2001 set. The resummed calculations are always matched to the corresponding fixed-order results, i.e. LL is matched to LO, NLL to NLO and NNLL to NNLO. We find that the effect of the inclusion of the collinear term in Eq. (5) is very small, whereas the effect of the coefficient $A^{(3)}$ is completely negligible.

In Fig. 2 we present our results at the LHC, by plotting the $K$-factor, defined as the hadronic cross section as a function of $\mu_F$ and $\mu_R$, normalized to the LO result at $\mu_F = \mu_R = M_H$. In the left side of the figure the LO, NLO and NNLO bands are shown, defined varying $\mu_F = \mu_R$ between $0.5M_H$ and $2M_H$. In the right side of the figure the corresponding resummed results are plotted, the bands being now obtained setting $\mu_F = M_H$ and letting $\mu_R$ to range between $0.5M_H$ and $2M_H$. In both cases we have defined the bands in such a way to maximize them but avoiding completely independent scale variations such as $\mu_R = 0.5M_H$ and $\mu_F = 2M_H$, by which the ratio $\mu_F/\mu_R$ would be 4. Different definitions of the uncertainty bands are of course possible. We see that soft-gluon resummation gives a moderate effect, the NNLL effect being about $5 - 6\%$ with respect to NNLO for $M_H \lesssim 200$ GeV. In

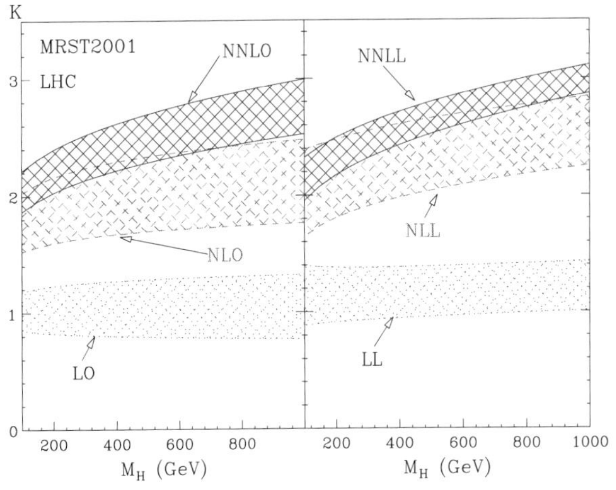


Figure 2. *Resummed K-factors at the LHC.*

Fig. 3 we report the analogous results at the Tevatron Run II. The bands are defined as in Fig. 2. Here the resummation effects are larger: going from NLO to NLL accuracy, the cross section increases by 25-30%. NNLL resummation increases the NNLO cross section by $\sim$ 12-15% when $M_H$ varies in the range 100-200 GeV. These results are not unexpected [6], since at the Tevatron the Higgs boson is produced closer to threshold and the effect of multiple soft-gluon emission is more important.

From these results we conclude that the theoretical predictions for Higgs boson production at hadron colliders are under control. A more detailed discussion of the present theoretical uncertainty will be given elsewhere [10].

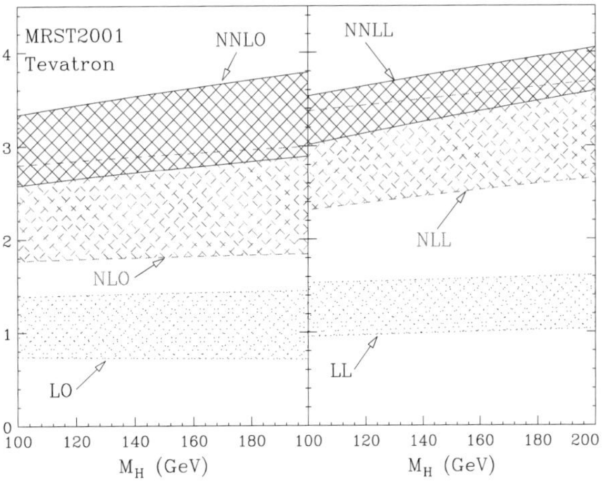


Figure 3. *Resummed K-factors at the Tevatron Run II.*

## REFERENCES


1. S. Dawson, Nucl. Phys. B **359** (1991) 283.
2. A. Djouadi, M. Spira and P. M. Zerwas, Phys. Lett. **B264** (1991) 440.
3. M. Spira, A. Djouadi, D. Graudenz and P. M. Zerwas, Nucl. Phys. B **453** (1995) 17.
4. S. Catani, D. de Florian and M. Grazzini, JHEP **0105** (2001) 025.
5. R. V. Harlander and W. B. Kilgore, Phys. Rev. D **64** (2001) 013015.
6. S. Catani, D. de Florian and M. Grazzini, JHEP **0201** (2002) 015.
7. R. V. Harlander and W. B. Kilgore, Phys. Rev. Lett. **88** (2002) 201801.
8. C. Anastasiou and K. Melnikov, hep-ph/0207004.
9. A. D. Martin, R. G. Roberts, W. J. Stirling and R. S. Thorne, Eur. Phys. J. C **23** (2002) 73, Phys. Lett. B **531** (2002) 216.
10. S. Catani, D. de Florian, M. Grazzini and P. Nason, in hep-ph/0204316 and preprint in preparation.
11. S. Catani, M. L. Mangano, P. Nason and L. Trentadue, Nucl. Phys. B **478** (1996) 273, and references therein.
12. M. Kramer, E. Laenen and M. Spira, Nucl. Phys. B **511** (1998) 523.
13. A. Vogt, Phys. Lett. B **497** (2001) 228.
14. J. F. Bennett and J. A. Gracey, Nucl. Phys. B **517** (1998) 241.
15. W. L. van Neerven and A. Vogt, Phys. Lett. B **490** (2000) 111.

ELSEVIER

Proceedings of ICHEP 2002, pp. 288–290
*S. Bentvelsen, P. de Jong, J. Koch and E. Laenen (Editors)*

31st INTERNATIONAL CONFERENCE ON HIGH ENERGY PHYSICS AMSTERDAM

# QCD corrections to associated $t\bar{t}h$ production at hadron colliders

S. Dawson[a], L. Orr[b], L. Reina[c] and D. Wackeroth[d]

[a]Physics Department, Brookhaven National Laboratory, Upton, NY 11973, USA

[b]Department of Physics and Astronomy, University of Rochester, Rochester, NY 14627, USA

[c]Physics Department, Floriday State University, Tallahassee, FL 32306, USA

[d]Department of Physics, SUNY at Buffalo, NY 14260, USA

We briefly present the status of QCD corrections to the inclusive total cross section for the production of a Higgs boson in association with a top-quark pair within the Standard Model at hadron colliders.

## 1. Introduction

The search for the Higgs boson of the Standard Model (SM) is one of the major tasks of the next generation of high-energy collider experiments. The direct limit on the SM Higgs boson mass, $M_h$, from LEP2 searches [1] and the indirect limit from electroweak precision data [2] strongly suggest the existence of a light Higgs boson, that may well be within the reach of a high-luminosity phase of the Fermilab Tevatron $p\bar{p}$ collider ($M_h \lesssim 180$ GeV) [3]. However, since the dominant SM Higgs production channels are plagued with low event rates and large backgrounds, the Higgs boson search requires the highest possible luminosity, and all possible production channels should be considered. The production of a SM Higgs boson in association with a top-quark pair, $p\bar{p} \to t\bar{t}H$, can play a role for almost the entire Tevatron discovery range [3,4]. Although the $t\bar{t}h$ event rate is small, the signature of such events is quite spectacular ($W^+W^-b\bar{b}b\bar{b}$). At the CERN LHC $pp$ collider, associated $t\bar{t}h$ production is one of the most promising processes for studying both the Higgs boson and the top quark. This process will provide a direct measurement of the top-Yukawa coupling [5,6]. As for any other hadronic cross section, the next-to-leading-order (NLO) QCD corrections are expected to be numerically important and are crucial in reducing the (arbitrary) dependence of the cross sections on the factorization and renormalization scales. In the following we briefly describe the status of NLO QCD predictions for associated $t\bar{t}h$ production at the Tevatron and the LHC.

## 2. QCD corrections to $t\bar{t}h$ production

The inclusive total cross section for $p\overset{(-)}{p} \to t\bar{t}h$ at $\mathcal{O}(\alpha_s^3)$ can be written as:

$$\sigma_{NLO}(p\overset{(-)}{p} \to t\bar{t}h) = \sum_{ij} \int dx_1 dx_2 \qquad (1)$$
$$\mathcal{F}_i^p(x_1,\mu)\mathcal{F}_j^{p(\bar{p})}(x_2,\mu)\hat{\sigma}^{ij}_{NLO}(x_1,x_2,\mu) \ ,$$

where $\mathcal{F}_i^{p,\bar{p}}$ are the NLO parton distribution functions for parton $i$ in a proton/antiproton, defined at a generic factorization scale $\mu_f = \mu$, and $\hat{\sigma}^{ij}_{NLO}$ is the $\mathcal{O}(\alpha_s^3)$ parton-level total cross section for incoming partons $i$ and $j$, renormalized at an arbitrary scale $\mu_r$, that we choose $\mu_r = \mu_f = \mu$. The NLO parton-level total cross section $\hat{\sigma}^{ij}_{NLO}$ consists of the $\mathcal{O}(\alpha_s^2)$ Born cross section and the $\mathcal{O}(\alpha_s)$ corrections to the Born cross section, including the effects of mass factorization. It comprises virtual and real corrections to the parton-level $t\bar{t}h$ production processes, $q\bar{q} \to t\bar{t}h$ and $gg \to t\bar{t}h$, and the tree-level $qg$ initiated processes, $qg \to t\bar{t}hq$, which are of the same order in perturbation theory. The main challenges in the calculation of the $\mathcal{O}(\alpha_s)$ corrections come from the presence of pentagon diagrams in the virtual part with several massive external and internal particles, and from the computation of the real part in the presence of infrared (IR) singularities.


doi:10.1016/S0920-5632(02)02136-9

### 2.1. $t\bar{t}h$ production at the Tevatron

The NLO inclusive total cross section for $p\bar{p} \to t\bar{t}h$ at the Tevatron center-of-mass energy $\sqrt{s_H} =$ 2 TeV has been calculated independently by two groups [7–9]. The numerical results of both calculations have been compared and they are found to be in very good agreement. For $p\bar{p}$ collisions at $\sqrt{s_H}$ = 2 TeV, more than 95% of the tree-level cross section comes from the sub-process $q\bar{q} \to t\bar{t}h$ and the $gg$ initial state is numerically irrelevant. Therefore, in [8,9] when calculating $\sigma_{NLO}(p\bar{p} \to t\bar{t}h)$ of Eq. 1, we only included the $q\bar{q} \to t\bar{t}h$ channel, summed over all light quark flavors. In [7] the $\mathcal{O}(\alpha_s)$ corrections to both $t\bar{t}h$ production channels have been considered.

In [8,9] we have calculated the pentagon scalar integrals as linear combinations of scalar box integrals using the method of [10]. The real corrections are computed using the phase space slicing method, in both the double [11] and single [12–14] cutoff approach. This is the first application of the single cutoff phase space slicing approach to a cross section involving more than one massive particle in the final state. As discussed in detail in [9], the numerical results of both methods agree within the statistical errors. In [7] the dipole subtraction formalism has been used to extract the IR singularities of the real part. To find agreement between calculations based on either of the three very different treatments of the real IR singularities represents a powerful check of the NLO calculations.

As illustrated in Fig. 1 for $M_h$ = 120 GeV, $\sigma_{NLO}(p\bar{p} \to t\bar{t}h)$ at $\sqrt{s_H}$ = 2 TeV shows a significantly reduced scale dependence as compared to the leading-order (LO) result and leads to increased confidence in predictions based on these results. Over the entire range of $M_h$ accessible at the Tevatron, the NLO corrections decrease the LO rate for $\mu < 2m_t + M_h$.

To determine the potential of the Tevatron to search for the SM Higgs boson, the predictions for all Higgs production processes have to be under good theoretical control. In Fig. 2 we show the scale dependence of the inclusive total cross sections at NLO (next-to-NLO in case of $gg \to h$) for the main SM Higgs production processes, including associated $t\bar{t}h$ production, in the $M_h$ range

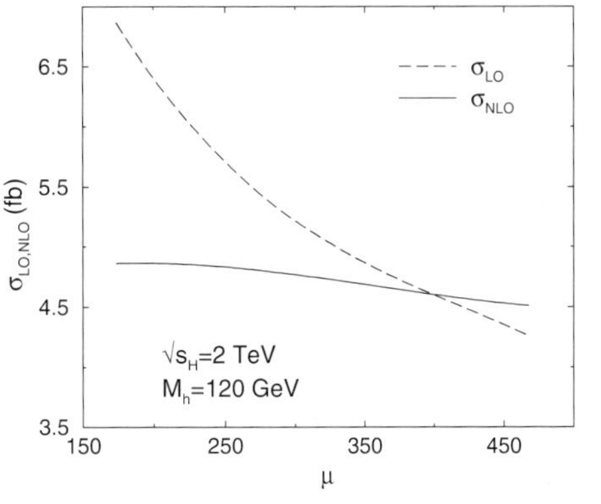


Figure 1. Dependence of $\sigma_{LO,NLO}(p\bar{p} \to t\bar{t}h)$ on the renormalization/factorization scale $\mu$, at the Tevatron ($\sqrt{s_H}$=2 TeV), for $M_h$=120 GeV [8,9].

accessible at the Tevatron. The $p\bar{p} \to b\bar{b}h$ cross section is only known at LO and thus not shown in Fig. 2. It is comparable in size to the $t\bar{t}h$ cross section for $\mu = m_t$ and $M_h$ =120 GeV. As illustrated in Fig. 2, the QCD NLO calculations for the $q\bar{q} \to Wh, Zh$ [15], $qq \to qqh$ [16] and $q\bar{q} \to t\bar{t}h$ [7–9] processes provide reliable predictions for the inclusive total cross sections at the Tevatron. However, the NLO corrections to the $gg \to h$ cross section [17,18] are large (up to $\sim$ 100%), and $\sigma_{NLO}$ still strongly depends on the factorization/renormalization scale. The QCD next-to-NLO (NNLO) $gg \to h$ cross section became available recently [19,20]. As illustrated in Fig. 2 [21], the NNLO corrections are crucial to obtain reliable predictions.

### 2.2. $t\bar{t}h$ production at the LHC

While the observation of the $t\bar{t}h$ process at the Tevatron will be quite challenging, at the LHC the $t\bar{t}h$ mode will play a crucial role in the 110 GeV$\leq M_h \leq$130 GeV mass region both for discovery and for precision measurements of the Higgs boson couplings. For determining $\sigma_{NLO}(pp \to t\bar{t}h)$ at the LHC, the $\mathcal{O}(\alpha_s)$ corrections to the subprocess $gg \to t\bar{t}h$ are crucial, since $pp$ collisions at $\sqrt{s_H}$=14 TeV are dominated by the $gg$ initial state. Results for $\sigma_{NLO}(pp \to t\bar{t}h)$ at $\sqrt{s_H}$ = 14 TeV have been provided in [7] and results of an independent calculation will be presented in [22]. A detailed comparison of the numerical results of

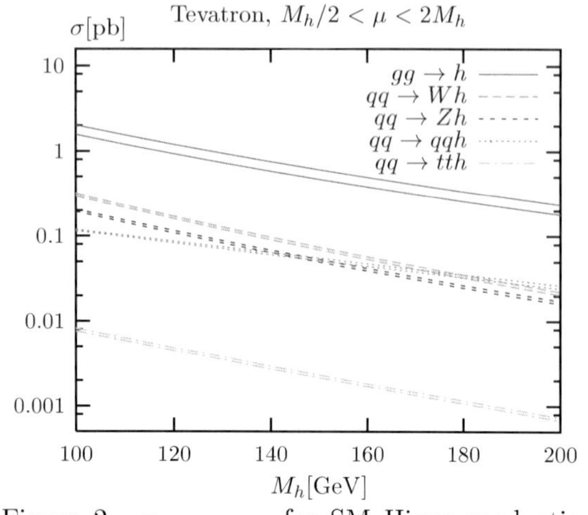


Figure 2. $\sigma_{NLO,NNLO}$ for SM Higgs production processes at the Tevatron ($\sqrt{s_H} = 2$ TeV) as a function of $M_h$. For $p\bar{p} \to t\bar{t}h$ the renormalization/factorization scale is varied between $m_t + M_h/2 < \mu < 2m_t + M_h$.

the two groups is presently under way.

Although the calculation of the $\mathcal{O}(\alpha_s)$ corrections to $gg \to t\bar{t}h$ is technically similar to the one for the $q\bar{q}$ initiated process, new challenges arise, e.g., the occurrence of spurious singularities in the calculation of the pentagon diagrams due to a vanishing Gram determinant at the phase space boundaries. The resulting numerical instabilities can be controlled by applying an extrapolation procedure from the numerically safe to the unsafe region [7]. However, we found that, also in the case of the $gg$ initiated subprocess, the occurrence of these singularities can be avoided by cancelling terms in the numerator against the propagators wherever possible, after interfering the pentagon amplitude with the Born-matrix element. We compared the numerical results of both treatments of these spurious singularities and found agreement within the statistical error. Details of the calculation will be presented elsewhere [22].

As is the case at the Tevatron, the scale dependence of $\sigma_{NLO}(pp \to t\bar{t}h)$ at $\sqrt{s_H} = 14$ TeV is strongly reduced compared to the LO result. At the LHC, the NLO QCD corrections enhance the LO cross section over the entire range of values for the renormalization/factorization scale shown in Fig. 1, unlike at the Tevatron where they mostly decrease the Born-level cross section.

## REFERENCES

1. LHWG Note/2002-01, July 2002.
2. LEPEWWG/2002-01, May 2002.
3. M. Carena and H. E. Haber, hep-ph/0208209.
4. J. Goldstein *et al.*, Int. J. Mod. Phys. A **16S1A**, 308 (2001); Phys. Rev. Lett. **86**, 1694 (2001).
5. A. Belyaev and L. Reina, JHEP **0208**, 041 (2002).
6. D. Zeppenfeld, hep-ph/0203123; and references therein.
7. W. Beenakker, S. Dittmaier, M. Krämer, B. Plümper, M. Spira and P. M. Zerwas, Phys. Rev. Lett. **87**, 201805 (2001).
8. L. Reina and S. Dawson, Phys. Rev. Lett. **87**, 201804 (2001).
9. L. Reina, S. Dawson and D. Wackeroth, Phys. Rev. D **65**, 053017 (2002).
10. Z. Bern, L. Dixon and D. A. Kosower, Phys. Lett. B **302**, 299 (1993) [Erratum-ibid. B **318**, 649 (1993)]; Nucl. Phys. B **412**, 751 (1994).
11. For a review see, e.g., B. W. Harris and J. F. Owens, Phys. Rev. D **65**, 094032 (2002).
12. W. T. Giele and E. W. Glover, Phys. Rev. D **46**, 1980 (1992).
13. W. T. Giele, E. W. Glover and D. A. Kosower, Nucl. Phys. B **403**, 633 (1993).
14. S. Keller and E. Laenen, Phys. Rev. D **59**, 114004 (1999).
15. T. Han and S. Willenbrock, Phys. Lett. B **273**, 167 (1991).
16. T. Han, G. Valencia and S. Willenbrock, Phys. Rev. Lett. **69**, 3274 (1992).
17. S. Dawson, Nucl. Phys. B **359**, 283 (1991).
18. A. Djouadi, M. Spira and P. M. Zerwas, Phys. Lett. B **264**, 440 (1991).
19. R. V. Harlander and W. B. Kilgore, Phys. Rev. Lett. **88**, 201801 (2002); W. B. Kilgore, these proceedings, hep-ph/0208143.
20. C. Anastasiou and K. Melnikov, hep-ph/0207004.
21. We thank the authors of [19] for providing the NNLO results.
22. S. Dawson, L. Orr, L. Reina, and D. Wackeroth, in preparation.

ELSEVIER

Proceedings of ICHEP 2002, pp. 291–293
*S. Bentvelsen, P. de Jong, J. Koch and E. Laenen (Editors)*

31st INTERNATIONAL CONFERENCE ON HIGH ENERGY PHYSICS AMSTERDAM

# Physics with forward protons at hadron colliders

V.A. Khoze[a] A.D. Martin[a], and M.G. Ryskin[a]

[a]Institute for Particle Physics Phenomenology,
University of Durham, DH1 3LE, UK

We emphasize the importance of tagging the outgoing forward protons to sharpen the predictions for New Physics at the LHC. We show that exclusive double-diffractive Higgs production, $pp \to p + H + p$, followed by the $H \to b\bar{b}$ decay, could play an important role in identifying a 'light' Higgs boson.

## 1. INTRODUCTION

Recently there has been much interest in studying events with tagged forward protons at present and forthcoming hadronic colliders, see, for example, [1,2]. This may not only allow the luminosity of the colliding protons to be monitored with high accuracy [3], but also can provide new ways to investigate the subtle issues of QCD and to search for the manifestations of the New Physics. As discussed in [1] the programme with the tagged forward protons is in many aspects complementary to both the standard physics at hadron colliders and to studies at a future linear collider.

The physics potential of high energy proton colliders can be significantly increased by studying exclusive double-diffractive-like processes of the type $pp \to p + M + p$. Here $M$ represents a system of invariant mass $M$, and the + signs denote the presence of rapidity gaps which separate the system $M$ from the protons. Such processes allow an exceptionally clean experimental environment to identify New Physics signals (such as the Higgs boson, SUSY particles, etc., see [1]). In such events we produce a colour-singlet state $M$ which is practically free from soft secondary particles. Moreover, if forward going protons are tagged we can reconstruct the 'missing' mass $M$ with good resolution, and so have an ideal means to search for new resonances and to study threshold behaviour phenomena. We have to pay a price for ensuring such a clean diffractive signal. In particular, the diffractive event rate is suppressed by the small probability, $\widehat{S}^2$, that the rapidity gaps survive soft rescattering effects between the interacting hadrons, which can generate secondary particles which populate the gaps, see, for example, [4,5] and references therein.

## 2. EXCLUSIVE HIGGS PRODUCTION

Double-diffractive Higgs production, $pp \to p + H + p$, at the LHC, is a good example to illustrate the pros and cons of exclusive processes. Let us assume a Higgs boson of mass $M_H = 120$ GeV and consider detection in the $b\bar{b}$ channel. It is possible to install proton taggers so that the 'missing mass' can be measured to an accuracy $\Delta M_{\rm missing} \simeq 1$ GeV [2]. Then the exclusive process will allow the mass of the Higgs to be measured in two independent ways. First the tagged protons give $M_H = M_{\rm missing}$ and second, via the $H \to b\bar{b}$ decay, we have $M_H = M_{b\bar{b}}$, although now the resolution is much poorer with $\Delta M_{b\bar{b}} \simeq 10$ GeV. The existence of matching peaks, centered about $M_{\rm missing} = M_{b\bar{b}}$, is a unique feature of the exclusive diffractive Higgs signal. Besides its obvious value in identifying the Higgs, the mass equality also plays a key role in reducing background contributions. Another advantage of this exclusive process, with $H \to b\bar{b}$, is that the leading order $gg \to b\bar{b}$ background subprocess is suppressed by a $J_z = 0$ selection rule [6,7]. The disadvantage is that, to ensure the survival of the rapidity gaps, the predicted $H \to b\bar{b}$ cross section is low, $\sigma \simeq 2$ fb, corresponding to a soft survival factor $\widehat{S}^2 = 0.02$. It is estimated that there is a factor two uncertainty in this prediction [2].


doi:10.1016/S0920-5632(02)02137-0

For an integrated luminosity of 30 $fb^{-1}$ the number of signal (background) events for this method of Higgs detection at the LHC is expected to be 11 (4). These include a factor 0.6 for the efficiency associated with proton tagging, 0.6 for $b$ and $\bar{b}$ tagging and 0.5 for the $b, \bar{b}$ jet polar angle cut, $60^\circ < \theta < 120^\circ$ (necessary to reduce the $b\bar{b}$ QCD background) [2].

There exists a huge spread of predictions of the cross sections for diffractive Higgs production, see, for example, [8–11], which can differ from [1,2] by orders of magnitude. A critical comparison between these predictions, and an explanation of their differences with our results, is given in [12].

A way to check experimentally the reliability of the predictions of exclusive production is to measure the much larger cross section for an analogous process: double-diffractive central production of a pair of high-$E_\perp$ jets [13]. Some of the existing approaches overshoot the current CDF dijet data [14] by a few orders of magnitude; others are just normalized to the experimental rates in order to account for the survival effects. However, the latter procedure is not unambiguous, since there is no direct way of using the dijet overshoot factors to correct the expectations for Higgs production. The perturbative approach [13] predicts about 1 nb [6] for the exclusive central production of dijets with $E_\perp > 7$ GeV, corresponding to the CDF kinematics [14] to be compared with the observed exclusive dijet bound of less than 3.7 nb. Moreover, in the case of central inelastic production (when secondaries are allowed in some central rapidity interval), our estimates give about 40 nb (with a factor two uncertainty) for the cross section in the CDF kinematical range, while the observed value [14] is $43.6 \pm 4.4 \pm 21.6$ nb. Bearing in mind that the accuracy of the theoretical predictions for $E_\perp > 7$ GeV jets at the Tevatron energy is far from perfect, we find this preliminary comparison quite encouraging.

Another valuable check of the calculations of the soft survival factor $\widehat{S}^2$ (which originates from the non-perturbative sector) is the description of diffractive dijet production at the Tevatron [15] in terms of the diffractive structure functions measured at HERA [5]. The remarkably good agreement of these predictions with the CDF measurements is a confirmation that our calculations of $\widehat{S}^2$ [4,5,13] are trustworthy. Moreover, the new fit to the H1 diffractive data [16] makes the agreement with the CDF results even better[1].

Details of the calculation of exclusive Higgs production are given in Ref. [1]. The main sources of the $b\bar{b}$ background are, at leading order, caused by gluon jets being misidentified as a $b\bar{b}$ pair, by a $J_z = 2$ admixture due to non-forward protons and by a $J_z = 0$ contribution arising from $m_b \neq 0$. Also there is a background contribution from $b\bar{b}g$ events in which the emitted gluon is approximately collinear with a $b$ jet. These backgrounds were considered in detail in Ref. [2], leading to a prediction of the signal-to-background ratio of about 3. Note that in [2] only the $gg \to b\bar{b}g$ hard subprocess was considered at NLO, and radiation for the spectator, screening gluon was not discussed. However, this latter process is numerically small because of the additional suppression of colour-octet $b\bar{b}$ production around 90°; rotational invariance around the $b$ quark direction causes the cross section to be proportional to $\cos^2\theta$ in the $b\bar{b}$ c.m. frame.

The cross sections for inclusive and central inelastic diffractive Higgs production are larger than for exclusive production. However, for these non-exclusive processes it is hard to suppress the QCD $b\bar{b}$ background and the signal-to-background ratio is small. Second, we cannot improve significantly the accuracy of the measurement of the mass of the Higgs boson by tagging the forward protons and measuring the missing mass.

Recall that, at medium and high luminosity at the LHC, the recorded events will be plagued by overlap interactions in the same bunch crossing. For example, at the medium luminosity of $10^{33}$ $cm^{-2}s^{-1}$, an average of 2.3 inelastic events are expected for each bunch crossing. Hence the rapidity gaps occurring in one interaction may be populated by particles created in an accompanying interaction. It is, however, possible to use

[1]Another probe of the calculations of $\widehat{S}^2$ could come from the experimental studies of the central $Z$ production by $WW$ fusion, see, for instance, [17].

detector information to locate the vertices of the individual interactions and, in principle, to identify hard scattering events with rapidity gaps. For the exclusive and central inelastic processes, the use of proton taggers makes it much more reliable to select the rapidity gap events.

## 3. CONCLUSION

The Physics menu for LHC studies with tagged forward protons looks quite attractive and promising [1]. In particular, the exclusive $pp \to p+H+p$ process has the advantage that the signal exceeds the background. The favourable signal-to-background ratio is offset by a low event rate, caused by the necessity to preserve the rapidity gaps so as to ensure an exclusive signal. Nevertheless, as shown in [2], the signal for a 'light' Higgs has reasonable significance in comparison to the standard $H \to \gamma\gamma$ and $t\bar{t}H$ search modes. Moreover, the advantage of the matching Higgs peaks, $M_{\rm missing} = M_{b\bar{b}}$, cannot be overemphasized.

We stress that the predicted value of the exclusive cross section can be checked experimentally. All the ingredients, except for the NLO correction to the $gg \to H$ vertex, are the same for our signal as for exclusive double-diffractive dijet production, $pp \to p + \text{dijet} + p$, where the dijet system is chosen in the same kinematic domain as the Higgs boson, that is $M(jj) \sim 120$ GeV. Therefore by observing the larger dijet production rate, we can confirm, or correct, the estimate of the exclusive Higgs signal.

## ACKNOWLEDGEMENTS

We thank Albert De Roeck, Aliosha Kaidalov and Risto Orava for valuable discussions, and the EU, PPARC and the Leverhulme Trust for support.

## REFERENCES

1. V.A. Khoze, A.D. Martin and M.G. Ryskin, Eur. Phys. J. **C23** (2002) 311.
2. A. De Roeck, V.A. Khoze, A.D. Martin, R. Orava and M.G. Ryskin, `hep-ph/0207042`, Eur. Phys. J. **C** (in press).
3. V.A. Khoze, A.D. Martin, R. Orava and M.G. Ryskin, Eur. Phys. J. **C19** (2001) 313.
4. V.A. Khoze, A.D. Martin and M.G. Ryskin, Eur. Phys. J. **C18** (2000) 167.
5. A.B. Kaidalov, V.A. Khoze, A.D. Martin and M.G. Ryskin, Eur. Phys. J. **C21** (2001) 521.
6. V.A. Khoze, A.D. Martin and M.G. Ryskin, `hep-ph/0006005`.
7. V.A. Khoze, A.D. Martin and M.G. Ryskin, Eur. Phys. J. **C19** (2001) 477, erratum **C20** (2001) 599.
8. J.-R. Cuddell and O.F. Hernandez, Nucl. Phys. **B471** (1996) 471.
9. B.E. Cox, J.R. Forshaw and B. Heinemann, Phys. Lett. **B540** (2002) 263.
10. M. Boonekamp, A. De Roeck, R. Peschanski and C. Royon, `hep-ph/0205332`.
11. R. Enberg, G. Ingelman, A. Kissavos and N. Timneanu, Phys. Rev. Lett. **89** (2002) 081801.
12. V.A. Khoze, A.D. Martin and M.G. Ryskin, `hep-ph/0207313`, Eur. Phys. J. **C** (in press).
13. V.A. Khoze, A.D. Martin and M.G. Ryskin, Eur. Phys. J. **C14** (2000) 525.
14. CDF Collaboration: T. Affolder et al., Phys. Rev. Lett. **85** (2000) 4215.
15. CDF Collaboration: T. Affolder et al., Phys. Rev. Lett. **84** (2000) 5043.
16. F.-P. Schilling, these proceedings.
17. H. Chehime and D. Zeppenfeld, Phys. Rev. **D47** (1993) 3898;
V.A. Khoze, M.G. Ryskin, W.J. Stirling and P.M. Williams, `hep-ph/0207365`.

Proceedings of ICHEP 2002, pp. 294–296
S. Bentvelsen, P. de Jong, J. Koch and E. Laenen (Editors)

31st INTERNATIONAL CONFERENCE ON HIGH ENERGY PHYSICS AMSTERDAM

# Top quark pair production and decay at hadron colliders: Predictions at NLO QCD including spin correlations

W. Bernreuther[a*], A. Brandenburg[b†], Z.G. Si[a*], P. Uwer[c]

[a]Institut f. Theoretische Physik, RWTH Aachen, 52056 Aachen, Germany

[b]DESY-Theorie, 22603 Hamburg, Germany

[c]Institut f. Theoretische Teilchenphysik, Universität Karlsruhe, 76128 Karlsruhe, Germany

We present results at NLO QCD for hadronic production and decay of top quark pairs, taking into account $t\bar{t}$ spin correlations.

The interactions of the top quark have not been precisely tested so far. The expected data samples of Run II of the Tevatron collider (eventually about $10^4$ $t\bar{t}$ pairs per annum) and the LHC ($\sim 10^7$ $t\bar{t}$ pairs p.a.) will make possible detailed investigations of top quark production and decay. In particular, one may search for possible deviations from the QCD production mechanism and the V-A structure of the top decay by analysing energy and angular distributions of the $t\bar{t}$ decay products. Such studies will be feasible since top quarks do not hadronize due to their very short lifetime, and perturbation theory gives a reliable description of top quark production and decay. To search for new physics effects, precise theoretical predictions for top quark pair production and decay at hadron colliders within the Standard Model are needed. We consider here in particular the following processes:

$$h_1 h_2 \to t\bar{t} + X \to \begin{cases} \ell^+\ell'^- + X \\ \ell^+ j_{\bar{t}} + X \\ \ell^- j_t + X \\ j_t j_{\bar{t}} + X, \end{cases} \tag{1}$$

where $h_{1,2} = p, \bar{p}$; $\ell = e, \mu, \tau$, and $j_t$ ($j_{\bar{t}}$) denote jets originating from hadronic $t$ ($\bar{t}$) decays. An observable that is very sensitive to the dynamics of top quark production and decay in the above reactions is the double differential angular distribution of the top decay products, e.g. for the dilepton channel:

$$\frac{1}{\sigma}\frac{d^2\sigma}{d\cos\theta_+ d\cos\theta_-} = \frac{1}{4}(1 - C\cos\theta_+\cos\theta_-). \tag{2}$$

In (2), $\theta_\pm$ are the angles between the $\ell^\pm$ direction of flight in the $t(\bar{t})$ rest frame with respect to arbitrary axes $\hat{\mathbf{a}}$ ($\hat{\mathbf{b}}$) which will be specified below. (Terms linear in $\cos\theta_\pm$ are absent in (2) for our choices of $\hat{\mathbf{a}}$, $\hat{\mathbf{b}}$ due to parity invariance of QCD.)

The calculation of the distribution (2) at next-to-leading order (NLO) QCD simplifies enormously in the leading pole approximation (LPA), which amounts to expanding the full amplitudes for (1) around the complex poles of the $t$ and $\bar{t}$ propagators. Only the leading pole terms are kept in this expansion, i.e. one neglects terms of order $\Gamma_t/m_t \approx 1\%$. Within the LPA, the radiative corrections can be classified into factorizable and non-factorizable [1] contributions. Here we consider only the factorizable corrections. In this case the coefficient in the distribution (2) factorizes:

$$C = \kappa_+\kappa_- D. \tag{3}$$

In (3), $D$ is the $t\bar{t}$ double spin asymmetry

$$D = \frac{N(\uparrow\uparrow) + N(\downarrow\downarrow) - N(\uparrow\downarrow) - N(\downarrow\uparrow)}{N(\uparrow\uparrow) + N(\downarrow\downarrow) + N(\uparrow\downarrow) + N(\downarrow\uparrow)}, \tag{4}$$

*supported by BMBF contract 05 HT1 PAA 4

†Speaker at the conference. Supported by a Heisenberg fellowship of D.F.G.


doi:10.1016/S0920-5632(02)02140-0

where $N(\uparrow\uparrow)$ denotes the number of $t\bar{t}$ pairs with $t$ ($\bar{t}$) spin parallel to the reference axis $\hat{\mathbf{a}}$ ($\hat{\mathbf{b}}$) etc. The directions $\hat{\mathbf{a}}$ and $\hat{\mathbf{b}}$ can thus be identified with the spin quantization axes for $t$ and $\bar{t}$, and $D$ reflects the strength of the correlation between the $t$ and $\bar{t}$ spins for the chosen axes. Further, $\kappa_\pm$ is the spin analysing power of the charged lepton in the decay $t(\bar{t}) \to b(\bar{b})\ell^\pm\nu(\bar{\nu})$ defined by the decay distribution

$$\frac{1}{\Gamma}\frac{d\Gamma}{d\cos\vartheta_\pm} = \frac{1 \pm \kappa_\pm \cos\vartheta_\pm}{2}, \qquad (5)$$

where $\vartheta_\pm$ is the angle between the $t$ ($\bar{t}$) spin and the $\ell^\pm$ direction of flight. From [2] we obtain

$$\kappa_+ = \kappa_- = 1 - 0.015\alpha_s, \qquad (6)$$

which means that the charged lepton serves as a perfect analyser of the top quark spin. For hadronic decays $t \to bq\bar{q}'$ one has a decay distribution analogous to (5), and the spin analysing power of jets can be defined. To order $\alpha_s$ [3] (and using $\alpha_s(m_t) = 0.108$),

$$\kappa_b = -0.408 \times (1 - 0.340\alpha_s) = -0.393, \qquad (7)$$
$$\kappa_j = +0.510 \times (1 - 0.654\alpha_s) = +0.474, \qquad (8)$$

where $\kappa_j$ is the analysing power of the least energetic non-b-quark jet. The results for $C$ given below are for the dilepton channel. The corresponding results for the single lepton channel are obtained by replacing $\kappa_+$ (or $\kappa_-$) in Eq. (3) by $\kappa_b$ or $\kappa_j$. The loss of analysing power using hadronic final states is overcompensated by the gain in statistics and in efficiency to reconstruct the $t$ ($\bar{t}$) rest frames.

To compute $D$ at NLO QCD, the differential cross sections for the following parton processes are needed to order $\alpha_s^3$, where the full information on the $t$ and $\bar{t}$ spins has to be kept:

$$q\bar{q} \to t\bar{t},\ t\bar{t}g;\ gg \to t\bar{t},\ t\bar{t}g;\ q(\bar{q})g \to t\bar{t}q(\bar{q}). \qquad (9)$$

Results at NLO QCD for the $\overline{\text{MS}}$ subtracted parton cross sections $\hat{\sigma}$ for the above processes with $t\bar{t}$ spins summed over have been obtained in [4], [5], while

$$\hat{\sigma}D = \hat{\sigma}(\uparrow\uparrow) + \hat{\sigma}(\downarrow\downarrow) - \hat{\sigma}(\uparrow\downarrow) - \hat{\sigma}(\downarrow\uparrow) \qquad (10)$$

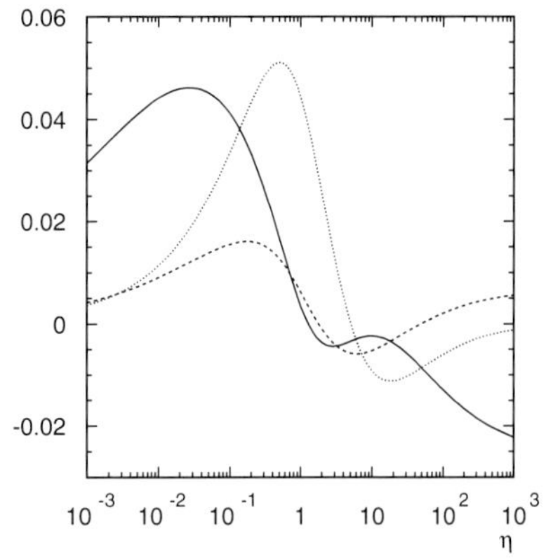


Figure 1. Dimensionless scaling functions $g^{(0)}(\eta)$ (full), $g^{(1)}(\eta)$ (dotted), and $\tilde{g}^{(1)}(\eta)$ (dashed) that determine $\hat{\sigma}D$ in the helicity basis for the $gg$ initial state.

has been computed for different spin quantization axes in [6], [7]. This combination of spin-dependent cross sections can be written as follows

$$\hat{\sigma}D = \frac{\alpha_s^2}{m_t^2}\left[g^{(0)}(\eta) + 4\pi\alpha_s\mathcal{G}^{(1)}\right], \qquad (11)$$

$$\text{with} \quad \mathcal{G}^{(1)} = g^{(1)}(\eta) + \tilde{g}^{(1)}(\eta)\ln\left(\frac{\mu^2}{m_t^2}\right), \qquad (12)$$

where $\eta = \hat{s}/(4m_t^2) - 1$, and we use a common scale $\mu$ for the renormalization and factorization scale. As an example, Fig. 1 shows, for the parton process $gg \to t\bar{t}(g)$, the three functions that determine $\hat{\sigma}D$ for the helicity basis, i.e. $\hat{\mathbf{a}}$ ($\hat{\mathbf{b}}$) is chosen to be the $t$ ($\bar{t}$) direction of flight. Apart from the helicity basis we also consider the beam basis, where $\hat{\mathbf{a}}$ and $\hat{\mathbf{b}}$ are equal to the direction of one of the hadron beams in the laboratory frame, and the so-called off-diagonal basis [8], which is defined by the requirement that $\hat{\sigma}(\uparrow\downarrow) = \hat{\sigma}(\downarrow\uparrow) = 0$ for the process $q\bar{q} \to t\bar{t}$ at tree level.

Table 1 lists our results [9] for the coefficient $C$ in the double differential distribution (2) for these different choices of spin quantization axes, using the parton distribution functions (PDFs) of [10]. One sees that at the Tevatron, the spin correlations are largest in the beam and off-diagonal basis. The QCD corrections reduce the LO results by about 10%. For the LHC, the best choice of the above three bases is the helicity basis, and the QCD corrections are small in that case.

Table 1
Coefficient $C$ defined in Eq. (2) to leading and next-to-leading order in $\alpha_s$ for the parton distribution functions of [10] and $\mu = m_t = 175$ GeV.

| | $p\bar{p}$, $\sqrt{s} = 2$ TeV | | $pp$, $\sqrt{s} = 14$ TeV | |
|---|---|---|---|---|
| | LO | NLO | LO | NLO |
| $C_{\text{hel.}}$ | −0.456 | −0.389 | 0.305 | 0.311 |
| $C_{\text{beam}}$ | 0.910 | 0.806 | −0.005 | −0.072 |
| $C_{\text{off.}}$ | 0.918 | 0.813 | −0.027 | −0.089 |

Table 2
Upper part: Dependence on $\mu$ of the correlation coefficients $C$, computed at NLO with the PDF of ref.[10]. Lower part: Correlation coefficients $C$ at NLO for $\mu = m_t$ and different sets of parton distribution functions: GRV98[11], CTEQ5[10], and MRST98 (c-g)[12].

| | Tevatron | | | LHC |
|---|---|---|---|---|
| $\mu$ | $C_{\text{hel.}}$ | $C_{\text{beam}}$ | $C_{\text{off.}}$ | $C_{\text{hel.}}$ |
| $m_t/2$ | −0.364 | 0.774 | 0.779 | 0.278 |
| $m_t$ | −0.389 | 0.806 | 0.813 | 0.311 |
| $2m_t$ | −0.407 | 0.829 | 0.836 | 0.331 |
| PDF | $C_{\text{hel.}}$ | $C_{\text{beam}}$ | $C_{\text{off.}}$ | $C_{\text{hel.}}$ |
| GRV98 | −0.325 | 0.734 | 0.739 | 0.332 |
| CTEQ5 | −0.389 | 0.806 | 0.813 | 0.311 |
| MRST98 | −0.417 | 0.838 | 0.846 | 0.315 |

Table 2 shows the dependence of the NLO results for $C$ on the scale $\mu$ (upper part) and on the choice of the PDFs (lower part). At the Tevatron the spread of results for different PDFs is larger than the scale uncertainty. This rather strong dependence on the PDFs is largely due to the fact that the contributions from $q\bar{q}$ and $gg$ initial states contribute to $C$ with opposite signs. This may offer the possibility to constrain the quark and gluon content of the proton by a precise measurement of the double angular distribution (2).

In all results above we used $m_t = 175$ GeV. A variation of $m_t$ from 170 to 180 GeV changes the results for the Tevatron, again for $\mu = m_t$ and PDFs of [10] as follows: $C_{\text{hel.}}$ varies from −0.378 to −0.397, $C_{\text{beam}}$ from 0.790 to 0.817, and $C_{\text{off.}}$ from 0.797 to 0.822. For the LHC, $C_{\text{hel.}}$ changes by less than a percent.

The results above have been obtained without imposing any kinematic cuts. Standard cuts on the top quark transverse momentum and rapidity only have a small effect on $C$: For the Tevatron, demanding $|\mathbf{k}^T_{t,\bar{t}}| > 15$ GeV and $|r_{t,\bar{t}}| < 2$ leads to the following results: $C_{\text{hel.}} = -0.386$, $C_{\text{beam}} = 0.815$, $C_{\text{off.}} = 0.823$. For the LHC, when imposing the cuts $|\mathbf{k}^T_{t,\bar{t}}| > 20$ GeV and $|r_{t,\bar{t}}| < 3$, we find $C_{\text{hel.}} = 0.295$.

In summary, $t\bar{t}$ spin correlations are large effects that can be studied at the Tevatron and the LHC by measuring double angular distributions both in the dilepton and single lepton decay channels. The QCD corrections to these distributions have been computed and are under control. Spin correlations are suited to analyse in detail top quark interactions, search for new effects, and may help to constrain the parton content of the proton.

## REFERENCES


1. W. Beenakker, F.A. Berends, A.P. Chapovsky, Phys. Lett. B **454** (1999) 129.
2. A. Czarnecki, M. Jezabek, J.H. Kühn, Nucl. Phys. B **351** (1991) 70.
3. A. Brandenburg, Z.G. Si, P. Uwer, Phys. Lett. B **539** (2002) 235.
4. P. Nason, S. Dawson, R.K. Ellis, Nucl. Phys. B **303** (1988) 607; Nucl. Phys. B **327** (1989) 49.
5. W. Beenakker, H. Kuijf, W.L. van Neerven, J. Smith, Phys. Rev. D **40** (1989) 54; W. Beenakker *et al.*, Nucl. Phys. B **351** (1991) 507.
6. W. Bernreuther, A. Brandenburg, Z.G. Si, Phys. Lett. B **483** (2000) 99.
7. W. Bernreuther, A. Brandenburg, Z.G. Si, P. Uwer, Phys. Lett. B **509** (2001) 53.
8. G. Mahlon, S. Parke, Phys. Lett. B **411** (1997) 173.
9. W. Bernreuther, A. Brandenburg, Z.G. Si, P. Uwer, Phys. Rev. Lett. 87 (2001) 242002.
10. H.L. Lai *et al.* [CTEQ Collaboration], Eur. Phys. J. C **12** (2000) 375.
11. M. Glück, E. Reya, A. Vogt, Eur. Phys. J. C **5** (1998) 461.
12. A.D. Martin, R.G. Roberts, W.J. Stirling, R.S. Thorne, Eur. Phys. J. C **4** (1998) 463.

ELSEVIER

Proceedings of ICHEP 2002, pp. 297–299
*S. Bentvelsen, P. de Jong, J. Koch and E. Laenen (Editors)*

31st INTERNATIONAL CONFERENCE ON HIGH ENERGY PHYSICS AMSTERDAM

# QCD Physics in Atlas at the Large Hadron Collider

James Proudfoot*

Argonne National Laboratory, High Energy Physics Division,
9700 S. Cass Avenue, Argonne Illinois, 60439 USA

## 1. INTRODUCTION

The Large Hadron Collider (LHC) is a proton-proton collider with a 14 TeV center of mass energy. The design luminosity is $10^{34}$ $cm^{-1}s^{-1}$ with beam collisions separated by 25 ns. The initial operation for physics will take place at a luminosity of $10^{33}cm^{-1}s^{-1}$ and it is expected that the integrated luminosity delivered in the first year will be 10 $fb^{-1}$. This integrated luminosity will result in very large event samples for most processes, for example: ~$10^8$ leptonic W decays, $10^4$ $\gamma$'s with $E_t$>500GeV and $10^4$ jets with $E_t$>1TeV. As a result of the high statistics event samples, the understanding of most QCD processes at 14TeV will be systematics limited after the first year of running.

The Atlas detector [1] is a general purpose detector designed to be sensitive to the many physics processes which are expected at the LHC. It contains high performance tracking using silicon detectors and a transition radiation tracker in a 2 Tesla solenoidal magnetic field, a high resolution electromagnetic calorimeter based on lead-liquid argon, a hadron calorimeter based on steel-scintillator and Cu/W-liquid argon, and a large instrumented air-core toroid magnet system for muon measurement. The basic performance characteristics of these systems are given in Table 1.

| Detector Sub-System | Performance |
|---|---|
| Tracking (Si + Transition Radiation Tracker) | $\sigma/p_t$~$5.10^{-4}$ $p_t$(GeV) + 0.01 |
| EM Calorimeter (Pb Liquid Argon) | $\sigma/E$~ $10\%/\sqrt{E}$ (GeV) |
| Hadron Barrel Calorimeter (Steel + Scintillator) | $\sigma/E$~ $50\%/\sqrt{E}$ (GeV)+0.03 |
| Hadron Endcap Calorimeter (Cu/W + Liquid Argon) | $\sigma/E$~ $60\%/\sqrt{E}$ (GeV)+0.03 |
| Muon (Instrumented Air Toroid) | $\sigma/p_t$ ~10% at 1 TeV |

**Table 1. Basic performance characteristics of the Atlas detector for the LHC.**

## 2. THE KINEMATIC REACH OF THE LHC

The increased kinematic reach of the LHC is indicated in Fig. 1 [2], which compares the jet inclusive cross-section at the Fermilab Tevatron with the LHC. The increased reach in $E_t$ is almost a factor of 10. QCD phenomenology can make predictions for inclusive jet cross-sections as well as other observables in this unexplored region of x and $Q^2$ and $E_t$. The overall consistency with QCD will provide an extensive test of the ***predictive*** power of this theory. In addition, QCD processes form the majority of the background processes in many of the searches, which will be carried out at the LHC. Some expectations for QCD physics with the Atlas detector are discussed below. These and other physics issues are discussed in more detail in reference [3].

## 3. "ENGINEERING" ISSUES

Many detector response and physics characteristics impact the overall performance of the Atlas detector for QCD physics. One physics issue is shown in Fig. 2, which shows the Monte Carlo predictions [3] of several models for the charged particle multiplicity in

*Work supported in part by the US Department of Energy, High Energy Physics Division, under Contract W-31-109-Eng-38.


doi:10.1016/S0920-5632(02)02141-2

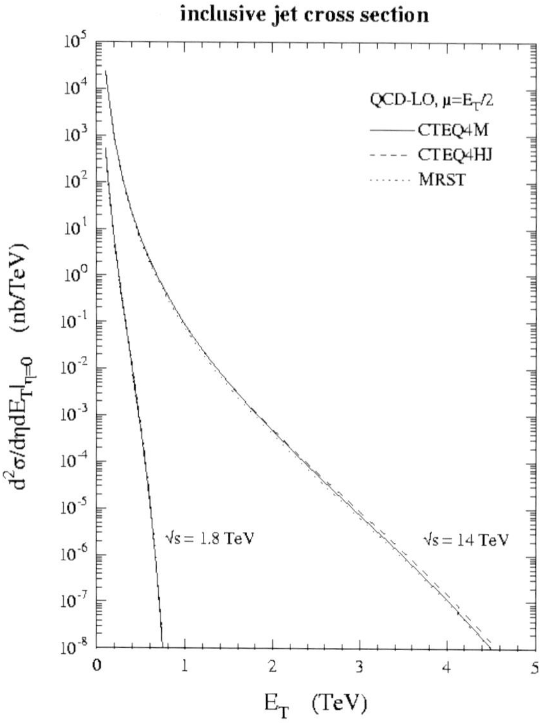


**Fig. 1. Jet inclusive cross-section at the Tevatron compared to the predicted cross-section for the LHC.**

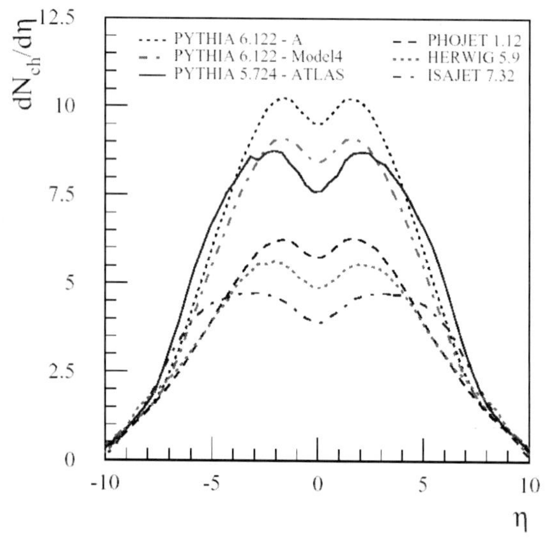


**Figure 2. Monte Carlo predictions for charged particle multiplicity distributions, for several models as described in [3].**

minimum bias events. The prediction varies from 5 to 10 particles per unit rapidity in the central region. The underlying event, resulting from overlapping minimum bias events, is one of the energy corrections in jet measurement contributing significantly to jet $E_t$ systematics. Therefore, this measurement is both an interesting test of (non-perturbative) QCD which is accessible at low luminosity, and a necessary "engineering" measurement for jet reconstruction.

A Geant simulation of the calorimeter predicts a jet energy resolution of ~2% for high $E_t$ jets. There are many systematics to jet measurement including: the jet energy scale which is expected to be ~1% from *in situ* calibration; calorimeter non-linearity (<3%); and effects such as the jet reconstruction algorithm used and energy from the underlying event. These are discussed in detail in the Atlas Physics TDR [3].

## 4. DETERMINATION OF PARTON DISTRIBUTION FUNCTIONS (PDF's)

In QCD phenomenology, the production cross-section for any process is factorized into the product of the parton-parton cross-section and the parton distribution function (or pdf). These pdf's are determined by fitting the cross-sections for several processes and from many experiments. The sensitivity of the jet $E_T$ cross-section to different sets of pdf's was studied in the Atlas Physics TDR [3]. At high $E_T$, the variation in the cross-section is ~20%. Once data from the LHC is available, the pdf's will also be directly measured by including results from the LHC in the global fits. For example, inclusive photon production provides input to the gluon distribution function since the two most important parton level processes are $qg \rightarrow q\gamma$ and $qq \rightarrow g\gamma$. Measurements of photons with $E_t$ from 40-500 GeV will provide information on the gluon density for $5.10^{-4} < x < 0.2$. Another good example of sensitivity to pdf's is the asymmetry in $W^+$ *vs.* $W^-$ production at high rapidity [3], as shown in Fig. 3. At high rapidity, W production is dominated by valence quark annihilation with sea quarks and therefore, the $W^+$ production cross-section is larger than that for $W^-$. This process will provide sensitivity to both sea and valence quarks with $x_1x_2 \sim 3x10^{-5}$.

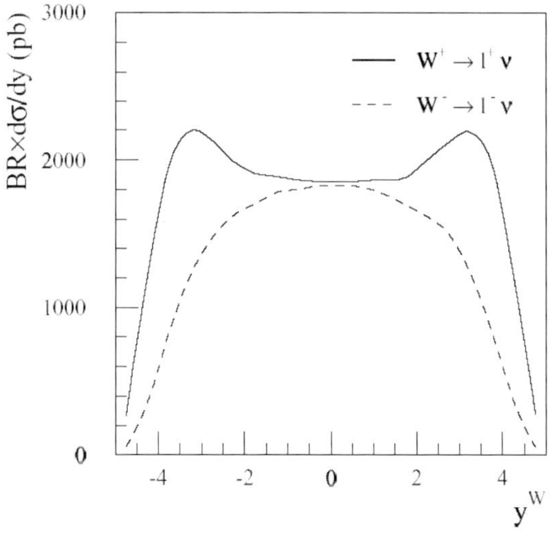


**Figure 3. Pseudo-rapidity distributions for $W^+$ and $W^-$ production in pp collisions at 14 TeV.**

## 5. MEASUREMENT OF THE STRONG COUPLING CONSTANT, $\alpha_s$.

In general there can be a strong correlation between the value of $\alpha_s$ and fitted pdf's, in the results from global fits. A study was carried out by Atlas to evaluate a method which allows the possibility of decoupling these two measurements [4]. The approach taken is to fit the inclusive jet cross-section as a function of $E_t$ and $\alpha_s(E_t)$ in two bins of pseudo-rapidity and study its sensitivity to different sets of pdf's. By inverting this fit, one can then determine $\alpha_s(E_t)$. An example of this analysis is shown in Fig. 4. The results indicate that $\alpha_s$ can be determined with an overall precision of ~10%, in which the contribution to the uncertainty from our knowledge of pdf's is ~3%.

## 6. SUMMARY

The LHC will provide a unique opportunity to test the predictive power of QCD, probing regions in x and $Q^2$ well beyond those studied to date. Almost any process one can think of will contribute to this understanding: from a measurement of the total cross section and inclusive cross sections of b's, t's, and W's to searches for physics beyond the Standard Model. In addition, a measurement of $\alpha_s(E_t)$ out to $E_t$ ~3TeV with a precision of ~10% should be possible.

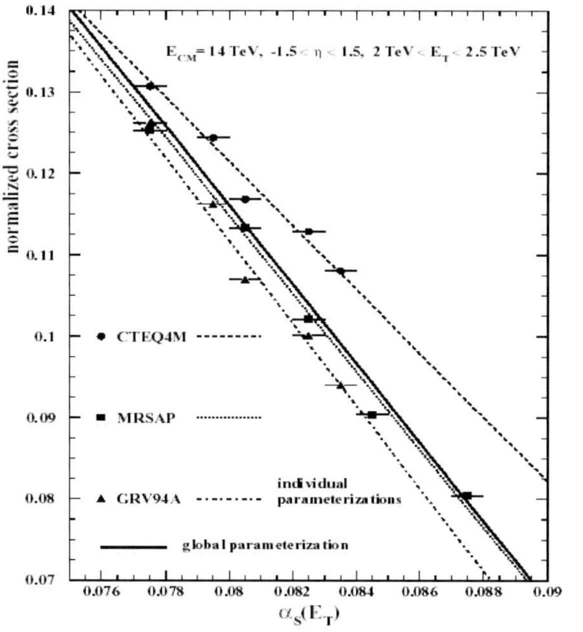


**Figure 4. An example fit of $\alpha_s(E_t)$ used in the determination of the precision with which $\alpha_s$ may be measured at the LHC.**

## 7. ACKNOWLEDGMENTS

I could not have prepared this presentation without the work of my colleagues in the Atlas physics groups and the many contributors to the Atlas Physics Technical Design Report. For this I give my thanks. I would like to thank, in particular, Fabiola Gianotti, Stefan Tapprogge, Joey Huston, and Bob Blair for informative discussions on QCD physics.

## 8. REFERENCES

1. ATLAS Technical Proposal, CERN/LHCC 94-43.
2. J. Huston, private communication.
3. ATLAS Detector and Physics Performance Technical Design Report, CERN/LHCC 99-14 and 99-15 (1999).
4. Determination of $\alpha_s$ using jet cross-section parameterizations at hadron colliders, H. Stenzel, ATL-PHYS-2001-003.

ELSEVIER

Proceedings of ICHEP 2002, pp. 300–303
*S. Bentvelsen, P. de Jong, J. Koch and E. Laenen (Editors)*

31st INTERNATIONAL CONFERENCE ON HIGH ENERGY PHYSICS AMSTERDAM

# Jet production in deep inelastic scattering at HERA

M.R. Sutton[a] on behalf of the ZEUS and H1 collaborations

[a] Denys Wilkinson Laboratory,
University of Oxford, Keble Road, Oxford OX1 3RH, United Kingdom

A small number of recent results from the wealth of precision HERA data on high transverse energy jet production in deep inelastic scattering are reviewed. These measurements are confronted with predictions from next-to-leading order (NLO) Quantum Chromodynamics and allow the extraction of the strong coupling constant, $\alpha_s$. Detailed studies of the underlying subprocess illustrate some areas of current interest.

## 1. INTRODUCTION

The production of high transverse energy jets in the hadronic final state of deep inelastic scattering (DIS) at HERA provides a unique laboratory for the study of the perturbative QCD subprocess. The large kinematic region available at HERA allows the detailed study of the hadronic final state with a variation of the hard scale, provided by the virtuality, $Q^2$, of the exchanged boson, or the jet transverse energy, $E_T$, over many orders of magnitude.

At sufficiently high transverse energies, the relative contribution from long distance, non-perturbative effects such as hadronisation is expected to be small. The high precision data now available at these scales enable more meaningful comparison with next-to-leading order (NLO) parton-level calculations, allowing questions relevant to the description of the QCD hard subprocess, such as the possible size of higher-order corrections and the choice of appropriate scale, to be addressed.

In the region of the $Q^2$ and $x_{\rm Bj}$ kinematic plane covered by jet measurements at HERA, the parton distributions, particularly the quark densities, are well constrained by fits [1] to inclusive charged- and neutral-current cross sections [2,3]. In the lower $Q^2$ region, the jet cross section is directly sensitive to the gluon distribution in the proton and so may be able to provide complementary information on the gluon density.

Where the hadronic final state is sufficiently well described by the NLO predictions, extractions of the QCD coupling, $\alpha_s$, are possible.

This paper presents a few of the most significant recent results from the large number of precision jet measurements now available from HERA, illustrating a number of areas of current interest.

## 2. JET PRODUCTION IN NEUTRAL CURRENT DIS

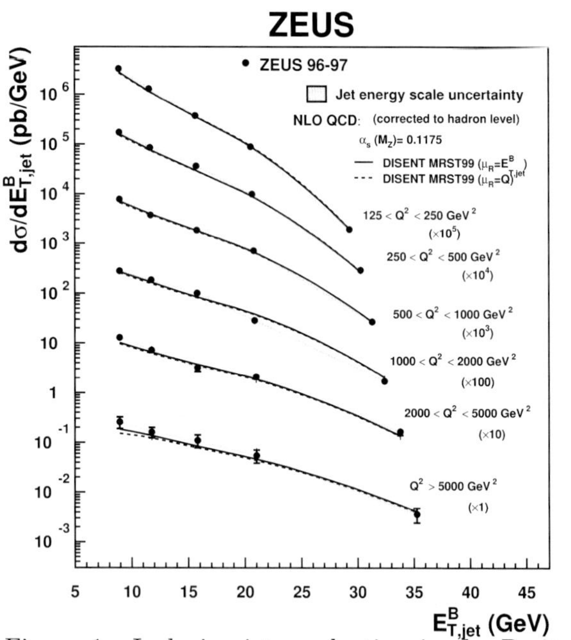


Figure 1. Inclusive jet production in the Breit frame from ZEUS.

The inclusive production of jets with high transverse energy in the Breit frame is an intrinsically $\mathcal{O}(\alpha\alpha_s)$ process, directly sensitive both to the QCD subprocess and the proton gluon density. The large variation possible in both $Q$ and


doi:10.1016/S0920-5632(02)02142-4

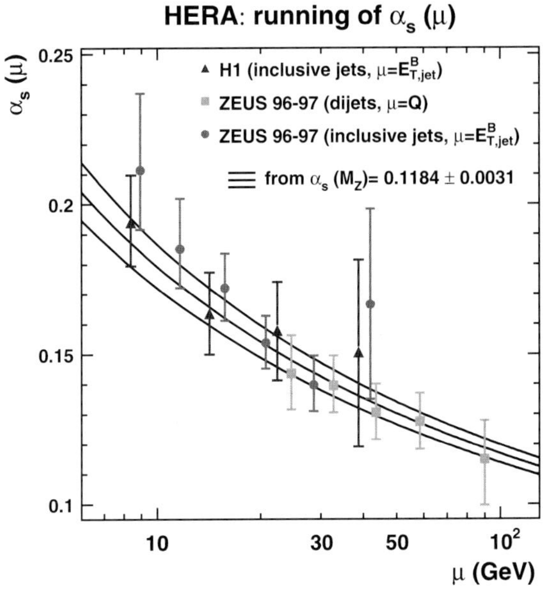


Figure 2. The strong coupling constant extracted from HERA jet data.

$E_T$ raises the question of which provides the most appropriate hard scale.

The cross section for inclusive jet production in neutral-current (NC) DIS at high virtuality, $Q^2 > 125$ GeV$^2$, is shown in Fig. 1, for jets with $E^{\mathrm{B}}_{T,\mathrm{jet}} > 8$ GeV and pseudorapidity in the Breit frame, $\eta^{\mathrm{B}}_{\mathrm{jet}}$, in the range $-2 < \eta^{\mathrm{B}}_{\mathrm{jet}} < 1.8$. Also shown is the NLO QCD prediction from DISENT [4] evaluated with both renormalisation scales, $\mu_R = Q$ and $\mu_R = E^{\mathrm{B}}_{T,\mathrm{jet}}$. The renormalisation scale uncertainty estimated using $\mu_0^2/2 < \mu_R < 2\mu_0$ is reasonably small and calculations at both scales are in reasonable agreement with the data over many orders of magnitude. For low $Q^2$ and low $E^{\mathrm{B}}_{T,\mathrm{jet}}$ there is a suggestion that both calculations fall below the data.

Given the level of agreement between the data and the NLO calculation, the strong coupling constant, $\alpha_s$, can be extracted. This is shown, for the HERA jet data, in Fig. 2. Clear running of $\alpha_s$ within a single environment is evident. The central value extracted by H1 for boson virtualities in the range $150 < Q^2 < 5000$ GeV$^2$ is

$$\alpha_s(M_Z) = 0.1186 \pm 0.0030(\mathrm{exp.})^{+0.0039}_{-0.0045}(\mathrm{theor.})$$

while that from ZEUS, for $Q^2 > 500$ GeV$^2$, is

$$\alpha_s(M_Z) = 0.1212 \pm 0.0017(\mathrm{stat.})\;{}^{+0.0023}_{-0.0031}(\mathrm{syst.})^{+0.0028}_{-0.0027}(\mathrm{theor.}).$$

In these measurements the largest single uncertainties result from the renormalisation scale dependence.

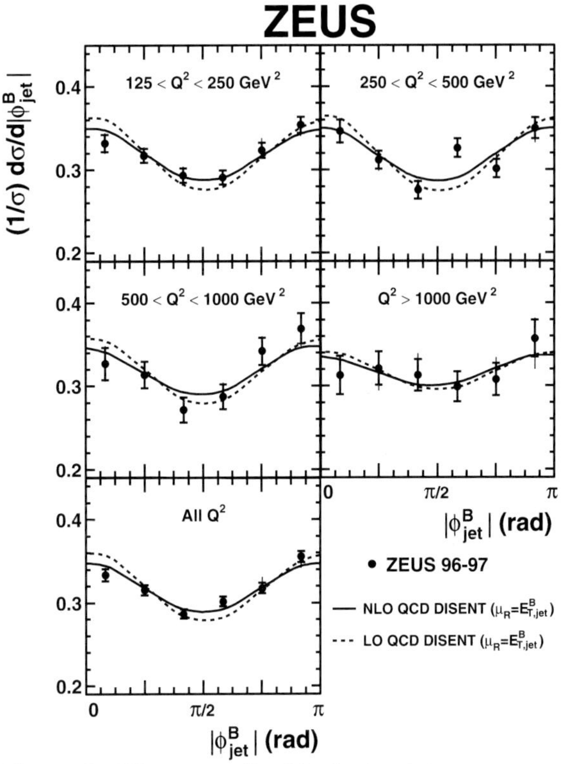


Figure 3. The normalised inclusive-jet cross section with respect to the angle, $|\phi^{\mathrm{B}}_{\mathrm{jet}}|$.

The azimuthal angle of the jet, $\phi^{\mathrm{B}}_{\mathrm{jet}}$, in the Breit frame with respect to the lepton scattering angle is shown in Fig. 3 for several regions of $Q^2$. Perturbative QCD predicts the cross section to have the form

$$\frac{\mathrm{d}\sigma}{\mathrm{d}\phi^{\mathrm{B}}_{\mathrm{jet}}} = A + B\cos\phi^{\mathrm{B}}_{\mathrm{jet}} + C\cos 2\phi^{\mathrm{B}}_{\mathrm{jet}}.$$

Since the observed azimuthal dependence arises predominantly from the boson-gluon-fusion process, as $Q^2$ increases the predicted asymmetry decreases. However, due to the large statistical uncertainty, the expected $Q^2$ dependence is not conclusively observed in the data.

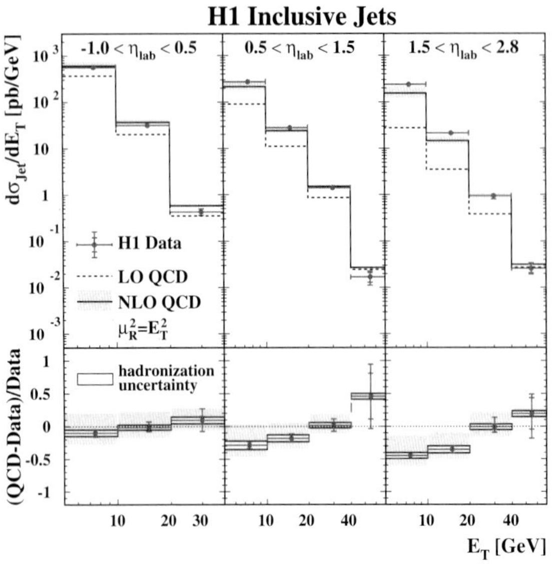


Figure 4. Inclusive jet production in NC DIS for lower $Q^2$.

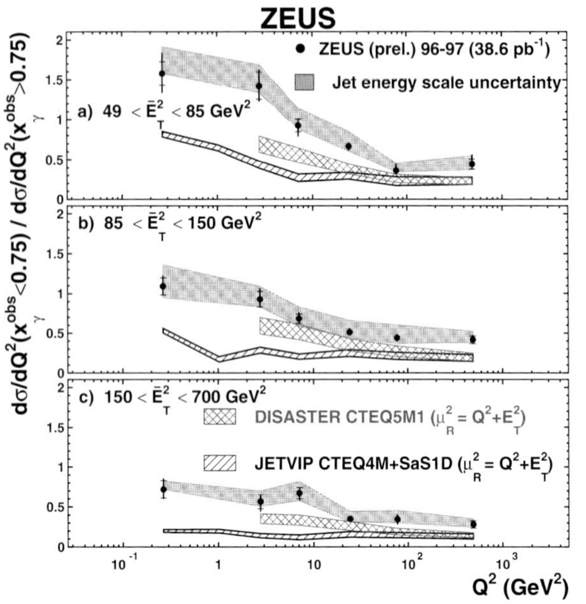


Figure 5. The ratio of resolved-enhanced dijet production to direct-enhanced dijet production.

The cross section for inclusive jet production in the Breit frame for the lower $Q^2$ region, $5 < Q^2 < 100$ GeV$^2$, is shown in Fig. 4 compared to the predictions of NLO QCD. In the forward region, $1.5 < \eta_{\rm lab} < 2.8$, the theory lies significantly below the data for low jet transverse energies, although the NLO corrections are large.

For this forward region, the agreement between data and the NLO calculation is reasonable towards higher $Q^2$ [5]. At lower $Q^2$, the data lie significantly above the calculation. In this region, the normal DGLAP evolution of the proton parton distributions might be expected to break down, and alternative evolution schemes such as BFKL, take over. However the large uncertainties in the DGLAP predictions preclude any firm conclusions on the need for BFKL effects in this data.

### 2.1. Virtual photon structure

At low $Q^2$, when $Q^2 < E_T^2$, the virtual photon can either interact directly, or fluctuate into a hadronic state, interacting via a resolved parton carrying a fraction $x_\gamma < 1$ of the photon's momentum.

Figure 5 shows the ratio of the cross section for resolved-enhanced dijet events, where $x_\gamma < 0.75$, to direct-enhanced events with $x_\gamma > 0.75$ as a function of $Q^2$. Also shown are NLO predictions from DISENT, which includes no virtual photon structure, and JetVIP [6], which is currently the only NLO calculation to include virtual photon structure. The predictions of both DISENT and JetVIP fail to describe the cross-section ratios for $Q^2 < \bar{E}_T^2$, where the resolved component is expected to be larger. This discrepancy decreases with increasing $Q^2$ and $\bar{E}_T^2$.

An alternative picture is available using $k_T$-unordered parton-evolution schemes, such as CCFM evolution. These allow the two highest transverse energy partons to come anywhere along the ladder between the proton and photon vertices, and so provide a qualitatively similar picture, but without explicit photon structure. Results from H1 [7] suggest that predictions from the CCFM Monte Carlo, CASCADE [8], are in reasonable accord with the data.

### 2.2. Multi-jet production

The large luminosity now collected by the ZEUS and H1 collaborations, enable measurements of intrinsically higher order processes. Three-jet production in DIS is an $\mathcal{O}(\alpha_s{}^2)$ process at leading-order, and an NLO calculation is available [9].

Figure 6 shows the three-jet cross section as

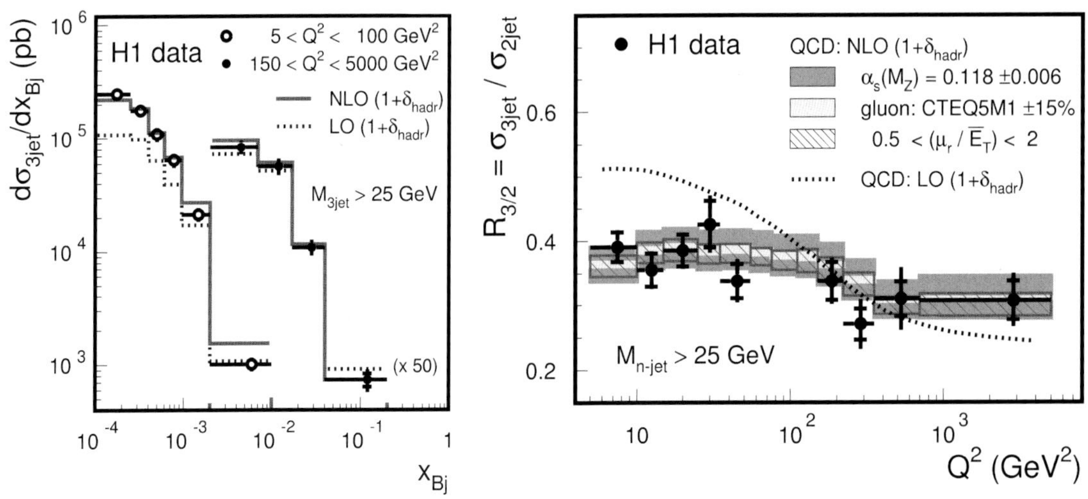


Figure 6. Three-jet production in NC DIS.

a function of $x_{\mathrm{Bj}}$ together with the ratio of the three-jet to two-jet cross-sections, $R_{3/2}$. Reasonable agreement is observed between the data and NLO calculation, with the three-to-two jet rate showing reduced uncertainty from the gluon in the proton. The future prosects for the extraction of $\alpha_s$ from this type of analysis are good, and should particularly benefit from the larger event samples expected with the upgraded HERA machine.

## 3. JETS IN CHARGED CURRENT DIS

Recently, results on single- and dijet inclusive production in charged current (CC) DIS events in the laboratory frame [10] have been compared with NLO QCD predictions [11] with a flavour changing matrix element. Although the data are still statistically limited, there is a possible discrepancy between data and the NLO predictions in the central pseudorapidity region.

## 4. SUMMARY

A small selection of recent results on jet production in DIS are presented and compared with the predictions of perturbative QCD At lower $Q^2$, some discrepancy exists, notably in the forward direction, where theoretical uncertainties are large, so that improved calculations are required to match the precision of the data. At higher $Q^2$, the data are well described by NLO QCD calculations, and represent a significant triumph of our understanding of QCD.

## REFERENCES


1. A.D. Martin *et al.*, Eur. Phys. J **C23** (2002) 73; CTEQ Collab., Preprint hep-ph/0201195 (2002); ZEUS Collab. DESY Red Report DESY-02-105, Submitted to Phys. Rev. **D**.
2. ZEUS Collab., Eur. Phys. J **C21** (2001) 443; H1 Collab., Eur. Phys. J **C21** (2001) 33.
3. ZEUS Collab., Phys. Lett. **B539** (2002) 197; H1 Collab., Eur. Phys. J. **C19** (2001) 269.
4. S. Catani and M.H. Seymour; Nucl. Phys. **B485** (1997) 291; erratum-ibid **B510** (1997) 503.
5. H1 Collab., Conf. Paper 1010, 31st Int. Conf. on High-Energy Phys. ICHEP02, Amsterdam, The Netherlands 2002.
6. B. Pötter, Comp. Phys. Comm. **133** (1997) 105.
7. H1 Collab., Conf. Paper 1009, 31st Int. Conf. on High-Energy Phys. ICHEP02, Amsterdam, The Netherlands 2002.
8. H. Jung, Comp. Phys. Comm.**143** (2002) 100.
9. Z. Nagy and Z. Trócsányi, Phys. Rev. Lett. **87** (2001) 082001.
10. ZEUS Collab., Conf. Paper 845, 31st Int. Conf. on High-Energy Phys. ICHEP02, Amsterdam, The Netherlands 2002.
11. E. Mirkes and D. Zeppenfeld, Phys. Lett. **B380** (1996) 205.

ELSEVIER

Proceedings of ICHEP 2002, pp. 304–307
*S. Bentvelsen, P. de Jong, J. Koch and E. Laenen (Editors)*

31[st] INTERNATIONAL CONFERENCE ON HIGH ENERGY PHYSICS AMSTERDAM

# Jets and Prompt Photons in Photoproduction at HERA

Günter Grindhammer[a*]

[a]Max-Planck-Institut für Physik, Werner-Heisenberg-Institut, München

Recent measurements of jet and prompt photon production by the H1 and ZEUS experiments are reviewed. The results are compared to next-to-leading order QCD calculations. No large deviations from the data are observed using proton and photon parton densities derived from other processes.

## 1. Introduction

The dominant hard processes in *ep* collisions at HERA are the direct and resolved production of jets in photoproduction. To leading order (LO) in the strong coupling $\alpha_s$ a quasi-real photon interacts either directly with a parton from the proton or a parton from the resolved photon interacts with a parton from the proton. These processes lead to final states containing two hard partons, which then hadronize into jets of particles. A rarer but nonetheless interesting process occurs when a so-called prompt photon is produced instead of a gluon. Experimentally (theoretically) jets are defined by finding clusters of charged particles and calorimetric energy depositions (partons) using the inclusive $k_\perp$ algorithm [1].

In QCD the jet-jet or jet-photon cross section may be written symbolically as $\sigma_{\mathrm{JJ}(\gamma)} \sim f_{\gamma/e}(y) \otimes f_{i/\gamma}(x_\gamma, \mu) \otimes f_{j/p}(x_p, \mu) \otimes \sigma_{ij}(\theta^\star, \mu)$ with the parton density functions (pdfs) of the photon being replaced by unity for direct processes. The scale $\mu$ represents the QCD renormalization and factorization scales. The variable $x_p$ ($x_\gamma$) is the fractional momentum of the parton from the proton (photon), while $\cos\theta^\star$ is the scattering angle of the jets in the dijet center of mass system.

We review here the latest measurements of inclusive jets, dijets, multijets, and prompt photons performed by the H1 and ZEUS collaborations at HERA. The main aim of these measurements is to provide stringent tests of our calculational abilities in QCD, specifically the adequacy of next to LO (NLO) matrix elements and the current parameterizations of proton and photon pdfs, obtained from deep-inelastic scattering measurements made at HERA, and other processes at other colliders.

*Presented on behalf of the H1 and ZEUS collaborations

## 2. NLO calculations

The data are compared to the NLO calculations of various groups [2–4]. Typically, the following settings are used: 5 flavors, $\alpha_s(M_Z) = 0.118$, $\mu = \mu_{\mathrm{R}} = \mu_{\mathrm{F}} = \mu_\gamma = \mu_p = E_{\mathrm{T,jet}}$ or $1/2\sum_{\mathrm{partons}} E_T$, proton pdfs CTEQ5M [5] and MRST99 [6], photon pdfs GRV-HO [7] and AFG-HO [8].

The band of uncertainty shown in the following figures is due to scale dependence and is obtained by varying the scale up and down by a factor of two, which typically leads to a 10 to 20% change in the cross section.

In order to compare data and NLO calculations, the latter are usually corrected for the effects of hadronization, which at high transverse jet energy $E_{\mathrm{T,jet}}$, are typically less than 10%.

## 3. Inclusive jet production

Preliminary results on inclusive jet production, demanding high $E_{\mathrm{T,jet}}$, i.e. $E_{\mathrm{T,jet}} > 17$ GeV (ZEUS [10]) or 21 GeV (H1 [9]), have become available. These measurements are experimentally the least restrictive in terms of phase space and theoretically most "safe". The NLO calculations are in good agreement with the data within the two dominant errors, the jet energy scale (1 to 2%) and the QCD scale uncertainty. Here we


doi:10.1016/S0920-5632(02)02143-6

only show in Fig. 1 the ratio of the scaled invariant jet cross section $(E_{\mathrm{T,jet}})^4(E_{\mathrm{jet}}d^3\sigma/d(p_{\mathrm{jet}})^3)$ as a function of $x_T = 2E_{\mathrm{T,jet}}/W_{\gamma p}$ for two mean values of $W_{\gamma p}$ of 180 and 255 GeV. In the naive par-

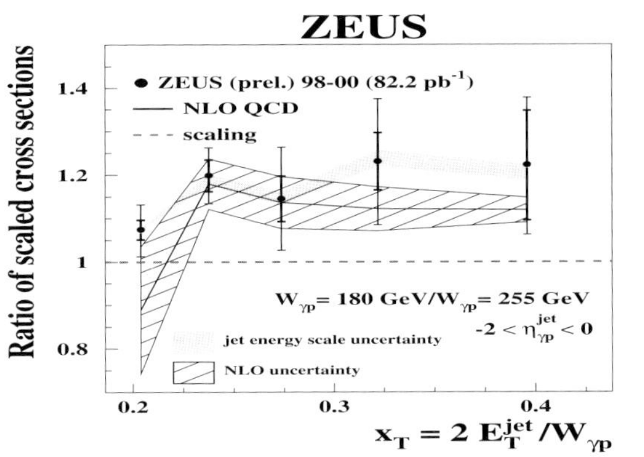


Figure 1. The ratio of the scaled invariant inclusive jet cross sections as a function of $x_T$.

ton model the scaled cross section as a function of $x_T$ is independent of $W_{\gamma p}$ and the ratio therefore should be unity. The photoproduction results show clear scaling violations in agreement with expectations from NLO QCD. In $p\bar{p}$ collisions scaling violations in jet production have been observed at the TEVATRON [11]. There, runs with different beam energies are needed while at HERA different $W_{\gamma p}$ are available in the same run, leading to smaller systematic errors.

## 4. Dijet production

The advantage of requiring at least two hard jets per event lies in being able to determine the $x_p$, $x_\gamma$ and $\cos\theta^\star$ distributions by measuring $E_{\mathrm{T,jet}}$ and $\eta_{\mathrm{jet}}$ for both jets, using the following LO expressions: $x_p = \sum_{1,2} E_{\mathrm{T,jet}} \exp(\eta_{\mathrm{jet}})/2E_p$, $x_\gamma = \sum_{1,2} E_{\mathrm{T,jet}} \exp(-\eta_{\mathrm{jet}})/2yE_e$, and $\cos\theta^\star = \tanh(\eta_{\mathrm{jet1}} - \eta_{\mathrm{jet2}})/2$. By selecting events with $x_\gamma < 0.75$ or $> 0.75$, for example, one can divide the dijet events into samples in which resolved or direct processes dominate. Many dijet distributions have been measured by H1 [12] and ZEUS [13] and compared to NLO calculations. The shapes of the $\cos\theta^\star$ distributions for example agree with the expectations from the dominant propagators, i.e. gluon propagator for resolved and quark propagator for direct processes. The $x_p$ distribution in the range $0.05 < x_p < 0.6$ for $x_\gamma > 0.8$ was measured by H1. They note that the data at the highest $x_p$ are sensitive to the gluon density of the proton, which contributes almost 40% of the cross section there. Currently, at high $x_p$, the NLO predictions agree well with the data, which are however statistics limited. This will change with the high luminosity of HERA II.

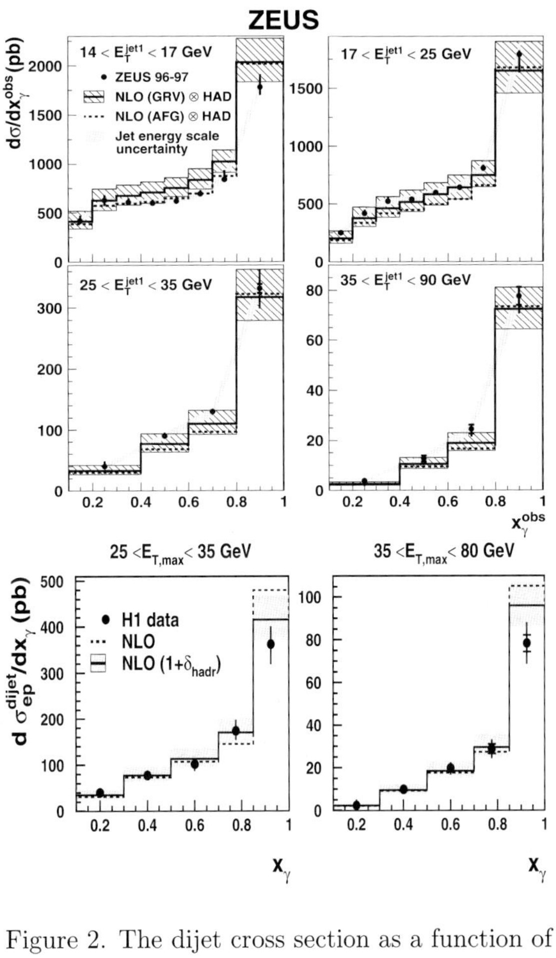


Figure 2. The dijet cross section as a function of $x_\gamma$ in different regions of $E_{\mathrm{T,jet1}}$.

While both experiments conclude that overall the data are described by NLO QCD, ZEUS find that neither the GRV-HO nor the AFG-HO photon pdfs fully describe all features of the data. This is in contrast to H1 which find GRV-HO to be quite adequate. The ZEUS and H1 data in question and the NLO calculations are shown in Fig. 2.

There are small differences in the phase space selected by the two experiments, ZEUS requires for the hardest jet $E_{T,jet1} > 14$ GeV and for the second hardest jet $E_{T,jet2} > 11$ GeV, while H1 demands 25 GeV and 15 GeV, respectively. These asymmetric cuts on $E_{T,jet}$ are needed to avoid an infra-red sensitivity of the NLO calculation in the case of symmetric cuts.

The seemingly contradictory conclusions by the two experiments can be at least partly understood by studying the significantly different sensitivity of the cross section to the cut on $E_{T,jet2}$ between data and NLO calculations, which is shown in Fig. 3. With the $E_{T,jet2}$ cut

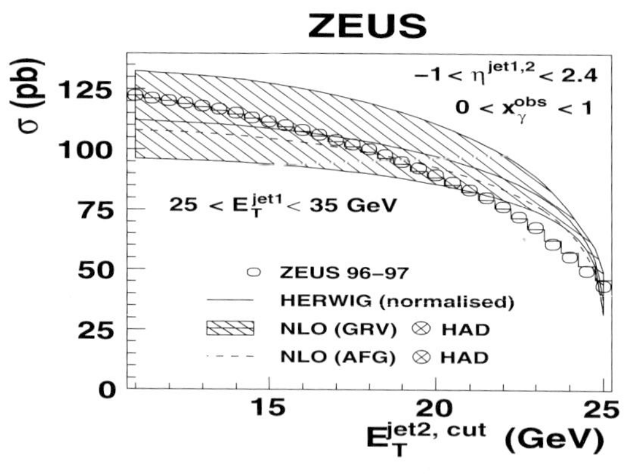


Figure 3. The dijet cross section as a function of $E_{T,jet2}$ for one range of $E_{T,jet1}$ compared to NLO predictions and to the Herwig MC model.

chosen by H1 the NLO calculation lies closer to the data than with the choice of ZEUS. It is also interesting to observe that the HERWIG Monte Carlo model (MC), incorporating LO matrix elements matched to parton showers, describes the $E_{T,jet2}$ dependence of the data quite well (it is normalized to the data in the lowest $E_{T,jet2}$ point). This success of HERWIG seems to indicate that the NLO calculations could be improved significantly by including soft perturbative effects.

## 5. Multijet production

Events with 4 jets resulting from a hard interaction are of $O(\alpha_s^3)$ in LO. In resolved events they may arise from additional interactions between partons of the proton and photon remnants and these are referred to as multi-parton interactions (mpi). The ZEUS collaboration has looked for mpi in events with 4 final state jets [14] $(1 + 2 \rightarrow 3 + 4 + 5 + 6)$ requiring low $x_\gamma$. For simplicity, the 4 jets are mapped onto three by combining the two jets of lowest invariant mass into one jet. The remaining three pseudo-jets are relabeled 3', 4', and 5' in order of decreasing energy. The resulting "3-jet" system can then be described by the usual two angles, $\cos\theta_3$ and $\psi_3$, and energy fractions $X_{3'} = 2E_{3'}/M_{4J}$ and $X_{4'} = 2E_{4'}/M_{4J}$. The jets 3 and 4 are required to have $E_{T,jet} > 6$ GeV and the jets 5 and 6 $E_{T,jet} > 5$ GeV. The resulting $x_\gamma$ distribution shows a significant enhancement at low $x_\gamma$, which can be better described by the PYTHIA MC including mpi. As soon as one requires, in addition, a high invariant mass of the jet system, the enhancement disappears, and the inclusion of mpi has no noticeable effect. The $\cos\theta_3$ distribution, with $\theta_3$ the angle between the beam direction and jet 3', turns out to be sensitive to whether mpi are modeled by adding a soft underlying event, as done in HERWIG, or as true mpi as done in Jimmy (interfaced to HERWIG) or in PYTHIA. The data clearly favor the latter.

## 6. Prompt photon production

Prompt photons are produced in direct and resolved interactions, with a hard photon emerging instead of a gluon jet. Since the properties of a hard photon can be better measured than those of a gluon jet, this process is expected to probe most sensitively our understanding of QCD at the parton level. Of course photons are also produced radiatively and in fragmentation and most copiously in the decays of $\pi^0$'s, and $\eta$'s. In order to suppress this background one looks for energetic and isolated photons. For the latest preliminary results by H1 [15], the requirements for prompt photon candidates are: $5 < E_T^\gamma < 10$ GeV, $-1 < \eta^\gamma < 0.9$, and an isolation cone with $E_T^{\rm cone}/E_T^\gamma < 0.1$. This isolation criterion was chosen to be able to compare to measurements from ZEUS [16]. Other criteria, more appropriate for theoretical calculations, have been suggested but not yet tried. The prompt photon fractions are determined in a likelihood analysis using shower shape observables for prompt photons, and for photons from $\pi^0$ and $\eta$ decays.

The cross section as a function of $E_{\rm T}^\gamma$ [15] is found to be in good agreement with the NLO prediction and with the shape, but not the normalization, predicted by PYTHIA. In Fig. 4 the cross

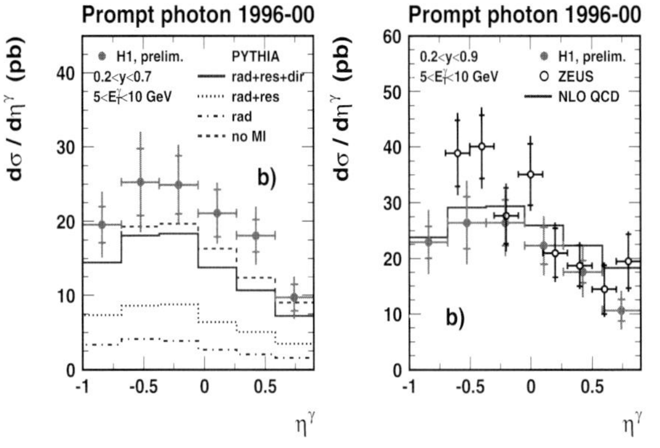


Figure 4. The prompt photon cross section of H1 as a function of $\eta^\gamma$ compared to NLO predictions, the PYTHIA MC and ZEUS data.

section as a function of $\eta^\gamma$ is shown. On the right, the NLO prediction can be seen to describe the data within the rather large errors. The ZEUS data agree with the H1 data and NLO calculations, but tend to be systematically above both at low $\eta^\gamma$. The figure on the left also includes comparisons to PYTHIA predictions which include radiative, resolved and direct contributions, with and without mpi effects. In PYTHIA the effects of mpi tend to reduce the cross section at positive $\eta^\gamma$. Taken at face value one expects that "adding" this effect to the NLO calculation would lower this prediction also.

## 7. Summary

The results reported here, and many more which can be found in the references, demonstrate that we have achieved a good understanding of jets in $\gamma p$ at HERA. NLO calculations with photon pdfs from $\gamma\gamma^\star$ scattering are able to describe the jet and prompt photon data. The data can also provide information on the gluon density in the photon and proton at large $x$ and scales.

Since the experimental uncertainties are now often smaller than the theoretical uncertainties, we need help from our theoretical friends to ensure that the calculations match and challenge the experimental precision achieved. HERA II, with its expected significantly higher luminosity, will allow advances to be made at the high $E_{\rm T,jet}$ and high $x$ frontiers.

## REFERENCES

1. S.D. Ellis and D.E. Soper, Phys. Rev. D **48** (1993) 3160.
2. S. Frixione and G. Ridolfi, Nucl. Phys. B **507** (1997) 315.
3. M. Klasen et al., Eur. Phys. J. direct C1 (1998) 1.
4. M. Fontannaz, et al., Eur. Phys. J. C **21** (2001) 303.
5. H.L. Lai et al., Eur. Phys. J. C **12** (2000) 375.
6. A.D. Martin et al., Eur. Phys. J. C **14** (2000) 133.
7. M. Glück et al., Phys. Rev. D **45** (1992) 3986.
8. P. Aurenche et al., Z. Phys. C **64** (1994) 621.
9. H1 Coll., ABS 1010 submitted to this Conf.
10. ZEUS Coll., ABS 843 submitted to this Conf.
11. F. Abe et al. [CDF Coll.], Phys. Rev. Lett. **70** (1993) 1376;
    B. Abbott et al. [D0 Coll.], Phys. Rev. Lett. **86** (2001) 2523.
12. C. Adloff et al. [H1 Coll.], Eur. Phys. J. C **25** (2002) 13.
13. S. Chekanov et al. [ZEUS Coll.], Eur. Phys. J. C **23** (2002) 615.
14. ZEUS Coll., ABS 849 submitted to this Conf.
15. H1 Coll., ABS 1007 submitted to this Conf.
16. J. Breitweg et al. [ZEUS Coll.], Phys. Lett. B **472** (2000) 175.

ELSEVIER

Proceedings of ICHEP 2002, pp. 308–311
*S. Bentvelsen, P. de Jong, J. Koch and E. Laenen (Editors)*

31st INTERNATIONAL CONFERENCE ON HIGH ENERGY PHYSICS AMSTERDAM

# Photon and Jet Physics at the Collider Detector at Fermilab

J. R. Dittmann[a] for the CDF collaboration

[a]Fermi National Accelerator Laboratory, M.S. 318, P.O. Box 500, Batavia, IL, 60510–0500, U.S.A.

We summarize recent Run 1 photon and jet measurements from $p\bar{p}$ collisions at $\sqrt{s} = 0.63$ TeV and 1.8 TeV using data collected at the Collider Detector at Fermilab (CDF). First Run 2 results at $\sqrt{s} = 1.96$ TeV are also presented together with predictions of the kinematic reach accessible with 15 fb$^{-1}$ of Run 2 data. Data are compared to the predictions of Quantum Chromodynamics (QCD).

## 1. Photon Physics at CDF

The point-like coupling of photons to the hard interaction in high energy collisions offers a direct way to test the predictions of perturbative QCD. At lowest order, prompt photon production is dominated by the Compton process $gq \to \gamma q$, which is sensitive to the gluon distribution of the proton. The ratio of prompt photon cross sections measured at different center-of-mass energies provides a precise test of the QCD matrix elements, as many experimental uncertainties cancel and the theoretical uncertainties due to the gluon distribution are reduced.

A recent CDF measurement of the isolated prompt photon cross section at $\sqrt{s} = 0.63$ and 1.8 TeV [1] reveals discrepancies between data and QCD that are difficult to explain with conventional theoretical uncertainties such as scale dependence and parton distribution parameterizations. Figure 1 highlights deviations from NLO QCD predictions [2] using the CTEQ5M parton distribution functions with a renormalization scale equal to the photon $P_T$. At low $P_T$, the data exhibit a steeper slope than theory, and at high $P_T$ the $\sqrt{s} = 1.8$ TeV data and theory disagree by an overall normalization factor. The experimental results are reinforced by measurements of DØ [3] and UA2 [4]. One proposed explanation for the discrepancy at low $P_T$ is an inadequate description of the initial-state parton shower in the NLO QCD calculation which could give a recoil $k_T$ to the photon-jet system [5]. Including an additional $k_T$ recoil by adding simple Gaussian smearing to the NLO QCD calculation yields better agreement with the data [1].

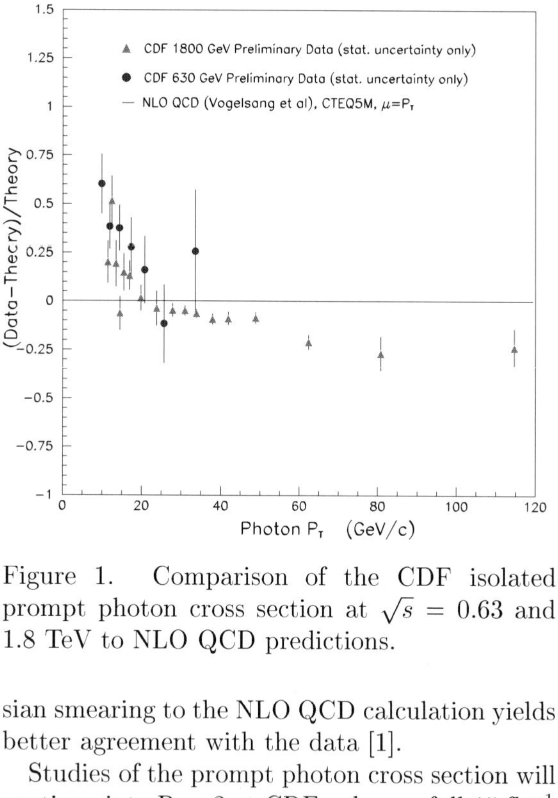


Figure 1. Comparison of the CDF isolated prompt photon cross section at $\sqrt{s} = 0.63$ and 1.8 TeV to NLO QCD predictions.

Studies of the prompt photon cross section will continue into Run 2 at CDF, where a full 15 fb$^{-1}$ of data are anticipated. In addition, other Run 2 photon analyses such as diphoton production and photon + heavy flavor production will play a leading role in tests of QCD and searches for new phenomena. In diphoton events, for example, where the final state kinematics can be completely reconstructed, the diphoton mass can be measured with good resolution out to nearly 600 GeV/$c^2$.


doi:10.1016/S0920-5632(02)02144-8

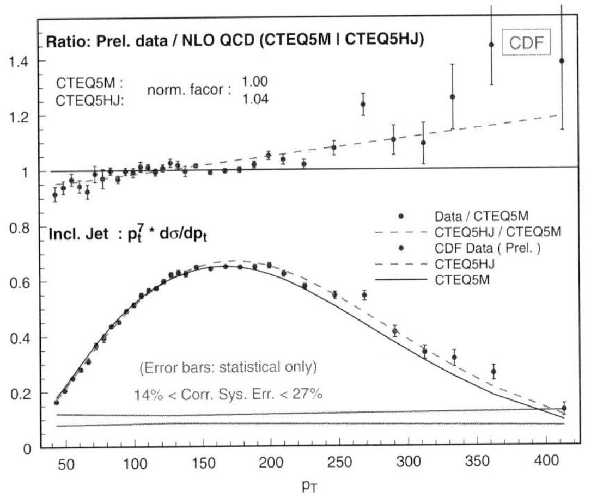


Figure 2. Ratio of the CDF inclusive jet cross section to NLO QCD theory using the CTEQ5M parton distribution functions. Predictions using CTEQ5HJ (dashed line) agree well with the data.

## 2. Jet Physics at CDF

### 2.1. Inclusive jet cross section

The inclusive jet cross section at CDF provides a fundamental test of QCD for jet energies ranging from 40–465 GeV, where the cross section changes by over 7 orders of magnitude. At the highest jet $E_T$, this measurement probes a distance scale of $\sim 10^{-17}$ cm and has traditionally been used to search for new physics.

Recent CDF Run 1B results [6] support the Run 1A observation of an excess in the jet cross section at large jet $E_T$ compared to NLO QCD predictions using parton distribution functions like MRST and CTEQ4M. Consistency between data and QCD can be attained, however, by virtue of a large uncertainty in the gluon distribution at high $x$. Figure 2 illustrates the level of agreement using CTEQ5HJ, where extra weight is given to high-$E_T$ jet data. Measurements of forward jets at DØ [7] further constrain the high-$x$ gluon distribution, yielding a more "HJ-like" central fit for CTEQ6 that is consistent with both CDF and DØ inclusive jet data [8].

A large sample of Run 2 data will determine whether the high-$x$ gluon distribution completely explains the inclusive jet cross section at high $E_T$. For Run 2, the change in the center-of-mass energy from $\sqrt{s} = 1.8$ to 1.96 TeV increases the rate of high-$E_T$ jets by a factor of 2–3, extending the reach beyond 650 GeV in the central region. In addition, the new CDF endplug calorimeter will permit jet measurements in the forward region. These improvements, together with a suite of new jet algorithms, offer a rich program of study for jet physics in Run 2.

### 2.2. Studies of the underlying event

Precise jet measurements at hadron-hadron colliders require a good model of the "underlying event" which consists of multiple parton scattering, beam-beam remnants, and (part of) the initial and final-state radiation. Underlying event energy must be subtracted from jet energies in order to compare jet cross sections to NLO QCD. In addition, studies of the underlying event are worthwhile because they probe the interface between perturbative and non-perturbative QCD. Two complementary Run 1 analyses provide a quantitative study of the underlying event from very soft to very hard collisions.

In one analysis [9], charged particle tracks with $p_T > 0.5$ GeV/c and $|\eta| < 1$ are identified in minimum bias and jet events. A simple algorithm is then applied to the list of tracks to locate "charged particle jets" with cone size $R = 0.7$ in the $\eta$–$\phi$ plane. Tracks pointing to a region "transverse" to the direction of the highest-$p_T$ jet ($60° < |\Delta\phi| < 120°$) are used to study the underlying event. Figure 3 illustrates the charged track multiplicity in the transverse region vs. the leading jet $p_T$, which extends up to 50 GeV/c. Comparisons to PYTHIA 6.206 reveal that the default parameters underestimate the underlying event energy in the transverse region. The parameters that control the amount of initial-state radiation in PYTHIA can be tuned, however, to yield good agreement with data.

In the second analysis, jets are selected with $E_T$ ranging from 50–300 GeV. Charged tracks are observed in two cones at the same $\eta$ as the leading jet, but at $\pm 90°$ away in $\phi$. Of the two cones, the one with the greater (lesser) $\Sigma p_T$ is defined as the max (min) cone. The $p_T$ in the max cone increases with leading jet $E_T$, whereas the min cone $p_T$ stays constant at a level similar to that found in minimum bias events at $\sqrt{s} = 1.8$ TeV.

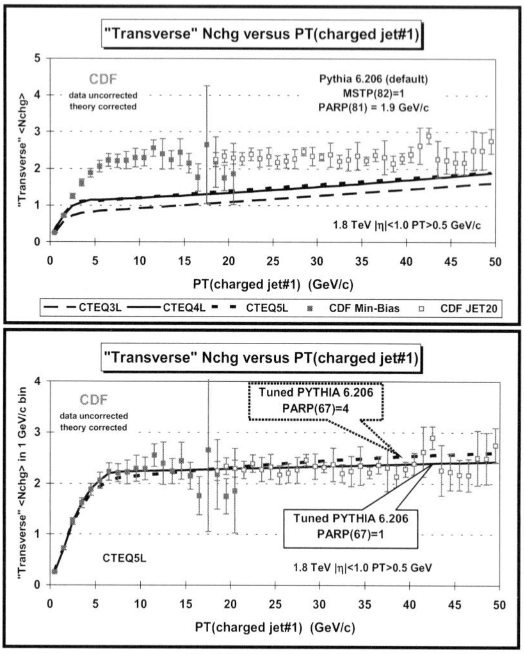


Figure 3. Mean number of charged tracks in the "transverse" region vs. the leading charged jet $p_T$ from Run 1 minimum bias and jet data. The data are compared to PYTHIA 6.206 using default parameters (top) and tuned parameters (bottom).

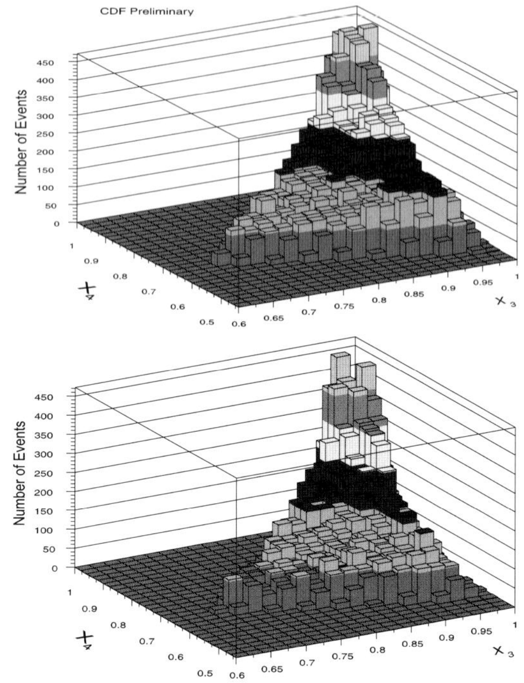


Figure 4. Event density in the $X_3$–$X_4$ plane for three-jet events in Run 1B data (top) and NLO QCD (bottom). The theoretical cross section is calculated using CTEQ3 with $\alpha_s = 0.116$.

Although HERWIG agrees well with the data without any tuning, the parameters in PYTHIA that control the regularization scale for multiple parton scattering and the impact parameters option can be tuned to better fit the data.

### 2.3. Three-jet production cross section

In a new measurement of inclusive three-jet production at CDF, features of three-jet events in 86 $pb^{-1}$ of Run 1B data are compared to NLO QCD predictions. Jets are identified using the CDF iterative cone algorithm (JETCLU) with cone size $R = 0.7$ in the $\eta$–$\phi$ plane. Events are selected that have $\geq$3 jets (each with $E_T^{\text{jet}} > 20$ GeV and $|\eta^{\text{jet}}| \leq 2.0$), $\Sigma E_T(3\,\text{jets}) > 320$ GeV, and a separation of $\Delta R > 1.0$ between jets. These events are boosted into the three-jet rest frame, and the energies of the three leading jets are corrected, unsmeared, and numbered such that $E_3 > E_4 > E_5$. The three-jet mass $m_{3\text{jet}}$ is calculated together with the Dalitz variables $X_i = 2E_i/m_{3\text{jet}}$.

Figure 4 compares the three-jet cross section vs. $X_3$ and $X_4$ to NLO QCD predictions [10]. A detailed, bin-by-bin comparison of data and theory indicates good agreement. The total measured three-jet production cross section, using the full kinematically-allowed Dalitz plane, is 466 $\pm$ 2 (stat) $^{+206}_{-71}$ (syst) pb, which is consistent with the NLO QCD prediction of 402 $\pm$ 3 pb.

### 2.4. Jet shapes in inclusive dijet events

The internal structure of jets is mainly dictated by multi-gluon emission from the primary parton which involves higher-order QCD processes. The study of this structure thus provides a stringent test of the models used to simulate parton showers. Here we present preliminary jet shape measurements using a sample of inclusive dijet events in 16 $pb^{-1}$ of Run 2 CDF data.

Jets are identified using JETCLU with cone size $R = 0.7$. Events are selected that contain at least two jets with $E_T^{\text{jet}} > 30$ GeV and $|\eta^{\text{jet}}| < 2.3$.

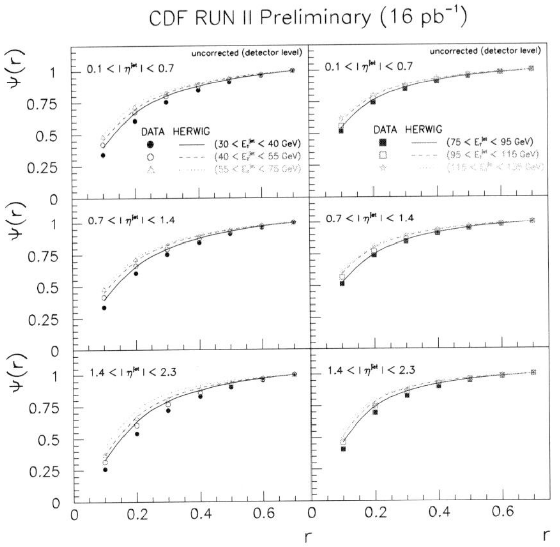


Figure 5. Integrated jet shapes, $\Psi(r)$, for various bins of $E_T^{\rm jet}$ and $\eta^{\rm jet}$, measured using the calorimeter. The data are compared to HERWIG Monte Carlo predictions.

The integrated jet shape, $\Psi(r)$, is defined as the average fraction of the jet's transverse energy that lies inside an inner cone of radius $r$ concentric to the jet cone:

$$\Psi(r) = \frac{1}{N_{\rm jet}} \Sigma_{\rm jets} \frac{E_T(0,r)}{E_T(0,R)} \qquad (r < R) \qquad (1)$$

A plot of $\Psi(r)$ for jets with 30 GeV $< E_T^{\rm jet} <$ 135 GeV and $0.1 < |\eta^{\rm jet}| < 0.7$ is shown in Figure 5. Jets become narrower as $E_T^{\rm jet}$ increases, but no significant dependence on $\eta^{\rm jet}$ is observed. The data are compared to HERWIG predictions with initial and final-state radiation and interactions between spectator partons ("underlying event"). Data and theory agree very well, although HERWIG predicts jets that are too narrow at low $E_T^{\rm jet}$ and high $\eta^{\rm jet}$. This disagreement may be due to the modeling of the underlying event.

The CDF Central Outer Tracker (COT) provides an independent method to reconstruct the jet shape, thereby validating the calorimeter-based measurement. For a given jet, tracks that point within the jet cone are identified and $\Psi(r)$ is calculated based on the $p_T$ of the tracks similar to Eq. 1. Figure 6 compares the integrated jet shapes for calorimeter towers and tracks in a region with good COT coverage ($0.1 < |\eta^{\rm jet}| < 0.7$). Excellent agreement between the calorimeter and tracking measurements is observed.

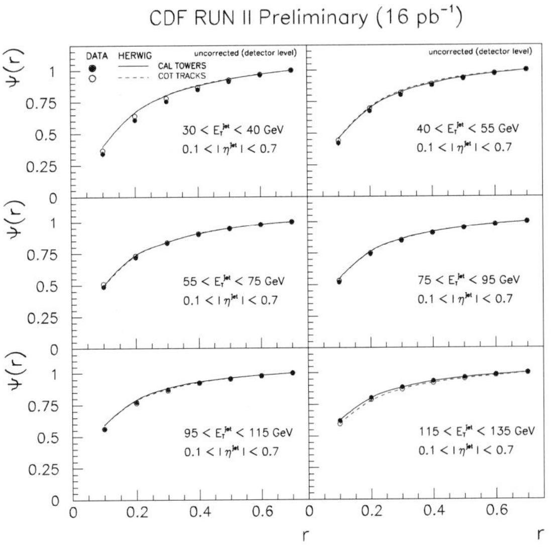


Figure 6. Measured integrated jet shapes, $\Psi(r)$, using calorimeter towers (solid circles) or COT tracks (open circles).

## REFERENCES


1. D. Acosta *et al.*, Phys. Rev. D **65**, 112003 (2002).
2. M. Gluck *et al.*, Phys. Rev. Lett. **73**, 388 (1994).
3. B. Abbott *et al.*, Phys. Rev. Lett. **84**, 2786 (2000).
4. J. Alitti *et al.*, Phys. Lett. **B 263**, 544 (1991).
5. L. Apanasevich *et al.*, Phys. Rev. D **59**, 074007 (1999).
6. T. Affolder *et al.*, Phys. Rev. D **64**, 032001 (2001).
7. B. Abbott *et al.*, Phys. Rev. Lett. **86**, 1707 (2001).
8. M. Zielinski, these proceedings.
9. T. Affolder *et al.*, Phys. Rev. D **65**, 092002 (2002).
10. W. Kilgore and W. Giele, hep-ph/0009193.

Proceedings of ICHEP 2002, pp. 312–315
*S. Bentvelsen, P. de Jong, J. Koch and E. Laenen (Editors)*

31st INTERNATIONAL CONFERENCE ON HIGH ENERGY PHYSICS AMSTERDAM

# Jet and Photon Physics at DØ

Marek Zieliński*
University of Rochester, Rochester, NY 14627
E-mail: marek@fnal.gov

We report several new DØ measurements related to issues in QCD. Because of its large reach in $x$-$Q^2$ and due to a thorough analysis of correlated systematic errors, the rapidity-dependent inclusive jet cross section has had a significant impact on the extraction of parton distribution functions. The ratios of jet and photon production at $\sqrt{s} = 1800$ and 630 GeV, although in general agreement with theory, indicate that not all aspects of these processes are fully understood. Studies of multiple jet production at moderate and low $E_T$ values have provided means to explore QCD scales and to tune the related aspects of event generators. A definitive signal for diffractive production of W bosons has been established, providing a new probe of the quark structure of the Pomeron. And, following on a great effort to bring up the experiment after a long shut-down and extensive upgrades, DØ is accumulating and analyzing new data at 2 TeV. The preliminary results for inclusive jet $p_T$ and dijet mass spectra are just the first steps in exploiting the opportunities that Run II of the Tevatron will provide.

Within the framework of perturbative QCD (pQCD), hard-scattering processes are described by a convolution of partonic cross sections with the parton distribution functions (PDFs). Thus, measurements of high $p_T$ interactions constrain PDFs, test current phenomenology, and provide opportunities to search for deviations from pQCD (*e.g.*, parton compositeness).

The hard-scattered quarks and gluons in the final state fragment into jets of particles that subsequently interact in the detectors. Experimentally, jets are usually defined by their energy deposition in multiple calorimeter towers. The jet results reported here employ the iterative fixed-radius cone algorithm, with $R = 0.7$ in $\eta$-$\phi$ space. While Run-I jets consist of $E_T$-weighted sums of calorimeter towers, in Run II jets are defined by 4-momentum sums of the towers inside the cone, and midpoints between such jets are used as additional seeds for the algorithm in order to ameliorate sensitivity to infrared divergence of gluon radiation [1]. Production of objects at low and moderate $p_T$ is sensitive to fragmentation, multiple parton interactions, and to the remnants of the colliding particles not participating in the hard collision. Studies of these effects help develop phenomenological descriptions of such soft physics and in the tuning of event generators.

The high-$x$ behavior of PDFs has been discussed intensely for the past several years. DØ measurement [2] of the inclusive jet cross section in 5 different $\eta$ regions, shown in Fig. 1, has provided powerful new information. The large coverage of parton phase-space results in unprecedented sensitivity to PDFs, and the evaluation of the correlated errors among the 90 experimental points permits a more rigorous incorporation of the data into recent global determinations of PDFs by the CTEQ [3] and MRST [4] groups. DØ jet data have been included in these fits without extra weight applied to their contribution to the total $\chi^2$; the resulting gluon PDFs show a significant enhancement at large $x$ over previously available best fits. The data are described very well by NLO QCD calculations using these new PDFs (see Fig. 2 for the CTEQ6M result).

The ratios of jet and photon measurements at $\sqrt{s}$ of 630 and 1800 GeV have been studied to investigate the scaling of the cross section with energy. Various experimental and theoretical uncertainties tend to cancel in the ratio. The ratio of the dimensionless central jet cross sections vs $x_T = 2E_T/\sqrt{s}$ is shown in Fig. 3. The 10-15% disagreement in normalization between data

*For the DØ Collaboration.


doi:10.1016/S0920-5632(02)02162-X

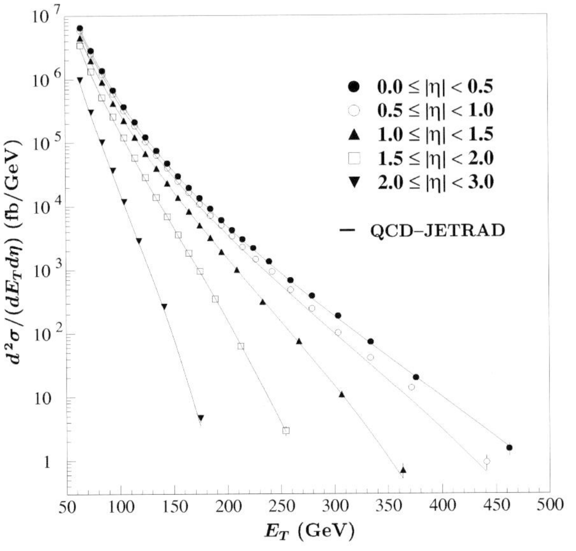


Figure 1. The inclusive jet cross section vs $E_T$, for 5 bins of $|\eta|$, compared to JETRAD calculation with the CTEQ4M PDF.

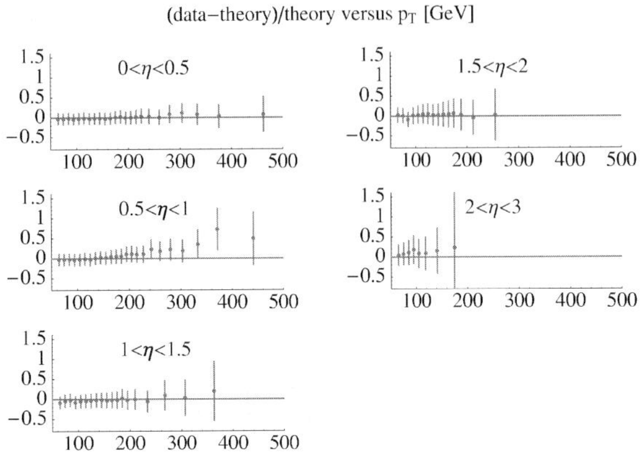


Figure 2. Comparison of jet data to NLO calculation using CTEQ6M PDF; $\chi^2 = 65$ for 90 points.

and theory, as well as some difference in shape, can not be accommodated through varying the choice of PDFs or of the pQCD scales [5]. Other explanations have been proposed, *e.g.*, invoking low-$E_T$ non-perturbative effects, or using different scales at the two $\sqrt{s}$. The analogous ratio for direct-photon production [6] is shown in Fig. 4. The data are in overall agreement with NLO QCD, with possible deviation at lowest $x_T$ values. However, current sensitivity is not sufficient to either prove or disprove the proposed presence of dynamic parton $k_T$ effects in photon production at these energies [7]. Run II measurements should shortly provide additional information to help clarify these issues.

The measurement [8] of the ratio of inclusive three-jet to two-jet production, $R_{32}$ vs $H_T = \sum E_T^{\text{jet}}$, reflects the rate of gluon emission in the jet production. The data were compared to predictions from JETRAD using different renormalization scales for the third jet (Fig. 5). No evidence was found to indicate the need to introduce such a separate scale. The common scale $\approx 0.3H_T$ describes the data satisfactorily.

DØ has recently performed a study of multiple jet production at low $E_T \approx 20$ GeV [9]. This process is expected to be sensitive to soft gluon emission and multiple-parton scattering effects. The data were compared to the predictions from PYTHIA 6.127 and HERWIG 5.9 (and with JETRAD for certain aspects). A good description of the results has been obtained after tuning parameters in the generators (PARP(83)=0.32 in PYTHIA and PTMIN=3.7 in HERWIG), as shown in Fig. 6, but the "off-the-shelf" versions with the default values disagree with the data.

DØ is completing an investigation of diffractive production of W and Z bosons using Run I data [10]. Since electroweak bosons couple only to quarks (at lowest order), this process facilitates examination of the quark structure of the Pomeron [11]. By using the rapidity-gap technique, a definitive signal for diffractive W boson production was established with 91 candidate events; diffractive Z boson production was observed for the first time, with only 9 candidate events. These data provide preliminary measurements of the diffractive component $R_W = 0.89^{+0.20}_{-0.19}\%$ and $R_Z = 1.44^{+0.62}_{-0.16}\%$ of total W and Z boson production, respectively. For Run II, DØ has installed Roman-pot detectors to tag protons and antiprotons in the final state. In particular, this will allow a direct measurement of the quark-gluon structure of the Pomeron, and to study for

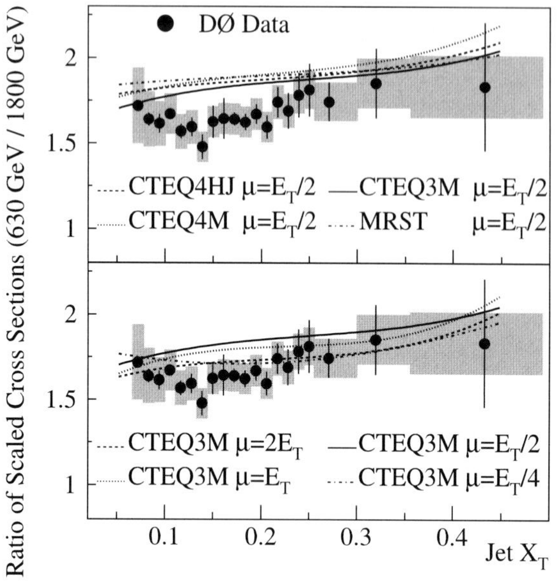


Figure 3. Ratio of jet cross sections at $\sqrt{s} = 630$ and at 1800 GeV, vs $x_T$, compared to calculations with different PDFs and variations of pQCD scales.

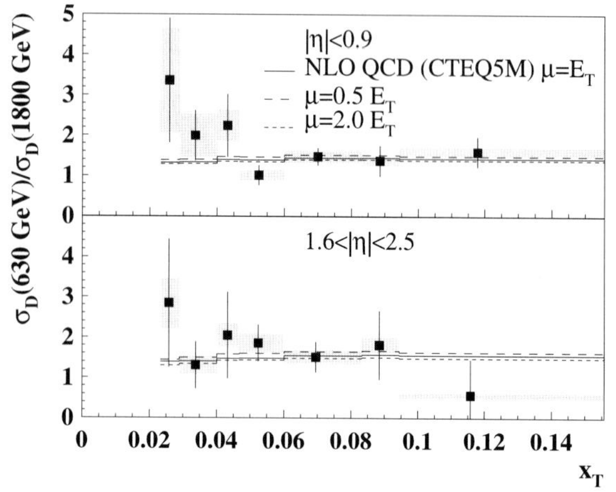


Figure 4. Ratio of photon cross sections at $\sqrt{s} =$ 630 and at 1800 GeV, vs $x_T$.

the first time the double Pomeron exchange.

After extensive upgrades of virtually all subsystems of the detector, DØ is collecting physics-quality data from Run II of the Fermilab Tevatron. High-$p_T$ jet production at $\sqrt{s} = 2$ TeV is the highest-priority measurement for the ultimate determination of the gluon PDF, and the search for new physics at shortest distance scales. In addition to the much larger integrated luminosity expected in Run II, jet measurements at $\sqrt{s} = 2$ TeV also benefit from a cross section increase by factors $\geq 2$ (at largest $p_T$) relative to Run I. Figures 7 and 8 show the preliminary jet inclusive and dijet mass spectra derived from $\approx 6$ pb$^{-1}$ of data taken during March–May 2002. Although the analyses are well under way, the displayed results have not yet been fully corrected for experimental effects, so comparisons to theory would be premature. The limited $\eta$ range is due to the fact that Level-1 jet triggers were constrained to $|\eta| < 0.8$ during that period (this range has now been expanded significantly).

The large kinematic reach of jet measurements in Run II is already apparent. Improvements in the precision of the results are anticipated thanks to the upgraded detectors, and the NNLO calculations are expected to be available shortly. We are entering an important period for studies of QCD.

## REFERENCES


1. U. Baur, R.K. Ellis, D. Zeppenfeld, editors, Fermilab-Pub-00/297.
2. B. Abbott *et al.*, Phys. Rev. Lett. 86 (2001) 1707.
3. J. Pumplin, D.R. Stump, J. Huston, H.L. Lai, P. Nadolsky and W.K. Tung, JHEP 0207 (2002) 012.
4. A.D. Martin, R.G. Roberts, W.J. Stirling and R.S. Thorne, Eur. Phys. J. C23 (2002) 73.
5. B. Abbott *et al.*, Phys. Rev. Lett. 86 (2001) 2523.
6. V. Abazov *et al.*, Phys. Rev. Lett. 87 (2001) 251805.
7. L. Apanasevich *et al.*, Phys. Rev. D59 (1999) 074007.
8. B. Abbott *et al.*, Phys. Rev. Lett. 86 (2001) 1955.
9. V. Abazov *et al.*, hep-ex/0207046.
10. DØ collaboration, paper in preparation.
11. P. Bruni and G. Ingelman, Phys. Lett. B311 (1993) 317.

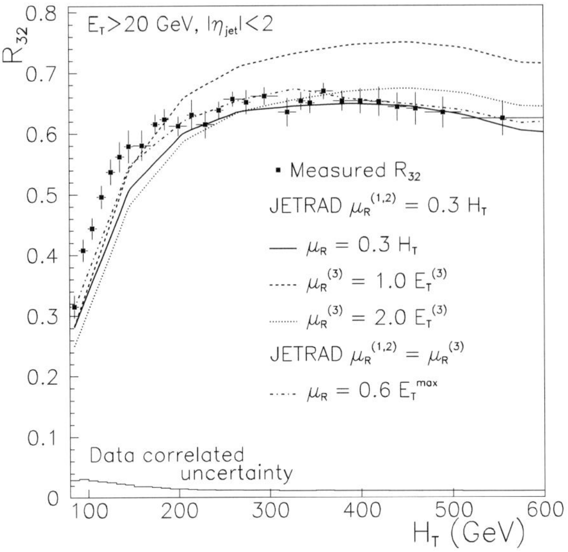


Figure 5. Ratio of $\geq$ 3-jet to $\geq$ 2-jet production vs $H_T$, compared to JETRAD calculations with various scales.

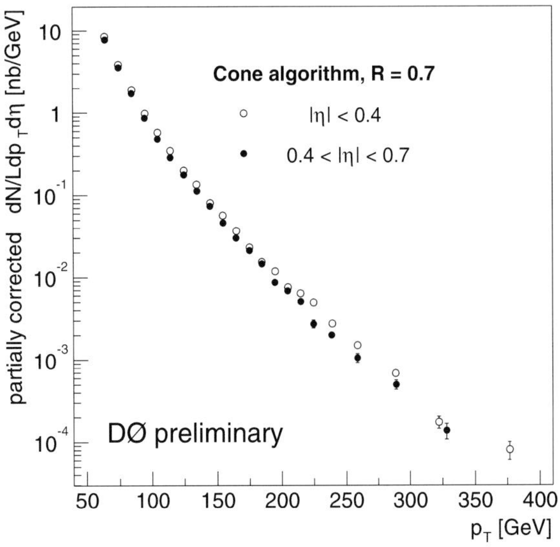


Figure 7. Inclusive jet-$p_T$ spectra in Run II data.

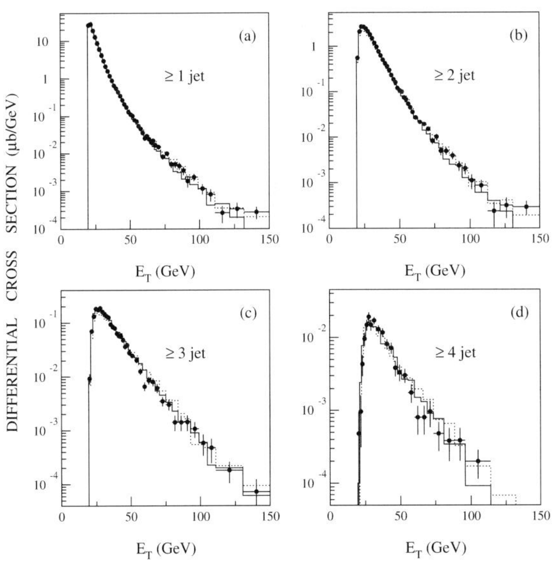


Figure 6. Leading jet $E_T$ for multiple inclusive jet production, compared to predictions from PYTHIA (solid) and HERWIG (dotted).

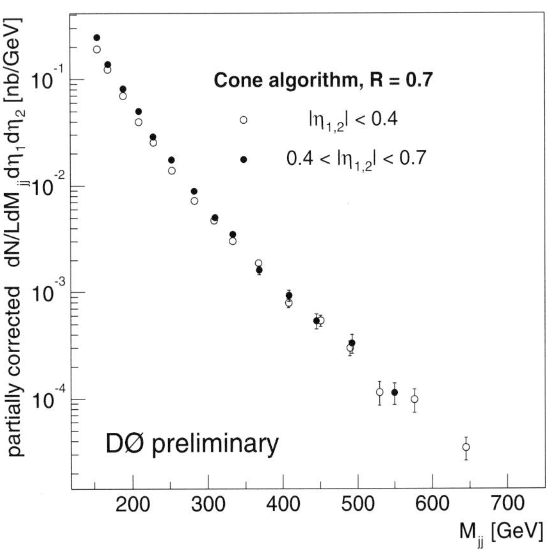


Figure 8. Dijet mass spectra in Run II data.

Proceedings of ICHEP 2002, pp. 316–318
*S. Bentvelsen, P. de Jong, J. Koch and E. Laenen (Editors)*

31st INTERNATIONAL CONFERENCE ON HIGH ENERGY PHYSICS AMSTERDAM

# Jet Measurements at DØ using a $k_T$ Algorithm

Ursula Bassler *on behalf of the DØ-collaboration* [a]

[a]LPNHE; 4, place Jussieu, 75252 Paris Cedex 05, France

A first implementation of the $k_T$ jet algorithm for hadron collider data has been developed by the DØ collaboration. Results on the measurement of the inclusive jet cross-section, sub-jet multiplicities in quark and gluon jets and a preliminary measurement of thrust cross-sections have been obtained using the Run I $p\bar{p}$ data collected between 1994−1995 at $\sqrt{s} = 1800$ GeV and $\sqrt{s} = 630$ GeV with an integrated luminosity of 90 pb$^{-1}$ and 0.6 pb$^{-1}$ respectively .

Jet reconstruction at hadron colliders is usually done using the cone algorithm, although the $k_T$ algorithm [1], initially developed for $e^+e^-$ data at LEP [2], has already been successfully used at HERA [3] and is now for the first time implemented by DØ for $p\bar{p}$ data. The $k_T$ algorithm is theoretically favored, as it is infrared save to all orders and presents fewer split-merge ambiguities than the cone algorithm.

The DØ implementation applies the $k_T$ algorithm to objects obtained after pre-clustering the energy deposited in the calorimeter in towers of $\eta \times \varphi = 0.2 \times 0.2$, reducing the number of objects to about 200. For all pre-clusters,

$$d_{ii} = k_{T,i}^2 \quad d_{ij} = min(k_{T,i}^2, k_{T,j}^2)\delta R/D^2 \qquad (1)$$

are computed, with $D$ being a resolution parameter. Using $D = 1$, NLO prediction are within 1% of the predictions for the cone algorithm with a cone size of 0.7. The resulting quantities $d_{ii}$ and $d_{ij}$ are ordered and if the smallest quantity is $d_{ii}$, the object is considered a jet, else the objects $i$ and $j$ are merged with a new value for $d_{ii}$ to be considered. A detailed description of the implementation of the $k_T$ algorithm is given in [4].

## 1. Inclusive Jet Cross-Section

The inclusive jet cross-section had been measured for jets with $60 < E_T < 450$ GeV and is steeply falling over seven orders of magnitude [5]. For this measurement, the jet $E_T$ has been corrected to the particle level. The total error is about 14(27)% at 60(450) GeV, largely dominated at high $E_T$ by the uncertainty on the jet energy scale.

A comparison of the measured cross-section with NLO-computation using the JETRAD [6] program with the CTEQ4HJ [7] proton structure function parameterization is shown in figure 1, where a marginal agreement is observed at low $E_T$. Using the CTEQ4M or the MRST [8] structure function parametrizations worsens the agreement at high $E_T$. A Comparison of the cross-section measurement using the cone algorithm with a cone size of 0.7 [9] to the corresponding JETRAD computation is shown on the same figure. Whereas for the cone algorithm a good agreement with JETRAD is observed, the behavior differs from the $k_T$ algorithm at low $E_T$ and

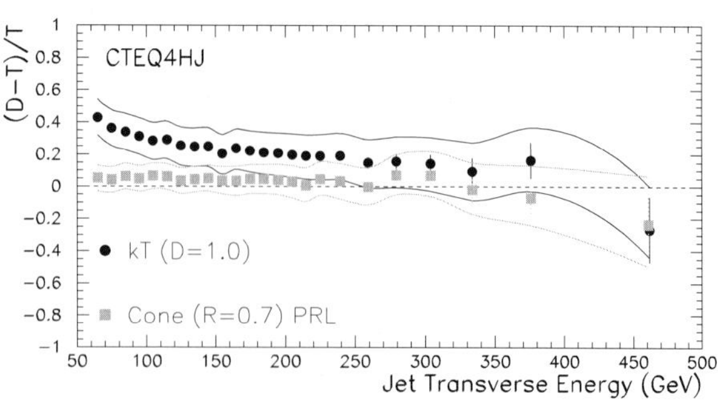


Figure 1. Comparison of Data-Theory/Theory for the inclusive jet cross-section as a function of the jet $E_T$ for jets identified with the $k_T$ and the cone algorithm. Each measurement is compared to its own NLO prediction.


doi:10.1016/S0920-5632(02)02164-3

the ratio of the measured cross-section over the computed cross-section is up to 45% higher when measured with the $k_T$ algorithm than with the cone algorithm.

The reasons of this discrepancy are studied comparing cone and $k_T$ jets which are matching in $\eta \times \varphi$. At 60(200) GeV $k_T$ jets are 7(3)% more energetic than cone jets in the data sample, after offset corrections to remove energy from pile-up, noise and multiple interactions have been applied. No dependence on the instatanious luminosity associated to a given sub-sample has been observed. This energy difference is also visible using events simulated with HERWIG [10], although the effect is smaller, 2(1.5)% at 60(200) GeV. As the cross-section is steeply falling, the difference in the jet energy allows to account for the difference in cross-section: when correcting the energy of cone jets by the difference observed with $k_T$ jets, the cross-section measurements are in agreement.

The origins of this energy difference are partially due to hadronization effects, energy losses outside the cone for the cone algorithm and the merging of nearby partons into a particle jet with the $k_T$ algorithm. In the simulation, the particle jets spectra is 5% higher (lower) at low $E_T$ compared to jets on the parton level when identified with the $k_T$ (cone) algorithm.

## 2. Sub-jet Multiplicity in Quark and Gluon Jets

Differences in the substructure of quark and gluon jets are expected from QCD. As the ratio of the color factor for radiated gluons from gluons/quarks is 9/4, the multiplicity ratio of objects in gluon/quark jets is expected to be comparable within the asymptotic limit of the color factor. Besides, particles in a gluon jet are expected to be softer than in a quark jet.

The fraction of gluon jets $f_g$ at fixed $E_T$ varies with $\sqrt{s}$ and has been estimated from HERWIG to be 59% at 1800 GeV and 33% at 630 GeV for jets with $55 < E_T < 100$ GeV. Assuming the sub-jet multiplicity independent of $\sqrt{s}$, the quark and gluon multiplicity can be extracted solving

$$M = f_g M_g + (1 - f_g) M_q \qquad (2)$$

at the two different center of mass energies [4].

To measure the sub-jet multiplicity, the $k_T$ algorithm is applied to all sub-clusters in already identified jets, by computing $d_{ij}$ with the resolution parameter $D = 0.5$ and merging only objects with $d_{ij} < y_{cut} E_T^{jet}$ using $y_{cut} = 10^{-3}$. Figure 2 shows the fraction of quark and gluon jets as a function of their sub-jet multiplicity. As expected, gluon jets show a higher multiplicity than quark jets. Good agreement can be found when comparing the result to a full simulation using the HERWIG event generator. Resummation calculations from Forshaw and Seymour [11] are qualitatively in agreement, but with a higher mean value. This effect may be due to not taking into account the pre-clustering.

The measured ratio $R = \langle M_g \rangle - 1 / \langle M_q \rangle - 1$ is

$$R = 1.84 \pm 0.15 \text{ (stat.)} \pm^{0.22}_{0.18} \text{ (syst.)} \qquad (3)$$

with the dominant uncertainty being the jet energy scale and the estimation of $f_g$, which depends on the parton density function used. The measurement in agreement with the expectation from HERWIG of $R = 1.91$ and a measurement in $e^+e^-$ data from ALEPH with $R = 1.7 \pm 0.1$ [2]. The obtained value is lower than the naive

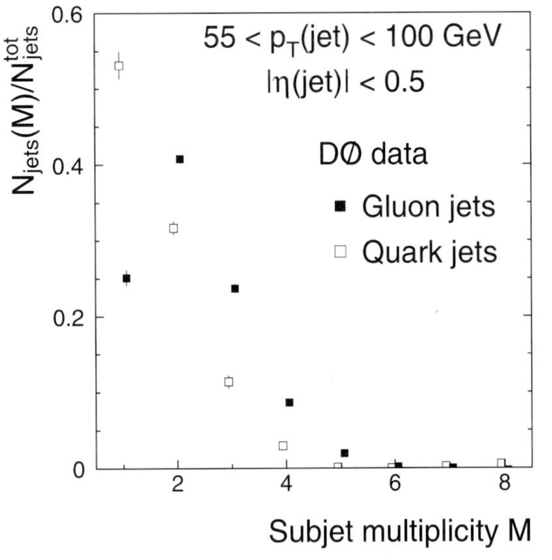


Figure 2. Number of quark (gluon) jets as a function of the sub-jet multiplicity.

expectation of $R = 2.25$ from QCD, due to the effect from higher order contributions.

## 3. Thrust Cross-Section

The thrust variable characterizes the sphericity of an event and is defined as:

$$T^T = max_{\hat{n}} = \frac{\sum_i |\vec{p}_i \cdot \hat{n}|}{\sum_i |\vec{p}_i|} \quad (4)$$

To reduce effects from the underlying event, pile-up and multiple interactions in hadron colliders, the transverse thrust from the two leading jets

$$T_2^T = max_{\hat{n}} = \frac{\sum_{i\leq 2} |\vec{p}_{ti} \cdot \hat{n}|}{\sum_{i\leq 2} |\vec{p}_{ti}|} \quad (5)$$

with $\sqrt{2}/2 < T_2^T < 1$ has been measured [12].

The measurement of the cross-section as a function for $1 - T$ is done in four bins of $H_3^T = \sum_{i\leq 3} |\vec{p}_{ti}|$ between 160 and 700 GeV, $H_3^T$ being proportional to $Q^2$ on the parton level. Figure 3 shows the cross-section in the lowest $H_3^T$ bin ($160 < H_3^T < 260$ GeV) on a linear(logarithmic) and logarithmic(linear) abscissa (ordinate) compared to a calculation using JETRAD.

For $\sqrt{2}/2 \leq T_2^T \leq \sqrt{3}/2$ the leading order is $\mathcal{O}(\alpha_s^4)$ and the available $\mathcal{O}(\alpha_s^3)$ calculation can not be used for comparisons. In the limit $(1 - T) \to 1$ the emission of soft and colinear gluons is contributing as logarithmic terms in $1 - T$ and resummation is needed to take into account this effect.

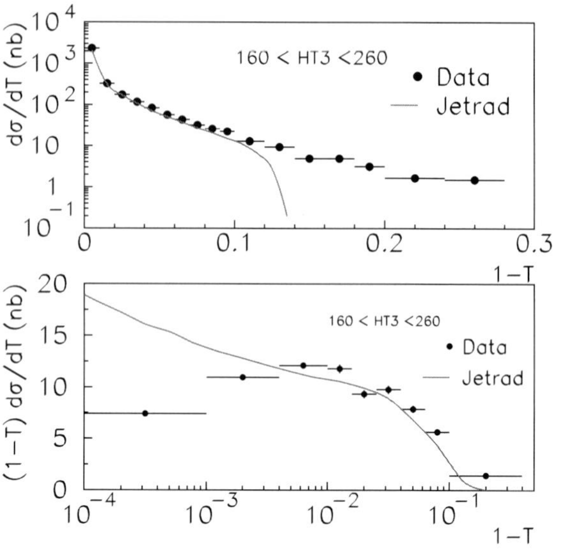


Figure 3. Comparison of Data-Theory/Theory for the cross-section as a function of the transverse thrust for the two leading jets in bins of scalar $E_T$ of the three leading jets. The upper (lower) distributions show the measurements as a function of $1 - T$ on a linear (logarithmic) scale.


## Acknowledgements

Many thanks to Daniel Elvira, Marek Zielinski, John Krane and Christophe Royon for their help during the preparation of the talk, the careful reading and comments to the proceedings.


## REFERENCES


1. S. Catani, Y.L. Dokshitzer and B.R. Webber, Phys.Lett. **B285** (1992) 1291. S. Catani et al., Nucl.Phys. **B406** (1993) 187. S.D. Ellis and D.E. Soper, Phys.Rev. **D48** (1993) 3160.
2. for example: D. Buskulic et al. (Aleph Collab.), Phys.Lett. **B346** (1995) 389.
3. C. Adloff et al. (H1 Collab.), Nucl.Phys. **B545** (1999) 3. J. Breitweg et al. (ZEUS Collab.), Eur.Phys.J. **C1** (1998) 109.
4. V.M. Abazov et al. (DØ Collab.), Phys.Rev. **D65** (2002) 052008.
5. V.M. Abazov et al. (DØ Collab.), Phys.Lett. **B525** (2002) 211.
6. W.T. Giele, E.W.N. Glover and D.A. Kosover, Phys.Lett. **73** (1994) 2019.
7. H.L. Lai et al. (CTEQ Collab.), Phys.Rev. **D55** (1997) 1280.
8. A.D. Martin et al., Eur.Phys.J. **C4** (1998) 463.
9. V.M. Abazov et al. (DØ Collab.), Phys.Rev. **D64** (2001) 032003.
10. G. Marchesini et al., Comp.Phys.Comm. **67** (1992) 465.
11. J.R. Forshaw and M.H. Seymour, JHEP **99909** (1999) 009.
12. DØ Collab., contributed paper to ICHEP-02 421.

ELSEVIER

Proceedings of ICHEP 2002, pp. 319–321
*S. Bentvelsen, P. de Jong, J. Koch and E. Laenen (Editors)*

31st INTERNATIONAL CONFERENCE ON HIGH ENERGY PHYSICS AMSTERDAM

# $\gamma\gamma$ Production of Heavy Flavour at LEP

S. Roth [a]

[a]III. Physikalisches Institut
Rheinisch-Westfälische Technische Hochschule Aachen
D-52056 Aachen, Germany

The measurements of the experiments ALEPH, DELPHI, L3 and OPAL on heavy quark production in two-photon collisions are presented. The total and differential cross sections for inclusive charm production have been measured at LEP 2 energies. The gluon content of the photon is important to describe the data. The total cross section of inclusive bottom production is three times larger than the QCD prediction.

## 1. INTRODUCTION

At LEP energies heavy flavour production in two-photon collisions is dominated by two processes, the direct and the single-resolved process which are shown in Figure 1. Due to the high mass of the charm and the bottom quark the production of heavy flavour can be calculated in pertubative QCD, where the resolved part also depends on the gluon content of the photon.

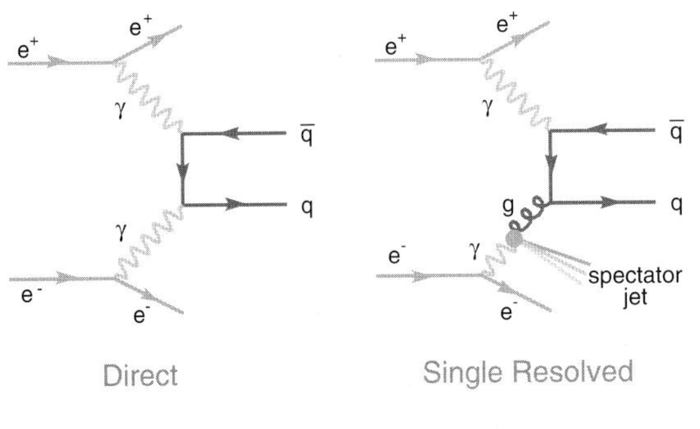


Figure 1. The main processes contributing to heavy-flavour production in two photon collisions at LEP

Charm production is about two orders of magnitude larger than the bottom cross section due to the smaller quark mass and the higher electric charge. It therefore allows to study differential cross sections and to seperate the direct and resolved part of the reaction, which are predicted to contribute equally at LEP 2 energies.

## 2. INCLUSIVE CHARM PRODUCTION

### 2.1. Inclusive $D^*$ Production

Inclusive charm production has been measured by all four LEP experiments using fully reconstructed $D^*$ mesons via the decay $D^{*\pm} \to D^0\pi^\pm$ [1,2]. Due to the small mass difference between $D$ and $D^*$ of 145 MeV one finds a slow pion of only 6 MeV kinetic energy and the reconstructed $D - D^*$ mass difference provides a clear signal with small background. Up to four decay modes of the $D^0$ are used.

Direct and single-resolved events can be distinguished by the fact that in resolved processes part of the energy of the $\gamma\gamma$ system is carried away by the spectator jet. The distribution for $x_T$ from the ALEPH experiment is shown in Figure 2, where $x_T = 2p_T^{D^*}/W_{\text{vis}}$ is the ratio of the transverse momentum of the $D^*$, which is a measure of the invariant mass of the $c\bar{c}$ system, and the visible invariant mass $W_{\text{vis}}$, which is a measure for the invariant mass of the $\gamma\gamma$ system. The relative contribution if the direct and the resolved part fitted to the data [1] is in agreement with the QDC prediction calculated by S. Frixione et al. [3].


doi:10.1016/S0920-5632(02)02166-7

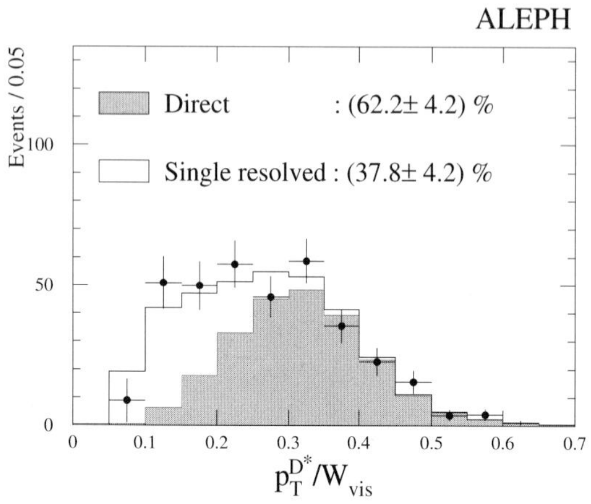


Figure 2. Separation of direct and resolved contributions for the two-photon production of charm quarks.

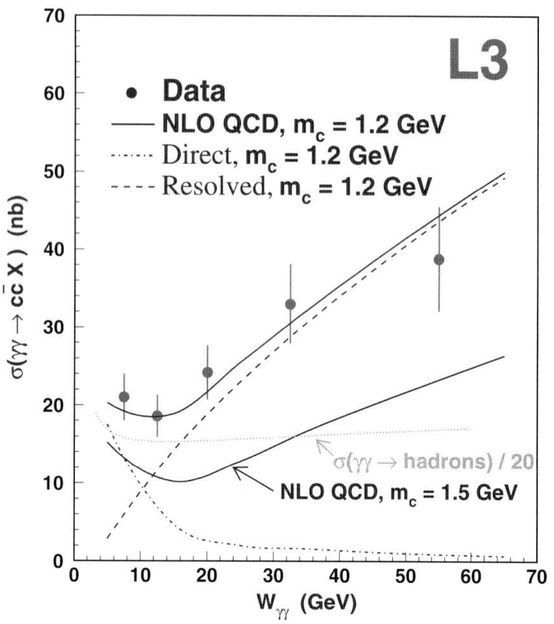


Figure 3. Dependence of the charm production cross section, $\sigma(\gamma\gamma \rightarrow c\bar{c})$, on the the two-photon center-of-mass energy, $W_{\gamma\gamma}$.

### 2.2. Semileptonic charm decays

The charm final state in two-photon collisions may also be selected using semileptonic charm decays with a high-$p_t$ lepton. The analysis performed by the L3 collaboration [6] yields a total charm cross section in agreement with the one using the $D^*$ tag.

The L3 experiment also measured the charm cross section as function of the two-photon centre-of-mass energy [5]. The charm quarks are identified by their semileptonic decay into electrons using a high statistics electron sample of 2434 events discussed in Reference [6]. Figure 3 shows $\sigma(\gamma\gamma \rightarrow c\bar{c}X)$ in the interval 5 GeV $< W_{\gamma\gamma} <$ 70 GeV together with NLO QCD calculations. In this calculations the charm mass is fixed to a rather low value of 1.2 GeV. The renormalization and factorization scales are set to $m_c$ and $2m_c$ , respectively and the GRS-HO photon parton density function has been used. Using these input parameters the QCD calculations reproduce the total rate and energy dependence measured in data. A calculation with the more realistic $m_c$ = 1.5 GeV results in a 50 % lower prediction of the cross-section.

In Figure 3 the data are also compared with the measured cross section of hadron production in two-photon collisions, scaled with a factor 1/20. A steeper rise with energy is observed for $\sigma(\gamma\gamma \rightarrow c\bar{c})$ compared to $\sigma(\gamma\gamma \rightarrow$ hadrons). The Pomeron/Reggeon ansatz suggesting a parametrization of the form $\sigma_{\rm tot} = As^{\epsilon} + Bs^{-\eta}$ describes the data well.

## 3. INCLUSIVE BOTTOM PRODUCTION

Leptons from the decay of heavy flavour show large transverse momentum with respect to the closest jet. In the analysises from L3 and OPAL [6,7] one exploits the $p_T$ spectrum of the leptons to measure the charm and bottom production in two-photon collisions.

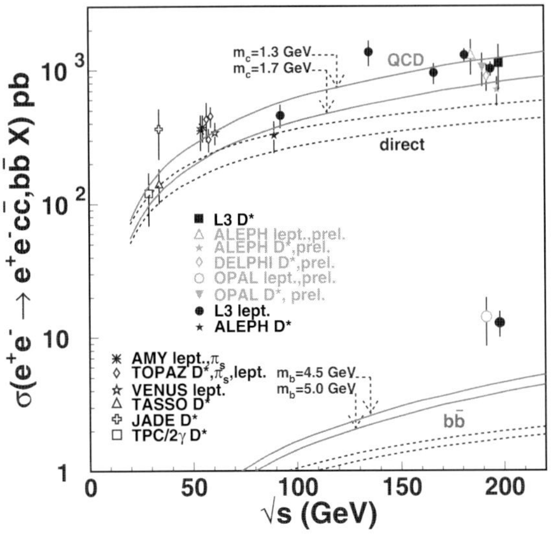


Figure 4. Total cross section of heavy-flavour production in two-photon collisions.

The total cross section of charm and bottom production in two-photon collisions is summarized in Figure 4. The measurements are compared to NLO calculations. Whereas the charm cross section is in agreement with theory, the calculations underestimate the measured b cross section by a factor three corresponding to five standard deviations.

## 4. EXCLUSIVE BOTTOM PRODUCTION

The ALEPH experiment performed a search for the still undiscovered $\eta_b$ pseudoscalar meson using an integrated luminosity of 699 $pb^{-1}$ [8]. One candidate event was found in the final state with six charged particles, in agreement with the expected background of about one event. An upper limit on the branching ratio $\Gamma_{\gamma\gamma} \times BR(\eta_b \to$ 6 charged particles) $< 132$ eV can be set. L3 started to search for the $\eta_b$ [9] and found three candidates compatible with background.

## REFERENCES


1. ALEPH Collab., *Measurement of the inclusive $D*\pm$ production in $\gamma\gamma$ collisions at LEP2*, ALEPH Note 2002-022, submitted to ICHEP02 Amsterdam.
2. L3 Collab., *Inclusive $D^*$ Production in Two-Photon Collisions at LEP*, Phys. Lett. B. **535** (2002) 59.
3. S. Frixione, M. Krämer, and E. Laenen, Nucl. Phys. B **571**, 169 (2000).
4. M. Glück, E. Reya and I. Schienbein, Phys. Rev. D **46**, 169 (1999).
5. L3 Collab., *Measurements of the Charm Production Cross Section in $\gamma\gamma$ Collisions at LEP*, Phys. Lett. B **514**, 19 (2001).
6. L3 Collab., *Measurements of the Cross Sections for Open Charm and Beauty Production in $\gamma\gamma$ Collisions at $\sqrt{s}$ = 189 - 209 GeV*, L3 Note 2761, submitted to ICHEP02 Amsterdam.
7. OPAL Collab., *Measurement of Open Beauty Production in Photon-Photon Collisions at $\sqrt{s_{ee}}$ = 189 - 202 GeV*, OPAL Note PN455, submitted to ICHEP02 Amsterdam.
8. ALEPH Collab., *Search for $\gamma\gamma \to \eta_b$ in $e^+e^-$ collisions at LEP*, CERN-EP/2002-009, submitted to Phys. Lett. B.
9. L3 Collab., *Study of $\eta_b$ in Two-Photon Collisions with the L3 Detector at LEP*, L3 Note 2773, submitted to ICHEP02 Amsterdam.

ELSEVIER

Proceedings of ICHEP 2002, pp. 322–324
S. Bentvelsen, P. de Jong, J. Koch and E. Laenen (Editors)

31st INTERNATIONAL CONFERENCE ON HIGH ENERGY PHYSICS AMSTERDAM

# Improved measurements of the $b$ quark mass at LEP

M.J. Costa [a], P. Bambade [b], J. Fuster [a] and P. Tortosa [a].

[a]IFIC, València

[b]LAL, Orsay

An improved determination of the $b$-quark mass at the $M_Z$ scale, $m_b(M_Z)$, has been obtained by the DELPHI and OPAL experiments at LEP. The normalized three-jet rate of $b$-quarks with respect to light quarks, $R_3^{b\ell}$, has been measured with different jet clustering algorithms and compared with the massive NLO calculations to extract $m_b(M_Z)$. For DELPHI, the increase in statistics and a better understanding of the flavour tagging, hadronization process and the gluon splitting rates leads to a precision of 380 MeV. The effect of the $b$-quark mass on the four-jet rate has also been studied by DELPHI.

## 1. Introduction

The radiation of gluons by quarks does not only depend on the strong coupling constant but also on the mass of the emitting quark. Massive quarks radiate less gluons than light quarks in a similar way to the suppression of bremsstrahlung for muons compared to electrons. At LEP energies, mass effects are negligible for inclusive obervables since they are suppressed by $m_b^2/m_Z^2 \leq 0.3\%$. But for more exclusive observables, like jet cross sections, mass effects can be enhanced considerably for small enough values of the jet resolution parameter, $y_c$, as they go like $m_b^2/m_Z^2/y_c$. Such mass effects have already been observed at LEP and SLC [1–5].

The observable used by DELPHI and OPAL to quantify these effects can be written as,

$$R_n^{b\ell}(y_c) = \frac{[\Gamma_n(y_c)/\Gamma_{tot}]^{Z\to b\bar{b}}}{[\Gamma_n(y_c)/\Gamma_{tot}]^{Z\to \ell\bar{\ell}}} \qquad (1)$$

being $[\Gamma_n(y_c)/\Gamma_{tot}]^{Z\to b\bar{b}}$ and $[\Gamma_n(y_c)/\Gamma_{tot}]^{Z\to \ell\bar{\ell}}$ the normalized $n$-jet cross sections for $b$- and $\ell$-quarks events ($\ell \equiv uds$ for DELPHI and $\ell \equiv udsc$ for OPAL).

The definition of the event flavour is slightly different for the two experiments; DELPHI considers the flavour of the quarks coupled to the $Z$ boson while OPAL considers $b$-quark events as those containing a $b$ quark in the final state.

DELPHI has measured the above observable for $n = 3$ jets using Durham and Cambridge jet clustering algorithms and compared with NLO theoretical predictions of [6,7], and $n = 4$ jets for Durham comparing with the LO calculations of [8]. OPAL has measured it for $n = 3$ jets and for the following jet finder algorithms: Jade and its variants E, E0, P and P0, Durham, Geneva and Cambridge and compared with the NLO calculations [9] for all algorithms but Cambridge.

When comparing with the theoretical predictions two possibilities are avaiable: extract $m_b(M_Z)$ assuming $\alpha_s$ universality or take the result obtained at threshold for the $b$-quark mass and test $\alpha_s$ flavour independence.

## 2. Experimental strategy

A sample of $Z^0$ hadronic events was first selected from the data collected at the $Z^0$ peak. A flavour tagging method was then used to separate the $b$ and $\ell$ quark events [10–12].

A complete simulation of the detector was used to correct for detector and acceptance effects and kinematic biases introduced in the tagging procedure. In order to correct for the hadronization process the models implemented in PYTHIA 6.131 [13] and Herwig 6.1 [14] generators were used by DELPHI and the model of JETSET 7.4 [15] by OPAL. All Monte Carlos were previously tuned to data.


doi:10.1016/S0920-5632(02)02167-9

Different sources of theoretical and systematic uncertainties (as those associated with imperfections on the physics and detector modelling and the hadronization process) were taken into account.

In order to reduce the hadronization model uncertainty an additional cut was applied by DELPHI for the case of $R_3^{b\ell}$ on $b$ tagged events, $x_E^b(\text{jet}) \geq 0.55$, where $x_E^b(\text{jet}) = E_{b-\text{jet}}/E_{\text{beam}}$ and this error was reduced from 6‰(7‰) to 2‰(3‰) for Durham (Cambridge). The error due to the $b$ mass parameter ambiguity in the generator has been considered in DELPHI for the first time and it is the dominant source of error.

## 3. Results

### 3.1. DELPHI

The $R_3^{b\ell}$ obtained at parton level, as a function of the $y_c$ for Durham and Cambridge, is shown in figure 1. The LO and NLO theoretical curves in terms of the pole and the running mass are also shown in the plot, for $M_b = 4.8$ GeV$/c^2$ and $m_b(M_Z) = 3$ GeV$/c^2$. From the values obtained at $y_c = 0.02$ and $y_c = 0.009$ for Durham and Cambridge respectively the measured $m_b(M_Z)$ for each algorithm was,

$$2.99^{+0.25}_{-0.27}\ (\text{st})^{+0.34}_{-0.37}\ (\text{syst}) \pm 0.28\ (\text{th})\ \text{GeV}/c^2$$

$$2.82 \pm 0.19\ (\text{st})^{+0.31}_{-0.33}\ (\text{syst}) \pm 0.06\ (\text{th})\ \text{GeV}/c^2$$

Both results are compatible within their statistical error. The measurement performed with Cambridge is more precise than the one obtained with Durham, with a total error of $^{+370}_{-390}$ MeV instead of $^{+500}_{-540}$ MeV.

If instead the $b$ quark mass measured at threshold is taken as an input [16], $\alpha_s$ flavour independence is verified with a precision of less than 1%. For the case of Cambridge,

$$\alpha_s^b/\alpha_s^\ell = 1.001 \pm 0.004(\text{st}) \pm 0.007(\text{syst}) \pm 0.002(\text{th})$$

Figure 2 shows the result obtained for $R_4^{b\ell}$ as a function of the $y_c$ at hadron level, together with the LO calculations for two different values of the $b$ mass, before (lines) and after (bands) correcting for hadronization. As can be observed, the data lies between the two LO curves for a mass of 3 and 4.8 GeV$/c^2$.

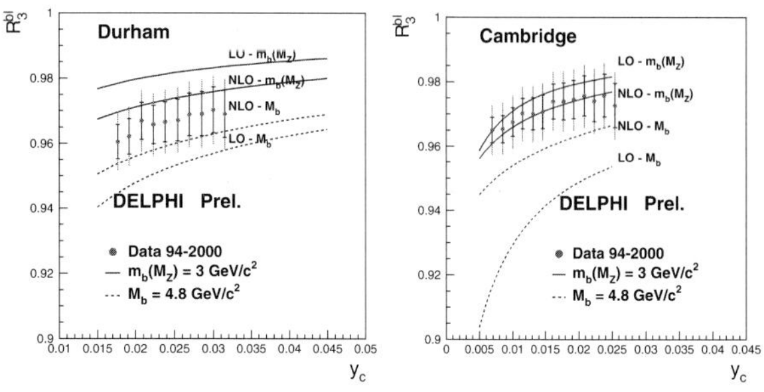


Figure 1. $R_3^{b\ell}$ as a function of $y_c$ measured by DELPHI at parton level.

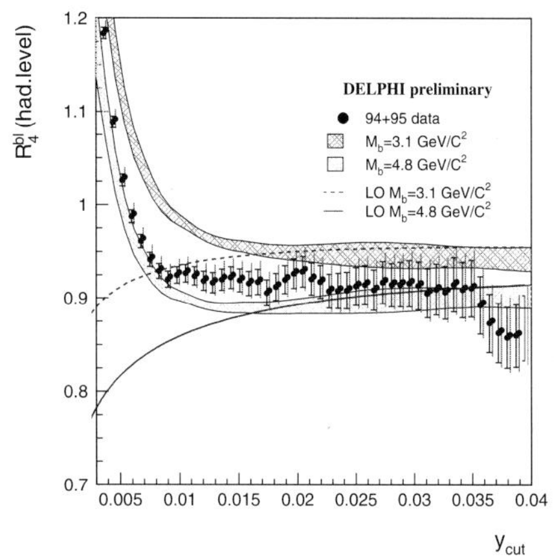


Figure 2. $R_4^{b\ell}$ as a function of $y_c$ measured by DELPHI at hadron level.

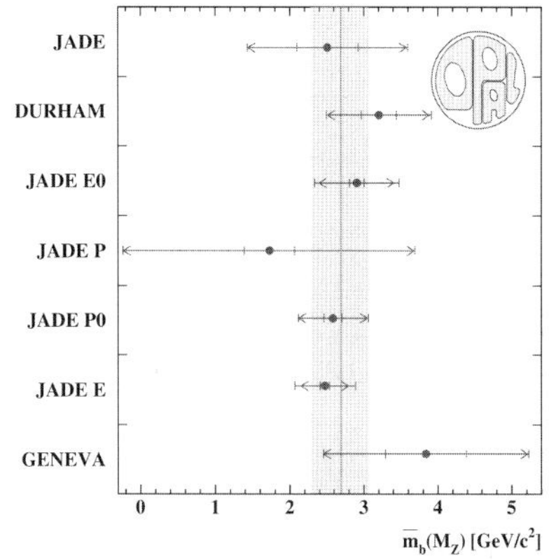


Figure 3. $m_b(M_Z)$ obtained for each jet finder algorithm by OPAL.

### 3.2. OPAL

The results obtained for $m_b(M_Z)$ for the seven different jet clustering algorithms are shown in figure 3. By minimizing the $\chi^2$ of the correlated determinations a single result for the $b$ quark mass $m_b(M_Z)$ was found to be,

$$2.67 \pm 0.03\ (\mathrm{st})^{+0.29}_{-0.37}\ (\mathrm{syst}) \pm 0.19\ (\mathrm{th})\ \mathrm{GeV/c^2}$$

## 4. Conclusions

Figure 4 shows the different results obtained for $m_b(M_Z)$ by LEP and SLC. As can be seen, all results are compatible and agree with measurements performed at threshold extrapolated to $\sqrt{s} = M_Z$.

A net change in the value of the running $b$-quark mass between the scales $M_Z$ and $M_\Upsilon/2$ is observed with almost 4 standard deviations. The $m_b(M_Z)$ determinations are systematics limited mainly because of the modelling of the hadronization.

The $b$-quark mass effects have also been observed by DELPHI with the four-jet observable $R_4^{b\ell}$, which has the advantage of having a higher suppression than $R_3^{b\ell}$, compensating the loss of statistics. A comparison with the LO theoretical predictions has been made but NLO computations would be needed in order to distinguish between the pole and the running mass.

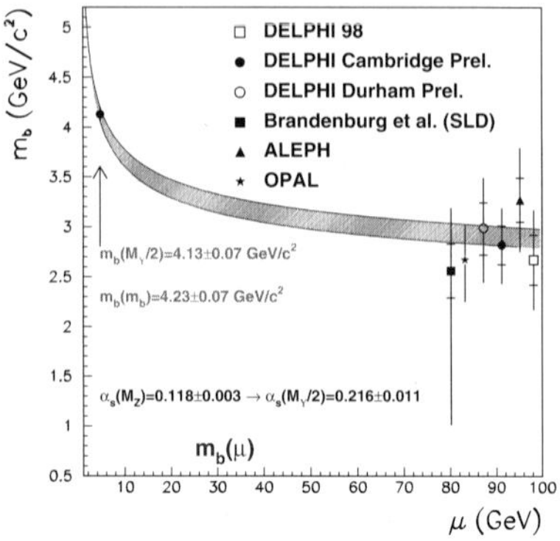


Figure 4. The running of $m_b(\mu)$. The shaded area corresponds to the band associated to $m_b(\mu)$ when running $m_b(\mu)$ at $M_\Upsilon/2$ up to the $M_Z$ scale using the QCD RGE and $\alpha_s(M_Z) = 0.118 \pm 0.003$.

## REFERENCES

1. DELPHI Coll., P.Abreu et al., Phys. Lett. **B418** (1998) 430-442;
S. Martí i García, J. Fuster and S. Cabrera, Nucl. Phys. B (Proc. Suppl.) 64 (1998) 376.
2. SLD Coll., K. Abe et al., Phys. Rev. **D59** (1999) 12002.
3. A. Brandenburg, P.N Burrows, D. Muller, N. Oishi, P Uwer, Phys. lett. **B468** (1999) 168.
4. OPAL Coll., G. Abbiendi et al., Eur. Phys. Jour. **C11** (1999) 643.
5. ALEPH Coll., R. Barate et al., Eur. Phys. Jour. **C18** (2000) 1.
6. G. Rodrigo, M. Bilenky, A. Santamaría, Phys. Rev. Lett. **79** (1997) 193.
7. M. Bilenkii, S. Cabrera, J. Fuster, S. Martí i García, G. Rodrigo, A. Santamaria, Phys. Rev. **D60** (1999) 114006.
8. A. Ballestrero et al., Yellow Report CERN-2000-009 of the LEP2 Monte Carlo Workshop, [hep-ph/0006259]
9. W.Bernreuther, A. Brandenburg, P.Uwer, Phys. Rev. Lett. **79** (1997) 189.
A. Brandenburg and P. Uwer, Nucl. Phys. **B515** (1998) 279.
10. DELPHI Coll. P. Abreu et al., Z. Phys. **C65** (1995) 555;
G.V. Borisov, C. Mariotti, Nucl. Instr. Meth. **A372** (1996) 181.
11. G. Borisov, Nucl. Instrum. Meth. **A417** (1998) 384.
12. R. Akers et al., Z. Phys. **C65** (1995) 17.
R. Akers et al., Z. Phys. **C63** (1994) 181.
13. T. Sjöstrand et al., hep-ph/0108264.
14. G. Marchesini et al. Computer Phys. Commun. 67 (1992) 465.
G. Corcella et al., JHEP 0101 (2001) 010 [hep-ph/0011363]; hep-ph/0201201.
15. T. Sjöstrand. Computer Physics Commun. **82** (1994) 74.
16. V. Giménez, L.Giusti, G.Martinelli and F.Rapuano, JHEP 0003:018 (2000).

ELSEVIER

Proceedings of ICHEP 2002, pp. 325–328
*S. Bentvelsen, P. de Jong, J. Koch and E. Laenen (Editors)*

31st INTERNATIONAL CONFERENCE ON HIGH ENERGY PHYSICS AMSTERDAM

# Charm Production at HERA

Christoph Grab[a]
representing the H1 and ZEUS collaborations

[a] Institute for Particle Physics, ETHZ,
8093 Zürich, Switzerland

Selected recent measurements of charm quark production in lepton-proton collisions performed at HERA by the experiments H1 and ZEUS are reviewed. The experimental results are compared in detail with theoretical predictions, based on QCD calculations in both leading order and next-to-leading order perturbative approaches. No large deviations from the expectations were observed within the theoretical uncertainties.

Results on inelastic leptoproduction of $J/\psi$ mesons are directly compared with predictions based on NRQCD/factorisation approaches. A search for colour octet processes predicted within this approach is conducted in many regions of phase space, but no unambiguous evidence has been found.

## 1. Introduction

One of the main interests in studying heavy flavours (HF) at the HERA $e-p$ collider is the investigation of the production mechanisms and detailed comparisons with theoretical predictions, in particular with perturbative QCD calculations, and extended studies of non-perturbative phenomena.

The kinematics of HF production is described by three independent variables, the centre-of-mass energy $\sqrt{s}$, the photon four-momentum squared $q^2 = -Q^2$ and either one of the scaling variables $y = (q \cdot P/l \cdot P)$, the inelasticity of the $ep$-interaction, or Bjorken-$x$, $x = Q^2/(2P \cdot l)$. Here $P$ and $l$ denote the four-momentum of the proton and the electron[1], respectively. The $\gamma - p$ centre-of-mass energy squared is given by $W_{\gamma p}^2 = W^2 \approx y \cdot s - Q^2$.

The results presented here are based upon data recorded by the H1 [1] and ZEUS [2] collaborations during 1996 through 2000, with an $e-p$ centre-of-mass energy $\sqrt{s} = 300$ GeV and 318 GeV, in the kinematic regimes of photoproduction ($Q^2 \approx 0$) and deep inelastic scattering (DIS, at large $Q^2$).

[1]Hereafter, a reference to electron implies a reference to either electron or positron. Most of the results shown were obtained in positron-proton collisions.

## 2. Open Charm Production

### 2.1. Production Mechanism and Theories

The description of open charm production is based on perturbative QCD (pQCD). At HERA charmed quarks are produced predominantly by the direct *photon gluon fusion* (BGF) process $\gamma g \to c\bar{c}$, where a real or virtual photon emitted by the electron and a gluon from the proton generate a $c\bar{c}$ pair. In photoproduction, resolved photon interactions contribute as well and lead to the notion of a photon structure function. However, beyond this LO *only the sum* of direct and resolved processes is a well-defined quantity. The heavy quarks then hadronise and are either detected as "open states", i.e., in final states with charmed hadrons, or alternatively as "hidden states", such as $J/\Psi$ .

One of the QCD calculations in NLO of interest here, is the *massive approach* [3], which is a fixed order (FO in $\alpha_s$) pQCD calculation with a massive charm quark. The massive calculations are expected to break down for large scales $p_T \gg m_c$ because no large-$p_T$ resummation is done.

Recently, for the photoproduction regime, an extended approach has become available. This so-called "matched calculation" (FONLL) [4] complements the FO-calculations by adding next-to-leading log (NLL) correction terms.


doi:10.1016/S0920-5632(02)02174-6

The available Monte Carlo simulation programs are based on the LO processes (BGF and resolved in some cases). Traditionally, they are based on DGLAP-evolution equations and leading-log parton showers, such as the implementations AROMA, HERWIG and PYTHIA.

## 2.2. Inclusive $D$-meson Production

$D$ mesons are identified in the H1 and ZEUS detectors by exclusive reconstruction of the relevant decay chains $D^{*+} \rightarrow D^0\pi^+$, $D^0 \rightarrow K^-\pi^+$ or $K^-\pi^+\pi^-\pi^+$, $D_s \rightarrow \phi\pi$ and $D^+ \rightarrow K^-\pi^+\pi^+$ in the *visible* kinematic regions limited in $p_t(D)$ and rapidity $\hat{y}(D)$.

The single differential production cross section distributions measured by ZEUS [5] are shown in Fig. 1 for $Q^2 < 1$ GeV$^2$, $| \hat{y}(D) |< 1.6$, $1.9 < p_t(D) < 20$ GeV and $130 < W < 285$ GeV.

Overlaid are the predictions by the NLO QCD calculations[2], in FO [3] (solid) and in FONLL [4] (dotted). In general the ZEUS data are reasonably well reproduced by both FO and FONLL, which are very similar in shape. However, the predictions tend to be below the data, slightly more so in the case of FONLL. There remain some discrepancies in the forward region, the clarification of which requires more studies.

H1 studied the inclusive production of $D$-mesons in DIS [6] based on about 48 pb$^{-1}$ of data taken in 1999 and 2000. Inclusive production cross sections are measured for the vector $D^{*+}$ and for the pseudoscalar mesons $D^0$, $D_S$ and, for the first time at HERA, also $D^+$. The long lifetimes of the pseudoscalar mesons lead to a separation of their production and decay vertices, which is exploited to distinguish signal and background processes and to substantially improve the signal qualities. This is illustrated in Fig. 2(top), which shows the $D^+$ signal with and without a vertex separation significance requirement. The resulting single differential distribution for the $D^+$-channel are presented in Fig. 2 for the visible kinematic range of $2 < Q^2 < 100$ GeV$^2$, $0.05 < y < 0.7$, $|\eta_{D^*}| < 1.5$ and $p_{t\,D^*} > 2.5$ GeV.

[2]The renormalisation scale $\mu_r$, the hadron and photon factorization scales $\mu_{Fh}$ and $\mu_{F\gamma}$ are customarily given relative to the charm quark mass $m_c$. Here the values are: $\mu = m_T = \sqrt{m_c^2 + p_t^2}$ ; and $\mu_r = \mu; \mu_{Fh} = \mu_{F\gamma} = 2\mu$.

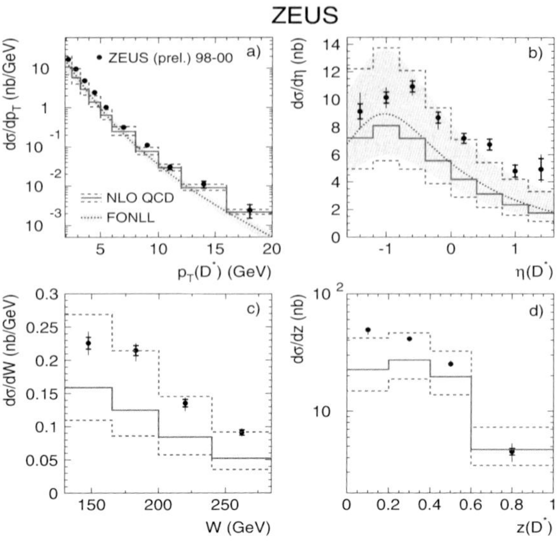


Figure 1. Single differential $D^*$ photoproduction cross sections compared with FO and FONLL QCD calculations; central choice of QCD predictions are given by solid (FO) and dotted (FONLL) histograms, with the bounds indicated separately for FO (dashed) and FONLL (band).

The LO QCD predictions (Aroma MC) overlaid in Fig. 2 (dark band) show rather good agreement with the measurements.

## 2.3. Studies of fragmentation

The question of the universality of non-perturbative fragmentation effects has been addressed by both ZEUS and H1. ZEUS studied the distribution of the energy fraction $z = (E+p_L)[D]/(E+p_L[jet])$ in $D^*$-meson photoproduction [7]. Within the framework of the LO simulation (in Pythia) complemented by a Peterson fragmentation parametrisation, a direct comparison allowed the extraction of an optimal fragmentation parameter $\epsilon_{Pete} = (0.064 \pm 0.006^{+0.011}_{-0.008})$, that compares pretty well with values measured in $e^+e^-$. H1 measured several differential cross section distributions [6] that allowed the determination of ratios of various D-meson quark and spin states, and subsequent comparisons with the numbers observed at $e^+e^-$-colliders.

All these measurements support the hypothesis that the charm fragmentation process is universal, e.g., the results from $e^+e^-$ can be applied to

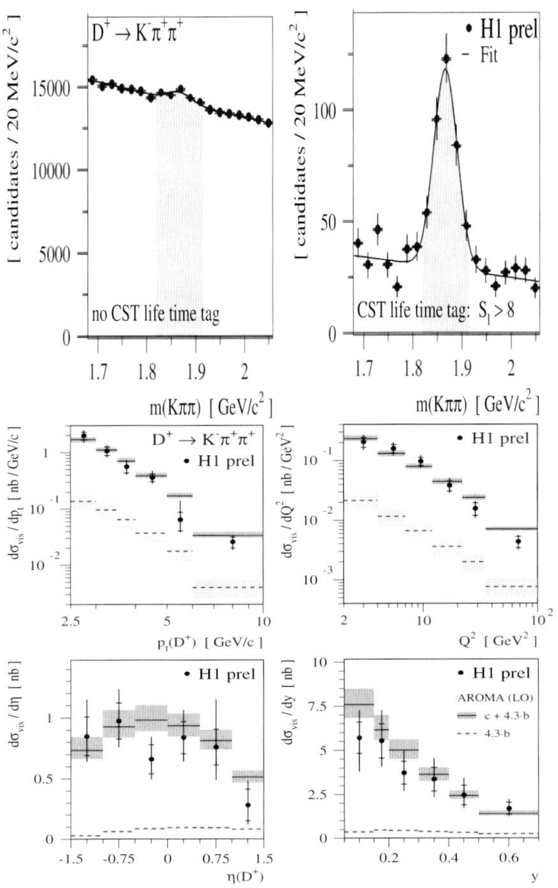


Figure 2. H1 data: $D^+$ signal signal enhancement (top) by a vertex separation significance requirement; differential $D^+$ production cross sections (bottom: solid dots), compared with the LO predictions (dark band).

*ep*-processes.

### 3. Inelastic Charmonium Production

One of the main interests in the leptoproduction of $J/\Psi$ mesons is a clarification of its production process.

Experimentally, inelastic production processes are selected by means of the *inelasticity* variable $z \equiv \frac{q \cdot P_\psi}{q \cdot P}$, where $P_\psi$ is the four-momentum of the $J/\psi$ meson. The $J/\psi$ mesons are identified by their decays to lepton pairs.

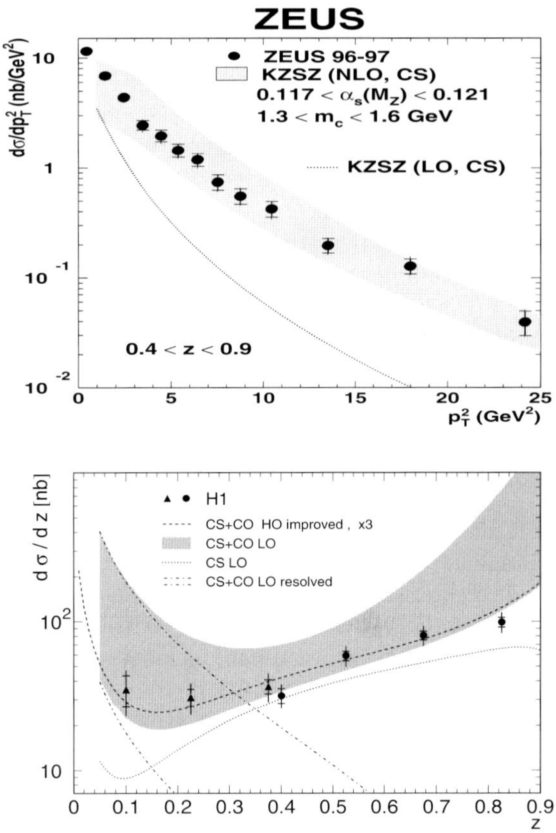


Figure 3. Inelastic $J/\psi$ photoproduction cross-sections measured as functions of $p_t^2(J/\Psi\ )$ by ZEUS (top) and $z$ by H1 (bottom). Overlaid are the predictions based on and NLO (top) and LO (bottom) QCD calculations.

In the theoretically favoured NRQCD/ factorization approach [9], the cross section is described by $\sigma(\gamma p \to J/\psi X) = \sum \langle \mathcal{O}[n] \rangle \cdot \hat{\sigma}(\gamma p \to c\bar{c}[n] X)$, where $\langle \mathcal{O}[n] \rangle$ are the NRQCD long distance matrix elements (LDME) and $\hat{\sigma}$ the short distance coefficients, calculable in pQCD, with $n$ being the quantum numbers of the colour state, ${}^{2S+1}L_J$.

This NRQCD-approach contains both the traditional colour-singlet model [10,11] (CSM) and the colour-octet model [12] (COM) where the transition from the coloured $c\bar{c}$ state produced in the hard sub-process to the colour-singlet final state proceeds non-perturbatively.

Since CO-contributions $\langle \mathcal{O}[8] \rangle$ are believed to be universal, they can be determined with data at e.g. the Tevatron [13] and applied to HERA data. Thus a measure of these CO-contributions at HERA provides a very sensitive test for the consistency of the model.

The $p_T^2$-dependence of the cross sections have been measured by both H1 [15,16] and ZEUS [17]. As an example the results by ZEUS are shown in fig 3 (top) and compared with both LO and NLO predictions [10,11,14] within the CSM. The LO contribution alone is too low, whereas the NLO calculations describe the data pretty well within the (rather large) theoretical uncertainties, indicated by the band.

Differential cross section $d\sigma/dz$ measured by H1 are shown in Fig. 3 (bottom) for a large z-range ($z > 0.1$), and compared with LO theoretical calculations within the NRQCD/factorization framework [14]. Direct photon contributions as well as resolved processes are included and shown separately. At $z$ above 0.3 direct photon contributions dominate and the CO contribution is dominant at $z \geq 0.6$. In order to describe the shape, the addition of CO contributions helps, as does a specific choice of quite small LDMEs, however the normalisations are still too low calling for large k-factors.

The number of parameters entering these comparisons is large, and thus the presently available measurements do not allow one to distinguish between different models yet.

## 4. Summary and Conclusions

The production of charm quarks in $ep$-collisions at HERA has been studied by identifying charm in the form of $D$- and $J/\psi$-mesons. In the overall picture, the hard nature of the charm production process appears to be firmly established, and thus yields reasonable predictions for the total production rates. The shape of differential distributions is in general reasonably well predicted by the pQCD calculations, with possible exceptions at large $\eta$-values, although the predictions tend to be below the data.

In the case of inelastic $J/\Psi$ meson production, the NLO calculations based on NRQCD/factorization with colour-singlet contributions alone describe the data well within errors, thus exhibiting no indications of colour-octet contributions.

## REFERENCES

1. H1 collab., I. Abt, *et al*, DESY 96-001 (1996).
2. ZEUS collab., J. Breitweg, *et al*, DESY 95-193 (1995).
3. S. Frixione, *et al*, *Phys. Lett. B* **348**, 633 (1995).
4. M. Cacciari, S. Frixione and P.Nason, *JHEP* **0103**, 006 (2001).
5. ZEUS collab., contribution 786 to ICHEP conference, Amsterdam 2002.
6. H1 collab., contribution 1015 to ICHEP conference, Amsterdam 2002.
7. ZEUS collab., contribution 778 to ICHEP conference, Amsterdam 2002.
8. ZEUS collab., contribution 781 to ICHEP conference, Amsterdam 2002.
9. G.T. Bodwin *et al*, *Phys. Rev. D* **51**, 1125 (1995), erratum *Phys. Rev. D* **55** 5853 (1997).
10. M. Krämer, *et al*, *Phys. Lett. B* **348**, 657 (1995).
11. M. Krämer, *Nucl. Phys. B* **459**, 3 (1996).
12. M. Cacciari, M. Krämer, DESY 96-005 (1995).
13. F. Abe *et al*, CDF collab., *Phys. Rev. Lett.* **69**, 3704 (1992).
14. M. Krämer, Prog. Part. Nucl. Phys. **47**, 1 (2001);
15. H1 collab., C. Adloff, *et al*, DESY-02/59, 2002.
16. H1 collab., C. Adloff, *et al*, DESY-02/60, 2002.
17. ZEUS collab., contribution 775 to ICHEP conference, Amsterdam 2002.

Proceedings of ICHEP 2002, pp. 329–331
S. Bentvelsen, P. de Jong, J. Koch and E. Laenen (Editors)

# Latest measurements of beauty quark production at HERA

V. Chiochia [a]

[a]DESY - Deutsches Elektronen-Synchrotron,
Notkestrasse 85, 22607 Hamburg, Germany

The latest results of beauty quark production measurements at HERA are presented. New measurements have been obtained both in the photoproduction and the deep inelastic scattering regimes. The results were compared with the NLO QCD calculations.

## 1. Introduction

Production of heavy quarks is a fascinating tool to understand the nucleon structure as well as the underlying QCD parton dynamics. In addition production of $b$ quarks is a source of background for many searches for new physics at existing and future colliders. The high $b$ mass ($M_b \gg \Lambda_{QCD}$) should provide a solid basis for a perturbative calculation. Nevertheless, so far discrepancies between the experimental cross section measurements and the NLO QCD calculation have been found.

The experimental procedure often relies on the measurement of the transverse momentum $p_T^{rel}$ of the muon produced in the semi-leptonic decay with respect to the axis of the closest jet. This spectrum is harder for $b$ quarks than for $c$ quarks and, therefore, allows a statistical separation of the signal from the background. Additional information such as lifetime measurements provided by silicon detectors can also be used. An alternative tag is given by the coincidence of $D^*$ mesons and muons which provide sensitivity to the region of low transverse $b$ quark momenta and is less affected by background.

The recent experimental results obtained at the HERA collider are reviewed in both the photoproduction ($Q^2 \sim 0$ GeV$^2$) and the Deep Inelastic Scattering (DIS: $Q^2 > 1$ GeV$^2$) kinematic regimes.

## 2. Open Beauty in Photoproduction

When the exchanged photon has small virtuality ($Q^2$) the time-scale of the interaction is such that its hadronic structure can be revealed. Photoproduction at HERA can therefore be similar to the processes at hadron colliders and supplies complementary information. Both ZEUS and H1 already published results on $b$ photoproduction [1–3]. The ZEUS experiment has now measured the differential cross sections of beauty photoproduction using events with at least two jets and a muon in the final state [4]. The luminosity used is almost three times larger than in the previous measurements. The fraction of events from $b$ decays has been extracted using the $p_T^{rel}$ method. The kinematic region is defined by $Q^2 < 1$ GeV$^2$, $0.2 < y < 0.8$, $p_T^{Jet1(2)} > 7(6)$ GeV, $|\eta^{Jet1(2)}| < 2.5$, $p_T^{\mu} > 2.5$ GeV and $-1.6 < \eta^{\mu} < 2.3$ [1]. Figure 1 shows a comparison between the measured differential cross section and a NLO QCD calculation for different regions of the muon pseudorapidity. The QCD prediction was calculated using the program FMNR [5]. The hadronization is modeled by a Peterson function and the spectrum of the semi-leptonic muon momentum was extracted from PYTHIA [6]. The bands around the NLO prediction show the results obtained by varying the $b$ quark mass as well as the renormalization and factorization scales. The measured cross sections are a factor 1.4 larger than the central NLO prediction but compatible with it within the experimental and theoretical uncertainties.

In addition, a dijet cross section ($ep \rightarrow b\bar{b}X \rightarrow jet\ jet\ X$) has been determined using PYTHIA to extrapolate to the unmeasured part

[1] $\eta = -\ln(\tan\theta/2)$ is the pseudorapidity, where $\theta$ is the polar angle measured with respect to the proton beam direction.


doi:10.1016/S0920-5632(02)02178-3

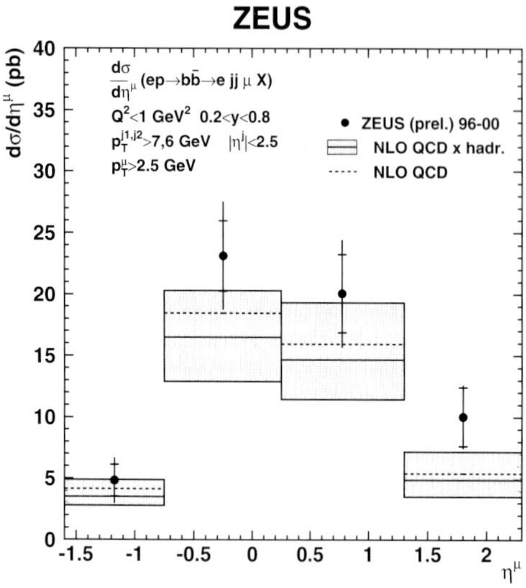


Figure 1. Differential cross section of beauty photoproduction as a function of the muon pseudorapidity compared with NLO QCD calculations.

of the muon kinematics and to correct for the branching ratio. For this measurement a different data sample has been used with looser cuts on the transverse momentum of the muon at large pseudorapidities. The result is $\sigma^{dijet} = 733 \pm 61 \pm 104$ pb while the NLO QCD prediction is $381^{+117}_{-78}$ pb, corresponding to a data excess of a factor of two.

## 3. Open Beauty in DIS

First results in the DIS region have been already released by the H1 collaboration [7]. Thanks to the high luminosity in the new ZEUS measurement [8] differential distributions have been measured for the first time in DIS. Events were selected by requiring the presence of at least one muon in the final state and at least one jet in the Breit frame[2]. A total visible cross section of $\sigma^{vis} = 38.7 \pm 7.7^{+6.1}_{-5.0}$ pb was measured for the reaction $ep \to eb\bar{b}X \to e\ jet\ \mu\ X$ in the kinematic region defined by: $Q^2 > 2$ GeV$^2$, $0.05 < y < 0.7$, $p^\mu > 2$ GeV, $30° < \theta^\mu < 160°$ and one jet in the Breit frame with $E_T^{Breit} > 6$ GeV and $-2 < \eta^{Lab} < 2.5$. Also for this measurement the $p_T^{rel}$ method was applied. This result has been compared with a NLO QCD calculation implemented in the HVQDIS program [9], after folding the $b$ quark momentum spectrum with a Peterson fragmentation function and subsequently with a spectrum of the semi-leptonic muon momentum extracted from RAPGAP [10]. The NLO QCD prediction is $28.1^{+5.3}_{-3.5}$ pb which agrees with the measured value within the errors. The differential cross section as a function of $Q^2$ compared to the NLO calculation is shown in Fig. 2. The

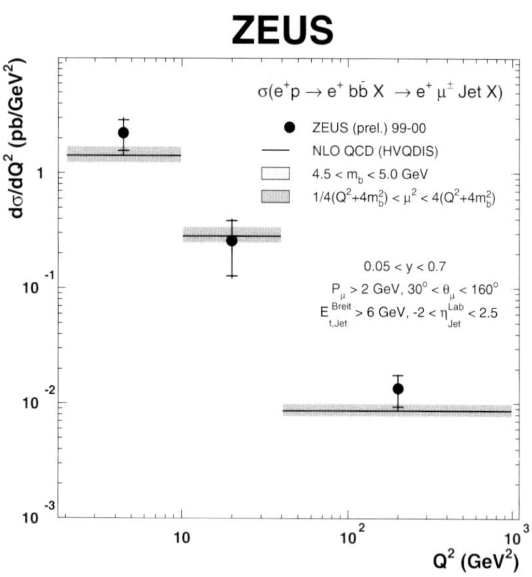


Figure 2. Differential beauty cross section as a function of $Q^2$ compared with the NLO QCD calculations.

prediction of the Monte Carlo program CASCADE [11], which implements a calculation based on the CCFM evolution equations [12] and uses a $k_T$-dependent gluon density, is 35 pb which is in very good agreement with the measurement. In addition, the simulation gives a good description of the measured differential cross sections.

## 4. $D^* - \mu$ correlations

The separation of charm and beauty contributions to the signal can also be performed by

[2] In the Breit frame, defined by $\vec{\gamma} + 2x\vec{P} = \vec{0}$, where $\vec{\gamma}$ is the momentum of the exchanged photon, $x$ is th Bjorken scaling variable and $\vec{P}$ is the proton momentum, a purely space-like photon and a proton collide head-on.

exploiting the charge and angle correlations of the $D^*$ meson and of the muon in the reaction $ep \to eb\bar{b}X \to eD^*\mu X$. Of particular interest is the configuration in which the muon and the $D^*$ originate from the same parent $B$-meson yielding unlike charge sign $D^* - \mu$ pairs produced in the same hemisphere.

Using this strategy and performing a likelihood fit on the kinematic distributions H1 has extracted the cross sections of beauty and charm production in the kinematic region defined by $p_T^{D^*} > 1.5$ GeV, $|\eta^{D^*}| < 1.5$, $p_T^{\mu} > 1$ GeV, $|\eta^{\mu}| < 1.74$ and $0.05 < y < 0.75$ [13]. The measured values, which confirm previous results, are respectively $\sigma_{vis}^b = 380 \pm 120 \pm 130$ pb and $\sigma_{vis}^c = 720 \pm 115 \pm 245$ pb which are well above LO+parton shower Monte Carlo expectations.

A similar analysis has been conducted by ZEUS [14] whose selection has been optimized for decays of $b$ quarks. The beauty cross section, measured in a slighty different phase space ($p_T^{D^*} > 1.9$ GeV, $|\eta^{D^*}| < 1.5$, $p_T^{\mu} > 1.4$ GeV, $-1.75 < \eta^{\mu} < 1.3$), is $\sigma_{vis}^b = 214 \pm 52^{+96}_{-84}$ pb. The result is in good agreement with the the H1 measurement once the same cuts are applied. In order to compare the measured cross section with NLO QCD predictions a photoproduction-enriched sample has been selected by applying the cuts $Q^2 < 1$ GeV$^2$ and $0.05 < y < 0.85$. Furthermore, the measurement is restricted in a $b$ quark rapidity range $\zeta^b < 1$ where the distributions of the Monte Carlo program used to extrapolate agree with the respective `FMNR` spectra within ±15%. The result for the extrapolated cross section is $\sigma_{\gamma p \to b(\bar{b})X} = 15.1 \pm 3.9^{+3.8}_{-4.7}$ nb while the NLO prediction of `FMNR` for this reaction is only $5.0^{+1.7}_{-1.1}$ nb.

## 5. Conclusions and outlook

The understanding of the $b$ quark production mechanism is an outstanding puzzle in QCD. A set of new visible cross sections of beauty production, defined close to the detector acceptance, have been measured at HERA. The results are about a factor 1.4 higher than the NLO QCD predictions but consistent within the experimental and theoretical uncertainties both in the photoproduction and the DIS regions. An excess of measured cross sections over the NLO QCD prediction is observed when attempts are made to extrapolate cross sections to regions which are not directly measured by the detectors.

The HERA collider is now starting a new phase of operation at higher luminosities. Together with the enhanced $b$-tagging capabilities of the new H1 and ZEUS vertex detector and tracking triggers more precise and differential measurements can be expected within the next five years.

**Acknowledgments**

I would like to thank my collegues of the H1 and ZEUS collaborations for their work and their suggestions to this talk.

## REFERENCES

1. C. Adloff *et al.* [H1 Collaboration], Phys. Lett. B **467** (1999) 156 [Erratum-ibid. B **518** (2001) 331].
2. H1 Coll., contrib. paper no. 979/982 to ICHEP2000, Osaka 2000
3. J. Breitweg *et al.* [ZEUS Collaboration], Eur. Phys. J. C **18** (2001) 625.
4. ZEUS Coll., contrib. paper no. 785.
5. S. Frixione, M. L. Mangano, P. Nason and G. Ridolfi, Phys. Lett. B **348** (1995) 633.
6. T. Sjostrand, Comput. Phys. Commun. **82** (1994) 74; E. Norrbin and T. Sjostrand, Eur. Phys. J. C **17** (2000) 137.
7. H1 Coll., contrib. paper no. 1013.
8. ZEUS Coll., contrib. paper no. 783.
9. B. W. Harris and J. Smith, Phys. Rev. D **57** (1998) 2806.
10. H. Jung, Comput. Phys. Commun. **86** (1995) 147.
11. H. Jung and G. P. Salam, Eur. Phys. J. C **19** (2001) 351; H. Jung, Comput. Phys. Commun. **143** (2002) 100.
12. M. Ciafaloni, Nucl. Phys. B **296** (1988) 49; S. Catani, F. Fiorani and G. Marchesini, Phys. Lett. B **234** (1990) 339; G. Marchesini, Nucl. Phys. B **445** (1995) 49.
13. H1 Coll., contrib. paper no. 1016.
14. ZEUS Coll., contrib. paper no. 784.

ELSEVIER

Proceedings of ICHEP 2002, pp. 332–337
*S. Bentvelsen, P. de Jong, J. Koch and E. Laenen (Editors)*

31st INTERNATIONAL CONFERENCE ON HIGH ENERGY PHYSICS AMSTERDAM

# Heavy Quark Production at CDF

Ch. Paus[a] for the CDF Collaboration

[a]Massachusetts Institute of Technology, 77 Massachusetts Avenue, Cambridge, MA 02139, USA

The contribution summarizes the latest results from CDF on heavy quark production. Results from top, bottom and charm production are included. Some new analysis using Run I (1991-1994) data have become available. More importantly there are a number of results using Run II data which began in April 2001. The data indicate the potential of CDF for bottom and charm production physics in the near future.

## 1. INTRODUCTION

The experiments at the Tevatron are, until the LHC turns on, in the unique position to study all quarks which are considered heavy: charm, bottom, and top. In the following a number of recent results from CDF are presented which are obtained from data of the Run I and Run II data taking periods. For heavy quark physics at hadron machines like the Tevatron the triggering system is crucial. Most of the collisions are uninteresting with relatively low momenta going in the transverse direction and originate from light quarks. Interesting events have to be filtered out carefully due to limited bandwidth.

During the Run I data taking period (1991-4) CDF took a total of 110 $pb^{-1}$ of data. These data are used extensively for analysis and even today new results of heavy quark production are being obtained. For $b$ physics analyses data originating from muon and electron triggers are used.

The Run II data taking period started in April 2001 and CDF has used up to 15 $pb^{-1}$ for the analyses presented here. The most exciting new feature of the CDF detector with respect to the Run I setup is the trigger for large impact parameter tracks [1]. For the first time at a hadron machine a level-2 trigger is able to measure the track impact parameter fast and accurately enough to allow a trigger based on these measurements. This opens a new window of opportunity for heavy flavor physics based on purely hadronic decays. This trigger will be essential to measure $B_s$ mixing, which is one of the most important measurements targeted by CDF in the near term future. Probably only the fully reconstructed final state leads to sufficiently precise proper time measurements to capture the rapid $B_s$ oscillations.

Heavy quark production is predicted by QCD and the larger the mass, the better the agreement with perturbative calculations. One of the main goals in measuring heavy quark production is to test the validity of QCD calculation schemes. The three quarks cover an interesting spectrum from the top quark mass of about 175 GeV/$c^2$ to the charm quark mass of about 1.5 GeV/$c^2$.

The text summarizes recent results for top and bottom production in the next two sections. Both are based on Run I data. The next section, number 4, is devoted to Run II data, first showing the detector performance in the leptonic triggers and then some first results on charm decays using the new hadronic trigger.

## 2. RECENT RESULTS FROM TOP PRODUCTION

Gluon radiation is an important aspect of top quark production. Not only is the phenomenon interesting in itself but it also constitutes a large component of the uncertainty of the top quark mass measurement. While the total uncertainty on the top quark mass is about $\pm 5$ GeV/$c^2$ in Run I, the uncertainty introduced by hard gluon radiation amounts to $\pm 2 - 3$ GeV/$c^2$ depending on the particular top decay channel.

Hard gluon radiation can in good approximation be separated into gluon radiation in production which means before the top quark decays, and gluon radiation in decay which happens dur-


doi:10.1016/S0920-5632(02)02182-5

ing or after the top decay. While jets reconstructed from the former type should not be included in the $t\bar{t}$ system the latter one has to be included to obtain the correct top quark mass.

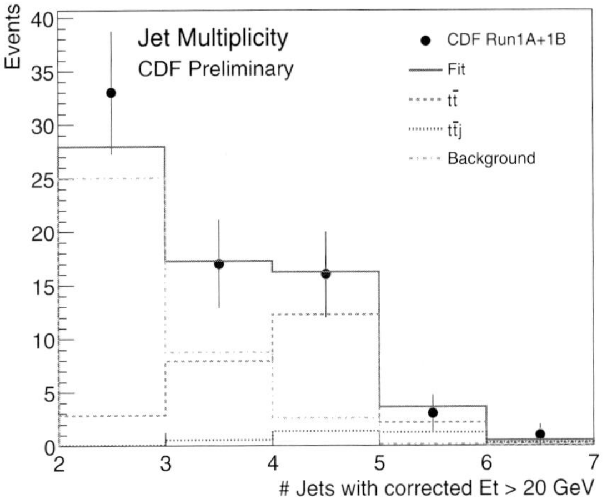


Figure 1. Fit of the Monte Carlo jet multiplicity templates to the data.

The $t\bar{t}$ is reconstructed in the Run I data as a lepton plus missing energy and at least two jets [2]. At least one of the jets is required to have a $b$-tag which means that it contains tracks originating from a secondary vertex. For the analysis the events are classified as $t\bar{t}$ plus 0, 1, 2 *etc.* jets. Jets are formed based on a cone algorithm [2] using different transverse energy thresholds: $E_T > 20$ GeV or $E_T > 25$ GeV. The jet multiplicity is then fit to Monte Carlo templates. The Monte Carlo uses Pythia [3], where gluon radiation in production is included via MADGRAPH matrix elements [4]. The processes $qq/gg \to tt$, $qq/gg \to ttg$, $qg \to ttq$, and $qg \to ttgg$ are included. Gluon radiation in the top decay is provided by Pythia itself. Using templates of the jet multiplicities of the signal Monte Carlo and the standard background simulation [2] allows one to determine the contribution of the different gluon radiation diagrams. A fit of the jet multiplicity templates for jets with transverse energy larger $E_T > 20$ GeV to the data is shown in Figure 1.

Table 1
Fraction of events originating from background, $t\bar{t}$ or $t\bar{t}j$ production.

| Contribution | Fit ($E_T > 20$ GeV) |
|---|---|
| $f_{tt}$ | $0.36\ ^{+0.13}_{-0.12}$ |
| $f_{ttj}$ | $0.051\ ^{+0.084}_{-0.059}$ |
| $f_{bg}$ | $0.523\ ^{+0.057}_{-0.056}$ |

For events with only 2 jets the background dominates while the fraction of top-pair events without radiation increases until jet multiplicities of 4. The contribution of top-pairs with one hard gluon radiation reaches its maximum at 4 to 5 jets. The result of the fit is the fraction of $t\bar{t}$, and $t\bar{t}j$ events: $f_{tt}$ and $f_{ttj}$, respectively, and is listed in Table 1.

The large uncertainty on the fraction of $t\bar{t}j$ events indicates that the number of such events is not yet sufficient to make a significant measurement of this process. A fit including an additional contribution of two hard gluons from production is obviously hopeless and not further discussed here. The fractions are converted into the cross section ratios as:

$$\sigma_{tt}/\sigma_{ttX} = f_{tt}/(1 - fbg)\,,$$
$$\sigma_{ttj}/\sigma_{ttX} = f_{ttj}/(1 - fbg)\,.$$

Limits at the 90% confidence level are also set on the cross section fractions:

$$\sigma_{ttj}/\sigma_{ttX}(E_T = 20\ \text{GeV}) < 0.48\,,$$
$$\sigma_{ttj}/\sigma_{ttX}(E_T = 25\ \text{GeV}) < 0.55\,.$$

Although the measurement lacks statistics, a method to determine the hard gluon radiation in top production has been established.

## 3. RECENT RESULTS FROM BOTTOM PRODUCTION

Bottom production is huge at the Tevatron. The total production cross section is on the order of about 100 $\mu$b as compared to 1 nb at the $\Upsilon(4s)$ resonance or 7 nb at the $Z$ resonance. Nevertheless $b$ physics is challenging at the Tevatron since the light quark inelastic cross section is another 3 orders of magnitude larger.

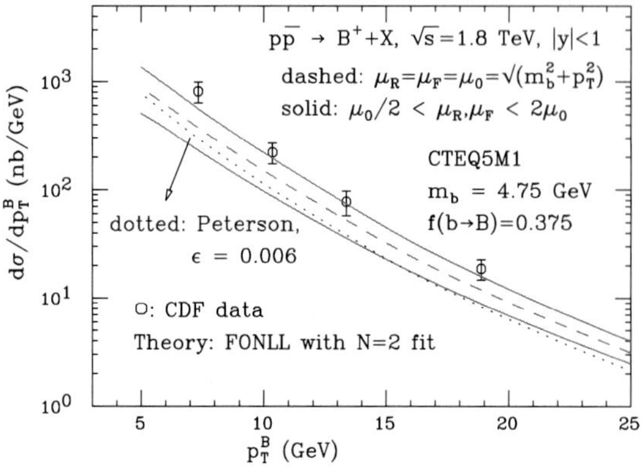


Figure 2. Differential production cross section of $b\bar{b}$ production at 1.8 TeV.

### 3.1. Single Inclusive Cross Section

A measurement of the single inclusive $b$ production is performed using the exclusive signal $B^+ \to J/\psi K^+$ with $J/\psi \to \mu^+\mu^-$ [5]. Since the $b$ quark is not measured directly the $B$ meson measurement has to be unfolded for various effects like fragmentation, structure functions, momentum scales and $b$ quark mass. When cross section was compared to the predictions in the original publication it comes out too high by a factor of 2.9. Recent developments in the theory sector [6] shows that the Peterson fragmentation function [8] with an $\epsilon$ parameter value of $0.006 \pm 0.002$ is not appropriate to use in conjunction with next-to-leading order, NLO, programs [7]. This is due to the effect that the fragmentation parameter $\epsilon$ has been determined using leading-log calculations which are commonly used in shower Monte Carlo programs like Pythia [3].

The authors of Reference [6] suggest a more correct implementation of the fragmentation function to be used in connection with the NLO calculation. This new calculation is referred to as FONLL by the authors. Using the new scheme for the implementation of the fragmentation the cross section prediction is only a factor of 1.7 higher than the measured cross section. The situation is nicely summarized in Figure 2. The data points include statistical and systematic errors. The dashed line is the central value of the FONLL calculations while the solid lines indicate the theoretical uncertainties introduced by the variation of the momentum scales of renormalization and factorization. Also shown as a dotted line is the cross section prediction when using Peterson fragmentation with a parameter of $\epsilon = 0.006$. The new calculation does not exhibit a large discrepancy to the measurements within the quoted uncertainties.

### 3.2. Correlations of $b\bar{b}$ Production

Recently CDF has started an effort to measure the correlations of $b\bar{b}$ production. Similar to the single inclusive cross section $b$ hadrons are used to identify the $b$ quark. Again an unfolding process has to be applied, only this time there are two $b$ hadrons, which makes the process much more complicated. From the measurement of $b\bar{b}$ correlations it is possible to distinguish contributions of different production mechanisms. The lowest order production mechanism is expected to produce $b\bar{b}$ pairs basically back-to-back while for the higher order production mechanisms the distribution of the angle between the two quarks is almost flat.

So far CDF has only results for correlations of $b$ hadrons which are not yet unfolded for detector effects. Two analyses have been performed.

In the first analysis an inclusive $J/\psi$ sample is used where the second $b$ hadron is identified by its semileptonic decay, a high $p_T$ lepton. The ratio of the numbers of events with the lepton in the same hemisphere with the $J/\psi$, over the number of events with the lepton in the opposite hemisphere is measured to be:

$$R(\text{same/opposite}) = 0.52 \pm 0.20$$

where the uncertainty is statistical and systematic. The measurement indicates that a large fraction of $b\bar{b}$ events have small angles between the two quarks and thus the data has contributions from higher order production mechanisms. The measurement suffers from low statistics, but $b$ hadrons are cleanly identified.

In the second analysis $b$ hadrons are identified by tagging secondary vertices in jets. This method has the advantage that there are many more events but at the cost of larger background.

The $\Delta\phi$ angle between the two jets is shown in Figure 3, and compared to the Pythia prediction at the lowest order. While data and Monte Carlo agree quite well for large opening angles it does not describe a significant fraction of events with small opening angle.

In summary, these first measurements indicate that there is a significant fraction of higher order production mechanisms present in the data, which can hopefully be disentangled in the near future.

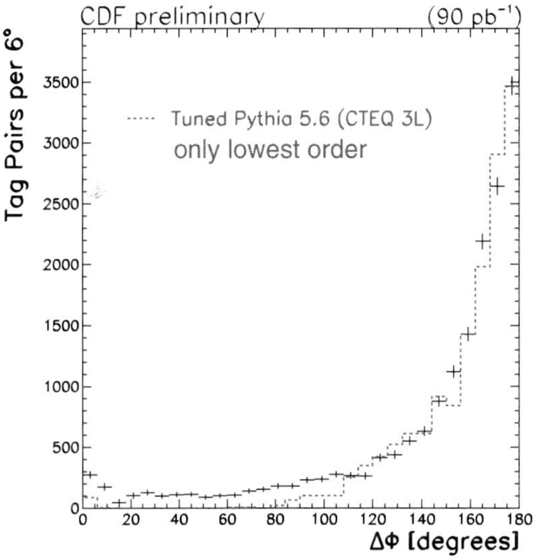


Figure 3. Smaller angle between the two $b$ jets, $\Delta\phi$ in the transverse plane. Dots with error bars are data.

## 4. SEARCH FOR $\eta_b$

An $\eta_b$ search is performed in about 80 $pb^{-1}$ of Run I data. The $\eta_b$ is searched for in its decay to two $J/\psi$ mesons, where the $J/\psi$ decays to a muon pair. The dataset is based on the dimuon $J/\psi$ trigger which constitutes the first $J/\psi$. Muons are identified by the trigger starting at about 2.2 GeV with a noticeable turn-on curve. For the second $J/\psi$ the momentum cuts are significantly relaxed and the fourth muon need not be identified by the muon system.

The $\eta_b$ mass is expected to lie in a window from 9.36 GeV to 9.46 GeV according to various NRQCD predictions [9]. The analysis turns up 7 events in the given mass window with a background contribution of 1.8 events. The mass distribution is shown in Figure 4, where the expected mass window is indicated by the arrows. Observing 7 events with an expected background of 1.8 translates into a probability of 1.5% for this observation to be due to background; a 2.2 standard deviation effect. Assuming the events come from the signal the mass of the $\eta_b$ is determined to be:

$$m(\eta_b) = 9446 \pm 6 \text{ MeV}/c^2 .$$

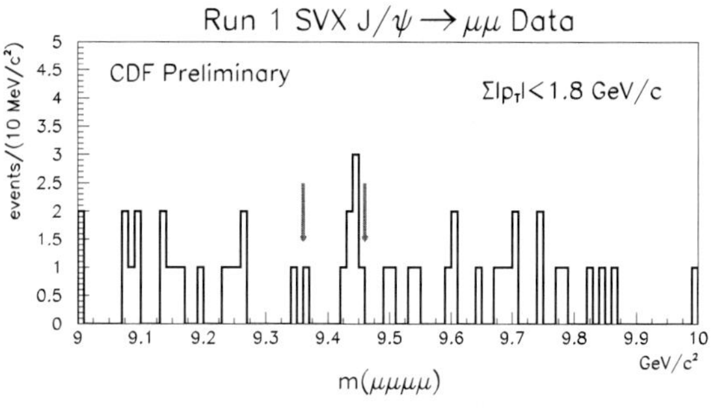


Figure 4. Mass distribution of the two $J/\psi$ mesons decaying to dimuons. The arrows indicate the region for the expected $\eta_b$ signal.

Run II data will allow us to significantly enlarge the data sample and if the signal is true a definite observation is soon possible.

## 5. A GLIMPSE AT RUN II DATA

Since April 2001 CDF has started its Run II data taking period with a significantly upgraded detector [10]. The most striking upgrade in terms of $b$ physics is certainly the trigger which allows us to trigger on high $p_T$ displaced tracks at the Level-2 trigger [1]. Also important are the upgrades of the silicon detector with higher vertexing precision and larger coverage, and the upgrade of the muon system with lower trigger thresholds and larger coverage.

### 5.1. Detector Performance

The performance of the CDF detector is already by now close to what has been achieved in the Run I data taking period. Since the $J/\psi$ trigger delivers a large signal and is well understood from Run I it is used to demonstrate CDF performance. The momentum scale calibration performed on this signal [11] allows us to measure $B$ meson masses. The masses of $B^+$, $B^0$ and $B_s^0$ mesons have been measured in their most prominent $J/\psi$ decay modes $B^+ \to J/\psi K^+$, $B^0 \to J/\psi K^{*0}$ ($K^{*0} \to K^+\pi^-$) and $B_s^0 \to J/\psi\phi$ ($K^{*0} \to K^+K^-$):

$$m(B^+) = 5280.6 \pm 1.7 \pm 1.1 \text{ MeV}/c^2$$
$$m(B^0) = 5279.8 \pm 1.9 \pm 1.4 \text{ MeV}/c^2$$
$$m(B_s^0) = 5360.3 \pm 3.8^{+2.1}_{-2.9} \text{ MeV}/c^2$$

where the first uncertainty is the statistical and the second the systematic one. All measurements are in good agreement with the world averages and the measurements are statistically limited. The $B_s^0$ mass measurement is the second most precise single measurement and will soon be improved significantly. The $B_s^0$ mass distribution containing 14.4 signal events after background subtraction is shown in Figure 5.

Also the vertex resolution is quite well understood already. Therefore the inclusive $b$ hadron lifetime has been measured again using the inclusive $J/\psi$ sample. Furthermore, the $B^+$ lifetime has been measured in its decay $B^+ \to J/\psi K^+$. The results are:

$$c\tau_{incl} = 458 \pm 10 \pm 11 \ \mu\text{m}$$
$$c\tau_{B^+} = 446 \pm 43 \pm 13 \ \mu\text{m}$$

where the first uncertainty is the statistical and the second the systematic one. Both measurements are in excellent agreement with the world average and the systematic uncertainties indicates the good understanding of the resolutions functions. In all analyses so far neither silicon hits with $z$ information nor the innermost silicon layer is used, because they are still being commissioned. A significant improvement in momentum

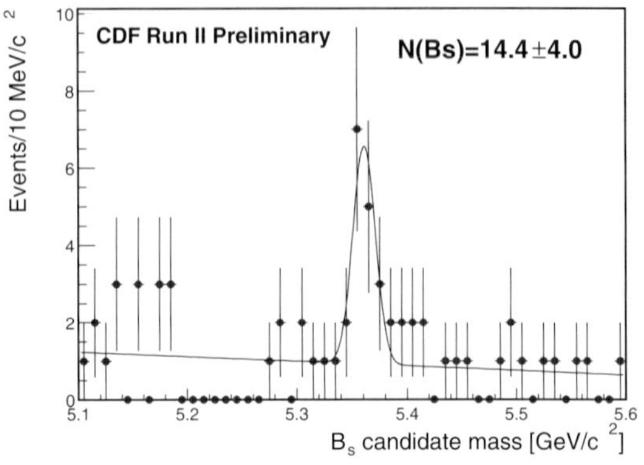


Figure 5. Mass distribution of the $B_s^0$ meson candidates in their decay $B_s^0 \to J/\psi\phi$ ($K^{*0} \to K^+K^-$).

and vertexing resolution is expected once they are available.

### 5.2. Charm Physics

The commissioning of the displaced track trigger has enabled CDF to trigger on hadronic decay products of longer, lived particles like $b$ and $c$ hadrons. Although the lifetime of $b$ hadrons is longer the huge production cross section of charm quarks makes this trigger especially rich in $D$ mesons. This has opened a completely new window of opportunities for CDF: charm physics. In about 5 pb$^{-1}$ of data CDF sees about 30k $D^0 \to K^-\pi^+$, 6k $D^{*+} \to D^0(K^-\pi^+)\pi^+$, 27k $D^+ \to K^-\pi^+\pi^+$ and 1k $D_s^+ \to \phi(K^+K^-)\pi^+$. This is an impressive collection of $D$ mesons taking into account that the integrated luminosity will improve by more than 2 orders of magnitude.

We start a measurement of the charm production cross section. The first step is to identify background from secondary charm. This exercise is also necessary for doing $b$ physics with this trigger where the background and the signal are reversed.

To separate prompt from secondary charm the impact parameter of the fully reconstructed $D$ meson is used. Direct charm should point back to the primary vertex corresponding to a zero impact parameter. Secondary charm on the contrary originates from a $b$ hadron which has a long lifetime and has other decay products apart from the $D$ meson. Thus the $D$ does not necessarily point back to the primary vertex and its impact parameter distribution is rather wide. The detector resolution smears out both prompt and secondary charm impact parameter distributions.

The detector resolution function, $R$, is measured from the impact parameter distributions from $K_s \to \pi^+\pi^-$ decays, which are predominantly prompt. This function should be the same for all decays. The inclusive impact parameter distributions of the $D$ mesons is than fitted to the function:

$$F(d_0) = f_B \int F_B(x) R(d_0 - x) dx + \\ + (1 - f_B) R(d_0) \,,$$

where $f_B$ and $F_B$ are the fraction of secondary $D$ mesons and the impact parameter distribution of the secondary $D$ mesons. The variable $x$ integrates over all possible values of the impact parameter. For the fit the un-smeared impact parameter distribution for secondary $D$ mesons, $F_B$, is obtained from Monte Carlo. The impact parameter distribution is shown in Figure 6 with the fit superimposed indicating the contribution of the secondary $D$ mesons. As a comparison the resolution function for prompt $D$ mesons is assumed to be a simple Gaussian. The result of both fits for the fraction of $D$ mesons which originate from the decay of a $b$ hadron are quoted below, in parenthesis the numbers when using the simple Gaussian resolution function:

| | | |
|---|---|---|
| $D^0$: | 16.43±0.65 % | (23.14±0.59 %) |
| $D^{*+}$: | 11.41±1.37 % | (20.00±1.19 %) |
| $D^+$: | 11.26±0.53 % | (17.29±0.53 %) |
| $D_s$: | 34.80±2.75 % | (37.84±2.60 %) |

The difference bbetween both models give an impression of the uncertainty of this measurement. In general we conclude that the charm sample has contamination from secondary $D$ mesons between 10% and 40%. The differences for the various $D$ mesons are due to the differing production fractions, lifetimes and the different number of decay products which have different probablities to pass the trigger.

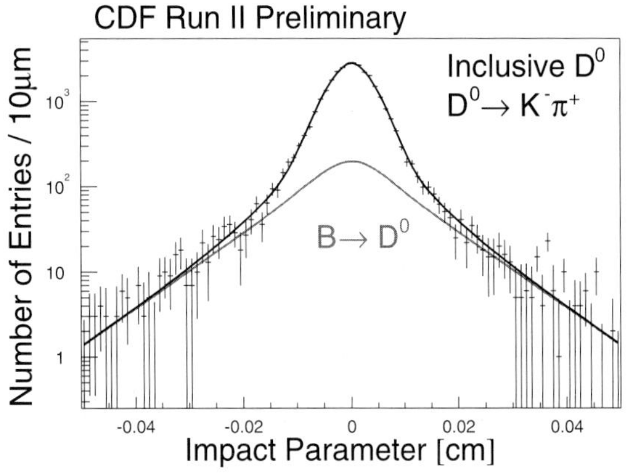


Figure 6. Impact parameter distribution for $D^0$ mesons in its decay $D^0 \to K^-\pi^+$. The tails are mostly due to secondary $D^0$ decays while the center is mostly due to prompt $D^0$ decays. Overlayed is the fit described in the text.

## REFERENCES


1. A. Cerri, in these proceedings.
2. The CDF collaboration, T. Affolder *et al.*, Phys. Rev. **D63** (2001) 032003.
3. H. U. Bengston and T. Sjöstrand, Comp. Phys. Comm. **43**, (1987) 43; T. Sjöstrand, Comp. Phys. Comm. **82** (1994) 74.
4. D. Rainwater, private communiction.
5. The CDF Collaboration, D. Acosta *et al.*, Phys. Rev. **D65** (2002) 052005.
6. M. Cacciari and P. Nason, Phys. Rev. Lett **89** (2002) 122003.
7. P. Nason, S. Dawson and R.K. Ellis, Nucl. Phys. **B303** (1988) 607; Nucl. Phys. **B327** (1989) 49.
8. C. Peterson *et al.*, Phys. Rev. **D27** (1983) 105.
9. Most recent only: A. Spitz *et al.*, hep/lat-9906009, June 1999; T. Manke, Phys. Rev. **D62** (2000) 114508; D.E. Groom *et al.*, Eur. Phys. J. **C15** (2000) 1.
10. The CDF II Collaboration, *The CDF II Detector Technical Design Report*, FERMILAB-Pub-96/390-E (1996).
11. F. Bedeschi, in these proceedings.

ELSEVIER

Proceedings of ICHEP 2002, pp. 338–340
*S. Bentvelsen, P. de Jong, J. Koch and E. Laenen (Editors)*

31[st] INTERNATIONAL CONFERENCE ON HIGH ENERGY PHYSICS AMSTERDAM

# Correlations between $D$ and $\overline{D}$ Mesons in High Energy Photoproduction

Erik E. Gottschalk[a]*,

[a]Fermilab,
P.O. Box 500, Batavia, IL 60510, USA

Over 7000 events containing a fully reconstructed $D\overline{D}$ pair have been extracted from data recorded by the FOCUS photoproduction experiment at Fermilab. Preliminary results from a study of correlations between $D$ and $\overline{D}$ mesons are presented. Correlations are used to study perturbative QCD predictions and investigate non-perturbative effects. We also present a preliminary result on the production of $\psi(3770)$.

## 1. Introduction

Quantum Chromodynamics (QCD) is widely accepted as the correct theoretical description of the strong interaction, and strong-interaction properties involving hard processes can be calculated using perturbative QCD computations. These computations are applicable to heavy quark production, and are applicable to the study of charm production in particular. One area of theoretical interest is the study of correlations between $D$ and $\overline{D}$ mesons.

$D\overline{D}$ correlations are used to study perturbative QCD predictions and are used to investigate non-perturbative effects. In this paper we focus on two distributions. The first is $\Delta\phi$, which is defined as the angle between the $D$ and $\overline{D}$ in the plane transverse to the beam. The second, $p_t^2(D\overline{D})$, is the transverse momentum squared for the $D\overline{D}$ pair. These distributions are important for studies of heavy-quark production in QCD [1]. At leading order in QCD the $c$ and $\overline{c}$ quarks are produced back-to-back with $\Delta\phi = \pi$ and $p_t(c\overline{c}) = 0$. By including NLO corrections the distributions broaden. The data, which tend to disagree with NLO calculations, suggest significant contributions from non-perturbative effects.

Results presented in this paper (most are preliminary) are based on the data recorded by the FOCUS experiment during the 1996-1997 fixed-target run at Fermilab. More than one million fully reconstructed charm-particle decays were recorded. Charm particles were produced by the interaction of photons (180 GeV average energy) with a segmented BeO target. For our study of $D\overline{D}$ correlations we used measurements from several detectors in the FOCUS spectrometer: the vertex detector, multiwire proportional chambers, and Cerenkov counters. FOCUS ran with an upgraded version of the E687 spectrometer[2]. The vertex detector included 16 planes of silicon-strip detectors. Four of the planes were interleaved with the BeO target elements[3], and 12 were located downstream of the target. Charged particles were reconstructed in the spectrometer using five multiwire proportional chambers. The momentum is determined by measuring the deflection of a particle's trajectory in two analysis magnets of opposite polarity. Particle identification is performed by using measurements from three multicell threshold Cerenkov counters[4].

*On behalf of the FOCUS collaboration: J.Link, M.Reyes, P.M.Yager (**UC DAVIS**); J.Anjos, I.Bediaga, C.Gobel, J.Magnin, A.Massafferri, J.M. de Miranda, I.M.Pepe, A.C. dos Reis, (**CPBF, Rio de Janeiro**); S.Carrillo, E.Casimiro, E.Cuautle, A.Sánchez-Hernández, C.Uribe, F.Vasquez (**CINVESTAV, México City**); L.Agostino, L.Cinquini, J.P.Cumalat, B.O'Reilly, J.E.Ramirez, I.Segoni (**CU Boulder**); J.N.Butler, H.W.K.Cheung, G.Chiodini, I.Gaines, P.H.Garbincius, L.A.Garren, E.E.Gottschalk, P.H.Kasper, A.E.Kreymer, R.Kutschke (**Fermilab**); L.Benussi, S.Bianco, F.L.Fabbri, A.Zallo (**INFN Frascati**); C.Cawlfield, D.Y.Kim, K.S.Park, A.Rahimi, J.Wiss (**UI Champaign**); R.Gardner, A.Kryemadhi (**IU Bloomington**); K.H.Chang, Y.S.Chung, J.S.Kang, B.R.Ko, J.W.Kwak, K.B.Lee (**Korea University, Seoul**); K.Cho, H.Park (**Kyungpook National University, Taegu**); G.Alimonti, S.Barberis, A.Cerutti, M.Boschini, P.D'Angelo, M.DiCorato, P.Dini, L.Edera, S.Erba, M.Giammarchi, P.Inzani, F.Leveraro, S.Malvezzi, D.Menasce, M.Mezzadri, L.Moroni, D.Pedrini, C.Pontoglio, F.Prelz, M.Rovere, S.Sala (**INFN and Milano**); T.F.Davenport III (**UNC Asheville**); V.Arena, G.Boca, G.Bonomi, G.Gianini, G.Liguori, M.M.Merlo, D.Pantea, S.P.Ratti, P.Vitulo (**INFN and Pavia**); H.Hernandez, A.M.Lopez, H.Mendez, L.Mendez, E.Montiel, D.Olaya, A.Paris, J.Quinones, C.Rivera, W.Xiong, Y.Zhang (**Mayaguez, Puerto Rico**); J.R.Wilson (**USC Columbia**); T.Handler, R.Mitchell (**UT Knoxville**); D.Engh, M.Hosack, W.E.Johns, M.Nehring, P.D.Sheldon, K.Stenson, E.W.Vaandering, M.Webster (**Vanderbilt**); M.Sheaff (**Wisconsin, Madison**)


doi:10.1016/S0920-5632(02)02185-0

## 2. Charm-pair analysis

FOCUS has acquired a large sample of 7066 ± 119 $D\overline{D}$ events. Each event has two $D$ mesons, a $D$ and a $\overline{D}$. Both $D$ mesons are fully reconstructed. The decay modes that are considered for the charm-pair analysis are $D^0 \to K^-\pi^+$, $D^+ \to K^-\pi^+\pi^+$, $D^0 \to K^-\pi^+\pi^+\pi^-$, and charge conjugate modes. By reconstructing both the $D$ and the $\overline{D}$ in an event, FOCUS is able to study $D\overline{D}$ correlations.

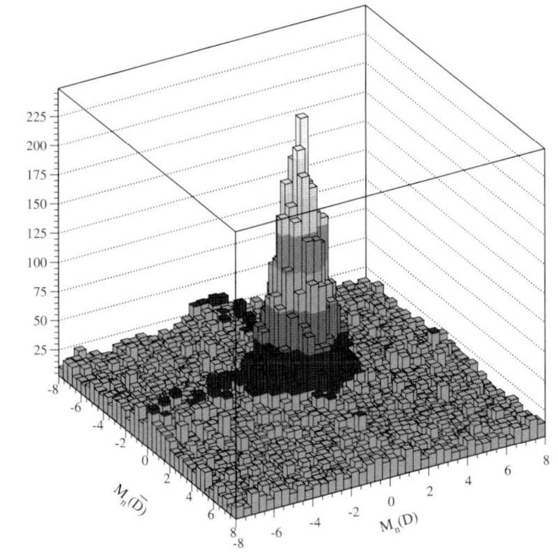


Figure 1. Normalized $D$ invariant mass vs. normalized $\overline{D}$ invariant mass distribution.

Figure 1 shows the $D\overline{D}$ signal that is obtained after applying a candidate-driven algorithm and various event-selection cuts to the data. The algorithm is used to reconstruct the $D$ and $\overline{D}$ decay vertices, as well as the primary interaction vertex. Events that survive the vertex reconstruction phase of the analysis are subjected to particle-identification cuts, which are based on measurements from the three threshold Cerenkov counters. After applying these cuts, we impose cuts that are based on the significance of detachment ($L/\sigma_L$) between the primary interaction vertex and the $D$ and $\overline{D}$ decay vertices in each event. The cuts for $L/\sigma_L$ range from $L/\sigma_L > 1$ to $L/\sigma_L > 4$ depending on the decay mode, whether a $D$ meson can be associated with a $D^*$ decay, and whether the $D$-decay vertex is located between target elements (for which background levels are lower) or in target material. Details of the analysis are being prepared for publication.

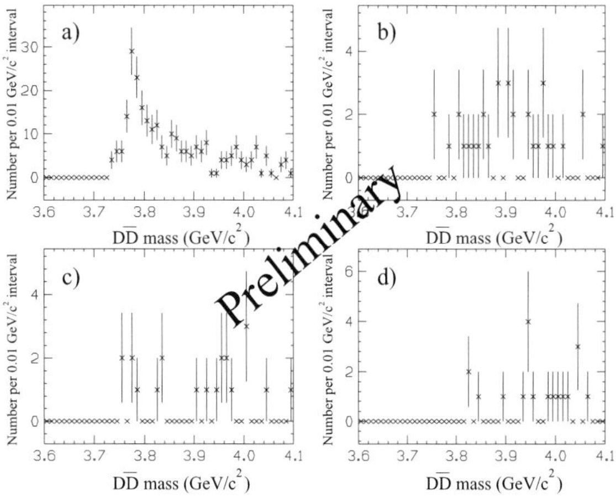


Figure 2. Invariant mass distribution of the $D\overline{D}$ pair for data and PYTHIA 6.203 for different charge combinations (see text) of the $D$ and $\overline{D}$ mesons. Distributions are shown for a) data with $Q_{D\overline{D}} = 0$, b) data with $Q_{D\overline{D}} \neq 0$, c) PYTHIA 6.203 with $Q_{D\overline{D}} = 0$, and d) PYTHIA 6.203 with $Q_{D\overline{D}} \neq 0$.

For our analysis of charm-pair events we require an additional event-selection cut to make comparisons between data and QCD production models. This cut is based on NPRIM, which is defined as the total number of charged tracks assigned to the primary interaction vertex plus two. Adding a value of two accounts for the reconstructed $D$ and $\overline{D}$ mesons associated with the primary vertex. The motivation for the additional cut, which is NPRIM > 2, is a discrepancy that is observed for events with NPRIM = 2 when comparing data to Monte Carlo events generated by PYTHIA[5].

The discrepancy seems to arise from an apparent signal for $\psi(3770)$, which is present in the data and not in PYTHIA. Figure 2a) shows a possible signal for $\psi(3770)$ for events with NPRIM = 2 and $Q_{D\overline{D}} = 0$, where $Q_{D\overline{D}}$ is defined as the sum of the charge of the $D$ and $\overline{D}$ mesons.

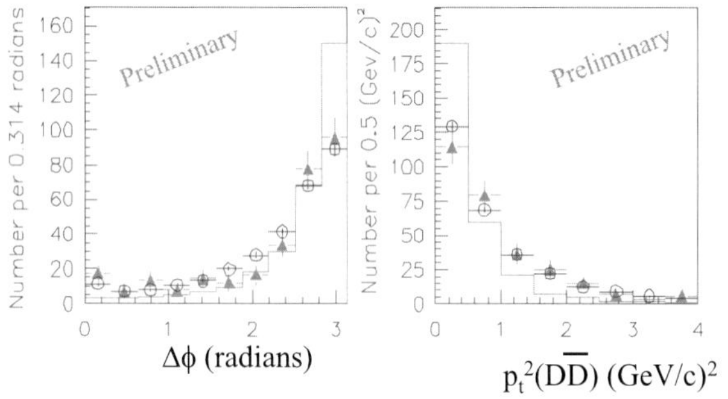


Figure 3. Comparison of $\Delta\phi$ and $p_t^2(D\overline{D})$ distributions for FOCUS data with an NPRIM > 2 cut (open circles with error bars), published E687 data (filled triangles with error bars), and PYTHIA 5.6 (solid histogram).

## 3. Charm-pair correlations

For our study of charm-pair correlations we compare $\Delta\phi$ and $p_t^2(D\overline{D})$ distributions for FOCUS data to published E687 data[6] and Monte Carlo events generated by two different versions of PYTHIA. The FOCUS data include the NPRIM > 2 cut to remove the observed discrepancy associated with the production of $\psi(3770)$. Figure 3 shows good agreement between FOCUS and E687 data for both distributions. The figure also shows that the data do not agree with PYTHIA version 5.6[7], which was the Monte Carlo generator that was used for the E687 publication[6]. The PYTHIA distributions are more sharply peaked compared to the data.

Figure 4 shows FOCUS data compared to a more recent version of PYTHIA[5]. Here the

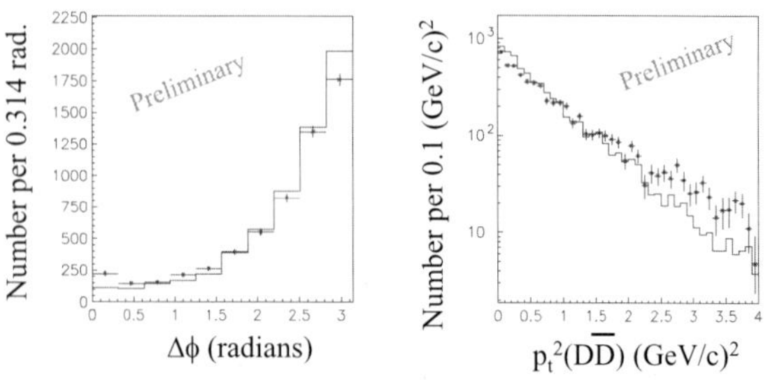


Figure 4. Comparison of $\Delta\phi$ and $p_t^2(D\overline{D})$ distributions for FOCUS data with an NPRIM > 2 cut (data points with error bars) and PYTHIA 6.203 (solid histogram).

agreement between data and PYTHIA version 6.203 has improved significantly. One of the explanations for the improvement (other explanations are being investigated) is the larger value for intrinsic $k_T$ in version 6.203 ($< k_T^2 > = 1$ GeV$^2$), which introduces a larger transverse momentum for initial partons.

In summary, FOCUS has extracted a large sample of $D\overline{D}$ events for correlation studies and comparisons to QCD models. A detailed description of the analysis and results is being prepared for publication.

## REFERENCES

1. S. Frixione, M.L. Mangano, P. Nason, G. Ridolfi, Adv. Ser. Direct. High Energy Phys. 15 (1998) 609.
2. P.L.Frabetti et al., Nucl.Instr.Meth. A320 (1992) 519.
3. J.M.Link et al., hep-ex/0204023 (2002), accepted for publication in Nucl.Instr.Meth.
4. J.M.Link et al., Nucl.Instr.Meth. A484 (2002) 270.
5. T. Sjostrand et al., Comput. Phys. Commun. 135 (2001) 238.
6. P.L.Frabetti et al., Phys. Lett. B308 (1993) 193.
7. H.U. Bengtsson, T. Sjostrand, Comput. Phys. Commun. 46 (1987) 43.

ELSEVIER

Proceedings of ICHEP 2002, pp. 341–344
*S. Bentvelsen, P. de Jong, J. Koch and E. Laenen (Editors)*

31st INTERNATIONAL CONFERENCE ON HIGH ENERGY PHYSICS AMSTERDAM

# Measurement of the $b\bar{b}$ production cross section in proton-nucleus collisions at HERA – B

Peter Kreuzer[a], for the HERA – B Collaboration

[a]University of California, Los Angeles,
CERN 32-R-B09, 1211 Geneva 23, Switzerland

Using the HERA – B detector, the $b\bar{b}$ production cross section has been measured in 920 GeV proton collisions on carbon and titanium targets. The $b\bar{b}$ production was tagged via inclusive bottom quark decays into $J/\psi$ by exploiting the longitudinal separation of $J/\psi \to l^+l^-$ decay vertices from the primary proton-nucleus interaction. Both $e^+e^-$ and $\mu^+\mu^-$ channels have been reconstructed and the combined analysis yields the cross section $\sigma(b\bar{b}) = 32^{+14}_{-12}(\text{stat})\,^{+6}_{-7}(\text{sys})$ nb/nucleon.

## 1. Introduction

The motivation of this measurement is to add an experimental constraint to the description of heavy quark hadroproduction in proton-nucleus (pA) collisions, which has been a subject of great attention in recent years [1–3]. Only two similar measurements have been reported so far [4,5]. Currently both experimental and theoretical uncertainties are large. A new measurement by HERA – B is relevant, specially because it covers the yet non-explored negative Feynman-x region of the produced $J/\psi$.

The HERA – B detector for this measurement is described in the corresponding publication [6]. The spectrometer has a large geometrical coverage and uses pA interactions produced on carbon and titanium target wires, in the halo of the 920 GeV HERA proton beam. A fast and multilevel dilepton trigger system, as well as a high resolution Silicon Vertex Detector System, are optimal for the isolation and reconstruction of both exclusive and inclusive $B$-meson decays. The most straightforward channel to measure the $b\bar{b}$ production cross section is the inclusive reaction $pA \to b\bar{b}\,X$, with $b\bar{b} \to J/\psi\,Y$ ("$b \to J/\psi$" in the following) and $J/\psi$ into dilepton.

The $b \to J/\psi$ production cross section per nucleon, for the measured range ($-0.25 \le x_F \le 0.15$) containing a fraction $f$ of prompt $J/\psi$, can be expressed as:

$$\Delta\sigma(b\bar{b}) = \sigma_r\, f\, \frac{N_B}{N_P}\, \frac{1}{\varepsilon_R\, \varepsilon_B^{\Delta z}\, \mathrm{Br}(b\bar{b} \to J/\psi X)} \qquad (1)$$

The measurement consists in selecting a clean prompt $J/\psi$ sample ($N_P$), followed by a detached vertex selection, in order to isolate $b \to J/\psi$ events ($N_B$) from the prompt $J/\psi$ background. The efficiency terms in Eq. 1 are determined from simulation: $\varepsilon_R$ is the relative trigger and $J/\psi$ reconstruction efficiency of $b \to J/\psi$ events with respect to prompt $J/\psi$ events, while $\varepsilon_B^{\Delta z}$ is the efficiency of the detached vertex selection. The branching ratio $\mathrm{Br}(b\bar{b} \to J/\psi X)$ in hadroproduction is assumed to be the same as in $Z$ decays, i.e. $2 \cdot (1.16 \pm 0.10)\%$ [7]. The normalization is done relative to the prompt $J/\psi$ production cross section $\sigma_r = 314 \pm 7(\text{stat}) \pm 31(\text{sys})$ nb/nucleon. The latter is obtained by averaging and rescaling the known E789 [8] and E771 [9] results to the HERA – B kinematics [6]. This normalization method minimizes the systematic errors related to detector and trigger efficiencies, which mostly cancel out in the ratio $\varepsilon_R$, and it removes the dependence on the absolute luminosity.

## 2. Trigger and Prompt $J/\psi$ Selection

The analyzed data sample was acquired in a short physics run during the HERA – B com-


doi:10.1016/S0920-5632(02)02194-1

missioning period in summer 2000, by triggering on dilepton signatures. Electron candidates were initiated in the Electromagnetic "shashlik" Calorimeter (ECAL) by requiring a transverse energy $E_T > 1.0$ GeV, while muon candidates were based on a double coincidence in the MUON system (cathod pad chambers). The First Level Trigger required two lepton candidates of the same type and the Second Level Trigger, which is a software filter running on a farm of 240 PCs, confirmed the lepton pairs by applying a simplified Kalman filter to the data in the Main Tracker and the Vertex Detector System. The full trigger chain gave a reduction of the initial interaction rate of 5 MHz to a final output rate of 20 Hz. A total of ≈900,000 dielectron and ≈450,000 dimuon candidates were acquired under these conditions. More detailed descriptions of various detector and trigger components are given in [6,10].

The $J/\psi$ selection requires events with two fully reconstructed trigger tracks, submitted additionally to specific lepton identification cuts and to a dilepton vertex requirement ($\chi^2 < 5$). $J/\psi \to e^+e^-$ decays are isolated from the large background contributions [1] via two identification criteria: the ratio of the ECAL cluster energy $E$ to the momentum $p$ from tracking ($E/p$) and the search for accompanying bremsstrahlung photons, emitted upstream of the magnet (i.e. maintaining the original electron direction). Figure 1 shows the signal purity for various bremsstrahlung requirements (and a 1 $\sigma$ $E/p$ cut). A good knowledge of these two criteria, confirmed by simulation, allows us to infer the number of prompt $J/\psi$ present in a sample with looser identification cuts, and hence preserve reasonable statistics for the detached vertex analysis. This leads to the normalization factor $N_P = 5710 \pm 380(\text{stat}) \pm 280(\text{sys})$ prompt $J/\psi \to e^+e^-$ decays. The dimuon selection is made using likelihood cuts in the MUON and RICH systems. The cuts are chosen to give the best signal significance over the dimuon background. The resulting spectrum is shown in Fig. 2, yielding $N_P = 2880 \pm 60$ prompt $J/\psi \to \mu^+\mu^-$ decays. The differences in shape between the dielectron and dimuon invariant mass spectra are due to differences both in background levels and in trigger selection cuts.

[1]Pions/hadrons overlapping energetic neutral showers.

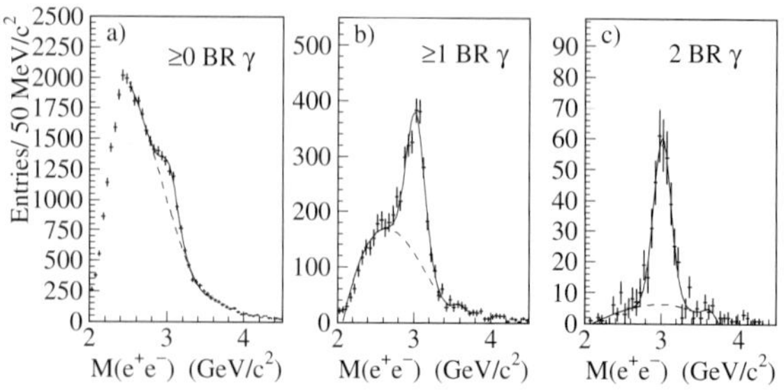


Figure 1. The $e^+e^-$ invariant mass with cuts on the number of identified bremsstrahlung photons. The fits (solid lines) assume a Gaussian $J/\psi \to e^+e^-$ signal and a polynomial background shape.

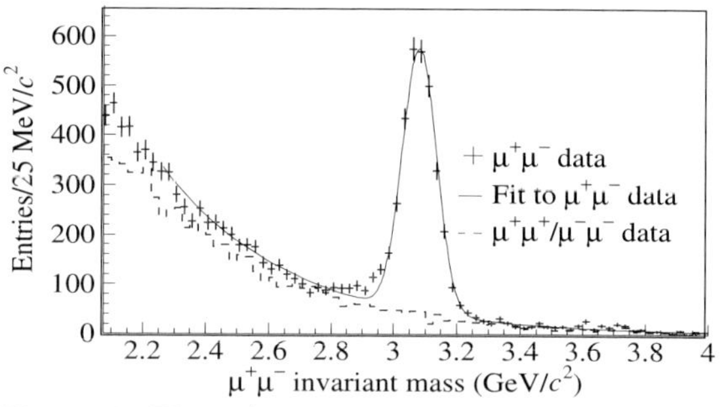


Figure 2. The $\mu^+\mu^-$ inv. mass after the $J/\psi$ selection. The fit assumes a Gaussian $J/\psi \to \mu^+\mu^-$ signal and an exponential background shape.

## 3. Detached $b \to J/\psi \to l^+l^-$ Selection

The achieved dilepton vertexing resolution (≈ 500 $\mu m$ along the beam direction and ≈ 40 $\mu m$ in the transverse plane) is sufficiently small compared to the average $b$ decay length (8000 $\mu m$) to allow a clean separation of $b \to J/\psi$ and prompt $J/\psi$ decays. The detached vertex cuts are applied on both the decay length ($\Delta z$), defined as the distance along the beam axis between the $J/\psi$ decay vertex and the closest target wire, and on the impact parameter. Several kinds of impact

parameter cuts are applied, in order to optimize the efficiency, namely: the impact parameter of the tracks to the wire ($I_w$) and to the production vertex ($I_v$) (or to any other track in the event ($I_{iso}$)); the latter two are used to reject events containing lepton tracks compatible with tracks coming from other primary vertices.
The resulting mass versus decay length distribution in the dielectron sample is shown in Fig. 3: the detached vertex cuts are $\Delta z > 0.5$ cm, $I_w > 200\,\mu$m OR $I_{iso} > 250\,\mu$m, resulting in 8 events upstream of the primary interaction region (pure combinatorial background) and 19 downstream events. Among the downstream events, 10 are found in the $J/\psi$ mass window (2.8 GeV/$c^2$ < $m_{e^+e^-}$ < 3.3 GeV/$c^2$). The events are projected

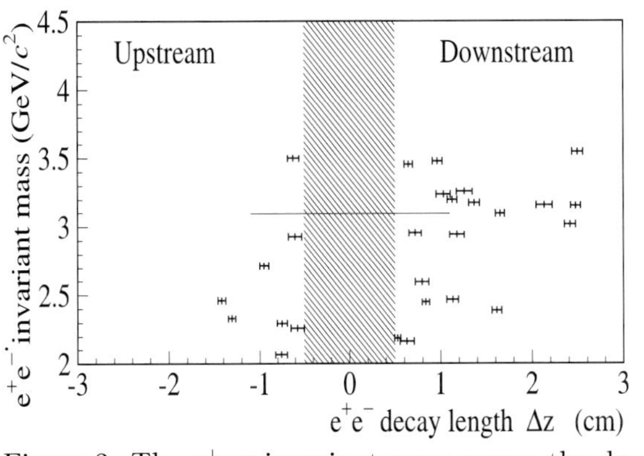


Figure 3. The $e^+e^-$ invariant mass versus the decay length for the selected detached events. The shaded region is removed by the $\Delta z$ cut. Events clustered around the $J/\psi$ mass (horiz. line) with large $\Delta z$ are observed in the downstream sample.

onto the invariant mass axis in Fig. 4: an unbinned likelihood fit is performed on the downstream $e^+e^-$ events, where the yields are left as free parameters, while the $b \to J/\psi \to e^+e^-$ signal shape is taken from simulation and the background shape is a combination of the simulated main physical contribution (double semileptonic bottom quark decays) and the pure combinatorial events. The result of the fit yields $8.6^{+3.9}_{-3.2}$ $b \to J/\psi \to e^+e^-$ events. No prompt $J/\psi$ background is expected from simulation while the number of observed downstream events ($\approx 11.4$ ) is compatible with background expectations. Several checks have been made to verify the stability of the signal against the change in types and values of the cuts: a $J/\psi$ signal with significance greater than 2 $\sigma$ is always observed in the downstream part of the spectrum, as opposed to the upstream part. Moreover, the $b$ assignment of the selected candidates is confirmed by an unbinned maximum likelihood fit on their decay length measurements, as opposed to the result obtained on the "non-physical" upstream events. Finally, the efficiency factors (Eq. 1) obtained from simulation are $\varepsilon_R \cdot \varepsilon_B^{\Delta z} = 0.44 \pm 0.02$.
A similar analysis with slightly looser cuts (but smaller input statistics) was applied on the dimuon sample, leading to $1.9^{+2.2}_{-1.5}$ $b \to J/\psi \to \mu^+\mu^-$ events and $\varepsilon_R \cdot \varepsilon_B^{\Delta z} = 0.41 \pm 0.01$. The results from both channels are combined below in a common cross section measurement.

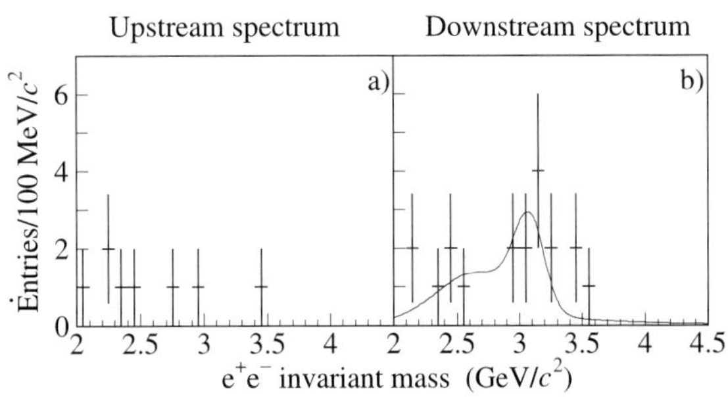


Figure 4. The $e^+e^-$ invariant mass spectra after the detached event selection. The downstream curve (b) shows the result of the unbinned likelihood fit (see text for details).

## 4. Cross Section Measurement

Using the selected $b \to J/\psi \to l^+l^-$ candidates and the computed efficiency terms from Eq. 1, we apply a four parameter likelihood maximization ($\Delta\sigma(b\bar{b})$, $\mu^+\mu^-$ background slope, $e^+e^-$ and $\mu^+\mu^-$ background yields), resulting in the following combined $b\bar{b}$ production cross section:

$$\sigma(b\bar{b}) = 32^{+14}_{-12}(\text{stat})\ ^{+6}_{-7}(\text{sys})\ \text{nb/nucleon.} \qquad (2)$$

where the quoted statistical uncertainty has been estimated directly from the likelihood fit.

The main sources of systematic uncertainties are (a) external, including the reference prompt $J/\psi$ cross section (11%) and the branching ratio $\text{Br}(b\bar{b} \to J/\psi X)$ (9%); (b) related to the low statistics of this analysis, namely the prompt $J/\psi \to e^+e^-$ counting (5%), the sensitivity to variations in the background shapes of the invariant mass fits and to the detached vertex cut values, for which we assign conservatively 13% and $^{+10}_{-24}$% to the $e^+e^-$ and $\mu^+\mu^-$ channels, respectively. The last contribution (c) concerns the Monte Carlo (MC) simulation tool, namely the trigger and detector simulation (5%), the prompt $J/\psi$ production model (3.5%) and the $b\bar{b}$ production model (5%); the latter is also used to extrapolate the measured $b\bar{b}$ cross section to the full kinematic range (Eq. 2). We refer the reader to [6] for a detailed description of the HERA - B MC simulation tool, which is similar to the model used in the E789 analysis [4,8], upgraded according to the latest theoretical developments.

## 5. Conclusions

In a sample of ≈1.35 million dilepton triggered events, acquired in a physics run during the HERA - B commissioning period in summer 2000, $8.6^{+3.9}_{-3.2}$ $b \to J/\psi \to e^+e^-$ and $1.9^{+2.2}_{-1.5}$ $b \to J/\psi \to \mu^+\mu^-$ events have been identified, leading to the $b\bar{b}$ production cross section result given in Eq. 2. It is the third measurement [6] at fixed target energies in pA collisions; it is found to be compatible with the existing measurements [4,5] and with the most recent QCD predictions [2,3] beyond next-to-leading order, as shown in Fig. 5. In the field of $b\bar{b}$ production, this agreement is actually one of the best available up to now [11], in the peculiar environment of production near threshold, compared to the collider experiment (production at high c.m.s. energies). In future running periods, HERA - B will produce $O(10^3)$ higher $B$ statistics, reducing the measurement error to $\leq 15\%$ (systematic limited), hence providing the most precise $b\bar{b}$ production cross section measurement at fixed target energies.

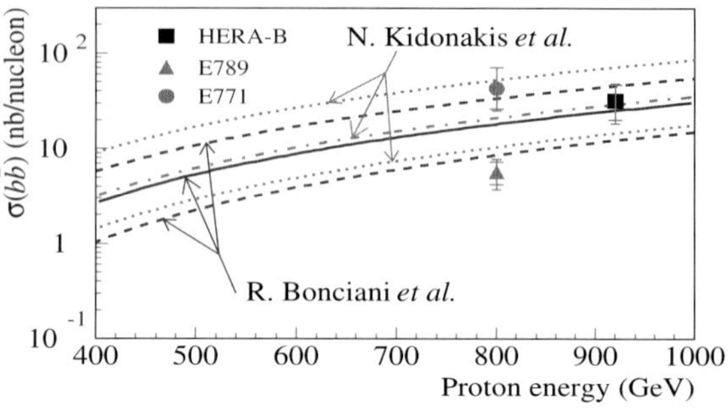


Figure 5. The comparison of the HERA - B $\sigma(b\bar{b})$ value with previous measurements and with the updated theoretical predictions of R. Bonciani *et al.* [2] and N. Kidonakis *et al.* [3].

## Acknowledgments

The author wishes to thank the organizers of the Conference for the kind invitation and the HERA - B collaborators for their help.

## REFERENCES

1. P. Nason, Proc. of the XX International Symposium on Lepton and Photon Interactions at High Energies, hep-ph/0111024; S. Frixione *et al.*, Adv. Ser. Direct. High Energy Phys. **15** (1998), 609.
2. R. Bonciani *et al.*, Nucl. Phys. **B529** (1998) 424.
3. N. Kidonakis *et al.*, Phys. Rev. **D64** (2001) 114001.
4. D. M. Jansen *et al.*, Phys. Rev. Lett. **74** (1995) 3118.
5. T. Alexopoulos *et al.*, Phys. Rev. Lett. **82** (1999) 41.
6. I. Abt *et al.*, hep-ph/0205106, in Press on Eur. Phys. J. **C**.
7. D. E. Groom *et al.*, Eur. Phys. J. **C15** (2000) 1.
8. M. H. Schub *et al.*, Phys. Rev. **D52** (1995) 1307.
9. T. Alexopoulos *et al.*, Phys. Rev. **D55** (1997) 3927.
10. E. Hartouni *et al.*, HERA - B Design Report, DESY-PRC-95-01 (1995).
11. G. P. Salam, hep-ph/0207147.

Proceedings of ICHEP 2002, pp. 345–348
S. Bentvelsen, P. de Jong, J. Koch and E. Laenen (Editors)

31st INTERNATIONAL CONFERENCE ON HIGH ENERGY PHYSICS AMSTERDAM

# Observation of double $c\bar{c}$ production in $e^+e^-$ annihilation

H. Palka[a]

[a]H. Niewodniczanski Institute of Nuclear Physics,
ul. Radzikowskiego 152, 31-342 Krakow, Poland

The production of prompt charmonia from the $e^+e^- \to q\bar{q}$ continuum is studied with the Belle detector at the KEKB asymmetric energy collider. The $J/\psi$ and $\psi(2S)$ cross-sections and momentum spectra are measured. The $J/\psi$ production angle distribution and polarization are determined. The system recoiling against the $J/\psi$ is studied. We find that a surprisingly high fraction ($\sim 60\%$) of prompt $J/\psi$'s originates from double $c\bar{c}$ production.

The production of prompt heavy quarkonia provides an interesting environment to study the interplay between perturbative QCD and non-perturbative effects. The process is poorly understood and already the measurements at the Tevatron have provided many surprises [1]. An effective field theory, called non-relativistic QCD (NRQCD) [2], provides a calculational framework for direct heavy quarkonium production. In NRQCD, a $c\bar{c}$ pair is produced at short distances in a colour-singlet or a colour-octet state, and then evolves nonperturbatively into an observed charmonium state. Further experimental information is needed to establish the applicability of NRQCD to charmonium production.

At $\sqrt{s} \approx 10.6$ GeV the following processes are expected [3–7] to contribute to $J/\psi$ production:

$$e^+e^- \to \gamma^* \to J/\psi^{(1)}gg \tag{1}$$

$$e^+e^- \to \gamma^* \to J/\psi^{(8)}g \tag{2}$$

$$e^+e^- \to \gamma^* \to J/\psi^{(1)}c\bar{c} \tag{3}$$

$$e^+e^- \to \gamma^* \to J/\psi^{(8)}q\bar{q} \tag{4}$$

The superscripts (1,8) in the $J/\psi$ symbol denote the colour state of the parent $c\bar{c}$ pair. According to LO calculations of Refs [4–7] the process (1) dominates the $J/\psi$ cross-section in the whole momentum range, the colour-octet process (2) is dominant at the endpoint and processes (3-4) give small contributions (O(10%) or smaller). The production angle distribution of the $J/\psi$ has been discussed in [3,4] and its polarization in [6].

It is interesting to note that in the Lund model of hadronization, the process (3) would be the default $J/\psi$ production mechanism, since the other three processes involve non-trivial colour rearrangements [8]. The colour reconnection effects, if they exist, may lead to an enhacement of processes (1), (2) and (4).

We report measurements of prompt charmonium production in data recorded in the Belle experiment at KEKB [9] at the $\Upsilon(4S)$ and in the continuum 60 MeV below the resonance. Our first analysis [10], which focused on inclusive prompt charmonium production, was based on a data set corresponding to integrated luminosities of 29.4 fb$^{-1}$ at the resonance and 3.0 fb$^{-1}$ in the continuum. The study of the recoil system to the $J/\psi$ [11] was performed for a larger data corresponding to 41.8 fb$^{-1}$ and 4.4 fb$^{-1}$ respectively.

The Belle detector is a large-solid-angle spectrometer with excellent particle tracking and identification capabilities [12].

We use hadronic events separated from QED, $\tau\tau$, two-photon and beam-gas interaction backgrounds by the selection criteria described in Ref. [10]. Candidate $J/\psi$ mesons are reconstructed using the leptonic decays $J/\psi \to \mu^+\mu^-$ and $e^+e^-$. We reconstruct $\psi(2S) \to J/\psi\pi^+\pi^-$ decays by combining $J/\psi$ candidates with $\pi^+\pi^-$ pairs and reconstruct $\chi_{c1}$ and $\chi_{c2}$ via their radiative decays to the $J/\psi$.

$J/\psi$ mesons from $B\overline{B}$ events are removed by requiring the momentum of the meson in the $e^+e^-$


doi:10.1016/S0920-5632(02)02198-9

center of mass (CM) system, $p^*$, to be greater than $2.0\,\mathrm{GeV}/c$. QED processes are suppressed by requiring the total charged multiplicity ($N_{\mathrm{ch}}$) to be greater than four. The process $e^+e^- \to \psi(2S)\gamma$ is partially removed by rejecting events with a photon of CM energy $E^* > 3.5\,\mathrm{GeV}$.

We combine off-resonance data with the $\Upsilon(4S)$ data for $p^* > 2$ GeV/$c$ to extract charmonium production cross-sections and momentum distributions. The full set of the measured partial cross-sections ($p^* > 2$ GeV/$c$) can be found in [10]. The total prompt $J/\psi$ cross-section in a full momentum range is $\sigma(e^+e^- \to J/\psi X) = 1.47 \pm 0.10 \pm 0.13$ pb. Our results on the $J/\psi$ total and partial cross-section are smaller than those of BaBar [13]. NRQCD predictions for $\sigma(e^+e^- \to J/\psi_{\mathrm{direct}}\, X)$ range from 0.8 to 1.7 pb [5–7]. The most recent analysis [7] obtains a total cross-section of $0.8 - 1.1$ pb, attributing 0.3 pb to the colour-singlet and $0.5 - 0.8$ pb to the colour-octet mechanism.

Momentum distributions are shown in Fig. 1. The $J/\psi$ distribution vanishes $\sim 300$ MeV/$c$ below the kinematical limit (4.84 GeV/$c$) while the $\psi(2S)$ momentum distribution extends to the end-point (4.65 GeV/$c$). The $J/\psi$ momentum distribution agrees qualitatively with the predicted shape of the color-singlet $J/\psi\, c\bar{c}$ component [4,6]. The predicted color-singlet $J/\psi\, gg$ and color-octet $J/\psi\, g$ momentum distributions [4,6] extend to the highest momenta where we do not observe the signal. Some NRQCD calculations also predict a dramatic rise in the cross-section at the end-point due to color-octet $e^+e^- \to J/\psi\, g$ [3,4]. No such signal is observed.

The distributions of the prompt $J/\psi$ CM production angle $\theta^*$, and the helicity angle $\theta_H$ have also been studied in different momentum intervals. The distributions are fitted with the parameterizations $1 + A\cos^2\theta^*$ and $1 + \alpha\cos^2\theta_H$; the results are summarized in Table 1. No statistically significant $p^*$ dependence is seen for either $A$ or $\alpha$ parameters. A large positive $A$ at all momenta for direct $J/\psi$ production is expected only for the color-singlet $J/\psi\, c\bar{c}$ mechanism [4]. However its contribution to the cross-section is thought to be small ($\sim 10\%$) [4,6]. The leading color octet process $J/\psi\, g$ is expected to yield $A \approx +1$ at the endpoint [3,4], although as noted above we do not observe these events in the $p^*$ distribution. BaBar finds $A = 0.05 \pm 0.22$ for $p^* < 3.5$ GeV/$c$ and $A = 1.5 \pm 0.6$ for $p^* > 3.5$ GeV/$c$ [13] and they interprete these measurements as evidence for colour-octet mechanism.

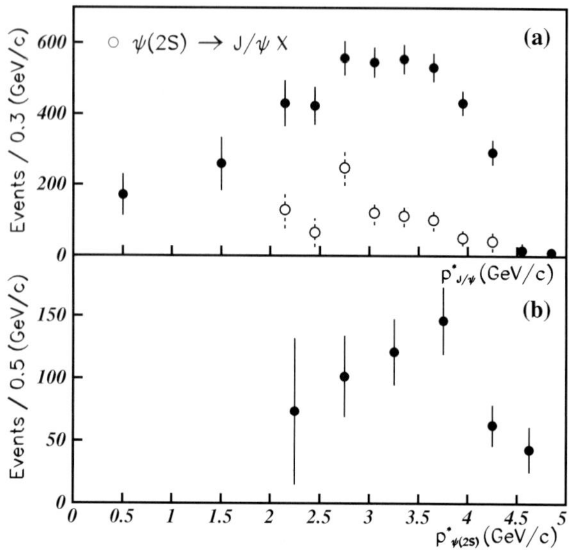


Figure 1. CM momentum distributions of prompt charmonia, corrected for efficiency : (a) $J/\psi$ (filled points) and $J/\psi$ mesons from $\psi(2S) \to J/\psi\, X$ (open points); (b) $\psi(2S)$.

The measured value of $\alpha \sim -0.4$ indicates partial longitudinal polarization of prompt $J/\psi$'s. A significant longitudinal polarization of direct $J/\psi$ mesons ($\alpha < -0.4$) is expected only for the color-singlet process $J/\psi\, gg$ [6].

In a subsequent analysis [11] we have studied the mass spectrum of a system recoiling against the $J/\psi$. The recoil mass is defined as $M_{\mathrm{recoil}} = \sqrt{\left(\sqrt{s} - E^*_{J/\psi}\right)^2 - p^{*\,2}_{J/\psi}}$. Distributions of the recoil mass are shown in Fig. 2$a$, for both $J/\psi$ signal and sidebands (background) regions. A clear threshold near $2\,m_c$ can be seen.

To avoid model dependence, we do not perform an acceptance correction. As examples, the Monte Carlo efficiencies for both $e^+e^- \to J/\psi\, q\bar{q}$

Table 1
Results of the fits to $J/\psi$ angular distributions.

| $p^*_{J/\psi}$ [GeV/$c$] | $A$ | $\alpha$ |
|---|---|---|
| $2.0-2.6$ | $0.3^{+0.5}_{-0.4}$ | $-0.4 \pm 0.2$ |
| $2.6-3.4$ | $1.1^{+0.4}_{-0.3}$ | $-0.4 \pm 0.1$ |
| $3.4-4.9$ | $1.1^{+0.4}_{-0.3}$ | $-0.2 \pm 0.2$ |
| $2.0-3.4$ | $0.7 \pm 0.3$ | $-0.5 \pm 0.1$ |
| $2.0-4.9$ | $0.9 \pm 0.2$ | $-0.4 \pm 0.1$ |

($q = u, d, s$) and $J/\psi\, c\bar{c}$ events are shown in the inset: the variation with $M_{\rm recoil}$ is smooth. For recoil masses smaller than $2.8\,{\rm GeV}/c^2$, no significant $J/\psi$ signal is observed.

In order to search for resonance structures in the region $2m_c \leq M_{\rm recoil} < 2m_D$, we apply a mass constrained fit to the $J/\psi$ candidates. This improves the $M_{\rm recoil}$ resolution by a factor close to two. The resulting recoil mass spectrum in the data is presented in Fig. 2*b*: a clear peak is observed around $3\,{\rm GeV}/c^2$. Since $e^+e^- \to \gamma^* \to J/\psi\, J/\psi$ is forbidden by charge conjugation symmetry, we interpret this peak as $e^+e^- \to J/\psi\, \eta_c$. Additional peaks at recoil masses consistent with the $\chi_{c0}$ [14] and $\eta_c(2S)$ [15] mass are also seen. A fit to the recoil mass spectrum finds an $\eta_c$ yield $N_{\eta_c} = 67^{+13}_{-12}$(sigificance $6.7\sigma$) at a mass $M = (2.962 \pm 0.013)\,{\rm GeV}/c^2$, with $\chi_{c0}, \eta_c(2S)$ yields $N_{\chi_{c0}} = 39^{+14}_{-13}(3.3\sigma)$, $N_{\eta_c(2S)} = 42^{+15}_{-13}(3.4\sigma)$ at masses $M_{\chi_{c0}} = (3.403 \pm 0.014)\,{\rm GeV}/c^2$ and $M_{\eta_c(2S)} = (3.622 \pm 0.012)\,{\rm GeV}/c^2$ respectively.

To study the $J/\psi\, c\bar{c}$ mechanism in the region $M_{\rm recoil} \geq 2m_D$, we search for fully reconstructed $D^{*+}$ and $D^0$ decays in events with a $J/\psi$ meson. We reconstruct $D^{*+} \to D^0\pi^+$ using five $D^0$ decay modes: $K^-\pi^+$, $K^-K^+$, $K^-\pi^-\pi^+\pi^+$, $K_S\pi^+\pi^-$ and $K^-\pi^+\pi^0$. To improve the $M_{D^0\pi^+}$ resolution $D^0$ candidates are refitted to the nominal $D^0$ mass. For the $J/\psi\, D^0$ study we use only the cleanest $D^0$ decay modes $D^0 \to K^-\pi^+$ and $K^-K^+$.

To remove the remaining $B\overline{B}$ background due to semileptonic $B$ decays which contribute to the background under the $J/\psi$ peak and lead to a large $D^{*+}(D^0)$ signal, we require that either the $D$ or one of the leptons from the $J/\psi$ candidate have a momentum above the kinematic limit for

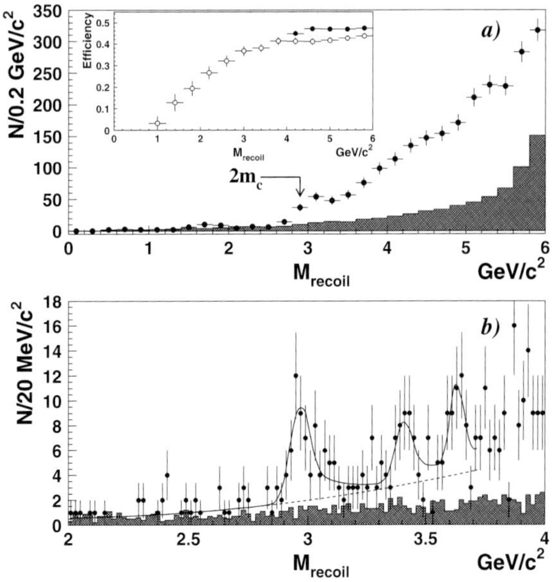


Figure 2. a) The $M_{\rm recoil}$ distribution for the $J/\psi$ signal region (points) and scaled sidebands (shaded histogram). The inset: efficiency for $e^+e^- \to J/\psi\, q\bar{q}$ (open circles: $q = u, d, s$; closed circles: $q = c$). b) The $M_{\rm recoil}$ distribution after refitting the $J/\psi$ with a mass constraint. The curve represents the fit described in the text.

$B$ decays: $p^*_D > 2.6\,{\rm GeV}/c$ or $p^*_{\ell^\pm} > 2.6\,{\rm GeV}/c$.

Scatter plots of the dilepton mass versus the charmed meson mass together with the $M_{D^0\pi^+}$ and $M_{K^-\pi^+}$ projections are shown in Figs. 3*a*-*d*. The projections are fit with a Gaussian for the $D$ meson signal and a polynomial for the background. The $J/\psi$ sideband is fit simultaneously, and used to estimate the combinatorial $(\ell^+\ell^-)\,D$ contribution to the $D$ meson yield in the signal region. We find significant yields: $N_{D^{*+}} = 10.1^{+3.6}_{-3.0}$ (significance $5.3\sigma$) and $N_{D^0} = 14.9^{+5.4}_{-4.8}$ $(3.7\sigma)$. The overall efficiencies for $J/\psi\, D^{*+}$ and $J/\psi\, D^0$ are calculated to be $\epsilon_{J/\psi\, D^{*+}} = (4.1 \pm 1.0) \cdot 10^{-4}$ and $\epsilon_{J/\psi\, D^0} = (3.7 \pm 0.8) \cdot 10^{-4}$ respectively. Using these values, we find cross-sections $\sigma(e^+e^- \to J/\psi\, D^{*+}X) = (0.53\,^{+0.19}_{-0.15} \pm 0.14)$ pb and $\sigma(e^+e^- \to J/\psi\, D^0X) =$

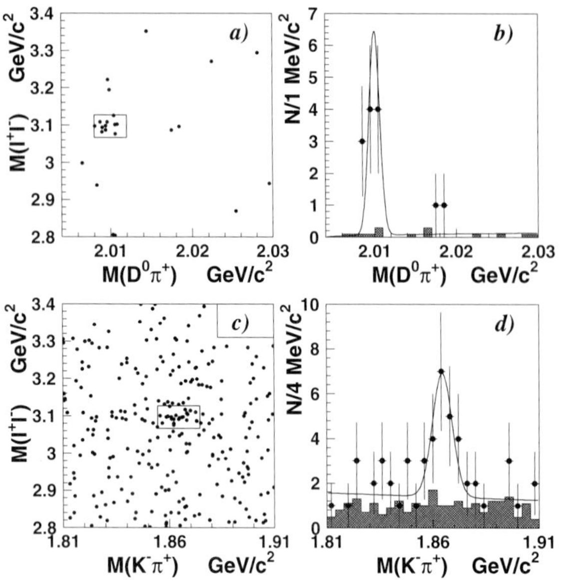


Figure 3. Associated production of $J/\psi$ and charm mesons, a) $M(l^+l^-)$ vs $M(D^0\pi^+)$; b) projection onto the $M(D^0\pi^+)$ axis; c) $M(l^+l^-)$ vs $M(K^-\pi^+(K^+K^-))$; d) projection onto the $M(K^-\pi^+(K^+K^-))$ axis. Points with error bars show the $J/\psi$ signal region and the shaded histograms show the scaled sidebands. The curves represent the fit described in the text.

$(0.87\,^{+0.32}_{-0.28} \pm 0.20)$ pb.

According to the Lund model, $c\bar{c}$ fragmentation produces charmed mesons at the rate of 0.53 per event for $D^{*+}$, and 1.18 per event for $D^0$, where both numbers include feed-down from higher states (in particular, $D^{*+} \to D^0\pi^+$) [16]. Assuming that these rates apply to $c\bar{c}$ fragmentation in $e^+e^- \to J/\psi\, c\bar{c}$, we calculate $\sigma(e^+e^- \to J/\psi\, c\bar{c})$ and find $(1.01\,^{+0.36}_{-0.30} \pm 0.26)$ pb and $(0.74\,^{+0.28}_{-0.24} \pm 0.19)$ pb based on our $D^{*+}$ and $D^0$ measurements, respectively. We average the results and obtain $(0.87\,^{+0.21}_{-0.19} \pm 0.17)$ pb. Using our measurement of the inclusive prompt $J/\psi$ cross-section we infer that a large fraction of prompt $J/\psi$ events: $\sigma(e^+e^- \to J/\psi\, c\bar{c})/\sigma(e^+e^- \to J/\psi\, X) = 0.59\,^{+0.15}_{-0.13} \pm 0.12$, is due to the $e^+e^- \to J/\psi\, c\bar{c}$ process.

In summary, our results favour $e^+e^- \to J/\psi\, c\bar{c}$ as the leading mechanism for prompt $J/\psi$ production at $\sqrt{s} \approx 10.6$ GeV.

## REFERENCES


1. For a recent review see, M. Kramer, Prog. Part. Nucl. Phys. **47**, 141 (2001); and references therein.
2. G.T. Bodwin, E. Braaten and G.P. Lepage, Phys. Rev. D **51**, 1125 (1995).
3. E. Braaten and Yu-Qi Chen, Phys. Rev. Lett. **76**, 730 (1996).
4. P. Cho and A.K. Leibovich, Phys. Rev. D **54**, 6690 (1996).
5. F. Yuan, C.-F. Qiao and K.-T. Chao, Phys. Rev. D **56**, 321 (1997); **56**, 1663 (1997).
6. S. Baek, P. Ko, J. Lee, and H.S. Song, J. Kor. Phys. Soc. **33**, 97 (1998); hep-ph/9804455.
7. G.A. Schuler, Eur. Phys. J. C **8**, 273 (1999).
8. T. Sjöstrand, private communication.
9. KEKB B Factory Design Report, KEK Report 95-7, 1995, (unpublished); Y.Funakoshi *et al.* Proc. 2000 European Accelerator Conference, Vienna 2000.
10. K. Abe *et al.* (Belle Collaboration), Phys. Rev. Lett. **88**, 052001 (2002).
11. K. Abe *et al.* (Belle Collaboration), Phys. Rev. Lett. **89**, 142001 (2002).
12. A. Abashian *et al.* (Belle Collaboration), Nucl. Instr. and Meth. **A479**, 117 (2002)
13. B. Aubert *et al.* (BABAR Collaboration), Phys. Rev. Lett. **87**, 162002 (2001).
14. D.E. Groom *et al.*, Eur. Phys. J. **C15**, 1 (2000).
15. S.-K. Choi, S.O. Olsen *et al.* (Belle Collaboration), Phys. Rev. Lett. **89**, 102001 (2002).
16. T. Sjöstrand, PYTHIA 5.7 / JETSET 7.4, Comp. Phys. Commun. **82**, 74 (1994). The $D^{*+}/D^0$ ratio is constant over $\sqrt{s}$ except for a decrease near $D^{*+}$ mass threshold: the effect on the result is negligible.

ELSEVIER

Proceedings of ICHEP 2002, pp. 349–351
*S. Bentvelsen, P. de Jong, J. Koch and E. Laenen (Editors)*

31st INTERNATIONAL CONFERENCE ON HIGH ENERGY PHYSICS AMSTERDAM

# Isolated Prompt Photon Production in Photon-Photon Collisions at $\sqrt{s}_{ee} = 183$–$209$ GeV

T. Schörner-Sadenius[a] on behalf of the OPAL collaboration.

[a]CERN, Divison EP, CH-1211 Geneva 23

An OPAL measurement of isolated prompt photon production in photon-photon collisions at center-of-mass energies between 183 and 209 GeV is presented. The total cross-section for isolated prompt photons with transverse momenta $p_T$ of more than 3 GeV and with pseudorapidities $|\eta| < 1$ is measured to be $\sigma = 0.26 \pm 0.04(\text{stat.}) \pm 0.03(\text{syst.})$ pb. Differential cross-sections as functions of $p_T$ and $|\eta|$ are compared to the predictions of an NLO QCD calculation and of the PYTHIA Monte Carlo program.

## 1. INTRODUCTION

Processes involving prompt photons in photon-photon collisions show features which make them ideal objects for detailed studies: First, at least in leading order (LO), prompt photons are restricted to resolved processes, i.e. to processes in which at least one of the incoming photons has fluctuated into hadrons, see figure 1. Therefore, prompt photons can be used to investigate the hadronic substructure of the photon. Second, QCD predictions for prompt photon production usually do not suffer from large next-to-leading order (NLO) corrections or from large scale dependencies. This makes prompt photons good candidates for tests of QCD predictions. In addition, the event signatures of prompt photons are usually less subject to hadronisation corrections than are dijet or hadronic signatures which are also used to extract the photonic substructure. However, the cross-section for prompt photon production is typically smaller by a factor $1/\alpha_{EM}$ than the dijet cross-section.

The data used in this analysis were collected with the OPAL detector at LEP in the years 1997–2000 at center-of mass energies $\sqrt{s}_{ee}$ between 183 and 209 GeV. Since over this energy range the variation of the cross-section for prompt photon production is expected to be smaller than the statistical error of the data, all energies are analysed together. The data correspond to an integrated luminosity of about 639 $\text{pb}^{-1}$.

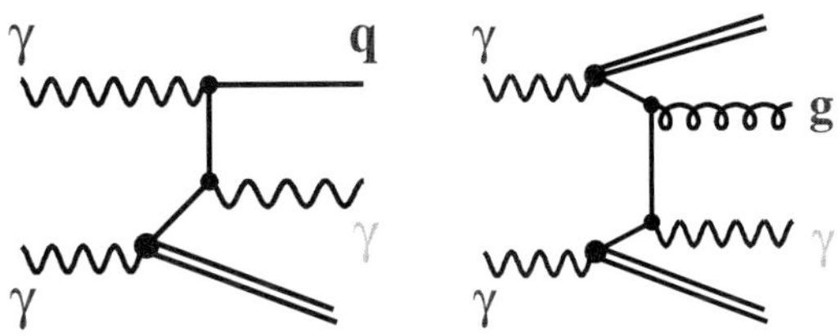


Figure 1. Feynman diagrams for single-resolved (left) and double-resolved (right) prompt photon production.

The OPAL detector stopped data taking in 2000. Its most important component for this analysis was an electromagnetic lead-glass calorimeter with about 23 radiation lengths over an acceptance of $|\cos\theta| < 0.98$ ($\theta$ is the polar angle measured from the beam lines).

## 2. MEASUREMENT

The event selection aims at keeping 'untagged' photon-photon collisions events, i.e. events where none of the beam leptons deposits a significant amount of energy in the forward detectors. The cuts applied correspond to a maximum allowed squared momentum transfer $Q^2$ of 4.5 $\text{GeV}^2$. Background from $e^+e^-$ annihilation events is mainly rejected by demanding the total visible


doi:10.1016/S0920-5632(02)02199-0

invariant mass $W_{vis}$ to be less then 30% of the center-of-mass energy $\sqrt{s}$ and by requiring that not more then 25% of $\sqrt{s}$ may be deposited in the calorimeters. Once untagged photon-photon collision events are selected, isolated photon candidates are searched for by looking for clusters of lead-glass blocks within the central rapidity region $|\eta| < 1$ with a transverse momentum $p_T$ of at least 3 GeV.

In order to reject photons from particle decays or fragmentation processes, the isolation criterion suggested by Frixione [1] was applied which has the advantage of retaining infrared safety. According to this criterion, particles are allowed less and less energy the closer they are to the prompt photon candidate:

$$\sum_i E_{T,i} \cdot \Theta(\delta - R_{i,\gamma}) < 0.2 \cdot E_{T,\gamma} \frac{1 - \cos\delta}{1 - \cos R}. \quad (1)$$

This criterion has to be fulfilled for all $\delta < R$ ($R = 1$). $R_{i,\gamma}$ is the distance between particle $i$ and the photon candidate in $\eta - \phi$ space; $E_{T,i}$ and $E_{T,\gamma}$ are the transverse energies of particle $i$ and the photon candidate, respectively. $\Theta$ is the familiar step function. Events with more than one candidate photon are rejected.

Background events from either real photons (for example from particle decays) or from particles faking a photon signature (for example antineutron annihilations) contaminate the data sample. However, the fraction of background events can be determined using an analysis of the shape of the clusters in the electromagnetic calorimeter. For this purpose, two variables $f$ and $\sigma$ describing the shape of the cluster in terms of the energy distribution and its spatial spread are defined, and the two-dimensional distribution of these variables as found in the data is fitted with a superposition of signal and background Monte Carlo samples. The only free parameters in this fit are taken to be the fraction of real isolated prompt photons $a$ and the fraction $b$ of photons coming from the decays of neutral pions. Other background source fractions are fixed to their Monte Carlo values. The result of the fit is $a = 0.85 \pm 0.08$(stat.) and $b = 0.11 \pm 0.08$(stat.).

In order to derive the total prompt photon cross-section, the ratio $r_{s/d}$ of events with one

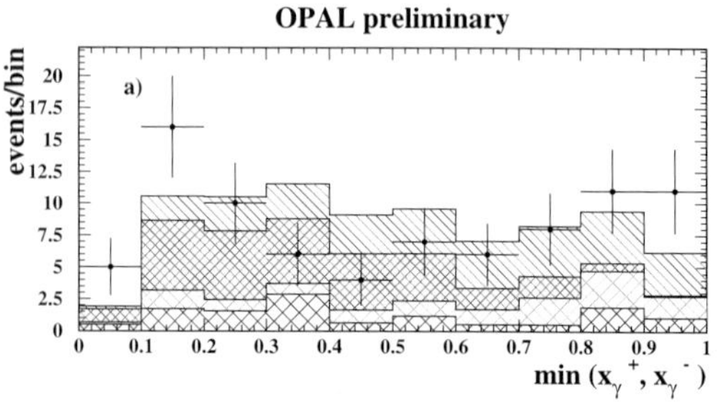


Figure 2. The distribution of the observable $min(x_\gamma^+, x_\gamma^-)$. The data are shown as points; the different histograms show the contributions from signal and background Monte Carlos.

resolved incoming photon to events with two resolved incoming photons has to be determined in order to correct for possibly different selection efficiencies for these two event classes. This is again done with a fit of the relevant signal and background Monte Carlo distributions to the data for two variables, $min(x_\gamma^+, x_\gamma^-)$ and $x_T$. The only free parameter in this fit is the above mentioned ratio $r_{s/d}$ governing the relative contributions of single- and double-resolved events. $x_T$ is defined as $x_T \equiv 2 \cdot p_T / W_{vis}$. The variables $x_\gamma^\pm$ are defined only for a subset of the data in which a cone jet finding algorithm ($R = 1$, $E_{T,min} = 3$ GeV) reconstructed one jet:

$$x_\gamma^\pm \equiv \frac{(E_\gamma \pm p_{z,\gamma}) + (E_{jet} \pm p_{z,jet})}{\sum_{hadrons,\gamma}(E \pm p_z)}. \quad (2)$$

They specify the fraction of the incoming photons' momenta that entered the hard scattering. Since (in LO) only resolved processes contribute to the prompt photon cross-section, the minimum of the $x_\gamma^\pm$ is expected to be distributed between 0 and 1, see figure 2. The two variables $min(x_\gamma^+, x_\gamma^-)$ and $x_T$ give different values for $r_{s/d}$ ($0.59 \pm 0.12$ and $0.41 \pm 0.08$, respectively), which are averaged.

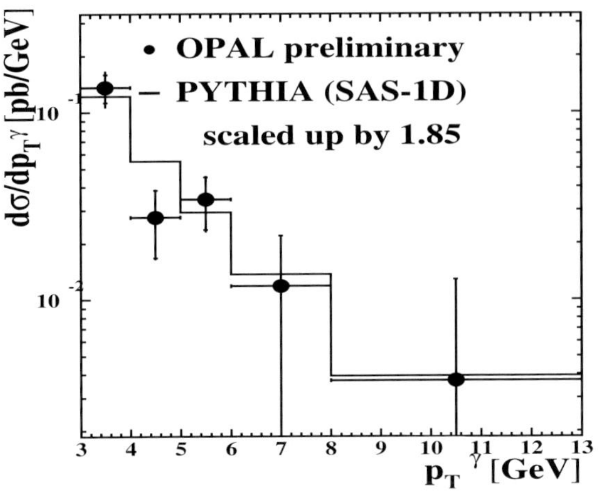


Figure 3. Differential isolated prompt photon cross-section as function of $p_T$.

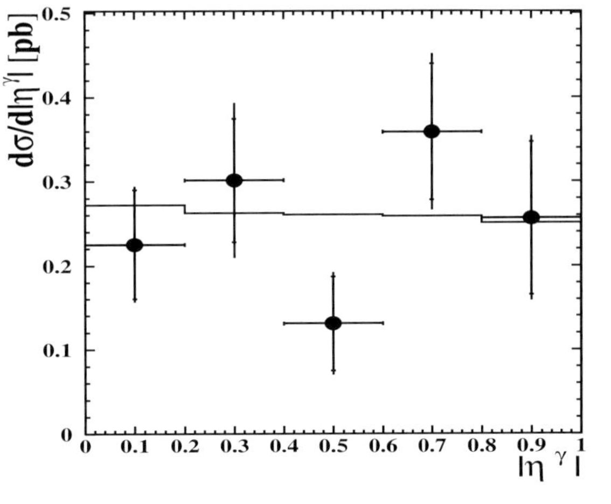


Figure 4. Differential isolated prompt photon cross-section as function of $|\eta|$.

## 3. RESULTS

The total cross-section $\sigma$ for isolated prompt photons can be derived according to

$$\sigma = \frac{r_{s/d} \cdot N}{\epsilon_s \cdot \mathcal{L}} + \frac{(1 - r_{s/d}) \cdot N}{\epsilon_d \cdot \mathcal{L}}, \qquad (3)$$

where $\mathcal{L}$ is the integrated luminosity, $\epsilon_{s/d}$ are the single- and double-resolved selection efficiencies (0.51 and 0.61, respectively) as estimated using Monte Carlo, and $N = 92.8$ is the number of prompt photon candidates after background subtraction. The number of background events to be subtracted was taken from the fit procedure described above and from Monte Carlo in the case of final state radiation (FSR). Inserting all numbers leads to a value of $\sigma = 0.26 \pm 0.04(\text{stat.}) \pm 0.03(\text{syst.})$ pb. The largest single systematic error of about 7% stems from the 3% uncertainty in the electromagnetic calorimeter energy scale.

Distributions differential in $|\eta|$ and $p_T$ have been studied and compared to predictions of an NLO QCD calculation by Fontannaz *et al.* [2] and of the PYTHIA Monte Carlo program [3] using the SaS-1d photon parton density distributions. PYTHIA can describe the shape of the data, but has to be scaled by a factor of 1.85 in order to match the normalisation of the data (figs. 3,4). The NLO calculation can describe the data both in shape and in normalisation.

## 4. SUMMARY

Isolated prompt photon production in photon-photon collisions has been studied using data collected with the OPAL detector at LEP at center-of-mass energies between 183 and 209 GeV. The total cross-section was measured to be $\sigma = 0.26 \pm 0.04(\text{stat.}) \pm 0.03(\text{syst.})$ pb. The differential cross-sections in $p_T$ and $|\eta|$ can be described in shape and normalisation by an NLO QCD calculation; the PYTHIA Monte Carlo program fails to reproduce the correct normalisation.

## REFERENCES


1. S. Frixione, Phys. Lett. **B429** (1998) 369.
2. M. Fontannaz *et al.*, Eur. Phys. J. **C23** (2002) 503.
3. T. Sjöstrand, Comp. Phys. Comm. **82** (1994) 74.

ELSEVIER

Proceedings of ICHEP 2002, pp. 352–354
*S. Bentvelsen, P. de Jong, J. Koch and E. Laenen (Editors)*

31st INTERNATIONAL CONFERENCE ON
HIGH ENERGY PHYSICS AMSTERDAM

# Parton dynamics and identified particles at HERA

H. Jung
Department of Physics, Lund University, Sweden
for the H1 and ZEUS Collaborations

Multi-particle production in the central and forward (proton) rapidity region is studied in photoproduction and deep inelastic scattering (DIS). The measurements in the forward region are a factor of $\sim 2$ larger than QCD predictions based on the DGLAP evolution equations. The central rapidity region is rather well described. Searches for instanton induced processes in DIS are presented.

## 1. Introduction

The production and emission of partons in *ep* interactions at high energies is extremely difficult to describe when there are more than 3 partons in the final state. In particular at high energies (small $x$) and in the rapidity region close to the outgoing proton the effects of new parton dynamics (BFKL [1]) beyond the standard DGLAP approach [2] are expected to become important. The measurements of multi-parton final states can be compared to hadron level Monte Carlo programs like PYTHIA [3], HERWIG [4] and RAPGAP [5] which implement the DGLAP approach. The Lund Color Dipole Model, as implemented in ARIADNE [6], reproduces some of the features of the BFKL [1] evolution equation. The CCFM evolution equation [7], as implemented in CASCADE [8], is based on the concept of color coherence,and reproduces the BFKL and DGLAP evolution equation in their asymptotic limits.

One of the long standing questions in high energy collisions is whether significant deviations from the successful DGLAP evolution equation can be observed at HERA energies.

To answer this question, we first study particle production in the central region of rapidity (in the laboratory frame) in photoproduction, where the DGLAP evolution equation is believed to be valid. Then we study jet and $\pi^0$ production in a region of phase space, in which the effects of evolution according to the BFKL equation are expected to be large, but contributions from DGLAP evolution are suppressed. In a further step, multi-particle production is studied, again in the central rapidity region, with the aim of searching for signatures of parton scattering in the background of an instanton field.

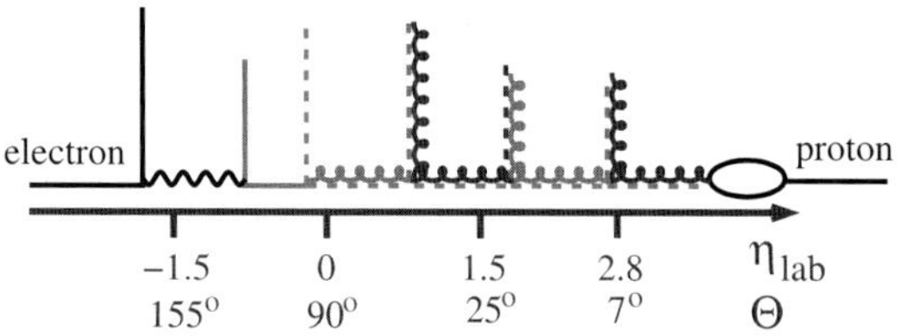


Figure 1. *Schematic diagram of an ep interaction indicating the angular and rapidity ranges in the HERA laboratory frame.*

## 2. Multi-particle and Jet production

The production of $K_s^0$ mesons in photoproduction has been studied by ZEUS [9] in events containing at least two jets with transverse energies of $E_T > 10(9)$ GeV (with $-2.4 < \eta^{jets} < 2$, cf. Fig. 1). In Fig. 2 the differential cross section $d\sigma(ep \to dijet + K_s^0 + X)/d\eta_{K_s^0}$ is shown. Both Monte Carlo Programs, PYTHIA and HERWIG, based on DGLAP, provide a good description of the data in a rapidity range not too far away from the scattered electron (central rapidities).

However, new effects are expected to become visible in a region of rapidity closer to the proton. H1 [10] has measured the cross section for a class of deep inelastic events ($5 < Q^2 < 75$ GeV$^2$) with a jet with $p_t > 3.5$ GeV close to the proton direction ($7^o < \Theta_{jet} < 20^o$) carrying a large fraction of


doi:10.1016/S0920-5632(02)02323-X

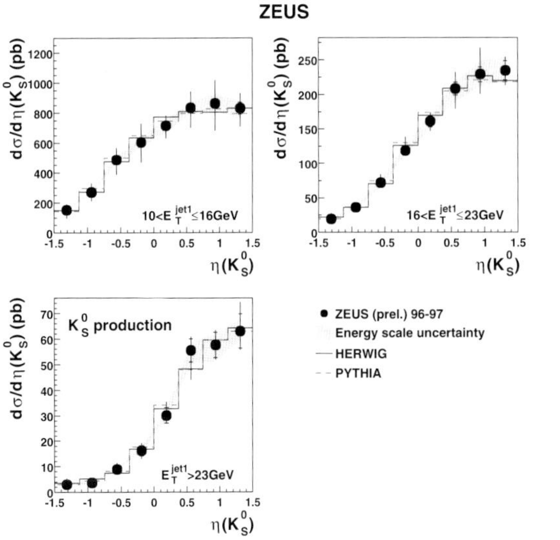


Figure 2. *The differential cross section $d\sigma/d\eta_{K^0}$ for $p_t(K^0) > 0.3$ GeV in different regions of the transverse energy of the leading jet. Also shown are Monte Carlo predictions which have been normalized to the measured cross section*

the protons momentum $x_{jet} = E_{jet}/E_p > 0.035$. In addition, $0.5 < p_{t\ jet}^2/Q^2 < 2$ is required to suppress contributions from DGLAP type evolution where the phase space for BFKL type evolution in $x$, from the large $x_{jet}$ to the small $x_{Bj}$, is enhanced. This cross section for forward jet production as a function of $x_{Bj}$ and $p_{t\ jet}$ is shown in Fig. 3 and 4. The measurements are up to a factor of two larger than the prediction based on $\mathcal{O}(\alpha_s)$ (and also $\mathcal{O}(\alpha_s^2)$) QCD calculations in the DGLAP approach. Using a hadron level Monte Carlo model incorporating resolved virtual photon processes [5] in addition to the usual direct photon processes allows a reasonable description of the data. ARIADNE [6], which simulates higher order QCD radiation without strong ordering of the transverse momenta of the emitted partons, also describes the measurements well. CASCADE [8], which is based on angular ordering coming from color coherence (CCFM), predicts a rate of forward jet events which is a little too high.

H1 [11] also measured single particles ($\pi^0$) as opposed to jets, allowing to reach angles closer to the proton direction to be investigated. The analysis is restricted to the kinematic range $2 < Q^2 <$

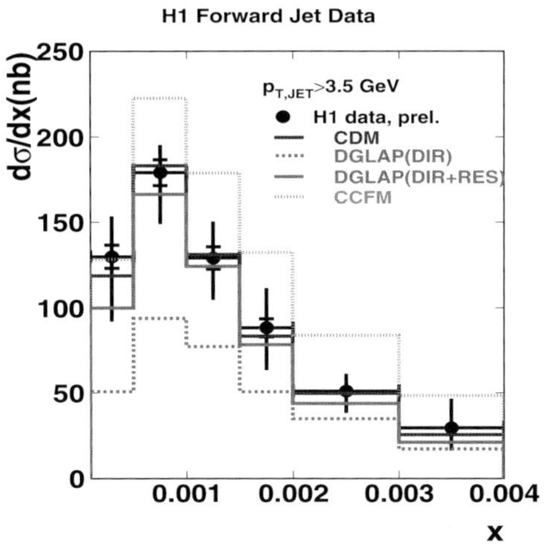


Figure 3. *The cross section for forward jet production at the hadron level, as a function of $x_{Bj}$ for $p_{t\ jet} > 3.5$ GeV. Also shown are the predictions from various Monte Carlo calculations.*

70 GeV$^2$, $5^o < \Theta_\pi < 25^o$, $x_\pi = E_\pi/E_p > 0.01$ and $p_{T,\pi}^* > 2.5$ GeV, defined in the hadronic center of mass system. The differential cross section as a function of $x_{Bj}$ for different regions in $Q^2$ is shown in Fig. 5. The prediction of DGLAP based Monte Carlos is well below the data, whereas a reasonable description is obtained in an approach with an additional resolved virtual photon contribution. CASCADE is below the data at small values of $x_{Bj}$ and, as in the case of the forward jets, above the measurement for $x_{Bj} > 0.001$.

Since the deviations of the measurements from the DGLAP predictions are related to the non-Abelian structure of QCD and the gluon splitting function at small $x$, it is interesting to search also for signatures coming from the complicated QCD - vacuum, e.g. instantons [12]. If a boson-gluon fusion process happens in DIS in the background of the instanton field, as simulated in the QCDINS program [13], many partons are produced in the central rapidity region. H1 [14] has performed an involved search for such processes using various methods to "control" the standard QCD background. Monte Carlo predictions for particle production in the signal region still differ, thus only upper limits on instanton production could be derived, consistent with calculations coming from lattice QCD.

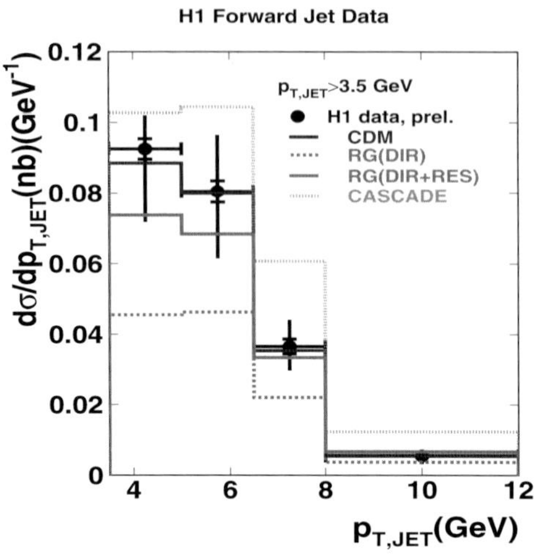


Figure 4. *The cross section for forward jet production at the hadron level, as a function of $p_{t\ jet} > 3.5$ GeV. Also shown are the predictions from various Monte Carlo calculations.*

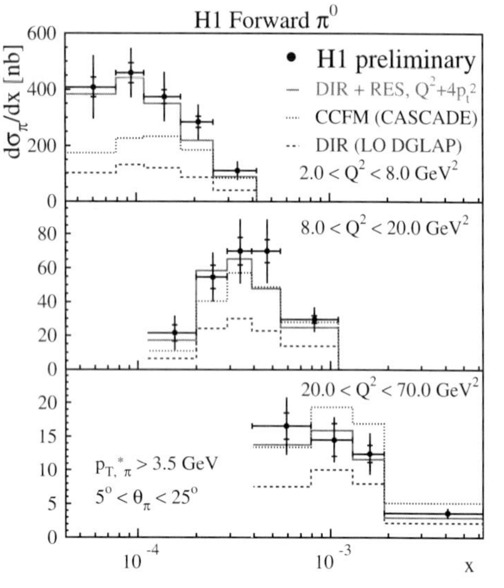


Figure 5. *Inclusive $\pi^0$ production cross section as a function of $x_{Bj}$ in 3 different regions of $Q^2$.*

## REFERENCES


1. E. Kuraev, L. Lipatov, V. Fadin, *Sov. Phys. JETP* **44** (1976) 443; E. Kuraev, L. Lipatov, V. Fadin, *Sov. Phys. JETP* **45** (1977) 199; Y. Balitskii, L. Lipatov, *Sov. J. Nucl. Phys.* **28** (1978) 822.
2. V. Gribov, L. Lipatov, *Sov. J. Nucl. Phys.* **15** (1972) 438 and 675; L. Lipatov, *Sov. J. Nucl. Phys.* **20** (1975) 94; G. Altarelli, G. Parisi, *Nucl. Phys.* **B 126** (1977) 298; Y. Dokshitser, *Sov. Phys. JETP* **46** (1977) 641.
3. T. Sjöstrand, *Comp. Phys. Comm.* **135** (2001) 238.
4. G. Marchesini et al., *Comp. Phys. Comm.* **76** (1992) 465, hep-ph/9912396.
5. H. Jung, *The RAPGAP Monte Carlo for Deep Inelastic Scattering, version 2.08*, Lund University, 2002, `www.quark.lu.se/~hannes/rapgap/`.
6. L. Lönnblad, *Comp. Phys. Comm.* **71** (1992) 15.
7. M. Ciafaloni, *Nucl. Phys.* **B 296** (1988) 49. S. Catani, F. Fiorani, G. Marchesini, *Phys. Lett.* **B 234** (1990) 339. S. Catani, F. Fiorani, G. Marchesini, *Nucl. Phys.* **B 336** (1990) 18. G. Marchesini, *Nucl. Phys.* **B 445** (1995) 49.
8. H. Jung, G. Salam, *Eur. Phys. J.* **C 19** (2001) 351, hep-ph/0012143. H. Jung, *Comp. Phys. Comm.* **143** (2002) 100, hep-ph/0109102, DESY 01-114
9. ZEUS Collaboration, M. Derrick et al., $K_s^0$ and $\Lambda$ Production in Photoproduction at HERA, 2002, contributed paper to ICHEP 2002.
10. H1 Collaboration, C. Adloff et al., Forward Jet production at HERA, 2002, contributed paper to ICHEP 2002.
11. H1 Collaboration, C. Adloff et al., Forward $\pi^0$ Production in DIS at HERA, 1999, contributed paper to ICHEP 2002.
12. A. Ringwald, F. Schrempp, (1999), hep-ph/9909338.
13. A. Ringwald, F. Schrempp, *Comput. Phys. Commun.* **132** (2000) 267, hep-ph/9911516.
14. H1 Collaboration, C. Adloff et al., (2002), hep-ex/0205078.

ELSEVIER

Proceedings of ICHEP 2002, pp. 355–357
*S. Bentvelsen, P. de Jong, J. Koch and E. Laenen (Editors)*

31st INTERNATIONAL CONFERENCE ON HIGH ENERGY PHYSICS AMSTERDAM

# Measurements of $\alpha_s$ from event shapes and the four–jet rate

G. Dissertori[a]

[a]Institute for Particle Physics, ETH Zürich,
ETH Hönggerberg HPK E28, CH-8093 Zürich, Switzerland

New results from measurements of the strong coupling constant $\alpha_s(M_Z)$ at LEP are presented. In particular, a new LEP combination of results based on event–shape variables has become available, where a new method for the estimation of the theoretical uncertainty has been implemented. Furthermore, two other analyses are quoted, based on power corrections and the four-jet rate.

## 1. Introduction

The strong coupling constant $\alpha_s$ is the only free parameter of the QCD Lagrangian and many measurement have been performed at various energy scales in different processes in order to determine it precisely [1]. The measurements at LEP have contributed significantly to this effort.

Two years after the closure of LEP still new results from QCD analyses become available [2]. These results are based on data from LEP1, where the centre-of-mass energy was around 91 GeV and the data sample consisted of several million events, practically background free, as well as from LEP2, with energies from 133 GeV up to 206 GeV. Here the data samples are much smaller, of the order of several thousand events, and the backgrounds (mainly fully hadronic W-pair events) can amount up to about 15%. In addition, at LEP2 a special selection has to be applied in order to eliminate the so-called *radiative return events*, where a hard photon is radiated in the initial state, reducing the effective centre-of-mass energy to the Z mass. This leads to a significant loss in statistics. Therefore the $\alpha_s$ measurements at LEP2 have a much larger statistical uncertainty. However, systematic uncertainties related to non-perturbative corrections or unknown higher order terms are reduced because of the larger energy scale. This motivates the combination of all measurements performed by the four LEP experiments ALEPH, DELPHI, OPAL and L3, in order to obtain a precise test of the running of the coupling, as well as a competitive measurement at LEP2 compared to LEP1, and finally the best possible overall combination. Such a combination has now been performed by the LEP QCD working group (LEPQCDWG) for measurements based on event–shape variables.

## 2. Event–shape variables

Event–shape variables are a classical tool to measure the strong coupling constant. They are sensitive to the topology of an event, therefore to gluon radiation and thus to the coupling strength. The most thoroughly tested variables are thrust, heavy jet mass, C–parameter, total and wide jet broadening and the $y_3$–distribution, where $y_3$ is the resolution parameter where a three–jet event is clustered into a two–jet event (Durham clustering algorithm). For these variables the perturbative QCD predictions exist at next–to–leading order (NLO), $\mathcal{O}(\alpha_s^2)$. In addition, large logarithms $L = \ln(1/y)$ have been resummed at next–to–leading logarithmic (NLL) accuracy to all orders in $\alpha_s$, where $y$ is an event–shape variable. Only recently some missing NLL terms have been calculated for the $y_3$–distribution [3]. Several matching schemes have been proposed over the years in order to combine the NLO and NLL calculations, such as the logR and R matching schemes [4]. The LEPQCDWG adopts so–called *modified matching schemes*, where the logarithmic term is replaced by $L \to \tilde{L} = \frac{1}{p}\ln\left[(1/y)^p - (1/y_{\max})^p + 1\right]$ . This ensures a physical behaviour of the prediction at


doi:10.1016/S0920-5632(02)02204-1

the phase space boundary $y_{\max}$. The choice of the power $p$ is arbitrary, in the sense that different choices only give different subleading contributions, whereas the NLL terms remain unchanged. The modified logR matching scheme corresponds to $p = 1$, whereas $p = 2$ is called the *second degree modification*. As mentioned

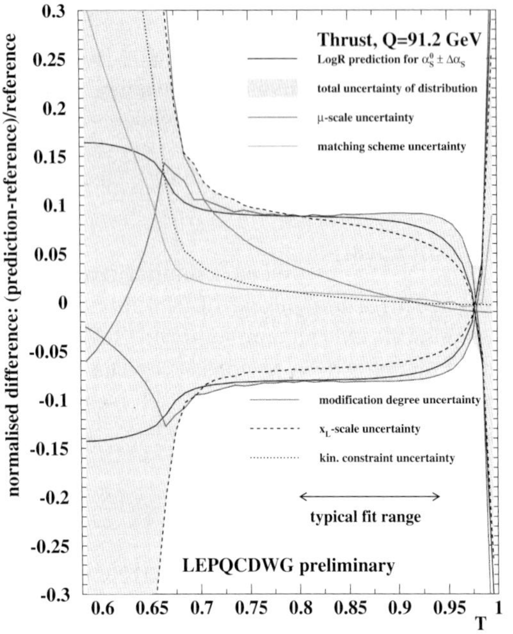


Figure 1. Uncertainty band for the thrust distribution.

earlier, the measurements at LEP1 are basically background free, whereas at LEP2 the W background is very difficult to suppress, in particular in the multi–jet region of the event–shape distribution which is also sensitive to hard gluon radiation. In order to avoid large systematic uncertainties related to the background subtraction, a careful choice of the fit range in the event–shape distribution is needed, over which the theoretical prediction is adjusted to the data. At the other end of the spectrum (the two–jet limit) the hadronization corrections become sizeable, and correspondingly the fit range has to be adapted. These corrections are obtained from phenomenological models implemented in Monte Carlo generators (JETSET, HERWIG, ARIADNE). The difference in the results obtained with different generators are quoted as systematic uncertainties due to hadronization, of the order of 1-2%.

The dominant systematic uncertainty is related to missing higher order terms in the perturbative calculations. In order to avoid a very bad fit quality when changing the renormalization scale, and in particular to reduce the multitude and arbitrariness of prescriptions used by the LEP experiments, the LEPQCDWG has proposed a new method, the so–called *uncertainty band method*. First a reference prediction is chosen, such as the modified logR matching scheme, together with a certain $\alpha_s$ value. Then various changes are applied to the prediction, keeping the same $\alpha_s$ value: The renormalization scale $\mu$ is varied in the range $0.5 < \mu/\sqrt{s} < 2$, the modified R matching scheme is applied, the second degree modification is tested, the logarithmic terms are changed to $L' = \ln(1/(x_L y))$ for $2/3 \leq x_L \leq 3/2$, which introduces different subleading terms, and the kinematic boundary $y_{\max}$ is varied over a range given by the difference between NLO and parton shower predictions. The difference of the distributions thus obtained to the reference distribution defines an uncertainty band (Fig. 1). In the next step again the reference prediction is taken, but now $\alpha_s(M_Z)$ is varied such that the changes in the prediction reproduce the spread given by the uncertainty band over the fit range. The necessary variations in $\alpha_s(M_Z)$ define the theoretical uncertainty.

The LEPQCDWG has combined all measurements based on event–shape distributions at LEP1 and LEP2 for the six variables mentioned above. Correlations between variables, energies and experiments have been taken into account. The results are $\alpha_s(M_Z) = 0.1197 \pm 0.0002_{\text{stat}} \pm 0.0008_{\text{exp}} \pm 0.0010_{\text{had}} \pm^{0.0048}_{0.0047}{}_{\text{theo}}$ for LEP1 alone, $\alpha_s(M_Z) = 0.1196 \pm 0.0005_{\text{stat}} \pm 0.0010_{\text{exp}} \pm 0.0007_{\text{had}} \pm^{0.0043}_{0.0044}{}_{\text{theo}}$ for LEP2 alone, and $\alpha_s(M_Z) = 0.1198 \pm 0.0003_{\text{stat}} \pm 0.0009_{\text{exp}} \pm 0.0008_{\text{had}} \pm 0.0046_{\text{theo}}$ for all LEP energies combined (including measurements at an effective centre-of-mass energy below $M_Z$, obtained by selecting events with hard initial state radiation). Excellent consistency between the LEP1 and LEP2 measurements is found. Taking the individual measurements at the various energy points, perfect agreement with the expected running of the strong coupling is observed (Fig. 2).

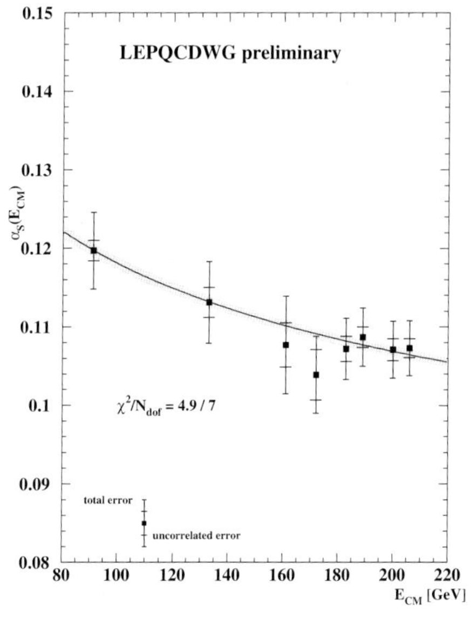


Figure 2. Combined measurements of $\alpha_s$ from event–shape variables at different LEP energies, compared to the expected running.

In an analysis by DELPHI the energy dependence of the mean values of event–shape variables has been analysed [5], but now the hadronization corrections are replaced by a power law as proposed in [6]. It results in a rather precise determination of $\alpha_s(M_Z) = 0.1184 \pm 0.0033_{\text{tot}}$. A more detailed discussion of power law studies can be found in [7].

## 3. Four–jet rate

Recently, the ALEPH collaboration has published a new determination of $\alpha_s$ [8], based on a measurement of the four-jet rate (Durham algorithm) at LEP1. The difference to the event–shape variables is that here the perturbative expansion starts at $\mathcal{O}(\alpha_s^2)$. The full NLO corrections are known, as well as the resummation of NLL terms [9]. The quoted result is extremely precise, namely $\alpha_s(M_Z) = 0.1170 \pm 0.0001_{\text{stat}} \pm 0.0009_{\text{exp}} \pm 0.0003_{\text{had}} \pm 0.0008_{\text{theo}}$. This is mainly due to the larger sensitivity, since the leading term is of $\mathcal{O}(\alpha_s^2)$. It is observed that although the $\chi^2$ of the fit changes substantially when varying the renormalization scale over a wide range, and also the location of the sharp minimum in this $\chi^2$ varies after changes in the fitted prediction, the actually fitted $\alpha_s(M_Z)$ value remains remarkably stable.

## 4. Summary

New combined results for measurements of $\alpha_s(M_Z)$ at LEP have been presented, based on event–shape distributions and a new prescription for the theoretical uncertainty. Furthermore, it has turned out that the measurement of the four-jet rate at LEP allows for a very precise determination of $\alpha_s(M_Z)$.

## 5. Acknowledgements

I would like to thank the members of the LEP-QCDWG for providing me with the relevant numbers and plots.

## REFERENCES

1. S. Bethke, *J. Phys.* **G26** (2000) R27.
2. ALEPH Collab., Note 2002-012 CONF 2002-002, ICHEP02 abs296; L3 Collab., *Phys. Lett.* **B536** (2002) 217; DELPHI Collab., Note 2002-050-CONF-584, ICHEP02 abs229; OPAL Collab., Note PN512, ICHEP02 abs368; Note PN473, ICHEP02 abs370.
3. A. Banfi *et al.*, *JHEP* **0201** (2002) 018.
4. S. Catani *et al.*, *Nucl. Phys.*, **B407** (1993) 3.
5. DELPHI Collab., Note 2002-050-CONF-584, ICHEP02 abs229.
6. Yu.L. Dokshitzer and B.R. Webber, *Phys. Lett.* **B352** (1995) 451.
7. P.A. Movilla Fernandez, these proceedings.
8. ALEPH Collab., CERN-EP-2002-029, accepted for publication in *Eur. Phys. J. C.*
9. Z. Nagy and Z. Trócsányi, *Phys. Rev. Lett.* **79** (1997) 3604; *Phys. Rev.* **D59** (1999) 014020; L. J. Dixon and A. Signer, *Phys. Rev. Lett.* **78** (1997) 811.

Proceedings of ICHEP 2002, pp. 358–360
*S. Bentvelsen, P. de Jong, J. Koch and E. Laenen (Editors)*

31st INTERNATIONAL CONFERENCE ON HIGH ENERGY PHYSICS AMSTERDAM

# Measurement of the QCD $\beta$-Function using Renormalisation Group Invariant Perturbation Theory

Klaus Hamacher[a]

[a]Fachbereich Physik, Bergische Univ. Wuppertal, Gaußstraße 20, D 42097 Wuppertal, Germany

It is shown that RGI (for Renormalisation Group Improved or Invariant) perturbation theory describes the energy dependence of 7 event shape distribution means. Consistent values of the strong coupling are obtained, $\langle\alpha_s(M_Z)\rangle = 0.121 \pm 0.002_{RMS}$, without need for Power or hadronisation corrections. The scatter of the results is much smaller than for standard Monte Carlo or Power correction methods. In particular no artificial theory error is required. The non-pert. parameters of $\overline{\mathrm{MS}}$ Power correction models can be predicted using RGI theory.

## 1. Introduction

The measurement of the energy evolution of the strong coupling, $\alpha_s$, as governed by the $\beta$-function of QCD is an experimental task of fundamental importance. It can, in principle, be performed by studying the evolution of event shapes (e.g. Thrust, $T$) with CMS energy in $e^+e^-$-annihilation. However, event shapes may be obscured by effects of the non-perturbative (NP) hadronisation vanishing as inverse powers of the energy, so-called "Power corrections" (PC). The size of the PC's depends on the borderline of pert. and NP physics. Thus it depends on the pert. calculation and the renormalisation scheme employed. PC's in general reduce if more complete calculations are used.

In RGI pert. theory UV divergent terms are resummed by demanding the observable to fulfil the Renormalisation Group Equation (RGE). This leads to vanishingly small PC's for event shape means and allows for a direct measurement of the $\beta$-function.

## 2. The Data on Event Shape Mean Values

The CMS energy dependence of 7 mean values $\langle y\rangle$ of event shape distributions as measured by DELPHI in the energy range 45–202 GeV has been studied. The data below $M_Z$ stem from events with hard $\gamma$ radiation. The explicit dependence of event shapes on hadron masses has been avoided and the data has been corrected for effects of B hadron decays. Also data of previous low energy experiments have partly been used. For a details and complete references see [1].

## 3. Comparing to Power Correction Models

The tube model is the simplest PC model. It assumes that the average transverse momentum produced in the hadronisation of a jet is constant and independent of energy. Hence the phase-space of jet fragmentation is a tube in transverse momentum – rapidity space. This implies that the angular spread of the particles due to hadronisation as well as the hadronisation correction to the event shape reduces $\propto 1/E$. Assuming independence of pert. radiation and hadronisation one obtains the following "naive" form for $\langle y\rangle$:

$$\begin{aligned}\langle y\rangle &= \langle y\rangle_{pert} + \langle y\rangle_{had}\\ &= A\alpha_s(\mu) + B(x_\mu)\alpha_s^2(\mu) + C_1/E\ .\end{aligned}$$

Here $A$ and $B$ are the LO and NLO coefficients of the $\mathcal{O}(\alpha_s^2)$ pert. expansion and $C_1$ is a NP parameter to be determined from data. The so called Dokshitzer–Webber (DW) PC model [2] is more elaborated. It parameterises the (analytical) behaviour of the strong coupling in the IR energy regime below a matching scale $\mu_I$ by its average value $\alpha_0(\mu_I) = \langle\alpha_s\rangle_{\mu<\mu_I}$. For mean values one obtains again a power term:


doi:10.1016/S0920-5632(02)02206-5

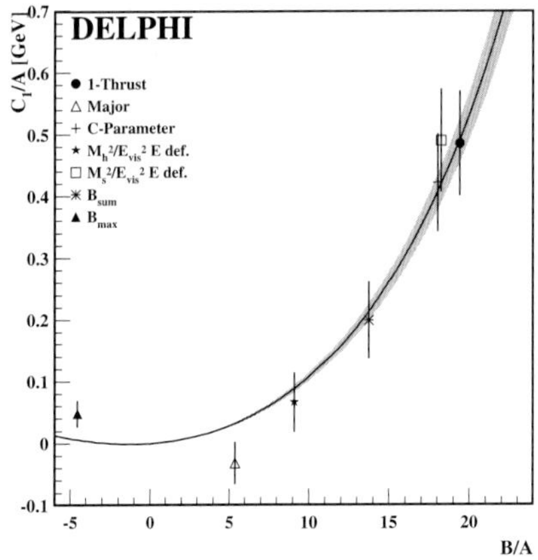


Figure 1. Correlation of the NP parameter $C_1$ with the pert. parameter $B$. The band is the RGI prediction for $\Lambda = 220\,\mathrm{MeV}$.

$$\langle y\rangle = \langle y_{pert}\rangle + c_y\mathcal{P} \qquad \text{with}$$
$$\mathcal{P} = \frac{4C_F}{\pi^2}\mathcal{M}\frac{\mu_I}{E_{\rm cm}}\cdot\left[\alpha_0 - \alpha_s - f(\frac{\mu}{\mu_I})\alpha_s^2\right].$$

$c_y$ is a parameter calculated for resummable observables only. The DW model describes the data on mean values well[1,3]. The fit results for $\alpha_s$ agree reasonably, while the $\alpha_0$ values agree within the expected uncertainty of $\sim 20\%$. Also the naive form of power corrections describes the data for all 7 observables. Here a strong correlation of the size of the NP term $C_1$ and the NLO pert. coefficient $B$ was observed, however (see Fig.1). This correlation indicates that the presumed NP terms to large extend parameterise uncalculated higher order contributions.

## 4. Comparing to RGI Pert. Theory

In RGI perturbation theory [4] the observable itself (normalised to its first order perturbative coefficient, i.e. $R = \langle y\rangle/A$) is used as expansion parameter. This choice avoids any dependence on a renormalisation scheme. The energy dependence of observables depending on a single energy scale can then be predicted by demanding the observables to fulfil the RGE:

$$Q\frac{dR}{dQ} = -\frac{\beta_0}{2}R^2(1+\rho_1 R+\rho_2 R^2+\ldots)$$
$$= \frac{\beta_0}{2}\rho(R) \qquad \text{with} \qquad \rho_1 = \frac{\beta_1}{2\beta_0}.$$

Here the parameters $\rho_i$ are scheme invariant, moreover $\beta_0$ and $\beta_1$ are universal and equal to the leading coefficients of the QCD $\beta$-function. The RGE can be integrated leading to:

$$\frac{\beta_0}{2}\ln\frac{Q}{\Lambda_R} = \frac{1}{R} - \rho_1\ln\left(1+\frac{1}{\rho_1 R}\right) + \ldots$$

This expression is identical to the evolution equation of $\alpha_s$. The observable dependent scale parameter $\Lambda_R$ can be transformed into $\Lambda_{\overline{MS}}$ using an exact expression (see [5]). It is important to note that the RGI theory resums UV divergent terms and is numerically equivalent to the Effective Charge renormalisation scheme. PC's can be included by using a modified $\beta$-function:

$$\rho(R) \to \rho(R) - \frac{2K_0}{\beta_0}R^{-\frac{\beta_1}{\beta_0^2}}e^{-\frac{2}{\beta_0 R}},$$

$K_0$ measures the size of the NP contribution.

RGI provides a very good description of the 7 shape means studied. For an example see Fig. 3. No significant power terms are required, contrast to the $\overline{\rm MS}$ case. A contribution of $\mathcal{O}(3\%)$ near $M_Z$ may be present, however. For pure RGI (as well as for RGI plus PC's) the $\alpha_s$ values from the 7 observables are consistent with each other yielding $\langle\alpha_s(M_Z)\rangle = 0.121 \pm 0.002_{RMS}$. Advantageously no additional theory error is required in contrary to the standard analyses showing RMS spreads of typically 5%.

The success of RGI supposes to predict the NP terms of the power models by setting both predictions equal and solving for the power model

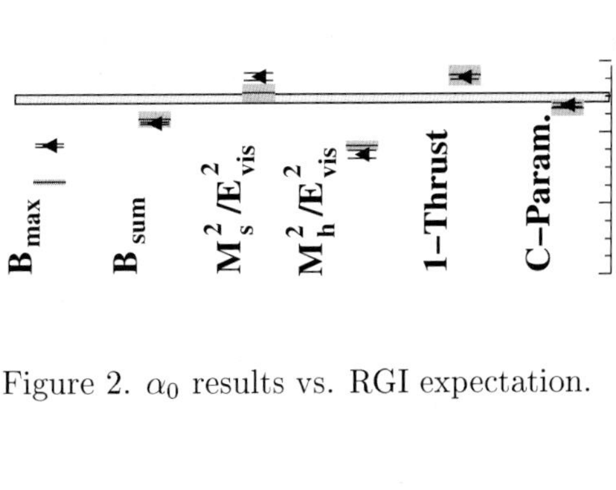


Figure 2. $\alpha_0$ results vs. RGI expectation.

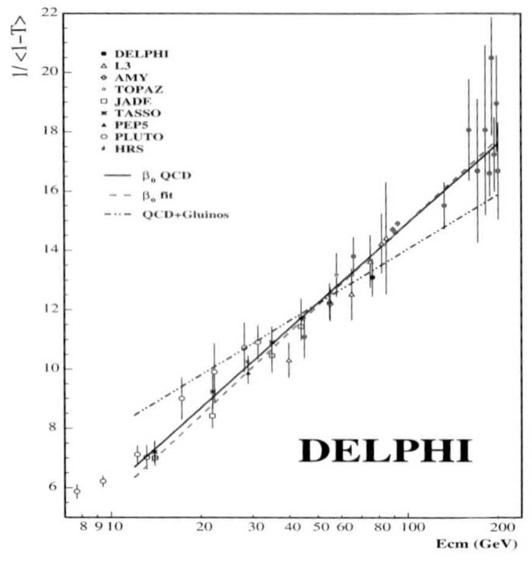


Figure 3. Fit of the $\beta$-function to $\langle 1-T\rangle(E_{CM})$.

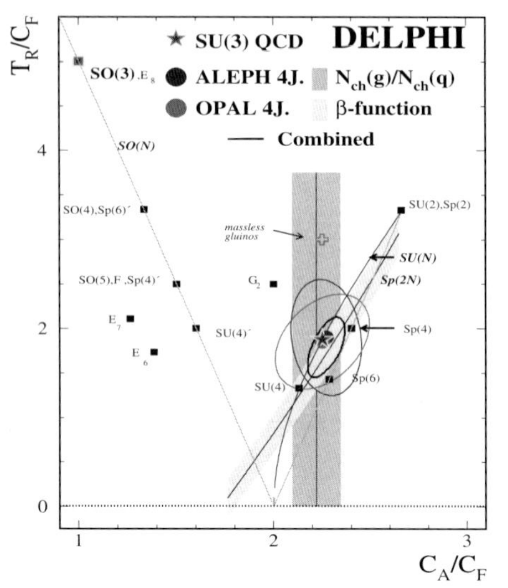


Figure 4. Colour factor measurements as obtained from various experiments.

parameters. This leads to the curve shown in Fig. 1 for the naive model and to the results shown in Fig. 2 for the DW model. In both cases the agreement with the data is very good. In the DW case it is even better than for any $\alpha_0$ value presumed universal.

## 5. Measuring the $\beta$-Function

The RGI theory imposes the RGE but does not fix the $\beta$-function. As also power terms appear vanishingly small, RGI allows for a direct measurement of the $\beta$-function of the observable $R$. As $\beta_0$ and $\beta_1$ are universal this is widely equivalent to a measurement of the $\beta$-function of $\alpha_s$. Rewriting the RGE:

$$\frac{dR^{-1}}{d\ln Q} = \frac{\beta(R)}{R^2} = \frac{\beta_0}{2}\cdot\left(1+\frac{\beta_1}{2\beta_0}\cdot R+\ldots\right)$$

it is seen that $\beta_0$ is approximately equal to the slope of $R^{-1}$ vs. $\ln Q$. As to be expected the fits yield consistent results from the 7 observables. In order to increase the precision low energy data on $\langle 1-T\rangle$ has been included yielding:

$$\frac{dR^{-1}}{d\ln Q^2} = 8.7 \pm 0.35 \qquad \text{or equivalently}$$

$$\beta_0 = 7.86 \pm 0.32 \qquad n_f = 4.75 \pm 0.44$$

The last line was obtained by either correcting for $\beta_1$ or assuming the QCD expressions for $\beta_{0/1}$, respectively. This result is superiour to the current determination from the world data on $\alpha_s$. The error includes statistical and systematic errors of the data and of the b-mass correction. The uncertainty due to possible PC's is small. This result agrees well with the expectation $\beta_0 = 7\frac{2}{3}$, $n_f = 5$ and excludes hypothetical light gluinos as perhaps introduced in [6][1]. Together with other measurements [8] it strongly restricts the colour factors of strong interactions to those of $SU(3)_c$ (see Fig. 4).

## REFERENCES


1. DELPHI Coll., contrib. paper 583.
2. Y.Dokshitzer et al., Nucl.Phys. **B** 511, 396.
3. P. Movilla-Fernandez, these proceedings.
4. A.Dhar V.Gupta Phys.Rev. **D 29**, 2822 .
5. J.M. Campbell et al., PRL 81, 1568.
6. E. Berger, these proceedings.
7. O. Passon, PhD thesis, WUB–DIS–2002/6.
8. see G. Dissertory and M.Siebel, these proc.


[1] A refined fit [7] shows that gluinos up to masses of 20 GeV are excluded by the data

ELSEVIER

Proceedings of ICHEP 2002, pp. 361–363
*S. Bentvelsen, P. de Jong, J. Koch and E. Laenen (Editors)*

31st INTERNATIONAL CONFERENCE ON HIGH ENERGY PHYSICS AMSTERDAM

# $\alpha_S$ and Power Corrections from JADE

P.A. Movilla Fernández[a]

[a] Max-Planck-Institut für Physik, Föhringer Ring 6, D-80805 München, Germany

Re-analysed JADE data were used to determine $\alpha_S$ at $\sqrt{s} = 14\text{-}44\,\text{GeV}$ on the basis of resummed calculations for event shapes and hadronisation models tuned to LEP data. The combined result is $\alpha_S(M_{Z^0}) = 0.1194^{+0.0082}_{-0.0068}$ which is consistent with the world average. Event shapes have also been used to test power corrections based on an analytical model and to verify the gauge structure of QCD. The only non-perturbative parameter $\alpha_0$ of the model was measured to $\alpha_0(2\,\text{GeV}) = 0.503\,^{+0.066}_{-0.045}$ and is found to be universal within the total errors.

## 1. INTRODUCTION

The re-analysis of $e^+e^-$ annihilation data collected with the JADE detector at the PETRA collider (1978-1986) has been shown to be a valuable effort [1–4] since the characteristic energy evolution of Quantum Chromodynamics (QCD) becomes more manifest towards decreasing centre-of-mass energies $\sqrt{s}$. Recently, data at energies down to $\sqrt{s} = 14\,\text{GeV}$ could be employed in state-of-the-art QCD studies due to the successful resurrection of the original JADE software.

Since the PETRA shutdown, significant progress has been made in the theoretical calculations of event shape observables. In the following, we present an $\alpha_S$ analysis at $\sqrt{s} = 14\text{-}44\,\text{GeV}$ based on the most complete perturbative calculations for event shapes [5,6] available so far. We included recently analysed data in the energy region $\sqrt{s} = 14\text{-}22\,\text{GeV}$ for which the calculations are applied for the first time. Furthermore, power corrections based on an analytical model by Dokshitzer, Marchesini, and Webber [7] (DMW) were investigated as a promising approach to describe non-perturbative effects in event shapes. Besides $\alpha_S$, the model depends only on one additional free parameter. Also the consistency of power corrections with the gauge structure of QCD was tested.

## 2. EVENT SHAPES

From multihadronic data samples, the distributions of *thrust* $(1-T)$, *heavy jet mass* $(M_\text{H})$, *total* and *wide jet broadening* ($B_\text{T}$ and $B_\text{W}$), *C parameter* and the *differential 2-jet rate* $y_{23}$ in the Durham scheme are calculated (cf. [1]). The data are corrected for the limited acceptance and resolution of the detector and for initial state photon radiation. Since electroweak decays of the heavy b-hadrons fake hard gluon radiation particularly at $\sqrt{s} \leq 22\,\text{GeV}$, we take the contribution $e^+e^- \to b\bar{b}$ as an additional background to be subtracted from the distributions.

The data have been used to assess the performance of various QCD event generators tuned to LEP data at $\sqrt{s} = M_{Z^0}$ (cf. [8]). The parton shower and string fragmentation model implemented in PYTHIA/JETSET is found to be well capable of describing event shapes down to 14 GeV. The quality of the models ARIADNE (colour dipole scheme) and HERWIG (cluster fragmentation), however, is more moderate, and COJETS (independent fragmentation) is clearly disfavoured. Obviously, the model parameters of these generators need a re-tune at lower $\sqrt{s}$.

## 3. DETERMINATION OF $\alpha_S$

The determination of $\alpha_S$ is based on a combination of an exact QCD matrix element calculation $\mathcal{O}(\alpha_S^2)$ [5] intended to describe the 3-jet region of phase space and a next-to-leading-logarithmic approximation (NLLA) [6] valid in the 2-jet region where multiple radiation of soft and collinear gluons from a system of two hard back-to-back partons dominate. We perform $\chi^2$-fits of the theoretical predictions corrected for


doi:10.1016/S0920-5632(02)02209-0

Table 1
Preliminary $\alpha_S$ results from JADE.

| $\sqrt{s}$ [GeV] | $\alpha_S(\sqrt{s})$ | fit+exp. | had. | hi. ord. | tot. |
|---|---|---|---|---|---|
| 14.0 | 0.1704 | $\pm 0.0051$ | $^{+0.0141}_{-0.0136}$ | $^{+0.0143}_{-0.0091}$ | $^{+0.0206}_{-0.0171}$ |
| 22.0 | 0.1513 | $\pm 0.0043$ | $\pm 0.0101$ | $^{+0.0101}_{-0.0065}$ | $^{+0.0144}_{-0.0121}$ |
| 34.8 | 0.1431 | $\pm 0.0019$ | $\pm 0.0073$ | $^{+0.0091}_{-0.0060}$ | $^{+0.0118}_{-0.0096}$ |
| 38.3 | 0.1397 | $\pm 0.0040$ | $\pm 0.0054$ | $^{+0.0084}_{-0.0056}$ | $^{+0.0108}_{-0.0087}$ |
| 43.8 | 0.1306 | $\pm 0.0037$ | $\pm 0.0056$ | $^{+0.0068}_{-0.0044}$ | $^{+0.0096}_{-0.0080}$ |

hadronisation effects. For the main results, we use the $\ln(R)$-scheme [6] for the perturbative prediction with the renormalisation scale factor $x_\mu \equiv \mu/\sqrt{s} = 1$ and PYTHIA for the estimation of non-perturbative contributions. We generally observe stable fits and good agreement with the data at each $\sqrt{s}$ with $\chi^2$/d.o.f. $\simeq$ 0.2-2.0. In case of $B_W$, a significant excess of the theory over the data in the 3-jet region of the distributions is present.

Experimental errors are under control for all data samples. Expectedly, hadronisation uncertainties increase rapidly towards 14 GeV. The individual results agree with each other within 1-2 standard deviations of the fit and experimental errors. For each $\sqrt{s}$, the $\alpha_S$ values from the six observables are combined using the weighted mean method of Ref.[1] (Tab. 1). The total errors are dominated by higher order uncertainties. At 14 and 22 GeV, hadronisation uncertainties are of the same order as the QCD scale ambiguities.

The $\alpha_S$ results obtained here and in similar analyses at higher energies based on resummed event shapes (Fig. 1) agree well with the QCD expectation for the running coupling [9]. A $\chi^2$-fit taking statistical and experimental errors into account yields $\alpha_S(M_{Z^0}) = 0.1213 \pm 0.0006$ with $\chi^2$/d.o.f. $= 8.3/11$. Even considering the total errors, the unphysical hypothesis $\alpha_S =$ const. is disfavoured by a fit probability of $\approx 10^{-5}$.

## 4. TEST OF POWER CORRECTIONS

The DMW model [7] describes non-perturbative effects to event shapes as contributions from gluon radiation at low energy scales, assuming that the physical strong coupling $\alpha_S(\mu)$ remains finite in the energy region around the Landau pole where simple perturbative evolution

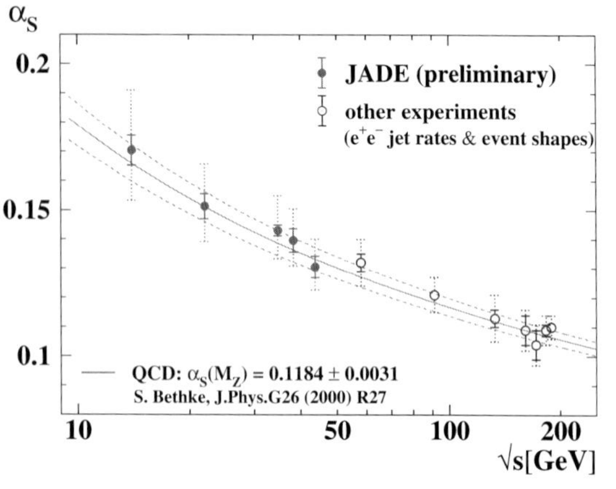


Figure 1. $\alpha_S(\sqrt{s})$ derived using $\mathcal{O}(\alpha_S^2)$+NLLA predictions for event shapes in e$^+$e$^-$ annihilation.

of $\alpha_S$ breaks down. This leads to the introduction of a parameter $\alpha_0(\mu_I) = 1/\mu_I \int_0^{\mu_I} d\mu\, \alpha_S(\mu)$ that absorbs all non-perturbative details of $\alpha_S(\mu)$ up to an arbitrary infrared matching scale $\mu_I$. The principle structure of power corrections is a shift $\mathcal{P}D_\mathcal{F}$ of the perturbative spectrum away from the 2-jet region, with $\mathcal{P} \propto \alpha_0/\sqrt{s}$ and $D_\mathcal{F}$ depending on $\mathcal{F}$. In case of $B_T$ and $B_W$, the shift is superimposed by a squeeze $D_\mathcal{F} \propto \ln 1/\mathcal{F}$. For mean values one obtains also an additive correction. The perturbative part is $\mathcal{O}(\alpha_S^2)$+NLLA for the distributions and $\mathcal{O}(\alpha_S^2)$ for the means.

The prediction, in particular the universality of $\alpha_0(\mu_I)$, has been tested by global fits to the hadron level data from this analysis and from other experiments e.g. at LEP/SLC up to $\sqrt{s} =$ 189 GeV (cf. [2]), with $\alpha_S(M_{Z^0})$ and $\alpha_0(2\,\text{GeV})$ as only free parameters. The major features of the distributions are reproduced well, thus supporting the $1/\sqrt{s}$ evolution of power corrections in event shapes. However, we observe discrepancies for the distributions of the less inclusive quantities $M_H$ and $B_W$ in particular at $\sqrt{s} < M_{Z^0}$.

For $y_{23}$ the leading power correction is known to be quadratic in $1/\sqrt{s}$. This expectation has been verified by means of the new JADE distributions at $\sqrt{s} = 14$ and 22 GeV using a simple additive power correction ansatz.

As shown in Fig. 2, there is a reasonable agreement between the individual results within the

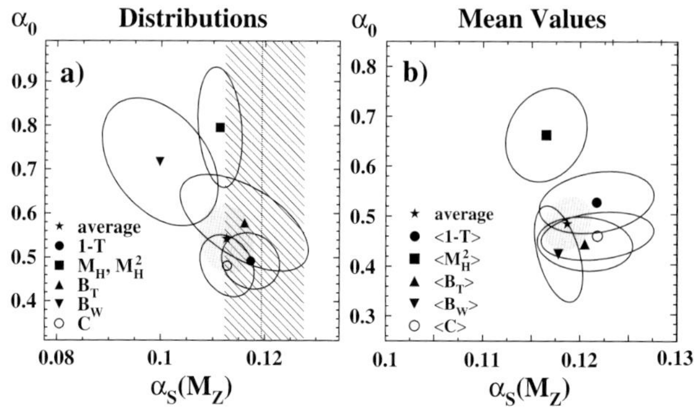


Figure 2. $\alpha_S(M_{Z^0})$ and $\alpha_0(2\,\mathrm{GeV})$ and one standard deviation errors of DMW-fits to distributions a) and means b). The hatched band represents the combined $\alpha_S(M_{Z^0})$ derived from the "conventional" analysis.

total uncertainties. However, the $\alpha_0$ results from the $M_H$ and $B_W$ distributions are large compared to the results from the other observables. This observation may be related to the non-inclusiveness of these variables. The $\alpha_S$ values from power corrections to the distributions are systematically smaller than the results based on MC corrections. This is due to the different amounts of squeeze of the perturbative spectrum predicted by both types of model, particularly in case of $B_W$.

Combining the results for the mean values and the distributions taking correlations between the systematic errors into account yields $\alpha_S(M_{Z^0}) = 0.1175^{+0.0031}_{-0.0021}$ and $\alpha_0(2\,\mathrm{GeV}) = 0.503^{+0.066}_{-0.045}$. The scatter of the $\alpha_0$ values is mostly covered by the theoretical uncertainty of the Milan factor [7].

## 5. STUDY OF QCD COLOUR FACTORS

The DMW ansatz has been exploited to extract the QCD colour factors $C_A$, $C_F$, and $n_f$ [3]. Various global fits to the event shape spectra trying alternative sets of the free model parameters support the $SU(3)$ symmetry group. The most stable and precise measurements are provided by $1-T$ and $C$. Combining the corresponding results for these variables with $\alpha_0(2\,\mathrm{GeV})$ and $n_f$ fixed, and $\alpha_S(M_{Z^0})$, $C_F$, and $C_A$ free, one finds $C_F = 1.29 \pm 0.18$ and $C_A = 2.84 \pm 0.24$. This is in good agreement with the QCD expectation while some other gauge symmetry groups are excluded.

## 6. CONCLUSIONS

Resummed QCD theory combined with LEP tuned hadronisation models fits event shape data well down to $\sqrt{s} = 14\,\mathrm{GeV}$ and allow consistent determinations of $\alpha_S$. The combined result evolved to the $Z^0$ mass scale is $\alpha_S(M_{Z^0}) = 0.1194^{+0.0082}_{-0.0068}$ which is substantially more precise than former PETRA measurements and also in good agreement with the world average value [9].

Power corrections $\propto 1/\sqrt{s}$ generally reproduce the overall event shape spectra, except for the distributions of the less inclusive variables ($M_H$ and $B_W$) at $\sqrt{s} < M_{Z^0}$. The results for $\alpha_0$ support the DMW prediction of universality within 25%. Using power corrections, the gauge structure of QCD has been verified with uncertainties competitive e.g. with traditional 4-jet angular correlation analyses. Potential biases from hadronisation models are reduced within this approach.

Thus, exploiting JADE data significantly improves the verification of QCD on the basis of $e^+e^-$ annihilation.

## REFERENCES

1. P.A. Movilla Fernández *et al.*, *Eur. Phys. J.* **C1** (1998) 461; O. Biebel *et al.*, *Phys. Lett.* **B459** (1999) 326.
2. P.A. Movilla Fernández *et al.*, *Eur. Phys. J.* **C22** (2001) 1.
3. S. Kluth *et al.*, *Eur. Phys. J.* **C21** (2001) 199.
4. P. Pfeifenschneider *et al.*, *Eur. Phys. J.* **C17** (2000) 19; M. Blumenstengel *et al.*, *Phys. Lett.* **B517** (2001) 37.
5. R.K. Ellis, D.A. Ross and A.E. Terrano, *Nucl. Phys.* **B178** (1981) 421.
6. S. Catani *et al.*, *Nucl. Phys.* **B407** (1993) 3; S. Catani *et al.*, *Phys. Lett.* **B295** (1992) 269; G. Dissertori and M. Schmelling, *Phys. Lett.* **B361** (1995) 167; S. Catani and B.R. Webber, *Phys. Lett.* **B427** (1998) 377; Y.L. Dokshitzer *et al.*, *JHEP* **01** (1998) 11.
7. Yu.L. Dokshitzer and B.R. Webber, *Phys. Lett.* **B404** (1997) 321; Yu.L. Dokshitzer *et al.*, *JHEP* **05** (1998) 003; Yu.L. Dokshitzer *et al.*, *Eur. Phys. J. direct* **C3** (1999) 1.
8. A description of the MC programs and tunes used here is given e.g. in: I.G. Knowles *et al.*, "QCD Event Generators" (1995), hep-ph/9601212.
9. S. Bethke, *J. Phys.* **G26** (2000) R27.

# Session 6

# QCD: Soft Interactions

*Session conveners:*
*W. Kittel, University of Nijmegen*
*M. Diehl, RWTH Aachen*
*A. Kaidalov, ITEP, Moscow*

ELSEVIER

Proceedings of ICHEP 2002, pp. 367–369
*S. Bentvelsen, P. de Jong, J. Koch and E. Laenen (Editors)*

31st INTERNATIONAL CONFERENCE ON HIGH ENERGY PHYSICS AMSTERDAM

# Perturbative QCD with Quark and Gluon Condensates

Paul Hoyer[a*]

[a]Department of Physical Sciences and Helsinki Institute of Physics
P.O. Box 64, FIN-00014 Helsinki University, Finland

The quark $\langle\bar{\psi}\psi\rangle$ and gluon $\langle F_{\mu\nu}F^{\mu\nu}\rangle$ vacuum expectation values are non-vanishing at low orders in $\alpha_s$ when the perturbative ground state includes quark and gluon pairs. This offers possibilities of studying the effects of a non-trivial vacuum using formally exact perturbative methods. The QCD Feynman rules are modified only for the free, on-shell quark and gluon propagators at zero four-momentum. Gauge and lorentz invariance is maintained, while chiral symmetry is spontaneously broken by the ground state.

## 1. THE GROUND STATE

The perturbative expansion is the basis of most comparisons of gauge field theories with data, and is used to establish general properties like analyticity and factorization of scattering amplitudes. The expansion is determined by the lagrangian $\mathcal{L}$ together with

(a) The renormalization scheme ($Q^2 \to \infty$)

(b) The perturbative ground state ($Q^2 \to 0$)

While the dependence on the short-distance regularization (a) has been thoroughly studied, I find it surprising that little attention is paid to the starting point (b) of the expansion. The essential difference between QED and QCD is believed to lie in the nature of the QCD vacuum. The poor approximation provided by the empty perturbative vacuum of the quark and gluon condensates is a potential reason that perturbative QCD (PQCD) fails to describe long-distance phenomena.

It may be argued that PQCD is in any case inadequate at long distance due to the large size of the strong coupling $\alpha_s$. While this possibility cannot be excluded, it is not a sufficient reason for neglecting to consider alternative choices of the perturbative vacuum. PQCD with a non-empty ground state could serve as a model to improve our understanding of vacuum effects (including confinement). Furthermore, analyses of data [1] indicate that the strong coupling freezes at a value $\alpha_s(Q^2 = 0) \simeq 0.5$. At a more qualitative level, the hadron spectrum shares many features with the hydrogen atom. Thus quark degrees of freedom and PQCD appear to be relevant in the confinement region [2] – but attention needs to be paid to the choice of ground state.

The perturbative expansion is based on using free $|in\rangle$ and $\langle out|$ states at asymptotic times, $t = \pm\infty$. Formally, any choice of asymptotic state that has an overlap with the true ground state is allowed, since the state will relax to the ground state during its evolution to finite $t$. At zeroth order in $\alpha_s$ the free asymptotic state is an eigenstate of the hamiltonian and thus constitutes the perturbative ground state at all times.

In path integral derivations of the perturbative expansion the choice of an empty perturbative vacuum is usually made implicitly. The dependence on the wave function of the asymptotic state can be made explicit by first considering a finite time interval, $-T \le t \le T$, and then taking the $T \to \infty$ limit [3]. Thus the generating functional of a free scalar field (in 0+1 dimensions, to keep the notation simple)

$$Z_0[J] = \int \mathcal{D}[\phi] \exp\left\{i \int dt\,[\mathcal{L}_0(\phi) + J(t)\phi(t)]\right\}$$

*Address until 1 August 2002: Nordita, Blegdamsvej 17, DK-2100 Copenhagen, Denmark. Research supported in part by the European Commission under contract HPRN-CT-2000-00130.


doi:10.1016/S0920-5632(02)02212-0

becomes

$$Z_0[J] = \int \mathcal{D}[\phi(t)] \exp\left\{ \frac{-i}{2} \int_{-T}^{T} dt \right.$$
$$\phi(t) \left[ \frac{\partial^2}{\partial t^2} + m^2 \right] \phi(t) + i \int_{-T}^{T} dt\, J(t)\phi(t)$$
$$\left. - \frac{C_s}{2} \left[ \phi^2(-T) + \phi^2(T) \right] \right\} \quad (1)$$

where the path integral is over $\phi(-T \le t \le T)$. The wave functions at $t = \pm T$ should not include interactions and thus are gaussian. The ground (empty) state of this harmonic oscillator system has $C_s = m$. Other choices of $C_s$ can be seen as superpositions of states with particle pairs through an expansion in powers of $(C_s - m)\phi^2$.

In the $T \to \infty$ limit the free generating functional (1) becomes

$$Z_0[J] = \exp\left\{ -\frac{1}{2} \int_{-\infty}^{\infty} \frac{dE}{2\pi} J(-E) \left[ \frac{i}{E^2 - m^2 + i\varepsilon} \right. \right.$$
$$\left. \left. + \left( \frac{C_s}{m} - 1 \right) \pi\delta(E^2 - m^2) \right] J(E) \right\} \quad (2)$$

Thus only the on-shell part of the free propagator is modified when $C_s \neq m$. This is natural since on-shell propagation is required to feel the boundary conditions at $t = \pm\infty$. It is also important to note that the generating functional of the interacting theory,

$$Z[J] = \exp\left[ iS_{int}\left( \frac{\delta}{i\delta J} \right) \right] Z_0[J] \quad (3)$$

defines the feynman rules in terms of standard interaction vertices and the free propagator of Eq. (2). Hence the perturbative expansion is modified *only* by the additional term $\propto C_s - m$ in the free, on-shell propagator.

The above method can be trivially extended to 3+1 dimensions. The freedom in choosing asymptotic states allows the addition of pairs with nonvanishing relative three-momentum [4]. Lorentz symmetry is restored due to the relaxation of the asymptotic state to the (lorentz invariant) ground state. In this case the perturbative expansion will not, however, be lorentz invariant order by order. Adding pairs only for massless fields with $\vec{p} = 0$ retains lorentz invariance order by order in $\alpha_s$ and is thus an attractive option [5].

Similar considerations show [3,5] that the PQCD expansion is modified only through the on-shell quark and gluon propagators,

$$S_q^{AB}(p) = \delta^{AB} \left[ \frac{i\not{p}}{p^2 + i\varepsilon} + C_q (2\pi)^4 \delta^4(p) \right] \quad (4)$$

$$D_g^{ab,\mu\nu}(p) = -g^{\mu\nu}\delta^{ab} \left[ \frac{i}{p^2 + i\varepsilon} + C_g (2\pi)^4 \delta^4(p) \right] \quad (5)$$

The constants $C_q$ and $C_g$ determine the density of $q\bar{q}$ and gluon pairs in the perturbative ground state and are the only parameters allowed by this generalization of PQCD. The corresponding modification of the ghost propagator is irrelevant since the ghost-gluon vertex is proportional to the ghost momentum.

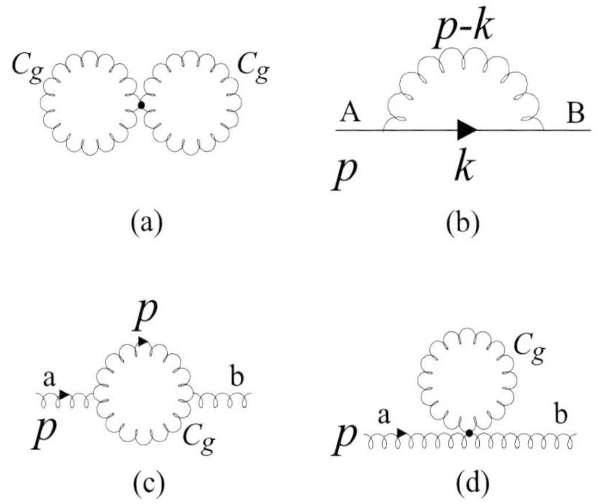


Figure 1. (a) The only diagram which contributes to $\langle F_{\mu\nu}F^{\mu\nu}\rangle$ at $\mathcal{O}(g^2)$. The gluon condensate term $\propto C_g$ in (5) is taken in both loops. (b) Quark propagator correction. (c,d) Gluon propagator corrections.

## 2. PHYSICAL EFFECTS

The modifications (4), (5) of the quark and gluon propagators change the analytic structure of scattering amplitudes. This reflects scattering on the on-shell partons in the perturbative ground state. Such scattering also affects quarks and gluons propagating in the condensates of the true QCD vacuum. PQCD with quark and gluon pairs in the perturbative vacuum may thus serve as a model of QCD vacuum effects in feynman diagrams. It is of particular interest to find the eigenstates of propagation in the perturbative vacuum, which can serve as external states in the S-matrix.

The quark propagator (4) gives a finite quark condensate already at zeroth order in perturbation theory,

$$\langle\overline{\psi}(x)\psi(x)\rangle = -4NC_q \tag{6}$$

where $N$ is the number of colors. Thus chiral symmetry is spontaneously broken by the asymptotic state.

In the case of the $\langle F_{\mu\nu}F^{\mu\nu}\rangle$ expectation value the $\mathcal{O}\left(g^0\right)$ contribution vanishes since the free part of $F_{\mu\nu} = \partial_\mu A_\nu - \partial_\nu A_\mu - gf_{abc}A^b_\mu A^c_\nu$ is proportional to the gluon momentum. Furthermore, at $\mathcal{O}\left(g^2\right)$ the gluon loop integrals vanish (when dimensionally regularized) due to the absence of a mass or momentum scale. This leaves only the two-loop diagram involving a 4-gluon coupling (Fig. 1a), with the $C_g$ part of the propagator taken in both loops. The result is [5]

$$\langle F_{\mu\nu}(x)F^{\mu\nu}(x)\rangle = 12g^2N(N^2-1)C_g^2 \tag{7}$$

The gluon condensate thus arises specifically from the non-abelian 4-gluon interaction.

The quark propagator correction shown in Fig. 1b gets a contribution from both the quark condensate term $C_q$ in (4) ($k=0$) and from the gluon propagator modification $C_g$ of (5) ($k=p$). Only the former breaks chiral symmetry and thus generates a proper 'constituent' quark mass [3],

$$M_q = \left(4g^2C_F|C_q|\right)^{1/3} \tag{8}$$

which then gets a correction from the gluon condensate term $C_g$.

The sum of the gluon propagator corrections shown in Figs. 1c,d has the transverse structure required by gauge invariance. At order $g^2$ the gluon mass

$$M_g^2 = -2g^2NC_g \tag{9}$$

is tachyonic [4,5] if $C_g > 0$.

The present framework allows a perturbative demonstration of how the pion (goldstone) mass

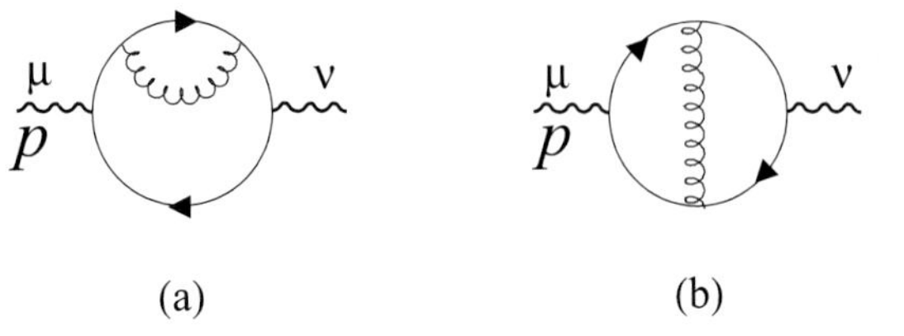


Figure 2. Corrections to the color singlet (axial) vector current. A third diagram where the gluon couples to the antiquark is not shown.

vanishes in spite of the constituent quark mass generated by corrections like Fig. 1b. The vanishing of the pion mass is a consequence of the general relation

$$\langle 0|J_5^\mu(x)|\pi(p)\rangle = -ip^\mu f_\pi \exp(-ip\cdot x) \tag{10}$$

and the conservation of the axial vector current, $\partial_\mu J_5^\mu(x) = 0$ when $m_q = 0$. The $\mathcal{O}\left(g^2\right)$ quark mass correction to the divergence of the axial vector propagator shown in Fig. 2a is exactly cancelled by the quark interaction term of Fig. 2b [3].

**Acknowledgments.** I am grateful for helpful discussions with Stéphane Peigné.

## REFERENCES

1. P. A. Movilla Fernandez, S. Bethke, O. Biebel and S. Kluth, Eur. Phys. J. C **22** (2001) 1 [arXiv:hep-ex/0105059].
2. Yu. L. Dokshitzer, Plenary talk at ICHEP 98, *Proc. Vancouver 1998, High energy physics, Vol. 1, 305-324* [hep-ph/9812252].
3. P. Hoyer, arXiv:hep-ph/0203236.
4. P. Hoyer and J. Rathsman, JHEP **0105** (2001) 020 [arXiv:hep-ph/0011209].
5. A. Cabo, S. Penaranda and R. Martinez, Mod. Phys. Lett. A **10** (1995) 2413;
   M. Rigol and A. Cabo Montes de Oca, Phys. Rev. D **62** (2000) 074018 [arXiv:hep-th/9909057];
   A. Cabo Montes de Oca and M. Rigol Madrazo, Eur. Phys. J. C **23** (2002) 289.

ELSEVIER

Proceedings of ICHEP 2002, pp. 370–374
*S. Bentvelsen, P. de Jong, J. Koch and E. Laenen (Editors)*

31st INTERNATIONAL CONFERENCE ON HIGH ENERGY PHYSICS AMSTERDAM

# Multiplicities, fluctuations and correlations*

Edward K.G. Sarkisyan[a b †]

[a]EP Division, CERN, CH-1211 Geneva 23, Switzerland

[b]Physics Department, University of Antwerpen, Universiteitsplein 1, B-2610 Wilrijk, Belgium

The recent results on hadron multiplicities in heavy and light quark fragmentation, multiplicity local fluctuations and multiparticle correlations submitted to the Conference are reviewed.

## 1. Introduction

One of the most important observables in particle production processes is *multiplicity*, i.e. the number of particles (mostly, hadrons) produced in the collision [1]. The multiplicity dependence on the energy scale, species of particles, event flavour content are among the main predictions of the theory of strong interactions, quantum chromodynamics (QCD) [1,2]. On the other hand, the multiplicity is used to select or to describe events, e.g. as a trigger for specific processes, as an input for kinematic variables' spectra. The distribution of multiplicity, its mean value and multiplicity fluctuations are the essential characteristics of the collision dynamics. However, the multiplicity distribution tells us just about the averaged, integrated numbers, while deeper information comes from the moments of the distribution, which measure particle *correlations*, i.e. probe the interaction dynamics [3].

Here, I report on the multiplicity flavour dependence[3], on the analyses of local multiplicity fluctuations and multiparticle correlations[4]. These studies provide us with details of strong interactions and allow us to estimate the level of the applicability of QCD, based on the partonic picture, to the production of hadrons which are the experimentally observed objects.

## 2. Definitions and notations [1,3]

The multiplicity distribution, or the density $\rho_n$, of multiplicity $n$ of particles with kinematic variables $p_1, p_2, \ldots, p_n$ is defined by the inclusive probability spectrum,

$$\rho_n(p_1, p_2, \ldots, p_n) = \frac{1}{N_{\rm ev}} \frac{dn(p_1, p_2, \ldots, p_n)}{dp_1 dp_2 \cdots dp_n} \,,$$

where $N_{\rm ev}$ is the number of events.

As it follows from this formula, the single particle distribution $\rho_1(p_1)$ gives an *average* multiplicity, $\langle n \rangle = \int \rho_1(p_1) dp_1$, while integration of the $q$-particle densities leads to the unnormalised $q$th order *factorial moments*,

$$f_q = \int \rho_1(p_1, p_2 \ldots p_n) dp_1 dp_2 \cdots dp_n$$
$$= \langle n(n-1)\cdots(n-q+1) \rangle \equiv \langle n^{[q]} \rangle \,. \quad (1)$$

The *normalised* factorial moments are then defined as $F_q = f_q / \langle n \rangle^q$.

The $q$-particle densities give us a way to study particle correlations described by the $q$-particle correlation functions, (factorial) *cumulants*, $C_q(p_1, \ldots, p_q)$. The cumulants vanish whenever one of their arguments is statistically independent, i.e. these functions measure *genuine* $q$-particle correlations.

The cumulants are constructed from the multiplicity densities, e.g.

$$C_1(p_1) = \rho_1(p_1),$$

*Talk given at the 31st International Conference on High Energy Physics (ICHEP 2002), Amsterdam, The Netherlands, 24 – 31 July 2002

†Electronic address: `sedward@mail.cern.ch`

[3]Other aspects of multiplicity studies such as multiplicity energy dependence, 3-jet multiplicity, multiplicities of quark and gluon jets etc. are reviewed by M. Siebel [4].

[4]Bose-Einstein correlation studies are reviewed by Š. Todorova-Nová [5].


doi:10.1016/S0920-5632(02)02123-0

$$C_2(p_1,p_2) = \rho_2(p_1,p_2) - \rho_1(p_1)\rho_1(p_2),$$
$$C_3(p_1,p_2,p_3) = \rho_3(p_1,p_2,p_3) - \sum_{(3)} \rho_1(p_1)\rho_2(p_2,p_3) + 2\rho_1(p_1)\rho_1(p_2)\rho_1(p_3)\,. \quad (2)$$

These functions, being properly normalised, are used to study *multi*particle correlations in different kinematic variables.

In studies of local fluctuations and correlations, i.e those in phase-space bins, one uses the *bin-averaged* factorial moments and cumulants. The phase space is divided into $M$ bins of equal size, so that $\langle n_m \rangle$ is the number of particles in the $m$th bin. Then, the normalised bin-averaged factorial moment is defined via Eq. (1) averaged over bins,

$$F_q(M) = \frac{1}{M}\sum_{m=1}^{M} \frac{f_q(m)}{\langle n_m \rangle^q} \equiv \frac{1}{M}\sum_{m=1}^{M} \frac{\langle n_m^{[q]} \rangle}{\langle n_m \rangle^q}. \quad (3)$$

These factorial moments have been extensively used to extract the *non-statistical* (non-Poissonian) fluctuations in many types of collisions [3]. Such fluctuations lead to the power-law scaling of factorial moments as a function of $M$ called the *intermittency* phenomenon.

In analogy with Eq. (3), Eqs. (2) are averaged to the bin-averaged normalised cumulants,

$$K_q(M) = \frac{1}{M}\sum_{m=1}^{M} \frac{\langle k_q^{(m)} \rangle}{\langle n_m \rangle^q}, \quad (4)$$

where $k_q^{(m)}$ are the Mueller moments, the functions of unnormalised moments $f_q(m)$, Eq. (1),

$$k_1 = \langle n_m \rangle, \quad k_2^{(m)} = \langle n_m^{[2]} \rangle - \langle n_m \rangle^2,$$
$$k_3^{(m)} = \langle n_m^{[3]} \rangle - 3\,\langle n_m^{[2]} \rangle \langle n_m \rangle + 2\,\langle n_m \rangle^3. \quad (5)$$

## 3. Hadronisation of heavy and light quarks

In this Section, I consider the recent results on hadron multiplicities from fragmentation of heavy and light quarks [6–8]. The study of the quark content in multiparticle production provides one of the basic tests of QCD. The results from LEP are of a special interest since they cover a wide centre-of-mass energy region and can be directly compared with QCD which is mostly predictable at the asymptotic energies [2].

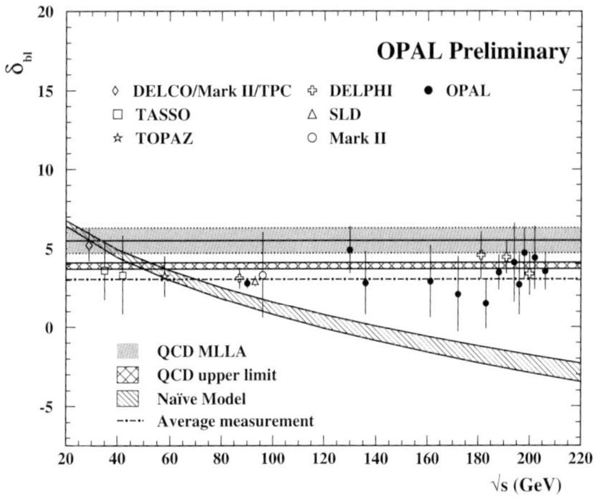


Figure 1. The difference in the mean charged multiplicities, $\delta_{\mathrm{bl}}$, between heavy and light quark pairs as a function of centre-of-mass energy. The dashed-dotted line is the combined result from all measurements, see Ref. [8] for more details.

In Refs. [6] and [8], a study of the fragmentation of heavy b-quark and light quarks (l = u, d, s) is performed. The measurements of the difference in charge particle multiplicities, $\delta_{\mathrm{bl}} = \langle n_{\mathrm{b\bar{b}}} \rangle - \langle n_{\mathrm{l\bar{l}}} \rangle$, for $\mathrm{b\bar{b}}$ and $\mathrm{l\bar{l}}$ events in $e^+e^-$ annihilation at the centre-of-mass energies above the $Z^0$ peak are carried out. The findings are compared to the theoretical predictions of QCD and to a more phenomenological (the so-called naïve) model (for review, see Ref. [2]). The QCD calculations predict energy-independent behaviour of the multiplicity difference $\delta_{\mathrm{bl}}$, while in the naïve model one expects the decrease with increasing energy. The latter is connected with the assumption that the hadron multiplicity accompanying the heavy hadrons in $\mathrm{b\bar{b}}$ events is the same as the multiplicity in $\mathrm{l\bar{l}}$ events at the energy left to the system once the heavy quarks have fragmented.

The difference between the heavy and light mean quark-pair multiplicities obtained by DELPHI at 206 GeV is $\delta_{\mathrm{bl}} = 4.50 \pm 1.05(\mathrm{stat}) \pm 0.52(\mathrm{syst})$ [8], while OPAL finds $\delta_{\mathrm{bl}} = 3.44 \pm 0.40(\mathrm{stat}) \pm 0.79(\mathrm{syst})$ in the energy range of 130 – 206 GeV [8]. The difference in the values is connected with some differences in the data processing procedure. In the meantime, the results of the energy-dependence is found to be the same for the both studies: the mean multiplicity difference is energy-independent and favours the QCD predictions while it is inconsistent with the flavour-independent naïve model (Fig. 1). Also in agreement with QCD is the ratio between light quark multiplicities, $\langle n_{\rm i}\rangle : \langle n_{\rm j}\rangle \sim 1$, where i,j ={u,d,s}, as obtained by OPAL at the $Z^0$ peak energy [7]. To note is that the $\langle n_{\rm u}\rangle$ and $\langle n_{\rm d}\rangle$ are highly statistically anti-correlated (−90%) which is due to their fractions in K-mesons.

## 4. Scaling of local fluctuations in hadronic $Z^0$ decays

During the last decades, the phase-space local multiplicity fluctuations are actively studied in many reactions, from leptonic to nuclear collisions, and the intermittency scaling of the fluctuations has been established [3].

All the studies show that the intermittency is more pronounced in high dimensions, while in one dimension (e.g., in rapidity) the effect is diluted by projection onto one dimension. This leads to flattenning of the factorial moments. Such a behaviour is well understood from a QCD parton shower which is a three-dimensional branching and naturally leads to the fractality. The best area to study the effect and its connection with QCD is given by $e^+e^-$ annihilation, where such investigations has been performed earlier [3,9,10] and new analysis is carried out by L3 [11].

The L3 analysis gives us further hints about the intermittency origin, which is currently far from understanding. The new study employs the fact that so far the fluctuations in many dimensions were studied with the same number $M$ of bins in each dimension/direction. This leads to self-*similar* fractals, i.e. the fluctuations in any direction are assumed to be the same, or *isotropic*. In case when the dynamics in different directions are not equivalent, the fluctuations become *anisotropic* and this tells us about the self-*affinity* [12]. A self-affine behaviour has been observed in hadronic interactions [12], while a self-similar scaling has been qualitatively confirmed in $e^+e^-$ annihilation [10]. In terms of factorial moments, Eq.(3), the 3-dimensional moments are expected to exhibit the intermittency property in $e^+e^-$ collisions when phase-space is partitioned isotropically, while this scaling in hadronic interactions only occurs for anisotropic partitioning.

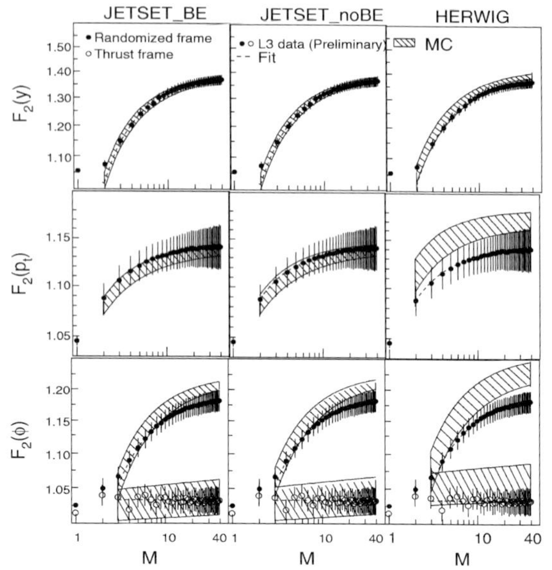


Figure 2. The 1-dimensional 2nd-order factorial moments, Eq. (3), as a function of the number $M$ of bins in the randomised and the thrust frames compared to the Monte Carlo predictions with and without Bose-Einstein correlations. The fit is given by the first f-la of Eqs. (7). The behaviour in the $q\bar{q}$ frame is similar to that in the randomised frame. See text and [11] for details.

In order to quantitatively study the observed isotropic fluctuations, the method of the Hurst exponent is used [12]. The Hurst exponent,

$$H_{ab} = \ln M_a / \ln M_b, \qquad (6)$$

is obtained from the fit of the one-dimensional second-order factorial moments,

$$F_2(M_a) = A_a + B_a\, M_a^{-\gamma_a} \;\Rightarrow\; H_{ab} = \frac{1+\gamma_b}{1+\gamma_a}. \qquad (7)$$

Here $a$, $b$ are the directions of the $(a,b)$ plane. For the isotropic dynamical fluctuations, $H_{ab} = 1$, while $H_{ab} \neq 1$ if the fluctuations are anisotropic.

L3 uses rapidity $y$, azimuthal angle $\varphi$ and transverse momentum $p_t$ – the variables often used in multiparticle studies – in the analysis. The variables are defined with respect to the

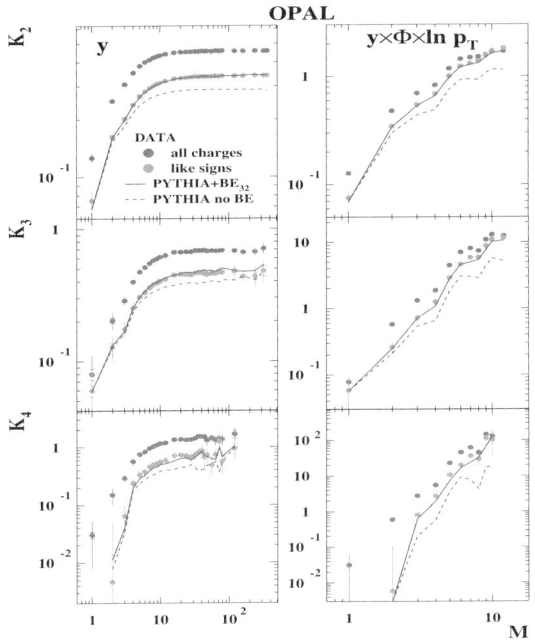


Figure 3. The all-charge and like-sign cumulants, Eq. (4), as a function of the number $M$ of bins in rapidity and in three dimensions. The like-sign cumulants are compared to the Monte Carlo predictions with and without Bose-Einstein correlations. See text and Ref. [13] for more details.

thrust axis, as well as the other two frames are considered: the randomised and the $q\bar{q}$ frames. In the former one, the $\varphi$-angles are randomly chosen being uniformly distributed in $[0, 2\pi]$, while in the latter case one uses Monte Carlo to correct the original thrust axis to the $q\bar{q}$ axis.

The 3-dimensional factorial moments are found to exhibit the intermittency scaling when phase-space is isotropically partitioned. This confirms earlier indications for isotropic fluctuations in $e^+e^-$ annihilations [9,10]. No sensitive differences in the different frames as well as no disagreements between the data and Monte Carlos are observed.

Fig. 2 shows the 2nd-order factorial moments in one dimension, from which the quantitative test of the isotropy of the fluctuations is made using Eqs. (7). The calculations of the Hurst exponent give: $H_{ab} \approx 1$, $a, b = \{y, \varphi, p_t\}$, i.e. due to the definition of Eq. (6) one concludes about the equivalence of the directions and about the isotropy of the fluctuations. The fluctuations in $\varphi$ are nearly absent in the thrust frame. In all frames the data are well described by Jetset, while somewhat less well by Herwig. No sensitivity to Bose-Einstein correlations is seen. From its study, L3 also concludes that there is the dependence on the QCD dynamics which serves to decrease the fluctuations in the thrust frame.

The 2-jet sub-samples are also analysed using the Durham jet algorithm. The fluctuations in the 2-jet events show the self-affine (anisotropic) fluctuations. This looks like the hard gluon emission leads to the isotropy.

## 5. Multiparticle correlations

The multiparticle correlation studies are performed for $e^+e^-$ annihilations at the $Z^0$ energy [13] and for $\bar{p}p$ annihilation at the incident momentum of 22.4 GeV/$c$ [14].

Genuine multiparticle correlations in $e^+e^-$ annihilations are studied by OPAL [13]. To extract the correlations, the method of the bin-averaged normalised cumulants is used, see Eqs. (4)–(5). In addition to the earlier analysis [10] of the cumulants of all charged particles ("all-charge cumulants"), OPAL studies the cumulants of same-charge particles ("like-sign cumulants"). The investigation is performed in rapidity $y$, azimuthal angle $\varphi$ and transverse momentum $p_t$, the same variables as in the above described L3 analysis, but the sphericity axis is used as a reference axis.

Fig. 3 shows the all-charge and like-sign cumulants calculated in one (rapidity) and three dimensions. Even in one dimension, positive genuine correlations of groups of two, three, four particles are present. The cumulants exhibit a scaling behaviour, although in rapidity the saturation already appears at the moderate bin sizes. The like-sign cumulants increase faster and drive the all-charge ones at small bins while the unlike-sign cumulants are almost constant. This points to the likely influence of Bose-Einstein correlations.

The comparison between the data and Monte Carlo like-sign cumulants shows that the model describes well the data when Bose-Einstein correlations are implemented. The same is true for the all-charge cumulants (not shown here). This

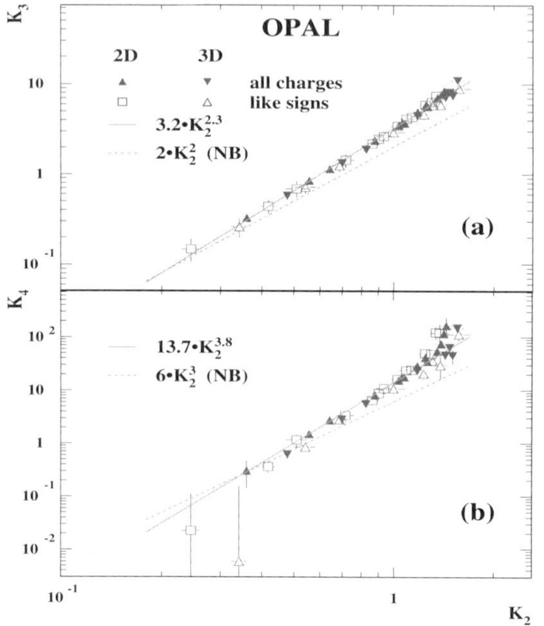


Figure 4. The Ochs-Wosiek plot in two and three dimensions for all-charge and like-sign cumulants. The dashed line shows the Negative Binomial prediction. The solid line is a fit, $\ln K_q = a_q + r_q \ln K_2$. See text and [13] for more details.

is not the case of the L3 results where no difference is found for the models with and without Bose-Einstein correlations, see Fig. 2. To note is that the OPAL data is described well in one dimension as well as in three dimensions.

From Fig. 3, one can see that Bose-Einstein correlations which are implemented in the model as the correlations of *two* identical particles and are pair-wise adjusted, well also describe the cumulants of $q > 2$. This suggests to consider the interdependence of the 2nd- and higher-order cumulants, Fig. 4. The fit akin to the Ochs-Wosiek one [3] gives the same parameter values for the all-charge and like-sign cumulants, while disagrees with the Negative Binomial predictions.

The multiparticle fluctuations are also studied in [14] using the Serpukhov fixed target $\bar{\text{p}}$p annihilation data at 22.4 GeV/$c$. The differential spectra of $k$ particles in pseudorapidity bins are analysed relative to the similar background spectra. The dips in the distributions of such ratios are found to be independent on the number $k$ of particles, $\sim 1.8$. The number of clusters is estimated to be about 2-3 with two particles per cluster in average. It is found that the data in the non-annihilation channel is similar to that from the inelastic pp collisions at 69 GeV/$c$ from the same accelerator. These observations are treated in frames of different mechanisms, e.g. a model with mesons emitted from intermediate nucleon isobars is suggested.

I am thankful to the ICHEP02 Organising Committee, to convenors of the QCD Session, and to my colleagues at CERN and particularly in OPAL for giving me the opportunity to make this presentation and for their kind help and support.

## REFERENCES

1. I.M. Dremin, J.W. Gary, Phys. Rep. 349 (2001) 301.
2. V.A. Khoze, W. Ochs, Int. J. Mod. Phys. A12 (1997) 2949.
3. E.A. De Wolf, I.M. Dremin, W. Kittel, Phys. Rep. 270 (1996) 1.
4. M. Siebel, Fragmentation of quarks and gluons, these Proceedings.
5. Š. Todorova-Nová, Bose-Einstein correlations at LEP and HERA, these Proceedings.
6. P. Abreu, A. De Angelis, DELPHI 2002-052 CONF 586 (2002), ICHEP02 abs. 232; DELPHI Collaboration, P. Abreu et al., Phys. Lett. B479 (2000) 118.
7. OPAL Collaboration, G. Abbiendi et al., Eur. Phys. J. C19 (2001) 257.
8. OPAL Collaboration, Physics Note PN511 (2002), ICHEP02 abs. 373.
9. DELPHI Collaboration, P. Abreu et al., Nucl. Phys. B386 (1992) 471.
10. OPAL Collaboration, G. Abbiendi et al., Eur. Phys. J. C11 (1999) 239.
11. L3 Collaboration, L3 Note 2758 (2002), ICHEP02 abs. 494.
12. Liu Feng, Liu Fuming, Liu Lianshou, Phys. Rev. D59 (1999) 114020 and refs. therein.
13. OPAL Collaboration, G. Abbiendi et al., Phys. Lett. B523 (2001)35, ICHEP02 abs.319.
14. E.G. Boos et al., ICHEP02 abs. 37.

ELSEVIER

Proceedings of ICHEP 2002, pp. 375–380
*S. Bentvelsen, P. de Jong, J. Koch and E. Laenen (Editors)*

31st INTERNATIONAL CONFERENCE ON HIGH ENERGY PHYSICS AMSTERDAM

# Fragmentation of Quarks and Gluons

M. Siebel[a]

[a]University of Wuppertal, Gaußstraße 20,
D-42097 Wuppertal, Germany
DELPHI-Collaboration

Experimental results on the hadronisation process under special consideration of the differences between quark and gluon fragmentation are presented.

## 1. Introduction

The fragmentation process, i.e. the formation of colour singlet hadrons out of quarks and gluons is still lacking a concise theoretical description. Therefore, the phenomenological results of experimental studies of the fragmentation features are an interesting topic of research. In this article several experimental results are presented, beginning with the properties of whole events, followed by studies concerning properties of quark and gluon jets and concluding with the investigation of the production of several explicitly identified particles.

## 2. Multiplicity

One of the most obvious results of fragmentation is a large hadronic multiplicity in the final state. In 2000 the LEP storage ring has been run with a center of mass energy up to 208GeV. The DELPHI-experiment has collected $164\text{pb}^{-1}$ with a mean center of mass energy of 206GeV. From the integrated $\xi_E$-distribution measured at this energy the charged multiplicity has been calculated. The main background for this measurement are events with initial state radiation or events with intermediate states with two gauge-bosons. After appropriate cuts 3047 events are left with expected 682 remaining background events. From this the charged multiplicity has been measured as

$$N_{ch}(206\text{GeV}) = 28.03 \pm 0.22_{\text{stat.}} \pm 0.27_{\text{sys.}} \quad .$$

The spread $D$ of the multiplicity distribution and the position of the maximum of the $\xi_p$-distribution have been measured as $D(206\text{GeV}) = 8.97 \pm 0.16_{\text{stat.}} \pm 0.13_{\text{sys.}}$ and $\xi_p^*(206\text{GeV}) = 4.21 \pm 0.04_{\text{stat.}} \pm 0.02_{\text{sys}}$. The value found for $N_{ch}$ agrees well with the NLLA extrapolation of the multiplicities found at lower center of mass energies. The ratio $N_{ch}/D$ gained at 206GeV is also found in agreement with measurements at lower energies thus confirming the hypothesis of Koba-Nielsen-Olesen scaling which predicts energy independence of this ratio. [1]

Using impact parameter techniques also the multiplicity of events with initial $b$-quarks has been investigated. With an unfolding procedure the multiplicity of $b\bar{b}$-events has been found to be $\langle n \rangle_{b\bar{b}}(206\text{GeV}) = 28.72 \pm 0.70_{\text{stat.}} \pm 0.32_{\text{sys.}}$ thus giving the difference in multiplicity between $b$-quark and light-quark events as $\delta_{bl}(206\text{GeV}) = 4.50 \pm 1.05_{\text{stat.}} \pm 0.52_{\text{sys.}}$. The value is in agreement with values found at lower center of mass energies. This confirms the MLLA expectation of an energy independent $\delta_{bl}$ due to coherent gluon radiation and the so-called dead cone effect. Therefore, a flavour independent fragmentation model which predicts a decrease of $\delta_{bl}$ with energy is disfavoured. [2]

In another analysis of DELPHI, the charged multiplicity of three jet events from $Z$-decays has been investigated. A MLLA-calculation in the framework of the colour-dipole model gives the multiplicity of a hadronic three jet event as

$$N_{q\bar{q}g} = N_{q\bar{q}}(L,\kappa) + \frac{1}{2} N_{gg}(\kappa) \quad , \qquad (1)$$


doi:10.1016/S0920-5632(02)02124-2

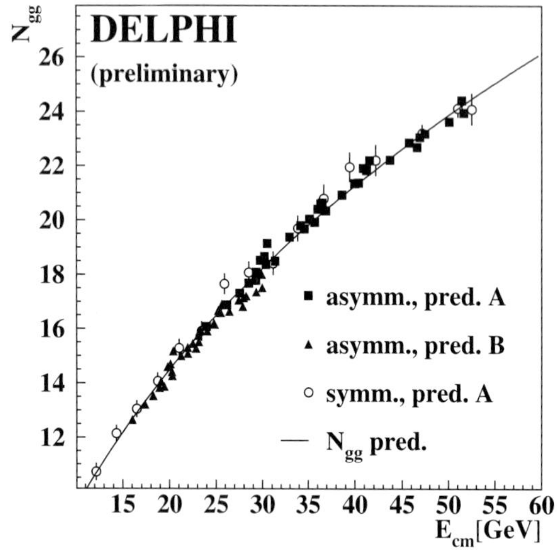


Figure 1. The multiplicity of a two-gluon colour-singlet system in dependence of an effective center of mass energy as obtained from the three jet event multiplicities

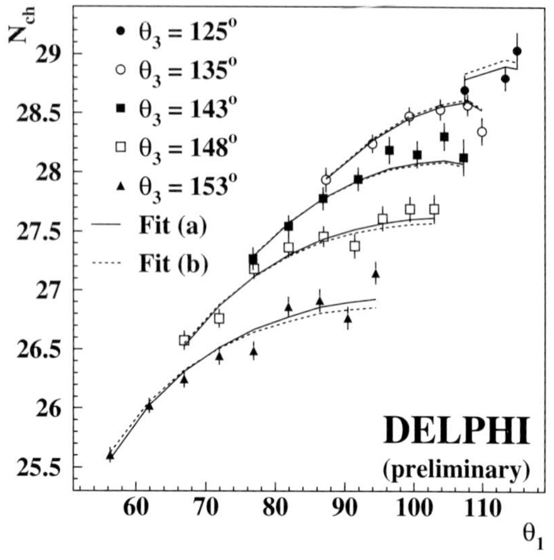


Figure 2. The charged multiplicity of three jet events as function of the event topology

where L denotes the logarithmic scale for the $q\bar{q}$-system and $\kappa$ represents the logarithmic $p_t$-like scale for the gluon. The $q\bar{q}$-contribution depends also on the gluon scale, as the resolution of a gluon jet restricts the phase space of the $q\bar{q}$-pair. Coherence effects are part of the calculation and are reflected by the choice of scales. Two different choices for the scales have been given, differing in the normalisation of the gluon-$p_t$ and the choice of the invariant mass of the $q\bar{q}$-system as either the energy for the quark contribution (formulation a) or the event energy (formulation b). This corresponds to different definitions of the gluonic contribution to the total event multiplicity. Assuming massless kinematics, the dynamical scales can be calculated from the event topology. Solving Eqn. 1 for $N_{gg}$, it is possible to extract the multiplicity of two-gluon colour-singlet systems from the measured three jet multiplicity. The function $N_{q\bar{q}}$ has been taken from an NLLA calculation fitted to the multiplicities measured in $e^+e^+$-annihilation at various $\sqrt{s}$. As no $b$-tagging procedure has been applied to the three jet event sample, whereas the data used to determine $N_{q\bar{q}}$ are corrected for the varying $b$-quark contributions, an additional offset $N_0$ has been introduced to Eqn. 1 and fitted to the data. This is justified, as the additional multiplicity due to $b$-quarks is constant over a wide energy range, as stated above. The extracted gluon multiplicities are shown in Fig. 1. Due to the different gluon scale variables, the two formulations result in gluon multiplicities for different energy ranges.

The energy dependence of $N_{gg}$ is related to $N_{q\bar{q}}$ via the colour factor ratio $C_A/C_F$:

$$\left.\frac{dN_{gg}(L')}{dL'}\right|_{L+c_g-c_q} = \frac{C_A}{C_F}\frac{dN^h_{q\bar{q}}(L)}{dL}\left(1-\frac{\alpha_0 c_r}{L}\right) \quad (2)$$

where $L$ again denotes a logarithmic energy scale, $c_g, c_q, c_r, \alpha_0$ are calculated constants. $C_F$ and $C_A$ are the $SU(3)_c$ Casimir operators. Integrating Eqn. 2 gives a prediction for $N_{gg}$ which is shown as a solid line in Fig. 1. The constant of integration left free by Eqn. 2 has been fixed by a direct measurement of $N_{gg}$ from $\chi_{b2}(2s)$-decays by CLEO. The gluon multiplicities extracted from 3 jet events are in good agreement with the prediction and a previous measurement using only symmetric 3 jet events.

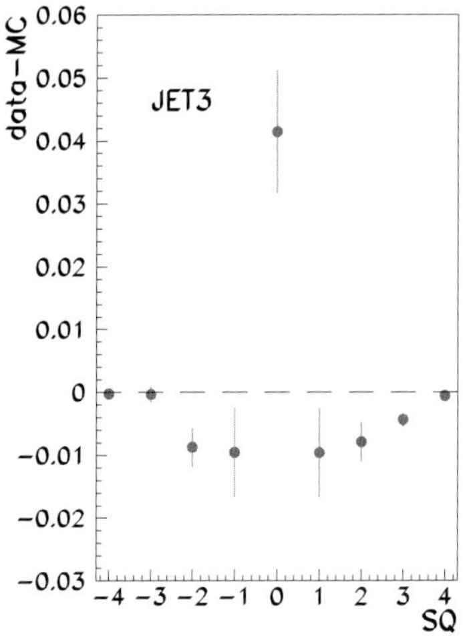


Figure 3. The differences between data and Monte-Carlo simulation for the probabilities of leading systems of specified charge for the gluon enriched sample

Alternatively, inserting the integrated $N_{gg}$ from Eqn. 2 into Eqn. 1, a prediction for the topology dependence of the three jet event multiplicity is given, which depends on the colour factor ratio $C_A/C_F$. A fit of the colour factor ratio to the measured three jet multiplicities yields $2.277 \pm 0.02 \pm 0.05$ and $2.093 \pm 0.05 \pm 0.08$ for formulations a and b of the prediction. Both values are in agreement with the QCD expectation of 2.25. The values found for the offset $N_0$ are $\simeq 0.8$ for formulation a and $\simeq 0.4$ for formulation b, while a value $\simeq 0.62$ is expected. The fits are shown in Fig. 2. [3]

## 3. Leading systems

Different from the hadronisation in quark jets, the colour blanching in gluon jets can proceed not only via colour triplet neutralisation, i.e. the production of $q\bar{q}$-pairs, but also via the process of colour octet neutralisation, i.e. the production of additional gluons. The existence of this process is necessary for the production of particles with gluons as valence particles. A proposed signature of this process is an abundance of electrically neutral leading systems in gluon jets. Here, the leading system of a jet is defined by particles with rapidity $y > 2$ and a subsequenting rapidity gap. This is realised by demanding that no particles in the jet have a lower rapidity. A sample enriched with gluon jets is obtained by selecting the least energetic jet of each three jet event. This sample has a gluon jet fraction of $\sim 70\%$. For comparison a quark jet enriched sample is obtained by selecting the most energetic jet of each three jet event. The quark jet purity of this sample is $\sim 90\%$. The probability for the occurrence of a leading system with a given charge is for both jet samples compared to the expectations from a Monte-Carlo simulation which does not include the octet neutralisation mechanism. In Fig. 3 the probability differences between data and simulation are shown as function of the charge of the leading system for the gluon jet sample. A significant higher occurrence of neutral leading systems in the data than in the simulation can be seen to an extend of more than four standard deviations. The same difference for the quark jet sample on the other hand is consistent with zero. This can be interpreted as a signature of the colour octet neutralisation mechanism.

To investigate the production of particles with valence gluons, the mass spectra of the neutral leading systems have been compared to the Monte-Carlo expectations. The quark jet enriched sample is found in good agreement with the simulation. Also no significant deviations from the simulation are found in the gluon jet enriched sample, although deviations are larger than in the quark jet sample and the surplus of neutral leading systems is concentrated below masses of 2GeV. [4]

## 4. Identified particles

Detailed information about the fragmentation process can be obtained by looking at the properties of different identified particles. Analyses on this topic have been performed by SLD in hadronic $Z$ decays from $e^+e^-$-annihilations. A veto against secondary vertices leaves only decays of the $Z$ into the light quarks $u\bar{u}/d\bar{d}/s\bar{s}$. The identification of the particles is done via a Ring Imaging Cherenkov Counter (CRID).

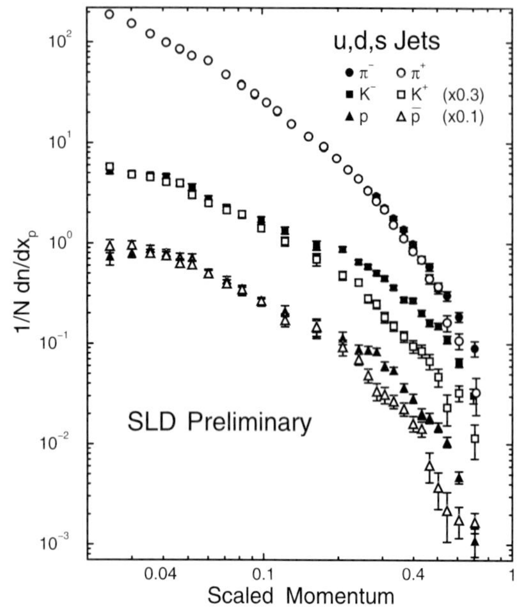


Figure 4. Production rate of $p$, $\pi^-$, $K^-$ and their anti-particles

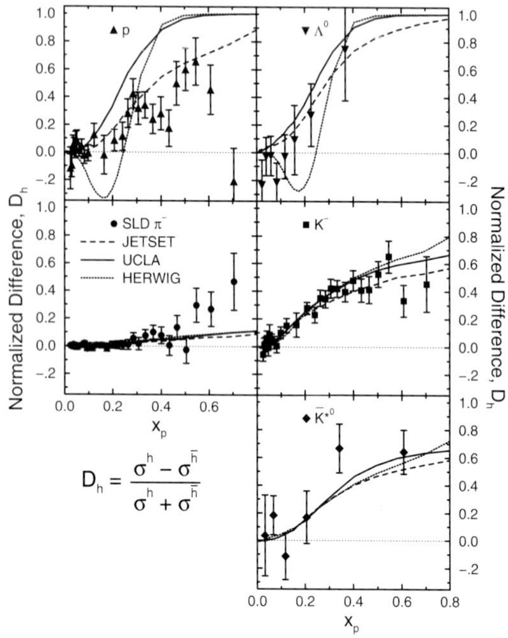


Figure 5. Normalised difference between the production rates of particles and their anti-particles

Due to the polarisation of the electron beam and the large production asymmetry of the $Z$-boson, the direction of the outgoing quark can statistically be distinguished from the direction of the anti-quark. Demanding an angle between the thrust axis and the beam pipe which is small enough to avoid the region perpendicular to the incoming beams where the asymmetry is low, the selected hemispheres have a quark purity of 0.73.

In Fig. 4 the distribution of the scaled momenta for pions, kaons and protons are plotted for the hemisphere of the quark. At high momenta there are e.g. more protons than anti-protons produced in quark jets. This can be interpreted as evidence for the production of protons as baryonic leading particles. Analogously more $K^-$ with high momentum are produced in the quark direction than $K^+$ due to the production of leading kaons in events with an initial $s\bar{s}$-pair, while the case of a leading $K^+$ with a $\bar{s}$-quark produced in the hadronisation process is less frequent. For pions the difference is smaller but a significantly higher production rate of $\pi^+$ over $\pi^-$ can still be seen above a scaled momentum of $\sim 0.3$. A more concise view of this behaviour can be seen in Fig. 5. Here the normalised difference between the production rates of a particle and its anti-particle, $D_h = (R_h^q - R_{\bar{h}}^q)/(R_h^q + R_{\bar{h}}^q)$, is shown. The significant overproduction of protons over anti-protons can be seen as a positive deviation from zero above $x_p \sim 0.2$. Also for pions a positive difference can be seen. Assuming equal leading pion production in $u$ and $d$ jets, a difference of 0.27 would be expected, while the expected difference is rather compatible with $0.27 \cdot D_p$. The positive deviation for kaons is well above this value for large $x_p$. Concluding, leading particle production has been observed in the obtained data without any model related assumptions and the fraction of produced leading pions has found to be similar to the fraction for leading protons. The higher value found for leading kaon production indicates additionally a suppression of leading kaon production in events with initial $u$- or $d$-quarks. [5]

Another property of the hadronisation process is the correlated production of particles. A corre-

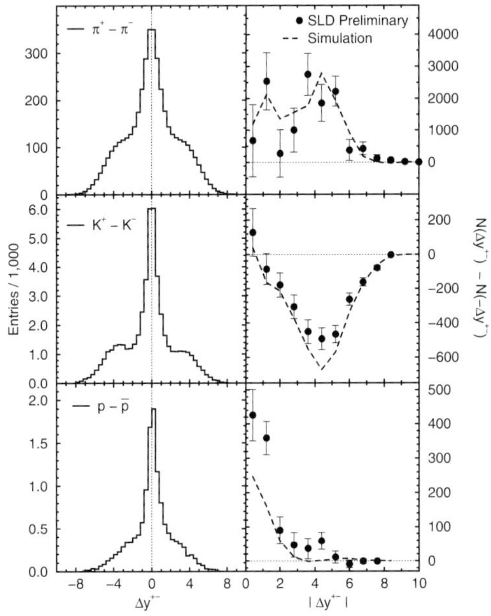


Figure 6. Distribution of the signed rapidity differences for $\pi$, $K$ and $p$

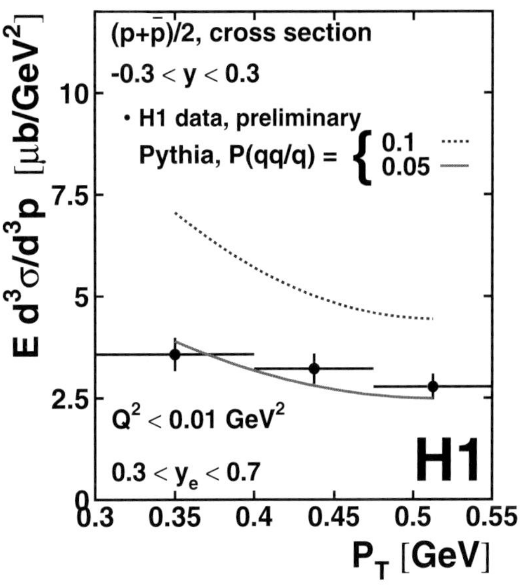


Figure 7. Differential photo-production cross section for protons

lation study by SLD is done via the difference in rapidity $y$ of the produced particles. The rapidity has been given a sign defined by the direction of a quark. The left column in plot Fig. 6 shows the differences of the signed rapidities for the identified positively charged particles and their antiparticles $\Delta y^{+-} = y(h^+) - y(h^-)$. So the sign of the rapidity difference is given by the direction of the quark and the charge of the hadron. Therefore, changing the sign of the rapidity difference corresponds to the C-conjugation of the quark. Thus any asymmetry in this distribution marks a difference between the fragmentation of a quark and an anti-quark and can give evidence of leading particle production.

The asymmetries in these distributions are shown in the right column of Fig. 6, where the difference between the distribution for positive and negative values of $\Delta y^{+-}$ is plotted against $|\Delta y^{+-}|$. For $K^+K^-$ correlations there is a large asymmetry at high rapidity differences reflecting the production of kaons as leading particles in opposite jets of $s\bar{s}$-events. A similar, but weaker effect appears in the plots for pion and proton pairs, too. The large asymmetry at low values of $|y^{+-}|$ in the proton plot can be understood as the close-by production of $p\bar{p}$-pairs with the proton preferably flying in the direction of the initial quark. This can be interpreted as evidence for local baryon number conservation and baryon number ordering. [6]

The final state of fragmentation has also been investigated in the framework of electron-proton collisions. In an analysis of the H1 experiment at HERA, the photo-production of protons has been studied. In photo-production processes the proton collides with a quasi-real photon radiated off the electron. This requires a low value of $Q^2$, so that the kinematic region for this analysis is restricted to $Q^2 < 0.01\text{GeV}^2$ and $0.3 < y < 0.7$. The produced protons are identified using the specific energy loss $dE/dx$ in the central tracker. In Fig. 7 the differential production cross section for protons is shown as function of $p_t$, it is found to be slightly decreasing with $p_t$. The dashed line indicates the expectations from PYTHIA Monte-Carlo simulation. One important simulation parameter for the production of baryons is the probability of di-quark production. In $e^+e^-$-

| $ep \to eDX$ | $D^+$ | $D^0$ | $D_s^+$ | $D^{*+}$ |
|---|---|---|---|---|
| $\sigma_{\rm vis}(nb)$ | 2.16 | 6.53 | 1.67 | 2.90 |
| stat. error | $\pm 0.19$ | $\pm 0.49$ | $\pm 0.41$ | $\pm 0.20$ |
| syst. error | $^{+0.46}_{-0.35}$ | $^{+1.06}_{-1.30}$ | $^{+0.54}_{-0.54}$ | $^{+0.58}_{-0.44}$ |

Table 1
Measured proton production cross sections

experiments the simulation described the measurements best with an di-quark production ratio of $P(qq/q) = 0.1$. Obviously, this curve overestimates the cross section significantly. Using a di-quark production ratio of half the size, the simulation describes the overall shape as well as the absolute value of the measured cross section reasonably. [7]

In another analysis by H1 the production of $D$ mesons has been studied. The $D$ mesons have been reconstructed along the decay modes $D^+ \to K^-\pi^+\pi^+$, $D^0 \to K^-\pi^+$, $D_s^+ \to \phi\pi^+ \to (K^+K^-)\pi^+$ and $D^{*+} \to D^0\pi^+ \to (K^-\pi^+)\pi^+$, with C-conjugation implied. For this analysis the kinematic region extended over $2\text{GeV}^2 \leq Q^2 \leq 100\text{GeV}^2$ and $0.05 \leq y \leq 0.7$ for the event as well as $p_t(D) \geq 2.5\text{GeV}/c$ and $|\eta(D)| \leq 1.5$ for the reconstructed $D$. The production cross section and their errors for the different mesons are given in Tab. 1. Also the differential cross sections have been studied and found to be in good agreement with the LO Monte-Carlo simulation. The fragmentation factors $f(c \to D)$ have been determined from the cross sections and the fraction of produced vector mesons $P_V$ has been calculated from the $f(c \to D)$ with two different approaches leading to $P_V = 0.693 \pm 0.045 \pm 0.004 \pm 0.009$ and $P_V = 0.613 \pm 0.061 \pm 0.033 \pm 0.008$ compatible with the value $P_V = 0.601 \pm 0.032$ found in $e^+e^-$-annihilation. Attention should be paid to the circumstance that the production of e.g. tensor mesons did not enter the calculation. [8]

## 5. Summary

Several features of the fragmentation process of quarks and gluons have been studied. The multiplicity of hadronic events has been measured at high energies. At lower energies the multiplicity of three jet events has been investigated regarding the gluon and quark specific contribution, giving a measurement of the ratio of the effective colour charges $C_A/C_F$ of gluons and quarks. Evidence for the existence of colour octet neutralisation processes has been found in the overproduction of neutral leading systems in gluon jets. The study of identified particles gave insight into the production of leading particles and local baryon number conservation. In electron proton scattering the production of photons and $D$-Mesons have been studied.

I would like to thank the organizers of the ICHEP'02 for their efforts to make the conference the inspiring experience it has been.

## REFERENCES


1. DELPHI Collaboration, DELPHI 2002-051 CONF 585, ABS231[1]
2. DELPHI Collaboration, DELPHI 2002-052 CONF 586, ABS232[1]
3. DELPHI Collaboration, DELPHI 2002-060 CONF 594, ABS247[1]
4. DELPHI Collaboration, DELPHI 2002-053 CONF 587, ABS233[1]
5. SLD Collaboration, SLAC-PUB-9287, ABS970[1]
6. SLD Collaboration, SLAC-PUB-9288, ABS971[1]
7. H1 Collaboration, ABS1002[1]
8. H1 Collaboration, ABS1015[1]


[1]contributed paper to ICHEP 2002, Amsterdam, 23-31 July 2002

Proceedings of ICHEP 2002, pp. 381–384
S. Bentvelsen, P. de Jong, J. Koch and E. Laenen (Editors)

# Statistical Models and Thermalization

J.Manjavidze[a]* and A.Sissakian[a]

[a]Joint Institute for Nuclear Research, Dubna, Russia

The status of thermodynamical approach is discussed. This approach is quiet popular in the heavy ion collision physics. It is argued that the "principle of vanishing of correlations" must be used for quantitative estimations of the rate of thermalization. Existing theoretical models and corresponding generator of events are unable to predict the thermalization phenomenon. A brief review of attendant papers is given.

## 1. Introduction

The question of *applicability of thermodynamical description in hadron physics* seem crucial. Its usage in hadron dynamics is reasonable since the inelastic collisions is the typical energy dissipation processes. This point of view was explored intensively by Fermi and Landau [1] and later by other authors [2].

It must be noticed that *if the system can be described by the thermodynamical "rough" parameter, then the system is equilibrium over it.* This important observation is a theoretical basis of our approach.

So, if only the temperature is sufficient for complete definition of the energy spectrum, then the system certainly is in the thermal equilibrium. But, generally speaking, it can be outside of equilibrium relative to other parameters. Generally, the thermalized state is characterized by uniform distribution of the energy perturbation over all degrees of freedom.

The ordinary statistical models predict the mean multiplicity proportional to the its threshold value, i.e. the statistical models assume the complete thermalization.

But the experimental data on the mean multiplicity, $\bar{n}(s) \sim \ln\sqrt{s}$, where $\sqrt{s}$ is the total CM energy. This indicates that the hadron interactions final state is far from equilibrium. It is impossible to use in this case the notions of thermodynamics.

The point is that the governing hadron dynamics non-Abelian symmetry constrain prevents the thermalization at the early stage of the process. This is why the multiple production dynamics is so complicated.

Therefore, it is very important to investigate: (i) the existence of phenomenological indications of statistics and (ii) the quantitative necessary and sufficient indication that the thermalized state is produced.

These two questions are the subject of the present paper. The introduction into the thermodynamics of hadron inelastic collisions one may find in [3]. We hope that the thermodynamical approach to the hadron dynamics is applicable at least for the case of very high multiplicities (VHM).

## 2. Phenomenological indications of statistics

One should mention the growing interest in the statistical approach in the inelastic hadron collisions. The ideological basis of it was originated by the number of authors [4]. In a series of later papers, part of which were presented at this Conference, the phenomenological basis of this approach is justified.

It should be noted that most of the presented papers are devoted to the description of the heavy ion central collisions.

— It was noted that the analysis of particle production yields in central Au-Au collisions at RHIC based on the statistical thermal model is in good agreement with the experimental data,

*Permanent address: Inst. of Physics, Tbilisi, Georgia.


doi:10.1016/S0920-5632(02)02125-4

see [5].

— It was shown that the improved statistical model allows one to conclude that the chemical equilibrium is reached in Pb-Pb collisions at SPS energies, see [6].

— The statistical model application indicates that the production of different particles and their momentum spectra are governed by one parameter, see [7].

### 3. Structure of the phase space

In spite of the progress of heavy ion collisions statistical description, it seems reasonable to be critical and to find the quantitative sign of its applicability.

It is useful for this purpose to have a transparent picture of the multiple production kinematics to classify various mechanisms of particle production.

One may consider the phase space projection on the longitudinal, $(p_{\|})$, vs. transverse, $(p_{\perp})$, momentum plane, see Fig.1.

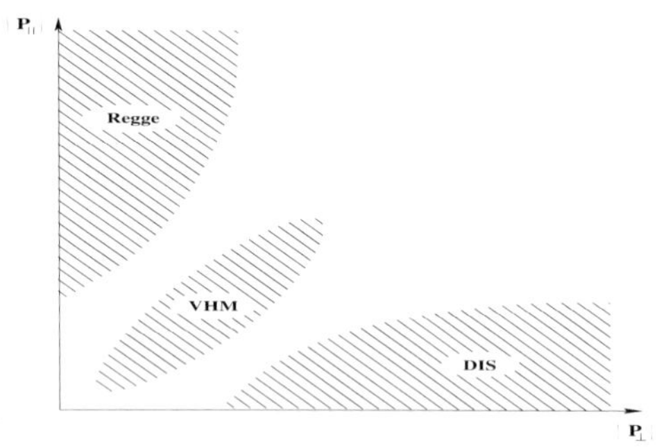


Figure 1. The $(p_{\|}, p_{\perp})$ plane

One may extract two domain, well described by the dynamical models:

— The multiperipheral-type models belong to the "Regge" domain. This kinematic is characterized by small and independent from initial energy and multiplicity the mean value of transverse momentum;

— The region of applicability of the LLA ideologies of pQCD belongs to the "Deep Inelastic Scattering" (DIS) domain. In this case the transverse momentum of produced hadrons is larger than the longitudinal one;

— In this case the transverse and longitudinal momenta of produced particles are comparable to each other and one can expect in this case the equilibrium among these degrees of freedom. The final state of inelastic heavy ion-ion collision belongs, most likely, to the intermediate region.

So, one may assume that the place of thermodynamics lies between the "Regge" and "DIS" domains. It is the "VHM" domain. To describe this region, a new perturbation theory, "topological QCD", was developed, see [8].

### 4. Necessary and sufficient condition of thermalization

The second step toward statistical description actually is a search of the necessary and sufficient condition of thermalization.

We can argue that if the inequality:

$$|K_l(E,n)|^{2/l} << K_2(E,n), \; l = 3, 4, ..., \qquad (1)$$

is satisfied then the statistical description actually occurs. Therefore, to have a right to introduce, for example, the temperature and to use the statistical models, one should check the experimental statistics from the point of view of this inequality.

If this condition (1) works, then it is possible to describe the system using thermodynamics.

One can observe here the analogy with the principle of "vanishing of correlations", offered by N.N. Bogolyubov for statistical physics, see [9].

It should be stressed that the conclusion (1) is general for hadron dynamics

### 5. The scenario of transition to thermalized state

We may offer a scenario as the thermalized state achieved with increasing multiplicities in hadron(ion) collisions. The Fig.2 illustrates it in terms of multiplicity distribution. So,

— Multiperipheral picture applicable up to $n \sim \bar{n}^2$ and the "hard" processes take place at larger values of $n$.

— The LLA ideology of pQCD is applicable up to $n \sim n_h$.

— **B** is the VHM region, where thermalization

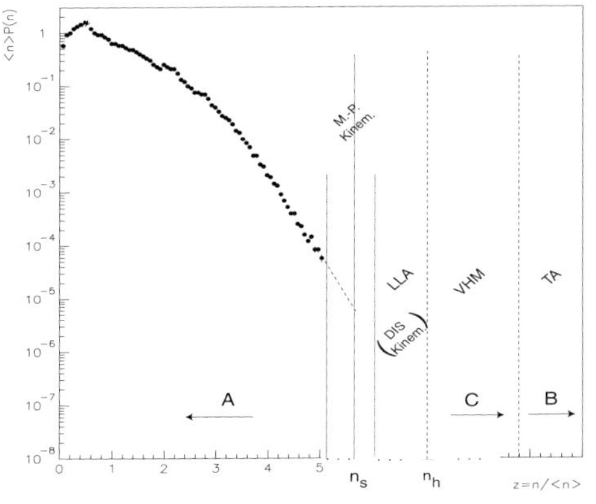


Figure 2. The multiplicity distribution. Dots from E-735 (FNAL) data.

occurs and all the secondary particles momentum are comparable to each other.

— **C** is the domain of the "ideal gas approximation".

## 6. Prediction of generators of events

It is interesting to investigate prediction of the generators of events.

— Fig.3 presents prediction of the generator PYTHIA for LHC energy. As follows from Fig.3, one may conclude that the dynamical models underlying the PYTHIA can not predict thermalization. This conclusion confirms our general prediction that no thermalization may occur in the "Regge" and "DIS" domains.

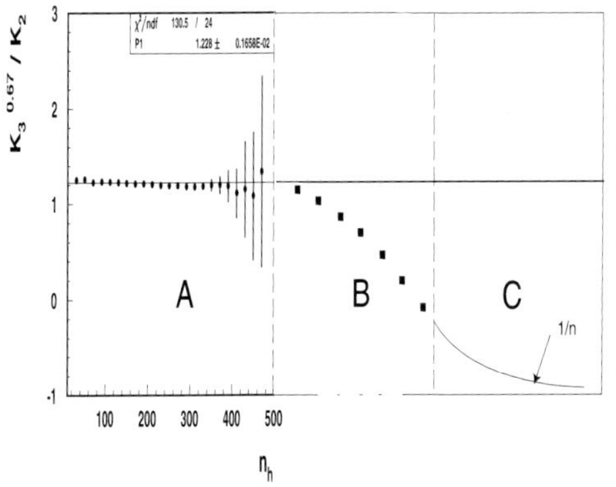


Figure 3. PYTHIA prediction for $|K_3|^{3/2}/|K_2|$ in the domain $A$.

— Fig.4 shows the prediction of the generator HIJING. It shows tendency to equilibrium since it takes re-scattering into account.

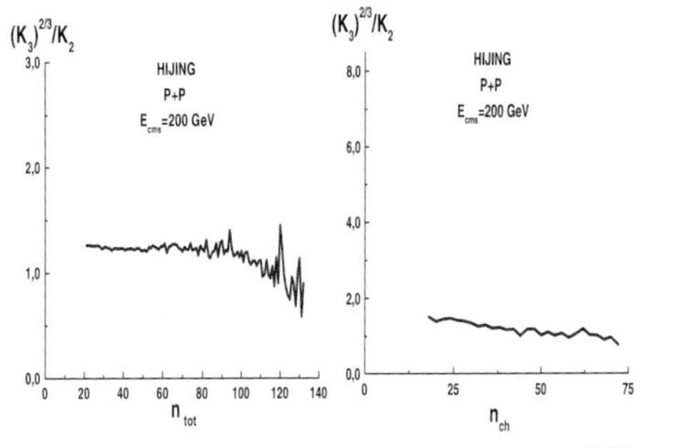


Figure 4. HIJING prediction for $|K_3|^{3/2}/|K_2|$

This shows that, presumably, the effect of thermalization may be seen in the heavy ion collisions at smaller multiplicities.

## 7. Toward the experiment

The problem of thermalization seems as a priority program in description of inelastic hadron reactions. For this purpose one must measure the ratio $|K_3|^{2/3}/|K_2|$, see (1).

It is important investigate the ratio of the mean values of the produced particles momentum $(\bar{p}_{\parallel}(E,n)/\bar{p}_{\perp}(E,n))$. If our conclusion about of dominance of central collisions in the VHM region then this ratio must tend to $\pi/4$ from above.

In conclusion, one may introduce also the "chemical potential" [3]

$$\mu(E,n) = - < \epsilon > \ln \frac{\sigma_n(E)}{\sigma_{tot}(E)} \quad (2)$$

if "the ratio $K_3$ to $K_2$" less than one. This interpretation of $\mu$ will allow to observe the first order phase transitions.

## 8. Conclusion

A few notes as conclusion.

— Definite indications of thermalization phenomena exist in the heavy ion collisions.

— Besides, some resent publications on $p\bar{p}$ collisions contains an interesting attempt of experimental observation of the colored plasma-like state which can be described by the "rough" parameters like temperature [10]

— The elaboration of quantitative estimations in the VHM kinematical region is outside of the LLA of pQCD abilities. Moreover, the ordinary dynamical models can not predict even the tendency to equilibrium.

— It is important that we have found the necessary and sufficient condition of thermalization and its $S$-matrix interpretation. This gave us a possibility to show that the thermalization must occur, at least, in a deep asymptotics over multiplicity.

— A new strong coupling perturbation theory, the "topological QCD", was offered. It is important that it includes the pQCD as the definite approximation.

— The experimental side of VHM physics is discussed intensively from various points of view by the ATLAS, CDF, STAR Collaborations.

— From the practical point of view, the tQCD "fast" generator of VHM events is the most important problem. The experimental side of this issue includes the problems of the trigger and absence of any experience in the analysis of the VHM events.

And at the very end, we would like to notice that only the statistical approach can give a chance of complete description of inelastic hadron collisions.

## Acknowledgement

Authors would like to thank Yu. A. Budagov, L.L. Enkovski, L.Gutay, V. G. Kadyshevski, E. A. Kuraev, E. M. Levin, L. N. Lipatov, V. A. Matveev, V. A. Nikitin, A. G. Olchevski, N. A. Russakovich, A. N. Tavkhelidze, E. Sarkisyan-Grinbaum for valuable discussions and important comments. We would like also to thank Yu. Kulchitsky, N. I. Shubitidze and V. Uzhinsky for realization of the Monte Carlo simulation programs.

## REFERENCES

1. E.Fermi, Progr. Theor. Phys., 4 (1950) 570, Phys. Rev., 81 (1950) 115, *ibid* 92 (1953) 452; L.D.Landau, Izv. AN SSSR, 17 (1953) 85
2. R. Hagedorn (1965), I.M. Dremin and V.A. Nechitailo, hep-ph/0207068; E.L. Feinberg, Phys. Rep., 56 (1972) 237; E. Shuryak, hep-ph/0205031
3. J.Manjavidze and A.Sissakian , Phys. Rep., 346 (2001)
4. J.Schwinger, J.Math.Phys., A9 (1994) 2363; L.Keldysh, Sov.Phys. JETP, 20 (1964) 1018; A.J.Niemi and G.Semenoff, Ann.Phys. (NY), 152 (1984) 105; P.Carruthers and F.Zachariasen, Phys.Rev., D13 (1986) 950; J.Manjavidze, El. Part. At. Nucl., 30 (1999) 124
5. F.Becattini, et al, hep-ph/0002267; hep-ph/00110221; hep-ph/0206203; P.Braun-Munzinger, et al., nucl-th/9903010; U.Heinz and P.F.Kolb, hep-ph/0204061
6. U.Henz, Nucl.Phys., A661 (1999) 140c; P.Braun-Munzinger, et al., hep-ph/0105229; H.Oeschler, nucl-ex/0011007; Zhong-Dao Lu, hep-ph/0207029; R.Baier et al., hep-ph/0204211
7. J.B.Elliot et al., Phys. Rev. Lett., 85 (2000) 1194; C.Tsallis, Lect. Notes in Phys. LNP 560 (2000), G.A.Kozlov, New J. Phys., 4 (2002) 23; D.Kharzeev, hep-ph/0204015
8. J.Manjavidze, Th. Math. Phys., 41 (2000) 5710, J.Manjavidze and A.Sissakian, J. Math. Phys., 42 (2001) 641, 42 (2001) 4158; J.Manjavidze and A.Sissakian, Th. Math. Phys., 123 (2000) 776, 130 (2002) 153
9. N.N.Bogolyubov, Studies in Statistical Mechanics, (North-Holland, Amsterdam, 1962)
10. A.Erwin, L.Cuty et al., Phys. Lett., B528 (2002) 43

ELSEVIER

Proceedings of ICHEP 2002, pp. 385–390
*S. Bentvelsen, P. de Jong, J. Koch and E. Laenen (Editors)*

31st INTERNATIONAL CONFERENCE ON HIGH ENERGY PHYSICS AMSTERDAM

# Bose-Einstein Correlations at LEP & HERA

Š.Todorova-Nová[a]

[a]FzU Prague, Czech Republic, and HEFIN Nijmegen, The Netherlands

Latest measurements of Bose-Einstein correlations at LEP and HERA colliders are reviewed.

## 1. Introduction

The enhanced production of 'close' pairs, or multiplets, of identical bosons is readily observed in high energy interactions. The effect is considered to be the outcome of quantum-mechanical interference (Bose-Einstein symmetrization) in the particle production mechanism.

The basic variable used in correlation measurements is n-particle density

$$\rho(Q) = \frac{1}{N_{ev}} \frac{dN(multiplets)}{dQ} \quad (1)$$

expressed in terms of n-particle momentum transfer $Q$

$$Q = \sqrt{M^2 - n^2 m^2}, \quad (2)$$

where $M$ stands for the mass of the multiplet, and $m$ is the mass of the bosons forming the multiplet. The correlation function is defined with respect to an uncorrelated reference sample as

$$R(Q) = \frac{\rho(Q)}{\rho_{ref}(Q)}. \quad (3)$$

The choice of reference sample is not unique, the following options are frequently found (not necessarily giving identical results): simulated samples without BE modelling, mixing of tracks from different events, or distributions of non-identical bosons.

Furthermore, the measured correlation function depends on many factors, reflecting the phase-space dependence (event and track selection, detector acceptance) and the sensitivity of the measurement (amount of secondary tracks in the sample, (mis)identification of particles, fraction of particles from the decay of long-lived resonances).

The correlation function carries information about the relative size of BE correlations, and the space-time properties of the particle source. Historically (Goldhaber et al.,1959), the particle correlations are treated in close analogy with the Hanbury Brown and Twiss (HBT) effect (intensity interferometry in astronomy), assuming the particle source is static, uniform, and gaussian. The correlation function is fitted with

$$C(Q) = 1 + \lambda \cdot \exp^{-(R\ Q)^2} \quad (4)$$

and parameters $R$, $\lambda$ are interpreted as radius of the source and the measure of incoherence of the particle emission. This is most probably an oversimplification: in particle physics, we deal with rapidly expanding systems with strong momentum-position correlations, and different event topologies are integrated over. Also, the HBT interpretation restricts the BE symmetrization to the system of final particles without taking into account the particle production stage (production amplitudes), thus obscuring the notion of (in)coherence.

The limitation of the above mentioned parameterization is also highlighted by the observation of a non-gaussian shape of the correlation function, and its asymmetry in multidimensional measurements. Some of these observations will be discussed below.

## 2. BE correlations in DIS

The ZEUS collaboration studied the dependence of the 2-particle correlations on the virtuality of the exchanged photon ($Q^2_\gamma$) in deep inelastic ep scattering [1], see Fig.1.


doi:10.1016/S0920-5632(02)02207-7

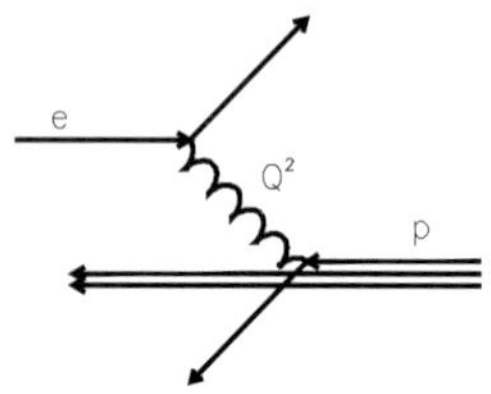


Figure 1. Kinematics of DIS events.

Data from the 1998-2000 period (83.0±1.2 $pb^{-1}$), with electron(positron) beam energy 27.5 GeV and proton beam energy 920 GeV, were used for this measurement. 2-particle densities were calculated for pairs of like-sign(±,±) and unlike-sign(+,-) charged particles (without further identification). The quantity $\xi = \rho(\pm,\pm)/\rho(+,-)$ was calculated and corrected for the resonance contribution using simulated data $R = \xi^{data}/\xi^{MC,no\ BE}$. The double ratio R was then fitted by the formula

$$R = \alpha(1 + \beta Q_{12})(1 + \lambda e^{-r^2 Q_{12}^2}), \qquad (5)$$

where $Q_{12}$ stands for 2-particle momentum transfer, and the $\beta$ term takes into account a long-distance non-BE contribution. The $K_0^s$ and $\rho^0$ regions were excluded from the fit.

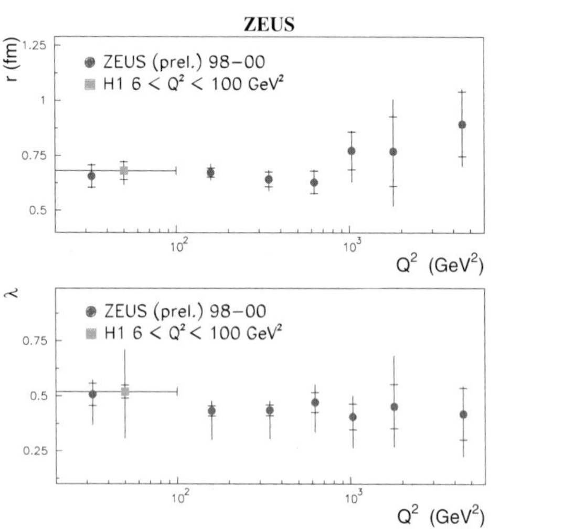


Figure 2. $Q^2$ dependence of the fitted parameters.

The $Q^2$ dependence of the fitted parameters $r, \lambda$ is shown in Fig.2. No significant change was observed in the range $25GeV^2 < Q^2 < 5000GeV^2$.

The fit for data with $Q^2 > 110\ GeV^2$ gives

$$r = 0.671 \pm 0.016(stat)^{+0.030}_{-0.032}(syst)\ fm,$$
$$\lambda = 0.431 \pm 0.012(stat)^{+0.042}_{-0.130}(syst),$$

in agreement with H1 and LEP measurements [2, 3].

The analysis was repeated in two dimensions using the longitudinally co-moving system [4], where the longitudinal direction was defined by the $\gamma^* p$ axis. The double ratio R was decomposed into the transverse and longitudinal components (see Fig.3) and fitted by

$$R = \alpha(1 + \beta_T Q_T + \beta_L Q_L)(1 + \lambda e^{-r_T^2 Q_T^2 - r_L^2 Q_L^2}) \qquad (6)$$

The values obtained for the data with $Q^2 > 110\ GeV^2$ are

$$r_T = 0.58 \pm 0.05(stat)^{+0.03}_{-0.12}(syst)\ fm,$$
$$r_L = 0.87 \pm 0.06(stat)^{+0.09}_{-0.01}(syst)\ fm,$$
$$\lambda = 0.35 \pm 0.0(stat)^{+0.03}_{-0.05}(syst),$$

indicating an *elongation* of the emitting region. The ratio $r_T/r_L = 0.67 \pm 0.08(stat)$ is consistent with similar LEP measurements [5].

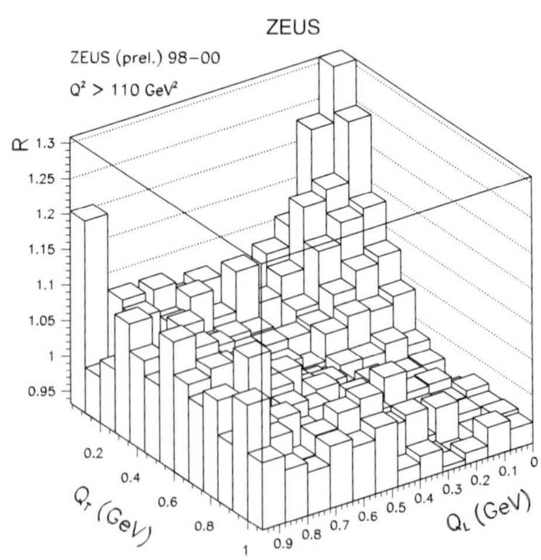


Figure 3. 2-dimensional 2-particle correlations.

## 3. BE correlations of Neutral and Charged Pions in Hadronic Z Decays

A first measurement of 2-particle correlations of neutral pions at LEP was reported by the L3 collaboration [6], and followed recently by another measurement from OPAL [7]. Both measurements are based on LEP1 data.

The reconstruction of pairs of $\pi^0$s proceeds as follows: $\pi^0$ candidates are selected in a given mass window of pairs of reconstructed photons, then paired and fitted, for each bin of $Q$, by a 2-dimensional function (with 18 parameters) estimating the fraction of true $\pi^0\pi^0$ pairs, with respect to the non-$\pi^0\pi^0$ background. The efficiency and purity of reconstruction in the L3 analysis, estimated using simulated data, is 17% and 54% for single $\pi^0$ with energy 0.2-6 GeV. The efficiency of reconstruction of $\pi^0\pi^0$ pairs range from 1% (at $Q$=0.3GeV) to 4% (at $Q$=2GeV). To obtain the correlation function, the $Q$ distribution is corrected for limited efficiency and acceptance with simulated data (Jetset with BE), and divided by a reference MC sample (without BE) at the generator level, with pions selected in the same kinematic range as in the data. To facilitate the comparison between neutral and charged pions, similar selection cuts and a similar correction procedure are applied on charged pions. The resulting correlation functions are shown in Fig.4. Parameters of the fit(Eq.5) of these functions in the interval 0.3 GeV $< Q <$ 2 GeV (the region below 0.3 GeV is excluded due to large systematic uncertainties) are summarized in Table 1.

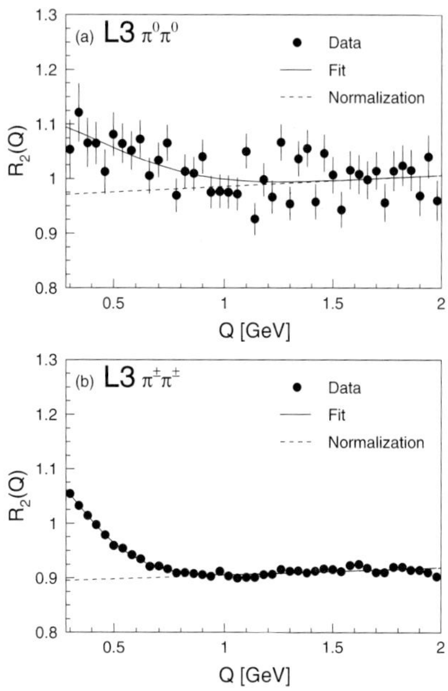

Figure 4. 2-particle correlation function for neutral(a) and charged(b) pairs of pions.

Table 1
Fitted parameters of the 2-particle correlation function for neutral and charged pions [6].

| | $\lambda$ | R(fm) |
|---|---|---|
| $\pi^0\pi^0$ | $0.16 \pm 0.05 \pm 0.07$ | $0.31 \pm 0.07 \pm 0.07$ |
| $\pi^\pm\pi^\pm$ | $0.29 \pm 0.01 \pm 0.03$ | $0.46 \pm 0.01 \pm 0.01$ |

The L3 data indicate both a weaker correlation and a smaller size of the source for neutral pions as compared to charged pions, at the level of 1.5 $\sigma$. The weakness of correlations of neutral pions can be partly explained by the bigger contribution of resonance decays. The difference in the effective size of source is in qualitative agreement with the predictions of the string fragmentation model. However, as it is pointed out in the OPAL analysis [7], the fraction of $\pi^0\pi^0$ pairs produced directly in the fragmentation is very small with respect to resonance decay contribution. Since the OPAL selection cuts and reconstruction efficiency for $\pi^0\pi^0$, as well as the choice the reference sample, differ from those described above, the interested reader is kindly invited to consult the references [6,7] for more details.

## 4. Genuine Three-Particle BE Correlations in Hadronic Z Decays

Three-particle correlations and their relation to the size of two-particle correlations were studied by L3 [8] using the LEP1 data from the 1994 data-taking period.

Under the assumption of totally incoherent particle emission and no momentum dependence of the density of the source, one can derive the relation[9]

$$\omega(Q_{12}, Q_{23}, Q_{31}) = \frac{R_3^{genuine} - 1}{2\sqrt{\Pi_{i \neq j}(R_2(Q_{ij}) - 1)}}, \quad (7)$$

where $\omega = \cos(\phi) = \cos(\sum_{i \neq j} \phi_{ij})$. For incoherent sources, $\omega \to 1$ is expected for $Q_{ij} \to 0$.

The genuine 3-particle correlation function is obtained from the 3-particle correlation function after subtraction of the contribution from 2-particle correlations:

$$R_3^{genuine} = R_3 + R_{1,2} + 1 = \frac{\rho_3}{\rho_0} + \frac{\sum \rho_1 \rho_2}{\rho_0} - 2 + 1,$$

where 3-particle densities $\rho_0, \rho_{1,2}$, corresponding to the uncorrelated sample, resp. sample containing only 2-particle correlations, are obtained via a mixing technique, where single tracks, respectively pairs of tracks from one event, are combined with tracks from another event. In the same way, the 2-particle correlation function $R_2 = \rho_2/\rho_0$ is measured. The measured correlation functions are corrected for detector resolution, limited acceptance, efficiency and pion purity with additional multiplicative factors, derived from simulated data. In addition, the distributions are corrected for Coulomb repulsion.

Table 2
$\omega$ values measured in various interactions.

| interaction | | experiment | $\omega$ |
|---|---|---|---|
| $e^+e^-$ | | L3 | $\approx 1$ |
| $hh$ | | UA1 | $> 1$ |
| | | NA22 | $\approx 1$ |
| $AA$ | $SPb$ | NA44 | $0.2 \pm 0.2$ |
| | $PbPb$ | NA44 | $0.85 \pm 0.22$ |
| | $PbPb$ | WA98 | $0.61 \pm 0.18$ |

The measured and fitted genuine 3-particle correlation function $R_3^{genuine}$ is shown in Fig.5. Since the enhancement of the correlation function is stronger than Gaussian, the fit is repeated using a so-called Edgeworth expansion [10]. From $R_3^{genuine}$ and $R_2$, the $\omega$ is calculated (Fig.6). In both parameterizations, the data are consistent with a fully incoherent pion production mechanism.

The result is compared with similar measurements performed in various types of interactions in Table 2. The spread of measured values does not provide a clear answer to the question of incoherence in the particle production. It should be noted that some of assumptions used to derive the formalism are probably not fulfilled. If so, a satisfactory explanation for the observed relation between 2- and 3-particle correlations is to be found yet.

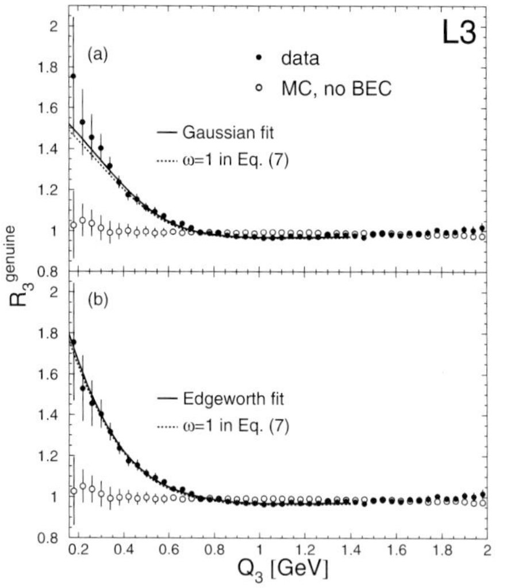


Figure 5. Genuine 3-particle correlations fitted with Gaussian(a) and Edgeworth expansion(b).

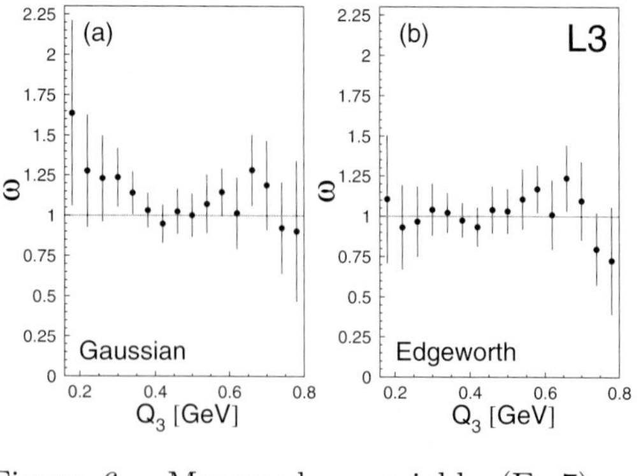


Figure 6. Measured $\omega$ variable (Eq.7), using Gaussian(a) and Edgeworth expansion(b).

## 5. Inter-W Correlations in $e^+e^- \to W^+W^-$ events at LEP

The size of the BE correlations between particles coming from independent hadronic systems can be investigated at LEP2 in the hadronic decay of a pair of W bosons. The life-time of the W boson is small in comparison with the size of the hadronization region, thus allowing a sizable phase-space overlap of the two hadronic systems. Moreover, the possibility to use the mixing of the hadronic parts of real measured semileptonic (semihadronic) WW events for the construction of an uncorrelated reference sample provides an absolute, model independent (except for background subtraction), measurement of inter-W correlations.

Following the method described in [11], L3[12] and DELPHI[13] analysed their data at $\sqrt{s}$ = 189 - 209 GeV. Comparison of charged 2-particle densities from hadronic WW events (after background subtraction) and from mixed (reference) WW events is done via the subtraction or ratio

$$\Delta\rho = \rho^{WW}(hadr) - \rho^{WW}(mixed), \quad (8)$$

$$D = \rho^{WW}(hadr)/\rho^{WW}(mixed). \quad (9)$$

Fig.7 shows the difference $\Delta\rho$ for like-sign and unlike-sign pairs of tracks as measured by L3. The comparison of the ratio $D$ obtained by DELPHI and L3 is shown in Fig.8.

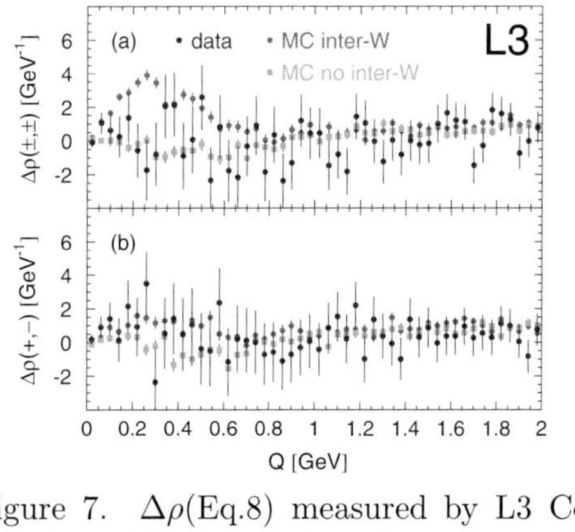


Figure 7. $\Delta\rho$(Eq.8) measured by L3 Coll.[12], together with expectations from the model with and without inter-W correlations.

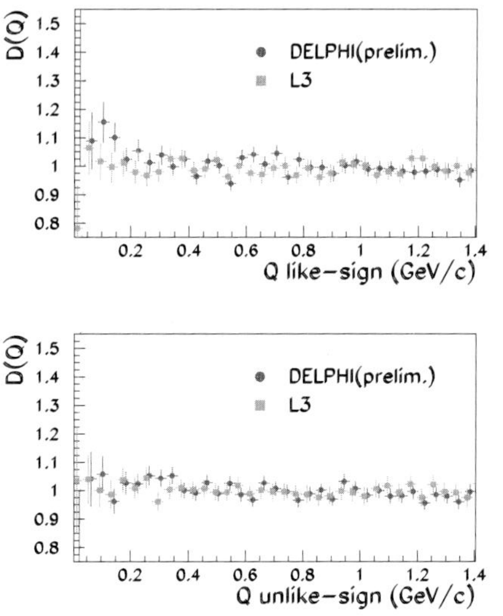


Figure 8. $D$(Eq.9) measured by DELPHI and L3.

The $D$ and $\Delta\rho$ variables provide information about the presence of inter-W correlations without relying on BE modelling. While the results of the two experiments are not inconsistent with each other, the L3 data are consistent with the absence of inter-W correlations, while DELPHI reports an effect at the level of 2.8 $\sigma$ for like-sign pairs, and 2.3 $\sigma$ for unlike-sign pairs.

A model dependent way of combination of LEP results, including the ALEPH measurement [15], is presented in Fig.9. Expressing the measured size of the correlation as a fraction of the tuned[1] model, the direct measurement of inter-W correlations allows to reduce the estimated systematics due to the BE effect in the measurement of the W mass (35 MeV $\to \sim$ 15 MeV).

## 6. Modelling

The interpretation of the available experimental data requires an adequate modelling of the BE effect. Unfortunately, a satisfactory quantum-mechanical description of the phenomenon is not available yet. The incorporation of space-time properties of the emission source and calculation

[1] PYBOEI model [14] with parameters adjusted to fit $Z^0$, resp. single W data, extended to WW events.

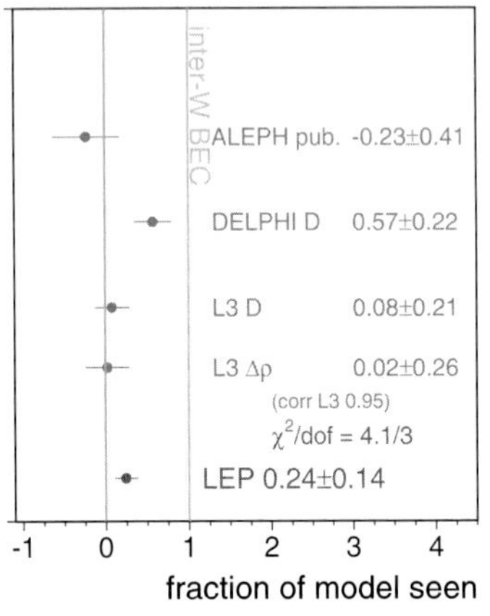


Figure 9. Model dependent combination of LEP results.

of BE interference terms is rather complicated, despite the fact that the relevant information is intrinsically present in some of the most successful hadronization models (string fragmentation). A pioneering work in BE modelling was done by B.Andersson and his collaborators [16], opening the way for explanation of some observed properties of particle correlations (elongation of the source, quantitative predictions of the size of correlations et cetera).

Since the vast majority (if not all) BE models currently on the market adjust the simulated data to the observed distributions with an ad-hoc parameterization, their predictive power is very limited. Even if some technical problems encountered in BE simulation, like spread of weights in events with high multiplicity, presence of negative weights, or difficulties with energy/momentum flow in case of reshuffling of particles can be minimized, one should nevertheless keep in mind that the crucial part of BE modelling, e.g. symmetrization of particle production amplitude, is missing. This is also the case of the modelling [17], which is based on the functional form of BE interference terms derived in [16], but stops short of calculation (and integration over the phase space) of these terms for particle pairs individually. Even so, the authors of [17] could demonstrate the stability of event properties with respect to BE reweighting, and rather limited effect on the reconstructed W mass, of particular interest for the W community at LEP.

## REFERENCES

1. ZEUS Coll.: One- and two-dimensional Bose-Einstein correlations in DIS, abstract 839.
2. H1 Coll., Z.Phys.C 75, 437(1997).
3. DELPHI Coll., Phys.Lett.B 286, 201(1992); ALEPH Coll., Z.Phys.C 54, 75(1992); L3 Coll., Phys.Lett.B 493, 233(2000); O.Smirnova, Nucl.Phys.Proc.Suppl.86,16(2000).
4. T.Csörgo, S.Pratt: Proceedings of the Budapest Workshop on Relativistic Heavy Ion Physics, p.75 (1991).
5. L3 Coll.,Phys.Lett.B 458, 517(1999); DELPHI Coll., Phys.Lett.B 471, 460(2000); OPAL Coll.,Eur.Phys.J.C 16, 185(2000).
6. L3 Coll.: Bose-Einstein Correlations of Neutral and Charged Pions in Hadronic Z Decays, Phys.Lett.B 524 (2002) 55.
7. OPAL Coll.: Bose-Einstein Correlations of $\pi^0\pi^0$ Pairs From Hadronic $Z^0$ Decays, OPAL Physics Note PN-513, September 5, 2002.
8. L3 Coll.: Measurement of Genuine Three-Particle Bose-Einstein Correlations in Hadronic Z Decay, CERN-EP/2002-035.
9. B.Lörstad, Int.J.Mod.Phys.A 4 (1989) 2861; V.L.Lyuboshitz, Sov.J.Nucl.Phys.553 (1991) 514.
10. F.Y.Edgeworth,Trans.Cambridge Phil.Soc.20 (1905) 36.
11. S.V.Chekanov, E.A.De Wolf, W.Kittel, E.Phys.J.C 6 (1999) 403.
12. L3 Coll., *Measurement of Bose-Einstein Correlations in $e^+e^- \to W^+W^-$ Events at LEP*, L3 Note 2751, CERN-EP/2002-034.
13. DELPHI Coll., *Correlations between particles coming from different Ws in $e^+e^- \to W^+W^-$ events*, DELPHI 2002-032-CONF-566.
14. L. Lönnblad and T. Sjöstrand, Eur.Phys.J. C2 (1998) 165-180
15. ALEPH Coll., *Bose-Einstein Correlations in W-pair decays*, Phys.Lett.B478 (2000) 50-64.
16. B.Andersson et al., Phys.Lett.B 169 (1986) 364; Phys.Lett.B 421 (1998) 283.
17. V.Kartvelishvili, R.Kvatadze, Phys.Lett.B 514(2001) 7.

ELSEVIER

Proceedings of ICHEP 2002, pp. 391–393
*S. Bentvelsen, P. de Jong, J. Koch and E. Laenen (Editors)*

31st INTERNATIONAL CONFERENCE ON HIGH ENERGY PHYSICS AMSTERDAM

# Scattering and Resonances in $QCD_2$*

Yitzhak Frishman[a] and Marek Karliner[b]

[a]Department of Particle Physics, Weizmann Institute of Science, 76100 Rehovot, Israel

[b]School of Physics and Astronomy, Raymond and Beverly Sackler Faculty of Exact Sciences, Tel Aviv University, Tel Aviv, Israel

Extending previous works on the spectrum of $QCD_2$, we now investigate the 2D analogue of meson-baryon scattering. We use semi-classical methods, perturbing around classical soliton solutions. In the case of one flavor, we find that the effective potential is reflectionless. In the case of several flavors, the method yields a potential which depends on the momentum of the incoming particle. In this case there is both transmission and reflection. In both cases no resonances appear.

## 1. INTRODUCTION

In 1+1 dimensions it is often possible to obtain analytic solutions. Thus instead of studying an approximate effective action in 3+1 dimensions, one can analyze the exact effective action in 1+1 dimensions.

For many purposes it is useful to have a strong analytic grip on the 1+1 dimensional analogues of the problems in 3+1 dimensions.

This line of thought has been pioneered by 't Hooft [2].

The Baryon spectrum of $QCD_2$ for general $N_f$ and $N_c$ was computed in Refs. [3],[4].

The $\bar{q}q$ content of baryons was calculated in Ref. [5]. The physical picture of baryons composed of constituent quark solitons was obtained in Ref. [6].

Here we will compute the meson-baryon scattering in $QCD_2$ at strong coupling, following the techniques of [7] applied to the bosonized action [8].

## 2. GENERAL FORMULATION

QCD in 1+1 dimensions, bosonized in the scheme

$$[\underbrace{SU(N_c)}_{h}]_{N_F} \times [\underbrace{U(N_F)}_{g}]_{N_c}$$

is given by the action[8]

$$S[g,h,A_+A_-] = N_c S[g] + N_f S[h]$$
$$-(1/2e_c^2)\int d^2x\,\mathrm{Tr}_c\,F_{\mu\nu}F^{\mu\nu}$$
$$+(N_F/2\pi)\int d^2x\,\mathrm{Tr}_c\left[i\left(A_+ h\partial_- h^\dagger + A_- h^\dagger \partial_+ h\right)\right]$$
$$-(N_F/2\pi)\int d^2x\,\mathrm{Tr}_c\left(A_+ h A_- h^\dagger - A_- A_+\right)$$
$$+m'^2 N_{\widetilde{m}}\int d^2x\,\mathrm{Tr}\left(g\,h + h^\dagger\,g^\dagger\right)$$

where

$m'^2 = m_q\,C\,\widetilde{m}$,

$\widetilde{m}$ is to be fixed, $C = \frac{1}{2}e^\gamma \approx 0.891$, and

$$S[u] \equiv S_{WZW}[u] = (1/8\pi)\int d^2x\,\mathrm{Tr}\left(\partial_\mu u \partial^\mu u^\dagger\right)$$
$$+(1/12\pi)\int_B d^3y\,\epsilon_{ijk}$$
$$\times\,\mathrm{Tr}\left[\left(u^{-1}\partial_i u\right)\left(u^{-1}\partial_j u\right)\left(u^{-1}\partial_k u\right)\right]$$

In the strong coupling limit $e_c/m_q \to \infty$ [8],

$$S_{\mathit{eff}} = N_c S[g] + m^2 N_m \int d^2x\left(\mathrm{Tr}\,g + \mathrm{Tr}\,g^\dagger\right)$$

$$m = \left[N_c C m_q\left(\frac{e_c\sqrt{N_f}}{\sqrt{2\pi}}\right)^{\Delta_c}\right]^{\frac{1}{1+\Delta_c}}$$

$$\Delta_c = \frac{N_c^2-1}{N_c(N_c+N_F)}$$

Equation of motion, as coefficient of $(\delta g)g^\dagger$,

$$(N_c/4\pi)\partial_+\left[(\partial_- g)\,g^\dagger\right] + m^2\left(g - g^\dagger\right) = 0$$

We adopt the approach of Ref. [7] for deriving the meson-baryon scattering $S$-matrix.

Expanding in small fluctuations around a given static classical solution

$$g = e^{-i\Phi_c(x)}\,e^{-i\tilde{\delta}\phi(x,t)}$$
$$\approx e^{-i\Phi_c(x)} - ie^{-i\Phi_c(x)}\tilde{\delta}\phi(x,t)$$

Equations of motion,

*Presented by Y. Frishman. Based on Ref. [1] which contains additional material and a complete set of references.


doi:10.1016/S0920-5632(02)02213-2

$\frac{N_c}{4\pi}\,\partial_+\left[e^{-i\Phi_c(x)}\left(\partial_-\tilde{\delta}\phi(x,t)\right)e^{i\Phi_c(x)}\right]$
$+m^2\left[e^{-i\Phi_c(x)}\tilde{\delta}\phi(x,t)+\tilde{\delta}\phi(x,t)e^{i\Phi_c(x)}\right]$
$=0$

Choose $\Phi_c(x)$ to have only the 11 entry non-zero, which we denote as $\phi_c(x)$. Then

$\phi_c(x)'' - \frac{8\pi}{N_c}m^2\sin\phi_c = 0$

$\phi_c(x) = 4\text{arctg}\,(e^{\mu x})\,, \qquad \mu = m\sqrt{\frac{8\pi}{N_c}}$

## 3. ONE FLAVOR

As in the nonabelian case, we expand in small fluctuations around the soliton [7]. Take $\delta\phi$ to commute with $\Phi_c$, in particular same entry in matrix.

Denote this case by $\delta\phi_A$, where the subscript "A" stands for "Abelian".

Then

$\delta g = -i\delta\phi_A(x)e^{-i\phi_c(x)}$

$\Box\delta\phi_A + \mu^2(\cos\phi_c)\delta\phi_A = 0$

$\cos\phi_c = \left[1-(2/\cosh^2\mu x)\right]$

Get

$\mathcal{L}_{eff} = \frac{1}{2}\left(\partial_\mu\delta\phi_A\right)^2 - \frac{1}{2}V(x)\left(\delta\phi_A\right)^2$

$V(x) = \mu^2\cos\phi_c(x) = \mu^2\left[1-(2/\cosh^2\mu x)\right]$

Take

$\delta\phi_A(x,t) = e^{-i\omega t}\chi_A(x)$

Then

$-\omega^2\chi_A - \chi_A'' + V(x)\chi_A = 0$

When $x\to\pm\infty$, the potential $\to\mu^2$, and so

$\chi_A''(\pm\infty)+\omega^2\,\chi_A(\pm\infty) = \mu^2\,\chi_A(\pm\infty)$

Take

$\chi_A(x)\longrightarrow_{|x|\to\infty} e^{\pm ikx}$

which results in

$\omega^2 = k^2+\mu^2$

It turns out that for the particular potential above, there is no reflection at all [9], i.e. the wavefunction for $x\to-\infty$ contains only the incoming wave. The transmission $T$ is

$T = e^{i\delta}$

$\text{ctg}\,\frac{1}{2}\delta = k/\mu$

$\delta$ varies smoothly and decreases monotonically from $\delta=\pi$ at $k=0$ to $\delta=0$ at $k=\infty$. Thus there is no resonance.

For actions that lead to solitons with potentials of no-reflection, see our paper, Ref. [1].

## 4. MULTI-FLAVOR CASE

$\Box\tilde{\delta}\phi - i\left(\partial_+\Phi_c\right)\left(\partial_-\tilde{\delta}\phi\right) + i\left(\partial_-\tilde{\delta}\phi\right)\left(\partial_+\Phi_c\right)$
$+\frac{1}{2}\mu^2\left[\tilde{\delta}\phi e^{-i\Phi_c(x)} + e^{i\Phi_c(x)}\tilde{\delta}\phi\right] = 0$

The equation for $\tilde{\delta}\phi_{ij}$ with $i,j\neq 1$ is like for the free case

$\Box\tilde{\delta}\phi_{ij} + \mu^2\tilde{\delta}\phi_{ij} = 0\,, \qquad i \text{ and } j\neq 1$

whereas the $i=1, j=1$ matrix element is like in the abelian case

$\Box\tilde{\delta}\phi_{11} + \mu^2\left(\cos\phi_c(x)\right)\tilde{\delta}\phi_{11} = 0$

with no reflection and no resonance.

So in order to proceed beyond these results, we need to consider $\tilde{\delta}\phi_{1j}$, $j\neq 1$, or $\tilde{\delta}\phi_{i1}$, $i\neq 1$. As $\tilde{\delta}\phi$ is hermitean, it is sufficient to discuss one of the above.

Thus we take

$\tilde{\delta}\phi_{1j} = e^{-i\omega t}u_j(x) \qquad j\neq 1$

resulting in

$u_j''(x) - i\phi_c'(x)u_j'(x) +$
$\left[\omega^2+\omega\phi_c'(x) - \frac{1}{2}\mu^2\left(1+e^{i\phi_c(x)}\right)\right]u_j(x) = 0$

Define

$u_j \equiv e^{\frac{i}{2}\phi_c}\,v_j$

Then

$v_j'' + \left[\omega^2+\omega\phi_c' - \frac{1}{2}\mu^2\left(1+\cos\phi_c\right)+\frac{1}{4}\left(\phi_c'\right)^2\right]v_j = 0$

Using

$\frac{1}{2}\left(\phi_c'\right)^2 = \mu^2\left(1-\cos\phi_c\right)$

we get

$v_j'' + \left[\omega^2+\omega\phi_c' - \mu^2\cos\phi_c\right]v_j = 0$

or

$-v_j'' - \omega^2 v_j + V(x)v_j = 0$

where

$V(x) = -\omega\phi_c' + \mu^2\cos\phi_c$
$= \mu^2 - 2\mu^2\left[\frac{(\omega/\mu)}{\cosh\mu x} + \frac{1}{\cosh^2\mu x}\right]$

with $\omega = \sqrt{k^2+\mu^2}$ as before. Note that the potential depends on the momentum of the incoming particle.

## 5. NUMERICAL RESULTS

$T$ is the transition amplitude and $R$ is the reflection amplitude, with

$|T|^2 + |R|^2 = 1$

We take

$v_j(x) = Te^{ikx}, \; x \to +\infty$
$v_j(x) = e^{ikx} + Re^{-ikx}, \; x \to -\infty$

Since the potential is symmetric, the symmetric and anti-symmetric scattering amplitudes don't mix, yielding two independent phase shifts $\delta_S$ and $\delta_A$, respectively. This leads to

$T = \frac{1}{2}\left(e^{i\delta_S} + e^{i\delta_A}\right)$
$R = \frac{1}{2}\left(e^{i\delta_S} - e^{i\delta_A}\right)$

Define

$\delta_\pm = \frac{1}{2}\left(\delta_S \pm \delta_A\right)$

Then

$T = e^{i\delta_+} \cos\delta_-$
$R = ie^{i\delta_+} \sin\delta_-$

Note that $R/T$ is purely imaginary.
The transmission and reflections probabilities are

$|T|^2 = \cos^2\delta_-$
$|R|^2 = \sin^2\delta_-$

The numerical results for the transmission probability $|T|^2$ and for the phase of $T$, $\delta_+$ are presented in Fig. 1.

For comparison and as an extra check we also plot the WKB result for $\delta_+$.

Note that no resonance appears.

Note also that the asymptotic value of the phase shift is $\pi$. This can also be obtained from a WKB calculation, which becomes exact at infinite energies.

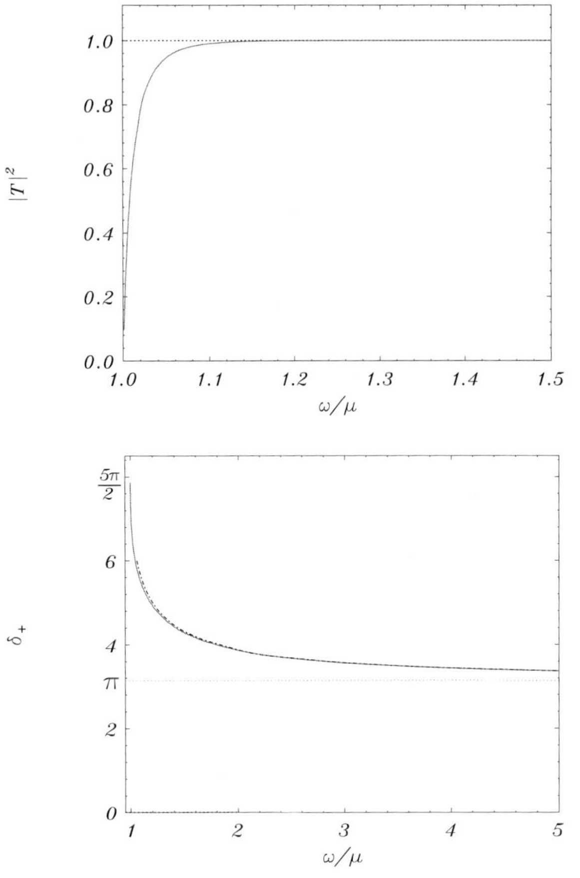


Figure 1. *Upper plot: transmission probability* $|T|^2$*; lower plot: phase of* $T$, $\delta_+$ *(continuous line). Also shown is the approximate result for* $\delta_+$ *from WKB (dot-dashed line).*

## ACKNOWLEDGMENTS

The research of one of us (M.K.) was supported in part by a grant from the United States-Israel Binational Science Foundation (BSF), Jerusalem and by the Einstein Center for Theoretical Physics at the Weizmann Institute.

## REFERENCES


1. Y. Frishman and M. Karliner, Phys. Lett. B **541**, 273 (2002) [arXiv:hep-ph/0206001].
2. G. 't Hooft, Nucl. Phys. B **75**, 461 (1974).
3. G. D. Date, Y. Frishman and J. Sonnenschein, Nucl. Phys. B **283**, 365 (1987);
4. Y. Frishman and J. Sonnenschein, Nucl. Phys. B **294**, 801 (1987).
5. Y. Frishman and M. Karliner, Nucl. Phys. B **344**, 393 (1990).
6. J. R. Ellis, Y. Frishman, A. Hanany and M. Karliner, Nucl. Phys. B **382**, 189 (1992) [arXiv:hep-ph/9204212].
7. M. P. Mattis and M. Karliner, Phys. Rev. D **31**, 2833 (1985).
8. Y. Frishman and J. Sonnenschein, Phys. Rept. **223**, 309 (1993) [arXiv:hep-th/9207017].
9. L.D. Landau and E.M. Lifshitz, Quantum Mechanics: non-relativistic theory, 3rd ed., §23 and §25, Pergamon Press, 1977.

ELSEVIER

Proceedings of ICHEP 2002, pp. 394–399
*S. Bentvelsen, P. de Jong, J. Koch and E. Laenen (Editors)*

31st INTERNATIONAL CONFERENCE ON HIGH ENERGY PHYSICS AMSTERDAM

# Colour Reconnection Effects in WW Production at LEP

E. Bouhova-Thacker[a]

[a]Department of Physics, Lancaster University, Lancaster LA1 4YB, United Kingdom

Preliminary results from the search for colour reconnection effects by the four LEP experiments and a first combination of their measurements of the inclusive particle flow distributions in four-jet events are presented. Data are compared with predictions of various models with and without colour reconnection. The extreme scenario of the SKI model, where almost all events are reconnected, is ruled out and a limit on the model parameter, controlling the fraction of reconnected events, is set.

## 1. $WW$ decays at LEP2

The LEP2 collider at CERN, Geneva ran for five years above the $W^+W^-$ threshold. During this time each of the four experiments, ALEPH, DELPHI, L3 and OPAL, collected ~10000 $WW$ pairs in the energy range 161 - 209 GeV. About 46% of the $WW$ pairs decay via the fully hadronic channel ($q\bar{q}q\bar{q}$), 44% via the semi-leptonic ($\ell\nu q\bar{q}$) and 10% via the fully leptonic ($\ell\nu\ell\nu$) channels. The separation of the $W$ decay vertices at LEP2 energies is ~0.1 fm, which is small compared to the QCD hadronisation scale, ~1 fm, and there is no reason to assume that the pair hadronise independently. The FSI leading to colour exchange between the decay products of the two $W$ bosons is known as colour reconnection (or rearrangement).

Colour reconnection (CR) may affect the multiplicities and the particle kinematics and can lead to an apparent shift in the $W$ mass measured in the fully hadronic channel. Colour reconnection effects in the perturbative phase are expected to be small [1]. On the other hand, significant interference is expected in the hadronisation process, which can be estimated only in the context of specific models.

## 2. Models of Colour Reconnection

The colour reconnection models studied by the LEP experiments are those implemented in the JETSET/PYTHIA, ARIADNE and HERWIG Monte Carlo programs. The Sjöstrand-Khoze (SK) models [1] are based on the Lund string fragmentation picture. In these models colour reconnection occurs for overlapping or crossing strings stretched between two $q\bar{q}$ systems. The models cannot be tuned to $Z^0$ data. In the SKI scenario, the strings have a significant transverse extension, comparable to hadronic dimensions, and the probability of reconnection depends on the volume of overlap, $V$, i.e. $P^{\mathrm{reco}} = 1 - \exp(-k_I V)$, where $k_I$ is a free parameter. In the SKII scenario, the strings have negligible thickness and a unit reconnection probability upon their first crossing. In SKII′ reconnection is only allowed to occur at the first string crossing which reduces the total string length of the system.

The Generalised Area Law (GAL) model [2], implemented in JETSET, is also based on the Lund string fragmentation model. The reconnection probability for this model is given by $P^{\mathrm{reco}} = R_0(1 - \exp(-b\Delta A))$, where $\Delta A$ is the area difference between the two configurations, with and without CR (in energy-momentum coordinates), $b$ is a phenomenological parameter of the order 0.6 $\mathrm{GeV}^{-2}$ and $R_0$ is a tunable parameter of the order $1/N_C^2 = 0.11$.

The ARIADNE Monte Carlo program utilises the dipole cascade model [3]. Colour reconnection is implemented at the end of the parton shower if it leads to a reduction in the string length and is suppressed by a factor of $1/N_C^2$ [4]. Multiple reconnections per event are permitted. They may occur within a single $q\bar{q}$ system (AR1 model). In the AR2 model, applicable to $q\bar{q}q\bar{q}$ events, recon-


doi:10.1016/S0920-5632(02)02215-6

nections are only allowed for gluons with energies less than 2 GeV. The AR3 model, which imposes no such restriction, is disfavoured on both theoretical and experimental grounds [5].

An alternative to string fragmentation is the cluster model implemented in HERWIG. After showering the gluons are split non-perturbatively into $q\bar{q}$ pairs and colour singlet clusters are formed, which then decay into a small number of hadrons. Colour reconnection would change the cluster formation if the cluster size can be reduced [6]. The multiplicity of the decayed cluster depends on the cluster mass, leading to an increase in the multiplicity predicted by the HERWIG CR model with respect to the non-reconnected scenario. In string-based models, on the other hand, colour reconnection tends to minimise the string length, which is correlated with the string energy, and leads to a decrease in multiplicity.

The SKI model predicts a $W$ mass shift of up to 250 MeV depending on the reconnection probability. Unless otherwise specified, results are quoted for the extreme case of maximal reconnection probability (referred to as SKI 100%). From theoretical considerations the fraction of reconnected events is expected to be in the range of 30 to 60%. All other models under study predict a much smaller shift, typically of the order of a few tens of MeV.

## 3. Multiplicity studies - an ALEPH update

The effects of colour reconnection on $WW$ charged track multiplicity was among the first signatures probed by the LEP experiments. ALEPH has submitted updated results [7], which include all data in the range 189 - 207 GeV. In the absence of colour reconnection the difference between the charged track multiplicity in fully hadronic ($4q$) and twice the multiplicity in semi-leptonic ($2q$) events in data should be consistent with zero. The latest ALEPH result, as well as some earlier results [8] from the other three experiments, are shown in Table 1. The first uncertainty is statistical, the second systematic. The results do not show any evidence for colour reconnection effects.

ALEPH also compare the observed data multiplicity to model predictions (see Table 2). In order to account for any selection biases, the double difference $\Delta_{cr} = [4q-2(2q)]_{data} - [4q-2(2q)]_{MC}$ is constructed. All models are compatible with the data, due to the large uncertainty in the measurements. The largest deviation between data and Monte Carlo is observed in the SKI 100% and AR2 models.

The charge particle multiplicities in fully hadronic events as a function of the rescaled momentum $X_p = \frac{2p}{\sqrt{s}}$ for data and Monte Carlo are shown in Figure 1. The largest colour reconnection effects are expected in the soft momentum region ($p < 1$ GeV). No significant effects are observed and the uncertainty on the data points in that region does not allow a distinction between the different models to be made.

The bottom plot shows a comparison of different models to JETSET, which has the best fit to data in the full momentum range. HERWIG pro-

Table 1
Difference in the charged particle multiplicity in fully hadronic and twice the multiplicity in semi-leptonic events.

| Experiment | $\sqrt{s}$ (GeV) | $4q - 2(2q)$ |
|---|---|---|
| ALEPH* | 189-207 | $0.31 \pm 0.23 \pm 0.10$ |
| DELPHI | 183-189 | $-0.26 \pm 0.60 \pm 0.38$ |
| L3 | 183-189 | $-0.29 \pm 0.26 \pm 0.30$ |
| OPAL | 183-202 | $0.07 \pm 0.39 \pm 0.37$ |

*Not corrected for event selection and $p_T$ cut of 200 MeV.

Table 2
The double difference $\Delta_{cr}$ for some Monte Carlo models without and with colour reconnection.

| MC Model | $\Delta_{cr}$ | significance |
|---|---|---|
| JETSET | $0.14 \pm 0.23 \pm 0.1$ | $0.6\sigma$ |
| HERWIG | $0.14 \pm 0.23 \pm 0.1$ | $0.6\sigma$ |
| ARIADNE | $0.10 \pm 0.23 \pm 0.1$ | $0.4\sigma$ |
| SKI 100% | $0.35 \pm 0.23 \pm 0.1$ | $1.4\sigma$ |
| SKII | $0.20 \pm 0.23 \pm 0.1$ | $0.8\sigma$ |
| SKII′ | $0.24 \pm 0.23 \pm 0.1$ | $1.0\sigma$ |
| HERWIG CR | $-0.15 \pm 0.23 \pm 0.1$ | $-0.6\sigma$ |
| AR2 | $0.36 \pm 0.23 \pm 0.1$ | $1.4\sigma$ |
| GAL | $0.12 \pm 0.23 \pm 0.1$ | $0.5\sigma$ |

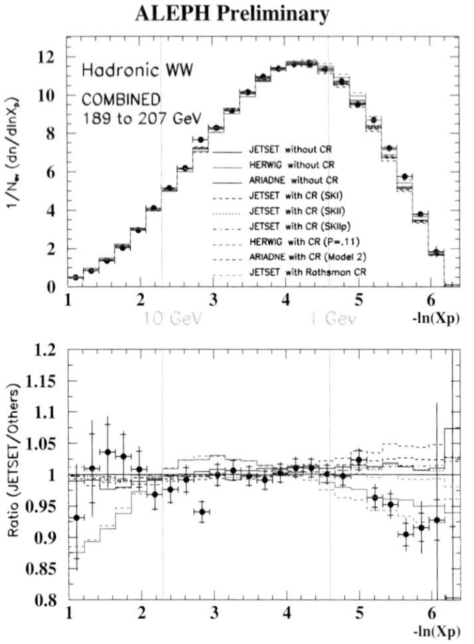


Figure 1. $\xi$ distributions for fully hadronic $WW$ decays for data and different MC models with and without colour reconnection, $\xi = -\ln(2p/\sqrt{s})$.

vides a relatively better description of data for low momentum tracks, but does not give a very good fit for higher momenta. The string fragmentation models, both with and without colour reconnection, do not describe data in the soft region very well. The same conclusions follow from semi-leptonic events and from studies at the $Z^0$ peak [9].

## 4. The particle flow method

Fully hadronic $WW$ events are reconstructed as four-jet events with two jet pairings with invariant mass close to the nominal $W$ mass. The particle flow method [10] aims to study the effects of colour reconnection on the distribution of particles in the regions between jets from the same $W$ and between jets from different $W$ bosons. Two types of selection are used by the experiments. The topological one selects events where the angles between the two jets of a $W$ decay (regions A and B in Figure 2) are in the range 100-140° and the angles between jets from different $W$ bosons are $< 100°$. This results in a rather low efficiency (~14%), but in most events (~90%) the correct assignment of the two reconstructed jets to a $W$ is made. The $W$ mass analysis selection uses the standard analysis of an experiment for measuring the $W$ mass, which has higher efficiency (40-80%). Each experiment's analysis [11] varies slightly in the exact method of resolving the ambiguities in constructing the four planes.

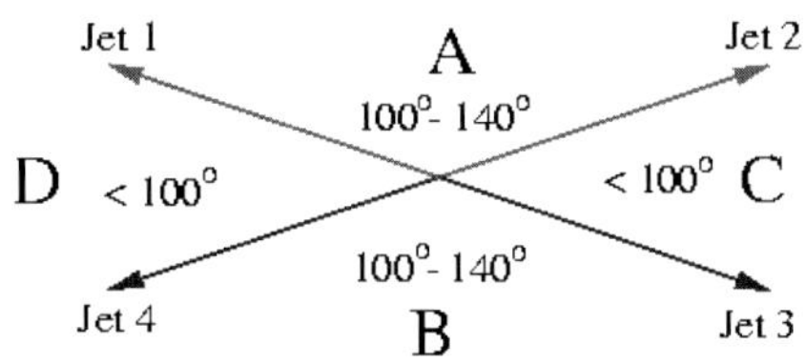


Figure 2. Schematic drawing of the topological selection. The four jets do not lie in a single plane. A and B define the regions between the jets from the same $W$ boson.

The next step is to project all particles onto the planes (Figure 3) and add together the distributions in the four planes to form the particle flow, ensuring that each particle enters the distribution only once. As the angles between the jets vary from event to event, a normalised particle flow (Figure 4) is constructed by rescaling the angles to the angle between the two jets defining the corresponding plane, $\phi_r = \phi/\phi_{j\,j+1}$.

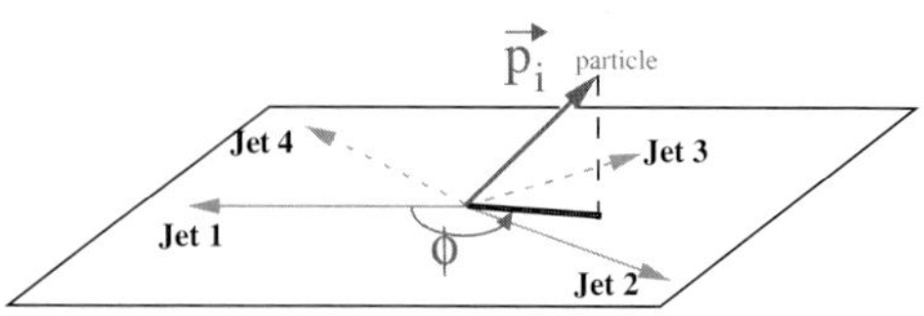


Figure 3. Projection of particles onto the four planes defined by pairs of reconstructed jets.

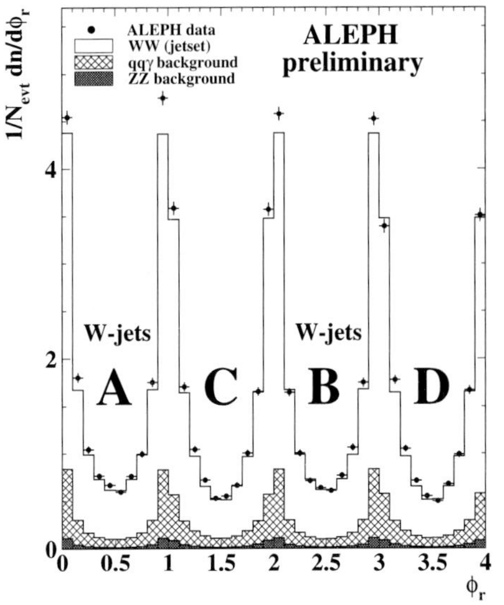


Figure 4. Normalised particle flow distribution.

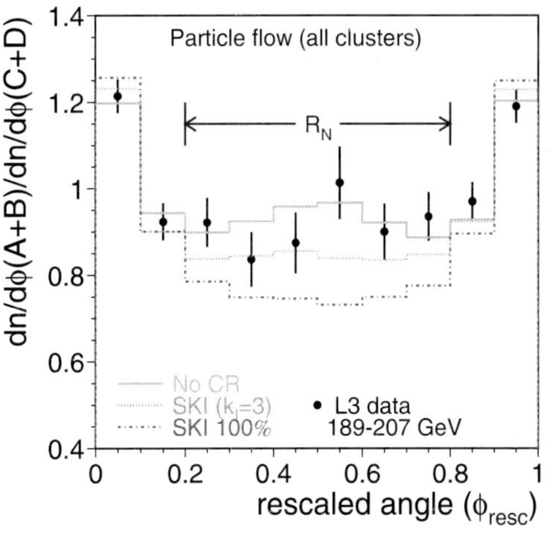


Figure 5. The ratio $R$ for data and the SKI model with and without colour reconnection.

In order to compare the distributions in the regions between jets from the same $W$ to the distributions in the regions between jets from different $W$ bosons, the ratio $R = \frac{A+B}{C+D}$ is formed. Colour reconnection leads to an increase in the particle flow in the region between jets from different $W$ bosons as can be seen from Figure 5 for the case of the SKI model. To extract quantitative information, the particle flow is integrated in the most sensitive region before taking the ratio. The observable $R_n$ is then defined as

$$R_n(0.2-0.8) = \frac{\int_{0.2}^{0.8}(\frac{dn_A}{d\phi_r} + \frac{dn_B}{d\phi_r})d\phi_r}{\int_{0.2}^{0.8}(\frac{dn_C}{d\phi_r} + \frac{dn_D}{d\phi_r})d\phi_r}.$$

The values for the statistical sensitivity, defined as $|R_n(\text{model noCR}) - R_n(\text{model CR})|/(\sigma_{R_n})_{stat}$, where $(\sigma_{R_n})_{stat}$ is the statistical error on the full data sample of each experiment, are shown in Table 3. The $R_n$ observable is sensitive to the predictions of the SKI model but the reconnected and non-reconnected HERWIG and ARIADNE models are only $1\sigma$ apart and therefore it would not be possible to distinguish between them using this method. The values are calculated using Monte Carlo samples, generated specifically for the purposes of LEP combination with ALEPH tuning at the $Z^0$ peak.

The experiments measure $R_n$ at several centre-of-mass energies which are combined by determining the energy dependence from non-reconnected JETSET samples and extrapolating all measurements to a single centre-of-mass energy, where a statistically weighted average is taken.

Table 3
Statistical sensitivity of $R_n$ using the full data samples (189 - 208 GeV).

| Experiment | SKI | HERWIG CR | AR2 |
|---|---|---|---|
| ALEPH | 5.9 | 1.0 | 0.9 |
| DELPHI | 3.2 | 0.1 | 0.1 |
| L3 | 5.0 | 0.0 | 0.2 |
| OPAL | 6.3 | 1.2 | 0.4 |

The values of $R_n$ for data for the four experiments are given in Table 4, where the first uncertainty is statistical, the second systematic and the third is from the extrapolation process. Due to the differences in the analysis methods and the types of events, which are selected, these values cannot be compared directly.

Table 4
Statistically weighted average values of $R_n$ in data estimated at 189 GeV.

| Experiment | $R_n$ |
|---|---|
| ALEPH | $1.095 \pm 0.014 \pm 0.006 \pm 0.006$ |
| DELPHI | $0.900 \pm 0.031 \pm 0.015 \pm 0.012$ |
| L3 | $0.844 \pm 0.022 \pm 0.021 \pm 0.002$ |
| OPAL | $1.257 \pm 0.025 \pm 0.020 \pm 0.003$ |

## 5. LEP combination procedure and results

The combination of the $R_n$ results [12] is performed with the help of common samples of events, processed by each experiment. For each model $m$, the ratio $r = \frac{R_n^{data}}{R_n^{m,noCR}}$ is formed, where $R_n^{data}$ and $R_n^{m,noCR}$ are the values of $R_n$ for data and for the common sample without CR of that model. The $r$ values of the four experiments for the model are then combined with weights, given by $w^m = (R_n^{m,CR} - R_n^{m,noCR})^2/(\sigma_{stat}^2 + \sigma_{syst}^2)$, where $R_n^{m,CR}$ is the $R_n$ value for the common sample with CR, $\sigma_{stat}$ and $\sigma_{syst}$ are the total statistical and systematic uncertainties.

Figure 6 shows the $r$ measurements of each experiment and the combined LEP result obtained with JETSET SKI 100%. The solid line is the prediction of JETSET without colour reconnection (at $2\sigma$ of the total error) and the dashed line is the combined LEP result for the prediction of SKI 100%, which is $5.2\sigma$ away from data.

The reconnection probability in the SKI model is governed by an arbitrary, free parameter, $k_I$. By comparing data with the model predictions for different values of $k_I$, it is possible to determine the value which is most consistent with data. The LEP combination procedure is repeated several times for each value of $k_I$ and the data preferred reconnection probability is found to be 0.49 ($k_I = 1.18$). The 68%CL lower and upper limits on $P^{\text{reco}}$ are 0.25 and 0.65, respectively. The upper limit corresponds to a 90 MeV shift in the measured mass of the $W$ in fully hadronic events.

The $r$ values obtained with the HERWIG models (Figure 7) with and without colour reconnection differ from the measured data value by $2.6\sigma$ and $3.3\sigma$, respectively. The AR2 model is $2.1\sigma$ away from data, while the non-reconnected ARIADNE model is $2.9\sigma$ away (Figure 8). The dashed lines in the plots are the combined LEP predictions of the CR models. The observed deviations of the $r$ values from all non-reconnected models may indicate a possible systematic effect in the description of particle flow for four-jet events.

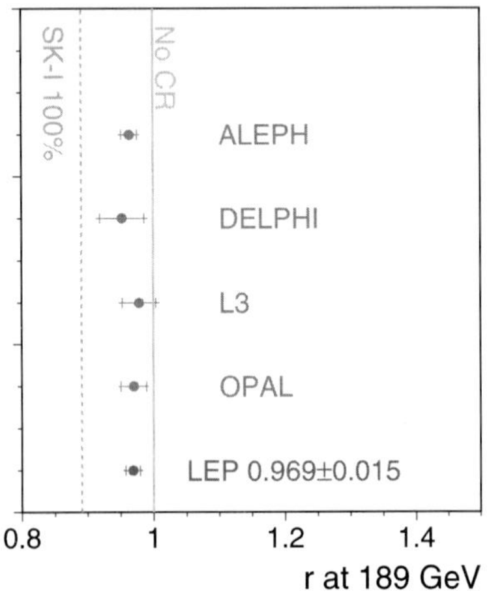


Figure 6. Combined particle flow results, testing the SKI model of colour reconnection with maximal fraction of reconnected events.

Table 5 shows the deviation of data $R_n$ from the predictions of some other models of colour reconnection, estimated using the experiment's own Monte Carlo samples.

Table 5
Deviation of data $R_n$ from the predictions of the SKII and GAL models of colour reconnection.

| Experiment | SKII | SKII′ | GAL* |
|---|---|---|---|
| ALEPH | $-2.01\sigma$ | $-2.32\sigma$ | $-1.65\sigma$ |
| L3 | $-0.16\sigma$ | – | – |
| OPAL | $-1.05\sigma$ | $-1.18\sigma$ | – |

$^*R_0 = 0.004$, from a global fit to $Z^0$ data

## 6. Conclusions

The study of $WW$ multiplicity and momentum distributions does not allow a distinction between the various models of colour reconnection

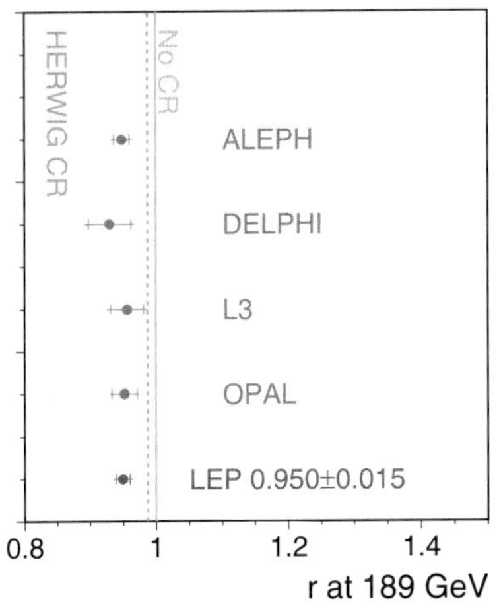


Figure 7. Combined particle flow results, testing the HERWIG model of colour reconnection.

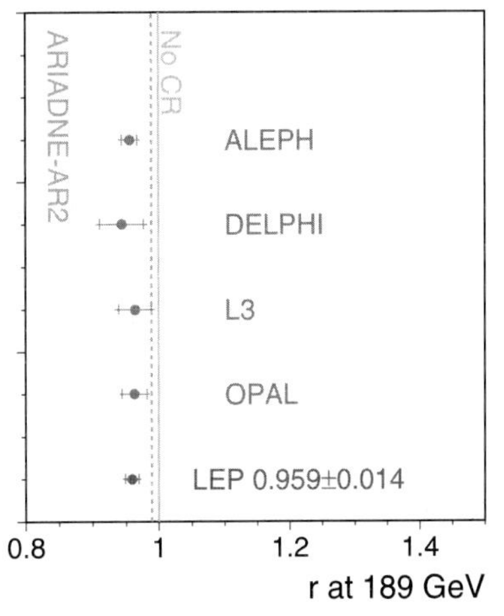


Figure 8. Combined particle flow results, testing the AR2 model of colour reconnection.

to be made. Apart from some earlier successes in the exclusion of more extreme CR models, not discussed here, the measurements are limited by statistics and large systematic uncertainties.

The particle flow distribution studies show that the observable $R_n$ is sensitive to colour reconnection in the SKI model. The combined LEP data disfavour by more than $5\sigma$ the extreme version of the model with maximal reconnection probability, however the no colour reconnection scenario is not ruled out. The combined data favours the model where about half of the events at 189 GeV are reconnected.

The values obtained with HERWIG and ARIADNE models differ from the measured data value by more than $2\sigma$. The differences between the results from the reconnected and non-reconnected scenarios for these models are small and do not allow the particle flow method to distinguish between them.

## REFERENCES


1. T. Sjöstrand, V.A. Khoze, Z. Phys. **C62** (1994) 281.
2. J. Rathsman, Phys. Lett. **B 452** (1999) 364.
3. G. Gustafson, Phys. Lett. **B 175** (1986) 453; G. Gustafson, U. Pettersson, Nucl. Phys. **B 306** (1988) 746; B. Andersson et al., Z. Phyz. **C43** (1989) 625.
4. L. Lönnblad, Z.Phys. **C70** (1996) 107.
5. OPAL Collaboration, G. Abbiendi et al., Eur. Phys. J.. **C 11** (1999) 217.
6. G. Marchesini et al., hep-ph/9607393; G. Corcella et al., CERN-TH/2000-284.
7. ALEPH Collaboration, ALEPH 2002-015, CONF 2002-004, ICHEP02 ABS343.
8. taken from M. Pearson's talk at Lake Louise Winter Institute, Workshop on Fundamental Interactions, 17-23 February 2002, proceedings to be published by World Scientific.
9. ALEPH Collaboration, CERN-PPE-96-186, Phys. Rept. **294** (1998) 1.
10. D. Duchesneau, preprint LAPP-EXP 2000-02.
11. ALEPH Collaboration, ALEPH 2002-020, CONF 2002-009, ICHEP02 ABS341; DELPHI Collaboration, DELPHI 2002-047, CONF 581, ICHEP02 ABS226; L3 Collaboration, L3 Note 2748, ICHEP02 ABS460; OPAL Collaboration, OPAL PN506, ICHEP02 ABS390.
12. The LEP Collaborations and the LEP W Working Group, LEPEWWG/FSI/2002-01, ICHEP02 ABS385.

Proceedings of ICHEP 2002, pp. 400–402
*S. Bentvelsen, P. de Jong, J. Koch and E. Laenen (Editors)*

ELSEVIER

31st INTERNATIONAL CONFERENCE ON HIGH ENERGY PHYSICS AMSTERDAM

# Analytic Amplitudes for Hadronic Forward Scattering : COMPETE update

B. Nicolescu[a], J. R. Cudell[b], V. V. Ezhela[c], P. Gauron[a], K. Kang[d], Yu. V. Kuyanov[c], S. B. Lugovsky[c], E. Martynov[e], E. A. Razuvaev[c], N. P. Tkachenko[c]

[a]LPNHE (Unité de Recherche des Universités Paris 6 et Paris 7, Associée au CNRS), Université Pierre et Marie Curie, Tour 12 E3, 4 Place Jussieu, 75252 Paris Cedex 05, France

[b]Institut de Physique, Bât. B5, Université de Liège, Sart Tilman, B4000 Liège, Belgium

[c]COMPAS group, IHEP, Protvino, Russia

[d]Physics Department, Brown University, Providence, RI, U.S.A.

[e]Bogolyubov Institute for Theoretical Physics, 03143 Kiev, Ukraine

We consider several classes of analytic parametrizations of hadronic scattering amplitudes, and compare their predictions to all available forward data ($pp, \bar{p}p, \pi p, Kp, \gamma p, \gamma\gamma, \Sigma p$). Although these parametrizations are very close for $\sqrt{s} \geq 9$ GeV, it turns out that they differ markedly at low energy, where a universal Pomeron term $\sim \ln^2 s$ enables one to extend the fit down to $\sqrt{s} = 4$ GeV. We present predictions on the total cross sections and on the ratio of the real part to the imaginary part of the elastic amplitude ($\rho$ parameter) for present and future $pp$ and $\bar{p}p$ colliders, and on total cross sections for $\gamma p \to$ hadrons at cosmic-ray energies and for $\gamma\gamma \to$ hadrons up to $\sqrt{s} = 1$ TeV.

Analytic parametrizations of forward ($t = 0$) hadron scattering amplitudes is a well-established domain in strong interactions.

However, in the past, the phenomenology of forward scattering hadquite a high degree of arbitrariness : i) Excessive attention was paid to $pp$ and $\bar{p}p$ scattering ; ii) Important physical constraints were often mixed with less general or even ad-hoc properties ; iii) The cut-off in energy, defining the region of applicability of the high-energy models, differed from one author to the other ; iv) The set of data considered by different authors was often different ; v) No rigorous connection was made between the number of parameters and the number of data points ; vi) No attention was paid to the necessity of the stability of parameter values ; vii) The experiments were performed in the past in quite a chaotic way : huge gaps are sometimes present between low-energy and high-energy domains or inside the high-energy domain itself.

The COMPETE (COmputerized Models and Parameter Evaluation for Theory and Experiment) collaboration has cured as much as possible the above discussed arbitrariness.

The $\chi^2/dof$ criterium is not able, by itself, to cure the above difficulties : new indicators have to be defined. Once these indicators are defined[1], it is possible to estimate the overall performance of each model, and to establish a ranking : the highest the numerical value of the rank the better the model under consideration.

The final aim of the COMPETE project is to provide our community with a periodic cross assessments of data and models via computer-readable files on the Web [2].

We consider the following exemplar cases of the imaginary part of the scattering amplitudes :

$$ImF^{ab} = s\sigma_{ab} = P_1^{ab} + P_2^{ab} + R_+^{ab} \pm R_-^{ab} \qquad (1)$$

where :
- the $\pm$ sign in formula (1) corresponds to antipar-


doi:10.1016/S0920-5632(02)02219-3

ticle (resp. particle) - particle scattering amplitudes.
- $R_\pm$ signify the effective secondary-Reggeon $((f, a_2), (\rho, \omega))$ contributions to the even (odd)-under-crossing amplitude

$$R_\pm(s) = Y_\pm \left(\frac{s}{s_1}\right)^{\alpha_\pm}, \tag{2}$$

where $Y$ is a constant residue, $\alpha$ - the reggeon intercept and $s_1$ - a scale factor fixed at 1 GeV$^2$ ;
- $P_1(s)$ is the contribution of the Pomeron Regge pole

$$P_1^{ab}(s) = C_1^{ab} \left(\frac{s}{s_1}\right)^{\alpha_{P_1}}, \tag{3}$$

$\alpha_{P_1}$ is the Pomeron intercept $\alpha_{P_1} = 1$, and $C^{ab}$ are constant residues.
- $P_2^{ab}(s)$ is the second component of the Pomeron corresponding to three different $J$-plane singularities :

a) a Regge simple - pole contribution

$$P_2^{ab}(s) = C_2^{ab} \left(\frac{s}{s_1}\right)^{\alpha_{P_2}}, \tag{4}$$

with $\alpha_{P_2} = 1 + \epsilon$, $\epsilon > 0$, and $C_2^{ab}$ constant ;

b) a Regge double-pole contribution

$$P_2^{ab}(s) = s\left[A^{ab} + B^{ab} \ln\left(\frac{s}{s_1}\right)\right], \tag{5}$$

with $A^{ab}$ and $B^{ab}$ constant ;

c) a Regge triple-pole contribution

$$P_2^{ab}(s) = s\left[A^{ab} + B^{ab} \ln^2\left(\frac{s}{s_0}\right)\right], \tag{6}$$

where $A^{ab}$ and $B^{ab}$ are constants and $s_0$ is an arbitrary scale factor.

We consider all the existing forward data for $pp, \bar{p}p, \pi p, Kp, \gamma\gamma$ and $\Sigma p$ scatterings. The number of data points is : 904, 742, 648, 569, 498, 453, 397, 329 when the cut-off in energy is 3, 4, 5, 6, 7, 8, 9, 10 GeV respectively. A large number of variants were studied. All definitions and numerical details can be found in Ref. 1.

The 2-component Pomeron classes of models are RRPE, RRPL and RRPL2, where by RR we denote the two effective secondary-reggeon contributions, by P - the contribution of the Pomeron Regge-pole located at $J = 1$, by E - the contribution of the Pomeron Regge-pole located at $J = 1 + \epsilon$, by L - the contribution of the component of the Pomeron, located at $J = 1$ (double pole), and by L2 - the contribution of the component of the Pomeron located at $J = 1$ (triple pole). We also studied the 1-component Pomeron classes of models RRE, RRL and RRL2.

The highest rank are get by the RRPL2$_u$ models (see Table 1), corresponding to the $\ln^2 s$ behaviour of total cross sections first proposed by Heisenberg 50 years ago [3]. The $u$ index denotes

Table 1
Ranking of the the 21 models having nonzero area of applicability. The number between paranthesis, in the Model Code column, denotes the number of free parameters.

| Model Code | Rank |
|---|---|
| RRPL2$_u$(19) | 230 |
| RRP$_{nf}$L2$_u$(21) | 222 |
| RRL$_{nf}$(19) | 222 |
| (RR$_c$)$^d$ PL2$_u$(15) | 204 |
| (RR)$^d$ P$_{nf}$L2$_u$(19) | 194 |
| [R$^{qc}$ L$^{qc}$]R$_c$(12) | 184 |
| (RR$_c$)$^d$P$^{qc}$L2$_u$(14) | 181 |
| (RR)$^d$P$^{qc}$L2$_u$(16) | 180 |
| RR$_c$ L2$^{qc}$(15) | 180 |
| (RR)$^d$ P$_{nf}$L2(20) | 178 |
| (RR)$^d$ PL2$_u$(17) | 174 |
| RRPL(21) | 173 |
| RR$_c$ L$^{qc}$(15) | 172 |
| RRL2$^{qc}$(17) | 170 |
| [R$^{qc}$ L2$^{qc}$]R$_c$(12) | 170 |
| RRL$^{qc}$(17) | 162 |
| RRPE$_u$(19) | 158 |
| [R$^{qc}$L$^{qc}$]R(14) | 155 |
| RRL2(18) | 152 |
| RR$_c$PL(19) | 142 |
| RRL(18) | 133 |

Table 2
Predictions for $\sigma_{tot}$ and $\rho$, for $\bar{p}p$ (at $\sqrt{s} = 1960$ GeV) and for $pp$ (all other energies). The central values and statistical errors correspond to the preferred model RRPL2$_u$.

| $\sqrt{s}$ (GeV) | $\sigma$ (mb) | $\rho$ |
|---|---|---|
| 100 | $46.37 \pm 0.06$ | $0.1058 \pm 0.0012$ |
| 200 | $51.76 \pm 0.12$ | $0.1275 \pm 0.0015$ |
| 300 | $55.50 \pm 0.17$ | $0.1352 \pm 0.0016$ |
| 400 | $58.41 \pm 0.21$ | $0.1391 \pm 0.0017$ |
| 500 | $60.82 \pm 0.25$ | $0.1413 \pm 0.0017$ |
| 600 | $62.87 \pm 0.28$ | $0.1416 \pm 0.0018$ |
| 1960 | $78.27 \pm 0.55$ | $0.1450 \pm 0.0018$ |
| 10000 | $105.1 \pm 1.1$ | $0.1382 \pm 0.0016$ |
| 12000 | $108.5 \pm 1.2$ | $0.1371 \pm 0.0015$ |
| 14000 | $111.5 \pm 1.2$ | $0.1361 \pm 0.0015$ |

Table 3
Predictions for $\sigma_{tot}$ for $\gamma p \to hadrons$ for cosmic-ray photons. The central values and the statistical errors are as in Table 2.

| $p^{\gamma}_{lab}$ (GeV) | $\sigma$ (mb) |
|---|---|
| $0.5 \cdot 10^6$ | $0.243 \pm 0.009$ |
| $1.0 \cdot 10^6$ | $0.262 \pm 0.010$ |
| $0.5 \cdot 10^7$ | $0.311 \pm 0.014$ |
| $1.0 \cdot 10^7$ | $0.333 \pm 0.016$ |
| $1.0 \cdot 10^8$ | $0.418 \pm 0.022$ |
| $1.0 \cdot 10^9$ | $0.516 \pm 0.029$ |

the *universality property* : the coupling $B$ of the $\ln^2 s$ term is the same in all hadron-hadron scatterings and $s_0$ is the same in all reactions.

We note that the familiar RRE Donnachie-Landshoff model is *rejected* at the 98% C.L. when models which achieve a $\chi^2/dof$ less than 1 for $\sqrt{s} \geq 5$ GeV are considered.

The predictions of the best RRPL2$_u$ model, adjusted for $\sqrt{s} \geq 5$ GeV, are given in Tables 2-4.

The uncertainties on total cross sections, including the systematic errors due to contradictory data points from FNAL (the CDF and E710/E811 experiments, respectively), can reach 1.9% at RHIC, 3.1% at the Tevatron, and 4.8% at the LHC, whereas those on the $\rho$ parameter are respectively 5.4%, 5.2%, and 5.4%. The global

Table 4
Predictions for $\sigma_{tot}$ for $\gamma\gamma \to hadrons$. The central values and the statistical errors are as in Table 2.

| $\sqrt{s}$ (GeV) | $\sigma$ ($\mu$ b) |
|---|---|
| 200 | $0.546 \pm 0.027$ |
| 300 | $0.610 \pm 0.035$ |
| 400 | $0.659 \pm 0.042$ |
| 500 | $0.700 \pm 0.047$ |
| 1000 | $0.840 \pm 0.067$ |

picture emerging from fits to all data on forward observables supports the CDF data and disfavors the E710/E811 data at $\sqrt{s} = 1.8$ TeV.

Any significant deviation from the predictions based on model RRPL2$_u$ will lead to a re-evaluation of the hierarchy of models and presumably change the preferred parametrisation to another one. A deviation from the "allowed region" would be an indication that strong interactions demand a generalization of the analytic models discussed so far, *e.g.* by adding Odderon terms, or new Pomeron terms, as suggested by QCD.

## REFERENCES


1. J. R. Cudell, V. V. Ezhela, P. Gauron, K. Kang, Yu. V. Kuyanov, S. B. Lugovsky, B. Nicolescu, and N. P. Tkachenko, Phys. Rev. **D65** (2002) 074024 ; see also 2002 Review of Particle Physics, K. Hagiwara et al. Phys. Rev. **D66** (2002) 010001-9.
2. See the preliminary version of a Web interface at the adress http:// sirius.ihep.su/~kuyanov/OK/eng/intro.html.
3. W. Heisenberg, Zeit. Phys. **133** (1952) 65 (in German).
4. J. R. Cudell, V. V. Ezhela, P. Gauron, K. Kang, Yu. V. Kuyanov, S. B. Lugovsky, E. Martynov, B. Nicolescu, E. A. Razuvaev, and N. P. Tkachenko, hep-ph/0206172, Phys. Rev. Lett. (in press).

ELSEVIER

Proceedings of ICHEP 2002, pp. 403–407
*S. Bentvelsen, P. de Jong, J. Koch and E. Laenen (Editors)*

31st INTERNATIONAL CONFERENCE ON HIGH ENERGY PHYSICS AMSTERDAM

# Inclusive Diffraction at HERA

F.-P. Schilling[a]* (on behalf of the H1 and ZEUS collaborations)

[a]DESY, Notkestr. 85, D-22603 Hamburg, Germany

New precision measurements of inclusive diffractive deep-inelastic $ep$ scattering interactions, performed by the H1 and ZEUS collaborations at the HERA collider, are discussed. A new set of diffractive parton distributions, determined from recent high precision H1 data, is presented.

## 1. INTRODUCTION

One of the biggest challenges in our understanding of QCD is the nature of colour singlet exchange or *diffractive* interactions. The electron-proton collider HERA is an ideal place to study hard diffractive processes in deep-inelastic $ep$ scattering (DIS). In such interactions, the point-like virtual photon probes the structure of colour singlet exchange, similarly to inclusive DIS probing proton structure.

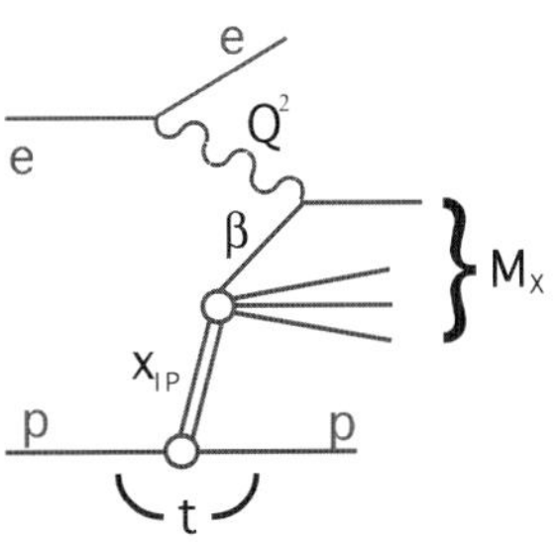


Figure 1: Illustration of a diffractive DIS event.

At HERA, around 10% of low $x$ events are diffractive [1]. Experimentally, such events are identified by either tagging the elastically scattered proton in *Roman pot* spectrometers $60-100$ m downstream from the interaction point or by asking for a large rapidity gap without particle production between the central hadronic system and the proton beam direction.

A diagram of diffractive DIS is shown in Fig. 1. A virtual photon coupling to the beam electron interacts diffractively with the proton through the exchange of a colour singlet and produces a hadronic system $X$ with mass $M_X$ in the final state. If the 4-momenta of the incoming (outgoing) electron and proton are labeled $l$ ($l'$) and $p$ ($p'$) respectively, the following kinematic variables can be defined: $Q^2=-q^2=-(l-l')^2$, the photon virtuality; $\beta = Q^2/q.(p-p')$, the longitudinal momentum fraction of the struck quark relative to the diffractive exchange; $x_{I\!P} = q.(p-p')/q.p$, the fractional proton momentum taken by the diffractive exchange and $t=(p-p')^2$, the 4-momentum squared transferred at the proton vertex. Bjorken-x is given by $x = x_{I\!P}\beta$. For the measurements presented here typical values of $x_{I\!P}$ are $< 0.05$. $y = Q^2/sx$ denotes the inelasticity, where $s$ is the $ep$ CMS energy.

A diffractive *reduced cross section* $\sigma_r^{D(4)}$ can be defined via

$$\frac{d^4\sigma^{ep\rightarrow eXp}}{dx_{I\!P}\ dt\ d\beta\ dQ^2} = \frac{4\pi\alpha^2}{\beta Q^4}\left(1-y+\frac{y^2}{2}\right)\sigma_r^{D(4)}(x_{I\!P},t,\beta,Q^2)\ , \qquad (1)$$

which is related to the diffractive structure functions $F_2^D$ and the longitudinal $F_L^D$ by

$$\sigma_r^D = F_2^D - \frac{y^2}{2(1-y+\frac{y^2}{2})}F_L^D\ . \qquad (2)$$

Except at the highest $y$, $\sigma_r^D = F_2^D$ to a very good approximation. If the outgoing proton is not detected, the measurements are integrated over $t$: $\sigma_r^{D(3)} = \int \mathrm{d}t\ \sigma_r^{D(4)}$.

*e-mail address: fpschill@mail.desy.de


doi:10.1016/S0920-5632(02)02222-3

## 2. FACTORIZATION PROPERTIES OF DIFFRACTIVE DIS

The proof [2] that QCD hard scattering factorization is valid for diffractive DIS justifies the expression of $\sigma_r^{D(4)}$, at fixed $x_{I\!P}$ and $t$, as a convolution of diffractive proton parton distributions (dpdf's) $f_i^D$ and partonic cross sections $\hat{\sigma}^{\gamma^* i}$:

$$\sigma_r^{D(4)} \sim \sum_i \hat{\sigma}^{\gamma^* i}(x.Q^2) \otimes f_i^D(x, Q^2, x_{I\!P}, t) \ . \quad (3)$$

The $f_i^D$ should obey the DGLAP evolution equations and the $\hat{\sigma}^{\gamma^* i}$ are the same as for standard DIS. In consequence, the framework of NLO QCD can be applied to diffractive DIS in a similar way as to inclusive DIS.

An extra assumption which is often made is that the $(x_{I\!P}, t)$ dependence of $\sigma_r^D$ can be factored out (Regge factorization) into a *Pomeron flux factor* $f_{I\!P}(x_{I\!P}, t)$:

$$f_i^D(x_{I\!P}, t, \beta, Q^2) = f_{I\!P}(x_{I\!P}, t) \ f_i^{I\!P}(\beta, Q^2) \ , \quad (4)$$

which is often parameterised using Regge phenomenology as

$$f_{I\!P}(x_{I\!P}, t) = x_{I\!P}^{1-2\alpha_{I\!P}(t)} e^{bt} \ , \quad (5)$$

where $\alpha_{I\!P}(t) = \alpha_{I\!P}(0) + \alpha'_{I\!P} t$ is the pomeron trajectory with intercept $\alpha_{I\!P}(0)$. The $f_i^{I\!P}(\beta, Q^2)$ are *Pomeron pdf's*. Although there is no firm basis in QCD for Eq. 4, it is approximately consistent with present data.

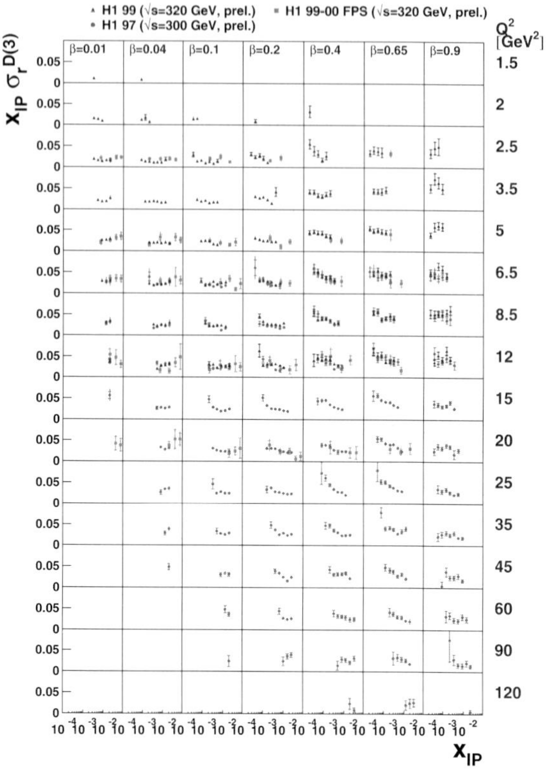


Figure 2. Compilation of new measurements of the diffractive reduced cross section $\sigma_r^D$ by H1.

## 3. EXPERIMENTAL RESULTS

### 3.1. The New Datasets

The H1 collaboration has performed several new measurements of diffractive DIS, presented in the form of the reduced cross section $\sigma_r^{D(3)}$, which are summarised in Fig. 2. Two high precision measurements in the kinematic range $1.5 < Q^2 < 120$ GeV$^2$ (labeled 'H1 99' [3] and 'H1 97' [4]) are based on the rapidity gap method. With respect to previous H1 data [5], a factor of typically 5 more data were analysed and the kinematic range was extended significantly to lower $Q^2$ and $\beta$. Another new measurement using a forward proton spectrometer covers $2.5 < Q^2 < 20$ GeV$^2$ ('H1 99-00 FPS', [6]). Good agreement between the different data sets and measurement techniques is observed.

The ZEUS collaboration has presented new data [7] on $F_2^{D(3)}$ for $2.2 < Q^2 < 80$ GeV$^2$, shown in Fig. 3. Diffractive events are selected by reconstructing a low value of $M_X$ in the detector. The data correspond to a factor 2 increase in integrated luminosity with respect to previous results [8]. ZEUS has also measured (Fig. 4, [9]) the diffractive cross section in the transition region $0.03 < Q^2 < 0.6$ GeV$^2$ between DIS and quasi-real photon-proton interactions ($Q^2 \sim 0$). Also shown in Fig. 4 is a fit based on a dipole model [10] in which the photon fluctuates into $q\bar{q}$ and $q\bar{q}g$ states.

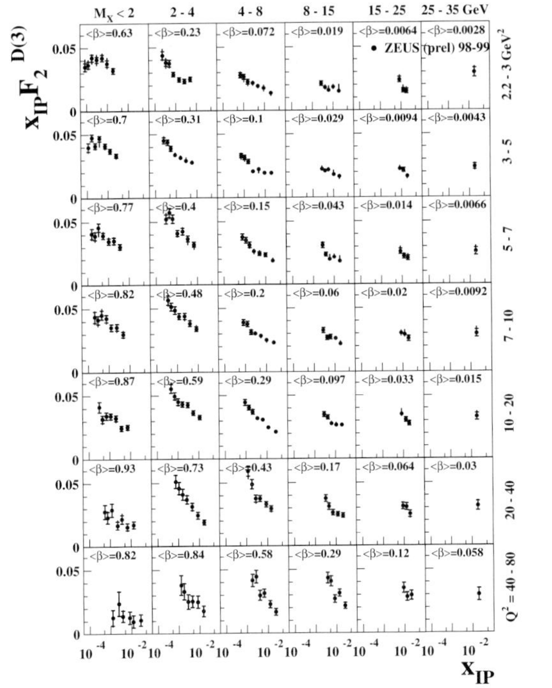


Figure 3. New measurement of the structure function $F_2^{D(3)}$ by ZEUS.

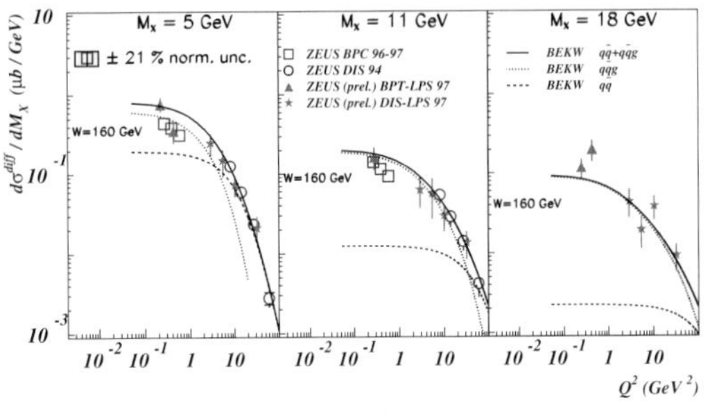


Figure 4. New measurement of the diffractive cross section at low $Q^2$ by ZEUS.

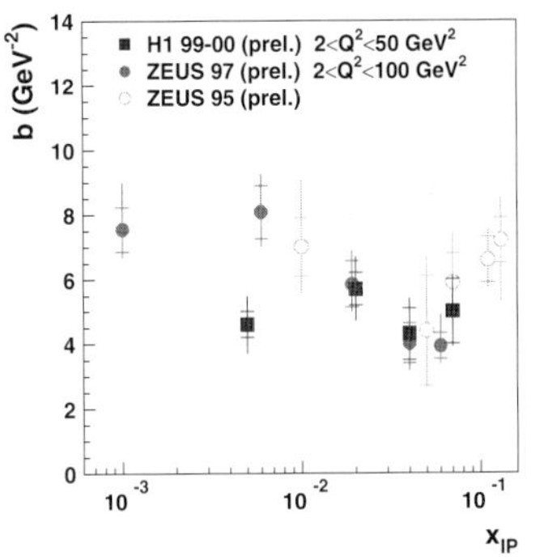


Figure 5. The slope parameter $b$ from fits to $\frac{d\sigma}{dt} \sim e^{bt}$ at different values of $x_{I\!P}$.

### 3.2. $t$ and $\Phi_{ep}$ Dependences

Both collaborations have measured the $t$ dependence of the cross section using their forward proton spectrometers. In the measured range $-0.45 \lesssim t < -|t|_{min}$, an exponential dependence $\frac{d\sigma}{dt} \sim e^{bt}$ is observed and the slope parameter $b$ is determined for different values of $x_{I\!P}$ (Fig. 5). Models based on Regge phenomenology predict an increasing steepness of the $t$ dependence with energy (*shrinkage*): $b = b_0 + 2\alpha' \log \frac{1}{x_{I\!P}}$. As can be seen in Fig. 5, the data at low $x_{I\!P}$ are so far inconclusive.

ZEUS has studied [11] the cross section dependence on the azimuthal angle $\Phi_{ep}$ between the electron and proton scattering planes, which is sensitive to the longitudinal cross section $\sigma_L^D$ via the interference between transverse and longitudinal photon induced contributions, which introduces an asymmetry in $\Phi_{ep}$. The measured asymmetries are small and, within the present statistical uncertainties, compatible with zero.

### 3.3. $x_{I\!P}$ Dependence

By making fits to the $x_{I\!P}$ dependence of the diffractive DIS data using Eqs. 4,5, values for the effective pomeron intercept $\alpha_{I\!P}(0)$ are determined. Results for measurements at different $Q^2$ are shown in Fig. 6 and compared with the *soft pomeron* value of 1.08 as well as with a parameterisation based on inclusive DIS data, obtained from fits of the form $F_2(x, Q^2) \sim cx^{-\lambda(Q^2)}$ where $\lambda = \alpha_{I\!P}(0) - 1$. The diffractive data suggest an increase of $\alpha_{I\!P}(0)$ with $Q^2$ and are significantly higher than the soft pomeron value for $Q^2 \gtrsim 10$ GeV. However, the naive expectation that the diffractive cross section grows twice as fast with energy as the inclusive one, correspond-

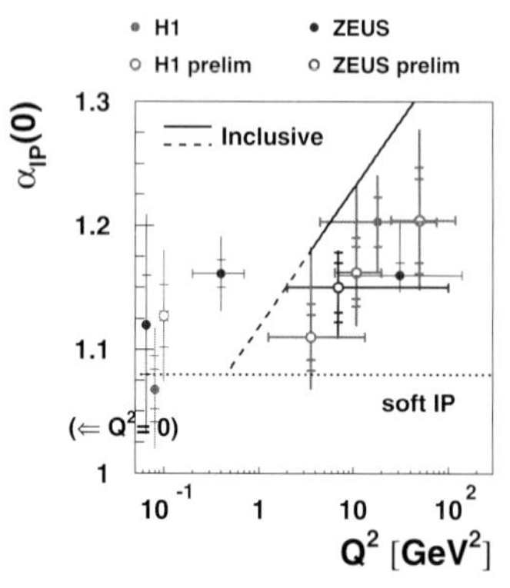


Figure 6. Measurements of the effective pomeron intercept $\alpha_{I\!P}(0)$ for different values of $Q^2$, compared with a parameterisation from inclusive DIS.

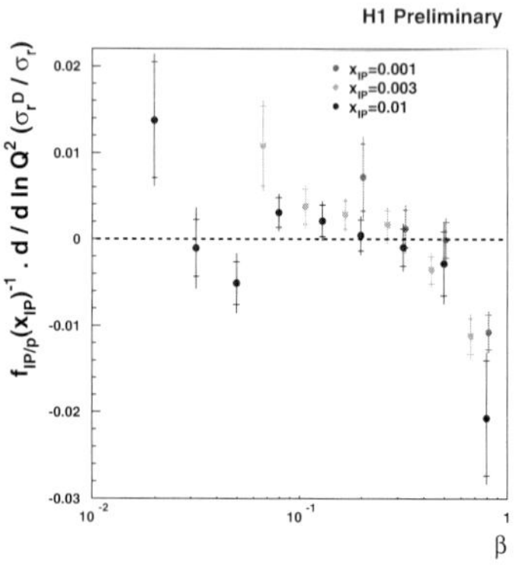


Figure 7. Logarithmic $Q^2$ derivatives of the ratio $\sigma_r^D/\sigma_r$ from H1, divided by $f_{I\!P}(x_{I\!P})$, shown as a function of $\beta$ for different $x_{I\!P}$.

ing to $\alpha_{I\!P}^{dif.}(0) = \alpha_{I\!P}^{incl.}(0)$, is hard to reconcile with the present data at high $Q^2$.

### 3.4. Ratio of Diffractive to Inclusive Cross Section

H1 has studied [4] the logarithmic $Q^2$ dependence of the ratio of the diffractive to the inclusive DIS reduced cross sections, presented as the quantity $f_{I\!P}(x_{I\!P})^{-1}$ d/dln$Q^2(\sigma_r^D/\sigma_r)$ in Fig. 7 for different $\beta$ and $x_{I\!P}$ values. Taking out the factorising $x_{I\!P}$ dependence, the same behaviour is observed at different $x_{I\!P}$: At low $\beta$, the logarithmic $Q^2$ dependences of $\sigma_r^D$ and $\sigma_r$ are very similar, indicating that the ratio of the diffractive to the total proton gluon density is approximately constant in this region. By contrast, at the highest $\beta \gtrsim 0.5$, the logarithmic $Q^2$ derivative becomes negative, suggestive of the presence of $Q^2$-suppressed higher twist contributions to $\sigma_r^D$ for $\beta \to 1$. However, this behaviour can also be explained by DGLAP evolution due to the the kinematic limit $x = x_{I\!P}$ ($\beta = 1$) for gluon radiation in the case of $\sigma_r^D$.

## 4. $\beta$ AND $Q^2$ DEPENDENCES OF $\sigma_r^D$ AND NLO QCD FIT TO H1 DATA

H1 has presented [4] a new NLO DGLAP QCD fit to the diffractive reduced cross section data, from which diffractive parton densities are determined. The shape of the dpdf's is assumed to be independent of $x_{I\!P}$ (Regge factorization), in agreement with the data, as can be seen in Figs. 8 and 9. The precisely measured $\beta$ and $Q^2$ dependences of the data are compared with the fit result, dividing out the $x_{I\!P}$ dependence, parameterised as stated in Eq. 5.

The diffractive exchange is parameterised by a light flavour singlet and a gluon distribution at a starting scale $Q_0^2 = 3$ GeV$^2$ and evolved to higher $Q^2$ using the standard NLO DGLAP equations. For the first time in diffraction, the experimental and model uncertainties are propagated to obtain error bands for the dpdf's.

The result of the fit is presented in Fig. 10. The diffractive pdf's remain large up to large fractional momenta $z$ (or $\beta$) and are dominated by the gluon distribution. In total 75% of the exchange momentum is carried by gluons at $Q^2 = 10$ GeV$^2$. To test factorization, the dpdf's can be used for updated comparisons with diffractive final state data from HERA and the TEVATRON [4,12].

## 5. CONCLUSIONS

Several new high precision data sets of inclusive diffractive DIS have recently become available. In the case of the rapidity gap method, the data are

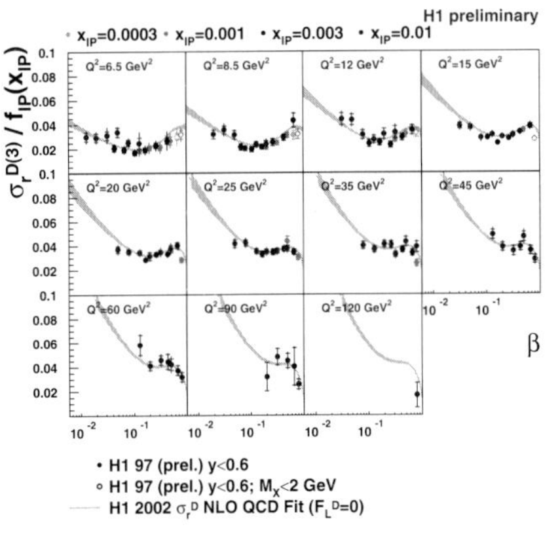


Figure 8. $\beta$ dependence of $\sigma_r^D$ from H1, scaled by $f_{IP}(x_{IP})^{-1}$ and compared with the NLO QCD fit.

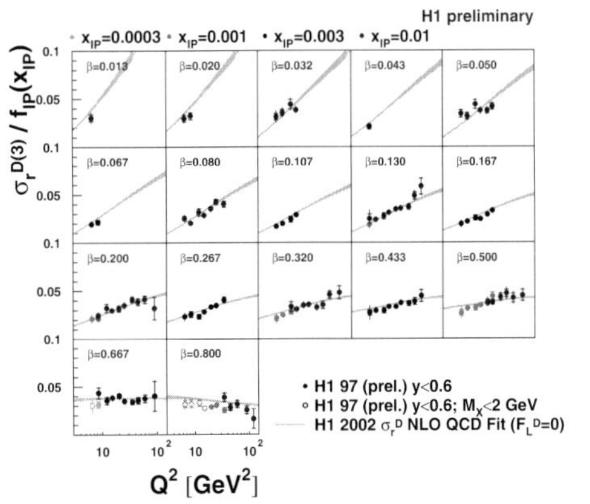


Figure 9. $Q^2$ dependence of $\sigma_r^D$ from H1, scaled by $f_{IP}(x_{IP})^{-1}$ and compared with the NLO QCD fit.

now systematically limited at low $Q^2$. These data are very useful to constrain the free parameters of QCD motivated models (e.g. dipole models) for diffractive DIS and to distinguish between different approaches. In particular, they have been used for the determination of a new generation of diffractive parton distributions – including their uncertainties – in the framework of NLO QCD, enabling tests of QCD factorization as applied to diffraction by comparisons with final state cross sections.

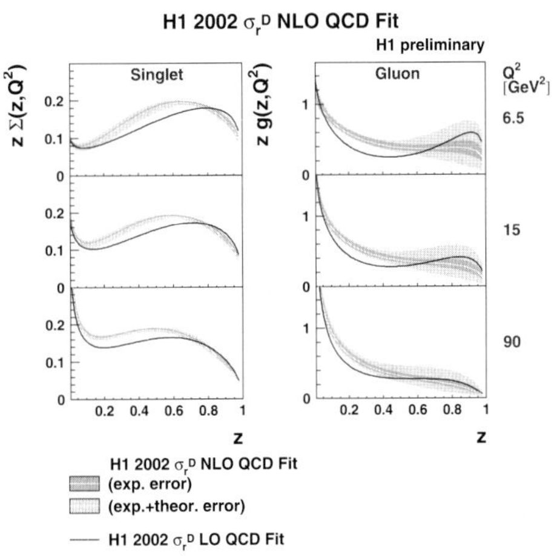


Figure 10. Diffractive singlet (left) and gluon (right) distributions from the H1 NLO QCD fit.

**Acknowledgements**

I thank my colleagues from H1 and ZEUS for their work reflected in this contribution.

## REFERENCES

1. ZEUS Coll., M. Derrick *et al.*, *Phys. Lett.* **B 315** (1993) 481; H1 Coll., T. Ahmed *et al.*, *Nucl. Phys.* **B 429** (1994) 477.
2. J. Collins, *Phys. Rev.* **D 57** (1998) 3051, err.-ibid. **D 61** (2000) 19902.
3. H1 Coll., paper 981 subm. to ICHEP 2002.
4. H1 Coll., paper 980 subm. to ICHEP 2002.
5. H1 Coll., C. Adloff *et al.*, *Z. Phys.* **C 76** (1997) 613.
6. H1 Coll., paper 984 subm. to ICHEP 2002.
7. ZEUS Coll., paper 821 subm. to ICHEP 2002.
8. ZEUS Coll., J. Breitweg *et al.*, *Eur. Phys. J.* **C 6** (1999) 43.
9. ZEUS Coll., paper 822 subm. to ICHEP 2002.
10. J. Bartels, J. Ellis, H. Kowalski, M. Wüsthoff, *Eur. Phys. J.* **C 7** (1999) 443.
11. ZEUS Coll., paper 823 subm. to ICHEP 2002.
12. A. Savin, these proceedings.

Proceedings of ICHEP 2002, pp. 408–410
S. Bentvelsen, P. de Jong, J. Koch and E. Laenen (Editors)

31st INTERNATIONAL CONFERENCE ON HIGH ENERGY PHYSICS AMSTERDAM

# Diffractive charm and jet production at HERA

Alexander A. Savin [a]*

[a]University of Wisconsin-Madison,
1150 University Ave., Madison WI 53706-1390, USA

A new high precision inclusive measurement of the diffractive production of $D^{*\pm}(2010)$ mesons in deep inelastic scattering (DIS) in the kinematic region $Q^2 > 1.5$ GeV$^2$, $0.02 < y < 0.7$, $x_{IP} < 0.035$ as well as differential dijet cross sections in low-$|t|$ diffractive photoproduction (PHP) process in the kinematic region $Q^2 < 0.01$ GeV$^2$, $165 < W < 240$ GeV$^2$, $x_{IP} < 0.03$ are presented. Diffractive parton densities extracted using a NLO DGLAP QCD fit are used for comparisons with diffractive DIS and PHP dijet and open charm cross sections at HERA and the Tevatron, thus testing the factorization properties of hard diffraction.

## 1. INTRODUCTION

A factorization theorem has been proven [1] for diffractive DIS and direct PHP processes. The proof establishes the universality of diffractive parton densities (dpdf) [2,3] for the class of processes to which the theorem applies. It means that QCD factorization in diffraction can be tested by taking the diffractive pdfs and predicting diffractive final state observables such as dijet and charm cross sections at HERA or diffractive dijet production at the Tevatron.

A particular motivation for studying diffractive DIS charm production is that one expects charm production to be a substantial fraction of the diffractive DIS cross section. Moreover, since open charm production is dominated by the photon-gluon fusion process, it also provides a good test of the normalization of the diffractive gluon density.

Theoretical pQCD-based models are expected to describe the data since high jet $p_t$ or charm-quark mass provides a sufficient hard scale.

## 2. DIJET CROSS SECTION

The RAPGAP MC program [4] is used to obtain the predictions. The renormalisation and factorization scales are set to $\mu^2 = Q^2 + p_t^2 + m_q^2$.

Figure 1 compares the new H1 LO dpdfs prediction [2] as well as predictions based on the previous H1 QCD fits [5] (H1 fit 2 and fit 3) with the H1 DIS dijet data [6] ($40 < Q^2 < 120$ GeV$^2$).

The prediction based on the recent QCD fit is generally below those for the previous fits due to the reduced diffractive gluon distribution. This leads to an improved description of the $D^*$ cross section (not shown), but a worse description for the dijet cross section.

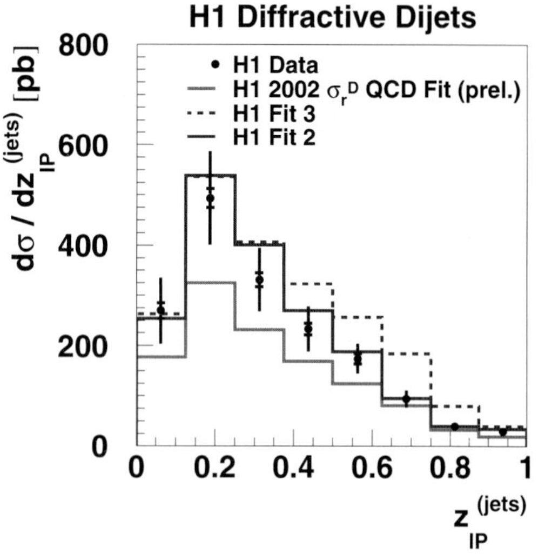


Figure 1. H1 measurements of diffractive DIS dijet cross section differential in the estimator $Z_{IP}^{jets}$ of the fractional momentum from the diffractive exchange.

*On behalf of the H1 and ZEUS collaborations


doi:10.1016/S0920-5632(02)02223-5

Applying the old dpdfs to predict diffractive dijet production at the Tevatron leads to an overestimation of the observed rate by an order of magnitude. The new prediction is slightly closer to the Tevatron data but the overall level of disagreement is the same [2].

The transition from DIS to hadron-hadron scattering can be studied in PHP, where the processes in which an almost real photon participates directly in the hard scattering are expected to be similar to DIS. Processes where the photon is first resolved into partons which then initiate the hard scattering resemble hadron-hadron scattering.

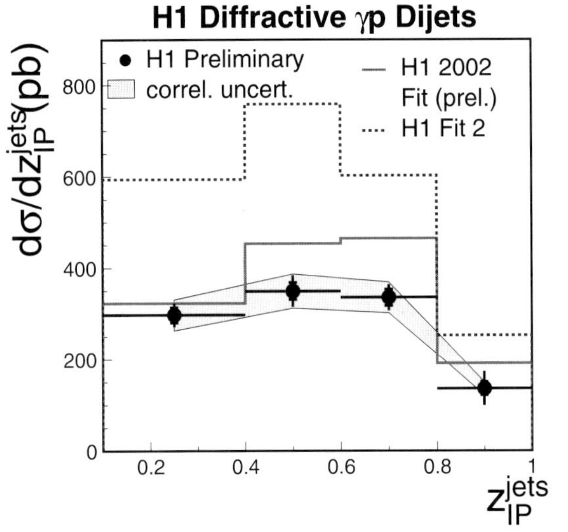


Figure 2. H1 measurements of diffractive PHP dijet [7]. Two predictions of the RAPGAP MC model with LO QCD fits are shown.

Figure 2 shows the cross section differential in the estimator $Z_{IP}^{jets}$. The RAPGAP prediction based on the H1 QCD fit 2 overestimates the normalization of the data by a factor $\approx$2. The new H1 2002 QCD fit overestimates the cross section by a factor $\approx$1.3, but is compatible with the data given the uncertainty of the gluon density resulting from the fit. Figure 3 shows the prediction of the H1 fit 2 with the contribution from direct photon processes displayed separately.

The same suppression of both resolved and direct photon processes is needed by an overall scale factor 1.8 ± 0.45(exp.) compare to the DIS dijet process. Although this factor is subject to considerable additional uncertainties, it is much smaller than the suppression factor $\approx$10 for single diffractive dijet production at the Tevatron.

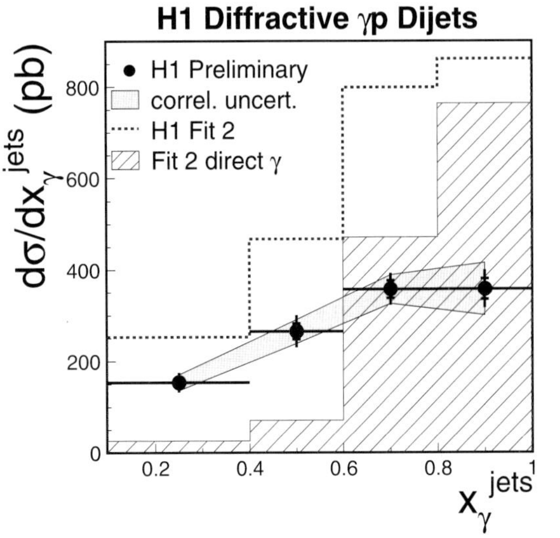


Figure 3. H1 measurements of diffractive PHP dijet differential in $x_\gamma^{jets}$, the longitudinal fraction momenta carried by the partons from the photon. The direct photon contributions (boson-gluon fusion and QCD compton) are indicated by the hatched histogram.

## 3. DIS $D^{*\pm}$(2010) PRODUCTION

Figures 4 and 5 show the measured by ZEUS differentioal $D^{*\pm}$(2010) cross sections [8]. Predictions of three models are compared to the data: (1) "saturation" model by Golec-Biernat and Wüsthoff [9], as implemented in the SATRAP MC generator; (2) two-gluon-exchange model of Bartels et al. (BJLW) [10]; (3) resolved Pomeron model, as implemented in the NLO fits to HERA data performed by Alvero et al.(ACTW) [3].

The SATRAP model decribes the data, both in shape and normalization. The $c\bar{c} + c\bar{c}g$ BJWL predictions, obtained for $m_c$=1.45 GeV and $\mu = \sqrt{p_{c,t}^2 + 4m_c^2}$ and value of cutoff on the gluon transverse momentum, $K_{T,g}^{out} = 1.5$ GeV, describes the shapes of the distributions reasonably well ex-

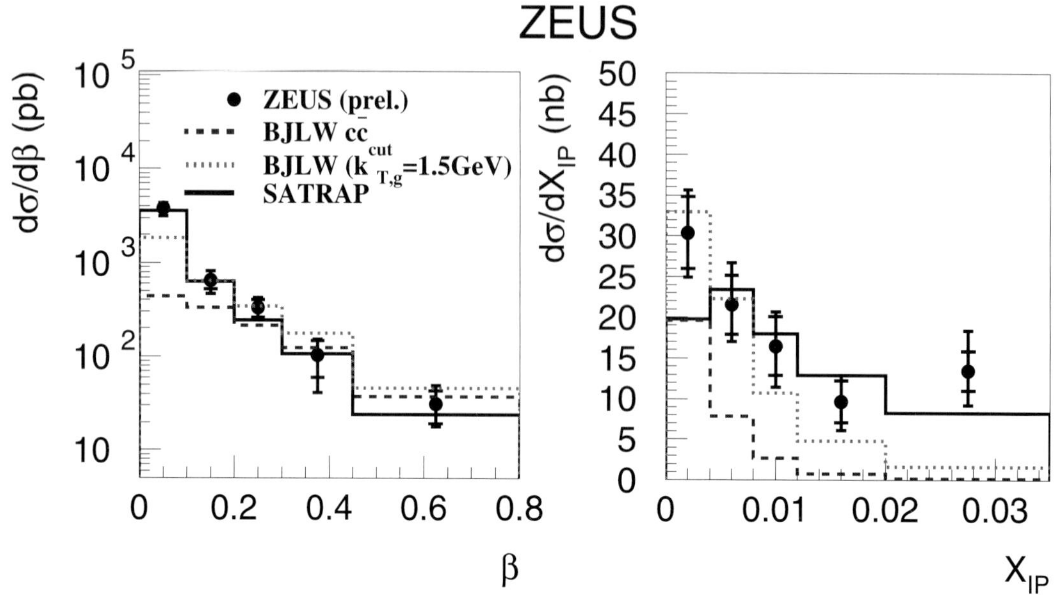


Figure 4. ZEUS measurement of differential cross section for $D^{*\pm}(2010)$ production. with respect to the kinematic variables $\beta$, the fraction of the Pomeron momentum carried by the parton participating in the hard interaction and $X_{IP}$, the fraction of the proton momentum carried by the Pomeron. The histograms correspondent to the different models described in the text.

cept for $x_{IP}$ (the reasons are discussed in [10]).

The ACTW NLO QCD prediction (gluon-dominated fit B) is also in reasonable agreement with the measured differential distributions supporting the notion of hard scattering factorization for difractive lepton-hadron processes.

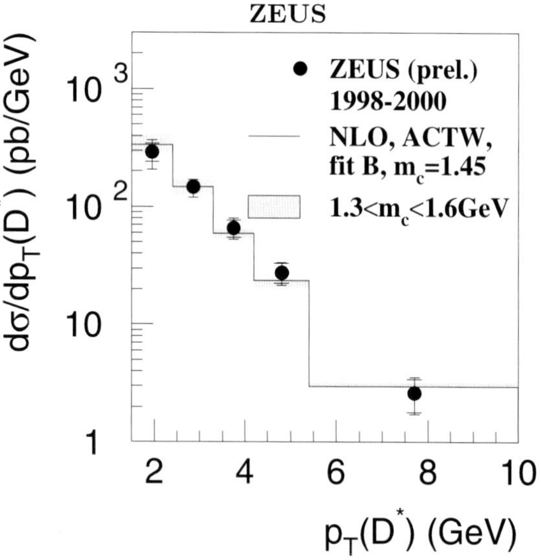


Figure 5. ZEUS measurement of differential cross section for D* production. The histogram shows the prediciotn from NLO ACTW model. The shaded area shows the effect of varying the charm-quark mass.

## REFERENCES


1. J.C. Collins, Phys.Rev. D57 (1998) 3051;
2. H1 Coll, Abstract 980 31 ICHEP, Amsterdam, July 2002;
3. L.Alvero et al., Phys.Rev. D59 (1999) 074022;
4. H. Jung, Comp.Phys.Commun. 86(1995)1973;
5. H1 Coll.,C.Adloff et al.,Z.Phys. C76(1997)613;
6. H1 Coll.,C.Adloff et al.,Eur.Phys.J. C20(2001)29;
7. H1 Coll., Abstract 987 31 ICHEP, Amsterdam, July 2002;
8. ZEUS Coll., Abstract 780 31 ICHEP, Amsterdam, July 2002;
9. K.Golec-Biernat and M.Wüsthoff, Phys.Rev. D59(1999)014017;
10. J.Bartels, H.Jung and M.Wüsthoff, Eur.Phys.J. C11(1999)111.

ELSEVIER

Proceedings of ICHEP 2002, pp. 411–413
*S. Bentvelsen, P. de Jong, J. Koch and E. Laenen (Editors)*

31st INTERNATIONAL CONFERENCE ON HIGH ENERGY PHYSICS AMSTERDAM

# Leading Baryons at HERA

W. B. Schmidke[a] on behalf of the H1 and ZEUS Collaborations

[a]Columbia University, Nevis Labs, Irvington on Hudson, N. Y., USA

H1 and ZEUS results on leading baryon production in *ep* scattering are presented. The data are shown to be consistent with particle exchange models of baryon production, and the pion structure function is extracted.

## 1. INTRODUCTION

Events with a baryon carrying a large fraction of the proton beam energy have been observed in *ep* scattering at HERA [1–3]. The dynamical mechanisms for their production are not completely understood. They may be the result of hadronization of the proton remnant, conserving baryon number in the final state. Exchange of virtual particles is also expected to contribute. In this picture, the target proton fluctuates into a virtual meson-baryon state. The virtual meson scatters with the projectile lepton, leaving the fast forward baryon in the final state. Neutron production occurs through the exchange of isovector particles, notably the $\pi^+$ meson. For proton production isoscalar exchange also contributes, including diffraction mediated by pomeron exchange.

Both the H1 and ZEUS experiments at HERA have installed forward detectors to measure leading baryons. The detectors measure the fraction of the beam energy carried by the baryon, $x_L$, and the transverse momentum transferred to the baryon, $p_T$. Here we report some results on leading baryon production, in both photoproduction (PHP), where the exchanged photon virtuality $Q^2$ is nearly zero, and in deep inelastic scattering (DIS), where $Q^2$ is greater than a few GeV$^2$. The baryon $x_L$ spectra are inconsistent with conventional hadronization models but in agreement with the meson exchange picture. Differences between the PHP and DIS $x_L$ spectra are consistent with exchange models that incorporate rescattering of the baryon. Using the exchange formalism, the structure function of the pion, $F_2^\pi$, is extracted.

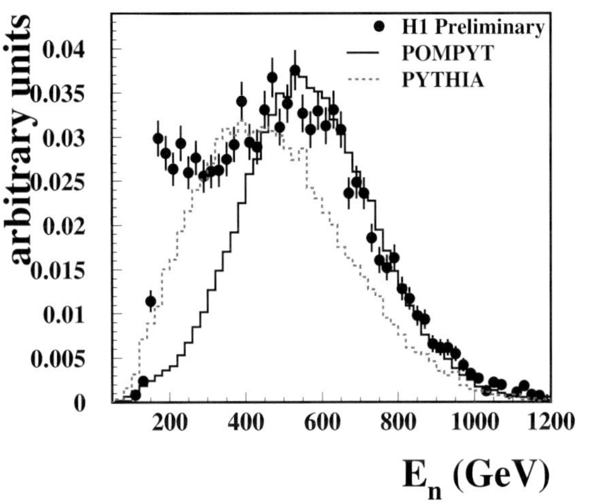


Figure 1. Energy spectrum of leading neutrons in PHP events.

## 2. LEADING BARYON SPECTRA

Figure 1 shows the energy spectrum of leading neutrons in PHP events [4]. Also shown are the predictions of two models of leading neutron production. In the PYTHIA model the neutrons are the result of hadronization of the proton remnant. This model peaks at a lower neutron energy than does the data. The POMPYT model produces the neutrons through virtual $\pi^+$ exchange, with the $\pi^+$ then scattering with the almost real photon. This model reproduces the data for $E_n > 450$ GeV.


doi:10.1016/S0920-5632(02)02227-2

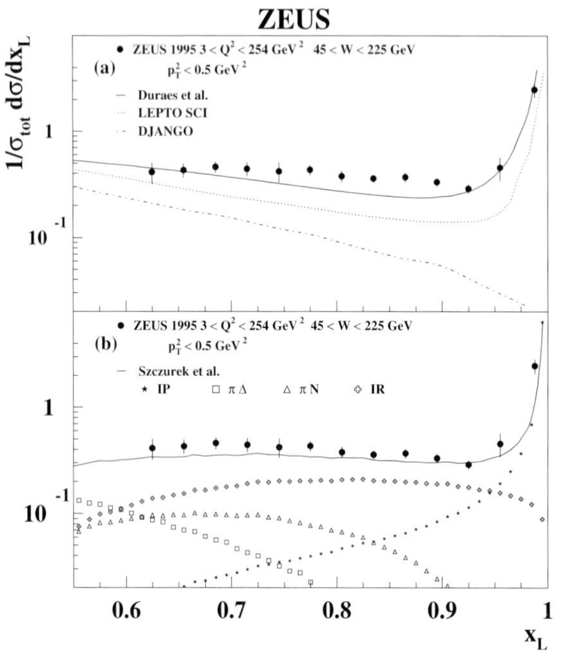


Figure 2. $x_L$ spectrum of leading protons in DIS events.

Figure 2 shows the $x_L$ spectrum of leading protons in DIS events [3]. In the upper plot the data are compared to the predictions of models where the protons result from hadronization of the incoming proton remnant. They all predict a rate of leading protons which is less than that observed. The lower plot displays the predictions of an exchange model. The dotted curves show the contributions of the different exchanges. Their sum, shown as the solid curve, is in good agreement with both the shape and normalization of the data.

## 3. RESCATTERING

In the particle exchange picture, the leading baryon spectra are determined by a factor which describes the flux of virtual mesons in the incoming proton. The proton fluctuates into a baryon with momentum fraction $x_L$ and a meson with momentum fraction $1-x_L$. Assuming vertex factorization, this flux is independent of the type of scattering the meson undergoes with the projectile lepton. Thus in the simplest picture the baryon $x_L$ spectra should be independent of $Q^2$. However, for decreasing $Q^2$, the virtual photon size is larger. If this size is comparable to the virtual baryon-meson separation, the baryon may also scatter with the photon. In this case it will emerge with lower $x_L$ and higher $p_T$ and may escape detection.

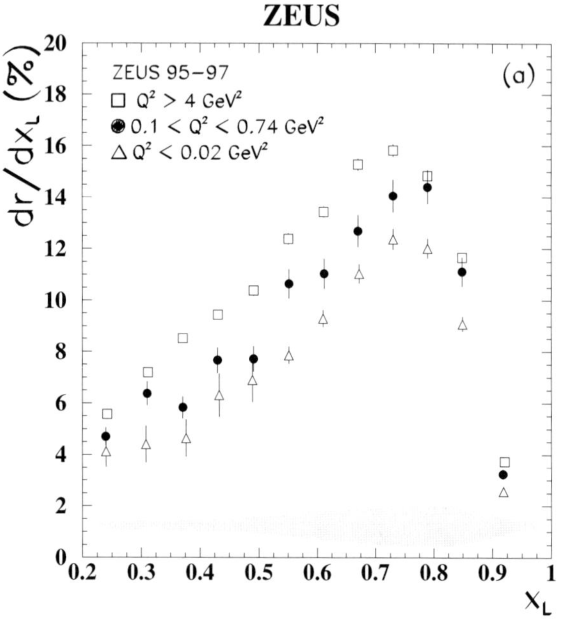


Figure 3. $x_L$ spectra of leading neutrons for different ranges of $Q^2$.

Figure 3 displays the leading neutron $x_L$ spectra for PHP, DIS, and events with an intermediate range of $Q^2$ [2]. Fewer neutrons are detected for decreasing $x_L$ and $Q^2$. Figure 4 shows the ratio of the PHP spectrum to the DIS spectra for different ranges of $Q^2$ [5]. The depletion of neutrons in PHP with decreasing $x_L$ is apparent. The predictions of two rescattering models are also shown [6,7]. In the model of D'Alesio and Pirner the rescattering for PHP decreases with increasing $x_L$ because the mean $n$–$\pi$ separation is increasing according to the pion exchange model. This model qualitatively describes the rise in the ratio above $x_L = 0.55$, although it underestimates

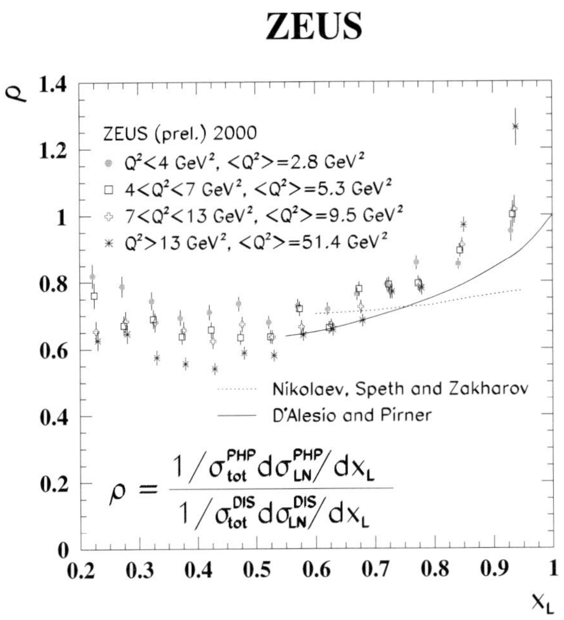


Figure 4. Ratio of neutron $x_L$ spectra PHP/DIS for different ranges of $Q^2$.

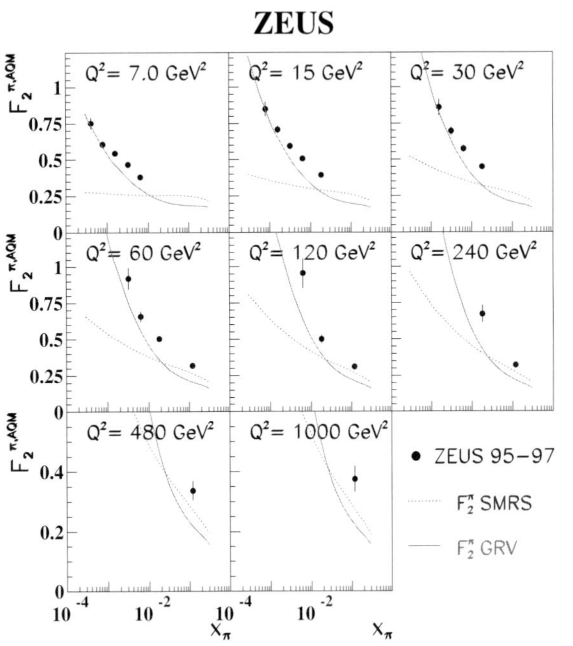


Figure 5. $F_2^\pi$ extracted from DIS events with a leading neutron. Here $x_\pi = x/(1 - x_L)$.

the magnitude.

## 4. PION STRUCTURE FUNCTION

In the same way that the inclusive DIS reaction is used to measure the proton structure function $F_2^p(x, Q^2)$, where $x$ is the Bjorken scaling variable, the semi-inclusive reaction $ep \to e'nX$ can be used to measure the neutron structure function $F_2^{LN}(x, Q^2, x_L, p_T)$. In the $x_L$ range where $\pi^+$ exchange dominates this can be related to the pion structure function:

$$F_2^{LN}(x, Q^2, x_L, p_T) = f(x_L, p_T) F_2^\pi(x/(1-x_L), Q^2).$$

Here $f(x_L, p_T)$ is the flux of virtual $\pi^+$ in the proton. Available $\pi$ exchange models have different forms for the flux. The range of choices results in an uncertainty of up to a factor of two in the normalization. For the flux based on the additive quark model, $F_2^\pi$ extracted from the measured $F_2^{LN}$ is shown in Fig. 5 [2]. Also shown are two parameterizations based on fixed target hadronic data at lower $Q^2$ and higher $x$. The present data discriminates between the two parameterizations shown.

## REFERENCES


1. ZEUS Coll., M. Derrick et al., Phys. Lett. B 384 (1995) 388; H1 Coll., C. Adloff et al., Eur. Phys. J. C 6 (1999) 587; ZEUS Coll., J. Breitweg et al., Nucl. Phys. B 596 (2000) 3; H1 Coll., C. Adloff et al., Nucl. Phys. B 619 (2001) 3.
2. ZEUS Coll., S. Chekanov et al., Nucl. Phys. B 637 (2002) 3.
3. ZEUS Coll., Abstract 833, XXXI International Conference on High Energy Physics, Amsterdam, Netherlands, 2002.
4. H1 Coll., Abstract 988, XXXI International Conference on High Energy Physics, Amsterdam, Netherlands, 2002.
5. ZEUS Coll., Abstract 824, XXXI International Conference on High Energy Physics, Amsterdam, Netherlands, 2002.
6. N.N. Nikolaev, J. Speth and B.G. Zakharov, Preprint KFA-IKP(TH)-1997-17 (hep-ph/9708290) (1997).
7. U. D'Alesio and H.J. Pirner, EPJ A 7 (2000) 109.

ELSEVIER

Proceedings of ICHEP 2002, pp. 414–419
*S. Bentvelsen, P. de Jong, J. Koch and E. Laenen (Editors)*

31st INTERNATIONAL CONFERENCE ON HIGH ENERGY PHYSICS AMSTERDAM

# Diffractive production of vector mesons at HERA

Alessia Bruni[a], on behalf of the H1 and ZEUS Collaborations

[a]Istituto Nazionale di Fisica Nucleare, Bologna,
Via Irnerio 46, I-40126 Bologna, Italy

An extended study of vector meson production in $ep$ interactions has been performed by the H1 and ZEUS Collaborations at the HERA collider. In this report, recent results are summarised and the dynamical picture of the diffractive interaction emerging in perturbative QCD is highlighted.

## 1. Introduction

The diffractive production of vector mesons $ep \rightarrow e\ V\ Y$, with $V = \rho^0$, $\omega$, $\phi$, $J/\psi$, $\psi'$ or $\Upsilon$ and with $Y$ being either the scattered proton or a low mass hadronic system, has been extensively investigated at HERA. The vector mesons are identified via their decays to two oppositely charged particles.

In QCD based models, the diffractive production of vector mesons is viewed as a sequence of processes: the virtual photon $\gamma^*$ fluctuates in a $q\bar{q}$ pair (dipole) which subsequently interacts elastically with the proton, and eventually forms a meson bound state. At high energies of the $\gamma^* p$ system, these successive processes are clearly separated in time (factorization). The transverse size of the $q\bar{q}$ pair depends on the photon virtuality $Q^2$ and on the quark masses; for $Q^2 > O(10)$ GeV$^2$ or for heavy charm quarks, it is considerably smaller than the proton size. For these distances the strong coupling is small and perturbative QCD (pQCD) can be applied.

HERA offers the unique opportunity to study the dependences of diffractive processes on several momentum scales: the mass of the vector meson $m_V$, the centre-of-mass energy of the $\gamma p$ system $W$, the photon virtuality $Q^2$ and the four-momentum transfer squared at the proton vertex $t$. Strong interactions can be investigated in the transition from the hard to the soft regime, where the confinement of quarks and gluons occurs. At small values of the Bjorken variable $x$, where the density of partons is large, the study of the vector meson production allows the investigation of whether unitarity limits have been reached.

The presented results cover a range in $Q^2$ up to 100 GeV$^2$, centre-of-mass energies $20 < W < 290$ GeV and values of $|t|$ up to 20 GeV$^2$.

## 2. $W$ dependence of the cross section

The dependence of the cross section on $W$ is shown in Fig. 1 for exclusive photoproduction of the different vector mesons together with the total photoproduction cross section. The $W$ dependence of the cross section for light vector meson ($\rho^0$, $\omega$, $\phi$) production is similar to that of the total cross section and is as expected for the exchange of the universal soft pomeron of hadronic physics. The regime clearly changes as the mass of the vector meson increases: the $J/\psi$ cross section has a steeper rise with $W$, which is a signature for a hard process. In pQCD models, the cross section is proportional to the square of the gluon density $G(x, Q^2)$ in the proton, $\sigma_L \propto \alpha_S^2(Q^2)/Q^6\ [xG(x,Q^2)]^2$. The measured $W$ dependence can be described by assuming a rapid rise of the gluon density in the proton at small $x$, since $W^2 \propto 1/x$.

The $J/\psi$ photoproduction data [1], shown also in Figure 2, are precise enough to distinguish between different parton-density function sets but no discrimination can be made at present due to the large theoretical uncertainties arising from the effects of higher-twist corrections, missing higher-order terms and skewed parton distributions. The $W$ dependence of the $J/\psi$ cross sec-


doi:10.1016/S0920-5632(02)02236-3

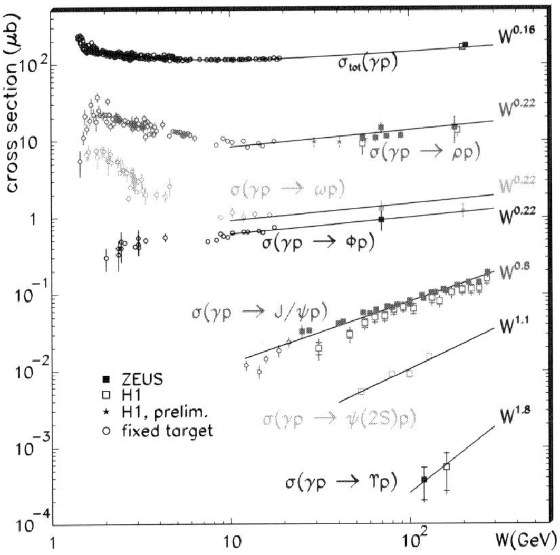


Figure 1. Compilation of cross sections for exclusive photoproduction of vector mesons as a function of $W$. The lines indicate a power law dependence, $W^{\delta}$. The total photoproduction cross section is also indicated.

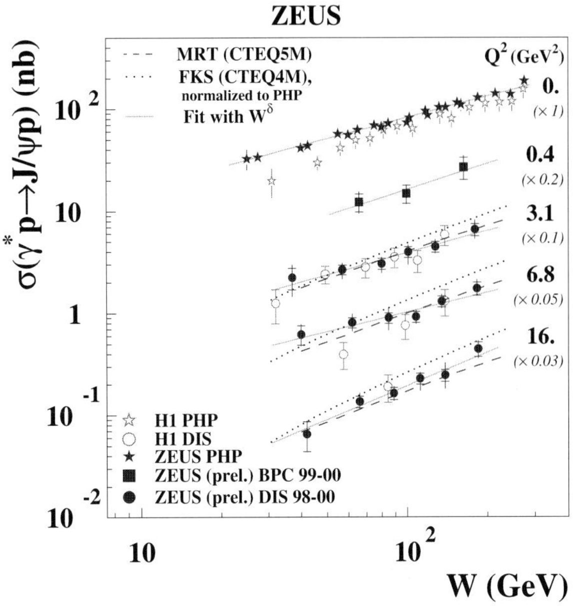


Figure 2. $J/\psi$ cross section as a function of $W$ for several values of $Q^2$. The curves show fits to the form $W^{\delta}$ as well as two pQCD predictions [3].

tions does not change for $Q^2$ up to 16 GeV$^2$ [2]. The already "hard" behaviour observed in photoproduction does not become steeper as $Q^2$ rises. The data are well described by pQCD calculations [3].

Figure 3 shows the different behaviour of the $\rho^0$ meson with a $W$ slope which increases with $Q^2$ [4,5].

The slope, $b$, of the exponential fall-off of the differential cross section $d\sigma/dt \propto e^{bt}$, that determines the size of the interaction region, has also been investigated [5,6]. As shown in Figure 4, $b$ shrinks for $\rho^0$ mesons with increasing $Q^2$, which is further evidence of a point-like interaction in the presence of a hard scale, while no change is observed for the $J/\psi$ meson.

The influence on the cross section of the meson wave function has been addressed [7]; the energy dependence of the $\psi(2S)$ and the $J/\psi$ is similar, but the production of the $\psi(2S)$ is suppressed. This is expected since the $\psi(2S)$ wave function has a node which leads to a larger expectation value for the $\psi(2S)$ radius than that characterising the ground state.

## 3. $Q^2$ dependence of the cross section

The QCD prediction for the longitudinal part of the cross section, $\sigma_L$, contains the factor $1/Q^6$ which is tamed by the $Q^2$ dependence of $\alpha_S$ and of the gluon density. For the transverse part, $\sigma_T$, a steeper $Q^2$ dependence is predicted [8]. Figure 5 shows the $Q^2$ dependence of the $\rho^0$ cross sections $\sigma_L$ and $\sigma_T$ with a fit to the form $1/(Q^2 + m_\rho^2)^n$ [9] .

## 4. Helicity studies

Sensitive tests of the dynamics of diffraction are provided by the study of the helicity struc-

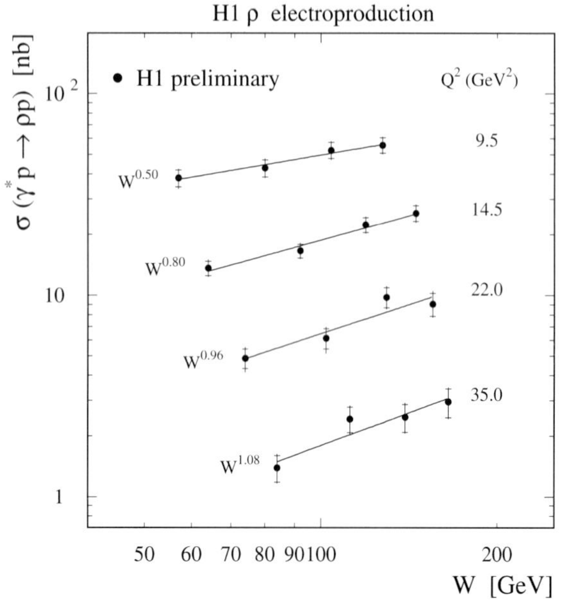


Figure 3. $\rho^0$ cross section as a function of $W$ for several values of $Q^2$. The lines indicate a power law dependence $W^\delta$.

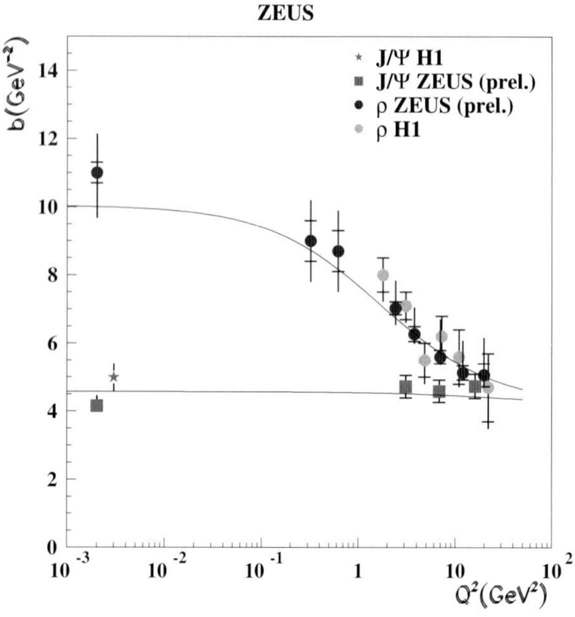


Figure 4. The $b$-slope parameter, $d\sigma/dt \propto e^{bt}$, as a function of $Q^2$ for $\rho^0$ and $J/\psi$ production.

ture of the interaction and its $t$ dependence. If $s$-channel helicity conservation (SCHC) holds, the vector meson retains the helicity of the photon. In QCD based models, the orbital momentum of the $q\bar{q}$ pair can be modified through the transfer of momentum carried by gluons. In such models, helicity flips between photon and meson are possible. Helicity analyses of $\rho^0$ electroproduction [10] have been recently extended [5,11] to higher value of $Q^2$, $2.5 < Q^2 < 60$ GeV$^2$, and $-t < 3$ GeV$^2$. As shown in Figure 6, a significant helicity non-conservation, consistent with a single flip amplitude increasing with $t$, is observed for the spin density matrix elements $r^5_{00} + 2r^5_{11}$. The measurement of the spin density matrix elements $r^1_{00} + 2r^1_{11}$, shown in Figure 7, confirms that the single flip amplitude is significantly larger than the double flip amplitude in the present kinematic domain. The data are compatible with a diffractive mechanism based on the exchange of two gluons [12]; non-conservation of $s$-channel helicity occurs when the longitudinal momentum of the photon is shared asymmetrically between the quark and the antiquark.

## 5. High $t$

Photoproduction of $\rho^0$, $\omega$, $\phi$ and $J/\psi$ has been studied [13] at $W \simeq 100$ GeV for $-t < 12$ GeV$^2$. The production of $J/\psi$ mesons has also been investigated [14] in the region $50 < W < 160$ GeV for $-t < 18$ GeV$^2$.

The $t$-dependence of the differential cross section is described by a power like behaviour $d\sigma/dt \propto (-t)^{-n}$ with $n$ decreasing with increasing mass of the vector meson, as shown in Figure 8. The data are described by a BFKL-like behaviour [15,16].

The cross-section ratios for the $\phi$ and $J/\psi$ with respect to the $\rho^0$ increase with $-t$. This observation suggests that the vector meson production

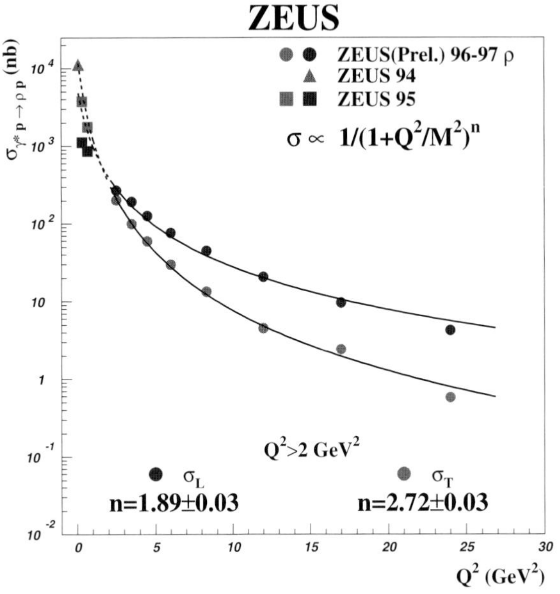


Figure 5. $Q^2$ dependence of the cross section for exclusive $\rho^0$ production; $\sigma_L$ and $\sigma_T$ are shown separately.

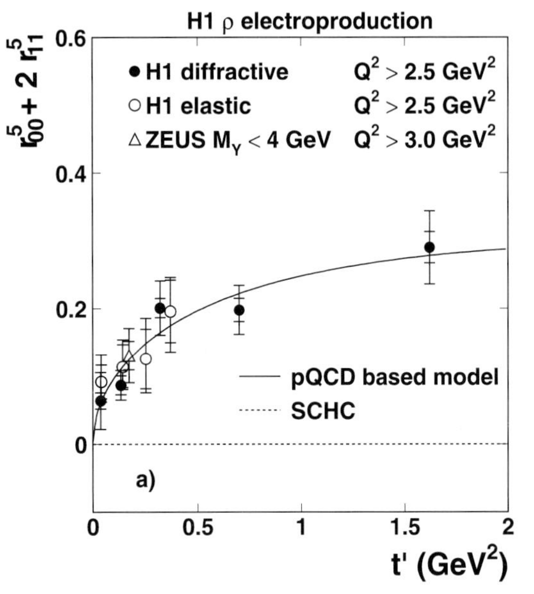


Figure 6. Measurement of the spin density matrix elements $r^5_{00}+2r^5_{11}$ as a function of $t' = |t|-|t|_{min}$. For SCHC a null result is expected, independent of $t'$. The full curve represents the expectations of a pQCD model [12].

mechanism may become flavour independent at high $-t$ values.

The $W$ dependence of the photoproduction of vector mesons at different $t$ values has also been investigated. Results are shown for the $J/\psi$ meson in Figure 9. The effective slope of the Pomeron trajectory, $\alpha'$, that parametrises the $W$ dependence as a function of $t$, is $\alpha' = -0.02 \pm 0.05(stat.)^{+0.04}_{-0.08}(syst.)$ for the $\rho^0$ meson and $\alpha' = -0.06 \pm 0.12(stat.)^{+0.05}_{-0.09}(syst.)$ for the $\phi$ meson. These values are lower than those measured in the exclusive photoproduction of vector mesons at lower $-t$ and suggest that the Pomeron trajectory, $\alpha$, is not linear with $t$.

## 6. Test of SU(4) universality

According to SU(4) flavour symmetry, the ratio of the cross sections for diffractive vector meson production should depend only on the charges of the quarks composing the vector mesons. This leads to an expectation of $\sigma_\rho : \sigma_\omega : \sigma_\phi : \sigma_{J/\psi} = [\frac{1}{\sqrt{2}}(u\bar{u} + d\bar{d})]^2 : [\frac{1}{\sqrt{2}}(u\bar{u} - d\bar{d})]^2 : [s\bar{s}]^2 : [c\bar{c}]^2 = 9 : 1 : 2 : 8$. A universal behaviour is expected when these cross sections are plotted as a function of the suitable scale. The cross sections for the light vector mesons, weighted by the appropriate SU(4) factors and plotted as a function of $Q^2 + m_V^2$, line up on a universal curve [17]. The new precise measurements of the $J/\psi$ cross section [1] are about 40% above this curve indicating that the momentum scale $Q^2 + M^2$ is a good variable for comparing the light vector mesons, but that it is not adequate for $J/\psi$ mesons.

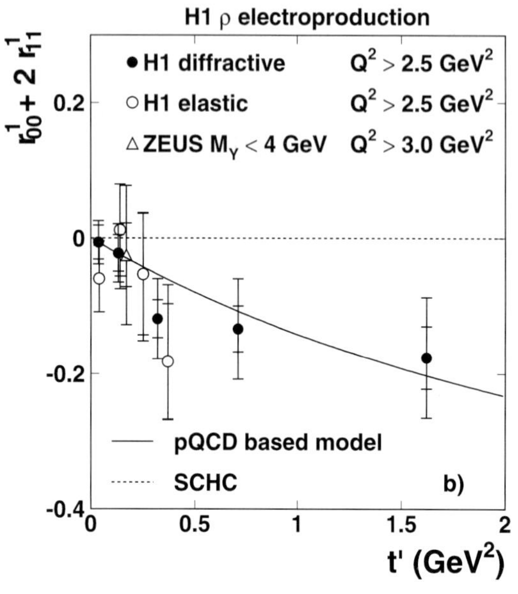


Figure 7. Measurement of the spin density matrix elements $r^1_{00}+2r^1_{11}$ as a function of $t' = |t|-|t|_{min}$. For SCHC a null result is expected, independent of $t'$. The full curve represents the expectations of a pQCD model [12].

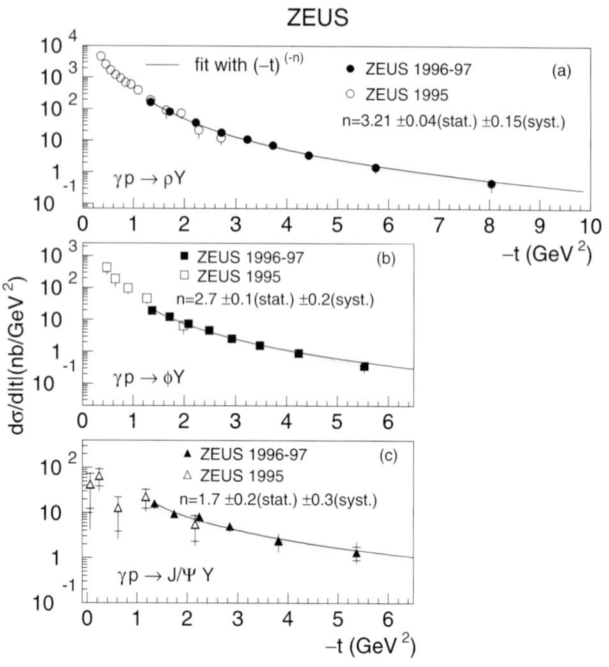


Figure 8. The differential cross section $d\sigma/dt$ in the range $80 < W < 120$ GeV for $\rho^0$, $\phi$ and $J/\psi$. The lines are the results of the fit to the function $A(-t)^{-n}$.

## 7. Test of the factorization of the proton vertex

The ratio of proton-dissociative and exclusive $\rho^0$ electroproduction has been investigated [6] in the kinematic region $Q^2 < 20$ GeV$^2$ and $|t| < 0.4$ GeV$^2$. The ratio, shown in Figure 10, is independent of $Q^2$, supporting the hypothesis of the factorization of the proton vertex: the probability of proton dissociation is independent of the photon virtuality.

## 8. Conclusions

Studies of vector meson production at HERA show that, in the presence of a hard scale given by the mass of the vector meson, $Q^2$ or $t$, the interaction can be described by perturbative QCD. Quantitative comparisons between data and calculations require further improvements on the theory side and would benefit from measurements of increased precision.

## 9. Acknowledgements

It is pleasure to thank my colleagues in H1 and ZEUS for stimulating discussions. A special thank to Halina Abramowicz for the help in preparing this talk.

## REFERENCES

1. ZEUS Coll., S. Chekanov et al., Eur. Phys. J. C24 (2002) 3, 345.
2. ZEUS Coll., *Exclusive electroproduction of $J/\psi$ mesons at HERA*, contr. paper 813 to ICHEP 2002.
3. L. Frankfurt, W. Koepf and M. Strikman,

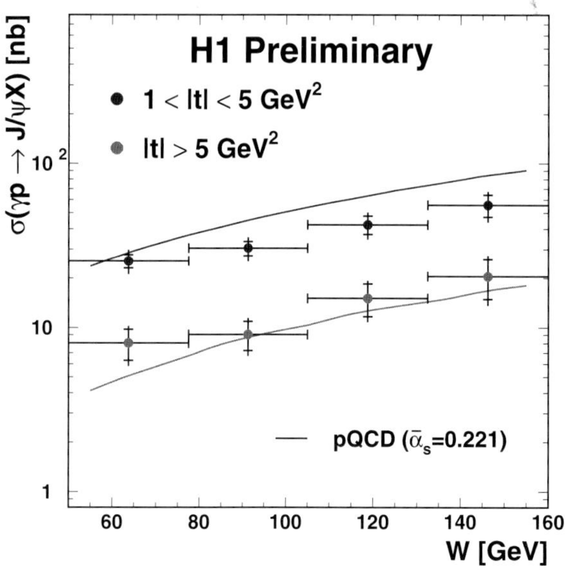


Figure 9. The differential cross section $d\sigma/dt$ in the range $50 < W < 160$ GeV for $J/\psi$ photoproduction. The lines are pQCD predictions [15].

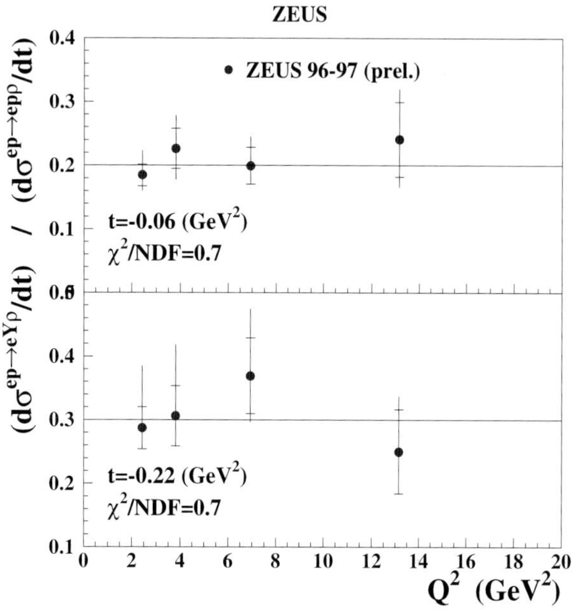


Figure 10. Ratio of proton-dissociative and exclusive $\rho^0$ electroproduction processes as a function of $Q^2$, for two values of $t$. The lines are the results of the fit to a constant function.

Phys. Rev. D57 (1998) 512;
A.D. Martin, M.G. Riskin and T. Teubner, Phys. Rev. D26 (1999) 14022.

4. H1 Coll., *Photoproduction of $\rho^0$ mesons with a leading proton*, contr. paper 991 to ICHEP 2002.
5. H1 Coll., *Elastic electroproduction of $\rho^0$ mesons at high $Q^2$*, contr. paper 989 to ICHEP 2002.
6. ZEUS Coll., *Exclusive and proton-dissociative electroproduction of $\rho^0$ mesons at HERA*, contr. paper 818 to ICHEP 2002.
7. H1 Coll., C. Adloff et al., Phys. Lett. B541 (2002) 251.
8. S.J. Brodsky et al., Phys. Rev. D50 (1994) 3134;
A.D. Martin, M.G. Ryskin and T. Teubner, Phys. Rev. D55 (1997) 4329;
A.D. Martin, M.G. Ryskin and T. Teubner, Phys. Rev. D56 (1997) 3007.
9. A. Kreisel for ZEUS Coll., Proceedings of DIS2001, Bologna, Italy.
10. ZEUS Coll., J. Breitweg et al., Eur. Phys. J. C12 (2000) 393;
H1 Coll., C. Adloff et al., Eur. Phys. J. C13 (2000) 371.
11. H1 Collab., C. Adloff et al., Phys. Lett. B539 (2002) 25.
12. D.Yu. Ivanov and R. Kirschner, Phys. Rev. D58 (1998) 114026.
13. ZEUS Coll., S. Chekanov et al., Preprint DESY-02-072, submitted to Eur. Phys. J.
14. H1 Coll., *Diffractive $J/\psi$ vector meson production at high $|t|$ at HERA*, contr. paper 993 to ICHEP 2002.
15. J.Bartels et al., Phys. Lett. B375 (1996) 301.
16. J.R.Forshaw and G.Poludniowski, Preprint hep-ph/0107068 (2001)
17. ZEUS Coll., *Energy dependence of exclusive vector-meson production in ep interactions at HERA*, contr. paper 820 to ICHEP 2002.

ELSEVIER

Proceedings of ICHEP 2002, pp. 420–423
*S. Bentvelsen, P. de Jong, J. Koch and E. Laenen (Editors)*

31st INTERNATIONAL CONFERENCE ON HIGH ENERGY PHYSICS AMSTERDAM

# Deeply Virtual Compton Scattering at HERA

J. Volmer[a]

[a]DESY-Zeuthen, Platanenallee 6, D-15738 Zeuthen, Germany

Deeply virtual Compton scattering has recently been studied by three HERA experiments, H1, ZEUS and HERMES, covering a broad range of kinematic regimes. We present cross section measurements of the two collider experiments in the kinematic region $2<Q^2<100$ GeV$^2$ and $30<W<140$ GeV, and compare them to QCD-based calculations. HERMES measurements of azimuthal asymmetries and their kinematical dependences are presented for $Q^2>1$ GeV$^2$ and $2<W<7$ GeV.

## 1. Introduction

The internal structure of the nucleon has long been studied with inclusive deep-inelastic lepton scattering. From these studies, the spin-dependent and spin-independent nucleon structure functions and corresponding parton distribution functions have been extracted. Recently, the formalism of generalized parton distributions (GPD) [1]-[3] was introduced. In this framework, non-forward degrees of freedom are probed in exclusive processes such as meson production or deeply virtual Compton scattering (DVCS), the electroproduction of a single real photon on the nucleon, which has been studied at DESY [4]-[6] and Jefferson Lab [7]. This opens the way to access the distribution of partons in the nucleon perpendicular to its momentum [8], and to access the contribution of the orbital angular momenta of the quarks to the nucleon spin [9].

## 2. DVCS at HERA

### 2.1. DVCS at H1 and ZEUS

The HERA storage ring at DESY provides an 820-920 GeV proton beam and a 27.6 GeV electron or positron beam to the two collider experiments, H1 and ZEUS. At collider energies, the leading order handbag diagram (Fig. 1 a) is accompanied by a sizable contribution of colour singlet two-gluon exchange (Fig. 1 b). In both cases, the final state is indistinguishable from that of the Bethe-Heitler (BH) process, in which the real photon is radiated off the lepton. The BH process constitutes a background to the DVCS process. Since it is purely electromagnetic, its amplitude can be calculated exactly. This contribution is subtracted from the combined DVCS+BH data by means of a Monte Carlo simulation that is cross checked with a BH-dominated sample of the experimental data [4,5].

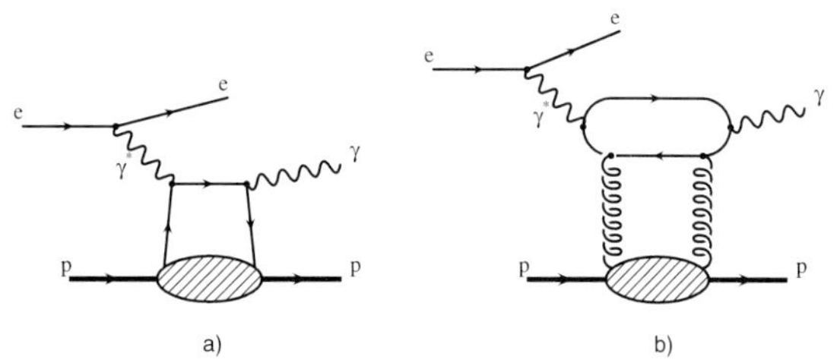


Figure 1. a) leading order handbag diagram, b) NLO colour singlet two-gluon exchange.

The cross section from the H1 experiment, shown in Figs. 2 and 3, was measured in the 1997 running period with an 820 GeV proton energy. The sample corresponds to an integrated luminosity of 8 pb$^{-1}$. The data cover a kinematic range in four-momentum transfer squared $Q^2$ from 2 to 20 GeV$^2$, invariant mass of the photon-nucleon system $W$ from 30 to 120 GeV and the square of the momentum transfer to the proton $|t|<1$ GeV$^2$, over which the cross section is integrated. The data in Figs. 2 and 3 have been scaled to central values of $W$=75 GeV and $Q^2$=4.5 GeV$^2$, respectively. Here, also theoretical predictions for DVCS at H1 kinematics from a GPD-based model (FFS, [10]), as well as from a colour dipole model (DD, [11]), are shown. In both models, the expo-


doi:10.1016/S0920-5632(02)02240-5

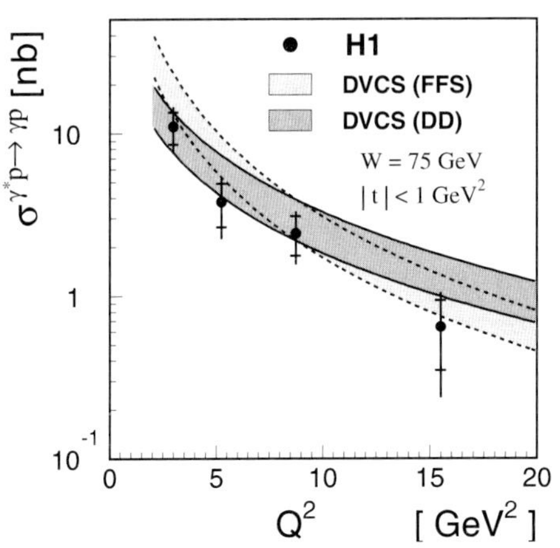


Figure 2. $Q^2$ dependence of DVCS cross section for the H1 data [4], GPD predictions (light shade) [10], a colour dipole model (dark shade) [11].

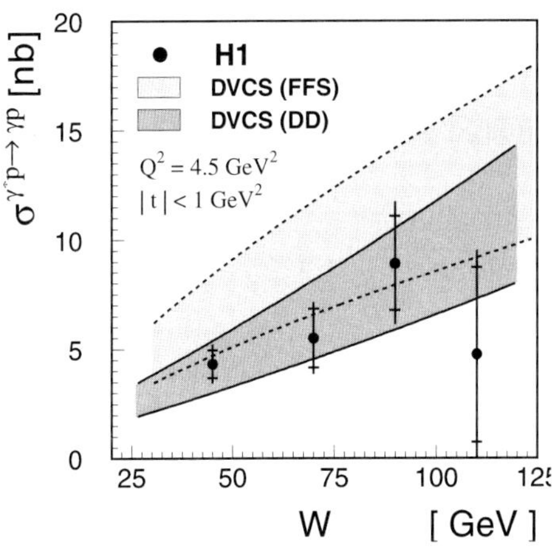


Figure 3. $W$ dependence of DVCS cross section for the 1997 H1 data [4], overlays as in Fig. 2.

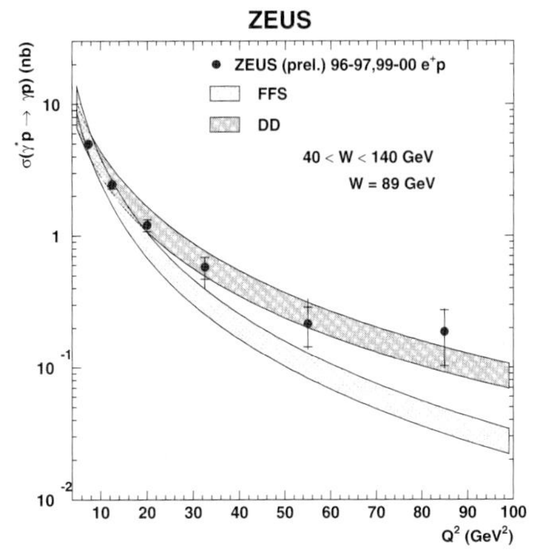


Figure 4. $Q^2$ dependence of DVCS cross section for the ZEUS positron data, overlays as in Fig. 2.

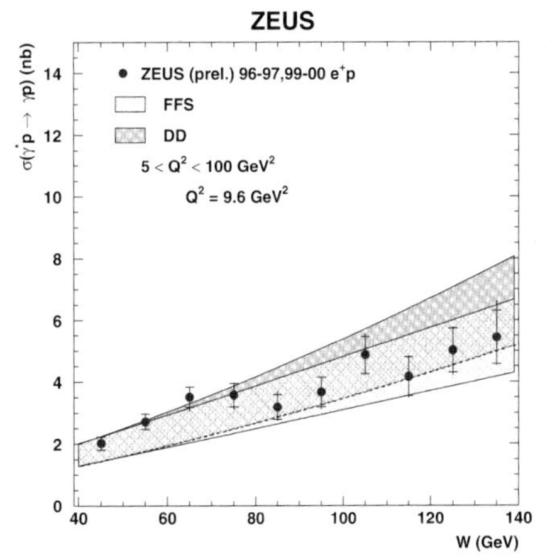


Figure 5. $W$ dependence of DVCS cross section for the ZEUS positron data, overlays as in Fig. 2.

nential $t$-dependence factorizes from the $Q^2$ and $W$ dependences. Since the $t$-dependence of the cross section was not measured, upper and lower bounds are determined by taking values of the $t$-slope parameter $b$ of 5 and 9 $\text{GeV}^{-2}$. A newer GPD-based approach can be found in [12].

The ZEUS results are based on data taken between 1996 and 2000, and, for the first time, also make use of the HERA electron beam. The integrated luminosities are 95 $\text{pb}^{-1}$ for $e^+p$ and 17 $\text{pb}^{-1}$ for $e^-p$ collisions. The data were analyzed in the kinematical region defined by $5<Q^2<100$ $\text{GeV}^2$ and $40<W<140$ GeV, with no cut on $t$. The data were found to agree for both lepton charges. The $e^+p$ data shown in Figs. 4 and 5 are scaled to central values of $W = 89$ GeV and $Q^2 = 9.6$ $\text{GeV}^2$, respectively. The data are overlaid with theoretical predictions from the same models as the H1 data. Due to the larger kinematical range and integrated luminosity, the ZEUS data have enough leverage and statistical accuracy to discriminate between different theoretical predictions and can be used to constrain the predicted $Q^2$ dependence of theoretical models. Here, the ZEUS data are seen to favour the

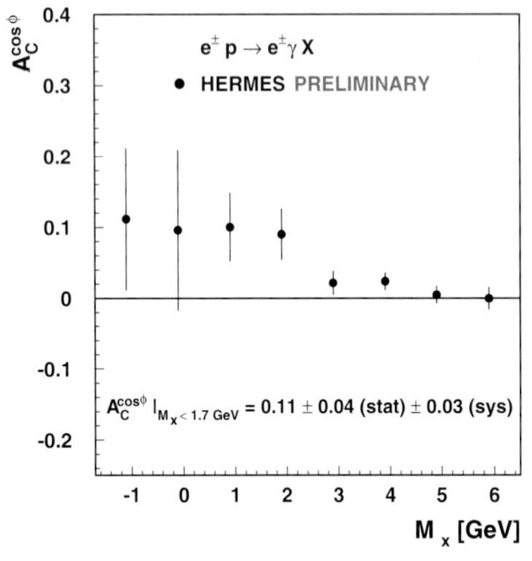


Figure 6. The $\cos\phi$ moment of the beam charge asymmetry from HERMES data from 1998 and 2000.

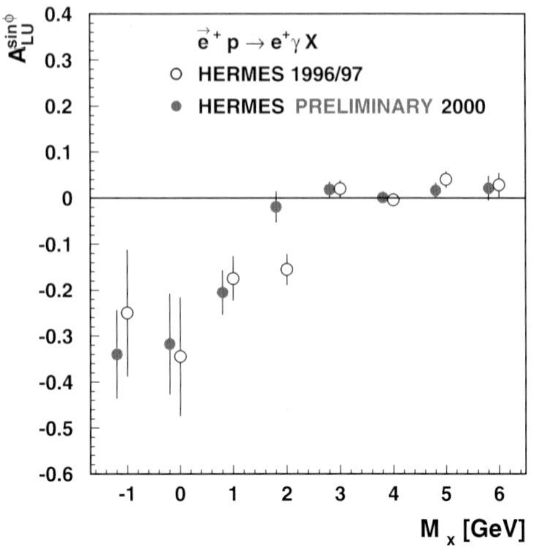


Figure 7. The $\sin\phi$ moment of the beam spin asymmetry from HERMES data from 1996/1997 (empty symbols) [6] and 2000 (filled symbols).

less steep slope in $Q^2$ of the colour dipole model. The $W$-dependence of the data is described well in both cases, reflecting the steep rise in gluon density at lower $x_B$, the momentum fraction of the struck parton in the nucleon.

The ZEUS and H1 data cannot be compared directly due to the different kinematic ranges. However, if the observed functional behaviour in $Q^2$ is used to scale the H1 data to the ZEUS central kinematics, the two data sets agree.

### 2.2. DVCS at HERMES

The fixed target experiment HERMES makes use of the 27.6 GeV polarized HERA positron beam incident on a storage cell that is filled with unpolarized gas (hydrogen, deuterium, or heavier nuclei). At HERMES kinematics, the BH amplitude largely dominates the DVCS amplitude. However, the DVCS and BH amplitudes interfere, giving rise to a single-spin asymmetry, which depends on beam charge, helicity, and the azimuthal angle $\phi$ between the lepton scattering plane and the reaction plane of the real photon and recoil proton. By extracting the $\sin\phi$ and $\cos\phi$ moments of the exclusive data, access is obtained to certain combinations of DVCS amplitudes [13].

The $\sin\phi$ moment of the beam spin asymmetry is calculated as

$$A_{\mathrm{LU}}^{\sin\phi} = \frac{2}{N}\sum_{i=1}^{N}\frac{\sin\phi_i}{(P_l)_i},$$

where $P_l$ is the beam polarization for each event. The $\cos\phi$ moment is determined similarly from the beam charge asymmetry with an unpolarized beam.

The HERMES forward spectrometer detects the scattered lepton and the real photon, but not the recoiling proton. Therefore the exclusivity of an event is measured through the missing mass ($M_x$) of the event, which in the exclusive case is equal to the proton mass. Examples are shown in Figs. 6 and 7, in which the $\cos\phi$ moment of the beam charge asymmetry for the 1998 and 2000 data, and the $\sin\phi$ moment of the beam spin asymmetry for the 1996/1997 data [6] and the 2000 data are shown. Asymmetries are observed at values of $M_x$<1-2 GeV, in the exclusive region. At higher missing masses, where non-exclusive processes dominate, asymmetries vanish. The two regions overlap slightly, because the resolution in $M_x$ is limited by the energy resolution of the electromagnetic calorimeter. The $\sin\phi$ measurements were confirmed at Jefferson

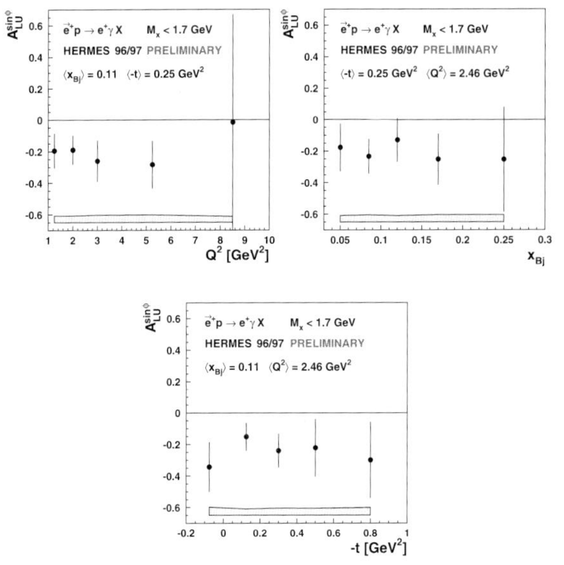


Figure 8. The $Q^2$, $x_{\rm B}$ and $t$-dependences of the $\sin\phi$ moment of the beam spin asymmetry from HERMES data from 1996 and 1997.

Lab [7]. GPD-based predictions can be found in Refs. [12,13].

From the exclusive sample ($M_x$<1.7 GeV) of the 1996/97 data the kinematic dependences of the beam spin asymmetry were extracted as functions of $Q^2$, $x_B$ and $t$ (preliminary results shown in Fig. 8). In the case of $t$, as with $M_x$, the reconstruction relies on the photon energy measured in an electromagnetic calorimeter. Therefore some events are reconstructed at $-t$<0.

## 3. Outlook

The HERA run II starting in 2002 will be used by all three experiments to improve the DVCS measurements. In the 2001 shutdown, spin rotators were installed before and after the collider experiments H1 and ZEUS, thus allowing for spin-dependent measurements. It is planned to run alternately with electron and positron beams.

The H1 experiment will install a Very Forward Proton Spectrometer in 2003 [14], allowing it to measure $|t|$ at small values of $W$. In addition, the precise tracking data from the Backward Silicon Tracker will be used for a measurement of azimuthal asymmetries. The measurement of differences in the cross section for the available beam polarizations and charges will be the main goal of the ZEUS DVCS analysis. The HERMES experiment will improve the exclusivity of its DVCS sample, as well as the $t$-resolution, by installing a large acceptance recoil detector around the HERMES target [15].

## Acknowledgements

I would like to thank Rainer Stamen, Paul R. Newman, Iwona Grabowska-Bołd and Yuji Yamazaki for useful and enlightening discussions.

## REFERENCES

1. D. Müller *et al.*, Fortschr. Phys. **42** (1994) 101.
2. A. V. Radyushkin, Phys. Lett. **B 385** (1996) 333.
3. X. Ji, Phys. Rev. **D 55** (1997) 7114.
4. C. Adloff *et al.*, H1 collaboration, Phys. Lett. B 517 (2001) 47-58.
5. ZEUS Collaboration, proceedings of the International Europhysics Conference on High Energy Physics, Budapest, 2001.
6. A. Airapetian *et al.*, Phys. Rev. Lett. 87 (2001) 182001.
7. S. Stepanyan *et al.*, Phys. Rev. Lett. 87 (2001) 182002.
8. J. P. Ralston, B. Pire, CPHT-S038-0901 (2001), hep-ph/0110075.
9. X. Ji, Phys. Rev. Lett. **78** (1997) 610.
10. L. L. Frankfurt, A. Freund and M. Strikman, Phys. Rev. **D**58 (1998) 114001.
11. A. Donnachie and H. G. Dosch, Phys. Lett. B 502 (2001) 74.
12. A. Freund and M. McDermott, Eur. Phys. J. C 23 (2002) 651, Phys. Rev. **D**65 (2002) 091901; A. Freund, M. McDermott and M. Strikman, hep-ph/0208160.
13. K. Goeke, M. V. Polyakov, M. Vanderhaeghen, Progr. Part. Nucl. Phys. Vol. 47, No. 2, 401 (2001), hep-ph/0106012.
14. L. Favart *et al.*, Proposal DESY-PRC-01/00 and H1-5/00-582.
15. R. Kaiser *et al.*, Technical Design Report, DESY-PRC 02-01, HERMES 02-003.

Proceedings of ICHEP 2002, pp. 424–427
*S. Bentvelsen, P. de Jong, J. Koch and E. Laenen (Editors)*

# Rapidity gaps in $\bar{p}p$, $ep$ and $e^+e^-$ collisions

K. Hatakeyama[a] for the CDF, L3, ZEUS and H1 collaborations

[a] The Rockefeller University,
1230 York Avenue, New York, NY, 10021, U.S.A.

Results on rapidity gaps in $\bar{p}p$ collisions obtained by the CDF collaboration, in $ep$ collisions by the ZEUS and H1 collaborations, and in $e^+e^-$ collisions by the L3 collaboration are presented.

## 1. Rapidity gaps between jets in $\bar{p}p$ and $ep$ collisions

Events with a rapidity gap, defined as a pseudorapidity ($\eta$) region devoid of particles, between the two highest $E_T$ jets have been studied in $\bar{p}p$ collisions by the CDF and DØ collaborations at the Tevatron, and in $ep$ collisions by the ZEUS and H1 collaborations at HERA. Such events are shown schematically in Fig. 1. The rapidity gap between the jets is generally thought to be due to the exchange of a color singlet (CS) object with vacuum quantum numbers.

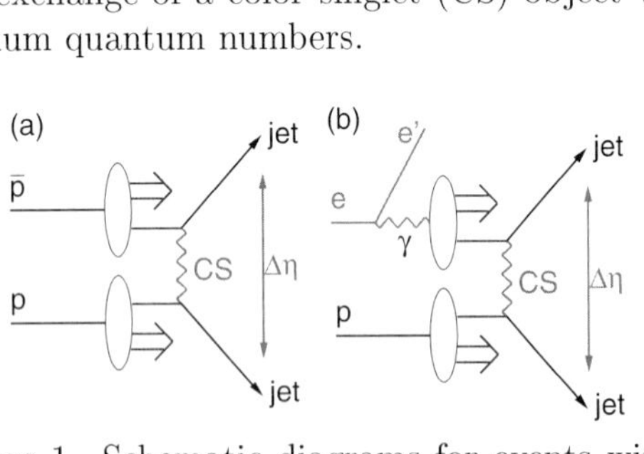


Figure 1. Schematic diagrams for events with a rapidity gap between jets produced in (a) $\bar{p}p$ and (b) $ep$ collisions.

The CDF [1] and DØ [2] collaborations measured the fraction of events due to color singlet exchange (CSE) at $\sqrt{s}$ = 1800 and 630 GeV. CDF measured the CSE fraction in dijet events with $1.8 < |\eta^{jet1,2}| < 3.5$ and $\eta^{jet1}\eta^{jet2} < 0$ to be $1.13 \pm 0.12(stat) \pm 0.11(syst)$ % for jets with $E_T^{jet1,2} > 20$ GeV at $\sqrt{s} = 1800$ GeV, and $2.7 \pm 0.7(stat) \pm 0.6(syst)$ % for jets with $E_T^{jet1,2} > 8$ GeV at $\sqrt{s} = 630$ GeV. The CSE fraction at 630 GeV is larger than that at 1800 GeV by $R[630/1800] = 2.4 \pm 0.7(stat) \pm 0.6(syst)$, which is consistent with the DØ result of $R[630/1800] = 3.2 \pm 1.2$.

Dijet events in photoproduction in which there is a large pseudorapidity separation $\Delta\eta > 2.5$ between the two highest $E_T$ jets were studied with the H1 detector [3]. The events containing 2 or more jets with $E_T^{jet1} > 6$ GeV, $E_T^{jet2} > 5$ GeV, $\eta^{jet1,2} < 2.65$ and $\Delta\eta > 2.5$ are used in this analysis. In the distribution of $E_T^{gap}$, the total transverse energy in the $\eta$ region between the two highest $E_T$ jets, the data show an excess above standard photoproduction predictions from the PYTHIA 5.7 and HERWIG 6.1 Monte Carlo (MC) generators at low values of $E_T^{gap}$.

Rapidity gap events are defined as events in which $E_T^{gap}$ is less than $E_T^{cut}$, where $E_T^{cut}$ lies between 0.5 and 2.0 GeV. The ratio of these events to the inclusive dijet events, the gap fraction, is shown as a function of $\Delta\eta$ in Fig. 2 for 4 different choices of $E_T^{cut}$. The MC distributions decrease with increasing $\Delta\eta$; however, the data distributions are approximately flat with $\Delta\eta$. Although HERWIG is closer to the data than PYTHIA, there is a noticeable excess in the data at large $\Delta\eta$ for all $E_T^{cut}$.

In Fig. 3, the gap fraction for $E_T^{gap}$ is shown along with predictions of HERWIG and PYTHIA with CSE models added. The measured gap fraction is reasonably well described with the addition of a CSE component.

A similar analysis was performed recently by the ZEUS collaboration. It was found that the gap fraction is larger at high $x_\gamma^{OBS}$, which is the fraction of the photon energy contributing to the production of the two jets. To further investigate the $x_\gamma^{OBS}$ dependence, the gap fraction is shown


doi:10.1016/S0920-5632(02)02245-4

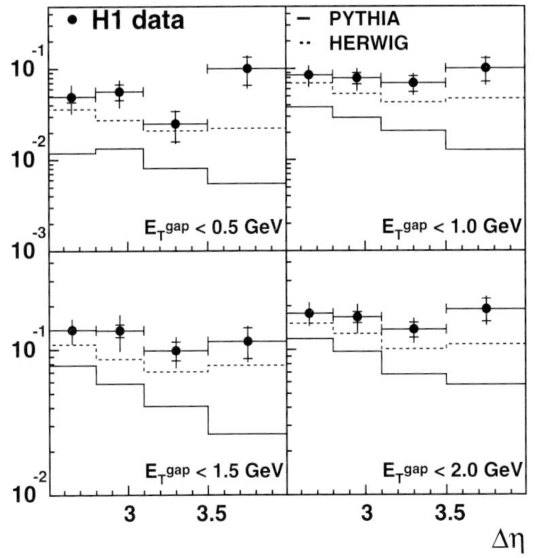


Figure 2. Gap fraction as a function of $\Delta\eta$ for 4 different values of $E_T^{cut}$.

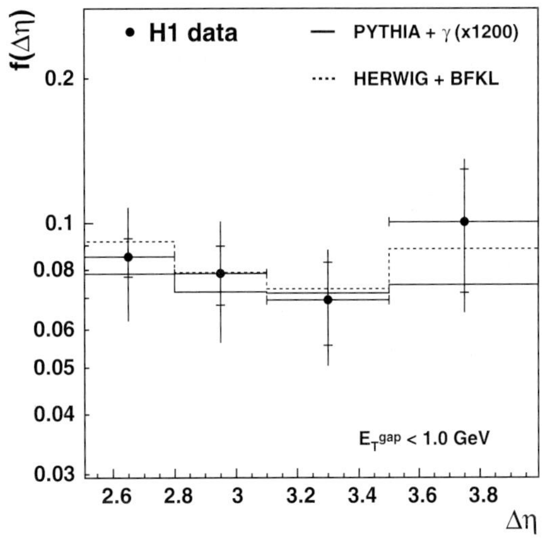


Figure 3. Gap fraction as a function of $\Delta\eta$ in data, and predictions of PYTHIA and HERWIG with 2 different CSE models added.

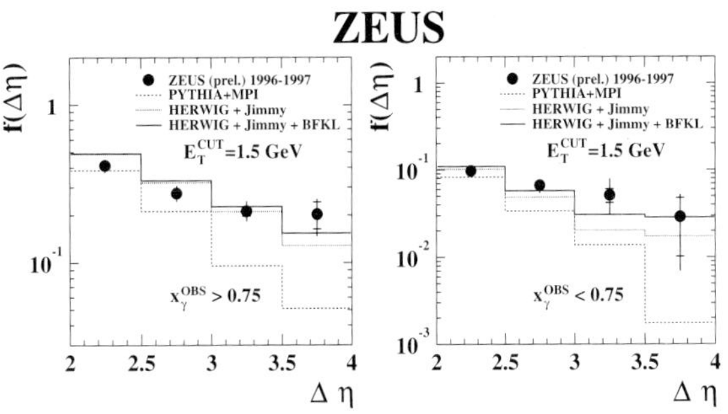


Figure 4. Gap fraction as a function of $\Delta\eta$ for (left) high $x_\gamma^{OBS}$ and (right) low $x_\gamma^{OBS}$.

as a function of $\Delta\eta$ separately for the high and low $x_\gamma^{OBS}$ regions in Fig. 4. At high $x_\gamma^{OBS}$, the MC models do not describe the data well, especially at low $\Delta\eta$. At low $x_\gamma^{OBS}$, the data are somewhat better described by the model including a BFKL CS component.

## 2. Rapidity gaps in hadronic $Z$ decays in $e^+e^-$ collisions

Events with a large rapidity between jets have been observed at the Tevatron and HERA as described in Sec. 1. By crossing symmetry, similar rapidity gaps are expected in 3-jet hadronic $Z$ decays. The L3 collaboration searched for rapidity gaps in symmetric 3-jet hadronic $Z$ decays produced in $e^+e^-$ collisions at LEP by exploiting differences in the 3-jet topology between color single exchange (CSE) and color octet exchange (COE) [4]. In COE, color flow is present between the $qg$ and $\bar{q}g$ gaps and is inhibited by destructive interference in the $q\bar{q}$ gap, while in CSE, color flow occurs only in the $q\bar{q}$ gap, as shown in Fig. 5. In the gaps where color flow is present, more particles and smaller angles between particles are expected than in the gaps where color flow is absent. Some asymmetry variables which are based on angles between particles in the inter-jet regions are found to be sensitive to differences between CSE and COE. By comparing the asymmetry distributions measured in the data and those expected from the COE and CSE models, one can estimate the CSE fraction present in the data.

The study revealed that the data can be explained by the COE model alone. The upper limit of the fraction of CSE events is estimated to be between 7 and 9 % at 95 % confidence level.

## 3. Multigap diffraction in $\bar{p}p$ collisions

Inclusive (soft) single diffractive (SD) and double diffractive (DD) events, which contain a rapidity gap and/or a leading (high longitudinal momentum) outgoing $p$ ($\bar{p}$), as illustrated in Fig. 6, have been previously studied by CDF [5,6]. Here,

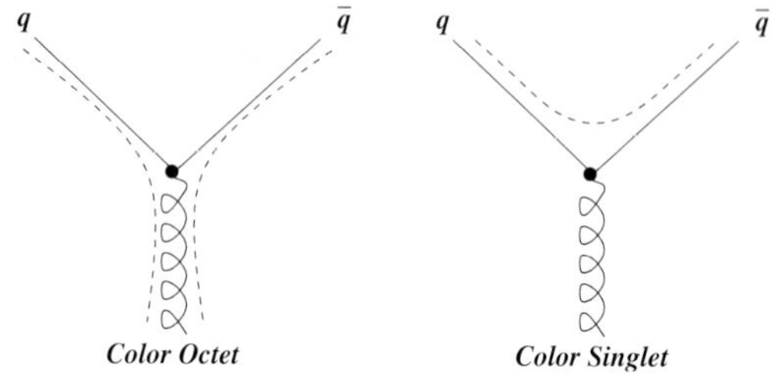


Figure 5. Color flow in COE (left) and CSE (right) shown by dotted lines.

new results are presented from studies of two types of events containing two rapidity gaps, namely double pomeron exchange (DPE) events and single ⊕ double diffractive (SDD) events, shown schematically in Fig. 6.

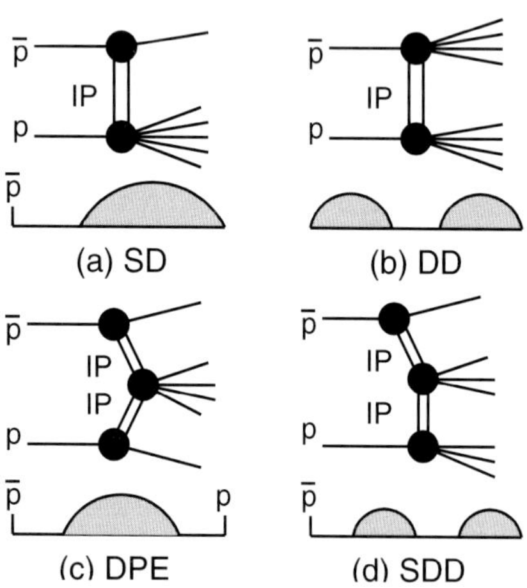


Figure 6. Diagrams and event topologies for (a) single diffraction (SD), (b) double diffraction (D-D), (c) double pomeron exchange (DPE) and (d) single ⊕ double diffraction (SDD).

Soft (low $p_T$) events have been traditionally treated in Regge theory, where rapidity gaps are formed due to Pomeron ($I\!P$) exchange. In [5,6], the measured SD and DD cross sections were found to be lower than Regge predictions by approximately an order of magnitude. Studying DPE and SDD events would help to further understand the mechanism of rapidity gap formation.

The data used in these analyses are the inclusive SD data samples ($\bar{p}p \to \bar{p}X$) collected at $\sqrt{s} = 1800$ and 630 GeV by triggering on a

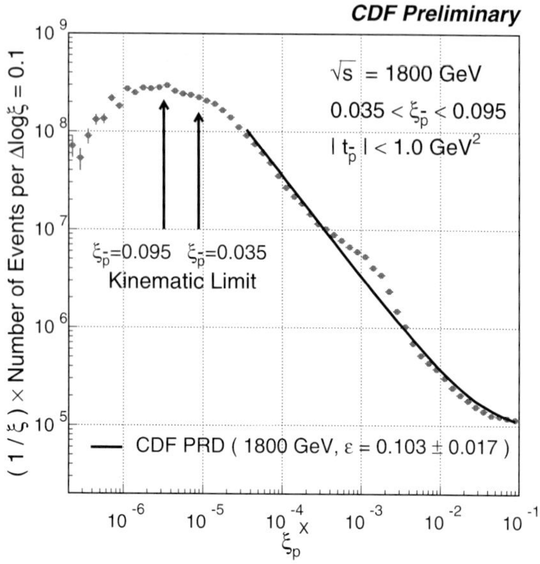


Figure 7. $\xi_{\bar{p}}^X$ distribution at $\sqrt{s} = 1800$ GeV for events with a leading $\bar{p}$ of $0.035 < \xi_{\bar{p}}^{RP} < 0.095$. The solid line is the distribution obtained in SD.

leading $\bar{p}$ detected in a Roman Pot spectrometer (RPS). The $\bar{p}$ momentum loss $\xi_{\bar{p}}$ and 4-momentum transfer squared $t_{\bar{p}}$ are determined from RPS information. In the DPE analysis, the $\xi_p$ is measured from the calorimeter and beam-beam counter (BBC) information using the formula,

$$\xi_p^X = \frac{M_X^2}{\xi_{\bar{p}} \cdot s} = \frac{\sum_i E_T^i \exp(\eta^i)}{\sqrt{s}} \tag{1}$$

where the sum is carried out over all the particles except the leading $p$. For each BBC hit, $\eta$ and $E_T$ values are chosen randomly from a flat $\eta$ distribution over the hit BBC $\eta$-segment and from the expected $E_T$ distribution, respectively. Fig. 7 shows the $\xi_p^X$ distribution for $\sqrt{s} = 1800$ GeV. The bump at $\xi_p^X \sim 10^{-3}$ is due to central calorimeter noise and is reproduced in MC simulations. The solid line representing the distribution obtained in SD is in good agreement with that measured in DPE.

The fraction of the DPE events with $\xi_p < 0.02$ in the inclusive SD events with a leading $\bar{p}$ of $0.035 < \xi_{\bar{p}} < 0.095$ and $|t_{\bar{p}}| < 1.0$ [0.2] GeV$^2$ was found to be $0.197 \pm 0.001(stat) \pm 0.010(syst)$ [$0.168 \pm 0.001(stat)^{+0.015}_{-0.020}(syst)$] at $\sqrt{s} = 1800$ [630] GeV.

These measured values are compared with predictions from Regge theory based on factorization, $I\!P$-flux renormalization for both exchanged Pomerons [7] and gap probability renormalization [8] in Fig. 8. The measured values are consistent with predictions of the gap probability renormalization model.

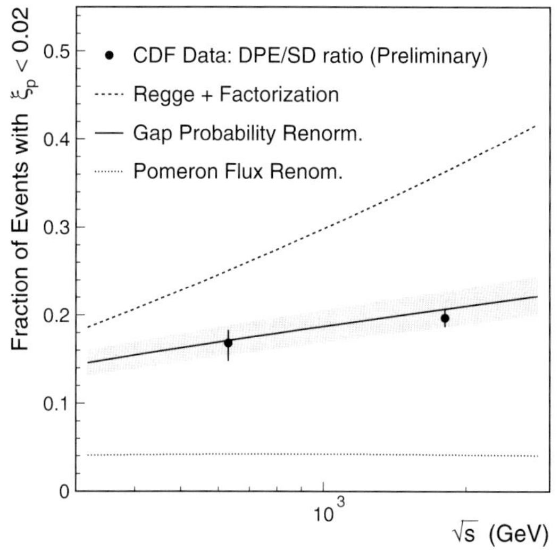


Figure 8. Measured ratio of DPE to SD rates (points) compared with predictions based on Regge theory (dashed), $I\!P$-flux renormalization for both exchanged Pomerons (dotted) and gap probability renormalization (solid line).

In the SDD analysis, the $\xi_{\bar{p}}$ region is restricted to $0.06 < \xi_{\bar{p}} < 0.09$, corresponding to a diffractive mass of $\sqrt{s'} = \sqrt{s\xi_{\bar{p}}} = 440\text{-}540[155\text{-}190]$ at $\sqrt{s} = 1800\,[630]$ GeV, so that the diffractive cluster $X$ of the process $\bar{p}p \to \bar{p}X$ covers almost the entire CDF calorimeter coverage of $-4.2 < \eta < 4.2$. CDF searched for *experimental* gaps overlapping $\eta = 0$, defined by no tracks or calorimeter towers above thresholds chosen to minimize calorimeter noise contributions.

The fraction of events with a rapidity gap $\Delta\eta > 3$ within the diffractive cluster $X$, corrected for detector acceptance and extrapolated to include gaps not overlapping with $\eta = 0$, is found to be $0.246 \pm 0.001(stat) \pm 0.042(syst)$ $[0.184 \pm 0.001(stat) \pm 0.043(syst)]$ for $0.06 < \xi_{\bar{p}} < 0.09$ and $|t_{\bar{p}}| < 1.0\,[0.2]$ GeV$^2$ at $\sqrt{s} = 1800\,[630]$ GeV. These fractions are presented in Fig. 9 along with the measured DD to total cross section ratios [6], predictions from Regge theory and factorization, and predictions based on the gap probability renormalization model [8]. The data are in good agreement with the predictions of the gap probability renormalization model.

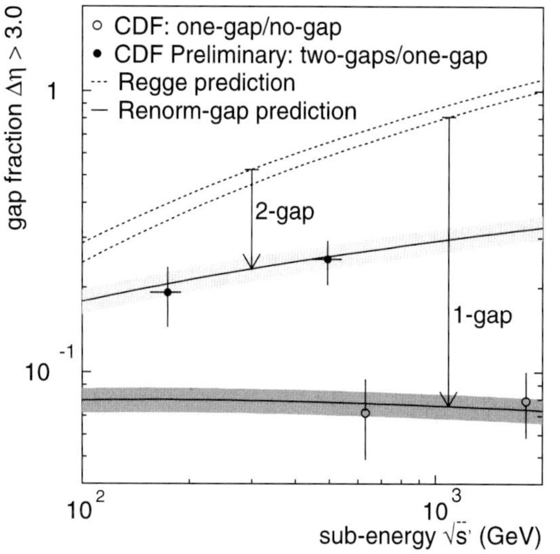


Figure 9. Measured ratios of SDD to SD rates (points) and DD to total rates (circles) as a function of $\sqrt{s'}$, the $I\!Pp$ and $\bar{p}p$ system energies, respectively. The dashed lines are predictions from Regge theory and the solid lines from the gap probability renormalization model.

## REFERENCES

1. F. Abe *et al.*, Phys. Rev. Lett. **81** (1998) 5278.
2. B. Abbott *et al.*, Phys. Lett. B **440** (1998) 189.
3. C. Adloff *et al.*, Eur. Phys. J. C **24** (2002) 517.
4. S. Banerjee, hep-ph/0112073; J. H. Field, hep-ex/0205004.
5. F. Abe *et al.*, Phys. Rev. D **50** (1994) 50.
6. T. Affolder *et al.*, Phys. Rev. Lett. **87** (2001) 141802.
7. K. Goulianos, Phys. Lett. B **358** (1995) 379; **363** (1995) 268.
8. K. Goulianos, hep-ph/0110240; hep-ph/0203141.

Proceedings of ICHEP 2002, pp. 428–433
*S. Bentvelsen, P. de Jong, J. Koch and E. Laenen (Editors)*

ELSEVIER

31st INTERNATIONAL CONFERENCE ON HIGH ENERGY PHYSICS AMSTERDAM

# Two Photon Physics

Valery P. Andreev[a]

[a]Department of Physics and Astronomy, University of California at Los Angeles, 405 Hilgard Avenue, Los Angeles, CA 90095-1547, USA

A review of the recent experimental results on two photon physics at LEP is presented. The total cross section for $\gamma\gamma \rightarrow hadrons$ is measured for two-photon centre-of-mass energies up to 160 GeV. The QCD predictions are tested with measurements of the total hadronic cross section for virtual $\gamma^*\gamma^*$ collisions, with heavy flavour ($c\bar{c}$ and $b\bar{b}$) production and inclusive particle production in collisions of quasi-real photons . Exclusive pair production of hadrons in $\gamma\gamma$ collisions is also discussed.

## 1. Introduction

A quasi-real photon interacts as a hadron due to quantum fluctuations into vector mesons or into quark-antiquark pairs. The $\gamma\gamma$ cross section

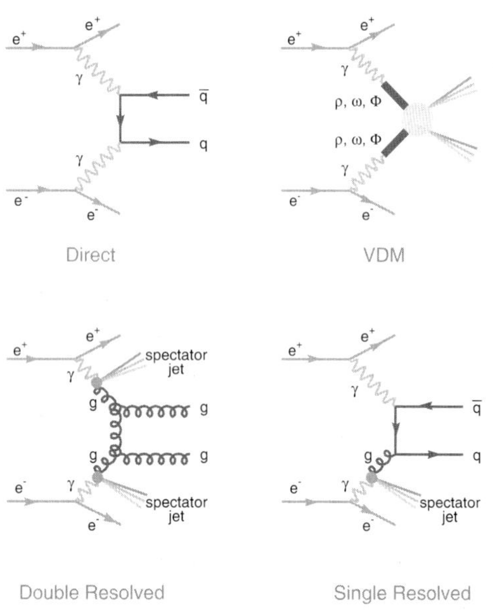


Figure 1. Hadronic final state in two-photon collisions.

is composed, Figure 1, by a "soft (VDM)" interaction, which can be described by hadronic phenomenology like the Regge poles, a point like two-photon reaction, $\gamma\gamma \rightarrow q\bar{q}$, which is exactly calculable in QED (direct component) and a "hard component" (resolved or anomalous or QCD component), which implies the knowledge of quark and gluon density in photon. The "hard component" of two-photon interactions can be accessed in measurements of high-$p_t$ single particle production, the heavy flavour production and scattering of highly virtual photons. Then a hard physical scale is given by either $p_t$ or by c(b)-quark mass or by large virtuality $Q^2$ of the photon. The LEP $e^+e^-$ collider is an ideal place to study two photon physics. The background from the annihilation events is normally very low, especially at LEP2 energies ($\sqrt{s} > 160$ GeV).

## 2. Total hadronic $\gamma\gamma$ cross section

The cross section of $\gamma\gamma \rightarrow hadrons$ process has been measured at LEP by the L3 [1] and the OPAL [2] for $\gamma\gamma$ centre-of-mass energies up to 160 GeV. The energy of a photon-photon collision is determined from the hadronic final state. The hadronic final state is not always fully contained in a detector acceptance. Unfolding procedure is applied to go from the measured $W_{vis}$ to the true $\gamma\gamma$ centre-of-mass energy $W_{\gamma\gamma}$. Figure 2(left) shows the $\sigma_{\gamma\gamma}$ measurements by the L3 and OPAL collaborations. The measurements agree when the same generator is used in the analysis, but there is a systematic difference of $\simeq$ 20% between the two Monte Carlo models used in analysis. This is mostly due to a different modelling of diffractive events. A parametrisation of


doi:10.1016/S0920-5632(02)02254-5

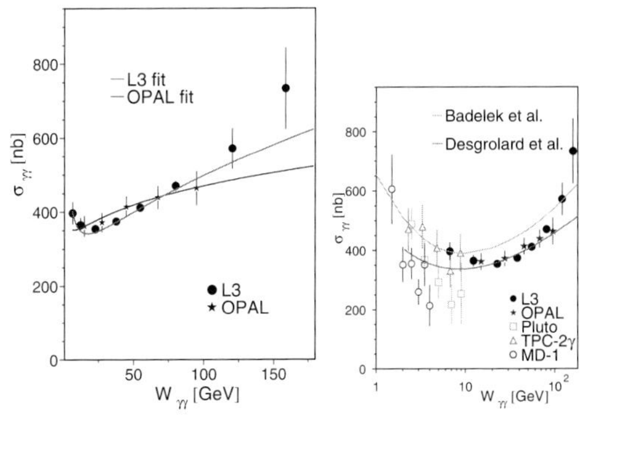


Figure 2. Total cross section for hadron production in two-photon collisions.

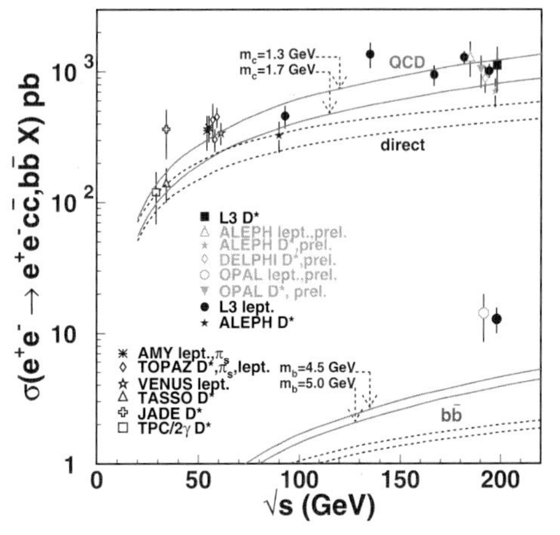


Figure 3. Charm and beauty production in two-photon collisions.

the form [3] $\sigma_{\text{tot}} = A\, s^{\epsilon} + B\, s^{-\eta}$ describes well the energy behaviour of all total hadron-hadron cross sections.

The two-photon cross section at high energies can be estimated from proton-proton and photoproduction total cross sections assuming factorization for the Pomeron term: $\sigma_{\gamma\gamma} \approx \sigma^2_{\gamma\text{p}}/\sigma_{\text{pp}}$. The predictions thus obtained are consistent with the two-photon measurement. Several QCD inspired models [4] are also in a qualitative agreement with the data, Figure 2 (right).

## 3. Charm and beauty production

The production of heavy quarks in two-photon collisions consists mainly of charm quarks. Because of their smaller electric charge and larger mass, the production of b-quarks is expected to be suppressed by more than two orders of magnitude relative to the production of charm quarks. The resolved photon cross section is dominated by the photon-gluon fusion diagram $\gamma\text{g} \to \text{c}\bar{\text{c}}, \text{b}\bar{\text{b}}$. At LEP energies, the direct and resolved processes are predicted to give comparable contributions to the cross section [5]. Measurements of charm production in two-photon collisions were done at LEP by ALEPH[6], DELPHI[7], L3[9] and OPAL[8] collaborations. Beauty production has been measured by L3[10] for the first time in gamma-gamma collisions.

### 3.1. Charm production

Charm particles in the final state were identified by the reconstruction of charged $D^*$ meson decays or using tagging by electrons and muons from semileptonic charm decays. The total inclusive charm cross sections at LEP are plotted in Figure 3 together with previous measurements. The data are compared to the theory predictions of Ref.[5]. The dashed line corresponds to the direct process, NLO QCD calculation, while the solid line shows the QCD prediction for the sum of the direct and the resolved processes calculated to NLO accuracy. The data require significant gluon content in the photon. The cross sections of charm production with a $D^{*\pm}$ tag is in agreement with the lepton tag measurements.

In Figure 4 the measurements of the differential cross section $\text{d}\sigma/dP_T^{D^*}$ are compared to NLO QCD calculations [12] based on a massive quark matrix elements. The measurements are in agreement with each other and with QCD calculations at high transverse momenta. At low $\text{P}_\text{T}$ the theory underestimates the data.

The L3 collaboration measured the cross sections $\sigma(\text{e}^+\text{e}^- \to \text{e}^+\text{e}^-\text{c}\bar{\text{c}}\text{X})$ and $\sigma(\gamma\gamma \to \text{c}\bar{\text{c}}\text{X})$ in the interval $5 \leq W_{\gamma\gamma} \leq 70$ GeV. Figure 5 shows the $\sigma(\gamma\gamma \to \text{c}\bar{\text{c}}\text{X})$ as function of $W_{\gamma\gamma}$ at $\sqrt{s} = 189 - 202$ GeV and the NLO QCD calculations[12]. In the calculations the charm mass, $m_\text{c}$, is fixed to 1.2 GeV. The calculation with $m_\text{c} = 1.5$ GeV results in about 50% lower cross section val-

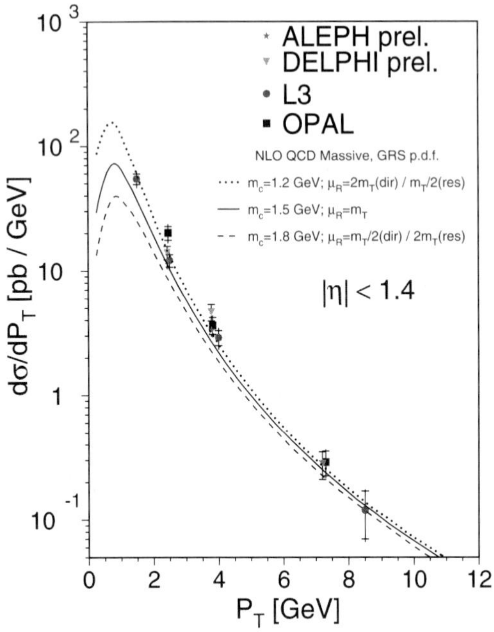


Figure 4. The differential $D^*$ cross section $d\sigma/dP_T^{D^*}$.

ues, except for the first point, where it is lower by 25%. The dotted curve is the measured by L3 total cross section $\sigma(\gamma\gamma \to \text{hadrons})$ [1] scaled by factor 1/20. A steeper rise with energy is observed as compared to hadron-hadron cross sections and to $\sigma(\gamma\gamma \to hadrons)$.

Using a sample of single-tagged events OPAL measured $F_{2,c}^{\gamma}$ photon structure function at $\langle Q^2 \rangle = 20$ GeV$^2$[14]. The HERWIG Monte Carlo event generator has been used in the study. Within the limited statistics of $55 \pm 11$ $D^{*\pm}$ mesons the HERWIG model is in agreement with the measured cross-sections. The data suggest a cross-section which is above the purely point-like component, i.e. the hadron-like component of $F_{2,c}^{\gamma}$ is needed at low x.

The inclusive $J/\Psi$ production in $\gamma\gamma$ collisions has been measured by the DELPHI collaboration [15]. The number of observed events is $36 \pm 7$ yielding a cross-section of $45.3 \pm 18.9$ pb.

### 3.2. Beauty production

Leptons from $b$ semi-leptonic decays are more energetic than from charm semi-leptonic decays and non-charm two-photon processes. To select $b\bar{b}$ events L3 apply cuts on the lepton momen-

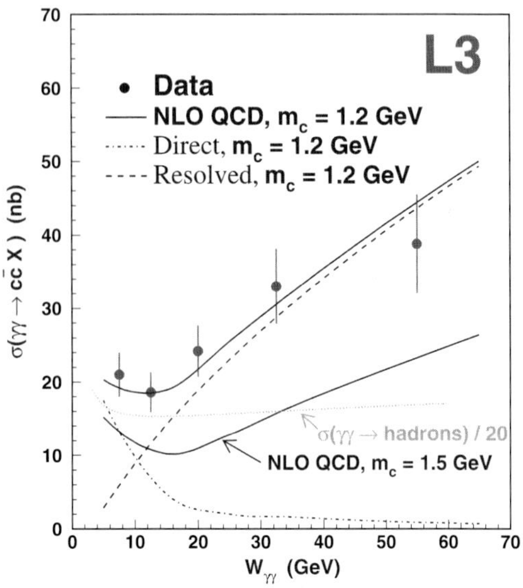


Figure 5. Cross section $\sigma(\gamma\gamma \to c\bar{c})$ versus $W_{\gamma\gamma}$ by L3.

tum and transverse momentum with respect to the closest jet defined by excluding the lepton from the jet. The previous L3 results[10] were updated [13] to include the full LEP2 data sample. After all cuts are applied, 435 muon and 219 electron event candidates remain. The beauty purity is 42% and 46%, respectively. The beauty selection efficiency is about 2.2% for muons and 1.25% for electrons. The preliminary beauty production cross section in $\gamma\gamma$ collisions has been measured by L3 to be $\sigma^{ee \to eebbX} = 12.8 \pm 1.7$ (st) $\pm$ 2.3 (sys) pb. The preliminary result by OPAL using muon tag is $\sigma^{ee \to eebbX} = 14.2 \pm 2.5$ (st) $\pm$ 5.0 (sys) pb [11].

The measured $b$ cross section lie about 5 standard deviations above QCD prediction, Figure 3. This is particularly interesting as measurements of beauty production in $p\bar{p}$ collisions by CDF [16] and DØ [17] as well as in ep collisions by H1 [18] and ZEUS [19] are a factor ~2–3 higher than NLO QCD predictions. Altogether the measurements compose a challenge for for the QCD theory.

## 4. Total hadronic $\gamma^*\gamma^*$ cross section

The cross-section measurement for scattering of two virtual photons is considered as "golden" process to test the BFKL dynamics which is proposed in [20]. Double-tag events in two-photon collisions are studied at LEP by ALEPH [22], L3 [23] and OPAL [24] collaborations. The cross-section of $\gamma^*\gamma^*$ collisions is measured at an average photon virtuality of $\langle Q^2 \rangle = 15 \div 18$ GeV$^2$. The $\gamma^*\gamma^*$ interaction can be seen as the interaction of two $q\bar{q}$ pairs scattering off each other via multiple gluon exchange. For the highly virtual two-photon process, with $Q_1^2 \simeq Q_2^2$, the calculation can be verified without phenomenological inputs as non-perturbative QCD contributions are expected to vanish [21]. In this scheme the cross-section for the collision of two virtual photons is $\sigma_{\gamma^*\gamma^*} \approx \sigma_0 \left( \frac{W_{\gamma^*\gamma^*}}{\sqrt{Q_1^2 Q_2^2}} \right)^{\alpha_P - 1}$, where $\alpha_P$ is the "hard Pomeron" intercept.

In Figure 6 the measured cross section $e^+e^- \rightarrow e^+e^-\gamma^*\gamma^* \rightarrow e^+e^- hadrons$ is plotted together with prediction for direct (QPM) process and with the prediction of QCD Monte Carlo model implemented in PHOJET generator [25]. The cross section is plotted as function of variable $Y = ln(s/s_0) \simeq ln(W_{\gamma^*\gamma^*}/Q_1Q_2)$. The QPM cross section lies below the data, while QCD prediction is in a reasonable agreement with the data. The value of the cross section at $5 < Y < 7$ exceeds the Monte Carlo prediction by about 3.5 standard deviations.

The two-photon cross-section is extracted from the $e^+e^- \rightarrow e^+e^- hadrons$ cross-section, $\sigma_{ee}$, by using only the transverse photon luminosity function, $\sigma_{ee} = L_{TT} \cdot \sigma_{\gamma^*\gamma^*}$. The two-photon cross-section $\sigma_{\gamma^*\gamma^*}$ is shown in Figure 7 as a function of $Y$. The predictions of the LO and NLO BFKL are also shown in Figure 7. The LO BFKL prediction is too high.

## 5. Single particle production

The production of charged hadrons, $\pi_0$ and $K_S^0$ mesons has been measured by L3 in [26,27].

Figure 8 shows the $p_T$ distribution for inclusive $\pi_0$ and $K_S^0$ mesons as well as $\pi^\pm$ and $K^\pm$ compared to NLO QCD calculations [28]. The spectra of different hadrons are consistent with each other. The high $p_T$ part of the $\pi$ meson spectrum is in excess of the NLO QCD prediction. The shape of the angular (pseudorapidity) spectrum is well reproduced by the QCD.

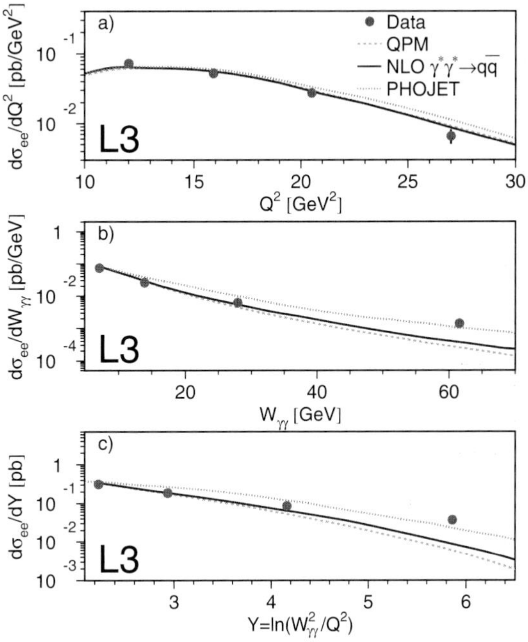


Figure 6. Cross section for $ee \rightarrow eeq\bar{q}$ in double-tag events by L3.

## 6. Exclusive final states

The study of exclusive meson and baryon pair production provides another important testing ground of QCD. The following processes have been measured at LEP: $\gamma\gamma \rightarrow \pi^+\pi^-, K^+K^-$ by ALEPH [29], $\gamma\gamma \rightarrow p\bar{p}$ by L3 [30] and OPAL [31], $\gamma\gamma \rightarrow \Lambda\bar{\Lambda}, \Sigma\bar{\Sigma}$ by L3 [32].

Cross section $\sigma(\gamma\gamma \rightarrow \pi^+\pi^-, K^+K^-)$ as function of $W_{\gamma\gamma}$ is plotted in Figure 9 compared to the theory curve from [33]. The measurement for pions is a factor 2 higher than the prediction. The pseudorapidity spectra for both pions and kaons agree well in shape with QCD prediction.

The process $\gamma\gamma \rightarrow$ *baryon antibaryon* is sensitive to the quark structure of the baryon. The results on measurement of the cross section of pair production of protons and $\Lambda$ baryons are plotted in Figure 10 compared to the three-quark calculation [34] and to the alternative quark-diquark

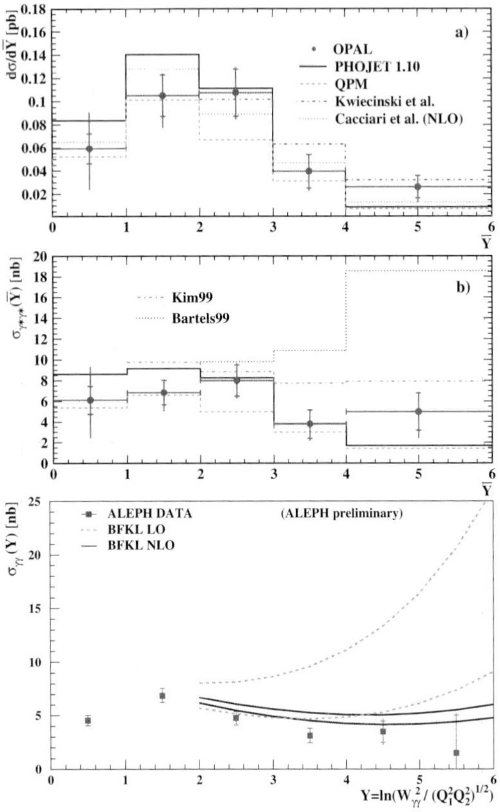


Figure 7. Two-photon cross-section for $\gamma^*\gamma^* \rightarrow q\bar{q}$ by OPAL and ALEPH.

model [35]. The absolute predictions of quark-diquark model using the standard distribution amplitude (Standard DA) reproduce the data. The asymptotic DA [36] and DZ-DA [37] as well as three-quark model are excluded.

## 7. Conclusion

Two Photon Physics studies at LEP have substantially improved (extended) the corresponding previous measurements. Several measurements are in disagreement with the perturbative QCD calculations at NLO accuracy.

## REFERENCES

1. L3 Coll., M. Acciari *et al*, Phys. Lett. **B 408** (1997) 450; Phys. Lett. **B 519** (2001) 33.

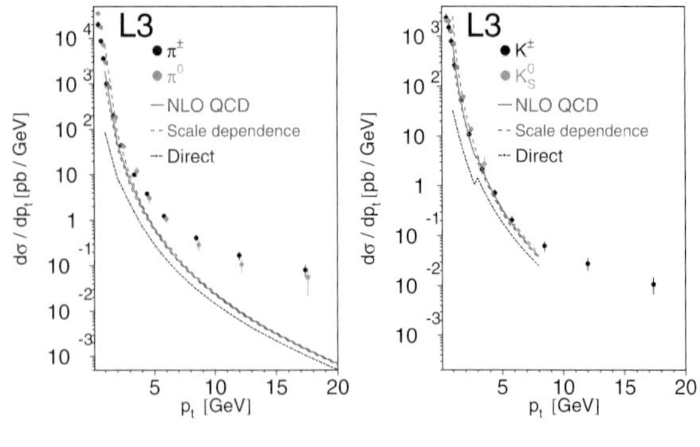


Figure 8. The inclusive $\pi^{\pm}$, $\pi_0$ (left) and $K^{\pm}$, $K^0_S$ (right) differential cross sections measured by L3 compared with NLO QCD calculations [28].

2. OPAL Coll., K. Ackerstaff *et al*, Eur. Phys. J **C 14** (2000) 199.
3. A. Donnachie and P.V. Landshoff, Phys. Lett. **B 296** (1992) 227.
4. B. Badelek, J. Kwiecinski and A.M. Stasto , hep-ph 9903248 (1999).
A. Donnachie, H.G. Dosch and M. Rueter, Phys. Rev. D **59**, 074011 (1999).
P.Desgrolard, M. Giffon, E. Martynov and E. Predazzi, Eur. Phys. J. **C9** (1999) 623.
A. Corsetti, R.M. Godbole and G. Pancheri, Phys. Lett. **B 435** (1998) 441.
5. M. Drees, M. Krämer, J. Zunft and P.M. Zerwas, Phys. Lett. B **306**, 371 (1993).
6. ALEPH Coll., D. Buskulic *et al*, Phys. Lett. B **355**, 595 (1995); the ALEPH note 2002-022, CONF 2002-011, July 8, 2002.
7. DELPHI Coll., M. Chapkin *et al*, DELPHI note 2002-083-CONF-617, July 3, 2002.
8. OPAL Coll., G. Abbiendi *et al*, Eur. Phys. J. **C16** (2000) 579.
9. L3 Coll., M. Acciarri *et al*, Phys. Lett. B **503**, 37 (2001); Phys. Lett. B **514**, 19 (2001); Phys. Lett. B **535**, 59 (2002).
10. L3 Coll., M. Acciarri *et al*, Phys. Lett. B **503**, 37 (2001).
11. OPAL Physics Note PN455, August 29, 2000.
12. S.Frixione, M.Krämer,E. Laenen, Nucl. Phys. B **571**, 169 (2000); priv. communication.
13. L3 note 2761, June 28, 2002.
14. OPAL Coll., G. Abbiendi *et al*, Phys. Lett.

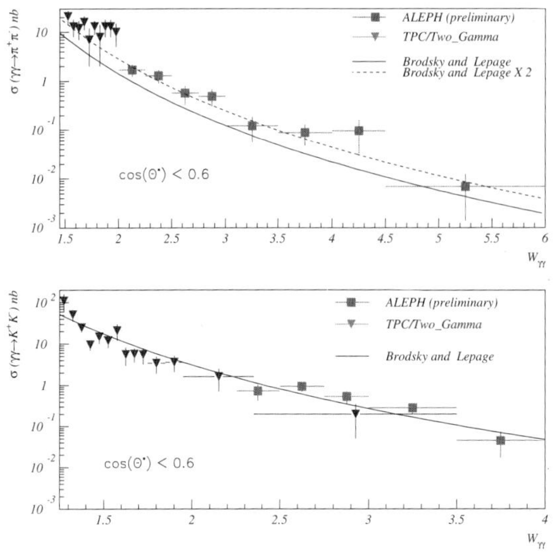


Figure 9. Cross section $\sigma(\gamma\gamma \to \pi^+\pi^-, K^+K^-)$ versus $W_{\gamma\gamma}$ by ALEPH.

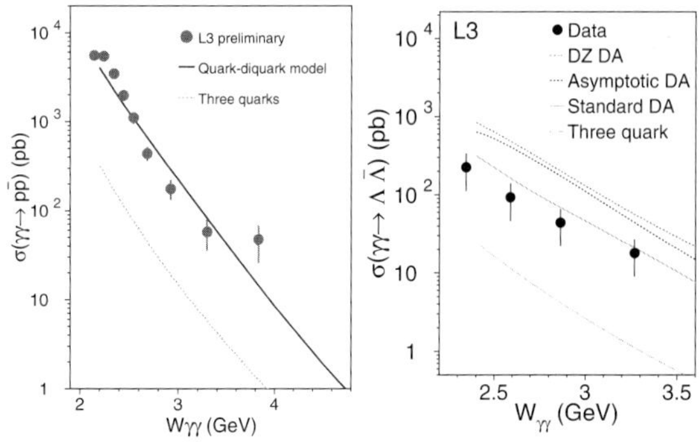


Figure 10. Measurements of the $\gamma\gamma \to p\bar{p}$ and $\gamma\gamma \to \Lambda\bar{\Lambda}$ by L3. The measurements are compared to the calculations of Reference [35].


**B 539** (2002) 13.

15. DELPHI note 2002-083-CONF-617.
16. CDF Coll., F. Abe *et al*, Phys. Rev. Lett. **71**, 500,2396,2537 (1993); Phys. Rev. Lett. **75**, 1451 (1995);Phys. Rev. Lett. **79**, 572 (1997).
17. DØColl., S. Abachi *et al*, Phys. Rev. Lett. **74**, 3548 (1995).
18. H1 Coll. , C. Adloff *et al*, Phys. Lett. B **467**, 156 (1999).
19. ZEUS Coll. , O. Deppe *et al*, Nucl. Phys. B. (Proc. Suppl.), **82** 206-211, 2000.
20. E.A. Kuraev, L.N. Lipatov and V.S. Fadin, Sov. Phys. JETP **45** (1977) 199; Ya.Ya. Balitski and L.N. Lipatov, Sov. J. Nucl. Phys. **28** (1978) 822.
21. S.J. Brodsky *et al*, Phys. Rev. **D 56** (1997) 6957;
    J. Bartels *et al*, Phys. Lett. **B 389** (1996) 742; hep-ph/9710500.
22. ALEPH Note 2002-026, CONF 2002-015, July 12, 2002.
23. L3 Coll., M. Acciari *et al.* Phys. Lett. **B 453** (1999) 333; Phys. Lett. **B 531** (2002) 39.
24. OPAL Coll., G. Abbiendi *et al*, Eur. Phys. J. **C24** (2002) 17.
25. R. Engel, Z. Phys. C **66**, 203 (1995);
    R. Engel and J. Ranft, Phys. Rev. D **54**, 4246 (1996).
26. L3 note 2750, June 26, 2002.
27. L3 Coll., M. Acciarri *et al*, Phys. Lett. B **524**, 44 (2002).
28. T. Kleinwort, G. Kramer, Nucl. Phys. B **477**, 3 (1996); Phys. Lett. B **370**, 141 (1996); M. Klasen, T. Kleinwort, G. Kramer, private communication.
29. ALEPH Note 2002-023, CONF 2002-012, June 27, 2002.
30. L3 note 2762, July 1, 2002.
31. OPAL Coll., G. Abbiendi *et al*, CERN-EP/2002-056,July 2002, Subm. Eur. Phys. J.C.
32. L3 Coll., M. Acciarri *et al*, Phys. Lett. B **536**, 24 (2002).
33. S.J. Brodsky and G.P. Lepage, Phys. Rev. D **24**, 1808 (1981).
34. G.Farrar,E. Maina,F. Neri, Nucl. Phys. B **259**, 702 (1985);Nucl. Phys. B **263**, 746 (1986).
35. C.F. Berger, B. Lechner and W. Schweiger, Fizika **B8** (1999) 371.
36. P. Kroll, M. Schürmann and W. Schweiger, Z. Phys. A **338**, 339 (1991).
37. Z. Dziembowski, Phys. Rev. D **37**, 2030 (1988).

NH
ELSEVIER

Proceedings of ICHEP 2002, pp. 434–436
*S. Bentvelsen, P. de Jong, J. Koch and E. Laenen (Editors)*

31st INTERNATIONAL CONFERENCE ON HIGH ENERGY PHYSICS AMSTERDAM

# Consequences of $t$-channel unitarity for $\gamma^{(*)}p$ and $\gamma^{(*)}\gamma^{(*)}$ amplitudes.

J.R. Cudell*, E. Martynov†‡ and G. Soyez§
Université de Liège, Bât. B5-a, Sart Tilman, B4000 Liège, Belgium

We show that $t$-channel unitarity constraints make it possible to obtain the photon-photon cross sections from the photon-proton and proton-proton cross sections, for $Q^2 < 150\ \text{GeV}^2$. In order to do so, one must postulate the existence of double-pole or triple-pole singularities in the complex $j$ plane.

It is an old result [1] that one can relate the amplitudes describing three elastic processes $aa \to aa$, $ab \to ab$, $bb \to bb$. The trick is to continue these to the crossed channels $a\bar{a} \to a\bar{a}$, $a\bar{a} \to b\bar{b}$, $b\bar{b} \to b\bar{b}$, where they exhibit discontinuities because of the $a$ and $b$ thresholds. One then obtains a nonlinear system of equations, which can be solved. Working in the complex $j$ plane above thresholds ($t > 4m_a^2$, $4m_b^2$), and defining the matrix

$$T_0 = \begin{pmatrix} A_{aa\to aa} & A_{ba\to ba} \\ A_{ab\to ab} & A_{bb\to bb} \end{pmatrix} \tag{1}$$

one obtains

$$T_0 = \frac{D}{1 - RD} \tag{2}$$

with $R_{km} = 2i\sqrt{\frac{t-4m_k^2}{t}}\delta_{km}$ and $D = T_0^\dagger$. The latter is made of the amplitudes on the other side of the cut. For any $D$, equation (2) is enough to derive factorization: the singularities of $T_0$ can only come from the zeroes of

$$\Delta = \det(1 - RD). \tag{3}$$

Taking the determinant of both sides of eq. (2), we obtain in the vicinity of $\Delta = 0$

$$A_{aa\to aa}A_{bb\to bb} - A_{ab\to ab}A_{ba\to ba} = \frac{C}{\Delta}, \tag{4}$$

where $C$ is regular at the zeroes of $\Delta$. As the l.h.s. is of order $1/\Delta^2$ we obtain the well-known factorization properties from eqs. (2) and (4):
- The elastic hadronic amplitudes have common singularities;
- At each singularity in the complex $j$ plane, these amplitudes factorise.

These equations are used to extract relations between the residues of the singularities, which can be continued back to the direct channel.

We have extended [2] the above argument including all possible thresholds, both elastic and inelastic. The net effect is to keep the structure (2), but with a matrix $D$ that includes multi-particle thresholds. Furthermore, we have shown that one does not need to continue the amplitudes from one side of the cuts to the other, but that the existence of complex conjugation for the amplitudes is enough to derive (2).

Hence there is no doubt that the factorization of amplitudes in the complex $j$ plane is correct, even when continued to the direct channel. For isolated simple poles one obtains the usual factorization relations for the residues.

If $A_{pq}(j)$ has coinciding simple and double poles (*e.g.* colliding simples poles at $t = 0$),

$$A_{pq} = \frac{S_{pq}}{j-z} + \frac{D_{pq}}{(j-z)^2}, \tag{5}$$

one obtains the new relations

$$\begin{aligned} D_{11}D_{22} &= (D_{12})^2, \\ D_{11}^2 S_{22} &= D_{12}(2S_{12}D_{11} - S_{11}D_{12}). \end{aligned} \tag{6}$$

In the case of triple poles

$$A_{pq} = \frac{S_{pq}}{j-z} + \frac{D_{pq}}{(j-z)^2} + \frac{F_{pq}}{(j-z)^3}, \tag{7}$$

the relations become

$$F_{11}F_{22} = (F_{12})^2,$$

*J.R.Cudell@ulg.ac.be
†E.Martynov@guest.ulg.ac.be
‡on leave from Bogolyubov Institute for Theoretical Physics, Kiev.
§G.Soyez@ulg.ac.be


doi:10.1016/S0920-5632(02)02257-0

$$\begin{aligned} F_{11}^2 D_{22} &= F_{12}(2D_{12}F_{11} - D_{11}F_{12}), \\ F_{11}^3 S_{22} &= F_{11}F_{12}\left(2S_{12}F_{11} - S_{11}F_{12}\right) \\ &+ D_{12}F_{11}\left(D_{12}F_{11} - 2D_{11}F_{12}\right) \\ &+ D_{11}^2 F_{12}^2. \end{aligned} \quad (8)$$

Although historically one has used $t$-channel unitarity to derive factorization relations in the case of simple poles, it is now clear [3] that a soft pomeron pole is not sufficient to reproduce the $\gamma^*$ data from HERA [4]. However, it is possible, using multiple poles, to account both for the soft cross sections and for the DIS data [5]. We shall see later that relations (6, 8) enable us to account for the DIS photon-photon data from LEP.

For photons, two theoretical possibilities exist: i) The photon cross sections are zero for any fixed number of incoming or outgoing photons [7]. In this case, it is impossible to define an S matrix, and one can only use unitarity relations for the hadronic part of the photon wave function. Because of this, photon states do not contribute to the threshold singularities, and the system of equations does not close. The net effect is that the singularity structure of the photon amplitudes is less constrained. $\gamma p$ and $\gamma\gamma$ amplitudes must have the same singularities as the hadronic amplitudes, but extra singularities are possible: in the $\gamma p$ case, these may be of perturbative origin, but must have non perturbative residues. In the $\gamma\gamma$ case, these singularities have their order doubled. It is also possible for $\gamma\gamma$ to have purely perturbative additional singularities.
ii) It may be possible to define collective states in QED for which an S matrix would exist. In this case, we obtain the same situation for on-shell photons as for hadrons. However, in the case of DIS, virtual photons come as external states. Because they are virtual, they do not contribute to the $t$-channel discontinuities, and hence the singularity structure for off-shell photons is as described in i).

In the following, we shall explore the possibility that no new singularity is present for on-shell photon amplitudes, and show that it is in fact possible to reproduce present data using pomerons with double or triple poles at $j = 1$.

For a given singularity structure, a fit to the $C = +1$ part of proton cross sections, and to $\gamma^{(*)}p$ data enables one, via relations (2), to predict the $\gamma^{(*)}\gamma^{(*)}$ cross sections. Hence we have fitted [2] $pp$ and $\bar{p}p$ cross sections and $\rho$ parameters, as well as DIS data from HERA [4].

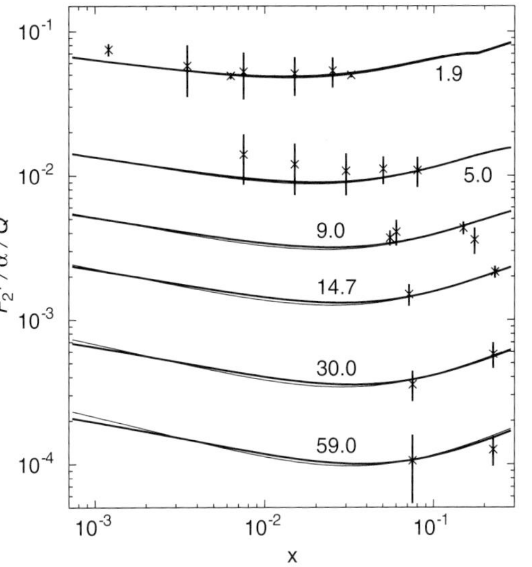


Figure 1. Fits to $F_2^\gamma$. The thick and thin curves correspond respectively to the triple-pole and to the double-pole cases. The data are from [6].

The general form of the parametrisations which we used is given, for total cross sections of $a$ on $b$, by the generic formula $\sigma_{ab}^{tot} = (R_{ab} + H_{ab})$. The first term, from the highest meson trajectories ($\rho$, $\omega$, $a$ and $f$), is parametrized via Regge theory as

$$R_{ab} = Y_{ab}^{+}\,(\tilde{s})^{\alpha_+ - 1} \pm Y_{ab}^{-}\,(\tilde{s})^{\alpha_- - 1} \quad (9)$$

with $\tilde{s} = 2\nu/(1\ \mathrm{GeV}^2)$. Here the residues $Y_+$ factorise. The second term, from the pomeron, is parametrized either as a double pole [8]

$$\begin{aligned} H_{ab} &= D_{ab}(Q^2)\Re e\left[\log\left(1 + \Lambda_{ab}(Q^2)\tilde{s}^{\,\delta}\right)\right] \\ &+ C_{ab}(Q^2) + (\tilde{s} \to -\tilde{s}) \end{aligned} \quad (10)$$

or as a triple pole

$$H_{ab} = t_{ab}(Q^2)\left[\log^2 \frac{\tilde{s}}{d_{ab}(Q^2)} + c_{ab}(Q^2)\right]. \quad (11)$$

The details of the form factors entering (10, 11) can be found in [2]. Such parametrisations give $\chi^2/dof$ values less than 1.05 in the region $\cos(\vartheta_t) \geq \frac{49}{2m_p^2}$, $\sqrt{2\nu} \geq 7$ GeV, x $\leq 0.3$, $Q^2 \leq 150$ GeV$^2$.

What is really new is that these forms can be extended to photon-photon scattering, using relations (6, 8). The total $\gamma\gamma$ cross section is well reproduced and the de-convolution using PHOJET is preferred. For photon structure functions, one needs to add one singularity at $j = 0$ corresponding to the box diagram [9], but otherwise the $\gamma\gamma$ amplitude is fully specified by the factorization relations. We see in Fig. 1 that DIS data are well reproduced by both parametrisations.

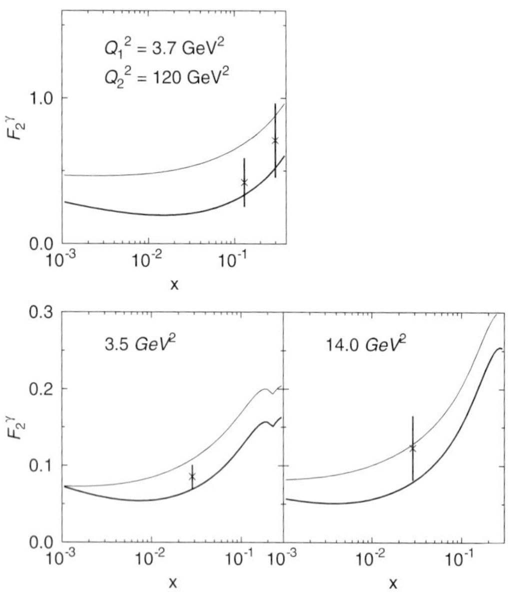


Figure 2. Fits to $F_2^\gamma$ for nonzero asymmetric values of $P^2$ and $Q^2$ and for $P^2 = Q^2$. The curves are as in Fig. 1. The data are from [6].

Even more surprisingly, it is possible to reproduce the $\gamma^*\gamma^*$ cross sections when both photons are off-shell, as shown in Fig. 2. This is the place where BFKL singularities may manifest themselves, but as can be seen such singularities are not needed.

In conclusion, we have shown that it is possible to reproduce soft data (*e.g.* total cross sections) and hard data (*e.g.* $F_2$ at large $Q^2$) using a common $j$-plane singularity structure, provided the latter is more complicated than simple poles. Furthermore, we have shown that it is then possible to predict $\gamma\gamma$ data using $t$-channel unitarity. How to reconcile such a simple description with DGLAP evolution, or BFKL results, remains a challenge.

## REFERENCES


1. V.N. Gribov and I. Ya. Pomeranchuk, Phys. Rev. Lett. **8**, 4343 (1962); J. Charap and E.J. Squires, Phys. Rev. 127, 1387 (1962).
2. J.R. Cudell, E. Martynov and G. Soyez, arXiv:hep-ph/0207196.
3. A. Donnachie and P. V. Landshoff, Phys. Lett. B **437** (1998) 408; Phys. Lett. B **533**, 277 (2002).
4. H1 collaboration: C. Adloff et al. , Eur. Phys. J. C **21**, 33, (2001); ZEUS collaboration: S.Chekanov et al. , Eur. Phys. J. C **21** (2001), 443-471. and references therein.
5. P. Desgrolard and E. Martynov, Eur. Phys. J. C **22**, 479 (2001) [arXiv:hep-ph/0105277]; J. R. Cudell and G. Soyez, Phys. Lett. B **516**, 77 (2001) [arXiv:hep-ph/0106307]; COMPETE collaboration, J. R. Cudell *et al.*, Phys. Rev. D **65** (2002) 074024 [arXiv:hep-ph/0107219].
6. L3 Collaboration: M. Acciarri *et al.*, Phys. Lett. B **519** (2001) 33; OPAL Collaboration: G. Abbiendi *et al.*, Phys. Lett. B **533** (2002) 207, and references therein.
7. F. Bloch and A. Nordsiek, Phys. Rev. **52** (1937) 54.
8. P. Desgrolard, A. Lengyel and E. Martynov, JHEP **0202** (2002) 029 [arXiv:hep-ph/0110149].
9. V. M. Budnev, I. F. Ginzburg, G. V. Meledin and V. G. Serbo, Phys. Rept. **15** (1974) 181.

ELSEVIER

Proceedings of ICHEP 2002, pp. 437–439
*S. Bentvelsen, P. de Jong, J. Koch and E. Laenen (Editors)*

31st INTERNATIONAL CONFERENCE ON HIGH ENERGY PHYSICS AMSTERDAM

# Charge and Spin Asymmetries from Pomeron-Odderon Interference *

P. Hägler [a], B. Pire[b], L. Szymanowski[bc], O.V. Teryaev [d]

[a] Institut für Theoretische Physik, Universität Regensburg, D-93040 Regensburg, Germany and CTP, MIT, Cambridge, MA 02139, USA

[b]CPhT, École Polytechnique, F-91128 Palaiseau, France

[c]Sołtan Institute for Nuclear Studies, Hoża 69, 00-681 Warsaw, Poland

[d] Bogoliubov Laboratory of Theoretical Physics, JINR, 141980 Dubna, Russia

We study Pomeron-Odderon interference effects giving rise to charge and single-spin asymmetries in diffractive electroproduction of a $\pi^+ \; \pi^-$ pair. We calculate these asymmetries in the Born approximation of QCD in a kinematical domain accessible to HERA experiments.

## 1. Introduction

Hadronic reactions at low momentum transfer and high energies (for charge-even exchange) are described in QCD in terms of QCD-Pomeron described by the BFKL equation [1]. The charge-odd exchange is less well understood although the corresponding BKP equations [2] have attracted much attention recently [3–6], thus reviving the relevance of phenomenological studies of the Odderon exchange pointed out years ago in Ref. [7]. Unfortunately, the recent studies of specific channels where the QCD Odderon contribution is expected to be singled out have turned out to be very disappointing. In particular, recent experimental studies at HERA of exclusive $\pi^0$ photoproduction [8] indicate a very small cross section for this process which stays in contradiction with theoretical predictions based on the stochastic vacuum model [9].

The general feature of all meson production processes is that scattering amplitude describing Odderon exchange enters quadratically in the cross section. This observation lead to the suggestion in Ref. [10], that the study of observables where Odderon effects are present at the amplitude level - and not at the squared amplitude level - is mandatory to get a convenient sensitivity to a rather small normalization of this contribution. This may be achieved by means of charge asymmetries, as for instance in open charm production [10]. Since the final state quark-antiquark pair has no definite charge parity both Pomeron and Odderon exchanges contribute to this process. Another example [11] is the charge asymmetry in soft photoproduction of two pions. Bearing in mind perturbative QCD (pQCD) description, we calculated the "hard" analogs of these asymmetries [12], and supplemented them by the single spin asymmetries, which may be studied at HERA with polarized lepton beam [13].

## 2. Method and Results

We consider the process $e^-(p_e)N(p_N) \rightarrow e^-(p'_e)\pi^+(p_+)\pi^-(p_-)N'(p'_N)$. The application of pQCD for the calculation of a part of this process is justified by the presence of a hard scale: the squared mass $-Q^2 = -(p_e - p'_e)^2$ of the virtual photon, $Q^2$ being of the order of a few $\mathrm{GeV}^2$. The amplitude of this process (Fig.1) includes the convolution of a perturbatively calculable hard subprocess with two non-perturbative inputs, the 2-pion generalized distribution amplitude (GDA) and the Pomeron-Odderon (P/O) proton impact

*This work is supported in part by the TMR, IHRP and INTAS Programmes of EU, Contracts No. FMRX-CT98-0194, No. HPRN-CT-2000-00130 and INTAS-00-00587.


doi:10.1016/S0920-5632(02)02259-4

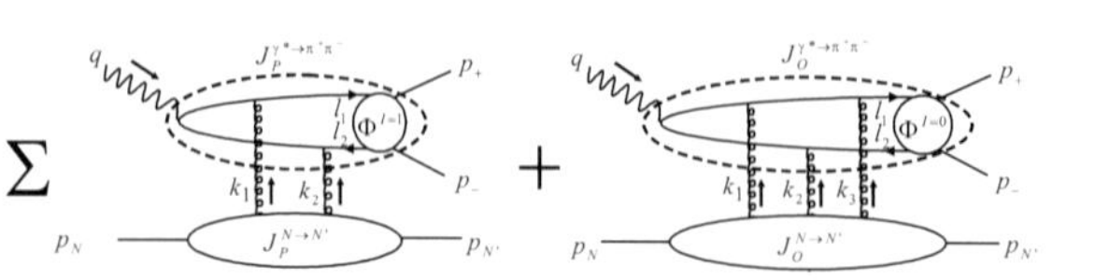

Figure 1. Pomeron and Odderon amplitudes within QCD factorization

factors. Since the $\pi^+\pi^-$ system is not a charge parity eigenstate, the GDA includes two charge parity components and allows for a study of the corresponding interference term. The relevant GDA is just given by the light cone wave function of the two pion system [14].

### 2.1. Charge asymmetry

We define the forward-backward or charge asymmetry $A(Q^2, t, m_{2\pi}^2, y, \alpha)$ by

$$\frac{\sum\limits_{\lambda=+,-} \int \cos\theta \, d\sigma(s, Q^2, t, m_{2\pi}^2, y, \alpha, \theta, \lambda)}{\sum\limits_{\lambda=+,-} \int d\sigma(s, Q^2, t, m_{2\pi}^2, y, \alpha, \theta, \lambda)} \quad (1)$$

as a weighted integral over polar angle $\theta$ of the relative momentum of two pions. Although this asymmetry depend on the full set of the kinematical variables, different dependencies, due to factorization, come from different sources. The most clean one are the dependencies on $Q^2$, coming from the hard subprocess, and on dipion mass $m_{2\pi} = \sqrt{(p_+ + p_-)^2}$, coming from GDA. The specific form of $m_{2\pi}$ dependence is explained by the fact, that the phase of GDA[14] should add to the phase shift between Pomeron and Odderon. The $Q^2$ dependence, which is a subject to corrections from BFKL, BKP and ERBL evolution, is due to the different $Q^2$ dependence of Pomeron and Odderon coefficient functions (perturbative impactfactors), the latter having the extra propagators, leading to the decrease of asymmetry with $Q^2$. The typical scale of $\alpha_s$ is determined by the gluons transverse momenta $\vec{k_i}$ (see Fig.1) and we checked that the result is not changed substantially if "hard" $\alpha_S(Q^2)$ is used for coefficient function and "soft" $\alpha_S \sim 0.5$ is used for non-perturbative proton impactfactors. The latter objects, together with the coefficient

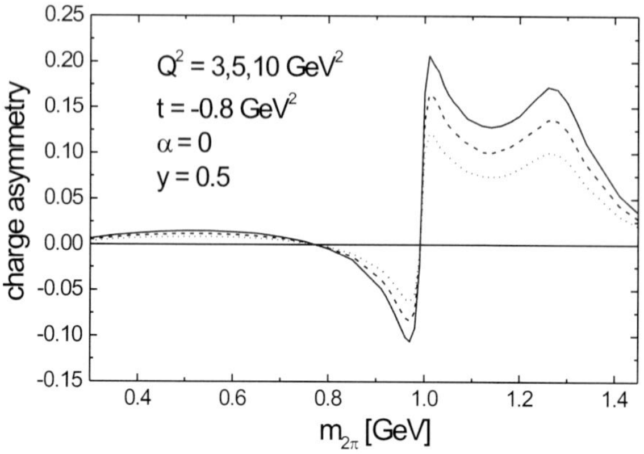


Figure 2. Dependence of charge asymmetry on $m_{2\pi}$ and $Q^2$

functions, define the dependence of asymmetry on $t = (p_N - p'_N)^2$. As to the dependence on the standard leptonic variables $y$ (the lepton energy fraction carried by the virtual photon) and $\alpha$ (the angle between lepton and photon scattering planes), it is determined by the relative size of the virtual photon helicity amplitudes and allows to estimate the promising kinematical region [13].

### 2.2. Spin Asymmetry

The single spin asymmetry $A_S(Q^2, t, m_{2\pi}^2, y, \alpha)$ is defined by

$$\frac{\sum\limits_{\lambda=+,-} \lambda \int \cos\theta \, d\sigma(s, Q^2, t, m_{2\pi}^2, y, \alpha, \theta, \lambda)}{\sum\limits_{\lambda=+,-} \int d\sigma(s, Q^2, t, m_{2\pi}^2, y, \alpha, \theta, \lambda)} \quad (2)$$

and requires to fix the lepton beam polarization $\lambda$. Contrary to charge asymmetry, this effect is proportional to the imaginary, rather than to real part of the interference term. As the Pomeron amplitude is imaginary and the Odderon one is real the relative phase between them is the maximal one for the emergence of single spin asymmetries [15]. The effect should be therefore maximal for zero relative phase between isoscalar and isovector GDA's, providing a complementary probe. This complementarity can be seen from the dependence of spin asymmetry on $m_{2\pi}$ and $Q^2$(Fig.2) The smaller numerical value

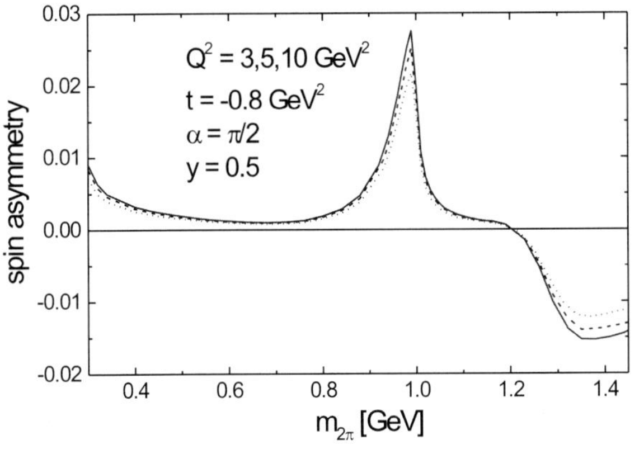


Figure 3. Dependence of spin asymmetry on $m_{2\pi}$ and $Q^2$

of spin asymmetry is to a large extent due to the kinematical factor $\sqrt{t}$. This asymmetry may be therefore important in the region of large $t$, which would become a relevant QCD scale and allow to have smaller $Q^2$.

## 3. Conclusions

We found that a sizable charge asymmetry may be a useful tool to look for QCD Odderon contribution at HERA. The spin asymmetry is smaller, but it can be important at larger $t$ and smaller $Q^2$. Note finally, that the numerical value of our predictions depend on the adopted model for proton impact factors, so the observation of the predicted effects with the different magnitude and/or $t-$dependence might be considered as an indirect experimental determination of these important non-perturbative objects.

## REFERENCES

1. E.A. Kuraev, L.N. Lipatov and V.S. Fadin, Sov. JETP **44** (1976) 443 ; *ibid* **45** (1977) 199 ; Ya.Ya. Balitsky and L.N. Lipatov, Sov. J. Nucl. Phys. **28** (1978) 822 ; L.N. Lipatov, Sov. Phys. JETP **63** (1986) 904
2. J. Bartels, Nucl. Phys. B **151** (1979) 293; *ibid* B **175** (1980) 365;
J. Kwiecinski and M. Praszalowicz, Phys. Lett. B **94** (1980) 413.
3. L.N. Lipatov, Phys. Lett. B **309** (1993) 394 ; JETP Lett. **59** (1994) 596 ; Sov. Phys. JETP Lett. **59** (1994) 571;
L.D. Fadeev and G.P. Korchemsky, Phys. Lett. B **342** (1995) 311
4. R.A. Janik and J. Wosiek, Phys. Rev. Lett. **82** (1999) 1092
5. J. Bartels, L.N. Lipatov and G.P. Vacca, Phys. Lett. B **477** (2000) 178
6. G. P. Korchemsky, J. Kotanski and A. N. Manashov, Phys. Rev. Lett. **88**, 122002 (2002).
7. L. Lukaszuk and B. Nicolescu, Lett. Nuovo Cim. **8** (1973) 405
8. J. Olsson (for the H1 Collab.), hep-ex/0112012
9. E.R. Berger, A. Donnachie, H.G. Dosch, W. Kilian, O. Nachtmann and M. Reuter, Eur. Phys. J. C **9** (1999) 491
10. S.J. Brodsky, J. Rathsman and C. Merino, Phys. Lett. B **461** (1999) 114
11. I.F. Ginzburg, I.P. Ivanov and N.N. Nikolaev, hep-ph/0207345 and Ref. therein.
12. P. Hagler, B. Pire, L. Szymanowski and O. V. Teryaev, Phys. Lett. B **535**, 117 (2002) [Erratum-ibid. B **540**, 324 (2002)]
13. P. Hagler, B. Pire, L. Szymanowski and O. V. Teryaev, hep-ph/0207224.
14. M. Diehl, T. Gousset, B. Pire and O.V. Teryaev, Phys. Rev. Lett. **81** (1998) 1782
15. O.V. Teryaev, *T-odd diffractive distributions*, in Proceedings of IX Blois Workshop, p.211-216.

ELSEVIER

Proceedings of ICHEP 2002, pp. 440–442
*S. Bentvelsen, P. de Jong, J. Koch and E. Laenen (Editors)*

31st INTERNATIONAL CONFERENCE ON HIGH ENERGY PHYSICS AMSTERDAM

# Odderon and Pomeron Physics in Multi-Photon Final States at HERA

T. Berndt[a*]

[a]Kirchhoff Institut für Physik — Universität Heideldberg
Im Neuenheimer Feld 227, 69120 Heidelberg, Germany

In this contribution the first search at HERA for Odderon induced reactions is presented and contrasted with cross section measurements for Pomeron induced processes. The searches are performed in the channels $\gamma p \to \pi^0 N^*$, $\gamma p \to f_2(1270)X$ and $\gamma p \to a_2^0(1320)X$, where $N^*$ denotes an excited nucleon state. The rates found are compatible with the background alone, and upper limits were derived which are below the expectations of a specific non-perturbative QCD model. In contrast to these findings the cross sections for the Pomeron mediated processes — $\gamma p \to \omega p$ and $\gamma p \to \omega\pi^0 X$ — were found to be in perfect agreement with earlier measurements and expectations from Regge theory with soft Pomeron exchange. The mean $\gamma p$ centre-of-mass energies were $\langle W \rangle =$ 200 GeV and 215 GeV, respectively.

## 1. Introduction

With the advent of HERA diffractive physics experienced a remarkable renaissance as it turned out that roughly ten per cent of the physics observed was diffractive in nature. In particular diffractive photoproduction of vector mesons could be analysed at hitherto unexplored energies.

Diffractive vector meson photo-(and electro)production is thought to proceed via Pomeron exchange, where according to Vector Meson Dominance (VMD) the photon fluctuates into a (virtual) $q\bar{q}$ pair that subsequently undergoes a hadronic scattering with the proton, where the $q\bar{q}$ pair is turned into a real vector meson by the interaction with the Pomeron[1].

According to [2] there should — aside the vacuum exchange of the Pomeron — also be a hadronic exchange with *odd* $C$-parity, which should manifest itself in a finite difference in the total cross sections for $pp$ and $p\bar{p}$ scattering, or respectively in a non-vaninishing $\Delta\rho_{pp}^{p\bar{p}}$ at asymptotic energies. Due to its odd $C$-parity this exchange became known as the so-called Odderon.

In a series of papers [3] predictions for Odderon-induced meson production for $\pi^0$-mesons and $f_2$ and $a_2^0$ mesons are given. The calculations are performed in the Stochastic Vaccum Model (SVM)[8]. In these calculations the proton is required to be excited into an $N^*$ nucleonic state. The $\pi^0$ analysis presented in the following takes this explicitely into account, in contrast to the $f_2$ and $a_2^0$ analyses which are inclusive in the sense that the baryonic final state is ignored completely.

## 2. Event Selection

The analyses presented here are based on data samples taken in the years 1996 ($\omega$, $\omega\pi^0$, $f_2$ and $a_2^0$), 1999 and 2000 (both for the exclusive $\pi^0$). The integrated luminosities are 4.5 pb$^{-1}$ for 1996 and 30.6 pb$^{-1}$ for the years 1999 and 2000 combined, respectively.

Electrons of 27.6 GeV were brought to collision with protons of 820 GeV (920 GeV for the years 1999 and 2000). Photoproduction events were selected by demanding the scattered electron to be detected under very low angles in a small angle electron detector 33 m downstream the electron beam, limiting the values for the virtuality $Q^2$ and inelasticity $y$ to $Q^2 < 0.01$ GeV$^2$ and $0.3 < y < 0.7$, respectively.

Depending on the meson analysed the num-

*Supported by the Bundesministerium für Bildung, Wissenschaft, Forschung und Technologie, Germany, under contract number 7HD27P.


doi:10.1016/S0920-5632(02)02260-0

ber of photons in the final state varies from two to five: $\pi^0 \to \gamma\gamma$, $\omega \to \pi^0\gamma \to 3\gamma$, $f_2 \to \pi^0\pi^0 \to 4\gamma$, $a_2^0 \to \pi^0\eta \to 4\gamma$, and $\omega\pi^0 \to (\pi^0\gamma)\pi^0 \to 5\gamma$, respectively. By counting the number of photons in the final state, one uniquely determines the $C$-parity of the produced system and thus of the hadronic exchange. These photons are detected in the backward calorimeter(s) of the H1-detector[4], since the mesons are produced with only little transverse but large longitudinal momentum. For 1996 the analyses were restricted to the backward electromagnetic calorimeter (SpaCal)[5] only, and for the exclusive $\pi^0$-measurement of the years 1999 and 2000 the very low angle calorimeter (VLQ)[6] and the Forward Neutron Calorimeter (FNC)[7] were included additionally.

To ensure that the events selected were indeed exclusive a variant of energy-momentum conservation was utilised, viz. 50(49) GeV$< \Sigma <$ 60 GeV, where $\Sigma := \sum_i E^{(i)} - p_z^{(i)}$, and the sum runs over all photons the SpaCal, the VLQ and the scattered electron. The number in brackets refers to the exclusive pion analysis. If $\Sigma$ is found to be equal (within resolution) twice the beam energy of 55 GeV the whole final state is detected.

For the mesons decaying via $\pi^0$'s and $\eta$'s appropriate mass windows for the respective $\gamma\gamma$-masses around the their nominal values of 135 MeV and 547 MeV were chosen.

For the $\omega$ it was further required that the interaction occurred elastically by imposing cuts limiting the mass of possible proton-excitations to less than 1.6 GeV, in contrast to the exclusive $\pi^0$ were it was explicitely demanded that the proton was excited into a state ($N^*$) decaying into a leading neutron, which was detected in the forward neutron calorimeter. For the remaining analyses no constraints on the outgoing baryonic final state were imposed.

## 3. Monte Carlo Models

In order to describe the processes under study DiffVM[10] and a derivative thereof called OPIUM[11] [2] were used. In addition to the processes based on Pomeron exchange combined with VMD as implemented in DiffVM, OPIUM is capable to generate pseudoscalar and tensor mesons as described in[3]. The background consisting of inclusive $\gamma p$-interactions and mesons decaying via charged pions, was studied using Pythia[9].

[2]OPIUM is an acronym for *Odderon, Pomeron Induced Unified Meson maker*

## 4. Results

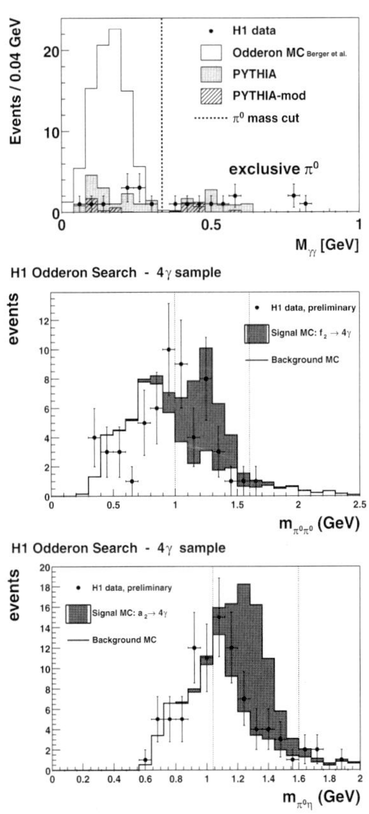


Figure 1. The top part shows the invariant mass of exclusive $\gamma\gamma$-pairs, the middle part shows the invariant $\pi^0\pi^0$-mass and in the lowest the mass of $\pi^0\eta$-pairs is shown.

Figure 1 shows the $\pi^0$-[12], $f_2$- and $a_2^0$-searches, respectively. Since the distributions observed are compatible with the background alone, 95% confidence level limits on the cross sections are given. The predictions of the SVM are give in brackets for comparison.

$$\begin{array}{rcll} \sigma_{\text{acc}}(\gamma p \to \pi^0 N^*) & < & 49\,\text{nb} & (> 200\,\text{nb}) \\ \sigma(\gamma p \to f_2 X) & < & 16\,\text{nb} & (21\,\text{nb}) \\ \sigma(\gamma p \to a_2^0 X) & < & 96\,\text{nb} & (190\,\text{nb}) \end{array}$$

The upper part of figure 2 shows the invariant $\pi^0\gamma$-mass spectrum from which a cross section for elastic $\omega$-photoproduction of

$$\sigma(\gamma p \to \omega p) = (1.3 \pm 0.2(\text{stat}) \pm 0.2(\text{syst}))\,\mu\text{b}$$

was derived. This result agrees well with expectations from the VDM combined with Regge extrapolations. In addition the diffractive slope $b$, characterising the differential cross section $\mathrm{d}\sigma/\mathrm{d}|t| \propto \exp(-b|t|)$ was extracted: $b = (10.2 \pm 1.1(\text{stat}) \pm 2.0(\text{syst})\,\text{GeV}^{-2}$ which is in accordance with other measurements.

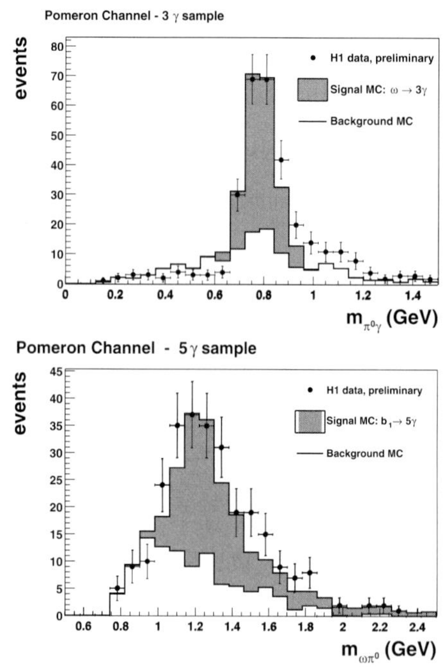


Figure 2. Three (top) and five (bottom) photon invariant mass spectra, made up of $\pi^0\gamma$- for the $\omega$ and $\omega\pi^0$ final states.

The lower part of the figure shows the invariant mass of exclusive $\omega\pi^0$-pairs for which a cross section of

$$\sigma(\gamma p \to \omega\pi^0 X) = (1.0 \pm 0.2(\text{stat}) \pm 0.2(\text{syst}))\,\mu\text{b}$$

was found. This result matches the expectation of resonant $b_1$ photoproduction, as shown by the filled histogram, although due to the lack of a spin parity analysis it cannot be firmly stated that the observed signal indeed stems from the decay $b_1 \to \omega\pi^0$. Subtracting a non-resonant contribution of $\sim 0.2\,\mu$b as taken from PYTHIA, one arrives at a cross section for resonant $\omega\pi^0$ photoproduction of $\sim 0.8\,\mu$b, which can be compared to a Regge extrapolation of low energy measurements of $b_1$-mesons that yields an expectation of $0.7\,\mu$b.

## REFERENCES

1. G. A. Schuler, T. Sjöstrand, *Nucl. Phys.*, **B407**, 539, (1993).
2. L. Łukaszuk, B. Nicolescu, *Nuov.Ciem.Lett.*, **8**, 405, (1973).
3. E. Berger et al., *Eur. Phys. J.*, **C9**, 491, (1999).
   E. Berger et al., *Eur. Phys. J.*, **C14**, 6731, (2000).
4. I. Abt et al., *Nucl. Instr. Meth.*, **A386**, 310 and 348, (1997).
5. T. Nicholls et al., *Nucl. Inst. Meth*, **A374**, 149, (1996).
6. M. Keller et al., *Nucl. Instr. Meth*, **A409**, 604, (1998).
7. T. Nunnemann, *Measurement of Leading Neutron Production in Deep-Inelastic ep Scattering at HERA*, PhD thesis, MPIH-V7-1999, Heidelberg, (1999).
8. H. G. Dosch et al., *Phys. Rev.*, **D50**, 1992, (1994).
9. T. Sjöstrand, *Comp. Phys. Comm.*, **82**, 74, (1994); private communication (2001).
10. B. List and A. Mastroberardino, *DIFFVM - A Monte Carlo Generator for Diffractive Processes in ep Scattering*, available as: `http://www.desy.de/~heramc/mclist.html`.
11. W. Kornelis, *private communication*, (2000).
12. C. Adloff et al., *Phys. Lett.*, **B554**, 35, (2002).

ELSEVIER

Proceedings of ICHEP 2002, pp. 443–445
*S. Bentvelsen, P. de Jong, J. Koch and E. Laenen (Editors)*

31st INTERNATIONAL CONFERENCE ON HIGH ENERGY PHYSICS AMSTERDAM

# Photoproduction of vector mesons in the Soft Dipole Pomeron model

E. Martynov[a*], E. Predazzi[b †] and A. Prokudin[b ‡]

[a]Institut de physique Bat B5-a Université de Liège, Sart Tilman B-4000 Liège, Belgique

[b]Dipartimento di Fisica Teorica, Università Degli Studi Di Torino, Via Pietro Giuria 1, 10125 Torino, Italy

Exclusive photoproduction of all vector mesons by real and virtual photons is considered in the Soft Dipole Pomeron model. It is emphasized that being the Pomeron in this model a double Regge pole with intercept equal to one, we are led to rising cross-sections but the unitarity bounds are not violated. It is shown that all available data for $\rho, \omega, \varphi, J/\psi$ and $\Upsilon$ in the region of energies $1.7 \leq W \leq 250$ GeV and photon virtualities $0 \leq Q^2 \leq 35$ $\mathrm{GeV}^2$ , including the differential cross-sections in the region of transfer momenta $0 \leq |t| \leq 1.6$ $\mathrm{GeV}^2$, are well described by the model.

## 1. Introduction

A new precise measurement of $J/\psi$ exclusive photoproduction by ZEUS [1] opens a new window in our understanding of the process and allows us to give more accurate predictions for future experiments. One of the key issues of the dataset [1] is the diffractive cone shrinkage observed in $J/\psi$ photoproduction which leads us to consider it a soft rather than pure QCD process so that we can apply the Soft Dipole Pomeron exchange [2] model.

The basic diagram is depicted in Fig.1; $s$ and $t$ are the usual Mandelstam variables, $Q^2 = -q^2$ is the virtuality of the photon.

It is well known that high-energy representation of the scattering amplitude may be expressed as a sum over the the appropriate Regge poles in the complex $j$ plane [3]

$$A(s,t)_{s\to\infty} \approx \sum_i \eta_i(t)\beta_i(t)(\cos\theta_t)^{\alpha_i(t)}, \tag{1}$$

where $\eta_i(t)$ is the signature factor and $\theta_t$ is the scattering angle in the $t$ channel.

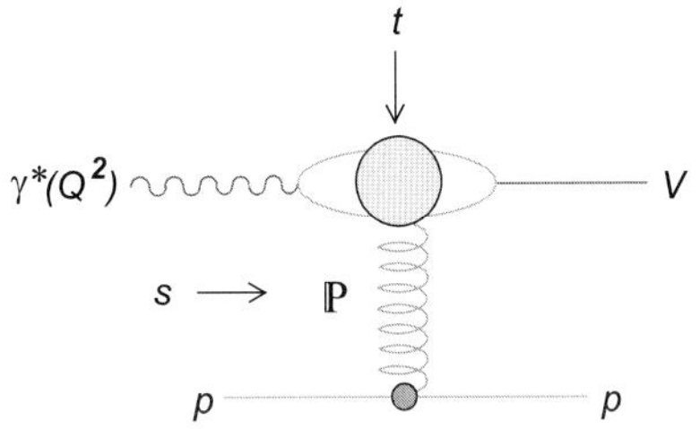


**Fig.1** Photoproduction of a vector meson.

## 2. Soft Dipole Pomeron (SDP) model

We utilize the variable $z \sim \cos\theta_t$

$$z = \frac{2(W^2 - M_p^2) + t + Q^2 - M_V^2}{\sqrt{(t + Q^2 - M_V^2)^2 + 4M_V^2 Q^2}} \tag{2}$$

where $W^2 = (p+q)^2 \equiv s$, $M_V$ is the vector meson mass, $M_p$ is the proton mass.

Following to [4] we take into account $SU(4)$ symmetry relations by introducing coefficients $N_V$ and writing the amplitude as $A_{\gamma p\to Vp} = N_V A_{Vp\to Vp}$, where

$$N_\rho = \frac{3}{\sqrt{2}}\,; N_\omega = \frac{1}{\sqrt{2}}\,; N_\phi = 1\,; N_{J/\psi} = 2\,. \tag{3}$$

The amplitude of the process $Vp \to Vp$ may be written in the following form [4]

$$A(z,t;M_V^2,\tilde{Q}^2) = \mathcal{P} + \mathcal{R} \tag{4}$$

**Permanent address:* Bogolyubov Institute for Theoretical Physics, Kiev, Ukraine
†Sezione INFN di Torino, Italy
‡*Permanent address:* Institute For High Energy Physics, Protvino, Russia

[4]Due to the lack of place we do not give here the explicit expressions for residue functions $g(t, M_V^2, Q^2)$ and the val-


doi:10.1016/S0920-5632(02)02261-2

where $\tilde{Q}^2 = Q^2 + M_V^2$.

For Pomeron contribution $\mathcal{P}$ we use the so called dipole Pomeron which gives a very good description of all hadron-hadron total cross sections [2],[6]. Specifically, $\mathcal{P}$ is given by [5,7]

$$\begin{aligned}\mathcal{P} &= ig_0(t; M_V^2, \tilde{Q}^2)(-iz)^{\alpha_{\mathcal{P}}(t)-1} \\ &+ig_1(t; M_V^2, \tilde{Q}^2)ln(-iz)(-iz)^{\alpha_{\mathcal{P}}(t)-1}.\end{aligned} \tag{5}$$

where the first term is a single $j$-pole contribution and the second one is a contribution of the double $j$-pole. We use a nonlinear pomeron trajectory

$$\alpha_{\mathcal{P}}(t) = 1 + \gamma(\sqrt{4m_\pi^2} - \sqrt{4m_\pi^2 - t}\,), \tag{6}$$

where $m_\pi$ is the pion mass. Attempts to add linear, $\alpha'_{\mathcal{P}}t$, term in (6) were failed because fit leads to $\alpha'_{\mathcal{P}} = 0$. We would like to stress that in this model

$$\alpha_{\mathcal{P}}(0) = 1. \tag{7}$$

Thus the model does not violate asymptotically the Froissart-Martin bound.

Contribution of the secondary reggeon ($\mathcal{R} = f, \pi$) is written in the standard form

$$\mathcal{R} = ig_{\mathcal{R}}(t; M_V^2, \tilde{Q}^2)(-iz)^{\alpha_{\mathcal{R}}(t)-1}. \tag{8}$$

with linear trajectory $\alpha_{\mathcal{R}}(t) = \alpha_{\mathcal{R}}(0) + \alpha'_{\mathcal{R}}\, t$.

For $\rho$ and $\varphi$ meson photoproduction we write the scattering amplitude as the sum of a Pomeron and $f$ contribution. For $\omega$ meson photoproduction $\pi$-reggeon exchange is added (see also the discussion in [8]). Only Pomeron contributes to $J/\psi$ amplitude.

In the integrated elastic cross section

$$\sigma_{el}^{\gamma p \to Vp} = 4\pi \int_{t_-}^{t_+} dt |A^{\gamma p \to Vp}|^2\,, \tag{9}$$

the upper and lower limits

$$\begin{aligned}2t_\pm =& \pm\lambda(W^2, -Q^2, M_p^2)\lambda(W^2, M_V^2, M_p^2)/W^2 \\ &-(W^2 + Q^2 - M_V^2 - 2M_p^2) \\ &+(Q^2 + M_p^2)(M_V^2 - M_p^2)/W^2,\end{aligned} \tag{10}$$

$$\lambda^2(x, y, z) = x^2 + y^2 + z^2 - 2xy - 2yz - 2zx, \tag{11}$$

ues of free parameters obtained in the fit to the data. One can find all details in [5]

are determined by the kinematical condition $-1 \leq \cos\theta_s \leq 1$ where $\theta_s$ is the scattering angle in the s-channel of the process. Thus when $W \to W_{threshold}$ we have $t_- \to t_+$ and the elastic cross section goes to zero. It allows us to describe effectively the behaviour of cross sections even at lowest energies.

For $Q^2 \neq 0$ we write

$$\sigma = \sigma_T + \sigma_L = \sigma_T\{1 + R(Q^2)\}, \quad R(0) = 1 \tag{12}$$

where $\sigma_T$ is defined by (9). For the ratio $R(Q^2)$ we considered a few possibilities of its behaviour at $Q^2 \to \infty$ (see details in [5]) and found that they lead to comparable description of cross-sections [5].

## 3. Fit to the data and results

Comparing the model with experimental data we made a fit and determined adjustable parameters in the two steps.

**1.** Fit to the data at $Q^2 = 0$, 12 free parameters. The data on differential and integrated cross-sections of $\rho, \omega, \varphi$ and $J/\psi$ mesons (357 points) were used. We found $\chi^2/d.o.f = 1.49$ for $W > 1.7$ GeV (the values of parameters as well as the partial $\chi^2$s one can see in [5]). However if we put $W_{min} = 4$ GeV$^2$ then $\chi^2/d.o.f. = 0.85$. The resulting description of $\sigma$ for all mesons and $d\sigma/dt$ for $J/\psi$ is shown in Figs. 2-3; for another mesons the figures are given in [5].

**2.** Fit to the data at $0 < Q^2 < 35$ GeV$^2$ and $W > 4$ GeV, 3 free parameters. Only the data on $\rho$ cross-section (283 points) were used in the minimization. For the choice of $R$, which better corresponds to the data on $\sigma_L/\sigma_T$, we obtained $\chi^2/d.o.f. = 1.53$. The results for $J/\psi$ cross-sections at various $Q^2$ is presented in the Fig.5 and Fig.6. (For other mesons see figures in [5]).

[5] We did not fit the model to the available data on $\sigma_L/\sigma_T$, however we found that they indicate qualitatively a slower growth than $R \propto Q^2$. A more precise data are needed to make a more definite conclusion about $R$

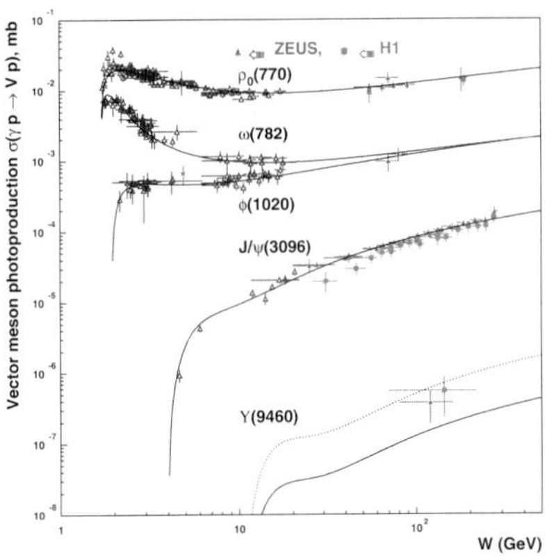


**Fig.2** Photoproduction of vector mesons in the SDP model. Two curves for $\Upsilon$-meson cross-section (calculated, not fitted) correspond to the two choices of the parameter $N_\Upsilon$ (see details in [5]).

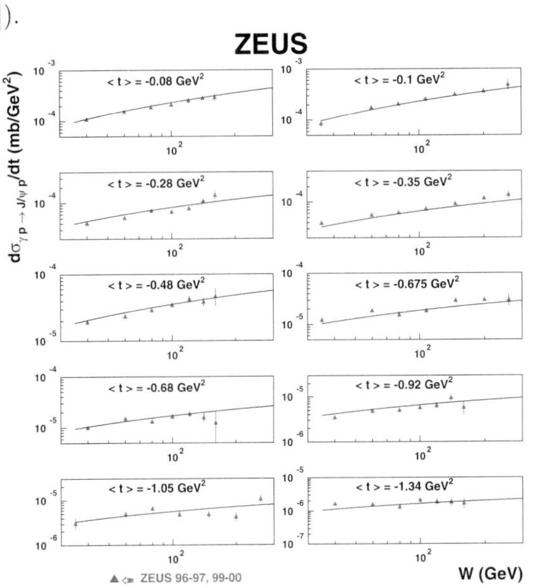


**Fig.3** Differential cross-section of $J/\psi$ photoproduction in the SDP model as function of $W^2$.

## 4. Conclusion

SDP model well describes the available data on vector meson photoproduction at all energies and transferred momenta without contribution of hard Pomeron (with $\alpha_{\mathcal{P}}(0) > 1$). The integrated cross-sections are rising because Pomeron is a double $j$-pole in the given model.

The new ZEUS data [1] quite definitely point towards the nonlinear Pomeron trajectory.

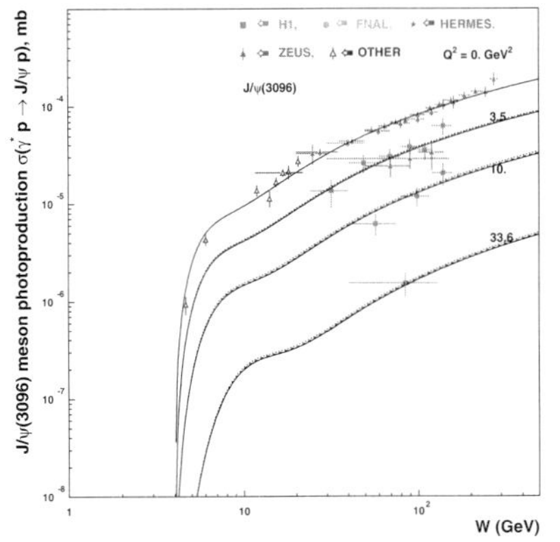


**Fig.4** Cross-section of $J/\psi$ photoproduction in the SDP model at various $Q^2$.

## REFERENCES

1. ZEUS Collaboration: S. Chekanov *et al.*, Euro Phys. J 24 (2002) 345.
2. P. Desgrolard, M. Giffon, A. Lengyel and E.S. Martynov, Nuovo Cim. 107 (1994) 637; P. Desgrolard and E. Martynov, Euro Phys. J 22 (2001) 479.
3. For a recent update on the Regge theory see: V. Barone and E. Predazzi *"High-Energy Particle Diffraction"*, Springer-Verlag Berlin Heidelberg (2002).
4. J. Nemchik, N.N. Nikolaev, E. Predazzi and B.G. Zakharov, Zeit. Phys. C 75 (1997) 71.
5. E. Martynov, E. Predazzi and A. Prokudin, hep-ph/0207272.
6. J.R. Cudell *et al.*, COMPETE Collaboration, Phys. Rev. D 65 (2002) 074024.
7. L.L. Jenkovszky, E.S. Martynov and F. Paccanoni, hep-ph/9608384; R. Fiore, L.L. Jenkovszky, F. Paccanoni and A. Papa, hep-ph/0110405; E. Martynov, E. Predazzi and A. Prokudin, hep-ph/0112242.
8. A. Donnachie and P. V. Landshoff, Phys. Lett. B 185 (1987) 403; ibid, 348 (1995) 213.

ELSEVIER

Proceedings of ICHEP 2002, pp. 446–448
*S. Bentvelsen, P. de Jong, J. Koch and E. Laenen (Editors)*

31st INTERNATIONAL CONFERENCE ON
HIGH ENERGY PHYSICS AMSTERDAM

# The $Q^2$ Dependence of the Generalized GDH Integrals for Proton, Deuteron, and Neutron.

A.P. Nagaitsev
Joint Institute for Nuclear Research,
P.O. 141980, Dubna, Russia

The Gerasimov–Drell–Hearn (GDH) [1] sum rule connects an energy–weighted integral of the difference of the helicity–dependent real–photon absorption cross sections with the anomalous contribution $\kappa$ to the magnetic moment of the nucleon:

$$\int_{\nu_0}^{\infty}[\sigma_{\frac{1}{2}}(\nu)-\sigma_{\frac{3}{2}}(\nu)]\frac{d\nu}{\nu}=-\frac{2\pi^2\alpha}{M^2}\kappa^2, \qquad (1)$$

where $\sigma_{\frac{1}{2}(\frac{3}{2})}$ is the photo-absorption cross section for total helicity of the photon-nucleon system equal to $\frac{1}{2}$ ($\frac{3}{2}$), $\nu$ is the photon energy in the target rest frame, $M$ is the nucleon mass, $\alpha$ the electromagnetic fine–structure constant and $\nu_0$ is the photoabsorption threshold. The sum rule holds for any type of target. It is valid for protons, neutrons or nuclei. The GDH sum rule is derived from the Compton forward–scattering amplitude by applying the general physics principles of Lorentz and gauge invariance and is non–perturbative in nature.

The generalisation of the GDH integral to non–zero photon virtuality $Q^2$ allows one to connect the static ground state properties of the nucleon with its helicity structure as measured in inelastic scattering in the resonance and deep inelastic regions. One such generalisation reads [2,3]:

$$I_{GDH}(Q^2)=\int_{\nu_0}^{\infty}[\sigma_{\uparrow\downarrow}(\nu,Q^2)-\sigma_{\uparrow\uparrow}(\nu,Q^2)]\frac{d\nu}{\nu}=$$

$$\frac{8\pi^2\alpha}{M}\int_0^{x_0}\frac{g_1(x,Q^2)-\gamma^2 g_2(x,Q^2)}{K}\frac{dx}{x}, \qquad (2)$$

where $g_1$ and $g_2$ are the spin–dependent structure functions of the nucleon, $\gamma^2=Q^2/\nu^2$, $x=Q^2/2M\nu$, $x_0=Q^2/2M\nu_0$ and $K$ is the virtual–photon flux factor. The subscript $\uparrow\uparrow$ ($\uparrow\downarrow$) refers to the orientation of the target spin parallel (anti–parallel) to the positron beam polarisation. In the leading–twist approximation Eq. 2 simplifies to $I_{GDH}(Q^2)=\frac{16\pi^2\alpha}{Q^2}\Gamma_1$. The first moment of the the spin structure function $g_1$, $\Gamma_1=\int_0^1 g_1(x)dx$, is predicted to have at large $Q^2$ only a logarithmic $Q^2$ dependence from QCD evolution.

Examining the generalised GDH integral provides a way to study the transition from polarised real–photon absorption ($Q^2=0$) on the nucleon to polarised deep inelastic lepton scattering (DIS). It constitutes an observable that allows the study of the transition from the non–perturbative regime at low $Q^2$ to the perturbative regime at high $Q^2$ .

The HERMES polarised data taken with hydrogen and deuterium targets in 1997-2000 were used to extract the $Q^2$ dependence of the generalised GDH integrals for proton, deuteron and neutron. The $Q^2$-range 1.2 GeV$^2 < Q^2 <$ 12.0 GeV$^2$ was divided into six bins. The full range in photon–nucleon invariant mass squared $W^2$ (1.0 GeV$^2 < W^2 < 45$ GeV$^2$) was separated into a resonance region (1.0 GeV$^2 < W^2 < 4.2$ GeV$^2$) and a DIS region (4.2 GeV$^2 < W^2 < 45.0$ GeV$^2$).

The longitudinal cross–section asymmetry $A_1$ for the absorption of virtual photons was calculated as

$$A_1=\frac{A_{\|}}{D}-\eta A_2, A_{\|}=\frac{N^{\uparrow\downarrow}L^{\uparrow\uparrow}-N^{\uparrow\uparrow}L^{\uparrow\downarrow}}{N^{\uparrow\downarrow}L_P^{\uparrow\uparrow}+N^{\uparrow\uparrow}L_P^{\uparrow\downarrow}}. \qquad (3)$$

$N$ is the number of detected scattered positrons, $L$ is the integrated luminosity corrected for dead time and $L_p$ is the integrated luminosity corrected for dead time and weighted by the prod-


doi:10.1016/S0920-5632(02)02135-7

uct of the beam and target polarisations, $\eta$ is the kinematic factor, $D$ is the photon depolarisation factor. $A_2$ is related to longitudinal–transverse photon–nucleon interference. In the DIS region $A_2$ can be parametrised as $A_2 = cMx/\sqrt{Q^2}$, where $c = 0.05$ (0.53) for the deuteron (proton). In the resonance region no data are available for the deuteron and $A_2 = 0$ was chosen, while for the proton a constant value of $A_2 = 0.06$ at $Q^2 = 3$ GeV$^2$ was adopted [4]. Radiative effects for both targets were calculated using the codes described in Ref. [5]. They were found not to exceed 7% (4%) of the asymmetry $A_1$ for the deuteron (proton). The generalised GDH integral Eq. (2) can be re-written in terms of the longitudinal virtual-photon asymmetry $A_1$ and the unpolarised structure function $F_1$:

$$I_{GDH}(Q^2) = \frac{8\pi^2\alpha}{M}\int_0^{x_0} \frac{A_1(x,Q^2)F_1(x,Q^2)}{K}\frac{dx}{x}, \quad (4)$$

where $K$ is the virtual-photon flux factor chosen in the Gilman convention [6]: $K = \nu\sqrt{1+\gamma^2}$.

Following the formalism described in Ref. [7], the GDH integral $I^n_{GDH}$ for the neutron was calculated from the GDH results on the deuteron $I^d_{GDH}$, as obtained in this analysis, and those on the proton $I^p_{GDH}$, reanalysed following the procedure detailed in Ref. [8]:

$$I^n_{GDH} = \frac{I^d_{GDH}}{1-1.5\omega_d} - I^p_{GDH}. \quad (5)$$

Here $\omega_d = 0.058 \pm 0.010$ [9] is the probability of the deuteron to be in a D-state. It has been shown in Ref. [7] that although the uncertainties in the structure functions in the integrand of Eq. (4) may be large, the resulting uncertainty for the integral $I^n_{GDH}$ does not exceed 3%.

Figure 1 presents the Q$^2$ dependence of the generalised GDH integrals obtained in the resonance region. The results is compared with predictions of the Aznauryan model [10] based on a Q$^2$ evolution of nucleon resonance amplitudes. The resonance part of the GDH integral decreases with increasing Q$^2$ and becomes small for Q$^2$ > 3 GeV$^2$. The model describes the neutron data well over the whole $Q^2$ region and also the proton and deuteron data for Q$^2$ > 2.5 GeV$^2$, but it underestimates these data for lower Q$^2$.

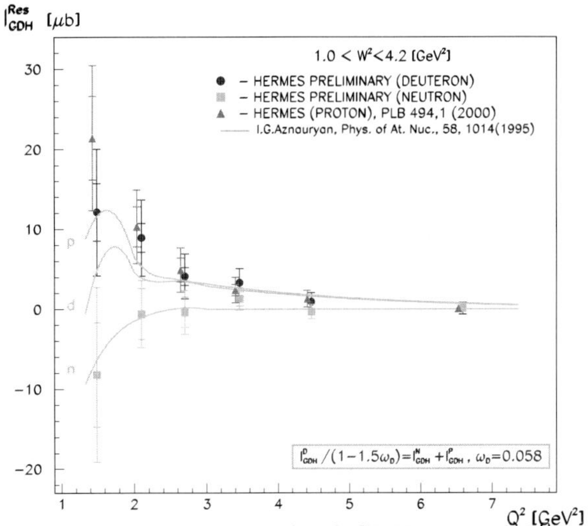


Figure 1. The generalised GDH integral $I^{Res}_{GDH}$ for the proton, neutron and deuteron, shown as a function of $Q^2$ for the resonance region. The inner(outer) error bars show the statistical (sum of statistical and systematic errors) uncertainties. The curves are from Ref. [10].

Figure 2 shows the Q$^2$ dependence obtained for the full generalised GDH integrals. The curves are predictions of the Soffer-Teryaev model [11] based on $Q^2$ evolution of $g_1$ and $g_2$. The proton data are very well described by this model, the deuteron data agree with the prediction within the total uncertainies, the neutron data are slightly underestimated, but still in agreement within experimental uncertainies.

The $Q^2$-behaviour of $I_{GDH}$ for proton and neutron can be more clearly studied when the GDH integral is multiplied by $Q^2/(16\pi^2\alpha)$, to test the Q$^2$ dependence expected from leading twist, as shown in Figure 3. Any contributions from resonance form factors or higher–twist contributions should become visible as a deviation from a constant line. Within the statistical and systematic uncertainties of the present measurement no deviation from a leading–twist behaviour is seen.

In summary the Generalised Gerasimov–Drell–Hearn integrals for proton, deuteron and neutron have been measured for the first time in both the resonance and DIS regions for the Q$^2$ range from

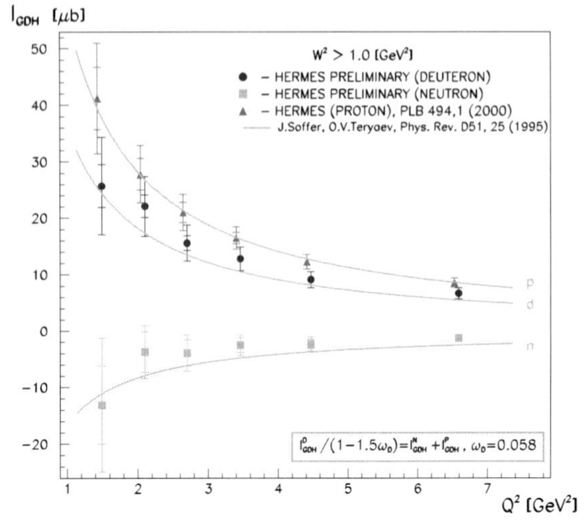


Figure 2. The full generalised GDH integral $I^d_{GDH}$ for the proton, neutron and deuteron, shown as a function of $Q^2$. The inner (outer) error bars show the statistical (sum of statistical and systematic errors) uncertainties. The curves are from Ref. [11].

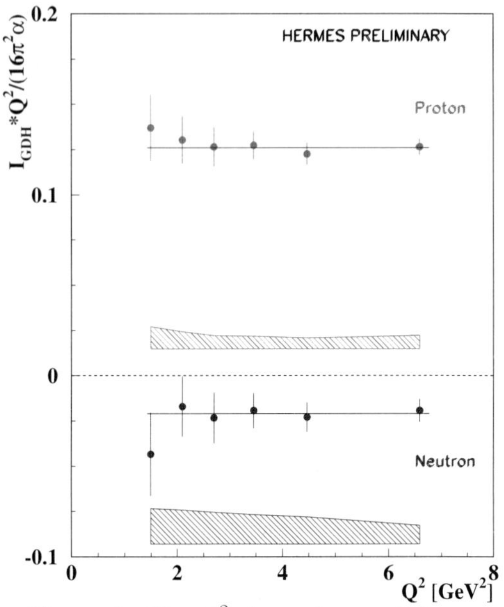


Figure 3. The $Q^2$–dependence of the generalised GDH integrals for the proton (top) and neutron(bottom) after the leading–twist dependence, $(16\pi^2\alpha)/Q^2$, has been divided out. The error bars represent the statistical errors. The systematic uncertainties are represented by the respective error bands. The lines are straight line fits to the data.

1.2 to 12.0 GeV$^2$. The contribution to the GDH integral from the resonance region for neutron is in agreement with predictions of the Aznauryan model [10] based on the $Q^2$ evolution of nucleon resonance helicity amplitudes. For proton and deuteron the model underestimates the data for $Q^2$ <2.5 GeV$^2$. Above $Q^2$=3 GeV$^2$ the DIS contributions to the generalised GDH integrals are dominant for all targets. The generalised GDH integral for the proton is very well described by the Soffer-Teryaev prediction and for deuteron and neutron it is in agreement within the uncertainties of the measurements. Within the statistical and systematic uncertainties the data show no indication for large non-leading twist effects.

## REFERENCES

1. S.B. Gerasimov, Sov. J. Nucl. Phys. **2** (1966) 430; S.D. Drell and A.C. Hearn, Phys. Rev. Lett. **16** (1966) 908.
2. R. Pantförder, PhD Thesis, Universität Bonn (1998), BONN-IR-98-06, hep-ph/9805434 and reference therein.
3. D. Drechsel, S.S. Kamalov and L. Tiator, Phys. Rev. **D 63** (2001) 114010.
4. E143 Collaboration, K. Abe *et al.*, Phys. Rev. **D 58** (1998) 112003.
5. I.V. Akushevich and N.M. Shumeiko, J. Phys. **G 20** (1994) 513; I. Akushevich *et al.*, Comput. Phys. Commun. **104** (1997) 201.
6. F.J. Gilman, Phys. Rev. **167** (1968) 1365.
7. C. Ciofi degli Atti *et al.*, Phys. Lett. **B 376** (1996) 309.
8. HERMES Collaboration, A. Airapetian *et al.*, Phys. Lett. **B 494** (2000) 1.
9. M. Lacombe *et al.*, Phys. Lett. **B 101** (1981) 139.
10. I. G. Aznauryan, Phys. of At. Nucl. **58** (1995) 1014 and private communication.
11. J. Soffer and O.V. Teryaev, Phys. Rev. **D 51** (1995) 25; Phys. Rev. Lett. **70** (1993) 3373.

Proceedings of ICHEP 2002, pp. 449–451
*S. Bentvelsen, P. de Jong, J. Koch and E. Laenen (Editors)*

# Detect Gluon Polarisation at BNL-RHIC Through $J/\psi$ Pair Production

Jiro Kodaira[a*] and Cong-feng Qiao[a†]

[a]Department of Physics, Hiroshima University Higashi-Hiroshima 739-8526, Japan

Our recent study shows that the double spin asymmetry of $J/\psi$ pair production is measurable in polarised $p-p$ collision at RHIC in near future. And hence it enables us to extract the polarised gluon distribution function $\Delta G(x)$.

Intrigued by the measurement of the European Muon Collaboration in polarised-target experiment [1], an enormous amount of researches have been carried out on the nucleon spin structure. The unpolarised deep inelastic scattering (DIS) experiments indicate clearly that gluons share a large portion of the parent proton's momentum. However, what portion of proton spin is carried by gluons is still an open question. Although there are some efforts [2] to parameterise the polarised gluon distributions, it is still quite unclear which of these parameterisations represents the reality of nature correctly. To explore this issue, commonly believe that the study of processes other than DIS is necessary. It is now expected that the polarised proton-proton collisions at BNL relativistic heavy-ion collider(RHIC) will provide copious experimental data to unveil the polarised parton distributions.

Up to now, three main kinds of schemes in measuring the gluon polarisation are proposed, which are thought to be feasible at RHIC technically. Those are: **P**rompt Photon Production, **J**et production, and **H**eavy Flavor Production. There are advantages and disadvantages in each of these scenarios. For detailed discussions on this issue, see recent review paper of Ref. [3].

Quarkonium production and decays have long been taken as an ideal means to investigate the nature of QCD and other phenomena. Due to the approximately non-relativistic nature, the description of heavy quark and antiquark system stands as one of the simplest applications of QCD. The very clean signals of quarkonium leptonic decays enable the experimental detection with a high precision, and therefore quarknoium plays a unique role in investigating other phenomena.

Nevertheless, theoretical description for quarkonium production is still premature, although the heavy quarkonium physics has been investigated for more than twenty years. In explaining the high-$p_T$ $J/\psi$ surplus production discovered by CDF group [4] at the Fermilab Tevatron, the color-octet scenario [5] was proposed based on a novel effective theory, the nonrelativistic QCD(NRQCD) [6]. Having achieved the first-step success in explaining the CDF data, however, color-octet mechanism also encounters difficulties in confronting with other phenomena [7]. Due to a recent discovery [8], the extent of importance of color-octet contributions in Charmonium production remains to be unfixed.

During the past years, a series of efforts have been made on detecting polarised parton distributions through quarkonium production processes [9–13]. Most of these investigations are not directly applicable to RHIC physics, and they are spoiled by a large amount of uncertainties. In recently, it is found [14] that double heavy quarkonium production in polarised proton-proton collision provides an ideal means to detect the polarised gluon distributions at RHIC, and which may at least play a supplemental role to the presently proposed program in this end. In literatures [15,16], there were discussions of detecting the gluon polarisation by means of double $J/\psi$ production, numerically, whereas with main emphasis not on RHIC physics.

*E-mail: kodaira@theo.phys.sci.hiroshima-u.ac.jp
†E-mail: qiao@theo.phys.sci.hiroshima-u.ac.jp


doi:10.1016/S0920-5632(02)02138-2

The double quarkonium production has everal advantages in reducing the theoretical uncertainties in investigating the gluon polarisation. (1) By considering double production, the relativistic corrections and color-octet uncertainties are suppressed, especially without or with a lower transverse momentum cut [17]. (2) The total contribution from the higher excited states are also doubly suppressed. (3)The higher order QCD correction can be properly controlled by applying a suitable $p_T$ cut for the Charmonium system. (4) Since the prevailing partonic process is the gluon-gluon fusion into double quarkonium, it stands as a very sensitive method in measuring the gluon polarisation. Note that at RHIC, the $q\bar{q}$ initiated process is negligible.

When calculating the helicity-dependent matrix elements, we we take the ghost free expression for the gluon state with helicity $\lambda = \pm$ ,

$$\epsilon_\mu(k_1, \lambda)\, \epsilon^\star_\nu(k_1, \lambda) = \frac{1}{2}\left[-g_{\mu\nu} + \frac{k_{1\mu}k_{2\nu} + k_{1\nu}k_{2\mu}}{k_1 \cdot k_2} + i\lambda\, \epsilon_{\mu\nu\alpha\beta}\frac{k_{1\alpha}k_{2\beta}}{k_1 \cdot k_2}\right] . \quad (1)$$

As usual, the measurable double spin asymmetry $A$, for $J/\psi$ pair production, is defined as

$$A = \frac{d\sigma(p_+p_+ \to J/\psi J/\psi) - d\sigma(p_+p_- \to J/\psi J/\psi)}{d\sigma(p_+p_+ \to J/\psi J/\psi) + d\sigma(p_+p_- \to J/\psi J/\psi)} = \frac{Ed\Delta\sigma/d^3p}{Ed\sigma/d^3p} , \quad (2)$$

where $p_+$ and $p_-$ denote the helicity projection of the incident protons being positive and negative, respectively. In terms of gluon densities and partonic cross sections, this asymmetry reads as

$$A = \frac{\int dx_1 dx_2 d\Delta\hat{\sigma}\Delta G(x_1, Q^2)\Delta G(x_2, Q^2)}{\int dx_1 dx_2 d\hat{\sigma} G(x_1, Q^2) G(x_2, Q^2)} , \quad (3)$$

where $\Delta G(x, Q^2) = G_+(x, Q^2) - G_-(x, Q^2)$ and $G(x, Q^2) = G_+(x, Q^2) + G_-(x, Q^2)$ are the polarised and unpolarised gluon distributions defined at the scale $Q^2$. The unpolarised (polarised) partonic cross section $\hat{\sigma}$ ($\Delta\hat{\sigma}$) is defined as

$$\hat{\sigma} = \frac{1}{4}\sum_{\lambda,\lambda'} \hat{\sigma}\,(\lambda,\, \lambda'),\ \Delta\hat{\sigma} = \frac{1}{4}\sum_{\lambda,\lambda'} \lambda\lambda'\, \hat{\sigma}\,(\lambda,\, \lambda').(4)$$

The expressions of corresponding partonic differential cross sections can be found in Refs. [14,17,18].

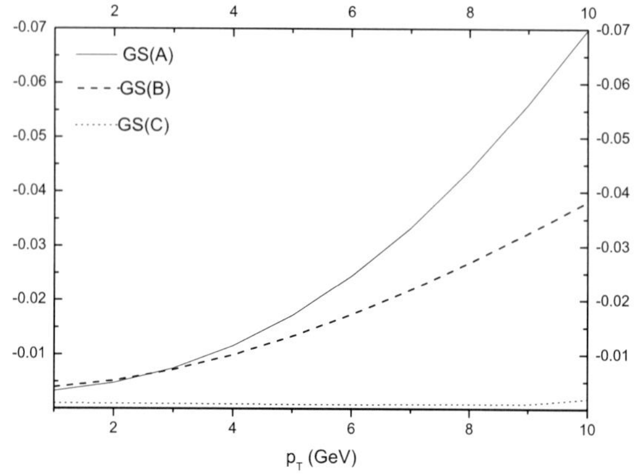


Figure 1. Spin asymmetry versus transverse momentum for different sets of GS polarised parton distribution at colliding energy $\sqrt{s} = 500$ GeV.

In doing numerical calculation, the scale $Q^2$ of the parton distribution function and the strong coupling constant is taken to be the transverse momentum of $J/\psi$ for the $p_T$ distributions. Whereas, in the calculation of the angular distribution of the spin asymmetries and the integrated cross sections, the scale is taken to be $Q^2 = m^2$.

In Figure 1 the double spin asymmetry versus the transverse momentum of $J/\psi$ is presented for different sets of Gehrmann and Stirling(GS) parameterisation [2]. Figure 2 shows the angular distributions of the asymmetry in the parton center-of-mass frame. For consistency the MRST parameterisation [19] for unpolarised gluon distribution is used in drawing these figures.

TABLE I. Total cross sections for $J/\psi$ pair production at RHIC with $\sqrt{s} = 500$ GeV, evaluated with different parton distributions.

| | $\sigma^{tot}_{\mu^+\mu^-}$ | $\sigma_{\mu^+\mu^-}(\lvert p_T\rvert > 1$ GeV$)$ |
|---|---|---|
| CTEQ5L[20] | 11.8 pb | 7.3 pb |
| MRST [19] | 6.5 pb | 4.3 pb |
| GRV [21] | 7.4 pb | 4.7 pb |

The total cross sections with different sets of

parton parameterisation are summarised in table I. The discrepancies among these results are not very large as expected. Here, the notation $\sigma_{\mu^+\mu^-}$ means that the branching ratio of $B(\psi \to \mu^+\mu^-) = 0.0588$, as the practical measuring mode to reconstruct the charmonium state, is included. From the predicted cross sections, we see that with the integrated luminosity of 800 $pb^{-1}$ in the future run of RHIC, there will be thousands of $J/\psi$ pair events to be detected, which can certainly give us some information on the gluon polarisation within the hadron.

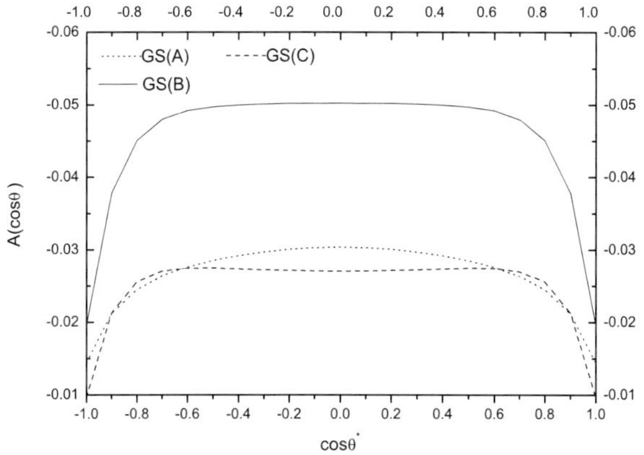


Figure 2. Angular differential asymmetry distribution of the $J/\psi$ pair in the parton centre-of mass frame at RHIC with colliding energy $\sqrt{s} = 500$ GeV.

In summary, it is found that the $J/\psi$ pair production at RHIC may stand as an independent means of observing the gluon spin distributions within the hadron. The asymmetry, rather than cross sections, eliminates large amount of uncertainties which come from the non-perturbative hadronization. In the future run of RHIC, that is to say with colliding energy of 500 GeV and accumulated luminosity 800 $pb^{-1}$, thousands of $J/\psi$ pair events could be detected and the gluon polarisation could be measured. With the expected upgrade of RHIC in future, the concerned process would show up to be more important in chasing the goal of uncovering the nucleon spin structures.

The work of J.K. was supported in part by the Monbu-kagaku-sho Grant-in-Aid for Scientific Research No.C-13640289. The work of C-F.Q. was supported by the Grant-in-Aid of JSPS.

## REFERENCES

1. European Muon Collab., J. Ashman *et al.*, Phys. Lett. **B206**, 364 (1988).
2. T. Gehrmann and W.J. Stirling, Phys. Rev. D53, 6100 (1996); M. Glück, E. Reya and W. Vogelsang, Phys. Lett. **B359**, 201 (1995).
3. G. Bunce, N. Saito, J. Soffer and W. Vogelsang, Ann. Rev. Nucl. Part. Sci. **50**, 525 (2000).
4. CDF Collaboration, F. Abe *et al.*, Phys. Rev. Lett. **69**, 3704 (1992); CDF Collaboration, F. Abe *et al.*, Phys. Rev. Lett. **79**, 572 (1997); CDF Collaboration, F. Abe *et al.*, Phys. Rev. Lett. **79**, 578 (1997).
5. E. Braaten and S. Fleming, Phys. Rev. Lett. **74**, 3327 (1995).
6. G.T. Bodwin, E. Braaten, and G.P. Lepage, Phys. Rev. D**51**, 1125 (1995); **55**, 5853(E) (1997).
7. For recent reviews see: e.g., M. Krämer, Hep-ph/0106120.
8. Cong-Feng Qiao, Hep-ph/0202227.
9. A.V. Batunin and S.R. Slabospitsky, Phys. Lett. **B188**, 269 (1987);
10. J.L. Cortes and B. Pire, Phys. Rev. D**38**, 3586 (1988).
11. M.A. Doncheski and R.W. Robinett, Phys. Lett. **B248**, 188 (1990).
12. T. Morii, S. Tanaka, and T. Yamanishi, Phys. Lett. **B372**, 165 (1996)
13. R.L. Jaffe and D. Kharzeev, Phys. Lett. **B455**, 306 (1999).
14. J. Kodaira and C.-F. Qiao, Hep-ph/0207318.
15. S.P. Baranov and H. Jung, Z. Phys. C**66**, 467 (1995).
16. T. Gehrmann, Phys. Rev. D**53**, 5310 (1996).
17. Cong-Feng Qiao, Phys. Rev. D**66** (2002) 057504.
18. B. Humbert and P. Mery, Z. Phys. C**20**, 83 (1983); Phys. Lett. **B124**, 256 (1983).
19. A.D. Martin, R.G. Roberts, W.J. Stirling and R.S Thorne, Eur. Phys. J. C**14**, 133 (2000).
20. CTEQ Collaboration, H.L. Lai *et al.*, Eur. Phys. J. C12, 375 (2000).
21. M. Glück, E. Reya and A. Vogt, Z. Phys. C**67**, 433 (1995).

Proceedings of ICHEP 2002, pp. 452–454
*S. Bentvelsen, P. de Jong, J. Koch and E. Laenen (Editors)*

# Transversity and Meson Photoproduction

Gary R. Goldstein[a] * Leonard Gamberg[b]

[a]Department of Physics and Astronomy, Tufts University, Medford, MA 02155, USA

[b]Physics Department, Penn State University, Berks-Lehigh Valley College, Reading, PA 19610, USA

Both meson photoproduction and semi-inclusive deep inelastic scattering can potentially probe transversity in the nucleon. We explore how that potential can be realized dynamically. The role of rescattering in both exclusive and inclusive meson production as a source for transverse polarization asymmetry is examined. We use a dynamical model to calculate the asymmetry and relate that to the transversity distribution of the nucleon.

The leading twist transversity distribution $h_1(x)$ [1] and its first moment, the tensor charge, are as fundamental to our understanding of the spin structure of the nucleon as are the helicity distribution and the axial vector charge. Unlike $g_1(x)$, though, the chiral odd $h_1(x)$ cannot be accessed in deep inelastic scattering. However, $h_1(x)$ can be probed when at least two hadrons are present, *e.g.* Drell Yan [2] or semi-inclusive deep inelastic scattering (SIDIS). In the latter process at leading twist, the effect of quark transversity can be measured via the azimuthal asymmetry in the fragmenting hadron's momentum and spin distributions. For spinless hadrons the so-called Collins asymmetry [3] depends on the transverse momentum of quark distributions in the target and fragmentation functions for outgoing hadrons [4]. Including transverse momentum leads to an increase in the number of leading twist distribution and fragmentation functions and can involve $T$-odd quark functions [5].

Non-zero transverse single spin asymmetries (SSA) have been measured at HERMES and SMC in semi-inclusive pion electroproduction [6]. These data could point to the essential role played by quark transverse momenta and $T$-odd distributions. Recently, further insight into transversity has come from the interpretation of deeply virtual Compton scattering (DVCS) [7] where the quark target helicity flip amplitudes, written in terms of the generalized parton distributions (GPD) $H_T^a(x,\xi,t)$, reduce in the forward limit to the ordinary transversity distribution, $\delta q^a(x)$. Angular momentum conservation in these amplitudes requires that helicity changes are accompanied by a transfer of 1 or 2 units of orbital angular momentum, highlighting the essential role played by the $k_\perp$ generalizations of the quark transversity distribution. The $t \to 0$ limit of the associated form factor is the tensor charge.

This interdependence of transversity on quark *orbital* angular momentum and $k_\perp$ is more general than the GPD analysis of transversity. This behavior arises in ref. [8] where we study the vertex function associated with the tensor charge. Again, angular momentum conservation results in the transfer of orbital angular momentum $\ell = 1$ carried by the dominant $J^{PC} = 1^{+-}$ mesons to compensate for the non-conservation of helicity across the vertex. Transverse momentum dependence arises from the axial vector mesons that dominate the tensor coupling. [2] These mesons are in the $(35 \otimes \ell = 1)$ multiplet of the $SU(6) \otimes O(3)$ symmetry group that best represents the mass symmetry among the low lying mesons. Along with axial vector dominance this symmetry results in the isoscalar and isovector contribution to the tensor charge

$$\delta u - \delta d = \frac{5}{6}\frac{g_A}{g_V}\frac{M_{a_1}^2}{M_{b_1}^2}\frac{\langle k_\perp^2\rangle}{M_N M_{b_1}},\ \delta u + \delta d = \frac{3}{5}\frac{M_{b_1}^2}{M_{h_1}^2}\delta q^v.$$

Each depends on two powers of the average intrinsic quark momentum $\langle k_\perp^2\rangle$, because the tensor

*Research suported in part by a grant from the US Department of Energy DE-FG02-29ER40702.

[2] $C$-odd mesons – $h_1(1170)$, $h_1(1380)$, $b_1(1235)$


doi:10.1016/S0920-5632(02)02139-4

couplings involve helicty flips that require kinematic factors of 3-momentum transfer.

The $k_\perp$ dependence can be understood on fairly general grounds from the kinematics of the exchange picture in exclusive pseudoscalar meson photoproduction. For large $s$ and relatively small momentum transfer $t$ simple combinations of the four helicity amplitudes involve definite parity exchanges. The four independent helicity amplitudes can have the minimum kinematically allowed powers,

$$f_1 = f_{1+,0+} \propto k_\perp^1,\ f_2 = f_{1+,0-} \propto k_\perp^0,$$
$$f_3 = f_{1-,0+} \propto k_\perp^2,\ f_4 = f_{1-,0-} \propto k_\perp^1.$$

However, in single hadron exchange (or Regge pole exchange) parity conservation requires

$$f_1 = \pm f_4 \quad \text{and} \quad f_2 = \mp f_3$$

for even/odd parity exchanges. These pair relations, along with a single hadron exchange model, force $f_2$ to behave like $f_3$ for small $t$. This introduces the $k_\perp^2$ factor into $f_2$. However for a nonzero polarized target asymmetry to arise there must be interference between single helicity flip and non-flip and/or double flip amplitudes. Thus this asymmetry must arise from rescattering corrections (or Regge cuts-eikonalization or loop corrections ) to single hadron exchanges. That is, one of the amplitudes in

$$P_y = \frac{2Im(f_1^* f_3 - f_4^* f_2)}{\sum_{j=1\ldots4} |f_j|^2}$$

must acquire a different phase. Rescattering reinstates $f_2 \propto k_\perp^0$ by integrating over loop $k_\perp$, which effectively introduces a $\langle k_\perp^2 \rangle$ factor [9]. This is true for the *inclusive process* as well, where only one final hadron is measured; a relative phase in a helicity flip three body amplitude is required.

Recently a rescattering approach was applied to the calculation of SSA in pion electroproduction, $ep \to e\pi X$, using a QCD motivated quark-diquark model of the nucleon [10] (BHS). In Ref. [11,12] the rescattering effect is interpreted as giving rise to the $T$-odd Sivers $f_{1T}^\perp$ function; the number density of upolarized quarks in a transversly polarized target. This function arises at leading twist in the SSA [13] in conjunction with the $T$-even unpolarized fragmentation function. Being $T$-odd, this asymmetry vanishes at tree level. The important lesson beyond the model calculation, is that, theoretically, final state interactions are essential for producing nonzero SSA's. Furthermore, the phenomenological determination of quark spin distributions can be disentangled from measurements of SSA's.

We have extended this approach to calculate the transversity distributions and corresponding SSA in SIDIS to access *transversity*. Collins [3] considered one such process, the production of pions from transversly polarized quarks in a transversely polarized target. The corresponding SSA involves the convolution of the transversity distribution function and the $T$-odd fragmentation function, $h_1(x) \star H_1^\perp(z)$ [13,14]. The analogous transversity distribution function can be defined through the light-cone quark distribution with gauge link indicated,

$$s_T^i \Delta f_T(x, k_\perp) = \frac{1}{2}\sum_n \int \frac{d\xi^- d^2\xi_\perp}{(2\pi)^3} e^{-i(\xi^- k^+ - \vec{\xi}_\perp \vec{k}_\perp)}$$
$$\langle P|\overline{\psi}(\xi^-, \xi_\perp)|n\rangle\langle n| \left(-ie_1 \int_0^\infty A^+(\xi^-, 0)d\xi^-\right)\gamma^+$$
$$\gamma^i\gamma^5\psi(0)|P\rangle + \text{h.c.},$$

where $e_1$ is the charge of the struck quark and $n$ represents intermediate diquark states. In the quark-diquark model the final state interaction contribution can be evaluated by integrating over $q^\mu$, the gluon momentum (similar to the calculation in refs. [10–12]). We obtain

$$s_T^i \Delta f_T(x, k_\perp) = \frac{e_1 e_2 g^2}{2(2\pi)^4} \frac{1-x}{\Lambda(k_\perp^2)}$$
$$\times \left\{ \left( S_T^i \left[ \left( m + xM \right)^2 + k_\perp^2 \right] + 2k_\perp^i \mathbf{S}_T \cdot \mathbf{k}_\perp \right) \right.$$
$$\times \frac{1}{k_\perp^2 + \Lambda(0)^2 + \lambda_g^2} \left( \ln\frac{\Lambda(k_\perp^2)}{\Lambda(0)} + \ln\frac{k_\perp^2 + \lambda_g^2}{\lambda_g^2} \right)$$
$$\left. - \left( S_T^i k_\perp^2 + 2k_\perp^i \mathbf{S}_T \cdot \mathbf{k}_\perp \right) \frac{1}{k_\perp^2} \ln\frac{\Lambda(k_\perp^2)}{\Lambda(0)} \right\},$$

$$\Lambda(k_\perp^2) = \mathbf{k}_\perp^2 + x(1-x)\left(-M^2 + \frac{m^2}{x} + \frac{\lambda^2}{1-x}\right).$$

The (Abelian) gluon mass (usually chosen at $\lambda_g \approx 1\ GeV$) is indicative of $\chi SB$ scale and appears here to regulate the IR divergence.

The first part has the same nucleon spin dependent structure as a tree level model calculation [15] - it is leading twist and a combination of $h_{1T}(x,k_\perp)$ and $h_{1T}^\perp(x,k_\perp)$. The second part has a different structure than tree level - it appears as a rescattering effect only. It is IR finite and, in this model, is proportional to the one loop result for $f_{1T}^\perp$ [11] and $\mathcal{P}_y$ in BHS. The ratio of $h_{1T}(x,k_\perp)$ to $h_{1T}^\perp(x,k_\perp)$ will differ from the tree level. Integrating over $k_\perp$ leaves $h_1(x)$. This one loop contribution constitutes the next order term in an eikonalization.

When combined with a measure of transversely polarized quarks, the fragmentation function $H_1^\perp(z)$, the integrated $h_1(x)$ ($h_{1T}(x)$ and the first moment of $h_{1T}^\perp(x)$) will contribute to the observable weighted meson azimuthal asymmetry from a transversely polarized nucleon [13,14]. Weighting by powers of $k_\perp$ gives asymmetries in $\sin(n\phi_{meson})$.

The $T$-odd structure function $h_1^\perp(x,k_\perp)$ is of more interest both theoretically, since it vanishes at tree level, and experimentally, since its determination does not necessarily involve polarized nucleons [13]. Repeating the calculation above *without nucleon polarization* leads to the result

$$h_1^\perp(x,k_\perp) = \frac{e_1e_2g^2}{2(2\pi)^4}\frac{(m+xM)(1-x)}{\Lambda(k_\perp^2)} \times \varepsilon_{+-\perp j}k_{\perp j}\frac{1}{k_\perp^2}\ln\frac{\Lambda(k_\perp^2)}{\Lambda(0)}.$$

This is again proportional to the $f_{1T}^\perp$ result. It is a leading twist, IR finite result. Being $T$-odd it will appear in SIDIS observables along with $T$-odd fragmentation functions. Many examples of such observables have been proposed, including SSA's, angular distributions of the final hadron and its polarization [13,16,17].

In summary, the Spin-flavor symmetry relates tensor charges to axial charges when supplemented with axial vector dominance. Secondly, axial vector dominance produces a $\langle k_\perp^2\rangle$ factor that appears in rescattering models in meson photoproduction. Particularly, transversely polarized nucleon asymmetries in exclusive and inclusive $\pi$ and $\eta$ photoproduction are interference phenomena that require rescattering to be nonzero. This is true in SIDIS as well. The exchange picture for photproduction merges with the struck quark perspective when rescattering is effective.

The spectator model provides a testing ground for these notions and yields simple relations. Are they too simple? We are exploring several related issues. How to specify asymmetries precisely that will focus on the interesting distribution functions? What are the $k_\perp$ dependences from different models? Are there more realistic intermediate states (that can carry spin information)?

Acknowledgement: We appreciate communications with S. Brodsky and D.S. Hwang.

## REFERENCES


1. X. Artu and M. Mekhfi, Z. Phys. C45 (1990) 669; R. L. Jaffe and X. Ji, Phys. Rev. Lett. 67 (1991) 552; Nucl. Phys. B375 (1992) 527.
2. J. Ralston and D. E. Soper, Nucl. Phys. B152 (1979) 109.
3. J. C. Collins, Nucl. Phys. B394 (1993) 169.
4. A. M. Kotzinian, Nucl. Phys B441 (1995) 234.
5. R. D. Tangerman and P. J. Mulders, Phys. Lett. B352 (1995) 129; Phys. Rev. D51 (1995) 3357; Nucl. Phys. B461 (1996) 197.
6. A. Airapetian *et al.*, Phys. Rev. Lett. 84 (2000) 4047 (2000); A. Bravar, Nucl. Phys. Proc. Suppl. 79 (1999) 520.
7. P. Hoodbhoy and X. Ji, Phys. Rev.D58 (1998), 054006; M. Diehl, Eur. Phys. J. C19 (2001) 485.
8. L. Gamberg and G. R. Goldstein, Phys. Rev. Lett. 87 (2001) 242001.
9. G. R. Goldstein and J. F. Owens, Phys. Rev. D7 (1973) 865; Nucl. Phys. B71 (1974) 461.
10. S. Brodsky *et al.*, Phys. Lett. B530 (2002) 99.
11. X. Ji and F. Yuan, hep-ph/0206057, 2002; A.V. Belitsky, *et al.*, hep-ph/0208038.
12. J. C. Collins, Phys. Lett. B536 (2002) 43.
13. D. Boer and P. J. Mulders, Phys. Rev. D57 (1998) 5780.
14. A. M. Kotzinian and P. J. Mulders, Phys. Lett. B406 (1997) 373.
15. R. Jakob, P.J. Mulders and J. Rodriques, Nucl. Phys. A626 (1997) 937.
16. L. Gamberg and G. R. Goldstein, in preparation.
17. D. Boer, R. Jakob and P.J. Mulders, Nucl. Phys. 564 (2000) 471; M. Radici, A. Bianconi and R. Jakob, Phys. Rev. D65 (2002) 074037.

## Session 7

# CP Violation and the CKM Matrix

*Session conveners:*
*P. Kluit, NIKHEF*
*M. Ciuchini, University of Roma III/INFN*

ELSEVIER

Proceedings of ICHEP 2002, pp. 457–461
*S. Bentvelsen, P. de Jong, J. Koch and E. Laenen (Editors)*

31st INTERNATIONAL CONFERENCE ON HIGH ENERGY PHYSICS AMSTERDAM

# The final measurement of $\epsilon'/\epsilon$ by NA48

G. Unal[a] *

[a]Laboratoire de l'Accélérateur Linéaire
IN2P3-CNRS et Université Paris-Sud,
B.P.34, 91898 Orsay Cedex, France

The direct CP violation parameter $\text{Re}(\epsilon'/\epsilon)$ has been measured from the decay rates of neutral kaons into two pions using the NA48 detector at the CERN SPS. The 2001 running period was devoted to collecting additional data under varied conditions compared to earlier years (1997-99). The 2001 data yield the result: $\text{Re}(\epsilon/\epsilon') = (13.7 \pm 3.1) \times 10^{-4}$. Combining this result with that published from the 1997,98 and 99 data, an overall value of $\text{Re}(\epsilon'/\epsilon) = (14.7 \pm 2.2) \times 10^{-4}$ is obtained from the NA48 experiment.

## 1. INTRODUCTION

### 1.1. CP violation in the neutral kaon system

CP violation has been discovered in the neutral kaon system in 1964 [1]. The main component of the effect [2] occurs in the mixing between $K_0$ and $\overline{K_0}$. The physical states $K_S$ and $K_L$ deviate from pure CP = $\pm$ 1 eigenstates, with the mixing described by the parameter $\epsilon$. Direct CP violation can occur in kaon decays to two pions through the interference of amplitudes with different isospins [3]. This is described by the parameter $\epsilon'$. The quantity which can be measured experimentally is the double ratio R of the decay widths:

$$\begin{aligned} R &= \frac{\Gamma(K_L \to \pi^0\pi^0)/\Gamma(K_S \to \pi^0\pi^0)}{\Gamma(K_L \to \pi^+\pi^-)/\Gamma(K_S \to \pi^+\pi^-)} \\ &\approx 1 - 6 \times Re(\epsilon'/\epsilon) \end{aligned}$$

In the Standard Model, CP violation arises from the irreducible complex phase in the CKM matrix [4]. Direct CP violation is predicted by the Standard Model, with typical computations for $\text{Re}(\epsilon'/\epsilon)$ ranging from $\approx$ -10 to 30 $\times 10^{-4}$ [5].

From the data taken in 1997-98-99, NA48 published [6] a result $(15.3 \pm 2.6) \times 10^{-4}$. KTeV [7] published a result based on the data taken in 1996-1997 $(20.7 \pm 2.8) \times 10^{-4}$. These recent results showed the existence of direct CP violation in the neutral kaon system.

We report here on the measurement of $\text{Re}(\epsilon'/\epsilon)$ performed using the 2001 data sample, recorded in somewhat different experimental conditions by the NA48 experiment.

After the 1999 data-taking period, the drift chambers of the experiment were damaged by the implosion of the beam tube. The data taking in 2001 took place with rebuilt drift chambers. Thanks to the possibility of a better SPS duty cycle, the data could be taken at a 30% lower beam intensity, allowing the insensitivity of the result to intensity-related effects to be checked, and the statistics for the final $\epsilon'/\epsilon$ measurement by NA48 to be completed. The statistics accumulated during the 2001 data-taking period is roughly half of the total statistics accumulated in the 1998 and 99 periods. The details of the analysis of this data set can be found in [8].

### 1.2. NA48 Method

The measurement of $R$ proceeds by counting the number of events in each of the four decay modes. The experiment is designed to exploit cancellations of systematic effects contributing symmetrically to different components of the R. Data are collected simultaneously in the four decay modes, cancelling the absolute fluxes and

*on behalf of the NA48 Collaboration (Cagliari, Cambridge, CERN, Dubna, Edinburgh, Ferrara, Firenze, Mainz, Orsay, Perugia, Pisa, Saclay, Siegen, Torino, Warsaw, Wien)


doi:10.1016/S0920-5632(02)02145-X

minimising the sensitivity of the measurement to accidental activity and variations of the detection efficiency. The $K_L$ and $K_S$ decays are provided by two nearly collinear beams. To minimise the difference in acceptance due to the large difference in average decay lengths, only $K_L$ decays occurring in the region also populated by $K_S$ are used. The $K_L$ events are furthermore weighted as a function of the proper lifetime in order to equalise the decay vertex distribution of $K_L$ and $K_S$ events. With this procedure, the acceptance correction cancels at first order in R, thus minimising the systematic uncertainty from Monte Carlo modelling of the experiment.

## 2. BEAMS AND DETECTOR

### 2.1. Beams

The neutral beams are derived from 400 GeV/$c$ protons extracted from the CERN SPS. Because of the different mean decay lengths of $K_L$ and $K_S$, two different production targets are used, located 126 m and 6 m upstream of the beginning of the decay region. For each SPS pulse (5.2 s every 16.8 s), $\approx 2.4\times10^{12}$ protons hit the $K_L$ production target. Three stages of collimation are used to define the $K_L$ beam, at a production angle of 2.4 mrad. Part of the non-interacting protons impinge on a bent silicon mono-crystal. A small fraction undergoes channelling and produces a proton beam of $\approx 5\times10^7$ protons per pulse transported to the $K_S$ production target, where a second neutral beam is derived, at a production angle of 4.2 mrad. The $K_S$ beam enters the fiducial decay region 6.8 cm above the $K_L$ beam. The beams converge with an angle of 0.6 mrad and the axes of the two beams cross at the position of the electromagnetic calorimeter. To distinguish $K_L$ and $K_S$ decays, the protons directed to the $K_S$ target are detected by an array of scintillation counters which comprise the tagging detector. The time of each proton is recorded and is compared with the time of the decay as measured in the main detector. The presence (absence) of a proton in coincidence with the event defines the event as $K_S$ ($K_L$). The $K_S$ beam traverses an anti-counter (AKS), formed by a set of scintillation counters following a 3 mm thick iridium crystal. This detector provides an accurate definition of the beginning of the decay region where it is located by vetoing $K_S$ decays occurring upstream.

### 2.2. Main detector

Charged pion decays are measured by a magnetic spectrometer comprised of four drift chambers and a dipole magnet giving a momentum kick of 265 MeV/$c$. The momentum resolution is $\sigma_p/p = 0.48\% \oplus 0.009\% \times p$ ($p$ in GeV/$c$). Two plastic scintillator hodoscope planes are located after the spectrometer. They are used to determine the event time of the charged events for the tagging procedure.

A quasi-homogeneous liquid krypton electromagnetic calorimeter with a projective tower readout is used to measure the photons from $\pi^0\pi^0$ events. The 13212 readout cells have each a cross-section of $\approx 2\times2$ cm$^2$. The energy resolution is $\sigma(E)/E = 0.032/\sqrt{E} \oplus 0.09/E \oplus 0.0042$, where $E$ is in GeV. The spatial resolution is better than 1 mm above 25 GeV. This detector is also used to measure the time of the photons for the tagging procedure.

Located after the electromagnetic calorimeter are a iron-scintillator hadron calorimeter, followed by a muon counter consisting of three planes of scintillators sandwiched between 80 cm thick iron walls.

### 2.3. Trigger

The electromagnetic calorimeter is used to trigger on $\pi^0\pi^0$ events. At the trigger level, the calorimeter data are reduced to $x$ and $y$ projections which are used to reconstruct the total energy as well as to estimate the decay position of the event. The efficiency of the $\pi^0\pi^0$ trigger is $(99.901 \pm 0.015)\%$ and is $K_S$-$K_L$ symmetric. A two level trigger is used for $\pi^+\pi^-$ decays. At first level, the hodoscope is placed in coincidence with a total energy condition defined using both calorimeters. The second level trigger uses information from the drift chambers to perform a fast event reconstruction. The efficiency of the $\pi^+\pi^-$ trigger is $(98.697 \pm 0.017)\%$. This is higher by 0.9% than for the 1998-99 data thanks mostly to the lower beam intensity. The efficiency is measured separately for $K_S$ and $K_L$ and the double ratio is corrected for the small difference.

## 3. EVENT RECONSTRUCTION AND SELECTION

$K \to \pi^0\pi^0$ decays are selected using only data from the LKr calorimeter. The details of the reconstruction and of the cuts applied can be found in [8] and [6]. From the measured energies and impact point positions on the calorimeter of the four showers, the decay vertex position along the beam axis is computed assuming that their invariant mass is the kaon mass. The invariant mass of the two photons pairs are then computed (the resolution is better than 1 MeV/$c^2$) and compared to the nominal $\pi^0$ mass, constructing a $\chi^2$ variable. To reject the residual background from $K_L \to 3\pi^0$ events, a cut on $\chi^2$ is applied. This residual background fraction is $(5.6 \pm 2.0) \times 10^{-4}$ in the $K_L$ sample, while the $K_S$ sample is background free. The background from $K_S$ decays produced by scattering of beam particles in the collimators in the $K_L$ beam is $(8.8 \pm 2.0) \times 10^{-4}$.

$K \to \pi^+\pi^-$ decays are reconstructed from tracks in the spectrometer. In the $\pi^+\pi^-$ mode, both the longitudinal and the transverse decay vertex position can be reconstructed, allowing a clean identification of $K_S$ and $K_L$ decays. A cut is applied on the ratio of the two track momenta to remove asymmetric decays in which one of the tracks could be close to the beam tube where the Monte Carlo modelling is more critical. To reject background from semileptonic $K_L$ decays, events with tracks consistent with being either an electron (from the E/p ratio of the energy deposited in the LKr calorimeter over the track momenta) or a muon (using hits in the muon counters) are rejected. Kinematical cuts are also applied on the $\pi^+\pi^-$ invariant mass (the resolution is typically 2.5 MeV/$c^2$) and on the reconstructed transverse momentum. The residual background in the $K_L$ sample is $(14.2 \pm 3.0) \times 10^{-4}$, while it is negligible in the $K_S$ sample.

The fiducial ranges in kaon energy $E_K$ and in proper time $\tau$ used to count events are chosen to be $70 < E_K < 170$ GeV and $0 < \tau < 3.5\ \tau_S$, where $\tau = 0$ is defined at the position of the AKS counter and $\tau_S$ is the $K_S$ mean lifetime. For $K_L$ events, the decay time cut is applied on reconstructed $\tau$, while for $K_S$ events the cut at $\tau = 0$ is applied using the AKS to veto decays occurring upstream. The determinations of the kaon energy, the decay vertex and the proper time in the $\pi^0\pi^0$ mode rely on measurements of the photon energies and positions with the calorimeter. The uniformity of the calorimeter response is optimised using $K_{e3}$ decays and checked using $\pi^0$ and $\eta$ decays produced during special ("$\eta$") runs in which a $\pi^-$ beam strikes two thin targets located near the beginning and the end of the fiducial decay region. The absolute energy scale is adjusted, with a 0.03% accuracy, using $K_S \to \pi^0\pi^0$ decays, such that the reconstructed AKS position matches the true value. It is also checked using data from the $\eta$ runs (the $\eta$ mass is taken from [9]). Taking all uncertainties into account (including non-linearities in the energy response), the total systematic uncertainty on R from the reconstruction of $\pi^0\pi^0$ events is found to be $\pm 5.3 \times 10^{-4}$ [8]. The uncertainty on R from the reconstruction of $\pi^+\pi^-$ events is $\pm 2.8 \times 10^{-4}$.

A decay is labelled $K_S$ is a coincidence is found within a $\pm$ 2 ns interval between its event time and a proton time measured with the tagger. Figure 1 shows the time distribution for $K_S$ and $K_L$ decays to $\pi^+\pi^-$ which have been identified by their vertex positions. From this, the misidentification probabilities in the $\pi^+\pi^-$ mode can be deduced: $\alpha_{SL} = (1.12 \pm 0.03) \times 10^{-4}$ for the $K_S$ tagging inefficiency and $\alpha_{LS} = (8.115 \pm 0.010)\%$ for the $K_L$ mistagging as $K_S$ due to an accidental coincidence between the event and a proton. As the same tagging procedure is used for $\pi^0\pi^0$ and $\pi^+\pi^-$ events, R is only sensitive to differences in misidentification probabilities between the $\pi^0\pi^0$ and $\pi^+\pi^-$ modes. Several methods have been developed to derive these differences directly from the data [8]. The results are $\Delta\alpha_{SL} = (0 \pm 0.5) \times 10^{-4}$ and $\Delta\alpha_{LS} = (3.4 \pm 1.4) \times 10^{-4}$ corresponding respectively to an uncertainty on R of $\pm 3.0 \times 10^{-4}$ and a correction of $(6.9 \pm 2.8) \times 10^{-4}$. The origin of $\Delta\alpha_{LS}$ is related to a higher loss of $\pi^+\pi^-$ events from accidental activity in the beam and can be predicted using a technique of overlaying "random" events (taken proportionally to the beam intensity) with $\pi\pi$ decays to estimate these losses. The agreement between the observed and the predicted $\Delta\alpha_{LS}$ values is illustrated in Figure 2.

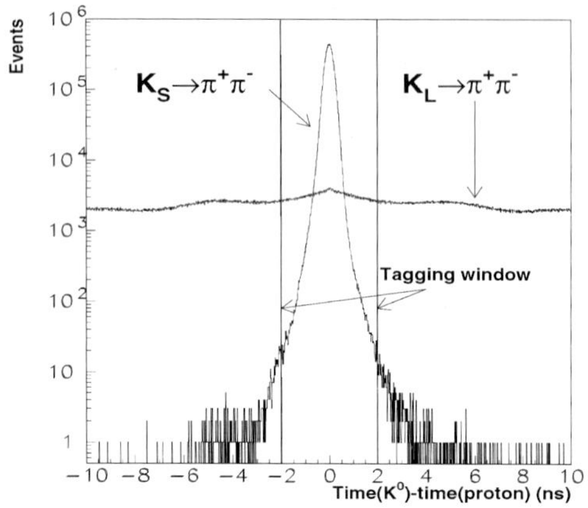


Figure 1. Time coincidence for $K_S$ and $K_L$ $\pi^+\pi^-$decays, identified by their reconstructed vertex position.

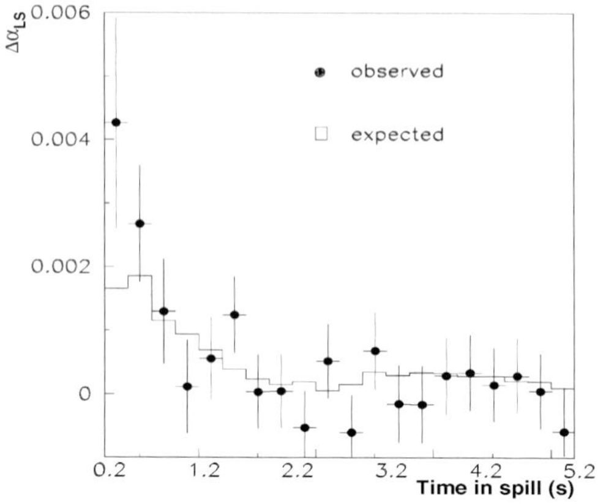


Figure 2. Measured compared to predicted values of $\Delta\alpha_{LS}$ as a function of the time during the spill.

## 4. ACCIDENTAL EFFECTS

The 2001 data were taken at lower beam intensity to reduce the uncertainties related to accidental effects. The overlap of extra particles related to kaon decays in the high intensity $K_L$ beam with a good event may result in the loss of the event. The effect on R is minimised by the simultaneous data collection in the four channels. The possible residual effect on R can be separated into two components:

1) A difference between the beam intensities seen by $K_S$ and $K_L$ events: $\Delta R = \Delta\lambda \times \Delta I/I$, where $\Delta\lambda$ is the difference between the event losses in the $\pi^+\pi^-$ and $\pi^0\pi^0$ modes and $\Delta I/I$ the difference between the $K_L$ beam intensity seen by $K_L$ and $K_S$ events. $\Delta\lambda$ is minimised by applying to all events the recorded dead time conditions. The largest one is an "overflow" condition in the drift chamber readout, which rejects 11% of the $\pi\pi$ events (this is significantly smaller than the 20% loss in the 1998-99 sample thanks to the lower beam instantaneous intensity and a lower noise in the drift chambers). $\Delta\lambda$ is estimated mainly from the overlay technique and is found to be $(1.0\pm0.5)\%$. $\Delta I/I$ is measured from the rate of out-of-time LKr clusters and tracks in good $K_S$ and $K_L$ decays. This is illustrated in Figure 3. Beam monitors, which have been improved for the 2001 data-taking period, are also used as cross-check. The result is $\Delta I/I = (0\pm1)\%$. The uncertainty on R from this effect is thus $\pm1.1\times10^{-4}$. This is significantly better than for the 1998-99 sample, thanks to the lower intensity and to better beam intensity monitors.

2) A difference in illumination between $K_S$ and $K_L$ decays coupled to a variation of the event loss with the impact points of the $K^0$ decay products. This effect is also computed from the overlay samples. No effect is found and the uncertainty on R is $\pm3.0\times10^{-4}$, from the statistics of the overlay samples.

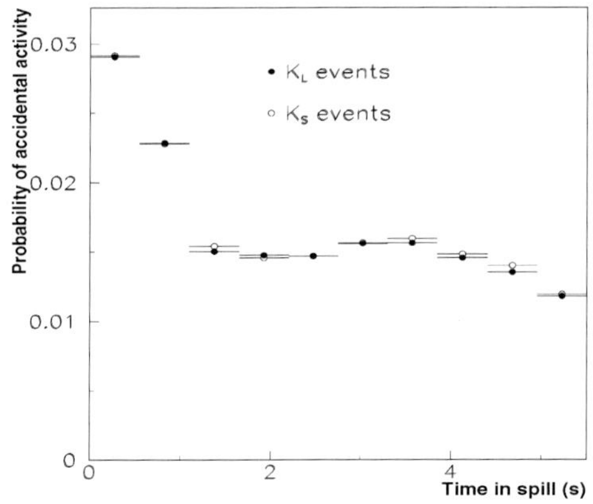


Figure 3. Probability of accidental activity in the LKr in the $\approx 150$ ns readout window as a function of the time during the spill, for $K_S$ and $K_L$ decays to $\pi^+\pi^-$.

## 5. RESULT AND CONCLUSIONS

Table 1 shows the summary of the corrections and systematic uncertainties on R for 2001 data. The residual acceptance correction is related to the small angle between the $K_L$ and $K_S$ beams and is computed using a large statistics Monte Carlo sample. As it is mostly given by the beam geometry, it does not rely on a detailed simulation of the detector. As expected from the design of the experiment, all corrections are small. Some systematic uncertainties are directly given by the statistics of the control samples used to study them. From the 2001 sample (comprising $1.55 \times 10^6$ $K_L \to \pi^0\pi^0$ events, $2.16 \times 10^6$ $K_S \to \pi^0\pi^0$, $7.14 \times 10^6$ $K_L \to \pi^+\pi^-$ and $9.61 \times 10^6$ $K_S \to \pi^+\pi^-$), the result $R = 0.99181 \pm 0.00147 \pm 0.00110$ is obtained (where the first error is statistical and the second systematic). The corresponding value of $Re(\epsilon'/\epsilon)$ is $Re(\epsilon'/\epsilon) = (13.7 \pm 2.5 \pm 1.8) \times 10^{-4}$. The agreement between this new and the earlier results is particularly significant since they were obtained from data taken at different average beam intensities. Taking into account the correlated systematic uncertainty of $\pm 1.4 \times 10^{-4}$, the final combined result from the NA48 experiment is $Re(\epsilon'/\epsilon) = (14.7 \pm 2.2) \times 10^{-4}$.

Table 1
Corrections and systematic uncertainties on R

| | | in $10^{-4}$ | |
|---|---|---|---|
| $\pi^+\pi^-$ trigger (stat) | | +5.2 | ±3.6 |
| AKS inefficiency | | +1.2 | ±0.3 |
| Reconstruction | of $\pi^0\pi^0$ | — | ±5.3 |
| | of $\pi^+\pi^-$ | — | ±2.8 |
| Background | to $\pi^0\pi^0$ | −5.6 | ±2.0 |
| | to $\pi^+\pi^-$ | +14.2 | ±3.0 |
| Beam scattering | | −8.8 | ±2.0 |
| Accidental tagging (stat) | | +6.9 | ±2.8 |
| Tagging inefficiency | | — | ±3.0 |
| Acceptance | statistical | +21.9 | ±3.5 |
| | systematic | | ±4.0 |
| Accidentals | intens. | — | ±1.1 |
| | illum. (stat) | | ±3.0 |
| $K_S$ in time activity | | — | ±1.0 |
| Total | | +35.0 | ±11.0 |

## REFERENCES


1. J.H.Christenson et al., Phys.Rev.Lett.13 (1964), 138.
2. L.Wolfenstein, Phys.Rev.Lett.13 (1964) 562.
3. T.T.Wu and C.N.Yang, Phys.Rev.Lett.13 (1964) 380.
4. M.Kobayashi and T.Maskawa, Prog.Theor.Phys.49 (1973) 652.
5. For a review, see for instance J. Donoghue in F.Costantini et al. (ed) Proc. Int. Conference on Kaon Physics, Frascati Physics Serie 26, 2001, p.93 and references therein.
6. A.Lai et al., Eur.Phys.Jour.C22 (2001) 231
7. A.Alavi-Harati et al., hep-ex/0208007, submitted to Phys.Rev.D
8. J.R.Batley et al., Phys.Lett.B544 (2002) 97
9. A.Lai et al., Phys.Lett.B533 (2002) 196

ELSEVIER

Proceedings of ICHEP 2002, pp. 462–464
*S. Bentvelsen, P. de Jong, J. Koch and E. Laenen (Editors)*

31st INTERNATIONAL CONFERENCE ON HIGH ENERGY PHYSICS AMSTERDAM

# $\epsilon'/\epsilon$ and the Electroweak Penguin Contribution

V. Cirigliano[a] * J.F. Donoghue[b] † E. Golowich[b‡], and K. Maltman[c] §

[a]Departament de Física Teõrica, IFIC, Universitat de València - CSIC, Apt. Correus 2085, E-46071 València, Spain

[b]Physics Department, University of Massachusetts, Amherst MA 01003 United States of America

[c]Department of Mathematics and Statistics, York University, 4700 Keele St., Toronto ON M3J 1P3 Canada

Our dispersive sum rule calculation of the electroweak penguin contribution to $\epsilon'/\epsilon$ is reviewed. A more recent analysis based on the finite-energy sum rule approach is described. Finally, a new determination of the electroweak penguin contribution to $\epsilon'/\epsilon$ is presented.

## 1. THE STANDARD MODEL AND $\epsilon'/\epsilon$

I shall describe ongoing work [1] to evaluate the electroweak penguin (EWP) contribution to $\epsilon'/\epsilon$.

After many years of intense effort, experimentalists have obtained the accurate determinations $[\epsilon'/\epsilon]_{\rm PDG} = (18. \pm 4.) \cdot 10^{-4}$ as cited in the latest PDG listings [2] or the updated (2002) value cited in Tony Barker's talk $[\epsilon'/\epsilon]_{2002} = (16.6 \pm 1.6) \cdot 10^{-4}$ [3]. The Standard Model prediction for $\epsilon'/\epsilon$ has two main components, from the QCD penguin operator $\langle\langle \mathcal{Q}_6\rangle\rangle \equiv \langle(\pi\pi)_{I=0}|\mathcal{Q}_6|K^0\rangle$ and from the EWP operator $\langle\langle \mathcal{Q}_8\rangle\rangle \equiv \langle(\pi\pi)_{I=2}|\mathcal{Q}_8|K^0\rangle$. Since the physics underlying such matrix elements is nonperturbative, their evaluation is a very difficult matter. Indeed, included in this class of matrix elements are those associated with the notorious $\Delta I = 1/2$ rule. Although such K-to-$2\pi$ matrix elements are investigated in lattice-QCD simulations, let us instead pursue an analytic approach.

### 1.1. Using Chiral Symmetry

A method which avoids the uncontrollable uncertainties of models is to employ the orderly expansion procedure of chiral perturbation theory (ChPT) [4],

$$\begin{aligned} \langle\langle \mathcal{Q}\rangle\rangle &= \overline{\langle\langle \mathcal{Q}\rangle\rangle} + \mathcal{O}(p^2) + \ldots \\ \langle\langle \mathcal{Q}_6\rangle\rangle &= 0 + \mathcal{O}(p^2) + \ldots \\ \langle\langle \mathcal{Q}_8\rangle\rangle &\simeq \overline{\langle\langle \mathcal{Q}_8\rangle\rangle}\,[1 - 0.3 + \ldots] \end{aligned} \quad (1)$$

where the superbar indicates evaluation in the chiral limit. The above three cases correspond respectively to a general matrix element, the QCD penguin and the EWP. Although the $\mathcal{Q}_6$ matrix element is *zero* in the chiral limit and so experiences all its contributions away from chiral corrections, the situation is better for $\mathcal{Q}_8$. Not only does it have a nonzero leading chiral component, its first chiral corection has been estimated to be 30% as large and of opposite sign [6]. Using a method of dispersive sum rules [5], we performed the following determination of the leading chiral component [7],

$$[\epsilon'/\epsilon]_{\rm EWP} = (-22. \pm 7.) \cdot 10^{-4} \quad . \quad (2)$$

The 32% uncertainty is sufficiently small to allow the conclusion that the EWP contribution is large in magnitude (roughly that of the experimental value) but opposite in sign.

*Research supported in part by MCYT, Spain (Grant No. FPA-2001-3031) and by ERDF funds from the European Commission.

†Research supported in part by the National Science Foundation under Grant PHY-9801875.

‡Speaker at ICHEP02. Research supported in part by the National Science Foundation under Grant PHY-9801875.

§Research supported in part by the Natural Sciences and Engineering Research Council of Canada.


doi:10.1016/S0920-5632(02)02146-1

### 1.2. Working in the Chiral Limit

To motivate the discussion to come, we briefly review the dispersive sum rule approach. The first step is to take the chiral limit,

$$\overline{\langle\langle \mathcal{Q}_8 \rangle\rangle}_\mu = -\langle \mathcal{O}_8 \rangle_\mu / F_\pi^{(0)3} + \ldots \ , \quad (3)$$

where $F_\pi^{(0)}$ is the pion decay constant as evaluated in the chiral limit and $\langle \mathcal{O}_8 \rangle_\mu$ is a vacuum matrix element evaluated at renormalization scale $\mu$. It is commonplace to work at $\mu = 2$ GeV. In Eq. (3), we have neglected the contribution from an additional vacuum matrix element $\langle \mathcal{O}_1 \rangle_\mu$, estimated to be a 2.5% effect.

The key to determining $\langle \mathcal{O}_8 \rangle_\mu$ is to employ the V-A correlator $\Delta\Pi$ and spectral function $\Delta\rho$,

$$\mathcal{I}m \ \Delta\Pi(s) = \pi \ \Delta\rho(s) \ , \quad (4)$$

where $\Delta\rho$ is measured directly in hadronic decays of the $\tau$ lepton. Our published work is based on the ALEPH determination [8] of $\Delta\rho$, although we are currently incorporating the OPAL database [9] as well. After much theoretical analysis, we were able to derive the sum rule [5,7]

$$\langle \mathcal{O}_8 \rangle_\mu = \int_0^\infty ds \ \mu^2 \frac{s^2}{s+\mu^2} \Delta\rho(s) + \ldots \ . \quad (5)$$

It is the numerical evaluation of this sum rule which led to the chiral result given in Eq. (2).

## 2. CAN WE DO BETTER?

The difficulty with Eq. (5) is that one needs to know $\Delta\rho(s)$ for *all* energies, but data supplies its value only for $s \leq m_\tau^2$. Clever use of chiral sum rules [10,11] allowed us to overcome this limitation, but at a cost. The chiral sum rules introduce additional dependence on imprecisely known (chiral limit) quantities like $F_\pi^{(0)3}$ and $\Delta m_\pi^{(0)2}$. In the following, we indicate how one can do better.

### 2.1. The OPE for $\Delta\Pi$

For large space-like momenta $Q^2 \gg \Lambda_{\rm QCD}^2$ and through $\mathcal{O}(\alpha_s^2)$ in QCD counting, $\Delta\Pi(Q^2)$ can be represented via the operator product expansion (OPE),

$$\Delta\Pi(Q^2) \sim \sum_d \frac{1}{Q^d} \left[ a_d(\mu) + b_d(\mu) \ln \frac{Q^2}{\mu^2} \right] \ , \quad (6)$$

where $d = 2, 4, 6, 8, 10, \ldots$.

The $d = 6$ OPE coefficient $a_6$ is of special interest because it is related to $\langle \mathcal{O}_8 \rangle_\mu$,

$$a_6(\mu) = 2 \left[ 2\pi \langle \alpha_s \mathcal{O}_8 \rangle_\mu + \ldots \right] \ . \quad (7)$$

It is clear from Eq. (7) that an attractive alternative for finding $\langle \mathcal{O}_8 \rangle_\mu$ is to determine the OPE coefficient $a_6(\mu)$. The finite energy sum rule (FESR) approach is naturally structured to do just that. Moreover, since only knowledge of $\Delta\rho$ in the data region is required, the prospect of an improved determination is bright.

## 3. FINITE ENERGY SUM RULES

There are several ways to implement the FESR machinary. Here, we integrate $w(q^2)\Delta\Pi(q^2)$ over a keyhole contour of radius $s_0$ to obtain

$$\int_{s_{\rm th}}^{s_0} ds \ w(s) \ \Delta\rho(s) - 2F_\pi^2 w(m_\pi^2) = -\frac{1}{2\pi i} \oint_{|s|=s_0} ds \ w(s) \ \Delta\Pi(s) \ . \quad (8)$$

Throughout we work with polynomial weights $w(s)$. Next we use the OPE form for $\Delta\Pi(q^2)$ on the circular part of the contour to obtain the still-exact relation,

$$\int_{s_{\rm th}}^{s_0} ds \ w(s) \ \Delta\rho(s) - 2F_\pi^2 w(m_\pi^2) = -\frac{1}{2\pi i} \oint_{|s|=s_0} ds \ w(s) \ \Delta\Pi_{\rm OPE}(s) - R[s_0, w] \ , \quad (9)$$

where the quantity $R[s_0, w]$ is defined as

$$R[s_0, w] \equiv -\frac{1}{2\pi i} \oint_{|s|=s_0} ds \ w(s) \times \left[ \Delta\Pi_{\rm OPE}(s) - \Delta\Pi(s) \right] \ . \quad (10)$$

Physically, $R[s_0, w]$ indicates the presence of duality violation since as $s_0$ is lowered one expects use of the OPE to become inadequate. To test for the presence of $R[s_0, w]$, we carry out our FESR studies using seven different values of $s_0$ in the range 2 GeV$^2 \leq s_0 \leq m_\tau^2$. This important aspect of our work is discussed at length in Ref. [1].

### 3.1. Choosing the Weights

We base our construction of the polynomial weights upon the following three criteria:

*(i) Pinched Weights*: Because the OPE is known to fail along the $\mathcal{R}e\, s \geq 0$ axis [12], we employ only 'pinched' weights $w(y) = (1-y)^2 p(y)$, where $y = s/s_0$ and $p(y)$ is a polynomial. An earlier study [13] has demonstrated that the suppression thus produced suffices to circumvent the OPE breakdown.

*(ii) Maximum Statistical Signal*: To maximize the statistical signal, we construct weights which avoid large cancellations in the spectral integrals (due to the shape of the spectral function $\Delta\rho$).

*(iii) Simplicity*: We maintain a relatively simple structure in our equation set by allowing the presence of no more than two of the $\{a_d\}$ with $d > 4$ on the OPE side of a given FESR relation.

## 4. PRELIMINARY RESULTS

First we display the FESR determination of OPE coefficients,

| $a_n$ | Value |
|---|---|
| $a_6$ | $(-5.2 \pm 0.7) \cdot 10^{-3}$ GeV$^6$ |
| $a_8$ | $(-3.0 \pm 3.2) \cdot 10^{-3}$ GeV$^8$ |
| $a_{10}$ | $(37 \pm 12) \cdot 10^{-3}$ GeV$^{10}$ |
| $a_{12}$ | $(-129 \pm 37) \cdot 10^{-3}$ GeV$^{12}$ |

next the *leading* chiral component for $[\epsilon'/\epsilon]_{\rm EWP}$ as obtained (i) in Ref. [7], (ii) in the FESR determination of $a_6$ and (iii) via an improved sum rule approach not described here,

| | |
|---|---|
| $(-22. \pm 7.) \cdot 10^{-4}$ | Sum Rule (old) |
| $(-16.2 \pm 3.4) \cdot 10^{-4}$ | FESR via$a_6$ |
| $(-16.5 \pm 3.7) \cdot 10^{-4}$ | Sum Rule (new) |

By 'an improved sum rule approach', I mean one that uses as input the OPE coefficients obtained in our FESR study. This allows the sum rule to be evaluated at lower renormalization scales than in Ref. [7] and leads ultimately to reduced uncertainties [1].

The different methodologies are seen to yield results which are in statistical agreement. Observe that the 13.5% error in $a_6$ has translated into a 21% error in $[\epsilon'/\epsilon]_{\rm EWP}$. We expect an even more accurate determination by the time we have completed our analysis.

## REFERENCES


1. V. Cirigliano, J. F. Donoghue, E. Golowich and K. Maltman, work in progress.
2. K. Hagiwara *et al* (Particle Data Group), Phys. Rev. **D66** (2002) 010001.
3. T. Barker, *Recent results from KTeV*, Parallel talk CP-1-1 presented at ICHEP02.
4. For an example of a higher order contribution in the ChPT expansion as it pertains to the VV correlator, see E. Golowich and J. Kambor, "Two loop analysis of vector current propagators in chiral perturbation theory," Nucl. Phys. B **447**, 373 (1995) [arXiv:hep-ph/9501318].
5. J. F. Donoghue and E. Golowich, "Dispersive calculation of $B_7^{(3/2)}$ and $B_8^{(3/2)}$ in the chiral limit," Phys. Lett. B **478**, 172 (2000) [hep-ph/9911309].
6. V. Cirigliano and E. Golowich, "Analysis of $\mathcal{O}(p^2)$ corrections to $\langle\pi\pi|Q_{7,8}|K\rangle$", Phys. Lett. B **475**, 351 (2000) [hep-ph/9912513].
7. V. Cirigliano, J. F. Donoghue, E. Golowich and K. Maltman, "Determination of $\langle(\pi\pi)_{I=2}|\mathcal{Q}_8|K^0\rangle$ in the chiral limit," Phys. Lett. B **522**, 245 (2001) [hep-ph/0109113].
8. R. Barate *et al.* [ALEPH Collaboration], "Measurement of the spectral functions of vector current hadronic tau decays," Z. Phys. C **76**, 15 (1997); "Measurement of the spectral functions of axial-vector hadronic tau decays and determination of $\alpha_s(m_\tau^2)$," Eur. Phys. J. C **4**, 409 (1998).
9. K. Ackerstaff *et al.* [OPAL Collaboration], Eur. Phys. J. C **7**, 571 (1999) [arXiv:hep-ex/9808019].
10. S. Weinberg, Phys. Rev. Lett. **18** (1967) 507.
11. T. Das, G.S. Guralnik, V.S. Mathur, F.E. Low and J.E. Young, Phys. Rev. Lett. **18** (1967) 759.
12. E. C. Poggio, H. R. Quinn and S. Weinberg, "Smearing The Quark Model," Phys. Rev. D **13**, 1958 (1976).
13. K. Maltman, "Constraints on hadronic spectral functions from continuous families of finite energy sum rules," Phys. Lett. B **440**, 367 (1998) [hep-ph/9901239].

Proceedings of ICHEP 2002, pp. 465–468
*S. Bentvelsen, P. de Jong, J. Koch and E. Laenen (Editors)*

# $|V_{ub}|$ measurements and related results from Belle

Youngjoon Kwon[a]*, for the Belle Collaboration

[a] Institute of Physics and Applied Physics, Yonsei University
Seoul 120-749, Republic of Korea

In this paper, we present preliminary measurements of $|V_{ub}|$ from exclusive semileptonic decay $B^0 \to \pi^- \ell^+ \nu_\ell$ and hadronic decay $B^0 \to D_s^+ \pi^-$ using data sample of asymmetric $e^+e^-$ collision recorded at $\Upsilon(4S)$ resonance with Belle detector at KEKB collider. Also presented are preliminary measurement of $\mathcal{B}(B^+ \to \omega \ell^+ \nu_\ell)$ and upper limits on $B^+ \to \ell^+ \nu_\ell$ ($\ell = e,\ \mu$).

## 1. INTRODUCTION

$V_{ub}$ is one of the smallest elements of the CKM matrix and plays a crucial role to explain CP violation in $B$ meson decays via KM mechanism[1]. For this reason, a precise measurement of $V_{ub}$ is one of the most important goals of the B-factory experiments such as Belle and BaBar. In this paper, we present recent results from the Belle experiment on the measurements of $|V_{ub}|$ using exclusive semileptonic and hadronic $B$ decays, and other related results.

This paper is organized as follows. In the next section, we will show recent Belle results on exclusive semileptonic $B$ decays such as $B^0 \to \pi^- \ell^+ \nu_\ell$[2], and $\omega \ell^+ \nu$. In addition to branching fractions, we will also show the measurement of $q^2$ distributions of $B^0 \to \pi^- \ell^+ \nu_\ell$ decays. From these measurements, we obtain preliminary measurements of $|V_{ub}|$. In the next two sections, $B$ decay modes containing a $b \to u$ decay vertex will be presented. In Section 3, we present results on hadronic $B$ decays $B \to D_s \pi$. In Section 4, we present a search for $W$-annihilated decay mode $B^+ \to \ell^+ \nu_\ell$. The results presented in this paper are based on the integrated luminosity of 60 fb$^{-1}$ on $\Upsilon(4S)$ resonance energy, unless explicitly stated otherwise.

## 2. EXCLUSIVE SEMILEPTONIC $B$ DECAYS

Semileptonic $B$ decays $B \to X_u \ell^+ \nu_\ell$ where $X_u$ denotes any meson containing a $u$ quark provide the best opportunity for measuring $|V_{ub}|$ because strong interaction effects are much simplified due to the appearance of two leptons in the final state. Analyzing the decays inclusively where we sum up all $X_u$ final states is less sensitive to theoretical modelling of the decays, but experimentally challenging because of the large background from $B \to X_c \ell^+ \nu_\ell$ decays. On the other hand, analyzing exclusive decay modes such as $B^0 \to \pi^- \ell^+ \nu_\ell$ provides experimentally cleaner measurements, but is subject to large systematic error from theoretical uncertainties in the form factor determination.

### 2.1. $B^0 \to \pi^- \ell^+ \nu_\ell$ and $V_{ub}$

Measuring the exclusive decay $B^0 \to \pi^- \ell^+ \nu_\ell$ is complicated because of the undetected neutrino in the final state. To extract information about undetected neutrino, we use a method that has been developed by CLEO[3], which exploit the knowledge of initial state of the $e^+e^- \to \Upsilon(4S) \to B\bar{B}$ reactions and near $4\pi$ coverage ("*hermeticity*") of the detector. We select events which are consistent with the hermeticity assumption by requiring that $|\sum Q| < 2$, only one lepton being identified in the event, and $17^\circ < \theta_{\rm miss} < 150^\circ$. Moreover, the missing mass of the event is required to be consistent with 0. Then, we attribute

*email: kwon@phya.yonsei.ac.kr


doi:10.1016/S0920-5632(02)02147-3

the missing energy-momentum 4-vector to that of the undetected neutrino.

In order to suppress background from $b \to c\ell\nu$ decays, we require $1.2 < p_\ell < 2.8$ GeV/$c$, $|p_\ell| + |p_\pi| > 3.1$ GeV/$c$, and $|\cos\theta_{B-\pi\ell}| < 1$, where the cosine of the angle $\theta_{B-\pi\ell}$ between the $B$ meson and the $\pi\ell$ system is calculated from directly measured quantities. To suppress the background from $q\bar{q}$ continuum process, we require $R_2 < 0.35$, $|\cos\theta_T| < 0.8$ and $q^2_{\rm obs} > 1.5$ GeV/$c^2$. These requirements are employed to make the best use of the event shape difference between the signal events and the continuum backgrounds.

From the measured 4-momenta of the signal lepton and the charged pion, and the missing 4-momentum which we attribute to the undetected neutrino, we calculate the energy difference $\Delta E (= E_{beam} - (E_\pi + E_\ell + E_\nu))$ and the beam-energy-constrained mass $M_{bc}$ for which we require $M_{bc} > 5.25$ GeV/$c^2$.

The signal yield is obtained by fitting $\Delta E$ and lepton momentum distributions simultaneously. Figure 1 shows the $\Delta E$ distribution after the final event selection using the 60 fb$^{-1}$ event sample of $B\bar{B}$ decays recorded by the Belle detector. Superimposed on the plot are the projections of the fitted signal and background components. The background components included in the fit are continuum, $b \to c$ and $B \to X_u\ell\nu$ other than the signal mode.

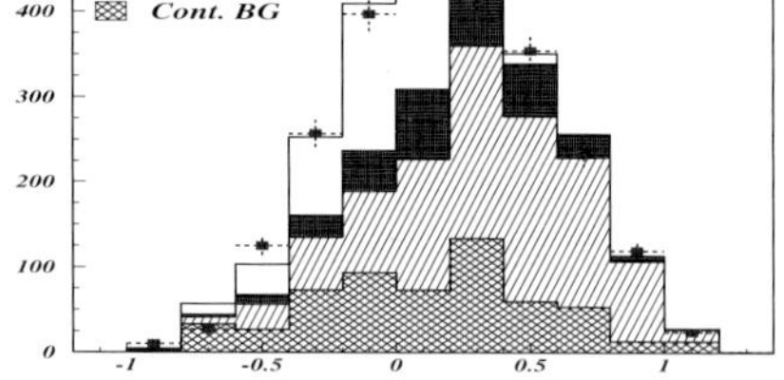


Figure 1. The $\Delta E$ distribution of $B^0 \to \pi^-\ell^+\nu_\ell$ after final event selection. Projections of the fitted signal and background components are superimposed.

In order to calculate $\mathcal{B}(B^0 \to \pi^-\ell^+\nu_\ell)$ from measured yield, we determine the signal efficiency using two models based on lattice QCD[4] and light-cone sum rule (LCSR)[5]. We also use the same two models to extract $|V_{ub}|$ by $\mathcal{B}(B^0 \to \pi^-\ell^+\nu_\ell) = \gamma_\pi |V_{ub}|^2 \tau_B$, where $\gamma_\pi$ is a factor determined by each model and $\tau_B$ is the $B$ meson lifetime. The results for detection efficiency and calculation of $\gamma_\pi$ are summarized in Table 1. We obtain $\mathcal{B}(B^0 \to \pi^-\ell^+\nu_\ell) = (1.33 \pm 0.11 \pm 0.21) \times 10^{-4}$, where we take an average of branching fractions based on the two models. The difference is included in the systematic uncertainty. Using the world average value for $\tau_B$[9], we obtain: $|V_{ub}| = (3.11 \pm 0.13 \pm 0.24 \pm 0.56) \times 10^{-3}$ based on lattice QCD, and $|V_{ub}| = (3.58 \pm 0.15 \pm 0.28 \pm 0.63) \times 10^{-3}$ based on LCSR. The errors are statistical, systematic, and theoretical, respectively. Since each model describes different region of $q^2$ better than the other (lattice QCD is in general good for large $q^2$ region, while LCSR is good for small $q^2$), we do not take an average of the two results.

Table 1
Detection efficiency and $\gamma_\pi$ calculated by two different methods: lattice QCD (UKQCD group) and LCSR.

| Model | UKQCD[4] | LCSR[5] |
|---|---|---|
| effi. (%) | 2.9 | 3.1 |
| $\gamma_\pi$ | $9^{+3+2}_{-2-2}$ | $7.3 \pm 2.5$ |

Attributing missing energy-momentum to that of a single neutrino, we can calculate $q^2$ for each event and obtain $q^2$ distribution after subtracting background using Monte-Carlo (MC) and off-resonance data. In order to account for kinematic effects due to various event selection requirements, we unfold the $q^2$ distribution considering detection efficiency and smearing effect. Figure 2 shows the unfolded $q^2$ distribution. Superimposed on the plot are theoretical distributions $d\Gamma/dq^2$ based on UKQCD and LCSR calculations. Eventually, we expect to be able to discriminate good models from poor ones based on the $q^2$ distribution.

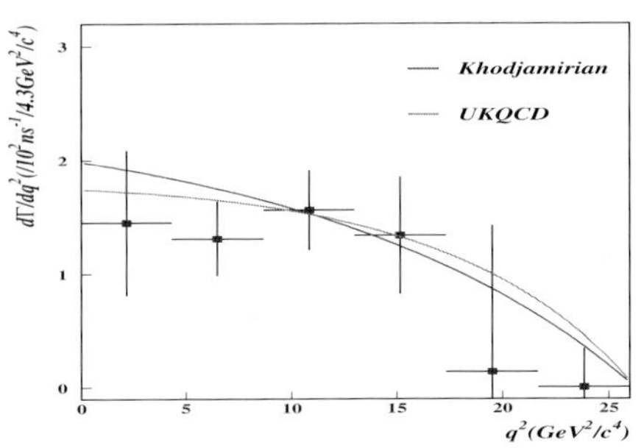


Figure 2. The unfolded $q^2$ distribution of $B^0 \to \pi^- \ell^+ \nu_\ell$. Superimposed curves are $d\Gamma/dq^2$ based on UKQCD and LCSR calculations.

### 2.2. Evidence for $B^+ \to \omega \ell^+ \nu_\ell$

The exclusive decays $B \to \pi \ell \nu$ and $B \to \rho \ell \nu$ have previously been observed by CLEO [3,6], but $B \to \omega \ell \nu$ decay has been unmeasured so far. To search for this decay, we use a similar method as in $B \to \pi \ell \nu$ analysis, *i.e.* using the detector hermeticity and attributing missing 4-momentum to the undetected neutrino.

After selecting events with only one identified lepton, we require $p_\ell > 2.2$ GeV/$c$. To enhance the probability that the signal neutrino is the only undetected particle in an event, we require $|M^2_{\rm miss}| < 3$ (GeV/$c^2$)$^2$ and the missing momentum must lie within the EM calorimeter acceptance of Belle ($17° < \theta_{\rm miss} < 150°$) . To suppress continuum background, $R_2 < 0.4$ is required. In addition, events with a uniform energy distribution around the lepton direction are selected by studying the energy distribution pattern across nine cones of equal polar angle around the lepton axis using a Fisher discriminant [7].

We reconstruct $\omega$ via its decay to $\pi^+\pi^-\pi^0$. The $\pi^0$ candidates are formed from pairs of photons with $E_\gamma > 30$ MeV, $120 < m(\gamma\gamma) < 150$ MeV/$c^2$ and $p_{\pi^0} > 200$ MeV/$c$. To suppress combinatoric background, we further require $p_\omega > 300$ MeV/$c$. In addition, the Dalitz amplitude $|\vec{p_{\pi^+}} \times \vec{p_{\pi^-}}|$ (calculated in the $\omega$ rest frame) is required to be larger than 75% of its maximum value.

The 4-momenta of the lepton and the $\omega$ are combined with the missing 4-momentum to form a $B$. Again, we apply the requirement $|\cos\theta_{B-\omega\ell}| < 1$, and calculate $M_{bc}$ and $\Delta E$ for the $B$ candidates. The signal window is defined as $5.25 < M_{bc} < 5.29$ GeV/$c^2$ and $|\Delta E| < 0.3$ GeV. The $\omega$ yield in the signal window is extracted by fitting the $\pi^+\pi^-\pi^0$ invariant mass distribution using a Gaussian signal function and a linear background. Background from continuum events is estimated using off-resonance data of 6 fb$^{-1}$ integrated luminosity. To estimate combinatoric background from $B\bar{B}$ decays, a sideband is defined as $5.10 < M_{bc} < 5.19$ GeV/$c^2$ and $|\Delta E| < 0.3$ GeV. It was checked with the MC sample that this sideband contains the same amounts of background as the signal window, within statistical errors.

The fit to the $\pi^+\pi^-\pi^0$ mass yields $73\pm15$ ($14\pm 11$) events in the signal window (side-band), resulting in an excess of 59 events which corresponds to $\mathcal{B}(B^+ \to \omega \ell^+ \nu_\ell) = (1.4 \pm 0.4 \pm 0.3) \times 10^{-4}$ where the errors are statistical and systematic, respectively. The biggest source of systematic error is in the estimation of background, due to limited statistics of off-resonance event sample.

## 3. $B^0 \to D_s^+ \pi^-$ AND $|V_{ub}|$

The decays $B \to D_s^{(*)} X_u$ occur dominantly via the same spectator decay as $B \to X_u \ell^+ \nu$; the only difference is that the virtual $W^+$ decays to $c\bar{s}$ and subsequently hadronizes to $D_s^{(*)+}$. In case of $B^0 \to D_s^+ \pi^-$, the final state particles can be fully reconstructed, hence one need not worry about missing neutrino. Converting the branching fraction of this decay to $|V_{ub}|$, however, is not easy because of two uncertain factors: the absolute branching fraction of $D_s^+ \to \phi \pi^+$, which is currently known to $\pm25\%$ fractional error, and the theoretical uncertainty in the form-factors of $B^0 \to D_s^+ \pi^-$ decay. According to a recent work by Kim *et al.*[8], such uncertainties might be alleviated by taking the ratio, for example, $\Gamma(B \to D_s^+ \pi^-)/\Gamma(B \to D_s^+ D^-)$. Assuming a generalized factorization scheme, they point out that form-factor uncertainty for $|V_{ub}|$ is around 10%.

Using a Belle data sample of 79 $fb^{-1}$ on-resonance integrated luminosity, we have searched for $B^0 \to D_s^+\pi^-$ decay in three $D_s^+$ decay modes: $D_s^+ \to \phi\pi^+$, $K_S^0K^+$, and $\bar{K}^{*0}K^+$. Signal yield is obtained by fitting the 2-dimensional $\Delta E$ vs. $M(D_s)$ distribution. Figure 3 shows the $\Delta E$ distribution of $B \to D_s\pi$ combining the three $D_s$ modes. The points with errors are the $M(D_s)$ signal yield in each $\Delta E$ bin, and the curve is the fit result. The signal yield is $10.1^{+4.4}_{-3.7}$ which gives an evidence of $B^0 \to D_s^+\pi^-$ decay with $3.6\sigma$ significance. We calculate the branching fraction $\mathcal{B}(B^0 \to D_s^+\pi^-) = (2.4^{+1.0}_{-0.8} \pm 0.7)\times 10^{-3}$. The dominant systematic error comes from the uncertainty in $\mathcal{B}(D_s^+ \to \phi\pi^+)$.

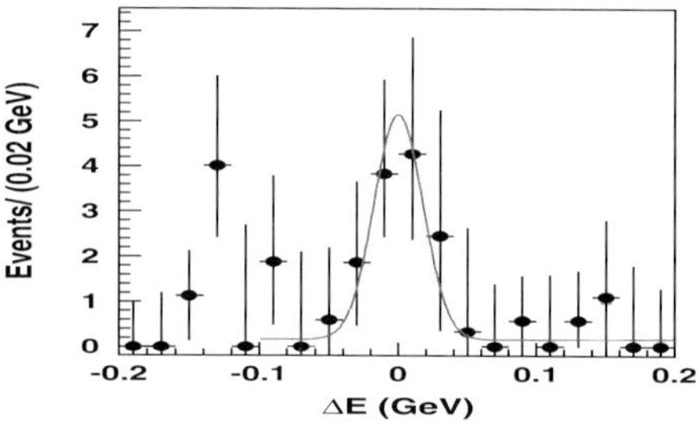


Figure 3. The $\Delta E$ distribution of $B \to D_s\pi$. The points with errors are the $M(D_s)$ signal yield in each $\Delta E$ bin, and the curve is the fit result.

From the result of Kim *et al.*[8] and the world average of $\mathcal{B}(B^0 \to D_s^+D^-)$[9], we calculate the ratio:

$$\frac{\Gamma(B \to D_s^+\pi^-)}{\Gamma(B \to D_s^+D^-)} = (3.0^{+1.8}_{-1.6}) \times 10^{-3} .$$

Consequently, using the world average of $|V_{cb}|$, we obtain $|V_{ub}| = (3.5^{+1.0}_{-0.9}) \times 10^{-3}$.

## 4. SEARCH FOR $B^+ \to \ell^+\nu_\ell$ DECAYS

In the Standard Model (SM), $B^+ \to \ell^+\nu_\ell$ decays occur via $W$-annihilation diagram. The decay rate is proportional to square of the charged lepton mass ("*helicity suppression*"). Therefore, $B^+ \to e^+\nu_e$ and $B^+ \to \mu^+\nu_\mu$ are highly suppressed in the SM compared to $B^+ \to \tau^+\nu_\tau$, which makes $B^+ \to e^+\nu_e$ and $B^+ \to \mu^+\nu_\mu$ very good places to search for new physics effects.

Experimentally, $B^+ \to \ell^+\nu_\ell$ ($\ell = e$ and $\mu$) events have a striking signature of a nearly monochromatic lepton in the $B$ rest frame. To exploit this feature in our search for $B^+ \to \ell^+\nu_\ell$, we analyze the momentum spectrum of charged leptons satisfying $2.2 < p_\ell^* < 2.9$ GeV/$c$ where $p_\ell^*$ is measured in the $B$ rest frame. The 4-momenta of the signal $B$ and opposite $B$ are determined by measuring the 4-momenta of all charged and neutral particles in an event except for the signal lepton. To suppress background, we use the beam-constrained mass and energy difference of the opposite $B$. The continuum background events are suppressed using event-shape variables.

The signal yield in each mode is obtained by fitting the $p_\ell^*$ distributions. The signal and background components are determined with MC event samples. We do not see any evidence for signals in either $B^+ \to e^+\nu_e$ or $B^+ \to \mu^+\nu_\mu$ and set the following upper limits with 90% confidence level:$\mathcal{B}(B^+ \to e^+\nu_e) < 5.4 \times 10^{-6}$ and $\mathcal{B}(B^+ \to \mu^+\nu_\mu) < 6.8 \times 10^{-6}$.

## REFERENCES

1. M. Kobayashi and T. Maskawa, Prog. Theor. Phys. **49** 652 (1973).
2. Throughout this paper, a charge-conjugate mode is also implied.
3. J.P. Alexander *et al.* (CLEO Collab.) Phys. Rev. Lett. **77** 5000 (1996).
4. UKQCD Collab., Phys. Lett. B **486**, 111 (2000).
5. Khodjamirian *et al.*, Phys. Rev. D **62**, 114002 (2000).
6. B. H. Behrens *et al.* (CLEO Collab.), Phys. Rev. D **61**, 052001 (2000).
7. R. A. Fisher, Annals of Eugenics **7**, 179 (1936).
8. C.S. Kim *et al.*, Phys. Rev. D **63** 094506 (2001).
9. K. Hagiwara *et al.*, Phys. Rev. D **66** 010001 (2002).

Proceedings of ICHEP 2002, pp. 469–471
*S. Bentvelsen, P. de Jong, J. Koch and E. Laenen (Editors)*

# Measurements of the CKM Matrix Element $|V_{ub}|$ with *BABAR*

L.H. Wilden for the *BABAR* Collaboration[a]

[a]Institut für Kern- und Teilchenphysik, TU Dresden,
Zellescher Weg 19, 01069 Dresden, Germany

A preliminary *BABAR* measurement of the inclusive partial branching fraction $b \to ue\nu$ is presented. We find $\Delta\mathcal{B}(b \to ue\nu\ , 2.3 \ldots 2.6\ \text{GeV}) = (0.152 \pm 0.014_{stat.} \pm 0.014_{syst.}) \times 10^{-3}$. In addition, a preliminary measurement of the branching fraction $B \to \rho e\nu$ is presented and the CKM matrix element $|V_{ub}|$ is determined from this measurement. We find $\mathcal{B}(B^0 \to \rho^- e^+ \nu) = (3.39 \pm 0.44_{stat.} \pm 0.52_{syst.} \pm 0.60_{theo.}) \times 10^{-4}$, and $|V_{ub}| = (3.69 \pm 0.23_{stat.} \pm 0.27_{syst.}\ {}^{+0.40}_{-0.59\ theo.}) \times 10^{-3}$.

## 1. INTRODUCTION

The matrix element $|V_{ub}|$ can be determined by measuring the semileptonic $b \to ue\nu$ rate, either inclusively or exclusively. Experimentally, the main difficulty for the observation of $b \to ue\nu$ signal events is the large background from $b \to ce\nu$ events. The analysis will be sensitive in a lepton energy range where $b \to ce\nu$ decays are suppressed for kinematical reasons. In the exclusive approach, additional kinematic constraints allow to include a somewhat larger lepton energy range and to reduce extrapolation uncertainties. To extract exclusive branching fractions and $|V_{ub}|$ requires the use of hadronic form-factors which have to be obtained from theory. In the exclusive analysis presented here we use five different form factor calculations: two quark models ISGW2 [1] and Beyer/Melikhov [2], a lattice calculation by the UKQCD group [3], a model based on light cone sum rules (LCSR [4]), and one based on heavy quark and $SU(3)$ symmetries (Ligeti/Wise [5]).

## 2. INCLUSIVE ANALYSIS

### 2.1. Data set

The data were collected with the *BABAR* detector [6] at the PEP-II $e^+e^-$ storage ring [7]. The integrated luminosity of the sample is 20.6 fb$^{-1}$ taken at the $\Upsilon(4S)$ mass, corresponding to 22.6 million $B\bar{B}$ meson pairs. An additional 2.6 fb$^{-1}$ of data were taken 40 MeV below the $\Upsilon(4S)$ resonance.

### 2.2. Analysis strategy

We measure the partial semileptonic branching fraction $\Delta\mathcal{B}$ in the electron center-of-mass (CM) energy interval $2.2 < E_e^{CM} < 2.6$ GeV. The continuum is subtracted by fitting a polynomial to the off-resonance data. Backgrounds from $B \to X_c e\nu$, electrons from $J/\Psi \to e^+e^-$ decays, cascade decays $B \to X_c \to e$, and misidentified hadrons are subtracted using Monte Carlo simulation.

### 2.3. Results

The electron momentum spectrum is shown in Fig. 1. The partial $b \to ue\nu$ branching fraction is

$$\Delta\mathcal{B} = (0.152 \pm 0.014_{stat.} \pm 0.014_{syst.}) \times 10^{-3}\ . \quad (1)$$

Using input from CLEO's $b \to s\gamma$ measurement [8] this translates into

$$\mathcal{B}(b \to ue\nu) = (2.05 \pm 0.27_{exp.} \pm 0.46_{f_u}) \times 10^{-3}\ , \quad (2)$$

$$|V_{ub}| = (4.43 \pm 0.29_{exp.} \pm 0.25_{OPE} \pm 0.50_{f_u} \pm 0.35_{s_\gamma}) \times 10^{-3}\ . \quad (3)$$

The errors are experimental, from the Operator Product Expasion (OPE) [9], from the uncertainty on the fraction of events in the measured electron energy range ($f_u$), and from theoretical uncertainties related to the $b \to s(u)\gamma$ shape functions ($s_\gamma$).


doi:10.1016/S0920-5632(02)02148-5

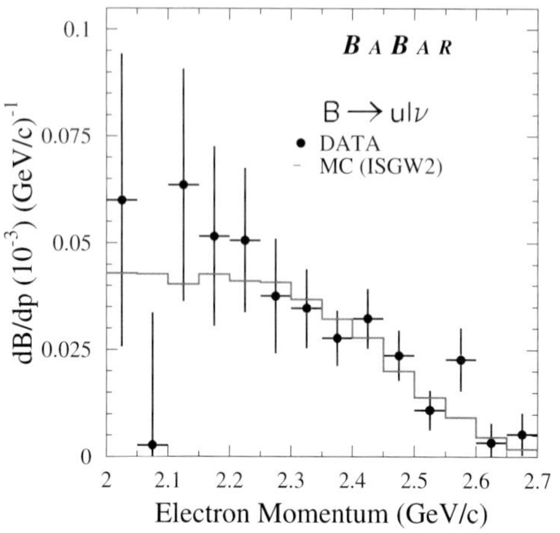


Figure 1. Inclusive electron momentum spectrum

## 3. EXCLUSIVE ANALYSIS

### 3.1. Data set

The integrated luminosity of the data sample is 50.5 $\text{fb}^{-1}$ taken at the $\Upsilon(4S)$ mass, corresponding to 55.2 million $B\bar{B}$ meson pairs. An additional 8.7 $\text{fb}^{-1}$ of data were taken 40 MeV below the $\Upsilon(4S)$ resonance.

### 3.2. Analysis strategy

We reconstruct the modes $B^+ \to \rho^0 e^+ \nu$, $B^0 \to \rho^- e^+ \nu$, $B^+ \to \omega e^+ \nu$, $B^+ \to \pi^0 e^+ \nu$, and $B^0 \to \pi^- e^+ \nu$ (with $\rho^0 \to \pi^+\pi^-$, $\rho^\pm \to \pi^0\pi^\pm$ and $\omega \to \pi^0\pi^+\pi^-$). The charge conjugate decays are implied throughout. The analysis is optimized for $B \to \rho e \nu$ decays; the $\pi$ and $\omega$ modes are included because of the crossfeeds into the $\rho$ modes. Isospin and quark model relations are used to effectively measure only two branching fractions, one for $B \to \rho e \nu$ and one for $B \to \pi e \nu$. Three kinematic variables are used: the electron energy $E_e^{\text{CM}}$; the invariant hadronic mass $M_{\pi\pi(\pi)}$ (for the $\rho$ and $\omega$ modes), and the difference between the reconstructed and expected $B$ meson energy ($\Delta E \equiv E_{\text{hadron}} + E_e + |\vec{p}_{\text{miss}}|c - E_{\text{beam}}$) in the center-of-mass system. Two electron energy regions are considered: $2.0 \leq E_e^{\text{CM}} < 2.3$ GeV (LOLEP), and $2.3 \leq E_e^{\text{CM}} < 2.7$ GeV (HILEP). The HILEP region is most sensitive to the signal because the $b \to c e \nu$ events are almost completely suppressed; the largest background source here is from continuum $e^+e^- \to q\bar{q}$ events. Real data taken below the $\Upsilon(4S)$ mass is used for the continuum subtraction. In the LOLEP region, $b \to c$ decays dominate and provide the normalization of the background at higher electron energies.

### 3.3. Fit method

We have performed a binned maximum-likelihood fit of the two-dimensional distributions ($M_{\pi\pi(\pi)}$, $\Delta E$) simultaneously in the two electron energy ranges (LOLEP, HILEP) and the five signal modes. The fit has contributions from the signal modes, other $b \to u e \nu$ decays, and $b \to c e \nu$ decays. Monte Carlo simulation provides the shapes of the distributions. The continuum, and a small contribution from misidentified electrons are determined using data. The decays $B \to D^{(*)} e \nu$ have been simulated using heavy quark effective theory (HQET [10]). The modes $B \to D^{(*)} \pi e \nu$ are simulated according to the Goity-Roberts model [11]. Resonant $b \to u e \nu$ downfeed modes are implemented according to the ISGW2 model. Non-resonant $b \to u e \nu$ modes are implemented according to a model by Fazio and Neubert [12]. The fit is performed five times using different form-factors for the signal modes $B^+ \to \rho^0 e^+ \nu$, $B^0 \to \rho^- e^+ \nu$, and $B^+ \to \omega e^+ \nu$.

### 3.4. Systematic errors

The systematic error includes contributions from uncertainties in the detector simulation ($\approx$ 8%), background modeling ($\approx$ 11%) and various other sources ($\approx$ 9%). The largest single systematic error comes from the uncertainty in the shape of the downfeed background.

### 3.5. Results

The isospin-constrained branching fraction results for the five different form-factors are shown in Fig. 2. The CKM matrix element $|V_{ub}|$ can be obtained from the branching fraction using

$$|V_{ub}| = \sqrt{\frac{\mathcal{B}(B^0 \to \rho^- e^+ \nu)}{\tilde{\Gamma}_{\text{thy}} \tau_{B^0}}}, \qquad (4)$$

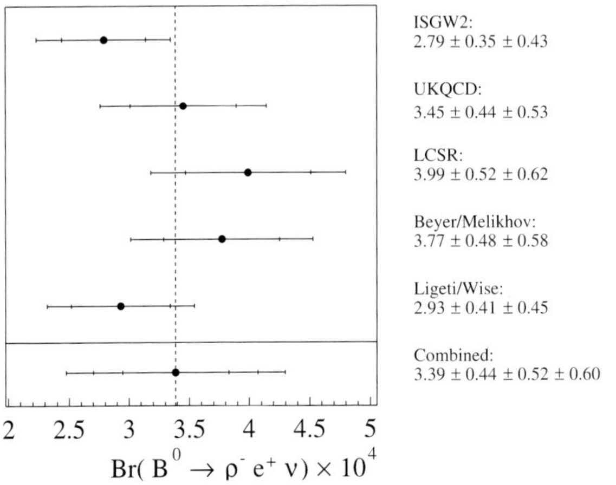


Figure 2. The $B^0 \to \rho^- e^+ \nu$ branching fraction results for different form-factors. The errors are statistical, systematic, and theoretical (in case of the combined result), successively combined in quadrature. The combined result is the average of the 5 form-factors, the theoretical error is half of the full spread observed.

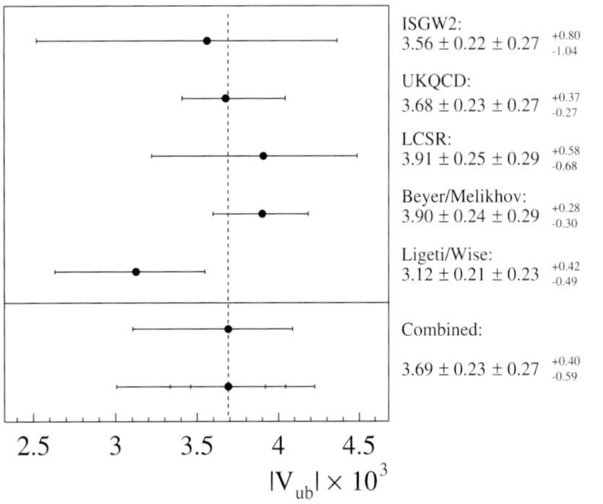


Figure 3. $|V_{ub}|$ determined using different form-factors. The results for each form-factor are drawn with theoretical error bars only. The combined result is described in the text.

where $\tilde{\Gamma}_{\rm thy}$ is the predicted form-factor normalization. The calculations quote errors on $\tilde{\Gamma}_{\rm thy}$ between 15% and 50%. We use $\tau_{B^0} = 1.548 \pm 0.032$ ps [13], and the branching fractions are taken separately for each form-factor as listed in Fig. 2. The combined central value is determined by taking the weighted mean of all form-factor results. The statistical and systematic errors of the combined result are determined by taking the mean of the relative errors of each individual result. The theoretical error is taken to be one half of the full spread of all fit results (including theoretical errors). The results for each form-factor and the combined result is shown in Fig. 3.

## REFERENCES


1. D. Scora and N. Isgur, CEBAF Preprint No. CEBAF-TH-94-14.
2. M. Beyer and D. Melikhov, *Phys. Lett.* **B436**, 344 (1998).
3. L. Del Debbio *et al.*, *Phys. Lett.* **B416**, 392 (1998).
4. P. Ball and V.M. Braun, *Phys. Rev.* **D58**, 094016 (1998).
5. Z. Ligeti and M.B. Wise, *Phys. Rev.* **D53**, 4937 (1996).
6. BABAR Collaboration, B. Aubert *et al.*, *Nucl. Instr. and Methods* **A479**, 1 (2002).
7. PEP II, SLAC-418, LBL-5379 (1993).
8. CLEO Collaboration, A. Bornheim *et. al.*, hep-ex/0202019 (2002)
9. J. Chay, H.Georgi, and B. Grinstein, *Phys. Lett.* **B247**, 399 (1990).
10. I.I. Bigi, M. Shifman, and N.G. Uraltsev, *Annu. Rec. Nucl. Part. Sci.* **47**, 591 (1997).
11. J.L. Goity and W. Roberts, *Phys. Rev.* **D51**, 3459 (1995).
12. F. Fazio and M. Neubert, *$B \to X_u \ell \nu$ decay distributions to order $\alpha_s$*, hep-ph/9905351v2.
13. Particle Data Group, D.E. Groom *et al.*, *Eur. Phys. Jour.* C **15**, 1 (2000).

NH
ELSEVIER

Proceedings of ICHEP 2002, pp. 472–474
*S. Bentvelsen, P. de Jong, J. Koch and E. Laenen (Editors)*

31st INTERNATIONAL CONFERENCE ON HIGH ENERGY PHYSICS AMSTERDAM

# Subleading Shape Functions and the Determination of $V_{ub}$

Thomas Mannel [a]

[a]Institut für Theoretische Teilchenphysik, Universität Karlsruhe, D–76128 Karlsruhe, Germany

The endpoint region of the inclusive lepton energy spectrum in $B \to X_u \ell \bar{\nu}$ can be used for a determination of $V_{ub}$ by comparing it to the endpoint region of the inclusive photon spectrum of $B \to X_s \gamma$. We consider the subleading twist contributions to this $V_{ub}$ determination.

A precise determination of the CKM matrix element $|V_{ub}|$ is one of the most important measurements in the current quest to overconstrain the unitarity triangle to test the standard model. It is currently believed that the theoretically cleanest determination of $|V_{ub}|$ comes from the inclusive rate for $B \to X_u \ell \bar{\nu}$, which may be calculated model-independently via an operator product expansion in terms of local operators [1]. Such a measurement is hampered by the experimental necessity to suppress the overwhelming charm background, which usually requires tight kinematic cuts. With a few exceptions [2], such cuts invalidate the local OPE usually used to calculate the inclusive branching ratio, and a twist expansion has to be performed instead [3]. At leading order in this twist expansion, and ignoring perturbative corrections, the lepton energy spectrum in inclusive $B \to X_u \ell \bar{\nu}$ decay is given by

$$\frac{d\Gamma}{dE_\ell} = \frac{2\Gamma_0}{m_b} \int\limits_{-\infty}^{m_b - 2E_\ell} d\omega\, f(\omega)\,,\ \Gamma_0 = \frac{G_F^2 |V_{ub}|^2 m_b^5}{192\pi^3}\,, \quad (1)$$

where $f(\omega)$ is the nonperturbative light cone structure function of the $B$ meson

$$f(\omega) = \frac{1}{2m_B} \langle B | \bar{h}\, \delta(in \cdot D + \omega)\, h | B \rangle\,. \quad (2)$$

Since this function is a property of the $B$ meson, it is process independent and can thus be measured elsewhere. The best way to measure this structure function is from the photon energy spectrum of the inclusive decay $B \to X_s \gamma$ [4]. Up to perturbative and subleading twist corrections, this spectrum is directly proportional to the structure function,

$$\frac{d\Gamma}{dE_\gamma} = \frac{G_F^2 |V_{tb} V_{ts}^*|^2 \alpha |C_7^{\rm eff}|^2 m_b^5}{32\pi^4} f(E_\gamma)\,. \quad (3)$$

Thus, combining data on $B \to X_s \gamma$ with data from $B \to X_u \ell \bar{\nu}$, one can eliminate the dependence on the structure function and therefore determine $|V_{ub}|$ with no model dependence at leading order [3–5]. The relation is

$$\left| \frac{V_{ub}}{V_{tb} V_{ts}^*} \right|^2 = \frac{3\alpha}{\pi} |C_7^{\rm eff}|^2 \frac{\Gamma_u(E_c)}{\Gamma_s(E_c)} + \cdots \quad (4)$$

where the elipses denote terms of order $1/m_b$ and $\alpha_s$ and

$$\Gamma_u(E_c) \equiv \int_{E_c}^{m_B/2} dE_\ell \frac{d\Gamma_u}{dE_\ell}$$

$$\Gamma_s(E_c) \equiv \frac{2}{m_b} \int_{E_c}^{m_B/2} dE_\gamma (E_\gamma - E_c) \frac{d\Gamma_s}{dE_\gamma}\,. \quad (5)$$

In this talk I describe the work done in collaboration with C. Bauer and M. Luke [6,7] (see also [8]), where we calculate the subleading twist corrections to Eq. (4). While perturbative corrections to this relation have been studied in detail [4,5], much less is known about nonperturbative corrections suppressed by powers of $\Lambda_{\rm QCD}/m_b$. In [6] we introduced a formalism which allows us to parametrize these subleading twist corrections in terms of subleading shape functions. The subleading twist corrections to the photon energy spectrum in $B \to X_s \gamma$ read

$$\frac{m_b}{\Gamma_0^s} \frac{d\Gamma}{dE_\gamma} = (4E_\gamma - m_b) F(m_b - 2E_\gamma) + \frac{1}{m_b} \left[ h_1(m_b - 2E_\gamma) + H_2(m_b - 2E_\gamma) \right] \quad (6)$$

where

$$\Gamma_0^s = \frac{G_F^2 |V_{tb} V_{ts}^*|^2 \alpha |C_7^{\rm eff}|^2 m_b^5}{32\pi^4} \quad (7)$$


doi:10.1016/S0920-5632(02)02151-5

and the functions $F$, $h_1$ and $H_2$ are defined in (14) below.

The subleading twist corrections to the lepton energy spectrum in $B \to X_u \ell \bar{\nu}$ may be calculated in much the same way as the photon spectrum for $B \to X_s \gamma$. The matching calculation is performed by expanding full QCD matrix elements to subleading twist and identifying the result with non-local operators introduced in [6]. The calculation is standard and uses the optical theorem to obtain the lepton energy spectrum from the imaginary part of a forward scattering matrix element. This matrix element is rewritten using $p_b^\mu = m_b v^\mu + k^\mu$ and $Q^\mu = m_b v^\mu - p_\ell^\mu$ and the light cone vectors $n^\mu$, $\bar{n}^\mu$ ($n^2 = \bar{n}^2 = 0$, $n \cdot \bar{n} = 2$, $2v = n + \bar{n}$), defined by $p_\ell^\mu = E_\ell \bar{n}^\mu$. The endpoint region is defined by

$$\bar{n} \cdot Q = m_b \quad n \cdot Q = m_b - 2E_\ell \sim O(\Lambda_{\rm QCD})$$

and the matrix element is expanded in powers of $1/m_b$, assuming $k^\mu \sim O(\Lambda_{\rm QCD})$.

The leading twist contribution is the first term of this expansion. It can be written as a convolution of a Wilson coefficient function $C_0(\omega)$ with an operator $O_0(\omega)$

$$\int d\omega\, C_0(\omega) O_0(\omega) = \int d\omega\, C_0(\omega) \bar{h} \delta(in \cdot D + \omega) h \quad (8)$$

and one finds at tree level

$$C_0(\omega) = 2\,\Gamma_0\, \theta(m_b - 2E_\ell - \omega)\,, \quad (9)$$

leading to Eq. (1).

At subleading twist there are five nonlocal operators contributing to $B$ meson decays [6]

$$\begin{aligned} O_1^\mu(\omega) &= \bar{h}_v \left\{ iD^\mu, \delta(in \cdot D + \omega) \right\} h_v \\ P_2^{\mu\alpha}(\omega) &= \bar{h}_v s^\alpha \left[ iD^\mu, \delta(in \cdot D + \omega) \right] h_v \\ O_T(\omega) &= i \int d^4y \, \frac{dt}{2\pi}\, e^{-i\omega t} T\left( \bar{h}_v(0) h_v(t) O_{1/m}(y) \right) \\ O_3(\omega) &= \int d\omega_1 d\omega_2 \, \frac{\delta(\omega - \omega_1) - \delta(\omega - \omega_2)}{\omega_1 - \omega_2} \\ &\quad \left[ h_v \delta(in \cdot D + \omega_2)(iD_\perp)^2 \delta(in \cdot D + \omega_1) h_v \right] \\ P_4^\alpha(\omega) &= g \int d\omega_1 d\omega_2 \, \frac{\delta(\omega - \omega_1) - \delta(\omega - \omega_2)}{\omega_1 - \omega_2} \\ &\quad \left[ h_v s^\alpha \delta(in \cdot D + \omega_2) \epsilon_\perp^{\mu\nu} G_{\mu\nu}^\perp \delta(in \cdot D + \omega_1) h_v \right], \end{aligned}$$

where $s^\alpha = P_+ \gamma^\alpha \gamma^5 P_+$ and $O_{1/m}$ are the subleading contributions to the Lagrangian.

To match onto these operators we expand the full theory calculations to one order higher in $k^\mu / m_b$ and $n \cdot Q / m_b$. The result for the coefficients for the subleading twist operators $O_1^\mu(\omega)$ and $O_3(\omega)$ is

$$C_1^\mu(\omega) = \frac{\Gamma_0}{m_b} n^\mu \theta(m_b - 2E_\ell - \omega) \quad (10)$$

$$C_3(\omega) = -\frac{2\Gamma_0}{m_b} \theta(m_b - 2E_\ell - \omega)\,. \quad (11)$$

The result for the the coefficient $C_3(\omega)$ agrees with the prediction from reparameterization invariance, which relates this coefficient to the leading order Wilson coefficient to all orders in perturbation theory [9]. For the operators $P_2^{\mu\alpha}(\omega)$ and $P_4^\alpha(\omega)$ we obtain the matching coefficients

$$D_2^{\mu,\alpha}(\omega) = -\frac{\Gamma_0}{m_b} i \epsilon_\perp^{\mu\alpha} \theta(m_b - 2E_\ell - \omega) \quad (12)$$

$$D_4^\alpha(\omega) = \frac{3\Gamma_0}{m_b} n^\alpha \theta(m_b - 2E_\ell - \omega)\,. \quad (13)$$

Finally we have to parametrize the matrix elements of these operators between $B$ meson states. We define

$$\begin{aligned} \langle H | O_0(\omega) | H \rangle &= 2m_B\, f(\omega) \\ \langle H | O_1^\mu(\omega) | H \rangle &= -4m_B\, \omega f(\omega)\, (v^\mu - n^\mu) \\ \langle H | O_3(\omega) | H \rangle &= 2m_B\, G_2(\omega) \\ \langle H | O_T(\omega) | H \rangle &= 2m_B\, t(\omega) \\ \langle H | P_2^{\alpha,\mu}(\omega) | H \rangle &= -2m_B\, i\epsilon_\perp^{\mu\alpha} h_1(\omega) \\ \langle H | P_4^\alpha(\omega) | H \rangle &= 4m_B\, (v^\alpha - n^\alpha) H_2(\omega) \end{aligned} \quad (14)$$

Due to reparametrization invariance and the general structure of the $1/m_b$ corrections we find that we only can have the combination

$$F(\omega) = f(\omega) + \frac{1}{2m_b} \left[ t(\omega) - 2\, G_2(\omega) \right]\,. \quad (15)$$

which is a universal function for any decay rate. Putting all this information together, we therefore find for the lepton energy spectrum in $B \to X_u \ell \bar{\nu}$

$$\frac{d\Gamma}{dE_\ell} = \frac{4\,\Gamma_0}{m_b} \int d\omega\, \theta(m_b - 2E_\ell - \omega) \left[ F(\omega) \left( 1 - \frac{\omega}{m_b} \right) - \frac{1}{m_b} h_1(\omega) + \frac{3}{m_b} H_2(\omega) \right] \quad (16)$$

up to terms of order $\Lambda_{\rm QCD}^2/m_b^2$. We can now use these results to get the $O(\Lambda_{\rm QCD}/m_B)$ corrections to the determination of $|V_{ub}|$ given in Eq. (4). We find for the observables defined in Eq. (5)

$$\Gamma_u(E_c) = \frac{4\Gamma_0}{m_b}\int_{E_c}^{m_B/2} dE \int_{-\infty}^{m_b-2E} d\omega \left[F(\omega)\left(1-\frac{\omega}{m_b}\right) - \frac{1}{m_b}h_1(\omega) + \frac{3}{m_b}H_2(\omega)\right] \quad (17)$$

$$\Gamma_s(E_c) = \frac{2\Gamma_0^s}{m_b}\int_{E_c}^{m_B/2} dE \int_{-\infty}^{m_b-2E} d\omega \left[F(\omega)\left(1-\frac{2\omega}{m_b}\right) + \frac{1}{m_b}h_1(\omega) + \frac{1}{m_b}H_2(\omega)\right] \quad (18)$$

where $\Gamma_0^s$ and $\Gamma_0$ are defined in Eqs. (7) and (1). This finally allows us to obtain the correction to the determination of $V_{ub}$

$$\left|\frac{V_{ub}}{V_{tb}V_{ts}^*}\right| = \left(\frac{3\alpha}{\pi}|C_7^{\rm eff}|^2\frac{\Gamma_u(E_c)}{\Gamma_s(E_c)}\right)^{\frac{1}{2}}(1+\delta(E_c)) \,, \quad (19)$$

where $\delta(E_c)$ can be calculated form (17) and (18).

To get a quantitative estimate for these corrections, we have to model the subleading structure functions. Using a simple model for the leading order structure function given in [10], we take the subleading functions as the derivative of the leading-order shape function, normalized such that they reproduce the known first moments.

$$f_{\rm mod}(\omega) = \frac{32}{\pi^2\Lambda}\left(1+\frac{\omega}{\Lambda}\right)^2 \exp\left[-\frac{4}{\pi}\left(1+\frac{\omega}{\Lambda}\right)^2\right] \theta\left(1+\frac{\omega}{\Lambda}\right) \quad (20)$$

where $\Lambda = m_B - m_b$. This gives

$$F_{\rm mod}(\omega) = f_{\rm mod}(\omega) + \frac{\lambda_1 - 9\lambda_2}{6\,m_b} f'_{\rm mod}(\omega) \quad (21)$$

$$h_{1\,\rm mod} = \lambda_2\, f'_{\rm mod}(\omega) \quad H_{2\,\rm mod} = -\lambda_2\, f'_{\rm mod}(\omega)$$

and allows us to compute the correction $\delta(E_c)$, which is plotted in Fig. 1.

The dashed line in Fig. 1 indicates a typical cut, and the corrections in this case are of the order of 15% and hence are sizable. Since this result is obtained using a model, the conservative estimate is to assign an uncertainty of this size to the determination of $V_{ub}$. However, it has been argued [11] that the situation may be not as bad, since the first moments of the subleading functions are known and that the situation may be improved including this information.

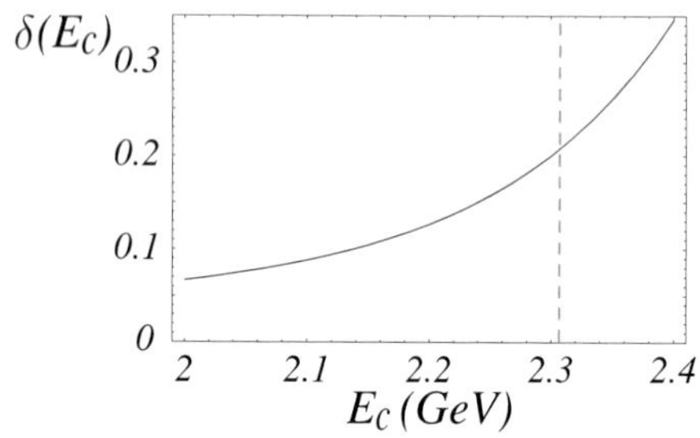


Figure 1. $\delta(E_c)$ as a function of $E_c$.

## REFERENCES


1. J. Chay, H. Georgi and B. Grinstein, Phys. Lett. B247 (1990) 399; M.A. Shifman and M.B. Voloshin, Sov. J. Nucl. Phys. 41 (1985) 120; I.I. Bigi *et al.*, Phys. Lett. B293 (1992) 430 [Erratum-ibid. B 297, 430 (1992)]; I.I. Bigi *et al.*, Phys. Rev. Lett. 71 (1993) 496; A.V. Manohar and M.B. Wise, Phys. Rev. D49 (1994) 1310.
2. C. W. Bauer, Z. Ligeti and M. E. Luke, Phys. Lett. B **479**, 395 (2000); Phys. Rev. D **64**, 113004 (2001)
3. I. I. Bigi, M. A. Shifman, N. G. Uraltsev and A. I. Vainshtein, Int. J. Mod. Phys. **A9**, 2467 (1994); M. Neubert, Phys. Rev. D **49**, 3392 (1994).
4. M. Neubert, Phys. Rev. **D49**, 4623 (1994).
5. A. K. Leibovich, I. Low and I. Z. Rothstein, Phys. Rev. D **61**, 053006 (2000).
6. C. W. Bauer, M. E. Luke and T. Mannel, `arXiv:hep-ph/0102089`.
7. C. W. Bauer, M. Luke and T. Mannel, Phys. Lett. B **543**, 261 (2002)
8. A. K. Leibovich, Z. Ligeti and M. B. Wise, Phys. Lett. B **539**, 242 (2002)
9. M. E. Luke and A. V. Manohar, Phys. Lett. B **286**, 348 (1992) F. Campanario and T. Mannel, Phys. Rev. D **65**, 094017 (2002)
10. T. Mannel and M. Neubert, Phys. Rev. D **50**, 2037 (1994)
11. M. Neubert, Phys. Lett. B **543**, 269 (2002)

ELSEVIER

Proceedings of ICHEP 2002, pp. 475–480
*S. Bentvelsen, P. de Jong, J. Koch and E. Laenen (Editors)*

31st INTERNATIONAL CONFERENCE ON HIGH ENERGY PHYSICS AMSTERDAM

# A Review of $|V_{ub}|$ and $|V_{cb}|$ Determinations

M. Battaglia[a]

[a]CERN, EP Division,
CH 1211 Geneva, Switzerland

## 1. INTRODUCTION

This paper reviews the status of experimental determinations of the $|V_{ub}|$ and $|V_{cb}|$ elements in the Cabibbo-Kobayashi-Maskawa (CKM) mixing matrix. Such measurements, with small and well understood uncertainties, represent a key goal for the present heavy flavour physics program.

Because $|V_{ub}|$, the smallest quark mixing element, provides a bound on the upper vertex of one of the triangles representing the unitarity property of the CKM matrix and $|V_{cb}|$ represent the overall normalisation, these studies play a crucial role in examining these unitarity constraints and the fundamental questions which relate to it. Comparing the direct determinations of the CP violating phases with the values predicted by the Standard Model (SM), from the size of these and other CKM elements, may reveal sources of CP violation arising from New Physics beyond the SM.

The statistical interpretation of these results will depend, not only on the level of accuracy achieved, but also on the possibility of properly defining probability density functions for their values. Theories, phenomenological models and experimental techniques are making constant progresses to address these requirements. They define a comprehensive program based on studies of both inclusive and exclusive methods where signal purities, control of backgrounds and of theory uncertainties and the possibility to extract the main theory parameters directly from the data are the main issues. Within the present and foreseen experimental accuracies, theory uncertainties dominate the error budget, making the test of the underlying assumptions an absolute priority.

The extraction of the CKM elements from inclusive s.l. decay is based on Heavy Quark Theory implemented through an Operator Product Expansion (OPE). This expansion provides a consistent theory framework for computing inclusive rates and partially integrated spectra. The main input parameters are the heavy quark masses $m_b$, $m_c$ [or $\bar{\Lambda}$], the kinetic energy of the $b$-quark inside the heavy hadron $\mu_\pi^2$ [or $\lambda_1$] and the coefficients governing the $1/m_b^3$ corrections. An important recent development has been the possibility to extract the non-perturbative parameters from moments of spectral distributions in $b \to X_c \ell \bar{\nu}$ and $b \to X_s \gamma$ decays. Also the shape function, which encodes the non-perturbative effects relating the parton level dynamics to the physical distributions, could be constrained using the photon energy spectrum in $b \to X_s \gamma$ radiative decays.

It is important to observe at this point that the OPE predictions rely on an implicit assumption of quark-hadron duality. While this is expected to hold best for s.l. decays, the levels of accuracies achieved have increased the awareness of the community for the need of direct, experimental tests of the underlying assumptions. Performing several determinations with different methods and verifying their consistency will add trust on the level of accuracy achieved.

Different kinematics and detector configurations available at symmetric $\Upsilon(4S)$ colliders, asymmetric $B$-factories and at colliders operating at the $Z^0$ peak and the large recorded data sets have brought an healthy competition of complementary approaches to this program.


doi:10.1016/S0920-5632(02)02153-9

Table 1
Some characteristics of $B \to D^*\ell\nu$ analyses

| | ALEPH [8] | BELLE [9] | CLEO [10] | DELPHI [11] | DELPHI [7] | OPAL [12] | OPAL [12] |
|---|---|---|---|---|---|---|---|
| Technique | Excl | Excl | Excl | Incl | Excl | Excl | Incl |
| $D^*$ Stats | 579 | 766 | 2683 | 6495 | 1688 | 1708 | |
| $\frac{b \to D^*\ell\nu}{b \to D^*}$ | 0.86 | 0.85 | 0.90 | 0.72 | 0.70 | 0.77 | |
| $\delta w$ | 0.07 | 0.04 | 0.03 | 0.13 | 0.11 | 0.10 | 0.12 |
| $\Delta\epsilon$ | 1.0 | 0.3 | 0.13/1. | 1.0 | 1.0 | 1.0 | 1.0 |

## 2. $|V_{cb}|$ DETERMINATIONS

### 2.1. Inclusive Determinations

The $|V_{cb}|$ value can be obtained from the measured $b$-quark s.l. decay partial width. Experimentally this is obtained by measuring the inclusive s.l. branching fraction of $B$ hadrons and their lifetime(s). These are presently known to a few percent and a fraction of percent respectively. This brings the experimental uncertainty on $|V_{cb}|$ to about 1%. New results from BABAR and BELLE [1] have added further accuracy, while confirming the central value determined at LEP and CESR. Issues on the model uncertainties of the LEP results are also better understood [2]. After averaging the $\Upsilon(4S)$ and $Z^0$ results and correcting for the $b \to u$ contribution, the $B$ s.l. decay width is $\Gamma(b \to X_c\ell\nu) = 0.434 \times (1 \pm 0.018) \times 10^{-10}$ MeV.

Theoretical expressions for the s.l. decay width depend on several parameters and include perturbative and non-perturbative QCD corrections. The uncertainty on the heavy quark masses, through $m_b$ and $m_b - m_c$, has traditionally been a major source of uncertainties, due to the $m_b^5$ dependence of the partial width. There has been a significant progress in evaluating the $b$ quark mass from spectroscopy and the mass definition itself has been refined by moving from pole masses to short-distance masses, which are free from renormalon ambiguities. Several determinations have been reported and $\bar{m}_b(m_b)$=(4.21±0.08) GeV, or $m_b$(1 GeV)=(4.57±0.09) GeV, represents a generally accepted estimate. Extracting $|V_{cb}|$ in this way has a ±3% relative uncertainty from $m_b$ and $\mu_{pi}^2$, 4% from perturbative and higher order corrections. More recently the analysis of spectral moments in $b \to s\gamma$ and $b \to X_c\ell\nu$ has provided means to extract the heavy quark masses and other non-perturbative parameters directly from the data. This allows to turn these uncertainties from theory parameters, whose ranges and statistical meaning are a point of controversy, into experimental uncertainties. Using a determination of $\Lambda$ and $\lambda_1$ from the CLEO moment analysis [3] and taking the $1/m_b^3$ and perturbative corrections as theoretical systematics, brings these to 2%, to get $|V_{cb}| = (40.7 \pm 0.6 \pm 0.8) \times 10^{-3}$. There have been further moments determinations by DELPHI [4] and BABAR [5]. Further improvement is expected by using the determinations of higher moments of hadron and lepton spectra to extract the $1/m_b^3$ coefficients.

### 2.2. Exclusive Determinations

Determining $|V_{cb}|$ using exclusive s.l. $B \to D^{(*)}\ell\bar{\nu}$ decays has several advantages. Heavy Quark Effective Theory (HQET) relates the differential decay rate $d\Gamma/dw$ to $|V_{cb}|$ by means of a form factor $\mathcal{F}(w)$ for which the normalisation $\mathcal{F}(\infty) = 1$ at zero recoil is known for $m_Q \to \infty$. The $D^{*+} \to D^0\pi^+$ decay can be reconstructed either through the exclusive $D^0 \to K\pi$, $K\pi\pi^0$, $K\pi\pi\pi$ modes or an inclusive method which profits of the small phase space available to the $\pi^+$ and uses the mass difference $\Delta M = M_{X\pi} - M_X$, where $X$ is a partially reconstructed $D^0$ meson. The first method gives higher signal purities and better $w$ resolution, while the inclusive reconstruction provides with higher statistics. One advantage in performing this measurement at the $Z^0$ peak has been the flat acceptance vs. $w$, since the slow $\pi^+$ momentum is enhanced by the large $B$ boost. At the $\Upsilon(4S)$, the reduced charged pion acceptance can be compensated in part by adding the $D^{*+} \to D^+\pi^0$ mode for which the

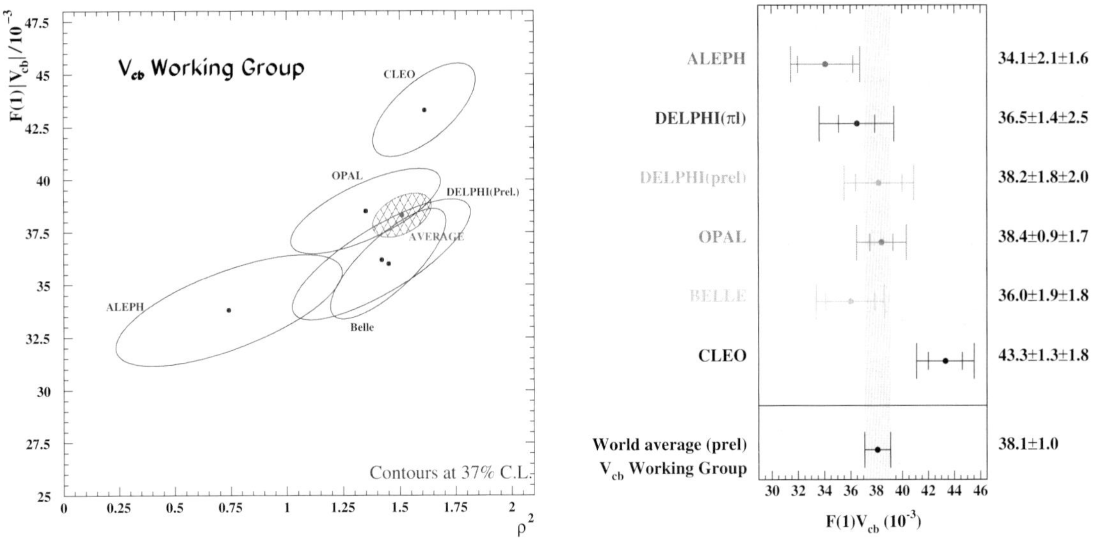


Figure 1. Results on exclusive determinations of $|V_{cb}|$: $1\sigma$ contours in the $\mathcal{F}(1) \times |V_{cb}|$ - $\rho^2$ plane for individual determinations (left) and summary of results and average value for $\mathcal{F}(1) \times |V_{cb}|$ (right).

$\pi^0$ efficiency is significantly more stable. An important source of correlated systematics is represented by the $b \to D^{**}\ell\nu$ background, which is still quite poorly known. Recent analyses fit the $D^*X\ell\nu$ fraction directly on data by using angular distributions or track topology. A new DELPHI study [7] finds BR($b \to D^{*+}X\ell\bar{\nu}$) = ($0.64 \pm 0.08 \pm 0.09$)%. The functional dependence of $d\Gamma/dw$ towards $w = 1$ is not known, but costraints are provided by dispersion relations [6]. An expansion around $w = 1$ in terms of the form factor slope $\rho$ is adopted for the extraopolation at zero recoil. The results for the measured product of $\mathcal{F}(1) \times |V_{cb}|$ are summarised in Figure X. These have been averaged accounting for correlations between the product $|V_{cb}| \times \mathcal{F}(1)$ and the $\rho^2$ parameter. In order to extract $|V_{cb}|$, the $\mathcal{F}(1)$ corrections due to finite $b$-quark mass need to be estimated. Computations have been performed using quark models, sum rules and lattice QCD. Results agree. Quark models, which rely on a quark-hadron duality assumption are not used and the adopted average [13] is $\mathcal{F}(1)$=0.91±0.04. This gives $|V_{cb}|$=(41.9±1.1±1.9)$\times 10^{-3}$. There has been a thorough review of the error budget for the lattice result and an effort to define the result in terms of a full probability density function [14], useful as input into fits of the unitarity triangle.

First results have been contributed by CDF [15] for the Tevatron Run II data making use of the Lepton + Si Vertex Trigger. The 960 fully reconstructed 960 $B \to D^*\ell\nu$ extrapolate to $\simeq 200k$ events for a Run II luminosity of 2 fb$^{-1}$. There is a real potential for extracting $|V_{cb}|$ precisely from such a large data set, which will crucially depend on the achievable $w$ resolution. In addition, large samples of $D^{**}$ will be available for measuring masses, widths and branching fractions and performing helicity analyses.

## 3. $|V_{ub}|$ DETERMINATIONS

Contrary to the case of $|V_{cb}|$, where the backgrounds are benign and the main issues are related to the experimental resolution and the theory inputs, the central issue in the determination of $|V_{ub}|$ is the extraction of the $b \to X_u\ell\bar{\nu}$ signal in presence of a $\simeq 75$ times larger $b \to X_c\ell\bar{\nu}$ backgrounds. This requires to know and control differential rates in regions of phase space where the $b \to X_c$ transitions are suppressed. The first determination of $|V_{ub}|$ was in fact obtained by measuring the yield of leptons produced above the kinematical limit for $b \to X_c \ell \bar{\nu}$ decays. However, the accuracy was limited by the model dependence. A recent review can be found in [16]

New paths, capable of leading to significant improvements in the control of these model uncertainties, have been undertaken. The first is the study of inclusive s.l. $B$ decays, where the signal is isolated by selections based on the lepton energy, hadronic mass, $\ell\nu$ mass or a combination of them.

The second is the determination of the branching fractions of exclusive decays, such as $B \to \pi$, $\rho \ell \bar{\nu}$ , which has reached a good statistical accuracy thanks to the large data sets collected at $\Upsilon(4S)$ facilities. Owing to the divergent directions taken by these developments, the present picture has been enriched by new data of good experimental accuracy, which should by affected by largely uncorrelated systematic uncertainties.

### 3.1. Inclusive Determinations

There are three main regions, available in the s.l. $b$ decay decay phase space. to enhance signal $b \to X_u \ell \bar{\nu}$: i) the lepton energy end-point region: $E_\ell > \frac{M_B^2 - M_D^2}{2M_B}$, ii) the low hadronic mass region: $M_X < M_D$ and iii) the high $q^2$ region: $M_{\ell\nu}^2 = q^2 > (M_B - M_D)^2$. They select $\simeq$15%, 70% and 20% of the inclusive charmless s.l. rate, respectively.

The lepton endpoint region has originally been affected by large systematics due to the uncertainties in extrapolating from a small fraction of selected events, at the edge of the phase space, to the full rate. However, it has been known since some time that, at leading twist, the same shape function corrects the parton level $b \to X_s \gamma$ photon spectrum and the $b \to X_u \ell \bar{\nu}$ lepton spectrum. CLEO first determined the shape function parameters from the photon spectrum, using different ansatz for its functional form [20]. This allowed to derive the fraction of $b \to u$ s.l. transitions yielding a lepton with energy 2.2 GeV< $E_\ell$ <2.6 GeV. A new preliminary analysis has now been presented by BaBar [17], which uses the same CLEO shape function result, but restricts the acceptance to 2.3 GeV< $E_\ell$ <2.6 GeV. The near future will provide with more accurate determinations of the $E_\gamma$ spectrum, allowing even better determinations. However, it will be important to understand the effects of subleading corrections which, although suppressed by $\Lambda_{QCD}/m_b$ factors, have recently been recognised to be numerically enhanced [18].

The use of an upper cut on the reconstructed hadronic mass, $M_X$, recoiling against the lepton, was first attempted by DELPHI, exploiting the favourable kinematics of LEP. This method has an advantage in the larger fraction of decay rate accepted but has to deal with a rather limited signal-to-background ratio. The shape function sensitivity is smaller compared to the end-point method, as it has been verified using the range of parameters determined by CLEO. At LEP, two inclusive analyses [19,24] have been performed, using a Neural Network (NN) discriminant, based on a selection of kinematic variables providing separation of the $b \to X_u \ell \bar{\nu}$ signal.

A recent development has been the introduction of a combined $M_X - q^2$ analysis to reduce the theory uncertainties, while restricting to a part of the phase space where the $b \to c$ background is suppressed. CLEO has presented the first experimental attempt to implement this method. The analysis is based on a full fit to $q^2/(E_\ell + E_\nu)^2$, $M_X$ and $\cos\theta_{W\ell}$. Models are needed to extract the sample composition and to relate the regions of higher sensitivity and theoretically safer to the inclusive charmless s.l. branching fraction.

With the severe selections needed to extract the signal, the effects of experimental cuts trimming inclusive distributions must be understood. An important question to assess is to which degree the inclusive analyses are evenly probing the selected phase space. The NN analyses are biased towards the region of large $E_\ell$ and low $M_X$ where the signal-to-background ratio is more favourable. While the resulting uncertainty is accounted by the range of models tested, for example in the ALEPH analysis, it is desirable to test more unbiased methods. DELPHI has shown that the $M_X$ analysis has a reasonably uniform sensitivity on the $M_X$-$E_\ell$ plane. The recent CLEO analysis has been repeated for different sets of $M_X$-$q^2$ selections, finding compatible $|V_{ub}|$ results.

Table 2
Some characteristics of inclusive $|V_{ub}|$ determinations.

| | ALEPH [19] | BABAR [17] | CLEO [20] | CLEO [21] | DELPHI [22] | L3 [23] | OPAL [24] |
|---|---|---|---|---|---|---|---|
| Technique | NN | $E_\ell$ | $E_\ell$ | $M_X - q^2$ | $M_X$ | $\pi - \ell$ | NN |
| $F_u$ | | 0.07 | .13 | 0.15 | 0.68 | | |
| Purity | 0.06 | 0.27 | 0.28 | | 0.09 | 0.18 | 0.04 |
| $\sigma_{b\to c}$ | 15% | 3% | 7% | 6% | 12% | 22% | 15% |
| $\sigma_{th}$ | 9% | 13% | 13% | 14% | 10% | 10% | 10% |

### 3.2. Exclusive Determinations

The exclusive reconstruction of charmless s.l. decays offers the comfortable signal-to-background ratios that the inclusive analyses lack. However, since here there is no equivalent to the HQET predictions for the form factors at zero recoil which make the $B \to D^*\ell\bar{\nu}$ channel so attractive, the form factors represent the main limitation in the accuracy of the exclusive $V_{ub}$ determinations. These uncertainties stem from three sources: i) the signal efficiency, ii) the subtraction of the cross-feed background and iii) the rate normalisation. Form factors have been computed with a variety of techniques. Typically, Light-Cone Sum Rules give 15-20% accuracy and are subject to quark-hadron duality uncertainties. Quenced Lattice calculations give a comparable precision, but their applicability is mostly confined to the high $q^2$ region. Progresses with Lattice calculations is expected to reduce these uncertainties significantly in the coming years.

At present, the statistical accuracies are continuously improved thanks to the larger data sets and improved reconstruction techniques. Larger statistics also allow to reduce model systematics. CLEO has reported the preliminary results of a $B^0 \to \pi^-\ell^+\nu$ analysis where the signal rates are extracted in three regions of $q^2$ independently [25]. This reduces the form factor sensitivity from the first two sources. The statistical accuracy of the data is also becoming sensitive to the details of the $q^2$ dependence of the signal yield predicted by different form factor computations. Results have been presented by both CLEO and BELLE for the $\pi\ell\nu$ channel. This makes possible to discriminate between models and the CLEO fit shows that the ISGW II model, with a $\chi^2$ probability of less than 1%, is not reliable for extracting $|V_{ub}|$ with this channel.

The $\pi\ell\nu$ mode, which involves a single form factor, is free from non-resonant backgrounds and has reduced feed-down uncertainties, will most probably enjoy the role of golden channel for performing the exclusive determination of $|V_{ub}|$. The $\rho$, $\omega$, $\eta\ell\nu$ channels will also play an important role in adding confidence, if they will result in compatible measurements.

An overall summary of the results presently available is given in Figure 2. At present no attept is made to average these measurements. As the understanding of the model dependence improves and parameter ranges can be consistently defined for different analyses, averages for the inclusive and exclusive determinations will become possible. While there is a remarkable agreement between the measurements obtained with a variety of different techniques, it will be interesting to monitor possible systematic shifts between the inclusive and exclusive measurements.

Other paths may also become available to the large data sets of the $B$ factories. An inclusive techniques, which uses a direct ratio of $b \to s\gamma$ and $b \to u\ell\nu$ spectra has been recently proposed [27]. The decay $B \to D_s\pi$ is also sensitive to the size of $|V_{ub}|$. BELLE has measured $\Gamma(B^0 \to D_s^+\pi^-)/\Gamma(B^0 \to D_s^+D^-) = (3.0^{+1.8}_{-1.6}) \times 10^{-3}$ which can be related to $|V_{ub}|^2/|V_{cb}|^2$.

## 4. CONCLUSIONS

A large set of determinations of $|V_{ub}|$ and $|V_{cb}|$ exploiting different techniques and affected by different assumptions and sources of uncertainties has become available. The determination of $|V_{cb}|$ is reaching a precision better than 1%. New approaches start allowing a determination of most

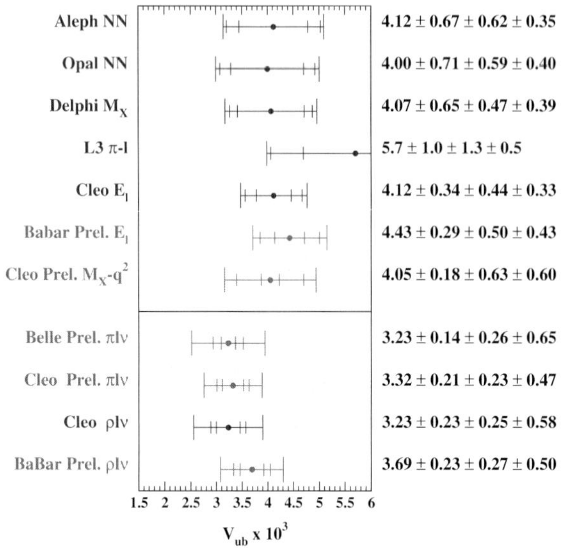

Figure 2. Summary of inclusive and exclusive determinations of $|V_{ub}|$.

of the input parameters directly from the data. Tests of OPE through moment analyses are also becoming available. The increasing ingenuity of experimentalists and phenomenologists in devising methods to measure $|V_{ub}|$ has brought individual accuracies to approach 15%.

The overall consistency is encouraging for the perspectives of bringing $|V_{ub}|$ and $|V_{cb}|$ towards the domain of precision physics. Averaging inclusive and exclusive results can improve numerical accuracy due to partly uncorrelated uncertainties. A deeper understanding of the sources of systematic uncertainties and thorough checks of consistency to the level of the anticipated total accuracy will be required to validate the new frontier of accuracy

I gratefully acknowledge contributions by M. Artuso, E. Barberio, I. Bigi, M. Calvi, K. Ecklund, P. Gambino L. Gibbons, Y. Kwon, T. Iijima, V. Lubicz, V. Luth, M. Luke, A. Oyanguren, P. Roudeau, K. Schubert, M. Shapiro, A. Stocchi, A. Tricomi, N. Uraltsev, and A. Warburton.

## REFERENCES

1. K. Abe *et al.* (BELLE Coll.), BELLE-CONF-0257 (2001).
2. A. Tricomi, these proceedings.
3. R.A. Briere *et al.* (CLEO Coll.), CLEO CONF 02-10, arXiv:hep-ex/0209024.
4. D. Bloch *et al.*, DELPHI 2002-070-CONF 604; M. Battaglia *et al.*, DELPHI 2002-071-CONF-605.
5. B. Aubert *et al.* BABAR Coll., SLAC-PUB-9312, arXiv:hep-ex/0207184.
6. C.G. Bloyd, B. Grinstein and R. Lebed, Phys. Lett B353 (1995) 306.
7. A. Oyanguren *et al* (DELPHI Coll.), DELPHI 2002-074 CONF608.
8. D. Buskulic *et al.* (ALEPH Coll.), Phys. Lett. B395 (1997) 373.
9. K. Abe *et al.* (BELLE Coll.), Phys. Lett. B526 (2002) 247.
10. R. A. Briere *et al.* (CLEO Coll.), Phys. Rev. Lett. 89 (2002) 081803.
11. P. Abreu *et al.* (DELPHI Coll.), Phys. Lett. B510 (2001) 55.
12. G. Abbiendi *et al.* (OPAL Coll.), Phys. Lett. B482 (2000) 15.
13. M. Artuso and E. Barberio, in K. Hagiwara *et al.* Phys. Rev. D66 (2002) 010001.
14. A. S. Kronfeld, *et al.*, arXiv:hep-ph/0207122.
15. CDF Coll., private communication.
16. M. Battaglia and L. Gibbons, in K. Hagiwara *et al.* Phys. Rev. D66 (2002) 010001.
17. B. Aubert *et al.* (BABAR Coll.), SLAC-PUB-9282, arXiv:hep-ex/0207081.
18. A. K. Leibovich, Z. Ligeti and M. B. Wise, Phys. Lett. B539 (2002) 242.
19. R. Barate *et al.* (ALEPH Coll.), Eur. Phys. J. C6 (1999) 555.
20. D. Cronin-Hennessy *et al.* (CLEO Coll.), Phys. Rev. Lett. 87 (2001) 251808.
21. A. Bornheim *et al.* (CLEO Coll.), CLEO CONF 02-08, arXiv:hep-ex/0207064.
22. P. Abreu *et al.* (DELPHI Coll.), Phys. Lett. B478 (2000) 14.
23. M. Acciarri *et al.* (L3 Coll.), Phys. Lett. B436 (1998) 174.
24. G. Abbiendi *et al.* (OPAL Coll.), Eur. Phys. J. C21 (2001) 399
25. N.E. Adam *et al.*, (CLEO Coll.), CLEO-CONF 02-09
26. B. Aubert *et al.* (BABAR Coll.), SLAC-PUB-9305, arXiv:hep-ex/0207080.
27. U. Aglietti, M. Ciuchini and P. Gambino, Nucl. Phys. B637 (2002) 427.

ELSEVIER

Proceedings of ICHEP 2002, pp. 481–484
*S. Bentvelsen, P. de Jong, J. Koch and E. Laenen (Editors)*

31st INTERNATIONAL CONFERENCE ON HIGH ENERGY PHYSICS AMSTERDAM

# $B_d$ mixing measurements with the BaBar detector

C. Voena
(for the BaBar collaboration)
Università "La Sapienza" and INFN Roma, P.le Aldo Moro 2 00185 Rome, Italy

The $B^0\bar{B}^0$ oscillation frequency $\Delta m_d$ has been measured with the BaBar detector at the PEP-II asymmetric $B$ factory with different experimental techniques. The discussion here is focused on the recent simultaneous measurement of $\Delta m_d$ and $\tau_{B^0}$ with exclusively reconstructed $B^0 \to D^{*-}l^+\nu_l$ decays, based on 23 million $B\bar{B}$ pairs collected by BaBar. The measurements of $\Delta m_d$ with fully reconstructed hadronic decays and with dilepton events are also reviewed. The average BaBar result is $\Delta m_d = 0.500 \pm 0.008 \pm 0.006\ ps^{-1}$.

## 1. Introduction

Particle-antiparticle oscillations or mixing, has been observed in the neutral $B\bar{B}$ meson system almost fifteen years ago [1]. This quantum-mechanical behavior is originated by the flavor eigenstates $B^0$ and $\bar{B}^0$ not being Hamiltonian eigenstates. The frequency of the oscillation is the mass difference between the mass eigenstates, $\Delta m_d$. In the Standard Model, $B^0\bar{B}^0$ mixing occours through second-order weak diagrams involving the exchange of up-type quarks, with the top quark contributing the dominant amplitude. A measurement of $\Delta m_d$ is therefore sensitive to the value of the Cabibbo-Kobayashi-Maskawa matrix element $V_{td}$ [2]. The oscillation frequency $\Delta m_d$ has been measured with both time-integrated and time-dependent techniques [3]. Asymmetric $B$ factory experiments like BaBar can perform high statistics time dependent measurements of $\Delta m_d$.

## 2. Measurements of $\Delta m_d$ with the BaBar detector

The BaBar detector [4] collects data at the PEP-II asymmetric $e^+e^-$ collider operated at or near the $\Upsilon(4S)$ resonance. $B\bar{B}$ pairs from $\Upsilon(4S)$ decay move along the high-energy beam direction ($z$) with a nominal Lorentz boost $<\beta\gamma>= 0.55$. Therefore, the two $B$ decays vertices are separated by about 260 $\mu m$ on average. The two $B$ mesons are produced in a coherent $P$-wave state and their proper decay-time difference $\Delta t$ distribution is governed by the following probabilities to observe *mixed*(-) or *unmixed*(+) events:

$$Prob(B^0\bar{B}^0 \to B^0\bar{B}^0, B^0B^0 or \bar{B}^0\bar{B}^0) \propto e^{-\frac{|\Delta t|}{\tau_{B^0}}}(1 \pm cos\Delta m_d\Delta t). \quad (1)$$

Therefore, a measurement of $\Delta t$ together with the identification of the $b$-flavor of both $B$ mesons at their time of decay, allows to observe the oscillations and to extract $\Delta m_d$.

In the following sections the simultaneous measurement of $\Delta m_d$ and $\tau_{B^0}$ with exclusively reconstructed $B^0 \to D^{*-}l^+\nu_l$ decays [5], the measurement of $\Delta m_d$ with fully reconstructed hadronic $B^0$ decays [6] and with inclusively reconstructed dilepton events [7], will be discussed.

## 3. Measurement of $\Delta m_d$ and $\tau_{B^0}$ with exclusively reconstructed $B^0 \to D^{*-}l^+\nu_l$ decays

The analysis is based on a sample of approximately 14,000 exclusively reconstructed $B^0 \to D^{*-}l^+\nu_l$ decays selected from 23 million $B\bar{B}$ pairs recorded in the years 1999-2000 by BaBar. The purity of the sample is 65-89% depending on the decay mode of the $\bar{D}^0$ from the $D^{*-}$.

One of the two $B$ produced by the $\Upsilon(4S)$ decay is reconstructed in the semileptonic mode and the charge of the final-state particles identifies the flavor of the $B$. $D^{*-}$ candidates are recon-


doi:10.1016/S0920-5632(02)02156-4

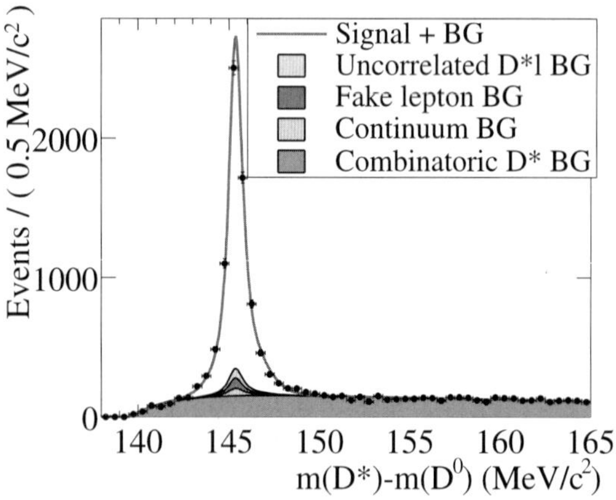


Figure 1. $\delta m$ distribution for events passing all selection criteria for $B^0 \rightarrow D^{*-}l^+\nu_l$ decays with a muon candidate. The points correspond to the data. The curve is the result of a fit. The shaded distributions correspond to the four types of background (BG) described in the text.

structed using the decay $D^{*-} \rightarrow \bar{D}^0\pi^-$, while $\bar{D}^0$ candidates are reconstructed in the modes $K^+\pi^-$, $K^+\pi^-\pi^0$, $K^+\pi^-\pi^+\pi^-$ and $K^0_s\pi^+\pi^-$. $D^{*-}$ candidates are then combined with oppositely charged high-energy electrons or muons in the event and the $D^{*-}l^+$ pair is required to pass kinematic cuts that enhance the contribution of $B^0 \rightarrow D^{*-}l^+\nu_l$ decays. The $B$ mass and energy cannot be reconstructed because of the presence of the neutrino, thus the distribution $\delta m$ of the difference between the $D^{*-}$ and the $D^0$ masses is used to select $B$ candidates. The distribution of $\delta m$ for events passing the selection criteria in the muon sample is shown in Figure 1.

There are two types of background with respect to the $\delta m$ distribution: combinatoric background and peaking background. Combinatoric background is due to events with a mis-reconstructed $D^{*-}$ and does not peak in the $\delta m$ distribution. Peaking background is due to $B\bar{B}$ events with mis-identified leptons or uncorrelated true leptons, and continuum events. The $\delta m$ distribution of the peaking background is the same as the distribution of the signal. Several control samples are selected to characterize the various backgrounds in both the fraction and the time distribution.

All the charged tracks in the event, except the reconstructed tracks from the $D^{*-}l^+$ pair, are used to identify the flavor of the other $B$ (referred to as $B_{tag}$). There are five types of tagging categories. The first two tagging categories rely on the presence of a prompt lepton or charged kaons in the event, whose charge is correlated with the $b$-flavor of the decaying B. The other three categories exploit a variety of inputs (e.g. slow pions, momentum of the track with the maximum center-of-mass momentum) with a neural network technique.

The difference $\Delta t$ between the two $B$ decay times is determined from the measured separation $\Delta z = z_{D*l} - z_{tag}$ along the beam axis between the $D^{*-}l^+$ vertex and the $B_{tag}$ vertex. The measured $\Delta z$ is converted into $\Delta t$ with the known $\Upsilon(4S)$ boost according to the relation $\Delta z = c\beta\gamma\Delta t$ which neglects the small $B$ momentum in the $\Upsilon(4S)$ frame. The resolution on the $D^{*-}l^+$ vertex is about 70 $\mu m$ while the resolution on the $B_{tag}$ vertex is about 160 $\mu m$.

Each tagging category $i$ has a probability $w_i$ of incorrectly assigning the flavor of the $B_{tag}$ and there is a limited precision on the $\Delta t$ measurement. These two experimental complications affect the $\Delta t$ distribution of Equation 1 which becomes:

$$Prob(B^0\bar{B}^0 \rightarrow B^0\bar{B}^0, B^0B^0 or \bar{B}^0\bar{B}^0) \propto$$
$$\mathcal{R}(\delta t;\hat{a}) \otimes e^{-\frac{|\Delta t|}{\tau_{B0}}}(1 \pm (1-2w_i)cos\Delta m_d\Delta t).$$

The function $\mathcal{R}(\delta t;\hat{a})$ is the resolution function which parametrizes the response to $\Delta t$ of the detector, $\delta t = \Delta t_{meas}$ - $\Delta t_{true}$ and $\hat{a}$ is a set of parameters. The final $\Delta t$ distribution also includes terms for each relevant background source.

The oscillation frequency $\Delta m_d$ and the lifetime $\tau_{B^0}$ are determined simultaneously with an unbinned maximum likelihood fit to the measured $\Delta t$ distribution. Note that other mixing measurements fix $\tau_{B^0}$ to the world average and

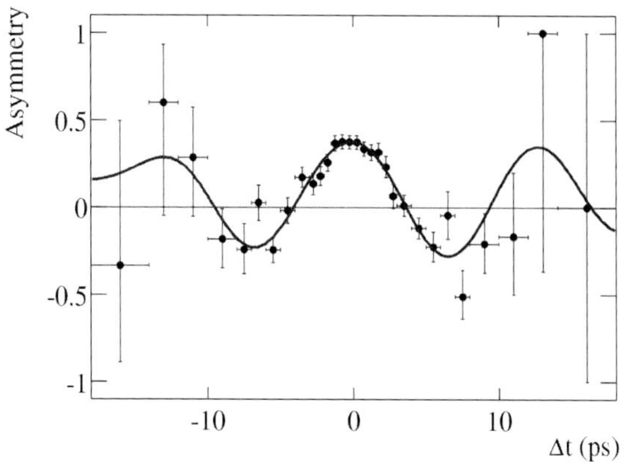


Figure 2. Mixing asymmetry plot for a 80% pure $B^0 \to D^{*-}l^+\nu_l$ sample. The dots are the data and the curve is the projection of the fit result.

this is the source of the dominant systematic error. Also the resolution, the fraction of charged $B$, the mistag and the background parameters are floated in the fit. The results are: $\Delta m_d = 0.492 \pm 0.018(stat) \pm 0.013(syst)\ ps^{-1}$ and $\tau_{B^0} = 1.523^{+0.024}_{-0.023}(stat) \pm 0.022(syst)\ ps$. The correlation between $\Delta m_d$ and $\tau_{B^0}$ is -0.22. A correction is applied to both $\Delta m_d$ and $\tau_{B^0}$ which takes into account selection and fit biases. The uncertainty on such a correction is the dominant systematic error. Other BaBar measurements for $\tau_{B^0}$ can be found in [8] and [9]. Figure 2 shows the mixing asymmetry defined as the difference between the number of unmixed and mixed events over their sum as a function of $\Delta t$.

## 4. Measurement of $\Delta m_d$ with fully reconstructed $B^0$ hadronic decays

The analysis is based on ~6300 $B^0$ selected from 32 million $B\bar{B}$ pairs. The $B^0$ are reconstructed in the flavor eigenstates $D^{*-}\pi^+$, $D^{*-}\rho^+$, $D^{*-}a_1^+$, $J/\Psi K^{*0}$ and the purity of the selected sample is about 86%. $B^0$ candidates are selected using the difference between the energy of the candidate and the beam energy $\sqrt{s}/2$ in the center of mass frame, and the beam energy substituted mass, calculated from $\sqrt{s}/2$ and the reconstructed momentum of the $B$.

The tagging algorithm and the tagging vertex reconstruction technique are described in the previous section. $\Delta m_d$ is extracted with an unbinned maximum likelihood fit to the $\Delta t$ distribution (obtained from $\Delta z$ as described in the previous section) where also all the mistag probabilities and the resolution parameters are floated. The result is $\Delta m_d = 0.516 \pm 0.016(stat) \pm 0.010(syst)\ ps^{-1}$, the dominant systematic error being the uncertainty on $\tau_{B^0}$ which is fixed in the fit.

## 5. Measurement of $\Delta m_d$ with inclusive dilepton events

The analysis is based on 23 million $B\bar{B}$ pairs. The measurement technique consists in the identification of events containing two high energy leptons from semileptonic decays of $B$ mesons. The flavor of the $B$ mesons at the time of their decay is determined by the charge of the leptons. About 99000 events are selected, $\sim 55$ % of them being $B^+B^-$ events which are not removed by the event selection criteria. Another non negligible background is due to leptons from the $b \to c \to l$ decay chain ($\sim 13$ %) which are also the main source of wrongly tagged events. The difference of the $z$ coordinates of the two $B$ decay vertices $\Delta z$ is determined using the two lepton tracks and a beam spot constraint. $\Delta t$ is then obtained from $\Delta z$ as described in section 3. A binned maximum likelihood fit is used to extract $\Delta m_d$ together with the resolution parameters, the fraction of charged $B$ and some of the background fractions and parameters. The result is $\Delta m_d = 0.493 \pm 0.012(stat) \pm 0.009(syst)\ ps^{-1}$. The dominant systematic error is due to the uncertainty on $B^0$ and $B^+$ lifetimes which are fixed in the fit.

## 6. Conclusions

The oscillation frequency $\Delta m_d$ of the $B^0\bar{B}^0$ system has been measured by the BaBar experiment with different experimental techniques. The combined BaBar result is $\Delta m_d = 0.500 \pm 0.008 \pm 0.006\ ps^{-1}$. The corresponding precision is 2% to be compared with the world average precision which is 1.2 %.

## REFERENCES


1. UA1 Collaboration, C.Albajar *et al.*, Phys.Lett. **B186**, 247 (1987); ARGUS Collaboration, H. Albrecht *et al.*, Phys. Lett. **B192**, 245 (1987).
2. N.Cabibbo, Phys.Rev.Lett. **10**, 531 (1963); M.Kobayashi and T.Maskawa, Prog. Th. Phys. **49**, 652 (1973).
3. D.E. Groom *et al.*, Eur. Phys. Jour. **C15**, 1 (2000);
4. The BaBar Collaboration, B. Aubert *et al.*, Nucl. Instrum. Meth. **A479**, 1 (2002).
5. The BaBar Collaboration, B. Aubert *et al.*, hep-ex/0207071, 2002.
6. The BaBar Collaboration, B. Aubert *et al.*, Phys. Rev. Lett. **88**, 221802 (2002).
7. The BaBar Collaboration, B. Aubert *et al.*, Phys. Rev. Lett. **88**, 221803 (2002).
8. The BaBar Collaboration, B. Aubert *et al.*, Phys. Rev. Lett. **89**, 011802 (2002).
9. The BaBar Collaboration, B. Aubert *et al.*, Phys. Rev. Lett. **87**, 201803 (2001).

ELSEVIER

Proceedings of ICHEP 2002, pp. 485–488
*S. Bentvelsen, P. de Jong, J. Koch and E. Laenen (Editors)*

31st INTERNATIONAL CONFERENCE ON HIGH ENERGY PHYSICS AMSTERDAM

# Measurement of $\Delta m_d$ with Belle

F. Ronga[a]

[a]IPHE, University of Lausanne, 1015 Lausanne, Switzerland

Four different measurements of $\Delta m_d$ with the Belle detector at KEKB are described here. Two inclusive and two exclusive methods were used on a 31 million $B\bar{B}$ sample. We obtain an average of $0.506 \pm 0.006 \pm 0.007$ ps$^{-1}$. A test for $CPT$ violation was also performed, allowing us to set new limits on the $CPT$ violation parameters.

## 1. Introduction

In the Standard Model, $B^0$-$\overline{B}^0$ oscillations occur through second-order weak interactions, mainly through internal loops containing virtual $t$ quarks. The mixing parameter $\Delta m_d$, the mass difference between the two mass eigenstates, is therefore related to the $V_{tb}$ and $V_{td}$ CKM matrix elements. The measurement of $\Delta m_d$ can thus in principle provide a mean to extract $|V_{td}|$. $\Delta m_d$ also plays a role in the parameterization of the $CP$ asymmetries in $B^0$ decays. A precise measurement of $\Delta m_d$ is therefore needed for such $CP$ violation measurements (see [1]).

In this article, we present four different measurements of $\Delta m_d$ from the time distribution of opposite-flavor (OF – $B^0\overline{B}^0$) and same-flavor (SF – $B^0B^0$, $\overline{B}^0\overline{B}^0$) neutral $B$ decays at the $\Upsilon(4S)$ resonance. The theoretical time-dependent probabilities for observing OF and SF states are given by:

$$\begin{aligned} \mathcal{P}^{\rm OF}(\Delta t) &= \frac{e^{-|\Delta t|/\tau_{B^0}}}{4\tau_{B^0}}[1+\cos(\Delta m_d\,\Delta t)] \\ \mathcal{P}^{\rm SF}(\Delta t) &= \frac{e^{-|\Delta t|/\tau_{B^0}}}{4\tau_{B^0}}[1-\cos(\Delta m_d\,\Delta t)] \end{aligned} \quad (1)$$

where $\tau_{B^0}$ is the $B^0$ lifetime and $\Delta t$ is the proper time difference between the two $B$ meson decays.

The analysis was performed on data collected with the Belle detector [2] at the KEKB storage ring. A 8.0 GeV electron beam and a 3.5 GeV positron beam are collided at a 22 mrad angle, resulting in a Lorentz boost of the center-of-mass of $\beta\gamma = 0.425$ in the laboratory frame. Since $B$ mesons are nearly at rest in the $\Upsilon(4S)$ frame, the proper time difference $\Delta t$ is approximated by $\Delta z/\beta\gamma c$, $\Delta z$ being the (signed) distance between the decay vertices of the two $B$ mesons along the beam axis. The flavor of the $B$ mesons is determined using flavor-specific decays.

## 2. Dilepton measurement

In this method [3], semi-leptonic decays $B \to X^- l^+ \nu_l$ from both $B$ mesons are used to tag the $B$ flavor with the sign of the lepton. Two fast leptons are searched for. The decay vertex position of the $B$ meson is determined from the interception of the lepton track with the profile of the interaction point. This selection has been applied on 29.4 fb$^{-1}$ of on-resonance (31.9 million $B\bar{B}$ pairs) and 3.0 fb$^{-1}$ of off-resonance data. 49,838 SF and 230,881 OF events satisfied all the cuts.

A binned maximum likelihood fit was then performed simultaneously on the OF and SF $\Delta z$ distributions. The signal, which consists of primary leptons coming from charged $B$ mesons (OF only, $\Delta m_d = 0$) or neutral $B$ mesons, is described by the theoretical distributions (Eq. 1) convolved with a detector resolution function. This resolution function is estimated from the $\Delta z$ distribution of events compatible with $J/\psi \to \ell^+\ell^-$ decays, where the true $\Delta z$ is equal to zero.

The backgrounds are classified into eight categories depending on their sources: $B^+B^-$, $B^0B^0$ or $\overline{B}^0\overline{B}^0$, $B^0\overline{B}^0$, and continuum, each of them contributing to both SF and OF events. The $\Delta z$ distribution for each category was estimated using Monte Carlo data. The amount of contin-


doi:10.1016/S0920-5632(02)02161-8

uum background was normalized using the off-resonance data. We labeled the six other categories as correctly-tagged (CT) and wrongly-tagged (WT) backgrounds.

A detailed comparison of the $\Delta z$ distribution of $J/\psi$ event samples between the data and Monte Carlo showed that the width of the real data distribution is wider than for the Monte Carlo by 50 $\mu$m. The Monte Carlo background distributions were thus smeared by this amount in order to compensate this discrepancy.

Signal and background efficiencies, as well as $\Delta m_d$ and $f^+/f^0$, the ratio between the branching ratios of $\Upsilon(4S)$ to charged and neutral $B$ mesons, are extracted from the fit. The result is shown on Fig. 1; we obtain $\Delta m_d = 0.503 \pm 0.008 \pm 0.009\ \mathrm{ps}^{-1}$ and $f^+/f^0 = 1.01 \pm 0.03 \pm 0.08$. The main systematic errors are summarized in Table 1.

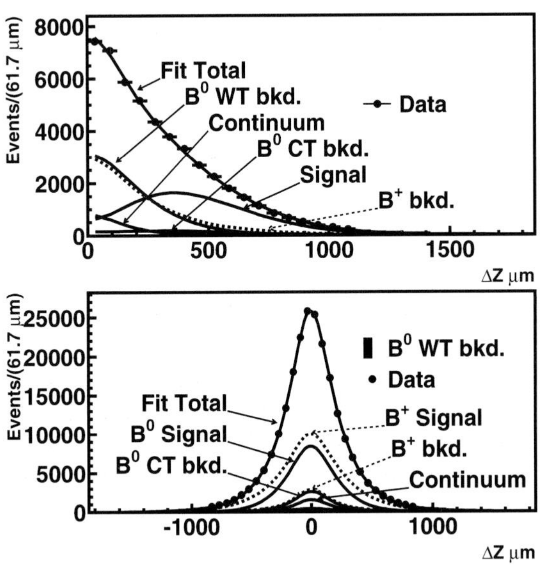


Figure 1. Result of the fit for the dilepton measurement. The various signal and background contributions are shown for same flavor (top) and opposite flavor (bottom) events.

Table 1
Main systematic errors: uncertainty on the $B$ meson lifetimes, error on the signal resolution function, error due to finite MC statistics and error on the smearing of the MC resolution.

| Source | $\Delta m_d$ [ps$^{-1}$] | $f_+/f_0$ |
|---|---|---|
| $\tau_{B^0}/\tau_{B^+}$ | $\pm 0.005$ | $\pm 0.07$ |
| $\Delta z$ resol. | $\pm 0.005$ | $\pm 0.02$ |
| MC stat. | $\pm 0.004$ | $\pm 0.01$ |
| MC $\Delta z$ resol. | $\pm 0.003$ | $\pm 0.02$ |

A fit including a possible $CPT$ violation has also been performed. In order to do so, the $\cos(\Delta m_d\, \Delta t)$ term in Eq. 1 is replaced by:

$$|\cos\theta|^2 + (1 - |\cos\theta|^2)\cos(\Delta m_d\, \Delta t) - 2\mathrm{Im}(\cos\theta)\sin(\Delta m_d\, \Delta t), \quad (2)$$

where $\theta$ is complex, and $CPT$ is violated if $\theta \neq \frac{\pi}{2}$. We obtain $\Delta m_d = 0.503 \pm 0.08\ \mathrm{ps}^{-1}$ and $f_+/f_0 = 1.02 \pm 0.03$ (stat. only), in agreement with the previous results. We also get $\mathrm{Re}(\cos\theta) = 0.00 \pm 0.12 \pm 0.01$ and $\mathrm{Im}(\cos\theta) = 0.03 \pm 0.01 \pm 0.02$. This allows us to set the following limits (at 90% CL) on the mass and decay width differences of the flavor eigenstates [4]:

$$\left|\frac{m_{B^0} - m_{\overline{B}^0}}{m_{B^0}}\right| < 1.2{\cdot}10^{-14}, \quad \left|\frac{\Gamma_{B^0} - \Gamma_{\overline{B}^0}}{\Gamma_{B^0}}\right| < 0.09$$

## 3. $D^*\pi$ partial reconstruction

Another measurement [5] of $\Delta m_d$ uses partially reconstructed $B^0 \to D^{*-}\pi_f^+$ decays. The $D^{*-}$ information is extrapolated from the soft pion of $D^{*-} \to \overline{D}^0\pi_s^-$, and then combined with the fast pion $\pi_f^+$ and the beam information to reconstruct the $B$ meson. The flavor is given by the charge of the fast pion. The other side is tagged by simply looking for a fast lepton.

Contributions of flat (combinatorial) background and signal are estimated from the $D^0$ missing mass. The signal shape comes from the Monte Carlo. The wrong-sign combination of $\pi_f$ and $\pi_s$ gives the flat background shape. Finally, the peaked background contribution (from

$B^0 \to D^{*-} \ell^+ \nu$ or $D^{**} \pi$) is fixed to the Monte-Carlo. From 4889 candidates in 31.3 million $B\bar{B}$ pairs, the fit yields $3433 \pm 81$ signal and $1466 \pm 37$ background events in the signal region (where the calculated $D^0$ mass is greater than 1.85 GeV/c$^2$). This corresponds to a purity of about 70%.

An unbinned maximum likelihood fit is then performed on the $\Delta z$ distributions. The signal probability density function (PDF) is a convolution of the theoretical probability (Eq. 1) and a triple-gaussian resolution function estimated from $J/\psi$ events (analogous to the previous analysis). The background distributions include prompt (zero lifetime) and finite lifetime components (determined from the $D^0$ missing mass side-band), as well as $B^0$-$\overline{B}^0$ mixing. The fit yields: $\Delta m_d = 0.505 \pm 0.017 \pm 0.020$ ps$^{-1}$. Figure 2 shows the mixing asymmetry $(N_{\rm OF} - N_{\rm SF})/(N_{\rm OF} + N_{\rm SF})$.

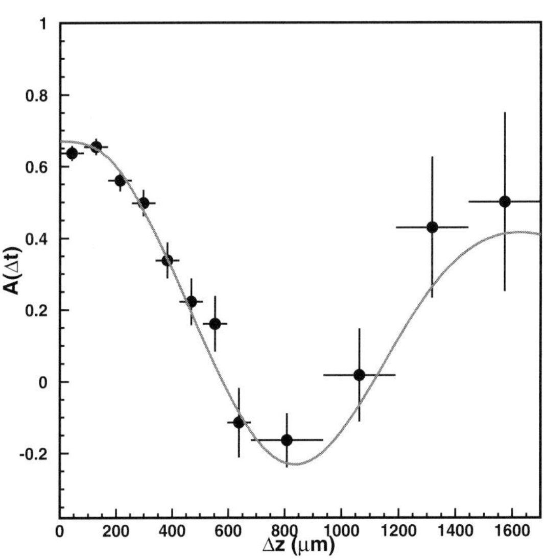


Figure 2. Mixing asymmetry as a function of $\Delta z$ (points) and fit result (curve) for the $D^* \pi$ partial reconstruction.

## 4. $D^* \ell \nu$ full reconstruction

This method [6] fully reconstructs $B^0 \to D^{*-} \ell^+ \nu$, with $D^{*-} \to \overline{D}^0 \pi^-$ and $\overline{D}^0 \to K^+\pi^-$, $K^+\pi^-\pi^0$ or $K^+\pi^-\pi^+\pi^-$. The flavor of the other $B$ meson is tagged using the algorithm of Ref. [1].

The cosine of the angle between the reconstructed $D^* \ell$ system and the $B$ meson momenta in the $\Upsilon(4S)$ frame is used to define the signal region ($|\cos\theta^*| < 1.1$) and a "$D^{**}$ dominated region" ($\cos\theta < -2$). Other backgrounds consist of fake $D^*$ resonances, random $D^* \ell$ combinations, and continuum. From 31.3 million $B\bar{B}$ pairs, we get 16,397 candidates, with a purity of 80.4%.

An unbinned maximum likelihood fit is then performed on the $\Delta z$ distributions. It includes the various types of backgrounds and their normalizations, as well as the detector resolution (determined from the untagged $\Delta z$ distribution). The fit is simultaneously performed on the signal region and the $D^{**}$ dominated region in order to determine background parameters. The result is displayed in Fig. 3. We obtain a $\Delta m_d$ value of $0.494 \pm 0.012 \pm 0.015$ ps$^{-1}$.

## 5. Hadronic modes

Finally, the exclusive reconstruction of $B$ mesons decaying into flavor specific hadronic states has been used to measure $\Delta m_d$ [8]. The reconstructed modes include $\overline{B}^0 \to D^+(K^-\pi^+\pi^+)\pi^-$, $D^{*+}\pi^-$ or $D^{*+}\rho^-(\pi^-\pi^0)$, with $D^{*+} \to D^0\pi^+$ and $D^0 \to K^-\pi^+$, $K^-\pi^+\pi^0$ or $K^-\pi^+\pi^-\pi^+$. The other side is tagged using the flavour tagging algorithm as in the previous analysis (see [1]). The signal region is defined by constraints on $\Delta E = E_B^* - E_{\rm beam}^*$ and $M_{\rm bc} = \sqrt{{E_{\rm beam}^*}^2 - {p_B^*}^2}$, where $E_B^*$ and $p_B^*$ are the center-of-mass energy and momentum of the fully reconstructed $B$ candidate, and $E_{\rm beam}^*$ is the center-of-mass energy of the beam. 6660 candidates from 31.3 million $B\bar{B}$ pairs satisfy the selection, with a purity of 80% in average.

This analysis uses an unbinned maximum likelihood fit to extract $\Delta m_d$. The $\Delta z$ distribution is described using a sum of three contributions: the signal (Eq. 1), combinatorial backgrounds (exponential and prompt components), and an "out-

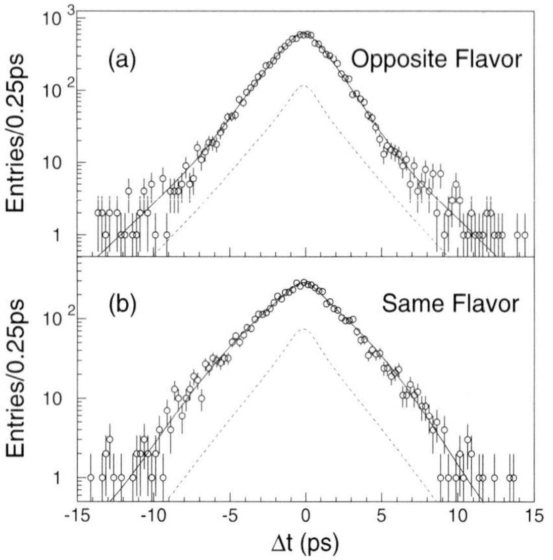


Figure 3. Fit result of the $\Delta t$ distributions of (a) OF events and (b) SF events, for the $D^* l \nu$ analysis. The solid lines are the result of the unbinned maximum likelihood fit. The dashed lines show the background contribution.

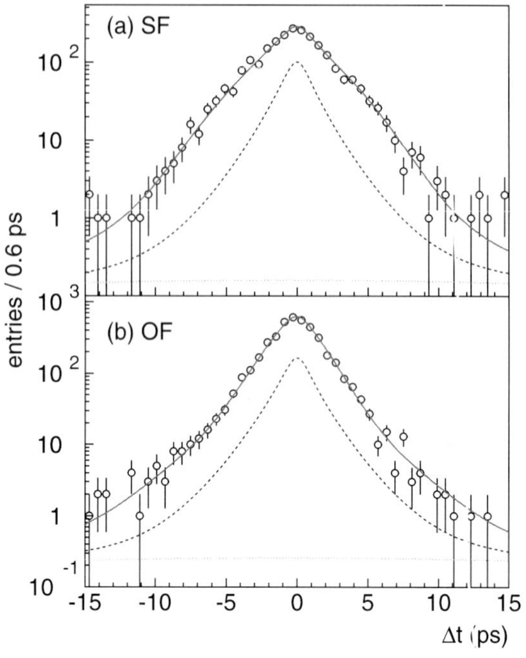


Figure 4. Distributions of $\Delta t$ for (a) SF and (b) OF hadronic events with the fitted curves superimposed. The dashed, dotted and solid curves show the background, outliers, and the sum of backgrounds and signal, respectively.

lier" component taking into account background events with very large $\Delta t$ [7]. The description of the signal resolution function is given in [7]. A sum of two gaussians is used to describe the background resolution function. The result of the fit is displayed on Fig. 4. A $\Delta m_d$ value of $0.528 \pm 0.017 \pm 0.011$ ps$^{-1}$ is obtained.

## 6. Conclusion

Four measurements of $\Delta m_d$ have been performed on about 31 million $B\bar{B}$ pairs using inclusive and exclusive reconstructions. The Belle average value is: $\Delta m_d = 0.506 \pm 0.09 \pm 0.07$ ps$^{-1}$. These errors take into account statistical and sytematical correlations between the measurements. In addition, new limits on $CPT$ violation parameters have been set.

## REFERENCES


1. K. Abe *et al.* (Belle Collab.), Phys. Rev. D **66** (2002) 032007.
2. A. Abashian *et al.* (Belle Collab.), Nucl. Instr. and Meth. **A479** (2002) 117.
3. K. Abe *et al.* (Belle Collab.), BELLE-CONF-0205.
4. A. Mohapatra, M. Satpathy, K. Abe, and Y. Sakai, Phys. Rev. **D58** (1998) 036003.
5. K. Abe *et al.* (Belle Collab.), BELLE-CONF-0204.
6. K. Hara, M. Hazumi *et al.* (Belle Collab.), hep-ex/0207045 (submitted to Phys. Rev. Lett.).
7. K. Abe *et al.* (Belle Collab.), Phys. Rev. Lett. **88** (2002) 171801.
8. T. Tomura *et al.* (Belle Collab.), Phys. Lett. **B542** (2002) 207.

Proceedings of ICHEP 2002, pp. 489–492
*S. Bentvelsen, P. de Jong, J. Koch and E. Laenen (Editors)*

ELSEVIER

31st INTERNATIONAL CONFERENCE ON HIGH ENERGY PHYSICS AMSTERDAM

# $B_s^0$ Oscillation Results

Stéphane Willocq[a]

[a]Department of Physics, University of Massachusetts,
Amherst, MA 01003, USA

We review new studies of the time dependence of $B_s^0$–$\overline{B_s^0}$ mixing by the ALEPH, DELPHI and SLD Collaborations, with an emphasis on the different analysis methods used. Combining all available results yields a preliminary lower limit on the oscillation frequency of $\Delta m_s > 14.4$ ps$^{-1}$ at the 95% C.L.

## 1. INTRODUCTION

Studies of the time dependence of $B_d^0$–$\overline{B_d^0}$ and $B_s^0$–$\overline{B_s^0}$ mixing continue to play an important role in the exploration of both the Cabibbo-Kobayashi-Maskawa (CKM) quark mixing matrix and the phenomenon of CP violation. The $B_d^0$ oscillation frequency is related to the poorly known CKM element $V_{td}$ according to [1]

$$\Delta m_d = \frac{G_F^2}{6\pi^2} m_{B_d} m_t^2 F(m_t^2/m_W^2) f_{B_d}^2 B_{B_d} \times \eta_{QCD} \left|V_{tb}^* V_{td}\right|^2 \quad (1)$$

and has recently been precisely measured by the BaBar and Belle collaborations—the current world average is $\Delta m_d = 0.503 \pm 0.006$ ps$^{-1}$ [2]. However, this measurement cannot be translated into a precise determination of $|V_{td}|$ due the 15-20% theoretical uncertainty in $f_{B_d}\sqrt{B_{B_d}}$ (see Ref. [3] for a review of Lattice QCD calculations). Uncertainties are reduced for the ratio

$$\frac{\Delta m_s}{\Delta m_d} = \frac{m_{B_s}}{m_{B_d}} \xi^2 \left|\frac{V_{ts}}{V_{td}}\right|^2, \quad (2)$$

where the quantity $\xi \equiv (f_{B_s}\sqrt{B_{B_s}})/(f_{B_d}\sqrt{B_{B_d}})$ is estimated to be $1.18 \pm 0.04^{+0.12}_{-0.00}$ from Lattice QCD [3]. This implies that the ratio $|V_{ts}/V_{td}|$ can be determined with an uncertainty smaller than that for $|V_{td}|$ [4]. In the Wolfenstein parameterization of the CKM matrix, we have $\Delta m_d \propto |V_{td}|^2 \simeq A^2\lambda^6[(1-\rho)^2 + \eta^2]$ and $\Delta m_s \propto |V_{ts}|^2 \simeq A^2\lambda^4$, where $\lambda = 0.2237 \pm 0.0033$ and $A = 0.819 \pm 0.040$ [5], but $\rho$ and $\eta$ are not well known. Studies of $B_d^0$ and $B_s^0$ oscillations thus provide some of the strongest constraints on the CKM unitarity triangle parameters $\rho$ and $\eta$.

Experimental studies of $B_s^0$ oscillations require two main ingredients: (i) reconstruction of the $B_s^0$ decay and its proper time, (ii) determination of the $B_s^0$ or $\overline{B_s^0}$ flavor at both production and decay to classify the decay as either 'mixed' (if the tags disagree) or 'unmixed' (otherwise). The significance for a $B_s^0$ oscillation signal can be approximated by [6]

$$S = \sqrt{\frac{N}{2}}\, f_s\, [1 - 2\,w]\, e^{-\frac{1}{2}(\Delta m_s \sigma_t)^2}, \quad (3)$$

where $N$ is the total number of decays selected, $f_s$ is the fraction of $B_s^0$ mesons in the selected sample, $w$ is the probability to incorrectly tag a decay as mixed or unmixed (i.e. the mistag rate) and $\sigma_t$ is the proper time resolution. The proper time resolution depends on both the decay length resolution $\sigma_L$ and the momentum resolution $\sigma_p$ according to $\sigma_t^2 = (\sigma_L m_B/p)^2 + (t\,\sigma_p/p)^2$. Based on the Wolfenstein parameterization, we see that $\Delta m_s/\Delta m_d \propto 1/\lambda^2$, which is of order of 20 (the other Wolfenstein parameters are of order 1). Therefore, $B_s^0$ oscillations are expected to be much more rapid than $B_d^0$ oscillations. The ability to resolve such rapid oscillations thus requires excellent decay length and momentum resolution, and benefits from having a low mistag rate and a high $B_s^0$ purity.


doi:10.1016/S0920-5632(02)02163-1

## 2. RECONSTRUCTION METHODS

The study of the time dependence of $B_s^0$–$\overline{B_s^0}$ mixing has been performed with different analysis techniques, ranging from fully inclusive to fully exclusive reconstruction of $B_s^0$ decay candidates. The study of $B_s^0$ oscillations is more chalenging than that of $B_d^0$ oscillations due to two main differences. First, only about 10% of $b$ quarks fragment into $B_s^0$ mesons, as compared to about 40% into $B_d^0$ mesons. Second, the $B_s^0$ oscillation frequency is expected to be at least a factor of 20 larger than that for $B_d^0$ oscillations. To address this, sophisticated analyses have been developed with an emphasis on lowering the mistag rate, increasing the $B_s^0$ purity and, especially, improving the proper time resolution, all of which affect the sensitivity to $B_s^0$ oscillations.

The production flavor tag combines a number of different individual tags. The single most powerful tag exploits the large polarized forward-backward asymmetry in $Z^0 \to b\,\bar{b}$ decays. This tag is available at SLD thanks to the large electron beam polarization ($P_e = 73\%$). A left- (right-) handed incident electron tags the quark produced in the forward hemisphere as a $b$ ($\bar{b}$) quark. This method yields a mistag rate of 28% with nearly 100% efficiency. Tags used in all analyses rely on charge information from the event hemisphere opposite that of the $B_s^0$ candidate: (i) charge of lepton from the direct transition $b \to \ell^-$, (ii) momentum-weighted jet charge, (iii) secondary vertex charge, (iv) charge of secondary vertex kaon from the dominant transition $b \to c \to s$, (v) charge dipole of secondary vertex (SLD only). Other tags from the same hemisphere as the $B_s^0$ candidate are also used: (i) unweighted (or weighted) jet charge, and (ii) charge of fragmentation kaon accompanying the $B_s^0$ meson. These various tags are combined on an event-by-event basis to yield an average mistag rate of approximately 22% at SLD and 27-29% at LEP.

The analyses differ in the way the $B_s^0$ decay is reconstructed and thus in the way the decay flavor is determined. Three general classes can be identified: inclusive, semi-exclusive and fully exclusive. Inclusive analyses benefit from the large available statistics but suffer from low $B_s^0$ purity, whereas more exclusive analyses benefit from higher purity and resolution but suffer from the lack of statistics (this is particularly true for the fully exclusive analyses). Several analyses are discussed below to highlight these differences.

Inclusive analyses have been performed by ALEPH, DELPHI, OPAL and SLD. The SLD charge dipole analysis is the most sensitive fully inclusive method [7]. It aims to reconstruct the $b$-hadron decay chain topology. This method takes full advantage of the superb decay length resolution of the SLD CCD pixel vertex detector to separate secondary tracks (from the $B$ decay point) from tertiary tracks (from the $D$ decay point). The decay length resolution is parameterized by the sum of two Gaussians with $\sigma_L = 78\,\mu$m (60% fraction) and $304\,\mu$m (40%), whereas the momentum resolution is parameterized with $\sigma_p/p = 0.07$ (60%) and 0.21 (40%). A "charge dipole" $\delta Q$ is defined as the distance between secondary and tertiary vertices signed by the charge difference between them such that $\delta Q > 0$ ($\delta Q < 0$) tags $\overline{B^0}$ ($B^0$) decays. The average decay flavor mistag rate is estimated to be 22% and is mostly due to decays producing two charmed hadrons. A sample of 11,462 decays is selected with a $B_s^0$ purity estimated to be 16% (higher than the production rate of 10% due to the fact that only neutral decays are selected).

The most sensitive of all analyses is the ALEPH inclusive lepton analysis [8], which selects semileptonic $B$ decays. In this analysis, a $D$ meson is reconstructed inclusively based on topological and kinematical properties of the decay, and a resultant $D$ track is vertexed with the lepton and the $b$-hadron direction (from the jet direction) to form a $B$ decay vertex. The average decay length and momentum resolutions are $\sigma_L = 251\,\mu$m (75% fraction) and $718\,\mu$m (25%), $\sigma_p/p = 0.064$ (60%) and 0.20 (40%). Fairly loose selection criteria are used at the various stages of the analysis to obtain a high statistics sample of 74,026 events. The analysis relies on several neural network algorithms to perform the following tasks: production flavor tagging, $b\bar{b}$ event selection, direct ($b \to \ell$) lepton selection, and $B_s^0$ fraction enhancement. To maximize sensitivity to $B_s^0$

oscillations, the analysis incorporates all the information event by event, including estimates of the decay length and momentum resolution.

Semi-exclusive analyses have been performed by ALEPH, CDF, DELPHI, OPAL and SLD. $B_s^0$ decays are partially reconstructed in the modes $B_s^0 \rightarrow D_s^- \ell^+ \nu_\ell X$ and $B_s^0 \rightarrow D_s^- h^+ X$, where $h$ represents any charged hadron (or system of several hadrons) and the $D_s^-$ meson decay is either fully or partially reconstructed in the modes $D_s^- \rightarrow \phi\pi^-$, $K^{*0}K^-$, $K^0K^-$, $\phi\pi^-\pi^+\pi^-$, $\phi\ell^-\overline{\nu_\ell}$, etc.

The most sensitive semi-exclusive analysis performed by DELPHI selects 436 $D_s^-\ell^+$ events [9]. The small statistics is compensated by the high $B_s^0 \rightarrow D_s^-\ell^+\nu_\ell$ purity, estimated to be $\sim 53\%$, and the good decay length and momentum resolution, $\sigma_L = 200\,\mu$m (82% fraction) and 670 $\mu$m (16%), $\sigma_p/p = 0.07$ (82%) and 0.16 (16%). Analyses selecting $D_s^-h^+$ final states benefit from higher statistics but are less sensitive than those selecting $D_s^-\ell^+$ states because of lower $B_s^0$ purity and worse proper time resolution. The SLD $D_s$+Tracks analysis [10] combines fully reconstructed $D_s$ mesons with either a lepton or one (or more) charged hadron(s). It contributes especially at large values of $\Delta m_s$ thanks to a $B_s^0$ purity of 40% and the best available decay length resolution: $\sigma_L = 50\mu$m (60% fraction) and $151\mu$m (40%).

Finally, fully exclusive analyses have been performed by ALEPH [8] and DELPHI [11] via the (all charged particles) modes $B_s^0 \rightarrow D_s^-\pi^+$, $D_s^-a_1^+$, $\overline{D^0}K^-\pi^+$, and $\overline{D^0}K^-a_1^+$ (last two for DELPHI only), where the $D_s^-$ and $\overline{D^0}$ are fully reconstructed. The decays $B_s^0 \rightarrow D_s^{*-}\pi^+$, $D_s^{*-}a_1^+$ and $D_s^{(*)-}\rho^+$ are also reconstructed by adding one or more photons to the above final states (ALEPH only) or by including the "satellite" mass region below the $B_s^0$ mass peak. The number of decay candidates is 80 for ALEPH and 44 for DELPHI with signal purities of approximately 36% and 50%, respectively. The main advantage of the exclusive method is its excellent proper time resolution with a negligible contribution from momentum resolution ($\sim 0.5\%$). As a result, unlike all other methods, $\sigma_t$ does not grow

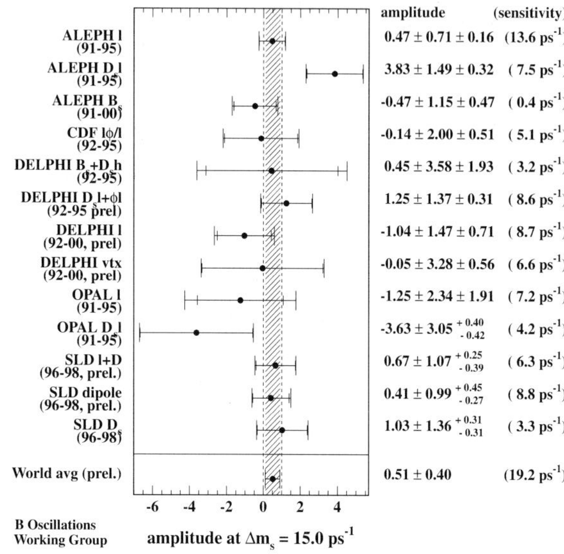


Figure 1. Measurements of the $B_s^0$ oscillation amplitude for $\Delta m_s = 15$ ps$^{-1}$.

significantly with increasing proper time $t$ and thus the oscillation amplitude is not damped as $t$ increases. Due to limited statistics, this method is not competitive with respect to the inclusive and semi-exclusive methods. However, this is the method of choice for future studies of $B_s^0$ oscillations at hadron colliders.

## 3. RESULTS

Studies of the time dependence of $B_s^0$–$\overline{B_s^0}$ mixing are carried out with the amplitude method, which is equivalent to a normalized Fourier transform [6]. The oscillation amplitude $A$ is expected to be $A = 0$ ($A = 1$) for frequencies sufficiently far from (close to) the true oscillation frequency. All available measurements of the oscillation amplitude at $\Delta m_s = 15$ ps$^{-1}$ are summarized in Figure 1. Also shown are the sensitivities for each analysis to set a 95% C.L. lower limit on $\Delta m_s$ ($\Delta m_s$ value at which $1.645\,\sigma_A = 1$).

The measured oscillation amplitudes are combined [2], taking statistical and systematic correlations into account, to obtain the world average amplitude spectrum shown in Figure 2. The

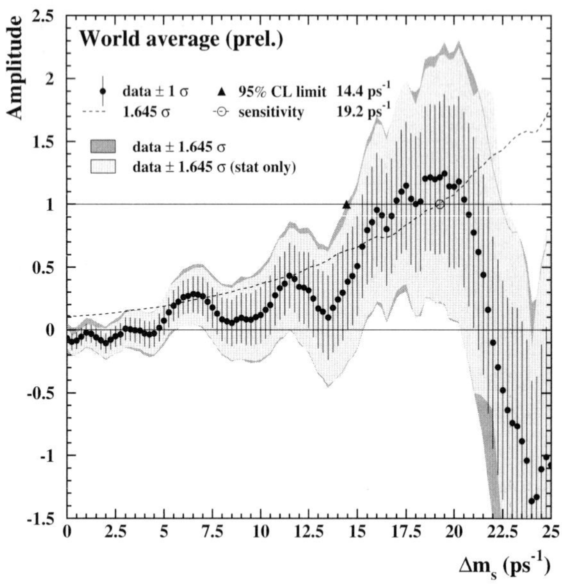


Figure 2. World average $B_s^0$ oscillation amplitude as a function of $\Delta m_s$. Values of the frequency for which $A + 1.645\,\sigma_A < 1$ are excluded at the 95% C.L.

combination also corrects for the different input parameters used ($\Delta m_d$, $b$-hadron lifetimes and production rates). Furthermore, the amplitude statistical uncertainties measured by the inclusive analyses are adjusted to take into account the $B_s^0$ production rate of $(9.3 \pm 1.1)\%$. The rise in statistical error as $\Delta m_s$ increases comes from the fact that an increasingly smaller fraction of the data sample has sufficient proper time resolution to resolve more rapid oscillations; the better the resolution, the smaller the rise. The preliminary combined amplitude spectrum excludes mixing ($A = 1$) for $\Delta m_s < 14.4$ ps$^{-1}$ at the 95% C.L., whereas the sensitivity is 19.2 ps$^{-1}$. The significance of the deviation from $A = 0$ near $\Delta m_s = 17.5$ ps$^{-1}$ is 2.3 $\sigma$. It is interesting to note that, while the sensitivity has been increasing steadily over the past 8 years, the limit has remained near 15 ps$^{-1}$ for the past 3 years. In August 1999, the limit was $\Delta m_s > 14.3$ ps$^{-1}$ but the sensitivity was only 14.7 ps$^{-1}$.

Many of the LEP and SLD analyses have been or are being finalized. One will thus have to wait for the next generation of experiments to measure the $B_s^0$ oscillation frequency. Prospects for such a measurement during Run 2 of the Fermilab Tevatron are excellent.

I wish to thank D.Abbaneo, P.Kluit, F.Parodi, O.Schneider, and A.Stocchi for their help with the preparation of the talk.

## REFERENCES

1. See, for example, A.J. Buras, W. Slominski, H. Steger, Nucl. Phys. B345, 369 (1984).
2. B Oscillations Working Group 2002 `http://www.cern.ch/LEPBOSC/`.
3. L. Lellouch, *in these proceedings.*
4. There is some debate as to whether the uncertainty due to the extrapolation of Lattice QCD calculations to light quark masses is properly accounted for, see for example, A.S. Kronfeld and S.M. Ryan, hep-ph/0206058, 2002.
5. M. Ciuchini *et al.*, JHEP 0107, 013 (2001).
6. H.G. Moser and A. Roussarie, Nucl. Instr. Methods A 384, 491 (1997).
7. K. Abe *et al.* (SLD Collaboration), SLAC-PUB-9285, 2002.
8. A. Heister *et al.* (ALEPH Collaboration), CERN-EP/2002-016.
9. P. Kluit *et al.* (DELPHI Collaboration), DELPHI 2002-073 CONF 607.
10. K. Abe *et al.* (SLD Collaboration), SLAC-PUB-9286, 2002. To appear in Phys. Rev. D.
11. P. Abreu *et al.* (DELPHI Collaboration), Eur. Phys. J. C18, 229 (2000).

ELSEVIER

Proceedings of ICHEP 2002, pp. 493–498
*S. Bentvelsen, P. de Jong, J. Koch and E. Laenen (Editors)*

31st INTERNATIONAL CONFERENCE ON HIGH ENERGY PHYSICS AMSTERDAM

# Lattice results relevant to the CKM matrix determination

Damir Becirevic[a]

[a]Dipartimento di Fisica, Università di Roma "La Sapienza", Piazzale Aldo Moro 2, I-00185 Rome, Italy

A brief discussion of the recent lattice studies of the quantities that are relevant to the Standard Model description of CP-violation in the hadronic sector is presented.

## 1. Introduction

The most popular way to estimate the amount of CP-violation in the Standard Model is to study the shape of the CKM unitarity triangle involving the third generation of quark flavors. The standard unitarity triangle analysis (UTA) [1], which requires the least theoretical assumption, comprises the $K^0 - \overline{K^0}$ and $B_d^0 - \overline{B}_d^0$ mixing amplitudes, as well as the SU(3) breaking ratio in the $B^0 - \overline{B^0}$ system. Additional constraint comes from information on $|V_{ub}/V_{cb}|$ ratio, which will be discussed towards the end of this review. The three "golden" mixing amplitudes over-constrain the CKM-triangle through a hyperbola and circles in the $(\bar{\rho}, \bar{\eta})$-plane which emerge from the comparison of the theoretical predictions with the experimental determinations of $|\varepsilon_K|$, $\Delta m_d$ and the limit on $\Delta m_s/\Delta m_d$. Besides the CKM parameters, the theoretical expressions contain the Wilson coefficients and the hadronic matrix elements. The Wilson coefficients encode the information about physics at high energy scales, they are computed in perturbation theory and are known at NLO accuracy (for a review see e.g. [2]). The hadronic matrix elements, instead, describe the low energy QCD dynamics and their computation requires a theoretical control of non-perturbative QCD. To be specific, the following matrix elements need to be computed:

$$\begin{aligned}
\langle \bar{K}^0 | O_{sd}(\mu) | \bar{K}^0 \rangle &= \frac{8}{3} m_K^2 f_K^2 B_K(\mu) \,, \\
\langle \bar{B}_d^0 | O_{bd}(\mu) | \bar{B}_d^0 \rangle &= \frac{8}{3} m_{B_d}^2 f_{B_d}^2 B_{B_d}(\mu) \,, \quad (1) \\
\langle \bar{B}_s^0 | O_{bs}(\mu) | \bar{B}_s^0 \rangle &= \frac{8}{3} m_{B_s}^2 f_{B_s}^2 B_{B_s}(\mu) \,,
\end{aligned}$$

where the operator $O_{Qq} = (\bar{Q}q)_{V-A}(\bar{Q}q)_{V-A}$. On the r.h.s. of eq. (1) we recognize the pseudoscalar decay constants $f_K$ and $f_{B_{d(s)}}$, and the familiar "bag" parameters, $B_K(\mu)$ and $B_{B_{d(s)}}(\mu)$. That is where the lattice QCD enters the stage and provides us with a completely non-perturbative way to calculate these (and other) hadronic quantities. The following two points should be emphasized: [1]

- ⊗ Lattice QCD is not a model but the approach allowing the computation of various Green functions from first principles of QCD: the only parameters entering the computation are those that appear in the QCD lagrangian, namely the quark masses and the (bare) strong coupling;

- ⊗ QCD is solved numerically on the lattice by using the Monte-Carlo methods which induce the statistical errors in the results. The central limit theorem applies and ensures that those errors fall as $1/\sqrt{N_{\rm conf}}$, with $N_{\rm conf}$ being a number of independent SU(3) gauge field configurations on which a given Green function has been calculated. Thus, in principle, computation of the hadronic quantities on the lattice can be made to an arbitrary accuracy.

In practice, however, various approximations have to be made, each introducing some systematic uncertainty into the final results. Yet all those approximations are improvable by: (a) increasing the computing power so that, one by one,

[1]For details about formulating QCD on the lattice, please see e.g. [3].


doi:10.1016/S0920-5632(02)02165-5

all the approximations can be removed from the computation; (b) improving the approach theoretically so that the hadronic quantities, computed on the lattice, converge faster to their values in the continuum limit.

Now I return to the lattice determination of the "golden" hadronic quantities entering the standard UTA. The present situation is discussed at length in ref. [4], and the actual results are:

$$\begin{aligned} \hat{B}_K &= 0.86(6)(14)\,, \\ \hat{B}_{B_d} &= 1.34(12)\,, \\ f_{B_d} &= 203(27)\,\left(^{+00}_{-20}\right)\ \text{MeV}\,, \\ f_{B_s}/f_{B_d} &= 1.18(4)\,\left(^{+12}_{-00}\right)\,, \\ \xi &= (f_{B_s}\sqrt{B_{B_s}})/(f_{B_d}\sqrt{B_{B_d}}) = 1.18(4)\,\left(^{+12}_{-00}\right)\,. \end{aligned} \qquad (2)$$

The above quantities are completely dominated by the systematic uncertainties, some of which are discussed in what follows.

## 2. $K^0 - \overline{K^0}$ mixing

Results of the precision lattice computations are presented as $B_K^{\overline{\rm MS}}(2\ \text{GeV})$, in the quenched approximation, i.e. the quark loops in the background gauge field configurations are being neglected, $n_f = 0$. In the real physical world, however, one needs $n_f = 3$. Numerically, the conversion from the $\overline{\rm MS}$(NDR) scheme to the renormalization group invariant (RGI) form, $\hat{B}_K$, is almost completely independent of $n_f$ (see discussion in ref. [5]). By using $\Lambda_{\overline{\rm MS}}^{n_f=3} = 338(40)$ MeV [6], I obtain [2]

$$\hat{B}_K = 1.382(15)\cdot B_K^{\overline{\rm MS}}(2\ \text{GeV})\,. \qquad (3)$$

A map of the precision quenched lattice QCD determinations of $\hat{B}_K$ is shown above the dashed line in fig. 1, where the asterisks denote the results presented this year. The most accurate is the value obtained by using the so-called *staggered* fermions [7], $\hat{B}_K = 0.87(6)$. The advantage of that action is that it preserves the chiral symmetry, but the drawback is that it breaks the flavor symmetry. With Wilson fermions, instead, the flavor symmetry is preserved but the

[2]For a comparison, with $\Lambda_{\overline{\rm MS}}^{n_f=0} = 250(25)$ MeV, one has $\hat{B}_K = 1.388(8)\cdot B_K^{\overline{\rm MS}}(2\ \text{GeV})$.

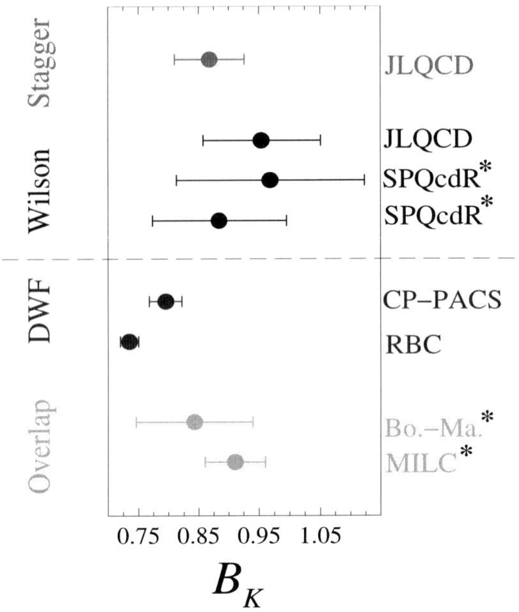


Figure 1. Quenched lattice results for the $\hat{B}_K$ parameter from 4 different lattice actions. The values above/below the dashed horizontal line are obtained after/without extrapolating to the continuum limit. The second result by SPQcdR is obtained by using the method proposed in ref. [11].

chiral symmetry is sacrificed. The loss of the chirality induces the additive renormalization of the operator $O_{sd}$, which is more involved than the (usual) multiplicative one. The non-perturbative method to subtract these mixing has been first implemented in ref. [8] by using the quark Ward identities, and then in ref. [9] by using the so-called RI/MOM method [10]. The problem of spurious mixing has been alleviated by a judicious application of the Ward identity on the parity violating operator which does not suffer from the spurious mixing problem on the lattice [11]. That proposal has been implemented in the high statistics lattice simulation and the result, in the continuum limit, reads $\hat{B}_K = 0.88(11)$. The results below the horizontal line in fig. 1 do not refer to the continuum limit. Their major significance lies in the fact that they are obtained by using the actions that satisfy both the chiral and the flavor symmetry. Computations with these actions are much more involved and require huge computational resources. This year, first results for $\hat{B}_K$, obtained by using the so-called overlap fermions, have been reported by the MILC [12] and by the Boston-Marseille [13] collaborations. Even though it is early to draw conclusions, the observed good agreement with the values above the horizontal line in fig. 1 is encouraging. Somewhat lower are the values obtained by using the so-called domain wall fermions (DWF). Although their statistical errors are very small, in my opinion, their systematic errors are not fully realistic (see discussion in ref. [5]). Nevertheless, it should be stressed that the results for $\hat{B}_K$ obtained by using DWF agree –within the error bars– with other lattice determinations. The remaining systematics is almost entirely due to quenching. Although the first (low statistics) lattice studies suggest that the unquenching is not changing the value of $B_K$ [16], we follow ref. [17] and assume that the quenching error is of the order of 15%. Thus it is of upmost importance, and challenging task for the lattice community, to unquench $B_K$.

## 3. $B^0_{s(d)} - \overline{B^0}_{s(d)}$ mixing

In this case, besides the light quark, one also has to deal with the heavy quark on the lattice. Since the Compton wavelength of the realistic $b$-quark is smaller than the presently attainable lattice resolution, one has to adopt some other strategy in order to compute the properties of the heavy-light hadrons. Four ways to deal with that problem have been implemented in the quenched computation of the decay constant $f_{B_{s(d)}}$, and/or the bag-parameter $\hat{B}_{B_{s(d)}}$.

⊙ Lattice QCD: one can afford to work with propagating heavy quarks of masses around the physical charm quark mass and then one has to *extrapolate* to the $b$-quark by means of standard heavy quark scaling laws [35];

⊙ Heavy quark effective theory (HQET) on the lattice is very difficult to extend beyond the static limit ($m_b \to \infty$). The long standing problem to renormalize the axial current non-perturbatively has been solved recently (see eg. [18]). The static results will therefore be possible to use in constraining the extrapolations of the results obtained in the full theory, i.e. with propagating heavy quarks;

⊙ Non-relativistic QCD (NRQCD), is the approach in which the $1/m_b$-corrections are included in both the lagrangian and the operators. Besides difficulties in renormalizing the operators in such a theory on the lattice, a particularly problematic is the fact that the expansion is in $1/(am_b)$ ($a$ being the lattice spacing), so that the continuum limit ($a \to 0$) cannot be taken;

⊙ Fermilab (FNAL) approach consists in using the lattice QCD with propagating quarks and "sending" the heavy quark mass over the lattice cut-off, followed by a redefinition of the hadron masses and reinterpretation of the theory in terms of a $1/m_b$-expansion. In some cases the approach is believed to be plagued by "renormalon shadows" [19].

None of the enumerated approaches is fully satisfactory on its own and, at present, they all should be used to check the consistency of the obtained results. For example, the results for $f_B$, as obtained by using various approaches, agree quite impressively (see ref. [4]). Nevertheless, the systematics (*within the quenched approximation*) cannot be further reduced until the fully QCD based lattice method is devised to compute the heavy-light quantities directly on the lattice. Otherwise, one should wait for the next generation of parallel computers to do the job. A candidate method for the precision computation of $f_B$ has been proposed this year in ref. [20], in which the step scaling function has been employed to make a controllable $1/(Lm_b)$-expansion, with $L$ being the side of the lattice box. From an exploratory study, $f_B = 170(11)(23)$ MeV has been quoted. The authors promise to reduce the error bars very soon and to respond to the questions raised in ref. [21].

The effect of unquenching has been extensively studied this year by the MILC collaboration [22]. They conclude that $f_B^{n_f=2}$ is about 15% larger than $f_B^{n_f=0}$. Moreover, in their preliminary study with $n_f = 3$, they show that the ratio $f_{B_s}/f_{B_d}$ remains essentially unchanged when one switches from $n_f = 0 \to 2 \to 3$ [23]. That brings me to the important estimate of the hadronic parameter $\xi$ (see eq. (2)). The SU(3) light flavor breaking effect in the bag parameter is equal to $B_{B_s}/B_{B_d} = 1.00(4)$ [4,21], and the whole effect of $\xi$ being different from unity comes from $f_{B_s}/f_{B_d}$.

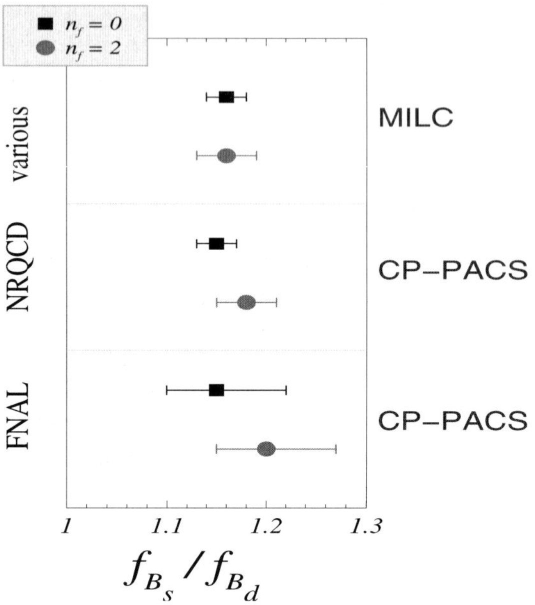


Figure 2. Quenched and unquenched ($n_f = 2$) lattice estimates for the SU(3) breaking ratio $f_{B_s}/f_{B_d}$, as obtained by three different methods indicated on the left. For more details see refs. [22] (MILC) and [24,25] (CP-PACS). No shift due to the chiral log term is displayed.

### 3.1. Chiral logs in $f_{B_s}/f_{B_d}$

Even if one gets a good handle over the heavy quark, there is still an extrapolation in the light quark mass to be made. With respect to the physical strange quark mass, the light quarks that are directly accessible from the lattice simulations lie in the range $1/2 \lesssim m_q/m_s \lesssim 3/2$, in which one observes that the heavy-light decay constant changes nearly linearly under the variation of the light quark mass. In other words, $f_{B_s}$ is accessed directly, whereas an extrapolation in the light quark mass is necessary in order to reach $f_{B_d}$. It has been argued recently that if in the standard (linear) extrapolation one also includes the chiral logarithmic terms, the SU(3) breaking ratio may get shifted from $f_{B_s}/f_{B_d} = 1.16$, by $+0.16$, which is a 100% error on the net SU(3) breaking [26]. In ref. [21] the shift can be as large as $+0.24$. There is an ambiguity while implementing the chiral logs in extrapolation. It is not clear at what mass scale the chiral logs dominate over the other, higher order, terms in the chiral expansion. If one sets it for the light meson masses around 600 MeV, the extrapolated value will be strongly shifted upwards (w.r.t. the result of linear extrapolation) [26]. The shift is roughly halved if the chiral logs are introduced for the light pseudoscalar mesons $\sim 350$ MeV. To avoid that ambiguity, ref. [27] proposes to study the double ratio

$$R = \frac{f_{B_s}/f_{B_d}}{f_K/f_\pi}\,, \tag{4}$$

in which almost all the chiral log corrections cancel, and one can make the linear extrapolations with a small error. From the available lattice data with $n_f = 2$, the NLO chiral correction in the ratio (4) seems to cancel too. From that consideration, ref. [27] concludes that $f_{B_s}/f_{B_d} \simeq f_K/f_\pi$, where the latter ratio is known very well from the experiments, $f_K/f_\pi = 1.22(1)$. Therefore, the shift is indeed present but it is not as spectacular as previously thought. Similar conclusion has been reached by the MILC collaboration [22]. They showed that if the chiral logarithms are consistently included in the entire analysis of the lattice data, the shift in $f_{B_s}/f_{B_d}$ is much smaller, namely $+0.04$.

Notice that the second errors in $f_{B_d}$, $f_{B_s}/f_{B_d}$ and $\xi$, quoted in eq. (2) (see also ref. [21]), reflect a guesstimate of the impact of the chiral logs on extrapolation in the light quark mass. As we just explained, those errors cannot be too large, and I personally believe that through the strategies such as those proposed in refs. [22] and [27], the lattice community will be able to cut down on those errors very soon.

## 4. $|V_{cb}|$ from $B \to D^{(*)}\ell\nu$

The extraction of $|V_{cb}|$ from the exclusive semileptonic modes, $B \to D^{(*)}\ell\nu$, requires the computation of the corresponding form factors. The cleanest method for their extraction has been proposed and implemented in ref. [28]. Interested reader is referred to those articles for more details. It would be very nice if the other lattice approaches adopt that strategy and compute the relevant form factors through the double ratios described in ref. [28].

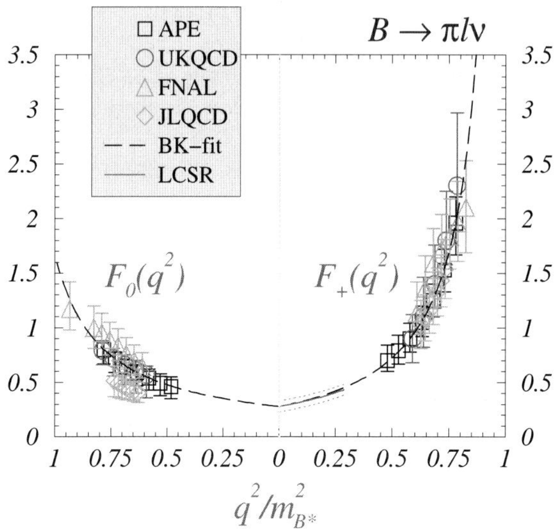


Figure 3. Recent (quenched) lattice results for the two form factors are plotted head-to-head, since they satisfy the condition $F_+(0) = F_0(0)$. For illustration a fit to the BK-form (6) is shown as well as the prediction obtained by using the light cone QCD sum rules (LCSR) [33].

## 5. $|V_{ub}|$ from $B \to \pi(\rho)\ell\nu$

In recent years several new lattice studies of the $B \to \pi$ form factors appeared [29–32]. The present situation with lattice results is depicted in fig. 3, where we plot the two form factors that parametrize the hadronic matrix element

$$\langle \pi(p')|V_\mu|B(p)\rangle = c_\mu^+ F_+(q^2) + c_\mu^0 F_0(q^2)\,, \quad (5)$$

with $c_{+/0}$ being the known kinematic factors. The range of the transfer momenta available from this decay is huge: $0 \le q^2 \le (m_B - m_\pi)^2 = 26.4\ \mathrm{GeV}^2$. The lattice can be used only for large $q^2$'s (smail recoil), and therefore the assumptions are necessary if we are to cover the entire $q^2$-range. In ref. [34] a simple parametrization has been proposed which includes various symmetry constraints onto the shapes of the form factors [35,36], as well as the kinematic condition $F_+(0) = F_0(0)$. It is often referred to as BK and it reads

$$F_+(q^2) = \frac{c\,(1-\alpha)}{(1-q^2/m_{B^*}^2)(1-\alpha q^2/m_{B^*}^2)}\,,$$
$$F_0(q^2) = \frac{c\,(1-\alpha)}{1-q^2/(\beta m_{B^*}^2)}\,, \quad (6)$$

thus consisting of three parameters only: $c$, $\alpha$ and $\beta$. Results of the fit of the lattice data to this form are given in table 1. We notice a pleasant agreement with the results obtained by LCSR [33], which is believed to be the method of choice in the low $q^2$-region. We also see a good consistency of the results obtained by using three different approaches. The agreement is especially good for the form factor $F_+(q^2)$, which is the only one entering the decay rate (for $\ell = e, \mu$). A bit less good is the agreement for the form factor $F_0(q^2)$. It is worth stressing that so far all the computations are performed in the quenched approximation. Recently, in ref. [37], it has been pointed out that the quenching effects may be significant, particularly for the form factor $F_0(q^2)$. That conclusion has been reached by confronting the corresponding expressions derived in the quenched and in the full ChPT. The remaining step, in my opinion, is to unquench the $B \to \pi$ form factors.

For the recent progress in taming the $B \to \rho\ell\nu$ and $B \to K^*\gamma$ on the lattice please see refs. [38, 39].

## 6. Conclusion

In this talk I discussed the status of the lattice QCD computation of the quantities that are relevant to the standard determination of the shape of the CKM unitarity triangle (see ref. [40] for the most recent update of the UTA). Quite a bit of progress has been made in gaining more control over the systematic uncertainties in the quenched lattice studies. A lot of work is still needed, especially in unquenching the current estimates.

Table 1
Results by various lattice groups are fitted to the form (6). For comparison, the LCSR results are presented in the same form (last line). Note that instead of the parameter $c$, we quote $F(0) = c\,(1-\alpha)$.

| $F(0)$ | $\alpha$ | $\beta$ | ref. |
|---|---|---|---|
| $0.30(^{+6}_{-5})(^{+4}_{-9})$ | $0.46(^{+09}_{-10})(^{+37}_{-05})$ | $1.27(^{+14}_{-11})(^{+04}_{-12})$ | UKQCD [29] |
| $0.28(6)(5)$ | $0.45(17)(^{+06}_{-13})$ | $1.20(13)(^{+15}_{-00})$ | APE-I [30] |
| $0.26(5)(4)$ | $0.40(15)(9)$ | $1.22(14)(^{+15}_{-00})$ | APE-II [30] |
| $0.33(^{+2}_{-3})$ | $0.34(^{+9}_{-3})$ | $1.31(^{+3}_{-9})$ | FNAL [31] |
| $0.23(^{+4}_{-3})$ | $0.58(^{+12}_{-09})$ | $1.28(^{+12}_{-20})$ | JLQCD [32] |
| $0.28(5)$ | $0.32(^{+21}_{-07})$ | – | KRWWY [33] |

Besides the "golden" quantities for the CKM triangle analysis, other quantities, such as the quark masses, decay constants, form factors, light cone wave functions, ... are computed on the lattice. Accurate determinations of these quantities is very useful for theoretical studies of the non-leptonic heavy-to-light decays. Finally, I would like to mention the intensive activity, within the lattice community, in taming the $K \to \pi\pi$ decay amplitudes which I did not have space nor time to cover in this talk. For a status report please see [41].

**Acknowledgments**

I am grateful to all members of the SPQcdR collaboration for their advice in preparing this talk. The correspondence with L. Lellouch, S. Sharpe and N. Yamada, as well as with C. Maynard, T. Onogi and J. Simone are kindly acknowledged. Finally, I thank the organizers for the invitation to this exciting and enjoyable conference.

## REFERENCES

1. M. Ciuchini *et al.*, JHEP **0107** (2001) 013; A. Hocker *et al.*, Eur. Phys. J. C **21** (2001) 225; A. J. Buras *et al.*, hep-ph/0207101.
2. G. Buchalla *et al.*, Rev. Mod. Phys. **68** (1996) 1125
3. I. Montvay and G. Münster, "Quantum Fields On A Lattice," *Cambridge, UK: Univ. Pr. (1994) 491 p.*.
4. L. Lellouch, this volume.
5. D. Becirevic, "$K^0 - \overline{K^0}$ mixing", presented at "QCD-02", Montpellier, France, 2-9 Jul 2002.
6. S. Bethke, hep-ex/0211012.
7. S. Aoki *et al.* [JLQCD Collaboration], Phys. Rev. Lett. **80** (1998) 5271.
8. S. Aoki *et al.* [JLQCD Collaboration], Phys. Rev. D **60** (1999) 034511.
9. D. Becirevic *et al.* [SPQcdR Collaboration], hep-lat/0209136.
10. A. Donini *et al.*, Eur. Phys. J. C **10** (1999) 121.
11. D. Becirevic *et al.*, hep-lat/0005013.
12. T. DeGrand [MILC], hep-lat/0208054.
13. L. Giusti, hep-lat/0211009.
14. A. Ali Khan *et al.* [CP-PACS Collaboration], Phys. Rev. D **64** (2001) 114506.
15. T. Blum *et al.* [RBC], hep-lat/0110075.
16. G. Kilcup *et al*, Nucl. Phys. PS. **53** (1997) 345; D. Becirevic,D. Meloni,A. Retico, JHEP **0101** (2001) 012.
17. S. R. Sharpe, Nucl. Phys. PS **53** (1997) 181.
18. R. Sommer, hep-lat/0209162.
19. G. Martinelli and C. T. Sachrajda, Nucl. Phys. B **478** (1996) 660; C. W. Bernard, Nucl. Phys. PS **94** (2001) 159.
20. M. Guagnelli *et al.*, Phys. Lett. B **546** (2002) 237.
21. N. Yamada, hep-lat/0210035.
22. C. Bernard *et al.* [MILC], hep-lat/0206016.
23. C. Bernard *et al.* [MILC], hep-lat/0209163.
24. A. Ali Khan *et al.* [CP-PACS Collaboration], Phys. Rev. D **64** (2001) 054504.
25. A. Ali Khan *et al.* [CP-PACS Collaboration], Phys. Rev. D **64** (2001) 034505.
26. A. S. Kronfeld and S. M. Ryan, Phys. Lett. B **543** (2002) 59; N. Yamada *et al.* [JLQCD], Nucl. Phys. P.S. **106** (2002) 397.
27. D. Becirevic *et al.*, hep-ph/0211271.
28. S. Hashimoto *et al.*, Phys. Rev. D **61** (2000) 014502, ibid **66** (2002) 014503.
29. K. C. Bowler *et al.* [UKQCD Collaboration], Phys. Lett. B **486** (2000) 111.
30. A. Abada *et al.* [APE Collaboration], Nucl. Phys. B **619** (2001) 565.
31. A. X. El-Khadra *et al.*, Phys. Rev. D **64** (2001) 014502.
32. S. Aoki *et al.* [JLQCD Collaboration], Phys. Rev. D **64** (2001) 114505.
33. A. Khodjamirian *et al.*, Phys. Rev. D **62** (2000) 114002.
34. D. Becirevic and A. B. Kaidalov, Phys. Lett. B **478** (2000) 417.
35. N. Isgur and M. B. Wise, Phys. Rev. D **42** (1990) 2388.
36. J. Charles *et al.*, Phys. Rev. D **60** (1999) 014001; C. W. Bauer *et al.*, Phys. Rev. D **63** (2001) 114020.
37. D. Becirevic *et al.*, hep-lat/0210048.
38. A. Abada *et al.*, hep-lat/0209116.
39. F. Mescia, see at http://www-zeuthen.desy.de/nic/euihp02/talks/mescia.pdf
40. A. Stocchi, hep-ph/0211245.
41. N. Ishizuka, hep-lat/0209108.

ELSEVIER

Proceedings of ICHEP 2002, pp. 499–502
*S. Bentvelsen, P. de Jong, J. Koch and E. Laenen (Editors)*

31[st] INTERNATIONAL CONFERENCE ON HIGH ENERGY PHYSICS AMSTERDAM

# Determination of the unitarity triangle parameters

F. Parodi[a]

[a]Dipartimento di Fisica, Università di Genova and INFN, Via Dodecaneso 33, 16146 Genova, Italy

I describe the present determination of the Unitarity Triangle parameters in the Standard Model; I discuss new strategies for future determinations and alternative sets of fundamental parameters. Implication of the present data on Minimal Flavour Violating models are shown.

## 1. CKM matrix and Unitarity Triangle

The determination of the Cabibbo-Kobayashi-Maskawa (CKM) matrix that parametrizes the weak charged current interactions of quarks is one of the important targets of particle physics.

The CKM matrix is a unitary matrix with four independent parameters (3 parameters and one phase). Many parametrization of the CKM matrix have been proposed in the literature; in the following the generalisation of the Wolfenstein parametrization introduced in [1] will be used:

$$|V_{us}|, \qquad |V_{cb}|, \qquad \bar{\varrho}, \qquad \bar{\eta} \tag{1}$$

where $(\bar{\varrho}, \bar{\eta})$, defined as

$$\bar{\varrho} = \varrho(1 - \lambda^2/2), \qquad \bar{\eta} = \eta(1 - \lambda^2/2), \tag{2}$$

determine the apex of the Unitarity Triangle (Figure 1) defined by the unitarity relation

$$V_{ud}V_{ub}^* + V_{cd}V_{cb}^* + V_{td}V_{tb}^* = 0 \tag{3}$$

Among the four independent parameters in (1) $|V_{us}|$ and $|V_{cb}|$ are determined using the semileptonic K and B decays, respectively, while $\bar{\varrho}$ and $\bar{\eta}$ are determined using several indirect measurements.

The presently available experimental constraints come from $K_d^0$-$\bar{K}^0$ mixing ($\epsilon_K$), $B_d^0$-$\bar{B}_d^0$ and $B_s^0$-$\bar{B}_s^0$ mixing ($\Delta m_d$ and $\Delta m_s$), charmless B decays and the CP asymmetry in $B_d^0 \to J/\psi K_S$.

In practice, theoretical expressions for these constraints also involve several additional parameters as quark masses, decay constants of B mesons, bag-factors,... The values of these parameters are constrained by other measurements (ex: top quark mass) or using theoretical expectations.

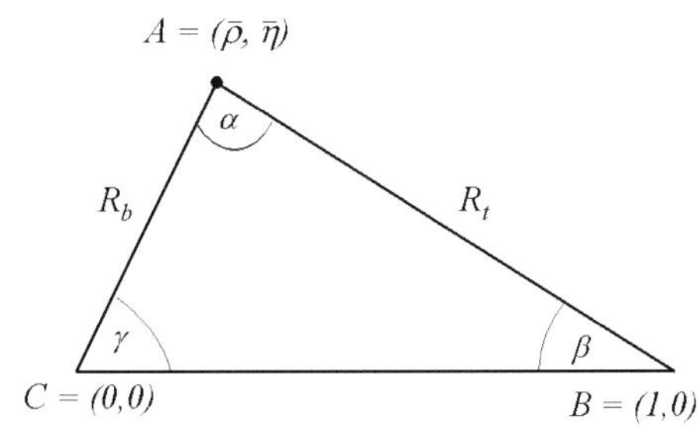


Figure 1. *Unitarity Triangle*

## 2. Different approaches to the UT analysis

Since several years there has been intense activity to constrain the apex of UT from the best knowledge of the experimental and theoretical inputs. All the recent methods [2–4] try to define the "correct" way to deal with experimental and theoretical uncertainties; differences arise from the statistical treatment of those uncertainties that could be reduced to two main statistical approaches: Bayesian and Frequentist.

The main concern are systematical uncertainties on theoretical parameters that are not of statistical origin, usually quoted with a maximal range of variation.

A Bayesian analysis [2] assume for those parameters a uniform probability density function in that range while a Frequentist analysis (Rfit)


doi:10.1016/S0920-5632(02)02168-0

| Parameter | Value | Gaussian $\sigma$ | Uniform half-width | Ref. |
|---|---|---|---|---|
| $\lambda$ | 0.2210 | 0.0020 | - | [5] |
| $\|V_{cb}\|$(excl.) | $42.1 \times 10^{-3}$ | $2.1 \times 10^{-3}$ | - | [6] |
| $\|V_{cb}\|$(incl.) | $40.4 \times 10^{-3}$ | $0.7 \times 10^{-3}$ | $0.8 \times 10^{-3}$ | [6] |
| $\|V_{cb}\|$(ave.) | $40.6 \times 10^{-3}$ | $0.8 \times 10^{-3}$ * | | |
| $\|V_{ub}\|$(excl.) | $32.5 \times 10^{-4}$ | $2.9 \times 10^{-4}$ | $5.5 \times 10^{-4}$ | [5] |
| $\|V_{ub}\|$(incl.) | $40.9 \times 10^{-4}$ | $4.6 \times 10^{-4}$ | $3.6 \times 10^{-4}$ | [5] |
| $\|V_{ub}\|$(ave.) | $36.3 \times 10^{-4}$ | $3.2 \times 10^{-4}$ * | | |
| $\|V_{ub}\|/\|V_{cb}\|$(ave.) | 0.089 | 0.008* | | |
| $\Delta M_d$ | 0.503 ps$^{-1}$ | 0.006 ps$^{-1}$ | - | [7] |
| $\Delta M_s$ | $>$ 14.4 ps$^{-1}$ at 95% C.L. | sensitivity 19.2 ps$^{-1}$ | | [7] |
| $m_t$ | 167 $GeV$ | 5 $GeV$ | - | [8] |
| $f_{B_d}\sqrt{\hat{B}_{B_d}}$ | 235 $MeV$ | 33 $MeV$ | $^{+0}_{-24}$ $MeV$ | [9] |
| $\xi = \frac{f_{B_s}\sqrt{\hat{B}_{B_s}}}{f_{B_d}\sqrt{\hat{B}_{B_d}}}$ | 1.18 | 0.04 | $^{+0.12}_{-0.00}$ | [9] |
| $\hat{B}_K$ | 0.86 | 0.06 | 0.14 | [9] |
| sin $2\beta$ | 0.734 | 0.054 | - | [10] |

Table 1
*Values of the relevant quantities entering into the expressions of $|V_{ub}/V_{cb}|$, $\Delta M_d$ and $\Delta M_s$. In the third and fourth columns the Gaussian and the flat part of the uncertainty are given, respectively. Here $m_t \equiv \overline{m}_t(m_t)$.*

[3] use this range as penalty function for the global $\chi^2$ of the fit (the fit is insensitive to variations of the parameters in their ranges). This different interpretation of the inputs is the main reason for the differences in the final results.

Another Frequentist analysis [4], presented at this conference, choose not relying on any ranges for the systematical uncertainties implying that, with the present theoretical uncertainties, a global allowed region for the apex of the UT cannot be determined. The compatibility among the constraints is evaluated, in this analysis, in the theoretical parameters space.

Recently [5] a comparison work between the Bayesian analysis and Rfit, with same inputs, has started. The results obtained [5] show that the width of the selected 95% C.L. regions for the relevant parameters differs by less than 50% (the regions are larger in the Rfit analysis); the difference decrease when increasing the C.L.

The results presented in the following are obtained using the Bayesian analysis. [2].

## 3. Present determination of the UT

The measurements and the theoretical determinations entering in the UT analysis are listed in Table 1

The region selected by present data in the $(\bar{\varrho}, \bar{\eta})$ plane is shown in Figure 2 and the results for the relevant UT parameters are:

| | | |
|---|---|---|
| $\bar{\eta}$ | $0.357 \pm 0.027$ | (0.305-0.411) |
| $\bar{\rho}$ | $0.173 \pm 0.046$ | (0.076-0.260) |
| $\sin 2\beta$ | $0.725\ ^{+0.035}_{-0.031}$ | (0.660-0.789) |
| $\sin 2\alpha$ | $-0.09 \pm 0.25$ | (-0.54-0.40) |
| $\gamma$ | $(63.5 \pm 7.0)^o$ | $(51.0\text{-}79.0)^o$ |
| $\Delta m_s$ | $(18.0^{+1.7}_{-1.5})$ ps$^{-1}$ | (15.4-21.7) ps$^{-1}$ |

Results in brackets represent the 95% probability intervals.

### 3.1. Consistency check of the CP violation picture in the SM

As four constraints are used to determine the values of two parameters, it is possible to relax, in turn, one (or more) of these constraints.

An interesting exercise consists in removing the constraints coming from experimental observations of CP violation ($\epsilon_K$ and $sin2\beta$) and comparing the selected region in the $(\bar{\varrho}, \bar{\eta})$ plane with these two constraints (Figure 3). The result of this comparison can be quantified by comparing the value of $sin2\beta$ from the measurement of the CP asymmetry in $J/\psi K_S$ decays with the value

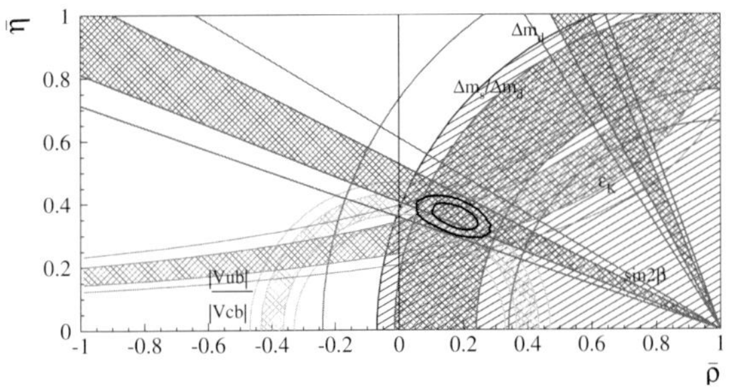


Figure 2. *The 68% and 95% probability regions for $\bar{\varrho}$ and $\bar{\eta}$ are compared with uncertainty bandes for the constraints.*

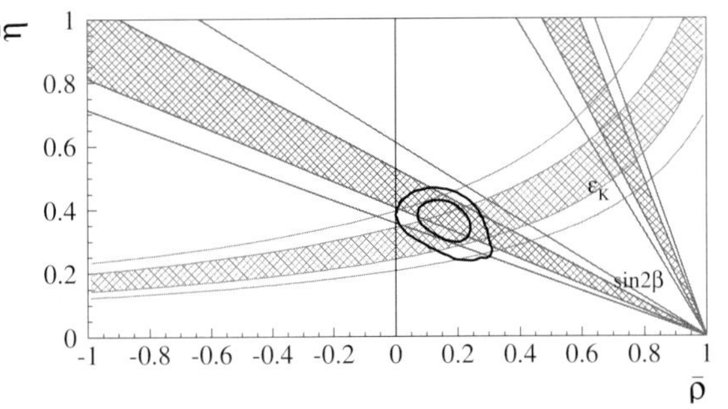


Figure 3. *The 68% and 95% probability regions for $\bar{\varrho}$ and $\bar{\eta}$, using the constraints given by the measurements of $|V_{ub}|\,/\,|V_{cb}|$, $\Delta m_d$ and $\Delta m_s$, are compared with the constraints from* $\sin 2\beta$ *and* $\epsilon_K$.

determined from the "sides" measurements:

$$\begin{array}{lll} sin2\beta & = & 0.725^{+0.055}_{-0.065} \quad \text{Sides only} \\ sin2\beta & = & 0.734 \pm 0.054 \quad \text{a}(J/\psi\ K_S)\ [10] \end{array}$$

The spectacular agreement of these values shows the consistency of the Standard Model in describing the CP violation phenomena in terms of one single parameter $\bar{\eta}$ with the description of the strong interaction through the HQET, OPE, Lattice QCD...

Using all the available constraints (but $sin2\beta$) the indirect determination of $sin2\beta$ turns out to be $0.715 \pm 0.055$.

## 4. Future Strategies for the UT

The determination of the apex of the UT triangle (Figure 1) only requires the measurement of two out of five parameters ($\alpha$, $\beta$, $\gamma$, $R_t$, $R_b$).

The following question is addressed in [11]: given the same relative errors on $R_b$, $R_t$, $\alpha$, $\beta$ and $\gamma$ which of the pairs gives the most accurate determination of $(\bar{\varrho}, \bar{\eta})$ ?

After dividing the parameters in the two classes with the lowest correlation ($(R_b, \beta)$ and $(R_t, \alpha, \gamma)$) it has been found [11] the following hierarchy (in decreasing order of precision):

1) $(\gamma, \beta)$, $(\gamma, R_b)$
2) $(\alpha, \beta)$, $(\alpha, R_b)$
3) $(R_t, \beta)$, $(R_t, R_b)$, $(R_b, \beta)$.

This analysis shows how important is the determination of $R_b$ and $\gamma$. On the other hand the strategy involving $R_t$ and $\beta$ will be most probably the cleanest one before the LHC experiments as the angle $\gamma$ cannot be precisely measured until then. The strategies involving $\alpha$ are in the second best class. However to determine $\alpha$ with good relative precision appears to be very difficult in view of the penguin pollution that could be substantial.

### 4.1. Present available strategies

At present concrete results can be obtained only for the strategies $(R_t, \beta)$, $(R_b, R_t)$ and $(R_b, \beta)$ as no direct measurements of $\gamma$ and $\alpha$ are available ($\beta$ is derived from $\sin 2\beta$, $R_t$ is extracted from $\Delta M_d$ and $\Delta M_d/\Delta M_s$ and $R_b$ is extracted from $|V_{ub}/V_{cb}|$).

The results for $\bar{\rho}$ and $\bar{\eta}$ for the three strategies in question are presented in Table 2

Once $\Delta M_s$ has been measured and the error on $\sin 2\beta$ reduced, the strategy $(R_t, \beta)$ will be leading among these three.

The most obvious candidate for the fundamental set in the quark flavour physics for the coming years appears to be

$$|V_{us}|, \qquad |V_{cb}|, \qquad R_t, \qquad \beta \tag{4}$$

having the advantage of containing, compared with the set in (1), only directly measurable quantities.

| Strategy | $\bar{\rho}$ | $\bar{\eta}$ |
|---|---|---|
| $(R_t, \beta)$ | $0.157\ ^{+0.056}_{-0.054}$ | $0.367\ ^{+0.036}_{-0.034}$ |
| | (0.047-0.276) | (0.298-0.439) |
| $(R_t, R_b)$ | $0.161^{+0.055}_{-0.057}$ | $0.361\ ^{+0.041}_{-0.045}$ |
| | (0.043-0.288) | (0.250-0.438) |
| $(R_b, \beta)$ | $0.137\ ^{+0.135}_{-0.135}$ | $0.373\ ^{+0.049}_{-0.063}$ |
| | (-0.095-0.357) | (0.259-0.456) |

Table 2
*Results for $\bar{\rho}$ and $\bar{\eta}$ for the three indicated strategies using the present knowledge summarized in Table 1. In bracket the 95% probability region is also given.*

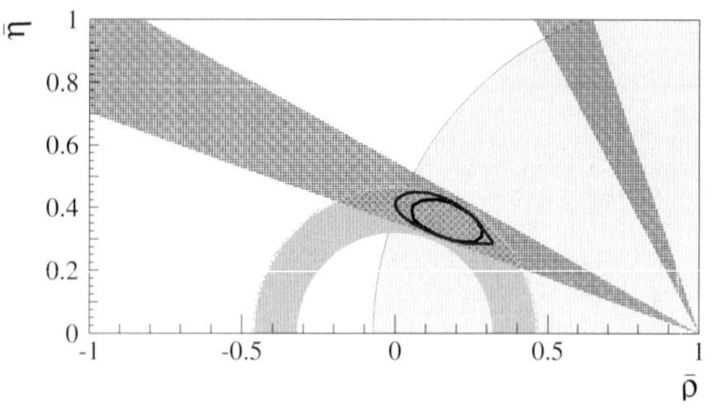


Figure 4. *The plot shows the allowed 95% region in the $(\bar{\varrho}, \bar{\eta})$ plane, consistent with UUT. The narrower region corresponds to the allowed 95% region consistent with SM.*

## 5. Implications on Minimal Flavour Violation Models

The simplest class of extensions of the SM are the "Minimal Flavour Violation" (MFV) models in which only the SM operators in the effective weak Hamiltonian are relevant and in which flavour violation is entirely governed by the CKM matrix. The interesting virtue of the MFV models is that with respect to $B^0_{d,s} - \bar{B}^0_{d,s}$ mixings and the CP-violating parameter $\varepsilon_K$, they all can be parametrized by a single function $F_{tt}$.

The presently available quantities that do not depend on the new physics parameters within the MFV-models and therefore can be used to determine the UUT are $R_t$ from $\Delta M_d/\Delta M_s$, $R_b$ from $|V_{ub}/V_{cb}|$, and $\sin 2\beta$ from the CP asymmetry in $B^0_d \to \psi K_S$.

Using these constraints it has been found [11] the allowed universal region for $(\bar{\varrho}, \bar{\eta})$ in the MFV models shown in Figure 4.

It should be stressed that any MFV model that is inconsistent with the broader allowed region in Figure 4 is ruled out. This analysis shows that little room is left for MFV models that in their predictions differ from the SM as the most ranges within the SM and MFV models do not significantly differ from each other.

## REFERENCES


1. A.J. Buras, M.E. Lautenbacher and G. Ostermaier, Phys. Rev. **D 50** (1994) 3433.
2. M. Ciuchini, G. D'Agostini, E. Franco, V. Lubicz, G. Martinelli, F. Parodi, P. Roudeau, A. Stocchi, JHEP 0107(2001) 013 (hep-ph/0012308), updated in contribution 657 to this conference.
3. A. Höcker, H. Lacker, S. Laplace and F. Le Diberder, Eur. Phys. J. **C21** (2001) 225.
4. G. P. Dubois-Felsmann D. G. Hitlin, F. C. Porter and G. Eigen, CALT 68-2396 June 2002, contribution 959 to this conference.
5. Results presented at the CKM Unitarity Triangle Workshop, CERN Feb 2002. http://ckm-workshop.web.cern.ch/ckm-workshop/
6. M. Artuso and E. Barberio, hep-ph/0205163
7. LEP Working group on oscillations : http://lepbosc.web.cern.ch/LEPBOSC/ combined_results/amsterdam_2002/
8. F. Abe *et al.*, CDF Collaboration, Phys. Rev. Lett. **74** (1995) 2626;
S. Abachi *et al.*, D0 Collaboration, Phys. Rev. Lett. **74** (1995) 2632.
9. L. Lellouch, these proceedings.
10. Y. Nir, these proceedings and reference therein.
11. A.J. Buras, F. Parodi and A. Stocchi, hep-ph/0207131 and references therein, contribution 661 to this conference.

ELSEVIER

Proceedings of ICHEP 2002, pp. 503–506
*S. Bentvelsen, P. de Jong, J. Koch and E. Laenen (Editors)*

31st INTERNATIONAL CONFERENCE ON HIGH ENERGY PHYSICS AMSTERDAM

# $CP$ asymmetry in $B \to$ Vector PseudoScalar decays

T. Aushev*, for the Belle Collaboration
Institue for Theoretical and Experimental Physics

We report the new results of a study of $B^0$ meson decays into the vector pseudoscalar final states using the data collected by the Belle detector at KEKB $e^+e^-$ collider. We have measured the branching ratios for the $B^0 \to J/\psi\pi^0, D^\pm D^{*\mp}, \phi K^0$ and $K^+K^-K_S$ decay modes. A search for $CP$-violation have been performed in $B^0 \to J/\psi\pi^0$, $\phi K_S$ and $K^+K^-K_S$ decays.

## 1. Introduction

The decays $B^0 \to J/\psi\pi^0$ and $B^0 \to D^\pm D^{*\mp}$ are mediated by the Cabibbo suppressed $\bar{b} \to \bar{c}c\bar{d}$ transition. The tree diagram which is believed to be the leading contribution to this transition results in the same weak phase as $B^0 \to J/\psi K_{S(L)}$. However, unlike $B^0 \to J/\psi K_{S(L)}$ the Standard Model penguin contribution can lead to another weak phase, thus a precision measurement of time-dependent $CP$-asymmetry in these modes can measure an asymmetry parameter different to $\sin 2\phi_1$. Theoretically the penguin contribution is estimated to be small and deviation from the measured $CP$ violation parameter in the $B^0 \to J/\psi\pi^0$ and $B^0 \to D^\pm D^{*\mp}$ decays should agree with $\sin 2\phi_1$ within 10%. This difference is beyond the present accuracy...

The decay $B^0 \to \phi K_S$ is expected to be a pure penguin $\bar{b} \to \bar{s}s\bar{s}$ transition and $CP$ violation in this channel can be reliably predicted to be related to $\sin 2\phi_1$. This channel is suitable for New Physics searches. On the one hand new particles can contribute to the penguin loop and change the weak phase in the $CP$ asymmetry. On the other hand within the Standard Model this weak phase is well known to be equal to those measured in $B^0 \to J/\psi K_S$ decay. Thus any significant difference in $\sin 2\phi_1$ measured in $B^0 \to \phi K_S$ and $B^0 \to J/\psi K_S$ decays can be a hint on New physics contribution.

The study of the decays of $B$ meson into three kaons can significantly broaden the understanding of $B$ decay mechanisms and also provide additional possibilities for $CP$-violation measurements in $\bar{b} \to \bar{s}s\bar{s}$ transitions.

## 2. Data sample and particle reconstruction

The data used for this analysis were taken with the Belle detector [2] at the KEKB asymmetric $e^+e^-$ (3.5 on 8.0 GeV) collider [3]. The integrated luminosity used for this analysis is 78 fb$^{-1}$ collected at the $\Upsilon(4S)$ resonance ($\sqrt{s} =$ 10.58 GeV) that corresponds to 85 million $B\bar{B}$ pairs. The Belle detector is a large-solid-angle magnetic spectrometer that consists of a three-layer silicon vertex detector (SVD), a 50-layer central drift chamber (CDC), a mosaic of aerogel threshold Čerenkov counters (ACC), time-of-flight scintillation counters (TOF), and an array of CsI(Tl) crystals (ECL) located inside a superconducting solenoid coil that provides a 1.5 T magnetic field. An iron flux-return located outside of the coil is instrumented to detect $K_L$ mesons and to identify muons (KLM).

Charged tracks are reconstructed in the CDC and their impact parameters are precisely determined using the SVD. Charged hardon identification is accomplished based on the combined information from the ACC, TOF and CDC $dE/dx$ systems. Electron identification is based on a combination of $dE/dx$ measurements, the ACC response and information about the shape, energy deposit and position of the associated shower in the ECL. Muons are identified by requiring an association between KLM hits and an extrapolated track. Photons are reconstructed in the ECL

*aushev@iris1.itep.ru


doi:10.1016/S0920-5632(02)02171-0

as showers that are not associated with charged tracks.

## 3. Study of the $B^0 \to J/\psi\pi^0$

$J/\psi$ candidates are reconstructed via their decay into $\ell^+\ell^-$ with selection criteria identical to those used in [1]. $\pi^0$ meson candidates are reconstructed from pairs of photons that have an invariant mass within $\pm3\sigma$ of the nominal $\pi^0$ mass. $B$ meson candidates are identified using the beam constrained mass $M_{bc} \equiv \sqrt{E_{beam}^2 - P_B^2}$ and the energy difference $\Delta E \equiv E_B - E_{beam}$, where $E_B(P_B)$ is the energy(momentum) of the $B$ candidate and $E_{beam}$ is the center-of-mass (CM) beam energy. The signal yield is found to be $55.5 \pm 8.6$. This number is obtained from the fit of $M_{bc}$ distribution presented on Fig. 1*a*. The branching fraction for this decay mode is calculated to be $(2.0 \pm 0.3 \pm 0.2) \times 10^{-5}$. The details of the systematics study can be found elsewhere [5].

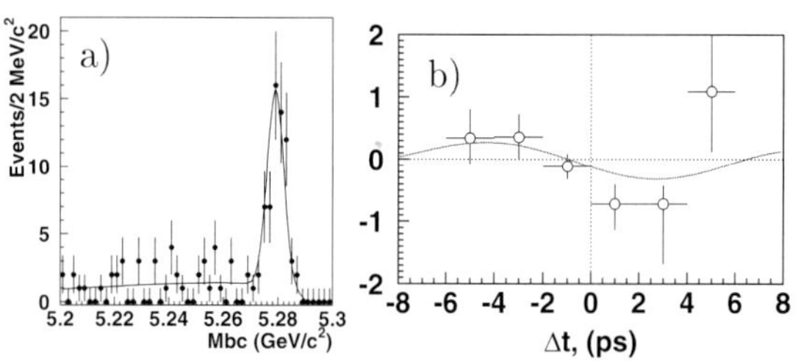


Figure 1. a) The beam-constrained mass distribution; the curve represents the result of the fit with a Gaussian to describe the signal and the ARGUS function to parameterize the background. b) the distribution of the time-dependent $CP$-asymmetry for the $B^0 \to J/\psi\pi^0$ decay mode; the curve shows the results of the fit.

For extracting the $CP$ asymmetry, the standard Belle tagging and vertexing procedures [1] are used. The probability density function used for the $CP$ fit is defined as:

$$\frac{1}{4\tau_{B^0}} e^{-\frac{|\Delta t|}{\tau_{B^0}}} \{1 + q[S\sin(\Delta m\Delta t) + A\cos(\Delta m\Delta t)]\},$$

where $A$ accounts for the direct $CP$ violation and $S = -\xi_f'' \sin 2\phi_1''$. The values of $S$ and $A$ found by the $CP$ fit are presented in Table 2. Fig. 1*b* shows the time-dependent asymmetry between the decay rates into $J/\psi\pi^0$ for $B$ mesons tagged as $B^0$ and $\overline{B}^0$. The results of the $CP$-fit is shown on this figure as the solid line.

## 4. Study of the $B^0 \to D^+D^{*-}$

For the $B^0 \to D^+D^{*-}$ decay mode we have updated the branching fraction published in [6]. In order to increase the statistics the partial reconstruction technique is used. $D^+$ mesons, fully reconstructed in its $K^-\pi^+\pi^+$ decay mode, is then combined with $\pi^-_{slow}$ originated from from the decay of $D^{*-} \to D^0\pi^-_{slow}$. The direction of the $\pi^-_{slow}$ from the $D^{*-}$ decay well approximates the direction of the mother particle due to the small energy release in the $D^{*-}$ decay. Thus the distribution of CM angle $\alpha$ between $\pi^-_{slow}$ and $D^+$ is peaked at $\alpha \sim 180^o$ for the signal events while background exhibits a flat behavior as estimated by the Monte Carlo simulation.

The distribution of $\cos\alpha$ is shown on Fig. 2. The number of signal events is obtained from the fit of this distribution with the exponential function, which is found to parameterize well the signal shape, and a second polynomial function to describe the background. The number of signal events found from the fit is $102 \pm 23$. Correcting the efficiencies we calculate the Branching ratio for $B^0 \to D^+D^{*-}$ to be equal to $(1.25 \pm 0.28)10^{-3}$. The systematic error is under the study now.

## 5. Study of the $B^0 \to \phi K_S$ and $B^0 \to K^+K^-K_S$

$K_S$ candidates are reconstructed from $\pi^+\pi^-$ pairs with a common vertex that is displaced from the interaction point and requiring an invariant mass within $\pm3\sigma$ of the nominal $K_S$ mass. As in the case of the $B^0 \to J/\psi\pi^0$ decay mode the $B$ meson is identified using $\Delta E$ and $M_{bc}$ distributions. The numbers of signal events are obtained from the fit of $\Delta E$ distribution (Fig. 3 and are found to be $39.4^{+7.5}_{-6.5}$ and $149 \pm 15$ for the $\phi K_S$ and

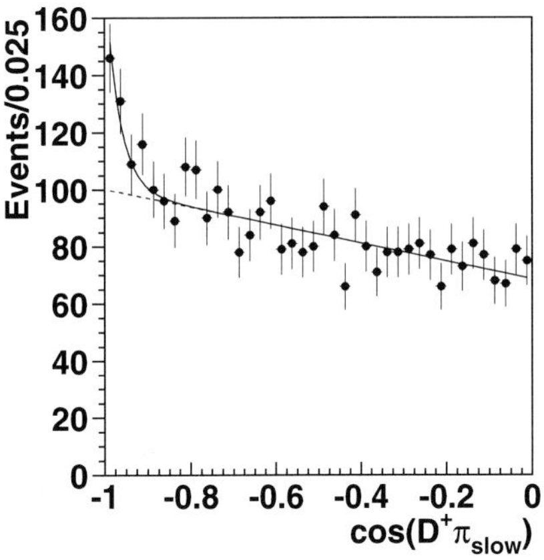


Figure 2. The distribution of $\cos\alpha$; the curve represents the fit described in the text.

$K^+K^-K_S$ respectively. The branching fractions measured for these decay are $(1.00^{+0.19+0.09}_{-0.17-0.13}) \times 10^{-5}$ and $(2.93 \pm 0.34 \pm 0.40) \times 10^{-5}$.

The results are listed in the Table 1. The details of the analysis can be found elsewhere [7,8].

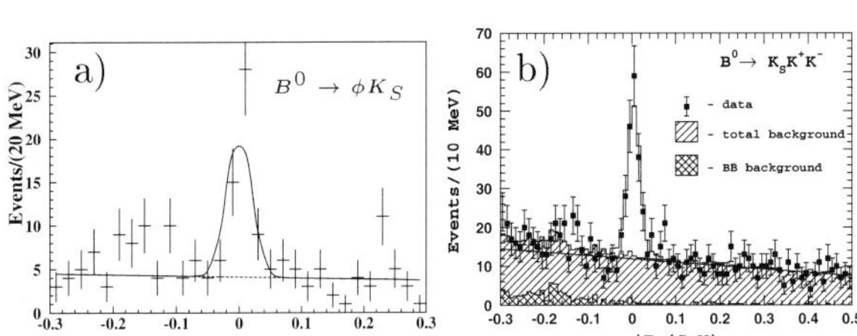


Figure 3. $\Delta E$ distribution for the decay mode a) $\phi K_S$ and b) $K^+K^-K_S$. The fit function is Gaussian as a signal function and first order polynomial as a background function.

As in the case of $J/\psi\pi^0$ the same $CP$ fit procedure is used for these decay modes. The fit results of $CP$-fit are presented on the Fig. 4 and listed in the Table 2.

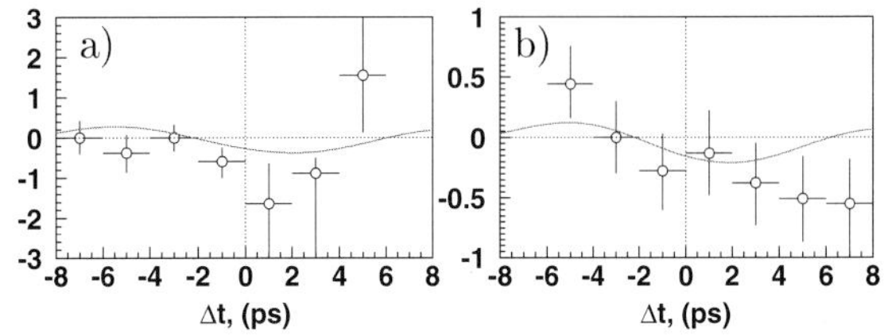


Figure 4. Time-dependent $CP$-asymmetry for the decay mode a) $\phi K_S$ and b) $K^+K^-K_S$. Solid lines present the $CP$-fit results.

Table 1
Summary of Branching fraction measurements

| Mode | # events | $\mathcal{B}, \times 10^{-5}$ |
|---|---|---|
| $J/\psi\pi^0$ | $55.5 \pm 8.6$ | $2.0 \pm 0.3 \pm 0.2$ |
| $D^\pm D^{*\mp}$ | $102 \pm 23$ | $125 \pm 28$ |
| $\phi K^0$ | $39.4^{+7.5}_{-6.5}$ | $1.00^{+0.19+0.09}_{-0.17-0.13}$ |
| $K^+K^-K^0$ | $149 \pm 15$ | $2.93 \pm 0.34 \pm 0.40$ |

## 6. Summary

The recent results of the branching ratios measurements of $B^0$ into $B^0 \to J/\psi\pi^0, D^\pm D^{*\mp}, \phi K^0$ and $K^+K^-K_S$ are presented using the data sample of $89 fb^{-1}$ collected by the Belle detector at KEKB $e^+e^-$ collider. Branching fractions for the $B^0 \to J/\psi\pi^0, D^\pm D^{*\mp}, \phi K^0$ and $K^+K^-K_S$ decays have been measured. $CP$-violation parameter $\sin 2\phi_1$ have been measured in $B^0 \to J/\psi\pi^0$, $\phi K_S$ and $K^+K^-K_S$ decays. The summary of the branching ratio measurements are listed in Table 1.

The $CP$ fit results are listed in the Table 2.

## REFERENCES

1. Belle Collaboration, K. Abe *et al.*, Phys. Rev. D **66**, 032007 (2002).
2. Belle Collaboration, A. Abashian *et al.*, Nucl. Instr. and Meth. A**479**, 117-232 (2002).
3. KEKB B Factory Design Report, KEK Report 95-1, 1995, unpublished.
4. ARGUS collaboration, H. Albrecht *et al.*,

Table 2
Summary of the $CP$-fit results

| Decay mode | "$\sin 2\phi_1$" | A |
|---|---|---|
| $J/\psi\pi^0$ | $+0.93 \pm 0.49 \pm 0.08$ | $-0.25 \pm 0.39 \pm 0.06$ |
| $\phi K_S$ | $-0.73 \pm 0.64 \pm 0.18$ | $-0.56 \pm 0.41 \pm 0.12$ |
| $K^+K^-K_S$ | $+0.52 \pm 0.46 \pm 0.11^{+0.27}_{-0.03}$ | $-0.42 \pm 0.36 \pm 0.09^{+0.03}_{-0.22}$ |


Phys. Lett. B**229**, 304 (1989); B**241**, 278 (1990).
5. Belle Collaboration, K. Abe *et al.*, BELLE-CONF-0209.
6. Belle Collaboration, K. Abe *et al.*, Phys. Rev. Lett. **89**, 122001 (2002).
7. Belle Collaboration, K. Abe *et al.*, BELLE-CONF-0232.
8. Belle Collaboration, K. Abe *et al.*, BELLE-CONF-0225.

ELSEVIER

Proceedings of ICHEP 2002, pp. 507–511
*S. Bentvelsen, P. de Jong, J. Koch and E. Laenen (Editors)*

31st INTERNATIONAL CONFERENCE ON HIGH ENERGY PHYSICS AMSTERDAM

# Measurement of sin2$\beta$ with *BABAR*

Douglas M. Wright[a]
(representing the *BABAR* Collaboration)

[a]Lawrence Livermore National Laboratory, P.O. Box 808, Livermore, CA 94566, USA

We present updated results on time-dependent *CP*-violating asymmetries in neutral $B$ decays to several *CP* eigenstates containing charmonium. In the Standard Model, the amplitude of these asymmetries is proportional to sin2$\beta$. We measure sin2$\beta$ = 0.741 ± 0.067 (stat) ± 0.034 (syst) from a data sample of about 88 million $\Upsilon(4S) \rightarrow B\overline{B}$ decays collected with the *BABAR* detector at the PEP-II asymmetric-energy $B$ Factory at SLAC. We have also measured *CP*-violating asymmetries in open charm, and penguin modes sensitive to sin2$\beta$, which provide important consistency tests of the Standard Model.

## 1. INTRODUCTION

In the Standard Model of electroweak interactions, *CP* violation arises as a consequence of a complex phase in the three-generation Cabibbo-Kobayashi-Maskawa (CKM) quark-mixing matrix. Observations of *CP* violation in $B^0$ decays to charmonium were reported last year by the *BABAR* [1] and Belle [2] collaborations. The PEP-II collider at SLAC has since delivered an additional 63 fb$^{-1}$, thereby approximately tripling the *BABAR* data sample near the $\Upsilon$(4S) resonance. Here we report a more precise measurement [3], as well as updated and new results from non-charmonium $B^0$ decays, using a sample of about 88 million $B\overline{B}$ decays. Changes in the analysis with respect to the previously published result [1] include the processing of all data with a uniform event reconstruction, a new flavor-tagging algorithm, and the addition of the decay mode $B^0 \rightarrow \eta_c K^0_S$. The *BABAR* detector and the measurement technique are described in detail in Refs. [4] and [5].

We reconstruct a sample of neutral $B$ mesons ($B_{CP}$) decaying to *CP* eigenstates and examine each event for evidence that the recoiling $B$ meson decayed as a $B^0$ or $\overline{B}^0$ (flavor tag). The proper-time distribution for such events can be expressed in terms of a complex parameter $\lambda$ that depends on both the $B^0$-$\overline{B}^0$ oscillation amplitude and the amplitudes describing $\overline{B}^0$ and $B^0$ decays to the *CP* final state. The decay rate $\mathrm{f}_+(\mathrm{f}_-)$, when the tagging meson is a $B^0(\overline{B}^0)$, is given by

$$\mathrm{f}_\pm(\Delta t) = \frac{e^{-|\Delta t|/\tau_{B^0}}}{4\tau_{B^0}}\left[1\pm\frac{2\,\mathcal{I}m\,\lambda}{1+|\lambda|^2}\sin(\Delta m_d\Delta t) \mp\frac{1-|\lambda|^2}{1+|\lambda|^2}\cos(\Delta m_d\Delta t)\right],(1)$$

where $\Delta t = t_{\rm rec} - t_{\rm tag}$ is the difference between the proper decay times of the reconstructed $B$ meson ($B_{\rm rec}$) and the tagging $B$ meson ($B_{\rm tag}$), $\tau_{B^0}$ is the $B^0$ lifetime, and $\Delta m_d$ is the $B^0$-$\overline{B}^0$ oscillation frequency.

In the Standard Model, $\lambda = \eta_f e^{-2i\beta}$ for charmonium modes (e.g. $B^0 \rightarrow J/\psi K^0_S$), where $\eta_f$ is the *CP* eigenvalue of the final state $f$, and $\beta \equiv \arg[-V_{\rm cd}V^*_{\rm cb}/V_{\rm td}V^*_{\rm tb}]$. The time-dependent *CP*-violating asymmetry then becomes

$$\begin{aligned} A_{CP}(\Delta t) &\equiv \frac{\mathrm{f}_+(\Delta t) - \mathrm{f}_-(\Delta t)}{\mathrm{f}_+(\Delta t) + \mathrm{f}_-(\Delta t)} \\ &= -\eta_f \sin 2\beta \sin(\Delta m_d\,\Delta t). \qquad (2)\end{aligned}$$

## 2. TAGGING AND VERTEXING

A measurement of $A_{CP}$ requires a determination of the experimental $\Delta t$ resolution and the fraction $w$ of events in which the tag assignment is incorrect. This mistag fraction reduces the observed *CP* asymmetry by a factor $1-2w$. Mistag fractions and $\Delta t$ resolution functions are determined from a sample of neutral $B$ mesons that


doi:10.1016/S0920-5632(02)02170-9

Table 1
Tagging efficiency, mistag fraction and quality factor for each category (measured from data).

| Category | $\varepsilon$ (%) | $w$ (%) | $Q$ (%) |
|---|---|---|---|
| `Lepton` | $9.1 \pm 0.2$ | $3.3 \pm 0.6$ | $7.9 \pm 0.3$ |
| `Kaon I` | $16.7 \pm 0.2$ | $10.0 \pm 0.7$ | $10.7 \pm 0.4$ |
| `Kaon II` | $19.8 \pm 0.3$ | $20.9 \pm 0.8$ | $6.7 \pm 0.4$ |
| `Inclusive` | $20.0 \pm 0.3$ | $31.5 \pm 0.9$ | $2.7 \pm 0.3$ |
| All | $65.6 \pm 0.5$ | | $28.1 \pm 0.7$ |

decay to flavor eigenstates ($B_{\rm flav}$) consisting of the channels $D^{(*)-}h^+(h^+ = \pi^+, \rho^+$, and $a_1^+$) [1] and $J/\psi K^{*0}(K^{*0} \to K^+\pi^-)$.

We use multivariate algorithms to identify signatures of $B$ decays that indicate the flavor of $B_{\rm tag}$. Based on the output of a multi-level neural network, we assign each event to one of four hierarchical, mutually exclusive tagging categories. The `Lepton` category contains events with an identified lepton. Events with kaon and soft pion candidates (but not identified leptons) are assigned to the `Kaon I` or `Kaon II` category. All other events are assigned to the `Inclusive` category or excluded from further analysis based on the estimated mistag probability. The tagging efficiencies $\varepsilon_i$ for the four tagging categories are measured from data and are summarized in Table 1. The figure of merit for tagging is the effective tagging efficiency $Q \equiv \sum_i \varepsilon_i(1-2w_i)^2$. This algorithm improves $Q$ by about 7% (relative) over the algorithm used in Ref. [5].

The time interval $\Delta t$ between the two $B$ decays is calculated from the measured separation $\Delta z$ between the decay vertices of $B_{\rm rec}$ and $B_{\rm tag}$ along the collision ($z$) axis. We employ constraints from the beam spot location and the $B_{\rm rec}$ momentum to improve vertex reconstruction efficiency and resolution. The fraction of events satisfying the vertex quality requirements is 95%. The r.m.s. $\Delta t$ resolution for practically all accepted events is 1.1 ps.

[1]Charge conjugation is implied throughout this paper, unless explicitly stated.

## 3. CHARMONIUM MODES

### 3.1. Event sample

We reconstruct $B^0$ decays to the following final states containing charmonium: $\eta_f = -1$ modes $J/\psi K_S^0$, $\psi(2S)K_S^0$, $\chi_{c1}K_S^0$, $\eta_c K_S^0$; $\eta_f = +1$ mode $J/\psi K_L^0$; and $J/\psi K^{*0}(K^{*0} \to K_S^0\pi^0)$. This last mode, $J/\psi K^{*0}$, is not a pure $CP$ eigenstate, but is a combination of even ($L$=0, 2) and odd ($L$=1) orbital angular momenta final states. In this analysis we do not separate the angular components, so the measured $CP$ asymmetry is reduced by a factor $1-2R_\perp$, where $R_\perp$ is the fraction of the $L$=1 component. We have measured $R_\perp = (16.0 \pm 3.5)\%$ [6], which gives an effective $\eta_f = 0.65 \pm 0.07$ for the $J/\psi K^{*0}$ mode after acceptance corrections.

The event selection criteria for the $B_{CP}$ sample are described in Ref. [3]. The selection includes two main discriminating variables constructed from the measured energy and momentum of the final state: the difference $\Delta E$ between the decay energy and the beam energy in the center-of-mass frame, and the beam-energy substituted mass $m_{\rm ES} = \sqrt{(E_{\rm beam}^{\rm cm})^2 - (p_B^{\rm cm})^2}$. Since it is not well measured, we determine the $K_L^0$ momentum in the $J/\psi K_L^0$ mode by applying one of the kinematic constraints (effectively $m_{\rm ES}$), leaving only one ($\Delta E$) of the two discriminating variables for the selection of this mode. Figure 1 shows the $m_{\rm ES}$ distribution for modes containing a $K_S^0$ or $K^{*0}$ and the $\Delta E$ distribution for the $J/\psi K_L^0$ candidates.

To determine numbers of events and purities, a signal region $5.270\ (5.273) < m_{\rm ES} < 5.290\ (5.288)\,{\rm GeV}/c^2$ is used for modes containing $K_S^0$ ($K^{*0}$). In the $J/\psi K_L^0$ mode, the $\Delta E$ resolution the signal region is defined by $|\Delta E| < 10\,{\rm MeV}$. The signal region contains 2641 events that satisfy the tagging and vertexing requirements. In Table 2 we list the number of events and the signal purity for the tagged $B_{CP}$ candidates.

### 3.2. Measurement of sin2$\beta$

We determine sin2$\beta$ with a simultaneous unbinned maximum likelihood fit to the $\Delta t$ distributions of the tagged $B_{CP}$ and $B_{\rm flav}$ samples. In this fit, the $\Delta t$ distributions of the $B_{CP}$ sample are described by Eq. 1 with $|\lambda| = 1$, i.e. with the assumption of no direct *CP* violation. The $\Delta t$ distributions of the $B_{\rm flav}$ sample evolve according to the known frequency for flavor oscillation in $B^0$ mesons. The observed amplitudes for the *CP* asymmetry and flavor oscillation are reduced by the same factor, $1 - 2w$, due to flavor mistags.

There are 34 free parameters in the fit: sin2$\beta$ (1), the average mistag fractions $w$ and the differences $\Delta w$ between $B^0$ and $\overline{B}^0$ mistag fractions for each tagging category (8), parameters for the signal $\Delta t$ resolution (8), and parameters for background time dependence (6), $\Delta t$ resolution (3), and mistag fractions (8). We fix $\tau_{B^0} = 1.542\,\text{ps}$ and $\Delta m_d = 0.489\,\text{ps}^{-1}$ [7]. The fit to the $B_{CP}$ and $B_{\rm flav}$ samples yields

$$\sin 2\beta = 0.741 \pm 0.067\ (\text{stat}) \pm 0.034\ (\text{syst}).$$

Figure 2 shows the $\Delta t$ distributions and the raw asymmetry $(N_{B^0} - N_{\overline{B}^0})/(N_{B^0} + N_{\overline{B}^0})$ versus $\Delta t$ for the $\eta_f = -1$ and $\eta_f = +1$, overlaid with the projection of the likelihood fit result.

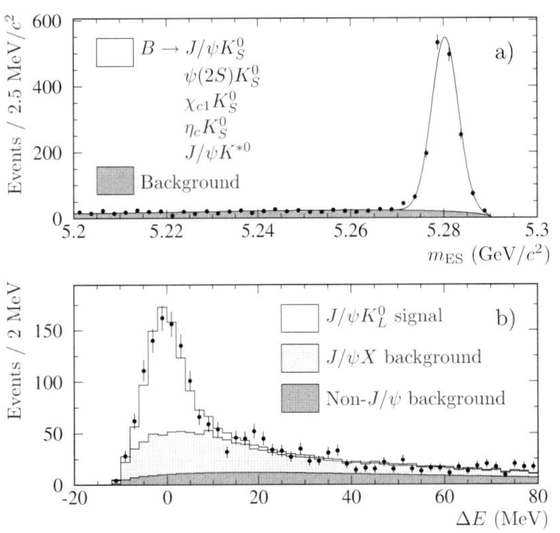


Figure 1. Distributions for $B_{CP}$ candidates satisfying the tagging and vertexing requirements: a) all final states except $J/\psi K^0_L$, and b) $J/\psi K^0_L$.

Table 2
Number of events in the signal region after tagging and vertexing requirements, signal purity, and results of the *CP* asymmetry fits.

| Sample | $N_{\rm tag}$ | $P(\%)$ | sin2$\beta$ |
|---|---|---|---|
| $(J/\psi, \psi(2S), \chi_{c1}, \eta_c)K^0_S$ | 1506 | 94 | $0.76 \pm 0.07$ |
| $J/\psi K^0_L$ $(\eta_f = +1)$ | 988 | 55 | $0.72 \pm 0.16$ |
| $J/\psi K^{*0}(K^{*0} \to K^0_S\pi^0)$ | 147 | 81 | $0.22 \pm 0.52$ |
| Full *CP* sample | 2641 | 78 | $0.74 \pm 0.07$ |
| $(J/\psi, \psi(2S), \chi_{c1}, \eta_c)K^0_S$ only $(\eta_f = -1)$ | | | |
| $J/\psi\,(K^0_S \to \pi^+\pi^-)$ | 974 | 97 | $0.82 \pm 0.08$ |
| $J/\psi\,(K^0_S \to \pi^0\pi^0)$ | 170 | 89 | $0.39 \pm 0.24$ |
| $\psi(2S)\ (K^0_S \to \pi^+\pi^-)$ | 150 | 97 | $0.69 \pm 0.24$ |
| $\chi_{c1}K^0_S$ | 80 | 95 | $1.01 \pm 0.40$ |
| $\eta_c K^0_S$ | 132 | 73 | $0.59 \pm 0.32$ |
| `Lepton` category | 220 | 98 | $0.79 \pm 0.11$ |
| `Kaon I` category | 400 | 93 | $0.78 \pm 0.12$ |
| `Kaon II` category | 444 | 93 | $0.73 \pm 0.17$ |
| `Inclusive` category | 442 | 92 | $0.45 \pm 0.28$ |
| $B^0$ tags | 740 | 94 | $0.76 \pm 0.10$ |
| $\overline{B}^0$ tags | 766 | 93 | $0.75 \pm 0.10$ |
| $B_{\rm flav}$ sample | 25375 | 85 | $0.02 \pm 0.02$ |
| $B^+$ sample | 22160 | 89 | $0.02 \pm 0.02$ |

The dominant sources of systematic error are the uncertainties in the background content, the assumed parameterization of the $\Delta t$ resolution function, and possible differences between the $B_{\rm flav}$ and $B_{CP}$ mistag fractions. Most systematic errors are determined with data and will continue to decrease with additional statistics.

The large $B_{CP}$ sample allows a number of consistency checks, including separation of the data by decay mode, tagging category, and $B_{\rm tag}$ flavor. The results of fits to these subsamples are found to be statistically consistent. Fits to the control samples of non-*CP* decay modes (the $B_{\rm flav}$ sample and $B^+$ mesons decaying to the final states $J/\psi K^{(*)+}$, $\psi(2S)K^+$, $\chi_{c1}K^+$, $\eta_c K^+$, and $\overline{D}^{(*)0}\pi^+$) indicate no statistically significant asymmetry, as expected. The results of these checks are given in Table 2.

This measurement of sin2$\beta$ improves upon the precision of each of the previous measurements [1,2] by a factor of two. The measured value is in good agreement with the expectation implied by the measurements and theoretical esti-

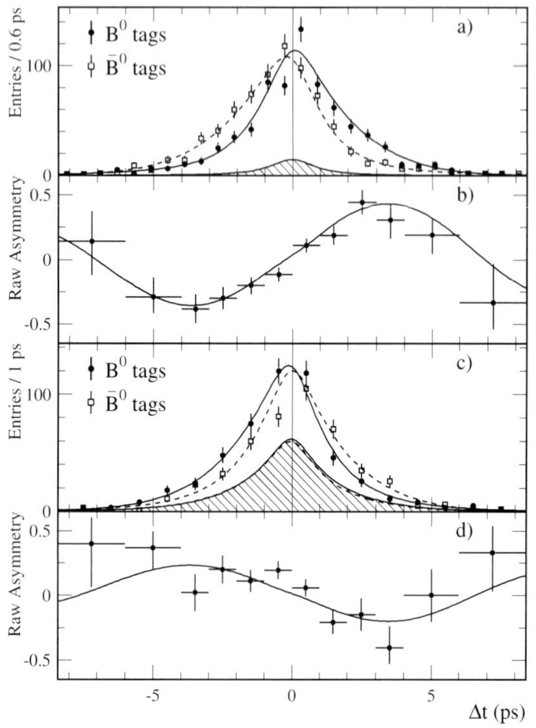


Figure 2. Number of tagged events, raw asymmetry, and fit projections as functions of $\Delta t$ for $\eta_f = -1$ (a & b) and $\eta_f = +1$ (c & d) candidates. The shaded regions represent the background contributions.

mates of the magnitudes of CKM matrix elements in the context of the Standard Model, and provides a precise and model-independent constraint on the position of the apex of the Unitarity Triangle.

### 3.3. Direct *CP* violation

We also measure the parameter $|\lambda|$ in Eq. 1 from a fit to the $\eta_f = -1$ sample, which has high purity and requires minimal assumptions on the effect of backgrounds. This parameter is sensitive to the difference in the number of $B^0$- and $\overline{B}^0$-tagged events. In order to account for differences in reconstruction and tagging efficiencies for $B^0$ and $\overline{B}^0$ mesons, we incorporate five additional free parameters in this fit. We obtain $|\lambda| = 0.948 \pm 0.051$ (stat) $\pm 0.030$ (syst). We simultaneously fit the coefficient of the sine term in Eq. 1 and find it to be $0.759 \pm 0.074$ (stat). These results are consistent with the no direct *CP* violation hypothesis ($|\lambda| = 1$) and with our $\sin 2\beta$ measurement.

## 4. OPEN CHARM MODE

Within the context of the Standard Model, the open charm decay $B^0 \to D^{*+}D^{*-}$ is expected to have a time-dependent *CP*-violating asymmetry [8]. Up to corrections due to theoretically uncertain penguin diagram contributions [9], the Cabibbo-suppressed tree diagram produces an asymmetry sensitive to $\sin 2\beta$. Factorization models predict that the penguin-induced correction is small [10], so comparisons with the *CP* asymmetry in charmonium modes provides a valuable test of these models and a consistency check on the Standard Model.

The selection of $B^0 \to D^{*+}D^{*-}$ candidates is described in Ref [11] and results in a sample of 102 tagged events with a purity of 82%. The $B^0 \to D^{*+}D^{*-}$ mode consists of three partial wave contributions with two different *CP* parities: even ($S$- and $D$-waves) and odd ($P$-wave). Since *CP*-even and odd components can be separated by a one-dimensional angular analysis in the transversity basis [12], we parameterize the tagged $B^0$ decay rate in terms of the transversity angle and fit for the complex *CP*-even (odd) parameters $\lambda_+(\lambda_\perp)$, and $R_\perp$ the fraction of signal events with *CP*-odd parity.

From a time-integrated fit and corrected for acceptance, we measure $R_\perp = 0.07 \pm 0.06(\text{stat}) \pm 0.03(\text{syst})$. Since the *CP*-odd fraction is small, we fix the *CP*-odd parameters in the time-dependent fit to those measured from the charmonium analysis, assuming the penguin diagram contribution is negligible, i.e. $|\lambda_\perp| = 1$ and $\text{Im}(\lambda_\perp) = -0.741$. With a suitably large statistical sample, one can fit the *CP*-odd and even parameters simultaneously.

Performing a time-dependent fit including both cosine and sine terms in Eq. 1 yields the preliminary result [11]

$$\begin{aligned}\mathrm{Im}(\lambda_+) &= 0.31 \pm 0.43(\mathrm{stat}) \pm 0.13(\mathrm{syst}) \\ |\lambda_+| &= 0.98 \pm 0.25(\mathrm{stat}) \pm 0.09(\mathrm{syst}).\end{aligned}$$

If the penguin contribution can be neglected, then $|\lambda_+| = 1$ and $\mathrm{Im}(\lambda_+) = -\sin 2\beta$. By making this assumption and refitting, we find that the change in the likelihood corresponds to a 2.7 standard deviation from the result of the fit in charmonium ($\sin 2\beta = 0.741$).

## 5. PENGUIN MODES

Since the tree diagram is Cabibbo-suppressed, the decay $B^0 \to J/\psi\pi^0$ ($\eta_f = +1$) has comparable tree and penguin contributions. The tree diagram has the same weak phase as charmonium modes containing neutral kaons (e.g. $B^0 \to J/\psi K^0_S$); however, a portion of the penguin diagram contributes a different weak phase. We therefore perform a time-dependent *CP*-violating asymmetry fit for the coefficients ($S_{J/\psi\pi^0}$ and $C_{J/\psi\pi^0}$) of the sine and cosine terms in Eq. 1.

We select 49 tagged events with a purity of 59% and measure the preliminary result [13]

$$\begin{aligned}C_{J/\psi\pi^0} &= 0.38 \pm 0.41\ (\mathrm{stat}) \pm 0.09\ (\mathrm{syst}), \\ S_{J/\psi\pi^0} &= 0.05 \pm 0.49\ (\mathrm{stat}) \pm 0.16\ (\mathrm{syst}).\end{aligned}$$

In the absence of the penguin contribution, we expect $C_{J/\psi\pi^0} = 0$ and $S_{J/\psi\pi^0} = -\sin 2\beta$.

The charmless decay $B^0 \to \phi K^0_S$ ($\eta_f = -1$) is dominated by gluonic penguin diagrams and within the Standard Model should have the same weak phase as the charmonium modes. The predicted deviation of the effective $\sin 2\beta$ measured in this mode, compared to the Standard Model parameter, is smaller than 4% [14]. A measurement of $\sin 2\beta$ in $\phi K^0_S$ therefore probes new physics participating in penguin loops.

From the analysis described in Ref. [15], we find 66 tagged candidates with a purity of 50%. Setting $|\lambda| = 1$ and fitting only the coefficient $S_{\phi K^0_S}$ of the sine term in Eq. 1, we determine $S_{\phi K^0_S} = -0.19^{+0.52}_{-0.50}(\mathrm{stat}) \pm 0.09(\mathrm{syst})$. In the absence of new physics, $S_{\phi K^0_S} = \sin 2\beta$ at the few percent level.

In these first measurements from *BABAR* on the *CP*-violating asymmetry in penguin modes sensitive to $\sin 2\beta$, the statistical precision of the results do not yet provide a meaningful test of the Standard Model; however, high statistics in the future provide a good opportunity to challenge the theory.

## 6. ACKNOWLEDGMENTS

We are grateful for the excellent luminosity and machine conditions provided by our PEP-II colleagues.

## REFERENCES

1. *BABAR* Collaboration, B. Aubert *et al.*, Phys. Rev. Lett. **87**, 091801 (2001).
2. BELLE Collaboration, K. Abe *et al.*, Phys. Rev. Lett. **87**, 091802 (2001).
3. *BABAR* Collaboration, B. Aubert *et al.*, SLAC-PUB-9293, hep-ex/0207042, to appear in Phys. Rev. Lett.
4. *BABAR* Collaboration, B. Aubert *et al.*, Nucl. Instr. and Methods **A479**, 1 (2002).
5. *BABAR* Collaboration, B. Aubert *et al.*, Phys. Rev. **D66**, 032003 (2002).
6. *BABAR* Collaboration, B. Aubert *et al.*, Phys. Rev. Lett. **87**, 241801 (2001).
7. Particle Data Group, K. Hagiwara *et al.*, Phys. Rev. **D66**, 010001 (2002).
8. R. Aleksan, *et al.*, Phys. Lett. B **317**, 173 (1993).
9. A.I. Sanda and Z.Z. Xing, Phys. Rev. D **56**, 341 (1997).
10. X.Y. Pham and Z.Z. Xing, Phys. Rev. B **458**, 375 (1999).
11. *BABAR* Collaboration, B. Aubert *et al.*, SLAC-PUB-9299, hep-ex/0207072.
12. I. Dunietz, *et al.*, Phys. Rev. D **43**, 2193 (1991).
13. *BABAR* Collaboration, B. Aubert *et al.*, SLAC-PUB-9298, hep-ex/0207058.
14. Y. Grossman and M.P. Worah, Phys. Lett. B **395**, 241 (1997); Y. Grossman, G. Isidori, and M.P. Worah, Phys. Rev. D **58**, 057505 (1998).
15. *BABAR* Collaboration, B. Aubert *et al.*, SLAC-PUB-9297, hep-ex/0207070.

ELSEVIER

Proceedings of ICHEP 2002, pp. 512–515
*S. Bentvelsen, P. de Jong, J. Koch and E. Laenen (Editors)*

31st INTERNATIONAL CONFERENCE ON HIGH ENERGY PHYSICS AMSTERDAM

# Measurements of time dependent CP asymmetry in $B \to VV$ decays with BELLE

Ryosuke Itoh[a]
Representing the Belle Collaboration

[a]Institute of Particle and Nuclear Studies, High Energy Accelerator Research Organization (KEK), 1-1 Oho, Tsukuba, Ibaraki 305-0801, Japan

A study of CP violation in $B \to J/\psi K^*(K_S^0\pi^0)$ decays by time dependent angular analysis is discussed. Status of time independent analyses for other $B \to VV$ decays is also reported. The data used for the analyses are taken with the Belle detector at KEK.

## 1. Introduction

There are three helicity states in $B \to VV$ decays. Although one of the states is a pure CP even state, CP even and odd states are mixed in other states and the observed CP asymmetry is diluted. By studying decay angles of final state particles, it is possible to project out each CP state in a statistical way. A theoretical expression of differential time-angular decay rate of $B \to VV$ decays consists of components corresponding to three helicity states and also to the interference between them. Each term is expressed as a product of an angular term and an amplitude term. The amplitude term contains CP violating phase(s) as a function of $\Delta t$, the decay time difference between two $B$ mesons from an $\Upsilon(4S)$ decay. A fit of the expression for the measured decay angles and $\Delta t$ gives the determination of the CP violating phase(s). The interference terms are also rich sources of various interesting physics such as the $\cos 2\phi_1$ measurement in $B^0 \to J/\psi K^{*0}(K_S^0\pi^0)$ decays, and the simultaneous determination of $r$ and $\sin(2\phi_1+\phi_3)$ in $B^0 \to D^{*-}\rho^+$ decays where $r$ is the ratio of doubly-Cabibbo-suppressed mode to Cabibbo favored mode contributing to the decay.

In this talk, a study of time dependent angular analysis for $B \to J/\psi K^*(K_S^0\pi^0)$ is discussed. Also reported is status of angular analyses for $B^0 \to D^{*+}D^{*-}$ and $B^0 \to D^{*-}\rho^+$. The data used in the analyses are taken with the Belle detector[1] in the KEKB accelerator[2]. The KEKB accelerator consists of two separate rings for electrons and positrons. The energies are set at 3.5 GeV for positrons and 8.0 GeV for electrons, respectively. These two beams are collided at the interaction point to produce $\Upsilon(4S)$ in motion for the measurement of $\Delta t$. The peak luminosity of the machine has reached $7.4 \times 10^{33} cm^{-2} sec^{-1}$ which is the current world record. The total integrated luminosity up to now is $89.1 fb^{-1}$. The Belle detector is placed at the interaction point of the KEKB accelerator. It is a general-purpose detector with a wide angle coverage featuring a particle identification using the Aerogel Cherenkov Counter assembly. The data sample used in the analyses corresponds to an integrated luminosity of 78 $\text{fb}^{-1}$ taken on the $\Upsilon(4S)$ resonance.

## 2. Time dependent angular analysis for $B^0 \to J/\psi K^*(K_S^0\pi^0)$[3]

Candidate $B^0$ mesons are reconstructed by selecting events with a $J/\psi$ identified from a pair of oppositely-charged leptons and a $K^*$ from a pair of $K_S^0$ and $\pi^0$ candidates. The beam-constrained mass ($M_{bc}$), which is the invariant mass of a reconstructed $J/\psi$ and $K^*$ calculated taking the energy to be the beam energy, is required to be in the range $5.27 - 5.29$ GeV/$c^2$ and the energy difference between $B$ candidate and the beam energy ($\Delta E$) to satisfy $-50$ MeV $< \Delta E <$ 30 MeV. To eliminate slow $\pi^0$ backgrounds, the angle of the kaon with respect to the $K^*$ direction in


doi:10.1016/S0920-5632(02)02176-X

the $K^*$ rest frame, $\cos\theta_{K^*}$, is required to satisfy $\cos\theta_{K^*} < 0.8$. When an event contains more than one candidate passing the above requirements, the combination for which $\Delta E$ is closest to zero is selected. The number of events remaining after the selection is 104.

The time dependent angular distribution of the decay is described as[4]:

$$\frac{d^4\Gamma(\theta_{tr},\phi_{tr},\theta_{K^*},\Delta t)}{d\cos\theta_{tr}d\phi_{tr}d\cos\theta_{K^*}d\Delta t} = \frac{e^{-|\Delta t|/\tau_{B^0}}}{2\tau_{B^0}}\sum_{i=1}^{6} g_i(\theta_{tr},\phi_{tr},\theta_{K^*})a_i(\Delta t) \quad (1)$$

where $g_i$ are the angular terms and $a_i$ are amplitude terms. $\tau_B$ is the lifetime of a $B^0$ meson. The $g_i$ are expressed as:

$$g_1 = 2\cos^2\theta_{K^*}(1-\sin^2\theta_{tr}\cos^2\phi_{tr}) \quad (2)$$
$$g_2 = \sin^2\theta_{K^*}(1-\sin^2\theta_{tr}\sin^2\phi_{tr}) \quad (3)$$
$$g_3 = \sin^2\theta_{K^*} sin^2\theta_{tr} \quad (4)$$
$$g_4 = \frac{-1}{\sqrt{2}}\sin 2\theta_{K^*}\sin^2\theta_{tr}\sin 2\phi_{tr} \quad (5)$$
$$g_5 = \sin^2\theta_{K^*}\sin 2\theta_{tr}\sin\phi_{tr} \quad (6)$$
$$g_6 = \frac{1}{\sqrt{2}}\sin 2\theta_{K^*}\sin 2\theta_{tr}\cos\phi_{tr} \quad (7)$$

where $\theta_{tr}$, $\phi_{tr}$ and $\theta_{K^*}$ are the decay angles defined in the transversity basis[5]. The $a_i$ are

$$a_1 = |A_0|^2(1+\eta\sin 2\phi_1\sin\Delta m\Delta t) \quad (8)$$
$$a_2 = |A_\parallel|^2(1+\eta\sin 2\phi_1\sin\Delta m\Delta t) \quad (9)$$
$$a_3 = |A_\perp|^2(1-\eta\sin 2\phi_1\sin\Delta m\Delta t) \quad (10)$$
$$a_4 = Re(A_\parallel^* A_0)(1+\eta\sin 2\phi_1\sin\Delta m\Delta t) \quad (11)$$
$$a_5 = \eta Im(A_\parallel^* A_\perp)\cos\Delta m\Delta t - \eta Re(A_\parallel^* A_\perp)\cos 2\phi_1\sin\Delta m\Delta t \quad (12)$$
$$a_6 = \eta Im(A_0^* A_\perp)\cos\Delta m\Delta t - \eta Re(A_0^* A_\perp)\cos 2\phi_1\sin\Delta m\Delta t \quad (13)$$

where $\Delta m$ is the $B^0-\overline{B^0}$ mixing parameter. $\eta$ is +1 for $B^0$ while -1 for $\overline{B^0}$. $A_0$, $A_\parallel$ and $A_\perp$ are the complex decay amplitudes for three helicity states in the transversity basis. Two CP violation parameters, $\sin 2\phi_1$ and $\cos 2\phi_1$, appear in the formula. The values of these parameters are determined by fitting the formula to the measured angles and $\Delta t$, taking into account the detection efficiency and background. The fit is done using an unbinned maximum likelihood method. The probability density function for an event is defined as

$$\begin{aligned} PDF &= f_{sig}(M_{bc})\epsilon(\theta_{tr},\phi_{tr},\theta_{K^*}) \\ &\times \frac{d^4\Gamma(\cos\theta_{tr},\phi_{tr},\cos\theta_{K^*},\Delta t)}{d\cos\theta_{tr}d\phi_{tr}d\cos\theta_{K^*}d\Delta t} \\ &+ \frac{e^{-|\Delta t|/\tau_{B^0}}}{2\tau_{B^0}} \\ &\times \{\sum_i f_{cf}^i(M_{bc})ADF_{cf}(\theta_{tr},\phi_{tr},\theta_{K^*}) \\ &\quad + f_{nr}(M_{bc})ADF_{nr}(\theta_{tr},\phi_{tr},\theta_{K^*})\} \\ &+ \delta(\Delta t)f_{combi}(M_{bc})ADF_{combi}(\theta_{tr},\phi_{tr},\theta_{K^*}). \quad (14) \end{aligned}$$

$f_{sig}$, $f_{cf}$, $f_{nr}$ and $f_{combi}$ are fractions of the signal, cross feed, non-resonant production, and combinatorial background as a function of $M_{bc}$, respectively, while $ADF_{cf}$, $ADF_{nr}$ and $ADF_{combi}$ are corresponding angular shape functions. $\epsilon$ is a three dimensional detection efficiency function for the signal. The determinations of these functions and the decay amplitudes are described in ref.[6]. Phases in the amplitudes are chosen so as to conserve $s$-quark helicity[7]. The $\tau_B$ and $\Delta m$ are fixed at the PDG values in the fit. The flavor tagging procedure[8] gives the flavor of tag-side $B$ meson $q$ and the probability of wrong tag $w$. To account for the effect of the wrong tagging in the data, $\eta$ is replaced with $-q(1-2w)$ in the PDF.

Each term in the PDF is then convolved with appropriate resolution functions[9] separately for the signal, the backgrounds with a $B^0$ lifetime and the combinatorial background with a $\delta(\Delta t)$ function shape. The parameters in the resolution functions are calculated event by event. From the fit to the data, we obtain $\sin 2\phi_1 = 0.13 \pm 0.51 \pm 0.06$, and $\cos 2\phi_1 = 1.40 \pm 1.28 \pm 0.19$. The systematic errors include uncertainties in the resolution parameters, wrong tagging fractions, decay amplitudes, background fractions and shapes, and others. The distributions of $\Delta t$ measured for the samples tagged as $q = +1$ and $-1$ are shown in Figure 1 together with the predictions by the PDF with obtained CP parameter values.

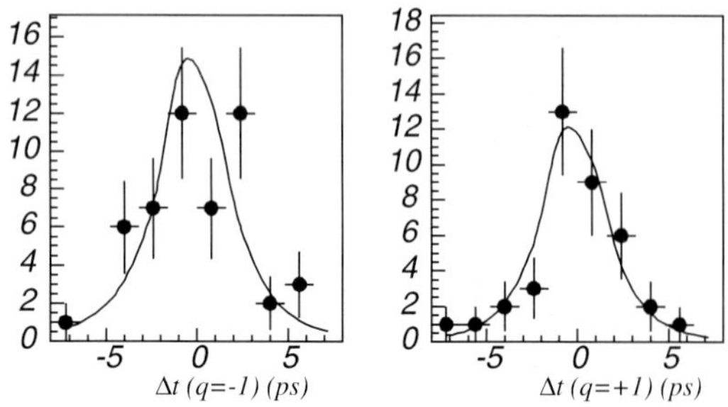


Figure 1. $\Delta t$ distributions for samples tagged as $q = +1$ and $-1$. The lines show the predictions by the PDF with determined CP parameter values.

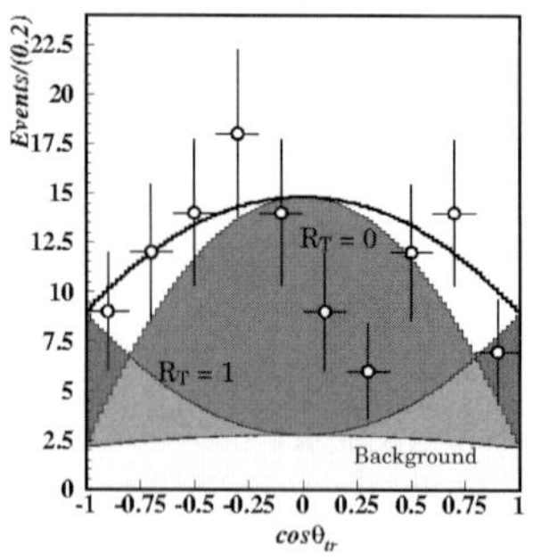


Figure 2. Distribution of $\cos\theta_{tr}$ for $B^0 \to D^{*+}D^{*-}$ candidates.

## 3. Time independent angular analyses

The time dependent angular analysis is applicable for other $B \to VV$ decays such as $B^0 \to D^{*+}D^{*-}$, $D^{-*}\rho^+$ and $\rho^+\rho^{-1}$ for the determination of CP violating phases including those other than $\phi_1$. Time dependent analyses for these modes are still in preparation and the status of time independent(integrated) analyses is reported here. The measurement of the decay amplitudes by the time independent angular analysis is essential for the determinations of the CP violating phases.

### 3.1. $B^0 \to D^{*+}D^{*-}$

The reconstruction of a $D^*$ is performed using a $D\pi$ decay with a subsequent $D$ decay into $K\pi, K\pi\pi, K\pi\pi\pi, KK$ and $KK\pi$. A $D$ meson is identified when the invariant mass of decay particles is within 3 to 6 $\sigma$ from the nominal $D$ mass depending on the decay mode. A $D^*$ is reconstructed by calculating the mass difference $M(D\pi) - M(D)$ and those satisfy the difference within $3.0MeV(D^0)$ or $2.25MeV(D^+)$ are identified as $D^*$. The candidate $B$ mesons are selected by applying $M_{bc} > M_{B^0} - 3\sigma$ and $|\Delta E| < 40MeV$.

The distribution of the transversity angle[5] is measured in the selected $B^0$ candidates as shown in Fig. 2. The distribution can be expressed as a function of the fraction of CP-odd component ($R_T$) as

$$\frac{d\Gamma}{d\cos\theta_t r} = \frac{3}{4}(1 - R_T)\sin^2\theta_{tr} + \frac{3}{2}R_T\cos^2\theta_{tr}. \quad (15)$$

The predictions given by this formula for cases with $R_T = 0$ and $R_T = 1$ are also shown in the figure. As seen, the distribution of the data is close to the prediction with $R_T = 0$.

### 3.2. $B^0 \to D^{*-}\rho^+$

The reconstruction of a $D^{*-}$ is performed in a similar manner as described above. A $\rho^+$ is reconstructed from a pair of $\pi^+$ and $\pi^0$ candidates. The $M_{bc}$ of a pair of reconstructed $D^*$ and $\rho$ is required to be in the range 5.27 - 5.29 GeV/$c^2$ to and $\Delta E$ to satisfy $-100 < |\Delta E| < 50MeV$.

The time independent angular distribution is studied using the helicity basis angles[4]. The decay rates as a function of each of three projected angles (two helicity angles $\theta_i$ where $i = 1, 2$, and an angle between two decay planes $\chi$) can be expressed as

$$\frac{d\Gamma}{d\cos\theta_i} = \frac{4\pi}{3}|A_0|^2\cos^2\theta_i + \frac{2\pi}{3}(|A_\perp|^2 + |A_\parallel|^2)\sin^2\theta_i, \quad (16)$$

$$\frac{d\Gamma}{d\chi} = (\frac{4}{9}|A_0|^2 + \frac{8}{9}|A_\perp|^2)\sin^2\chi + (\frac{4}{9}|A_0|^2 + \frac{8}{9}|A_\parallel|^2)\cos^2\chi + \frac{8}{9}Im(A_\parallel^* A_\perp)\sin 2\chi \quad (17)$$

[1]The angular analysis for $B^+ \to \rho^+\rho^0$ is covered in other talk[10].

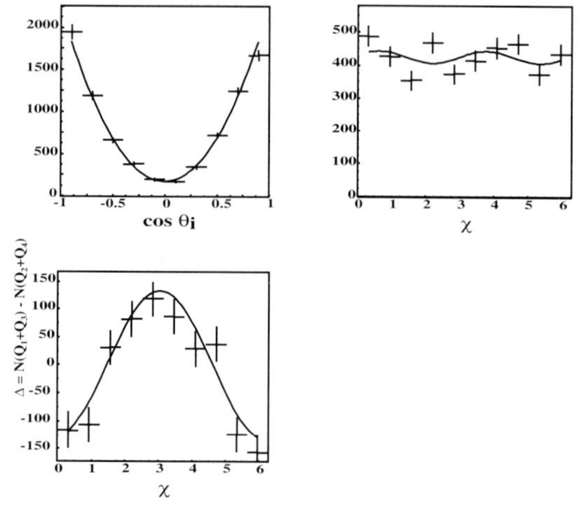


Figure 3. Distributions of angles used in the angular analysis for $B^0 \to D^{*-}\rho^+$ candidates. The distributions are corrected for the acceptance and background contaminations.

where $A_0$, $A_\parallel$ and $A_\perp$ are the decay amplitudes in the transversity basis. In addition, we also used $\chi$-differential decay rates projected in 4 quadrants of a $\theta_1 - \theta_2$ plane (angular regions from 0 to $\pi/2$ and $\pi/2$ to $\pi$ for each of them) to observe the effects in interference terms. The difference in the decay rates between two diagonally-summed quadrants ($\Delta$) can be written as

$$\frac{d\Delta}{d\chi} = -\frac{8}{9\sqrt{2}} Im(A_0^* A_\perp)\sin\chi + \frac{8}{9\sqrt{2}} Re(A_0^* A_\parallel)\cos\chi. \tag{18}$$

Fig. 3 shows the distributions of $\cos\theta_i$ ($\theta_1$ and $\theta_2$ distributions are added), $\chi$, and $\Delta(\chi)$. The measurement of the decay amplitudes is now in progress by fitting these distributions with the theoretical formula given above considering effects of the detector acceptance and background contaminations.

## 4. Summary

A full time dependent angular analysis is performed for $B^0 \to J/\psi K^{*0}(K_S^0\pi^0)$ decays collected with the Belle detector at KEK B-factory. The decay angles in the transversity basis ($\theta_{tr}$, $\phi_{tr}$ and $\theta_{K^*}$) and the decay time difference of two $B$ mesons ($\Delta t$) are measured for each event with the determination of the flavor of tag-side $B$ meson. The values are fitted to the theoretical distribution using an unbinned maximum likelihood method considering the effects of detector acceptance, background contamination, $\Delta t$ resolution and wrong flavor tagging. From the fit, two CP violation parameters are determined to be $\sin 2\phi_1 = +0.13 \pm 0.51 \pm 0.06$, and $\cos 2\phi_1 = +1.40 \pm 1.28 \pm 0.19$. If we take the s-quark helicity conservation choice of the amplitude phases and the $\sin 2\phi_1$ value measured using other decay modes ($0.72 \pm 0.07 \pm 0.04$)[9], the obtained sign of $\cos 2\phi_1$ suggests the $2\phi_1$ to be in the range $0° < 2\phi_1 < 90°$ and the other choice of $2\phi_1 \geq 90°$ is not preferred although the current statistics is not enough to conclude this.

A time independent angular analysis for other decay modes, $B^0 \to D^{*+}D^{*-}$, $B^0 \to D^{*-}\rho^+$, and $B^+ \to \rho^+\rho^0$, are now being performed.

## REFERENCES

1. A.Abashian *et al.* (Belle Collaboration), Nucl. Inst. and Meth. **A479**, 117 (2002).
2. E. Kikutani ed., KEK Preprint 2001-157 (2001), to appear in Nucl. Inst. and Meth. **A**.
3. K.Abe *et al.* (Belle Collaboration), BELLE-CONF-0212 (2002), submitted to this conference (ABS700).
4. K.Abe, M.Satpathy and H.Yamamoto, "Time-dependent angular analysis of B decays", hep-ex/0103002 (2001).
5. I.Dunietz, H.Quinn, A.Snyder, W.Toki and H.J.Lipkin, Phys. Rev. D **43**, 2193 (1991).
6. K.Abe *et. al.* (Belle Collaboration), Phys. Lett. B **538**, 11 (2002).
7. M.Suzuki, Phys. Rev. D **64**, 117503 (2001).
8. K.Abe *et al.* (Belle Collaboration), Phys. Rev. D 66 , 032007 (2002).
9. K.Abe *et al.* (Belle Collaboration), hep-ex/0208025 (2002), to appear in Phys. Rev. D.
10. A.Gordon, "Charmless two-body B decays at Belle", presented at this conference.

ELSEVIER

Proceedings of ICHEP 2002, pp. 516–520
S. Bentvelsen, P. de Jong, J. Koch and E. Laenen (Editors)

31st INTERNATIONAL CONFERENCE ON HIGH ENERGY PHYSICS AMSTERDAM

# Study of *CP* Asymmetries in Charmless Hadronic *B* Decays: Toward a measurement of $\alpha$

James D. Olsen (Representing the *BABAR* Collaboration) [a]

[a]Department of Physics, Princeton University
Princeton, NJ, 08544

We present preliminary measurements of time-dependent *CP* asymmetries in neutral $B$ meson decays to $\pi^+\pi^-$ and $\pi^+\pi^-\pi^0$ final states, where the latter is measured in the region of the Dalitz plane dominated by the $\rho$ resonance. We also present preliminary measurements of the branching fraction for the decay $B^+ \to \pi^+\pi^0$ and an improved upper limit for $B^0 \to \pi^0\pi^0$, both of which are needed to extract the *CP* parameter $\alpha$ from the time-dependent *CP* asymmetry in the $\pi^+\pi^-$ channel. These results are obtained from a data sample of approximately 88 million $\Upsilon(4S) \to B\overline{B}$ decays collected with the *BABAR* detector at the PEP-II asymmetric-energy $B$ Factory at SLAC.

## 1. Introduction

The Standard Model of electroweak interactions describes *CP* violation as a consequence of a single complex phase in the three-generation Cabibbo-Kobayashi-Maskawa (CKM) quark-mixing matrix [1]. The agreement between the direct measurements of the *CP* parameter sin2$\beta$ reported at this conference [2,3], and the Standard Model prediction based on indirect constraints on the magnitudes of the elements of the CKM matrix, is strong evidence that the Kobayashi-Maskawa mechanism is the dominant source of *CP* violation in flavor-changing processes. One of the primary goals of the $B$-Factory experiments in the future will be to measure the remaining angles ($\alpha$ and $\gamma$) of the Unitarity Triangle in order to further test whether the Standard Model description of *CP* violation is correct.

In $B$ decays to $\pi^+\pi^-$ and $\pi^+\pi^-\pi^0$ final states, the time-dependent *CP* asymmetry is related to the angle $\alpha \equiv \arg\left[-V_{\rm td}V_{\rm tb}^*/V_{\rm ud}V_{\rm ub}^*\right]$. The final state $\pi^+\pi^-$ is a *CP* eigenstate and the decay rate distribution $f_+$ ($f_-$) when the second $B$ meson ($B_{\rm tag}$) is identified as a $B^0$ ($\overline{B}^0$) is given by [4]

$$f_\pm(\Delta t) = \frac{e^{-|\Delta t|/\tau}}{4\tau}[1 \pm S_{\pi\pi}\sin(\Delta m_d\Delta t) \mp C_{\pi\pi}\cos(\Delta m_d \Delta t)], \quad (1)$$

where $\tau$ is the mean $B^0$ lifetime, $\Delta m_d$ is the mixing frequency due to the eigenstate mass difference, and $\Delta t$ is the difference in proper decay times between the reconstructed meson and the tagging meson $B_{\rm tag}$. The parameters $S_{\pi\pi}$ and $C_{\pi\pi}$ are defined as

$$S_{\pi\pi} \equiv \frac{2\,\mathcal{I}m\lambda}{1+|\lambda|^2} \quad \text{and} \quad C_{\pi\pi} \equiv \frac{1-|\lambda|^2}{1+|\lambda|^2}, \quad (2)$$

and vanish in the absence of *CP* violation. If the decay proceeds purely through the $b \to u$ tree amplitude, the complex parameter $\lambda$ is given by

$$\lambda(B\to\pi^+\pi^-) = \left(\frac{V_{\rm tb}^*V_{\rm td}}{V_{\rm tb}V_{\rm td}^*}\right)\left(\frac{V_{\rm ud}^*V_{\rm ub}}{V_{\rm ud}V_{\rm ub}^*}\right). \quad (3)$$

In this case $C_{\pi\pi} = 0$ and $S_{\pi\pi} = \sin 2\alpha$. In general, the $b \to d$ gluonic penguin amplitude modifies both the magnitude and phase of $\lambda$, so that $C_{\pi\pi} \neq 0$ and $S_{\pi\pi} = \sqrt{1-C^2_{\pi\pi}}\sin 2\alpha_{\rm eff}$, where $\alpha_{\rm eff}$ depends on the magnitudes and relative strong and weak phases of the tree and penguin amplitudes. In this case, measurements of the branching fractions for the decays $B^+ \to \pi^+\pi^0$ and $B^0 \to \pi^0\pi^0$ can be used to extract $\alpha$ from a measurement of $\alpha_{\rm eff}$ [5,6].

Since the final state $\rho^+\pi^-$ is not a *CP* eigenstate, four possible decay paths must be considered: $B^0(\overline{B}^0) \to \rho^\pm\pi^\mp$. In this case, the decay


doi:10.1016/S0920-5632(02)02180-1

Table 1
Average tagging efficiency $\epsilon$, average mistag fraction $w$, mistag fraction difference $\Delta w = w(B^0) - w(\overline{B}^0)$, and effective tagging efficiency $Q = \epsilon(1-2w)^2$ for signal events in each tagging category. The quantities are measured in the $B_{\rm flav}$ sample.

| Category | $\epsilon$ (%) | $w$ (%) | $\Delta w$ (%) | $Q$ (%) |
|---|---|---|---|---|
| Lepton | $9.1 \pm 0.2$ | $3.3 \pm 0.7$ | $-1.6 \pm 1.3$ | $8.0 \pm 0.3$ |
| Kaon I | $16.6 \pm 0.2$ | $9.5 \pm 0.7$ | $-2.8 \pm 1.3$ | $10.7 \pm 0.4$ |
| Kaon II | $19.8 \pm 0.3$ | $20.6 \pm 0.8$ | $-5.3 \pm 1.3$ | $6.7 \pm 0.4$ |
| Inclusive | $20.1 \pm 0.3$ | $31.7 \pm 0.9$ | $-2.6 \pm 1.4$ | $2.7 \pm 0.3$ |
| Untagged | $34.4 \pm 0.5$ | | | |
| Total $Q$ | | | | $28.4 \pm 0.7$ |

rate distributions can be written as [7]

$$f_T^{\rho^\pm\pi^\mp}(\Delta t) = (1 \pm A_{CP}^{\rho\pi})\frac{e^{-|\Delta t|/\tau}}{4\tau}[1 + T((S_{\rho\pi} \pm \Delta S_{\rho\pi})\sin(\Delta m_d \Delta t) - (C_{\rho\pi} \pm \Delta C_{\rho\pi})\cos(\Delta m_d \Delta t))], \quad (4)$$

where $T$ indicates the flavor of $B_{\rm tag}$ ($B^0 = +1$, $\overline{B}^0 = -1$). The parameters $S_{\rho\pi}$ and $C_{\rho\pi}$ are analogous to $S_{\pi\pi}$ and $C_{\pi\pi}$ and are sensitive to $\alpha$ and direct *CP* violation, respectively. The time-integrated asymmetry $A_{CP}^{\rho\pi} \equiv \left[N_{\rho^+\pi^-} - N_{\rho^-\pi^+}\right] / \left[N_{\rho^+\pi^-} + N_{\rho^+\pi^-}\right]$ also measures direct *CP* violation, but independent of the flavor of $B_{\rm tag}$. The parameters $\Delta C_{\rho\pi}$ and $\Delta S_{\rho\pi}$ are intrinsically *CP* invariant. The asymmetry between $N(B^0_{\rho\pi} \to \rho^+\pi^-) + N(\overline{B}^0_{\rho\pi} \to \rho^-\pi^+)$ and $N(B^0_{\rho\pi} \to \rho^-\pi^+) + N(\overline{B}^0_{\rho\pi} \to \rho^+\pi^-)$ is described by $\Delta C_{\rho\pi}$, while $\Delta S_{\rho\pi}$ is sensitive to the strong phase difference between the amplitudes contributing to $B^0 \to \rho\pi$ decays.

The results reported here are described in more detail in separate papers [8] submitted to this conference.

## 2. Common Analysis Issues

At the $\Upsilon(4S)$, $B\overline{B}$ events can be distinguished from continuum $q\bar{q}$ production using two kinematic variables, the difference $\Delta E$ between the center-of-mass (CM) energy of the $B$ meson candidate and $\sqrt{s}/2$, and the beam-energy substituted mass $m_{\rm ES} = \sqrt{(s/2 + \mathbf{p}_i \cdot \mathbf{p}_B)^2/E_i^2 - \mathbf{p}_B^2}$, where $\sqrt{s}$ is the total CM energy, and the $B$ momentum $\mathbf{p_B}$ and the four-momentum of the initial state $(E_i, \mathbf{p_i})$ are defined in the laboratory frame. For signal decays $m_{\rm ES}$ peaks near the $B$ mass and $\Delta E$ peaks near zero when the correct mass hypothesis is assigned to all daughter particles. When there is ambiguity, as in the decay $B^0 \to h^+h^-$ with $h = \pi$ or $K$, we use the pion mass when calculating four-momenta, which results in a shift of $-45$ MeV in $\Delta E$ for each kaon in the decay.

Identification of charged tracks as pions or kaons is accomplished with the Cherenkov angle measurement $\theta_c$ from a detector of internally reflected Cherenkov light [9]. We use the information on $\theta_c$ directly in our maximum likelihood fits, which allows simultaneous measurement of $\pi^+(\pi^-/K^-)$, $\pi^0(\pi^+/K^+)$, and $\rho^+(\pi^-/K^-)$ decays. The typical separation between pions and kaons varies from $8\sigma_{\theta_c}$ at 2 GeV/$c$ to $2.5\sigma_{\theta_c}$ at 4 GeV/$c$.

The dominant background to charmless $B$ decays is from the process $e^+e^- \to q\bar{q}$ ($q = u, d, s, c$). To suppress this background, we use multivariate techniques that discriminate between the jet-like topology of $q\bar{q}$ events and the more spherical topology of $B\overline{B}$ events. For the analysis of $\pi^+\pi^-\pi^0$ decays we use a neural network, while the remaining analyses use a Fisher discriminant.

We use a multivariate technique [2] to determine the flavor of the $B_{\rm tag}$ meson. Separate neural networks are trained to identify primary leptons, kaons, soft pions from $D^*$ decays, and high-momentum charged particles from $B$ decays. Events are assigned to one of five mutually exclusive tagging categories based on the estimated mistag probability and the source of the tagging information. Table 1 summarizes the tagging performance measured in a data sample $B_{\text{flav}}$ of fully reconstructed neutral $B$ decays to $D^{(*)-}(\pi^+, \rho^+, a_1^+)$.

The time difference $\Delta t$ is obtained from the known boost of the $e^+e^-$ system and the measured distance between the $z$ positions of the $B$ candidate and the $B_{\text{tag}}$ decay vertices. A detailed description of the algorithm is given in Ref. [10]. We determine the parameters of the signal resolution function from a fit to the $B_{\text{flav}}$ sample (including events in all five tagging categories), while the form of the background $\Delta t$ distribution is obtained from a signal-free region in data.

## 3. Analysis of $B^0 \to \pi^+\pi^-$

We analyze the decays $B^0 \to \pi^+\pi^-$, $K^+\pi^-$, $K^+K^-$ simultaneously. Event yields are obtained from an unbinned extended maximum likelihood fit using $m_{\text{ES}}$, $\Delta E$, the Fisher discriminant, and $\theta_c$ measurements for each track. We include separate components for $K^+\pi^-$ and $K^-\pi^+$ in order to measure the direct $CP$ asymmetry $\mathcal{A}_{CP}^{K\pi}$. The parameters of the background PDFs are obtained directly from the data, while signal parameters are determined from a mix of GEANT4 Monte Carlo simulation and control samples in data. Figure 1 shows distributions of $m_{\text{ES}}$ and $\Delta E$ for subsamples of events enriched in $\pi\pi$ and $K\pi$ decays using probability ratios. The results are summarized in Table 2. The dominant sources of systematic error on the branching fraction measurements are from uncertainty in track and $\theta_c$ reconstruction efficiencies and imperfect knowledge of the PDF shapes. For $\mathcal{A}_{CP}^{K\pi}$ the systematic error is dominated by the $\theta_c$ PDF shape and possible charge bias in track reconstruction.

The parameters $S_{\pi\pi}$ and $C_{\pi\pi}$ are obtained from a second fit including tagging and $\Delta t$ information. The $\Delta t$ PDF for signal $\pi^+\pi^-$ decays is given by Eq. 1, modified to include $w_k$ and $\Delta w_k$ for each tagging category and convolved with the

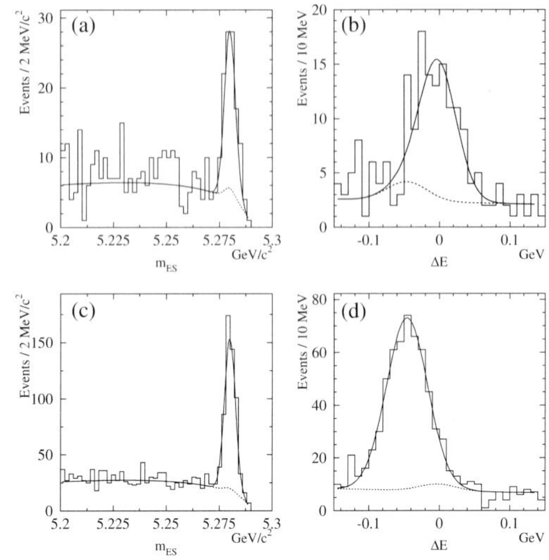


Figure 1. Distributions of $m_{\text{ES}}$ and $\Delta E$ for events enhanced in signal (a), (b) $\pi^+\pi^-$ and (c), (d) $K^\mp\pi^\pm$ decays. Solid curves represent projections of the maximum likelihood fit, dashed curves represent $q\bar{q}$ and $\pi\pi \leftrightarrow K\pi$ cross-feed background.

signal resolution function. The $\Delta t$ PDF for signal $K^+\pi^-$ events takes into account $B^0$–$\overline{B}^0$ mixing based on the charge of the kaon and the flavor of $B_{\text{tag}}$. As a means of validating the analysis technique, we determine $\tau$ and $\Delta m_d$ using signal $\pi^+\pi^-$ and $K^+\pi^-$ decays and find $\tau = (1.56 \pm 0.07)\,\text{ps}$ and $\Delta m_d = (0.52 \pm 0.05)\,\text{ps}^{-1}$. The fit yields

$$
\begin{aligned}
S_{\pi\pi} &= 0.02 \pm 0.34\,(\text{stat}) \pm 0.05\,(\text{syst}) \\
C_{\pi\pi} &= -0.30 \pm 0.25\,(\text{stat}) \pm 0.04\,(\text{syst}),
\end{aligned}
$$

where the correlation between $S_{\pi\pi}$ and $C_{\pi\pi}$ is $-10\%$. Systematic uncertainties on $S_{\pi\pi}$ and $C_{\pi\pi}$ are dominated by imperfect knowledge of the PDF shapes and possible fit bias. Figure 2 shows distributions of $\Delta t$ for events with $B_{\text{tag}}$ tagged as $B^0$ or $\overline{B}^0$, and the asymmetry as a function of $\Delta t$ for tagged events enhanced in signal $\pi\pi$ decays. We find no evidence for large mixing-induced, or direct $CP$ violation, as reported in [13].

Table 2

Summary of results for total detection efficiencies, fitted signal yields $N_S$, charge-averaged branching fractions $\mathcal{B}$, and $CP$ asymmetries $\mathcal{A}_{CP} \equiv [N_{\overline{B}^0} - N_{B^0}]/[N_{\overline{B}^0} + N_{B^0}]$ for various charmless two-body $B$ decays. Branching fractions are calculated assuming equal rates for $\Upsilon(4S) \to B^0\overline{B}^0$ and $B^+B^-$. The upper limit yields and branching fractions for $\pi^+\pi^0$ and $K^+K^-$ correspond to the 90% C.L. All results are consistent with similar measurements from other experiments [11].

| Mode | Efficiency (%) | $N_S$ | $\mathcal{B}(10^{-6})$ | $\mathcal{A}_{CP}$ |
|---|---|---|---|---|
| $\pi^+\pi^-$ | $38.0 \pm 0.8$ | $157 \pm 19 \pm 7$ | $4.7 \pm 0.6 \pm 0.2$ | |
| $\pi^+\pi^0$ | $26.1 \pm 1.7$ | $125^{+23}_{-21} \pm 10$ | $5.5^{+1.0}_{-0.9} \pm 0.6$ | $-0.03^{+0.18}_{-0.17} \pm 0.02$ |
| $\pi^0\pi^0$ | $16.5 \pm 1.7$ | $23^{+10}_{-9} (< 46)$ | $< 3.6$ | |
| $K^+\pi^-$ | $37.5 \pm 0.8$ | $589 \pm 30 \pm 17$ | $17.9 \pm 0.9 \pm 0.7$ | $-0.102 \pm 0.050 \pm 0.016$ |
| $K^+\pi^0$ | $21.5 \pm 1.5$ | $239^{+21}_{-22} \pm 6$ | $12.8^{+1.2}_{-1.1} \pm 1.0$ | $-0.09 \pm 0.09 \pm 0.01$ |
| $K^+K^-$ | $36.2 \pm 0.8$ | $1 \pm 8 (< 16)$ | $< 0.6$ | |

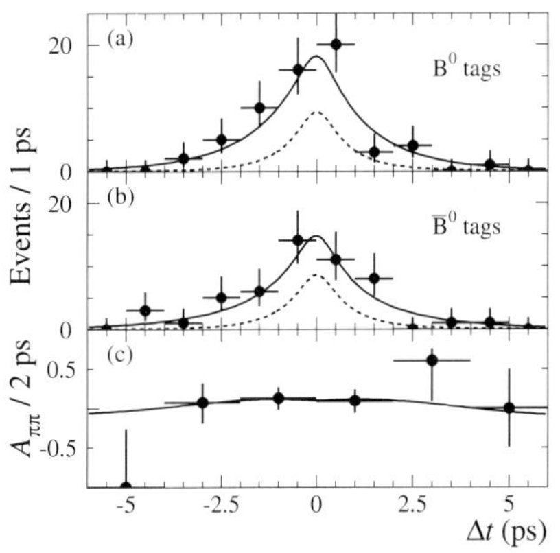


Figure 2. Distributions of $\Delta t$ for events enhanced in signal $\pi\pi$ decays with $B_{\rm tag}$ tagged as (a) $B^0$ ($N_{B^0}$) or (b) $\overline{B}^0$ ($N_{\overline{B}^0}$), and (c) the asymmetry $[N_{B^0} - N_{\overline{B}^0}] / [N_{B^0} + N_{\overline{B}^0}]$ as a function of $\Delta t$. Solid curves represent projections of the maximum likelihood fit, dashed curves represent the sum of $q\bar{q}$ and $K\pi$ background events.

### 3.1. Constraining $|\alpha_{\rm eff} - \alpha|$

The isospin relationship between the final states $\pi^+\pi^-$, $\pi^+\pi^0$, and $\pi^0\pi^0$ allows the determination of $\alpha$ given a measurement of $\alpha_{\rm eff}$. Assuming isospin invariance, the shift in $\alpha$ is bounded by the relation $\sin^2(\alpha_{\rm eff} - \alpha) < \mathcal{B}(B^0 \to \pi^0\pi^0)/\mathcal{B}(B^+ \to \pi^+\pi^0)$ [6], where the branching fractions are averaged over $B^0/\overline{B}^0$ and $B^+/B^-$. Details of the branching fraction measurements are reported elsewhere [8], and summarized in Table 2. Including correlations and systematic errors, we find $BR(B^0 \to \pi^0\pi^0)/\mathcal{B}(B^+ \to \pi^+\pi^0) < 0.61$ at 90% C.L., which corresponds to $|\alpha_{\rm eff} - \alpha| < 51°$.

## 4. Analysis of $B^0 \to \rho^+\pi^-$

A clean determination of $\alpha$ from the decay $B^0 \to \pi^+\pi^-\pi^0$ can only be accomplished with a full Dalitz analysis, including resonant and non-resonant contributions. However, a full analysis requires significantly more data than is currently available. We therefore restrict ourselves in this analysis to the regions of the Dalitz plane dominated by the $\rho$ resonance, and we analyze $\rho\pi$ and $\rho K$ decays simultaneously.

Due to the presence of a neutral pion, there is a significant background from other $B$ decay modes which can fake the $\rho\pi$ final state. We have investigated more than 80 decay modes and identified 20 charmless decays which have more than one event entering the final sample. These decays are parameterized and included in the maximum likelihood fit. We also include contributions from misreconstructed signal events and continuum $q\bar{q}$ background.

The fit technique is similar to the one used in the $\pi^+\pi^-$ analysis. We find $413^{+34}_{-33}$ $\rho\pi$ and $147^{+22}_{-21}$ $\rho K$ candidates and measure the following

*CP* parameters

$$S_{\rho\pi} = 0.16 \pm 0.25\,(\text{stat}) \pm 0.07\,(\text{syst}),$$
$$C_{\rho\pi} = 0.45^{+0.18}_{-0.19}\,(\text{stat}) \pm 0.09\,(\text{syst}),$$
$$A_{CP}^{\rho\pi} = -0.22 \pm 0.08\,(\text{stat}) \pm 0.07\,(\text{syst}),$$
$$A_{CP}^{\rho K} = 0.19 \pm 0.15\,(\text{stat}) \pm 0.11\,(\text{syst}),$$

and the *CP* invariant quantities

$$\Delta S_{\rho\pi} = 0.15 \pm 0.26\,(\text{stat}) \pm 0.05\,(\text{syst}),$$
$$\Delta C_{\rho\pi} = 0.38^{+0.19}_{-0.20}\,(\text{stat}) \pm 0.11\,(\text{syst}).$$

The dominant source of systematic error in this analysis is from uncertainty in the level of background from other $B$ decays, and in the parameterization of the PDFs for those decays. Figure 3 shows distributions of $\Delta t$ for events with $B_{\text{tag}}$ tagged as $B^0$ or $\overline{B}^0$, and the asymmetry as a function of $\Delta t$ for tagged events enhanced in signal $\rho\pi$ decays.

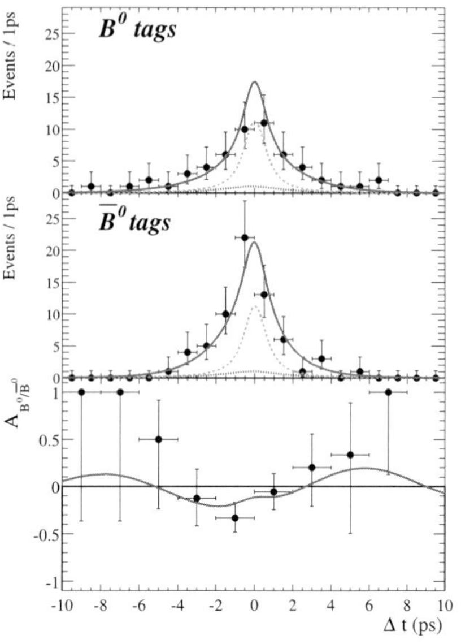


Figure 3. Distributions of $\Delta t$ for events enhanced in signal $\rho\pi$ decays with $B_{\text{tag}}$ tagged as (a) $B^0$ ($N_{B^0}$) or (b) $\overline{B}^0$ ($N_{\overline{B}^0}$), and (c) the asymmetry $[N_{B^0} - N_{\overline{B}^0}] / [N_{B^0} + N_{\overline{B}^0}]$ as a function of $\Delta t$. Solid curves represent projections of the maximum likelihood fit, dotted curves represent backgrounds from $B\overline{B}$ events, and dashed curves represent the sum of $q\bar{q}$ and $B\overline{B}$ background events.

## REFERENCES

1. N. Cabibbo, Phys. Rev. Lett. **10**, 531 (1963); M. Kobayashi and T. Maskawa, Prog. Th. Phys. **49**, 652 (1973).
2. *BABAR* Collaboration, B. Aubert *et al.*, SLAC-PUB-9293, hep-ex/0207042, to appear in Phys. Rev. Lett.
3. BELLE Collaboration, K. Abe *et al.*, hep-ex/0208025, submitted to Phys. Rev. D .
4. For a general review of the formalism see Y. Nir and H. Quinn, Ann. Rev. Nucl. Part. Sci. **42**, 211 (1992).
5. M. Gronau and D. London, Phys. Rev. Lett. **65**, 3381 (1990).
6. Y. Grossman and H.R. Quinn, Phys. Rev. D **58**, 017504 (1998)
7. R. Aleksan, I. Dunietz, B. Kayser, and F. Le Diberder, Nucl. Phys. B **361**, 141 (1991).
8. *BABAR* Collaboration, B. Aubert *et al.*, SLAC-PUB-9310, hep-ex/0207063; SLAC-PUB-9304, hep-ex/0207065; SLAC-PUB-9303, hep-ex/0207068.
9. The *BABAR* detector is described in B. Aubert *et al.*, Nucl. Instr. and Methods A **479**, 1 (2002).
10. *BABAR* Collaboration, B. Aubert *et al.*, SLAC-PUB-9060, hep-ex/0201020, to appear in Phys. Rev. D .
11. BELLE Collaboration, B.C.K. Casey *et al.*, hep-ex/0207090, submitted to Phys. Rev. D ; CLEO Collaboration, D. Cronin-Hennessy *et al.*, Phys. Rev. Lett. **85**, 515 (2000); CLEO Collaboration, S. Chen *et al.*, Phys. Rev. Lett. **85**, 525 (2000).
12. Particle Data Group, K. Hagiwara *et al.*, Phys. Rev. D **66**, 010001 (2002).
13. Belle Collaboration, K. Abe *et al.*, Phys. Rev. Lett. **89**, 071801 (2002).

ELSEVIER

Proceedings of ICHEP 2002, pp. 521–524
*S. Bentvelsen, P. de Jong, J. Koch and E. Laenen (Editors)*

31st INTERNATIONAL CONFERENCE ON HIGH ENERGY PHYSICS AMSTERDAM

# Measurements of direct $CP$ violation in two-body decays of $B$ meson with Belle

K. Suzuki[a]

[a]IPNS, KEK,
Oho 1-1, Tsukuba-City, Ibaraki, 305-0801 Japan

We report measurements of partial rate asymmetries in charge conjugate modes using two-body $B$ decays with flavor-specific final states at the Belle experiment. A time-integrated analysis is applied on data samples of $31-85$ million $B\overline{B}$ pairs. All asymmetries are null consistent within errors.

## 1. Introduction

The observation of mixing-induced or "indirect" $CP$ violation is a great achievement of $B$-factory experiments [1], and strongly supports the Kobayashi-Maskawa scheme [2]. On the other hand, decay rate asymmetry or "direct" $CP$ violation (DCPV) has not yet been observed in the $B$ meson system. DCPV can occur through interference of more than one diagram for a single decay. In the case that two diagrams contribute, the partial rate asymmetry ($\mathcal{A}_{CP}$) is expressed as

$$\mathcal{A}_{CP} = \frac{2r\sin\Delta\delta\sin\Delta\phi}{1+r^2+2r\cos\Delta\delta\cos\Delta\phi}, \quad (1)$$

where $r$ denotes the magnitude of ratio of two decay amplitudes and $\Delta\delta$ and $\Delta\phi$ represent the differences of strong and weak phases, respectively, between the two amplitudes. A sizable $\mathcal{A}_{CP}$ requires $r \sim \mathcal{O}(1)$ with $\Delta\delta \neq 0$ and $\Delta\phi \neq 0$. Decays in which $b \to u$ tree and $b \to s$ penguin diagrams contribute may exhibit such interference. Sensitivity to $\phi_3$ is possible in such $\mathcal{A}_{CP}$ measurements because $\Delta\phi = \phi_3$ in eq. 1, where $\phi_3 \simeq \arg(V_{ub}^*)$ denotes an internal angle of the unitarity triangle. On the other hand, when no asymmetry is expected in the Standard Model (SM) framework, a non-zero asymmetry would be an indication of new physics beyond the SM. Such an effect can be searched for in one-loop penguin diagrams. The search for DCPV is an important issue at $B$-factory experiments.

We report $\mathcal{A}_{CP}$ measurements in two-body $B$ decays with flavor-specific final states including $B \to K\pi^+$, $\pi^+\pi^0$, $\eta' K^+$, $\eta K^*$, $\omega K^+$, $\overline{D}^0 K^+$ and $K^*\gamma$. Throughout the text, the inclusion of charge conjugate modes are implied. Detailed descriptions can be found in refs. [3–6]. The results are based on $29-78$ fb$^{-1}$ data samples collected at the $\Upsilon(4S)$ resonance, corresponding to $31-85$ million $B\overline{B}$ pairs, by the Belle detector [7] at the KEKB $e^+e^-$ storage ring [8]. Applying time-integrated analysis, $\mathcal{A}_{CP}$ is measured as

$$\mathcal{A}_{CP}(f) = \frac{N(\overline{B}\to\overline{f}) - N(B\to f)}{N(\overline{B}\to\overline{f}) + N(B\to f)}, \quad (2)$$

where $N(B \to f)$ denotes the signal yield of $B^0$ or $B^+$ decaying into a flavor-specific final state $f$ and $N(\overline{B} \to \overline{f})$ denotes that of the charge conjugate mode.

## 2. Charmless hadronic decays

### 2.1. $B \to K\pi$ and $\pi^+\pi^0$

Tree-penguin interference can cause DCPV in $B \to K^+\pi^-$ and $K^+\pi^0$ decays while no large asymmetry is expected in $B^+ \to K_S^0\pi^+$ and $\pi^+\pi^0$ decays, which are dominated by a single penguin or tree contribution, respectively. In $B^+ \to K_S^0\pi^+$, the re-scattering process $B^+ \to (K^+\pi^0)_{\rm tree} \to K^0\pi^+$ could interfere with the original penguin within the SM framework.

Candidate $B$ mesons are reconstructed using charged tracks, $K_S^0(\to \pi^+\pi^-)$ mesons and $\pi^0(\to$


doi:10.1016/S0920-5632(02)02183-7

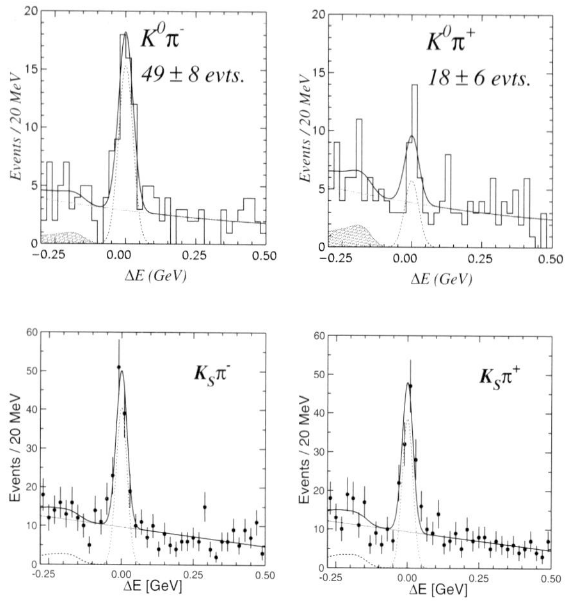


Figure 1. The $\Delta E$ fit results for $B^+ \to K_S^0\pi^+$ at 29 fb$^{-1}$ (upper) and at 78 fb$^{-1}$ (lower) divided into $B^-$ (left) and $B^+$ (right) samples.

$\gamma\gamma$) mesons. Signal candidates are identified using $m_{bc} \equiv \sqrt{E_{\rm beam}^2 - p_B^{*2}}$ and $\Delta E \equiv E_B^* - E_{\rm beam}$, where $E_{\rm beam} = 5.29$ GeV and $p_B^*$ and $E_B^*$ are the momentum and energy of the reconstructed $B$ meson in the center of mass frame. The dominant background comes from $e^+e^- \to q\bar{q}$ ($q = u$, $d$, $s$, $c$) continuum process while $b \to c$ decays are negligible. The $q\bar{q}$ background is suppressed based on the event topology and the angle of the $B$ flight direction with respect to the beam axis. The $h\pi^+$ and $hK^+$ ($h = \pi$, $K$) final states are discriminated using high momentum particle identification (PID) based on the light yield in the aerogel Čerenkov counter and $dE/dx$ measurements in the central drift chamber. Using a binned maximum likelihood fit, signal yields are extracted from $\Delta E$ distributions that consist of signal, $q\bar{q}$ background and other charmless $B$ decays (the bump in lower $\Delta E$ region).

An analysis on the complete set of $K\pi$ and $\pi^+\pi^0$ modes was performed using 29 fb$^{-1}$ of data. The results are summarized in Table 1. In

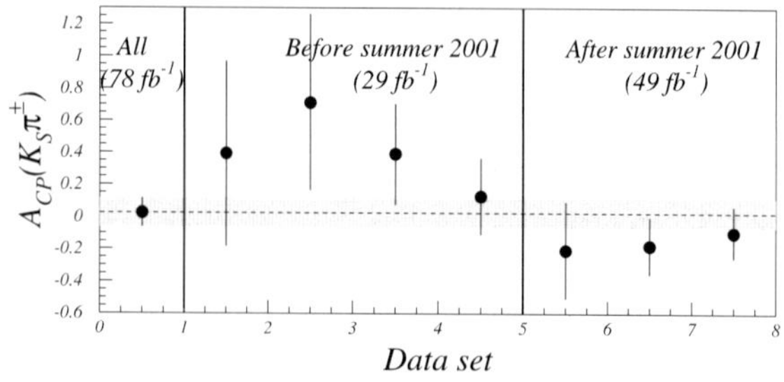


Figure 2. The $\mathcal{A}_{CP}(K_S^0\pi^+)$ in each data subsamples.

$B^+ \to K_S^0\pi^+$, a large asymmetry was found as shown in Figure 1 while other channels show null asymmetries within statistical errors.

To confirm the sizable $\mathcal{A}_{CP}(K_S^0\pi^+)$, the same analysis was redone using 78 fb$^{-1}$ of data. Preliminary results for the $B^+ \to K_S^0\pi^+$ decay are also shown in Figure 1 and given in Table 1 along with that for $B^0 \to K^+\pi^-$. In this data sample, $\mathcal{A}_{CP}(K_S^0\pi^+)$ is measured to be consistent with zero. Figure 2 shows $\mathcal{A}_{CP}(K_S^0\pi^+)$ as a function of the data sub-sample. We now conclude that the sizable $\mathcal{A}_{CP}(K_S^0\pi^+)$ observed with the 29 fb$^{-1}$ sample was a statistical fluctuation. For $\mathcal{A}_{CP}(K^+\pi^-)$, the statistical error is reduced to 6% while the centeral value is almost the same as the previous value.

## 2.2. $B \to \eta' K^+$, $\eta K^*$ and $\omega K^+$

Although $B \to \eta' K^+$ and $\eta K^*$ decays are strongly penguin dominated, a new physics contribution, indicated by the unexpectedly large branching fractions [4], may give a large asymmetry. In $B^+ \to \omega K^+$, which was first observed by Belle [5], tree-penguin interference can cause DCPV. The analysis is based on 42 fb$^{-1}$ of data for $B^+ \to \eta' K^+$ and 29 fb$^{-1}$ of data for $B \to \eta K^*$ and $\omega K^+$. Candidate $B$ mesons are reconstructed using the following intermediate decays where applicable: $\eta' \to \eta(\to \gamma\gamma)\pi^+\pi^-$ and $\rho^0\gamma$; $\eta \to \gamma\gamma$ and $\pi^+\pi^-\pi^0$; $K^* \to K^+\pi^-$, $K^+\pi^0$ and $K_S^0(\to \pi^+\pi^-)\pi^+$; and $\omega \to \pi^+\pi^-\pi^0$. The analysis is performed in a similar way to that described in section 2.1, where the time-of-flight information is also used for the PID. Signal yields

Table 1
Summary of $\mathcal{A}_{CP}$ results. Each result is based on the listed integrated luminosity ($\int \mathcal{L}dt$). The 90% confidence intervals of $\mathcal{A}_{CP}$ are also listed. The quoted errors of signal yields are statistical while the first and second errors of $\mathcal{A}_{CP}$ are statistical and systematic, respectively.

| Decay Mode | $N(\overline{B} \to \overline{f})$ | $N(B \to f)$ | $\mathcal{A}_{CP}$ | 90%C.L. | $\int \mathcal{L}dt$ [fb$^{-1}$] |
|---|---|---|---|---|---|
| $B^0 \to K^+\pi^-$ | $102.8 \pm 12.6$ | $115.0 \pm 13.3$ | $-0.06 \pm 0.09 {}^{+0.01}_{-0.02}$ | $[-0.21, 0.09]$ | 29 |
| | $235.4 {}^{+19.8}_{-9.1}$ | $270.2 {}^{+19.7}_{-18.9}$ | $-0.07 \pm 0.06 \pm 0.01$ | $[-0.18, 0.04]$ | 78 |
| $B^+ \to K^0\pi^+$ | $49.5 \pm 8.4$ | $18.6 \pm 6.3$ | $0.46 \pm 0.15 \pm 0.02$ | $[0.19, 0.72]$ | 29 |
| | $96.5 {}^{+12.4}_{-11.7}$ | $92.0 {}^{+12.1}_{-11.6}$ | $0.02 \pm 0.09 \pm 0.01$ | $[-0.14, 0.18]$ | 78 |
| $B^+ \to K^+\pi^0$ | $28.7 \pm 7.8$ | $30.1 \pm 7.7$ | $-0.02 \pm 0.19 \pm 0.02$ | $[-0.35, 0.30]$ | 29 |
| $B^+ \to \pi^+\pi^0$ | $24.2 \pm 8.4$ | $13.0 \pm 7.3$ | $0.30 \pm 0.30 {}^{+0.06}_{-0.04}$ | $[-0.23, 0.86]$ | 29 |
| $B^+ \to \eta' K^+$ | $139.7 {}^{+14.3}_{-13.5}$ | $144.0 {}^{+14.6}_{-13.8}$ | $-0.015 \pm 0.070 \pm 0.009$ | $[-0.13, 0.10]$ | 42 |
| $B^0 \to \eta K^{*0}$ | $11.6 {}^{+5.4}_{-4.3}$ | $14.6 {}^{+6.0}_{-4.8}$ | $0.17 {}^{+0.28}_{-0.25} \pm 0.01$ | $[-0.24, 0.70]$ | 29 |
| $B^+ \to \eta K^{*+}$ | $12.5 {}^{+6.1}_{-4.2}$ | $13.1 {}^{+6.1}_{-4.6}$ | $-0.05 {}^{+0.25}_{-0.30} \pm 0.01$ | $[-0.72, 0.38]$ | 29 |
| $B^+ \to \omega K^+$ | $7.3 \pm 3.5$ | $11.2 \pm 3.7$ | $-0.21 \pm 0.28 \pm 0.03$ | $[-0.70, 0.28]$ | 29 |
| $B^+ \to D_f K^+$ | $81.1 \pm 10.4$ | $80.6 \pm 10.1$ | $0.003 \pm 0.089 \pm 0.037$ | $[-0.15, 0.16]$ | 29 |
| $B^+ \to D_1 K^+$ | $14.7 \pm 4.6$ | $8.1 \pm 3.9$ | $0.29 \pm 0.26 \pm 0.05$ | $[-0.14, 0.73]$ | 29 |
| $B^+ \to D_2 K^+$ | $10.6 \pm 4.2$ | $16.4 \pm 5.5$ | $-0.22 \pm 0.24 \pm 0.04$ | $[-0.62, 0.18]$ | 29 |
| $B \to K^*\gamma$ | | | $-0.022 \pm 0.048 \pm 0.017$ | $[-0.106, 0.062]$ | 60 |
| $B^0 \to K^{*0}\gamma$ | $165 \pm 14$ | $186 \pm 15$ | $-0.061 \pm 0.061 \pm 0.022$ | | 60 |
| $B^+ \to K^{*+}\gamma$ | $95 \pm 10$ | $85 \pm 10$ | $0.053 \pm 0.083 \pm 0.016$ | | 60 |

are extracted by fitting the $m_{bc}$ and $\Delta E$ distributions simultaneously using an unbinned maximum likelihood fit. As summarized in Table 1, all results show null asymmetries.

## 3. Charmed hadronic decay: $B^+ \to D_{CP}K^+$

DCPV in $B^+ \to D_{CP}K^+$ decays, where $D_{CP}$ denotes a $D^0$ meson decaying into a $CP$ eigenstates, can arise from the interference of $b \to c$ and $b \to u$ tree processes. This interference provides a theoretically clean way to extract $\phi_3$ form branching fraction and asymmetry measurements of $B^+ \to D_iK^+$ and $D_i\pi^+$ decays, where $D_i$ represents $D_1$, $D_2$ and $D_f$ for the $CP$-even, $CP$-odd and flavor-specific final states, respectively [6].

The analysis is based on 29 fb$^{-1}$. Candidate $B$ mesons are reconstructed using $D_1 \to K^+K^-$ and $\pi^+\pi^-$ decays, $D_2 \to K^0_S\pi^0$, $K^0_S\eta(\to \gamma\gamma)$, $K^0_S\eta'(\to \eta\pi^+\pi^-)$, $K^0_S\omega(\to \pi^+\pi^-\pi^0)$ and $K^0_S\phi(\to K^+K^-)$ decays, and $D_f \to K^-\pi^+$ decay. In addition to the requirements described in section 2, vetoes of other $b \to c$ decays are also applied. Signal yields are extracted by fitting the $\Delta E$ distributions as described in section 2.1. The results are listed in Table 1. Although all results are null consistent within statistical errors, this is the first step in the program of $\phi_3$ extraction.

## 4. Radiative decay: $B \to K^*\gamma$

$B \to K^*\gamma$ decays are dominated by $b \to s\gamma$ electroweak penguins, which are very sensitive to contributions from non-SM particles. Since the SM predicts $\mathcal{A}_{CP}(K^*\gamma) < 1\%$, any large asymmetry would be an indication of non-SM effects.

The analysis is based on 60 fb$^{-1}$. Candidate $B$ mesons are reconstructed using a high energy primary photon, and $K^{*0} \to K^+\pi^-$, $K^{*+} \to K^0_S\pi^+$ and $K^+\pi^0$ decays. The analysis is similar to that described in section 2. Signal yields are extracted by fitting $m_{bc}$ distributions which consists of signal and backgrounds from $q\overline{q}$ and other charmless $B$ decays. Preliminary results are shown in Fig-

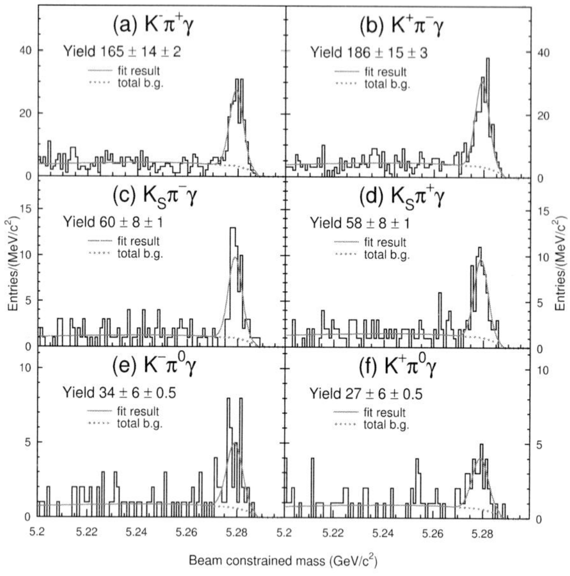


Figure 3. The $m_{bc}$ fit results for $B \to K^{*0}\gamma$ (top), $K^{*+}(\to K^0_S\pi^+)\gamma$ (middle) and $K^{*+}(\to K^+\pi^0)\gamma$ (bottom) decays divided into $\overline{B}$ (left) and $B$ (right) samples.

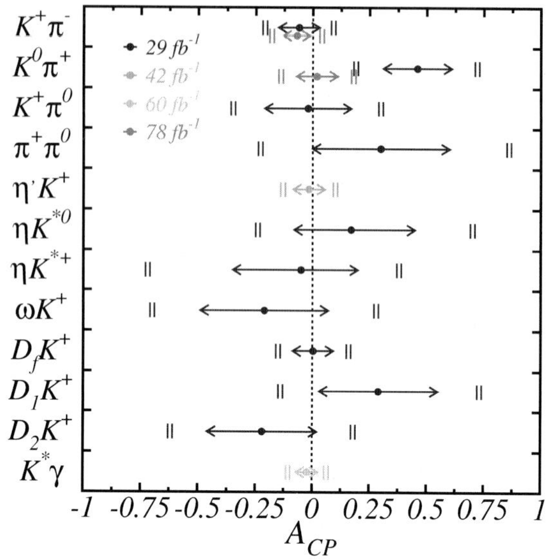


Figure 4. Summary of $\mathcal{A}_{CP}$ results based on $29 - 78$ fb$^{-1}$ data samples. The point, arrows and double bar show the $\mathcal{A}_{CP}$, statistical errors and 90% confidence intervals, respectively.

ure 3 and summarized in Table 1. The asymmetries are consistent with zero and agree with the SM at the current level of statistics.

## 5. Conclusion

Using time-integrated analysis based on $29-78$ fb$^{-1}$data samples, partial rate asymmetries are measured in the flavor-specific decays $B \to K\pi$, $\pi^+\pi^0$, $\eta' K^+$, $\eta K^*$, $\omega K^+$, $D_{CP}K^+$ and $K^*\gamma$ decays. All asymmetries are consistent with zero at the current level of statistics as shown in Figure 4 and summarized in Table 1. The sizable $\mathcal{A}_{CP}(K^0_S\pi^+)$ obtained with 29 fb$^{-1}$ has disappeared with 78 fb$^{-1}$ and, therefore, is interpreted as a statistical fluctuation. The statistical precision in several modes has achieved $6 - 10\%$. In a few years, much more interesting information, such as $\phi_3$, strong phases and new physics, can be expected.

## REFERENCES

1. Belle Collaboration, K. Abe *et al.*, Phys. Rev. Lett. **87**, 091802 (2001); hep-ex/0208025 (2002); BABAR Collaboration, B. Aubert *et al.*, Phys. Rev. Lett. **87**, 091801 (2001).
2. M. Kobayashi and T. Maskawa, Prog. Theor. Phys. **49**, 652 (1973).
3. Belle Collaboration, B. Casey *et al.*, hep-ex/0207090 (2002); Belle Collaboration, M. Nakao *et al.*, ICHEP 2002 contributed paper, BELLE-CONF-0239.
4. Belle Collaboration, K.F. Chen *et al.*, hep-ex/0207033 (2002); Belle Collaboration, C.H. Wang *et al.*, ICHEP 2002 contributed paper, BELLE-CONF-0229.
5. Belle Collaboration, R.-S. Lu *et al.*, hep-ex/0207019 (2002).
6. Belle Collaboration, K. Abe *et al.*, hep-ex/0207012 (2002).
7. Belle Collaboration, K. Abe *et al.*, Nucl. Inst. Meth. A **479**, 117 (2002).
8. E. Kikutani ed., KEK Preprint 2001-157 (2001).

ELSEVIER

Proceedings of ICHEP 2002, pp. 525–528
*S. Bentvelsen, P. de Jong, J. Koch and E. Laenen (Editors)*

31st INTERNATIONAL CONFERENCE ON HIGH ENERGY PHYSICS AMSTERDAM

# CP Asymmetries and Charmless Branching Fractions with *BABAR*

A. J. Bevan[a] *,

[a]Oliver Lodge Laboratory, University of Liverpool, Oxford Street, Liverpool L69 7ZE, UK

A summary of recent results of searches for direct CP Violation and measurements of charmless B-meson decay branching fractions with the *BABAR* detector is presented.

## 1. INTRODUCTION

Direct CP Violation (CPV) is manifest in processes where the amplitude for a B-meson to decay to a final state f, $|A_f|$, differs from that of its CP conjugate, $|\overline{A}_{\overline{f}}|$. In order for a process to be direct CP violating, one requires interference between decay amplitudes with different weak, $\phi_i$, and strong, $\delta_i$, phases (i.e. different CP-odd and CP-even phases) as can be seen in the following

$$|A_f| - |\overline{A}_{\overline{f}}| \propto \sum_{i,j} A_i A_j sin(\phi_i - \phi_j) sin(\delta_i - \delta_j) \quad (1)$$

Thus far direct CPV has only been observed in the neutral kaon system and is measured via the double ratio method. Recent experimental results of these measurements from the NA48 and KTeV collaborations show a non-zero $\epsilon'$ [1,2]. In contrast to the neutral kaon system, direct CPV in B-meson decays is expected to be large (few $\sim$ 80% depending on the mode).

In B-meson decays one can define a time integrated CP asymmetry, $A_{CP}$ , as

$$A_{CP} = \frac{BR(\overline{B} \to \overline{f}) - BR(B \to f)}{BR(\overline{B} \to \overline{f}) + BR(B \to f)} \quad (2)$$

where $BR(B \to f)$ is the branching ratio of a B-meson decaying to a final state $f$, and $BR(\overline{B} \to \overline{f})$ is the branching ratio for the CP conjugate process. For neutral B-meson decays one either calculates $A_{CP}$ and subsequently corrects for the effect of mixing when looking for direct CPV, or performs a time dependent analysis. Charged B-mesons do not mix and as such one can determine the B-flavour from decay products or tagging the other B-meson decay in the event. So one can use a time integrated asymmetry to look for direct CPV in charged B-meson decays.

Direct CPV is maximal when the interfering amplitudes are of comparable magnitude. As a result one is interested in rare or suppressed processes. Measurement of decays involving $b \to u$ suppressed tree and $b \to s, d$ penguin processes are be useful for direct CPV searches. The former can manifest large $A_{CP}$ [3] whereas the latter penguin dominated modes are predicted to have small $A_{CP}$. Standard model behaviour such as $A_{CP}$'s and branching ratios can be modified in the presence of new physics effects via loops [4–6].

## 2. Extracting signals

PEP-II is an asymmetric $e^+e^-$ collider located at the Stanford Linear Accelerator Center. The accelerator collides 9.0 $GeV$ electrons with 3.1 $GeV$ positrons at the $\Upsilon(4s)$ resonance to produce a $B\overline{B}$ pair nearly at rest in the center of mass. The *BABAR* detector is used to measure the decay products of these events. Both PEP-II and *BABAR* are described in detail elsewhere [7].

When measuring $A_{CP}$'s, one has to correct for flavour mixing of neutral B-mesons and detector effects such as charge asymmetries in triggering, reconstruction, event selection and final analysis. In addition to this, there is the formidable task of extracting a rare process from a large light quark continuum background coming from $u, d, s, c$ quarks.

The main variables used to discriminate signal

*representing the *BABAR* collaboration.


doi:10.1016/S0920-5632(02)02187-4

| Mode | $BR(\times 10^{-6})$ | $A_{CP}$ |
|---|---|---|
| $B^0 \to K^+\pi^-$ | $(17.9 \pm 0.9 \pm 0.7)$ | $-0.102 \pm 0.050 \pm 0.016$ |
| $B^+ \to K^+\pi^0$ | $(12.8^{+1.2}_{-1.1} \pm 1.0)$ | $-0.09 \pm 0.09 \pm 0.01$ |
| $B^+ \to \pi^+\pi^0$ | $(5.5^{+1.0}_{-0.9} \pm 0.6)$ | $-0.03^{+0.18}_{-0.17} \pm 0.02$ |
| $B^0 \to K^0\pi^0$ | $(10.4 \pm 1.5 \pm 0.8)$ | $0.03 \pm 0.36 \pm 0.09$ |
| $B^0 \to \pi^+\pi^-$ | $(4.7 \pm 0.6 \pm 0.2)$ | $0.30 \pm 0.25 \pm 0.04$ |
| ‡ $B^+ \to K^0\pi^+$ | $(17.5^{+1.8}_{-1.7} \pm 1.3)$ | $-0.17 \pm 0.10 \pm 0.02$ |
| $B^0 \to \pi^0\pi^0$ | $< 3.6$ (90% C.L.) | - |
| $B^0 \to K^+K^-$ | $< 0.6$ (90% C.L.) | - |
| ‡ $B^+ \to K^+K^0$ | $< 1.3$ (90% C.L.) | - |

Table 1
Branching ratios, upper limits and $A_{CP}$ measured for the decays $B \to hh$.

from continuum background in the analyses discussed here are $M_{ES}$, $\Delta E$, cosines of the angle between the direction of the B-candidate and the thrust or sphericity axis of the rest of the event [3] and fisher discriminants based on energy flow of the events. The beam energy substituted mass, $M_{ES}$ is defined as $M_{ES} = \sqrt{(E^{*2}_{BEAM} - P^2_B)}$, where $E^*_{BEAM}$ is the center of mass energy of the beam and $P_B$ is the reconstructed B-candidate momentum. The energy variable $\Delta E$ is given by $\Delta E = E_B - E^*_{BEAM}$, where $E_B$ is the energy of the B-candidate. There are two main fisher discriminants used in the different analyses presented here which are described in [8] and [9].

## 3. RESULTS

### 3.1. Charmless two body decays

The decays $B \to hh$, where $h = \pi, K$, can be used in the determination of the weak phases $\alpha$ and $\gamma$ and may also prove useful in searching for direct CPV [10]. Analysis of these decays is discussed in detail elsewhere [8]. The $A_{CP}$'s and branching ratios measured for these modes come from a data set comprising approximately $88 \times 10^6$ $B\overline{B}$ pairs as shown in Table 1 (entries denoted by ‡ use $56.2 \times 10^6$ $B\overline{B}$ pairs [11]). The $A_{CP}$ measured in $B^0 \to K^+\pi^-$ is the most precise one from *BABAR* in a mode where there is a non-zero asymmetry expected. The total uncertainty on this asymmetry is 5.2%. Also the $B^0 \to \pi^+\pi^-$ asymmetry quoted explicitly assumes $q/p = 1$.

| Mode | $BR(\times 10^{-6})$ |
|---|---|
| $B^+ \to K^+\pi^-\pi^+$ | $59.2 \pm 4.7 \pm 4.9$ |
| $B^+ \to K^+K^-K^+$ | $34.7 \pm 2.0 \pm 1.8$ |
| $B^+ \to \pi^+\pi^-\pi^+$ | $< 15$ (90% C.L.) |
| $B^+ \to K^+K^-\pi^+$ | $< 7$ (90% C.L.) |

Table 2
Branching ratios and upper limits of the decays $B^+ \to h^+h^-h^+$.

### 3.2. Charmless three body decays

The branching fractions of the decays $B^+ \to h^+h^-h^+$, where $h = \pi, K$, have been measured using a cut based analysis of the whole Dalitz plot [12]. The data sample analysed for these modes contain approximately $56.2 \times 10^6$ $B\overline{B}$ pairs. Once observed these decays are of interest as one can search for direct CPV and ultimately, with sufficient statistics, fit to the Dalitz plot in order to extract the weak angle $\gamma$ [13].

Significant signals are observed for the decays $B^+ \to K^+K^-K^+$ and $B^+ \to K^+\pi^-\pi^+$, whereas we place an upper limit on $B^+ \to \pi^+\pi^-\pi^+$ and $B^+ \to K^+K^-\pi^+$ as given in Table 2. The Dalitz plots and $M_{ES}$ distributions for the decays $B^+ \to K^+K^-K^+$ and $B^+ \to K^+\pi^-\pi^+$ are shown in Figure 1. One can see in the case of $B^+ \to K^+\pi^-\pi^+$ that there is a significant resonant contamination in the Dalitz plot. This comes from B-decays with charm in the final state, i.e. $J/\psi$, $\psi_{2s}$ and $D$-mesons. To remove these, $3\sigma$ cuts are applied to the data around the nominal resonance masses. This leaves behind a small peaking background under the signal which can be seen in the $M_{ES}$ distribution for $B^+ \to K^+\pi^-\pi^+$.

### 3.3. Charmless quasi-two-body decays

One can search for direct CPV in the suppressed tree and penguin dominated quasi-two-body decays $B \to \eta' K^{(*)}$ and $B \to \omega K$, $\omega\pi$ respectively. Decays with $\eta'$ in the final state are of interest as, depending on the final state one may have a large $A_{CP}$. Also, by virtue of a $b \to u$ transition one can try to extract information about the weak angle $\gamma$ from these decays. Experimentally one observes larger than expected branching

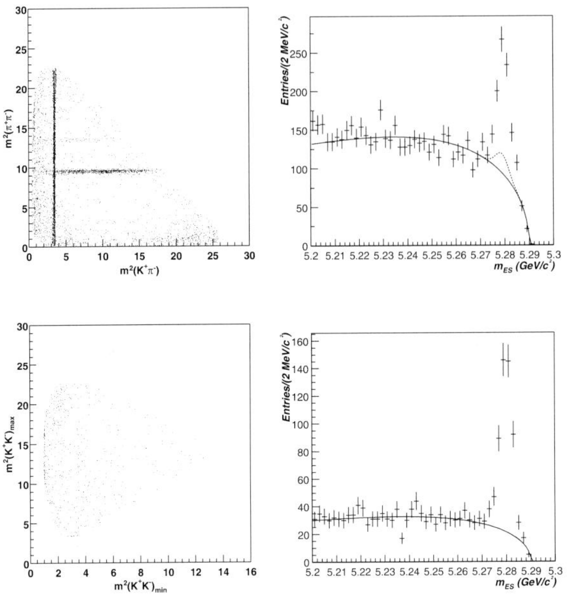


Figure 1. The (left) Dalitz plot and (right) $M_{ES}$ distributions for (top) $B^+ \to K^+\pi^-\pi^+$and (bottom) $B^+ \to K^+K^-K^+$.

| Mode | $A_{CP}$ |
|---|---|
| $B^+ \to \eta' K^+$ | $-0.11 \pm 0.11 \pm 0.02$ |
| $B^+ \to \omega\pi^+$ | $-0.01^{+0.29}_{-0.31} \pm 0.03$ |
| $B^+ \to \phi K^+$ | $-0.05 \pm 0.20 \pm 0.03$ |
| $B^+ \to \phi K^{\star +}$ | $-0.43^{+0.36}_{-0.30} \pm 0.06$ |
| $B^0 \to \phi K^{\star 0}$ | $0.00 \pm 0.27 \pm 0.03$ |

Table 3
Measured $A_{CP}$ 's for quasi-two-body charmless decays using $22.7 \times 10^6$ $B\overline{B}$ pairs

ratios for these modes. As such it is important to measure as many of these as possible in the hope of understanding the mechanism involved.

Decays with $\omega$ in the final state have a theoretically small $A_{CP}$ and give access to $\alpha$ and $\gamma$ [14]. Table 3 lists the measured $A_{CP}$ of these decays and the branching ratios can be found in Table 4. The decay $B^0 \to \omega K^0$ has been observed with a significance of $6.6\sigma$.

The penguin decays $B \to \phi K^{(\star)}$, $\phi\pi$ are interesting as the standard model $A_{CP}$ is a few percent, whereas new physics can alter this appreciably [5]. In addition, one can measure $sin2\beta$ which in the absence of new physics has the same value as that measured in charmonium modes. The measured $A_{CP}$'s and branching fractions for these decays are shown in Tables 3 and 4 respectively. Details of the quasi-two body analyses can be found in [15].

| Mode | $N(B\overline{B}) \times 10^6$ | $BR(\times 10^{-6})$ |
|---|---|---|
| $B^+ \to \eta' K^+$ | 61.5 | $67 \pm 5 \pm 5$ |
| $B^+ \to \eta' K^0$ | 61.5 | $46 \pm 6 \pm 4$ |
| $B^+ \to \eta' K^{\star 0}$ | 61.5 | $< 13$ (90% C.L.) |
| $B^+ \to \omega\pi^+$ | 22.7 | $6.6^{+2.1}_{-1.8} \pm 0.7$ |
| $B^0 \to \omega K^0$ | 61.5 | $5.9^{+1.7}_{-1.5} \pm 0.9$ |
| $B^+ \to \phi K^+$ | 61.5 | $9.2 \pm 1.0 \pm 0.8$ |
| $B^0 \to \phi K^0$ | 61.5 | $8.7^{+1.7}_{-1.5} \pm 0.9$ |
| $B^+ \to \phi K^{\star +}$ | 61.5 | $9.7^{+4.2}_{-3.4} \pm 1.7$ |
| $B^0 \to \phi K^{\star 0}$ | 61.5 | $8.7^{+2.5}_{-2.1} \pm 1.1$ |
| $B^+ \to \phi\pi^+$ | 61.5 | $< 0.56$ (90% C.L.) |

Table 4
Measured branching ratios for quasi-two-body charmless decays.

### 3.4. Other decays

One can search for direct CPV in decays other than the charmless ones described above, in particular we have looked for direct CPV in $B^0 \to K^\star\gamma$ and $B^- \to D^0_{(CP)}K^-$. The radiative penguin decay $B^0 \to K^\star\gamma$ provides a window to constrain new physics, as this can enhance the standard model $A_{CP}$ from $< 1\%$ to large values [6]. Using $22.7 \times 10^6$ $B\overline{B}$ decays we measure $A_{CP} = -0.044 \pm 0.076 \pm 0.012$ [16].

Given sufficient statistics, one can access $\gamma$ via the decay $B^- \to D^0_{(CP)}K^-$ [17]. Here the CP subscript refers to reconstruction of a D-meson decaying to a CP eigenstate. The measured $A_{CP}$ for this mode is also consistent with zero within errors; $0.17 \pm 0.23^{+0.09}_{-0.07}$ [18].

## 4. CONCLUSIONS

There is no signal for direct CPV in the $B$ system yet. The search for direct CPV in $B_d$ decays is well underway at *BABAR* but more data are required over the next few years to further these efforts. The most precise measurement of a CP asymmetry, where a non-zero value of $A_{CP}$ is predicted, has a total uncertainty of 5.2%. This uses $88 \times 10^6 B\overline{B}$ pairs which is almost our full data-set. In searching for direct CPV in radiative penguin decays *BABAR* has already started to constrain the parameter space of new physics models. As we increase our statistics over the next few years, the search for direct CPV will become more interesing as will measurements of the weak angles of the unitarity triangle and constraints on new physics models from these rare decays.

## 5. ACKNOWLEDGEMENTS

We are grateful for the extraordinary contributions of our PEP-II colleagues in achieving the excellent luminosity and machine conditions that have made this work possible.

## REFERENCES

1. The NA48 Collaboration, Phys. Lett. B **544** 97-112 (2002)
2. The KTeV collaboration, hep-ex/0208007 (Submitted to Phys. Rev. D)
3. P.F. Harrison, ed., Helen R. Quinn, ed., SLAC-R-0504 (Oct 1998)
4. Y.Y. Keum, hep-ph/0003155; B. Dutta et al., hep-ph/0207171
5. Y. Grossman et al., Phys. Rev. D **58** 057504 (1998) and M. P. Worah, Phys. Lett. B **395**, 241 (1997); R. Fleischer, Int. J. Mod. Phys. A **12**, 2459 (1997); R. Fleischer and T. Mannel, Phys.Lett. B **511** 240-250 (2001)
6. A. Kagan and M. Neubert, Phys Rev D **58** 094012 (1998)
7. The *BABAR* Collaboration, Nucl. Instrum. Meth. A**479** p1-116 (2002)
8. The *BABAR* Collaboration, hep-ex/0207055, hep-ex/0207063, hep-ex/0207065
9. The CLEO collaboration, Rhys. Rev. D **53** 1039-1050 (1996)
10. M. Benke et al., Nucl. Phys. B **606** 245-321 (2001)
11. The *BABAR* Collaboration, hep-ex/0206053
12. The *BABAR* Collaboration, hep-ex/0206004.
13. I. Bediaga et al., Phys. Rev. Lett. **81** 4067-4070 (1998); R. E. Blanco et al., Phys. Rev. Lett. **86** 2720-2723 (2001)
14. D. Atwood and A. Soni, Phys Lett B **516** 39-46 (2001)
15. The *BABAR* collaboration, Phys. Rev. Lett. **87** 221802 (2001); Phys. Rev. Lett. **87** 151801 (2001); Phys. Rev. D**65** 051101 (2002); A. Telenov talk at DPF 24-28 May 2002, Williamsburg, Virginia; P. Bloom talk at DPF 24-28 May 2002, Williamsburg, Virginia
16. The *BABAR* collaboration, Phys. Rev. Lett. **88** 101805 (2002)
17. D. Atwood et al., Phys. Rev. Lett. **78** 3257 (1997)
18. The *BABAR* collaboration hep-ex/0207087

## Session 8

# Heavy Quark Mesons and Baryons (including lattice calculations)

*Session conveners:*
*E. Barberio, CERN/SMU*
*S. Ryan, Trinity College, Dublin*

ELSEVIER

Proceedings of ICHEP 2002, pp. 531–534
*S. Bentvelsen, P. de Jong, J. Koch and E. Laenen (Editors)*

31st INTERNATIONAL CONFERENCE ON HIGH ENERGY PHYSICS AMSTERDAM

# First CDF II Heavy Flavor physics results with the Silicon Vertex Trigger

A. Cerri[a]*

[a]Physics Department, Lawrence Berkeley National Laboratory,
1 Cyclotron road, 94720 Berkeley, CA, USA

The renewed CDF II experiment recently re-started data taking. The data collected lately already provides insight in the heavy flavor capabilities of this $p\bar{p}$ experiment.
As a benchmark of these possibilities we present two preliminary results: the measurement of the branching fractions $BR\left(D^\circ \to \pi^+\pi^-\right)$ and $BR\left(D^\circ \to K^+K^-\right)$ relative to $BR\left(D^\circ \to K^\pm\pi^\mp\right)$ and the measurement of the invariant mass difference between $D_s^\pm \to \phi\pi^\pm, \phi \to K^+K^-$ and $D^\pm \to \phi\pi^\pm, \phi \to K^+K^-$. The preliminary results are $\frac{BR\left(D^\circ\to\pi^+\pi^-\right)}{BR\left(D^\circ\to K^\pm\pi^\mp\right)} = 3.37 \pm 0.20\,(stat) \pm 0.16\,(sys)$ %, $\frac{BR\left(D^\circ\to K^+K^-\right)}{BR\left(D^\circ\to K^\pm\pi^\mp\right)} = 11.17 \pm 0.48\,(stat) \pm 0.98\,(sys)$ % and $\Delta m\left(D_s - D^+\right) = 99.28 \pm 0.43\,(stat) \pm 0.27\,(sys)\ MeV/c^2$. All results are consistent with PDG expectations.

## 1. Introduction

A detailed description of the renewal of the CDF experiment commonly referred to as "CDF II" can be found elsewhere [1]. In this paper we will focus mainly on the physics results that the experiment has obtained so far exploiting the new Silicon Vertex Trigger based selections.
The Silicon Vertex Trigger is one of the major upgrades of CDF II: it consists of a real-time tracking device capable of reconstructing 2D silicon tracks with offline-quality resolution [1].
This allows the selection of low momentum heavy flavor enriched events: lifetime based selections like impact parameter cuts are in fact a well known tool for discriminating $b\bar{b}$ and $c\bar{c}$ physics events from the mostly prompt hadronic background typical of hadron colliders [2,3].
The results described here have to be considered preliminary benchmarks rather than full-fledged analyses: at the moment of writing this proceeding we already know that statistical and systematical errors will be better under control in the future. Nevertheless this is the first time that non-leptonic heavy flavor physics is collected efficiently at an hadron collider: this will be the key to competitivity in the beauty and charm sectors for CDF II.

*on behalf of the CDFII collaboration

## 2. Charm physics

The charm production cross section at the momentum range explorable with the hadronic B trigger (SVT trigger from now on) is as large as uncertain [4]. At this stage we know that the amount of promptly produced charm is quite sizeable (see figure 1). Also Cabibbo suppressed de-

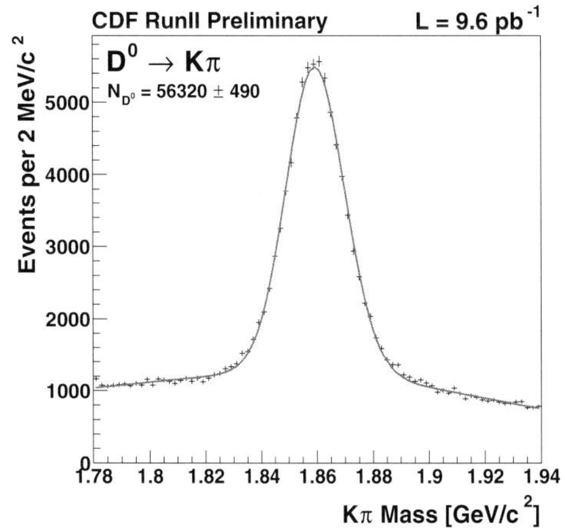


Figure 1. Offline reconstructed invariant mass distribution of the two trigger tracks. Note the huge collection rate of $D^\circ$ decays: this is so large that we can identify candidates in data samples collected in few hours.


doi:10.1016/S0920-5632(02)02192-8

cays to CP eigenstates (like $KK$ and $\pi\pi$) are collected at such huge rates with exactly the same trigger: this suggests the possibility of directly comparing branching fractions into the three possible different 2-body final states containing two charged tracks: $D^\circ \to K^\pm\pi^\mp$, $D^\circ \to K^+K^-$, $D^\circ \to \pi^+\pi^-$. Systematics are in fact minimized requiring a common set of selection and reconstruction cuts, while intrinsic differences coming from the slight difference in the physics of the final states can be taken into account properly through detector simulation, and then treated as an additional systematics uncertainty.
The interest in this measurement is three prong: (i) first of all it is a thermometer of the CDF II performances (ii) secondly, it is the first step towards interesting charm physics measurements such as the experimental determination of the $c\bar{c}$ parameter $y$ [5] and last but not least, (iii) it provides a very important control sample for the study of the detector performances (particle id in particular) for the reconstruction of two prong

| Systematic source | $\frac{D^\circ \to K^+K^-}{D^\circ \to K^\pm\pi^\mp}$ | $\frac{D^\circ \to \pi^+\pi^-}{D^\circ \to K^\pm\pi^\mp}$ |
|---|---|---|
| Background model | 7.9% | 1.4 % |
| Signal model | 1.3% | 1.3% |
| Trigger simulation | 1.9% | 2.5% |
| Beam $z$ profile | 0.7% | 0.5% |
| Material description | 0.5% | 0.5% |
| Lifetime difference | 2.2% | 2.2% |
| DCS Decays | 0.4% | 0.4% |
| Input $D^\circ$ spectra | 1.8% | 1.8% |
| $b/c$ ratio | 1.1% | - |
| Tracking | 0.4% | 1.7% |
| Total | 8.8% | 4.6% |

Table 1
Systematic uncertainty contribution for the measurement of the relative BR of $D^\circ \to K^+K^-$ and $D^\circ \to \pi^+\pi^-$ WRT $D^\circ \to K^\pm\pi^\mp$

charmless $B$ decays.
The relative branching fractions are measured fitting the invariant mass distributions of the candidates with suitable signal and background models. The measurement of the relative selection and reconstruction efficiency relies on detector and trigger simulation, with the uncertainty on the input spectra and on the bottom to charm relative contribution being treated as additional systematics.
In the final systematics evaluation also the effect of possible differences in the states lifetimes and the contribution from DCS decays are accounted. Even with the limited integrated luminosity used in this study ($\approx 10.5\,pb^{-1}$), CDF II is able to measure the relative branching fractions with an accuracy comparable to the best $c\bar{c}$ results available in literature [5]: for instance the CLEO II results [6] obtained with $10\,fb^{-1}$ of integrated luminosity at the $\Upsilon\,(4s)$ resonance. The preliminary results are:

$$\frac{BR(D^\circ \to \pi^+\pi^-)}{BR(D^\circ \to K^\pm\pi^\mp)} = 3.37 \pm 0.20\,(stat) \pm 0.16\,(sys)\%$$

$$\frac{BR(D^\circ \to K^+K^-)}{BR(D^\circ \to K^\pm\pi^\mp)} = 11.17 \pm 0.48\,(stat) \pm 0.98\,(sys)\%$$

Systematic uncertainty in our measurement is to date dominated by the background modeling systematics in the resonance fit for $D^\circ \to K^+K^-$ and by the trigger simulation systematics in $D^\circ \to \pi^+\pi^-$. The detailed breakdown of the systematics uncertainties is reported in table 1.

## 3. Charm and Charm-Strange signals

Analogously to what happens for the different decay modes of the $D^\circ$, also the systematics in the

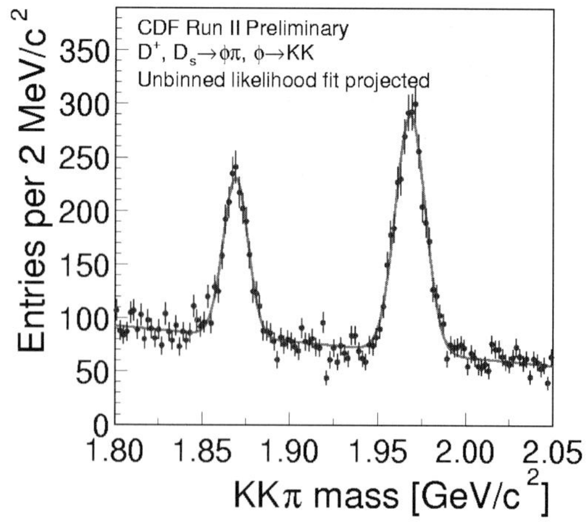


Figure 2. Offline reconstructed invariant mass distribution of the $\phi\pi$ candidates.

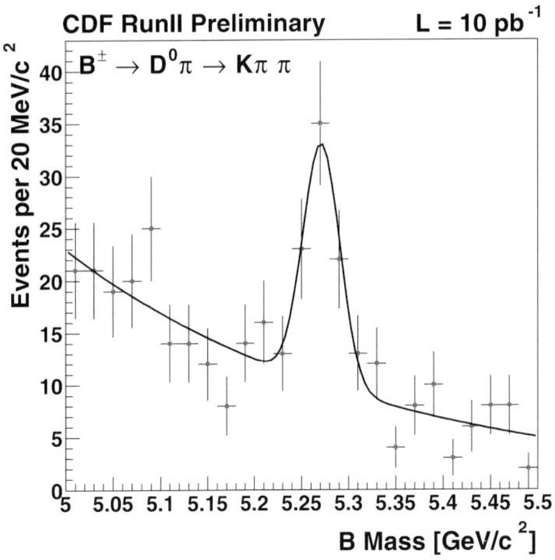


Figure 3. Invariant mass distribution for the $B^{\pm} \rightarrow D^{\circ}\pi^{\pm}$ decay, after appropriate selections.

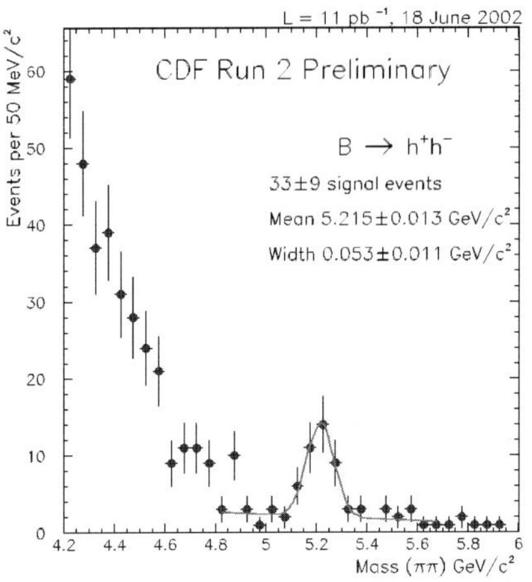


Figure 4. Invariant mass distribution for the $B \rightarrow h^+h^-$ decay, after appropriate selections.

extraction of the $D_s - D^+$ mass difference can be reduced selecting a configuration where the decay mode and the trigger selection are identical. This is the case of $D^{\pm}$, $D_s^{\pm} \rightarrow \phi\pi^{\pm}, \phi \rightarrow K^+K^-$. $D^+$ and $D_s$ are fitted in $11.6\,pb^{-1}$ of data, after correcting for charged tracks energy loss. This measurement gives a feeling of our ability in achieving precise mass measurements. The preliminary result is:

$$\Delta m = 99.28 \pm 0.43\,(stat) \pm 0.27\,(sys)\ MeV/c^2$$

The most important contributions to systematics are coming from selection/reconstruction cuts ($\chi^2$ for the $\phi\pi$ combined fit, transverse plane decay length, and helicity angle are the three main variables which could introduce biases related to the small kinematic differences between $D^+$ and $D_s$) and from the modeling of the resonances used to extract the mass differences. Tracking systematics are also important (uncertainty on the scale of the track parameters covariance matrix, detector material uncertainty, the magnetic field scale and tracking biases related to alignment all contribute significantly to the systematic error). A detailed breakdown of these contributions is reported in table 2

## 4. Conclusions

CDF II just begun data taking. Even with the small amount of luminosity integrated so far, our ability of performing a series of world class heavy flavor physics measurements is already evident. As a proof of principle, we report in figures 3 and 4 the signals $B^{\pm} \rightarrow D^{\circ}\pi^{\pm}$ and $B \rightarrow h^+h^-$ reconstructed from the hadronic SVT trigger.
As a preview of our ability, we reported in this paper the results of a preliminary measurement of $\frac{BR(D^{\circ}\rightarrow\pi^+\pi^-)}{BR(D^{\circ}\rightarrow K^{\pm}\pi^{\mp})}$, $\frac{BR(D^{\circ}\rightarrow K^+K^-)}{BR(D^{\circ}\rightarrow K^{\pm}\pi^{\mp})}$ and $\Delta m\,(D_s - D^+)$. In all cases the achieved resolution is close to the world averaged measurements and will be soon improved by better understanding of the detector and analysis systematics, together with larger statistics.

| Systematic source | Syst $[MeV/c^2]$ |
|---|---|
| $\chi^2$ | 0.06 |
| $L_{xy}$ | 0.09 |
| $\Theta_{hel}$ | 0.09 |
| Multiple cand. | 0.04 |
| kinematics | 0.004 |
| Tracking error scale | 0.03 |
| Alignment | 0.055 |
| Material | 0.015 |
| B field | 0.025 |
| fit | 0.22 |
| Total | 0.273 |

Table 2
Systematic uncertainty contribution for the measurement of the $D_s$-$D^+$ invariant mass difference.

## REFERENCES


1. CDF II TDR, FERMILAB-Pub-96/390-E
2. B physics at the Tevatron, FERMILAB Pub-01/197
3. Tevatron Measurements Related to CP violation, FERMILAB-CONF-00/272-E
4. See Ch. Paus, Heavy flavor production at CDF, this same proceeding
5. K. Hagiwara et al., Physical Review D66, 01000001 (2002)
6. Cleo Collab., Hep-ex/0111024

Proceedings of ICHEP 2002, pp. 535–538
*S. Bentvelsen, P. de Jong, J. Koch and E. Laenen (Editors)*

ELSEVIER

31st INTERNATIONAL CONFERENCE ON HIGH ENERGY PHYSICS AMSTERDAM

# b Fragmentation and Energy Correlation in $Z \to b\bar{b}$ Decays (LEP-1, SLD)

K. Harder[a]

[a]DESY, D-22603 Hamburg, Germany

Recent measurements of the fragmentation and hadronization of b quarks at LEP-1 and SLD are discussed. Hadronization models are compared with data, and a model-independent description of the B hadron energy spectrum is obtained. The results of all recent analyses are in good agreement. SLD investigates also the energy-dependent angular correlation of the B hadrons in events where two B hadrons are reconstructed.

The transition of isolated quarks to hadrons involves QCD interactions at an energy scale of the order of light hadron masses, which is close to $\Lambda_{QCD}$. It can therefore not be described using perturbative QCD techniques alone. In $e^+e^-$ annihilation reactions at high energies, the transition is usually described in terms of an initial phase of gluon radiation and $q\bar{q}$ creation at high energies, followed by the formation of hadrons. The former phase, the so-called fragmentation, can be described perturbatively. A successful description of the latter phase, hadronization, requires non-perturbative approaches.

Fragmentation and hadronization are predicted to be factorizable. Factorization is exploited in Monte Carlo event generators, where fragmentation and hadronization are usually treated separately. Fragmentation is modeled in terms of QCD matrix elements and/or parton showers. Hadronization usually involves either a cluster model (e.g. HERWIG [1]), or a string model (e.g. JETSET/PYTHIA [2]). The kinematics of hadrons created during hadronization is described with a few general parameters in the cluster model. The string model is more flexible: each new hadron can inherit an arbitrary fraction $z$ of the light cone momentum in the string. Specific models exist that predict probability density distributions $f(z)$ depending only on one or a few parameters. One of these parametrizations, the Lund symmetric function [3], is derived from basic symmetry considerations. The Bowler model [4] is a more consistent implementation of the same symmetry constraints. The UCLA model [5] is a modification of the default JETSET hadronization scheme with Lund symmetric parametrization, exploiting phase space considerations to reduce the number of parameters. Another category of hadronization models is based on phenomenological constraints. This involves the functions by Kartvelishvili et al. [6], Peterson et al. [7], and Collins-Spiller [8]. Attempts to describe heavy quark hadronization with a perturbative ansatz exist as well, e.g. in the BCFY model [9].

This article gives an overview of measurements of the B hadron energy spectrum by ALEPH [10], DELPHI [11] (preliminary results), OPAL [12] and SLD [13]. Another, though yet preliminary, analysis from SLD investigates the energy-dependent angular correlations of two B hadrons within the same event [14]. It will be discussed at the end of this article.

## 1. Hadronization of b quarks

Major prerequisite of any b fragmentation/hadronization analysis is a selection and reconstruction of B hadrons. ALEPH selects B mesons exclusively in $B \to D^{(*)}\ell\nu$ decays, where D and $D^*$ mesons are reconstructed in five channels. Electrons and muons are considered as lepton candidates. Missing energy in the event hemisphere containing the candidate tracks is used as estimator for the neutrino energy. DELPHI and OPAL identify charged and neutral decay products with artificial neural nets. Information like track impact parameters with respect to primary and secondary vertices, impact parameter uncertainties, transverse momentum with respect to


doi:10.1016/S0920-5632(02)02196-5

Table 1
Fit $\chi^2$ per degree of freedom for the hadronization model fits.

| | ALEPH | DELPHI (prelim.) | | | OPAL | SLD |
|---|---|---|---|---|---|---|
| | | $x_{wd}$ | $x_L$ | $z$ | | |
| Bowler | — | 35/8 | 43/8 | 1/2 | 67/44 | 17/15 |
| Lund | — | 42/8 | 53/8 | 2/2 | 75/44 | 17/15 |
| UCLA | — | — | — | — | — | 27/17 |
| Kartvelishvili et al. | 107/94 | — | — | 36/3 | 99/45 | 32/16 |
| Peterson et al. | 117/94 | 287/9 | 245/9 | 187/3 | 159/45 | 70/16 |
| BCFY | — | — | — | — | — | 105/16 |
| Collins/Spiller | 181/94 | — | — | 536/3 | 407/45 | 142/16 |

jet axis or B flight direction are used to estimate the probability that a track or cluster originates from a B hadron decay. This selection method allows an inclusive reconstruction of all B hadron species. The B hadron energy is either estimated directly by neural networks (DELPHI), or obtained from a weighted sum of all track and cluster energies, where the weights are transformed neural network outputs (OPAL). The DELPHI neural networks estimate also the energy of the primary B hadron (an excited B hadron created in the hadronization process and undergoing subsequent cascade decay to the weakly decaying B hadron) and $z$. SLD achieves an inclusive B hadron reconstruction using a significantly different technique. They use only charged B decay products. The difference between B flight direction, as obtained from primary and secondary vertex reconstruction, and the momentum sum of all tracks attached to the B decay vertex allows to put an upper limit on the missing B momentum. Monte Carlo studies show that the true missing momentum peaks sharply at its maximum allowed value, and therefore SLD approximates the true value with this maximum. The assumption of a B meson mass of the decaying particle is used as further constraint to obtain the B momentum.

Hadronization models are then compared to data by fitting their parameters to optimize the agreement between predicted and observed B hadron energy distributions. ALEPH, OPAL and SLD perform this fit at detector level, whereas DELPHI unfolds detector effects from $x_{wd}$, the

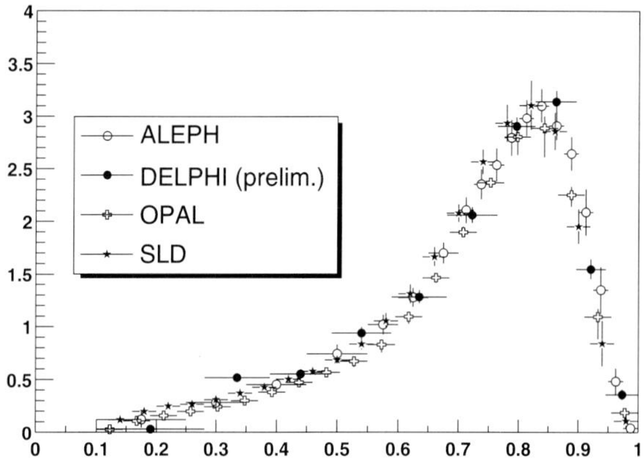


Figure 1. Comparison of the $x_{wd}$ spectra seen in the analyses discussed in this article.

scaled energy of the primary B hadron $x_L$ and $z$ and performs three independent fits on these three variables. The model parameter values themselves are strongly model-dependent; different values are found within the context of the different Monte Carlo setups of the LEP/SLD collaborations. However, the ranking of the models in terms of fit probability agrees between all experiments, as shown in Table 1. The string hadronization models are clearly favoured against the HERWIG cluster hadronization. Among the investigated hadronization parametrizations for the string model those seem to match the data best that are derived from intrinsic symmetries of this model.

To allow further models to be tested without having to repeat the full analysis, a model-independent description of the observed B hadron energy spectrum is given by all collaborations. This involves unfolding detector effects from the energy distribution. An overlay of the $x_{wd}$ spectra seen by all experiments is shown in Figure 1. The shape of the unfolded spectrum is then described in terms of moments of the energy distribution. The first non-trivial moment is the mean of the scaled energy distribution, $\langle x_{wd} \rangle$. The following $\langle x_{wd} \rangle$ values are obtained:

| | |
|---|---|
| ALEPH [10] | $0.7163 \pm 0.0061 \pm 0.0056$ |
| DELPHI [11] | $0.7153 \pm 0.0007^{+0.0049}_{-0.0052}$ |
| OPAL [12] | $0.7193 \pm 0.0016^{+0.0036}_{-0.0031}$ |
| SLD [13] | $0.709 \pm 0.003 \pm 0.003 \pm 0.002$ |

The first uncertainty is statistical, the second systematic. SLD separately quotes a model uncertainty. Good agreement is seen among the experiments. In comparison to earlier measurements, the $\langle x_{wd} \rangle$ value seems to have increased: the latest LEP average amounts to $0.702 \pm 0.008$ [15]. However, this is only due to a different method to evaluate $\langle x_{wd} \rangle$. Earlier analyses usually assumed a Peterson et al. shape of the hadronization parametrization. As it turns out now that the Peterson et al. function does not lead to a good description of the data, this assumption has been dropped, and most new analyses evaluate $\langle x_{wd} \rangle$ without any assumption on the shape. SLD averages over $\langle x_{wd} \rangle$ values obtained with a large number of different model assumptions.

The dominant systematic uncertainties in the calculation of $x_{wd}$ moments are mostly detector-related and thus not correlated among the analyses. Despite the relative smallness of the statistical error, it is therefore useful to combine the new LEP/SLD measurements. A preliminary fit to the results of the four analyses presented here was performed, resulting to the moments

$$\begin{aligned} \langle x_{wd} \rangle = D_2 &= 0.7151 \pm 0.0025 \\ D_3 &= 0.5426 \pm 0.0012 \\ D_4 &= 0.4268 \pm 0.0010 \\ D_5 &= 0.3440 \pm 0.0017 \end{aligned}$$

Correlations are not yet taken into account in this combination. The calculation of higher moments beyond $\langle x_{wd} \rangle$ is of vital importance for the transfer of these results to hadron colliders. As pointed out by Cacciari and Nason [16], a fit of hadron collider Monte Carlo fragmentation parameters to $x_{wd}$ moments leads to a more accurate description of the b production rates than a simple transfer of fragmentation function parameters measured in $e^+e^-$ annihilation.

Another convenient feature of $x_{wd}$ moments is that they provide simple means to separate perturbative and non-perturbative contributions. The factorization theorem predicts that the $x_{wd}$ moments $D_i$ can be expressed as product of purely perturbative moments $P_i$, which can be calculated, and a non-perturbative contribution $M_i$. DELPHI has demonstrated an application of this by dividing their measured $x_{wd}$ moments by a prediction of the perturbative part. Hadronization models can then be fitted directly to the series of non-perturbative moments. DELPHI finds that the results of this fit confirm the hadronization model fits to the shape of the energy spectrum.

## 2. Energy-dependent angular correlation of B hadrons

Perturbative QCD does not only predict the perturbative energy loss of quarks, expressed in the moments $P_i$ introduced above. It furthermore predicts a correlation between the energy of the two b quarks in one event and the angle between their momenta. Both quantities are affected by the amount of hard gluon radiation. Experimentally, this correlation is smeared out by the additional, non-perturbative hadronization energy loss of the two b quarks. However, as in the one-dimensional case, moments of these energy distributions should factorize into a perturbative and a non-perturbative contribution.

Simultaneous integration over both scaled B hadron energies $x_1$ and $x_2$ leads to two-dimensional moments which depend on the angle $\Phi$ between the flight directions of both hadrons:

$$D_{ij}(\Phi) = \int dx_1 \int dx_2 \; x_1^{i-1} x_2^{j-1} D(x_1, x_2, \Phi)$$

According to the factorization theorem, these

moments can be expressed as $D_{ij}(\Phi) = M_i M_j P_{ij}(\Phi)$, where the non-perturbative contributions $M_i$ and $M_j$ are the same as in the one-dimensional case, provided that the factorization into perturbative and non-perturbative contributions is done with the same scheme and same parameters.

The above relation between one-dimensional and two-dimensional moments is exploited in a preliminary analysis by SLD, who use their investigation of the hadronization of b quarks to extract information on the perturbative energy loss of b quarks. The moments $M_i$ are extracted from the measured one-dimensional moments $D_i$ assuming a specific LO prediction for $P_i$. They can then extract $P_{ij}(\Phi)$ from the energy and angular distribution of the B hadrons in events where two B hadrons are reconstructed. To reduce uncertainties due to residual non-perturbative effects further, these two-dimensional perturbative moments $P_{ij}(\Phi)$ are then normalized to $P_1 1(\Phi)$. The result is shown in Figure 2. Evidently the data agrees well with the perturbative prediction, which strongly supports the factorization approach. This method can furthermore be used to test specific perturbative calculations. Parameters like $\Lambda_{QCD}$ can be constrained and measured flavor-dependently, albeit much larger data samples, like those expected at a future linear $e^+e^-$ collider, would be necessary to obtain strong constraints.

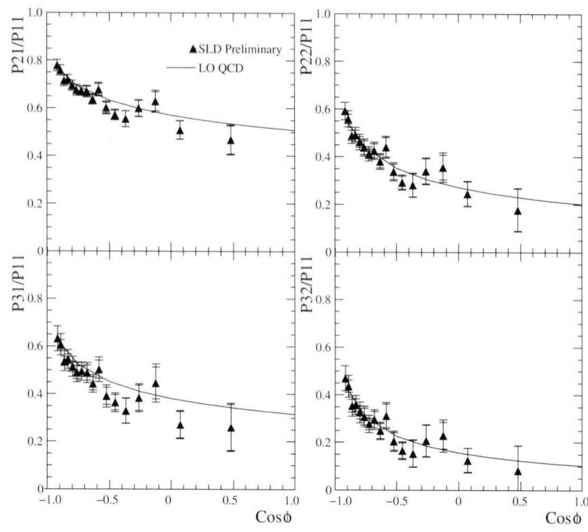


Figure 2. Normalized angle-dependent two-dimensional moments $P_{ij}(\Phi)/P_{11}(\Phi)$.

## 3. Acknowledgments

The preparation of this article would not have been possible without the efforts and contributions by T. Boccali (ALEPH), E. Ben Haim, U. Kerzel, L. Ramler, P. Roudeau (DELPHI), D. Dong, D. Muller, G. Nesom (SLD). The author wishes to express his gratefulness for the close and efficient collaboration.

## REFERENCES

1. G. Corcella et al., JHEP 0101 (2001) 010.
2. T. Sjöstrand, Comp. Phys. Comm. 82 (1994) 74.
3. B. Andersson et al., Z. Phys. C20 (1983) 317.
4. M.G. Bowler, Z. Phys. C11 (1981) 169.
5. C.D. Buchanan, S.B. Chun, Phys. Rev. Lett. 59 (1987) 1997.
6. V.G. Kartvelishvili et al., Phys. Lett. B78 (1978) 615.
7. C. Peterson et al., Phys. Rev. D27 (1983) 105.
8. P.D.B. Collins, T.P. Spiller, J. Phys. G11 (1985) 1289.
9. E. Braaten et al., Phys. Rev. D51 (1995) 4819.
10. ALEPH Collaboration, Phys. Lett. B512 (2001) 30.
11. DELPHI Collaboration, ICHEP 2002 Abstract 583.
12. OPAL Collaboration, CERN-EP-2002-051, submitted to Eur. Phys. J. C.
13. SLD Collaboration, Phys. Rev. D65 (2002) 092006.
14. SLD Collaboration, ICHEP 2002 Abstract 967.
15. ALEPH, CDF, DELPHI, L3, OPAL, SLD Collaborations, hep-ex/0009052.
16. M. Cacciari, P. Nason, Phys. Rev. Lett. 89 (2002) 122003.

ELSEVIER

Proceedings of ICHEP 2002, pp. 539–541
*S. Bentvelsen, P. de Jong, J. Koch and E. Laenen (Editors)*

31st INTERNATIONAL CONFERENCE ON HIGH ENERGY PHYSICS AMSTERDAM

# Determination of $\alpha_s$ and the charm and bottom quark masses from recent measurements of $R(s)$

J.H. Kühn[a] and M. Steinhauser[b]

[a] Institut für Theoretische Teilchenphysik, Universität Karlsruhe, D-76128 Karlsruhe, Germany

[b] II. Institut für Theoretische Physik, Universität Hamburg, D-22761 Hamburg, Germany

Recent experimental data for the total cross section $\sigma(e^+e^- \to \text{hadrons})$ are compared with the up-to-date theoretical prediction of perturbative QCD for those energies where perturbation theory is reliable. This allows for a determination of the strong coupling $\alpha_s^{(4)}(5\ \text{GeV}) = 0.235 \pm 0.047$. The precise data from the charm threshold region, when combined with the recent evaluation of moments with three loop accuracy, lead to a direct determination of the short distance $\overline{\text{MS}}$ charm quark mass $m_c(m_c) = 1.304(27)$. Applying the same approach to the bottom quark we obtain $m_b(m_b) = 4.191(51)$ GeV.

## 1. Introduction

During the past years new and more precise data for $\sigma(e^+e^- \to \text{hadrons})$ have become available in the low energy region between 2 and 10 GeV. At the same time increasingly precise calculations have been performed in the framework of perturbative QCD (pQCD), both for the cross section as a function of the center-of-mass energy $\sqrt{s}$, including quark mass effects, and for its moments which allow for a precise determination of the quark mass. A fresh look at the determination of $\alpha_s$ and the evaluation of the charm quark mass with the help of sum rules is thus an obvious task. We will concentrate on low moments as suggested by the ITEP group long ago [1]. This is a natural route to determine directly a short distance mass, say $m_c(m_c)$ or even $m_c(3\ \text{GeV})$ in the $\overline{\text{MS}}$ scheme [1,2]. In the present work we concentrate on the mass in the $\overline{\text{MS}}$ scheme, $m_c(\mu)$, and adopt $\mu = 3$ GeV as our default value, a scale characteristic for the present problem and sufficiently high to ensure convergence of the perturbative series. As already discussed in [1], fixed order pQCD is adequate, if only low moments are used in the analysis.

## 2. $\alpha_s$ from the continuum region

As stated in the Introduction, we distinguish two energy regions: first, the continuum region where pQCD and local duality are expected to give reliable predictions for the hadronic cross section and, second, the charm threshold region starting from the $D$ meson threshold at 3.73 GeV up to approximately 5 GeV, where the cross section exhibits rapid variations, plus the $J/\Psi$ and $\Psi'$ resonances. The former will be mainly sensitive to the value of $\alpha_s$. The latter will be used to evaluate moments and to determine the charm quark mass.

The most recent theoretical results for $R(s)$ based on NNLO pQCD, including quark mass effects, were employed in [3]. Assigning a fully correlated systematic error of 4.3% to the BES data [4] they find

$$\alpha_s^{(3)}(3\ \text{GeV}) = 0.369^{+0.047+0.123}_{-0.046-0.130}, \qquad (1)$$

from the BES data below 3.73 GeV and

$$\alpha_s^{(4)}(4.8\ \text{GeV}) = 0.183^{+0.059+0.053}_{-0.064-0.057}, \qquad (2)$$

from the point at 4.8 GeV. The uncertainties refer to the uncorrelated (statistical) and correlated errors, respectively. The same analysis is applicable to the MD-1 results. From their average value


doi:10.1016/S0920-5632(02)02202-8

$R((8.9\ \text{GeV})^2) = 3.578 \pm 0.021 \pm 0.140$ as quoted in their summary one obtains

$$\alpha_s^{(4)}(8.9\ \text{GeV}) = 0.193^{+0.017+0.127}_{-0.017-0.107}, \quad (3)$$

From the CLEO value $R((10.52\ \text{GeV})^2) = 3.56 \pm 0.01 \pm 0.07$ one obtains

$$\alpha_s^{(4)}(10.52\ \text{GeV}) = 0.186^{+0.008+0.061}_{-0.008-0.057}. \quad (4)$$

Combining the results of Eqs. (1), (2), (3) and (4) by assuming uncorrelated systematic errors and evolving them to a common scale of 5 GeV leads to

$$\alpha_s^{(4)}(5\ \text{GeV}) = 0.235^{+0.047}_{-0.047}. \quad (5)$$

Evolving this value up to $M_Z$

$$\alpha_s^{(5)}(M_Z) = 0.124^{+0.011}_{-0.014}, \quad (6)$$

good agreement with other determinations [5] is observed.

## 3. Charm quark mass determination from the threshold region

The optimal method for the determination of the charm quark mass is based on the direct comparison of theoretical and experimental moments of the charm quark contribution to the photon polarization function. In the limit of small momentum the latter can be cast into the form [6]

$$\Pi_c(q^2) = Q_c^2 \frac{3}{16\pi^2} \sum_{n\geq 0} \bar{C}_n z^n, \quad (7)$$

with $z = q^2/(4m_c^2)$ where $m_c = m_c(\mu)$ is the $\overline{\text{MS}}$ charm quark mass at the scale $\mu$. The perturbative series for the coefficients $\bar{C}_n$ up to $n = 8$ is known analytically [7,6] up to order $\alpha_s^2$. We define the moments through

$$\mathcal{M}_n^{\text{exp}} = \int \frac{\text{d}s}{s^{n+1}} R_c(s) = \mathcal{M}_n^{\text{th}} = \frac{9}{4} Q_c^2 \left(\frac{1}{4m_c^2}\right)^n \bar{C}_n. \quad (8)$$

The charm quark mass can be obtained from

$$m_c(\mu) = \frac{1}{2} \left(\frac{\bar{C}_n}{\mathcal{M}_n^{\text{exp}}}\right)^{1/(2n)}. \quad (9)$$

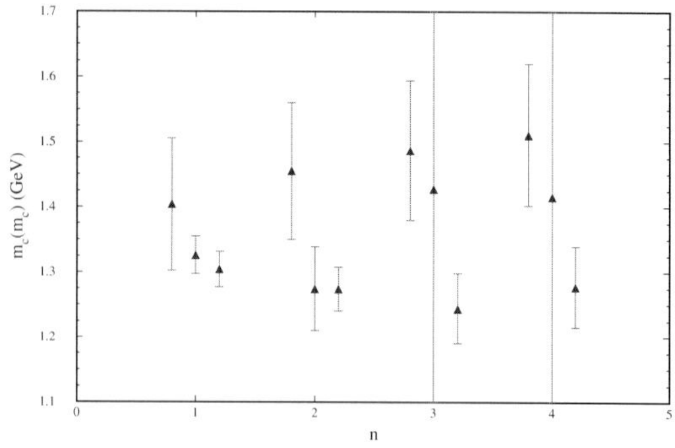


Figure 1. $m_c(m_c)$ for $n = 1, 2, 3$ and 4. For each value of $n$ the results from left to right correspond the inclusion of terms of order $\alpha_s^0$, $\alpha_s^1$ and $\alpha_s^2$ in the coefficients $\bar{C}_n$. Note, that for $n = 3$ and $n = 4$ the errors can not be determined with the help of Eq. (9) in those cases where only the two-loop corrections of order $\alpha_s$ are included into the coefficients $\bar{C}_n$ as the equation cannot be solved for $m_c(3\ \text{GeV})$.

The analysis of the experimental moments with $n = 1, \ldots, 4$ leads to the results displayed in Fig. 1.

The moment with $n = 1$ is evidently least sensitive to non-perturbative contributions from condensates, to the Coulombic higher order effects, the variation of $\mu$ and the parametric $\alpha_s$ dependence. Hence

$$m_c(m_c) = 1.304(27)\ \text{GeV}. \quad (10)$$

is adopted as the final result. Using the two- and three-loop relation [8–10] between the pole- and the $\overline{\text{MS}}$-mass this corresponds to

$$M_c^{(2-\text{loop})} = 1.514(34)\ \text{GeV}, \qquad M_c^{(3-\text{loop})} = 1.691(35)\ \text{GeV}. \quad (11)$$

## 4. The bottom quark mass

The same approach is also applicable to the determination of $m_b$. Again a significant improvement of the stability of the prediction by inclusion of the NNLO terms is observed. As final result one finds

$$m_b(m_b) = 4.191(51)\ \text{GeV}, \quad (12)$$

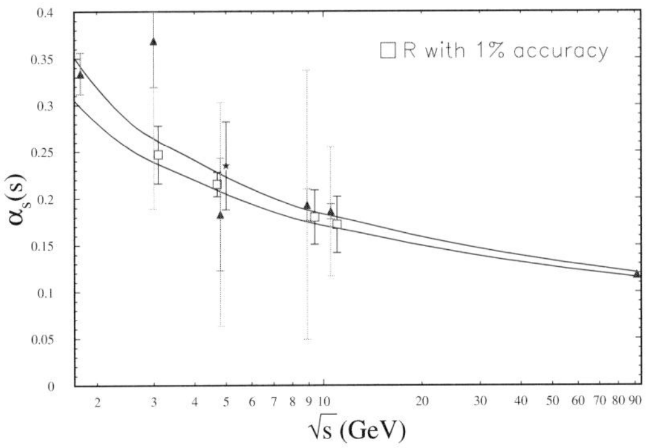


Figure 2. $\alpha_s$ as a function of $\sqrt{s}$. Results from $\tau$ and $Z$ decays and those extracted in Section 2 are shown. The two error bars on the data points indicate the statistical (inner) and systematical (outer) uncertainty. For the combined result (indicated by a star) at $\sqrt{s} = 5.0$ GeV only the error after adding the statistical and systematical uncertainty in quadrature is shown. For illustration $\delta\alpha_s$ reduced to the error as expected from $\delta R/R = 1\%$ are also shown (slightly displaced in order to make the presentation more visible).

well consistent with evaluations based on the analysis of bottonium and high spectral moments [11–16]. The $\overline{\text{MS}}$ result of Eq. (12) corresponds to a pole mass of [8–10]

$$M_b^{(2-\text{loop})} = 4.651(57)\text{ GeV},$$
$$M_b^{(3-\text{loop})} = 4.819(57)\text{ GeV}, \qquad (13)$$

using $\mathcal{O}(\alpha_s^2)$ and $\mathcal{O}(\alpha_s^3)$ accuracy, respectively.

## 5. Perspectives [17]

Precise measurements of the $R$-ratio in the charm- and bottom-threshold region are well within reach at CLEO. The values of $\alpha_s$ as derived from present measurements in Section 2 and the impact of a reduction of the relative error down to 1% are demonstrated in Fig. 2 [17].

The quark mass determinations are at present strongly affected by the relatively large error in the electronic widths of the narrow charmonium and bottomium resonances and by the poorly measured threshold region. More precise measurements might well lead to quark mass determinations with $\delta m_Q \approx 15$ MeV and 30 MeV for charm and bottom, respectively.

## REFERENCES


1. V. A. Novikov *et al.*, Phys. Rep. C **41** (1978) 1.
2. M. A. Shifman, A. I. Vainshtein, and V. I. Zakharov, Nucl. Phys. B **147** (1979) 385; Nucl. Phys. B **147** (1979) 448.
3. J.H. Kühn and M. Steinhauser, Nucl. Phys. B **619** (2001) 588, (E) ibid. B **640** (2002) 415.
4. Z. G. Zhao, private communication.
5. D. E. Groom et al., Eur. Phys. J. C **15** (2000) 1.
6. K. G. Chetyrkin, J. H. Kühn, and M. Steinhauser, Nucl. Phys. B **505** (1997) 40.
7. K. G. Chetyrkin, J. H. Kühn, and M. Steinhauser, Phys. Lett. B **371** (1996) 93; Nucl. Phys. B **482** (1996) 213.
8. N. Gray, D. J. Broadhurst, W. Grafe, and K. Schilcher, Z. Phys. C **48** (1990) 673.
9. K. G. Chetyrkin and M. Steinhauser, Phys. Rev. Lett. **83** (1999) 4001;
K. G. Chetyrkin and M. Steinhauser, Nucl. Phys. B **573** (2000) 617.
10. K. Melnikov and T. v. Ritbergen, Phys. Lett. B **482** (2000) 99.
11. J. H. Kühn, A. A. Penin, and A. A. Pivovarov, Nucl. Phys. B **534**, 356 (1998);
A. A. Penin and A. A. Pivovarov, Phys. Lett. B **435** (1998) 413; Nucl. Phys. B **549** (1999) 217.
12. K. Melnikov and A. Yelkhovsky, Phys. Rev. D **59** (1999) 114009.
13. M. Beneke and A. Signer, Phys. Lett. B **471** (1999) 233.
14. A. H. Hoang, hep-ph/0008102.
15. A. Pineda, JHEP **0106** (2001) 022.
16. A.A. Penin and M. Steinhauser, Phys. Lett. B **538** (2002) 335.
17. J.H. Kühn and M. Steinhauser, Report Nos.: TTP02–26, hep-ph/0209357.

ELSEVIER

Proceedings of ICHEP 2002, pp. 542–545
*S. Bentvelsen, P. de Jong, J. Koch and E. Laenen (Editors)*

31st INTERNATIONAL CONFERENCE ON HIGH ENERGY PHYSICS AMSTERDAM

# Inclusive $B$ Decays - Spectra, Moments and CKM Matrix Elements.

Daniel Cronin-Hennessy[a]

[a]University of Rochester, Laboratory for Elementary Particle Physics,
Ithaca, NY 14853-8001
hennessy@lns.cornell.edu

I present improved measurements of $|V_{cb}|$ and $|V_{ub}|$ from four recent CLEO analyses.

## 1. Introduction

The heavy quark limit of QCD[1] is potentially a very useful tool for relating measured inclusive properties in $B$ meson decay, such as semileptonic branching fractions, to fundamental CKM parameters like $|V_{cb}|$ and $|V_{ub}|$. The expressions for inclusive observables are expansions in inverse powers of the $B$ meson mass $M_B$ [2–4]. At order $1/M_B$, the non-perturbative parameter $\overline{\Lambda}$ enters, and at order $1/M_B^2$ two more parameters, $\lambda_1$ and $\lambda_2$ appear. Intuitively, these parameters may be thought of as the energy of the light quark and gluon degrees of freedom ($\overline{\Lambda}$), the average momentum-squared of the $b$ quark ($-\lambda_1$), and the energy of the hyperfine interaction of the spin of the $b$ quark with the light degrees of freedom ($\lambda_2/M_B$). The parameter $\lambda_2$ can be extracted directly from the $B^* - B$ mass splitting[3]. Measurements of inclusive observables have been used to extract values for $\overline{\Lambda}$ and $\lambda_1$ [5]. If these non-perturbative parameters are extracted with high precision they can be used in the HQET expression for the semi-leptonic width to obtain a precise measurement of $|V_{cb}|$.

There are several sources of uncertainty in the extraction of HQET parameters $\overline{\Lambda}$ and $\lambda_1$ from inclusive observables. One is the validity of underlying assumption of quark-hadron duality. A second is uncertainty that arises from the use of expressions calculated to finite order in the expansion. A third is the uncertainty in additional non-perturbative parameters introduced at order $1/M^3$. The size of the third order parameters are estimated to be less than approximately (0.5 GeV)$^3$ from dimensional considerations. We take this size as an estimate of their uncertainty and evaluate the uncertainty on the third order terms. Given that there is no clear prescription for estimating the uncertainties described above it is neccessary to extract the HQET parameters using as many observables as can be measured. The consistency of the results can then be used to ascertain the reliability in extracting a precise value of $|V_{cb}|$.

## 2. Extractions of $|V_{cb}|$

A recent CLEO analysis has measured the moments of the lepton energy distribution from $B \to X_c l\nu$[6] . The measured lepton energy distribution for electrons and muons are shown in figure 1. The moments are defined for $P_{lepton} > 1.5GeV/c$ as

$$\mathrm{R}_0 = \frac{\int_{1.7}(d\Gamma_{sl}/dE_l)dE_l}{\int_{1.5}(d\Gamma_{sl}/dE_l)dE_l}, \text{ and} \tag{1}$$

$$\mathrm{R}_1 = \frac{\int_{1.5}E_l(d\Gamma_{sl}/dE_l)dE_l}{\int_{1.5}(d\Gamma_{sl}/dE_l)dE_l}. \tag{2}$$

Using these equations and the combined electron and muon data we measure preliminary values for $R_0 = 0.6187 \pm 0.0014 \pm 0.0016$ and $R_1 = 1.7810 \pm 0.0007 \pm 0.0009$ where the uncertainties are statistical and systematic respectively. Including CLEO's previous measurements we now have a total of six moments to use in extraction of HQET parameters. Using the HQET expressions for these moments ($\overline{MS}$ scheme) we plot in figure 2 the allowed curves in the $\overline{\Lambda}$-$\lambda_1$space. The


doi:10.1016/S0920-5632(02)02205-3

preliminary lepton energy moments are shown in red and black. The ellipse represents the allowed $\overline{\Lambda}$-$\lambda_1$ pairs as presented in our publication of the first moment of the hadronic mass squared and the first moment of the photon energy spectrum[5]. All known experimental and theoretical uncertainties are included. The moments from the various measurements show excellent agreement. In figure 3 we show the curves in the $\overline{\Lambda}$-$m_b$ space for an alternative expansion scheme [7]. Extracting $|V_{cb}|$ using the first photon energy moment and the first hadronic mass squared moment yields $|V_{cb}| = 4.05 \pm 0.09 \pm 0.04 \pm 0.10$, which agrees well with the extracted value of $|V_{cb}| = 4.04 \pm 0.09 \pm 0.04 \pm 0.10$ in the $\overline{MS}$ scheme. The first uncertainty is experimental uncertainty from the semileptonic width, the second is the experimental uncertainty from the moments measurements and the third uncertainty is from theory. Details on the inputs to the extraction can be found in reference[5].

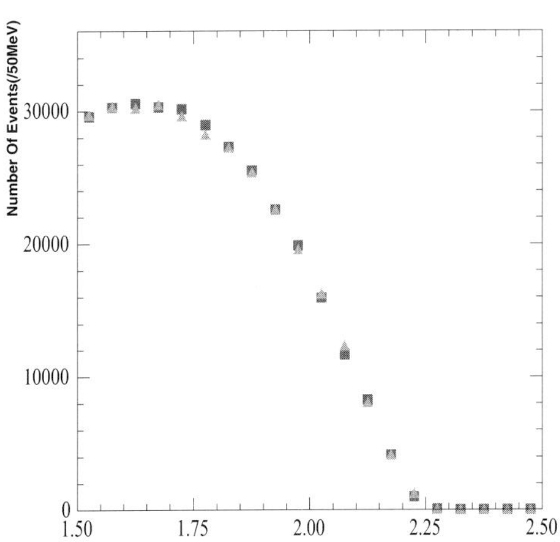


Figure 1. Corrected electron (green triangles) and muon (red squares) momentum spectra in the $B$-meson rest frame.

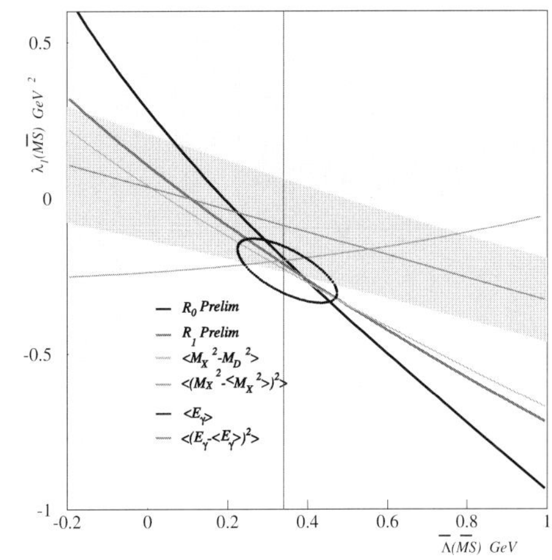


Figure 2. Allowed curves in $\overline{\Lambda}$-$\lambda_1$ space using moments measurements from the lepton energy and hadronic mass in $b \to cl\nu$ and photon energy in $b \to s\gamma$. Ellipse is from $1^{st}$ $M^2$ moment and $1^{st}$ $E_\gamma$ moment. The band shows the total uncertainty from $2^{nd}$ $E_\gamma$ moment.

## 3. Extractions of $|V_{ub}|$

The method of substituting experimental measurements for poorly known operator expectation values has reduced the uncertainties on the extraction of $|V_{cb}|$. Similar advances are starting to show promise for the extraction of $|V_{ub}|$. CLEO presents three new measurements used to extract values for $|V_{ub}|$. The first [8] uses the familiar lepton endpoint approach. This analysis measures the $b \to ul\nu$ yield in the region kinematically inaccessible to $b \to cl\nu$ production which dominates at lower lepton energy. Previously, large model dependence was introduced in extrapolating the $b \to ul\nu$ yield below the endpoint. The new measurement employs a technique to use the shape of the photon energy spectrum in $b \to s\gamma$ to derive the shape of the lepton energy spectrum in $b \to ul\nu$ [9]. Once the shape is derived the fraction of events above any energy cut can be found by integration. Recently, there has been

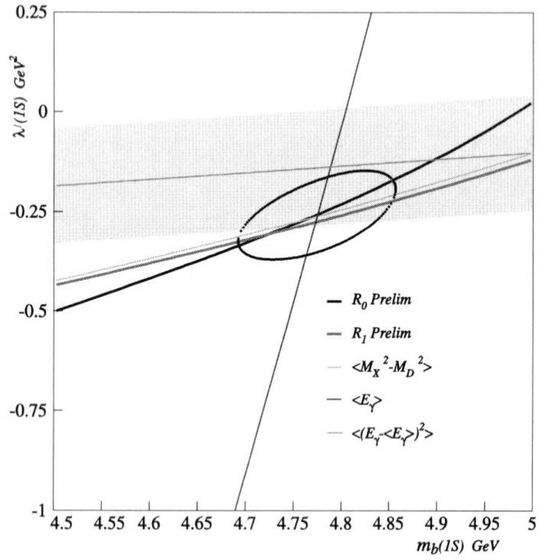


Figure 3. Allowed curves in $\overline{\Lambda}$-$\lambda_1$ space using moments measurements from the lepton energy and hadronic mass in $b \to cl\nu$ and photon energy in $b \to s\gamma$. Ellipse is from $1^{st}$ $M^2$ moment and $1^{st}$ $E_\gamma$ moment. The band shows the total uncertainty from $2^{nd}$ $E_\gamma$ moment.

criticism on the size of sub-leading correction to the convolution function used in the derivation of the lepton spectrum [10]. We have used reference [11] to include part of the sub-leading corrections which are shown in figure 4. At low values of the endpoint experimental uncertainties grow quickly while at high values the sub-leading corrections may be large. The results emphasize the need for careful choice of the endpoint cut.

A second inclusive analysis [12] performs a 3-D maximum likelihood fit to the variables $q^2$, $Cos(\theta_{Wl})$ and $M_X^2$ to discriminate the $b \to ul\nu$ from $b \to cl\nu$. The approach has the merit of including a much larger region of phase space than the endpoint analysis. Figure 5 shows the projection of the fit onto the $M_X^2$ axis (with $q^2 > 11.0 GeV^2$) with the $b \to u$ component (blue) prominent in the low mass region. Recent theoretical work [13] relates the partial branching fraction to $|V_{ub}|$. From our fit we find $|V_{ub}| = 4.05 \pm 0.18_{stat} \pm 0.58_{syst} \pm 0.25_{b \to cmodel} \pm 0.21_{b \to umodel} \pm 0.56_{theory}$. Currently the correlation between the endpoint and 3-D fit approaches have not been completely determined so that the reader is cautioned against averaging the results.

The final analysis that I present is an exclusive approach to extracting $|V_{ub}|$. CLEO has measured the branching ratio for $B \to \pi l\nu$[14]. Our preliminary measurement is $B(B^0 \to \pi^- l^+ \nu) = (1.376 \pm 0.180^{+0.116}_{-0.135} \pm 0.008 \pm 0.102 \pm 0.021) \cdot 10^{-4}$ where the errors are statistical, systematic, $\pi l\nu$ and $\rho l\nu$ form factors, and modeling other $B \to X_u l\nu$ decays. Figure 6 shows the branching fractions in three $q^2$ bins for the data and for three models. ISGW2 shows poor agreement with the data. The other models are used in the determination of the model dependent uncertainty. This is the first exclusive $b \to u$ analysis that uses the $q^2$ dependence to limit the choice of models used in the determination of the systematic uncertainties in the branching ratio measurement. The extracted value of $|V_{ub}| = (3.32 \pm 0.21^{+0.17+0.55}_{-0.19-0.39} \pm 0.12 \pm 0.07) \cdot 10^{-4}$

## 4. Summary

I have presented moments measurements from the lepton energy spectrum in $b \to cl\nu$ and have shown consistency among the six CLEO moments in the extraction of HQET parameters $\overline{\Lambda}$ and $\lambda_1$. Such demonstrations are necessary if we are to gain confidence in the reliability of the HQET framework which can allow precision measurements of $|V_{cb}|$ and $|V_{ub}|$. I have also presented three analysis which yield new values for the CKM matrix element $|V_{ub}|$. The extracted values of $|V_{ub}|$ are consistent among the measurements.

## REFERENCES

1. J. Chay, H. Georgi, and B. Grinstein, Phys. Lett. **B247**, 399 (1990); I. I. Bigi, N. G. Uraltsev, and A. I. Vainshtein, Phys. Lett. **B293**, 430 (1992); I. I. Bigi, M. A. Shifman, N. G. Uraltsev, and A. I. Vainshtein, Phys. Rev. Lett. **71**, 496 (1993); A. V. Manohar and M. B. Wise, Phys. Rev.D **49**, 1310 (1994); A. Falk, M. Luke, and M. Savage, Phys. Rev.

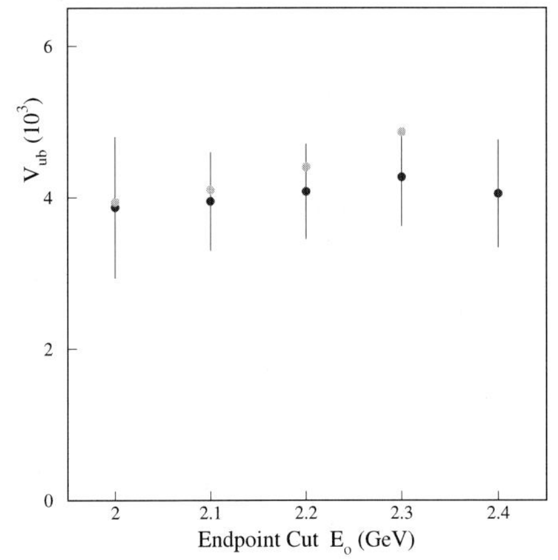


Figure 4. Blue points with error bars show $V_{ub}$ versus endpoint energy cut. Green points show new value after including sub-leading corrections

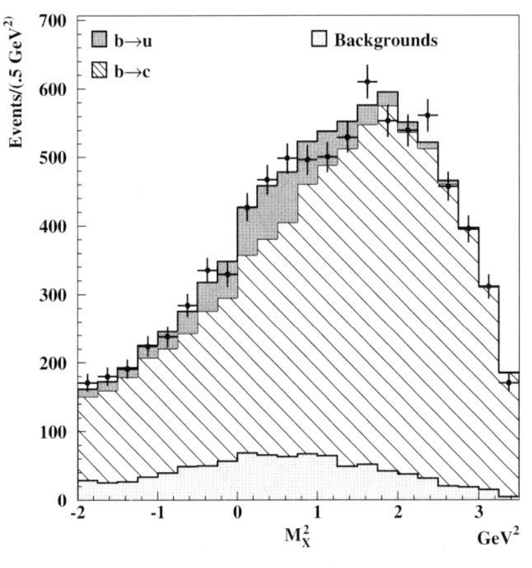


Figure 5. Projection of the 3-D maximum likelihood fit to the $B \to Xl\nu$ data onto the $M_X^2$ axis (with $q^2 > 11.0GeV^2$).

D **49**, 3367 (1994); A. Falk, hep-ph/9610363, (1996).

2. A. Falk, M. Luke, and M. Savage, Phys Rev. D **53**, 2491 (1996); *ibid*, **53**, 6316 (1996).
3. M. Gremm and A. Kapustin, Phys. Rev. D **55**, 6924 (1997).

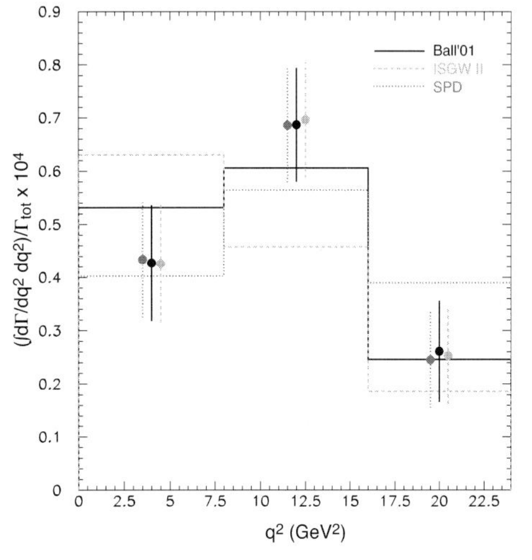


Figure 6. Branching fraction for $B^0 \to \pi^- l^+ \nu$ versus $q^2$.

4. M. Voloshin, Phys. Rev. D **51**, 4934 (1995).
5. D. Cronin-Hennessy *et al.*, Phys. Rev. Lett. **87**, 251808 (2001).
6. R. A. Briere *et al.*, ICHEP02 ABS932, (2002)
7. C.W. Bauer, Z. Ligeti, M. Luke, A. Manohar, hep-ph/0210027
8. A. Bornheim *et al.*, Phys. Rev. Lett. **88**, 231803 (2002).
9. M. Neubert, PR D49,3392(1994), hep-ph/9311325; A.L. Kagan and M. Neubert, Eur. Phys. J. C 7, 5(1999),hep-ph/9805303; F. DeFazio and M. Neubert, JHEP 06, 017 (1999), hep-ph/9905351 A.K. Leibovich, L. Low, and I.Z. Rothstein, PR D61, 053006(2000), hep-ph/9909404
10. C. W. Bauer, M. Luke, T. Mannel hep-ph/0205150; A. Leibovich, Z. Ligeti, and M.B. Wise, hep-ph/0205148
11. M. Neubert, hep-ph/0207002
12. N. E. Adam *et al.*, ICHEP02 ABS933, (2002)
13. C. W. Bauer, Z. Ligeti, and M. E. Luke, Phys. Rev D **64**, 113004 (2001)
14. A. Bornheim *et al.*, ICHEP02 ABS931, (2002)

ELSEVIER

Proceedings of ICHEP 2002, pp. 546–550
*S. Bentvelsen, P. de Jong, J. Koch and E. Laenen (Editors)*

31st INTERNATIONAL CONFERENCE ON HIGH ENERGY PHYSICS AMSTERDAM

# Studies of Inclusive Semileptonic $B$ Meson Decays with the BABAR Detector

Vera G. Lüth [a] representing the *BABAR* Collaboration

[a]Stanford Linear Accelerator Center
PO Box 20450, Stanford, CA 94309, USA

In this report the measurement of the inclusive branching fraction $\mathcal{B}(\overline{B} \to X\ell^-\overline{\nu})$ and the preliminary measurement of the first moment of the hadron mass $M_X$ in such decays are presented and their implications for the determination of $|V_{cb}|$ are discussed.

## 1. Introduction

For many years, semileptonic decays have been the topic of a variety of studies because they are theoretically simple at the parton level, sensitive to the coupling of quarks to the weak charged current, and allow us to probe the impact of strong interactions on the bound quark. Experimentally they are readily accessible because of the large branching fraction and clear signature in form of a high momentum lepton. The principal goal of studies of semileptonic $B$ meson decays is the determination of the CKM matrix elements $|V_{cb}|$ and $|V_{ub}|$.

The decay width for inclusive semileptonic $B$ decays to the charm states $X_c$ is

$$\Gamma^c_{SL} \equiv \mathcal{B}(\overline{B} \to X_c\ell^-\overline{\nu})/\tau_B = \gamma_c|V_{cb}|^2, \quad (1)$$

i.e., $|V_{cb}|$ can be determined from the branching fraction and the average $B$ lifetime, provided the factor $\gamma_c$ is known. The theoretical QCD parameter $\gamma_c$ requires both perturbative and nonperturbative input. In the framework of Heavy Quark Effective Theory the uncertainties in the estimate of $\gamma_c$ can be reduced by using information from inclusive measurements, for instance, the moments of the mass of hadrons $X_c$ and/or the moments of the lepton energy spectrum. Like the total decay rate, these inclusive observables can be calculated using expansions in powers of the strong coupling constant $\alpha_s(m_b)$ and in inverse powers of the $B$ meson mass, $m_B$, that include non-perturbative parameters. At order $1/m_B^2$, there are three parameters, $\bar{\Lambda}$, $\lambda_1$, and $\lambda_2$. From the $B^* - B$ mass splitting, we have $\lambda_2 = 0.128 \pm 0.010\,\mathrm{GeV}^2$.

In the following, we report two measurements performed with the *BABAR* detector [1] operating at the $\Upsilon(4S)$ resonance at the PEP-II energy asymmetric $e^+e^-$ storage ring [2] at SLAC, the semileptonic branching fraction [3] and the first moment of the hadron mass distribution as a function of the minimum lepton momentum [4].

## 2. Inclusive Semileptonic Branching Ratio

This measurement employs a method introduced by ARGUS [5] and later also used by CLEO [6], in which $B\overline{B}$ events are tagged by the presence of a high momentum lepton. We choose electrons with a momentum $p^*$ in the interval 1.4 to 2.3 GeV as a tag ($p^*$ is measured in the rest frame of the $\Upsilon(4S)$ resonance). A second electron in the event is taken as the signal lepton for which we require $p^* > 0.6\,\mathrm{GeV}/c$, so as to avoid very significant background from photon conversion and misidentified hadrons at lower momenta. Signal electrons are mostly from primary $B$ decays if they are accompanied by a tag electron of opposite charge (unlike-sign), those with a tag of the same charge (like-sign) originate predominantly from secondary decays of charm particles produced in the decay of the other $B$ meson.

The analysis is based on a data set corresponding to an integrated luminosity of $4.1\,\mathrm{fb}^{-1}$


doi:10.1016/S0920-5632(02)02210-7

collected at the $\Upsilon(4S)$ resonance, and $0.97\,\mathrm{fb}^{-1}$ recorded about 40 MeV below the $\Upsilon(4S)$ peak.

Electrons are identified using a likelihood-based algorithm, combining the track momentum with the energy, position, and shape of the shower measured in the calorimeter, the number and angle of Cherenkov photons, and the specific energy loss in the drift chamber. The electron identification efficiency has been measured with radiative Bhabha events as a function of momentum and polar angle yielding an average of $(92 \pm 2)\%$. These measurements are corrected for the fact that in $B\overline{B}$ events the efficiencies are lower by $(4 \pm 2)\%$ at low momenta and by $(2 \pm 1)\%$ above $p^* = 1.6$ GeV/$c$. The misidentification rates for pions, kaons, and protons are extracted from selected track samples. Per hadronic track these rate vary between 0.05% and 0.1%.

Three samples of electrons are selected: (1) the tag electrons, (2) unlike-sign and (3) like-sign pairs of a tag and signal electron candidate. Misidentified hadrons, non-$B\overline{B}$ (continuum) events, photon conversions, $\pi^0, \eta \to \gamma e^+e^-$ ("Dalitz") and $J/\psi\,\psi(2S) \to e^+e^-$decays contribute background to all three samples. The continuum background is derived from off-resonance spectra. Electrons from photon conversions and Dalitz decays $\pi^0, \eta \to \gamma e^+e^-$ are identified in the data by pairing them with oppositely charged tracks. The number of observed electrons from such pairs is corrected for the momentum and polar angle dependent pair finding efficiency.

A significant contamination to the unlike-sign sample arises from events in which the signal electron originates from a decay of the charm particle produced from the same $B$ meson as the tag. We reduce this background by imposing conditions on the opening angle $\alpha$ of $e^+e^-$ pairs, namely $\cos\alpha + p_e^* > 1.0$ and $\cos\alpha > -0.2$, where $p_e^*$ is measured in GeV/$c$. This selection reduces the background by a factor of 24, the loss in signal efficiency can be calculated on the basis of geometric acceptance. The selection also eliminates most of the $e^+e^-$ pairs from inclusive $B \to J/\psi X$ decays. The remaining electrons from sources other than primary $B$ decays are estimated from Monte Carlo simulations with uncertainties dominated by the current errors on the measured branching fractions.

Because of $B^0\overline{B}^0$ flavor oscillations, electrons from primary $B$ decays and $B \to \overline{D}X, \overline{D} \to e^-\nu_e Y$ secondary electrons contribute to both unlike- and like-sign pairs. These contributions can be separated bin-by-bin in momentum, taking into account the fraction of $B^0$ production at the $\Upsilon(4S)$ of $f_0 = 0.500 \pm 0.025$ and flavor mixing $\chi_0 = 0.174 \pm 0.009$ [7].

We arrive at $304{,}048 \pm 880_{stat} \pm 2{,}100_{syst}$ tag electrons, and $25,700 \pm 410_{stat}$ primary signal electrons in the momentum range from 0.6 to 2.3 GeV/$c$. To derive the branching fraction, the polar angle acceptance (84%), relative selection selection efficiencies ($98.0 \pm 0.50\%$), and losses due to bremsstrahlung $(2.20 \pm 0.35)\%$ are taken into account. Figure 1 shows the fully corrected momentum spectrum. Based on a fit of the measured spectrum to the sum of the spectra from the various exclusive decays we estimate the fraction of the spectrum below the momentum cut-off at 0.6 GeV/$c$ to be $6.1 \pm 0.9\%$.

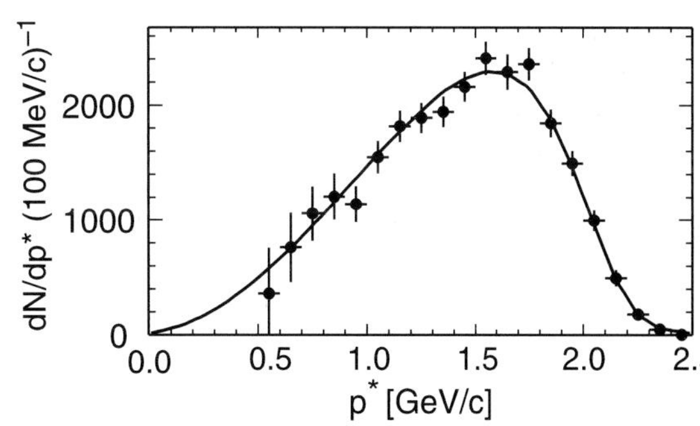


Figure 1. Momentum spectrum of electrons from $B \to X\ e\ \nu$ decays, corrected for efficiencies and for external bremsstrahlung. The errors represent the combined statistical and systematic uncertainties. The curve indicates a fit to the spectrum.

Thus the total semileptonic branching fraction

is measured to be

$$\mathcal{B}(B \to Xe\nu) = (10.87 \pm 0.18(stat) \pm 0.30(syst))\%. \quad (2)$$

Based on the calculations by Hoang *et al.* [12], we relate the decay rate to the CKM matrix element,

$$|V_{cb}| = 0.0419 \left( \frac{\mathcal{B}(B \to X_c\, e\, \nu)}{0.105} \frac{1.6\,\text{ps}}{\tau_B} \right)^{1/2}$$
$$(1.0 \pm 0.019_{(pert)} \pm 0.017_{(\lambda_1)} \pm 0.012_{(m_b)}).$$

Using the current world averages [7] of $\tau_B = (1.601 \pm 0.022)$ ps and $\mathcal{B}(B \to X_u\ e\ \nu) = (1.7 \pm 0.6) \times 10^{-3}$, we obtain $|V_{cb}| = 0.0423 \pm 0.0007(exp) \pm 0.0020(theory)$, where the individual contributions to the theoretical uncertainty have been added linearly.

## 3. Hadronic Mass Moments

We measure the first moment of the invariant mass $M_X$ distribution of the hadronic system recoiling against the charged lepton and the neutrino in semileptonic $B$ decays, $\langle M_X^2 - \overline{m}_D^2 \rangle$, where $\overline{m}_D = (m_D + 3m_{D^*})/4 = 1.975\ \text{GeV}/c^2$ is the spin-averaged $D$ meson mass.

The analysis is based on a sample of 55 million $B\overline{B}$ events, from which we select a subsample of 5,800 events (above a background of 3,600 events that are statistically subtracted using the energy-constrained $B$ mass distribution). In these events one $B$ meson is fully reconstructed in a hadronic decay mode ($B_{reco}$) and the semileptonic decay of the second $B$ ($B_{recoil}$) is identified by a high momentum electron or muon. The hadron system in the decay $B \to X\ell\nu$ is made up of hadrons and photons that are not associated with the $B_{reco}$ candidate. We exploit the available kinematic information of the full event, the $B_{reco}$ and the $B_{recoil}$, by performing a 2C kinematic fit that imposes four-momentum conservation, the equality of the masses of the two $B$ mesons, and forces $M_{miss}^2 = M_\nu^2 = 0$. The fit takes into account event-by-event the measurement errors of all individual particles and the measured missing mass. This leads not only to a significant improvement of the r.m.s. of the mass resolution of the $X$ system but also provides an almost unbiased estimator of the mean and a resolution that is largely independent of $M_{miss}^2$.

Figure 2 shows the resultant $M_X$ distribution of the selected events, for a minimum lepton momenta $P_{min}^* = 0.9\,\text{GeV}/c$. Different $B \to X_c l\nu$ decays contribute to this distribution. The dominant decays are $\bar{B} \to D^* l\nu$ and $\bar{B} \to Dl\nu$, but we also expect contributions from decays to higher mass charm states, $D^{**}$ resonances with a mass distribution $X_H^{reso}$ peaked near 2.4 GeV/$c^2$, and potentially non-resonant $D^*\pi$ final states for which we assume a broad mass distribution $X_H^{nreso}$ extending to the kinematic limit. The background is dominated by secondary semileptonic charm decays, contributions from lepton (primarily muon) misidentification are much smaller. The backgrounds decrease significantly for higher $P_{min}^*$.

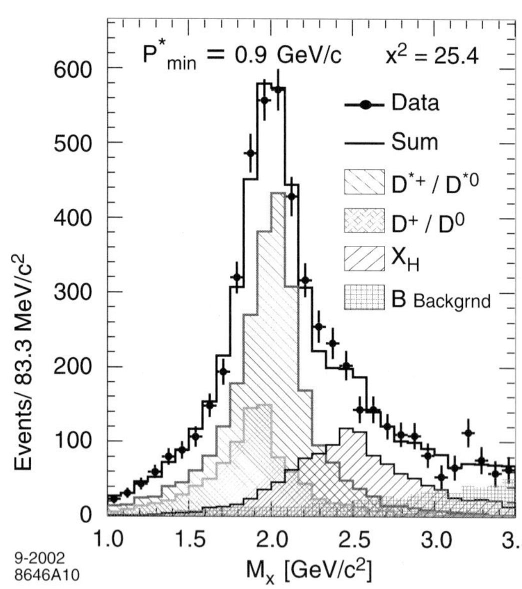


Figure 2. The measured $M_X$ distribution for $P_{min}^* = 0.9\,\text{GeV}/c$. The hatched histograms show the fitted contributions from $\bar{B} \to D^* l\nu$, $\bar{B} \to Dl\nu$ and $\bar{B} \to X_H l\nu$ decays, as well as the background distribution. The open histogram represents the sum of all the individual distributions.

A binned $\chi^2$-fit to the $M_X$ distribution is performed to determine the relative size of the three signal contributions, $f_{D*}$, $f_D$, and $f_{X_H}$, where $f_{X_H}$ refers to the sum of the resonant and non-resonant high mass charm states. Taking into account the true particle masses (the $D$ and $D^*$ masses are basically $\delta$ functions, and the mean of the $X_H$ contribution is taken from generated events) the first moments are calculated according to the following expression

$$\begin{aligned}\langle M_X^2 - \overline{m}_D^2 \rangle &= f_{D^*} \cdot (M_{D^*}^2 - \overline{m}_D^2) \\ &+ f_D \cdot (M_D^2 - \overline{m}_D^2) \\ &+ f_{X_H} \cdot \langle M_{X_H}^2 - \overline{m}_D^2 \rangle. \qquad (3)\end{aligned}$$

Figure 3 shows the first moment as a function of the minimum lepton momentum $P^*_{min}$. The increase at lower momenta is attributed to contributions from the high mass states, i.e. the non-resonant $X_H^{nreso}$ decays. The CLEO Collaboration has measured the same moment for $P^*_{min} = 1.5$ GeV/$c$, based on similar assumption about the high mass hadronic states. Their result [9] is compatible with the measurement presented here.

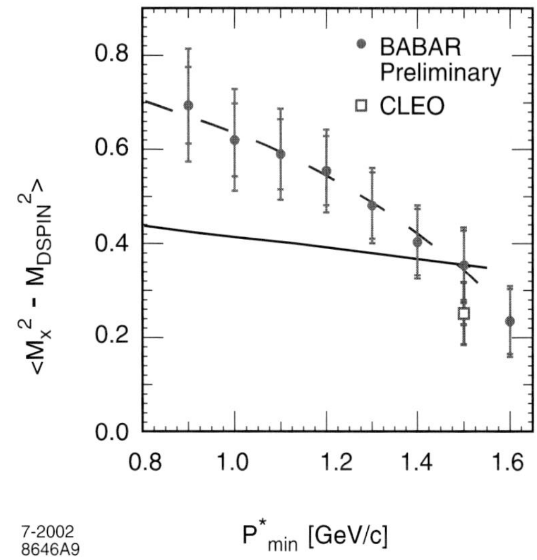


Figure 3. Measured mass moments as a function of the minimum lepton momentum, $P^*_{min}$. The errors indicate the sum of the statistical and systematic uncertainties, they are highly correlated. The dashed curve shows the best description of the data by the OPE expansion [10] with $\lambda_1$ and $\bar{\Lambda}$ as free parameters. For comparison, the solid line shows variation of the moments for the parameters $\lambda_1 = -0.17$ GeV$^2$ and $\bar{\Lambda} = 0.35$ GeV [11].

Extensive studies have been performed to assess the systematic uncertainties and potential biases in the moment measurement. These studies involve both changes in the event selection and variations of the corrections for efficiencies and resolution, and comparisons of data with Monte Carlo simulations. The leading systematic error is due to the uncertainty in the model for the higher mass states $X_H$. Other sizable errors are due to uncertainties in the Monte Carlo simulation of the detector resolution and efficiencies as well as in the background contributions.

As one of the many cross checks we use the relative contributions $f_i$ to determine branching fractions by correcting for acceptance and setting the total semi-leptonic branching fraction to 10.87%. The resultant partial branching fractions are fully compatible with previous measurements and independent of $P^*_{min}$. This is also the case when we split the $X_H$ contribution into a resonant and non-resonant part and allow both of them to vary independently.

Heavy Quark Effective Theory (HQET) calculations of the first mass moment $\langle M_X^2 - \overline{m}_D^2 \rangle$ have been carried out [10] using Operator Product Expansions (OPE) up to order $\alpha_s^2 \beta_0$ and $1/m_B^3$. These expansions contain the non-perturbative parameters $\bar{\Lambda}$ ($\mathcal{O}(m_B)$), $\lambda_1$ and $\lambda_2$ ($\mathcal{O}(m_B^2)$). The observed dependence of the moments on the minimum lepton momentum can be reproduced, as long as we adjust the values of the non-perturbative parameters.

If we restrict the data to $P^*_{min} = 1.5\,\mathrm{GeV}/c$ and use a recent, independently measured value of $\bar{\Lambda} = 0.35 \pm 0.08 \pm 0.10$ GeV [11], we obtain $\lambda_1 = -0.17 \pm 0.06 \pm 0.07\,\mathrm{GeV}^2$, a result that is in good agreement with the CLEO value of $\lambda_1 = -0.236 \pm 0.071 \pm 0.078\,\mathrm{GeV}^2$. However, if we take these values of $\lambda_1$ and $\bar{\Lambda}$ and calculate the moments as a function of $P^*_{min}$, we find a much smaller momentum dependence than the data indicate (see Figure 3). In contrast, we find a momentum dependent variation of $\lambda_1$,

$$\begin{aligned} &\lambda_1|_{P^*_{min}=0.9\,\mathrm{GeV}} - \lambda_1|_{P^*_{min}=1.5\,\mathrm{GeV}} \\ &= 0.05\,\mathrm{GeV}^2 - (-0.17\,\mathrm{GeV}^2) \\ &= (0.22 \pm 0.04 \pm 0.05)\ \mathrm{GeV}^2. \qquad (4) \end{aligned}$$

Here the first error is statistical and takes into account the correlation between the individual moment measurements, the second error is dominated by the uncertainty in $\bar{\Lambda}$.

In summary, if the assumption is correct that there are significant contributions from charm states with masses extending well beyond the resonance $D^{**}$, the first moment is expected to rise for lower lepton momenta, an effect that we observe and that is not described by OPE using other independently measured values of the parameter $\bar{\Lambda}$.

## 4. Conclusions and Outlook

In summary, we have used electrons in $\Upsilon(4S)$ events tagged by a high momentum electron to measure $\mathcal{B}(B \to X\ e\ \nu) = (10.87 \pm 0.18(stat) \pm 0.30(syst))\%$. This measurement is largely model independent. The result is in agreement with previous measurements [6,13], but the systematic uncertainties are reduced. The resulting measurement of $|V_{cb}|$ remains dominated by theoretical uncertainties.

The preliminary measurement of the first moment of the charmed hadron mass distribution assumes that there are sizable contributions from higher mass charm states, both resonant and nonresonanat. The data are fully compatible with this assumption. The resultant dependence of the moment on the lepton momentum is not compatible with other measurement of $\bar{\Lambda}$ and $\lambda_1$.

To resolve this problem, more precise measurements of various exclusive semileptonic branching fractions will be very important. In addition, we expect to measure higher mass moments and add moments of the lepton energy spectrum in the near future. Larger data samples will permit differential moment measurements and a better understanding of the correlated errors. Measurements of the inclusive photon spectrum will also add critical information on the nonperturbative effects that impact the translation of inclusive decays rates to $|V_{cb}|$ and $|V_{ub}|$.

## REFERENCES

1. B. Aubert *et al.*, BABAR Collaboration, Nucl. Instrum. Meth. **A479** (2002) 1.
2. PEP-II , SLAC-418, LBL-5379 (1993).
3. B. Aubert *et al.*, BABAR Collaboration, SLAC-PUB-9306, BABAR-PUB-02-011, e-Print Archive: hep-ex/0208018.
4. B. Aubert *et al.*, BABAR Collaboration, SLAC-PUB-9314, BABAR-CONF-02-029, Contribution to this Conference (ICHEP 2002), e-Print Archive: hep-ex/0207084.
5. H. Albrecht *et al.*, ARGUS Collaboration, Phys. Lett. **B318** (1993) 397.
6. B. Barish *et al.*, CLEO Collaboration, Phys. Rev. Lett. **76** (1996) 1570.
7. Review of Particle Properties, Eur. Phys. J. **C15** (2000) 1.
8. A. Hoang, Z. Ligeti and A. V. Manohar, Phys. Rev. Lett. **82** (1999) 277; similar expressions have been derived by I.I. Bigi, M. Shifman, and N. Uraltsev, *Ann. Rev. Nucl. Part. Science* **47** (1997) 591.
9. D. Cronin-Hennessy *et al.*, CLEO-Collaboration, Phys. Rev. Lett. **87** (2001) 251808.
10. A.F. Falk and M.E. Luke, Phys. Rev. **D57** (1998) 424., and also private communication.
11. S. Chen *et al.*, CLEO Collaboration, Phys. Rev. Lett. **87** (2001) 251.
12. A. Hoang, Z. Ligeti and A. V. Manohar, Phys. Rev. Lett. **82** (1999) 277; similar expressions have been derived by I.I. Bigi, M. Shifman, and N. Uraltsev, *Ann. Rev. Nucl. Part. Science* **47** (1997) 591.
13. M. Accierri *et al.*, L3-Collaboration, Eur. Phys. J. **C13** (2000) 47; G. Abbiendi *et al.*, OPAL-Collaboration, Eur. Phys. J. **C13** (2000) 225; P. Abreu *et al.*, DELPHI-Collaboration, Eur. Phys. J. **C20** (2001) 455; A. Heister *et al.*, ALEPH-Collaboration, Eur. Phys. J. **C22** (2002) 613.

ELSEVIER

Proceedings of ICHEP 2002, pp. 551–553
S. Bentvelsen, P. de Jong, J. Koch and E. Laenen (Editors)

31st INTERNATIONAL CONFERENCE ON HIGH ENERGY PHYSICS AMSTERDAM

# Study of spectral moments in semileptonic decays of the b hadron with the DELPHI detector at LEP.

M. Calvi [a] on behalf of the DELPHI Collaboration

[a]Dep. of Physics, University of Milano Bicocca and INFN, piazza della Scienza 3, I-20126 Milano, Italy

The measurement of the moments of hadronic mass spectrum and of lepton energy spectrum based on a sample of semileptonic decays of the b hadron selected from $Z^0 \to b\bar{b}$ events recorded with the DELPHI detector at LEP, are presented. These results are interpreted in terms of constraints on the quark masses and on the $b$-quark kinetic energy value.

## 1. INTRODUCTION

Moments of hadronic mass spectrum and of lepton energy spectrum are sensitive to the masses of the $b$ and $c$ quarks as well as to the non-perturbative parameters of the Heavy Quark Expansion. They allow to improve the determination of the $|V_{cb}|$ element of the CKM quark coupling matrix which can be measured from inclusive semileptonic B decays by using the relation:

$$\Gamma(b \to c\ell\nu) = |V_{cb}|^2 f(par.) = BR(b \to c\ell\nu)/\tau_b$$

The current experimental accuracy on the semileptonic branching ratio and B lifetime is about 1%, while the evaluation of the function $f$, based on Operation Product Expansion, brings an uncertainty of the order of several percent, making it the dominant error contribution.

Moreover, the comparison of these results with different measurements of the same parameters provides a test of the consistency of the theoretical predictions for inclusive semileptonic B decays and of the underlying assumptions.

Previous measurements of spectral moments have been performed at the $\Upsilon(4S)$ [1], this is the first measurement in $b$ hadron semileptonic decays at the $Z^0$. There are several advantages in the $Z^0$ kinematics, mainly the large boost acquired by the b quark ($E_B \sim 30$ GeV) which gives access in the laboratory frame to the low region of the lepton energy spectrum. The challenge in this case is the complete reconstruction of the B system. The lepton energy acceptance extending down to the lower end of the spectrum makes these results both easier to interpret and complementary to those obtained at the $\Upsilon(4S)$.

In this study two different formulations, using different mass definitions for deriving the constraints on the OPE parameters, are used.

## 2. MOMENTS OF HADRONIC MASS DISTRIBUTION IN $b$ HADRON SEMILEPTONIC DECAYS

The hadronic mass distribution of $\bar{B}^0_d \to D^{**}\ell\bar{\nu}$ events have been studied [2]. $D^{**}$ events have been reconstructed in the three channels: $D^0\pi^+$, $D^+\pi^-$, $D^{*+}\pi^-$ with the $D^0$, $D^+$ and $D^{*+}$ mesons fully reconstructed. Leptons have been required to have a momentum greater than 2 GeV/$c$ in the laboratory frame. The separation of the signal from the background has been achieved by means of a discriminant variable based on the topological properties of the secondary vertex such as the presence of additional charged particles in addition to the D, the lepton and the neutrino.

D$\pi$ candidates have been separated in "right sign" and "wrong sign", considering the charge correlation between the D and the pion, and the discriminant variable distributions have been fitted to the simulation expectation in the two samples separately. Only $D^{(*)}\pi^+\pi^-$ events, with one missing pion, can contribute to "wrong sign" candidates and no evidence of signal have been found in this sample. The following upper limits have been derived at 90% C.L.:


doi:10.1016/S0920-5632(02)02217-X

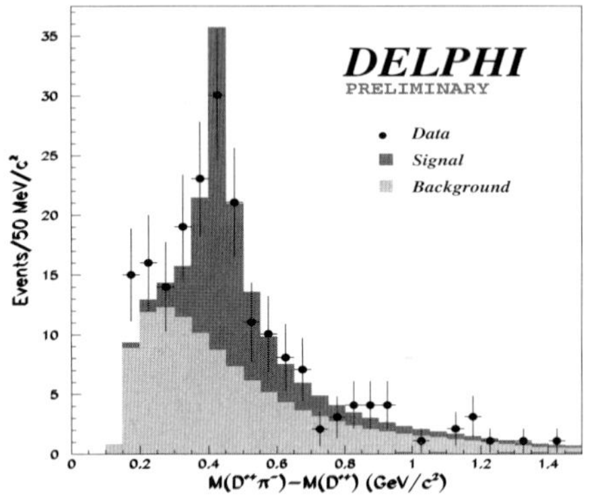


Figure 1. *$\Delta M$ distributions in the $D^{*+}\pi^-$ reconstructed decay channel.*

$BR(b \to D^0\pi^+\pi^-\ell^-\bar{\nu}) = BR(b \to D^+\pi^+\pi^-\ell^-\bar{\nu}) < 0.18\%$ and $BR(b \to D^{*+}\pi^+\pi^-\ell^-\bar{\nu}) < 0.17\%$.

Therefore only $D\pi$ states have been considered in the following analysis of the hadronic mass distribution of the $D^{**}$ states. A fit to the variable $\Delta M = m(D^{(*)}\pi) - m(D^{(*)})$ has been performed, considering the contributions of narrow and broad resonant states $D_0^{*+}$, $D_1^{*+}$, $D_1^+$ and $D_2^+$ as well as non resonant $D\pi$ states. An example of $\Delta M$ distribution is shown in Figure 1.

The total rate for $D^{**}$ production amounts to: $\bar{B}_d^0 \to D^{**}\ell\bar{\nu} = (2.6 \pm 0.5 \pm 0.6)\%$ and the broad $D_1^{*+}$ is the dominant contributing channel.

Moments of the $D^{**}$ mass distribution have been evaluated from the fitted mass distributions. To determine the moments of the complete hadronic mass distribution, in $b$-hadron semileptonic decays, the $b \to D\ell\bar{\nu}$ and $b \to D^*\ell\bar{\nu}$ components have been included using the relation $\langle m_H^n \rangle = p_D m_D^n + p_{D^*} m_{D^*}^n + p_{D^{**}} \langle m_{D^{**}}^n \rangle$ where $p_D$ and $p_{D^*}$ are the relative branching fractions derived from published results and $p_{D^{**}}$ is obtained by imposing the constraint $p_D + p_{D^*} + p_{D^{**}} = 1$ and using the above measurement.

The following preliminary DELPHI results have been obtained:

$$\langle m_H^2 - m_{\bar{D}}^2 \rangle = (0.534 \pm 0.041 \pm 0.074)\ GeV/c^2$$
$$\langle (m_H^2 - m_{\bar{D}}^2)^2 \rangle = (1.51 \pm 0.20 \pm 0.23)\ (GeV/c^2)^2$$
$$\langle (m_H^2 - \langle m_H^2 \rangle)^2 \rangle = (1.23 \pm 0.16 \pm 0.15)\ (GeV/c^2)^2$$
$$\langle (m_H^2 - \langle m_H^2 \rangle)^3 \rangle = (2.97 \pm 0.67 \pm 0.48)\ (GeV/c^2)^3$$

where the first uncertainty is statistic and the second is systematic.

## 3. MOMENTS OF LEPTON ENERGY SPECTRUM IN $b$ HADRON SEMILEPTONIC DECAYS

The moments of the lepton energy spectrum provide constraints similar to those of the hadronic mass. Further they offer an important consistency test of the theory.

An inclusive reconstruction of the semileptonic decays has been performed [3]. Muons and electrons with a momentum greater than 2.5 and 3.0 GeV/c, respectively, have been tagged in a sample of $Z^0 \to b\bar{b}$ events. Secondary vertices have been reconstructed using an iterative procedure. The B energy has been reconstructed adding to the charm vertex energy, the lepton energy and the neutrino energy, evaluated from the event missing energy. The B direction has been estimated from both the reconstructed B momentum and the B decay flight direction. Boosting the leptons in the B reconstructed rest frame provides a typical energy resolution of 250 MeV. In order to separate signal $b \to \ell$ events form $b \to c \to \ell$ and other backgrounds without introducing significant biases to the lepton energy distribution, two probabilistic variables have been defined based on charge correlation and event topology. The measured energy spectrum after background subtraction is shown in Figure 2. For this preliminary result only 1994 and 1995 statistics have been used, corresponding to a sample of 18,300 leptons. After unfolding the resolution smearing, the first, second and third moments have been calculated. The relevant corrections for distortions

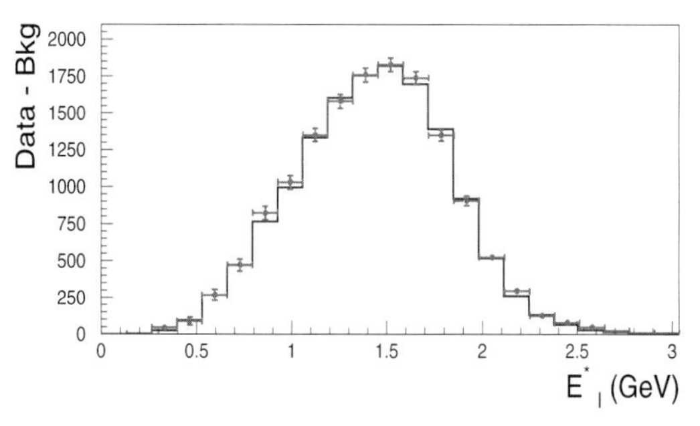


Figure 2. *Lepton energy spectrum after background subtraction. Points are data and histogram is the result of the fit.*

due to electromagnetic radiation, contamination of $b \to u\ell\nu$ decays and contribution of $B_s^0$ and $\Lambda_b$ hadrons in the events sample have been applied. The following preliminary DELPHI results have been obtained:

$$\langle E_\ell^* \rangle = (1.383 \pm 0.012 \pm 0.008)\ GeV$$
$$\langle (E_\ell^* - \langle E_\ell^* \rangle)^2 \rangle = (0.192 \pm 0.005 \pm 0.010)\ GeV^2$$
$$\langle (E_\ell^* - \langle E_\ell^* \rangle)^3 \rangle = (-0.029 \pm 0.005 \pm 0.005)\ GeV^3$$

where the first uncertainty is statistic and the second is systematic.

## 4. INTERPRETATION OF THE RESULTS

Two different approaches have been followed in order to obtain constraints on the non-perturbative parameters of OPE from the measured spectral moments. The first [4] is based on an expansion on the pole masses $m_b$ and $m_c$ and expresses the $b$-energy parameter as $\lambda_1$, while the second [5] uses running heavy quark masses $m_b(\mu)$ and $m_c(\mu)$ and the kinetic energy expectation value $\mu_\pi^2$, corresponding to $\lambda_1$.

Results obtained from the measured values of the first two moments of the hadronic mass spectrum and lepton energy spectrum have been found to be compatible. An exemplification is given in Figure 3 showing constraints extracted in the $\bar{\Lambda}$ - $\lambda_1$ plane which give:

$$\bar{\Lambda} = (\ 0.44 \pm 0.04 \pm 0.05 \pm 0.07)\ GeV$$
$$\lambda_1 = (-0.23 \pm 0.04 \pm 0.05 \pm 0.08)\ GeV^2$$

where the quoted uncertainties are statistic, systematic and related to the power corrections and $\alpha_s$ uncertainties, respectively.

Another way to express these results exploits the correlation in the expected values for the charm and beauty quark masses to extract the charm quark mass $m_c$ by using an independently determined value of $m_b$. Using the first two moments of the lepton energy spectrum and $m_b$=(4.60±0.05 ) GeV, with the second theoretical formulation we find:

$$m_c(1\ \text{GeV}) = (1.19{\pm}0.05{\pm}0.05{\pm}0.07{\pm}0.04)\ GeV$$

where the quoted uncertainties are statistic, systematic ± 0.050 GeV on $m_b$ and $1/m_b^3$ corrections, respectively.

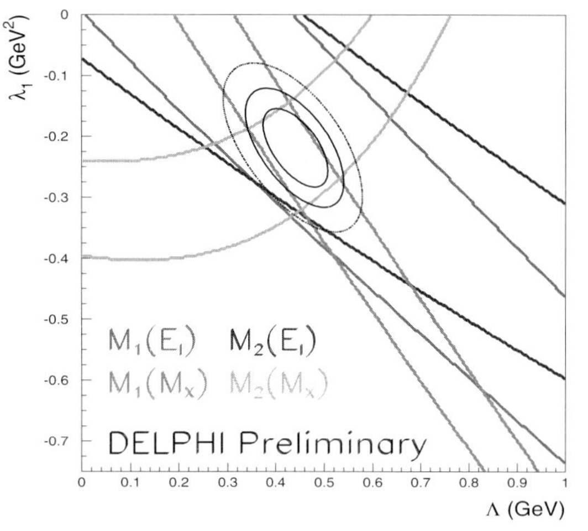


Figure 3. *The constraints on the $\bar{\Lambda}$ - $\lambda_1$ plane obtained from the combination of the first two moments of the lepton energy ($M_1(E_\ell)$,$M_2(E_\ell)$) and hadronic mass ($M_1(M_X)$,$M_2(M_X)$). The bands represents the regions selected by each moment within $\pm 1\sigma_{stat+syst}$. The ellipsis show the $1\sigma$, 68% and 90% C.L. respectively.*

## 5. ACKNOWLEDGMENT

I would like to thank M.Battaglia and P.Roudeau for their contributions to this work, N.Uraltsev and P.Gambino for contributing to the theoretical predictions used in this paper.

## REFERENCES


1. J. Bartelt *et al.* (CLEO Collaboration), CLEO-CONF 98-21. D. Cronin-Hennessy *et al.* (CLEO Collaboration), Phys. Rev. Lett. **87** (2001)251808
2. D. Bloch *et al.*, DELPHI 2002-070 CONF 604, contributed paper to this Conference.
3. M. Battaglia, M. Calvi and L. Salmi. DELPHI 2002-071 CONF 605, contributed paper to this Conference.
4. A. F. Falk and M. Luke, Phys. Rev. **D57** (1998)424. P.Gambino, private communication.
5. M.B.Voloshin Phys. Rev. **D51** (1995)4934. I. I. Bigi, M. A. Shifman and N. Uraltsev Ann. Rev Nucl.Part. Sci **47** (1997)591. N. Uraltsev, private communication.

Proceedings of ICHEP 2002, pp. 554–557
*S. Bentvelsen, P. de Jong, J. Koch and E. Laenen (Editors)*

31st INTERNATIONAL CONFERENCE ON HIGH ENERGY PHYSICS AMSTERDAM

# Strong interaction effects in semileptonic $B$ decays

Nikolai Uraltsev INFN, Sezione di Milano, Milan, Italy;
Department of Physics, University of Notre Dame du Lac, Notre Dame, IN 46556, U.S.A.;
St. Petersburg Nuclear Physics Institute, Gatchina, St. Petersburg 188300, Russia

A comprehensive approach for extracting $|V_{cb}|$ is described not relying on a $1/m_c$ expansion; it allows a percent accuracy without ad hoc assumptions about higher-order effects. An alternative to the $M_X^2$ variable is proposed improving convergence. Intrinsic hardness of moments with a cut on $E_\ell$ is discussed; it can be responsible for the behavior of $\langle M_X^2\rangle$ reported by BaBar. Consequences of the 'BPS' limit are addressed.

The heavy quark (HQ) expansion, a first-principle QCD application of the Wilsonian OPE to heavy quarks has yielded novel insights into the dynamics of heavy flavor hadrons. (For a review see [1] and references therein.) An important application of the HQ expansion is accurately extracting $|V_{cb}|$ and $|V_{ub}|$ from decay rates with little model dependence. This requires a genuine control over nonperturbative effects in $B$ decays.

A popular method to determine $|V_{cb}|$ uses the decay rate $B \to D^*\,\ell\nu$ near zero recoil. At this kinematic point the $B\to D^*$ formfactor $F_{D^*}(0)$ is unity when $m_b, m_c \to \infty$. Driven by the charm mass scale, power corrections are still significant: $F_{D^*} \simeq 0.9$ to order $1/m_Q^2$ [2], and the $1/m_c^3$ effects were estimated to be in the 3% range [3]. Relying on expansion in $1/m_c$ makes it difficult to overcome a 5% level of reliable accuracy here.

These estimates were supported by recent lattice studies which yielded surprisingly close central values, $F_{D^*} \simeq 0.88$ and $F_{D^*} \simeq 0.91$ to order $1/m_Q^2$ and $1/m_Q^3$, respectively [4]. Since the method is based on $1/m_Q$ expansion for both $b$ and $c$, an important issue is higher-order as well as exponential in $m_c$ terms. This sophisticated lattice approach will hopefully be refined. Presently a large fraction of the corrections to $F_{D^*}(0)=1$ is still added theoretically rather than emerges directly in the lattice simulations.

Extrapolating the decay amplitude to zero recoil introduces additional experimental uncertainty. (A parametrization relied upon is alleged to be rigorously derived from QCD. This is not true.) It can be reduced incorporating the model-independent inequalities for the slope of the IW function stemming from the set of the HQ sum rules; however this has not yet been implemented in the experimental analyses.

More extensive opportunities are provided by inclusive semileptonic $B$ decays. Theory-wise nonperturbative effects are controlled by the 10 year old QCD theorem [5] which established absence of the leading $\Lambda_{\rm QCD}/m_b$ power corrections to total decay rates. It applies to all sufficiently inclusive decay probabilities, not only semileptonic. As pointed out shortly after, the masses and relevant nonperturbative parameters can be determined from the $B$ decay distributions themselves [6,7]. Nowadays this strategy is being implemented in a number of experimental studies.

The new generation of data provides accurate measurements of many inclusive characteristics in $B$ decays. The details can be found in experimental talks [8–11]. It is also encouraging that proper theoretical formalism gradually finds its way into their analyses. Recent theoretical findings allow to shrink theoretical uncertainties – among them constraints from the exact HQ sum rules and the consequences of the proximity to the so-called 'BPS' regime signified by the hierarchy $\mu_\pi^2-\mu_G^2 \ll \mu_\pi^2$ suggested by experiment.

Present theory allows to aim a percent accuracy in $|V_{cb}|$. The low-scale running masses $m_b(\mu)$, $m_c(\mu)$, the expectation values $\mu_\pi^2(\mu)$, $\mu_G^2(\mu)$... are completely defined and can be determined from experiment with an in principle unlimited accu-


doi:10.1016/S0920-5632(02)02221-1

racy. Violation of local duality potentially limiting theoretical predictability, has been scrutinized and found to be negligibly small in total semileptonic $B$ widths [12]. Present-day perturbative technology allows computing $\alpha_s$-corrections to the Wilson coefficients. It is also understood how to treat higher-order power corrections.

Such accuracy can be achieved in a comprehensive approach where many observables are measured in $B$ decays to extract necessary 'theoretical' input parameters. Since $b\to c$ widths strongly depend on $m_b-m_c$, previous analyses to some extent relied on expansion in $1/m_c$ employing its relation to $\overline{M}_B-\overline{M}_D$. Reliability of the $1/m_c$ expansion is however questionable. Already terms $\frac{1}{m_c^2}>14\frac{1}{m_b^2}$, and even $\frac{1}{(m_b-m_c)^2}<\frac{1}{8m_c^2}$. On top of that there are indications that the nonlocal correlators affecting meson masses can be particularly large [13] – a pattern also observed in the 't Hooft model [14]. This expectation is supported by the pilot lattice study [15] suggesting a very large value of $\rho_{\pi\pi}^3+\rho_S^3$. On the other hand, non-local correlators are not measured in inclusive $B$ decays. A partial cure was suggested recently [13]: The proximity to the 'BPS' limit leads to much smaller power corrections for the mass relation applied to ground-state mass difference $M_B-M_D$ rather than for standard spin-averaged masses.

Fortunately, there is a way totally free from relying on charm mass expansion. It was put forward some time ago [16] to utilize the whole potential of the comprehensive approach and makes full use of a few key facts [7,6]:

- Total width to order $1/m_b^3$ is affected by a single new Darwin operator (the moments also weakly depend on $\rho_{LS}^3$).
- No nonlocal correlators ever enter *per se.*
- Deviations from the HQ limit in the expectation values are driven by the full scale $1/m_b$ (and are additionally suppressed by proximity to the BPS limit); they are negligible in practice.
- Exact sum rules and inequalities which hold for properly defined Wilsonian parameters.

Some of the HQ parameters like $\mu_G^2$ are known beforehand. Proper field-theoretic definition allows its accurate determination from the $B^*-B$ mass splitting: $\mu_G^2(1\,\text{GeV}) = 0.35^{+.03}_{-.02}\,\text{GeV}^2$ [13]. A priori less certain is $\mu_\pi^2$. However, the inequality $\mu_\pi^2>\mu_G^2$ valid for any definition of kinetic and chromomagnetic operators respecting the commutation relation $[D_j,D_k]=-ig_sG_{jk}$, essentially limits its range: $\mu_\pi^2(1\,\text{GeV})=0.45\pm0.1\,\text{GeV}^2$.

Running $b$ quark mass was accurately extracted from $\sigma(e^+e^-\to\Upsilon(nS))$ in the end of the 1990s: $m_b(1\,\text{GeV}) = 4.57\pm0.06\,\text{GeV}$ for the "kinetic" $m_b(\mu)$. However, considering all available constraints, I think that 4.57 GeV is on the lower side of the $m_b$ range centered rather around 4.63 GeV.

Often extracted from the data are the "HQET parameters" $(-\lambda_1,\overline{\Lambda})$ – they actually correspond to extrapolating the $\mu$-dependent quantities down to $\mu=0$. They are ill-defined and meaningful only as intermediate stage entries, however can often be translated into properly defined parameters. Say, in the context of the recent CLEO analyses

$$\begin{aligned}\overline{\Lambda}_{\text{HQET}} &\simeq \overline{\Lambda}(1\,\text{GeV}) - 0.255\,\text{GeV}\\ -\lambda_1 &\simeq \mu_\pi^2(1\,\text{GeV}) - 0.18\,\text{GeV}^2\ ;\end{aligned}\tag{1}$$

the central CLEO's values [8] thus correspond to $m_b(1\,\text{GeV})=4.62\,\text{GeV}$, $\mu_\pi^2(1\,\text{GeV})=0.43\,\text{GeV}^2$, surprisingly close to the theoretical expectations!

The utility of hadronic moments $\langle M_X^{2k}\rangle$ follows from the fact [7] that, at least if charm were heavy enough the first, second and third moments would more or less directly yield $\overline{\Lambda}$, $\mu_\pi^2$ and $\rho_D^3$.

Let me give an illustration. Leptonic moments, for instance, are approximated as

$$\begin{aligned}\langle E_\ell\rangle &= 1.38\,\text{GeV} + 0.38[(m_b-4.6\,\text{GeV}) -\\ &\quad 0.7(m_c-1.15\,\text{GeV})] + 0.03(\mu_\pi^2-0.4\,\text{GeV}^2)\\ &\quad - 0.09(\tilde\rho_D^3-0.12\,\text{GeV}^3)\ ,\\ \langle(E_\ell-\langle E_\ell\rangle)^2\rangle &= 0.18\,\text{GeV}^2 + 0.1[(m_b-4.6\,\text{GeV})-\\ &\quad 0.6(m_c-1.15\,\text{GeV})] + 0.045(\mu_\pi^2-0.4\,\text{GeV}^2)\\ &\quad - 0.06(\tilde\rho_D^3-0.12\,\text{GeV}^3)\ ,\\ \langle(E_\ell-\langle E_\ell\rangle)^3\rangle &= -0.033\,\text{GeV}^3 - 0.03\,[(m_b-\\ &\quad 4.6\,\text{GeV}) - 0.8(m_c-1.15\,\text{GeV})] + 0.024(\mu_\pi^2-\\ &\quad 0.4\,\text{GeV}^2) - 0.035(\tilde\rho_D^3-0.12\,\text{GeV}^3)\ .\end{aligned}\tag{2}$$

The moments depend basically on one and the same combination of masses $m_b-0.65m_c$.

The value of $|V_{cb}|$ extracted from $\Gamma_{\rm sl}(B)$ has the following dependence on the HQ parameters:

$$\frac{|V_{cb}|}{0.042}=1-0.65[(m_b-4.6\,\text{GeV})-0.61(m_c-1.15\,\text{GeV})]$$

$$+ 0.013\,(\mu_\pi^2 - 0.4\,\mathrm{GeV}^2) + 0.1(\tilde\rho_D^3 - 0.12\,\mathrm{GeV}^3)$$
$$+ 0.06(\mu_G^2 - 0.35\,\mathrm{GeV}^2) - 0.01(\rho_{LS}^3 + 0.15\,\mathrm{GeV}^3) =$$
$$1 - \tfrac{0.65}{0.38}\,[\langle E_\ell\rangle - 1.38\,\mathrm{GeV}] - 0.06\,(m_c - 1.15\,\mathrm{GeV})$$
$$- 0.07(\mu_\pi^2 - 0.4\,\mathrm{GeV}^2) - 0.05(\tilde\rho_D^3 - 0.12\,\mathrm{GeV}^3) -$$
$$0.08(\mu_G^2 - 0.35\,\mathrm{GeV}^2) - 0.005\,(\rho_{LS}^3 + 0.15\,\mathrm{GeV}^3); \quad (3)$$

a combination of the parameters has been replaced by the first lepton moment in Eq. (2), and sensitivity to $\mu_G^2$ and $\rho_{LS}^3$ is illustrated. We see that the precise value of charm mass is irrelevant, but reasonable accuracy in $\mu_\pi^2$ and $\tilde\rho_D^3$ is needed.

The first hadronic moment is

$$\langle M_X^2\rangle = 4.54\,\mathrm{GeV}^2 - 5.0\,[(m_b - 4.6\,\mathrm{GeV}) - 0.62\,(m_c - 1.15\,\mathrm{GeV})] - 0.66\,(\mu_\pi^2 - 0.4\,\mathrm{GeV}^2) + (\tilde\rho_D^3 - 0.12\,\mathrm{GeV}^3), \quad (4)$$

i.e., given by nearly the same combination $m_b - 0.7m_c + 0.1\mu_\pi^2 - 0.2\rho_D^3$ as the lepton moment. Not very constraining, it provides, however a highly nontrivial check of the HQ expansion.

The dependence expectedly changes for higher moments:

$$\langle (M_X^2 - \langle M_X^2\rangle)^2\rangle = 1.2\,\mathrm{GeV}^4 - 0.003(m_b - 4.6\,\mathrm{GeV}) - 0.68\,(m_c - 1.15\,\mathrm{GeV}) + 4.5\,(\mu_\pi^2 - 0.4\,\mathrm{GeV}^2) - 5.5\,(\tilde\rho_D^3 - 0.12\,\mathrm{GeV}^3)\,,$$
$$\langle (M_X^2 - \langle M_X^2\rangle)^3\rangle = 4\,\mathrm{GeV}^6 + (m_b - 4.6\,\mathrm{GeV}) - 3\,(m_c - 1.15\,\mathrm{GeV}) + 5\,(\mu_\pi^2 - 0.4\,\mathrm{GeV}^2) + 13\,(\tilde\rho_D^3 - 0.12\,\mathrm{GeV}^3)\,. \quad (5)$$

Ideally, they would measure the kinetic and Darwin expectation values separately. At the moment, however, we have only an approximate evaluation and informative upper bound on $\tilde\rho_D^3$.

Measuring the second and third hadronic moments is the real step in implementing the comprehensive program of extracting $|V_{cb}|$. It is crucial that this extraction carries no hidden assumptions, and at no point we rely on $1/m_c$ expansion.

The true '`hardness`' of the moments deteriorates when the cut on $E_\ell$ is imposed. As a result, say the extraordinary accuracy of CLEO's $R_0$–$R_2$ cannot be even nearly utilized by theory, whether or not the expressions we use make this explicit.[1]

For total widths the effective energy scale parameter is generally $\mathcal{Q} = m_b - m_c$. When OPE applies it is typically given by $\mathcal{Q} \lesssim \omega_{\max}$, with $\omega_{\max}$ the threshold energy at which the decay process disappears once $m_b$ is replaced by $m_b - \omega$. With the $E_\ell > E_{\min}$ cut then

$$\mathcal{Q} \simeq m_b - E_{\min} - \sqrt{E_{\min}^2 + m_c^2} \quad (6)$$

constituting only 1.25 GeV for $E_{\min} = 1.5\,\mathrm{GeV}$. This may explain the unexpected behavior of the first hadronic moment with respect to the cut on $E_\ell$ reported by BaBar earlier at this session. In $b \to s + \gamma$ decays one has $\mathcal{Q} \simeq m_b - 2E_{\min}$, once again a rather soft scale 1.2 GeV if the lower cut is set at $E_\gamma = 2\,\mathrm{GeV}$. For higher moments the hardness deteriorates further. A high premium should be placed for lowering the cuts [17].

Considering alternative kinematic variables will help to improve the theoretical precision for higher hadronic moments. Namely, it is advantageous to trade the traditional hadronic mass $M_X^2$ for the observable more closely corresponding to the quark virtuality $\Delta$, defined as

$$\mathcal{N}_X^2 = M_X^2 - 2\tilde\Lambda E_X\,, \qquad E_X = M_B - q_0 \quad (7)$$

using the $B$ restframe total hadronic energy, and $\tilde\Lambda$ a fixed mass parameter. Its preferred values are about $M_B - m_b(1\,\mathrm{GeV}) \simeq 650\,\mathrm{MeV}$. The higher moments $\langle(\mathcal{N}_X^2 - \langle\mathcal{N}_X^2\rangle)^2\rangle$, $\langle(\mathcal{N}_X^2 - \langle\mathcal{N}_X^2\rangle)^3\rangle$... should enjoy better theoretical stability.

The kinematic variable $\mathcal{N}_X^2$ is not well constrained inclusively at LEP experiments, however can be used in the $B$ threshold production at CLEO and $B$ factories.[2] This possibility should be carefully explored.

An intriguing theoretical environment opens up if $\mu_\pi^2(1\,\mathrm{GeV})$ is confirmed to be close enough to $\mu_G^2(1\,\mathrm{GeV})$. If $\mu_\pi^2 - \mu_G^2 \ll \mu_\pi^2$ it is useful to analyze strong dynamics expanding around the point $\mu_\pi^2 = \mu_G^2$ [13]. This is not just one point of a continuum in the parameter space, but a quite special 'BPS' limit where the ground state satisfies functional relations $\vec\sigma\vec\pi|B\rangle = 0$, remarkable in many respects. In some instances like the $B \to D$

[1] An instructive example of how naive analysis can miss such effects was given in [17], Sect. 5.

[2] I am grateful to experimental colleagues for discussing this point.

zero-recoil amplitude it extends the HQ symmetry to higher orders in $1/m_Q$. One practical application has been mentioned – the robust relation for $m_b - m_c$ via $M_B - M_D$. Exclusive $B \to D^*$ decay can also benefit from the proximity to BPS. The exact spin sum rules [18] yield a bound for the IW slope

$$\mu_\pi^2 - \mu_G^2 = 3\tilde{\varepsilon}^2(\varrho^2 - \tfrac{3}{4}), \quad 0.45\,\mathrm{GeV} \lesssim \tilde{\varepsilon} \lesssim 1\,\mathrm{GeV} \quad (8)$$

thus leaving only a small room for the slope of the $B \to D^*$ formfactor, excluding values exceeding 1.15–1.2. This would be a very constraining result for a number of experimental studies.

*In conclusion:* experiment has entered a new era of exploring $B$ physics at the nonperturbative level, with qualitative improvement in $|V_{cb}|$. The comprehensive approach will allow to reach a percent level of reliable accuracy in translating $\Gamma_{\rm sl}(B)$ to $|V_{cb}|$. Recent experiments have set solid grounds for dedicated future studies at $B$ factories. We already observe a nontrivial consistency between quite different measurements, and between experiment and QCD-based theory.

There are obvious lessons to infer. Experiment must strive to weaken the cuts in inclusive measurements used in extracting $|V_{cb}|$. Close attention should be paid to higher moments or their special combinations, as well as exploring complementary kinematic observables.

There are a number of directions where theory can be developed further. The analyses presented at the Conference should give theorists substantial motivation to refine it at least in:

- Calculating perturbative corrections to Wilson coefficients of subleading operators.
- Scrutiny of higher-order power corrections.
- A thorough study of alternative kinematic variables, for instance moments of $\mathcal{N}_X^2$.

To fully realize the physical information in the quest for the ultimate precision, a truly comprehensive analysis must implement all theoretical constraints on HQ parameters; the suitable framework uses well-defined running parameters having physical meaning. Heavy quark sum rules appear to yield strong constraints on the parameter space; it is important to study the question of their saturation. If a low $\mu_\pi^2$ around $0.45\,\mathrm{GeV}^2$ is confirmed by experiment, the BPS expansion will play an important role in analyzing nonperturbative effects – in particular, guide us through higher-order corrections.

**Acknowledgments:** It is a pleasure to thank M. Artuso, M. Battaglia, I. Bigi, M. Calvi, V. Luth and P. Roudeau for helpful discussions. This work was supported in part by the NSF under grant number PHY-0087419.

## REFERENCES

1. N. Uraltsev, **in** Boris Ioffe Festschrift 'At the Frontier of Particle Physics–Handbook of QCD', Ed. M. Shifman (World Scientific, Singapore, 2001), Vol. 3, p. 1577; hep-ph/0010328.
2. M. Shifman, N.G. Uraltsev and A. Vainshtein, Phys. Rev. **D51** (1995) 2217.
3. N.G. Uraltsev, Nucl. Instrum. & Methods **A 384** (1996) 17; I. Bigi, M. Shifman and N.G. Uraltsev, Ann. Rev. Nucl. Part. Sci. **47** (1997) 591.
4. J. N. Simone, HQ-2-3 (these proceedings), and the original papers cited therein.
5. I. Bigi, N. Uraltsev and A. Vainshtein, Phys. Lett. **B293** (1992) 430; B. Blok and M. Shifman, Nucl. Phys. **B399** (1993) 441 and 459.
6. I.I. Bigi et al., Int. Journ. Mod. Phys. **A9** (1994) 2467.
7. I.I. Bigi et al., Phys. Rev. **D52** (1995) 196.
8. CLEO collab., these proc.; hep-ex/0209024.
9. DELPHI collab., HQ-2-1 (these proceedings).
10. BaBar collab., these proc.; hep-ex/0207084.
11. M. Battaglia, CP-2-2 (these proceedings).
12. I. Bigi and N. Uraltsev, Int. J. Mod. Phys. **A16** (2001) 5201.
13. N. Uraltsev, Phys. Lett. **B545** (2002) 337.
14. R. Lebed and N. Uraltsev, Phys. Rev. **D62** (2000) 094011.
15. A. S. Kronfeld and J. N. Simone, Phys. Lett. **B490**, 228 (2000) 228 [Erratum-ibid. **B495**, (2000) 441].
16. N. Uraltsev, http://ckm-workshop.web.cern.ch /ckm-workshop/WorkingGroups/Presentations /Uraltsev/Nikolai.ps
17. I. Bigi and N. Uraltsev, hep-ph/0202175; Int. Journ. Mod. Phys. **A** (2002), to appear.
18. N. Uraltsev, Phys. Lett. **B501** (2001) 86.

ELSEVIER

Proceedings of ICHEP 2002, pp. 558–560
*S. Bentvelsen, P. de Jong, J. Koch and E. Laenen (Editors)*

31st INTERNATIONAL CONFERENCE ON HIGH ENERGY PHYSICS AMSTERDAM

# The $\mathbf{B} \to \mathbf{D}^{*}\ell\nu$ Form Factor at Zero Recoil and the Determination of $|\mathbf{V_{cb}}|$

J.N. Simone[a],S. Hashimoto[b],A.S. Kronfeld[a], P.B. Mackenzie[a] and S.M. Ryan[c]

[a]Fermi National Accelerator Laboratory, Batavia, IL 60510, USA

[b]High Energy Accelerator Research Organization (KEK),Tsukuba, 305-0801, Japan

[c]Department of Applied Mathematics, Trinity College, Dublin 2, Ireland

We describe a model independent lattice QCD method for determining the zero recoil form factor, $h_{A_1}(1)$, for the semileptonic decay $B \to D^{*}\ell\nu$. We find $h_{A_1}(1) = 0.913^{+0.024}_{-0.017} \pm 0.016^{+0.003}_{-0.014}{}^{+0.000}_{-0.016}{}^{+0.006}_{-0.014}$, where uncertainties, respectively, are from statistics and fit procedures, matching of lattice theory to QCD, residual lattice spacing dependence, spectator quark mass effects and effects due to working in quenched lattice QCD.

## 1. INTRODUCTION

The CKM matrix element $V_{cb}$ is an important ingredient in precision tests of the CKM picture of $CP$ violation. Its value is used in the "unitarity triangle" analysis of the CKM matrix to constrain the triangle apex in $\bar{\rho}$-$\bar{\eta}$ plane of Wolfenstein parameters.

The determination of $|V_{cb}|$ from the exclusive semileptonic decay $B \to D^{*}\ell\nu$ presents two important advantages: rates are readily measured in experimental studies[1] and heavy quark symmetry imparts important theoretical constraints on the hadronic form factors near zero recoil.

The differential decay rate, divided by a kinematic factor, has the zero recoil ($\omega = 1$) limit

$$\lim_{\omega\to 1} \frac{1}{(\omega^2-1)^{1/2}} \frac{d\Gamma(B \to D^{*}\ell\nu)}{d\omega} = \frac{G_F^2}{4\pi^3} m_{D^*}^3 (m_B - m_{D^*})^2 |V_{cb}|^2 |h_{A_1}(1)|^2 \quad (1)$$

where the hadronic dependence reduces to the single form factor $h_{A_1}(1)$.

Heavy quark symmetry constraints lead to the heavy quark expansion[2,3]

$$h_{A_1}(1) = \eta_A \left[ 1 - \frac{\ell_V}{(2m_c)^2} + \frac{2\ell_A}{4m_c m_b} - \frac{\ell_P}{(2m_b)^2} + \mathcal{O}\left(1/m_Q^3\right) \right] \quad . \quad (2)$$

The radiative corrections, $\eta_A$, are known to two loops[4]. The coefficients $\ell_X$ and the coefficients of higher-order terms in the expansion above all depend on long QCD distance effects.

Previous determinations of $h_{A_1}(1)$ rely on quark models or sum rule estimates for $\ell_V$, $\ell_A$ and $\ell_P$[5,6]. This paper summarizes a model independent method of determining $h_{A_1}(1)$ using lattice QCD to compute coefficients $\ell_X$ and much of the $1/m_Q^3$ corrections. The full description of the method and details of this calculation are found in Reference [7].

## 2. DOUBLE RATIO METHOD

We consider three double ratios of hadronic matrix elements for which the bulk of statistical and systematic errors are expected to cancel. From these ratios we extract three zero recoil form factors:

$$\frac{\langle D | \mathcal{V}_0 | B \rangle \langle B | \mathcal{V}_0 | D \rangle}{\langle D | \mathcal{V}_0 | D \rangle \langle B | \mathcal{V}_0 | B \rangle} \Rightarrow |h_+(1)|^2 \quad (3)$$

$$\frac{\langle D^* | \mathcal{V}_0 | B^* \rangle \langle B^* | \mathcal{V}_0 | D^* \rangle}{\langle D^* | \mathcal{V}_0 | D^* \rangle \langle B^* | \mathcal{V}_0 | B^* \rangle} \Rightarrow |h_1(1)|^2 \quad (4)$$

$$\frac{\langle D^* | \mathcal{A}_1 | B \rangle \langle B^* | \mathcal{A}_1 | D \rangle}{\langle D^* | \mathcal{A}_1 | D \rangle \langle B^* | \mathcal{A}_1 | B \rangle} \Rightarrow |H_{A_1}(1)|^2 \quad (5)$$

Form factor $h_+$ is one of two form factors contributing to $B \to D\ell\nu$ decays[8]. Note that the third ratio yields $H_{A_1}$ and not $h_{A_1}$.


doi:10.1016/S0920-5632(02)02225-9

Form factors $h_+$, $h_1$ and $H_{A_1}$ depend, respectively, upon parameters $\ell_P$, $\ell_V$ and $\ell_A$ and have quark mass dependence:

$$\frac{1 - |h_X(1)|}{\Delta^2} = \ell_x - \ell_x^{[3]} \left( \frac{1}{2m_c} + \frac{1}{2m_b} \right) + \ldots \quad (6)$$

where $\Delta = (\frac{1}{2m_c} - \frac{1}{2m_b})$. The bare charm and bottom quark masses are inputs in lattice QCD. We compute double ratios for a range of "charm" and "bottom" quark masses. After matching the lattice theory to HQET, we determine all the $\ell_x$ as well as the order $1/m_Q^3$ coefficients $\ell_x^{[3]}$ by studying the mass dependence of the form factors. These long-distance coefficients are combined as in Equation 2 to give our determination of $h_{A_1}(1)$.

## 3. RESULT AND UNCERTAINTIES

Our result computed in quenched QCD is

$$h_{A_1}(1) = 0.913^{+0.024}_{-0.017} \pm 0.016^{+0.003+0.000+0.006}_{-0.014-0.016-0.014} .$$

Uncertainties, respectively, are from statistics and fitting procedures, matching of lattice theory to QCD, residual lattice spacing dependence, spectator quark mass effects and effects due to working in quenched lattice QCD. We combine in quadrature uncertainties from setting $m_c$ and $m_b$, perturbative matching, order $\alpha_s(\bar{\Lambda}/2m_c)^2$ terms and order $\bar{\Lambda}/(8m_c^2 m_b)$ terms to obtain the uncertainty in matching to QCD.

We discuss sources of uncertainty and their effect upon $h_{A_1}(1)$.

- **Statistics and fitting** Statistical errors arise from Monte Carlo evaluations of quantum path integrals. Our results are based upon three ensembles, at different lattice spacings, each having between 200 and 350 configurations of the gluon field. Physics parameters are estimated in fits by $\chi^2$ minimization. Correlations among parameters are propagated throughout the entire analysis. The statistical error for $h_{A_1}(1)$ is combined with an uncertainty from effects due to variations in fitting procedures.
- **Setting $\mathbf{m_c}$ and $\mathbf{m_b}$** Bare charm and bottom quark masses are adjusted so that the spin-averaged $1S$ states in $D_s$ and $B_s$ systems match the experimental values. Uncertainties in quark masses result from statistical errors in the tuning procedures and systematic effects estimated by repeating the tuning using -onia systems.
- **Peturbative matching** Much of the renormalization matching lattice theory to QCD is captured nonperturbatively by the double ratios. The perturbative matching coefficients are known to one-loop order[9]. These are combined in the V-scheme with the one-loop results of Czarnecki and Melinkov to match the lattice ratios to HQET. We use a BLM scale for the QCD coupling. Pertubative corrections to ratios are small: no more than 1% over the range of quark masses in our analysis. We assign an overall matching uncertainty by combining estimates for non-BLM higher-order corrections, effects of quenching upon $\alpha_s$ and effects from the (re)definition of the quark mass.
- $\mathcal{O}(\alpha_s(\bar{\Lambda}/2m_c)^2)$ **terms** An uncertainty of this order arises from the adjustment of our action and currents in the Szymanzik approach to matching HQET. Additional uncertainties of this order arise from the definition of the renormalized quark mass.
- $\mathcal{O}(\bar{\Lambda}/(8m_c^2 m_b))$ **terms** Heavy quark symmetry relates $1/m_Q^3$ corrections for $h_{A_1}(1)$ to those for double ratios. For example, spin symmetry relates $B \to D^*$ and $B^* \to D^*$. Thus, the replacement $\ell_V^{\rm eff} = \ell_V - \ell_V^{[3]}/(2m_c)$ reproduces the $1/(m_c^3)$ correction for $h_{A_1}(1)$. Similar arguments recover the $1/(m_b^3)$ corrections. The replacement $\ell_A^{\rm eff} = \ell_A - \ell_A^{[3]} 1/4(1/m_c + 1/m_b)$ leaves a remaining uncertainty for $h_{A_1}(1)$ of the order $\frac{\bar{\Lambda}}{8m_c m_b}\left[\frac{1}{m_c} - \frac{1}{m_b}\right]$.
- **Dependence upon the light spectator quark** Spectator quarks in our lattice calculations have masses in the range $m_s/2$ to $m_s$. Results must be extrapolated to $m_d$ in order to obtain $h_{A_1}$. Due to the limited range of quark mass, lattice results are sufficient to determine only the linear (in $m_\pi^2$) terms in the chiral expansion for $h_{A_1}(1)$. The solid line with dotted 68% confidence limits contours in Figure 1 show this linear dependence upon $m_\pi^2$. We take the extrapolated value and uncertainty as our determination.

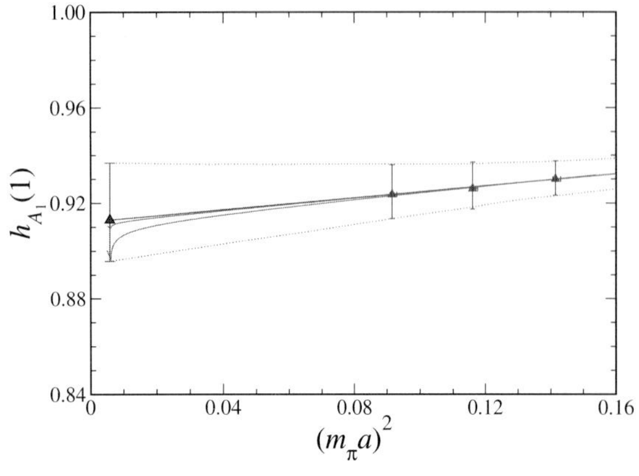


Figure 1. Extrapolation of $h_{A_1}(1)$ to the physical spectator quark mass. The solid straight line with dotted 68% confidence level contours are from an extrapolation linear in $m_\pi^2$. The (red) curve with the cusp at light quark masses indicates the expectation from the chiral expansion.

Randall and Wise[10] predict a $g^2_{D^* D\pi} \log(m_K^2/m_\pi^2)$ "chiral log" term in the expansion for $h_{A_1}(1)$. Our results are not sensitive to such a term since the range of spectator masses is limited. The predicted chiral behavior, using a phenomenological estimate for the $D^* \to D\pi$ coupling, is indicated in Figure 1 by the (red) curve having the downward cusp at the physical pion mass. We take the difference of this prediction and our extrapolation at physical $m_\pi^2$ as the uncertainty.

• **Lattice spacing dependence** We computed $h_{A_1}$ at three lattice spacings to test for lattice spacing dependence. Comparing results from our two finest $a = 0.075\,\text{fm}$ and $a = 0.109\,\text{fm}$ lattices, we conclude residual lattice spacing effects are smaller than the statistical errors. Our best $h_{A_1}(1)$ determination is the weighted average of these two results. The $a = 0.109\,\text{fm}$ result is greater than the average – we use it as an upper estimate for residual lattice spacing dependence. The result from the coarsest, $a = 0.170\,\text{fm}$, lattice has greater uncertainty due to heavy quark discretization effects. A linear (in $a$) extrapolation using all three lattice results gives a value about one-sigma below the average result from our two finer lattices. We take the extrapolated ($a = 0$) value as a lower estimate for residual discretization effects.

• **Quenching** Our results are computed in quenched QCD. The uncertainty due to quenching, like most other uncertainties, affects the deviation of $h_{A_1}(1)$ from one. Quenching affects both the short distance physics – the running of the QCD coupling and the long distance physics of hadronic matrix elements. We attribute uncertainties in $h_{A_1}$ to both long- and short-distance effects. Short distance effects are estimated by comparing the running of $\alpha_s$ in the $n_f = 0$ and $n_f = 3$ theories. We use lattice $f_B$ and $f_D$ decay constant studies as a guide to possible long distance quenching effects. For decay constants 10% quenching effects have been found. We estimate long-distance effects for $h_{A_1} - 1$ to be of similar magnitude.

## REFERENCES

1. The CLEO Collaboration, hep-ex/0203032; K. Abe, *et al.*, Phys. Lett., **B526** (2002) 247; P. Abreu, *et al.*, Phys. Lett., **B510** (2001) 55; D. Buskulic, *et al.*,Phys. Lett., **B395** (1997) 373; G. Abbienddi, *et al.*, Phys. Lett., **B482** (2000) 15.
2. A.F. Falk and M. Neubert, Phys. Rev., **D47** (1993) 2965.
3. T. Mannel, Phys. Rev., **D50** (1994) 428.
4. A. Czarnecki and K. Melinkov, Nucl. Phys., **B505** (1997) 65.
5. M. Neubert, Phys. Rept., **245** (1994) 259.
6. M. Shifman, *et al.*, Phys. Rev., **D51** (1995) 2217; *erratum* Phys. Rev., **D52** (1995) 3149.
7. S. Hashimoto, *et al.*, Phys. Rev., **D66** (2002) 014503.
8. S. Hashimoto, *et al.*, Phys. Rev., **D61** (2000) 014502.
9. J. Harada, *et al.*, Phys. Rev., **D65** (2002) 094514.
10. L. Randall and M. Wise, Phys. Lett., **B303** (1993) 135.

ELSEVIER

Proceedings of ICHEP 2002, pp. 561–563
*S. Bentvelsen, P. de Jong, J. Koch and E. Laenen (Editors)*

31st INTERNATIONAL CONFERENCE ON HIGH ENERGY PHYSICS AMSTERDAM

# Power corrections in heavy-to-light decays at large recoil energy *

M. Diehl[a]

[a]Institut für Theoretische Physik E, RWTH Aachen, 52056 Aachen, Germany

I briefly present recent work on QCD power corrections in heavy-to-light meson decays, using an effective field theory approach.

## 1. INTRODUCTION

The strong interaction physics in $B$ decays is characterized by the presence of several distance scales. The general idea of factorization in QCD is to separate the dynamics at short distances (which can be treated perturbatively) from the long-distance physics (encoded in hadronic matrix elements like decay constants or form factors). There are different techniques to achieve this separation. One is the detailed analysis of Feynman graphs, used for instance in [1] to establish QCD factorization for exclusive $B$ decays. Here I focus on an effective field theory approach, whose principal objects are operators and fields. This framework is very general and well suited to be applied to a wide class of processes.

As an explicit application I consider exclusive $B$ decays to light mesons (e.g. $B \to \pi \ell \nu$, $B \to K^* \ell^+ \ell^-$) in kinematics where the recoil energy of the light meson is large, $E \sim m_b$. In the limit $m_b \to \infty$ symmetry relations between the various form factors describing the $B \to \pi$ or $B \to K^*$ transition emerge [2], which greatly reduce hadronic uncertainties when analyzing these decays in the Standard Model and beyond. These relations receive corrections in inverse powers of $m_b$, which we have systematically studied in effective field theory [3]. They also receive corrections in $\alpha_s$ from the heavy quark decay vertex (which are part of the effective theory) and from hard interactions of the spectator quark in the $B$ meson (which have so far only been treated by direct calculation of the relevant Feynman graphs in QCD [4]).

*Based on work with A. Chapovsky, M. Beneke and Th. Feldmann [3]


doi:10.1016/S0920-5632(02)02228-4

## 2. EFFECTIVE FIELD THEORY

To set up the effective field theory appropriate for a given process one first identifies the relevant momentum regions in the corresponding Feynman graphs. One wants to keep in the theory only the nearly on-shell momentum modes that directly connect to the external hadrons. In the decays of our study (see Fig. 1) this is a heavy quark in the initial $B$ meson, a cluster of light quarks and gluons which move approximately collinear to each other and form the fast outgoing meson, and "ultrasoft" gluons in either the initial or final state. (At this point we exclude the hard spectator quark interactions mentioned above.) Modes with large momenta $k^\mu \sim m_b$ are "integrated out" from the theory: this is done in perturbation theory and provides the Wilson coefficients describing the shaded blob in the figure. The same happens to "soft" gluons with momenta $k^\mu \sim \sqrt{m_b \Lambda_{\text{QCD}}}$. The heavy quark and its interactions with ultrasoft gluons is described in conventional heavy-quark effective theory (HQET), and the dynamics of collinear and ultrasoft modes in the soft-collinear effective theory (SCET) pioneered by Bauer et al. [5].

For each momentum mode a separate field is introduced, e.g. $A_c^\mu(x)$ and $A_{us}^\mu(x)$ for collinear and ultrasoft gluons. For these fields one then must find the Lagrangian and current operators which reproduce the results of QCD to a given accuracy in a power-counting parameter $\lambda$, in our case given by $\sqrt{\Lambda_{\text{QCD}}/m_b}$. It proves useful to project out the large spinor components of fermion fields and to integrate out the small ones. In HQET this is the usual projection on $h_v(x) = \frac{1}{2}(1 + \not{v})\, Q_v(x)$, where $Q_v(x) = e^{i m_b (vx)} Q(x)$ has

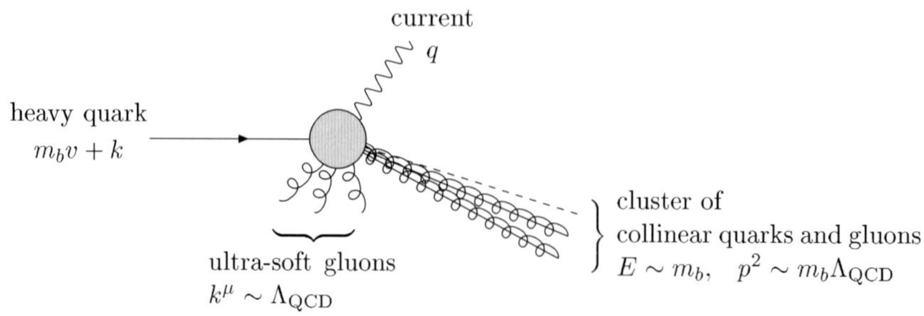


Figure 1. Heavy-quark decay into a cluster of collinear and ultrasoft particles.

the large phase of the full heavy quark field removed. For collinear quarks the large component is

$$\xi(x) = \tfrac{1}{4}\not{n}_-\not{n}_+\, q(x),$$

where $n_-$ and $n_+$ are lightlike reference vectors, $n_-$ giving the direction of the collinear particles and $n_+$ the opposite direction. In contrast to [5] we do not split off the large phase of collinear fields in our approach [3]. For ultrasoft quarks all spinor components are of same size and hence kept in the theory.

To decide whether an operator is relevant to a given accuracy in $\lambda$ one determines power counting rules for momentum components (collinear momenta are large in the $n_-$ direction and small in the others), for fields, their derivatives, etc. Finally, the effective theory "inherits" part of the gauge invariance of full QCD, and one has to define appropriate gauge transformations for the effective fields to implement this.

### 2.1. The effective Lagrangian

To obtain the effective Lagrangian for collinear and ultrasoft quarks and gluons one can start from the QCD quark Lagrangian $\mathcal{L} = \bar{q}i\not{D}q$, separate out the different momentum modes, and integrate out the small spinor components of the collinear quark field (using e.g. path integral methods). The result is

$$\mathcal{L}_c = \bar{\xi}\left(in_-D + i\not{D}_\perp W \frac{1}{in_+\partial} W^\dagger i\not{D}_\perp\right)\frac{\not{n}_+}{2}\,\xi$$

if one neglects ultrasoft quarks. Because of the inverse derivative $(n_+\partial)^{-1}$ the action is nonlocal. It is explicitly gauge invariant: the sum $A = A_c + A_{us}$ of gluon fields appears in the covariant derivative $D = \partial - igA$ and in the Wilson line

$$W(x) = P\exp\left\{ig\int_{-\infty}^{0} ds\; n_+A(x+sn_+).\right\}$$

One must still expand $\mathcal{L}_c$ in $\lambda$: the components of $A_c^\mu$ and $A_{us}^\mu$ have different scaling in $\lambda$, which requires a Taylor expansion of $D$ and $W$. Furthermore, $A_{us}(x)$ varies more slowly in $x$ than collinear fields with their large-momentum modes, so that $A_{us}(x)$ must be appropriately Taylor expanded in $x$.

To leading order in $\lambda$ ultrasoft quarks just appear in $\mathcal{L}_{us} = \bar{q}_{us}\, i\not{D}_{us}\, q_{us}$, but in higher orders they couple to collinear fields. The resulting Lagrangian including $O(\lambda)$ and $O(\lambda^2)$ corrections also contains ultrasoft Wilson lines $Z$, obtained from $W$ by setting $A_c = 0$.

### 2.2. The effective heavy-to-light current

Before decaying via an electroweak current a heavy quark can radiate collinear gluons, which puts it off shell by an amount of order $m_b^2$. The corresponding diagrams have no place in the effective theory and must be summed into an effective vertex. This leads one to match the QCD heavy quark field onto an effective field, $Q \to e^{-im_b(vx)}\, Q_{\text{eff}}(A_c, A_{us}, h_v)$. By explicitly summing diagrams or by solving the Dirac equation for $Q_{\text{eff}}$ in an external field $A = A_c + A_{us}$

one obtains $Q_{\text{eff}}$ as

$$WZ^\dagger Q_v - \frac{1}{\mathcal{V}^2 - 1}\left(\not{\mathcal{V}}\not{\mathcal{V}} WZ^\dagger - WZ^\dagger \not{\mathcal{V}}_{\text{us}} \not{\mathcal{V}}_{\text{us}}\right) Q_v$$

including corrections of order $\lambda$ and $\lambda^2$. Here $\mathcal{V}^\mu = v^\mu + iD^\mu/m_b$ is the reparametrization invariant "velocity" known from HQET.

Matching of the QCD light quark field onto an effective field, $q \to q_{\text{eff}}(A_c, A_{us}, \xi)$, is readily performed in SCET. The complete QCD current is matched as $\overline{q}\,\Gamma Q \to e^{-im_b(vx)}\,\overline{q}_{\text{eff}}\,\Gamma\, Q_{\text{eff}}$ and finally has to be Taylor expanded in $\lambda$ as in the case of the effective Lagrangian.

## 3. POWER CORRECTIONS TO FORM FACTORS

The symmetry relations for $E, m_b \to \infty$ between the transition form factors from $B$ to a light meson $L = \pi, \rho, K^*, \ldots$ receive corrections in powers of $\lambda = \sqrt{\Lambda_{\text{QCD}}/m_b}$, which can be systematically described in the effective theory of Sec. 2 by matching the QCD matrix elements $\langle L|\,\overline{q}\,\Gamma\, Q\,|B\rangle$ onto $\langle L|\,\overline{q}_{\text{eff}}\Gamma\, Q_{\text{eff}}\,|B\rangle$. Projecting $\overline{q}_{\text{eff}}$ and $Q_{\text{eff}}$ on their large and small spinor components one finds that

1. at leading order in $\lambda$ only the large components are relevant. This gives 1 (2) independent form factors for the transition to a pseudoscalar (vector) meson [2,4].

2. the $O(\lambda)$ corrections involve the small components of either $\overline{q}_{\text{eff}}$ or $Q_{\text{eff}}$, and one remains with 2 (5) independent form factors.

3. the $O(\lambda^2)$ corrections come from the small components of both fields. Hence no symmetry relations are left to this accuracy, and one has 3 (7) independent form factors as in full QCD.

There is a caveat concerning the corrections at order $\lambda$. The corresponding operators in the effective theory involve different parton configurations in the light meson than the leading-order operators, for instance with an additional transversely polarized gluon. There may well be further dynamical suppression of their matrix elements, bringing them down from $O(\lambda)$ to $O(\lambda^2)$. Whether this happens requires knowledge of the light meson wave functions which goes beyond the scope of SCET. We remark that these putative $O(\lambda)$ corrections affect the form factor relations used in the analysis of the forward-backward asymmetry in $B \to K^*\,\ell^+\ell^-$, see [6].

## 4. SUMMARY

We have performed a systematic analysis of power corrections to soft-collinear factorization in the effective field theory formalism. The interactions between collinear and ultrasoft quarks and gluons are described in SCET, whose effective Lagrangian was obtained including the power corrections of first and second order in $\lambda$. Heavy-to-light meson decays at large recoil energy can be described by the effective actions of SCET and HQET plus the effective heavy-to-light current, which was also obtained including $O(\lambda^2)$ corrections. Aspects of the $O(\lambda)$ corrections were previously investigated in [7].

Among the 2 (5) leading-order symmetry relations between the transition form factors for $B$ to pseudoscalar (vector) mesons, 1 (3) receive power corrections at order $\sqrt{\Lambda_{\text{QCD}}/m_b}$. Whether these are further suppressed dynamically remains to be investigated. To accuracy $\Lambda_{\text{QCD}}/m_b$ no form factor relations are left. Phenomenological analyses of the corresponding decays thus remain dependent on models for the relevant hadronic physics.

## REFERENCES

1. M. Beneke et al., Nucl. Phys. B 591 (2000) 313.
2. J. Charles et al., Phys. Rev. D 60 (1999) 014001.
3. M. Beneke et al., hep-ph/0206152, Nucl. Phys. B (in press).
4. M. Beneke and Th. Feldmann, Nucl. Phys. B 592 (2001) 3.
5. C. W. Bauer et al., Phys. Rev. D 63 (2001) 014006; Phys. Rev. D 63 (2001) 114020; Phys. Rev. D 65 (2002) 054022.
6. G. Burdman, Phys. Rev. D 57 (1998) 4254.
7. J. Chay and C. Kim, Phys. Rev. D 65 (2002) 114016; C. W. Bauer et al., hep-ph/0205289.

Proceedings of ICHEP 2002, pp. 564–568
*S. Bentvelsen, P. de Jong, J. Koch and E. Laenen (Editors)*

ELSEVIER

31st INTERNATIONAL CONFERENCE ON HIGH ENERGY PHYSICS AMSTERDAM

# Branching Fraction and Properties of $B$ meson decays to Charmonium at Belle

Yasushi Watanabe*, for the Belle collaboration
Graduate School of Science and Engineering, Tokyo Institute of Technology, 2-12-1, Meguro-ku, Tokyo, 152-8551 Japan

The latest Belle results on $B$ meson decays to charmonium are reported. The topics include the first observation of the $\eta_C(2S)$ meson, which is a charmonium state so far missing; an observation of factorization-forbidden decays: $B \to \chi_{C2}X$ and $\chi_{C0}K$; and search for more $CP$ eigenstates in both Cabibbo-favored and Cabibbo-suppressed decays.

## 1. Introduction

New physics frontier, "luminosity frontier" has been just opened; B factory experiments, Belle at KEKB accelerator (KEK) and BABAR at PEP-II (SLAC) have each accumulated close to 100 fb$^{-1}$ (the initial target luminosity) since their start in May 1999 and started producing important results.

Here I report the latest results from the Belle collaboration on $B$ meson decays to charmonium. The most important reason for interest in these decays is to measure a $CP$ violation parameter through a time dependent asymmetry of the golden mode $B^0(\overline{B}^0) \to J/\psi K_S$ and others [1]. Another reason is that this is the best place to search for missing charmonium states such as the $\eta_C(2^1S_0)$ and $h_C(1^1P_1)$ mesons, which are supposed to lie just below the $\psi(2S)$ level. These decay modes also provide a testing ground for theoretical ideas: both perturbative and non perturbative QCD, heavy quark effective theory, etc.

Before going into individual topics, it is appropriate to have an overview of analyses. Most of the results presented here are based on a data sample of 29.4 fb$^{-1}$ (~1/3 of the data taken so far) or 32M $B\overline{B}$ pairs. For most of the analyses, we look for signals by searching for an accumulation of $B$ meson candidates in the signal box of the beam-constrained mass ($M_{\rm bc}$) vs. the energy difference ($\Delta E$) and for a peak at the charmonium mass, where $M_{\rm bc}$ and $\Delta E$ are defined as:

$$M_{\rm bc} \equiv \sqrt{E^2_{\rm beam} - |\boldsymbol{p}_{\rm recon}|^2} \quad (1)$$

$$\Delta E \equiv E_{\rm recon} - E_{\rm beam} \ . \quad (2)$$

Here, $E_{\rm beam}$ is the beam energy and $E_{\rm recon}$ ($\boldsymbol{p}_{\rm recon}$) is the reconstructed energy (momentum) of a $B$ meson candidate, both defined in the center of mass system.

## 2. Observation of the $\eta_C(2S)$ meson

It has been suggested that a $B$ factory is the best place to search for missing charmonium states, thanks to its overwhelming luminosity and to considerable branching fractions of $B$ meson decays to charmonium [2].

The Crystal Ball group reported an evidence for the $\eta_C(2S)$ meson with a mass of $3594 \pm 5$ MeV/$c^2$ and width of 8 MeV in 1982 [3], but subsequent experiments have not supported it. The Belle collaboration has looked for the $\eta_C(2S)$ meson in $B$ decays and reported its first clear observation [4]. The property of the $\eta_C(2S)$ meson is expected to be similar to that of the $\eta_C(1S)$ meson, which decays to $K_SK^\pm\pi^\mp$ with a branching fraction of 5.5%. We looked for $B$ meson signals in $(K_SK^\pm\pi^\mp)K$ for various mass range of $K_SK^\pm\pi^\mp$ and found a clear signal at $3620 < M(K_SK^\pm\pi^\mp) < 3660$ MeV/$c^2$ as shown in Fig. 1.

Figure 2 shows a distribution of $M(K_SK^\pm\pi^\mp)$, where a clear peak corresponding to $\eta_C(2S)$ is

*email: watanabe@hp.phys.titech.ac.jp


doi:10.1016/S0920-5632(02)02230-2

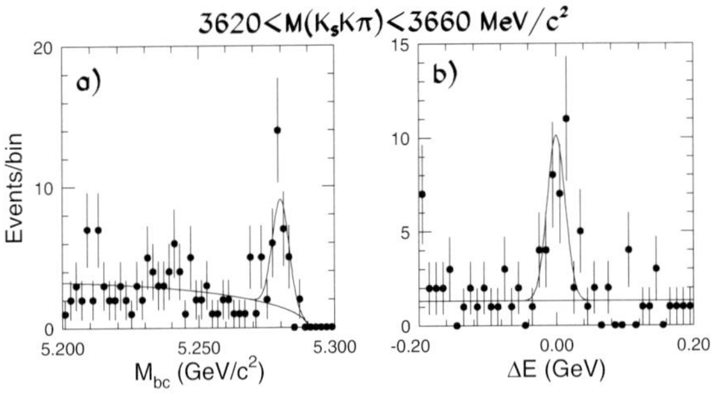


Figure 1. $B$ meson signals in $M_{bc}$ and in $\Delta E$ for $3620 < M(K_S K^{\pm}\pi^{\mp}) < 3660$ MeV$/c^2$.

seen. The mass is obtained to be $3654 \pm 6 \pm 8$ MeV$/c^2$ and the width less than 55 MeV at 90 % confidence, which are in agreement with theoretical expectations [2].

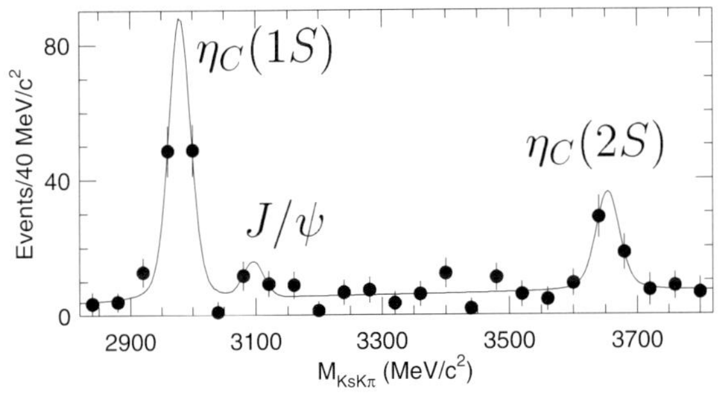


Figure 2. Invariant mass distribution of $K_S K^{\pm}\pi^{\mp}$ system.

## 3. Observation of Factorization-Forbidden Decays

The amplitude for a charmonium production in $B$ decays can be factorized as:

$$\mathcal{A}(B \to c\bar{c}X) = \frac{G_F}{\sqrt{2}} vC\langle X|\,(\bar{q}_a b_a)_{V-A}\,|B\rangle \times \langle c\bar{c}|\,(\bar{c}_b c_b)_{V-A}\,|0\rangle, \quad (3)$$

where $G_F$ is the Fermi coupling constant, $v$ is a product of the Cabibbo-Kobayashi-Maskawa (CKM) matrix elements, $C$ is a QCD factor and $\overline{q'}_a q_a$ stands for the color sum [5].

According to Eq. (3), $B$ meson decays to the $\chi_{C0}$ and $\chi_{C2}$ are forbidden while to the $\chi_{C1}$ is allowed. This is because the production of the $\chi_{CX}$ proceeds through the vector ($V$) current and the last matrix element in Eq. (3) is zero for the $\chi_{C0}$ and $\chi_{C2}$ because of the angular momentum conservation and the CVC (conserved vector current). Note that this naive factorization hypothesis may be easily violated through a soft gluon exchange, especially in such color-suppressed decays. Indeed we observe such factorization-forbidden decays with considerable branching fractions as discussed below.

### 3.1. $B \to \chi_{C2}(3556)X$ decay

The inclusive decay mode: $B \to \chi_{C2}(3556)X$ was looked for by first enriching data with $B\overline{B}$ events, and then by looking for the decay chain of the $\chi_{C2}(3556)$ into $J/\psi\gamma$ and the $J/\psi$ into $e^+e^-$ and $\mu^+\mu^-$ pairs [6]. Figure 3 shows the mass difference $M(l^+l^-\gamma) - M(l^+l^-)$, where clear peaks corresponding to the $\chi_{C1}$ (larger peak) and the $\chi_{C2}$ (smaller peak) are visible. In this measurement, it is crucially important to have a good energy resolution for low energy photons; we have achieved a photon energy resolution of $(2.61 \pm 0.04)\%$, through recalibration using the decay mode: $D^{*0} \to D^0\gamma$.

The branching fraction is measured to be:

$$\mathcal{B}(B \to \chi_{C2}X) = (1.80^{+0.23}_{-0.28} \pm 0.27) \times 10^{-3}. \quad (4)$$

By subtracting the cascade decay contribution of $\psi(2S) \to \chi_{C2}\gamma$, the branching fraction of the direct decay is:

$$\mathcal{B}_{\text{direct}}(B \to \chi_{C2}X) = (1.53^{+0.23}_{-0.28} \pm 0.27) \times 10^{-3}, (5)$$

which is substantial compared to a factorization-allowed $B$ decay into $\chi_{C1}$:

$$\mathcal{B}(B \to \chi_{C1}X) = (3.32 \pm 0.22 \pm 0.34) \times 10^{-3}. (6)$$

Figure 4 shows the momentum spectra of the $\chi_{C1}$ and $\chi_{C2}$, where a marked difference is seen between the two; that of $\chi_{C2}$ is suppressed near the end point compared to that of $\chi_{C2}$ even

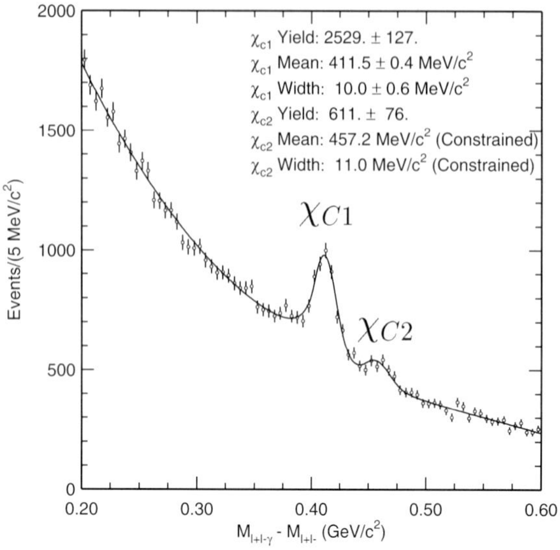


Figure 3. Distribution of $M(l^+l^-\gamma) - M(l^+l^-)$.

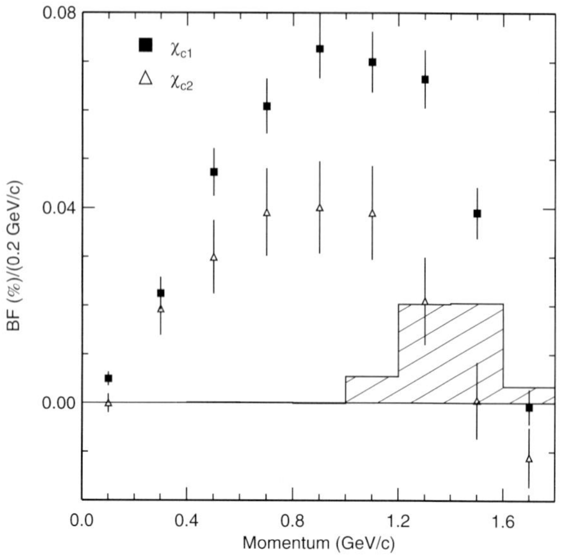


Figure 4. Momentum distribution of the $\chi_{C1}$ and $\chi_{C2}$. The hatched histograms correspond to a Monte Carlo estimation of $B \to \chi_{C2}K$ assuming the same branching fraction as that for the $\chi_{C1}$.

though the mass difference is small. This indicates that the exclusive decay mode, $B \to \chi_{C2}K$ is suppressed in contrast to $B \to \chi_{C1}K$.

### 3.2. $B^\pm \to \chi_{C0}(3417)K^\pm$ decay

The decay mode of $B^\pm \to \chi_{C0}(3417)K^\pm$ was looked for through $\chi_{C0} \to \pi^+\pi^-$ and $K^+K^-$ decays [7]. Clear peaks can be seen in an invariant mass distribution of both $\pi^+\pi^-$ and $K^+K^-$ as shown in Fig. 5. The peak for $K^+K^-$ is shifted slightly which may be due to an interference with the background. The branching fraction measured is:

$$\mathcal{B}(B^\pm \to \chi_{C0}K^\pm) = (0.60^{+0.21}_{-0.18} \pm 0.11) \times 10^{-3}, \quad (7)$$

which is also substantial compared to $\mathcal{B}(B^\pm \to \chi_{C1}K^\pm) = (1.0 \pm 0.4) \times 10^{-3}$ and to $\mathcal{B}(B^\pm \to J/\psi K^\pm) = (1.00 \pm 0.10) \times 10^{-3}$ [8].

## 4. Search for more $B$ Decays to $CP$ eigenstates

It is useful to find more $B$ Decays to $CP$ eigenstates for $CP$ studies in order to further increase the statistics and to study systematics. We have looked for decay modes into $CP$ eigenstates in both Cabibbo-favored and Cabibbo-suppressed decays. The obtained branching fractions are summarized in Table 1 together with those for other decay modes.

### 4.1. Cabibbo-Favored Decay Modes: ($b \to (\bar{c}c)s$)

The decay mode of $B \to J/\psi K_1(1270)$ has been observed for the first time through the decay mode of $K_1(1270) \to K\pi\pi$ [9]. The analysis is done in a data sample of 10.5 fb$^{-1}$.

$B \to J/\psi K^*_X(1430)$ has been found as shown in Fig. 6, where an invariant mass distribution of $K\pi$ is plotted; a clear peak at 1430 MeV/$c^2$ can be seen with some activities below. A large fraction of $K^*_X(1430)$ seems to be due to the $K^*_2(1430)$ meson.

Other decay modes studied include: the first observation of $B \to \eta_C K^*$, where the $K^*$ is identified through the decays $K\pi$; and an observation of $B \to \eta_C K$, through $\eta_C \to$

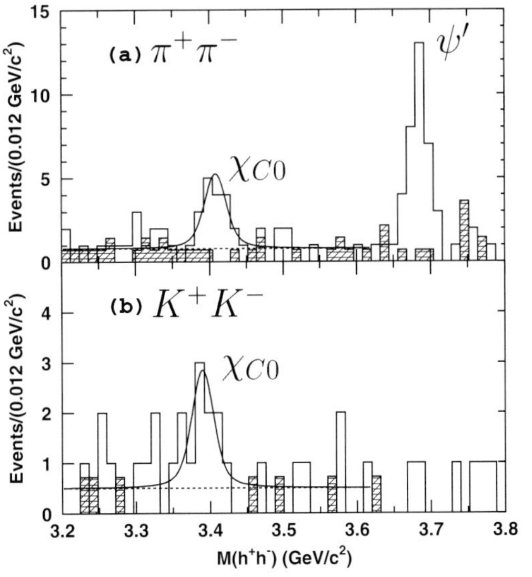


Figure 5. Invariant Mass distribution of (a) $\pi^+\pi^-$ and (b) $K^+K^-$ system. Hatched histograms correspond to events in the $\Delta E$ sidebands.

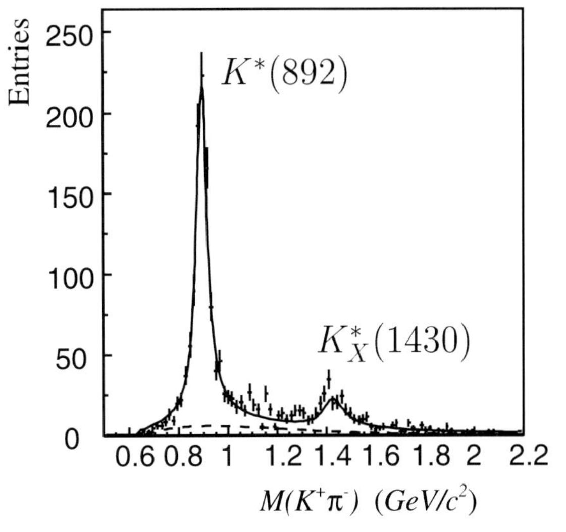


Figure 6. Invariant mass Distribution of $K\pi$.

$K_S K^-\pi^+, K^+K^-\pi^0, K^*K^-\pi^+$.

## 4.2. Cabibbo-suppressed Decay Modes: $(b \to (\bar{c}c)d)$

A search for Cabibbo-suppressed decay modes is made by analyzing more data sample. $B^0 \to J/\psi\pi^0$ is found by using almost all the data of 78.1 fb$^{-1}$. A clear accumulation of events is seen in the signal box of $M_{\rm bc}$ vs. $\Delta E$, as shown in Fig. 7, yielding a net signal of 52.3±8.6 events [9]. We have also looked for $B^0 \to J/\psi\rho^0$ analyzing a 60.1 fb$^{-1}$ data sample and found a healthy signal [9].

## 5. Branching Fractions of Exclusive Decays

Branching Fractions of Exclusive Decays obtained by Belle [11] are summarized and compared with the values in PDG(2000) [8] in Table 1. As can be seen, considerable number of modes are newly established and the values already existing are significantly improved; dominant errors are now systematic in most of the decay modes.

## 6. Summary

I reported the latest Belle results on $B$ meson decays into charmonium: the first observation of a missing charmonium state, the $\eta_C(2S)$ meson; an observation of factorization-forbidden decays: $B \to \chi_{C2}X$ and $\chi_{C0}K$ decays; and a search for more $B$ meson decays into $CP$ eigenstates.

I am grateful to the members of the Belle collaboration. Special thanks are due to T.Browder, F.Fang, H.Ishino, S.U.Kataoka, K.Miyabayashi, Y.Sakai, S.Schrenk and S.Uno.

## REFERENCES


1. B.Aubert *et al.*, BABAR collaboration, Phys. Rev. Lett. 87, 091801 (2001).
   K.Abe *et al.*, Belle collaboration, Phys. Rev. Lett. 87, 091802 (2001).
2. E.J.Eichten, K.Lane and C.Quigg, Fermilab-

Table 1
Branching Fractions of $B$ meson decays to Charmonium (*exclusive mode*)

| | $B^+$ $(\times 10^{-4})$ | | $B^0$ $(\times 10^{-4})$ | |
|---|---|---|---|---|
| mode | PDG(2000)[1)] | Belle | PDG(2000)[1)] | Belle |
| $\eta_C K$ | | $12.5 \pm 1.4^{+1.0}_{-1.2} \pm 3.8$[2)] | | $12.3 \pm 2.3^{+1.2}_{-1.6} \pm 3.8$ |
| $\eta_C K^*$ | | | | $16.2 \pm 3.2^{+2.4}_{-3.4} \pm 5.0$ |
| $J/\psi K$ | $10.0 \pm 1.0$ | $10.1 \pm 0.3 \pm 0.8$ | $8.9 \pm 1.2$ | $7.7 \pm 0.4 \pm 0.7$ |
| $J/\psi K^*$ | $14.8 \pm 2.7$ | $12.8 \pm 0.7 \pm 1.4$ | $15.0 \pm 1.7$ | $12.9 \pm 0.5 \pm 1.3$ |
| $J/\psi K_1(1270)$ | | $18.0 \pm 3.4 \pm 3.9$ | | $13.0 \pm 3.4 \pm 3.1$ |
| $J/\psi \pi$ | $0.51 \pm 0.15$ | $0.52 \pm 0.07 \pm 0.07$ | $< 0.58$ | $0.18 \pm 0.03 \pm 0.02$ |
| $J/\psi \eta$ | | | | $< 0.14$ |
| $J/\psi \rho$ | | | | $0.28 \pm 0.05 \pm 0.07$ |
| $\psi(2S)K$ | $5.8 \pm 1.0$ | $6.2 \pm 0.4 \pm 0.5$ | $< 8$ | $6.6 \pm 0.8 \pm 0.7$ |
| $\chi_{C1} K$ | $10 \pm 4$ | $6.1 \pm 0.6 \pm 0.6$ | $< 27$ | $3.1 \pm 0.9 \pm 0.4$ |
| $\chi_{C0} K$ | | $6.0^{+2.1}_{-1.8} \pm 1.1$ | | |

[1)]The world average compiled by PDG before the era of Belle and BABAR [8].
[2)]The first error is statistical, the second is systematic and the third when given is due to uncertainty from the branching fraction.


pub-02/104-T, hep-ph/0206018 (submitted to Phys. Rev. D) and references therein.
3. C. Edwards *et al.*, Crystal Ball collaboration, Phys. Rev. Lett. 48, 70 (1982).
4. S.-K Choi, S.Olsen *et al.*, Belle collaboration, Phys. Rev. Lett. 89, 102001 (2002).
5. J.H.Kühn, S.Nussinov and R.Rückl, Z. Phys. C5, 117 (1980).
6. K.Abe *et al.*, Belle collaboration, Phys. Rev. Lett. 89, 011803 (2002).
7. K.Abe *et al.*, Belle collaboration, Phys. Rev. Lett. 88, 031802 (2002).
8. D.E.Groom *et al.*, PDG group, Euro. Phys. J. C15, 1 (2000).
9. K.Abe *et al.*, Belle collaboration, Phys. Rev. Lett. 87, 161601 (2001).
10. K.Abe *et al.*, Belle collaboration, BELLE-CONF-209.
11. References not already given are in K.Abe *et al.*, Belle collaboration: BELLE-CONF-215 [$\eta_C K^*$]; and Phys. Lett. B538, 11 (2002) and ICHEP02-CONF-213 [$J/\psi K^{(*)}$]; BELLE-CONF-101 [$J/\psi \pi^+$].


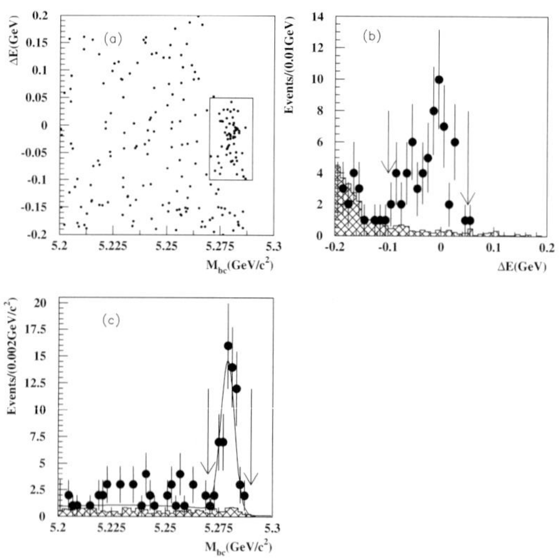


Figure 7. Distribution of $J/\psi \pi^0$ candidate events in $M_{\rm bc}$ vs. $\Delta E$ plane, and their projections. Hatched histograms correspond to events in side-bands of $M_{\rm bc}$ or $\Delta E$.

ELSEVIER

Proceedings of ICHEP 2002, pp. 569–573
*S. Bentvelsen, P. de Jong, J. Koch and E. Laenen (Editors)*

31st INTERNATIONAL CONFERENCE ON HIGH ENERGY PHYSICS AMSTERDAM

# $B$ Meson Decays to Open Charm and Charmonium at BaBar

E. W. Varnes[a]
(representing the BaBar collaboration)

[a]Department of Physics
P.O. Box 708
Princeton University
Princeton, NJ 08540

Since it began recording data in 1999, the BaBar detector at the SLAC B Factory has collected nearly 90 million $B$ meson pairs. Using this sample, we are able to make precise measurements of $B$ decays to open charm and charmonium. In particular, we have measured the rates for $B \to D\overline{D}K$, for the color-suppressed modes $B^0 \to D^0\pi^0$,$B^0 \to D^0\eta$ , and $B^0 \to D^0\omega$, for $B^+ \to \chi_c^0 K^+$, and for $B^0 \to D_s^{(*)+}D^{*-}$. We have also begun preparations for a measurement of the CKM angle $\gamma$ by studying the modes $B^+ \to D^0K^+$, $B^0 \to D_s^{(*)+}\pi^-$, and $B^0 \to D_s^{(*)+}K^-$.

## 1. Introduction

There are two primary motivations for studying $B$ decays to open charm and charmonium. The first is to improve the understanding of hadronic $B$ meson decay. In particular, we can test the decay rates to two-body modes predicted in some models.

The second motivation is to improve our understanding of CP violation in the $B$ meson system. Decays to charmonium states have already been used to perform a precision measurement of $\sin 2\beta$ [1]. A program aimed at measuring the other sides and angles of the Unitarity Triangle may reveal CP violation that cannot be explained by the standard model. Some open charm decays offer the promise of measuring the angle $\gamma$, and the first steps in the program are presented here.

## 2. Detector and Dataset

The BaBar detector is described in detail in [2]. It consists of a five-layer double-sided silicon tracker followed by a 40-layer drift chamber and Cherenkov particle identification system. A CsI(Tl) electromagnetic calorimeter provides precision measurement of electron and photon energies. All of these detectors are contained within the bore of a 1.5T solenoid. The flux return is instrumented with resistive plate chambers to allow detection of muons and neutral hadrons.

BaBar begin collecting data in 1999, and has recorded nearly 90 million $B$ meson pair events; some analyses discussed here use only a subset of this sample.

## 3. $B$ Meson Reconstruction

$B$ mesons are reconstructed by combining tracks or photons to form candidates for the intermediate mesons in the decay, constraining these mesons to their known masses, and then combining them to form $B$ candidates. Two kinematic variables distinguish signal from background. The first is $\Delta E \equiv E_B^* - E_{\rm beam}^*$, where $E_B$ is the $B$ candidate energy and $E_{\rm beam}$ is the beam energy, and the asterisks denote that these quantities are calculated in the center-of-mass frame. The second variable is $m_{ES} \equiv \sqrt{E_{\rm beam}^{*2} - p_B^{*2}}$, where $p_B$ is the $B$ candidate's momentum. True $B$ mesons will have $\Delta E$ near zero and $m_{ES}$ near the known $B$ mass.

Backgrounds from continuum quark pair production are reduced using variables that differentiate the two-jet structure typical of continuum events from the uniform energy distribution of $B\overline{B}$ events. Two of the most important variables


doi:10.1016/S0920-5632(02)02233-8

are the ratio of the second to zeroth Fox-Wolfram moments and the angle between the $B$ candidate direction and the thrust axis of the particles not taken as originating from the $B$ candidate.

## 4. Measurement of $\mathcal{B}(B \to D^{(*)}\overline{D}^{(*)}K)$

Measurements of the $B$ meson's semileptonic branching fraction $\mathcal{B}_{\rm sl}$ are lower than one would expect from a naive parton model. In addition, the average number of charm quarks produced during a decay $n_c$ is also lower than expected [3], and reconciling these two measurements is not straightforward.

Recently, more sophisticated models of $B$ decay have predicted a $\mathcal{B}_{\rm sl}$ consistent with experiment [4]. Also, in the past $n_c$ has been inferred under the assumption that the quark transition $b \to c\bar{c}s$ always produced either a $D_s$ or charmonium meson, so there may be a neglected contribution arising from light quark pair production from the vacuum, resulting in $D^{(*)}\overline{D}^{(*)}K$ final states.

The primary challenge in measuring $\mathcal{B}(B \to D^{(*)}\overline{D}^{(*)}K)$ is that there are a total of 10 (12) such modes for neutral (charged) $B$ mesons. The measurement presented here is the first in which all 22 $B$ decay modes are considered. $D^{*+}$ mesons are reconstructed via their decay to $D^0\pi^+$, and $D^{*0}$ mesons are reconstructed via both $D^0\pi^0$ and $D^0\gamma$. $D^0$ candidates are formed from $K^-\pi^+$, $K^-\pi^+\pi^0$, and $K^-\pi^+\pi^-\pi^+$, and $D^+$ candidates are formed from $K^-\pi^+\pi^+$. After selecting candidates with $\Delta E$ less than $2.5\sigma$ from its central value, a fit to the $m_{ES}$ distribution of all modes summed in a sample of 82 million $B\overline{B}$ pairs yields a signal of $823\pm57$ $B^0$ and $969\pm65$ $B^+$ events. To extract the branching fraction, a binned Poisson likelihood fit is performed for each submode. The fit accounts for the varying reconstruction efficiencies, background levels, and cross-feed among modes. Of the 22 $B$ decay modes, 11 have signals with greater than $4\sigma$ significance. The branching fraction for $B^0 \to D^*\overline{D}^{(*)}K$ is found to be $(4.3 \pm 0.3(\text{stat.}) \pm 0.6(\text{syst.}))\%$, and that for $B^+ \to D^*\overline{D}^{(*)}K$ is $(3.5\pm0.3(\text{stat.})\pm0.5(\text{syst.}))\%$. These values confirm that $D^*\overline{D}^{(*)}K$ modes do make a significant contribution to $n_c$. A more complete description of this analysis is available in [5].

## 5. Measurement of Color-suppressed Decay Rates

In the simplest versions of the factorization model, two-body $B$ decays are classified according to whether the quarks from the virtual $W$ boson decay hadronize as a single meson (color-allowed decay) or pair with the $c$ and spectator quarks in forming mesons (color-suppressed decay) [6]. The non-perturbative components of the decay amplitude are assumed to be the same for all decays in each class, and are denoted $a_1$ and $a_2$ for color-allowed and color-suppressed decays respectively. Within this framework, the rates for the color-suppressed decays $B \to D^0\pi^0$, $B \to D^0\eta$, and $B \to D^0\omega$ are expected to be between 0.3 and $0.7 \times 10^{-4}$.

We reconstruct these decays by forming $D^0$ mesons from $K^-\pi^+$, $K^-\pi^+\pi^0$, and $K^-\pi^+\pi^-\pi^+$, $\pi^0$ and $\eta$ mesons from photon pairs, and $\omega$ mesons from $\pi^+\pi^-\pi^0$. The $D^0\pi^0$ mode faces a substantial background from the color-allowed decay $B^+ \to D^0\rho^+$, which can mimic the signal if the $\pi^+$ originating from the $\rho^+$ is sufficiently soft. To reduce this background, any candidates to which a $\pi^+$ can be added to create a viable $D^0\rho^+$ candidates are excluded. In the $D^0\omega$ analysis, the vector nature of the $\omega$ results in its daughters having distinctive angular distributions, and this fact is exploited to reduce backgrounds.

The $\Delta E$ distributions for candidates in all three modes in a sample of 49 million $B$ meson pairs are shown in Fig. 1. The signals are extracted using an unbinned maximum likelihood fit to the $m_{ES}$ distributions, and corrected for the amount of peaking background predicted from Monte Carlo simulations. $291 \pm 31$ $D^0\pi^0$, $101 \pm 14$ $D^0\eta$, and $78 \pm 12$ $D^0\omega$ events are observed, which correspond to branching fractions of $(2.89 \pm 0.29(\text{stat.}) \pm 0.38(\text{syst.})) \times 10^{-4}$, $(2.41 \pm 0.39(\text{stat.}) \pm 0.32(\text{syst.})) \times 10^{-4}$, and $(2.48 \pm 0.40(\text{stat.}) \pm 0.32(\text{syst.})) \times 10^{-4}$ respectively. These values are significantly higher than the factorization model predicts, indicating that the phenomenology of $B$ decays requires a more detailed treatment. A more complete description of this analysis can be found in [7].

## 6. Measurement of $\mathcal{B}(B^+ \to \chi_{c0} K^+)$

Another prediction of factorization models is that decays such as $B^+ \to \chi_{c0} K^+$ are disallowed, since there is no color-singlet operator that can give rise to this final state [8]. However, if soft gluon interactions (i.e. color-octet operators) play a significant role, the decay can proceed [9]. By comparing the rate of $B^+ \to \chi_{c0} K^+$ to that for $B^+ \to \chi_{c1} K^+$ (which can proceed through both color-singlet and -octet operators) the importance of soft gluon interactions in these decays can be estimated.

We reconstruct $\chi_{c0}$ candidates via their decay to $\pi^+\pi^-$ and $K^+K^-$, with particle identification information used to assign the proper mass to the tracks. Rejection of continuum background is achieved by combining eleven variables into a Fisher discriminant. The most dangerous residual background is that arising from non-resonant $B^+ \to \pi^+\pi^- K^+$ or $K^+K^-K^+$, which have the same distribution in $m_{ES}$ and $\Delta E$ as the signal. To mitigate their impact, any candidates with a $K\pi$ pair having invariant mass near the $D^0$ mass, or a $KK$ pair near the $\phi$ mass, are excluded from the analysis. Monte Carlo simulation predicts that the remaining "exact match" backgrounds are negligible. This is confirmed by extracting the signal with two methods. In the first, candidates near the $\chi_{c0}$ mass are selected, and the signal taken from a fit to the $m_{ES}$ distribution. In the second method, the $\chi_{c0}$ mass window is widened, and a two-dimensional fit to the $m_{ES}$ and $\chi_{c0}$ mass distributions is used to extract the signal. The central value for the branching fraction is derived from the first method, and the difference between the two is taken as a systematic uncertainty.

The resulting preliminary branching fraction of $(2.4 \pm 0.7(\text{stat.} \oplus \text{syst.})) \times 10^{-4}$ is significantly non-zero, but smaller than that for $B^+ \to \chi_{c1} K^+$, implying that soft gluon exchange plays an important, but not dominant, role in $B$ decay to charmonium final states. Further details are available in [10].

## 7. Measurement of Rate and Polarization of $B^0 \to D_s^{(*)+} D^{*-}$)

An additional constraint on models of $B$ decay can be obtained from decays of the type $B^0 \to D_s^{(*)+} D^{*-}$. In this case, not only the decay rate but also the polarization of the vector-vector mode $B^0 \to D_s^{*+} D^{*-}$ can be compared to their predicted values. The factorization model predicts: $\mathcal{B}(B^0 \to D_s^+ D^{*-}) = (0.58 - 0.70)\%$, $\mathcal{B}(B^0 \to D_s^{*+} D^{*-}) = (2.32 - 2.45)\%$, and $\Gamma_L/\Gamma_{\text{tot}} = 52\%$ [6].

For this analysis, a full reconstruction of the $B$ meson is not done. Rather, only the $D_s^{(*)}$ and the soft pion from the $D^*$ decay are found. Applying kinematic constraints from the $B$ and $D^*$ masses and the beam energy allows the $B$ meson to be reconstructed up to an arbitrary angle about the slow pion direction. The variable used to distinguish signal from background is the missing mass, which peaks at the $D$ mass for signal. This technique is much more efficient than fully reconstructing the $D^*$ meson, so only 23 million $B$ meson pairs are required for this analysis.

The signal yields are determined by fitting a Gaussian signal and empirical background shape to the missing mass distribution, and the branching fractions are calculated using the signal efficiency and cross-feed determined by Monte Carlo simulation. The preliminary results are $\mathcal{B}(B^0 \to D_s^+ D^{*-}) = (1.03 \pm 0.14(\text{stat.}) \pm 0.13(\text{syst}) \pm 0.26(\text{ meson } \mathcal{B}))\%$ and $\mathcal{B}(B^0 \to D_s^{*+} D^{*-}) = (1.97 \pm 0.14(\text{stat.}) \pm 0.13(\text{syst}) \pm 0.26(\text{ meson } \mathcal{B}))\%$. These values are consistent within $2\sigma$ with the values predicted by factorization.

The polarization of the $D_s^{*+} D^{*-}$ mode is determined from the helicity angles of the slow pion from the $D^*$ decay and the photon from the $D_s^{*+}$ decay. Here the missing mass is constrained to the $D$ mass to fix the pion angle. The observed distributions are corrected for background and detector acceptance. To minimize the magnitude of these corrections, only the purest subsample, that in which the $D_s^+$ is reconstructed as $\phi\pi^+$, is used. The preliminary resulting polarization

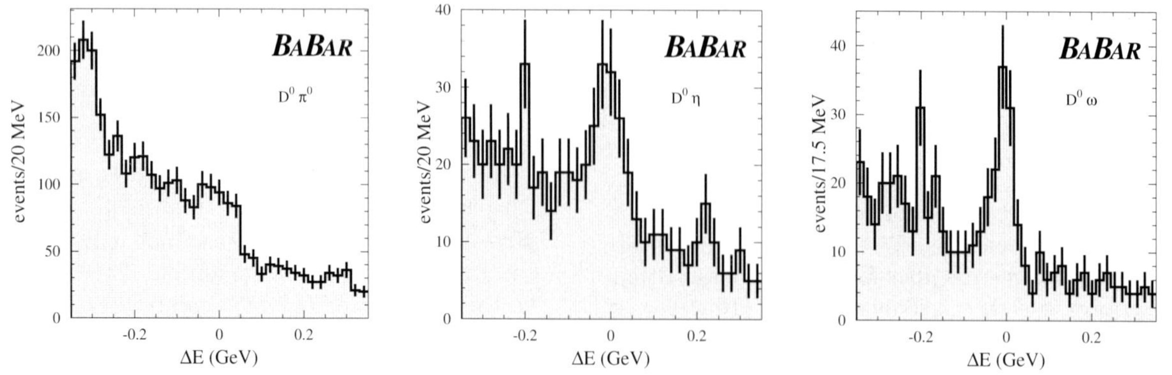


Figure 1. $\Delta E$ distributions for $B^0 \to D^0\pi^0$ (left), $B^0 \to D^0\eta$ (center), and $B^0 \to D^0\omega$ (right) candidates.

is $\Gamma_L/\Gamma_{\rm tot} = (51.9 \pm 5.0(\text{stat.}) \pm 2.8(\text{syst.}))\%$, in excellent agreement with the prediction from factorization. For further details on this analysis, see [11].

## 8. Measurement of Rate and CP Asymmetry for $B^+ \to D^0K^+$

Measurement of the CKM angle $\gamma$ in a manner analogous to that of the $\sin 2\beta$ measurement requires the study of $B_s$ decay, and thus will not be possible at BaBar for the foreseeable future. However, other techniques have been proposed, some of which depend on measuring the rate and direct CP asymmetry in the decay $B^+ \to D^0K^+$ [12].

For the rate analysis, $D^0$ mesons are reconstructed via their decays to $K^-\pi^+$, $K^-\pi^+\pi^0$, and $K^-\pi^+\pi^-\pi^+$. The primary challenge is that the $D^0K^-$ signal has the same distribution in $m_{ES}$ as the more common $D^0\pi^-$ mode. To separate the signal from this and other backgrounds, a three-dimensional unbinned maximum likelihood fit to the $m_{ES}$, $\Delta E$, and $K$ identification probability distributions is used. To partially cancel systematics, the relative rate of $D^0K^-$ to $D^0\pi^-$ is measured, and the preliminary result (based on a total extracted signal of $810 \pm 33$ events from 58 million $B$ meson pairs), is $(8.31 \pm 0.35(\text{stat.}) \pm 0.20(\text{syst.}))\%$.

For the CP asymmetry measurement, decays of the $D$ to a CP eigenstate must be used, and we choose the mode $D^0 \to K^+K^-$. Since the uncertainty for this measurement is dominated by statistics, the sample size is increased to 81 million $B$ pairs. Within this sample we observe $21.7 \pm 5.6$ $B^-$ and $15.3 \pm 5.6$ $B^+$ candidates, yielding a preliminary CP asymmetry of $0.17 \pm 0.23(\text{stat.})^{+0.09}_{-0.08}(\text{syst.})$, consistent with zero. Further information about these analyses can be found in [13].

## 9. Study of $B^0 \to D_s^{(*)+}\pi^-$ and $B^0 \to D_s^{(*)+}K^-$

Another way to obtain information about $\gamma$ is to measure the time-dependent CP asymmetry in $B^0 \to D^{(*)+}\pi^-$, for which the amplitude $\propto \sin(2\beta + \gamma)$. However, the measurement is not as simple that for $\sin 2\beta$ since the interference that leads to the asymmetry is between the Cabibbo-allowed $\overline{B}^0 \to D^{(*)+}\pi^-$ and Cabibbo-suppressed $B^0 \to D^{(*)+}\pi^-$ decays. This means not only that the asymmetry is small ($\mathcal{O}(10^{-2})$) but also that one must know the ratio of the Cabibbo-suppressed to Cabibbo-allowed rates $\lambda_{D\pi}$ in order to extract $\sin(2\beta + \gamma)$.

A step in this direction may be taken by measuring the rate of $B^0 \to D_s^{*+}\pi^-$, which in the factorization approximation is related via SU(3) to $\lambda_{D\pi}$. This relation only holds, however, to the extent that the $W$-exchange contribution to $B^0 \to D^{(*)+}\pi^-$ is small. This can be determined by measuring the rate for $B^0 \to D_s^{*+}K^-$, which proceeds only via $W$ exchange. The latter measurement is also of interest as no pure $W$ exchange $B$ decays have been previously observed.

Using 84 million $B$ pairs, evidence at the 3.3 and $3.5\sigma$ levels is seen for the decays $B^0 \to D_s^+\pi^-$ and $B^0 \to D_s^+K^-$ respectively, with branching fractions of $(3.2\pm0.9(\text{stat.})\pm1.0(\text{syst.}))\times10^5$ and $(3.2\pm1.0(\text{stat.})\pm1.0(\text{syst.}))\times10^5$. There is less evidence for $B^0 \to D_s^{*+}\pi^-$ and $B^0 \to D_s^{*+}\pi^-$, so we only quote upper limits on the branching fractions at 90% confidence level of $4.1\times10^{-5}$ and $2.5\times10{-}5$ respectively. The $m_{ES}$ distributions are shown in Fig. 2.

These results imply that the rate for $W$-exchange diagrams is less than 1% of that for the dominant $W$-emission diagram, and therefore that the correction to $\lambda_{D\pi}$ under the factorization assumption is negligible. In this model, then, $\lambda_{D\pi}$ is $0.020\pm0.005(\text{stat.}\oplus\text{syst.})$. The central question of whether factorization is a reasonable model for these decays will only be answered through an extensive program of measuring these and related decay rates. Further details of this analysis are available in [14].

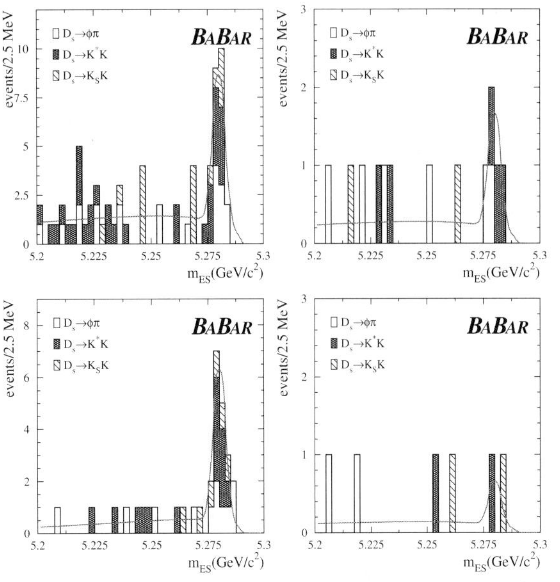


Figure 2. $m_{ES}$ distributions of candidates for $B^0 \to D_s^+\pi^-$ (upper left), $B^0 \to D_s^{*+}\pi^-$ (upper right), $B^0 \to D_s^+K^-$ (lower left) and $B^0 \to D_s^{*+}K^-$ (lower right).

## 10. Summary

$B$ meson decays to open charm and charmonium provide excellent constraints on models of hadronic $B$ decay, and also offer possibilities of providing additional information about CP violation in the $B$ system. The results presented here show that the average charm multiplicity in $B$ meson decay is consistent with expectations. It is also seen that the simplest factorization models cannot explain all hadronic $B$ decays, such as those to color-suppressed modes. Finally, we have demonstrated the ability to complete the first steps of a program to measure the CKM angle $\gamma$.

## REFERENCES

1. B. Aubert et al. (BaBar collaboration), submitted to Phys. Rev. Lett. (hep-ex/0207042).
2. B. Aubert et al. (BaBar collaboration), Nucl. Instrum. Meth. A479 (2002) 1.
3. D.E. Groom et al. (Particle Data Group), Eur. Phys. J 15 (2000) 1.
4. M. Neubert and C.T. Sachrajda, Nucl. Phys. B483 (1997) 339.
5. B. Aubert et al. (BaBar collaboration), hep-ex/0207086 (2002).
6. M. Neubert and B. Stech, in Heavy Flavors, 2nd Edition (1997).
7. B. Aubert et al.(BaBar collaboration), hep-ex/0207092 (2002).
8. J.H. Kühn and R. Rückl, Phys. Lett. B135 (1984) 477.
9. G.T. Bodwin et al., Phys. Rev. D46 (1992) 3703.
10. B. Aubert et al.(BaBar collaboration) , hep-ex/0207066 (2002).
11. B. Aubert et al.(BaBar collaboration) , hep-ex/0207079 (2002).
12. M. Gronau and D. Wyler, Phys. Lett. B265 (1991) 172.
13. B. Aubert et al.(BaBar collaboration) , hep-ex/0207087 (2002).
14. B. Aubert et al.(BaBar collaboration) , hep-ex/0207042 (2002).

ELSEVIER

Proceedings of ICHEP 2002, pp. 574–577
*S. Bentvelsen, P. de Jong, J. Koch and E. Laenen (Editors)*

31st INTERNATIONAL CONFERENCE ON HIGH ENERGY PHYSICS AMSTERDAM

# Charmless two-body decays

Ascelin Gordon*, for the Belle Collaboration
School of Physics, University of Melbourne, Victoria 3010 Australia

We report the latest results of a study of charmless two-body $B$ meson decays using data recorded by the Belle experiment at KEKB. Branching fractions or upper limits are provided for the complete set of $K\pi$, $\pi\pi$ and $K\overline{K}$ final states; the $B \to \rho\pi$ decays with final states $\rho^0\pi^+$, $\rho^\pm\pi^\mp$ and $\rho^0\pi^0$; we also provide a first measurement of the branching fraction of the vector-vector decay $B^+ \to \rho^+\rho^0$.

## 1. Introduction

This report summarises the latest results from the Belle collaboration on measuring branching fractions and upper limits for $B$ meson decays to charmless two-body modes. These include decays with final states $K\pi$, $\pi\pi$, and $K\overline{K}$ (which we collectively label $hh$) as well as the pseudo-two-body decays $\rho\pi$ and $\rho^+\rho^0$. Through investigating these modes, many of the parameters that describe $CP$ violation can be measured. The $hh$ modes alone contain enough information to measure all angles of the Unitarity Triangle [1]. The $\rho\pi$ and $\rho^+\rho^0$ decays also provide ways of measuring $\phi_2$ [2] and $\phi_3$ [3]. Partial-rate and time-dependent asymmetry measurements have been performed using some of the $hh$ modes; the results are reported elsewhere [4,5].

The results presented here are based on data taken with the Belle detector [6] at KEKB [7]. The $hh$ and $\rho\pi$ results are based on a 29 fb$^{-1}$ data sample while the $\rho\rho$ results use a 43 fb$^{-1}$ data sample.

## 2. Event Reconstruction

$B$ meson candidates are identified using the beam constrained mass $M_{bc} = \sqrt{E_{\text{beam}}^2 - p_B^2}$ and the energy difference $\Delta E = E_B - E_{\text{beam}}$. Here, $p_B$ and $E_B$ are the momentum and energy of a $B$ candidate in the centre-of-mass (CM) frame and $E_{\text{beam}}$ is the CM beam energy.

For charged particle identification (PID), three subsystems of the Belle detector are used. The central drift chamber $dE/dx$ measurements are combined with the responses of the aerogel threshold Čerenkov counters and time-of-flight scintillation counters to form likelihoods $\mathcal{L}_\pi$ or $\mathcal{L}_K$. We distinguish pions from kaons by applying selection requirements on the likelihood ratio $\mathcal{LR} = \mathcal{L}_\pi/(\mathcal{L}_\pi + \mathcal{L}_K)$. A similar likelihood ratio including calorimeter information is used to identify electrons. All charged tracks that originate from the interaction point and are not positively identified as electrons are considered as kaon or pion candidates.

Candidate $K_S^0$ mesons are reconstructed using pairs of oppositely charged tracks with a displaced vertex as well as a flight direction and invariant mass consistent with a $K_S^0$ originating from the interaction point. Neutral pion candidates are selected from $\gamma\gamma$ pairs by requiring that their invariant mass to be within about $3\sigma$ of the nominal $\pi^0$ mass. To reduce combinatorial backgrounds, mode-dependent momentum cuts are made on the $\gamma$ and $\pi^0$ momenta. Candidate $\rho$ mesons are reconstructed from pion pairs whose invariant mass is required to be within a mode-dependent window of about 150 MeV either side of the nominal $\rho$ mass.

All the modes presented here have large backgrounds from $e^+e^- \to q\overline{q}$ continuum events ($q = u, d, s, c$), which typically exhibit a two-jet-like structure. This background is suppressed by means of a likelihood ratio derived from two variables. One is a Fisher discriminant formed from six modified Fox-Wolfram [8] moments. The coefficients of the Fisher discriminant are determined

*email: gordon@physics.unimelb.edu.au


doi:10.1016/S0920-5632(02)02234-X

by optimizing the separation between the signal and continuum MC events. The other variable is $\cos\theta_B$, where $\theta_B$ is the angle between the $B$ flight direction and the electron beam direction. Probability density functions obtained from the Fisher discriminant and the $\cos\theta_B$ distribution are multiplied to form the likelihood functions for signal ($\mathcal{L}_{sig}$) and continuum ($\mathcal{L}_{cont}$) [9]. A requirement on the likelihood ratio $\mathcal{L}_{sig}/(\mathcal{L}_{sig}+\mathcal{L}_{cont})$ is then made which varies from mode to mode. As the $\rho^+\rho^0$ mode suffers from the largest continuum background, additional suppression is achieved by making an initial requirement of $|\cos\theta_{thr}| < 0.8$ ($\theta_{thr}$ is the angle between the thrust axis of the candidate tracks and that of the remaining tracks in the event).

## 3. Signal Yield Extraction

Figures 1 to 3 show the $\Delta E$ distributions in the $M_{bc}$ signal region for many of the modes analysed after all the selection criteria have been applied. Signal yields are obtained by performing maximum likelihood fits on the $\Delta E$ distributions, each using a signal function and one or more background functions. The signal functions are obtained from the MC and adjusted based on comparisons of $B \to D\pi$ decays in data and MC. Different $D$ meson decays are used to obtain the same final state particles as the modes being measured. The solid line in all plots shows the sum of the signal and background components. A Chebyshev polynomial with coefficients determined from the $M_{bc}$ sideband is used to model the continuum background for all fits. Backgrounds from other $B$ decays vary for the different modes. Below is a detailed description of $\Delta E$ fits for the modes analysed.

### 3.1. $B \to hh$

Apart from continuum, the $\Delta E$ fits for the $hh$ modes include up to two background components: crossfeeds from other misidentified $B \to hh$ decays and backgrounds from multi-body and radiative charmless $B$ decays. For charged particle final states, the signal is modeled with a Gaussian. For modes containing $\pi^0$s, the signal is modeled as the sum of a primary Gaussian and a secondary asymmetric Gaussian. The crossfeed component has the same shape as the signal component but is shifted by 45 MeV. Backgrounds from charmless $B$ decays are modeled by a smoothed MC histogram. For all final states except $K^+\pi^0$ and $\pi^+\pi^0$, the normalizations of the four components are the only free parameters in the fits. Due to the large overlap of the signal and crossfeed components in the $K^+\pi^0$ and $\pi^+\pi^0$ signals, we perform a simultaneous fit to the $K^+\pi^0$ and $\pi^+\pi^0$ $\Delta E$ distributions constraining the crossfeed to the expected values based on the PID fake rates.

### 3.2. $B \to \rho\pi$

For the $\rho^0\pi^+$ mode there are rare $B$ decays that are expected to contaminate the $\Delta E$ distribution. The modes are of the type $B^0 \to h^+h^-$ , $B \to \rho\rho$ and $B \to K\pi\pi$. These background modes are modeled by smoothed MC histograms. The normalisations of the $K\pi\pi$ and $h^+h^-$ components are fixed to their expected yields, which are calculated using efficiencies determined by MC and branching fractions measured by previous Belle analyses [10]. The $\rho\rho$ component is left to float.

For the $\rho\pi$ modes with a $\pi^0$, the signal component is modeled by a Crystal Ball (CB) function [11]. The parameterization for rare $B$ decays includes one component for the $B \to K\pi\pi^0$ modes [12], one for all the $B \to \rho\rho$ modes and a component for the background from the $b \to c$ transition. The normalisation of the $B \to \rho\rho$ component is left to float while the other components are fixed to their expected yields.

### 3.3. $B \to \rho\rho$

The $\Delta E$ fit for the $\rho^+\rho^0$ mode uses the sum of a Gaussian and a CB function to represent the signal. The only background contributions in the fit is from the $b \to c$ transition and is modeled by a smoothed histogram with a shape that is obtained from MC. All parameters are fixed except the overall normalisation.

## 4. Results

Table 1 summarizes the results of the $\Delta E$ fits, showing the number of events, signal yields, reconstruction efficiencies, statistical significance and branching fractions (BF) or upper limits

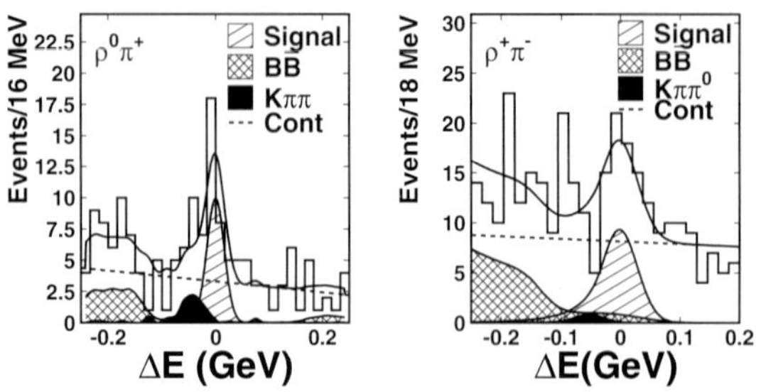


Figure 1. The $\Delta E$ distribution for the $B \to \rho^0\pi^+$ and $B^0 \to \rho^\pm\pi^\mp$ modes. The various backgrounds are shown with different hatchings.

Table 1
The signal yields, efficiencies, statistical significance, and branching fractions/upper limits from the $\Delta E$ fits for all modes. * See text for description of $\rho^+\rho^0$ efficiency.

| mode | $N_{sig}$ | $\epsilon$ (%) | S($\sigma$) | BF/UL[$\times 10^{-6}$] |
|---|---|---|---|---|
| $K^+\pi^-$ | 218 | 31 | 16.4 | $22.5 \pm 0.19 \pm 0.18$ |
| $K^+\pi^0$ | 59 | 30 | 6.4 | $13.0^{+0.25}_{-0.24} \pm 0.13$ |
| $K^0\pi^+$ | 67 | 20 | 7.6 | $19.4^{+0.31}_{-0.30} \pm 0.16$ |
| $K^0\pi^0$ | 20 | 14 | 2.8 | $8.0^{+0.33}_{-0.31} \pm 0.16$ |
| $\pi^+\pi^-$ | 51 | 16 | 5.4 | $5.4 \pm 0.12 \pm 0.05$ |
| $\pi^+\pi^0$ | 37 | 32 | 3.5 | $7.4^{+0.23}_{-0.22} \pm 0.09$ |
| $\pi^0\pi^0$ | 13 | 17 | 2.4 | $< 6.4$ |
| $K^+K^-$ | $0^{+3.2}_{-0}$ | 23 | - | $< 0.9$ |
| $K^+\bar{K}^0$ | $0^{+2.0}_{-0}$ | 13 | - | $< 2.0$ |
| $K^0\bar{K}^0$ | $0^{+2.9}_{-0.9}$ | 20 | - | $< 8.2$ |
| $\rho^0\pi^+$ | 24.3 | 9.6 | 4.4 | $8.0^{+2.3+0.7}_{-2.0-0.7}$ |
| $\rho^+\pi^-$ | 44.6 | 6.8 | 3.7 | $20.8^{+6.0+2.8}_{-6.3-3.1}$ |
| $\rho^0\pi^0$ | -4.4 | 8.5 | - | $< 5.3$ |
| $\rho^+\rho^0$ | 31 | * | 4.2 | $38.5 \pm 10.9^{+5.9+2.5}_{-5.4-7.5}$ |

(UL) obtained from each fit. The first error in the BF is statistical, the send systematic. The statistical significance is defined as $\sqrt{-2\ln(\mathcal{L}_0/\mathcal{L}_{max})}$, where $\mathcal{L}_{max}$ denotes the likelihood at the nominal signal yield and $\mathcal{L}_0$ is the likelihood with the signal yield fixed to zero. The systematic error in the fitting procedure is determined by varying the parameters of the fitting functions within their errors and measuring the change in the signal yield.

There is an added complication for the $\rho^+\rho^0$ mode: the final state is a vector-vector system so both longitudinal ($H_{11}$) and transverse ($H_{00}$) polarizations of the $\rho$ meson are possible. For these two polarizations the $\rho \to \pi\pi$ daughters have different momentum distributions giving the $H_{00}$ state a lower reconstruction efficiency of $1.81 \pm 0.04\%$ compared to the $3.27 \pm 0.05\%$ for the $H_{11}$ final state.

The fraction of events with each polarisation is obtained by performing a simultaneous fit to the background-subtracted $\rho^+$ and $\rho^0$ helicity-angle distributions using MC-determined expectations for the $H_{00}$ and $H_{11}$ helicity states. The fit results are $H_{00} = 0.86 \pm 0.41$, $H_{11} = 0.14 \pm 0.23$ which indicate that the $H_{00}$ state dominates, as expected [13]. We estimate the branching fraction error due to the uncertain mixture of helicity states by shifting the ratio of the $H_{00}$ state to the $H_{11}$ state by $\pm 1\sigma$, and assign the change in the branching fraction as the third error.

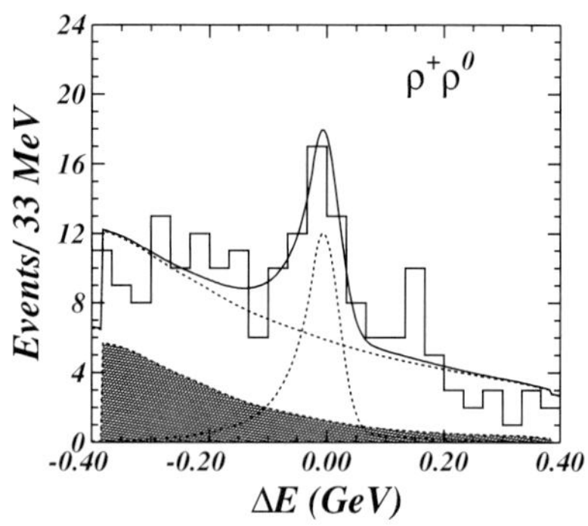


Figure 2. The $\Delta E$ distribution for the $B \to \rho^+\rho^0$ candidates. The hatched histogram represents the $b \to c$ background component.

## 5. Conclusion

The complete set of $hh$ modes has been investigated with significant signals seen in the $K\pi$, $\pi^+\pi^-$ and $\pi^+\pi^0$ modes. We place upper limits on $\pi^0\pi^0$ and the $K\overline{K}$ modes. For the $B \to \rho\pi$ modes, branching fractions $\mathcal{B}(B^+ \to \rho^0\pi^+) = (8.0^{+2.3+0.7}_{-2.0-0.7}) \times 10^{-6}$ and $\mathcal{B}(B^0 \to \rho^\pm\pi^\mp) = (20.8^{+6.0+2.8}_{-6.3-3.1}) \times 10^{-6}$ are obtained. No evidence is seen for the mode $B^0 \to \rho^0\pi^0$ and we pro-

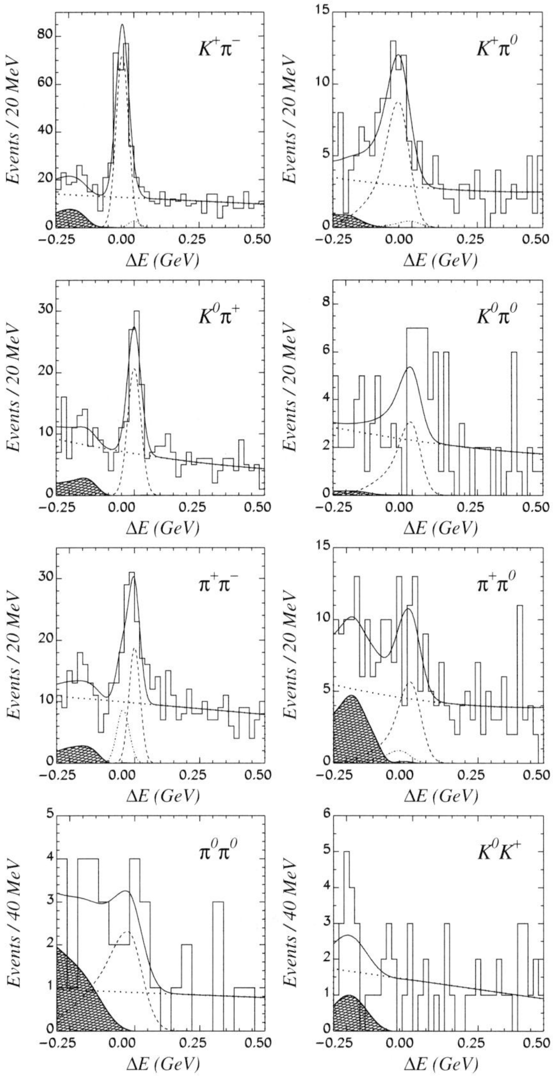


Figure 3. The $\Delta E$ distribution for the $B \to hh$ modes. The hatched histograms represents the charmless $B$ background component. For the $K^+\pi^-$ and $K^+\pi^0$ distributions, the crossfeed components from $\pi^+\pi^-$ and $\pi^+\pi^0$ are shown by dot-dashed curves centered 45 MeV above the signal components.

vide an upper limit of $5.3 \times 10^{-6}$ at 90% confidence level. We also provide a measurement of the first charmless $b \to u$ vector-vector mode $B^+ \to \rho^+\pi^0$ with a branching fraction $\mathcal{B}(B^+ \to \rho^+\rho^0) = (38.5 \pm 10.9^{+5.9+2.5}_{-5.4-7.5}) \times 10^6$.

## REFERENCES

1. H. Quinn, A. I. Sanda, Eur. Phys. Jour. C **15**, 628 (2000).
2. A.E. Snyder and H.R. Quinn, Phys. Rev. D **48**, 2139 (1993).
3. I. Bediaga, R.E. Blanco, C. Göbel, and R. Méndez-Galain, Phys. Rev. Lett. **81**, 4067 (1998).
4. Belle Collaboration, B. C. K. Casey, *et al.* submitted to PRD ; hep-ex/0207090
5. Belle Collaboration, K.Abe, *et al.* Phys. Rev. Lett. **89**, 071801 (2002).
6. Belle Collaboration, A. Abashian *et al.*, Nucl. Instr. and Meth. A **479**, 117 (2002).
7. E .Kikutani ed., KEK Preprint 2001-157 (2001), to appear in Nucl. Instr. and Meth. A.
8. G.C. Fox and S. Wolfram, Phys. Lett. **B** 41, 1581 (1978).
9. For more details see: Belle Collaboration, A. Gordon *et al.*, Phys. Lett. B **542** 183, (2002).
10. Belle Collaboration, A. Garmash *et al.*, Phys. Rev. D **65**, 092005 (2002).
11. J.E. Gaiser *et al.*, Phys. Rev. D **34**, 711 (1986).
12. Belle Collaboration, K. Abe *et al.*, BELLE-CONF-0115, submitted as a contribution paper to the 2001 International Europhysics Conference on High Energy Physics (EPS-HEP 2001).
13. R. Aleksan *et al.*, Phys. Rev. Lett. **356**, 95 (1995).

ELSEVIER

Proceedings of ICHEP 2002, pp. 578–581
*S. Bentvelsen, P. de Jong, J. Koch and E. Laenen (Editors)*

31st INTERNATIONAL CONFERENCE ON HIGH ENERGY PHYSICS AMSTERDAM

# Charmless Multi-body $B$ Decays

P. Chang for the Belle Collaboration [a]

[a]Department of Physics, National Taiwan University,
No. 1, Sec. 4, Roosevelt Rd., Taipei, Taiwan

We report on current studies of three body charmless hadronic $B$ meson decays with at least one kaon in the final state. The results are obtained using the data sample produced on the $\Upsilon(4S)$ resonance and collected with the Belle and CLEO detectors. The branching fractions of $B$ decays to three kaon final states are presented. And further study using an isospin analysis shows the dominance of CP-even component in the non-$\phi$ $K_S K^+ K^-$ final state. We observe the decays $B^0 \to K^0\pi^+\pi^-$ and $B^0 \to K^+\pi^-\pi^0$. In the same final states, significant signals of $B^0 \to K^*(892)^+\pi^-$ channel have been seen by both CLEO and Belle. However, $K^*(892)^+ \to K^+\pi^0$ candidates in the Belle analysis are not fully polarized, suggesting that other states contribute.

## 1. Introduction

Recently observations of large three-body charmless hadronic decays of $B$ mesons have been reported by the $B$ factory experiments[1]. Studies of these charmeless three-body decays have significantly broaden the understanding of $B$ meson decay mechanism and provide additional possibilities for CP violation searches. Experimentally, these three-body studies provide a better way to search for $B$ decays to pseudoscalar-vector final states where possible feed-down or quantum interferences from different states need to be taken into account. In this article, we report on studies of three-body $B$ meson decays with at least one kaon in the final state using the data samples collected by the Belle or CLEO detectors. Search for two-body states with intermediate vector resonances is also reported. The inclusion of charged conjugate states is always implied.

Detailed descriptions of the Belle and CLEO detectors can be found elsewhere[2,3]. Charged tracks in both analyses are required to satisfy criteria based on track fit quality and the distances closest to the interaction point. $K-\pi$ separation is achieved using the particle identification devices in both experiments. $K_S$ candidates are selected from pairs of charged tracks with their invariant massed consistent to the nominal $K_S$ mass and well-measured displaced vertices. $\pi^0$ candidates are identified from pairs of photons with the minimum momentum of 200 MeV$/c$ and 1 GeV$/c$ for the Belle and CLEO analyses, respectively. $B$ meson candidates are identified using the beam constrained mass $M_{bc} = \sqrt{E^2_{\text{beam}} - P^2_B}$ and the energy difference $\Delta E = E_B - E_{\text{beam}}$.

The dominant background in these analyses is the continuum $q\bar{q}$ events, which is suppressed using various event shape variables. Belle combines these shape variables into a fisher discriminant or likelihood ratio and $B$ candidates are obtained using the $\Delta E$ fit after applying a cut on fisher discriminant or likelihood ratio. CLEO applies a loose cut using the sphericity angle [4] and then obtain $B$ yields from a multi-dimension fit on $M_{bc}, \Delta E, dE/dX$ and fisher discriminant formed by all other shape variables. We reject candidates that are consistent with the exclusive $b \to c$ transitions $B \to D\pi$, where $D \to K\pi$, and $B \to \psi K$, where two muons from $\psi$ decays are misidentified as pions. Other $B$ decay backgrounds are studied using a large GEANT based Monte Carlo sample.

## 2. $B$ to Three Kaons

Figure 1 1 shows the $\Delta E$ distributions for all three-kaon final states using 78 fb$^{-1}$ data with the Belle detector. Significant signals show up in all four three-kaon channels. Their results of branching fraction measurements summarized in


doi:10.1016/S0920-5632(02)02239-9

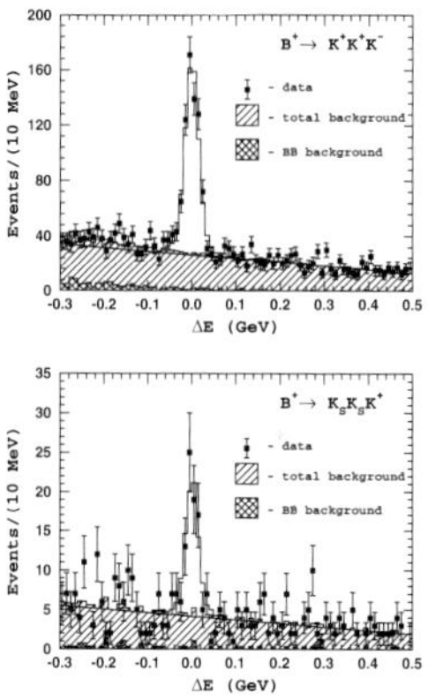


Figure 1. $\Delta E$ distributions for $B \to KKK$ decays. Points are data, open histograms show the fit results, the straight lines indicate the $q\bar{q}$ background and the dark shaded area represents the $B\bar{B}$ background expected from Monte Carlo simulations.

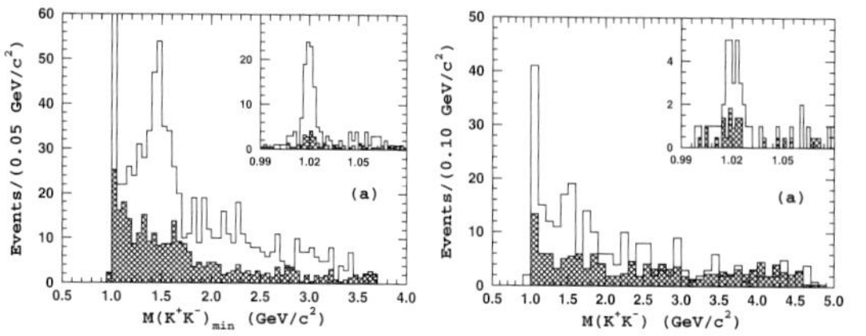


Figure 2. The $K^+K^-$ invariant mass distributions for the $K^+K^+K^-$ (left) and $K^0_SK^+K^-$ (right) channels. Open histograms are for the events in the $M_{bc} - \Delta E$ signal region while hatched histograms for events in the $\Delta E$ sideband. The insets in both plots are the blow-up in the low mass region.

Table 1 are obtained by normalizing yields to the observed $B^+ \to \overline{D}^0\pi^+, \overline{D}^0 \to K^+\pi^-$ and $B^0 \to D^-\pi^+, D^- \to K^0_S\pi^-$ signals. The reconstruction efficiencies for $K^+K^+K^-$ and $K^0K^+K^-$ modes are determined using a model parameterized by a $\phi K$ and a $f_X K$ intermediate states to take into account the non-uniform distributions over the Dalitz plot. The $f_X$ state is a hypothetical side scalar, which was observed in the previous Belle analysis [1]. Figure 2 shows the $K^+K^-$ invariant mass distributions, where similar $\phi$ and high mass $f_X$ signatures are observed in both $K^+K^+K^-$ and $K^0_SK^+K^-$ modes. For the $K^0_SK^0_SK^+$ and $K^0_SK^0_SK^0_S$ final states, MC simulated events with a simple phase-space model are used to determine their reconstruction efficiencies.

Large $B^0 \to K^0_SK^+K^-$ signals could be very useful to measure the CP violation parameter $\sin(2\phi_1)$ (also known as $\sin(2\beta)$) and the result with a significant deviation from the $b \to c\bar{c}K^0$ channel indicates new physics. However, the CP-parity of the $K_SK^+K^-$ final state, besides $\phi K_S$ state, is not fixed. One needs to measure their CP content before performing an time dependent CP search. Moreover, although $b \to s$ penguin processes are expected to be dominant, possible $b \to u$ tree contributions may introduce an additional weak phase in the $B^0 \to K_SK^+K^-$ amplitude and complicates the interpretation of any observed CP violation. A quantitative estimate of the size of the $b \to u$ tree amplitude can be achieved using the $B^+ \to K^+K^-\pi^+$ mode, where $W^+ \to \bar{d}u$ is the dominant contributor, and the following ratio:

$$F \equiv \frac{|A^{KKK}_{b\to u}|^2}{|A^{KKK}_{total}|^2} \sim \frac{\mathcal{B}(B^+ \to K^+K^-\pi^+)}{\mathcal{B}(B^+ \to K^+K^+K^-)} \times (\frac{f_K}{f_\pi})^2 \times \tan^2\theta_c,$$

where $f_K$ and $f_\pi$ are kaon and pion decay constants and $\theta_C$ is the Cabibbo angle. Using the results in Table 1, we obtain $F = 0.023 \pm 0.013$, indicating that $b \to u$ contribution is indeed small. A similar result is also measured using the branching fractions of $B^0 \to K^0_SK^+\pi^-$ and $B^0 \to K^0_SK^+K^-$

To check the CP content of the $K^0_SK^+K^-$ state, we perform an isospin analysis with the following relations:

$$\mathcal{B}(B^0 \to K^0K^+K^-) = \mathcal{B}(B^+ \to K^+K^+K^-) \times \frac{\tau_{B^0}}{\tau_{B^+}}; \quad (1)$$

$$\mathcal{B}(B^0 \to K^0K^+K^-) = \mathcal{B}(B^+ \to K^+K^0\overline{K}^0) \times \frac{\tau_{B^0}}{\tau_{B^+}}, \quad (2)$$

where the factor $\frac{\tau_{B^0}}{\tau_{B^+}}$ takes into account the width difference between charged and neutral $B$ mesons. We use Eq. 1 to check the isospin violating contribution in data:

$$R \equiv \frac{\mathcal{B}(B^0 \to K^0K^+K^-)}{\mathcal{B}(B^+ \to K^+K^+K^-)} \times \frac{\tau_{B^+}}{\tau_{B^0}} = 0.95 \pm 0.11 \pm 0.06,$$

which agrees with one unity within its statistical error. Then we decompose the $K^+K^0\overline{K}^0$ into the following form:

$$|K^+K^0\overline{K}^0> \; = \; \alpha\frac{|K^+K_SK_S> + |K^+K_LK_L>}{\sqrt{2}} + \beta|K^+K_SK_L>, \quad (3)$$

where $\alpha^2+\beta^2 = 1$ an $\alpha$ is the CP even amplitude. The parameter $\alpha^2$ can be determined using Eq. 2 and the information in Table 1: $\alpha^2 = 0.86 \pm 0.15 \pm 0.05$. If we remove $B^0 \to \phi K^0_S$ events by requiring $|M(K^+K^-) - M_\phi| > 15$ MeV, the value of $\alpha^2$ for the remaining events is: $\alpha^2_{non,\phi} = 1.04 \pm 0.19 \pm 0.06$, indicating that non-$\phi$ $K^0_SK^+K^-$ state is predominantly CP even.

## 3. $B^0 \to K^+\pi^-\pi^0$

Belle studies three-body $B$ decays to $K^+\pi^-\pi^0$ using 29.7 fb$^{-1}$ data and large signals are seen as shown in Fig. 3.a. Further investigation on the two-body intermediate states indicates the dominance of $K_X\pi$ states where $K_X \to K\pi$ with masses around 1 to 1.6 GeV/$c^2$ as reported in Ref. [1]. These high mass $K_X$ states cannot be clearly identified with limited statistics. The reconstruction efficiency of the inclusive $K^+\pi^-\pi^0$ signal is obtained by taking into account the observed yields in data for the $K_X\pi$ states as well as three Pseudoscalar-vector (PV) decays: $K^*(892)^0\pi^0, K^*(892)^+\pi^-$ and $\rho^+K^-$.

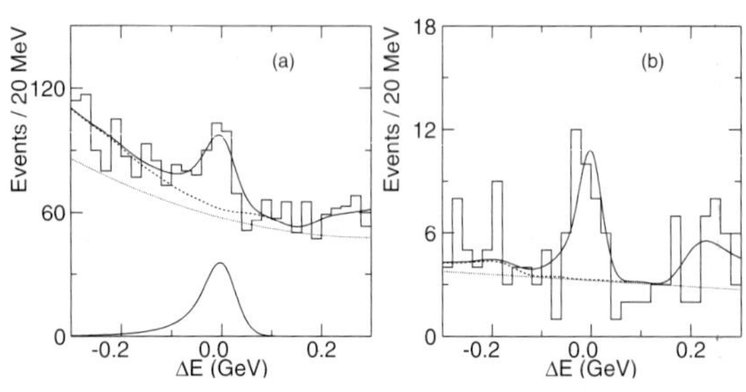


Figure 3. $\Delta E$ distributions for the inclusive $K^+\pi^-\pi^0$ (a) and $K^*(892)^+\pi^-$ (b) modes.

Belle also performs a search for the three PV modes and significant signals (see Fig. 3.b) are only observed in the $K^*(892)^+\pi^-$ channel. However, a study of angular distribution shows that $K^*(892)^+$ is not fully polarized, suggesting other states may feed cross to the $K^*(892)^+\pi^-$ channel. Using a hypothetical charged scalar $\kappa(797)^+\pi^-$ with $\Gamma(\kappa) = 410$ MeV/$c^2$ or a three-body phase-space decay to model the possible states, the true $K^{*+}\pi^-$ fraction is estimated by fitting the $K^+\pi^0$ mass and $\cos\theta_H$ distributions (see Fig. 4) with PDFs from a $K^{*+}\pi^-$ MC and one additional modeled state. Assuming no quantum interference between these two states, the $K^{*+}\pi^-$ fraction is $0.45^{+0.24}_{-0.23}$, which is similar to the $0.50^{+0.23}_{-0.22}$ obtained from the 3-body phase-space model. Since the significance of the true $K^{*+}\pi^-$ signal is less than 4 $\sigma$ after performing an angular analysis, an upper limit at 90% confidence level is given. The inclusive $K^+\pi^-\pi^0$ study and three PV decay searches are summarized in Table 1.

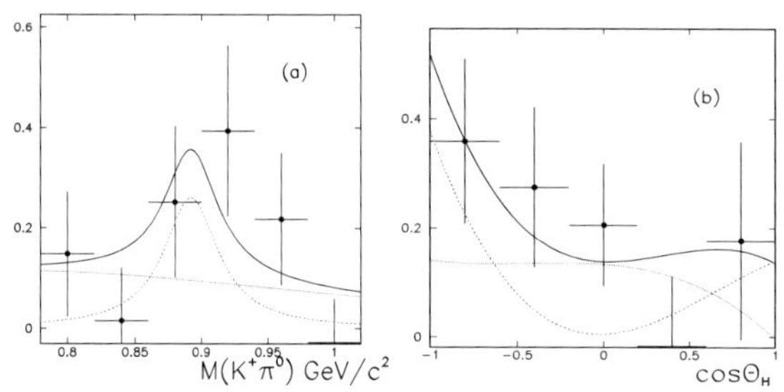


Figure 4. $\Delta E$ yields as functions of $M(K^+\pi^0)$ and $\cos\theta_H$. Solid lines are the combined fit curves for the $K^*(892)^+\pi^-$ (dashed line) and $\kappa^+(797)\pi^-$ (dotted line) components.

## 4. Three body study from CLEO

After a loose continuum suppression, CLEO searches for $B$ decays to three-body states $K^0_Sh^+\pi^-, K^+\pi^-\pi^0, K^0_Sh^+\pi^0$ using unbinned maximum likelihood fits to $M_{bc}, \Delta E, dE/dx$ and other event shape variables. The data sample consists of 9.1 fb$^{-1}$ on resonance data and 4.4 fb$^{-1}$ off resonance data. Significant $B^0 \to K^0_S\pi^+\pi^-$ events have been observed as shown in Fig. 5 and Table 1. The corresponding branching fraction agrees with early Belle result. A sizeable $K^+\pi^-\pi^0$ signal is also seen; however,

Table 1
Summary of studies of Multi-body $B$ decays

| Mode | Efficiency (%) | Signal Yield (events) | $\mathcal{B}$ or 90% CL UL ($10^{-6}$) | Data Sample ($\text{fb}^{-1}$ | Exp. |
|---|---|---|---|---|---|
| $K^+K^+K^-$ | $23.5 \pm 0.50$ | $565 \pm 30$ | $33.0 \pm 1.8 \pm 3.0$ | 78 | Belle |
| $K^0K^+K^-$ | $7.20 \pm 0.17$ | $149 \pm 15$ | $29.3 \pm 3.4 \pm 4.0$ | 78 | Belle |
| $K_SK_SK^+$ | $6.78 \pm 0.19$ | $66.5 \pm 9.3$ | $13.4 \pm 1.9 \pm 1.4$ | 78 | Belle |
| $K_SK_SK_S$ | $3.98 \pm 0.17$ | $12.2^{+4.5}_{-3.8}$ | $4.3^{+1.6}_{-1.4} \pm 0.7$ | 78 | Belle |
| $K^+K^-\pi^+$ | 13.8 | $93.7 \pm 23.2$ | $9.3 \pm 2.3 (< 13)$ | 78 | Belle |
| $K^0K^+\pi^-$ | 4.53 | $26.8 \pm 16.6 (< 48)$ | $8.4 \pm 5.2 (< 15)$ | 78 | Belle |
| $K^+\pi^-\pi^0$ | 11.6 | $172.8^{+30.5}_{-29.6}$ | $47.6^{+8.3}_{-7.7} \pm 5.2$ | 29.7 | Belle |
| $K^*(892)^0\pi^0$ | 10.3 | $0.6^{+7.6}_{-6.5}$ | $< 7.0$ | 29.7 | Belle |
| $K^*(892)^+\pi^-$ | 8.6 | $13.0^{+3.6+6.1}_{-3.3-5.9}$ | $13.0^{+3.9+2.0+6.9}_{-3.6-1.8-6.1} < 30.4$ | 29.7 | Belle |
| $\rho^+K^-$ | 6.4 | $22.5^{+11.8}_{-11.2}$ | $11.2^{+5.9+1.9}_{-5.6-1.8}$ | 29.7 | Belle |
| $K^0\pi^+\pi^-$ | 12 | $60.2^{+11.5}_{-10.6}$ | $50^{+10}_{-9} \pm 7$ | 9.1 | CLEO |
| $K^+\pi^-\pi^0$ | 19 | $43.0^{+14.5}_{-13.5}$ | $< 40$ | 9.1 | CLEO |
| $K^0\pi^+\pi^0$ | 6.8 | $20.3^{+10.1}_{-8.8}$ | $< 66$ | 9.1 | CLEO |

the statistical significance is less than 4 $\sigma$ and an upper limit is given (see Table 1). Note that CLEO result on the $K^+\pi^-\pi^0$ channel cannot be compared with Belle measurement directly because CLEO requires a minimum 1 GeV/$c$ $\pi^0$ momentum. Investigation on intermediate states shows that besides $K^*(892)^+\pi^-$ high mass $K_X^+\pi^-$ states also exist. CLEO then performs a Dalitz plot fits on both $K_S^0\pi^+\pi^-$ and $K^+\pi^-\pi^0$ topologies simultaneously with the branching fraction for each $K^+(X)\pi^-$ decay constrained to be equal in each subsample. Statistical significant signal ($4.6\sigma$) is observed for $B \to K^*(892)^+\pi^-$ with a yield of $12.6^{+4.6}_{-3.9}$ for $K^*(892)^+ \to K_S^0\pi^+$ and $6.1^{+2.2}_{-1.9}$ for $K^*(892)^+ \to K^+\pi^0$ with efficiencies 8.1% and 3.9%, respectively. The branching fraction is measured to be $\mathcal{B}(B \to K^*(892)^+\pi^-) = (16^{+6}_{-5} \pm 2) \times 10^{-6}$.

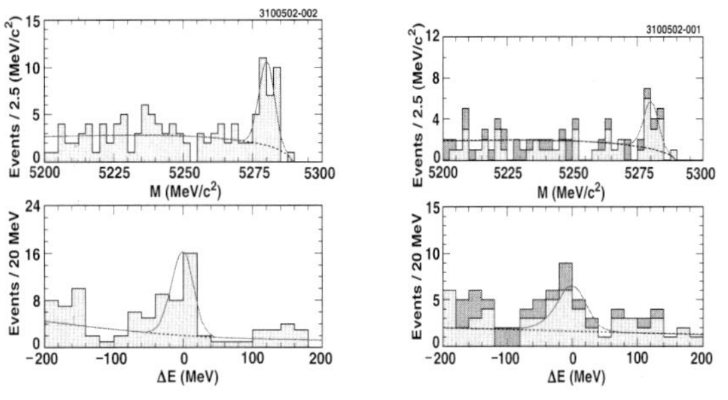


Figure 5. $M_{bc}$ and $\Delta E$ projections for $B \to K^0\pi^+\pi^-$ (left) and $B \to K^*(892)^+\pi^-$ (right), which includes $K^*(892)^+ \to K^0\pi^+$ (light shade) and $K^*(892)^+ \to K^+\pi^0$ (dark shade).

## REFERENCES


1. Belle Collaboration, A. Garmash *et al.*, Phys. Rev. D. **65** 092005 (2002); Belle Collaboration K. Abe *et al.*, BELLE-CONF-0115 (2001), contributed to the Proceedings of the XX International Symposium on Lepton and Photon Interactions at High Energies, July 23-28, 2001, Rome, Italy; H. C. Huang for the Belle Collaboration, hep-ex/0205062, to appear in the *Proceedings of the XXXVII Recontres de Moriond on Electroweak Interactions and Unified Theories, Les Arcs, France, 9-16, March 2002*; Belle Collaboration, K. Abe *et al.*, Phys. Rev. Lett. Phys. Rev. D **88** (2002) 1818103; BaBar Collaboration, B. Aubert *et al.*, hep-ex/0206004.
2. Belle collaboration, A. Abashian *et al.*, Nucl. Inst. and Meth. A 479 117 (2002) .
3. CLEO collaboration, Y. Kubota *et al.*, Nucl. Nucl. Inst. and Meth. A 320, 66 (1992); T. S. Hill, Nucl. Nucl. Inst. and Meth. A 418, 32 (1998).
4. S. L. Wu, Phys. Rep. C107, 59 (1984).

ELSEVIER

Proceedings of ICHEP 2002, pp. 582–586
S. Bentvelsen, P. de Jong, J. Koch and E. Laenen (Editors)

31st INTERNATIONAL CONFERENCE ON HIGH ENERGY PHYSICS AMSTERDAM

# Open charm $B$ decays

P. Krokovny[a], for the Belle Collaboration

[a]Budker Institute of Nuclear Physics, Novosibirsk, Russia

New results on $B$ decays to charm-related modes from the Belle and CLEO experiments are presented.

## 1. Introduction

Studies of charm-related $B$ decays provide information about the dynamics of $B$ meson decays. Results of these fields can be used to check predictions of theoretical models. Recently new results appeared from the $B$ factories at KEK and SLAC and from CLEO. This report covers the recent results on this subject from Belle [1] and CLEO [2].

The CLEO results are based on a 9.15 $\text{fb}^{-1}$ data sample collected at the center-of-mass (CM) energy of the $\Upsilon(4S)$ resonance, while the Belle results are obtained using various data samples from 29.2 $\text{fb}^{-1}$ to 78.7 $\text{fb}^{-1}$. Both group identify $B$ candidates by two kinematic variables: the energy difference, $\Delta E = (\sum_i E_i) - E_b$, and the beam constrained mass, $M_{\text{bc}} = \sqrt{E_b^2 - (\sum_i \vec{p}_i)^2}$, where $E_b = \sqrt{s}/2$ is the beam energy and $\vec{p}_i$ and $E_i$ are the momenta and energies of the decay products of the $B$ meson in the CM frame.

## 2. $B \to D\pi$ Isospin Analysis (CLEO Collaboration)

Precise measurements of the $B \to D\pi$ branching fractions can be used to extract of the strong phase difference $\delta_I$ between the $I = 1/2$ and $I = 3/2$ isospin amplitudes in the $D\pi$ system. Observation of the color-suppressed $\bar{B}^0 \to D^0\pi^0$ decay [3,4] completed the measurements of the $D\pi$ final states. Recently CLEO collaboration improved the accuracy in the $B^- \to D^0\pi^-$ and $\bar{B}^0 \to D^+\pi^-$ branching fractions [5], these results are presented here.

$D^0$ mesons are reconstructed using three decay channels $D^0 \to K^-\pi^+$, $K^-\pi^+\pi^-\pi^+$ and $K^-\pi^+\pi^0$. Charged $D$ mesons are reconstructed via the mode $K^-\pi^+\pi^+$. In each case, $D$ candidates are required to have an invariant mass within $3\sigma$ of the nominal $D$ mass. The fitted $M_{\text{bc}}$ distributions for each of the $D$ decay modes are presented in Fig. 1. The following branching fractions have been obtained: $\mathcal{B}(B^- \to D^0\pi^-) = (4.97 \pm 0.12 \pm 0.29 \pm 0.22) \times 10^{-3}$ and $\mathcal{B}(\bar{B}^0 \to D^+\pi^-) = (2.68 \pm 0.12 \pm 0.24 \pm 0.12) \times 10^{-3}$. Here the first error is statistical, the second is systematic and the third one is a separate systematic error due to the experimental uncertainty of the production fractions of charged and neutral $B$ mesons from $\Upsilon(4S)$ decays. Using the measurements of $\mathcal{B}(\bar{B}^0 \to D^0\pi^0)$ [3,4], $\cos\delta_I = 0.863^{+0.024+0.036+0.038}_{-0.023-0.035-0.030}$ has been obtained. The difference of $\delta_I$ from zero is $2.3\sigma$ indicating the presence of final state interactions in $B \to D\pi$ decays.

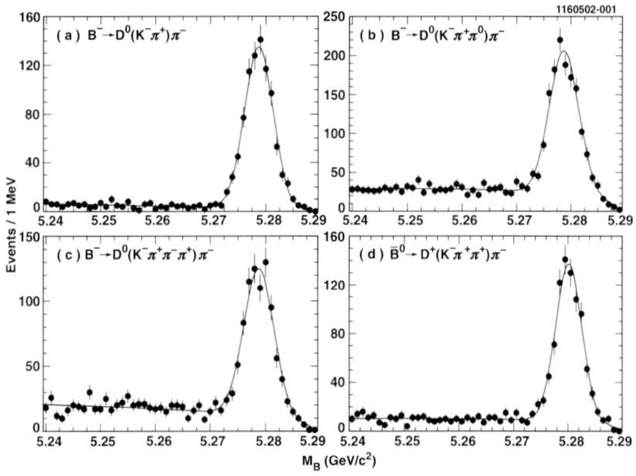


Figure 1. The $M_{\text{bc}}$ distributions for the $B \to D\pi$ candidates.


doi:10.1016/S0920-5632(02)02242-9

Table 1
The signal yields and branching fractions for the $\bar{B}^0 \to D_s^+ K^-$ and $B^0 \to D_s^+ \pi^-$ decay channels.

| Mode | $M(D_s)$ - $\Delta E$ yield | $M(D_s)$ yield | $\Delta E$ yield | $\mathcal{B}$ $(10^{-5})$ | Significance |
|---|---|---|---|---|---|
| $\bar{B}^0 \to D_s^+ K^-$ | $16.4^{+4.6}_{-3.9}$ | $15.0^{+4.5}_{-3.8}$ | $17.5^{+4.8}_{-4.2}$ | $4.6^{+1.2}_{-1.1} \pm 1.3$ | $6.4\sigma$ |
| $B^0 \to D_s^+ \pi^-$ | $10.1^{+4.4}_{-3.7}$ | $10.3^{+4.1}_{-3.4}$ | $9.5^{+4.5}_{-3.8}$ | $2.4^{+1.0}_{-0.8} \pm 0.7$ | $3.6\sigma$ |

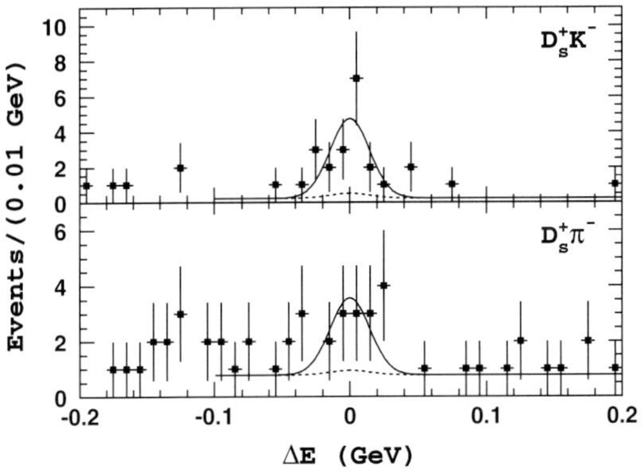


Figure 2. The $\Delta E$ spectra for the $\bar{B}^0 \to D_s^+ K^-$ (top) and $B^0 \to D_s^+ \pi^-$ (bottom) candidates.

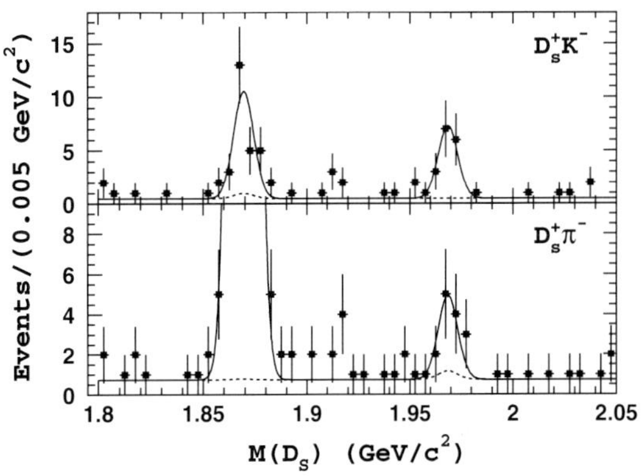


Figure 3. The $M(D_s)$ spectra for the $\bar{B}^0 \to D_s^+ K^-$ (top) and $B^0 \to D_s^+ \pi^-$ (bottom) candidates.

## 3. $B \to D_s \pi/K$ (Belle Collaboration)

The decay $B^0 \to D_s^+ \pi^-$ is expected to be dominated by a $b \to u$ transition, with no penguin contribution. Therefore, it can provide a way to determine the CKM matrix element, $|V_{ub}|$ [6]. The decay $\bar{B}^0 \to D_s^+ K^-$ can occur via $W$-exchange or final state rescattering and cannot be described by a spectator graph. The measurement of this decay mode can be used to estimate $W$-exchange or final state rescattering contributions in other $B$ decays. Only upper limits have been reported for these decays by CLEO [7]. Recently, the BaBar collaboration presented evidence for the $B^0 \to D_s^+ \pi^-$ decay [8].

Here Belle results [9] on a search for these decays are presented. The $D_s^+$ candidates are reconstructed in the $D_s^+ \to \phi\pi^+$, $\bar{K}^{*0}K^+$ and $K_S^0 K^+$ decay channels. $D_s^+$ candidates are combined with a charged kaon or pion to form a $B$ meson. To extract the signal, a binned maximum likelihood fit to the two-dimensional distribution of data in $M(D_s)$ and $\Delta E$ is performed. The $D_s^+$ signal is described by a two-dimensional Gaussian, while the background function includes three components: combinatorial (flat in $M(D_s)$ and $\Delta E$), $q\bar{q}$ events that peak in $M(D_s)$ and are flat in $\Delta E$, and $B$ decay events that peak in $\Delta E$ and are flat in $M(D_s)$. The levels of three background components are allowed to vary independently in the three reconstructed $D_s^+$ modes.

Figures 2 and 3 show the $\Delta E$ and $M(D_s)$ projections for events from the signal region, the fitted signal plus background combined shape by solid lines and background shape including the peaking background by dashed lines. In addition to the clear signals at the $D_s^+$ mass in Fig. 3, also seen are peaks at the $D^+$ mass, corresponding to the $\bar{B}^0 \to D^+\pi^-$ and $\bar{B}^0 \to D^+K^-$ decays. The fit results are given in Table 1. The results of one-dimensional fits to the $M(D_s)$ and $\Delta E$ distributions are also shown in Table 1 for comparison. A statistically significant signal ($6.4\sigma$) is observed for the $\bar{B}^0 \to D_s^+ K^-$ decay channel, while only evidence is reported for the $B^0 \to D_s^+ \pi^-$ decay.

Table 2
The branching fractions and resonance parameters for the $D^{(*)+}\pi^-\pi^-$ final states.

| Mode | $\mathcal{B}(B^- \to D_X\pi^-)\mathcal{B}(D_X \to D^{(*)+}\pi^-)$ | $M(D_X)$, MeV/$c^2$ | $\Gamma(D_X)$, MeV/$c^2$ |
|---|---|---|---|
| $B^- \to D_2^{*0}\pi^- \to D^+\pi^-\pi^-$ | $(3.5 \pm 0.3 \pm 0.5) \times 10^{-4}$ | $2460.7 \pm 2.1 \pm 3.1$ | $46.4 \pm 4.4 \pm 3.1$ |
| $B^- \to D_0^{*0}\pi^- \to D^+\pi^-\pi^-$ | $(5.5 \pm 0.5 \pm 0.8) \times 10^{-4}$ | $2290 \pm 22 \pm 20$ | $300 \pm 30 \pm 30$ |
| $B^- \to D_v^{*0}\pi^- \to D^+\pi^-\pi^-$ | $(1.4 \pm 0.3 \pm 0.2) \times 10^{-4}$ | [10] | [10] |
| $B^- \to D_1\pi^- \to D^{*+}\pi^-\pi^-$ | $(6.2 \pm 0.5 \pm 1.1) \times 10^{-4}$ | $2423.9 \pm 1.7 \pm 0.2$ | $26.7 \pm 3.1 \pm 2.2$ |
| $B^- \to D_2^{*0}\pi^- \to D^{*+}\pi^-\pi^-$ | $(2.0 \pm 0.3 \pm 0.5) \times 10^{-4}$ | [13] | [13] |
| $B^- \to D_1^{*0}\pi^- \to D^{*+}\pi^-\pi^-$ | $(4.1 \pm 0.5 \pm 0.8) \times 10^{-4}$ | $2400 \pm 30 \pm 20$ | $380 \pm 100 \pm 100$ |

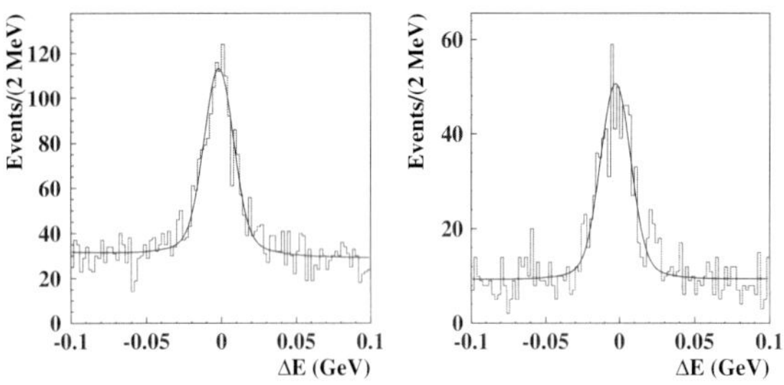


Figure 4. The $\Delta E$ distributions for the $B^- \to D^+\pi^-\pi^-$ (left) and $B^- \to D^{*+}\pi^-\pi^-$ (right) candidates.

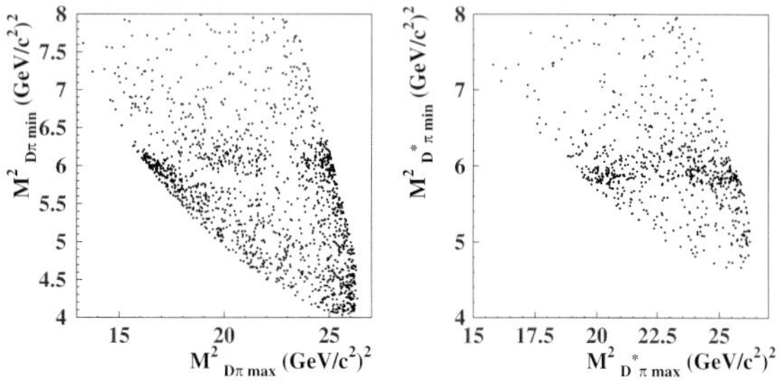


Figure 5. The Dalitz plot distributions for $B^- \to D^+\pi^-\pi^-$ (left) and $B^- \to D^{*+}\pi^-\pi^-$ (right) candidates.

## 4. $B^- \to D^{(*)+}\pi^-\pi^-$ (Belle Collaboration)

A study of charmed meson production in $B$ decays provides an opportunity to test predictions of Heavy Quark Effective Theory (HQET) and QCD sum rules. $B$ decays to $D^{(*)}\pi$ final states are the dominant hadronic $B$ decay modes and are measured quite well [10]. The large data sample accumulated in the Belle experiment allows to study production of $D$ meson exited states. $D^{**}$s are P-wave excitations of quark-antiquark systems that contain one charmed and one light ($u$,$d$) quark. The $B \to D^{**}\pi$ decays have been studied by Belle [11] using the $D^+\pi^-\pi^-$ and the $D^{*+}\pi^-\pi^-$ final states.

Figure 4 shows the $\Delta E$ distributions for the $B^- \to D^+\pi^-\pi^-$ and $B^- \to D^{*+}\pi^-\pi^-$ candidates. The following branching fractions are measured: $\mathcal{B}(B^- \to D^+\pi^-\pi^-) = (1.07 \pm 0.05 \pm 0.16) \times 10^{-3}$ and $\mathcal{B}(B^- \to D^{*+}\pi^-\pi^-) = (1.24 \pm 0.07 \pm 0.22) \times 10^{-3}$, without any assumption about the intermediate final states.

To study the dynamics of $B \to D^{(*)}\pi\pi$ decays, an analysis of the Dalitz plots shown in Fig. 5 is performed. The fit to the $D^+\pi^-\pi^-$ Dalitz plot includes three final states: $D_2^{*0}\pi^-$, $D_0^{*0}\pi^-$ and the contribution of the process with virtual $D^{*0}\pi$ production($D_v^{*0}\pi^-$). The results are presented in Table 2.

Two additional degrees of freedom should be taken into account in the $D^{*+}\pi^-\pi^-$ final state: the angle ($\alpha$) between the pion from the $D^{**}$ decay and the pion from the $D^*$ decay in the $D^*$ rest frame and the azimuthal angle ($\gamma$) of this particle relative to the plane of the $B^- \to D^{*+}\pi^-\pi^-$ decay. The fit to the $D^{*+}\pi^-\pi^-$ distribution includes three final states: $D_1\pi^-$, $D_2^{*0}\pi^-$ and $D_1^{*0}\pi^-$. The results are presented in Table 2.

Using these measurements the ratio of $D_2^{*0}$ branching fractions $h = \mathcal{B}(D_2^{*0} \to D^+\pi^-)/\mathcal{B}(D_2^{*0} \to D^{*+}\pi^-) = 1.77 \pm 0.49$, consistent with the world average $h = 2.3 \pm 0.6$ [10], is obtained. The measured ratio $R = \mathcal{B}(B^- \to D_2^{*0}\pi^-)/\mathcal{B}(B^- \to D_1^0\pi^-) = 0.89 \pm 0.14$ is lower than the CLEO measurement $1.8 \pm 0.8$ [12] (al-

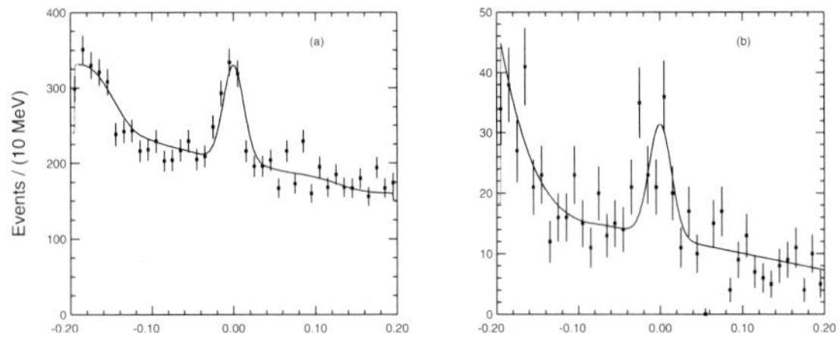


Figure 6. The $\Delta E$ distributions for (a) $\bar{B}^0 \to D^0\pi^+\pi^-$ and (b) $\bar{B}^{*0} \to D^0\pi^+\pi^-$ candidates.

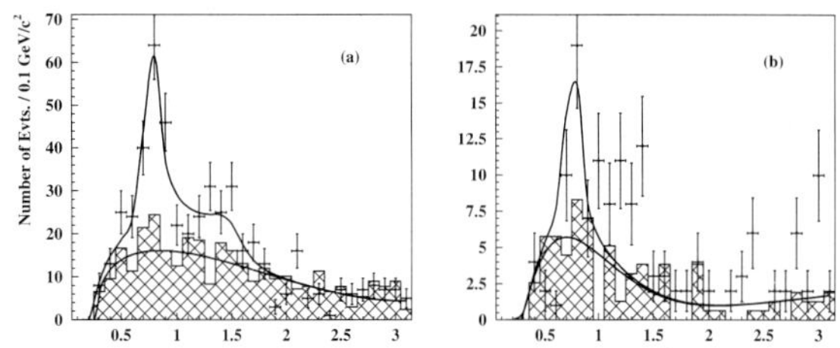


Figure 7. The $\pi^+\pi^-$ invariant mass spectra for the (a) $\bar{B}^0 \to D^0\pi^+\pi^-$ and (b) $\bar{B}^{*0} \to D^0\pi^+\pi^-$ candidates.

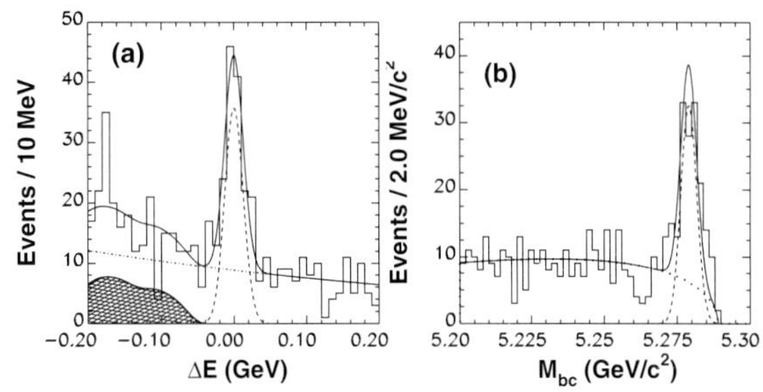


Figure 8. The (a) $\Delta E$, (b) $M_{\rm bc}$ distributions for the $B^- \to D^0K^{*-}$ candidates.

though the results are consistent within errors) but is still a factor of 3 larger than the factorization prediction [14]. Belle measurements show that the narrow resonances compose $(33 \pm 4)\%$ of the $D\pi\pi$ decays and $(66 \pm 7)\%$ of the $D^*\pi\pi$ decays. This result is inconsistent with the QCD sum rule prediction and may indicate a large contribution from a color suppressed amplitude.

## 5. $\bar{B}^0 \to D^{(*)0}\pi^+\pi^-$ (Belle Collaboration)

Exclusive hadronic decay rates provide important tests of models for $B$ meson decays. The $\bar{B}^0 \to D^{(*)0}\pi^+\pi^-$ decays provide a precision testing ground for factorization [15] and the possibility to search for resonant substructure in the final state. At present, only an upper limit $\mathcal{B}(\bar{B}^0 \to D^0\pi^+\pi^-) < 1.6 \times 10^{-3}$ [16] exists.

Here the Belle results [17] on a study of these decays are presented. $D^0$ candidates are reconstructed in the $K^-\pi^+$, $K^-\pi^+\pi^-\pi^+$ and $K^-\pi^+\pi^0$ final states. Figure 6 shows the $\Delta E$ distributions for the $\bar{B}^0 \to D^{(*)0}\pi^+\pi^-$ candidates. The following branching fractions are measured: $\mathcal{B}(\bar{B}^0 \to D^0\pi^+\pi^-) = (7.5 \pm 0.7 \pm 1.5) \times 10^{-4}$ and $\mathcal{B}(\bar{B}^0 \to D^{*0}\pi^+\pi^-) = (6.2 \pm 1.2 \pm 1.7) \times 10^{-4}$. Figure 7 shows the $\pi^+\pi^-$ invariant mass spectra for the selected events. These distributions are fitted by a sum of the $\rho^0$,$\sigma$ and $f_0(1370)$ resonances. The amplitudes and phases are free parameters in the fit. The branching fraction of $\mathcal{B}(\bar{B}^0 \to D^0\rho^0) = (3.0 \pm 1.2 \pm 0.4) \times 10^{-4}$ and the 90% CL upper limit $\mathcal{B}(\bar{B}^0 \to D^{*0}\rho^0) < 5.1 \times 10^{-4}$ are obtained.

## 6. $B^- \to D^0K^{*-}$ (Belle Collaboration)

The decay $B^- \to D^0K^{*-}$ can be used for $\phi_3$ determination [18]. This mode has been previously observed by the CLEO Collaboration [19].

$D^0$ mesons are reconstructed in the decay channels $D^0 \to K^-\pi^+$, $K^-\pi^+\pi^-\pi^+$ and $K^-\pi^+\pi^0$ using a $2\sigma$ mass window from the nominal $D^0$ mass. The $K^{*-}$ candidates are combined from the $K^0_S\pi^-$ pairs.

The $\Delta E$ and $M_{\rm bc}$ distributions are presented in Fig. 8(a) and (b) respectively. A signal of $114.4 \pm 13.5$ events with $10.9\sigma$ statistical significance is observed. The fit results are presented in Table 3. The measured branching fraction $\mathcal{B}(B^- \to D^0K^{*-}) = (5.4 \pm 0.6 \pm 0.8) \times 10^{-4}$ agrees well with the world average [10]. Figure 9 shows the $K^{*-}$ helicity and invariant mass distributions obtained by fitting the $M_{\rm bc}$ spectra in each bin.

Table 3
The signal yields and branching fractions for the $B^- \to D^0 K^{*-}$ decay channel.

| Decay mode | $\Delta E$ yield | $M_{\rm bc}$ yield | $\mathcal{B}$ $(10^{-4})$ |
|---|---|---|---|
| $D^0 \to K^-\pi^+$ | $52.5 \pm 8.1$ | $51.6 \pm 8.5$ | $6.1 \pm 0.9 \pm 0.8$ |
| $D^0 \to K^-\pi^+\pi^0$ | $36.1 \pm 6.8$ | $32.9 \pm 6.4$ | $5.1 \pm 0.9 \pm 0.7$ |
| $D^0 \to K^-\pi^+\pi^+\pi^-$ | $31.0 \pm 7.0$ | $29.4 \pm 6.8$ | $4.6 \pm 1.1 \pm 0.8$ |
| Weighted mean | | | $5.4 \pm 0.6 \pm 0.8$ |

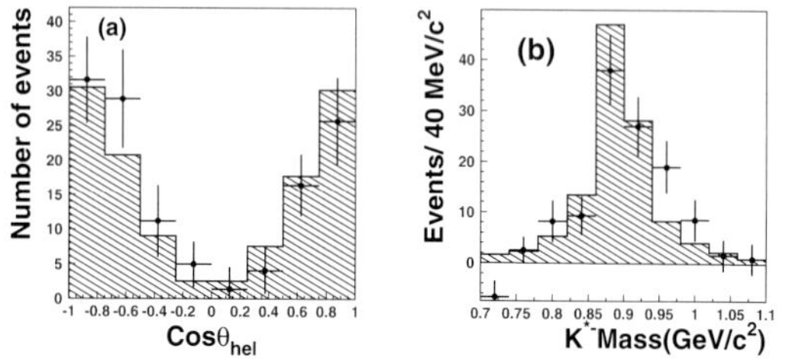


Figure 9. The (a) $K^{*-}$ helicity and (b) $K^{*-}$ invariant mass distributions for the $B^- \to D^0 K^{*-}$ signal region.

## 7. Conclusion

A nonzero strong phase $\delta_I$ is obtained using the CLEO and Belle measurements of the branching fractions $\mathcal{B}(B \to D\pi)$. The $2.3\sigma$ difference from zero indicates the presence of final state interactions in $B \to D\pi$ decays. The observation of the $\bar{B}^0 \to D_s^+ K^-$ decay by Belle emphasizes the importance of $W$ exchange or final state rescattering. In $B^- \to D^{(*)+}\pi^-\pi^-$ decays all four P-wave $D^{**}$ have been observed and their parameters have been measured by Belle. For the broad $D_0^{*0}$ and $D_1^{*0}$ states this is the first measurement.

## REFERENCES

1. Belle Collaboration, A. Abashian *et al.*, Nucl. Inst. and Meth. A **479**, 117 (2002).
2. CLEO Collaboration, Y. Kubota *et al.*, Nucl. Inst. and Meth. A **320**, 66 (1996); T. Hill, Nucl. Inst. and Meth. A **418**, 32 (1998).
3. Belle Collaboration, K. Abe *et al.*, Phys. Rev. Lett. **88**, 052002 (2002).
4. CLEO Collaboration, T. E. Coan *et al.*, Phys. Rev. Lett. **88**, 062001 (2002).
5. CLEO Collaboration, S. Ahmed *et al.*, Phys. Rev. D **66**, 031101 (2002).
6. C.S. Kim, Y. Kwon, J. Lee, W. Namgung, Phys. Rev. D **63**, 094506 (2001).
7. CLEO Collaboration, J. Alexander *et al.*, Phys. Lett. B **319**, 365 (1993).
8. BaBar Collaboration, B. Aubert *et al.*, Proc. of the Flavor Physics and CP Violation (FPCP), May 2002, Philadelphia, **hep-ex/0205102**.
9. Belle Collaboration, K. Abe *et al.*, BELLE CONF 02-48, Cont. paper for International conference for High Energy Physics (ICHEP 02), July 2002, Amsterdam; hep-ex/**0207077**.
10. K. Hagiwara *et al.* (Particle Data Group), Phys. Rev. D **66**, 010001 (2002).
11. Belle Collaboration, K. Abe *et al.*, BELLE CONF 02-35, Cont. paper for ICHEP 02, July 2002, Amsterdam.
12. CLEO Collaboration, G. Gronberg *et al.*, CLEO CONF 96-25, Proc. of the ICHEP 96, July 1996, Poland.
13. Fixed from the $D_2^{*0}$ parameters obtained in the $B^- \to D^+\pi^-\pi^-$ analysis.
14. M. Neubert, Phys. Lett. B **418**, 173 (1998).
15. C. Reader and S. Isgur, Phys. Rev. D **47**, 1007 (1993).
16. CLEO Collaboration, M.S. Alan *et al.*, Phys. Rev. D **50**, 43 (1994).
17. Belle Collaboration, K. Abe *et al.*, BELLE CONF 02-33, Cont. paper for ICHEP 02, July 2002, Amsterdam.
18. M. Gronau and D. Wyler, Phys. Lett. B **265**, 172 (1991).
19. CLEO Collaboration, R. Mahapatra *et al.*, Phys. Rev. Lett. **88**, 101803 (2002).
20. Belle Collaboration, K. Abe *et al.*, BELLE CONF 02-36, Cont. paper for ICHEP 02, July 2002, Amsterdam.

ELSEVIER

Proceedings of ICHEP 2002, pp. 587–592
*S. Bentvelsen, P. de Jong, J. Koch and E. Laenen (Editors)*

31st INTERNATIONAL CONFERENCE ON HIGH ENERGY PHYSICS AMSTERDAM

# Heavy Meson Spectra and quark masses from Lattice QCD

S. Collins[a]

[a]Department of Physics and Astronomy, University of Glasgow, Glasgow, G12 8QQ, UK

I review recent lattice results for bottom and charm quarkonia, heavy hybrids and the bottom quark mass.

## 1. Introduction

Lattice calculations of spectroscopy have been performed for a wide range of hadrons containing heavy quarks: $\Upsilon$, $b\bar{b}g$ hybrids, $B$ and $B_s$ mesons, $\Lambda_b$ baryons, $J/\psi$, $c\bar{c}g$, $D$ mesons, .... Here I concentrate on quarkonia and the $b$ quark mass. I omit results for $B$ and $D$ mesons due to the lack of new work, the current state of the art can be found in reference [1].

## 2. Simulating $b$ and $c$ quarks on the lattice

Traditionally simulating heavy quarks on the lattice was problematic; the standard relativistic quark action has discretisation terms which introduce errors of $O(m^2a^2, a^2\Lambda^2_{QCD})$, where $a$ is the lattice spacing, $m$ is the quark mass and $\Lambda_{QCD}$ is some typical soft QCD scale. Since the present restrictions on computer power limit $a^{-1}$ to the range $1-4$ GeV it is clear that the $b$ quark mass in lattice units $am_b \gg 1$, while for the charm quark $am_c \sim 1$; discretisation errors are significant for simulations in this range.

Several approaches have been developed to tackle this problem.

**NRQCD**: as many heavy quark systems are non-relativistic this approach abandons the relativistic quark action in favour of a non-relativistic one. The NRQCD action [2]

$$S = -\int d^4x Q^\dagger(D_t - D^2/(2M))Q + \dots \quad (1)$$

is an expansion in the heavy quark velocity squared, $v^2$ (for quarkonia). The series is truncated at the order which gives the desired accuracy. Each term in the action has a corresponding coefficient which must be determined to achieve this accuracy. As a low energy effective theory $am > 1$ and so the continuum limit cannot be taken. However, the rationale is to control all systematic errors at finite lattice spacing. NRQCD has been shown to work well for bottomonium [3], but has not been successful when applied to charmonium spectroscopy, for which $v^2 \sim .3$.

**Fermilab**: this approach [4] uses a generalised set of operators to encompass both the light and heavy quark regimes. Thus it can be applied to both charm and bottom quarks. A disadvantage is that the method is more involved and the perturbation theory more complicated.

**Standard action**: here, the standard $O(a^2)$ action is used and large discretisation errors are present. These are tackled by performing a careful continuum extrapolation of the results. This can be computationally expensive if many simulations are required. This approach has been applied to the charm quark.

**Asymmetric lattices**: a recent, novel, approach has been to use the standard relativistic quark action with asymmetric lattices, where $a_t \ll a_s$, such that $a_t m < 1$ but $a_s$ remains fairly coarse. The latter point allows the computational cost to be kept under control. Both charm and bottom quarks have been simulated using this method.

## 3. Accuracy of lattice simulations

This depends, of course, on how well the sources of error have been controlled and/or estimated. Typical sources of error in a simulation are finite statistics, finite lattice spacing, finite volume, chiral extrapolation, renormalisation etc. In most studies these errors have been investi-


doi:10.1016/S0920-5632(02)02247-8

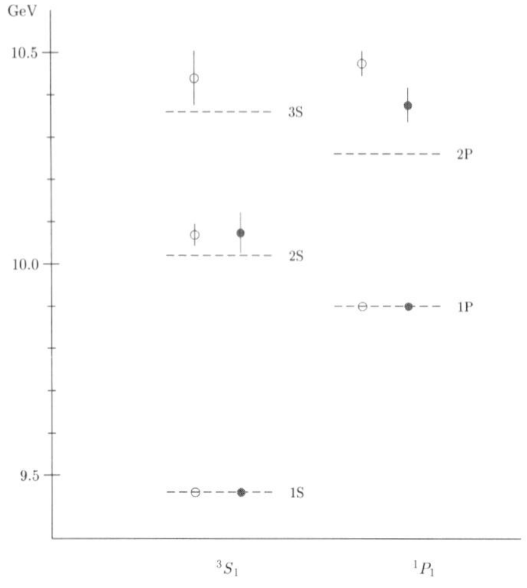


Figure 1. The gross $\Upsilon$ spectrum calculated by L. Marcantonio et al. [7] using NRQCD to $O(Mv^4)$ with $n_f = 0$ (open circles) and with $n_f = 2$ degenerate flavours of sea quarks for $m_{sea} \sim m_s$ (filled circles).

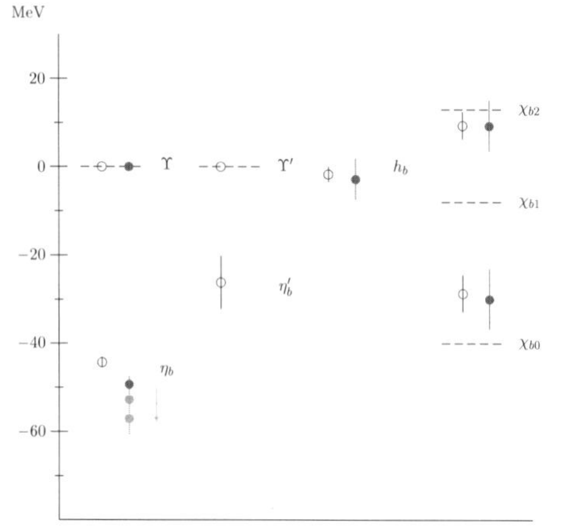


Figure 2. $\Upsilon$ fine structure from the same simulation as in figure 1. The grey filled symbols for the hyperfine splitting show the increase in the result when it is extrapolated to $m_{sea} = m_{u,d}$ and then to $n_f = 3$ from $n_f = 2$.

gated and in some cases are under control to the level of a few percent.

However, the uncertainty which still hangs over many simulations is the quenched approximation. Due to the high computational cost many simulations omit internal quark loops. One consequence of this is that when the lattice spacing is determined by comparing simulation results to experiment, different values are obtained depending on which quantity is used. When converting lattice results into physical units the predictions can vary by 10% due to the variation in $a$. With increasing computer power becoming available the first serious studies including sea quarks are being performed. Note must be taken of the number of sea quark flavours ($n_f$) and masses ($m_{sea}$) as they do not necessarily correspond to the real world.

Lattice calculations are from first principles: only the bare coupling and bare quark masses are inputed to the simulation. As mentioned above the lattice spacing which this coupling corresponds to is determined post simulation. Once this is done, which bare mass corresponds to the $b$ and $c$ quarks is also determined using experiment.

## 4. $\Upsilon$ spectrum

The $\Upsilon$ spectrum has been studied extensively using NRQCD [5]. The spectrum in figure 1 is representative of the results. The $O(Mv^4)$ action was used which leads to an expected accuracy of 1% in spin-independent splittings and 10% in fine structure splittings. This assumes that the coefficients in the action have been determined accurately. At the moment, however, only modified tree-level estimates are used.

The lattice spacing is set using the $^1P_1 - ^3S_1$ splitting and $am_b$ is fixed using the $^3S_1$ mass. The predictions are for the $2S$, $3S$ and $2P$ states. The quenched results agree with experiment, apart from the $2P$ state. For the latter the disagreement is reduced when sea quarks are partially included. In the quenched case the results have been shown to be reliable: the results scale when the lattice spacing is changed, and using potential models finite volume errors have been found to be within the statistical errors [6].

With $a$ set from the gross spectrum, one expects to see quenching effects in the fine structure, which is dominated by a much higher physical scale. Unfortunately, the statistical errors are

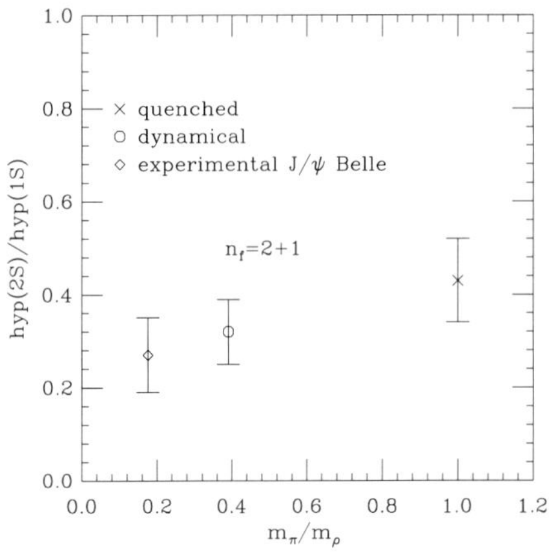


Figure 3. The ratio of $\Upsilon$ hyperfine splittings from A. Gray et al [8] as a function of $m_\pi/m_\rho$ corresponding to the $m_{sea}$ values. $n_f = 2+1$ refers to 2 light sea quark masses and 1 heavier sea quark around $m_s$ ($m_\pi/m_\rho \sim .7$).

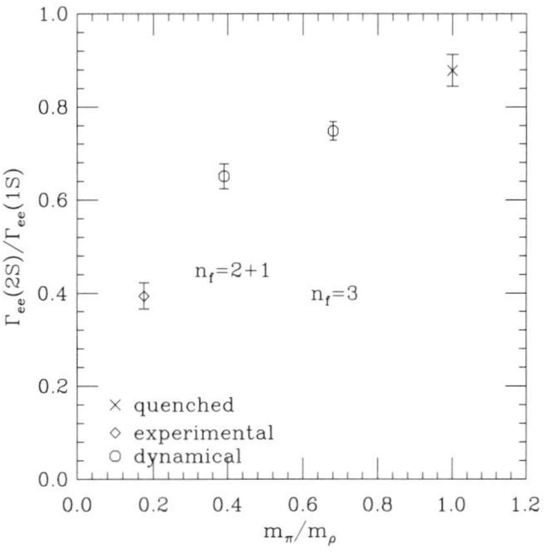


Figure 4. The ratio of $\Upsilon$ $1S$ and $2S$ leptonic widths from the same simulation as in figure 3.

too large to see a significant effect in the $P$ fine structure. For the $1S$ hyperfine splitting there is no experimental result to compare to, however, thoughout lattice calculations of spectra hyperfine splittings in the quenched approximation are significantly lower than the experimental values. We see a significant increase when sea quarks are partially included and a further increase when the results are extrapolated to the physically relevant $n_f$ and $m_{sea}$.

However, the fine structure splittings have been shown to depend significantly on the lattice spacing. This could be due to not determining the coefficients in the NRQCD accurately enough, and/or discretisation errors arising from the gluon action. A more recent calculation by A. Gray et al. (UKQCD and HPQCD collaboration) [8], figure 3, aims to improve on this. Work is in progress to calculate the coefficients to 1-loop in perturbation theory, in the mean time the ratio of $1S$ and $2S$ hyperfine splittings is taken to remove this uncertainty. Taking the ratio also significantly reduces the dependence of the splitting on the heavy quark mass and this suggests one can compare to the experimental result for $J/\psi$. Furthermore, an improved gauge action has been used which reduces the discretisation errors from this source, and which has been implemented with much lighter sea quark masses, directly at the physically relevant $n_f = 2+1$.

We see from the figure that the dynamical result is much closer to what the experimental value is likely to be than the quenched result, however, the statistical errors are still large. Figure 4 of the ratio of leptonic widths shows more clearly the improvement when introducing sea quarks. A significant deviation from experiment still remains but this calculation is, so far, only to leading order in the currents. Further work is in progress.

## 5. $b\bar{b}g$ hybrids

Hybrids are mesons containing gluon excitation. The $b$ quark is even less relativistic in a hybrid meson and so most simulations use NRQCD to $O(Mv^2)$, omitting spin-dependent terms. This means that states corresponding to the same gluon quantum numbers are degenerate:

$$(1^{+-})_g \quad \rightarrow \quad 1^{--}, O^{-+}, \mathbf{1^{-+}}, 2^{-+} \qquad (2)$$

$$(1^{-+})_g \quad \rightarrow \quad 1^{++}, \mathbf{O^{+-}}, 1^{+-}, \mathbf{2^{+-}} \qquad (3)$$

where the exotic quantum numbers (those not possible using quark model states) are shown in

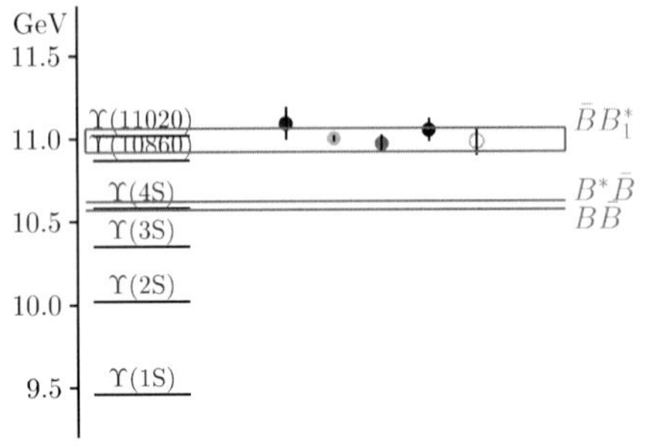


Figure 5. Recent results for the lowest hybrid, $1^{-+}$ compared to the experimental $\Upsilon$ spectrum and the thresholds for decay. Filled circles indicate results for $n_f = 0$, open circles $n_f = 2$.

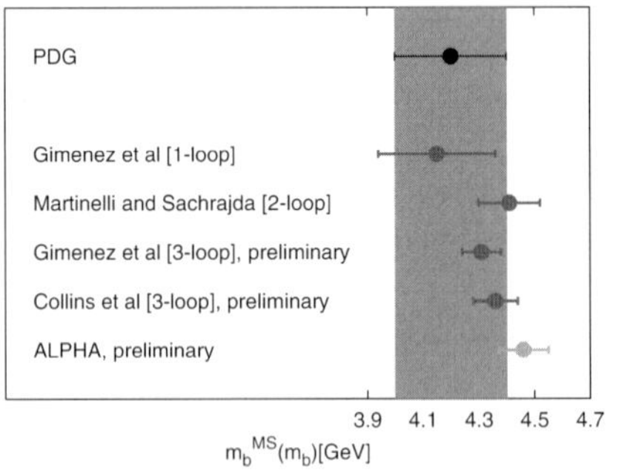


Figure 6. Quenched results for $m_b$ from 1- [12] and 2-loop [13] perturbation theory and the new preliminary results at 3-loop from Gimenez et al [14] and Collins et al [15]. The preliminary nonperturbative result from the ALPHA collaboration [10] is also included. The latest 3 results are all presented using the same method to determine $a$.

bold. Most work concentrates on the exotic states as they are easier to simulate: there is no mixing with quark model states (or any other states when in the quenched approximation). Whether these particles are seen in experiment depends on where they lie in relation to the relevant thresholds for decay. A strong decay to $\bar{B}B$ is not allowed by $CP$ arguments, $B^*$ and $\bar{B}$ in a relative $P$ wave is allowed but expected to be suppressed. Decay to $B_1^*\bar{B}$ is allowed and the most interesting threshold. In all simulations the $1^{-+}$ state has been found to be the lowest exotic. Figure 5 presents the most recent results [9].

It is difficult to extract a hybrid signal on the lattice as the correlators fall into noise very quickly as the hybrid propagates across the lattice in time. Asymmetric lattices are employed by CP-PACS, Juge et al and Liao et al [9] to overcome this problem; with $a_t \ll a_s$ the hybrid signal can be sampled more often before it is lost, but at reasonable computational cost. Figure 5 shows that this leads to much smaller statistical errors, however, the assymetric calculations are more involved and more checks are needed. Most groups have investigated finite volume and lattice spacing dependence and found that the predictions are not significantly affected.

Interestingly all results sit on the $\bar{B}B_1^*$ threshold. In the quenched approximation there is a significant variation in the prediction due to the uncertainty in $a$ (discussed in section 3). However, for a fixed method of determining $a$, potential models suggest unquenching decreases the exotic mass slightly. Although in this case one has to consider possible mixing with $b\bar{b}q\bar{q}$ molecules. For example, since the hybrid sits on the $\bar{B}B_1^*$ threshold there is likely to be strong mixing with a $b\bar{u}\bar{b}u$ state. While, the bottomonium spectrum is a clean system (not many states, all stable) to study, unfortunately, with no appropriate experiments, the hybrid is unlikely to be seen.

## 6. $b$ quark mass

The most accurate way to determine $m_b$ at the moment on the lattice is using the $B$ meson with the approximation of an infinitely heavy $b$ quark. This approximation leads to errors of only $10-50$ MeV in $m_b$, compared to several 100 MeV if one sticks to finite $b$ quark mass. Both perturbative and non-perturbative approaches are being followed. The nonperturbative method, developed by the ALPHA collaboration [10] and described in more detail in [18], has the advantage of avoiding the perturbative subtractions of power divergences inherent in the other approach. First preliminary results are appearing and the systematic errors must be fully understood.

$m_b$ is calculated, normally in the $\overline{MS}$ scheme at the scale $m_b$, using results for a binding energy $\Lambda_{bind}$ and the spin-averaged experimental $B$ mass, $M_{\overline{B}}$:

$$\overline{m_b}(\overline{m_b}) = Z_{cont}(\overline{m_b})(M_{\overline{B}} - \Lambda_{bind}) \quad (4)$$

where $\Lambda_{bind} = E_{sim} - E_0$, and $E_{sim}$ and $E_0$ are the energy of the $B$ meson in the simulation and heavy quark self energy respectively. $Z_{cont}$ is the continuum renormalisation factor relating the pole mass to the $\overline{MS}$ mass, known to $O(\alpha^3)$. $E_0$ is also known to $O(\alpha^3)$ in the static limit (compared to $O(\alpha)$ at finite $m_b$). Two groups using different numerical methods, Trottier et al. and Di Renzo et al., agree on the 3rd order coefficient, $c_3 = 86.6(5)$ and $86.2(6)(10)$ respectively [11], where

$$aE_0 = 2.12\alpha_L + 11.15\alpha_L^2 + c_3\alpha_L^3. \quad (5)$$

and $\alpha_L = g^2/(4\pi)$. In practice, both groups use physically motivated definitions of the strong coupling which lead to much more convergent perturbative series.

Figure 6 compares the results for $m_b$ to previous ones at 1- and 2-loop and the current world average quoted in the PDG [17]. The results converge nicely with the shift going from 1- through to 3-loops being only $1\sigma$ at most. There is also agreement with the nonperturbative result. All groups have studied systematics, including lattice spacing dependence when estimating their errors.

With the uncertainty in these calculations being less than 100 MeV one must consider the remaining uncertainties of working in the static limit and in the quenched approximation. As mentioned above the former is likely to be a very small correction, while preliminary studies with fixed methods to extract $a$, suggest quenching leads to $< 1\sigma$ decrease [15,16]. In the quenched approximation the shift in the prediction due to the variation in determinations of $a$ depends on the method. For the nonperturbative approach this can be a 10% effect [10], while for the perturbative method it is $< 5\%$.

The lattice results have become the most reliable and accurate determinations of $m_b$. A lattice world average of 4.38(9)(10) GeV is quoted by Lellouch [18], where the second error is from quenching.

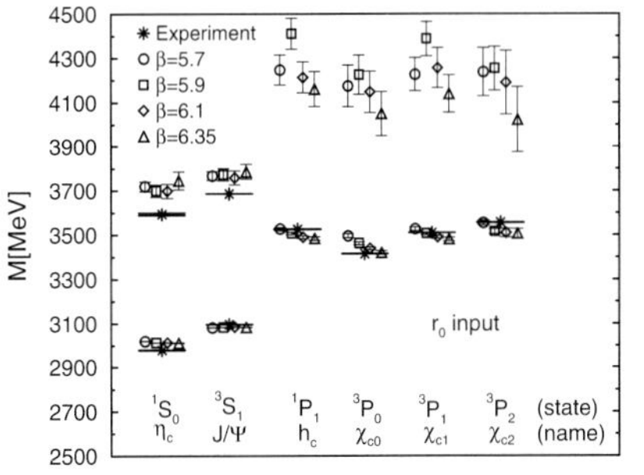


Figure 7. The quenched charmonium spectrum as calculated by Okamoto et al [21]. Results are shown from simulations at several different lattice spacings, where $\beta = 6/g^2$.

## 7. $J/\psi$ spectrum and $c\bar{c}g$ hybrids

In the past, there have not been many simulations of the charmonium spectrum. This reflects the difficulty in simulating with a quark mass which lies between the heavy and light regimes. However, recent work by Klassen [19] and Chen [20], using the standard relativistic action and asymmetric lattices has produced extensive calculations of the spectrum.

Following this method, the CP-PACS collaboration [21] have achieved high statistical accuracy for the lower lying $S$ and $P$ states at several lattice spacings, shown in figure 7. The lattice spacing dependence of the splittings is not severe. The assymetry is such that $2.9 < (a_t GeV)^{-1} < 8.44$, while $0.97 < (a_s GeV)^{-1} < 2.81$; the physical volume is kept fixed at 1.6 fm. Performing the continuum limit of these results we obtain figure 8.

Charmonium hybrids, $c\bar{c}g$, have also been calculated. Chen et al [22], using the same assymetric approach, at several lattice spacings, find the lowest hybrid $1^{-+}$ at $\sim 4.294(71)$ GeV. We see from figure 8 that the hybrid sits on the $\bar{D}D_1^*$ threshold.

There is some disagreement with experiment in the quenched spectrum, which must be resolved

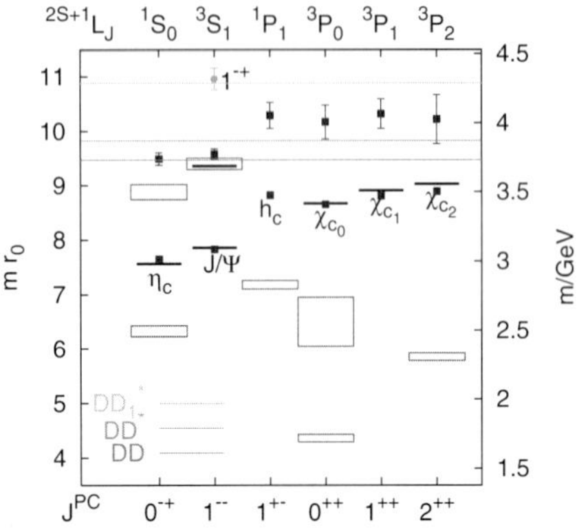


Figure 8. The continuum limit of the Okamoto et al [21] and Chen et al [22] results (squares) at $n_f = 0$. The lines indicate the experimental results and the thresholds for decay of the $1^{-+}$ hybrid. The boxes indicate the quenched lattice values for glueballs.

by including sea quarks. However, the presence of other flavour singlet particles, i.e. glueballs, with similar masses will make such a simulation challanging; mixing between these states and charmonium states must be considered. In addition, while implementing asymmetric lattices has produced very precise predictions in the quenched approximation, it will be difficult to implement this approach in the presence of sea quarks.

## 8. Conclusions and outlook

Heavy meson spectroscopy remains a very active area of research within the lattice community. In particular, recent progress in the quenched approximation is seen in terms of the extensive calculation of charmonium and $c\bar{c}g$ exotics, and precise predictions for the $b$ quark mass.

The next few years will be dominated by the exploration of sea quark effects. The results for the $\Upsilon$ hyperfine and leptonic widths suggest long standing discrepancies will be resolved. Mixing between quark model states, hybrids, molecules and glueballs will present a challenge.

The author acknowledges support from the European Community's Human Potential Programme under contract HPRN-CT-2000-00145, Hadrons/Lattice QCD.

## REFERENCES

1. J. Hein et al, Phys. Rev. D62:074503, 2000.
2. G. P. Lepage et al, Phys. Rev. D46:4052-4067, 1992.
3. C. Davies, "Schladming 1997, Computing particle properties", 1-64.
4. A. El-Khadra et al, Phys. Rev. D55:3933-3957, 1997.
5. C. Davies et al, Phys. Rev. D58:054505, 1998. N. Eicker et al, Phys. Rev. D57:4080-4090, 1998. T. Manke et al, Phys. Rev. D62:114508, 2000.
6. G. Bali and P. Boyle, Phys. Rev. D59:114504, 1999.
7. L. Marcantonio et al, Nucl. Phys. Proc. Suppl. 94:363-366, 2001.
8. A. Gray et al, hep-lat/0209022.
9. Going from left to right in figure 5 T. Manke et al, Phys. Rev. D57:3829-3832, 1998. T. Manke et al, Phys. Rev. Lett. 82:4396-4399, 1999. J. Juge et al, Phys. rev. Lett. 82:4400-4403, 1999. X. Liao et al, Phys. Rev. D65:074508, 2002. T. Manke et al, Phys. Rev. D64:097505, 2001
10. R. Sommer, talk at Lattice '02, Boston, USA.
11. H. Trottier et al, Phys. Rev. D65:094502, 2002. F. Di Renzo et al, Jhep 0102:020, 2001.
12. V. Gimenez et al, Phys. Lett. B393:124, 1997.
13. G. Martinelli et al, Nucl. Phys. B559:429, 1999.
14. V. Gimenez, private communication.
15. S. Collins et al, "Wien 2000, Quark confinement and the hadron spectrum", 325-327, 2000, and re-analysis by the author.
16. V. Gimenez et al, JHEP 0003:018, 2000.
17. Particle Data Group, http://pdg.lbl.gov/
18. L. Lellouch, plenary talk these proceedings.
19. T. Klassen, Nucl. Phys. Proc. Suppl. 73, 918, 1999.
20. P. Chen, Phys. Rev. D64:034509, 2001.
21. M. Okamoto et al, Phys. Rev. D65:094508, 2002.
22. P. Chen et al, Nucl. Phys. Proc. Suppl. 94:342-345, 2001.

ELSEVIER

Proceedings of ICHEP 2002, pp. 593–597
*S. Bentvelsen, P. de Jong, J. Koch and E. Laenen (Editors)*

31st INTERNATIONAL CONFERENCE ON HIGH ENERGY PHYSICS AMSTERDAM

# Studies of Radiative Penguin Decays at *BABAR*.

Colin P. Jessop (for the *BABAR*collaboration.)[a]

[a]Stanford Linear Accelerator Center, 2575 Sand Hill Road,
Menlo Park, CA 94025, U.S.A.

The electromagnetic radiative "penguin" decays $b \to s\gamma$, $b \to d\gamma$ are sensitive to physics beyond the Standard Model. We present recent studies made with the *BABAR* detector at the PEP-II asymmetric $e^+e^-$ storage ring.

## 1. Introduction

The radiative $b \to s\gamma$, $b \to d\gamma$ transitions occur by the electromagetic "penguin " loop diagram in the Standard Model. These decays are sensitive to the CKM matrix elements $V_{td}$ and $V_{ts}$ and to new virtual high mass particles such as a charged Higgs appearing in the loop causing deviations in the rate [1] and possibly introducing a new $CP$-violating phase. The Standard Model predictions [2] for the leading exclusive decays $B \to K^*\gamma$ ($b \to s\gamma$) and the as yet unobserved $B \to \rho\gamma$ ($b \to d\gamma$) decay suffer from large uncertainties compromising the sensitivity to both CKM matrix elements and new physics. However in the ratio $\mathcal{B}(B \to \rho\gamma)/\mathcal{B}(B \to K^*\gamma)$ some of these uncertainties cancel and a competitive measurement of $V_{ts}/V_{td}$ is possible. We present measurements of the exclusive decay, $\mathcal{B}(B \to K^*\gamma)$ [5], the CP-violating charge asymmetry, $A_{cp}(B \to K^*\gamma)$ and results of a search for $B \to \rho\gamma$ and $B \to \omega\gamma$ [6].

The predictions for the inclusive rates [3] are more precise as quark-hadron duality is believed to be a good approximation so that $B \to X_s\gamma = b \to s\gamma$. We present two measurements of $\mathcal{B}(B \to X_s\gamma)$ using both a semi-inclusive [8] and a fully inclusive technique [7]. The photon energy spectrum from the $B \to X_s\gamma$ measured with the semi-inclusive technique is also used to extract values for the HQET parameters $\overline{\lambda}$ and $\lambda_1$.

The data were collected with the *BABAR* detector [9] at the PEP-II asymmetric $e^+(3.1\,\mathrm{GeV})$ – $e^-(9\,\mathrm{GeV})$ storage ring. We use Monte Carlo simulations of the *BABAR* detector based on GEANT 4.0 [10] to optimize our selection criteria,to determine signal efficiencies, and in some cases to estimate backgrounds. These simulations take account of varying detector conditions and beam backgrounds.

The event selection for both the exclusive and inclusive analysis begins by requiring a high-energy radiative photon candidate. A photon candidate is defined as a localized energy maximum in the calorimeter. It must be isolated by from any other photon candidate or track and have a lateral energy profile consistent with a photon shower. We veto photons from a $\pi^0(\eta)$.

## 2. $\mathcal{B}(B \to K^*\gamma)$ and $A_{cp}(B \to K^*\gamma)$

The details of the event selection are described in [5]. The $K^*$ is reconstructed from $K^+$, $K^0_S$, $\pi^-$ and $\pi^0$ candidates through the four modes $K^{*0} \to K^+\pi^-$, $K^0_S\pi^0$ and $K^{*+} \to K^+\pi^0$, $K^0_S\pi^+$ and combined with a high energy photon to form a $B$ meson candidate. The dominant background is from continuum production from continuum $q\overline{q}$ production, where $q$ can be a $u$,$d$,$s$ or $c$ quark, with the high-energy photon originating from initial-state radiation or from $\pi^0$ and $\eta$ decays. The continuum background is produced above threshold and so is lorentz boosted to a "jet-like" topology in contrast to the signal which is produced isotropically. Event topology variables are used to suppress the background. Finally we construct from the $B$ candidates two independent variables $\Delta E$ and $m_{\mathrm{ES}}$ used to extract the signal yield. We define $\Delta E = E^*_B - E_{beam}$ and $m_{\mathrm{ES}} = \sqrt{E^{*2}_{beam} - p^{*2}_B}$ where $E^*_B$ and $p^*_B$ are the energy and momentum of the reconstructed


doi:10.1016/S0920-5632(02)02150-3

$B$, computed in the center of mass frame of the $e^+e^-$ and $E_{beam}$ is the precisely known beam energy. These variables exploit the fact that the beam energy is equal to the energy of the produced $B$ but more accurately known than the detector reconstructed energy measurement.

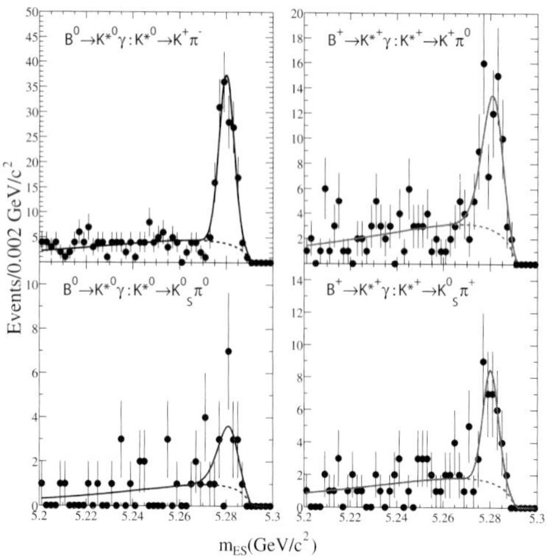


Figure 1. $m_{\rm ES}$ for the $B \to K^*\gamma$ candidates. The solid and dashed curves show respectively the fitted signal-plus-background and the background alone.

The analysis is performed on 22.7 million $B\overline{B}$ pairs. Figure 1 shows the $m_{\rm ES}$ distributions for the four decay modes. The signal is extracted from a maximum likelihood fit. The background is modeled by a threshold function [4]. The measured branching fractions and $CP-violating$ charge asymmetries are summarised in Table 1. The branching fractions are roughly a factor of two lower than the next-to-leading order Standard Model predictions and in agreement with previous measurements [11].

## 3. Search for $B \to \rho\gamma$ and $B \to \omega\gamma$

The analysis, described in detail in reference [6], is considerably more challenging than the $B \to K^*\gamma$ measurements. The signal rate is expected to be of order 50 times smaller than $B \to K^*\gamma$ ($\mathcal{B}[B^+ \to \rho^+\gamma] = (0.9-1.5)\times10^{-6}$ [2]) with significantly higher continuum backgrounds and additional backgrounds from $b \to s\gamma$ processes. The decay $B \to \rho\gamma$ is reconstructed with $\rho^0 \to \pi^+\pi^-$ and $\rho^+ \to \pi^+\pi^0$, while $B^0 \to \omega\gamma$ is reconstructed with $\omega \to \pi^+\pi^-\pi^0$. Continuum backgrounds are suppressed with event topology variables, vertexing and flavor tagging. The variables are combined into a neural net, trained on Monte Carlo, and validated with several data control samples. A dedicated pion selector has been developed for this analysis optimized to reject the kaon fake background from $b \to s\gamma$ processes. The analysis was performed "blind" on 84.4 million BB pairs. Figure 2 shows the $\Delta E$,$m_{\rm ES}$ dsitributions for the data after all cuts. The signal yield is estimated from a multi-dimensional maximum likelihood fit. The results are given in table 2. No signal is observed and we set 90 % confidence limits.

Table 1
The measured branching fraction $\mathcal{B}(B \to K^*\gamma)$ and $A_{CP}$ for each of the decay modes.

| Mode | $\mathcal{B}(B \to K^*\gamma)$ -±stat. ± sys. $\times10^{-5}$ | $A_{CP}$(signal) (±stat. ± sys.) |
|---|---|---|
| $K^+\pi^-$ | $4.24 \pm 0.41 \pm 0.22$ | $-0.049 \pm 0.094 \pm 0.012$ |
| $K^0_S\pi^0$ | $4.10 \pm 1.71 \pm 0.42$ | |
| $K^0_S\pi^+$ | $3.01 \pm 0.76 \pm 0.21$ | $-0.190 \pm 0.210 \pm 0.012$ |
| $K^+\pi^0$ | $5.52 \pm 1.07 \pm 0.38$ | $0.044 \pm 0.155 \pm 0.021$ |

Table 2
The signal yields and errors obtained from the signal extraction fit, and the 90 % Confidence Limit.

| Mode | Yield (Events) | 90% C.L. $\times10^{-6}$ |
|---|---|---|
| $B^0 \to \rho^0\gamma$ | $4.8 \pm 5.2$ | 1.4 |
| $B^+ \to \rho^+\gamma$ | $6.2 \pm 5.5$ | 2.3 |
| $B^0 \to \omega\gamma$ | $0.1 \pm 2.3$ | 1.9 |

The limits for the individual modes can be combined assuming isospin symettry to give $\mathcal{B}(B \to \rho\gamma) < 1.9 \times 10^{-6}$ at 90 % C.L.. In addition this limit can be compared to the measured $\mathcal{B}(B \to$

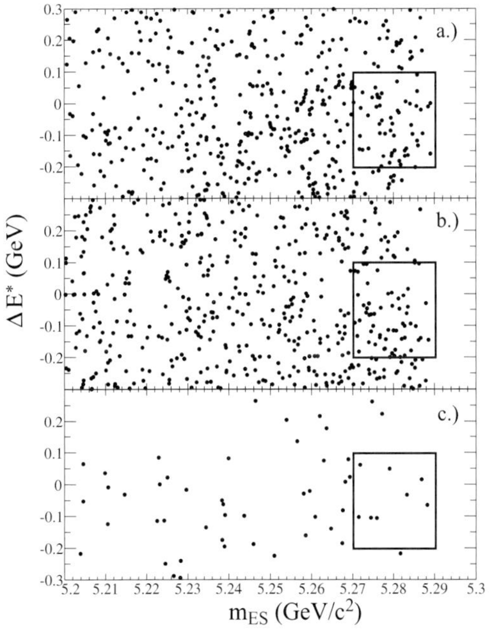


Figure 2. $\Delta E$ vs. $m_{\rm ES}$ for a.) $B^0 \to \rho^0\gamma$, b.) $B^+ \to \rho^+\gamma$ and c.) $B^0 \to \omega\gamma$ candidates. The boxes indicate the blinded region.

$K^*\gamma$) above to give $\mathcal{B}(B \to \rho\gamma)/\mathcal{B}(B \to K^*\gamma) < 0.047$ at 90 % C.L.

## 4. Measurement of $\mathcal{B}(B \to X_s\gamma)$ using a "semi-inclusive" technique

The fragmentation of the s-quark in the $b \to s\gamma$ transistion results in final states, $X_s$, containing kaons and pions proceeding through a complicated resonance structure. One technique to measure the inclusive rate is to attempt to reconstruct all possible $X_s$ states. However, we are not sensitive to states containing a $K^0_L$ while the prohibitively large combinatorics from high multiplicity modes and modes with $K^0_S \to \pi^0\pi^0$ limit the fraction of accessible modes to approximately 50 %. We recomstruct the $X_s$in the $K^+\pi^-$, $K^0_S\pi^0$, $K^+\pi^-\pi^0$, $K^0_S\pi^+\pi^-$, $K^+\pi^-\pi^+\pi^-$, $K^0_S\pi^+\pi^-\pi^0$ $K^+\pi^0$, $K^0_S\pi^+$, $K^+\pi^-\pi^+$, $K^0_S\pi^+\pi^0$, $K^+\pi^-\pi^+\pi^0$, $K^0_S\pi^+\pi^-\pi^+$ modes in 22.7 million $B\overline{B}$ pairs. The basic reconstruction technique is

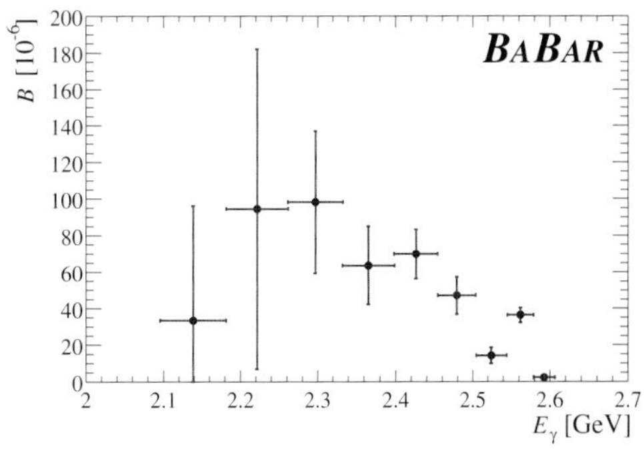


Figure 3. Branching fraction as a function of $E_\gamma$. The errors are purely statistical.

essentially the same as for the $B \to K^*\gamma$ analysis described above, except that there are additional backgrounds from $B\overline{B}$ decays and cross-feed between the modes. The analysis is described in detail in reference [8]. The $X_s$is divided into several $m_{X_s}$(dual to $E^*_\gamma$) bins and the signal yield extracted from a fit to $m_{\rm ES}$ for each of these bins. Figures 3 and 4 show the resultant $E^*_\gamma$and

$m_{X_s}$ spectra respectively. The first moment of the $E^*_\gamma$ spectra can be used to extract a measurement of the HQET parameter $\overline{\Lambda}(\alpha_s^2, 1/M_B^3)$ [13]. We find $< E_\gamma > |_{E_\gamma > 2.094 \text{ GeV}} = 2.35 \pm 0.04$ $(stat) \pm 0.04$ $(syst)$ GeV from which we derive $\overline{\Lambda} = 0.37 \pm 0.09$ $(stat) \pm 0.07$ $(syst) \pm 0.10$ $(model)$ GeV. The parameter $\overline{\Lambda}$ then provides information to help constrain the parameters in the fit to the $m_{X_s}$ spectrum shown in fig 4 used to extract $\mathcal{B}(b \to s\gamma) = 4.3 \pm 0.5$ $(stat) \pm 0.8$ $(syst) \pm 1.3$ $(model) \cdot 10^{-4}$ and $\lambda_1 = -0.24^{+0.03}_{-0.04}$ $(stat) \pm 0.02$ $(syst)^{+0.15}_{-0.21}$ $(model)$ $[\text{GeV}/c^2]^2$. consistent with standard model predictions [3] and previous measurements [12]. Note that this technique gives large systematic errors due to uncertainties in the fragmentation model and resonance structure of the $X_s$.

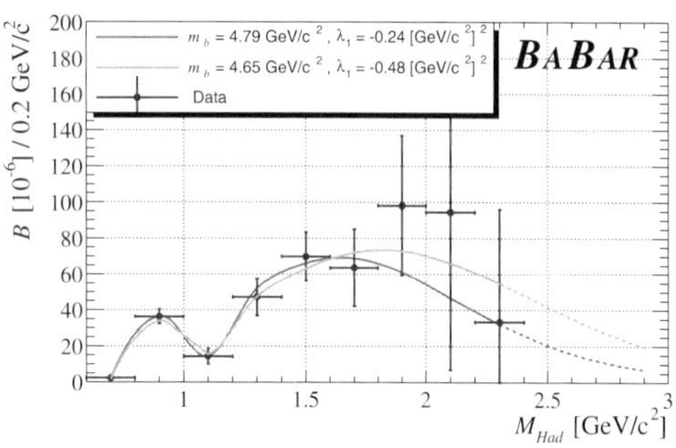


Figure 4. Superposition of the predicted spectrum for $m_b = 4.79$ GeV/$c^2$ and $m_b = 4.65$ GeV/$c^2$ on the observed hadronic mass spectrum.

## 5. Measurement of $\mathcal{B}(B \to X_s\gamma)$ using a "Fully-inclusive" technique

The semi-inclusive measurement of $\mathcal{B}(B \to X_s\gamma)$ offers a powerful technique to suppress backgrounds but the sensitivity to the details of the $X_s$fragmentation incurs large systematic uncertainties. A fully inclusive measurement sensitive to 100 % of the decay is highly desirable but complicated by the large backgrounds as shown in Figure 5. Here for pedagogical purposes we have selected a high energy photon using standard quality cuts from a sample of signal and continuum and $B\overline{B}$ Monte Carlo. To reduce

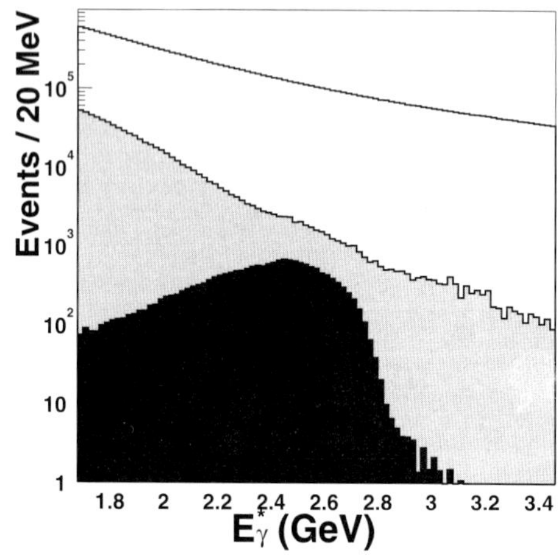


Figure 5. The energy distribution, in the $\Upsilon(4s)$ center of mass, of simulated photon candidates, after "photon quality" cuts. Shown are $B \to X_s\gamma$ signal (dark shading), $B\overline{B}$ background (grey shading) and continuum background (unshaded), all normalized to 54.6 Million BB.

these backgrounds while remaining insensitive to the $X_s$fragmentation we employ a lepton tagging technique. The analysis is described in detail in reference [7]. The tagging reduces the continuum background by 3 orders of magnitude but at a cost of low efficiency ($\approx 0.5\%$). Also a significant $B\overline{B}$ background remains which must be estimated by Monte Carlo. Figure 6 shows the $E^*_\gamma$ spectrum for data with the estimated background. We find $\mathcal{B}(B \to X_s\gamma) = 3.88 \pm 0.36(stat.) \pm 0.37(syst.) \pm^{0.43}_{0.23}$ $(model) \times 10^{-4}$. The dominant systematic uncertainty is from the modelling of the $B\overline{B}$ background which has been compared to data control samples. The measurement is consistent with both Standard Model expectations [3] and previous measurments [12]

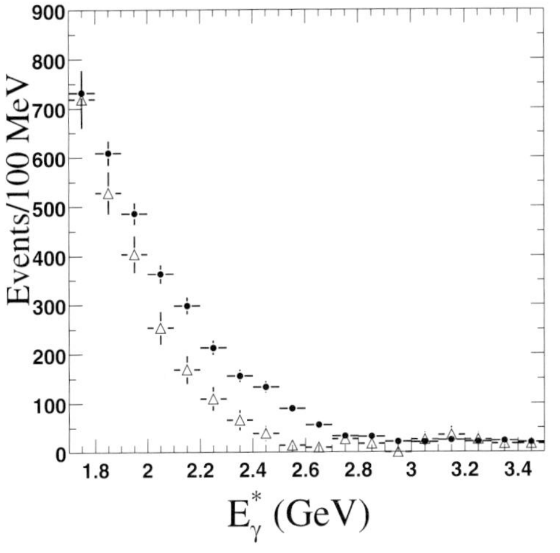


Figure 6. The $E^*_\gamma$ distribution of on-resonance data (solid points) compared to background expectation. All errors are statistical only.

## REFERENCES


1. See for example, S. Bertolini, F. Borzumati and A. Masiero, Nucl. Phys. B **294**, 321 (1987); H. Baer, M. Brhlik, Phys. Rev. D **55**, 3201 (1997); J. Hewett and J. Wells, Phys. Rev. D **55**, 5549 (1997); M. Carena *et al.*, Phys. Lett. B **499**, 141 (2001).
2. Z. Ligeti and M. B. Wise, Phys. Rev. D **60**, 117506 (1999;)M. Beneke, T. Feldmann and D. Seidel; Nucl. Phys. B **612**, 25 (2001) S. W. Bosch and G. Buchalla, Nucl. Phys. B **621**, 459 (2002); A. Kagan and M. Neubert, e-print hep-ph/0110078(2001); A. Ali and A. Y. Parkhomenko, Eur. Phys. J. C **23**, 89 (2002). B. Grinstein and D. Pirjol, Phys. Rev. D **62**, 093002 (2000).
3. K. Abel and Y. P. Yao, Phys. Rev. D **49**, 4945 (1994); A. Ali and C. Greub, Phys. Lett. B **361**, 146 (1995); C. Greub and T. Hurth, Phys. Lett. B **380**, 2934 (1997); C. Greub, T. Hurth and D. Wyler, Phys. Lett. B **361**, 385 (1996); K. Chetyrkin, M. Misiak and M. Munz, Phys. Lett. B **400**, 296 (1997); A. J. Buras, A. Kwiatkowski and N. Pott, Phys. Lett. B **414**, 157 (1997); A. Kagan and M. Neubert, Eur. Phy. Journal 7 5 1999 ; A. Kagan and M. Neubert, Phys. Rev. D **58**, 094012 (1998;)P. Gambino and M. Misiak, Nucl. Phys. B **611**, 338 (2001).
4. ARGUS Collaboration, H. Albrecht *et al.*, Z. Phys. C **48**, 543 (1990).
5. "Measurement of $B \to K^*\gamma$ Branching Fractions and Charged Asymmetries"
The BaBar Collaboration, B. Aubert *et al.* Phys. Rev. Lett. **88**, 101805 (2002).
6. "Search for the Exclusive Radiative Decays $B \to \rho\gamma$ and $B^0 \to \omega\gamma$."
The BaBar Collaboration, B. Aubert *et al.* e-print hep-ex/0207073.
7. "Determination of the Branching Fraction for Inclusive Decays $B \to X_s\gamma$."
The BaBar Collaboration, B. Aubert *et al.* e-print hep-ex/0207076.
8. "$B \to X_s\gamma$ Using a Sum of Exclusive Modes."
The BaBar Collaboration, B. Aubert *et al.* e-print hep-ex/0207074.
9. "The BaBar Detector"
The BaBar Collaboration, B. Aubert *et al.* Nucl. Instrum. Methods **A**, 479 (1)2002 .
10. "GEANT Detector Description and Simulation Tool", CERN Program Library Long Writeup W5013 (1994).
11. T.E. Coan *et al.* [CLEO Collaboration], Phys. Rev. Lett. **84**, 5283 (2000;)Y. Ushiroda *et al.* [BELLE Collaboration], contributed to BCP4, Ago Town, Japan, Feb 2001. hep-ex/0104045.
12. CLEO Collaboration, M.S. Alam *et al.*, Phys. Rev. Lett. **74**, 2885 (1995); ALEPH Collaboration, R. Barate *et al.*, Phys. Lett. B **429**, 169 (1998); BELLE Collaboration, K. Abe *et al.*, Phys. Lett. B **511**, 151 (2001); CLEO Collaboration, S. Chen *et al.*, Phys. Rev. Lett. **87**, 251807 (2001).
13. Z. Ligeti, M.E. Luke, A.V. Manohar, M.B. Wise, Phys. Rev. D **60**, 034019 (1999;)C. Bauer, Phys. Rev. D **57**, 5611 (1998).

Proceedings of ICHEP 2002, pp. 598–601
*S. Bentvelsen, P. de Jong, J. Koch and E. Laenen (Editors)*

# Electroweak Penguin Decays of $B$ Mesons

Shohei Nishida[a]

[a]Department of Physics, Kyoto University

We report measurements of the electroweak penguin decays of $B$ mesons at the Belle detector, based on a data sample containing 65 million $B\bar{B}$. We perform a precise measurement of $B \to K^*(892)\gamma$ and a search for the $b \to d\gamma$ decay. We also analyze the $b \to s\ell^+\ell^-$ process using exclusive and inclusive modes. We measure the inclusive $b \to s\ell^+\ell^-$ rate for the first time with the branching fraction of $(6.1 \pm 1.4\,^{+1.3}_{-1.1}) \times 10^{-6}$.

## 1. Introduction

The $b \to s\gamma$, $b \to d\gamma$ and $b \to s\ell^+\ell^-$ decays are flavor changing neutral current processes. In the Standard Model (SM), the lowest order diagram for these processes is a one loop electroweak penguin (or box) diagram. Due to the possible contribution of non-SM particles to the loop, these processes are sensitive to physics beyond the SM.

We report measurements of the electroweak penguin decays of $B$ mesons at Belle. Belle is a general purpose detector at the KEKB collider [1], which is an $e^+e^-$ asymmetric collider (3.5 GeV on 8 GeV) operated at the $\Upsilon(4S)$ resonance. Details of the Belle detector are described elsewhere [2]. The analysis is based on a data sample containing 65 million $B\bar{B}$.

## 2. $B \to K^*(892)\gamma$

In the exclusive mode $B \to K^*(892)\gamma$, the branching fraction itself does not provide a tight constraint on the New Physics, because of the large uncertainty in the theoretical prediction. However, the rate difference between charged and neutral modes or the partial rate asymmetry between CP conjugate mode can be a probe for the New Physics, since it is expected to be small in the SM.

We form $B$ meson candidates from a high energy photon candidate and a $K^*$ candidate, which is reconstructed from $K^+\pi^-$, $K_S\pi^+$, $K^+\pi^0$, and $K_S\pi^0$ with $|M(K\pi) - M_{K^*}| < 75$ MeV$/c^2$ [1]. The main background originates from continuum $q\bar{q}$ ($q = u, d, s, c$) production, including contributions from initial state radiation ($e^+e^- \to q\bar{q}\gamma$). To suppress this background, we use a likelihood ratio (LR) from the SFW and the cosine of $B$ meson flight direction ($\cos\theta_B$). Here, SFW is a Fisher discriminant formed from six modified Fox-Wolfram moments [3]. A small contribution from other $B$ decays such as $B \to K^*\pi^0$ and $B \to K^*\pi\gamma$ is estimated by Monte Carlo (MC).

The signal yield is extracted by fitting the beam constrained mass ($M_{\rm bc}$) distribution after applying the $\Delta E$ selection. The $M_{\rm bc}$ distributions and yields for each mode are shown in Fig. 1. The reconstruction efficiencies are estimated to be (13.42±0.84)%, (4.12±0.35)%, (2.40±0.24)% and $(0.69 \pm 0.08)$% respectively for $K^+\pi^-\gamma$, $K_S\pi^+\gamma$, $K^+\pi^0\gamma$ and $K_S\pi^0\gamma$. Then, we obtain the branching fraction to be

$$\mathcal{B}(B^0 \to K^{*0}\gamma) = (39.1 \pm 2.3 \pm 2.5) \times 10^{-6}$$
$$\mathcal{B}(B^+ \to K^{*+}\gamma) = (42.1 \pm 3.5 \pm 3.1) \times 10^{-6}.$$

We find no significant difference between the neutral and charged $K^*(892)\gamma$ modes.

The partial rate asymmetry is defined as

$$A_{\rm CP} = \frac{1}{1-2w}\frac{N(\bar{B} \to \bar{K}^*\gamma) + N(B \to K^*\gamma)}{N(\bar{B} \to \bar{K}^*\gamma) + N(B \to K^*\gamma)},$$

where $N$ is the signal yield, $w$ is the wrong tag fraction, $B$ ($K^*$) indicates $B^0$ or $B^+$ ($K^{*0}$ or $K^{*+}$), and $\bar{B}$ and $\bar{K}^*$ are their conjugates respectively. The wrong tag fraction is about 0.9% for

[1]The inclusion of the charge conjugate mode is implied unless otherwise stated.


doi:10.1016/S0920-5632(02)02195-3

the neutral mode, and negligible for the charged mode. We study an inclusive $K^*$ sample, but find no asymmetry in it. We assign 1.5% to the charge asymmetry systematic error. By separately performing $M_{\rm bc}$ fits for the six self-tagging modes, we obtain $A_{\rm CP}(K^*\gamma) = (-2.2 \pm 4.8 \pm 1.7)\%$, which is consistent with null asymmetry.

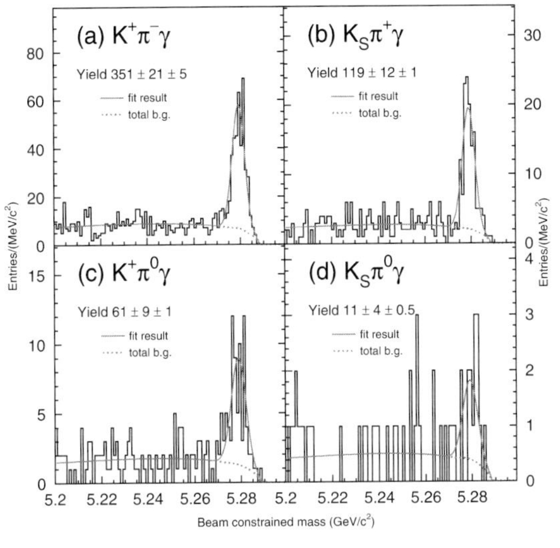


Figure 1. $M_{\rm bc}$ distributions for $B \to K^*(892)\gamma$

## 3. $B \to K_2^*(1430)\gamma$, $B \to K\pi\pi\gamma$

Other exclusive $b \to s\gamma$ modes are also studied[2] [4]. For the $K^+\pi^-\gamma$ final states, we find an excess around $M(K\pi) = 1.4$ GeV$/c^2$. By performing a unbinned likelihood fit on $M_{\rm bc}$, the cosine of the decay helicity angle ($\cos\theta_{\rm hel}$) and $M(K\pi)$, we find that the $B \to K_2^*(1430)\gamma$ component is dominant. We obtain

$$\mathcal{B}(B^0 \to K_2^*(1430)^0\gamma) = (15\,^{+6}_{-5} \pm 1) \times 10^{-6}.$$

We also observe the decay $B^+ \to K^+\pi^-\pi^+\gamma$ with a branching fraction of $(24 \pm 5\,^{+4}_{-2}) \times 10^{-6}$. From the $M(K\pi)$ and $M(\pi\pi)$ distributions, we find that the $B \to K^*\pi\gamma$ and $B \to K\rho\gamma$ processes are dominant. However, the statistics are not sufficient to disentangle the resonant structure for the $K\pi\pi$ system.

[2]The results in this section are based on a data sample containing 32 million $B\bar{B}$.

## 4. $B \to \rho\gamma$, $B \to \omega\gamma$

We search for the radiative decays $B \to \rho\gamma$ and $B \to \omega\gamma$. These modes proceed through the $b \to d\gamma$ process, and are useful to extract $|V_{td}/V_{ts}|$.

We reconstruct $\rho^0$ ($\rho^+$) from $\pi^+\pi^-$ ($\pi^+\pi^0$) requiring $|M(\pi\pi) - M_\rho| < 150$ MeV$/c^2$, while $\omega$ is reconstructed in $\pi^+\pi^-\pi^0$ with $|M(\pi^+\pi^-\pi^0) - M_\omega| < 15$ MeV$/c^2$. The selection criteria are similar to those for the $B \to K^*(892)\gamma$ analysis, but we must suppress the contamination from the $B \to K^*(892)\gamma$ background in the case of $B \to \rho\gamma$. For this, we apply a tight pion identification requirement. In addition, we calculate the invariant mass of the two pions by assigning a kaon mass to the pion with the higher momentum, and reject the event if the invariant mass is within 50 MeV$/c^2$ of the nominal kaon mass. After applying these requirements, the contribution of $B \to K^*(892)\gamma$ to $B^0 \to \rho^0\gamma$ is estimated to be $0.9 \pm 0.2$ events, and the contribution to $B^+ \to \rho^+\gamma$ becomes negligible.

The signal yield is obtained using an unbinned maximum likelihood fit to $M_{\rm bc}$ and $\Delta E$. The $M_{\rm bc}$ and $\Delta E$ distributions are shown in Fig. 2. We find no signal and set the upper-limits listed in Table 1.

Table 1
Measured signal yields, reconstruction efficiencies and 90% confidence level upper limits (UL) of branching fractions ($\times 10^{-6}$) for $B \to \rho\gamma$ and $B \to \omega\gamma$.

| mode | yield | efficiency | UL |
|---|---|---|---|
| $B^0 \to \rho^0\gamma$ | $2.1 \pm 4.0$ | $(6.3 \pm 0.5)\%$ | 2.6 |
| $B^+ \to \rho^+\gamma$ | $1.9 \pm 2.7$ | $(3.3 \pm 0.3)\%$ | 4.9 |
| $B^0 \to \omega\gamma$ | $1.5 \pm 1.6$ | $(4.5 \pm 0.5)\%$ | 3.1 |

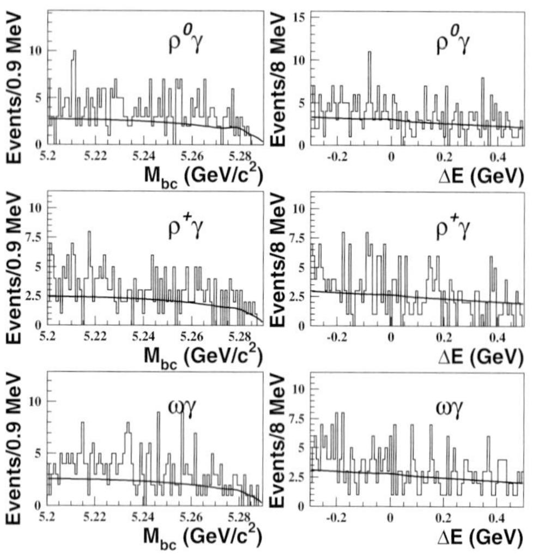


Figure 2. $M_{\rm bc}$ and $\Delta E$ distributions for $B \to \rho\gamma$ and $B \to \omega\gamma$. The fit results are overlaid.

## 5. $B \to K^{(*)}\ell^+\ell^-$

Next, we report on the measurement of the exclusive modes $B \to K^{(*)}\ell^+\ell^-$ ($\ell = e, \mu$), where $K^{(*)}$ is either $K^+$, $K^0_S$ or $K^*$ reconstructed from $K^+\pi^-$, $K^+\pi^0$ or $K^0_S\pi^+$.

The selection criteria are similar to those in Ref. [5]. To form a $B$ candidate, we combine a $K^{(*)}$ candidate with two leptons. Here, the invariant mass of the lepton pair is required to be greater than 0.14 GeV/$c^2$ in order to suppress background from photon conversions and $\pi^0$ Dalitz decay. To remove a contribution from $B \to J/\psi(\psi')K^{(*)}$, we reject an event if the invariant mass of the two leptons is consistent with the nominal $J/\psi$ ($\psi'$) mass. To suppress the $q\bar{q}$ continuum background, we form a LR from virtual calorimeter and $\cos\theta_B$.

The main background is semi-leptonic $B$ decays from $B\bar{B}$ events. The missing energy ($E_{\rm miss}$) of the event is useful to suppress the background, since we expect a large amount of missing energy due to neutrinos in semi-leptonic decays. Therefore, we form another LR from $E_{\rm miss}$ and $\cos\theta_B$.

We fit the $M_{\rm bc}$ distribution to extract the signal yield as shown in Fig. 3. Here, a small contribution from $B \to K^{(*)}h^+h^-$ is included, where $h$ is a hadron incorrectly identified as a muon. The shape and the normalization for this contribution is determined by reconstructing $B \to K^{(*)}h^+h^-$ without lepton identification and then multiplying the yield by momentum dependent muon mis-identification rate. We expect $0.32 \pm 0.03$ ($0.21 \pm 0.02$) events in $K\mu\mu$ ($K^*\mu\mu$) modes in the peak region.

The obtained results are listed in Fig. 2. We observe a clear signal for $B \to K\ell^+\ell^-$, while the signal for $B \to K^*\ell^+\ell^-$ is not significant. The measured branching fractions are consistent with the SM prediction [6]. We also find the dilepton invariant mass distribution in the peak region agrees with the prediction.

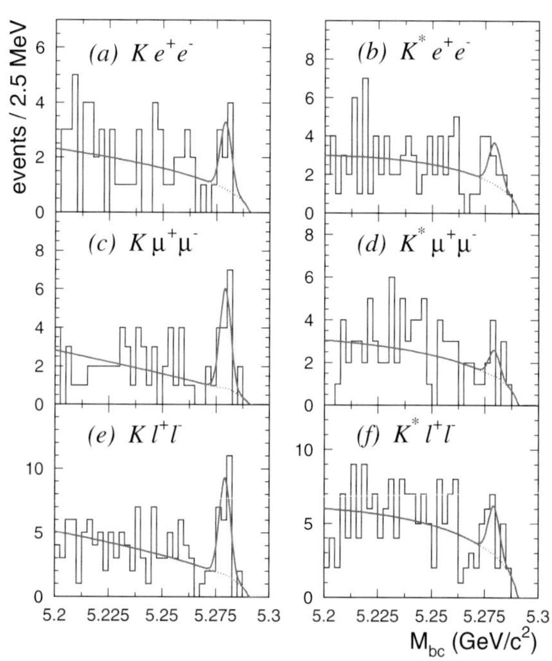


Figure 3. $M_{\rm bc}$ distributions and the fit results for $B \to K^{(*)}\ell^+\ell^-$.

## 6. $B \to X_s\ell^+\ell^-$

Finally, we present the inclusive analysis of $B \to X_s\ell^+\ell^-$ [7], where $X_s$ is a hadronic recoil

Table 2
Measured signal yields, reconstruction efficiencies and 90% confidence level upper limits (UL) of branching fractions ($\times 10^{-7}$) for $b \to s\ell^+\ell^-$ processes

| mode | yield | efficiency | $\mathcal{B}$ ($\times 10^{-7}$) | signif. |
|---|---|---|---|---|
| $B \to Ke^+e^-$ | $6.1^{+3.5}_{-2.8}{}^{+0.7}_{-0.8}$ | $(12.5 \pm 1.0)\%$ | $3.8^{+2.1}_{-1.7} \pm 0.6$ | 2.7 |
| $B \to K\mu^+\mu^-$ | $13.0^{+4.5}_{-3.8}{}^{+0.8}_{-0.9}$ | $(12.4 \pm 1.1)\%$ | $8.0^{+2.8}_{-2.3} \pm 0.8$ | 4.9 |
| $B \to K\ell^+\ell^-$ | $19.0^{+5.5}_{-4.8}{}^{+1.0}_{-1.2}$ | $(12.5 \pm 1.1)\%$ | $5.8^{+1.7}_{-1.5} \pm 0.6$ | 5.4 |
| $B \to K^*e^+e^-$ | $5.4^{+4.0}_{-3.3}{}^{+1.2}_{-1.3}$ | $(4.6 \pm 0.5)\%$ | $< 24$ | — |
| $B \to K^*\mu^+\mu^-$ | $3.0^{+3.3}_{-2.6}{}^{+1.0}_{-1.1}$ | $(6.5 \pm 0.7)\%$ | $< 12$ | — |
| $B \to K^*\ell^+\ell^-$ | $8.3^{+5.0}_{-4.3}{}^{+1.6}_{-1.7}$ | $(5.3 \pm 0.5)\%$ | $< 14$ | — |
| $B \to X_se^+e^-$ | $26 \pm 11^{+5}_{-4}$ | $(3.9 \pm 0.4 \pm 0.5)\%$ | $50 \pm 23^{+12}_{-11}$ | 3.4 |
| $B \to X_s\mu^+\mu^-$ | $37 \pm 10^{+7}_{-4}$ | $(3.6 \pm 0.4 \pm 0.4)\%$ | $79 \pm 21^{+20}_{-15}$ | 4.7 |
| $B \to X_s\ell^+\ell^-$ | $60 \pm 14^{+9}_{-5}$ | $(3.7 \pm 0.4 \pm 0.5)\%$ | $61 \pm 14^{+13}_{-11}$ | 5.4 |

system. We reconstruct $X_s$ from either a $K$ or $K_S$ combined with 0 to 4 pions, of which up to one $\pi^0$ is allowed. To suppress background from $B\bar{B}$ and $q\bar{q}$, in addition to SFW and a Fisher discriminant from total visible energy and missing mass, we apply a cut based on a LR formed from $\Delta E$ and $\cos\theta_B$. The LR is also used to choose the best candidate in one event. We require $M(\ell^+\ell^-) > 0.2$ GeV/$c^2$ and $M_{X_s} < 2.1$ GeV/$c^2$. The other selection criteria are similar to the exclusive analysis. We apply a $J/\psi$ ($\psi'$) veto and estimate the $B \to X_sh^+h^-$ contribution to be $2.6 \pm 0.2$ events.

The $M_{\rm bc}$ distributions are shown in Fig. 4. We observe a clear peak in $X_s\mu^+\mu^-$ mode. We also analyze the $X_s\mu^+e^-$ mode as a cross-check and find no peak in $M_{\rm bc}$. The resulting signal yields and branching fractions are listed in Table 2. This is the first measurement of the inclusive $B \to X_s\ell^+\ell^-$ branching fraction. Although the experimental error is still large, the background-subtracted $M(\ell^+\ell^-)$ and $M_{X_s}$ distributions agree with SM based theoretical predictions [8].

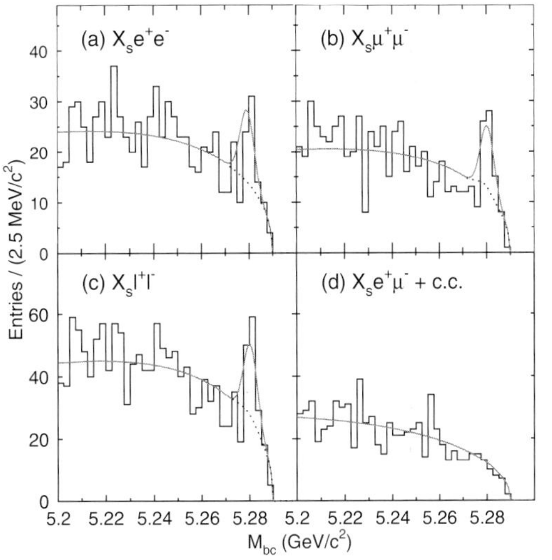


Figure 4. $M_{\rm bc}$ distribution for $B \to X_s\ell^+\ell^-$. Curves are fit results.

## REFERENCES

1. KEKB B Factory Design Report, KEK Report 95-1, 1995, unpublished.
2. A. Abashian *et al.* (Belle Collaboration), Nucl. Instr. and Meth. **A479**, 117 (2002).
3. K. Abe *et al.* (Belle Collaboration), Phys. Lett. **B511**, 151 (2001).
4. S. Nishida *et al.*, hep-ex/0205025.
5. K. Abe *et al.*, Phys. Rev. Lett. **88**, 021801 (2002).
6. A. Ali *et al.* Phys. Rev. **D61**, 074024 (2000).
7. J. Kaneko *et al.*, hep-ex/0208029.
8. A. Ali *et al.* Phys. Rev. **D55**, 4105 (1997); A. Ali *et al.*, hep-ph/0112300.

ELSEVIER

Proceedings of ICHEP 2002, pp. 602–606
S. Bentvelsen, P. de Jong, J. Koch and E. Laenen (Editors)

31st INTERNATIONAL CONFERENCE ON HIGH ENERGY PHYSICS AMSTERDAM

# Gluonic Penguin Decays of $B$ mesons with Belle

Kai-Feng Chen[a]

[a]Department of Physics, National Taiwan University,
No.1, Sec. 4, Roosevelt Road, Taipei, Taiwan

We present measurements of gluonic penguin decays of $B$ mesons with a large data sample recoreded at the $\Upsilon(4S)$ resonance with the Belle detector at KEKB. We have observed large signals for $B \to \eta' K$ and $B \to \eta K^*$. Upper limits for other related decays were obtained. We measured the $B^+ \to \omega K^+$ branching fraction and found evidence for $B^+ \to \omega\pi^+$. The branching fractions for $B \to \phi K$ and $B \to \phi K^*$ were also measured.

## 1. Two-body $B$ meson decays to $\eta'$ or $\eta$

Two-body charmless hadronic $B$ decays with $\eta'$ or $\eta$ mesons in the final state are of interest because of the unexpectedly large branching fractions The large branching fractions $B \to \eta' K$ and $B \to \eta K^*$ are still difficult to account in theoretical calculations [1–4] and might be a hint of new physics beyond the Standard Model (SM).

The difference between the $B^+ \to \eta' K^+$ ($B^+ \to \eta K^{*+}$) and $B^0 \to \eta' K^0$ ($B^0 \to \eta K^{*0}$) decays is also interesting. An additional SU(3)-singlet contribution or a interference between tree and penguin amplitudes may be required.

We report measurements of branching fractions based on a data sample collected at the $\Upsilon(4S)$ resonance with the Belle detector [5] at the KEKB [6] $e^+e^-$ storage ring. The integrated luminosity of this data set is 29.4 $fb^{-1}$ corresponding to 31.7 $\times 10^6$ $B\overline{B}$ pairs.

All primary charged tracks were required to pass track quality cuts based on the impart parameters relative to the interaction point (IP). The particle identification was based on a likelihood ratio, $KID = \mathcal{L}_K/(\mathcal{L}_K + \mathcal{L}_\pi)$, where $\mathcal{L}_K$ ($\mathcal{L}_\pi$) stands for the likelihood combination of subdectectors for charged kaons (pions).

The $\eta$ was reconstructed through $\eta \to \gamma\gamma$ and $\pi^+\pi^-\pi^0$, and the $\eta'$ are reconstructed by $\eta' \to \eta\pi^+\pi^-$ and $\rho\gamma$ decays. The $K^{*0}$ was formed by $K^-\pi^+$ and $K^+\pi^0$ and $\overline{K^0}\pi^+$ for the $K^{*+}$. The $\rho$ meson was reconstructed through $\rho \to \pi^+\pi^-$ and $\pi^+\pi^0$ with the $\pi\pi$ invariant mass between 570 and 970 $MeV/c^2$.

$B$ meson candidates were identified by the beam constrained mass ($M_{bc} = \sqrt{E_{beam}^2 - P_B^2}$) and the energy difference ($\Delta E = E_B - E_{beam}$), where the $E_{beam} = 5.29$ GeV is the beam energy at the rest frame.

Several event-shape variables were selected to distinguish $B$ events that are spherical from the non-$B$ continuum events which are jet-like. A Fisher discriminant [7] ($= \overrightarrow{\alpha} \cdot \overrightarrow{R}$) was used to optimize the separation power, where $\overrightarrow{R}$ is a vector consisting event shape variables and $\overrightarrow{\alpha}$ is the vector of Fisher coefficients. The Fisher and other non-correlated variables were combined in a likelihood ratio, $\mathcal{LR} = \mathcal{L}_s/(\mathcal{L}_s + \mathcal{L}_b)$, where $\mathcal{L}_s$ ($\mathcal{L}_b$) was the combined likelihood for signal (background).

The signal yield was obtained by an extended unbinned maximum liklihood fit on both $M_{bc}$ and $\Delta E$. The signal probability density functions (PDFs) were obtained from signal Monte Carlo (MC). The fitting results are shown in Table 1. Figure 1 and 2 show the projection plots for $B \to \eta' K$ and $B \to \eta K^*$ modes.

The systematic errors are due to the efficiency estimation and fitting. The main sources are from the reconstruction of low-momentum charged tracks and low energy photon finding, and the background suppression cut.

In summary, we have studied exclusive two-body charmless hadronic B decays with $\eta'$ or $\eta$ in the final state. The results confirmed the branching fractions from previous measurements.


doi:10.1016/S0920-5632(02)02190-4

Table 1
Summary of results for each channel. The branching fraction are in units of $10^{-6}$. $\Sigma$ is the statistical significance.

| Mode | This measurement | $\Sigma$ | CLEO | BaBar | Theory |
|---|---|---|---|---|---|
| $B^+ \to \eta' K^+$ | $77.9^{+6.2+9.3}_{-5.9-8.7}$ | 21.2 | $80^{+10}_{-9}\pm7$ | $70 \pm 8 \pm 5$ | 21-53 |
| $B^0 \to \eta' K^0$ | $68.0^{+10.4+8.8}_{-9.6\ -8.2}$ | 10.9 | $89^{+18}_{-16}\pm9$ | $42^{+13}_{-11} \pm 4$ | 20-50 |
| $B^+ \to \eta' K^{*+}$ | $< 85$ | - | $< 35$ | - | 1-3.7 |
| $B^0 \to \eta' K^{*0}$ | $< 25$ $(4.8^{+7.2}_{-5.9})$ | - | $< 24$ | $< 13$ $(4.0^{+3.5}_{-2.4})$ | 1-8.0 |
| $B^0 \to \eta' \rho^0$ | $< 14$ | - | $< 12$ | - | 0.1-11 |
| $B^0 \to \eta K^{*0}$ | $16.5^{+4.6}_{-4.2}\pm1.2$ | 5.0 | $13.8^{+5.5}_{-4.6}\pm1.6$ | $19.8^{+6.5}_{-5.6} \pm 1.7$ | 0.1-15.2 |
| $B^+ \to \eta K^{*+}$ | $26.5^{+7.8}_{-7.0}\pm3.0$ | 4.7 | $26.4^{+9.6}_{-8.2}\pm3.3$ | $22.1^{+11.1}_{-9.2} \pm 3.3$ | 0.2-16.2 |
| $B^0 \to \eta \rho^0$ | $< 2.7$ | - | $< 15$ | - | 4-7 |
| $B^+ \to \eta \rho^+$ | $< 6.2$ | - | $< 10$ | - | 0.1-6.5 |
| $B^+ \to \eta K^+$ | $< 7.7$ $(5.2^{+1.7}_{-1.5})$ | - | $< 6.9$ | $< 6.4$ | 0.2-5.0 |
| $B^+ \to \eta \pi^+$ | $< 8.2$ $(5.3^{+2.0}_{-1.7})$ | - | $< 5.7$ | $< 5.2$ | 1.9-7.4 |

## 2. $B^\pm \to \omega K^\pm$ and $\omega\pi^\pm$

We report measurements of charmless $B^- \to \omega\pi^-$ and $B^- \to \omega K^-$ decays. The $B^- \to \omega K^-$ mode was first reported by CLEO collaboration in 1998 [8], but superseded by nonobseravtion with a larger data set [9]. This result was supported by BaBar in 2001 [10]. The $B^- \to \omega\pi^-$ channel had also been reported by the CLEO and BaBar collaborations. However, we find a significant signal in the $B^- \to \omega K^-$ mode. The data used in this analysis were based on an integrated luminosity of 29.4 fb$^{-1}$ data sample.

Charged tracks were required to be inconsistent with electrons or muons, and were identified as kaons or pions according to a $K/\pi$ likelihood ratio. Neutral pions were reconstructed from pairs of photons, each consisting of energy greater than 50 MeV, with the $m_{\gamma\gamma}$ within a $\pm3\sigma$ ($\sigma$ = 5.4 MeV/c$^2$) mass window around the $\pi^0$ mass. $\omega$ meson candidates were obtained from $\pi^+\pi^-\pi^0$ combinations with invariant mass within $\pm30$ MeV/c$^2$ of the nominal $\omega$ mass. The momentum of candidate $\pi^0$'s under $\omega$ was required to $> 350$ MeV/c.

A charged kaon or a pion track was selected to combine with an $\omega$ candidate to form a $B^\pm$ candidate. $B$ mesons were identified by the $M_{bc}$ and $\Delta E$. The resolutions were 3 MeV/c$^2$ for $M_{bc}$ and 23 MeV for $\Delta E$ from Monte Carlo simulations.

Continuum $e^+e^- \to q\bar{q}$ processes were the dominant background in this analysis. A single likelihood ratio combining the Fisher discriminant, $B$-meson direction, and the helicity angle was used.

Background from charmed and charmless $B$ decays such as $B \to \omega K^*$, $B \to \omega\rho$, and non-resonant $B^- \to K^-\pi^+\pi^-\pi^0$ were studied by Monte Carlo simulation with an effective luminosity up to 20 times larger than current data amount. These backgrounds were negligible in the signal region.

The signal yields were extracted by an unbinned maximum likelihood fit to $M_{bc}$ and $\Delta E$. The PDFs were determined by Monte Carlo simulation. The yields and branching fractions are summarized in Table 2. The $M_{bc}$ and $\Delta E$ projections of candidate events are shown in Figure 3.

The systematic error of $\omega$ detection efficiency was determined from detailed studies of charged particle tracking, $\pi^0$ detection, and particle identification. The combined uncertainty of the efficiency was 10.1%.

We made several consistency checks of our analysis. The $\Delta E$ distributions without KID requirement confirmed the $\omega K^-$ signals were larger than $\omega\pi^-$. We also examine the $\omega$ properties to con-

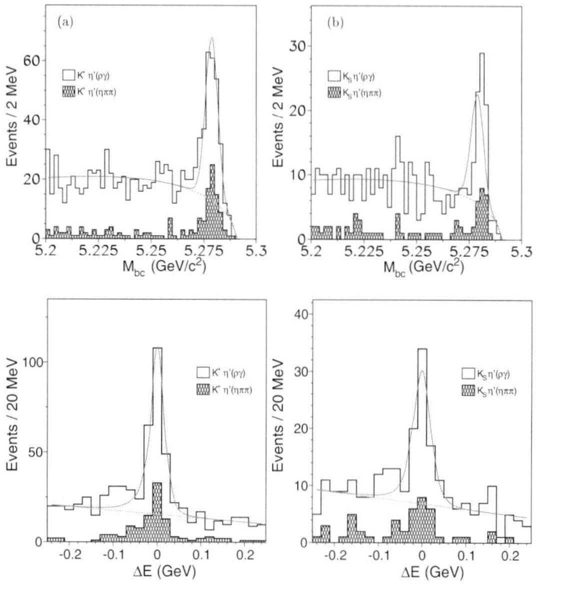


Figure 1. Projections on $M_{bc}$ and $\Delta E$ for $\eta' K^+$ and $\eta' K_S$ events. The shaded area represents the contribution of $\eta' \to \eta\pi^+\pi^-$ decays.

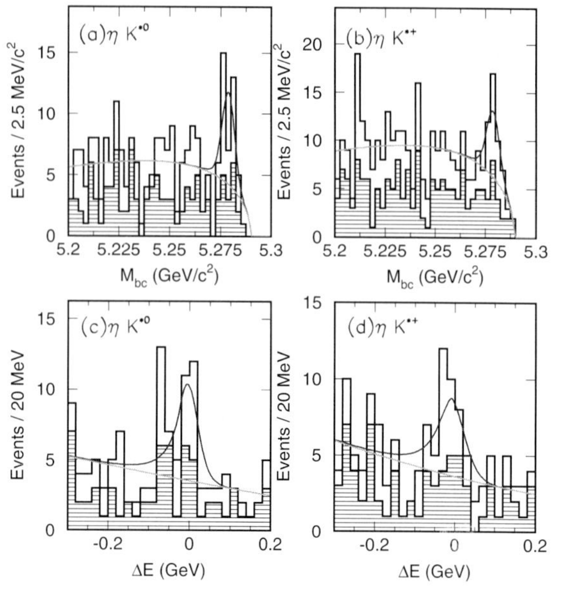


Figure 2. Projections on $M_{bc}$ and $\Delta E$ for $\eta K^{*+}$ and $\eta K^{*0}$ events. The shaded area represents $\eta \to \pi^+\pi^-\pi^0$ decays for the $\eta K^{*0}$ mode, and $K^{*+} \to K^+\pi^0$ for the $\eta K^{*+}$ mode.

firm the $\omega K^-$ signal candidates. The $\omega$ invariant mass spectrum and the cosine of the helicity angle of the $B$ candidate events are also shown in Figure 4.

### 3. $B \to \phi K$ and $\phi K^*$

$B$ meson decays involving $b \to ss\bar{s}$ transitions, such as $B \to \phi K$ are forbidden in the first order of SM. Such decays are generated by second order loop such as penguin diagrams, leading to a $b \to s$ flavour changing neutral current (FCNC) transition.

We report measurements of $B \to \phi K$ and $B \to \phi K^*$ with a data sample having an integrated luminosity of $78.1 fb^{-1}$ corresponding to $85.5 \times 10^6$ $B\overline{B}$ pairs.

The $\phi$ mesons were reconstructed from two charged kaons with KID requirement and the mass window cut: $|m_{KK} - m_\phi| < 10$ MeV/c$^2$. The $K_s \to \pi^+\pi^-$ candidates were formed from two oppositely charged pions within a $|m_{\pi\pi} - m_{K^0}| < 15$ MeV/c$^2$ mass window. Neutral pions were reconstructed from pairs of photons and a minimum $\gamma$ energy of 100 MeV was required for the $\pi^0$ daughters. The $K^*$ candidates are reconstructed in three decay channels, $K^{*0} \to K^+\pi^-$, $K^{*+} \to K^+\pi^0$, and $K^{*+} \to K_S\pi^0$. A $\pm 50$ MeV mass window cut was performed on the $K^*$ candidates.

Candidate B meson were reconstructed from a candidate $\phi$ meson and a $K^\pm$, $K_S$, or a $K^*$, and identified by the kinematical variables, $\Delta E$ and $M_{bc}$.

Background was studied using the events outside of the signal window. The event shape variables were used to suppress the continuum background. A likelihood ratio method was used to combine all the information and a single cut was made.

An extended maximum likelihood fit to $M_{bc}$

Table 2
Results for $B^- \to \omega K^-$ and $B^- \to \omega\pi^-$ decays. $\Sigma$ is the statistical significance. $\epsilon$ is the detection efficiency. The upper limit is defined at 90% confidence level.

| Mode | Signal Yield | $\Sigma$ | $\epsilon(\%)$ | $\mathcal{B}$ ($\times10^{-6}$) | $UL$ ($\times10^{-6}$) |
|---|---|---|---|---|---|
| $\omega K^-$ | $19.7^{+5.8+0.7}_{-4.7-0.5}$ | $6.4\sigma$ | 6.3 | $9.9^{+2.7}_{-2.4} \pm 1.0$ | - |
| $\omega\pi^-$ | $10.6^{+4.3+0.4}_{-4.5-0.6}$ | $3.3\sigma$ | 7.7 | $4.3^{+2.0}_{-1.8} \pm 0.5$ | $< 8.2$ |

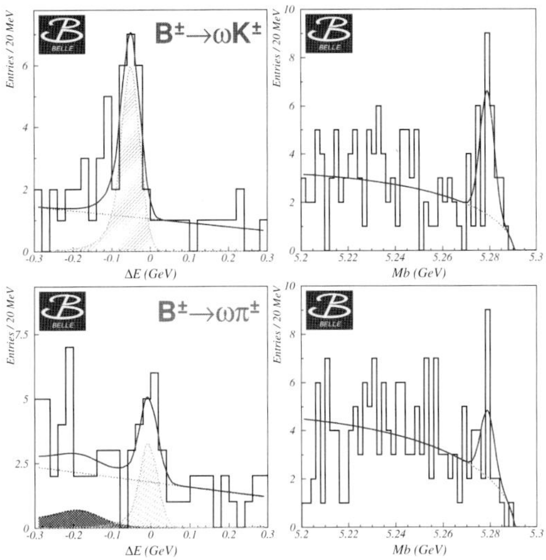


Figure 3. The $M_{bc}$ and $\Delta E$ projection plots of $B^- \to \omega K^-$ and $B^- \to \omega\pi^-$ modes.

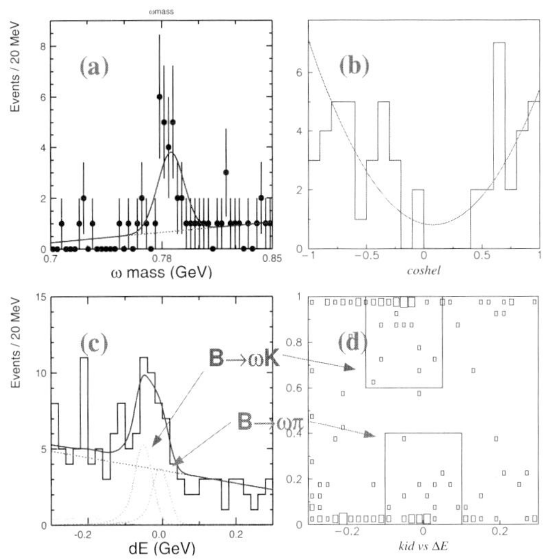


Figure 4. (a) The $\pi^+\pi^-\pi^0$ invariant mass spectrum. (b) The cosine of the helicity angle of $B \to \omega h$ modes. (c) The $\Delta E$ distribution without KID information. (d) The Belle KID versus $\Delta E$ spectrum.

and $\Delta E$ was used to extract the signal yields. Signal PDFs for $M_{bc}$ and $\Delta E$ are represented by Gaussians which were determined from Monte Carlo and verified with control sample. The background PDFs were determined from sideband events. The yields and branching fractions are summarized in Table 3. The $M_{bc}$ and $\Delta E$ projections of candidate events are shown in Figure 5 and 6.

The systematic errors due to background suppression were estimated by varing the cut on the $\mathcal{LR}$ and also by comparing to the results of sequential cuts analysis. The efficiency uncertainty of polarization of the $K^*$ in the $B \to \phi K^*$ mode was included. The non-resonant contribution was estimated from the yields of the $\phi$ and $K^*$ sidebands and included as an systematic error.

## REFERENCES


1. A. Ali, G. Kramer, and C.-D. Lu, Phys. Rev. D **58**, 094009 (1998).
2. Y.-H. Chen, H.-Y. Cheng. B. Tseng, and K.-C. Yang, Phys. Rev. D **60**,094014 (1999); H.-Y. Cheng and K.C. Yang, Phys. Rev. D **62**, 054029 (2000).
3. E. Kou and A.I. Sanda, Phys. Lett. B **525**, 240 (2002).

Table 3
Results for $B \to \phi K$ and $B \to \phi K^*$ decays.

| Mode | Signal Yield | $\epsilon(\%)$ | $\Sigma$ | $\mathcal{B}$ $(\times 10^{-5})$ |
|---|---|---|---|---|
| $B^\pm \to \phi K^\pm$ | $158.7^{+15.3}_{-14.6}$ | 17.3 | 16.7 | $1.07^{+0.10}_{-0.10}{}^{+0.09}_{-0.16}$ |
| $B^0 \to \phi K^0$ | $39.4^{+7.5}_{-6.8}$ | 4.6 | 8.7 | $1.00^{+0.19}_{-0.17}{}^{+0.09}_{-0.13}$ |
| $B^0 \to \phi K^{*0}$ | $31.6^{+7.9}_{-7.1}$ | 4.6 | 5.7 | $0.80^{+0.20}_{-0.18}{}^{+0.08}_{-0.11}$ |
| $B^\pm \to \phi K^{*\pm}$ | $28.0^{+7.4}_{-6.4}$ | – | – | $1.12^{+0.33}_{-0.29}{}^{+0.13}_{-0.17}$ |
| $\to K^\pm \pi^0$ | $9.0^{+4.8}_{-4.1}$ | 1.2 | 2.6 | $0.88^{+0.47}_{-0.40}{}^{+0.10}_{-0.13}$ |
| $\to K^0_S \pi^\pm$ | $19.0^{+5.6}_{-4.9}$ | 1.4 | 5.5 | $1.36^{+0.47}_{-0.41}{}^{+0.15}_{-0.20}$ |

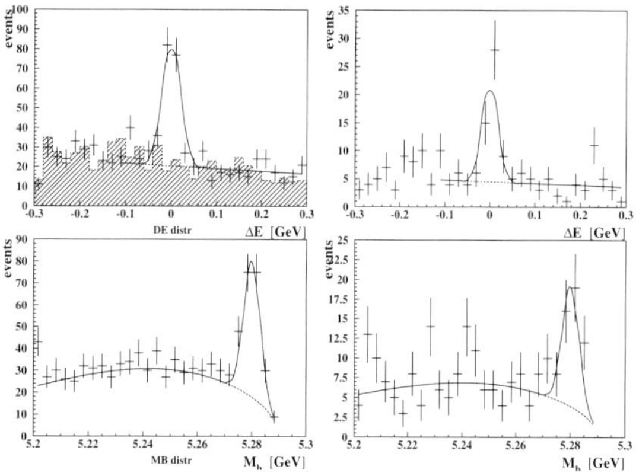


Figure 5. The $M_{bc}$ and $\Delta E$ projection plots for the $B^+ \to \phi K^+$ (left) and $B^0 \to \phi K_S$ (right) modes.

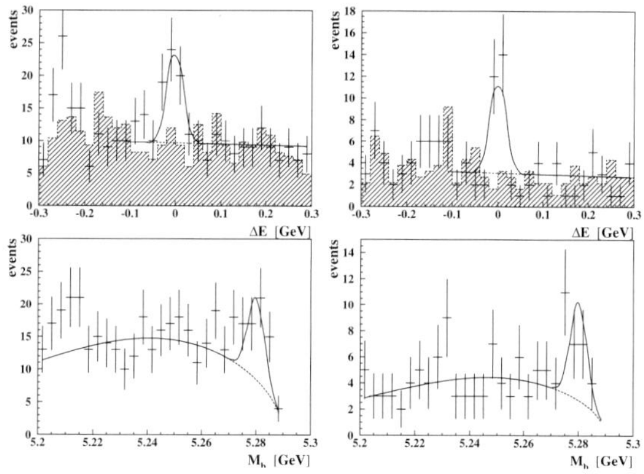


Figure 6. The $M_{bc}$ and $\Delta E$ projection plots for the $B^0 \to \phi K^{*0}$ (left) and $B^+ \to \phi K^{*+}(K^{*+} \to K_S \pi^+)$ (right) modes.

4. X.-G. He, W.-S. Hou and K.-C. Yang, Phys. Rev. Lett. **83**, 1100 (1999); W.-S. Hou and K.-C. Yang, Phys. Rev. D **61**, 073014 (2000).
5. Belle Collaboration, A. Abashian *et al.*, Nucl. Inst. Meth. **A 479** (2002) 117.
6. KEKB B Factory Design Report, KEK Report 95-1, 1995, unpublished.
7. R. A. Fisher, Annals of Eugenics, **7**, 179 (1936).
8. CLEO Collaboration, T. Bergfeld *et al.*, Phys. Rev. Lett. **81**, 272 (1998).
9. CLEO Collaboration, C.P. Jessop *et al.*, Phys. Rev. Lett. **85**, 2881 (2000).
10. BaBar Collaboration, B. Aubert *et al.*, Phys. Rev. Lett. **87**, 221802 (2001).

Proceedings of ICHEP 2002, pp. 607–609
*S. Bentvelsen, P. de Jong, J. Koch and E. Laenen (Editors)*

31st INTERNATIONAL CONFERENCE ON HIGH ENERGY PHYSICS AMSTERDAM

# Excited b states

Christian Weiser[a]

[a]Institut für Experimentelle Kernphysik, Universität Karlsruhe, Postfach 6980, D-76128 Karlsruhe

The status of excited B-meson and b-baryon states is reviewed and new, preliminary analyses on $B^{**}_{u,d}$, $B^{**}_s$ and $\Sigma^{(*)}_b$ are presented.

## 1. INTRODUCTION

Heavy Quark Effective Theory (HQET) [1] predicts the multiplet structure for mesons with $l = 0$ and $l = 1$ shown in Table 1. The orbitally excited states, commonly labelled $B^{**}$, are grouped in doublets with same $j_q$ and are degenerate in mass in the heavy quark limit $m_Q \to \infty$. The dominant decay mode of $B^{**}_{u,d}$ mesons is into the $B^{(*)}\pi$ final state. For $B^{**}_s$ mesons, isospin conservation forbids $B^{**}_s \to B^{(*)}_s\pi$, thus $B^{**}_s \to B^{(*)}K$ should dominate if above threshold. The $j_q = 1/2$ (3/2) states are expected to be broad (narrow) because the strong decay proceeds via s-wave (d-wave) transitions. Several theoretical predictions for the masses of $B^{**}$ mesons exist [2–10]. Some of the models predict the mass of the narrow doublet ($j_q = 3/2$) to be above the broad doublet ($j_q = 1/2$), whereas others predict the opposite, called 'spin-orbit-inversion'.

$B^{**}_{u,d}$ mesons have been identified by all four LEP Collaborations [11–17] and CDF [18] using different analysis methods (Table 1). The

Table 1
The $l = 0$ and $l = 1$ states of $B_{u,d}$ mesons. Other final states, e.g. $B^{(*)}\pi\pi$, are also expected.

| $l$ | $j_q$ | $J^P$ | state | dom. decay mode | |
|---|---|---|---|---|---|
| 0 | $\frac{1}{2}$ | $0^-$ | B | | |
| | | $1^-$ | $B^*$ | $B\gamma$ | |
| 1 | $\frac{1}{2}$ | $0^+$ | $B^*_0$ | $B\pi$ | (s-wave) |
| | | $1^+$ | $B^*_1$ | $B^*\pi$ | (s-wave) |
| 1 | $\frac{3}{2}$ | $1^+$ | $B_1$ | $B^*\pi$ | (d-wave) |
| | | $2^+$ | $B^*_2$ | $B^{(*)}\pi$ | (d-wave) |

Table 2
Results on $B^{**}_{u,d}$ masses and production rates from analyses using inclusive (row 1-4), partial (5) and exclusive B reconstruction (6). Some numbers have been recomputed using updated values for input parameters and a similar set of assumptions.

| | $M(B^{**})[MeV/c^2]$ | $\sigma_{B^{**}_{u,d}}/\sigma_{b-jet}$ |
|---|---|---|
| OPAL [11] | $5712 \pm 11$ | $0.22 \pm 0.05$ |
| DELPHI [17] | $5734 \pm 5 \pm 17$ | $0.26 \pm 0.02 \pm 0.05$ |
| ALEPH [13] | $5734 \pm 3 \pm 16$ | $0.14 \pm 0.01 \pm 0.03$ |
| L3 [14] | $B^*_2 : 5768 \pm 5 \pm 6$ | $0.32 \pm 0.03 \pm 0.06$ |
| CDF [18] | $B_1 : 5710 \pm 20$ | $0.22 \pm 0.05 \pm 0.02$ |
| ALEPH [15] | $B^*_2 : 5739^{+8}_{-11}{}^{+6}_{-4}$ | $0.25 \pm 0.07^{+0.05}_{-0.04}$ |

results show some disagreement on the production rates. Furthermore, OPAL has measured the fraction of $B^{**}$ decays with a $B^*$ in the final state [16]: $BR(B^{**} \to B^*\pi(X)) = 0.85^{+0.26}_{-0.27}(\text{stat.}) \pm 0.12(\text{syst.})$. Up to now, the composition of the signals could not been clearly resolved, and no final conclusions can be drawn on the mass hierarchy. In the past, OPAL [11] and DELPHI [17] had reported excesses at the level of 2-3 standard deviations in the $B^{(*)}K$ spectrum, interpreted as stemming from $B^{**}_s$ decays. Furthermore, DELPHI [19] had found evidence for $\Sigma_b$ and $\Sigma^*_b$ production[1] in the decay channel $\Sigma^{(*)}_b \to \Lambda_b\pi$.

## 2. New DELPHI Analyses

The new DELPHI analyses presented here [20] profit from the reprocessed LEP I data sample us-

[1] $\Sigma_b$ and $\Sigma^*_b$ are b-baryon states ($bqq, q = u, d$) with $S_{qq} = 1$ coupling with the b-quark spin to total spin $J = 1/2$ ($\Sigma_b$) and $J = 3/2$ ($\Sigma^*_b$).


doi:10.1016/S0920-5632(02)02179-5

ing strongly improved reconstruction algorithms and analysis techniques. Special emphasis has been put on a better control of the background to reduce the modelling uncertainty. The main physics motivation is, to try to answer some of the open questions concerning the composition of the signals, the production rates and the mass hierarchy of the two doublets. Furthermore, the situation of earlier results on $B_s^{**}$ and $\Sigma_b^{(*)}$ production which have not been published should be clarified.

Two independent analysis lines have been set up following different approaches. Common to both is the inclusive reconstruction of the weakly decaying b hadron which is then combined with a pion or kaon from the primary vertex. The quantity analysed is the Q value: $Q = m(B^{(*)}\pi/K) - m(B^{(*)}) - m(\pi/K)$. Correlations between the charge of the $\pi/K$ and the flavour or charge of the b quark are used to define samples enriched in $B_d^{**}$ and $B_u^{**}$ and samples enriched and depleted in signal. One analysis, called 'Soft Analysis' in the following, uses relative soft cuts tuned to produce similar background shapes for the signal enriched and depleted samples. The background is then extracted from the data itself by a simultaneous fit to the signal enriched/depleted samples. The other analysis, 'NN Analysis', is heavily based on the use of artificial Neural Networks for B momentum reconstruction, charge and flavour tags, vertex assignment, selection of the pion/kaon from the decay of the excited b hadron etc. to boost the performance to its limit and to get signal samples with high purity.

### 2.1. The $B_{u,d}^{**}$ Analyses

Figure 1 shows the Q value distributions of the 'soft' analysis for the $B_d^{**}$ enriched sample. Fitting the signal part with a single Gaussian gives a width compatible with what is expected from detector resolution, suggesting an interpretation of the signal as stemming mainly from narrow states. It also prefers a low mass splitting within the $j_q = 3/2$ doublet and a small contribution from the decay $B_2 \rightarrow B\pi$ which is shifted by $m_{B^*} - m_B \approx 45$ MeV/$c^2$ in the Q value distribution. The averaged values for $B_d^{**}$ and $B_u^{**}$ states are: $Q = 298 \pm 4(\text{stat.}) \pm 12(\text{syst.})$

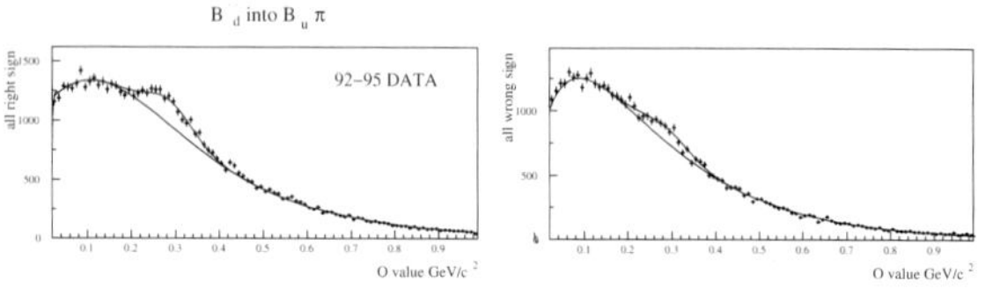


Figure 1. The 'right-sign' (left) and 'wrong-sign' (right) spectra for the $B_d^{**}$ sample from the 'soft' analysis. The signal has been fitted with a single Gaussian.

MeV/$c^2$, $\sigma = 47 \pm 3(\text{stat.}) \pm 5(\text{syst.})$ MeV/$c^2$, $(\sigma(B_d^{**})+\sigma(B_u^{**}))/\sigma_{b-jet} = 0.098 \pm 0.007(\text{stat.}) \pm 0.012(\text{syst.})$. Given the interpretation above, this production cross section very likely corresponds to the rate of narrow states only. To search for additional broad states, the parameters of the Gaussian for the narrow states have been fixed to the values obtained before and two additional parameters, $Q$ and $\Gamma$ for a mass dependent Breit-Wigner function, are introduced. The relative contribution of narrow and broad states has been fixed, e.g. according to state counting (1:1) or spin counting (2:1). A broad component with a $Q$ value at about 100 MeV/$c^2$ above the narrow states and a width of $\Gamma \approx 250$ MeV/$c^2$ significantly improves the fit quality (Figure 2). Figure 2 also shows the spectra obtained with the 'NN' analysis working at much higher purity. A narrow component compatible with detector resolution is found there, too, with the mean Q value

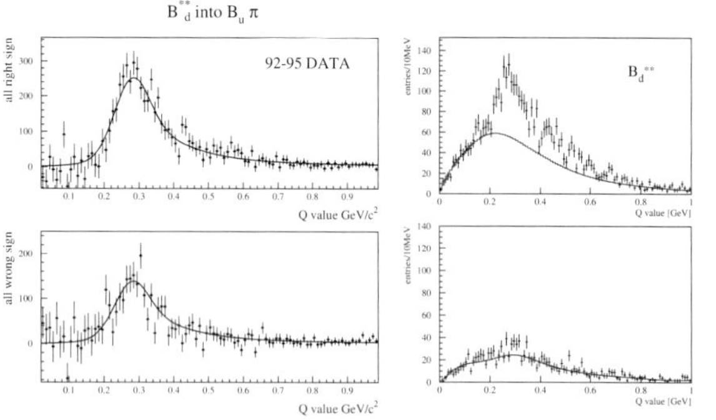


Figure 2. Left: the background subtracted $B_d^{**}$ spectra from the 'soft' analysis using an additional signal component for broad states (the ratio narrow/broad has been fixed to 1). Right: the $B_d^{**}$ spectra from the 'NN' analysis.

being in good agreement with the 'soft' analysis. However, this component alone does not give a good description of the spectrum. Additional contributions, also preferably at higher masses, are needed. So, the data slightly seem to favour models predicting spin-orbit-inversion. However, final conclusions can't be drawn yet and more stringent test of models have to be performed.

### 2.2. $B_s^{**}$ and $\Sigma_b^{(*)}$

The $B_s^{**}$ analyses are very similar to the $B_{u,d}^{**}$ analyses, the main difference being the replacement of the pion from the primary vertex with an identified kaon. Figure 3 shows the results from both analyses. In the NN analysis some activity at $Q \approx 70$ MeV/$c^2$ can be seen which has to be investigated further. An upper limit on the production rate of a narrow resonance compatible with detector resolution up to $Q = 150$ MeV/$c^2$ can be set: $\sigma(B_s^{**})/\sigma_{b-jet} < 0.015$ @ 95% CL.

For the $\Sigma_b^{(*)}$ analyses, the inclusively reconstructed B-meson is replaced with a $\Lambda_b$. Figure 4 shows the resulting spectra from both analyses. No significant signal is found and an upper limit on the production rate of a narrow resonance up to $Q = 150$ MeV/$c^2$ can be set: $\sigma(\Sigma_b^{(*)})/\sigma_{b-jet} < 0.015$ @ 95% CL. These results supersede and do not confirm the old unpublished results from DELPHI on $B_s^{**}$ and $\Sigma_b^{(*)}$ production.

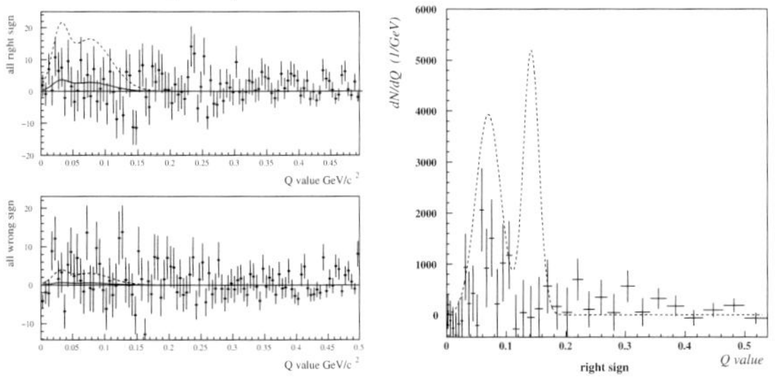


Figure 3. The background subtracted spectra for $B_s^{**}$ from the 'soft' (left) and 'NN' (right) analysis. The dashed lines indicate the expectation from the model of [4] with a production rate of 3% (left) or the expectation from the previous DELPHI result [17] corresponding to a production rate of 3.7% for narrow states.

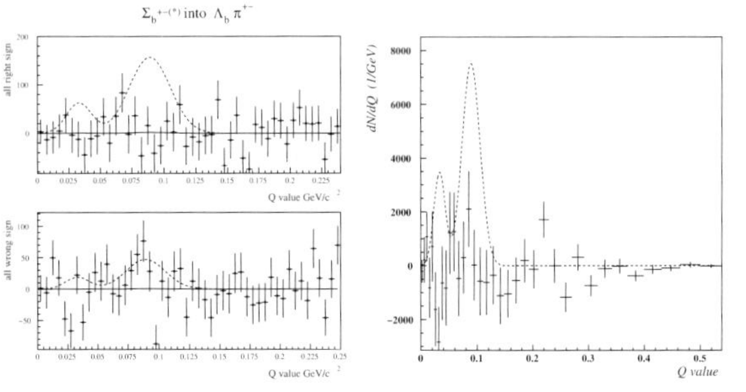


Figure 4. The background subtracted spectra for $\Sigma_b^{(*)}$ from the 'soft' (left) and 'NN' (right) analysis. The dashed lines indicate the expectation if the Q values would be as measured in the old DELPHI analysis [19] with corresponding production rates of 3% (left) and 4.8% (right).

## REFERENCES


1. N. Isgur, M.B. Wise; *Phys. Lett.* **B232** (1989) 113 and others
2. S. Godfrey, N. Isgur; *Phys. Rev.* **D32** (1985) 189
3. S. Godfrey, R. Kokoski; *Phys. Rev.* **D43** (1991) 1679
4. E.J. Eichten, C.T. Hill, C. Quigg; *Phys. Rev. Lett.* **71** *(1993) 4116 and FERMILAB-CONF-94/118-T*
5. *S.N. Gupta, J.M. Johnson;* Phys. Rev. **D51** *(1995) 168*
6. *A.F. Falk, T. Mehen;* Phys. Rev. **D53** *(1996) 231*
7. *N. Isgur;* Phys. Rev. **D57** *(1998) 4041*
8. *D. Ebert, V.O. Galkin, R.N. Faustov;* Phys. Rev. **D57** *(1998) 5663*
9. *A.H. Örsland, H. Högaasen;* Eur. Phys. J. **C9** *(1999) 503*
10. *M. Di Pierro, E. Eichten;* Phys. Rev. **D64** *(2001) 114004*
11. *OPAL Collaboration, R. Akers et al.;* Z. Phys. **C66** *(1995) 27*
12. *DELPHI Collaboration, P. Abreu et al.;* Phys. Lett. **B345** *(1995) 598*
13. *ALEPH Collaboration, D. Buskulic et al.,* Z. Phys. **C69** *(1996) 393*
14. *L3 Collaboration, M. Acciarri et al.;* Phys. Lett. **B465** *(1999) 323*
15. *ALEPH Collaboration, R. Barate et al.;* Phys. Lett. **B425** *(1998) 215*
16. *OPAL Collaboration, G. Abbiendi et al.;* Eur. Phys. J. **C23** *(2002) 437*
17. *DELPHI Collaboration; DELPHI Note 95-105 PHYS 540; EPS-HEP'95* **Ref. #0563**
18. *CDF Collaboration, T. Affolder et al.;* Phys. Rev. **D64** *(2001) 072002*
19. *DELPHI Collaboration; DELPHI Note 95-105 PHYS 542; EPS-HEP'95* **Ref. #0565**
20. *DELPHI Collaboration, Z. Albrecht et al.; DELPHI Note 2002-079 CONF-613;* contribution to this conference

Proceedings of ICHEP 2002, pp. 610–612
S. Bentvelsen, P. de Jong, J. Koch and E. Laenen (Editors)

# Search for $\tau \to \mu\gamma$ Decays at *BABAR*

J. M. Roney[a]
(representing the BABAR Collaboration)

[a]Department of Physics and Astronomy,
University of Victoria,
P.O. Box 3055 STN CSC, Victoria, B.C., Canada

Using data collected with the *BABAR* detector between 1999 and 2001, we describe a search for the neutrinoless decay $\tau \to \mu\gamma$. This data sample includes data collected both on and off the $\Upsilon(4S)$ resonance and corresponds to $56 \times 10^6$ produced tau pair events. The search has an efficiency of $5.2 \pm 0.1(MCstat) \pm 0.5(sys)\%$ and an expected background rate of $7.8 \pm 1.4$ events. We select 13 events in the final sample. As there is no evidence for a signal in this data, we set an upper limit of $\mathcal{B}(\tau \to \mu\gamma) < 2.0 \times 10^{-6}$@90%CL.

The decay $\tau \to \mu\gamma$ is an anticipated lepton-number violating process in supersymmetric models[1–3], left-right supersymmetric models[4] and in supersymmetric string unified models[5]. For some ranges of model parameters, decay rates as high several parts per million are expected for this decay[3,5], even in light of the current experimental limit on the related $\mu \to e\gamma$ decay[6]. Assuming that neutrinos have mass, the standard model branching ratio is expected at the $\mathcal{O}(10^{-34})$ level. The current best published limit is $B(\tau \to \mu\gamma) < 1.1 \times 10^{-6}$ from the CLEO experiment using 12.6 million $\tau$ pairs[7]. The BELLE collaboration has presented an unpublished preliminary result with an upper limit on the branching ratio of $1.0 \times 10^{-6}$[8].

This analysis uses $56 \times 10^6$ $e^+e^- \to \tau^+\tau^-$ events from the 1999-2001 *BABAR* data set, corresponding to an integrated luminosity of $56.4\,\text{fb}^{-1}$ on the peak of the $\Upsilon(4s)$ resonance (10.58 GeV) and $6.4\,\text{fb}^{-1}$ below the resonance. The *BABAR* detector is described in detail elsewhere[9]. It consists of a precision silicon vertex tracker, drift chamber, ring-imaging Cherenkov detector, and CsI(Tl) calorimeter surrounded by a superconducting solenoid operating with a 1.5 T field. The steel flux return of the magnet is instrumented for muon and neutral hadron detection.

The general signature of the $\tau \to \mu\gamma$ signal is the presence of an isolated $\mu$ and $\gamma$ which have an invariant mass consistent with that of the $\tau$(1.777 GeV) and with an energy in the centre-of-mass system (CMS) consistent with the beam energy in the CMS (5.29 GeV). In the current analysis, the rest of the particles in the event are required to have properties that are consistent with being produced in a generic 1-prong $\tau$ decay. The analysis is performed using a blinded mass-energy region in the data which corresponds to a $3\sigma$ error ellipse centred on the expected peak of the distributions.

Background sources arising from non-$\tau^+\tau^-$ sources, the most problematic of which are radiative $\mu^+\mu^-$ pair events, are reduced to small levels. This leaves the generic 1-prong $\tau$ decays as the major source of background. The dominant, and irreducible, background is from radiative muonic decays of the tau lepton, $\tau^+ \to \mu^+\nu\bar{\nu}(\gamma)$, in which the two neutrinos have very little energy.

The $e^+e^-$ environment lends itself very well to the use of a mass of the $\mu\gamma$ system calculated from the kinematic fit employing the beam energy constraint on the energy of the system, denoted as $m_{\text{EC}}$. The energy variable used is the difference between the measured $\mu\gamma$ energy and the beam energy, denoted by $\Delta E$. The resolutions of the core of these distributions, which represent those events with well reconstructed photons and tracks, are $\sigma_{\Delta E} = 88\,\text{MeV}$ and $\sigma_{m_{\text{EC}}} = 19\,\text{MeV}$.

A cut-based approach is used, in which the


doi:10.1016/S0920-5632(02)02173-4

background in the signal region is estimated from extrapolations using side-band data and verified with the Monte Carlo simulation. The determination of selection criteria in this analysis was based on Monte Carlo simulation of background and signal as well as on data in side-band regions further away from the signal-box in a region referred to as the 'Grand Sideband'. These regions are depicted in Figure 1.

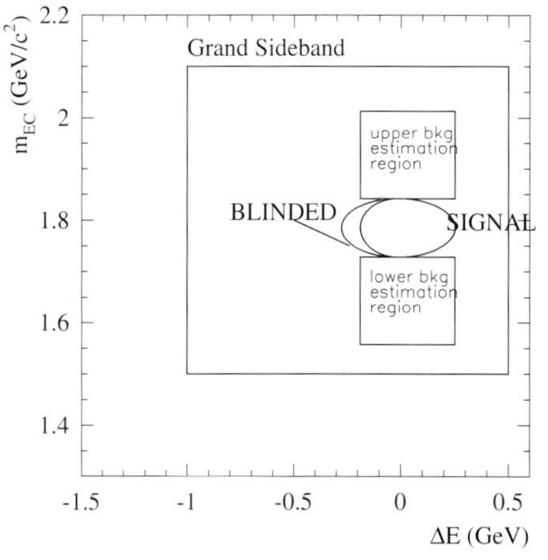


Figure 1. Regions of the $m_{\rm EC}$ $-\Delta E$ plane shown schematically overlaying the distribution of the simulated signal. The signal region is the assymetric ellipse within the elliptical blinded region.

In order to be selected, the event must have two oppositely charged particles separated in the CMS by at least 90°. The reconstructed track associated with each particle is required to fall within the fiducial region: $0.775 < \cos\theta_{LAB} < 0.094$, where $\theta_{LAB}$ is the angle between the track and the electron beam direction in the LAB frame. One of the charged particles is required to be consistent with being a muon using information from the instrumented flux return and calorimeter. A photon with at least 400 MeV of energy in the CMS must be detected within 90° of the $\mu$ in the CMS. The other charged particle is required to be inconsistent with a muon hypothesis. In order to further reduce the $\mu^+\mu^-$ backgrounds, the non-$\mu$ particle is required to be consistent with coming from either the $\tau \rightarrow e\nu\overline{\nu}$ or $\tau \rightarrow h \geq 1\pi^0\nu$ decays, which have branching ratios of 18% and 37% respectively. The total measured energy of all particles, both charged and neutral, in the CMS hemisphere opposite the $\mu$ is required to be less than 80% of the beam energy in the CMS. Electron identification uses information from the calorimeter and drift chamber, including E/p and dE/dx energy loss information, and information from the ring-imaging Cherenkov detector. The $\tau \rightarrow h \geq 1\pi^0\nu$ decays are associated with a hemisphere whose total invariant mass is at least 300 MeV and which has a missing mass squared between -0.5 GeV $^2$ and +0.5 GeV $^2$.

The efficiency of the selection is determined using a signal simulation incorporated into the KORALB[10] event generator employing TAUOLA[11] for the generic decays. Simulated events are processed through a complete simulation of the detector[12] using the same reconstruction software used for the real data. Small corrections to the efficiencies predicted by the simulation are obtained from independent control samples of data.

After all the selection criteria have been applied, before applying the signal-box cut, there remain background sources in the Grand Sideband region that, from generic tau Monte Carlo simulation, consist of $\tau \rightarrow \mu\nu\overline{\nu}$ (85.9%), $\tau \rightarrow \pi\nu$ and $\tau \rightarrow K\nu$ (10.6%) and $\tau \rightarrow \rho\nu$ (3.5%). Comparisons between data and Monte Carlo indicate excellent agreement between the overall rates of these events (observed/expected $= 1.022 \pm 0.069 \pm 0.025$) as well as shapes of distributions, such as the spectra of the $\mu$ and $\gamma$ (Figure 2). Such comparisons give confidence that the non-$\tau$ background is expected to be very small in the final sample. As these events have nearly an identical signature as the signal, these studies confirm estimations of the efficiencies obtained from the simulation. The overall selection efficiency is $5.2 \pm 0.1 \pm 0.5\%$. The systematic error is dominated by statistics in comparisons between data and Monte Carlo.

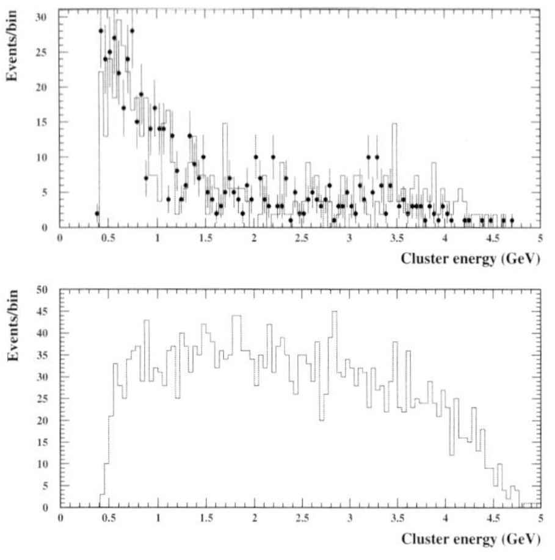


Figure 2. Distributions of the photon energy in the Grand Sideband, after all other requirements have been applied. The data (points) and $\tau$ Monte Carlo (histogram) are in the top plot and the $\tau \to \mu\gamma$ signal Monte Carlo in the bottom plot. The $\tau^+\tau^-$ Monte Carlo is normalized based on the absolute luminosity.

The background, estimated by interpolating between the $m_{\mathrm{EC}}$ sideband regions in the data, is expected to be $7.8 \pm 1.4\%$. Note that if 8 events were to be observed, one would expect a limit on the branching ratio of $9.5\times10^{-7}$.

Upon unblinding, 13 events are observed in the signal region (Figure 3). Of these, 9 had electrons identified in the non-$\mu\gamma$ hemisphere; 5 were $\tau \to h \geq 1\pi^0\nu$ and one was in both categories. Various distributions were examined and no pathological grouping of events is evident: they are consistent with both signal and $\tau^+\tau^-$ background. This yields a limit of 11.5 event upper limit on the number of signal events @90%CL when the systematic errors on the background are included. The 11.5 event upper limit translates into a limit: $\mathcal{B}(\tau \to \mu\gamma) < 2.0 \times 10^{-6}$@90%CL. The probability of a background of $7.8\pm1.4$ events fluctuating up to 13 observed events in the absence of a signal is 7.6% if one includes the systematic errors.

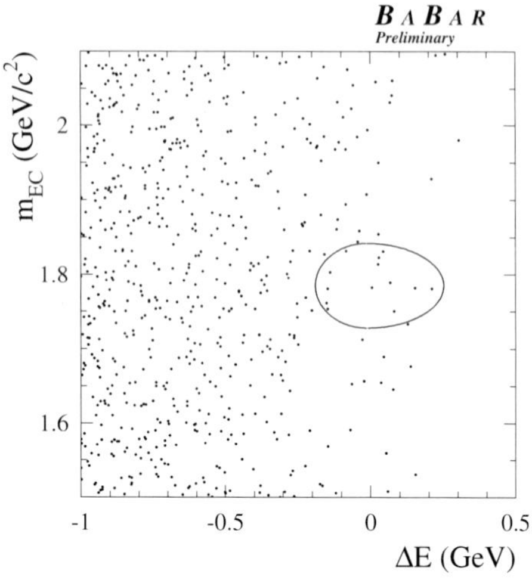


Figure 3. $m_{\mathrm{EC}}$ vs $\Delta E$ distributions for events surviving in the data, including the 13 events in the signal region.

## REFERENCES


1. R. Barbieri and L. J. Hall, Phys. Lett. B **338**, 212 (1994)
2. J. Hisano *et al*, Phys. Lett. B **357**, 579 (1995)
3. J. Hisano and D. Nomura, Phys. Rev. D **59**, 116005 (1999)
4. K. S. Babu, B. Dutta and R. N. Mohapatra, Phys. Lett. B **458**, 93 (1999)
5. S. F. King and M. Oliveira, Phys. Rev. D **60**, 035003 (1999)
6. M. L. Brooks *et al.* [MEGA Collaboration], Phys. Rev. Lett. **83**, 1521 (1999)
7. S. Ahmed *et al.* [CLEO Collaboration], Phys. Rev. D **61**, 071101 (2000)
8. Belle Collaboration, BELLE-CONF-0118.
9. B. Aubert *et al.* [BABAR Collaboration], Nucl. Instrum. Meth. A **479**, 1 (2002)
10. S. Jadach and Z. Was, Comput. Phys. Commun. **85**, 453 (1995).
11. Z. Was, Nucl. Phys. Proc. Suppl. **98**, 96 (2001)
12. K. Amako [Geant4 Collaboration], Nucl. Instrum. Meth. A **453**, 455 (2000).

Proceedings of ICHEP 2002, pp. 613–615
*S. Bentvelsen, P. de Jong, J. Koch and E. Laenen (Editors)*

31st INTERNATIONAL CONFERENCE ON HIGH ENERGY PHYSICS AMSTERDAM

# The b semileptonic branching ratio in $Z^0 \to b\bar{b}$ decays

Alessia Tricomi[a] *

[a]Dipartimento di Fisica e Astronomia, Università di Catania, and INFN Catania, Via S. Sofia 64, I-95100 Catania, Italy

A review of the latest results on the inclusive semileptonic branching fractions, BR($b \to \ell^- \bar{\nu}$ X) and BR($b \to c \to \ell^+ \nu$ X), from the LEP experiments is reported.

## 1. Introduction

The measurement of the inclusive semileptonic branching fraction of b hadrons, BR($b \to \ell^- \bar{\nu}$ X), provides the opportunity to test and improve our knowledge of the dynamics of heavy hadrons. Within the context of the Heavy Quark Theory its value can also be used together with the average lifetime of the b mesons to extract the value of the $|V_{cb}|$ CKM matrix element [1–3]. Besides, both the direct semileptonic b decay rate, BR($b \to \ell$), and the cascade decay rate, BR($b \to c \to \ell$), are important input parameters for many heavy flavour analyses.

The large statistics of b hadrons collected at LEP1 and the peculiar kinematical signatures of b semileptonic decays (e.g. high $p$ or $p_\perp$ leptons), offer one of the best enviroments for the determination of the semileptonic branching ratios of the b hadrons.

## 2. Measurement of $b \to \ell^- \bar{\nu}$ X

Measuring BR($b \to \ell^- \bar{\nu}$ X) involves identifying lepton candidates in samples enriched in $Z^0 \to b\bar{b}$ decays; b-tagging is performed with algorithms that exploiy the long lifetime and heavy mass of b quarks. A double-tag technique is tipically used to measure the tagging purity directly from the data. Events are divided in two hemispheres by the plane perpendicular to the thrust axis and lepton candidates are searched for in the hemisphere opposite to the one used for b tagging.

Usually BR($b \to \ell$) and BR($b \to c \to \ell$) are measured simultaneously, since the latter constitutes the main background for the direct decays and also is an important input to many heavy flavour measurements.

In order to measure the two branching ratios, the different sources of prompt leptons in heavy flavour decays need to be disentangled, as well as separated from from non-prompt leptons (K and $\pi$ decays in flight) and background (hadrons misidentified as leptons) produced in events of all flavours. Several strategies have been used by the four LEP Collaborations in order to separate the different sources of lepton candidates and extract the signal.

The direct and cascade decay branching ratios can be measured from a fit to the $p$ and $p_\perp$ spectra of the reconstructed leptons. Since this kind of analyses rely on the models used to describe the fragmentation of the b quarks and the lepton energy spectrum in the b hadron decay, the related uncertainties lead to the largest systematic errors in all these measurements. Several different approaches have been used by the LEP experiments in order to reduce the dependence on the modelling. The most recent results from ALEPH [4] is summarized here. Results from DELPHI [5], L3 [6] and OPAL [7] are summarized in Fig. 3.

### 2.1. The ALEPH analyses

The ALEPH Collaboration has presented a new measurement of the inclusive b branching ratio. It combines an analysis based on a fit to the lepton $p_\perp$ spectrum (Fig. 1) with one based on the charge correlation between the lepton candidate and a charge estimator (hemisphere charge or

*On behalf of the LEP Collaborations


doi:10.1016/S0920-5632(02)02169-2

high $p_\perp$ lepton charge) measured in the opposite hemisphere. The two methods have different sen-

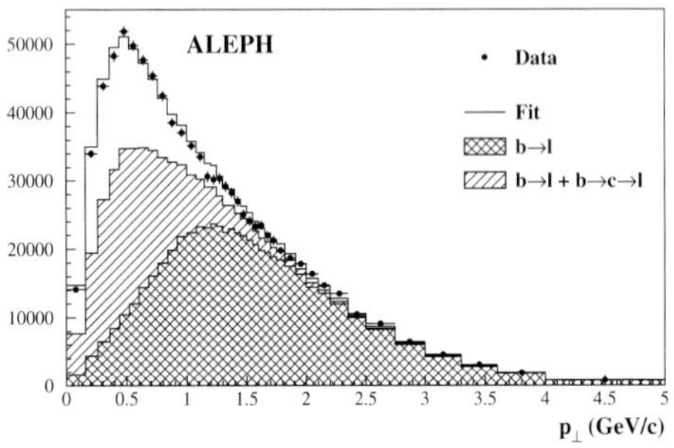


Figure 1. Comparison between the lepton $p_\perp$ spectrum measured in the data and the fit result. The expected contribution fromb $\to \ell$ as well as the sum of b $\to \ell$ and b $\to$ c $\to \ell$ are shown.

sitivities to the different systematic effects. The transverse momentum analysis is largely dependent on the modelling of the lepton spectrum: a novel approach has been used to estimate the size of the associated uncertainty. The strategy commonly adopted by the LEP experiments [8] is to reweight the lepton momentum in the b hadron rest frame accordingly to the ACCMM model, while the systematic uncertainties are estimated using ISGW and ISGW** models. In this analysis, instead, ALEPH exploits the available measurements of the B meson decay rates into the different charm hadron final states (D, D* and D** mesons, as well as non resonant $D^{(*)}\pi$). The leptons coming from each of these components have a different momentum spectrum so that the shape of the inclusive B→ $X\ell\nu$ spectrum depends on the branching fractions of $B^0$ and $B^+$ into the various charmed species. The fractions of D, D*, $D_1$ and $D_2^*$ in the simulation are reweighted according to the latest measured values [9,10]. The broad D** states are assumed to be equal to the sum of the narrow $D_1$ and $D_2^*$ states, and the non-resonant $D^{(*)}\pi$ decays account for the rate needed in order to add up the measured inclusive $D^{(*)}X$ branching ratio. The systematic uncertainty is evaluated by varying the measured branching ratios within their estimated errors. Additionally, in each case the B meson energy spectrum measured in [11] is used, thus taking the correlation between the two analyses into account.

In the charge correlation analysis the model dependence is significantly reduced but the main source of systematics comes from the uncertainty on the rate of b $\to W^- \to \bar{c} \to \ell^-\bar{\nu}$ X decays. Since no information about the lepton $p_\perp$ is used in this analysis, the $p_\perp$ spectrum can be extracted from the data (Fig. 2).

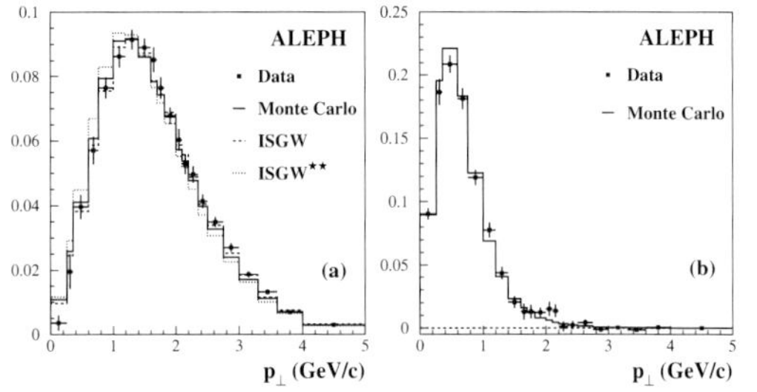


Figure 2. Comparison between the lepton $p_\perp$ spectra measured in the data with the charge correlation analysis and that expected from the simulation. (a) b $\to \ell$; data are compared with the simulation corrected as described in the text, ISGW and ISGW** models. (b) b $\to$ c $\to \ell$. All the spectra are normalized to unit area.

The results obtained with the two methods are compatible and can be averaged; the charge correlation analysis has the largest weight, due to the smaller systematic error.

The ALEPH final result, together with all the other LEP measurements for the BR(b $\to \ell$), is shown in Fig. 3. The LEP average quoted is obtained from the LEP Electroweak Working Group fit [12]. This value leads to $|V_{cb}| = (40.7 \pm 0.6 \pm 0.8) \times 10^{-3}$ [3].

In Fig. 4 all the BR(b $\to$ c $\to \ell$) measurements made at LEP are summarized.

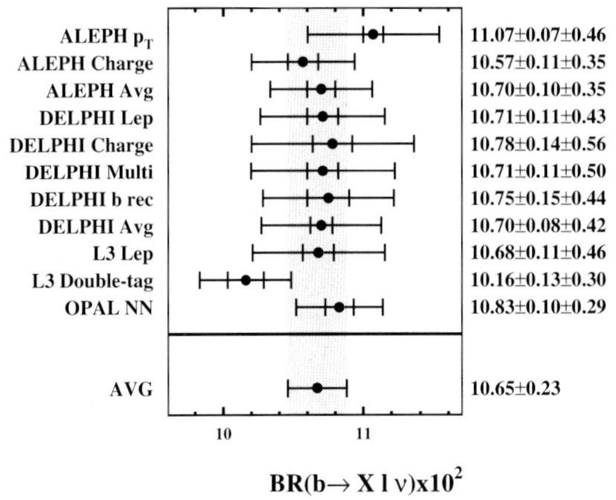


Figure 3. BR(b $\rightarrow \ell$) results from the LEP experiments. The average value is the result of the LEP EWWG global fit.

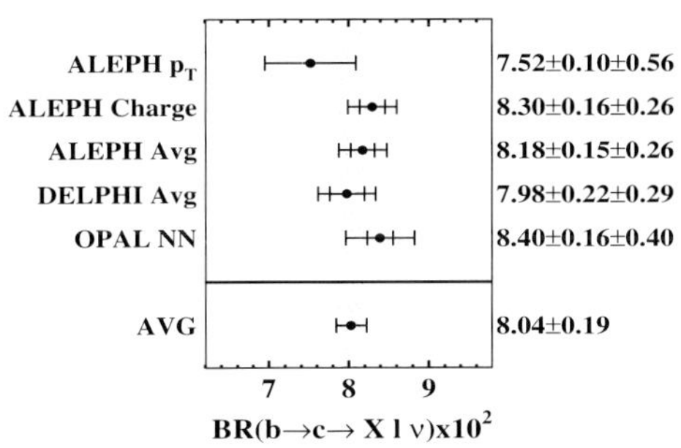


Figure 4. BR(b $\rightarrow$ c $\rightarrow \ell$) results from the LEP experiments. The average value is the result of the LEP EWWG global fit.

## 3. Conclusions

The LEP Collaborations have presented to this Conference the final results for the BR(b $\rightarrow \ell$)=$(10.65 \pm 0.23) \times 10^{-2}$ and BR(b $\rightarrow$ c $\rightarrow \ell$)=$(8.04 \pm 0.19) \times 10^{-2}$.

The discrepancies between BR(b $\rightarrow \ell$) measurements at the Z and $\Upsilon$(4S) energies shown in the past are reduced to about $1.2\sigma$. A comparison of LEP results with the one obtained at the $\Upsilon$(4S) and HQET predictions are shown in Fig. 5.

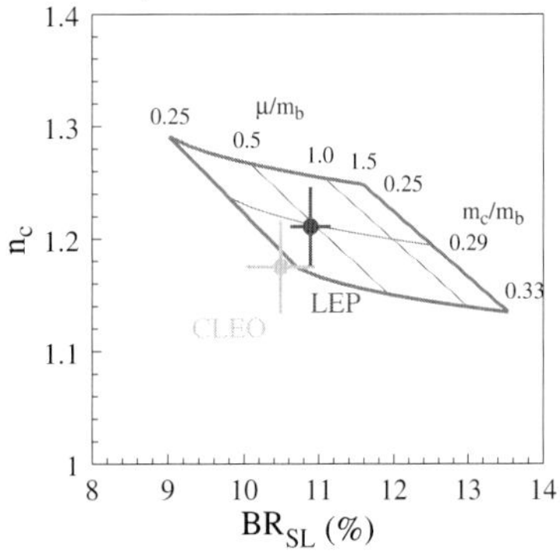


Figure 5. Charm multiplicity per b hadron decay $n_c$ *vs* BR(b $\rightarrow \ell$) from Z and $\Upsilon$(4S) results compared to the HQET prediction.

## Acknowledgements

I would like to thank the LEP EWWG for providing their results. A special thank to D. Abbaneo, M. Battaglia, E. Barberio, N. Marinelli, F. Palla and R. Tenchini for fruitful discussions and assistance.

## REFERENCES

1. A.H. Hoang, Z. Ligeti, A.V. Manohar, Phys. Rev. Lett.**82**, 277 (1999). Phys. Rev. D **59**, 074017 (1999).
2. I.I. Bigi, M. Shifman, N. Uraltsev, Ann. Rev. Nucl. Part. Sci. **47**, 591 (1997).
3. M. Battaglia, these proceedings.
4. ALEPH Collaboration, Eur. Phys. J. C **22**, 613 (2002).
5. DELPHI Collaboration, Eur. Phys. J. C **20**, 455 (2001).
6. L3 Collaboration, Z. Phys. C **71**, 379 (1996); Eur. Phys. J. C **13**, 47 (2000).
7. OPAL Collaboration, Eur. Phys. J. C **13**, 225 (2000).
8. The LEP Experiments: ALEPH, DELPHI, L3 and OPAL, Nucl. Instrum. Meth. **A378**, 101 (1996).
9. The ALEPH, CDF, DELPHI, L3, OPAL and SLD Collaborations, SLAC-PUB/8492, CERN-EP/2000-096.
10. The Particle Data Group, Review of Particle Physics, Eur. Phys. J. C **15**, 1 (2000).
11. ALEPH Collaboration, Phys. Lett. B **512**, 30 (2001).
12. LEP Electroweak Working Group, Summer 2002 results, *http://lepewwg.web.cern.ch.*

NH
ELSEVIER

Proceedings of ICHEP 2002, pp. 616–619
*S. Bentvelsen, P. de Jong, J. Koch and E. Laenen (Editors)*

31st INTERNATIONAL CONFERENCE ON HIGH ENERGY PHYSICS AMSTERDAM

# Evidence for the Flavor-Changing Neutral Current Decays $B \to K\ell^+\ell^-$ and $B \to K^*\ell^+\ell^-$

Jeffrey D. Richman[a] (Representing the *BABAR* Collaboration)

[a]Department of Physics, University of California
Santa Barbara, CA, 93106

We present preliminary results from a search for the rare, flavor-changing neutral current decays $B \to K\ell^+\ell^-$ and $B \to K^*\ell^+\ell^-$, where $\ell^+\ell^-$ is either an $e^+e^-$ or $\mu^+\mu^-$ pair. The data sample comprises $(84.4 \pm 0.9) \times 10^6$ $\Upsilon(4S) \to B\overline{B}$ decays (77.8 fb$^{-1}$) collected with the *BABAR* detector at the PEP-II $B$ Factory. For $B \to K\ell^+\ell^-$, we observe a signal with greater than $4\sigma$ significance and obtain $\mathcal{B}(B \to K\ell^+\ell^-) = (0.78^{+0.24+0.11}_{-0.20-0.18}) \times 10^{-6}$ (averaged over $\ell = e$ and $\mu$). For $B \to K^*\ell^+\ell^-$, we observe an excess of events over background with less than $4\sigma$ significance. We obtain $\mathcal{B}(B \to K^*\ell^+\ell^-) = (1.68^{+0.68}_{-0.58} \pm 0.28) \times 10^{-6}$ and the 90% C.L. upper limit $\mathcal{B}(B \to K^*\ell^+\ell^-) < 3.0 \times 10^{-6}$.

## 1. Introduction

The flavor-changing neutral current decays $B \to K\ell^+\ell^-$ and $B \to K^*(892)\ell^+\ell^-$, where $\ell^\pm$ is a charged lepton, are highly suppressed in the Standard Model, with branching fractions predicted to be of order $10^{-7} - 10^{-6}$ [1–3]. The dominant contributions arise at one-loop level and include both electroweak penguin and box diagrams (Fig. 1). Besides probing Standard Model loop effects, these rare decays are important because their rates and kinematic distributions are sensitive to new, heavy particles—such as those predicted by supersymmetric models—that can appear virtually in the loops [1–3].

Searches for $B \to K^{(*)}\ell^+\ell^-$ decays have been performed by *BABAR* [4], Belle [5], CDF [6], and CLEO [7]. Using a sample of 29 fb$^{-1}$, the Belle Collaboration has observed a signal for the decay $B \to K\ell^+\ell^-$ with branching fraction $\mathcal{B}(B \to K\ell^+\ell^-) = (0.75^{+0.25}_{-0.21} \pm 0.09) \times 10^{-6}$, averaged over electron and muon modes. The recently published *BABAR* [4] result was based on a 20.7 fb$^{-1}$ sample and yielded the 90% C.L. upper limits $\mathcal{B}(B \to K\ell^+\ell^-) < 0.51 \times 10^{-6}$ and $\mathcal{B}(B \to K^*\ell^+\ell^-) < 3.1 \times 10^{-6}$. In the present study [8], we update our result with a total sample of 77.8 fb$^{-1}$ accumulated with the *BABAR* detector at the $\Upsilon(4S)$. This sample is nearly four times that used in the original *BABAR* analysis, and it contains $(84.4 \pm 0.9) \times 10^6$ $B\overline{B}$ events.

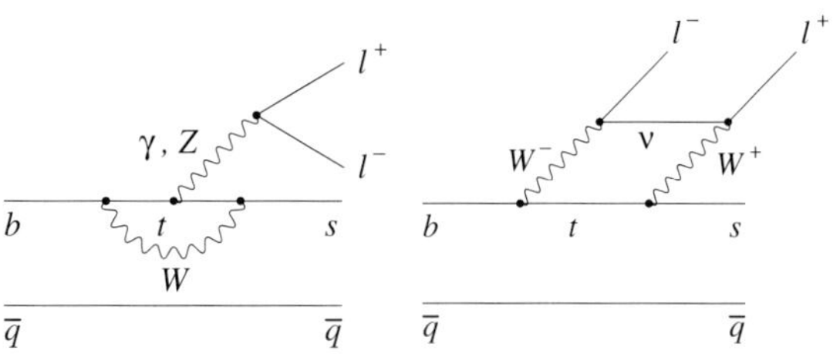


Figure 1. Standard Model processes for $B \to K^{(*)}\ell^+\ell^-$.

## 2. Theoretical predictions

Table 1 summarizes theoretical predictions for the branching fractions for $B \to K^{(*)}\ell^+\ell^-$. Both $B \to K^*e^+e^-$ and $B \to K^*\mu^+\mu^-$ have substantial enhancements at very low $q^2 = m^2_{\ell^+\ell^-}$ due to the pole at $q^2 = 0$ associated with the EM penguin that produces $B \to K^*\gamma$. This enhancement is larger in the electron mode, leading to a larger predicted rate for $B \to K^*e^+e^-$ than for $B \to K^*\mu^+\mu^-$.


doi:10.1016/S0920-5632(02)02160-6

Table 1
Theoretical predictions for $B \to Kl^+l^-$ and $B \to K^*l^+l^-$ based on the Standard Model.

| Authors | $\mathcal{B}(B \to Kl^+l^-)$ /$10^{-6}$ | $\mathcal{B}(B \to K^*\mu^+\mu^-)$ /$10^{-6}$ | $\mathcal{B}(B \to K^*e^+e^-)$ /$10^{-6}$ |
|---|---|---|---|
| Ali *et al.* (2000) | $0.57^{+0.17}_{-0.10}$ | $1.9^{+0.5}_{-0.4}$ | $2.3^{+0.7}_{-0.5}$ |
| Ali *et al.* (2001) [NNLO] | $0.35 \pm 0.12$ | $1.19 \pm 0.39$ | $1.58 \pm 0.49$ |
| Aliev *et al.* (1997) | $0.31 \pm 0.09$ | 1.4 | |
| Colangelo *et al.* (1996) | 0.3 | 1.0 | |
| Faessler *et al.* (2002) | 0.55 | 0.81 | |
| Geng and Kao (1996) | 0.5 | 1.4 | |
| Melikhov *et al.* (1998) | 0.44 | 1.15 | 1.50 |
| Zhong *et al.* (2002) | $0.69^{+0.28}_{-0.25}$ | $1.98^{+0.66}_{-0.71}$ | $2.01^{+0.65}_{-0.73}$ |

Calculation of the decay amplitude for $B \to K^{(*)}\ell^+\ell^-$ is complicated by strong interactions associated with the hadronic current for the $B \to K$ or $B \to K^*$ transition. The Wilson coefficients relevant to these decays have recently been calculated to NNLO accuracy [9]. The hadronic current, however, also depends on form factors, which parametrize non-perturbative QCD interactions and are functions of $q^2$. Part of the spread in the theoretical predictions is due to progress in calculating the Wilson coefficients, but much of it is due to uncertainties in the form factors.

A recent calculation by Ali *et al.* [3] uses NNLO Wilson coefficients, as well as a set of light-cone QCD form factors that have lower values than those used earlier. The form factor choice was motivated by the measured value for the $B \to K^*\gamma$ rate, which is sensitive to the magnetic moment form factor $T_1(q^2 = 0)$. Since $T_1$ also plays a role in the $B \to K^*\ell^+\ell^-$ amplitude, this approach partially ties the overall form factor scale to data. This calculation gives $B \to K^{(*)}\ell^+\ell^-$ predictions that are lower than than those in the earlier Ali *et al.* paper [1].

## 3. Data sample and event selection

We search for the following decays: $B^+ \to K^+\ell^+\ell^-$, $B^0 \to K^0_S\ell^+\ell^-$, $B^+ \to K^{*+}\ell^+\ell^-$, and $B^0 \to K^{*0}\ell^+\ell^-$, where $K^{*0} \to K^+\pi^-$, $K^{*+} \to K^0_S\pi^+$, $K^0_S \to \pi^+\pi^-$, and $\ell$ is either an $e$ or $\mu$. Throughout this paper, charge-conjugate modes are implied. The event selection and fitting procedure are described in a paper [8] submitted to this conference. Because $B$ mesons are only produced in our experiment via the process $e^+e^- \to \Upsilon(4S) \to B\overline{B}$, it is useful to define two kinematic variables that serve to distinguish between signal and background:

$$m_{\rm ES} = \sqrt{E_{\rm b}^{*2} - (\sum_i \mathbf{p}_i^*)^2}$$

$$\Delta E = \sum_i \sqrt{m_i^2 + \mathbf{p}_i^{*2}} - E_{\rm b}^*,$$

where $E_{\rm b}^*$ is the beam energy in the $e^+e^-$ rest (cm) frame, $\mathbf{p}_i^*$ is the cm frame three-momentum of daughter particle $i$ of the $B$ meson candidate, and $m_i$ is the mass hypothesis for particle $i$. Note that $m_{ES}$, the beam-energy substituted mass, requires no particle identification of the daughter particles. For signal events, $m_{\rm ES}$ peaks at the $B$ meson mass with a resolution of about 2.5 MeV/$c^2$. The quantity $\Delta E$ measures the energy difference between the candidate system of $B$ daughter particles and the beam energy in the cm frame; $\Delta E$ peaks near zero for signal events.

Backgrounds in this analysis can be classified as peaking or non-peaking. Peaking backgrounds produce peaks similar to those of the signal processes in $m_{\rm ES}$ and $\Delta E$. Each of the eight signal channels has peaking background from $B \to J/\psi(\to \ell^+\ell^-)K^{(*)}$ and $B \to \psi(2S)(\to \ell^+\ell^-)K^{(*)}$ decays that must be carefully understood and vetoed. On the other hand, these backgrounds provide nearly ideal control samples for each channel, allowing us to perform comparisons between data and Monte Carlo and checks of event selection efficiencies on samples that are nearly identical to the signal modes. Other peaking backgrounds include decays such as $B^- \to K^-\pi^+\pi^-$ (including $D^0 \to K^-\pi^+$), in which two of the hadrons are misidentified as leptons (usually muons). To suppress this type of background, we use strict criteria for kaon and lepton identification.

Non-peaking peakgrounds arise from random combinations of particles found in hadronic events. These include (1) $B\overline{B}$ backgrounds with either two real leptons or one real lepton and one hadron misidentified as a lepton and (2) continuum processes, especially $c\bar{c}$ events with a pair of $D \to K^{(*)}\ell^+\nu$ decays or events with hadrons faking leptons. In the $B \to K\ell^+\ell^-$ modes, the dominant non-peaking background is from continuum events. Much of this background can be suppressed using event-shape variables that discriminate between the jet-like topology of continuum events and the much more spherical topology of signal (and other $B\overline{B}$ events). In $B \to K^*\ell^+\ell^-$ channels, the $B\overline{B}$ background is substantially larger than that from the continuum. Much of the $B\overline{B}$ background is due to events with leptons from $B$ or $D$ semileptonic decay. Such events are characterized by large amounts of missing energy associated with the two neutrinos and can be suppressed on this basis.

To prevent bias, the analysis is performed using a "blind" technique in which all event selection criteria are defined on samples independent of both the signal region in the $m_{\rm ES}$ vs. $\Delta E$ plane, and the sideband region used to determine the background.

## 4. Fits, yields, and results

We extract the signal and background yields in each channel using a two-dimensional extended unbinned maximum likelihood fit in the region defined by $m_{\rm ES} > 5.2$ GeV/$c^2$ and $|\Delta E| < 0.25$ GeV. The two-dimensional signal shapes, including the effects of radiation on the $\Delta E$ distribution and the correlation between $m_{\rm ES}$ and $\Delta E$, are obtained by parameterizing a GEANT4 Monte Carlo simulation of the signal. Small shifts in the mean values of $m_{\rm ES}$ and $\Delta E$ are applied to the signal shapes based on studies of these quantities in the charmonium control samples. The non-peaking background normalization is allowed to float, as are other parameters describing the background shape. Thus, while we use Monte Carlo samples to motivate our background parametrizations, the parameter values themselves are not fixed from the Monte Carlo simulation. In each channel we also include a term to describe the estimated amount of peaking background. Table 2 summarizes the results for each channel.

Systematic errors arise from uncertainties on the efficiency and on the number of $B\overline{B}$ pairs in the data sample (both of which are multiplicative errors) and on the yields extracted from the fits (additive systematic errors). The additive systematic errors arise from three sources: (1) uncertainties in the signal shapes, (2) uncertainties in the non-peaking background shapes, and (3) uncertainties in the amount of peaking backgrounds. Note that because the background shapes and yields float in the fit, much of the uncertainty associated with the background is incorporated in the statistical uncertainty on the signal yield.

We also perform combined fits in which the branching fractions in the four $B \to K\ell^+\ell^-$ channels are constrained to each other, as are the branching fractions for the four $B \to K^*\ell^+\ell^-$ channels (Fig. 2). From these fits we obtain $\mathcal{B}(B \to K\ell^+\ell^-) = (0.78^{+0.24+0.11}_{-0.20-0.18}) \times 10^{-6}$. The significance of this signal, based purely on the statistical errors, is $5.6\sigma$. When systematic uncertainties are applied, this significance remains above $4\sigma$. For this reason, we conclude that we have observed a $B \to K\ell^+\ell^-$ signal. In the $B \to K^*\ell^+\ell^-$ combined fit, we also observe an excess of events over background, but the statistical significance is less than $4\sigma$. We obtain a central value of $\mathcal{B}(B \to K^*\ell^+\ell^-) = (1.68^{+0.68}_{-0.58} \pm 0.28) \times 10^{-6}$ and a 90% C.L. upper limit $\mathcal{B}(B \to K^*\ell^+\ell^-) < 3.0 \times 10^{-6}$, where extracted yield corresponds to the electron mode. The central values are within the ranges expected in the Standard Model, although both the experimental and theoretical uncertainties are still large.

Table 2
Results from individual fits to $B \to K^{(*)}\ell^+\ell^-$ modes. The columns are, from left to right: (1) the fitted signal yield; (2) the contribution of the background to the error on the signal yield, expressed as an effective background yield; (3) the signal efficiency, $\epsilon$ (not including the branching fractions for $K^*$, $K^0$, and $K_S^0$ decays); (4) the multiplicative systematic error on the selection efficiency, $(\Delta\mathcal{B}/\mathcal{B})_\epsilon$; (5) the additive systematic error from the fit, $(\Delta\mathcal{B})_{\rm fit}$; and (6) the branching fraction central value ($\mathcal{B}$).

| Mode | Signal Yield | Effective Bkgnd. | $\epsilon$ (%) | $(\Delta\mathcal{B}/\mathcal{B})_\epsilon$ (%) | $(\Delta\mathcal{B})_{\rm fit}$ $(/10^{-6})$ | $\mathcal{B}$ $(/10^{-6})$ |
|---|---|---|---|---|---|---|
| $B^+ \to K^+e^+e^-$ | $14.4^{+5.0}_{-4.2}$ | 2.2 | 17.5 | $\pm6.8$ | $^{+0.14}_{-0.21}$ | $0.98^{+0.34+0.16}_{-0.28-0.22}$ |
| $B^+ \to K^+\mu^+\mu^-$ | $0.5^{+2.3}_{-1.3}$ | 2.2 | 9.2 | $\pm6.6$ | $^{+0.09}_{-0.08}$ | $0.06^{+0.30+0.09}_{-0.17-0.08}$ |
| $B^0 \to K^0e^+e^-$ | $1.3^{+2.6}_{-1.7}$ | 1.5 | 18.6 | $\pm7.9$ | $\pm0.14$ | $0.24^{+0.49+0.14}_{-0.32-0.15}$ |
| $B^0 \to K^0\mu^+\mu^-$ | $3.6^{+2.9}_{-2.1}$ | 0.9 | 9.4 | $\pm7.7$ | $\pm0.24$ | $1.33^{+1.07}_{-0.78} \pm 0.26$ |
| $B^0 \to K^{*0}e^+e^-$ | $10.6^{+5.2}_{-4.3}$ | 5.5 | 10.6 | $\pm7.6$ | $^{+0.46}_{-0.47}$ | $1.78^{+0.87+0.48}_{-0.72-0.49}$ |
| $B^0 \to K^{*0}\mu^+\mu^-$ | $3.4^{+3.9}_{-2.8}$ | 4.6 | 6.1 | $\pm9.3$ | $\pm0.38$ | $0.99^{+1.14}_{-0.82} \pm 0.39$ |
| $B^+ \to K^{*+}e^+e^-$ | $0.3^{+3.7}_{-2.3}$ | 6.3 | 10.3 | $\pm9.5$ | $^{+0.69}_{-0.72}$ | $0.15^{+1.87+0.69}_{-1.16-0.72}$ |
| $B^+ \to K^{*+}\mu^+\mu^-$ | $3.6^{+3.9}_{-2.5}$ | 3.4 | 5.2 | $\pm11.1$ | $\pm1.80$ | $3.61^{+3.91}_{-2.51} \pm 1.84$ |

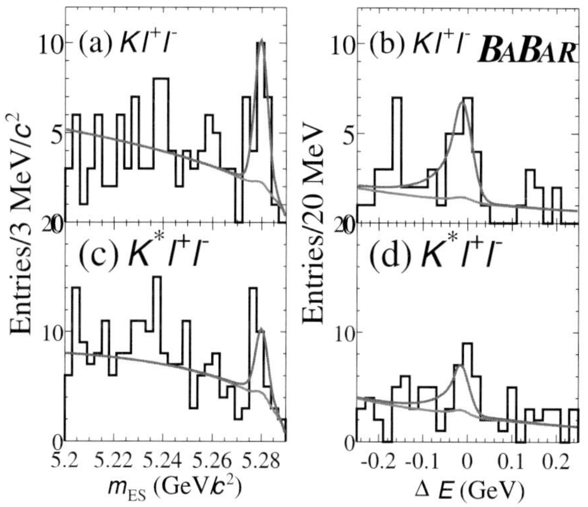


Figure 2. Summed distributions for the four $B \to K\ell^+\ell^-$ and the four $B \to K^*\ell^+\ell^-$ channels, showing the projections of the combined fits.

## REFERENCES

1. A. Ali *et al.*, Phys. Rev. D **61**, 074024 (2000); P. Colangelo *et al.*, Phys. Rev. D **53**, 3672 (1996); D. Melikhov, N. Nikitin, and S. Simula, Phys. Rev. D **57**, 6814 (1998).
2. T.M. Aliev *et al.*, Phys. Lett. B **400**, 194 (1997); M. Beneke, Th. Feldmann, and D. Seidel, Nucl. Phys. B **612**, 25 (2001); G. Burdman, Phys. Rev. D **52**, 6400 (1995); N.G. Deshpande and J. Trampetic, Phys. Rev. Lett. **60**, 2583 (1988); A. Faessler *et al.*, hep-ph/0205287 (2002); C. Greub, A. Ioannissian, and D. Wyler, Phys. Lett. B **346**, 149 (1995); C.Q. Geng and C.P. Kao, Phys. Rev. D **54**, 5636 (1996).
3. A. Ali, E. Lunghi, C. Greub, and G. Hiller, Phys. Rev. D **66**, 034002 (2002).
4. *BABAR* Collaboration, B. Aubert *et al.*, Phys. Rev. Lett. **88**, 241801 (2002).
5. Belle Collaboration, K. Abe *et al.*, Phys. Rev. Lett. **88**, 021801 (2002).
6. CDF Collaboration, T. Affolder *et al.*, Phys. Rev. Lett. **83**, 3378 (1999).
7. CLEO Collaboration, S. Anderson *et al.*, Phys. Rev. Lett. **87**, 181803 (2001).
8. *BABAR* Collaboration, B. Aubert *et al.*, *BABAR*-CONF-02/023, submitted to this conference; hep-ex/0207082.
9. C. Bobeth, M. Misiak, J. Urban, Nucl. Phys. B 574, 291 (2000); H.H. Asatrian, H.M. Asatrian, C. Greub, and M. Walker, Phys. Rev. D **65**, 074004 (2002).

ELSEVIER

Proceedings of ICHEP 2002, pp. 620–623
*S. Bentvelsen, P. de Jong, J. Koch and E. Laenen (Editors)*

31st INTERNATIONAL CONFERENCE ON HIGH ENERGY PHYSICS AMSTERDAM

# Leptonic $B$ Decays at *BABAR*

Thomas B. Moore[a]

[a]University of Massachusetts, Amherst MA 01003-4525 USA

We present results on searches for the rare $B$ decays $B^0 \to \ell^+\ell^-$ ($\ell = e$ or $\mu$) and $B^+ \to K^+\nu\bar{\nu}$ in $\Upsilon(4S)$ decays. The data used in these analyses were collected with the *BABAR* detector at the PEP-II $e^+e^-$ storage ring during the years 2000 and 2001. The $B^0 \to \ell^+\ell^-$ search was performed using an integrated luminosity of 54.4 fb$^{-1}$ accumulated at the $\Upsilon(4S)$ resonance (about 60 million $B\bar{B}$ pairs) while the $B^+ \to K^+\nu\bar{\nu}$ search was performed with 50.7 fb$^{-1}$ (about 56 million $B\bar{B}$ pairs). Both analyses use an additional 6.4 fb$^{-1}$ of data accumulated about 40 MeV below resonance. We see no evidence for a signal in either the $B \to \ell^+\ell^-$ or $B \to K^+\nu\bar{\nu}$ decay modes and set the following upper limits at the 90% CL: $\mathcal{B}(B^0 \to e^+e^-) < 3.3 \times 10^{-7}$, $\mathcal{B}(B^0 \to \mu^+\mu^-) < 2.0 \times 10^{-7}$, $\mathcal{B}(B^0 \to e^\pm\mu^\mp) < 2.1 \times 10^{-7}$, and $\mathcal{B}(B^+ \to K^+\nu\bar{\nu}) < 9.4 \times 10^{-5}$.

## 1. $B^0 \to \ell^+\ell^-$

In the Standard Model (SM), the rare leptonic $B$ decays $B^0 \to \ell^+\ell^-$, where $\ell$ refers to $e$ or $\mu$, are expected to proceed through box and penguin diagrams. These decays are highly suppressed since they involve a $b \to d$ transition and require an internal quark annihilation within the $B$ meson which further suppresses the decay by a factor of $(f_B/M_B)^2 \approx 2 \times 10^{-3}$ relative to the electroweak "penguin" $b \to d\gamma$ decays. In addition, the decays are helicity suppressed by a factor of $(m_\ell/m_B)^2$. The expected branching ratios are $1.9 \times 10^{-15}$ and $8.0 \times 10^{-11}$ for $B^0 \to e^+e^-$ and $B^0 \to \mu^+\mu^-$ respectively. $B^0$ decays to leptons of different flavors violate lepton flavor conservation and are forbidden in the SM. They are allowed, however, in extensions to the SM with non-zero neutrino mass. To date, $B^0 \to \ell^+\ell^-$ decays have not been observed. The previous best upper limits are from the CLEO [1] and BELLE [2] collaborations.

Since these processes are highly suppressed in the SM, they are potentially sensitive probes of new physics. Significant enhancements are possible through interaction with a charged Higgs in multi-Higgs doublet models [3], in models with an extra vector-like down-type quark [4], in models containing leptoquarks [5], and in supersymmetric models with R-parity violation.

The presence of two charged high-momentum leptons provides a very clean signature for the three decay modes under consideration. In the center-of-momentum (CM) frame we require two oppositely charged high momentum leptons from a common vertex consistent with the decay of a $B^0$ meson. Since the signal events contain two $B$ mesons and no additional particles, the energy of each $B$ in the CM frame must be equal to the $e^+$ or $e^-$ beam energy. We therefore define

$$m_{ES} = \sqrt{(E^*_{beam})^2 - (p^*_{B^0})^2}\,, \tag{1}$$

$$\Delta E = E^*_{B^0} - E^*_{beam} \tag{2}$$

where $E^*_{beam}$ is the beam energy and $p^*_{B^0}$ and $E^*_{B^0}$ are the reconstructed momentum and energy of the $B^0$ candidate in the CM frame. For signal events, the beam energy-substituted mass $m_{ES}$ peaks at $m_B$. The quantity $\Delta E$ is used to determine whether a candidate system of particles has total energy consistent with the beam energy in the CM frame. Table 1 lists the $m_{ES}$ and $\Delta E$ resolution for each decay mode. Note that the resolution is degraded for the $B^0 \to e^+e^-$ mode and, to a lesser extent, the $B^0 \to e^\pm\mu^\mp$ mode due to bremsstrahlung by the electrons. We define a signal box in the $m_{ES}$-$\Delta E$ plane whose dimensions are optimized for each mode to produce the best upper limit on the branching ratio. The various signal box dimensions are given in table 1. We also define a *Grand Sideband* (GSB) as 5.20


doi:10.1016/S0920-5632(02)02159-X

Table 1
The $m_{ES}$ and $\Delta E$ resolution for each decay channel. Also shown are the dimensions of the signal box for each channel which have been optimized to produce the best upper limit.

| | $B^0 \to e^+e^-$ | $B^0 \to \mu^+\mu^-$ | $B^0 \to e^\pm\mu^\mp$ |
|---|---|---|---|
| $\sigma(m_{ES})$ [MeV/c$^2$] | 3.0±0.2 | 2.6±0.1 | 2.7±0.1 |
| $\sigma(\Delta E)$ [MeV] | 29.3±0.9 | 24.7±0.3 | 26.8±0.4 |
| Signal Box $m_{ES}$ range [GeV/c$^2$] | 5.273-5.285 | 5.274-5.285 | 5.274-5.284 |
| Signal Box $\Delta E$ range [GeV] | −0.105-0.050 | −0.050-0.050 | −0.070-0.050 |

$< m_{ES} < 5.26$ GeV/c$^2$ and -0.40 $< \Delta E <$ 0.40 GeV in which to study backgrounds.

Suppression of background from non-resonant $q\bar{q}$ production is provided by a series of topological requirements. We employ restrictions on the overall magnitude of the event thrust $|T|$ and on the magnitude of the cosine of the angle $\theta_T$ defined as the angle between the thrust axis of the particles that form the reconstructed $B^0$ candidate and the thrust axis of the remaining tracks and neutral clusters in the event. We require $|T| < 0.9$ and $\cos\theta_T < 0.84$. The efficiencies for the full selection are given in table 2. The systematic error on the efficiency is determined by a comparison of the control sample $B^0 \to J/\psi K^0_s$ with $J/\psi \to e^+e^-$ for $B^0 \to e^+e^-$ and $J/\psi \to \mu^+\mu^-$ for $B^0 \to \mu^+\mu^-$. These comparisons indicate the dominant uncertainty on the signal efficiency to be the resolution and scale of $\Delta E$, contributing 4.4% (2.6%) for the $B^0 \to e^+e^-$ ($B^0 \to \mu^+\mu^-$) channel.

We estimate the background in the signal box, $N_{BG}$, by assuming it is described by the ARGUS function[6] in $m_{ES}$ and an exponential function in $\Delta E$. These shapes are fit to the data in the sidebands and extrapolated into the signal box using the number of events in the GSB to determine the normalization. For the $B^0 \to e^+e^-$ channel, the background is dominated by pairs of true electrons from $c\bar{c}$ and two-photon events. For the $B^0 \to \mu^+\mu^-$ channel, about half of the total background is due to misidentified hadrons in combination with a real muon. For the $B^0 \to e^\pm\mu^\mp$ channel, the background is composed of real electrons and fake muons with two-photon processes contributing strongly. The expected background is roughly 0.5-0.6 events for all channels. In the signal box we observe 1 event in the $B^0 \to e^+e^-$ mode and no events in the other two modes as shown in figure 1. We therefore find no evidence for a signal and set upper limits at the 90% confidence level. For the purpose of setting upper limits, all observed events are assumed to be signal. These results are summarized in table 2.

## 2. $B \to K^+\nu\bar{\nu}$

The flavor changing neutral current decay $B^+ \to K^+\nu\bar{\nu}$ proceeds via $Z^0$ penguin and $W$ box diagrams in the SM and is, therefore, highly suppressed. The predicted branching ratio for $B \to K^+\nu\bar{\nu}$ summed over all neutrino species is $\mathcal{B}(B^+ \to K^+\nu\bar{\nu}) = 3.8^{+1.2}_{-0.6} \times 10^{-6}$ [7]. Since the decay is highly suppressed it can be used to probe the indirect effects of new particles and interactions emerging in the loop diagrams. New physics models which may enhance the rate of $B^+ \to K^+\nu\bar{\nu}$ include minimal supersymmetry(SUSY), multi-Higgs doublet models, leptoquarks, and SUSY models with R parity [7,8]. Moreover, the SM prediction for the inclusive process $b \to s\nu\bar{\nu}$ is nearly free from strong interaction effects and has very small theoretical uncertainty. The best previous experimental limit from the CLEO collaboration [9] is $\mathcal{B}(B^+ \to K^+\nu\bar{\nu}) < 2.4 \times 10^{-4}$ at 90% CL.

Since the signal mode contains two neutrinos there are only minimal kinematic constraints that can be applied. Therefore, the identification of the signal mode demands the reconstruction and removal of the decay products of the other $B$ in the event. The particles not used in the reconstruction of the companion or *tag* $B$ may then be compared with the signature expected for $B^+ \to K^+\nu\bar{\nu}$. The low multiplicity of the signal decay greatly reduces the combinatorial back-

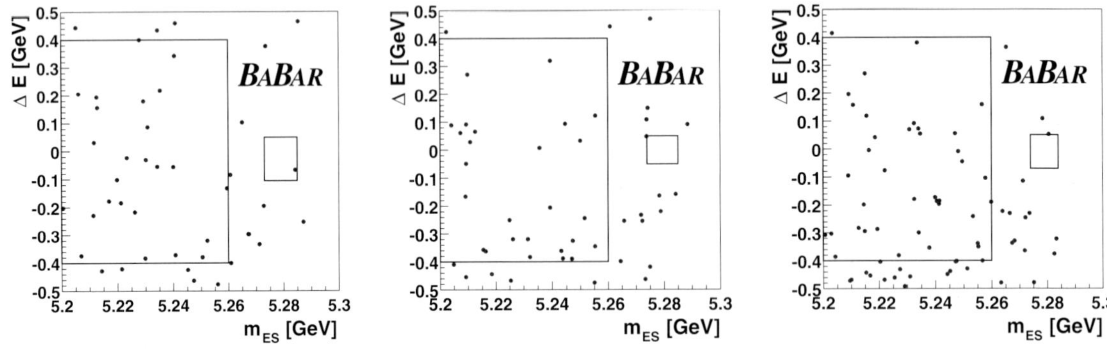


Figure 1. Distribution from data of $m_{ES}$ vs $\Delta E$ for $B^0 \to e^+e^-$ (left), $B^0 \to \mu^+\mu^-$ (middle), and $B^0 \to e^\pm\mu^\mp$ (right).

Table 2
Results of the $B \to \ell^+\ell^-$ search. Shown are the number of events in the Grand Sideband, $N_{GSB}$, number of events in the signal box, $N_{SigBox}$, number of background events expected, $N_{BG}$, total signal efficiency, and 90% CL upper limit on the branching ratio.

| mode | $N_{GSB}$ | $N_{SigBox}$ | $N_{BG}$ | efficiency[%] | Upper Limit (90% CL) |
|---|---|---|---|---|---|
| $B \to e^+e^-$ | 25 | 1 | 0.60±0.24 | 19.3±0.4 | $3.3\times10^{-7}$ |
| $B \to \mu^+\mu^-$ | 26 | 0 | 0.49±0.19 | 18.8±0.3 | $2.0\times10^{-7}$ |
| $B \to e^\pm\mu^\mp$ | 37 | 0 | 0.51±0.17 | 18.3±0.4 | $2.1\times10^{-7}$ |

ground in the tag reconstruction, allowing the use of decay modes that would not be sufficiently clean under other circumstances. Therefore, the companion $B$ is reconstructed in the decay mode $B^- \to D^0\ell^-\bar{\nu}X$ with the $D^0$ reconstructed in the $K^-\pi^+$, $K^-\pi^+\pi^-\pi^+$ and $K^-\pi^+\pi^0$ modes. The $X$ system is kinematically constrained to be either nothing or a soft transition pion or photon from a higher mass charm state. This method results in roughly 0.5% of $B^-$ decays being reconstructed as tags.

The following kinematic requirements are imposed on the tag $B$ candidate: $p^*_{D^0} > 0.5$ GeV/c and $m_{D\ell} > 3$ GeV/c$^2$ where $p^*_{D^0}$ is the CM frame momentum of the $D^0$ candidate and $m_{D\ell}$ is the invariant mass of the $D^0\ell^-$ combination. We also require $-2.5 < \cos\theta_{B,D\ell} < 1.1$ where,

$$\cos\theta_{B,D\ell} = \frac{2E_B E_{D\ell} - m_B^2 - m_{D\ell}^2}{2|\vec{p}_B||\vec{p}_{D\ell}|}. \quad (3)$$

Here $E_B$ and $|\vec{p}_B|$ are the known energy and momentum of the $B$ in the $\Upsilon(4S)$ frame. This requirement on $\cos\theta_{B,D\ell}$ is the most important for restricting the kinematics of the $D^0\ell^-$ to be consistent with a semileptonic $B$ decay.

After the tag $B$ is selected, there must be exactly one remaining charged track in the event whose charge is opposite that of the tag lepton and it must satisfy the particle identification criteria for charged kaons. The kaon from the decay $B^+ \to K^+\nu\bar{\nu}$ has a stiff momentum spectrum in the $\Upsilon(4S)$ rest frame while the background peaks at small momenta. Therefore, we require $p^*_K > 1.5$ GeV/c. We also require the angle between the kaon and lepton to satisfy $-0.9 < \cos\theta^*_{K,\ell} < 0.8$ in the CM frame since $e^+e^- \to q\bar{q}$ and $e^+e^- \to \tau^+\tau^-$ backgrounds tend to peak in the forward and backward directions.

The $B^+ \to K^+\nu\bar{\nu}$ signal leaves very little neutral energy (calorimeter energy not associated with a charged track) in the detector and does not contain any neutral hadrons. We therefore require that the remaining neutral energy detected in the electromagnetic calorimeter, $E_{left}$, be less than 0.5 GeV and that there be no neutral clusters consistent with neutral hadrons located in the muon system. A signal box is defined by the requirements $E_{left} < 0.5$ GeV and $|m_D - m_D^{fit}| < 3\sigma_D^{fit}$ where $m_D$ is the nominal $D^0$ mass and $m_D^{fit}$ and $\sigma_D^{fit}$ are the reconstructed $D^0$ mass and resolution respectively. The expected background in the signal box from the simulation is 2.2 events.

Systematic errors are evaluated by comparing data with the simulation in two $E_{left}$ vs $(m_D - m_D^{fit})$ sideband regions. In addition to the sideband samples, we use events in which both the $B^+$ and $B^-$ are reconstructed in the tag mode $B^- \to D^0 \ell^- \bar{\nu} X$. These "double tagged" events are used to quantify the uncertainty in the efficiency of our signal selection. Systematic errors on the total number of $\Upsilon(4S)$ events, tagging efficiency, kaon selection and momentum, $E_{left}$ and neutral hadron multiplicity have all been studied. The total relative uncertainty on the selection efficiency is found to be 8.7%.

As shown in figure 2, the signal region contains two events, consistent with the 2.2 background events predicted by the simulation. For the purpose of setting a limit, each candidate is assumed to be signal. We therefore find $\mathcal{B}(B^+ \to K^+ \nu\bar{\nu}) < 9.4 \times 10^{-5}$ at 90% CL.

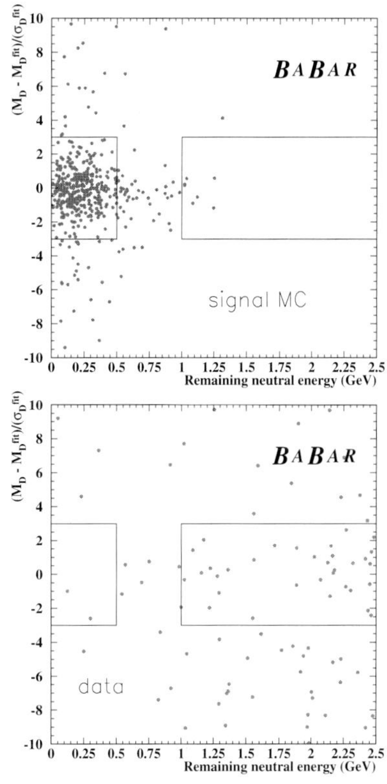


Figure 2. The distribution of events in the $(m_D - m_D^{fit})$ vs $E_{left}$ plane for signal Monte Carlo (top) and data (bottom).

## 3. CONCLUSIONS

We see no evidence for a signal in either the $B \to \ell^+\ell^-$ or $B \to K^+\nu\bar{\nu}$ decay modes and set the following upper limits at the 90% CL: $\mathcal{B}(B^0 \to e^+e^-) < 3.3 \times 10^{-7}$, $\mathcal{B}(B^0 \to \mu^+\mu^-) < 2.0 \times 10^{-7}$, $\mathcal{B}(B^0 \to e^\pm\mu^\mp) < 2.1 \times 10^{-7}$, and $\mathcal{B}(B^+ \to K^+\nu\bar{\nu}) < 9.4 \times 10^{-5}$.

## REFERENCES


1. CLEO Collaboration, T. Bergfeld *et al.*, *Phys. Rev.* D **62**, 091102 (2000)
2. Belle Collaboration, K. Abe *et al.*, BELLE-CONF-0127, (2001)
3. J.L. Hewitt, S. Nandi, and T.G. Rizzo, *Phys. Rev.* D **39**, 250 (1989)
4. Y. Nir and D. Silverman, *Phys. Rev.* D **42**, 1477 (1990)
5. J. Pati and A. Salam, *Phys. Rev.* D **10**, 275 (1974)
6. H. Albrecht *et al. Phys. Lett.* B **241**, 278 (1990)
7. G. Buchalla, G. Hiller, and G. Isidori, *Phys. Rev.* D **63**, 014015 (2001)
8. Y. Grossman, Z. Ligeti, and E. Nardi, *Nucl. Phys.* B **465**, 369 (1996)
9. T.E. Browder *et al.*, *Phys. Rev. Lett.* **86**, 2950 (2001)

ELSEVIER

Proceedings of ICHEP 2002, pp. 624–630
*S. Bentvelsen, P. de Jong, J. Koch and E. Laenen (Editors)*

31st INTERNATIONAL CONFERENCE ON HIGH ENERGY PHYSICS AMSTERDAM

# Next-to-leading Order Calculations of the Radiative and Semileptonic Rare $B$ Decays in the Standard Model and Comparison with Data

A. Ali[a*]

[a]Deutsches Elektronen-Synchrotron DESY,
Notkestraße 85, D-22603 Hamburg, Federal Republic of Germany

We review some selected rare $B$ decays calculated in next-to-leading order accuracy in the Standard Model (SM). These include the radiative decays $B \to (X_s, K^*, \rho)\gamma$ and the semileptonic decays $B \to (X_s, K, K^*)\ell^+\ell^-$, for which new data from the BABAR and BELLE collaborations have been presented at this conference. SM is in agreement with the current measurements within the theoretical and experimental errors. The impact of rare $B$-decays on the CKM phenomenology is quantitatively discussed.

## 1. INTRODUCTION AND OVERVIEW

First measurements of rare $B$ decays $B \to K^*\gamma$ [1] and $B \to X_s\gamma$ [2] were reported by CLEO in 1993 and 1995, respectively.. Since then, a wealth of data on these decays has become available through the subsequent work of the CLEO collaboration [3,4], the ALEPH collaboration at LEP [5], and more recently from the experiments at the B factories, BABAR [6] and BELLE [7], which now dominate this field. We quantify the impact of $B \to X_s\gamma$ measurement on the CKM phenomenology.

Concerning the exclusive decays $B \to K^*\gamma$, the NLO calculations of the decay widths were completed last year [8–10]. Radiative decays provide a cleaner test of the underlying theoretical framework of perturbative QCD factorization which has been invoked initially to calculate the two-body exclusive non-leptonic $B$ decays, such as $B \to \pi\pi$ [11]. Quantitative *rapport* of theory and experiment in these decays is on hold at present due to the imprecise knowledge of the form factor, as discussed later in this talk.

A big leap forward has been reported in the experimental study of the Cabibbo-suppressed decays $B \to \rho\gamma$ and $B \to \omega\gamma$ at this conference [6,7]. These decays which enact $b \to d\gamma$ transition at the quark level yield complementary information on the CKM parameters and hence are potentially very important. We work out the constraints following from the current upper limits on the $B \to \rho\gamma$ branching ratios [6,7]. Apart from these, the Cabibbo-suppressed radiative decays $B \to \rho\gamma$ allow us to study isospin- and CP-asymmetries, which in the SM have a sensitive dependence on the angle $\alpha$ of the unitarity triangle [12,8,9]. We present the SM expectations for the decay rates and asymmetries, which are also potentially of interest in searching for physics beyond-the-SM [12,13].

The semileptonic decays $B \to (X_s, K, K^*)\ell^+\ell^-$ ($\ell^\pm = e^\pm, \mu^\pm$) are the new additions to the flavour changing neutral current FCNC decays, where B factories have made first experimental inroads [7,14]. Of particular interest is the inclusive decay $B \to X_s\ell^+\ell^-$, whose measurement has been reported at this conference by the BELLE collaboration [7]. These decays have been calculated in the SM to the same degree of theoretical precision as their radiative counterpart $B \to X_s\gamma$, and take into account the explicit $O(\alpha_S)$ corrections to the dilepton invariant mass distribution [15,16], the forward-backward asymmetry of the charged lepton [17,18] and the leading power corrections in

*Present address: CERN (-TH-), CH-1211 Geneva 23.


doi:10.1016/S0920-5632(02)02158-8

$1/m_b$ [19,20] and $1/m_c$ [21]. Precise measurements of the dilepton invariant mass spectrum and the forward-backward asymmetry would allow to extract the (effective) Wilson coefficients, providing precision tests of the SM and enabling searches for physics beyond-the-SM [22].

What concerns the exclusive decays $B \to (K, K^*)\ell^+\ell^-$, important theoretical progress has been made recently [23,10], using ideas based on the large energy expansion [24]. In particular, both the dilepton invariant mass spectrum and the forward-backward asymmetry in $B \to K^*\ell^+\ell^-$ have been calculated in the lower region of the dilepton invariant mass ($s < m^2_{J/\psi}$) to the same accuracy as is the case for $B \to X_s\ell^+\ell^-$, albeit using perturbative QCD factorization. This framework has also been used to carry out a helicity analysis of the decays $B \to K^*\ell^+\ell^-$ and $B \to \rho\ell\nu_\ell$ [25]. Data on $B \to (X_s, K^*, K)\ell^+\ell^-$ are compared with the SM estimates [26,27] and the two are found to be compatible with each other.

## 2. Inclusive Decay $B \to X_s\gamma$

The theoretical framework for analyzing the transition $B \to X_s\gamma$ is provided by the effective Hamiltonian

$$\mathcal{H}_{\text{eff}} = -\frac{4G_F}{\sqrt{2}} V^*_{ts} V_{tb} \sum_{i=1}^{8} C_i(\mu) O_i, \tag{1}$$

where $G_F$ is the Fermi coupling constant and $V_{ij}$ are the CKM matrix elements. The effective Hamiltonian is obtained from the SM by integrating out all the particles that are much heavier than the $b$-quark. The Wilson coefficients $C_i(\mu)$ play the role of effective coupling constants for the interaction terms $O_i$. The generic structure of the operators $O_i$ can be seen, for example, in Ref. [27].

Perturbative calculations are used to determine the Wilson coefficients in a given renormalization scheme, such as the $\overline{\text{MS}}$ scheme, at a renormalization scale, typically $\mu_b \sim O(m_b)$,

$$C_i(\mu_b) = C_i^{(0)}(\mu_b) + \frac{\alpha_s(\mu_b)}{4\pi} C_i^{(1)}(\mu_b) + \ldots . \tag{2}$$

Here, $C_i^{(n)}(\mu_b)$ depend on $\alpha_s$ only via the ratio $\eta \equiv \alpha_s(\mu_0)/\alpha_s(\mu_b)$, where $\mu_0 \sim m_W$. In the leading order (LO) calculations, everything but $C_i^{(0)}(\mu_b)$ is neglected in Eq. (2). In NLO, one takes in addition $C_i^{(1)}(\mu_b)$ into account.

It has become customary to quote the theoretical branching ratios for $B \to X_s\gamma$ decay with a cut on the photon energy $E_\gamma$. In the $\overline{\text{MS}}$ scheme, the results in the SM (in units of $10^{-4}$) are [28,29]:

$$\begin{aligned} \mathcal{B}(\bar{B} \to X_s\gamma\,; E_\gamma > 1.6\text{ GeV}) &= (3.57 \pm 0.30), \\ \mathcal{B}(\bar{B} \to X_s\gamma\,; E_\gamma > \tfrac{1}{20} m_b) &= (3.70 \pm 0.31). \end{aligned} \tag{3}$$

The present world averages (also in units of $10^{-4}$)

$$\begin{aligned} \mathcal{B}(\bar{B} \to X_s\gamma\,; E_\gamma > 1.6\text{ GeV}) &= \left(3.28\,^{+0.41}_{-0.36}\right), \\ \mathcal{B}(\bar{B} \to X_s\gamma\,; E_\gamma > \tfrac{1}{20} m_b) &= \left(3.40\,^{+0.42}_{-0.37}\right) \end{aligned} \tag{4}$$

result from the following four measurements: BABAR [6,30], CLEO [3], BELLE [31], and ALEPH [5]. They are in good agreement with the NLO results in the $\overline{\text{MS}}$ scheme. However, there is a residual theoretical uncertainty on the SM estimates related to the scheme-dependence of the quark masses. This can be judged from the theoretical branching ratios in the $\overline{\text{MS}}$ scheme for which one gets $\mathcal{B}(\bar{B} \to X_s\gamma) = (3.73 \pm 0.31) \times 10^{-4}$ [28], and in the pole quark mass scheme yielding $\mathcal{B}(\bar{B} \to X_s\gamma) = (3.35 \pm 0.30) \times 10^{-4}$ [32]. Reducing this theoretical uncertainty, which currently represents an error of $O(10\%)$, requires calculation of the next-to-next-to-leading order contribution to the decay width (i.e., explicit $O(\alpha_s)^2$ improvements), which is a formidable but doable task in a concerted theoretical effort [33]. Anticipated experimental precision at the $B$ factories will make it mandatory to undertake this heroic effort.

The transition $b \to s\gamma$ is not expected to yield useful information on the CKM-Wolfenstein parameters $\bar{\rho}$ and $\bar{\eta}$, which define the apex of the unitarity triangle of current interest. The test of CKM-unitarity for the $b \to s$ transitions in rare $B$-decays lies in checking the relation $\lambda_t \simeq -\lambda_c$, which holds up to corrections of order $\lambda^2$. This is implicit in the theoretical decay rates quoted above. Alternatively, one can drop the explicit use of the CKM unitarity and calculate the various contributions in the $b \to s\gamma$ amplitude from the current knowledge of $\lambda_c \simeq |V_{cb}| = (41.0 \pm 2.1) \times 10^{-3}$ and $\lambda_u$ from PDG [34], which then yields $\lambda_t \simeq V_{ts} = -(47 \pm 8) \times 10^{-3}$ [35]. This is consistent with $\lambda_t = -\lambda_c$, but less precise. This is due to the strong mixing of the operator $O_2$ with $O_7$ under QCD renormalization, which enhances the coefficient of the $\lambda_c$ term in the $b \to s\gamma$ amplitude in comparison with the coefficient of the $\lambda_t$ term, thereby reducing the precision on $\lambda_t$.

## 3. EXCLUSIVE DECAYS $B \to (K^*, \rho)\gamma$

To compute the branching ratios for $B \to V\gamma$ ($V = K^*, \rho$) reliably, one needs to calculate the explicit $O(\alpha_s)$ improvements to the lowest order decay widths. This requires the calculation of the renormalization group effects in the appropriate Wilson coefficients in the effective Hamiltonian [32], an explicit $O(\alpha_s)$ calculation of the matrix elements involving the hard vertex corrections [36–38], annihilation contributions [39–41], which are more important in the decays $B \to \rho\gamma$ but have also been worked out for the $B \to K^*\gamma$ decays [42], and the so-called hard-spectator contributions involving (virtual) hard gluon radiative corrections off the spectator quarks in the $B$-, $K^*$-, and $\rho$-mesons [8–10]. We discuss the decays $B \to K^*\gamma$ and $B \to \rho\gamma$ in turn.

### 3.1. $B \to K^*\gamma$ Decays

The branching ratio for the decays $B \to K^*\gamma$ can be written as [8–10]:

$$\mathcal{B}(B \to K^*\gamma) = \tau_B \frac{G_F^2 \alpha |V_{tb}V_{ts}^*|^2}{32\pi^4} m_{b,\mathrm{pole}}^2 M_B^3 \times \left[\xi_\perp^{(K^*)}(0)\right]^2 \left(1 - \frac{m_{K^*}^2}{M^2}\right)^3 \left|C_7^{(0)\mathrm{eff}} + A^{(1)}(\mu)\right|^2 .$$

Here, $M_B$ is the $B$-meson mass and $\xi_\perp^{(K^*)}(0)$ is the form factor in the large energy effective theory. It differs from the corresponding form factor in QCD, $T_1^{K^*}(0)$, by $O(\alpha_s)$ terms calculated in Ref. [23]. Numerically, this relation can be expressed as $T_1^{K^*}(0) \simeq 1.08\xi_\perp^{(K^*)}$. The effect of the $O(\alpha_s)$ corrections is encoded in the function $A^{(1)}(\mu)$ and its numerical value can be expressed in terms of a $K$-factor [8–10]:

$$K = \frac{\left|C_7^{(0)\mathrm{eff}} + A^{(1)}(\mu)\right|^2}{\left|C_7^{(0)\mathrm{eff}}\right|^2}, \quad \text{with} \quad 1.5 \leq K \leq 1.7.$$

This yields a charge-conjugate averaged branching ratio

$$\langle \mathcal{B}(B \to K^*\gamma) \rangle \simeq (7.2 \pm 1.1) \times 10^{-5} \times \left(\frac{m_{b,\mathrm{pole}}}{4.65\ \mathrm{GeV}}\right)^2 \left(\frac{\xi_\perp^{(K^*)}(0)}{0.35}\right)^2 ,$$

where the default values of the most sensitive parameters are indicated. The value given for $\xi_\perp^{(K^*)}(0)$ is based on using the light-cone QCD sum rules. Varying the parameters within reasonable ranges (see, for example, [8]) and adding the errors in quadrature, one gets

$$\langle \mathcal{B}(B \to K^*\gamma) \rangle \simeq (7.2 \pm 2.7) \times 10^{-5} , \tag{5}$$

to be compared with the corresponding experimental measurement [4,6,7]:

$$\langle \mathcal{B}(B \to K^*\gamma) \rangle \simeq (4.22 \pm 0.28) \times 10^{-5} . \tag{6}$$

Given the large theoretical errors, there is agreement between theory and experiment. Since there exists good agreement between the SM and experiment in the inclusive decay $B \to X_s\gamma$, one could use the measured branching ratios for $B \to K^*\gamma$ to determine the form factor $T_1^{K^*}(0)$, which in the stated theoretical context yields $T_1^{K^*}(0) = 0.27 \pm 0.04$. This is to be compared with the light-cone QCD sum rule result $T_1^{K^*}(0) = 0.38 \pm 0.05$ [43,26], and with $T_1^{K^*}(q^2 = 0) = 0.32^{+0.04}_{-0.02}$ [44], obtained by using the Lattice UK-QCD simulations combined with a LC-QCD-sum-rule inspired input for extrapolation from $q^2 \gg 0$, where the Lattice simulations are actually done, to the physical value $q^2 = 0$. A preliminary result on $T_1^{K^*}(0)$ using Lattice-QCD has also been reported at this conference [45]. Due to the spread in the theoretical estimates it is too early to draw a quantitative conclusion on the precise value of the form factor, and hence on the perturbative QCD factorization method underlying the theoretical estimates in $B \to K^*\gamma$.

### 3.2. $B \to \rho\gamma$ Decays

The effective Hamiltonian for the radiative decays $B \to (\rho, \omega)\gamma$ can be seen for the SM in [8]. These transitions involve the CKM matrix elements in the first and third column, with the unitarity constraints taking the form $\sum_{u,c,t} \xi_i = 0$, with $\xi_i = V_{ib}V_{id}^*$. All three matrix elements are of order $\lambda^3$, with $\xi_u \simeq A\lambda^3(\bar{\rho} - i\bar{\eta})$, $\xi_c \simeq -A\lambda^3$, and $\xi_t \simeq A\lambda^3(1 - \bar{\rho} - i\bar{\eta})$. This equation leads to the same unitarity triangle as studied through the constraints $V_{ub}/V_{cb}$, $\Delta M_{B_d}$ (or $\Delta M_{B_d}/\Delta M_{B_s}$).

We shall concentrate here on $B \to \rho\gamma$ decays. As the absolute values of the form factors in this decay and in $B \to K^*\gamma$ decays discussed earlier are quite uncertain, it is advisable to calculate instead the ratios of the branching ratios

$$R^{\pm}(\rho\gamma/K^*\gamma) \equiv \frac{\mathcal{B}(B^{\pm} \to \rho^{\pm}\gamma)}{\mathcal{B}(B^{\pm} \to K^{*\pm}\gamma)}, \tag{7}$$

$$R^{0}(\rho\gamma/K^*\gamma) \equiv \frac{\mathcal{B}(B^{0} \to \rho^{0}\gamma)}{\mathcal{B}(B^{0} \to K^{*0}\gamma)}. \tag{8}$$

They have been calculated in the NLO accuracy and the result can be expressed as [8]:

$$\begin{aligned} R^{\pm}(\rho\gamma/K^*\gamma) &= \left|\frac{V_{td}}{V_{ts}}\right|^2 [\mathrm{PS}]\ \zeta^2(1 + \Delta R^{\pm})\ , \\ R^{0}(\rho\gamma/K^*\gamma) &= \frac{1}{2}\left|\frac{V_{td}}{V_{ts}}\right|^2 [\mathrm{PS}]\ \zeta^2(1 + \Delta R^{0})\ , \end{aligned}$$

where $[\mathrm{PS}] = (M_B^2 - M_\rho^2)^3/(M_B^2 - M_{K^*}^2)^3$ represents a kinematic factor, which is 1 to a good approximation, and $\zeta = \xi_\perp^\rho(0)/\xi_\perp^{K^*}(0)$, with $\xi_\perp^\rho(0)(\xi_\perp^{K^*}(0))$ being the form factors (at $q^2 = 0$) in the effective heavy quark theory for the decays $B \to \rho(K^*)\gamma$. Noting that in the SU(3) limit one has $\zeta = 1$, we take $\zeta = 0.76 \pm 0.10$ - a range which straddles various theoretical estimates of this quantity [39,43,46,47]. The functions $\Delta R^{\pm}$ and $\Delta R^0$, appearing on the r.h.s. of the above equations and whose parametric dependence is suppressed here for ease of writing, encode both the $O(\alpha_s)$ and annihilation contributions. They have been evaluated as [8,13]: $\Delta R^{\pm} = 0.055 \pm 0.130$ and $\Delta R^0 = 0.015 \pm 0.110$, where the errors also include the effect of varying the angle $\alpha$ in the currently allowed region.

There are two more observables of interest, namely the isospin-violating ratio $\Delta(\rho\gamma)$ defined as

$$\Delta(\rho\gamma) \equiv \frac{\Gamma^{\pm}(B \to \rho\gamma)}{2\,\Gamma^0(B \to \rho\gamma)} - 1\ , \tag{9}$$

and the CP asymmetry in the decay rates, which for $B^{\pm} \to \rho^{\pm}\gamma$ decays is defined as

$$A_{CP}^{\pm}(\rho\gamma) \equiv \frac{\mathcal{B}(B^- \to \rho^-\gamma) - \mathcal{B}(B^+ \to \rho^+\gamma)}{\mathcal{B}(B^- \to \rho^-\gamma) + \mathcal{B}(B^+ \to \rho^+\gamma)}. \tag{10}$$

The expected values of the observables in the $B \to \rho\gamma$ system defined in Eqs. (7) - (10) and $A_{CP}^0(\rho\gamma)$, the CP asymmetry in the neutral modes, have been recently updated in Ref. [13]. Including the errors on the current determination of the CKM parameters, the results are:

$$\begin{aligned} R^{\pm}(\rho\gamma/K^*\gamma) &= 0.023 \pm 0.012\ , \\ R^{0}(\rho\gamma/K^*\gamma) &= 0.011 \pm 0.006\ , \\ \Delta(\rho\gamma) &= 0.04^{+0.14}_{-0.07}\ , \\ A_{CP}^{\pm}(\rho\gamma) &= 0.10^{+0.03}_{-0.02}\ , \\ A_{CP}^{0}(\rho\gamma) &= 0.06 \pm 0.02\ . \end{aligned} \tag{11}$$

At this conference, the BABAR collaboration has reported a significant improvement on the upper limits of the branching ratios for the decays $B^0(\bar{B}^0) \to \rho^0\gamma$ and $B^{\pm} \to \rho^{\pm}\gamma$. Averaged over the charge conjugated modes, the current 90% C.L. upper limits are [6]:

$$\mathcal{B}(B^0 \to \rho^0\gamma) < 1.4 \times 10^{-6}\ , \tag{12}$$

$$\mathcal{B}(B^{\pm} \to \rho^{\pm}\gamma) < 2.3 \times 10^{-6}\ , \tag{13}$$

$$\mathcal{B}(B^0 \to \omega\gamma) < 1.2 \times 10^{-6}\ . \tag{14}$$

They have been combined, using isospin weights for $B \to \rho\gamma$ decays and assuming $\mathcal{B}(B^0 \to \omega\gamma) = \mathcal{B}(B^0 \to \rho^0\gamma)$, to yield the improved upper limit [6]

$$\mathcal{B}(B \to \rho\gamma) < 1.9 \times 10^{-6}\ . \tag{15}$$

Together with the current measurements of the branching ratios for $B \to K^*\gamma$ decays by BABAR, this yields a 90% C.L. upper limit [6]

$$R(\rho\gamma/K^*\gamma) \equiv \frac{\mathcal{B}(B \to \rho\gamma)}{\mathcal{B}(B \to K^*\gamma)} < 0.047\ . \tag{16}$$

Thus, we see that the current experimental upper limit on the ratio $R(\rho\gamma/K^*\gamma)$ is typically a factor 2 away from the central values of the SM. This can be seen graphically in Fig. 1, where we show the fit of the unitarity triangle resulting from the various direct and indirect measurements, including the CP asymmetry $a_{\psi K_s}$, reported by the BABAR and BELLE collaboration at this conference, with the current world average being $a_{\psi K_s} = 0.734 \pm 0.054$ [48]. The (almost concentric) constraints in the $(\bar{\rho}, \bar{\eta})$ plane from the measured value $\Delta M_{B_d} = 0.503 \pm 0.006\ \text{ps}^{-1}$ [49], and the two competing bounds $\Delta M_{B_s} > 14.4\ \text{ps}^{-1}$ [49], and $R(\rho\gamma/K^*\gamma) < 0.047$ [6] are also shown. The two set of curves for the constraints following from $\Delta M_{B_d}$ and $\Delta M_{B_s}$, shown with and without the chiral logs, reflect the current uncertainty inherent in Lattice calculations due to chiral extrapolations from the region $m_s$ to $m_d$ [50]. We see that the quantity $R(\rho\gamma/K^*\gamma)$ is not yet competitive with the other two constraints from either $\Delta M_{B_d}$ or $\Delta M_{B_s}$. However, this sure will change as the B factory experiments will soon reach the SM-sensitivity on the decay mode $B \to \rho\gamma$, and hence on $R(\rho\gamma/K^*\gamma)$.

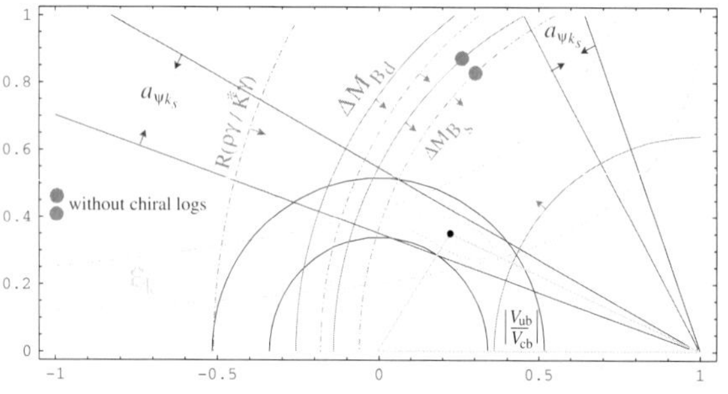


Figure 1. *Unitary triangle fit in the SM and the resulting 95% C.L. contour in the $\bar{\rho}$ - $\bar{\eta}$ plane. The impact of the $R(\rho\gamma/K^*\gamma) < 0.047$ constraint is also shown. (From Ref. [13].)*

## 4. THE DECAYS $B \to (X_s, K, K^*)\ell^+\ell^-$

To discuss the FCNC semileptonic decays in the SM, one has to extend the operator basis given in Eq. (1) by adding two semileptonic operators:

$$O_9 = \frac{e^2}{g_s^2}(\bar{s}_L\gamma_\mu b_L)\sum_\ell \underbrace{(\bar{\ell}\gamma^\mu\ell)}_{\text{V}},$$

$$O_{10} = \frac{e^2}{g_s^2}(\bar{s}_L\gamma_\mu b_L)\sum_\ell \underbrace{(\bar{\ell}\gamma^\mu\gamma_5\ell)}_{\text{A}}, \qquad (17)$$

where $e$ ($g_s$) is the QED (QCD) coupling constant and the subscript $L$ refers to the left-handed components of the fermion fields. There are additional non-local contributions, which can be added to $C_9$, defining an *effective* Wilson coefficient, which is a misnomer as it is actually a function of the dilepton mass squared ($\hat{s} = s/M_B^2$): $C_9^{\text{eff}}(\hat{s}) = C_9\eta(\hat{s}) + Y(\hat{s})$, where $\eta(\hat{s}) = (1 + O(\alpha_s))$ includes the real gluon bremsstrahlung and virtual contributions to the matrix element of the operator $O_9$ [51], and $Y(\hat{s})$ contains perturbative charm loops and Charmonium resonances $(J/\psi, \psi', ...)$.

Concentrating on the perturbative contribution only, which is expected to be the dominant contribution away from the resonances, and including the leading power corrections in $1/m_b$ and $1/m_c$, the dilepton invariant mass distribution for $B \to X_s\ell^+\ell^-$ can be written as [27]:

$$\begin{aligned}\frac{d\Gamma(b \to s\ell^+\ell^-)}{d\hat{s}} &= \left(\frac{\alpha_{em}}{4\pi}\right)^2 \frac{G_F^2 m_{b,pole}^5 \left|\lambda_{ts}\right|^2}{48\pi^3} \\ &\times (1-\hat{s})^2 \left[(1+2\hat{s})\left(\left|\widetilde{C}_9^{\text{eff}}\right|^2 + \left|\widetilde{C}_{10}^{\text{eff}}\right|^2\right) G_1(\hat{s})\right. \\ &+ 4\,(1+2/\hat{s})\left|\widetilde{C}_7^{\text{eff}}\right|^2 G_2(\hat{s}) + 12\text{Re}\left(\widetilde{C}_7^{\text{eff}}\widetilde{C}_9^{\text{eff}*}\right) \\ &\times G_3(\hat{s}) + G_c(\hat{s})\Big]\,, \end{aligned} \qquad (18)$$

where the functions $G_i(\hat{s})$ $(i = 1, 2, 3)$ encode the $1/m_b$ corrections [19,20] and the function $G_c(\hat{s})$ the corresponding $1/m_c$ corrections [21]. Their explicit forms as well as of the other effective coefficients in the above equation can be seen in Ref. [27]. As noted in Ref. [16], inclusion of the explicit $O(\alpha_s)$ corrections reduces the branching ratio for $B \to X_s\ell^+\ell^-$, compared to the partial

$O(\alpha_s)$-corrected result [19] and its scale dependence is considerably reduced.

Data from the BELLE collaboration on the dilepton invariant mass spectrum has been analyzed using Ref. [27], which incorporates all the explicit $O(\alpha_s)$ and power corrections mentioned above. The hadron invariant mass spectrum in $B \to X_s \ell^+ \ell^-$, calculated in the heavy quark effective theory (HQET) and in a phenomenological Fermi motion model [52], has been used to estimate the effect of the experimental cuts on $M_{X_S}$. The results are given in Table 1, and are to be understood with a cut on the dilepton invariant mass $M_{\ell^+\ell^-} > 0.2$ GeV. Given the current experimental errors, the agreement with the SM-based estimates is reasonably good.

The corresponding results for the exclusive decays $B \to K \ell^+ \ell^-$, averaged over $\ell = e, \mu$, and for $B \to K^* \ell^+ \ell^-$, with $\ell = e, \mu$, presented at this conference [7,14] are also given in Table 1 and compared with the SM-based estimates [26,27]. One notes the comparatively larger theoretical uncertainty in the exclusive decay rates due to the form factors. However, within the theoretical and experimental errors, SM accounts well the current data on all the inclusive and exclusive rare $B$ decays. This can be used to constrain models of physics beyond-the-SM [27].

## REFERENCES


1. R. Ammar *et al.* (CLEO Collaboration), Phys. Rev. Lett. **71**, 674 (1993).
2. M. S. Alam *et al.* (CLEO Collaboration), Phys. Rev. Lett. **74** (1995) 2885.
3. S. Chen *et al.* (CLEO Collaboration), Phys. Rev. Lett. **87** (2001) 251807 [hep-ex/0108032].
4. T. E. Coan *et al.* (CLEO Collaboration), Phys. Rev. Lett. **84** (2000) 5283 [hep-ex/9912057].
5. R. Barate *et al.* (ALEPH Collaboration), Phys. Lett. B **429** (1998) 169.
6. C. Jessop (BABAR Collaboration), these proceedings.
7. S. Nishida (BELLE Collaboration), these proceedings.
8. A. Ali and A. Y. Parkhomenko, Eur. Phys. J. C **23** (2002) 89 [hep-ph/0105302].
9. S. W. Bosch and G. Buchalla, Nucl. Phys. B **621** (2002) 459 [hep-ph/01060].
10. M. Beneke, T. Feldmann and D. Seidel, Nucl. Phys. B **612** (2001) 25 [hep-ph/0106067].
11. M. Beneke, G. Buchalla, M. Neubert and C. T. Sachrajda, Phys. Rev. Lett. **83** (1999) 1914 [hep-ph/9905312].
12. A. Ali, L. T. Handoko and D. London, Phys. Rev. D **63** (2000) 014014 [hep-ph/0006175].
13. A. Ali and E. Lunghi, DESY-02-089, hep-ph/0206242.
14. J. Richman (BABAR Collaboration), these proceedings.
15. C. Bobeth, M. Misiak and J. Urban, Nucl. Phys. B **574**, 291 (2000) [hep-ph/9910220].
16. H. H. Asatrian, H. M. Asatrian, C. Greub and M. Walker, Phys. Lett. B **507**, 162 (2001) [hep-ph/0103087].
17. A. Ghinculov, T. Hurth, G. Isidori and Y. P. Yao, [hep-ph/0208088].
18. H. M. Asatrian, K. Bieri, C. Greub and A. Hovhannisyan, [hep-ph/0209006].
19. A. Ali, G. Hiller, L. T. Handoko and T. Morozumi, Phys. Rev. D **55**, 4105 (1997) [hep-ph/9609449].
20. A. F. Falk, M. E. Luke and M. J. Savage, Phys. Rev. D **49**, 3367 (1994) [hep-ph/9308288].
21. G. Buchalla, G. Isidori and S. J. Rey, Nucl. Phys. B **511**, 594 (1998) [hep-ph/9705253].
22. A. Ali, G. F. Giudice and T. Mannel, Z. Phys. C **67**, 417 (1995) [hep-ph/9408213].
23. M. Beneke and T. Feldmann, Nucl. Phys. B **592**, 3 (2001) [hep-ph/0008255].
24. J. Charles, A. Le Yaouanc, L. Oliver, O. Pene and J. C. Raynal, Phys. Rev. D **60**, 014001 (1999) [hep-ph/9812358].
25. A. Ali and A. S. Safir, [hep-ph/0205254].
26. A. Ali, P. Ball, L. T. Handoko and G. Hiller, Phys. Rev. D **61**, 074024 (2000) [hep-ph/9910221].
27. A. Ali, E. Lunghi, C. Greub and G. Hiller, Phys. Rev. D **66**, 034002 (2002) [hep-ph/0112300].
28. P. Gambino and M. Misiak, Nucl. Phys. B **611** (2001) 338 [hep-ph/0104034].
29. A.J. Buras, A. Czarnecki, M. Misiak and

Table 1
SM predictions ($O(\alpha_s)$ and leading power corrected) and comparison with data (in units of $10^{-6}$)

| Decay Mode | Theory (SM) [26,27] | BELLE [7] | BABAR [14] |
|---|---|---|---|
| $B \to K\ell^+\ell^-$ | $0.35 \pm 0.12$ | $0.58^{+0.17}_{-0.15} \pm 0.06$ | $0.78^{+0.24+0.11}_{-0.20-0.18}$ |
| $B \to K^* e^+ e^-$ | $1.58 \pm 0.52$ | $< 5.1$ | $1.68^{+0.68}_{-0.58} \pm 0.28$ |
| $B \to K^* \mu^+ \mu^-$ | $1.2 \pm 0.4$ | $< 3.0$ | $< 3.0$; weighted $e^+e^-$, $\mu^+\mu^-$ |
| $B \to X_s \mu^+ \mu^-$ | $4.15 \pm 0.70$ | $7.9 \pm 2.1^{+2.0}_{-1.5}$ | – |
| $B \to X_s e^+ e^-$ | $4.2 \pm 0.7$ | $5.0 \pm 2.3^{+1.2}_{-1.1}$ | – |
| $B \to X_s \ell^+ \ell^-$ | $4.18 \pm 0.70$ | $6.1 \pm 1.4^{+1.3}_{-1.1}$ | – |

The inclusive measurements and the corresponding SM rates are given with a cut on the dilepton mass, $M_{\ell^+\ell^-} > 0.2$ GeV.

J. Urban, Nucl. Phys. B **631** (2002) 219 [hep-ph/0203135].
30. B. Aubert *et al.* [BaBar Collaboration], [hep-ex/0207076].
31. K. Abe *et al.* (BELLE Collaboration), Phys. Lett. B **511** (2001) 151 [hep-ex/0103042].
32. K. G. Chetyrkin, M. Misiak and M. Münz, Phys. Lett. B **400**, 206 (1997) [Erratum-ibid. B **425**, 414 (1998)] [hep-ph/9612313].
33. M. Misiak, talk presented at the DESY-Theory Workshop, Hamburg, September 24-26, 2002, and private communication.
34. K. Hagiwara *et al.* (Particle Data Group Collaboration), Phys. Rev. D **66** (2002) 010001.
35. A. Ali and M. Misiak, in Proceedings of the CERN-CKM Workshop, Geneva, Feb. 13-16, 2002.
36. J. M. Soares, Nucl. Phys. B **367**, 575 (1991).
37. C. Greub, H. Simma and D. Wyler, Nucl. Phys. B **434**, 39 (1995) [Erratum-ibid. B **444**, 447 (1995)] [hep-ph/9406421].
38. C. Greub, T. Hurth and D. Wyler, Phys. Rev. D **54**, 3350 (1996) [hep-ph/9603404].
39. A. Ali and V. M. Braun, Phys. Lett. B **359**, 223 (1995) [hep-ph/9506248].
40. A. Khodjamirian, G. Stoll and D. Wyler, Phys. Lett. B **358**, 129 (1995) [hep-ph/9506242].
41. B. Grinstein and D. Pirjol, Phys. Rev. D **62**, 093002 (2000) [hep-ph/0002216].
42. A. L. Kagan and M. Neubert, Phys. Lett. B **539**, 227 (2002) [hep-ph/0110078].
43. P. Ball and V. M. Braun, Phys. Rev. D **58**, 094016 (1998) [hep-ph/9805422].
44. L. Del Debbio, J. M. Flynn, L. Lellouch and J. Nieves [UKQCD Collaboration], Phys. Lett. B **416**, 392 (1998) [hep-lat/9708008].
45. D. Becirevic, these proceedings.
46. S. Narison, Phys. Lett. B **327** (1994) 354 [hep-ph/9403370].
47. D. Melikhov and B. Stech, Phys. Rev. D **62** (2000) 014006 [hep-ph/0001113].
48. Y. Nir, hep-ph/0208080.
49. A. Stocchi, these proceedings.
50. L. Lellouch, these proceedings.
51. M. Jezabek and J. H. Kühn, Nucl. Phys. B **320**, 20 (1989).
52. A. Ali and G. Hiller, Phys. Rev. D **58**, 074001 (1998) [hep-ph/9803428].

NH ELSEVIER

Proceedings of ICHEP 2002, pp. 631–635
*S. Bentvelsen, P. de Jong, J. Koch and E. Laenen (Editors)*

31st INTERNATIONAL CONFERENCE ON HIGH ENERGY PHYSICS AMSTERDAM

# $D^0$ Mixing, Lifetime Differences, and Hadronic Decays of Charmed Hadrons

D.C.Williams[a] Representing the *BABAR* Collaboration

[a] Santa Cruz Institute of Particle Physics, University of California,
1156 High Street, Santa Cruz, CA 94040, USA

Presented are preliminary results from the *BABAR* Collaboration on charm physics, including limits on $D^0$ mixing from the ratio of lifetimes from $K^-\pi^+$, $K^+K^-$, and $\pi^+\pi^-$ final states (using 47.8 $\text{fb}^{-1}$ of data) and branching ratio, and Dalitz amplitude analyses of the decays $D^0 \to K^0K^-\pi^+$, $\overline{K^0}K^+\pi^-$, and $\overline{K^0}K^+K^-$ (using 22 $\text{fb}^{-1}$ of data). The status of direct measurements of $D^0$ mixing using wrong-sign $D^0 \to K^+\pi^-$ decays is also discussed.

## 1. INTRODUCTION

Because of a large charm cross section, $e^+e^-$ colliders such as PEP-II that run near the $\Upsilon(4S)$ resonance produce copious amounts of charm hadrons. For example, there are approximately 120 million charm pair events contained in the current sample of 91 $\text{fb}^{-1}$ data collected with the *BABAR* experiment. This data, currently the largest sample of charm available to any single experiment, is being used by the *BABAR* Collaboration for a series of charm studies.

Presented in this paper are summaries of two recent measurements: a limit on $D^0$ mixing from the ratio of lifetimes and Dalitz plot analyses of $D^0$ decays to $K_S$ and two charged mesons. In addition, recently discovered issues involving the measurement of $D^0$ mixing using wrong-sign $K^+\pi^-$ decays are also discussed.

The *BABAR* detector, a general purpose, solenoidal, magnetic spectrometer, is described in more detail elsewhere [1].

## 2. $D^0$ MIXING

Like the $K^0$ and $B^0$ meson systems, $D^0$ mixing can occur if two $D^0$ states can be distinguished by either different widths $\Gamma_1$ and $\Gamma_2$ or different masses $M_1$ and $M_2$.[1] That is, if either of the quantities

$$x = \frac{1}{2}\frac{M_1 - M_2}{\Gamma_1 + \Gamma_2} \quad \text{or} \quad y = \frac{\Gamma_1 - \Gamma_2}{\Gamma_1 + \Gamma_2} \tag{1}$$

are non-zero, $D^0$ mesons will mix. Both $x$ and $y$ are expected to be small ($10^{-3}$) within the Standard Model [2]. A value of $x$ whose magnitude $|x|$ is significantly larger than $|y|$ would be evidence of physics beyond the Standard Model.

### 2.1. Lifetime Ratios

The parameter $y$ may be measured [3] using the $D^0$ lifetime $\tau_{K\pi}$ measured for a sample decaying into the Cabibbo-favored, *CP*-mixed $K^-\pi^+$ final state[2] and the lifetime $\tau_{hh}$ from a sample decaying into the Cabibbo-suppressed, *CP*-even $K^-K^+$ or $\pi^-\pi^+$ final states:

$$y = \frac{\tau_{K\pi}}{\tau_{hh}} - 1\ . \tag{2}$$

Since all three of these final states have similar topologies, many systematic uncertainties in the $D^0$ lifetimes cancel in their ratio, making Eq. 2 a particularly sensitive measurement.

To identify $D^0$ candidates, pairs of oppositely charged tracks are combined in a fit which assumes a common point of origin. Charged $K$ and $\pi$ mesons are separated from other tracks using an algorithm that employs information from a

[1] The convention is that the parameters $\{\Gamma_1, M_1\}$ would be associated with the *CP*-even state if *CP* is conserved.

[2] In this paper, statements involving $D^0$ mesons and their decay modes are intended to apply in addition to their charged conjugates.


doi:10.1016/S0920-5632(02)02155-2

Cerenkov detector (DIRC) and $dE/dx$ measured in the tracking chambers. The $D^0$ momentum $p^*$ in the center-of-mass is required to be greater than 2.5 GeV/$c^2$ to remove contamination from $B$ decays. To further reduce backgrounds, the $D^0$ is matched to a $\pi^+$ to form a $D^{*+}$ candidate with a value of $\Delta m = m_{D^0} - m_{D^{*+}}$ within 2 to 3 MeV/$c^2$ of the peak, depending on the quality of $\pi^+$ track.

The lifetime of each $D^0$ sample is calculated using a unbinned maximum likelihood fit. The likelihood is divided into signal and background functions. The signal function is an exponential convoluted by a resolution function consisting of three Gaussians, the first two being proportional to the event-by-event decay time error and the third designed to catch outliers. The background distribution is similar, except that it includes a zero lifetime component. The probability that an event belongs to the signal is determined by comparing the reconstructed $D^0$ mass to a fit to the $D^0$ mass distribution. An example lifetime fit is shown in Fig. 1.

The values of $y$ extracted from the lifetimes of the three $D^0$ samples are listed in Table 1 using 47.8 fb$^{-1}$ of data collected during 2001–2002. Systematic uncertainties are calculated from the Monte Carlo by applying variations in event selection and event weights that correspond to the level of understanding of the detector response.

Table 1
Preliminary results for $y$. The first error is statistical; the second, systematic.

| Decay | y (%) |
|---|---|
| $D^0 \to K^+K^-$ | $1.6 \pm 1.2\ ^{+0.6}_{-0.7}$ |
| $D^0 \to \pi^+\pi^-$ | $1.0 \pm 1.7\ ^{+1.2}_{-1.4}$ |
| Combined | $1.4 \pm 1.0\ ^{+0.6}_{-0.7}$ |

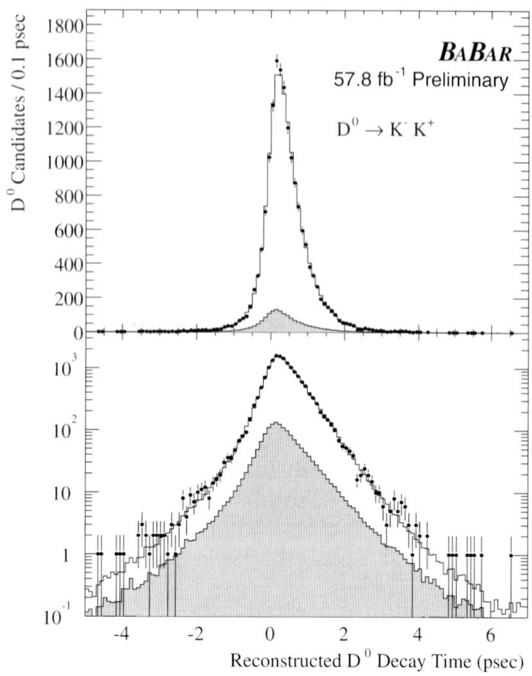


Figure 1. The fit to the reconstructed $D^0$ lifetime for the $K^-K^+$ decay mode for all events including the $D^0$ mass sidebands. The white histogram represents the result of the unbinned maximum likelihood fit described in the text. The gray histogram is the portion assigned to the background.

## 2.2. Wrong Sign $K^+\pi^-$ Decays

The selection of wrong-sign $D^0 \to K^+\pi^-$ decays is similar to the method described in the previous section. The $\Delta m$ distribution of the resulting sample is shown in Fig. 2 and consists of approximately 600 signal events.

If *CP* is violated, the decay time distribution for $D^0$ mesons does not need to match that of their anti-particles $\overline{D^0}$. Each of these time distributions depend upon the parameters $|x'^{\pm}|$ and $y'^{\pm}$ [4]:

$$|x'^{\pm}| = |a_M (x' \cos\varphi y' \pm \sin\varphi)| \quad (3)$$

$$y'^{\pm} = a_M (y' \cos\varphi \mp x' \sin\varphi) , \quad (4)$$

where the upper (lower) sign corresponds to $D^0$ ($\overline{D^0}$) mesons, $\{x', y'\}$ are the mixing parameters $\{x, y\}$ rotated by an unknown strong phase, and $a_m = 1$ and $\varphi = 0$ in the limit of *CP* conservation. It has recently been recognized that, as a consequence of Eqs. 3 and 4, there is an eight-fold ambiguity in the values of $\{x', y', \varphi\}$. In particu-

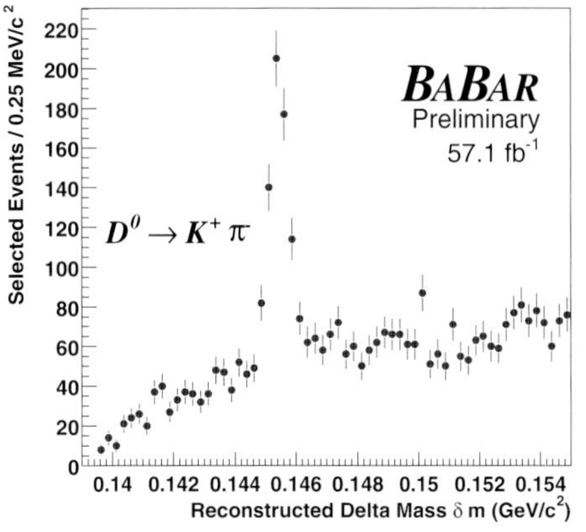


Figure 2. The $\Delta m$ distribution of the $D^0 \to K^+\pi^-$ (wrong-sign) sample.

lar, the existence of one ambiguity $\{x', -y', \varphi+\pi\}$ prevents the measurement of the sign of $y'$.

## 3. THREE BODY $D^0$ DECAYS

The Dalitz plot analysis is the most complete method of studying the dynamics of three-body charm decays. Some of the cleanest such samples are those involving a $K_S$ and two charged mesons: $D^0 \to K_S h^+ h^-$, where $h$ is either a pion or kaon.

To identify these $D^0$ decays, $K_S$ candidates are selected by combining all positive and negative tracks. False $K_S$ candidates are removed by requiring a flight distance of at least 0.4 cm. Particle identification is used to select a clean sample of $K^+$ candidates. A $D^*$ tag with a value of $\Delta m$ within $2\sigma$ of the peak is used to remove background and distinguish between $D^0$ and $\overline{D^0}$.

The selection efficiencies for the four $D^0$ decay modes are determined by a sample of Monte Carlo events in which each decay mode is generated according to uniform phase space (such that the Dalitz plot is uniformly populated). The Dalitz distribution of events selected from these samples are used to calculate the acceptance of each event in the data. Background levels are estimated by fitting the $D^0$ mass distribution to a Gaussian on top of a linear background. Listed in Table 2 are the resulting branching ratios, based on a data sample of 22 fb$^{-1}$ of data collected during 2001.

Table 2
Branching ratios relative to $D^0 \to \overline{K^0}\pi^+\pi^-$. The first error is statistical; the second, systematic. All numbers are preliminary.

| Channel | Branching Ratio (%) |
|---|---|
| $D^0 \to K^0 K^- \pi^+$ | $8.32 \pm 0.29 \pm 0.56$ |
| $D^0 \to \overline{K^0} K^+ \pi^-$ | $5.68 \pm 0.25 \pm 0.41$ |
| $D^0 \to \overline{K^0} K^- K^+$ | $16.30 \pm 0.37 \pm 0.27$ |

An unbinned maximum likelihood fit is applied to the Dalitz distribution of $D^0 \to K^0 K^- \pi^+$, $\overline{K^0} K^+ \pi^-$, and $\overline{K^0} K^+ K^-$ in order to determine the relative amplitudes and phases of intermediate resonant and nonresonant (NR) states. Using a method employed by ARGUS [5] and CLEO [6], each resonant state is represented by an amplitude that is a product of a complex Breit-Wigner function and an angular function. The likelihood function consists of the sum of these amplitudes, normalized by the selection efficiency, and with a separate term for a flat contribution from background, the relative size of which is determined by the sidebands in a fit to the $D^0$ mass distribution.

The parameters of the resonant states are held fixed in the fit, with the exception of the width of the $\phi(1020)$ meson which is allowed to vary in order to approximately account for detector resolution. The $f_0(980)$ and $a_0(980)$ are described using coupled-channel Breit-Wigner functions with parameters taken from the CERN/WA76 [7] and the Crystal Barrel [8] experiments, respectively. The parameters of the $K_0^*(1430)$ ($m = 1.435$ GeV/$c^2$, $\Gamma = 279$ GeV/$c^2$) are taken from a recent reanalysis of LASS data [9]. The NR contribution is represented by a constant term.

The fraction and relative phase $\phi$ of each contribution to the fits are listed in Tables 3–5. The Dalitz plots and their projections are shown in Fig. 3. Systematic errors in the fractions are evaluated by varying the assumptions in the fits including removing the Blatt-Weisskopf terms and assuming a uniform acceptance.

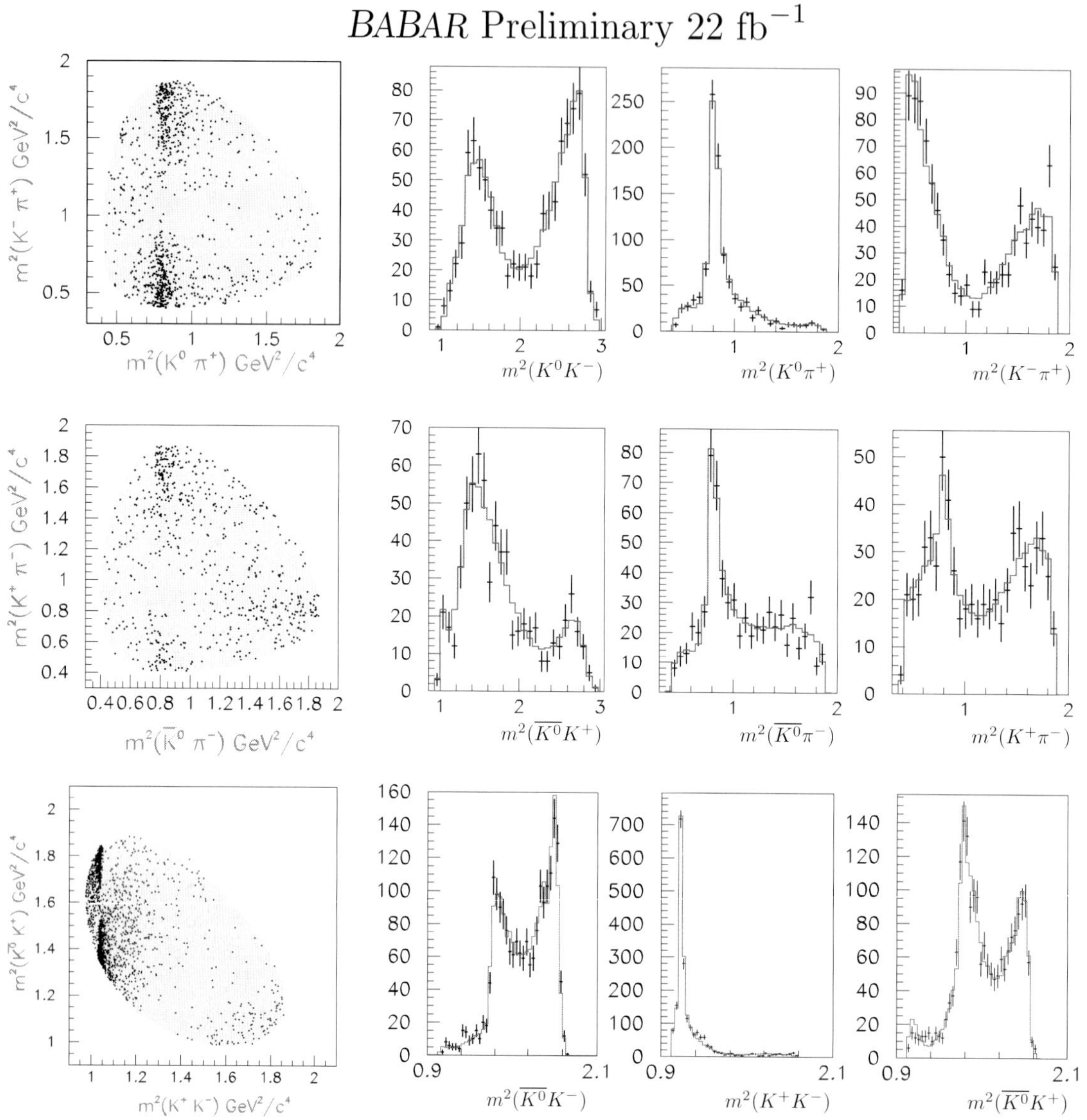


Figure 3. Dalitz plots and projections for (top row) $D^0 \to K^0K^-\pi^+$, (middle row) $D^0 \to \overline{K^0}K^+\pi^-$, and (bottom row) $D^0 \to \overline{K^0}K^+K^-$. On the left are the Dalitz plots, with the kinematic region indicated in yellow. On the right are the three $m^2$ projections for (points) data and (histogram) the Dalitz plot fit.

Table 3
Preliminary results from the Dalitz plot analysis of $D^0 \to K^0 K^- \pi^+$. The first error is statistical; the second, systematic.

| Final state | Fraction (%) | $\phi$ (degrees) |
|---|---|---|
| $\overline{K_0^{*0}}(1430)\ K^0$ | $4.8 \pm 1.4 \pm 1.6$ | $52 \pm 27$ |
| $\overline{K_1^{*0}}(892)\ K^0$ | $0.8 \pm 0.5 \pm 0.1$ | $175 \pm 22$ |
| $\overline{K_1^{*0}}(1680)\ K^0$ | $6.9 \pm 1.2 \pm 1.0$ | $-169 \pm 16$ |
| $\overline{K_2^{*0}}(1430)\ K^0$ | $2.0 \pm 0.6 \pm 0.1$ | $51 \pm 18$ |
| $K_0^{*+}(1430)\ K^-$ | $13.3 \pm 3.5 \pm 3.9$ | $-41 \pm 25$ |
| $K_1^{*+}(892)\ K^-$ | $63.6 \pm 5.1 \pm 2.6$ | 0 |
| $K_1^{*+}(1680)\ K^-$ | $15.6 \pm 3.0 \pm 1.4$ | $-178 \pm 10$ |
| $K_2^{*+}(1430)\ K^-$ | $13.8 \pm 2.6 \pm 7.9$ | $-52 \pm 7$ |
| $a_0^-(980)\ \pi^+$ | $2.9 \pm 2.3 \pm 0.7$ | $-100 \pm 13$ |
| $a_0^-(1450)\ \pi^+$ | $3.1 \pm 1.9 \pm 0.9$ | $31 \pm 16$ |
| $a_2^-(1310)\ \pi^+$ | $0.7 \pm 0.4 \pm 0.1$ | $-149 \pm 27$ |
| NR | $2.3 \pm 0.5 \pm 5.6$ | $-136 \pm 23$ |
| Sum | $130 \pm 8$ | |

## ACKNOWLEDGEMENTS

We are grateful for the excellent luminosity and machine conditions provided by our PEP-II colleagues. This work is supported by DOE and NSF (USA), NSERC (Canada), IHEP (China), CEA and CNRS-IN2P3 (France), BMBF (Germany), INFN (Italy), NFR (Norway), MIST (Russia), and PPARC (United Kingdom). Individuals have received support from the Swiss NSF, A. P. Sloan Foundation, Research Corporation, and Alexander von Humboldt Foundation.

## REFERENCES

1. The *BABAR* Collaboration, B. Aubert *et al.*, Nucl. Instr. Methods **A479** (2002) 1.
2. F. Buccella, M. Lusignoli, and A. Publiese, Phys. Lett. **B379** (1996) 249; F. Buccella *et al.*, Phys. Rev. **D51** (1995) 3478.
3. E. Golowich and S. Pakvasa, Phys. Lett. **B505** (2001) 94.
4. K. Hagiwara *et al.*, Phys. Rev. **D66** (2002) 010001.
5. H. Albrecht *et al.*, Phys. Lett. **B308** (1993) 435.
6. S. Kopp *et al.*, Phys. Rev. **D63** (2001) 092011.
7. T.A. Armstrong *et al.*, Z. Phys. **C51** (1991) 351.
8. A. Abele *et al.*, Phys. Rev. **D57** (1998) 3860.
9. D. Aston *et al.*, Nucl. Phys. **B296** (1998) 493. D. Aston and B. Dunwoodie, private communication.

Table 4
Preliminary results from the Dalitz plot analysis of $D^0 \to \overline{K^0} K^+ \pi^-$. The first error is statistical; the second, systematic.

| Final state | Fraction (%) | $\phi$ (degrees) |
|---|---|---|
| $K_0^{*0}(1430)\ \overline{K^0}$ | $26.0 \pm 16.1 \pm 3.3$ | $-38 \pm 22$ |
| $K_1^{*0}(892)\ \overline{K^0}$ | $2.8 \pm 1.4 \pm 0.5$ | $-126 \pm 19$ |
| $K_1^{*0}(1680)\ \overline{K^0}$ | $15.2 \pm 11.9 \pm 0.5$ | $161 \pm 9$ |
| $K_2^{*0}(1430)\ \overline{K^0}$ | $1.7 \pm 2.5 \pm 0.2$ | $53 \pm 38$ |
| $K_0^{*-}(1430)\ K^+$ | $2.4 \pm 8.2 \pm 1.0$ | $-142 \pm 115$ |
| $K_1^{*-}(892)\ K^+$ | $35.6 \pm 7.7 \pm 2.3$ | 0 |
| $K_1^{*-}(1680)\ K^+$ | $5.1 \pm 5.7 \pm 1.1$ | $124 \pm 27$ |
| $K_2^{*-}(1430)\ K^+$ | $1.0 \pm 1.0 \pm 0.2$ | $-26 \pm 38$ |
| $a_0^+(980)\ \pi^-$ | $15.1 \pm 12.5 \pm 0.6$ | $-160 \pm 42$ |
| $a_0^+(1450)\ \pi^-$ | $2.2 \pm 2.7 \pm 1.2$ | $148 \pm 25$ |
| NR | $36.6 \pm 25.8 \pm 2.7$ | $-172 \pm 13$ |
| Sum | $144 \pm 37$ | |

Table 5
Preliminary results from the Dalitz plot analysis of $D^0 \to \overline{K}^0 K^+ K^-$. The first error is statistical; the second, systematic.

| Final state | Fraction(%) | $\phi$ (degrees) |
|---|---|---|
| $\overline{K^0}\ \phi$ | $45.4 \pm 1.6 \pm 1.0$ | 0 |
| $\overline{K^0}\ a_0^0(980)$ | $60.9 \pm 7.5 \pm 13.3$ | $109 \pm 5$ |
| $\overline{K^0}\ f_0(980)$ | $12.2 \pm 3.1 \pm 8.6$ | $-161 \pm 14$ |
| $a_0(980)^+\ K^-$ | $34.3 \pm 3.2 \pm 6.8$ | $-53 \pm 4$ |
| $a_0(980)^-\ K^+$ | $3.2 \pm 1.9 \pm 0.5$ | $-13 \pm 15$ |
| NR | $0.4 \pm 0.3 \pm 0.8$ | $40 \pm 44$ |
| Sum | $155 \pm 9$ | |

Proceedings of ICHEP 2002, pp. 636–640
*S. Bentvelsen, P. de Jong, J. Koch and E. Laenen (Editors)*

ELSEVIER

31st INTERNATIONAL CONFERENCE ON HIGH ENERGY PHYSICS AMSTERDAM

# Meson lifetimes, decays, mixing and CPV in FOCUS

S.Malvezzi [a] *

[a]I.N.F.N, Milan, Via Celoria 16
20133 Milan, Italy

In recent years charm physics has encountered renovated interest; the excellent quality of the data and high statistics allow for unprecedented sensitivity and sophisticated studies never possible before. Investigation of decay dynamics in both the hadronic and semileptonic sectors reveals manifestation of quantum mechanical interference and phase effects that allow probing FSI and possible CPV effects. Lifetime measurements at a level of 1% or better can shed light onto the non-spectator contribution to the charm decays, and mixing could represent a possible window to new physics. It is in this context that the results selected from the Focus experiment here have to be accommodated.

## 1. The $D^+, D_s \to K^-K^+K^+$ channels

We present the first observation of the singly Cabibbo suppressed and the first clear observation of the doubly Cabibbo suppressed decays of the $D_s$ and $D^+ \to K^-K^+K^+$ [1]. With a yield of $31.4 \pm 7.4$ and $65.5 \pm 15.0$ events respectively we obtained a branching ratio of $\frac{\Gamma(D_s \to K^-K^+K^+)}{\Gamma(D_s \to K^-K^+\pi^+)} = (8.62 \pm 2.04 \pm 2.07) \times 10^{-3}$ and $\frac{\Gamma(D^+ \to K^-K^+K^+)}{\Gamma(D^+ \to K^-\pi^+\pi^+)} = (9.49 \pm 2.17 \pm 0.22) \times 10^{-4}$.

The DCSD is particularly interesting since it requires either the intervention of resonances coupling to $\pi\pi$ and $KK$ simultaneously or weak-annihilation. Unfortunately only a complete Dalitz analysis could point towards a likely interpretation, however the statistics is still too poor.

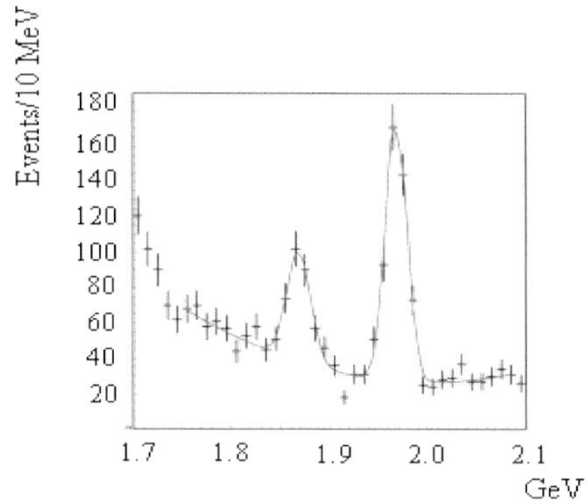


Figure 1. $K\pi\pi$ invariant mass plot.

In contrast, the statistics is sufficient to perform a Dalitz analysis of other suppressed or heavily suppressed modes.

## 2. Preliminary amplitude analysis of some three-body decays

Focus has indeed reconstructed the singly and doubly Cabibbo suppressed $K^+\pi^+\pi^-$ decay of the $D_s$ and $D^+$ (fig. 1). The signal yield of about 500 events of the former allows for the first Dalitz analysis of this decay; the good signal-to-noise ratio ($\sim 1$) of the latter, whose yield is

*On behalf of the FOCUS collaboration: J.Link, M.Reyes, P.M.Yager (**UC DAVIS**); J.Anjos,I.Bediaga, C.Gobel, J.Magnin, A. Massafferri, J.M. de Miranda, I.M.Pepe, A.C. dos Reis, (**CPBF, Rio de Janeiro**); S.Carrillo, E.Casimiro, A.Sánchez-Hernández, C.Uribe, F.Vasquez (**CINVESTAV, México City**); L.Agostino, L.Cinquini, J.P.Cumalat, B.O'Reilly, J.E.Ramirez, I.Segoni (**CU Boulder**); J.N.Butler, H.W.K.Cheung, G.Chiodini, I.Gaines, P.H.Garbincius, L.A.Garren, E.Gottschalk, P.H.Kasper, A.E.Kreymer, R.Kuschke (**Fermilab**); L.Benussi, S.Bianco, F.L.Fabbri, A.Zallo (**INFN Frascati**); C.Cawlfield, D.Y.Kim, K.S.Park, A.Rahimi, J.Wiss (**UI Champaign**); R.Gardner, A.Kryemadhi (**IU Bloomington**); K.H.Chang, Y.S.Chung, J.S.Kang, B.R.Ko, J.W.Kwak, K.B.Lee (**Korea University, Seoul**); K.Cho, H.Park (**Kyungpook National University, Taegu**); G.Alimonti, S.Barberis, A.Cerutti, M.Boschini, P.D'Angelo, M.DiCorato, P.Dini, L.Edera, S.Erba, M.Giammarchi, P.Inzani, F.Leveraro, D.Menasce, M.Mezzadri, L.Moroni, D.Pedrini, C.Pontoglio, F.Prelz, M.Rovere, S.Sala (**INFN and Milano**); T.F.Davenport III (**UNC Asheville**); V.Arena, G.Boca, G.Bonomi, G.Gianini, G.Liguori, M.M.Merlo, D.Pantea, S.P.Ratti, C.Riccardi, P.Vitulo (**INFN and Pavia**); H.Hernandez, A.M.Lopez, H.Mendez, L.Mendez, E.Montiel, D.Olaya, A.Paris, J.Quinones, C.Rivera, W.Xiong, Y.Zhang (**Mayaguez, Puerto Rico**); J.R.Wilson (**USC Columbia**); T.Handler, R.Mitchell (**UT Knoxville**); D.Engh, M.Hosack, W.E.Johns, M.Nehring, P.D.Sheldon, K.Stenson, E.W.Vaandering, M.Webster (**Vanderbilt**); M.Sheaff (**Wisconsin, Madison**)


doi:10.1016/S0920-5632(02)02149-7

Table 1
$D^+$ and $D_s \to K^+\pi^+\pi^-$ decay fit fractions and phases.

| $D^+ \to K^+\pi^+\pi^-$ | | | $D_s \to K^+\pi^+\pi^-$ | |
|---|---|---|---|---|
| channel | fit fraction | phase | fit fraction | phase |
| $NR$ | $9 \pm 5$ % | $(-6 \pm 16)^\circ$ | $18 \pm 4\%$ | $(34 \pm 7)^\circ$ |
| $\bar{K}^{*0}(892)$ | $43 \pm 7$ % | $(208 \pm 16)^\circ$ | $22 \pm 3\%$ | $(163 \pm 7)^\circ$ |
| $K(1410)$ | $12 \pm 8$ % | $(133 \pm 23)^\circ$ | $14 \pm 5\%$ | $(-10 \pm 7)^\circ$ |
| $K_2(1430)$ | $6 \pm 3$ % | $(48 \pm 27)^\circ$ | | |
| $K^*(1430)$ | | | $14 \pm 6$ % | $(68 \pm 7)^\circ$ |
| $K^*(1680)$ | $22 \pm 10$ % | $(2 \pm 20)^\circ$ | | |
| $\rho(770)$ | $51 \pm 10$ % | $0\,(fixed)$ | $40 \pm 4$ % | $0\,(fixed)$ |
| $f_0(980)$ | $9 \pm 5$ % | $(73 \pm 31)^\circ$ | | |
| $f_2(1270)$ | | | $2 \pm 1$ % | $(33 \pm 21)^\circ$ |
| $\rho(1450)$ | $10 \pm 5$ % | $(-113 \pm 15)^\circ$ | $8 \pm 2$ % | $(219 \pm 14)^\circ$ |

of about 200 events, reveals clear resonant decay substructures (fig. 2). The results of the Dalitz fits are shown is Table 1 [2]. Both the $D_s$ and $D^+$ are dominated by $\rho$ and $\bar{K}^{*0}$ in a relatively real phase-shift configuration indicating a non-prominent role of FSI. The scenario is quite different for the $D^+ \to K^-K^+\pi^+$ (fig 3). The Dalitz plot, fig. 4, is highly structured, dominated by $\phi$ and $K^{*0}$ with a visible accumulation of events in the $K\pi$ high-mass region.

The striking asymmetry in the $\bar{K}^{*0}$ lobes can indeed be interpreted as an interference term between the $\bar{K}^{*0}$ itself and a broad, nearly constant, scalar. When a complete Dalitz analysis is carried out a 60% $K^*(1430)$ fraction is returned by the fit, with a relatively imaginary phase-shift, $\sim 62^\circ$, w.r.t. the $\bar{K}*^{0}$, suggesting a major contribution of the FSI effects.

## 3. CPV and Dalitz analysis

The amplitude analysis also allows for a new method of looking at CPV effects. A Dalitz analysis gives the full observation of a three-body decay, providing access to the coefficients and phases of the various channels. CP effects are intimately connected to phase phenomena; the comparison between the relative phases of the CP-conjugate states for the various resonant channels can address CPV effects. Each measured phase in the Dalitz plot can be interpreted as a sum of two components, one CP conserving, the other CP violating. Under CP conjugation the first does not switch sign, while the second does. The $D^+ \to K^-K^+\pi^+$ channel is a good candidate to observe CP effects. This decay is Cabibbo suppressed and thus contains two diagrams contributing to the final state, spectators and penguins; it is also dominated by strong effects, as discussed above, and thus satisfies the two necessary requirements for detecting a CP asymmetry.

The study proceeds through a complete Dalitz analysis of the $D^+/D^-$ split samples. The very preliminary results on our $D^+/D^-$ coefficients and phases, shown in fig. 5, do not point to any CPV manifestation. The relative phases are measured with an accuracy of about 5–10°. The errors are affected by a large correlation among the various resonant components necessary to fit the Dalitz plot. We believe that the more robust formalism that we are adopting now, namely the K-matrix approach, will allow us to perform a good fit with fewer resonances, thus ameliorating the accuracy of our phase measurements.

## 4. New results in the semileptonic sector

Decays in the semileptonic sector also reveal the presence of quantum mechanical effects. Of particular relevance is the study of

[2] For a description of the adopted formalism see Frabetti *et al.* Phys. Lett. B331 (1994) 217; S. Malvezzi HQ96 proceedings, Frascati Series Vol. VII, (pp. 183-195)

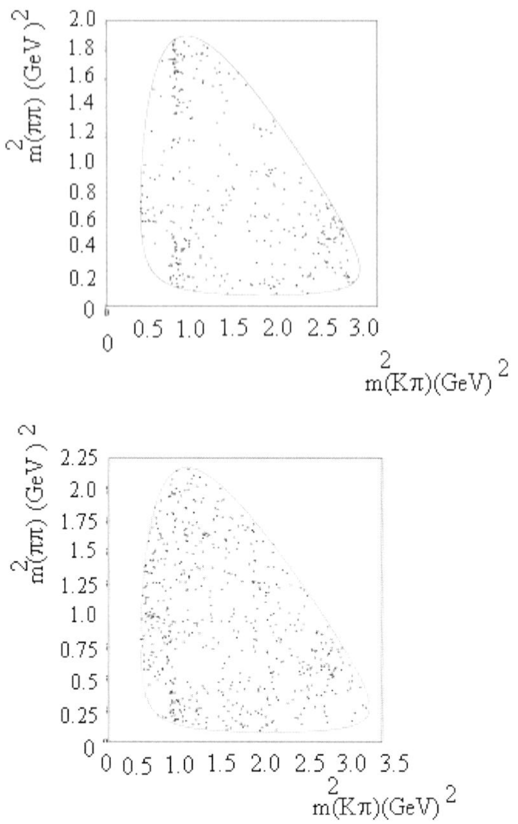


Figure 2. $D^+$ and $D_s \to K^+\pi^-\pi^+$ Dalitz plot.

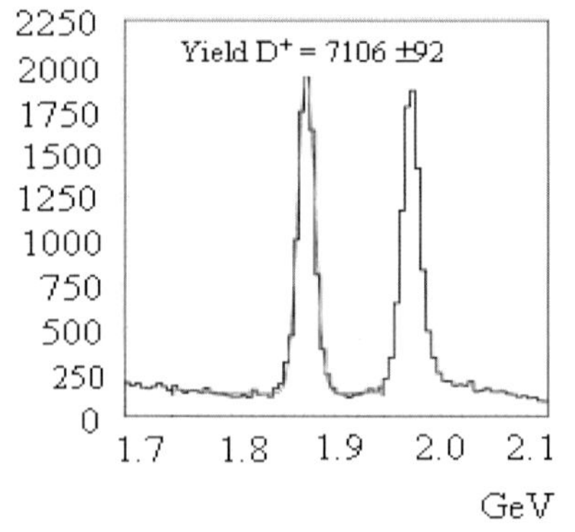


Figure 3. KK$\pi$ invariant mass plot.

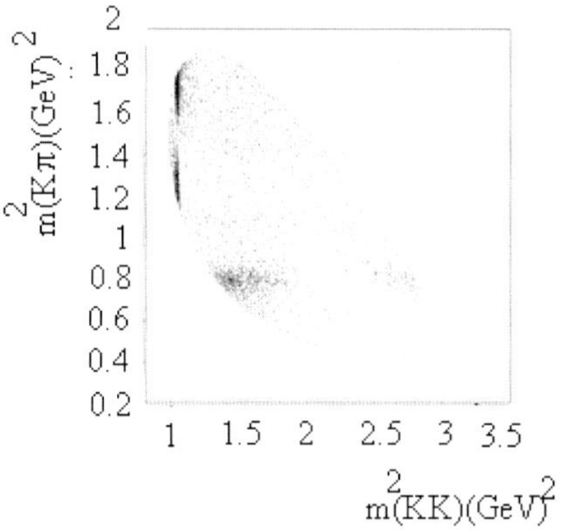


Figure 4. KK$\pi$ Dalitz plot.

$D^+ \to K^-\pi^+\mu^+\nu$ [2]. The Focus mass spectrum appears, as expected, dominated by $\bar{K}^{*0}(892)$; yet an unexpected forward-backward asymmetry was exposed (fig. 7) in the $\cos\theta_V$ variable, defined in fig. 6. The asymmetry is striking below the pole and essentially absent above the pole: it thus suggests quantum-mechanical interference between the Breit-Wigner amplitude describing the $\bar{K}^{*0}(892)$ and a broad or nearly constant $s$-wave amplitude. We tried to describe this behaviour of the decay distribution for $D^+ \to K^-\pi^*\mu^+\nu$ through a simple model. We wrote it in terms of the three helicity-basis form factors: $H_+, H_0, H_-$ and added the $s$-wave amplitude modelled as a constant, with modulus $A$ and phase $\delta$. Angular momentum conservation restricts the $s$-wave contribution to the $H_0$ piece that describes the amplitude for having the virtual $W^+$ in a zero helicity state relative to its momentum vector in the $D^+$ rest-frame.

The shape of the $\cos\theta_V$ term versus $m_{K\pi}$ turns out to be a strong function of the interfering $s$-wave amplitude phase $\delta$. Fig. 8 shows how the $\cos\theta_V$ distribution (properly weighted) is consistent with a constant $s$-wave amplitude of the form $0.36\exp(i\pi/4)\,(\mathrm{GeV})^{-1}$, although, at this stage of the analysis, the solution is not unique; alterna-

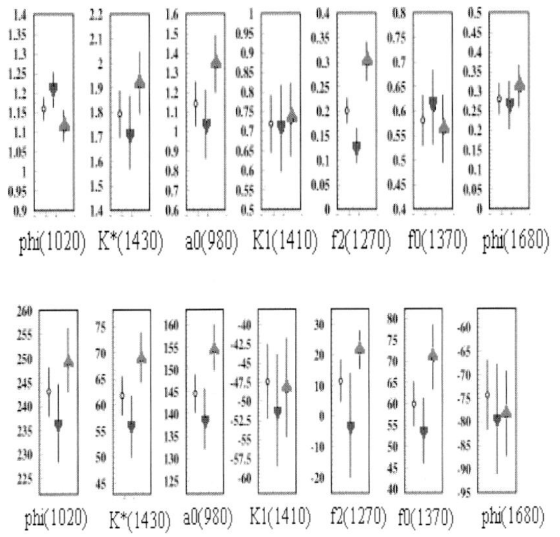


Figure 5. Amplitude coefficient and relative phases (in degrees) of all the resonant contributions of the $D^+ \to K^- K^+ \pi^+$ decay: the circles are for the full sample, upward triangles for $D^-$ and downward triangles for $D^+$.

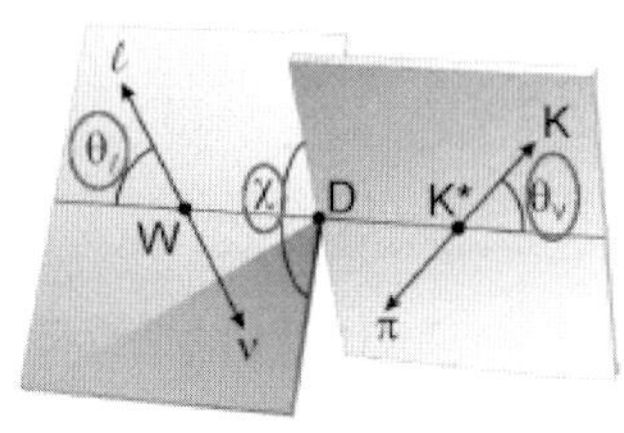


Figure 6. The definition of the kinematic variables for the $D^+ \to \bar{K}^{*0} \mu^+ \nu$.

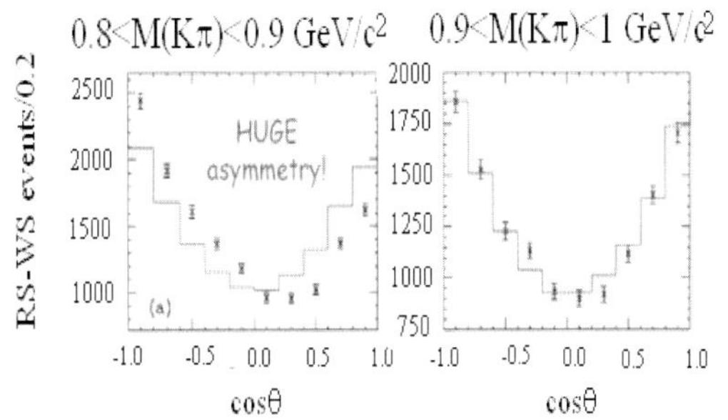


Figure 7. The $\cos\theta_V$ distribution.

tive modellings of the $s$-wave amplitude, as shown in fig. 8, could also fit the data.

A more precise determination can be obtained through a complete analysis of the form factor ratios $r_v$ and $r_2$ [3]. We have found $r_v = 1.504 \pm 0.057 \pm 0.039$ and $r_2 = 0.875 \pm 0.049 \pm 0.064$. For the $s$-wave we obtained an amplitude modulus of $A = 0.330 \pm 0.022 \pm 0.015\,\mathrm{GeV}^{-1}$ and a phase $\delta = 0.68 \pm 0.07 \pm 0.05\,\mathrm{rad}$, in reasonable agreement with our very informal previous estimate of $A = 0.36 \exp(i\pi/4)\,\mathrm{GeV}^{-1}$. The inclusion of the $s$-wave amplitude dramatically improved the quality of our form factor fits but created only minor shifts in the resulting form factor values. A measurement of the branching ratio $\frac{\Gamma(D^+ \to \bar{K}^{*0} \mu^+ \nu)}{\Gamma(D^+ \to K^- \pi^+ \pi^+)} = 0.602 \pm 0.01(\mathrm{stat.}) \pm 0.021(\mathrm{syst.})$ was then performed including interference with this broad scalar resonance. To complete the semileptonic study, we performed an analysis of the $D_s^+ \to \phi \mu^+ \nu$ final state. The $\frac{\Gamma(D_s^+ \to \phi \mu^+ \nu)}{\Gamma(D_s^+ \to \phi \pi^+)}$ was measured to be $0.54 \pm 0.033(\mathrm{stat.}) \pm 0.048(\mathrm{syst.})$ [4].

## 5. Charm meson lifetimes

The very high statistics of our photoproduced samples of charm mesons allows for a very precise measurements of $D^0$ and $D^+$ lifetimes. Using about 210000 and 110000 events for the $D^0$ and $D^+$ respectively, we obtained the following values: $409.6 \pm 1.1(\mathrm{stat.}) \pm 1.5(\mathrm{syst.})$ fs for the $D^0$ and $1039.4 \pm 4.3(\mathrm{stat.}) \pm 7.0(\mathrm{syst.})$ fs for the $D^+$ [5]. Our results significantly decrease the errors on the current world average values. Our determination of the ratio $\tau(D^+)/\tau(D^0) = 2.538 \pm 0.023$ to better than 1% and the inclusive semileptonic branching ratios $D^+ \to eX = 17.2 \pm 1.9\%$ and

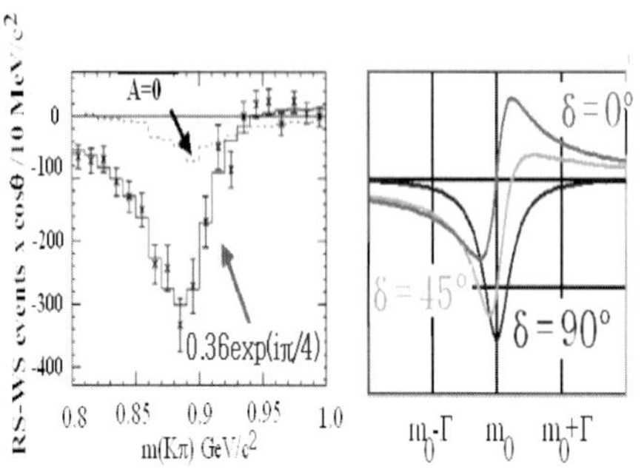


Figure 8. Asymmetry distribution in $K\pi$ invariant mass. The dashed line is the MC with no $s$-wave amplitude, the solid line is the MC with an $s$-wave amplitude of $0.36\exp(i\pi/4)\,(\mathrm{GeV})^{-1}$. Different modellings of the $s$-wave are also shown.

$D^0 \to eX = 6.87 \pm 0.28$ confirm that the total decay-width difference must be due to the hadronic sector. Focus is also studying the $D_s$ lifetime; a measurement with the accuracy of 1% or better will allow us to gauge the role of the annihilation/exchange diagrams and the possible interference effects with the spectator process in the charm decay [6]. Focus has analysed 50% of its statistics in the $D_s^+ \to \phi\pi^+$ channel for a sample of $5668 \pm 95$ events. We obtained a determination of $\tau(D_s) = 506 \pm 8\,\mathrm{fs}$, to be compared with the PDG02 value of $490 \pm 9\,\mathrm{fs}$, and a ratio $\tau(D_s)/\tau(D^0) = 1.23 \pm 0.02$. A statistical improvement by a factor of four is expected from Focus by completing the $D_s \to \phi\pi$ analysis and including the $D_s \to \bar{K}^{*0}K$ mode.

## 6. New mixing results

The mixing effects are parametrized by two dimensionless amplitudes $x = \Delta M/\Gamma$ and $y = \Delta\Gamma/2\Gamma$ with $\Delta M$ mass difference between the two mass eigenstates, $\Delta\Gamma$ the width difference and $\Gamma$ the average width. Recent measurements in the charm sector hint at a possible mixing effect ($R_{\mathrm{mix}} = 1/2(x^2 + y^2)$), at the edge of experimental sensitivity. Focus has already published its lifetime difference result [7] $y_{CP} = \frac{\tau(D^0\to K^-\pi^+)}{\tau(D^0\to K^-K^+)} - 1 = (3.42 \pm 1.39 \pm 0.74)\%$ and is now working on the hadronic and semileptonic decays. In particular, the experiment has recently "opened the box" of a blind analysis of the $D^0 \to K^+\mu^-\nu_\mu$ channel obtaining a very preliminary results of $R_{\mathrm{mix}} < 0.0012$ at 95 % C.L., which includes the statistical error only. This new Focus semileptonic measurement along with the preliminary DCSD $D^0 \to K^+\pi^-$ [8], can be translated into $x', y'$: $x' = x\cos\delta + y\sin\delta; y' = y\cos\delta - x\sin\delta$, where $\delta$ is the unknown strong phase between the DCS decay and the CF $D^0 \to K^-\pi^+$, The allowed $x', y'$ regions are shown in fig. 9. Focus is constraining the CLEO result and pointing at a strong relative phase.

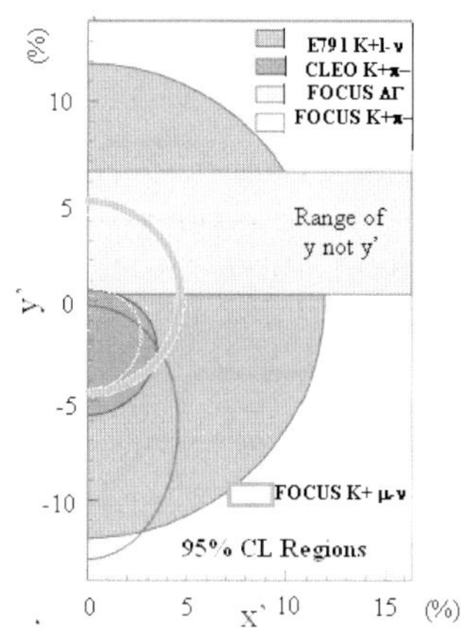


Figure 9. The $x'$–$y'$ $D$-mixing allowed regions.

## REFERENCES


1. J. Link *et al.*, Phys. Lett. B541, 227 (2002).
2. J. Link *et al.*, Phys. Lett. B535, 43 (2002).
3. J. Link *et al.*, Phys. Lett. B544, 89 (2002).
4. J. Link *et al.*, Phys. Lett. B541, 243 (2002).
5. J. Link *et al.*, Phys. Lett. B537, 192 (2002).
6. I.I. Bigi and N.G. Uraltsev, Z. Phys. C62, 623 (1994).
7. J. Link *et al.*, Phys. Lett. B485, 62 (2000).
8. hep-ex/010693.

ELSEVIER

Proceedings of ICHEP 2002, pp. 641–644
*S. Bentvelsen, P. de Jong, J. Koch and E. Laenen (Editors)*

31st INTERNATIONAL CONFERENCE ON HIGH ENERGY PHYSICS AMSTERDAM

# $B$ decays to baryonic final states and $\Omega_c^0$ at Belle

Ruslan Chistov*, for the Belle Collaboration
Institute for Theoretical and Experimental Physics, Moscow, Russia

We report on the results for the baryonic $B$ decays using data collected with the Belle experiment at $e^+e^-$ asymmetric collider KEKB. We studied the exclusive $B$ decays to 2-, 3- and 4-body final states with charmed baryons and observed for the first time $\bar{B}^0 \to \Sigma_c(2455)^{++}\bar{p}\pi^-$ decay. First observations and measured branching fractions are presented for the decay modes $\bar{B} \to D^{(*)0}p\bar{p}$, $B^+ \to p\bar{p}K^+$ and $B^0 \to p\bar{\Lambda}\pi^-$. The upper limits for two-body charmless baryonic $B$ decays to $p\bar{p}, \Lambda\bar{\Lambda}$ and $p\bar{\Lambda}$ are improved. We also provide the measurement of $\Omega_c^0$ mass and semileptonic $\Omega_c^0$ decay at Belle.

## 1. Introduction

Up to now, only a few $B \to$ *baryons* decay modes have been measured experimentally. Previous results of a searching for exclusive modes[1] and the measured momentum spectrum of charmed baryons from $B$ decays[2] suggest the dominance of multybody final states. With the large statistics of $B$ decays accumulated by the Belle experiment we are able to investigate further $B \to$ *baryons* processes. New information will help to understand better the dinamics of $B$ decays and provide the test of theoretical approaches describing these decays.

In this report the $B^0 \to p\bar{\Lambda}\pi^-$ analysis is based on a 60 fb$^{-1}$ data sample. Others use a 32.6 fb$^{-1}$ data sample. The selection criteria and the details of particular analysis can be found through the references in corresponding sections.

## 2. Study of Exclusive $B$ Decays to Charmed Baryons

We studied $B$ decays to $\Lambda_c^+\bar{p}$, $\Lambda_c^+\bar{p}\pi^-$ and $\Lambda_c^+\bar{p}\pi^+\pi^-$ final states[3]. The $\Lambda_c^+$ was reconstructed in $pK^-\pi^+$ mode. Figure 1 presents the $M_{bc}$ and $\Delta E$ distributions for $B$ candidates. The measured branching fractions are: $\mathcal{B}(\bar{B}^0 \to \Lambda_c^+\bar{p}) = (0.12^{+0.10}_{-0.07} \pm 0.02 \pm 0.03)(< 0.31@90\%\text{CL})\times 10^{-4}$, $\mathcal{B}(B^- \to \Lambda_c^+\bar{p}\pi^-) = (1.87^{+0.43}_{-0.40} \pm 0.28 \pm 0.49) \times 10^{-4}$ and $\mathcal{B}(\bar{B}^0 \to \Lambda_c^+\bar{p}\pi^+\pi^-) = (11.0^{+1.2}_{-1.2} \pm 1.9 \pm 2.9) \times 10^{-4}$.

We investigated the resonant substracture by selecting events from the $B$ signal region($|\Delta E| <$ 0.030 GeV and $M_{bc} >$ 5.270 GeV) and look for $\Lambda_c^+\pi^\pm$ mass distributions (see Figure 2). After the background subtraction we made the following measurements: $\mathcal{B}(\bar{B}^0 \to \Sigma_c(2455)^{++}\bar{p}\pi^-) = (2.38^{+0.63}_{-0.55} \pm 0.41 \pm 0.62) \times 10^{-4}$ (first observation), $\mathcal{B}(\bar{B}^0 \to \Sigma_c(2520)^{++}\bar{p}\pi^-) = (1.63^{+0.57}_{-0.51} \pm 0.28 \pm 0.42) \times 10^{-4}$ (first evidence), $\mathcal{B}(\bar{B}^0 \to \Sigma_c(2455)^0\bar{p}\pi^+) < 1.59\times 10^{-4}$ @90% CL, $\mathcal{B}(\bar{B}^0 \to \Sigma_c(2520)^0\bar{p}\pi^+) < 1.21\times 10^{-4}$ @90% CL, $\mathcal{B}(B^- \to \Sigma_c(2455)^0\bar{p}) < 0.93 \times 10^{-4}$ @90% CL and $\mathcal{B}(B^- \to \Sigma_c(2520)^0\bar{p}) < 0.46 \times 10^{-4}$ @90% CL.

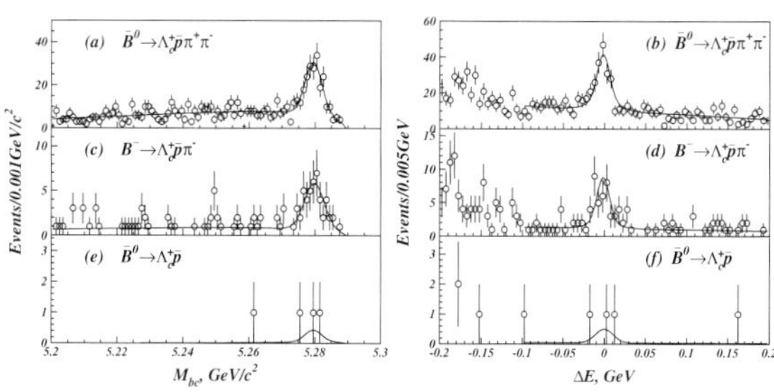


Figure 1. The $M_{bc}$ ((a), (c), (e)) and $\Delta E$ ((b), (d), (f)) distributions for 4-, 3- and 2-body baryonic $B$ decays respectively.

*email: chistov@iris1.itep.ru


doi:10.1016/S0920-5632(02)02152-7

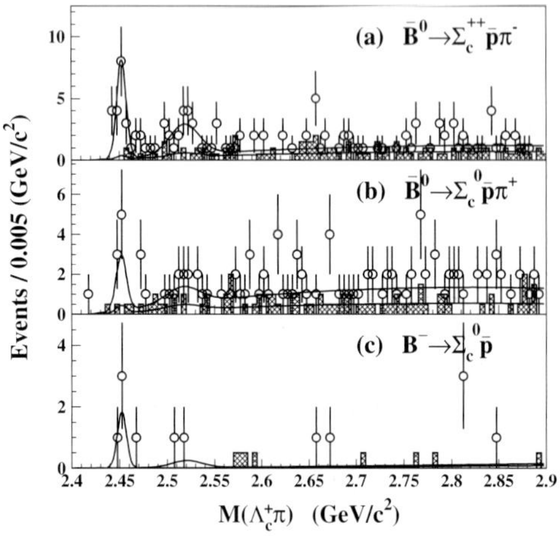


Figure 2. (a): $\Lambda_c^+\pi^+$ and (b): $\Lambda_c^+\pi^-$ invariant mass distribution for $B \to \Lambda_c^+\bar{p}\pi^+\pi^-$ events, (c): $M(\Lambda_c^+\pi^-)$ for $B \to \Lambda_c^+\bar{p}\pi^-$ events.

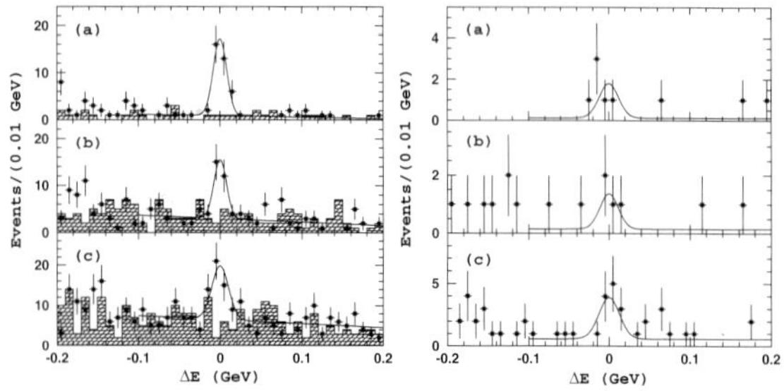


Figure 3. The $\Delta E$ distribution for the $\bar{B}^0 \to D^0 p\bar{p}$ (left) and $\bar{B}^0 \to D^{*0} p\bar{p}$ (right) candidates, where: (a) $D^0 \to K^-\pi^+$, (b) $D^0 \to K^-\pi^+\pi^+\pi^-$, (c) $D^0 \to K^-\pi^+\pi^0$

## 3. Observation of $\bar{B}^0 \to D^{(*)0} p\bar{p}$

For this decay we used the following $D^0$ modes: $K^-\pi^+$, $K^-\pi^+\pi^+\pi^-$ and $K^-\pi^+\pi^0$. For the $D^{*0}$ we used its decay to $D^0\pi^0$. Figure 3 presents the $\Delta E$ distribution for $\bar{B}^0 \to D^{(*)0} p\bar{p}$ for each $D^0$ submode. The superimposed curves are the result of a simultaneous fit which gives the branching fractions for $\bar{B}^0 \to D^0 p\bar{p}$ and $\bar{B}^0 \to D^{*0} p\bar{p}$ decays to be $(1.18 \pm 0.15 \pm 0.16) \times 10^{-4}$ and $(1.20^{+0.33}_{-0.29} \pm 0.21) \times 10^{-4}$ respectively**??**. These are the first observations for color-suppressed baryonic $B$ decays.

## 4. Observation of $B^+ \to p\bar{p}K^+$

To reconstruct this decay we suppress continuum and apply the charm veto against the decays $B^+ \to J/\psi(\eta_c/\psi'/\chi_{c[0,1]})K^+$ followed by $(c\bar{c}) \to p\bar{p})$. The $\Delta E$ distribution for $B^+ \to p\bar{p}K^+$ candidates is presented in Figure 4. The fit gives $42.8^{+10.0}_{-9.6}$ events for signal with $5.6\sigma$ significance. This is a first observation of charmless baryonic $B$ decay. The branching fraction for this decay was extracted to be $(4.3^{+1.1}_{-0.9} \pm 0.5) \times 10^{-6}$[5]. Figure 5 gives the invariant mass distribution for $p\bar{p}$ pair from this decay. The underlying histogram shows the expected $M(p\bar{p})$ distribution from non-resonant $B^+ \to p\bar{p}K^+$ MC simulation. We conclude that this distribution is inconsistent with the observed one from data. The search for the related modes $B^0 \to p\bar{p}K_s^0$ and $B^+ \to p\bar{p}\pi^+$ gives no signal and upper limits were set to be $< 7.2 \times 10^{-6}$ @90% and $< 3.7 \times 10^{-6}$ @90% respectively.

## 5. Observation of $B^0 \to p\bar{\Lambda}\pi^-$

The strategy for this analysis is similar to that in the previous section and the details can be found elsewhere[6]. The $\Delta E$ distribution is shown in Figure 6. The fit gives for signal $24.8^{+6.6}_{-5.9}\pm$ events with $5.1\sigma$ significance. This is the first observation of charmless hyperonic $B$ decay. The branching fraction for this decay was extracted to be $(3.56^{+.86}_{-0.74} \pm 0.43) \times 10^{-6}$. The background subtracted $p\bar{\Lambda}$ invariant mass spectrum for this decay peaks at a low masses (see Figure 7) and inconsistent with the expected $M(p\bar{\Lambda})$ distribution from non-resonant $B^+ \to p\bar{\Lambda}\pi^-$ MC simulation. For the related modes $B^0 \to p\bar{\Lambda}K^-$ and $B^0 \to p\bar{\Sigma}^0\pi^-$ upper limits were set to be $< 1.8 \times 10^{-6}$ @90% and $< 4.1 \times 10^{-6}$ @90% respectively.

## 6. Search for Charmless Two-Body $B^0 \to p\bar{p}$, $B^0 \to \Lambda\bar{\Lambda}$ and $B^+ \to p\bar{\Lambda}$ Decays

Here we present the results of a search for two-body charmless baryonic $B$ decays. In the Standard Model these decays are expected to proceed

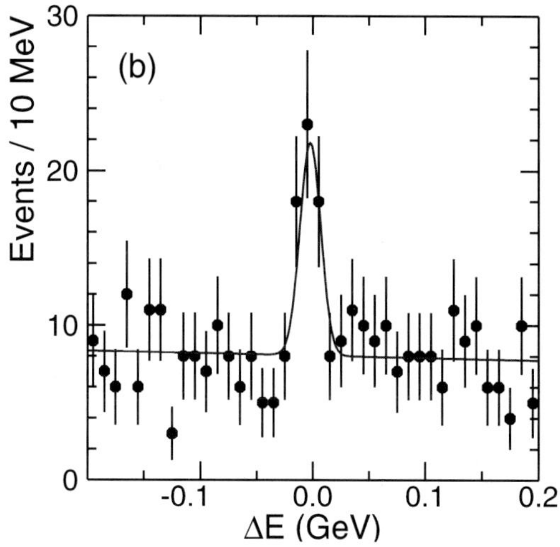


Figure 4. The $\Delta E$ distribution for the $B^+ \to p\bar{p}K^+$ candidates.

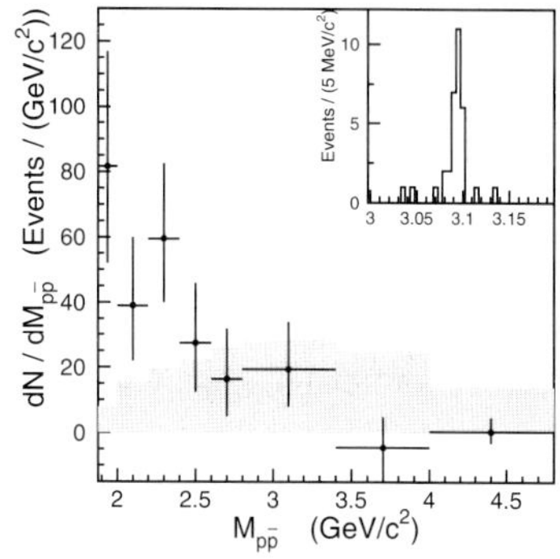


Figure 5. Background subtracted $p\bar{p}$-invariant mass for $B^+ \to p\bar{p}K^+$ decay. The superimposed dashed histogram is expected phase-space distribution (see text).

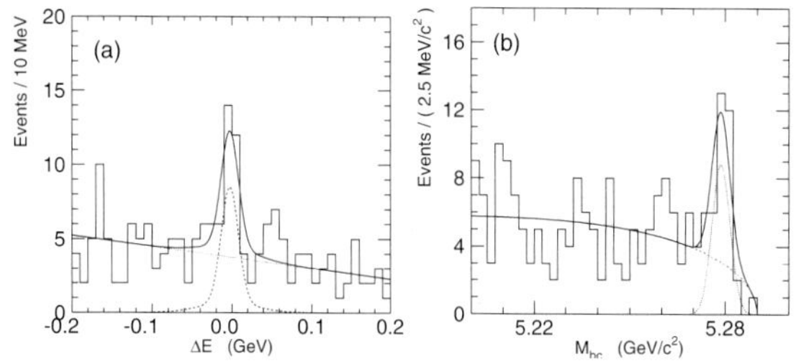


Figure 6. (a): The $\Delta E$ and (b): $M_{bc}$ distributions for the $B^0 \to p\bar{\Lambda}\pi^-$ candidates.

via tree $b \to u$ and penguin $b \to s$ and $b \to d$ transitions. We found no signals and set upper limits( @90% CL): $\mathcal{B}(B^0 \to p\bar{p}) < 1.2 \times 10^{-6}$, $\mathcal{B}(B^0 \to \Lambda\bar{\Lambda}) < 1.0 \times 10^{-6}$ and $\mathcal{B}(B^+ \to p\bar{\Lambda}) < 2.2 \times 10^{-6}$[7]. These limits are improved in comparison with the previously obtained[1].

## 7. Study of $\Omega_c^0$ at Belle

The $\Omega_c^0$ is the least well understand ground state charmed baryon. The measurement of $\Omega_c^0$ mass and its decay modes will test theoretical models which provide quantitative predictions for the properties of heavy baryons.

### 7.1. Measurement of $\Omega_c^0$ Mass

We reconstructed the $\Omega_c^0$ in its decay to $\Omega^-\pi^+$. The invariant mass distribution for the selected $\Omega^-\pi^+$ pairs satisfying the criterion of $x_p(\Omega^-\pi^+) > 0.5$ is presented in Figure 8. The fit gives for $\Omega_c^0$ signal $37.6 \pm 6.9$ events at mass of $2693.7 \pm 1.3^{+1.1}_{-1.0}$ MeV. Our measurement of $\Omega_c^0$ mass is consistent with the CLEO measurement $M(\Omega_c^0) = 2694.6 \pm 2.6 \pm 1.9$ MeV [1].

We measured the product of the production cross section and branching fraction $\sigma(e^+e^- \to \Omega_c^0 X) \times \mathcal{B}(\Omega_c^0 \to \Omega^-\pi^+)$ in the whole momentum interval and in the interval $x_p(\Omega^-\pi^+) > 0.5$ to be $24.5^{+20.0}_{-10.7} \pm 3.4$ fb and $17.3 \pm 3.2 \pm 1.7$ fb respectively.

### 7.2. Observation of $\Omega_c^0 \to \Omega^- l^+ \nu$

The decay $\Omega_c^0 \to \Omega^- e^+ \nu$ has previously been seen by CLEO[8]. Here we present our observation of the $\Omega_c^0$ semileptonic decay in both electron and muon modes. By requiring the electron or muon in the event with the $\Omega^-$ and applying the cut $x_p(\Omega^- l^+) > 0.5$, we observed the excess of $\Omega^-$ signal for rihgt-sign $\Omega^\mp l^\pm$ combinations over wrong-sign $\Omega^\mp l^\mp$ one. The value of $\sigma(e^+e^- \to \Omega_c^0 X) \times \mathcal{B}(\Omega_c^0 \to \Omega^- l^+ \nu)$ was ex-

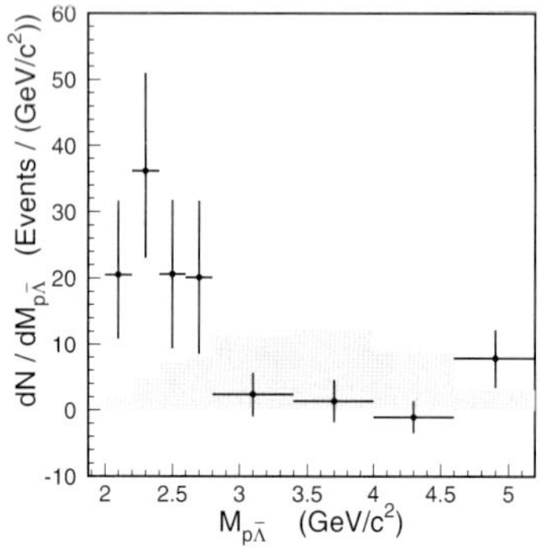


Figure 7. Background subtracted $p\bar{\Lambda}$ invariant mass for $B^0 \to p\bar{\Lambda}\pi^-$ decay. The superimposed dashed histogram is expected phase-space distribution (see text).

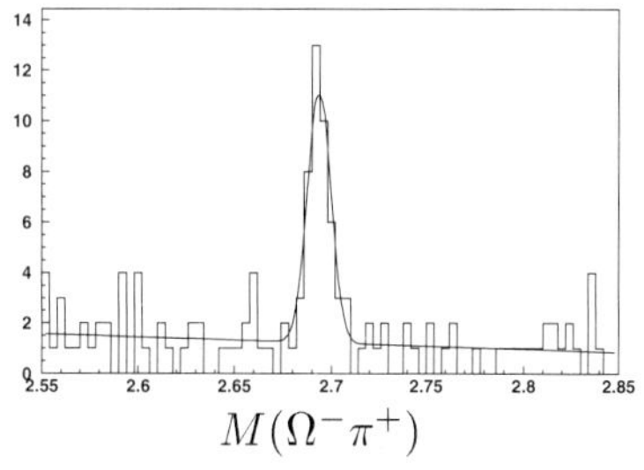


Figure 8. The $\Omega^-\pi^+$ invariant mass spectrum after the requirement $x_p(\Omega^-\pi^+) > 0.5$

tracted to be $20.4 \pm 5.4 \pm 2.7$ fb. The ratio of $\mathcal{B}(\Omega_c^0 \to \Omega^-\pi^+)/\mathcal{B}(\Omega_c^0 \to \Omega^- l^+\nu)$ has been measured to be $0.83 \pm 0.28 \pm 0.14$ which is in marginally agreement with the CLEO result[8]. The details of our analysis for $\Omega_c^0$ mass measurement and $\Omega_c^0$ semileptonic decay can be found elsewhere[9].

## 8. Conclusions

In summary, we studied the exclusive $B$ decays to charmed baryons and observed for the first time the exclusive decay $\bar{B}^0 \to \Sigma_c(2455)^{++}\bar{p}\pi^-$ with the branching fraction of $(2.38^{+0.63}_{-0.55} \pm 0.41 \pm 0.62) \times 10^{-4}$. We also obtain the first evidence for $\bar{B}^0 \to \Sigma_c(2520)^{++}\bar{p}\pi^-$ decay and measured its branching fraction to be $(1.63^{+0.57}_{-0.51} \pm 0.28 \pm 0.42) \times 10^{-4}$. The color-suppressed $\bar{B}^0 \to D^0 p\bar{p}$ and $\bar{B}^0 \to D^{*0} p\bar{p}$ decays have been observed for the first time and the branching fractions were extracted to be $(1.18 \pm 0.15 \pm 0.16) \times 10^{-4}$ and $(1.20^{+0.33}_{-0.29} \pm 0.21) \times 10^{-4}$ correspondingly. For the charmless baryonic and hyperonic $B$ decays we observed $B^+ \to p\bar{p}K^+$ and $B^0 \to p\bar{\Lambda}\pi^-$ decays and measured $\mathcal{B}(B^+ \to p\bar{p}K^+) = (4.3^{+1.1}_{-0.9} \pm 0.5) \times 10^{-6}$, $\mathcal{B}(B^0 \to p\bar{\Lambda}\pi^-) = (3.56^{+.86}_{-0.74} \pm 0.43) \times 10^{-6}$. For decays $B^0 \to p\bar{p}$, $B^0 \to \Lambda\bar{\Lambda}$ and $B^+ \to p\bar{\Lambda}$ we improved upper limits for their branching fractions. Our measurements of $\Omega_c^0$ mass and productions properties were obtained to be consistent with the previous CLEO measurements. We also observed semileptonic decay $\Omega_c^0 \to \Omega^- l^+\nu$. The ratio $\mathcal{B}(\Omega_c^0 \to \Omega^-\pi^+)/\mathcal{B}(\Omega_c^0 \to \Omega^- l^+\nu)$ has been measured to be $0.83 \pm 0.28 \pm 0.14$.

## REFERENCES


1. K. Hagiwara *et al.*, Phys. Rev. D 66 (2002) 010001
2. H.Albrecht *et al.*(ARGUS Collaboration), Phys.Lett.**B210** 263 (1988); G. Crawford *et al.*(CLEO Collaboration), Phys.Rev.D **45**, 752-770, (1992); B. Barish *et al.*(CLEO Collaboration), Phys. Rev. Lett. **79**, 3599-3603, (1997).
3. N.Gabyshev, H.Kichimi, *et al.*(Belle Collaboration), submitted to Phys.Rev.Lett.; hep-ex/0208041.
4. K.Abe, *et al.*(Belle Collaboration), Phys.Rev.Lett. **89**, 151802 (2002).
5. K.Abe, *et al.*(Belle Collaboration), Phys.Rev.Lett. **88**, 181803 (2002).
6. K.Abe, *et al.*(Belle Collaboration), BELLE-CONF-0237.
7. K.Abe, *et al.*(Belle Collaboration), Phys.Rev. D **65**, 091103 (2002).
8. R. Ammar *et al.*(CLEO Collaboration), CLNS 02/1782, CLEO 02-5.
9. K.Abe, *et al.*(Belle Collaboration), BELLE-CONF-0253.

ELSEVIER

Proceedings of ICHEP 2002, pp. 645–647
*S. Bentvelsen, P. de Jong, J. Koch and E. Laenen (Editors)*

31st INTERNATIONAL CONFERENCE ON HIGH ENERGY PHYSICS AMSTERDAM

# Charm and bottom baryon masses in the 1/N expansion

E. Jenkins[a*]

[a]Department of Physics, University of California, San Diego,
9500 Gilman Drive, La Jolla, California 92093, USA

The masses of heavy quark baryons are studied in an expansion in $1/N_c$, $SU(3)$ flavor symmetry breaking, and heavy-quark symmetry breaking. Very accurate model-independent mass relations are obtained for charm and bottom baryons.

## 1. INTRODUCTION

The $1/N_c$ expansion has proven useful for studying the spin-flavor properties of baryons containing light quarks. A spin-flavor symmetry for baryons is defined in large-$N_c$ QCD. Away from the $N_c \to \infty$ limit, $1/N_c$ corrections which break large-$N_c$ baryon spin-flavor symmetry can be classified in terms of the spin and flavor symmetries which remain for finite $N_c$. Since explicit $SU(3)$ flavor symmetry breaking $\epsilon \sim m_s/\Lambda_\chi$ is comparable to $1/N_c = 1/3$ for QCD baryons, a combined expansion in $1/N_c$ and $\epsilon$ is necessary to explain the symmetry-breaking pattern. The $1/N_c$ expansion has had considerable phenomenological success for $qqq$ baryons; $1/N_c$ suppression factors clearly are present in experimental data.

For baryons containing a single heavy quark $Q$ in HQET, there is spin-flavor symmetry in the large-$N_c$ limit as well as in the heavy quark limit [1]. The spin-flavor symmetry of heavy quark baryons with one heavy quark flavor contains a $SU(6)_\ell$ symmetry of the light degrees of freedom in the $N_c \to \infty$ limit, and a $SU(2)_Q$ spin symmetry of the heavy quark in both the $m_Q \to \infty$ and $N_c \to \infty$ limits. The spin-flavor symmetry of the light degrees of freedom is broken by corrections suppressed by factors of $1/N_c$ and $SU(3)$ flavor symmetry breaking, whereas the heavy-quark spin symmetry is broken by terms suppressed by factors of $1/N_c$ and heavy quark symmetry breaking $\delta_Q \sim \Lambda_{\rm QCD}/m_Q$. Note that $SU(3)$ flavor symmetry and heavy quark spin symmetry are better symmetries for baryons than for mesons because violation of spin-flavor symmetry is suppressed by additional factors of $1/N_c$ for baryons.

Heavy quark symmetry for two heavy quark flavors $Q = c$ and $Q = b$ in HQET generalizes to heavy-quark spin-flavor symmetry $SU(4)_Q$, which relates the heavy quark spin-flavor properties of charm and bottom hadrons. Again, heavy quark spin-flavor symmetry is a better symmetry for baryons than for mesons because heavy quark spin-flavor symmetry violation is accompanied by factors of $1/N_c$.

The lowest-lying spin-flavor representation for $Qqq$ baryons consists of a completely symmetric spin-flavor representation of two light quarks combined with a single heavy quark with $J_Q = \frac{1}{2}$. Under light-quark spin and flavor, this representation decomposes into a $J_\ell = 0$ flavor $\bar{\mathbf{3}}$, which consists of the isosinglet $\Lambda_Q(Qud)$ and the isodoublet $\Xi_c(Qsq)$ with $J = J_\ell + J_Q = \frac{1}{2}$, and a $J_\ell = 1$ flavor $\mathbf{6}$, which consists of $\Sigma_Q$, $\Xi'_Q$, $\Omega_Q$ with $J = \frac{1}{2}$ and $\Sigma^*_Q$, $\Xi^*_Q$, $\Omega^*_Q$ with $J = \frac{3}{2}$.

The mass hierarchy of the lowest-lying charm and bottom baryon masses was predicted in a combined expansion in $1/N_c$, $\epsilon$ and $\delta_Q$ in Ref. [1]. Here, the theoretical hierarchy is compared with experiment for charm baryon masses, and the predicted pattern is seen. Bottom baryons are predicted to obey the same mass hierarchy with $1/m_c$ replaced by $1/m_b$.

*Supported in part by the Department of Energy under grant DOE-FG03-97ER40546.


doi:10.1016/S0920-5632(02)02154-0

## 2. $1/N_c$ Expansion

In the $1/m_Q$ expansion of HQET, the mass of a hadron containing a single heavy quark is given by

$$M(H_Q) = m_Q + \bar{\Lambda} - \frac{\lambda_1}{2m_Q} - d_H \frac{\lambda_2}{2m_Q} + \cdots, \quad (1)$$

where

$$\begin{aligned} \lambda_1 &= \langle H_Q(v)|\bar{Q}_v(iD)^2 Q_v|H_Q(v)\rangle, \quad (2) \\ d_H\lambda_2 &= \frac{1}{2} Z_Q \langle H_Q(v)|\bar{Q}_v g G_{\mu\nu}\sigma^{\mu\nu} Q_v|H_Q(v)\rangle, \end{aligned}$$

are the matrix elements of the $1/m_Q$-suppressed operators in the heavy hadron, $d_H = -4\,(J_\ell \cdot J_Q)$, and corrections of order $1/m_Q^2$ have been neglected. Eq. (1) can be applied to mesons and to baryons, but the values of $\bar{\Lambda}$, $\lambda_1$ and $\lambda_2$ will be different in the two cases.

The spin-$\frac{1}{2}$ **3**, and spin-$\frac{1}{2}$ and $\frac{3}{2}$ **6** baryons have masses given by the $1/m_Q$ expansions

$$\begin{aligned} T_Q &= m_Q + \bar{\Lambda}_T - \frac{\lambda_{1T}}{2m_Q} + \cdots, \\ S_Q &= m_Q + \bar{\Lambda}_S - \frac{\lambda_{1S}}{2m_Q} - \frac{4\lambda_{2S}}{2m_Q} + \cdots, \quad (3) \\ S_Q^* &= m_Q + \bar{\Lambda}_S - \frac{\lambda_{1S}}{2m_Q} + \frac{2\lambda_{2S}}{2m_Q} + \cdots, \end{aligned}$$

respectively. Large-$N_c$ spin-flavor symmetry implies that the hadronic matrix elements of these baryons are equal in the $N_c \to \infty$ limit. For finite $N_c$, the matrix elements have expansions in terms of $1/N_c$ operators given by

$$\begin{aligned} \bar{\Lambda} &= N_c \mathbf{1} + \frac{J_\ell^2}{N_c}, \\ \frac{\lambda_1}{2m_Q} &= \frac{1}{m_Q} N_Q + \frac{1}{N_c^2}\frac{1}{m_Q} N_Q J_\ell^2, \quad (4) \\ -d_H \frac{\lambda_2}{2m_Q} &= \frac{1}{N_c}\frac{1}{m_Q}\,(J_\ell \cdot J_Q), \end{aligned}$$

where $N_Q$ is the heavy quark number operator which is equal to 1 for baryons containing a single heavy quark, **1** is the unit operator for baryons, and $J_\ell^i$ and $J_Q^i$ are the spins of the light degrees of freedom and the heavy quark, respectively. In Eq. (4), it is to be understood that each $1/N_c$ operator is accompanied by an unknown, dimensionful, $\mathcal{O}(1)$ coefficient, which has been suppressed for simplicity. Eq. (4) makes a number of interesting predictions. For instance, $\bar{\Lambda}_T$ and $\bar{\Lambda}_S$ are equal at leading order $N_c$ in the $1/N_c$ expansion. However, at order $1/N_c$, the two matrix elements are not equal, but are split by a contribution which is order $1/N_c^2$ relative to the leading $\mathcal{O}(N_c)$ term. Similar remarks apply for the $\lambda_1$ matrix elements.

The generalization of Eq. (4) to include $SU(3)$ flavor symmetry and its breaking is provided in Ref. [1]. At the time of this work, the $\Xi_c'$ mass had not been measured, and the $1/N_c$ analysis was used to successfully predict $\Xi_c' = 2580.8 \pm 2.1$ MeV [2], to be compared with the subsequent experimental value $\Xi_c' = 2576.5 \pm 2.3$ MeV. This theoretical prediction and its precision required the $1/N_c$ expansion.

Today the masses of all singly charmed baryons are measured except for the spin-$\frac{3}{2}$ $\Omega_c^*$. The most suppressed mass combination

$$\frac{1}{4}\left[(\Sigma_c^* - \Sigma_c) - 2\,(\Xi_c^* - \Xi_c') + (\Omega_c^* - \Omega_c)\right], \quad (5)$$

which is suppressed by $\delta_Q \epsilon^2/N_c^3$ relative to the $\mathcal{O}(N_c)$ baryon mass, can be used to extract $\Omega_c^* = 2770.7 \pm 5.9$ MeV. Thus, it is possible to evaluate the charm baryon mass hierarchy and compare with theory. For bottom baryons, only the $\Lambda_b^0$ mass is measured. Using heavy quark spin-flavor symmetry, it is possible to predict all of the other bottom baryon masses in terms of the charm baryon masses [2].

Figure 1 plots seven of the eight independent mass combinations of the lowest-lying charm baryon spin-flavor multiplet. [The mass splitting Eq. (5) is not plotted since it was used to determine the $\Omega_c^*$ mass.] The first mass combination is the leading order $N_c\Lambda + m_Q$ baryon mass where $Q = c$. The remaining mass splittings are order $\Lambda$ times $\frac{1}{N_c}$, $\epsilon$, $\frac{\epsilon}{N_c}$, $\frac{\epsilon^2}{N_c}$, $\frac{\delta_Q}{N_c}$ and $\frac{\epsilon\delta_Q}{N_c^2}$, respectively. Fig. 1 shows that the $1/N_c$ splitting is comparable to the $\epsilon$ splitting, and that the $\delta_Q/N_c$ splitting is a bit larger than the $\epsilon/N_c$ splitting for $Q = c$. The most suppressed splittings are consistent with the predicted $\frac{\epsilon^2}{N_c}$ and $\frac{\epsilon\delta_Q}{N_c^2}$ hierarchy as well.

It is possible to determine the $1/m_Q$-dependent contributions to the charm baryon mass splittings which do not violate heavy quark spin symme-

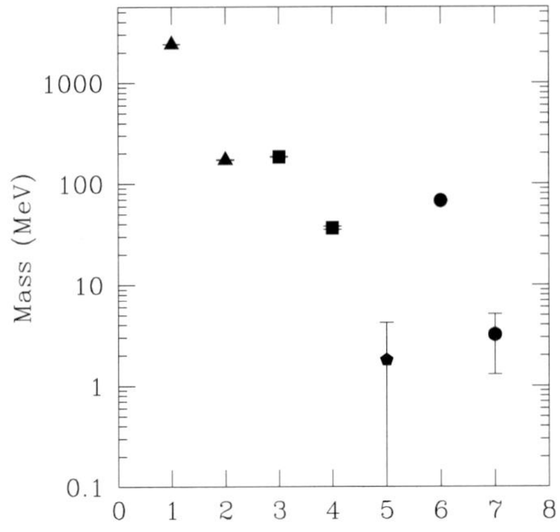


Figure 1. Mass hierarchy of charm baryons.

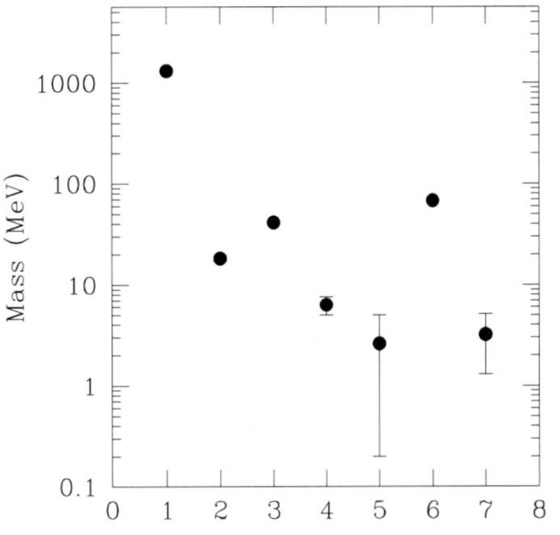


Figure 2. $1/m_Q$-dependent mass splittings for charm baryons.

try by comparison with the analogous mass splittings for baryons containing no heavy quark. The $1/N_c$ expansions of $qqq$ and $Qqq$ baryon masses are given in the flavor symmetry limit by

$$\begin{aligned} M_{(qqq)} &= N_c \mathbf{1} + \frac{1}{N_c} J_\ell^2, \\ M_{(Qqq)} &= N_c \mathbf{1} + N_Q m_Q + \frac{1}{N_c} J_\ell^2 \\ &+ \frac{1}{N_c^2 m_Q} N_Q J_\ell^2 + \frac{1}{N_c m_Q} \left( J_\ell \cdot J_Q \right), \end{aligned} \quad (6)$$

which shows that the $N_Q$ and the $N_Q J_\ell^2$ splittings can be extracted by making this comparision.

Figure 2 plots the $1/m_Q$-dependent portion of the first five mass splittings of Fig. 1 together with the two heavy quark spin-violating mass splittings (points 6 and 7) of Fig. 1. The first mass (point 1) yields $m_Q$ or the charm quark mass at leading order in $1/m_Q$, whereas the six other mass splittings are order $\Lambda$ times the the dimensionless suppression factors $\frac{\delta_Q}{N_c^2}$, $\frac{\epsilon\delta_Q}{N_c}$, $\frac{\epsilon\delta_Q}{N_c^2}$, $\frac{\epsilon^2\delta_Q}{N_c^2}$, $\frac{\delta_Q}{N_c}$, and $\frac{\epsilon\delta_Q}{N_c^2}$, respectively. It is interesting to note, for example, that the two $\frac{\epsilon\delta_Q}{N_c^2}$ splittings (points 4 and 7) are in good agreement, showing that the same heavy quark symmetry-violating parameter $\delta_Q$ is governing heavy quark spin-conserving and spin-violating mass splittings.

## 3. CONCLUSIONS

The $1/N_c$ hierarchy of the $1/N_c$ expansion is evident in the masses of $Qqq$ baryons as well as $qqq$ baryons. The $1/N_c$ expansion, together with $SU(3)$ flavor violation and heavy-quark symmetry violation, gives a quantitative understanding of spin-flavor symmetry breaking for heavy quark baryons. An intricate pattern of spin-flavor symmetry breaking is predicted since $1/N_c$, $\epsilon$, and $\delta_Q$ for $Q = c$ are comparable in magnitude. The same $1/N_c$ hierarchy is expected to appear for bottom baryon masses, and it is possible to predict the bottom baryon mass splittings in terms of charm baryon mass splittings. Heavy quark spin-flavor symmetry is a better symmetry for $Qqq$ baryons than for heavy quark mesons because violation of the spin-flavor symmetry is suppressed by factors of $1/N_c$ as well as $1/m_Q$.

## REFERENCES

1. E. Jenkins, Phys. Rev. D 54 (1996) 4515.
2. E. Jenkins, Phys. Rev. D 55 (1997) R10.

NH
ELSEVIER

Proceedings of ICHEP 2002, pp. 648–651
*S. Bentvelsen, P. de Jong, J. Koch and E. Laenen (Editors)*

31st INTERNATIONAL CONFERENCE ON
HIGH ENERGY PHYSICS AMSTERDAM

# Charmed Baryons and Heavy Quark spectroscopy results from FOCUS

Cristina Riccardi[a]*,

[a]Dept. of Physics and I.N.F.N. of Pavia,
via Bassi 6, 27100 Pavia, Italy

Using data taken at the Fermilab FOCUS experiment(FNAL-E831), we have measured the lifetimes of the single charm baryons: $\Lambda_c^+$,$\Xi_c^+$,$\Xi_c^0$. A preliminary lifetime measurement of the $\Omega_c^0$ charm baryon will be presented. We also report our measurements of branching ratios for some $\Lambda_c^+$ decay modes and in particular the first measurement of the relative branching ratio for the Cabibbo suppressed decay mode $\Lambda_c^+ \to \Sigma^+ K^*(892)$ with respect to the $\Lambda_c^+ \to \Sigma^+\pi^+\pi^-$ mode and an upper limit for the Cabibbo suppressed decay $\Lambda_c^+ \to \Sigma^- K^+\pi^+$.
Finally our search of the double charm baryons will be presented.

## 1. Introduction

In the single charm baryon decays sector, the W-exchange mechanism and the interference effects, which can be constructive and destructive, are expected to play a large role and they can produce significant differences in the lifetimes. Besides that the semileptonic widths are expected not universal for charm baryons. An improvement of the precision for the lifetime measurements of the $\Xi_c^+$, $\Xi_c^0$ and $\Omega_c^0$ to the 10% level is requested for a better knowledge of the different contributions[1][2]. More precise measurements of some exclusive decay modes may also provide a further investigation of the W-exchange contribution[3]. Furthermore useful inputs for a better understanding of the $\Xi_c^+$ lifetime may be provide by the measurements of exclusive Cabibbo suppressed decay modes of the $\Lambda_c^+$ baryon[4].

The studies presented in this paper are based on the full data sample collected by the FOCUS experiment during the Fermilab 1996-1997 fixed-target run. Over one million fully reconstructed charm particles have been produced by the interaction of photons with average energy of 180 GeV in a segmented BeO target. FOCUS used an upgraded version of the E687 spectrometer[5]. The vertex detector was formed by 16 layers of silicon microstrip detector (four interleaved with the target segments[6] and 12 downstream the target) which provide an excellent proper time resolution of about 40 fs for most charm baryon decays. The momentum of charged particles is determined by measuring their deflection in two analysis magnets of opposite polarity with five stations of multiwire proportional chambers. Particle identification is performed by three multicell threshold Cerenkov counters[7], electromagnetic calorimeters and muon counters.

Several techniques to reconstruct the decays of the hyperons have been developed for an extensive search of charm baryon decays. The implemented algorithms provide a fully reconstruction of the $\Lambda^0 \to p\pi^-$, $\Xi^- \to \Lambda^0\pi^-$, $\Omega^- \to \Lambda^0 K^-$ decays and a partial reconstruction of the $\Sigma^- \to n\pi^-$, $\Sigma^+ \to n\pi^+$, $\Sigma^+ \to p\pi^0$, $\Xi^- \to \Lambda^0\pi^-$, $\Omega^- \to \Lambda^0 K^-$ decays where the neutral particle is undetected. A detailed description of these techniques is reported in [8].

*On behalf of the FOCUS collaboration: J.Link, M.Reyes, P.M.Yager (**UC DAVIS**); J.Anjos,I.Bediaga, C.Gobel, J.Magnin, A. Massafferri, J.M. de Miranda, I.M.Pepe, A.C. dos Reis, (**CPBF, Rio de Janeiro**); S.Carrillo, E.Casimiro, A.Sánchez-Hernández, C.Uribe, F.Vasquez (**CINVESTAV, México City**); L.Agostino, L.Cinquini, J.P.Cumalat, B.O'Reilly, J.E.Ramirez, I.Segoni (**CU Boulder**); J.N.Butler, H.W.K.Cheung, G.Chiodini, I.Gaines, P.H.Garbincius, L.A.Garren, E.Gottschalk, P.H.Kasper, A.E.Kreymer, R.Kuschke (**Fermilab**); L.Benussi, S.Bianco, F.L.Fabbri, A.Zallo (**INFN Frascati**); C.Cawlfield, D.Y.Kim, K.S.Park, A.Rahimi, J.Wiss (**UI Champaign**); R.Gardner, A.Kryemadhi (**IU Bloomington**); K.H.Chang, Y.S.Chung, J.S.Kang, B.R.Ko, J.W.Kwak, K.B.Lee (**Korea University, Seoul**); K.Cho, H.Park (**Kyungpook National University, Taegu**); G.Alimonti, S.Barberis, A.Cerutti, M.Boschini, P.D'Angelo, M.DiCorato, P.Dini, L.Edera, S.Erba, M.Giammarchi, P.Inzani, F.Leveraro, S.Malvezzi, D.Menasce, M.Mezzadri, L.Moroni, D.Pedrini, C.Pontoglio, F.Prelz, M.Rovere, S.Sala (**INFN and Milano**); T.F.Davenport III (**UNC Asheville**); V.Arena, G.Boca, G.Bonomi, G.Gianini, G.Liguori, M.M.Merlo, D.Pantea, S.P.Ratti, P.Vitulo (**INFN and Pavia**); H.Hernandez, A.M.Lopez, H.Mendez, L.Mendez, E.Montiel, D.Olaya, A.Paris, J.Quinones, C.Rivera, W.Xiong, Y.Zhang (**Mayaguez, Puerto Rico**); J.R.Wilson (**USC Columbia**); T.Handler, R.Mitchell (**UT Knoxville**); D.Engh, M.Hosack, W.E.Johns, M.Nehring, P.D.Sheldon, K.Stenson, E.W.Vaandering, M.Webster (**Vanderbilt**); M.Sheaff (**Wisconsin, Madison**)


doi:10.1016/S0920-5632(02)02157-6

## 2. Charm Baryon lifetimes

In FOCUS the lifetime measurements are performed using a binned maximum likelihood fit method. The lifetime fit variable is the *reduced* proper time $t'$. This variable is defined as $t' = \frac{\ell - N\sigma_\ell}{\beta\gamma c}$, where $\ell$ is the detachment between primary and secondary vertex, $\sigma_\ell$ is the resolution on $\ell$ and $N$ is the minimum detachment cut to tag the charm particle through its lifetime. The fit is performed on the data which lie in the signal region. The distortions from a pure exponential behaviour expected for the signal events are corrected using a Montecarlo correction function which reproduces the acceptance and selection efficiency effects. Furthermore the background events in the signal region are modeled by the $t'$ distributions of the events in the sideband regions of the invariant mass distribution. In the likelihood function a term is also included to tie the sideband background level to the one observed in the signal region. The fit parameters are the lifetime and the background level.

FOCUS performed a high statistics lifetime measurement of the $\Lambda_c^+$ baryon using a very clean sample of about 8000 $\Lambda_c^+ \to pK^-\pi^+$ decays. A detailed study of the systematic sources has been made and in particular an extensive search of effects related to the acceptance and efficiency corrections. This very high accuracy measurement results to be $204.6 \pm 3.4 \pm 2.5$ fs[9].

The $\Xi_c^+$ lifetime has been studied with a total sample of $532 \pm 30$ candidates of four decay modes: $\Xi_c^+ \to \Xi^-\pi^+\pi^+$, $\Xi_c^+ \to \Sigma^+K^-\pi^+$, $\Xi_c^+ \to pK^-\pi^+$, $\Xi_c^+ \to \Lambda^0K^-\pi^+\pi^+$, reconstructed through 8 different topologies. The fitted lifetime is $439 \pm 22 \pm 9$ fs[10], where the systematic error is dominated by effects related to the fit conditions.

In addition to the longest-lived charm baryons, FOCUS measured also the $\Xi_c^0$ and $\Omega_c^0$ lifetimes.

The study on the $\Xi_c^0$ baryon has been performed using a total sample of 110±17 events reconstructed through the two decay modes: $\Xi_c^0 \to \Xi^-\pi^+$ ($97 \pm 17$ events)and $\Xi_c^0 \to \Omega^-K^+$ ($21 \pm 8$ events). Since the proper time resolution (about 40 fs for the $\Xi^-\pi^+$ mode and 80 fs for the $\Omega^-K^+$ mode) is comparable to the measured lifetime, an extensive investigation of possible resolution effects has been conducted. The measurement has been performed with a convolved binned likelihood method [11] to check the fitting technique. In this method the exponential decay is convolved with the smearing due to the time resolution and the proper decay time is fit instead of the reduced one. The performance of the fitting method has been also checked using a mini Montecarlo study with a wide range of input lifetimes. In addition other studies on acceptance and production effects have been made. The FOCUS measurement of the lifetime results $118^{+14}_{-12} \pm 5$[12] fs and it represents a big improvement on the previous existing measurements reducing the error from 20% to 10%.

A preliminary measurement of the $\Omega_c^0$ lifetime[13] has been obtained analizing the two decay modes: $\Omega_c^0 \to \Xi^-K^-\pi^+\pi^+$ ($38 \pm 9$ events) and $\Omega_c^0 \to \Omega^-\pi^+$ ($23 \pm 7$ events). The invariant mass plots

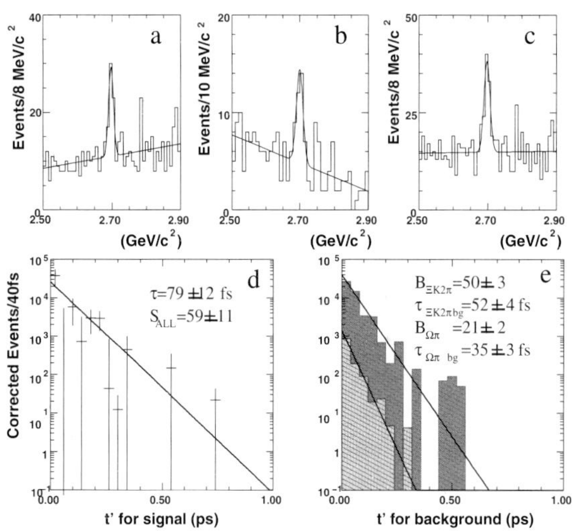


Figure 1. Invariant mass distribution for the $\Omega_c^0 \to \Xi^-K^-\pi^+\pi^+$ candidates (a), the $\Omega_c^0 \to \Omega^-\pi^+$ candidates (b) and the total sample (c). The t' distributions for the $\Omega_c^0$ signal events (d) and the background events (e).

of the $\Omega_c^0$ candidates are shown in figures 1a-b-c, where a total sample of 64±14 events is found at a mass value of 2697.5±2.2(stat.) MeV which is consistent with the previous and recent high statistics measurements [14][15]. Figures 1d-e show respectively the reduced proper time distribution of the signal events and the one of the background events lying into the sideband regions of the invariant mass distributions. Systematics studies, including investigations of the time resolution effects, has been performed similar to the ones made for the $\Xi_c^0$ measurement. A preliminary result for the $\Omega_c^0$ lifetime is $79 \pm 12 \pm 9$ fs. The lifetime measurements provided by FOCUS

Table 1
Charm baryon lifetimes (in fs). The FOCUS measurement of the $\Omega_c^0$ lifetime value is preliminary.

| Charm Baryon | FOCUS | PDG[14] |
|---|---|---|
| $\Lambda_c^+$ | $204.6 \pm 3.4 \pm 2.5$[9] | $200 \pm 6$ |
| $\Xi_c^+$ | $439 \pm 22 \pm 9$[10] | $442 \pm 26$ |
| $\Xi_c^0$ | $118^{+14}_{-12} \pm 5$[12] | $98^{+23}_{-15}$ |
| $\Omega_c^0$ | $79 \pm 12 \pm 9$ | $64 \pm 20$ |

for all the charm baryons are compared in table with the PDG world averages, which already include the FOCUS $\Lambda_c^+$ and $\Xi_c^+$ lifetime values. The PDG average of the ratio $\tau(\Xi_c^+)/\tau(\Lambda_c^+)$ results to be $2.21 \pm 0.15$, while the expected ratio should range from 1.2 to 1.7.

## 3. Measurements of the branching ratios of the $\Lambda_c^+$ decay modes

Some $\Lambda_c^+$ decay modes with a charged $\Sigma$ in the final state have been investigated. The FOCUS largest samples are $1666 \pm 60$ events for the mode $\Lambda_c^+ \to \Sigma^+\pi^+\pi^-$ and $523 \pm 35$ events for the mode $\Lambda_c^+ \to \Sigma^-\pi^+\pi^+$, when the selection criteria are the same and optimized for both modes. The invariant mass distributions of these modes are shown in fig. 2. A preliminary measurement of their relative branching ratio is $\Gamma(\Lambda_c^+ \to \Sigma^-\pi^+\pi^+)/\Gamma(\Lambda_c^+ \to \Sigma^+\pi^+\pi^-) = 0.422 \pm 0.033$,

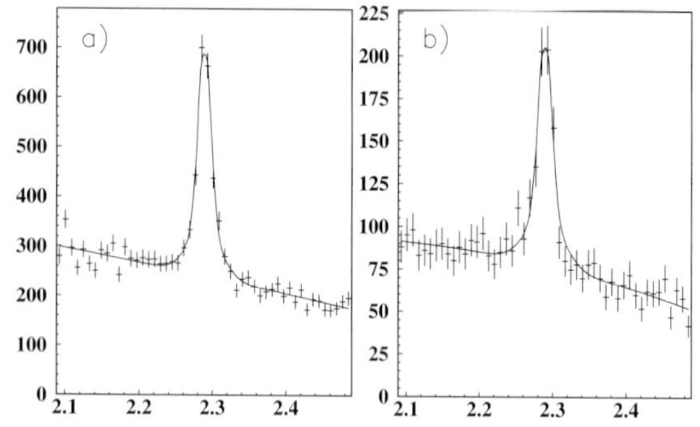


Figure 2. Invariant mass distributions for the $\Lambda_c^+ \to \Sigma^+\pi^+\pi^-$ (a) and $\Lambda_c^+ \to \Sigma^-\pi^+\pi^+$ (b) decay candidates.

where the error is only statistical. This result is consistent with the unique previous measurement provided by the E687 experiment which is $0.53 \pm 0.15 \pm 0.07$[?].
The W-exchange dominated decay $\Lambda_c^+ \to \Sigma^+K^+K^-$ has been analized using a total sample of $105 \pm 15$ events[17]. We've measured the resonant contribution to these decays through the modes: $\Lambda_c^+ \to \Sigma^+\phi$ ($57 \pm 10$ events) and $\Lambda_c^+ \to \Xi^{*0}(1690)K^+$ ($34 \pm 8$ events). The events of each mode have been selected in the resonance ($\phi$ or $\Xi^{*0}(1690)$) signal region of the corresponding ($K^+K^-$ or $\Sigma^+K^-$) invariant mass distribution. The contamination of the non-resonant or other resonant components has been rejected using a sideband subtraction technique or using mass cuts. Almost all of the $\Sigma^+K^+K^-$ decays occur through the resonant modes $\Xi^{*0}(1690)K^+$ and $\Sigma^+\phi$ and no evidence of a non-resonant contribution has been found.
The branching ratios of the investigated decays have been measured relative to the $\Lambda_c^+ \to \Sigma^+\pi^+\pi^-$ decays to minimize the systematic sources related to the $\Sigma^+$ reconstruction. The results are: $\Gamma(\Lambda_c^+ \to \Sigma^+K^+K^-)/\Gamma(\Lambda_c^+ \to \Sigma^+\pi^+\pi^-) = (7.1 \pm 1.1 \pm 1.1)\%$, $\Gamma(\Lambda_c^+ \to \Sigma^+\phi)/\Gamma(\Lambda_c^+ \to \Sigma^+\pi^+\pi^-) = (8.7 \pm 1.6 \pm 0.6)\%$, $\Gamma(\Lambda_c^+ \to \Xi^*(\Sigma^+K^-)K^+)/\Gamma(\Lambda_c^+ \to$

$\Sigma^+\pi^+\pi^-) = (2.2 \pm 0.6 \pm 0.6)\%$, $\Gamma(\Lambda_c^+ \to \Sigma^+K^+K^-)_{NR}/\Gamma(\Lambda_c^+ \to \Sigma^+\pi^+\pi^-) < 2.8\%$ at the 90% CL. These measurements are consistent

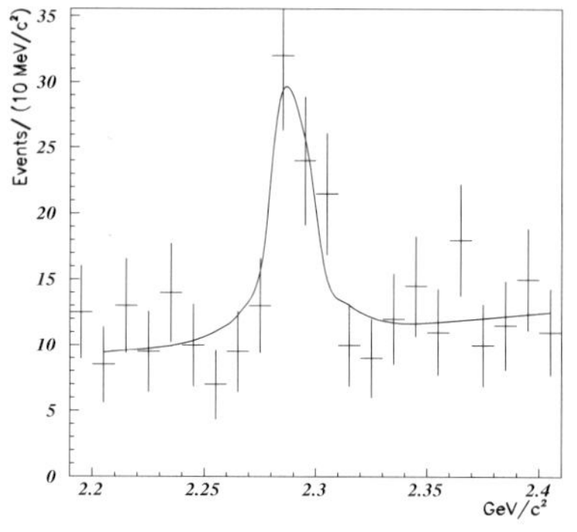


Figure 3. Invariant mass distribution for the $\Lambda_c^+ \to \Sigma^+K^{*0}$ decay candidates.

with the recent ones by BELLE [18].
FOCUS have also studied two Cabibbo-suppressed decays: $\Lambda_c^+ \to \Sigma^+K^{*0}(892)$ and $\Lambda_c^+ \to \Sigma^-K^+\pi^+$. The analysis of the $\Lambda_c^+ \to \Sigma^+K^{*0}$ decays has been performed in a similar way as described for the previous resonant decays. More details can be found in [17]. A signal of $49 \pm 10$ events for the $\Lambda_c^+ \to \Sigma^+K^{*0}$ mode (see fig.3) has been observed; the first measurement of its branching ratio relative to the $\Lambda_c^+ \to \Sigma^+\pi^+\pi^-$ mode results $(7.8 \pm 1.8 \pm 1.3)\%$. Because no evidence of the $\Lambda_c^+ \to \Sigma^-K^+\pi^+$ decay has been found, the relative branching ratio $\Gamma(\Lambda_c^+ \to \Sigma^-K^+\pi^+)/\Gamma(\Lambda_c^+ \to \Sigma^+K^{*0}) < 35\ \%$ at the 90% CL has been set. The non-resonant component of the $\Lambda_c^+ \to \Sigma^+K^+\pi^-$ mode has been also checked finding no evidence of it. A possible explanation of this suppression may be related to the need of creating a quark pair in order to get the three body decays.

## 4. Double Charm Baryons search

In the year 2000 FOCUS has made an extensive search for the decays of the double charm baryons $\Xi_{cc}^+$ and $\Xi_{cc}^{++}$. A sample of $19444 \pm 262$ $\Lambda_c^+ \to pK^-\pi^+$ decays has been used to investigate 7 $\Xi_{cc}$ decay modes, while a sample of about 1.2 million of the reconstructed charm mesons ($D^+ \to K\pi\pi, D^0 \to K\pi, K\pi\pi\pi$) has been used to study 14 $\Xi_{cc}$ decay modes. For more details about this analysis see [19]. FOCUS found no evidence of double charm baryons.

## REFERENCES


1. G.Bellini, I.I.Y.Bigi, P.J.Dornan, Phys. Rept. 289 (1997) 1.
2. I.I.Y.Bigi, Proc. of Heavy quarks at fixed target, St. Goar 1996, 131, hep-ph/9612293.
3. Y.Kohara, Nuovo Cim.A 111 (1998) 67 and references therein; M.A.Ivanov et al., Phys.Rev.D 57 (1998) 5632 and references therein.
4. B.Guberina and H.Stefancic, hep-ph/0202080 (2002).
5. P.L.Frabetti et al., Nucl.Instr.Meth. A320 (1992) 519.
6. J.M.Link et al., hep-ex/0204023 (2002), submitted to Nucl.Instr.Meth.
7. J.M.Link et al., Nucl.Instr.Meth. A484 (2002) 270.
8. J.M.Link et al., Nucl.Instr.Meth. A484 (2002) 174.
9. J.M.Link et al., Phys.Rev.Lett. 88 (2002) 161801.
10. J.M.Link et al., Phys.Lett. B523 (2001) 53.
11. P.L.Frabetti et al., Phys.Lett. B357 (1995) 678.
12. J.M.Link et al., Phys.Lett. B541 (2002) 211.
13. J.Kwak, Preliminary measurement of the $\Omega_c^0$ lifetime in FOCUS, (2002) http://dpf2002.velopers.net/talks_pdf/215talk.pdf.
14. K. Hagiwara et al., Phys.Rev. D66 (2002) 010001
15. R.Chistov, B decays to baryon final states and $\Omega_c^0$ from Belle, in this proceedings.
16. P.L.Frabetti et al., Phys.Lett. B328 (1994) 193.
17. J.M.Link et al., Phys.Lett. B540 (2002) 25.
18. K.Abe et al., Phys.Lett. B524 (2002) 33.
19. FOCUS internal report, www.hep.vanderbilt.edu/~stenson/xicc/xicc_focus.html.

Proceedings of ICHEP 2002, pp. 652–654
*S. Bentvelsen, P. de Jong, J. Koch and E. Laenen (Editors)*

ELSEVIER

31st INTERNATIONAL CONFERENCE ON HIGH ENERGY PHYSICS AMSTERDAM

# First Observation of a Family of Double-Charm Baryons

J. S. Russ[a]
on behalf of the SELEX Collaboration*

[a]Carnegie Mellon University
Pittsburgh, PA 15213 USA

The SELEX experiment (E781) at Fermilab has candidates for high mass states decaying to $\Lambda_c^+ K^- \pi^+$ and $\Lambda_c^+ K^-$ pi$^+\pi^+$, Cabibbo-allowed decay modes of doubly-charmed baryons $\Xi_{cc}^+$ and $\Xi_{cc}^{++}$. The masses are consistent with theoretical considerations, but the spectroscopy is surprising. Limited lifetime information suggests that $\tau_{\Xi_{cc}^{++}} \sim \tau_{\Xi_{cc}^+}$.

## 1. Introduction

The existence of baryons with two and three charm quarks is expected from our present understanding of hadronic structure. In the double-charm system one expects a J=1/2 ground state iso-doublet, termed $\Xi_{cc}^{+,++}$ in PDG notation [1]. Most predictions for the masses of the J=1/2 states and the J=3/2 hyperfine excitations expect the ground state near 3.6 GeV/c$^2$ and a hyperfine split of 60 MeV/c$^2$ [2].

The production mechanism for double-charm states is not clear. Perturbative pictures treat production as successive $c\bar{c}$ pair production with enough spatial overlap to form hadrons. Such cross sections are small compared to single-charm production.

## 2. Features of the Selex spectrometer

The SELEX experiment at Fermilab is a 3-stage magnetic spectrometer [3]. The negative 600 GeV/c Fermilab Hyperon Beam had about equal fluxes of $\pi^-$ and $\Sigma^-$. The positive beam was 92% protons. For charm momenta in a range of 100-500 GeV/c mass resolution is constant and primary (secondary) vertex resolution is typically 270 (560) $\mu$m. A RICH detector labelled all particles above 25 GeV/c [4]. Details of single-charm analyses involving $\Lambda_c^+ \to pK^-\pi^+$ reconstructions can be found in [5,6]. The double-charm search discussed here began with the sample of 1630 $\Lambda_c^+$ events used in the lifetime analysis [5].

## 3. Double-charm Analysis

This topological search for double-charm baryons asks for a decay vertex lying between the primary vertex and the observed $\Lambda_c^+$ decay vertex. A Cabibbo-allowed $\Xi_{cc}^+$ decay can give a final-state $\Lambda_c^+$, a $K^-$, and a $\pi^+$, shown schematically in Fig. 1.

We reconstruct total charge states Q=1 (neutral 2-prong vertex) and Q=2 (positive 3-prong vertex) in separate reconstructions. Few of the tracks from the intermediate vertex are RICH-identified. We call the negative track a kaon in the right-sign reconstruction. The same events have a wrong-sign reconstruction, calling the negative track a pion.

Event selection cuts used here were taken without change from previous single-charm studies. For short-lived states, $L_1/\sigma_1 \geq 1$ and the $\Lambda_c^+$ momentum vector must point back to the primary vertex within a $\chi^2$ cut of 4. We have varied the

*SELEX Collaboration: Ball State University, Bogazici University, Carnegie Mellon University, Centro Brasileiro de Pesquisas Fisicas (Rio de Janeiro), Fermi National Accelerator Laboratory, Petersburg Nuclear Physics Institute, IHEP (Beijing), IHEP (Protvino), ITEP (Moscow), Moscow State University, Max-Planck-Institut für Kernphysik (Heidelberg), Tel Aviv University, Universidad Autónoma de San Luis Potosí, University of Bristol, Universidade Federal da Paraiba, University of Iowa, University of Michigan–Flint, University of Rome “La Sapienza”, INFN, University of São Paulo, University of Trieste.


doi:10.1016/S0920-5632(02)02172-2

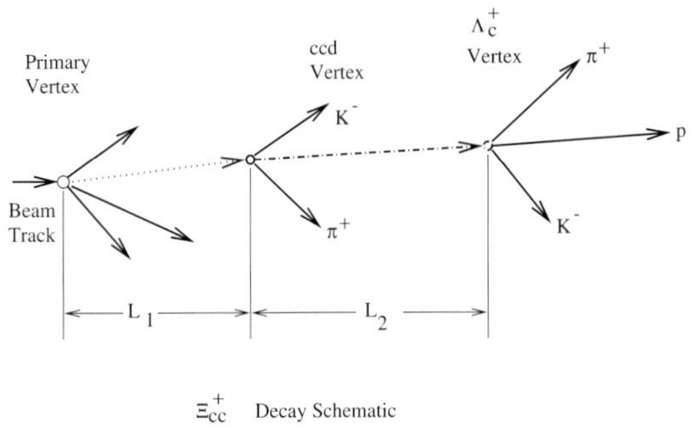


Figure 1. Schematic of $\Xi_{cc}^+ \to K^-\pi^+\Lambda_c^+$

cuts and observe that no signal significance depends critically on any cut value. The cuts were checked with simulation studies using several production models for double charm.

### 3.1. Q = 1 reconstruction

For single-charged baryons, the $K^-\pi^+\Lambda_c^+$ mass distribution is shown in Fig. 2. Fig. 2(c) shows a $\Xi_{cc}^+$ candidate at 3520 MeV/c$^2$, consistent with most model calculations. The peak is narrow but consistent with simulation. The general agreement between right-sign (a,c) and wrong-sign (b) average levels and fluctuations in Fig. 2 confirms that most events are combinatoric background. The signal region in (c) contains 22 events with a background of 6.1 ± 0.51 events, for a single-bin significance of 6.3 $\sigma$. The probability of such an excess is less than $10^{-6}$ for a single bin. We searched for a peak in the interval 3.2-4.3 GeV/c$^2$, or 110 bins. The probability of such a fluctuation anywhere in the search interval is $< 1.1 \times 10^{-4}$. SELEX has reported this as the first observation of a doubly-charmed baryon [7].

Using the reduced proper decay time $t^* = M(L - L_{min})/pc$ we find that the lifetime < 33 fs at 90% confidence. Here $L_{min} = \sigma_1$ the error on the vertex separation $L_1$. The SELEX proper time resolution is about 20 fs. With constructive interference between the two c-quark decay amplitudes along with the W-exchange amplitudes, this state could have a lifetime shorter than the $\Xi_c^0$.

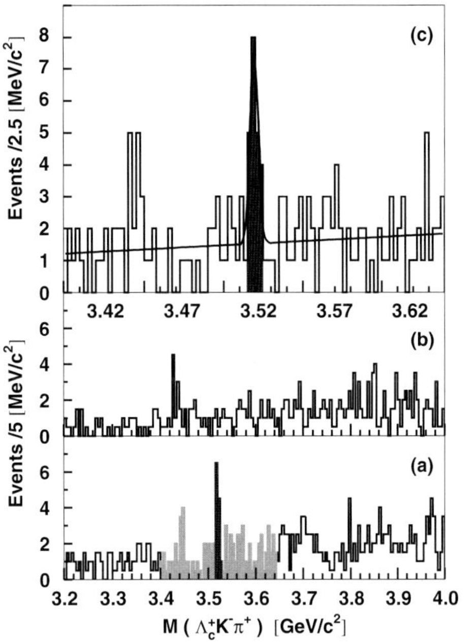


Figure 2. (a) The $\Lambda_c^+ K^-\pi^+$ mass distribution in 5 MeV/$c^2$ bins. The shaded region is shown in more detail in (c). (b) The wrong-sign combination $\Lambda_c^+ K^+\pi^-$ mass distribution in 5 MeV/$c^2$ bins. (c) The signal (shaded) region (22 events) and sideband mass regions (140 events) in 2.5 MeV/$c^2$ bins. The fit is a Gaussian plus linear background.

### 3.2. Q=2 Reconstruction

For double-charged baryons, we look for an isospin partner of the $\Xi_{cc}^+(3520)$. The $K^-\pi^+\pi^+\Lambda_c^+$ mass distribution in the vicinity of 3520 MeV/c$^2$ is shown in Fig. 3, along with the wrong-sign background.

There is a $\Xi_{cc}^{++}$ candidate at 3460 MeV/c$^2$. The observed width (6 ± 1 MeV/c$^2$) matches simulation. Events outside the signal region show a strong preference for the center-of-mass (CM) angle of the negative track to be near 180 degrees. Simulation indicates that a cut to remove such events should have very little effect on the signal

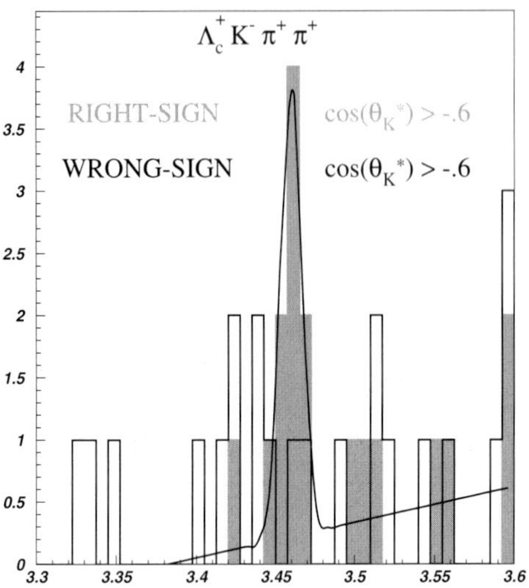


Figure 3. $\Xi_{cc}^+ \to K^-\pi^+\pi^+\Lambda_c^+$ mass distribution in 7.5 MeV/c$^2$ bins. Signal events are shaded. Wrong-sign background is shown as open histogram boxes.

region for a phase-space decay distribution. That is indeed the case in the data. With the selection shown, we find 9 events in the peak, compared to an expected background of 1 event. The Poisson probability that there is an excess of 8 events or more anywhere on the plot is $10^{-5}$. There are too few events to attempt a lifetime analysis. The $\Xi_{cc}^{++}$ candidates have a raw average proper time (uncorrected for acceptance) comparable to that for the $\Xi_{cc}^+$ candidates, suggesting that $\tau_{\Xi_{cc}^{++}} \sim \tau_{\Xi_{cc}^+}$.

### 3.3. Production

Both $\Xi_{cc}^+$ and $\Xi_{cc}^{++}$ states are produced only by baryon beams in SELEX data. There are no signal candidates from the pion beam. Simulation studies suggest that the double-charm states may account for as much as 40% of the $\Lambda_c^+$ sample seen in this experiment, a surprisingly high fraction. This situation is reminiscent of the discovery of the $\Xi_c^+$ baryon in the WA62 experiment at CERN, using a 135 GeV hyperon beam [8]. The FOCUS photoproduction experiment at Fermilab has looked for these states using their $\Lambda_c^+$ events and sees no signal peaks [9]. If SELEX is correct, the hadroproduction mechanism is unusual.

## 4. Summary

SELEX has introduced two statistically-compelling new high-mass states that decay into a final state $\Lambda_c^+$, $K^-$ and one or two $\pi^+$, as expected for double-charm baryon decays. The 3520 MeV/c$^2$ state satisifies all expectations for being a $\Xi_{cc}^+$ state [7]. Its lifetime is shorter than 30 fs at 90% confidence. The 3460 MeV/c$^2$ state has the decay characteristics of a $\Xi_{cc}^{++}$ state. It is difficult to understand the 60 MeV/c$^2$ mass difference between the Q=1 and Q=2 states if they are members of the ground state isodoublet. However, any other interpretation is also problematic.

## REFERENCES

1. Particle Data Group, D.E. Groom, *et al.*, Eur. Phys. J.**C15**,1 (2000)
2. see references 111-124 in hep/ph0201071, B Physics at the Tevatron.
3. J. Russ, *et al.* , *in Proceedings of the 29th International Conference on High Energy Physics,* edited by A. Astbury *et al.* (World Scientific, Singapore, 1999) VolII, p. 1259;hep-ex/9812031.
4. J. Engelfried, *et al.*, Nucl.Instr.and Methods A**431**, 53, 1999
5. A. Kushnirenko *et al.* Phys. Rev. Lett. **86**, 5243 (2001), hep-ex/0010014.
6. F. Garcia *et al.* Phys. Lett. **B528**, 49 (2002), hep-ex/0109017.
7. M. Mattson *et al.* Phys. Rev. Lett. **89**, 112001 (2002)
8. S. Biagi, *et al.*, Phys. Lett. **B122**, 455 (1983)
9. C. Riccardi, these proceedings. See also http://www.hep.vanderbilt.edu/~stenson/xicc/xicc_focus.html

# Session 9

# Hadron Spectroscopy and Exotics (experiment and theory)


*Session conveners:*
*V. Burkert, JLab, Newport News*
*J. Urheim, University of Minnesota*

ELSEVIER

Proceedings of ICHEP 2002, pp. 657–661
*S. Bentvelsen, P. de Jong, J. Koch and E. Laenen (Editors)*

31[st] INTERNATIONAL CONFERENCE ON
HIGH ENERGY PHYSICS AMSTERDAM

# Review of baryon spectroscopy

S. Krewald[a]

[a]Institut für Kernphysik, Forschungszentrum Jülich, D-52425 Jülich, Germany

There is a mismatch between the number of experimentally established baryon resonances and predictions based on naive quark models. New lattice results support phenomenological analyses which suggest a $non - q^3$ structure for the Roper. Meson production experiments in the vicinity of the Roper resonance may help to resolve the issue.

## 1. Introduction

The investigation of baryon resonances has lead to the concepts of quarks and color. At present, baryon resonances intrigue us by the so-called *missing resonances problem* : at energies above 2 GeV, the number of experimentally established resonances is less than what is predicted by theories based on a triple quark structure of the resonances[1,2]. On the other hand, for smaller energies there are resonances which appear not to fit into $q^3$ multiplets. The Roper resonance $N^*(1440)$, the first excited state of the nucleon showing the quantum numbers of the nucleon, has challenged theorists for a long time by its relatively low mass which is less than that of the first excited state with negative parity, the $N^*(1535)$ [2]. There is the exciting possibility, however, that instanton motivated effective quark interactions or color singlet interactions reduce the mass of the Roper[3,4].

## 2. Excited baryons on the lattice

During the last decade, lattice groups have started to investigate excited baryons. By a suitable choice of interpolating fields, the lattice can distinguish between resonances of positive and negative parity, while contaminations due to backward propagation are removed by a boundary condition in the time direction. The numerical results cannot yet be compared directly to the experimental masses because unrealistically large quark masses have to be adopted on those lattices accessible to present numerical technology. In Fig.1, the masses of the first and the second nucleons of both positive and negative parity are shown as a function of the pion mass[5]. While

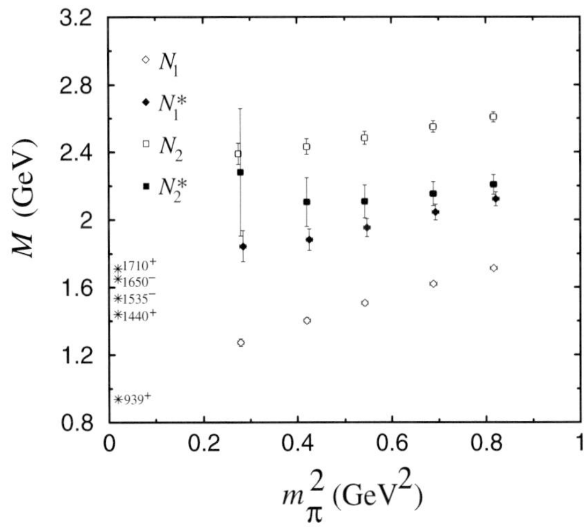


Figure 1. Masses of the $1/2^+$ and $1/2^-$ nucleons as a function of the square of the pion mass. The empirical masses of the lowest nucleon states of either parity are indicated by the asterisks[5].

the lattice results for the nucleon are pointing towards the experimental mass, it is obvious that the mass of the excited state of positive parity remains above the one of the states of negative


doi:10.1016/S0920-5632(02)02175-8

parity. It is unlikely that the recently developped chiral extrapolation methods would lead to qualitative changes [6]. The lattice calculations have employed interpolating fields consistent with a triple quark structure of the excited states and thus appear to confirm the level ordering of the excited baryons obtained by the naive quark models[2]. During the preparation of this manuscript, a calculation by the Kentucky group has appeared which finds a decreasing mass of the Roper[7]. One has to ask whether the new statistical weighting procedure emphasizes configurations beyond triple quark states which to some extent are included even in quenched calculations.

## 3. The $N^*(1440)$ in pion-nucleon reactions

A remarkable difference between the $N^*(1440)$ and the $N^*(1520)$ is seen when one compares the phase shifts and inelasticities of the pion-nucleon partial wave amplitudes, see Fig.2.

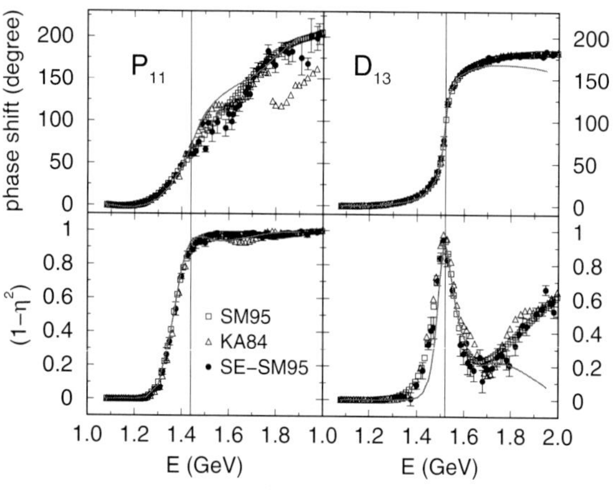


Figure 2. Phase shifts and inelasticities in the pion-nucleon partial waves $P_{11}$ and $D_{13}$ compared with the Höhler and the Arndt analyses. The vertical lines corrrespond to the suggested resonance masses. The solid curves are the results of Ref.[8].

The $N^*(1520)$ causes a change in the $D_{13}$ partial wave up to $180^0$ with a resonant phase at 1520 MeV, which is also the maximum of the inelasticity. After passing the resonant phase of $90^0$, the amplitude goes back to being almost elastic. The situation is completely different for the $N^*(1440)$. Here the phase shift increases slowly which corresponds to a very broad resonance, but the inelasticity opens rapidly and remains inelastic over a very large energy range. Furthermore, the suggested resonance position of $m_R = 1440$ MeV does not correspond to $\delta = 90^0$. The strong deviation from a Breit-Wigner like resonance suggests an investigation of final state interaction effects. In Ref.[8], the pion-nucleon reaction has been treated by coupling the reaction channels $\pi N$, $\eta N$, $\rho N$, $\sigma N$, and $\pi\Delta$ within a meson-exchange model. Thus those channels relevant for two-pion decays have been incorporated. The model provides a theory for the physical background due to meson-baryon interactions. The background may show some structure because the exchange of mesons in the t-channels can act as a long-range attraction between meson and baryon and thus lead to a somewhat localized enhancement of strength. In order to reproduce resonances with a relatively narrow width, such as the $\Delta_{33}$, one has to introduce pole diagrams. A pole diagram is interpreted as an underlying triple quark structure. It was found that all pion-nucleon resonances below 2 GeV required pole diagrams within this formalism, with one exception: the $N^*(1440)$, see Fig.2.

In Fig.3, a simplified (separable )version of the model is shown. If one includes only the $\pi N$ and the $\pi\Delta$ channel, one can fit the phase shift of the $P_{11}$ resonance, but not the inelasticities, even if one allows unrealistically large coupling strengths between the $\pi N$ and the $\pi\Delta$ channel (dashed line). On the other hand, as soon as the $\sigma N$ channel is considered, there is no problem in reproducing the early onset of the inelasticity (solid line). This points to a dominant role of the $\sigma N$ reaction channel in the description of the Roper. Note that the $\sigma$ represents the correlated two-pion state with scalar-isoscalar quantum numbers. A dominant role of the $\sigma N$ channel would imply a large two-pion decay of the Roper resonance. Let's therefore look next into the ex-

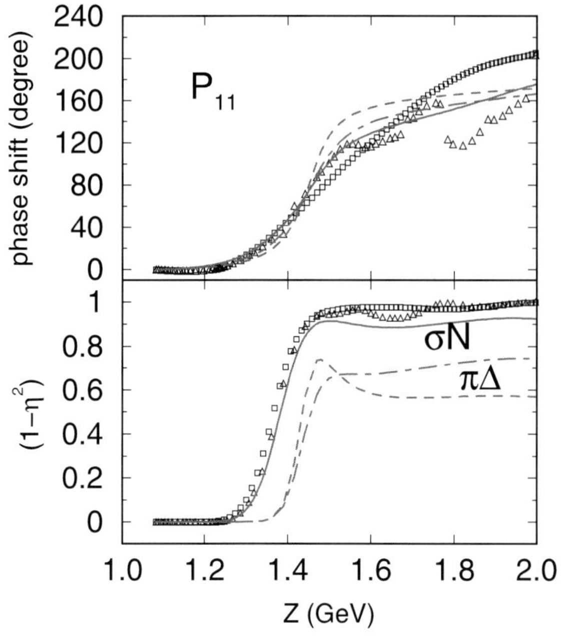


Figure 3. Phase shifts and inelasticities of the $P_{11}$ partial wave in a simplified model. Dashed lines: only $\pi N$ and $\pi\Delta$ channels; solid line: only $\pi N$ and $\sigma N$ channels are coupled.

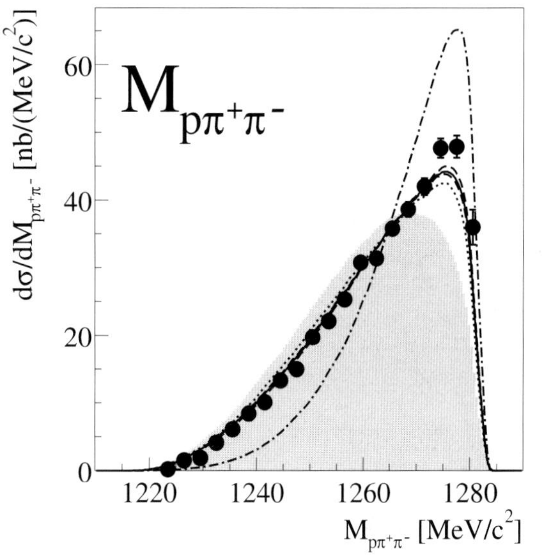


Figure 4. Cross sections for two pion production in the proton-proton reaction as function of the invariant mass $M_{p\pi\pi}$ [9].

perimental analyses of the decay properties of the Roper.

## 4. Two meson production in hadronic reactions

Double pion production in proton collisions $pp \rightarrow pp\pi^+\pi^-$ has been measured recently at CELSIUS[9], see Fig. 4.

The two-pion production proceeds via the intermediate excitation of the Roper resonance and of the $\Delta_{33}$. A theory for this reaction has been worked out by L. Alvarez-Ruso et al.[10]. Close to threshold, the intermediate excited baryon decays into relative s-waves. Since the $\sigma$-correlated two pions are in a relative s-wave, the decay of the Roper into an $\sigma N$ channel (dotted line) is hard to distiguish from the phase space (the shaded area). Fortunately, there is some admixture of the $\pi\Delta$ channel (dashed-dotted line) which causes strong deviations. The measured angular distributions therefore allow to discriminate between reaction channels involving the $\Delta_{33}$ and the Roper. The data show that a direct decay of the Roper into a $\sigma N$ channel (dotted line) is by far the dominant one. There is room for a 25% admixture of the decay of the Roper into the $\pi\Delta$ channel.

An independent confirmation of this result may be obtained via the $p\alpha$ reaction, which shows the Roper as a bump in the cross section, in contrast to pion-nucleon scattering[11]. The analysis of those data is compatible with two components of the Roper.

Two-pion production in the pion-proton reaction has been investigated theoretically by various groups. Close to threshold, chiral perturbation theory has been applied to predict the total cross sections and angular distributions[12,13]. The low energy constants required in those calculations have been estimated by the assumption of "resonance saturation". As one moves away from the threshold, one is forced to reconsider the relevant resonances explicitly. This has been done by a unitarization of a tree-level model[14].

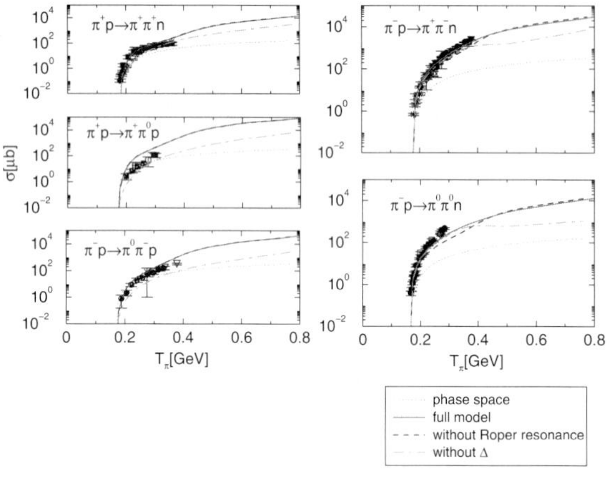


Figure 5. The total cross sections for pion induced two pion production on the proton are shown as functions of the kinetic energy of the incident pion. solid curve: full calculation, dashed : without Roper; dash-dotted: without $\Delta$; dotted: phase space

So far, total cross sections have been worked out. The $\Delta 33$ resonance has a very strong effect on the total cross sections. The Roper may be studied best in the reaction $\pi^- p \to \pi^0 \pi^0 n$, see Fig.5. The Crystal Ball Collaboration is investigating these and other reactions[15].

## 5. The $N^*(1440)$ in electromagnetic reactions

The electroexcitation of the Roper has been worked out in a non-relativistic constituent quark model by Cano and Gonzalez[16]. The helicity amplitude $A_{1/2}$ is found to have a zero for $Q^2 = 0.4GeV^2$. This zero can be pushed up to about $Q^2 = 1.0GeV^2$ depending on the parameterization adopted, see Fig.6.

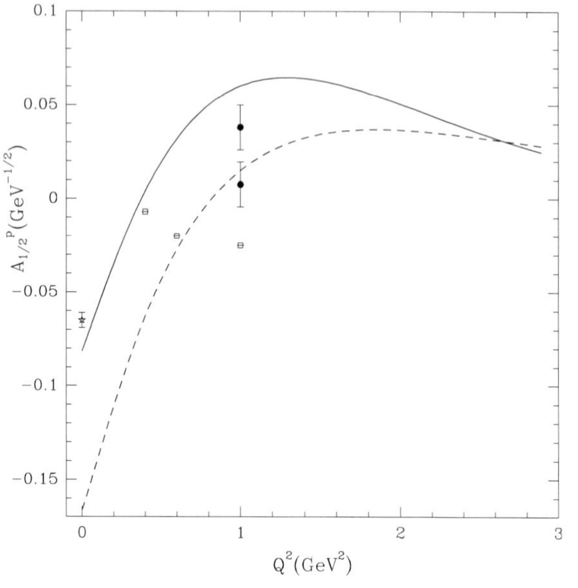


Figure 6. Helicity production amplitude of the Roper assuming a quark model[16].

The experimental data we possess has been taken more than twenty years ago and is somewhat ambiguous, as is discussed in Ref.[17]. A preliminary estimate based on the meson-exchange model [8] is shown in Fig.7. Here it is assumed that the $\sigma N$ channel is dominant.

This implies that the incoming photon attaches to the proton, since there is no coupling to the effective $\sigma$ meson. The helicity amplitude now remains negative for momentum transfers well beyond $Q^2 = 1GeV^2$. The reason is that in this model the helicity amplitude closely follows the electromagnetic form factor of the proton. Should this feature survive more elaborate investigations, one might have a possiblity to distinguish between a triple quark structure and other explanations for the Roper resonance by the electroproduction data.

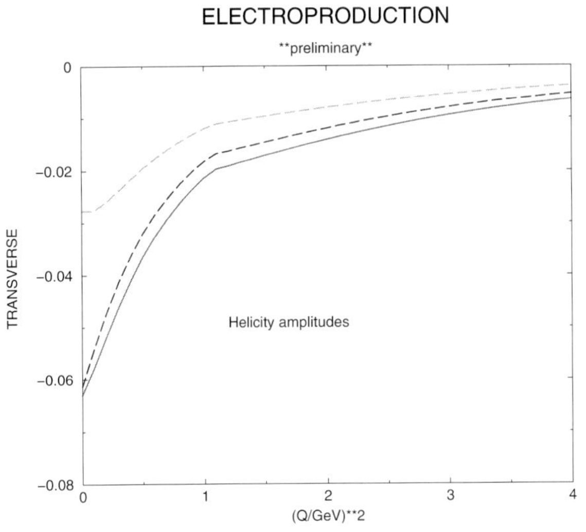


Figure 7. Helicity production amplitude of the Roper assuming dominance of the $\sigma N$ reaction channel (solid line); preliminary.

## 6. Conclusions

The field of baryon resonances continues to provide challenging puzzles, as the experimental studies start to investigate resonances beyond the lowest multiplets and their decays. In the case of the Roper, the near future will bring a large amount of new experimental data.

## 7. Acknowledgements

Thanks is given to S.Schneider and J.Speth for the collaboration concerning the Jülich meson-exchange model. Discussions with H. Clement, P. Morsch, D.B. Leinweber and A.W. Thomas are gratefully acknowledged.

## REFERENCES

1. N. Isgur and G.Karl, Phys. Rev. D19 (1979) 2653.
2. S. Capstick, nucl-th/008028.
3. U.Loring et al., Eur. Phys. J. A10(2001)309.
4. L.Y.Glozman et al., Phys. Rev. D58 (1998) 094030.
5. W.Melnitchouk et al., hep-lat/0202022.
6. D.B.Leinweber et al., Phys. Rev. Lett. 86 (2001) 5011.
7. F.X.Lee et al., hep-lat/0208070.
8. O.Krehl et al., Phys.Rev.C62 (2000)025207.
9. W. Brodowski et al., Phys. Rev. Lett. 88 (2002) 192301.
10. L. Alvarez-Ruso et al., Nucl.Phys. A633 (1998) 519.
11. H.P. Morsch and P.Zupranski, Phys. Rev. C61 (1999)024002.
12. V. Bernard, N. Kaiser, Ulf-G.Meißner, Nucl. Phys. A619 (1997) 261
13. N. Fettes, V. Bernard, Ulf-G.Meißner, Nucl. Phys. A669 (2000) 269
14. S.Schneider et al., to be publ.
15. J.R.Comfort, $\pi N$ Newsletter 16 (2002)31.
16. F.Cano and P.Gonzalez, Phys. Lett. B431 (1998) 270.
17. Z. Lee, V.Burkert, and Z.Li, Phys. Rev. D46 (1992) 70.

ELSEVIER

Proceedings of ICHEP 2002, pp. 662–667
*S. Bentvelsen, P. de Jong, J. Koch and E. Laenen (Editors)*

31st INTERNATIONAL CONFERENCE ON HIGH ENERGY PHYSICS AMSTERDAM

# Topics in Meson Spectroscopy *

Stephen Godfrey[a] [b]

[a]Department of Physics, Carleton University, Ottawa K1S 5B6 CANADA

[b]DESY, Deutsches Elektronen-Synchrotron, D22603 Hamburg, GERMANY

In this mini-review I discuss three topics in meson spectroscopy. The production of heavy quarkonium states, S-wave scattering below 1 GeV, and exotic hybrid meson production. This is not intended to be a comprehensive review, just an overview of several topics of current interest.

## 1. INTRODUCTION

Meson spectroscopy is an extremely broad subject, covering $\pi\pi$ scattering with $\sqrt{s} < 1$ GeV to the heaviest $b\bar{b}$ states with masses greater than 10 GeV [1,2]. The goal at both ends of the energy scale is to understand QCD in the nonperturbative regime. It is impossible to do justice to such a large body of work in a short talk so I restrict myself to a few selected topics. Many other aspects of meson spectroscopy are covered in other contributions where the interested reader can obtain a more balanced view of recent developments.

In a very short period of time, the subject of heavy quarkonium has come to the forefront. This is the result of two recent developments; the CLEO collaboration is taking data at the $\Upsilon(3S)$ and $\Upsilon(2S)$ which will exceed the world's existing data by a large factor and the B-factories have observed a number of charmonium states in B-decay [3–5] culminating in the discovery of the $\eta_c'$ in $B \to \eta_c' K$ by the Belle collaboration [3]. These measurements have led to the hope that some of the missing charmonium states may also be observed in B decay. Simultaneously, there has been great progress in Lattice QCD [6] so that we have progress on both the experimental and theoretical fronts.

On a second front there continues to be vigorous discussion about the structure seen in $\pi\pi$ S-wave scattering below 1 GeV. My impression is that a consensus is emerging among workers in the field with respect to general features. However, there continues to be considerable debate about the details. Low energy S-wave scattering is discussed in Section 3.

Finally, one of the most important topics in meson spectroscopy is the issue of gluonic excitations. To highlight this topic I briefly describe a proposal that should make a definitive statement about the existence of these objects and map out their properties.

## 2. HEAVY QUARKONIUM

The discovery of a D-wave $b\bar{b}$ state by the CLEO collaboration [7] and the discovery of the $\eta_c'$ in B-decay by the BELLE collaboration [3] represent important advances in heavy quarkonium physics. With continued running by CESR on the $\Upsilon(2S)$ and $\Upsilon(3S)$, large samples of B-mesons being collected by the B-factories, and future measurements at CLEO-c/CESR-c, I anticipate further discoveries in the near future. At the same time, advances in Lattice QCD calculations of the heavy quarkonium spectra are leading to quantitative predictions which need to be tested against experiment [6]. Taken together we should expect a much better understanding of heavy quarkonium from first principles QCD.

Numerous $b\bar{b}$ states are expected to exist below the $B\bar{B}$ decay threshold. However, only spin triplet states have been observed in the $b\bar{b}$ system and only 2 spin-singlet states have been ob-

*This work was partially funded by the Natural Sciences and Engineering Research Council of Canada.


doi:10.1016/S0920-5632(02)02177-1

served in $c\bar{c}$. Given that there is a wide variation of the spin dependent splittings predicted by various calculations, the observation of additional quarkonium states, including the missing spin-singlet states, would pose an important test of the calculations (comparisons are given in Ref. [8–10]). Similarly, various calculations predict different values for the mass of $b\bar{b}$ D-wave cog (the spin weighted mass of the triplet states). Observation of these states would be a powerful discriminator of the various calculations.

CESR/CLEO has just completed a high statistics data taking run at the $\Upsilon(3S)$ and will soon be running at the $\Upsilon(2S)$. We expect that analysis of the data will yield a rich spectroscopy. To give direction to searches we use the quark model to estimate radiative widths and branching ratios.

### 2.1. D-Wave States

There are several strategies to search for D-wave mesons [8]. The first is via direct scans in $e^+e^-$ collisions to produce the $^3D_1$. However, given the small coupling of the $^3D_1(b\bar{b})$ to the photon and the effects of line smearing this approach would require substantial statistics. A second approach is via the electromagnetic cascades $\Upsilon(3S) \to \gamma\chi_b' \to \gamma\gamma{}^3D_J$. Using quark model estimates of both E1 dipole transitions and decays to final state hadrons [11] we estimate the number of events for $4\gamma$ cascades that proceed via various intermediate states [8]. These are given in Table 1. We expect $\sim$ 38 events per $10^6$ $\Upsilon(3S)$'s to be produced via $^3D_J$ states. Backgrounds can be reduced using the $\Upsilon \to e^+e^- + \mu^+\mu^-$ final states as event tags. A substantial background will be $4\gamma$ cascades proceeding via the $2^3S_1$ state which will also produce $\sim$ 38 events per $10^6$ $\Upsilon(3S)$'s. These events can be separated from the $1^3D_J$ intermediate state using the fact that the energies of the intermediate photons are different. The CLEO collaboration successfully employed this strategy for the first observation of a triplet $\Upsilon(1D)$ state which was announced at this conference [7].

### 2.2. $\eta_b(nS)$ States

$\eta_b(nS)$ can be produced via magnetic dipole (M1) transitions $\Upsilon(nS) \to \eta_b(n'S) + \gamma$ [9]. The rates for magnetic dipole transitions in quarkonium ($Q\bar{Q}$) bound states are given in the nonrelativistic approximation by

$$\Gamma(^3S_1 \to ^1S_0 + \gamma) = \frac{4}{3}\alpha\frac{e_Q^2}{m_Q^2}|\langle f|j_0(kr/2)|i\rangle|^2\omega^3 (1)$$

where $\alpha$ is the fine-structure constant, $e_Q$ is the quark charge in units of $|e|$ ($-1/3$ for $Q = b$), $m_Q$ is the quark mass (which we take equal to 4.8 GeV/$c^2$) and $k$ ($\omega$) is the emitted photon's momentum (energy). The available phase space for allowed transitions (the principle quantum number is unchanged) is small leading to a small partial width. In contrast, the hindered transitions (the principal quantum number changes)

Table 1
Predicted numbers of $4\gamma$ $e^+e^-$ events corresponding to $3S \xrightarrow{\gamma} 2P \xrightarrow{\gamma} 1D \xrightarrow{\gamma} 1P \xrightarrow{\gamma} 1S \to e^+e^-$ or $3S \xrightarrow{\gamma} 2P \xrightarrow{\gamma} 2S \xrightarrow{\gamma} 1P \xrightarrow{\gamma} 1S \to e^+e^-$ per $10^6$ $\Upsilon(3S)$ decays. From Ref. [8].

| $2^3P_J$ state | Next state | $1^3P_J$ state | Events |
|---|---|---|---|
| $2^3P_2$ (87) | $1^3D_3$ | $1^3P_2$ | 7.8 |
| | $1^3D_2$ | $1^3P_2$ | 0.3 |
| | | $1^3P_1$ | 2.7 |
| | $1^3D_1$ | $1^3P_2$ | 0.0 |
| | | $1^3P_1$ | 0.1 |
| | | $1^3P_0$ | 0.0 |
| | $2^3S_1$ | $1^3P_2$ | 4.1 |
| | | $1^3P_1$ | 8.8 |
| | | $1^3P_0$ | 0.4 |
| $2^3P_1$ (100) | $1^3D_2$ | $1^3P_2$ | 2.5 |
| | | $1^3P_1$ | 20.1 |
| | $1^3D_1$ | $1^3P_2$ | 0.1 |
| | | $1^3P_1$ | 3.3 |
| | | $1^3P_0$ | 0.4 |
| | $2^3S_1$ | $1^3P_2$ | 7.5 |
| | | $1^3P_1$ | 15.9 |
| | | $1^3P_0$ | 0.7 |
| $2^3P_0$ (123) | $1^3D_1$ | $1^3P_2$ | 0.0 |
| | | $1^3P_1$ | 0.3 |
| | | $1^3P_0$ | 0.0 |
| | $2^3S_1$ | $1^3P_2$ | 0.3 |
| | | $1^3P_1$ | 0.7 |
| | | $1^3P_0$ | 0.0 |

Table 2
Predictions for branching ratios of M1 transitions between $n^3S_1$ and $n'^1S_0$ $b\bar{b}$ levels taking into account relativistic corrections. From Ref. [9] using wavefunctions from Ref. [2].

| | Transition | $\mathcal{B}(10^{-4})$ |
|---|---|---|
| $\Upsilon(3S)$ | $\to 3^1S_0$ | 0.10 |
| $(\Gamma_{tot} = 52.5$ keV) | $\to 2^1S_0$ | 4.7 |
| | $\to 1^1S_0$ | 25 |
| $\Upsilon(2S)$ | $\to 2^1S_0$ | 0.21 |
| $(\Gamma_{tot} = 44$ keV) | $\to 1^1S_0$ | 13 |
| $\Upsilon(1S)$ | $\to 1^1S_0$ | 2.2 |
| $(\Gamma_{tot} = 26.3$ keV) | | |

have large available phase space. In the nonrelativistic limit the overlap integral for hindered transitions is zero due to the orthogonality of the initial and final wavefunctions. However, relativistic corrections from the hyperfine interaction leads to differences in the $^3S_1$ and $^1S_0$ wavefunctions resulting in a non-zero overlap. The branching ratios for $\Upsilon(nS) \to \eta_b(n'S)\gamma$ are given in Table 2. Note that the radiative widths are quite sensitive to model dependent details. From these predictions we expect $\eta_b$'s to be produced at a substantial rate. Another possible production channel for $\eta_b$'s is the process $\Upsilon(3S) \to h_b(^1P_1)\pi\pi \to \eta_b + \gamma + \pi\pi$ [12]. The BR for $\Upsilon(3S) \to h_b(^1P_1)\pi\pi$ is expected to be 0.1-1 % while the BR for $h_b(^1P_1) \to \eta_b + \gamma$ is estimated to be $\sim 50\%$.

### 2.3. Production of $^1P_1$ States

We consider two types of cascades to produce the $^1P_1$ $c\bar{c}$ and $b\bar{b}$ states [10]. The first starts with an $\Upsilon(3S)$ or $\psi(2S)$ produced in $e^+e^-$ annihilation which then underdoes an M1 transition to the $\eta(2S)$ state followed by an E1 transition to the $h(^1P_1)$ state and a final E1 transition to the $\eta(1S)$ state:

$$\Upsilon(3S) \to \eta_b(2S) + \gamma \to h_b + \gamma\gamma \to \eta_b + \gamma\gamma\gamma \quad (2)$$

$$\psi(2S) \to \eta_c(2S) + \gamma \to h_c + \gamma\gamma \to \eta_c + \gamma\gamma\gamma \quad (3)$$

The second cascade consists of the single pion emission transition from the initial $^3S_1$ state followed by an E1 transition to the $\eta(1S)$:

$$\Upsilon(3S) \to h_b + \pi \to \eta_b + \gamma + \pi \quad (4)$$

$$\psi(2S) \to h_c + \pi \to \eta_c + \gamma + \pi \quad (5)$$

To determine if either of these cascades is promising we need the appropriate branching ratios and hence the partial widths. In the quark model they are given by:

$$\Gamma(^1S_0 \to {}^1P_1 + \gamma) = \frac{4}{3}\alpha\, e_Q^2\, \omega^3\, |\langle {}^1P_1|r|{}^1S_0\rangle|^2 \quad (6)$$

so that $\Gamma[\eta_b(2S) \to h_b(1P) + \gamma] = 2.3$ keV. In addition we need the hadronic widths for the $\eta_b(2S)$. There are considerable theoretical uncertainties which to a large extent can be reduced by relating ratios of theoretical expressions to observed hadronic widths. For the $\eta_b(2^1S_0)$ we take

$$\Gamma[^1S_0 \to gg] = \frac{27\pi}{5(\pi^2-9)\alpha_s} \times \Gamma[^3S_1 \to ggg] \quad (7)$$

where 1st order QCD corrections are included in our numerical results but are not explicitly shown. This gives $\Gamma[\eta_b(2^1S_0) \to gg] = 4.1 \pm 0.7$ MeV. Combining the resulting branching ratios for $\mathcal{B}[\Upsilon(3S) \to \eta_b' + \gamma]$ and $\mathcal{B}[\eta_b' \to h_b + \gamma]$ gives $\mathcal{B}[\Upsilon(3S) \to \eta_b'\gamma \to h_b\gamma\gamma] = 2.6 \times 10^{-7}$. This would result in only 0.3 events per $10^6$ $\Upsilon(3S)$'s. A similar exercise yields $B[\psi(2S) \to \eta_c'\gamma \to h_c\gamma\gamma] = 10^{-6}$ or 1 event per $10^6$ $\psi'$'s.

A more promising approach utilizes the hadronic decay $\Upsilon(3S) \to \pi 1^1P_1$ which is estimated to have a BR of around 0.1% [13]. Again we need to estimate the radiative and hadronic widths of the $^1P_1$ states:

$$\Gamma[^1P_1 \to {}^1S_0 + \gamma] = \frac{4}{9}\alpha\, e_Q^2\, \omega^3\, |\langle {}^1S_0|r|{}^1P_1\rangle|^2 \quad (8)$$

giving $\Gamma[h_b(1P) \to \eta_b(1S) + \gamma] = 37$ keV and

$$\Gamma[^1P_1 \to ggg] = \frac{5}{2n_f} \times \Gamma[^3P_1 \to q\bar{q}g] \quad (9)$$

giving $\Gamma[h_b(1P) \to ggg] = 50.8$ keV. Combining the branching ratios in this decay chain yields

$$\mathcal{B}[\Upsilon(3S) \to 1^1P_1 + \pi \to 1^1S_0 + \gamma\pi] = 4 \times 10^{-4} \quad (10)$$

which would yield 400 events per $10^6$ $\Upsilon(3S)$'s. Similarly one obtains

$$\mathcal{B}[\psi(2S) \to 1^1P_1 + \pi \to 1^1S_0 + \gamma\pi] = 3.8 \times 10^{-4} \quad (11)$$

In both cases it appears that the $1^1P_1$ should be produced in sufficient numbers to be observed.

There is also the possibility that charmonium states can be observed in $B$ decay. This was recently highlighted by the Belle observation of the $\eta_c(2S)$ in $B \to \eta_c(2S)K \to KK_sK^-\pi^+$ [3]. Belle had previously reported the observation of the $\chi_{c0}$, $\chi_{c2}$ [4] and the $\chi_{c1}$ had been observed by both the BaBar and Belle collaborations[5,4]. It had long been proposed that the $h_c$ could be seen in $B$ decays [14] and the recent Belle result has led to a renewal of interest in this process [15].

## 3. S-WAVES BELOW 1 GEV

The subject of the S-waves below 1 GeV garners some of the most vigorous discussion in hadron spectroscopy [16]. Thus, the discussion below reflects my impression of this subject which may not be held by all workers in this field. Despite this note of caution I believe that there is a growing consensus on the general features of S-wave scattering in this kinematic region, if not on all specific details.

The phenomenology of the $0^{++}$ sector reflects the many different components of the Fock space: conventional $q\bar{q}$ mesons, $q\bar{q}q\bar{q}$ states, glueballs, meson-meson molecules, and threshold effects. It is clearly a challenge to understand this sector. One starts by analyzing the raw data to produce phase shift plots and then one attempts to intrepret these results in terms of underlying models.

Much of the information on low energy S-wave $\pi\pi$ scattering comes from measuring $\pi^-p \to \pi^+\pi^-n$ at CERN [17] and $\pi^-p \to \pi^0\pi^0n$ at BNL [18]. Unfortunately the experiments provide fewer observables than are needed to unambiguously describe the partial waves. One must therefore make assumptions in the PWA. In particular the role of nucleon spin is ignored and the role of the $a_1$ exchange amplitude is generally neglected assuming the dominance of pion exchange. These assumptions result in ambiguities in the extraction of phase shifts from the PWA.

Kaminski, Lesniak, and Rybicki studied this problem [20] constraining the allowed solutions by imposing unitarity and using the non-observation of inelastic scattering below the $f_0(980)$. They further related the $\pi^+\pi^-$ to the $\pi^0\pi^0$ amplitudes to obtain $\pi\pi$ phase shifts. Including these ingredients they presented what they believe to be a unique physical solution.

Furthermore, by separating the pseudoscalar from the pseudovector exchange amplitudes they obtain the cross sections shown in Fig. 1. The conclusion is that the $a_1$ contributions are significantly different from zero. This is an important point which should be kept in mind in future experiments. High statistics enables the differentiation between different t-channel exchange particles which results in different couplings to final states. It is a very useful bit of information in understanding the properties of the produced excited mesons.

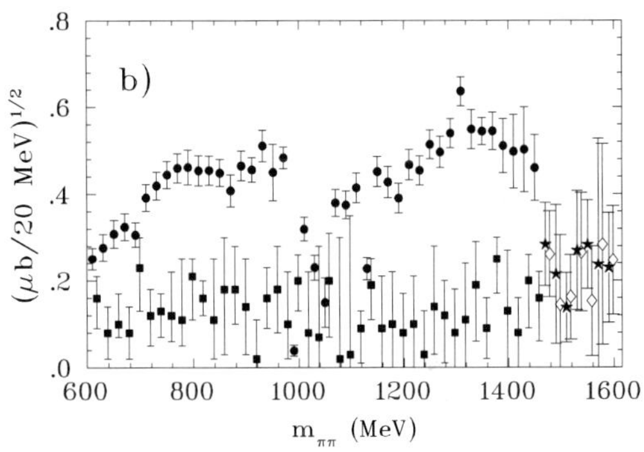


Figure 1. Moduli of the pseudoscalar (circles) and pseudovector amplitudes (squares) for the $\pi\pi$ scattering "down-flat" solution of Ref. [20].

Given this information the next question is how to describe these phase shifts in terms of true resonances and from the dynamics between $K\bar{K}$, $K\pi$ and $\pi\pi$. This was discussed in contributions by Van Beveren and Rupp [21] and by Furman and Lesniak [22]. It is clear that one must look at the meson-meson scattering and extract the phase shifts not the resonance positions. To do this requires a multichannel approach which includes resonances (from constituent $q\bar{q}$ channels) and meson-meson interactions. An interesting and important consequence is that the pole parameters are dependent on the coupling strengths.

As a concrete example of this approach I show results for the coupled channel calculation of Furman and Lesniak [22]. They constructed a coupled channel of two $a_0$ resonances decaying into the $\pi\eta$ and $K\bar{K}$ mesons using the coupled channel Lippman-Schwinger equation:

$$\langle p|T|q\rangle = \langle p|V|q\rangle + \int \frac{d^3 s}{(2\pi)^3} \langle p|V|s\rangle\langle s|G|s\rangle\langle s|T|q\rangle \quad (12)$$

where $T$, $V$, and $G$ are $2 \times 2$ matrices with $G$ the matrix of propagators. The resulting Argand diagrams are shown in Fig. 2. These plots show significantly different widths in the two channels which agrees with E852 and Crystal Barrel observations. The $I = 0$ sector was also studied in this approach.

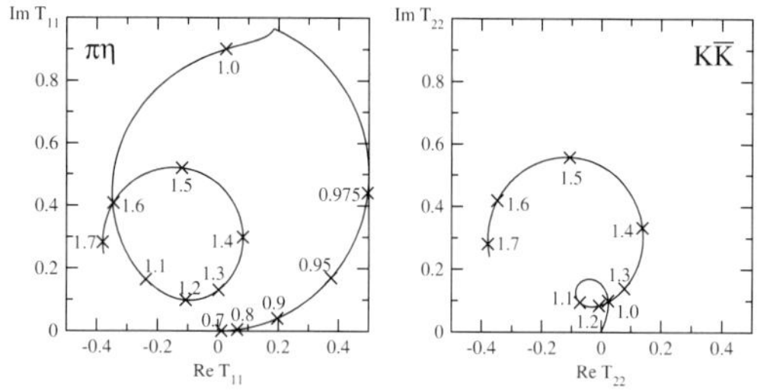


Figure 2. Argand diagrams for $\pi\eta$ and $K\bar{K}$ elastic scattering amplitudes. Numbers denote effective mass in units of GeV. From Ref. [22].

What these multichannel approaches by Van Beveren and Rupp [21] and by Furman and Lesniak [22] and by others show is that one can obtain a good description of the low S-wave spectrum in this approach. My impression is that workers in this field have come to a consensus on the general description of S-wave scattering although there continues to be considerable debate on the details and interpretation of what these results mean.

## 4. EXOTIC HYBRID MESONS

The most important qualitative question in "soft QCD" is if and how glue manifests itself in the hadron spectrum. Lattice QCD and hadron models predict glueballs, objects with no constituent quark content, and hybrids, hadrons with quark content and an explicit gluon degree of freedom. It is expected that the most definitive evidence for a non $q\bar{q}$ state is to discover a state with $J^{PC}$ quantum numbers inconsistent with the quark model, so called exotics. Because the lowest lying exotic glueballs are quite massive this directs the search to exotic hybrid mesons.

The emerging picture of hybrid mesons is that of transverse excitations of the flux tube. For the light quarks this leads to the lowest mass hybrids of mass $\sim 1.9$ GeV with doubly degenerate quantum numbers. In particular, states are predicted with $J^{PC} = 0^{+-}$, $1^{-+}$, and $2^{+-}$ which are inconsistent with the quark model and would therefore provide the smoking gun for new hadronic physics. There have been a number of sightings of such states [1]. For example the Brookhaven E852 experiment has seen evidence for the $\hat{\rho}(1^{-+})$ exotic in the reaction $\pi^- p \to \pi^+\pi^-\pi^- p$ [19].

Here I briefly mention an ambitious proposal to search for and map out exotic gluonic excitations at the Jefferson Laboratory in Virginia. The proposal takes advantage of the fact that exotic hybrid mesons have their quark spins aligned in an $S = 1$ state. In standard hadroproduction the incoming beam consists of $\pi$ mesons whose quark spins are in an $S = 0$ state. Since exotic hybrids have quark spins in an $S = 1$ state it is believed that hadroproduction of exotic hybrids is suppressed. In contrast, via vector meson dominance, a photon beam consists of quarks in an $S = 1$ state so it is believed that exotic hybrids are more readily produced in diffractive photoproduction by "plucking" the gluon flux tube. To this end there is a proposal to upgrade CEBAF to 12 GeV and produce photons via coherent bremsstrahlung with a spectrum that peaks at 9 GeV. This project would add considerably to our knowledge of hybrid mesons and our understanding of confinement.

## 5. SUMMARY

Over the last while, there have been significant developments in heavy quarkonium physics. We expect this progress to continue with a high likelihood that the CESR collaboration should find evidence for the $2^1S_0$ $1^1S_0$, $1^1P_1$ $b\bar{b}$ states. It is also possible that they will find evidence for the $3^1S_0$, $1^3D_1$ and $1^3D_3$ states. At the same time we expect that in the near future the various B-factories will discover some of the missing charmonium states in B-decay. Beyond this, the CLEO-c/CESR-c project would provide an additional avenue to study the charmonium system. Taken together, this would represent the largest advances in heavy quarkonium spectroscopy in two decades. These results would provide an important benchmark against which to measure the results of lattice QCD and the various models in the literature.

The situation with S-wave scattering at low energies is converging to a consensus on the overall details. Good agreement between theory and experiment is achieved by describing low energy scattering in the framework of a multichannel quark model. Nevertheless, there is still vigorous debate about the details and the interpretation.

Finally, we mention hybrid meson production as an important tool in understanding QCD in the low $Q^2$ regime and the nature of confinement. In the near future we look forward to results from, for example, the COMPASS experiment at CERN and photoproduction results at HERA. In the longer term we expect that the GlueX experiment at Jefferson Lab will provide a major leap forward in our understanding of gluon dynamics in the regime of Soft QCD.

## REFERENCES


1. For a recent review see S. Godfrey and J. Napolitano, Rev. Mod. Phys. **71**, 1411 (1999). See also F. Close, hep-ph/0110081, LP 01, Rome, Italy, 23-28 Jul 2001.
2. S. Godfrey and N. Isgur, Phys. Rev. **D32**, 189 (1985).
3. BELLE collaboration (S.K. Choi *et al.*), Phys. Rev. Lett. **89**, 102001 (2002), Erratum *ibid*, **89**, 129901 (2002) [hep-ex/0206002].
4. Belle Collaboration (K. Abe *et al.*), Phys. Rev. Lett. **88**, 031802 (2002); hep-ex/0107050; hep-ex/0202028.
5. BaBar Collaboration, (B. Aubert *et al.*), Phys. Rev. **D65**, 032001 (2002).
6. See e.g. C.T. Davies *et al.* (UKQCD Collaboration), Phys. Rev. **D58**, 054505 (1998); T. Manke *et al.*, Phys. Rev. **D62**, 114508 (2000); M. Okamoto *et al.*, Phys. Rev. **D65**, 094508 (2002).
7. CLEO Collaboration (S.E. Csorna *et al.*), [hep-ex/0207060].
8. S. Godfrey and J. Rosner, Phys. Rev. **D64**, 097501 (2001).
9. S. Godfrey and J. Rosner, Phys. Rev. **D64**, 074011 (2001).
10. S. Godfrey and J. Rosner, Phys. Rev. **D66**, 014012 (2002).
11. W. Kwong and J.L. Rosner, Phys. Rev. **D38**, 279 (1988).
12. Y.-P. Kuang and T.-M. Yan, Phys. Rev. **D24**, 2874 (1981); M. B. Voloshin, Yad. Fiz **43**, 1571 (1986) [Sov. J. Nucl. Phys. **43**, 1011 (1986)].
13. M. B. Voloshin, Sov. J. Nucl. Phys. **43**, 1011 (1986); M. B. Voloshin and V. I. Zakharov, Phys. Rev. Lett. **45**, 688 (1980).
14. G. Bodwin *et al.*, Phys. Rev. **D46**, R3703 (1992); F. Close *et al.*, Phys. Rev. **D57**, 5653 (1998).
15. M. Suzuki, Phys. Rev. **D66**, 037503 (2002); Y.F. Gu, Phys. Lett. **B538**, 6 (2002); E.J. Eichten, K. Lane, and C. Quigg, [hep-ph/0206018].
16. F. Close and N. Tornqvist [hep-ph/0204205].
17. G. Grayer *et al.*, Nucl. Phys. **B75**, 189 (1974); H. Becker *et al.*, Nucl. Phys. **B150**, 301 (1979); *ibid*, **B151**, 46 (1979).
18. J. Gunter *et al.*, (E852 Collab.) Phys. Rev. **D64**, 07003 (2001).
19. E852 Collaboration (G.S. Adams et al.) Phys. Rev. Lett.81, 5760 (1998)
20. R. Kaminski, L. Lesniak, K. Rybicki, Eur. Phys. J. **C4**, 1 (2002).
21. E. van Beveren and G. Rupp, hep-ph/0207022 and references therein.
22. Furman and Lesniak, [hep-ph/0203255].

ELSEVIER

Proceedings of ICHEP 2002, pp. 668–670
*S. Bentvelsen, P. de Jong, J. Koch and E. Laenen (Editors)*

31st INTERNATIONAL CONFERENCE ON HIGH ENERGY PHYSICS AMSTERDAM

# Measurement of high lying nucleon resonances and search for a missing state in double charged pion electroproduction off the proton

E.Golovach[a], V.Burkert[b], V.Mokeev[c,a], M.Ripani[d], M.Anghinolfi[d], M.Battaglieri[d], A.Boluchevskii[e], G.Fedotov[a], R.De Vita[d], B.Ishkhanov[a,e], M.Osipenko[a,d], G.Ricco[d], V.Sapunenko[b], N.Shvedunov[e], M.Taiuti[d] and CLAS collaboration.

[a]Nuclear Physics Institute, Moscow State University, Vorob'evy gory, 119992 Moscow. Russia.

[b]Thomas Jefferson National Accelerator Facility, Newport News, VA 23606. USA.

[c]Christopher Newport University, Newport News, Virginia 23602, USA.

[d]Istituto Nazionale di Fisica Nucleare, Via Dodecanneso 33, I-16146 Genova. Italy.

[e]Physics Department, Moscow State University, 119992 Moscow. Russia.

We have analyzed data taken by the CLAS collaboration in double charged pion electroproduction off the proton. We have used an isobar model specifically developed for the analysis of the $N\pi\pi$ channel. This approach relates baryon resonance parameters to the experimental cross section, allowing the extraction of electromagnetic and some strong couplings from the fit to the data. A remarkable feature of the CLAS data is a resonance structure around 1.7 GeV which has not been seen in double pion photoproduction. The structure appears at all $Q^2$ values from 0.5 to 1.5 GeV$^2$ and it can not be reproduced with conventional N$^*$ parameters.

## 1. Introduction

Studies of double charged pion production by virtual photons offer promising opportunities to investigate the structure of high lying nucleon excitations that decay preferably into multi-pion final states. The virtual photon cross-section exhibits resonance structures at hadron invariant masses $W$ above 1.6 GeV, not seen in experiments with real photons [1]. Moreover, this exclusive channel can be used to search for missing baryon states predicted by the quark model [2–4], which may have low probability for decay to a single pion final state, and a much larger probability for two pion decays.

Here we discuss the results of an analysis of CLAS data on two pion production by virtual photons at $W$ below 2.1 GeV and $Q^2$ below 1.5 GeV$^2$ [5]. Our goals were: a) to extract information on high lying $N^*$ ($M > 1.6$ GeV) photocouplings and b) to search for signals from missing baryon states.

## 2. The approach for data analysis

A phenomenological approach [6–8] has been developed to describe double charged pion production, relating N$^*$ photo and strong couplings to the measured cross-sections and asymmetries, and allowing extraction of N$^*$ parameters from fits to the data.

Nucleon resonance excitations (all good established states below 2.0 GeV masses with observed two pion decays) are described by a Breit-Wigner ansatz [6]. The strong decay couplings are taken from hadronic data analysis [10,11] while the photocoupling values are treated as free parameters determined by a fit to the data. Non-resonant mechanisms in the $\pi\Delta$ sub-channel are described by Reggetized gauge-invariant Born terms [6]. A diffractive ansatz is used to account for non-resonant mechanisms in the $\rho p$ sub-channel [7,8]. Our approach provides a reasonable description of world data [1,9] in the kinematical region of interest to us ($W < 3.5$ GeV $Q^2$,$-t \sim$ few GeV$^2$).


doi:10.1016/S0920-5632(02)02181-3

## 3. The results of CLAS data analysis.

The measured integrated cross-section on double charged pion production is presented in Fig 1. The analysis is based on $(\pi^+\pi^-)$, $(\pi^+p)$ invariant mass and $\pi^-$ angular distributions, measured in 25 MeV $W$ bins and three $Q^2$ bins centered at 0.65, 0.95 and 1.3 GeV$^2$.

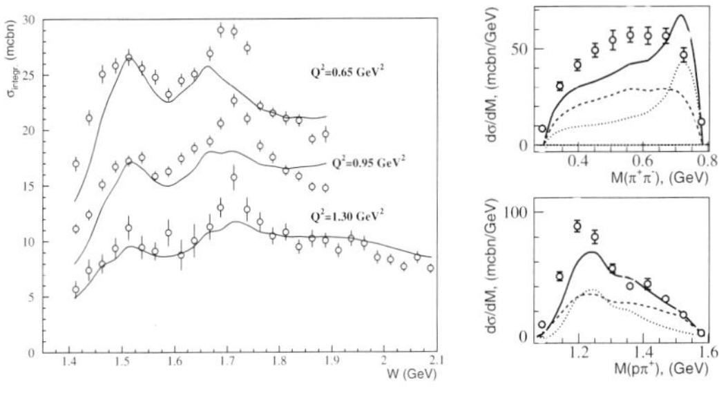


Figure 1. Left: total cross section for $\gamma_v p \to p\pi^+\pi^-$ as a function of W. Data from CLAS for three $Q^2$ bins (see text) Error bars are statistical only. The curves represent our initial step calculation. Right: $\frac{d\sigma}{dM_{\pi^+\pi^-}}$ and $\frac{d\sigma}{dM_{p\pi^+}}$ from CLAS $Q^2$=0.8–1.1(GeV/c)$^2$) and $W$=1.7–1.725 GeV. Dotted lines include all resonances, the dashed lines include only the non-resonant part, and the solid lines are the full calculation.

First, we tried to describe the CLAS data [5] with $N^*$ photocouplings taken from an interpolation of the world data in the framework of the Single Quark Transition Model (SQTM) [12]. Quark-model predictions [12] for the $D_{15}(1675)$ and $D_{13}(1700)$ states were used, since these states can not be excited in the SQTM. The calculation reproduces gross features of the integrated cross-sections, but fails to describe the structure around 1.7 GeV (Fig 1). The calculated $(\pi^+\pi^-)$ invariant mass distributions at $W$ around 1.7 GeV exhibit a sharp $\rho$ meson production peak, contrary to the smooth behavior of the experimental data (right panel on Fig. 1). We found that the major contribution to this $\rho$ peak arises from the $P_{13}(1720)$.

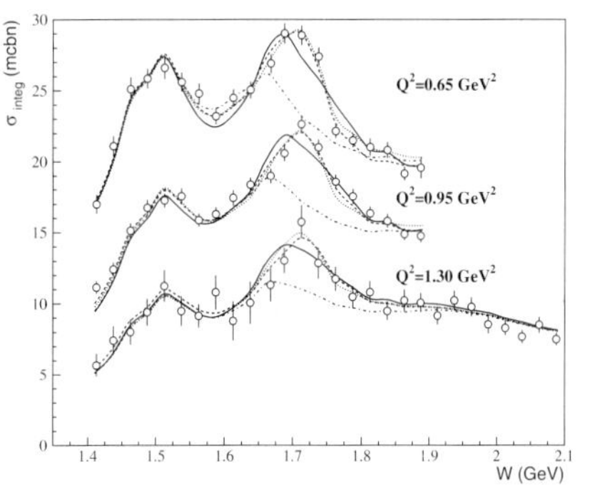


Figure 2. Integrated cross section from CLAS data [5] at three $Q^2$ intervals and results of the fit (see text).

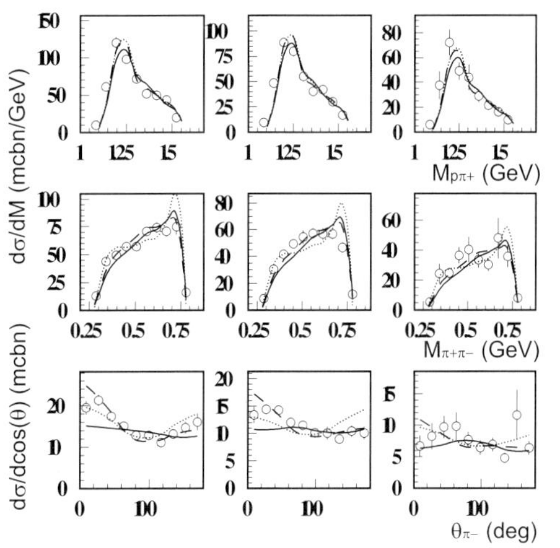


Figure 3. $\frac{d\sigma}{dM_{\pi^+\pi^-}}$ $\frac{d\sigma}{dM_{p\pi^+}}$ and $\frac{d\sigma}{d\cos\theta_{\pi^-}}$ from CLAS at $W$=1.7–1.725 GeV at three $Q^2$ intervals (left to right) and the results of three fits (see text).

Next, we attempted to eliminate the above mentioned discrepancies by the following means: a reduction of the $P_{13}(1720)$ photocoupling by more than a factor of two; implementation of a sizeable Roper excitation, and simultaneous variations of the photocouplings, considering uncertainties in experimental data and in the SQTM fit, while keeping the $N^*$ strong couplings inside the uncertainties obtained in the analysis [10]. We obtained a reasonable description of the data apart from the structure around 1.7 GeV. (dash-dotted lines on Fig 2) without changing the $N^*$ photocouplings by more than 30% from the initial values from the SQTM [12]. This suggests that the single quark transition is a relevant mechanism for $N^*$ excitation at $Q^2 < 1.5$ GeV$^2$. Considering resonances that could contribute to the structure in the data around 1.7 GeV, the $D_{13}(1700)$, $P_{11}(1710)$ and $P_{13}(1720)$ states are the most likely candidates. Accordingly, we performed three separate fits, in which the strong couplings and $A_{1/2}$ and $A_{3/2}$ were varied for a) $D_{13}(1700)$ b)$P_{11}(1710)$&$D_{13}(1700)$, keeping $P_{13}(1720)$ strong couplings inside of the uncertainties of its published parameters [10,11] and for c) $D_{13}(1700)$&$P_{13}(1720)$ outside of the the uncertainties [10,11] The results are shown in Figs. 2 and 3 by solid (a), dotted (b), and dashed (c) lines, respectively. All fits reproduce the integrated cross-section reasonably well (Fig 2), but fits (a) and (b) fail to reproduce the $\pi^-$ angular and/or the $(\pi^+\pi^-)$ invariant mass distributions ($\chi^2$/dof are 5.2 and 4.3 respectively).

The best result is achieved with fit (c) with $\chi^2/\text{dof} = 3.4$ (dashed line). However, the strong decay parameters of $P_{13}(1720)$ that emerged from our fit are completely different from the values available from analysis of data with hadron probes (see Table 1). Since such a drastic change in the $P_{13}(1720)$ strong coupling is not confirmed in any other experiment, we studied the possibility of fitting the structure at 1.7 GeV by using a missing baryon state with decay widths and quantum numbers determined from a fit to the data, keeping the decay widths of the conventional $P_{13}(1720)$ at the values from [10,11]

A missing state with quantum numbers $P_{I3}(1720)$ (where the index $I$ stands for isospin 1/2 or 3/2) provides a description of the data indistinguishable within errors in the data to the fit (c). New precise data on the strong decay branching fractions of the $P_{13}(1720)$ state are necessary to distinguish between these two ways to describe the structure at 1.7 GeV in double charge pion production by virtual photons, which has been observed for the first time by the CLAS Collaboration.

| | $\Gamma_{\Delta\pi}/\Gamma_{tot}$ | $\Gamma_{\rho N}/\Gamma_{tot}$ |
|---|---|---|
| PDG | absent | 70–85% |
| Our fit (c) | 63±13% | 19±9% |

Table 1
$\Delta\pi$ and $\rho N$ branching fractions for $P_{13}(1720)$.

## REFERENCES


1. ABBHHM Collaboration, Phys. Rev. 175 (1968) 1669.
2. R.Koniuk, N.Isgur, Phys. Rev. D21 (1980) 1868.
3. M.M.Giannini, Rep. Progr. Phys. 54 (1990) 453.
4. S.Capstick, W.Roberts, Phys.Rev. D48 (1994) 4570.
5. V.Burkert, BARYON2002 conference, (to be published in the conference proceedings); M.Ripani, Proceedings of the Workshop NSTAR2001, ed D.Dreshsel, L.Tiator, 429.
6. M.Ripani, e.a. Nucl. Phys. A672 (2000) 220.
7. M.Ripani, e.a. Phys. of Atom. Nucl. 63 (2000) 76.
8. V.Mokeev, e.a. Phys. of Atom. Nucl. 64 (2001) 1292.
9. K.Wacker, e.a. Nucl. Phys. B144 (1978) 269.
10. D.M.Manley, E.M.Salesky Phys. Rev. D43 (1992) 4002.
11. T.P.Vrana, e.a. Phys. Rep. 328 (2000) 181.
12. V.D.Burkert, Czech. Journ. of Phys. 46 (1996) 627.

ELSEVIER

Proceedings of ICHEP 2002, pp. 671–673
*S. Bentvelsen, P. de Jong, J. Koch and E. Laenen (Editors)*

31st INTERNATIONAL CONFERENCE ON HIGH ENERGY PHYSICS AMSTERDAM

# Light Quark Spectroscopy Results from FOCUS and E687

Stefano Bianco [a] on behalf of the FOCUS and E687 Collaborations *

[a] Laboratori Nazionali di Frascati dell'INFN - via E. Fermi 40, Frascati I-00044

Using the FOCUS spectrometer (experiments 687 and 831 at Fermilab) we confirm the existence of a diffractively photoproduced enhancement in $K^+K^-$ at 1750 MeV/c$^2$ with nearly 100 times the statistics of previous experiments. We also observe a narrow dip structure at 1.9 GeV/c$^2$ in a study of diffractive photoproduction of the $3\pi^+3\pi^-$ final state.

In this paper we present results on the spectroscopy of high-energy light meson photoproduction from the FOCUS (E831) and E687 experiments at Fermilab. Although both experiments were focused on charm physics, a very large sample of diffractively photoproduced light-meson events was also recorded. Details on the detector can be found in Ref. [1], Ref. [2], and references therein.

## 1. Observation of a 1750 MeV/c$^2$ Structure in the Diffractive Photoproduction of $K^+K^-$

The data for this analysis was collected by the Wideband photoproduction experiment FOCUS during the Fermilab 1996–1997 fixed-target run. In FOCUS, a forward multi-particle spectrometer is used to measure the interactions of high energy photons on a segmented BeO target. A photon beam is derived from the bremsstrahlung of secondary electrons and positrons with an $\approx$ 300 GeV endpoint energy produced from the 800 GeV/$c$ Tevatron proton beam. The FOCUS detector is a large aperture, fixed-target spectrometer with excellent vertexing, particle identification, and reconstruction capabilities for photons, $\pi^0$'s, and $K_S$.

Our sample of $K^+K^-$ events, selected using the criteria described in detail in Ref. [1], shows a clear $\phi(1020)$ signal dominating the spectrum (Fig. 1). The diffractive component of the production of the $\phi(1020)$ shows up as a peak in the $p_T$ spectrum. Cutting around this peak by requiring $p_T < 0.15$ GeV/$c$, we select a diffractive sample of $K^+K^-$ events, in which a clear enhancement appears in the mass spectrum near 1750 MeV/c$^2$ (Fig. 2.a). Figure 2.b confirms that the enhancement appears at only low $p_T$. Plotting the $p_T$ spectra in the 1750 region (1640–

*Co-authors are P.L. Frabetti (**INFN and Bologna**), J.M. Link, V.S. Paolone, M. Reyes, P.M. Yager (**UC DAVIS**); J.C. Anjos, I. Bediaga, C. Göbel, J. Magnin, A. Massafferri, J.M. de Miranda, I.M. Pepe, A.C. dos Reis, F. Simão (**CPBF, Rio de Janeiro**); S. Carrillo, E. Casimiro, H. Mendez, A.Sánchez-Hernández,, C. Uribe, F. Vasquez (**CINVESTAV, México City**); L. Agostino, L. Cinquini, J.P. Cumalat, C. Dallapiccola, J.F. Ginkel, J.E. Ramirez, B. O'Reilly, I. Segoni, E.W. Vaandering (**CU Boulder**); J.N. Butler, H.W.K. Cheung, G. Chiodini, S. Cihangir, I. Gaines, P.H. Garbincius, L.A. Garren, E. Gottschalk, S.A. Gourlay, D.J. Harding, P.H. Kasper, A.E. Kreymer, R. Kutschke, P. Lebrun, S. Shukla, M. Vittone (**Fermilab**); R. Baldini-Ferroli, L. Benussi, F.L. Fabbri, A. Zallo (**INFN Frascati**); C. Cawlfield, R. Culbertson, R. Greene, D.Y. Kim, K.S. Park, A. Rahimi, J. Wiss (**UI Champaign**); R. Gardner, A. Kryemadhi (**Indiana** ); Y.S. Chung, J.S. Kang, B.R. Ko, J.W. Kwak, K.B. Lee, S.S. Myung (**Korea University, Seoul**); K.Cho, H.Park (**Kyungpook National University, Taegu**); G. Alimonti, S. Barberis, A. Cerutti, G. Bellini, M. Boschini, D. Brambilla, B. Caccianiga, A. Calandrino, L. Cinquini, P. D'Angelo, M. DiCorato, P. Dini, L. Edera, S. Erba, M. Giammarchi, P. Inzani, F. Leveraro, S. Malvezzi, D. Menasce, E. Meroni, M. Mezzadri, L. Milazzo, L. Moroni, D. Pedrini, L. Perasso, C. Pontoglio, F. Prelz, M. Rovere, A. Sala, S. Sala D. Torretta (**INFN and Milano**); D. Buchholz, D. Claes, B. Gobbi (**Nortwestern**); J.M. Bishop, N.M. Cason, C.J. Kennedy, G.N. Kim, T.F. Lin, D.L. Puseljic, R.C. Ruchti, W.D. Shephard, J.A. Swiatek, Z.Y. Wu (**Notre Dame**); T.F. Davenport III (**UNC Asheville**); V. Arena, G. Boca, G. Bonomi, C. Castoldi, G. Gianini, G. Liguori, M. Merlo, D. Pantea, S.P. Ratti, C. Riccardi, P. Torre, L. Viola, P. Vitulo (**INFN and Pavia**); H. Hernandez, A.M. Lopez, L. Mendez, A. Mirles, E. Montiel, D. Olaya, J. Quinones, C. Rivera, Y. Zhang (**Mayaguez, Puerto Rico**); N. Copty, M. Purohit, J.R. Wilson (**USC Columbia**); K. Cho, T. Handler (**UT Knoxville**); D. Engh, W.E. Johns, M. Hosack, M.S. Nehring, M. Sales, P.D. Sheldon, K. Stenson, M.S. Webster (**Vanderbilt**); M. Sheaff (**Wisconsin, Madison**); Y. Kwon (**Yonsei University, Korea**).


doi:10.1016/S0920-5632(02)02184-9

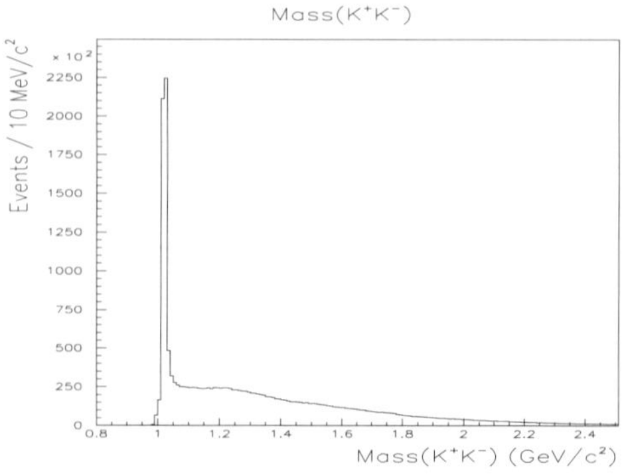


Figure 1. The $K^+K^-$ sample with no cut on $p_T$.

1860 MeV/c$^2$ ) and in the two sideband regions (1500–1600 MeV/c$^2$ and 1900–2100 MeV/c$^2$ ), it is seen that the 1750 region has a peak in the $p_T$ spectrum in nearly the same place as the $\phi(1020)$ peak, but the sideband regions have significantly smaller $p_T$ peaks, indicating that the background under the X(1750) signal is largely non-diffractive. Fitting the 1750 MeV/c$^2$ mass region with a non-relativistic Breit-Wigner distribution and a quadratic background, we find

Yield $= 11,700 \pm 480$ Events

M $= 1753.5 \pm 1.5 \pm 2.3$ MeV/c$^2$

$\Gamma = 122.2 \pm 6.2 \pm 8.0$ MeV/c$^2$

There is a region near 1600 MeV/c$^2$ where there is some discrepancy in our fit to the $K^+K^-$ mass spectrum. The residuals show that the statistical significance of this discrepancy is not strong (Fig. 2.d). Finally, we searched for the X(1750) enhancement in K*K and we find

$$\frac{\Gamma(X(1750) \to \overline{K}^{*0} K^0 \to K^-\pi^+K_S + c.c.)}{\Gamma(X(1750) \to K^+K^-)} < 0.065 \text{ at } 90\% \text{ C.L.}$$

$$\frac{\Gamma(X(1750) \to K^{*+}K^- \to K_S\pi^+K^- + c.c.)}{\Gamma(X(1750) \to K^+K^-)} < 0.183 \text{ at } 90\% \text{ C.L.}$$

The two relative branching ratios were measured to be $-0.083 \pm 0.081$ and $0.065 \pm 0.072$, respectively. Because of the large discrepancies in

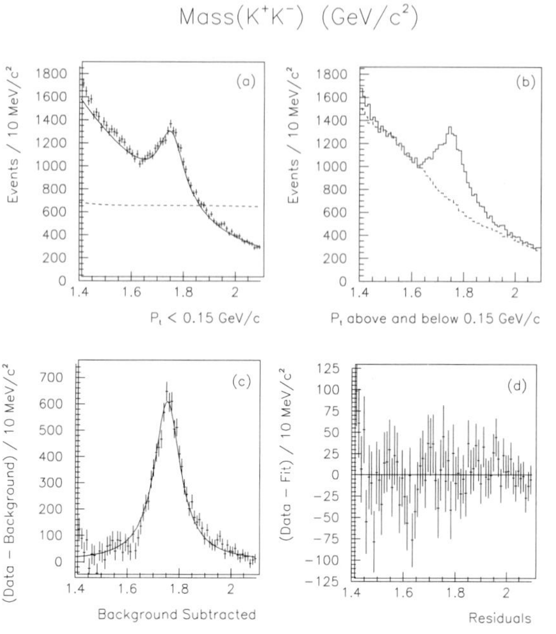


Figure 2. (a) The $K^+K^-$ mass spectrum with the requirement that $p_T < 0.15$ GeV/$c$. The spectrum is fit with a non-relativistic Breit-Wigner distribution and a quadratic background. The dotted line is the Monte Carlo efficiency on a scale from 0 to 100%. (b) The solid line is the $K^+K^-$ mass spectrum with the requirement that $p_T < 0.15$ GeV/$c$. The dotted line is the $K^+K^-$ mass spectrum with $p_T > 0.15$ GeV/$c$ scaled to the size of the low $p_T$ spectrum for comparison. (c) The data and fit after subtracting the quadratic polynomial background shape. (d) The data minus the fit.

mass and relative branching fractions to $K^+K^-$ and $K^*K$, we do not believe it is reasonable to identify the X(1750) (which should be assigned the photon quantum numbers $J^{PC} = 1^{--}$) with the $\phi(1680)$. In fact, because the mass of the X(1750) is significantly higher than all known vector mesons, the most massive of which are the $\omega(1650)$, $\phi(1680)$, and $\rho(1700)$, an interpretation claiming the X(1750) is some combination of interfering vector mesons also seems highly unlikely. The interpretation of the X(1750) remains uncertain.

## 2. A Narrow Dip in A Study of Diffractive Photoproduction

The Fermilab experiment 687, FOCUS predecessor, collected data during the 1990/91 fixed-target runs at the Wideband Photon beamline at Fermilab. We report on a study of the diffractive photoproduction of the $3\pi^+3\pi^-$ final state and the observation of a narrow dip in the mass spectrum at 1.9 GeV/c$^2$. Pions are produced in photon interactions in the Be target. The data acquisition trigger requires a minimum energy deposition in the hadron calorimeters located behind the electromagnetic calorimeters and at least three charged tracks outside the pair region. Details of event selection, analysis cuts and fitting strategies are reported in [2]. The invariant mass distribution of diffractively produced $3\pi^+3\pi^-$ final states shows a dip structure at about 1.9 GeV/c$^2$. No evidence for structures is shown in the incoherent ($P_T^2 > 0.040$GeV/c$^2$ subsample (Fig.3a). Fig. 3b) shows the coherent mass distribution after correcting for efficiency and acceptance, and unfolding the spectrometer mass resolution. The dip structure has been characterized by a two-component fit, adding coherently a relativistic Breit-Wigner resonance to a diffractive continuum contribution. Fit values show consistent evidence for a narrow resonance at $M_r = 1.911 \pm 0.004 \pm 0.001$ GeV/c$^2$ with a width $\Gamma = 29 \pm 11 \pm 4$ MeV/c$^2$, where the errors quoted are statistic and systematic, respectively. Such a resonance could be assigned the photon quantum numbers ($J^{PC} = 1^{--}$) and G=+1, I=1 due to the final state multiplicity. There is little understanding of the specific mechanism responsible for this destructive interference.

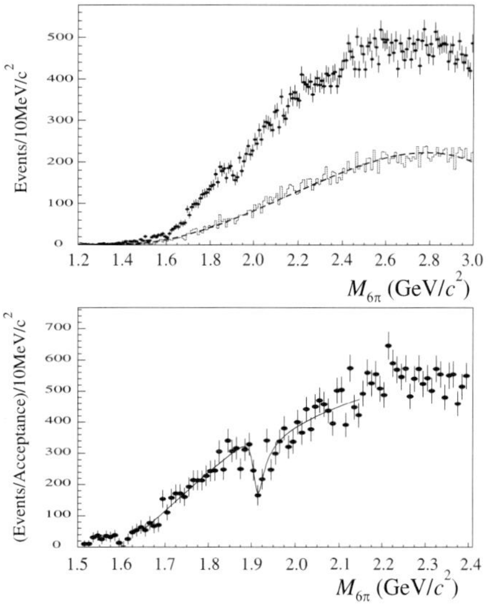


Figure 3. a) Distribution of $3\pi^+3\pi^-$ invariant mass in the 1.2–3.0 GeV/c$^2$ mass range: coherent plus incoherent contribution. Dotted distribution: incoherent contribution. b) Acceptance-corrected distribution of $3\pi^+3\pi^-$ invariant mass for diffractive events. The mass resolution has been unfolded.

## 3. Conclusions

We have reported on new results from fixed target photoproduction at Fermilab, which show how a high-resolution multiparticle spectrometer, coupled to a high-energy photon beam, constitutes a tremendously effective tool to explore the light vector meson spectroscopy and dynamics. More results will be coming soon.

## REFERENCES

1. J. M. Link *et al.* (FOCUS Coll.), Phys. Lett. B **545** (2002) 50 [arXiv:hep-ex/0208027].
2. P. L. Frabetti *et al.* (E687 Coll.), Phys. Lett. B **514** (2001) 240 arXiv:hep-ex/0106029.

Proceedings of ICHEP 2002, pp. 674–676
*S. Bentvelsen, P. de Jong, J. Koch and E. Laenen (Editors)*

ELSEVIER

31st INTERNATIONAL CONFERENCE ON HIGH ENERGY PHYSICS AMSTERDAM

# Measurement of $\pi^+\pi^-$ atom lifetime at DIRAC

L. Afanasyev[a]
on behalf of DIRAC Collaboration

[a]Joint Institute for Nuclear Research, Dubna, Russia

The DIRAC experiment aims to measure the lifetime of $\pi^+\pi^-$ atoms in the ground state with 10% precision, using the 24 GeV/$c$ proton beam of the CERN Proton Synchrotron. As the value of the above lifetime of order $10^{-15}$s is dictated by a strong interaction at low energy, the precise measurement of this quantity enables to determine a combination of S-wave pion scattering lengths to 5%. Pion scattering lengths have been calculated in the framework of chiral perturbation theory with high precision. Thus the accurate measurement of these values would submit the understanding of chiral symmetry breaking of QCD to a crucial test. The results on the $\pi^+\pi^-$ atom detection and the preliminary estimation of their lifetime are presented.

## 1. Introduction

Pionium ($A_{2\pi}$) is the hydrogen-like atom consisting of $\pi^+$ and $\pi^-$ mesons. Its lifetime is dominated by the charge-exchange process $\pi^+\pi^- \to \pi^0\pi^0$ and decay width is proportional to the square of the difference between the isoscalar and isotensor S-wave $\pi\pi$-scattering lengths $a_0 - a_2$. The predicted value for the $A_{2\pi}$ lifetime in the ground state, according to the Chiral Perturbation Theory at leading and next-to-leading order in isospin breaking, is of the order of $3 \cdot 10^{-15}$ s [1–3].

The goal of the DIRAC Experiment [4] is to measure the $A_{2\pi}$ lifetime with 10% accuracy in order to determine $|a_0 - a_2|$ with 5% precision in a model independent way [5]. The determination of the $S$-wave $\pi\pi$-scattering lengths will yield the size of the two-flavour quark condensate [6].

## 2. Experimental method

The production cross section of $A_{2\pi}$ in the high-energy inclusive processes have been obtained by L.Nemenov [5]. Relativistic $A_{2\pi}$ produced in hadron-nucleus interaction moves in the target material before their decay. The electro-magnetic interactions with the atoms of the medium might, however, lead to the pionium excitation or breakup (ionization). From the knowledge of the atomic interaction cross sections, for a given target material and thickness, one can calculate the breakup probability $P_{br}$ for arbitrary values of the atom momentum and lifetime [7].

"Atomic pairs" produced from the $A_{2\pi}$ breakup have very small relative momentum in their c.m. system $Q < 3$ MeV/$c$ that allows to observe they over the background of pions pairs produced in the free state. Moreover the number of the produced $A_{2\pi}$ can be obtained through the number of free pions pairs. Thus comparing the ratio of detected "atomic pairs" to the number of produced $A_{2\pi}$ with the calculated dependence of $P_{br}$ on $\tau$, one can determine the atom lifetime.

## 3. The DIRAC apparatus

The DIRAC setup is located at CERN, at the ZT8 beam area of the PS East Hall. It became operational at the end of 1998 and has been collecting data since the summer of 1999, using the 24 GeV/$c$ proton beam extracted from the PS accelerator.

DIRAC is a fixed-target experiment and its apparatus is designed to detect charged pion pairs with high resolution over the pair relative momentum (see Fig.1). It consists of a straight detector section between the target station and the analyzing magnet (2.3 Tm bending power), and of a double arm spectrometer downstream the magnet.


doi:10.1016/S0920-5632(02)02186-2

A multi-level trigger system has been designed to select "atomic pairs" from the overwhelming background of uncorrelated $\pi^+\pi^-$ pairs. The Cherenkov, preshower and muon detectors ensure rejection of $e^+e^-$ and $\mu^+\mu^-$ pairs at trigger level. Pairs of $p\pi^-$ are identified by offline analysis of time-of-flight measurements.

## 4. Experimental results

The number of ionized $A_{2\pi}$ is measured directly from the excess of $\pi^+\pi^-$ pairs with very low relative momentum ($Q < 3$ MeV/$c$) compared to the expected number of "free" pairs simulated using pairs of accidental coincidences. The $Q$-distribution of $\pi^+\pi^-$ pairs obtained with the platinum, titanium and nickel targets in 1999–2002 are presented on the Fig.2. The first preliminary result on the lifetime of $\pi^+\pi^-$ atom obtained with the Ni 2000 data are given on the Fig.3.

The DIRAC experiment is collecting statistics to achieve the final goal of measuring the lifetime of pionium with 10% accuracy. A preliminary analysis of data collected two last years with three targets shows the excess of $\pi^+\pi^-$ pairs coming from the breakup of $A_{2\pi}$. By now the total number of the observed "atomic pairs" amounts 5500. A preliminary result in the $A_{2\pi}$ lifetime measurement averaged for Ni and Ti targets is $3.6^{+0.9}_{-0.7} \times 10^{-15}$ s (statistical error only amounts 22%). The accurate determination of systematic errors requires a background measurement. The 10% accuracy on the atom lifetime value will require running beyond 2002.

## REFERENCES


1. A. Gall et al., Phys.Lett.B 462 (1999) 335.
2. J. Gasser et al., Phys.Lett.B 471 (1999) 244-250.
3. G. Colangelo et al., Phys.Lett.B 488 (2000) 261-268.
   G. Colangelo et al., Nucl.Phys.B 603 (2001) 125-179.
4. B. Adeva et al., Proposal to the SPSLC, CERN/SPSLC 95-1, SPSLC/P 284, (1994).
5. L.L. Nemenov, Sov.J.Nucl.Phys. 41 (1985) 629.
6. N.H. Fuchs et al., Phys.Lett.B 269 (1991) 183; Phys. Rev. D 47 (1993) 3814.
7. L.G.Afanasyev and A.V.Tarasov: Yad.Fiz. 59 (1996) 2212; Phys.Atom.Nucl. 59 (1996) 2130.


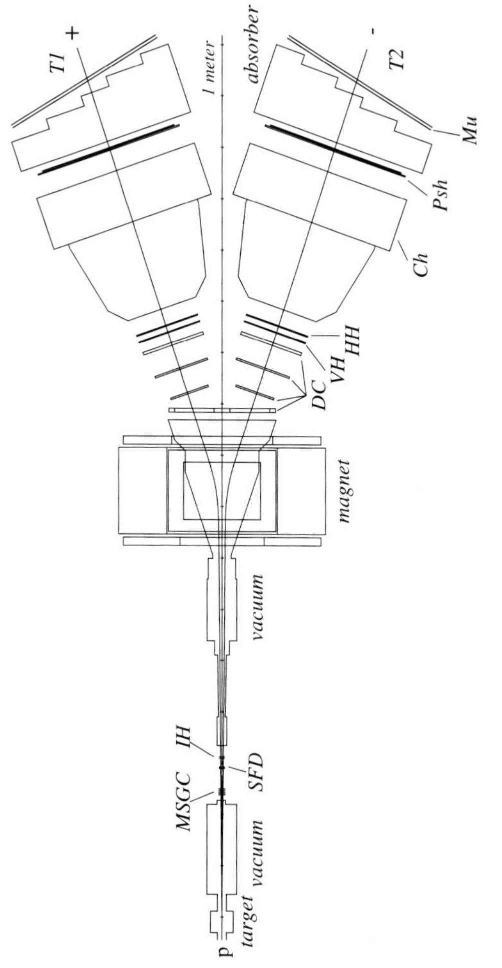


Figure 1. Schematic top view of the DIRAC spectrometer. Moving from the target station toward the magnet there are: four MicroStrip Gas Chambers (*MSGC*), two Scintillating Fiber Detectors (*SFD*) and two Ionization Hodoscopes (*IH*). Downstream the dipole magnet, on each arm of the spectrometer, are located: four modules of Drift Chambers (*DC*), the Vertical and Horizontal Scintillation Hodoscopes (*VH, HH*), the Cherenkov counter (*Ch*), the Preshower detector (*PSh*) and, behind the iron absorber, the Muon detector (*Mu*).

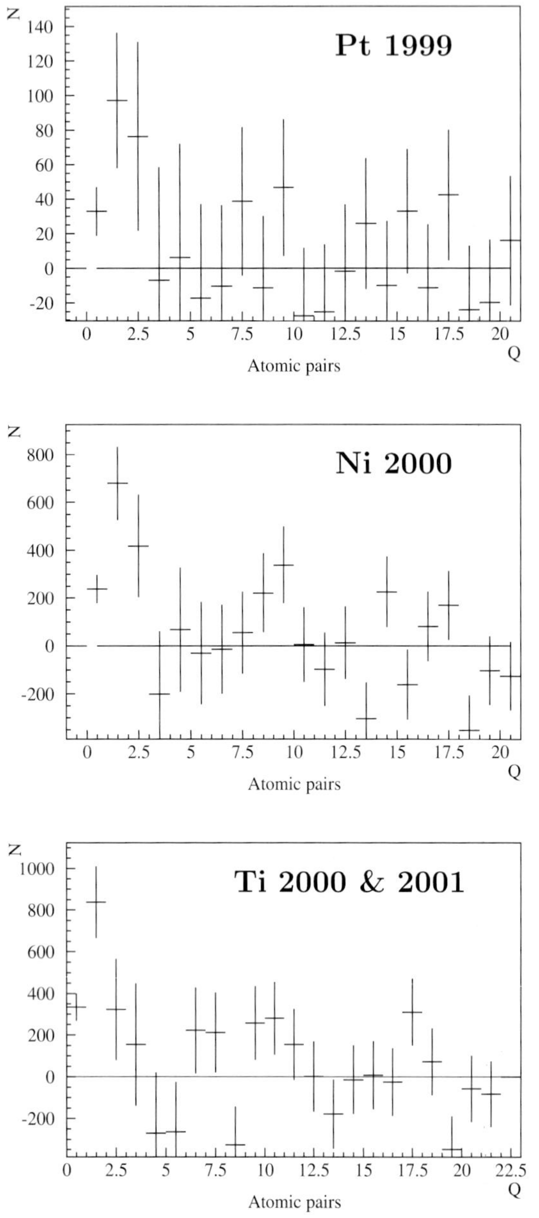


Figure 2. Distributions of the $\pi^+\pi^-$ pair excess over relative momentum for data collected with the platinum, titanium and nickel targets in 1999–2002. The "atomic pairs" are observed as the peaks at low $Q$.

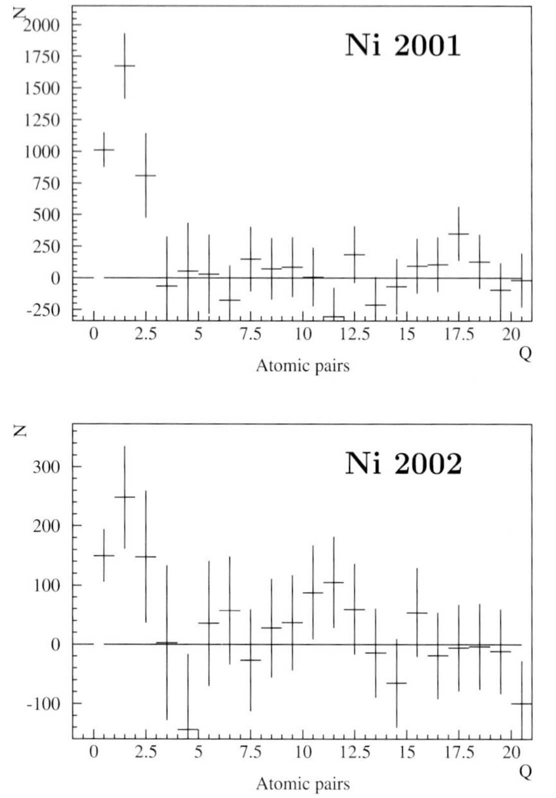


Figure 2. Continue

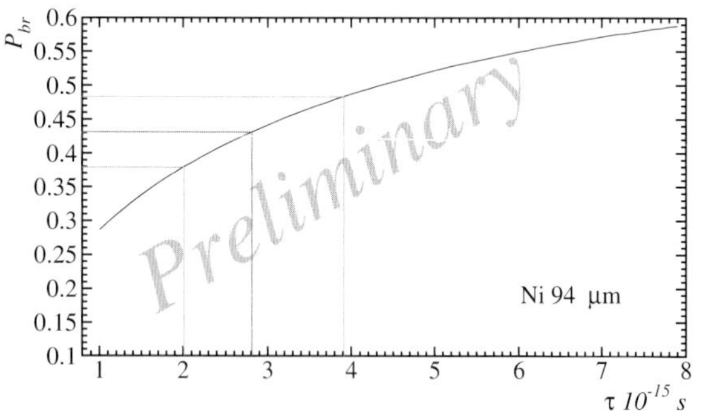


Figure 3. The calculated probability of pionium breakup as a function of its lifetime for the Ni target of 94 $\mu$m thick (curve). The measured probability of the pionium breakup obtained form Ni 2000 data is shown with its statistical errors as the horizontal lines. The crossing gives the lifetime value of $\tau = (2.8^{+1.1}_{-0.8}) \cdot 10^{-15}$ s

ELSEVIER

Proceedings of ICHEP 2002, pp. 677–680
*S. Bentvelsen, P. de Jong, J. Koch and E. Laenen (Editors)*

31st INTERNATIONAL CONFERENCE ON HIGH ENERGY PHYSICS AMSTERDAM

# Studies of $\phi$ meson radiative decays with KLOE

The KLOE Collaboration*
Presented by P.Gauzzi[a†]

[a]Dipartimento di Fisica, Università degli Studi di Roma "La Sapienza" and INFN Sezione di Roma, P.le A.Moro 2, 00185 - Rome (Italy)

A sample of $5.3 \times 10^7 \phi$ mesons, produced at the Frascati $\phi$-factory DAΦNE, has been used by the KLOE Collaboration to study the $\phi$ radiative decays. The decays $\phi \to \eta\pi^0\gamma$ and $\phi \to \pi^0\pi^0\gamma$ have been exploited to study the scalar mesons $a_0(980)$ and $f_0(980)$. Furthermore a new determination of the $\eta - \eta'$ mixing angle has been obtained from the measurement of the ratio of the decay rates of $\phi \to \eta'\gamma$ to $\phi \to \eta\gamma$.

## 1. Introduction

The Frascati $\phi$-factory DAΦNE is an $e^+e^-$ collider working at the peak of the $\phi(1020)$ resonance. The peak cross section is $\sigma_\phi \approx 3300$ nb. The KLOE experiment has collected an integrated luminosity of 25 $pb^{-1}$, corresponding to $7.5 \times 10^7\phi$, during the 2000 data taking, and 190 $pb^{-1}$, corresponding to $5.7 \times 10^8\phi$, during the 2001 one. The results reported in this paper have been obtained from the analysis of a sample of 16 $pb^{-1}$ from the year 2000 data. The analysis of the 2001 data is in progress. More details on the analyses presented here can be found in refs.[1].

## 2. $\phi \to \mathbf{S(0^{++})}\gamma$ (S=$f_0$, $a_0$)

The nature of the scalar mesons ($J^{PC}=0^{++}$) $f_0(980)$ and $a_0(980)$ is not well established. They are not easily interpreted as ordinary $q\bar{q}$ mesons, belonging to the $^3P_0$ nonet, alternative hypotheses have been proposed: $q\bar{q}q\bar{q}$ states[2] or $K\bar{K}$ bound states[3]. According to the theory the branching ratios of $\phi \to f_0\gamma$ and $\phi \to a_0\gamma$, as well as the $f_0$ and $a_0$ mass shapes, are sensitive to the structure of these particles[4].

Concerning the $f_0$, the decay $f_0 \to \pi^0\pi^0$ has been measured, by selecting events without charged tracks and with exactly five prompt photons in the detector. The same final state has been used to study the decay $\phi \to a_0\gamma$, with $a_0 \to \eta\pi^0$ and $\eta \to \gamma\gamma$. These two processes have been previously measured by the experiments SND and CMD-2 at VEPP-2M in Novosibirsk[5]. In addition another final state, consisting of two charged pions and five prompt photons, was also used to study the decay $\phi \to a_0\gamma$, corresponding to $a_0 \to \eta\pi^0$ and $\eta \to \pi^+\pi^-\pi^0$. This last decay chain has been observed by KLOE for the first time.

The processes that are expected to contribute to the five photon final state, besides the signals, are: (i) $\phi \to \sigma(500)\gamma$ with $\sigma \to \pi^0\pi^0$ ($\sigma(500)$ is

*The KLOE Collaboration: A. Aloisio, F. Ambrosino, A. Antonelli, M. Antonelli, C. Bacci, G. Bencivenni, S. Bertolucci, C. Bini, C. Bloise, V. Bocci, F. Bossi, P. Branchini, S. A. Bulychjov, R. Caloi, P. Campana, G. Capon, G. Carboni, M. Casarsa, V. Casavola, G. Cataldi, F. Ceradini, F. Cervelli, F. Cevenini, G. Chiefari, P. Ciambrone, S. Conetti, E. De Lucia, G. De Robertis, P. De Simone, G. De Zorzi, S. Dell'Agnello, A. Denig, A. Di Domenico, C. Di Donato, S. Di Falco, A. Doria, M. Dreucci, O. Erriquez, A. Farilla, G. Felici, A. Ferrari, M. L. Ferrer, G. Finocchiaro, C. Forti, A. Franceschi, P. Franzini, C. Gatti, P. Gauzzi, S. Giovannella, E. Gorini, F. Grancagnolo, E. Graziani, S. W. Han ,M. Incagli, L. Ingrosso, W. Kluge, C. Kuo, V. Kulikov, F. Lacava, G. Lanfranchi, J. Lee-Franzini, D. Leone, F. Lu, M.Martemianov, M. Matsyuk, W. Mei, L. Merola, R. Messi, S. Miscetti, M. Moulson, S. Müller, F. Murtas, M. Napolitano, A. Nedosekin, F. Nguyen, M. Palutan, E. Pasqualucci, L. Passalacqua, A. Passeri, V. Patera, E. Petrolo, L. Pontecorvo, M. Primavera, F. Ruggieri, P. Santangelo, E. Santovetti, G. Saracino, R. D. Schamberger, B. Sciascia, A. Sciubba, F. Scuri, I. Sfiligoi, T. Spadaro, E. Spiriti, G. L. Tong, L. Tortora, E. Valente, P. Valente, B. Valeriani, G. Venanzoni, S. Veneziano, A. Ventura, G. Xu, G. W. Yu

†E-mail: Paolo.Gauzzi@roma1.infn.it


doi:10.1016/S0920-5632(02)02189-8

the other scalar meson expected in the mass region below 1 GeV, recently observed at FNAL by the E791 Collaboration[6]), (ii) $\phi \to \rho^0\pi^0$ with $\rho \to \pi^0\gamma$, $\eta\gamma$, (iii) $e^+e^- \to \omega\pi^0$ with $\omega \to \pi^0\gamma$. Some contamination is expected from events with three ((iv) $\phi \to \eta\gamma$, $\eta \to \gamma\gamma$ and $\phi \to \pi^0\gamma$) and seven ((v) $\phi \to \eta\gamma$, $\eta \to \pi^0\pi^0\pi^0$) prompt photons, wrongly reconstructed as five photon ones. A kinematic fit is performed, by requiring the 4-momentum conservation to reject events from (iv) and (v), then the best photon pairing is searched for, by associating pairs of photons to $\pi^0$ and $\eta$, to discriminate between $\phi \to \pi^0\pi^0\gamma$ and $\phi \to \eta\pi^0\gamma$. The $\omega\pi^0$ background is reduced by rejecting events with $\pi^0\gamma$ invariant mass compatible with the $\omega$ mass.

The final sample for $\phi \to \pi^0\pi^0\gamma$ consists of

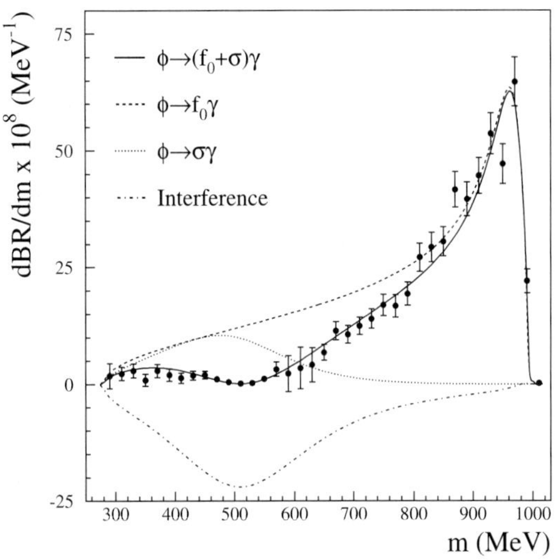


Figure 1. Differential decay rate for $\phi \to \pi^0\pi^0\gamma$; the solid line is the result of Fit B, individual contributions are also shown.

3102 events with a 20% contamination (evaluated by Monte Carlo (MC)). The total detection efficiency for the signal is 40%. By normalizing to the number of $\phi$ produced the following branching ratio can be obtained: $\mathrm{Br}(\phi \to \pi^0\pi^0\gamma) = (1.09 \pm 0.03 \pm 0.05) \times 10^{-4}$ in agreement with the Novosibirsk measurements[5], with much smaller uncertainties.

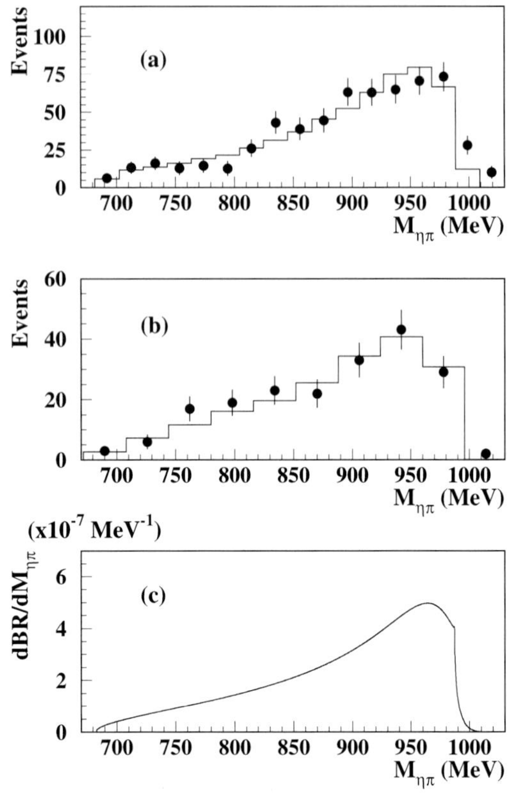


Figure 2. $\eta\pi^0$ invariant mass spectra: (a) five photon sample; (b) $\pi^+\pi^-$ plus five photon sample. (c) Theoretical curve obtained from the combined fit.

For $\phi \to \eta\pi^0\gamma$, 916 events have been selected with 30% contamination (MC), and 32% overall efficiency. By normalizing to the number of $\phi$ produced, KLOE obtains $\mathrm{Br}(\phi \to \eta\pi^0\gamma) = (8.51 \pm 0.51 \pm 0.57) \times 10^{-5}$, also in agreement with the Novosibirsk measurements[5].

The final state of $\phi \to \eta\pi^0\gamma$ and $\eta \to \pi^+\pi^-\pi^0$ consists of two charged tracks coming from the interaction point and five prompt photons; this signature is unique among the possible final states, so that the background comes from events with two tracks and three ($\phi \to \eta\gamma$, $\eta \to \pi^+\pi^-\pi^0$), four ($e^+e^- \to \omega\pi^0$, $\omega \to \pi^+\pi^-\pi^0$; $\phi \to K_L K_S$, $K_S \to \pi^0\pi^0$ and $K_L \to \pi l \nu$) or six ($\phi \to K_L K_S \to \pi^+\pi^-\pi^0\pi^0\pi^0$, with prompt $K_L$ decay) prompt photons. The analysis scheme is similar to that of the five photon sample: 197 events have been selected with $4 \pm 4$ background events and with 19% efficiency. From this decay chain the $\mathrm{Br}(\phi \to \eta\pi^0\gamma) = (7.96\pm0.60\pm0.47)\times10^{-5}$ is extracted, in good agreement with the result of the five photon sample.

In order to disentangle the various contributions, the data have been fitted to theoretical mass spectra obtained from the following model: (a) the coupling of $\phi$ to $S(0^{++})\gamma$ is assumed to occur through a charged kaon loop[4], (b) the scalar propagator with finite width corrections is used, (c) the $\phi \to \rho^0\pi^0$ parametrization is taken from VDM calculations[7].

Table 1
Fit results.

| $f_0$ | Fit A | Fit B |
|---|---|---|
| $\chi^2/\mathrm{ndf}$ | 109.5/34 | 43.2/33 |
| $M_{f_0}$ (MeV) | $962 \pm 4$ | $973 \pm 1$ |
| $g^2_{f_0KK}/(4\pi)$ ($\mathrm{GeV}^2$) | $1.29 \pm 0.14$ | $2.79 \pm 0.12$ |
| $g^2_{f_0KK}/g^2_{f_0\pi\pi}$ | $3.22 \pm 0.29$ | $4.00 \pm 0.14$ |
| $g_{\phi\sigma\gamma}$ | - | $0.060 \pm 0.008$ |
| $a_0$ | | |
| $\chi^2/\mathrm{ndf}$ | 27.2/25 | |
| $g^2_{a_0KK}/(4\pi)$ ($\mathrm{GeV}^2$) | $0.40 \pm 0.04$ | |
| $g_{a_0\eta\pi}/g_{a_0KK}$ | $1.35 \pm 0.09$ | |
| $\mathrm{Br}(\phi \to \rho^0\pi^0 \to \eta\pi^0\gamma)$ | $(0.5 \pm 0.5) \times 10^{-5}$ | |

Two different fits have been tried on the $\pi^0\pi^0$ mass spectrum, after background subtraction: without (Fit A) and with (Fit B) $\phi \to \sigma(500)\gamma$; the $\sigma(500)$ has been parametrized as a fixed width Breit-Wigner ($M_\sigma = 478$ MeV, $\Gamma_\sigma = 324$ MeV[6]) and a point-like $\phi - \sigma\gamma$ coupling has been assumed. The free parameters of the fits are: the $f_0$ mass, the coupling $g^2_{f_0KK}/(4\pi)$, the ratio $g^2_{f_0KK}/g^2_{f_0\pi\pi}$, the $\mathrm{Br}(\phi \to \rho^0\pi^0 \to \pi^0\pi^0\gamma)$ and $g_{\phi\sigma\gamma}$ (only for Fit B).

The parameter values are listed in tab.1, the $\phi \to \rho^0\pi^0$ contribution turns out to be negligible for both fits. Fit A gives a large $\chi^2$, while Fit B shows a good agreement with the experimental data. In fig.1 the differential decay rate is plotted, together with the various contributions from Fit B; the depletion of the mass spectrum below 700 MeV is produced by a large negative interference between $f_0$ and $\sigma$.

For $\phi \to \eta\pi^0\gamma$, a combined fit of the two mass spectra, after background subtraction, has been performed, by fixing their relative normalization to the ratio $\mathrm{Br}(\eta \to \gamma\gamma)/\mathrm{Br}(\eta \to \pi^+\pi^-\pi^0)$. The free parameters are: the coupling $g^2_{a_0KK}/(4\pi)$, the ratio $g_{a_0\eta\pi}/g_{a_0KK}$ and $\mathrm{Br}(\phi \to \rho^0\pi^0 \to \eta\pi^0\gamma)$ (tab.1). The mass of the $a_0$ is not free, it has been fixed to the PDG value $M_{a_0}$=984.8 MeV. The results are shown in fig.2; by integrating the theoretical curve the $a_0$ contribution can be evaluated: $\mathrm{Br}(\phi \to a_0\gamma \to \eta\pi^0\gamma) = (7.4\pm0.7)\times10^{-5}$. In tab.2 the fit results are compared to the predictions of the $K^+K^-$ loop model for some possible structures of the scalar mesons. The $f_0$ parameters are compatible with the $q\bar{q}q\bar{q}$ model, while the $a_0$ ones disagree with that model.

## 3. $\phi \to \eta'\gamma/\phi \to \eta\gamma$

The mass eigenstates $\eta$ and $\eta'$ are related to the SU(3) octet-singlet states through the pseudoscalar mixing angle. Even in the case for two mixing angles which appears in extended chiral perturbation theory, as well as from phenomenological analyses, it has been argued that the two mixing parameters in the quark flavour basis are equal, apart from small terms which violate the OZI rule[8]. It is then possible to parametrize the mixing with only one angle $\varphi_P$, which can be extracted from the ratio $R$=$\mathrm{Br}(\phi \to \eta'\gamma)/\mathrm{Br}(\phi \to \eta\gamma)$[9].

Moreover the large $\mathrm{Br}(B{\to}K\eta')$ value observed[10], also raises interest about the gluonium contents of the $\eta'$[11]. This gluonium contents can be tested with a precise measurement of $\mathrm{Br}(\phi \to \eta'\gamma)$.

The same final state, $\pi^+\pi^-$ and three prompt photons, has been choosen for both decays. This final state proceeds through the chains: $\phi \to \eta\gamma$, $\eta \to \pi^+\pi^-\pi^0$, $\pi^0 \to \gamma\gamma$ ($\mathrm{Br}{\approx}\ 3 \times 10^{-3}$) and $\phi \to \eta'\gamma$, $\eta' \to \eta\pi^+\pi^-$, $\eta \to \pi^0\pi^0$ ($\mathrm{Br}{\approx}\ 2\times10^{-5}$). $\phi \to \eta\gamma$ decays are easily selected with small background; 50210 events have been counted with 37% efficiency. Since this decay chain is about 100 times more probable than the $\eta'$ decay, it is also the main source of background for $\phi \to \eta'\gamma$. After all background removal, $120\pm12\pm5$ $\phi \to \eta'\gamma$ events have been selected, with 23% efficiency.

Thus $R = (4.70\pm0.47\pm0.31)\times10^{-3}$ is obtained.

Table 2
KLOE results compared to $K^+K^-$ loop predictions[4].

| $f_0$ | KLOE | $s\bar{s}(u\bar{u}+d\bar{d})/\sqrt{2}$ | $(u\bar{u}+d\bar{d})/\sqrt{2}$ | $s\bar{s}$ |
|---|---|---|---|---|
| $g^2_{f_0KK}/(4\pi)$ (GeV$^2$) | $2.79 \pm 0.12$ | 2.3 | 0.15 | 0.3 |
| $g_{f_0\pi\pi}/g^2_{f_0KK}$ | $0.50 \pm 0.01$ | 0.3 – 0.5 | 2 | 0.5 |
| $Br(\phi \to \pi^0\pi^0\gamma) \times 10^4$ | $1.08 \pm 0.07$ | $\sim 1$ | $\sim 0.15$ | $\sim 0.2$ |
| $a_0$ | | $s\bar{s}(u\bar{u}-d\bar{d})/\sqrt{2}$ | $(u\bar{u}-d\bar{d})/\sqrt{2}$ | |
| $g^2_{a_0KK}/(4\pi)$ (GeV$^2$) | $0.40 \pm 0.04$ | 2.3 | 0.15 | |
| $g_{a_0\eta\pi}/g_{a_0KK}$ | $1.35 \pm 0.09$ | 0.91 | 1.53 | |
| $Br(\phi \to a_0\gamma) \times 10^4$ | $0.74 \pm 0.07$ | $\sim 2$ | $\sim 0.2$ | |

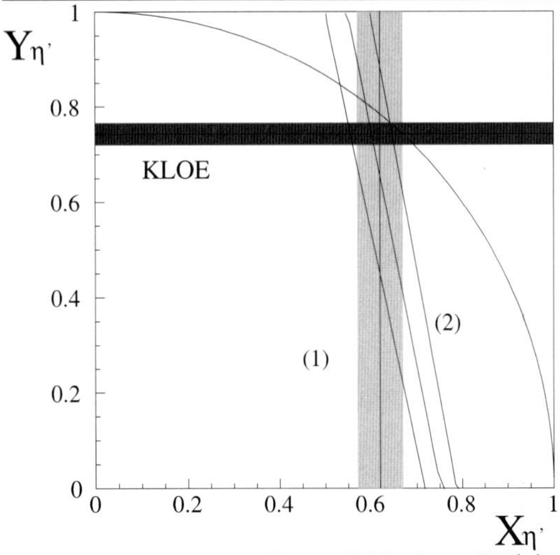


Figure 3. Bounds on $X_{\eta'}$ and $Y_{\eta'}$ from SU(3) calculations and experimental branching fractions.

From this ratio the mixing angle in the quark flavour basis can be extracted: $\varphi_P = (41.8^{+1.9}_{-1.6})^\circ$, that is equivalent to $\theta_P = (-12.9^{+1.9}_{-1.6})^\circ$ in the octet-singlet basis, in the picture with only one mixing angle. By using the current PDG value $Br(\phi \to \eta\gamma) = (1.297 \pm 0.003)\%$, it follows $Br(\phi \to \eta'\gamma) = (6.10 \pm 0.61 \pm 0.43) \times 10^{-5}$; this is the most accurate determination up to now.
The above values have been obtained neglecting any gluonium contents of $\eta$ and $\eta'$. If one allows for gluonium:

$$|\eta\rangle = X_\eta \left|u\bar{u} + d\bar{d}\right\rangle \sqrt{2} + Y_\eta \left|s\bar{s}\right\rangle + Z_\eta \left|glue\right\rangle$$
$$|\eta'\rangle = X_{\eta'} \left|u\bar{u} + d\bar{d}\right\rangle \sqrt{2} + Y_{\eta'} \left|s\bar{s}\right\rangle + Z_{\eta'} \left|glue\right\rangle$$

A consistency check can be performed: if $Z_{\eta'} = 0$, then $|Y_{\eta'}| = \cos\varphi_P$; other constraints on $X_{\eta'}$ and $Y_{\eta'}$ can be obtained from (1) $\Gamma(\eta' \to \rho\gamma)/\Gamma(\omega \to \pi^0\gamma)$ and (2) $\Gamma(\eta' \to \gamma\gamma)/\Gamma(\pi^0 \to \gamma\gamma)$. Then in the $X_{\eta'}$, $Y_{\eta'}$ plane (fig.3) the overlap region of the three allowed bands is consistent within one standard deviation with no gluonium contents in the $\eta'$, in fact: $X^2_{\eta'} + Y^2_{\eta'} = 0.94^{+0.06}_{-0.09}$.

## REFERENCES


1. A. Aloisio et al.(KLOE Collaboration), Phys. Lett. B536 (2002) 209; Phys. Lett. B537 (2002) 21; Phys. Lett. B541 (2002) 45.
2. R.L. Jaffe, Phys. Rev. D15 (1977) 267.
3. J. Weinstein, N. Isgur, Phys. Rev. Lett. 48 (1982) 659.
4. N.N. Achasov, V.I. Ivanchenko, Nucl. Phys. B315 (1989) 465.
5. M.N. Achasov et al., Phys. Lett. B485 (2000) 349; Phys. Lett. B479 (2000) 53; R.R. Akhmetshin et al., Phys. Lett. B462 (1999) 380.
6. E.M. Aitala et al., Phys. Rev. Lett. 86 (2001) 770.
7. N.N. Achasov, V.V. Gubin, Phys. Rev. D63 (2001) 094007.
8. H. Luetwyler, Nucl. Phys. Proc. Suppl. 64 (1998) 223, R. Escribano, J.M.Frère, Phys. Lett. B459 (1999), 288.
9. A.Bramon et al., Eur. Phys. J. C7 (1999) 271; T.Feldmann, Int. J. Mod. Phys. A15 (2000) 159.
10. S.J. Richichi et al., Phys. Rev. Lett. 85 (2000) 520; B. Aubert et al., Phys. Rev. Lett. 87 (2001) 221802; K. Abe et al., Phys. Lett. B517 (2001) 309.
11. E. Kou, Phys. Rev. D63 (2001) 054027.

NH
ELSEVIER

Proceedings of ICHEP 2002, pp. 681–683
*S. Bentvelsen, P. de Jong, J. Koch and E. Laenen (Editors)*

31st INTERNATIONAL CONFERENCE ON
HIGH ENERGY PHYSICS AMSTERDAM

# Possible Hints and Search for Glueballs in Charmless Rare B Decays

George W.S. Hou[a]*

[a]Department of Physics, National Taiwan University
Taipei, Taiwan 10764, R.O.C.

We suggest possibility for glueball search in 3-body charmless B decays, and compare hints in these channels with other possibilities, such as $p\bar{p}$ annihilation, $pp$ central production, and $\Upsilon$ decay.

## 1. HINT FOR $\xi$ IN B DECAY?

With 30 $\text{fb}^{-1}$ of data, the Belle experiment has reported the observation of various 3-body charmless $B \to Kh^+h^-$ decays, where $h$ is $\pi$, $K$ [1] or $p$ [5]. We point out [3], as shown in Fig. 1(a), (c) and (e), possible hints of some structure at $m_{hh} \sim 2.2$–$2.3$ GeV, or the "$\xi$" candidate for a $2^{++}$ glueball. Admittedly, the "evidence", if at all, are rather weak. We shall discuss in Sec. 2 the underlying curve for the $B \to p\bar{p}K$ decay, giving some support for a "bump" structure at 2.2–2.4 GeV bin. Fig. 1(c) for the $K_S\pi^+\pi^-$ mode is very low in statistics, and there is no indication in the corresponding $K^+\pi^+\pi^-$ and $K_SK^+K^-$ modes, as shown in Fig. 1(b) and (d). Fig. 1(e) and (f) are the high mass $m_{K^+K^-}$ spectra of the $B \to K^+K^+K^-$ transition, where $B \to K\phi$ is observed, with a broad $K^+K^-$ spectrum at 1.2–1.6 GeV. These spectra, complicated by identical particle effects, are not yet understood.

We give some reasoning in Sec. 3 for why glueball production in charmless $B$ decays may be interesting, and in Sec. 4 we gather a few other processes where a hint for $\xi$ production can be entertained.

## 2. UNDERLYING 3-BODY DYNAMICS

Stimulated by observation of $\overline{B} \to D^*p\bar{n}$ decay, we proposed to account for the process by factorized currents. The current induced $\overline{B} \to D^*$ transition is standard, but for the "current-produced" $p\bar{n}$ pair, we invoke isospin relations to tap into proton and neutron EM form factor data. Although axial vector current data is lacking, we find [4] that the vector current alone accounts for 60% of $\overline{B} \to D^*p\bar{n}$ rate.

Armed with this success, when Belle reported the first ever charmless baryonic $B$ decay, $B \to p\bar{p}K$ [2], we etried understand it (and other modes) within the factorization framework. The extension [5] seemed to work, giving rise to *threshold enhancement* in $m_{p\bar{p}}$ spectra, in agreement with experiment. Such threshold enhancement is

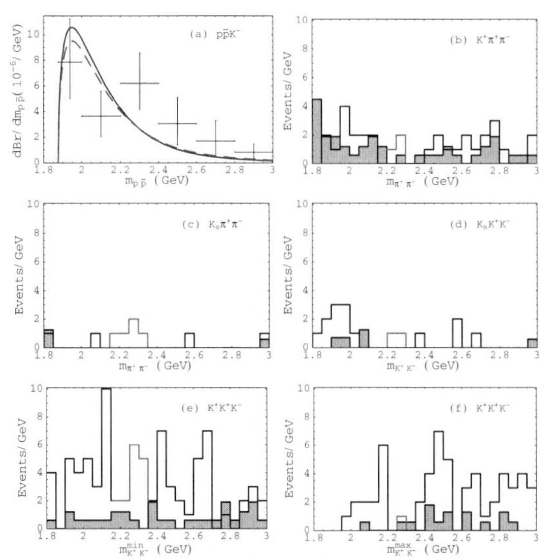


Figure 1. Spectra for $B \to$ (a) $p\bar{p}K$; (b) $K\pi\pi$ ($m_{K\pi} > 2$ GeV); (c) $K_S\pi\pi$; (d) $K_SK\bar{K}$; and (e), (f) for $KK\bar{K}$ vs. $m^{\min}_{K\bar{K}}$, $m^{\max}_{K\bar{K}}$ ($m^{\min}_{K\bar{K}} > 1.1$ GeV).

*The work reported here is done in collaboration with C.K. Chua and S.Y. Tsai, and supported y the NSC.


doi:10.1016/S0920-5632(02)02191-6

rooted in the baryon form factor structure. The idea is further confirmed by the recent Belle observation of $\overline{B} \to D^{(*)}K^-K^{0(*)}$, where one utilizes [6] kaon EM form factor data.

## 3. CHARMLESS B DECAYS & GLUEBALL SEARCH

Why are charmless $B$ decays potentially interesting for glueball search?

First, glueballs turn out to be relatively heavy, probably $> 1.5$ GeV. Thus, for flavored meson decays, they are only accessible in the $B$ system. Lattice calculations give [7] the $0^{++}$ and $2^{++}$ glueball masses at 1.5–1.7 GeV and $\sim 2.2$ GeV, respectively, with the latter coinciding well with the "$\xi$" particle hinted by BES data [8].

Second, it has always been a curiosity, at least to this author, why the gluonic penguin, $b \to sg^*$, i.e. a QCD-induced $b \to s$ weak "loop" transition, evolves only into $sq\bar{q}$ final states. The parton level $sgg$ final state suffers cancellation via "Low's low energy theorem" and becomes considerably smaller. This is in contrast to the equivalent presence of quarks and gluons in hadrons, or the equivalence of $N_C = 3$ vs. the number of light quarks. On the other hand, we were quite surprised in 1997 to learn the enormous strength of $B \to \eta' K$, and especially $B \to \eta' X_s$, decays. The latter, at $10^{-3}$ level, is a sizable fraction of the expected $b \to sq\bar{q}$ which is at $10^{-2}$ level.

It was pointed out in this context that the large inclusive $B \to \eta' X_s$ rate is likely related to the glue-rich nature of the $\eta'$. An intriguing possibility is that $\eta'$ is radiated preferentially from the virtual $g^*$ in the $b \to sg^*$ transition via $g^* \to g\eta'$, which is an extension of the QCD gluon anomaly. This picture [9] is so far the only one that can give the effective 3-body $m_{X_s}$ spectrum first observed by CLEO, which is now confirmed by BaBar. Caution has to be taken, however, that a straightforward application of the $g^* \to g\eta'$ idea would lead to energetic $\eta'$ mesons from $\Upsilon \to ggg + \eta'$ decay, another glue rich environment, which does not seem [10] to be observed by experiment.

Keeping the caution in mind, let us visualize $\xi$ production in analogy with $\eta'$, i.e. $g^* \to g\xi$ (or, via $(c\bar{c})_8 \to g\xi$, where $c\bar{c}$ is in color octet). How the $sg\bar{q}$ system evolves into a kaon is of no concern, for we now consider an effective $B \to \xi K$ transition, much in analogy with $B \to \eta' K$. A numeric exercise shows that, if $\xi \to p\bar{p}$ is of order $\sim$ %, and $B \to \xi K$ is comparable to $B \to \eta' K \sim 10^{-4}$, then $B \to p\bar{p}K$ could be at $10^{-6}$ order and could show up in the spectrum of Fig. 1(a). Whether and how $\xi$ would show up in the $\pi\pi$ or $K\overline{K}$ spectra of the remainder of Fig. 1 depends on $\xi \to \pi\pi$, $K\overline{K}$ branching ratios, which are controversial, like the existence of $\xi$ itself.

We suggest the $B$ factories to combine modes in their analysis.

## 4. OTHER EVIDENCE FOR $\xi$ AND POSSIBLE COMPETITION

The $\xi$ state has a rather checkered history, to say the least. Early on, it was seen by MARK III, but not DM2. In the 1990's, the BES experiment claimed [8] observation in quite a few channels. In particular, the $p\bar{p}$ channel stimulated a lot of interest in experiments at LEAR. But a multitude of null search results made $p\bar{p}$ and $pp$ experiments very skeptic of the very existence of $\xi$ itself. Note that the null results could simply be explained by $\xi$ having many open channels to decay to, with each channel having branching ratio considerably below % level.

In the course of our studies, however, we encountered a few curiosities, which could be counted as possible hints for $\xi$.

### 4.1. Glueball-filtered Central Production?

An empirical "glueball filter" [11] has been argued and applied to centrally produced $pp \to p(\pi^+\pi^-\pi^+\pi^-)p$ by the WA102 collaboration, to data with proton beams at 450 GeV. The glueball filter works as a "$dP_T$" cut. For $dP_T > 0.5$ GeV, the $\pi^+\pi^-\pi^+\pi^-$ invariant mass spectrum shows a prominent $f_1(1285)$ peak, which is definitely $q\bar{q}$. One sees a small bump at 1500 MeV. As $dP_T$ is lowered, i.e. for 0.2 GeV $< dP_T <$ 0.5 GeV, the significance of $f_1(1285)$ drops while the 1500 MeV structure is enhanced. For $dP_T <$ 0.2 GeV, the $f_1(1285)$ state becomes less significant than the 1500 MeV structure, which is a possible candidate for the $0^{++}$ glueball.

In the plot [11] that demonstrates this "glueball filter" phenomenon, we note one curiosity. For $dP_T < 0.2$ GeV, in contrast to the other two regions, a one bin *spike* pops up at $\sim 2400$ MeV, with statistical significance $> 5.5\sigma$! The claimed $\xi$ width is very narrow, of order 20 MeV, and the WA102 experiment has sufficient resolution. The mass may seem high, but it may in part be due to the plots in Ref. [11] are perhaps based on raw data. A similar spike in a published WA102 paper [12] appears at $\sim 2200$ MeV.

### 4.2. $p\bar{p} \to \phi\phi$?

As stated, the strongest doubt for $\xi$ comes from LEAR experiments, which claim nonobservation in $p\bar{p} \to K_S K_S$, $\phi\phi$, $\pi^0\pi^0$ and $\eta\eta$ channels. It has come to our attention, however, that some hint may exist in the $\phi\phi$ channel.

In an unpublished work [13], a partial wave analysis of $p\bar{p} \to \phi\phi$ finds evidence for a 70 MeV "resonance", in both the cross section *and* the phase (relative to other channels) for the $2^+_{00}$ channel. It appears to us that there may be some underlying structure in addition to the Breit-Wigner resonance. If so, the resonance part could be narrower. To my eye, it could well be 30 MeV. Reanalysis seems to be in progress.

### 4.3. $\Upsilon \to \gamma p\bar{p}$?

The CLEO experiment has searched [14] for $\Upsilon \to \gamma\pi^+\pi^-$, $\gamma K^+K^-$, $\gamma p\bar{p}$. Using 61.3 $\mathrm{fb}^{-1}$ data, 1, 1, 2 events were found in the $\xi$ window, respectively, against estimated background of 0.12, 0.21, 0.28 events. CLEO allowed for larger background and stopped short of claiming evidence. But it now has 1.3 $\mathrm{fb}^{-1}$ data on the $\Upsilon$, a factor of 20 increase. This means that 10–40 events (scaling up from BES branching ratios) could pop up in these channels once the data is analyzed, and could establish the $\xi$ state.

### 4.4. Disappearance of $J/\psi \to \gamma\xi \to \gamma p\bar{p}$?

If $\xi$ becomes established in $\Upsilon \to \gamma p\bar{p}$ by CLEO would be rather cute. But the rug seems pulled from under the original source ...

The BES experiment has just shown [15] its new result on $J/\psi \to \gamma p\bar{p}$, with 7 times the data of Ref. [8]. The $\xi$ state has all but disappeared, but one has a sizable peak right at threshold.

Glueball search is a hazardous occupation ...

## 5. Conclusion

We have pointed out that charmless $B$ decays could be a hunting ground for glueballs, because i) the $B$ meson is heavy enough; ii) strong penguins are gluon induced; iii) glue-rich channels such as the $\eta'$ modes seem rather enhanced.

A possible hint is pointed out for the $\xi$, or $2^{++}$ glueball candidate, in the $B \to Kh^+h^-$ spectra; but it is far from conclusive. We have further collected a few other possible hints. Alas, the rug seems pulled from underneath: BES no longer seems to see $\xi$ in $J/\psi \to \gamma p\bar{p}$.

Still, with luck, glueballs may (or may not!) emerge in some of these channels within a year.

## REFERENCES


1. A. Garmash *et al.* (Belle Collab.), Phys. Rev. D **65**, 092005 (2002).
2. K. Abe *et al.* (Belle Collab.), Phys. Rev. Lett. **88**, 181803 (2002).
3. C.K. Chua, W.S. Hou and S.Y. Tsai, Phys. Lett. **B544**, 139 (2002).
4. C.K. Chua, W.S. Hou and S.Y. Tsai, Phys. Rev. D **65**, 034003 (2002).
5. C.K. Chua, W.S. Hou and S.Y. Tsai, Phys. Rev. D **66**, 054004 (2002).
6. C.K. Chua *et al.*, hep-ph/0209164.
7. C.J. Morningstar and M.J. Peardon, Phys. Rev. D **60**, 034509 (1999).
8. J.Z. Bai *et al.* (BES Collab.), Phys. Rev. Lett. **76**, 3502 (1996).
9. W.S. Hou and B. Tseng, Phys. Rev. Lett. **80**, 434 (1998).
10. A.L. Kagan, AIP Conf. Proc. **618**, 310 (2002).
11. F.E. Close and A. Kirk, Phys. Lett. **B397**, 333 (1997).
12. D. Barberis *et al.* (WA102 Collab.), Phys. Lett. **B471**, 440 (2000).
13. A. Palano, talk at *Workshop on Hadron Spectroscopy*, Frascati, 1999.
14. G. Masek *et al.* (CLEO Collab.), Phys. Rev. D **65**, 072002 (2002).
15. S. Jin, following article, this proceedings.

ELSEVIER

Proceedings of ICHEP 2002, pp. 684–687
*S. Bentvelsen, P. de Jong, J. Koch and E. Laenen (Editors)*

31st INTERNATIONAL CONFERENCE ON HIGH ENERGY PHYSICS AMSTERDAM

# Radiative $J/\psi$ Decays and Searches for Glueballs

Shan JIN[a]

[a]Institute of High Energy Physics, P.O.Box 918 (1), Beijing 100039, P.R. China

Latest results from BES II, CLEO and L3 on some interesting states in the radiative $J/\psi$ decays and two-photon collisions are reviewed in this talk. Partial wave analyses are performed in $J/\psi \to \gamma K\overline{K}$, $\gamma\pi^+\pi^-$ and near threshold enhancement is observed in $J/\psi \to \gamma p\bar{p}$ at the BES II experiment. No resonance is observed in $K_S^0 K_S^0$ final state in the mass region around 2.2 GeV in the two-photon collisions at the CLEO and L3 experiments.

This talk contains three parts: 1) Partial Wave Analyses (PWA) of $J/\psi \to \gamma K\overline{K}$, $\gamma\pi^+\pi^-$ at BES II. 2) Observation of an enhancement near $p\bar{p}$ threshold in $J/\psi \to \gamma p\bar{p}$ at BES II. 3) Anti-search for glueball candidates in two-photon collisions at CLEO and L3.

## 1. PWA of $J/\psi \to \gamma K\overline{K}$, $\gamma\pi^+\pi^-$ at BES II (preliminary)

The spin-parity of the structure around 1.7 GeV has a long history of uncertainty. Partial wave analyses (PWA) are applied to $J/\psi \to \gamma K^+K^-$ and $\gamma K_S^0 K_S^0$ channels, based on $5.8 \times 10^7 J/\psi$ events collected at the BES II detector [1]. Fig. 1 shows the invariant mass spectra of $K^+K^-$ and $K_S^0K_S^0$, where $f_2'(1525)$ and the structure around 1.7 GeV are clearly seen in both cases.

The amplitudes are fitted to relativistic covariant tensor expressions, and a maximum likelihood method is applied in the fit. Global and slice fits are performed to the 1-2 GeV $KK$ mass region. For the slice fit, a 40MeV bin width is adopted. Fig. 2 shows $0^{++}$ and $2^{++}$ intensity distributions as a function of the invariant mass of $K^+K^-$ and $K_S^0K_S^0$. The dots with error bars are the efficiency corrected data points, solid curves stand for the coherent superposition of the individual Breit-Wigner resonance fits and dashed histograms represent the global fit results. Both $K^+K^-$ and $K_S^0K_S^0$ D wave intensities

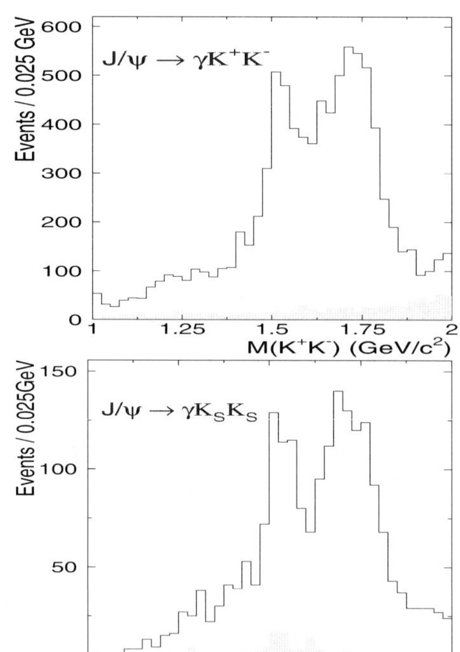


Figure 1. $K^+K^-$ *and* $K_s^0K_s^0$ *invariant mass spectra in* $J/\psi \to \gamma K^+K^-$ *and* $J/\psi \to \gamma K_S^0K_S^0$ *at BES II*


doi:10.1016/S0920-5632(02)02193-X

show a clear $f_2'(1525)$ with the mass and width of $M = 1518 \pm 6$ MeV and $\Gamma = 84^{+28}_{-24}$ MeV. There is evidence for $f_2(1270)$ and a weak $2^{++}$ intensity in the mass region around 1.7 GeV. While $0^{++}$ is found to be dominant in 1.7 GeV region, which is well determined by a Breit-Wigner resonance. The mass and width of the $0^{++}$ component are:

$$M = 1703^{+8}_{-10} MeV, \quad \Gamma = 163^{+27}_{-22} MeV$$

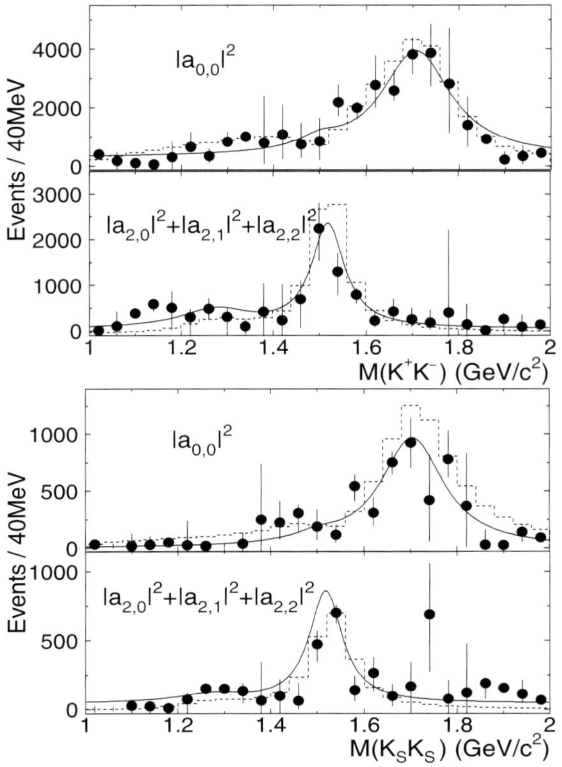


Figure 2. *PWA fit results of $K^+K^-$ and $K^0_S K^0_S$ at BES II*

Fig. 3 is the invariant mass spectrum of $\pi^+\pi^-$ in $J/\psi \to \gamma\pi^+\pi^-$.

Besides the well known $f_2(1270)$, a shoulder in the high mass region of $f_2(1270)$ and a bump near 1.7 GeV are seen. PWA is performed in the 1-2 GeV $\pi^+\pi^-$ mass region. Fig. 4 shows the

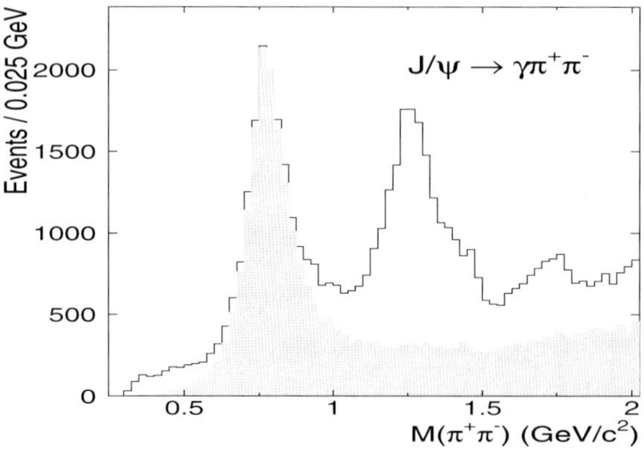


Figure 3. *$\pi^+\pi^-$ invariant mass spectrum in $J/\psi \to \gamma\pi^+\pi^-$ at BES II*

$0^{++}$ and $2^{++}$ intensity distributions, where the definition of the symbols are the same as in $K\overline{K}$ case.

A strong $2^{++}$ component is well determined at around 1.27 GeV, and two $0^{++}$'s are observed at around 1.4 and 1.7 GeV. If the $0^{++}$ at around 1.7 GeV is considered as the same one as in $K\overline{K}$ and the mass and width are fixed to be the values obtained from $K\overline{K}$ analyses, *i.e.*

$$M = 1703 MeV, \quad \Gamma = 163 MeV,$$

the fit results give the masses and widths of $2^{++}$ and another $0^{++}$ as:

$$M = 1266 \pm 6 MeV, \quad \Gamma = 170^{+37}_{-30} MeV$$

$$M = 1383^{+20}_{-18} MeV, \quad \Gamma = 274^{+59}_{-128} MeV$$

The preliminary results indicate that the relative ratio $BR(f_J(1710) \to \pi\pi)/BR(f_J(1710) \to K\bar{K})$ is around 30%.

## 2. Near threshold structure in $J/\psi \to \gamma p\bar{p}$

Based on the BESII $J/\psi$ data, $J/\psi \to \gamma p\bar{p}$ process is studied. The invariant mass spectrum of $p\bar{p}$ is shown in Fig. 5. In addition to the $\eta_c$ signal, there is a clear enhancement near $2m_p$. An

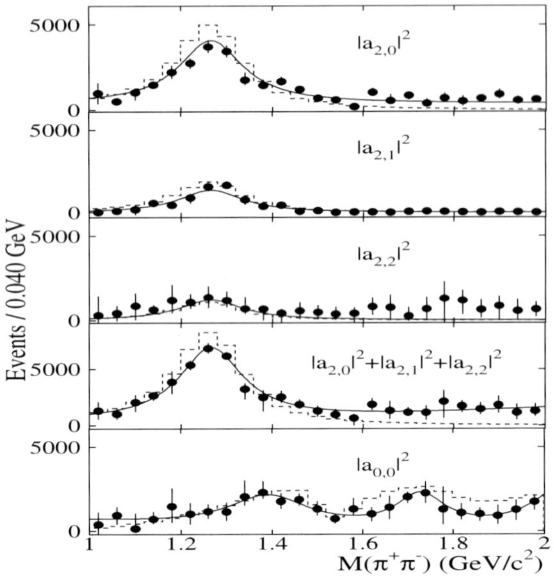


Figure 4. *PWA fit results of $\pi^+\pi^-$ at BES II*

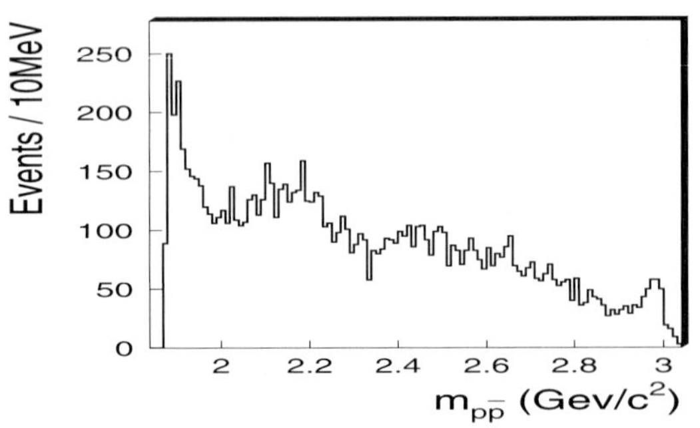


Figure 5. *$p\bar{p}$ mass spectrum at BES II*

S-wave Breit-Wigner function, as shown below, is used to fit the near threshold enhancement.

$$BW \sim \frac{M_0\Gamma_0(q/q_0)}{(M^2 - M_0^2)^2 + (M_0\Gamma_0(q/q_0))^2}$$

Where, $q$ is the momentum of daughter particle, $q_0$ is the momentum of daughter particle at peak. Fig. 6 shows the fit curve. The preliminary fit gives (statistical errors only):

$$M = 1894.1 \pm 1.4 MeV, \quad \Gamma = 48.0 \pm 5.9 MeV$$

with 16 $\sigma$ statistical significance.

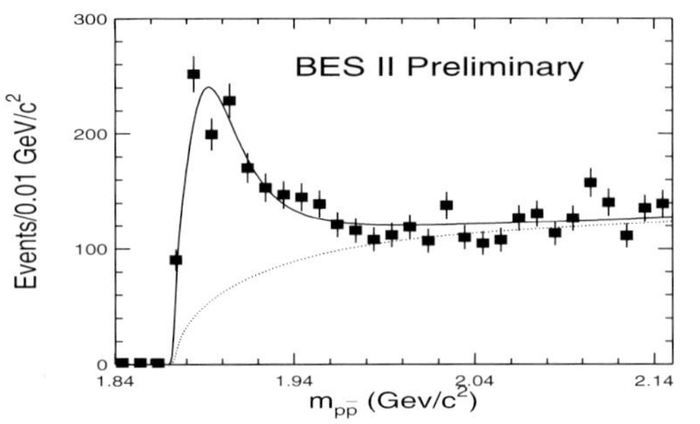


Figure 6. *S-wave BW fit to $p\bar{p}$ near threshold enhancement at BES II*

## 3. Anti-search for glueball candidates in two-photon collisions at CLEO and L3

The two-photon width of a glueball $\Gamma_{\gamma\gamma}$ is expected to be very small since gluons do not couple directly to photons, so a state which is observed in a gluon rich environment but not in two-photon fusions has the typical signature of a glueball. Both CLEO and L3 experiments studied the process $\gamma\gamma \to K_S^0 K_S^0$ [2,3], where the background is made very small by cutting on second vertex and mass of $K_S^0$.

Fig.7 shows $K_S^0 K_S^0$ mass distribution in the region around 2.2 GeV at CLEO. A statistically insignificant excess of $15 \pm 11$ events is found in the $\xi(2230)$ region, with a mass of 2228 MeV and a width of 31 MeV, corresponding to an upper limit of 29.9 events at 95% confidence level and yielding

$$\Gamma_{\gamma\gamma} B(\xi(2230) \to K_S^0 K_S^0) < 1.1eV, \quad (95\% C.L.)$$

assuming spin J=2.

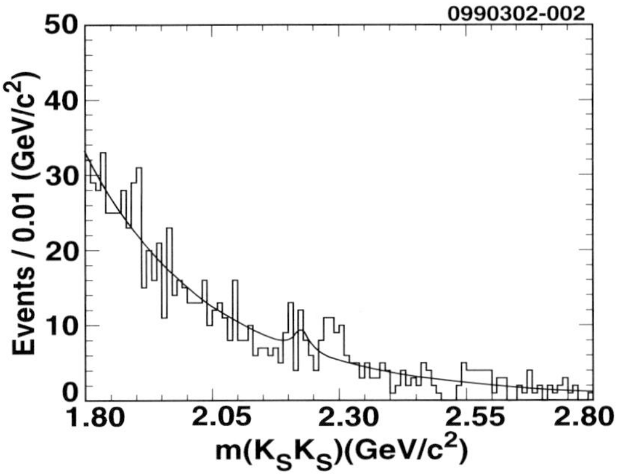


Figure 7. *$K_S^0 K_S^0$ mass distribution observed in data around $\xi(2230)$ mass region at CLEO*

The $K_S^0 K_S^0$ mass spectrum at L3 is presented in Fig.8 showing three distinct peaks over a low background. Despite their large two-photon widths, the $f_2(1270)$ and the $a_2^0(1320)$ tensor mesons produce a small signal in the $K_S^0 K_S^0$ final state due to their destructive interference. The spectrum is dominated by the formation of the $f_2'(1525)$ tensor meson. A signal for the formation of the $f_J(1710)$ is presented while no resonance is observed in the mass region of the $\xi(2230)$. The spin-parity analysis shows that the structure in the $f_J(1710)$ is dominated by $J^{PC} = 2^{++}$, which might be a different state from what is observed at BES II. For the $\xi(2230)$, an upper limit

$$\Gamma_{\gamma\gamma} B(\xi(2230) \to K_S^0 K_S^0) < 1.4eV, \quad (95\% C.L.)$$

is obtained under the hypothesis of a pure helicity-two state.

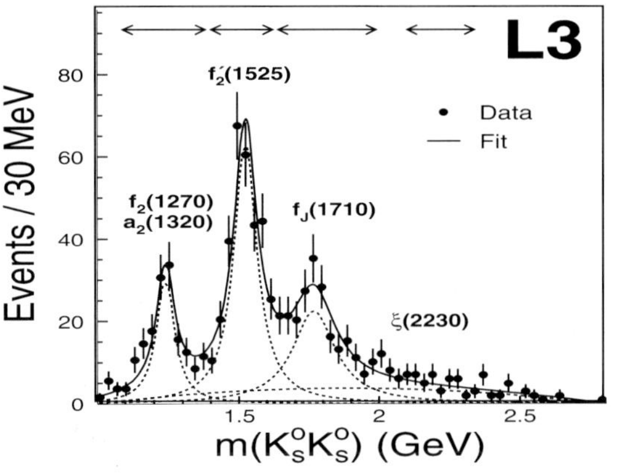


Figure 8. *The $K_S^0 K_S^0$ mass spectrum at L3. The arrows correspond to the $f_2(1270)/a_2^0(1320)$, the $f_2'(1525)$, the $f_J(1710)$ and the $\xi(2230)$ mass regions.*

## 4. Summary

PWA of $J/\psi \to \gamma K\overline{K}$, $\gamma\pi^+\pi^-$ at BES II show that the $f_J(1710)$ is dominated by $J^{PC} = 0^{++}$. A signifcant enhancement near $p\bar{p}$ mass threshold is observed in $J/\psi \to \gamma p\bar{p}$ at the BES II experiment. No resonance is observed in $K_S^0 K_S^0$ final state in the mass region around 2.2 GeV in the two-photon collisions at the CLEO and L3 experiments.

From the very preliminary results at BES II, no clear signal of the $\xi(2230)$ has been observed in the 58 million $J/\psi$ data so far. All possible problems are still being checked at BES II.

## REFERENCES

1. BES Collaboration,J.Z. Bai*et al*, Nucl. Instr. Meth., **A458** 627 (2001).
2. CLEO Collaboration, K. Benslama *et al*, hep-ex/0204019, CLNS 02/1780, CLEO 02-03.
3. L3 Collaboration, M. Acciarri *et al*, CERN-EP/2000-135.

ELSEVIER

Proceedings of ICHEP 2002, pp. 688–691
*S. Bentvelsen, P. de Jong, J. Koch and E. Laenen (Editors)*

31st INTERNATIONAL CONFERENCE ON HIGH ENERGY PHYSICS AMSTERDAM

# E835 results on $\chi_c$ states of charmonium

Roberto Mussa[a] *

[a]INFN, Sezione di Torino, via Pietro Giuria 1, 10125 Torino , Italy

The experiment E835 at FNAL has done an extensive set of studies on charmonium triplet P states. Precision measurements on mass and width of the $\chi_{c0}$ resonance, not studied by the predecessor experiment E760, are reported. Results on both partial widths and higher multipoles of the radiative transitions $\chi_c \to J/\psi\gamma$ are given, together with the $\gamma\gamma$ widths of $\chi_{c0}$ and $\chi_{c2}$. The first clear evidence of charmonium from $p\bar{p}$ annihilations in hadronic decay modes is described, and preliminary results on $BR(\chi_{c0} \to \pi^o\pi^o, \eta\eta) * BR(\chi_{c0} \to p\bar{p})$ are given.

## 1. Introduction

The charmonium system has been a rich source of both experimental and theoretical information on the dynamics of $q\bar{q}$ bound states. The first precision measurements were done in $e^+e^-$ annihilations experiments, where only vector states can be formed directly. In $e^+e^-$ colliders, the production of states with positive C-parity is allowed only via radiative decays and/or via photon-photon fusion processes. In both cases the resolution of the detector is limiting the precision of mass and width measurements.

The $p\bar{p}$ annihilation allows the direct formation of all charmonium states via 2 or 3 perturbative gluons; the resonant signal is usually extracted from the huge hadronic background by selecting charmonium decay modes to EM final states. This limitation can be also overcome by studying scattering at large angles in the CM frame, as well as OZI violating processes (e.g. $\phi\phi$). Here we also report on the first evidence of charmonium detection in hadronic final states, namely $\pi^o\pi^o$ and $\eta\eta$.

In 1984, the experiment R704 at Cern ISR pioneered the study of charmonium states in $p\bar{p}$ annihilations [1], and the experiment E760 at the FNAL $\bar{p}$ accumulator has brought the technique to a mature stage. The scans done by E760 (1989-91) at the $\chi_{c1}$ and $\chi_{c2}$ resonances [2] gave the most accurate measurements available so far on the masses and widths of these two states.

## 2. The E835 Experiment

The experiment is located in the Fermilab Antiproton Accumulator, where a stochastically cooled ($\Delta p/p \sim 10^{-4}$) beam intersects a variable density internal jet target, in order to keep a constant instantaneous luminosity (L$\sim 2 * 10^{31}$) throughout each stack.The E835 detector, described in detail in [3], is a major upgrade of the E760 detector. It is a non-magnetic cylindrical spectrometer with full azimuthal coverage and polar angle acceptance from 2 to 70 degrees in the lab frame. It consists of a lead-glass EM calorimeter divided into a barrel and a forward section. The inner part of the barrel is instrumented with a multicell threshold Čerenkov counter, triggering hodoscopes and charged tracking chambers. In order to withstand the ~3 MHz interaction rate, all channels are instrumented with multihit TDCs. The experiment collected 143 pb$^{-1}$ of data in Run I (Oct.1996-Sept.1997) and 113 pb$^{-1}$ in Run II(Jan.-Nov.2000).

During Run I, E835 collected large samples of $\chi_{c1}$ and $\chi_{c2}$ events, mostly on the resonance peaks, to study the angular distribution of their radiative decay, as well as the $\chi_{c2} \to \gamma\gamma$ partial width. A failure in the beam orbit measurement system prevented to use the Run I data samples to improve our knowledge on masses and widths of these states.

For both E760 and E835-I, the transition en-

*on behalf of the E835 Collaboration: Fermilab, Ferrara, Genoa, Irvine, Minnesota, Northwestern, Torino


doi:10.1016/S0920-5632(02)02197-7

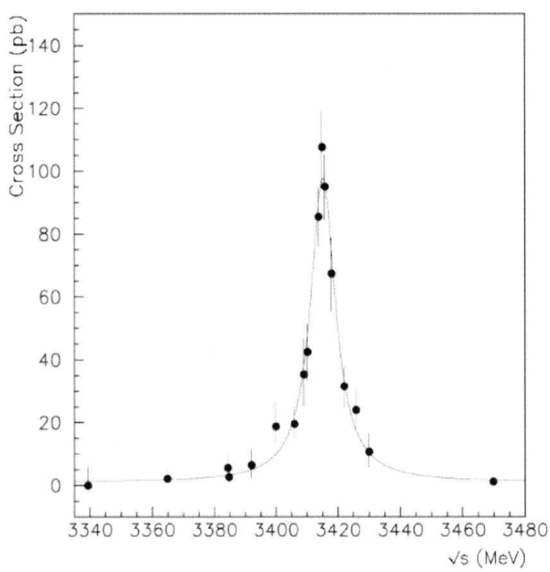


Figure 1. Cross section of $p\bar{p} \to J/\psi\gamma \to e^+e^-\gamma$

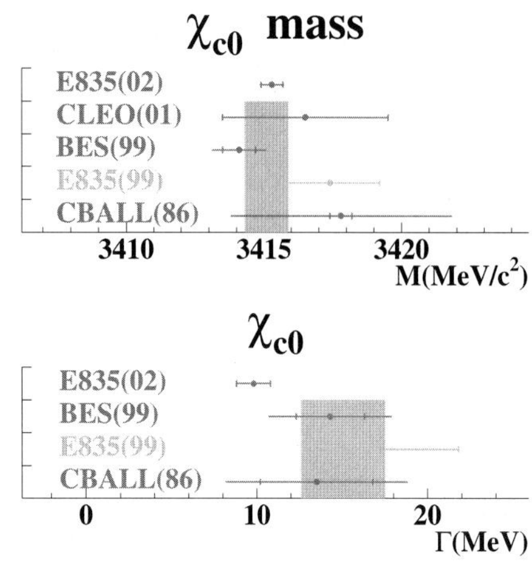


Figure 2. $\chi_{c0}$ mass and width results.

ergy of the accumulator was close enough to the $\chi_{c0}$ mass to prevent stable running with large stacks in this energy region. Nevertheless, a few stacks were decelerated to the $\chi_{c0}$ region at the end of Run I, yielding an unexpectedly high rate of $J/\psi\gamma$ events. The Accumulator underwent a major upgrade between 1997 and 2000, shifting the transition energy upwards and allowing a smooth running at the $\chi_{c0}$.

## 3. Masses and total widths

The 32 pb$^{-1}$ of data taken in scanning the $\chi_{c0}$ during Run II yield the most accurate measurement of the excitation parameters of this resonance. Figure 1 shows the experimental cross section of $J/\psi\gamma$ events for the entire scan, and Fig.2 compares the E835-II measurement [4] of mass and width with former experiments [5–8]: the bands show the PDG 2002 values[11].

Scans at the $\chi_{c1}$ and $\chi_{c2}$ were also made, to cross-check the energy calibration. The results are consistent with E760 [2], as shown in Table 1.

## 4. Radiative transitions $\chi_c \to J/\psi\gamma$

According to a non relativistic calculation, the radiative widths $\Gamma_{rad}(\chi_c)$ are supposed to scale with $E_\gamma^3$, and deviations from this simple scaling rule allow to study the relativistic corrections [9] to these predictions, as well as the O($\alpha_s$) corrections on the orbital wavefunctions. The determination of the radiative width cannot be extracted by a single measurement , as $\psi'$ factories determine BR($\psi' \to \chi_c\gamma$)*BR($\chi_c \to J/\psi\gamma$) from the double radiative decay $\psi' \to J/\psi\gamma\gamma$, $\gamma\gamma$ analyses measure the product $\Gamma_{rad}(\chi_c)$*BR($\chi_c \to \gamma\gamma$) (only the $\chi_{c2}$ has been seen in this process), and $p\bar{p}$ experiments measure $\Gamma_{rad}(\chi_c) * BR(\chi_c \to p\bar{p})$. A set of cross-checks is mandatory, to reach < 10% accuracies on these widths. The global refitting [10,11] of all the $\psi' \to \gamma\chi_c$ and $\chi_c \to J/\psi\gamma$ branching fractions hints to larger systematic errors on the double cascades. The rightmost column in Table 1 shows the E760 and E835-II results on $\Gamma(\chi_c \to J/\psi\gamma) * BR(\chi_c \to p\bar{p})$.

The radiative transitions $\chi_{c1,2} \to J/\psi\gamma$ are dominated by the electric dipole term (E1). The higher multipoles, arising in the relativistic treatment of the transitions, can be studied by measuring the E1-M2 and E1-E3 interference terms in the angular distributions of the radiative decays. Previous measurements of the M2/E1 ratio were done by E760 [12] and Crystal Ball[13]. E835 results[14] are compared with them in Fig.3.

Table 1
Parameters of $\chi_c$ states from E760 and E835-II

| R | Expt. | Mass(MeV/c$^2$) | $\Gamma$(MeV) | $\Gamma(J/\psi\gamma) * BR(p\bar{p})$(eV) |
|---|---|---|---|---|
| $\chi_{c0}$ | | | | |
| | E835-II | 3415.4±0.4±0.2 | 9.9±1.0±0.1 | 26.6±2.5±1.4 |
| $\chi_{c1}$ | | | | |
| | E760 | 3510.53±0.04±0.12 | 0.88±0.11±0.08 | 21.8±1.5±2.2 |
| | E835-II | 3510.62±0.02±0.17 | 0.88±0.06±0.09 | 19.0±0.7±0.9 |
| $\chi_{c2}$ | | | | |
| | E760 | 3556.15±0.07±0.12 | 1.98±0.17±0.07 | 28.2±1.5±2.0 |
| | E835-II | 3556.10±0.09±0.17 | 1.93±0.19±0.09 | 25.3±1.3±1.0 |

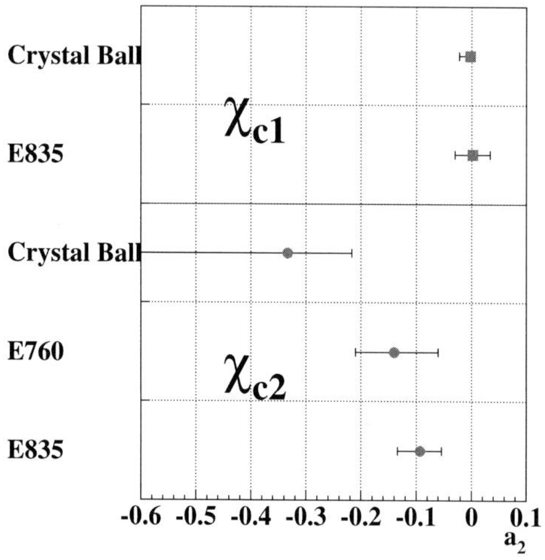


Figure 3. $a_2 = M2/E1$ results

## 5. Two photon widths

Using the data taken at $\chi_{c0}$ and $\chi_{c2}$ energies during Run I, the E835 collaboration has published[15] the value of the ratio $R_{\gamma\gamma}(\chi_{cJ}) = BR(\chi_{cJ} \to \gamma\gamma)/BR(\chi_{cJ} \to J/\psi\gamma)$:

$$R_{\gamma\gamma}(\chi_{cJ}) = \begin{cases} (1.45 \pm 0.74\%) * 10^{-2} & \text{for J=0} \\ (0.99 \pm 0.18\%) * 10^{-3} & \text{for J=2} \end{cases}$$

A substantial contribution to the background on this process comes from the feeddown of $\pi^0\gamma$ and $\pi^o\pi^o$ with one or two undetected photons: the cross section of both reactions is measured with our detector and a MonteCarlo simulation of the showers is used to calculate the feeddown. The $\gamma\gamma$ width is obtained from $\Gamma_{tot}(\chi_c) * R_{\gamma\gamma}(\chi_c) * BR(\chi_c \to J/\psi\gamma)$. Fig.4 shows all the latest measurements [16–20] of the partial width $\chi_{c2} \to \gamma\gamma$, corrected using the new PDG value [11] for $BR(\chi_{c2} \to J/\psi\gamma) = 18.7 \pm 2.0\%$.

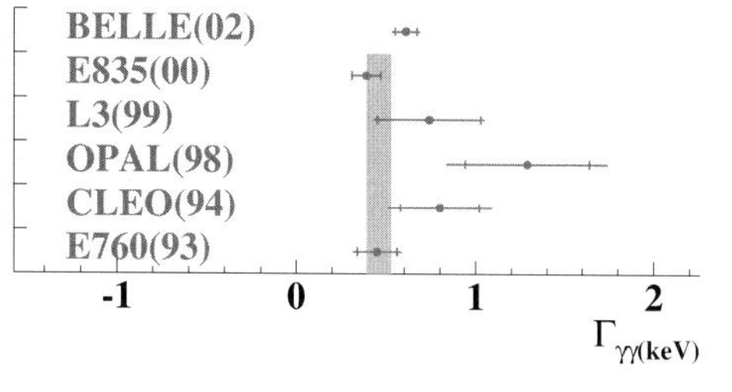


Figure 4. Results on $\Gamma(\chi_{c2} \to \gamma\gamma)$.

The discrepancy between published $p\bar{p}$ and $e^+e^-$ results on this quantity was probably due to larger systematic errors on the measure of $BR(\chi_{c2} \to J/\psi\gamma)$. The analysis of the large sample of $\chi_{c0}$ events taken in Run II is under way. This will allow a 20% measurement of the ratio $R_{\gamma\gamma}(\chi_{c0})$. The large error on both $BR(\chi_{c0} \to p\bar{p})$ and $BR(\chi_{c0} \to J/\psi\gamma)$, limits the precision on the evaluation of $\Gamma(\chi_{c0} \to \gamma\gamma)$ from our result.

## 6. Recent evidence of $p\bar{p} \to \chi_{c0} \to \pi^o\pi^o, \eta\eta$

Other than in the radiative decay modes, the E835-II data taking has yielded the observation of a resonant $\chi_{c0}$ signal in two more decay channels, $\pi^o\pi^o$, $\eta\eta$. Despite the large continuum cross-section in the hadronic reactions , the observation of the charmonium signal is made possible by the interference between resonant and non-resonant amplitudes at $\theta_{CM} = 90^o$. The differential cross section of the $p\bar{p} \to \pi^o\pi^o$ reaction can be written as :

$$\frac{d\sigma}{dz} = |\frac{-A_R}{x+i} + A_I(x,z)e^{i\delta_I(z)}|^2 + |A_{NI}(x,z)|^2$$

where $x$=2($\sqrt{s}$-M)/$\Gamma$ and $z$=cos($\theta_{CM}$). The helicity one amplitude $A_{NI}$ dominates the cross-section at large cos$\theta_{CM}$, does not interfere with the resonant amplitude, and goes to zero at cos$\theta_{CM}$=0. The interference effect magnifies by a factor 20 the resonant signal, and explains the skewness of the cross-section around the peak, as shown in Fig.6. The fit to the data allows to calculate the product of branching ratios:

$$BR(\chi_{c0} \to \pi^o\pi^o)*BR(\chi_{c0} \to p\bar{p}) = (5.1\pm0.8)*10^{-7}$$

An analogous interference signal was observed in the $\eta\eta$ channel and yields:

$$BR(\chi_{c0} \to \eta\eta)*BR(\chi_{c0} \to p\bar{p}) = (4.0\pm1.2)*10^{-7}$$

The results are preliminary and a study of statistical errors is under way.

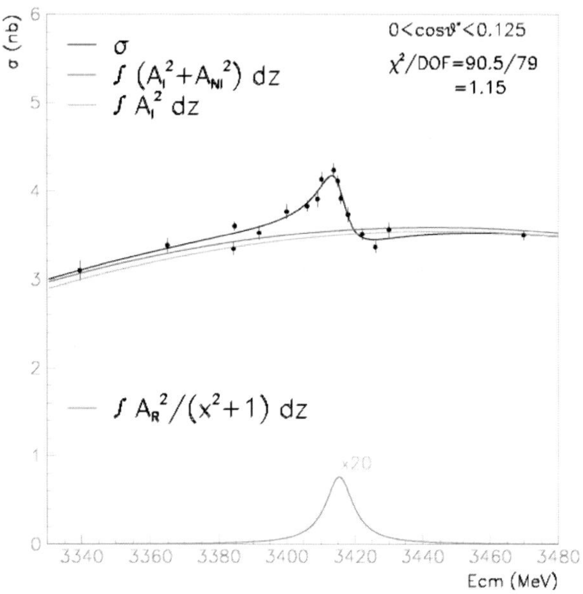


Figure 5. The cross section $\sigma(p\bar{p} \to \chi_{c0} \to \pi^o\pi^o)$ for $|cos(\theta_{CM})| < 0.125$.

## REFERENCES


1. C. Baglin *et al.* [R704 Collaboration], Phys. Lett. B **172** (1986) 455.
2. T. A. Armstrong *et al.* [E760 Collaboration], Nucl. Phys. B **373** (1992) 35.
3. M. Ambrogiani *et al.* [E835 Collaboration], Phys. Rev. D **60** (1999) 032002.
4. S. Bagnasco *et al.* [Fermilab E835 Collaboration], Phys. Lett. B **533** (2002) 237.
5. M. Ambrogiani *et al.* [E835 Collaboration], Phys. Rev. Lett. **83** (1999) 2902.
6. J. Gaiser *et al.*, Phys. Rev. D **34** (1986) 711.
7. J. Z. Bai *et al.* [BES Collaboration], Phys. Rev. D **60** (1999) 072001
8. B. I. Eisenstein *et al.* [CLEO Collaboration], Phys. Rev. Lett. **87** (2001) 061801
9. R. McClary and N. Byers, Phys. Rev. D **28** (1983) 1692.
10. C. Patrignani, Phys. Rev. D **64** (2001) 034017
11. K. Hagiwara *et al.* [Particle Data Group Collaboration], Phys. Rev. D **66** (2002) 010001.
12. T. A. Armstrong *et al.* [E760 Collaboration], Phys. Rev. D **48** (1993) 3037.
13. M. Oreglia *et al.*, Phys. Rev. D **25** (1982) 2259.
14. M. Ambrogiani *et al.* [E835 Collaboration], Phys. Rev. D **65** (2002) 052002.
15. M. Ambrogiani *et al.* [E835 Collaboration], Phys. Rev. D **62** (2000) 052002.
16. T. A. Armstrong *et al.* [E760 Collaboration], Phys. Rev. Lett. **70** (1993) 2983.
17. V. Shelkov *et al.* [CLEO Collaboration], Phys. Rev. D **50** (1994) 4265.
18. K. Ackerstaff *et al.* [OPAL Collaboration], Phys. Lett. B **439** (1998) 197
19. M. Acciarri *et al.* [L3 Collaboration], Phys. Lett. B **453** (1999) 73.
20. K. Abe *et al.* [Belle Collaboration], Phys. Lett. B **540** (2002) 33

ELSEVIER

Proceedings of ICHEP 2002, pp. 692–694
*S. Bentvelsen, P. de Jong, J. Koch and E. Laenen (Editors)*

31st INTERNATIONAL CONFERENCE ON
HIGH ENERGY PHYSICS AMSTERDAM

# Recent $\psi(2S)$ and $\eta_c$ Results from BES

Frederick A. Harris[a]
for the BES Collaboration

[a]The University of Hawaii, Honolulu, Hawaii 96822, USA

We report measurements of branching fractions for $\psi(2S)$ decays into $\omega\pi^+\pi^-$, $b_1\pi$, $\omega f_2(1270)$, $\omega K^+K^-$, $\omega p\bar{p}$, $\phi\pi^+\pi^-$, $\phi f_0(980)$, $\phi K^+K^-$, and $\phi p\bar{p}$ final states based on a data sample of $(4.02 \pm 0.22) \times 10^6$ $\psi(2S)$ events collected with the BESI detector at the Beijing Electron-Positron Collider.

Cross sections for $e^+e^- \to e^+e^-$, hadrons, $\pi^+\pi^- J/\psi$, and $\mu^+\mu^-$ have been measured in the vicinity of the $\psi(2S)$ resonance using the BESII detector. The $\psi(2S)$ total width; partial widths to hadrons, $\pi^+\pi^- J/\psi$, and muons; and corresponding branching fractions have been determined to be $\Gamma_t = 264 \pm 27$ keV; $\Gamma_h = 258 \pm 26$ keV, $\Gamma_\mu = 2.44 \pm 0.21$ keV, and $\Gamma_{\pi^+\pi^- J/\psi} = 85.4 \pm 8.7$ keV; and $B_h = (97.79 \pm 0.15)\%$, $B_{\pi^+\pi^- J/\psi} = (32.3 \pm 1.4)\%$, $B_\mu = (0.93 \pm 0.08)\%$, respectively.

Decays of $J/\psi \to \gamma\eta_c$ are used to determine preliminary values for the mass and width of the $\eta_c$ using a sample of 58 M $J/\psi$ events: $M_{\eta_c} = (2977.4 \pm 0.9 \pm 2.6)$ MeV and $\Gamma_{\eta_c} = (15.7 \pm 3.8 \pm 10.0)$ MeV.

## 1. Introduction

The Beijing Spectrometer (BES) is a general purpose solenoidal detector at the Beijing Electron Positron Collider (BEPC). BEPC operates in the center of mass energy range from 2 to 5 GeV with a luminosity at the $J/\psi$ energy of approximately $5 \times 10^{30}$ cm$^{-2}$s$^{-1}$. BES (BESI) is described in detail in Ref. [1], and the upgraded BES detector (BESII) is described in Ref. [2].

## 2. Hadronic $\psi(2S)$ decays

Both $J/\psi$ and $\psi(2S)$ decays to light hadrons are expected to proceed dominantly via $\psi \to ggg$, with widths that are proportional to the square of the $c\bar{c}$ wave function at the origin [3]. This yields the expectation that

$$Q_X = \frac{B(\psi(2S) \to X_h)}{B(J/\psi \to X_h)} \approx \frac{B(\psi(2S) \to e^+e^-)}{B(J/\psi \to e^+e^-)} = (11.6 \pm 2.2)\%$$

It was first observed by MarkII [4] that the vector-pseudoscalar $\rho\pi$ and $K^*\overline{K}$ channels are suppressed with respect to the 12% expectation - the "$\rho\pi$ puzzle". BES finds a $\rho\pi$ suppression factor of $\sim 60$; this and many other BES $\psi(2S)$ branching ratio results can be found in Refs. [5–9].

Here, we report measurements of branching fractions for $\psi(2S)$ decays involving an $\omega$ or a $\phi$, including $\omega\pi^+\pi^-$, $b_1\pi$, $\omega f_2(1270)$, $\omega K^+K^-$, $\omega p\bar{p}$, $\phi\pi^+\pi^-$, $\phi f_0(980)$, $\phi K^+K^-$, and $\phi p\bar{p}$ final states, based on a data sample of $(4.02\pm0.22)\times10^6$ $\psi(2S)$ events collected with the BESI detector at the Beijing Electron-Positron Collider. Events are selected using particle identification and kinematic fitting. As an example, the $K^+K^-$ invariant mass distribution for candidate $\psi(2S) \to \phi\pi^+\pi^-$ events is shown in Fig. 1, where a clear $\phi$ peak can be seen. We obtain the branching ratios and $Q_X$ values shown in Table 1. The branching fractions for $b_1\pi$ and $\omega f_2(1270)$ update previous BES results, while those for other decay modes are first measurements. The ratios of $\psi(2S)$ and $J/\psi$ branching fractions are smaller than the expected 12% rule by a factor of six for $\omega f_2(1270)$, by a factor of two for $\omega\pi^+\pi^-$, $\omega p\bar{p}$, and $\phi K^+K^-$, while for other studied channels the ratios are consistent with expectation within errors. For more detail on this analysis, see Ref. [10].


doi:10.1016/S0920-5632(02)02200-4

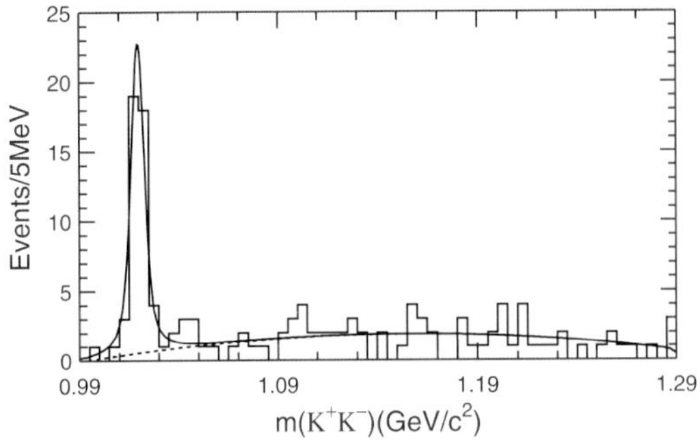


Figure 1. The $K^+K^-$ invariant mass distribution for candidate $\psi(2S) \to \phi\pi^+\pi^-$ events.

### 3. BES $\psi(2S)$ Scan Results

In 1999, after the $R$-scan [11], BES did a careful $\psi(2S)$ scan. The purpose was to improve the accuracies of the $\psi(2S)$ parameters: the total width $(\Gamma_t)$, and partial widths into hadrons $(\Gamma_h)$, $\mu^+\mu^-$ $(\Gamma_\mu)$, and $\pi^+\pi^- J/\psi$ $(\Gamma_{\pi^+\pi^- J/\psi})$, and the corresponding branching fractions, $B(h)$, $B(\mu)$, and $B(\pi^+\pi^- J/\psi)$. $B(\pi^+\pi^- J/\psi)$ and $B(\mu)$ are important because these decays are used to identify $\psi(2S)$ in B decays $(B \to \psi(2S)K^0_S)$.

A total of 24 energy points between 3.67 and 3.71 GeV were scanned. The total integrated luminosity was 760 nb$^{-1}$. We assume $\Gamma_t = \Gamma_h + \Gamma_\mu + \Gamma_e + \Gamma_\tau$, along with lepton universality: $\Gamma_e = \Gamma_\mu = \Gamma_\tau/0.38847$. The cross sections versus scan point energy and fit curves are shown in Fig. 2, and the preliminary fit results are given in Table 2. We obtain a first measurement of $\Gamma_{\pi^+\pi^- J/\psi}$, and $B(h)$, $B(\mu)$, and $B(\pi^+\pi^- J/\psi)$ have improved precision compared to the PDG values [12]. The value for $\Gamma_t$ agrees with a previous BES value of (252±37) keV [13] within errors. A complete description of this work can be found in Ref. [14].

### 4. $\eta_c$ Parameters

The mass and width of the $\eta_c$ are rather poorly known; the confidence level for the PDG weighted average mass is only 0.001 [12]. Previously we

Table 1
Branching fractions of $\psi(2S)$ and $Q_X$ values for $\psi(2S)$ and $J/\psi$ hadronic decays. The $B_{J/\psi}$ are taken from the PDG [12]. To determine $B(\phi f_0(980))$, we use $B_{f_0 \to \pi^+\pi^-} = 0.521 \pm 0.016$ (PDG'96).

| Channel X | $B_{\psi(2S)\to X}$ $(10^{-4})$ | $Q_X(\%)$ (%) |
|---|---|---|
| $\omega\pi^+\pi^-$ | $4.8 \pm 0.6 \pm 0.7$ | $6.7 \pm 1.7$ |
| $b_1^\pm\pi^\mp$ | $3.2 \pm 0.6 \pm 0.5$ | $11 \pm 3$ |
| $\omega f_2(1270)$ | $1.1 \pm 0.5 \pm 0.2$<br>$< 1.5$ | $2.4 \pm 1.3$ |
| $\omega K^+K^-$ | $1.5 \pm 0.3 \pm 0.2$ | $20 \pm 8$ |
| $\omega p\bar{p}$ | $0.8 \pm 0.3 \pm 0.1$ | $6.0 \pm 2.8$ |
| $\phi\pi^+\pi^-$ | $1.5 \pm 0.2 \pm 0.2$ | $18 \pm 5$ |
| $\phi f_0(980)$<br>$\times(f_0 \to \pi^+\pi^-)$ | $0.6 \pm 0.2 \pm 0.1$ | |
| $\phi f_0(980)$ | $1.1 \pm 0.4 \pm 0.1$ | $33 \pm 15$ |
| $\phi K^+K^-$ | $0.6 \pm 0.2 \pm 0.1$ | $7.3 \pm 2.6$ |
| $\phi p\bar{p}$ | $0.12 \pm 0.06 \pm 0.02$<br>$< 0.26$ | $27 \pm 16$<br>$< 58$ |

measured the $\eta_c$ mass using the BESI 4.02 M $\psi(2S)$ sample and obtained $M_{\eta_c} = (2975.8 \pm 3.9 \pm 1.2)$ MeV [15]. We also used 7.8 M BESI $J/\psi$ events and obtained $M_{\eta_c} = (2976.6 \pm 2.9 \pm 1.3)$ MeV [16]. For the two data sets combined, $M_{\eta_c} = (2976.3 \pm 2.3 \pm 1.2)$ MeV and the total width $\Gamma_{\eta_c} = (11.0 \pm 8.1 \pm 4.1)$ MeV [16].

Here, the mass and width have been determined using our 58 M $J/\psi$ sample. We use the channels $J/\psi \to \gamma\eta_c$, with $\eta_c \to p\bar{p}$, $K^+K^-\pi^+\pi^-$, $\pi^+\pi^-\pi^+\pi^-$, $K^\pm K^0_S\pi^\mp$, and $\phi\phi$. Events are selected using particle identification and kinematic fitting. Combining the five decay channels, we obtain a *preliminary* result: $M_{\eta_c} = (2977.4 \pm 0.9 \pm 2.6)$ MeV and $\Gamma_{\eta_c} = (15.7 \pm 3.8 \pm 10.0)$ MeV, to be compared to the current PDG values: $M_{\eta_c} = (2979.7 \pm 1.5)$ MeV and $\Gamma_{\eta_c} = (16.0^{+3.6}_{-3.2})$ MeV [12]. The systematic errors here are conservative and should be smaller in the final analysis.

Table 2
$\psi(2S)$ scan results and comparison with the PDG2002 [12]. The numbers in parenthesis denote the relative errors. $\Gamma_\mu$ value given using the assumption $\Gamma_e = \Gamma_\mu$.

| Value | BES | PDG2002 |
|---|---|---|
| $\Gamma_t$(keV) | $264 \pm 27$ (10.1%) | $300 \pm 25$ (8.3%) |
| $\Gamma_h$(keV) | $258 \pm 26$ (10.1%) | |
| $\Gamma_{\pi\pi J/\psi}$(keV) | $85.4 \pm 8.7$ (10.1%) | |
| $\Gamma_\mu$(keV) | $2.44 \pm 0.21$ (8.8%) | |
| $\mathcal{B}_h$(%) | $97.79 \pm 0.15$ (0.16%) | $98.10 \pm 0.30$ (0.31 %) |
| $\mathcal{B}_{\pi\pi J/\psi}$(%) | $32.3 \pm 1.4$ (4.4%) | $30.5 \pm 1.6$ (5.2 %) |
| $\mathcal{B}_\mu$(%) | $0.93 \pm 0.08$ (8.5%) | $0.7 \pm 0.09$ (12.9 %) |

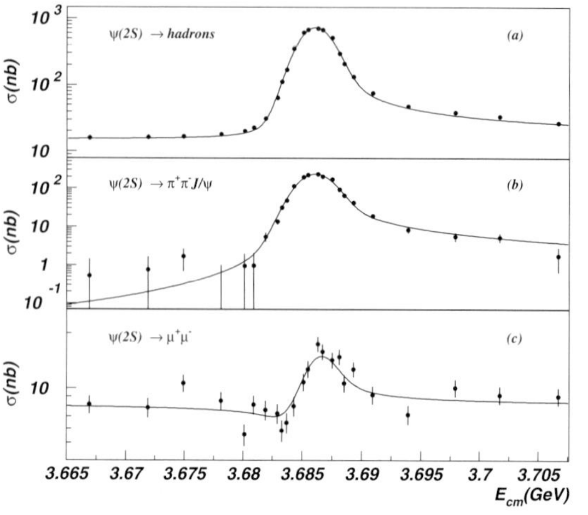


Figure 2. The cross section for (a) $e^+e^- \to$ hadrons, (b) $e^+e^- \to \pi^+\pi^- J/\psi$, and (c) $e^+e^- \to \mu^+\mu^-$ versus center-of-mass energy. The solid curves represent the results of the fit to the data.

## 5. Summary

BES has measured decay modes of the $\psi(2S)$ to final states containing $\omega$ and $\phi$ particles in order to provide more information on the 12 % Rule. The $\psi(2S)$ scan results improve the precision of the $\psi(2S)$ resonance parameters and measure $\Gamma_{\pi^+\pi^- J/\psi}$ for the first time. Finally using our 58 M $J/\psi$ sample, we obtain preliminary values for the $\eta_c$ mass and width.

## REFERENCES


1. J. Z. Bai *et al.*, (BES Collab.), Nuc. Inst. Meth. **A344**, 319 (1994).
2. J. Z. Bai *et al.*, (BES Collab.), Nuc. Inst. Meth. **A458**, 627 (2001).
3. T. Appelquist and H.D. Politzer Phys. Rev. Lett. **34**, 43 (1975); and A. De Rujula and S. L. Glashow, ibid, page 46.
4. M.E.B. Franklyn *et al.*, (MarkII Collaboration), Phys. Rev. Lett. **51**, 963 (1983).
5. Y. S. Zhu, Proceedings of the 28th International Conference on High Energy Physcis, ed. Z. Adjuk and A. K. Wroblewski, World Scientific, 507 (1997).
6. J. Z. Bai *et al.*, (BES Collab.), Phys. Rev. Lett. **81**, 5080 (1998).
7. J.Z. Bai *et al.*, (BES Collab.), Phys. Rev. **D58**, 097101 (1998).
8. J. Z. Bai *et al.*, (BES Collab.), Phys. Rev. Lett. **83**, 1918 (1999).
9. J.Z. Bai *et al.*, (BES Collab.), Phys. Rev. **D63**, 032002 (2001).
10. J.Z. Bai *et al.*, (BES Collab.), submitted to Phys. Rev. D, hep-ex/0209080.
11. J. Z. Bai *et al.*, (BES Collab.), Phys. Rev. Lett. **84**, 594 (2000); J. Z. Bai *et al.*, (BES Collab.), Phys. Rev. Lett. **38** (2002) 101802.
12. K. Hagiwara *et al.*, Phys. Rev. **D66**, 010001 (2002).
13. J. Z. Bai *et al.*, (BES Collab.), Phys. Rev. **D65**, 052004-1 (2002).
14. J. Z. Bai *et al.*, (BES Collab.), submitted to Phys. Lett., hep-ph/0209354.
15. J. Z. Bai *et al.*, (BES Collab.), Phys. Rev. **D60**, 72001 (1999).
16. J.Z. Bai *et al.*, (BES Collab.), Phys. Rev. **D 62**, 72001 (2000).

ELSEVIER

Proceedings of ICHEP 2002, pp. 695–697
*S. Bentvelsen, P. de Jong, J. Koch and E. Laenen (Editors)*

31st INTERNATIONAL CONFERENCE ON HIGH ENERGY PHYSICS AMSTERDAM

# $\eta_c(2980)$ formation in two-photon collisions at LEP energies

A. Oblakowska-Mucha[a], B. Muryn[a], G. Polok[b], T. Szumlak[a]

[a]Faculty of Nuclear Physics and Techniques, UMM, PL-30055 Kracow, Poland

[b]Institute of Nuclear Physics, PL-31342 Krakow, Poland

$\eta_c(2980)$ production in $\gamma\gamma$ interactions has been detected via its decays into $K_s^0 K^\pm \pi^\mp$, $K^+K^-K^+K^-$ and $K^+K^-\pi^+\pi^-$ in the data taken with the DELPHI detector at LEP1 and LEP2 energies. The two-photon radiative width averaged over all observed decay channels is $\Gamma_{\gamma\gamma} = 13.9 \pm 2.0$ (stat.) $\pm$ 1.4(syst.) $\pm$ 2.7 (BR) keV. No direct decay channel $\eta_c \to \pi^+\pi^-\pi^+\pi^-$ has been observed. An upper limit $\Gamma_{\gamma\gamma} < 3.8$ keV at 95% confidence level has been evaluated for that decay mode.

## 1. Introduction

An estimations of the $\eta_c$ partial width, $\Gamma_{\gamma\gamma}(\eta_c)$ based on different models and corrections provided values from 3 to 14 keV, see [1] and references therein. An even bigger discrepancy is observed between values obtained by numerous experimental groups [2]. The results for $\Gamma_{\gamma\gamma}(\eta_c)$ range from 4 keV to 27 keV.

In this paper we report on the production and decays of the $\eta_c$ resonance using data collected by the DELPHI detector during the period 1994-1999 corresponding to a range of centre-of-mass energies from 90 GeV up to 202 GeV and an integrated luminosity of $\mathcal{L} = 531$ pb$^{-1}$. The aim of this analysis was to determine the radiative width of the $\eta_c$ resonance separately for each decay channel, using the production process: $e^+e^- \to e^+e^-\eta_c(2980)$ in no-tag mode.
The following exclusive final states have been analysed:

$$\eta_c \to K_s^0(\pi^+\pi^-)K^\pm\pi^\mp \tag{1}$$

$$\eta_c \to K^+K^-K^+K^- \tag{2}$$

$$\eta_c \to K^+K^-\pi^+\pi^- \tag{3}$$

$$\eta_c \to \pi^+\pi^-\pi^+\pi^- \tag{4}$$

## 2. Detector

A general description of the DELPHI detector can be found elsewhere [3]. The main features relevant to this analysis are particle tracking and identification. Due to the low momenta of the decay products, their identification is based on measurement of ionization losses (dE/dx) in the Time Projection Chamber (TPC).

## 3. General Data Selection

Candidates for the $\eta_c(2980)$ decay channels (1)-(4) were selected by requiring:

- exactly four charged particle tracks with zero total charge,
- the square of the total transverse momentum, $(\Sigma\vec{p_t})^2$, of charged particles to be less than 0.04 $(\mathrm{GeV/c})^2$,
- the total detected energy of charged particles to be less than 10 GeV,
- each of the track should fulfil standard DELPHI quality cuts [3]
- each track to pass through the TPC,
- no particles identified as electrons or muons by the standard lepton identification algorithms,

Additional criteria which are specific to particular channels are discussed in the next section.

## 4. Analysis

In order to calculate the acceptance and detection efficiency, a Monte Carlo generation program has been used, with the full kinematics of a system produced in $\gamma\gamma$ interactions [4]. The Monte Carlo generated events were passed through the


doi:10.1016/S0920-5632(02)02203-X

standard DELPHI detector simulation procedure [3] and the trigger simulation.
The efficiency for each decay mode as a function of the invariant mass was calculated bin-by-bin and is shown in Fig.1.

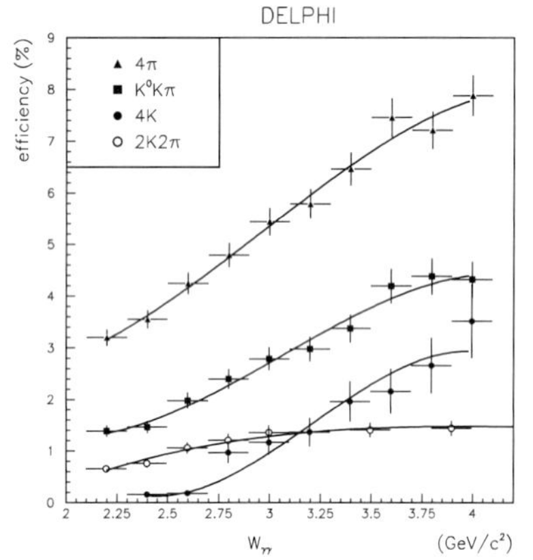


Figure 1. The average luminosity-weighted efficiencies for different $\eta_c$ decay final states as a function of the corresponding invariant mass. In the decay to $K^0$, its branching fraction to $\pi^+\pi^-$ has been taken into account.

It should be noted that particle identification was essential for all the channels analysed and was based on dE/dx energy loss measurements [3] since this is the only method providing an identification probability.

**4.1.** $\eta_c \rightarrow K_s^0 K^\pm \pi^\mp$

For this channel the $K_s^0 \rightarrow \pi^+\pi^-$ decay is identified using an algorithm which takes pairs of oppositely charged particle tracks, intersecting them and determining a secondary vertex. It was also required that one of the remaining two particles should be identified as a kaon.

**4.2.** $\eta_c \rightarrow K^+K^-K^+K^-$

Additional requirements for this decay channel are that at least three particles must be identified as charged kaons and there are no secondary vertices. Only kaons with the probability of identification greater then 0.5 were considered.
A scatter plot (not shown) of the invariant mass of $K^+K^-$ combinations does not indicate any intermediate $\phi\phi$ state.

**4.3.** $\eta_c \rightarrow K^+K^-\pi^+\pi^-$

In order to select events resulting from the process (3) it was required that one of the particles was identified as a kaon with probability $\geq$0.5 and two of the three remaining particles should satisfy selection criteria for pion identification with probability $\geq$0.5.
The intermediate states of $\eta_c \rightarrow K^+K^-\pi^+\pi^-$ decay via one or two $K^{*0}(892)$ have not been observed.

**4.4.** $\eta_c \rightarrow \pi^+\pi^-\pi^+\pi^-$

The observation of an $\eta_c \rightarrow \pi^+\pi^-\pi^+\pi^-$ decay mode reported by numerous experimental groups remains controversial. This decay has been found by MARK III, DM2, TASSO [2] and references therein. (where the last one did not distinguish between the $4\pi$ and the $\rho^0\rho^0$ decay channels). Other, more recent experiment did not confirm it [6] providing only an upper limit.
In addition to the general selection and the stringent cut on the total transverse momentum, it was also required that all particles were pions with the single track identification probability $\geq$ 0.5.
In this decay channel of $\eta_c$ an upper limit of 18 events at the 95% confidence level has been calculated.
A search for the intermediate decay mode, $\rho^0\rho^0$ was also performed. These events were attributed to non-resonant vector meson production.

## 5. Results

In experiment one measures directly the invariant mass distribution of the $\gamma\gamma$ system, $\frac{\Delta N}{\Delta W}$. Given the detector efficiency $\mathcal{E}_f$, the integrated luminosity $\mathcal{L}$, and flux $L_{\gamma\gamma}$ of the two interacting photons parametrised by well known equivalent photon approximation formula, the invariant mass distribution can be converted into two-photon cross section multiplied by corresponding branching ratio:

$$\sigma_{\gamma\gamma\to\eta_c}(W_{\gamma\gamma})\cdot BR(\eta_c\to f) = \frac{\Delta N_{(e^+e^-\to e^+e^-\eta_c\to e^+e^-f)}}{\Delta W_{\gamma\gamma}\ \cdot\ \mathcal{L}\ \cdot \mathcal{E}_f(W_{\gamma\gamma})\ \cdot L_{\gamma\gamma}(W_{\gamma\gamma})} \quad (5)$$

The efficiency was calculated dividing bin-by-bin the simulated invariant mass distribution for events that passed all the cuts in mass interval 2.5-4.0 GeV/c$^2$ by the invariant mass distribution for the generated events.

In order to determine the value of the $\eta_c$ radiative width, the $\gamma\gamma$ invariant mass cross-section has been fitted to the Breit Wigner distribution describing the $\eta_c$ production cross-section convoluted with a Gaussian mass resolution $G(W_{\gamma\gamma},\sigma)$ together with a background parametrisation expressed in terms of polynomial function of the third order $P_3(W_{\gamma\gamma})$:

$$\sigma_{\gamma\gamma\to\eta_c}(W_{\gamma\gamma})\cdot BR(\eta_c\to f) = G(W_{\gamma\gamma},\sigma)\otimes [BW(\Gamma_{\gamma\gamma},M_{\eta_c},\Gamma_{tot},W_{\gamma\gamma})\ +\ P_3(W_{\gamma\gamma})] \quad (6)$$

According to (6) the fit determines the product of the radiative width and the branching ratio, the mass of the resonance and the experimental mass resolution $\sigma$. All these fitted parameters have been determined separately for each data sample.

The final $\sigma_{\gamma\gamma}$ plots are average distributions from different samples. The resulting cross-sections multiplied by the corresponding branching ratios for $\eta_c \to K^0_s K^\pm\pi^\mp$, $\eta_c \to K^+K^-K^+K^-$ and $\eta_c \to K^+K^-\pi^+\pi^-$ are presented in Fig.2.

A weighted mean of the radiative width of the first three channels is calculated and provides the result:

$$\Gamma_{\gamma\gamma} = 13.9 \quad \pm 2.0\ (\text{stat.}) \quad \pm 1.4\ (\text{syst.}) \quad \pm 2.7\ (\text{BR}) \quad \text{keV}$$

The product $\Gamma_{\gamma\gamma}\cdot BR$ in in the analysis of the four charged kaon decay channel is in agreement, within the large errors, with the result of the ARGUS Collaboration [5].

The $\eta_c$ decay into $\pi^+\pi^-\pi^+\pi^-$ has not been detected. This striking result is supported also by other experiments [6].

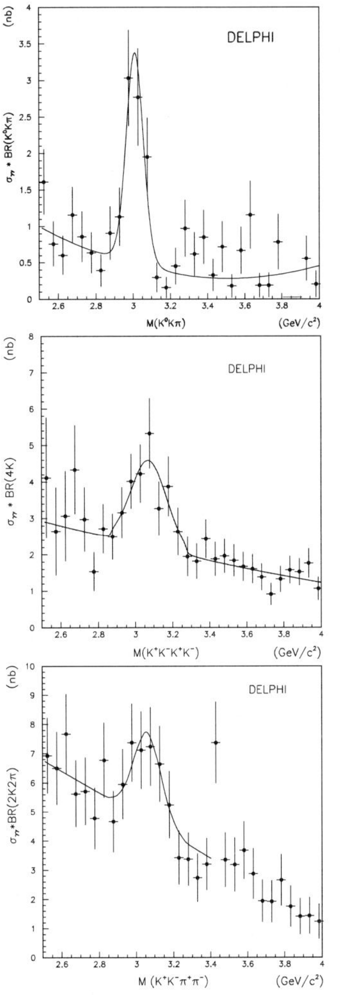


Figure 2. $\sigma(\gamma\gamma\to\eta_c)\cdot BR(\eta_c\to final)$

## REFERENCES

1. E.S.Ackleh and T.Barnes, Phys. Rev. **D45** (1992) 232; L.J.Reinders, H.Rubinstein and S.Yazaki, Phys. Rep. **127-1** (1985) 1.
2. D.E. Groom et al,(Particle Data Group) Eur. Phys. Jour. **C15** (2000) 1.
3. DELPHI Collab., P. Aarnio et al., Nucl. Instr. and Meth. **A303** (1991) 233.
4. H.Krasemann and J.A.M.Vermaseren, Nucl. Phys. **B184** (1981) 265.
5. ARGUS Collab., H.Albrecht et al., Phys. Lett. **B338** (1994) 390.
6. L3 Collab., M.Acciari et al., Phys. Lett. **B461** (1999) 155.

ELSEVIER

Proceedings of ICHEP 2002, pp. 698–702
*S. Bentvelsen, P. de Jong, J. Koch and E. Laenen (Editors)*

31st INTERNATIONAL CONFERENCE ON HIGH ENERGY PHYSICS AMSTERDAM

# Bottomonium studies via $\Upsilon(3S)$ decays. First observation of $\Upsilon(1D)$ states.

T. Skwarnicki[a]

[a]Department of Physics
Syracuse University
Syracuse, NY 13244, USA

*(Representing the CLEO Collaboration)*

The CLEO III experiment has recently accumulated a large statistics sample of $4.73 \cdot 10^6$ $\Upsilon(3S)$ decays. We present the first evidence for the production of the triplet $\Upsilon(1D)$ states in the four-photon cascade, $\Upsilon(3S) \to \gamma\chi_b(2P)$, $\chi_b(2P) \to \gamma\Upsilon(1D)$, $\Upsilon(1D) \to \gamma\chi_b(1P)$, $\chi_b(1P) \to \gamma\Upsilon(1S)$, followed by the $\Upsilon(1S)$ annihilation to $e^+e^-$ or $\mu^+\mu^-$. We have also analyzed two-photon cascades between the triplet $S$ and triplet $P$ states and report improved measurements of the photon energies and transition rates. We have searched for $\Upsilon(3S) \to \gamma\eta_b(1S)$ hindered M1 photon transition in inclusive photon spectrum. No evidence for such a transition is found. We set upper limits on the branching ratio, which rule out many previously published phenomenological estimates of the rate for this transition.

## 1. INTRODUCTION

Long-lived $b\bar{b}$ states are especially well suited for testing QCD via lattice calculations [1] and effective theories of strong interactions, such as potential models [2]. The narrow triplet $S$ states, $\Upsilon(1S)$, $\Upsilon(2S)$ and $\Upsilon(3S)$, were discovered in 1977 in proton-nucleus collisions at Fermilab. Later, they were better resolved and studied at various $e^+e^-$ storage rings. Six triplet $P$ states, $\chi_b(2P_J)$ and $\chi_b(1P_J)$ with $J = 0, 1, 2$ were discovered in radiative decays of the $\Upsilon(3S)$ and $\Upsilon(2S)$ in 1982 and 1983, respectively. There have been no observations of new narrow $b\bar{b}$ states since then despite the large number of such states predicted below the open flavor threshold (see Fig. 1).

The CLEO III experiment has recently accumulated $4.73 \cdot 10^6$ $\Upsilon(3S)$ decays, which constitutes roughly a ten-fold increase in $\Upsilon(3S)$ statistics compared to the CLEO II data set [3], and roughly a four-fold increase compared to the integrated CUSB data set [4]. Thanks to the good granularity and the large solid angle of the CLEO CsI(Tl) calorimeter, the CLEO III detection efficiencies for the analyses presented here are $2-5$ times larger than in the CUSB detector.

We have analyzed the $\Upsilon(3S)$ data sample for two- and four-photon cascade transitions between the $\Upsilon(3S)$ and either the $\Upsilon(2S)$ or the $\Upsilon(1S)$, with the final state $b\bar{b}$ state annihilating to $e^+e^-$ or $\mu^+\mu^-$. The transitions in these cascades are of electric dipole type (E1) and occur between spin-triplet states ($S = 1$) by changing one unit of orbital angular momentum ($S \leftrightarrow P$, $P \leftrightarrow D$) - see Fig. 1. The experimental backgrounds in this type of analysis are small. We also search for the hindered magnetic dipole (M1) transition of the $\Upsilon(3S)$ to $\eta_b(1S)$. Since a single virtual photon cannot be produced in the $\eta_b(1S)$ annihilation, this state decays to two gluons which produces hadrons. Thus, in this analysis we select hadronic events and look for a monochromatic peak in the inclusive photon spectrum. This type of analysis suffers from large experimental backgrounds.

## 2. SEARCH FOR $\eta_b$

The experimental signature for $\Upsilon(3S) \to \gamma\eta_b$ M1 transitions is a monoenergetic peak in the photon energy distribution for hadronic events.


doi:10.1016/S0920-5632(02)02262-4

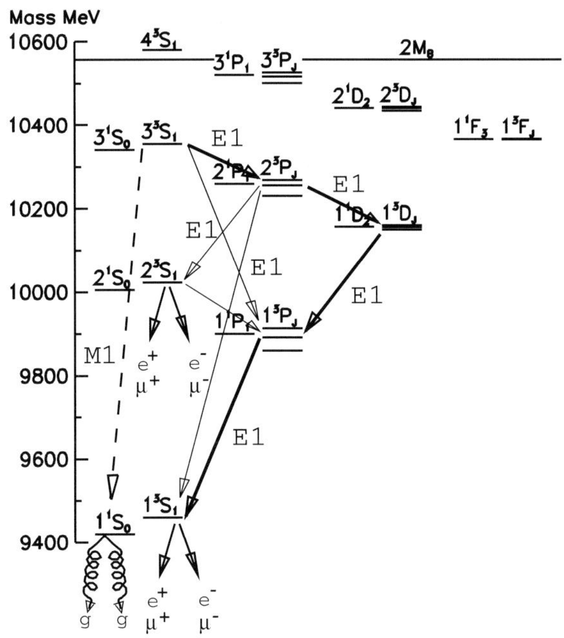

Figure 1. $b\bar{b}$ mass levels as predicted by one of the potential models. Electric dipole (E1) and magnetic dipole (M1) transitions studied in this work are indicated.

Since the mass of the $\eta_b$ is unknown, we inspect the 880 – 1000 MeV photon energy range which covers the $\eta_b$ mass range predicted by various theoretical calculations. To minimize the width of the expected peak we use only the barrel part of the electromagnetic calorimeter ($|\cos\theta| < 0.804$) and require that the photon is isolated from other energy depositions in the calorimeter. Since the main background is due to $\pi^0$ decays, we remove photons which, when paired with a second photon, have an invariant mass within 2.5 standard deviations of the known $\pi^0$ mass. The remaining background is still predominantly $\pi^0$ mesons where the companion photon in the $\pi^0$ decay is not reconstructed. Our signal selection efficiency is $(46 \pm 3)\%$ (the error is systematic and is dominated by modeling of $\eta_b$ decays). The resulting photon energy spectrum is shown in Fig. 2. We fit this spectrum using a third order polynomial for the background, three overlapping Gaussians to represent the Doppler broadened E1 spectral lines from $\chi_b(2P_J) \to \gamma\Upsilon(1S)$ $(J = 2, 1, 0)$, and a signal peak due to $\Upsilon(3S) \to \gamma\eta_b(1S)$. The

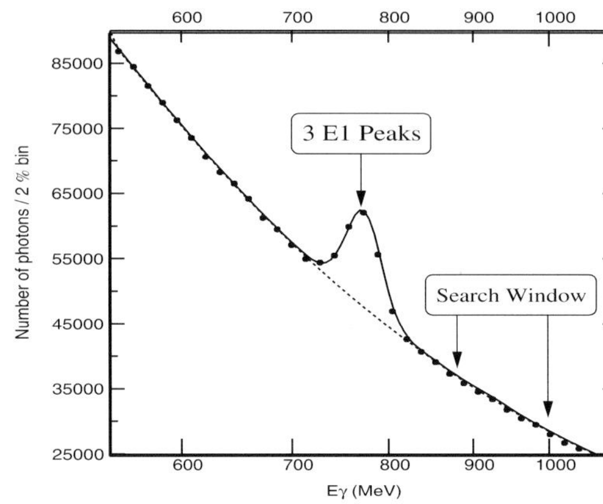


Figure 2. The inclusive photon energy spectrum, in logarithmic energy scale, showing the search window for the M1 transition $\Upsilon(3S) \to \gamma\eta_b(1S)$ and the nearby $\chi_b(2P) \to \gamma\Upsilon(1S)$ peak due to the three overlapping E1 lines. The overall production rate for the E1 lines is about $2 \cdot 10^{-2}$.

maximum signal yield found in the search window corresponds to only 1.5 standard deviations above the expected background. Since we do not find evidence for the signal we set an upper limit on its production rate (see Fig. 3). Predictions for this rate are also shown. The matrix element for any hindered M1 transition is zero at lowest order, thus relativistic and finite-size corrections must be invoked. Since most of the existing calculations can be ruled out by our results, such corrections are clearly not well understood.

A more detailed description of these results can be found in Ref. [6].

## 3. TWO PHOTON CASCADES

The large photon background from $\pi^0$ mesons produced in gluonic annihilation of the $b\bar{b}$ system can be completely eliminated by selecting events consistent with $b\bar{b}$ annihilation to $e^+e^-$ or $\mu^+\mu^-$ via a virtual photon. Only triplet $S$ states have substantial dilepton widths, thus this method of background suppression can be employed for two- and four-photon cascades between the $\Upsilon(3S)$ and either the $\Upsilon(2S)$ or the $\Upsilon(1S)$. The background in such analyses comes from radiative Bhabha scattering and $\mu$–pair production. Since the Bhabha cross-section is much larger, the dimuon

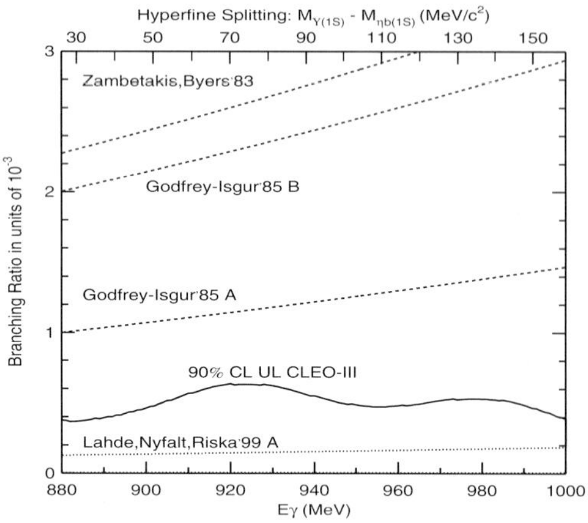


Figure 3. Upper limit on $\mathcal{B}(\Upsilon(3S) \to \gamma\eta_b(1S))$ at 90% C.L. for various values of photon energy (the bottom axis) or hyperfine mass splitting (the top axis). The model predictions are taken from Ref. [5]. They are extrapolated to various singlet-triplet splitting values via the phase space factor $(E_\gamma^3)$.

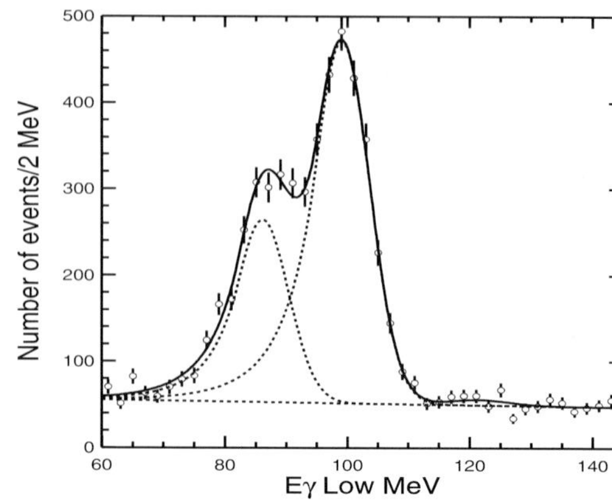


Figure 4. Energy of the less energetic photon in $\Upsilon(3S) \to \gamma\gamma l^+l^-$ events. The two large peaks correspond to $\Upsilon(3S) \to \gamma\chi_b(2P_2)$ and $\Upsilon(3S) \to \gamma\chi_b(2P_1)$ transitions. There is also an insignificant peak at an energy corresponding to $\Upsilon(3S) \to \gamma\chi_b(2P_0)$. The solid line represents the fit. The dashed lines show the fitted background (linear function) and the two individual photon lines on top of the background.

sample has a better signal-to-background ratio.

In this section, we present results for two photon cascades. We select events with exactly two photons and two high momentum leptons. The total momentum of all reconstructed particles must be consistent with zero. We also require that the mass of the two leptons to be consistent with either the $\Upsilon(1S)$ or $\Upsilon(2S)$ mass. Even better identification of these states is obtained using the recoil mass against the two photons. The average detection efficiency is about 32%. The energy spectrum of the lower energy photon is shown in Fig. 4. From the three photon lines expected from $\Upsilon(3S) \to \gamma\chi_b(2P_{2,1,0})$ transitions, only the $J = 2$ ($1286 \pm 68$ events) and $J = 1$ ($2726 \pm 180$ events) lines are clearly visible. The third $J = 0$ line is suppressed by the larger hadronic width of this state, which results in a small radiative branching fraction. A fit to this photon energy spectrum yields $E_{J=2} = (86.09 \pm 0.30 \pm 0.29)$ MeV, $E_{J=1} = (99.08 \pm 0.17 \pm 0.34)$ MeV. These results are consistent with, but more precise than, previous determinations [7].

From the fitted amplitudes, we compute product branching ratios for the entire decay sequence. These measurements are much improved over, and in good agreement with the previous CLEO II results [3] (see Fig. 5). The CUSB results [4] for the transitions via the $J = 1$ state appear to be systematically lower than the CLEO results. The numerical results and a more detailed description of this analysis can be found in Ref. [8].

We also observe a signal for the radiative cascade via the $\chi_b(1P)$ states. Since different spin states cannot be resolved, and the two photons in the cascade have similar energies, we extract the signal from the recoil mass spectrum (see Fig. 6) with the requirement that the photon energies are within 100 MeV of each other. We observe $167 \pm 19$ events, corresponding to a branching ratio of $\mathcal{B}(\Upsilon(3S) \to \gamma\chi_b(1P))\times\ \mathcal{B}(\chi_b(1P) \to \gamma\Upsilon(1S))\times\ \mathcal{B}(\Upsilon(1S) \to l^+l^-) = (5.2 \pm 0.5 \pm 0.5) \cdot 10^{-5}$. Hereafter, $\mathcal{B}(\Upsilon(1S) \to l^+l^-)$ denotes $[\mathcal{B}(\Upsilon(1S) \to e^+e^-) + \mathcal{B}(\Upsilon(1S) \to \mu^+\mu^-)]/2$. While the $\Upsilon(3S) \to \gamma\chi_b(2P)$ rates are well predicted by various potential models, the predictions for the $\Upsilon(3S) \to \gamma\chi_b(1P)$ rates are model dependent and only a few calculations agree with the results presented here (see Ref. [8]).

Among $\gamma\gamma l^+l^-$ final states we also search for $\pi^0$ and $\eta$ transitions between the triplet $S$-states. No evidence for these transitions are found and

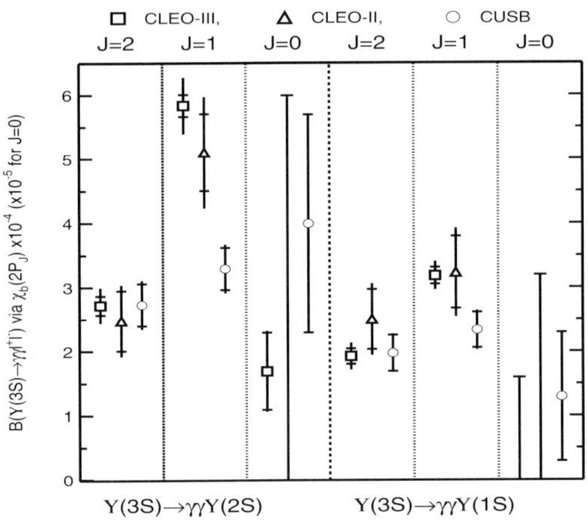


Figure 5. Product branching ratios, $\mathcal{B}(\Upsilon(3S) \to \gamma\chi_b(2P_J))\times$ $\mathcal{B}(\chi_b(2P_J) \to \gamma\Upsilon(2,1S))\times$ $\mathcal{B}(\Upsilon(2,1S) \to l^+l^-)$, in units of $10^{-4}$ ($10^{-5}$) for $J = 2, 1$ ($J = 0$) obtained in this analysis (squares), CLEO II (triangles) and CUSB (circles). The vertical error bars indicate the total errors. The dashes on the error bars indicate the statistical errors alone. The bars with no central value give the 90% C.L. upper limits.

the following 90% C.L. upper limits are set: $\mathcal{B}(\Upsilon(3S) \to \pi^0\Upsilon(1S)) < 1.7 \cdot 10^{-4}$, $\mathcal{B}(\Upsilon(3S) \to \eta\Upsilon(1S)) < 0.9 \cdot 10^{-3}$ and $\mathcal{B}(\Upsilon(3S) \to \pi^0\Upsilon(2S)) < 1.2 \cdot 10^{-3}$.

## 4. FOUR PHOTON CASCADES

The main interest in the $\gamma\gamma\gamma\gamma l^+l^-$ events is to observe production of the triplet $\Upsilon(1D)$ states (see Fig. 1). The production rate was predicted by Godfrey and Rosner to be $3.76 \cdot 10^{-5}$ [9]. Potential sources of background include four-photon cascades via the $\Upsilon(2S)$, instead of the $\Upsilon(1D)$, as well as the dipion transition $\Upsilon(3S) \to \pi^0\pi^0\Upsilon(1S)$. To address the latter, we try all combinations of 4 photons into two $\pi^0$s and pick the one that minimizes the chi-squared variable, $\chi^2_{\pi^0\pi^0}$, containing the two $\pi^0$ mass deviation terms. We identify the dipion transitions by requiring $\chi^2_{\pi^0\pi^0} < 6$. We observe $737 \pm 28$ such events, with an efficiency of 14%. The measured branching ratio, $\mathcal{B}(\Upsilon(3S) \to \pi^0\pi^0\Upsilon(1S))\times$ $\mathcal{B}(\Upsilon(1S) \to l^+l^-) = (5.7 \pm 0.2 \pm 0.4) \cdot 10^{-4}$, is in good agreement with the previous CLEO II

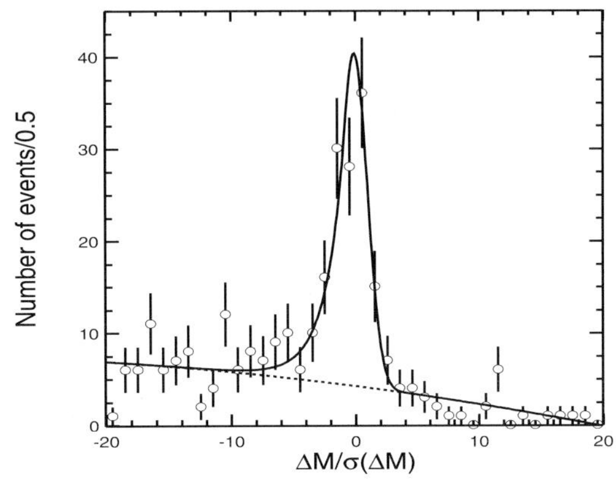


Figure 6. The standard deviation of the recoil mass against the two photons from the $\Upsilon(1S)$ mass in the $\Upsilon(3S) \to \gamma\chi_b(1P)$ analysis. Only dimuon events are shown here, which have the smaller background. The solid line represents the fit. The dashed line shows the estimated background.

[10] and CUSB [4] measurements. By requiring $\chi^2_{\pi^0\pi^0} > 6$ and every two-photon mass to be at least two standard deviations away from the nominal $\pi^0$ mass, we reduce the dipion background to less than 1 event. To suppress the photon cascades via the $\Upsilon(2S)$ we consider all possible ways of assigning the observed photons to the four transitions for all possible values of $J_{2P}$ and $J_{1P}$. For each combination we sum up in quadrature the standard deviations of the measured photon energies from the energies calculated using the known masses of the intermediate $b\bar{b}$ states. The minimal value of this chi-squared variable, $\chi^2_{2S}$, is required to be at least 12. After this requirement only about 2 background events are expected from this source. Having accounted for the backgrounds, we now try all possible ways of assigning the observed photons to the four transitions while going through an $\Upsilon(1D)$ state for all possible values of $J_{2P}$ and $J_{1P}$. Since the masses of the $\Upsilon(1D)$ states are not known we minimize the corresponding chi-squared variable, $\chi^2_{1D}$, for each event with respect to this mass. The distribution of $\chi^2_{1D}$ is shown in Fig. 7 for data (top) and for signal Monte Carlo (bottom). A narrow peak at zero is observed, as expected for $\Upsilon(1D)$ events. No other process can produce

such a narrow peak in the $\chi^2_{1D}$ distribution. By fitting this distribution we disentangle the signal and background contribution. We observe 40.7±6.8 signal events, with an efficiency of about 13.2% and a statistical significance of 9.7 standard deviations. The measured production rate, $\mathcal{B}(\Upsilon(3S) \to \gamma\chi_b(2P))\times$ $\mathcal{B}(\chi_b(2P) \to \gamma\Upsilon(1D))\times$ $\mathcal{B}(\Upsilon(1D) \to \gamma\chi_b(1P))\times$ $\mathcal{B}(\chi_b(1P) \to \gamma\Upsilon(1S))\times$ $\mathcal{B}(\Upsilon(1S) \to l^+l^-) = (3.3 \pm 0.6 \pm 0.5) \cdot 10^{-5}$, is in good agreement with the predictions by Godfrey and Rosner (see above).

The distribution of the $\Upsilon(1D)$ mass minimizing the $\chi^2_{1D}$ and of the recoil mass against the two lowest energy photons in the event indicate that our data are dominated by production of a state with a mass of $10162.2 \pm 1.6$ MeV. From the experimental point of view this could be either $\Upsilon(1D_2)$ or $\Upsilon(1D_1)$ state. The production rate for the spin 2 state is predicted to be about 6 times larger and its mass is predicted to be only 0.5-1.0 MeV lower than the center-of-gravity mass for the $1D$ triplet [9]. Accepting these theoretical inputs, we determine the center-of-gravity mass to be $10163 \pm 2$ MeV, in agreement with most potential model calculations, but in disagreements with others [9].

This is the first observation of the $\Upsilon(1D)$ states. A more detailed description of this analysis can be found in Ref. [11].

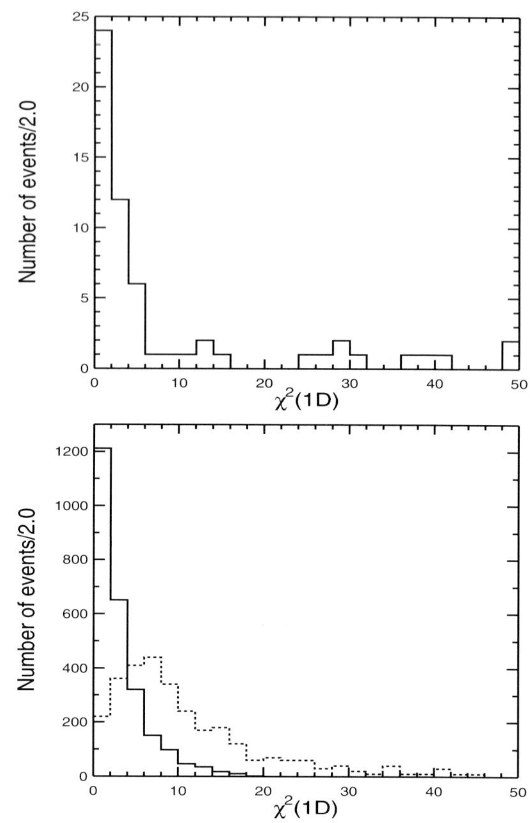


Figure 7. $\chi^2_{1D}$ distribution for the data (top) and for the Monte Carlo (bottom). The signal Monte Carlo is represented by the solid histogram. The background Monte Carlo for the cascades via the $\Upsilon(2S)$ state is shown by the dashed histogram. The latter distribution was scaled up by a factor of 10 to be visible on the scale of the signal Monte Carlo.

## REFERENCES

1. See e.g. C. T. Davies *et al.* (UKQCD Collaboration), Phys. Rev. **D58** (1998) 054505; T. Manke *et al.*, Phys. Rev. **D62** (2000) 114508; M. Okamoto *et al.*, Phys. Rev. **D65** (2002) 094508.
2. For a review see e.g. D. Besson and T. Skwarnicki, Ann. Rev. Nucl. Part. Sci. **43** (1993) 333.
3. G. Crawford *et al.* (CLEO II), Phys. Lett. **294B** (1992) 139.
4. U. Heintz *et al.* (CUSB), Phys. Rev. **D46** (1992) 1928.
5. S. Godfrey and J.L. Rosner, Phys. Rev. **D64** (2001) 074011 and references therein. See also S. Godfrey, these proceedings.
6. A.H. Mahmood *et al.* (CLEO III), CLEO-CONF-02-05, ICHEP-02-ABS-947, hep-ex/0207057.
7. K. Hagiwara *et al.* (Particle Data Group) Phys. Rev. **D66** (2002) 010001.
8. D. Cinabro *et al.* (CLEO III), CLEO-CONF-02-07, ICHEP-02-ABS-949 hep-ex/0207062.
9. S. Godfrey and J. L. Rosner, Phys. Rev. **D64**, 097501 (2001) and references therein. See also S. Godfrey, these proceedings.
10. F. Butler *et al.* (CLEO II), Phys. Rev. **D49** (1994) 40.
11. S.E. Csorna *et al.* (CLEO III), CLEO-CONF-02-06, ICHEP-02-ABS-948 hep-ex/0207060.

ELSEVIER

Proceedings of ICHEP 2002, pp. 703–705
*S. Bentvelsen, P. de Jong, J. Koch and E. Laenen (Editors)*

31st INTERNATIONAL CONFERENCE ON HIGH ENERGY PHYSICS AMSTERDAM

# Search for $\eta_b$ in two-photon collisions with L3 detector at LEP

C. Palomares[a]

[a]CERN,
Division EP, 1211 Meyrin, Geneva-23.

A search for the pseudoscalar meson $\eta_b$ is performed in two-photon interactions using the L3 detector at LEP. The data sample corresponds to a total integrated luminosity of 610 $pb^{-1}$ collected at the centre-of-mass energies from 189 GeV to 209 GeV. Preliminary results of the analysis in the $\eta_b$ mass region are presented. Several candidate events are found in different final states.

## 1. INTRODUCTION

Two-photon collisions are well suited for the study of pseudoscalar mesons, for which $J^{PC} = 0^{-+}$. For instance, the $c\bar{c}(1S)$ state, $\eta_c$ meson, has been studied at LEP and its two-photon width is found to be $\Gamma_{\gamma\gamma} = 6.9 \pm 1.7 \pm 2.1$ keV [1]. The high energy and luminosity make LEP a good environment to search also for $\eta_b$, the $b\bar{b}$ ground state, not yet observed.

Many theoretical estimates exist from potential models and lattice QCD of the mass difference, $\Delta m$, between $\eta_b$ an $\Upsilon(1S)$ ($m_\Upsilon = 9.46$ GeV) [2]. The predictions lie between 30 and 150 MeV. The same models predict the partial decay width of the $\eta_b$ into two photons, $\Gamma_{\gamma\gamma}(\eta_b)$, to be around 500 eV.

Here, a search is presented using the L3 data sample collected at the centre-of-mass energies, $\sqrt{s}$, from 189 GeV to 209 GeV, corresponding to a total integrated luminosity of 610 $pb^{-1}$. The $\eta_b$ resonance is produced at LEP via the interaction of two quasi-real photons. The $e^+e^- \to e^+e^-\eta_b$ cross-section depends on the two-photon luminosity function and on the form factors used. In the present analysis we use an infinite pole form factor, $F(Q^2)=1$, compatible with our measurements of the $\eta_c$ production, leaving the two-photon width, $\Gamma_{\gamma\gamma}(\eta_b)$, as the only unknown parameter, linearly proportional to the total cross-section:

$$\sigma(e^+e^- \to e^+e^-\eta_b) = \kappa\Gamma_{\gamma\gamma}(\eta_b) \quad (1)$$

This gives an expected production cross-section $\sigma(e^+e^- \to e^+e^-\eta_b)$ of 0.26 pb at $\sqrt{s} = 197$ GeV for $\Gamma_{\gamma\gamma}(\eta_b) = 0.5$ keV. The $\eta_b$ production is studied through the decay into four and six charged particles only or associated with a $\pi^0$. These particles are considered to be pions or/and kaons. In the last case, because of G-parity conservation, the decay products must include a $K^+K^-$ pair.

Searches have already been conducted by CUSB and CLEO Collaborations in the cascade decay of the $\Upsilon(3S)$ [3]. The CLEO Collaboration finds an upper limit on the product of branching ratios $BR(\Upsilon(3S) \to \pi^+\pi^- h_b) \times BR(h_b \to \gamma\eta_b)$ of about 0.1% at 90% C.L for the $\eta_b$ mass range from 9.32 GeV to 9.46 GeV. In tow-photon collision, a first investigation by ALEPH [4] observes a candidate event in the channel $K^0_S K^-\pi^+\pi^-\pi^+$ with a mass of $9.30 \pm 0.02 \pm 0.02$ GeV, in agreement with the total expected background of about one event. The following upper limits on $\Gamma_{\gamma\gamma}(\eta_b)$ at 95% confidence level are obtained:

$$\Gamma_{\gamma\gamma} \times BR(\eta_b \to 4\text{charged particles}) \quad < 48 \text{ eV}$$
$$\Gamma_{\gamma\gamma} \times BR(\eta_b \to 6\text{charged particles}) \quad < 132 \text{ eV}$$

For each channel, a continuum mass spectrum and the $\eta_b$ resonance are generated with the EGPC Monte Carlo. The background from $e^+e^- \to e^+e^-\tau^+\tau-$ is generated by the JAMVG program. All Monte Carlo events are passed through the L3 detector simulation.

## 2. DATA SELECTION

The $e^+e^- \to e^+e^-\eta_b$ events are selected by requiring four or six tracks with charge balance


doi:10.1016/S0920-5632(02)02263-6

and two isolated electromagnetic clusters in case of a $\pi^0$ decay.

The charged particles are measured in the central tracker in the polar angle $13^o \leq \theta \leq 167^o$ and are selected by a track trigger with a threshold on the track transverse momentum of 150 MeV. The photons are detected by the electromagnetic calorimeter in the barrel region, $42^o \leq \theta \leq 138^o$, and in the forward and backward regions, $11.6^o \leq \theta \leq 38^o$. A photon is an electromagnetic cluster with energy greater than 80 MeV and no nearby tracks in a cone of 0.2 rad half angle. The reconstructed $\pi^0$ mass is required to be between 105 GeV and 165 GeV (Fig 1).

To ensure that no final state particle of the resonance decay has escaped detection, the squared vectorial sum of the transverse momenta of all detected particles, $\sum \vec{p_t}^2$, is required to be smaller than 0.1 $GeV^2$.

The $\gamma\gamma \to \tau^+\tau^-$ background is reduced in the four track events by requiring the invariant mass of 3 charged particles to be greater than 1.9 GeV. In addition, events are excluded if a photon conversion is detected, i.e., when, assigning the electron mass to an opposite charge pair, the effective mass of the pair is smaller than 50 MeV.

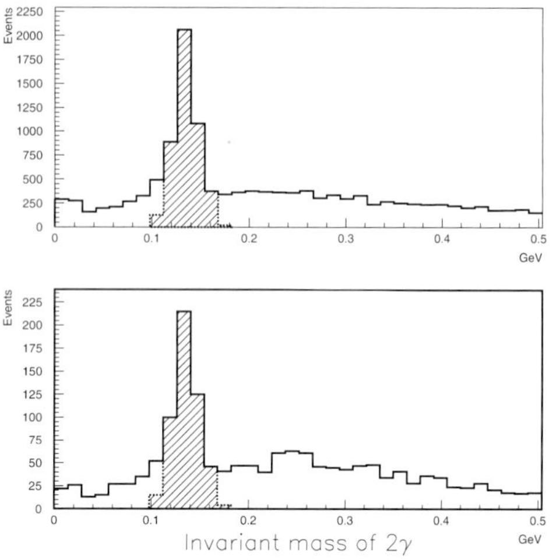


Figure 1. Invariant mass of the $\pi^0$-decay two photon candidates for the preselected data. The upper histogram corresponds to the 4 tracks $+\pi^0$ channel and the second plot includes the 6 tracks $+\pi^0$ preselected events. In both plots, the dashed histograms contain the accepted events.

## 3. RESULTS

The mass spectrum, $W_{\gamma\gamma}$, of the selected events is presented in Fig 2. In the signal region, chosen between 9 and 10 GeV, three candidates are found. The mass spectrum is well represented by two-photon cross-section exponentially decreasing with $W_{\gamma\gamma}$. The main background, $e^+e^- \to e^+e^-\tau^+\tau^-$, is estimated with a MC sample six times larger than the data. This contamination is also presented in Fig 2.

The selection efficiency for each channel in the signal region together with the number of expected background events and the selected candidates are given in Table 1.

A display of the candidate event with 4 charged particles in the final state is shown in Fig. 3.

### 3.1. Mass of the Candidates

In Table 1 the values of the reconstructed mass of the candidates are listed together with the mass resolution. These masses are determined assuming that all tracks are pions if no $\pi^0$ is present and that there is a $K^+K^-$ pair otherwise. The ionization loss measurement, $dE/dx$, has a good $\pi/K$ separating power only for tracks with a momentum below 0.5 GeV. As the momenta of the decay particles of the $\eta_b$ extend from 0.15 to 3 GeV the $dE/dx$ has a weak separating power. The mass resolution due to detector effects is about 300 MeV, while the mass shift due to $\pi/K$ misidentification is 100 MeV. The average reconstructed mass of the three candidates is $9.7 \pm 0.3$ GeV.

### 3.2. Interpretation of the Results

The observation (3 selected events) is found to be compatible with the background expectation (1.3±1 events) within less than 2 sigmas. If we assume the production of the $\eta_b$ resonance in the invariant mass range from 9 GeV to 10 GeV, upper

Table 1
Selection results in the signal region, 9 GeV $< W_{\gamma\gamma} <$ 10 GeV. The reconstructed mass for each $\eta_b$ candidate is also shown.

| Channel | Selected Events | Eff(%) | Expected background | Reconstructed mass |
|---|---|---|---|---|
| 4 tracks | 1 | 4.9 | $0.25 \pm 0.06$ | $9.9 \pm 0.3$ |
| 4 tracks+$\pi^0$ | 0 | 1.9 | $0.09 \pm 0.01$ | |
| 6 tracks | 1 | 3.6 | $0.65 \pm 0.06$ | $9.4 \pm 0.3$ |
| 6 tracks +$\pi^0$ | 1 | 0.9 | $0.29 \pm 0.07$ | $9.9 \pm 0.3$ |

limits at 95% confidence level for $\Gamma_{\gamma\gamma}(\eta_b) \times$ BR are calculated for each channel. The results are:

$\Gamma_{\gamma\gamma}(\eta_b)\times$ BR(4 tracks) $< 0.3$ keV
$\Gamma_{\gamma\gamma}(\eta_b)\times$ BR(4 tracks+$\pi^0$) $< 0.5$ keV
$\Gamma_{\gamma\gamma}(\eta_b)\times$ BR(6 tracks) $< 0.4$ keV
$\Gamma_{\gamma\gamma}(\eta_b)\times$ BR(6 tracks+$\pi^0$) $< 1.4$ keV

Assuming that the $\eta_b$ branching ratio for each channel is the same, the combined upper limit for $\Gamma_{\gamma\gamma}(\eta_b)\times$ BR(6 and 4 tracks) is 0.2 keV.

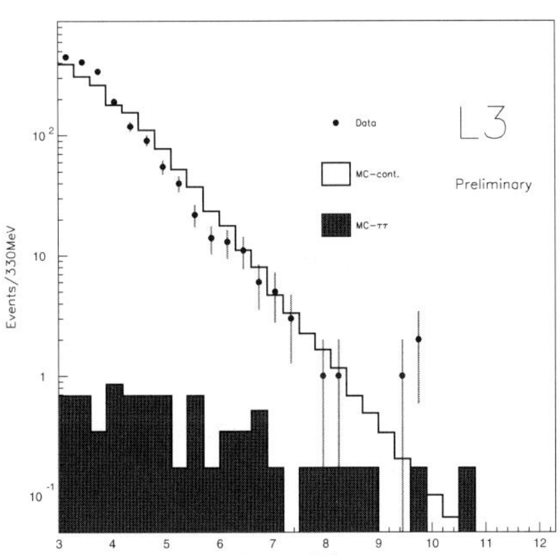


Figure 2. Mass spectrum of selected events. The dots represent the data, the open histogram is the expected mass distribution for the exclusive channels. The shaded histogram corresponds to the $e^+e^- \to e^+e^-\tau^+\tau^-$ background.

## REFERENCES


1. L3 Collaboration, Phys. Lett. **B461** (1999) 155
2. A.X. El-Khadra, hep-ph//9508266; G.S. Bali, K. Schilling and A. Wachter, Phys. Rev. **B56** (1997) 2566; S. Narison, Phys. Lett. **B387** (1996) 162; T. Barnes, hep-ph/0103142; N. Brambilla, Y. Sumino and A. Vairo, Phys. Lett. **B513** (2001) 381.
3. CUSB Collaboration, Phys. Rev. Lett. **66** (1991) 3113; CLEO Collaboration, Phys. Rev. **D49** (1994) 40.
4. ALEPH Collaboration, Phys. Lett. **B530** (2002).


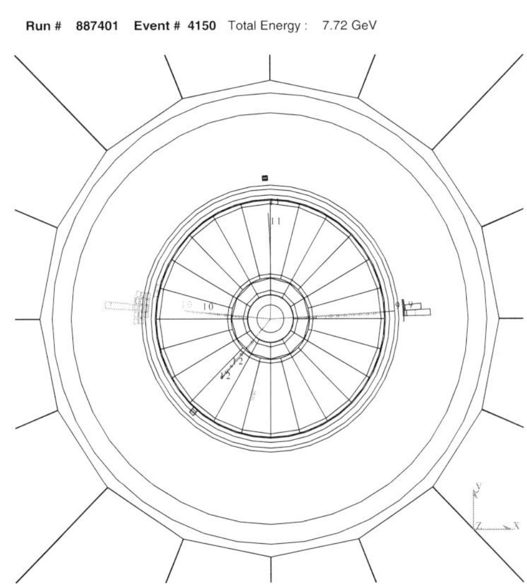


Figure 3. A view in the detector transverse plane of the $\eta_b \to \pi^+\pi^-\pi^+\pi^-$ candidate event.

## Session 10

# Beyond the Standard Model (experiment and theory)

*Session conveners:*
*G. Weiglein, University of Durham*
*S. Mele, CERN*

ELSEVIER

Proceedings of ICHEP 2002, pp. 709–712
*S. Bentvelsen, P. de Jong, J. Koch and E. Laenen (Editors)*

31st INTERNATIONAL CONFERENCE ON HIGH ENERGY PHYSICS AMSTERDAM

# SUSY Particles Searches at LEP and Interpretations within the MSSM

Giacomo Sguazzoni[a]

[a]European Laboratory for Particle Physics (CERN), CH-1211 Geneva 23, Switzerland

Searches for R-parity conserving supersymmetric particles have been performed in $e^+e^-$ data collected by LEP detectors, at centre-of-mass energies up to 209 GeV, corresponding to an integrated luminosity of $\sim 3.1\,\text{fb}^{-1}$. The results and their interpretation in the context of MSSM frameworks are briefly reviewed.

## 1. INTRODUCTION

Theories with Supersymmetry (SUSY) are the most promising extensions of the Standard Model (SM) [1]. The simplest version is the Minimal Supersymmetric Model (MSSM), which contains the minimal number of additional particles. The scalar fermions or *sfermions*, $\tilde{f}_L$ and $\tilde{f}_R$, are the partners of the left- and right-handed SM fermions and mix to form the mass eigenstates. The mixing angle $\theta_{\tilde{f}}$ is so defined that $\tilde{f} = \tilde{f}_L \cos\theta_{\tilde{f}} + \tilde{f}_R \sin\theta_{\tilde{f}}$ is the lightest sfermion. In general mixing is relevant for the third family, while $\tilde{f} \equiv \tilde{f}_R$ otherwise. The SM gauge boson states have fermionic super-partners or *gauginos*. The MSSM higgses are arranged into two doublets and their fermionic super-partners are the *higgsinos*. Neutral gauginos and higgsinos mix into four mass eigenstates, the *neutralinos* $\chi$, $\chi_2$, $\chi_3$, $\chi_4$ ($M_{\chi_4} > M_{\chi_3} > M_{\chi_2} > M_\chi$). The charged gauginos and higgsinos mix into two mass eigenstates, the *charginos* $\chi^\pm$ and $\chi_2^\pm$ ($M_{\chi_2^\pm} > M_{\chi^\pm}$).

The lepton and baryon number conservation translates into the "R-parity" conservation. The LSP (Lightest Supersymmetric Particle) is stable and must be also neutral and weakly interacting to fit the cosmological observations. Within the MSSM the LSP is the lightest neutralino $\chi$ or, less likely, the sneutrino, $\tilde{\nu}$. At LEP the sparticles are pair produced and the decay brings to final states containing at least one LSP.

The success of LEP searches is also due to the impressive performance of the accelerator that provided $e^+e^-$ collisions at centre-of-mass energies between 161 and 209 GeV, and an integrated luminosity of $\sim 775\,\text{pb}^{-1}$ per experiment. The results, based on the full high-energy data sample, are presented in the form of 95% C.L. exclusion domains in the space of the relevant parameters, since no excess has been observed. When available, the LEP SUSY Working Group combinations, based on the outcomes from ALEPH, DELPHI, L3 and OPAL (ADLO), are reported [2].

## 2. PRIMARY SIGNALS

Except few pathological cases, sparticle pair production leads to the typical acoplanar particles topology due to missing energy ($\not{E}$) and momentum ($\not{P}$) from escaping LSP's. The energy of the visible system is related to the mass difference between the sparticle $\tilde{P}$ and the LSP ($\Delta M = M_{\tilde{P}} - M_{\text{LSP}}$). The acoplanar topologies studied cover each type of visible final state (leptons, hadronic jets, $\gamma$'s).

### 2.1. Acoplanar leptons

The analyses for slepton signals ($e^+e^- \to \tilde{\ell}^+\tilde{\ell}^-$, $\tilde{\ell} \to \ell\chi$) search for acoplanar leptons by using the powerful lepton and tau identification of LEP detectors [3–6]. The LEP combined cross section upper limits range from 10 to 60 fb [2]. The resulting mass lower limits are $100\,\text{GeV}/c^2$, $94\,\text{GeV}/c^2$ and $86\,\text{GeV}/c^2$ for $\tilde{e}_R$, $\tilde{\mu}_R$ and $\tilde{\tau}_R$ respectively, valid for $\Delta M > 10\,\text{GeV}/c^2$, as shown in Figure 1.

### 2.2. Acoplanar jets

The production of a squark pair results into an acoplanar jet topology. These hadronic events can be selected by using event variables and re-


doi:10.1016/S0920-5632(02)02264-8

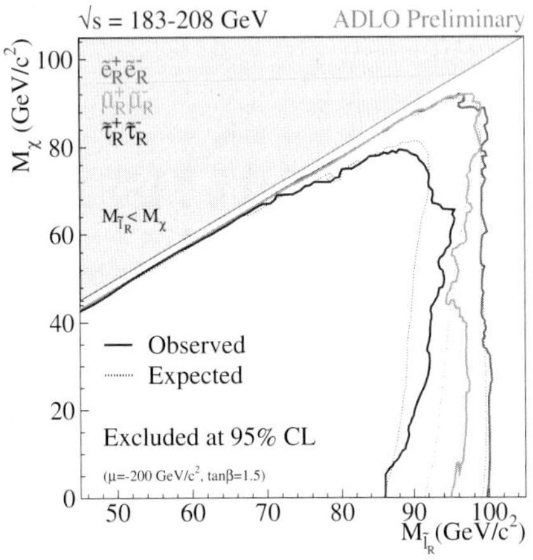


Figure 1. Slepton mass exclusion plot from the LEP SUSY Working Group.

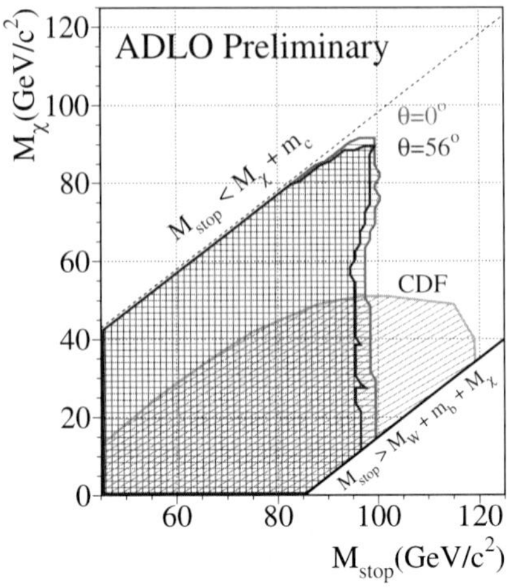


Figure 2. Stop mass exclusion plot from the LEP SUSY Working Group in case of $\tilde{t} \to c\chi$ decay for minimal ($\theta_{\tilde{t}} = 56°$) and maximal production cross section ($\theta_{\tilde{t}} = 0°$). The CDF result is also shown.

quiring $\not{E}$ and $\not{P}$ [4,5,7,8]. In case of $e^+e^- \to \tilde{t}\bar{\tilde{t}}$, $\tilde{t} \to c\chi$, the mass lower limit is $94\,\mathrm{GeV}/c^2$ for $\Delta M > 10\,\mathrm{GeV}/c^2$ and any mixing, as visible in Figure 2. Further specialized selections are used for other squark processes: b-tagging is effective for $e^+e^- \to \tilde{b}\bar{\tilde{b}}$, $\tilde{b} \to b\chi$, allowing a limit of $92\,\mathrm{GeV}/c^2$ to be set ($\Delta M > 10\,\mathrm{GeV}/c^2$, any $\theta_{\tilde{b}}$); leptons are required in case of $e^+e^- \to \tilde{t}\bar{\tilde{t}}$, $\tilde{t} \to b\ell\tilde{\nu}$, leading to a mass lower limit of $95\,\mathrm{GeV}/c^2$ ($\Delta M > 10\,\mathrm{GeV}/c^2$, any $\theta_{\tilde{t}}$). The stop decay $\tilde{t} \to b\chi f_u \bar{f}_d$, recently recognized as relevant, leads to a multi-body final state topology addressed by a dedicated ALEPH selection [7]. As an example, assuming the decay $\tilde{t} \to b\chi W^*$, the result is $M_{\tilde{t}} > 77\,\mathrm{GeV}/c^2$ ($\Delta M > 10\,\mathrm{GeV}/c^2$, any $\theta_{\tilde{t}}$). ALEPH analyses also consider the case in which a stop quasi-degenerate with the LSP acquires a sizeable lifetime and hadronizes [9]. This scenario has been excluded searching for long-lived heavy hadrons and an absolute stop mass lower limit of $63\,\mathrm{GeV}/c^2$ has been set for any $\theta_{\tilde{t}}$, any branching ratio and any $\Delta M$ [7].

### 2.3. Other topologies with $\not{E}$ and $\not{P}$

Topologies with two or more visible fermions in the final state plus $\not{E}$ and $\not{P}$ are expected in case of charginos and neutralinos production [4, 5,10,11]. The processes are of the type $e^+e^- \to \chi_{i>1}\chi$ and $e^+e^- \to \chi_{i>1}\chi_{j>1}$ with $\chi_{i>1} \to \chi f\bar{f}$, and $e^+e^- \to \chi^+\chi^-$ with $\chi^\pm \to \chi f_u f'_d$. Cross section upper limits of $\sim 0.1$–$0.3\,\mathrm{pb}$ are obtained by the LEP-wide outcome of dedicated selections [2].

Topologies with photon(s) can be very powerful in detecting new phenomena [5,6,12,13]. Within the MSSM this case applies when heavier neutralinos are assumed to decay radiatively: $e^+e^- \to \chi_2\chi_2$ and $e^+e^- \to \chi_2\chi$ with $\chi_2 \to \chi\gamma$. In this hypotheses the cross section upper limits range between 10 fb and 0.1 pb depending on the process [2].

## 3. INTERPRETATION

The negative results of the search for sparticle production can be translated into constraints on the parameter space in the context of specific SUSY models. Such a method allows the exclusions to be extended to sparticles otherwise not accessible, either because invisible, as the LSP, either because too heavy to be produced [4–6,10].

A widely accepted framework is the constrained MSSM (CMSSM). The unification of masses and couplings at the GUT scale allow the EW scale phenomenology to be set by few parameters: $\tan\beta$, the ratio of the vacuum expectation values of the two Higgs doublets; $\mu$, the Higgs sec-

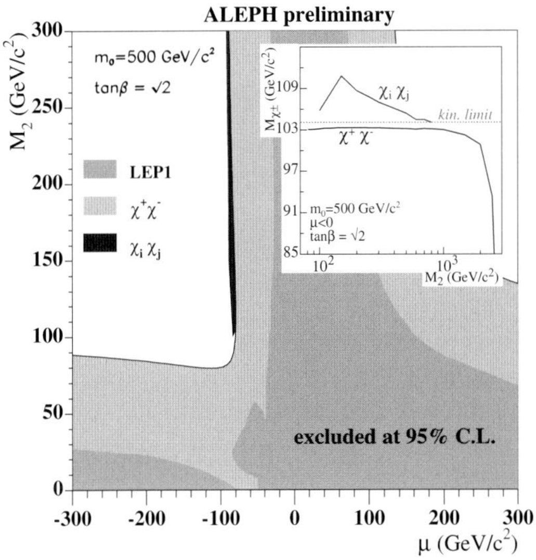


Figure 3. Excluded domains in the $M_2$ vs. $\mu$ plane for $\tan\beta = \sqrt{2}$ and $m_0 = 500\,\mathrm{GeV}/c^2$. The upper-right plot shows the corresponding chargino mass lower limits for $\mu < 0$.

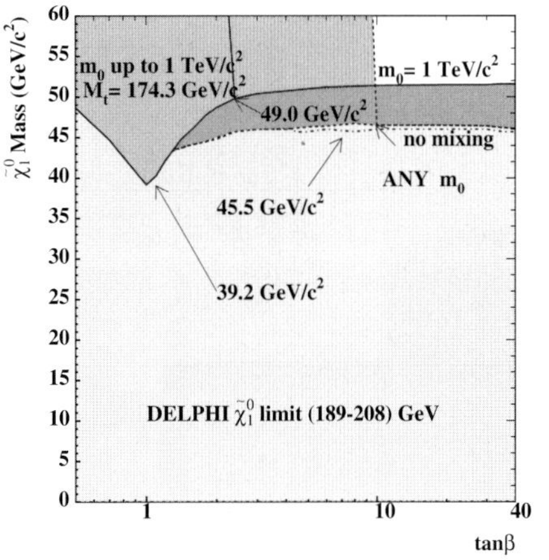


Figure 4. The $\chi$-LSP limit vs. $\tan\beta$: large $m_0$ (solid curve) and any $m_0$ (dashed curve) in case of no-mixing; any $m_0$ and $A_\tau = A_\mathrm{b} = A_\mathrm{t} = 0$, i.e. mixing in the third family (dash-dotted curve). The steep lines show the impact of Higgs boson searches for two $m_0$ scenarios [4].

tor mass parameter; $M_2$, the EW scale common gaugino mass; $m_0$, the GUT scale common scalar mass; the trilinear couplings $A_\mathrm{f}$, that enter in the prediction of the sfermion mixing and are generally set to fit the no-mixing hypothesis.

### 3.1. LSP limit

The negative outcome of charginos and neutralinos searches can be used to exclude regions in the $(\mu, M_2)$ plane, as shown, as an example, in Figure 3 in which the sleptons are assumed to decouple (i.e. large $m_0$). The upper-right plot of Figure 3 shows how neutralino searches allow chargino exclusions to be improved for small $\tan\beta$ and $\mu < 0$. If the sleptons are lighter (small $m_0$ values), the chargino and neutralino cross sections decrease for the enhancement of negative-interfering slepton-exchange diagrams. The consequent loss of sensitivity is recovered by slepton searches in such a way that lower mass limits on gauginos and other sparticles as $\tilde{\mathrm{e}}_\mathrm{R}$ or $\tilde{\nu}$ could be set [4–6,14]. Among these, the most important is the LSP limit, i.e. the mass lower limit on $\chi$, shown in Figure 4 as a function of $\tan\beta$ for coupling and decoupling sleptons. The LSP mass lower limits from LEP experiments fall between $36.3\,\mathrm{GeV}/c^2$ and $39.6\,\mathrm{GeV}/c^2$ and are set for $\tan\beta \sim 1$.

The LEP mass lower limits on the Higgs boson mass $m_\mathrm{h}$ can be also used to further exclude small $\tan\beta$ ranges. Roughly, this just derives from the MSSM tree-level relation $m_\mathrm{h} < m_\mathrm{Z}|\cos\beta|$. However, the details of the exclusion depend on $M_2$, $m_0$ and the stop mass because of the large radiative corrections to $m_\mathrm{h}$. Adding the Higgs constraints the LSP mass lower limit substantially improves (up to $\sim 45\,\mathrm{GeV}/c^2$) and moves towards $\tan\beta \sim 4$, as shown in Figure 4.

### 3.2. Impact of mixing in the third family

The robustness of the LSP limit has been checked with respect to the mixing effects in the third family, neglected in the above discussion. A stau getting light for mixing may be mass degenerate with the LSP, making the chargino decays into staus difficult to detect [4,10]. Dedicated selections for $\chi^\pm \to \tilde{\tau}\nu_\tau \to \tau\chi\nu_\tau$ with soft taus, $\mathrm{e^+e^-} \to \chi_2\chi$ and $\mathrm{e^+e^-} \to \chi_2\chi_2$ with $\chi_2 \to \tau\tau\chi$, and for chargino production in association with an ISR photon ($\mathrm{e^+e^-} \to \chi^+\chi^-\gamma$) allow to solve this problem. As shown in Figure 4, the LSP

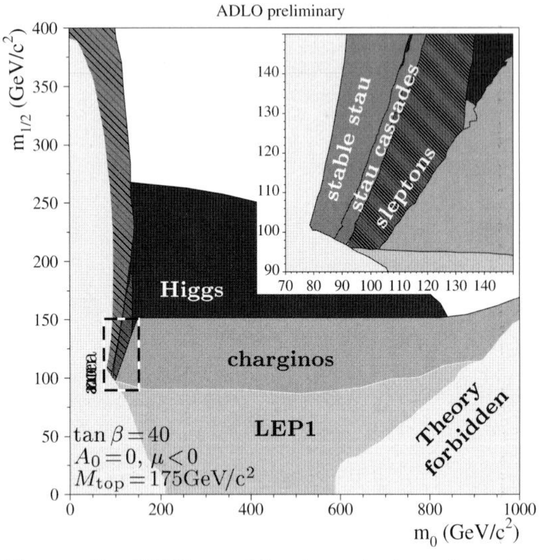


Figure 5. LEP combined exclusion domains in the mSUGRA $m_{1/2}$ vs. $m_0$ plane for $\tan\beta = 40$, $A_0 = 0$ and $\mu < 0$. A peculiar area is zoomed to show the interplay between selections.

limits reported above have been demonstrated to hold by using this studies, extended also considering the mixing configurations for $\tilde{\tau}$, $\tilde{\mathrm{t}}$ and $\tilde{\mathrm{b}}$ that can be explored by setting $A_\tau$, $A_{\mathrm{t}}$ and $A_{\mathrm{b}}$ to zero.

### 3.3. mSUGRA

The results have been also interpreted within an even more constrained version of the CMSSM, usually referred to as Minimal Supergravity (mSUGRA). The relevant parameters are: $\tan\beta$, the sign of $\mu$ and $m_0$; $m_{1/2}$, the GUT scale common gaugino mass, that replaces $M_2$; $A_0$, the GUT scale common trilinear coupling.

On top of LEP1 exclusions and theory-forbidden regions, small $m_0$ and $m_{1/2}$ areas are constrained from sleptons and gaugino searches, respectively. Higgs boson searches are also effective, even in the large $\tan\beta$ range. As an example, Figure 5 illustrates $m_{1/2}$ vs. $m_0$ excluded domains for $\tan\beta = 40$, $\mu < 0$ and $A_0 = 0$. The zoomed area focuses on the pathological region in which, for the mixing, $\tilde{\tau}$ and $\chi$ are almost degenerate and the selections for stau-cascades and stable staus have to be used. The resulting mSUGRA LSP mass lower limits lie between 52 and $59\,\mathrm{GeV}/c^2$, depending on the top mass, and turn out to be $\sim 8$–$9\,\mathrm{GeV}/c^2$ lower if $A_0$ is allowed to assume values other than zero [2].

## 4. CONCLUSION

Despite the negative outcome, LEP has substantially contributed to the study of supersymmetric scenarios. Stringent constraints on sparticle parameters have been set by direct search and by interpretation studies within widely accepted frameworks. This huge experience is an important part of the LEP legacy and it will result very useful for SUSY searches at future experiments.

## ACKNOWLEDGMENTS

I would like to thank G. Ganis for the careful reading of the manuscript. Special thanks also to P. Azzurri for his kind hospital-ity.

## REFERENCES

1. K. Hagiwara *et al.*, Phys. Rev. D **66** (2002) 010001; S.P. Martin, hep-ph/9709356 (1997); N. Polonsky, hep-ph/0108236 (2001).
2. `http://lepsusy.web.cern.ch/lepsusy/`
3. ALEPH Collaboration, Phys. Lett. B **526** (2002) 206.
4. DELPHI Collaboration, DELPHI Note 2002-027-CONF-561 (2002).
5. `http://l3.web.cern.ch/l3/`
6. OPAL Collaboration, OPAL Physics Note PN470 (2002).
7. ALEPH Collaboration, Phys. Lett. B **537** (2002) 5.
8. OPAL Collaboration, Phys. Lett. B **545** (2002) 272.
9. ALEPH Collaboration, Phys. Lett. B **488** (2000) 234.
10. ALEPH Collaboration, ALEPH/2002-028 CONF/2002-017 (2002).
11. OPAL Collaboration, OPAL Physics Note PN503 (2002).
12. ALEPH Collaboration, CERN-EP-2002-033.
13. DELPHI Collaboration, DELPHI Note 2002-077-CONF-611 (2002).
14. ALEPH Collaboration, Phys. Lett. B **544** (2002) 73.

ELSEVIER

Proceedings of ICHEP 2002, pp. 713–716
*S. Bentvelsen, P. de Jong, J. Koch and E. Laenen (Editors)*

31st INTERNATIONAL CONFERENCE ON HIGH ENERGY PHYSICS AMSTERDAM

# LEP searches in R-parity Violation scenarios

Theodora D. Papadopoulou

Physics Department, National Technical University of Athens
GR-157 80 Zografou, Athens, Greece

An overview of searches from LEP experiments ( ALEPH, DELPHI, L3 and OPAL) is presented on pair and single production of supersymmetric particles under the assumption that R-parity is violated via a single dominant coupling using the data collected at LEP2. The observed canditate events in the data are in agreement with the Standard Model expectation, and limits on the production cross sections and the masses of the supersymmetric particles are derived.

## 1. INTRODUCTION

R-parity [1] is a multiplicative symmetry in supersymmetric models connected to matter parity with a quantum number defined as:

$$R_p = (-1)^{3B+L+2S} \quad (1)$$

where $S$ is the spin and $L(B)$ are the Lepton (Baryon) numbers. For ordinary particles $R_p$ is $+1$, while it is $-1$ for their supersymmetric partners. R-parity conservation implies that sparticles can only be produced in pairs and then decay in cascade to the lightest supersymmetric particle (LSP), which is stable.

The most general superpotential with the minimal field content of the supersymmetric standard model contains the trilinear terms:

$$W_{\not R_p} = \lambda_{ijk} L_i L_j \bar{E}_k + \lambda'_{ijk} L_i Q_j \bar{D}_k + \lambda''_{ijk} \bar{U}_i \bar{D}_k (2)$$

where $L(Q)$ is the lepton(quark) doublet superfields and $\bar{D}, \bar{U}, (\bar{E})$ are the down-like and up-like quark(lepton) singlet superfields, $\lambda, \lambda', \lambda''$ are Yukawa couplings and i,j,k=1,2,3 are the generation indices [2]. Subsets of terms of Eq(2) are allowed in phenomenological models in which R-parity is violated. Bilinear terms are not considered in this report.

These three types of trilinear terms, two violating L conservation, and one violating B conservation contain 45 R-parity Violating Couplings, if Spontaneous $\not R_p$ is considered there are three additional terms. Hierarchies in $\not R_p$ Couplings are expected as for Yukawa Couplings generating fermion masses. One assumes that a single coupling dominates all others. Low energy measurements provide upper limits for the $\lambda, \lambda', \lambda''$ Couplings [3].

$\not R_p$ opens a new scenario in supersymmetric searches complementary to the $R_pC$ ones. The phenomenological consequences of the $\not R_p$ are that the LSP is allowed to decay to standard fermions, single sparticle production is allowed via a $\Delta L \neq 0$ or a $\Delta B \neq 0$ operator (not simultaneously present to suppress fast proton decays) and the golden signature is the multilepton or multijet events in excess.

## 2. RESEARCH STRATEGIES

The Minimal Supersymmetric Standard Model (MSSM) scheme with the assumption that the gaugino masses are unified at the Grand Unified Theories (GUTs) scale is the theoretical framework of the majority of searches . Relevant parameters for these $\not R_p$ searches are the $M_1$, $M_2$, the $U(1)$ and $SU(2)$ gaugino mass at the electroweak scale (with $M_1 \approx 0.5 M_2$), $m_0$, the scalar common mass at the GUT scale, $\mu$, the mixing mass term of the Higgs doublets at the electroweak scale and $\tan\beta$, the ratio of the vacuum expectation values of the two Higgs doublets.

Two types of supersymmetric particle decays are considered. First, the *direct decay* of the sparticles into two standard fermions or the neutralino


doi:10.1016/S0920-5632(02)02267-3

(chargino) decay via a sparticle virtual exchange to standard particles through $\not{R}_p$ vertex. Second, the *indirect decay* of the sparticle cascade decay through $R_p$ conserving vertex to on-shell sparticles down to the LSP, which then decays via $\not{R}_p$ vertex.

In all searches for $\not{R}_p$, the LSP lifetime is a crucial parameter, since the analyses performed are mainly sensitive only if the LSP has a negligible lifetime. LEP analyses are sensitive in case of a LSP mass greater than 10 $GeV/c^2$ and a coupling value $\sim 10^{-5} < \lambda < 10^{-2}$ up to 1. Otherwise the LSP decay vertex is a displaced vertex or it is outside the detector.

The variety of final states to be explored is extreme on LEP2 $\not{R}_p$ searches and many channels have to be combined. The optimization of the signal selection on various topologies follows a variety of methods; sequential cuts, lepton identification and lepton isolation criteria, neural network methods or jet algorithms and b tagging, depending on the topology.

## 3. DATA SAMPLES

The total integrated luminosity collected by LEP2 experiments during the five last years (1996-2000) at center of mass energies up to 209 $GeV$ was $\sim$ 714 $pb^{-1}$ per experiment. To evaluate background contaminations, different contributions coming from the Standard Model processes were considered by using various generators.

Signal events were generated with the SUSYGEN program 2.2(3.0) [4]. Faster simulation programs are also used in some cases.

## 4. REVIEW OF ANALYSES RESULTS

No evidence for a significant signal is observed in all the $\not{R}_p$ searches. Many limits at 95 % CL are derived on cross-sections, couplings and sparticle masses and exclusion plots in the MSSM regions are produced.

A short review on the important recent updated analyses results from all LEP2 experiments (ALEPH, DELPHI, L3 and OPAL) is given below.

### 4.1. Single sneutrino production

One spectacular manifestation of $\not{R}_p$ could be the presence of the sneutrino resonance. This process probes masses up to the centre of mass energy. Assuming $\lambda_{ijk}$ coupling dominant, $\tilde{\nu}_\mu$ or $\tilde{\nu}_\tau$ resonant production is possible. Analysis of the final states with 2 leptons and missing energy ($E_{miss}$), 4 or 6 leptons (with or without $E_{miss}$ and leptons plus jets is performed by DELPHI [5]. Upper limits on the $\lambda_{121}$ or $\lambda_{131}$ couplings at the level of 1 to $3 \times 10^{-3}$, depending on $M_{\tilde{\nu}}$, are derived for $\tan\beta = 1.5$ or 30 .

A search for single $\tilde{\nu}_e, \tilde{\nu}_\mu, \tilde{\nu}_\tau$ production through the process $e\gamma \to \tilde{\nu}_j l_k$ is also performed by ALEPH [6] at the centre of mass energies $\sqrt{s} = 189$ to 209 $GeV$. The 95 % CL upper limits obtained on $\lambda$ couplings, assuming a $\tilde{\nu}$ mass of 100 $GeV/c^2$, are in the range $\sim 7.10^{-3} - 3.10^{-2}$.

### 4.2. Gaugino pair production

Pair production of supersymmetric particles in MSSM with $\not{R}_p$ is identical to the pair production with $R\_p$ conservation, since the trilinear couplings are not present in the production vertex.

The neutralino ($\tilde{\chi}_i^0$) and chargino ($\tilde{\chi}_k^\pm$) pair productions were considered for different values of $\tan\beta$, $m_0$, $\mu$ and $M_2$, via a dominant $\lambda$, $\lambda'$ or $\lambda''$ coupling by LEP2 Collaborations. The main final topologies analyzed via a dominant $LL\bar{E}$ coupling include 2, 4 and 6 leptons plus ($E_{miss}$), via a dominant $LQ\bar{D}$ coupling include 2, 4 jets and lepton(s) plus ($E_{miss}$) and via a $\bar{U}\bar{D}\bar{D}$ coupling include multijets and leptons plus ($E_{miss}$). An extensive presentation of the results can be found in the references [7] to [10].

Table 2 includes the gaugino pair production mass limits, at 95 % CL, in $GeV/c^2$, extracted from searches of all LEP Collaborations (ADLO).

### 4.3. Slepton/Sneutrino pair production

The direct decay of a pair of sleptons/sneutrinos via a dominant $LL\bar{E}$ coupling to four charged lepton final states gives a clear signal with high efficiency. The study of a 4-jet final state gives limits via a $\lambda'$ coupling. Table 1 contains mass limits of slepton/sneutrino pair production from direct and indirect decays.

### New lower mass Limits (GeV/$c^2$) ADLO

| sfermion | λ (dd) A | λ (dd) L | λ (dd) O | λ′(dd) A | λ′(dd) O | λ′′(dd) L | λ (id) A | λ (id) D | λ (id) L | λ (id) O | λ′(id) A | λ′(id) O | λ′′(id) A | λ′′(id) D | λ′′(id) L |
|---|---|---|---|---|---|---|---|---|---|---|---|---|---|---|---|
| $\tilde{e}_R(\tilde{e}_L)$ | 96 | 69 | **(89)** | | (89) | 96 | 96 | 95 | 79 | **99** | 93 | **92** | 94 | 92 | 96 |
| $\tilde{\mu}_R(\tilde{\mu}_L)$ | 87 | 61 | (74) | (81) | (77) | 86 | 96 | 90 | 87 | **92** | 90 | **87** | 85 | 87 | 86 |
| $\tilde{\tau}_R,(\tilde{\tau}_L)$ | 87 | 61 | (74) | | (76) | 75 | 95 | 90 | 86 | **90** | 76 | - | - | - | 75 |
| $\tilde{\nu}_e$ | 100 | 95 | **90** | | **90** | 99 | 98 | 98 | 99 | **95** | 91 | **88** | 88 | - | 99 |
| $\tilde{\nu}_\mu$ | 90 | 65 | **76** | 79 | **75** | 70 | 89 | 85 | 78 | **81** | 78 | - | 65 | - | 70 |
| $\tilde{\nu}_\tau$ | - | 65 | **76** | | **75** | 70 | 89 | 85 | 78 | **81** | 78 | - | 65 | - | 70 |
| $\tilde{t}$ $(\tilde{t}_L)$ | - | - | - | (97) | - | - | (91) | (92) | - | - | (85) | - | (71.5) | (87) | - |
| $\tilde{b}$ $(\tilde{b}_L)$ | - | - | - | - | | - | (90) | - | - | - | (80) | - | (71.5) | (78) | - |
| $\tilde{u}_R(\tilde{u}_L)$ | - | - | - | - | | A (82.5) L 80 (87) | - | | | | - | | L 79 (87) | | |
| $\tilde{d}_R(\tilde{d}_L)$ | - | - | - | - | | A (77) L 56 (86) | - | | | | - | | L 55 (86) | | |

Table 1
Sparticle mass limits in $GeV/c^2$ from $\not{R}_p$ pair production searches of SUSY particles from the **ADLO** Collaborations. (Letter A is used for **ALEPH**, D for **DELPHI**, L for **L3** and O for **OPAL** ).
Abbreviations used: **( dd )** for direct decays , **( id )** for indirect decays .
The limits from the left-handed sparticles are inside parenthesis.
The various validity conditions used for gauginos and details from analyses performed are included in Ref. [7] to [10].

Table 2
Chargino and neutralino pair production mass limits at 95 % CL in $GeV/c^2$

| Gaugino | exp | $\lambda$ | $\lambda'$ | $\lambda''$ |
|---|---|---|---|---|
| $\tilde{\chi}_1^0$ | L | 40.2 | - | 39.9 |
| | D | 39.5 | - | 38.0 |
| $\tilde{\chi}_2^0$ | L | 84.0 | - | 80.0 |
| $\tilde{\chi}_3^0$ | L | 107.2 | - | 107.2 |
| $\tilde{\chi}_1^{\pm}$ | A | 103 | 103 | 103 |
| | D | 103 | - | 102.5 |
| | L | 103.0 | - | 102.7 |

### 4.4. Squark pair production

Squarks cannot decay directly via a $\lambda$ coupling but they can decay indirectly via the lightest neutralino. Searches for squarks mostly in the case of indirect decays via various couplings are performed and limits on the production cross-sections and on the masses are included in the Ref. [7] to [10]. A summary of limits is given in Table 1. Fig. 1 shows the exclusion contour in the ( $m_{\tilde{t}}$ , $m_{\tilde{\chi}_1^0}$) plane , at 95 % CL, for a stop with minimum mixing via a dominant $\bar{U}\bar{D}\bar{D}$ coupling from DELPHI [7] searches.

### 4.5. Other searches

Preliminary results on the LSP mass limits under $\not{R}_p$ via a dominant $\lambda, \lambda', \lambda''$ coupling by using direct/indirect decays are reported in Ref. [11] and single top quark production searches in the framework of $\not{R}_p$ in Ref. [12].

## 5. CONCLUSIONS

$\not{R}_p$ has inspired new interesting scenarios of SUSY searches that are complementary with the $R_pC$ ones . There is no evidence for SUSY with $\not{R}_p$ so far at LEP . Limits on the production cross-sections, the masses and couplings of $\not{R}_p$ are set at 95 % CL. $\not{R}_p$ limits are comparable with the $R_pC$ ones.

## 6. As an epilogue

SUSY with $\not{R}_p$ predicts very promising signatures especially for couplings with $\Delta L \neq 0$ .
All hopes are shifted towards future colliders !

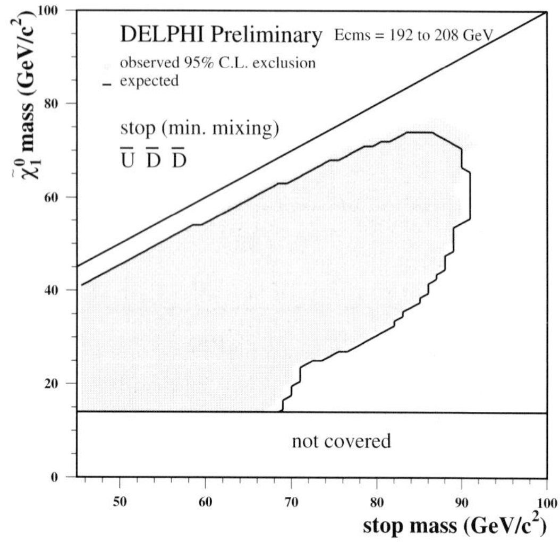


Figure 1. Exclusion contour, at 95 % CL, for a stop mass with minimum mixing as a function of the $\tilde{\chi}_1^0$ mass via a dominant $\bar{U}\bar{D}\bar{D}$ coupling.

## REFERENCES

1. P.Fayet, Phys. Lett. B69 (1977) 489; G.Farrar and P.Fayet, Phys. Lett. B76 (1978) 575.
2. H.Nilles, Phys. Rep. 110 (1984) 1; H.Haber and G.Kane, Phys. Rep. 117 (1985) 75.
3. V.Barger et al., Phys .Rev. D40 (1989) 2987.
4. S.Katsanevas, P.Morawitz, Computer Phys. Comm. 112 (1998) 227
5. DELPHI Collab., ICHEP2002 ABS223.
6. ALEPH Collab., ICHEP2002 ABS618.
7. DELPHI Collab., ICHEP2002 ABS164.
8. ALEPH Collab., ICHEP2002 ABS276.
9. L3 Collab., ICHEP2002 ABS409.
10. OPAL Collab., ICHEP2002 ABS513.
11. ALEPH Collab., ICHEP2002 ABS294.
12. DELPHI Collab., ICHEP2002 ABS225.

ELSEVIER

Proceedings of ICHEP 2002, pp. 717–719
*S. Bentvelsen, P. de Jong, J. Koch and E. Laenen (Editors)*

31st INTERNATIONAL CONFERENCE ON HIGH ENERGY PHYSICS AMSTERDAM

# Supersymmetry breaking and the SUSY flavour problem

O. Vives[a]

[a]Department of Physics, University of Oxford,
1 Keble Road, Oxford, OX1 3NP, UK.

Nonuniversality in the soft breaking terms is a generic feature in softly broken Supersymmetry. In particular, we show here that trilinear couplings are always nonuniversal in a theory that explains the Yukawa hierarchy in the frame of a spontaneously broken flavour symmetry. We analyse the constraints from low energy experiments and we find that large SUSY contributions are expected in the $\mu \to e\gamma$ transition in the near future with SUSY masses around the TeV scale.

## 1. INTRODUCTION

Our knowledge of the Supersymmetry breaking sector in the Minimal Supersymmetric extension of the Standard Model (MSSM) is still very limited due to the reduced experimental information available. Only after the discovery of SUSY particles and the measurement of the Supersymmetric spectrum we will be able to explore in detail and improve our understanding on this fundamental piece of the MSSM. Nevertheless, we have already today a lot of useful information on this sector from experiments looking for indirect effects of SUSY particles in low-energy experiments [1]. In fact, it was readily realized at the beginning of the SUSY phenomenology era that large contributions to Flavour Changing Neutral Currents (FCNC) and $CP$ violation phenomena were expected in Supersymmetric theories with a generic soft breaking sector. The absence of these effects much beyond the level most theorists considered reasonable became to be known as the SUSY flavour and $CP$ problems. These problems are closely related to the flavour structure of the soft breaking terms and therefore to the source of flavour itself which remains the most intriguing open question in high energy physics.

The negative result in the search for indirect effects of SUSY provides indeed very strong constraints on the flavour structure of the unknown soft breaking terms. In fact, the main available solutions to the Supersymmetric flavour problem are either exact universality of the soft breaking terms or alignment to the Yukawa matrices. However, as we will show here, in any model where we intend to explain at the same time the strongly hierarchical structure of the Yukawa matrices, these two solutions are not perfect and the soft breaking terms are necessarily nonuniversal and not aligned with the Yukawa couplings.

## 2. NONVANISHING F-TERMS

It is well-known that in Superstring models the SUSY soft breaking terms are generically nonuniversal. Only in some particular limits these terms can be, in principle, completely flavour blind. In the dilaton domination SUSY breaking scenario [2], only the dilaton field has a non vanishing F-term that contributes to SUSY breaking. In this case due to the universality of dilaton couplings, all soft breaking terms would be completely universal. However, we show here that this limit is never strictly possible and as we will see, other fields necessarily acquire a non vanishing F-term and therefore also contribute to SUSY breaking. This is due to the fact that in a theory with broken local Supersymmetry, i.e. Supergravity, any field that acquires a vacuum expectation value (vev) obtains at the same time a non-vanishing F-term.

The breaking of local Supersymmetry is indicated by a non zero gravitino mass $m_{3/2} = e^{K/2}\langle W\rangle \neq 0$ (in units of $M_{Pl} = 1$), which is


doi:10.1016/S0920-5632(02)02272-7

equivalent to a vev of the superpotential. Moreover, the full supergravity F-terms, for fields with canonical K'ahler potential, are given by $F^i = e^{K/2}\left(\frac{\partial W}{\partial \phi_i} + \phi^{i*} W\right)$. Therefore, if a certain field $\theta$ gets a non-vanishing vev in the presence of broken Supergravity, the natural size for its F-term will be given by the second term as $m_{3/2}\langle\theta\rangle$. However, it is still possible that the first term partially reduces the size of the F-term. In fact, we calculate the vev of the $\theta$ filed and the corresponding F-term through the minimization of the scalar potential,

$$V = \left[\sum_i (K^{-1})_i^j F^i F_j - 3e^K |W|^2\right] \qquad (1)$$

with $(K^{-1})_i^j$ the inverse of the Kähler metric. In particular, we are interested in the case of fields that obtain small vevs in units of the Plank mass and are charged under a some new symmetry. This symmetry can be either a continuous or a discrete symmetry. If the symmetry is continuous, $W$ only includes interaction terms of $\theta$ with matter fields which are not relevant in the minimization of the scalar potential since these matter fields will have zero vevs. A second possibility is to have a discrete symmetry, $Z_p$, which allows for a superpotential of the kind $W = A + B\theta^p$, where $A$ represents the vev of the hidden sector superpotential. Even in this case, it is possible to show that the F-term is still $m_{3/2}\langle\theta\rangle$ [3].

However, it is clear that this F-term is in principle small due to the $\langle\theta\rangle \ll 1$, when compared with the dominant F-terms of order $m_{3/2}$. So, does this F-term have any observable effect?? The answer to this question is positive if we consider a $\theta$ field that generates the hierarchy in the Yukawa couplings through an effective superpotential,

$$W = W^{hid} + \left(\frac{\theta}{M}\right)^{\alpha(i,j)} H_a Q_{Li} q^c_{Rj} + \ldots \qquad (2)$$

In this conditions, it is easy to see that the trilinear terms are nonuniversal and are given by,

$$A_{ij} = A_0 + \alpha_{ij}\; m_{3/2}, \qquad (3)$$

where $A_0$ includes the flavour independent contribution from the dilaton. Here, the smallness of $F^\theta$ is not effective because it is reabsorbed in the $\theta$ dependence of the Yukawas themselves and we have a completely nonuniversal trilinear couplings due to the different $\alpha_{ij}$ powers. Therefore, in any framework explaining the hierarchy in the Yukawa textures in terms nonrenormalizable couplings with vevs of scalar fields the trilinear couplings are necessarily nonuniversal as in Eq. (3).

## 3. PHENOMENOLOGICAL EFFECTS

As an example we take a generic Yukawa texture that fit the experimental data [4],

$$\frac{M}{m_3} = \left(\begin{array}{ccc} 0 & b\epsilon^3 & c\epsilon^3 \\ b'\epsilon^3 & d\epsilon^2 & a\epsilon^2 \\ f\epsilon^m & g\epsilon^n & 1 \end{array}\right), \qquad (4)$$

with $\epsilon_d = \sqrt{m_s/m_b} = 0.15$ and $\epsilon_u = \sqrt{m_c/m_t} = 0.05$. at the unification scale, and $a, b, b', c, d, g, f$ coefficients $\mathcal{O}(1)$ and complex in principle. In this texture there are still two undetermined elements that would account for the unobservable right-handed quark mixings. These are the $(3,1)$ and $(3,2)$ elements that can have an arbitrary power of $\epsilon$. If we assume that the hierarchy in the Yukawa matrices is generated by different powers in the vevs of scalar fields, $\epsilon_a = \langle\theta_a\rangle/M$, we can immediately calculate the nonuniversality in the trilinear terms from Eq. (3). In fact, after going to the basis of diagonal Yukawa matrices through superfield rotations, the so-called SCKM basis, we can see that large off-diagonal terms necessarily remain in the trilinear couplings. The Mass Insertions (MI) formalism is very useful to investigate the effects of nonuniversality in physical observables at $M_W$. The left–right MI are defined in the SCKM basis as $(\delta_{LR})_{ij} = (m^2_{LR})_{ij}/m^2_{\tilde{f}}$, with $m^2_{\tilde{f}}$ the average sfermion (squark or slepton) mass at the electroweak scale. Once we calculate the values of these MI in the model, we can directly compare with the phenomenological bounds from low energy experiments [5].

Therefore, we can immediately estimate the value of the LR MI at the electroweak scale. The stronger constraints are obtained from gluino mediated contributions in the down sector. For instance the flavour changing $LR$ MI are given by,

$$(\delta^d_{LR})_{21} \simeq \frac{m_{3/2}\ m_b}{19\ m^2_{3/2}} b'\ \epsilon^3_d \simeq b'\ 8 \times 10^{-6} \qquad (5)$$

using an average squark mass of $m_{\tilde{q}} \simeq 4.2 m_{3/2} \simeq$ 500 GeV which corresponds to $m_{3/2} \simeq 120$ GeV and a value of $\epsilon_d \simeq 0.15$ which correctly fits the down quark masses and mixings [4]. The strongest bound on this element comes from $\varepsilon'/\varepsilon$ and corresponds to the imaginary part of this element. Still, we can see that in the presence of a sizeable phase $\varepsilon'/\varepsilon$ naturally receives a sizeable Supersymmetric contribution [6,7].

On the other hand the MI corresponding to the $b \to s\gamma$ decay are,

$$(\delta^d_{LR})_{32} \simeq \frac{m_{3/2}\ m_b\ g\ n\ \epsilon^n_d}{19\ m^2_{3/2}} \simeq 2.2 \times 10^{-3}\ g\ n\ \epsilon^n_d \qquad (6)$$

again with $m_{3/2} \simeq 120$ GeV. Although this estimate is already below the phenomenological bound of $(\delta^d_{LR})_{32} \leq 1.6 \times 10^{-2}$ [5], for any value of $n$.

The situation in the leptonic sector is even more interesting. Assuming a similar texture in the charged lepton sector, as required by Grand Unification, we can calculate the nonuniversality in the LR transitions.

$$(\delta^e_{LR})_{12} \simeq \frac{m_{3/2}\ m_\tau}{5.5\ m^2_{3/2}} (b'\ \epsilon^3_d) \simeq b'\ 8.7 \times 10^{-6} \qquad (7)$$

where we take an average slepton mass of $m_{\tilde{l}} =$ 280 GeV, which again corresponds approximately to $m_{3/2} \simeq 120$ GeV. This MI must be compared with the bound from neutralino mediated contributions to the $\mu \to e\gamma$ decay,

$$(\delta^e_{LR})_{12} \leq 7 \times 10^{-7} \left(\frac{280}{100}\right)^2 = 5.5 \times 10^{-6} \qquad (8)$$

Clearly the estimate in Eq. (7) is still too large when compared with this experimental bound. Therefore this bound necessarily requires a larger value of the slepton mass. For an average slepton mass of 320 GeV our estimate would be just below the MI bound (notice that this would correspond to an average squark mass of 600 GeV). Assuming a quark-lepton unification 'a la' Georgi-Jarskog would predict a large $\mu \to e\gamma$ branching ratio very close to the current experimental bounds.

In summary, we have shown that nonuniversality in the soft breaking terms is a generic feature in softly broken SUSY theories. In particular trilinear couplings are always nonuniversal in a theory the explains the Yukawa hierarchy in the frame of a spontaneously broken flavour symmetry. We have analyzed the constraints from low energy experiments and we have concluded that large SUSY contributions are expected in the $\mu \to e\gamma$ transition in the near future with SUSY masses around the TeV scale.

## REFERENCES


1. A. Masiero and O. Vives, Ann. Rev. Nucl. Part. Sci. **51** (2001) 161 [arXiv:hep-ph/0104027].
2. A. Brignole, L. E. Ibanez and C. Munoz, arXiv:hep-ph/9707209.
3. G.G. Ross and O. Vives, work in progress
4. R. G. Roberts, A. Romanino, G. G. Ross and L. Velasco-Sevilla, Nucl. Phys. B **615** (2001) 358 [arXiv:hep-ph/0104088];
   G. G. Ross and L. Velasco-Sevilla, arXiv:hep-ph/0208218.
5. F. Gabbiani, E. Gabrielli, A. Masiero and L. Silvestrini, Nucl. Phys. B **477**, 321 (1996) [arXiv:hep-ph/9604387].
6. A. Masiero and H. Murayama, Phys. Rev. Lett. **83**, 907 (1999) [arXiv:hep-ph/9903363].
7. S. Khalil, T. Kobayashi and O. Vives, Nucl. Phys. B **580** (2000) 275 [arXiv:hep-ph/0003086].

ELSEVIER

Proceedings of ICHEP 2002, pp. 720–723
*S. Bentvelsen, P. de Jong, J. Koch and E. Laenen (Editors)*

31st INTERNATIONAL CONFERENCE ON HIGH ENERGY PHYSICS AMSTERDAM

# Recent developments in Anomaly Mediated Supersymmetry Breaking

I. Jack[a], D.R.T. Jones[a*]

[a] Dept. of Mathematical Sciences,
University of Liverpool, Liverpool L69 3BX, UK

We describe some recent developments in the Anomaly Mediated Supersymmetry Breaking scenario. In particular we focus on resolutions of the tachyonic slepton puzzle based on extending the MSSM by an anomaly free $U_1$ symmetry, so that exact RG invariance is preserved.

## 1. Introduction

The usual assumption of the CMSSM is that at gauge unification we have

| | |
|---|---|
| GAUGINO MASSES ($M_i$) | $\to M_0$ |
| SOFT $\phi^3$ TERMS ($h$) | $\to AY^{ijk}$ |
| $\phi^*\phi$ MASSES ($m^2$) | $\to m_0^2$ |

(here $Y^{ijk}$ are the Yukawa couplings).

THERE IS NO CONVINCING THEORETICAL BASIS FOR THIS.

A persuasive alternative is provided by[1] Anomaly Mediated Supersymmetry Breaking (AMSB):

$$\begin{aligned} M_i &= m_{\frac{3}{2}}\beta_{g_i}/g_i \\ h &= -m_{\frac{3}{2}}\beta_Y \\ m^2 &= \frac{1}{2}m_{\frac{3}{2}}m^*_{\frac{3}{2}}\mu\frac{d}{d\mu}\gamma \end{aligned} \tag{1}$$

Here $\gamma$ is the matter multiplet anomalous dimension and $m_{\frac{3}{2}}$ is the gravitino mass. The above relations are precisely RG invariant, and so can be evaluated at any scale; running from the unification scale is not necessary.

### 1.1. The Gaugino sector

An elementary consequence of Eq. (1) is that at low energies we have (approximately)

$$M_1 : M_2 : M_3 = 0.3 : 0.1 : 1, \tag{2}$$

*Work supported by a PPARC Senior Fellowship.

to be compared with the usual assumption that $M_1 = M_2 = M_3$ at gauge unification, which gives

$$M_1 : M_2 : M_3 = 0.1 : 0.3 : 1. \tag{3}$$

Thus in the AMSB scenario, there is likely to be an approximately degenerate triplet of light winos (a chargino and a neutralino). The characteristic phenomenology (involving the characteristic decay of the chargino to the neutralino and a charged pion) has been explored in a number of papers.

Another interesting consequence[2] is that in both AMSB and the CMSSM there is a sum rule relating the chargino and neutralino masses:

$$\begin{aligned} \Delta M^2 &= 2\sum_{i=1,2}(M_{\chi_i^{\pm}})^2 - \sum_{j=1\cdots 4}(M_{\chi_j^0})^2 \\ &= f(g_i)M_3^2 + 4M_W^2 - 2M_Z^2 \end{aligned} \tag{4}$$

In AMSB models $\Delta M^2$ is negative while in the CMSSM it is positive; see Fig. 1 (taken from Ref. [2], which presents sum rules and IR focus point analysis in a range of AMSB scenarios).

### 1.2. The slepton mass problem

Direct application of the AMSB solution to the MSSM leads, unfortunately, to negative (mass)$^2$ sleptons. The solution to this which has been most popular is the replacement $m^2 \to \hat{m}^2$ where

$$(\hat{m}^2)^i{}_j = (m^2)^i{}_j + m_0^2\delta^i{}_j. \tag{5}$$

Here $m^2$ is the basic AMSB solution from Eq. (1) and $m_0^2$ is constant. For example, de Campos et al[3] applied this solution to an extension of the


doi:10.1016/S0920-5632(02)02273-9

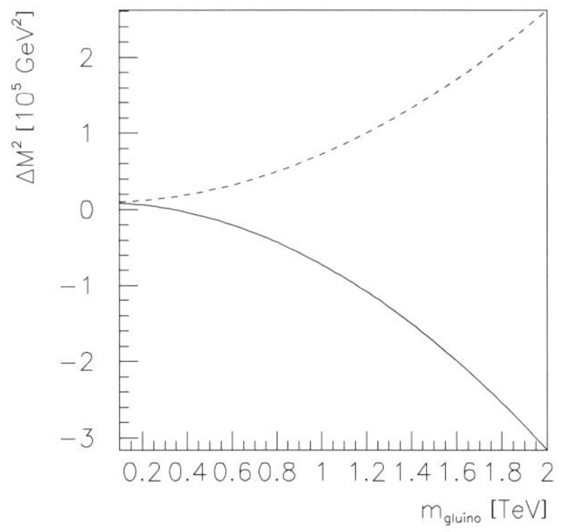


Figure 1. $\Delta M^2$ v. gluino mass for the CMSSM (dashed line) and AMSB (solid line).

MSSM to incorporate bilinear R-parity violation, i.e.

$$W \to W + \sum_i \lambda_i L_i H_2, \quad (6)$$

where $W$ is the MSSM superpotential,

$$W = H_2 Q Y_t t^c + H_1(QY_b b^c + LY_\tau \tau^c) + \mu H_1 H_2. \quad (7)$$

This results in interesting phenomenology; for example the mixing between neutral gauginos and neutrinos induces neutrino masses.

Eq. (5) has a major defect, however, which is that it is not RG invariant. If instead we have

$$(\hat{m}^2)^i{}_j = (m^2)^i{}_j + m_0^2 \sum_{a=1}^{\mathcal{N}} k_a (Y_a)^i{}_j \quad (8)$$

then $\hat{m}^2$ *is* RG invariant as long as each $Y_a$ corresponds to a $U_1$ invariance of the superpotential $W$ and also has vanishing mixed anomaly with each MSSM gauge group factor[4]. In the MSSM there are two possible flavour-blind linearly independent sets of such $Y_a$; the hypercharge gauge group $U_1^Y$ and another, which could be chosen[5] to be $U_1^{B-L}$, or a linear combination of $U_1^Y$ and $U_1^{B-L}$, which we will call $U_1'$. A

Table 1
Table of $U_1$ and $U_1'$ hypercharges.

| | $Q$ | $L$ | $t^c$ | $b^c$ | $\tau^c$ | $H_1$ | $H_2$ |
|---|---|---|---|---|---|---|---|
| $Y$ | $\frac{1}{6}$ | $-\frac{1}{2}$ | $-\frac{2}{3}$ | $\frac{1}{3}$ | 1 | $-\frac{1}{2}$ | $\frac{1}{2}$ |
| $Y'$ | $\frac{7}{3}$ | $-7$ | $\frac{5}{3}$ | $-\frac{19}{3}$ | 3 | 4 | $-4$ |

Flavour-blind case.

set of possible $U_1'$ charges (chosen so as to satisfy $\mathrm{Tr}(YY') = 0$) and the corresponding $U_1$ ones are shown in Table 1. It is easy to show that for $k' < 0$ and $-3 > k/k' > -14$, the contributions to both slepton mass terms from the $k_a$ terms in Eq. (8) are positive. In Ref. [4] it was shown that with, for example, $\zeta_1 = km_0^2 = 0.2\mathrm{TeV}^2$ and $\zeta_2 = k'm_0^2 = -0.02\mathrm{TeV}^2$, electroweak symmetry breaking occurs and a distinctive sparticle spectrum is obtained.

## 2. AMSB and the FN mechanism

Although the flavour-blind scenario described in section 1 has the advantage that flavour changing neutral currents (FCNCs) are naturally suppressed, it is interesting to explore a marriage between the $U_1$-based solution to the AMSB slepton mass problem and the Froggatt-Nielsen (FN) mechanism by replacing $U_1^{B-L}$ with a generation-dependent $U_1$ symmetry: in the hope of providing an explanation for the mass hierarchies and also a more natural framework for the introduction of neutrino masses.

At first sight this is unattractive because evidently generation-dependent contributions proportional to $Y_a$ in Eq. (8) will give rise to off-diagonal squark and slepton masses when we rotate to the quark and lepton mass-diagonal bases. However it was shown in Ref. [6] that the following set of textures provide a solution to this conundrum:

$$Y_t \sim \begin{pmatrix} \lambda^8 & \lambda^4 & 1+\mathcal{O}(\lambda^2) \\ \lambda^8 & \lambda^4 & 1+\mathcal{O}(\lambda^2) \\ \lambda^8 & \lambda^4 & 1+\mathcal{O}(\lambda^2) \end{pmatrix},$$

$$Y_b \sim \begin{pmatrix} \lambda^4 & \lambda^2 & 1+\mathcal{O}(\lambda^2) \\ \lambda^4 & \lambda^2 & 1+\mathcal{O}(\lambda^2) \\ \lambda^4 & \lambda^2 & 1+\mathcal{O}(\lambda^2) \end{pmatrix},$$

$$Y_\tau \sim \begin{pmatrix} \lambda^4 & \lambda^2 & 1 \\ \lambda^4 & \lambda^2 & 1 \\ \lambda^4 & \lambda^2 & 1 \end{pmatrix} \quad (9)$$

In $Y_\tau$ the third column entries are $O(1)$; likewise in $Y_{t,b}$ except that in these cases we require them to differ from each other at $O(\lambda^2)$ only. Imposing this leads to a CKM matrix of the form

$$CKM \sim \begin{pmatrix} 1 & 1 & \lambda^2 \\ 1 & 1 & \lambda^2 \\ \lambda^2 & \lambda^2 & 1 \end{pmatrix} \quad (10)$$

which is not of the form of the standard Wolfenstein parametrisation, but does reproduce its most significant feature, which is the smallness of the couplings to the third generation.

Textures of this "democratic" form manifestly correspond to generation-independent charge assignments for the left-handed fields; however the Yukawa matrices are diagonalised by rotations (to a good approximation) of *the left handed fields only*. Thus looking again at Eq. (8) we see that FCNC problems are avoided: for LH fields (which are rotated) because their charges are flavour-independent, and for RH fields (whose charges differ) because they are not rotated.

We assume the texture pattern is produced via higher dimensional terms such as $H_2 Q_i t_j^c (\frac{\theta_t}{M_U})^{a_{ij}}$ where $M_U$ represents the scale of new physics. We assume that each Yukawa matrix $Y_{t,b,\tau}$ gains its texture from the vev of a *particular* $\theta$-charge and that the vevs of the $\theta$-charges are approximately the same.

A set of charges leading to automatic cancellation of mixed anomalies (for arbitrary charges $h_2$ and $t_1^c$) is shown in Table 2. It is easy to verify that with these charges we obtain the democratic texture of Eq. (9) if we use $\theta$-charges $\theta_t = -1$, $\theta_b = \theta_\tau = 2$.

It is straightforward to show that in order to resolve the slepton mass problem we then require

$$3t_1^c + 4h_2 < 24 \quad \text{or} \quad 3t_1^c + 4h_2 > 32, \quad (11)$$

For the particular choices $h_2 = 12$, $t_1^c = -7/2$, $m_{\frac{3}{2}} = 40\text{TeV}$, $\tan\beta = 5$ and $\text{sign}\,\mu = -1$, we show in Fig. 2 the triangular region in the $\zeta_{1,2}$ plane which leads to an acceptable electroweak vacuum. For a typical point in the neutralino

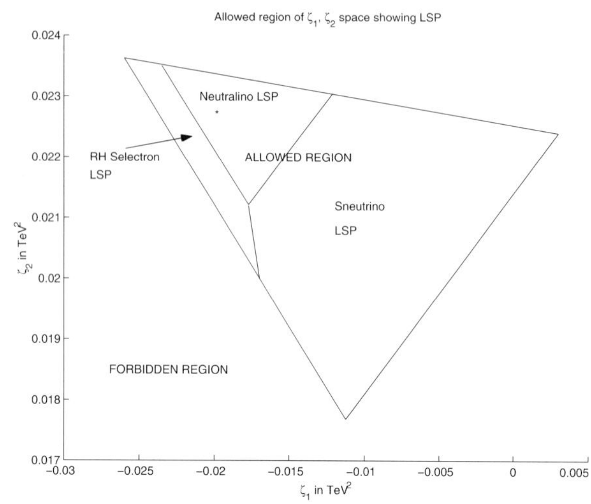


Figure 2. Allowed values of $\zeta_{1,2}$ for $\tan\beta = 5$, $m_{\frac{3}{2}} = 40\text{TeV}$ and $\text{sign}\,\mu = -1$.

LSP region ($\zeta_1 = -0.02\text{TeV}^2$, $\zeta_2 = 0.0227\text{TeV}^2$), we obtain the following spectrum:

$$\begin{aligned} m_{\tilde{t}_{1,2}} &= 869, 484,\ m_{\tilde{b}_{1,2}} = 825, 1082, \\ m_{\tilde{\tau}_{1,2}} &= 148, 442,\ m_{\tilde{u}_L,\tilde{c}_L} = 931,\ m_{\tilde{u}_R} = 908, \\ m_{\tilde{c}_R} &= 856,\ m_{\tilde{d}_L,\tilde{s}_L} = 934,\ m_{\tilde{d}_R} = 998, \\ m_{\tilde{s}_R} &= 1042,\ m_{\tilde{e}_L,\tilde{\mu}_L} = 149,\ m_{\tilde{e}_R} = 117, \\ m_{\tilde{\mu}_R} &= 323,\ m_{\tilde{\nu}_e,\tilde{\nu}_\mu} = 126,\ m_{\tilde{\nu}_\tau} = 125 \\ m_{h,H} &= 122, 166, m_A = 161,\ m_{H^\pm} = 181, \\ m_{\tilde{\chi}^\pm_{1,2}} &= 112, 575,\ m_{\tilde{\chi}_{1\cdots 4}} = 111, 369, 579, 579 \\ m_{\tilde{g}} &= 1007 \end{aligned} \quad (12)$$

where all masses are given in GeV. The squarks $\tilde{t}_1, \tilde{b}_1$ and $\tilde{\tau}_1$ couple more strongly to $t_L$, $b_L$ and $\tau_L$ respectively, though (for our chosen $\tan\beta$) the $\tilde{t}_{1,2}$ mixing is of course substantial.

## 3. Massive Neutrinos

We could introduce Dirac masses for the neutrinos preserving $U_1^{B-L}$, and hence maintain the flavour-blind scenario described in Section 1. This is obviously unappealing however, as it provides no explanation for the smallness of neutrino

Table 2
Table of $U_1$ and $U_1'$ hypercharges.

| $Q_i$ | $b_1^c$ | $b_2^c$ | $b_3^c$ | $t_2^c$ | $t_3^c$ |
|---|---|---|---|---|---|
| $8 - t_1^c - h_2$ | $2h_2 + t_1^c - 16$ | $2h_2 + t_1^c - 12$ | $2h_2 + t_1^c - 8$ | $t_1^c - 4$ | $t_1^c - 8$ |
| $L_i$ | $\tau_1^c$ | $\tau_2^c$ | $\tau_3^c$ | $h_1$ | |
| $3t_1^c + 3h_2 - 24$ | $16 - 2h_2 - 3t_1^c$ | $20 - 2h_2 - 3t_1^c$ | $24 - 2h_2 - 3t_1^c$ | $-h_2$ | |

Froggatt-Nielsen case.

masses. So we adopt instead the seesaw mechanism:

$$W \to W + \frac{1}{2}\nu^c M_{\nu^c}\nu^c + H_2 L Y_\nu \nu^c \tag{13}$$

where $Y_\nu$ is $3 \times n_\nu$, $n_\nu$ being the number of RH neutrinos. In fact we can still preserve the flavour-blind solution if we assign zero $U_1'$ charge to all the $\nu^c$, and generate the $Y_\nu$ matrix via the FN mechanism. We eschew this for two reasons; firstly because it seems unnatural to thus generate *only* $Y_\nu$ (and not $Y_{t,b,\tau}$); and secondly because we would then have to consider (or somehow exclude) R-parity violation since clearly, for example the $LH_2$ bilinear is allowed if $H_2 L \nu^c$ is (with zero charge for $\nu^c$). We prefer therefore to use a more exotic assignment of charges such as described in Section 2 with a view to simultaneously producing an acceptable neutrino spectrum with a $U_1'$ which (without further assumption) naturally suppresses the R-parity violation sector.

The simplest such possibility we have found is to begin with the charge assignments given in Table 2. Then with two RH neutrinos with $U_1'$ charges $\pm q_\nu$ and a FN field $\theta_\nu$ such that

$$\begin{aligned} L_i + h_2 + q_\nu + n q_{\theta_\nu} &= 0 \\ L_i + h_2 - q_\nu + m q_{\theta_\nu} &= 0, \end{aligned} \tag{14}$$

it is easy to show that if we choose, for example, $n = 2$ and $m = 1$ and $q_{\theta_\nu} = -9$ then (with $h_2 = 12$, $t_1^c = -7/2$ as above) no renormalisable $R$-parity violating interactions can be generated using the available $\theta$-charges.

The resulting pattern of neutrino masses and mixings can accommodate the currently favoured pattern, i.e. $m_1 \ll m_2 \ll m_3$ (with $m_1 = 0$) and a mixing matrix of the form, for example,

$$U \approx \begin{pmatrix} 2/\sqrt{6} & 1/\sqrt{3} & 0 \\ -1/\sqrt{6} & 1/\sqrt{3} & 1/\sqrt{2} \\ 1/\sqrt{6} & -1/\sqrt{3} & 1/\sqrt{2} \end{pmatrix} \tag{15}$$

While, however, these patterns can indeed be obtained they are not a specific prediction of the model in its present form. It would be interesting if a marriage of the AMSB and texture paradigms that naturally predicted an appropriate hierarchy and form of $U$ could be found. There is already a considerable literature devoted to building neutrino mass models; for recent reviews of the experimental and theoretical situations see for example [7].

## REFERENCES


1. L. Randall and R. Sundrum, *Nucl. Phys.* B557 (1999) 79;
   G.F. Giudice, M.A. Luty, H. Murayama and R. Rattazzi, *JHEP* 9812 (1998) 27;
   I. Jack and D.R.T. Jones, *Phys. Lett.* B465 (1999) 148.
2. K. Huitu, J. Laamanen and P.N. Pandita, Phys. Rev. D65 (2002) 115003.
3. F. de Campos et al, *Nucl. Phys.* B623 (2002) 47.
4. I. Jack and D.R.T. Jones, *Phys. Lett.* B482 (2000) 167.
5. N. Arkani-Hamed et al, *JHEP* 0102 (2001) 041.
6. I. Jack, D.R.T. Jones and R. Wild, *Phys. Lett.* B535 (2002) 193.
7. C. Giunti, hep-ph/0209103;
   S.F. King, hep-ph/0208266

ELSEVIER

Proceedings of ICHEP 2002, pp. 724–727
*S. Bentvelsen, P. de Jong, J. Koch and E. Laenen (Editors)*

31st INTERNATIONAL CONFERENCE ON HIGH ENERGY PHYSICS AMSTERDAM

# Exotic SUSY scenarios

R.M. Godbole[a] *

[a]CERN, Theory Division, CH-1211, Geneva-23, Switzerland.

In this talk I discuss some scenarios which involve small extensions of the ideas usually considered in the Minimal Supersymmetric Standard Model (MSSM). I present results of a study of the implication of nonuniversal gaugino masses (NGM) for the invisible decays of the lightest scalar and the correlation of the same with the relic density of the lightest supersymmetric particle (LSP) in the Universe. Further I discuss SUSY with $\not{R}_p$ . Decay of $\tilde{\chi}_1^0$ in case of dominant $\not{R}_p$ and $\not{L}$, trilinear $\lambda$ and $\lambda'$ couplings and also that of $\tilde{\chi}_1^0, \tilde{\chi}_1^+$ caused by the bilinear $\not{R}_p$ ,$\not{L}$ couplings are discussed. The effect of the latter for the trilepton signal at the Tevatron is presented. I end with a discussion of signals of Heavy Majorana Neutrinos(HMN) at the LHC.

## 1. Introduction

In this talk I summarise salient features of the results of four different investigations; small departures from the standard assumptions of the SM and the MSSM, being the common feature to all of them. These are based on four abstracts[1–4] submitted to this meeting. Supersymmetry and models for nonzero $\nu$ masses indeed form a very big part of all the current Beyond the Standard Model (BSM) discussions. In this talk I refer to the effects of relaxing two of the assumptions normally made: that of universal gaugino masses[1] and $R_p$ conservation[2,3]. $\not{R}_p$ SUSY provides one of the most economical way of generating non-zero $\nu$ masses. In the last abstract[4] a new aspect of the collider sigantures of the heavy majorana neutrino, an important ingradient of all the models of $\nu$ masses not involving $\not{R}_p$ SUSY, is presented.

## 2. Non Universal Gaugino Masses

MSSM assumes universal masses for the $SU(3), SU(2)$ and $U(1)$ gauginos at the high scale. This implies $M_1 \simeq 0.5 M_2$ at the electroweak scale, where $M_1, M_2$ are the $U(1), SU(2)$ gaugino masses. However, this assumption need not be true even in the very restrictive mSUGRA model wherein the nonuniversality is possible for a nonminimal kinetic term for the gauge superfields. Non-universal gaugino masses are expected also in models with anomaly-mediated SUSY breaking (AMSB) or moduli dominated SUSY breaking. In general, therefore, we can expect $M_1 = r M_2$ with $r \neq 0.5$ at the EW scale. We studied the effect of such a scenario on the 'invisible' decays of the $h$. A ratio $r$ between the two gaugino masses at the EW scale, needs

$$M_1 = 2 r M_2 \tag{1}$$

at the GUT scale. Most of the abovementioned models normally imply values of $r > 1$. Taking a phenomenological approach, however, we consider two values of $r < 1$ viz. $0.1, 0.2$. As a result of such a nonuniversality, for a given $\tilde{\chi}_1^{\pm}$ mass, the mass of the $\tilde{\chi}_1^0$ is smaller than that in the universal case. Hence it is possible to have $h \to \tilde{\chi}_1^0 \tilde{\chi}_1^0$ while respecting all the LEP constraints, in spite of the theoretical upper limit of $\sim 130$ GeV on the mass of the lightest $h$ among the SUSY higgses. Of course it is necessary to rework[5] all the LEP constraints for the nonuniversal case. B.R. ($h \to \tilde{\chi}_1^0 \tilde{\chi}_1^0$) is maximised for moderate $\tan\beta$, small $\mu$ and $M_2$. Unlike the case of universal gaugino masses, we found large regions of $M_2$–$\mu$ plane where B.R. for the 'invisible', $\tilde{\chi}_1^0 \tilde{\chi}_1^0$ decay mode can be as large as 0.65 *even* after all the LEP constraints are imposed, as shown in Fig. 2. Such large 'invisible' branching ratio for the $h$ de-

*Permanent Address: Centre for Theoretical Studies, Indian Institute of Science, Bangalore, 560 012, India.


doi:10.1016/S0920-5632(02)02274-0

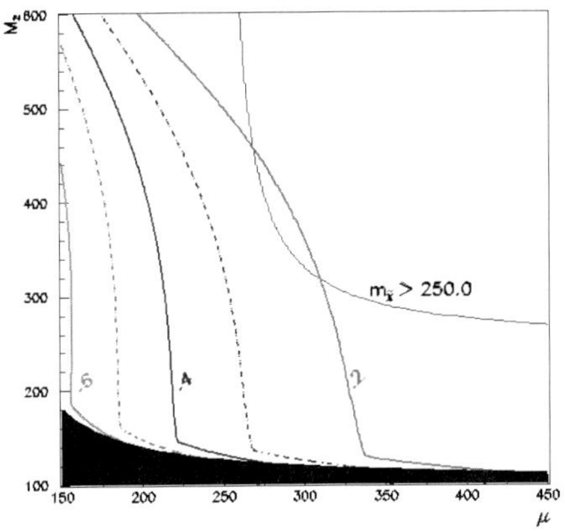


Figure 1. Contours of $B.R.(h \to \tilde{\chi}_1^0\tilde{\chi}_1^0)$ for $r = 0.1$ along with the LEP constraints on the $\tilde{\chi}_1^+$ mass indicated by the black region. $\tan\beta = 5$ and $m_h = 125$ GeV.

creases that into the $\gamma\gamma$ and $b\bar{b}$ channels, which respectively provide the best possible signature for the $h$ produced 'inclusively' or in association with a $W/Z/t\bar{t}$. If we define,

$$R_{\gamma\gamma} = \frac{B.R.(h \to \gamma\gamma)_{SUSY}}{B.R.(h \to \gamma\gamma)_{SM}}$$

and $R_{b\bar{b}}$ similarly, we find that

$$R_{\gamma\gamma} = R_{b\bar{b}} = 1.0 - B.R.(h \to \tilde{\chi}_1^0\tilde{\chi}_1^0).$$

Thus B.R. $(h \to \tilde{\chi}_1^0\tilde{\chi}_1^0) \sim 0.3 - 0.4$ can mean loss of the signal for the lightest Higgs at the LHC.

A light $\tilde{\chi}_1^0$ also has implications for the relic density of the neutralinos in the universe, as the latter is decided by $\sigma(\tilde{\chi}_1^0\tilde{\chi}_1^0 \to f^+f^-)$. For a light $\tilde{\chi}_1^0$, the $Z/h$ mediated $s$ channel process contributes to the annihilation. If the $\tilde{l}_R$ is light the cross-section also receives contribution from the $t$-channel $\tilde{l}_R$ exchange. There is a clearly a correlation between the expected 'invisible' branching ratio for the $h$ and the relic density of the $\tilde{\chi}_1^0$ as the same couplings are involved. Fig. 2 shows results obtained for $r = 0.1$ with $\tan\beta = 5$ and $m_{\tilde{l}_R} \sim 100$ GeV[1] using the micrOMEGAs[6] program to calculate the relic density. These show that the requirement of an acceptable relic density does constrain the $M_2 - \mu$ plane quite substantially. However, there exist large regions

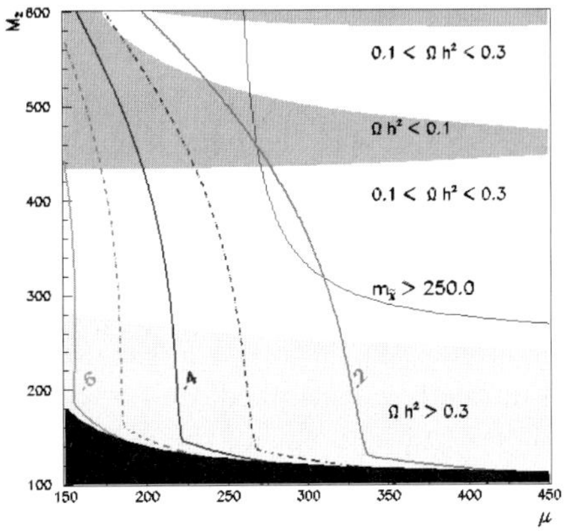


Figure 2. Contours of $B.R.(h \to \tilde{\chi}_1^0\tilde{\chi}_1^0)$ for $r = 0.1$ along with the DM and LEP constraints. The white region corresponds to $0.1 < \Omega h^2 < 0.3$, for $m_0 = 94$ GeV, $\tan\beta = 5$ and $m_h = 125$ GeV. The black region is the LEP excluded region. The lightly (heavily) shaded region corresponds to $\Omega h^2 > 0.3$ ($< 0.1$).

of this plane where the 'invisible' branching ratio of $h$ is as large as 0.5–0.6, even for 'large' ($\sim 200$ GeV) $\tilde{l}_R$, consistent with the LEP constraints and with an acceptable relic density. The small mass of the $\tilde{\chi}_1^0$ in this case means that the trilepton signal at the Tevatron will be qualitatively different than in the universal case. Thus this scenario can be tested at the Tevatron via the trilepton events. As a matter of fact, one can also obtain a model independent limit on $M_1$ and hence on the $\tilde{\chi}_1^0$ mass, by considerations of relic density.

## 3. Decaying $\tilde{\chi}_1^0$ in $R\!\!\!/_p$ violating theories:

$B, L$ are symmetries of the SM but not of the MSSM; i.e. it is possible to write terms in the Lagrangian which are $L\!\!\!/, B\!\!\!/$ but respect gauge invariance as well as supersymmetry. Nonzero $\nu$ masses can be generated in $R\!\!\!/_p$ Supersymmetric theories, without introducing any new fields; at the tree level via the bilinear $R\!\!\!/_p$ terms in the superpotential and quantum one or two loop level via the trilinear $\lambda, \lambda'$ couplings[7]. A large number of these couplings are constrained by low enery processes, cosmological arguments as well as explicit collider searches[8]. Some of the strongest constraints on the $\Delta L = 1$ processes come from

$\nu$ masses. As said earlier, since the heavy scale involved in the mass generation mechanism is decided by the sparticle masses, this leads to constraints on the size of the $R\!\!\!/_p$ couplings and hence to clear predictions of the $L\!\!\!/$ ,$B\!\!\!/$ signals in collider processes involving these sparticles. Since some of the constraints do depend on the details of the specific SUSY model, it is important to study the direct effect of the *very same* couplings that are so indirectly constrained, at the colliders. Such studies can help clarify the picture for model building.

The main constraints on $L\!\!\!/$ couplings $\lambda, \lambda'$ with more than one third generation index come from models for $\nu$ masses and not from collider experiments. Probes of these $L\!\!\!/$ couplings at the colliders involve studying the physics of third generation fermions. The third generation sfermions are also likely to give rise to larger virtual effects as they are expected to be lighter than those of first two generations. For $\tilde{\chi}_1^0, \tilde{\chi}_1^+$ with masses of interest at the $LHC$ and $NLC$, final states with third generation fermions including $t$ are possible. One thus needs a study of the $R\!\!\!/_p$ decays of $\tilde{\chi}_1^0, \tilde{\chi}_1^\pm$ retaining effects of the mass of the third generation fermions, for $L$ violating coupling. An interesting example is the resonant or non-resonant production of the $\tilde{\tau}$ via the $L\!\!\!/$ , $\lambda'_{333}$ coupling in $pp \to t\bar{b}\tilde{\tau}$, similar to the case of $t\bar{b}H^+$ production via the Yukawa coupling. Even for $\lambda'_{333}$ as small as 0.01 and $m_{\tilde{t}} > m_t$, the rates are appreciable. The $\tilde{\tau}$ so produced will have both $R_p$ conserving decay $\tilde{\tau} \to \tau\tilde{\chi}_1^0, \tilde{\tau} \to \nu_\tau\tilde{\chi}_1^-$ for $m_t > m_{\tilde{\tau}}$ and the $R_p$ violating one $\tilde{\tau} \to b\bar{t}$ for $m_t < m_{\tilde{\tau}}$. The final state composition will depend on the $R\!\!\!/_p$ decays of the $\tilde{\chi}_1^0$ as well. Production of $\tilde{\tau}$ through $R\!\!\!/_p$ couplings and its decay via the *same* will give rise to $pp \to t\bar{b}\tilde{\tau}X \to t\bar{b}t\bar{b}X$ which is the same as the expected final state for the $H^\pm$ production. Thus the $\tilde{\tau}$ in this case can fake the $H^\pm$ signal. $R_p$ conserving decays of the $\tilde{\tau}$ and $R\!\!\!/_p$ decays of the $\tilde{\chi}_1^0$ can also produce,

$$pp \to t\bar{b}\tilde{\tau}X \quad \to (2t)(2b)(2\tau)X \\ \qquad\qquad \to tb\ (2\bar{b})\ \tau\nu_\tau X. \qquad (2)$$

In the latter case the final state has the telltale signature of $L\!\!\!/$ , the like sign fermion pairs.

We calculated[2] the three body decays of the LSP for dominant $L\!\!\!/$ trilinear couplings and obtained explicit expressions for the most general case of complex mass matrices, including the effect of the mass of the third generation fermions. Of course the results depend on the $L$–$R$ mixing in the sfermion sector, gaugino-higgsino mixing and $\tan\beta$ as well as on the nature of the particular $L\!\!\!/$ couplings which is large. One can have the 'massive' and 'massless' modes with/without a $t$ in the final state. Fig. 3 shows the results for the case with $\lambda'_{333}$ dominant, for two different choices of $L$–$R$ mixing in the squark sector and different composition of the $\tilde{\chi}_1^0$. The massive decay mode has a large width for low $\tan\beta$ and large higgsino-gaugino mixing. Otherwise the massless mode is larger, though the massive mode is nonnegligible. Even for smaller but dominant $R\!\!\!/_p$ , $\lambda$ and $\lambda'$ couplings, with more than one third generation index, the three body $R\!\!\!/_p$ decays of $\tilde{\chi}_1^0, \tilde{\chi}_1^+$ can have important phenomenological consequences for new particle searches at the future colliders.

In the above discussion the bilinear $R\!\!\!/_p$ terms were assumed to be zero. In the analysis of Ref. [3], particular case of nonzero $\kappa_3$ is considered. The $R\!\!\!/_p$ and $R_p$ conserving decays of the $\tilde{\chi}_1^0, \tilde{\chi}_1^\pm$ in this case are given by Eq.3 and 4 respectively.

$$\begin{aligned} \tilde{\chi}_1^0 &\to \nu_\tau Z^* \to \nu_\tau f\bar{f}, \ \ \tilde{\chi}_1^0 \to \tau W^\pm \to \tau f\bar{f}' \\ \tilde{\chi}_1^\pm &\to \tau Z^* \to \tau f\bar{f}, \ \ \tilde{\chi}_1^\pm \to \nu_\tau W^\pm \to \nu_\tau f\bar{f}', \end{aligned} \qquad (3)$$

and

$$\tilde{\chi}_2^0 \to \tilde{\chi}_1^0\ell^+\ell^-, \tilde{\chi}_1^\pm \to \tilde{\chi}_1^0\ell\nu_\ell. \qquad (4)$$

It is clear the $R\!\!\!/_p$ decays will enhance the number of leptons in the final state and give very clear signals. These authors analysed, in an mSUGRA picture, the multilepton signals at the Tevatron. With the current mass bounds on various supersymmetric particles, the multilepton signal offer of the best chance of the discovery of SUSY at the Tevatron. Hence such an analysis is important. This one shows that the reach in the multilepton channel for the $R\!\!\!/_p$ case is much better than for the $R_p$ conserving case at large values of $m_0$. For small $m_0$ the situation is reversed due to slepton mediated $\tilde{\chi}_2^0$ decay.

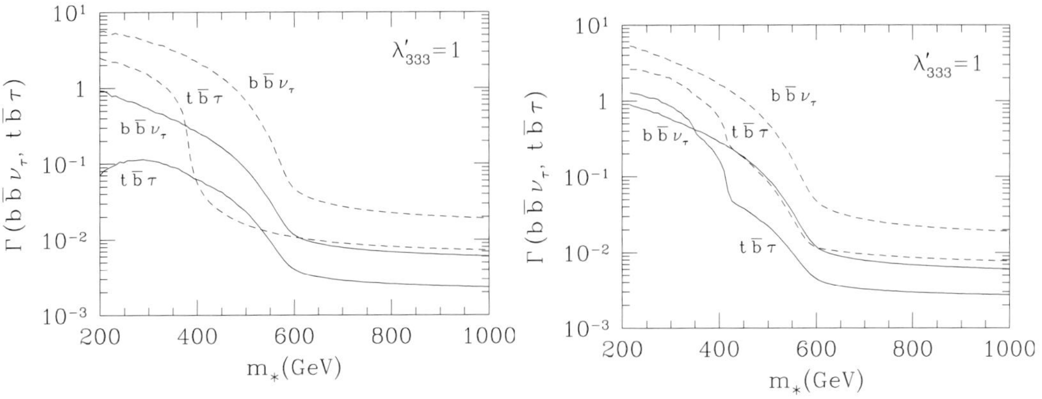


Figure 3. $\tilde{\chi}_1^0$ widths for no (moderate) $L$–$R$ squark mixing in left (right) panel. Dshed (solid) lines for a wino(bino) like $\tilde{\chi}_1^0$ . $A_t - \mu \cot\beta = 150$ GeV, $A_b - \mu\tan\beta = 2000$ for nonzero mixing.

## 4. Signals for the Heavy Majorana Neutrino

In this work[4] the authors consider models of $\nu$ masses which have an isosinglet neutrino $N$ which mixes with the ordinary light one but wherein the mass and the mixing of the $N$ are not linked together, unlike in the usual models with the See-Saw mechanism. This mixing is then treated as a phenomenological parameter constrained by the LEP data. It is possible in these models to have $N$ mass 'naturally' in the range: $100 < M_N < 1000$ GeV[9]. The issue of characteristic $\not{L}$ signals for the HMN via Like Sign Dilepton (LSD) events. was revisited in this investigation. They considered both the fusion contribution from

$$q_i + \bar{q}_j \rightarrow q_k + \bar{q}_l + W^{+*} + W^{+*} \rightarrow q_k + \bar{q}_l + l^+ + l^+ \quad (5)$$

where the $N$ is exchanged in $t$ channel and also from the resonant production of $N$ via

$$q_i + \bar{q}_j \rightarrow W^{+*} \rightarrow l^+ N \rightarrow l^+ + l^+ + W^- \rightarrow l^+ l^+ q_k \bar{q}_l . \quad (6)$$

Explicit formulae were obtained in helicity formalism and errors in the earlier calculations[10], traced to dropping of the ghost diagrams, were corrected. The predicted LSD cross-section increases by about 20% as a result of this.The reach of LHC just from rates is $M_N = 200 - 250$ GeV. Detailed simulations will be needed to make these conclusions firmer, as the cuts can then be optimised.

## REFERENCES

1. G. Bélanger, F. Boudjema, R.M. Godbole, F. Contrant and A. Semenov, Abstract 103, *Phys. Lett.* B **519** (2001) 93, hep-ph/0106275.
2. F. Borzumati, R. Godbole, J.L. Kneur and F. Takayama Abstract 176, *JHEP* **07** (2002) 037, hep-ph/0108244.
3. F. de Campos, O.J.P. Eboli, M.B. Magro, W. Porod, D. Restrepo and J.W.F. Valle, Abstract 115 at ICHEP02.
4. O. Panella, M. Cannoni, C. Carimalo and Y.N. Srivastava, Abstract 962: *Phys. Rev.* D **65** (2002) 035005.
5. G. Bélanger, F. Boudjema F. Donato, R. Godbole and S. Rosier-Lees, *Nucl. Phys.* B**581** (2000) 3, hep-ph/0002039.
6. G. Bélanger, F. Boudjema, A. Pukhov and A. Semenov, "micrOMEGAs: A program for calculating the relic density in the MSSM," hep-ph/0112278.
7. For a recent discussion see, for example, F. Borzumati and J. S. Lee, arXiv:hep-ph/0207184.
8. For a recent summary, see for example, Ref. [2].
9. J. Gluza and M. Zralek, Phys. Rev. D **55** (1997) 7030.
10. F. M. Almeida, Y. A. Coutinho, J. A. Martins Simoes and M. A. do Vale, *Phys. Rev.* D **62** (2000) 075004.

ELSEVIER

Proceedings of ICHEP 2002, pp. 728–730
*S. Bentvelsen, P. de Jong, J. Koch and E. Laenen (Editors)*

31st INTERNATIONAL CONFERENCE ON HIGH ENERGY PHYSICS AMSTERDAM

# Little Higgs Models: New Approaches to the Hierarchy Problem

J. G. Wacker[a*]

[a]Theory Group, Department of Physics
University of California,
Berkeley, CA 94720 USA

In this note we present a review of the little Higgs models that stabilize the electroweak by realizing the Standard Model Higgs as a pseudo-Goldstone boson.

## 1. Introduction

The experimental results of the past decade indicate that the chiral Lagrangian describing EWSB works to one loop accuracy and irrelevant operators (such as $|D\omega|^4$, where $\omega$ are the Goldstone bosons eaten by $W^\pm$ and $Z^0$) are suppressed by a scale $\Lambda \gg 1$ TeV. This strongly indicates perturbative physics at 1 TeV and a physical Higgs boson $H = (h^0 + v)\omega$. However, this linear sigma model is unstable to quantum corrections leading to the hierarchy problem, meaning that the Standard Model is in an incomplete description of physics parametrically above the weak scale. To date, only the MSSM provides a solution to the hierarchy problem and weakly coupled physics at 1 TeV.

In this note we review models where the Higgs is a pseudo-Goldstone boson, $\Sigma = e^{i\sigma/f}$ [1], for early attempts see [2]. Approximate global symmetries, $\Sigma \to L\Sigma R^\dagger$ keep the Higgs vev small relative to the cut-off, $\Lambda \sim 4\pi f$. The global symmetries are broken by couplings, however, the structure of the symmetry breaking prevents large corrections to the Higgs vev. At 100 GeV the model is a two Higgs doublet model with a Higgs potential identical to the MSSM. There is a mini-desert up to 1-3 TeV where new particles emerge that cancel the quadratic divergences of the Standard Model. The novelty of this set of models is that particles of the *same* spin cancel the respective quadratic divergences. Above the TeV scale there is another mini-desert until the 10-30 TeV scale where the description of the model becomes strongly coupled and new physics emerges. However, the Standard Model is insensitive to this physics.

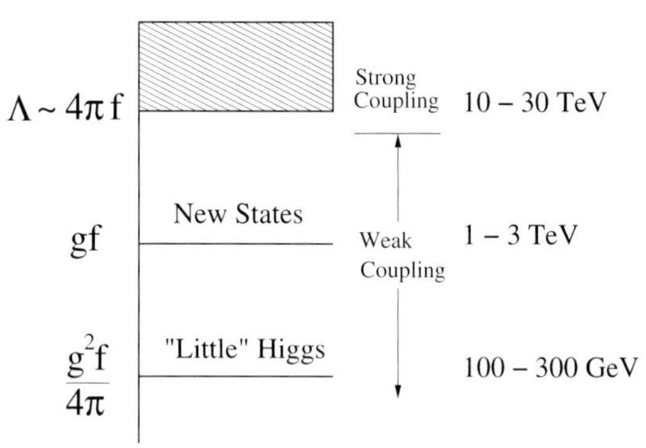


## 2. Little Higgs Models

Little Higgs models are gauged non-linear sigma model with Lagrangians that can be divided into four pieces:

$$\mathcal{L} = \mathcal{L}_{\text{Gauge}} + \mathcal{L}_{\text{nl}\sigma\text{m Kin}} + \mathcal{L}_{\text{Plaq}} + \mathcal{L}_{\text{Yuk}} \quad (1)$$

The first term is the standard kinetic term for gauge bosons. The second term is the nl$\sigma$m kinetic terms that give mass to some of the gauge bosons. There is the possibility of adding a scalar potential, often called the plaquette potential, that gives mass to some of the nl$\sigma$m fields. Finally there are the Yukawa couplings between the nl$\sigma$m fields and fermions. To study the low energy dynamics of the theory we want to integrate

*Theory Group, Harvard University


doi:10.1016/S0920-5632(02)02281-8

out all the states that are classically massive and have an effective Lagrangian in terms of the classically massless fields. Some nl$\sigma$m can be represented by theory spaces where points denote gauge groups, lines are nl$\sigma$m fields, and faces are plaquette potentials. This was studied in depth in [6]. The classically massless scalars are the little Higgs and are associated with non-contractible loops in theory space. The little Higgs has a mass generated radiatively, but the size of the radiative corrections are parametrically smaller than the TeV scale.

In the limit of vanishing couplings the nl$\sigma$m fields become exact Goldstone bosons under the symmetry $\Sigma \to L\Sigma R^{\dagger}$. This leaves the little Higgs exactly massless, but also without non-derivative interactions. Coupling constants break these chiral symmetries and allow masses for the little Higgs to be generated radiatively. Little Higgs models always require at least two couplings to communicate sufficient chiral symmetry breaking to generate a mass radiatively, and therefore quadratic divergences first arise at two loop order. To ensure that there are no one loop quadratic divergences to the the little Higgs mass there are simple rules for theory space models [4,6]. The first rule is that if all nl$\sigma$m fields are bi-fundamentals, then there are no one loop gauge quadratic divergences. If no plaquette contains any nl$\sigma$m field more than once, then there one loop quadratic divergences to the little Higgs mass from the scalar potential.

Having removed the one loop quadratic divergences, there are always finite contributions to the little Higgs mass. We can estimate the size of these finite contributions by studying the theory beneath the scale of the classically massive modes. In this low energy thoery there are a quadratic divergences and they are cut-off by the mass of the first heavy states and become finite corrections to the mass squared of the little Higgs. The mass of the heavy modes is roughly $M_{\rm H} \sim gf$. This means that the correction to the little Higgs mass from the low energy quadratic divergence is:

$$\delta m_{\rm LH}^2 \sim \frac{g^2}{16\pi^2} M_{\rm H}^2 = \frac{g^2}{16\pi^2} g^2 f^2 = \left(\frac{g^2}{16\pi^2}\right)^2 \Lambda^2. \quad (2)$$

This is the same size as the two loop quadratic divergence in the full theory. Thus, even if we remove higher order quadratic divergences, we can not push the cut-off higher than two loop factors above the EW scale, setting $\Lambda \sim 10-30$ TeV. This still allows for a seperation of scales $m_{\rm LH} \ll M_{\rm H} \ll \Lambda$. Furthermore, this means that the states that stabilizes the electroweak scale is invisible to roughly 1 TeV.

## 3. The Minimal Moose

The minimal theory space model[2] that has the requisite properties for EWSB has two gauge group and four nl$\sigma$m fields and is illustrated in Fig. 1. The two gauge groups are taken to be $SU(3)$ and $SU(2) \times U(1)$ with the $U(1)$ embedded as the $\mathbf{T_8}$ generator of $SU(3)$. There are two plaquettes added:

$$\begin{aligned} \mathcal{L}_{\rm Plaq.} &= \lambda_1 f^4 \,{\rm Tr}\, X_1 X_2^{\dagger} X_3 X_4^{\dagger} + \text{ h.c.} \\ &\quad + \lambda_2 f^4 \,{\rm Tr}\, X_2 X_3^{\dagger} X_4 X_1^{\dagger} + \text{ h.c.} \end{aligned} \quad (3)$$

This gauged nl$\sigma$m breaks the $SU(3) \times [SU(2) \times U(1)]$ gauge symmetry down to the diagonal $[SU(2) \times U(1)]$ that is identified as the electroweak gauge group. The plaquettes give mass to one linear combination of nl$\sigma$m fields and leave three massless. One is eaten by the massive $SU(3)$ gauge bosons while two are physical and massless and are the little Higgs of the theory.

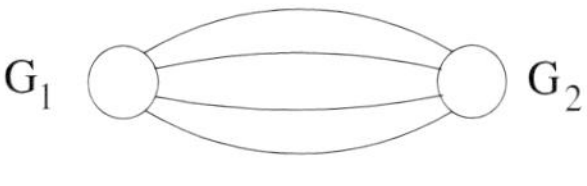


Figure 1. Minimal theory space model.

Since all the link fields are bi-fundamentals under the two gauge groups and the plaquettes contain each link field only once, there are no one loop quadratic divergences to the little Higgs masses.

The scalar potential for little Higgs, $\Sigma_{1,2}$, is of the form ${\rm Tr}\,\Sigma_1\Sigma_2\Sigma_1^{\dagger}\Sigma_2^{\dagger}$ or in terms of the linearized modes:

$$V(\sigma_1, \sigma_2) = \lambda_{\rm eff}\,{\rm Tr}\,[\sigma_1, \sigma_2]^2 + \cdots \quad (4)$$

[2]see [7,8] for other classes of models

where $\lambda_{\text{eff}}^{-1} = \lambda_1^{-1} + \lambda_2^{-1}$. Written in terms of the doublets, the potential is:

$$V(h_1, h_2) = \lambda_{\text{eff}}\left(|h_1|^2 - |h_2|^2\right)^2 \quad (5)$$

This is identical to the Higgs potential inside the MSSM except that the coefficient is undetermined. Hence, the mass of the Higgs is undetermined, although we do expect $\lambda_{\text{eff}} \sim \mathcal{O}(1)$, leading to a physical Higgs mass in the 100 – 300 GeV range.

The Standard Model fermions are charged under the $SU(2) \times U(1)$ gauge group. It is convenient to write the fermions as triplets under the global $SU(3)$, i.e. $Q = (q, 0)$ and $U^c = (0, 0, u^c)$. They are coupled to the little Higgs, $\Sigma = \exp(i(\sigma_1 \pm \sigma_2))$, through operators such as:

$$yfQ\Sigma U^c \sim yq(h_1 \pm ih_2)u^c + \cdots . \quad (6)$$

This operator induces a one loop quadratically divergent contribution to the little Higgs mass. For everything but the top quark, the quadratic divergence is negligeable. For the top quark, the one loop quadratic divergence must be dealt with. By adding an additional massive vector-like fermion, $\chi$ and $\chi^c$, we can promote the quark doublet, $q$, to quark triplet under the global $SU(3)$, $Q = (q, \chi)$. By preserving the global symmetry of the coupling, we remove the one loop quadratic divergence to the top quark. The quadratically divergent part of the one loop Coleman-Weinberg effective potential is:

$$V_{\text{eff}} = -\frac{y^2\Lambda^2 f^2}{16\pi^2} \operatorname{Tr} P_R \Sigma P_L \Sigma^\dagger \quad (7)$$

where $P_R = (0, 0, 1)$ from the $U^c$ coupling and $P_L = (1, 1, 0)$ if without the vector-like fermion $\chi$ and becomes the identity matrix with with $\chi$. When $P_L$ is the identity matrix, we preserve the $SU(3)$ symmetry and the effective potential is independent of $\Sigma$.

One loop log divergences from the top sector drives one linear combination of the Higgs to acquire a mass and the orthogonal combination are stabilized by one loop log divergences from the gauge and scalar sectors [3,7]. This leads to phenomenologically acceptable EWSB driven radiatively much as in the MSSM. We refer the reader to [7] for more details.

## 4. Conclusions and Outlook

We have presented new models of EWSB that stabilize the weak scale to radiative corrections and have weakly coupled physics at 1 TeV. At the 100 GeV scale it is a 2HDM model with possibly additional scalar singlets and $SU(2)$ triplets. At 1 TeV there are additional vector bosons that cancel the SMs guage $\Lambda^2$ divergences, additional scalars that cancel the Higgs quartic interaction $\Lambda^2$ divergence and additional coloured fermions that cancel the top's $\Lambda^2$ divergence. This description is valid up to 10 – 30 TeV where the nl$\sigma$m becomes strongly coupled and new physics emerges.

There is a tremendous amount of physics that is still to be done: precision studies [9], experimental signature [3], exploring other little Higgs models [7,8], and UV completing the theory above 10 – 30 TeV.

## REFERENCES

1. N. Arkani-Hamed, A. G. Cohen and H. Georgi, Phys. Lett. B **513**, 232 (2001)
2. D. B. Kaplan and H. Georgi, Phys. Lett. B **136**, 183 (1984).
3. N. Arkani-Hamed, A. G. Cohen, T. Gregoire and J. G. Wacker, arXiv:hep-ph/0202089.
4. N. Arkani-Hamed, A. G. Cohen, E. Katz, A. E. Nelson, T. Gregoire and J. G. Wacker, arXiv:hep-ph/0206020.
5. N. Arkani-Hamed, A. G. Cohen and H. Georgi, JHEP **0207**, 020 (2002)
6. T. Gregoire and J. G. Wacker, arXiv:hep-ph/0206023.
7. N. Arkani-Hamed, A. G. Cohen, E. Katz and A. E. Nelson, JHEP **0207**, 034 (2002)
8. I. Low, W. Skiba and D. Smith, arXiv:hep-ph/0207243.
9. R. S. Chivukula, N. Evans and E. H. Simmons, arXiv:hep-ph/0204193.

ELSEVIER

Proceedings of ICHEP 2002, pp. 731–734
*S. Bentvelsen, P. de Jong, J. Koch and E. Laenen (Editors)*

31st INTERNATIONAL CONFERENCE ON HIGH ENERGY PHYSICS AMSTERDAM

# Deconstructing Dimensional Deconstruction

Kenneth Lane
Department of Physics
Boston University
590 Commonwealth Avenue
Boston, MA 02215, USA*

Dimensional deconstruction (DD) abstracts from higher dimensional models features of related 4–dimensional ones. DD was proposed in Refs. [1–3] as a scheme for constructing models of *naturally* light composite Higgs boson. These are models in which—*without fine–tuning of parameters*—the composite Higgs's mass $M$ and vacuum expectation value $v$ are much lighter than its binding energy scale $\Lambda$. We review the basic idea of DD. It is easy to arrange $M \ll \Lambda$. We show, however, that DD fails to give $v \ll \Lambda$ in a model that is supposed to contain a naturally light composite Higgs [4].

## 1. WHAT IS DIMENSIONAL DECONSTRUCTION?

There has been considerable interest lately in a new approach to model–building called "dimensional deconstruction" (DD). In the beginning, there were two views of DD. The one we discuss in this paper is due to Arkani-Hamed, Cohen and Georgi (ACG) [1,2]. It is based on the fact that certain renormalizable, asymptotically free 4d field theories look, *for a limited range of energies*, like $d > 4$–dimensional theories in which the extra dimensions are compactified and discretized (on a periodic lattice). Here, the extra dimensions are a mirage. The other view is that of Hill and his collaborators [3] who assume the extra dimensions are real. They discretize the extra dimensions too—to regulate the theory. Both Arkani-Hamed *et al.* and Hill *et al.* use features of the higher dimensional model to deduce the form, magnitude, and sensitivity to high–scale ($\Lambda$) physics of phenomenologically important operators such as mass terms (generically, $M$), self–interactions ($\lambda$), and vevs ($v$) of light *composite* Higgs bosons (LCH) [5]. Their LCH models aim for $M \simeq v \simeq$ 100–200 GeV and $\Lambda \simeq 10\,\mathrm{TeV}$, relevant to electroweak symmetry breaking.

The simplest DD example is the $d = 5$ "moose ring" model [1] depicted in Fig. 1. This shows the full content ("UV–completed") of the model at the high energy scale $\Lambda$. It contains $N$ strong $SU(n)$ and weak $SU(m)$ (coupling $g^2/4\pi \ll 1$) gauge groups, with matter fields that are the massless chiral fermions

$$\psi_{Lk} \in (n, m, 1)\,,\ \ \psi_{Rk} \in (n, 1, m) \quad \text{of } (SU(n)_k,\, SU(m)_k,\, SU(m)_{k+1})\,. \quad (1)$$

The index $k$ is periodically identified with $k + N$. As $g \to 0$, these fermions have a large chiral symmetry, $[SU(m)_L \otimes SU(m)_R]^N$. At $\Lambda$, the strong $SU(n)$ interactions cause them to condense, creating $N$ sets of $m^2 - 1$ *composite* Goldstone bosons (GBs), $\pi_k^a$ with $k = 1, \dots, N$ and $a = 1, \dots, m^2 - 1$. Their decay constant $f \simeq \Lambda/4\pi$.

Below $\Lambda$, this is a nonlinear sigma model, with fields $U_k = \exp(i\pi_k^a t_a/f) \equiv \exp(i\pi_k/f)$ interacting with the weakly–coupled $SU(m)_k$ gauge fields $A_{k\mu} = A_{k\mu}^a t_a$. They transform as $U_k \to W_k U_k W_{k+1}^\dagger$ with $W_k \in SU(m)_k$. This low energy theory is represented by the "condensed moose" obtained from Fig. 1 by erasing the $SU(n)$ squares and linking the $SU(m)_k$ and $SU(m)_{k+1}$ circles by $U_k$.

Now, $N-1$ gauge boson multiplets eat $N-1$ sets of GBs and acquire the masses $\mathcal{M}_k = 2gf\sin(k\pi/N)$ for $k = 1, \dots, N$. The massless

*Email: lane@bu.edu


doi:10.1016/S0920-5632(02)02282-X

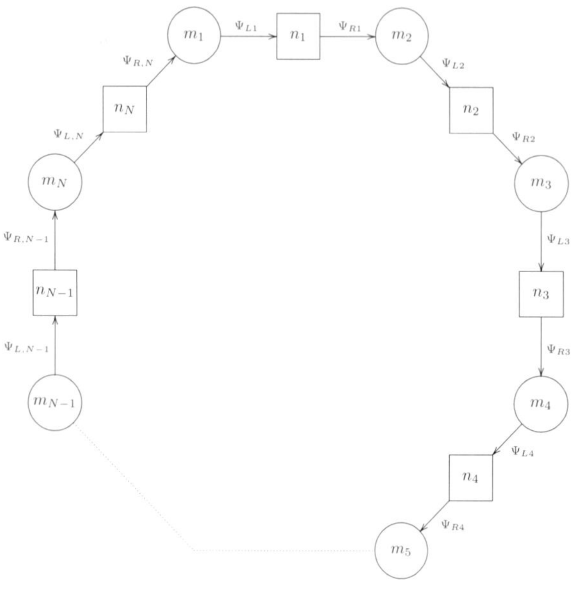


Figure 1. The full moose for the ring model of Ref. [1], showing its UV completion. Strong gauge groups are labeled by $n_1, \ldots, n_N$ and weak gauge groups by $m_1, \ldots, m_N$.

gauge field $A^a_\mu = (A^a_{1\mu} + \cdots + A^a_{N\mu})/\sqrt{N}$ couples with strength $g/\sqrt{N}$ and the uneaten GB is $\pi^a = (\pi^a_1 + \cdots + \pi^a_N)/\sqrt{N}$. In the unitary gauge, the 4d theory below $\Lambda$ is described by uniform link variables $U_k = \exp(i\pi^a t_a/\sqrt{N}f)$ plus the massless and massive gauge fields.

Alternatively, at energies well below $gf$, this looks like a 5d gauge theory: The fifth dimension is compactified on a discretized circle, represented exactly by the *condensed* moose. For $k \ll N$, there is a Kaluza–Klein tower of gauge excitations with masses $\mathcal{M}_k = 2\pi gfk/N$ [1]. The circumference of the circle is $R = Na$ where the lattice spacing $a = 1/gf$ and the 5–dimensional gauge coupling is $g_5^2 = g^2 a$. The fifth component of the gauge boson $A^a_5 = g\pi^a/\sqrt{N}$. The geometrical connection is clear: $\pi^a$ is the zero mode associated with rotation about the circle of $SU(m)$ groups in four dimensions and it corresponds to the fifth–dimensional gauge freedom associated with $A^a_5$. At higher energies, $\sim f$ or $\Lambda$, the fifth dimension is deconstructed as the underlying asymptotically free 4d theory appears.

## 2. WHAT IS DD GOOD FOR?

But, $\pi^a$ is really a 4d pseudoGoldstone boson (PGB) whose symmetry is explicitly broken by the weak $SU(m)_k$ interactions. So it might be a candidate for the LCH of electroweak symmetry breaking. To be a truly *natural* LCH, its vev $v \ll \Lambda$ also. This requires its quartic couplings $\lambda \simeq M^2/v^2 = \mathcal{O}(1)$ or, at least, not $\ll 1$. The idea of DD is that the magnitude and $\Lambda$–dependence of $M^2$ and $\lambda$ can be deduced from the higher dimensional theory. Let's see.

Higher dimensional gauge invariance allows mass for $A_5$ from $|\mathcal{W}|^2$, where $\mathcal{W} = P\exp(i\int dx_5 A_5)$ is the nontrivial Wilson loop around the fifth dimension [2]. Since $|\mathcal{W}|^2$ is a nonlocal operator, it cannot be generated with a UV–divergent coefficient. On the discretized circle, $\mathcal{W} = \mathrm{Tr}[\Pi_{k=1}^N \exp(iaA_{5k})]$. In the 4d theory this is just the gauge–invariant $\mathrm{Tr}(U_1U_2\cdots U_N)$, and so this is what provides the mass for $\pi^a$. Standard power counting indicates that the strength of $|\mathrm{Tr}(U_1U_2\cdots U_N)|^2$ is $\Lambda^2 f^2(g^2/16\pi^2)^N$. This is correct only for $N = 1$. For $N \geq 2$ infrared singularities from the gauge boson masses at $g \to 0$ overcome this power counting. For $N = 2$, $M^2 \sim g^4 f^2 \log(\Lambda^2/\mathcal{M}_B^2) \sim g^4 f^2 \log(N^2/g^2)$ where $\mathcal{M}_B^2 \sim g^2 f^2/N^2$ is a typical $SU(m)$ gauge boson mass. For $N \geq 3$, $M^2 \sim g^4 f^2$. Thus, for $N \geq 2$ and $g^2/4\pi \sim 10^{-2}$, we have $M \ll \Lambda$, as desired.

DD predicts that $\pi^a$ will fail as an LCH because the quartic interactions of $A_5$ are derivatively coupled and/or induced by weak $SU(m)$ interactions. This is true for $\pi^a$ as well. Since $p/f \sim M/f \sim g^2$, all quartic couplings of $\pi^a$ are $\leq \mathcal{O}(g^4)$. So, in this model, DD is a reliable guide. To achieve larger $\lambda$, ACG applied DD to a 6d model with nonderivative PGB interactions [2].

## 3. THE 6*d* TOROIDAL MOOSE MODEL.

Consider a 4d theory described below its UV–completion scale $\Lambda$ by the condensed moose in

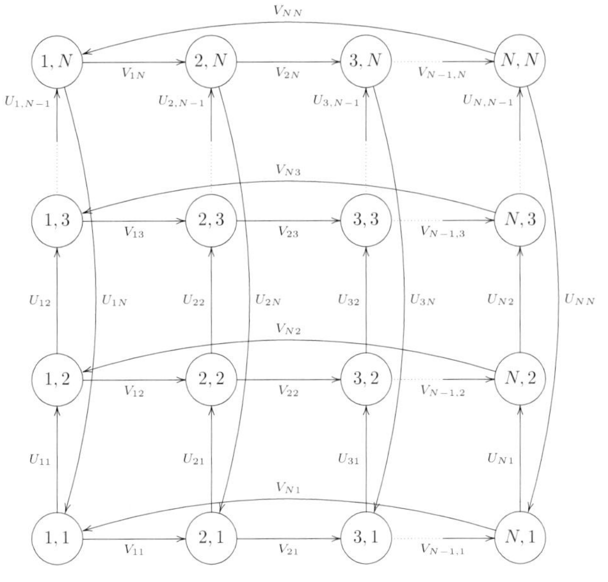

Figure 2. The condensed moose for the 6d toroidal model of Ref. [2].

Fig. 2. This resembles a discretized torus with $N \times N$ sites labeled periodically by integers $(k, l)$. Weakly–coupled $(g)$ gauge groups $SU(m)_{kl}$ at the sites are linked by nonlinear sigma model fields $U_{kl}$ and $V_{kl}$, transforming as

$$\begin{aligned} U_{kl} &= \exp\left(i\pi_{u,kl}/f\right) \to W_{kl}\, U_{kl}\, W^{\dagger}_{k,l+1}\,, \\ V_{kl} &= \exp\left(i\pi_{v,kl}/f\right) \to W_{kl}\, V_{kl}\, W^{\dagger}_{k+1,l}\,. \end{aligned} \quad (2)$$

The $\pi_{u,kl}$ and $\pi_{v,kl}$ comprise $2N^2$ $SU(m)$ adjoints of composite Goldstone bosons.

The $SU(m)_{kl}$ gauge bosons eat $N^2 - 1$ sets of GBs. The spectrum of massive gauge bosons, $\mathcal{M}^2_{kl} = 4g^2 f^2[\sin^2(k\pi/N) + \sin^2(l\pi/N)]$ is KK–like for small $k, l$. The massless gauge boson is $B^{\mu}_{NN} = N^{-1}\sum_{k,l} A^{\mu}_{kl}$ and its coupling is $g/N$. Among the $N^2 + 1$ leftover PBS, two that ACG proposed as light composite Higgses are

$$\pi_u = \frac{1}{N}\sum_{k,l} \pi_{u,kl}\,, \qquad \pi_v = \frac{1}{N}\sum_{k,l} \pi_{v,kl}\,. \quad (3)$$

These are the zero modes associated with going around the torus in the $U$ and $V$–directions.

What does DD predict for the masses and couplings of $\pi_{u,v}$? Viewing the condensed moose as the compactified and discretized dimensions 5,6 of a 6d gauge theory, the extra–dimensional gauge fields are $A^a_{5,6} = g\pi^a_{u,v}/N$. As before, DD predicts small $M^2_{\pi_{u,v}} \sim g^4 f^2 \log(N^2/g^2)$ for $N = 2$ and $g^4 f^2$ for $N \geq 3$.

In the 6d model, $A_{5,6}$ have moderately strong nonderivative interactions [2]. They come from the term $\mathrm{Tr}F^2_{56} = \mathrm{Tr}([A_5, A_6]^2) + \cdots = \lambda \mathrm{Tr}([\pi_u, \pi_v]^2) + \cdots$ which, in turn, arises from the "plaquette" Hamiltonian

$$\mathcal{H}_{\mathrm{P}} = \sum_{k,l} \lambda_{kl}\, f^4\, \mathrm{Tr}\left(U_{kl} V_{k,l+1} U^{\dagger}_{k+1,l} V^{\dagger}_{kl}\right) + \mathrm{h.c.} \quad (4)$$

Note that $\mathcal{H}_{\mathrm{P}}$ leaves $\pi_{u,v}$ massless.

In 6d, the quartic coupling may be shown to be $\lambda \equiv \frac{1}{2}\sum_{k,l}\lambda_{kl}/N^4 = g^2/2N^2$ [4]. Depending on the $N$–dependence of the Higgs masses, this may be large enough to give a Higgs vev comparable to $M_{\pi_{u,v}}$. In 4d, this prediction of DD fails. The strength of $\lambda$ depends entirely on the nature of the toroidal moose model's UV completion.

The most natural UV completion of this model is the analog of Fig. 1: At $\Lambda$, there are $2N^2$ massless fermions $\psi$ with strong $SU(n)$ interactions located midway between the weak $SU(m)$'s [4]. Then, the plaquette interaction arises only from weak gauge interactions. It is of $\mathcal{O}(g^4)$ and, so, $v^2 \sim M^2_{\pi_{u,v}}/\lambda \gg M^2_{\pi_{u,v}}$.

It is possible to find UV completions of the toroidal moose that yield larger $\lambda$. They involve elementary scalars and, therefore, supersymmetry to avoid unnatural fine–tuning of parameters [4,6]. More generally, one can construct sigma models whose symmetries are tailored to give an effective Lagrangian with *arbitrarily and separately tunable* $M^2$ and $\lambda$—at least at the one–loop level. This is the basis of an interesting new direction that has evolved from DD [7,8]. But this approach, called "little Higgs", has nothing to do with the original idea of deconstruction—that the strengths of a composite Higgs' mass and interactions may be deduced from corresponding terms in higher dimensional gauge theories. Finding a truly dynamical, natural way of UV–completing little Higgs models remains one of the greatest challenges to this new idea for electroweak symmetry breaking.

## REFERENCES


1. N. Arkani-Hamed, A. G. Cohen and H. Georgi, Phys. Rev. Lett. **86**, 4757 (2001); hep-ph/0104005.
2. N. Arkani-Hamed, A. G. Cohen and H. Georgi, Phys. Lett. **B513**, 232 (2001) and hep-ph/0105239 v4; N. Arkani-Hamed, *et al.*, JHEP 0208:020 (2002); hep-ph/0202089.
3. C. T. Hill, S. Pokorski and J. Wang, Phys. Rev. **D64**, 105005 (2001); hep-th/0104035; H. C. Cheng, C. T. Hill, S. Pokorski and J. Wang, Phys. Rev. **D64**, 065007 (2001); hep-th/0104179; H. C. Cheng, C. T. Hill and J. Wang, Phys. Rev. **D64**, 095003 (2001); hep-ph/0105323.
4. This talk is largely based on my paper, K. Lane, *A Case Study in Dimensional Deconstruction*, Phys. Rev. **D65**, 115001 (2002); hep-ph/0202093.
5. D. B. Kaplan and H. Georgi, Phys. Lett. **B136**, 183 (1984); S. Dimopoulos et al., Phys. Lett. **B136**, 187 (1984).
6. A. Cohen, H. Georgi, communications.
7. N. Arkani-Hamed, *et al.*, JHEP 0208:021 (2002), hep-ph/0206020 and JHEP 0207:034 (2002), hep-ph/0206021.
8. M. Schmaltz, *Beyond the Standard Model (theory)*, invited talk at the 31st International Conference on High Energy Physics, Amsterdam, The Netherlands, July 24-31, 2002.

ELSEVIER

Proceedings of ICHEP 2002, pp. 735–738
*S. Bentvelsen, P. de Jong, J. Koch and E. Laenen (Editors)*

31st INTERNATIONAL CONFERENCE ON HIGH ENERGY PHYSICS AMSTERDAM

# LEP searches for Higgs bosons in models beyond the Standard Model

Pierre Lutz[a*]

[a]Service de Physique des Particules, DSM/DAPNIA, CEA Saclay,
91191 GIF sur YVETTE CEDEX, France
e-mail : Pierre.Lutz@cern.ch

Various models for Higgs bosons beyond the Standard Model (SM) have been examined, most of them within the framework of general Two Higgs Doublet Models (2HDM). The large amount of LEP2 data (about 3000 $pb^{-1}$) allows precise analyses. Results are consistent with the expected SM background, no evidence for any signal has been found, and stringent limits on Higgs boson masses are set. Moreover, some "model independent" limits, eg. limits on cross-sections and/or branching fractions become available. On top of the classical models (MSSM, charged Higgs, invisible Higgs), some very interesting new results have been presented at this conference (a scenario violating CP in the MSSM, some bosonic decays of the charged Higgs, studies of a doubly-charged Higgs). Nevertheless, all combinations of the results of the 4 experiments are still to come.

## 1. Introduction

Two Higgs Doublet Models (2HDM) are the simplest extensions of the SM, with two complex scalar fields doublets, which give rise to 5 physical Higgs states : two neutral scalars h and H arising from the mixing (with an angle $\alpha$) of the real parts of the fields, one pseudo-scalar A which is the remaining of the imaginary part, and two charged scalars $H^+$ and $H^-$. At tree level, these models are described by 6 parameters (4 masses, $\alpha$ and the ratio of the vevs $\tan\beta = v2/v1$) and the decay channels are 2 by 2 complementary (hZ with hA, and HZ with HA) since

$$\sigma_{hZ} \propto \sin^2(\alpha-\beta) \quad \sigma_{hA} \propto \cos^2(\alpha-\beta)$$
$$\sigma_{HZ} \propto \cos^2(\alpha-\beta) \quad \sigma_{HA} \propto \sin^2(\alpha-\beta)$$

2HDM come in two types : in type I, one doublet couples to fermions, the other to bosons, while in type II, one doublet couples to "up" fermions and the other to "down" fermions. MSSM is a particular case of a type II 2HDM.

Most of the contributions are within these general models, and the next sections will describe them (MSSM, Charged Higgs and Invisible Higgs searches). Doubly-charged Higgs belong to other models since a field triplet is needed : section 5 is dedicated to their searches.

*on leave of absence from CNRS-IN2P3.

## 2. MSSM

Up to now, all searches were classically done for three benchmark scans, following the recommendations from [1]. The "no- or minimal-mixing" scenario has no mixing in the stop sector, the "maximum $m_h$" scenario was designed to yield the maximal value of the lightest neutral Higgs boson mass (around 130 GeV), while the "large $\mu$" scenario produces suppressed Higgs bosons decays to $b\bar{b}$.

For these scenarios, a preliminary combination of the 4 experiments has been produced in 2001 [2], Aleph [3] presented their final results for this conference, Delphi [4] and L3 [5] updated their first results and Opal [6] kept its 2001 preliminary results. A new combination is expected for the coming months.

Table 2 summarizes the situation in the maximum $m_h$ scenario.

In the no-mixing scenario, the max. $m_h$ is around 118 GeV. The previous combination (see Fig. 1) has two unexcluded regions. A first one, at high $\tan\beta$ corresponds to $m_h \sim m_A$ beyond the hA kinematic reach : it would have probably been covered by a mild upgrade of LEP energy. The second one, at low $\tan\beta$ and $m_{H+} \sim 74$ GeV may probably be covered in the next LEP com-


doi:10.1016/S0920-5632(02)02285-5

Table 1
Limits in max. $m_h$ scenario ; observed and (expected) limits in GeV/$c^2$

| | $m_h$ | $m_A$ | $\tan\beta$ excl. |
|---|---|---|---|
| Aleph | 89.8 (91.3) | 90.1 (91.6) | 0.7 - 2.3 |
| Delphi | 89.1 (89.4) | 90.0 (90.2) | 0.54 - 2.36 |
| L3 | 86.0 (88.4) | 86.5 (88.6) | 0.55 - 2.2 |
| Opal | 79.3 (85.1) | 80.6 (86.9) | 0.9 - 1.7 |
| LEP | 91.0 (94.6) | 91.9 (95.0) | 0.5 - 2.4 |

bination, thanks to dedicated $H^+ \to WA$ analyses (see section 3.2) and/or flavor independent searches.

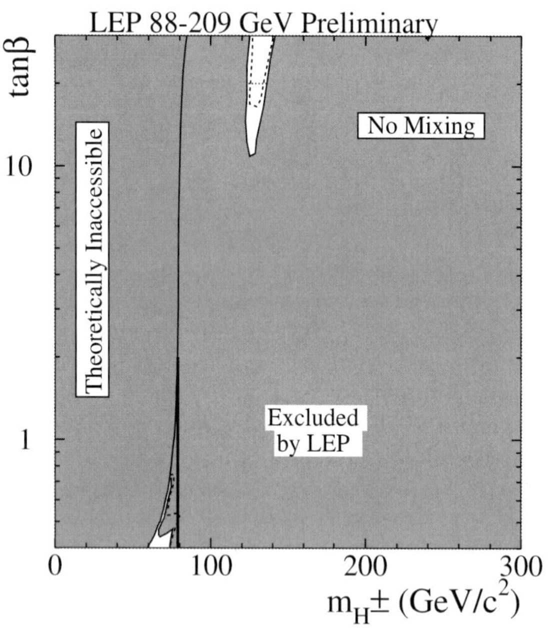


Figure 1. Excluded regions in the no-mixing scenario

With the help of flavor-independent searches, the large $\mu$ scenario was already excluded in 2001.

### 2.1. A CP violating scenario

In general, the SUSY breaking trilinear couplings have phases, inducing CP violation. Some new features appear : the Higgs mass eigenstates ($H_1$, $H_2$ and $H_3$) are no more CP eigenstates and the search channels are $H_1Z$, $H_2Z$ and $H_1H_2$ (two Higgsstrahlung processes have to be examined simultaneously). Furthermore, while it is still true that the lightest neutral Higgs boson ($H_1$) has a mass bounded to 130 GeV and decays mainly into $b\bar{b}$ (and $\tau\tau$), it might have escaped detection at LEP2 due to too low a coupling to the Z, and the $H_2$ boson decays dominantly into $H_1H_1$ as soon as kinematically allowed, which is the case for $\tan\beta > 10$ where the mass difference is large, needing dedicated analyses.

A benchmark (named CPX) has been designed [7], taking care of Electric Dipole Moment constraints, with parameters maximizing CP violation effects. Opal [8] presented the first analysis for this scenario at this conference, reinterpreting searches for SM or CP conserving MSSM and adding dedicated analyses for most of the channels not yet covered. Fig. 2 shows clearly that the most difficult region is for $\tan\beta$ between 4 and 10. It should be noted that the excluded tie in that region completely disappears if $m_{top}$ is taken at 179.3 GeV.

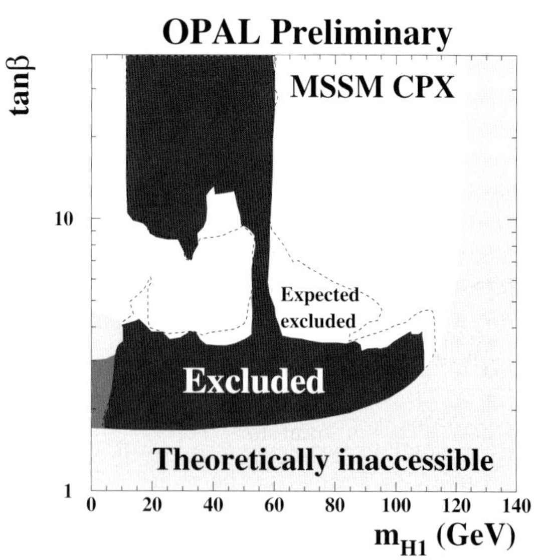


Figure 2. Excluded regions in the CP violating scenario

## 3. Charged Higgs boson

### 3.1. fermionic decays

BR($H^+ \to \tau^+\nu$)+BR($H^+ \to c\bar{s}$) = 1 was up to now the working assumption. The LEP combined limit (from summer 2001) [9] was $m_{H^+} > 78.6$ GeV for any BR, far from the kinematic limit due to the irreducible loss of sensitivity near $m_W$. Since then, Aleph [10], Delphi [11] and Opal [12] produced not only new limits (resp. 79.3, 74.3 and 75.5 GeV), but also, in a more "model-independent" way, limits in cross-sections : see Fig.3 as an example.

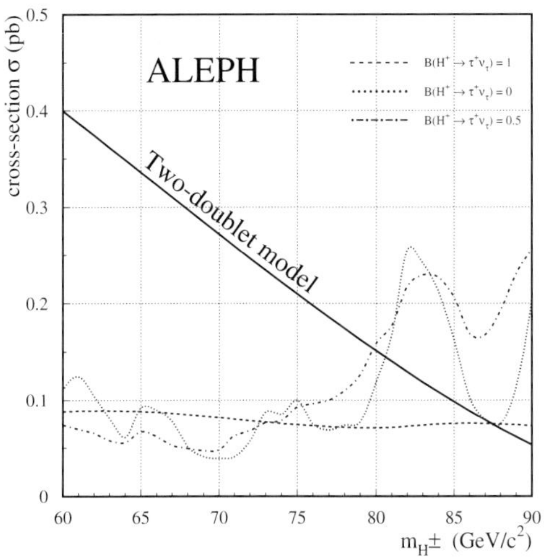


Figure 3. 95% CL limits in excluded cross-sections.

### 3.2. bosonic decays

In 2HDM type II, BR($H^+ \to AW^{\star +}$) becomes important at low $\tan\beta$. In 2HDM type I, this decay mode is dominant for $\tan\beta > 1$. OPAL [6] and DELPHI [11] search for the possible final states in that channel, assuming the A to decay into $b\bar{b}$, and thus for $m_A > 12$ GeV. Fig.4 is an example of what OPAL reaches when combining the fermionic searches with a simple counting analysis, while DELPHI, with a more complete analysis, including a mass reconstruction, is able to exclude a charged Higgs with masses up to 76.7 GeV for any $\tan\beta$ and $m_A$ in the range 12 to 70 GeV.

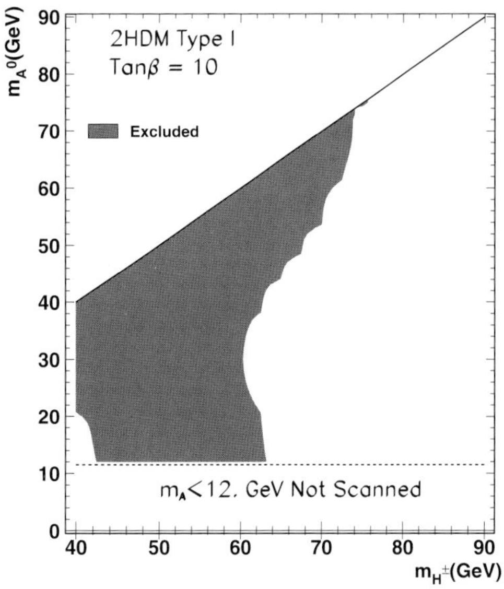


Figure 4. Example of exclusion domain obtained by OPAL in Type I 2HDM

## 4. Higgs boson decaying invisibly

In SUSY models, the Higgs boson might decay into neutralinos. Searching for acoplanar jets or leptons, compatible with a Z boson, covers this scenario, and all 4 experiments did such a search. The preliminary (summer 2001) LEP combination excludes this hypothesis up to 114.2 GeV for BR($h \to$ *invisible*) = 100%. DELPHI [13] updated their result at this conference, excluding such a Higgs boson up to 112.0 GeV for any BR($h \to$ *invisible*) when combining with the SM Higgs boson channels (see Fig.5).

## 5. Doubly charged Higgs boson

Doubly charged Higgs bosons appear naturally in left-right symmetric models in which the gauge symmetry is broken by triplet Higgs fields. Both pair- and single-production are possible at LEP, through their coupling to charged leptons. Furthermore, they may induce modifications on Bhabha scattering where a doubly charged Higgs would contribute via $t$-channel exchange. Indirect constraints (from Bhabha, but also from FCNC

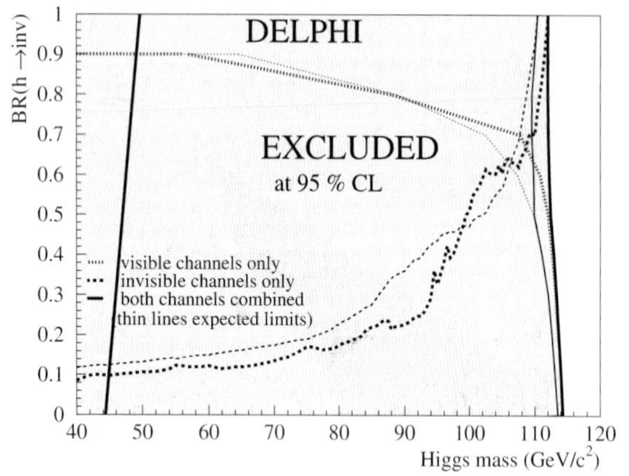


Figure 5. Exclusion domain for "invisible" Higgs search.

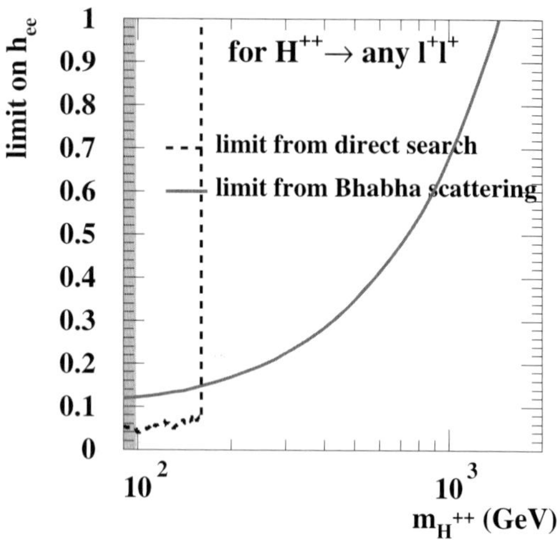


Figure 6. Limits on the Yukawa coupling $h_{ee}$: the shaded region is excluded by the OPAL pair production search.

and $\mu$ decay) make the $\tau\tau$ decay particularly interesting, and if the Yukawa coupling $h_{\tau\tau} > 10^{-7}$, the decay occurs at interaction point. Both OPAL [14] and DELPHI [15] presented results for pair-produced doubly charged Higgs bosons, OPAL for all possible decays but assuming decay at interaction point, and DELPHI only for $\tau\tau$ decays but for all lifetimes. Furthermore, triggered by an excess of events seen recently by H1, that can be explained by such an object, OPAL [16] looked also for single-production and indirect constraints from Bhabha scattering.

Concerning the analysis of pair-produced $H^{++}$, OPAL (DELPHI) set limits for $h_{\tau\tau} > 10^{-7}$ at $m_{H_L^{++}} > 99.0$ (99.6) GeV and $m_{H_R^{++}} > 98.5$ (99.1) GeV. Additionally, DELPHI sets a limit at $m_{H^{++}} > 97.3$ GeV for any lifetime. For the singly-produced $H^{++}$, OPAL does not observe any evidence for the existence of $H^{++}$, and sets an upper limit on the Yukawa coupling $h_{ee}$ at 0.08 for $m_{H^{++}} <160$ GeV. Furthermore, indirect constraints from Bhabha scattering are set for $m_{H^{++}} <2$ TeV. Fig.6 summarizes their results.

## REFERENCES

1. Carena et al., hep-ph/9912223.
2. LEP Higgs Working Group, LHWG note 2001-04 or hep-ex/0107030.
3. ALEPH Collaboration, CERN-EP/2001-095 and ICHEP02 251.
4. DELPHI Collaboration, Delphi note 2002-041-CONF-575 and ICHEP02 189.
5. L3 Collaboration, L3 note 2749 and ICHEP02 404.
6. OPAL Collaboration, OPAL Physics Note PN472 and ICHEP02 521.
7. Carena et al., hep-ph/0009212.
8. OPAL Collaboration, OPAL Physics Note PN505 and ICHEP02 397.
9. LEP Higgs Working Group, LHWG note 2001-05 or hep-ex/0107031.
10. ALEPH Collaboration, CERN-EP/2002-054 and ICHEP02 283.
11. DELPHI Collaboration, Delphi note 2002-038-CONF-572 and ICHEP02 178.
12. OPAL Collaboration, OPAL Physics Note PN509 and ICHEP02 393.
13. DELPHI Collaboration, Delphi note 2002-025-CONF-559 and ICHEP02 078.
14. OPAL Collaboration, CERN-EP/2001-082 and ICHEP02 266.
15. DELPHI Collaboration, Delphi note 2002-039-CONF-573 and ICHEP02 179
16. OPAL Collaboration, OPAL Physics Note PN502 and ICHEP02 1074.

ELSEVIER

Proceedings of ICHEP 2002, pp. 739–742
*S. Bentvelsen, P. de Jong, J. Koch and E. Laenen (Editors)*

31st INTERNATIONAL CONFERENCE ON HIGH ENERGY PHYSICS AMSTERDAM

# Flavour independent searches and anomalous production and decay of Higgs bosons at LEP

J. Alcaraz [a] [b], on behalf of the LEP Collaborations

[a]CIEMAT, Research Centre for Energy, Environment and Technology, Avda. Complutense 22, 28040-Madrid, Spain

[b]CERN, European Laboratory for Particle Physics, 1211, Genève 23, Switzerland

We report on extended searches for neutral Higgs bosons in $e^+e^-$ collisions at LEP. Scenarios with two Higgs doublets, unusual decays, as well as model-independent deviations from the Standard Model have been investigated by the LEP Collaborations. No evidence for signals or deviations due to the presence of a light scalar boson have been observed so far. Limits in the context of the different hypotheses are presented.

## 1. Introduction

The search for the lightest Higgs particle is one of the main objectives of the Large Electron-Positron Collider (LEP) programme. In the framework of the Standard Model (SM), a LEP-combined lower mass limit of $m_H > 114.4$ GeV has been obtained [1]. The experimental indications of the existence of a scalar sector being so weak, it is important to look for any behaviour beyond the SM leading to visible signals in LEP data. The LEP Collaborations have carried out searches for neutral scalars using different and interesting approaches. The results of these studies are the subject of this report.

## 2. Two Higgs Doublet Models (2HDM)

The production and decay properties of the lightest Higgs boson are modified in theories containing more than one Higgs doublet. A typical and relevant example is the case of supersymmetric theories [2]. The phenomenology of neutral Higgs bosons in 2HDM scenarios depends strongly on the values of Higgs masses, on the parameter $\alpha$ (the rotation angle required to go from the original Higgs doublets $\Phi_1, \Phi_2$ to the two neutral scalar eigenstates) and on $\tan\beta = <\Phi_2^0> / <\Phi_1^0>$ (the ratio of vacuum expectation values).

Dedicated searches in some pathological regions of the 2HDM phase space have reported at this Conference [3,4]. An interesting example is the case in which the lightest Higgs particle is the pseudoscalar A and the lightest scalar, h, has a dominant decay into $h \to AA$. In the MSSM [2], this scenario corresponds to relatively low values of $\tan\beta$ in a region of phase space still unexcluded in the last LEP combination [1] (2001, Figure 1). The OPAL collaboration has presented at this Conference a dedicated search for A masses below the threshold of $b\bar{b}$ production [3]. The signal is assumed to be: $e^+e^- \to hZ \to AAZ$, with the subsequent decay of A into $c\bar{c}$, gg or $\tau^+\tau^-$. Since the A is highly boosted, events tend to follow a two-jet plus Z topology. No signal is observed, leading to the excluded region 45 GeV $< m_h <$ 85 GeV, $2 < m_A < 9.5$ GeV in the framework of the MSSM. The other LEP experiments also employ dedicated AAZ searches in order to exclude this problematic region [5].

The DELPHI collaboration has adopted a general approach to search for Higgs particles in 2HDM scenarios [6], providing parameter-independent limits for all the possible production mechanisms: $e^+e^- \to hZ$ (Higgsstrahlung), $e^+e^- \to Ah$ (pair production), $e^+e^- \to f\bar{f}h(A)$ (Yukawa production), and for many different de-


doi:10.1016/S0920-5632(02)02287-9

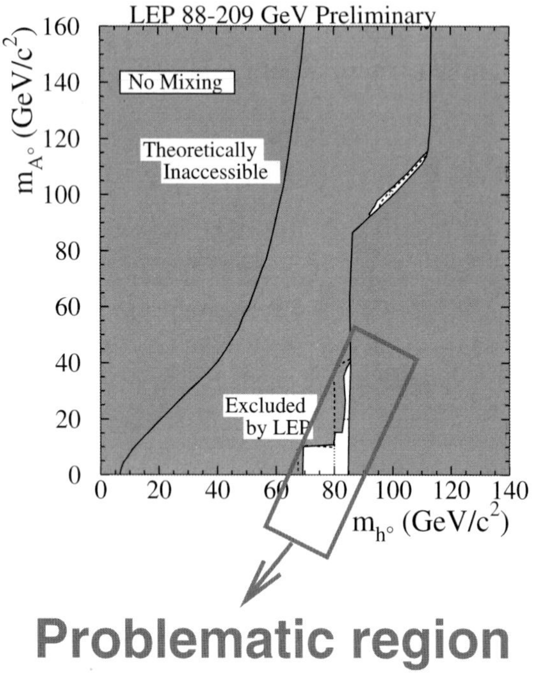


Figure 1. Unexcluded region in the $(m_H, m_A)$ plane from LEP combined MSSM searches [1] (2001).

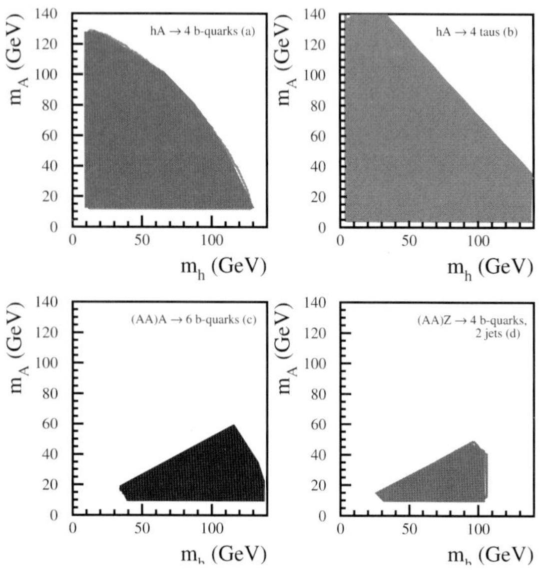


Figure 2. DELPHI excluded masses for different Higgs production mechanisms in the context of 2HDM models [6]. Decays into $b\bar{b}$ and $\tau^+\tau^-$ final states are considered. The excluded regions are given for the reference values described in [6]. These results are preliminary.

cay modes ($h \to f\bar{f}$, $h \to AA$, ...). Figure 2 shows the limits obtained for different production mechanisms with subsequent decays into $b\bar{b}$ and $\tau^+\tau^-$ final states. The excluded regions correspond to the reference values for the couplings described in [6].

### 3. Flavour-blind decay searches

Most of the LEP searches for the Higgs boson assume a dominant decay mode into bottom quarks. However, the $H \to b\bar{b}$ decay is not dominant in many extensions of the SM. One example is a 2HDM model with a relatively low $\tan\beta$ value and enhanced couplings to up-type quarks. Another example is a model with a large Higgs coupling to top quarks, leading to a larger branching decay fraction into gluons (involving loops with $Ht\bar{t}$ and $gt\bar{t}$ couplings). The signature in these scenarios would a dominant hadronic decay with undetermined flavour composition.

All four LEP Collaborations have searched for hadronic Higgs decays without specific flavour tagging. Under the assumption of a SM production cross section, every single experiment is typically sensitive to Higgs masses up to 110 GeV. The preliminary LEP combined limit is shown in Figure 3: $m_H > 113$ GeV [1]. A natural extension of this study is the inclusion of $\tau^+\tau^-$ decay channels. In this context, the ALEPH Collaboration excludes the existence of a Higgs boson with dominant hadronic or tauonic decays for masses smaller than 109.1 GeV [7].

### 4. Higgs deviations from the SM

The L3 Collaboration has followed an alternative strategy to look for deviations in the Higgs sector [8]. The study assumes that the Stan-

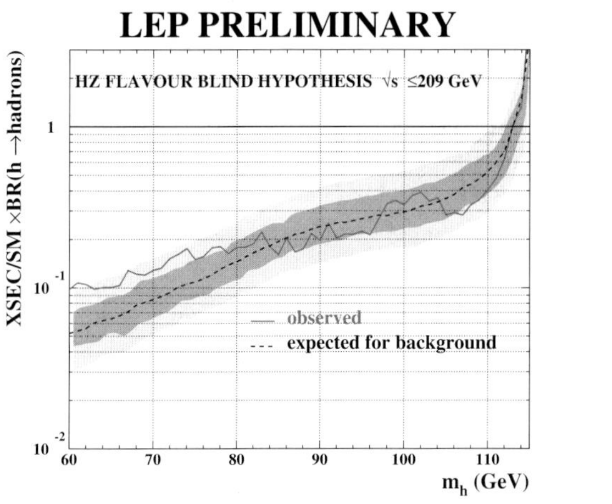


Figure 3. LEP limits as a function of the Higgs mass hypothesis for a Higgs boson with purely hadronic decays [1].

dard Model is a valid theory up to a scale of New Physics $\Lambda$, larger than the electroweak scale. Following an approach similar to the one used in the measurement of WWZ and WW$\gamma$ couplings, only the Lagrangians of lowest dimension (six) [9], not severely constrained by low energy experiments, are kept. The deviations from the SM are expressed in terms of five adimensional couplings: $\xi \equiv 1 + \delta$ [10], $\Delta g_1^Z$, $\Delta\kappa_\gamma$ [11], and $d$, $d_B$ [12]. These couplings modify in general the production and decay properties of the Higgs boson. One of the most striking effects would be an enhancement of the photonic couplings H$\gamma\gamma$ and HZ$\gamma$, giving access to a Higgs signal in the unexplored region $\sqrt{s} - m_Z < m_H < \sqrt{s}$ via the process $e^+e^- \to \{Z^*, \gamma^*\} \to H\gamma$.

The L3 Collaboration has set preliminary limits on the five anomalous couplings discussed above. Typical limits are: $| d | \lesssim 0.15$, $| d_B | \lesssim 0.05$, $| \Delta g_1^Z | \lesssim 0.4$ and $| \Delta\kappa_\gamma | \lesssim 0.5$ for Higgs masses up to 140 GeV. LEP limits [1] on the parameter $\xi$ are shown in Figure 4. Since large $d$ or $d_B$ couplings may lead to large photonic couplings of the Higgs, limits in the $(d, d_B)$ plane can be directly interpreted as upper limits in the $(\mathrm{Br}(H \to \gamma\gamma), \mathrm{Br}(H \to Z\gamma))$ plane. Figure 5 presents the experimentally excluded regions for Higgs masses in the range of interest for LHC searches.

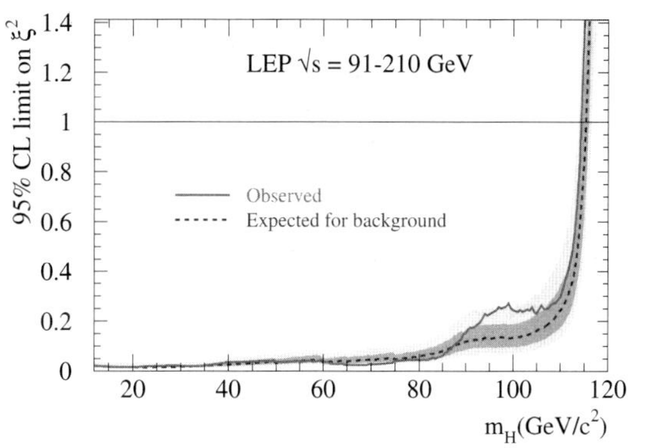


Figure 4. LEP limits on the anomalous coupling $\xi^2$ as a function of the Higgs mass hypothesis [1].

## 5. Relaxing Higgs decay assumptions

The OPAL Collaboration has carried out Higgs searches via the Higgsstrahlung process with minimal assumptions for the decay of the Higgs [14]. The basic idea is to perform an analysis in which the sensitivity is almost independent of the Higgs decay mode (hadronic, leptonic, photonic or invisible). Since the study must rely on a good identification of the Z decay products, $Z \to e^+e^-$ and $Z \to \mu^+\mu^-$ decays are considered. $Z \to \nu\bar{\nu}$ decays are also employed for the $H \to e^+e^-$ and $H \to \gamma\gamma$ modes in the case of a low mass Higgs hypothesis, due to the large backgrounds from final state radiation effects ($e^+e^- \to \{Z, \gamma^*\}^* \to f\bar{f}\gamma\gamma$ and $e^+e^- \to \{Z^*, \gamma^*\} \to f\bar{f}e^+e^-$). Under the assumption of a SM Higgsstrahlung production cross section, Higgs masses below 81 GeV are excluded.

A step beyond consists in relaxing the assumption that the Higgs width must be small compared to the experimental resolution (of the order

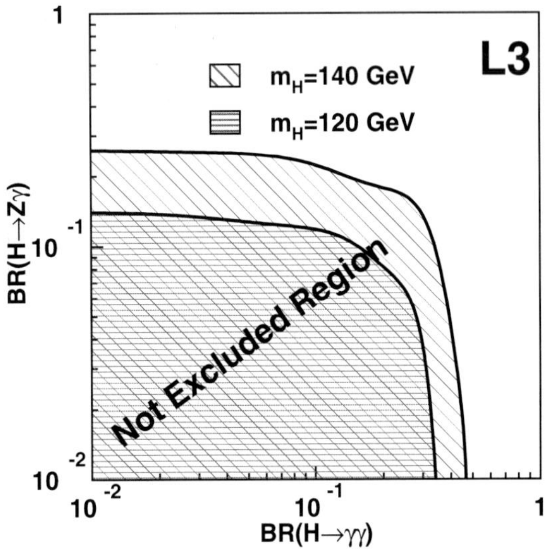


Figure 5. L3 limits on the branching fractions of a Higgs boson into final states containing photons [8]. Two mass hypotheses, $m_{\rm H} = 120, 140$ GeV, of interest for LHC searches, are shown.

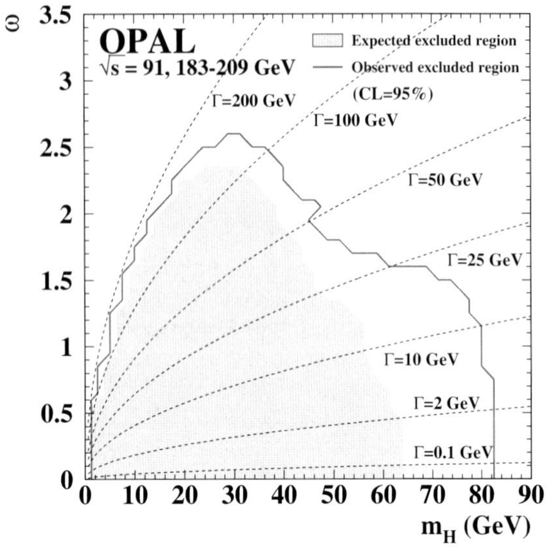


Figure 6. OPAL limits on the mass a wide Higgs boson in the context of the Stealthy Higgs Model [13].

of GeVs). A possible scenario of this type is the "Stealthy Higgs Model" [13], where many Higgs singlets, called "phions", couple significantly to the SM Higgs doublet. Since a Higgs singlet does not interact with matter, the relevant signature is a very wide scalar decaying predominantly to invisible particles (the phions). The OPAL Collaboration has searched for different scenarios with a "wide Higgs" [14]. In the context of the Stealthy Higgs Model with many phions, they obtain the limits shown in Figure 6 as a function of the additional Higgs width $\Delta\Gamma_{\rm H}$ ( $\Delta\Gamma_{\rm H} = \omega^2 v^2/(32\pi m_{\rm H})$, $v = 246$ GeV).

## REFERENCES


1. LEP Collaborations, http://cern.ch/lephiggs and references therein.
2. S. Dawson *et al.*, "Higgs Hunter's Guide", Addison Wesley, Menlo Park, 1989.
3. OPAL Collaboration, CERN-EP/2002-058.
4. OPAL Collaboration, Eur. Phys. J. C23 (2002) 397.
5. ALEPH Collaboration, Phys. Lett. B526 (2002) 191; DELPHI Collaboration, 2002-041-CONF-575 L3 Collaboration, CERN-EP-2002-066.
6. DELPHI Collaboration, 2002-035-CONF-569.
7. ALEPH Collaboration, CERN-EP-2002-02.
8. L3 Collaboration, Internal Note 2774.
9. W. Buchmüller and D. Wyler, Nucl. Phys. B268 (1986) 621; C.J.C. Burgess and H.J. Schnitzer, Nucl. Phys. B228 (1983) 424; C.N. Leung, S.T. Love and S. Rao, Z Phys. C31 (1986) 433.
10. B. Grządkowski and J. Wudka, Phys. Lett. B364 (1995) 49.
11. K. Hagiwara *et al.*, Nucl. Phys. B282 (1987) 253.
12. G.J. Gounaris, F.M. Renard and N.D. Vlachos, Nucl. Phys. B459 (1996) 51.
13. T. Binoth and J.J. van der Bij, Z. Phys. C75 (1997) 17; T. Binoth and J.J. van der Bij, hep-ph/9908256.
14. OPAL Collaboration, CERN-EP-2002-032.

ELSEVIER

Proceedings of ICHEP 2002, pp. 743–745
*S. Bentvelsen, P. de Jong, J. Koch and E. Laenen (Editors)*

31st INTERNATIONAL CONFERENCE ON HIGH ENERGY PHYSICS AMSTERDAM

# Fermiophobic Higgs Bosons at LEP

A. Rosca[a]

[a]DESY Zeuthen, Platanenallee 6, 15738 Zeuthen, Germany

This work describes the results of the searches for a Higgs boson decaying into gauge bosons carried out by the four LEP collaboration, ALEPH, DELPHI, L3 and OPAL. A lower bound of 109.7 GeV is set at 95% confidence level on the mass of a fermiophobic Higgs boson decaying into photons. This mass limit can be extended by considering the Higgs decay mode into weak bosons. Such a combination has been done by the L3 collaboration which extended the fermiophobic mass reach by 5 GeV.

## 1. Introduction

Photonic final states from the process $e^+e^- \to Zh \to Z\gamma\gamma$ do not occur in the Standard Model at the tree level, but may be present at a low rate due to charged weak boson and top quark loops. Very specific models can lead to an enhancement of the $h \to \gamma\gamma$ rate, such as the fermiophobic 2 Higgs doublet models of type I [1], where all fermions are assumed to couple to the same scalar field, and the couplings can thus be suppressed simultaneously by appropriate parameter choices.

Since there are several possible models which predict fermiophobic Higgs bosons, the LEP Higgs working group (LHWG) defined the so-called "benchmark fermiophobic model" where the production cross-section is the same as in the Standard Model, and the couplings of the Standard Model Higgs to fermions are removed, resulting in increased branching fractions into gauge bosons. A fermiophobic Higgs in the benchmark model decays predominantly via the process $h \to \gamma\gamma$ if its mass is below 90 GeV, while at higher masses it decays mostly via the processes $h \to WW^*, ZZ^*$.

## 2. Photonic Higgs decays

A fermiophobic Higgs can be produced in $e^+e^-$ collisions at the energies available at LEP by the Higgs-Strahlung mechanism in the s-channel, $e^+e^- \to Zh$, where the Higgs boson is radiated off an intermediate Z boson. For the case of photonic decays, this results in three categories of possible final states with different topologies, efficiencies and background rates:

1. the hadronic final state, involving two high energy photons and two jets from the hadronic decay of the Z;
2. the missing energy final state, involving two high energy acoplanar photons and two neutrinos;
3. the dilepton final state, involving two high energy photons and two high energy leptons.

For all the channels, the major background is the double radiative Z production.

The searches used data samples collected in $e^+e^-$ collisions at centre-of-mass energies between 88 and 209 GeV. The ALEPH [2], DELPHI

Table 1
Results of the $h \to \gamma\gamma$ search.

| Experiment | Data | Bkgd. | Mass limit (GeV) |
|---|---|---|---|
| ALEPH | 23 | 30.8 | 105.4 |
| DELPHI | 54 | 51.6 | 104.1 |
| L3 | 62 | 72.0 | 105.4 |
| OPAL | 124 | 135.2 | 105.5 |
| SUM | 263 | 289.6 | 109.7 |


doi:10.1016/S0920-5632(02)02289-2

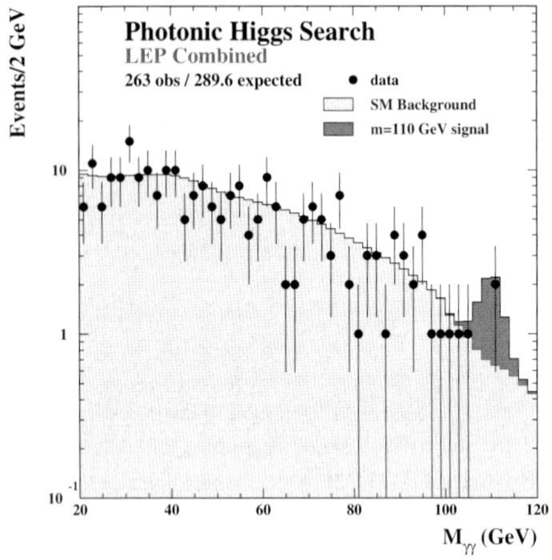


Figure 1. Distribution of the reconstructed di-photon invariant mass for all final states combined. The combined LEP data are shown together with the background and a Higgs boson signal with mass $m_{\mathrm{h}} = 110$ GeV.

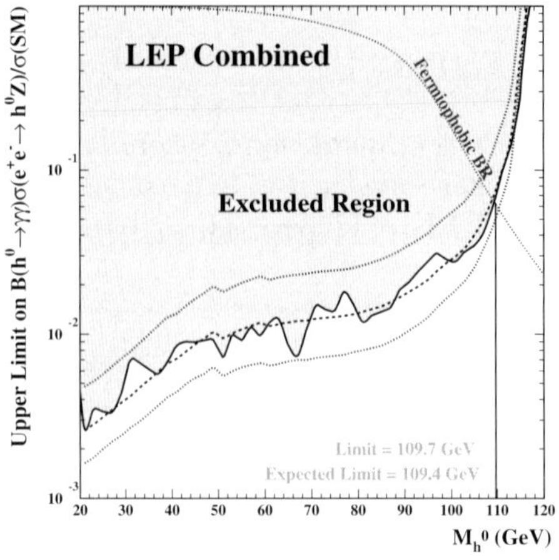


Figure 2. Excluded values at 95% confidence level of BR(h→ $\gamma\gamma$)×$\sigma$($e^+e^-$ → Zh)/$\sigma_{\mathrm{SM}}$ as a function of the Higgs mass, in the fermiophobic model. The expected (dashed line) 95% confidence limit and the theoretical prediction (dotted line) are also presented.

[3], L3 [4] and OPAL [5] analyses are described in journal articles or CERN preprints.

The results of the analyses are summarised in Table 1, where we list the number of observed events, the estimated background and the lower mass limit set at 95% confidence level for the benchmark fermiophobic Higgs boson.

The distribution of the di-photon invariant mass is shown in Figure 1 for the combined data from the four LEP experiments. No excess with respect to the Standard Model predictions is observed in the data. In the absence of a signal, the results are given in terms of upper limits on the BR(h→ $\gamma\gamma$) at 95% confidence level, assuming a Standard Model production cross section for the Higgs boson. The LEP combined limits are shown in Figure 2.

## 3. Decay into weak bosons

In the case Higgs decays via the process h→ WW* or ZZ*, and considering all the Z and W decay modes, there are nine possible channels. From these the L3 [7] collaboration analysed the following six: Zh→ $\mathrm{q\bar{q}q\bar{q}q\bar{q}}$, Zh→ $\mathrm{q\bar{q}q\bar{q}}\ell\nu$, Zh→ $\mathrm{q\bar{q}}\ell\nu\ell\nu$, Zh→ $\nu\nu\mathrm{q\bar{q}q\bar{q}}$, Zh→ $\nu\nu\mathrm{q\bar{q}}\ell\nu$ and Zh→ $\ell^+\ell^-\mathrm{q\bar{q}q\bar{q}}$, covering a total of 92% of the h→ WW* branching fraction. Of these the $\mathrm{qqqq}\ell\nu$ final state has the largest sensitivity.

The background in these channels comes from a variety of Standard Model processes, including W pair production, Z pair production and $\mathrm{q\bar{q}}$ events.

The number of selected events in 336.4 $\mathrm{pb}^{-1}$ of data collected at center-of-mass energies from 199 GeV up to 209 GeV is 566 and 568.3 events are expected from the Standard Model processes. Since no indication for a signal was found in the data, the negative search result is given in terms of upper limits on the branching fraction BR(h→ WW*) as a function of the Higgs mass. The limits are shown in Figure 3. A fermiophobic Higgs boson decaying to weak bosons is excluded for 83.8 GeV $< m_{\mathrm{h}} <$ 104.2 GeV with a region between 88.9 GeV $< m_{\mathrm{h}} <$ 89.4 GeV which can be excluded only at 93% confidence level.

Model-independent fermiophobic results can be derived by scanning the relative branching fractions of h→ $\gamma\gamma$ and h→ WW*, assuming

$$\mathrm{BR}(\mathrm{h}\to\gamma\gamma)+\mathrm{BR}(\mathrm{h}\to\mathrm{WW}^*)+\mathrm{BR}(\mathrm{h}\to\mathrm{ZZ}^*)=1.$$

The search results are presented as excluded regions in the plane [BR(h→ $\gamma\gamma$),$m_{\mathrm{h}}$] and are shown in Figure 4. The solid line crossing the excluded area gives the predicted value of BR(h→ $\gamma\gamma$) for the benchmark fermiophobic model. In the benchmark model, the 95% confidence level for a fermiophobic Higgs mass is set at 108.3 GeV, with an expected limit of 110.7 GeV. This result represents a significant extension of the mass limit of 105.3 GeV obtained by the L3 collaboration from the search in the photonic decay mode. The 95% confidence level mass limit, valid for any value of BR(h→ $\gamma\gamma$) and assuming a Standard Model production cross section is set at 107 GeV.

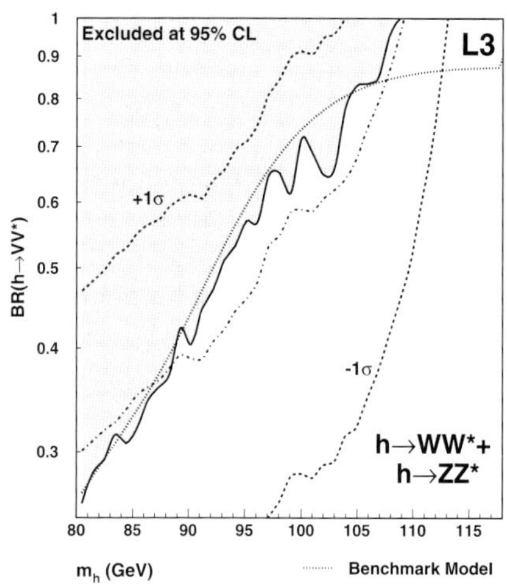


Figure 3. Excluded values at 95% confidence level of BR(h→ WW*) as a function of the Higgs mass in the fermiophobic model. The expected 95% confidence level limit (dot-dash line) and the theoretical prediction (dotted line) are also presented.

**Acknowledgements**

I would like to thank the members of the four LEP collaborations and the LEP Higgs working group for providing me with plots and their help in preparing this talk.

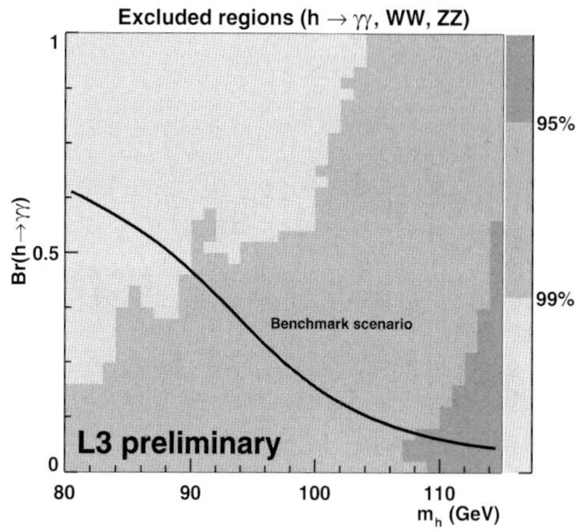


Figure 4. Excluded area at 95% confidence level in the (BR(h→ $\gamma\gamma$),$m_{\rm h}$) plane. The solid line gives the prediction of the benchmark fermiophobic model.

## REFERENCES

1. A.G. Akeroyd, Phys. Lett. **B368** (1996) 89; L. Brucher, R. Santos, Eur. Phys. J. **C12** (2000) 87.
2. ALEPH collaboration, CERN-EP-2002-044 (26 June 2002), submitted to Phys. Lett. B.
3. DELPHI 2002-086 PHYS 921 (July 2002); DELPHI 2001-073 CONF 501 (June 2001).
4. L3 collaboration, Phys. Lett. **B534** (2002) 28; Phys. Lett. **B489** (2000) 115.
5. OPAL collaboration, Phys. Lett. **B544** (2002) 44.
6. LHWG Note 2002-02, ALEPH 2002-019 CONF 2002-008, DELPHI 2002-087-CONF-620, L3 Note 2765, OPAL Technical Note TN720 (12 July 2002).
7. L3 collaboration, L3 Note 2685 (27 June 2001).

ELSEVIER

Proceedings of ICHEP 2002, pp. 746–748
*S. Bentvelsen, P. de Jong, J. Koch and E. Laenen (Editors)*

31st INTERNATIONAL CONFERENCE ON HIGH ENERGY PHYSICS AMSTERDAM

# Search for MSSM Higgses at the Tevatron

A. Connolly [a*]
For the CDF and D0 Collaborations

[a]Lawerence Berkeley National Laboratory, 50B-5239, One Cyclotron Rd., Berkeley, CA 94720 U.S.A.

We present an overview of searches for MSSM Higgs at the Tevatron, concentrating on searches probing the high $\tan\beta$ region. We discuss the search for $A/H \rightarrow \tau\tau$ which is soon to be completed in the Run I data and review the new tau triggers implemented by CDF and D0 in Run II, which will greatly impact this analysis. We also present the results of a Run I search for $A/Hbb \rightarrow bbbb$ performed by CDF and highlight expected improvements in this channel by both experiments in Run II.

## 1. MOTIVATION

The Higgs mechanism breaks electroweak symmetry in the Standard Model, giving mass to particles through its couplings. Current data from electroweak precision measurements points to a light Higgs ($M_{Higgs} < 199$ GeV @ 95% CL [1]). However, the Higgs has never been definitively observed ($M_{Higgs} > 114$ GeV at 95% CL [2]).

A Standard Model Higgs suffers from the so-called hierarchy problem. The theory needs fine-tuned parameters to accomodate a light Higgs mass. Supersymmetry offers a solution to this problem, through a symmetry between fermions and bosons.

The Minimal Supersymmetric Standard Model (MSSM) contains two Higgs doublets, leading to five physical Higgs bosons: Two neutral CP-even states (h and H), one neutral CP-odd (A), and two charged states ($H^+$ and $H^-$). At tree-level, the masses are governed by two parameters, often taken to be $m_A$ and $\tan\beta$ [3]. When $\tan\beta >> 1$, A is nearly degenerate with one of the CP-even states (denoted $\phi$). Where $m_A \leq 130$ GeV ($m_A \geq 130$), $m_A \cong m_h$ ($m_A \cong m_H$).

In this same large $\tan\beta$ region, the cross sections for some production mechanisms such as $pp \rightarrow A(\phi)$ and $pp \rightarrow A(\phi)b\bar{b}$ are enhanced by factors of $\tan\beta^2(\sec\beta^2)$. For example, with $\sqrt{s} = 2$ TeV, $\tan\beta = 30$ and $m_A = 100$ GeV, the cross sections for $pp \rightarrow A$ and $pp \rightarrow \phi$ are each of order 10 pb[4]. The cross section for $pp \rightarrow A/\phi b\bar{b}$ is smaller, but within the same order of magnitude. In the same region, the branching ratios to $A/\phi \rightarrow b\bar{b}$ and $\tau\tau$ dominate, at $\sim 90\%$ and $\sim 10\%$ respectively, independent of mass.

Due to their similar masses, cross-sections and branching ratios in the high $\tan\beta$ region, we search for both A and $\phi$ simultaneously. At the Tevatron, we search for $pp \rightarrow A/\phi \rightarrow \tau\tau$ (the $b\bar{b}$ final state is expected to be overwhelmed by dijet background) and $pp \rightarrow A/\phi b\bar{b} \rightarrow b\bar{b}b\bar{b}$.

## 2. SEARCH FOR $pp \rightarrow A/\phi \rightarrow \tau^+\tau^-$

This search is underway at CDF. The dominant issues for this analysis are: tau identification, di-tau mass reconstruction, irreducible background from $Z \rightarrow \tau\tau$, and event loss at the trigger level.

Wherever not specified, we use the benchmark case of $m_A = 95$ GeV and $\tan\beta = 40$ to quote efficiencies and cross-sections.

### 2.1. Tau Identification

Compared to QCD jets, taus are highly collimated, leaving narrow jets with low track and photon multiplicity, and low mass.

In CDF, when selecting taus, one typically requires a jet with high visible $E_T$ containing a high $p_T$ track. The jet is required to be isolated in a $10^o - 30^o$ annulus around the high $p_T$ track. The visible energy in a $10^o$ cone is required to satisfy low track and photon multiplicity requirements

*Current address is Fermi National Accelerator Laboratory, P.O. Box 500, M.S. 318, Batavia, IL 60510 U.S.A.


doi:10.1016/S0920-5632(02)02291-0

and to reconstruct a mass $m < 1.8$ GeV. Additionally, a requirement is made on the charge of the tracks in the $10^o$ cone when appropriate. In Run I, CDF acheived fake rates in the range 1.2 - 0.7 % for jet $E_T$ between 20 and 200 GeV[5].

### 2.2. Ditau Mass Reconstruction

The full mass of a ditau system may be reconstructed [6] if the neutrinos are assumed to travel in the same direction as their parent taus, by solving the following system of equations:

$$\not{E}_x^{meas} = \not{E}_x^{\tau 1} + \not{E}_x^{\tau 2} \quad (1)$$

$$\not{E}_y^{meas} = \not{E}_y^{\tau 1} + \not{E}_y^{\tau 2} \quad (2)$$

where $\not{E}_{x,y}^{meas}$ are the x and y components of the measured event missing energy, and $\not{E}_{x,y}^{\tau 1}$ and $\not{E}_{x,y}^{\tau 2}$ denote the missing energy from each tau.

Equations 1 and 2 do not give a meaningful solution when the taus are back-to-back in the transverse plane. Therefore, we require that $|\sin\Delta\phi| > 0.3$, where $\Delta\phi$ is the azimuthal angle between the tau candidates.

When the solution to Equations 1 and 2 gives $\not{E}^{\tau 1} < 0$ or $\not{E}^{\tau 2} < 0$, the event is thrown out, causing about 50% of the Higgs signal to be lost. However, 97% of W+jets events are rejected in this way, which would otherwise be a formidable background.

We generate $A/\phi \to \tau\tau$ events in Pythia 6.203 with $m_A = 95$ GeV and $\tan\beta = 40$. After simulation of the Run I CDF detector, a ditau mass distribution is reconstructed with a mean value of 93.7 GeV with an RMS of 24.1 GeV.

### 2.3. Irreducible Background

The dominant *reducible* backgrounds to this analysis are QCD, $Z \to ee$, and W+jets. $Z \to \tau\tau$ is an irreducible background, but Higgs events are more efficient for this search than Z's for a couple of reasons.

First, in the high $\tan\beta$ region, $A/\phi$'s have a high branching ratio to taus (9%) compared to Z's (3.7%). Second, an $A/\phi$ is typically produced with a stiffer $p_T$ than a Z. This means that the requirement $|\sin\Delta\phi| > 0.3$, which is nearly equivalent to $p_T^{A/\phi/Z} > 15$ GeV, is $\sim 30\%$ more efficient for Higgs events than Z events.

### 2.4. Triggers

Since there was no $\tau$ trigger in Run I at CDF, the analysis uses a lepton trigger requiring $p_T >$ 18 GeV, seeking events with one leptonic and one hadronic decay. Since only half of the signal events decay in this way, and of these, only 20% contain a lepton which satisfies the $p_T$ requirement within the acceptance, the signal rate is greatly diminished at the trigger level.

This major loss at the trigger level is problematic, since the cross section drops by a factor of 4 from $m_A = 95$ GeV to $m_A = 120$ GeV, before the mass reconstruction, with an RMS of 24 GeV, can discriminate from $Z \to \tau\tau$. Therefore, in Run II, CDF and D0 are both implementing triggers designed for tau physics. Lowered $p_T$ thresholds and new decay modes available will greatly increase the acceptance for this search.

In Run II, CDF and D0 both have lepton + track triggers and $\tau + \not{E}_T$ triggers. In addition, both experiments are triggering on events with two hadronic taus. CDF's trigger is calorimeter-based, while D0's is track-based.

The Run I search for $A/\phi \to \tau\tau$ is still work in progress, and the Run II analysis is also in the works.

## 3. SEARCH FOR $pp \to A/\phi b\bar{b} \to b\bar{b}b\bar{b}$

CDF performed this search in Run I. Both experiments expect to improve on the analysis in Run II.

### 3.1. Run I search

The Run I search at CDF [7]utilized a 4-jet trigger requiring $\Sigma E_T > 125$ GeV. Three b-tags were required based on displaced vertices, and the b jets were required to be separated in azimuthal angle, $\Delta\phi > 1.9$. To optimize sensitivity, the $E_T$ cuts on the jets varied with mass hypothesis. For mass hypothesis below 120 GeV (above 120 GeV), the second and third b-tagged jets (first and second jets) ordered in $E_T$ were chosen for the mass reconstruction. The search is performed in mass windows dependedent on mass hypothesis.

The product of branching ratio and acceptance ranged from 0.2 to 0.6% in the mass range 70 and 300 GeV. For a mass hypothesis of 70 GeV,

5 events were observed with 4.6 ± 1.4 expected. Only these 5 events appear in the higher mass windows. No excess above predicted is observed. Figure 1 shows the $m_A - \tan\beta$ region excluded.

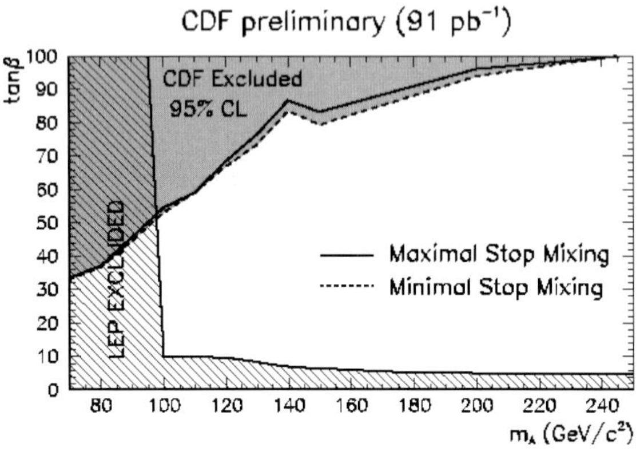


Figure 1. Region of the $m_A - \tan\beta$ region exluded by the CDF search.

### 3.2. Run II Improvements

At CDF, studies of $Z \to b\bar{b}$ events show an improved dijet mass resolution after correcting for muons, $\not{E}_T$, and nonlinearities in the hadronic calorimeter. Separate studies of QCD jets using similar techniques show a 30% improvement in jet resolution.

B-tagging in Run II at CDF will be improved with the new ability to reconstruct three-dimensional tracks. Extended coverage from $|\eta| < 1$ (Run I) to $|\eta| < 2$ means improve b-jet and lepton acceptance. Additionally, new triggers will also recover acceptance, including a displaced track trigger, and an improved multijet trigger.

With a new silicon detector, D0 will also be performing this search in Run II, expecting a 12% dijet mass resolution. Both experiments perform a study of their expected sensitivity to $pp \to A/\phi b\bar{b} \to b\bar{b}b\bar{b}$ in Run II, and obtain similar results[8]. We present the D0 study here.

D0 also uses a multijet trigger, requiring four jets, each with $E_T > 15$ GeV. To maximize sensitivity, mass dependent $E_T$ cuts are made on the jets. At least 3 b tags are required. All mass combinations are plotted, and a $2.5\sigma$ $b\bar{b}$ mass window is used. With $2fb^{-1}$, D0 concludes that the Tevatron is expected to exclude $m_A < 160$ GeV for $\tan\beta = 40$ at 95% CL, and a $5\sigma$ discovery for $m_A < 115$ GeV for the same $\tan\beta$.

## 4. CONCLUSIONS

Run I results of the search for $A/\phi \to \tau\tau$ at CDF are to be completed soon, and a first glimpse of Run II data is on the way.

The Run I search for $pp \to A/\phi b\bar{b} \to b\bar{b}b\bar{b}$ derives lower mass limits for $\tan\beta$ in excess of 35. In Run II with both experiments searching for this decay mode, the Tevatron is expected to exclude (or make a discovery in) a significant region of MSSM parameter space. Both experiments are optimistic about improvements from triggers, jet resolution, and b-tagging to make this search even stronger than the current projections.

## REFERENCES

1. K. Hagiwara et al., Physical Review D66, 010001 (2002).
2. ALEPH Collaboration, DELPHI Collaboration, L3 Collaboration, OPAL Collaboration and LEP Higgs Working Group, LHWG Note/2001-03,hep-ex/0107029.
3. For a review of the MSSM, see H.P. Nilles, Phys. Rept. 110:1,1984, H. Haber, G. Kane, Phys. Rept. 117:75,1985.
4. M. Spira, Nucl.Instrum.Meth.A389:357-360,1997
5. CDF Collaboration, Phys.Rev.Lett.79:357-362, 1997
6. CMS Technical Proposal, CERN/LHCC 94-38, December 1994, pp. 191-192.
7. CDF Collaboration, Phys.Rev.Lett.86(2001) 4472-4478.
8. M.Carena, J.S.Conway, H.E.Haber, J.D.Hobbs, et al, Fermilab-Conf-00/279-T and SCIPP-00/37,hep-ph/0010338.

ELSEVIER

Proceedings of ICHEP 2002, pp. 749–752
*S. Bentvelsen, P. de Jong, J. Koch and E. Laenen (Editors)*

31st INTERNATIONAL CONFERENCE ON HIGH ENERGY PHYSICS AMSTERDAM

# Searches at the Tevatron

W. Orejudos[a], for the CDF and D0 Collaborations

[a]Lawrence Berkeley National Laboratory,
1 Cyclotron Rd., Berkeley, CA, 94720 USA

Results of searches performed by CDF and D0 are presented. Most of the results are based on data taken during the 1994-95 data taking period (Run I), but some preliminary results from the current data taking period (Run II) are included.

## 1. Introduction

Run II of the Tevatron accelerator began in Spring 2001. The D0 and CDF experiments are currently recording collisions at a center-of-mass energy $\sqrt{s}$ of 1.96 GeV, an increase of ~10% with respect to the Run I center-of-mass energy. Both D0 and CDF have undergone substantial upgrades since the end of Run I and both have begun to present preliminary results based on Run II data. Both experiments have also continued to work on new analyses based on the Run I data.

## 2. Searches for new gauge bosons

CDF has previously looked for $W'$ in the decay $W' \to \ell\nu$ [1]. A new Run I analysis [2] looks for the decay $W' \to$ tb, providing sensitivity in the case that the decay to leptons is suppressed. The analysis requires transverse energy $E_T > 20$ GeV, an electron (muon) with $E_T > 20$ GeV ($p_T > 20$ GeV/$c$) and 2 or 3 jets ($j$) with $E_T > 15$ GeV, where one of the jets has a high probability of coming from a b-quark. The $\ell\nu jj$ invariant mass, shown in Fig. 1, is formed by fixing the $\ell\nu$ mass to the mass of the $W$. No excess over the background expectation is seen in this distribution, and the $W'$ mass $M_{W'}$ is excluded in the region 225 GeV/$c^2$ < $M_{W'}$ < 536 (566) GeV/$c^2$ for $M_{W'} >> M_{\nu_R}$ ($M_{W'} < M_{\nu_R}$). The D0 experiment has looked for $W'$ using the dijet final state and has excluded the region 400 GeV/$c^2$ < $M_{W'}$ < 625 GeV/$c^2$ [3].

CDF has looked for $Z'$ production in the Drell-Yan channel [4]. The Run I analysis requires two isolated, high $E_T$ ($p_T$) oppositely charged electrons (muons) and has set a lower limit on the $Z'$ mass of 690 GeV/$c^2$. At Run IIa (the first ~2 fb$^{-1}$ of Run II data), $Z'$ masses up to 1000 GeV/$c^2$ will be probed [5].

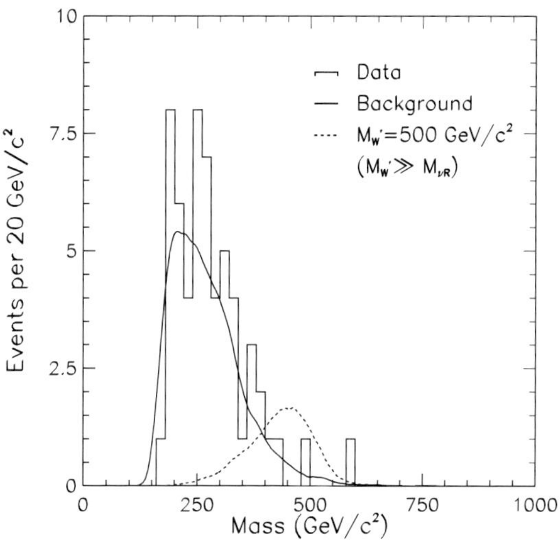


Figure 1. The $\ell\nu jj$ invariant mass distribution for the CDF $W' \to$ tb analysis. The histogram shows the data, the solid line the background expectation and the dashed line the expectation for a $W'$ with a mass of 500 GeV/$c^2$.


doi:10.1016/S0920-5632(02)02292-2

## 3. Searches for compositeness and technicolor

The Drell-Yan final state has also been used to search for other types of new physics. A limit on the mass of color singlet technirhos ($\rho_{T1}$) and techniomegas ($\omega_{T1}$) of 207 GeV/$c^2$ has been set by D0 at Run I using the $e^+e^-$ final state [6]. At Run IIa this search should be sensitive up to 410 GeV/$c^2$ [7]. D0 has also used this final state to set a lower limit on the quark-lepton compositeness scale of $\Lambda^- > 4.2$ ($\Lambda^+ > 3.3$) TeV assuming that the contact interaction and the Standard Model Lagrangian interact constructively (destructively) [8]. The corresponding reach at Run IIa is $\Lambda^- > 10$ ($\Lambda^+ > 6.5$) TeV [9].

D0 has searched for quark compositeness using the dijet final state. Limits of $\Lambda^- > 2.4$ ($\Lambda^+ > 2.7$) TeV have been set with the Run I data [10]. The same final state has been used by D0 to set a lower limit of 775 GeV/$c^2$ on the mass of excited quarks [3]. The Run IIa reach for this search is 940 GeV/$c^2$ [11].

Additional searches for technicolor have been performed in a variety of final states at Run I. CDF has searched for $\rho_{T1}$ in the decay $\rho_{T1} \to W\pi_{T1} \to \ell\nu b\bar{b}$ [12]. The $\rho_{T1}$ mass $M_{\rho_{T1}}$ has been excluded in the region 170 GeV/$c^2$ < $M_{\rho_{T1}}$ < 200 GeV/$c^2$, assuming that the $\pi_{T1}$ mass $M_{\pi_{T1}}$ is equal to about one-half of $M_{\rho_{T1}}$. At Run IIa CDF will probe the region 160 GeV/$c^2$ < $M_{\rho_{T1}}$ < 240 GeV/$c^2$ [7]. CDF has searched for $\omega_{T1}$ in the decay $\omega_{T1} \to \gamma\pi_{T1} \to \gamma b\bar{b}$ [13]. The $\omega_{T1}$ mass $M_{\omega_{T1}}$ has been excluded in the region 240 GeV/$c^2$ < $M_{\omega_{T1}}$ < 310 GeV/$c^2$ for $M_{\pi_{T1}} = 120$ GeV/$c^2$, and 140 GeV/$c^2$ < $M_{\omega_{T1}}$ < 290 GeV/$c^2$ has been excluded for $M_{\pi_{T1}}$ = 60 GeV/$c^2$. CDF has also searched for color octet technirhos ($\rho_{T8}$) in the decay $\rho_{T8} \to \pi_{LQ}\pi_{LQ} \to b\nu b\nu$ [14]. A limit on the $\rho_{T8}$ mass $M_{\rho_{T8}}$ of $M_{\rho_{T8}} > 600$ GeV/$c^2$ has been set. At Run IIa CDF will be sensitive to values of $M_{\rho_{T8}}$ up to 850 GeV/$c^2$ [15].

CDF has searched for particles decaying to a $t\bar{t}$ final state [16]; this would be the favored decay mode of a leptophobic $Z'$ found in certain models of topcolor [17]. The analysis of Run I data looked for $t\bar{t}$ events with one decay $t \to b\ell\nu$ and another decay $t \to bqq'$. A fit was performed for each event to get the mass of the $t\bar{t}$ system, $M_{t\bar{t}}$. The distribution of $M_{t\bar{t}}$ was found to be consistent with the background expectation, and a lower limit on the leptophobic $Z'$ mass $M_{Z'}$ of 780 (480) GeV/$c^2$ was set for a natural width $\Gamma_{Z'}$ of $0.04M_{Z'}$ ($0.012M_{Z'}$). At Run IIa CDF will be sensitive to leptophobic $Z'$ masses up to 1100 GeV/$c^2$ for $\Gamma_{Z'} = 0.04M_{Z'}$ [15].

## 4. Search for heavy, long-lived charged particles

A search for CHAMPS, heavy charged particles long-lived enough to leave the detector, has been performed by CDF at Run I [18]. The analysis looks for high $p_T$ charged particles with a large ionization energy loss. No evidence of a signal was found and a lower limit of 220 (190) GeV/$c^2$ has been set on the CHAMP mass in the context of a long-lived up (down)-type 4$^{th}$ generation quark. CDF is also developing an analysis [19] for Run II which will use the new time-of-flight (TOF) detector installed during the upgrade. The analysis requires a high $p_T$ charged particle with a large time-of-flight. A total of $2.2 \pm 0.8$ events from background sources are expected in the $16.5 \pm 1.7$ pb$^{-1}$ of Run II data that have been analyzed thus far.

## 5. Final states with photons

CDF has searched for new physics in the Run I diphoton sample [20]. Two isolated photons, each with $E_T > 22$ GeV, are required in the analysis. The main background comes from jets which fake photons. No evidence for new physics was found in this sample. A subset of this sample was used to search for a Higgs which is produced in association with a $W$ or $Z$ and which decays to two photons [20]. Events with a $W$ or $Z$ are selected by asking for an electron (muon) with $E_T > 20$ GeV ($p_T > 20$ GeV/$c$), or for missing $E_T > 20$ GeV, or for two jets with $E_T > 15$ GeV that form an invariant mass between 40 and 130 GeV/$c^2$. No excess above the Standard Model expectation is found in the resulting sample and a lower limit on the mass of a bosophilic Higgs of 82 GeV/$c^2$

is set.

CDF has also performed a model independent search for new physics with photons in the final state [21]. Several final states were defined, based on the presence of leptons, photons, and missing $E_T$. One such final state is defined by the presence of a photon with $E_T > 25$ GeV, an electron (muon) with $E_T > 25$ GeV ($p_T > 25$ GeV/$c$), missing $E_T > 25$ GeV, and an azimuthal angle between the photon and lepton which is less than 150°. Most of the events which produce this final state are expected to come from $(W/Z)\gamma$ production or from events with a real lepton and a jet which fakes a photon. A total of $7.6 \pm 0.7$ events are expected in the Run I data while 16 events are observed.

## 6. Search for large extra dimensions

In theories of large extra dimensions [22], gravitons may be detected at energies currently accessible by the Tevatron experiments. Searches by D0 and CDF can be divided into two categories: searches in the diphoton or dilepton final state for evidence of graviton exchange and searches in final states with large missing $E_T$ for direct graviton production.

In graviton exchange diagrams a virtual graviton is produced which couples to a $\gamma\gamma$ or $\ell^+\ell^-$ final state. D0 has combined their Run I diphoton and dielectron samples and performed a search [23] by requiring two electromagnetic objects with $E_T > 45$ GeV. In addition, the missing $E_T$ is required to be less than 25 GeV. The main backgrounds in this search come from Drell-Yan and $\gamma\gamma$ processes, as well as events in which two jets fake a photon or electron. The analysis looks at the angular distributions of the electromagnetic objects as well as their invariant mass. No deviation from the Standard Model expectation is observed, and a lower limit on the effective Planck scale of $M_S > 1.1$ (1.0) TeV is set in the Hewett convention [24], assuming constructive (destructive) interference between the Kaluza-Klein and Standard Model contributions.

In direct graviton production the graviton escapes the detector, producing missing $E_T$. A gluon or a photon can also be produced in the event. D0 has performed a search at Run I for the graviton-gluon final state [25]. The analysis looks for events that have a single jet with $E_T > 150$ GeV, missing $E_T > 150$ GeV, and no other jets with $E_T > 50$ GeV. The main backgrounds come from the processes $Z \to \nu\bar{\nu}$ + jets, $W \to \tau\nu$ + jets, and from QCD and cosmics rays. A total of $38.0 \pm 9.6$ events are expected from background processes, while 38 are actually observed. The lack of observation of a signal leads to a limit of $M_S > 886$ (617) GeV for 2 (7) large extra dimensions.

CDF has searched for direct graviton production at Run I in the graviton-photon final state [26]. The analysis looks for events that have a photon with $E_T > 55$ GeV, missing $E_T > 45$ GeV, and no jets with $E_T > 15$ GeV in the event. The main backgrounds come from $Z \to \nu\bar{\nu}\gamma$ and from cosmic rays. A total of $11.0 \pm 2.2$ events are expected from the background sources, and 11 events are actually observed. Again, no signal is observed and a limit of $M_S > 549$ (602) GeV is set for 4 (8) large extra dimensions.

D0 has also performed a search for graviton exchange using $9.85 \pm 0.38$ pb$^{-1}$ of data from Run II [27]. Two electromagnetic objects with $E_T > 25$ GeV are required in the analysis. The missing $E_T$ is required to be less than 30 GeV. No evidence of a signal is seen, and a limit of $M_S > 0.82$ TeV is set in the Hewett convention, assuming constructive interference between the Kaluza-Klein and Standard Model contributions. More data needs to be analyzed in order to improve on the Run I result.

## 7. Leptoquark search at Run II

D0 has looked for first generation leptoquark pair production in the Run II data [27]. Each leptoquark would decay to an electron and a quark; D0 looks for this final state by requiring two electromagnetic clusters with $E_T > 25$ GeV and at least two jets with $E_T > 20$ GeV. The jets are required to not form an invariant mass near that of the $Z$. The main backgrounds to this search come from Drell-Yan production and from jets faking electrons. Fig. 2 shows the scalar sum of

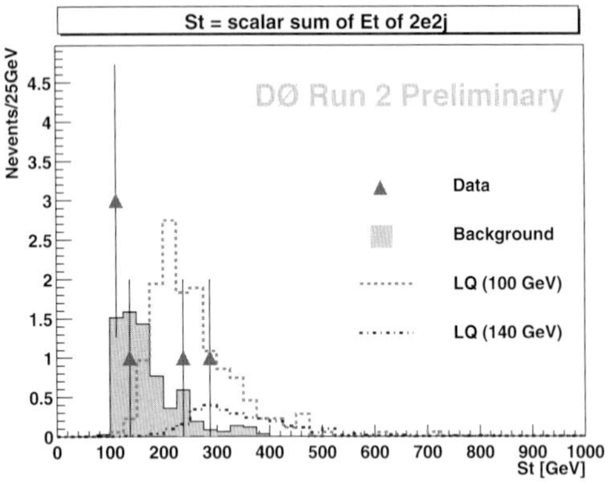


Figure 2. Scalar sum of the two electromagnetic clusters and the two highest $E_T$ jets. The solid histogram is the Standard Model expectation , the points are data and the dashed (dashed-dotted) lines show the expectation for a leptoquark mass of 100 (140) GeV/$c^2$.

the two electromagnetic clusters and the 2 highest $E_T$ jets. No evidence of a signal was seen in the $8.0 \pm 0.8$ pb$^{-1}$ of Run II data that was analyzed, leading to a lower limit on the leptoquark mass of 113 GeV/$c^2$. More data needs to be analyzed in order to improve on the Run I result [28].

## 8. Conclusion

In recent years the D0 and CDF experiments have continued to perform searches for new physics using the Run I data. Run II is now underway and preliminary search results are starting to appear. The next few years promise to be a very exciting time for D0 and CDF as both experiments continue to search the Run II data for hints of new physics.

## REFERENCES


1. T. Affolder *et al.*, CDF Collaboration, Phys. Rev. Lett. **87**, 231803 (2001).
2. D. Acosta *et al.*, CDF Collaboration, hep-ex/0209030 (2002).
3. C. Hebert for the D0 Collaboration, Presentation given at the APS 2002 Spring Meeting, Albuquerque, NM, April 20-24 2002.
4. F. Abe *et al.*, CDF Collaboration, Phys. Rev. Lett. **79**, 2191 (1997).
5. R. Blair *et al.*, CDF Collaboration, FERMILAB-PUB-96/390-E (1996).
6. V.M. Abazov *et al.*, D0 Collaboration, Phys. Rev. Lett. **87**, 061801 (2001).
7. K. Lane, hep-ph/0006143 (2000).
8. B. Abbott *et al.*, D0 Collaboration, Phys. Rev. Lett. **82**, 4769 (1999).
9. K. Cheung and R. Harris, hep-ph/9610382 (1996).
10. B. Abbott *et al.*, D0 Collaboration, Phys. Rev. Lett. **82**, 2457 (1999).
11. R. Harris, hep-ph/9609319 (1996).
12. T. Affolder *et al.*, CDF Collaboration, Phys. Rev. Lett. **84**, 1110 (2000).
13. T. Affolder *et al.*, CDF Collaboration, Phys. Rev. Lett. **83**, 3124 (1999).
14. T. Affolder *et al.*, CDF Collaboration, Phys. Rev. Lett. **85**, 2056 (2000).
15. T.L. Barlow *et al.*, hep-ph/0201243 (2002).
16. T. Affolder *et al.*, CDF Collaboration, Phys. Rev. Lett. **85**, 2062 (2000).
17. T. Affolder *et al.*, CDF Collaboration, FERMILAB-FN-687 (1999).
18. A. Connolly for the CDF Collaboration, FERMILAB-CONF-99/092-E (1999).
19. CDF Collaboration, internal note 5997 (2002).
20. T. Affolder *et al.*, CDF Collaboration, Phys. Rev. D **64**, 092002 (2001).
21. D. Acosta *et al.*, CDF Collaboration, Phys. Rev. D **66**, 012004 (2002).
22. N. Arkani-Hamed, S. Dimopoulos, and G. Dvali, Phys. Lett. B **429**, 263 (1998).
23. B. Abbott *et al.*, D0 Collaboration, Phys. Rev. Lett. **86**, 1165 (2001).
24. J.L. Hewett, Phys. Rev. Lett **82**, 4765 (1999).
25. S. Hagopian for the D0 Collaboration, hep-ex/0205048 (2002).
26. D. Acosta *et al.*, CDF Collaboration, FERMILAB-PUB-02/119-E.
27. D0 Collaboration, internal note 3995 (2002).
28. B. Abbott *et al.*, D0 Collaboration, Phys. Rev. Lett. **79**, 4321 (1997).

ELSEVIER

Proceedings of ICHEP 2002, pp. 753–755
*S. Bentvelsen, P. de Jong, J. Koch and E. Laenen (Editors)*

31st INTERNATIONAL CONFERENCE ON HIGH ENERGY PHYSICS AMSTERDAM

# Light Bottom Squark Phenomenology

Edmond L. Berger [a*]

[a]High Energy Physics Division, Argonne National Laboratory, Argonne, IL 60439, USA

Agreement of theoretical calculations with the observed production rate of bottom quarks at hadron colliders is improved by the introduction of a contribution from pair-production of light gluinos, of mass 12 to 16 GeV, having two-body decays into bottom quarks and light bottom squarks with mass $\simeq$ 2 to 5.5 GeV. Predictions are made for hadronic and radiative decays of the Upsilon states. In the limit of large $\tan\beta$, the dominant decay mode of the light scalar Higgs boson is into a pair of light bottom squarks that materialize as jets of hadrons.

## 1. INTRODUCTION

The cross section for bottom-quark $b$ production at hadron collider energies exceeds the central value of predictions of next-to-leading order (NLO) perturbative quantum chromodynamics (QCD) by about a factor of two or three [1]. The NLO contributions are large, and a combination of further higher-order effects in production and/or fragmentation may eventually reduce the discrepancy [2]. In Ref. [3], my collaborators and I propose a contribution from physics beyond the standard model (SM). In this paper, I summarize the proposal in Ref. [3] of light gluinos $\tilde{g}$ and light bottom squarks $\tilde{b}$, as well as subsequent work [4–8]. In this scenario the $\tilde{b}$ is the lightest supersymmetric (SUSY) particle, and the masses of all other SUSY particles are arbitrarily heavy, i.e., of order the electroweak scale or greater. The lifetime of the $\tilde{b}$ is assumed to be less than the cosmological time scale so that these squarks make no contribution to the dark matter density. References to an extensive body of other recent theoretical papers can be found in [7]. Various experimental constraints and phenomenological implications are examined in Ref. [5].

There are important restrictions on couplings of the $\tilde{b}$ from precise measurements of $Z^0$ decays. A light $\tilde{b}$ would be ruled out unless its coupling to the $Z^0$ is very small. The squark couplings to the $Z^0$ depend on the mixing angle $\theta_b$. The lowest-order (tree-level) coupling to the $Z^0$ can be arranged to be small [9] if $\sin^2\theta_b \sim 1/6$. The couplings of the heavier bottom squark $\tilde{b}_2$ survive. A careful phenomenological analysis is needed of expected $\tilde{b}_2$ decay signatures, along with an understanding of detection efficiencies and expected event rates, before one knows the admissible range of its masses consistent with LEP data. In the first paper of Ref. [10] it is argued that one-loop contributions may render the light $\tilde{g}$ and light $\tilde{b}$ scenario inconsistent with data, unless the mass of the $\tilde{b}_2$ is less than about 125 GeV. The possibility that $\tilde{b}_2$ lies in this mass range is not excluded. In the second paper, the mass bound is relaxed to about 180 GeV, and in the third, the constraint is further relaxed if $CP$-violating phases are present.

## 2. HADRON COLLIDERS

The light gluinos are produced in pairs via standard QCD subprocesses, dominantly $g+g \to \tilde{g}+\tilde{g}$ at Tevatron and Large Hadron Collider (LHC) energies. The $\tilde{g}$ has a strong color coupling to $b$'s and $\tilde{b}$'s and, as long as its mass satisfies $m_{\tilde{g}} > m_b + m_{\tilde{b}}$, the $\tilde{g}$ decays promptly to $b + \tilde{b}$. The magnitude of the $b$ cross section, the shape of the $b$'s transverse momentum $p_{Tb}$ distribution, and the CDF measurement [11] of $B^0 - \bar{B}^0$ mixing are three features of the data that help to estab-

*I thank B. Harris, C.-W. Chiang, L. Clavelli, J. Jiang, D. E. Kaplan, J. Lee, Z. Sullivan, T. Tait, and C. Wagner for their collaboration. Work in the High Energy Physics Division at Argonne is supported by the U.S. Department of Energy, Division of High Energy Physics, Contract W-31-109-ENG-38.


doi:10.1016/S0920-5632(02)02293-4

lish the preferred masses of the $\tilde{g}$ and $\tilde{b}$. Values of $m_{\tilde{g}} \simeq 12$ to 16 GeV are chosen because the resulting $\tilde{g}$ decays produce $p_{Tb}$ spectra that are enhanced primarily in the neighborhood of $p_{Tb}^{\min} \simeq m_{\tilde{g}}$ where the data show the most prominent enhancement above the QCD expectation. Larger values of $m_{\tilde{g}}$ yield too little cross section to be of interest, and smaller values produce more cross section than seems tolerated by the ratio of like-sign to opposite-sign leptons from $b$ decay.

After the contributions of the NLO QCD and SUSY components are added, the magnitude of the bottom-quark cross section and the shape of the integrated $p_{Tb}^{\min}$ distribution are described well. The SUSY process produces $b$'s in a four-body final state. Nevertheless, the angular correlations between $b$'s in the SUSY case are nearly indistinguishable from those of QCD once experimental cuts are applied. The energy dependence of the $b$ cross section is a potentially important constraint on models in which new physics is invoked. Since the assumed $\tilde{g}$ mass is larger than the mass of the $b$, the $\tilde{g}$ pair process will turn on more slowly with energy than pure QCD production of $b\bar{b}$ pairs. The new physics contribution will depress the ratio of cross sections at 630 GeV and 1.8 TeV from the pure QCD expectation. An explicit calculation with CTEQ4M parton densities and the $b$ rapidity selection $|y| < 1$, yields a pure NLO QCD prediction for the ratio of 0.17 +/- 0.02 for $p_{Tb}^{\min} = 10.5$ GeV, and 0.16 +/- 0.02 after inclusion of the $\tilde{g}$ pair contribution. Either of these numbers is consistent with data [12].

### 2.1. Like/Unlike-sign B's and Leptons

If, as in many scenarios, the $\tilde{g}$ is a Majorana particle, its decay yields both quarks and antiquarks. Pair production of Majorana gluinos and subsequent decay to $b$'s will generate $bb$ and $\bar{b}\bar{b}$ pairs, as well as the $b\bar{b}$ final states that appear in QCD production. Therefore, a "gold-plated" prediction is production of $B^+B^+$ and $B^-B^-$ pairs. For the cuts chosen in current hadron collider experiments, an equal number of like-sign and opposite-sign $b$'s is expected from the SUSY mechanism, leading to an increase of like-sign leptons in the final state after semi-leptonic decays of the $b$ and $\bar{b}$ quarks. This increase could be confused with an enhanced rate of $B^0 - \bar{B}^0$ mixing.

Time-integrated mixing analyses of lepton pairs observed at hadron colliders are interpreted in terms of the quantity $\bar{\chi}$. The CDF measurement [11] of $\bar{\chi}_{\text{eff}} = 0.131 \pm 0.02 \pm 0.016$ is marginally larger than the world average value $\bar{\chi} = 0.118 \pm 0.005$ [13], assumed to be the contribution from the pure QCD component only. After the contribution from new physics is included, the predictions are $\bar{\chi}_{\text{eff}} = 0.17 \pm 0.02$ for $m_{\tilde{g}} =$ 14 GeV, and $\bar{\chi}_{\text{eff}} = 0.16 \pm 0.02$ with $m_{\tilde{g}} = 16$ GeV. The calculated $\bar{\chi}_{\text{eff}}$ is consistent with the data within uncertainties if $m_{\tilde{g}} > 12$ GeV. The published result is based on an analysis of only 20% of the run-I sample and only the $\mu\mu$ final state. It would be valuable to extend the analysis to the full sample in both the $e\mu$ and $\mu\mu$ modes.

With $\sigma_{\tilde{g}\tilde{g}}/\sigma_{\text{qcd}} \sim 1/3$, the mixing data and the magnitude and $p_T$ dependence of the $b$ production cross section can be satisfied.

## 3. Υ DECAY

If $m_{\tilde{b}}$ is less than half the mass of one of the Upsilon states, then $\Upsilon$ decay to a pair of bottom squarks might proceed with sufficient rate for observation or exclusion of a light $\tilde{b}$. The rate for $\Upsilon(nS) \to \tilde{b}\tilde{b}^*$ is computed in Ref. [4] as a function of the masses of the $\tilde{b}$ and the $\tilde{g}$, and $\chi_{bJ}$ decays are treated in Ref. [6]. The data sample is largest at the $\Upsilon(4S)$. For a fixed $\tilde{g}$ mass of 14 GeV, the branching fraction into a pair of $\tilde{b}$'s is about $10^{-4}$ for $m_{\tilde{b}} = 4.85$ GeV. A large sample may be available from the CLEO, BaBar, and BELLE experiments. Direct observation of $\Upsilon(nS)$ or $\chi_b$ decay into $\tilde{b}$'s requires an understanding of the ways that $\tilde{b}$'s may manifest themselves, discussed in Refs. [4–6]. Possible baryon-number-violating R-parity-violating decays of the $\tilde{b}$ lead to $u + s$; $c + d$; and $c + s$ final states.

It is possible that the $\tilde{b}$ is relatively stable and, hence, bound states of a bottom squark and bottom antisquark (sbottomonium) could exist. These bound states could be produced in radiative decays of bottomonium states, such as $\Upsilon \to \tilde{S}\gamma$, where $\tilde{S}$ is the $S$-wave bound state of a $\tilde{b}\tilde{b}^*$ pair. In Ref. [8], a calculation is presented of the radiative decay of the $\Upsilon(nS)$ states into a bound state of $\tilde{b}$'s. Predictions are provided of the branching fraction as a function of the masses $m_{\tilde{b}}$ and $m_{\tilde{g}}$. Branching fractions as large as several times $10^{-4}$ are obtained for SUSY particle masses in the range suggested by the analysis of the $b$ cross section. Provided that a bound state can be formed, the resonance search by the CUSB Collaboration [14] raises the allowed lower bounds on $m_{\tilde{b}}$ and $m_{\tilde{g}}$. Discovery of the $\tilde{S}$ bound states may be possible with the high-statistics 2002 CLEO-c data set, or a larger range of $\tilde{b}$ and $\tilde{g}$ masses may be disfavored [8].

## 4. HIGGS BOSON DECAY

Current strategies for discovery and measurement of the properties the neutral scalar Higgs particle $h$ with $m_h < 135$ GeV rely heavily on the presumption that the principal branching fractions are close to those predicted in the SM or in the usual minimal supersymmetric standard model (MSSM). For masses in this range, the decay width of the SM Higgs boson is dominated by its decay into bottom quarks, $b\bar{b}$. In Ref. [7], my collaborators and I show explicitly that these assumptions are not warranted if there are non-standard light particles such as $\tilde{b}$'s in the spectrum. We analyze the possibility that the $h$ decays into new particles that manifest themselves as hadronic jets without necessarily significant bottom or charm flavor content. As an example of this possibility, we present the case of a light $\tilde{b}$, with mass smaller than about 10 GeV.

We work in the decoupling limit in which the mass of the pseudo-scalar Higgs boson ($m_A$) is large compared to $m_Z$, and we assume that the ratio of Higgs vacuum expectation values $\tan\beta$ is large. No assumption is made about the gluino mass; a light $\tilde{g}$ is not required. Under these conditions, the dominant decay of $h$ is into a pair of light $\tilde{b}$'s. The total decay width of the $h$ becomes several orders of magnitude larger than the width for decay into $b$'s. Branching fractions into SM decay channels are reduced from their SM values by a factor proportional to $\tan^{-2}\beta$. For values of the branching ratio $BR(h \to \tilde{b}\tilde{b}^*)$ larger than two to five times that into bottom quarks, the large QCD jet backgrounds will make observation of the $h$ very difficult in Tevatron and LHC experiments. Because they rely principally on the production process $e^+e^- \to hZ^0$, experiments at proposed $e^+e^-$ linear colliders remain fully viable for direct observation of the $h$ and measurement of its mass and some of its branching fractions [7].

## REFERENCES


1. CDF Collaboration, F. Abe *et al.*, *Phys. Rev. Lett.* **71**, 500 (1993); *ibid* **79**, 572 (1997); *ibid* **79**, 572 (1997); *ibid* **75**, 1451 (1995); D. Acosta *et al.*, Phys.Rev. D65, 052005 (2002); D0 Collaboration, B. Abbott *et al.*, *Phys. Lett. B* **487**, 264 (2000) and *Phys. Rev. Lett.* **85**, 5068 (2000).
2. M. Cacciari and P. Nason, Phys. Rev. Lett. **89**, 122003 (2002).
3. E. L. Berger, B. W. Harris, D. E. Kaplan, Z. Sullivan, T. M. Tait, and C. E. M. Wagner, Phys. Rev. Lett. **86**, 4231 (2001) [arXiv:hep-ph/0012001].
4. E. L. Berger and L. Clavelli, Phys. Lett. B **512**, 115 (2001) [arXiv:hep-ph/0105147].
5. E. L. Berger, arXiv:hep-ph/0201229, Int. J. Mod. Phys. A (in press).
6. E. L. Berger and J. Lee, Phys. Rev. D **65**, 114003 (2002) [arXiv:hep-ph/0203092].
7. E. L. Berger, C. W. Chiang, J. Jiang, T. M. Tait, and C. E. M. Wagner, arXiv:hep-ph/0205342, Phys. Rev. D (in press).
8. E. L. Berger, G. T. Bodwin, and J. Lee, arXiv:hep-ph/0206115.
9. M. Carena, S. Heinemeyer, C. E. M. Wagner, and G. Weiglein, Phys. Rev. Lett. **86**, 4463 (2001).
10. J. Cao, Z. Xiong, and J. M. Yang, Phys. Rev. Lett. **88**, 111802 (2002); G. C. Cho, arXiv:hep-ph/0204348; S. W. Baek, Phys. Lett. B **541**, 161 (2002).
11. CDF Collaboration, F. Abe *et al.*, *Phys. Rev. D* **55**, 2546 (1997).
12. CDF Collaboration, D. Acosta *et al.*, Phys. Rev. D **66**, 032002 (2002).
13. Particle Data Group, *Eur. Phys. J. C* **15** 1, (2000).
14. P. Franzini *et al.*, Phys. Rev. D **35**, 2883 (1987).

Proceedings of ICHEP 2002, pp. 756–758
*S. Bentvelsen, P. de Jong, J. Koch and E. Laenen (Editors)*

31st INTERNATIONAL CONFERENCE ON HIGH ENERGY PHYSICS AMSTERDAM

# Higgs Decays into Leptons

Chung Kao[a]*

[a]Department of Physics and Astronomy, University of Oklahoma, Norman, Oklahoma 73019, USA

We discuss the prospects for the discovery of neutral Higgs bosons ($\phi^0 = H^0, h^0, A^0$) in multi-Higgs doublet models via their decays into leptons at hadron colliders. A major focus will be searching for Higgs bosons produced with bottom quarks via Higgs decays into muon pairs ($pp \to b\bar{b}\phi^0 \to b\bar{b}\mu\bar{\mu} + X$) at the CERN Large Hadron Collider within the framework of the minimal supersymmetric model.

**1. Introduction.** In this article we discuss the prospects of discovering neutral Higgs bosons via their decays into lepton pairs of the same flavor as well as final states with lepton flavor violation. In particular, we present a summary of our results for the minimal supersymmetric Higgs bosons produced with bottom quarks via Higgs decays into muon pairs at the CERN Large Hadron Collider (LHC) [1].

In a two Higgs doublet model such as the minimal supersymmetric model (MSSM) with Yukawa interactions of Model II, there are two Higgs doublets $\phi_1$ and $\phi_2$ coupling to the $t_3 = -1/2$ and $t_3 = +1/2$ fermions, respectively. After spontaneous symmetry breaking, there remain five Higgs bosons: two charged Higgs bosons $H^\pm$, two neutral CP-even scalars $H^0$ (heavier) and $h^0$ (lighter), and a neutral CP-odd pseudoscalar $A^0$. The couplings of $\phi^0 b\bar{b}$ and $\phi\ell\bar{\ell}$ ($\phi^0 = H^0, h^0, A^0$) are enhanced by factors of $1/\cos\beta$.

For large $\tan\beta \equiv v_2/v_1$, the $\tau\bar{\tau}$ decay mode [2,7] is a promising discovery channel for the $A^0$ and the $H^0$ at the CERN LHC. It has also been suggested that Higgs bosons might be observable via their $b\bar{b}$ decays [3] in a large region of the $(m_A, \tan\beta)$ plane. However, simulations for the ATLAS detector concluded that detection of the $b\bar{b}$ channel would be difficult [4].

The LHC discovery potential of the muon pair channel for MSSM Higgs bosons was demonstrated by Kao and Stepanov [5,6], and was later confirmed by the ATLAS collaboration [2,7]. In the minimal supergravity model, the significance of $pp \to \phi^0 \to \mu\bar{\mu} + X$ is greatly improved at large $\tan\beta$ [8] because $m_A$ and $m_H$ become small from the evolution of the renormalization group equations with large $b\bar{b}\phi^0$ couplings. However, it is very challenging to search for Higgs decays into muon pairs in the Standard Model (SM) [9].

Furthermore, Higgs decays might lead to final states with lepton flavor violation [10]. In the MSSM, radiative corrections via gaugino-sfermion loops can lead to lepton flavor violating Higgs decays with a branching ratio approximately $B(\phi^0 \to \tau\mu) \sim 4 \times 10^{-4}$ which would be a $3\sigma$ signal at the LHC. In an $E_6$−inspired multi-Higgs model with an abelian flavor symmetry, lepton flavor violating effects also arise in Higgs decays ($\phi^0 \to \tau\mu$) that might be observable at the Tevatron and the LHC [10].

In this article, we focus on the prospects of discovering the MSSM Higgs bosons ($\phi^0 = H^0, h^0, A^0$) produced with bottom quarks[2] via Higgs decays into muon pairs ($pp \to b\bar{b}\phi^0 \to b\bar{b}\mu\bar{\mu} + X$) at the LHC. The Higgs signal and the SM background are evaluated with realistic cuts.

**2. Higgs Decays into Muons.** We calculate the cross section at the LHC for $pp \to b\bar{b}\phi^0 + X$ ($\phi^0 = H^0, h^0, A^0$) from two subprocesses $gg \to b\bar{b}\phi^0$ and $q\bar{q} \to b\bar{b}\phi^0$. It turns out that the contribution from quark-antiquark annihilation ($q\bar{q} \to b\bar{b}\phi^0$) is negligible.

*E-mail address: Kao@physics.ou.edu

[2]Recently, it was suggested that the production of a Higgs boson associated with a bottom quark ($gb- > \phi^0 b$) might allow for the suppression of backgrounds [11].


doi:10.1016/S0920-5632(02)02294-6

It is a good approximation to evaluate the next to the leading order (NLO) cross section of $pp \to \phi^0 b\bar{b} + X$ with the leading order (LO) contribution multiplied by a K factor[3] of 0.8 [13] and using the pole mass $M_b = 4.7$ GeV. The results of Ref. [12] are applied to compute the NLO rates for $pp \to \phi^0$. We take a K factor of 1.5 for the contribution from $gg \to H^0, h^0$ for $\tan\beta < 6$ and a K factor of 1.3 for $\tan\beta \geq 6$. The contribution from $gg \to A^0$ is calculated with the LO contribution multiplied by a K factor of 1.7 for $\tan\beta < 6$ and a K factor of 1.3 for $\tan\beta \geq 6$.

The cross section for $pp \to b\bar{b}\phi^0 \to b\bar{b}\mu\bar{\mu} + X$ is evaluated with the Higgs production cross section $\sigma(pp \to b\bar{b}\phi^0 + X)$ multiplied by the branching fraction of the Higgs decay into muon pairs $B(\phi^0 \to \mu\bar{\mu})$. When the $b\bar{b}$ mode dominates Higgs decays, the branching fraction for $A^0 \to \mu\bar{\mu}$ is approximately $3 \times 10^{-4}$ for $m_A = 100$ GeV.

**3. The Discovery Potential at the LHC.** The dominant physics backgrounds to the final state of $b\bar{b}\mu\bar{\mu}$ come from $gg \to b\bar{b}\mu\bar{\mu}$ and $q\bar{q} \to b\bar{b}\mu\bar{\mu}$ as well as $gg \to b\bar{b}W^+W^-$ and $q\bar{q} \to b\bar{b}W^+W^-$ followed by the decays of $W^\pm \to \mu^\pm\nu_\mu$. We have applied the same acceptance cuts and efficiencies of $b$-tagging and mistagging as those of the ATLAS collaboration [7].

We show the invariant mass distribution of muon pairs in Fig. 1 for $pp \to b\bar{b}A^0 + X \to b\bar{b}\mu\bar{\mu} + X$ with $\tan\beta = 10$ and 50, the SM processes of $pp \to b\bar{b}\mu\bar{\mu} + X$, and contributions from $pp \to b\bar{b}W^+W^- + X$. Also shown are the muon pair invariant mass distributions for the inclusive final state of $pp \to A^0 \to \mu\bar{\mu} + X$ and its dominant background from the Drell-Yan process $pp \to Z, \gamma \to \mu\bar{\mu} + X$. There are several interesting aspects to note from this figure: (a) including two $b$ quarks in the final state greatly improves the signal to background ratio with realistic efficiencies of b tagging and jet mistagging at the ATLAS and the CMS detectors, (b) the SM subprocess of $gg \to b\bar{b}\mu\bar{\mu}$ and $q\bar{q} \to b\bar{b}\mu\bar{\mu}$ make the major contribution to the physics background for $M_{\mu\bar{\mu}} \lesssim 100$ GeV, but $gg \to b\bar{b}W^+W^-$ and $q\bar{q} \to b\bar{b}W^+W^-$ become the dominant background for higher muon pair invariant mass.

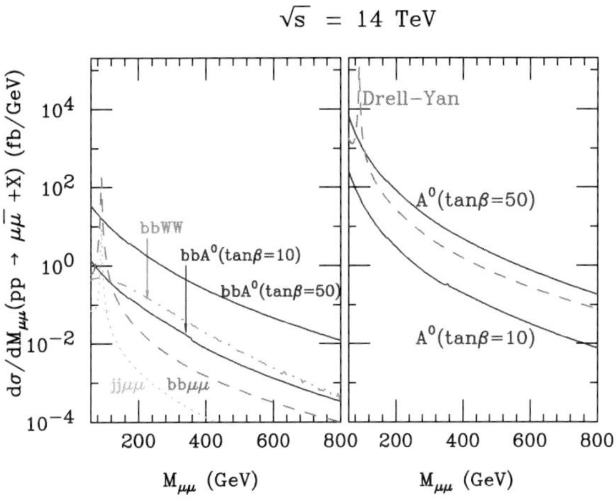


Figure 1. The invariant mass distribution for $pp \to b\bar{b}A^0 + X \to b\bar{b}\mu\bar{\mu} + X$ at $\sqrt{s} = 14$ TeV with $M_{\mu\mu} = m_A$. Also shown are the SM backgrounds from $pp \to b\bar{b}\mu\bar{\mu} + X$ (dashed), $jj\mu\bar{\mu} + X$ (dotted), and $b\bar{b}W^+W^- + X$ (dot-dashed). The panel on the right shows the invariant mass distribution for the inclusive final state $pp \to A^0 \to \mu\bar{\mu} + X$ and the Drell-Yan background (dashed).

To study the discovery potential of $pp \to b\bar{b}\phi^0 + X \to b\bar{b}\mu\bar{\mu} + X$ at the LHC, we calculate the background from the SM processes of $pp \to b\bar{b}\mu\bar{\mu} + X$ in the mass window of $m_\phi \pm \Delta M_{\mu\bar{\mu}}$ where $\Delta M_{\mu\bar{\mu}} \equiv 1.64[(\Gamma_\phi/2.36)^2 + \sigma_m^2]^{1/2}$ [7], $\Gamma_\phi$ is the total width of the Higgs boson, and $\sigma_m$ is the muon mass resolution. We take $\sigma_m$ to be 2% of the Higgs boson mass [7]. The CMS mass resolution will be better than 2% of $m_\phi$ for $m_\phi \lesssim 500$ GeV [5,6].

The $5\sigma$ discovery contours for the MSSM Higgs bosons at $\sqrt{s} = 14$ TeV for an integrated luminosity of $L = 30$ fb$^{-1}$ and $L = 300$ fb$^{-1}$ are shown in Fig. 2. We have chosen $M_{\rm SUSY} = m_{\tilde{q}} = m_{\tilde{g}} = m_{\tilde{\ell}} = \mu = 1$ TeV.

[3]The K factor is defined as $K = \sigma_{\rm NLO}/\sigma_{\rm LO}$, where $\sigma_{LO}$ is evaluated with LO parton distribution functions and 1-loop evolution of the strong coupling ($\alpha_s$) and $\sigma_{NLO}$ is evaluated with NLO parton distribution functions and 2-loop evolution of $\alpha_s$.

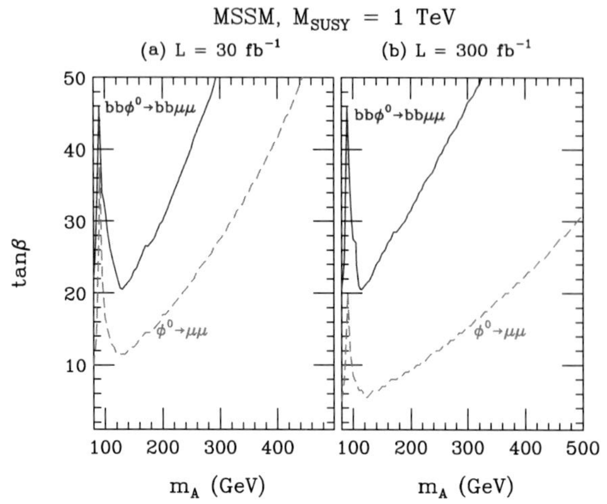


Figure 2. The $5\sigma$ contours at the LHC for an integrated luminosity ($L$) of 30 fb$^{-1}$ and 300 fb$^{-1}$ in the $m_A$ versus $\tan\beta$ plane. The signal includes $\phi^0 = A^0$ and $h^0$ for $m_A < 125$ GeV, but $\phi^0 = A^0$ and $H^0$ for $m_A \geq 125$ GeV. The discovery region is above the $5\sigma$ contour.

We have presented results for $M_{\rm SUSY} = 1$ TeV. If $M_{\rm SUSY}$ is smaller, the discovery region of $A^0, H^0 \to \mu\bar{\mu}$ will be slightly reduced for $m_A \gtrsim 250$ GeV, because the the Higgs bosons can decay into SUSY particles and the branching fraction of $\phi^0 \to \mu\bar{\mu}$ is suppressed [5,8].

**4. Conclusions** The muon pair decay mode is a promising channel to discover the MSSM Higgs bosons at the LHC. The associated final state of $b\bar{b}\phi^0 \to b\bar{b}\mu\bar{\mu}$ could discover the $A^0$ and the $H^0$ at the LHC with an integrated luminosity of 30 fb$^{-1}$ if $m_A \lesssim 300$ GeV. At a higher luminosity of 300 fb$^{-1}$, the discovery region in $m_A$ is not expanded much, because the harder $p_T$ cut on $b$ quarks reduces the Higgs production cross section.

The inclusive final state of $\phi^0 \to \mu\bar{\mu}$ could allow the discovery of the $A^0$ and the $H^0$ at the LHC with an integrated luminosity of 30 fb$^{-1}$ if $m_A \lesssim 450$ GeV. At a higher luminosity of 300 fb$^{-1}$, the discovery region in $m_A$ is significantly extended to $m_A \lesssim 650$ GeV for $\tan\beta = 50$.

The discovery of both $\phi^0 \to \tau\bar{\tau}$ and $\phi^0 \to \mu\bar{\mu}$ will allow us to understand the Higgs Yukawa couplings with the leptons. The discovery of the associated final state of $b\bar{b}\phi^0 \to b\bar{b}\mu\bar{\mu}$ will provide information about the Yukawa couplings of $b\bar{b}\phi^0$ and an opportunity to measure $\tan\beta$. Although the discovery region of the $\mu\bar{\mu}$ mode is smaller than the $\tau\bar{\tau}$ channel, the $\mu\bar{\mu}$ channel allows a precise reconstruction of the Higgs boson masses.

**Acknowledgments.** I am very grateful to Sally Dawson and Duane Dicus for an inspiring and enjoyable collaboration. This research was supported in part by the U.S. Department of Energy under Grant No. DE-FG03-98ER41066.

## REFERENCES

1. S. Dawson, D. Dicus and C. Kao, to be published in Phys. Lett. B, arXiv:hep-ph/0208063.
2. E. Richter-Was, D. Froidevaux, F. Gianotti, L. Poggioli, D. Cavalli and S. Resconi, Int. J. Mod. Phys. A **13**, 1371 (1998).
3. J. Dai, J. F. Gunion and R. Vega, Phys. Lett. B **345** (1995) 29; Phys. Lett. B **387** (1996) 801; C. Balazs, J. L. Diaz-Cruz, H. J. He, T. Tait and C. P. Yuan, Phys. Rev. D **59** (1999) 055016.
4. E. Richter-Was and D. Froidevaux, Z. Phys. C **76** (1997) 665.
5. C. Kao and N. Stepanov, Phys. Rev. D **52** (1995) 5025.
6. CMS Technical Proposal, CERN/LHCC 94-38 (1994).
7. ATLAS Detector and Physics Performance Technical Design Report, CERN/LHCC 99-14/15 (1999).
8. V. D. Barger and C. Kao, Phys. Lett. B **424** (1998) 69.
9. T. Plehn and D. Rainwater, Phys. Lett. B **520**, 108 (2001); T. Han and B. McElrath, Phys. Lett. B **528**, 81 (2002).
10. J. L. Diaz-Cruz, arXiv:hep-ph/0207030.
11. J. Campbell, R. K. Ellis, F. Maltoni and S. Willenbrock, arXiv:hep-ph/0204093.
12. M. Spira, A. Djouadi, D. Graudenz and P.M. Zerwas, Nucl. Phys. **B453** (1995) 17.
13. B. Plumper, DESY-THESIS-2002-005.

ELSEVIER

Proceedings of ICHEP 2002, pp. 759–763
*S. Bentvelsen, P. de Jong, J. Koch and E. Laenen (Editors)*

31st INTERNATIONAL CONFERENCE ON HIGH ENERGY PHYSICS AMSTERDAM

# Low Scale Gravity and Extra Dimensions at HERA, LEP and Tevatron

Gregorio Bernardi [a]

[a]LPNHE, University of Paris VI; 4, place Jussieu, 75252 Paris Cedex 05, France

Searches for Low Scale Gravity and Extra Dimensions at HERA, LEP and the Tevatron are reviewed.

In spite of his many successes, the Standard Model (SM) is plagued by some intriguing characteristics, like the hierarchy problem, i.e. the smallness of the ratio of the electroweak scale to the Planck scale ($M_{Pl}$). The conjecture [1] that the existence of $n$ extra spatial dimensions (ED) could remove this problem is a new approach which can be tested experimentally at high energy accelerators. In this hypothesis, the SM particles are confined to our usual 3+1 dimensional space, while quantum gravity permeates the whole space.

The mass scale ($M_S$) for which cross section calculations are available [2–5] is expected to be comparable to the Planck mass($M_D$) in $D = 4+n$ dimensions, which is related to $M_{Pl}$ by $M_{Pl}^2 = M_D^{n+2}\ R^n$, $R$ being the compactification radius of the extra dimensions, each of them supposed to be a torus. For $M_D = 1$ TeV, this translates in values of $R = 8 \times 10^{12}$ m, 0.7 mm, 3 nm, $6 \times 10^{-12}$ m, for $n$ respectively equal to 1,2,3,4, i.e. only the case of one single extra dimension is clearly ruled out experimentally, by the non-observation of deviations from Newton's law on large distances. Recent Cavendish-type experiments exclude $R > 0.19$ mm, which translates as $M_S > 1.9$ TeV for $n = 2$.

In collider experiments, extra dimensions can be searched for in two different ways: either in an "indirect" way via interference effects of SM processes with graviton ($G$) exchange interactions which modify the production cross-sections, or by direct $G$ production, which appears experimentally as missing energy in the detector. The interference effects have been searched for at HERA, LEP and the Tevatron, while direct production has been investigated only at LEP in the $\gamma G$ and $ZG$ production channels and at the Tevatron in the $\gamma G$ and gluon($g$)$G$ production channels.

## 1. Indirect ED Searches at HERA

At HERA, the H1 and ZEUS experiments have searched for ED using the complete statistics taken until 1999, corresponding to 16 (112) pb$^{-1}$ in $e^-p$ ($e^+p$) scattering. The interference of the $G$ exchange with the SM $\gamma$ and $Z$ processes leads to an effective contact type interaction with a coupling $\eta_G = \lambda/M_S^4$, where the parameter $\lambda$ depends on the details of the theory and is expected to be of order unity. However, by convention, negative couplings are also allowed and limits are derived assuming $\lambda = \pm 1$. The interference term behaves oppositely for electrons and positron scattering, as illustrated in fig. 1, which shows the ZEUS results [6], excluding $M_S < 0.82(0.81)$ TeV for $\lambda = +(-)1$. Similarly, the H1 experiment excludes $M_S < 0.83(0.79)$ TeV [7]

## 2. Indirect ED Searches at LEP

At LEP, the 4 collaborations ALEPH, DELPHI, L3 and OPAL have searched for virtual graviton effects in difermion and diboson production, the most sensitive channels being $e^+e^- \to e^+e^-$ and $e^+e^- \to \gamma\gamma$. The expected cross section in presence of ED in the $\gamma\gamma$ channel is given by $(\frac{d\sigma}{d\cos\theta})^{SM+ED} = (\frac{d\sigma}{d\cos\theta})^{SM} - \alpha s\frac{\lambda}{M_S^4}(1 + cos^2\theta) + \frac{s}{8\pi}\frac{\lambda}{M_S^8}(1 - cos^4\theta)$, where $s$ is the center of mass energy and $\theta$ the photon scattering angle. A similar formula exists for the $e^+e^-$ final state, but with more complicated coefficients due to the presence of $s$ and $t$ channel in this latter case. The $e^+e^- \to e^+e^-$ cross section at different center of mass energies, from which the limits


doi:10.1016/S0920-5632(02)02295-8

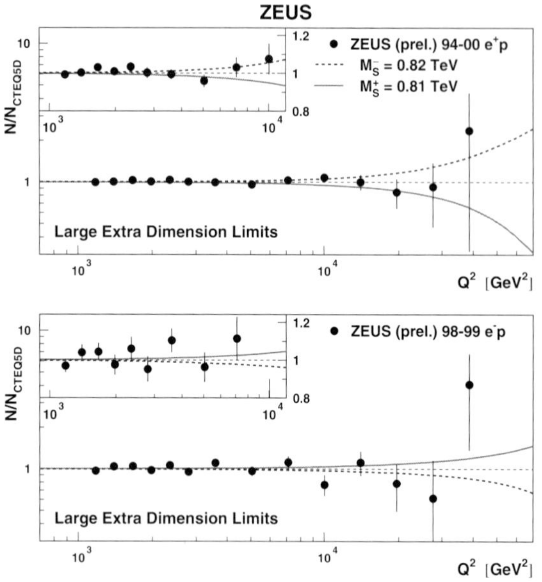


Figure 1. Comparison of Data/Theory for $e^+p$ and $e^-p$ cross-section as a function of the transfer momentum squared $Q^2$. Also shown are the SM prediction and the deviations expected from ED with mass scales for $\lambda = \pm 1$ of 0.81 and 0.82 TeV.

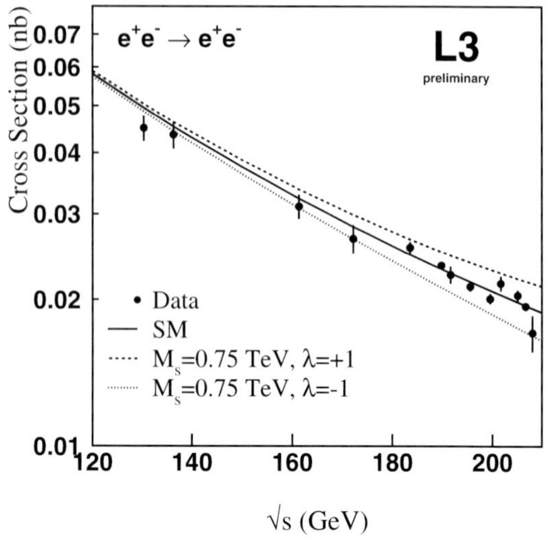


Figure 2. $e^+e^- \to e^+e^-$ cross-section measured at different center of mass energies by the L3 experiment. The SM prediction and the expected deviations due to ED with $M_S = 0.75$ TeV and $\lambda = \pm 1$ are also shown.

can be derived is shown as an example in fig. 2, as obtained by L3.The limits obtained by the 4 LEP experiments are given in table 2 [8–11]. The preliminary combination of all the LEP results on $e^+e^- \to e^+e^-$ differential cross-section is shown in fig. 3, excluding $M_S > 1.20(1.09)$ TeV for $\lambda = +1$ (–1) [12].

| Collab. | $e^+e^-(\lambda = +1/-1)$ | $\gamma\gamma$ ($\lambda = +1/-1$) |
|---|---|---|
| ALEPH | 0.81 / 1.04 TeV | 0.80 / 0.85 TeV |
| DELPHI | – | 0.77 / 0.70 TeV |
| L3 | 1.04 / 1.05 TeV | 0.84 / 0.99 TeV |
| OPAL | 1.00 / 1.15 TeV | 0.81 / 0.96 TeV |

Table 1
Limits on $M_S$ obtained from virtual graviton effects at LEP in $e^+e^- \to e^+e^-$ and $e^+e^- \to \gamma\gamma$.

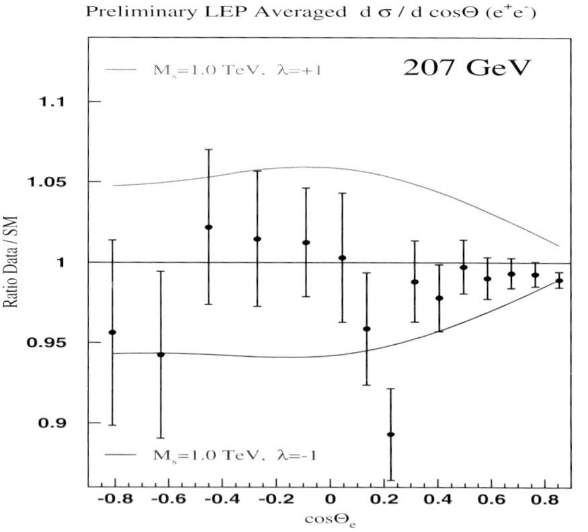


Figure 3. Ratio Data/SM of the averaged LEP results in $\frac{d\sigma(e^+e^-)}{d\cos\theta}$.The expected deviations due to ED with $M_S = 1$ TeV and $\lambda = \pm 1$ are also shown.

## 3. Indirect ED Searches at the Tevatron

At hadronic colliders the virtual effects of the graviton in dilepton production in the $s$ channel can come not only from $q\bar{q}$ annihilation but also from gluon fusion. For diphoton production there is interference between the $t$ channel SM production and the $s$ channel production mediated by a graviton. Here also the dielectron cross section can be written as a sum of 3 terms: the SM contribution, the interference term weighted by $\eta \equiv F/M_S^4$, and the Kaluza-Klein term weighted by $\eta^2$. Throughout this paper, we take the definition $F = \frac{2\lambda}{\pi}$ [3], except when we derive the limits as a function of the number $n$ of extra dimensions, in which case we have $F \equiv \log(M_S^2/s)$ for n=2 and $F \equiv \frac{2}{n-2}$ for $n > 2$ [4].

The DØ experiment performs the search by combining the dielectron and diphoton final states and looking for deviations in the distributions of the polar angle of these electromagnetic objects in the helicity frame and in their invariant mass. The good agreement with the SM in the 1992-1996 data sample (corresponding to $\sim$ 120 $pb^{-1}$), allows to place limits at $M_S > 1.1(1.0)$ TeV for $\lambda = 1(-1)$, i.e. $M_S > 1$ TeV for $n$ between 2 and 6.

The CDF collaboration has made an exclusive search on a similar statistics, separating the dielectron and diphoton candidates, and fitting the corresponding invariant mass distributions, as shown in fig. 4 for the diphoton sample. The limits obtained are $M_S > 0.80, 0.81(0.90, 0.83)$ TeV from the diphoton,dielectron samples and $\lambda = 1(-1)$, giving a combined limit of 0.85 (0.94) TeV. On the first data taken this year after the a 5 years long upgrade of the machine and the detectors, DØ has repeated its search, based on a smaller statistics ($\sim$ 9 $pb^{-1}$) but at a larger center of mass energy (2 TeV vs 1.8 before). No signal has been found and the preliminary limit has been set at 0.82 TeV. This result, based on a few months of data taking shows the potential of the Run II which is expected to accumulate a statistics larger by 2 orders of magnitude compared to Run I over the next 6 years.

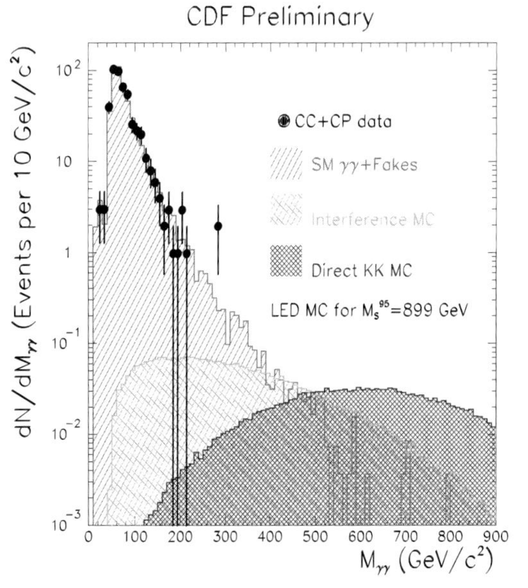


Figure 4. Comparison of Data and SM expectation for the diphoton invariant mass spectrum as obtained by CDF on 110 $pb^{-1}$ of Tevatron Run I data. The SM-ED interference and the direct ED effects are also shown for $M_S = 0.9$ TeV.

## 4. Direct Searches of ED at LEP

If a graviton is emitted in an $e^+e^-$ collision, it can propagates in the 4+n dimensions, and the energy momentum will not be conserved in our usual space. The direct searches look for missing energy, which can be interpreted in terms of a missing mass due to an escaping particle. At LEP the most favourable process is $e^+e^- \to \gamma G$. The differential cross-section for this process depends on both the gravity scale $M_D$ and the number $n$ of ED and is given in ref. [2].

In presence of ED, the distributions of the missing mass and polar angle $\theta$, would be enhanced at high missing mass and large $\cos\theta$. As shown for instance from ALEPH in fig. 5 this has not been observed allowing to set limits up to $M_S = 1.36$ for $n = 2$ in the HLZ formalism [4]. The results of the 4 LEP experiments are summarized in table 4, and the variation of these limits as a function of $n$ are shown for DELPHI in fig. 6.

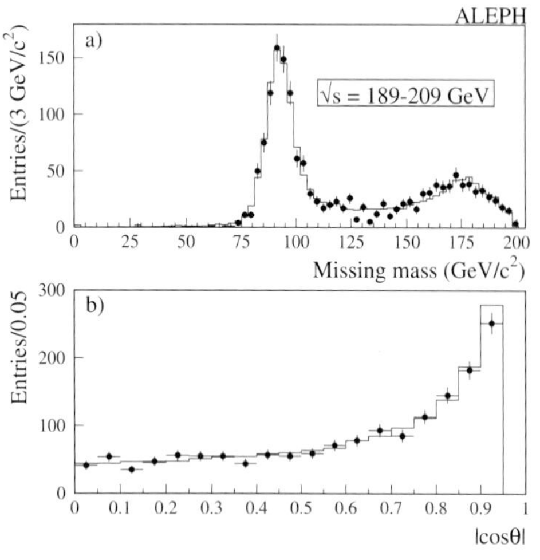


Figure 5. Distributions from the $\gamma + E_T^{miss}$ ALEPH sample :a) invariant mass of the system recoiling against the photon candidate; b) $\gamma$ polar angle. The effects of ED would be visible at large missing mas and polar angle.

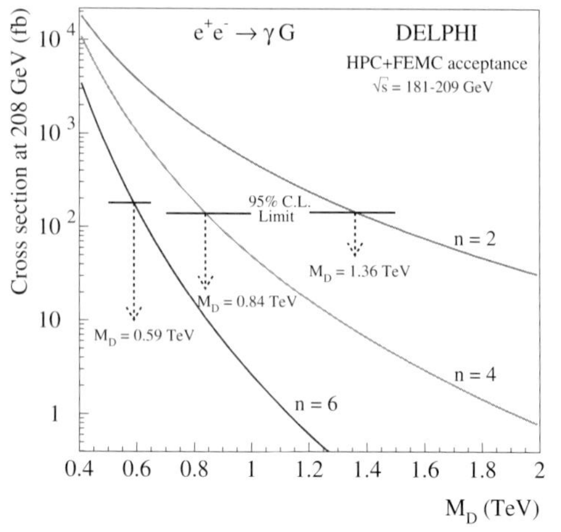


Figure 6. Cross section limits for $e^+e^- \to \gamma G$ production at $\sqrt{s} = 208$ GeV as obtained by DELPHI, compared to the expected ED cross section for $n = 2, 4, 6$ as a function of $M_D$.

| Collab. | $n = 2$ | $n = 4$ | $n = 6$ |
|---|---|---|---|
| ALEPH | 1.26 TeV | 0.77 TeV | 0.57 TeV |
| DELPHI | 1.36 TeV | 0.84 TeV | 0.59 TeV |
| L3 | 1.02 TeV | 0.67 TeV | 0.51 TeV |
| OPAL | 1.09 TeV | 0.71 TeV | 0.53 TeV |

Table 2
Limits on $M_S$ obtained from direct graviton production at LEP in the $e^+e^- \to \gamma G$ process.

## 5. Direct Search for ED at the Tevatron

At the Tevatron another channel can be searched for: the production of gluon-graviton, leading to a monojet + missing transverse ($E_T^{miss}$) energy topology. DØ has performed such a search on 80 pb$^{-1}$ of Run I data, asking for large missing and leading jet transverse energies ($> 150 GeV$). The measured distribution of ($E_T^{miss}$) and the expected effect of ED with $M_D$= 1 TeV are shown in fig. 7. The data/SM agreement allows to derive limits displayed in fig. 8, and compared to the LEP limits, showing that this channel is more sensitive when the number of ED is large ($\geq 5$).

A search in the $\gamma G$ channel has been performed by CDF on 87 pb$^{-1}$ of Run I data, asking for large missing and photon transverse energies ($>$ 45 GeV). After all cuts the 11 observed events are compatible with the expected SM background (11±2), leading to a limit on $M_D$ of 0.55,0.58,0.60 TeV for $n = 4, 6, 8$, as shown in fig. 8.

## 6. Prospects

At present the limits obtained at the 3 highest energy colliders are comparable, of the order of 1 TeV, and the searches have been made in a large variety of channels reinforcing the confidence in this limit. In future, while waiting for the start of the LHC, this domain will mostly be explored at the Tevatron, with a sensitivity up to about 2 (3) TeV with the full luminosity of the Run II a(b). Eventually with the advent of the LHC and an hypothetical linear collider this domain will be explored up to about 8 TeV, allowing to draw

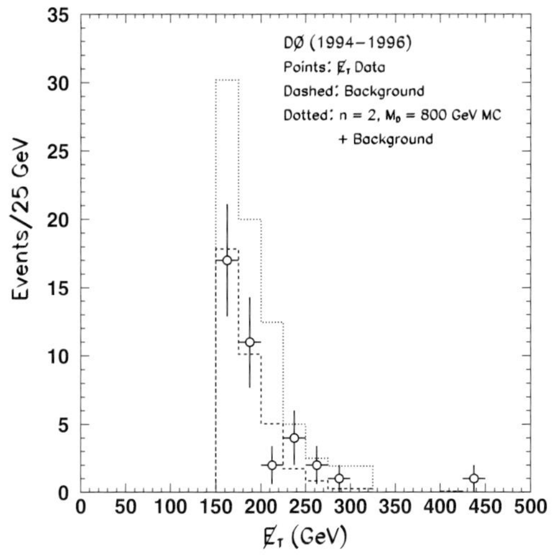


Figure 7. Distribution of $E_T^{miss}$ as obtained by DØ in 1992-1996 Data, compared to SM expectations and SM+ED expectations for $n = 2$ and $M_D = 800$ GeV.

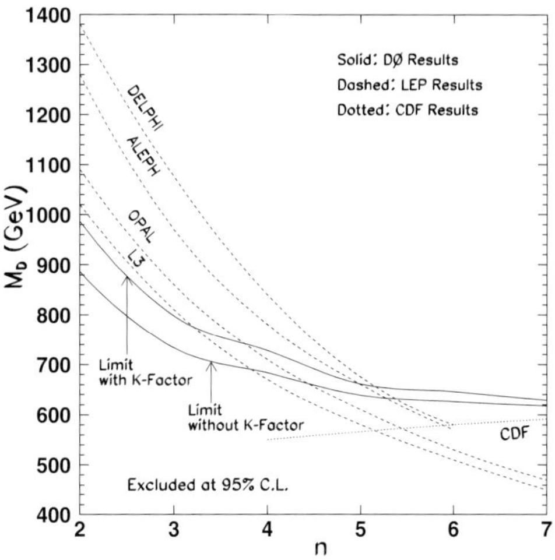


Figure 8. 95 % C.L. limits in the $\gamma G$ (LEP,CDF) and $gG$ (DØ) processes as a function of $M_D$ and $n$. The DØ limit is shown with and without the application of a K factor of 1.3 to the expected $gG$ cross-section.

definite conclusions on the possibilities that ED can solve the hierarchy problem.

## REFERENCES


1. N. Arkani-Hamed, S. Dimopoulos, G. Dvali, Phys. Lett. B429, 263 (1998);
2. G. Giudice, R. Rattazzi, and J. Wells, Nucl. Phys. B544, 3 (1999), and revised version hep-ph/9811291 v2.
3. J. Hewett, Phys. Rev. Lett. 82, 4765 (1999).
4. T. Han, J. Lykken, and R. Zhang, Phys. Rev. D 59, 105006 (1999), and revised version.
5. K. Agashe and N.G. Deshpande, Phys. Lett. B 456, 60 (1999).
6. ZEUS Collab., Eur.Phys.J. **C1** (1998) 109.
7. H1 Collab.,Contributed paper to ICHEP-2002, Abstract 979.
8. ALEPH Collab., CERN-EP/2002-033 (2002), subm. to Eur. Phys. J. C.
9. DELPHI Collab., DELPHI 2002-077 CONF 611, Contributed paper to ICHEP-2002, Abstract 592.
10. OPAL Collab., OPAL Phys. Note PN471 (2002), Contributed paper to ICHEP-2002, Abstract 509; OPAL Collab., CERN-EP/2002-060 (2002), subm. to Eur. Phys. J. C.
11. L3 Collab., CERN-EP/2001-096 (2001), subm. to Eur. Phys. J. C.; L3 Collab., L3 Note 2759 (2002), Contributed paper to ICHEP-2002, Abstract 462.
12. J. Holt, LEP-EWeak W.G., Private communication.
13. DØ Collab., Phys. Rev. Lett. 86, 1156 (2001); DØ Collab., contributed paper to ICHEP-2002, abstract 556.
14. CDF Collab., FERMILAB-CONF-01/363-E; CDF Collab., hep-ex/0205057 (2002), subm. to Phys. Rev. Lett.

ELSEVIER

Proceedings of ICHEP 2002, pp. 764–766
*S. Bentvelsen, P. de Jong, J. Koch and E. Laenen (Editors)*

31st INTERNATIONAL CONFERENCE ON HIGH ENERGY PHYSICS AMSTERDAM

# Collider implications of models with extra dimensions

C. Macesanu[a], C.D. McMullen[b] and S. Nandi[a*]

[a]Department of Physics, Oklahoma State University,
Stillwater, Oklahoma, 78074, USA

[b]Department of Physics, Penn State Altoona University,
Altoona, PA, 16601, USA

We discuss the collider signals of large extra dimensions in which gravity as well as the SM particles propagate into the extra dimensions. These signals arise either from the production of Kaluza-Klein excitations of the SM particles and their subsequent decay, or from their off-shell exchanges. Depending on the scenario, the dominant signals are two high $p_T$ jets + missing energy, two high $p_T$ photons + missing energy and soft leptons, or a combination of photon + jet and missing energy. For the scenario in which only the gauge bosons propagate into the extra dimensions, Tevatron Run II (LHC) can observe such signals up to a compactification scale of about 2 TeV (7 TeV), while for the case of universal extra dimensions, the corresponding limits are about 600 GeV (3 TeV) respectively.

## 1. Introduction

Recent developements [1] in string theory have led to the construction of models in which the compactification scale of the extra dimensions can be as large as inverse TeV [2] or even submilimeter [3]. In this later case, due to experimental constraints, only gravity can be allowed to propagate into the extra dimensions, while for the inverse TeV scale, both gravity and the Standard Model particles can be in the bulk. The phenomenological implications for the models with submilimeter extra dimensions have been extensively discussed in the past few years. In this talk, we will concentrate on the implications of the inverse TeV scale extra dimensions. In this case, there are two major possible scenarios: only the SM gauge bosons propagate into the extra dimensions, or all the SM particles propagate in the bulk (also called the Universal Extra Dimensions scenario). A hybrid scenario [4] in which the model accomodates sub-mm (in which only gravity propagates) as well as inverse TeV scale extra dimensions (where SM particles also propagate) has additional interesting features. Such a model gives rise to deviations from Newton's law of gravity at sub-mm distances and several interesting astrophysical phenomena. It also leads to new physics signals at high energy colliders due to the production of KK excitations of SM particles.

## 2. Only gauge bosons in the bulk

In this scenario, only the SM gauge bosons propagate into the extra dimensions. The SM fermions are confined to the usual 3+1 dimensional wall, the $D_3$-brane. We will consider the case of only one extra dimension of inverse TeV size. The collider signals will be similar for more than one extra dimension. The 5th dimension is compactified on an $S^1/Z_2$ orbifold (with $y \to -y$ symmetry) and the physical space is $M_4 \times S^1/Z_2$, where $M_4$ is the Minkowski space. The five dimensional gauge fields can be expanded as:

$$A^a_\mu = \frac{1}{\sqrt{\pi R}}\left[A^a_{\mu,0} + \sqrt{2}\sum_{n=1}^{\infty} A^a_{\mu,n}\cos(\frac{ny}{R})\right]$$
$$A^a_4 = \frac{\sqrt{2}}{\sqrt{\pi R}}\sum_{n=1}^{\infty} A^a_{4,n}\sin(\frac{ny}{R}). \qquad (1)$$

Here $A^a_{\mu,n}$ are the $n$th KK excitations of the gauge bosons with masses $m_n \equiv n/R$. The collider sig-

*Invited talk presented by S. Nandi at the ICHEP 2002, Amsterdam, The Netherlands, July 24-31, 2002


doi:10.1016/S0920-5632(02)02296-X

nals will mainly come from the production of the KK excitations of the gluons ($g_n$) and their subsequent decays, or through the off-shell exchange of the KK gluons in the usual SM processes. In this model, we can have new couplings such as $\bar{q}qg_n, g g_n g_n, g_n g_m g_p$ etc. (The detailed interactions and the Feynman rules for all the allowed couplings can be found in [5]). The new collider signals arise from the processes $q\bar{q} \to g^* \to q\bar{q}$ (dijet signal); $q\bar{q} \to g^* g \to q\bar{q}g$ (3-jet signal) and $q\bar{q} \to g^* g^* \to (q\bar{q})(q\bar{q})$ (4-jet signal). (Here $g^*$ stands for a general KK excitation of the gluon). Since these jets are coming from the decay of a very heavy particle, they carry much higher $p_T$ than the jets produced by the usual SM processes. The high $p_T$ dijet signal is observable over the usual QCD background, and gives the reach for the compactification scale of about 2 TeV at the Tevatron Run II and 7 TeV at the LHC [5].

## 3. Universal extra dimensions

In this scenario, all SM particles propagate in the bulk [6]. Again, for simplicity, we consider the case of only one extra dimension of inverse TeV size. With compactifiation on an $S^1/Z_2$ orbifold, the decomposition of the 5-dimensional gauge fields is the same as above (Eq. (1)). In 5 dimensions, the fermion fields are vectorlike; in order to obtain the chiral fermions of the SM, there are needed two sets of fermionic fields, with different properties under the $Z_2$ parity. Thus, we introduce an SU(2) doublet, whose left-handed fields are even and right-handed fields are odd under $Z_2$ parity:

$$Q = \frac{1}{\sqrt{\pi R}}\left\{Q_L + \sqrt{2}\sum_{n=1}^{\infty}\left[Q_L^n \cos\left(\frac{ny}{R}\right) + Q_R^n \sin\left(\frac{ny}{R}\right)\right]\right\} \tag{2}$$

and an SU(2) singlet with opposite properties:

$$q = \frac{1}{\sqrt{\pi R}}\left\{q_R + \sqrt{2}\sum_{n=1}^{\infty}\left[q_R^n \cos\left(\frac{ny}{R}\right) + q_L^n \sin\left(\frac{ny}{R}\right)\right]\right\}. \tag{3}$$

With these definitions, we obtain the correct chiral nature of the SM fermions (which are build from the zero modes of the above fields). For the KK exciations we then obtain two towers for each SM fermionic field: $q_n^\bullet = Q_L^n + Q_R^n$ and $q_n^\circ = q_L^n + q_R^n$ (ignoring small mixing due to Yukawa interactions). The masses of the fermionic excitations are given by $m_n^2 = (n/R)^2 + m_{SM}^2$. The Feynman rules for the allowed couplings in this model can be found in [7]. The most interesting feature of this model is the KK number conservation at tree level. As a consequence, excitations of gluons and quarks can only be pair produced at colliders, which results in significantly lower limits (around 350 GeV) set on the compactification scale $1/R$ from current experiments [6,7].

### 3.1. Decay mechanisms

There are two possible decay scenarios for KK excitations in the UED models.

(i) Loop effects: In first order, the masses of the KK excitations of the gluons and light quarks are essentialy degenerate at each level. However, loop corrections due to virtual particles propagating in the bulk, as well as boundary terms due to the breaking of translational invariance at the orbifold boundaries will remove this degeneracy. With some assumptions (which include universality for the boundary terms as well as the introduction of a cutoff scale $\Lambda$) it is possible to compute the corrections to the KK masses at all levels [8]. Taking $\Lambda = 10$ TeV, the relative corrections to the first KK level masses are as follows: $\delta m_{g^*} \simeq 30\%$, $\delta m_{q^*} \simeq 20\%$ and $\delta m_{W^*,Z^*,l^*} \leq 10\%$. The lightest KK particle (LKP) will then be the $\gamma^*$, and, due to KK number conservation, it will be stable, with all other first level excitations decaying to it. In this scenario, the decay modes for the quark and gluon excitations produced at a hadron collider will be:

$$g_1 \to q\bar{q_1^\bullet}, q\bar{q_1^\circ}\ ;\quad q_1^\circ \to q\gamma^*\ ;\ \text{and}$$

$$q_1^\bullet \to q\ Z_1^* \to q\ l\ l_1^\bullet \to q\ l\ l\ \gamma^*\ , \text{Br.} \sim 33\%,\ \text{or}$$

$$q_1^\bullet \to q\ W_1^* \to q\ l'\ l_1^\bullet \to q\ l'\ l\ \gamma^*\ , \text{Br.} \sim 65\% \tag{4}$$

The final states in this scenario contains two $\gamma^*$, plus soft jets and leptons. If this is the only decay mode, then the observable signal will be soft leptons plus large missing energy (from the LKP).

Then, the Tevatron Run II reach is about 300 GeV, while the LHC reach is about 1.5 TeV [9].
(ii)Gravity mediated decays (fat brane scenario): In this scenario, gravity-mediated interactions break KK number conservation, allowing decays such as

$$g^* \to gG_n,\ q^* \to qG_n,\ \gamma^* \to \gamma G_n \quad (5)$$

where $G_n$ is the KK excitation of the graviton. We assume here that gravity propagates in $N$ extra dimensions of submilimeter size (such that the above decays are kinematically allowed), while the SM particles can propagate only a small distance -inverse TeV size- along one of these dimensions, i.e., the SM matter lives on a fat $D_3$ brane [10]. In this case, one obtains a form factor describing the superposition of the matter wave function with the graviton wave function along the fifth dimension, which gives rise to the couplings which allow the decays in Eq. (5).

### 3.2. Collider signals for the fat brane scenario

In UED models, KK excitations of quarks and gluon are pair produced at hadron colliders. The final state signal will depend on which of the above decay mechanism, (i) or (ii), dominates. If $\Gamma_{(ii)} \gg \Gamma_{(i)}$, then the final state signal will be two jets plus missing energy due to the escaping graviton excitations. The transverse momenta of the observable jets will be quite large, as will be the missing transverse energy (for an analysis, see [7]). This will allow elimination of the SM backgrounds with suitable $p_T$ and $\not{E}_T$ cuts. The Tevatron Run II reach in this mode is about 500 GeV, while the LHC reach is about 3 TeV [7].

If the decay mechanism (i) dominates over (ii), i.e. $\Gamma_{(i)} \gg \Gamma_{(ii)}$, then the KK excitations produced at colliders will first cascade decay to the LKP as in (4). However, due to gravitational interaction the LKP is not stable anymore and it will decay: $\gamma^* \to \gamma G_n$. The signal in this case will the be two hard photons plus large $\not{E}_T$ (and some soft leptons and jets). Collider reach in this case is around 550 GeV at the Tevatron Run II and 3 TeV at the LHC [11].

If we are in a region of the parameter space where $\Gamma_{(i)} \simeq \Gamma_{(ii)}$, then the signal will be a mixture of the modes discussed above, that is $\gamma\gamma$, $\gamma$+jet, 2 jets, plus large $\not{E}_T$ in the final states. For details, see [11].

## 4. Conclusions

Models with extra dimensions in the inverse TeV range can lead to interesting phenomenology at high energy colliders. Depending on the scenario, the signals may be high $p_T$ dijets plus missing $E_T$, or even high $p_T$ diphotons plus $\not{E}_T$. Tevatron reach is from $\sim$ 400 GeV to 2 TeV, while LHC reach is from $\sim$ 1.5 TeV to around 7 TeV, depending on the model (the higher limits are for the case when only the SM gauge bosons propagate in the extra dimensions). For low compactification scales, LHC may observe the second level KK excitations, which will be a clear signal for the existence of extra dimensions.

**Acknowledgements**

This work was supported in part by the U.S. Department of Energy Grant Numbers DE-FG03-98ER41076 and DE-FG02-01ER45684.

## REFERENCES

1. E. Witten, Nucl. Phys. **B471**, 135 (1996); J. D. Lykken, Phys. Rev. D **54**, 3693 (1996).
2. I. Antoniadis, Phys. Lett. **B246**, 377 (1990).
3. N. Arkani-Hamed, S. Dimopoulos and G. Dvali, Phys. Lett. **B429**, 263 (1998).
4. J. Lykken and S. Nandi, Phys. Lett. **B485**, 224 (2000).
5. D.A. Dicus, C.D. McMullen and S. Nandi, Phys. Rev. **D65**, 076007 (2002).
6. T. Appelquist, H.-C. Cheng and B.A. Dobrescu, Phys. Rev. **D64**, 035002 (2001).
7. C. Macesanu, C.D. McMullen and S. Nandi, Phys. Rev. **D66**, 015009 (2002).
8. H.C. Cheng, K.T. Matchev and M. Schmaltz, Phys. Rev. **D66**, 036005 (2002).
9. H.C. Cheng, K.T. Matchev and M. Schmaltz, Phys. Rev. **D66**, 056006 (2002).
10. A. DeRujula, A. Donini, M.B. Gavela and S. Rigolin, Phys. Lett. **B482** 195 (2000); T.G. Rizzo, Phys. Rev. **D64**, 095010 (2001).
11. C. Macesanu, C.D. McMullen and S. Nandi, Phys. Lett. **B546**, 253 (2002).

ELSEVIER

Proceedings of ICHEP 2002, pp. 767–769
*S. Bentvelsen, P. de Jong, J. Koch and E. Laenen (Editors)*

31st INTERNATIONAL CONFERENCE ON HIGH ENERGY PHYSICS AMSTERDAM

# Black Holes at Future Colliders

G. Landsberg[a]

[a]Brown University, Department of Physics, 182 Hope St., Providence, RI 02912, USA
E-mail: landsberg@hep.brown.edu

One of the most dramatic consequences of low-scale ($\sim 1$ TeV) quantum gravity is copious production of mini black holes at future accelerators. Hawking radiation of these black holes is constrained mainly to our (3+1)-dimensional world and results in rich phenomenology. We discuss a sensitive probe of the dimensionality of extra space, as well as an exciting possibility of finding new physics in the decays of black holes.

## 1. BLACK HOLE PHENOMENOLOGY

It has been pointed out in the past [1] that an exciting consequence of TeV-scale quantum gravity [2] is the possibility of production of black holes (BHs) at the high-energy accelerators. Recently, this phenomenon has been quantified for the case of TeV-scale particle collisions [3,4], resulting in a mesmerizing prediction that future colliders would produce mini black holes at an enormous rate (e.g., $\sim 1$ Hz at the LHC for the fundamental Planck scale $M_P = 1$ TeV), thus becoming black-hole factories.

In this talk we review the phenomenology of the black hole production and decay in high-energy collisions and point out exciting ways of studying quantum gravity and searching for new physics using large samples of black holes accessible at future colliders.

Black holes are well understood general-relativistic objects when their mass $M_{\rm BH}$ far exceeds the fundamental (higher dimensional) Planck mass $M_P \sim$TeV. As $M_{\rm BH}$ approaches $M_P$, the BHs become "stringy" and their properties complex. In what follows, we will ignore this obstacle and estimate the properties of light BHs by simple semiclassical arguments, strictly valid for $M_{\rm BH} \gg M_P$. We expect that this will be an adequate approximation, since the important experimental signatures rely on two simple qualitative properties: (i) the absence of small couplings and (ii) the "democratic" nature of BH decays, both of which may survive as average properties of the light descendants of BHs. We will focus on the production and sudden decay of Schwarzschild black holes. The Schwarzschild radius $R_S$ of an $(4+n)$-dimensional black hole has been derived in Ref. [5], assuming that all $n$ extra dimensions are large ($\gg R_S$).

As we expect unknown quantum gravity effects to play an increasingly important role for the BH mass approaching the fundamental Planck scale, following the prescription of Ref. [3], we do not consider BH masses below the Planck scale. It is expected that the BH production rapidly turns on, once the relevant energy threshold $\sim M_P$ is crossed. (At lower energies, we expect BH production to be exponentially suppressed due to the string excitations or other quantum effects.)

Consider two partons with the center-of-mass energy $\sqrt{\hat{s}} = M_{\rm BH}$ moving in opposite directions. Semiclassical reasoning suggests that if the impact parameter is less than the (higher dimensional) Schwarzschild radius, a BH with the mass $M_{\rm BH}$ forms. Therefore the total cross section can be estimated from geometrical arguments and is of order $\pi R_S^2$.

Using the expression for the Schwarzschild radius [5], we derive the following parton level BH production cross section [3]:

$$\sigma(M_{\rm BH}) \approx \pi R_S^2 = \frac{1}{M_P^2}\left[\frac{M_{\rm BH}}{M_P}\left(\frac{8\Gamma\left(\frac{n+3}{2}\right)}{n+2}\right)\right]^{\frac{2}{n+1}}.$$

In order to obtain the production cross section in $pp$ collisions at the LHC, the parton level


doi:10.1016/S0920-5632(02)02297-1

cross section is convoluted with parton distribution functions (PDF) [3,4]. (The dependence of the cross section on the choice of PDF is $\sim 10\%$.) The total production cross section for $M_{\rm BH} > M_P$ at the LHC ranges between 15 nb and 1 pb for the Planck scale between 1 TeV and 5 TeV, and varies by $\sim 10\%$ for $n$ between 2 and 7.

Once produced, the TeV BHs quickly evaporate via Hawking radiation [6] with a characteristic temperature [5]

$$T_H = M_P \left( \frac{M_P}{M_{\rm BH}} \frac{n+2}{8\Gamma\left(\frac{n+3}{2}\right)} \right)^{\frac{1}{n+1}} \frac{n+1}{4\sqrt{\pi}} = \frac{n+1}{4\pi R_S}$$

of $\sim 100$ GeV [3,4]. The average multiplicity of particles produced in the process of BH evaporation is given by [3,4] and is of the order of half-a-dozen for typical BH masses accessible at the LHC. Since gravitational coupling is flavor-blind, a BH emits all the $\approx 120$ SM particle and antiparticle degrees of freedom with roughly equal probability. Accounting for the color and spin, we expect $\approx 75\%$ of particles produced in the BH decays to be quarks and gluons, $\approx 10\%$ charged leptons, $\approx 5\%$ neutrinos, and $\approx 5\%$ photons or $W/Z$ bosons, each carrying hundreds of GeV of energy. Consequently, if there exist new particles with the mass $\sim 100$ GeV, they would be produced in the decays of BHs with the probability similar to that for the SM species. For example, a sufficiently light Higgs boson is expected to be emitted in BH decays with $\sim 1\%$ probability. This has exciting consequences for searches for new physics at the LHC and beyond, as the production cross section for any new particle via this mechanism is: (i) large and (ii) depends only weakly on particle mass, in contrast with the exponentially suppressed direct production mechanism.

A relatively large fraction of prompt and energetic photons, electrons, and muons are expected in the high-multiplicity BH decays, which would make it possible to select pure samples of BH events, which are also easy to trigger on [3,4]. At the same time, only small fraction of the particles produced in the BH decays are undetectable gravitons and neutrinos, so most of the BH mass is radiated in the form of visible energy.

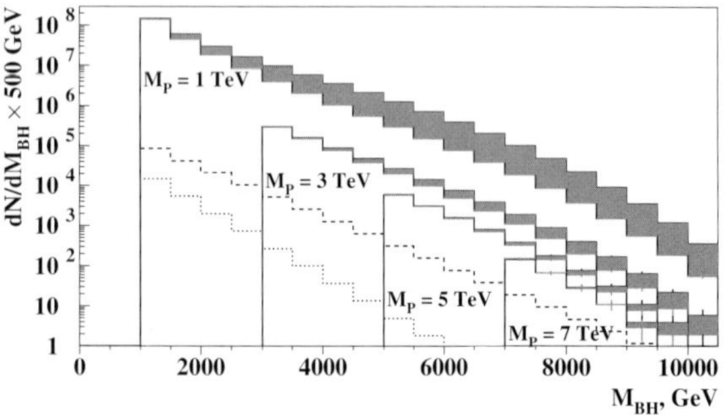


Figure 1. Number of BHs produced at the LHC in the electron or photon decay channels, with 100 fb$^{-1}$of integrated luminosity, as a function of the BH mass. The shaded regions correspond to the variation in the number of events for $n$ between 2 and 7. The dashed line shows total SM background (from inclusive $Z(ee)$ and direct photon production). The dotted line corresponds to the $Z(ee)+X$ background alone. From Ref. [3].

In Fig. 1 we show the number of BH events tagged by a presence of an energetic electron or photon among the decay products in 100 fb$^{-1}$ of data collected at the LHC, together with the SM backgrounds, as a function of the BH mass [3]. The LHC reach extends up to $M_P \approx 9$ TeV ($n =$ 2–7), where one would expect to see a handful of BH events with negligible background.

A sensitive test of properties of Hawking radiation can be obtained by measuring the relationship between the mass of the BH (reconstructed from the total energy of all the decay products) and its Hawking temperature (measured from the energy spectrum of the decay products). We can use the measured $M_{\rm BH}$ vs. $T_H$ dependence to determine both the fundamental Planck scale $M_P$ and the dimensionality of space $n$. This is a multidimensional equivalent of the Wien's law. Note that determination of $n$ can be done in largely model-independent way via taking a logarithm of both parts of the expression for Hawking temperature: $\log(T_H) = -\frac{1}{n+1}\log(M_{\rm BH}) + {\rm const}$, where the constant does not depend on the BH mass, but only on $M_P$ and on detailed properties of the bulk space [3]. Therefore, the slope of a straight-line fit to the $\log(T_H)$ vs. $\log(M_{\rm BH})$ data offers a direct way of determining the dimensionality of space. The reach of this method at the LHC is discussed in detail in Ref. [3]. Note that this method is fundamentally different from other ways of determining $n$, e.g. by studying a monojet signature or a virtual graviton exchange processes, also predicted by theories with large extra dimensions, but sensitive to the volume of extra space, rather than $n$.

Finally, as an example of a radically new way of looking for new physics in decay of black holes [7], we study the discovery potential of the BH sample collected at the LHC for a SM-like Higgs boson with the mass of 130 GeV, still allowed in low-scale supersymmetry models, but very hard to establish at the Fermilab Tevatron. We consider decay of the Higgs boson into a pair of jets, dominated by the $b\bar{b}$ final state (57%), with an additional 10% contribution from the $c\bar{c}$, $gg$, and hadronic $\tau\tau$ final states. We assume a 1% probability of the Higgs boson emission in the BH evaporation and take into account the geometry and resolution of a typical LHC detector.

The simplest way to look for the Higgs boson in the BH decays is to reconstruct the dijet invariant mass spectrum for all possible combinations of jets found in the final state. With this method, the $5\sigma$ discovery of a 130 GeV Higgs boson may be possible with $\mathcal{L} \approx 2$ pb$^{-1}$ (first day), 100 pb$^{-1}$ (first week), 1 fb$^{-1}$ (first month), 10 fb$^{-1}$ (first year), and 100 fb$^{-1}$ (full LHC sample) for the fundamental Planck scale of 1, 2, 3, 4, and 5 TeV, respectively, even with incomplete and poorly calibrated LHC detectors. The integrated luminosity required is significantly lower than that for direct Higgs boson production, if the Planck scale is below 4 TeV. While the studies were done for a particular Higgs boson mass, the dependence on the mass is small. Moreover, this approach is applicable to searches for other new particles with the masses $\sim 100$ GeV, e.g. low-scale supersymmetry. Light slepton or top squark searches via this technique may be particularly fruitful.

## 2. CONCLUSIONS

To conclude, black hole production at the LHC may be one of the early signatures of TeV-scale quantum gravity. It has three distinct advantages: (i) large cross section; (ii) hard, prompt, charged leptons and photons in the final state; and (iii) little missing energy. Large samples of black holes accessible at the LHC would allow for precision determination of the parameters of the bulk space and may result in the discovery of new particles in the BH evaporation.

If the models with large extra dimensions are realized in nature, the production of BH in the lab is just years away. That would mark an exciting transition for astroparticle physics, truly unifying it with cosmology.

## 3. ACKNOWLEDGEMENTS

I would like to thanks the ICHEP 2002 organizers for the kind invitation and an exciting conference. I am indebted to Savas Dimopoulos, my coauthor on the first black hole paper [3], for a number of stimulating discussions and support. This work has been supported partially by the U.S. Department of Energy under Grant No. DE-FG02-91ER40688 and by the Alfred P. Sloan Foundation.

## REFERENCES

1. P.C. Argyres, S. Dimopoulos, and J. March-Russell, Phys. Lett. **B441**, 96 (1998); T. Banks and W. Fischler, JHEP **9906**, 014 (1999); R. Emparan, G.T. Horowitz, and R.C. Myers, Phys. Rev. Lett. **85**, 499 (2000).
2. N. Arkani-Hamed, S. Dimopoulos, and G. Dvali, Phys. Lett. B **429**, 263 (1998); I. Antoniadis, N. Arkani-Hamed, S. Dimopoulos, and G. Dvali, Phys. Lett. B **436**, 257 (1998); N. Arkani-Hamed, S. Dimopoulos, and G. Dvali, Phys. Rev. D **59**, 086004 (1999).
3. S. Dimopoulos and G. Landsberg, Phys. Rev. Lett. **87**, 161602 (2001).
4. S.B. Giddings and S. Thomas, Phys.Rev. D **65**, 056010 (2002).
5. R.C. Myers and M.J. Perry, *Ann. Phys.* **172**, 304 (1986).
6. S.W. Hawking, Commun. Math. Phys. **43**, 199 (1975).
7. G. Landsberg, Phys. Rev. Lett. **88**, 181801 (2002).

Proceedings of ICHEP 2002, pp. 770–773
S. Bentvelsen, P. de Jong, J. Koch and E. Laenen (Editors)

# Searches for New Physics at the LHC

D.R. Tovey [a] on behalf of the ATLAS and CMS collaborations

[a]Department of Physics and Astronomy, University of Sheffield, Hounsfield Road, Sheffield S3 7RH, UK.

The considerable centre-of-mass energy and luminosity provided by the Large Hadron Collider will ensure a discovery reach for new particles which extends well into the multi-TeV region. ATLAS and CMS have carried out many studies of the implications of this capability for Beyond the Standard Model physics, focussing in particular on models incorporating supersymmetry and extra dimensions. In this paper we summarise some key results of these studies.

## 1. Introduction

The ATLAS [1] and CMS [2] experiments at the Large Hadron Collider will give access to an unprecedented range of processes at $p_T$ scales and event rates far in excess of those generated by earlier colliders. Many rare Standard Model (SM) processes will be studied with high statistics while at the same time searches will be carried out for particles predicted to exist by theories extending the SM. The need to stabilise the mechanism for breaking electroweak (EW) symmetry in the SM requires that at least some of these new particles have masses of order the EW scale ($\sim$ 100 GeV). Hence there is every reason to believe that they should be observable at the LHC.

In this paper we review the results of some of the studies of 'Beyond the Standard Model' physics at the LHC which have been carried out by ATLAS and CMS, concentrating on those involving supersymmetry (SUSY) or extra dimensions. For reasons of space we do not discuss results of recent studies of compositeness or technicolour models but further details of this and other work can be found in Refs. [3–5].

## 2. Supersymmetry

Supersymmetry is one of the oldest and best motivated of the theories predicting the existence of physics beyond the Standard Model [6]. SUSY models typically require the existence of at least one SUSY partner for each SM fermion and vector boson, together with a considerably enlarged Higgs sector. With the exception of spin these SUSY particles ('sparticles') possess the same quantum numbers as their SM counterparts. The fact that we have not observed such states in nature implies that SUSY is broken by a mechanism which causes sparticles to acquire masses greater than SM states. The precise nature of this mechanism is curently unknown but candidates include gravity-mediation (e.g. minimal Supergravity or mSUGRA models [7]) as well as gauge mediation (GMSB models) and anomaly mediation (AMSB models). A further feature of many SUSY models is the absolute conservation of a multiplicative quantum number known as R-parity at each SUSY vertex, which causes SUSY states to be pair produced and forces the Lightest Supersymmetric Particle (LSP) to be neutral and stable. Missing transverse energy generated by the escape of two such LSPs from each SUSY event provides the classic signature for R-parity conserving SUSY at hadron colliders.

Inclusive searches for R-parity conserving SUSY using the generic $E_T^{\rm miss}$ signature will be carried out by ATLAS and CMS from the outset of LHC operation. In addition to cuts on $E_T^{\rm miss}$, cuts can be imposed on the multiplicity and $p_T$ of jets and leptons in order to reduce SM background from W+jet, Z+jet and $t\bar{t}$ processes and mismeasured QCD events. Studies have shown that within the mSUGRA framework both ATLAS and CMS should be sensitive to SUSY mod-


doi:10.1016/S0920-5632(02)02298-3

els with gluino and squark masses $\sim$ 2 TeV after one year of low luminosity running (10 $\text{fb}^{-1}$ integrated luminosity), rising to $\sim$ 2.5 TeV after one year of high luminosity running (100 $\text{fb}^{-1}$) [3,8–11]. Following discovery of a SUSY signal using this signature, the mass scale of the produced sparticles can then be measured in a model-independent fashion with an ultimate error $\lesssim$ 10 % (mSUGRA models) by fitting to the distribution of a variable based upon the scalar sum of jet $p_T$ and $E_T^{\text{miss}}$ in jet + $E_T^{\text{miss}}$ + 0 lepton events [9,11–13]. The normalisation of this distribution also provides a measure of the total SUSY production cross-section and together these two pieces of information can be used to constrain the SUSY breaking machanism [9,13].

Further information regarding the SUSY model and its parameters can be obtained by measuring SUSY particle masses. In R-parity conserving models the presence of two invisible LSPs prevents the reconstruction of invariant mass peaks, however combinations of sparticle masses can be measured by searching for edges in dilepton invariant mass spectra [3,12]. In mSUGRA models the most prominent such edge arises from $\tilde{\chi}_2^0$ decays, with its position depending on the masses of the $\tilde{\chi}_2^0$ and its SUSY decay products, such as the $\tilde{\chi}_1^0$ and light sleptons. By selecting events in the region of the dilepton edge and forming invariant mass combinations of (di)leptons with jets it is also possible to gain additional sensitivity to the masses of left-handed squarks. Additional observation of thresholds in such invariant mass spectra then provides sufficient constraints for the full sparticle mass spectrum to be reconstructed [14]. Similar procedures have been developed for mass reconstruction when $\tilde{\chi}_2^0$ decays to the light higgs are kinematically accessible [3].

Work by CMS has investigated the origin and observability of dilepton edges over the mSUGRA $m_0$ - $m_{1/2}$ plane [15]. When kinematically allowed the $\tilde{\chi}_2^0$ decay proceeds via two sequential two-body processes involving slepton exchange:

$$\tilde{\chi}_2^0 \to \tilde{l}^{+(-)} l^{-(+)} \to \tilde{\chi}_1^0 l^+ l^- \tag{1}$$

while elsewhere a direct three body decay to the $\tilde{\chi}_1^0$ is favoured:

$$\tilde{\chi}_2^0 \to \tilde{\chi}_1^0 l^+ l^- \tag{2}$$

These two cases can be distinguished experimentally by measuring the $p_T$ asymmetry between the two leptons [11]. In certain very limited regions of parameter space the $\sigma$.BR of both decay channels can be appreciable and in this case two edges are observed, giving additional constraints on the sparticle mass spectrum (Fig. 1). In addition, measurements of the cross-section for events in the most prominent edge can be compared with the height of the Z peak arising from higher gaugino decays in order to constrain the value of $\tan(\beta)$ [15].

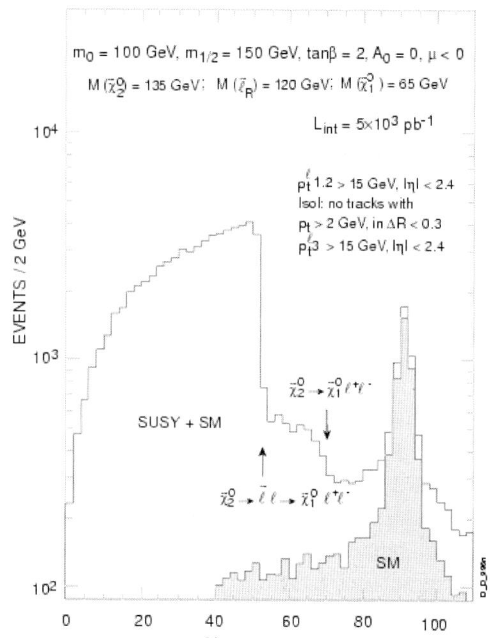


Figure 1. *Dilepton invariant mass spectrum of events generated with an mSUGRA model with parameters shown. The left and right hand edges arise from two and three body decays of* $\tilde{\chi}_2^0$ *respectively. Taken from Ref. [11].*

## 3. Extra Dimensions

Models incorporporating extra space-time dimensions (ED) have recently become extremely popular with phenomenologists. An attractive feature of many of these models is that the effective Planck scale is brought down to the EW scale by power-law running of the gauge coupling constants, thereby removing the gauge hierarchy problem. A by-product of this running is that we expect quantum gravity effects to begin to manifest themselves at the TeV scale, possibly within the range accessible to the LHC. Striking potential signatures can include graviton production, the observation of graviton resonances contributing to Drell-Yan processes and new gauge and scalar bosons. The precise characteristics of these signatures depend strongly on the size, 'shape' and number of extra dimensions.

ATLAS and CMS have investigated three different classes of ED models incorporating 'Large' [16], 'warped' [17] or $\text{TeV}^{-1}$ scale [18] extra dimensions. 'Large' ($>>$ $\text{TeV}^{-1}$) extra dimension models such as that proposed by Arkani-Hamed, Dimopoulos and Dvali [16] typically assume the existence of $\delta$ EDs of size $R$ in which only gravitons can propagate. The observed Planck scale $M_{Pl}$ following compactification is then related to the fundamental scale of gravity $M_D$ by the relation

$$M_{Pl}^2 \sim R^\delta M_D^{2+\delta} \tag{3}$$

If $M_D$ is near the EW scale then the gauge hierarchy problem is neatly solved. Such models predict the existence of an infinite number of Kaluza-Klein (KK) excitations of the graviton of very small mass splitting propagating in the $\delta$ dimensional bulk space-time. This leads to possible discovery signatures based on searches for KK graviton + quark (monojet) or photon production, in which the graviton is emitted into the bulk leaving a large amount of $E_T^{\text{miss}}$. An ATLAS study of the monojet signature [19] has shown that after one year of high luminosity running (100 $\text{fb}^{-1}$) ATLAS should be able to confirm ADD models with $M_D < 9.1$, 7.0 or 6.0 TeV for 2, 3 or 4 EDs respectively. An alternative approach searching for associated production of KK gravitons with dileptons and diphotons [20] has given equivalent reaches of 8.1, 7.9 and 7.4 TeV, indicating that potential signals may be observable in more than one channel.

Other ED models construct the universe from the usual four dimensions together with large EDs (as in ADD type models) and a number of smaller $\text{TeV}^{-1}$ scale EDs [18]. The SM fermions propagate only in the usual 4D space-time while the SM gauge bosons propagate additionally in the $\text{TeV}^{-1}$ scale EDs and gravitons propagate throughout (as in ADD). The new dynamics associated with the $\text{TeV}^{-1}$ scale EDs leads to 4D KK excitations of the SM bosons. With 100 $\text{fb}^{-1}$ peaks in dilepton invariant mass spectra arising from excitation of the first KK modes of the photon and $Z^0$ can be detected for compactification scales $M_c < 5.8$ TeV and excesses above SM Drell-Yan tails for $M_c < 9.5$ TeV [21].

A third class of ED models studied by ATLAS and CMS [22–24] is the 'warped' ED scenario of Randall and Sundrum [17]. In these models the EW scale is generated from the Planck scale through warping of one small ED, rather than the presence of 'large' flat EDs (as in ADD). The universe is here envisaged to consist of two 3-branes bounding a warped 5D bulk. The SM gauge fields propagate on one of the 3-branes while gravitons are localised on the other. This again leads to the existence of graviton KK modes, although in this case, unlike in ADD, their mass splittings are likely to be at least of order the EW scale. Direct searches for new resonances in difermion and digluon invariant mass spectra have recently been studied in this context [22] and have been found to permit observation of KK gravitons with multi-TeV masses. An equally important finding is that these states can be determined to be spin-2 rather than spin-1 for masses up to 1.72 TeV, allowing $Z'$ interpretations of potential signals to be tested (Fig. 2).

A further novel aspect of warped ED models is their prediction of the existence of a new scalar (the radion $\phi$) corresponding to radial excitations of the compactified bulk fifth dimension. Mechanisms introduced to stabilise this dimension [25,26] also generate a non-zero, EW scale mass for the radion. The radion can also mix

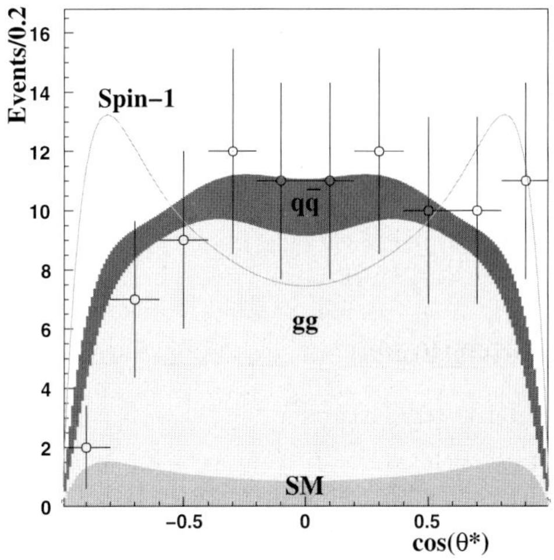


Figure 2. *Angular distribution of $q\bar{q}$ and gg events arising from excitation of a KK graviton resonance of mass $m = 1.5$ TeV for 100 $fb^{-1}$ of data. Taken from Ref. [22].*

strongly with the SM higgs scalar leading to modified phenomenology. Studies of searches using existing higgs search strategies have shown that radion signals could be identified with high significance ($S/\sqrt{B} \sim 100$) with 100 $\mathrm{fb}^{-1}$ data given 200 GeV $< m_\phi <$ 600 GeV and a scale for new physics $\Lambda_\phi = 1$ TeV [23]. With only 30 $\mathrm{fb}^{-1}$ of data limits of between 2.2 TeV and 1.0 TeV could be set on $\Lambda_\phi$ for values of $m_\phi$ between 300 GeV and 600 GeV using the $\phi \to hh \to b\bar{b}\gamma\gamma$ and $\phi \to hh \to b\bar{b}\tau\tau$ channels, assuming no higgs-radion mixing [23].

**Acknowledgements**

The author wishes to thank Georges Azuelos, Frank Paige and Giacomo Polesello for discussions. He also wishes to thank members of the ATLAS and CMS collaborations for the work described in this paper.

## REFERENCES

1. ATLAS Collaboration, *Technical Proposal*, CERN/LHCC/94-43.
2. CMS Collaboration, *Technical Proposal*, CERN/LHCC/94-38.
3. ATLAS Collaboration, *Detector and Physics Performance Technical Design Report*, CERN/LHCC/99-15.
4. see *e.g.* O. Cakir *et al.*, ATLAS NOTE ATL-PHYS-2002-014; A.K. Gupta *et al.*, CMS-NOTE-1999/075.
5. see *e.g.* G. Azuelos *et al.*, ATLAS NOTE ATL-PHYS-99-020.
6. H.P. Nilles, *Phys. Rep.* **111** (1984) 1; H.E. Haber and G.L. Kane, *Phys. Rep.* **117** (1985) 75.
7. see *e.g.* L. Alvarez-Gaume *et al.*, *Nucl. Phys.* **B221** (1983) 495.
8. H. Baer *et al.*, *Phys. Rev.* **D52** (1995) 2746; *ibid. Phys. Rev.* **D53** (1996) 6241.
9. D.R. Tovey, ATLAS SCIENTIFIC NOTE SN-ATLAS-2002-020.
10. S. Abdullin and F. Charles, *Nucl. Phys.* **B547** (1999) 60; M. Dzelalija *et al.*, *Mod. Phys. Lett.* **A15** (2000) 465.
11. S. Abdullin *et al.*, *J. Phys.* **G28** (2002) 469.
12. I. Hinchliffe *et al.*, *Phys. Rev.* **D55** (1997) 5520 9.
13. D.R. Tovey, *Phys. Lett.* **B498** (2001) 1.
14. C.G. Lester *et al*, *JHEP* **0009** (2000) 004.
15. D. Denegri, *et al.*, *Phys. Rev.* **D60** (1999) 035008.
16. N. Arkani-Hamed *et al.*, *Phys. Lett.* **B429** (1998) 263.
17. L. Randall and R. Sundrum, *Phys. Rev. Lett.* **83** (1999) 3370.
18. see *e.g.* I. Antoniadis and M. Quiros, *Phys. Lett.* **B392** (1997) 61.
19. L. Vacavant and I. Hinchliffe, *J. Phys.* **G27** (2001) 1839.
20. V. Kabachenko *et al.*, ATLAS NOTE ATL-PHYS-2001-012.
21. G. Azuelos and G. Polesello, hep-ph/0204031.
22. B.C. Allanach *et al.*, *JHEP* **0009** (2000) 019.
23. G. Azuelos *et al.*, ATLAS SCIENTIFIC NOTE (in preparation).
24. P. Traczyk and G. Wrochna, CMS NOTE 2002/003; M.C. Lemaire and J.-P. Pansart, CMS NOTE 2002/020.
25. W.D. Goldberger and M.B. Wise, *Phys. Rev. Lett* **83** (1999) 4922.
26. G.F. Giudice *et al.*, *Nucl. Phys.* **B595** (2001) 250.

ELSEVIER

Proceedings of ICHEP 2002, pp. 774–777
*S. Bentvelsen, P. de Jong, J. Koch and E. Laenen (Editors)*

31st INTERNATIONAL CONFERENCE ON HIGH ENERGY PHYSICS AMSTERDAM

# Top quark Kaluza-Klein mode mixing in the Randall-Sundrum bulk Standard Model and Constraint from $\Delta\rho$

C. S. Kim[a*], J. D. Kim[a†], J. Song[b]

[a]Department of Physics & IPAP, Yonsei University, Seoul 120-749, Korea

[b]School of Physics, Korea Institute for Advanced Study, Seoul 130-012, Korea

The Randall-Sundrum (RS) scenario with all the standard model (SM) fermions and gauge bosons in the bulk is phenomenologically studied. Even though the simple assumption of universal bulk fermion mass $m_\psi$ leads to the same Kaluza-Klein (KK) mass spectrum for all the SM fermions and thus suppresses new contributions to Flavor-Changing-Neutral-Current (FCNC) and the $\rho$ parameter, large Yukawa coupling of the top quark generates its KK mode mixing and breaks the degeneracy: unacceptably large contribution to $\Delta\rho$ occurs. With a different bulk fermion mass to SU(2) singlet bottom quark, we demonstrate that there exists some parameter space to satisfy the $\Delta\rho$ constraint.

## 1. Introduction

Extra dimensional models have drawn a lot of interest since it can leave distinct phenomenological signatures at future colliders. In the original RS scenario, the SM fields are assumed to be confined to our brane. Phenomenological signatures come from KK gravitons with electroweak scale masses and couplings to matter, characterized by $\Lambda_\pi$. However, the small size of the RS-bulk allows that the SM fields may also be in the bulk. In Ref. [1], it is demonstrated that placing the SM gauge fields in the RS-bulk while confining the fermions to our brane is strongly constrained by the current precision electroweak data so that $\Lambda_\pi$ is pushed up to about 100 TeV. This is disfavored as a solution of the gauge hierarchy problem. If both the SM gauge and fermion fields are in the bulk [2], their phenomenological signatures are very sensitive to the bulk fermion mass $m_\psi$.

In the early study of the RS-bulk SM, Yukawa interactions with the Higgs field have been ignored due to small quark masses compared to the KK mass scale. Then a simple assumption of universal bulk fermion mass suppresses their contributions to FCNC as well as the $\rho$ parameter [2,3]. This is because the common $m_\psi$ leads to the same KK mass spectra for all the SM fermions. The degeneracy of the up-type (and down-type) quark KK modes operates the GIM cancellation [4] KK-level by level: With the minimal flavor violation assumption that at the tree level the flavor mixing comes only through the CKM matrix, FCNC is suppressed as in the SM. Since two constituents of SU(2)–doublet have also the same KK mass, their contribution to the $\rho$ parameter vanishes. However, Yukawa interaction with the Higgs field mixes the fermion KK tower members, which can be substantial for the top quark [2,5]. Recently it has been shown that the large mixing in the top quark KK sector leads to unacceptably large contribution to the $\rho$ parameter and raises $\Lambda_\pi$ above 100 TeV [6]. To accommodate those electroweak precision data, a 'mixed' scenario was proposed with the third generation fermions on the TeV brane but the other generations in the bulk. However, the first excited KK mode of gauge bosons should be heavier than 11 TeV due to the strong constraints from precision measurements: It is hard to probe the new physics effects at LHC. In addition, the obvious discrimination of fermions according to genera-

*Supported in part by Grant No. R02-2002-000-00168-0 from BRP of the KOSEF, in part by Grant No. 2001-042-D00022 of the KRF.

†Supported by Grant No. 2001-042-D00022 of the KRF.


doi:10.1016/S0920-5632(02)02208-9

tion may lead to potentially dangerous FCNC due to the absence of GIM mechanism.

It is worthwhile to keep the original framework where all the fermions are in the bulk, and to question other unsubstantiated assumptions. Relaxing the universal bulk fermion mass assumption, we assign a different bulk fermion mass $m'_\psi$ to the SU(2)–singlet bottom quark field, and see whether there exists some parameter space to accommodate the $\rho$ constraint[7]. We show that this $m'_\psi$ allows some limited parameter space where the degeneracy between the top and bottom quark KK modes is retained, suppressing their contribution to the $\rho$ parameter.

## 2. Extended Bulk SM in the RS scenario

### 2.1. Original set-up

Let us review the KK solution of a bulk fermion with arbitrary Dirac bulk mass in the RS scenario[3,8,9], which causes a subtle problem when discussing the bulk SM. The five-dimensional action of a Dirac fermion $\Psi$ with the bulk mass $m_\psi$ is

$$S = \int \mathrm{d}^4x \int \mathrm{d}\phi \sqrt{-G}\left\{ E_{\underline{A}}^{A}\left[\frac{i}{2}\bar{\Psi}\gamma^{\underline{A}}(D_A - \overleftarrow{D_A})\Psi\right] - m_\psi \mathrm{sign}(\phi)\bar{\Psi}\Psi\right\}.$$

With the KK expansion of $\Psi$

$$\Psi_{L,R}(x,\phi) = \sum_{n=0}^{\infty} \psi_{L,R}^{(n)}(x)\,\frac{e^{2\sigma(\phi)}}{\sqrt{r_c}}\,\hat{f}_{L,R}^{(n)}(\phi)\,, \qquad (1)$$

Eq. (1) becomes the action for a tower of massive Dirac fermions

$$S = \sum_{n=0}^{\infty}\int \mathrm{d}^4x\left\{\bar{\psi}^{(n)}\, i\partial\!\!\!/\, \psi^{(n)} - M_f^{(n)}\,\bar{\psi}^{(n)}\,\psi^{(n)}\right\}. (2)$$

Note that $Z_2$-symmetric action constrains $\bar{\Psi}\Psi = \bar{\Psi}_L\Psi_R + \bar{\Psi}_R\Psi_L$ to be $Z_2$-odd. If $\widehat{f}_L^{(n)}$ is a $Z_2$-even function $\chi^{(n)}$ then $\widehat{f}_R^{(n)}$ should be a $Z_2$-odd function $\tau^{(n)}$ and vice versa. The order one parameter $\nu \equiv m_\psi/k$ determines the $\chi^{(n)}$ and $\tau^{(n)}$ as some combination of Bessel functions.

Let us discuss the physical implication of the parameter $\nu$. Note that the canonically re-scaled zero mode of $Z_2$-even bulk fermion is proportional to $e^{(1/2+\nu)kr_c|\phi|}$. For $\nu \ll -1/2$ the fermion bulk wave functions are localized toward the Planck brane: The magnitudes of its gauge couplings with KK gauge bosons are quite small. Numerically for $\nu \lesssim -0.5$ the couplings are too small to be probed at high energy colliders[3,10]. For $\nu \gg 1$ the SM fermions become localized closer to the TeV brane: The model approaches the RS model with the gauge fields only in the bulk, which is phenomenologically disfavored[1]. To be specific, if $\nu \gtrsim -0.3$, the large contribution to the precision electroweak data pushes $M_A^{(1)}$ up to about 6 TeV, beyond the direct production at any planned collider. Therefore, we consider the parameter space of $\nu$ between $-0.5$ and $-0.3$.

Another subtle point when placing the SM fermions in the $\mathrm{AdS}_5$ bulk is that the fermion field contents should be doubled. In the SM, a fermion field with left-handed chirality and that with right-handed chirality belong to different representations of a gauge group. In the RS-bulk SM, however, if a fermion which belongs to a specific representation of a gauge group has a single chirality, the bulk wave function cannot be determined. For each generation, we introduce four five-dimensional Dirac fields, an SU(2)–doublet fermion field $Q = (q_u, q_d)^T$ and two SU(2)–singlet fermion fields, $u$ and $d$, with weak hypercharges $Y = 1/6$, $2/3$, and $-1/3$ respectively.

Since the SM fermion should correspond to the KK zero mode, we assign $Z_2$-even wave function $\chi^{(n)}$ to the left-handed SU(2)–doublet and the right-handed SU(2)–singlet, $u(x,\phi)$ and $d(x,\phi)$. The charged current interactions, mediated by the bulk $W$ boson, connect $q_u$ and $q_d$:

$$S \supset \int d^4x \frac{g}{\sqrt{2}}\sum_{l=0}^{\infty}\left[\sum_{n,m=0}^{\infty}\bar{q}_{uL}^{(n)} W\!\!\!\!/^{+(l)} q_{dL}^{(m)}\left\{C_{nml}^{\bar{f}f'W}\right\} + \sum_{n,m=1}^{\infty}\bar{q}_{uR}^{(n)} W\!\!\!\!/^{+(l)} q_{dR}^{(m)}\left\{H_{nml}^{\bar{f}f'W}\right\}\right] + h.c.,$$

where $g = g_5/\sqrt{2\pi r_c}$. $C_{nml}^{\bar{f}f'W}$ and $H_{nml}^{\bar{f}f'W}$ denote the couplings of the $m$-th and the $n$-th fermion states to the $l$-th $W$ boson in the unit of the SM coupling. It is defined by

$$C_{nml}^{\bar{f}f'W} = \sqrt{2\pi}\int_{-\pi}^{\pi} d\phi\, e^{\sigma}\chi^{(n)}(\phi)\chi^{(m)}(\phi)\chi_A^{(l)}(\phi),$$

$$H^{\bar{f}f'W}_{nml} = \sqrt{2\pi}\int_{-\pi}^{\pi} d\phi\, e^{\sigma}\tau^{(n)}\tau^{(m)}\chi^{(l)}_{A}.$$

## 2.2. Extended Model

Now every SM fermion possesses its KK tower. We remind the reader that in the RS background the fermion KK mass spectrum is determined by the bulk fermion mass $m_\psi$. A simple assumption of universal bulk fermion mass results in the same KK mass spectrum for all the SM fermions. However, there is another mass source, Yukawa interaction. This Yukawa mass relates SU(2)–doublet with singlet (e.g., $m_Y \overline{q}_{uL} u_R$). It is to be compared to the KK mass which is the coupling of $\overline{q}_{uL} q_{uR}$ and $\overline{u}_L u_R$. This difference results in mixing among the fermion KK modes. Since quark masses are much smaller than the KK mass scale, this mixing effect has been neglected in the early study. One exception is the top quark. Its heavy mass yields substantial mixing. The mass shift of the top quark KK mode from the rest up-type quark KK modes invalidates the GIM cancellation; FCNC becomes inevitable. More severe problem happens if precision measurements are taken into account. In particular, the $\rho$ parameter becomes dangerous since it does not follow the decoupling theorem since its quantum correction increases with the squared mass difference between the $T = 1/2$ and $T = -1/2$ fermions. As each top quark KK mass deviates from the corresponding $b$ quark KK mass, their additive contribution yields disastrous and unacceptable value of $\Delta\rho$. The problem does not ameliorate but worsens as we add more and more KK states[6].

In Ref. [6], a 'mixed' scenario was proposed such that the third generation fermions are confined on the TeV brane while the other two generations propagate in the bulk. Instead, here we keep the original set-up but relax the unsubstantiated assumption of the universal bulk fermion mass[7]. For the simplest extension, we assume that the SU(2)–singlet bottom quark field has different bulk fermion mass $m'_\psi$, and see whether this introduction of another parameter can accommodate the $\Delta\rho$ constraint without a new hierarchy. Then the five dimensional action for the third generation quarks becomes

$$S = \int d^4x \int d\phi\, \sqrt{-G}[E^A_a i(\bar{Q}\gamma^a \mathcal{D}_A Q + \bar{t}\gamma^a \mathcal{D}_A t + \bar{b}\gamma^a \mathcal{D}_A b) - \text{sign}(\phi)\big(m_\psi\{\bar{Q}Q + \bar{t}t\} + m'_\psi \bar{b}b\big)].$$

The KK mass terms become

$$\mathcal{L} = -\sum_{n=1}^{\infty} k_{EW}[x^{(n)}_f(\nu)\{\bar{q}^{(n)}_{tL} q^{(n)}_{tR} + \bar{q}^{(n)}_{bL} q^{(n)}_{bR} + \bar{t}^{(n)}_L t^{(n)}_R\} + x^{(n)}_f(\nu')\bar{b}^{(n)}_L b^{(n)}_R] + h.c., \quad (3)$$

where we introduced additional dimensionless parameter $\nu' = m'_\psi/k$. As explicitly shown in Eq. (3), the KK masses of fermions depend on the bulk fermion masses, $m_\psi$ and $m'_\psi$.

The five-dimensional action for Yukawa interaction with the confined Higgs field is

$$S_{ffH} = -\int d^5x\sqrt{-G}[\frac{\lambda^b_5}{k}\overline{Q}(x,\phi)\cdot H(x)b(x,\phi) + \frac{\lambda^t_5}{k}\epsilon^{ab}\overline{Q}(x,\phi)_a \cdot H(x)_b t(x,\phi) + h.c.]\delta(\phi - \pi),$$

where $\lambda^{b,t}_5$ is the five-dimensional Yukawa coupling. Spontaneous symmetry breaking shifts the Higgs field as $H^0 \to v_5 + H^{'0}$ with a VEV of the order Planck scale.

## 3. Constraints from the $\rho$ parameter

The $\rho$ parameter is defined by the difference between the $W$ and $Z$ boson self-energy functions re-scaled by each mass. It has been known to play a special role among precision measurements since it is sensitive to heavy fermions beyond SM. With the Higgs mass below 1 TeV, the current electroweak precision data constrain $\Delta\rho < 2 \times 10^{-3}$ at 95% CL (with $\Delta\rho \equiv \rho - \rho_{\rm SM}$), which restricts new SU(2)–doublet fermion mass spectrum to satisfy [11]

$$\sum_i \Delta m^2_i \leq (115\,\text{GeV})^2. \quad (4)$$

Note that for $|M^{(n)}_t - M^{(n)}_b| \ll M^{(n)}_t$, we have

$$\Delta m^2 \approx \sum_n (M^{(n)}_t - M^{(n)}_b)^2. \quad (5)$$

Thus different mass spectra for the top and bottom quark KK modes lead to dangerous contributions to $\Delta\rho$. First let us demonstrate that

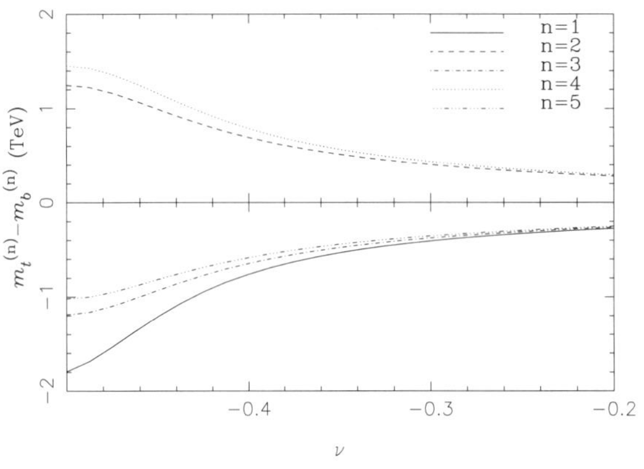


Figure 1. The KK mass difference between the top and bottom quarks when $\nu = \nu'$ and $k_{EW} = 1$ TeV.

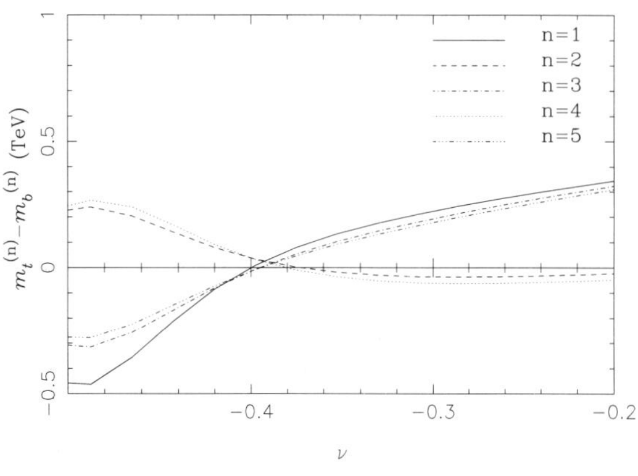


Figure 2. The KK mass difference between the top and bottom quarks when $\nu' = -0.6$ and $k_{EW} = 1$ TeV.

the RS-bulk SM with the universal bulk fermion mass assumption ($\nu = \nu'$) cannot satisfy this $\Delta\rho$ constraint. In Fig. 1, we show top and bottom KK mass differences up to the fifth KK excitation states as a function of parameter $\nu$, which is too large to satisfy Eq. (4).

Now let us allow $\nu \neq \nu'$ and see whether there exists a parameter space to satisfy Eq. (4). In Fig. 2, we shows, with the fixed $\nu' = -0.6$, the top-bottom quark mass differences for the first five excited modes. We find that for example the $\nu = -0.39$ case with $\nu' = -0.6$ gives vanishing mass differences below 20 GeV for the first five KK excited states. A remarkable point is that the $\Delta m$ decreases for higher KK modes; the contribution of higher KK modes becomes less important.

In summary, the minimal RS-bulk SM with a common bulk fermion mass has disastrous contribution to the $\rho$ parameter due to quite large mass shifts between the top and bottom quark KK modes. We relax the universal bulk fermion mass, and let the SU(2)–singlet bottom quark field have a different bulk fermion mass $m'_\psi$. It is shown that for example if $m'_\psi/k \simeq -0.6$ and $m_\psi/k \simeq -0.4$, the degeneracy of the top and bottom quark KK mode is good enough to suppress the new contribution to $\Delta\rho$.

## REFERENCES


1. H. Davoudiasl, J. L. Hewett and T. G. Rizzo, Phys. Lett. B **473**, 43 (2000).
2. H. Davoudiasl, J. L. Hewett and T. G. Rizzo, Phys. Rev. D **63**, 075004 (2001).
3. T. Gherghetta and A. Pomarol, Nucl. Phys. B **586**, 141 (2000).
4. S. L. Glashow, J. Iliopoulos, and L. Maiani, Phys. ReV. D **2**, 1285 (1970).
5. F. del Aguila, M. Pérez-Victoria and J. Santiago, Phys. Lett. B**492**, 98 (2000); D. E. Kaplan and T. M. Tait, JHEP **0111**, 051 (2001).
6. J. L. Hewett, F. J. Petriello and T. G. Rizzo, arXiv:hep-ph/0203091.
7. C. S. Kim, J. D. Kim and Jeonghyeon Song, arXiv:hep-ph/0204002.
8. Y. Grossman and M. Neubert, Phys. Lett. B **474**, 361 (2000).
9. S. Chang, J. Hisano, H. Nakano, N. Okada and M. Yamaguchi, Phys. Rev. D **62**, 084025 (2000).
10. H. Davoudiasl, J. L. Hewett and T. G. Rizzo, Phys. Lett. B **493**, 135 (2000).
11. D. E. Groom *et al.* [Particle Data Group Collaboration], Eur. Phys. J. C **15** (2000) 1.

ELSEVIER

Proceedings of ICHEP 2002, pp. 778–780
*S. Bentvelsen, P. de Jong, J. Koch and E. Laenen (Editors)*

31st INTERNATIONAL CONFERENCE ON HIGH ENERGY PHYSICS AMSTERDAM

# LEP search for single top production and new fermions

M. Antonelli[a]

[a] Laboratori Nazionali di Frascati dell'INFN,
Via E. Fermi 40,
I-00044 Frascati (Roma), Italy
E-mail: Mario.Antonelli@lnf.infn.it

LEP searches for different possible extensions of the fermion sector of the Standard Model are summarised in this report. The results are based on the final LEP data sample consisting of about 600 $\mathrm{pb}^{-1}$ collected at centre-of-mass energies ranging from 130 GeV up to 209 GeV. No deviation from the Standard Model expectation is observed and constraints on the parameter space of the new physics models have been set at the 95% CL.

## 1. Search for single top production

Because of the large mass of the top quark [1], at LEP2 only single top production is possible. This could occur via flavour changing neutral currents (FCNC) in the reactions[1]

$$e^+e^- \to \bar{\mathrm{t}}\mathrm{c}, \bar{\mathrm{t}}\mathrm{u}\,.$$

In the Standard Model (SM), such a process is forbidden at tree level and can only proceed via loops with cross sections $< 10^{-9}$ fb [2]. Extensions of the Standard Model could lead to enhancements of FCNC single top production and to measurable effects as proposed, for example, in Refs. [3–4]. It is customary to parametrise the FCNC transitions in terms of anomalous vertices whose strengths are described by the parameters $k_Z$ and $k_\gamma$ for Z and $\gamma$ exchange, respectively, using the formalism of Ref. [4].

At the Tevatron, the CDF Collaboration has performed a search for FCNC in top decays $\mathrm{t} \to \gamma\mathrm{c(u)}$ and $\mathrm{t} \to \mathrm{Zc(u)}$ in $\mathrm{p}\bar{\mathrm{p}}$ collisions at a centre-of-mass energy of 1.8 TeV. They obtain upper limits at the 95% confidence level (CL) on the branching fractions [5]: $\mathrm{Br}(\mathrm{t} \to \gamma\mathrm{c} + \mathrm{t} \to \gamma\mathrm{u}) < 3.2\%$ and $\mathrm{Br}(\mathrm{t} \to \mathrm{Zc} + \mathrm{t} \to \mathrm{Zu}) < 33\%$

The processes $e^+e^- \to \bar{\mathrm{t}}\mathrm{c}$, $\bar{\mathrm{t}}\mathrm{u}$ ($\bar{\mathrm{t}} \to \bar{\mathrm{b}}\mathrm{W}^-$) are characterised by a multijet final state with one b jet. The event properties depend significantly on the W decay mode. Two separate analyses have been designed for the leptonic and the hadronic decays of the W. Typical efficiencies are about 4-7% for the leptonic selection and 10-15% for the hadronic selection. The background, dominated by WW and $\mathrm{q}\bar{\mathrm{q}}$ processes, has an expected cross section of 10-20 fb and 50-100 fb for the leptonic and hadronic selections respectively. In total, 293 candidates were observed in the LEP data, in agreement with the 289.4 events expected from SM backgrounds.

Table 1
$e^+e^- \to \bar{\mathrm{t}}\mathrm{c}$, $\bar{\mathrm{t}}\mathrm{u}$ cross section upper limits

| $<\sqrt{s}>$ (GeV) | $\overline{\sigma}_{95}^{exp.}$ (pb) | $\sigma_{95}^{obs.}$ (pb) |
|---|---|---|
| 189 | 0.11 | 0.11 |
| 192 | 0.33 | 0.38 |
| 196 | 0.20 | 0.36 |
| 200 | 0.21 | 0.21 |
| 202 | 0.35 | 0.30 |
| 205 | 0.25 | 0.22 |
| 207 | 0.18 | 0.17 |

LEP combined 95% CL upper limits on the signal cross section have been derived [6]. The expected and observed signal cross section upper limits are shown in Table 1 for the differ-

[1] Throughout this paper, the notation $\bar{\mathrm{t}}\mathrm{q}$ is used for both $\bar{\mathrm{t}}\mathrm{q}$ and $\bar{\mathrm{q}}\mathrm{t}$ (q=c,u).


doi:10.1016/S0920-5632(02)02211-9

ent centre-of-mass energies for a top mass of 174 GeV/$c^2$. The reduction of Br($\bar{t} \to \bar{b}W$) due to possible FCNC top decays has been taken into account.

The negative results of these searches are translated into limits on the top quark couplings $k_Z$ and $k_\gamma$ [4]. The region of the ($k_Z$,$k_\gamma$) plane excluded at 95% CL is shown in Fig. 1.

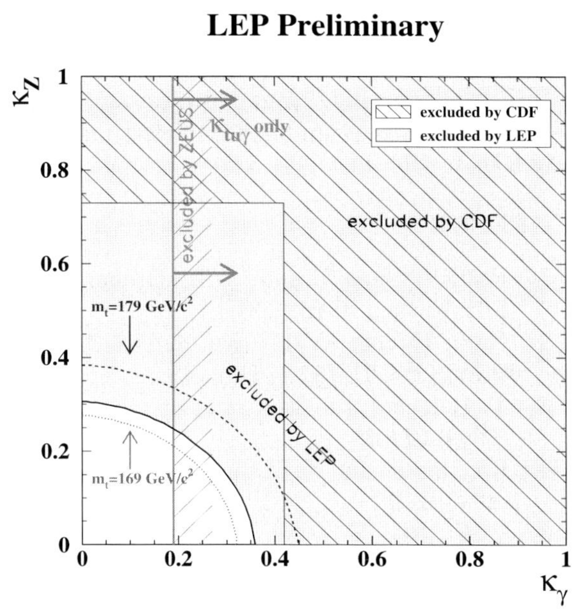


Figure 1. Exclusion curves at 95% CL in the ($k_Z$, $k_\gamma$) plane for $m_t$ = 169, 174, 179 GeV/c$^2$. The region excluded by CDF and by ZEUS is also shown.

This exclusion translates into branching fraction limits Br($t \to Zc$) + Br($t \to Zu$) < 6.7, 8,1 and 12.2% for $k_\gamma = 0$ and $m_t$ = 169, 174, and 179 GeV/$c^2$.

## 2. Search for heavy and excited leptons

Many theories beyond the Standard Model predict the existence of heavy new leptons. Excited leptons arise naturally in models predicting a sub–structure in the fermionic sector. New leptons couple to gauge bosons through their SU(2)×U(1) charges and can be pair produced at LEP. Single production of excited leptons is also possible through their mixing with SM leptons.

New leptons can decay into SM leptons by radiating a $\gamma$, Z or W. In the case of the charged heavy lepton, the decay $L^\pm \to L^0W$ is also possible.

The different combinations of several production mechanisms and decay modes give rise to many, very different experimental signatures, from final states with only $\gamma$s to isolated leptons or jets. In most of the topologies, there is a significant amount of missing energy. Heavy Stable Charged Particle (HSCP) topology is also a signal for heavy charged lepton production if the mixing between heavy leptons and SM leptons is small and the decay $L^\pm \to L^0W$ is kinematically forbidden.

No signal of heavy or excited lepton production has been observed and lower limits on the heavy letpon masses have been set.

A search for heavy–lepton pair production has been performed by L3 [7]. The lower limits on the neutral heavy lepton masses are shown in Table 2. Lower limits on the charged heavy lepton masses exceed 100 GeV/$c^2$ independently of the model.

Table 2
Lower mass limits for neutral heavy leptons in GeV/$c^2$

| decay mode | Model | Dirac | Majorana |
|---|---|---|---|
| | Seque. | 101.3 | 89.5 |
| $L^0 \to eW$ | Mirror | 102.6 | - |
| | Vector | 100.8 | 89.5 |
| | Seque. | 101.5 | 90.7 |
| $L^0 \to \mu W$ | Mirror | 102.7 | - |
| | Vector | 101.0 | 90.7 |
| | Seque. | 90.3 | 80.5 |
| $L^0 \to \tau W$ | Mirror | 99.3 | - |
| | Vector | 90.3 | 80.5 |

The lower limits on the masses of pair–

Table 3
Lower mass limits on the excited lepton masses in GeV/$c^2$

| | $e^*$ | $\mu^*$ | $\tau^*$ | $\nu_e^*$ | $\nu_\mu^*$ | $\nu_\tau^*$ |
|---|---|---|---|---|---|---|
| $f = f'$ | 103.0 | 103.1 | 102.2 | 102.0 | 102.4 | 95.3 |
| $f = -f'$ | 98.0 | 98.0 | 98.0 | 102.7 | 102.8 | 102.8 |

produced excited leptons obtained by DELPHI[8] are shown in Table 3. Two scenarios are possible for the constants $f$, $f'$ associated to the gauge group SU(2)×U(1): $f = f'$ ( Br($\nu^* \to \nu\gamma$)=0), and $f = -f'$ ( Br($e^* \to e\gamma$)=0).

The searches for single production of excited leptons extend the mass reach up to $\sqrt{s}$. The LEP combined limits [6] on the ratio of the coupling ($f$) to the compositness scale ($\Lambda$) are shown in Figure 2.

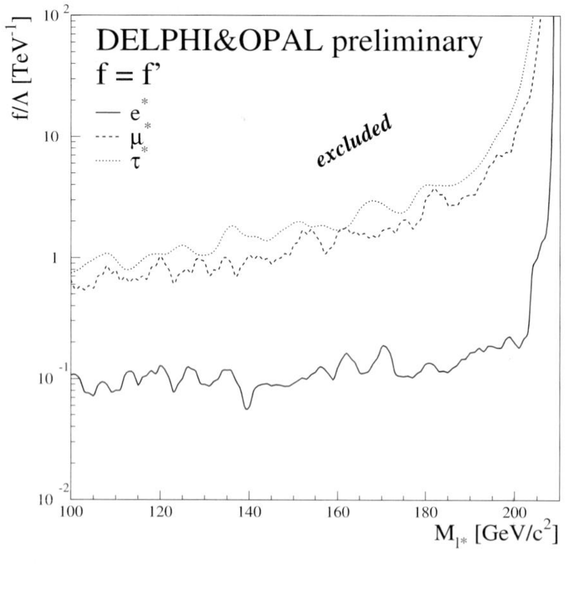


Figure 2. Upper limits on the ratio of the coupling $f$ to the compositness scale $\Lambda$ as a function of the excited lepton mass.

The more restricitve limit in the case of excited electrons is due to the t–channel contribution to the production cross section. The existence of excited electrons can also be probed indirectly through the t-channel excited electron contribution to the $e^+e^- \to \gamma\gamma$ process. The combined LEP results exclude a compositness scale below 200 GeV for $f = 1$ up to excited electron masses of 600 GeV/$c^2$ [9].

## REFERENCES

1. Particle Data Group, D.E. Groom *et al.*, Eur. Phys. J. **C15** (2000) 1.
2. C.-S. Huang, X.-H. Wu and S.-H. Zhu, Phys. Lett. **B 452** (1999) 143.
3. T. Han and J.L. Hewett, Phys. Rev. **D 60** (1999) 074015.
4. R.D. Peccei and X. Zhang, Nucl. Phys. **B 337** (1990) 269.
5. CDF Collaboration, *Search for Flavor-Changing neutral Current Decays of the Top Quark in $p\bar{p}$ Collisions at $\sqrt{s}$ = 1.8 TeV* , Phys. Rev. Lett. **80**(1998) 2525.
6. LEP Exotica Working Group, http://lepexotica.web.cern.ch/LEPEXOTICA/notes/notes.html.
7. L3 Collaboration, *Search for Heavy Neutral and Charged Leptons in $e^+e^-$ Annihilation at LEP*, Phys. Lett. **B 517** (2001) 75.
8. DELPHI Collaboration, *Search for excited leptons with the DELPHI detector at LEP* , DELPHI note 2002-062-CONF-596.
9. LEP2 diphoton group, http://lepewwg.web.cern.ch/LEPEWWG/lep2/photons/welcome.html

ELSEVIER

Proceedings of ICHEP 2002, pp. 781–784
*S. Bentvelsen, P. de Jong, J. Koch and E. Laenen (Editors)*

31st INTERNATIONAL CONFERENCE ON HIGH ENERGY PHYSICS AMSTERDAM

# Minireview on other signatures of new physics at LEP and HERA

M.Kobel[a]

[a]Universität Bonn,
Nußallee 12, 53115 Bonn, Germany
representing the LEP and HERA collaborations

We review the search at HERA and LEP for physics beyond the Standard Model, other than leptoquarks, excited leptons, and large extra dimensions, which are covered elsewhere in these proceedings. The search methods are twofold: On the one hand there are indirect methods, via deviations of precision measurements from Standard Model predictions, like the search for extra Z Bosons, noncommutative geometry, or contact interactions. On the other hand explicit final states are searched for, which do not exist in the Standard Model, like those arising from decays of doubly charged Higgs Bosons or Technipions.

## 1. EXTRA Z′ BOSONS

Several Grand Unified Theories predict additional heavy Gauge Bosons, which would manifest themselves through their interference with Standard Model processes. In $E_6$ and left-right symmetric (LR) models the Z′ generally is a mixture of two Gauge Bosons, and its couplings are predicted as a function of the mixing angles $\Theta_6$ and $\alpha_{LR}$, respectively. In the "Sequential Standard Model" (SSM) the Z′ is assumed to have the same couplings as the Z Boson. In both cases experimental data can be used to obtain lower limits on its mass within the given model. Vice versa, assuming a Z′ mass, model independent limits on its couplings can be set.

The latter approach has been used by L3 [1] for setting limits on the Z′ leptonic couplings $v'_\ell$ and $a'_\ell$ using measurements of leptonic cross-sections and asymmetries at LEP 2, and assuming lepton universality. The resulting limits in Fig. 1 show no hint for the existence of an extra Z′ boson.

The combined 2-fermion cross-section and asymmetry measurements of the LEP collaborations [2] achieve a precision of about 1% for hadrons and 3% for leptons at the highest LEP 2 center-of-mass energies, and are the most sensitive testing grounds for possible contributions of Z′ Bosons. Precision measurements on the Z resonance suggest negligible ZZ′ mixing, which is

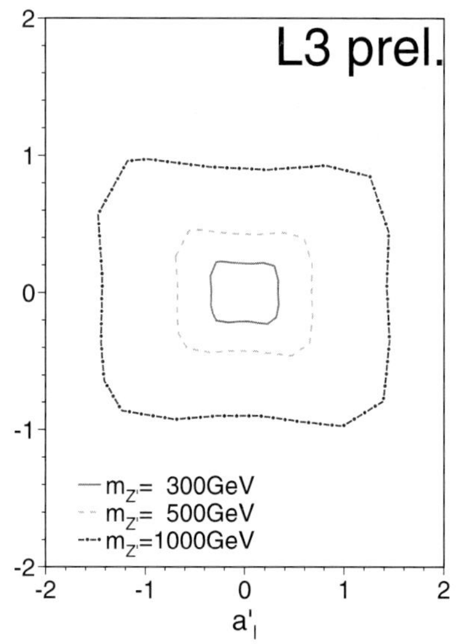


Figure 1. *Model independent 95% C.L. upper bounds (L3, preliminary) on $Z'\ell^+\ell^-$ couplings for various Z′ masses, assuming zero Z-Z′ mixing.*

therefore set to zero in all fits. For an SSM Z′, LEP finds a lower mass limit of 1787 GeV, compared to 690 GeV from CDF. The limits for the $E_6$ and LR models, shown in Fig. 2 as function of the mixing parameters, range from 340 GeV to 900 GeV, compared to typically 600 GeV from


doi:10.1016/S0920-5632(02)02214-4

CDF. Limits for special parameter values, like the superstring-motivated $\eta$ model, are also shown.

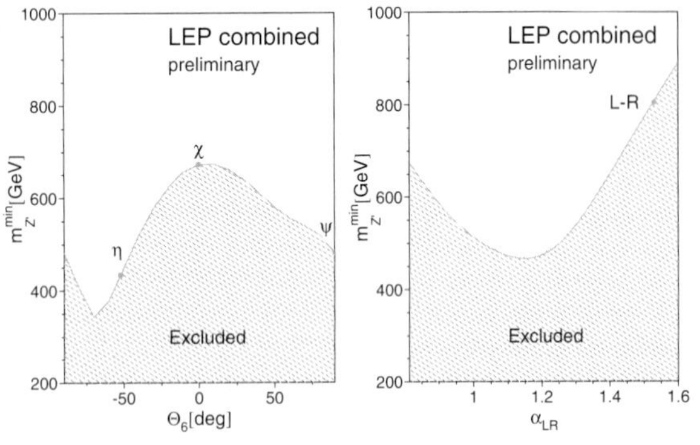


Figure 2. *Preliminary 95% CL LEP-combined limits on Z′ masses as function of the mixing parameters in the E6 and LR Model, respectively.*

## 2. NON-COMMUTATIVE QED

Recently, it has been realized that the quantisation of strings in the presence of background fields may go along with a non-commutative space-time geometry of "fuzzy" coordinates [3,4], which obey an uncertainty relation with a typical scale $\Lambda_{\rm NC}$. This scale might naturally assumed to be close to the Planck Scale, but could conceivably be lowered to the TeV region in scenarios with large extra dimensions. Non-commutative space-time, currently consistently formulated only for QED and particles with integer electric charge (NCQED), would manifest itself in a preferred direction, fixed in space. This would result in time-dependent modifications of differential cross-sections due to the earth rotation, as the laboratory frame moves through space .

OPAL[5] has searched for such effects in the process $e^+e^- \rightarrow \gamma\gamma$, using more than 5000 events at a mean center-of-mass energy of about 197 GeV. No statistical significant modulation of differential cross-sections in $\phi$ and $\theta$ was observed for any assumption for the orientation of the preferred NCQED direction. Independent of this orentation, a lower limit of $\Lambda_{\rm NC} > 142$ GeV could be set on the 95% CL, corresponding to a fuzzy space scale of less than $1.4 \times 10^{-18}$m.

## 3. CONTACT INTERACTIONS

New physics at scales well above the typical momentum transfers of the experiments can generally be described by an effective Lagrangian in terms of contact interactions, just like Fermi's low-energy formulation of weak interactions in terms of $G_F \sim g_W^2/m_W^2$. In the most general case the effective Langrangian of the 4-Fermion contact interaction reads

$$\mathcal{L}_{\rm eff} = \frac{g_{\rm new}^2}{\Lambda_{\rm new}^2} \sum_{i,j={\rm L,R}} \eta_{ij} (\bar{\rm g}_i \gamma_\mu {\rm g}'_i)(\bar{\rm f}_j \gamma^\mu {\rm f}'_j) \qquad (1)$$

where $g_{\rm new}^2/4\pi$ is taken to be 1 by convention and f,$g_{i,j}$ are left-($i,j$=L) or right-($i,j$=R) handed fermion spinors which take part in the contact interaction according to the set of parameter choices for $\eta_{ij} = \pm 1$ or 0.

Possible effective couplings for Fermion-Boson interactions are $G_6 \propto \Lambda_6^{-2}$ for ee$\gamma$ vertices and $G_7 \propto \Lambda_7^{-3}$ or $G_8 \propto \Lambda_8^{-4}$ for ee$\gamma\gamma$ vertices , according to [6].

### 3.1. Contact interactions in $e^+e^- \rightarrow \gamma\gamma$

All four LEP experiments have measured cross-sections for the process $e^+e^- \rightarrow \gamma\gamma$ for center-of-mass energies ranging from 183 GeV to 208 GeV. The overall average of the cross-section ratio over its Standard Model prediction, $0.982 \pm 0.010$, severely constrains contributions of new physics, which would increase the cross-section $\propto G_6^2 s, G_7^2 s^2$, and $G_8^2 s^2 m_e^2$, respectively. From a fit to all center-of-mass energies 95% CL lower limits of $\Lambda_6 > 1595$ GeV, $\Lambda_7 > 831$ GeV, and $\Lambda_8 > 23$ GeV are obtained, the latter being much smaller due to its $m_e^2/\Lambda^2$ suppression.

### 3.2. Four-Fermion Contact Interactions

Four-fermion contact interactions have been studied at HERA for eeuu and eedd (f,g=e; f′,g′=u,d) and at LEP (g,g′=e) for all eeff vertices.

As an example Fig. 3 shows the deep inelastic $e^+p$ scattering data of the ZEUS experiment [7], normalized to the Standard Model prediction, as a function of $Q^2$. Indicated are predicted deviations for the so-called VV model, where all $\eta_{ij} = \pm 1$ for the eeuu and eedd vertices. Depending on the model for the set of $\eta_{ij}$, ZEUS

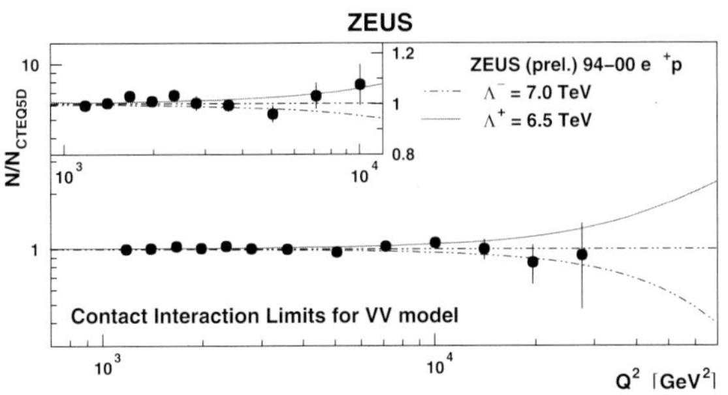


Figure 3. *Preliminary ZEUS data on deep inelastic $e^+p$ scattering, compared with 95% CL limits for the effective contact interaction mass scale in the VV model, depending on the sign of the $\eta$ parameters.*

and H1 [8] find lower limits for $\Lambda_{ij}$ ranging from 2 TeV to 7 TeV. For some models they can be translated into limits on the ratio of certain Leptoquark couplings $\lambda_{LQ}$ and masses $m_{LQ}$ according to $g_{new}/\Lambda_{new} = \lambda_{LQ}/m_{LQ}$.

LEP has separately fitted contact interactions for eeee, ee$\mu\mu$, ee$\tau\tau$, eeqq, eecc, and eebb vertices, and combinations of these, from measurements of 2-Fermion cross-sections and asymmetries [2]. The corresponding limits range up to $\Lambda > 20$ GeV, as is shown in the selection of limits in Table 1.

## 4. DOUBLY CHARGED $H^{\pm\pm}$ BOSONS

A special interpretation of eeee contact interactions is the search for doubly charged Higgs Bosons, which appear in several extensions of the Standard Model, and would e.g. modify the Bhabha scattering process via $t$-channel exchange. OPAL [9] has combined indirect and direct searches deriving limits on the ee$H^{\pm\pm}$ coupling $h_{ee}$ versus the mass of the $H^{\pm\pm}$. The exploitation of the contact interactions according to $g_{new}/\Lambda_{new} = h_{ee}/m_{H^{\pm\pm}}$, together with a fit to the $\theta$ distribution, shows no significant deviation from the Standard Model prediction. A direct search is looking for single $H^{\pm\pm}$ production in the process $e^+e^- \rightarrow e^+e^+ H^{--} \rightarrow e^+e^+\ell^-\ell^{-\prime}$ and its charge conjugate. No lepton flavor ID is applied in order to cover all possible $H^{\pm\pm}$ decays to lep-

Table 1: *Combined LEP Limits on scales of new physics in TeV, for several models of contact interactions, given by the $\eta$ parameters. The following assumptions on affected vertices are considered, from left to right: both ee$\mu\mu$ and ee$\tau\tau$, only ee$\tau\tau$, all eeqq vertices, either eeuu or eecc, only eebb. Limits for other combinations can be found in [2]. The signs of $\Lambda$ correspond to the sign of the $\eta$ parameters.*

| Model | $\eta_{LL}$ | $\eta_{RR}$ | $\eta_{LR}$ | $\eta_{RL}$ | $\ell^+\ell^-$ | | $\tau^+\tau^-$ | | $q\bar{q}$ | | up-type | | $b\bar{b}$ | |
|---|---|---|---|---|---|---|---|---|---|---|---|---|---|---|
| | | | | | $\Lambda_-$ | $\Lambda_+$ | $\Lambda_-$ | $\Lambda_+$ | $\Lambda_-$ | $\Lambda_+$ | $\Lambda_-$ | $\Lambda_+$ | $\Lambda_-$ | $\Lambda_+$ |
| LL | ±1 | 0 | 0 | 0 | 9.8 | 13.3 | 9.1 | 8.6 | 3.7 | 6.0 | 6.7 | 10.2 | 9.1 | 12.3 |
| RR | 0 | ±1 | 0 | 0 | 9.3 | 12.7 | 8.7 | 8.2 | 5.5 | 3.9 | 5.7 | 8.3 | 2.2 | 8.1 |
| VV | ±1 | ±1 | ±1 | ±1 | 16.0 | 21.7 | 14.2 | 14.5 | 8.1 | 5.3 | 9.6 | 14.3 | 9.4 | 14.1 |
| AA | ±1 | ±1 | ∓1 | ∓1 | 15.1 | 17.2 | 14.0 | 11.3 | 5.1 | 8.8 | 8.0 | 11.5 | 11.5 | 15.3 |
| LR | 0 | 0 | ±1 | 0 | 8.6 | 10.2 | 2.2 | 7.9 | 5.1 | 4.3 | 4.2 | 2.3 | 3.1 | 5.5 |
| RL | 0 | 0 | 0 | ±1 | 8.6 | 10.2 | 2.2 | 7.9 | 7.2 | 9.3 | 3.5 | 2.8 | 7.0 | 2.4 |
| V0 | ±1 | ±1 | 0 | 0 | 13.5 | 18.4 | 12.7 | 11.8 | 5.1 | 6.0 | 8.7 | 13.4 | 10.8 | 14.5 |
| A0 | 0 | 0 | ±1 | ±1 | 12.4 | 14.3 | 9.8 | 10.8 | 8.0 | 3.9 | 4.9 | 2.8 | 6.3 | 3.9 |

tons, including those violating lepton flavor. The limits of the direct search are stronger than the indirect ones for 100% leptonic branching ratio of the $H^{\pm\pm}$, i.e. masses below the threshold for $W^{\pm}W^{\pm}$ decay. A summary of the OPAL limits, together with past and future limits from other experiments is shown in Fig. 4.

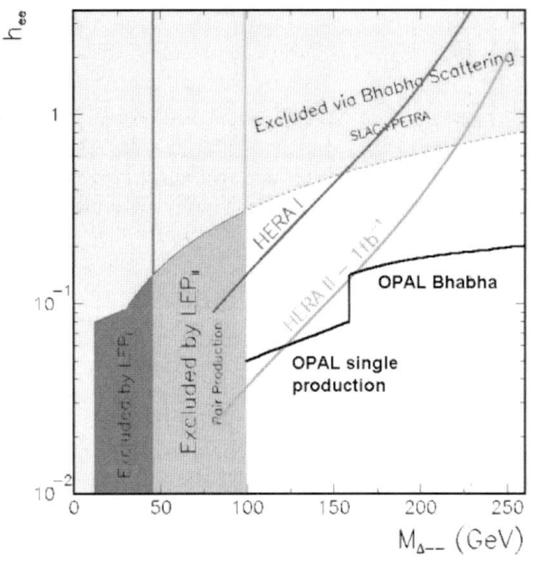


Figure 4. *Preliminary OPAL limits on the doubly charged Higgs Boson couplig to electrons, $h_{ee}$ as function of its mass. Indicated are present limits from $e^+e^-$ machines, and expected future limits from HERA [10].*

## 5. TECHNICOLOR

In case no Higgs Bosons exist, there must be another mechanism of Electroweak Symmetry Breaking. Candidates for this are various extended technicolor models. OPAL [11] has searched for technipions $\pi_T$, the lightest bound states within the Technicolor Strawman Model (TCSM)[12], via the processes $e^+e^- \rightarrow \pi_T^+\pi_T^-$ and $\gamma\pi_T^0$. No excess was observed above the background, which mainly comes from Boson pair production processes. Independent of the TCSM parameters, a lower mit of $m_{\pi_T} > 62$ GeV was derived, similar to the Moriond 2001 result from DELPHI [13]. Excluded contours in the $m_{\pi_T}$-$m_{\rho_T}$ plane are shown in Fig. 5.

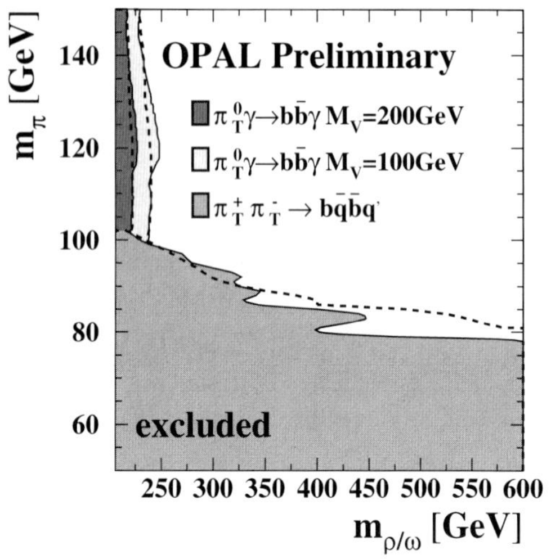


Figure 5. *Preliminary OPAL limits on technimeson masses within the TCSM model.*

## REFERENCES


1. L3 Note 2759, June 2002.
2. The LEP Electroweak Working Group, $f\bar{f}$ subgroup, LEP2FF/02-03, October 2002
3. A.Connes, M.R.Douglas, and A.Schwarz, J.High Energy Phys. 02 (1998) 3.
4. N.Seiberg and E.Witten, J.High Energy Phys. 09 (1999) 32.
5. OPAL Physics Note PN 500, July 2002.
6. O.J.P.Eboli, A.A.Natale, and S.F.Novaes, Phys.Lett.B271 (1991) 274.
7. ZEUS Collaboration, contributed paper 602 to EPS HEP01 conference, July 2001.
8. J. Scheins et al., H1 note H1prelim-02-062, presented at La Thuile, March 2002.
9. OPAL Physics Note 502, July 2002.
10. Y.Sirois, presented at SUSY02, Hamburg, June 2002.
11. OPAL Physics Note 485, July 2002.
12. K.Lane, Phys.Rev.D60 (1999) 75007.
13. DELPHI 2001-028, Conf 469(2001).

ELSEVIER

Proceedings of ICHEP 2002, pp. 785–788
S. Bentvelsen, P. de Jong, J. Koch and E. Laenen (Editors)

31st INTERNATIONAL CONFERENCE ON HIGH ENERGY PHYSICS AMSTERDAM

# Searches for signatures of physics beyond the Standard Model with high-$P_T$ leptons at HERA

Jochen Dingfelder[a]*

[a]Physikalisches Institut, Universität Heidelberg, Philosophenweg 12, 69120 Heidelberg, Germany

The latest HERA results on searches for events with high-$P_T$ leptons in $ep$ collisions are presented. The full HERA I data set, corresponding to 115 pb$^{-1}$ (H1) and 130 pb$^{-1}$ (ZEUS), is used. Searches for $W$ production and for signatures of new physics including single top production via flavor changing neutral currents have been performed in events with high-$P_T$ leptons and missing transverse momentum. The production of multi-lepton events at high transverse momenta has been studied. Some 2- and 3-electron events are observed at high di-electron masses, for which the Standard Model expectation is low. A possible interpretation in terms of doubly charged Higgs boson production is ruled out.

## 1. ISOLATED LEPTONS AND $W$ PRODUCTION

The H1 and ZEUS collaborations at HERA have investigated events with leptons which have the topology of a $W$ produced in $ep$ collisions at $\sqrt{s} \approx 320$ GeV ($ep \to eWX$). The latest measurements use a data sample corresponding to three times the luminosity of the previously published results [1,2]. The events are required to contain an isolated lepton with high transverse momentum, $P_T^\ell$, substantial missing transverse momentum, $P_T^{miss}$, and may also have a hadronic final state $X$ with transverse momentum $P_T^X$. The H1 selection is optimized for $W$-like topologies [3], while ZEUS carries out an inclusive selection, with a higher expectation from non-$W$ background processes, as well as a selection designed to identify $W$ bosons at high $P_T^X$ [4].

At low values of $P_T^X$ the observed event yields are in agreement with the Standard Model (SM) prediction. At high $P_T^X$, ZEUS also sees good agreement with the SM in the electron and muon channels, while H1 observes an excess of events. The hadronic transverse momentum distribution of the H1 events is shown in Fig. 1. The observed H1 and ZEUS event rates at high $P_T^X$ are summarized in tables 1 and 2. ZEUS has recently carried out a search in the tau channel [5], looking for hadronic decays of the tau. At $P_T^X > 25$ GeV two events are observed for $0.12 \pm 0.02$ expected. They have similar characteristics as the outstanding H1 electron and muon events.

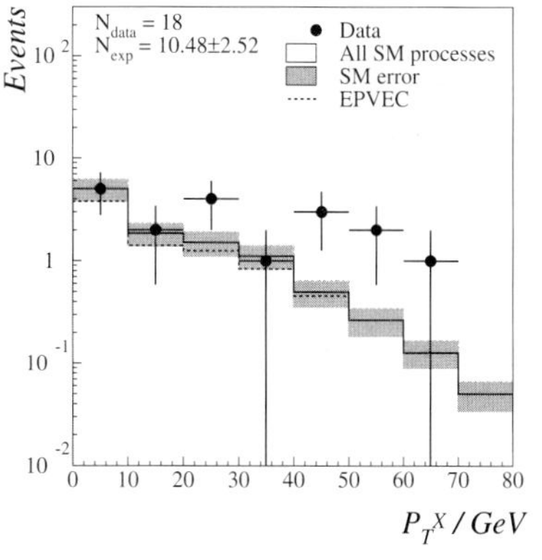


Figure 1. Hadronic transverse momentum distribution of the H1 isolated lepton events (electrons and muons combined).

## 2. SINGLE TOP PRODUCTION

A possible interpretation of the excess of isolated lepton events at high $P_T^X$ could be the production of top quarks in flavor changing neutral current (FCNC) processes with a coupling $\kappa_{tu\gamma}$ of the top quark to a $u$-quark and a photon.

*On behalf of the H1 and ZEUS collaborations.


doi:10.1016/S0920-5632(02)02216-8

Table 1
Observed and expected lepton event yields at high $P_T^X$ for the H1 analysis.

| H1 preliminary $e^+$p (101.6 pb$^{-1}$) | Electrons obs./exp. (W) | Muons obs./exp. (W) |
|---|---|---|
| $P_T^X > 25$ GeV | 4 / 1.29 ± 0.33 (1.05) | 6 / 1.54 ± 0.41 (1.29) |
| $P_T^X > 40$ GeV | 2 / 0.41 ± 0.12 (0.40) | 4 / 0.58 ± 0.16 (0.53) |

Table 2
Observed and expected lepton event yields at high $P_T^X$ for the ZEUS analysis.

| ZEUS preliminary $e^\pm$p (130.5 pb$^{-1}$) | Electrons obs./exp. (W) | Muons obs./exp. (W) | Taus obs./exp. (W) |
|---|---|---|---|
| $P_T^X > 25$ GeV | 1 / 1.14 ± 0.06 (1.10) | 1 / 1.29 ± 0.16 (0.95) | 2 / 0.12 ± 0.02 (0.10) |
| $P_T^X > 40$ GeV | 0 / 0.46 ± 0.03 (0.46) | 0 / 0.50 ± 0.08 (0.41) | 1 / 0.06 ± 0.01 (0.05) |

In top quark decays ($t \to bW$) a high hadronic transverse momentum is expected. The search for top quarks has been carried out in both the leptonic and hadronic decay modes of the $W$ boson ($W \to \ell\nu$, $W \to q\bar{q}'$).

### 2.1. Leptonic channel

To search for top candidates, the lepton events that pass the $W$ selection are required to have $P_T^X$ >25 (40) GeV for H1 (ZEUS) [4,6]. ZEUS observes no event for an expectation of 1.0. To enhance a possible top signal, H1 applies an additional cut on the transverse mass of the lepton-neutrino system ($M_T > 10$ GeV) and requires a positive lepton charge (since $\bar{t}$ production is strongly suppressed). After this selection, H1 sees 5 events for 1.8 ± 0.5 expected. The mass of the lepton-neutrino-X system can be computed after reconstructing the neutrino kinematics, applying a $W$ mass constraint. For each event two solutions are possible. Several of these masses are compatible with the top mass distribution.

### 2.2. Hadronic channel

In the hadronic decay channel of the top quark, events with at least three high-$P_T$ jets are selected. In addition, mass windows are set around the di-jet mass closest to the $W$ mass and the 3-jet mass [4,6]. Both experiments see agreement with the Standard Model. ZEUS observes 19 events for 20.0 expected, H1 observes 14 events and expects 19.6 ± 7.8. The 3-jet mass distribution of the ZEUS events is shown in Fig. 2. Although no sign of top quark production is seen in the hadronic channel, these results do not rule out a possible top interpretation of the H1 isolated lepton excess. Given the H1 result in the hadronic channel, one would expect less than 5.4 events in the leptonic channel at 95% CL.

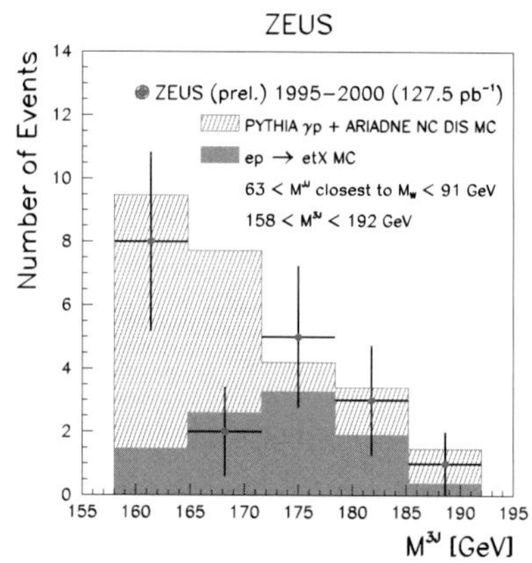

Figure 2. 3-jet mass distribution of the top candidates in the ZEUS hadronic channel compared to the SM prediction and the top simulation (arbitrary normalization).

### 2.3. Limit on the anomalous coupling $\kappa_{tu\gamma}$

Combining the results of the top searches in the leptonic and hadronic channels, limits on the FCNC coupling of the top quark to a $u$-quark and a photon are obtained by ZEUS ($\kappa_{tu\gamma} < 0.19$) and H1 ($\kappa_{tu\gamma} < 0.22$). The HERA limits extend into a region of parameter space not excluded by the LEP and TeVatron collider experiments (Fig. 3).

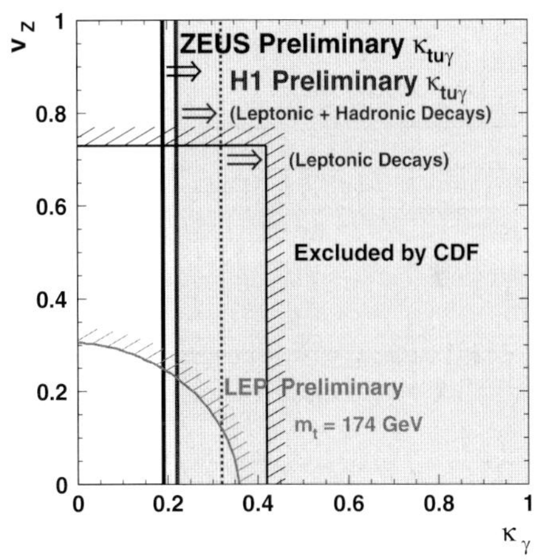


Figure 3. Present exclusion limits on the anomalous $tu\gamma$ and $tuZ$ couplings obtained by HERA, LEP and TeVatron experiments.

## 3. MULTI-LEPTON EVENTS

Studies of events containing two or more leptons in the final state have been carried out at HERA. The dominant SM contribution to multi-lepton production comes from photon-photon interactions, where one photon is radiated from the incoming electron, the other from the proton. The transverse momentum spectrum of the produced leptons ($P_T^\ell$) falls off steeply, making multi-lepton events with high $P_T^\ell$ a possible place to search for new physics. Additional experimental backgrounds are neutral current and QED Compton interactions in which a photon or a hadron is misidentified as an electron.

The selection of multi-electron events mainly requires two central electron candidates in the polar angle range $20° < \theta_e < 150°$ ($17° < \theta_e < 164°$) for H1 (ZEUS). The transverse momentum of at least one electron must be above 10 GeV. Events with two central electrons are classified as di-electrons, while events containing an additional third electron candidate within $5° < \theta_e < 175°$ are assigned to a 3-electron sample. More details are given in [7,8]. The selections for the di-muon analyses can be found in [8,9].

In both the electron and muon analyses, overall agreement is observed by H1 and ZEUS between the observed multi-lepton event yields and the expectation from SM processes. In the multi-electron analyses, however, some outstanding events are observed with high electron transverse momenta and invariant masses, $M_{12}$, of the highest-$P_T$ electron pair. H1 sees three 2-electron and three 3-electron events with masses $M_{12} > 100$ GeV for a SM expectation of $0.25 \pm 0.05$ and $0.23 \pm 0.04$ events, respectively. ZEUS sees two di-electron events with $M_{12} > 100$ GeV for $0.77 \pm 0.08$ expected and no 3-electron event for $0.37 \pm 0.04$ expected (see table 3 and Fig. 4). The higher ZEUS rate for the $\gamma\gamma$ contribution to the SM expectation is due to the extended polar angle range of the electrons compared to H1.

In the muon channel no excess above the SM prediction is seen by H1 or ZEUS in a smaller data sample of 105 pb$^{-1}$ (ZEUS) and 71 pb$^{-1}$ (H1). In particular, no event is seen with a di-muon mass above 100 GeV. Taking into account the difference in integrated luminosities and selection efficiencies for the electrons and muons, the two channels are compatible for high masses.

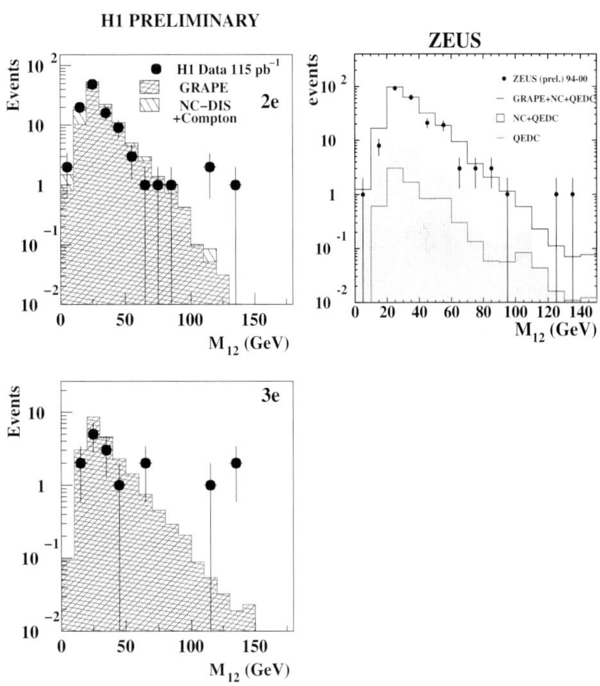


Figure 4. Invariant mass of the highest-$P_T$ electron pair in the H1 and ZEUS multi-electron selection. (Left: H1 2e and 3e events, right: ZEUS 2e and 3e events combined).

Table 3
Multi-electron event yields for di-electron masses $M_{12}$ above 100 GeV.

| Selection | H1 DATA | SM expectation | ZEUS DATA | SM expectation |
|---|---|---|---|---|
| 2e $M_{12} > 100$ GeV | 3 | $0.25 \pm 0.05$ | 2 | $0.77 \pm 0.08$ |
| 3e $M_{12} > 100$ GeV | 3 | $0.23 \pm 0.04$ | 0 | $0.37 \pm 0.04$ |

## 4. SEARCH FOR DOUBLY CHARGED HIGGS BOSONS

The search for doubly charged Higgs bosons carried out by H1 is motivated by the high mass multi-electron events discussed above. The search is performed in the framework of models with a Higgs triplet coupling to leptons of the $i^{th}$ generation via Yukawa couplings $h_{ii}$. The Higgs signal is searched for in the electronic and muonic decay modes ($H^{\pm\pm} \to e^{\pm}e^{\pm}$ or $\mu^{\pm}\mu^{\pm}$). Details of the selection can be found in [10]. Only one of the six high mass events in the electron channel is a doubly charged Higgs candidate. No candidate is found in the muon channel. Thus a limit is set on the coupling $h_{ee}$ as a function of the $H^{\pm\pm}$ mass. It is shown in Fig. 5 together with the preliminary bounds recently obtained by the OPAL collaboration [11] at LEP. The results of the H1 Higgs search and the new strong bounds from OPAL rule out an interpretation of the multi-electron excess as doubly charged Higgs decays.

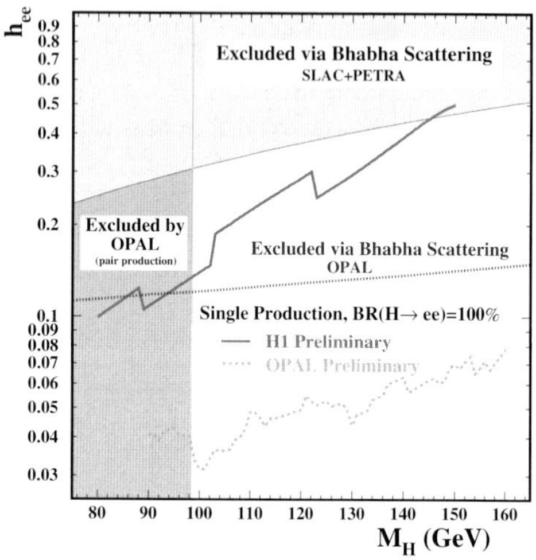


Figure 5. Exclusion limits on the coupling $h_{ee}$ as a function of the doubly charged Higgs mass.

## 5. OUTLOOK

HERA is currently starting a phase of operation at higher luminosity. The goal is to achieve luminosities a factor of $\sim$ five greater than possible to date at HERA. The new data will shed more light on the H1 and ZEUS results for the production of high-$P_T$ leptons with missing $P_T$, and the multi-electron events observed at high masses.

## REFERENCES


1. C. Adloff et al. (H1 Collaboration), *Eur. Phys. J.* **C 5** (1998) 575.
2. J. Breitweg et al. (ZEUS Collaboration), *Phys. Lett.* **B 471** (2000) 411.
3. H1 Collaboration, contributed paper 1022, ICHEP02, Amsterdam.
4. ZEUS Collaboration, contributed paper 650, EPS2001, Budapest.
5. ZEUS Collaboration, contributed paper 909, ICHEP02, Amsterdam.
6. H1 Collaboration, contributed paper 1024, ICHEP02, Amsterdam.
7. H1 Collaboration, contributed paper 1019, ICHEP02, Amsterdam.
8. ZEUS Collaboration, contributed paper 910, ICHEP02, Amsterdam.
9. H1 Collaboration, contributed paper 1021, ICHEP02, Amsterdam.
10. H1 Collaboration, contributed paper 1020, ICHEP02, Amsterdam.
11. OPAL Collaboration, contributed paper 1074, ICHEP02, Amsterdam.

Proceedings of ICHEP 2002, pp. 789–791
*S. Bentvelsen, P. de Jong, J. Koch and E. Laenen (Editors)*

31st INTERNATIONAL CONFERENCE ON HIGH ENERGY PHYSICS AMSTERDAM

# Excited fermions and other searches at HERA

L. Bellagamba[a]

[a]INFN Bologna
on behalf of the H1 and ZEUS Collaborations

This paper reports a search in *ep* collisions for events with isolated $\tau$ leptons and large missing transverse momentum and for heavy excited fermions. The analysis of events with isolated $\tau$ followed the strategy used in the search for isolated electrons and muons. Two candidate events were found, while $0.12 \pm 0.02$ were expected from the Standard Model. The search for excited electrons and neutrinos led to direct constraints on the couplings and masses of these new states, which are the best ever reported for masses beyond 200 GeV.

## 1. Introduction

Events with isolated leptons and large missing transverse momentum in *ep* collisions are an interesting signature for processes beyond the Standard Model (SM). The background due to SM processes is expected to be small and due almost completely to $W$ boson production. Both HERA collaborations have previously reported results on the electron and muon channels which led to constraints on single-top production mediated by an anomalous FCNC coupling [1–4]. A search in the $\tau$ channel, performed by the ZEUS collaboration, is reported.
HERA is also an ideal place to look for excited fermion states that would be a clear signal of fermion compositness. Such new states could be produced directly via t-channel exchange of a gauge boson followed by decay to a known fermion and a gauge boson. The results of a search for excited electrons and neutrinos are presented and compared to results obtained at LEP.

## 2. Search for isolated $\tau$

ZEUS data, corresponding to $\sim 130\,\mathrm{pb}^{-1}$, were used in this analysis [5]. The search for $\tau$ leptons is based on their hadronic decays, with one charged particle detected in the central tracking detector. The narrow jets with low charged-particle multiplicity that characterized $\tau$ decay, were used to distinguish them from quark- and gluon-induced jets [6]. The internal jet structure was parametrized by six variables related to the jet shape. The separation between signal and background was studied using Monte Carlo (MC) events, $W \to \tau\nu_\tau$ for the signal and charged current (CC) DIS for the background. In order to separate signal from background the six variables were combined to form a discriminant D. Fig. 1 shows the tau discriminant distribution for the MC signal and background events and for CC DIS data. The event selection was as used in

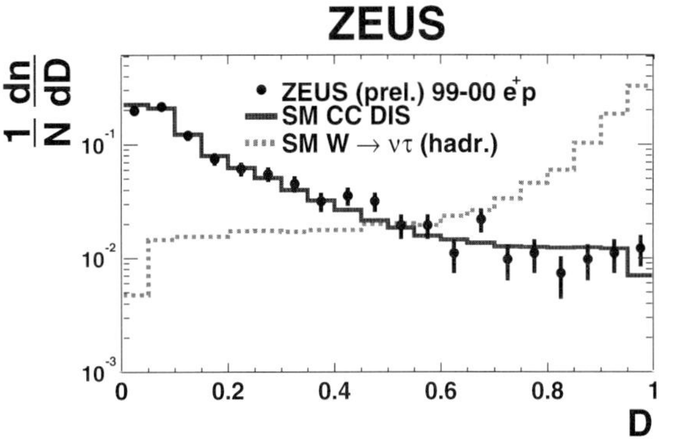


Figure 1. Distribution of the tau discriminant D for an inclusive selection of CC DIS events. The histograms are normalized to the total number of events N.

the previous ZEUS searches for events with isolated leptons and large missing $P_t$ [3,4] and required an isolated track with transverse momen-


doi:10.1016/S0920-5632(02)02218-1

tum $p_T > 5$ GeV and calorimeter transverse momentum $P_T^{CAL} > 20$ GeV. At least one jet with transverse energy $E_T > 5$ GeV within the range $-1.0 < \eta < 2.5$ was also required. Events with identified electrons and a topology characteristic of neutral current (NC) DIS events were rejected. In order to select $\tau$ candidates, $D > 0.95$ was required for events whose isolated tracks were not identified as muons or electrons and, to further reject CC DIS, the isolation requirement between the track and the closest jet was tightened. At this stage three events remain while $0.23 \pm 0.06$ are expected from the SM. Figure 2 shows the hadronic transverse momentum ($P_T^X$) distributions of the data and MC events that survive the $\tau$ selections. The final cuts have been designed

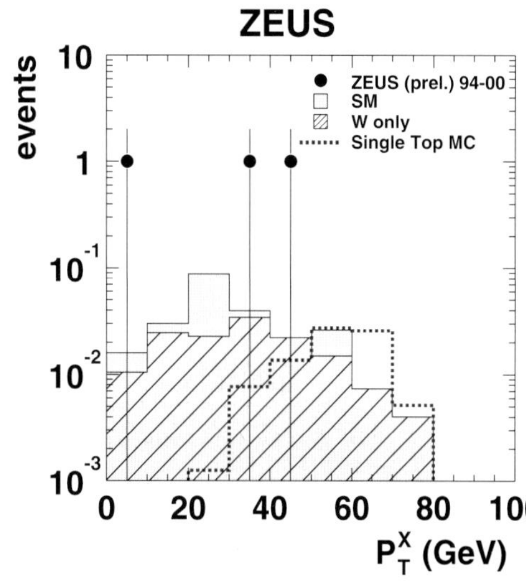


Figure 2. Distribution of the hadronic transverse momentum, $P_T^X$. The data (points) are compared to the SM expectations (shaded histogram). The dashed histogram represents the contribution from $W$ production. The dashed line represents the single-top signal MC normalized to the cross section limit obtained by the previous ZEUS search [4].

using the single-top MC as a template. A cut at $P_T^X > 25$ GeV gave the best expected upper limit assuming no signal. Two events remained in the data, while $0.12 \pm 0.02$ are expected from the SM, mainly from direct $W$ production. The observed excess cannot be explained by single-top production, since this process is already strongly constrained by the electron and the muon channel results which have a much higher sensitivity.

## 3. Excited fermions

In this section a search for excited electrons and neutrinos is presented [7–9]. At HERA excited electrons (neutrinos) could be singly produced through t-channel $\gamma$ and $Z$ ($W$) boson exchange. Their production cross-section and partial decay width have been calculated using an effective Lagrangian [10,11] which depends on a compositness scale $\Lambda$ and on form factors $f$ and $f'$, reduced here to constants, describing the relative coupling strengths of the excited leptons to the $SU(2)_L$ and $U(1)_Y$ gauge bosons, respectively. Usually these coupling parameters are assumed to be of comparable strength and the relationships $f = +f'$ and $f = -f'$ are imposed. In this case the production cross sections and partial decay widths depend only on the excited lepton mass and the ratio $f/\Lambda$. The decay channels $e^*(\nu^*) \to e(\nu)\gamma$, $e^*(\nu^*) \to \nu(e)W$ and $e^*(\nu^*) \to e(\nu)Z$ have been searched for, with subsequent hadronic decays of $W$ and $Z$ (the ZEUS collaboration also considered the decay $Z \to \nu\bar{\nu}$ in the $e^*$ case). The signature of such processes is very clear: a highly energetic electron and/or large missing transverse momentum in addition to a peak in the invariant mass of the lepton and the gauge boson. Such clear signatures allow the efficient selection of the signal. For the $e^*$ decay in $e\gamma$ channel, the efficiency was in the range $70-80\%$, while in the other cases it was typically $30-60\%$ for masses in the range $130-250$ GeV. The most important sources of background are NC and CC DIS and $W$ production processes. Since no excess over the SM expectations was found, limits at 95% confidence level were set on the $e^*$ and $\nu^*$ masses as a function of $f/\Lambda$. Figure 3 shows the HERA and LEP $e^*$ results for $f = +f'$ while figure 4 shows H1 and L3 [12] $\nu^*$ results (the ZEUS collaboration has similar results [7]) for $f = -f'$. The HERA limits extends the excluded region to higher masses than reached in previous direct searches.

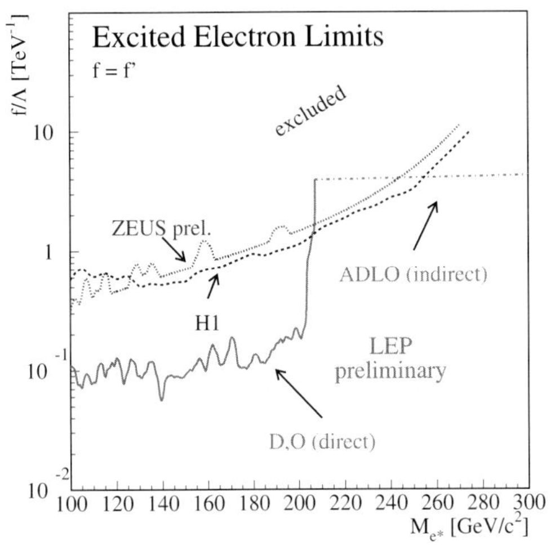


Figure 3. Exclusion limit on $f/\Lambda$ at 95% confidence level as a function of the $e^*$ mass for $f = +f'$. Exclusion limits are reported for H1 ($\sim 120$ pb$^{-1}$ of 1994-2000 $e^{\pm}p$ data), ZEUS ($\sim 47$ pb$^{-1}$ of 1994-1997 $e^+p$) and LEP experiments.

## 4. Conclusions

A search was presented for isolated $\tau$s and excited fermions at HERA. The isolated $\tau$ search extends the previous searches performed by the H1 and the ZEUS collaborations for high-$p_t$ isolated electrons and muons in events with large missing transverse momentum. Two events were observed while $0.12 \pm 0.02$ were expected from the SM. The observed excess is unlikely to be explained by single-top production, since this process is already strongly constrained by the analysis of the electron and muon channels.

Limits on $f/\Lambda$ as a function of mass were presented for excited electrons and neutrinos. The HERA limits extend the excluded region to higher masses than reached in previous direct searches.

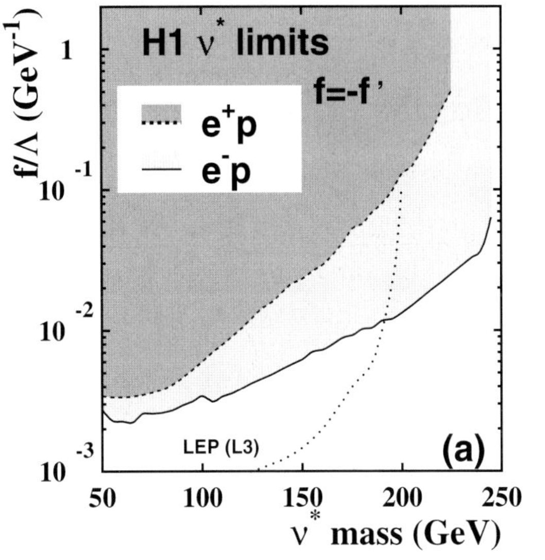


Figure 4. Exclusion limit on $f/\Lambda$ at 95% confidence level as a function of the $\nu^*$ mass for $f = -f'$. Exclusion limit reported by H1 refers to $e^-p$ data collected in 1998-1999 ($\sim 15$ pb$^{-1}$) and to $e^+p$ data collected in 1994-1997 ($\sim 37$ pb$^{-1}$).

## REFERENCES


1. H1 Coll., Eur. Phys. J. **C 5**, 575 (1998).
2. H1 Coll., paper 512, EPS01, Budapest, Hungary, July 12-18, 2001.
3. ZEUS Coll., Phys. Lett. **B 471**, 411 (2000).
4. ZEUS Coll., paper 650, EPS01, Budapest, Hungary, July 12-18, 2001.
5. ZEUS Coll., paper 909, ICHEP02, Amsterdam, The Holland, July 24-31, 2002.
6. C. N. Nguyen, Diploma Thesis, Hamburg University, Hamburg (Germany), DESY-THESIS-2002-024, 2002.
7. ZEUS Coll., DESY-01-132, hep-ex/0109018.
8. H1 Coll., Phys. Lett. **B 525**, 9 (2002).
9. H1 Coll., DESY-02-096, hep-ex/0208041.
10. K. Hagiwara, D. Zeppenfeld and S. Komamiya, Z. Phys. **C 29**, 115 (1993).
11. U. Baur, M. Spira and P. M. Zerwas, Phys. Rev. **D 42**, 815 (1990).
12. L3 Coll., Phys. Lett. **B 502**, 37 (2001).

ELSEVIER

Proceedings of ICHEP 2002, pp. 792–795
*S. Bentvelsen, P. de Jong, J. Koch and E. Laenen (Editors)*

31st INTERNATIONAL CONFERENCE ON
HIGH ENERGY PHYSICS AMSTERDAM

# Minireview of Leptoquark Searches

P. Bruce Straub [a]

[a]Yale University
DESY, F1/ZEUS
Notkestrasse 85
22607 Hamburg, Germany

Direct and indirect leptoquark searches at colliders are reviewed.

## 1. Introduction

Leptoquarks are hypothetical bosons which couple to a lepton and a quark via a Yukawa coupling (denoted $\lambda$). In the Standard model, both quarks and leptons occur in left-handed $SU(2)$ doublets and right-handed $SU(2)$ singlets. The symmetry between quarks and leptons leads to the cancellation of triangle anomalies which make the SM renormalizable. Leptoquarks appear in theories[1] in which this symmetry is more fundamental.

Leptoquarks (LQs) are color triplets, which would be pair produced in either $q\bar{q}$ or $gg$ interactions at $p\bar{p}$ or $pp$ colliders. Because they carry electroweak charge, they would also be pair produced in $e^+e^-$ or $\mu^+\mu^-$ collisions. Only standard model gauge couplings are involved in pair production; therefore the cross sections depend neither on the quark-lepton-LQ Yukawa coupling nor on the quark and lepton generations to which the leptoquark couples. In contrast, leptoquarks would be singly produced via the Yukawa coupling in a lepton-quark collision. Searches at electron-proton colliders are sensitive only to LQs which couple to electrons and the sensitivity to LQs which couple to second and third generation quarks is far below that of first-generation LQs. Leptoquarks are usually (but not always) assumed to be *generation diagonal.* Models in which LQs couple to more than one generation of quarks or leptons would induce flavor-changing neutral currents[2,3] or lepton flavor violation[2,4] respectively.

The model of Buchmüller, Rückl and Wyler[5] (BRW), in which leptoquarks couple to a single generation of SM fermions via chiral Yukawa couplings which are invariant under $SU(3)\times SU(2)\times U(1)$ is often used to classify possible leptoquark species. In the BRW model baryon and lepton numbers ($B$ and $L$) are conserved; there exist ten possible leptoquark species characterized by the chirality of the coupling, the the spin ($J = 0$ or 1), the weak isospin ($T$ =0, 1/2, or 1), and the fermion number, $F = 3B + L = 0$ or 2. Four of the species could have separate couplings to right- and left-handed leptons. A leptoquark which coupled first generation quarks to both left- and right-handed electrons would mediate $\pi \rightarrow e\nu$ decay[6]. The agreement of the measured decay rate with SM predictions motivates the conventional assumption that only one of the couplings can be nonzero. In the most commonly used notation[7] $F = 0$ scalar ($S$) and vector ($V$) species are denoted $S^L_{1/2}$, $S^R_{1/2}$, $\tilde{S}^L_{1/2}$, $V^L_0$, $V^R_0$, $\tilde{V}^R_0$ and $V^L_1$, while the $F = 2$ species are $S^L_0$, $S^R_0$, $\tilde{S}^R_0$, $S^L_1$, $V^L_{1/2}$, $V^R_{1/2}$, and $\tilde{V}^L_{1/2}$. The sub- and superscripts indicate the lepton chirality and the weak isospin respectively. The species $\tilde{S}$ and $\tilde{V}$ differ by two units of hypercharge from $S$ and $V$ respectively. Partial widths for scalar and vector LQs with mass $M_{LQ}$ are given by $\lambda^2 M_{LQ}/16\pi$ and $\lambda^2 M_{LQ}/24\pi$ respectively.

## 2. Leptoquark Searches the Tevatron

The Tevatron experiments have performed searches for first, second, and third generation


doi:10.1016/S0920-5632(02)02220-X

| | | $M_{LQ}$ Limit (GeV) | | |
|---|---|---|---|---|
| Gen. | $\beta$ | Scalar | Vector | Ref. |
| 1 | 1 | 225 | 246,345 | D0[8–10] |
| | 1/2 | 204 | 233,337 | |
| 1 | 1 | 213 | | CDF[12] |
| 1 | 1 | 242 | | CDF+D0[13] |
| 2 | 1 | 200 | 275,325 | D0[14,15] |
| | 1/2 | 180 | 260,310 | |
| 2 | 1 | 202 | | CDF[16,17] |
| | 1/2 | 160 | | |
| | 0 | 123 | 171,222 | |
| 1,2 | 0 | 98 | 200,298 | D0[11] |
| 3 | 0 | 94 | 148,209 | D0[18] |
| 3 | 1 | 99 | 170,217 | CDF[19,17] |
| | 0 | 148 | 199,250 | |

Table 1
Limits on leptoquarks from the Tevatron. The first two columns give the generation and $\beta$, the branching ratio to a charged lepton and a quark. The next two columns give the lower bounds on $M_{LQ}$ (95% CL). Vector LQs limits are given for both low and high cross section models.

leptoquarks, including separate analyses for decays to charged leptons and to neutrinos. The results are summarized in table 1. Limits are more restrictive for LQs which decay to $eq$ and $\mu q$ than those decaying to $\nu q$. CDF has used their silicon vertex detector to tag $b$ and $c$ quarks[17] while D0 identifies $b$-jets by associated muons[18]. For scalar LQs, NLO cross section calculations[20] are used in limit setting. In vector LQ cross section calculations[21] (done at LO), the coupling of the LQ and gluon fields is model dependent and limits for several different models[8,19] are quoted.

## 3. Leptoquark Searches at HERA

At the HERA electron-proton collider, first-generation leptoquarks can be singly produced via fusion of the beam electron with valence quarks in the proton. Event by event, the electron-jet and neutrino-jet final states resulting from LQ production (or exchange of virtual LQs) are indistinguishable from Neutral and Charged Current Deep Inelastic Scattering (DIS). If $M_{LQ}$ is below the center of mass energy, ($\sqrt{s} = 318\,\text{GeV}$) and the width of the LQ state is small, LQs would be produced as an $s$-channel resonance with cross section proportional to $\lambda^2$. Leptoquark production and decay would give rise to a a peak in the spectrum of lepton-jet invariant mass ($M_{\ell j}$), while the DIS background appears as a continuum. The lepton-jet invariant mass is related to the fraction $x$ of the proton momentum carried by the interacting quark by $M_{\ell j}^2 = xs$. A LQ signal is distinguished from the DIS background by the angular distribution of the final-state leptons. Define $\theta^*$ as the angle between the incident electron and the final state lepton in the LQ rest frame and $y = 1 - \cos\theta^*$. The distribution of $y$ for scalar LQ events is flat while vector LQs have a $1 - y^2$ distribution. In contrast, the NC-DIS background follows a $1/y^2$ distribution (for $\gamma$ exchange). Both H1[22,23] and ZEUS[24] have searched for LQs by computing a $\lambda$-dependent likelihood for the 2-dimensional distributions of electron-jet and neutrino-jet events in $M_{\ell j}$ and $y$ (or $\cos\theta^*$). This method is also sensitive to the changes in the observed $(M_{\ell j}, \cos\theta^*)$ distribution resulting from $u$- and $s$-channel LQ diagrams with $M_{LQ} \gtrsim \sqrt{s}$. Example of the resulting limit curves are shown in Fig. 1. In table 2, the ZEUS lower limits on $M_{LQ}$ for $\lambda = 0.1$ are shown. Both H1[25] and ZEUS[26] have also obtained limits for LQs with $M_{LQ} > \sqrt{s}$ by examining the $Q^2$ spectrum of NC-DIS-like events ($Q^2$ is the negative square of the 4-momentum transfer between the incident $e$ and the hadronic system). Limits obtained in these analyses are consistent with those in the analyses described above. In addition ZEUS[27] and H1[23] have searched by LQs which couple to two different lepton generations, thereby inducing lepton flavor violation.

## 4. Leptoquark Searches at LEP

Three sorts of leptoquark searches have been performed at LEP. Leptoquarks with $M_{LQ} < \sqrt{s}/2$ would be pair produced via electroweak coupling. Using data with $\sqrt{s}$ from 189 to 209 GeV OPAL[28] searched for scalar and vector LQs from all three generations, obtaining lower limits on $M_{LQ}$ ranging from 72 to 102 GeV. Single LQ

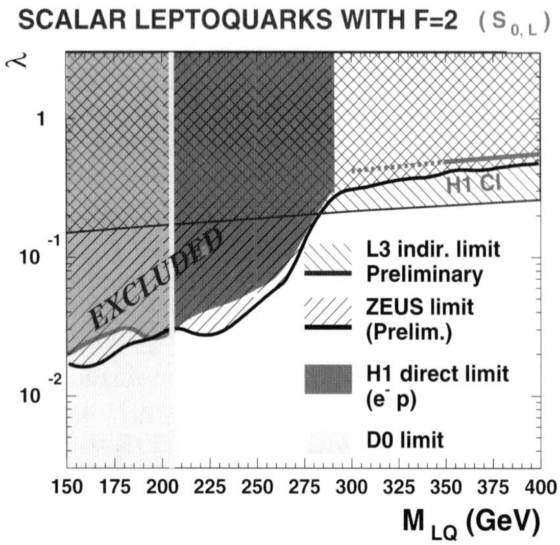


Figure 1. Limits on the $S_0^L$ leptoquark in the $(M_{LQ}, \lambda)$ plane. The excluded region lies above the curves for H1, L3, and ZEUS and to the left of the line for D0.

production can occur via the interaction of an electron with a radiated photon which resolves into $q\bar{q}$. Limits in the $(M_{LQ}, \lambda)$ plane for various LQ species obtained by OPAL[29] range from 189 to 209 GeV for $\lambda = \sqrt{4\pi\alpha_{\rm em}}$, where $\alpha_{\rm em}$ is the fine structure constant. See [30] for DELPHI results. The most interesting LQ searches[31–33] performed at LEP look for indirect effects of $t$-channel ($F = 0$) or $u$-channel ($F = 2$) LQ exchange in the reaction $e^+e^- \rightarrow q\bar{q}$. L3 limits on $\lambda$ vs. $M_{LQ}$ for scalar LQs are shown in Fig. 2. Table 2 shows combined LEP limits[34] on LQs which couple to $e$ and quarks of different generations. These limits were obtained using the cross section for $e^+e^- \rightarrow$ hadrons for $\sqrt{s}$ between 130 and 207 GeV. For LQs which couple to $ec$ and $eb$, the the $c\bar{c}$ and $b\bar{b}$ fractions $R_c$ and $R_b$, and the forward-backward asymmetries were also used.

| species | LEP | | | ZEUS |
|---|---|---|---|---|
| | 1st | 2nd | 3rd | 1st |
| $S_0^L$ | 655 | 762 | NA | 274 |
| $S_0^R$ | 520 | 625 | NA | 272 |
| $\tilde{S}_0^R$ | 202 | 209 | 454 | 249 |
| $S_{1/2}^L$ | 178 | 215 | NA | 281 |
| $S_{1/2}^R$ | 232 | 185 | - | 281 |
| $\tilde{S}_{1/2}^L$ | - | - | 226 | 270 |
| $S_1^L$ | 361 | 408 | 1036 | 274 |
| $V_0^L$ | 917 | 968 | 765 | 268 |
| $V_0^R$ | 165 | 147 | 167 | 271 |
| $\tilde{V}_0^R$ | 489 | 478 | NA | 279 |
| $V_{1/2}^L$ | 303 | 165 | 208 | 248 |
| $V_{1/2}^R$ | 227 | 466 | 499 | 273 |
| $\tilde{V}_{1/2}^L$ | 176 | 101 | NA | 272 |
| $V_1^L$ | 659 | 687 | 765 | 283 |

Table 2
Lower limits on $M_{LQ}$ in GeV for different LQ species which couple to first-generation leptons. The first column gives the LQ species; the next three columns give the LEP combined limits for first, second, and third generation quarks with $\lambda = \sqrt{4\pi\alpha_{\rm em}}$. A dash indicates that no limit could be set and NA indicates an LQ which couples only to top quarks, which would not be visible at LEP. The last column gives the ZEUS limits for $\lambda = 0.1$.

## 5. Summary

For first generation leptoquarks, the Tevatron experiments have set limits on scalars (coupling to $eq$) of $M_{LQ} > 242$ GeV and corresponding vector LQ limits in the range from 233 to 345 GeV, depending on model assumptions. The LEP experiments have set lower limits on $M_{LQ}$ (which are approximately proportional to $\lambda$) which range from 165 to 917 GeV for $\lambda = \sqrt{4\pi\alpha_{\rm em}} \sim 0.31$. The HERA experiments have set lower mass limits in the range of $\sim 250$ to 280 GeV for $\lambda = 0.1$. In addition, searches at the Tevatron and LEP have constrained LQs coupling to leptons and quarks of the second and third generations.

## REFERENCES


1. J.C. Pati and A. Salam, Phys. Rev. **D19**, 275 (1974); H. Georgi and S. Glashow, Phys. Rev. Lett. **32**, 438 (1974); J.L. Hewett and T.G. Rizzo, Phys. Rep. **183**, 193 (1989).
2. S. Davidson, D. Bailey, and B. Campbell, Z.

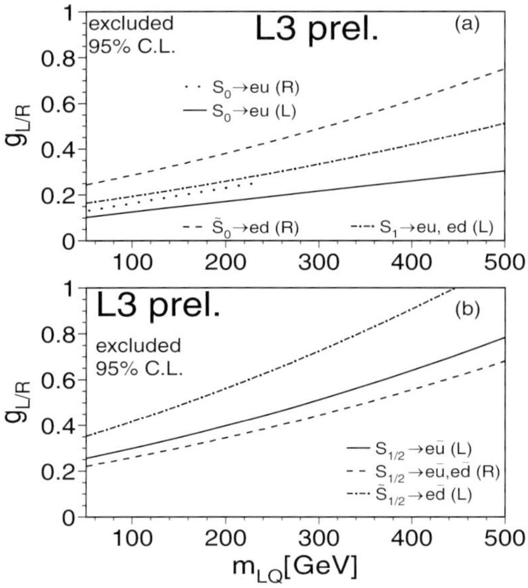


Figure 2. Scalar leptoquark limits in the $(M_{LQ}, \lambda)$ plane from L3[31]. The area above the curves is excluded.


Phys. **C61**, 613 (1994)

3. M. Leurer, Phys. Rev. **D49**, 333 (1994).
4. E. Gabrielli, Phys. Rev. **D62**, 055009 (2000)
5. W. Buchmüller, R. Rückl, and D. Wyler, Phys. Lett. **B191**, 442 (1987).
6. W. Buchmüller and D. Wyler, Phys. Lett. **B177**, 377 (1986).
7. B. Schrempp, Proc. Workshop on Physics at HERA, W. Buchmüller and G. Ingelman (eds.), Vol. 2, 1034, Hamburg (1991).
8. Abazov *et al.* D0 Coll., Phys. Rev. **D64**, 092004 (2001).
9. Abbott *et al.* D0 Coll., Phys. Rev. Lett. **79**, 4321 (1997).
10. Abbott *et al.* D0 Coll., Phys. Rev. Lett. **80**, 2051 (1998).
11. Abazov *et al.* D0 Coll., Phys. Rev. Lett. **88**, 191801 (2002).
12. Abe *et al.* CDF Coll., Phys. Rev. Lett. **79**, 4327 (1997).
13. CDF and D0 Leptoquark Limit Combination Working Group, `hep-ex/9810015`.
14. Abbott *et al.* D0 Coll., Phys. Rev. Lett. **84**, 2088 (2000).
15. Abbott *et al.* D0 Coll., Phys. Rev. Lett. **83**, 2896 (1999).
16. Abe *et al.* CDF Coll., Phys. Rev. Lett. **81**, 4806 (1998).
17. Affolder *et al.* CDF Coll., Phys. Rev. Lett. **85**, 2056 (2000).
18. Abbott *et al.* D0 Coll., Phys. Rev. Lett. **81**, 38 (1998).
19. Abe *et al.* CDF Coll., Phys. Rev. Lett. **78**, 2906 (1997).
20. M. Krämer, T. Plehn, M. Spira, and P.M. Zerwas, Phys. Rev. Lett. **79**, 341 (1997).
21. J. Blümlein, E. Boos, and A. Kryukov, Z. Phys. **C76**, 137 (1997).
22. ICHEP 2002 paper 1027 (H1 preliminary note 02-064); Adloff *et al.* H1 Coll., Phys. Lett. **B523**, 234 (2001).
23. Adloff *et al.* H1 Coll., Eur. Phys. J. **C11**, 447 (1999); (Erratum-ibid. **C14**, 553 (2000).
24. ICHEP 2002 paper 907; Breitweg *et al.* ZEUS Coll., Phys. Rev. **D63**, 052002 (2001); Breitweg *et al.* ZEUS Coll., Eur. Phys. J. **C16**, 253 (2000).
25. ICHEP 2002 paper 979 (H1 preliminary note 02-062); Adloff *et al.* H1 Coll., Phys. Lett. **B479**, 358 (2000).
26. ZEUS Coll., International Europhysics Conference on High Energy Physics, paper 602, Budapest 2001; Breitweg *et al.* ZEUS Coll., Eur. Phys. J. **C14**, 239 (2000).
27. ICHEP 2002 paper 906; Chekanov *et al.* ZEUS Coll., Phys. Rev. **D65**, 092004 (2002).
28. ICHEP 2002 paper 516 (OPAL Physics Note PN498); Abbiendi *et al.* OPAL Coll., Eur. Phys. J. **C13**, 15 (2000)
29. Abbiendi *et al.* OPAL Coll., Phys. Lett. **B526**, 233 (2002)
30. Abreu *et al.* DELPHI Coll., Phys. Lett. **B446**, 62 (1999)
31. ICHEP 2002 paper 462 (L3 Note 2759); Acciari *et al.* L3 Coll., Phys. Lett. **B489**, 81 (2000)
32. Abbiendi *et al.* OPAL Coll., Eur. Phys. J. **C6**, 1 (1999)
33. Barate *et al.* ALEPH Coll., Eur. Phys. J. **C12**, 183 (2000)
34. C. Goy personal communication, $f\bar{f}$ subgroup of LEP Electroweak Working Group, (`http://lepewwg.web.cern.ch/LEPEWWG/lep2`).

Proceedings of ICHEP 2002, pp. 796–798
*S. Bentvelsen, P. de Jong, J. Koch and E. Laenen (Editors)*

31st INTERNATIONAL CONFERENCE ON HIGH ENERGY PHYSICS AMSTERDAM

# Minireview on lepton flavour violation at $e^+e^-$ colliders

Guoliang Tong[a*]

[a]The Institute of High Energy Physics, Chinese Academy of Sciences, Beijing 100039,China

In this paper I review on recent results of searching for lepton flavor violation at $e^+e^-$ colliders: $J/\psi \to e\mu$ at BES and $\tau \to \mu\gamma$, $B^0 \to l^+l^-$, $B^- \to K^-\nu\overline{\nu}$ at BaBar and $e^+e^- \to e\mu, \mu\tau, e\tau$ at LEP with $\sqrt{S} = 189 - 209GeV$ by OPAL.

## 1. Introduction

In the standard model of electroweak theory, the lepton flavor symmetries are conserved, but speculated to be violated in many extensions of the standard model, such as supersymmetric standard models, left-right symmetric models and models where electroweak symmetry is broken dynamically.

Recent Super-Kamiokande and SNO experimental results[1] indicate strongly that neutrinos have masses, mix with each others. Consequently the lepton flavor symmetries or lepton number symmetry are broken. There have been a lot of studies both experimentally and theoretically on searching for the lepton flavor violating processes. At present we have various bounds listed in the particle data book from $\mu$, $\tau$ and Z decays.

The theoretical studies on the lepton flavor violation in decays of Charmonium, Bottomonium systems as well as $\phi$, Z are given in Ref.[2].

In this paper we report the results of searching for $J/\psi \to e\mu$ at BESII[3] first time, and we also review on recent results of searching for LVF at BaBar and OPAL.

## 2. Search for LFV in $J/\psi \to e\mu$ at BES

The results presented here are based on 58 milion $J/\psi$ events collected with the BESII detector at the Beijing Electron-Positron Collider(BEPC)[4].

The electron and $\mu$ particle have been identified with the information of the ratio of momentum (from Main Draft Chamber-MDC)/energy (from Shower Counter) and the $\mu$ counter.Cosmic rays were rejected with time information (fromTOF).The electron and $\mu$ particle in the candidates must be back to back, came from the $e^+e^-$ interacting point , and without isolated photon. Beside Monte Carlo simulation, the pion and kaon sample selected from $J/\psi \to \rho^\pm\pi^\mp$ and $J/\psi \to K^{*\pm}K^\mp$ are also used to study the background from $J/\psi \to \pi^+\pi^-$ and $K^+K^-$. Main possible background come from $J/\psi \to e^+e^-, \mu^+\mu^-, \pi^+\pi^-, K^+K^-$ and $ee \to ee(\gamma)$. Finally four $e\mu$ candidates are observed and 5.23 events for the Standard Model background are expeted. Our results are given in Table 1.

Table 1

| | Background,candidates and upper limit |
|---|---|
| $J/\psi \to ee$ | 2.55 |
| $J/\psi \to \mu\mu$ | 0 |
| $J/\psi \to \pi\pi$ | 1.29 |
| $J/\psi \to KK$ | 0.24 |
| $ee \to ee(\gamma)$ | 1.15 |
| Total backgrounds | 5.23 |
| $J/\psi \to e\mu$ | 4 |
| The upper limit | $1.35 \times 10^{-6}(90\%CL)$ |

*E-mail: tonggl@mail.ihep.ac.cn


doi:10.1016/S0920-5632(02)02229-6

We obtain a upper limit on BR($J/\psi \to e\mu$) with 58M $J/\psi$ sample at BES with the below fomular[5],

$$Br < \lambda(N_{OB})/[N_T \epsilon_{J/\psi \to e\mu}] \quad (1)$$

The upper limit in Table 1 is a conservative estimation(set background to zero).

Now we are working on the analysis of $J/\psi \to \mu\tau, e\tau$.

## 3. New results on rare leptonic B decays and $\tau \to \mu\gamma$ at BaBar

The results presented here are based on the data collected with the BaBar detector[6]. The upper limits on the BR of $B^0 \to l^+l^-$ and $B^- \to K^-\nu\bar{\nu}$ at 90% CL are listed in Table 2.

Table 2

| Mode | The upper limits |
|---|---|
| $B(B^- \to K^-\nu\bar{\nu})$ | $9.4 \times 10^{-5}$ |
| $B(B^0 \to e^+e^-)$ | $3.3 \times 10^{-7}$ |
| $B(B^0 \to \mu^+\mu^-)$ | $2.0 \times 10^{-7}$ |
| $B(B^0 \to e^\pm\mu^\mp)$ | $2.1 \times 10^{-7}$ |
| Luminosity | $54.4 fb^{-1}$ |

A search for the lepton number violating decay $\tau \to \mu\gamma$ is based on 56 Million $e^+e^- \to \tau^+\tau^-$ events produced in BaBar. In this search, $\tau \to e\nu\nu$ (BR=18%) and $h \geq 1\pi^0\nu$ (37%) are as the tag. Fig.1 shows the plot of $m_{EC}$ vs $\Delta E$[7]. Finally, 13 $\mu\gamma$ candidates are observed with an expected SM background of $7.8 \pm 1.4$ events, the upper limit on the BR($\tau \to \mu\gamma$) at 90% CL is: $BR(\tau \to \mu\gamma) < 2.0 \times 10^{-6}$.

## 4. Search for LFV in $e^+e^-$ collisions at $\sqrt{S} = 189 - 209 GeV$ at OPAL

The OPAL detector is described in detail in ref.[8]. In this search, a single $e\mu$ candidate has been found at $\sqrt{S} = 189 GeV$, the effects of the selection criteria are summarised in Table 3.

Table 3: Selected data events versus SM expectation[9]. "No - $l^+l^-$" stands for the cut against events having a pair of electron($e^+e^-$) or a pair of muon($\mu^+\mu^-$) candidates in the final state.

| | $e^+e^-$ | $\mu^+\mu^-$ | $\tau^+\tau^-$ | Other | Total background | Data |
|---|---|---|---|---|---|---|
| Selected | 20745 | 1359 | 564 | 24 | 22683 | 23164 |
| $\Sigma E_{ECAL} < 1.6 E_{beam}$ | 2275 | 1359 | 520 | 23 | 4185 | 4201 |
| No - $l^+l^-$ | 559 | 57 | 67 | 21 | 704 | 713 |
| $e\mu$ Candidates | 0 | 0 | 0.015 | 0.004 | 0.019 | 1 |
| $e\tau$ Candidates | 4.010 | 0.017 | 0.520 | 0.004 | 5.01 | 5 |
| $\mu\tau$ Candidates | 0.017 | 5.901 | 8.400 | 0 | 14.3 | 11 |

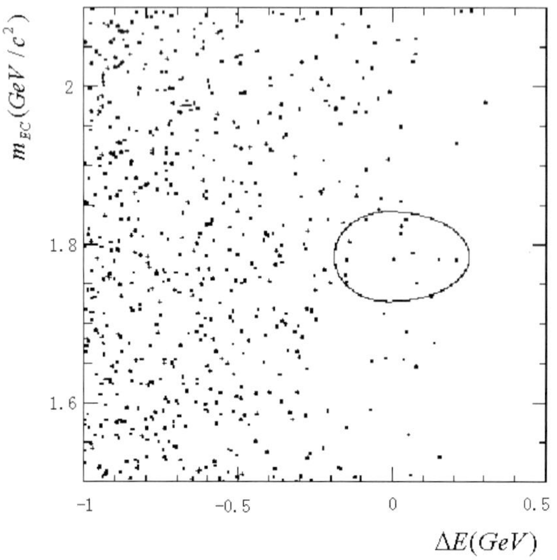


Figure 1. $m_{EC}$ vs $\Delta E$ (BaBar)

About the one $e\mu$ candidate, the most likely SM interpretation is $e^+e^- \rightarrow \tau^+\tau^-\gamma$, where at least one of the leptons from the tau gives rise to a fake electron or muon. There is also a cluster in the forward detector, which is probably from the gamma produced in initial state radiation.

Results of the present search are quantified in terms of 95% CL upper limits of the prodiction cross section of LFV events, assuming uniform final state angular distribution. The upper limits are displayed in Table 4 for different intervals of $\sqrt{S}$.

Table 4
95% confidence level upper limits on $\sigma(e^+e^- \rightarrow e\mu, e\tau, \mu\tau)$ as a function of $\sqrt{S}$

| Channel | $e\mu$ | $e\tau$ | $\mu\tau$ |
|---|---|---|---|
| $\sqrt{S}(GeV)$ | $\sigma[fb]$ | $\sigma[fb]$ | $\sigma[fb]$ |
| 189 | 58 | 95 | 115 |
| $192 \leq \sqrt{S} \leq 196$ | 62 | 144 | 116 |
| $200 \leq \sqrt{S} \leq 209$ | 22 | 78 | 64 |

## REFERENCES


1. Y.Fukuda *et al.* (Super -Kamiokande Collaboration), Phys. Rev. Lett., 1998:811562; Q.R.Ahmad *et al.* (SNO Collaboration), Phys. Rev. Lett.,37,071301(2001)
2. Xinmin Zhang, Invited talk given at the National Conference on High Energy Physics, Chengde, P.R. China, 1998 April; S. Nussinov, R.D. Peccei and Xinmin Zhang, Phys. Rev. D63, 016003 (2000); Xinmin Zhang, "Searching for new physics in $J/\psi$ decays" Hep-ph/0010105; Jose Bordes, Hong-Mo Chan, Sheung Tsun Tsou, Phys.Rev.D63:016006,2001; Taifu Feng, Wu-Jun Huo, Chongxing Yue and Xinmin Zhang (in preparation).
3. "Searching for lepton flavor violation in $J/\psi \rightarrow e\mu$ decays"; BES Collaboration (in preparation)
4. M.H.Ye and Z.P.Zheng, in *Proceedings of the XIVth International Symposium on Lepton and Photon Interactions*, Stanford, California, 1989, edited by M.Riordan(World Scientific, Singapore, 1990), p.122.
5. G.J. Feldman and R.D. Cousins, Phys. Rev. D, 1998,57:3873.
6. BABAR Collaboration, B.Aubert *et al.*,Nucl. Instrum. Methods Phys. Res. Sect. A 479, 1 (2001).
7. M. Roney, Search for $\tau \rightarrow \mu\gamma$ at BABAR, this Conference.
8. OPAL Collaboration, K.Ahmet *et al.*, Nucl. Inst. Meth. A305 (1991) 275; S. Anderson *et al.*, Nucl. Inst. Meth. A403 (1998) 326; B.E. Anderson *et al.*, IEEE Trans. on Nucl. Science 41 (1994) 845; G. Aguillon *et al.*, Nucl. Inst. Meth. A416 (1998) 266.
9. OPAL Collaboration, G. Abbiendi *et al.*, Phys. Lett. B 519 (2001) 23-32.

ELSEVIER

Proceedings of ICHEP 2002, pp. 799–802
*S. Bentvelsen, P. de Jong, J. Koch and E. Laenen (Editors)*

31st INTERNATIONAL CONFERENCE ON HIGH ENERGY PHYSICS AMSTERDAM

# $B$ Physics and Supersymmetry

L. Silvestrini[a]

[a]INFN, Sez. di Roma, Dip. di Fisica, Univ. di Roma "La Sapienza",
P.le A. Moro 2, I-00185 Roma, Italy

In this talk, I briefly review a selection of SUSY effects in $B$ physics. First, I consider models with Minimal Flavour Violation. Then I discuss SUSY models with new sources of flavour violation in squark mass matrices, analyzing present constraints and possible developments with forthcoming data on $b \to s$ and $b \to d$ transitions.

## 1. INTRODUCTION

Flavour changing neutral currents (FCNC) and CP-violating processes are very sensitive probes of new physics. The GIM suppression of FCNC amplitudes is generally absent beyond the Standard Model (SM), giving possibly large enhancements of FCNC processes over the SM predictions. Furthermore, CP violation in weak decays in the SM is entirely governed by a single phase in the CKM matrix, resulting in a strong correlation between all CP-violating processes. In general, this correlation is lost beyond the SM. Until now, all experimental data are fully compatible with the SM, as can be clearly seen from the standard analysis of the Unitarity Triangle (UT) [1]. One is then left with two basic questions: *i)* How can the SM be extended without spoiling the impressive consistency of the SM UT fit? *ii)* Can we still hope to see some hint of new physics in low-energy FCNC and CP-violating phenomena? While these two questions can be considered in full generality, I will focus in the rest of this talk on Supersymmetry (SUSY), which is at present the most successful extension of the SM with respect to the consistency with precision electroweak data. SUSY can modify the SM predictions on FCNC and CP-violating processes in three different ways: 1) With additional loop contributions still proportional to elements of the CKM matrix. A typical example of this kind of contributions is given by stop and chargino loops. This effect is present in any SUSY extension of the SM. 2) With additional loop contributions governed by new sources of flavour and CP violation [2]. When these new sources of FCNC are present, typically the largest contributions arise from gluino exchange. This effect is absent in SUSY models with Minimal Flavour Violation (MFV) (see Sec. 2). 3) With additional tree-level contributions. These can only arise in models with R-parity violation, and typically affect FCNC processes in a dramatic way.

Clearly, models in which only the first kind of contributions are present tend to agree much better with experimental data than more general models. On the other hand, the presence of contributions of the second or third kind usually generates larger deviations from SM predictions on yet unobserved FCNC and CP violating processes. In this respect, $B$ physics is the next frontier of testing SUSY in weak decays: $B$ factories and the Tevatron will provide us with data on a large variety of FCNC and CP-violating $B$ decays, and $B$ physics is playing an ever-increasing role in the UT fit. In the following, I will discuss in some detail SUSY effects in $B$ physics in models with R-parity conservation.[1]

## 2. $B$ PHYSICS IN MODELS WITH MFV

Let me start by considering models with MFV, defined by the following requirements. First, I assume that all sfermion masses are flavour diagonal and real at the electroweak scale, and all gaugino and Higgs mass parameters are real. This ensures

[1]For a recent example of the large effects in $B$ physics generated by R-violating couplings, see [3].


doi:10.1016/S0920-5632(02)02232-6

that the CKM matrix is the only source of flavour and CP violation, so that only the first kind of contributions discussed above can arise. Second, I assume that $\tan\beta$ is not too large ($\tan\beta \leq 10$), so that no new operator is generated in the effective Hamiltonian for $B$ decays. This implies that the SM analysis of perturbative and non-perturbative strong interaction effects can also be applied in this model. Finally, I assume that all squarks are degenerate except for the stop. This ensures that SUSY only modifies the contribution of the top quark in the SM. In particular, this means that SUSY contributions cancel from all quantities that in the SM do not depend on the top quark mass. A typical example is given by the ratio of the mass difference of $B_d$ and $B_s$ mesons. Indeed, a Universal Unitarity Triangle (UUT) can be constructed in these models, independent of SUSY contributions [4], once a sufficient number of top-mass independent quantities becomes available. At present, this is not the case and one needs to perform the full UT analysis to determine the CKM parameters for any given value of SUSY masses. The relevant SUSY parameters for the analysis are stop masses and mixing angle, the mass of the lightest chargino, the $\mu$ parameter, $\tan\beta$ and the charged Higgs mass. Potentially large contributions to $b \to s\gamma$ arise; once the constraints from $b \to s\gamma$ and the lower bounds on Higgs and SUSY masses from direct searches are taken into account, the UT fit in these models is indistinguishable from the SM one [5,6]. Deviations at the level of $20-30\%$ are possible in rare $B$ decays [6]. This is a property shared also by non-SUSY MFV models [7]. In the near future, $b \to s\gamma$ and $b \to s\ell^+\ell^-$ will be the most sensitive probes of SUSY models with MFV [8].

If one allows $\tan\beta$ to be very large, huge effects are possible in $B_s \to \mu^+\mu^-$ [9], increasing the BR from $BR(B_s \to \mu^+\mu^-)_{SM} \sim 4\cdot 10^{-9}$ to $10^{-6}$, behind the corner of $BR(B_s \to \mu^+\mu^-)_{EXP} < 2\cdot 10^{-6}$ [10]. This large enhancement of $B_s \to \mu^+\mu^-$ is in general correlated to a decrease of the $B_s - \bar{B}_s$ mass difference $\Delta M_s$, cutting out a large part of parameter space [11]. Once a measurement of $\Delta M_s$ becomes available, this correlation will be very useful in determining the prediction for $B_s \to \mu^+\mu^-$ in this model. It is also interesting to notice that, in minimal supergravity models, a correlation can be established between contributions to $B_s \to \mu^+\mu^-$ and $(g-2)_\mu$. In the region of parameter space favoured by the anomalous magnetic moment of the muon, one can obtain enhancements of one or two orders of magnitude of $BR(B_s \to \mu^+\mu^-)$ over the SM prediction [12].

## 3. $B$ PHYSICS IN GENERAL SUSY

We now turn from MFV to SUSY models with arbitrary sfermion mass matrices. In general, FCNC and CP-violating processes impose stringent constraints on off-diagonal sfermion mass terms [13]. To study these models in full generality, it is convenient to parameterize FCNC amplitudes in terms of $(\delta^d_{ij})_{AB}$, the ratio of the off-diagonal squark mass term $(\Delta^d_{ij})_{AB}$ connecting squarks of flavour $i$ and $j$ and helicities $A$ and $B$ over the average squark mass $m^2_{\tilde{q}}$. Let us first consider $(\delta^d_{13})_{AB}$, the mass insertion that induces $b \leftrightarrow d$ transitions. Constraints on this parameter from the $B_d - \bar{B}_d$ mass difference $\Delta M_d$ and from the time-dependent CP asymmetry in $B_d \to J/\Psi K$ decays $a_{\Psi K}$ have been recently studied in [14]. This analysis includes NLO QCD corrections [15] and lattice matrix elements [16]. As an example of the constraints one obtains from this kind of analysis, I report in Fig. 1 the allowed regions in the $\mathrm{Abs}(\delta^d_{13})_{AB} - \mathrm{Arg}(\delta^d_{13})_{AB}$ plane ($AB = LL, LR$), for $m_{\tilde{q}} = 500$ GeV. Other results and details of the analysis can be found in [14]. A similar analysis including also chargino contributions has been very recently carried out in [17]. Effects of $(\delta^d_{13})_{LR}$ can also be tested using $B \to \rho\gamma$ decays [18].

A similar analysis can be carried out for $(\delta^d_{23})_{AB}$, the mass insertions that generate $b \leftrightarrow s$ transitions [19]. In this case, however, at present we only have a lower bound on $\Delta M_s$ and the precise measurement of $BR(B \to X_s\gamma)$. The latter quantity is very effective in constraining $(\delta^d_{23})_{LR}$, while its impact on $(\delta^d_{23})_{LL}$ is quite limited. In Fig. 1 I report preliminary results on the allowed regions in the $\mathrm{Re}(\delta^d_{23})_{AB} - \mathrm{Im}(\delta^d_{23})_{AB}$ plane, for $AB = LL, LR$, for the present lower bound $\Delta M_s > 14$ ps$^{-1}$. To illustrate the possible

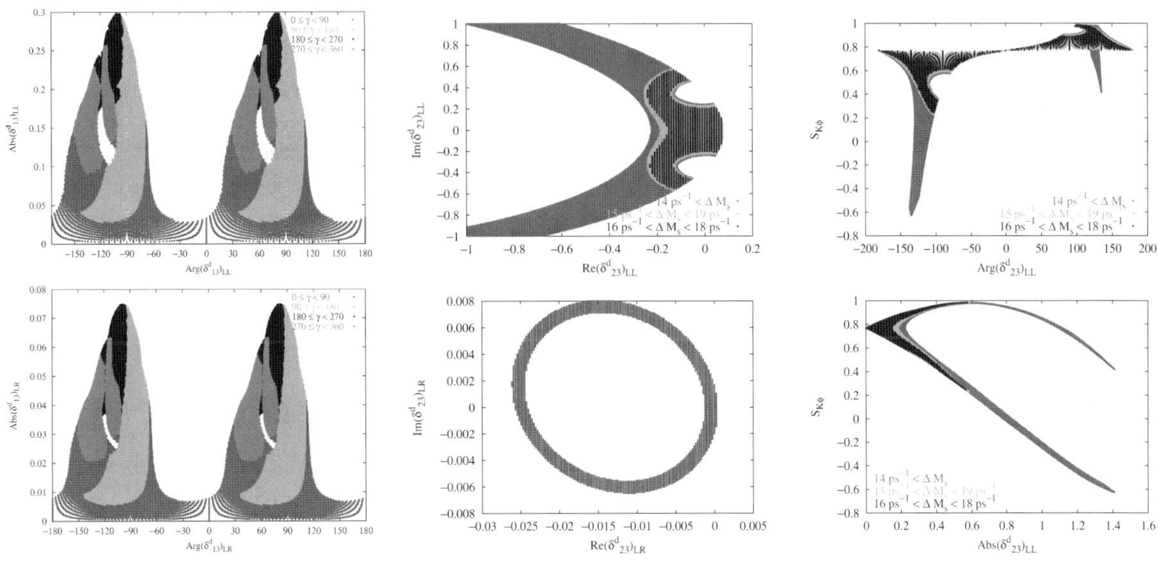


Figure 1. Left: Allowed values of $\mathrm{Abs}(\delta^d_{13})_{AB}$ as a function of $\mathrm{Arg}(\delta^d_{13})_{AB}$ for $AB = LL, LR$ and $m_{\tilde{q}} = 500$ GeV (from [14]). Different colours denote values of $\gamma$ belonging to different quadrants. Center: Allowed regions in the $\mathrm{Re}(\delta^d_{23})_{AB}$ – $\mathrm{Im}(\delta^d_{23})_{AB}$ plane for $AB = LL, LR$ and $m_{\tilde{q}} = 250$ GeV (from [19]). Results are preliminary. Right: $S_{\phi K_s}$ in the presence of $(\delta^d_{23})_{LL}$ for $m_{\tilde{q}} = 250$ GeV (from [19]). Results are preliminary.

impact of a future measurement of $\Delta M_s$, in Fig. 1 I also report the allowed regions for 15 $\mathrm{ps}^{-1} < \Delta M_s < 19$ $\mathrm{ps}^{-1}$ and 16 $\mathrm{ps}^{-1} < \Delta M_s < 18$ $\mathrm{ps}^{-1}$. While the constraints on $(\delta^d_{23})_{LR}$ are completely dominated by $BR(B \to X_s\gamma)$, one can see clearly that a measurement of $\Delta M_s$ will drastically reduce the allowed values of $(\delta^d_{23})_{LL}$. However, given the errors on hadronic matrix elements, the allowed range for $(\delta^d_{23})_{LL}$ cannot shrink too much when the experimental error on $\Delta M_s$ is reduced.

Clearly, $(\delta^d_{23})_{AB}$ enter many other interesting $b \to s$ decays. Let us first consider $B \to \phi K_s$. The first measurements of $a_{\phi K_s}$ by the BaBar and Belle collaborations display a $2.7\sigma$ deviation from the observed value of $a_{\Psi K}$ [20], while in the SM both quantities should measure $\sin 2\beta$ with negligible hadronic uncertainties [21] (here $\beta$ is one of the angles of the UT). In Fig. 1 I report $S_{\phi K_s}$ (the coefficient of the $\sin \Delta M_d t$ term in $a_{\phi K_s}$) as a function of $(\delta^d_{23})_{LL}$ for different values of $\Delta M_s$ [19]. At present, SUSY effects can account for the observed central value, but this may change with future data on $\Delta M_s$. It is interesting to notice that direct CP violation can also occur in this channel [22]. Another interesting process in which effects of $(\delta^d_{23})_{AB}$ can be seen is $b \to s\ell^+\ell^-$. Here large deviations from the SM in the asymmetries can be generated even for values of the BR close to the SM expectations [23].

## 4. CONCLUSIONS

For reasons of space, in this talk I was able to discuss only a few of the many interesting implications of SUSY in $B$ physics. However, already from these selected topics it is clear that the richer the flavour structure of superpartners is, the more probable it is to discover indirect signs of SUSY in $B$ physics. In any case, forthcoming data in this field will certainly help us learn more on SUSY extensions of the SM.

## ACKNOWLEDGMENTS

I am much indebted to M. Ciuchini for his kind and patient help in preparing the figures.

## REFERENCES

1. M. Ciuchini *et al.*, JHEP **0107** (2001) 013 [arXiv:hep-ph/0012308]; A. Hocker, H. Lacker, S. Laplace and F. Le Diberder, Eur. Phys. J. C **21** (2001) 225 [arXiv:hep-ph/0104062]; F. Parodi, talk at this conference.
2. L. J. Hall, V. A. Kostelecky and S. Raby, Nucl. Phys. B **267** (1986) 415.
3. S. Bar-Shalom, G. Eilam and Y. D. Yang, arXiv:hep-ph/0201244; B. Dutta, C. S. Kim and S. Oh, arXiv:hep-ph/0208226.
4. A. J. Buras, P. Gambino, M. Gorbahn, S. Jager and L. Silvestrini, Phys. Lett. B **500** (2001) 161 [arXiv:hep-ph/0007085].
5. S. Bertolini, F. Borzumati, A. Masiero and G. Ridolfi, Nucl. Phys. B **353** (1991) 591; E. Gabrielli and G. F. Giudice, Nucl. Phys. B **433** (1995) 3 [Erratum-ibid. B **507** (1997) 549] [arXiv:hep-lat/9407029]; P. L. Cho, M. Misiak and D. Wyler, Phys. Rev. D **54** (1996) 3329 [arXiv:hep-ph/9601360]; M. Misiak, S. Pokorski and J. Rosiek, Adv. Ser. Direct. High Energy Phys. **15** (1998) 795 [arXiv:hep-ph/9703442]. M. Ciuchini, G. Degrassi, P. Gambino and G. F. Giudice, Nucl. Phys. B **534** (1998) 3 [arXiv:hep-ph/9806308].
6. A. J. Buras, P. Gambino, M. Gorbahn, S. Jager and L. Silvestrini, Nucl. Phys. B **592** (2001) 55 [arXiv:hep-ph/0007313].
7. A. J. Buras, F. Parodi and A. Stocchi, arXiv:hep-ph/0207101.
8. G. D'Ambrosio, G. F. Giudice, G. Isidori and A. Strumia, arXiv:hep-ph/0207036.
9. K. S. Babu and C. F. Kolda, Phys. Rev. Lett. **84** (2000) 228 [arXiv:hep-ph/9909476]; P. H. Chankowski and L. Slawianowska, Phys. Rev. D **63** (2001) 054012 [arXiv:hep-ph/0008046]; C. Bobeth, T. Ewerth, F. Kruger and J. Urban, Phys. Rev. D **64** (2001) 074014 [arXiv:hep-ph/0104284]; C. S. Huang, W. Liao, Q. S. Yan and S. H. Zhu, Phys. Rev. D **63** (2001) 114021 [Erratum-ibid. D **64** (2001) 059902] [arXiv:hep-ph/0006250]; G. Isidori and A. Retico, JHEP **0111** (2001) 001 [arXiv:hep-ph/0110121].
10. F. Abe *et al.* [CDF Collaboration], Phys. Rev. D **57** (1998) 3811.
11. A. J. Buras, P. H. Chankowski, J. Rosiek and L. Slawianowska, arXiv:hep-ph/0207241.
12. A. Dedes, H. K. Dreiner and U. Nierste, Phys. Rev. Lett. **87** (2001) 251804 [arXiv:hep-ph/0108037].
13. F. Gabbiani and A. Masiero, Nucl. Phys. B **322** (1989) 235 ; E. Gabrielli, A. Masiero and L. Silvestrini, Phys. Lett. B **374** (1996) 80 [arXiv:hep-ph/9509379]; F. Gabbiani, E. Gabrielli, A. Masiero and L. Silvestrini, Nucl. Phys. B **477** (1996) 321 [arXiv:hep-ph/9604387];
14. D. Becirevic *et al.*, Nucl. Phys. B **634** (2002) 105 [arXiv:hep-ph/0112303].
15. M. Ciuchini *et al.*, Nucl. Phys. B **523** (1998) 501 [arXiv:hep-ph/9711402]; A. J. Buras, M. Misiak and J. Urban, Nucl. Phys. B **586** (2000) 397 [arXiv:hep-ph/0005183].
16. D. Becirevic *et al.*, JHEP **0204** (2002) 025 [arXiv:hep-lat/0110091].
17. E. Gabrielli and S. Khalil, arXiv:hep-ph/0207288.
18. A. Ali and E. Lunghi, arXiv:hep-ph/0206242.
19. M. Ciuchini, E. Franco, G. Martinelli, A. Masiero and L. Silvestrini, work in progress.
20. B. Aubert *et al.* [BABAR Collaboration], arXiv:hep-ex/0207070; K. Abe [Belle Collaboration], arXiv:hep-ex/0207098.
21. Y. Grossman, G. Isidori and M. P. Worah, Phys. Rev. D **58** (1998) 057504 [arXiv:hep-ph/9708305].
22. M. Ciuchini and L. Silvestrini, arXiv:hep-ph/0208087.
23. E. Lunghi, A. Masiero, I. Scimemi and L. Silvestrini, Nucl. Phys. B **568** (2000) 120 [arXiv:hep-ph/9906286]. A. Ali and E. Lunghi, Eur. Phys. J. C **21** (2001) 683 [arXiv:hep-ph/0105200]. E. Gabrielli and S. Khalil, Phys. Lett. B **530** (2002) 133 [arXiv:hep-ph/0201049]. G. Isidori and A. Retico, arXiv:hep-ph/0208159.

ELSEVIER

Proceedings of ICHEP 2002, pp. 803–806
*S. Bentvelsen, P. de Jong, J. Koch and E. Laenen (Editors)*

31st INTERNATIONAL CONFERENCE ON HIGH ENERGY PHYSICS AMSTERDAM

# New physics searches at a Linear Collider with polarized beams

G. Moortgat-Pick[a] *, A.Bartl[b], K. Hidaka[c], T. Kernreiter[b], H.Liivat[d], R.-K.Loide[d], I. Ots[d], W. Porod[e], R.Saar[d], H. Uibo[d]

[a]Inst. für Theoretische Physik, Universität Hamburg, D–22607 Hamburg, Germany

[b]Institut für Theoretische Physik, Universität Wien, A-1090 Vienna, Austria

[c]Dept. of Physics, Tokyo Gakugei University, Koganei, Tokyo 184-8501, Japan

[d]Institute of Physics, University of Tartu, Riia 142, 51014 Tartu, Estonia

[e]Inst. für Theor. Physik, Universität Zürich, CH-8057 Zürich, Switzerland

A future $e^+e^-$ Linear Collider has a large physics potential for the discovery of new physics beyond the Standard Model and precision studies of the Standard Model itself. It is well suited to complement and extend the physics program of the LHC concerning the precise determination of the underlying theory. The use of polarized beams at a LC will be one of the decisive tools.

## 1. Introduction

A future Linear Collider (LC) designed with high luminosity of about $\mathcal{L} = 3.4 \cdot 10^{34}\mathrm{cm}^{-2}\mathrm{s}^{-1}$ at $\sqrt{s} = 500$ GeV and $\mathcal{L} = 5.8 \cdot 10^{34}\mathrm{cm}^{-2}\mathrm{s}^{-1}$ at $\sqrt{s} = 800$ GeV will provide precision studies in the energy range between LEP and the TeV range [1]. Concerning the searches for New Physics it complements and extends the physics program of the Hadron Colliders, Tevatron and the future Large Hadron Collider (LHC)[1]. The LC has large potential for the discovery of new particles and is unique concerning the precision analysis of new physics (NP) as well as of the Standard Model (SM) [1]. An important tool of a LC is the use of polarized beams [2]. Already in the base line design it is foreseen to use the electron beams polarized to around 80% via a strained photocathode technology. In order to generate also polarized positrons the use of a helical undulator is favoured, producing polarized photons which generate via pair production polarized positrons of about 40% (with full intensity of the $e^+$ beam) to 60% (with 55% intensity).

In the next sections we summarize the new physics searches at a LC starting from high precision studies of the SM as a motivation.

*Talk given by G. Moortgat-Pick, gudrid@mail.desy.de
[1]contact Georg Weiglein, see http://www.ippp.dur.ac.uk/~georg/lhclc/

## 2. High precision analysis of the SM

One possibility to see hints for new physics, even if new particles are not yet found, are electroweak precision tests with an unprecedented accuracy.

### 2.1. Triple gauge couplings in $e^+e^- \to Z\gamma$

The study of triple gauge couplings is very promising. In [3] a method has been worked out to derive complete analytical expressions for the spin polarization and alignment of the $Z$ boson in $e^+e^- \to Z\gamma$ including the contributions from $ZZ\gamma$ and $Z\gamma\gamma$ anomalous couplings.

### 2.2. Anomalous couplings in $e^+e^- \to W^+W^-$

Searches for deviations from the SM have already been studied, comparing polarized cross sections in a general parametrization including anomalous gauge couplings with expected precision measurements at the LC [1]. It turns out that for this purpose the polarization of the beams is very powerful: e.g. the polarization of $P(e^-) = \pm 80\%$ (together with $P(e^+) = \mp 60\%$) improves the sensitivity up to a factor 1.8 (2.5).

### 2.3. Transversely polarized beams in $e^+e^- \to W^+W^-$

A promising possibility to study electroweak symmetry breaking is the use of transversely po-


doi:10.1016/S0920-5632(02)02235-1

larized $e^+e^-$ beams which projects out $W_L^+W_L^-$ [4]. The asymmetry with respect to the azimuthal angle of this process focusses on the $LL$ mode. This asymmetry is very pronounced at high energies reaching about 10%. The advantage of this observable is that at high energies this asymmetry peaks at larger angles and not in beam direction where the analysis might be difficult. One has to note, however, that for the use of transverse beams the polarization of both beams is needed. The effect does not occur if only one beam is polarized since the cross section is given by:

$$\sigma = (1 - P_{e^-}^L P_{e^+}^L)\sigma_{unp} + (P_{e^-}^L - P_{e^+}^L)\sigma_{pol}^L + P_{e^-}^T P_{e^+}^T \sigma_{pol}^T. \quad (1)$$

### 2.4. GigaZ

A very spectacular method for probing the SM is foreseen as an upgrade of a LC: the GigaZ option. Here $e^+e^- \to Z \to f\bar{f}$ is studied at the pole with very high luminosity of about $10^9$ Z's within three month. With this option the effective electroweak leptonic mixing angle can be measured via the left–right asymmetry of this process with an unprecedented accuracy, see Table 1 [1].

### 2.5. Sensitivity to CP violation in the SM

In [6] electroweak dipole form factors of the $\tau$ lepton have been analysed with regard to CP violation. CP–odd triple product correlations have been studied and sensitivity bounds for the real and imaginary parts of these form factors have been set. Using polarized beams a LC could be sensitive to these form factors up to $O(10^{-19})$ ecm.

## 3. Revealing the structure of Susy

Supersymmetry is widely regarded as the best motivated extension of the SM. However, since the SM particles and their Susy partners are not mass degenerate, Susy has to be broken, which leads even for its minimal version, the Minimal Supersymmetric Standard Model (MSSM), to about 100 free parameters. In specific scenarios of Susy breaking one has much less parameters: 5 in mSUGRA, 4 in AMSB and 5 in GMSB.

In order to exactly pin down the structure of the underlying model it is unavoidable to extract the parameters without assuming a particular breaking scheme. The LC with its clear signatures is well suited to determine the general parameters with high precision and to test fundamental Susy assumptions as e.g. the equality of quantum numbers or of couplings of the particles and their Susy partners. It turns out that the use of polarized beams plays a decisive role in this context.

### 3.1. Stop mixing angle in $e^+e^- \to \tilde{t}_1\tilde{t}_1$

As demonstrated in [1] the mass and the mixing angle of $\tilde{t}$ can be extracted with high precision via the study of polarized cross sections for light stop production. At a high luminosity LC and with $P(e^-) = 80\%$ and $P(e^+) = 60\%$ an accuracy of $\delta(m_{\tilde{t}_1}) \approx 0.8$ GeV and $\delta \cos\theta_{\tilde{t}} \approx 0.008$ could be reachable.

### 3.2. Quantum numbers in $e^+e^- \to \tilde{e}^+_{L,R}\tilde{e}^-_{L,R}$

Susy transformations associate $e^-_{L,R} \leftrightarrow \tilde{e}^-_{L,R}$ and the antiparticles $e^+_{L,R} \leftrightarrow \tilde{e}^+_{R,L}$. In order to prove this association of scalar particles to chiral quantum numbers the use of polarized beams is necessary [7]. The process occurs via $\gamma$ and $Z$ exchange in the s–channel and via $\tilde{\chi}^0_i$ exchange in the t–channel. Only in the t–channel the SM particle is directly coupled to its scalar partner and in order to test the association of quantum numbers one has to projects out the t–channel exchange.

With completely polarized $e^-_L e^+_L$ only $\tilde{e}^-_L\tilde{e}^+_R$ contributes. Due to their $L, R$ coupling character $\tilde{e}_L$, $\tilde{e}_R$ can be discriminated via their decays and can be identified via their charge. One has to note that a polarized $e^+$ beam is necessary. Even completely polarized $e^-$ would not be sufficient, since otherwise the s-channel exchange could not be switched off. In reality partially polarized beams of $P_{e^-} = -80\%$ and $P_{e^+} = -60\%$ can be sufficient to probe this association. With this polarization $\tilde{e}^-_L\tilde{e}^+_R$ dominates e.g. by a factor of 3, Fig. 1.

### 3.3. Gaugino/higgsino sector

In [8] and references therein a strategy has been worked out to determine accurately the MSSM parameters $M_1$, $\Phi_{M_1}$, $M_2$, $\mu$, $\Phi_\mu$ and moderate $\tan\beta$ via the study of polarized rates in $e^+e^- \to \tilde{\chi}^\pm_i\tilde{\chi}^\mp_j$ and $e^+e^- \to \tilde{\chi}^0_i\tilde{\chi}^0_j$. Even if only

$\tilde{\chi}_1^+\tilde{\chi}_1^-$, $\tilde{\chi}_1^0\tilde{\chi}_2^0$ were accessible, it would be sufficient for determining the MSSM parameters [8]. With light chargino pair production and polarized initial beams one can unambiguously determine the mixing angle but $M_2$, $\mu$, $\tan\beta$ still depend on $m_{\tilde{\chi}_2^\pm}$. If one furthermore includes now the light neutralino system and studies the polarized cross sections $e^+e^- \to \tilde{\chi}_1^0\tilde{\chi}_2^0$ in the complex $M_1$ plane, one can simultaneously determine the parameter $M_1$ and predict $m_{\tilde{\chi}_2^\pm}$.

Once the parameters are determined one can efficiently test whether the gauge couplings $g_{Bee}$ and $g_{Wee}$ are identical to the Yukawa couplings $g_{\tilde{B}e\tilde{e}}$ and $g_{\tilde{W}e\tilde{e}}$, respectively, by studying the polarized cross sections with a variable ratio of $g_{Bee}/g_{\tilde{B}e\tilde{e}}$ and $g_{Wee}/g_{\tilde{W}e\tilde{e}}$ and comparing it with experimental values [8].

### 3.4. Case of high $\tan\beta$: $\tau$ polarization

In case of high $\tan\beta > 10$ the chargino and neutralino sector is insensitive to this parameter. However, one could then determine $\tan\beta$ from another sector whose particles are relatively light in many scenarios: the $\tilde{\tau}$ sector.

The polarization of $\tau$'s from $\tilde{\tau}_i \to \tau\tilde{\chi}_1^0$ is sensitive to $\tan\beta$. In case of a sufficient higgsino admixture in the $\tilde{\chi}_1^0$ it is even possible to determine high $\tan\beta$ as well as $A_\tau$, without any assumptions on the Susy breaking mechanism [9]: after determining the $\tilde{\tau}$ mixing angle via the polarization asymmetry $A_{pol} = (\sigma_L - \sigma_R)/(\sigma_L + \sigma_R)$ in $e^+e^- \to \tilde{\tau}_1\tilde{\tau}_1$, one can determine $\tan\beta$ from the $\tau$ polarization in the decay $\tilde{\tau} \to \tau\tilde{\chi}_1^0$, see Fig.2. Even for high $\tan\beta \approx 40$ one can reach an accuracy of about 10%. Measuring now $m_{\tilde{\tau}_2}$ leads to the determination of $A_\tau$.

### 3.5. CP–phases in the $\tilde{\tau}$ sector

In the $\tilde{\tau}_i$ system CP–violating phases can occur in the parameters $A_\tau$, $\mu$ and $M_1$. These phases have a decisive influence on masses, couplings and various CP–violating observables. The influence of the phases could be quite strong even on CP–conserving decay branching ratios (BR's) and hence one should take it into account when determining the MSSM parameters at the LHC and LC. In this context the BR's also might be a sensitive probe for the CP–phases [10].

### 3.6. Extended Susy models

In case of R–parity violating Susy non–standard couplings could occur which produce a scalar particle in the s–channel: $e^+e^- \to \tilde{\nu} \to e^+e^-$. The process gives a significant signal over the background. Since it requires both left–handed $e^-$ and $e^+$ beams, it can be easily analysed and identified by the use of beam polarization ([2] and references therein): here simultaneously polarized beams enhance the signal by about a factor of 10.
Introducing in the MSSM an additional Higgs singlet leads to the (M+1)SSM with one additional neutralino. Since the mass spectra of the four light neutralinos could be similar to those in the MSSM a distinction between the models might be difficult via spectra and rates alone. However, polarization effects might then indicate the different coupling structure in the (M+1)SSM [11].

## 4. Other kind of New Physics

Another approach to resolve the hierachy problem is the introduction of large extra dimensions. At a LC the process $e^+e^- \to \gamma G$ is promising and it has been worked out that running on two different $\sqrt{s}$ one can determine the number of extra dimensions [1]. The use of polarized beams in this context enlarges on one hand the sensitivity to the new scale $M_*$ and suppresses on the other hand the main background $e^+e^- \to \nu\nu\gamma$ significantly. The ratio $S/\sqrt{B}$ is enhanced by a factor of about 2.1 (4.4) if $P(e^-) = +80\%$ (and $P(e^+) = -60\%$) is used.

## 5. Summary

It has been shown that a LC in the TeV range is well suited for new physics searches and in particular for its precise determination. The use of polarized beams plays a decisive role[2]. In this context the use of simultaneously polarized $e^-$, $e^+$ beams has several advantages: enhancing considerably the accuracy for SM precision tests at GigaZ, providing higher sensitivity to non–standard couplings and supplying efficient tools to identify quantum numbers of the new particles

[2]for updates see POWER group (POlarization at Work in Energetic Reactions), forum for machine physicists, experimentalists, theorists: http://www.desy.de/ ~ gudrid/POWER/

Table 1
Ew. precision parameters at different colliders [5]

| Accuracy | LEP/Tev. | Tev./LHC | LC | GigaZ |
|---|---|---|---|---|
| $\delta M_W$ | 34 MeV | 15 MeV | 15 MeV | 6 MeV |
| $\delta \sin^2 \theta_{eff}$ | 0.00017 | 0.00017 | 0.00017 | 0.00001 |
| $\delta m_t$ | 5 GeV | 2 GeV | 0.2 GeV | 0.2 GeV |
| $\delta m_h$ | – | 0.2 GeV | 0.05 GeV | 0.05 GeV |

and to determine the underlying parameters of the theory.

G.M.–P. was partially supported by the Graduiertenkolleg 'Zukünftige Entwicklungen in der Teilchenphysik' of the University of Hamburg, Project No. GRK 602/1. This work was also supported by the EU TMR Network Contract No. HPRN-CT-2000-00149. The group from IP, University of Tartu was partly supported by ESF grant 4510.

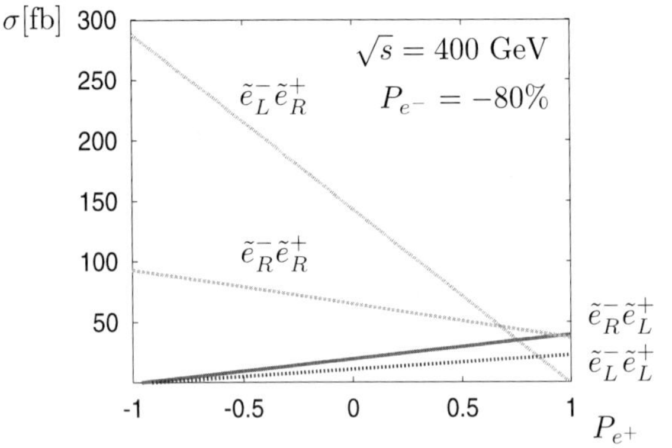


Figure 1. Test of selectron quantum numbers in $e^+e^- \to \tilde{e}^+_{L,R}\tilde{e}^-_{L,R}$ with fixed electron polarization $P(e^-) = -80\%$ and variable positron polarization $P(e^+)$. For $P(e^-) = -80\%$ and $P(e^+) < 0$ both pairs $\tilde{e}_L^- \tilde{e}_R^+$ and $\tilde{e}_R^- \tilde{e}_R^+$ still contribute. For $P(e^+) = -60\%$ $\tilde{e}_L^- \tilde{e}_R^+$ dominates by more than a factor 3 [7].

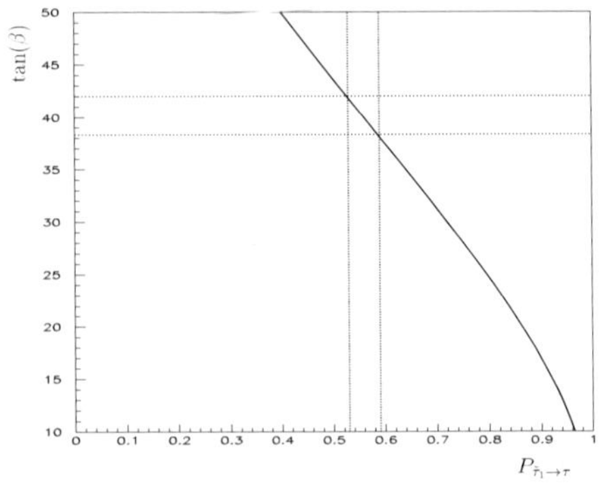


Figure 2. Measuring the $\tau$ polarization $P_{\tilde{\tau}_1 \to \tau}$ from the decay $\tilde{\tau}_1 \to \tau \tilde{\chi}_1^0$ for an already measured mixing angle $\cos\theta_{\tilde{\tau}}$ leads in an example of [9] to an accurate determination of high $\tan\beta$: $\tan\beta = 40 \pm 2$.

## REFERENCES

1. J. A. Aguilar-Saavedra *et al.*, ECFA/DESY LC Physics Working Group Collaboration, [hep-ph/0106315].
2. G. Moortgat-Pick and H. M. Steiner, Eur. Phys. J. directC **6** (2001) 1 [hep-ph/0106155].
3. I. Ots, H. Uibo, H. Liivat, R. K. Loide and R. Saar, Nucl. Phys. B **588** (2000) 90 and *Possible $Z\gamma\gamma$ and $ZZ\gamma$ coupling in $e^+e^- \to Z\gamma$ in a general relativistic density matrix formalism*, Contrib. to ICHEP02, Amsterdam.
4. J. Fleischer, K. Kolodziej and F. Jegerlehner, Phys. Rev. D **49** (1994) 2174.
5. J. Erler, S. Heinemeyer, W. Hollik, G. Weiglein and P. M. Zerwas, Phys. Lett. B **486** (2000) 125.
6. B. Ananthanarayan, S. D. Rindani and A. Stahl, hep-ph/0204233.
7. C. Blöchinger, H. Fraas, G. Moortgat-Pick and W. Porod, Eur. Phys. J. C **24** (2002) 297.
8. S. Y. Choi, J. Kalinowski, G. Moortgat-Pick and P. M. Zerwas, Eur. Phys. J. C **22** (2001) 563 ; S. Y. Choi, J. Kalinowski, G. Moortgat-Pick and P. M. Zerwas, Eur. Phys. J. C **23** (2002) 769 [hep-ph/0202039].
9. E. Boos, U. Martyn, G. Moortgat–Pick, M. Sachwitz, S. Vologdin, P.M. Zerwas, in prep..
10. A. Bartl, K. Hidaka, T. Kernreiter and W. Porod, hep-ph/0207186 and Phys. Lett. B **538** (2002) 137.
11. S. Hesselbach, F. Franke, H. Fraas, hep-ph/0003272; F. Franke and S. Hesselbach, Phys. Lett. B **526** (2002) 370.

ELSEVIER

Proceedings of ICHEP 2002, pp. 807–810
*S. Bentvelsen, P. de Jong, J. Koch and E. Laenen (Editors)*

31st INTERNATIONAL CONFERENCE ON HIGH ENERGY PHYSICS AMSTERDAM

# Sfermion Precision Measurements at a Linear Collider

A. Freitas*, B. Ananthanarayan, A. Bartl, G. Blair, C. Blöchinger, E. Boos, A. Brandenburg, A. Datta, A. Djouadi, H. Fraas, J. Guasch, S. Hesselbach, K. Hidaka, W. Hollik, T. Kernreiter, M. Maniatis, A. v. Manteuffel, H.-U. Martyn, D. J. Miller, G. Moortgat-Pick, M. Mühlleitner, U. Nauenberg, H. Nowak, W. Porod, J. Solà, A. Sopczak, A. Stahl, M. M. Weber, P. M. Zerwas

SUSY Collaboration of ECFA/DESY and International Linear Collider Workshops

At prospective $e^{\pm}e^{-}$ linear colliders, the large cross-sections and clean signals of scalar fermion production — in particular for the scalar leptons — allow very precise measurements of their masses and couplings and the determination of their quantum numbers. Various methods are proposed for extracting these parameters from the data at the sfermion thresholds and in the continuum. At the same time, NLO radiative corrections and non-zero width effects have been calculated in order to match the experimental accuracy. The substantial mixing expected in the third generation opens up additional opportunities. Techniques are presented for determining potential CP-violating phases and for extracting $\tan\beta$ from the stau sector, in particular at high values. The consequences of possible large mass differences in the stop and sbottom system are explored in dedicated analyses.

## 1. INTRODUCTION

If supersymmetry is detected at a future collider, its exploration will require two main tasks. On one hand, to establish supersymmetry experimentally, it is necessary to accurately test the validity of supersymmetry relations, i.e. whether the quantum numbers and couplings of the superpartners agree with the corresponding parameters of the Standard Model. On the other hand, the pattern of the supersymmetry breaking needs to be explored. In particular, it is important to determine the breaking parameters with high precision in order to reconstruct the underlying breaking mechanism, which eventually involves extrapolations to high energy scales. A high-energy $e^+e^-$ linear collider [1] is particularly suited for precision measurements of supersymmetric particle properties. In order to match the experimental accuracy, sufficiently precise and reliable theoretical predictions for the production cross-sections and decays of the superpartners are required.

This report is focused on recent developments in precision analyses of scalar fermions at a linear collider. Among the most important issues are the determination of their quantum numbers, couplings and masses, and furthermore the mixing parameters and potential CP-violating phases.

*afreitas@mail.desy.de


doi:10.1016/S0920-5632(02)02237-5

## 2. SPARTICLE COUPLINGS

The chiral quantum numbers of the selectrons can be unambiguously determined by studying the selectron production cross-sections in $e^+e^-$ annihilation with polarized beams [2]. In the t-channel neutralino exchange contributions, the quantum numbers of the produced selectron and spositron are directly related to the polarization of the incoming electron and positron beams.

This is of relevance for one of the most fundamental predictions of supersymmetry, the equivalence between gauge couplings and their supersymmetric Yukawa counterparts. For instance, the usual gauge coupling between a vector boson $V$ and a fermion $f$, $g(Vff)$, is is related to the Yukawa coupling between the gaugino partner $\tilde{V}$ of the vector boson, the fermion $f$ and the sfermion $\tilde{f}$, $\hat{g}(\tilde{V}f\tilde{f})$. Within supersymmetric theories both kinds of (bare) couplings are required to be identical, $g = \hat{g}$.

This relation can be tested in the electroweak sector by measuring the production cross-sections of scalar leptons. The Yukawa couplings $\hat{g}$ can best be probed in the production of selectrons, as a result of the t-channel neutralino exchange contributions. Simulations have shown that the U(1) and SU(2) couplings $\hat{g}'$ and $\hat{g}$ can be determined to better than one percent by measuring the slepton cross-sections at a linear collider [3], i.e. $\delta\hat{g}'/\hat{g}' \approx 0.2\%$, $\delta\hat{g}/\hat{g} \approx 0.7\%$.

Clearly the experimental precision requires the inclusion of radiative corrections in the theoretical predictions for the slepton cross-sections. In Ref. [4] the pair production of smuons in $e^+e^-$ and the production of selectrons in $e^+e^-$ and $e^-e^-$ collisions have been calculated at next-to-leading order. The $\mathcal{O}(\alpha)$ corrections for the most important decay channels of the sleptons, $\tilde{l}^\pm \to l^\pm \tilde{\chi}_i^0$ and $\tilde{l}^\pm \to \overset{(-)}{\nu_l} \tilde{\chi}_i^\pm$ have been considered in Ref. [5]. The effects of the electroweak corrections were found to be sizeable, of the order of 5–10%. They include important effects from supersymmetric particles in the virtual corrections, in particular non-decoupling logarithmic contributions, e.g. terms $\propto \log m_{\tilde{f}}/m_{\mathrm{weak}}$ from fermion-sfermion-loops.

The equality of gauge and Yukawa couplings in the $SU(3)_C$ gauge sector can be tested at a linear collider by investigating the associated production of quarks $q$ and squarks $\tilde{q}$ with a gluon $g$ or gluino $\tilde{g}$. While the processes $e^+e^- \to q\bar{q}g$ and $e^+e^- \to \tilde{q}\bar{\tilde{q}}g$ are sensitive to the strong gauge coupling of quarks and squarks, respectively, the corresponding Yukawa coupling can be probed in $e^+e^- \to q\bar{\tilde{q}}\tilde{g}$. Recently, the complete $\mathcal{O}(\alpha_s)$ corrections to all three processes in the MSSM have been calculated [6]. The NLO contributions enhance the cross-section in the peak region by roughly 20% with respect to the LO result. Furthermore, the scale dependence is reduced by a factor of about six when the NLO corrections are included.

## 3. MASS DETERMINATION

Both the mass of scalar fermions and the unobserved LSP can be determined by measuring the end points of the kinematical energy distributions of the visible decay fermions. However the efficiency of this method is affected by initial-state radiation (ISR) and beamstrahlung effects and cuts necessary to reduce the backgrounds.

A novel approach for determining the selectron masses [7] takes advantage of the fact that selectrons can be produced in mixed pairs, $e^+e^- \to \tilde{e}_R^+\tilde{e}_L^-$, $\tilde{e}_L^+\tilde{e}_R^-$. Because of the mass difference between L- and R-selectrons, the minima and maxima of the final state lepton energy distributions originating from an $\tilde{e}_R^+\tilde{e}_L^-$ pair will be different for the final electron and positron. When subtracting the normalized electron and positron distributions, the relevant kinematical edges will therefore remain visible while the Standard Model background and a large part of supersymmetric backgrounds are canceled. In particular, the contributions from the production of selectrons with same chirality, $\tilde{e}_R^+\tilde{e}_R^-$ and $\tilde{e}_L^+\tilde{e}_L^-$, are cancelled while the signal from mixed pairs, $\tilde{e}_R^+\tilde{e}_L^-$ and $\tilde{e}_L^+\tilde{e}_R^-$ is preserved. In order to disentangle $\tilde{e}_R^+\tilde{e}_L^-$ pairs from $\tilde{e}_L^+\tilde{e}_R^-$ pairs, the subtracted electron and positron distributions for left- and right-handed $e^-$ beam polarization are again subtracted. The final distribution exhibits the characteristic kinematical edges while being unaffected by backgrounds. With this method it is possible to determine the masses of the selectrons and the neutralino LSP with a precision of 200–500 MeV for an integrated luminosity of $\int L = 1000$ fb$^{-1}$.

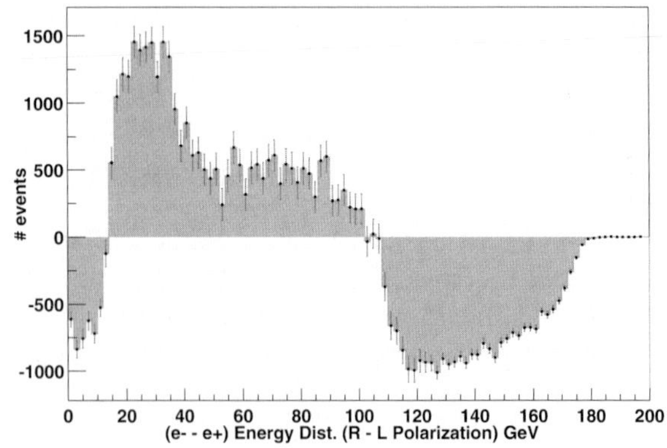

Figure 1. Difference of the subtracted normalized electron and positron energy distributions for 80% L- and R-polarized $e^-$ beam.

Alternatively, the sfermion masses can be extracted from the measurement of the pair production cross-section near threshold. The steep and distinct rise of the excitation curve allows a very precise determination of the sfermion masses in threshold scans. Since the expected experimental accuracy is of $\mathcal{O}(100\ \mathrm{MeV})$, it is necessary to incorporate effects beyond leading order in the theoretical predictions [9]. The non-zero widths of the sfermions, which considerably affect the cross-sections near threshold, must be included in a gauge-invariant manner. Beamstrahlung, ISR and the Coulomb correction due to photon exchange between the produced sfermions also play an important role. The production of smuons, staus and squarks proceeds via s-channel

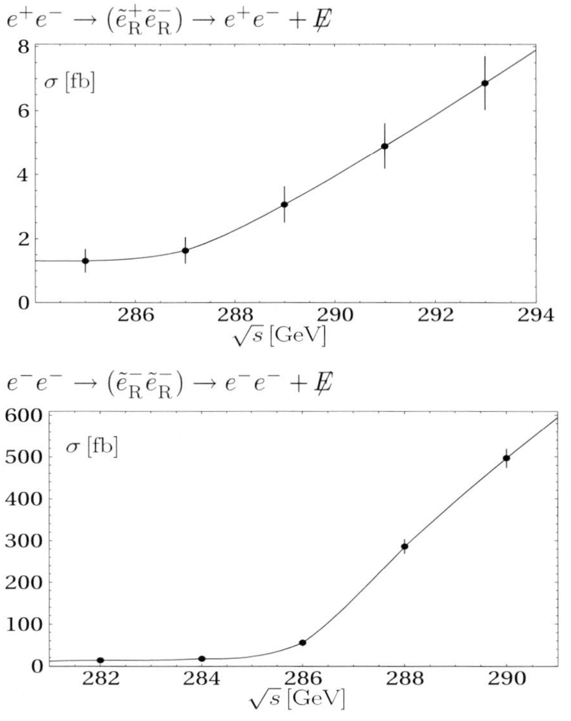


Figure 2. Threshold excitation curves for R-selectron pair production. The error bars indicate the statistical error for a luminosity per scan point of $\int L = 10\ \text{fb}^{-1}$ in $e^+e^-$ and $\int L = 1\ \text{fb}^{-1}$ in $e^-e^-$ mode, respectively (SPS1 scenario).

gauge-boson exchange, so that the sfermions are produced in a P-wave with a characteristic rise of the excitation curve $\sigma \propto \beta^3$, where $\beta = \sqrt{1 - 4m_{\tilde{f}}/s}$ is the sfermion velocity. Due to the exchange of Majorana neutralinos in the t-channel, selectrons can also be produced in S-waves ($\sigma \propto \beta$), namely for $\tilde{e}^\pm_R \tilde{e}^\mp_L$ pairs in $e^+e^-$ annihilation and $\tilde{e}^-_R \tilde{e}^-_R, \tilde{e}^-_L \tilde{e}^-_L$ pairs in $e^-e^-$ scattering.

The selectron cross-sections for both collider modes are shown in Fig. 2 including five equidistant scan points. Using four free parameters, the mass, width, normalization and flat background contribution, it is possible to fit the excitation cross-section in a model-independent way. As an example, results for the production of R-selectrons are given in Tab. 1. The numbers have been obtained including ISR and the theoretical refinements mentioned above, as well as backgrounds from both SM and MSSM sources,

Table 1
Expected precision for the determination of selectron masses and widths from threshold scans [3]. Values for SPS1 scenario.

| Process | Lumn. [fb$^{-1}$] | Mass [GeV] | Width [MeV] |
|---|---|---|---|
| $e^+e^- \to (\tilde{e}^+_R \tilde{e}^-_R) \to e^+e^- + \not{E}$ | $5 \times 10$ | $143.0^{+0.21}_{-0.19}$ | $150^{+300}_{-250}$ |
| $e^-e^- \to (\tilde{e}^-_R \tilde{e}^-_R) \to e^-e^- + \not{E}$ | $5 \times 1$ | $142.95^{+0.048}_{-0.053}$ | $200^{+50}_{-40}$ |

which have been reduced by appropriate cuts.

These high-precision measurements are necessary for reconstructing the soft-breaking parameters at the GUT and Planck scales [10]. As an example, in [11] a method is proposed to probe a nonuniversality between the left and right slepton masses at the GUT scale if only both selectrons and the light chargino are accessible at a linear collider. Inspired by the one-loop RGE for slepton masses in mSUGRA-type models the quantity $\Delta = m^2_{\tilde{e}_R} - m^2_{\tilde{e}_L} + m^2_{\tilde{\chi}^\pm_1}/\big(2\alpha_2^2(m_{\tilde{\chi}^\pm_1})\big) \times \big[\frac{3}{11}\big(\alpha_1^2(m_{\tilde{e}}) - \alpha_1^2(M_{\text{GUT}})\big) - 3\big(\alpha_2^2(m_{\tilde{e}}) - \alpha_2^2(M_{\text{GUT}})\big)\big]$ is defined which is strongly correlated with the slepton mass nonuniversality $\delta m^2 \equiv m^2_{\tilde{e}_R}(M_{\text{GUT}}) - m^2_{\tilde{e}_L}(M_{\text{GUT}})$. The method allows to detect a nonuniversality if $|\delta m^2| \gtrsim 5 \cdot 10^3\ \text{GeV}^2$. Using additional information on $M_2$ this improves to $|\delta m^2| \gtrsim 2 \cdot 10^3\ \text{GeV}^2$, i.e. $\delta m/m \gtrsim 1.6\%$ for $m \approx 250$ GeV.

In general, only sfermions with masses up to the beam energy, $m_{\tilde{f}} < \sqrt{s}/2$, can be investigated using the pair production process. Below the pair production threshold, sfermions can only be produced in association with their partner fermions and charginos or neutralinos. With sufficiently high luminosity at a future linear collider ($\int L = 500\ \text{fb}^{-1}$), it is possible to probe sfermions with masses that exceed the beam energy by a few tens of GeV [12].

## 4. THIRD GENERATION SFERMIONS

In contrast to the first two generation sfermions, large mixings are expected between the left- and right-chiral components of the third generation sfermions due to the large Yukawa coupling. The mixing effects are thus sensitive to the Higgs parameters $\mu$ and $\tan\beta$ as well as the trilinear couplings $A_f$.

For instance, the production and decay of staus is interesting for the determination of $\tan\beta$, in particular at high values, by examining the polarization of the taus in the decay $\tilde{\tau}^-_{1,2} \to \tau^-_{L,R}\tilde{\chi}^0_1$. In a recent study within the MSSM [13] it was found that $\tan\beta$ in the range of 30–40 can be determined with an error of about 10% when the $\tau$ polarization is measured for both $\tilde{\tau}_1$ and $\tilde{\tau}_2$.

The parameters $\mu$ and $A_\tau$ as well as the neutralino parameter $M_1$ may carry CP-violating phases. By using measurements of stau masses, cross-sections and branching ratios, it is possible to extract these parameters including their phases from the stau system. Very high accuracy on all involved parameters, with errors of the order of a few percent, can be obtained [14] if the decay of the heavy staus into heavy Higgs bosons is kinematically allowed.

CP-violating parameters in the stau sector can also give rise to electric and weak dipole moments of the taus. At a linear collider with high luminosity and polarization of both $e^+$ and $e^-$ beams it is possible to probe CP-violating tau dipole form factors up to the level of $\mathcal{O}(3) - \mathcal{O}(5) \cdot 10^{-19}e$cm [15]. While this precision would improve the current experimental bounds by three orders of magnitude, it is still about one order of magnitude beyond the expectations from supersymmetric models with CP-violation.

The study of scalar top quarks is of particular interest since due to large mixing the lighter stop mass eigenstate is likely to be the lightest squark. As a consequence, the decay channel into $t\tilde{\chi}^0_1$ may be kinematically forbidden so that other dedicated search strategies are necessary. The most promising decay modes in this case are $\tilde{t}_1 \to c\tilde{\chi}^0_1$ and $\tilde{t}_1 \to b\tilde{\chi}^\pm_1$. A simulation including full-statistics SM backgrounds and a detailed investigation of quark tagging and SUSY backgrounds for the SPS5 scenario[2] [8] has been performed [16]. Using the decay channel $\tilde{t}_1 \to c\tilde{\chi}^0_1$, the mass of the lighter stop and the mixing angle can be constrained up to $\Delta m_{\tilde{t}_1} \approx 1$ GeV and $\Delta\cos\theta_{\tilde{t}_1} \approx 0.0125$ with a luminosity of 1000 fb$^{-1}$.

For stops and sbottoms with large mass differences, bosonic decay modes, $\tilde{t}_1 \to \tilde{b}_1 + (H^+ \text{ or } W^+)$, $\tilde{b}_1 \to \tilde{t}_1 + (H^- \text{ or } W^-)$, may become important. Due to the large Yukawa couplings and mixings of $\tilde{t}$ and $\tilde{b}$ these decay modes can dominate in a wide range of MSSM parameters [17]. Large radiative SUSY-QCD corrections are taken into account by SUSY-QCD running parameters for the stop/sbottom masses and the trilinear couplings. As a consequence, the precise investigation of these decay modes is important for stop/sbottom searches and the determination of the underlying MSSM parameters.

[2]The SPS5 scenario is a dedicated "light-stop" scenario

## REFERENCES


1. E. Accomando *et al.*, Phys. Rept. 299 (1998) 1.
2. C. Blöchinger, H. Fraas, G. Moortgat-Pick and W. Porod, Eur. Phys. J. C **24** (2002) 297.
3. A. Freitas, Ph.D. thesis, Hamburg (2002), DESY-THESIS-2002-023.
4. A. Freitas, A. v. Manteuffel and P. M. Zerwas, in preparation.
5. J. Guasch, W. Hollik and J. Solà, Phys. Lett. B 510 (2001) 211; hep-ph/0207364.
6. M. M. Weber, Ph.D. thesis, Hamburg (2001), DESY-THESIS-2002-003; M. Maniatis, Ph.D. thesis, Hamburg (2002), DESY-THESIS-2002-007; A. Brandenburg, M. Maniatis and M. M. Weber, in *Proc. of SUSY02*, hep-ph/0207278.
7. M. Dima *et al.*, Phys. Rev. D 65 (2002) 071701.
8. B. C. Allanach *et al.*, hep-ph/0202233.
9. A. Freitas, D. J. Miller and P. M. Zerwas, Eur. Phys. J. C 21 (2001) 361; A. Freitas and D. J. Miller, hep-ph/0111430.
10. G. A. Blair, W. Porod and P. M. Zerwas, Phys. Rev. D 63 (2001) 017703; P. M. Zerwas *et al.*, these proceedings.
11. H. Baer, C. Balazs, S. Hesselbach, J. K. Mizukoshi and X. Tata, Phys. Rev. D 63 (2001) 095008.
12. A. Datta and A. Djouadi, hep-ph/0111466; A. Datta, A. Djouadi and M. Mühlleitner, hep-ph/0204354.
13. E. Boos *et al.*, in preparation.
14. A. Bartl, K. Hidaka, T. Kernreiter and W. Porod, Phys. Lett. B 538 (2002) 137; hep-ph/0207186.
15. B. Ananthanarayan, S. D. Rindani and A. Stahl, hep-ph/0204233.
16. H. Nowak and A. Sopczak, contributed paper to this conference, ABS614.
17. K.Hidaka and A.Bartl, Phys.Lett.B501 (2001) 78.

ELSEVIER

Proceedings of ICHEP 2002, pp. 811–814
*S. Bentvelsen, P. de Jong, J. Koch and E. Laenen (Editors)*

31st INTERNATIONAL CONFERENCE ON HIGH ENERGY PHYSICS AMSTERDAM

# Reconstruction of Fundamental SUSY Parameters

P.M. Zerwas, G.A. Blair, S.Y. Choi, A. Freitas, J. Kalinowski, H.U. Martyn, G. Moortgat-Pick, W. Porod
ECFA/DESY SUSY Collaboration
Deutsches Elektron–Synchrotron DESY, D-22603 Hamburg, Germany.

We summarize methods and expected accuracies in determining the basic low-energy SUSY parameters from experiments at prospective $e^+e^-$ linear colliders in the TeV energy range, combined with LHC results. In a second step we demonstrate how the fundamental supersymmetric theory can be reconstructed at high scales near the grand unification or Planck scale. These analyses have been carried out for minimal supergravity and a string effective theory.

Standard particle physics is characterized by energy scales of order 100 GeV. However the roots for all the phenomena we observe experimentally in this range, may go as deep as the Planck length of $10^{-33}$ cm, equivalently to energies as high as the Planck scale around $\Lambda_P \sim 10^{19}$ GeV or the grand unification [GUT] scale. Supersymmetry [SUSY] provides us with a stable bridge between these two vastly different energy regions.

A rich corpus of information on physics near the Planck scale may become available from extrapolating fundamental parameters, measured with high precision at laboratory energies, to high scales. Such extrapolations extend over 13 to 16 orders of magnitude. Despite this huge distance, the extrapolations can be carried out in a stable way in supersymmetric theories. To this purpose renormalization group (RG) techniques can be exploited, by which parameters are transported from low to high scales based on nothing but measured quantities in laboratory experiments. This procedure, very successful for the three gauge couplings, can be expanded to a large ensemble of supersymmetry parameters [1], the soft SUSY breaking parameters, i.e. gaugino and scalar masses, as well as trilinear couplings. We have analyzed this procedure for two examples, minimal supergravity [2] with a naturally very high degree of regularity near the grand unification scale, and superstring based effective field theories [3]. This RG bottom-up approach makes use of the low-energy measurements to the maximum extent possible and it reveals the quality with which the fundamental theory at the high scale can be reconstructed in a transparent way.

High-quality experimental data are necessary in this context, which should become available in a unique way by future lepton colliders [4].

## 1. Minimal Supergravity

Supersymmetry is broken in mSUGRA in a hidden sector and the breaking is transmitted to our eigenworld by gravity [2]. This mechanism suggests, yet does not enforce, the universality of the soft SUSY breaking parameters at a scale which we will identify with the unification scale in this study for simplicity.

The typical form of the mass spectrum in mSUGRA scenarios can be exemplified by the Snowmass point SPS#1A [5], slightly modified for illustrative purpose by increasing the scalar mass parameter, cf. Fig. 1. Masses can best be obtained in threshold scans at $e^+e^-$ colliders [6]. The excitation curves for chargino production in S-waves [7] rise steeply with the velocity of the particles near the threshold and thus are very sensitive to their masses, cf. Fig. 2(a); the same is true for selectrons in mixed pairs in $e^+e^-$ and for diagonal pairs in $e^-e^-$ collisions. Other scalar fermions and neutralinos are produced generally in P-waves, with a somewhat less steep threshold behavior proportional to the third power of the velocity [8]. Additional information, in particular on the lightest neutralino $\tilde{\chi}^0$, can be extracted from decay spectra, cf. Fig. 2(b). A selection of parameters combined from TESLA, LHC and CLIC measurements is collected in Tab. 1. Mixing parameters must be obtained from measurements of cross sections, in particular from the production of chargino pairs and neutralino pairs [7], both in diagonal or mixed form,


doi:10.1016/S0920-5632(02)02238-7

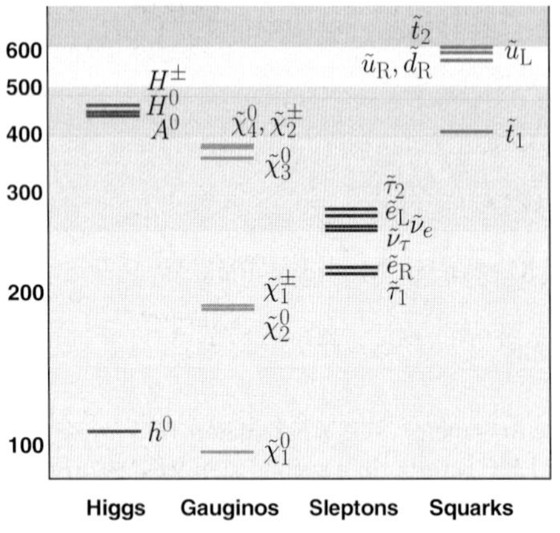


Figure 1. *The spectrum in the mSUGRA scenario mod.SPS#1A defined by* $M_0$ = 200 GeV, $M_{1/2}$ = 250 GeV, $A_0$ = −100 GeV, $\tan\beta$ = 10, *and* $\text{sign}(\mu) = (+)$.

$e^+e^- \to \tilde\chi_i^+\tilde\chi_j^-$ [$i,j$ = 1,2] and $\tilde\chi_i^0\tilde\chi_j^0$ [$i,j$ = 1,. . .,4]. The production cross sections for charginos are simple binomials of $\cos 2\phi_{L,R}$, the mixing angles rotating current to mass eigenstates. Using polarized beams, the cosines can be determined in a model-independent way [7].

Based on this high-precision information, the fundamental SUSY parameters can be extracted at low energy in analytic form. To lowest order:

$$\begin{aligned}
|\mu| &= M_W[\Sigma + \Delta[\cos 2\phi_R + \cos 2\phi_L]]^{1/2} \\
\text{sign}(\mu) &= [\Delta^2 - (M_2^2 - \mu^2)^2 - 4m_W^2(M_2^2 + \mu^2) \\
&\quad -4m_W^4 \cos^2 2\beta]/8m_W^2 M_2|\mu| \sin 2\beta \\
M_2 &= M_W[\Sigma - \Delta[\cos 2\phi_R + \cos 2\phi_L]]^{1/2} \\
|M_1| &= \left(\textstyle\sum_i m_{\tilde\chi_i^0}^2 - M_2^2 - \mu^2 - 2M_Z^2\right)^{1/2} \\
|M_3| &= m_{\tilde g} \\
\tan\beta &= \left[\frac{1 + \Delta(\cos 2\phi_R - \cos 2\phi_L)}{1 - \Delta(\cos 2\phi_R - \cos 2\phi_L)}\right]^{1/2} \qquad (1)
\end{aligned}$$

where $\Delta = (m_{\tilde\chi_2^\pm}^2 - m_{\tilde\chi_1^\pm}^2)/(4M_W^2)$ and $\Sigma = (m_{\tilde\chi_2^\pm}^2 + m_{\tilde\chi_1^\pm}^2)/(2M_W^2) - 1$. The signs of $M_{1,3}$ with respect to $M_2$ will follow from measurements of the cross sections for $\tilde\chi\tilde\chi$ and gluino production processes. In practice the one-loop corrections have been used to improve on the accuracy.

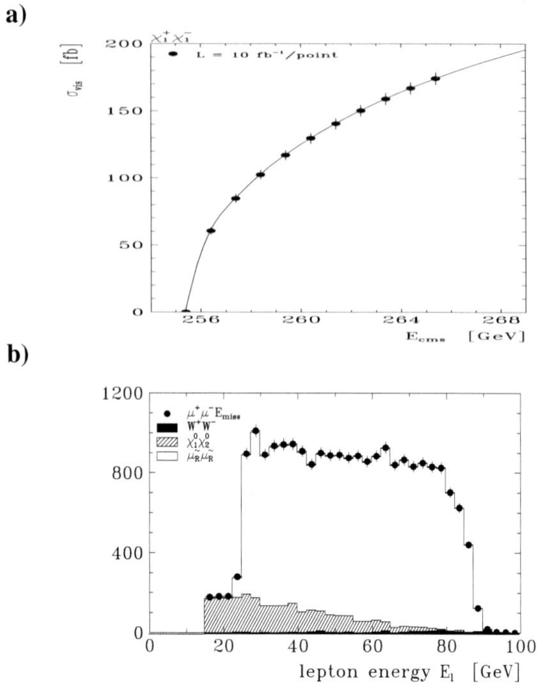


Figure 2. *a) Threshold excitation in chargino pair production, b)* $\mu$ *energy sprectrum in* $\tilde\mu_R$ *decays [6].*

Accuracies expected for the various parameters in *mod.*SPS#1A are typically at the per-mille level for LC measurements, while the others turn out to be in the per-cent range.

The evolution to the high scale is governed by solutions of the renormalization group equations [9]:

| | | |
|---|---|---|
| gauge couplings | : | $\alpha_i = Z_i\,\alpha_U$ |
| gaugino masses | : | $M_i = Z_i\,M_{1/2}$ |
| scalar masses | : | $M_{\tilde j}^2 = M_0^2 + c_j M_{1/2}^2 + \sum_{\beta=1}^2 c'_{j\beta}\Delta M_\beta^2$ |
| trilinear couplings | : | $A_k = d_k A_0 + d'_k M_{1/2}$ |

The $Z$ transporter is given by $Z_i^{-1} = 1 + b_i\,\alpha_U/(4\pi) \times \log(M_U/M_Z)^2$ with $b[SU_3, SU_2, U_1] = -3,\ 1,\ 33/5$. The $c$, $d$ coefficients as well as the shifts $\Delta M_\beta^2$ depend on the high-energy parameters to be calculated, so that complicated implicit equations emerge – partly with low sensitivity in the final solutions. In practice the transport equations have been solved to two loops.

Examples for the gaugino masses and the scalar

| | Mass + Errors | |
|---|---|---|
| $\tilde{\chi}_1^{\pm}$ | $183.05 \pm 0.15$ | LC |
| $\tilde{\chi}_2^{\pm}$ | $383.28 \pm 0.28$ | LC |
| $\tilde{\chi}_1^0$ | $97.86 \pm 0.20$ | LC |
| $\tilde{\chi}_2^0$ | $184.65 \pm 0.30$ | LC |
| $\tilde{e}_R$ | $224.82 \pm 0.15$ | LC |
| $\tilde{e}_L$ | $269.09 \pm 0.28$ | LC |
| $\tilde{u}_R$ | $572 \pm 10$ | LHC+CLIC |
| $\tilde{u}_L$ | $589 \pm 10$ | LHC+CLIC |
| $\tilde{g}$ | $598 \pm 10$ | LHC |
| $h^0$ | $113.38 \pm 0.05$ | LHC+LC |
| $A^0$ | $435.5 \pm 1.5$ | CLIC |

Table 1
*A sample of masses and expected accuracies for the mod.SPS#1A point. [Masses in* GeV.*]*

| | Ideal | Experimental Error |
|---|---|---|
| $M_U$ | $2 \cdot 10^{16}$ | $1.6 \cdot 10^{14}$ |
| $\alpha_U^{-1}$ | 24.361 | 0.007 |
| $M_{\frac{1}{2}}$ | 250 | 0.08 |
| $M_0$ | 200 | 0.09 |
| $A_0$ | -100 | 1.8 |
| $\mu$ | 358.23 | 0.21 |
| $\tan\beta$ | 10 | 0.1 |

Table 2
*Comparison of the ideal parameters with the experimental expectations for the mod.SPS#1A point. [All mass parameters are given in units of GeV.]*

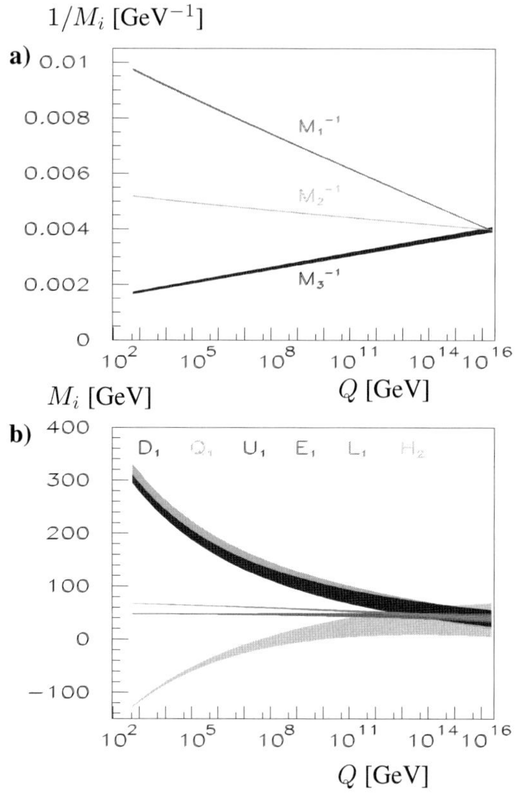


Figure 3. *Evolution, from low to high scales, of a) gaugino mass parameters, b) first-generation sfermion mass parameters and the Higgs mass parameter $M^2_{H_2}$ for the mSUGRA mod.SPS#1A point. [The widths of the bands indicate the $1\sigma$ CL.]*

masses of the first two generations are depicted in Fig. 3. The mass squared of the Higgs field $H_2$ crosses to negative values, when approached from the universality point, at a scale of order $10^4$ TeV, in accordance with radiative symmetry breaking in mSUGRA. In an overall-fit, Tab. 2, based on the universality hypothesis *per se*, we observe accuracies of order per-mille in the gaugino sector while being order per-cent in the scalar sector.

## 2. String Effective Field Theory

For a string effective field theory in four dimensions based on orbifold compactification of the 10-dimensional heterotic superstring [3], SUSY breaking is generated non-perturbatively, mediated by a Goldstino field that is the superposition of the dilaton field $S$ and the moduli fields $T$ [for simplicity assumed to be of identical structure]. The gaugino and scalar mass parameters can be expressed to leading order by the gravitino mass $m_{3/2}$, the vacuum values $\langle S \rangle$, the mixing parameter $\sin\theta$, and the modular weights $n_j$:

$$\begin{aligned} M_i &= -g_i^2 m_{3/2} \langle S \rangle \sqrt{3} \sin\theta + \cdots \\ M_{\tilde{j}}^2 &= m_{3/2}^2 \left(1 + n_j \cos^2\theta\right) + \cdots \end{aligned} \qquad (2)$$

while in next-to-leading order the vacuum value $\langle T \rangle$ and the Green-Schwarz parameter $\delta_{\rm GS}$ are included. The system is completed by the relations between the universal gauge coupling $\alpha(M_{\rm string})$ at the string scale $M_{\rm string}$ and the [slightly non-universal] gauge couplings $\alpha_i(M_{\rm GUT})$ at the SU(5) unification scale

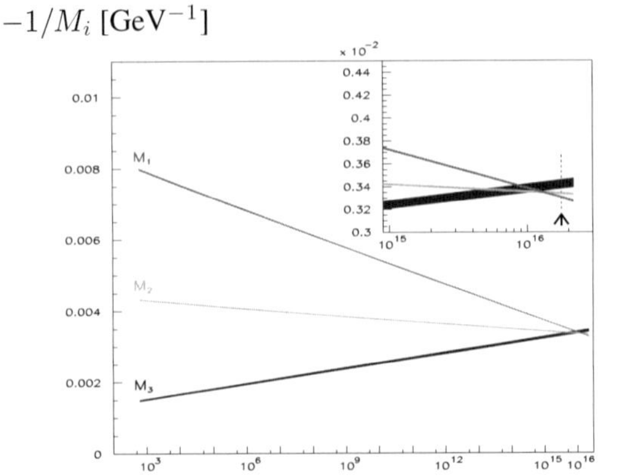


Figure 4. *Gaugino mass parameters for the heterotic string [the insert expands on the breaking of universality at the GUT scale] for the point defined in Tab. 3. [The widths of the bands indicate the* $1\,\sigma$ *CL.]*

| Parameter | Ideal | Reconstructed |
|---|---|---|
| $m_{3/2}$ | 180 | $179.9 \pm 0.4$ |
| $\langle S \rangle$ | 2 | $1.998 \pm 0.006$ |
| $\langle T \rangle$ | 14 | $14.6 \pm 0.2$ |
| $\sin\theta$ | 0.949 | $0.948 \pm 0.001$ |
| $g_s^2$ | 0.5 | $0.501 \pm 0.002$ |
| $\delta_{GS}$ | 0 | $0.1 \pm 0.4$ |
| $n_L$ | -3 | $-2.94 \pm 0.04$ |
| $n_E$ | -1 | $-1.00 \pm 0.05$ |
| $n_Q$ | 0 | $0.02 \pm 0.02$ |
| $n_U$ | -2 | $-2.01 \pm 0.02$ |
| $n_D$ | +1 | $0.80 \pm 0.04$ |
| $n_{H_1}$ | -1 | $-0.96 \pm 0.06$ |
| $n_{H_2}$ | -1 | $-1.00 \pm 0.02$ |
| $\tan\beta$ | 10 | $10.00 \pm 0.13$ |

Table 3
*Comparison of the experimentally reconstructed values with the ideal fundamental parameters in a specific example for a string effective field theory. [Masses in* GeV.*]*

$M_{\rm GUT}$: $\alpha_i^{-1}(\rm M_{GUT}) = \alpha^{-1}(M_{string}) + \Delta\alpha_i^{-1}[n_j]$. The gauge coupling at $M_{\rm string}$ is related to the dilaton field, $g_s^2 = 1/\langle S \rangle$.

The evolution of gaugino masses is shown in Fig. 4, and the relevant parameters derived from the evolution of couplings and masses are collected in Tab. 3. It turns out that the ideal values, from which the experimental input observables were derived, can indeed be extracted from the data collected at the high-energy hadron- and lepton-colliders.

The bottom-up approach of evolving the parameters from the low-energy scale to the high scale by means of renormalization group techniques provides us with a transparent picture in a region where gravity is linked to particle physics, and string theory becomes effective directly. This – truly exciting – observation has been exemplified for minimal supergravity theories, and for a string effective field theory based on orbifold compactification of the heterotic string.

## REFERENCES

1. G.A. Blair, W. Porod and P.M. Zerwas, Phys. Rev. D 63 (2001) 017703, and in preparation.
2. A.H. Chamseddine, R. Arnowitt, and P. Nath, Phys. Rev. Lett. 49 (1982) 970.
3. M. Cvetič, A. Font, L.E. Ibáñez, D. Lüst and F. Quevedo, Nucl. Phys. B 361 (1991) 194; A. Brignole, L.E. Ibáñez and C. Muñoz, Nucl. Phys. B 422 (1994) 125 [Erratum-ibid. B 436 (1995) 747]; A. Love and P. Stadler, Nucl. Phys. B 515 (1998) 34; P. Binetruy, M.K. Gaillard and B.D. Nelson, Nucl. Phys. B 604 (2001) 32.
4. E. Accomando *et al.*, ECFA/DESY LC Working Group, Phys. Rep. 299 (1998) 1. J.A. Aguilar-Saavedra *et al.*, ECFA/DESY LC Physics Working Group, "TESLA Technical Design Report Part III: Physics at an $e^+e^-$ Linear Collider," hep-ph/0106315.
5. B.C. Allanach *et al.*, Eur. Phys. J. C 25 (2002) 113.
6. G.A. Blair and U. Martyn, Proceedings, LC Workshop, Sitges 1999, hep-ph/9910416.
7. S.Y. Choi *et al.*, Eur. Phys. J. C 14 (2000) 535; S.Y. Choi, J. Kalinowski, G. Moortgat-Pick and P.M. Zerwas, Eur. Phys. J. C 22 (2001) 563 [Addendum-ibid. C 23 (2002) 769].
8. A. Freitas *et al.*, these Proceedings.
9. S. Martin and M. Vaughn, Phys. Rev. D50, 2282 (1994); Y. Yamada, Phys. Rev. D 50, 3537 (1994); I. Jack, D.R.T. Jones, Phys. Lett. B 333 (1994) 372.

ELSEVIER

Proceedings of ICHEP 2002, pp. 815–817
*S. Bentvelsen, P. de Jong, J. Koch and E. Laenen (Editors)*

31st INTERNATIONAL CONFERENCE ON HIGH ENERGY PHYSICS AMSTERDAM

# Physics at a Photon Collider

Stefan Söldner-Rembold[a]

[a] Fermi National Accelerator Laboratory, P.O. Box 500, Batavia, USA

A Photon Collider will provide unique opportunities to study the SM Higgs boson and to determine its properties. MSSM Higgs bosons can be discovered at the Photon Collider for scenarios where they might escape detection at the LHC. As an example for the many other physics topics which can be studied at a Photon Collider, recent results on Non-Commutative Field Theories are also discussed.

## 1. Introduction

The Photon Collider option of a Linear Collider (LC) is based on laser photon back-scattering on high energy electrons. The maximum photon energy is 205 GeV for a laser with $\lambda = 1.06\mu$m and an electron beam energy of 250 GeV. A high degree of polarisation with opposite helicities of the electron and the laser photon is crucial for obtaining a peaked spectrum of high energy polarised photons close to the maximum energy. In this case the high energy part of the $\gamma\gamma$ spectrum is dominated by the spin-0 configuration which is important to enhance the signal and suppress the background for Higgs production. Alternatively, e$\gamma$ interactions are also possible. The technical aspects of the photon collider are discussed in [1,2].

## 2. Higgs Production

Neutral Higgs bosons are produced in the scattering of two photons as a $s$-channel resonance through a loop. In this loop all charged particles contribute which obtain their mass from electroweak symmetry breaking. The two-photon partial width of the Higgs boson is therefore sensitive to physics beyond the SM.

For Higgs bosons decaying predominantly into b$\overline{\text{b}}$, the main source of background are $\gamma\gamma \rightarrow \text{Q}\overline{\text{Q}}$ processes (Q=c,b). The spin-0 component of these processes is suppressed in Leading Order (LO) by a factor $m_Q^2/s$. Since this suppression is only valid in LO, a realistic background simulation should be based on a next-to-leading (NLO) calculation.

Several analyses of H $\rightarrow$ b$\overline{\text{b}}$ decays for a SM Higgs boson in the mass range from 120 GeV to 160 GeV have been performed [3,4]. All analyses exploit the kinematic differences between the $s$-channel signal and the $t$-channel background by cutting on visible energy and angular distributions. The results are shown in Fig. 1.

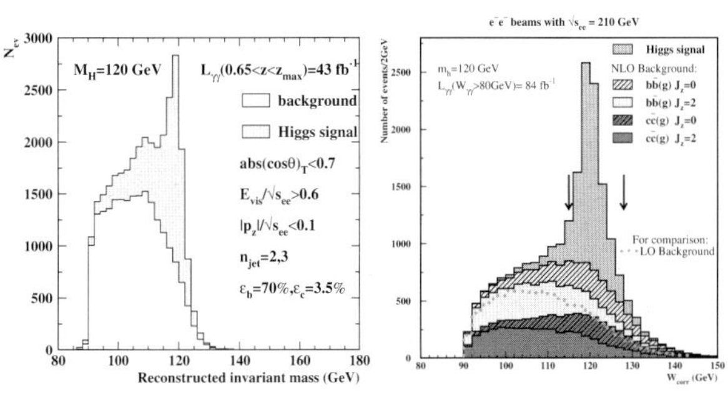


Figure 1. Reconstructed invariant mass for a Higgs boson mass of 120 GeV with the full NLO background simulation. The detector response is simulated with SIMDET. Left plot from [3] and right plot from [4].

Before b-tagging the ratio of background from $\gamma\gamma \rightarrow$ c$\overline{\text{c}}$ events to $\gamma\gamma \rightarrow$ b$\overline{\text{b}}$ events is approximately 16 due to the quark charges. Excellent b-tagging is therefore required to suppress the charm background. The need to minimize the


doi:10.1016/S0920-5632(02)02244-2

radius of the beam pipe is one of the main challenges for a Photon Collider, since the beam pipe has to accomodate the optical system for producing back-scattered photons.

The analyses presented here have assumed that $b\bar{b}$ events are tagged with 70% efficiency and $c\bar{c}$ events with 3.5% efficiency. For a luminosity corresponding to roughly one year of running, a statistical uncertainty of about 2% for a Higgs mass of 120-140 GeV for the two photon width measurement can be achieved. The uncertainty increases to about 10% for a Higgs mass of 160 GeV.

At Higgs boson masses above 160 GeV decays into WW and ZZ pairs become important. In this case the interference between signal and $\gamma\gamma \to$ WW background needs to be taken into account. The interference gives access to an additional observable, the phase $\phi_{\gamma\gamma}$ of the $\gamma\gamma \to$ H amplitude. The combined precision of phase and partial width determination gives sufficient precision to distinguish the SM from SM-like 2HDM (II) scenarios [5]. The results of a detector simulation are shown in Fig. 2.

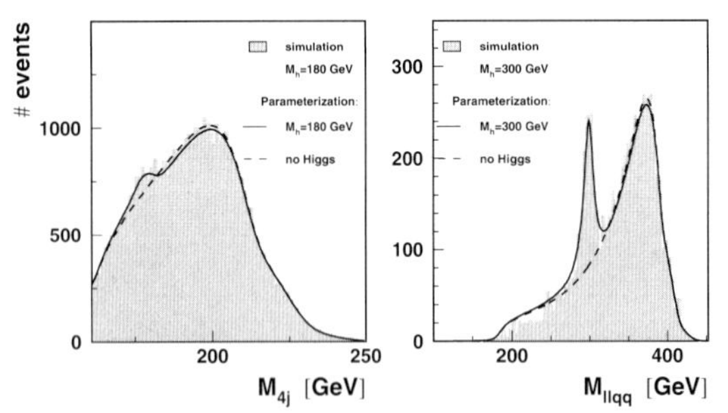


Figure 2. Reconstructed invariant mass for $\gamma\gamma \to WW$ events with a SM Higgs mass of 180 GeV (left) and for $\gamma\gamma \to$ ZZ events with a SM Higgs mass of 300 GeV (right) [5].

The neutral MSSM Higgs Boson H,A for masses above 200 GeV and for moderate $\tan\beta \approx 7$ might escape detection at the LHC. In this parameter region, where decays into $b\bar{b}$ are the dominant SM decays up to Higgs masses around 550 GeV, the Photon Collider can discover the neutral MSSM Higgs Bosons [6]. In contrast to the $e^+e^-$ option of the LC, the Photon Collider can produce these Higgs Boson with masses up to about 80% of $\sqrt{s_{ee}}$. Cross-sections for signal and background are shown in Fig. 3.

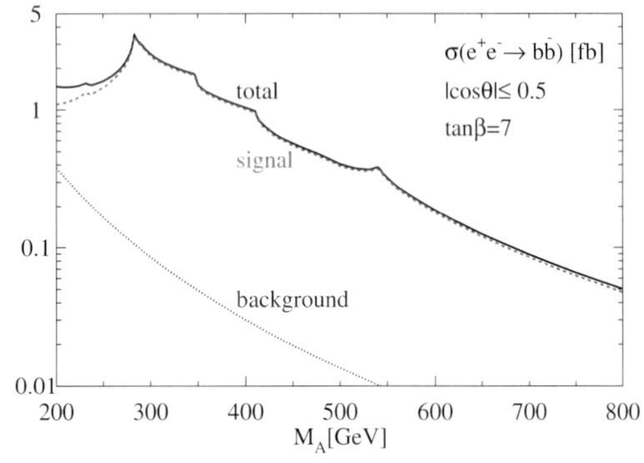


Figure 3. Cross-section for the process $\gamma\gamma \to A \to b\bar{b}$ and for the background $\gamma\gamma \to b\bar{b}$. A mass window of $\pm 3$ GeV has been applied, 100% polarisation is assumed, and only the two-jet configuration is considered [6].

Many other Higgs scenarios have been studied for the Photon Collider option, adding to the discovery potential: For example, the measurement of CP properties of the Higgs bosons A,H in $t\bar{t}$ decays [7] or the production of charged Higgs bosons in the process $\gamma\gamma \to H^+H^-$ [8].

## 3. Non-Commutative Field Theories

One of many other interesting topics which can be studied at a $\gamma\gamma$ or at an $e\gamma$ collider are non-commutative quantum field theories (NC-QFT) with non-commuting (NC) space-time operators [9]. The additional cubic coupling ($\gamma\gamma\gamma$) contributing to the process $\gamma\gamma \to f\bar{f}$ is shown in Fig. 4.

The parameter $\Lambda_{NC}$ characterises the threshold where NC effects become relevant. The current limit from $e^+e^-$ scattering is $\Lambda_{NC} > 142$ GeV at 95% confidence level [10].

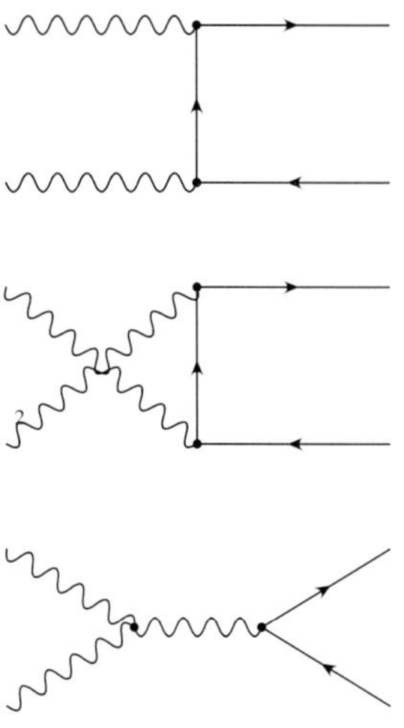

Figure 4. Diagrams contributing to fermion pair production ($\gamma\gamma \to f\bar{f}$) in NCQFT.

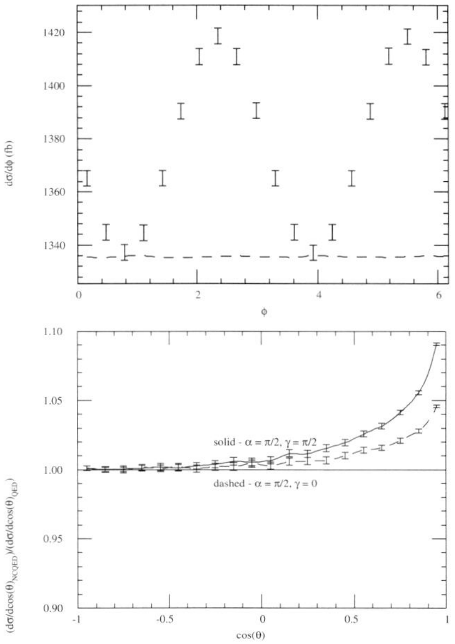


Figure 5. $e\gamma \to e\gamma$ scattering at $\sqrt{s}_{ee} = 500$ GeV a) Differential cross-section $d\sigma/d\phi$ for the SM (dashed line) and for $\Lambda_{NC} = 500$ GeV ($\alpha = \gamma = \pi/2$). The expected statistical uncertainties are also shown. b) Ratio of the differential cross-sections $d\sigma/d\cos\theta$ for NCQFT and for the SM.

A theoretical analysis has been performed of the processes $\gamma\gamma \to f\bar{f}$ and $e\gamma \to e\gamma$ for $L_{ee} = 500\ fb^{-1}$. A transverse momentum greater than 10 GeV and a polar angle in the range $10^\circ < \theta < 170^\circ$ has been required for the final state particles.

In Fig. 5 the NCQFT effects on the angular distributions of the final state photons in $e\gamma \to e\gamma$ scattering are shown. The parameters $\alpha$ and $\gamma$ are related to the Maxwell field tensor [9]. Significant deviations from the SM can be observed.

## 4. Conclusion

Studies of various Higgs scenarios show that a Photon Collider has a unique potential for Higgs boson physics over a wide range of masses and model parameters. Excellent b-tagging and good energy resolution are very important for the Photon Collider to suppress background. The $e\gamma$ option of the Photon Collider is complementary to the $e^+e^-$ Linear Collider in discovering effects from Non-Commutative Field Theories. Many other topics which can be studied at a Photon Collider (e.g. SUSY, Leptoquarks, QCD) had to be omitted in this short summary.

## REFERENCES

1. J. Gronberg, these proceedings.
2. ECFA/DESY Photon Collider Working Group, hep-ex/0108012 (ABS 770).
3. G. Jikia, S. Söldner-Rembold, Nucl. Inst. and Meth. A472 (2001) 133; Nucl. Phys. Proc. Suppl. 82 (2000) 373 (ABS 812).
4. P. Nieżurawski, A.F. Żarnecki and M. Krawczyk, hep-ph/0208234 (ABS 665).
5. P. Nieżurawski, A.F. Żarnecki and M. Krawczyk, hep-ph/0207294 (ABS 155).
6. M. Mühlleitner et al., Phys. Lett. B508 (2001) 311; D. Asner et al., hep-ph/0110320.
7. E. Asakawa, hep-ph/0101234.
8. D. Asner et al., hep-ph/0208219.
9. S. Godfrey, M. Doncheski, hep-ph/0111147; Phys. Rev. D65 (2002) 015005 (ABS 625).
10. OPAL Physics Note PN500 (ABS 889).

## Session 11

# Computational Quantum Field Theory (methods)

*Session conveners:*
*J. Vermaseren, NIKHEF*
*E. Laermann, University of Bielefeld*

ELSEVIER

Proceedings of ICHEP 2002, pp. 821–824
*S. Bentvelsen, P. de Jong, J. Koch and E. Laenen (Editors)*

31st INTERNATIONAL CONFERENCE ON HIGH ENERGY PHYSICS AMSTERDAM

# Foam: A General purpose Monte Carlo Cellular Algorithm*

S. Jadach[a]

[a]Henryk Niewodniczanski Institute of Nuclear Physics,
ul. Radzikowskiego 152, 31-342 Cracow, Poland

A general-purpose, self-adapting Monte Carlo (MC) algorithm implemented in the program `Foam` is described. The high efficiency of the MC, that is small maximum weight or variance of the MC weight is achieved by means of dividing the integration domain into small cells. The cells can be $n$-dimensional simplices, hyperrectangles or a Cartesian product of them. The grid of cells, "foam", is produced in the process of the binary split of the cells. The next cell to be divided and the position/direction of the division hyperplane is chosen by the algorithm which optimizes the ratio of the maximum weight to the average weight or (optionally) the total variance. The algorithm is able to deal, in principle, with an arbitrary pattern of the singularities in the distribution.

## Introduction and motivation

For the problem of function minimalization of the arbitrary user-function one applies MINUIT or other program from the NAG library. One finds also many *general purpose* programs for integration of "arbitrary" integrand. The above *general-purpose* tools work, in principle, for a very wide range of the user-functions. For *multi-dimensional Monte Carlo simulation* problem, that is for the problem of generating randomly points according to $n$-dimensional distribution, there is precious little examples of the *general purpose* Monte Carlo Simulators (GPMCS), that is programs which work (in principle) for an arbitrary integrand.

Inevitably the GPMCS has to work in 2 stages: "exploration" and "generation". During exploration GPMCS is "digesting" the entire shape of the $n$-dimensional distribution $\rho(x_1, x_2, ...x_n)$ to be generated and memorizes it as efficiently as possible using all CPU power and memory available. Obviously, for the memorized $\rho'(x_1, x_2, ...x_n)$ a method of the MC generation of the points $\vec{x}$ *exactly* according to $\rho'(\vec{x})$, has to be available. The quality of the distribution of the weight $w = \rho/\rho'$ for events in the *generation* (small variance, good ration of maximum to average, etc.) is determined by the algorithm of the *exploration*. In other words, the quality of the "target weight distribution" in the latter generation is determining/driving the algorithm of the former exploration.

At the time of writing up this contribution a detailed description of the algorithm and the program is available in ref. [1]. The original, less advanced version was described in ref. [2]. The other old and new efforts in the area of general purpose MC methods see refs. [3–9]

## Cellular algorithm of the `Foam`

The most obvious method to minimize the variance (or maximum weight) of the target weight distribution in generation is to split integration domain into many cells[2], such that $\rho(\vec{x})$ is approximated by constant $\rho'(\vec{x})$ within each cell. This is a *cellular class* of general purpose MC algorithms[3] which always involves a kind of cellular exploration of the user defined distribution.

The obvious questions are: what kind (shape) of cells to use and how to cover the space with cells? In `Foam` algorithm/program three types of cells are used: pure *simplices*, pure *hyperrectangles* and *Cartesian products of them.* Cells of this kind can be parametrized rather easily and the related computer memory consumption is not ex-

*Work partly supported by the European Community's Human Potential Programme under contract HPRN-CT-2000-00149 "Physics at Colliders", by Polish Government grants KBN 5P03B09320 and 2P03B00122, and by NATO grant PST.CLG.977751.

[2]This general idea is at least 40 years old.

[3]"Stratified sampling", used in the literature, has a narrower meaning.


doi:10.1016/S0920-5632(02)02249-1

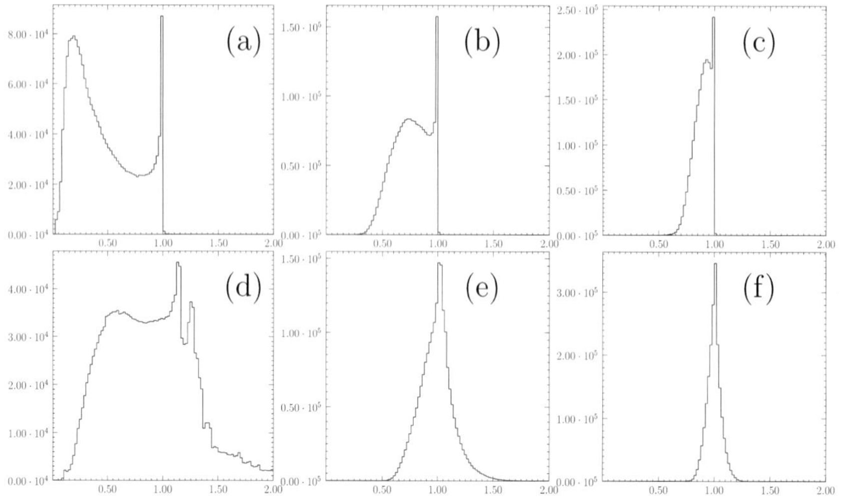


Figure 1. Weight distribution of the Foam with the maximum weight optimization (a-c) compared to the variance optimization (d-f). Number of cells is 200, 2000 and 20000 correspondingly.

cessive.

The system of cells can be created all at once (like in VEGAS [3]) or in an evolutionary way, by the "splitting process". The Foam algorithm does the *binary split* of cells. Next cell to be split is chosen by examining the "target weight distribution", that is of the event generation in the latter phase. The binary split provides automatically the *full coverage* of the space, simply because the primary "root cell" is identical with the entire integration domain and there is never any mismatch between parent cell and two daughter cells.

Another important issue is the question of the variance reduction versus maximum weight reduction. In construction of the Foam algorithm we have put most effort on the minimization of the ratio of the maximum weight to the average weight $w_{\max}/\langle w\rangle$. This parameter is essential, if we want to transform $w$-ted events into $w = 1$ events, at the latter stage of the MC generation. Minimizing maximum weight is not the same as minimizing variance $\sigma = \sqrt{\langle w^2\rangle - \langle w\rangle^2}$. Usually minimizing $w_{\max}$ is more difficult. In Foam minimizing variance is also implemented and optionally available. It can be useful in the case when a MC with $w$-ted events is acceptable. In fig. 1 we show two examples of the weight distribution evolution in the Foam, when adding more and more cells.

In the Foam algorithm we define two auxiliary distributions $\rho'(x)$ and $\rho_{loss}(x)$ related to integrand $\rho(x)$. Both are constructed together step by step, simultaneously with the construction of the foam of cells, in the exploration process. When our ultimate aim is to minimize $w_{\max}$, we define

$$\rho'(x) \equiv \max_{y\in Cell_I} \rho(y), \quad \text{for} \quad x \in Cell_I,$$
$$R_{loss} = \int d^n x\, [\rho'(x) - \rho(x)] = \int d^n x\, \rho_{loss}(x).$$

here $\rho_{loss}$ is the difference between the "ceiling distribution" $\rho'$ and the actual distribution $\rho$ from which it is derived. The rejection rate in the final MC run is just proportional to $\int \rho_{loss}(x)$ by construction, i.e. the rejection rate $= R_{loss}/R$. Also, $\rho_{loss}(x)$ has a clear geometrical meaning, see below. When our principal aim is to minimize the ratio of the variance to the average of the weight, $\sigma/\langle w\rangle$, in the final MC generation, we are led to the following definition:

$$\rho'(x) \equiv \sqrt{\langle \rho^2\rangle_I}, \quad \text{for} \quad x \in Cell_I,$$
$$\rho_{loss}(x) \equiv \sqrt{\langle \rho^2\rangle_I} - \langle \rho\rangle_I, \quad \text{for} \quad x \in Cell_I.$$

The average $\langle ...\rangle_I$ is over the $I$-th cell[4]. The ratio $\sigma/\langle w\rangle$ in the final MC generation is a monotonous

[4]See ref. [1] for a details of derivation.

ascending function of $R_{loss} = \int \rho_{loss}(x)dx^n$ – hence, minimization of $R_{loss}$ is equivalent to minimization of $\sigma/\langle w\rangle$.

The basic rules governing binary split of a cell is that each split of a cell: $\omega \to \omega' + \omega''$ should decrease the total $R_{loss}$ and the decrease should be as big as possible. How to get the best total decrease $\Delta R_{loss}$?
(1) For each next cell split we choose a cell with the biggest $R_{loss}$.
(2) Position/direction of a plane dividing a parent cell into two daughter cells is chosen to get the smallest total $R_{loss}$.
The actual procedure relies on the small MC exercise done for each single cell during its exploration, in which events are generated with the flat distribution, weighted with $\rho$ and projected onto $n$ (simplical case) or $n(n+1)/2$ (hyperrectangular case) of the cells. Resulting projection histograms are then analysed and the best "division geometry" is found, for which the estimate of $\Delta R_{loss}$ is calculated. How the geometry of the division algorithm looks like? In case of a $n$-dimensional simplex defined by $n+1$ vertices, a pair of vertices $x_i$ and $x_j$ is chosen and a new vertex $Y$ is put somewhere on the line in between: $Y = \lambda x_i + (1-\lambda)x_j, \quad 0 < \lambda < 1$. Two daughter simplices are defined with two list of vertices: $(x_1, ..., x_{i-1}, Y, x_{i+1}, ..., x_n, x_{n+1})$ and $(x_1, x_{j-1}, Y, x_{j+1}, ..., x_n, x_{n+1})$. Now, how do we choose a pair $(i,j)$ and the value of $\lambda$? A short sample of the MC events (100-1000) is generated $\in$ cell. Each MC point is projected (see fig. 2) $X \to Y$ onto an edge $(i,j), i \neq j$. For each $(i,j)$ the $dN/d\lambda$ is histogrammed, and its "loss" functional $R_{loss}$ is estimated. The optimal edge $(i,j)$ with the biggest loss is selected. For the optimal edge the cell division ration $\lambda$ is deduced from its histogram ($\lambda$ is always a rational number, $n/N_{bin}$).

In fig. 3 we show several examples of the evolution of the 2-dimensional foam of cells in case of simplices (triangles) and rectangles for up to 2500 cells.

The important practical question is whether hyperrectangular cells or simplical cells are beter as a building blocks. There is no simple answer. Simplices are limited to low dimensions $n < 6$ because calculation of increasingly large number of determinants leads to much CPU time consumption. In addition, for simplices memory consumption is $\sim 16n$ Bytes/Cell while for hyperrectangles memory consumption can be limited to below 50Bytes/Cell *independently* of $n$. See ref. [1] for details. Experimenting with many testing functions has also shown that quite often hyperrectangles provide final MC efficiency as good or sometimes even better than for simplices.

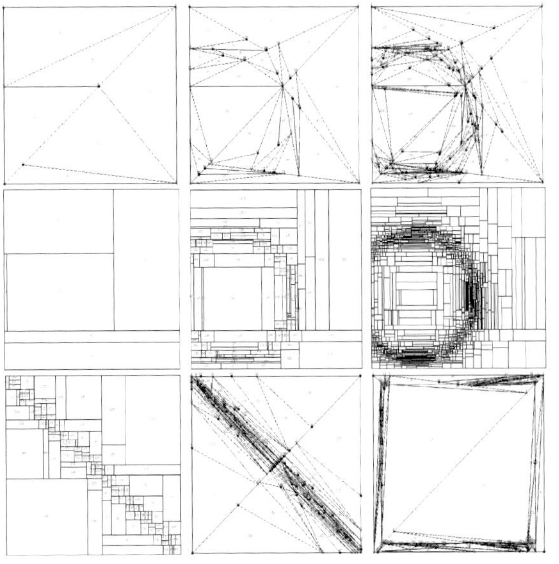

Figure 3. Examples of the 2-dimensional foam. Number of cells from 10 to 2500.

The method of cell division relying on the analysis of the projection histograms requires many function calls in the cell exploration. The overall efficiency of the final MC is best for large number of cells. That points towards necessity of the large number of the function calls during the exploration, which may lead to excessive CPU time consumption. The following efficient trick was found to limit number of the function calls in the cell exploration: During MC exploration of a given new cell we continuously monitor the number of accumulated effective $W=1$ events, $N_{eff} = \frac{(\sum w_i)^2}{\sum w_i^2}$, and stop exploration when $N_{eff}/n_{bin} > 25$, where $n_{bin}$ is the number

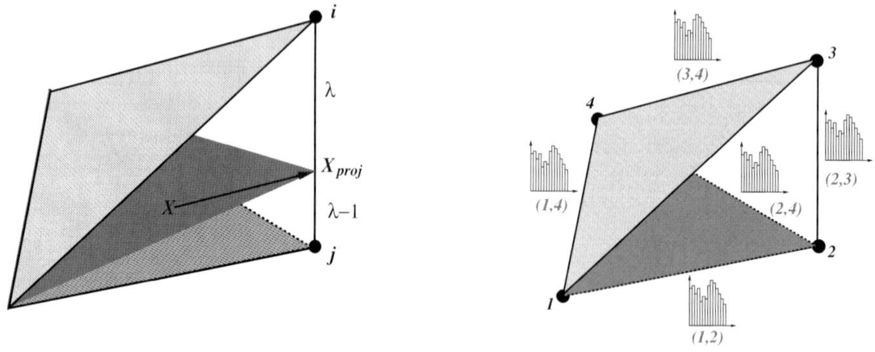


Figure 2. Geometry of the split of the 3-dimensional simplex cell.

of bins in each histogram used to estimate the best division direction/edge and parameter. In this way the increase of $N_{samp}$ is not wasted for cells inside which the integrand is already varying very little.

## Programs

At present the main development version of the `Foam` implementation is available in the C++ language. Two other versions written in Fortran 77 are also available. They are functionally equivalent to version in C++[5].

## Conclusions

`Foam` algorithm is a versatile *adaptive, general-purpose* type of an algorithm based on the *cellular division* of the integration domain. The geometry of the "foam of cells" is rather simple, cells of *simplical and/or hyperrectangular* shape are constructed in the process of a binary split. It works in principle for arbitrary distribution – *no assumption of factorizability* as in `VEGAS` of Ref. [3]. Encoding cells with *memory-efficient methods* allows up to $\sim 10^6$ cells to be built in the computer memory of a typical desktop computer.

## REFERENCES

1. S. Jadach, "Foam: A general purpose cellular Monte Carlo event generator," `physics/0203033`.
2. S. Jadach, "Foam: Multi-dimensional general purpose Monte Carlo generator with self-adapting symplectic grid," *Comput. Phys. Commun.* **130** (2000) 244–259, `physics/9910004`.
3. G. P. Lepage, "A new algorithm for adaptive multidimensional integration," *J. Comput. Phys.* **27** (1978) 192.
4. S. Kawabata *Comput. Phys. Commun.* **88** (1995) 309.
5. T. Ohl, "Vegas revisited: Adaptive monte carlo integration beyond factorization," *Comput. Phys. Commun.* **120** (1999) 13–19, `hep-ph/9806432`.
6. G. I. Manankova, A. F. Tatarchenko, and F. V. Tkachov, "MILXy way: How much better than VEGAS can one integrate in many dimensions?." A Contribution to AINHEP-95, Pisa, Italy, Apr 3-8, 1995 (extended version).
7. E. de Doncker and A. Gupta, "Multivariate integration on hypercubic and mesh networks," *Parallel Computing* **24** (1998) 1223–1244.
8. E. de Doncker, A. Gupta, and R. Zanny, "Large scale parallel numerical integration," *Journal of Computational and Applied Mathematics* **112** (1999).
9. E. de Doncker, K. Kaugars, and A. Gupta, "User manual of PARINT1.1." http://www.cs.wmich.edu/~parint/.

[5] Future development will be restricted to C++.

Proceedings of ICHEP 2002, pp. 825–827
*S. Bentvelsen, P. de Jong, J. Koch and E. Laenen (Editors)*

ELSEVIER

31st INTERNATIONAL CONFERENCE ON HIGH ENERGY PHYSICS AMSTERDAM

# Project SANC (former CalcPHEP): Support of Analytic and Numeric calculations for experiments at Colliders

A. Andonov[a] *, D. Bardin[a*], S. Bondarenko[a*], P. Christova[a*], L. Kalinovskaya[a*], G. Nanava[a*], G. Passarino[b†]

[a]Laboratory of Nuclear Problems, Joint Institute for Nuclear Research, Dubna, Russia

[b]Department of Theoretical Physics, University of Torino, INFN, Torino, Italy

The project, aimed at the theoretical support of experiments at modern and future accelerators — TEVATRON, LHC, electron Linear Colliders (TESLA, NLC, CLIC) and muon factories, is presented. Within this project a four-level computer system is being created, which must automatically calculate, at the one-loop precision level the pseudo- and realistic observables (decay rates and event distributions) for more and more complicated processes of elementary particle interaction, using the principle of knowledge storing.

It was already used for a recalculation of the EW radiative corrections for Atomic Parity Violation [1] and complete one-loop corrections for the process $e^+e^- \to t\bar{t}$ [2–4]; for the latter an, agreement up to 11 digits with FeynArts and the other results is found. The version of SANC that we describe here is capable of automatically computing the decay rates and the distributions for the decays $Z(H,W) \to f\bar{f}$ in the one-loop approximation.

## 1. SANC PROJECT AND ITS ROOTS

The main goal of this project is the creation of a software product, accessible via Internet, for the automatic computation of pseudo- and realistic observables with a one-loop precision for various processes of elementary particle interactions, such as: $1 \to 2$, $1 \to 3$, $2 \to 2$, $1 \to 4$, $2 \to 3$, etc. [5].

It has two roots: 1) Codes aimed at a theoretical support of HEP experiments, such as TOPAZO [6], ZFITTER [7] and similar ones.
2) Numerous FORM2-codes, written by the authors of ref. [8] while they were working on it.

It is supposed that the main software products of the project should be the Monte-Carlo event generators that are being created in collaboration with S. Jadach and Z. Was from INF (Krakow, Poland) and B.F.L. Ward from University of Tennessee (Knoxville, USA). On top of these generators, for some processes the semi-analytic codes are also provided, the latter being used both for the cross-checks of MC generators and for fits of inclusive observables to the theory predictions.

## 2. BASIC NOTIONS

### 2.1. The SM Lagrangian in the $R_\xi$ gauge

All the calculations start from the SM Lagrangian in the $R_\xi$ gauge. It depends on 25 input parameters (IPS), fields, and on three gauge parameters:

$$\mathcal{L} = \mathcal{L}(\text{IPS: 25 parameters, fields}, \xi_{_A}, \xi_{_Z}, \xi), \quad (1)$$

where the fields are: fermions, vector bosons, physical Higgs field, $H$, and unphysical fields, $\phi^0$, $\phi^\pm$, $Y^A$, $Y^Z$, $X^\pm$ [8]. We give the example of Feynman Rules for vector boson propagators:

$$A: \quad \frac{1}{p^2}\left\{\delta_{\mu\nu} + \left(\xi_{_A}^2 - 1\right)\frac{p_\mu p_\nu}{p^2}\right\},$$
$$Z: \quad \frac{1}{p^2+M_{_Z}^2}\left\{\delta_{\mu\nu} + \left(\xi_{_Z}^2 - 1\right)\frac{p_\mu p_\nu}{p^2+\xi_{_Z}^2 M_{_Z}^2}\right\},$$
$$W^\pm: \frac{1}{p^2+M_{_W}^2}\left\{\delta_{\mu\nu} + \left(\xi^2 - 1\right)\frac{p_\mu p_\nu}{p^2+\xi^2 M_{_W}^2}\right\}.$$

### 2.2. Scalar functions and reduction

At present, SANC knows how to deal with up to third-rank tensorial reduction to the usual scalar

*Supported by INTAS grant N$^o$ 00-00313
†Supported by the European Union under contract HPRN-CT-2000-00149 and by MIUR under contract 2001023713006.


doi:10.1016/S0920-5632(02)02250-8

functions: $A_0$, $B_0$, $C_0$ and $D_0$; and to the auxiliary scalar functions: $a_0$, $b_0$, $c_0$ and $d_0$, which are due to particular form of photonic propagator in $R_\xi$ gauge. **A new fortran library** for numerical calculations of all these functions but one ($D_0$) is created and thoroughly tested by means of comparison with the other codes. For the $D_0$ function, we use an old coding of `TOPAZO` [6].

### 2.3. Amplitude basis, scalar form factors, helicity amplitudes

`SANC` computes the one-loop covariant amplitude ($\mathcal{A}$) of a process parametrized in a certain basis by a certain number of scalar form factors (SFF). Next, it computes helicity amplitudes (HA) in terms of SFF. As an example, we present three covariant amplitudes of the decays $B(Q) \to f(p_1)\bar{f}(p_2)$, where $B = H, Z, W$:

$$H \to f\bar{f} : \mathcal{A} \propto I\mathcal{F}_S, \qquad (2)$$

1 structure (S-basis), 1 SFF, 1 HA;

$$Z \to f\bar{f} : \mathcal{A} \propto i\gamma_\mu\gamma_6\mathcal{F}_{\mathrm{L}} + i\gamma_\mu\mathcal{F}_{\mathrm{Q}} + m_f D_\mu\mathcal{F}_{\mathrm{D}},$$

3 structures (LQD-basis), 3 SFF, 3 HA;

$$W \to u\bar{d} : \mathcal{A} \propto i\gamma_\mu\gamma_6\mathcal{F}_{\mathrm{L}} + i\gamma_\mu\gamma_7\mathcal{F}_{\mathrm{R}} + m_u D_\mu\gamma_6\mathcal{F}_{\mathrm{LD}} + m_d D_\mu\gamma_7\mathcal{F}_{\mathrm{RD}},$$

4 structures (LRD-basis), 4 SFF, 4 HA,

where $D_\mu = (p_1 - p_2)_\mu$. Besides SFF, the HA depend on kinematical factors and coupling constants. For instance, for $Z$ decay three HA are:

$$\mathbf{A}^{\mathbf{Z}}_{\mathbf{0}\pm\pm} = \frac{g m_f}{c_W}\left[a_f\mathcal{F}_{\mathrm{L}} + \delta_f\mathcal{F}_{\mathrm{Q}} + \frac{1}{2}a_f\beta_f^2 M_Z^2\mathcal{F}_{\mathrm{D}}\right],$$
$$\mathbf{A}^{\mathbf{Z}}_{\pm\pm\mp} = \frac{g M_Z}{\sqrt{2}c_W}\left[a_f(1\mp\beta_f)\mathcal{F}_{\mathrm{L}} + \delta_f\mathcal{F}_{\mathrm{Q}}\right], \qquad (3)$$

where $\beta_f^2 = 1 - 4\dfrac{m_f^2}{M_Z^2}$,

and $\delta_f = v_f - a_f = -2Q_f s_W^2$, $a_f = I_f^{(3)}$.

We note that the number of SFF and the number of independent non-zero HA always coincide. In general, a $2f \to 2f$ process with four different external fermion masses is described by 16 SFF and 16 independent HA. In case of $e^+e^- \to t\bar{t}$ process, one has six SFF and six independent HA if the electron mass is ignored [2–4].

## 3. STATUS OF THE PROJECT

### 3.1. Four levels of computer system

Here we explain what the `SANC` system does at its four levels in the case of calculation of pseudo-observables (decay rates, event distributions) for the simplest decays: $H(Z,W) \to f_1\bar{f}_2$.

**Level 1:** the chain of calculations *from* $\mathcal{L}_{SM}$ *to the ultraviolet-free helicity amplitudes* is realized. (All codes are written in `FORM3` [9]). It contains four sublevels, which calculate:

- the scalar form factors;
- the *soft* and *hard* photonic contributions to the decay rates (analytic chain);
- the helicity amplitudes for basic processes;
- the helicity amplitudes for accompanying bremsstrahlung processes;

**Level 2:** an `s2n.f` software generates automatically the `fortran` codes for $\Gamma^{(1)} = \Gamma^{\mathrm{Born}} + \Gamma^{\mathrm{Virt}} + \Gamma^{\mathrm{Soft}} + \Gamma^{\mathrm{Hard}}$;

**Level 3:** an infrared rearrangement (or exponentiation) procedure should work here (it is still at the stage of development);

**Level 4:** a Monte Carlo event generator works out. For the time being, we have here a 'manually written' `fortran` code.

### 3.2. Some keywords, characterizing SANC

`SANC` is an **Internet-based** system (address *http://brg.jinr.ru/*) the use of which is supposed to be as simple as the usual surf on Internet. It is a **database-based** system. This means that there is a storage of source codes written in several languages, which talk to one another, being placed into a homogeneous environment written in `JAVA` (linker). It uses the **precomputation or intermediate access principle** which means that:

1) All the relevant one-loop diagrams are precomputed and stored. This greatly saves the CPU.

2) It has several intermediate 'entries' to bypass CPU consuming computations. The user may also access its final product, a MC generator. From the other side, the full chain of calculation 'from the Lagrangian to realistic distribution' can be worked out in real time upon the user request. This option is intended to demonstrate the self-reproducibility of the full chain of calculation at any time.

## 4. First versions of SANC

- **v0.01, 3/01** realizes a part of analytic calculations of level 1 for the decays $H$ $(Z,W) \to f_1 \bar{f}_2$ (demonstration of viability);
- several versions, **v0.02c/d**, were tested on the road toward realization of levels 2,4 for decays $H(Z,W) \to f_1 \bar{f}_2$;
- the current version, **v0.03** (summer'02), realizes the full chain of calculations, returns numbers and distributions for these decays at the one-loop level of precision (demonstration of workability);
- the next version, **v0.10** (at work), should contain all FORM3 codes of level 1 for the processes $2f \to 2f$ and the decays $F \to 3f$.

### 4.1. Comparison with other codes

A lot of comparisons with the other codes were made. For instance, we compared numbers with those of the ZFITTER code for all channels of $e^+e^- \to f\bar{f}$ with light fermion masses and found the following agreement: within **8–9 digits** for the SFF; within **7–8 digits** for the one-loop differential cross-sections $d\sigma^{(1)}/d\cos\vartheta$; within **6–7 digits** for the total cross section and $\sigma^{FB}$.

Several comparisons were made for the $e^+e^- \to t\bar{t}$ process. First is an internal one, between two `fortran` codes: `s2n.f` and `eeffLib`. The latter is a 'manually written' code aimed at benchmarking of `s2n.f` software. For the SFF and for the complete one-loop differential cross-section $d\sigma^{(1)}/d\cos\vartheta$ the numbers agree within **12–13 digits**. A comparison between `s2n.f` and `FeynArts` [10] for the one-loop cross-section *without soft photons* revealed an agreement within **11 digits**. Finally, we compared `s2n.f` numbers with those of the `topfit` code [11,12] for the differential one-loop cross-section *with soft photons* and found an **8-digit** agreement.

## 5. CONCLUDING REMARKS

SANC clearly is a long-term project. Its first phase is realized in *demonstration version 0.10.*

Its second phase assumes the creation of a complete software product, accessible via an Internet-based environment, and realizing the full chain of calculations at the one-loop level of precision including *processes 2 → 3* and *decays 1 → 4*.

## REFERENCES

1. D. Y. Bardin, P. Christova, L. Kalinovskaya and G. Passarino, *Eur. Phys. J.* **C22** (2001) 99–104.
2. D. Bardin, L. Kalinovskaya and G. Nanava, "An electroweak library for the calculation of EWRC to $e^+e^- \to f\bar{f}$ within the CalcPHEP project", CERN-TH/2001-308, hep-ph/0012080.
3. A. Andonov, D. Bardin, S. Bondarenko, P. Christova, L. Kalinovskaya and G. Nanava, "Further study of the $e^+e^- \to f\bar{f}$ process with the aid of CalcPHEP system", CERN-TH/2002-068, hep-ph/0202112.
4. A. Andonov, D. Bardin, S. Bondarenko, P. Christova, L. Kalinovskaya and G. Nanava, "Update of one loop corrections for $e^+e^- \to f\bar{f}$, first run of CalcPHEP system", hep-ph/0207156.
5. D. Bardin *et al.*, "Project CalcPHEP: Calculus for precision high energy physics", hep-ph/0202004.
6. G. Montagna, O. Nicrosini, F. Piccinini and G. Passarino, *Comput. Phys. Commun.* **117** (1999) 278.
7. D. Bardin, M. Bilenky, P. Christova, M. Jack, L. Kalinovskaya, A. Olshevsky, S. Riemann and T. Riemann, *Comput. Phys. Commun.* **133** (2001) 229-395.
8. D. Y. Bardin and G. Passarino, *The standard model in the making: Precision study of the electroweak interactions.* Oxford, UK: Clarendon (1999) 685 p.
9. J. A. M. Vermaseren, "New features of FORM", math-ph/0010025.
10. T. Hahn, *Nucl. Phys. Proc. Suppl.* **89** (2000) 231–236.
11. J. Fleischer, T. Hahn, W. Hollik, T. Riemann, C. Schappacher and A. Wertenbach, "Complete electroweak one-loop radiative corrections to top-pair production at TESLA: – a comparison", hep-ph/0202109.
12. J. Fleischer, J. Fujimoto, T. Ishikawa, A. Leike, T. Riemann, Y. Shimizu and A. Wertenbach, "One-loop corrections to the process $e^+e^- \to t\bar{t}$ including hard bremsstrahlung", hep-ph/0203220.

ELSEVIER

Proceedings of ICHEP 2002, pp. 828–830
*S. Bentvelsen, P. de Jong, J. Koch and E. Laenen (Editors)*

31[st] INTERNATIONAL CONFERENCE ON
HIGH ENERGY PHYSICS AMSTERDAM

# Using Coulomb gauge for next-to-leading order numerical calculations

Michael Krämer [a*] and Davison E. Soper [b†]

[a]School of Physics, University of Edinburgh, Edinburgh EH9 3JZ, Scotland

[b]Institute of Theoretical Science, University of Oregon, Eugene, OR 97403, USA

Calculations of observables in quantum chromodynamics can be performed using a method in which all of the integrations, including integrations over virtual loop momenta, are performed numerically. We use this method to perform next-to-leading order calculations for event shape variables in electron-positron annihilation in Coulomb gauge. The use of a physical gauge provides the potential to go beyond a purely order $\alpha_s^2$ calculation by including, for instance, renormalon or parton showering effects.

## 1. Introduction

QCD calculations at next-to-leading order can be done in a style in which all of the integrations over loop three-momenta are done numerically. In this method, in particular, the integrations for virtual loop diagrams are performed numerically instead of analytically. The method works well in Feynman gauge [1–3] when applied to three-jet-like observables in electron-positron annihilation (*eg.* the three jet cross section).

Feynman gauge is the simplest gauge from a calculational point of view. However, it is unphysical in that unphysical gluon polarizations propagate into the final state. The contributions from unphysical polarizations cancel when one sums over graphs. However, one may want to go beyond a pure next-to-leading order calculation by incorporating, in an approximate way, some effects at all orders in $\alpha_s$. (We have in mind, for instance, renormalon and parton showering effects.) For such applications, one must approximate, and the presence of unphysical degrees of freedom propagating over long distances makes it difficult to see what approximations to apply. The remedy is simple: do the calculation in a physical gauge, such as Coulomb gauge. In this talk, we review some key points from this calculation, which was described in Ref. [4]. The corresponding computer code can be found at [5].

## 2. The numerical method

Before describing the calculation in Coulomb gauge, we need to outline the numerical method itself, which works in any gauge. We present a very brief sketch only since the details can be found in Refs. [1–3].

We wish to calculate an observable $\mathcal{I}$ with the following structure

$$\mathcal{I} = \frac{1}{\sigma_0\left(\sqrt{S}\right)} \sum_n \frac{1}{n!} \int d\vec{P}_1 \cdots d\vec{P}_n \times \frac{d\sigma}{d\vec{P}_1 \cdots d\vec{P}_n} \mathcal{S}_n(\vec{P}_1, \ldots, \vec{P}_n) \quad (1)$$

The quantity $d\sigma/[d\vec{P}_1 \cdots d\vec{P}_n]$ is the cross section to make $n$ massless partons. The functions $\mathcal{S}_n$ are the measurement functions that define the observable. They have the property of infrared safety. We are concerned with three-jet-like quantities, which means that $\mathcal{S}_2 = 0$, so that the smallest value of $n$ that contributes is $n = 3$. We work in next-to-leading order perturbation theory, so that there are at most four partons in the final state. Thus the sum over $n$ runs over $n = 3$ and $n = 4$.

*This work was supported in part by the European Union under contract HPRN-CT-2000-00149, and by the British Particle Physics and Astronomy Research Council.

†This work was supported in part by the U.S. Department of Energy and by the British Particle Physics and Astronomy Research Council.


doi:10.1016/S0920-5632(02)02251-X

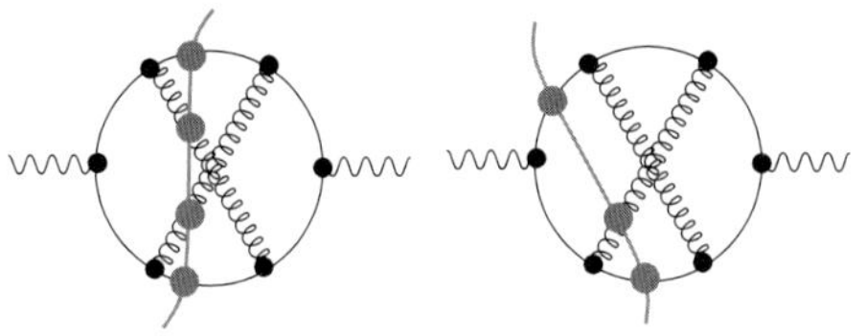

Figure 1. Two cuts of one of the Feynman diagrams that contribute to $e^+e^- \to hadrons$.

After suitable manipulations, the contribution $\mathcal{I}^{(2)}$ to $\mathcal{I}$ proportional to $\alpha_s^2$ can be expressed as an integral of the form

$$\mathcal{I}^{(2)} = \sum_G \int d\vec{l}_1\, d\vec{l}_2\, d\vec{l}_3 \sum_C \mathcal{J}(G,C;l) \times g(G,C;l+i\kappa(G,C;l)). \tag{2}$$

Here there is a sum over graphs $G$ (of which one is shown in Fig. 1) and there is a sum over the possible cuts $C$ of a given graph. There is an integration over three loop momenta. We take care of the ultraviolet divergences by simple renormalization subtractions on the integrand. We make certain deformations $l \to l + i\kappa$ on the integration contours so as to keep away from poles of the form $1/[E_F - E_I \pm i\epsilon]$, where $E_F$ is the energy of the final state and $E_I$ is the energy of an intermediate state. There is a jacobian $\mathcal{J}$ associated with this contour deformation. Then the integrals are all convergent and we calculate them by Monte Carlo numerical integration.

The functions $g$ include the measurement functions. The calculation is organized so that the numerical integration sends random "events" to a routine that represents the measurement function, with each event supplied with an appropriate weight (which can be positive or negative).

## 3. The gluon propagator in Coulomb gauge

In part, doing the calculation described above in Coulomb gauge means substituting the Coulomb gauge Feynman rules for the Feynman gauge Feynman rules in the algebraic calculation of the numerator functions for the graphs. However, the self-energy functions require a separate treatment that is adapted to the gauge choice. See Refs. [6,7] for a discussion on the subtleties of Coulomb gauge. We present here a sketch of the result for the gluon loop contribution to the gluon self-energy.

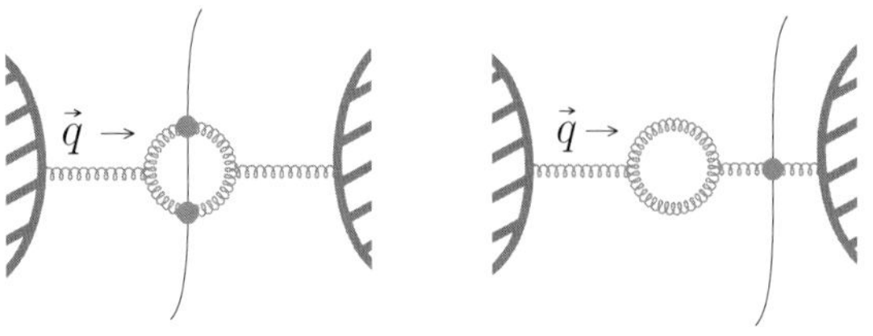

Figure 2. Diagrams with a real gluon self-energy and a cut gluon self-energy.

Consider self-energy diagrams of the form shown in Fig. 2. After some analysis, we obtain a result of the following form:

$$\mathcal{I}[\text{real}] + \mathcal{I}[\text{virtual}] = \frac{1}{2|\vec{q}|}\int_0^\infty \frac{d\bar{q}^2}{\bar{q}^2}\int_0^1 dx \int_{-\pi}^{\pi}\frac{d\phi}{2\pi}\Big\{ \mathcal{M}_g^{\mu\nu}(\bar{q}^2,x,\phi)\, R_{\mu\nu}(\bar{q}^2,x,\phi) + \mathcal{W}_g^{\mu\nu}(\bar{q}^2,x,\phi)\, R_{\mu\nu}^{(0)}\Big\}. \tag{3}$$

Here $\mathcal{I}[\text{real}]$ and $\mathcal{I}[\text{virtual}]$ represent the contributions of the two graphs of Fig. 2. There is an integration over the loop momentum $\vec{l}$ in the self-energy subdiagram. We choose certain coordinates $\{\bar{q}^2, x, \phi\}$ for this loop momentum. Of particular interest is the variable $\bar{q}^2$, which is the virtuality of the gluon pair if the gluon energies are determined so that the gluon lines are on-shell. The function $R_{\mu\nu}(\bar{q}^2,x,\phi)$ represents the factors associated with the rest of the cut self-energy graph, while $R_{\mu\nu}^{(0)}$ represents the factors associated with the rest of the virtual self-energy graph. The function $R_{\mu\nu}(\bar{q}^2,x,\phi)$ approaches $R_{\mu\nu}^{(0)}$ when $\bar{q}^2 \to 0$. We calculate $\mathcal{M}_g^{\mu\nu}(\bar{q}^2,x,\phi)$ and $\mathcal{W}_g^{\mu\nu}(\bar{q}^2,x,\phi)$. The $\bar{q}^2 \to 0$ behavior of these functions is important:

$$\mathcal{M}^{\mu\nu} \sim D(q)^{\mu\nu}\,\frac{\alpha_s}{2\pi}\,\frac{1}{2}\,\tilde{P}_{g/g}(x),$$

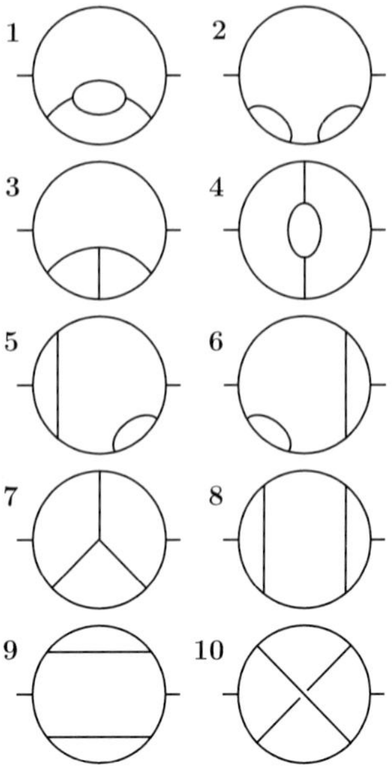


Figure 3. The ten possible graph topologies for $e^+e^- \to hadrons$.

$$\mathcal{W}^{\mu\nu} \quad \sim \quad -D(q)^{\mu\nu}\,\frac{\alpha_s}{2\pi}\,\frac{1}{2}\,\tilde{P}_{g/g}(x), \tag{4}$$

where $\tilde{P}_{g/g}(x)$ is the usual parton splitting function for $g \to gg$. Because of this behavior, the two terms multiplying $1/\vec{q}^{\,2}$ in Eq. (3) cancel each other as $\vec{q}^{\,2} \to 0$, so that the integration is convergent and can be performed numerically.

## 4. Results

We have tested the program by computing the coefficients of $(\alpha/\pi)^2$ in the moments of the thrust distribution,

$$\mathcal{I}_n = \frac{1}{\sigma_0(\alpha_s/\pi)^2}\int_0^1 dt\;(1-t)^n\,\frac{d\sigma^{(2)}}{dt}. \tag{5}$$

There are ten possible graph topologies, as illustrated in Fig. 3. (Fig. 1 is an example of topology 10.) When we calculate an $\mathcal{I}_n$ for one of the graph topologies, we get very different results depending on whether we do the calculation in Coulomb or in Feynman gauge. For example, for $n = 2$ and topology 1, we get $-0.327$ in Coulomb gauge and 0.113 in Feynman gauge. However, the sum of all the graphs is gauge invariant: $1.561 \pm 0.003 \pm 0.006$ in Coulomb gauge and $1.565 \pm 0.002 \pm 0.007$ in Feynman gauge for $n = 2$.

## 5. Conclusions

Numerical calculations in at next-to-leading order for three jet quantities in electron positron are completely practical in a physical gauge. This gives one the potential to go beyond a purely order $\alpha_s^2$ calculation by including, for instance, renormalon or parton showering effects, which should be simplest when expressed in a gauge in which only physically polarized gluons propagate over long distances.

## REFERENCES

1. D. E. Soper, Phys. Rev. Lett. **81**, 2638 (1998) [hep-ph/9804454].
2. D. E. Soper, Phys. Rev. D **62**, 014009 (2000) [hep-ph/9910292].
3. D. E. Soper, Phys. Rev. D **64**, 034018 (2001), [hep-ph/0103262].
4. M. Krämer and D. E. Soper, Phys. Rev. D (to be published), [hep-ph/0204113].
5. D. E. Soper, *beowulf* Version 2.0, http://zebu.uoregon.edu/~soper/beowulf/.
6. N.H. Christ and T. D. Lee, Phys. Rev. D **22**, 939 (1980).
7. G. Leibbrandt and J. Williams, Nucl. Phys. **B475**, 469 (1996) and subsequent papers.

ELSEVIER

Proceedings of ICHEP 2002, pp. 831–833
*S. Bentvelsen, P. de Jong, J. Koch and E. Laenen (Editors)*

31st INTERNATIONAL CONFERENCE ON HIGH ENERGY PHYSICS AMSTERDAM

# WHIZARD: Complete simulations for electroweak multi-particle processes

W. Kilian[a]

[a]Institut für Theoretische Teilchenphysik, University of Karlsruhe, 76128 Karlsruhe, Germany

We describe the generic Monte-Carlo package WHIZARD which interfaces symbolic programs for automatic matrix element calculation with an adaptive multi-channel integration and event generation algorithm. The program generates leading-order cross sections and distributions together with unweighted event samples for a large set of multi-particle processes. As an application, new results for eight-fermion processes are briefly discussed.

## 1. ELECTROWEAK PROCESSES

Exploring the mechanism of electroweak symmetry breaking (EWSB) and mass generation will be a primary goal for the next generation of collider experiments. We know that the particles associated with EWSB include scalars (which manifest themselves either as longitudinal components of massive vector bosons or as physical Higgs bosons), heavy fermions, and potentially new degrees of freedom such as pseudo-Goldstone bosons, superpartners, etc. Essentially all of those are unstable and decay either directly or indirectly into light Standard Model particles. Their production and decay processes at colliders typically result in final states of high multiplicity which cannot be handled exactly by analytical methods.

Since precise measurements of the form factors and self-interactions of EWSB particles at future colliders will be both feasible and necessary, theoretical predictions and Monte-Carlo simulations cannot rely on approximations but need to incorporate complete matrix elements and an accurate description of multi-particle phase space in order to be useful.

## 2. THE SETUP

The large number of different multi-particle processes that will be analyzed in experiments at the Tevatron, LHC and future lepton colliders can only be handled by generic program packages. Traditionally, universal Monte Carlo generators [1–3] have been implemented as libraries of pre-calculated matrix elements, interfaced to robust phase space sampling algorithms. However, this approach becomes less practical as the number of different partonic processes increases exponentially with the number of external particles. Simultaneously, the complex singularity structure of the multi-particle phase space makes brute-force approaches inefficient.

The WHIZARD [4] program system is designed to overcome these limitations. Matrix elements are calculated only when needed. The user sets up an arbitrary list of processes he is interested in, which is automatically fed into external programs that analytically calculate tree-level matrix elements in the form of Fortran 90 computer code. Currently, the matrix element generators O'Mega [5], MadGraph [6] and CompHEP [7] are interfaced[1] and can be selected on a process-by-process basis. Together they cover the full Standard Model. Once set up, the process code is compiled and linked with the other WHIZARD modules to produce a stand-alone Monte Carlo generator. Alternatively, one can generate a subroutine library to be included in a larger simulation and analysis framework.

The package uses the standard `configure` and `make` tools of UNIX systems. Internally, the calling of external programs and compilation is controlled by PERL scripts. All physical and non-physical parameters are defined in input files,

[1]The MadGraph and CompHEP versions included in the WHIZARD system are somewhat out-of-date compared to the respective current versions.


doi:10.1016/S0920-5632(02)02324-1

to be used at runtime by the Monte Carlo executable. Results are given on screen in command-line mode and additionaly written to log files. Generated events can be output in several formats, including human-readable ASCII and binary STDHEP formats.

## 3. PHASE SPACE

While the idea of full automatization has been pursued before (e.g., [7,8]), `WHIZARD` is aimed at complex multi-particle processes in particular. The modular structure allows to select the most efficient version of the matrix element available. The phase-space integration is performed by the adaptive multi-channel integration algorithm `VAMP` [9]. Building up on the well-known `VEGAS` algorithm, each dimension of the integration space is divided in initially equidistant bins which are adapted to the variance of the integrand after each of several sampling runs. Simultaneously, the weights of individual channels are adapted according to the total per-channel variance. The final version of the integration grids is used for generating an arbitrary number of weight-one events for a realistic Monte-Carlo simulation.

The possible parameterizations are derived from the denominator structure of the dominant Feynman diagrams contributing to a given process. The selection is is done automatically, allowing for user intervention where it is desired. While process-dependent default cuts are suggested by the program to regulate infinities, these can be overrun by arbitrary user-defined cuts, and reweighting factors can be applied, to account for any particular needs.

## 4. FURTHER MODULES

The properties of a partonic initial state are described by structure functions. For hadronic collisions, `WHIZARD` uses the standard PDF library. For high-energy lepton collisions beamstrahlung is implemented by an interface to `CIRCE` [10]. Furthermore, the leading-logarithmic ISR and effective-photon approximation spectra are included in the `WHIZARD` code. The structure functions can be concatenated to simulate, e.g., resolved photons emitted from electrons, whenever this makes sense.

Table 1
Six- and eight-fermion matrix elements

| $e^-e^+ \to$ | $\nu_e\bar{\nu}_e ud\bar{d}\bar{u}$ | $\nu_e\bar{\nu}_e b\bar{b}ud\bar{d}\bar{u}$ | factor |
|---|---|---|---|
| diagrams | 618 | 22,958 | 37.1 |
| fusions | 305 | 1,371 | 4.5 |
| helicities | 32 | 128 | 4 |
| fus.× hel. | | | 18 |

Hadronization of colored particles is supported by calling appropriate libraries [2]. On the partonic level, an analysis module is included in the package which allows for histogramming distributions in various kinematical variables and displaying the results in the form of LaTeX/POSTSCRIPT output.

Polarization of both the initial and final state is fully supported as long as the selected matrix element generator allows for it.

## 5. RESULTS

In the present version, `WHIZARD` is able to calculate cross sections and distributions and to generate unweighted event samples for Standard-Model processes with up to 6–8 fermions or bosons in the final state. The efficiency depends on the complexity of the problem: Processes with a dominant resonance structure as they occur in the analysis of EWSB are handled more efficiently than QCD-like massless parton cascades, since in the former case, the number of integration channels can be kept significantly smaller, resulting in a lower number of degrees of freedom in the `VAMP` grid adaptation algorithm.

In a systematic survey, `WHIZARD`'s six-fermion cross sections have recently been verified to agree within statistical errors with the six-fermion integration package `LUSIFER` [11]. Similar comparisons have been made for other process classes, verifying the correctness of the matrix-element and phase-space generation algorithms.

Eight-fermion processes will be of interest once the next generation of colliders becomes operational. To test the applicability `WHIZARD` to this

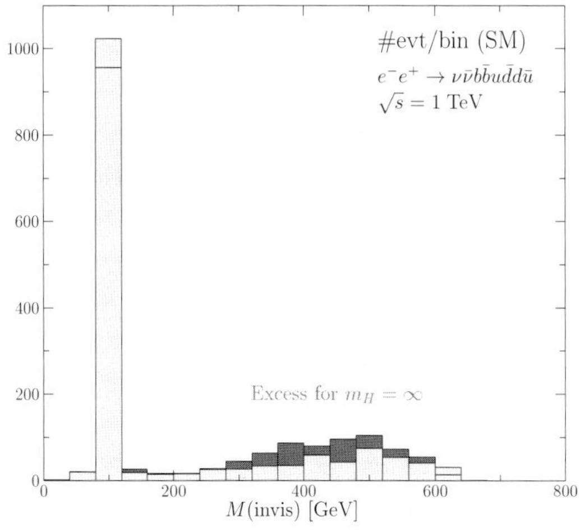


Figure 1. Eight-fermion distribution

problem, we have considered a process which contains $W^+W^- \to t\bar{t}$ as a signal,

$$e^+e^- \to \nu_e\bar{\nu}_e b\bar{b}ud\bar{d}\bar{u},$$

with the complete matrix element generated by `O'Mega`. Comparing this to $W^+W^- \to W^+W^-$ scattering,

$$e^+e^- \to \nu_e\bar{\nu}_e ud\bar{d}\bar{u},$$

one expects the matrix element evaluation alone to consume two orders of magnitude more CPU time, if the traditional Feynman-diagrammatic approach is adopted. The `O'Mega` algorithm, where the amplitude is recursively constructed from the fusions of off-shell wave functions, reduces this by one order of magnitude, cf. Tab.1. However, the increased complexity of the eight-fermion phase space calls for larger event samples during the adaptation, so the step from six to eight fermions generically results in a factor 100 increase in CPU time. Nevertheless, this can still be handled, and unweighted event samples corresponding to 1 $\text{ab}^{-1}$ can be obtained in reasonable time using current PC technology. A sample result, the distribution of the invisible invariant mass for two values of the Higgs mass, is shown in Fig. 1.

## 6. IMPROVEMENTS

While `WHIZARD`'s phase space sampling works well for most processes up to six fermions in the final state, the efficiency is insufficient in some cases where many identical particles are present, e.g., $e^+e^- \to e^+e^-e^+e^-e^+e^-$. This problem can be solved by an improved treatment of massless splittings and by accounting for permutation symmetries, which will be included in the version of fall 2002. In addition, this version will support the new standard Monte Carlo interface [12] along with a number of minor improvements. Furthermore, new versions of the matrix element generators will allow to generate full matrix elements in several SM extensions, including the MSSM.

## REFERENCES

1. `HERWIG`, G. Corcella *et al.*, JHEP 0101 (2001) 010.
2. `PYTHIA`, T. Sjöstrand, Comput. Phys. Commun. 135 (2001) 238.
3. `PANDORA`, M. Peskin, SLAC–PUB–8290, hep-ph/9910519.
4. `WHIZARD`, W. Kilian, LC–TOOL–2000–039.
5. `O'Mega`, T. Ohl, in: Proc. ACAT 2000, Fermilab, IKDA-2000-30, hep-ph/0011243; M. Moretti, Th. Ohl, and J. Reuter, LC-TOOL-2001-040
6. `MadGraph`, T. Stelzer and W.F. Long, Comput. Phys. Commun. 81 (1994) 357.
7. `CompHEP`, A. Pukhov, *et al.*, Preprint INP MSU 98-41/542, `hep-ph/9908288`.
8. `GRACE`, F. Yuasa, *et al.*, Prog. Theor. Phys. Suppl. 138 (2000) 18.
9. `VAMP`, T. Ohl, Comput. Phys. Commun. 120 (1999) 13.
10. `CIRCE`, T. Ohl, Comput. Phys. Commun. 101 (1997) 269.
11. `LUSIFER`, S. Dittmaier and M. Roth, DESY–02–081, hep-ph/0206070.
12. E. Boos *et al.*, Proc. Les Houches 2001, hep-ph/0109068.

ELSEVIER

Proceedings of ICHEP 2002, pp. 834–836
*S. Bentvelsen, P. de Jong, J. Koch and E. Laenen (Editors)*

31st INTERNATIONAL CONFERENCE ON HIGH ENERGY PHYSICS AMSTERDAM

# The technique of inverse Mellin transform for processes occurring in a background magnetic field

Guey-Lin Lin[a]*

[a]Institute of Physics, National Chiao-Tung University
Hsinchu 300, Taiwan

We develop the technique of inverse Mellin transform for processes occurring in a background magnetic field. We show by analyticity that the energy (momentum) derivatives of a field theory amplitude at the zero energy (momentum) is equal to the Mellin transform of the absorptive part of the amplitude. By inverting the transform, the absorptive part of the amplitude can be easily calculated. We apply this technique to calculate the photon polarization function in a background magnetic field.

## 1. INTRODUCTION

The analytic properties of scattering amplitudes in quantum field theories are well known. They are described by the so called cutting rules[1,2]. With the cutting rules, the absorptive part of a scattering amplitude can be calculated, while the dispersive part of the amplitude is obtained from the former by the Kramers-Kronig relation. The above procedures can be easily implemented for any quantum field theory with a trivial vacuum. For such a vacuum, the energy-momentum relation of an asymptotic state is simply $E^2 = \mathbf{p}^2 + m^2$ with $m$ the mass of the asymptotic state. On the other hand, for processes occurring in a background magnetic field, the energy-momentum relations of asymptotic states receive corrections from the magnetic-field effects. For instance, the energy-momentum relation of an electron (positron) in the background magnetic field is given by

$$E^2_{n,s_z} = m_e^2 + p_z^2 + eB\left(2n + 1 + 2s_z\right), \qquad (1)$$

where the background magnetic field is taken to be along the $+z$ direction, $s_z$ is the electron spin projection, and $n$ labels the Landau levels. Due to the above energy quantization caused by a background magnetic field, one expects the absorptive part of a physical amplitude contains an infinite number of contributions distinguished by their corresponding Landau levels. Hence the calculation of the absorptive part is becoming rather involved, and certainly the calculation of the dispersive part is even more intricate.

In this talk, we present a new approach for computing the absorptive part of a physical amplitude in the background magnetic field. The idea is based upon the analyticity. For illustration, we take the one-loop photon polarization function as an example, with a background magnetic field along the $+z$ direction. Since the energy of the internal electron in the photon polarization function is given by Eq. (1), the value of the photon longitudinal momentum $q_\parallel^2 \equiv q_0^2 - q_z^2$ determines the threshold for the absorptive part. Therefore the photon polarization function is an analytic function of the photon longitudinal momentum $q_\parallel^2$ except on the positive real axis. To obtain the photon polarization function for an arbitrary $q_\parallel^2$, it suffices to know the function's power-series expansion in $q_\parallel^2$ at $q_\parallel^2 = 0$, because the analytic continuation can map the function from the neighborhood of $q_\parallel^2 = 0$ to any value of $q_\parallel^2$. A powerful way to perform this analytic continuation is through the inverse Mellin transform[3,4]. With the inverse Mellin transform, one calculates the absorptive part of the photon polarization function with the knowledge of the above-

*This work is supported in part by the National Science Council of Taiwan under the grant number NSC90-2112-M009-023.


doi:10.1016/S0920-5632(02)02224-7

mentioned power-series expansion. We note that it is straightforward to obtain such a power series because the neighborhood of $q_{\|}^2 = 0$ is free of resonant singularities caused by the creation of electron-positron pairs. Knowing the absorptive part, the dispersive part of the photon polarization at any $q_{\|}^2$ is calculable by the Kramers-Kronig relation.

We organize this presentation as follows: In Section 2, we illustrate the technique of inverse Mellin transform using the vacuum QED as an example. In Section 3, we apply this technique to the photon polarization function in a background magnetic field. A short conclusion is given in Section 4.

## 2. VACUUM QED

The vacuum polarization tensor in QED is written as

$$i\Pi^{\mu\nu}(q) = (q^2 g^{\mu\nu} - q^\mu q^\nu) i\Pi(q^2). \tag{2}$$

We are interested in calculating the one-loop finite part $\bar{\Pi}(q^2) = \Pi(q^2) - \Pi(0)$. We observe that $\bar{\Pi}(q^2)$ satisfies the sum rule [3]:

$$\frac{1}{n!}\left(\frac{d^n}{d(q^2)^n}\bar{\Pi}(q^2)\right)_{q^2=0} = \frac{1}{\pi}\int_{M^2}^{\infty} dq^2 \mathrm{Im}\bar{\Pi}(q^2)(q^2)^{-(n+1)}, \tag{3}$$

where $M^2$ is the threshold energy for the absorptive part $\mathrm{Im}\bar{\Pi}(q^2)$. Although the value for $M^2$ is not specified at this moment, the procedure of analytic continuation will automatically generate it. Let us rewrite the sum rule in dimensionless variables:

$$\frac{1}{n!}\left(\frac{d^n}{dt^n}\bar{\Pi}(t)\right)_{t=0} = \frac{1}{\pi}\int_0^1 du \mathrm{Im}\bar{\Pi}(u)u^{n-1}, \tag{4}$$

where $t = q^2/M^2$ and $u = 1/t$. It is easily seen that the derivatives of $\bar{\Pi}$ at $t = 0$ is proportional to the Mellin transform of $\mathrm{Im}\bar{\Pi}$. Hence $\mathrm{Im}\bar{\Pi}$ at any value of $t$ is obtainable by the inverse Mellin transform:

$$\mathrm{Im}\bar{\Pi}(u) = \frac{1}{2i}\int_C ds\, a(s)u^{-s}, \tag{5}$$

where $a(s)$ is the analytic continuation of $a(n)$ which appears in the power series expansion

$$\bar{\Pi}(t) = \sum_{n=0}^{\infty} a(n)t^n. \tag{6}$$

For the current case, $a(n)$ is given by

$$a(n) = -\frac{2\alpha}{\pi}\frac{\Gamma^2(n+2)}{n\Gamma(2n+4)}\left(\frac{M^2}{m_e^2}\right)^n. \tag{7}$$

Applying Eq. (5), we obtain

$$\begin{aligned}\mathrm{Im}\bar{\Pi}(u) = & -\frac{\alpha}{3}\sqrt{1-\frac{4m_e^2}{M^2}u}\left(1+\frac{2m_e^2}{M^2}u\right) \\ & \times\ \Theta\left(1-\frac{4m_e^2}{M^2}u\right).\end{aligned} \tag{8}$$

We like to remark that, with $u = 1/t = M^2/q^2$, $\mathrm{Im}\bar{\Pi}(u)$ is independent of the undetermined threshold scale $M^2$, and it agrees with the known result obtained by applying the cutting rules. Having obtained $\mathrm{Im}\bar{\Pi}(u)$, one can calculate $\mathrm{Re}\bar{\Pi}(u)$ by the Kramers-Kronig relation.

## 3. QED IN A BACKGROUND MAGNETIC FIELD

The photon polarization function in a background magnetic field is given by the following proper time representation[5,6]:

$$\begin{aligned}\Pi_{\mu\nu}(q) = & -\frac{e^3 B}{(4\pi)^2}\int_0^{\infty} ds \int_{-1}^{+1} dv \times \\ & \{e^{-is\phi_0}[T_{\mu\nu}N_0 - T_{\|,\mu\nu}N_{\|} + T_{\perp,\mu\nu}N_{\perp}] \\ & -e^{-ism_e^2}(1-v^2)T_{\mu\nu}\},\end{aligned} \tag{9}$$

where the photon momentum has been decomposed into $q_{\|}^{\mu} = (\omega, 0, 0, q_z)$ and $q_{\perp}^{\mu} = (0, q_x, q_y, 0)$; while $T_{\mu\nu} = (q^2 g_{\mu\nu} - q_\mu q_\nu)$, $T_{\|,\mu\nu} = (q_{\|}^2 g_{\|\mu\nu} - q_{\|\mu}q_{\|\nu})$, and $T_{\perp,\mu\nu} = (q_{\perp}^2 g_{\perp\mu\nu} - q_{\perp\mu}q_{\perp\nu})$. The phase $\Phi_0$ and the functions $N_0$, $N_{\|}$ and $N_{\perp}$ are given by

$$\phi_0 = m_e^2 - \frac{1-v^2}{4}q_{\|}^2 - \frac{\cos(zv)-\cos(z)}{2z\sin(z)}q_{\perp}^2 \tag{10}$$

with $z = eBs$, and

$$N_0 = \frac{\cos(zv) - v\cot(z)\sin(zv)}{\sin(z)},$$

$$N_{\parallel} = -\cot(z)\left(1 - v^2 + \frac{v\sin(zv)}{\sin(z)}\right) + \frac{\cos(zv)}{\sin(z)},$$
$$N_{\perp} = -\frac{\cos(zv)}{\sin(z)} + \frac{v\cot(z)\sin(zv)}{\sin(z)} + \frac{2\,(\cos(zv) - \cos(z))}{\sin^3(z)}. \quad (11)$$

The two independent eigenmodes of the above polarization tensor are $\epsilon^{\mu}_{\parallel}$ and $\epsilon^{\mu}_{\perp}$ which are respectively parallel and perpendicular to the plane spanned by the photon momentum $\mathbf{q}$ and the magnetic field $\mathbf{B}$. They obey the dispersion equations $q^2 + \Pi_{\parallel} = 0$ and $q^2 + \Pi_{\perp} = 0$ respectively with $\Pi_{\parallel,\perp} = \epsilon^{\mu}_{\parallel,\perp}\Pi_{\mu\nu}\epsilon^{\nu}_{\parallel,\perp}$. It turns out that $\Pi_{\parallel}$ and $\Pi_{\perp}$ are proportional to $N_{\parallel}$ and $N_{\perp}$ respectively. We shall not discuss the contribution by $N_0$ since it does not correspond to an independent eigenmode.

For simplicity, we shall only focus on the calculation of $\Pi_{\parallel}$. In particular, we only illustrate the procedure of calculating the partial contribution $\Pi^A_{\parallel}$ given by

$$\Pi^A_{\parallel} = -\frac{\alpha\omega^2\sin^2\theta}{4\pi}\int_0^{\infty} dz \int_{-1}^{+1} dv \times \exp(-is\phi_0)\frac{\cos(zv)}{\sin(z)}, \quad (12)$$

where the integrand $\cos(zv)/\sin(z)$ is taken from the second term of $N_{\parallel}$, and $\theta$ is the angle between $\mathbf{q}$ and $\mathbf{B}$.

The scalar function $\Pi^A_{\parallel}$ is easy to calculate only for $q^2_{\parallel} < 4m_e^2$, since, in this momentum region, the contour rotation $s \to -is$ is allowed[6]. The oscillating trigonometric functions are then rotated into hyperbolic functions. Similar to the previous section, one writes $\Pi^A_{\parallel}$ as a power series in $q^2_{\parallel}$:

$$\Pi^A_{\parallel} = -\frac{\alpha\omega^2\sin^2\theta}{4\pi}\sum_{n=0}^{\infty} a(n, q^2_{\perp})\left(\frac{q^2_{\parallel}}{4m_e^2}\right)^n. \quad (13)$$

Following Eq. (5), we calculate $\mathrm{Im}\Pi^A_{\parallel}$ by

$$\mathrm{Im}\Pi^A_{\parallel}(v) = \frac{i\alpha\omega^2\sin^2\theta}{8\pi}\int_C ds\, a(s, q^2_{\perp})v^{-s}, \quad (14)$$

with $a(s, q^2_{\perp})$ the analytic continuation of $a(n, q^2_{\perp})$, and $v = 4m_e^2/q^2_{\parallel}$. We arrive at

$$\mathrm{Im}\Pi^A_{\parallel} = \frac{2\alpha eB\omega^2\sin^2\theta}{q^2_{\parallel}}\sum_{l_1=1,l_2=0}^{\infty} T_{l_1,l_2}(q^2_{\perp}) \times \frac{\Theta(1 - \frac{4\lambda eB}{q^2_{\parallel}} + \frac{4\rho^2e^2B^2}{q^4_{\parallel}})}{\sqrt{1 - \frac{4\lambda eB}{q^2_{\parallel}} + \frac{4\rho^2e^2B^2}{q^4_{\parallel}}}}, \quad (15)$$

where $\lambda = l_1 + l_2 + m_e^2/eB$, $\rho = l_1 - l_2$, and the step function indicates the threshold for creating an electron-positron pair occupying the $l_1$-th and $l_2$-th Landau levels. The detailed form of $T_{l_1,l_2}(q^2_{\perp})$ is given in Ref. [4]. With $\mathrm{Im}\Pi^A_{\parallel}$ determined, $\mathrm{Re}\Pi^A_{\parallel}$ can once more be calculated by the Kramers-Kronig relation.

## 4. CONCLUSION

In this talk, we have demonstrated the usefulness of analyticity for computing field theory amplitudes. We observed that a field theory amplitude in the neighborhood of zero energy (momentum) determines the amplitude at the arbitrary energy (momentum). The latter can be calculated from the former by the inverse Mellin transform and the Kramers-Kronig relation. We also showed that such an approach is very appropriate for a process occurring in a background magnetic field, where an infinite number of thresholds occur in the amplitude of the process.

## REFERENCES


1. R. E. Cutkosky, J. Math. Phys. 1 (1960) 429.
2. M. Veltman, Physica 29 (1963) 186.
3. W.-F. Kao, G.-L. Lin, and J.-J. Tseng, Phys. Lett. B495 (2000) 105.
4. W.-F. Kao, G.-L. Lin, and J.-J. Tseng, Phys. Lett. B522 (2001) 257; Erratum-ibid. B541 (2002) 411.
5. J. Schwinger, Phys. Rev. 82 (1951) 664.
6. W.-y. Tsai and T. Erber, Phys. Rev. D 10 (1974) 492.

ELSEVIER

Proceedings of ICHEP 2002, pp. 837–839
*S. Bentvelsen, P. de Jong, J. Koch and E. Laenen (Editors)*

31st INTERNATIONAL CONFERENCE ON HIGH ENERGY PHYSICS AMSTERDAM

# Lattice study of the Coleman–Weinberg mass in the SU(2)-Higgs model *

F. Csikor[a] Z. Fodor[a] and P. Kovács[a]

[a]Institute for Theoretical Physics, Eötvös University,
P.O. Box 32, H-1117 Budapest, Hungary

Radiative symmetry breaking is a well known phenomenon in perturbation theory. We study the problem in a non-perturbative framework, i.e. lattice simulations. The example of the bosonic sector of the SU(2)-Higgs model is considered. We determine the minimal scalar mass which turns out to be higher than the mass value given by 1-loop continuum perturbation theory.

## 1. INTRODUCTION

Symmetry breaking is a fundamental concept in particle physics. The most popular mechanism is the Higgs mechanism. Also radiative symmetry breaking (Coleman-Weinberg (CW) mechanism [1]) is well known. The CW mass is obtained when the tree level mass parameter is zero, i.e. the Higgs mass is induced by radiative corrections. The value of the CW mass is near to the smallest possible Higgs mass.

Connected to the CW mechanism is the problem of vacuum stability, i.e. the value of the minimal Higgs mass e.g. in the Standard Model. Clearly the latter problem would be more interesting to study in a non-perturbative framework. However, putting a chiral top quark on the lattice is difficult, therefore in this contribution we study symmetry breaking on the lattice only in the bosonic sector of the SU(2)-Higgs model. Still this is a difficult two scale problem, since the CW mass is an order of magnitude smaller than the W mass.

## 2. LATTICE FORMULATION

The lattice action in standard notation is:

$$S[U,\varphi] = \beta \sum_{pl} \left(1 - \frac{1}{2}\mathrm{Tr}\, U_{pl}\right)$$
$$+ \sum_x \left\{ \frac{1}{2}\mathrm{Tr}\,(\varphi_x^+ \varphi_x) + \lambda \left[\frac{1}{2}\mathrm{Tr}\,(\varphi_x^+ \varphi_x) - 1\right]^2 \right.$$
$$\left. - \kappa \sum_{\mu=1}^{4} \mathrm{Tr}\,(\varphi_{x+\hat{\mu}}^+ U_{x,\mu} \varphi_x) \right\}, \qquad (1)$$

where $U_{x,\mu}$ denotes the SU(2) gauge link variable, $U_{pl}$ the path-ordered product of the four $U_{x,\mu}$ around a plaquette; $\varphi_x$ stands for the Higgs field. The continuum bare parameters are given: $g^2 = 4/\beta$, $\lambda_c = \lambda/4/\kappa^2$, $\varphi_c = \sqrt{2\kappa/a}\varphi$ and $a^2 m_0^2 = (1-2\lambda)/\kappa - 8$. We choose $g^2$ =0.5, which gives renormalized gauge couplings close to the Standard Model value.

The CW mechanism is an inherently perturbative notion. Therefore, on the lattice we want to determine the smallest Higgs mass possible (which is not far from the CW mass). The lattice action is bounded from below for positive $\lambda$, while it is not for $\lambda < 0$. Thus $\lambda$=0 is the interesting point relevant to the minimal and CW masses. Since the Higgs mass is a monotonous function of $\lambda/\kappa^2$, the limit $\lambda \to 0$ gives the smallest Higgs mass possible. Our aim is to determine the Higgs mass in this region and try to interpret the results in terms of the CW mechanism.

## 3. SIMULATIONS

The lattice simulation goes in the standard way as described e.g. in [2]. It is essential to simulate along a line of constant physics (LCP). For CW we define the LCP with $\lambda$=0, $g_R^2$=fixed ($g_R$ is the renormalized gauge coupling, determined from a fit to the static potential) and the hopping parameter $\kappa$ tuned to the finite temperature phase

*Presented by F. Csikor


doi:10.1016/S0920-5632(02)02226-0

Figure 1. Higgs mass squared as a function of $\lambda/\kappa_{cr}^2$. The masses corresponding to the three highest values of $\lambda/\kappa_{cr}^2$ are from [2,3], while the other point are new simulations.

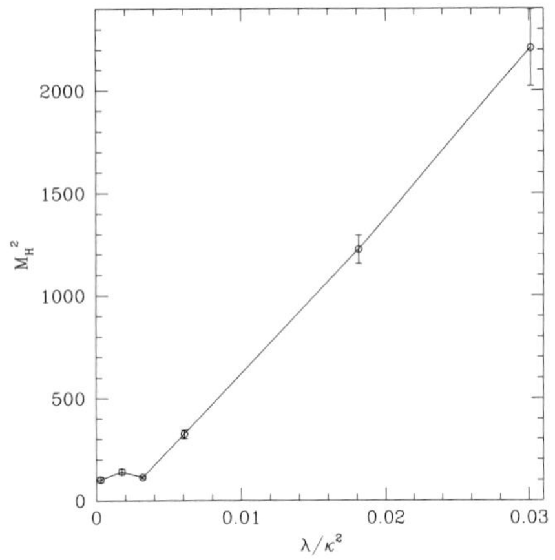


transition point.

We simulate on $L_t \cdot L_s^3$ hot lattices ($L_t << L_s$) and perform an $L_s \to \infty$ extrapolation. Next we extrapolate to the continuum limit ($L_t \to \infty$). The physical lattice sizes for the various $L_t$'s are kept constant by choosing appropriate lattice extensions. The extrapolations to $L_s = \infty$ were carefully performed in order to eliminate finite size effects.

In practice we take $\lambda$ very small ($5 \cdot 10^{-6}$, $2.875 \cdot 10^{-5}$, $5.25 \cdot 10^{-5}$) and fix $\beta = 8$. We found $\kappa_{cr}$ by *constrained simulation*, which is an appropriate method for strong phase transitions.

The physical scale is found for each $\lambda$, $\beta$, $\kappa_{cr}$ by performing simulations at $T$=0 and fixing $M_W$=80 GeV. The quantity we extrapolate is $R_{HW} = M_H/M_W$.

Note that our LCP is not necessarily close to an LCP defined by the perturbative renormalization group running of $\lambda$ and $\beta$. In particular this is the case for the smallest $\lambda$ value given above.

## 4. RESULTS

In one loop perturbation theory one gets

$$M_H^2 = v^2 \cdot (8\lambda_c + 12B), \quad B = \frac{9g^4}{1024\pi^2} \qquad (2)$$

The CW case corresponds to $\lambda_c = -B/2$. The minimal Higgs mass is for $\lambda_c = -B$.

Our results for the Higgs mass squared as a function of $\lambda/\kappa_{cr}^2$ are given in Fig. 1. It is convenient to plot the Higgs mass squared as a function of $\lambda/\kappa_{cr}^2$, since (at tree level) this is proportional to the continuuum quartic coupling. Similar plots for higher Higges masses have been presented in [4]. The straight line part of the curve on the right hand side of the figure obviously corresponds to the perturbative result. Even the slope is in reasonable aggreement with the perturbative prediction. Clearly, for small values of $\lambda/\kappa_{cr}^2$ the straight line has to flatten out, since otherwise we would obtain negative Higgs mass squares. This is what is seen on the figure. It is easy to extrapolate to $\lambda/\kappa_{cr}^2$=0. We get $M_{H,min}^2 = 115.78 \pm 4.95$ GeV$^2$ to be compared to the perturbative value: 22.80 GeV$^2$ (91.20 GeV$^2$) corresponding to the minimal Higgs mass (CW) cases. The lattice result is clearly much larger than the perturbative value.

One would expect that the perturbative and nonperturbative results are close to each other, since the quartic coupling is small in this region. However, it is clear that the LCP choosen brings in a true nonperturbative feature in the lattice calculation. The CW LCP is namely very far from the renormalization group LCP for small $\lambda$. Still, the large value of the smallest nonperturbative Higgs mass is a surprising result. We have carefully studied the finite size effects and believe that the discrepancy is caused most probably by the fact that our simulations are still far from the continuum limit.

## 5. ACKNOWLEDGEMENT

This work was partially supported by Hungarian Science Foundation Grants No. OTKA-T34980/T29803/-M37071/OM-MU-708

## REFERENCES


1. S. Coleman and E. Weinberg, Phys. Rev. D7 (1973) 1888; for a review see e.g. M. Sher, Phys. Reports 179 (1989) 173.
2. F. Csikor, Z. Fodor, J. Hein, A. Jaster and I. Montvay, Nucl.Phys. B474 (1996) 421.
3. Z. Fodor, J. Hein, K. Jansen, A. Jaster and I. Montvay, Nucl. Phys. B439 (1995) 147.
4. Y. Aoki, Phys. Dev. D56 (1997) 3860.

ELSEVIER

Proceedings of ICHEP 2002, pp. 840–842
*S. Bentvelsen, P. de Jong, J. Koch and E. Laenen (Editors)*

31st INTERNATIONAL CONFERENCE ON HIGH ENERGY PHYSICS AMSTERDAM

# Lattice QCD data versus Chiral Perturbation Theory: the case of $M_\pi$

S. Dürr[a]

[a]DESY, Platanenallee 6, 15738 Zeuthen, Germany

I present a selection of recent lattice data by major collaborations for the pseudo-Goldstone boson masses in full ($N_f = 2$) QCD, where the valence quarks are chosen exactly degenerate with the sea quarks. At least the more chiral points should be consistent with Chiral Perturbation Theory for the latter to be useful in extrapolating to physical masses. Perspectives to reliably determine NLO Gasser-Leutwyler coefficients are discussed.

## 1. INTRODUCTION

In lattice QCD physical observables like masses and decay constants are extracted from the fall-off pattern of correlation functions; for instance

$$\begin{aligned} C(x) &= \langle Q(x)^\dagger Q(0)\rangle \\ &= \frac{1}{Z}\int DU\,D\bar{q}Dq\; e^{-S_g-S_f}\; Q(x)^\dagger Q(0) \quad (1) \end{aligned}$$

with $Q(x) = \bar{u}(x)\Gamma d(x)$ and $\Gamma \in \{\gamma_5, \gamma_4\gamma_5\}$ allows to determine $M_\pi$ and $F_\pi$. Like in the continuum, the fermions are integrated out and (1) reads

$$\begin{aligned} &\langle \bar{d}(x)\Gamma u(x)\; \bar{u}(0)\Gamma d(0)\rangle = \\ &\quad \frac{1}{Z}\int DU\,\det(D+m_{\rm sea})e^{-S_g} \times \qquad (2) \\ &\quad \mathrm{Tr}\,\Gamma(D+m_{\rm val})^{-1}_{x0}\Gamma(D+m_{\rm val})^{-1}_{0x} \end{aligned}$$

where Tr goes over colour and spinor indices. In this form, the quark masses $m_{\rm sea}, m_{\rm val}$ should be equal, since they stem from the very same term $S_f = \int dx\, \bar{q}(D+m)q$ in the original form, eqn. (1).

However, state-of-the-art algorithms for the determinant $\det(D+m)$ and the Green's function $S(x,0) = ((D+m_{\rm val})^{-1})_{x0}$ get parametrically slow as the quark mass is taken light. Since this effect is much more pronounced for the determinant, most groups are restricted to a regime where the *sea*-quark mass for the two light flavours (the one in the determinant) is about half of the physical strange mass, while the *valence*-quark mass for the $u/d$-quarks (the one in the propagator) may be pushed down to $\sim 1/4$ of the physical $m_s$.

With these data at hand, one has to *extrapolate* both in $m_{\rm sea}$ and $m_{\rm val}$ down to the physical $u$- and $d$-masses (which we take degenerate); see Fig 1. Since the extrapolation range is large, it is crucial to know the functional form against which the data may be matched. Partially quenched chiral perturbation theory (PQ-XPT) provides such a framework [1], but its limited range of applicability (around the twofold chiral limit) might cause a problem; it is a priori not clear whether the bulk of the existing data lies in this range.

## 2. ELEMENTS OF XPT

The mandatory test is most conveniently done along the diagonal line in Fig. 1, which is the domain of (ordinary) XPT [2,3]. This is the effective

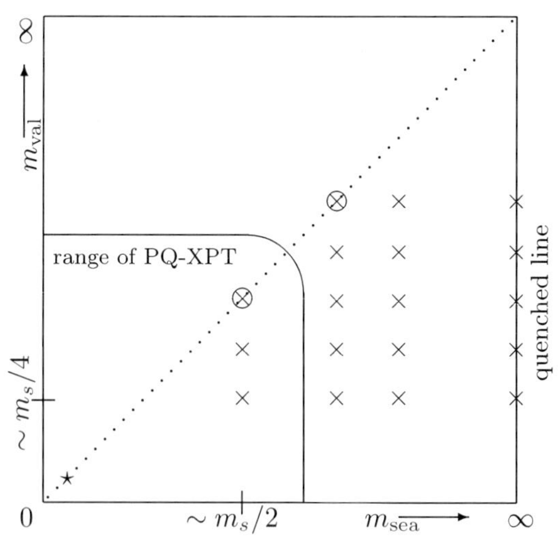


Figure 1. Schematic view of partially quenched data taking, with crosses indicating measurements in the $(m_{\rm sea}, m_{\rm val})$ plane for degenerate u/d-quarks. While the real-world pion ($\star$) is on the unitary line (diagonal), substantial savings in CPU-time stem from taking the detour through the sector with $m_{\rm val} < m_{\rm sea}$.


doi:10.1016/S0920-5632(02)02231-4

theory of QCD at energies small compared to the intrinsic scale, $4\pi F_\pi$, which faithfully reproduces the structure in the chiral limit. Its Green's functions are dominated by pseudo-Goldstone contributions and computed in a systematic expansion in the external momenta and the quark mass, the latter counted as $m \sim p^2$ [2].

The leading part of the Lagrangian, $L^{(2)}$, contains two low-energy constants $(B, F)$ which are finite dimensionful quantities[1] and the quark mass matrix $M = \text{diag}(m_d, m_u\, [, m_s])$. The $O(p^2)$ suppressed part, $L^{(4)}$, contains about a dozen so-called Gasser-Leutwyler (GL) coefficients, denoted $l_i$ for 2 flavours [2] and $L_i$ if the strange quark is included [3]. In the dimensionally regulated theory, the $l_i$ or $L_i$ consist of a divergent part ($\propto \epsilon^{-1}$, where $\epsilon = 4-d$) and a finite part ($\propto \epsilon^0$) which is scale-dependent. After some reshuffling of finite contributions in $\overline{\text{MS}}$ style, the latter part (denoted by $l_i^{\text{r}}(\mu)$ or $L_i^{\text{r}}(\mu)$) encodes for the QCD short distance contributions. In an arbitrary Green's function calculated at order $O(p^4)$, the divergence from the loop compensates exactly the divergent part of the GL-coefficient (see Fig. 2), and an important check of the finite part is that the scale-dependent (finite) integral $\bar{I}(\mu)$ and the renormalized "constants" $l_i^{\text{r}}(\mu)$ (or $L_i^{\text{r}}(\mu)$) add up to form an exactly *scale-independent* expression.

What we are interested in is the prediction for the squared pseudo-Goldstone boson mass in 2-flavour QCD versus the sum of the degenerate valence quark masses, i.e. $M_\pi^2$ versus $2m$. For the latter, we adopt ($\overline{\text{MS}}$, 2 GeV) conventions, which implies that the leading-order (LO) constant $B$ is also quoted in $\overline{\text{MS}}$ scheme at $\mu \sim 2$ GeV, since the product $mB$ is scheme- and scale-independent (as is $F$). An important simplification stems from

[1] $F$ is the pion decay constant in the chiral limit $\sim 87$ MeV, while $B = -\Sigma/F^2$ with $\Sigma$ the chiral (one-flavour) condensate (in the 2- or 3-flavour theory) in the chiral limit.

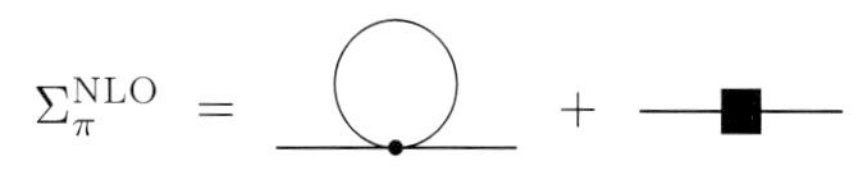

Figure 2. Contributions to the pion self-energy at NLO: 1-loop graph with a vertex from $L^{(2)}$ (little dot) and a counterterm from $L^{(4)}$ (fat box). The divergent parts ($\propto \epsilon^{-1}$) compensate each other, and in the finite parts ($\propto \epsilon^0$) the $\mu$-dependence cancels exactly.

the fact that the GL-coefficients $l_i^{\text{r}}(\mu)$ may be traded for the low-energy scales $\Lambda_i$, and in terms of these also the prediction at next-to-leading order (NLO) [2] in XPT takes a form in which the scale-independence of the physical result is explicit (see [4,5] for details and references):

$$M_\pi^2 \overset{\text{LO}}{=} 2mB \equiv: M^2 \tag{3}$$

$$M_\pi^2 \overset{\text{NLO}}{=} M^2 \Big(1 - \frac{M^2}{32\pi^2 F^2} \log(\frac{\Lambda_3^2}{M^2})\Big) \tag{4}$$

## 3. AIMING AT NLO GL COEFFICIENTS

Equations (3, 4) may be used in two ways: Either phenomenological values for $B, F, \Lambda_3$ are *plugged in* and the resulting curves are compared to lattice data, or the functional forms are *fitted* to the data, which amounts to a lattice determination of the QCD low-energy constants.

The first option is shown in Fig. 3, where quark masses determined via the axial (AWI) or vector (VWI) Ward-Takahashi identity [6,7] are renormalized by 1-loop perturbation theory [6,5]. The heaviest pion is $\sim$1.1 GeV, and one sees that it is still compatible with the LO graph, but not with the NLO or NNLO version. The reason behind is that XPT yields an *asymptotic series*: the LO result for $M_\pi$ may be good up to $\sim$1 GeV, while the NLO/NNLO forms are more precise at small quark masses, but break down earlier, e.g. the NLO functional form might be good up to (say) 600 MeV, to NNLO form up to (say) 400 MeV.

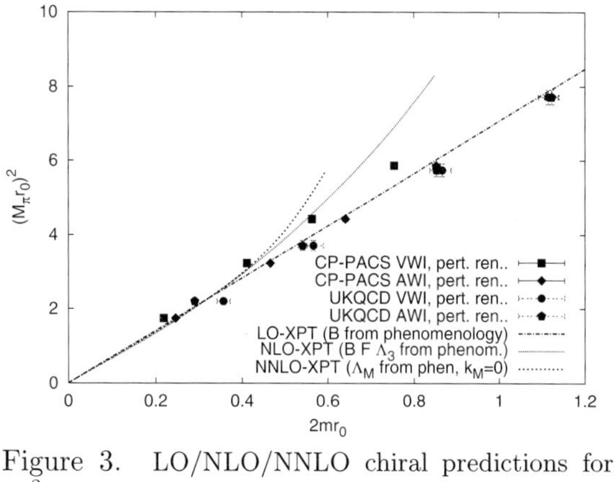


Figure 3. LO/NLO/NNLO chiral predictions for $M_\pi^2$ (with phenomenological values for the low-energy constants) compared to CP-PACS and UKQCD data, using the VWI and AWI definition of the quark mass.

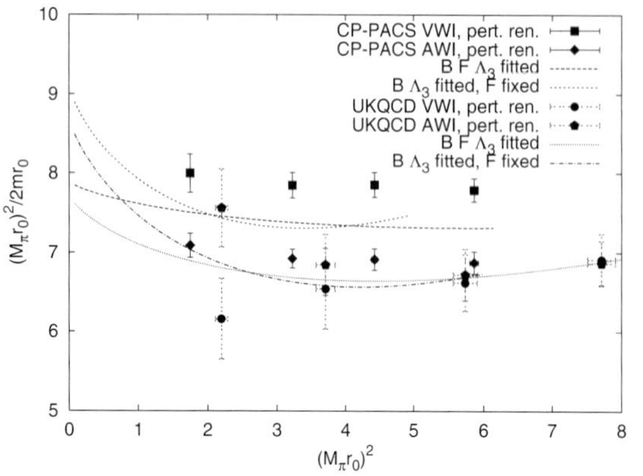

Figure 4. Fits of the NLO functional form (5) to the CP-PACS [6] and UKQCD [7] data, after dividing by $2m$. Either $F$ is kept fixed with the 3 most chiral points used or $B, F, \Lambda_3$ are fitted, using all 4 points.

The second option is likely premature: If the estimate that the NLO functional form is good up to $(M_\pi r_0)^2 \sim 2$ is correct [5] (with $r_0 \simeq 0.5\,\text{fm}$), then Fig. 1 depicts the current situation, and there is no room for using eqn. (4) to fit the data, since it has 3 parameters and in either set at best 1 data-point is in the permissible range. Still, one might simply go ahead and see how inconsistent results get, if one ignores that the bulk of the data is likely not in a regime where (PQ-)XPT applies.

For practical reasons, a factor $2m$ is taken out and parameters are made dimensionless with powers of $r_0$, i.e. the actual NLO form used is

$$\frac{(M_\pi r_0)^2}{2mr_0} = Br_0 - \frac{(M_\pi r_0)^2 Br_0}{32\pi^2 (F_\pi r_0)^2} \log \frac{(\Lambda_3 r_0)^2}{(M_\pi r_0)^2} \quad (5)$$

with the pertinent fits shown in Fig. 4 and the resulting parameters summarized in Tab. 1 [5].

Clearly, results should be taken with care, but

| | $Br_0$ | $F_\pi r_0$ | $\Lambda_3 r_0$ |
|---|---|---|---|
| "phen." [2,6] | 7.06 | 0.233 | 1.51 |
| CP-PACS | 9.10 | (0.233) | 3.03 |
| CP-PACS | 7.88 | 0.503 | 3.97 |
| UKQCD | 8.69 | (0.233) | 3.38 |
| UKQCD | 7.70 | 0.318 | 3.46 |

Table 1
Coefficients in the fits of the functional form (5) to the degenerate CP-PACS [6] and UKQCD [7] data with average VWI and AWI masses [5]. Constrained values in brackets, ranges such that #(d.o.f.) = 1.

it is reassuring to see that the agreement among the fitted values of the NLO GL constant $\Lambda_3$ is not substantially worse than for the LO constant $B$. The numerical values in the last column of Tab. 1 correspond to $\Lambda_3 \sim 1.4\,\text{GeV}$. This is consistent with the early estimate $\Lambda_3 \sim 0.6 \pm {}^{1.4}_{0.4}\,\text{GeV}$ [2], and hence supports the hypothesis that at least in two-flavour QCD the *chiral condensate is large* (see [4] for references to this topic). However, the difference between $F_\pi$ from the unconstrained fit and the physical value (cf. the central column in Tab. 1) reminds one that to date the bulk of the data is likely outside the regime of validity of (PQ-)XPT and accurate simulation results at smaller quark masses are needed (cf. [8]) before one may draw definite conclusions.

In summary, $M_\pi^2$ vs. $2m$ illustrates that lattice QCD is, in principle, an ideal tool to determine certain NLO GL coefficients. To date, most studies with $N_f = 2$ Wilson quarks are likely in a mass-range where there is only a "point-like" overlap with the regime where (PQ-)XPT holds. Once moderately lower masses are covered and with further improvements in renormalizing the bare data and controlling the lattice artefacts (see [5] for a discussion) the lattice will yield a competitive determination of those NLO GL coefficients which encode how low-energy QCD Green's functions depend on the quark mass.

## REFERENCES

1. C.W. Bernard and M.F. Golterman, Phys. Rev. D **49**, 486 (1994) [hep-lat/9306005]. S.R. Sharpe, Phys. Rev. D **56**, 7052 (1997); Errat. D **62**, 099901 (2000) [hep-lat/9707018].
2. J. Gasser and H. Leutwyler, Annals Phys. **158**, 142 (1984).
3. J. Gasser and H. Leutwyler, Nucl. Phys. B **250**, 465 (1985).
4. H. Leutwyler, Nucl. Phys. Proc. Suppl. **94**, 108 (2001) [hep-ph/0011049].
5. S. Dürr, hep-lat/0208051.
6. A. Ali Khan *et al.* [CP-PACS], Phys. Rev. D **65**, 054505 (2002) [hep-lat/0105015].
7. C.R. Allton *et al.* [UKQCD], Phys. Rev. D **65**, 054502 (2002) [hep-lat/0107021].
8. F. Farchioni *et al.*, hep-lat/0209142.

Proceedings of ICHEP 2002, pp. 843–846
S. Bentvelsen, P. de Jong, J. Koch and E. Laenen (Editors)

31st INTERNATIONAL CONFERENCE ON HIGH ENERGY PHYSICS AMSTERDAM

# Summary of a Talk on Peculiar Phases in a Supersymmetric Model at Finite Temperature

Moshe Moshe [a] * and Jean Zinn-Justin[b] †

[a]Physics Department, Technion, Haifa 32000,Israel

[b] CEA-Saclay, Service de Physique Théorique, F-91191 Gif-sur-Yvette Cedex, FRANCE
and
Institut de Mathématiques de Jussieu–Chevaleret, Université de Paris VII

Bosons and fermions behave differently at finite temperature as they interact with an heat bath, and consequently supersymmetry is softly broken. The $O(N)$ symmetric, supersymmetric model in d=3 studied here is known to demonstrate spontaneous breaking of scale invariance with no explicit breaking. This is observed in the spectrum by the appearance of bosonic and fermionic massless bound states at $N \to \infty$. We discuss here those properties of the peculiar phase structure seen at $T = 0$ which are also maintained at finite temperature.

Breakdown of supersymmetry at finite temperature has attracted attention since the early interest in supersymmetry and involved controversy on its consequences, appearance and phase structure [1–3]. Recently, there is a continued interest in the thermodynamics of supersymmetric Yang-Mills theory [4] and temperature effects on the minimal supersymmetric model [5].

We considered [6] an $O(N)$ invariant supersymmetric action in d=3

$$\mathcal{S}(\Phi) = \int \mathrm{d}^3x\,\mathrm{d}^2\theta\,\left[\tfrac{1}{2}\bar{\mathrm{D}}\Phi\cdot\mathrm{D}\Phi + NU(\Phi^2/N)\right] \quad (1)$$

$\Phi$ is an $N$-component vector superfield, $\Phi(\theta) = \varphi + \bar{\theta}\psi + \frac{1}{2}\bar{\theta}\theta F$. The action translates, in components form, for a generic super-potential, into:

$$\mathcal{S}=\int \mathrm{d}^3x\,[-\tfrac{1}{2}\bar{\psi}\not{\partial}\psi + \tfrac{1}{2}(\partial_\mu\varphi)^2 - \tfrac{1}{2}U'(\varphi^2/N)\bar{\psi}\psi - U''(\varphi^2/N)(\bar{\psi}\varphi)(\varphi\psi)/N + \tfrac{1}{2}\varphi^2U'^2(\varphi^2/N)]$$

Large $N$ methods can be applied by introducing two new superfields:
$L(\theta) = M + \bar{\theta}\ell + \frac{1}{2}\bar{\theta}\theta\lambda \qquad R(\theta) = \rho + \bar{\theta}\sigma + \frac{1}{2}\bar{\theta}\theta s$
and adding an extra term to the action:

$$\mathcal{S}(\Phi, L, R) = \int \mathrm{d}^3x\,\mathrm{d}^2\theta \;\; L(\theta)[\Phi^2(\theta) - NR(\theta)] \quad (2)$$

After integrating out $N-1$ superfields of $\Phi$, keeping only the superfield $\Phi_1 \equiv \phi$

$$\mathcal{Z} = \int [\mathrm{d}\phi][\mathrm{d}R][\mathrm{d}L]\mathrm{e}^{-\mathcal{S}_N(\phi,R,L)}$$

The large N action is $\sim N$ and thus, three super-saddle point equations will result from:

$$\mathcal{S}_N = \textstyle\int \mathrm{d}^3x\,\mathrm{d}^2\theta[\frac{1}{2}\bar{\mathrm{D}}\phi\mathrm{D}\phi + NU(R) + L(\phi^2 - NR)] + \frac{1}{2}N\mathrm{Str}\,\ln[-\bar{\mathrm{D}}\mathrm{D} + 2L]. \quad (3)$$

The *Action density* $\mathcal{E} = \mathcal{S}_N/\mathrm{vol}$ simplifies into

$$\mathcal{E}/N = M^2\frac{\varphi^2}{2N} + \frac{1}{24\pi}(m - |M|)^2(m + 2|M|) \quad (4)$$

and one notes that $\mathcal{E}$ is positive and has an absolute minimum at $m_\varphi \equiv m = |M| = m_\psi$ ( a supersymmetric ground state ).

**$\Phi^4$ super-potential in $d = 3$:**

$$U(R) = \mu R + \tfrac{1}{2}uR^2$$

Gap equations (saddle point equations) reduce to

$$M = \mu - \mu_c + u\frac{\varphi^2}{N} - \frac{u}{4\pi}|M|, \quad M\varphi = 0 \quad . \ (5)$$

Inserting $M^2(m,\varphi)$ from Eq. (5) into $\mathcal{E}(M,m,\varphi)/N$ in Eq.(4) one obtains the action density $\mathcal{E}(M(m,\varphi),m,\varphi)$ from which the phase structure of the theory is clearly seen and it is summarized in Fig.1 (see also Figs. 2 and 3)

*e-mail: moshe@physics.technion.ac.il Supported in part by the Israel Science Foundation grant number 193-00 and the VPR fund for promotion of research.
†e-mail:zinn@spht.saclay.cea.fr


doi:10.1016/S0920-5632(02)02241-7

When the coupling constant has the special value $u = u_c = 4\pi$. one finds in the spectrum bound states of massless fermion and boson $O(N)$ singlets [7]. This can be seen by considering the poles in the $\langle LL \rangle$ propagator. The $\langle LL \rangle$ part in the action and the $\langle LL \rangle$ propagator are given by:

$$\frac{-N}{2u}\int \mathrm{d}^3x\,\mathrm{d}^2\theta(L-\mu)^2+\tfrac{1}{2}(N-1)\mathrm{Str}\,\ln\left(-\bar{\mathrm{D}}\mathrm{D}+2L\right)$$

$$\Delta_L^{-1}=\frac{-N}{4\pi|M|}\left[1+\{M+|M|\frac{u_c}{u}\}\delta^2(\theta'-\theta)\right]\mathrm{e}^{i\bar{\theta}\not{\partial}\theta'}$$

Corresponds, at $M<0$, to a super-particle of mass $2M(1-u_c/u)$

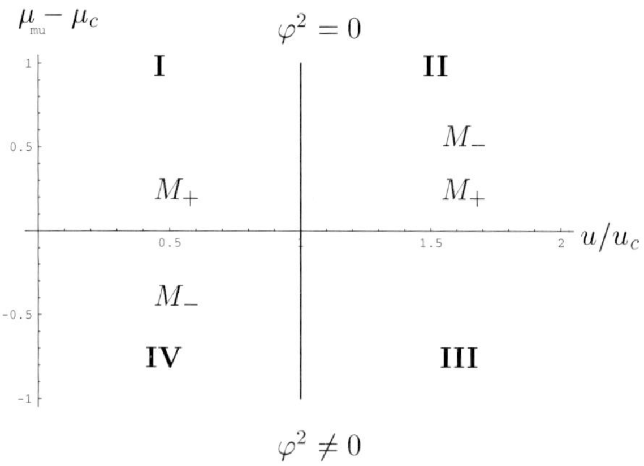


Figure 1. Summary of the phases of the model in the $\{\mu-\mu_c, u\}$ plane. Here $m_\varphi = m_\psi = |M_\pm| = (\mu-\mu_c)/(u/u_c\pm 1)$. The lines $u=u_c$ and $\mu-\mu_c=0$ are lines of first and second order phase transitions.

At $u=u_c$ the bound state mass of vanishes. E.g. the $\psi\varphi$ scattering amplitude $T_{\psi\varphi,\psi\varphi}(p^2)$, in the limit $p^2\to 0$ satisfies [7]

$$T_{\psi\varphi,\psi\varphi}(p^2)\sim\frac{2u}{N}\left[1+\frac{u}{4\pi}\frac{M}{|M|}+\frac{u}{2\pi}\frac{i\not{p}}{|M|}\right]^{-1}$$
$$\to-\frac{4\pi i}{N}\frac{|M|}{\not{p}}\quad\text{for } M<0 \text{ and } u\to u_c$$

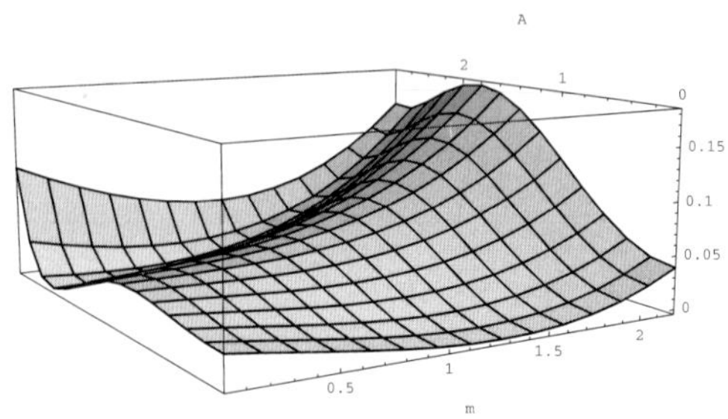


Figure 2. **Region IV** : The action density $\frac{1}{N}\mathcal{E}(m,\varphi)$ as a function of the boson mass $(m)$ and $A$, where $A^2=\varphi^2/u_c$. Here $\mu-\mu_c=-1$, $u/u_c=0.2$. As seen here there are two distinct degenerate phases. One is an ordered phase ($\varphi\neq 0$) with a massless boson and fermion, the other is a symmetric phase ($\varphi=0$) with a massive ($m=|M_-|$) boson and fermion.

### Large $N$ supersymmetric non-linear $\sigma$

We also studied [6] the supersymmetric non-linear $\sigma$-model in $d$ dimensions, $2\le d\le 3$.

$$\mathcal{S}(\Phi)=\frac{1}{2\kappa}\int \mathrm{d}^dx\,\mathrm{d}^2\theta\,\bar{\mathrm{D}}\Phi\cdot\mathrm{D}\Phi \qquad (6)$$

One finds in dimension $d=3$ the action density:

$$\mathcal{E}/N=\frac{1}{2N\kappa}M^2\varphi^2+\frac{1}{24\pi}(m-|M|)^2(m+2|M|)$$

and in dimension $d=2$ we have $M=m=\Lambda\mathrm{e}^{-2\pi/\kappa}$.

$$\mathcal{E}/N=\frac{1}{8\pi}\left[m^2-M^2+2M^2\ln(M/m)\right]. \qquad (7)$$

### $O(N)$ susy model at finite temperature

Using the propagator at finite temperature for the bosons and fermions

$$G_{B/F}(T)=\int^{\Lambda}\frac{\mathrm{d}^{d-1}k}{(2\pi)^{d-1}}\frac{1}{\omega(k)}\left(\frac{1}{2}\pm\frac{1}{\mathrm{e}^{\omega(k)/T}\mp 1}\right)$$

with $\omega(k)=\sqrt{k^2+m_T^2}$ and $\omega(k)=\sqrt{k^2+M_T^2}$ respectively, the action density at finite $T$ is given at large $N$ by [6]

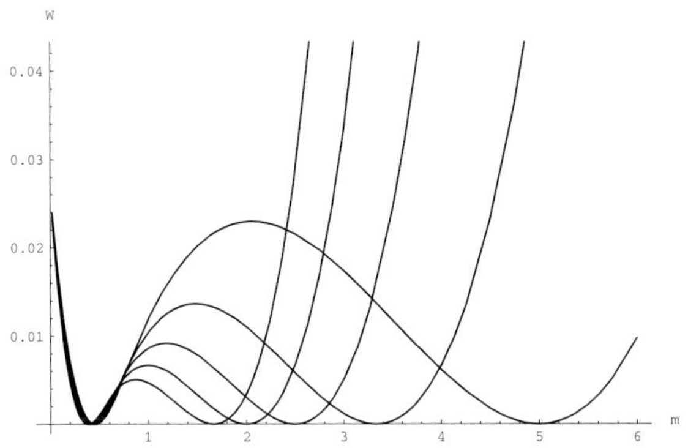


Figure 3. The action density $\frac{1}{N}\mathcal{E}(m,\varphi=0)$ in region **II** of Fig. (1). Here $\mu-\mu_c=1$ (sets the mass scale) and $u/u_c$= varies between 1.6 and 1.2. There are two massive, degenerate $O(N)$ symmetric SUSY vacua with $\varphi=0$ and possible large hierarchy of masses $|M_-/M_+|>>1$ (as $u\to u_c$ ) with $m_\psi=m_\varphi=M_+=(\mu-\mu_c)/(u/u_c+1)$ and $m_\psi=m_\varphi=-M_-=(\mu-\mu_c)/(u/u_c-1)$.

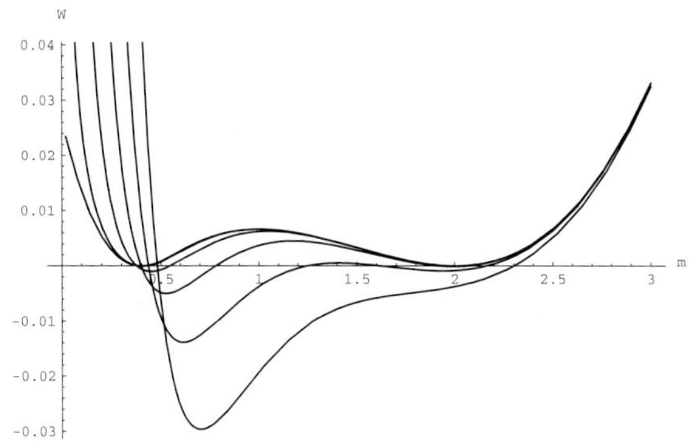


Figure 4. The action density $\frac{1}{N}\mathcal{F}(m\equiv m_\varphi,\varphi,T)$ at $\varphi=0$ as a function of the boson mass ($m$) at different temperatures. Here $\mu-\mu_c=1$(this sets the mass scale) , $u/u_c=1.5$ and $T$ varies between $T=0--0.5$. At $T=0$ two degenerate $O(N)$ symmetric phases exist with a light $m=M_+$ and heavier $m=M_->M_+$ massive boson (and fermion). As the temperature increases, the light mass phase is stronger affected as its entropy increases faster and becomes the only ground state.

$$\begin{aligned}\mathcal{F}/N=&M_T^2\frac{\varphi^2}{2N}+\frac{1}{24\pi}(m_T-|M_T|)^2(m_T+2|M_T|)\\&+T\int\frac{\mathrm{d}^2k}{(2\pi)^2}\{\ln(1-\mathrm{e}^{-\beta\omega_\varphi})-\ln(1+\mathrm{e}^{-\beta\omega_\psi})\\&+\frac{T\lambda}{4\pi}\ln(1-\mathrm{e}^{-\beta m_T})\}.\end{aligned}\quad(8)$$

Eq. (8) is the finite temperature version of Eq. (4). $M_T$ is the expectation value of $M(x)$ at finite temperature $\beta^{-1}=T$ and satisfy a gap equation which replaces at $T\neq0$ Eq.(5), namely

$$M_T=\mu-\mu_c+u\frac{\varphi^2}{N}-\frac{u}{2\pi}T\ln(2\sinh(\tfrac{1}{2}\beta m_T))$$

It is important to note that the r.h.s. of this gap equation diverges when $m_T\to0$. The finite temperature system in three dimensions has the property of a statistical system in two dimensions. In two dimensions a spontaneous breaking of a continuous symmetry is impossible, due to the IR behaviour of a system with potential massless Goldstone particles. Therefore, the $O(N)$ symmetry is never broken, $\varphi=0$

**Peculiar transitions can occur in this system** as seen e.g. in Fig. (4) that describes the change in the ground state free energy of the system as the temperature is changed when the parameters are chosen in region **(II)** $\mu-\mu_c\geq0$ and $u\geq u_c$: ( of Fig. (1)). If the system was initially at $T=0$ in the ground state with a boson mass $m_\varphi=M_-$, it will eventually go, as $T$ is increased, into the state with mass $m_\varphi=M_+$, namely, into the lower mass ground state. This is due to the entropy negative contribution to the free energy forcing the system to favors the lowest mass state. On the other hand, a system that started at a high temperature in the state of low mass $m_\varphi=M_+$ will stay in this state as the system is cooled and will never roll back into the $m=M_-$ high mass phase. Favoring the low-

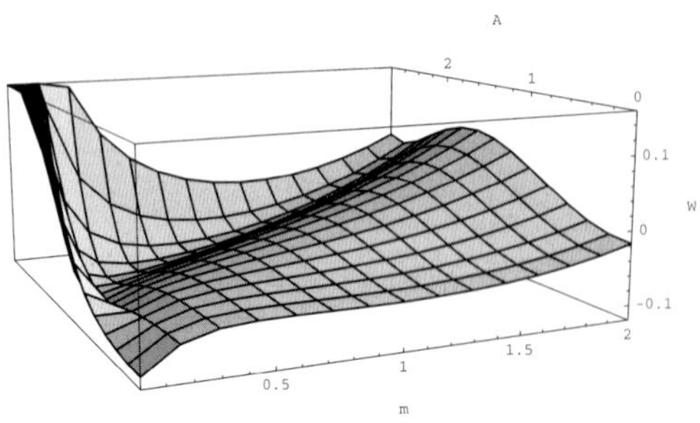


Figure 5. Same as Fig. (2) but the temperature has been increased from $T = 0$ (in Fig. (2)) to $T = 0.7$ (here). $\frac{1}{N}\mathcal{F}(m \equiv m_\varphi, \varphi, T)$ as a function of the boson mass ($m$) and $A$, where $A^2 = \varphi^2/u_c$. Here $\mu = -1$ and $u/u_c = 0.2$. A non-degenerate $O(N)$ symmetric ground state ($\varphi = 0$) appears with a very small boson mass (the non-zero mass is not seen here due to the limited resolution of the plot).

est mass phase as the temperature increases is a general effect that will occur in any physical system that is initially (at $T = 0$) mass degenerated. Here, supersymmetry imply that the $M_+$ and $M_-$ vacua are at the same energy $E = 0$ at $T = 0$.

Similarly, in region **(IV)** $\mu - \mu_c \leq 0$ and $u \leq u_c$: here one finds at $T = 0$ two distinct degenerate phases. One is an ordered phase ($\varphi \neq 0$) with a zero boson and fermion mass, the other is a symmetric phase ($\varphi = 0$) with a massive ($m = M_-$) boson and fermion (as seen in Fig. 2). As the temperature increases, the ordered phase with the broken $O(N)$ symmetry ($m = 0, \varphi^2 \neq 0$) disappears into a $\varphi^2 = 0$ symmetric phase and a very small mass ground state $m_\varphi = m \geq 0$ is created (as seen in Fig. 5 and in Fig. 2). Eventually, also the other $O(N)$ symmetric vacuum ($m = M_-, \varphi^2 = 0$) goes into the small mass $O(N)$-symmetric ground state.
As in **II**, a system that was initially, at $T = 0$, in the $O(N)$-symmetric phase ($m_\varphi = M_-, \varphi^2 = 0$) will decay into the smaller mass state when the temperature is raised. But upon cooling the system in the small mass phase will roll into the ordered phase ($m_\varphi = m = 0, \varphi^2 \neq 0$) at $T = 0$. The system will never roll back into the symmetric phase ($m_\varphi = M_-, \varphi = 0$).

**Susy $O(N)$ non-linear $\sigma$ model at $T \neq 0$**
As mentioned above, the $d = 3$ finite temperature theory phase structure is analogous to a two-dimensional theory. Therefore, the $O(N)$ symmetry remains unbroken and $\varphi = 0$. The free energy density in the symmetric phase is given by [6]:

$$\frac{1}{N}\mathcal{F}=\frac{(M_T^2-m_T^2)}{2}\left(\frac{1}{\kappa}-\frac{1}{\kappa_c}\right)-\left(\frac{m_T^3-|M_T|^3}{12\pi}\right) +T\int\frac{\mathrm{d}^2k}{(2\pi)^2}\left\{\ln[1-\mathrm{e}^{-\beta\omega_\varphi}]-\ln[1+\mathrm{e}^{-\beta\omega_\psi}]\right\}$$

**Acknowledgements**
MM would like to thank W.A. Bardeen and J. Feinberg for useful discussions. The warm hospitality of the Saclay and Fermilab theory groups is also appreciated.

## REFERENCES

1. L.Girardelo, M.T. Grisaru, P. Salomonson *Nucl. Phys.* B178 (1981) 331 .
2. A.Das and M.Kaku *Phys.Rev* D18 (1978) 4540; A. Das *Physics A* 158 (1989) 1.
3. A. Riotto and G. Senjanović *Phys. Rev. Lett.* 79 (1997) 349 and references therein
4. A. Fotopoulos and T.R. Taylor *Phys. Rev.* D59 (1999) 061701 and references therein
5. J.R. Espinosa *Nucl. Phys.* B475 (1996) 273 ; M. Laine *Nucl. Phys.* B481 (1996) 43 , Erratum. ibid. B548 (1999) 637
6. M. Moshe and J. Zinn-Justin, Supersymmetric Models at Finite Temperature(hep-th 0209045 to be published in *Nucl. Phys.* B648 issue (1,2) (2003) ); Review on Quantum Field Theory in the Large N limit (Saclay-Technion preprint, to be published in *Physics Reports* (2003) ).
7. W.A. Bardeen, K. Higashijima and M. Moshe, *Nucl. Phys.* B250 (1985) 437

## Session 12

# String Theory and Mathematical Quantum Field Theory

*Session conveners:*
*S. Vandoren, Utrecht University*
*J. Louis, University of Halle*

ELSEVIER

Proceedings of ICHEP 2002, pp. 849–854
*S. Bentvelsen, P. de Jong, J. Koch and E. Laenen (Editors)*

31st INTERNATIONAL CONFERENCE ON
HIGH ENERGY PHYSICS AMSTERDAM

# Regge amplitudes from AdS/CFT duality

R. Peschanski [a]

[a]CEA/DSM/SPhT,Unité de recherche associée au CNRS,
CE-Saclay, F-91191 Gif-sur-Yvette Cedex,France;
email: pesch@spht.saclay.cea.fr

String theory has long ago been initiated by the quest for a theoretical explanation of the observed high-energy "Regge behaviour" of strong interaction amplitudes, but this 35-years-old puzzle is still unsoved. We discuss how modern tools like the AdS/CFT correspondence give a new insight on the problem.

## 1. Introduction

It is well-known that string theory started from the proposal of scattering amplitudes which may grasp the two major structures of soft interaction phenomenology in a condensed form : resonances and Regge poles. Two types of amplitudes were proposed for four-point amplitudes, see Fig.1. The Veneziano amplitude corresponds to *Reggeon* exchanges with non-vacuum quantum numbers and the Shapiro-Virasoro amplitude corresponds to *Pomeron* exchange with vacuum quantum numbers. As soon after demonstrated, they correspond to respectively open and closed bosonic string theory amplitudes at tree-level.

However, despite many efforts during years, no widely recognized progress have been done in the string theory of strong interaction amplitudes, and after the discovery of QCD as the gauge field theory of quarks and gluons, there remained little place for it. Indeed, major theoretical obstacles have been raised, for instance:

- The conformal anomaly of string theories in Minkowski $D$-dimensional space leads to the limitation $D = 26, 10$ for bosonic and super strings.
- Zero mass gauge and gravitational fields appear in the string spectra of asymptotic states, if not tachyonic states.

To these (non exhaustive list of) difficulties, new questions have to be added after the discovery of QCD, for instance:

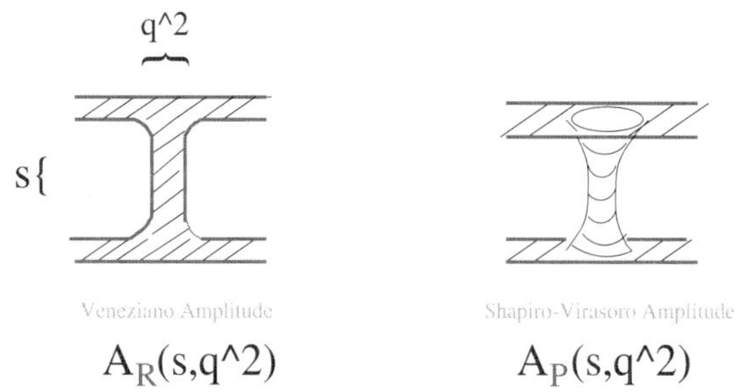


Figure 1. Reggeon and Pomeron String Amplitudes.

- Where are "hard" interactions recovered in a string theory framework?
- Can we elaborate a suitable string Theory which could coherently describe the properties of gauge fields?

To these pending questions, the recently proposed AdS/CFT correspondence between certain string and gauge field theories may give new and reliable answers. Among these questions, the understanding of Regge amplitudes in terms of the AdS/CFT duality have been the subject of an approach which I will now describe.

## 2. String/Gauge fields Duality

The AdS/CFT correspondence [1] has many interesting formal and physical facets. Concerning the aspects which are of interest for our problem, it allows one to find relations between gauge


doi:10.1016/S0920-5632(02)02243-0

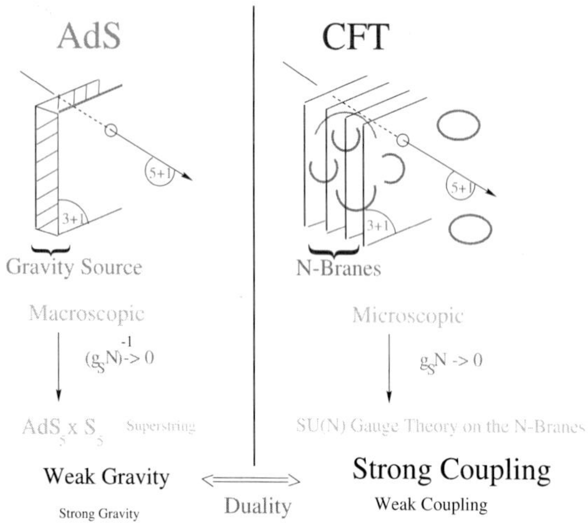


Figure 2. $\mathrm{AdS}_5/\mathrm{CFT}_4$ duality correspondence.

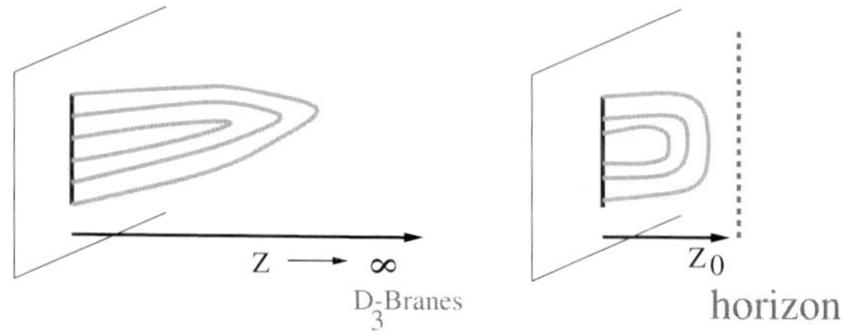


Figure 3. Exemple of minimal surfaces with Wilson lines boundary.

field theories at strong coupling and string gravity at weak coupling in the limit of large number of colours ($N_c \to \infty$), see Fig.2. It can be examined quite precisely in the $\mathrm{AdS}_5/\mathrm{CFT}_4$ case which conformal field theory corresponds to $SU(N)$ gauge theory with $\mathcal{N}=4$ supersymmetries.

Some existing extensions to other gauge theories with broken conformal symmetry with less or no supersymmetries will be valuable for our approach, since they lead to confining gauge theories which are more similar to QCD. Indeed, one important question is to examine to what extent confinement plays a rôle in the Reggeization of amplitudes.

However, note that the appropriate string gravity dual of QCD has not yet been identified, and thus we are forced to restrict for the moment our use of AdS/CFT correspondence to features which are expected to be a general feature of confining theories duals.

Let us schematically recall the canonical derivation of the $\mathrm{AdS}_5$ background. One starts from the (super)gravity classical solution of a system of $N$ $D_3$-branes in a $10-D$ space of the (type IIB) superstrings. The metrics solution of the (super)Einstein equations read

$$ds^2 = f^{-1/2}(-dt^2 + \sum_{1-3} dx_i^2) + f^{1/2}(dr^2 + r^2 d\Omega_5) ,$$

where the first four coordinates are on the brane and $r$ corresponds to the coordinate along the normal.

$$f = 1 + \frac{R^4}{r^4} \ ; \qquad R = 4\pi g_{YM}^2 \alpha'^2 N ,$$

and $g_{YM}^2 N$ is the ‘t Hooft-Yang-Mills coupling and $\alpha'$ the string tension. Then one introduce the “Maldacena limit”, where one sits near-by the branes while in the same time going to weak string coupling limit. the space-time is thus distorted due to the (super) gravitational field of the branes. One writes

$$\frac{\alpha'(\to 0)}{r(\to 0)} \to z \ , \ R \ fixed \ \Rightarrow g_{YM}^2 N \sim \frac{1}{\alpha'^2} \to \infty \ .$$

By reorganizing the two parts of the metrics one obtains

$$ds^2 = \frac{1}{z^2}(-dt^2 + \sum_{1-3} dx_i^2 + dz^2) + R^2 d\Omega_5 ,$$

which corresponds to the $\mathrm{AdS}_5 \times S_5$ background structure, $S_5$ being the 5-sphere. More detailed analysis shows that the isometry group of the 5-sphere is the geometrical dual of the $\mathcal{N}=4$ supersymmetries.

In order to illustrate the way how one formulates the AdS/CFT correspondence, let us consider the example of the vacuum expectation value (*vev*) of Wilson lines in a configuration parallel to the time direction of the branes. This configuration allows a determination of the potential between colour charges [2].

One writes

$$\langle e^{iP\int_C \vec{A}\cdot\vec{dl}} \rangle = \int_\Sigma e^{-\frac{Area(\Sigma)}{\alpha'}} \approx e^{-\frac{Area_{min}}{\alpha'}} \times Fluct. ,$$

where $C$ is the Wilson line contour near the $D_3$ branes and $\Sigma$ the surface in $AdS$-space with $C$ as the boundary, see Fig.3. The minimal area approximation is the *vev* evaluation classical $\alpha' \to 0$ limit which can eventually be improved by calculating the fluctuation determinant around the minimal surface. In Fig.3, we have sketched Two cases: the $AdS_5$ "conformal" one and a a confining case, $AdS_{BH}$, where a black-hole (BH) in the AdS bulk determines a characteristic horizon scale $R_0$ breaking conformal invariance (see Witten [1]).

The *vev* results can be summed up as follows:

$$AdS_5 : \langle Wilson\ Lines \rangle = e^{TV(L)} \sim e^{\#T/L}$$
$$AdS_{BH} : \langle Wilson\ Lines \rangle = e^{TV(L)} \sim e^{\#TL/R_0^2} ,$$

where, the potential behaviour is as expected for respectively conformal (perimeter law) and confining (area law) cases. Note that there is interesting information in the coupling dependent numbers here denoted by # .

## 3. Supergravity Duals of Scattering Amplitudes

Interestingly enough, high energy amplitudes in gauge field theories can be related to other related configurations of minimal surfaces [3]. At high energy, fast moving colour sources propagate along linear trajectories in coordinate space thanks to the eikonale approximation. An analytic continuation from Minkowski to Euclidean $\mathcal{R}^4$ space allows one to find a geometrical interpretation in terms of a well-defined minimal surface problem. Let us consider for illustration different applications.

### 3.1. "Quark" elastic scattering

Calling "(anti)quarks" the colour sources in the (anti)fundamental representation of SU(N), the high-energy elastic quark-(anti)quark amplitude can be written [4]

$$A(s, q^2) = 2is \int d\vec{l}\, e^{i\vec{q}\cdot\vec{l}}\, \langle W_1 W_2 \rangle_{L=|\vec{l}|}^{\chi=\log s/m^2} ,$$

where $\vec{l}$ is the impact parameter between the two trajectories, conjugated to the momentum transfer $\vec{q}$, $\chi$ the total rapidity interval. Performing an analytic continuation to the Euclidean space:

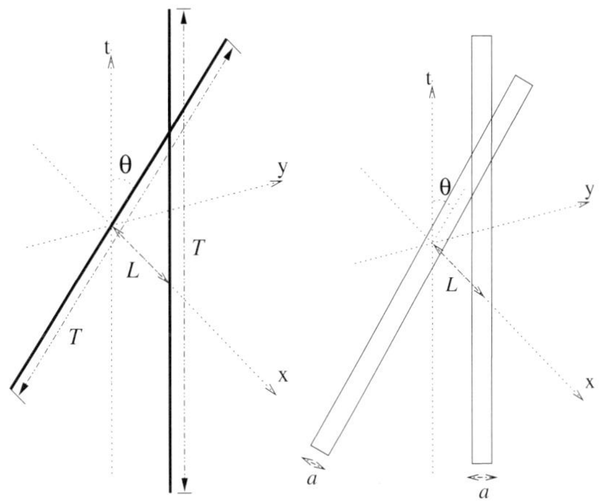

Figure 4. Wilson lines for "quark" and dipole elastic scattering in $\mathcal{R}^4$.

$$\chi \to i\theta \quad ; \quad t_{Mink} \to -it_{Eucl} ,$$

the Wilson line *vev* can be expressed as a minimal surface problem whose boundaries are two straight lines with an angle $\theta$ in $\mathcal{R}^4$, see Fig.4. In flat space, with the same boundary conditions, the minimal surface is the helicoid. One thus realizes that the problem can be formulated as a minimal surface problem whose mathematical formal solution is a *generalized helicoidal* manifold embedded in Euclidean AdS Spaces.

### 3.2. Dipole elastic scattering

For elastic scattering of colourless states, it is interesting to consider QCD dipoles, which are known to be a good toy model. Their propagation in coordinate space within the eikonale approximation can be represented by elongated Wilson loops near both right and left moving light-cone directions. Using the same analytic continuation framework, one has to compute the Wilson loop correlator in the configuration displayed in Fig.4.

The minimal area solution with the corresponding boundary conditions is difficult to find in analytic form, necessary for the continuation to Minkowski space. Some approximation schemes may be used [2]. A quite general and intringuing

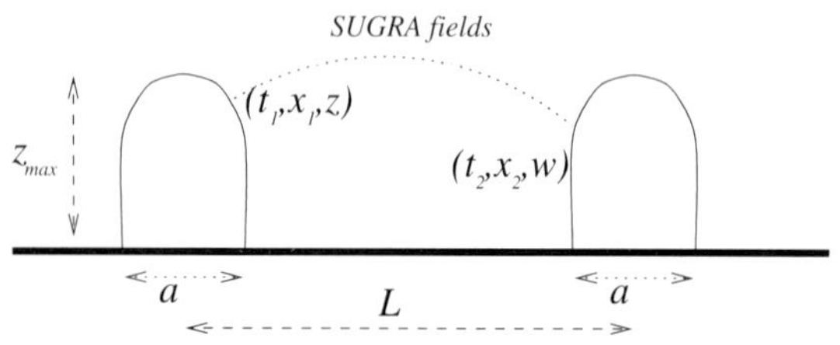


Figure 5. Dipole scattering at large impact parameter.

feature is the existence of a geometrical transition between small and large impact parameter, corresponding to the realization of disconnected minimal surfaces, see Fig.5.

### 3.3. Dipole inelastic scattering

The application of AdS/CFT correspondence for the two previous exemples is not so easy, even if partial results are encouraging. For "quark" elastic scattering, an infra-red time-like cut-off is to be introduced due to the colour charges of the quarks which implies a regularization scheme and a complication of the geometrical aspects. For dipole elastic scattering, there is no need for a cut-off but the geometry of the minimal surface is complicated. Inelastic scattering of dipoles allows one to circumvent these difficulties. Indeed, the helicoidal geometry remains valid due to the eikonale approximation for the "spectator quarks" while the "exchanged quarks" define a trajectory drawn on the helicoid, see Fig.6. This trajectory plays the rôle of a dynamical time-like cut-off which takes part in the minimization procedure.

Adopting the "world-line" path integral scheme of Feynman [5], one may write the inelastic amplitude in terms of a Wilson loop *vev*:

$$\int \mathcal{D}\tau \ \langle W(1\to 3'\to 4'\to 2'\to 1')\rangle_{\mathcal{A},[\tau]} \ e^{-2m\mathcal{L}[\tau]} ,$$

where $\tau$ parametrizes the boundary trajectories and $\mathcal{L}$ is their total length. Using the AdS-CFT correspondence in the same framework as previously, one may formally integrate over the gauge degrees of freedom and write

$$\langle W(1\to 3'\to 4'\to 2'\to 1')\rangle_{\mathcal{A},[\tau]} = e^{-\frac{Area[\tau]}{2\pi\alpha'}} \times Fluct.$$

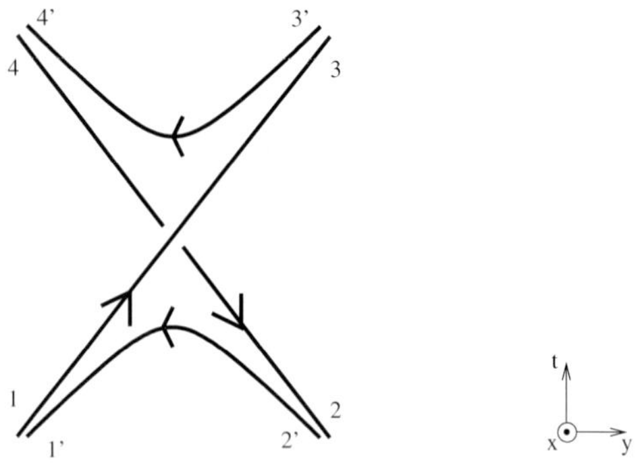


Figure 6. Wilson lines for inelastic dipole scattering

Note that the remaining minimization in $\tau$ runs now on both the area and its boundary.

## 4. Reggeization from the geometry of its AdS dual

To give a practical exemple of calculation we focus on the configuration of Wilson lines of Fig.6 in the context of a confining theory, dual to the $AdS_{BH}$ case with a limiting horizon, see Fig.3. The $AdS_{BH}$ metrics, as well as the canonical $AdS_5$ one, are caracterized by a singularity at $z = 0$ which implies a rapid growth in the $z$ direction towards the $D_3$ branes, then stopped near the horizon at $z_0$. Thus, to a good approximation, and for large enough impact parameter (compared to the horizon distance), the main contribution to the minimal area is from the metrics in the bulk near $z_0$ which is nearly flat. Hence, near $z_0$, the relevant minimal area can be drawn on a classical helicoid . Parametrizing the helicoid:

$$\begin{aligned} t &= \tau \cos\theta\sigma/L \\ y &= \tau \sin\theta\sigma/L \\ x &= \sigma \\ z &\sim z_0 \end{aligned}$$

the solution of the amplitude boils down to an Euler-Lagrange minimization over $\tau$, namely

$$\begin{aligned} A_{\mathcal{R}}(s, L^2) \ &\propto \ \frac{1}{s} \lim_{\alpha'\to 0} \int \mathcal{D}\tau \, e^{-\frac{1}{2\pi\alpha'} Area[\tau]} \times \\ &\times \ e^{-2m\mathcal{L}[\tau]} \times Fluct. , \end{aligned}$$

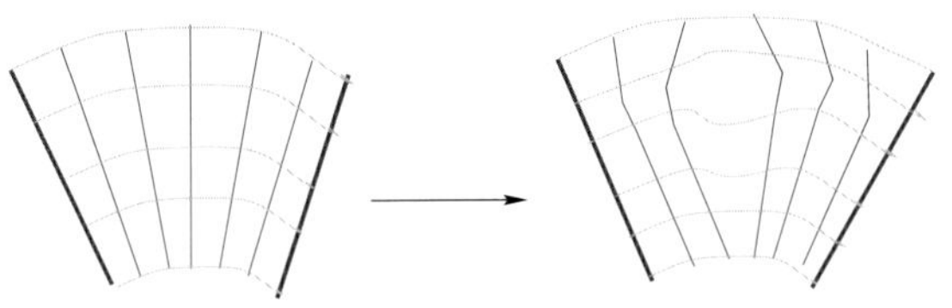

Figure 7. Fluctuations around the minimal surface

where $Area[\tau]$ is the section of an helicoid bounded by the quark trajectories having total length $\mathcal{L}[\tau]$. Note that the kinematical factor $1/s$ in front comes from an infinite-dimensional fermionic spin factor along the quark trajectories corresponding to a bosonized representation [5]. Its non-trivial calculation makes use of the 3-dimensional embedding of the quark trajectories on the helicoid surface [2].

After some technical steps and analytic continuation back to Minkowski space, the resulting amplitude reads:

$$A_{\mathcal{R}}(s,q^2) = \int d\vec{l}\, e^{i\vec{q}\cdot\vec{l}}\, e^{-\frac{L^2}{4\alpha'_{eff}\chi}} \propto s^{-\alpha'_{eff}q^2} ,$$

corresponding to a linear Regge trajectory with intercept 0 and slope $\alpha'_{eff}$ related to the quark potential calculated within the same AdS/CFT framework.

A semi-classical correction comes from the fluctuations near the minimal surface sketched in Fig.7. It can be shown to be intimately related to a contribution to the quark potential similar to the well-known Lüscher term [6]. One finally finds

$$A_{\mathcal{R}}(s,q^2) \propto s^{\frac{n_\perp}{24}-\alpha'_{eff}q^2} ,$$

where $n_\perp$ is the number of transverse zero-modes (zero-mass transverse excitations of the string) in the AdS dual theory. It gives a shift in the intercept of the reggeon trajectory.

The value of this intercept depends on the particular realization of the AdS/CFT duality. In known examples (see [2]) it is quoted to be $n_\perp = 7$ *or* 8, , to be compared with the ordinary Lüscher term having $n_\perp = 2$.. Note that fermionic d.o.f. are not expected to remain massless in an AdS background, and thus to give contributions with opposite sign.

It is worth comparing the resulting *inelastic* dipole amplitude with the one obtained for dipole *elastic* scattering in the same framework. With approximations required by to the non-trivial geometry of dipole loops one finds:

$$A_{\mathcal{P}}(s,q^2) \sim s^{1+\frac{n_\perp}{96}-\frac{\alpha'_{eff}}{4}q^2} ,$$

with the same values of $\alpha'_{eff}$ and $n_\perp$. Reggeon and Pomeron Regge trajectories are thus linear and related to the quark potential. Note the factor $1/4$ in the exponent of $A_{\mathcal{P}}$ with respect to $A_{\mathcal{R}}$ which has some consistency with actual amplitudes, since it gives phenomenologically consistent Pomeron and (average) dominant Reggeon trajectories [2].

## 5. Conclusion: Reggeization and confinement

Lattice calculations, which is the only presently known way to evaluate directly QCD observables at strong coupling, are not able to compute high-energy amplitudes. hence, an interesting output of the application of AdS/CFT correspondence to high energy amplitudes at strong coupling is to discuss the relation between Reggeization and confinement, using the description in the dual theory.

When comparing $AdS_5$ duality - which corresponds to a conformal, non-confining gauge theory - with $AdS_{BH}$ duality, which leads to reggeization, the difference ultimately comes from the different metrics in the bulk and hence from the minimal surfaces for the same boudary conditions. Taking into account their different geometry, see e.g. Fig.3, one finds after analytic continuation and in the large energy ($\chi \to \infty$) limit:

$$Area^{AdS}_{min} \sim \lim_{\chi\to\infty} \frac{L}{L/\chi} \;;\; Area^{BH}_{min} \sim \lim_{\chi\to\infty} L \times \frac{L}{\chi} .$$

As expected, the $AdS_5$ case leads to a $L$-invariant value and, after Fourier transformation, to a high-energy amplitude with a $q^2$ independent energy exponent (or flat Regge trajectory). On the other hand, the $AdS_{BH}$ case leads to a linear Regge trajectory after Fourier transformation. Hence confinement appears as an essential ingredient for the reggeized structure of high-energy amplitudes. We expect this result not to be dependent on the precise geometrical $AdS_{BH}$ setting and thus to indicate a quite general property of confining theories.

As a conclusion, let us discuss the list of problems for which the AdS/CFT framework give new insights on the 35-years-old puzzle of high-energy amplitudes at strong gauge coupling. As usual this also adds some new problems in this context! Among the new insights:

- *String dimensionality:* $D = 4+1+5$ : Extra-dimensions play an important dynamical rôle.
- *Gravitation:* It is decoupled. The (classical) Pomeron intercept is $1 + \epsilon$ instead of 2 for the graviton. $\epsilon$ is related to a Lüsher term.
- *Regge trajectories:* They come out linear, with slopes and intercepts related to the quark potential including a Lüscher term.

Among the problems:

- *High-energy phenomenology:* Many aspects, like the Flavor/Spin dependences, remain to be studied.
- *Approximations:* The dual gauge theory is not specified, and the exact minimal surface in the bulk metrics to be determined.
- *Dual of QCD?* In the present framework, the confining scale $R_0$ has no relation with $\Lambda_{QCD}$.
- *Unitarity:* A more complete investigation requires the study of multi-leg amplitudes.
- *Deeper general problems:* The formulation of string theory in AdS backgrounds and last but not least, a proof of the AdS/CFT conjecture.

## 6. Acknowledgement

The whole approach of high-energy amplitudes described in this review paper comes from a collaboration with Romuald Janik from Cracow University.

## REFERENCES

1. J. Maldacena, *Adv. Theor. Math. Phys.* **2** (1998) 231; S.S. Gubser, I.R. Klebanov and A.M. Polyakov, *Phys. Lett.* **B428** (1998) 105; E. Witten, *Adv. Theor. Math. Phys.* **2** (1998) 253; , *Adv. Theor. Math. Phys.* **2** (1998) 505. O. Aharony, S.S. Gubser, J. Maldacena, H. Ooguri and Y. Oz, *Phys.Rept.* **323** (2000)183.
2. J. Maldacena, *Phys. Rev. Lett.* **80** (1998) 4859; S.-J. Rey and J. Yee, hep-th/9803001; JJ. Sonnenschein and A. Loewy, *JHEP* **0001** (2000) 042. Y. Kinar, E. Schreiber, J. Sonnenschein and N. Weiss, *Nucl. Phys.* **B583** (2000) 76.
3. R.A. Janik and R. Peschanski, *Nucl. Phys.* **B586** (2000) 163 , *Nucl. Phys.* **B565** (2000) 193 , *Nucl. Phys.* **B565** (2000) 193; R.A. Janik, *Phys. Lett.* **B500** (2001) 118. See also: M. Rho, S.-J. Sin and I. Zahed, *Phys. Lett.* **B466** (1999) 199.
4. O. Nachtmann, hep-ph/9609365. E. Meggiolaro, *Z. Phys.* **C76** (1997) 523.
5. R.P. Feynman, *Phys. Rev.* **80** (1950) 440 A. M. Polyakov, *Mod. Phys. Lett.* **A3** (1988) 325. G. P. Korchemsky, *Int. J. Mod. Phys.* **A7** (1992) 339; , *Phys. Lett.* **B232** (1989) 334.
6. M. Lüscher, K. Symanzik and P. Weisz, *Nucl.Phys.* **B173** (1980) 365. O. Alvarez, *Phys.Rev.* **D24** (1981) 440.

ELSEVIER

Proceedings of ICHEP 2002, pp. 855–860
*S. Bentvelsen, P. de Jong, J. Koch and E. Laenen (Editors)*

31st INTERNATIONAL CONFERENCE ON HIGH ENERGY PHYSICS AMSTERDAM

# (Non-)Abelian Gauged Supergravities in Nine Dimensions

E. Bergshoeff[a], T. de Wit[a], U. Gran[a], R. Linares[a] and D. Roest[a]

[a]Centre for Theoretical Physics, University of Groningen,
Nijenborgh 4, 9747 AG Groningen, The Netherlands

We construct five massive deformations of the unique nine–dimensional $N = 2$ supergravity, each with two parameters. All of these deformations have a higher–dimensional origin via Scherk–Schwarz reduction and correspond to gauged supergravities. The gauge groups we encounter are $SO(2), SO(1,1)^+, \mathbb{R}, \mathbb{R}^+$ and the unique two–dimensional non–Abelian Lie group $C\mathbb{R}^1$, which consists of scalings and translations in one dimension.

We make a systematic search for two classes of vacuum solutions: maximally symmetric solutions with constant scalars and half-supersymmetric domain wall solutions. In the first category we find explicit solutions in the form of (non-supersymmetric) de Sitter space solutions. In the second category we find precisely the three classes of domain wall solutions that were given in an earlier work.

## 1. Introduction

The procedure of gauging a global symmetry includes the replacement of the ordinary derivative by a covariant derivative:

$$\partial_\mu \longrightarrow D_\mu = \partial_\mu + gA_\mu \,. \tag{1}$$

Here $A_\mu$ is the gauge field and $g$ is the gauge coupling constant which acts as a deformation parameter of the ungauged theory. In the case of Einstein gravity with scalars one can consider as an independent deformation the addition of a scalar potential $V(\varphi)$:

$$R + (\partial\varphi)^2 \rightarrow R + (\partial\varphi)^2 + m^2 V(\varphi) \,. \tag{2}$$

In the supersymmetric case, i.e. the case of *gauged* supergravity, the two deformations are not independent. Supersymmetry relates the two deformation parameters:

$$g = m \,. \tag{3}$$

Due to the scalar potential the Minkowski spacetime is no longer a maximally supersymmetric vacuum solution of the gauged supergravity. Instead we will search for other vacuum solutions, like, e.g., non-supersymmetric de Sitter space solutions. A natural class of *half-supersymmetric* vacuum solutions that makes use of the scalar potential is the set of domain wall solutions. Recently, domain wall solutions of supergravity theories have attracted attention in view of their relevance for a supersymmetric Randall-Sundrum scenario [1,2], the domain wall/QFT correspondence [3,4] and applications to cosmology [5,6]. In all these applications the properties of the domain walls play a crucial role and these properties are determined by the details of the scalar potential.

Motivated by this we studied general domain wall solutions in D=9 dimensions [7]. We took D=9 because on the one hand this case shares some of the complexities of the lower-dimensional cases, on the other hand the scalar potential for this case is simple enough to study the corresponding domain wall solutions in full detail. The supergravity theory we considered in [7] was obtained by a generalized Scherk-Schwarz (SS) reduction of D=10 IIB supergravity. This is not the most general possibility in D=9. In this talk we will present a systematic search for massive deformations of the unique D=9, N=2 supergravity theory. All deformations we find correspond to *gauged* supergravities. The hope is that the D=9 case will teach us something about the more complicated situation in D $<$ 9 dimensions. The results presented in this talk are taken from [8].


doi:10.1016/S0920-5632(02)02246-6

## 2. Massive deformations of D=9, $N = 2$ Supergravity

The field content of the unique $D = 9$, $N = 2$ massless supergravity theory is given by $(i = 1, 2)$

$$e_\mu{}^a, \phi, \varphi, \chi, A_\mu, A_\mu^{(i)}, B_{\mu\nu}^{(i)}, C_{\mu\nu\rho}, \psi_\mu, \lambda, \tilde{\lambda}\,. \tag{4}$$

The massless 9–dimensional theory has four global scaling symmetries, with parameters $\alpha, \beta, \gamma$ and $\delta$, respectively. The scaling weights of all these symmetries are given in Table 1.

It turns out that only three out of the four scaling symmetries given in Table 1 are linearly independent. There is a relation

$$\tfrac{4}{9}\alpha - \tfrac{8}{3}\beta = \gamma + \tfrac{1}{2}\delta\,. \tag{5}$$

The massless N=2, D=9 theory also has an $SL(2,\mathbb{R})$ symmetry whose explicit rules can be found in [8].

We now turn to massive deformations of the 9D theory. To obtain these deformations we will apply a SS reduction which can be best illustrated by an example. Consider a single scalar field:

$$\hat{\mathcal{L}} = \sqrt{-\hat{g}}\,(\partial\hat{\varphi})^2\,, \tag{6}$$

which is invariant under the $\mathbb{R}$–symmetry $\varphi \to \varphi + c$. In the SS procedure one gives the field a dependence on the compactification coordinate $z$ which is governed by a global symmetry, in this case the $\mathbb{R}$–symmetry:

$$\hat{\varphi}(x, z) = \varphi(x) + mz\,. \tag{7}$$

Using the standard reduction rules the Lagrangian reduces to

$$\mathcal{L} = \sqrt{-g}\left((D\varphi)^2 + m^2\right)\,, \tag{8}$$

where $D_\mu\phi = \partial_\mu\varphi - mA_\mu$ with $A_\mu$ being the Kaluza-Klein vector.

Applying the above outlined SS dimensional reduction we obtain a number of massive deformations in nine dimensions, as illustrated in Figure 1. By employing the different global symmetries of 11D, IIA and IIB supergravity we obtain

Table 1: *The scaling weights of the 9 dimensional supergravity fields.*

| | $e_\mu{}^a$ | $e^\phi$ | $e^\varphi$ | $\chi$ | $A_\mu$ | $A_\mu^{(1)}$ | $A_\mu^{(2)}$ | $B_{\mu\nu}^{(1)}$ | $B_{\mu\nu}^{(2)}$ | $C_{\mu\nu\rho}$ | $\psi_\mu$ | $\lambda$ | $\tilde{\lambda}$ |
|---|---|---|---|---|---|---|---|---|---|---|---|---|---|
| $\alpha$ | $\frac{9}{7}$ | 0 | $\frac{6}{\sqrt{7}}$ | 0 | 3 | 0 | 0 | 3 | 3 | 3 | $\frac{9}{14}$ | $-\frac{9}{14}$ | $-\frac{9}{14}$ |
| $\beta$ | 0 | $\frac{3}{4}$ | $\frac{\sqrt{7}}{4}$ | $-\frac{3}{4}$ | $\frac{1}{2}$ | $-\frac{3}{4}$ | 0 | $-\frac{1}{4}$ | $\frac{1}{2}$ | $-\frac{1}{4}$ | 0 | 0 | 0 |
| $\gamma$ | 0 | $-2$ | 0 | 2 | 0 | 1 | $-1$ | 1 | $-1$ | 0 | 0 | 0 | 0 |
| $\delta$ | $\frac{8}{7}$ | 0 | $-\frac{4}{\sqrt{7}}$ | 0 | 0 | 2 | 2 | 2 | 2 | 4 | $\frac{4}{7}$ | $-\frac{4}{7}$ | $-\frac{4}{7}$ |

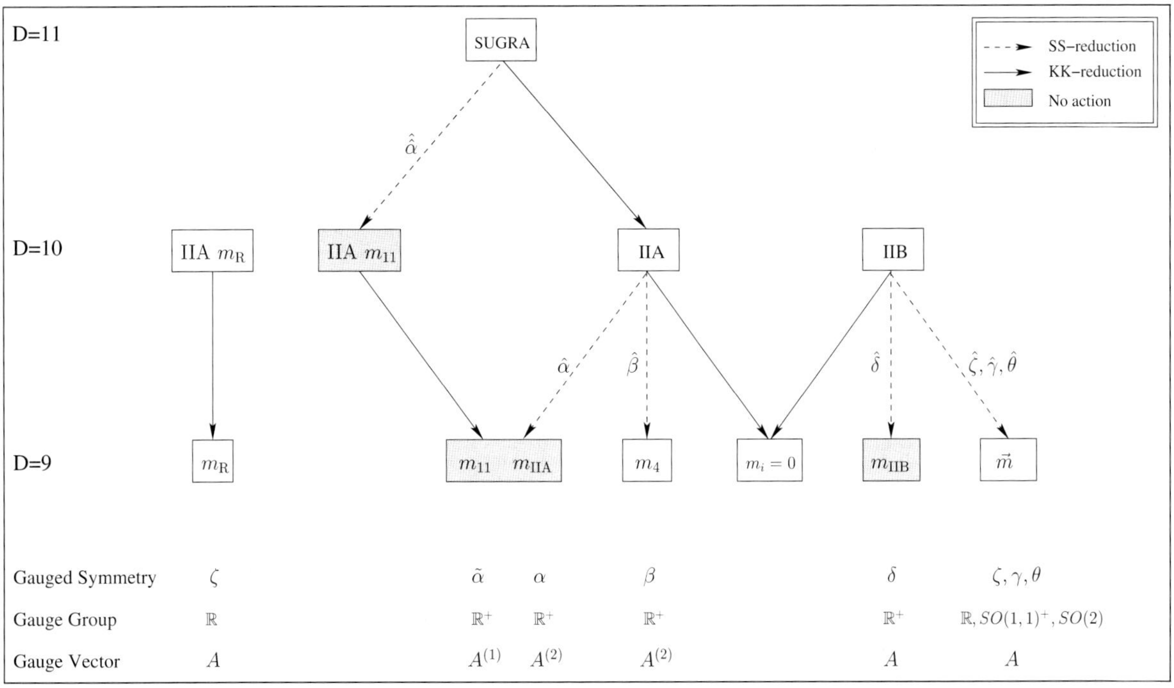


Figure 1. *Overview of all reductions discussed in this talk. These cases can all be interpreted as gauged supergravities, with gauged symmetry and corresponding gauge field as given in the Figure. Mass parameters in the same box, such as $m_{11}, m_{IIA}$ or $m_1, m_2, m_3$, form a multiplet under $SL(2, \mathbb{R})$.*

seven deformations of the unique $D = 9$ supergravity.

Note that the different massive deformations can be related. Symmetries of the massless theory become field redefinitions in the massive theory that only act on the massive deformations. This means that the mass parameters transform under such transformations: they have a scaling weight under the different scaling symmetries and fall in multiplets of $SL(2,\mathbb{R})$. In Table 2 the multiplet structure of the massive deformations under $SL(2,\mathbb{R})$ is given. The mass parameter $\tilde{m}_4$ is defined as the S-dual partner of $m_4$ and can not be obtained by a SS reduction of IIA supergravity. All these deformations correspond to a

Table 2
*This tables indicates the different multiplets that the D=9 mass parameters form under $SL(2,\mathbb{R})$.*

| mass parameters | $SL(2,\mathbb{R})$ |
|---|---|
| $(m_1, m_2, m_3)$ | triplet |
| $(m_4, \tilde{m}_4)$ | doublet |
| $(m_{11}, m_{\rm IIA})$ | doublet |
| $m_{\rm IIB}$ | singlet |

gauging of a 9D global symmetry. In particular, it is always the symmetry that is employed in the SS reduction Ansatz that becomes gauged upon reduction. The corresponding gauge vector is always provided by the metric, i.e. is the Kaluza–Klein vector of the dimensional reduction. In all but one case this is the complete story and one finds an Abelian gauged supergravity. It turns out that there is one exception, i.e. the case with $m_4 \neq 0$, where we find a *non-Abelian* gauge symmetry. The (non-semi-simple) gauge group is $C\mathbb{R}^1$, the group of scalings and translations of the real line. Further details of the different massive deformations can be found in [8].

## 3. Combining Massive Deformations

We next try to combine the different massive deformations we found above. Requiring that the fermionic field equations transform under supersymmetry to a complete set of bosonic field equations restricts us to five cases, each containing two nonzero mass parameters:

- **Case 1** with $\{m_{\rm IIA}, m_4\}$: this combination can also be obtained by Scherk-Schwarz reduction of IIA employing a linear combination of the symmetries $\hat{\alpha}$ and $\hat{\beta}$, guaranteeing its consistency. It is also a gauging of both this symmetry and (for $m_4 \neq 0$) the parabolic subgroup of $SL(2,\mathbb{R})$ in 9D, giving the non-Abelian gauge group $C\mathbb{R}^1$.

- **Case 2,3,4** with $\{\vec{m}, m_{\rm IIB}\}$: as in the case with $m_{\rm IIB} = 0$ and only $\vec{m}$ this combination contains three different, inequivalent cases depending on $\vec{m}^2$ (depending crucially on the fact that $m_{\rm IIB}$ is a singlet under $SL(2,\mathbb{R})$):

  - **Case 2** with $\{\vec{m}, m_{\rm IIB}\}$ and $\vec{m}^2 = 0$.
  - **Case 3** with $\{\vec{m}, m_{\rm IIB}\}$ and $\vec{m}^2 > 0$.
  - **Case 4** with $\{\vec{m}, m_{\rm IIB}\}$ and $\vec{m}^2 < 0$.

  All these combinations can also be obtained by Scherk-Schwarz reduction of IIB employing a linear combination of the symmetries $\hat{\delta}$ and (one of the subgroups of) $SL(2,\mathbb{R})$, guaranteeing its consistency. All cases (assuming that $m_{\rm IIB} \neq 0$) correspond to the gauging of an Abelian non-compact symmetry in 9D. Only in the special case $\vec{m}^2 < 0, m_{\rm IIB} = 0$ corresponds to a SO(2)–gauging.

- **Case 5** with $\{m_4 = -\frac{12}{5} m_{\rm IIA}, m_2 = m_3\}$: this case can be understood as the generalized dimensional reduction of Romans' massive IIA theory, employing the $\mathbb{R}^+$ symmetry that is not broken by the $m_{\rm R}$ deformations: $\hat{\beta} - \frac{5}{12}\hat{\alpha}$. It gauges both this linear combination of $\mathbb{R}^+$'s and the parabolic subgroup of $SL(2,\mathbb{R})$ in 9D, giving a non-Abelian gauge group provided $m_4 \neq 0$.

All five cases are gauged theories and have a higher–dimensional origin. Both case 1 and case 5 have a non-Abelian gauge group provided $m_4 \neq 0$.

## 4. Solutions

We have constructed a variety of gauged supergravities with 32 supersymmetries. They all have in common that there is a scalar potential. Our next goal is to make a systematic search for solutions that are based on this scalar potential. In the next Subsections we will search for two types of solutions: (i) 1/2 BPS domain wall (DW) solutions and (ii) maximally symmetric solutions with constant scalars, i.e. de Sitter (dS), Minkowski (Mink) or anti–de Sitter (AdS) solutions.

### 4.1. 1/2 BPS Domain Wall Solutions

In [7] we already made a systematic search for half-supersymmetric Domain Wall (DW) solutions of the gauged supergravities corresponding to the cases 2, 3 and 4 (with $m_{\rm IIB} = 0$). Due to a one-to-one relationship with 7-branes in D=10 dimensions [9] we could even make a systematic investigation of the quantization of the mass parameters by using the results of [10,11].

We now want to investigate whether the five massively deformed supergravities we found in the previous Section allow new half-supersymmetric DW solutions. Since we are looking for 1/2 BPS solutions it is convenient to solve the Killing spinor equations, which are obtained by setting the supersymmetry variation of the gravitino and dilatinos to zero. The projector[1] for a DW is given by $\frac{1}{2}(1 \pm \gamma_y)$, where $y$ denotes the transverse direction. We find that, in order to make a projection operator in the Killing spinor equations, we are forced to set all mass parameters to zero except for $\vec{m}$, which corresponds to cases 2, 3 and 4 of Section 3 with $m_{\rm IIB} = 0$. This is a consistent combination of masses and we obtain three classes of domain wall solution which were discussed in detail in [7].

To summarize, we find that there are no new codimension-one 1/2 BPS solutions to the D=9 supergravity theories we obtained in the previous Sections, as compared to the three classes of domain wall solutions given in [7].

### 4.2. Maximally Symmetric Solutions with Constant Scalars

The second category of vacuum solutions we consider are the solutions with all three scalars constant. This is a consistent truncation in two cases which both have two mass parameters. In this truncation one is left with the metric only satisfying the Einstein equation with a cosmological term

$$R_{\mu\nu} - \tfrac{1}{2} g_{\mu\nu} R = -\Lambda g_{\mu\nu}\,, \tag{9}$$

with $\Lambda$ quadratic in the two mass parameters. Depending on the sign of this term one thus has anti-de Sitter, Minkowski or de Sitter geometry.

We find that solutions with constant scalars are possible in the following massive supergravities:

- **D=10** with $\{m_{11}\}$ has $\Lambda = +36 {m_{11}}^2 e^{-3\hat{\phi}/2}$, which gives rise to de Sitter$_{10}$ [12], breaking all supersymmetry. The D=11 origin of this solution is Mink$_{11}$ written in a basis where the $x$–dependence is of the required form [12]:

$$ds^2 = e^{2m_{11}x}\big(-dt^2 + e^{2m_{11}t} dx_9^2 + dx^2\big)\,. \tag{10}$$

- **D=9, Case 1** with $\{m_{\rm IIA} = -\frac{2}{3} m_4\}$ has $\Lambda = +\frac{63}{4} {m_4}^2 e^{\phi - 3\varphi/\sqrt{7}}$, which gives rise to De Sitter$_9$, breaking all supersymmetry. This case follows from the reduction of Mink$_{10}$ by using a combination of IIA scale symmetries that leave the dilaton invariant so that. This particular scale symmetry allows a SS reduction of a configuration with a zero dilaton sothat, after reduction, one is left with a non-trivial metric field only.

- **D=9, Case 4** with $\{m_{\rm IIB}, m_3\}$ has $\Lambda = +28 {m_{\rm IIB}}^2 e^{4\varphi/\sqrt{7}}$, which gives rise to de Sitter$_9$ for non-vanishing $m_{\rm IIB}$. This case follows from the reduction of Mink$_{10}$ by using a combination of IIB scale symmetries

[1] From a general analysis of the possible projectors in 9 dimensions, we find that there is a second projector given by $\frac{1}{2}(1 \pm i\gamma_t)$. This projector would give a *Euclidean* DW, i.e. a DW having time as a transverse direction. Note that such a Euclidean DW can never be 1/2 BPS since if there existed a Killing spinor it would square to a Killing vector in the *transverse* direction, i.e. time, which is not an isometry of the euclidean DW.

that leave the dilaton invariant. Note that for vanishing $m_{\rm IIB}$ this reduces to $\text{Mink}_9$, despite the presence of $m_3$ [13]. For either $m_{\rm IIB}$ or $m_3$ non-zero this solution breaks all supersymmetry.

## 5. Conclusions

We have constructed five different D=9 massive deformations with 32 supersymmetries, each containing two mass parameters. All these five theories have a higher–dimensional origin via SS reduction from D=10 dimensions. Furthermore, the massive deformations gauge a global symmetry of the massless theory. The gauge group we have obtained are the Abelian groups $SO(2), SO(1,1)^+, \mathbb{R}, \mathbb{R}^+$ and the unique two–dimensional non-Abelian Lie group $C\mathbb{R}^1$ of scalings and translations on the real line.

We have analyzed the possibility of combining massive deformations to obtain more general massive supergravities that are not gauged or do not have a higher–dimensional origin. Our analysis shows that the only possible combinations are the five two–parameter deformations, which are all gauged and can be uplifted. We have not made a systematic search for massive D=9 supergravities that are not the combination of gaugings and we cannot exclude that there are more possibilities. This requires a separate calculation.

Finally, not all gauged supergravities we constructed are necessarily the leading terms in a low-energy approximation to (compactified) superstring theory. The deformations with $\vec{m} \neq 0$ or $m_{\rm IIA} = \frac{1}{12}m_4 \neq 0$ definitely correspond to a sector of (compactified) string theory. The other cases are less clear sofar [8].

## REFERENCES

1. L. Randall and R. Sundrum, A large mass hierarchy from a small extra dimension, Phys. Rev. Lett. 83 (1999) 3370–3373.
2. L. Randall and R. Sundrum, An alternative to compactification, Phys. Rev. Lett. 83 (1999) 4690–4693.
3. H. J. Boonstra, K. Skenderis and P. K. Townsend, The domain wall/QFT correspondence, JHEP 01 (1999) 003.
4. K. Behrndt, E. Bergshoeff, R. Halbersma and J. P. van der Schaar, On domain wall/QFT dualities in various dimensions, Class. Quant. Grav. 16 (1999) 3517–3552,
5. R. Kallosh, A. D. Linde, S. Prokushkin and M. Shmakova, Gauged supergravities, de Sitter space and cosmology, Phys. Rev. D65 (2002) 105016
6. P. K. Townsend, Quintessence from M-theory, JHEP 11 (2001) 042.
7. E. Bergshoeff, U. Gran and D. Roest, Type IIB seven-brane solutions from nine-dimensional domain walls, Class. Quant. Grav. 19 (2002) 4207–4226.
8. E. Bergshoeff, T. de Wit, U. Gran, R. Linares, and D. Roest, (Non-)Abelian Gauged Supergravities in Nine Dimensions, hep-th/0209205.
9. P. Meessen and T. Ortín, An Sl(2,Z) multiplet of nine-dimensional type II supergravity theories, Nucl. Phys. B541 (1999) 195–245.
10. O. DeWolfe, T. Hauer, A. Iqbal and B. Zwiebach, Uncovering the symmetries on (p,q) 7-branes: Beyond the Kodaira classification, Adv. Theor. Math. Phys. 3 (1999) 1785–1833.
11. O. DeWolfe, T. Hauer, A. Iqbal and B. Zwiebach, Uncovering infinite symmetries on (p,q) 7-branes: Kac-Moody algebras and beyond, Adv. Theor. Math. Phys. 3 (1999) 1835–1891.
12. I. V. Lavrinenko, H. Lu and C. N. Pope, Fibre bundles and generalised dimensional reductions, Class. Quant. Grav. 15 (1998) 2239–2256.
13. J. Gheerardyn and P. Meessen, Supersymmetry of massive D = 9 supergravity, Phys. Lett. B525 (2002) 322–330.

ELSEVIER

Proceedings of ICHEP 2002, pp. 861–866
*S. Bentvelsen, P. de Jong, J. Koch and E. Laenen (Editors)*

31st INTERNATIONAL CONFERENCE ON HIGH ENERGY PHYSICS AMSTERDAM

# What are Instantons made of?

C. Ford[a]*

[a]Instituut-Lorentz for Theoretical Physics,
Niels Bohrweg 2, 2300 RA Leiden, The Netherlands

It is well known that certain finite temperature instantons, or calorons, can be viewed as bound states of several monopole constituents. This suggests that, in general, periodic instantons can be constructed out of 'elementary' constituents. It is argued that recent studies of doubly periodic instantons lend further support to this supposition.

## 1. INTRODUCTION

Over 25 years ago 't Hooft and Mandelstam advanced a convincing, albeit qualitative, 'dual superconductivity' picture of confinement in QCD. Here a magnetic monopoles were supposed to play the role of the Cooper pairs in the BCS theory. The ensuing magnetic Meißner effect would then enforce the confinement of quarks. Unfortunately, efforts to put this picture onto a more quantitative footing have been hampered by the lack of a proper definition of such monopole degrees of freedom. Following 't Hooft, one can identify gauge singularities of particular gauge fixings (abelian gauges) with magnetic monopoles. However, the monopole content of a Yang-Mills potential depends on the choice of gauge fixing. Is there a gauge independent way to formulate monopoles in pure Yang-Mills theory? While there is, as yet, no definitive answer, some progress has been made with regard to instantons. In particular, certain finite temperature instantons, or calorons, can be viewed as bound states of several monopole constituents [1,2]. This identification does not hinge on a particular gauge choice as the monopoles are clearly visible as peaks in the action density.

The monopole constituent picture is most transparent within a formalism due to Nahm. Here calorons, that is self-dual gauge potentials on $S^1 \times R^3$ are mapped to self-dual potentials on a circle. In the one-instanton sector $SU(N)$ calorons are mapped to simple abelian $U(1)$ potentials. Moreover these potentials are *piecewise constant.* If one executes the Nahm transform on a BPS monopole a constant potential is produced; the spatial components determine the position of the monopole. Thus each interval for which the Nahm potential is constant can be identified with a BPS monopole constituent. However, in the higher topological charge sectors, the Nahm potentials are non-abelian and, in general, not piecewise constant. Therefore, one cannot simply read off the monopole content from the Nahm potential. A similar situation occurs for multiply periodic instantons, where the spacetime is taken to be $T^n \times R^{4-n}$, $n > 1$. Under a generalisation of Nahm's transform, multiply periodic instantons are mapped to self-dual potentials on the *dual torus* $\tilde{T}^n$ which is the torus obtained by inverting the periods of $T^n$. As for calorons, charge-one instantons are mapped to abelian potentials, but these are never piecewise constant. Some recent results regarding the $n = 2$ case, i.e. doubly periodic instantons, suggests that these objects also are built from elementary constituents [3]. These constituents are not static; they can be thought of as overlapping instanton cores. Though for sufficiently high temperature they merge to form monopole-like structures. In this contribution these results are reviewed and possible extensions briefly outlined.

*Supported through a European Community Marie Curie Fellowship, contract HPMF-CT-2000-00841.


doi:10.1016/S0920-5632(02)02248-X

## 2. THE NAHM TRANSFORM

The Nahm transform is easiest to formulate on the four torus $T^4$ [4]. So before we specialise to $T^2 \times R^2$ it is instructive to start with the $T^4$ version. Consider a self-dual anti-hermitian $SU(N)$ potential, $A_\mu(x)$, on a euclidean four-torus with topological charge $k$. A gauge field on $T^4$ is understood to be an $R^4$ potential which is periodic (modulo gauge transformations) with respect to $x_\mu \to x_\mu + L_\mu$ $(\mu = 0, 1, 2, 3)$, the $L_\mu$ being the four periods of the torus. The next, apparently trivial, step is to turn the $SU(N)$ instanton into a $U(N)$ instanton by adding a constant $U(1)$ potential, $A_\mu(x) \to A_\mu(x) - iz_\mu$, where the $z_\mu$ are real numbers. We can regard the $z_\mu$ as coordinates of the *dual torus* , $\tilde{T}^4$, since the shifts $z_\mu \to z_\mu + 2\pi/L_\mu$ can be effected via *periodic* $U(1)$ gauge transformations.

Now consider the $U(N)$ Weyl operator

$$D_z(A) = \sigma_\mu D^\mu_z(A), \qquad (1)$$

with $D^\mu_z(A) = \partial^\mu + A^\mu(x) - iz^\mu$, and $\sigma_\mu = (1, i\tau_1, i\tau_2, i\tau_3)$, the $\tau_i$ being Pauli matrices. Provided certain mathematical technicalities are met $D^\dagger_z(A) = -\sigma^\dagger_\mu D^\mu_z(A)$ has $k$ square-integrable zero modes $\psi^i(x; z)$ with $i = 1, 2, ..., k$. The Nahm potential is defined as

$$\hat{A}^{ij}_\mu(z) = \int_{T^4} d^4x\, \psi^{i\dagger}(x; z) \frac{\partial}{\partial z^\mu} \psi^j(x; z). \qquad (2)$$

Here the zero modes are taken to be orthonormal. Remarkably, $\hat{A}(z)$ is a $U(k)$ instanton on the dual torus with topological charge $N$. If we execute a second Nahm transformation on $\hat{A}$, the original $SU(N)$ instanton will be recovered.

Formally, one can obtain the $T^2 \times R^2$ Nahm transform by taking two of the periods, say $L_0$ and $L_3$, to be infinite. Given an $SU(N)$ instanton periodic with respect to $x_1 \to x_1 + L_1$, $x_2 \to x_2 + L_2$ its Nahm transform is

$$\hat{A}^{ij}_\mu(z) = \int_{T^2 \times R^2} d^4x\, \psi^{i\dagger}(x; z) \frac{\partial}{\partial z^\mu} \psi^j(x; z), \qquad (3)$$

when $\mu = 1, 2$, and

$$\hat{A}^{ij}_\mu(z) = \int_{T^2 \times R^2} d^4x\, \psi^{i\dagger}(x; z) i x_\mu \psi^j(x; z), \qquad (4)$$

if $\mu = 0, 3$. Here the $\psi^i(x; z)$ with $i = 1, ..., k$ are orthonormal zero modes of $D^\dagger_z(A)$. Note that we may gauge $z_0$ and $z_3$ to zero. The Nahm potential is a self-dual $U(k)$ gauge field with $N$ singularities in $\tilde{T}^2$ (periods $2\pi/L_1$ and $2\pi/L_2$).

## 3. CHARGE ONE INSTANTONS

One can regard $\hat{A}_1(z)$ and $\hat{A}_2(z)$ as the components of a two dimensional gauge potential, and combine $\hat{A}_0(z)$ and $\hat{A}_3(z)$ into a 'Higgs' field $\Phi(z) = \frac{1}{2}(\hat{A}_0(z) - i\hat{A}_3(z))$. The next step is to seek solutions of the dimensionally reduced self-duality (or Hitchin) equations. In the one-instanton sector this is rather straightforward since the corresponding Nahm potential is abelian. What is more tricky, however, is to execute the second Nahm transformation to recover the instanton itself. The simplest case corresponds to zero Higgs. This means that the associated instanton will be radially symmetric: local gauge invariants such as the action density depend on $x_1$, $x_2$ and $r = \sqrt{x_0^2 + x_3^2}$ only. With this restriction the self-duality equations are just

$$\hat{F}_{y\bar{y}} = \partial_y \hat{A}_{\bar{y}} - \partial_{\bar{y}} \hat{A}_y = 0. \qquad (5)$$

Here we have used complex coordinates; $y = z_1 + iz_2$, $\bar{y} = z_1 - iz_2$, $\partial_y = \frac{1}{2}(\partial_{z_1} - i\partial_{z_2})$, $\partial_{\bar{y}} = \frac{1}{2}(\partial_{z_1} + i\partial_{z_2})$, $\hat{A}_y = \frac{1}{2}(\hat{A}_1(z) - i\hat{A}_2(z))$ and $\hat{A}_{\bar{y}} = \frac{1}{2}(\hat{A}_1(z) + i\hat{A}_2(z))$. Consider the ansatz

$$\hat{A}_y = \partial_y \phi, \qquad \hat{A}_{\bar{y}} = -\partial_{\bar{y}} \phi, \qquad (6)$$

which gives $\hat{F}_{y\bar{y}} = -2\partial_y \partial_{\bar{y}} \phi$. Then, in order to satisfy (5), $\phi$ must be harmonic except at two singularities, for the simplest gauge group, $SU(2)$. A suitable $\phi$ satisfies

$$\left(\partial^2_{z_1} + \partial^2_{z_2}\right) \phi(z) = 2\pi\kappa \left[\delta^2(z + \omega) - \delta^2(z - \omega)\right] \qquad (7)$$

where $\kappa$ is a constant and $\pm\omega$ are the positions of the two singularities (we have used translational invariance to shift the 'centre of gravity' of the singularities to the origin). The delta functions should be read as periodic (with respect to $z_1 \to z_1 + 2\pi/L_1$ and $z_2 \to z_2 + 2\pi/L_2$). Physically, the Nahm potential describes two Aharonov-Bohm fluxes of strength $\kappa$ and $-\kappa$ threading the dual torus. They must have equal and opposite strength to ensure a periodic $\hat{A}$. We may assume that $\kappa$ lies between 0 and 1 since it is possible via a (singular) gauge transformation to shift $\kappa$ by an integer amount (under such a transformation the *total* flux through $\tilde{T}^2$ remains zero). Note that $\phi(z)$ is only defined up to a constant. This is usually fixed by setting the integral of $\phi$ over the dual to be zero. This renders $\phi(z)$ an odd function, $\phi(-z) = -\phi(z)$. It is simple to determine Fourier or theta-function representations of $\phi$. The Nahm potential can also be obtained via the ADHM construction [5]. Here the flux strength $\kappa$ is related to the ADHM 'size', $\lambda$, of an instanton centred at $x_\mu = 0$,

$$\kappa = \frac{\pi\lambda^2}{L_1 L_2}. \tag{8}$$

The $SU(2)$ instanton we seek, $A_\mu(x)$, is the Nahm transform of the abelian potential $\hat{A}(z)$

$$A_\mu^{pq}(x) = \int_{\tilde{T}^2} d^2z\, \psi^{p\dagger}(z;x) \frac{\partial}{\partial x^\mu} \psi^q(z;x), \tag{9}$$

where the $\psi^p(z;x), p = 1, 2$ are orthonormal zero modes of

$$\frac{D_x^\dagger}{2i} = \begin{pmatrix} \frac{1}{2}\bar{x}_\perp & \partial_y + \partial_y\phi - \frac{i}{2}\bar{x}_{||} \\ \partial_{\bar{y}} - \partial_{\bar{y}}\phi - \frac{i}{2}x_{||} & \frac{1}{2}x_\perp \end{pmatrix}. \tag{10}$$

In addition to the complex coordinates $y$, $\bar{y}$ on $\tilde{T}^2$ we have introduced two sets of complex coordinates for $T^2 \times R^2$; in the 'parallel' directions $x_{||} = x_1 + ix_2$, $\bar{x}_{||} = x_1 - ix_2$, and in the 'transverse' non-compact directions $x_\perp = x_0 + ix_3$, $\bar{x}_\perp = x_0 - ix_3$.

When $x_\perp = 0$ the Weyl equation decouples and the two zero modes have a simple form [5]

$$\psi^1(z;x) = \begin{pmatrix} 0 \\ e^{-\phi(z)}G_+(z-\omega) \end{pmatrix}, \tag{11}$$

and

$$\psi^2(z;x) = \begin{pmatrix} e^{\phi(z)}G_-(z+\omega) \\ 0 \end{pmatrix}, \tag{12}$$

where $G_\pm(z)$ are periodic Green's functions satisfying

$$\left(-i\partial_y - \frac{1}{2}\bar{x}_{||}\right) G_+(z) = \frac{1}{2}\delta^2(z), \tag{13}$$

and

$$\left(-i\partial_{\bar{y}} - \frac{1}{2}x_{||}\right) G_-(z) = \frac{1}{2}\delta^2(z). \tag{14}$$

Like $\phi(z)$, $G_\pm(z)$ have theta function representations. The zero modes have square-integrable singularities at both fluxes. For $z$ near $\omega$, $|\psi^1|^2 \sim |y-\omega_1-i\omega_2|^{2(\kappa-1)}$ while $|\psi^2|^2 \sim |y-\omega_1-i\omega_2|^{-2\kappa}$. When $z$ approaches the other flux at $-\omega$ the singularity profiles are reversed, with $|\psi^1|^2 \sim |y+\omega_1+i\omega_2|^{-2\kappa}$ and $|\psi^2|^2 \sim |y+\omega_1+i\omega_2|^{2(\kappa-1)}$. There are two values of $x_{||}$ (in each copy of $T^2$) for which the above discussion does not apply. When $x_{||} = 0$ the Green's function's $G_\pm$ do not exist. This does not mean there are no zero modes, indeed one can see that

$$\psi^1(z;x=0) = \begin{pmatrix} 0 \\ e^{-\phi(z)} \end{pmatrix}, \tag{15}$$

and

$$\psi^2(z;x=0) = \begin{pmatrix} e^{\phi(z)} \\ 0 \end{pmatrix}, \tag{16}$$

are solutions of the $x_{||} = 0$ Weyl equation. $\psi^1$ has the expected square-integrable singularity at $z = -\omega$, but for $z = \omega$, $\psi^1(z; x = 0)$ is zero. On the other hand $\psi^2(z, x = 0)$ diverges at $z = \omega$ but not at $z = -\omega$. The finiteness of $G_-(2\omega)$ was used to derive the quoted asymptotics. But if it is *zero*, $\psi^1$ will lose its singularity at $z = -\omega$. Using standard complex variables theory or the explicit theta function representations one can see that there is exactly one value of $x_{||}$ in $T^2$ where this occurs, $\pi x_{||} = -iL_1L_2(\omega_1 + i\omega_2)$. Do these 'soft' points represent the locations of some kind of constituents?

Notice that the singularity profiles of $\psi^1$ and $\psi^2$ are exchanged under the replacement $\kappa \to 1 - \kappa$. This suggests that the constituents are exchanged under this mapping. That is, if there are indeed lumps at the two points, then $\kappa \to 1 - \kappa$ swaps the two lumps. The identity $F_{\mu\nu}(x_{||}, x_\perp, \kappa) = V^{-1}(x)F_{\mu\nu}\left(-x_{||} + \frac{L_1L_2}{i\pi}(\omega_1 + i\omega_2), -x_\perp, 1-\kappa\right)V(x)$ formalises this ideas. Here $V(x)$ is some $U(2)$ gauge transformation.

We have seen that any lumps centred at the constituent points have a simple exchange property. To actually see the lumps we must compute the field strengths. We can do this explicitly in the $x_\perp = 0$ slice, since here we have the exact zero modes (11,12) Before we insert these modes into (9), they must be normalised. This can be done by dividing both modes by $\sqrt{\rho}$, where

$$\rho = \int_{\tilde{T}^2} d^2z\, e^{2\phi(z)} |G_-(z+\omega)|^2. \tag{17}$$

In fact, one can write two components of the gauge potential solely in terms of this normalisation factor

$$A_{x_{||}} = -\frac{1}{2}\tau_3 \partial_{x_{||}} \log \rho, \quad A_{\bar{x}_{||}} = \frac{1}{2}\tau_3 \partial_{\bar{x}_{||}} \log \rho, \tag{18}$$

so that

$$F_{x_{||}\bar{x}_{||}} = \tau_3 \partial_{x_{||}} \partial_{\bar{x}_{||}} \log \rho. \tag{19}$$

The computation of the other components is more involved since the derivatives of the zero modes with respect to $x_\perp$ are required. Here we just quote the results

$$F_{x_{||}\bar{x}_\perp} = 2\pi i(\tau_1 + i\tau_2)\,\kappa\rho\, \partial^2_{x_{||}} \frac{\nu}{\rho}. \tag{20}$$

where

$$\nu = \int_{\tilde{T}^2} d^2z\, G_+(\omega - z) e^{2\phi(z)} G_-(z+\omega). \tag{21}$$

The other components are fixed by self-duality, i.e. $F_{x_\perp \bar{x}_\perp} + F_{x_{||}\bar{x}_{||}} = 0$ and $F_{x_{||}x_\perp} = 0$. Note that $\rho$ is dimensionless, real and periodic (with respect to $x_1 \to x_1 + L_1$, $x_2 \to x_2 + L_2$), while $\nu$ is dimensionless, complex and periodic up to constant phases. Both $\rho$ and $\nu$ diverge at $x_{||} = 0$, but the field strengths should not. A careful analysis of $\rho$ and $\nu$ in the neighbourhood of $x_{||} = 0$ shows that the field strengths are well defined at this point.

Since we have all field strengths within the $x_\perp = 0$ slice we can compute the action density

$$-\frac{1}{2}\mathrm{Tr} F_{\mu\nu}F^{\mu\nu} = 16\left(\partial_{x_{||}}\partial_{\bar{x}_{||}} \log \rho\right)^2 + 128\pi^2\kappa^2\rho^2 \left|\partial^2_{x_{||}} \frac{\nu}{\rho}\right|^2. \tag{22}$$

We have provided integral representations of $\rho$ and $\nu$ where the integrands are expressible in terms of standard functions. Using these results, we have made numerical plots of the action density for various values of $\kappa$ and $\omega$, and for the periods we have taken $L_1 = L_2 = 2\pi$, i.e. a square torus. The most symmetric case, $\kappa = \frac{1}{2}$ and $\omega_1 = \omega_2 = \frac{1}{4}$ is plotted in Fig. 1. Two equal sized lumps are observed, one at $x_{||} = 0$, the other at $x_{||} = (1+i)\pi$. Decreasing $\kappa$ at fixed $\omega$ reduces the size of the $x_{||} = 0$ constituent but *increases* its contribution to the action. Conversely, the second constituent becomes larger but contributes less to the action density. These two effects tend to (rather quickly) flatten the second peak as $\kappa$ is reduced. Here we provide plots with $\kappa = \frac{7}{16}$ (Fig. 2) and $\kappa = \frac{3}{8}$ (Fig. 3), again with $\omega_1 = \omega_2 = \frac{1}{4}$. Already at $\kappa = \frac{3}{8}$ we see that the $x_{||} = (1+i)\pi$ constituent is more of a plateau than a peak. If $\kappa$ is decreased much further the first peak will dominate completely; this peak will be essentially that of a BPST instanton on $R^4$ with a scale parameter given by the ADHM formula (8). Indeed, if we take (8) at face value, our two constituents are BPST instanton cores, one with scale $\lambda = \sqrt{\kappa L_1 L_2/\pi}$ at $x_{||} = x_\perp = 0$, the other with scale $\lambda = \sqrt{(1-\kappa)L_1 L_2/\pi}$ at the second soft point. When $L_1 = L_2$ at least one $\lambda$ is of the order of the size of the torus in line with the strong overlap seen in the plots.

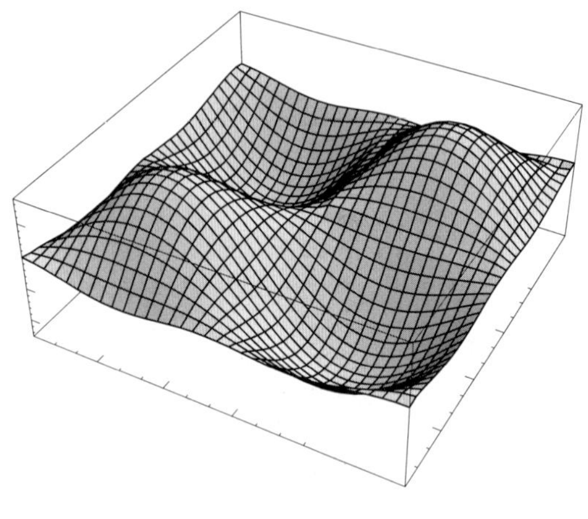

Figure 1. Plot of action density for $x_\perp = 0$, $\kappa = \frac{1}{2}$ and $\omega_1 = \omega_2 = \frac{1}{4}$.

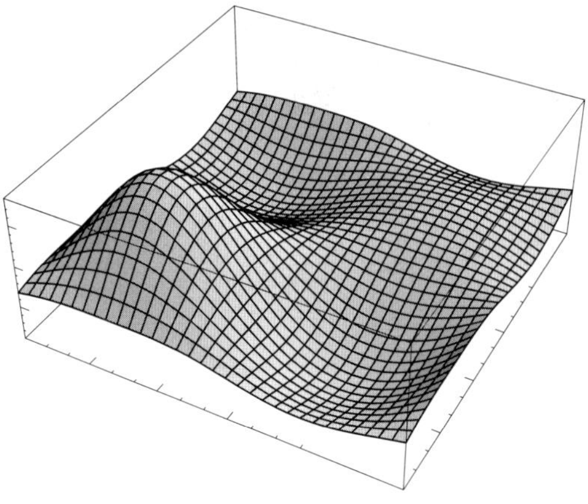

Figure 2. Plot of action density for $x_\perp = 0$, $\kappa = \frac{7}{16}$ and $\omega_1 = \omega_2 = \frac{1}{4}$.

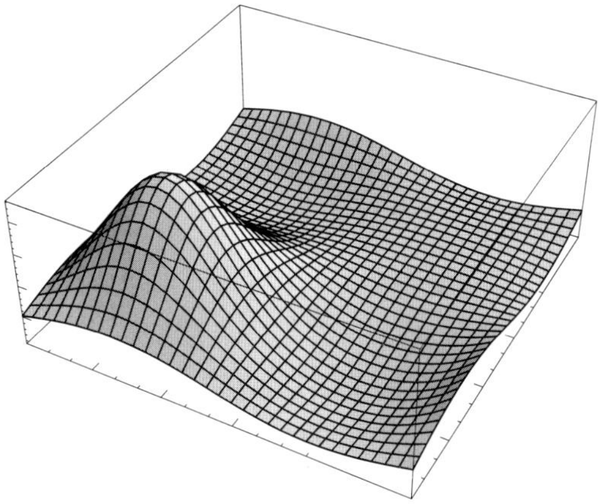

Figure 3. Plot of action density for $x_\perp = 0$, $\kappa = \frac{3}{8}$ and $\omega_1 = \omega_2 = \frac{1}{4}$.

The peaks are best resolved at $\kappa = \frac{1}{2}$, when the constituents are identical. The overlap can be avoided by taking one period, say $L_2$, smaller than the other, and then taking a sufficiently large constituent separation in the $x_1$ direction. However, then the size of one or both cores will exceed $L_2$, tending to suppress or smooth out the dependence on $x_2$. In this case the constituents become *periodic monopoles.* Actually, it is possible to see some hint of this monopole limit even for a square torus; taking $\omega_1 = \frac{1}{4}$, $\omega_2 = 0$, $\kappa = \frac{1}{2}$ with $L_1 = L_2 = 2\pi$ the two constituents (at $x_{||} = 0$ and $x_{||} = i\pi$) overlap so strongly that they merge into a monopole worldline, in that the dependence of the field strengths on $x_2$ is very weak.

The $\kappa = \frac{1}{2}$ case has another interesting feature. If we choose the constituent locations so that they are separated by half periods the charge one instanton can be 'cut' to yield a charge $\frac{1}{2}$ instanton. Doubly periodic charge $\frac{1}{2}$ instantons have also been found in lattice simulations [6]. Many qualitative features reported are in accord with our findings. In particular, we agree that these objects are single lumps with radial symmetry, and that the action density is never zero in the $x_\perp = 0$ plane.

## 4. WHAT ARE INSTANTONS MADE OF?

The very special double periodic instantons considered here have constituents that look like instanton cores. Their locations are identified with the 'soft points' in $T^2 \times R^2$ where one of the Nahm zero modes has a weakened singularity profile. These results as well as the monopole constituent description of charge-one calorons suggest the following general picture: doubly periodic charge $k$ instantons comprise $NK$ constituents located at the soft points of the Nahm equations. Interpreting these constituents as instanton cores begs the question why this leads to a topological charge $k$ rather than $Nk$. However, the constituents always overlap strongly to deliver the correct instanton number. In the high temperature limit, i.e. when one period, say $L_2$, is much smaller than the other, the cores will overlap with the periodic copies of themselves to form

$x_2$-independent monopole like structures.

Is this viable only for doubly periodic instantons or can it be extended to $T^n \times R^{4-n}$ with $n \neq 2$. The $n = 3$ case seems to be the best bet since here the Nahm zero modes are also expected to have (square-integrable) singularities, and so the notion of soft zero modes makes perfect sense. One additional complication of $T^3 \times R^3$ is the role of open boundary conditions (associated with the disconnected boundary). Indeed, it has been suggested that open boundary conditions increase the number of singularities in the Nahm potential [7] (in the $SU(2)$ case from 2 to 4) which might affect the counting of soft modes. For the four torus the distinction between soft and generic zero modes doesn't exist since here all gauge fields and associated zero modes are smooth. This does not mean that one cannot have constituents, rather the soft mode definition needs to be modified or replaced altogether.

Finally, we should not overlook the caloron case. Here the Nahm potentials do not have associated zero modes unless source terms are added to the Weyl equation; these sources can be determined via ADHM-based methods. However, these 'zero modes' are not divergent, but they have cusps at the $N$ jumping points of the Nahm potential. Perhaps there are points in $S^1 \times R^3$ for which one of the cusps of a given zero mode disappears. If so we would have an exact analogue of the soft points for the calorons.

## Acknowledgements

Much of this work was done in collaboration with J. M. Pawlowski. The author thanks F. Bruckmann and P. van Baal for helpful discussions.

## REFERENCES

1. K. M. Lee and C. h. Lu, Phys. Rev. D **58** (1998) 025011 [hep-th/9802108].
2. T. C. Kraan and P. van Baal, Nucl. Phys. B **533** (1998) 627 [hep-th/9805168].
3. C. Ford and J. M. Pawlowski, Phys. Lett. B **540** (2002) 153 [hep-th/0205116].
4. P. J. Braam and P. van Baal, Commun. Math. Phys. **122** (1989) 267.
5. C. Ford, J. M. Pawlowski, T. Tok and A. Wipf, Nucl. Phys. B **596** (2001) 387 [hep-th/0005221].
6. A. Gonzalez-Arroyo and A. Montero, Phys. Lett. B **442** (1998) 273 [hep-th/9809037].
7. P. van Baal, Phys. Lett. B **448** (1999) 26 [hep-th/9811112].

ELSEVIER

Proceedings of ICHEP 2002, pp. 867–869
*S. Bentvelsen, P. de Jong, J. Koch and E. Laenen (Editors)*

31st INTERNATIONAL CONFERENCE ON HIGH ENERGY PHYSICS AMSTERDAM

# Construction of Supersymmetric Nonlinear Sigma Models on Noncompact Calabi-Yau Manifolds with Isometry

Kiyoshi Higashijima[a*], Tetsuji Kimura[a] and Muneto Nitta[b†]

[a]Department of Physics, Graduate School of Science, Osaka University, Toyonaka, Osaka 560-0043, Japan

[b]Department of Physics, Purdue University, West Lafayette, IN 47907-1396, USA

We propose a class of $\mathcal{N}=2$ supersymmetric nonlinear sigma models on the noncompact Ricci-flat Kähler manifolds, interpreted as the complex line bundles over the hermitian symmetric spaces. Kähler potentials and Ricci-flat metrics for these manifolds with isometries are explicitly constructed by using the techniques of supersymmetric gauge theories. Each of the metrics contains a resolution parameter which controls the size of these base manifolds, and the conical singularity appears when the parameter vanishes.

## 1. Introduction

Nonlinear sigma models (NL$\sigma$Ms) in two dimensions are interesting for several reasons. They help us to understand various non-perturbative phenomena in four dimensional gauge theories such as confinement or dynamical mass generation. They also provide the description of strings propagating in the curved space-time. In the latter case, the consistency of strings requires the conformal invariance of the NL$\sigma$Ms. In the case of superstrings, furthermore, NL$\sigma$Ms have to possess the $\mathcal{N}=2$ world sheet superconformal symmetry. Although all two-dimensional NL$\sigma$Ms with $\mathcal{N}=2$ supersymmetry have $\mathcal{N}=2$ superconformal symmetry at the tree level, they suffer from superconformal anomaly at the quantum level. Since these anomalies are proportional to the Ricci curvatures $R_{ab}$ of the background (target) manifolds, $\mathcal{N}=2$ supersymmetric NL$\sigma$Ms taking values on the target manifolds with vanishing Ricci curvatures, provide the possible candidates of the consistent superstrings propagating in the curved space-time. They are also candidates of finite quantum field theories in two dimensions. $\mathcal{N}=2$ supersymmetry in two dimensions requires the target manifolds of NL$\sigma$Ms are Kähler manifolds. The Kähler manifolds with vanishing Ricci curvatures are called the Calabi-Yau manifolds. Therefore, it is important to study the $\mathcal{N}=2$ supersymmetric NL$\sigma$Ms defined on the Calabi-Yau manifolds.

To find the metric of the Ricci-flat manifolds, we have to solve the Einstein equation $R_{ab}=0$ for the vacuum. If the manifold has enough symmetries, we can reduce the partial differential equation to an ordinary differential equation, which is usually easy to solve. In compact Calabi-Yau manifolds, however, there is no isometry and not a single explicit metric is known in this case. In some noncompact Calabi-Yau manifolds, the number of isometries is sufficient to reduce the Einstein equation to an ordinary differential equation. In this article, we report our recent progress to obtain the explicit metric for certain class of the noncompact Calabi-Yau manifolds [1–3].

## 2. NL$\sigma$Ms with $\mathcal{N}=2$ Supersymmetry in Two Dimensions

$\mathcal{N}=2$ supersymmetry requires that both scalars $A^a(x)$ and fermions $\psi^a(x)$ are complex fields. The scalar fields $A^a(x)$ $(a=1,\cdots,N)$ are coordinates of the target manifold, whose

*Work supported in part by the Grant-in-Aid for Scientific Research
†Work supported by the U.S. Department of Energy under grant DE-FG02-91ER40681 (Task B)


doi:10.1016/S0920-5632(02)02252-1

metric is given by the Kähler potential $K(A,\bar{A})$ which characterizes the theory:

$$g_{a\bar{b}}(A,\bar{A}) = \frac{\partial^2 K(A,\bar{A})}{\partial A^a \partial \bar{A}^{\bar{b}}} = K,_{a\bar{b}}(A,\bar{A}). \quad (1)$$

The lagrangian for $\mathcal{N}=2$ supersymmetric NL$\sigma$M is given by

$$\mathcal{L}(x) = g_{a\bar{b}}\partial_\mu A^a \partial^\mu \bar{A}^{\bar{b}} + i g_{a\bar{b}} \bar{\psi}^{\bar{b}} (\not{D}\psi)^a + \frac{1}{4} R_{a\bar{b}c\bar{d}} \psi^a \psi^c \bar{\psi}^{\bar{b}} \bar{\psi}^{\bar{d}}, \quad (2)$$

where the covariant derivative and the Riemann curvature are defined by

$$(D_\mu \psi)^a = \partial_\mu \psi^a + \partial_\mu A^b \Gamma^a{}_{bc} \psi^c \quad (3)$$

$$R_{a\bar{b}c\bar{d}} = \partial_c \partial_{\bar{d}} g_{a\bar{b}} - g^{e\bar{f}} \partial_c g_{a\bar{f}} \partial_{\bar{d}} g_{e\bar{b}}, \quad (4)$$

with the connection being

$$\Gamma^c{}_{ab} = g^{c\bar{d}} g_{b\bar{d},a} = g^{c\bar{d}} K,_{ab\bar{d}}\,. \quad (5)$$

Since all geometrical quantities are derived from the Kähler potential $K$, all we have to do is to assume an appropriate symmetry of the scalar function $K$. Symmetries of a scalar function is far simpler than the symmetries of the tensor $g_{a\bar{b}}$.

## 3. Noncompact Calabi-Yau Manifolds

Although the NL$\sigma$M defined by (2) is conformally invariant at the tree level, it suffers from the conformal anomaly proportional to the Ricci curvature at the quantum level[4]. To obtain the conformally invariant theory at the quantum level, we impose the Ricci-flatness condition

$$R_{a\bar{b}} = 0 \quad (6)$$

to the target manifold.

In order to reduce an Einstein equation to an ordinary differential equation of single variable, we assume our target manifold $M_1$ has a symmetry, which carries each point of the manifold to a certain point on a line, which is parameterized by a single variable. In other word, any point of our manifold $M_1$ can be reached from a point on a line by a suitable action of some symmetry group $G$. If any point of the manifold can be reached from a single point, say the origin, by appropriate action of a symmetry group $G$, the manifold is a homogeneous coset space $G/H$ where $H$ denotes a subgroup of $G$. The local structure of our manifold is, therefore, the product of a homogeneous space $M = G/H$ and a line. Since $\mathcal{N}=2$ supersymmetry requires the manifold $M_1$ to be a Kähler manifold, we assume that the homogeneous space $M = G/H$ is *Kähler* by itself and the line is a *complex* line. Furthermore, we assume the Kähler coset space $M = G/H$ is a compact *Einstein manifold*, which satisfies the Einstein equation with a positive cosmological constant $h > 0$

$$R_{m\bar{n}} = h g_{m\bar{n}}, \quad (7)$$

where we use the indices $m, n$ for the compact Kähler manifold $M = G/H$ whose complex dimension is $d = dim(G/H)/2$. The complex dimension of the Ricci-flat manifold $M_1$ is $D = 1 + d$.

Our total space $M_1$ is parameterized by the complex coordinates $\phi_m$ $(m = 1,\cdots,d)$ and a complex number $\rho$ that represents the complex line. We denote the Kähler potential of the compact Kähler-Einstein manifold $M$ by $\Psi(\phi,\bar{\phi})$. Our assumption for the Kähler potential of the Ricci-flat manifold $M_1$ is as follows:

$$K(A,\bar{A}) = K(X), \quad (8)$$

where

$$X = \log|\rho|^2 + h\Psi(\phi,\bar{\phi}). \quad (9)$$

Now, if we notice

$$R_{a\bar{b}} = -g^{c\bar{d}} R_{a\bar{b}c\bar{d}} = -\partial_a \partial_{\bar{b}} \log\det(g_{c\bar{d}}) \quad (10)$$

the Ricci-flat condition (6) takes a very simple form

$$\det(g_{c\bar{d}}) = |\mathrm{hol.}|^2, \quad (11)$$

where hol. is a holomorphic function of the coordinates $\phi$ and $\rho$.

Components of the metric for the whole space $M_1$

$$g_{a\bar{b}} = \begin{pmatrix} g_{\rho\bar{\rho}} & g_{\rho\bar{n}} \\ g_{m\bar{\rho}} & g_{m\bar{n}} \end{pmatrix}, \quad (12)$$

are given by

$$g_{\rho\bar{\rho}} = K'' \frac{\partial X}{\partial \rho} \frac{\partial X}{\partial \bar{\rho}}, \quad (13)$$

$$g_{\rho\bar{n}} = K''\frac{\partial X}{\partial \rho}\frac{\partial X}{\partial \bar{\phi}^{\bar{n}}}, \quad (14)$$

$$g_{m\bar{n}} = K''\frac{\partial X}{\partial \phi^m}\frac{\partial X}{\partial \bar{\phi}^{\bar{n}}} + K'\frac{\partial^2 X}{\partial \phi^m \partial \bar{\phi}^{\bar{n}}}, \quad (15)$$

where the prime denotes the differentiation with respect to the argument $X$. Using equations

$$\partial X/\partial \rho = 1/\rho (\rho \neq 0), \quad \frac{\partial^2 X}{\partial \phi^m \partial \bar{\phi}^{\bar{n}}} = h g_{m\bar{n}}, \quad (16)$$

we can express the determinant of the metric of $M_1$ in terms of the determinant of the metric of $M = G/H$

$$\det g_{a\bar{b}} = g_{\rho\bar{\rho}} \cdot \det(g_{m\bar{n}} - g_{\rho\bar{\rho}}^{-1} g_{m\bar{\rho}} g_{\rho\bar{n}}) \quad (17)$$

$$= \frac{1}{|\rho|^2} K''(K')^d \cdot \det(h g_{m\bar{n}}). \quad (18)$$

Since we have assumed that the Kähler coset space is an Einstein space, we can easily calculate the determinant of its metric. By substituting the expressions of the Ricci curvature (10) and the metric $g_{m\bar{n}}$ in terms the Kähler potential $\Psi$ (1) to the Einstein equation (7), we obtain

$$R_{m\bar{n}} = -\partial_m \partial_{\bar{n}} \log\det(g_{k\bar{l}}) = h\frac{\partial^2 \Psi(\phi, \bar{\phi})}{\partial \phi^m \partial \bar{\phi}^{\bar{n}}}, \quad (19)$$

which can be integrated to give a convenient formula for the determinant of the metric in Kähler-Einstein manifold $M$

$$\det(g_{m\bar{n}}) = e^{-h\Psi}|\mathrm{hol.}|^2. \quad (20)$$

Combining the formula (18), (20) and (11), we obtain the Ricci-flatness condition for the Kähler potential of the manifold $M_1$

$$\det(g_{a\bar{b}}) = e^{-X} K''(K')^d |\mathrm{hol.}|^2 = (const.) \cdot |\mathrm{hol.}|^2,$$

from which we find an ordinary differential equation for the Kähler potential:

$$e^{-X}\frac{d}{dX}(K')^D = a, \quad (21)$$

where $a$ is a constant.

The equation (21) is easily solved to yield

$$K' = \left(\lambda e^X + b\right)^{\frac{1}{D}}, \quad (22)$$

where $\lambda$ is a constant related to $a$ and $D$, and $b$ is an integration constant, interpreted as a resolution parameter. To calculate the metric (13)(14)(15) of the Ricci-flat manifold $M_1$, the Kähler potential itself is not necessary, though it is easily obtained by integrating (22).

Our method can be applied to any Kähler-Einstein coset manifold $M$, however, we need the Kähler potential $\Psi$ of $M$ to obtain the explicit expression for the metric of the noncompact Calabi-Yau manifold $M_1$. Fortunately, the general prescription to calculate $\Psi$ is given by Ito, Kugo and Kunitomo[5]. In this respect, our results gives the explicit realization of the method of Page and Pope, who constructed Ricci-flat manifolds by introducing a complex line as a fibre over a Kähler-Einstein manifold[6].

Here, we further require that the Kähler coset space $M$ is a symmetric space, called the hermitian symmetric space, then the Kähler potential can be explicitly obtained by the gauge theory technique[2,7]. Therefore, we find a new class of Ricci-flat manifolds corresponding to various hermitian symmetric spaces with $h = C_2(G)/2$. For $M = CP^{N-1}$, $M_1$ reduces to the manifold obtained by Calabi[8].

## REFERENCES

1. K. Higashijima, T. Kimura and M. Nitta, Phys. Lett. **B515** (2001) 421, hep-th/0104184.
2. K. Higashijima, T. Kimura and M. Nitta, Phys. Lett. **B518** (2001) 301, hep-th/0107100; Nucl. Phys. **B623** (2002) 133, hep-th/0108084; Ann. Phys. **296** (2002) 347, hep-th/0110216.
3. K. Higashijima, T. Kimura and M. Nitta, to appear in Nucl. Phys. B, hep-th/0202064.
4. L. Alvarez-Gaumé and P. Ginsparg, Comm. Math. Phys. **102** (1985) 311.
5. K. Itoh, T. Kugo and H. Kunitomo, Nucl. Phys. **B263** (1986) 295; Prog. Theor. Phys. **75** (1986) 386.
6. D. N. Page and C. N. Pope, Class. Quantum Grav. **4** (1987) 213.
7. K. Higashijima and M. Nitta, Prog. Theor. Phys. **103** (2000) 635, hep-th/9911139.
8. E. Calabi, Ann. Scient. Ec. Norm. Sup. **12** (1979) 269.

Proceedings of ICHEP 2002, pp. 870–871
*S. Bentvelsen, P. de Jong, J. Koch and E. Laenen (Editors)*

31st INTERNATIONAL CONFERENCE ON HIGH ENERGY PHYSICS AMSTERDAM

# Virtual black holes in generic 2D dilaton theories

D. Grumiller[a]*

[a]Institut für Theoretische Physik, Technische Universität Wien
Wiedner Hauptstr. 8-10, A-1040 Vienna, Austria

Recent years have seen remarkable progress in the quantum treatment of two dimensional models of gravity, so called Generalized Dilaton Theories (GDTs), which include, most prominently, the Schwarzschild Black Hole (BH) and the string inspired dilaton BH. In this note some of the latest results for GDTs with matter are summarized.

Ever since the seminal paper of Callan, Giddings, Harvey and Strominger [1] GDTs have served as a link between string theory, black hole physics and (due to their close relation to Poisson-$\sigma$ models [2]) non-commutative geometry. For a recent review on GDTs with matter cf. [3].

It turned out that even in the presence of matter an exact path integration of all geometric quantities is possible for all GDTs, proceeding along well established paths of quantum field theory. The effective theory obtained in this way solely depends on the matter fields in which it is nonlocal and non-polynomial. Already at the level of the (nonlocal) vertices of matter fields, to be used in a systematic perturbative expansion in terms of Newton's constant, a highly nontrivial and physically intriguing phenomenon can be observed, namely the so-called "virtual black hole" (VBH). This notion originally has been introduced by S. Hawking [4], but in our recent approach the VBH for spherically reduced gravity (SRG) emerges naturally in Minkowski signature space-time, without the necessity of additional *ad hoc* assumptions.

The first discussion in [5] (SRG with in $D = 2$ minimally coupled scalars) was too simple to yield an interesting $S$-matrix, unless mass terms were added. The situation had improved for non-minimally coupled scalars where the lowest order $S$-matrix indeed exhibited interesting features: forward scattering poles, monomial scaling with energy, CPT invariance, and pseudo-self-similarity in its kinematic sector [6].

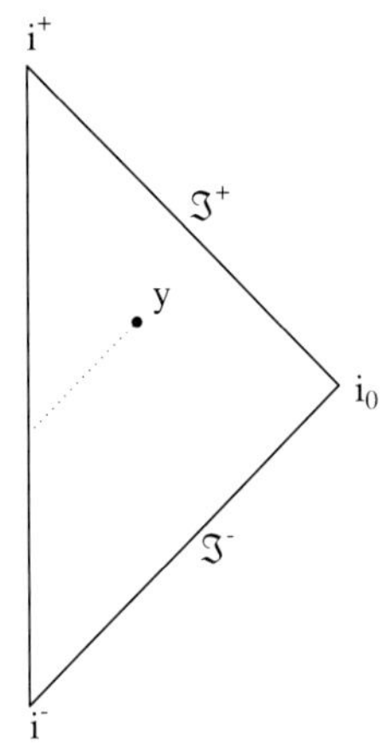


Figure 1. CP diagram of the VBH

It was possible to reconstruct geometry self-consistently from a (perturbative or, if available, exact) solution of the effective theory. For the simplest case of four-point tree-graph scattering the corresponding Carter-Penrose (CP) diagram is presented in Fig. 1. It is non-local in the sense that it depends not only on one set of coordinates but on two. This was a consequence of integrat-

*This work has been supported by project P-14650-TPH of the Austrian Science Foundation (FWF). The participation at the conference has been sponsored by the Austrian Ministry of Education, Science and Culture (I am especially grateful to Mag. Neuhäuser for his support). Last but not least I would like to thank J. Koch and S. Vandoren for their patience and understanding regarding my administrative problems.


doi:10.1016/S0920-5632(02)02253-3

ing out geometry non-perturbatively. For each choice of $y$ (one of the two sets of coordinates) it is possible to draw an ordinary CP-diagram. The non-trivial part of our effective geometry (i.e. the VBH) is concentrated on the light-like cut. For SRG the ensuing line-element has the following (Bondi-Sachs) form

$$(ds)^2 = 2drdu + \left(1 - \frac{2m}{r} - ar + d\right)(du)^2, \quad (1)$$

with $m$, $a$ and $d$ being localized[2] on the cut $u = u_0$ with compact support $r < r_0$. These quantities depend on the second set of coordinates $u_0$, $r_0$.

One should not take the effective geometry at face value – this would be like over-interpreting the role of virtual particles in a loop diagram. It is a nonlocal entity and one still has to "sum" (read: integrate) over all possible geometries of this type in order to obtain the nonlocal vertices and the scattering amplitude. Nonetheless, the simplicity of this geometry and the fact that all possible configurations are summed over are nice features of this picture.

Qualitatively it is clear what has to be done in order to obtain the $S$-matrix[3]: Take all possible VBHs of Fig. 1 and sum them coherently with proper weight factors and suitably attached external legs of scalar fields. This had been done quantitatively in a straightforward but rather lengthy calculation for gravitational scattering of s-waves in the framework of SRG, the result of which yielded the lowest order tree-graph $S$-matrix for ingoing modes with momenta $q, q'$ and outgoing ones $k, k'$,

$$T(q, q'; k, k') = -\frac{i\kappa\delta\,(k + k' - q - q')}{2(4\pi)^4|kk'qq'|^{3/2}} E^3\tilde{T}, \quad (2)$$

with the total energy $E = q + q'$, $\kappa = 8\pi G_N$,

$$\tilde{T}(q, q'; k, k') := \frac{1}{E^3}\Bigg[\Pi \ln \frac{\Pi^2}{E^6} + \frac{1}{\Pi} \sum_{p \in \{k,k',q,q'\}} p^2 \ln \frac{p^2}{E^2}\left(3kk'qq' - \frac{1}{2}\sum_{r \neq p}\sum_{s \neq r,p}\left(r^2 s^2\right)\right)\Bigg], \quad (3)$$

and the momentum transfer function $\Pi = (k + k')(k - q)(k' - q)$. The interesting part of the scattering amplitude is encoded in the scale independent factor $\tilde{T}$.

It is possible to generalize the VBH phenomenon to arbitrary GDTs with matter as well as most of its properties (for instance, the CP-diagram, CPT invariance and the role played in the $S$-matrix) [8].

[2] The localization of "mass" and "Rindler acceleration" on a light-like cut is not an artifact of an accidental gauge choice, but has a physical interpretation in terms of the Ricci-scalar. Certain parallels to Hawking's Euclidean VBHs can be observed, but also essential differences. The main one is our Minkowski signature which we deem to be a positive feature.

[3] The idea that BHs must be considered in the $S$-matrix together with elementary matter fields has been put forward some time ago [7]. The approach [6] reviewed here, for the first time allowed to derive (rather than to conjecture) the appearance of the BH states in the quantum scattering matrix of gravity.

## REFERENCES


1. C. G. Callan, S. B. Giddings, J. A. Harvey and A. Strominger, Phys. Rev. D **45** (1992) 1005 [arXiv:hep-th/9111056].
2. P. Schaller and T. Strobl, Mod. Phys. Lett. A **9** (1994) 3129 [arXiv:hep-th/9405110].
3. D. Grumiller, W. Kummer and D. V. Vassilevich, "Dilaton gravity in two dimensions," to appear in Phys. Rept. [arXiv:hep-th/0204253].
4. S. W. Hawking, Phys. Rev. D **53** (1996) 3099 [arXiv:hep-th/9510029].
5. D. Grumiller, W. Kummer and D. V. Vassilevich, Nucl. Phys. B **580** (2000) 438 [arXiv:gr-qc/0001038].
6. P. Fischer, D. Grumiller, W. Kummer and D. V. Vassilevich, Phys. Lett. B **521** (2001) 357 [Erratum-ibid. B **532** (2002) 373] [arXiv:gr-qc/0105034]; D. Grumiller, PhD thesis [arXiv:gr-qc/0105078]; Int. J. Mod. Phys. A **17** (2002) 989 [arXiv:hep-th/0111138]; Class. Quant. Grav. **19** (2002) 997 [arXiv:gr-qc/0111097].
7. G. 't Hooft, Int. J. Mod. Phys. A **11** (1996) 4623 [arXiv:gr-qc/9607022].
8. D. Grumiller, W. Kummer and D. V. Vassilevich, "Virtual Black Holes in Generalized Dilaton Theories and Their Special Role in String Gravity," submitted to Eur. Phys. J. C [arXiv:hep-th/0208052].

## Session 13

# R&D for Future Accelerators and Detectors

*Session conveners:*
*Y. Kuno, Osaka University*
*T. Behnke, DESY Hamburg*

ELSEVIER

Proceedings of ICHEP 2002, pp. 875–877
*S. Bentvelsen, P. de Jong, J. Koch and E. Laenen (Editors)*

31st INTERNATIONAL CONFERENCE ON
HIGH ENERGY PHYSICS AMSTERDAM

# The Status of Run II at Fermilab

K. Gounder, [a] *

[a]Beams Division, Fermi National Accelerator Laboratory,
P.O. Box 500, Batavia, IL 60510. U.S.A.

The Run II at Fermilab is progressing steadily. In the Run IIa scheme, 36 antiproton bunches collide with 36 proton bunches at CDF and D0 interaction regions in the Tevatron at 980 GeV beam energy. The current status and performance of the Fermilab Accelerator Complex is reviewed. The plans for Run IIb along with the Antiproton Source upgrade and incorporation of the Recycler Ring in the accelerator chain are outlined. The prospects of achieving Run II integrated luminosity goal of 15 $fb^{-1}$ are discussed.

## 1. INTRODUCTION

The Accelerator Complex at Fermilab has delivered an integrated luminosity of 80 $pb^{-}$ as part of collider Run IIa through August 2002. The performance has been improving steadily in the midst of many upgrades to the accelerator complex. The major components of the complex are: (1) The proton source consists of Linac and Booster; (2) The Antiproton Source containing the Debuncher and Accumulator; (3) The Tevatron as the collider storage ring for protons and antiprotons; (4) The Main Injector for accelerating and transfering beams between the beam sources and the Tevatron; (5) The Recycler Ring to be incorporated shortly in the accelerator complex as an antiproton storage facility. After a brief over view of the present collider operations in the next section, the present status of Run II is discussed in section 3. Details of the Recycler Ring commissioning and upgrades for improving the luminosity in Run IIb are outlined in sections 4 and 5.

## 2. OVERVIEW OF OPERATIONS

There are two important operational modes for the Fermilab accelerator complex during the collider runs [1], [2]: (1) Stacking - collecting antiprotons in the Accumulator; (2) Tevatron Shots - loading the Tevatron with protons and antiprotons for colliding stores.

During the stacking mode, the proton beam is delivered from the Booster to antiproton target via the Main Injector (MI) every 2.4 seconds. About 84 8 GeV 53MHz bunches of protons from the Booster are accelerated to 120 GeV in the Main Injector before delivery to the Antiproton target. The resulting 8 GeV antiprotons from the production target are collected, debunched and stochastically cooled in the Debuncher. Then the beam is transferred to the Accumulator where they are stacked and cooled into a 'core'.

The Tevatron shots start with 7 53 MHz bunches of proton beam in the Booster. They are brought to the Main Injector, accelerated to 150 GeV and coalesced into one bunch and injected into the Tevatron. This is repeated until 36 bunches separated by 396 nanoseconds are loaded into the Tevatron central orbit. In the next step, antiproton beam made up of sets of 7 bunches with 53 MHz structure from the Accumulator are transferred to the Main Injector and accelerated to 150 GeV. This beam is coalesced into four bunches and injected into the Tevatron. This is continued nine times till 36 bunches of antiprotons are circulating in the Tevatron in separated helical orbits with the proton bunches. Then the whole beam is accelerated to 980 GeV and are brought into collision at the interaction regions at B0 and D0 after a low beta squeeze. A collider store normally lasts for about 1 luminosity

*Fermilab is operated by Universities Research Association, Inc. under contract with the U.S. Department of Energy


doi:10.1016/S0920-5632(02)02255-7

lifetime that is ~ 12-14 hours.

## 3. RUN IIa: PRESENT STATUS

The luminosity history of Run II from the beginning through August of 2002 is shown in Figure 1. Except for the periods of shutdowns, the weekly integrated luminosity is climbing steadily. The total integrated luminosity is displayed by the curve marked with diamonds reaching about $80pb^{-1}$. The important parameters characterizing the performance of the Run and the goals set for them during Run II are listed in Table 1. These numbers indicate that much progress has to be made even though considerable gains have been realized during the past several months. The major issues facing the performance have been: (1) The transverse emittance of the antiprotons from the Accumulator [3] - almost twice that during Run I for similar size of the core intensities. This reduces the transfer efficiency of the antiprotons to the Tevatron significantly; (2) The reduction of antiproton lifetime in the Tevatron via excessive emittance growth by the long range beam-beam interaction at 150 GeV.

The problem of large transverse emittance of the antiprotons from the accumulator has been solved recently by employing two lattices - one for stacking and the other for shots. Hence as of now, the beam blow-up along the transfer lines is being addressed. To reduce the long range beam-beam effects, efforts are being made to remove restrictive apertures for maximum size possible helical orbits. Other modifications for improvements such as installing more octupoles, longitudinal and transverse dampers for reduced chromaticity are in the works to be installed during the next shutdown.

## 4. RECYCLER COMMISSIONING

The Recycler Ring is located in the Main Injector tunnel about ~ 2 meters above the Main Injector beam pipe [4]. This machine will serve as an antiproton storage ring by recycling the residual antiproton stores from the Tevatron. This Ring will improve the stacking rate of the Accumulator by rapid transfer of the antiproton stores

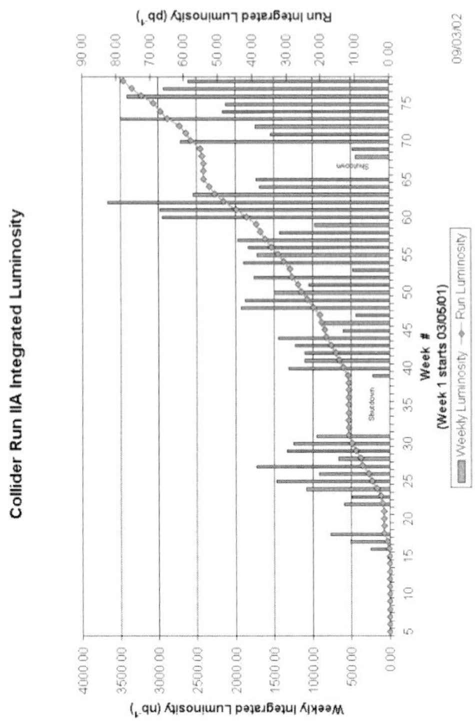


Figure 1. The Integrated Luminosity history for Run II through August, 2002.

to the Recycler. Thus the Recycler is designed to improve the overall Run II luminosity by a factor of 2-3.

The Recycler commissioning is in full swing. Both protons and antiprotons are being circulated with $> 85\%$ efficiency. Presently, the antiproton lifetime with stochastic cooling is $>100$ hours even while the Main Injector is ramping. The challenging issues facing the Recycler commissioning and the proposed solutions are:

(a) To further improve the beam lifetime and curb emittance growth, the number of Ion pumps are being doubled around the ring. Combined with baking of the beam pipe longer at $> 100^0$C, the lifetime is expected to improve by a factor of 3.

(b) To improve the orbit control, new trims and

Table 1
Status on luminosity parameters as of 7/15/2002

| | Highest Luminosity to 12/15/01 | Highest Luminosity to date | Run IIa Goals |
|---|---|---|---|
| Maximum $\bar{p}$ stackrate (E10/hr) | 10 | 11.4 | 18 |
| Maximum $\bar{p}$ stacksize (E10) | 115 | 144 | 165 |
| $\bar{p}$ transfer Eff. | 0.23 | 0.49 | 0.80 |
| $\bar{p}$/bunch at low $\beta$ (E9) | 7.6 | 14.1 | 33.0 |
| p/bunch at low $\beta$ (E9) | 115 | 211 | 270 |
| Emitt. at low $\beta$ ($\pi$-mm-mr) | 16.0 | 16.3 | 17.5 |
| Peak luminosity (E31 $cm^{-2}sec^{-1}$) | 0.84 | 2.12 | 8.6 |

more diagnostics are being installed. A new and improved design of beam position monitoring devices (BPM) is being developed.

(c) Due to initial design limitations and poor matching, the beam extraction and injection apertures are small. Studies to rectify the matching problems as well as new transfer line designs are being done to improve the relevant apertures.

(d) More beam pipe sheilding is added to reduce the effect of the Main Injector ramping. Also ramping power supplies have been installed.

The present expectation is that the Recycler Ring will be commissioned and fully integrated into the Fermilab Collider Complex by the middle of 2003. Without a functioning Recycler Ring, peak luminosities in the Tevatron greater than 1E32 are probably very difficult to achieve.

## 5. OUTLOOK: FUTURE PLANS

As the number of protons per bunch in the Tevatron stores is limited by the beam-beam interaction, the luminosity can be improved by increasing the number of antiprotons per bunch at collisions. But the collider detectors will experience increased difficulty in reconstructing the interesting events due to multiple collisions per bunch crossing (optimum $< 5$). Therefore, the increment of the number of antiprotons per bunch should be also accompanied by increasing the number of bunches in the store. This leads to the possibility of operating the Tevatron at 132 nano second bunch separation mode with 140 proton bunches and 100 antiproton bunches.

To improve the antiproton production rate, major steps will include the following: (a) Slip stacking in the Main Injector to deliver twice the amount of protons on the production target, (b) The design of Lithium lens used to collect produced antiprotons have to be improved to increase the operating gradient from 750 $\rightarrow$ 900 T/m, (c) The transverse collection aperture in the debuncher should be improved to 30$\pi$-mm-mr from 16$\pi$-mm-mr, (d) The antiproton cooling has to be increased from 2-4 GHz $\rightarrow$ 4-8 GHz.

The 132 nsec operation of the Tevatron produces additional challenges to be overcome. The long range beam-beam effects can be $\sim$ 2 times worse while the proton beam instability could be worse by a factor 3-4! A crossing angle must be introduced at the interaction regions to avoid parasitic head-on collisions. The collider detector backgrounds could increase by a factor of 4.

## REFERENCES


1. J. Marriner et. al., "Run II Handbook", at http://www-bd.fnal.gov/runII/index.html
2. M. Church, "Tevatron Run II Performance and Plans", EPAC2002, Paris, June 2002.
3. M. Church and J. Marriner, *Ann. Rev. Nucl. Part. Sci. 43:253-95 (1993)*.
4. S. Mishra, "Status of Fermilab Recycler Ring", EPAC2002, Paris, June 2002.

Proceedings of ICHEP 2002, pp. 878–881
S. Bentvelsen, P. de Jong, J. Koch and E. Laenen (Editors)

ELSEVIER

31st INTERNATIONAL CONFERENCE ON HIGH ENERGY PHYSICS AMSTERDAM

# The BTeV experiment

Marina Artuso [a] * and Jianchun Wang [a]

[a]Department of Physics
Syracuse University, Syracuse, NY 13244, USA

BTeV is an experiment being constructed at the Fermilab $p\bar{p}$ collider, designed to study mixing, CP violation, and rare decays of beauty and charm hadrons. The detector covers the forward region (10-300 mrad) and includes a state of the art silicon pixel detector providing real time information to the first level detached vertex trigger, a Ring Imaging Cherenkov detector, providing excellent charge particle identification, a $PbWO_4$ electromagnetic calorimeter, and a muon system.

## 1. INTRODUCTION

The BTeV experiment is designed to study beauty and charm physics at the Fermilab Tevatron collider. Its goals are an exhaustive search for physics beyond the Standard Model (SM) as well as a precise determination of the Standard Model quark mixing parameters. The key measurements include CP violation, flavor oscillations, and rare decays of hadrons containing $b$ or $c$ quarks.

The CP violation in the SM originates from quark mixing with complex terms in the Cabibbo Kobayashi Maskawa (CKM) matrix. The constraints imposed by the unitarity of the CKM matrix can be visualized as 6 triangles. The most commonly used arises from the orthogonality of the $d$ and $b$ columns: $V_{ud}^* V_{ub} + V_{cd}^* V_{cb} + V_{td}^* V_{tb} = 0$, which defines the the CKM phases $\alpha, \beta$ and $\gamma$ with the constraint that $\alpha + \beta + \gamma = \pi$. Other independent angles are $\chi$ and $\chi'$ where:

$$\chi = arg\left(-\frac{V_{cs}^* V_{cb}}{V_{ts}^* V_{tb}}\right), \quad \chi' = arg\left(-\frac{V_{ud}^* V_{us}}{V_{cd}^* V_{cs}}\right).$$

While $\alpha, \beta$ and $\gamma$ may be relatively large, the angle $\chi$ is small, and $\chi'$ is even smaller. One goal of the BTeV experiment is to measure the CKM phases: $\alpha, \beta, \gamma$ and $\chi$.

We usually measure a trigonometric function of the angles. For example, the recently measured asymmetry in the decay $B^\circ \to J/\psi K_s$ [1] determines $\sin(2\beta)$. There is a 4-fold ambiguity generated by the conversion from $\sin(2\beta)$ to $\beta$. This ambiguity can be reduced by finding $\cos(2\beta)$.

Inconsistencies in determinations of CKM parameters using different physics processes may also reveal new physics. One good candidate is the decay mode $B^\circ \to \phi K_s$. The CP asymmetry in the SM for this mode is the same as for $B^\circ \to J/\psi K_s$. In models containing new physics both the mixing amplitude and the decay amplitude can be modified by new phases. However, the $J/\psi K_s$ decay, being tree level, does not usually pick up any new phase, while the loop level $\phi K_s$ decay is likely to have a new phase. Then the CP asymmetry in $B^o \to J/\psi K_s$ becomes proportional to $\sin(2\beta + \theta_M)$, where $\theta_M$ identifies a new physics induce phase in $B$ mixing, while in $B^o \to \phi K_s$ we have $\sin(2\beta + \theta_M + \theta_A)$, where $\theta_A$ is an additional phase in the loop diagram in the decay $B \to \phi K_s$. Thus, a measurement of the difference in CP asymmetries between these two modes would definitively demonstrate new physics and measure the decay phase $\theta_A$ of the new physics [2].

The angle $\chi$ can be extracted by measuring the time dependent CP violation asymmetry using CP eigenstates in $B_s$ decay modes. Silva and Wolfenstein [3], [4] show that the $\chi$ measurement

*on behalf of the BTeV collaboration


doi:10.1016/S0920-5632(02)02256-9

is a critical check to the SM by seeing if

$$\sin\chi = \left|\frac{V_{us}}{V_{ud}}\right|^2 \frac{\sin\beta\ \sin\gamma}{\sin(\beta+\gamma)} \quad .$$

BTeV can measure $\chi$ in the decay modes $B_s \to J/\psi\eta^{(\prime)}$, where $\eta \to \gamma\gamma$ and $\eta' \to \rho^\circ\gamma$, or $\pi^+\pi^-\eta$. Since $\chi$ is very small ($\approx 0.03$), we need several years of data collection to achieve the precision required by this test.

## 2. THE BTeV SPECTROMETER

The BTeV detector is an one-arm spectrometer that covers the angular region between 10-300 mrad with respect to the beam as shown in Fig. 1. It fully exploits two advantages of the "forward" direction: the correlation in the direction of the $b\bar{b}$ pair produced, and the boost that allows an easier identification of detached vertices. This provides efficient flavor tagging and sensitivity to a great variety of heavy flavor decays.

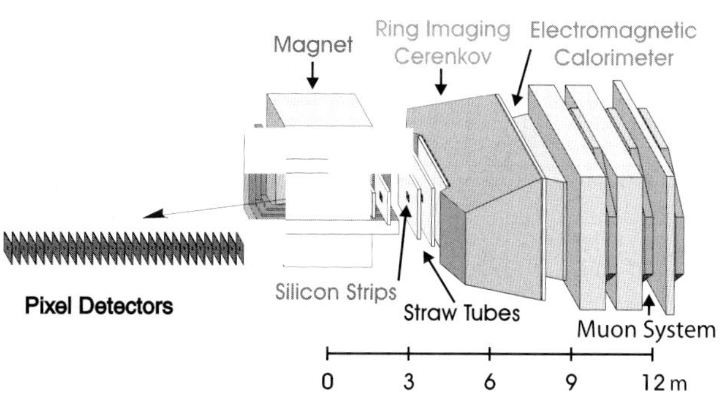


Figure 1. Layout of the BTeV spectrometer.

The spectrometer consists of a precision vertex detector composed of silicon pixel planes, a forward tracking system comprised of silicon strips near the beam and straw tube chambers at larger radius, a Ring-Imaging Cherenkov Detector (RICH), an $PbWO_4$ electromagnetic calorimeter, a muon system, and a deadtimeless trigger and DAQ system. Here we will give a brief description on some key sub-systems. More information can be found elsewhere [5].

The pixel vertex detector resides in a secondary vacuum separated from the beam by a thin RF shield, inside an 1.6 Tesla dipole magnetic field [6]. It provides accurate vertex information for offline analysis, and delivers very precise space points to the vertex trigger. The pixel vertex detector consists of 30 stations along the beam direction. Each station includes two planes with othogonal orientation of the pixel arrays, composed of elongated cells, whose dimensions are $50 \times 400\mu m^2$. The pixel technology is chosen to provide excellent signal-to-noise, spatial resolution, high speed and low occupancy. A beam test of prototype pixel detectors had been carried on in 1999-2000 at Fermilab. The resolution obtained is excellent, better than the 9 $\mu$m requirement [7]. A simulation package describes the detector performance quite well [8].

The BTeV Ring Imaging Cherenkov detector (RICH) is designed to separate $\pi/K/p$ in a momentum range of 3 to 70 GeV [9]. It is essential to CP violation studies, providing separation of specific final states, such as $K^+\pi^-$ from $\pi^+\pi^-$, and flavor tagging. The RICH detector is also a fine supplement to the electromagnetic calorimeter and the muon detector in lepton identification. It increases significantly the reconstruction efficiency in interesting modes like $B^\circ \to J/\psi K_s$. We use freon ($C_4F_{10}$) gaseous radiator to generate Cherenkov light in the optical frequency range. The light is focused by mirrors onto Hybrid Photo-Diode (HPD) tubes at upstream. To separate kaons from protons below the threshold of gaseous radiator, a liquid radiator ($C_5F_{10}$) is used and the light is focused onto the side of the vessel and detected by a photomultiplier tube array.

BTeV uses radiation hard lead tungsten scintillation crystals ($PbWO_4$) to detect photons and electrons [10]. The crystals are 220 mm long (25 $X_0$) and have very small transverse cross-section ($\approx 28 \times 28mm^2$), providing excellent segmentation. The light is collected by photomultiplier tubes. Sample crystals were tested in a beam at Protvino, Russia. The energy and position resolutions were demonstrated to be excellent [11].

With a nominal luminosity of $2 \times 10^{32}\text{cm}^{-2}\text{s}^{-1}$, the Tevatron delivers $2 \times 10^{11}$ $b$-hadron per year. It is an ideal place for study of $b$-physics. It also poses a severe challenge to the data acquisition (DAQ) system. Only a small portion of inelastic events are from heavy quarks, and the interaction rate is very high at the bunching spacing of 132 ns.

A sophisticated detached vertex algorithm is the heart of the BTeV trigger scheme, using the unique event topology to distinguish "B" events from undesired backgrounds [12]. It uses hits provided by the pixel vertex detector and parallel pipelines insure that the data are held until the Level 1 algorithm is completed, thus it is deadtimeless. At Level 2 and Level 3, there is much more time to implement a more refined reconstruction. With this robust system, the total event rate is reduced from 7.6 MHz to 2 - 4 kHz. The trigger efficiency for a typical beauty decays is greater than 50%.

## 3. PHYSICS REACH

We use GEANT-based Monte Carlo simulation to determine our sensitivity to the processes that we plan to study. A full pattern recognition is done at trigger level. The algorithm is demonstrated to have high efficiency and very few false tracks. For "offline" analysis, pattern recognition is not included since it has very little impact because of the excellent segmentation of the pixel detector. The charged tracks are fitted with a Kalman filter. Realistic shower reconstruction, and the fine details of the charged particle identification system are implemented in our simulation. a The results on CKM parameters for $10^7$s running at nominal luminosity are shown in Table 3, showing that this detector has the ability to explore the whole $b$ decay phenomenology relevant to precise and comprehensive Standard Model test.

The BTeV data samples will be large enough to identify new physics signatures. In Table 2 we list some of the event samples relevant to new physics studies with one "snow mass" year data collection. Also shown in comparison are samples from $e^+e^-$ $B$ factories that has 500 fb$^{-1}$ accumulated data, approximately the expected amount when BTeV starts. BTeV will easily surpass the $e^+e^-$ $B$ factories and also be optimal for the important CP violation measurements that need to be made in $B_s$ modes.

Table 1
BTeV sensitivity to CKM angles with a data sample accumulated in 1 Snowmass year ($10^7$ s).

| Decay modes | Events | S/B | Parameter | Error or (Value) |
|---|---|---|---|---|
| $B^\circ \to \pi^+\pi^-$ | 14,600 | 3 | Asy | 0.030 |
| $B_s \to D_s^+ K^-$ | 7,500 | 7 | $\gamma$ | $8^\circ$ |
| $B^\circ \to J/\psi K_s$ | 168,000 | 10 | $\sin(2\beta)$ | 0.017 |
| $B_s \to D_s^+ \pi^-$ | 59,000 | 3 | $x_s$ | (75) |
| $B^- \to \overline{\overline{D}}^\circ(K^+\pi^-)K^-$ | 170 | 1 | | |
| $B^- \to D^\circ(K^+K^-)K^-$ | 1,000 | $> 10$ | $\gamma$ | $13^\circ$ |
| $B^- \to K_s\pi^-$ | 4,600 | 1 | | $< 4^\circ$ + |
| $B^\circ \to K^+\pi^-$ | 62,100 | 20 | $\gamma$ | th. err. |
| $B^\circ \to \rho^+\pi^-$ | 5,400 | 4.1 | | |
| $B^\circ \to \rho^\circ\pi^\circ$ | 780 | 0.3 | $\alpha$ | $\sim 4^\circ$ |
| $B_s \to J/\psi\eta$ | 2,800 | 15 | | |
| $B_s \to J/\psi\eta'$ | 9,800 | 30 | $\sin(2\chi)$ | 0.024 |

Table 2
Reconstructed events in new physics modes for BTeV and $e^+e^-$ $B$-factories

| Decay mode | BTeV ($10^7$ s) | | | $B$-factories (500 fb$^{-1}$) | | |
|---|---|---|---|---|---|---|
| | Yield | Tagged | S/B | Yield | Tagged | S/B |
| $B_s \to J/\psi\eta^{(\prime)}$ | 12,650 | 1,645 | $> 15$ | 0 | | |
| $B^- \to \phi K^-$ | 6,325 | 6,325 | $> 10$ | 700 | 700 | 4 |
| $B^o \to \phi K_s$ | 1,150 | 115 | 5.2 | 250 | 75 | 4 |
| $B^o \to K^{*o}\mu^+\mu^-$ | 2,530 | 2,530 | 11 | ~50 | ~50 | 3 |
| $B_s \to \mu^+\mu^-$ | 6 | 0.7 | $> 15$ | 0 | | |
| $B^o \to \mu^+\mu^-$ | 1 | 0.1 | $> 10$ | 0 | | |

The LHCb experiment will be the main competitor to BTeV. Compared to LHCb, BTeV has a more robust trigger system that use a detached vertex algorithm in the first event selection level. In the "golden" mode $B_s \to D_s K^-$ for $\gamma$ measurement, although LHCb has initial advantages in $b$ cross-section, the yield and signal to background ratio are compatible between the two experiments. BTeV includes a much better electromagnetic calorimeter. Thus it is superior in $\alpha$ measurement that relies on $B^\circ \to \rho^+\pi^-, \rho^\circ\pi^\circ$ decay modes.

## 4. CONCLUSIONS

The BTeV experiment is presently in a stage of intense and successful R& D. Several detector components have been already tested with beams and more extensive test beam runs are planned for the year 2003. Components located in intense radiation environment has been irradiated and proven to be able to withstand the expected doses. Progress has been made on the trigger and DAQ. We have received IT funding from the NSF to develop fault-tolerant, fault-adaptive software to control the trigger system in real time. This is named "The Real Time Embedded Systems" (RTES) project.

In summary, quoting the recent unanimous recommendation from the Fermilab program advisory committee, "... BTeV has designed and prototyped an ambitious trigger that will use $B$ decay displaced vertices as its primary criterion. This capability, together with BTeV's excellent electromagnetic calorimetry and particle ID and enormous yields, will allow this experiment to study a broad array of $B$ and $B_s$ decays. BTeV has a broader physics reach than LHCb and should provide definitive measurements of CKM parameters and the most sensitive tests for new physics in the flavor sector."

## 5. ACKNOWLEDGEMENTS

The authors would like to thank the many BTeV collaborators who have contributed to the analysis and R& D advances described in this paper. This work was supported by National Science Foundation.

## REFERENCES

1. B. Aubert *et al.* [BaBar], Phys. Rev. D **66**, 032003 (2002); K. Abe *et al.* [Belle], Phys. Rev. D **66**, 071102 (2002).
2. Y. Nir, arXiv:hep-ph/9911321.
3. J. P. Silva and L. Wolfenstein, Phys. Rev. D **55**, 5331 (1997).
4. R. Aleksan, B. Kayser and D. London, Phys. Rev. Lett. **73**, 18 (1994).
5. J. C. Wang, Int. J. Mod. Phys. A **16S1C**, 1062 (2001).
6. S. Kwan [BTeV Collaboration], Nucl. Instrum. Meth. A **408**, 235 (1998).
7. J. A. Appel *et al.*, Nucl. Instrum. Meth. A **485**, 411 (2002).
8. M. Artuso and J. C. Wang, Nucl. Instrum. Meth. A **465**, 115 (2000).
9. T. Skwarnicki [BTeV Collaboration], Nucl. Instrum. Meth. A **408**, 204 (1998).
10. S. R. Menary [BTeV Collaboration], Nucl. Instrum. Meth. A **446**, 253 (2000).
11. V. Batarin *et al.* [BTeV Electromagnetic Calorimeter Group], hep-ex/0209055 (2002).
12. P. Lebrun, Int. J. Mod. Phys. A **16S1C**, 1153 (2001).

Proceedings of ICHEP 2002, pp. 882–884
*S. Bentvelsen, P. de Jong, J. Koch and E. Laenen (Editors)*

31st INTERNATIONAL CONFERENCE ON HIGH ENERGY PHYSICS AMSTERDAM

# Status of CDF Silicon Tracking

S. Nahn[a] for the CDF Collaboration

[a]Yale University Physics Department
PO Box 208121, New Haven, CT 06520-2181

A snapshot of the status of the CDF Run II Silicon Detector dated July 2002 is presented, with a summary of commissioning issues since the start of Run II, current performance of the detector, and the use of the data in both the trigger and offline reconstruction.

## 1. The CDF Silicon Detector in Run II

Coincident with the Tevatron Run II upgrade for higher $\sqrt{s}$ and increased luminosity, CDF replaced the Run I silicon detector with an entirely new detector with nearly double the coverage of both the luminous region and pseudorapidity ($\eta$), the ability to do three dimensional tracking, and enhanced impact parameter resolution for better B tagging [1,2]. The 8 layer, 704 ladder, 722432 channel detector that resulted is comprised of three subdetectors: SVXII, ISL, and L00 ("Layer Zero Zero"). SVXII is 360 double-sided ladders in a layout of six 15 cm axial sections $\times$ twelve 30° $\phi$ slices $\times$ five radial layers between 2.5 and 10.6 cm. ISL covers the area between SVXII and the CDF wire tracker (COT), with 296 double-sided ladders at radii of 20 or 28 cm, in total *1.9 m* long, providing silicon hits out to $|\eta| \leq 2$. L00 is a single-sided layer of 48 ladders mounted directly on the beampipe, 1.5 cm from the beamline, which enhances the impact parameter resolution. The $r-\phi$ and $r-z$ views of the detector are shown in Figure 1.

The three subdetectors share the same readout system, starting with the SVX3D chip, a custom designed ASIC with a 128 channel $\times$ 46 capacitor analog storage ring, which makes it possible to acquire data in a "deadtimeless" fashion, integrating charge on one capacitor while reading out another. The SVX3D chip also features common mode noise suppression, sparsification, and other settings used to optimize the charge collection. A "wedge" of up to 5 ladders are serviced by one

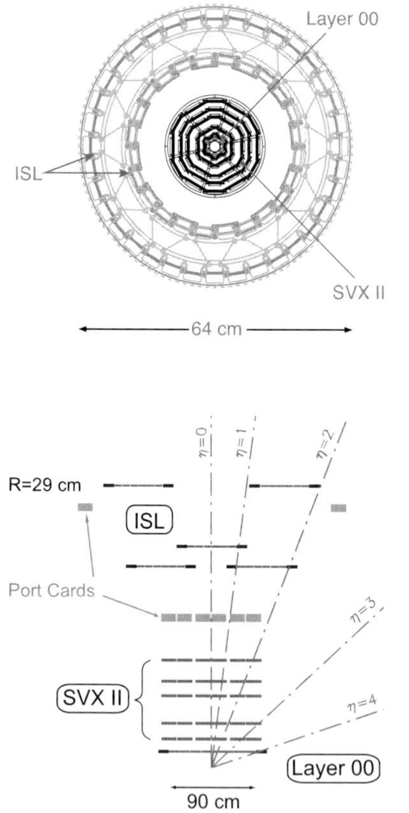


Figure 1. $r-\phi$ and $r-z$ cartoons of the Silicon detector, showing all sub-components. Note the compression of the $z$ scale.


doi:10.1016/S0920-5632(02)02258-2

"portcard" mounted near the detector which provides a $1 \rightarrow 5$ fanout of electrical control signals and converts the data from electrical to optical for long distance transmission, as well as bringing in the analog, digital, and bias voltages to the chips and ladders. Behind this is a VME based data acquisition system which controls the entire detector and presents data to the Level 2 and Level 3 trigger systems. All detectors are incorporated in the same power distribution, cooling, interlock, and radiation protection systems.

## 2. Silicon Commissioning Experience

Commissioning commenced with the onset of Run II in early 2001. Since then, progress has been steady in getting the detector to a robust operating point, but seriously hampered by a series of problems, from which lessons for future detectors can be drawn. The first serious problem was an issue with a vendor, overcome with relentless pursuit. The full detector was not outfitted with power supplies until late October, forcing "supply juggling" to ensure functionality and cabling during installation. Furthermore, the supply interlock mechanisms failed, such that the detector could be powered without cooling. This danger necessitated the shutdown of all Silicon operation until the problem was solved. The second set of problems arose from a lack of testing under realistic conditions. This manifested itself in a variety of ways, most notably the discovery of cooling lines blocked by epoxy in ISL, which have since been bored out with a laser, but also the presence of excess noise in L00, and mismatches in optical power of receivers and transmitters, both of which required much effort to work around. The solutions were developed as soon as the issues were were exposed by the true operating conditions, but could have been seen before. In contrast, the last set of impediments were *beam-induced* damage to SVX3D chips, and ladder and VME power supplies, the latter of which caused large sections of the detector to be turned off. The diagnosis of the chip damage occupied over a month during which the detector was shut off while scenarios were explored on the bench to understand what was damaged, since the damaged parts are inaccessible. Considerable delay may have been eliminated had the failure modes been well catagorized before installation. Having all pieces available, testing under operational conditions, and understanding in detail all failure modes are the important lessons to take away from the silicon commissioning experience.

## 3. Performance of the Silicon

Despite the litany of problems described in the previous section, by July 2002 over 90% of the silicon detector was powered, and more than 80% reading out quality data. For any ladder type and chip setting on average $\frac{Signal}{Noise} \geq 10$, with operating conditions set at 11-15 depending on ladder type. Figure 3 shows a histogram of a typical ladder's pedestal and signal ADC counts, exhibiting clear seperation between the guassian noise and Landau distribution for the signal. With high signal to noise, the single hit efficiencies can be kept $\geq 99\%$, maintaining high signal efficiency without prohibitive data volume. With tracks in the data both the global and relative alignment continue to improve. In addition, the cluster resolution of

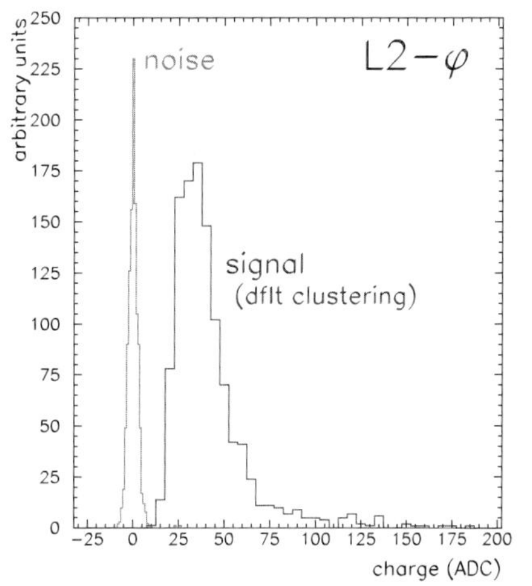


Figure 2. Pedestal and signal for a SVXII L2 $\phi$ side ladder

the detector is as expected, with for example an 11 $\mu m$ intrinsic resolution on 2 strip clusters for the $r - \phi$ measurement, as shown in Figure 3. In general, the raw silicon detector performance shows promise for physics in Run II.

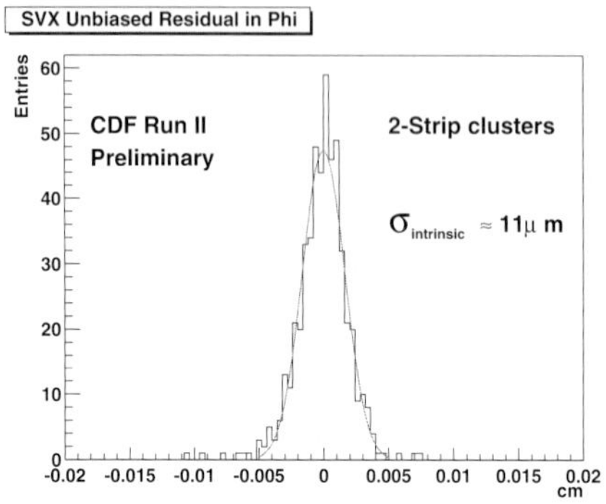


Figure 3. Two-Strip $\phi$ cluster resolution $\sim 11\ \mu m$

## 4. Tracking Applications

While performance benchmarks are necessary for validation, the real test of the detector is whether one can perform measurements. In addition to traditional offline tracking, the silicon detector is also utilized in the Level 2 trigger every $\sim 50\ \mu s$, where tracking information from the COT is combined with silicon hits and high impact parameter tracks are then used to trigger on heavy flavor physics. This allows CDF to trigger on lepton-less B decays, which both relieves the penalty of the small B leptonic branching fraction and enhances the fraction of fully reconstructed B decays in the data. The details and results of this application were presented to the conference and can be found in [3]. Of course, the detector is also used in offline tracking, and there are results from Run II both in "calibration" channels such as the $J/\psi$ mass (Figure 4) and B physics channels, as presented in [4]. The conclusion is that this detector is useful for collecting and analyzing interesting physics data.

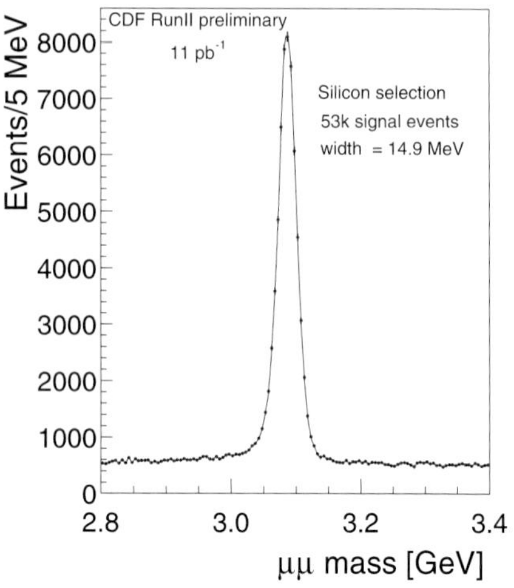


Figure 4. Calibration using the $J/\psi$ mass

## 5. Conclusion

With the first $\sim 35\ pb^{-1}$ of Run II data, the new CDF silicon detector has been commissioned, with some setbacks, and is currently taking data of a quality useful for competitive measurements. The detector will continue to improve as subtle problems are addressed, until the end of its useful lifetime, when the Run IIb detector will be installed [5].

## REFERENCES

1. CDF Collaboration, "The CDF II Detector Technical Design Report", FERMILAB-Pub-96/390-E.
2. CDF II Collaboration, "Proposal for Enhancement of the CDF II Detector: An Inner Silicon Layer and a Time of Flight Detector" FERMILAB-Proposal-909
3. A. Cerri, *these proceedings*
4. F. Bedeschi, *these proceedings*
5. M. Kruse, *these proceedings*

ELSEVIER

Proceedings of ICHEP 2002, pp. 885–887
*S. Bentvelsen, P. de Jong, J. Koch and E. Laenen (Editors)*

31st INTERNATIONAL CONFERENCE ON HIGH ENERGY PHYSICS AMSTERDAM

# Silicon Detector Upgrades for the Tevatron Run 2

M. C. Kruse [a]*

[a]Department of Physics, Duke University, P.O. Box 90305, Durham, North Carolina 27708-0305, USA

The current silicon devices being used by the DØ and CDF collaborations for the Tevatron Run 2a, which is expected to end in 2005 after accumulating about $2\,\mathrm{fb}^{-1}$ of data, will need to be replaced due to radiation damage for the following data collection period designated as Run 2b. We will discuss these silicon replacement plans, the more uniform design of the detectors between DØ and CDF, and the current status of their fabrication.

## 1. INTRODUCTION

The goal of the Tevatron Run 2a is to deliver about $2\,\mathrm{fb}^{-1}$ of integrated luminosity to the DØ and CDF detectors by 2005 [1]. Based on some recent improvements to the Tevatron performance, we expect to accumulate about $100\,\mathrm{pb}^{-1}$ by the end of 2002, and $400\,\mathrm{pb}^{-1}$ by the end of 2003. The radiation damage that will be incurred by the Run 2a DØ and CDF silicon detectors, both to the bulk silicon and the frontend readout electronics, will necessitate at least the partial replacement of these systems.

Following Run 2a, and after a short shutdown period, Tevatron running will resume as Run 2b, with the ultimate integrated luminosity goal of $15\,\mathrm{fb}^{-1}$ per experiment. Such an integrated luminosity is essential for the sensitivity to the discovery of a light Higgs boson at the Tevatron. The Higgs sector is the last cornerstone of the Standard Model yet to be directly explored, and our findings, or lack thereof, will have deep ramifications to our understanding of matter. A light Higgs boson ($M_H < 140\,\mathrm{GeV}/c^2$) decays predominantly to a $b\bar{b}$ pair. Therefore, critical to the discovery of a light Higgs is the detection of the secondary vertices from $b$-quark decays, accomplished with high efficiency through the use of silicon vertex detectors. In Run 2 we expect to be able to "tag" at least one $b$-quark jet from a Higgs decay with about 60% efficiency.

The silicon vertex detectors have been designed for both the CDF and DØ Run 2b upgrade projects, with construction and prototyping now well underway. The main factors governing the design of these detectors were the following.

- Greater radiation tolerance: both sensors and frontend electronics have been designed to operate for at least $15\,\mathrm{fb}^{-1}$.
- Simplicity of construction to meet the scheduled completion in 2005/2006.
- Retention or improvement of the tracking capabilities of the Run 2a silicon systems.
- Preservation of as much of the existing infrastructure as possible. This includes most of the DAQ systems, the external cooling systems, and the available space and structural support, present in both the CDF and DØ experiments.
- Given the infrastructure differences between the two experiments, exploit the areas where we can benefit from common design features. This includes common readout chips, and sensor specifications.

## 2. MECHANICAL LAYOUTS

Both the DØ and CDF Run 2b silicon detectors have a 6-layer geometry arranged in 2 barrels, each about 60 cm long, joined at $z = 0$. The Run 2a DØ silicon tracker [2] will be replaced in its entirety. The outer layers of the CDF silicon system, the ISL [3], will remain unchanged, with the present SVX 2a and Layer 00 devices [4] being replaced. The geometrical similarity is apparent in Figure 1, which shows the cross-sectional views of the replacement detectors.

The 6 layers for both detectors (labelled 0 to 5) vary in radii from about 2 cm to 16 cm. This

*Representing the CDF and DØ collaborations


doi:10.1016/S0920-5632(02)02265-X

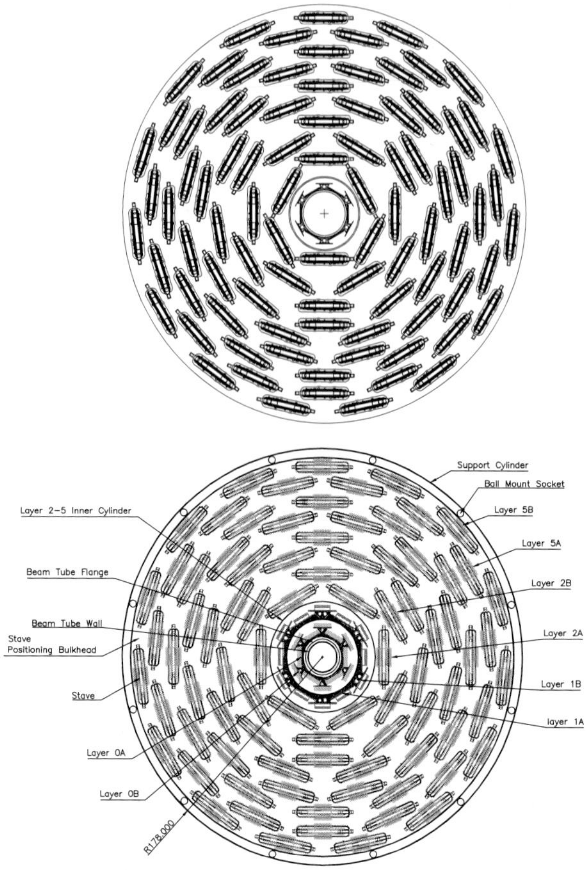


Figure 1. Transverse views of the mechanical layout of the CDF(top) and DØ (bottom) silicon detector upgrades.

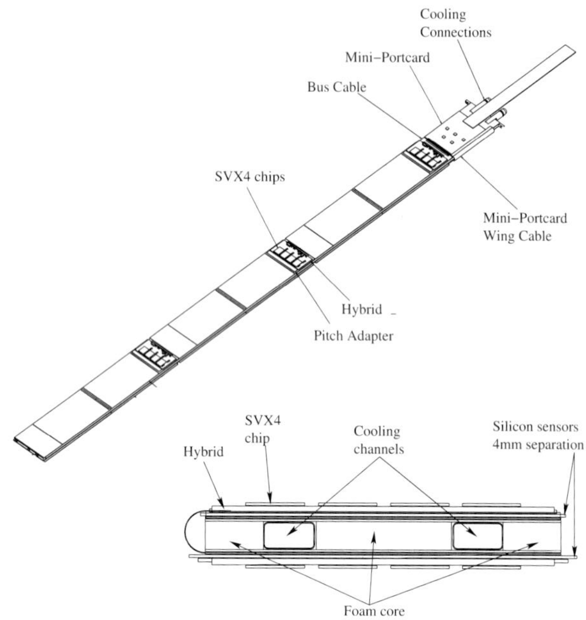


Figure 2. Isometric (top) and cross-sectional (bottom) views of the CDF *stave* design for all the outer layers.

fills more of the tracking volume with silicon sensors as opposed to the Run 2a designs, as a lot of the readout electronics have been pushed out to the ends of the barrels. The basic detector unit is the "stave". Details in the stave design differ between DØ and CDF but the more general design features are similar, in that they are composed of two layers of single-sided silicon sensors upon which the SVX4 readout chips are mounted (by way of "hybrid" circuit boards), and between which the cooling channels have been integrated. A schematic of the CDF stave is shown in Figure 2. There will be no $90°$ stereo sensors, however, layers 2 through 5 have either $2.5°$ or $1.2°$ small-angle-stereo sensors on one side of the staves. DØ will have the same stave design for layers 2 to 5, and CDF for layers 1 to 5. Therefore, both detectors will be entirely comprised of only 3 sensor types (axial and stereo for the outer layers, and axial for the inner layer(s)), greatly reducing the design complexity. The sensor pitches differ slightly for both experiments: DØ use a strip(readout) pitch of $30\mu m (60\mu m)$ for the outer layers (2 to 5), $29\mu m (58\mu m)$ for layer 1, and $25\mu m (50\mu m)$ for layer 0; and CDF uses pitches of $37.5\mu m (75\mu m)$ for the outer layers (1 to 5), and $25\mu m (50\mu m)$ for layer 0. The smaller pitch used for the DØ outer layers, means 5 SVX4 chips are required to read out the channels spanning the width of a sensor, as opposed to 4 readout chips for CDF.

The single-sided sensors can be operated at higher bias voltages than the double-sided Run

2a sensors which are limited to about 100V by the coupling of the applied voltages to each side. As the bulk silicon suffers radiation damage the voltages required to maintain full depletion increase, giving the single-sided sensors a considerably longer lifetime in high radiation environments. It is envisioned by the end of Run 2b bias voltages for the inner layers will need to be about 600V. The use of single-sided sensors also increases the simplicity of the construction. Initial tests on prototype sensors have been very encouraging, with measured leakage currents being about $0.15\mu A$ all the way to 1000 V (the specification being $0.20\mu A$). Construction of prototype staves is now underway.

## 3. DATA ACQUISITION

The frontend readout system has undergone the most significant changes from the Run 2a DAQ. The rest of DAQ system has largely remained intact.

Both CDF and DØ will use the SVX4 readout chip, replacing the current SVX3 and SVX2 chips respectively. The SVX4 chip is based on the SVX3 design with an analog front-end, digital back-end, a 42 cell pipeline, and 128 channels wide, but produced with 0.25 micron technology, giving it an intrinsic radiation hardness that is expected to withstand the total doses to the inner layers. The sub-micron technology has also resulted in the halving of the operating voltage to 2.5 V. Prototype SVX4 chips were received in June 2002 and are fully functional with a limited number of fixes required. Further studies, including noise tests are ongoing.

The SVX4 chips are mounted on, and serviced by, hybrid circuit boards, which are in turn mounted on the staves. The same hybrid substrate is used by both DØ and CDF, however, the hybrid designs are somewhat different. The hybrids use a BeO substrate, and are smaller than in Run 2a, significantly reducing their impact in the tracking volume. The one exception is the inner CDF layer, which uses the same hybrid design as for the current CDF Layer 00 detector, with alumina substrate and placed outside of the tracking volume.

CDF has two hybrid types, one with 2 SVX4 chips for layer 0, and a 4-chip version for the outer layers. As can be seen in Figure 2 each hybrid reads out a 2-sensor module. A total of 72 2-chip and 1080 4-chip hybrids are required.

DØ has four hybrid types and utilizes a double-ended design which reduces the total hybrid readout count. Outer layers (2 to 5) are read out by 10-chip hybrids (5 chips on each end), layer 1 by double-ended 6-chip hybrids, and layer 0 by single-ended 2-chip hybrids. A total of 672 10-chip, 72 6-chip, and 144 2-chip hybrids are required.

Circuit boards outside the tracking volume take signals from the hybrids and feed them to the rest of the DAQ system, similarly to the Run 2a system, however, with some of the hybrid and port-card functionality moved further from the tracking volume, and with no optical components.

## 4. CONCLUSIONS

Extending the Tevatron Run 2 to 15 $fb^{-1}$ opens up many exciting physics opportunities. The greater integrated luminosity also extends the radiation dose to the existing inner layers of the CDF and DØ silicon vertex detectors beyond their design limits, which has compelled the design and construction of silicon detector upgrades. Both the DØ and CDF upgrades have many similarities within the constraints of their existing infrastructure, the most notable being similar sensor specifications, and the use of the same SVX4 readout chip. Progress on many components has been excellent, for example, we have SVX4 chips working on hybrids, and working sensors have been delivered to both experiments. These devices will play a crucial role in extending our capabilities to answer some very important physics questions this decade.

## REFERENCES

1. K. Gounder, *these proceedings.*
2. G. Ginther, *these proceedings.*
3. A. Affolder, *et al.* (CDF collaboration), Nucl. Instrum. Meth. A461:216-218,2001
4. S. Nahn, *these proceedings.*

ELSEVIER

Proceedings of ICHEP 2002, pp. 888–890
*S. Bentvelsen, P. de Jong, J. Koch and E. Laenen (Editors)*

31st INTERNATIONAL CONFERENCE ON HIGH ENERGY PHYSICS AMSTERDAM

# Tools for simulation and analysis

A. Pfeiffer[a], M. G. Pia[b] (on behalf of the Anaphe and Geant4 Collaborations)

[a]CERN, CH-1123 Geneva, Switzerland

[b]INFN Sezione di Genova, Via Dodecaneso 33, 16146 Genova, Italy

Software plays an incrwasingly important role in the design and operation of large-scale HEP detectors, as well as in the analysis of experimental data. In the recent past, new software technologies have emerged in the HEP domain, and new software systems, in particular for detector simulation (Geant4) and data analysis (Anaphe), have been developed to address the challenging requirements of new-generation HEP experiments.

## 1. The challenge of HEP software

The software for high energy physics experiments faces an unprecedented challenge, determined by the complexity of the physics under study, of the detectors and of the environments where they operate.

The complexity of the physics of the LHC experiments, related to the high machine luminosity and the significant QCD background, is the source of their enormous computational requirements: 5-8 Petabytes of data will be generated each year, and their analysis will require $\approx 10$ Petabytes of disk storage and the equivalent of 200,000 of today's fastest PC processors.

The software for astroparticle experiments, which embrace a wide spectrum of scientific programmes, from underground neutrino physics and dark matter searches to astrophysics measurements on satellites to cosmic ray physics, is characterised by a similar variety of requirements. The physics models necessary for the simulation of such experiments, also accounting for the space radiation environment or underground backgrounds, span an energy range from the order of few eV to the PeV scale. Moreover, for some of these experiments, like, for instance, those on satellites, simulation is often mission critical, thus requiring high reliability and rigorous software engineering standards in addition to powerful functionality.

While facing the diversity of requirements from many different applications, the software for HEP has to deal with a growing trend to share functionalities across diverse fields: for instance, models to extend the simulation of electromagnetic interactions down to low energies ($< 1keV$), motivated originally from medical physics applications, have been successfully applied to astroparticle experiments, and are of interest for the precise simulation of LHC detectors.

The intrinsic complexity of the software for the new generation of HEP experiments is further complicated by the rapidly changing computing environment. The evolution of hardware, operating systems and software over the long time scale of LHC operation, as well as likely changes in the requirements of the experiments, imply that the software being built now must be able to anticipate changes.

The response to such challenges consists of the adoption of a rigorous software engineering, together with a collaborative effort to share resources effectively. The adoption of object oriented technology makes the software open to extension and evolution, and facilitates its maintenance. A component-based architecture allows both the specialisation of components and their easy integration with experimental frameworks, providing the flexibility to customise the software by means of alternative, interchangeable implementations. Computing facilities implemented as a global computational grid aim to integrate geographically distributed computing fabrics into a


doi:10.1016/S0920-5632(02)02266-1

worldwide virtual computing environment.

## 2. The Geant4 Simulation Toolkit

Geant4 [1] is an object oriented simulation toolkit designed for the new generation of high energy physics experiments; its current applications also cover astrophysics, nuclear physics, medical physics, space science and radiation background studies.

Software engineering plays a fundamental role in Geant4: advanced techniques are adopted in response to the requirements of functionality, modularity, extensibility and openness, resulting in a wide set of capabilities to model detectors and particle interactions with matter over a wide energy range.

Thanks to its modular architecture, Geant4 can be easily integrated into the complex software frameworks of large-scale HEP experiments, to collaborate with other software components such as data analysis.

### 2.1. Overview of Geant4 functionality

The Geant4 kernel handles the management of runs, events and tracking. The Event package provides an abstract interface to external physics event generators for the creation of the primary particles. Geant4 Tracking handles the propagation of a track, determined by physics interactions. It is completely general and common to all processes for any particle type, thus allowing great flexibility in the implementation of a variety of physics processes, as well as openness to further extensions. The management of particles is based on Particle Data Group compliant definitions and data, including their decay processes and modes.

The Geant4 Geometry package provides tools to describe the detailed geometrical structure of a detector; it also handles the equation of motion solvers in different fields and geometrical boundary conditions for the propagation of particles. An ISO STEP compliant solid modeller allows the exchange of models from CAD systems. Multiple solid representations, such as Constructive Solid Geometry or Boundary Represented Solids are available. Boolean operations on solids are supported.

The Geant4 Materials package allows the description materials consisting of a single element or a composition of elements, which in turn can consist of a single isotope or a mixture of isotopes.

The Hits and Digi domains provide the functionality to reproduce the readout structure of a detector and its electronic response, independently from the geometry used for tracking particles.

Geant4 Visualisation provides the capability to visualise detector geometry, particle trajectories, tracking steps, hits and text. Its design, based on abstract interfaces, makes Geant4 independent from any particular graphics system; at the same time it allows multiple implementations of drivers to interface the simulation with a variety of such systems. The User Interface domain adopts a similar approach, allowing the usage of a variety of user interfaces, from simple command-line driven ones to sophisticated GUIs.

A Fast Parameterisation facility is integrated with the full simulation, allowing independent and simplified detector descriptions and direct production of hits.

Extensive possibilities for interaction with the Geant4 system are offered to the user via a set of dedicated user-action classes.

### 2.2. Geant4 physics

One of the most important goals of Geant4 consists in making the design and implementation of the physics open and transparent. The Geant4 design makes physics data, models and assumptions transparently accessible to the user, rather than hard-coded in black-box packages, thus improving the verification and the reliability of simulation results. It also exposes the granular implementation of the physics, each component of which can be inspected at source code level. Thanks to object oriented technology, a variety of alternative or complementary physics models can be provided for the same physical process, with openness to further extensions.

Geant4 electromagnetic physics [2] provides a variety of implementations of electron, positron, photon, charged hadron and ion interactions. Photon processes include Compton and Rayleigh scattering, $\gamma$-conversion and the photoelectric effect. Electron/positron processes include bremsstrahlung, ionisation and $\delta$-ray production, positron annihilation and synchrotron radiation. The ionisation and energy loss are available for hadrons and ions as well. The multiple scattering process handles all charged particles, computing the mean path length correction and the mean lateral displacement. Low energy extensions [3] are implemented, down to 250 eV for photons and electrons, and to the energy corresponding to the ionisation potential of the material for hadrons and ions. The validity range of all the muon processes, based on theoretical models, scales up to the thousand PeV region, allowing the simulation of ultra-high energy and cosmic ray physics.

Geant4 can also handle the optics of scintillation and Cherenkov detectors and their associated light guides, including a set of dedicated processes for optical photons: refraction and reflection at medium boundaries, absorption and Rayleigh scattering.

The Geant4 Toolkit provides a rich set of physics models for the simulation of hadronic interactions [4]. The models follow three different basic approaches: data-driven, parameterisation-driven and theory-driven modelling; in the overall framework they offers both complementary and alternative options.

## 3. Analysis tools

### 3.1. Abstract Interfaces for Data Analysis

The AIDA [5] project represents an ongoing effort to define a set of common abstract interfaces for data analysis in high energy physics experiments; it encompasses histograms, ntuples, functions, vectors, fitting, plotting. The interfaces have been defined in C++ and Java.

Abstract interfaces, i.e. classes with pure virtual methods and only inheriting from other abstract interfaces, enforce a weak coupling among the components of a software system: the components may therefore evolve independently, thus reducing the maintenance overhead. This approach also provides a great flexibility, since different implementations of a component can be easily interchanged by using dynamically loadable libraries.

### 3.2. The Anaphe Analysis Toolkit

The Anaphe [6] toolkit provides an object oriented software environment for data analysis in high energy physics experiments, compliant with AIDA interfaces. The architecture of Anaphe is the result of a significant effort to minimise the dependencies: the system is characterised by a strict partitioning into independent packages, which communicate only through interfaces. This component approach facilitates the maintainability of the software, as well as the flexibility of customisation of user applications.

A range of commercial and public domain libraries is used to cover basic functionalities; on top of these libraries a set of HEP-specific class libraries for histogram management, fitting, plotting and ntuple-like data analysis has been developed. In some cases alternative implementations for the same interface are provided.

An interactive analysis and visualisation tool (Lizard) provides a flexible Python-based framework, which can be fully configured at run-time to load any of the Anaphe libraries or external modules.

## REFERENCES

1. S. Agostinelli et al., GEANT4: a simulation toolkit, SLAC-PUB-9350.
2. S. Chauvie et al., Geant4 electromagnetic physics, Proc. of the Monte Carlo 2000 Conference, Springer, 2000.
3. S. Chauvie et al., Geant4 low energy electromagnetic physics, Proceedings of CHEP2001, Beijing, 2001.
4. J.P. Wellisch, Hadronic shower models in Geant4: the frameworks, Comput.Phys. Commun. No. 140 (2001) 65.
5. G. Barrand et al., Abstract Interfaces for Data Analysis: Component Architecture for Data Analysis Tools, Proceedings of CHEP2001, Beijing, 2001.
6. B. Ferrero Merlino et al., Anaphe: OO Libraries and Tools for Data Analysis, Proceedings of CHEP2001, Beijing, 2001.

ELSEVIER

Proceedings of ICHEP 2002, pp. 891–894
*S. Bentvelsen, P. de Jong, J. Koch and E. Laenen (Editors)*

31st INTERNATIONAL CONFERENCE ON HIGH ENERGY PHYSICS AMSTERDAM

# Silicon Detectors at the LHC

C. Parkes[a]*

[a]Department of Physics and Astronomy, Kelvin Building,
University of Glasgow, Glasgow, G12 8QQ, Scotland, U.K.

The silicon systems of the Large Hadron Collider (LHC) detectors are briefly described. The complexity and diversity of the projects are illustrated by highlighting for discussion different components of the silicon systems in each experiment.

## 1. Introduction

The LHC detectors will have to function in a high radiation environment, fluences of up to $10^{15}$ 1 MeV neutron equivalents ($n_{eq}$)/$cm^2$ are expected over a 10 year period. They will have to cope with a 40 MHz collision rate and high particle multiplicities. In this environment, the detectors must provide efficient tracking and b-tagging capabilities.

These requirements are met by the silicon systems that are being constructed for the LHC detectors on a scale previously unprecedented in particle physics.

## 2. ALICE

ALICE, the heavy ion collider experiment of the LHC, has the highest expected particle multiplicities at the LHC: up to 8000 charged particles per unit of rapidity. In order to achieve a good two-track separation three dimensional hit information is required wherever possible.

The primary functions of the ALICE silicon system [1] are the determination of primary and secondary vertices necessary for the reconstruction of charm and hyperon decays, to provide identification (using dE/dx) for low momentum particles and to assist the ALICE time projection chamber in particle tracking.

The silicon system consists of six cylindrical layers of silicon detectors. The two inner layers, at radii of $4-7$ cm, and coping with multiplicities of up to 90 particles/$cm^2$ are comprised of 240 modules containing 9.8 million $50 \times 425$ $\mu$m $p^+ - \text{on} - n$ pixels .

The two central layers consist of silicon drift detectors, ALICE is the only experiment at the LHC to exploit this technology. The need for three dimensional information combined with the relatively modest central event rate ($\sim$40 Hz trigger) and radiation hardness requirements (12.7 krad in 10 years for the inner layer) at ALICE make this a suitable technology choice. A series of $p^+$ field strips are implanted on both sides of an $n$-type silicon wafer and connected by built-in voltage dividers. Hence, under an applied bias voltage, an electric drift field parallel to the wafer surface is produced. Electrons liberated by charged particles are drifted up to 35 mm in 4.5 $\mu$s and collected on the 256 $n^+$ anodes. The doping inhomogeneities in the neutron transmutation doped material cause the electrons to deviate significantly from their ideal trajectory and this must be compensated for. The deep sub-micron CMOS readout electronics samples the collected charge at 40 MHz with a 10 bit ADC. After corrections are applied a spatial precision of 25 $\mu$m in the drift co-ordinate and 33 $\mu$m in the perpendicular co-ordinate has been achieved over the full drift distance of 35 mm.

The outermost two layers of the silicon system, at radii of 39 and 43 cm, are comprised of 1770 modules containing 2.7 million 95 $\mu$m pitch double sided silicon strips. Analogue read-out, implemented in the 0.25 $\mu$m CMOS process, is used in order that the dE/dx information is preserved.

*email: Chris.Parkes@cern.ch


doi:10.1016/S0920-5632(02)02268-5

## 3. LHCb

LHCb, the dedicated b physics experiment at the LHC, has a forward geometry covering the range 15 to 300 mrad. The tracking system comprises a set of circular silicon sensors known as the vertex locator [2], a silicon strip station and three hybrid tracking stations. The silicon sensors will all be read out with radiation hard analogue electronics.

The hybrid tracking stations employ two different technologies: they have silicon sensors in the higher occupancy region close to the beam and straw drift tubes surrounding these. The silicon strip station and the hybrid layers will use the same sensor design and a similar layout. Each station comprises four layersof single sided 200 $\mu$m $p^+ - \text{on} - n$ pitch sensors. Two of these layers are rotated by a small stereo angle. Typically, two 11cm long sensors are connected together to form a ladder.

The vertex locator (VELO) is required to perform primary and b decay vertex reconstruction, with the displaced vertex capabilities being used in the second level trigger. Hence, good vertex resolution and fast tracking capabilities were strong considerations for the design. The VELO uses two sets of 42 4.2 cm outer radius semi-circular sensors, arranged over approximately one metre with the beam direction as their axis, covering the interaction point and the region downstream. The sensors are of two types, having strips that run either radially or circumferentially with pitches varying between 40 and 100 $\mu$m. The two sets of half disks are separated by 6 cm during injection of the LHC beam, but the active strips at the inner radius of the disks are brought to only 8 mm from the LHC beam when stable conditions are provided. Naturally this leads to an extreme radiation environment, a fluence of $10^{14}$ $\text{n}_{\text{eq}}/\text{cm}^2$ per year is anticipated. The performance of irradiated $p^+ - \text{on} - n$ and $n^+ - \text{on} - n$ prototype sensors, evaluated in a test beam, are compared in Figure 1. $n^+ - \text{on} - n$ sensors have been chosen as they remain fully efficient even when only 60 % of the detector thickness is depleted. A best resolution of 4 $\mu$m has been achieved with these sensors.

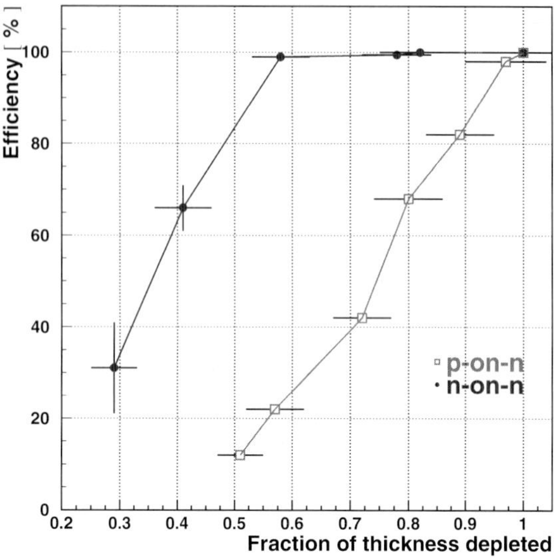


Figure 1. A comparison of the performance of irradiated LHCb VELO $p^+ - \text{on} - n$ and $n^+ - \text{on} - n$ prototype sensors.

## 4. ATLAS

The inner detector tracking system of ATLAS [3] consists of pixels, silicon strips and a straw tracker.

The pixel system has three forward disks per side and three barrel layers at radii of 5 to 12 cm. The detector is composed of 1500 modules and has 100 million 50 × 400 $\mu\text{m}^2$ $n^+ - \text{on} - n$ pixels. The modules are mounted on carbon frames, where their positions must be monitored and maintained at a 10 $\mu$m accuracy. The 15 kW of power dissipated in the 1.5 $\text{m}^2$ pixel system is removed by a coolant in the integrated cooling ducts. A new 7 m long insertable tubular layout has been designed which allows independent installation and removal of the pixels from the silicon strip layers. In addition to decoupling the construction schedule for the pixels and strips this will also facilitate upgrades, for example a change in the inner pixel layer which (due to radiation damage) is only anticipated to be operational for the first three to five years of the LHC. The pixels are readout with a In/SnPb bump-bonded front-end chip implemented in the radiation hard 0.25 $\mu$m CMOS process. A recent production run has been highly successful with 85 % of the chips passing the initial tests, and irradiated bump-bonded assemblies retaining high efficiency (98%), low noise (350 $e^-$) and the expected resolution of 13 $\mu$m.

Four barrel layers at radii of 27 to 52 cm and nine end cap disks per side of silicon strips surround the pixel detector. Single sided $p^+ - \text{on} - n$ strips are used throughout with a square sensor containing 80 $\mu$m pitch strips used in the barrel and wedge geometries used in the forward region. The sensors are readout with binary DMILL BiCMOS front-end chips. A module consists of two sets, mounted back to back, of two daisy-chained single sided sensors with a 40 mrad stereo angle between the planes. The effective strip length is around 12 cm.

The designs of all the LHC detectors are now well advanced and indeed production of the larger systems of ATLAS and CMS has already commenced. Indeed, in the case of ATLAS, 85 % of the silicon strip sensors have already been delivered, with less than 1 % of the manufacturer's supplied detectors failing the ATLAS tests. Module production is also now underway.

In order to successfully assemble such large systems a well structured testing and quality assurance plan is necessary. Extensive tests are performed on the components such as the sensors, hybrids and front-end chips before assembly, during assembly and on the completed modules. The scale of these tests is illustrated by listing the tests that will be performed on the final assembled ATLAS silicon module: 100 points are measured on each detector; Leakage current measurements up to 500V are taken and the stability checked over a 24 hour time period; a range of electrical tests are performed including the requirement that 99 % of the strips are efficient; the modules are required to withstand ten thermal cycles between -25°C and +40°C; finally the modules are operated, clocked and triggered, at the standard operating environment and temperature of -7°C for 24 hours.

## 5. CMS

CMS has by far the largest of the silicon systems, as they have adopted an entirely silicon solution to their tracking [4].

The initial barrel pixel system consists of two layers close to the interaction point at radii of 4.4 and 7.2 cm and two forward disks, with a $100 \times 150\mu\text{m}^2$ pixel size. The inner layer is expected to receive a fluence of about $3\times10^{14}\text{n}_{\text{eq}}/\text{cm}^2$ per year at full machine luminosity, with the readout chip being the most vulnerable part of the system and expected to withstand approximately two years of running. At a fluence of $6 \times 10^{14}$ $\text{n}_{\text{eq}}/\text{cm}^2$ the sensors are expected to be operated at less than half the full depletion voltage, hence the choice of $n^+ - \text{on} - n$ pixels.

One of the key differences of the CMS and ATLAS silicon strip detectors has already been mentioned: while the ATLAS system has 4,000 modules the CMS system has a staggering 15,000 modules to assemble. There are 10 detector layers in the barrel region and three inner disks and nine outer disks per side of the detector: a total of 206 $m^2$ of silicon. The CMS module design uses either two or four $p^+ - \text{on} - n$ sensors in a module for single sided or back-to-back layers. The strip pitches vary between 61 $\mu$m in the inner barrel layer at 21 cm and 140 $\mu$m in the outer layer at 115 cm.

Another significant difference is the choice of readout mode for the system. While ATLAS has selected the cheaper option of binary readout, the additional control and resolution of analogue readout has been chosen by CMS and the other two experiments. The larger data volume of the CMS analogue system is then readout of the 128 channel front-end chips and passed to a laser driver which transmits the signals in fibre optic cables over a 100m distance to the counting room.

An entirely automated system has been developed both for the module assembly and for the testing. The setup that is used at each of the seven laboratories assembling modules is shown in figure 2. The system uses video from a CCD video camera to automatically locate fiducial marks on each of the module components. Glue is dis-

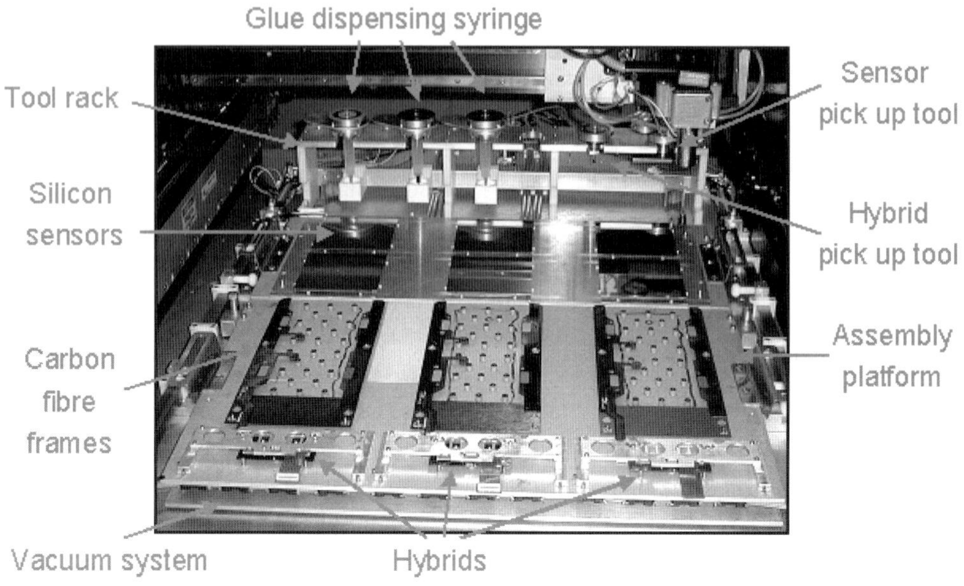


Figure 2. The features of the CMS automatic module assembly system are indicated.

pensed onto the carbon fibre frame and then the silicon sensors and hybrid are positioned by a robotic arm. Results show that two sensors in a module have an rms translational displacement of only 3 $\mu$m and a rotational accuracy of 1.5 mrad has been achieved. Three modules can be assembled simultaneously in around 25 minutes.

## 6. Closing Remarks

With the development and construction of the LHC a paradigm shift in silicon detector history is occurring. There are nine distinct subsystems, over 280 $m^2$ of silicon and 22,000 modules to assemble. The silicon systems of the LHC do not simply provide superior vertexing capabilities or provide the first space points for tracking, like their LEP forerunners, but rather their combined attributes of spatial precision, speed and radiation tolerance have gained these technologies their roles as primary components of the tracking system for the LHC experiments.

## 7. Acknowledgements

I would like to thank the many members of the silicon groups of the four LHC experiments who assisted in the preparation of my presentation and of this paper, particularly Ariella Cattai, Leonardo Rossi and Flavio Tosello.

## REFERENCES

1. ALICE Collab., *ALICE Technical Design Report of the Inner Tracking System* CERN LHCC 99-12
2. LHCb Collab., *VELO Technical Design Report* CERN/LHCC 2001-0011
3. C Gemme, ATLAS Pixel Collab., submitted to NIMA, proceedings of Vertex 2001, Brunnen, Switzerland 23-28 September 2001
   C Lacasta, ATLAS SCT Collab., submitted to NIMA, proceedings of 9th European Symposium on Semiconductor Detectors, Schloss Elmau, June 23 - 27, 2002
4. CMS Collab., *Addendum to the CMS Tracker TDR* CERN/LHCC 2000-016

ELSEVIER

Proceedings of ICHEP 2002, pp. 895–898
*S. Bentvelsen, P. de Jong, J. Koch and E. Laenen (Editors)*

31st INTERNATIONAL CONFERENCE ON HIGH ENERGY PHYSICS AMSTERDAM

# New developments for silicon detectors

M. Winter[a]*

[a]Institut de Recherches Subatomiques (IReS),
BP 28 CRO, F-67037 Strasbourg cedex, France

The success of future particle physics experiments will rely on vertex detectors of unprecedented performances. This observation has led to several new developements in Si detector technologies, aiming for very high granularity, very modest material budget, good to high radiation hardness and fast read-out time. A review of the on-going developements is provided, emphasising those aiming for a future $e^+e^-$ Linear Collider (LC).

## 1. Motivations for R&D on Si detectors

The need for High Energy Physics experiments to aim for ever increasing reaction energies, in order to cross new kinematical thresholds and access new phenomena, is at the origin of a sizeable effort of R&D for a new generation of Si pixel detectors. This is mainly to equip future vertex detectors operating in an environnement characterised by a large number of jets (typically $> 10$), a significant fraction of them containing b or c quarks or $\tau$ leptons. An example is provided by the reaction: $e^+e^- \to HA \to t\bar{t}t\bar{t} \to b\bar{b}b\bar{b}W^+W^-W^+W^-$, which may lead to at least 12 jets, 4 of them coming from b quarks and - eventually - 4 from c quarks. Moreover, gluon radiation tends to increase the number of jets and of b and c quarks.

Such extreme experimental conditions call for a vertex detector with unprecedented performances. The importance of the latter is further enhanced by the need to characterise (new) phenomena and particles. This ability is mandatory to disentangle various possible explanations for new observations and pin down the dynamics they express (e.g. electroweak breaking mechanism, supersymmetry, new gauge interactions, low scale quantum gravity).

Future vertex detectors must therefore allow reconstructing the flavour of each vertex in a "poly-jet" environnement with very high efficiency and purity (for charm quarks in particular). Pixels are mandatory in most of these cases, where it is necessary to assign each track to its vertex origin, to reconstruct the vertex mass, energy and electric charge as well as to establish the link between vertices (e.g. between tertiary and secondary ones), to help the reconstruction of energy flow, to identify the nature of electrons in jets (e.g. photon conversions or prompt electrons), a.s.o.

Experiments will therefore require very granular, ultra-light and poly-layer vertex detectors, installed very close to the interaction region. Among existing technologies, Charged Coupled Devices (CCD) offer performances which are closest to those ambitionned. They are however not fully satisfactory, as shown below.

The smallness of the cross-sections related to the processes of interest make very high luminosity collisions mandatory. This translates into harsh radiation conditions which degrade the detector performances, and into high background, which leads to high occupancy. CCDs are therefore not applicable in cases where the read-out speed or the radiation tolerance demanded are high and where only Hybrid Pixel Sensors (HPS) may be used. However, the latter are substantially less granular and thin than CCDs. This situation originated a wide panel of R&D activities, which aim to combine specific advantages of CCDs with those of HPSs, mainly in perspective of a Future $e^+e^-$ Linear Collider. Another class of developments aims for more futuristic objectives (e.g. Super-LHC), which require winning

*M.Caccia, Ch.Damerell, R.Richter and M.Trimpl are aknowledged for their help in preparing this contribution.


doi:10.1016/S0920-5632(02)02269-7

one or two additionnal orders of magnitude in read-out speed and radiation tolerance.

## 2. R&D on Charged Coupled Devices

CCDs are mainly developed by the LCFI collaboration [1] in perspective of a LC. This technology was retained for the baseline of the vertex detector considered in the Technical Design Report of the TESLA collider [2].

### 2.1. Principle of operation

The signal is generated in a typicaly 20 $\mu m$ thick, partially depleted, epitaxial layer. The signal electrons are collected through diffusion and drift, the latter occuring in the depleted part of the layer. Fig.1 illustrates the charge collection in a typical device consisting of a lightly doped epitaxial p-layer on a heavily doped $p^+$ substrate, covered with a $\sim 1$ $\mu m$ thick $n^+$ layer. The partial depletion of the $n^+/p$ region creates a potential minimum for electron storage just above the $n^+/p$ edge. The charge collected is clocked out through the sensor material.

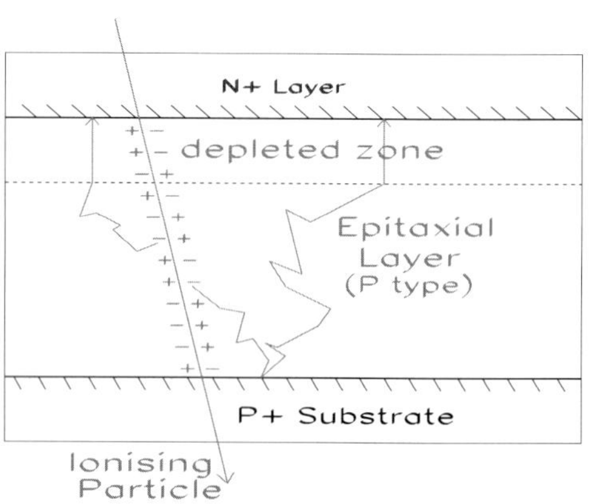


**Figure 1:** Schematic charge collection in a CCD.

CCDs were used very successfuly for the vertex detector of the SLD experiment. Made of about 307 million pixels, the detector was read out at a frequency of 5 MHz, translating into a (serial) frame read-out time of $\sim$ 200 ms. Based on a 20 $\mu m$ pitch, on layers representing 0.4 % radiation length only, and on a small beam pipe radius, it provided an unprecedented impact parameter resolution of $\sigma = 8 \oplus 33/p \cdot sin^{3/2}\theta$ $\mu m$. These performances need however substantial improvement to exploit the physics at a LC.

### 2.2. R&D objectives and achievements

The LC running conditions call for a much shorter frame read-out time and a significantly lighter material budget. Moreover, the radiation tolerance of the CCDs for the foreseen radiation levels needs to be assessed. The exact performance goals are not the same for all LC projects. For TESLA, which has the most demanding running conditions, the first layer of the detector, which has the highest occupancy, needs to be read out in less than 50 $\mu s$, making a clock frequency of at least 50 MHz mandatory. The material budget of each of the 5 layers (with radii ranging from 15 to 60 mm) of the detector should be kept below 0.1 % radiation length. The impact parameter resolution should therefore be better than $\sigma = 5 \oplus 10/p \cdot sin^{3/2}\theta$ $\mu m$. The detector should stand a few 100 kRads ionising radiation and fluences exceeding $10^{10}$ $n \cdot cm^{-2}$.

Several steps are being made to achieve these goals. The first column parallel CCD was designed, due for delivery in October. 50 MHz column parallel operation may already be possible with this prototype. The first column parallel read-out chip is in advanced design, and should be back from fabrication on a similar timescale. Successful operation of a commercial CCD was achieved, clocked at 50 MHz with a driving voltage of 3 V. This device will also allow radiation tests. Further prototyping aims in particular for rather high resistivity Si, allowing a larger depleted volume, and for low voltage two-phase clocking.

Trials have been made to achieve an unsupported ladder system, where the latter were thinned down to about 60 $\mu m$ and hold by springs at both ends. Up-to-now, these trials were semi-successful. While longitudinal stabilisation was satisfactory, the processed Si tends to have a transverse curvature which is difficult to control sufficiently well by the spring tension. Alternative solutions consisting of semi-supported options (e.g. Si/Be) are presently under study, which should yield better overall stability.

## 3. R&D on Hybrid Pixel Sensors

This pixel technology is best suited to LHC experiments, because of its proven 25 ns time stamping and sparse data scan capabilities and of its high hadron radiation tolerance. It does however not provide sufficient spatial resolution for the LC physics goals. The R&D activity is therefore mainly aiming for more granular and thinner devices. For the long term, effort is also made to further improve the HPS radiation tolerance in order to stand running conditions such as those foreseen at a very high luminosity LHC.

Substantially better granularity was already obtained by a Como-Cracow-Milano-Warsaw collaboration [3], which designed test structures exploiting the concept of interleaved (i.e. floating) pixels. Tests performed with one of them featuring 60/100 $\mu m$ implant width/pitch and 200 $\mu m$ read-out pitch showed that the charge sensing by interleaved pixels works efficiently, translating into an average single point resolution of $\sim$ 6.5 $\mu m$, i.e. $\sim$ 4 times better than the binary resolution associated to the implant pitch. New tests structures, having $\sim$ 25 $\mu m$ implant pitch, are being fabricated, which are expected to provide a single point resolution better than 3 $\mu m$.

## 4. Monolithic Active Pixel Sensors

The developement of these sensors (called MAPS) for charged particle tracking was initiated by an IReS-LEPSI (Strasbourg) collaboration in 1999. This novel technology allows to integrate signal processing (amplification, discrimination, etc.) in the detector substrate, which may be thinned down to a few tens of $\mu m$.

### 4.1. Principle of operation

The principle of detection is similar to that used in visible light CMOS cameras. The signal delivered by a charged particle is generated in a low resistivity, p- type, epitaxial layer of 5 - 15 $\mu m$ thickness (see fig.2). The electrons diffuse thermally and are collected by regularly implanted n wells, where the signal gets converted into an electrical pulse. Since the lightly doped epitaxial layer is embedded between two heavily doped media (i.e. p wells and substrate), electrons reaching the boundary regions get reflected by the potential barrier due to the doping profile.

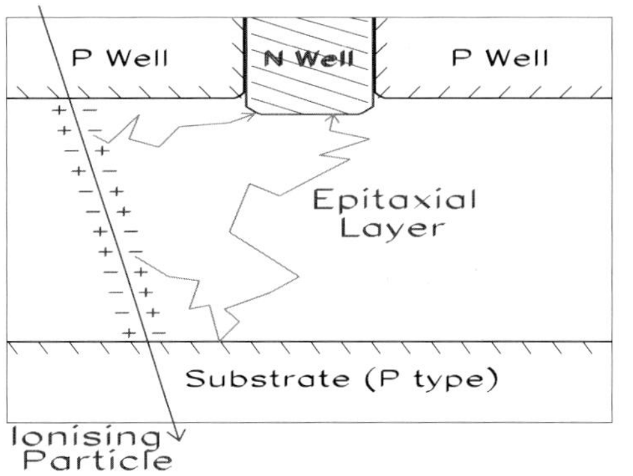

**Figure 2:** Schematic charge collection in a MAPS.

### 4.2. Present status [4]

The first prototypes, made of small arrays of a few thousands of pixels, demonstrated the viability of the technology and its high tracking performances (e.g. $>$ 99 % detection efficiency, single point resolution of 1.5 $\mu m$, double track resolution below 30 $\mu m$). Moreover, a prototype manufactured in a process without epitaxy, but based on a low doping substrate, exhibited a detection efficiency close to 100%, and a single point resolution of $\sim$ 4 $\mu m$. It was also shown that the sensors could stand fluences of $10^{12}$ to $10^{13}$ $\text{n}\cdot\text{cm}^{-2}$ without significant performance drop.

Important steps were achieved recently: the first real scale prototype was fabricated. Made of arrays of $10^6$ pixels (17x17 $\mu m^2$ wide), it reproduced the tracking performances of the small prototypes rather well (e.g. $>$ 99 % detection efficiency, single point resolution of 1.7 $\mu m$). Moreover, the average gain dispersion over the sensor surface was of the order of a few per-mill only.

Next R&D steps include the design of microcircuits providing integrated amplification, noise suppression and signal discrimination in a column parallel architecture aiming to meet the LC readout speed requirements. A first test structure of this kind was fabricated and is being tested.

## 5. DEPleted Field Effect Transistors

DEPFET (i.e. DEPleted Field Effect Transistor) pixel sensors offer performances which make them particularly attractive for X-ray imaging. Their application to charge particle tracking started recently, promoted by a MPI(München)-Bonn collaboration.

### 5.1. Principle of operation

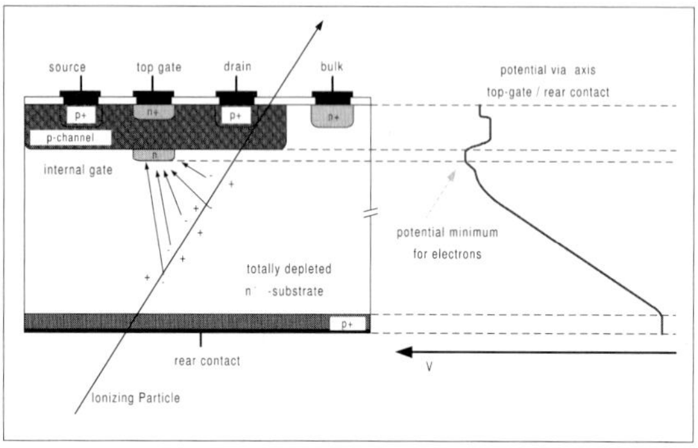


**Figure 3:** DEPFET basic operation principle.

The signal collection is illustrated on fig.3. A MOS or junction field effect transistor is integrated in a high resistivity, fully depleted, n-type substrate. By means of sidewards depletion and additional n-implants below the transistor, the signal electrons get collected by the n-implant underneath the transistor channel, for which it acts as an internal gate. The signal holes drift towards the rear contact of the sensor. The detection of a charged particle exploits the modulation in the transistor current induced by the change in the internal gate potential.

### 5.2. Present status [5]

A 64x64 (50 x 50 $\mu m^2$) pixel prototype was fabricated in 2000 for biomedical imaging and tested. A noise of 40 $e^-$ was observed, as well as a single point resolution of $\sim$ 10 $\mu m$, the frame read-out time being $\sim$ 1 ms. The first read-out chip achieving zero suppression was fabricated recently in 0.25 $\mu m$ TSMC technology. It is currently tested.

The next step of the developement will consist of a sizeable array of 30 x 30 $\mu m^2$ pixels, read-out with a row clock frequency of 50 MHz, and thinned down to $\sim$ 50 $\mu m$.

## 6. Other developments

Several other R&D activities are carried on, which aim for the long-term, and are therefore not described in this short review. This is the case for 3-D detectors, defect engineered Si, cryogenic detectors, Si-on-Insulator, amorphous Si, etc., for which more information can be found in [6].

## 7. Summary

Major future high energy physics projects will require pixel vertex detectors with unprecedented performances, allowing to tag with high efficiency and purity, b and c quarks as well as $\tau$ jets, in high jet multiplicity final states. Existing technologies (CCD, HPS) don't fulfil these requirements at a satisfactory level, a fact which originated various research lines, either to improve their performances or to explore new technologies (MAPS, DEPFET, SoI, 3-D, etc.). Substantial effort is also made to improve the radiation tolerance of Si against bulk damage (defect-engineering, cryogenic detectors, amorphous silicon, etc.), a crucial issue for future Hadron Colliders.

Given the importance of the material budget for the impact parameter resolution at the LC, the technology best suited to its physics goals (CCD, HPS, MAPS, DEPFET, ...) may not emerge before several years, since a quite detailed vertex detector design is needed for such a choice.

## REFERENCES

1. C.J.S.Damerell (for the LCFI collaboration), *A CCD based Vertex Detector for TESLA*, preprint LC-DET-2001-023;
2. Part III of the TESLA Technical Design Report, DESY 2001-011, March 2001;
3. M.Caccia et al., *Characterisation of Hybrid Pixel Detectors with capacitive charge division*, contribution to the Linear Collider Workshop 2000, FNAL, October 24-28, 2000;
4. Y.Gornushkin et al., proceedings of the 6th Conference on Position Sensitive Detectors, Leicester, September 9-13, 2002;
5. P.Fischer at al., *A DEPFET based pixel vertex detector for the detector at TESLA*, preprint LC-DET-2002-004;
6. F.Gianotti et al., *Physics potential and experimental challenges of the LHC luminosity upgrade*, preprint hep-ph/0204087.

ELSEVIER

Proceedings of ICHEP 2002, pp. 899–902
*S. Bentvelsen, P. de Jong, J. Koch and E. Laenen (Editors)*

31st INTERNATIONAL CONFERENCE ON HIGH ENERGY PHYSICS AMSTERDAM

# Recent Developments in Gaseous Tracking Detectors

S. Roth [a]

[a]III. Physikalisches Institut
Rheinisch-Westfälische Technische Hochschule Aachen
D-52056 Aachen, Germany

An overview of recent developments in gaseous tracking detectors is given focussing on the application of micro pattern devices. For trigger purposes in LHC-B a triple-GEM detector has been optimized for good time performance through a detailed investigation of the role played by electric fields and gas mixture. The design studies for a time-projection chamber at the detector of a $e^+e^-$ linear collider advise the usage of micro pattern devices. Ongoing R&D work for the construction of a TPC using either GEM foils or Micromegas is presented.

## 1. INTRODUCTION

In recent years the Micro Pattern Gas Detectors have been extensively studied and are currently used as tracking devices in various experiments. They have a good spatial resolution, show high rate capabilities and are robust against radiation damage.

However, little interest was devoted so far to the optimization of their time resolution, which is important for triggering applications. In addition, the charge transfer processes of micro pattern devices have to be better unterstood in order to use these devices in a TPC. Their transparency and their ion feedback suppression have to be optimised and the behaviour in high magnetic fields up to 4 T has to be studied.

## 2. A TRIPLE GEM DETECTOR FOR LHC-B

The triple-GEM detector for LHC-B [1] consists of three GEM foils sandwiched between a cathode and a anode plane. The anode is segmented in pads and connected to the readout electronics. The primary electrons, produced between the cathode and the first GEM are attracted by electric fields through the three GEM foils where they get multiplied. After the last GEM the charge is collected on the pads. A photograph of the triple-GEM detector used for the presented studies is shown in Figure 1.

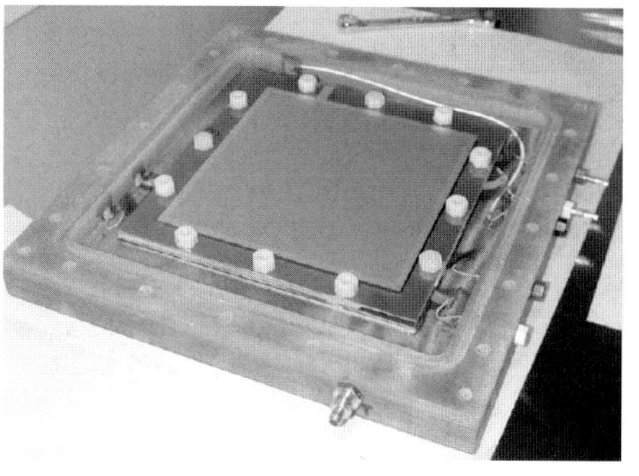

Figure 1. Photograph of the triple-GEM detector used for the test measurements

The triple-GEM detectors will be used in the inner part of the LHC-B muon system, where particle rates above 100 kHz/cm$^2$ are expected. One important task of the muon system is to give the Level 0 trigger for muons coming from b-flavoured hadron decays. To achieve the required performance one has to trigger muons inside the acceptance with 95 % efficiency and unambigous bunch-crossing information.

For leading-edge triggering the arrival of the first ionisation cluster defines the time of the


doi:10.1016/S0920-5632(02)02270-3

event. The time resolution therefore improves proportional to the number of ionisation clusters per unit length, $n$, and to the drift velocity, $v_{\mathrm{drift}}$. Therefore, a large average atomic number and high drift velocity gas mixture should be used.

The collection efficiency for primary electrons decreases for high drift fields, because the drift lines then tend to end on the upper surface of the first GEM foil. Therefore the detector should be operated using a gas with high drift velocity at moderate drift fields.

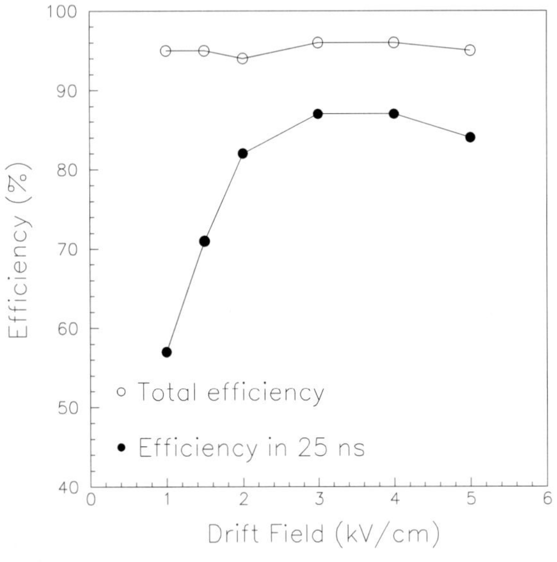


Figure 2. Efficiency as a function of the drift field

Efficiency measurements were performed at the PS hadron beam facility at CERN, which deliveres 2–4 GeV pions at a rate of $1\,\mathrm{kHz/cm^2}$. Using instead of the generic GEM gas $Ar/CO_2$ (70/30) a gas mixture of $Ar/CO_2/CF_4$ (60/20/20) improved the time resolution by a factor of two. A time distribution r.m.s. of about 6 ns was obtained. Figure 2 shows the measured efficiency versus the drift field using the ternary gas mixture. For low drift field an efficiency drop due to the low drift velocity can be observed, whereas at high drift fields the reduced GEM transparency limits the detector efficiency. For a single chamber an efficiency of about 85 % in a 25 ns time window could be achieved and make the triple-GEM detector a promising option for high rate particle triggering at LHC.

## 3. A TPC FOR TESLA

A Time Projection Chamber (TPC) is foreseen as the main tracker of the detector for the TESLA linear collider [2]. It should provide a momentum resolution of $\delta(1/p_t) < 2 \cdot 10^{-4}(\mathrm{GeV/c})^{-1}$ to exploit especially the physics of final states with high energy leptons. Additionally the specific ionisation $dE/dx$ should be measured with a precision better than 5% to allow particle identification in hadronic jets.

To reach these goals readout planes with the finest possible granularity are required. They must provide a single point resolution of 100 – 150 $\mu$m and their systematic distortions must be controlled to better than 10 $\mu$m over the whole radius of 1.6 m. Enough charge amplification has to be provided by the system to keep the signals well above the noise level of modern readout electronics. At the same time the produced ions have to be suppressed intrinsically, because active gating seems impossible in between a TESLA bunch spacing of only 337 ns. Finally one aims for a minimum of material in front of the calorimeters, especially in the endcap region.

A conventional TPC using a multi wire plane for charge amplification is limited by the $\vec{E} \times \vec{B}$ effects in the region close to the wires. In strong magnetic fields these effects result in a broadening of the electron cloud and a worsening of the resolution. Additionally, as the wires define a prefered direction, the reconstructed hit location depends on the projected angle between track and wires.

Using Gas Electron Multipliers (GEM) [3] or Micromegas [4] as the charge amplifying system could solve some of the drawbacks of wire planes. When using micro pattern devices for the TPC end plate the pads directly detect the amplified electron cloud which results in a fast and narrow charge signal. Also the slow ion tail is cut off since the ion cloud does not reach the induction region. These devices show no preferred direc-

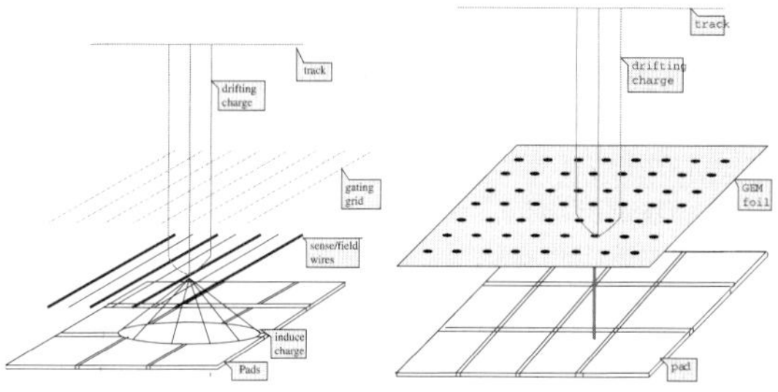


Figure 3. Scetch for the wire readout and the GEM readout

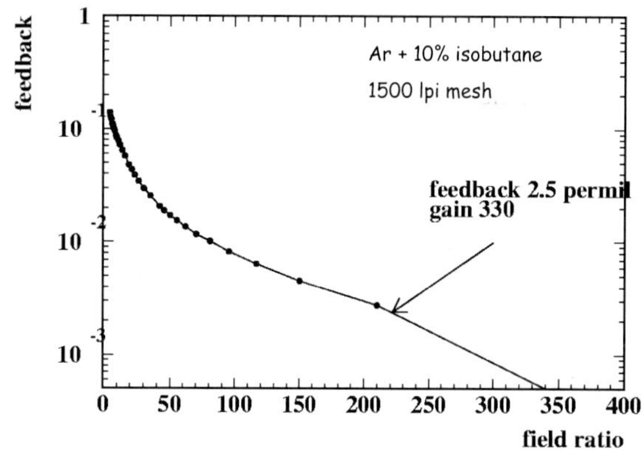


Figure 4. Ion feedback measured in Micromegas

tion, thus any $\vec{E}\times\vec{B}$ effects will be isotropic. And finally the back drift of the produced ions, the ion feedback, can be largely suppressed using highly asymmetric electric fields between drift and amplification region.

### 3.1. Ion feedback

The intrinsic suppression of ion feedback into the drift volume is one of the advantages of GEMs or Micromegas with respect to wires. First measurements for GEMs by groups from Aachen, Hamburg and Novosibirsk yield an ion feedback at the percent level for effective gains in the order of $10^4$. For Micromegas the ion feedback decreases proportional to the field ratio between drift and amplification region. A first measurment of ion feedback in Micromegas is shown in Figure 4. A feedback of only 0.25% at a gain of 330 was reached. At this operation point the absolut number of ions drifting back into the drift volume is already suppressed to the level of the primary ionisation.

Basic investigations on the charge transfers in multiple GEM structures are performed by the Aachen group [6]. They measure the collection and extraction efficiency for electrons and ions from a single GEM foil and compare the results with numerical calculations based on the finite elements program Maxwell 3D [7]. This program provides for a given set of external potentials the electrical field map for a unit cell of the GEM structure. Measurement and simulation are plotted in Figure 5 and are in good agreement. Additionaly a parametrisation of the numerical results has been established which can be used in future to optimize the working point of the GEM structures with respect to ion feedback and transparency.

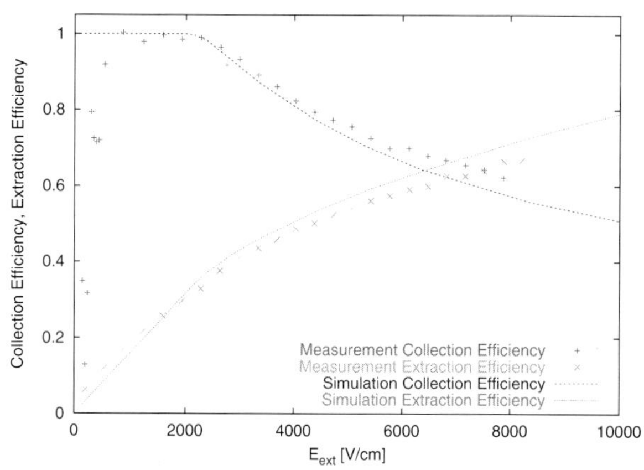


Figure 5. Collection and Extraction Efficiency in GEM structures

### 3.2. Influence of magnetic fields

The primary motivation for tests in high magnetic fields is to investigate whether there is a significant drop in collection efficiency for GEMs in high B-fields parallel to the electric fields as suggested by the Langevin Formula

$$\vec{v}_{Drift} \propto \hat{\vec{E}} + \omega\tau \; (\hat{\vec{E}} \times \hat{\vec{B}}) + \omega^2\tau^2 \; (\hat{\vec{E}} \cdot \hat{\vec{B}}) \; \hat{\vec{B}} \; ,$$

where $\hat{\vec{E}}$ and $\hat{\vec{B}}$ are unit vectors and $\omega$ is the cyclotron frequency. The last term, proportional to $B^2$, which gives the contribution along the magnetic field lines, could cause a drop in collection efficiency for high magnetic fields. When this term dominates, most electrons will no longer be collected into a GEM hole but will stay on drift lines perpendicular to the GEM surface and eventually reach the GEM surface. Those charges will be lost for the signal and consequently decrease the $dE/dx$ capabilities due to the loss in primary ionisation statistics.

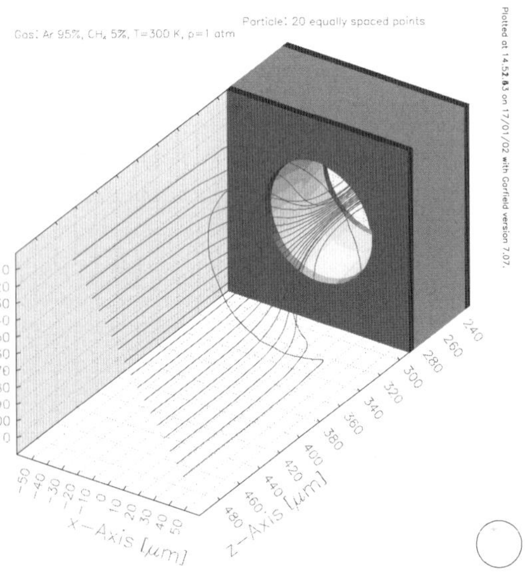


Figure 6. Garfield simulation of electron drift lines into a GEM hole at 4 T magnetic field

Figure 6 gives a first qualitative view of the electrons behaviour. It shows drift lines calculated with the simulation program Garfield [8] in the vicinity of a GEM hole pervaded by a 4 T magnetic field. A typical TPC gas mixture, $Ar/CH_4$ (95/5), was used. Each of the drift lines starts at a virtual straight ionisation track consisting of equally spaced electrons. In Figure 6 the effect of the Lorentz term is well visible for electrons: they follow a track spiraling into the GEM hole. In contrast there is no contribution to the drift velocity along the magnetic field lines. Hence we have no indication for a significant drop in the collection efficiency for primary electrons.

## REFERENCES


1. G. Bencivenni et al.,
*A triple GEM detector with pad readout for high rate charged particle triggering*,
submitted to ICHEP02 and to NIM A.
2. T. Behnke, S. Bertolucci, R.D. Heuer, R. Settles (editors),
*TESLA Technical design report, Part IV "A Detector for TESLA"*,
DESY-01-011.
3. F. Sauli,
*A New Concept for Electron Amplification in Gas Detectors*,
Nucl. Instr. and Meth. A 386 (1997) 531.
4. Y. Giomataris et al.,
*Micromegas: a High Granularity Position Sensitive Gaseous Detector for High Particle Flux Enviroments*,
Nucl. Instr. and Meth. A 376 (1996) 29.
5. The tracking group of the ECFA/DESY linear collider study,
*Linear Collider TPC R&D*,
submitted to ICHEP02.
6. M. Killenberg et al.,
*Modelling and Measurement of Charge Transfer in Multiple GEM Structures*,
submitted to ICHEP02.
7. MAXWELL, Commercial Finite Element Computation Package, Ansoft Co. Pittsburg, PA, USA.
8. R. Veenhof, Nucl. Instr. and Meth. A 419 (1998) 726.

Proceedings of ICHEP 2002, pp. 903–905
*S. Bentvelsen, P. de Jong, J. Koch and E. Laenen (Editors)*

ELSEVIER

31st INTERNATIONAL CONFERENCE ON HIGH ENERGY PHYSICS AMSTERDAM

# Calorimetry in CDF Run 2

R. D. Erbacher [a] for the CDF 2 Collaboration.

[a]Fermi National Accelerator Laboratory, P.O. Box 500, MS 318, Batavia, IL U.S.A.

The integrated calorimetry environment of CDF Run 2 comprises the Run 1 sampling scintillator calorimeters, electromagnetic preshower and shower maximum detectors, new scintillating-fiber endplug detectors, radioactive source calibration systems, dedicated trigger paths, and new custom front-end electronics. Together they form a general-purpose calorimetry system which was successfully commissioned in 2000–2001. The initial performance is described here, along with a glimpse of first CDF Run 2 data.

## 1. Introduction

The Tevatron Run 2 began in the spring of 2001 after completion of major upgrades to the accelerator complex. The improvements have provided an increase in energy from $\sqrt{s} = 1.8$ TeV to 1.96 TeV, a decrease in bunch spacing from 3.5 $\mu$s to 396 ns, and expected instantaneous luminosities reaching $5 \times 10^{32}$ cm$^{-2}$ s$^{-1}$ in Run 2b. To take advantage of the increased energies and collision rates, CDF underwent a major overhaul, with upgrades to almost every system [1]. With these improvements, CDF is poised to exploit the physics at the energy frontier for the remainder of the decade.

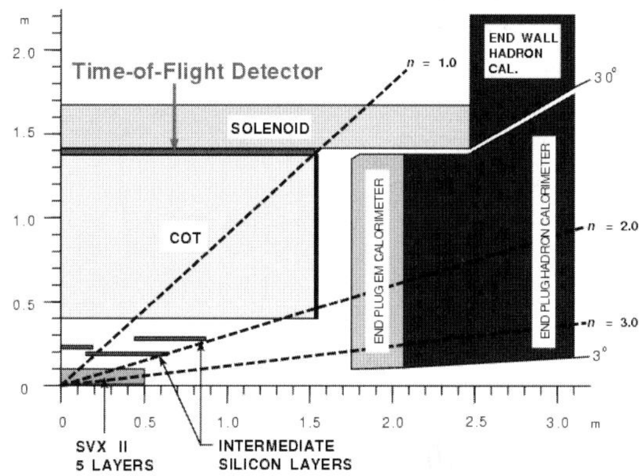


Figure 1. Cutaway quadrant of CDF2 without central calorimeters and muon chambers.

## 2. Calorimeter Endplug Upgrade

As part of the CDF Run 2 project [1], several portions of the calorimetry detectors were upgraded. While the original, central ($|\eta| < 1.1$), scintillator-based calorimeters [2] [3] were sufficient for the Run 2 operating conditions, the time response of the Run 1 gas endplug detectors was too slow for the decreased bunch spacing of Run 2, and they had to be replaced. The active elements of both the electromagnetic and hadronic portions of the new detectors [1] [4] are scintillator tiles read out by wavelength-shifting (WLS) fibers embedded in the scintillator. The new design has an improved sampling fraction, reduces forward gaps existing in Run 1, and utilizes technology similar to the central calorimeters, providing greater detector uniformity over the full solid angle to $|\eta| = 3.6$. Fig. 1 shows the new detectors positioned in the ends of CDF, and Fig. 2 shows the endplug support structure and photomultiplier tube (PMT) boxes.

### 2.1. Shower Maximum

The central shower maximum detectors [5] have had little problem with aging, so they remain for Run 2. These detectors consist of gas chambers with wire and strip elements. However, the slow gas chambers of the central preradiator and crack detectors do suffer from aging, and will be upgraded to more finely segmented scintillator detectors for Run 2b.

The plug shower maximum detector (PES) and preradiator were built into the new endplug detector. The PES consists of scintillating strips


doi:10.1016/S0920-5632(02)02271-5

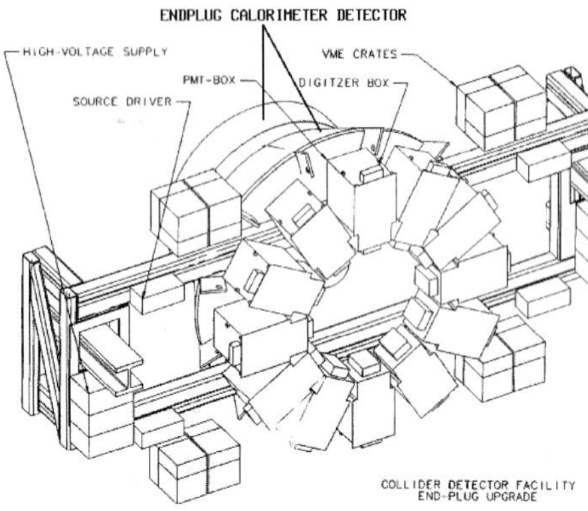


Figure 2. The new endplug calorimeter detector, its support structure, and the PMT boxes.

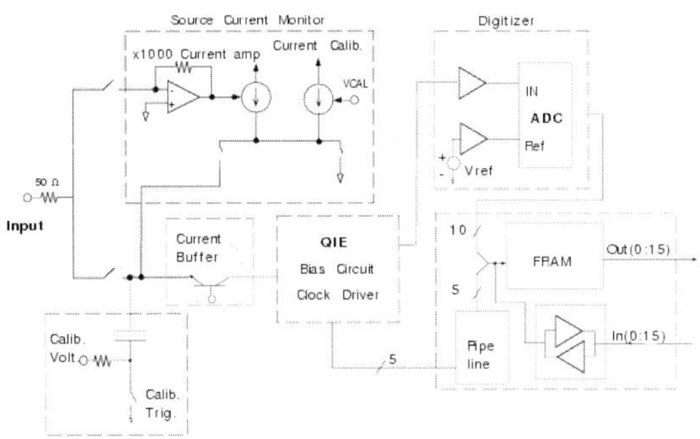


Figure 3. Simplified schematic of the CAFÉ card, containing a QIE chip and a 10-bit ADC.

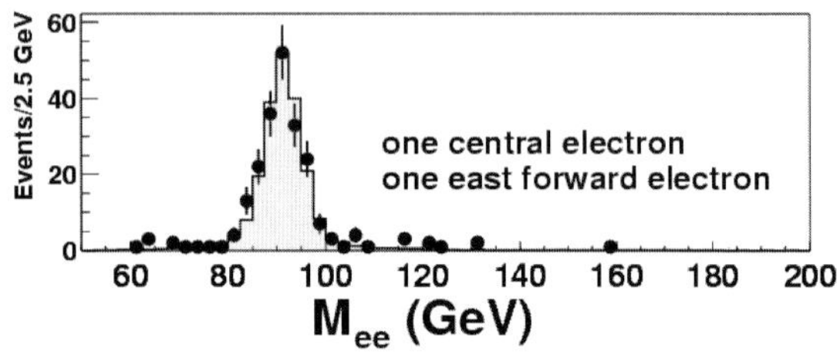


Figure 4. Preliminary CDF Run 2 Z→ee distribution using 10 $pb^{-1}$ of data, compared to MC (solid).

and WLS fibers in two layers located ~6 $X_o$ into the PEM. The energy scale is well-matched to the plug electromagnetic calorimeter. The position is determined to within 1.5 cm, and will be improved using the forward silicon layers.

## 3. Electronics Upgrade

The move to scintillator-based calorimetry in the forward regions of the CDF detector has allowed for a common design for the readout electronics [6] throughout the central, endwall, and endplug calorimeters. Signal size estimates for collider energies up to 2 TeV indicated that the maximum PMT signal could be a factor of two greater than that seen in Run 1, requiring greater dynamic range than the previous 16 bits. In addition, with bunch crossings every 396 ns, the readout required fast digitization of signals with a large dynamic range. The Fermilab QIE (Charge Integrating and Encoding) [7] chip provided a multi-range solution for signal readout.

The QIE is a custom, multi-ranging circuit used originally in the KTeV experiment. It was modified for use in CDF for both calorimeter (QIE6) and shower max (SMQIE) readout, and again for use in both MINOS and CMS. The QIE architecture consists of a binary-weighted splitter and eight current ranges. The QIE6 together with a 10-bit FADC provides the needed 18 bits of dynamic range. These components, along with a FRAM and calibration/charge injection circuitry, are mounted on single-channel CAFÉ card modules (Fig. 3). The CAFÉ daughter cards are in turn mounted onto front-end VME ADMEM (ADC+MEMory) [8] boards. The ADMEMs provide the Level 1 trigger with transverse energy sums using Xilinx FPGA's. They also use a pipelined Level 1 buffer 42 clock-cycles (5.5 $\mu$s) deep (which provides deadtimeless readout), and four event buffers for pre-decision Level 2 storage.

### 3.1. Performance

The new calorimeter readout is performing well. Pedestals are stable over long time periods, and RMS values are typically 1.5–2.5 counts (5–6 MeV or 10–15 fC). The QIE has been extremely reliable, which bodes well for future experiments. Rare comparator failures make up the majority of the QIE problems, and periodic failures of tantalum capacitors comprise the main problems on the ADMEMs.

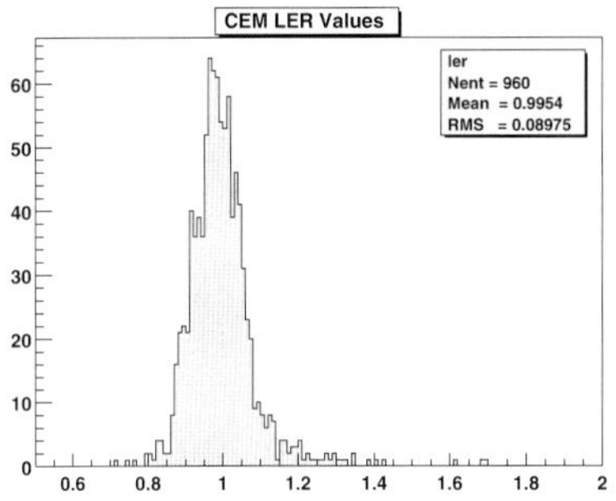


Figure 5. LER coeffs. for the EM calorimeter.

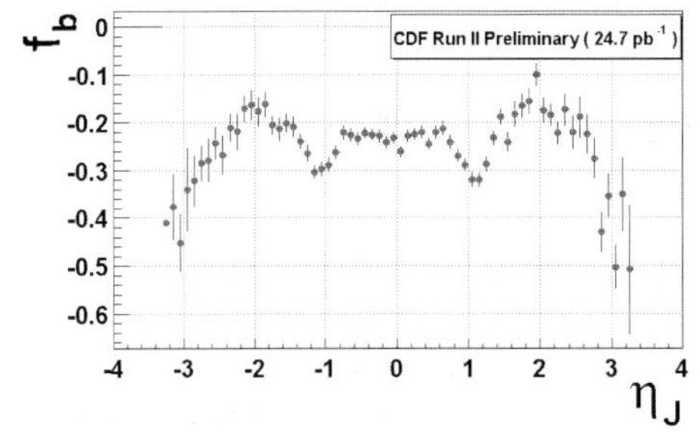


Figure 6. $f_b$, the fractional difference between $p_T$ of the photon and the jet, as a function of $\eta$.

## 4. Calibration

The absolute energy scales were determined during the original test beam calibrations. These scales are maintained by measuring the response to radioactive sources. During the upgrade the $^{137}$Cs source system was refurbished for the central detectors, and a $^{60}$Co system was installed into the endplug. The energy scales are verified using data. Fig. 4 shows a sample of Z bosons having one electron in the central detector and one in the endplug. By construction the Z peak is at the correct location. For the hadron detectors, MIP peaks from $J/\psi$ samples have shown that the central hadron calorimeter scale is within 4% of the Run 1 scale, and has been adjusted.

The relative energy scales are set using data. To correct for PMT gain and tower response variations, linear energy response (LER) coefficients are determined and applied to individual channels (Fig. 5). The responses are then tracked using a laser system for the hadron detectors, and LED/Xenon flashers for the EM detectors. The jet energy scales are corrected using dijet and

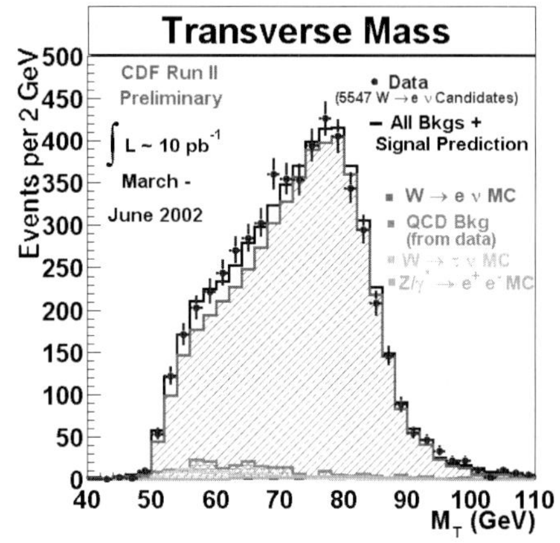


Figure 7. Preliminary CDF Run 2 W$\rightarrow e\nu$ sample using 10 pb$^{-1}$, including a MC prediction (non-filled plot) of signal plus all backgrounds.

photon-jet events by balancing the transverse momenta in individual events. For example, Fig. 6 shows $f_b = (p_T(\gamma) - p_T(jet))/p_T(\gamma)$ across the detector. The relative jet corrections are already quite good in the central detector region, and should be flat when the corrections are complete.

## 5. Conclusion

With the integrated calorimetry system in place and commissioned for Run 2, CDF can take full advantage of Tevatron collision data at the energy frontier. Analyses using the calorimeter, such as the initial W cross-section measurement [9] (Fig. 7) are well underway, establishing the tools for numerous exciting high-$p_T$ physics measurements to come.

## REFERENCES

1. CDF II Detector - Technical Design Report - Fermilab-Pub-96/390-E.
2. Balka,*et al.*(CDF), NIM **267** 272, 1988.
3. Bertolucci,*et al.*(CDF), NIM **267** 301, 1988.
4. Albrow,*et al.*(CDF), NIM **480** 524, 2002.
5. Nodulman, *et al.*(CDF), NIM **204** 351, 1983.
6. Article on the CDF2 front-end calorimeter electronics to be submitted to NIM in 2002.
7. Yarema, IEEE Trans.Nucl.Sci. **4** 750, 1993.
8. Shaw, IEEE Trans.Nucl.Sci. **47** 1834, 2000.
9. K. Bloom, these proceedings.

ELSEVIER

Proceedings of ICHEP 2002, pp. 906–908
*S. Bentvelsen, P. de Jong, J. Koch and E. Laenen (Editors)*

31st INTERNATIONAL CONFERENCE ON HIGH ENERGY PHYSICS AMSTERDAM

# The DØ Detector for Run II

Levan Babukhadia[a]* (for the DØ Collaboration)

[a]State University of New York, Stony Brook, New York 11794, U.S.A.

The general purpose DØ collider detector at the Fermilab Tevatron has undergone major upgrades for Run II. We describe the current status and performance of the DØ detector.

## 1. INTRODUCTION

The Fermilab Tevatron has undergone significant upgrades for Run II. Presently, it is delivering higher instantaneous luminosity than in Run I with a ten-fold decrease in the time between beam crossings. The Run IIa will provide an integrated luminosity of 2 $\text{fb}^{-1}$, about 20 times that of Run I. Also the center-of-mass energy of $p\bar{p}$ collisions has increased modestly, from 1.8 to 1.96 TeV.

To achieve the physics goals for Run II which include searches for the Higgs boson, supersymmetry, extra dimensions, and other new phenomena, as well as precision studies of weak bosons, top quark, QCD, and $B$-physics, the collider detectors at the Tevatron need superb electron, muon, and tau identification, excellent jet and missing transverse energy ($\not\!\!E_T$) reconstruction, and very good flavor tagging of jets through displaced vertices and leptons.

To meet these challenges in a high event rate environment, the DØ detector has undergone major upgrades [1] (Fig. 1a). The central tracking system has been replaced by a silicon microstrip tracker, surrounded by a scintillating fiber tracker, both immersed in a 2T magnetic field from a superconducting solenoid [2]. To provide fast energy and position measurements for the electron trigger and offline electron identification, scintillating fiber preshower detectors have been added in the central and forward regions. Calorimeter trigger and readout electronics have been upgraded and the muon system has been significantly improved. A new forward proton spectrometer (FPD) has been added for the diffractive and elastic physics program. To handle high event rates, an entirely new trigger and data acquisition system has been designed and implemented, see Fig. 1b.

## 2. STATUS AND PERFORMANCE

**Calorimeters.** The nearly compensating, uniform, and hermetic liquid-argon uranium calorimeters from Run I, with completely upgraded readout and trigger electronics, provide about 50k readout cells of electromagnetic (EM) and fine and coarse hadronic compartments with less than 0.1% dead or noisy channels. About 5k calorimeter towers are also used for fast trigger decisions. The system is fully commissioned.

For an *in situ* calibration of EM calorimeters the $Z \to ee$ signal is employed. Fig. 1c shows the $Z$ peak in a di-EM invariant mass distribution with close to expected resolution. On the other end of the spectrum, Fig. 1d shows the $J/\psi$ resonance. The calorimeter response to jets is calibrated using photon-jet data. Fig. 1e shows the distribution of $\not\!\!E_T$ in multijet events, spanning several orders of magnitude; the present $\not\!\!E_T$ resolution is about 7 GeV.

**The Muon system** consists of central and forward regions providing coverage up to $|\eta| = 2$ with three layers of scintillators and drift tubes, one inside and two outside the magnetized iron toroids. In the central region, there are over 6k cells of proportional drift tubes and 630 (360) scintillating counters inside (outside) the toroid. In the forward regions, there are about 6k 8-cell mini drift tubes in eight octants per layer and

*Current address: MS 357, Fermilab P.O. Box 500, Batavia, Illinois, U.S.A. Electronic mail: blevan@fnal.gov.


doi:10.1016/S0920-5632(02)02275-2

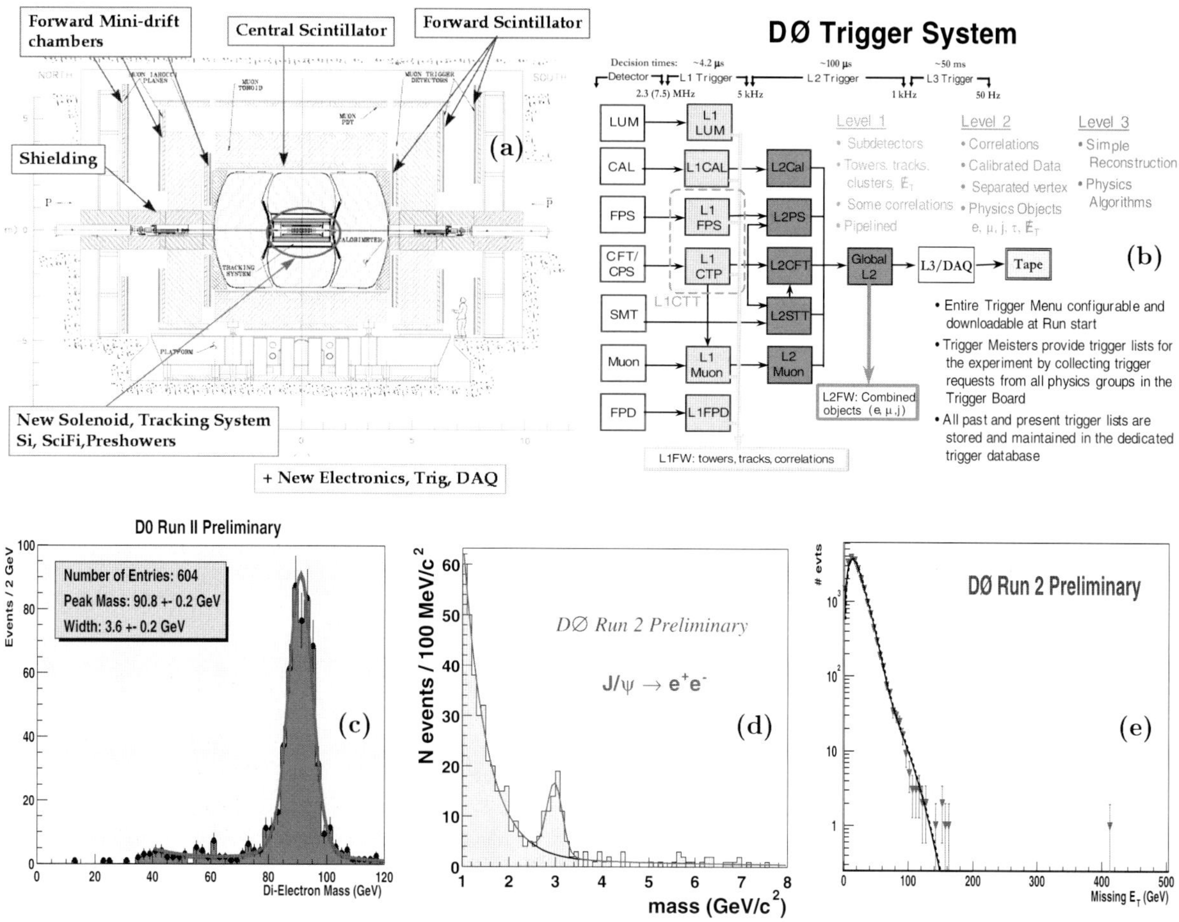


Figure DØ detector for Run II (a) and diagram of the trigger system (b). Invariant mass distribution of di-EM objects showing $Z$-boson (c) and $J/\psi$ (d) resonances. Distribution of missing transverse energy $\not{E}_T$ in multijet events (e).

over 4k forward scintillation counters or pixels. The muon system is fully commissioned.

Fig. 2a shows the invariant mass distribution of di-muons from the muon stand-alone system, with the $J/\psi$ resonance visible above background. Matching muons with central tracks significantly improves muon transverse momentum resolution and Figs. 2b and 2c show the $J/\psi$ and the $Z$-boson resonances, respectively. Timing cuts on muons reduce backgrounds from cosmic rays, and could aid in detection of slow moving particles.

**The FPD system** consists of two arms of 18 Roman pots in four quadrupole and two dipole "castles." Hits in scintillating fiber detectors installed in the Roman pots will be used for measuring fractional energy lost by the proton and scattering angle, and for triggering on elastic, diffractive, and double pomeron events. Currently the system is in commissioning. Several million elastic events have been recorded by FPD stand-alone system, and integration within DØ is ongoing.

**The DØ Trigger** is a three stage hardware and software system that reduces the raw event rate of about 7 MHz down to under 50 Hz for recording events to tape (Fig. 1b). The hardware triggers in Level 1 (L1) allow triggering on EM and hadronic calorimeter objects, tracks, and muons. In L2 more physics-like objects are used employing correlations and allowing triggering on separated vertices. In L3, physics algorithms are

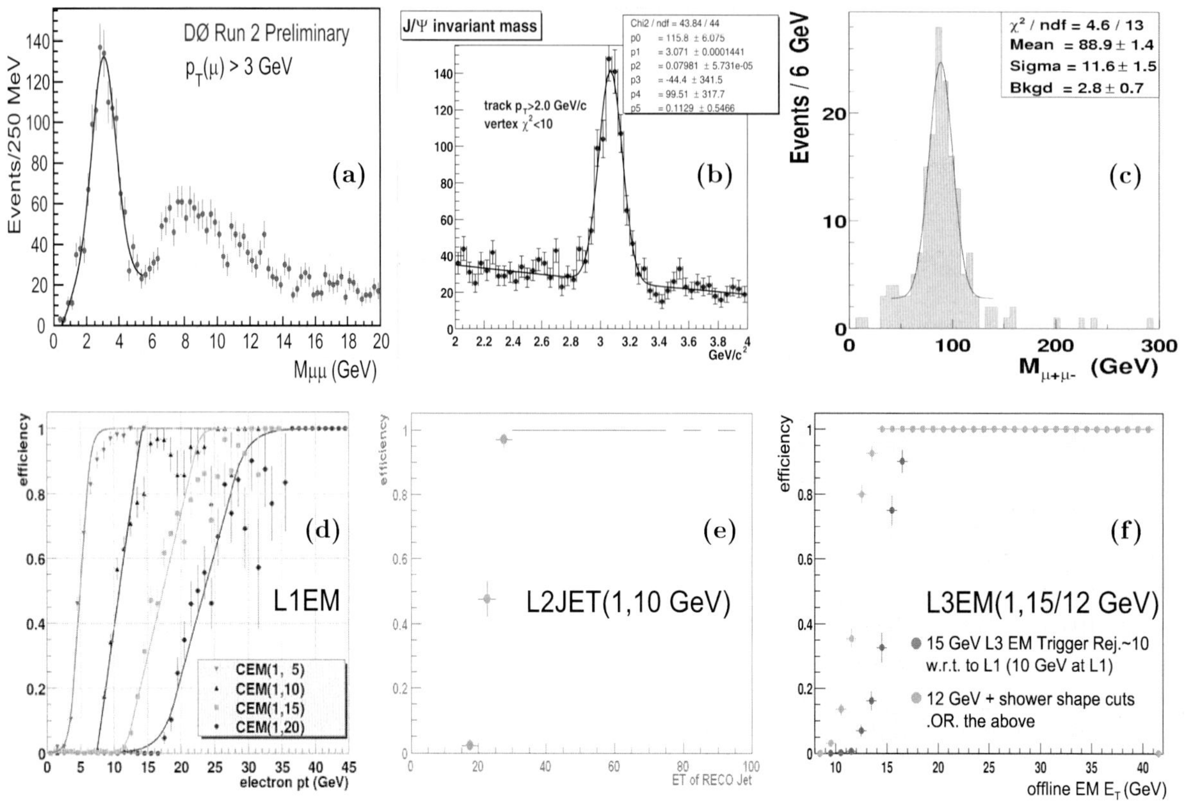


Figure 2. Invariant mass distributions of di-muons using: muon stand-alone system showing $J/\psi$ resonance (a) and muons matched with central tracks showing $J/\psi$ (b) and $Z$-boson (c) resonances. Efficiencies for triggering on electromagnetic and hadronic objects of different $E_T$ thresholds in the first (d), second (e), and third (f) stages of the DØ trigger system.

employed after fast event reconstruction.

Fig. 2d shows the efficiency "turn-on" for triggering on EM objects with various transverse energy ($E_T$) thresholds using the L1 calorimeter trigger. Similarly, Fig. 2e shows the efficiency for triggering on a L2 "jet" of $E_T = 10$ GeV. Finally, Fig. 2f shows the performance of L3 software filter to trigger on $E_T = 15$ GeV EM objects, also indicating that the $E_T$ threshold can effectively be lowered by utilizing EM shower shape requirements.

## 3. SUMMARY AND OUTLOOK

The DØ detector for Run II is operating and collecting physics data. There has been enormous progress over the past year in installation, integration, and commissioning of the detector, and understanding the data. Performance of the Run II DØ detector is very encouraging, all subdetectors are operating well, software and computing systems are working well, we are reconstructing electrons, muons, jets, $\not{E}_T$, $J/\psi$'s, $W$'s and $Z$'s. We are working hard to complete commissioning, improve calibration and alignment, optimize detector, trigger, and DAQ performance, and are on the way to exciting physics, with first physics results now appearing [3].

## REFERENCES


1. The DØ Experiment, www-d0.fnal.gov.
2. G. Ginther, these proceedings.
3. M. Narain, these proceedings.

ELSEVIER

Proceedings of ICHEP 2002, pp. 909–912
*S. Bentvelsen, P. de Jong, J. Koch and E. Laenen (Editors)*

31st INTERNATIONAL CONFERENCE ON HIGH ENERGY PHYSICS AMSTERDAM

# The ATLAS liquid argon calorimetry system

Pascal Perrodo[a]*

[a]LAPP, 9 chemin de Bellevue, BP110, 74941 Annecy-le-vieux CEDEX, FRANCE

The status of construction and recent test beam results for the ATLAS liquid argon calorimetry system (electromagnetic, hadronic and forward calorimeters) are reported as well as a new development on the first-level trigger system for electromagnetic clusters and isolated hadrons.

## 1. Introduction

The ATLAS detector is a multi-purpose detector which will be installed on the LHC collider at CERN. It has been designed to search for possible new physics phenomena in the energy range of ≈ 100GeV to ≈ 1TeV. The traditional signatures for new physics are based on high $P_T$ photons, leptons ($e^{\pm}, \mu^{\pm}, \tau^{\pm}$) and missing transverse energy. The calorimeters plays a crucial role in such a search.

The choice of liquid argon calorimetry has been made for its radiation hardness, stability and good calorimetry performances. A short description is given here pointing out the specificities of each detector and the evolution of the design with respect to the initial one [1]. Recent test beam results of series modules for the electromagnetic (EM), hadronic (HEC) and forward calorimeters (FCAL) are presented (more details can be found in [2]) as well as the status of the construction.

The calorimeter analog signals are used to build not only the first level trigger signal on jets, total and missing transverse energies but also to trigger on events with isolated electrons/photons or isolation hadrons (likely coming from hadronic $\tau^{\pm}$ decays) which are crucial for the LHC physics. A new development of a board to implement this trigger is presented.

## 2. The electromagnetic calorimeter

The EM calorimeters are required to measure the energy and position of $e^{\pm}$ and $\gamma$ with a good uniformity as well as a good $\gamma/\pi^0$ separation. The response has to be fast, and an accordion shaped geometry for lead absorbers and electrodes has been chosen which also insures a total hermiticity in $\phi$. Lots of care has been taken at the start of the production period with the honeycomb spacers to avoid high voltage shorts, with the bending of the kapton-copper electrodes, and the handling of the resistive ink which reduces the impact of sparks on electronics.

The ionisation signals are amplified and $\mathrm{CR}-\mathrm{RC}^2$ shaped, then sampled every 25ns before digitization. These samples are used after first level trigger decision to reconstruct the energy and the arrival time for each channel using optimal filtering techniques. The calibration is achieved by injecting exponentially decaying signals with very well known amplitudes through precision (0.1%) resistors in the $\ell$Ar. It has been observed recently that these resistors are sentitive to possible HV sparks. A protection diode has therefore been installed on each channel.

The EM calorimeters (barrel and endcap) are currently half produced and assembled. The insertion of this first half into the cryostats should occur beginning of 2003.

Seven series modules have been tested with a beam at CERN. A substantial effort has been made to understand the shape and the normalisation of the calibration and physics signals with respect to the electrical characteristics of the detec-

*on behalf of the ATLAS liquid argon and first-level trigger groups


doi:10.1016/S0920-5632(02)02276-4

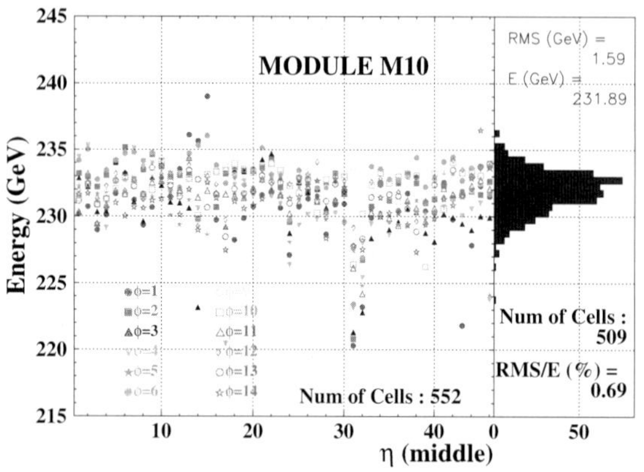


Figure 1. Uniformity of energy response for one electromagnetic barrel module (M10).

tor (inductances, ground returns and cross talk). This has led to an absolute electronic calibration and the possibility to compute meaningful calibrated optimal filtering coefficients [3]. From this the uniformity in position of the energy response has been extracted: over one full series module a promissing figure of 0.7% is obtained for the uniformity after clustering and time corrections (see Fig. 1) when 0.7% is expected over the whole EM calorimeter.

Two HV lines supply a region $\Delta\eta = 0.2, \Delta\phi = 0.2$. When one is missing, energy has to be multiplied by 2. However at the boundary with a fully supplied region a non linear behaviour ($\approx 5\%$) has been observed. It has been fully reproduced by simulation and is now understood as a geometrical effect of the accordion. This proves that such a local possible HV problem can be fully corrected.

The position resolution has been studied for the endcap calorimeter. For the front and middle layers (granularities $\Delta\eta = 0.025/6$, $\Delta\eta = 0.025$) the resolution has a sampling terms of $(2.69 \pm 0.05)\text{mm}/\sqrt{E}$ (E in GeV), and $(5.40 \pm 0.20)\text{mm}/\sqrt{E}$ respectively. This can be converted into an angular resolution of $50\text{mrad}/\sqrt{E}$ in agreement with the ATLAS requirements. Similarly the position resolution can also be measured from the EM barrel middle compartment for various beam energies. A $3.4\text{mm}/\sqrt{E}$ sampling term with 0.32mm constant term is extracted at $\eta = 0.47$.

Finally preliminary results on $\gamma/\pi^0$ separation have been obtained using events recorded from a photon beam. After kinematical $\gamma$ pairing and cuts using the first strip compartment a promissing rejection factor (due to the first calorimeter compartment with fine $\Delta\eta = 0.025/8$ segmentation) between 2.6 and 3 (as a function of the photon's $P_T$) is obtained for a 90% photon's efficiency when the figure of 3.0 is expected especially for the $H \rightarrow \gamma\gamma$ search.

## 3. The Hadronic Endcap Calorimeter

The HEC calorimeter is designed for forward jet measurements in association with the EM endcap calorimeter. It requires a fast response in a high radiation environment. It consists of flat copper absorber plates interleaved with electrodes organised on the principle of the electrostatic transformer (EST). This increases the signal amplitude without increasing the capacitance and the required HV. The readout is achieved with cryogenics AsGa preamplifiers in order to optimize the signal-to-noise ratio, followed in the warm by pre-shapers instead of preamplifiers like for the EM and then the same chain as for EM. The calibration is performed in the cold in a similar way as the EM. Currently 90% of the HEC calorimeters modules are produced and tested. A first wheel of 32 modules is completed (see Fig. 2).

Beam tests have been performed on 24 over 134 HEC modules at CERN. First of all the calibration and physics signals have been simulated according to an electrical model of the readout circuits. The agreement found is within 1%. Measurements performed at various $\ell$Ar temperatures have been used to extract the ionisation drift time from the physics pulse shape and the relative amplitude variation after shaping is found to be $2\%/^o$K in agreement with ionisation models and similar measurements performed with the EM. The drift velocity and the ionisation cur-

Figure 2. View of the first endcap wheels (almost) assembled.

rent have also been measured as a function of the high voltage in the range $[1, 10]$kV/cm and are in good agreement with the models. The modules have been scanned in 15 points with $e^{\pm}$ and $\pi^{\pm}$ beams. The sampling and constant terms for electrons are found to be $(21.4 \pm 0.2)\%$ and $(0.3 \pm 0.2)\%$ with a spatial uniformity better than 1%. These two terms have been checked to be stable in position over the calorimeter within the errors and are compatible with the simulation expectations. The $\mu^{\pm}$ to noise ratio has been measured to be $\approx 5$ and the $e/\mu$ ratio is found to be $0.93 \pm 0.04$ in agreement with simulations. Finally the $\pi^{\pm}$ resolutions as a function of the beam energy give sampling and constant terms estimated to be $(70.6 \pm 1.5)\%$ and $(5.8 \pm 0.2)\%$ with an intrinsic ratio $e/h = 1.49 \pm 0.10$ in fair agreement with the GCALOR simulation program.

## 4. The Forward Calorimeter

The FCAL calorimeter extends the detector acceptance from $|\eta| = 3.2$ to 4.9. It plays an important role for forward jet tagging, vital for essential physics channels at LHC [5]. It has to cope with a constant high particle flux from minimum bias events and a very high radiation level. The technological choice consists of successive wheels in copper then tungsten for the absorber. The sensitive volumes are thin cylindrical $\ell$Ar gaps (tube/rod assemblies). A consequence is a fast drift time and a large cell capacitance. In order to adapt the signal to the same readout electronics than the EM an 2:1 autotransformer is used. Currently one half of the FCAL detector is completed and being assembled at CERN, the other half is expected for june 2003.

FCAL prototype modules have been tested in a beam at CERN in 1998. Energy resolution and linearity for $e^{\pm}$ and $\pi^{\pm}$ beams as well as the $\mu^{\pm}$ to noise ratio have been studied [6]. The $e^{\pm}$ linearity is measured to be within 1%. For $\pi^{\pm}$ beam the possibility to apply weights with a radial dependence has been studied and it has been shown that the noise can be improved by $\approx 10\%$ and the sampling term reduced from $89\%/\sqrt{E}$ to $80\%/\sqrt{E}$. The radial dependence of the weights shows well the suppression of the central part, mainly due to $\pi^0$'s, the enhancement of the peripheral zone and the suppression at large radius corresponding to the presence of coherent noise. This weighting method looks interesting for the FCAL calorimeter as the forward jets are very narrow. So the lateral shower extension for jets is mainly governed by the hadronic shower phenomenon and not by the jet opening.

## 5. Isolated clusters at first level trigger

Triggering at first level on isolated electromagnetic showers $(e^{\pm}, \gamma)$ and isolated hadrons most likely coming from $\tau^{\pm}$ is crucial for LHC physics as these particles are among the main signatures for possible new phenomena.

The calorimeter-based first-level trigger signals are formed from the analog signals after shaping. They are aligned in time, summed into trigger towers $(\Delta\eta = 0.1, \Delta\phi = 0.1)$, converted into transverse energy, digitized and associated to a beam crossing number [7]. These digitized signals are used to form jet and missing energy clusters or for isolated e.m./hadron clusters.

The isolated e.m./hadron cluster trigger is based on a sliding window algorithm using $4 \times 4$ trigger towers in both electromagnetic and hadronic calorimeters. This window is divided into four regions: the e.m. core, the e.m isolation ring, the had. core and the had. isolation ring (see [7] p. 45). The e.m. trigger is achieved by requiring energy in two adjacent trigger towers in the e.m. core and no energy elsewhere. The isolated hadron trigger is on an other hand achieved by requiring energy in the the sum of two adjacent trigger towers in the e.m. core and all 4 trigger towers in the hadronic core, and no energy in both isolation rings.

The cluster trigger covers the phase space, $|\eta| \leq 2.5$, and all $\phi$. It consists of 4 crates, each containing 14 cluster Processing Modules (CPM). Each CPM (see Fig. 3) processes $4\times16$ windows using 8 FPGA's (Xilinx Virtex chip XCV1000E). Because of the slidling window algorithm a CPM needs information processed by the adjacent boards. This communication is achieved via the crate back plane using a 160 $\mathrm{Mbs}^{-1}$ serial data stream in order to limit the number of electrical connections.

Simulations of this implementation have been made. It has been shown that the cluster algorithm itself is evaluated within 25ns. Including the other steps required, the cluster trigger takes a total of 175 ns, to be compared with the total time available to the level-1 calorimeter trigger of about 600 ns. A complete CPM has now been produced and is under test.

## 6. Conclusion

The status on the construction of the ATLAS liquid argon calorimeters (barrel, endcap electromagnetic and hadronic and forward) has been reported. Almost half of the detector is up to now produced. Several test beam have been performed on the serie modules and give good results in agreement with expectations. On an other hand a board for triggering electromagnetic clusters and isolated hadrons at the first-level trigger has been developed. The performances in time are good within the acceptable first-level time window.


### Acknowledgements
I am indebted to the members of the ATLAS collaboration, to P. Loch, P. Schacht, L. Serin and especially J. Garvey for his carefull reading of the manuscript.

Figure 3. The CPM board for the electromagnetic and isolated hadron first-level trigger.

## REFERENCES


1. *ATLAS Liquid Argon Calorimeter Technical Design Report, CERN/LHCC/96-41.*
2. *International Conference on Calorimetry in High Energy Physics, Pasadena, March 25-29, 2002, see talks given by M. Fincke-Keeler, K.K. Joo, A. Kiryunin, S. Rodier and D. Zerwas.*
3. *ATLAS Coll., Performance of the barrel module 0 of the ATLAS electromagnetic calorimeter, to be published in NIM.*
4. *ATLAS Coll., Performance of the end-cap module 0 of the ATLAS electromagnetic calorimeter, to be published in NIM.*
5. *ATLAS Detector and Physics Performance, vol 2, CERN/LHCC/99-15.*
6. *A. Savine et al., in Proc. of the* 7th *Int. Conf. on Calorimetry in High Energy Physics, Tucson 1997.*
7. *ATLAS First-level Trigger Technical Design Report, CERN/LHCC/98-14.*

ELSEVIER

Proceedings of ICHEP 2002, pp. 913–921
*S. Bentvelsen, P. de Jong, J. Koch and E. Laenen (Editors)*

31st INTERNATIONAL CONFERENCE ON HIGH ENERGY PHYSICS AMSTERDAM

# Recent Developments in Crystal Calorimeters (featuring the CMS $PbWO_4$ Electromagnetic Calorimeter)

S. Gascon-Shotkin[a]
on behalf of the CMS ECAL Collaboration

[a]Institut de physique nucléaire de Lyon, Bât. Paul Dirac
4, rue Enrico Fermi, 69622 Villeurbanne Cedex, France

In the mass range of 110-150 GeV the favored process for Higgs boson detection via p-p collisions is via its decay into two photons, which demands a very high-resolution electromagnetic calorimeter. This physics goal plus the Large Hadron Calorimeter (LHC)-imposed design constraints of 25ns bunch spacing and a hostile radiation environment have led the Compact Muon Solenoid (CMS) collaboration to the choice of lead tungstate ($PbWO_4$) crystals. These factors plus the presence of a 4T magnetic field and the relatively low room-temperature scintillation photon yield of $PbWO_4$ make photodetection a real challenge, which CMS has met via the choice of devices providing gain amplification: Avalanche photodiodes (APD) in the central barrel region and vacuum phototriodes (VPT) in the forward and backward endcap regions.

In the past year the CMS electromagnetic calorimeter has entered the construction phase. We review progress in the areas of crystals, barrel and endcap photodetection devices, plans for detector calibration as well as the status of assembly and quality control. We also invoke relevant developments in other crystal calorimeters currently in operation or under development. Crystal calorimeters remain the medium of choice for precision energy and position measurements in high energy physics.

## 1. INTRODUCTION

Crystal calorimeters have historically been used in high energy physics for precision energy measurements of electromagnetically-interacting particles such as e, $\gamma$ and $\pi^0$, and to help in precision position measurements. Since Crystal Ball[1], the first large-scale apparatus (672 NaI crystals, 15.7 radiation lengths ($X_0$)) in 1977, whose goal was the study of the charmonium spectrum, crystal calorimeters have evolved significantly. Calorimeter size in number of crystals has increased by more than two orders of magnitude (e.g. the 12000-crystal L3 BGO calorimeter[2] at LEP (1989)), innovations in crystal composition (e.g. doping) and machining techniques have made possible increased crystal length such that nearly-perfect longitudinal shower containment (>25 $X_0$) has become possible, and the range of applications has become very wide, from physics of b-quark-containing hadrons and CP violation (CLEOII[3], with 7800 CsI(Tl) crystals; KTEV[4], with 3256 pure CsI crystals, the currently-running BaBar[5] and BELLE[6] experiments, with 6580 and 8636 CsI(Tl) crystals, respectively, and the future BTEV[7], with 23000 $PbWO_4$ crystals) to searches for new phenomena (L3 and the future CMS[8], described in this presentation) to physics of heavy ions (the future ALICE[9] at the LHC, with 18000 $PbWO_4$ crystals), all with accelerators, to astrophysics experiments such as the future gamma-ray telescope GLAST[10], which will use 1536 CsI(Tl) crystals.

The electromagnetic calorimeter[11](ECAL) currently under construction for the general-purpose CMS experiment at the LHC at CERN, scheduled to begin data collection in 2007, will consist of ~77,000 lead tungstate ($PbW0_4$) crystals, and thus will become by more than a factor of 5 the largest crystal calorimeter ever built. The $PbW0_4$ crystals have been chosen in order to optimize the probability of observing the Higgs boson in the mass range of 110-150 GeV via its decay into 2 photons, which demands a very high-


doi:10.1016/S0920-5632(02)02277-6

resolution ECAL and in particular maximal longitudinal containment of electromagnetic showers. To achieve a targeted mass resolution of $\sigma(M)/M<1\%$ given the LHC-imposed constraints of a 25ns bunch spacing and a hostile radiation environment (0.15-0.3 Gy/hour in the barrel and 0.3-15 Gy/hour in the endcap region), some light yield (that of $PbW0_4$ at room temperature is three orders of magnitude less than that of NaI) has been sacrificed in favor of a very short light decay time (~10ns) and radiation length (0.9cm).

The CMS ECAL barrel, which will be off-pointing by 3° in each of $\eta$ and $\phi$ , will consist of 61,200 crystals of ~26 $X_0$ in 17 truncated pyramidal shapes of slightly different dimensions, with an extent of $\eta$ <1.48. The two endcaps, each consisting of 2 "dees", will consist of a total of 16,000 identical crystals of ~25 $X_0$, having a slightly larger cross-section than the barrel crystals, and extending from 1.48< $\eta$ <3.0. In addition, a sampling preshower detector located in front of each endcap, comprising 3 $X_0$ of layers of lead and silicon ladders, will increase granularity in the forward/backward zones and thus aid in rejection of $\pi^0$s faking Higgs photons. The target parametrized calorimeter energy resolutions are

$$\sigma/E = \frac{2.7\%}{\sqrt{E(GeV)}} \oplus 0.55\% \oplus \frac{0.155(0.210)}{E(GeV)} \quad (1)$$

and

$$\sigma/E = \frac{5.7\%}{\sqrt{E(GeV)}} \oplus 0.55\% \oplus \frac{0.770(0.915)}{E(GeV)} \quad (2)$$

for barrel and endcap regions, respectively[1]. The targeted contributions to each term in the above resolution expression are detailed in[11].

## 2. CRYSTALS

The Bogoroditsk Techno-Chemical Plant (BCTP)[2] will furnish all of the CMS $PbWO_4$ crystals, although R&D (which began in 1994) was also undertaken and continues with the Shanghai Institute of Ceramics (SIC). To date 8700 production-ready barrel crystals have been received and 6000 characterized. Figure 1 shows longitudinal scintillation light transmission (in %) vs. wavelength for a typical production crystal, superimposed upon which is the distribution of longitudinal transmission at 420 nm (the scintillation emission peak value) for the 6000 measured crystals. Virtually all meet the CMS minimum specified value of 55%.

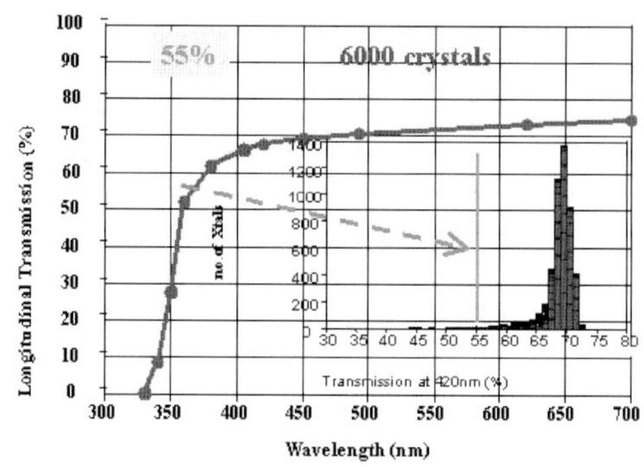


Figure 1. CMS experiment: Longitudinal scintillation light transmission in % vs. wavelength in nm for a typical production $PbWO_4$ crystal. Inset: Distribution of longitudinal transmission at 420 nm (minimum acceptable is 55%) for the 6000 measured production-quality crystals[12].

Light yield non-uniformity, defined as the percentage change in scintillation light yield per radiation length as a function of the distance from the photodetector, will contribute significantly to the calorimeter resolution via the constant term[11]. The targeted contribution of 0.3% corresponds to a maximal absolute value of frontal non-uniformity (FNUF) in the region of shower maximum of $3.5\%/X_0$. To achieve this goal

[1]The first (second) numbers in the numerators of the third terms of the above expressions refer to the expected "low-" ("high-") luminosity conditions at the LHC; where low (high) luminosity $\sim 10^{33(34)} cm^{-2}s^{-1}$.

[2]Tula, Russia

one face of each barrel crystal is depolished to a roughness of $0.34\mu$. The 6000 crystals measured to-date overwhelmingly satisfy the FNUF requirement as well as the minimal light yield specification of 8 photoelectrons/MeV[12][3].

Crystal radiation hardness will be crucial at the LHC. In $PbWO_4$ radiation does not affect the scintillation mechanism directly, but affects light yield through decreased transmission. The CMS crystals are sample-tested with a longitudinal irradiation of 10 hours at a typical LHC dose rate (0.15Gy/hour). Of 368 production-ready crystals thus tested so far, only 1% have suffered losses in transmission in excess of the permitted 6% (the average loss being $2.45\pm1.06\%$), and almost all of these have in addition been rejected by other selection criteria[12]. Improvements in radiation hardness since the beginning of the R&D period have been due to stoichiometric fine-tuning, optimization of crystal growth conditions and doping with yttrium and niobium. These last two factors have also contributed to improvements in scintillation light transmission.

The BTEV experiment, which will also use $PbWO_4$ crystals of a size comparable to the CMS endcap crystals, recently presented first testbeam results on radiation hardness[13]. Figure 2 shows signal loss in one crystal furnished by BCTP vs. dose rate and integrated dose from an 85-hour irradiation in a 27GeV electron beam at Protvino in April 2002. Also reported were the results of a cross-check performed on two CMS endcap crystals: After a low-dose rate (50 rad/hour at Protvino) irradiation, BTEV obtained comparable signal loss rates to those obtained by CMS on the same crystals after a high-dose-rate (20 krad/hour at the Hôpital Cantonal de Genève) irradiation.

Since the end of 2000, BCTP has been able to produce ingots large enough (65mm wide vs. 32mm) to allow the obtention of two barrel crystals per ingot instead of one. The "pair-produced" crystals seem optically indistinguishable from the singly-produced ones[12]. In the future, the fabrication of even larger ingots (75, 85mm wide) may be possible, thus allowing the

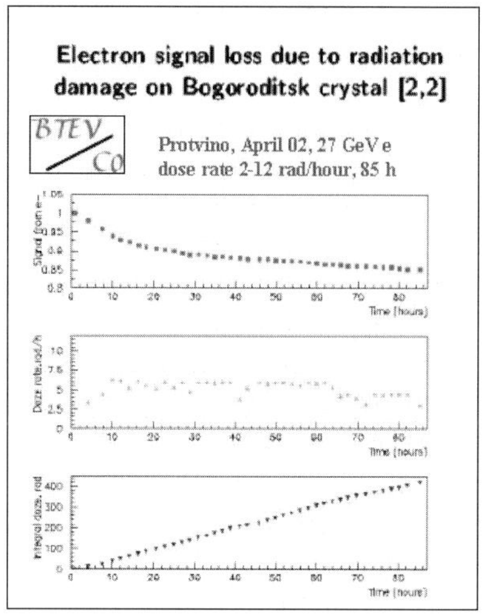


Figure 2. BTEV collaboration: Relative signal loss in one BCTP $PbWO_4$ crystal vs. dose rate and integrated dose from an 85-hour irradiation during test beam, April 2002.[13].

obtention of two endcap crystals or 4 barrel crystals per ingot.

## 3. PHOTODETECTORS

### 3.1. Avalanche photodiodes (APDs)

The CMS barrel avalanche photodiode (APD) is a gain-amplification-providing, magnetic-field-insensitive, fast, radiation-hard device and is the fruit of a 10-year R&D effort between the collaboration and Hamamatsu Photonics[4]. The principle of operation is shown in Figure 3: Scintillation photons enter the device and convert in a $p^{++}$- layer; these conversion electrons are then accelerated by a strong-gradient electric field in a p-layer. This field peaks at a p-n junction so as to induce avalanche multiplication of the electrons, which then drift under constant electric field in an n-region and are collected in an $n^{++}$-region as the photoelectric current. Figure 3 also shows a schematic of the APD[14], a key feature of which is the groove added to minimize surface leakage

[3]as measured with a photomultiplier

[4]Hamamatsu City, Japan. An R&D program was also undertaken with EG&G Corporation (Canada)

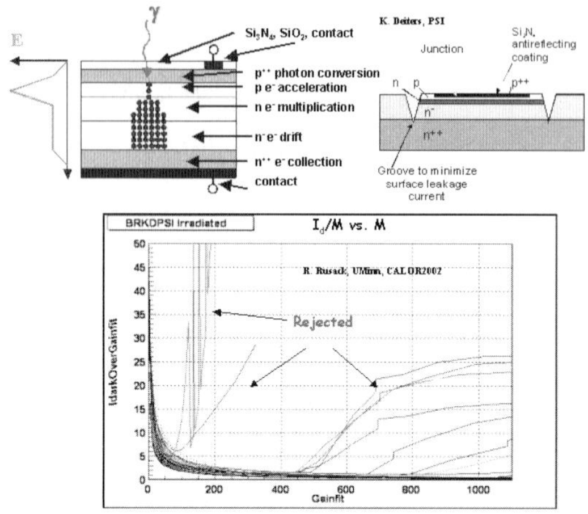


Figure 3. CMS experiment: Upper left: Principle of operation of a silicon avalanche photodiode (APD). Upper right: Schematic of the CMS barrel APD furnished by Hamamatsu Photonics[14]. Below: Dark current ($I_{dark}$) normalized to gain (M) vs. gain for a group of CMS APDs[15]

current.

Two APDs connected in parallel constitute the fundamental barrel photodetection device, known as a 'capsule', providing an active area per crystal of $2\times25mm^2=50mm^2$. The capsules influence the calorimeter energy resolution through all three terms in the parametrization: The active area and APD quantum efficiency ($\sim$75%) contribute to the stochastic term, the operating gain (M) of 50 (gains in excess of 1000 are possible) contributes to the constant term through intercalibration effects, and the APD capacitance ($\sim$80pF) and excess noise factor F (F=2 at M=50) contribute to the noise term (F also contributes to the part of the stochastic term represented by statistical fluctuations introduced by the gain amplification process.). The APDs are relatively charged-particle-insensitive, with an effective thickness $d_{eff}$ $\sim 6\mu$m.

Once glued to the crystal and mounted inside the calorimeter, the capsules become essentially inaccessible during the 10-year life of the experiment, giving rise to a maximum allowed APD failure rate of 1/1000. To fulfil this criterion all APDs are subjected to a predictive screening program involving a 500-krad pre-irradiation with a $Co^{60}$ gamma source followed by combined annealing and accelerated ageing under bias. A selection is then made based on the following quantities: Difference between breakdown and operating voltages, difference in breakdown voltage before and after the pre-irradiation, dark current ($I_{dark}$), $I_{dark}$/M vs. M and noise. Figure 3 shows $I_{dark}$/M vs. M curves for a group of APDs; those having a portion of the curve with positive slope are rejected[15]. The APDs are also sample-irradiated with a neutron source having a flux of $2\times10^{13}$neutrons/cm.

Each capsule is fully characterized[16] as to its gain, dark current, breakdown voltage, noise and pulse shape; the first three measurements are repeated after the capsule is glued to a crystal. Good correlations have been found between capsule measurements and Hamamatsu test results on the constituent APDs as well as with the post-gluing test results. Barrel photodetection performance continues to be checked in photovoltaic mode at each subsequent stage of calorimeter assembly.

The electromagnetic calorimeter of the ALICE experiment, which will study interactions from colliding heavy-ion beams at the LHC, will also be made of $PbWO_4$ crystals and will use the same APDs as the CMS ECAL. Figure 4 shows the parametrized energy resolution parameters obtained from measurements during the August-September 2001 test beam run, using 3X3 matrices of crystals furnished by BCTP and Apatity[5], with one APD per crystal; the values are close to those targeted[17]. The plot shows the distribution of the 3X3 energy sum from a 30-GeV electron beam; the lack of a high-energy tail attests to the insensitivity of the APD to the direct passage of charged particles.

### 3.2. Vacuum Phototriodes (VPTs)

Silicon photodetectors are not suitable for use in the CMS ECAL endcaps, where the radiation rates can be orders of magnitude higher than in

[5] Murmansk, Russia

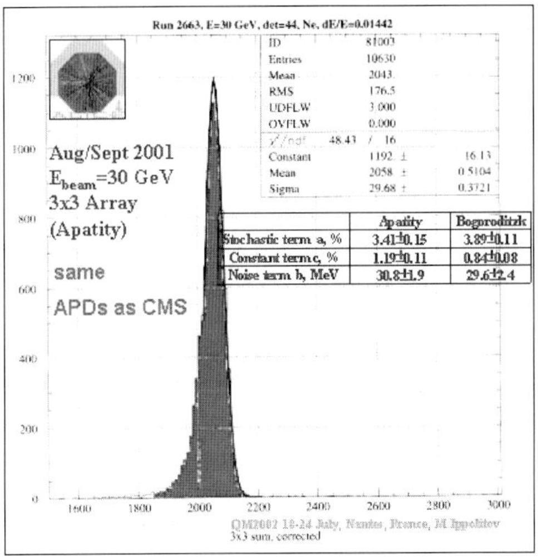


Figure 4. ALICE experiment: Distribution of the 3 X 3 Apatity $PbWO_4$ crystal energy sum from a 30-GeV electron beam, and parametrized energy resolution parameters (Apatity and BCTP crystals) from test beam, August-September 2001[17].

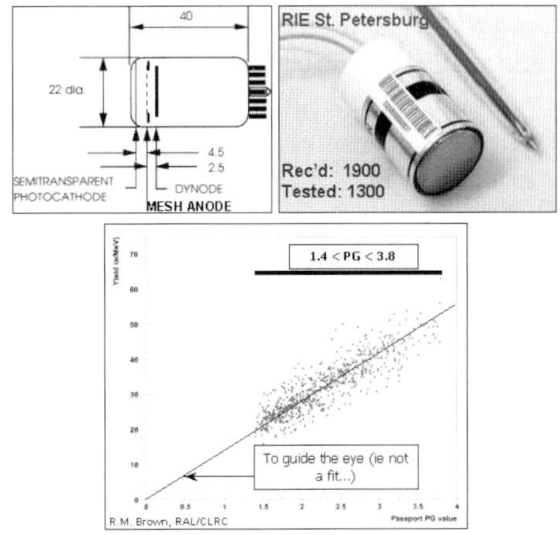


Figure 5. CMS experiment: Top: Schematic of the CMS endcap vacuum phototriode (VPT) furnished by RIE corporation. Bottom: Anode response at 1.8T under pulsed light vs. VPT net gain at 0T under continuous light for production VPTs (preliminary result)

the barrel. The CMS collaboration has instead chosen vacuum phototriodes (VPTs), single-stage photomultiplier tubes with fine metal grid anodes and UV windows (see Figure 5), which will be furnished by the RIE corporation[6]. With an active area of $280mm^2$/crystal, they will provide gains of between 8 and 10 in the CMS magnetic field of 4T. In particular, the VPT response is uniform for the range of angles their axes make with the magnetic field direction. They have a quantum efficiency of $\sim 20\%$ at 420nm and an excess noise factor of 3. The decrease in VPT response due to irradiation over the 10-year life of the experiment is anticipated to be less than 10%, which is deemed acceptable.

Figure 5 also shows the distribution of the anode response of production VPTs (expressed as a light yield) measured in a magnetic field of 1.8T under pulsed light versus the VPT net gain in zero magnetic field measured by the producer under continuous light (result preliminary). The correlation attests to a good understanding of the measurement schematics.

[6]St. Petersburg, Russia

The photoelectric currents from the APDs/VPTs will be processed by a 'light-to-light' on-detector readout chain consisting of only radiation-hard components[11,18,19].

## 4. CALORIMETER CALIBRATION

The dominant contribution to the energy resolution constant term for the CMS ECAL will be that from crystal intercalibration[11]. It is planned to perform the calibration of the CMS ECAL in four steps: First, by using the correlations between direct laboratory measurements performed during the detector construction and the same quantities calculated indirectly from test beam measurements[20], it is estimated that the final calibration constants can be predicted to within 6%. Second, a 'pre-calibration' of a significant fraction of the detector prior to installation will be performed in test beam with 50 and 120 GeV electrons, which should produce calibration constants with an estimated accuracy of within 2%[21]. Third, once data-taking starts, the crystals in a given ring in $\phi$ could be intercalibrated with minimum bias or jet trigger events in less than two hours, with a precision of 1-2%[22]. The process $Z \rightarrow e^+e^-$ would then be used to set the absolute energy scale for the $\phi$-rings and to intracalibrate them in less than one day, to 1%. Finally, electrons from $W \rightarrow e\nu$ events will be used for definitive periodic calibrations which will take 2 months to perform, with a final precision of within 0.5%[11].

The KTeV experiment, which collected data during 1996-1997 and 1999, has used a total of 600 million decays from the interaction $K_L \rightarrow \pi^{\pm}e^{\pm}\nu$ to establish the energy scale for each of the 3256 channels every 1-2 days with a precision of 0.03%[23]. Figure 6 (a) shows the ratio E/p for electrons/positrons from these decays, where E is measured with the calorimeter and p from tracks in a spectrometer. The crystal calorimeter resolution is obtained by subtracting the momentum resolution from the E/p resolution (calorimeter and momentum resolutions shown in Figure 6 (b)). The dependence of the mean E/p value on momentum, shown in Figure 6 (c), indicates an average energy nonlinearity of 0.5% in the energy range of 3 GeV<E<75 GeV. This nonlinearity, measured separately for each CsI crystal is used as a correction.

The BaBar experiment, running since 1999, uses three different processes to calibrate the 6580 CsI(Tl) crystals in three different energy regimes[24]: A neutron-activated fluid circulating through tubes in front of the crystals provides 6.13 MeV photons with which relative and absolute calibrations can be performed in a half-hour. Relative and absolute calibrations can be performed in 12 hours using Bhabha scattering events for the energy range 3-9 GeV. Finally, an absolute energy calibration in the energy range of 0.03-3 GeV is provided in 4 hours by pi-zero decays into two photons. Figure 7 shows the resolution Δ E/E vs. photon energy for the three calibration processes and in addition for two physics processes: $\eta \rightarrow \gamma\gamma$ and $\chi_c \rightarrow J/\Psi\gamma$, from which the two-term energy resolution parametrization and precisions shown are then extracted.

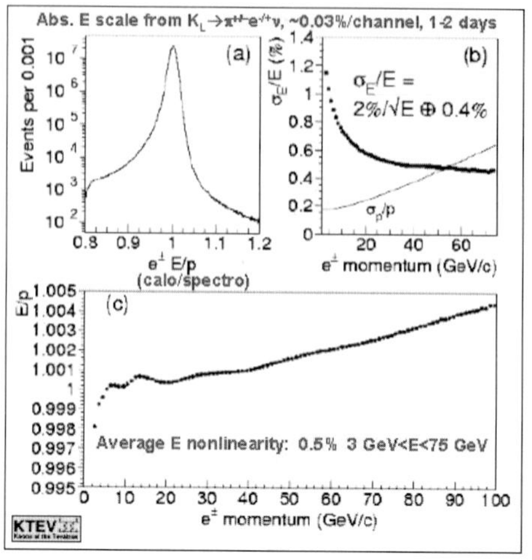


Figure 6. KTeV experiment: a) Ratio of energy E (from CSI crystal calorimeter) to track momentum p (from spectrometer) for electrons from the decay $K_L \rightarrow \pi^{\pm}e^{\pm}\nu$ b) Crystal calorimeter energy resolution vs. electron momentum and spectrometer momentum resolution c) E/p vs. electron momentum[23].

## 5. SCINTILLATION LIGHT LOSS MONITORING

Between calibrations with physics events, CMS will dynamically correct the calibration constant of each crystal for temporary losses in scintillation light transmission due to radiation[25]. Laser light of three frequencies will be distributed through a 2-level fanout system with cross-verification by precision radiation-hard PN diodes at each level, and eventually to each crystal via a radiation-hard optical fiber. Under radiation, the loss of the scintillation signal from the laser light follows that from particles according to a ratio R≡ $\Delta_{particle}/\Delta_{laser}$, considered to be sufficiently linear and uniform between crystals.

Figure 8 shows a demonstration of the laser monitoring system performed in test beam in 2000. The intercalibration constants for three neighboring crystals in a 30-crystal matrix after the monitoring corrections are shown as a function of time. Only the middle crystal has been irradiated and has undergone three successive exposures (0.57, 2.05 and 1.47 Gy, respectively) and

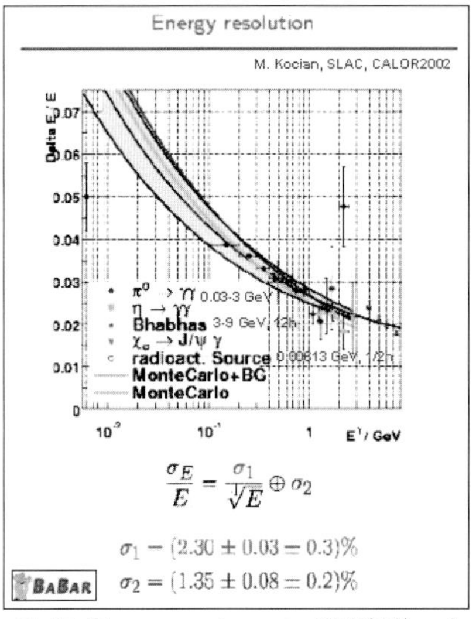


Figure 7. BaBar experiment: CSI(Tl) calorimeter energy resolution vs. photon energy for three calibration processes and two physics processes[24].

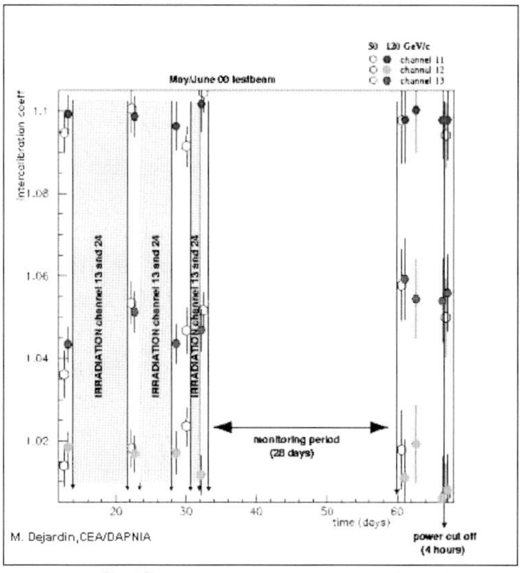


Figure 8. CMS experiment: Normalized intercalibration constants vs. time for three neighboring $PbWO_4$ crystals in a 30-crystal matrix, of which one (middle) has been irradiated, after laser monitoring corrections, during test beam May-June 2000.

recoveries. Within errors, the stability of the calibration constants is the same as for the two non-irradiated crystals.

## 6. THE CMS ECAL: ASSEMBLY STATUS AND QUALITY CONTROL

Each of the CMS ECAL barrel crystals undergoes fully automated testing of geometrical dimensions, scintillation light transmission, light yield and uniformity on a multipurpose system called ACCOS (Automatic Crystal COntrol System) [26]. CMS has received ~35,000 production Hamamatsu APDs and fabricated and characterized ~6500 barrel photodetection capsules. Of these, 4300 have been glued to crystals. Up to the module (400 or 500 crystals each) level, production and quality control activity proceeds in parallel at two distinct assembly centers, one at CERN and one at INFN ENEA/Casaccia (Rome); supermodule (1700 crystals) assembly, including FE electronics mounting, is performed at CERN.

The construction of the barrel began in September 2001 and is now well underway. In June 2002, the mechanical assembly of the first supermodule was completed. All construction and test activities are managed by and data input to the CRISTAL (Concurrent Repository & Information System for Tracking Assembly and production Lifecycles)[27] production management/quality control system, built on an object-oriented database.

For the endcaps, 1900 production VPTs have been received and 1300 tested.

## 7. CONCLUSION

Crystals remain the medium of choice for precision energy measurements in medium- and high-energy physics and astroparticle physics experiments. Construction of the CMS ECAL, the largest (in number of crystals) crystal calorimeter ever built, is now well under way, with as much as ~10% of some components already delivered and validated. A complete module with 100 channels equipped with near-final very-front-end electronics is currently in beam at CERN, representing the first large-scale system test. In summer 2003 the first supermodule (1/36th of the calorimeter) will be beam-tested with final front-end electronics.

## 8. ACKNOWLEDGEMENTS

The author wishes to thank all the members of the CMS ECAL collaboration, as well as the members of the KTeV, BaBar, ALICE and BTEV collaborations who provided material for this talk, and finally, the organizers and session convenors of ICHEP2002.

## REFERENCES


1. M. Oreglia et al., Phys. Rev. D 25 (1982) 2259
   R. Nernst et al., Phys. Rev. Lett. 54 (1985) 2195
2. L3 Collaboration, B. Adeva et al., "The Construction of the L3 Experiment", NIM A 289 (1990) 35
   L3 Technical Proposal to CERN-LEPC (May 1983)
   L3 Collaboration (O. Adriani et al.), "Results from the L3 Experiment at LEP", Physics Reports (236) 1,2 (1993)
3. Y. Kubota et al., Nucl. Inst. and Meth. A320, 66 (1992).
4. K. Arisaka et al., KTeV Design Report, FN-580.
5. B. Aubert et al., The BABAR detector, NIM A479 (2002)
6. A.Abashian et al. (The Belle Collaboration) [edited by S.Mori] KEK Progress Report 2000-4, to appear in NIM A
7. BTEV Proposal & TDR, http://www-btev.fnal.gov/public/hep/general/proposal /index.shtml
8. The Compact Muon Solenoid (CMS) Technical Proposal, CERN/LHCC 94-38, LHCC/P1 (1994).
9. ALICE - Technical Proposal for A Large Ion Collider Experiment at the CERN LHC, CERN/LHCC/95-71, December 1995.
   PHOS Technical Design Report, CERN/LHCC 99-4, ALICE TDR 2.5 March 1999
10. "Studying the High-Energy Gamma-Ray Sky with GLAST", Kamae, T., Ohsugi, T., Thompson, D.J., Watanabe, K. (on behalf of the GLAST Collaboration), 2000, AdSpR, 25, 905
    "GLAST: The Next-Generation High Energy Gamma-Ray Astronomy Mission", Gehrels, N., Michelson, P., 1999, Astroparticle Phys, 11, 277
11. The Electromagnetic Calorimeter Project, Technical Design Report, CERN/LHCC 97-33, CMS TDR 4 (1997).
12. E. Auffray, "Status on PWO Crystals from Bogoroditsk for CMS-ECAL", presentation at SCINT2001, Chamonix, Sept. 17. 2001
13. A. Matulenko, V. Mochalov, A. Vasiliev, BTEV Collaboration Meeting, Fermilab, July 17-18 2002
14. K. Deiters, "APDs for CMS and other Users", presentation at BEAUNE2002–3d International Conference on Photodetection, 17-21 June 2002
15. R. Rusack, "Avalanche Photodiodes for the CMS Lead Tungstate Calorimeter", presentation at CALOR2002, Caltech, 25-30 May 2002
16. P. Depasse et al., "A Validation Program for the CMS ECAL Photodetection Capsules", CMS IN-2000/013.
17. M. Ippolitov, "Development of the ALICE Electromagnetic Calorimeter PHOS", presentation at QM2002, Nantes, 18-24 July 2002
18. "Addendum to the CMS ECAL TDR–Changes to the CMS ECAL Electronics", CERN/LHCC-2002-027 CMS TDR 4-Addl, 19 Sept. 2002
19. G. Anelli et al.,"Radiation tolerant VLSI circuits in standard deep submicron CMOS technologies for the LHC experiments: practical design aspects", IEEE trans. nucl. science, vol.46, n.6, Dec.1999
    P. Jarron, "RD49 project: radiation tolerant techniques and strategies of electronics systems for the LHC experiments", presentation at 6th International Conference on Advanced Technology and Particle Physics, Villa Olmo, Como 5-9 October 1998
20. F. Cavallari et al., "Test-beam Results on the Performance of Two Matrices of $PbWO_4$

Crystals for the CMS ECAL and Comparison with Laboratory Measurements", CMS IN 2001/033.
21. CMS ECAL Precalibration Group (P. Bordalo et al.) "The Ecal Precalibration in H4. Sequences of Operations", CMS IN-2001/005. J. Bourotte et al.,"The ECAL Precalibration in H4: User Requirements Document Version 1.0", CMS NOTE 1999/024.
22. D. Futyan, "Intercalibration of ECAL Crystals in Phi Using Symmetry of Energy Deposition", CMS NOTE-2002/031.
23. A. Alavi-Harati et al. (the KTeV Collaboration), "Measurements of Direct CP Violation, CPT Symmetry, and Other Parameters in the Neutral Kaon System", submitted to Phys. Rev. D. (August 6, 2002).
24. M. Kocian, "The Electromagnetic Calorimeter of the BaBar Detector", presentation at CALOR2002, Caltech, 25-30 May 2002
25. CMS-Saclay Group, M. Déjardin et al., The CMS-ECAL Monitoring System: Data Generation, Acquisition And Processing At LHC, CMS IN-2002/012.
26. E. Auffray et al.,"Performance of ACCOS, an Automatic Crystal Quality Control System for PWO", CMS NOTE-1999/067.
27. E. Auffray et al.,"CMS ECAL User Requirements For CRISTAL2" , CMS IN-2002/045. J.-M. Le Goff et al., "Detector Construction Management and Quality Control: Establishing and Using a CRISTAL System", CMS NOTE-1998/033.

ELSEVIER

Proceedings of ICHEP 2002, pp. 922–926
*S. Bentvelsen, P. de Jong, J. Koch and E. Laenen (Editors)*

31st INTERNATIONAL CONFERENCE ON HIGH ENERGY PHYSICS AMSTERDAM

# Calorimetry at a future $e^+e^-$ collider

J. Cvach[a]*

[a]Centre for Particle Physics, Institute of Physics AS CR, Na Slovance 2, 18221 Praha 8, The Czech Republic

on behalf of the CALICE Collaboration

The requirements on calorimetry for future $e^+e^-$ linear colliders are formulated. Approaches and recent R&D results for electromagnetic and hadron calorimeters for the TESLA detector are given.

## 1. INTRODUCTION

The physics program for an electron–positron linear collider (LC) with the collision CMS energy between $90-800$ GeV is dominated by final states containing many jets. Among them, the production of heavy bosons $W's, Z's$ and probably Higgses followed by hadron and lepton decays are the most important. The resolution of calorimeters in the energy range of the future LC is mediocre. More promising method called here "analytic energy flow" (AEF) [1] relies on precision measurements of charged tracks in the tracker and photons in the fine grained electromagnetic calorimeter (ECAL). These particles represent about 90% of an event energy. The remaining 10% of energy with large fluctuations, of course, must be separated from charged track deposits in the fine grain hadron calorimeter. The behaviour of the detector with respect to jets is then the following. The strong magnetic field sweeps the charged particles away from the neutrals and the separation between photons and neutral hadrons is done in a calorimeter with a large interaction to radiation length ratio. The presence of a charged lepton is a signature of the presence of a neutrino.

To profit from the AEF new effective software algorithms are as important as new detectors. A complete set of generation and reconstruction programs exists in Europe [2] and in the United States [3]. Practically all approaches use the following steps:

- charged track–cluster association using the charged track extrapolation
- unassociated clusters in calorimeters are identified with neutral particles - photons or neutral hadrons

The AEF performance of program packages will be tested with the same events with $W^+W^-$, $Z\bar{Z}$, $W^+W^-\nu\bar{\nu}$, $Z\bar{Z}\nu\bar{\nu}$, $t\bar{t}$ final states. The goal is to reach the energy resolution for jets $\sigma/E \sim 30\%/\sqrt{E}$. The particle identification and reconstruction put requirements on calorimeters:

`cell size of the ECAL` is comparable to the Molière radius to resolve $\pi^0$ from $2\gamma$'s

`longitudinal segmentation of ECAL` is fine to allow high quality 3D shower reconstruction and to evaluate hypotheses for different types of particles - 10 segments are recommended

`cell size of the HCAL` is comparable to 1 $X_0$ to follow the core of the hadron shower made by electromagnetic component of the hadron cascade

Based on the above considerations and also on the previous detailed studies [4] an R&D program for calorimeters was formulated [5] and accepted by DESY PRC in autumn 2001. The main task of the R&D program is to choose the technology

*supported by the Ministry of Education of the Czech Republic under the project LN00A006.


doi:10.1016/S0920-5632(02)02278-8

and to optimize the calorimeter design. A prototype of the calorimeter (1 $m^3$) will be built and tested in beams in 2003-4. The results of the tests will help to tune and optimize the simulation and reconstruction software. This work will be done within CALICE Collaboration [6].

## 2. THE ECAL DESIGN

The requirement of ECAL is to identify photons down to energies of several hundreds MeV without collecting fake ones. To identify electrons, accurately measure photons and electrons in jets. To optimize the separation between photons, electrons and hadrons, the calorimeter should be far from the interaction point, dense and highly granular with large ratio of interaction $\lambda_I$ (cm) over radiation $X_0$ (cm) lengths. The

| material | $\lambda_I$ | $X_0$ | $\rho_M$ (cm) | $\lambda_I/X_0$ |
|---|---|---|---|---|
| Fe | 16.8 | 1.76 | 1.69 | 9.5 |
| Cu | 15.1 | 1.43 | 1.52 | 10.6 |
| W | 9.6 | 0.35 | 0.93 | 27.4 |
| Pb | 17.1 | 0.56 | 1.00 | 30.5 |

comparison for iron, copper, tungsten and lead is given in the table. The need of containing high energy electromagnetic showers leads to the length of $\sim$ 24 $X_0$. The tungsten radiator provides a very compact detecting medium. To reach the transversal size close to Molière radius $\rho_M$ the obvious solution is to use silicon diodes as sensing detectors. This solution is proposed in [4] and called a Si-W calorimeter. To keep a good energy resolution it should finely sample the showers in non projective geometry without cracks. By using 30 thin layers of sampling (0.4 $X_0$) followed by 10 coarser ones (1.2 $X_0$), the calorimeter is only 19 cm thick and has an energy resolution close to $10\%/\sqrt{E}$. The sensing pad size of 1 $cm^2$ is made on 4" Si wafers 0.5 mm thick of high resistivity of 5 k$\Omega$.cm. Because of large pad area and higher noise they are AC coupled to the front-end electronics mounted above pads. The mechanical structure is not sufficient for dissipation of heat and an active cooling is mandatory. This can be

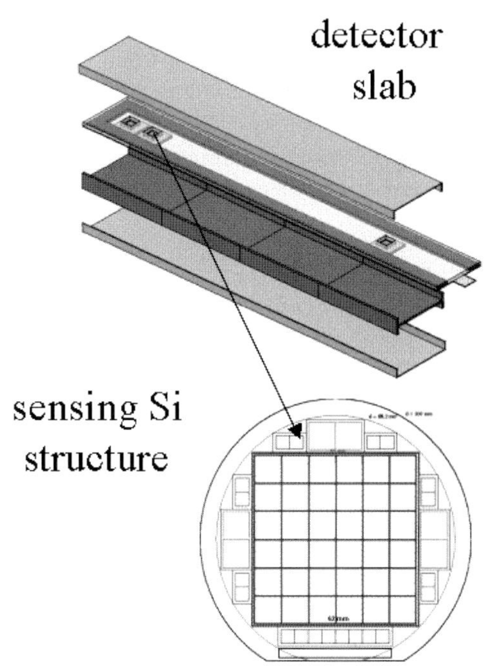


Figure 1. The smallest construction unit of ECAL is a detector slab 150 cm long which contains sensing Si diodes in arrays of 6 x 6 pads, each pad of 1 $cm^2$ area. It is supported by an H profile made of tungsten–carbon–epoxy molding and covered from top and bottom. Slabs are interleaved with W absorber plates and arranged into larger alveoli structure (not shown).

done by tiny pipes laying on sides of silicon diodes and running parallel the long sides of a detector slab (see Fig. 1). The front-end chip for the prototype is directly derived from a chip done for the Opera experiment.

The ongoing studies with prototyping concern the Si sensor design, cooling, the pick-up noise control, the front-end electronics, the building of the detector slabs and of mechanical structure.

## 3. THE HCAL DESIGN

As mentioned in the Introduction the main task of the hadron calorimeter HCAL is to identify properly clusters of neutral hadrons - mainly

$n$, $K_L^0$ and to measure their energy. The best compact hadron calorimeter [7] would use tungsten as radiator but mainly the price considerations incline to the solution with the stainless steel or copper (less favourable due to eddy currents caused by magnet quenching). The current design counts on stainless steel plates 2 cm thick with 6.5 mm gap for the sensing structure. The new role of the hadron calorimeter which acts as a coarse tracking device leads to small pads. As mentioned in the Introduction their size is governed by the radiation length of the calorimeter material which is of several cm's. There are two approaches to the HCAL performance. In the traditional one, the energy is given by the sum of analog signals from sensing cells. The second approach relies on the number of particle passages through the the sensitive cells of the calorimeter. Such discrete counting requires a high granular calorimeter. Number of binary counts is proportional to the deposited energy. This solution is called the digiHCAL contrary to the first solution which uses scintillator tiles of sizes 5x5 cm$^2$ and larger and is called here the tileHCAL. In both cases the calorimeter is 4.4 $\lambda$ deep in the barrel and 12.8 $\lambda$ in the end-cap.

### 3.1. The tileHCAL

The status of design of calorimeter can be found in [8]. Plastic scintillator plates 5 mm thick serve as active layers. The scintillation light produced in the tiles enters the attached wavelength shifter (WLS) fibres where it is re-emitted as green light at around 500 nm wavelength. About 5% of emitted light is captured in the double clad WLS fibre and guided to sensitive photo-detectors outside the calorimeter volume. Different types of photo-detectors are studied for optimal photo-cathode efficiency, gain and large signal to noise ratio. Possible choices are:

- avalanche photo-diode arrays
- silicon photomultiplier arrays (pixel or metal resistor photo-diodes)
- multi-anode photomultiplier arrays

Detectors differ in basic parameters which are given in the following table where $\epsilon$ gives a photo-conversion efficiency for green light. The photo-detectors should be linear in the signal range from a mip to maximal shower cell energy (0.03-15 GeV). This requires a dynamic range of 500. For

| detector | $\epsilon$ (%) | gain | bias volt.(V) |
|---|---|---|---|
| APD | 70-85 | 100-400 | 300 |
| Si-PM | 12-18 | $10^6$ | 30-50 |
| PM | 11-15 | $10^4$-$10^7$ | 500-2000 |

further questions of signal processing, calibration and monitoring see e.g. [9]. Actual R&D studies started in autumn 2001 and have come already to concrete results which will be used and further tested in the first prototype – minical – a tile calorimeter of size 21x21x75 cm$^3$ which will be exposed to cosmics and to 2 GeV electron beam at DESY. The best yield for 5x5 cm$^2$ tiles is obtained with the Bicron BC-408 scintillator viewed by a Y11(200) double clad WLS-fibre from Kuraray. From Russian scintillators the best is a polystyrole based scintillator from Protvino SC-306 with a LY $\sim$ 70% of BC-408. The effort is put to further improve this scintillator. The optimal fibre wrap is Tyvek but an improvement of several tens % was observed using special super-reflector foils from 3M (see Fig. 2). At present fibres are read by Hamamatsu multi-anode PMs with 4x4 channels. ADC spectra show sufficient LY up to 17 photoelectrons for a tile (see Fig. 2). The study of tile–fibre coupling showed that this result can be improved with two fibre loops in the groove in the tile. The LY is $1.5 - 1.7$ higher than in the simplest configuration with a fibre on the tile side. Complete results from the investigations will be soon published.

The final detector will be placed in 4 T magnetic field and photo-detectors insensitive to magnetic field must be used. The first results with 32 element Hamamatsu Si APD array S8550 are promising. A low noise charge preamplifier developed in CALICE collaboration is used. In parallel new developments with Si-PMs are done in Moscow Institutes.

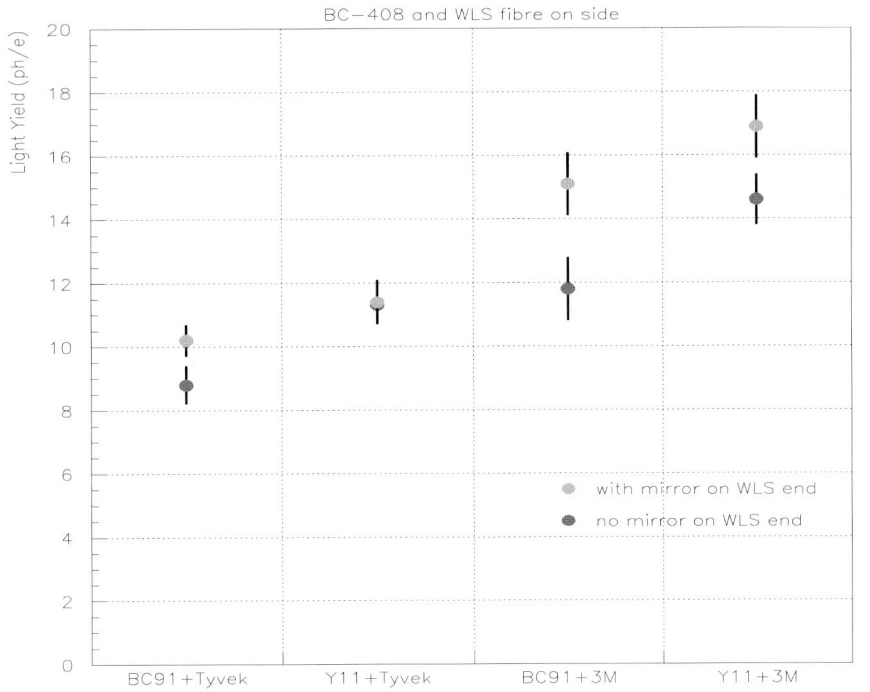


Figure 2. Light yield in photo-electrons from Bicron BC-408 tiles 5x5 $cm^2$ in side contact with a WLS 1 mm fibre with mirror (light circles) and without mirror (dark circles) on the fibre end. Double clad fibres from Kuraray Y11(200) and Bicron BC 91A are compared for two different tile reflectors: two layers of Tyvek paper and a super-reflector foil from 3M. The highest light yield is achieved with Y11(200) fibre and 3M foil. The effect of fibre mirror is modest. The light from fibre was read by a PM.

### 3.2. digiHCAL

For this option it is proposed to use gaseous detectors. Two candidates are considered [10]: thin resistive plate chambers (RPC) with glass sheets as resistive plates and short drift tube (SDT) chambers with 2–3 mm tube length. Both options have to work in the streamer/Geiger mode providing large output signals to simplify digital front-end electronics. Cosmic and beam tests were performed with small 80x80 $mm^2$ RPC prototypes in streamer mode. Commercially available glass sheets of 1 mm thickness were used for resistive plates with $\rho \sim 10^{13}$ Ω.cm. Gas gap was 1.2 mm. The gas mixture selected for the streamer operation was TFE:$N_2$:IB = 80:10:10. All prototypes had 16 pads of 1 $cm^2$ area with 1 mm gap between pads. The exposition in the 5 GeV positive beam showed 98% efficiency for 7.5 kV HV and 50–200 mV thresholds. At 200 mV threshold the hit multiplicity reached 1.3 level.

The same beam set up was used also for the SDT chamber prototype. The gas mixture selected for RPC showed low efficiency for SDT. It was found that the best properties had the mixture rich of isobutane Ar:TFE:IB = 10:10:80. At HV of 7.5 kV the hit multiplicity was reasonably low 1.05 and efficiency at 96%.

It seems that both the RPC and SDT chamber options are promising for the digital calorimeter. The RPC option has a low rate capability ($< 10$ Hz/$cm^2$) and relative large cell to cell signal overlap ($\sim 30\%$). The SDT chamber option has efficiency slightly lower. Non flammable gas

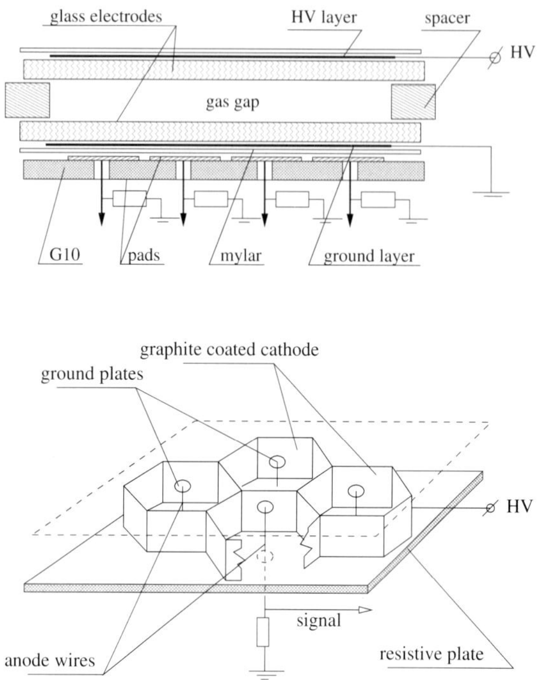


Figure 3. Scheme of RPC prototype (upper part) and a sketch of SDT chamber (lower part) for digiHCAL.

mixture is an open topic. Easy technology for the mass production (60 million channels for the final calorimeter) has to be found.

## 4. CONCLUSIONS.

To finalize the design of a calorimeter for a linear collider a considerable effort is needed both on the technical and software sides. The electromagnetic calorimeter as a sandwich of tungsten and silicon appears to be adequate from the point of view of physics, it remains to be fully proved technologically. For the hadron part a choice has to be made between two possibilities. For the decision full simulation and reconstruction as well as building of prototypes is foreseen. Due to the limited future access to high energy test beams a world wide effort is needed.

**Acknowledgments**

I wish to thank to all colleagues in the CALICE Collaboration who contributed to the ideas and results presented in this report. Especially I want to thank to V. Ammosov, T. Behnke, J.-C. Brient, E. Devitsin, V. Korbel, V. Morgunov, S. Němeček, S. Reiche, J. Repond, R. Settles, H. Videau, V. Vrba for many enlightening discussions and providing me with the material.

## REFERENCES

1. H. Videau and J-C. Brient, Proceedings of CALOR 2002, Pasadena, March 2002.
2. J-C. Brient, Programs MOKKA+REPLIC, http://polywww.in2p3.fr/tesla/calice_software.html;
   T. Behnke and V. Morgunov, Programs BRAHMS+SNARK, see also V. Morgunov, Proceedings of CALOR 2002, Pasadena, March 2002;
   P. Gay, MOKKA + programs for AEF.
3. N. Graph, T. Abe and M. Iwasaki, GEANT4 + reconstruction energy flow package, A. Maciel, GEANT4 + reconstruction energy flow package.
4. TESLA Technical Design Report, part IV, DESY 2001-011, March 2001, http://tesla.desy.de/new_pages/TDR_CD/start.html.
5. J-C. Brient et al., Proposal for R&D, DESY-PRC R&D-01/02 and Memorandum to the PRC, DESY, Hamburg 2001.
6. CALICE Collaboration: http://polywww.in2p3.fr/tesla/calice.html.
7. J.-C. Brient and H. Videau, preprint LLR 02-001, Lab. Leprince-Ringuet, Ecole Polytechnique.
8. V. Morgunov and V. Korbel, Proceedings of the Snowmass 2001 Conference, see http://www.slac.stanford.edu/econf/C0010630/forweb/ E310_morgunov1.pdf and E315_morgunov.pdf.
9. V. Korbel, Proceedings of CALOR 2002, Pasadena, March 2002.
10. V. Ammosov, Proceedings of INSTR-02 conference, Novosibirsk, February 2002.

ELSEVIER

Proceedings of ICHEP 2002, pp. 927–929
*S. Bentvelsen, P. de Jong, J. Koch and E. Laenen (Editors)*

31st INTERNATIONAL CONFERENCE ON HIGH ENERGY PHYSICS AMSTERDAM

# Photon Colliders

J. Gronberg[a]*

[a]Lawrence Livermore National Lab, Livermore CA, 94550, USA

A photon collider interaction region has the possibility of expanding the physics reach of a future TeV scale electron-positron collider. A survey of ongoing efforts to design the required lasers and optics to create a photon collider is presented in this paper.

## 1. Photon-Photon Interactions

The idea for creating a photon-photon collider by adding high power lasers to an electron linear collider was proposed by Ginzburg *et al.* [1] in the early 1980s. Such a machine opens up new possiblities for the study of elementary particles. The Higgs boson can be produced in photon-photon collisions through a loop diagram in which all particles with charge and mass contribute. This diagram provides an opportunity to search for new particles of arbitrarily high mass through their effects on the loop. Additionally, control of the laser polarization allows some control over the spin and charge-parity state of the initial two photon system. This provides a unique mechanism to study the properties of a Higgs boson. Two photon collisions also provide a mechanism of charged particle production that has only electromagnetic contributions at the tree level. This would alleviate the theoretical uncertainties in particle production commonly encountered at other machines. The physics potential of such a machine is the subject of on-going study.

## 2. Photon Collider Hardware

The basic idea of a photon collider interaction region is shown in Figure 1. As the electron beams converge to the interaction point (IP) they are intercepted by laser pulses a few millimeters from the IP. A large fraction of the incoming electrons, ~60%, undergo a Compton backscatter and produce a high energy photon. The Compton backscattering process has been studied in detail [2] and simulation programs to model and parameterize the luminosity exist [3–6].

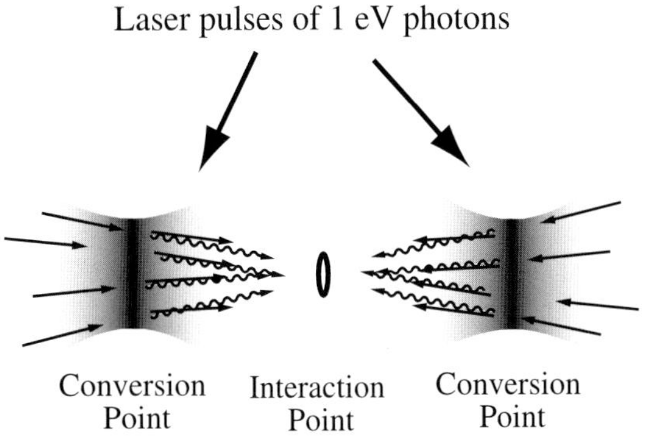


Figure 1. A schematic diagram of the photon collider interaction point. The high energy Compton backscattered photons focus to the same spot size as the electron beam.

In order to achieve a high luminosity of photon collisions, each electron bunch must be intercepted by a laser pulse. A laser pulse of 1 Joule compressed to 1 picosecond is required to achieve the ~60% Compton backscattering rate. This must then be multiplied by the repetion rate of proposed machines such as NLC [7] and TESLA [8] to compute the average laser power required. A single pass system that used each laser

*This work was performed under the auspices of the U.S. Department of Energy by the University of California, Lawrence Livermore National Laboratory under Contract No. W-7405-Eng-48.


doi:10.1016/S0920-5632(02)02279-X

pulse once would require around 20kW of average power. A formidable number which pushes the state of the art in laser engineering.

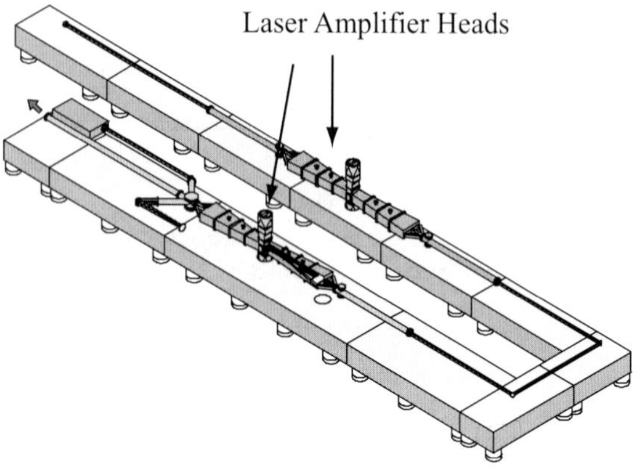


Figure 2. The layout of the MERCURY laser. Each laser head contains diode-pumped solid-state Yb:S-FAP crystals that are cooled by helium gas flow.

A design of the laser architecture for the NLC has been produced, based on the MERCURY [9] laser under construction at LLNL. The MERCURY laser, as shown in Figure 2, is designed to be a high average power solid-state laser producing 100 Joule pulses at one micron wavelength at a repetion rate of 10 Hz. An array of 12 of these lasers can be fired sequentially to match the 120 Hz repetition rate of the NLC. The single 100 Joule pulse can then be subdivided and time delayed using a series of optical splitters and delay lines to produce a train of 100 one Joule pulses separated by 2.8 ns. This would provide one laser pulse for every electron bunch in the machine. A system of focusing mirrors in the interaction region (IR) allow each laser pulse to be used twice, providing laser pulses for each arm of the linear accelerator. The layout of the focusing optics is shown in Figure 3.

While the technical solution proposed for NLC could be adapted to TESLA, the 5 Hz repetition rate of TESLA does not use the MERCURY laser efficiently. To use this architecture, TESLA would require 2-3 times more average power and the development of high average power electro-optical switches. An alternative architecture [10] which takes advantage of the large inter-bunch spacing of TESLA is under development at the Max Born Institute. A single laser pulse travels around a ring colliding with all of the bunches, as shown in Figure 4. This greatly reduces the required laser power, but requires high precision in the construction of the optical cavity.

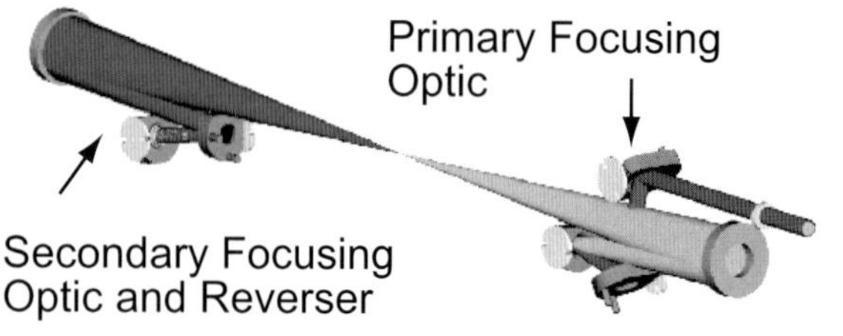


Figure 3. The arrangement of focusing optics which are contained within the accelerator's beam pipe. This is a two-pass system in which each laser pulse collides with two electron bunches, one from each arm of the accelerator.

## 3. Future Work

The MERCURY laser has begun commissioning at lower pulse energy and repetition rate. It is expected to be operating at full power by fall of 2003. A prototyping effort for the TESLA optical ring is proposed for the Max Born Institute.

While the conceptual designs for a photon collider are maturing, a photon collider of this type has never been demonstrated. The existence of the Stanford Linear Collider offers the opportunity [11] for a proof-of-principle demonstration of the integration of the photon collider optics into a real IR with the ability to produce and observe actual photon-photon luminosity. Such a demonstration would provide technical risk reduction for a photon collider experiment at a future TeV scale linear collider.

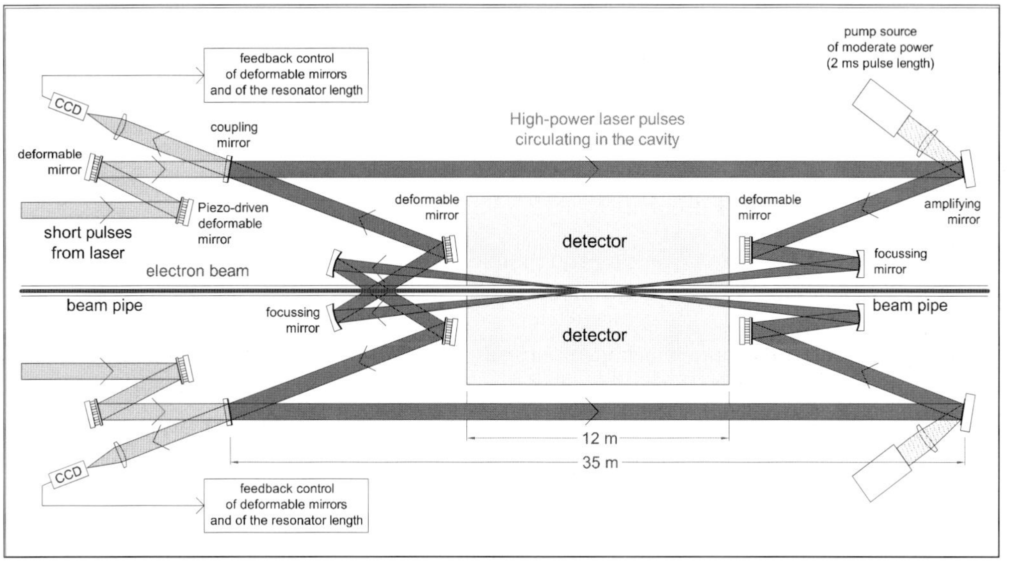


Figure 4. A diagram of the ring laser proposed for TESLA. The inter-bunch spacing of TESLA allows the laser pulse to travel around the outside of the detector before the arrival of the next electron bunch. Figure courtesy of K. Moenig, DESY.

## 4. Conclusion

No conceptual obstacles to the construction of a photon collider experiment have been discovered. The designs have reached the level of maturity where prototyping and systems demonstration are the next steps.

## REFERENCES


1. I. Ginzburg, G. Kotkin, V.Serbo and V. Telnov, *Pizma ZhETF* **34** (1981) 514; *JETP Lett.* **34** (1982) 491 (Preprint INF 81-50, Novosibirsk (1981) in English).
2. V. Telnov, *Nucl. Instr. and Meth.* **A 355** (1995) 3.
3. K. Yokoya, "A Computer Simulation Code for the Beam-Beam Interaction in Linear Colliders", KEK report 85-9, Oct. 1985.
4. A. Zarnecki, "CompAZ: parameterization of the photon collider luminosity spectra", hep-ex/0207021
5. V. I. Telnov, "A code for the simulation of luminosities and QED backgrounds at photon colliders", talk presented at Second Workshop of ECFA-DESY study, St. Malo, France, April 2002.
6. T. Ohl, "Circe Version 2.0: Beam Spectra for Simulating Linear Collider and Photon Collider Physics", WUE-ITP-2002-006, May 2002 (in preparation)
7. The NLC Design Group, "Zeroth Order Design Report for the Next Linear Collider", *LBNL-5424*, *SLAC-474*, *UCRL-ID-124161*, *UC-414* (1996).
8. The TESLA Collaboration, "TESLA Technical Design Report", *DESY-2001-011*, *ECFA-2001-209*, *DESY-TESLA-2001-23*, *DESY-TESLA-FEL-2001-05*, Mar. 2001.
9. L. Hackel, *et al.*, *UCRL-TB-136126-01-02.*
10. I. Will, T. Quast, H. Redlin and W. Sander, "A Laser System For The TESLA Photon Collider Based On An External Ring Resonator", *Nucl. Instrum. Meth.* A **472** (2001) 79.
11. http://www-project.slac.stanford.edu/lc/linx

Proceedings of ICHEP 2002, pp. 930–932
*S. Bentvelsen, P. de Jong, J. Koch and E. Laenen (Editors)*

ELSEVIER

31st INTERNATIONAL CONFERENCE ON
HIGH ENERGY PHYSICS AMSTERDAM

# A European Neutrino Factory Complex

S. Gilardoni[a]*, representing the European Neutrino Factory Working Groups

[a]Université de Genève, Département de physique nucléaire et corpusculaire
24, Quai Ernest-Ansermet 1211 Genève 4, Suisse

The Neutrino Factory is the ultimate machine to measure all neutrino mixing parameters, in particular $\theta_{13}$ and the leptonic CP violation, together with the sign of the mass splitting. This paper reviews the physics reach and the European accelerator design and R&D for a Neutrino Factory.

## 1. Introduction and Physics Motivation

Neutrino oscillations are the best explanation, at the same time, of the Solar Neutrino puzzle and for the azimuthal variation of the atmospheric muon neutrino flux. SuperKamiokande has provided from atmospheric neutrinos the mixing angle $\theta_{23}$ and the mass splitting $\Delta m^2_{23}$. The SNO measurements for the solar neutrinos clearly prefers the LMA solution, which fixes the range for $\theta_{12}$ and $\Delta m^2_{12}$. Still missing are the measurements of $\theta_{13}$, the sign of the mass splitting $\Delta m^2_{23}$ and the discovery of the leptonic CP violation. The best way to access $\theta_{13}$ is to look for $\nu_e$ transition into $\nu_\mu$, since the oscillation probability is proportional to $\theta^2_{13}$. High energy $\nu_e$ can be produced under controlled conditions by a machine which accelerates an intense muon beam whose decay $\mu^+ \rightarrow e^+ + \nu_e + \overline{\nu}_\mu$ will assure a naturally pure $\nu_e$ and $\overline{\nu}_\mu$ beam, namely a Neutrino Factory (N.F.). The signature of the $\nu_e \rightarrow \nu_\mu$ transition is given by the neutrino interactions producing $\mu^-$ (*wrong sign muons* of opposite sign of the muons used to generate the neutrino beam) at the far detector location. The typical neutrino energy considered is 20-50 GeV, and the maximum of the oscillation probability is at few thousand kilometres from the machine. At such distance the MSW effect is quite large for neutrinos around 10 GeV; this allows to determine the sign of $\Delta m^2_{23}$. In the N.F. the sign of muons can be changed to reach the *Holy Grail*, the leptonic CP violation, from the asymmetry:

$$A_{CP} = \frac{P(\nu_e \rightarrow \nu_\mu) - P(\overline{\nu}_e \rightarrow \overline{\nu}_\mu)}{P(\nu_e \rightarrow \nu_\mu) + P(\overline{\nu}_e \rightarrow \overline{\nu}_\mu)}$$

The final goal of the N.F. is to measure the mixing angles $\theta_{13}$ and $\theta_{23}$ with high precision and the CP violating $\delta$ phase at the few percent level[1].

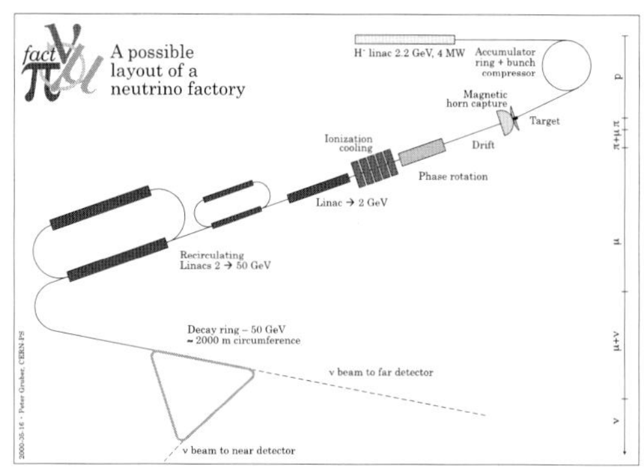


Figure 1. Layout of the CERN Neutrino Factory scheme

## 2. Machine description

The aim of a N.F. is the production of high energy, highly collimated (anti)electron and (anti)muon neutrino beams from muon decay. The goal is to achieve a neutrino flux of the order of $10^{21}\nu/year$. The typical scheme of the factory is the following (see fig.1): a high power proton

*E-mail:\Simone.Gilardoni@cern.ch Supported by the CERN Doctoral Student Programme. The author is grateful to K. Long for giving the talk in his place.


doi:10.1016/S0920-5632(02)02280-6

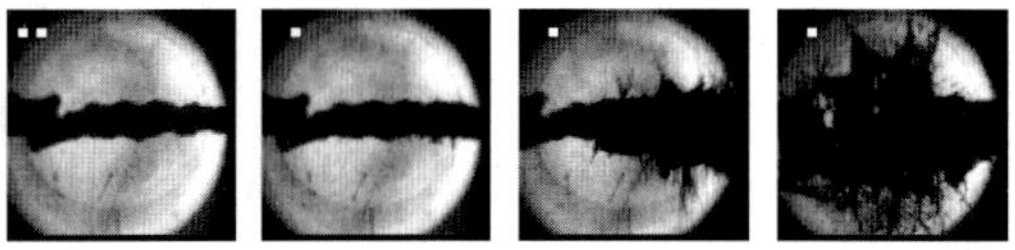

Figure 2. Mercury jet explosion due to protons impact. (Protons right to left, time left to right)

beam (4 MW) impinges on a mercury target to produce pions. Pions are focused by a magnetic lens (e.g. a horn) into a solenoidal decay channel. At the end of the channel muons produced by pion decay have an enormous energy spread (300%), an enormous divergence (>0.5 rad) and an enormous transverse dimension (radius $\approx$ 30 cm). This beam cannot be injected into a traditional accelerator without intolerable losses. The first step is the reduction of the muon energy spread via a series of RF (Radio Frequency) cavities. Then the transverse phase space is reduced by ionization cooling. The resulting beam can feed two recirculating linacs placed one after the other to reach the final energy of 50 GeV. The last step is to store the muons in a triangular decay ring, whose two longer sides are pointing to two locations placed at different distances.

### 2.1. SPL: Superconducting Proton Linac

The first element of the N.F. is the proton driver. In the CERN scheme $H^-$ ions are accelerated using a superconducting linac to 2.2 GeV[2]. The requested proton beam power is at least 4 MW because the flux of muons needed to meet the oscillation experiment goals is estimated around $10^{21}\mu/year$, and the rate proton/muon conversion is around 1%. This leads to $10^{23}$ protons per year, and at 2.2 GeV it equals to 4 MW. The SPL is composed by a room temperature low energy part (until 120 MeV), followed by a series of superconducting accelerating section to reach 2.2 GeV. The SPL injects the stripped $H^-$ into two rings, the first to accumulate protons and the second to compress them in time before the interaction with the target.

### 2.2. Target

Protons impinge on a target to produce pions. The main challenge of such system is to sustain 4 MW and a high proton pulse repetition rate (50 Hz) in a volume that cannot exceed the one of a pint of beer (L=30 cm and 1 cm radius), since a bigger target will decrease considerably the efficiency of the pion collection system. The baseline material chosen is mercury because it has a high Z, which means high pion production per unit of length, it is liquid at room temperature and it can be easily replaced at every proton shot, while a solid will not survive more than few proton pulses impacts. A number of experiments have been performed to measure the speed of the mercury explosion (see fig.2) induced by the proton energy deposition[3]. A major difficulty in evaluating the best target material comes from the large uncertainty on the pion production and interaction cross section for the range of energies between some hundred MeV and few GeV. The data taking of the Harp experiment will provide new measurements of those quantities with an error of few percent [4] and clarify which is the best material to be used for a given proton energy.

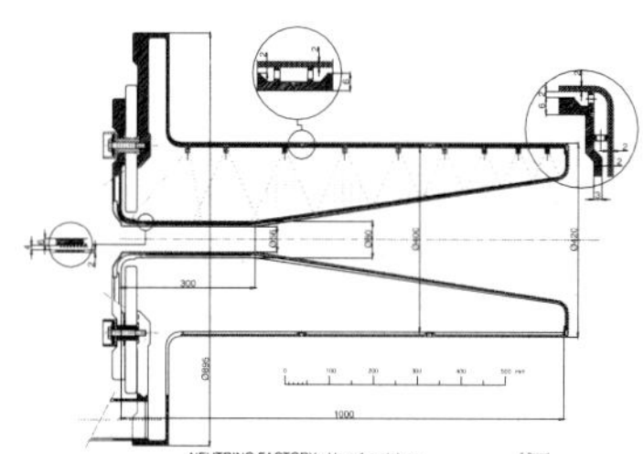


Figure 3. Design of N.F. horn prototype.

### 2.3. Pion collection system: Horn

Pions produced in the target are focused by a magnetic lens called horn. A horn consists of two concentric conductors which delimit a closed volume (see fig.3). The current of 300 kA pulsed at 50 Hz running in the conductors generates in that volume a toroidal magnetic field whose intensity decreases proportionally to the distance from the horn axis. Pions entering in the magnetic volume are bent by the field and focused in the direction of the decay channel section [5]. The sign of pion charge is selected by the current polarity, changing in this way the sign of the muons in the machine. The major limitation of the system comes from its short life-time, estimated around 6 weeks, mainly due to thermal and mechanical stresses induced by the high repetition rate.

### 2.4. Ionization cooling

After the focusing system pions are injected into a solenoidal channel where they decay into muons of a few hundred MeV but with an enormous energy spread. A series of RF cavities compress the muons around 200 MeV, but the transverse phase space dimensions (radius and angular divergence) are still too large to be accepted by a conventional accelerator: the beam has to be transversally cooled. Traditional techniques like stochastic cooling are too slow compared to the mean life-time of a muon, even for ultrarelativistic energies, so a faster process called ionization cooling is used. Ionization cooling involves three elements: solenoids, absorbers, RF cavities. The beam is confined by a series of solenoids whose waist is centered into an energy absorber of low Z material, like liquid hydrogen. Muons lose momentum in all direction passing through the material, while the RFs placed after the absorber restore only the longitudinal momentum, thus reducing the divergence of the beam. Liquid hydrogen is preferred for the absorber material to keep the beam heating due to multiple scattering as low as possible [6]. Ionization cooling was invented 20 years ago, but no experiment was ever tried in the past to demonstrate the feasibility of a section of a cooling device. The MICE experiment (see fig.4[7]) will test a part of the cooling channel of the N.F.. An International Collaboration has been already set up and a technical proposal will be submitted to RAL by the end of this year.

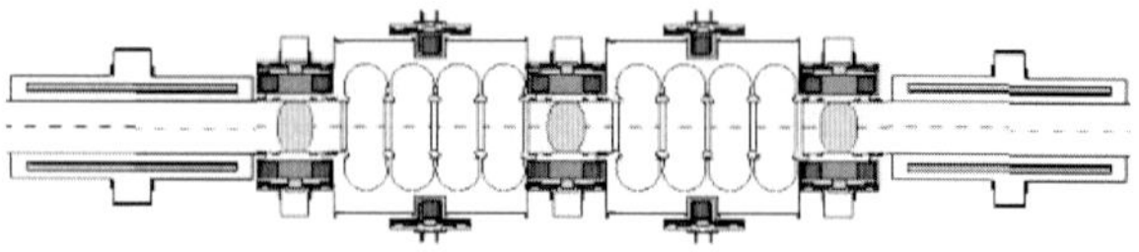

Figure 4. MICE experiment layout. Three $H_2$ absorbers are placed between pairs of focusing coils and separated by two series of RFs. The muon beam is measured after and the before the cooling section in the long solenoids.

### 2.5. Acceleration, storage ring, detector

After the cooling section, muons are injected in a series of two recirculating linacs and accelerated to the final energy of 50 GeV. Then they feed the triangular storage ring. This shape has two preferred decay directions pointing towards two experimental sites, one at short distance ($\approx$ 700 km), and the other at far distance ($\approx$ 3000 km). Two locations are already identified inside Europe, namely the Gran Sasso Laboratories in Italy and the Canaria Island in Spain. Two different detectors are proposed: an iron magnetized calorimeter or a large magnetized liquid argon detector. The magnetic field ($\approx$1 T) separates the wrong sign muon signal from the background.

## 3. Conclusions

The study of the N.F. complex is coordinated by the European Muon Coordination and Oversight Group[8], with the aim of producing a conceptual design report by the start time of LHC and JHF.

## REFERENCES


1. A. Blondel et al., Nucl. Instrum. Methods Phys. Res., A451(2000)102
2. M. Vretenar (ed.), CERN 2000-012.
3. A. Fabich et al., J. Nucl. Mater. (2002), t.b.p.
4. G. Catanesi et al., CERN-SPSC/2002-019
5. S. Gilardoni et al., Nufact02 proceedings, J. Phys. G (2002), t.b.p.
6. A. Lombardi et al., hep-ph/0008065
7. Mice LOI, http://hep04.phys.iit.edu/cooldemo
8. http:://muonstoragerings.cern.ch

ELSEVIER

Proceedings of ICHEP 2002, pp. 933–935
*S. Bentvelsen, P. de Jong, J. Koch and E. Laenen (Editors)*

31st INTERNATIONAL CONFERENCE ON HIGH ENERGY PHYSICS AMSTERDAM

# Status of R&D on Neutrino Factories and Muon Colliders

Gail G. Hanson[a]

[a]Department of Physics, University of California at Riverside, Riverside, California 92521-0413

Muon storage rings offer the possibility of intense neutrino beams to measure neutrino oscillation parameters. Muon colliders could be the best way to discover and study Higgs bosons in supersymmetry. The ultimate goal is a multi-TeV $\mu^+\mu^-$ collider. Recent and on-going progress in research and development on neutrino factories and muon colliders in the U.S. will be summarized, including targetry and ring coolers.

## 1. WHY MUON COLLIDERS?

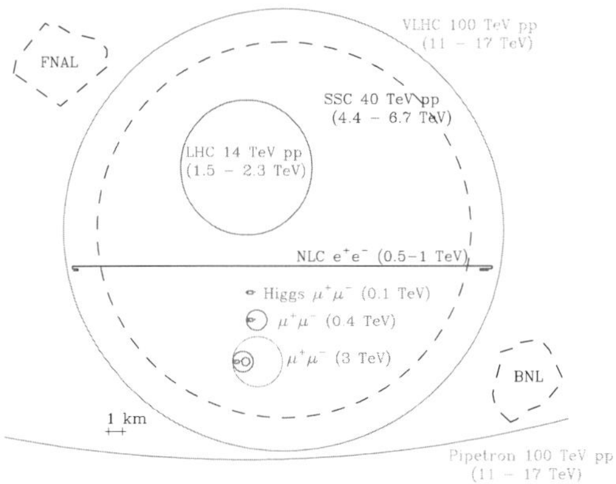


Figure 1. Comparison of the sizes of various future and proposed accelerators.

Muons are fundamental particles like electrons, so muon colliders have the same advantages as electron-positron colliders. The energy of the interaction is the full center-of-mass energy of the colliding particles, not of constituents. Synchrotron radiation by muons is less than that by electrons by $(m_e/m_\mu)^4 \approx 6 \times 10^{-10}$, which allows muon colliders to be much smaller than $e^+e^-$ colliders for the same rf power. Both of these effects allow $\mu^+\mu^-$ colliders to be smaller than hadron colliders of the same effective center-of-mass energy. A comparison[1] of the sizes of proposed accelerators is shown in Fig. 1. The muon beam can have a very narrow energy spread ($\sim 10^{-5}$).

## 2. NEUTRINO FACTORIES

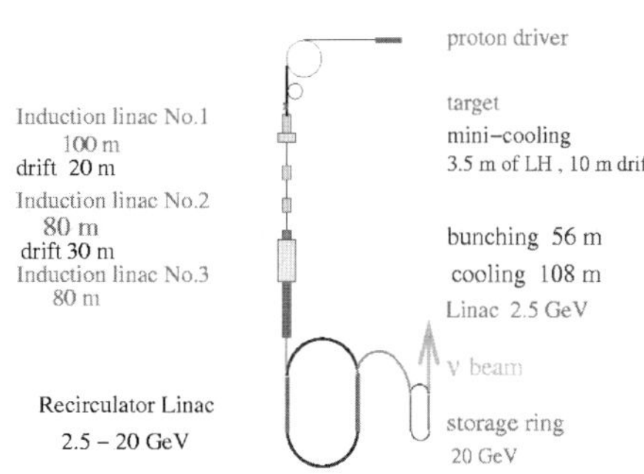


Figure 2. Schematic of a neutrino factory - Study II version.

Since muons decay ($\mu^- \rightarrow e^-\nu_\mu\bar{\nu}_e$), a muon storage ring produces an intense beam of neutrinos and can be thought of as a "neutrino factory." There is no other comparable source of $\nu_e$'s and $\bar{\nu}_e$'s. A muon storage ring can produce $10^{19}$ to $10^{21}$ muon decays per year. These intense neutrino beams can be used to study neutrino oscillations and possible $CP$ violation in the leptonic sector. Two feasibility studies were carried out in the U.S. [2,3]. Figure 2 shows a schematic of the Study II neutrino factory.

### 2.1. Target and Capture

The Study II target is a liquid mercury jet, which is being studied in BNL Experiment E951. Solenoidal magnets capture the produced pions. The mercury is contained in a pool. Liquid mercury can accommodate a proton beam power of


doi:10.1016/S0920-5632(02)02283-1

up to 4 MW. Experiment E951 has shown that the mercury jet does not break up until long after being struck by an incident proton beam.

### 2.2. Cooling

The ionization cooling of the muon beam is accomplished by passing the beam through liquid hydrogen absorbers and then restoring the longitudinal momentum in rf cavities. Solenoidal focusing is employed. The SFOFO cooling channel of Study II is shown in Fig. 3.

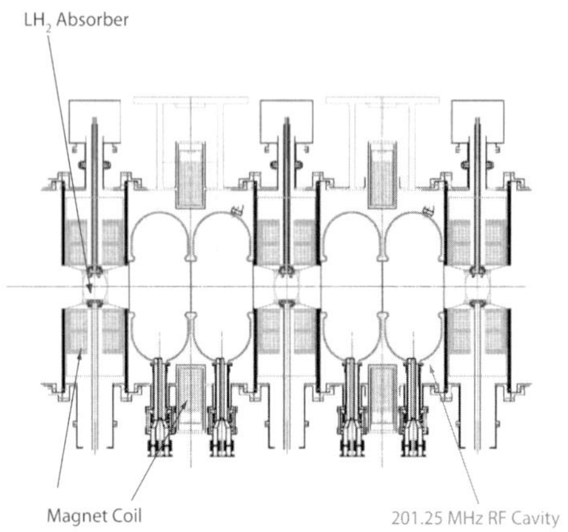


Figure 3. The solenoidal focusing (SFOFO) cooling channel of Feasibility Study II.

## 3. Muon Colliders

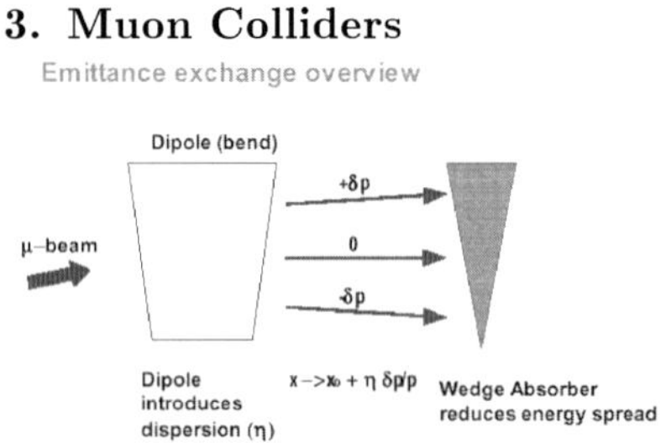


Figure 4. Overview of an emittance exchange section, in which longitudinal emittance is reduced by using a wedge absorber at nonzero dispersion. From D. Neuffer.

The cooling channel for the neutrino factory cools only in the transverse direction. For a muon collider the beam must be cooled in the longitudinal direction also. Emittance exchange is needed to transfer the transverse cooling to the longitudinal direction, as illustrated schematically in Fig. 4.

A ring cooler[4], shown in Fig. 5, has been shown in simulation to cool both longitudinally and transversely, also shown in Fig. 5. However, there is a problem with injection and extraction of the beam.

This problem has been solved in the RFOFO ring cooler[5], shown in Fig. 6, which uses a cell much like the Study II cooling channel. The solenoids are tilted for bending. The performance of the RFOFO ring cooler is also shown in Fig. 6.

Another type of ring cooler uses quadrupole focusing[5], as shown in Fig. 7.

With several ring coolers the longitudinal emittance can be reduced to the level needed for a muon collider. Another factor of $\sim 10$ is needed in each of the transverse directions. For this the use of a lithium lens is under study.

There has been a tremendous amount of progress in R&D on cooling. Ring coolers can increase the performance and may reduce the cost of a neutrino factory, thus allowing the development of a better design.

## REFERENCES

1. C. M. Ankenbrandt *et al.*, Muon Collider Collaboration, "Status of Muon Collider Research and Development and Future Plans," *Phys. Rev. ST Accel. Beams* **2**, 081001 (1999).
2. N. Holtkamp and D. Finley, eds., "A Feasibility Study of a Neutrino Source Based on a Neutrino Storage Ring," Fermilab-Pub-00/108-E (2000).
3. S. Ozaki, R. B. Palmer, M.S. Zisman, and J. C. Gallardo, eds., "Feasibility Study-II of a Muon-Based Neutrino Source," BNL-52623, June 2001; http://www.cap.bnl.gov/mumu/studyii/FS2-report.html.
4. V. Balbekov, "Ring Cooler Update," NuMu Note # 189 (2001).
5. R. B. Palmer, "Ring Coolers: Status and Prospects," talk at the 4th International Workshop on Neutrino Factories Based on Muon Storage Rings – NuFact'02, London, U.K., July 1-6, 2002.

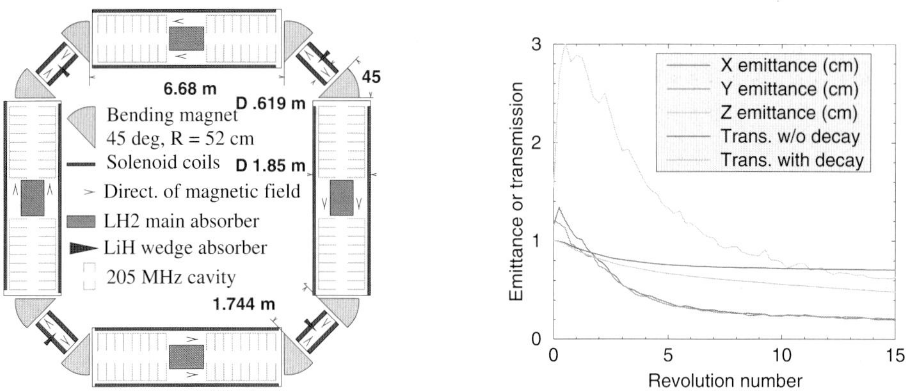


Figure 5. Balbekov ring cooler and its simulated performance.

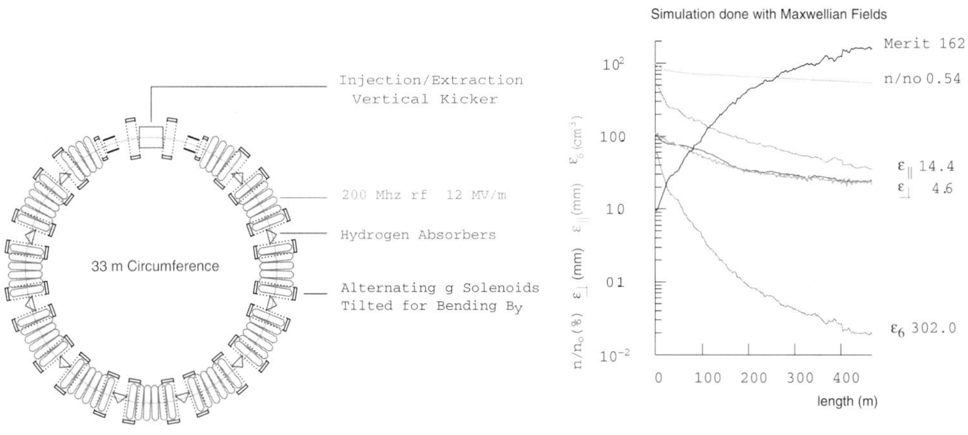


Figure 6. RFOFO ring cooler and its simulated performance.

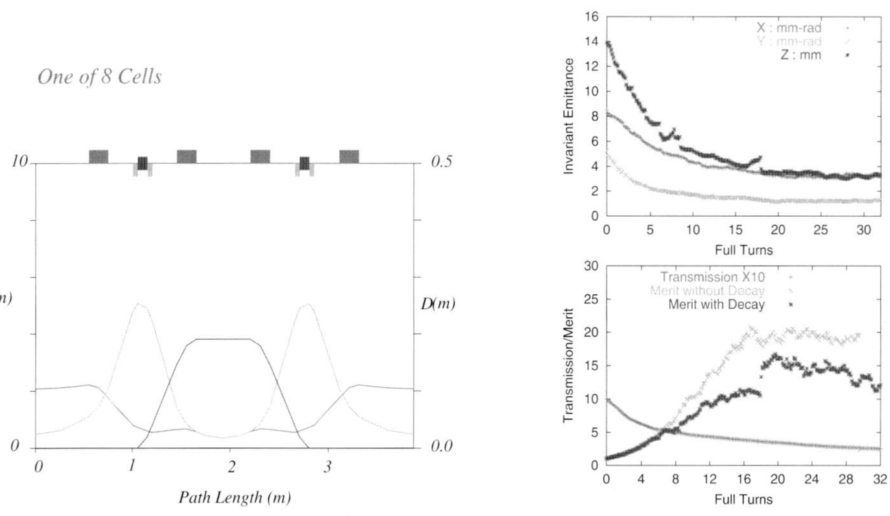


Figure 7. Section of quadrupole ring cooler and simulated performance of the quadrupole ring cooler.

ELSEVIER

Proceedings of ICHEP 2002, pp. 936–938
*S. Bentvelsen, P. de Jong, J. Koch and E. Laenen (Editors)*

31st INTERNATIONAL CONFERENCE ON HIGH ENERGY PHYSICS AMSTERDAM

# The BEPCII Project

Hesheng Chen[a]

[a]Institute of High energy Physics, Chinese Academy of Sciences, Beijing 100039, China

## 1. BEPC/BES and the Physics Results

Beijing Electron-Positron Collider (BEPC) was constructed successfully at the Institute of High Energy Physics (IHEP), Chinese Academy of Sciences, in 1988, with the beam energy range between 1 and 2.8 GeV. The Beijing Spectrometer (BES) is the only detector in BEPC, with the major physics goals of the Tau-Charm physics. BES started data taking from 1989. Now BEPC runs 5 months per year for the physics collision with the luminosity of $10^{31}$ $cm^{-2}s^{-1}$ at the beam energy of 1.89 GeV, and 3 months per year for the dedicated synchrotron radiation running with 140 mA at the beam energy of 2.2 GeV.

BES collected large data samples of $J/\Psi, \Psi', D_s$, and $\tau$. Many important results were obtained from BES, including the precision measurement of $\tau$ mass, the study of the decay property $J/\Psi$ and $\Psi'$, and the precision R-measurement between 2 - 5 GeV. The new R-values from BES made a relevant impact to the prediction of the Higgs mass in the Standard Model, the calculation of $\alpha(M_Z^2)$ and the g-2 experiment. These results draw great interests in the high energy physics community. During the last three running year, 58 M $J/\Psi$ hadron events and 14 M $\Psi'$ hadron events were collected. Many interesting physics results are expected to be published soon.

## 2. Future Development of BEPC: BEPCII

The recent developments in the high energy physics require the precision measurements in the Tau-Charm energy region to answer many important questions in the frontier of the particle physics. There are many Charmonium, charmed mesons and other resonances in the energy region. It is an ideal place to study the light hadron spectroscopy, the $J/\Psi$ family, and the excited baryons. This energy region is also the transition between continuum and resonances, perturbative and non-perturbative QCD, and could study many important issues in QCD. In addition, new types of the hadronic matter, such as glueball and quark-gluon hybrid are predicted in the energy region, the radioactive decays of the Charmonium are the best channels to search for them. The search for new phenomena, such as $D^0 - \bar{D}^0$ mixing and CP violation in the decays of the Charmonium and Tau, are also very interesting. To study on the precision measurements in these physics topics, high statistical data samples and small systematic error are the key issues.

BEPC runs at the threshold of the Charmonium family, the production cross sections are high, and the backgrounds are low. The pure initial states and the lower multiplicities provide optimum signal to noise ratios to study the physics. Thus BEPC has a unique advantage for many interesting physics studies, which cannot be done by B-factories or the fixed target experiments. To reach these interesting physics goals, IHEP decided to make a major upgrade in both the machine and the detector, called BEPCII / BESIII. Chinese Government approved the upgrade plan.

## 3. Machine Upgrade: Double Ring Collider

The earlier design of BEPC II was based on the multiple bunches and the bunch trains in a pretzel orbit at the existing BEPC storage ring, increasing the luminosity by one order of magnitude, similar to CESR. However, the great importance of the Tau - Charm physics and the interesting results from BEPC attract CESR to


doi:10.1016/S0920-5632(02)02284-3

join the competition. To extend the physics potential and to be more competitive, last year I-HEP modified the design of BEPC II to the double ring version, which significantly improved the performance with the calculated luminosity of $10^{33}\ cm^{-2}s^{-1}$ at 1.89 GeV beam energy, which is a two orders of magnitude improvement in the luminosity.

The double ring design requires a new storage ring to be built inside the existing storage ring at the BEPC tunnel. The two half new rings and two half old rings will cross at the two interaction points to form two identical rings, shown in the figure. Each ring can be filled with up to 93 bunches with 10 mA per bunch. The total beam current per ring is 0.9 A at the beam energy of 1.89 GeV. The beams collide at the south interaction point with a horizontal crossing angle of 11 mrad..

To reduce the beam length, the superconducting micro-beta quads will be installed at the interaction region. The superconducting cavities with a 499.8 MHz RF system will further reduce the bunch length and provide much higher RF power. The low impedance vacuum chambers with a coupling impedance of 0.4 $\Omega$ will be installed in the machine. The position ring will use antechamber. The designed beta function $\beta_y^*$ is 1.5 cm. The expect beam lifetime is 2.7 hours. The instrumentation and the control system will be also upgraded.

The upgrade of the Linac injector includes: increasing the injection frequency from 12.5 Hz into 50 Hz; increasing the current and the energy bombing on the position target; multiple packet injection. The designed positron injection rate is 50 mA per minute. The high power klystrons of 45 MW will replace the existing 30MW klystrons to provide the full energy injection up to 1.89 GeV. This is essential for the top-up injection to optimize the average luminosity of BEPCII.

To retain the dedicated synchrotron radiation running in the existing outer ring, a bridge will connect the two half outer rings at the north interaction point. At the collision point of the south interaction region, special dipole coils will be installed in the superconducting quads to keep the electron beam in the outer ring during the dedicated synchrotron radiation runs. Thus all insert devices and synchrotron radiation beam-lines will have no change. The beam current during dedicated synchrotron radiation running could be higher than 250 mA at 2.5 GeV, because the SC RF system provides much higher RF power .

The length of the BEPC tunnel is only 240m, and the straight section in the interaction region is rather shorter. In addition, one has to keep the synchrotron radiation insert devices in the double ring. Thus the design and the construction of BEPCII are big challenge, and more difficult than the two B-factories.

## 4. Detector Upgrade

The design goals of the Beijing Spectrometer upgrade (BES III) are

a) To improve the detector resolution, specially the measurement of photon, and improve the particle identification, to reduce the systematic errors, matching the high statistics of BEPCII;

b) To increase the acceptance of the detector;

c) To adapt to the high event rate (3 KHz at the $J/\Psi$ peak) and the short bunch spacing (8ns).

Starting from the interaction point, the BES III detector consists of Beryllium beam pipe, main drift chamber (MDC), time of flight counters (TOF), electromagnetic calorimeter (EMC), superconducting magnet, muon detector and luminosity monitor. The superconducting magnet with the field of 1.0 Tesla has a length of 3.2 meter, an inner radius of 1.05 meter, and an outer radius of 1.45 meter.

The main draft chamber consists of 47 layers with small cells to adapt to the high event rate of BEPCII, and helium-based gas to reduce the multiple scattering and. The filed wires are gold plated aluminum wires of 110 microns, and the sense wires are gold plated tungsten wires. The single wire resolution is expected to be better than 130 microns. The length of the MDC is 2400 mm, an inner radius of 59 mm and an outer radius of 810 mm. The expected momentum resolution is 0.6% at 1 GeV. The expected resolution of dE/dx is about 6%. The solid angle coverage for tracks going through all layers is $\cos\theta = 0.83$. To provide the space for the superconducting micro beta quads and to reduce the backgrounds, the end plates of the inner part of the MDC (r = 70 mm to 192 mm) will be step-shaped.

The barrel of the time of flight counters consists of two layers of plastic scintillator with 88 pieces in azimuth per layer. The length is 2400 mm, and the thickness is 50 mm per layer. The counters are readout by Hamamatsu R5924 at the both ends. The time resolution is better than 90 ps. This will provide a $K/\pi$ separation ($2\sigma$) up to 1.1 GeV. The endcap of the time of flight counters is single layer of plastic scintillator with 48 pieces in azimuth per layer. The inner radius is 410 mm and an outer radius 890 mm. and the thickness is 50 mm. The counters are readout by the ends near the beam pipe.

The CsI electromagnetic calorimeter consists of the barrel, covering up to $\cos\theta = 0.83$, and the two endcaps, covering up to $\cos\theta = 0.93$. The length of the CsI crystal is 24 cm, i.e. 13 radiation lengths. The cross sections are $4\ cm \times 4\ cm$ at the inner surface and $6\ cm \times 6cm$ at the outer surface readout by two photodiodes. The total number of the CsI crystals is about 10000. The total weight is about 22 tons. The expected resolution is better than 2.5% at 1 GeV. The position resolution is expected to be better than 5 mm/$\sqrt{E}$ (GeV).

The muon detector consists of 9 layer of RPC, interleaved with iron plates, which are also the magnet return yoke. The inner radius of the barrel muon detector is 1700mm, and the outer radius 2600mm. The total thickness of the iron absorber is 490mm. The total area of the RPC in the barrel is 1100 $m^2$, and in the endcaps 900 $m^2$. The coverage of the muon detector is up to $\cos\theta = 0.89$.

## 5. Expect Physics Results

BEPCII/BESIII will provide high statistical data samples for the Tau-Charm physics. Assume BEPCII runs 5 months per year for physics, the average luminosity is 30% of the peak luminosity, and the acceptance of 60-70%, each year one could expect to obtain 6000 M - 10000 M $J/\Psi$ events, or 2000 M $\Psi'$ events, or 10 M Tau pairs at 3.67 GeV. From the data sample, some examples of the expected physics results are:

a) precision measurement of CKM matrix, e.g. measure $V_{cd}$ and $V_{cs}$ from the ratio of the semi-leptonic decay and pure leptonic decay with accuracy of a few percents.

b) Precision R-value measurement with the accuracy of 2 - 3% between the CMS energy 2 - 4 GeV.

c) Search for glueballs, fit the overlapping resonances, and to determine the spin-parity.

d) Search for $^1P_1$.

## 6. Budget and Schedule

The estimated total cost of BEPCII, including both the machine and the detector, is about 640 M RMB (77 M US$). The Chinese Government agreed to provide 540 M RMB (65 M US$), covering most of the machine cost and about 3/4 of the detector cost. The rest will be the contributions from the Chinese Academy of Sciences and the international collaborations. The international contribution and collaboration are highly welcomed. The international participation and contributions to BEPCII / BES III will help IHEP to meet the technical challenges in the construction of BEPCII/BESIII and to make BESIII more competely.

The technical design of BEPCIII/BESIII is nearly finished. Several international reviews on the design were carried out during last one and half years. The funding paper work is under way. We expect to start the construction by the end of this year. We expect to finish the construction within 4 years. There will be a shutdown of 6 months starting from April 2004 to upgrade the linac and to remove the BESII detector from the interaction region. The installation of the two rings will be done from April to Dec. 2005. The machine tuning without the detector is scheduled at Jan. - May 2006. The BESIII detector will be moved into the interaction region June - Aug. 2006. The machine + detector tuning should be finished by the end of 2006, and the physics data taking will be started at the beginning of 2007.

The work was supported by the Chinese National Nature Science Foundation under the grant 19775050.

ELSEVIER

Proceedings of ICHEP 2002, pp. 939–941
*S. Bentvelsen, P. de Jong, J. Koch and E. Laenen (Editors)*

31st INTERNATIONAL CONFERENCE ON HIGH ENERGY PHYSICS AMSTERDAM

# TRI$\mu$P - Trapped Radioactive Isotopes: $\mu$icrolaboratories for fundamental Physics

K. Jungmann [a], G.P. Berg [a], P. Dendooven [a], O. Dermois[a], M.N. Harakeh[a], R. Hoekstra[a], R. Morgenstern[a], A. Rogachevskiy[a], M. Sanchez-Vega[a], R.G. Timmermans[a], L. Willmann[a], H.W. Wilschut[a]

[a]Kernfysisch Versneller Instituut, Rijksuniversiteit Groningen, Zernikelaan 25, 9747 AA Groningen, The Netherlands

At the Kernfysisch Versneller Instituut (KVI) a new facility (TRI$\mu$P) is under development. It aims for producing a spectrum of radioactive isotopes in direct, inverse kinematics fusion and evaporation or fragmentation reactions with heavy ion beams from the superconducting cyclotron AGOR. The products will be slowed down and trapped for accurate measurements of fundamental symmetries and interactions in physics. The scientific focus of the local KVI research groups includes precision studies of nuclear $\beta$-decays through $\beta$–neutrino (recoil nucleus) momentum correlations in weak decays and searches for permanent electric dipole moments in heavy atomic systems. This research programme offers a large potential for discovering new physics or to limit parameters in models beyond standard theory significantly. The scientific approach chosen in TRI$\mu$P can be regarded as complementary to such high energy physics. The TRI$\mu$P facility in Groningen will be open for users worldwide.

The new facility TRI$\mu$P (**T**rapped **R**adioactive **I**sotopes: $\mu$icro-laboratories for fundamental **P**hysics) at the Kernfysisch Versneller Instituut (KVI) in Groningen, The Netherlands, has been designed with the central goal to provide cold radioactive atoms and ions for precise measurements of fundamental symmetries and basic interactions in physics [1–3]. It exploits several major recent developments in different fields of physics, in particular the production of intense radioactive beams and very advanced techniques for cooling and trapping of charged and neutral atomic systems in Penning, Paul or laser traps.

The TRI$\mu$P facility design is based on three main radioactive isotope production mechanisms, i.e. direct, inverse kinematics fusion and evaporation or fragmentation reactions. For this purpose heavy ion beams from the superconducting AGOR cyclotron, are directed on fixed targets, the materials of which are chosen for optimal production rates. The foreseen reactions favor in general proton rich nuclei. The created isotopes of interest are separated from the primary beam and other reaction products in a novel designed combined fragment and (gas filled) recoil separator. This mass and momentum selective ion optical system consists of two pairs of dipole magnets for the primary particle selection and quadrupoles for accurate imaging. Two possible target positions are provided within the arrangement of magnets: one at its very entrance for fragmentation reactions and another one between the two dipole pairs for inverse kinematics reactions. Gas filling of the separator is essential for good imaging if the reaction products appear in a distribution of charge states. With appropriate gas densities concurrent electron capture and stripping processes result in an effective charge for the ions, which determines their trajectory in the ion optical system [4].

When exiting from the separator the product particles will have typically 1 MeV/c momentum. For later trapping they need to be slowed down. Moderation to eV energies is foreseen in a high pressure gas cell. Presently detailed measurements of relevant atomic physics processes during slow down in gases are under way at KVI to determine the design parameters of this essential device, where an as high as possible throughput needs to be achieved. At present little informa-


doi:10.1016/S0920-5632(02)02286-7

tion is available in the interesting range where the particle velocities are of order $\alpha \cdot c$, which is the order of valence electron velocities of the slower gas and where charge exchange and, in particular, neutralization cross sections are high.

Further cooling of the ions, which are expected to be singly charged upon exiting the gas cell slower, occurs in a low density buffer gas. Here the ions are longitudinally guided by an electrostatic potential gradient while radially confined by a radio frequency quadrupole field. They are collected in a Paul Trap the end voltage of which can be rapidly switched in order to allow this device to act as a beam buncher. After neutralization the atoms can be stored in atoms traps, e.g. a magneto-optical trap.

Research using trapped radioactive atoms and ions is pursued in many laboratories worldwide. It covers a wide range of physics topics in atomic, nuclear and particle physics [5]. The local researchers at KVI concentrate therefore, on two groups of experiments, which are optimal suited for the facility: precision measurements of nuclear $\beta$-decays and searches for permanent electric dipole moments in atoms.

In standard theory the structure of weak interactions is of the V-A type, where V and A are vector and axial-vector currents with opposite relative sign causing a left handed structure of the interaction and parity violation [6]. Other possibilities like scalar, pseudo-scalar and tensor type interactions would be clear signatures of new physics. In particular, right-handed currents would give rise to deviations from standard theory predictions. Such an observation could be expected to shine some light onto the mysteries behind parity violation.

The double differential decay probability $d^2W/d\Omega_e d\Omega_\nu$for a $\beta$-radioactive nucleus is related to the electron and neutrino momenta $\vec{p}$ and $\vec{q}$ through

$$\begin{aligned} \frac{d^2W}{d\Omega_e d\Omega_\nu} &\sim 1 + a\ \frac{\vec{p}\cdot\vec{q}}{E} + b\ \sqrt{1-(Z\alpha)^2}\ \ \frac{m_e}{E} \\ &+ < \vec{J} > \cdot \left[ A\ \ \frac{\vec{p}}{E} + B\ \ \vec{q} + D\ \ \frac{\vec{p}\times\vec{q}}{E} \right] \\ &+ < \vec{\sigma} > \cdot \left[ G\ \ \frac{\vec{p}}{E} + Q\ \ \vec{J} + R\ \ < \vec{J} > \times\ \frac{\vec{q}}{E} \right] \end{aligned} \qquad (1)$$

where $m_e$ is the $\beta$-particle mass, $E$ its energy, $\vec{\sigma}$ its spin, and $\vec{J}$ is the spin of the decaying nucleus. Among the coefficients in equation (1) D describes the correlation between the neutrino and $\beta$-particle momentum vectors for spin polarized nuclei and is time reversal violating in nature. It renders a particularly high potential for further restricting parameters in speculative models. The coefficient R relates is highly sensitive within a smaller set of models, since in this region there exist some already well established constraints, e.g., from searches for a permanent electric dipole moment (edm) of fundamental particles. Because the neutrino momentum cannot be determined directly in a meaningful way, the recoiling nucleus will detected instead. The neutrino momentum can be reconstructed from the kinematics of the process. Since the recoil nuclei have energies typically a few 10 eV, precise measurements can only be performed, if the decaying isotopes are suspended using sufficiently shallow potential wells. Such exist in magneto-optical traps, where atoms can be stored at temperatures below 1 mK. Since the atoms must have suitable spectral lines for optical trapping and since also the nuclear properties must be such that rather clean transitions can be observed, the isotopes of primary interest for the KVI experimenters are $^{20,21}$Na and $^{18,19}$Ne.

An edm of any fundamental particle violates both parity and time reversal symmetries [7]. With the assumption of a generally valid CPT invariance a permanent dipole moment also violates CP. CP violation, such as it is known from the K and B systems, induces edm's for all particles through higher order loops. Their values are for any known system at least 4 orders of magnitude below the present experimentally established limits. It should be noted that the known sources of CP violation are not sufficient in Sakharov's model for the baryon asymmetry, i.e. the dominance of matter over anti-matter in the universe [8]. New sources of CP violation need to be discovered. Indeed, a large number of speculative models foresees edm's which could be as large as the present experimental bounds. Historically the non-observation of any edm has ruled out more speculative models than any other experimental approach in all of particle physics [9].

Permanent electric dipole moments have been

searched for in various systems with different sensitivities [10–15]. Since the searched for, yet unknown, new fundamental forces may act completely different on various particles and systems, a comparison of experiments solely on the basis of the smallest limit in units of $e \cdot cm$ is not possible. Careful and detailed analysis is indispensable. In composed systems such as molecules or atoms fundamental particle dipole moments can be enhanced significantly [16]. Particularly in polarizable systems where large internal fields can exist. Ra atoms in excited states are of particular interest for edm searches, because of the rather close lying $7s7p^3P_1$ and $7s6d^3_2$ states the a significant enhancement has been predicted [17]. This gives a several orders of magnitude advantage over the $^{199}$Hg atom, the system which has given the best limits so far [15]. Further enhancements have been predicted in Ra isotopes with strongly octupole deformed nuclei. Speculative models would allow to observe an atomic edm in Ra already at an absolute value some three orders of magnitude below the present limit set by Hg!

From a technological point of view Ra atoms well accessible spectroscopically and a variety of isotopes can be produced in fusion and evaporation or in fission reactions. The advantage of an accelerator based Ra experiment is apparent, because a nuclear edm requires an isotope with spin. and all Ra isotopes with finite nuclear spin are relatively short-lived.

TRI$\mu$P at KVI is expected to offer new possibilities to study with high precision fundamental interactions in physics and fundamental symmetries in nature. The approach combines nuclear physics, atomic physics and particle physics in experimental techniques as well as in the conceptual approaches. The scientific approach chosen in TRI$\mu$P can be regarded as complementary to such high energy physics.

For TRI$\mu$P there exists also a large variety of possibilities for research with cold radioactive isotopes in connection with applied sciences. For example, cold polarized $\beta$-emitters could be the basis for extending the method of $\beta$-NMR, which is very successful in bulk material [18], to condensed matter surfaces, on which such atoms could be softly deposited.

At KVI a user facility is created which is open to the worldwide scientific community. TRI$\mu$P is jointly funded by FOM (Stichting voor Fundamenteel Onderzoek der Materie, Dutch funding agency) and the Rijksuniversiteit Groningen in the framework of a managed programme. TRI$\mu$P is expected to receive a 50% share of the AGOR beam time. The time planning foresees that the facility is set up by 2004 followed by an exploitation phase until 2013. First physics experiments are expected in 2004.

## REFERENCES


1. J.W. Turkstra, Hyperf.Int. **127** (2000) 533.
2. H.W. Wilschut et al., Abstr.Pap.Am. Chem.Soc. **222** (2001) 66.
3. K. Jungmann, Acta Physica Polonica **33** (2002) 2049.
4. M. Leino. Nucl. Instr. Meas. **126** (1997) 320.
5. "Atomic Physics at Accelerators: Laser Spectroscopy and Applications", Hyperf.Int. **127**, L. Schweikhard and H.-J. Kluge (eds.)
6. P. Herczeg, Prog. Part. and Nucl. Phys. **46** (2001) 413.
7. "CP Violation without Strangeness", I.B. Khriplovich, S.K. Lamoreaux, Springer, Berlin, 1997
8. M. Trodden, Rev.Mod.Phys. **71** (1999) 1463; A.D. Sakharov, JETP Lett. **5** (1967) 24.
9. N. Ramsey, at "Breit Symposium", Yale, 1999
10. C. Regan et al., Phys. Rev. Lett.**88** (2002) 071805; see also: E.A. Hinds and B.E. Sauer, Physics World **10** (1997) 37.
11. J. Bailey et al., Nucl. Phys. B**150**(1978) 1.
12. K. Ackerstaff et al., Z.Phys. C**74** (1997) 403.
13. H.M. Quiney et al., Phys. Rev. A**57**(1998) 920.
14. P.G. Harris et al., Phys. Rev. Lett. **82** (1999) 904.
15. M.V. Romalis et al. Phys. Rev. Lett. **86** (2001) 2505.
16. P.G.H. Sandars, Contemp.Phys. **42** (2001) 97.
17. V. Dzuba et al., Phys.Rev. A **63** (2001) 062101.
18. D. Fick, Hyperf.Int. **127** (2000) 463.

Proceedings of ICHEP 2002, pp. 942–944
*S. Bentvelsen, P. de Jong, J. Koch and E. Laenen (Editors)*

# The Detector for the New $\mu \rightarrow e\gamma$ Experiment MEG

T. Mori[a]*

[a]International Center for Elementary Particle Physics (ICEPP), the University of Tokyo,
7-3-1 Hongo, Bunkyo-ku, Tokyo 113-0033, Japan

Searches for muon's decays that violate lepton flavors are the key to unveiling new physics underlying the recently established neutrino oscillation phenomenology and an apparent unification of the three coupling constants at extremely high energies. A new experiment [1] approved at the Paul Scherrer Institut to search for $\mu \rightarrow e\gamma$ decays, the MEG experiment, is expected to be sensitive to such new physics. The status of the preparations for the experiment, in particular, the positron spectrometer magnet and the liquid xenon photon detector, are briefly described here.

## 1. MEG Physics

During the last several years possible signs of new physics beyond the Standard Model are finally beginning to emerge.

One of such signs is a possible supersymmetric grand unification of the three coupling constants (SUSY GUT). SUSY GUT generically predicts rather large lepton flavor violation (LFV) in the charged lepton sector [2]. In particular, muon's LFV decays $\mu \rightarrow e\gamma$ are predicted to occur at a branching ratio of $10^{-11}$ – $10^{-14}$, just below the current experimental limit of $1.2 \times 10^{-11}$ (90% C.L.) [3]. The recent measurement of the muon $(g-2)$ [4] and 'no observation' of SUSY particles and Higgs at LEP both suggest large $\tan\beta$ values, which would make the $\mu \rightarrow e\gamma$ branching ratio even higher.

Another sign of new physics is the neutrino oscillations [5]. The tiny neutrino masses may be explained by the see-saw mechanism caused by ultra heavy right-handed neutrinos. In supersymmetric theories these heavy particles induce large LFV, making $\mu \rightarrow e\gamma$ decays occur at a branching ratio as large as the current experimental limit, especially if the solar neutrino deficit is due to the MSW large mixing [6], as strongly indicated by the recent measurements by the SNO Collaboration [7].

A new experiment to search for $\mu \rightarrow e\gamma$ decays, MEG [1], has been approved by the Paul Scherrer Institut (PSI), where a DC muon beam with the world's highest intensity of more than $10^8 \mu$/sec is available. Its experimental set-up is shown in Fig. 1. The MEG Collaboration consist of some 40 physicists from Japan, Switzerland, Italy and Russia. In this presentation the development of the detectors of the MEG experiment, in particular, the positron spectrometer magnet and the liquid xenon photon detector, is described.

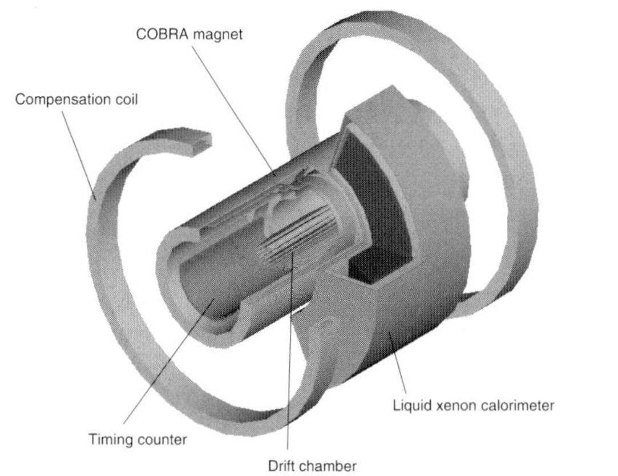


Figure 1. An experimental set-up of the MEG experiment

*representing the MEG Collaboration.


doi:10.1016/S0920-5632(02)02288-0

## 2. COBRA Magnet

The COBRA (COnstant Bending RAdius) positron spectrometer consists of a superconducting solenoidal magnet designed to form a special gradient field, in which monochromatic positrons from the target follow trajectories with a constant projected bending radius, independent of the emission angle over a wide angular range. This allows us to sharply define the absolute momentum window. The gradient field also sweeps away curling tracks quickly out of the tracking volume, thereby reducing the accidental pile-up of the Michel positrons.

Very thin superconducting cables have been developed to make the magnet as thin as 3.8 g/cm$^2$ so that the gamma rays traverse it without much loss. A He-free, simple and easy operation of the magnet is realized with a GM regrigerator.

The magnet is currently in production. All the superconducting cables were produced and are being wound (Fig 2). After careful tests towards the end of 2002, the magnet should be ready in early 2003 for beam line optimization.

Figure 2. The central coil of the COBRA magnet is being wound.

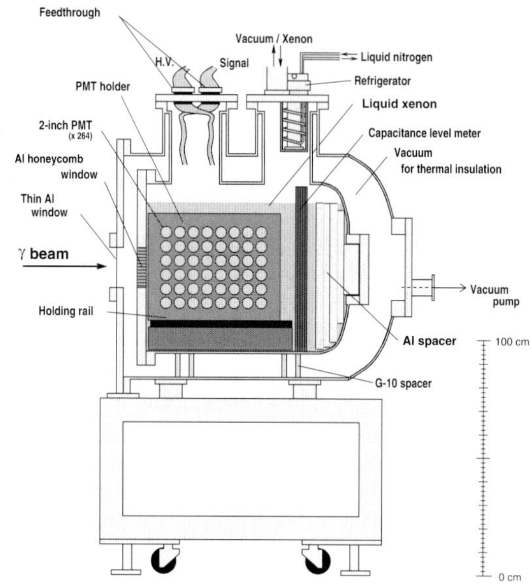


Figure 3. A prototype liquid xenon photon detector.

## 3. Liquid Xenon Detector

A newly invented liquid xenon scintillation detector of "mini-Kamiokande" type, placed outside the magnet, detects the penetrating gamma rays. The detector holds a 0.8m$^3$ volume of liquid xenon. Scintillation light emitted inside xenon are viewed from all sides by arrays of ~800 photomultiplier tubes (PMTs).

High light yield of liquid xenon (roughly 75% of NaI) and its uniformity are necessary ingredients for good energy resolution. The scintillation pulse from xenon is very fast and has a short tail, thereby minimizing the pile-up problem. Distributions of the PMT outputs enable a measurement of the incident position of the gamma ray with a few mm accuracy. The position of the conversion point can be also estimated with an accuracy that corresponds to timing resolution of about 50psec.

The major ambiguity of this innovative detector is absorption of the scintillation light inside xenon. Impurities in xenon, such as water, that absorbs UV light could possibly degrade the detector perfomance.

Prototype detectors have been built (Fig 3). PMTs sensitive to the UV scintillation light of xenon ($\lambda = 175$ nm) under the low temperature ($-100$C$^\circ$) and the pressure of liquid xenon has been developed in cooperation with Hamamatsu Photonics. A pulse-tube refrigerator that operates efficiently at the liquid xenon temperature has been developed at KEK.

After initial successful operations, cosmic rays and alpha sources placed inside the detector were measured. These measurements indicated clear light absorption both in the total light yield and in the light yield as a function of the distance from the light source. The detector was also irradiated with 10–40MeV gamma rays at the laser Compton back-scattering facility at the TERAS storage ring at the AIST laboratory, Tsukuba, Japan (Fig 4). While the light absorption jeopadized energy measurement, position measurements were less affected and a resolution as good as 2.5–3.0mm ($\sigma$) was obtained. According to the subsequent mass spectrometer measurement, the main inpurity was water.

To remove the water contamination, a system to circulate and purify xenon was installed to the detector. Light absorption was monitored with cosmic rays and the alpha sources. After one-month circulation, light absorption length improved from about 7 cm to 100 cm, which is long enough to achieve a 2 % (FWHM) energy resolution for 40–50 MeV gamma rays. It is planned to carry out another laser Compton gamma ray measurement at TERAS later in 2002 to demonstrate the performance.

A design of the final xenon detector is being finalized accordingly and its construction is planned in 2003.

## 4. Conclusion

With the recent results of the solar neutrino and muon $(g-2)$ measurements, the expectation for the MEG experiment is now higher than ever. The R&D and prototype studies indicate that the necessary performance for a possible discovery is in hand with a good chance of further improvements. It is foreseen that the experiment may start in 2004 with an aim to get the final result before LHC.

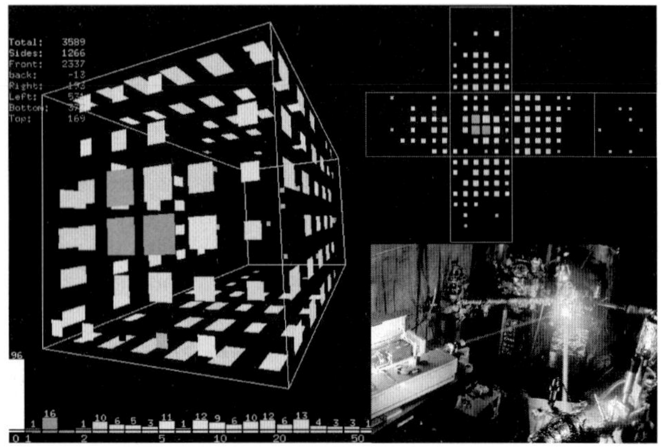


Figure 4. A 40 MeV gamma ray observed by the prototype detector at the TERAS beam test.

## REFERENCES


1. T. Mori *et al.*, "Search for $\mu \to e\gamma$ down to $10^{-14}$ branching ratio," PSI Proposal R-99-5 (1999). The MEG collaboration's web page is `http://meg.psi.ch`.
2. R. Barbieri and L.J. Hall, Phys. Lett. **B338** (1994) 212; R. Barbieri, L.J. Hall, and A. Strumia, Nucl. Phys. **B445** (1995) 219.
3. MEGA Collaboration, M.L. Brooks *et al.*, Phys. Rev. Lett. **83** (1999) 1521.
4. Y. Semertzidis, this conference; Muon (g-2) Collaboration, G.W. Bennett *et al.*, Phys. Rev. Lett. **89** (2002) 101804.
5. Super-Kamiokande Collaboration, Y. Fukuda *et al.*, Phys. Rev. Lett. **81** (1998) 1562; Phys. Lett. **B433** (1998) 9 and **B436** (1998) 33; Phys. Rev. Lett. **81** (1998) 1158.
6. J. Hisano and D. Nomura, Phys. Rev. **D59** (1999) 116005 and references therein.
7. SNO Collaboration, Q.R. Ahmad *et al.*, Phys. Rev. Lett. **87** (2001) 071301; Phys. Rev. Lett. **89** (2002) 011301; Phys. Rev. Lett. **89** (2002) 011302;

## LIST OF PARTICIPANTS

| | | | |
|---|---|---|---|
| Elena | Accomando | Paul Scherrer Institut | Switzerland |
| Ana | Achucarro | University of Leiden | The Netherlands |
| Wolfgang | Adam | HEPHY-Vienna | Austria |
| Ann Marie | Adams | Cornell University | United States of America |
| Gary | Adams | Rensselaer Polytechnic Institute | United States of America |
| Petar | Adzic | Vinca Institute of Nuclear Sciences | Yugoslavia |
| Leonid | Afanasyev | JINR-Dubna | Russian Federation |
| Manuel | Aguilar-Bentez | CIEMAT | Spain |
| Mohammad | Ahmady | Mount Allison University | Canada |
| Hiroaki | Aihara | University of Tokyo | Japan |
| Christopher | Ainsley | University of Cambridge | United Kingdom |
| Zygmunt | Ajduk | Warsaw University | Poland |
| Juan | Alcaraz | CERN / CIEMAT | Switzerland |
| Mark | Alford | Glasgow University | United Kingdom |
| Ahmed | Ali | DESY | Germany |
| Phil | Allport | University of Liverpool | United Kingdom |
| Christopher | Altman | Pierre Laclede Honors College / Starlab | United States of America |
| Attilio | Andreazza | Università di Milano | Italy |
| Valery | Andreev | CERN / UCLA | Switzerland |
| Roberta | Antolini | INFN-Assergi | Italy |
| Mario | Antonelli | INFN-Frascati | Italy |
| Giorgio | Apollinari | Fermilab | United States of America |
| Thomas | Appelquist | Yale University | United States of America |
| Firooz | Arash | Amir Kabir Univ. of Technology | Iran |
| Katsushi | Arisaka | UC Los Angeles | United States of America |
| Marina | Artuso | Syracuse University | United States of America |
| Alan | Astbury | University of Victoria | Canada |
| Jean-Jacques | Aubert | CNRS-IN2P3 | France |
| Jean-Eudes | Augustin | Universités de Paris 6&7 / IN2P3 | France |
| Tagír | Aushev | Moscow | Russian Federation |
| Markus | Axer | RWTH-Aachen | Germany |
| Kaladi | Babu | Oklahoma State Univeristy | United States of America |
| Levan | Babukhadia | SUNY at Stony Brook | United States of America |
| Alessandro | Bacchetta | Vrije Universiteit | The Netherlands |
| Sebastian | Bachmann | Universität Heidelberg | Germany |
| Ben | Bakker | Vrije Universiteit Amsterdam | The Netherlands |
| Vinod | Baldew | NIKHEF | The Netherlands |
| Gunnar | Bali | University of Glasgow | United Kingdom |

| | | |
|---|---|---|
| Paul Balm | NIKHEF | The Netherlands |
| Bruce Baller | Fermilab | United States of America |
| Myron Bander | University of California Irvine | United States of America |
| Shaouly Bar-Shalom | Technion | Israel |
| Elisabetta Barberio | CERN / SMU | Switzerland |
| Dmitri Bardin | JINR-Dubna | Russian Federation |
| Marcello Barisonzi | NIKHEF | The Netherlands |
| Tony Barker | University of Colorado | United States of America |
| Ursula Bassler | LPNHE-Paris | France |
| Marco Battaglia | CERN | Switzerland |
| Gunter Baum | University of Bielefeld | Germany |
| Ulrich Baur | State University of New York Buffalo | United States of America |
| Aurelio Bay | University of Lausanne | Switzerland |
| Andrew Bazarko | Princeton University | United States of America |
| Damir Becirevic | University of Rome 'La Sapienza' | Italy |
| Franco Bedeschi | INFN-Pisa | Italy |
| Ties Behnke | DESY | Germany |
| Anastasios Belias | Rutherford Appleton Lab | United Kingdom |
| Gideon Bella | Tel-Aviv University | Israel |
| Lorenzo Bellagamba | INFN-Bologna | Italy |
| James Bensinger | Brandeis University | United States of America |
| Stan Bentvelsen | NIKHEF | The Netherlands |
| Alberto Benvenuti | INFN-Bologna | Italy |
| Frits Berends | University of Leiden | The Netherlands |
| Arjun Berera | University of Edinburgh | United Kingdom |
| Edmond Berger | Argonne National Laboratory | United States of America |
| Douglas Bergman | Rutgers University | United States of America |
| Eric Bergshoeff | University of Groningen | The Netherlands |
| Karl Berkelman | Cornell University | United States of America |
| Zvi Bern | UC Los Angeles | United States of America |
| José Bernabeu | University of Valencia | Spain |
| Gregorio Bernardi | Université de Paris 7 | France |
| Thomas Berndt | Universität Heidelberg | Germany |
| Sergio Bertolucci | INFN / LNF-Frascati | Italy |
| Khamidbi Beshtoev | JINR-Dubna | Russian Federation |
| Adrian Bevan | University of Liverpool | United Kingdom |
| Helena Bialkowska | Institute for Nuclear Studies | Poland |
| Stefano Bianco | INFN / LNF-Frascati | Italy |
| Tomas Blazek | University of Southampton | United Kingdom |
| Freya Blekman | NIKHEF | The Netherlands |
| Ian Bloodworth | University of Birmingham | United Kingdom |
| Ken Bloom | University of Michigan | United States of America |
| Barry Blumenfeld | Johns Hopkins University | United States of America |
| Furio Bobisut | Padova University | Italy |
| Manfred Boehm | Universität Würzburg | Germany |

| | | | |
|---|---|---|---|
| Armin | Boehrer | Universität Siegen | Germany |
| Daniel | Boer | Vrije Universiteit Amsterdam | The Netherlands |
| David | Boersma | NIKHEF | The Netherlands |
| Peter | Bogdanov | RF Ministry for Atomic Energy | Russian Federation |
| Daniela | Bortoletto | Purdue University | United States of America |
| Francisco | Botella | Universidad de Valencia | Spain |
| Joseph | Boudreau | University of Pittsburgh | United States of America |
| Evelina | Bouhova-Thacker | Lancaster University | United Kingdom |
| Arnd | Brandenburg | DESY | Germany |
| Fernando | Brandt | Universidade de São Paulo | Brazil |
| Siegmund | Brandt | Universität Siegen | Germany |
| Francois | Brochard | CNRS | France |
| Ian | Brock | Universität Bonn | Germany |
| Tom | Browder | University of Hawaii | United States of America |
| Robert | Brown | Rutherford Appleton Laboratory | United Kingdom |
| Alessia | Bruni | INFN-Bologna | Italy |
| Andrzej | Buras | Technical University Munich | Germany |
| Toby | Burnett | University of Washington | United States of America |
| Gerd | Buschhorn | MPI für Physik München | Germany |
| Joel | Butler | Fermilab | United States of America |
| Paula | Butler | Fermilab | United States of America |
| Neil | Calder | Stanford Linear Accelerator Center | United States of America |
| Marta | Calvi | Università di Milano Bicocca | Italy |
| Tiziano | Camporeti | CERN | Switzerland |
| Vincento | Canale | INFN-Napoli | Italy |
| Roger | Cashmore | CERN | Switzerland |
| David | Cassel | Cornell University | United States of America |
| Gianluca | Cavoto | Princeton University | United States of America |
| Marcos | Cerrada | CIEMAT | Spain |
| Alessandro | Cerri | Lawrence Berkeley Nat. Lab. | United States of America |
| Laynam | Chang | Virginia Tech | United States of America |
| Ngee-Pong | Chang | City College of CUNY | United States of America |
| Paoti | Chang | National Taiwan University | Taiwan |
| Jérôme | Charles | CPT-Marseille | France |
| Claude | Charlot | LLR-Ecole Polytechnique | France |
| Vladimir | Chekelian | MPI für Physik München | Germany |
| Hesheng | Chen | IHEP Beijing | China |
| Kai-Feng | Chen | National Taiwan University | Taiwan |
| Hai-Yang | Cheng | Academia Sinica | Taiwan |
| Giovanni | Chiefari | Università e INFN-Napoli | Italy |
| Vincenzo | Chiochia | DESY | Germany |
| Ruslan | Chistov | Moscow | Russian Federation |
| Eung Jin | Chun | Korea Institute for Advanced Study | South-Korea |
| Jiri | Chyla | Institute of Physics | Czech Republic |
| Anna | Ciaccio | University of Roma 'Tor Vergata' | Italy |

| | | | |
|---|---|---|---|
| Marco | Ciuchini | INFN-Roma III | Italy |
| Robert | Clare | University of California Riverside | United States of America |
| Jacques | Colas | LAPP-Annecy | France |
| Tilmann | Colberg | Institut Kern-Teilchenphysik Dresden | Germany |
| Paula | Collins | CERN | Switzerland |
| Sara | Collins | University of Glasgow | United Kingdom |
| Amy | Connolly | Lawrence Berkeley Nat. Lab. | United States of America |
| John | Conway | Rutgers University | United States of America |
| Amanda | Cooper-Sarkar | Oxford University | United Kingdom |
| Fernando | Cornet | Universidad de Granada | Spain |
| Giuseppe | Costa | INFN-Milano | Italy |
| Maria Jose | Costa | Instituto de Fisica Corpuscular | Spain |
| Daniel | Cronin-Hennessy | University of Rochester | United States of America |
| Ferenc | Csikor | Eötvös University | Hungary |
| Jean-René | Cudell | Université de Liège | Belgium |
| Jaroslav | Cvach | Institute of Physics AS CR | Czech Republic |
| Bernard | D'Almagne | Laboratoire de l'Accelerateur Lineaire | France |
| Juan Carlos | D'Olivo | Instituto de Ciencias Nucleares, UNAM | Mexico |
| Samuel | Dagan | Tel-Aviv University | Israel |
| Sylvie | Dagoret-Campagne | LPNHE-Paris | France |
| Mogens | Dam | Niels Bohr Institute | Denmark |
| Michael | Danilov | ITEP | Russian Federation |
| Gavin | Davies | Imperial College London | United Kingdom |
| Jan | De Boer | Universiteit van Amsterdam | The Netherlands |
| Wim | De Boer | Universität Karlsruhe | Germany |
| Isabelle | De Bonis | LAPP-Annecy | France |
| Kees | De Jager | Jefferson Lab | United States of America |
| Paul | De Jong | NIKHEF | The Netherlands |
| Sijbrand | De Jong | University of Nijmegen | The Netherlands |
| Eddi | De Wolf | University of Antwerpen | Belgium |
| Els | De Wolf | NIKHEF / Universiteit van Amsterdam | The Netherlands |
| Pascal | Debu | CEA-Saclay | France |
| Regina | Demina | Kansas State University | United States of America |
| Malcolm | Derrick | DESY | Germany |
| Klaus | Desch | Inst. f. Experimentalphysik Hamburg | Germany |
| Nilendra | Deshpande | University of Oregon | United States of America |
| Adriano | Di Giacomo | Pisa University & INFN | Italy |
| Cristinel | Diaconu | CNRS / IN2P3 | France |
| Jorge | Dias De Deus | Instituto Superior Técnico | Portugal |
| Lorenzo | Diaz-Cruz | Universidad Autonoma de Puebla | Mexico |
| M. | Diehl | RWTH-Aachen | Germany |
| Mark | Dierckxsens | NIKHEF | The Netherlands |
| Hans | Dijkstra | CERN | Switzerland |
| Tim | Dijkstra | NIKHEF | The Netherlands |
| Jochen | Dingfelder | Physikalisches Institut Heidelberg | Germany |

| | | | |
|---|---|---|---|
| Günther | Dissertori | ETH-Zürich | Switzerland |
| Jay | Dittmann | Fermilab | United States of America |
| Eduardo | Do Couto e Silva | Stanford University / SLAC | United States of America |
| Brian | Dolan | National University Ireland Maynooth | Ireland |
| Jean | Dolbeau | PCC-Collège de France | France |
| Gabor | Domokos | Johns Hopkins University | United States of America |
| John | Donoghue | University of Massachusetts | United States of America |
| Jonathan | Dorfan | Stanford Linear Accelerator Center | United States of America |
| Jeroen | Dreschler | Universiteit van Amsterdam | The Netherlands |
| Dongsheng | Du | Institute of High Energy Physics, CAS | China |
| Stanislav | Dubnicka | Slovak Academy of Sciences | Slovak Republic |
| Anna Zuzana | Dubnickova | Comenius University | Slovak Republic |
| Gregory | Dubois-Felsmann | Caltech | United States of America |
| Guenter | Duckeck | Universität München | Germany |
| Silke | Duensing | Universiteit Nijmegen | The Netherlands |
| Pieter | Duinker | NIKHEF | The Netherlands |
| Stephan | Durr | DESY-Zeuthen | Germany |
| Claire | Earnshaw | CERN | Switzerland |
| Oscar | Eboli | Universidade de São Paulo | Brazil |
| Paula | Eerola | Lund University | Sweden |
| Gerald | Eigen | University of Bergen | Norway |
| Gad | Eilam | Israel Institute of Technology | Israel |
| Tord | Ekelof | Uppsala University | Sweden |
| Mohamed | El-Nagdy | Hewan University Cairo | Egypt |
| Stephen | Ellis | University of Washington | United States of America |
| Eckhard | Elsen | DESY | Germany |
| Ralph | Engel | Bartol Research Institute | United States of America |
| Jos | Engelen | NIKHEF | The Netherlands |
| Sarah | Eno | University of Maryland | United States of America |
| Robin | Erbacher | Fermilab | United States of America |
| Erez | Etzion | Tel-Aviv University | Israel |
| Harold | Evans | Columbia University | United States of America |
| Tim Oliver | Eynck | NIKHEF | The Netherlands |
| Bernard | Fabbro | CEA-Saclay | France |
| Svjetlana | Fajfer | University Ljubljana | Slovenia |
| Lutz | Feld | University of Freiburg | Germany |
| Gary | Feldman | Harvard University | United States of America |
| Jol | Feltesse | CEA | France |
| Tom | Ferbel | University of Rochester | United States of America |
| Thomas | Ferguson | Carnegie Mellon University | United States of America |
| Samir | Ferrag | LPNHE-Paris | France |
| Joana | Ferreira Montenegro | NIKHEF | The Netherlands |
| Antonio | Ferrer | IFIC-University of Valencia | Spain |
| Rick | Field | University of Florida | United States of America |
| Frank | Filthaut | University of Nijmegen | The Netherlands |

| | | | |
|---|---|---|---|
| Margret | Fincke-Keeler | University of Victoria | Canada |
| Gunter | Flugge | RWTH-Aachen | Germany |
| Chris | Ford | University of Leiden | The Netherlands |
| Brian | Foster | DESY / Bristol University UK | Germany |
| Paolo | Franzini | Università di Roma 'La Sapienza' | Italy |
| Ayres | Freitas | DESY | Germany |
| Klaus | Freudenreich | ETH-Zürich | Switzerland |
| Ulrich | Fricke | DESY | Germany |
| Jan | Friedrich | Technische Universität München | Germany |
| Yitzhak | Frishman | Weizmann Institute | Israel |
| Harald | Fritzsch | Universität München | Germany |
| Stefano | Frixione | LAPP-Annecy / INFN-Genova | France |
| Colin | Froggatt | University of Glasgow | United Kingdom |
| Andrea | Fuster | NIKHEF | The Netherlands |
| Emidio | Gabrielli | University of Helsinki | Finland |
| Karel | Gaemers | University of Amsterdam / NIKHEF | The Netherlands |
| Pauline | Gagnon | Indiana University | United States of America |
| Thomas | Gaisser | University of Delaware | United States of America |
| Cristina | Galea | Universiteit Nijmegen | The Netherlands |
| Paolo | Gambino | CERN | Switzerland |
| K.K. | Gan | The Ohio State University | United States of America |
| Robert | Garisto | Brookhaven National Laboratory | United States of America |
| Erika | Garutti | NIKHEF | The Netherlands |
| John | Garvey | University of Birmingham | United Kingdom |
| Suzanne | Gascon-Shotkin | Institut de Physique Nucléaire de Lyon | France |
| Paolo | Gauzzi | Rome University 'La Sapienza' | Italy |
| Colin | Gay | Yale University | United States of America |
| Ulrich | Gensch | DESY | Germany |
| Jean-Marc | Gerard | Institut de Physique Théorique | Belgium |
| C. | Geweniger | Kirchhoff-Institut für Physik | Germany |
| Giorgio | Giacomelli | University of Bologna | Italy |
| Walter | Giele | Fermilab | United States of America |
| Simone | Gilardoni | Université de Genève | Switzerland |
| Camille | Ginsburg | University of Wisconsin | United States of America |
| George | Ginther | University of Rochester | United States of America |
| Paolo | Giubellino | INFN-Torino | Italy |
| Carlo | Giunti | INFN-Torino | Italy |
| Leonid | Gladilin | Weizmann Institute | Israel |
| Douglas | Glenzinski | Fermilab | United States of America |
| Ferenc | Gluck | Johannes Gutenberg Universität Mainz | Germany |
| Rohini | Godbole | Indian Institute of Science | India |
| Stephen | Godfrey | Carleton University | Canada |
| Marianne | Goeger-Neff | Technische Universität München | Germany |
| Marvin | Goldberg | National Science Foundation | United States of America |
| Gary | Goldstein | Tufts University | United States of America |

| | | | |
|---|---|---|---|
| Evgeniy | Golovach | SINP / MSU Moscow | Russian Federation |
| Eugene | Golowich | University of Massachusetts | United States of America |
| Concha | Gonzalez-Garcia | CERN / IFIC-UV / Stony Brook | Switzerland |
| Ascelin | Gordon | University of Melbourne | Australia |
| Alfred | Goshaw | Duke University | United States of America |
| Errol | Gotsman | Tel-Aviv University | Israel |
| Erik | Gottschalk | Fermilab | United States of America |
| Krishnaswamy | Gounder | Fermilab | United States of America |
| Anna | Goussiou | Imperial College London | United Kingdom |
| Christoph | Grab | ETH-Zürich | Switzerland |
| Per | Grafström | CERN | Switzerland |
| Raphael | Granier de Cassagna | Laboratoire Leprince-Ringuet | France |
| Giacomo | Graziani | LAL-Orsay | France |
| Massimiliano | Grazzini | INFN-Firenze | Italy |
| Tim | Greenshaw | Liverpool University | United Kingdom |
| Christoph | Greub | University of Bern | Switzerland |
| Sjors | Grijpink | NIKHEF | The Netherlands |
| Günter | Grindhammer | MPI für Physik München | Germany |
| Benjamin | Grinstein | University of California San Diego | United States of America |
| Jean-Francois | Grivaz | LAL-Orsay | France |
| Jeff | Gronberg | Lawrence Livermore Nat. Lab. | United States of America |
| Eilam | Gross | Weizmann Institute | Israel |
| Yuval | Grossman | Technion | Israel |
| Daniel | Grumiller | TU-Vienna | Austria |
| Martin | Grünewald | RWTH-Aachen | Germany |
| Carlo | Gustavino | INFN / LNOS Assergi | Italy |
| Laszlo | Gutay | Purdue University | United States of America |
| Sharon | Hagopian | Florida State University | United States of America |
| Vasken | Hagopian | Florida State University | United States of America |
| Carter | Hall | Harvard University | United States of America |
| Yogiro | Hama | Universidade de São Paulo | Brazil |
| Klaus | Hamacher | Universität Wuppertal | Germany |
| Hideki | Hamagaki | University of Tokyo | Japan |
| Gautier | Hamel de Monchenault | CEA-Saclay | France |
| Gail | Hanson | University of California Riverside | United States of America |
| Kael | Hanson | University of Wisconsin Madison | United States of America |
| Kristian | Harder | University of Hamburg / DESY | Germany |
| Frederick | Harris | University of Hawaii | United States of America |
| John | Harris | Yale University | United States of America |
| Robert | Harris | Fermilab | United States of America |
| Kenichi | Hatakeyama | The Rockefeller University | United States of America |
| Richard | Hawkings | CERN | Switzerland |
| Mitsuo | Hayashi | Tokai University | Japan |
| Yoshinari | Hayato | High Energy Accelerator Research Org. | Japan |
| Beate | Heinemann | University of Liverpool | United Kingdom |

| | | | |
|---|---|---|---|
| Richard | Hemingway | Carleton University | Canada |
| Pierre | Henrard | Laboratoire de Physique Corpusculaire | France |
| Jose | Hernandez | DESY | Germany |
| Gregor | Herten | Universität Freiburg | Germany |
| Willem | Hesselink | Vrije Universiteit / NIKHEF | The Netherlands |
| Rolf | Heuer | Universität Hamburg | Germany |
| Keisho | Hidaka | Tokyo Gakugei University | Japan |
| Kiyoshi | Higashijima | Osaka University | Japan |
| Malte | Hildebrandt | Physik-Institut Zürich | Switzerland |
| Chris | Hill | University of California | United States of America |
| Aksel | Hiorth | Oslo | Norway |
| David | Hitlin | Caltech | United States of America |
| John | Hobbs | SUNY at Stony Brook | United States of America |
| Hallstein | Hogaasen | University of Oslo | Norway |
| Sven Olov | Holmgren | Stockholm University | Sweden |
| Walter | Hoogland | Universiteit van Amsterdam | The Netherlands |
| Jiri | Horejsi | Charles University | Czech Republic |
| Yutaka | Hosotani | Osaka University | Japan |
| Gerardus | 't Hooft | Universiteit Utrecht | The Netherlands |
| Paul | Hoyer | Nordita | Denmark |
| Josef | Hrubec | HEPHY-Vienna | Austria |
| Yuan | Hu | Universiteit Nijmegen | The Netherlands |
| Tao | Huang | IHEP Beijing | China |
| Todd | Huffman | Oxford University | United Kingdom |
| Lennaert | Huiszoon | NIKHEF | The Netherlands |
| Katri | Huitu | Helsinki Institute of Physics | Finland |
| Wouter | Hulsbergen | NIKHEF | The Netherlands |
| Olya | Hulsbergen-Igonkina | ITEP | Russia |
| K. | Hultqvist | Stockholm University | Sweden |
| Joey | Huston | Michigan State University | United States of America |
| Enzo | Iarocci | Istituto Nazionale di Fisica Nucleare | Italy |
| Ia | Iashvili | University of California Riverside | United States of America |
| Keiji | Igi | Kanagawa University | Japan |
| Jelena | Ilic | VINCA Institute of Nuclear Sciences | Yugoslavia |
| John | Iliopoulos | ENS Paris | France |
| Ryosuke | Itoh | KEK | Japan |
| Yoshitaka | Itow | University of Tokyo | Japan |
| Hiroyuki | Iwasaki | KEK | Japan |
| Joseph | Izen | University of Texas Dallas | United States of America |
| Judith | Jackson | Fermilab | United States of America |
| Stanislaw | Jadach | Institute of Nuclear Physics Cracow | Poland |
| Rainer | Jakob | University of Wuppertal | Germany |
| Goran | Jarlskog | Lund University | Sweden |
| Hassan | Jawahery | University of Maryland | United States of America |
| Fred | Jegerlehner | DESY | Germany |

| | | | |
|---|---|---|---|
| Elizabeth | Jenkins | University of California San Diego | United States of America |
| Rick | Jesik | Imperial College London | United Kingdom |
| Colin | Jessop | Stanford Linear Accelerator Center | United States of America |
| Stephane | Jezequel | LAPP-Annecy | France |
| Shan | Jin | IHEP Beijing | China |
| Erik | Johansson | Stockhom University | Sweden |
| Marvin | Johnson | Fermilab | United States of America |
| Julia | Jones | University of Liverpool | United Kingdom |
| Tim | Jones | University of Liverpool | United Kingdom |
| Isabel | Josa | CIEMAT | Spain |
| Hannes | Jung | Lund University | Sweden |
| Klaus | Jungmann | Rijksuniversiteit Groningen | The Netherlands |
| Vladimir | Kadyshevsky | JINR-Dubna | Russian Federation |
| Alexey | Kaidalov | ITEP-Moscow | Russian Federation |
| Peter | Kalmus | Queen Mary University of London | United Kingdom |
| Marc | Kamionkowski | California Institute of Technology | United States of America |
| Kyungsik | Kang | Brown University | United States of America |
| Chung | Kao | University of Oklahoma | United States of America |
| Herbert | Kapitza | DESY | Germany |
| Alexander | Kappes | Universität Bonn | Germany |
| Inga | Karliner | University of Illinois | United States of America |
| Vakhtang | Kartvelishvili | Lancaster University | United Kingdom |
| Yannis | Karyotakis | SLAC & LAPP-Annecy | United States of America |
| Matthias | Kasemann | Fermilab | United States of America |
| Richard | Kass | Ohio State University | United States of America |
| Nobu | Katayama | KEK | Japan |
| Ken | Kawarabayashi | Tokyo International University | Japan |
| Richard | Keeler | University of Victoria | Canada |
| Vladimir | Kekelidze | JINR-Dubna | Russian Federation |
| T.W. | Kephart | Vanderbilt University | United States of America |
| Avto | Kharchilava | University of Notre Dame | United States of America |
| Valeri | Khoze | IPPP-Durham University | United Kingdom |
| Maria Novella | Kienzle-Focacci | Université de Genève | Switzerland |
| Christian | Kiesling | MPI for Physics München | Germany |
| William | Kilgore | Brookhaven National Laboratory | United States of America |
| Wolfgang | Kilian | University of Karlsruhe | Germany |
| Choong Sun | Kim | Yonsei University | South-Korea |
| Shinhong | Kim | University of Tsukuba | Japan |
| Barry | King | University of Liverpool | United Kingdom |
| Wolfram | Kittel | University of Nijmegen / NIKHEF | The Netherlands |
| Robert | Klanner | DESY | Germany |
| Joshua | Klein | University of Pennsylvania | United States of America |
| Katarzyna | Klimek | Universität Hamburg | Germany |
| Reiner | Klingenberg | Universität Dortmund | Germany |
| Frans | Klinkhamer | Universität Karlsruhe | Germany |

| | | | |
|---|---|---|---|
| Peter | Kluit | NIKHEF | The Netherlands |
| Karl | Knoepfle | MPI für Kernphysik Heidelberg | Germany |
| Michael | Kobel | Universität Bonn | Germany |
| Justus | Koch | NIKHEF | The Netherlands |
| Els | Koffeman | NIKHEF | The Netherlands |
| Yoshio | Koide | University of Shizuoka | Japan |
| Hermann | Kolanoski | Humboldt-Universität Berlin | Germany |
| Karol | Kolodziej | University of Silesia | Poland |
| Sachio | Komamiya | University of Tokyo | Japan |
| Paul | Kooijman | NIKHEF | The Netherlands |
| Susumu | Koretune | Shimane Medical University | Japan |
| Andrey | Korytov | University of Florida | United States of America |
| Alexandre | Kourilin | JINR-Dubna | Russian Federation |
| Susan | Kovesi-Domokos | Johns Hopkins University | United States of America |
| Guennadi | Kozlov | JINR-Dubna | Russian Federation |
| Peter | Kreuzer | UC Los Angeles | United States of America |
| Siegfried | Krewald | Forschungszentrum Jülich | Germany |
| Hubert | Kroha | MPI für Physik München | Germany |
| Pavel | Krokovnyi | BINP-Novosibirsk | Russian Federation |
| Mark | Kruse | Duke University | United States of America |
| J.H. | Kuehn | Universität Karlsruhe | Germany |
| Moshe | Kugler | Weizmann Institute | Israel |
| Dietmar | Kuhn | Innsbruck University | Austria |
| Paul | Kuijer | NIKHEF | The Netherlands |
| Kresimir | Kumericki | Zagreb University | Croatia |
| Yoshitaka | Kuno | Osaka University | Japan |
| Masaaki | Kuroda | Meijigakuin University | Japan |
| Shawn | Kwang | University of Maryland | United States of America |
| Youngjoon | Kwon | Yonsei University | South-Korea |
| Eric | Laenen | NIKHEF / Universiteit Utrecht | The Netherlands |
| Edwin | Laermann | Universität Bielefeld | Germany |
| Remi | Lafaye | LAPP-Annecy / IN2P3 / CNRS | France |
| George | Lafferty | University of Manchester | United Kingdom |
| Giovanni | Lamanna | CERN | Switzerland |
| Greg | Landsberg | Brown University | United States of America |
| Rolf | Landua | CERN | Switzerland |
| Kenneth | Lane | Boston University | United States of America |
| Dankfried | Lanske | RWTH-Aachen | Germany |
| Louk | Lapikas | NIKHEF | The Netherlands |
| Achilleas | Lazopoulos | Universiteit Nijmegen | The Netherlands |
| Luca | Latronico | INFN / University of Pisa | Italy |
| Elliot | Leader | Imperial College London | United Kingdom |
| Gerhard | Leder | HEPHY-Vienna | Austria |
| Taekoon | Lee | KAIST | South-Korea |
| Juliet | Lee Franzini | Laboratori Nazionali di Frascati | Italy |

| | | | |
|---|---|---|---|
| Jacques | Lefrancois | LAL-Orsay | France |
| Laurent | Lellouch | CPT-Marseille | France |
| Jacques | Lemonne | Vrije Universiteit Brussel | Belgium |
| C.N. | Leung | University of Delaware | United States of America |
| Ling-Fong | Li | Carnegie Mellon University | United States of America |
| Weiguo | Li | IHEP Beijing | China |
| Peter | Limon | Fermilab | United States of America |
| Guey-Lin | Lin | National Chiao-Tung University | Taiwan |
| Frank | Linde | Universiteit van Amsterdam / NIKHEF | The Netherlands |
| Fu-Sin | Ling | Inst. Fund. Theory Gainsville | United States of America |
| Harry | Lipkin | Weizmann Institute of Science | Israel |
| Peter | Loch | University of Arizona | United States of America |
| Gregory | Loew | Stanford Linear Accelerator Center | United States of America |
| Kenneth | Long | Imperial College London | United Kingdom |
| Egidio | Longo | Università La Sapienza / INFN | Italy |
| Claude | Longuemare | CNRS-IN2P3 / University of Caen | France |
| Angel | Lopez | University of Puerto Rico – Mayaguez | Puerto Rico |
| Abdenour | Lounis | Inst. Subatomiques Strassbourg | France |
| Sherwin | Love | Purdue University | United States of America |
| Cai-Dian | Lu | IHEP Beijing | China |
| Gongru | Lu | Henan Normal University | China |
| Vera | Luth | Stanford Linear Accelerator Center | United States of America |
| Anne Marie | Lutz | Laboratoire de l'Accelerateur Linéaire | France |
| Pierre | Lutz | CEA-Saclay | France |
| Louis | Lyons | Oxford University | United Kingdom |
| Ernest | Ma | University of California Riverside | United States of America |
| Hong | Ma | Brookhaven National Laboratory | United States of America |
| Bruno | Machet | LPTHE-Paris | France |
| Erik | Maddox | NIKHEF | The Netherlands |
| Kalyana | Mahanthappa | University of Colorado | United States of America |
| Hanna | Mahlke-Krueger | Cornell University | United States of America |
| Luciano | Maiani | CERN | Switzerland |
| Giuseppe | Maiella | University of Naples 'Federico II' | Italy |
| Andrew | Maier | Universität Siegen | Switzerland |
| Aki | Maki | KEK | Japan |
| Naomi | Makins | Univ. of Illinois, Urbana-Champaign | United States of America |
| Alexandre | Malakhov | JINR-Dubna | Russian Federation |
| Luca | Malgeri | Université de Genève | Switzerland |
| Ehab | Malkawi | University of Science & Technology | Jordan |
| Sandra | Malvezzi | INFN-Milano | Italy |
| Joseph | Manjavidze | JINR-Dubna | Russian Federation |
| Italo | Mannelli | INFN / Scuola Normale Superiore | Italy |
| Uri | Maor | Tel-Aviv University | Israel |
| Chiara | Mariotti | CERN | Switzerland |
| Thomas | Markiewicz | Stanford Linear Accelerator Center | United States of America |

| | | | |
|---|---|---|---|
| Evgenij | Martynov | Université de Liège | Belgium |
| David | Mason | University of Oregon | United States of America |
| Graziano | Massaro | NIKHEF | The Netherlands |
| Konstantin | Matchev | University of Florida | United States of America |
| Jean Francois | Mathiot | Laboratoire Physique Romagnat | France |
| Giorgio | Matthiae | University of Roma 'Tor Vergata' | Italy |
| Christopher | Mauger | SUNY at Stony Brook | United States of America |
| Chris | Maxwell | University of Durham | United Kingdom |
| Robert | McCarthy | State University of New York | United States of America |
| Ronan | McNulty | University of Liverpool | United Kingdom |
| Robert | McPherson | University of Victoria | Canada |
| Michael | Medinnis | DESY | Germany |
| Terje | Meisler | NTNU Dept. of Physics Trondheim | Norway |
| Salvatore | Mele | CERN & INFN | Switzerland |
| Ezio | Menichetti | INFN-Torino | Italy |
| Marcel | Merk | NIKHEF | The Netherlands |
| Marco | Meschini | INFN-Firenze | Italy |
| Federica | Messer | SUNY at Stony Brook | United States of America |
| Wesley | Metzger | University of Nijmegen | The Netherlands |
| Maaike | Mevius | Universiteit Utrecht | The Netherlands |
| Andreas | Meyer | Universität Hamburg | Germany |
| Thomas | Meyer | Iowa State University | United States of America |
| Vanessa | Mexner | NIKHEF | The Netherlands |
| Douglas | Michael | Caltech | United States of America |
| Ernesto | Migliore | Università di Torino / INFN-Torino | Italy |
| Giora | Mikenberg | Weizmann Institute | Israel |
| Predag | Milenovic | VINCA Institute of Nuclear Sciences | Yugoslavia |
| Jose Guilherme | Milhano | Vrije Universiteit Amsterdam | The Netherlands |
| David | Miller | University College London | United Kingdom |
| Tadao | Mitsui | Research Center for Neutrino Science | Japan |
| William | Molzon | University of California | United States of America |
| Gerard | Montarou | Lab. Physique Corpusculaire Aubière | France |
| Teresa | Montaruli | Università di Bari | Italy |
| Thomas | Moore | SLAC | United States of America |
| Gudrid | Moortgat-Pick | University of Hamburg | Germany |
| Toshinori | Mori | University of Tokyo | Japan |
| Masahiro | Morii | Harvard University | United States of America |
| Moshe | Moshe | Technion | Israel |
| Pedro A. | Movilla Fernandez | Max-Planck-Institut für Physik | Germany |
| Piet | Mulders | Vrije Universiteit Amsterdam | The Netherlands |
| Michael | Murray | Texas A&M University | United States of America |
| Roberto | Mussa | INFN-Torino | Italy |
| Alexander | Nagaitsev | JINR-Dubna | Russian Federation |
| Yorikiyo | Nagashima | Osaka University | Japan |
| Steven | Nahn | Yale University | United States of America |

| | | | |
|---|---|---|---|
| Satyanarayan | Nandi | Oklahoma State University | United States of America |
| Meenakshi | Narain | Boston University | United States of America |
| Thomas | Naumann | DESY-Zeuthen | Germany |
| Uriel | Navenberg | University of Colorado | United States of America |
| Martin | Nedden | Humboldt Universität Berlin | Germany |
| Harry | Nelson | University of California Santa Barbara | United States of America |
| Peter | Nemethy | New York University | United States of America |
| Francesca | Nessi-Tedaldi | ETH-Zurich | Switzerland |
| Norbert | Neumeister | HEPHY-Vienna | Austria |
| Roman | Nevzorov | ITEP & DESY | Russian Federation |
| Catherine | Newman-Holmes | Fermilab | United States of America |
| Harvey | Newman | California Institute of Technology | United States of America |
| Paul | Newman | DESY / Birmingham University | Germany |
| Basarab | Nicolescu | CNRS / Université de Paris 6 | France |
| Carsten | Niebuhr | DESY | Germany |
| Ulrich | Nierste | Fermilab | United States of America |
| Yosef | Nir | Weizmann Institute | Israel |
| Shohei | Nishida | Kyoto University | Japan |
| Richard | Nisius | MPI-Munich | Germany |
| Seishi | Noguchi | Nara Women's University | Japan |
| Dennis | Nordstrom | American Physical Society | United States of America |
| Paul | Norman | University of Birmingham | United Kingdom |
| Yury | Novoseltsev | Institute for Nuclear Research of RAS | Russian Federation |
| Jeremy | Nowell | Imperial College London | United Kingdom |
| Mina | Nozar | Jefferson Lab | United States of America |
| Robert | Oakes | Northwestern University | United States of America |
| John | O'Fallon | U.S. Department of Energy | United States of America |
| Alexander | Oh | CERN | Switzerland |
| Sechul | Oh | Yonsei University | South-Korea |
| S.K. | Oh | Konkuk University | South-Korea |
| Benedict | Oh | Penn State University | United States of America |
| Tommy | Ohlsson | Royal Institute of Technology | Sweden |
| Alexandre | Olchevski | JINR-Dubna | Russian Federation |
| James | Olsen | Princeton University | United States of America |
| Stephen | Olsen | University of Hawaii | United States of America |
| William | Orejudos | Lawrence Berkeley Nat. Lab. | United States of America |
| Yona | Oren | Tel-Aviv University | Israel |
| Scott | Oser | University of Pennsylvania | United States of America |
| Per | Osland | University of Bergen | Norway |
| Andrei | Ostaptchouk | RWTH-Aachen | Germany |
| Ilmar | Ots | Inst. of Physics / University of Tartu | Estonia |
| Marco | Paganoni | University of Milan 'Bicocca' / INFN | Italy |
| Sandro | Palestini | CERN | Switzerland |
| Henryk | Palka | Institute of Nuclear Physics Cracow | Poland |
| Carmen | Palomares | CERN | Switzerland |

| | | | |
|---|---|---|---|
| Nikolaos | Papadopoulos | Universität Mainz | Germany |
| Theodora | Papadopoulou | Nat. Technical University of Athens | Greece |
| Joannis | Papavassiliou | Universidad de Valencia | Spain |
| Hyangkyu | Park | University of Michigan | United States of America |
| Chris | Parkes | University of Glasgow | United Kingdom |
| Fabrizio | Parodi | Genova University / INFN | Italy |
| Giampiero | Passarino | University of Torino | Italy |
| Antonio | Passeri | INFN-Roma | Italy |
| James | Paterson | Stanford Linear Accelerator Center | United States of America |
| Thomas | Patzak | Univ. de Paris 7 / Collège de France | France |
| Christoph | Paus | Massachusetts Institute of Technology | United States of America |
| Ken | Peach | Rutherford Appleton Laboratory | United Kingdom |
| Thomas | Peitzmann | Universität Münster | Germany |
| Antonio | Pellegrino | NIKHEF | The Netherlands |
| Jose | Pearrocha | Universitat de Valencia | Spain |
| Simon | Peeters | NIKHEF | The Netherlands |
| Emmanuelle | Perez | CEA-Saclay | France |
| Roberto | Perrino | INFN-Lecce | Italy |
| Pascal | Perrodo | LAPP-Annecy | France |
| Onne | Peters | NIKHEF | The Netherlands |
| Robi | Peschanski | CEA-Saclay | France |
| Bernard | Peyaud | CEA-Saclay | France |
| Maria Grazia | Pia | INFN-Genova | Italy |
| Fulvio | Piccinini | CERN | Switzerland |
| Fetze | Pijlman | Vrije Universiteit | The Netherlands |
| Philip | Pile | Brookhaven National Laboratory | United States of America |
| Krzysztof | Piotrzkowski | Université Catholique de Louvain | Belgium |
| Nikolai | Piskunov | JINR-Dubna | Russian Federation |
| Roberto | Pittau | University of Torino / INFN | Italy |
| Daniel | Pitzl | DESY | Germany |
| Wieslaw | Placzek | Jagiellonian University Cracow | Poland |
| Levon | Pogosian | Imperial College London | United Kingdom |
| Martin | Pohl | Université de Genève | Switzerland |
| Oliver | Pooth | RWTH-Aachen | Germany |
| Tracey | Pratt | University of Oxford | United Kingdom |
| Margherita | Primavera | INFN-Lecce | Italy |
| James | Proudfoot | Argonne National Laboratory | United States of America |
| Philipp | Puls | University of Vienna | Austria |
| Jovan | Puzovic | University of Belgrade | Yugoslavia |
| Cong-Feng | Qiao | Hiroshima University | Japan |
| Catarina | Quintans | LIP Lisbon | Portugal |
| Sergio | Ratti | INFN / University of Pavia | Italy |
| Frank | Raupach | RWTH-Aachen | Germany |
| Gerhard | Raven | NIKHEF / Vrije Universiteit | The Netherlands |
| Laura | Reina | Florida State University | United States of America |

| | | | |
|---|---|---|---|
| Andreas | Reischl | NIKHEF | The Netherlands |
| Christoph | Rembser | CERN | Switzerland |
| John | Renner Hansen | Niels Bohr Institute | Denmark |
| Cristina | Riccardi | University of Pavia | Italy |
| Jeffrey | Richman | University of California Santa Barbara | United States of America |
| Stefan | Ritt | Paul Scherrer Institut | Switzerland |
| Dieter | Roehrich | University of Bergen | Norway |
| Herbert | Rohringer | Institut für Hochenergiephysik | Austria |
| Ewa | Rondio | Soltan Inst. Nuclear Studies Warsaw | Poland |
| Mike | Roney | University of Victoria | Canada |
| Frederic | Ronga | IPHE Lausanne | Switzerland |
| Eduardo | Ros | IFIC Valencia | Spain |
| Aura | Rosca | DESY-Zeuthen | Germany |
| Carl | Rosenfeld | University of South Carolina | United States of America |
| Stefan | Roth | RWTH-Aachen | Germany |
| Igor | Roufanov | JINR-Dubna | Russian Federation |
| Alexandre | Rozanov | CPPM-Marseille / Univ. Mediterranée | France |
| Roy | Rubinstein | Fermilab | United States of America |
| James | Russ | Carnegie Mellon University | United States of America |
| Sinead | Ryan | Trinity College Dublin | Ireland |
| Rein | Saar | University of Tartu | Estonia |
| Gerard | Sajot | Institut des Sciences Nucléaires | France |
| Yoshihide | Sakai | KEK / IPNS | Japan |
| Makoto | Sakuda | KEK | Japan |
| David | San Segundo Bello | NIKHEF | The Netherlands |
| Heinz-Georg | Sander | Universität Mainz | Germany |
| Mario | Sannino | Genova University / INFN-Genova | Italy |
| Arcadi | Santamaria | IFIC / Universitat de Valencia | Spain |
| Mariusz | Sapinski | Groupe d'Astroparticules Montpellier | France |
| Edward | Sarkisyan-Grinbaum | CERN & Universiteit Antwerpen | Switzerland |
| Alexander | Savin | University of Wisconsin | United States of America |
| Antonello | Sbrizzi | NIKHEF | The Netherlands |
| Joseph | Scanio | University of Cincinnati | United States of America |
| Stefan | Schael | RWTH-Aachen | Germany |
| Sven | Schagen | NIKHEF | The Netherlands |
| Bert | Schellekens | NIKHEF / University Nijmegen | The Netherlands |
| Paolo | Schiavon | University of Trieste and INFN | Italy |
| Frank-Peter | Schilling | DESY | Germany |
| Peter | Schlein | UC Los Angeles | United States of America |
| Martin | Schmaltz | Boston University | United States of America |
| Michael | Schmelling | MPI for Nuclear Physics Heidelberg | Germany |
| William | Schmidke | Columbia University New York | United States of America |
| Walter | Schmidt-Parzefall | Inst. Experimental Physics Hamburg | Germany |
| M.P. | Schmidt | Yale University | United States of America |
| Uwe | Schneekloth | DESY | Germany |

| | | | |
|---|---|---|---|
| Howard | Schnitzer | Brandeis University | United States of America |
| Andre | Schoening | ETH-Zürich | Switzerland |
| Thomas | Schoerner-Sadenius | CERN | Switzerland |
| Rene | Scholte | NIKHEF | The Netherlands |
| Jaap | Schotanus | Nijmegen University | The Netherlands |
| Henning | Schröder | Universität Rostock | Germany |
| Peter | Schuler | DESY | Germany |
| Hans-Christian | Schultz-Coulon | Universität Dortmund | Germany |
| Gabriella | Sciolla | Massachusetts Institute of Technology | United States of America |
| Lalit | Sehgal | Institute of Theoretical Physics Aachen | Germany |
| Yannis | Semertzidis | Brookhaven National Laboratory | United States of America |
| Meltem | Serin | M.E.T.U. | Turkey |
| Giacomo | Sguazzoni | CERN | Switzerland |
| Qaisar | Shafi | University of Delaware | United States of America |
| Avinash | Sharma | GGSIP University | India |
| Vivek | Sharma | University of California San Diego | United States of America |
| Tino | Shawish Nyawelo | NIKHEF | The Netherlands |
| Tara | Shears | University of Liverpool | United Kingdom |
| Xiaoyan | Shen | IHEP Beijing | China |
| Edward | Shibata | Purdue University | United States of America |
| Kazunari | Shima | Saitama Institute of Technology | Japan |
| Martin | Siebel | Universität Wuppertal | Germany |
| Luca | Silvestrini | INFN-Roma | Italy |
| Maria Chiara | Simani | NIKHEF | The Netherlands |
| James | Simone | Fermilab | United States of America |
| Pekka | Sinervo | University of Toronto | Canada |
| Paul | Singer | Technion | Israel |
| Alexey | Sissakian | JINR-Dubna | Russian Federation |
| Torbjörn | Sjöstrand | Lund University | Sweden |
| Goran | Skoro | Belgrade University | Yugoslavia |
| Patrick | Skubic | University of Oklahoma | United States of America |
| Andris | Skuja | University of Maryland | United States of America |
| Jonivar | Skullerud | Universiteit van Amsterdam | The Netherlands |
| Tomasz | Skwarnicki | Syracuse University | United States of America |
| Gerard | Smadja | Institut de Physique Nucléaire de Lyon | France |
| Jan | Smit | Universiteit van Amsterdam | The Netherlands |
| Stewart | Smith | Princeton University | United States of America |
| Christopher | Smith | Université Catholique de Louvain | Belgium |
| Raimond | Snellings | NIKHEF | The Netherlands |
| Stefan | Söldner-Rembold | Fermilab | United States of America |
| Dongchul | Son | Kyungpook National University | South-Korea |
| Hee-Sung | Song | Seoul National University | South-Korea |
| Amarjit | Soni | Brookhaven National Laboratory | United States of America |
| Davison | Soper | University of Oregon | United States of America |
| Arnd | Specka | Ecole Polytechnique | France |

| | | | |
|---|---|---|---|
| Joachim | Spengler | DESY | Germany |
| Paraskevas | Sphicas | CERN / University of Athens | Switzerland |
| Hubert | Spiesberger | Universität Mainz | Germany |
| Alexandre | Spiridonov | DESY-Zeuthen | Germany |
| Michel | Spiro | CEA-Saclay | France |
| Yogendra | Srivastava | University of Perugia | Italy |
| Peter | Steinberg | Brookhaven National Laboratory | United States of America |
| Olaf | Steinkamp | Universität Zürich | Switzerland |
| Paul | Stevenson | Rice University | United States of America |
| Friedrich | Stinzing | Physikalisches Institut Erlangen | Germany |
| Achille | Stocchi | LAL-Orsay | France |
| Dominik | Stoeckinger | DESY | Germany |
| Sheldon | Stone | Syracuse University | United States of America |
| Bruce | Straub | Yale University | United States of America |
| Ulrich | Straumann | Universität Zürich | Switzerland |
| Jelper | Striet | Universiteit van Amsterdam | The Netherlands |
| Roberto | Stroili | Università di Padova & INFN | Italy |
| Ryszard | Stroynowski | Southern Methodist University | United States of America |
| Daniel | Stump | Michigan State University | United States of America |
| Hirotaka | Sugawara | KEK | Japan |
| Lawrence | Sulak | Boston University | United States of America |
| Gregory | Sullivan | University of Maryland | United States of America |
| Sham | Sumorok | Massachusetts Institute of Technology | United States of America |
| Peter | Suranyi | University of Cincinnatti | United States of America |
| Mark | Sutton | University of Oxford | United Kingdom |
| Kazuhito | Suzuki | KEK-IPNS | Japan |
| Shiro | Suzuki | Yokkaichi University | Japan |
| Yoichiro | Suzuki | Tokyo University | Japan |
| Andre | Sznajder | Universidade Rio de Janeiro | Brazil |
| Tomasz | Szumlak | University of Mining and Metallurgy | Poland |
| Anyes | Taffard | Liverpool University | United Kingdom |
| Fumihiko | Takasaki | KEK | Japan |
| Mossadek | Talby | CPPM-Marseille | France |
| Paul | Taras | Université de Montréal | Canada |
| Marek | Tasevsky | CERN | Switzerland |
| Frank | Taylor | Massachusetts Institute of Technology | United States of America |
| Pedro | Teixeira-Dias | University of London | United Kingdom |
| Jeff | Templon | NIKHEF | The Netherlands |
| Roberto | Tenchini | INFN-Pisa | Italy |
| Oleg | Teryaev | JINR-Dubna | Russian Federation |
| Thomas | Teubner | CERN | Switzerland |
| Jon | Thaler | University of Illinois | United States of America |
| Jim | Thomas | Lawrence Berkeley Nat. Lab. | United States of America |
| Jonathan | Thron | Argonne National Laboratory | United States of America |
| Henk | Tiecke | NIKHEF | The Netherlands |

| | | | |
|---|---|---|---|
| Maury | Tigner | Cornell University | United States of America |
| Charles | Timmermans | University of Nijmegen | The Netherlands |
| Jan | Timmermans | CERN | Switzerland |
| Sharka | Todorova | FOM-Genève | Switzerland |
| Katsuo | Tokushuku | KEK | Japan |
| P. | Tolun | M.E.T.U. | Turkey |
| Guoliang | Tong | Institute of High Energy Physics, CAS | China |
| Ezio | Torassa | INFN-Padova | Italy |
| Edwige | Tournefier | LAPP-Annecy / CNRS | France |
| Daniel | Tovey | University of Sheffield | United Kingdom |
| Stuart | Tovey | University of Melbourne | Australia |
| Fumihiko | Toyoda | Kinki University | Japan |
| Anders | Tranberg | Universiteit van Amsterdam | The Netherlands |
| Thomas | Trefzger | Universität Mainz | Germany |
| Alessia | Tricomi | University / INFN-Catania | Italy |
| Isabel | Trigger | CERN | Switzerland |
| B. | Trocme | IPN-Lyon | France |
| Boris | Tuchming | CEA-Saclay | France |
| Michal | Turala | Institute of Nuclear Physics Cracow | Poland |
| Maria Luisa | Turluer | CEA-Saclay | France |
| Vladis | Tvaskis | NIKHEF | The Netherlands |
| Michael | Tyndel | Rutherford Appleton Laboratory | United Kingdom |
| Guillaume | Unal | LAL-Orsay | France |
| Nikolai | Uraltsev | INFN-Milano | Italy |
| Jon | Urheim | University of Minnesota | United States of America |
| Ulrich | Uwer | Universität Heidelberg | Germany |
| Mark | Vagins | University of California Irvine | United States of America |
| Alice | Valkarova | Charles University | Czech Republic |
| Pierre | Van Baal | University of Leiden | The Netherlands |
| Eef | Van Beveren | Center for Theoretical Physics | Portugal |
| Rene | Van Dantzig | NIKHEF | The Netherlands |
| Martin | Van den Akker | Universiteit Nijmegen | The Netherlands |
| Johannes | Van den Brand | Vrije Universiteit / NIKHEF | The Netherlands |
| Paul | Van der Nat | NIKHEF | The Netherlands |
| Gerard | Van der Steenhoven | NIKHEF / Rijksuniversiteit Groningen | The Netherlands |
| Stefan | Vandoren | Universiteit Utrecht | The Netherlands |
| Hubert | Van Hecke | Los Alamos National Laboratory | United States of America |
| Jan-Willem | Van Holten | NIKHEF / Vrije Universiteit | The Netherlands |
| Pierre | Van Hove | CNRS Strasbourg | France |
| Miranda | Van Iersel | Vrije Universiteit | The Netherlands |
| Ger | Van Middelkoop | NIKHEF | The Netherlands |
| Willy | Van Neerven | University of Leiden | The Netherlands |
| Gerrit | Van Nieuwenhuizen | Massachusetts Institute of Technology | United States of America |
| Jan | Van Tilburg | NIKHEF | The Netherlands |
| Erich | Varnes | Princeton University | United States of America |

| | | |
|---|---|---|
| Alexei Vasiljev | Budker INP Novosibirsk | Russian Federation |
| Michael Vaughn | Northeastern University | United States of America |
| Jaap Velthuis | NIKHEF | The Netherlands |
| Andrea Venturi | CERN / INFN-Pisa | Switzerland |
| Giovanni Venturi | University of Bologna | Italy |
| Frans Verbeure | Universiteit Antwerpen | Belgium |
| Jos Vermaseren | NIKHEF | The Netherlands |
| Jan Visschers | NIKHEF | The Netherlands |
| Oscar Vives | University of Oxford | United Kingdom |
| Cecilia Voena | University 'La Sapienza' & INFN-Rome | Italy |
| Andreas Vogt | NIKHEF | The Netherlands |
| Jochen Volmer | DESY-Zeuthen | Germany |
| Helge Voss | Université de Lausanne | Switzerland |
| Marcel Vreeswijk | NIKHEF | The Netherlands |
| Jacob Wacker | University of California Berkeley | United States of America |
| Doreen Wackeroth | University of Rochester | United States of America |
| Albrecht Wagner | DESY | Germany |
| Hernan Wahlberg | Universiteit Utrecht | The Netherlands |
| Helmut Wahlen | Universität Wuppertal | Germany |
| Roland Waldi | Universität Rostock | Germany |
| Kameshwar Wali | Syracuse University | United States of America |
| Xuelei Wang | Henan Normal University | China |
| Andreas Warburton | Cornell University | United States of America |
| Bennie Ward | University of Tennessee / MPI-Munich | United States of America |
| David Wark | Rutherford Appleton Laboratory | United Kingdom |
| Harmen Warringa | Vrije Universiteit | The Netherlands |
| Yasushi Watanabe | Tokyo Institute of Technology | Japan |
| Harry Weerts | Michigan State University | United States of America |
| Georg Weiglein | University of Durham | United Kingdom |
| Marvin Weinstein | Stanford Linear Accelerator Center | United States of America |
| Christian Weiser | Universität Karlsruhe | Germany |
| Mike Whalley | University of Durham | United Kingdom |
| Jim Whitmore | U.S. National Science Foundation | United States of America |
| Daniel Wicke | Universität Wuppertal | Germany |
| Leo Wiggers | NIKHEF | The Netherlands |
| Richard Wigmans | Texas Tech University | United States of America |
| L.C.R. Wijewardhana | University of Cincinnati | United States of America |
| Bram Wijngaarden | Universiteit Nijmegen | The Netherlands |
| Frank Wilczek | Massachusetts Institute of Technology | United States of America |
| Leif Wilden | Technische Universität Dresden | Germany |
| David Williams | University of California Santa Cruz | United States of America |
| P.K. Williams | U.S. Department of Energy | United States of America |
| Stephane Willocq | University of Massachusetts | United States of America |
| Hans Wilschut | Kernfysisch Versneller Instituut KVI | The Netherlands |
| Marc Winter | IReS / IN2P3 | France |

| | | |
|---|---|---|
| Michael Witherell | Fermilab | United States of America |
| Stanley Wojcicki | Stanford University | United States of America |
| Steven Worm | Rutgers University | United States of America |
| Guy Wormser | LAL-Orsay | France |
| Joerg Wotschack | CERN | Switzerland |
| Martin Woudstra | NIKHEF | The Netherlands |
| Alison Wright | Nature | United Kingdom |
| Douglas Wright | Lawrence Livermore Nat. Lab. | United States of America |
| Jose Wudka | University of California | United States of America |
| Stephan Wynhoff | Princeton University | United States of America |
| Boleslaw Wysłouch | Massachusetts Institute of Technology | United States of America |
| Zhi-zhong Xing | IHEP Beijing | China |
| Tongzhou Xu | IHEP Beijing | China |
| Philip Yager | University of California Davis | United States of America |
| Sakue Yamada | KEK | Japan |
| Richard Yamamoto | Massachusetts Institute of Technology | United States of America |
| Taku Yamanaka | Osaka University | Japan |
| Masa Yamauchi | KEK | Japan |
| Gabriel Ybeles Smit | NIKHEF | The Netherlands |
| Motohiko Yoshimura | Institute for Cosmic Ray Research | Japan |
| Changzheng Yuan | Institute of High Energy Physics / CAS | China |
| Francisco Yumiceva | University of South Carolina | United States of America |
| Adriano Zallo | Laboratori Nazionali di Frascati | Italy |
| Aleksander Zarnecki | Institute of Experimental Physics | Poland |
| Peter Zerwas | DESY | Germany |
| Torsten Zeuner | Universität Siegen | Germany |
| Zhiqing Zhang | LAL-Orsay | France |
| Shouhua Zhu | Inst. Theoretical Physics Karlsruhe | Germany |
| Vadym Zhuravlov | JINR-Dubna | Russian Federation |
| Marek Zielinski | University of Rochester | United States of America |
| Daria Zieminska | Indiana University | United States of America |
| Andrzej Zieminski | Indiana University | United States of America |
| Renata Zukanovich Funchal | Universidade de São Paulo | Brazil |
| Jure Zupan | Institut Josef Stefan | Slovenia |
| Vishnu Zutshi | Northern Illinois University | United States of America |
| Fabio Zwirner | Univ. of Rome 'La Sapienza' / INFN | Italy |

## AUTHOR INDEX

# The Gregory Rift Valley and Neogene–Recent Volcanoes of Northern Tanzani

## Geological Society books refereeing procedures

The Society makes every effort to ensure that the scientific and production quality of its books matches that of its journals. Since 1997, all book proposals have been refereed by specialist reviewers as well as by the Society's Books Editorial Committee. If the referees identify weaknesses in the proposal, these must be addressed before the proposal is accepted.

Once the book is accepted, the Society Book Editors ensure that the volume editors follow strict guidelines on refereeing and quality control. We insist that individual papers can only be accepted after satisfactory review by two independent referees. The questions on the review forms are similar to those for *Journal of the Geological Society*. The referees' forms and comments must be available to the Society's Book Editors on request.

Although many of the books result from meetings, the editors are expected to commission papers that were not presented at the meeting to ensure that the book provides a balanced coverage of the subject. Being accepted for presentation at the meeting does not guarantee inclusion in the book.

More information about submitting a proposal and producing a book for the Society can be found on its web site: www.geolsoc.org.uk.

It is recommended that reference to all or part of this book should be made in the following way:

*From*: Dawson, J. B. 2008. *The Gregory Rift Valley and Neogene–Recent Volcanoes of Northern Tanzania.* Geological Society, London, Memoirs, **33**.

GEOLOGICAL SOCIETY MEMOIR NO. 33

# The Gregory Rift Valley and Neogene–Recent Volcanoes of Northern Tanzania

BY

J. B. DAWSON
University of Edinburgh, UK

2008
Published by
The Geological Society
London

# THE GEOLOGICAL SOCIETY

The Geological Society of London (GSL) was founded in 1807. It is the oldest national geological society in the world and the largest in Europe. It was incorporated under Royal Charter in 1825 and is Registered Charity 210161.

The Society is the UK national learned and professional society for geology with a worldwide Fellowship (FGS) of over 9000. The Society has the power to confer Chartered status on suitably qualified Fellows, and about 2000 of the Fellowship carry the title (CGeol). Chartered Geologists may also obtain the equivalent European title, European Geologist (EurGeol). One fifth of the Society's fellowship resides outside the UK. To find out more about the Society, log on to www.geolsoc.org.uk.

**The Geological Society Publishing House** (Bath, UK) produces the Society's international journals and books, and acts as European distributor for selected publications of the American Association of Petroleum Geologists (AAPG), the Indonesian Petroleum Association (IPA), the Geological Society of America (GSA), the Society for Sedimentary Geology (SEPM) and the Geologists' Association (GA). Joint marketing agreements ensure that GSL Fellows may purchase these societies' publications at a discount. The Society's online bookshop (accessible from www.geolsoc.org.uk) offers secure book purchasing with your credit or debit card.

To find out about joining the Society and benefiting from substantial discounts on publications of GSL and other societies worldwide, consult www.geolsoc.org.uk, or contact the Fellowship Department at: The Geological Society, Burlington House, Piccadilly, London W1J 0BG: Tel. +44 (0)20 7434 9944; Fax +44 (0)20 7439 8975; E-mail: enquiries@geolsoc.org.uk.

For information about the Society's meetings, consult *Events* on www.geolsoc.org.uk. To find out more about the Society's Corporate Affiliates Scheme, write to enquiries@geolsoc.org.uk

Published by The Geological Society from:
The Geological Society Publishing House, Unit 7, Brassmill Enterprise Centre, Brassmill Lane, Bath BA1 3JN, UK

(*Orders*: Tel. +44 (0)1225 445046, Fax +44 (0)1225 442836)
Online bookshop: www.geolsoc.org.uk/bookshop



**British Library Cataloguing in Publication Data**

A catalogue record for this book is available from the British Library.
ISBN 978-1-86239-267-0

Typeset by Techset Composition Ltd., Salisbury, UK
Printed by Antony Rowe Ltd, Chippenham, UK

**Distributors**

***North America***
For trade and institutional orders:
The Geological Society, c/o AIDC, 82 Winter Sport Lane, Williston, VT 05495, USA
*Orders*: Tel. +1 800-972-9892
Fax +1 802-864-7626
E-mail gsl.orders@aidcvt.com

For individual and corporate orders:
AAPG Bookstore, PO Box 979, Tulsa, OK 74101-0979, USA
*Orders*: Tel. +1 918-584-2555
Fax +1 918-560-2652
E-mail bookstore@aapg.org
Website http://bookstore.aapg.org

***India***
Affiliated East-West Press Private Ltd, Marketing Division, G-1/16 Ansari Road, Darya Ganj, New Delhi 110 002, India
*Orders*: Tel. +91 11 2327-9113/2326-4180
Fax +91 11 2326-0538
E-mail affiliat@vsnl.com

# Contents

# Acknowledgements

My first encounter with East African geology was as a geologist with the Tanganyika Geological Survey in 1960. My late wife Christine was a wonderful companion in the bush in the early days before family commitments kept her reluctantly at home. Amongst other things, before the arrival of the official Ordnance Survey 1:50 000 sheets, she made me topographic maps from aerial photographs that enabled mapping of the Oldoinyo Lengai and Monduli areas. I was well served by my Tanganyika field staff, particularly my assistant Salum Kichwa, and I had the stalwart company of my Survey colleague Ray Pickering during my first descent into the crater of Oldoinyo Lengai. The rigours of fieldwork in Maasailand were much eased by Francis and Helen Townsend, John Stephenson and Joe Fourie of the Maasai District Headquarters in Monduli. At Survey headquarters in Dodoma I learned much from discussion with Dan Harkin, Alec McKinlay, Eddie Haldemann, Trevor James and John Whittingham, and I was taught the rudiments of Up-Country Kiswahili by Giovanni Balletto (of 'No Picnic On Mt Kenya' fame).

In later work in Tanzania, I enjoyed the company of Thure Sahama, Colin Clark, Magnus Garson, Miroslav Krs, Harry Pinkerton, Gill Norton, Dave Pyle, Celia Nyamweru, Cassi Paslick, Dodie James, Aloyce Tesha, Roger Mitchell, Valerie Dennison, Tony Mariano and Barbara Scott Smith. On several expeditions, the logistics were expertly managed by Mike, Dave and Thad Peterson of Dorobo Safaris, Arusha, and their staff, especially Alex, Peter and Lomnyak.

Projects on Tanzanian rocks, in particular, have involved Joe Smith and Ian Steele at the University of Chicago; others involved were Dave Powell, Joy Baldwin, Tom Frisch, Peter Bowden, Bob Hutchinson, Noel Gale, Mitch Macintyre, Keith Bell, Colin Donaldson, Richard Batchelor, Ray Kanaris-Sotiriou, Adrian Jones, Arch Reid, Ian Ridley, Mike Rhodes, Mike Henderson, Jeff Cooper, Roy Cohen, Keith O'Nions, Tony Fallick, Peter Hill, John Craven, Richard Hinton, Cassi Paslick, Alex Halliday, Kevin Burton and Roger Mitchell.

On aspects of Tanzanian and East African geology, I have benefited from discussions and correspondence with Ray Macdonald, Laurie Williams, Basil King, Glyn Isaac, Jörg Keller, Charles Downie, Peter Wilkinson, Dick Hay, Roberta Rudnick, Andy Nyblade, Gail Ashley, Sue Brantley, Alan Woolley, Peter Wyllie, Cindy Ebinger, Adrian Foster, Sally Gibson and Martin Roberts. I also thank the people who kindly sent me reprints of their work and I am particularly grateful to Magnus Garson who provided data on Kwaraha and Hanang. I am grateful to Colin Bristow, Mikko Punkari and Dean Polley for providing the photographs for Figures 7.7, 7.15, 7.16 and 7.19.

The technical and secretarial staff in the Earth sciences departments in the universities of St Andrews, Sheffield, Edinburgh and Chicago are particularly thanked for their cheerful and unstinting support. The Natural Environment Research Council and NATO are thanked for several research grants, and the Carnegie Trust for the Universities of Scotland, the Institute of Mining and Metallurgy, and the Royal Society also provided support for fieldwork in Tanzania.

During the preparation of this memoir, Yvonne Cooper and Gerry White helped with photography and drafting respectively, and John Wexler and Marie Cope of the Graphics and Multimedia Resource Centre, University of Edinburgh, are thanked for assistance with image processing. The staff of the National Library of Scotland and the Special Collection of the Edinburgh University Library were of great help in accessing some of the older literature. I am grateful to Ray Macdonald, Martin Smith and Nick Rodgers for undertaking the task of reviewing the original manuscript and for their encouraging and constructive comments. Phil Leat was responsible for the overall editing and Sally Oberst and Angharad Hills gave much welcome advice during the production stage.

Finally, I would not have become involved in African geology without the influence of W. Q. Kennedy, former Professor of Geology at the University of Leeds. He established the Research Institute of African Geology in Leeds in 1956, and I was sent with my postgraduate colleague Pete Nixon to Basutoland in 1957 to study kimberlites. Tom Clifford, Magnus Garson, John Hepworth and John Vail, some of whose work is cited in this volume, were amongst the other members of the Institute. During our research, Kennedy engendered in us the enthusiasm for Africa that he himself had developed when, as a student at the University of Glasgow, he had been taught and inspired by J. W. Gregory.

# Dedication

This volume is dedicated to Christine, Sarah, Michael and Rebecca

# Chapter 1

# Introduction

There are only a few areas on Earth where continental plate break-up and the attendant magmatism are taking place at the present day. In this context, the East African Rift system takes pride of place as it is the most extensive, presently active, continental extension zone, the extension being accompanied by seismicity, crustal thinning and, in some sectors, magmatism. The reason for this spectacular fracturing of the African Plate has been the subject of much debate but there is general consensus that it is due, at least in part, to the presence of rising thermal plumes in the mantle beneath Africa. The extension is now held

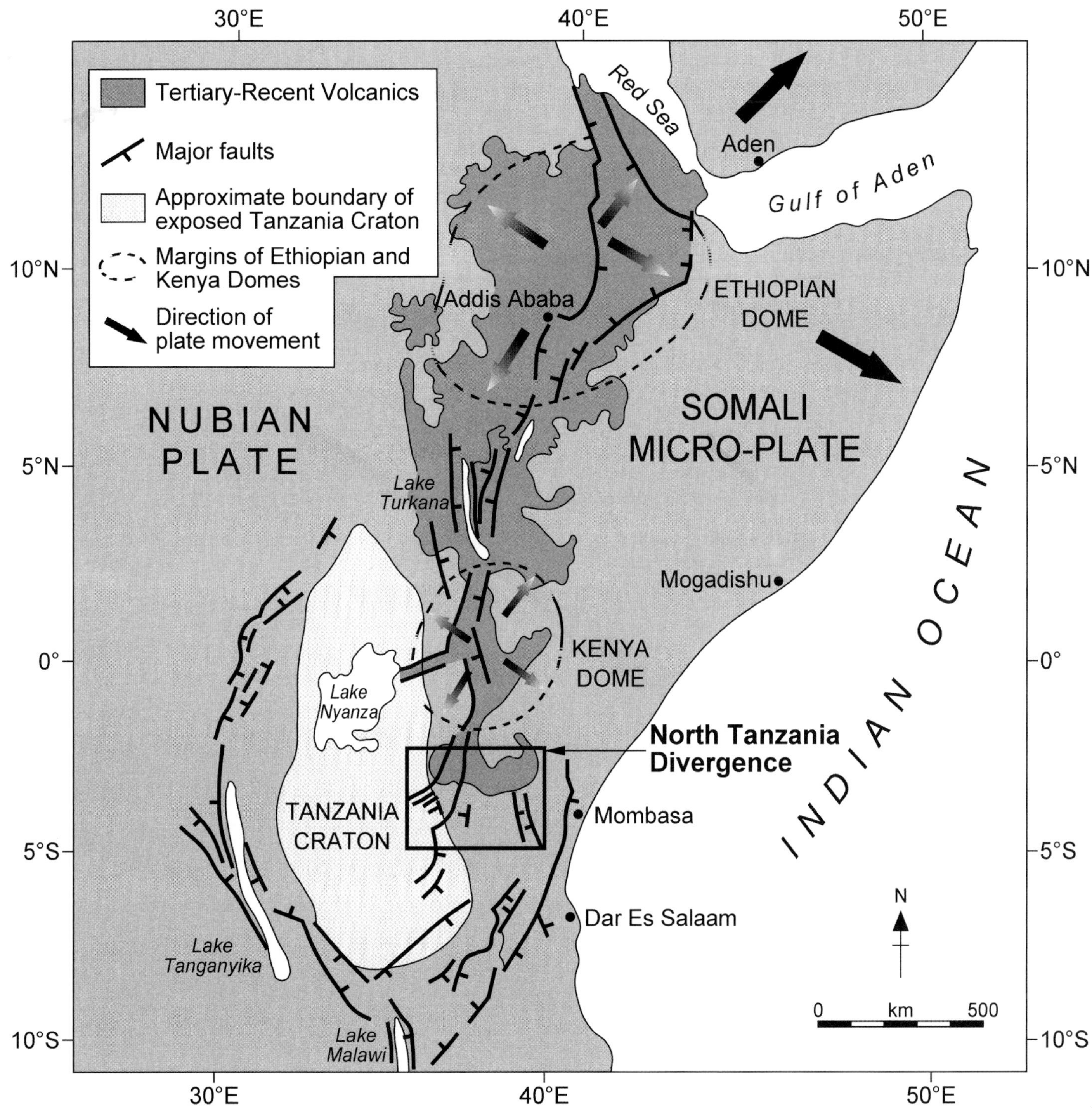


**Fig. 1.1.** Map of East Africa, showing the plate tectonic position of the North Tanzania Divergence relative to the Red Sea spreading centre, and the main structural units of the Ethiopian and Kenya domes and the Tanzania Craton.

J. B. Dawson, University of Edinburgh (*jbdawson@glg.ed.ac.uk*)

*From*: DAWSON, J. B. 2008. *The Gregory Rift Valley and Neogene–Recent Volcanoes of Northern Tanzania*. Geological Society, London, Memoirs, **33**, 1–2.
0435-4052/08/$15.00 © The Geological Society of London 2008. DOI: 10.1144/M33.1

to be due to the incipient separation caused by the plume-related eastward drift of the Somalia microplate away from the more stationary Nubian Plate (Fig. 1.1). The attendant fracturing extends from the Afar triple junction in the Red Sea in the north to at least the Zambezi River in the south. It splits into two branches around the Tanzania Craton which forms the topographic high of the East Africa Plateau, which itself is part of the larger highland area that covers much of southern Africa, referred to as the African Superswell (Nyblade & Robinson 1994).

The eastern branch of the fracture system, stretching from the Gulf of Aden, through Ethiopia and Kenya to northern Tanzania, mainly follows the north–south trend of the Mozambique Fold Belt and passes over the localized uplifts of the Ethiopia and Kenya Domes where the sub-parallel rift structures are best developed. The extension is greatest in the north, in the Afar area of Ethiopia, but reduces to little more than a few kilometres in northern Tanzania where the rifting stalls against the Tanzania Craton and splays out into what has been termed the North Tanzania Divergence.

Compared with the western branch of the system, this Eastern or Gregory Rift Valley is characterized by a greater amount of magmatism and, within this branch, the initiation of rift-associated volcanism has moved southwards over time. It began in Ethiopia, where magmatism started *c.* 30 Ma ago; commenced later in Kenya and was only initiated in northern Tanzania around 8 Ma. As a result northern Tanzania holds a special position in that there are examples of the volcanic rocks extruded at the onset of mantle melting. Further north, in Ethiopia and Kenya, where the rifting and volcanism are relatively mature, the products of early melting are largely buried by the larger volumes of volcanic rocks resulting from more extensive mantle melting.

The Tertiary–Recent volcanic area in northern Tanzania is a small appendage to the larger volcanic massifs in Ethiopia and Kenya to the north, but this belies the wealth of volcanic features and magma types to be found within this relatively small province. It contains features such as snow-capped Kilimanjaro, the highest mountain on the African continent; Ngorongoro, one of the largest calderas on Earth; and the active volcano Oldoinyo Lengai, which is unique for its extrusion of alkali carbonatite. Moreover, this young volcanic area contains an extraordinarily wide variety of volcanic rock types, ranging from ultramafic lavas through alkali basalts, peralkaline trachytes, phonolites and comendites to volcanic carbonatites. There is an ongoing investigation into the origin and evolution of this compositionally wide range of rocks which has been advanced by study of the upper-mantle xenoliths that provide insights into the nature of the upper mantle beneath the Gregory Rift. Xenolith-bearing volcanic rocks occur at a greater number of localities in northern Tanzania than in any other sector of the Gregory Rift Valley, and the chemical and isotopic compositions of the xenoliths have provided evidence for a complex mantle history which is partly reflected in the range in chemistry of the volcanic rocks referred to above.

In addition to the magmatism, a further significant consequence of the extension-related fracturing was the formation of elongate basins now filled with volcanic rocks, sediments, and fresh and saline lakes. The chronology of these basins has been a major tool in understanding the history of the rifting and, in the context of anthropology and human evolution, these basins have been of prime importance. Once again, Tanzania figures prominently. The discovery of remains of an ancient hominid, *Zinjanthropus boisie* ('Nutcracker Man'), in the basin sediments of the Olduvai Gorge stimulated further anthropological studies in the rift basins of Kenya and Ethiopia, and has resulted in East Africa being acknowledged as 'the Cradle of Mankind'.

The volcanic areas in Ethiopia and, particularly Kenya, have been more thoroughly studied and described than those in northern Tanzania. In its areal relationship between the rift faulting and volcanic activity, northern Tanzania broadly resembles Kenya and Ethiopia but, although the rifting and volcanicity form a geographical continuum with those in southern Kenya, there are significant differences with respect to the overall rift structure, the relatively young age of the volcanicity in Tanzania, and in contemporaneous magma types. These were reviewed briefly by Dawson (1992) and, in the following account, commentary will be made on the differences in the regional geology and geophysics.

The monograph begins with an account of observations of the East African landscape made during explorations in the mid- to late- 1800s; observations that ultimately led to the recognition of the fracture system that we now refer to as 'the rift valleys' (Chapter 2). The following section is a short review of the pre-1960 geological studies in the part of East Africa that is now known as Tanzania. Chapter 3 places the Gregory Rift Valley within its regional geological setting, and is followed by an account of the geophysical advances made during the past two decades. These advances, together with our knowledge of the regional distribution of rock units, have led to an integrated picture of the deeper structural setting of the rifting and magmatism (Chapter 4). Chapter 5 details the progressive development of the rift structures, particularly in the context of the influence of pre-existing structural elements, and this is followed by descriptions of the sedimentary basins that formed as the result of the rifting and the relevance of their sedimentary deposits for dating the rifting (Chapter 6). Descriptions of the northern Tanzania volcanoes form the next, most extensive section (Chapter 7), and the petrogenesis of the volcanic rocks is interpreted in the context of the known composition of mantle beneath the rift valley and the processes that impact on magmas whilst *en route* to the surface (Chapter 8).

Overall, it is the objective of this monograph to give a summary of the present knowledge of the interplay between the break-up of the northern Tanzania segment of the African Plate and its associated magmatism

## References

DAWSON, J. B. 1992. Neogene tectonics and volcanicity in the North Tanzania sector of the Gregory Rift Valley: contrasts with the Kenya sector. *Tectonophysics*, **204**, 81–92.

NYBLADE, A. A. & ROBINSON, S. W. 1994. The African Superswell. *Geophysical Research Letters*, **21**, 765–768.

# Chapter 2

# Discovery of the African rift valleys: early work on the Gregory Rift Valley and volcanoes in Northern Tanzania

## Discovery of the African rift valleys

The Austrian geomorphologist Eduard Suess was the first to recognize the importance of the African rift valleys. However, Suess never visited Africa, and he would never have been able to make his perceptive recognition of this aspect of the geomorphology of the African continent had it not been for earlier field explorers.

The discovery of the African rift valleys can be traced back to the middle part of the nineteenth century. At the time, little was known about the interior of much of Africa, partly because, due to uplift of the continental rim, easy access to much of the interior of Central Africa was precluded by major cataract systems near the mouths of many of the larger rivers (Congo, Niger, Zambesi).

The mainspring for the exploration of East Africa came from a largely unacknowledged source, the missionary community on the East African coast. In 1846, the Church Missionary Society of London established a mission station at Kisuludini in the Rabai Hills on the mainland across from Mombasa Island. The mission was started by Johann Krapf, who was soon joined by Johann Erhardt and Johann Rebmann. All three were Lutherans, trained at Basel; apparently the Society had difficulties recruiting British missionaries for the arduous mission life in East Africa (Oliver 1952). In addition to their pastoral duties, the missionaries travelled inland in attempts to find suitable sites for further mission stations and, during these journeys, they made observations on the geographical features they encountered. It was on such journeys that Rebmann first saw the snow on Kilimanjaro on 11 May 1848, and Krapf saw the snow-capped Mt Kenya in May 1849. In an attempt to find more information about possible routes to the interior, the missionaries (Erhardt in particular) questioned Arab traders who had ventured into the interior in search of slaves and ivory (Krapf 1860). This resulted in the production in 1855 of a map of the interior of East Africa drawn 'in true accordance with information received from natives'. This map is in the possession of the Royal Geographical Society in London, and is catalogued 'Tanzania General 16'. Amongst other features, the map shows three 'Snow Mountains': Kilimanjaro, 'Kignea' and Ndonyo Engai, the 'snow' on the last being in reality a capping of white carbonatite. Oldoinyo Lengai was specifically identified as an active volcano in a later account of Arab trade routes (Wakefield 1870).

At the time, little was known of the area beyond the coastal fringe, and the 1855 map became the stimulus to European exploration of the interior of East Africa that was the domain of the Sultan of Zanzibar. Victorian Britain, from its bases in South Africa and Egypt, had colonial interests in Central Africa and a particular interest in the source of the Nile. The 1855 map shows, well into the interior, a vast area of water termed the Sea of Unyamwezi, a large round body with a thin tail pointing to the south, the shape leading the map to be referred to colloquially as the 'Slug Map'. In retrospect, the body of the slug proved to be Lake Victoria/Nyanza, and the tail to be lakes Tanganyika and Nyasa/Malawi. The possibility that this body of water might be the long sought-after source of the Nile was not lost on the Royal Geographical Society of London which, under the presidency of the geologist Sir Roderick Murchison, sponsored Richard Burton and John Hanning Speke to explore what later became known as the Lake Regions.

Leaving Zanzibar on 16 June 1857, Burton and Speke followed the Arab caravan route from Bagamoyo on the coast and crossed the central plateau of Tanzania, arriving, via Tabora, at Ujiji on Lake Tanganyika on 13 February 1858 (Fig. 2.1). Burton noted the steep descent from the plateau to the lake and, during a journey to the north end of Lake Tanganyika was impressed by the length but narrowness of the lake (Burton 1860). This morphology was further confirmed for the southern part of the lake when it was circumnavigated by Henry Stanley in 1876 (Stanley 1878). On their return journey to the coast, Speke temporarily left a sick Burton with Arab hosts at Tabora to make a side-journey to the north where he discovered the Sea of Unyamwezi, i.e. Lake Nyanza which he named 'Victoria' and which subsequently did indeed prove to be the source of the Nile.

Further south, in 1859 David Livingstone travelled up the Zambesi and Shire rivers with a party that included his brother Charles and John Kirk (later to become the British Consul in Zanzibar). On 16 September 1859, they set eyes on Lake Nyasa and Livingstone noted 'at this point, the valley is about twelve miles wide' (Livingstone & Livingstone 1865, p.108). It subsequently transpired that the lake is as long and narrow as Lake Tanganyika. Then, in May 1864, Samuel Baker, having travelled south up the Nile from the Sudan, discovered Lake Albert Nyanza (Baker 1866), the narrow and elongate form of which became apparent after its circumnavigation in 1876 by Romolo Gessi, one of General Gordon's lieutenants operating from Khartoum (Gessi 1892). Thus, from the heart of Africa, a pattern of long, narrow lakes with a dominantly north–south trend was beginning to emerge.

Meanwhile, exploration into Kenya had not proceeded so rapidly, mainly due to the deterrent effect of the warlike Maasai. A proposal by the Scottish geographer Alexander Keith Johnston to explore Maasailand was agreed by the African Committee of the Royal Geographical Society, but the remit of the expedition in 1879–1880 was re-directed towards exploration of the country between the northern end of Lake Nyasa and the southern end of Lake Tanganyika. A member of the expedition was Joseph Thomson, who had studied geology at the University of Edinburgh under Archibald Geikie (J. B. Thomson 1896), and it fell to Thomson to assume leadership of the expedition after the death from dysentery of its leader, Johnston, and to record the results of the expedition (Thomson 1880, 1881). Thomson's 1880 paper, the first with any significant geological observations pertaining to the African rift valleys, contains geological cross-sections across the area, and records sheared and crushed rocks at the southern end of Lake Tanganyika. From this Thomson deduced 'such a condition indicates the existence of a great fault'. He also recorded the presence of the Rungwe volcanics at the northern end of Lake Nyasa which he tentatively, but mistakenly, correlated with the Stormberg volcanics of South Africa.

In addition to the interest in East Africa shown by Britain, France and Italy, by the 1880s, there had also been several German expeditions, both scientific and commercial, into what is now Tanzania. Chronologically and territorially some of these overlapped with British expeditions. One example is the journey by Count von der Decken who, in 1862, visited Kilimanjaro and Lake Jipe and, on a second visit, climbed Kilimanjaro to 14 000 feet and surveyed Mt Meru (Decken 1869).

J. B. Dawson, University of Edinburgh (*jbdawson@glg.ed.ac.uk*)
*From*: Dawson, J. B. 2008. *The Gregory Rift Valley and Neogene–Recent Volcanoes of Northern Tanzania*. Geological Society, London, Memoirs, **33**, 3–7.
0435-4052/08/$15.00 © The Geological Society of London 2008. DOI: 10.1144/M33.2

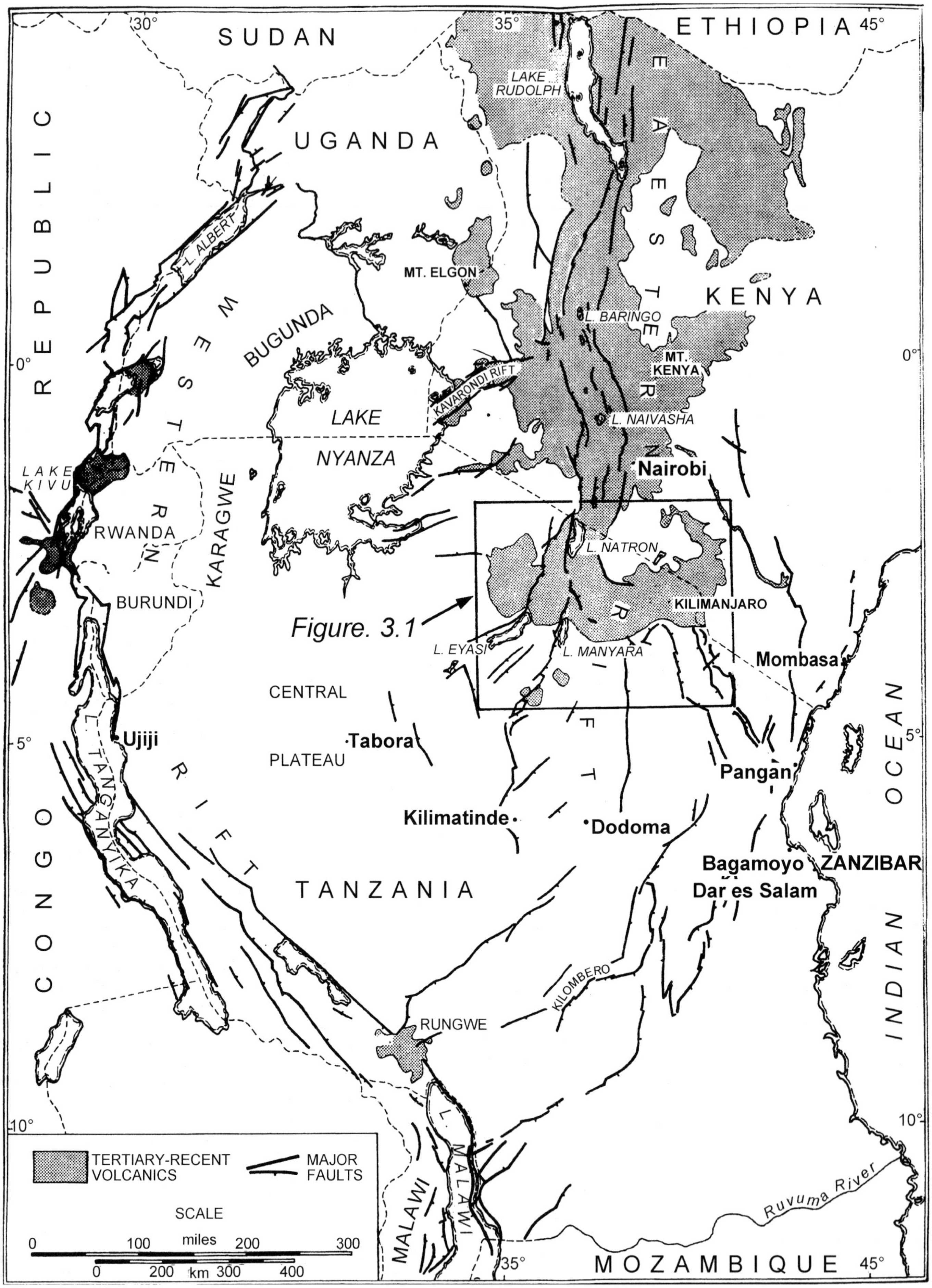


**Fig. 2.1.** Map of East Africa showing the rift system and localities mainly referred to in Chapter 2 (modified after King 1970).

Subsequent to his travels to Lakes Nyasa and Tanganyika, in 1883 Thomson at last obtained sponsorship from the Royal Geographical Society to explore Maasailand. His plan was to follow an Arab trade route, infrequently travelled because of the predations of the Maasai, to Lake Baringo, a body of water previously unseen by Europeans but whose surrounding country was fabled by the Arabs as a source of much ivory (Farler 1882). Leaving from Mombasa, near Kilimanjaro he had difficulties with the Maasai as the result of resentment caused by a battle between the Maasai and an earlier, large German expedition led by Gustav Fischer, which had then proceeded westwards towards Lake Manyara (referred to below). After placating the Maasai, Thomson's onward journey took him further north than the route taken by Fischer; he descended into the rift valley near Mt Longonot, continued to Lake Naivasha (the furthest point of the Fischer expedition) and, eventually reached Lake Baringo. *En route* he visited Lakes Elmenteita and Nakuru and noted that all the lakes lie within the elongate depression that we now recognize as the rift valley. After visiting the lower western slopes of Mount Kenya, where his hopes of climbing the mountain were dashed by hostile Maasai, he reached Lake Baringo on 10 November 1883. Thomson's description of the setting of the lake from the edge of the Laikipia Plateau is the first description of rift valley geomorphology: 'suddenly emerging from the dense forest ... at the edge of the meridional trough ... there was the mysterious Lake Baringo several thousand feet below ... Imagine if you can a trough or depression 3300 feet above the sea-level, and twenty miles broad, the mountains rising with a very great abruptness on both sides to a height of 9000 feet. In the centre of this depression lies a dazzling expanse of water. Round the lake ... a remarkable assemblage of straight lines, wall-like extensions and angular lines produces an impressive and quite unique landscape. It speaks eloquently, however, of igneous disturbances; for there you observe numerous earth movements, faults crossing each other at right angles, and other features, which are clearly not modelled by surface agents, all of them so recent in origin as to remain comparatively untouched by the hand of time, which seems to abhor anything approaching a straight line' (Thomson 1887, 228–230). Later on this journey, Thomson travelled west to Lake Nyanza, and then returned to Baringo via Mt Elgon, noting *en route* the major fault escarpments on the western side of the rift valley.

Between 1887 and 1888 the Hungarian, Count Samuel Teleki, accompanied by the Austrian Ludwig von Höhnel, travelled on a hunting and natural history expedition from Pangani on the East African coast across central Kenya to the southern borders of Ethiopia. It was von Höhnel who kept the records and made maps of the terrain crossed by the expedition (Höhnel 1890, 1894). In southern Ethiopia, the expedition discovered two large lakes, Basso Narok and Basso Ebor, which Teleki renamed respectively lakes Rudolph and Stephanie after the Prince Regent of the Austro-Hungarian Empire and his consort. Like lakes Tanganyika and Nyasa, Lake Rudolph in particular proved to be long and narrow, and von Höhnel had the perception to link the Rudolph depression with Thomson's 'meridional trough' in Kenya to the south. Von Höhnel was a friend of Eduard Suess who had been plotting the shapes of the various lakes as they were discovered; sitting in Vienna, he had never set foot in Africa. Drawing on the cumulative information from the early explorers, and with the collaboration of von Höhnel after his return to Vienna, Suess (1891) proposed, in a compendium on the geology of East Africa (Höhnel *et al.* 1891), that the chains of ribbon-like lakes lie in sunken valleys, which he termed 'grabens', resulting from major fractures running across the African continent from the Red Sea to the Zambesi. Later, in a more extended work, he proposed that these fractures linked northwards via the Red Sea to the Dead Sea depression (Suess 1885–1909). This has indeed proved to be the case, but the nature of the fracturing and the geological events attending it in East Africa had to await the observations made by J. W. Gregory during two journeys into central Kenya in 1893 and 1919 (Gregory 1894, 1896, 1921). Gregory made several fundamental observations about the Quaternary of East Africa, published in his classic '*The Great Rift Valley*' (Gregory 1896); these included observations on the former extension of glaciers on Mt Kenya, the former existence of great lake systems in areas that are now arid, and the subsidence of East African Plateau on the site of Lake Nyanza and the consequent interruption of major river systems. But it was in his earlier 1894 account that Gregory introduced the term 'rift valley', for which he will be remembered, defining it as 'a linear valley with parallel and almost vertical sides, which has fallen owing to a series of parallel faults'.

## Early geological work in northern Tanzania

As noted earlier, it was not only Britain that was showing an interest in East Africa in the latter part of the nineteenth century. Following the virtual annexation of part of the East African mainland by the Company for German Colonization under Carl Peters in 1884, a joint boundary commission under the direction of Kirk, the British Consul in Zanzibar, was set up in 1885, and in October 1890, apart from a thin coastal strip ten miles wide, the former mainland domains of the Sultan of Zanzibar were partitioned into British and German spheres of influence, with the boundary running from the coast south of Mombasa northwestwards towards Lake Nyanza/Victoria (Kirk 1887; Smith 1894).

Pakenham (1991) details the attendant and subsequent political and commercial manoeuvrings, but suffice to say that, even before the partitioning, and beginning in the 1860s, there had been an increasing number of German expeditions into what is now Tanzania. It is due to these expeditions that we have the first accounts of the Tanzania sector of the rift valley and many of the volcanoes of northern Tanzania. The visits of von der Decken to Kilimanjaro and Meru were mentioned earlier. In 1882–1883, Gustav Fischer, in what turned out to be a rival expedition to that of Thomson, travelled from Pangani to Kilimanjaro and Arusha, but then west to Lake Manyara and north up the rift valley, arriving at Lake Naivasha in Kenya before Thomson. In Njorowa Gorge, on the southern shore of Lake Naivasha, stands a pinnacle of peralkaline rhyolite. Its name, 'Fischer's Tower', perpetuated to the present day, commemorates his visit. Also during his northwards journey from Lake Manyara he became the first European to see the active volcano Oldoinyo Lengai (Fischer 1884–1885).

Other notable expeditions were those of Baumann in 1892–1893 (Baumann 1894); of Meyer, who went as far as the Western Rift Valley and discovered the Kivu volcanic province, and who also was the first European to reach the summit of Kilimanjaro in 1889 (Meyer 1893, 1900); and of Schöller, who crossed Maasailand to reach Uganda (Schöller 1901–1904). Although rocks collected on the Fischer, Baumann and Schöller expeditions were later described by Mügge (1886), Lenk (1894) and Kunzli (1901), respectively, as were rocks from Kilimanjaro by Finckh (1903, 1906), these expeditions were mainly concerned with the geography of the area. An exception was the first geophysical fieldwork that took place in 1899 and 1990, when Ernst Kohlschütter (the German astronomer who had been attached to the survey that had carried out the earlier boundary commission) made the first pendulum gravity readings across the Tanganyika Plateau and the Gregory Rift Valley (Kohlschütter 1901). His data were incorporated into a later synthesis of East African geology by Krenkel (1922).

More detailed geological observation arose in 1904 when Carl Uhlig led an expedition which surveyed the Manyara–Natron sector of the rift valley more thoroughly than had Fischer; Uhlig was the first to note the asymmetry of this sector of the rift

(Uhlig 1905, 1907*a*, *b*). During the expedition, Fritz Jaeger made the first recorded ascents of Meru and Oldoinyo Lengai in 1904 and also explored the Crater Highlands to the west of the rift valley (Jaeger 1911, 1913). Rocks collected by Jaeger were described by Finckh (1911). Oldoinyo Lengai continued to hold a fascination for the German geologists, and was climbed again by Uhlig and F. Th. Müller in 1910 (Uhlig & Jaeger 1942), by Hans Reck in 1911 (Reck 1914*c*, 1923) and by Shulze in 1915 (Reck & Shulze 1921).

In 1913, Reck made the earliest stratigraphic studies of the Olduvai Gorge to the west of the Crater Highlands (Reck 1914*a*, *b*), during which he collected many mammalian fossils that he took to Berlin. These fossils subsequently turned out to have important implications for East African anthropology. Geological fieldwork in German East Africa was interrupted by the 1914–1918 war but publications arising from pre-war work continued for some time afterwards, notable contributions being Reck's collection of photographs of Oldoinyo Lengai (Reck 1924), Erdmannsdorfer's descriptions of wollastonite-bearing igneous rocks from Oldoinyo Lengai (Erdmannsdorfer 1935), and a detailed volume on the Manyara–Natron area (Uhlig & Jaeger 1942).

Following the 1914–1918 war, the former German East Africa became Tanganyika Territory under a mandate from the League of Nations. Under the new British administration, the Tanganyika Geological Survey was established in Dodoma. Initially, because of their relatively low economic potential, the areas around the Gregory Rift Valley and the Neogene volcanoes had a low priority within the regional-geology mapping programme This was redressed in the 1950s and early 1960s and maps of most of these areas have now been published on a scale of 1:125 000. The individual maps, which are referred to in the appropriate places later in this volume, contain brief explanations of the geology.

In addition to the efforts of the Survey staff, geologists and geophysicists from academic institutions have carried out detailed studies, particularly in more recent years. These are mainly referred to later, but three significant contributions made prior to 1960 are mentioned here:

(1) In 1928 the anthropologist L. S. B. Leakey visited Berlin and, despite Reck's scepticism, recognized artefacts amongst the material gathered by Reck at Olduvai Gorge in 1913. This recognition resulted in a series of expeditions, beginning in the 1930s, to collect new material. The work arising from these expeditions has led to Olduvai being recognized as a world-famous hominid locality and, via Leakey's connection with the Coryndon Museum in Nairobi (now the National Museum of Kenya), was the impetus for anthropological research in Kenya and Ethiopia, as well as other parts of Tanganyika. In addition, *en route* to Olduvai on the 1934–1935 expedition, P. E. Kent recognized and later described the lake beds at Lake Manyara, and, on the same expedition, carried out the first mapping of the Laetoli Beds to the south of Olduvai. The dating of volcanic components in the strata at both Olduvai and Laetoli, in connection with elucidating the chronology of hominid evolution, has led to a clearer knowledge of the eruption history of some of the northern Tanzania volcanoes.

(2) Between 1930 and 1935, Bailey Willis, of the Carnegie Institute of Washington, travelled extensively in both the Western and Gregory Rifts, and a monograph detailing his observations contains the most exhaustive analysis to that date of the roles of extension v. compression in the formation of the rift valleys (Willis 1936). In contrast to Gregory, and on the basis of metamorphic rocks apparently overlying younger volcanic rocks on the west shore of Lake Natron, he concluded that thrusting and compression are the causes of the rift valley depression. It later transpired that the igneous rocks are intrusive into the schists.

(3) As part of the regional mapping programme, in 1953 and 1957 joint expeditions of the Tanganyika Geological Survey and the Department of Geology at the University of Sheffield, carried out geological and glaciological surveys of Kilimanjaro. The geological account (Downie & Wilkinson 1972) is one of the most comprehensive reports pertaining to a single African volcano, and the glaciological data (Downie 1964) provide a yardstick against which to measure glacier retreat over the past half-century.

On 9 December 1961, Tanganyika became independent and the highest point on Kilimanjaro, formerly known as Kaiser Wilhelm Point, was renamed Uhuru (Freedom) Peak on that date. After its unification with Zanzibar, on 29 October 1964 the name of Tanganyika merged with that of Zanzibar to become Tanzania.

## References


Baker, S. W. 1866. *Albert N'Yanza, Great Basin of the Nile*. MacMillan, London, 2 volumes.

Baumann, B. 1894. *Durch Masailand zur Nilquelle*. Reimer, Berlin.

Burton, R. F. 1860. *The Lake Regions of Central Africa*. Longman, Green, Longman & Roberts, London. 2 volumes.

Decken, C. C. & von der, 1869. *Reisen in Ost Afrika* (referred to in Thomson (1887, p.4) and Gregory (1921, p.53)).

Downie, C. 1964. Glaciations of Mount Kilimanjaro, Northeast Tanganyika. *Bulletin of the Geological Society of America*, **75**, 1–16.

Downie, C. & Wilkinson, P. 1972. *The Geology of Kilimanjaro*. University of Sheffield, Sheffield.

Erdmannsdorfer, O. 1935. Über wollastoniturtit und die Entstehungsweisen von alkaligesteiner. *Sitzungberichte der Heidelberg Akademie für Wissenschaften, Mathematike und Naturgeschichte*, **K 1.2**, 1–23.

Farler, J. P. 1882. Native routes in East Africa from Pangani to the Masai country and the Victoria Nyanza. *Proceedings of the Royal Geographical Society, New Series*, **4**, 730–753.

Finckh, L. 1903. Die trachydolerite des Kibo (Kilimandscharo) und die kenyte des Kenya. *Zeitschrift der Deutsches Geologische Gesellschaft*, **55**, Monatsberich 1–14.

Finckh, L. 1906. Die rhomberporphyre des Kilimanscharo. *In*: *Festschrift zum siebsigsten Geburtstag von Harry Rosenbusch*. Stuttgart, 372–398.

Finckh, L. 1911. Die von F. Jaeger in Deutsch–Ostafrika gesammelten Gesteine. *In*: Jaeger, F. (ed.) Das Hochland der Riesenkrater and die umliegender Hochländer Deutsch Ostafrika, Part 1. *Mitteilungen aus den Deutsches Schutzgebieten, Ergänzungsheft*, **4**, 71–85.

Fischer, G. A. 1884–1885. Reise in das Masailand. *Mitteilungen der Geographisches Gesellschaft Hamburg*, Pt 1 (1884), 36–99, Pts 2 and 3 (1885) 189–237 and 238–279.

Gessi, R. 1892. *Seven Years in the Sudan*. Sampson Low, Marston & Co., London.

Gregory, J. W. 1894. Contributions to the physical geography of British East Africa. *Geographical Journal*, **4**, 290–315, 408–424, 505–514.

Gregory, J. W. 1896. *The Great Rift Valley*. John Murray, London.

Gregory, J. W. 1921. *The Rift Valleys and Geology of East Africa*. Seeley, Service, London.

Höhnel, L. R. von, 1890. Ost-Äquatorial Afrika zwischen Pangani und dem neuentdecken Rudolf-See. *Petermann's Mitteilungen Ergänzungsheft*, **99**, 1–44.

Höhnel, L. R. von, 1894. The Discovery of Lakes Rudolf and Stefanie. A narrative of Count Teleki's exploring and hunting expedition in Eastern Equatorial Africa in 1887–1888. Longmans, Green, London, 2 volumes. Abridged from the German edition of 1892.

Höhnel, L. R., von, Rosiwal, A., Toula, F. & Suess, E. 1891. Beiträge zur geologischen Kenntniss der östlichen Afrika. *Denkschrifte der Königlich Akademie für Wissenschaften Wien, Math- Naturwissenschaften Klasse*, **58**, 447–584.

Jaeger, F. 1911. Das Hochland der Riesenkrater and die umliegender Hochländer Deutsch Ostafrika, Part 1. *Mitteilungen aus den Deutsches Schutzgebieten, Ergänzungsheft*, **4**, 1–133.

Jaeger, F. 1913. Das Hochland der Riesenkrater and die umliegender Hochländer Deutsch Ostafrika, Part 2. *Mitteilungen aus den Deutsches Schutzgebieten, Ergänzungsheft*, **8**, 1–213.

KING, B. C. 1970. Vulcanicity and rift tectonics in East Africa. *In*: CLIFFORD, T. N. & GASS, I. G. (eds) *African Magmatism and Tectonics*. Oliver & Boyd, Edinburgh, 263–283.

KIRK, J. 1887. Britain and Germany in East Africa; note. *In*: THOMSON, J. *Through Masailand: a Journey of Exploration among the Snow-clad Volcanic Mountains and Strange Tribes of Eastern Equatorial Africa*. Sampson Low, Marston, Searle & Rivington, London. 357–359.

KOHLSCHÜTTER, E. 1901. Über den Bau der Erdkrust in Deutsch-Ostafrika. *Vorläufige Mitteilungen, Nachrichte der Königlichen Gesellschaft der Wissenschaften zu Göttingen*, **1901**, 1–40.

KRAPF, J. L. 1860. *Travels, Researches and Missionary Labours during an Eighteen Years' Residence in Eastern Africa*. Trübner & Co., London.

KRENKEL, E. 1922. *Die Bruchzonen Ostafrikas: Tektonik, Vulkanismus, Erdbeben and Schwereanomalien*. Bornträger, Berlin.

KUNZLI, E. 1901. Die petrographische Ausbeute der Schoellerschen Expedition in Äquatorial Ostafrika. (Masailand). *Vierteljahrsheft der Naturforscheres Gesellschaft Zürich*, **46**, 128–172.

LENK, H. 1894. Über Gesteine aus Deutsch Ostafrika. *In*: BAUMANN, B. *Durch Masailand zur Nilquelle*. Reimer, Berlin. 263–294.

LIVINGSTONE, D. & LIVINGSTONE, C. 1865. *Narrative of an Expedition to the Zambesi and its Tributaries, and the Discovery of Lakes Shirwa and Nyasa 1858–1864*. John Murray, London.

MEYER, H. 1893. Die grossen Bruchspalten und Vulkane in Äquatorial-Afrika. *Deutsche Geographische Blätter*, **16**, 105–127.

MEYER, H. 1900. *Der Kilimandjaro*. Reimer, Berlin.

MÜGGE, O. 1886. Untersuchungen der von Dr. G. A. Fischer gesammelten gesteine des Masailand. *Neues Jahrbuch für Mineralogie*, **4**, 238–264.

OLIVER, R. 1952. *The Missionary Factor in East Africa*. Longman, Green & Co, London.

PAKENHAM, T. 1991. *The Scramble for Africa*. Weidenfelt & Nicolson, London.

RECK, H. 1914*a*. Erste vorläufige Mitteilungen über den Fund eines fossil Menschskelets aus Zentralafrika. *Sitzungsberichten der Gesellschaft Naturforschender Freunde*, **3**, 81–95.

RECK, H. 1914*b*. Zweite vorläufige Mitteilungen über fossile Tier- und Menschenfunde aus Oldoway in Zentralafrika. *Sitzungsberichten der Gesellschaft Naturforschender Freunde*, **3**, 305–318.

RECK, H. 1914*c*. Oldoinyo Lengai, ein tätigerVulkan in Gebiet der Deutschostafrikanischen Bruchstufe. *Branca Festschrift*, **7**, 373–409.

RECK, H. 1923. Über der Ausbruch des Oldoinyo Lengai in Deutschostafrika in Jahre 1917. *Zeitschrift für Vulkanologie*, **7**, 55–56.

RECK, H. 1924. L'Engai Bilder. *Zeitschrift für Vulkanologie*, **8**, 172–174.

RECK, H. & SCHULZE, G. 1921. Ein Beitrag zur Kenntnis des Baues an der jüngsten Veränderung des l'Engai-Vulkanes im nordlichen Deutsch-Ostafrika. *Zeitschrift für Vulkanologie*, **6**, 47–71.

SCHÖLLER, M. 1901–1904. *Mitteilungen über meine Reise nach Äquatorial-Ostafrika und Uganda, 1897–97*. Reimer, Berlin, Pt I (1901), Pt II (1904), Pt III 16 maps (1901).

SMITH, C. S. 1894. The Anglo-German boundary in East Equatorial Africa. Proceedings of the British Commission. *Geographical Journal*, **4**, 424–439.

STANLEY, H. M. 1878. *Through the Dark Continent*. Sampson Low, Marston, Searle & Rivington, London.

SUESS, E. 1891. Die Brüche des östlichen Afrika. Chapter 4. *In*: HÖHNEL, L. R. VON *ET AL*. *Denkschrifle der Königlich Akademie für Wissenschaften Wien, Math-Naturwissenschaften Klasse*, **58**, 555–584.

SUESS, E. 1885–1909. *Das Antlitz der Erde (The Face of the Earth*. Translated by H. Sollas & W. J. Sollas, 1904–1909). Clarendon Press, Oxford, 3 volumes.

THOMSON, J. 1880. Notes on the geology of East–Central Africa. *Nature*, **28**, 102–104.

THOMSON, J. 1881. *To the Central African Lakes and Back: the Narrative of the Royal Geographical Society's Central African Expedition 1879–80*. Sampson Low, Marston, Searle & Rivington, London, 2 volumes.

THOMSON, J. 1887. *Through Masailand: a Journey of Exploration among the Snow-clad Volcanic Mountains and Strange Tribes of Eastern Equatorial Africa*. Sampson Low, Marston, Searle & Rivington, London. (New and revised edition of the original 1884 version).

THOMSON, J. B. 1896. *Joseph Thomson – African Explorer*. Sampson Low, Marston & Co., London.

UHLIG, C. 1905. Bericht über die Expedition der Otto-Winter Stiftung nach dem Umgebungen des Meru. *Zeitschrift der Gesellschaft für Erdkund*, **1905**, 120–123.

UHLIG, C. 1907*a*. Vom Kilimandscharo zum Meru. *Zeitschrift der Gesellschaft für Erdkunde*, **1907**, 627–650.

UHLIG, C. 1907*b*. Der sogennante grosse Ostafrikanische Graben zwischen Magad (Natron See) und Lawa ya Mweri (Manyara See). *Geographisches Zeitung*, **15**, 478–505.

UHLIG, C. & JAEGER, F. 1942. Die Ostafrikanische Bruchstufe and die angrenzenden Gebiete zwischen die Seen Magad und Lawa ja Mweri sowie den Westfuss des Meru. *Deutsches Institut für Länderkunde, Wissenschaften Veröffentlichungen*, New Series **10**.

WAKEFIELD, T. 1870. Routes of native caravans from the coast to the interior of Eastern Africa, chiefly from information given by Sadi Bin Ahedi, a native of a district near Gazi, in Udigo, a little north of Zanzibar. *Journal of the Royal Geographical Society*, **40**, 303–339.

WILLIS, B. 1936. *East African Plateaus and Rift Valleys*. Carnegie Institution of Washington, Washington D.C.

# Chapter 3

# Regional geology

The present-day Gregory Rift Valley in northern Tanzania is an elongate north–south half graben flanked on its western side by a high, eastwards-facing escarpment; this differs from Kenya where there are also major faults on the eastern side of the rift valley. The words 'present-day' are used advisedly, as the present morphology is imposed upon an earlier, wider, volcano-infilled tectonic depression. As in Kenya, within and to either side of the rift valley there are extensive areas of Neogene and Quaternary volcanic rocks (Fig. 3.1). To the SW of the main volcanic area is the Eyasi half graben and, to the SE, the Pangani graben; these major features are infilled by volcanoes at their northeastern and northern ends, respectively.

The northern Tanzania volcanic province is younger overall than the Ethiopian, where magmatism was initiated at around 40 Ma, and Kenya provinces (Baker *et al.* 1972; George *et al.* 1998) suggesting that more recent mantle perturbations are responsible for the intra-plate magmatism and crustal fracturing in this part of Africa.

The province stands astride the surface interface between the Archaean rocks of the Tanzania Craton and the north–south-trending Mozambique orogenic fold belt (Fig. 3.1). The rocks of the craton have been divided into three major formations, the Dodoman (oldest), the Nyanzian and the Kavirondian (Quennell *et al.* 1956; Schlüter 1997) but, in brief, the craton is an amalgamation of several terranes comprising Archaean metasediments, including the so-called greenstone belts (some as old as >3 Ga), that were intruded by granites and migmatized during three short tectono-metamorphic events at around 2.9, 2.7 and 2.4 Ga (Pinna *et al.* 1996). A minor phase of granite intrusion was the latest event to affect this stable block at 1.85 Ga (Cahen *et al.* 1984).

The age, make-up and terminology of the 'Mozambique Belt' require some explanation. With the exception of Hanang, the most southerly of the northern Tanzanian volcanoes, which overlies cratonic rocks, the Tanzania Geological Survey maps show the volcanic province to be surrounded and underlain by metamorphic rocks assigned to the 'Usagaran', a major subduction-related, tectono-thermal event dated at around 2 Ga at the type locality in the Usagara Mountains at the southeastern margin of the Tanzania Craton (Quennell *et al.* 1956). Such rocks extend part-way across the Serengeti Plains to the west of the volcanic province, and over part of the Maasai Block to the east. However, metamorphic rocks in the Pare Mountains further to the east are dated at 645 $\pm$10 Ma (Muhongo & Lenoir 1994).

The term 'Mozambique orogenic belt' was originally used by Holmes (1951) to describe the north–south-trending fold belt stretching from Ethiopia to Mozambique, partly bounded on the west by the Tanzania Craton, and this fold belt was later included by Kennedy (1964) in the series of Africa-wide fold belts affected by the so-called 'Pan-African orogenic event' at around 650 Ma. Specifically in northern Tanzania, Möller *et al.* (1995, 1998) suggested that, whereas the Usagaran metamorphic event is still recognizable at the type locality, the effects of the Usagaran metamorphism were largely erased in northern Tanzania during the Pan-African event due to large-scale thrusting of fold belt metamorphic rocks westward over the eastern margin of the Tanzania Craton (Shackleton 1986). Hepworth (1972) had earlier advocated that the complex metamorphism and structure of a large tract of terrain, to the east of the mapped craton boundary but to the west of the Pare Mountains, should be assigned to an event earlier than the Pan-African, and Cahen *et al.* (1984) agreed that 'ancient assemblages are involved in the Mozambique Belt along its western margin and are recognizable over some distance into the belt'. This has been further confirmed by recent work that shows the Mozambique Belt in Central Tanzania to consist of reworked, ancient crust of Archaean (2970–2500 Ma) or Palaeoproterozoic (2124–1837 Ma) age (Sommer *et al.* 2003). Hence, the reality is that, with the Tanzania Craton acting as a foreland, the Mozambique Belt is a polycyclic orogenic complex involving basement and cover rocks of more than one orogenic event, with the last taking place around 650 Ma. Similarly, Key *et al.* (1989), Mosley (1993) and Hauzenberger *et al.* (2007) provide evidence for the polycyclic nature of the Mozambique Belt in Kenya.

Petrographically, the Usagaran has been divided into two major series: (i) the lower Masasi Series comprising mainly quartzo-feldspathic gneisses, charnockites and hornblende–biotite gneisses; and (ii) the higher Crystalline Limestone Series, comprising quartzites, dolomitic marbles, graphitic marble, graphite schists, mica schists, kyanite gneisses and rarer metabasites (Quennell *et al.* 1956). Both series are extensively veined by pegmatites. Particularly interesting units in the Crystalline Limestone Series are the tanzanite-bearing marbles of the Leletema Hills (south of Kilimanjaro), ruby corundum-Cr zoisite amphibolite in the Longido area (Game 1954), and the rocks of Mautia Hill [6°8′S 36°29′E], the type locality for yoderite ($[MgAl]_8SiO_4[O,OH]_{20}$) (McKie & Redford 1959), where yoderite–kyanite–talc schists are accompanied by piedmontite marbles.

The metamorphic grade of the Usagaran rocks ranges from biotite facies to upper amphibolite facies, whereas the Pan-African age rocks of the Pare Mountains are of granulite facies. However, granulite-facies granite (charnockite) and anorthosite also occur as the cores of mantled gneiss domes within the dominantly Usagaran metamorphic rocks at Loliondo and Longido near the Kenya border to the west and east of the rift valley, respectively.

As noted above, the northern Tanzania volcanic province stands astride the surface interface between the Archaean rocks of the Tanzania Craton and the north–south-trending Mozambique orogenic fold belt. Thus, on the basis of surface geology alone, it might appear that the volcanism and the rift structures are located within a tectonic crustal suture regime where there are mechanical and thermal contrasts between thick, cold, rigid Archaean lithosphere and thinner, anisotropic mobile-belt lithosphere. However, Smith (1994) and Smith & Mosley (1993) proposed that, in southern Kenya, due to the westward overthrusting by the Mozambique Belt rocks, the buried margin of the craton lies some 100 km to the east of the surface contact between the craton and the fold belt metamorphics. In support of this is a 1 km thick low-velocity P-wave layer, extending for 120 km westwards from the rift valley at Lake Magadi; the layer is interpreted as highly fractured and sheared Mozambique Belt rocks overlying the Tanzania Craton (Birt *et al.* 1997). In northern Tanzania, granulites dated at 2.0 Ga (Cohen *et al.* 1984) occur as xenoliths at the Lashaine volcano (see Dawson 2008, fig. 5.8) again some 100 km east of the mapped contact between the craton and the mobile belt. The age of the Lashaine granulites has not been reset due to tectonic downgrading, which suggests a stable, underlying structure i.e. the craton. Further, the suggestion of Archaean lithosphere below a thin layer of Proterozoic nappes is supported by heat flow values that, in central and northeastern Tanzania, are

J. B. Dawson, University of Edinburgh (*jbdawson@glg.ed.ac.uk*)
*From*: Dawson, J. B. 2008. *The Gregory Rift Valley and Neogene–Recent Volcanoes of Northern Tanzania*. Geological Society, London, Memoirs, **33**, 9–12.
0435-4052/08/$15.00 © The Geological Society of London 2008. DOI: 10.1144/M33.3

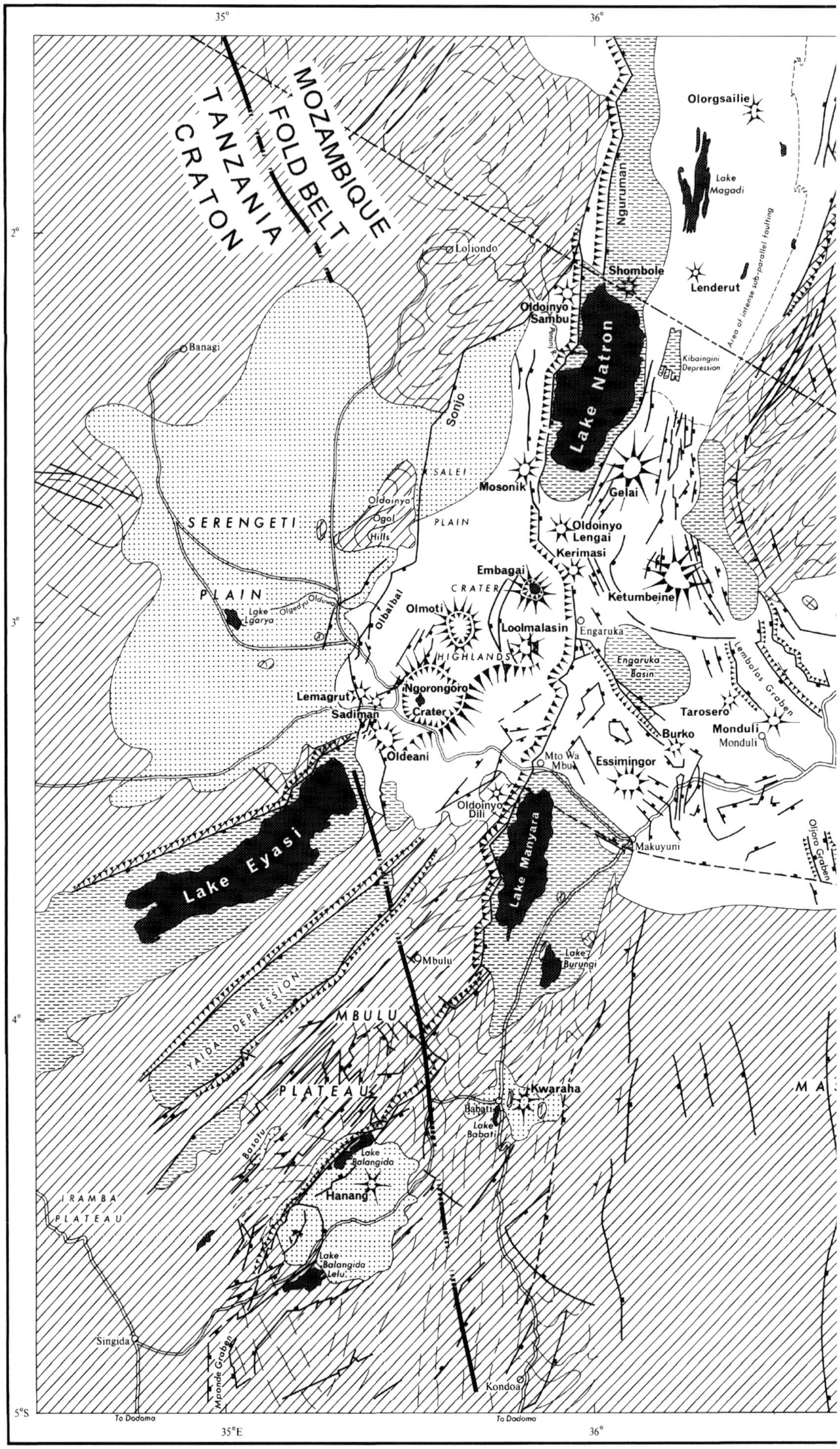


**Fig. 3.1.** Distribution of Neogene–Recent faults and volcanic rocks in northern Tanzania and adjoining parts of southern Kenya.

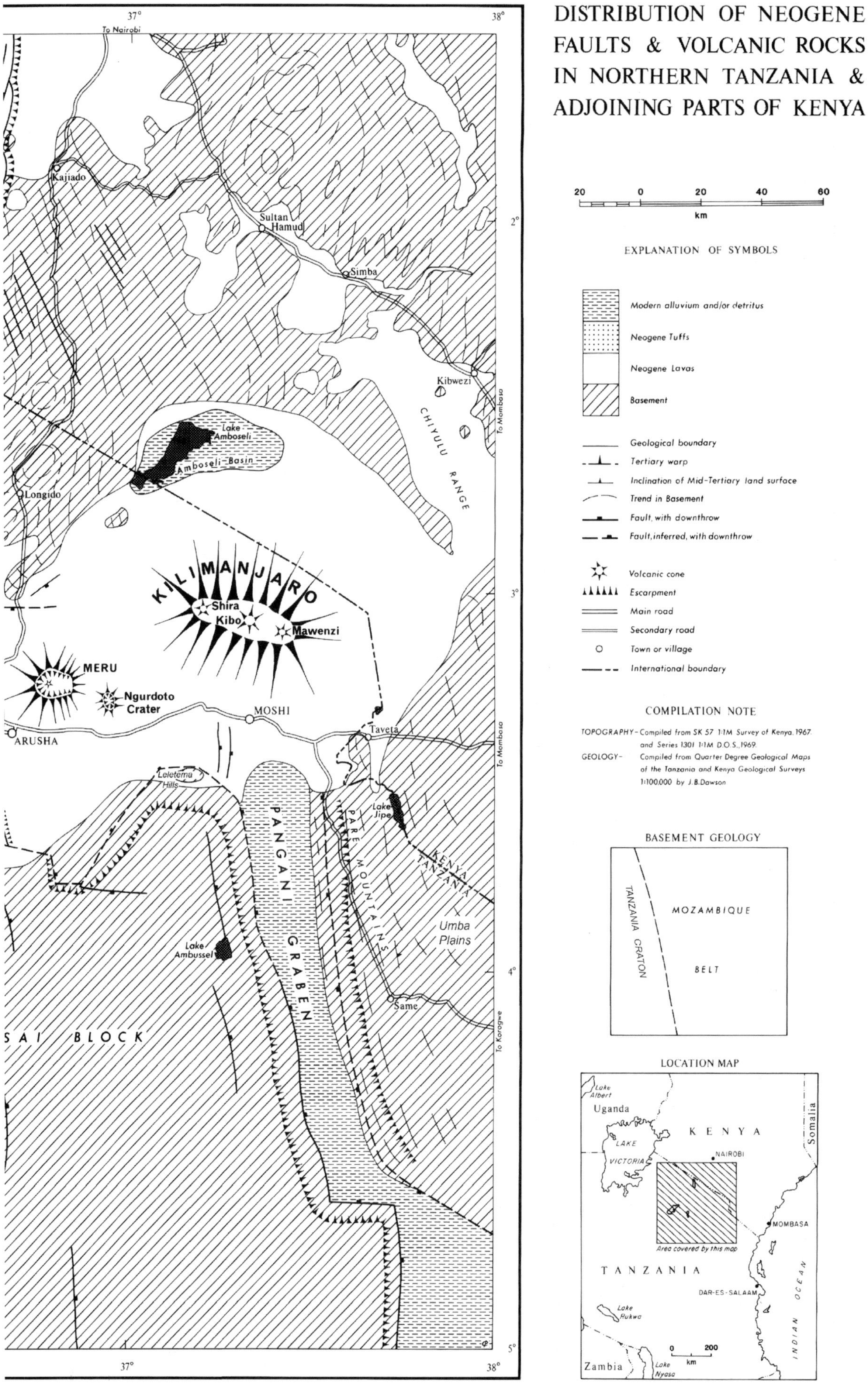


Fig. 3.1. (*Continued*)

those typical of a craton (see Chapter 4, Dawson 2008). Thus, it can be reasonably inferred that many of the Tanzanian volcanoes have erupted through Archaean cratonic lithosphere that is overlain by only a thin veneer of relatively young fold belt rocks. In the context of potential crustal contamination, this implication is important when considering the composition and isotope systematics of the volcanic rocks.

## References


BAKER, B. H., MOHR, P. A. & WILLIAMS, L. A. J. 1972. *Geology of the Eastern Rift System of Africa*. Geological Society of America, Special Publications, **36**.

BIRT, C. S., MAGUIRE, P. K. H., KHAN, M. A., THYBO, H., KELLER, G. R. & PATEL, J. 1997. The influence on pre-existing structures on the evolution of the southern Kenya Rift Valley – evidence from seismic and gravity studies. *Tectonophysics*, **278**, 211–242.

CAHEN, L., SNELLING, N. J., DELHAL, J. & VAIL, J. R. 1984. *The Geochronology and Evolution of Africa*. Clarendon Press, Oxford.

COHEN, R. S., O'NIONS, R. K. & DAWSON, J. B. 1984. Isotope geochemistry of xenoliths from East Africa: implications for the development of mantle reservoirs and their interaction. *Earth and Planetary Science Letters*, **68**, 209–220.

DAWSON, J. B. 2008. *The Gregory Rift Valley and Neogene–Recent Volcanoes of Northern Tanzania*. Geological Society, London, Memoirs, **33**.

GAME, P. M. 1954. Zoisite-amphibolite with corundum from Tanganyika. *Mineralogical Magazine*, **30**, 458–466.

GEORGE, R. M., ROGERS, N. W. & KELLEY, S. 1998. Earliest magmatism in Ethiopia: evidence for two mantle plumes in one flood basalt province. *Geology*, **26**, 923–926.

HAUZENBERGER, C. H., SOMMER, H., FRITZ, H., BAUERNHOFER, A., KRÖNER, A., HOINKES, G., WALLBRECHER, E. & THÖNI, M. 2007. SHRIMP U–Pb zircon and Sm–Nd garnet ages from the granulite-facies basement of SE Kenya: evidence for Neoproterozoic polycyclic assembly of the Mozambique Belt. *Journal of the Geological Society, London*, **164**, 189–201.

HEPWORTH, J. V. 1972. The Mozambique orogenic belt and its foreland in northeast Tanzania: a photogeologically-based study. *Journal of the Geological Society, London*, **128**, 461–500.

HOLMES, A. 1951. The sequence of pre-Cambrian orogenic belts in south and central Africa. *Proceedings of the 18th International Geological Congress, 1948*, **14**, 254–269.

KENNEDY, W. Q. 1964. The structural differentiation of Africa in the Pan-Africa (500 ± m.y.) tectonic episode. *8th Annual Report of the Research Institute of African Geology, University of Leeds, Session 1962–63*, **8**, 48–49.

KEY, R. M., CHARSLEY, T. J., HACKMAN, B. D., WILKINSON, A. F. & RUNDLE, C. C. 1989. Superimposed Upper Proterozoic collision-controlled orogenies in the Mozambique orogenic belt of Kenya. *Precambrian Research*, **44**, 197–225.

MCKIE, D. & REDFORD, A. J. 1959. Yoderite, a new hydrous magnesium iron aluminosilicate from Mautia Hill, Tanganyika. *Mineralogical Magazine*, **32**, 282–307.

MÖLLER, A., APPEL, P., MEZGER, K. & SCHENK, V. 1995. Evidence for a 2 Ga subduction zone: eclogites in the Usagaran belt of Tanzania. *Geology*, **23**, 1067–1070.

MÖLLER, A., MEZGER, K. & SCHENK, V. 1998. Crustal age domains and the evolution of the continental crust in the Mozambique Belt of Tanzania: a combined Sm–Nd, Rb–Sr and Pb–Pb evidence. *Journal of Petrology*, **39**, 749–783.

MOSLEY, P. N. 1993. Geological evolution of the late Proterozoic 'Mozambique Belt' of Kenya. *Tectonophysics*, **221**, 223–250.

MUHONGO, S. & LENOIR, J.-L. 1994. Pan-African granulite-facies metamorphism in the Mozambique Belt of Tanzania: U–Pb geochronology. *Journal of the Geological Society of London*, **151**, 343–347.

PINNA, P., COCHERIE, A., THIEBLEMONT, D., FEYBESSE, J.-L. & LAGNY, P. 1996. Évolution géodynamique du craton Est-Africain et déterminisme gitologique. *Chronique de la Recherche Minière*, **525**, 33–43.

QUENNELL, A. M., MCKINLAY, A. C. M. & AITKEN, W. G. 1956. *Summary of the Geology of Tanganyika. Part1. Introduction and stratigraphy*. Government Printer, Dar es Salaam.

SCHLÜTER, T. 1997. *Geology of East Africa*. Borntraeger, Berlin.

SHACKLETON, R. M. 1986. Precambrian collision tectonics in Africa. *In*: COWARD, M. P. & RIES, A. C. (eds) *Collision Tectonics*. Geological Society, London, Special Publications, **19**, 329–349.

SMITH, M. 1994. Stratigraphic and structural constraints on mechanisms of active rifting in the Gregory Rift, Kenya. *Tectonophysics*, **236**, 3–22.

SMITH, M. & MOSLEY, P. 1993. Crustal heterogeneities and basement influence on the development of the Kenya Rift, East Africa. *Tectonics*, **12**, 591–606.

SOMMER, H., KRÖNER, A., HAUZENBERGER, C., MUHONGO, S. & WINGATE, T. D. 2003. Metamorphic petrology and zircon geochronology of high-grade rocks from the central Mozambique Belt of Tanzania: crustal recycling of Archaean and Palaeoproterozoic material during the Pan-African orogeny. *Journal of Metamorphic Petrology*, **21**, 915–934.

# Chapter 4

## Geophysical evidence for the structure of the crust and upper mantle of the Tanzania Craton and the Gregory Rift Valley

The past three decades have seen an upsurge in geophysical fieldwork across the East African Rift Valley in Kenya (KRISP: the Kenya Rift International Seismic Project; Prodehl *et al.* 1994; Fuchs *et al.* 1997) and in Ethiopia (Project EAGLE: the Ethiopia Afar Geophysical Lithosphere Experiment; Yirgu *et al.* 2006). In northern Tanzania, in 1994, A. A. Nyblade and co-workers established an array of broadband receivers in northern Tanzania, the Tanzania Broadband Seismic Experiment, TBSE (Nyblade *et al.* 1996) that, together with interpretation of earlier commercially acquired data, have provided new insights into the present-day structure of the Tanzania Craton and the Mozambique belt. Although in many experiments data have also been acquired for the Western Rift, attention will be confined in the following section to observations on the northern Tanzania and southern Kenya sectors of the Gregory Rift Valley.

The interpretation of gravity data alone does not give a unique solution because Bouguer anomalies can reflect lateral changes in: (i) the density of the crust; (ii) the density of the upper mantle; (iii) the depth to the crust/mantle interface, the Moho; or (iv) a combination of these possible variables. Gravity measurements can, when combined with data from seismic experiments and measured rock densities, lead to integrated models for the structure of the crust and upper mantle, but it is relevant to discuss here the results obtained by individual disciplines.

### Gravity studies

Northern Tanzania lies on the southern margin of a broad, regional, negative Bouguer anomaly around $350 \pm 50$ km wide and with an amplitude of 800 g.u. relative to a background value of $-1200$ g.u. (Fig. 4.1). The anomaly, which is highest in Kenya and decreases southwestwards, lies slightly to the west of the present-day rift valley, strikes NNE and is superimposed on a wider 1000 km negative area (Fairhead 1976). This zone is isostatically compensated with respect to adjacent lithosphere, resulting in plateau uplift and high topographic relief. This is particularly apparent where the domal uplift in Kenya coincides with the most negative gravity values. The overall, regional, negative gravity anomaly has been interpreted as reflecting a zone of hot, low-density asthenosphere which has risen due to thinning of the lithosphere.

Along the axis of the rift valley in Kenya is a thin, positive (high-density) anomaly that has been attributed to either the injection of a single 10 km wide dyke (Fairhead 1976), or to a zone of dyke injection, the zone having a density ($\rho$: density in kg m$^{-3}$) of 2750 kg m$^{-3}$, compared with normal basement density of 2700 kg m$^{-3}$; the depth of the injection zone has been modelled as extending down to 22 km (Swain 1992).

In northern Tanzania, an elongate gravity low is almost continuous to the east and south side of the Sonjo–Eyasi fault (Fig. 4.1) and is interpreted as due to large thicknesses of low-density lavas that ponded against the fault escarpment (Fairhead 1976). Other large gravity lows occur in the Kitingiri Basin, to the west of the Eyasi Basin, and in the Mponde graben, south of Lake Balangida Lelu (Ebinger *et al.* 1997). Less intense, elongate gravity lows occur over the Natron and Manyara basins and at the north end of the Pangani graben. After removal of the regional negative Bouguer anomaly associated with the Kenya Dome, negative anomalies are associated with many of the individual volcanoes; as an example, the volcanic chain running from Burko to Meru, coincides with low density ($\rho$ 2100 kg m$^{-3}$) surface volcanics overlying Precambrian basement with $\rho$ 2670 kg m$^{-3}$. Within this, the Meru anomaly can be modelled as being due to a dense volcanic plug, 2 km in diameter with $\rho$ 2900 kg m$^{-3}$ extending to within 0.5 km of the surface (Fairhead 1980).

### Seismic experiments

During the late 1980s and early 1990s, seismic refraction profiles were obtained by the Kenya Rift International Seismic Project (KRISP) to investigate the crustal structure in Kenya, the main results having been reported by Prodehl *et al.* (1994) and Fuchs *et al.* (1997). Their results have been complemented in Tanzania by data from the Tanzania Broadband Seismic Experiment (TBSE) reported by Nyblade *et al.* (1996), Last *et al.* (1997), Ritsema *et al.* (1998) and Brazier *et al.* (2000). The following account draws largely on a review of these results by Nyblade (2002).

KRISP found that, along the rift axis, the depth to the Moho is 35 km in southern Kenya but shallows to 20 km northwards. Away from the rift, crustal thickness is 35–42 km for the Mozambique Belt, and 34–40 km for the Tanzania Craton. The KRISP results for the upper mantle give low Pn velocities of 7.5–7.8 km s$^{-1}$ under the rift axis, compared with 8.1–8.3 km s$^{-1}$ under the unrifted Mozambique Belt and Tanzania Craton. The change from low to high velocities is abrupt and coincides with the main border faults, indicating that thermal modification of the uppermost mantle in Kenya is confined to the rift. However, teleseismic results indicate a broader zone of thermal modification in the deeper parts of the mantle (Green *et al.* 1991; Slack *et al.* 1994).

For northern Tanzania, the TBSE results give depths to Moho of 36–39 km for the Mozambique Belt, and 37–42 km for the Tanzania Craton; mean crustal shear velocities are 3.74 and 3.79 km s$^{-1}$, respectively (Last *et al.* 1997). Further, Brazier *et al.* (2000) found Pn velocities of 8.3–8.35 km s$^{-1}$ beneath the Mozambique Belt and 8.4–8.45 km s$^{-1}$ beneath the Tanzania Craton down to depths of *c.* 200 km. Compared with Pn velocities in the range 8.0–8.6 km s$^{-1}$ in the upper mantle beneath other Archaean terranes (e.g. Finland and western Russia, King & Calcagnile 1976; the Canadian shield, Le Fevre & Helmberger 1989; the Kaapvaal craton, Simon *et al.* 2003), these velocities are at the high end of the range, suggesting denser (?colder) mantle.

### An integrated model and possible constraints

Most early models of the regional Bouguer anomaly lacked collaborative data for seismically determined crustal thicknesses or measured densities for crustal and mantle rocks from the region. Simiyu & Keller (1997) give an integrated seismic/gravity/rock-density model involving measured densities for non-volcanic crustal rocks. Gravity and seismic data suggest that, in southern Kenya away from the rift, the crust, with density 2850 kg m$^{-3}$ and a thickness of *c.* 35 km, overlies mantle lithosphere with density 3260 kg m$^{-3}$. For the anomaly within the rift, Simiyu & Keller (1997) proposed low, anomalous mantle densities of 3170–3120 kg m$^{-3}$ which are lower than the 3220 kg m$^{-3}$ of

J. B. Dawson, University of Edinburgh (*jbdawson@glg.ed.ac.uk*)
*From*: Dawson, J. B. 2008. *The Gregory Rift Valley and Neogene–Recent Volcanoes of Northern Tanzania*. Geological Society, London, Memoirs, **33**, 13–20.
0435-4052/08/$15.00 © The Geological Society of London 2008. DOI: 10.1144/M33.4

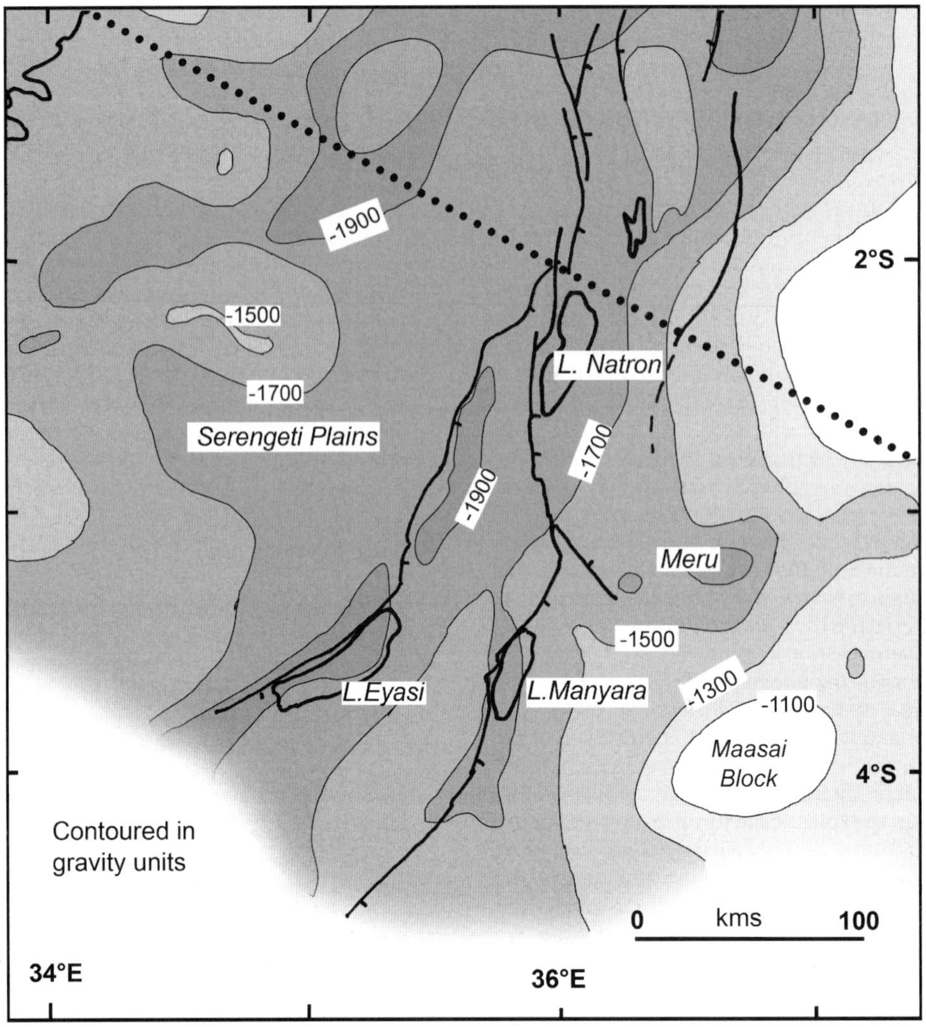


**Fig. 4.1** Bouguer anomaly map of northern Tanzania, contoured at 100 g.u. intervals (modified after Fairhead 1976).

Fairhead (1976). A north–south gravity profile from Lake Turkana to southern Tanzania (Simiyu & Keller 1997), shows that anomalous mantle is at a depth of around 35 km (i.e. at the depth of the Moho) from Lake Turkana to Nairobi but, further south, the depth to the anomaly increases rapidly from around 35 km at Lake Magadi to around 90 km south of Lake Manyara, whence it persists southwards. Moreover, this model requires that, south of Lake Magadi, the anomalous mantle underlies a wedge of normal mantle with density 3200 kg $m^{-3}$. This change in mantle structure just south of Lake Magadi coincides with the changes in tectonic style and contemporaneous magmatism noted by Dawson (1992), who also proposed thickening of the crust across a suture between the Tanzania Craton and the mobile belt, coincident with the rift valley.

Geophysical models must be assessed within possible geological and geochemical constraints. First, as discussed above, beneath a thin veneer of Mozambique Belt nappes, the Tanzania Craton most probably extends at depth some distance to the east of its surface contact with the mobile belt; thus the rift structures in southern Kenya and northern Tanzania do not coincide with the craton margin, as assumed in some gravity models (Tesha *et al.* 1997; Simiyu & Keller 1997). Second, an alternative interpretation to the proposed anomalously-light asthenosphere wedge is provided by xenoliths of metasomatized and veined mantle peridotite in the Pello Hill tuff cone just south of Lake Natron (Dawson & Smith 1988). In these lithosphere xenoliths, veins rich in mica and amphibole have metasomatized adjacent peridotite wall rock, and density varies from 3150 kg $m^{-3}$ (vein) to 3280 kg $m^{-3}$ (metasomatized peridotite) and 3350 kg $m^{-3}$ for unmodified peridotite; mixtures of these three lithosphere densities can give the 'asthenosphere' density of 3220 kg $m^{-3}$ required by the gravity models. In addition, Lee *et al.* (1999) give density 3280–3350 kg $m^{-3}$ for peridotite xenoliths from Labait. The measured density of 3350 kg $m^{-3}$ for both sets of unmodified peridotite is closer to the 3290 kg $m^{-3}$ of Fairhead (1976) than to the 3220 kg $m^{-3}$ of the Simiyu & Keller (1997) model.

The geochemical signature of basaltic rocks in the Magadi area of Kenya is another factor apparently at variance with some geophysical models. The rock compositions require them to have been derived by variable degrees of melting within the mantle garnet peridotite stability field (Le Roex *et al.* 2001). The inference is that the lavas must have formed within the lithosphere, rather than in the ambient asthenosphere or within a rising mantle plume. Hence, it is concluded that, in contrast to most gravity models, the lithospheric mantle must extend to at least 75 km beneath the axial part of the rift. However, this constraint is compatible with the gravity model of Byrne *et al.* (1997) who, having noted the continuity of two strong reflectors beneath the rift valley, argued that, although the lithosphere in the region has been thinned and stretched, it has not been disrupted or replaced by asthenosphere upwelling to depths as shallow as the Moho.

## Teleseismic results and inferences

Although the Pn wave data suggest no regional thermal anomalies in the craton and Mozambique Belt for the deeper parts of the mantle, broadband teleseismic experiments (utilizing major source regions for teleseismic earthquakes in the Hindu Kush/Pamir region and the Fiji/Tonga subduction zone) indicate mantle S-wave velocity variations on the 410 km and 660 km discontinuities in Tanzania (Ritsema *et al.* 1998; Nyblade *et al.* 2000; Nyblade 2002). The S-velocity variations derive from travel-time tomography. Higher-than-average velocities exist beneath the Tanzania Craton, whereas a low-velocity zone, with a vertical extent to depths of >400 km and laterally over a region *c.* 300 km, is found under the Mozambique Belt. As defined by the faster S-velocities, the lithospheric keel beneath the Tanzania Craton extends to a depth of *c.* 250 km (Priestley *et al.* 2006) which is in broad agreement with minimum depths of derivation of 180 km for peridotite xenoliths from the Lashaine, Labait and Igwisi Hills volcanoes (Chapter 8, Dawson 2008). In general, the 410 km and 660 km discontinuities are interpreted as being due to the transitions in $(MgFe)SiO_4$ (olivine being the dominant upper mantle phase at higher levels) from first the $\alpha$- to the $\beta$-form and second, at greater pressures, from $\gamma$-(MgFe)SiO4 to perovskite + magnesiowustite. The slopes of equilibrium phase boundaries indicate that, in regions of higher temperature, the 410 km discontinuity should be deflected downwards and the 660 km discontinuity should deflect upwards. In Tanzania, along a 4.5°S profile, where the rifting is in the Tanzania Craton, the 410 km discontinuity is depressed by 30–40 km over a zone 200–400 km wide, whereas the 660 km discontinuity is relatively flat. The topography of these discontinuities indicates that the lower-than-average velocities under this section of the rift result from elevated temperature, with the thinning of the zone between the two discontinuities corresponding to a temperature increase of 200–300 °C. The coincidence of the depressed 410 km discontinuity with the low-velocity zone indicates that the differences in the S-wave velocities between the craton and the Mozambique Belt are due to elevated temperature. Although on the basis of the S-wave parameters the thermal anomaly does not appear to extend beyond depths of 500–600 km, Rayleigh and Love wave tomography indicates the presence of a large thermal anomaly that can be interpreted as the presence of a plume head beneath the craton (Weeraratne *et al.* 2003; Sebai *et al.* 2006). The anomaly becomes diffuse at around 280 km depth suggesting that the Tanzania cratonic keel is a shallower structure than the West Africa, Congo and Kalahari cratons (Sebai *et al.* 2006).

In a review of seismic anisotropy beneath continental interiors in general, Fouch & Rondenay (2006) report evidence for multiple layers of seismic anisotropy in the area around the Tanzania Craton, which they attribute to structural complexity. Around the edges of the craton, shear waves have fast polarization directions that mimic both the shape of the craton and the seismically detected lithospheric keel, whilst beneath the craton itself there is more complexity in splitting behaviour. Lithospheric anisotropy is clearly present, and the complexity in some areas suggests a component of sub-lithospheric anisotropy. Kendall *et al.* (2006) suggested that, at least in Ethiopia, there is a coincidence between zones of seismic anisotropy and melt migration.

## Seismicity

The distribution of earthquake epicentres in East Africa shows that they are largely confined to the present-day rift valleys or their immediate surroundings, but there are more large-magnitude earthquakes in the Western Rift than in the Gregory Rift (Wohlenberg 1969; Kebede & Kulhanek 1991).

Båth (1975) lists 319 earthquakes of magnitude >5 in the region bounded by latitudes 2°N and 12°S and 28°E and 40°E during the period December 1910 to December 1973. In the area covering the Gregory Rift Valley and its surrounding area in northern Tanzania (2–5°S and 34–38°E), there were 51 earthquakes of magnitude >5, two being of magnitude 6.0 and 5.9 (Fig. 4.2). Another, of magnitude 7.2 on 7 May 1964, caused damage and casualties in the Mbulu area, and resulted in a change in the shape of Lake Babati, though no surface ruptures were observed (Whittingham 1964). Five seismological studies analysing this event (summarized by Nyblade & Langston 1995) largely agreed that the best estimate for the focal depth of this earthquake was 31 ± 3 km and that the focal mechanism indicated strike-slip faulting, unusual in an area of predominantly normal faulting. Many smaller earthquakes testify to continuous movement in the crust beneath the Gregory Rift. Rykounov *et al.* (1972) reported up to 30 events per day during 1968–69, all between magnitude 0.5 and 3.5 and all originating in the crust. Seismicity was relatively high around Ngorongoro, Oldoinyo Lengai and the southern end of Lake Manyara. Nyblade *et al.* (1996) reported some 2000 events in northeastern Tanzania between June 1994 and June 1995; most nucleated in the upper crust but several occurred at lower crustal depths of >15–20 km. The events were mainly concentrated in two swarms, one in the rifted area near the south end of Lake Manyara, and another, shallower swarm, to the south of Kondoa in an area with no significant surface expressions of rifting; the Kondoa area data are interpreted as reflecting southward propagation of the rift. More recent tremors in June and July 2007, with magnitudes varying between 4.0 and 5.94, had their centres in the area immediately to the east of Gelai and Ketumbeine, SE of Lake Natron (U.S.G.S. Earthquakes Hazards Program, Denver).

Foster & Jackson (1998) have analysed data for two Tanzanian earthquakes occurring in May 1990. Both had focal depths of *c.* 5 km and have epicentres just west of the Manyara basin boundary-escarpment, and were most probably related to the rifting. The east-dipping nodal plane of one event (900515a—magnitude 5.6) strikes north–south and is interpreted as the fault plane. The nodal plane of the second event (900515b) is east–west, not parallel to faults exposed at the surface but rather to the Archaean trend. On a continental scale, slip vectors analysed by Foster & Jackson (1998) for 84 major seismic events in eastern Africa broadly agree with a best-fit plate model that predicts an overall east–west extension between the Nubian and Somalia plates (Jestin *et al.* 1994) (see Dawson 2008, fig. 1.1), but the analysis shows that, on a regional scale, the extension is more NE–SW in Kenya and Tanzania, compared with NW–SE in Malawi and southern Africa. This concurs with NE–SW extension derived from analysis of fault directions in northern Tanzania (Dawson 1964; Ring *et al.* 2005).

For earthquakes generated below 20 km, Nyblade & Langston (1995) conclude that, to be brittle enough to deform seismogenically, the lower crust in East Africa must be cooler than predicted by some earlier models. The inference is that this is due to lower abundances of radiogenic elements, and hence the lithology must be of mafic composition, as predicted by Shudovsky *et al.* (1987). Although it is established that seismicity occurs throughout both the lower and upper crust, Foster & Jackson (1998) showed that: (i) parts of the East African crust have a seismogenic thickness of up to 35 km; and (ii) heat flow calculations suggest temperatures of *c.* 325–475 °C at 35 km depth, again requiring that the lower crust must have an anhydrous, mafic bulk composition. The corollary is that significantly more heat production must come from the upper crust, and it also raises the question of the formation of a mafic lower crust. Although magmatic underplating is a possible explanation (as suggested by Shudovsky *et al.* 1987), it is open to question whether this transfer of mafic material was achieved during the Cenozoic magmatism when the underplating might have been more closely restricted to the volcanogenic zones, or whether it was during earlier more widespread magmatism associated with the Proterozoic dyke swarms of the Mbulu Plateau (Nyblade & Langston 1995). Some geological constraint is provided by deep crustal granulite xenoliths in the Lashaine

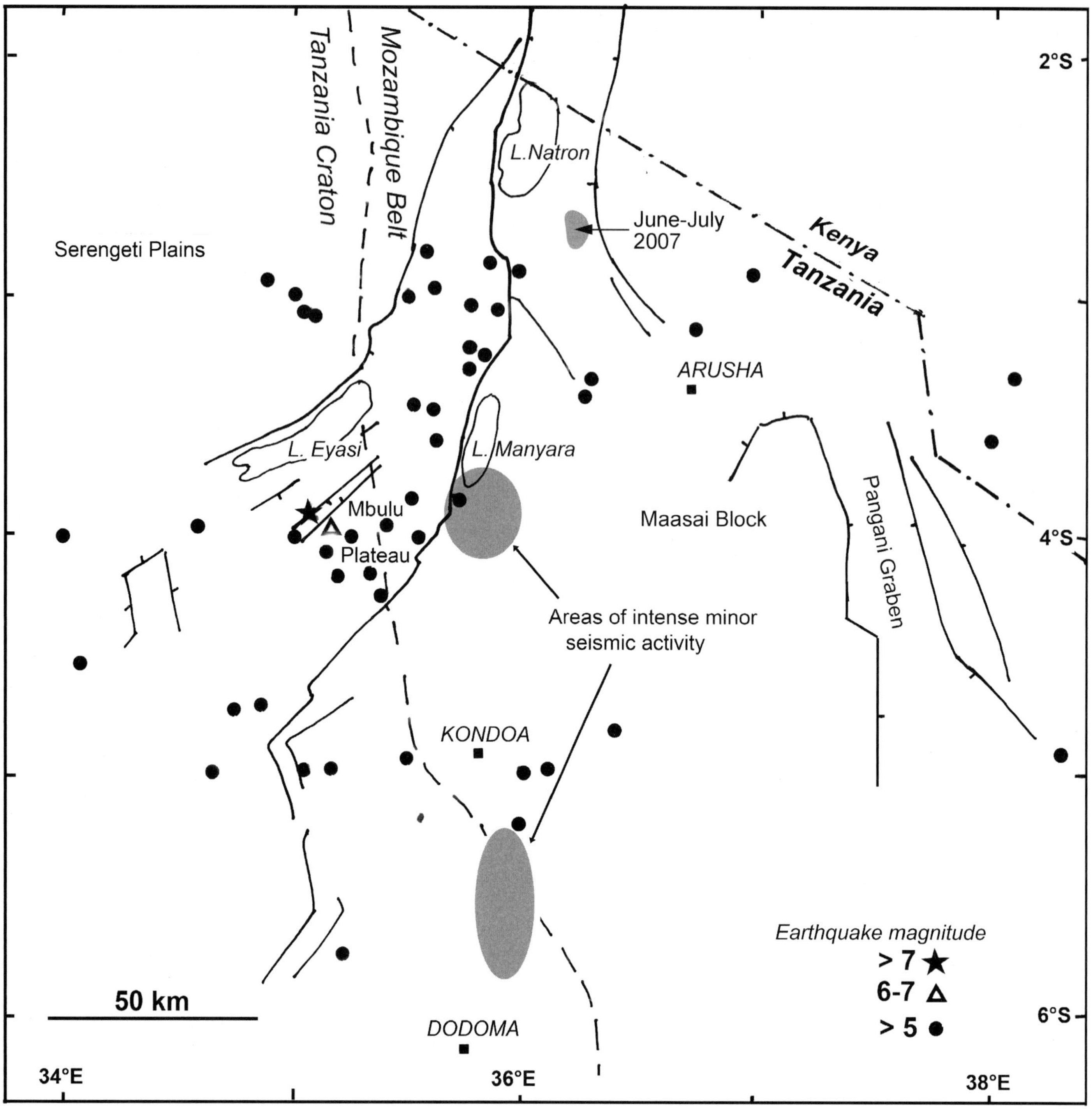


**Fig. 4.2** Epicentres of major (>5 magnitude) earthquakes in northern Tanzania between 1951 and 1965 (data from Båth 1975). Locations of the two swarms of minor earthquakes, occurring between 1994 and 1996, are from Nyblade *et al.* (1996).

tuff cone (Dawson 1964; Jones *et al.* 1983) which are of mafic to intermediate bulk composition (Dawson 1977); if they originated by underplating, their age of 2.0 Ga (Cohen *et al.* 1984) would favour a Palaeoproterozoic event.

Nyblade *et al.* (1996) drew attention to the fact that, in contrast to the large number of earthquakes in the Eyasi and Manyara areas, there are few events associated with the Pangani graben. One, in May 1989, in the Pare Mountains on the east side of the Pangani graben, had a source depth of 28 km and the focal mechanism for a normal fault with a WNW–ESE or east–west fault plane, neither nodal plane coinciding with the NNW orientation of the graben boundary faults. Le Gall *et al.* (2004), citing some seismic activity and an alleged small tuff eruption in the Pare Mountains in 1998, suggested that the Pangani graben is still tectonically active. However, in the context of possible directions for the further propagation of the Gregory Rift (i.e. southwestwards into the Tanzania Craton, southwards along the craton/mobile belt suture, or southeastwards via the Pangani graben to link up with the Davie Fracture zone offshore (Mougenot *et al.* 1986), the relatively sparse seismic activity appears to relegate the Pangani Rift to a minor role.

## Heat flow

In Kenya, heat flow on the western and eastern flanks of the rift (the latter in the Mozambique Belt) is 40–50 and

40–60 $mWm^{-2}$ respectively, whereas on the rift floor values range between 50 and 100 $mWm^{-2}$ (Wheildon *et al.* 1994).

In northern Tanzania, the average of eight heat flow measurements on the Tanzania craton is 34 $mWm^{-2}$ (range 21–47) (Nyblade *et al.* 1990), and a further measurement on the craton, south of Lake Nyanza, gave a heat flow of 33 $mWm^{-2}$ (Nyblade 1997), close to the average. Nyblade *et al.* (1990) give an average value of 47 $mWm^{-2}$ (range 39–62) for five measurements in the Mozambique Belt, these values being very similar to those for eastern Kenya. Although sites at Basotu (on the Mbulu Plateau), and Arkatan (near Makuyuni), are close to major faults associated with the most recent rifting, the heat flow recorded at these sites is none the less lower than values of between 51 and 87 $mWm^{-2}$ recorded on the Tanzanian offshore islands of Pemba and Zanzibar, which are generally held to be close to the eastern rifted margin of East Africa. Only one measurement of 62 $mWm^{-2}$, at Holili near Moshi, lies within the coastal heat flow range; this figure also distorts the average value for the Mozambique Belt. In detail, the heat flow at some sites in the mobile belt (e.g. Longido 44, Arkatan 39, and Olboloti (near Kondoa) 41 $mWm^{-2}$, respectively) is similar to, or even lower than, some sites on the craton (Basotu 46, and Kizaga 47 $mWm^{-2}$). In view of the probability that much of the western margin of the Mozambique Belt is underlain at quite shallow depths by cratonic rocks (Chapter 2), this convergence is perhaps not surprising. The data for world-wide Archaean cratons and adjacent Proterozoic terrains that have not been affected by later tectonism show low heat flow of around 41 $mW\ m^{-2}$ (Nyblade & Pollack 1993), comparable to that for the Tanzania Craton and fold belt. This comparison suggests that the absence of a broad, heat flow anomaly in Tanzania and also in Kenya away from the rift (which is also in agreement with the seismic evidence) indicating that the crust and uppermost mantle have not been subject to significant thermal modification during the Tertiary–Recent tectonism.

Moreover, geothermobarometric estimates of upper mantle garnet lherzolite xenoliths from Igwisi (40 $mWm^{-2}$; Dawson 1994) on the Tanzania Craton and at Lashaine (*c.* 45 $mWm^{-2}$; Henjes-Kunst & Altherr 1992) lie on a typical cratonic geothermal gradient, whereas xenoliths from the contemporaneous Labait volcano (*c.* 55 $mW\ m^{-2}$; Dawson *et al.* 1997), just south of Hanang and also lying on the craton, derive from hotter lithosphere. The Labait data suggest that a thermal anomaly, not yet expressed in surface heat flow, is present in the upper mantle beneath the southerly-propagating tip of the rift (Dawson *et al.* 1997; Dawson 1997).

The ambient thermal state of the sub-Tanzania mantle is exemplified by the peridotite xenoliths from Lashaine that mainly document temperatures of 1045 $\pm$ 50 °C at a pressure of 40 $\pm$ 4 GPa, plotting along a *c.* 45 $mWm^{-2}$ model geotherm. In contrast, xenoliths from the Chyulu Hills (50 km north of Kilimanjaro and of approximately the same age as Lashaine) plot along a 60 $mWm^{-2}$ geotherm (Henjes-Kunst & Altherr 1992). This points to differences in the thermal state of the upper mantle between southern Kenya and northern Tanzania, again underlining the structural and petrological differences between the two areas (Dawson 1992).

## Magnetic data

Aeromagnetic anomaly data give higher resolution images of shallow crustal structure than those provided by gravity data. Recent surveys show, across the Archaean area of the Mbulu Plateau, narrow NE- and NNW trending anomalies or broad east–west anomalies that correspond to dyke swarms and greenstone belts, respectively. In contrast, the Maasai Block is characterized by NE and north–south magnetic lineaments that, together with a relatively positive Bouguer anomaly, suggest that it is a discrete structural unit (Ebinger *et al.* 1997).

Magnetic signals are masked or subdued in the rift valley basins and on the Serengeti, indicating that the magnetic basement lies deeper. Using the 3D Euler deconvolution technique, which provides a means of estimating the depth and orientation of magnetic sources, Ebinger *et al.* (1997) found that shallow Euler solutions correspond to greenstone belts, lithological contacts and and shear zones on the exposed basement, whereas deeper solutions (>3 km) occur over basin gravity highs or areas of steep magnetic gradients. Intermediate solutions (1–3 km) correlate with the subsurface projection of Neogene normal faults bounding the rift valley basins, and also some basement structures. Analysis of clusters of Euler solutions indicate the following: (i) circular clusters of solutions up to 3 km deep correspond to most of the Neogene volcanoes (Lemagrut, Sadiman, Ngorongoro, Loolmalasin, Kerimasi, Oldoinyo Lengai, Essimingor, Tarosero, Meru & Hanang); however, Gelai and Kwaraha have no magnetic sources; (ii) a cluster of solutions coincides with a gravity high that separates the east and west Eyasi basins; depth to the magnetic basement in the east Eyasi basin is <2 km; (iii) the most continuous solution clusters in the Natron–Manyara–Balangida basins are at the subsurface projections of the southern Natron, central Manyara and Mponde graben border faults. Other aligned solutions correlate with the Engaruka Basin fault at its intersection with the Natron boundary fault, and also along faults on the east side of Lake Natron. Euler solution alignments also suggest east–west structures cross-cutting the rift beneath Lake Manyara, these solutions coinciding with a zone of high seismicity (Nyblade *et al.* 1996) and the inferred southern boundary of the late Tertiary tectonic depression (see Chapter 3, Dawson 2008); (iv) in the Eyasi basin, which overlies the mapped interface between the Tanzania Craton and the Mozambique Belt metamorphic rocks, there is no correlation with Euler solutions, which argues against a steep contact between the two units and against differences in magnetic properties between the two tectonic units; and (v) maximum depths for the rift valley basins from Euler deconvolution are: W. Eyasi 3.5 km, E. Eyasi <2 km, Natron 3.3 km. Manyara 2.9 km, Balangida <1 km.

## Magnetotelluric data

Magnetotelluric data collected in southern Kenya, along a west–east profile from Lake Nyanza via Lake Magadi to the Chyulu Hills, show a sharp break from low conductivity on the western flank of the rift valley to high conductivity in the rift valley immediately to the east of the Nguruman escarpment. This is interpreted as due to the high conductivity of sediments within the rift. In the area to the east of Lake Magadi, an eastward change towards lower conductivity is gradual, coinciding with the more gentle step faulting (Simpson *et al.* 1997). The magnetotelluric evidence does not support the hypothesis, derived from seismic data, that melts are present in the upper mantle beneath the rift (Simpson 2000). To date, no magnetotelluric data are available for the rift valley in northern Tanzania.

## Lithosphere strength

Estimates of the effective plate thickness (*Te*) can place some constraint on models of crust and upper mantle structure derived from gravity studies. Using a mean crustal density of 2800 $kg\ m^{-3}$ and 3200 $kg\ m^{-3}$ for the mantle, Ebinger *et al.* (1997) derived the following *Te*: for craton lithosphere: 64 $\pm$ 5 km; for Mozambique mobile belt lithosphere: 30 $\pm$ 3 km; for lithosphere beneath the Natron–Manyara–Balangida section of the rift: 23 km; and beneath the Eyasi rift: >30 km. The estimates for the craton and mobile belt are typical of Precambrian lithosphere, and are in agreement with the gravity and heat flow data for unmodified lithosphere away from the rift, but for perturbed lithosphere beneath the rift itself.

## Crustal extension and its causes

The East African rifting is generally accepted as a reflection of perturbation in the upper mantle and it is a matter for debate whether this perturbation, manifest in the more recent Tanzania magmatism and giving rise to a southerly-propagating lithosphere fracture system, might be linked to easterly movement of the Somalia microplate relative to the larger Nubian Plate or whether the African Plate as a whole is drifting north over a mantle plume, the buoyancy of which may have caused crustal stretching. Domal uplift, a potential cause of lithosphere extension, is small in northern Tanzania compared with that in Kenya and Ethiopia. Contrasting with the generally held view that the underlying cause for the extension arises in the mantle, Coblenz & Sandiford (1994) suggested that large extensional stresses in the elevated regions of Ethiopia and Kenya arise from lithospheric density variations; however, they did not address the fundamental question of why these regions are in fact elevated, an explanation for which is given in the African Superswell model of Nyblade & Robinson (1994).

Implicit in the concept of rifting is the belief that there must be horizontal extension, and various models have arrived at differing estimates. Plate tectonic modelling, based on poles of rotation and angular separation of the Red Sea and the Gulf of Aden gave estimates of crustal separation in Kenya of 30 km (McKenzie *et al.* 1970) whereas Royer *et al.* (2006), again using poles of rotation, proposed that eastward drift of the Somalia Plate has caused $129 \pm 62$ km extension at the northern tip of the eastern African Rift over the past 11 Ma. Calais *et al.* (2006), using both earthquake slip vector data and GPS (Global Positioning Systems) data and, assuming constant rates of extension since initiation, estimated extension in the Western Rift of 30 km over *c.* 7 Ma and of 8 to 0 km (varying from north to south) in the Eastern Rift over 5 Ma. These estimates compare with 15 km extension for the Western Rift calculated from surface fault geometries (Morley 1988), and 15 km for the Eastern Rift in Kenya (based on a width of 10 km for the width of dense intrusion obtained from interpretation of the gravity high, plus lateral extension of *c.* 5 km due to normal faulting (Baker & Wohlenberg 1971; Fairhead 1976). Although there is disparity in these estimates, there is consensus that extension is greatest at the northern end of the rift valley system and that the extension, at least in the case of Kenya, is too large to be entirely due to domal uplift (Baker & Wohlenberg 1971). Moreover, domal uplift is minimal in northern Tanzania compared with that in Kenya and Ethiopia.

In the specific case of northern Tanzania, inclusions of metasomatized granitic and basic crustal rocks (fenites) in the ejectamenta from the Oldoinyo Lengai volcano (Dawson 1962) show that crustal rocks still underlie the median axis of the rifted area, but the presence of dykes and veins in peridotite blocks from Pello Hill, just south of Lake Natron (Dawson & Smith 1988) provides evidence for brittle-fracture extension at mantle depths.

## Summary of geophysical work

On a regional scale, gravity and seismic studies in northern Tanzania show that, away from the rifted zones, the thickness of the crust of the Tanzania Craton and the Mozambique mobile belt is similar, with depths to Moho of 37–42 and 36–39 km, respectively. Mean crustal shear velocities are 3.74 and 3.79 km $s^{-1}$ respectively. The depth to the base of the lithosphere beneath the Tanzania Craton is 150–200 km and, combined with evidence that the effective elastic strength and heat flow is similar to Precambrian terrains world-wide, the inference is that, unlike the case in Kenya, the lithosphere has not been substantially thinned during Neogene tectono-thermal events. However, there is seismic evidence for an upper mantle thermal anomaly beneath the Tanzania Craton which, if plume-related, is a possible explanation for the isostatic uplift of the East African Plateau (Birt *et al.* 1995); this itself is part of the anomalously high plateau area extending over much of southern Africa, the African Superswell (Nyblade & Robinson 1994), which is attributed to upwelling mantle flow from the core–mantle boundary (Lithgow-Bertelloni & Silver 1998).

In the immediate vicinity of the rift, gravity lows coincide with volcanic centres and thick sediments in half graben basins. Seismicity is most pronounced in connection with faults cutting the Tanzania Craton, and minor earthquake swarms and centres to the south of the present rift structure indicate southwards propagation of the rift. Earthquake depths suggest that their epicentres are located in both the lower and the upper crust, and that the lower crust must be of anhydrous mafic composition. Some exceptionally deep events originate in the mantle which, by inference, must be atypically cold and brittle. Although heat flow measurements along the rift are not exceptionally high, there are numerous hot springs in the Natron, Manyara and Eyasi basins, and in the area to the south of the main rifting.

## References


Baker, B. H. & Wohlenberg, J. 1971. Structure and evolution of the Kenya Rift Valley. *Nature*, **229**, 538–542.

Båth, M. 1975. Seismicity of the Tanzania region. *Tectonophysics*, **27**, 358–379.

Birt, C. S., Maguire, P. H. K., Khan, M. A. & Thybo, H. 1995. A combined interpretation of the KRISP 94 seismic and gravity data; Evidence for a mantle plume beneath the East African plateau. *EOS, Transactions of the American Geophysical Union* (abstracts), **76**, F 607–608.

Brazier, N. A., Nyblade, A. A., Langston, C. A. & Owens, T. J. 2000. Pn wave velocities beneath the Tanzania craton and adjacent mobile belts, East Africa. *Geophysical Research Letters*, **27**, 2365–2368.

Byrne, G. F., Jacob, A. W. B., Mechie, J. & Dindi, E. 1997. Seismic structure of the upper mantle beneath the southern Kenya Rift from wide-angle data. *Tectonophysics*, **238**, 243–260.

Calais, E., Ebinger, C., Hartnady, C. & Nocquet, J. M. 2006. Kinematics of the East African Rift from GPS and earthquake slip vector data. *In*: Yirgu, G., Ebinger, C. Y. & Maguire, K. H. (eds) *The Afar Volcanic Province within the East African Rift System.* Geological Society, London, Special Publications, **259**, 9–22.

Coblentz, D. D. & Sandiford, M. 1994. Tectonic stresses in the African plate: Constraints on the ambient lithospheric stress state. *Geology*, **22**, 831–834.

Cohen, R. S., O'Nions, R. K. & Dawson, J. B. 1984. Isotope geochemistry of xenoliths from East Africa: implications for the development of mantle reservoirs and their interaction. *Earth and Planetary Science Letters*, **68**, 209–220.

Dawson, J. B. 1962. The geology of Oldoinyo Lengai. *Bulletin Volcanologique*, **24**, 349–387.

Dawson, J. B. 1964. *Monduli.* Tanzania Geological Survey, Quarter Degree Sheet **54**.

Dawson, J. B. 1977. Sub-cratonic and upper mantle models based on xenolith suites in kimberlite and nephelinitic diatremes. *Journal of the Geological Society, London*, **134**, 173–184.

Dawson, J. B. 1992. Neogene tectonics and volcanicity in the North Tanzania sector of the Gregory Rift Valley: contrasts with the Kenya sector. *Tectonophysics*, **204**, 81–92.

Dawson, J. B. 1994. Quaternary kimberlitic volcanism on the Tanzania craton. *Contributions to Mineralogy and Petrology*, **116**, 473–485.

DAWSON, J. B. 1997. A thermal anomaly in the upper mantle beneath the southerly-propagating tip of the East African rift valley: evidence from mantle xenoliths in northern Tanzania. *In:* JACOB, A. W. B., DELVAUX, D. & KHAN, M. A. (eds) *Lithospheric Structure, Evolution and Sedimentation in Continental Rifts*. Communications of the Dublin Institute for Advanced Studies, Ser.D., Geophysical Bulletin, **48**, 29–31.

DAWSON, J. B. 2008. *The Gregory Rift Valley and Neogene–Recent Volcanoes of Northern Tanzania*. Geological Society, London, Memoirs, **33**.

DAWSON, J. B. & SMITH, J. V. 1988. Veined and metasomatized peridotites from Pello Hill, Tanzania: evidence for anomalously light mantle beneath the Tanzania sector of the eastern Rift Valley. *Contributions to Mineralogy and Petrology*, **100**, 510–527.

DAWSON, J. B., JAMES, D., PASLICK, C. & HALLIDAY, A. M. 1997. Ultrabasic potassic low-volume magmatism and continental rifting in north–central Tanzania: association with enhanced heat flow. *Russian Geology and Geophysics*, **38**, 69–81.

EBINGER, C., POUDJOM DJOMANI, Y., MBEDE, E., FOSTER, A. & DAWSON, J. B. 1997. Rifting Archaean lithosphere: the Eyasi–Manyara–Natron rifts, East Africa. *Journal of the Geological Society, London*, **154**, 947–960.

FAIRHEAD, J. D. 1976. The structure of the lithosphere beneath the eastern rift, East Africa, deduced from gravity studies. *Tectonophysics*, **30**, 269–298.

FAIRHEAD, J. D. 1980. The structure of the cross-cutting volcanic chain of northern Tanzania and its relation to the East African rift system. *Tectonophysics*, **65**, 193–208.

FOSTER, A. & JACKSON, J. J. 1998. Source parameters of large African earthquakes: implications for crustal rheology and regional kinematics. *Geophysical Journal International*, **134**, 422–448.

FOUCH, M. J. & RONDENAY, S. 2006. Seismic anisotropy beneath stable continental interiors. *Physics of the Earth and Planetary Interiors*, **158**, 292–320.

FUCHS, K., ALTHERR, R., MULLER, S. & PRODEHL, C. (eds) 1997. Structure and dynamic processes in the lithosphere of the Afro-Arabian rift system. *Tectonophysics*, **278**, 1–352.

GREEN, W. V., ACHAUER, U. & MEYER, R. P. 1991. A three-dimensional seismic image of the crust and upper mantle beneath the Kenya rift. *Nature*, **354**, 199–203.

HENJES-KUNST, F. & ALTHERR, R. 1992. Metamorphic petrology of xenoliths from Kenya and northern Tanzania and implications for geotherms and lithospheric structures. *Journal of Petrology*, **33**, 1125–1156.

JESTIN, F., HUCHON, P. & GAULIER, J. M. 1994. The Somali plate and the East African rift system: present day kinematics. *Geophysical Journal International*, **116**, 637–654.

JONES, A. P., SMITH, J. V., DAWSON, J. B. & HANSEN, E. C. 1983. Metamorphism, partial melting and K-metasomatism of garnet–scapolite–kyanite granulite xenoliths from Lashaine, Tanzania. *Journal of Geology*, **91**, 143–165.

KEBEDE, F. & KULHANEK, O. 1991. Recent seismicity of the East African Rift system and its implications. *Physics of the Earth and Planetary Interiors*, **68**, 259–273.

KENDALL, J. M., PILIDOU, S., KEIR, D., BASTOW, I. D., STUART, G. W. & AYELE, A. 2006. Mantle upwellings, melt migration and the rifting of Africa: insights from seismic anisotropy. *In*: YIRGU, G., EBINGER, C. Y. & MAGUIRE, K. H. (eds) *The Afar Volcanic Province within the East African Rift System*. Geological Society, London, Special Publications, **259**, 55–72.

KING, D. W. & CALCAGNILE, G. 1976. P-wave velocities in the upper mantle beneath Fennoscandinavia and western Russia. *Journal of Geophysical Research of the Royal Astronomical Society*, **46**, 407–432.

LAST, R. J., NYBLADE, A. A., LANGSTON, C. A. & OWENS, T. J. 1997. Crustal structure of the East African Plateau from receiver functions and Rayleigh wave phase velocities. *Journal of Geophysical Research*, **102**, 24469–24483.

LE GALL, B., GERNIGON, L. ET AL. 2004. Neogene–Holocene rift propagation in central Tanzania: morphostructural and aeromagnetic evidence from the Kilombero area. *Geological Society of America, Bulletin*, **83**, 490–510.

LE ROEX, A. P., SPÄTH, A. & ZARTMAN, R. E. 2001. Lithospheric thickness beneath the southern Kenya Rift: implications from basalt geochemistry. *Contributions to Mineralogy and Petrology*, **142**, 89–106.

LEE, C. T. & RUDNICK, R. L. 1999. Compositionally-stratified cratonic lithosphere: petrology and geochemistry of peridotite xenoliths from the Labait volcano, Tanzania. *In*: GURNEY, J. J., GURNEY, J. L., PASCOE, M. D. & RICHARDSON, S. H. (eds) *Proceedings of the 7th International Kimberlite Conference, Cape Town, 1998*, Vol. 2, 503–521.

LE FEVRE, L. V. & HELMBERGER, D. V. 1989. Upper mantle P-velocity of the Canadian Shield. *Journal of Geophysical Research*, **94**, 17749–17765.

LITHGOW-BERTELLONI, C. & SILVER, P. S. 1998. Dynamic topography, plate driving forces and the African superswell. *Nature*, **395**, 269–272.

MCKENZIE, D. O., DAVIES, D. & MOLNAR, P. 1970. Plate tectonics of the Red Sea and East Africa. *Nature*, **226**, 243–248.

MORLEY, C. K. 1988. Variable extension in Lake Tanganyika. *Tectonics*, **7**, 785–801.

MOUGENOT, D., RECQ, M., VIRLOGEUX, P. & LEPVRIER, C. 1986. Seaward extension of the East African rift. *Nature*, **321**, 599–603.

NYBLADE, A. A. 1997. Heat flow across the East African Plateau. *Geophysical Research Letters*, **24**, 2083–2086.

NYBLADE, A. A. 2002. Crust and upper mantle structure in East Africa: implications for the origin of Cenozoic rifting and volcanism and the formation of magmatic rifted margins. *In*: MENZIES, M., KLEMPERER, S. L., EBINGER, C. J. & BAKER, J. (eds) *Volcanic Rifted Margins*. Geological Society of America, Special Papers, **362**, 15–26.

NYBLADE, A. A. & LANGSTON, C. A. 1995. East African earthquakes below 20 km and their implications. *Geophysical Journal International*, **121**, 49–62.

NYBLADE, A. A. & POLLACK, H. N. 1993. A global analysis of heat flow from Precambrian terrains: implications for the thermal structure of Archean and Proterozoic lithosphere. *Journal of Geophysical Research*, **98**, 12207–12218.

NYBLADE, A. A. & ROBINSON, S. W. 1994. The African Superswell. *Geophysical Research Letters*, **21**, 765–768.

NYBLADE, A. A., POLLACK, H. N., JONES, D. L., PODMORE, F. & MUSHAYANDEBVU, M. 1990. Terrestrial heat-flow in east and southern Africa. *Journal of Geophysical Research*, **95**, 17371–17384.

NYBLADE, A. A., BURT, C., LANGSTON, C. A., OWENS, T. J. & LAST, R. J. 1996. Seismic experiment reveals rifting of craton in Tanzania. *EOS, Transactions of the American Geophysical Union*, **77**, 519–521.

NYBLADE, A. A., OWENS, T. J., GURROLA, H., RITSEMA, J. & LANGSTON, C. A. 2000. Seismic evidence for a deep upper mantle anomaly beneath East Africa. *Geology*, **28**, 599–602.

PRIESTLEY, K., MCKENZIE, D. O. & DEBAYLE, E. 2006. The state of the upper mantle beneath southern Africa. *Tectonophysics*, **416**, 101–112.

PRODEHL, C., KELLER, G. R. & KHAN, M. A. 1994. Crustal and upper mantle structure of the Kenya Rift. *Tectonophysics*, **236**, 1–483.

RING, U., SCHWARTZ, H. L., BROMAGE, T. G. & SANAANE, C. 2005. Kinematic and sedimentological evolution of the Manyara Rift in northern Tanzania, East Africa. *Geological Magazine*, **142**, 355–368.

RITSEMA, J., NYBLADE, A. A., OWENS, T. J. & LANGSTON, C. A. 1998. Upper mantle seismic velocity structure beneath Tanzania, East Africa: implications for the stability of cratonic lithosphere. *Journal of Geophysical Research*, **103**, 21210–21213.

ROYER, J. Y., GORDON, R. G. & HORNER-JOHNSON, B. C. 2006. Motion of Nubia relative to Antarctica since 11 Ma: Implications for the Nubia–Somalia, Pacific–North America and Indi-Eurasia motion. *Geology*, **34**, 501–504.

RYKOUNOV, L. N., SEDOV, V. V., SAVRINA, L. A. & BOURMIN, V. J. 1972. Study of micro-earthquakes in the rift zones of East Africa. *Tectonophysics*, **15**, 123–130.

SEBAI, A., STUTZMAN, E., MONTAGNER, J. P., SICILIA, D. & BEUCLER, E. 2006. Anisotropic structure of the African upper mantle from Rayleigh and Love wave tomography. *Physics of the Earth and Planetary Interiors*, **155**, 48–62.

SHUDOFSKY, G. N., CLOETINGH, S., STEIN, S. & WORREL, R. 1987. Unusually deep earthquakes in East Africa constraints on the thermo-mechanical structure of a continental rift system. *Geophysical Research Letters*, **14**, 741–744.

SIMIYU, S. M. & KELLER, G. R. 1997. An integrated analysis of lithospheric structure across the East African plateau based on gravity anomalies and recent seismic studies. *Tectonophysics*, **278**, 291–313.

SIMON, R. E., WRIGHT, C., KWADIBA, M. T. O. & KGASWANE, E. M. 2003. Mantle structure and composition to 800-km depth beneath southern Africa and surrounding oceans from broadband body waves. *Lithos*, **71**, 353–367.

SIMPSON, F. 2000. A three-dimensional electromagnetic model of the southern Kenya Rift: Departure from two dimensionality as a possible consequence of a rotating stress field. *Journal of Geophysical Research*, **105**, 19 321–19 334.

SIMPSON, F., HAAK, V., KHAN, M. A., SAKKAS, V. & MEJU, M. 1997. The KRISP magnetotelluric survey of 1995: first results. *Tectonophysics*, **278**, 261–271.

SLACK, P. D. & DAVIES, P. M. & the KRISP Teleseismic Working Group 1994. Attenuation and velocity of P-waves in the mantle beneath the East African rift, Kenya. *Tectonophysics*, **236**, 331–358.

SWAIN, C. J. 1992. The Kenya rift axial high: a re-interpretation. *Tectonophysics*, **204**, 59–70.

TESHA, A. L., NYBLADE, A. A., KELLER, R. A. & DOSER, D. I. 1997. Rift localization in suture-thickened crust: evidence from Bouguer gravity anomalies in north-eastern Tanzania, East Africa. *Tectonophysics*, **278**, 315–328.

WEERARATNE, D. S., FORSYTH, D. W., FISCHER, K. M. & NYBLADE, A. A. 2003. Evidence for an upper mantle plume beneath the Tanzanian craton from Rayleigh wave tomography. *Journal of Geophysical Research*, **108** (B9), 2427, doi: 10.1029/2002JB002273.

WHEILDON, J., MORGAN, P., WILLIAMSON, K. H., EVANS, T. R. & SWANBERG, C. A. 1994. Heat flow in the Kenya rift zone. *Tectonophysics*, **236**, 131–149.

WHITTINGHAM, J. K. 1964. Report on the earth tremor of 7 May 1964. *Geological Survey of Tanzania, Unpublished Report* JKW/ **34**.

WOHLENBERG, G. J. 1969. Some remarks on the seismicity of East Africa between 4°N–12°S and 23°E–40°E. *Tectonophysics*, **8**, 562–577.

YIRGU, G., EBINGER, C. Y. & MAGUIRE, K. H. (eds) 2006. *The Afar Volcanic Province within the East African Rift System*. Geological Society, London, Special Publications, **259**.

# Chapter 5

# Tectonic development of the rift structures

During the Neogene–Recent there have been two major phases of crustal deformation in northern Tanzania. Each deformation event has been followed by a major pulse of volcanism arising from instabilities in the mantle.

The chronology of the development of the rift faulting has been made possible by the dating of volcanic rocks that erupted before and after particular episodes of faulting. Appendix 1 (Dawson 2008) lists the current dating results for the volcanic rocks. However, most of the dates were obtained many years ago by the K–Ar method and more refined stratigraphy will doubtless arise from Ar/Ar dating in the future.

## Early rift structure

Whereas in Kenya, there is evidence for an elongate pre-rift depression in which the earliest volcanics were deposited (Baker 1986), there is at present only sparse evidence for the presence of pre-volcanic sedimentary basins in northern Tanzania. The limited evidence comes from an exposure at the base of the rift escarpment at the north end of Lake Manyara, where a sedimentary formation, termed the Manyara Group, comprises a boulder conglomerate unconformably overlying metamorphic basement rocks. The conglomerate itself is overlain by siltstones, sandstones, waterlain tuffs and ashes with an intercalated basalt that has been dated at $4.86 \pm 0.24$ Ma. This is taken to indicate that, in this area, limited basin subsidence (? related to faulting) and basaltic volcanism had begun by around 4.9 Ma (Foster 1997; Foster *et al.* 1997). This shallow-basin formation follows the oldest volcanism in the area, the eruption of Essimingor lying 25 km to the east which has rocks dated between 8.1 and 3.2 Ma. In some respects this is analogous to western Kenya, where Miocene basin formation was preceded by the eruption of highly explosive volcanoes such as Mt Elgon and Napak (Baker *et al.* 1972).

The main evidence for the earliest phase of Tertiary crustal deformation in East Africa derives from studies of the shape and altitude of a widespread erosion surface of sub-Miocene age (Saggerson & Baker 1965; Baker *et al.* 1972). In Kenya this warped land surface rises to an elliptical dome >1500 m above sea level (a.s.l.) in the Mau and Aberdare Ranges (Fig. 5.1). This symmetrical dome is bisected by the narrow graben of the present-day rift valley. In northern Tanzania, due to the effect of the Tanzania Craton, the shape of the deformed land surface is less regular. The surface is at 1500 m a.s.l. around Longido and decreases in elevation both westwards towards the rift valley and southeastwards towards Kilimanjaro and the coast. On the Maasai Block, it is around 1200 m but decreases in height westward towards the rift valley, and also northwards and eastwards. The eastern margin of the Maasai Block is still at around 900 m where it forms the western side of the Pangani graben and the northerly drop, together with decrease in height from Longido, have combined to form the 'Kilimanjaro Depression' (Downie & Wilkinson 1972). The surface on the Pare Mountains, to the east of the Pangani graben, is at >1800 m above the steep, west-facing escarpment and decreases gently eastwards towards the Umba Plains (McConnell 1972). At the eastern margin of the Mbulu plateau, the surface is at >1800 m but it decreases in height northwestwards to 1200 m at Lake Eyasi. On the upthrow-side of the Eyasi scarp it is again at >1800 m, as further north in the Loliondo area; this section of the surface then slopes decreasingly in height westwards across the Serengeti Plains towards Lake Nyanza.

## The northern Tanzania late Tertiary depression

The earliest faulting in East Africa follows the deformation of the sub-Miocene land-surface. In Kenya, following a period of down-warping, the first effect was a southerly extension of a pre-existing fault (the Turkwell fault) to create an asymmetric graben, faulted on its western side, running from Lake Turkana to Lake Eyasi (Baker *et al.* 1972). Later faulting created the trough-shaped graben around 60 km wide that cuts across the updomed land surface in Kenya. In northern Tanzania, due to the effects of the Tanzania Craton, there is a wider area of faulting, termed the North Tanzania Divergence (Baker *et al.* 1972). Dawson (1992) has proposed that the earliest post-Miocene faulting gave rise to a wide, tectonic depression, bounded by faults or by downwarps of the sub-Miocene land-surface (Fig. 5.2). The depression is bounded to the south by the Maasai (or Kondoa) Block around which the area of tectonic disturbance bifurcates continuing to the SW as the NE-trending Eyasi half graben, and to the SE as the NNW-trending Pangani graben. The structure of the north-eastern margin of the depression is a downwarp with minor faulting, that has given rise to a series of benches, the highest coinciding with the unfaulted but tilted peneplain. The breaks between the benches are old, eroded fault escarpments first noted by Uhlig (1907*a*, *b*) and remarked upon by Willis (1936). The structure of the northern margin is speculative (?faulted or downwarped) as the structure is obscured by the later Meru and Kilimanjaro volcanics.

The north-western margin, cutting across the eastern edge of the Serengeti Plains, comprises the Sonjo–Eyasi fault, that runs NE to join up with a major fault that runs *en échelon* with the Nguruman fault north of the present-day Lake Natron (Dawson 2008, fig. 3.1); the Nguruman fault had formed by 7 Ma, creating a half graben (Baker 1986). An elongate gravity low, that is almost continuous to the east and south side of the Sonjo–Eyasi fault (Dawson 2008, fig. 4.1), has been interpreted as due to large thicknesses of low-density lavas that ponded along the foot of the escarpment (Fairhead 1976). The southwestern segment of this system is the fault giving rise to the escarpment on the north side of the Eyasi half graben that loses height westwards until it fades out in the Wembere basin; some minor NW-throwing opposing faults are present at the NE end of the Eyasi basin. To the SE of the Eyasi half graben is the Yaida graben that consists of two opposing, almost parallel faults between 8 and 5 km apart This lies where the sub-Miocene land surface is descending from the Mbulu Plateau into the Eyasi half graben, and results from fracturing of the hinge of the Eyasi downwarp.

The continuity and age of the Sonjo–Eyasi fault, as proposed by Baker *et al.* (1972), Dawson (1992) and Manega (1993), have been questioned, so it is necessary to consider the arguments and counter-arguments. Foster *et al.* (1997) propose that the age of the Eyasi section of the fault is possibly the same as that of the main rift escarpment faulting at around 1.2 Ma. To support this, they cite: (1) the steep, immature morphology of the Eyasi escarpment; and (2) that the fault cuts 3.1 Ma ankaramites from Lemagrut (and hence must be younger than 3.1 Ma). Counter-arguments are that: (a) the northeastern end of the Eyasi half graben is buried beneath the Crater Highland volcanoes which predate 1.2 Ma; (b) the upper part of the Eyasi escarpment is strongly eroded (Pickering 1961),

J. B. Dawson, University of Edinburgh (*jbdawson@glg.ed.ac.uk*)
*From*: Dawson, J. B. 2008. *The Gregory Rift Valley and Neogene–Recent Volcanoes of Northern Tanzania*. Geological Society, London, Memoirs, **33**, 21–31.
0435-4052/08/$15.00 © The Geological Society of London 2008. DOI: 10.1144/M33.5

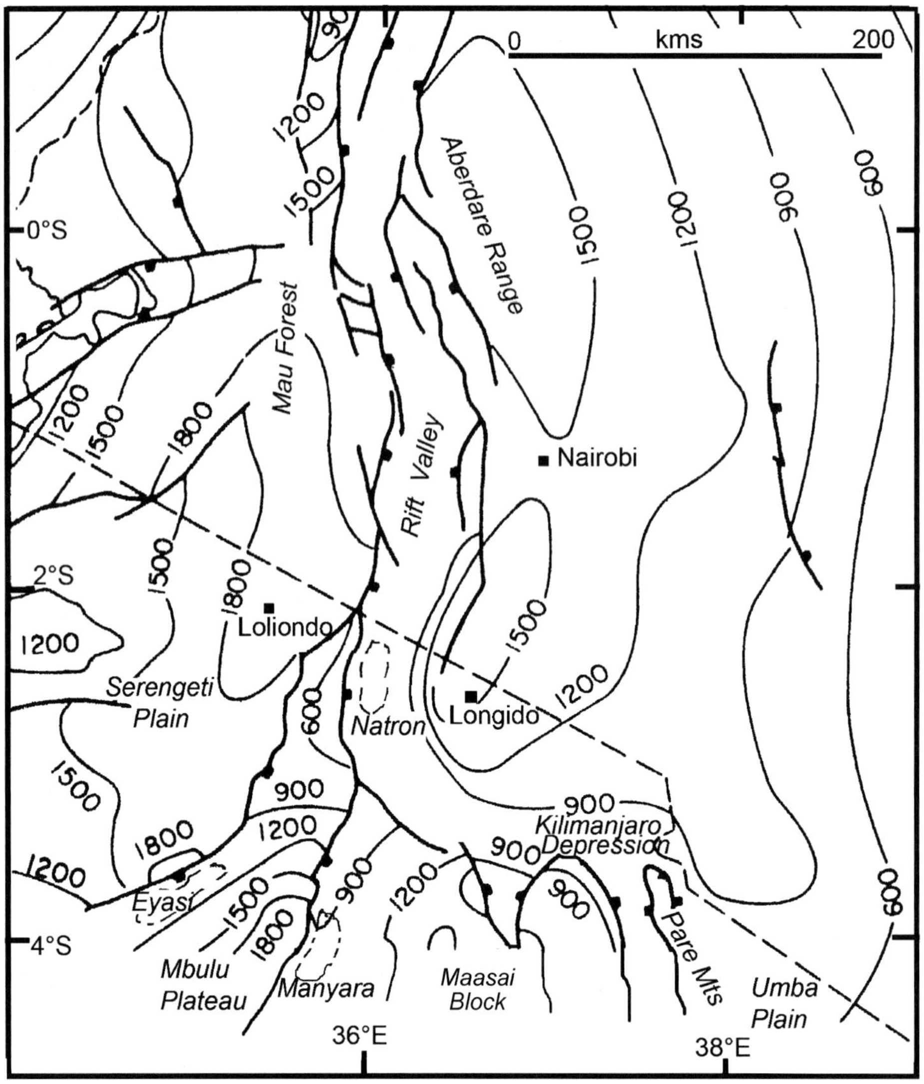


**Fig. 5.1.** Isobases of the sub-Miocene erosion surface in southern Kenya and northern Tanzania (modified after Baker *et al.* 1972).

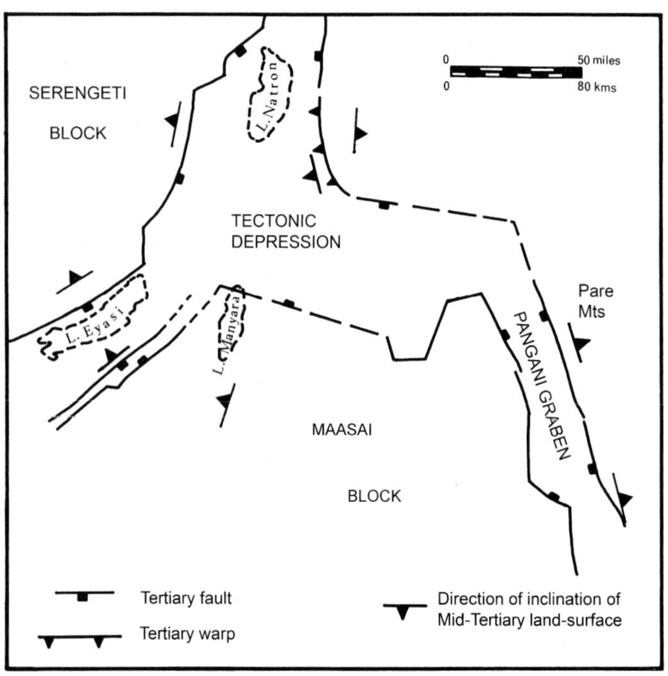


**Fig. 5.2.** The late Tertiary northern Tanzania tectonic depression. The positions of some of the large present-day lakes are shown for reference.

indicating more than one movement on the fault; and (c) the faulting that cuts the Lemagrut volcanics is of much less magnitude than the Eyasi fault. The collective field evidence points to at least two phases of movement on the fault (a possibility not discussed by Foster *et al.* 1997), with post-3.1 Ma reactivation along the Lemagrut section. Foster *et al.* (1997) also questioned the continuity of the Sonjo–Eyasi fault, claiming that the fault could not be traced south of the Oldoinyo Ogol Hills, and that such a fault would have impeded deposition of easterly-derived volcanoclastic sediments into the Olduvai Basin. The Tanganyika Geological Survey Quarter Degree Sheets 38 (Oldoinyo Ogol; Pickering 1958) and 52 (Endulen; Pickering 1964) show a line of minor escarpments (in this area not unreasonably interpreted as resulting from faults) continuing from the south side of the Oldoinyo Ogol Hills along the west side of the Olbalbal depression, to join up with faults cutting across the Lemagrut volcano; these eventually link up with the Eyasi fault. The Ogol Hills are a group of inselbergs rising above the flat peneplain of the Serengeti, and the apparent relatively minor displacement of the fault in the area immediately south of the Ogol Hills can be interpreted as due to topographic unevenness of the basement land-surface across which the fault cuts. Regarding the second point, some upper flows from the Crater Highlands volcanoes have ages of up to 3.8 Ma (Appendix 1, Dawson 2008). Volcanic rocks deeper in the sequences are currently unexposed, but must have been extruded before 3.8 Ma. The depositional history in the Olduvai Basin only begins at *c.* 2 Ma (Hay 1976; Walter *et al.*

1991) and thus there would have been ample time (a minimum of 1.8 Ma) for a fault-bounded depression to the east of the Olduvai Basin to have been infilled by volcanic flows and sediments from the Crater Highlands before the boundary scarp was overstepped. An analogy can be made with the Sonjo section of the fault further north, where lavas from Oldoinyo Sambu have infilled the depression to the east of the fault, and eventually overstepped onto the foot wall (Dundas & Awadalla, unpublished).

The southern margin of the post-Miocene depression can be inferred from both geophysical and field evidence. First, the direction of the contours on the regional Bouguer anomaly changes abruptly from a NNE direction in the area to the south of Lake Manyara and on the Masai Block, to east–west in the area to the east and north of Lake Manyara. The contours then trend eastwards on the south side of the chain of volcanoes running from Essimingor to Meru (see Dawson 2008, fig. 4.1); for greater detail see Fairhead (1976). Second, the geological evidence is that, at the western end of the inferred depression, the present-day *c.* 350 m high rift scarp on the western side of Lake Manyara exposes a major, abrupt break in lithology between Neogene basalts to the north of the Msasa River, and Precambrian gneisses to the south. This break is best explained by a fault with a downthrow of least 350 m with a northerly component, although its strike is not apparent. East of the Manyara Basin, at a latitude of *c.* 3°35′S, discontinuous outcrops show successions of Neogene basalt to the north in faulted contact with Maasai Block gneisses to the south, the faults trending east–west or WNW–ESE (Jones 1967). The faults run parallel to the contours on the regional Bouguer anomaly map. To the south of this line of faults, discontinuous thin lava flows overlie the undulose gneiss terrain. Further east, in the Shambarai area (to the south of Arusha), Guest (1953), quoting data from the Tanganyika Water Development Department, reported that, just to the north of basement outcrops, a borehole (B.H. 6/48) ‘penetrated lava to a depth of 1065 feet ... (which) ... estimates a downthrow of the basement of between, 1100 to 1300 feet’. Finally, the eastern end is well documented where the northern edge of the Maasai Block is downfaulted into the Kilimanjaro depression (Downie & Wilkinson 1972; Grainger 1968). Hence, despite burial by Pleistocene sediments in the Lake Manyara basin, and being obscured further east by overstepping Pleistocene lava flows, it is possible to infer a major pre-Pleistocene fault with a dominantly east–west or WNW–ESE trend along a latitude of approximately 3°35′S. North of this line, the basalt terrain is extensively faulted, indicative of a low-strength, brittle lithology; in the gneiss terrain to the south, the comparatively few faults are long (30–80 km). In the area between the well documented eastern end and the inferred western end, the inferred east–west-trending fault has suffered minor disruption by Pleistocene north–south faults of the Oljoro graben, SW of Arusha.

## Eruption of the first set of volcanoes: the Older Extrusives

Volcanoes erupted within the late Tertiary tectonic depression, infilling the depression until, at various points around the margins of the depression, lavas overstepped the boundary faults (Fig. 5.3). These volcanoes are described more fully in Chapter 6 (Dawson 2008). Sedimentary successions accumulated at various points within the depression.

## Plio-Pleistocene faulting

### *Late Pliocene faulting*

Minor faulting in the Upper Pliocene–Lower Pleistocene formed the Lembolos graben (Fig. 5.4), and affected the trachyte–phonolite volcanics on the Tarosero volcano, where faults cutting lavas on the lower slopes, dated at *c.* 2.2 Ma, are buried by later 1.98 Ma lavas in the summit area (MacIntyre *et al.* 1974).

In the Lake Natron area, the Peninj Group sediments were deposited in the shallow, faulted depression of proto-Lake Natron (Dundas & Awadalla, unpublished). Feldspar in the basal tuff of the group has given an age of 1.7 Ma (Manega 1993) which is the minimum age for the pre-sedimentation faulting.

An erosion bevel on the main rift escarpment on the west side of Lake Manyara, together with a corresponding nick-point on the Endabash River, has been interpreted as due to early, relatively minor faulting on the main Manyara Basin boundary fault (Orridge 1965). A distinct erosion bevel at the top of the otherwise sharp fault escarpment between Lake Manyara and Engaruka (Dawson personal observation) may also be due to pre-1.2 Ma faulting.

### *Major Pleistocene faulting*

A major episode of faulting at *c.* 1.0 $\pm$ 0.2 Ma has given rise to what is geomorphologically the present-day rift valley. The age of this faulting, based on dates obtained on the highest lavas in faulted lava sequences and also on extrusions up the main fault in the sector north of Lake Manyara (see Appendix 1, Dawson 2008), is between 1.2 and 0.9 Ma (MacIntyre *et al.* 1974; Foster *et al.* 1997). This age equates with the major faulting episode in the Kirikiti–Lengitoto area in Kenya in the period 1.3 to 0.9 Ma (Fairhead *et al.* 1972). This faulting has been superimposed on the volcanic and sedimentary successions previously extruded and deposited within, and around the margins of, the late Tertiary tectonic depression (Fig. 5.5). A reversal in the drainage in the Olduvai Basin took place at this time (*c.* 1.15 Ma; Hay 1976).

The main manifestation of this major faulting event is the high east-facing escarpment that runs from Lake Natron southwards to Lake Balangida Lelu (Fig. 5.6). In the north, the faulting exposes the volcanic rocks forming the eastern flanks of Oldoinyo Sambu and the volcanoes of the Crater Highlands and, further south, exposes basement rocks of the Mozambique fold belt and the Tanzania Craton.

Between the Kenya border and Lake Manyara the trend of the fault escarpment is generally north–south, but south of Manyara, it swings gradually westwards until at Lake Balangida the trend is NW–SE (Fig. 3.1). It forms an asymmetric half graben, having no major, eastern equivalent (unlike the rift valley in Kenya to the north), and the eastern margin of the present-day rift depression is formed by a series of small faults and downwarps that are the southern continuation of the Ngong–Turoka fault in Kenya.

A significant feature of the faulting was a disruption of older drainage directions, and the formation of a series of enclosed sedimentation basins at the foot of the main escarpment, these being, from north to south, the Natron, Engaruka, Manyara, Balangida and Balangida Lelu basins; of these, the Natron, Balangida and Balangida Lelu basins have small opposed faults facing the main escarpment. Before the onset of the Pleistocene drying-out event the Natron Basin most probably linked up with the Magadi Basin in Kenya and, prior to the eruption of Oldoinyo Lengai and Kerimasi, which form a topographic high, the Natron and Engaruka basins may have been continuous. To the west of the main rift, the asymmetric older Eyasi Basin lies at the foot of the SE-facing Eyasi escarpment.

### *The Natron Basin boundary fault system*

The boundary fault trends approximately north–south and, between the Kenya border and the south end of Lake Natron, the system is segmented (Fig. 5.7). The main Sambu segment extends from the border to the Peninj delta (where there are some east–west offsets), followed to the south by the Binini,

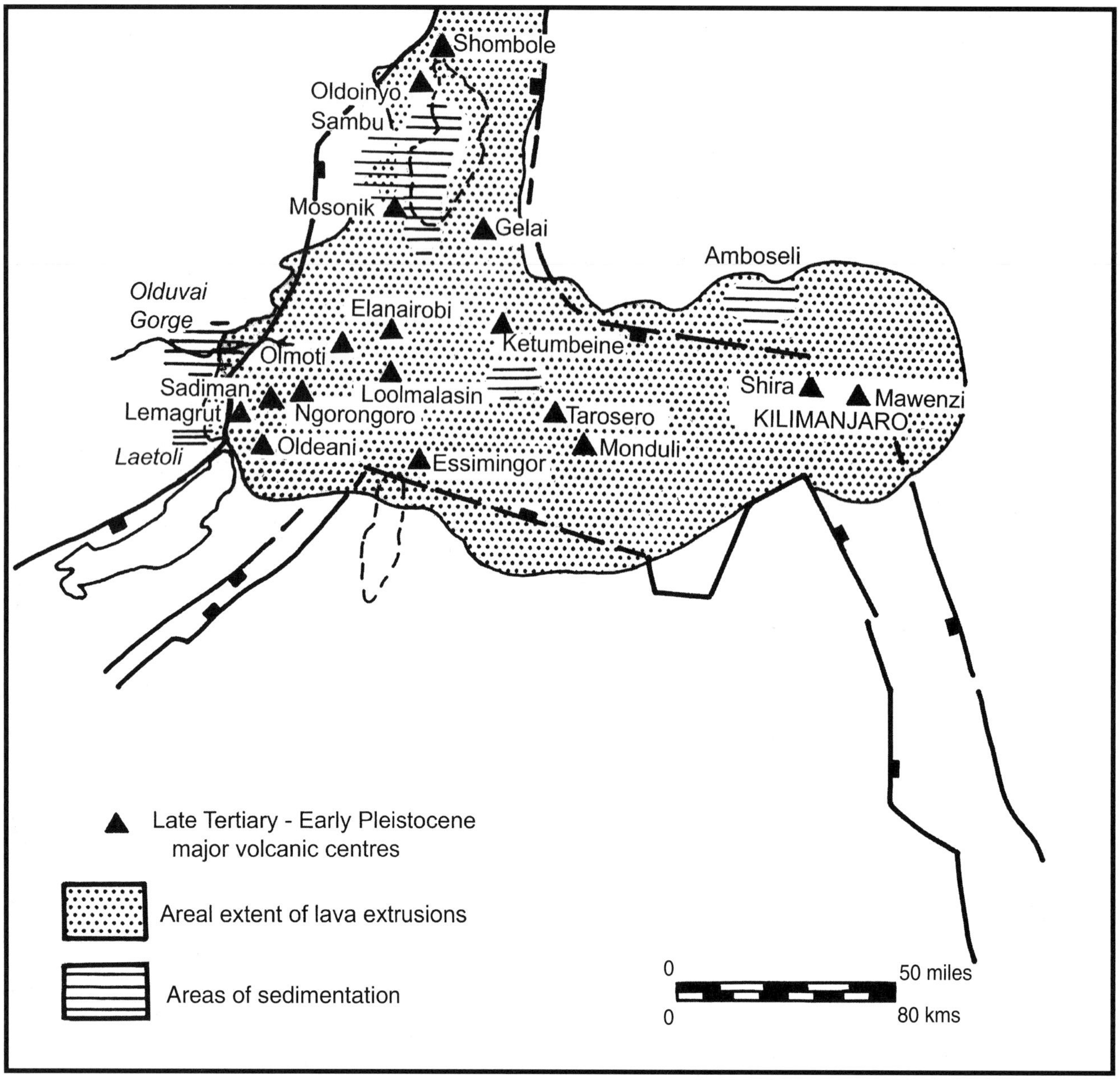


**Fig. 5.3.** Distribution of the Older Extrusive volcanic centres, and areas of Plio-Pleistocene sedimentation.

**Fig. 5.4.** The Matunginini fault escarpment, looking across the Lembolos graben from the SW (see Fig. 5.8). It forms the eastern boundary of the graben and exposes 250 m of horizontal basalts, the youngest of which is dated at 2.0 Ma.

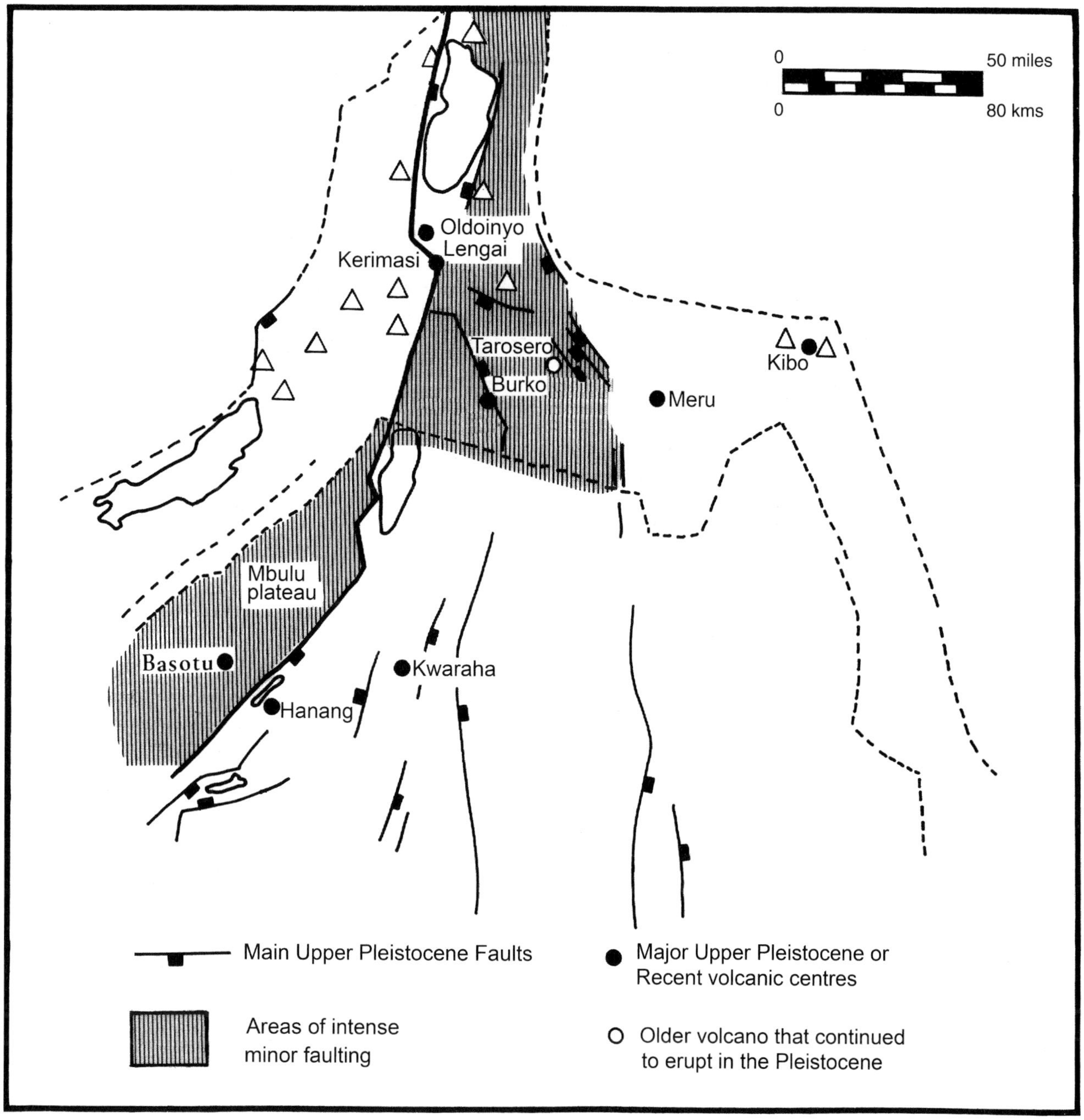


**Fig. 5.5.** Distribution of the main Upper Pleistocene faults and Younger Extrusive volcanoes.

Ngongolo, Mto wa Mbagai, and Sanjan fault segments. The height of the escarpment varies from 1200 m at Oldoinyo Sambu to zero where the Engaruka–Loliondo dirt road crosses the line of the escarpment, whence the height increases again southwards. The Sanjan Fault is the most westerly of these segments, and between it and the other segments is the Moinik valley, in which are extensive outcrops of sediments deposited in the proto-Lake Natron (see above). South of Lake Natron, the fault, with an escarpment 600 m high, runs north–south to a point just south of Oldoinyo Lengai, where a small SE-trending sector occurs before the fault is blanketed beneath the Kerimasi volcano. The fault reappears running north–south of Kerimasi, having been offset some 8 km to the east relative to its line between Lake Natron and Oldoinyo Lengai. It is not possible to ascertain whether this offset results from a continuity of the short visible SE-trending sector beneath Kerimasi, or whether the fault is segmented at this point. The escarpment height at Engaruka is *c*. 550 m.

South of Engaruka, the throw of the main rift fault is taken up by two fault systems: the main boundary fault of the Manyara Basin that continues southwards, and a diverging, lesser SE-trending fault, the Engaruka Basin fault, that forms the southwestern boundary of the Engaruka Basin (Fig. 5.8). The triangular block of terrain between these two faults, the Engaruka Block, is back-tilted southwestwards towards the Manyara boundary fault. This block itself, which is broken by many minor faults, decreases in height from its highest elevation (1220 m a.s.l.) just south of Engaruka southwards towards Lake Manyara (945 m a.s.l.).

Further to the SE, the Engaruka Basin fault is replaced by several closely spaced faults that form a SE-trending zone of parallel, elongate horsts and grabens (Fig. 5.8). Several faults on the Engaruka Block trend in this same dominant direction, and similar faults occur prominently in the area between and south of the Essimingor and Burko volcanoes. Also in this area, numerous small faults trending north–south or NNW–ESE combine with the SE-trending faults to form grabens, horsts and fault blocks,

**Fig. 5.6.** The Manyara Basin boundary fault escarpment, north of Lake Manyara (see Fig. 5.8). The escarpment exposes the eastern flanks of some Crater Highland volcanoes, and varies in height from *c.* 250 m at Manyara to 600 m at its northern end near Engaruka.

the last being tilted to the west or SW. The NNW–SSE trend is also apparent south of Meru in faults forming the Oljoro graben. Lavas in the wall of this graben have been dated at 2.49–2.3 Ma (Wilkinson *et al.* 1986).

The pronounced NW–SE fault direction is also found in the area NW of Meru, where SE-trending faulting cuts both the lower northern slopes of the Tarosero volcano, and is also the trend of the Tarosero, Matunginini and Matisiwi faults that form the Lembolos graben. The ages of these faults are *c.* 2.0 Ma, based on faulted basalt and trachyte (2.4–2.04) and fault-blanketing phonolite (1.98 Ma) on Tarosero, and also on a date of 2.0 Ma on the highest faulted basalt flow on the Matunginini escarpment (MacIntyre *et al.* 1974). Further, the Lembolos graben is buried beneath extrusive rocks from Meru, the oldest lavas of which are 1.5 Ma.

A minor fault direction in this area is WNW–ESE or east–west, shown by faults cutting across the southern slopes of Ketumbeine, by faults forming the southern boundary of the Ardai Basin, and a possible fault on the south side of the Monduli volcano. This trend is that of the inferred southern boundary of the late Tertiary tectonic depression.

Analyses of the fault directions in the Engaruka–Arusha area indicate that the dominant direction is NW–SE, with a minor one at NNW–SSE, indicating maximum extension was *c.* N50°E (Dawson 1964; Ring *et al.* 2005). This compares with a dominant NNE–WSW direction with maximum extension at *c.* N100°E in the Magadi area (Atmaoui & Hollnack 2003) which again emphasizes structural differences between southern Kenya and northern Tanzania.

In the area between Engaruka and Lake Manyara, and to the north and south of the main road between Makuyuni and Arusha, many of these relatively minor faults are the eruption sites of small flows and scoria cones (see Fig. 5.8). Most of these minor faults (e.g. Yayai and Elunata) have escarpments of $<$100 m.

### *Manyara Basin boundary fault system*

South of Engaruka, the Manyara Basin boundary fault has a SSW trend, though with minor indentations caused by small north–south sectors near Kitete and Mto wa Mbu (Fig. 5.8). The escarpment exposes layered lava flows derived from the Crater Highland volcanoes to the west, and continuity of these horizontal flows suggests no segmentation of the fault as far south as Mto wa Mbu, where a thin ramp between two sectors enables the Makuyuni–Ngorongoro road to ascend the escarpment. To the north of Lake Manyara itself, a minor basin at Erumkoko is the result of back-tilting of a small fault block westwards towards the main boundary fault.

Just south of Mto wa Mbu, at the gorge of the Msasa River, the lithology exposed in the entire escarpment changes abruptly from Neogene lavas to the north to basement gneisses further south. The highly variable height of the escarpment south from this point, the undulating topography of the gneiss terrain of the footwall, and the absence of horizontal formations, makes it difficult to ascertain whether the boundary fault is segmented but south of Lake Manyara the trace of the fault becomes saw-toothed due to SW-oriented sections being increasingly interspersed with north- or NW-trending sections. This pattern continues to around 4°15′S, where the SW-trend becomes dominant and continues well to the SW past Hanang (Dawson 2008, fig. 3.1). This change of pattern coincides with exposure in the fault escarpment of Archaean rocks, that also occupy the floor of the depression (Selby & Thomas 1966). This area, comprising the Balangida and Balangida Lelu basins, is still bounded on its northwestern side by the continuation of the Manyara boundary fault, but a variation here is that small, opposing, SW-trending faults form the southeastern margin of both basins.

Along its length, the height of this continuous boundary fault escarpment decreases from 500–600 m at Lake Manyara to no more than 200 m at Balangida Lelu. SW of Lake Balangida Lelu the fault escarpments fade out in the relatively subdued topography (escarpments of up to 100 m) of the north–south-trending Mponde graben (Fozzard 1958). Further south, minor scarps and an area of seismicity are interpreted to be the expressions of incipient faulting (Nyblade *et al.* 1996; Le Gall *et al.* 2004).

### *The Mbulu Plateau and the Maasai Block*

The plateau to the west of the Manyara and Balangida basins is cut by numerous SW-trending faults mainly showing only minor topographic expression. The age of most cannot be determined, though some at the northeastern end of the plateau cut lavas from the

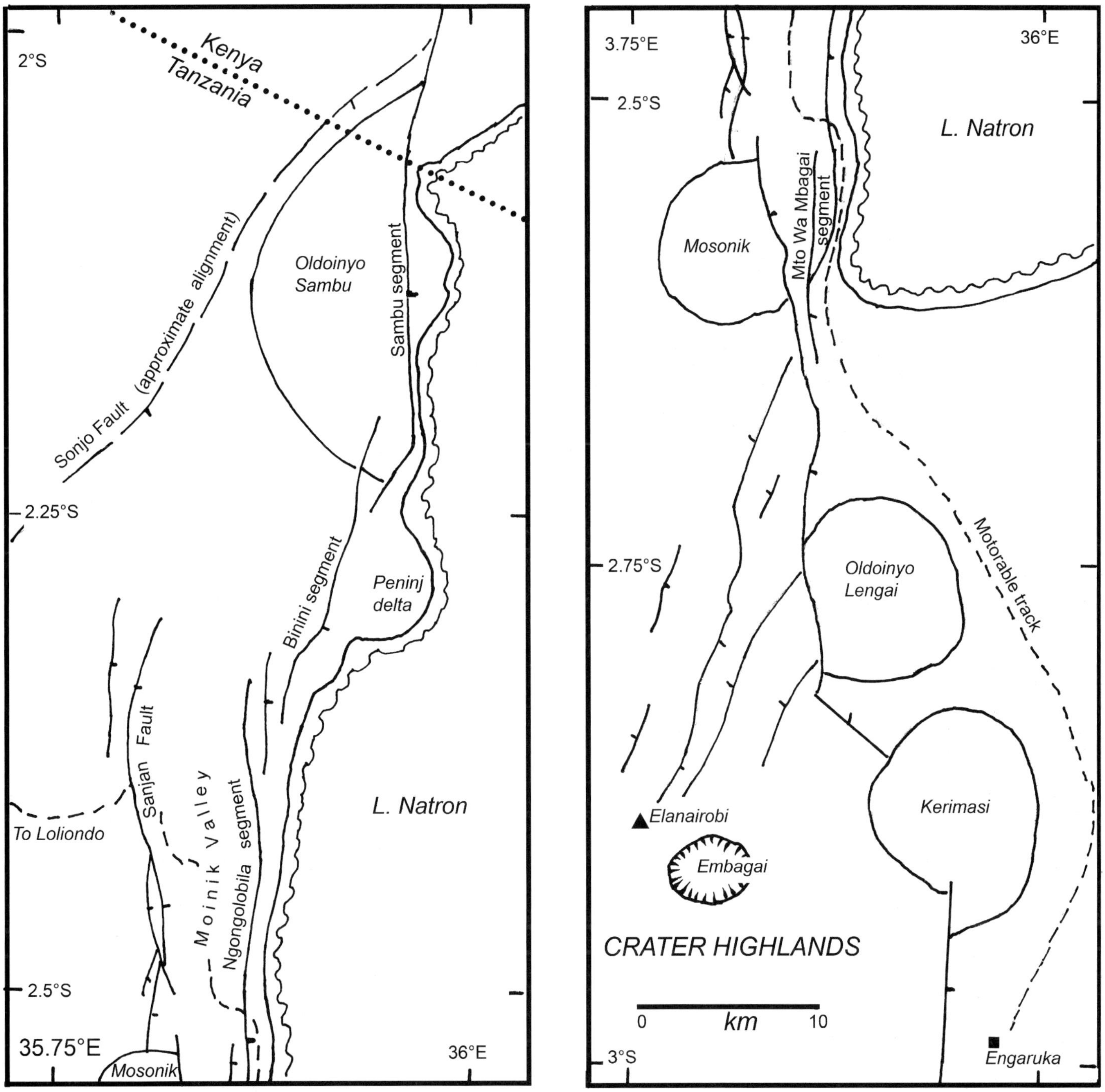


**Fig. 5.7.** The Natron Basin boundary fault system, showing segmentation particularly in the area immediately west of Lake Natron.

Crater Highland volcanoes and could thus be attributed to the post-1.2 Ma faulting episode, or to reactivation at that time.

To the east of the Manyara Basin, the Maasai Block (Fig 5.2) is cut by several north–south or NNW–SSE faults; the more westerly are the boundaries of large west-dipping fault blocks.

### *Faulting of the volcanoes in the Crater Highlands*

Between Lake Natron and Oldoinyo Lengai, several faults branch away from the Natron boundary fault to run SSW or SW across the northern slopes of Elanairobi (Fig. 5.7). They run parallel to, or are a continuation of, faults running in the same direction across the Crater Highlands, including one cutting across the northwestern rim of the Ngorongoro crater (Dawson 2008, fig. 7.5). Foster (1997) also identified a linear SW-trending feature along the western side of the Crater Highlands, based on a marked contrast in the drainage to the east and west of the feature. He proposed that this is the surface expression of a major west-dipping fault which he termed the Angata Salei Fault. Minor faults, related to reactivation of the Eyasi section of the Sonjo–Eyasi fault, cut north–south across Lemagrut and the western flank of Oldeani. Other north–south faults cut the eastern slopes of Oldeani, then link up with minor NE-trending faults on the south side of Lake Eyasi. Many minor faults on the lower slopes of the Ngorongoro and Olmoti volcanoes were formed at the time of the deepening of the Olbalbal depression in the late Pleistocene (Hay 1976). Numerous small faults occur on the footwall of the Manyara Basin boundary fault in the area between Mto wa Mbu and Karatu (Dawson 2008, fig. 7.5). These are exposed in deforested areas, and their apparent absence in contiguous areas to the north may be due to obscuring by forest.

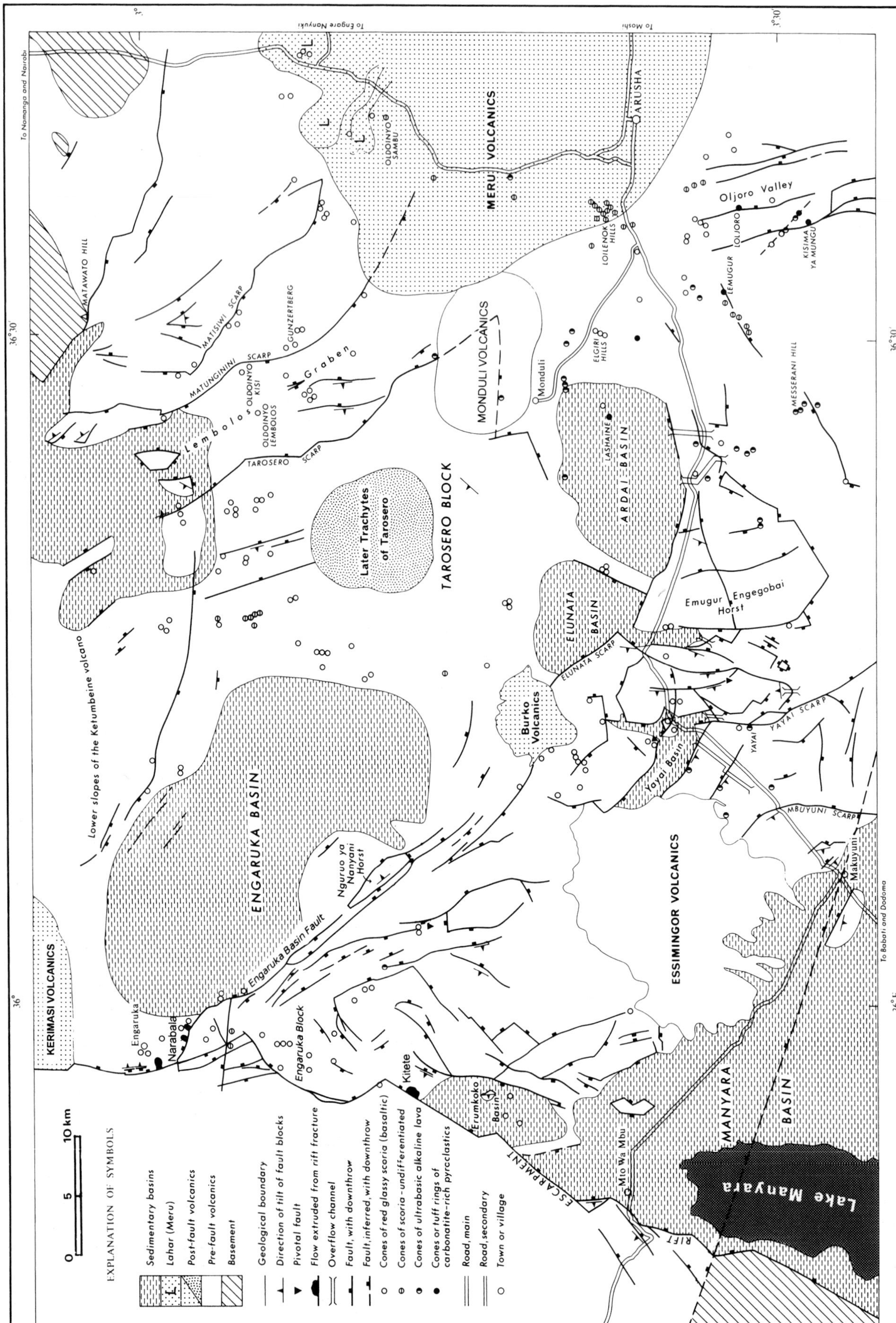


**Fig. 5.8.** Distribution of Pleistocene faults, post-fault volcanoes, minor volcanic features and sedimentary basins in the Engaruka–Monduli–Arusha area.

*The Eyasi half graben faulting*

As noted earlier, faulting occurred in the late Pliocene to form the half graben. Following the extrusion of the Crater Highlands volcanoes that filled the NE end of the graben, reactivation caused minor faults that cross Lemagrut and Oldeani to link up with faults in the Olbalbal graben. The reactivated part of the fault escarpment above Lake Eyasi created by this faulting is sharp and *c.* 500 m high. The maximum height on the escarpment is around 900 m (Kent 1941).

*The Kibangaini grid-faulted area*

The area between the northern flanks of the Gelai volcano and the Kenya border, around Lake Natron (Dawson 2008, fig. 3.1) is characterized by a style of faulting unlike that described above. The term 'grid-faulting' was introduced by Gregory (1921) to describe 'that condition by which the valley floor is divided into narrow strips which rise to different elevations and are bounded by near vertical walls' with the vertical displacement of the strips being no more than 20–30 m. The term was first used to describe this style of faulting in the Magadi area of Kenya, of which the Kibangaini area is a southerly continuation. An inference of the sharpness of the fault scarps, combined with their small throw, is that they must be of Recent age.

## Upper Pleistocene–Recent faulting

In addition to the Recent faulting in the Kibangaini area, the pyroclastics of Kerimasi were faulted at *c.* 0.37 Ma prior to the initial eruption of Oldoinyo Lengai, the youngest major centre (MacIntyre *et al.* 1974).

## Faulting: broader issues

*Single-stage or continuous faulting?*

Whereas the overall geochronology of the lava sequences indicates a major faulting episode at 1.0–1.2 Ma, it is open to question whether the whole throw on the faults results from a single event or whether the overall displacement is the accumulation of a number of smaller movements. Hay (1976) suggested that episodic faulting may be more apparent than real, citing the sediments of the Olduvai Gorge basin that have been faulted several times between 2.0 and 0.4 Ma. Some support for sporadic, cumulative faulting is also found in the southern part of the Kenya Rift just north of the Tanzanian border, where Isaac (1968) documented faulting during much of the deposition of the mid-Pleistocene Olorgesaile Formation. Recent dating of trachytes in the Magadi area also showed that faulting on the floor of the rift valley took place sporadically between 2.25 and 0.36 Ma (Dunningham 2005).

*Style of faulting*

There is no evidence for reverse faults. Petroleum industry seismic traverses have shown that listric faults form the boundaries to many half grabens in the Western Rift and in many of the rift-valley basins in Kenya (references in Morley 1999*a*).

No similar traverses have been made across the main Tanzania half graben. Although intuitively it might be expected that listric faults also bound the Tanzania basins, Birt *et al.* (1997) found no evidence in southernmost Kenya for the good reflections that should arise from major listric faults. In the case of minor faults on the western side of the Engaruka Basin (Fig. 5.8), one at least is pivotal, suggesting that it is vertical.

*Southern extension of the Gregory Rift?*

Until recently, there was only limited evidence that Neogene–Recent rifting was propagating further south than the Mponde graben (Dawson 2008, fig. 3.1). There are no Pleistocene volcanoes south of Hanang, but the swarms of minor earthquakes in the Kondoa area are mentioned above. In the Kilimatinde area (Dawson 2008, fig. 2.1), linking with the Mponde graben, are escarpments which Fozzard (1958) links to the Pleistocene faulting further north, noting that they follow shear- and shatter-zones in the basement migmatites.

Further south in Tanzania are older half graben structures, mainly bounded by NE–SW-trending faults and infilled with Karoo (Upper Carboniferous and Triassic) sediments. These have been recognized for many years (e.g. Haughton 1963) and in the Kilombero area (Dawson 2008, fig. 2.1), where uplifted basement blocks are associated with major half grabens, Whittingham (1963) documented fault movements in late Neogene times. More recently, Le Gall *et al.* (2004) proposed that, in agreement with seismicity associated with the faults, there is Neogene–Recent extensional overprinting of the older, Karoo-age fault zone along both rift-parallel and transverse faults in the Kilombero area. They further suggested that the Kilombero rift zone connects northwestward to the Manyara rift zone by way of an active transverse fault zone which cuts NW across the Tanzania Craton. Further south, they envisaged the rifting continuing via the Karoo-age Ruhuhu graben to the Tanganyika–Rukwa–Malawi rift, thereby linking the eastern and western arms of the overall East African rift system.

*Relationship of the rift valley faults to structural elements in the underlying basement rocks*

There is a long-standing tenet that the rift valleys in central and eastern Africa follow the trends of the Pan-African orogens that wrap round the ancient cratons. First proposed by Holmes (1951), and followed by others such as Dixey (1956) and McConnell (1972), this thesis is broadly correct for the western branch of the African rift system that follows the Kibaride and Ubendian fold belts round the western margin of the Tanzania Craton, and round the Zambian Craton (Clifford 1970; Morley 1999*b*). There is also some linearity in the Kenya sector of the Mozambique fold belt, though many departures from the general north–south trend have been noted (Baker *et al.* 1972). However, nowhere in the East African rift system is there such a departure from the overall trend of the fold belts as in the North Tanzania Divergence (Dawson 2008, fig. 1.1). Dawson (1992) and Nyblade & Brazier (2002) have proposed a key role for the rigid Tanzania Craton, against which the eastern branch of the rift has stalled.

The most critical analysis of the structure of the Tanzania Craton and Mozambique Belt in northern Tanzania (Hepworth 1972) recognizes different structural trends that are possibly relevant to the interpretation of the Tertiary–Recent rift fractures.

On the craton, Hepworth (1972) recognizes:

1. A strong, straight NE–SW trend which is followed by the present day Eyasi half graben, and thus termed the Eyasi trend. The underlying trend is that of a prominent Archaean dyke-swarm around 400 km long and 300 km wide, dated at 2.2 to 2.6 Ga (Vail 1970). Although Halls *et al.* (1987) claimed that this dyke swarm terminates at the mapped craton-Mozambique fold belt boundary, aeromagnetic data show faint NE–SW magnetic lineaments across (?beneath) the Mozambique Belt (Ebinger *et al.* 1997); the trend is unrelated to the Mozambique Belt.
2. A straight NW–SE trend, developed as shear-belts (the Bubu cataclasites) in the Archaean granite basement. This is the same trend as the Kondoan (see below) and may have been

superimposed on the craton during a later orogenesis in the Mozambique Belt.

3. A WNW–ESE trend characteristic of the Dodoman Archaean fold belt. Unlike the two other trends, it is gently curving and is perpetuated by trends in the Mozambique Belt.

Within the Mozambique fold belt are:

4. The most prominent trend, referred to as the Kondoan trend which is WNW–ESE or NW–SE, and oblique to the trend of the orogenic belt as a whole. This is the same direction as two important late-Proterozoic (635–550 Ma) mylonite and cataclasite shear zones, the Aswa–Nandi–Loita and Nyangea–Athi–Ikutha shear zones, that cut across the Mozambique Belt in central Kenya (Smith & Mosley 1993). On a continental scale, this trend coincides with the northern termination of the western rift valley (Morley 1999*b*) and, more locally, is the trend of the northwestern arm of Lake Magadi.
5. A less prominent NE–SW or NNE–SSW trend, termed the Parangan, which is dominant on the Mbulu Plateau and on the northern part of the Maasai Block; this is the same direction as the Eyasi trend.
6. Large, closed folds to the west of the Pare Mountains, perhaps due to the combined effects of NE–SW and NW–SE folds.
7. A NNW–SSE trend now picked out by the boundary faults of the Pangani graben.

In the case of the faults forming the late Tertiary depression, the Eyasi Basin and Yaida graben, the faults parallel the cratonic Eyasi trend. Some sectors of the Sonjo Fault, particularly those to NW of Lake Natron, follow the Eyasi trend but are interspersed with sectors running north–south, a direction also followed by the boundary fault continuing southwards from Kenya on the eastern side of the depression. The postulated southern boundary fault of the depression trends east–west. It lies along the northern edge of the Maasai Block which, on geophysical grounds, appears to be a distinct structural terrane.

In the case of the Pleistocene faults, most on the Mbulu Plateau, and those cutting across the Crater Highlands follow the Eyasi trend. The north–south sectors of the Natron and Manyara boundary fault systems follow the same trend as some cutting the Maasai Block which are a continuum of the main north–south trend in Kenya. As noted earlier, the group of faults in the Engaruka–Manyara–Arusha area are dominantly NW–SE. In Kenya, where this trend has also been recognized in induction transfer functions during the KRISP-94 magnetotelluric survey (Simpson *et al.* 1997), it has been linked to the continental-scale Proterozoic Aswa shear zone (Smith & Mosley 1993). In Tanzania this is the Bubu and Kondoan trend, although to suggest such an influence on the Pleistocene faults is at odds with the absence of such an influence on the orientation of the earlier late-Pliocene faulting. The orientation of these faults (Dawson 1964; Ring *et al.* 2005) is consistent with NE–SW extension rather than to basement structure control.

The linearity of the chain of volcanoes that trends west–east across the Tanzania volcanic province from Essimingor to Meru and Kilimanjaro has given rise to speculation as to their being either due a transform fault running normal to the rift trend (Fairhead 1980) or, in the case of the more easterly volcanoes of both Kenya and Tanzania (the so-called 'off-rift' volcanoes), to eruption up listric faults (Bosworth 1987). Analysis of the distribution of the northern Tanzania volcanoes, particularly with regard to distribution during each of the two main episodes of eruption during the evolution of the North Tanzania Divergence (Dawson 1992), does not support these suggestions. Some groups of volcanoes undoubtedly have a well-defined linearity, the best being the volcanoes of the Crater Highlands that follow the NE Eyasi trend, as do the explosion craters and maars of Basotu on the Mbulu Plateau. Another is the NW–SE line formed by the coeval Gelai–Kitumbeine–Tarosero–Monduli group, parallel to the downwarped eastern margin of the late Tertiary depression. Alternative explanations to Fairhead's (1980) Burko–Monduli–Meru Line are that Burko erupted along the line of the Engaruka Basin fault swarm, that Meru formed close to the junction of the Lembolos and Oljoro grabens, and that Kilimanjaro (an extension of the line) grew at the intersection of the Pangani graben with the late Tertiary depression.

To summarize, whereas there is evidence that the faulting and chains of volcanoes to the west of the main Natron–Manyara–Balangida boundary fault reflect the dominant Eyasi trend in the Archaean basement, to the east of the main fault the linkage of both faulting and volcanicity to structures in the basement is more tenuous. In this context, the deeper structure of the Maasai Block, that has the geophysical properties of a small discrete terrane (Ebinger *et al.* 1997), is still unknown.

## References


Atmaoui, N. & Hollnack, D. 2003. Neotectonics and extension direction of the southern Kenya Rift, Lake Magadi area. *Tectonophysics*, **364**, 71–83.

Baker, B. H. 1986. Tectonics and volcanism in the southern Kenya Rift Valley and its influence on rift sedimentation. *In*: Frostick, L. E., Renaut, R. W., Reid, I. & Tiercelin, J. J. (eds) *Sedimentation in the African Rifts*. Geological Society, London, Special Publications, **25**, 45–57.

Baker, B. H., Mohr, P. A. & Williams, L. A. J. 1972. *Geology of the Eastern Rift System of Africa*. Geological Society of America, Special Publications, **36**.

Bosworth, W. 1987. Off-axis volcanism in the Gregory rift, east Africa: Implications for models of continental rifting. *Geology*, **15**, 397–400.

Birt, C. S., Maguire, P. K. H., Khan, M. A., Thybo, H., Keller, G. R. & Patel, J. 1997. The influence on pre-existing structures on the evolution of the southern Kenya Rift Valley – evidence from seismic and gravity studies. *Tectonophysics*, **278**, 211–242.

Clifford, T. N. **1970**. The structural framework of Africa. *In*: Clifford, T. N. & Gass, I. G. (eds) *African magmatism and tectonics*. Oliver & Boyd, Edinburgh, 1–26.

Dawson, J. B. 1964. *Monduli*. Tanzania Geological Survey. Quarter Degree Sheet **54**.

Dawson, J. B. 1992. Neogene tectonics and volcanicity in the North Tanzania sector of the Gregory Rift Valley: contrasts with the Kenya sector. *Tectonophysics*, **204**, 81–92.

Dawson, J. B. 2008. *The Gregory Rift Valley and Neogene–Recent Volcanoes of Northern Tanzania*. Geological Society, London, Memoirs, **33**.

Dixey, F. 1956. The East African Rift System. *Colonial Geology and Mineral Resources, Supplement Series, Bulletin Supplement* **1**. H M Stationery Office, London.

Downie, C. & Wilkinson, P. 1972. *The Geology of Kilimanjaro*. University of Sheffield, Sheffield.

Dundas, D. L. & Awadalla, G. W. (unpublished). *Loliondo*. Unpublished report and map of Quarter Degree Sheet 16 and 27. Tanzania Geological Survey. Mapped 1966.

Dunningham, J. P. 2005. *Long-term Evolution of Normal Fault Systems: Controls on the Development and Evolution of Extensional Structures in the Neotectonic Kenyan Rift, East Africa*. PhD thesis, University of Edinburgh.

Ebinger, C., Poudjom Djomani, Y., Mbede, E., Foster, A. & Dawson, J. B. 1997. Rifting Archaean lithosphere: the Eyasi–Manyara–Natron rifts, East Africa. *Journal of the Geological Sociey, London*, **154**, 947–60.

Evans, A. L., Fairhead, J. D. & Mitchell, J. G. 1971. Potassium–argon ages from the volcanic province of northern Tanzania. *Nature*, **229**, 19–20.

Fairhead, J. D. 1976. The structure of the lithosphere beneath the eastern rift, East Africa, deduced from gravity studies. *Tectonophysics*, **30**, 269–98.

FAIRHEAD, J. D. 1980. The structure of the cross-cutting volcanic chain of northern Tanzania and its relation to the East African rift system. *Tectonophysics*, **65**, 193–208.

FAIRHEAD, J. D., MITCHELL, J. G. & WILLIAMS, L. A. J. 1972. New K/Ar determinations on Rift volcanics of S. Kenya and their bearing on the age of Rift faulting. *Nature*, **238**, 66–9.

FOSTER, A. 1997. *The Seismicity and Tectonics of Africa*. PhD thesis, University of Cambridge.

FOSTER, A., EBINGER, C., MBEDE, E. & REX, D. 1997. Tectonic development of the northern sector of the East African Rift System. *Journal of the Geological Society, London*, **154**, 689–700.

FOZZARD, P. M. H. 1958. The Mponde river area, Singida District: Quarter Degree Sheet 41 NW. *Records of the Geological Survey of Tanganyika*, **VIII**, 1–5.

GRAINGER, D. J. 1968. *Shambarai*. Tanzania Mineral Resources Division, Quarter Degree Sheet **71**.

GREGORY, J. W. 1921. *The Rift Valleys and Geology of East Africa*. Seeley, Service, London.

GUEST, N. J. 1953. *The Geology of Part of Northern Tanganyika Territory*. PhD thesis, University of Sheffield.

HALLS, H. C., BURNS, K. G., BULLOCK, S. J. & BATTERHAM, P. M. 1987. Mafic dyke swarms of Tanzania interpreted from aeromagnetic data. *In*: HALLS, H. C. & FAHRIG, W. (eds) *Mafic Dyke Swarms*. Geological Association of Canada, Special Papers, **34**, 173–218.

HAUGHTON, S. H. 1963. *The Stratigraphic History of Africa South of the Sahara*. Oliver & Boyd, Edinburgh.

HAY, R. L. 1976. *Geology of the Olduvai Gorge*. University of California Press, Berkeley.

HEPWORTH, J. V. 1972. The Mozambique orogenic belt and its foreland in northeast Tanzania: a photogeologically-based study. *Journal of the Geological Society, London*, **128**, 461–500.

HOLMES, A. 1951. The sequence of pre-Cambrian orogenic belts in south and central Africa. *Proceedings of the 18$^{th}$ International Geological Congress, 1948*, **14**, 254–269.

ISAAC, G. L. 1968. *The Acheulian Site Complex at Olorgesaile, Kenya: A Contribution to the Interpretation of Middle Pleistocene Culture in East Africa*. PhD thesis, University of Cambridge.

JONES, H. 1967. *Makuyuni*. Tanzania Mineral Resources Division. Quarter Degree Sheet **70**.

KENT, P. E. 1941. The Recent history and Pleistocene deposits of the plateau north of Lake Eyasi, Tanganyika. *Geological Magazine*, **78**, 173–184.

LE GALL, B., GERNIGON, L., *ET AL*. 2004. Neogene–Holocene rift propagation in central Tanzania: morphostructural and aeromagnetic evidence from the Kilombero area. *Geological Society of America, Bulletin*, **83**, 490–510.

MACINTYRE, R. M., MITCHELL, J. G. & DAWSON, J. B. 1974. Age of the fault movements in the Tanzanian sector of the East African rift system. *Nature*, **247**, 354–356.

MANEGA, P. C. 1993. *Geochronology, Geochemistry and Isotope Study of the Plio-Pleistocene Hominid Sites and the Ngorongoro Volcanic Highlands in Northern Tanzania*. PhD thesis, University of Colorado.

MCCONNELL, R. B. 1972. Geological development of the rift system of eastern Africa. *Bulletin of the Geological Society of America*, **83**, 2549–2572.

MORLEY, C. K. (ed.) 1999*a*. *Geoscience of Rift Systems – Evolution of East Africa*. American Association of Petroleum Geologists, Studies in Geology, **44**.

MORLEY, C. K. 1999*b*. Influence of pre-exisiting fabric on rift structure. *In*: MORLEY, C. K. (ed.) *Geoscience of Rift Systems – Evolution of East Africa*. American Association of Petroleum Geologists, Studies in Geology, **44**, 151–160.

MOUGENOT, D., RECQ, M., VIRLOGEUX, P. & LEPVRIER, C. 1986. Seaward extension of the East African rift. *Nature*, **321**, 599–603.

NYBLADE, A. A. & BRAZIER, R. A. 2002. Precambrian lithospheric controls on the development of the East African rift system. *Geology*, **30**, 755–758.

NYBLADE, A. A., BURT, C., LANGSTON, C. A., OWENS, T. J. & LAST, R. J. 1996. Seismic experiment reveals rifting of craton in Tanzania. *EOS, Transactions of the American Geophysical Union*, **77**, 519–521.

ORRIDGE, G. R. 1965. *Mbulu*. Tanzania Mineral Resources Division. Quarter Degree Sheet **69**.

PICKERING, R. 1958. *Oldoinyo Ogol*. Tanganyika Geological Survey. Quarter Degree Sheet **38**.

PICKERING, R. 1961. The geology of the country round Endulen. *Records of the Geological Survey of Tanganyika*, **XI**, 1–9.

PICKERING, R. 1964. *Endulen*. Tanzania Geological Survey. Quarter Degree Sheet **52**.

RING, U., SCHWARTZ, H. L., BROMAGE, T. G. & SANAANE, C. 2005. Kinematic and sedimentological evolution of the Manyara Rift in northern Tanzania, East Africa. *Geological Magazine*, **142**, 355–368.

SAGGERSON, E. P. & BAKER, B. H. 1965. Post-Jurassic erosion surfaces of East Africa and their deformation in relation to rift structure. *Journal of the Geological Society, London*, **121**, 51–72.

SELBY, J. & THOMAS, C. M. 1966. *Balangida Lelu*. Tanzania Mineral Resources Division. Quarter Degree Sheet **103**.

SIMPSON, F., HAAK, V., KHAN, M. A., SAKKAS, V. & MEJU, M. 1997. The KRISP magnetotelluric survey of 1995: first results. *Tectonophysics*, **278**, 261–271.

SMITH, M. & MOSLEY, P. 1993. Crustal heterogeneities and basement influence on the development of the Kenya Rift, East Africa. *Tectonics*, **12**, 591–606.

UHLIG, C. 1907*a*. Vom Kilimandscharo zum Meru. *Zeitschrift der Gesellschaft für Erdkunde*, **1907**, 627–650.

UHLIG, C. 1907*b*. Der sogennante grosse Ostafrikanische Graben zwischen Magad (Natron See) und Lawa ya Mweri (Manyara See). *Geographisches Zeitung*, **15**, 478–505.

VAIL, J. R. 1970. Tectonic controls of dykes and related irruptive rocks in eastern Africa. *In*: CLIFFORD, T. N. & GASS, I. G. (eds) *African Magmatism and Tectonics*. Oliver & Boyd, Edinburgh, 337–354.

WALTER, R. C., MANEGA, P. C., HAY, R. L., DRAKE, R. E. & CURTIS, G. H. 1991. Laser fusion $^{40}Ar/^{39}Ar$ dating of Bed I, Olduvai Gorge, Tanzania. *Nature*, **354**, 145–149.

WHITTINGHAM, J. K. 1963. The geological environment of the Kilombero valley. *Records of the Geological Survey of Tanganyika*, **X**, 17–27.

WILKINSON, P., MITCHELL, J. G., CATTERMOLE, P. J. & DOWNIE, C. 1986. Volcanic chronology of the Meru–Kilimanjaro region, Northern Tanzania. *Journal of the Geological Society of London*, **143**, 601–605.

WILLIS, B. 1936. *East African Plateaus and Rift Valleys*. Carnegie Institution of Washington, Washington D.C.

YIRGU, G., EBINGER, C. Y. & MAGUIRE, K. H. (eds) 2006. *The Afar Volcanic Province within the East African Rift System*. Geological Society, London, Special Publications, **259**.

# Chapter 6

# Rift-associated sedimentary basins

The warping of the mid-Tertiary land surface and the subsequent faulting gave rise to a series of half grabens. Due to the disruption of earlier river systems, inland drainage basins formed within the half grabens and, in these basins, lake sediments were deposited, often intercalated with lava flows and ash-fall deposits, the thickest sedimentary successions forming in depocentres immediately adjacent to the boundary faults.

The most prominent of the basins are along the foot of the east-facing Pleistocene escarpment, being, from north to south, Natron, Engaruka, Manyara, Balangida and Balangida Lelu (Fig. 6.1). Unlike the half graben rift basins in Lake Tangayika (Rosendahl 1987) and in parts of Kenya, the northern Tanzania basins are not enclosed within opposed faults, nor are they linked by accommodation zones. These basins, together with the Eyasi half graben basin, now contain saline lakes. Other sedimentation basins are those of Olduvai/Laetoli and Amboseli, and some small ones in the Monduli area.

## The Natron Basin

The Natron Basin in its early stages was probably joined to the Magadi Basin to the north. The maximum depth of the basin, estimated from Euler deconvolution (Ebinger *et al.* 1997), is around 3.3 km, which is similar to the 3.5 km depth to the basement derived from seismic refraction data for the Magadi Basin (Birt *et al.* 1997). Much of the basin is underlain by the Sambu volcanics and the Peninj Beds which were deposited in the palaeo-Lake Natron until *c.*1.2 Ma when the sedimentation was ended by the Natron boundary faulting.

*Palaeo-Lake Natron.* Sedimentation in the Natron Basin began at *c.* 1.7 Ma in a shallow, depression that may have been fault-generated. A succession of basalts and nephelinites from Oldoinyo Sambu and Mosonik, dated at 3.5–1.77 Ma (Isaac & Curtis 1974), is overlain by the sediments of the Peninj Group, from which they are separated by a thin layer of sediment and a weathered lava (the Hajaro lava) collectively termed the Pre-Peninj Beds. The Peninj Group, named after the Peninj River where they are best exposed between the Sanjan and Natron faults immediately to the west of the present-day Lake Natron, is a 100 m succession of sediments with interbedded lavas. The sediments contain hominid remains and artefacts (Leakey & Leakey 1964; Isaac 1965; Howell 1972; Dundas & Awadalla, unpublished). The group comprises the lower Humbu Formation, formed of deltaic freshwater beds deposited in a shallow depression. The sediments are sands with clays and reworked tuffs with one thin limestone horizon. The palaeomagnetism of the tuffs record the Matuyama Epoch, including the Olduvai reversal at 1.87–1.67 Ma (Thouveny & Taieb 1986). The overlying, more widespread, Moinik Formation was deposited in a saline lake that extended westwards across the Salei Plains towards the Oldoinyo Ogol escarpment (Isaac 1965; Dundas & Awadalla, unpublished). The sediments contrast strongly with those of the underlying Humbu Formation, the conditions probably being like those of the modern Lake Natron, with deposition of fine clays, laminated trachytic tuffs and evaporites. Cherts derived from the clays are widespread, and a chert nodule and a bedded chert with trona moulds have $\delta^{18}O$ values of 40.3 and 36.6‰, respectively (O'Neil & Hay 1973). The overall Peninj Group was deposited between 1.7 and 1.2 Ma (Dawson 2008, Appendix 1). From the tectonic aspect, the Peninj area provides important evidence for: (a) the formation of an internal drainage basin due to pre-Pleistocene faulting; and (b) an age of <1.2 Ma for the faulting that terminated the Peninj Group sedimentation and formed the Natron Basin boundary fault.

*Present-day Lake Natron.* This is surrounded by, and sits below, the High Natron Beds, a 7 m section of zeolitized clays and tuffs with irregular nodules of chert, some rimmed by magadiite $[NaSi_7O_{13}(OH_3)\cdot 3H_2O]$ and kenyaite $[Na_2Si_{22}O_{41}(OH)_8\cdot 6H_2O]$. Chert nodules and magadiite have $\delta^{18}O$ values of 32.4 to 34.6‰ (O'Neil & Hay 1973). The High Natron Beds are probably correlatable with the High Magadi Beds surrounding Lake Magadi, which were deposited during a lake highstand at around 9 ka (Butzer *et al.* 1972). There was an earlier highstand at around 130 ka during which freshwater stromatolite growth took place (Hilaire-Marcel *et al.* 1986).

The modern lake lies at an average height of 610 m a.s.l., is 22–35 km wide and 75 km long, with a present surface area of around 1059 $km^2$. The lake is shallow (only 3–4 m depth), and the evaporation rate is high, up to 20 mm per day. Its saline waters are rich in sodium carbonate/bicarbonate, NaCl and silica, and its pH is >9.5 (Howell 1972). The lake is fed by four permanent rivers, being from north to south: (1) the Engare Nyiro that flows south into the lake from Kenya; (2) the Peninj; (3) the Moinik rivers that enter the west side of the lake; and (4) the Engare Sero which enters the SW corner. All rise in the highlands to the west of the Natron Basin boundary fault. As well as volcanic detritus, the first three rivers contribute metamorphic debris from the basement hills to the west of the Nguruman and Sambu escarpments, as well as volcanic detritus, whereas the Engare Sero transports only volcanic debris from the Crater Highlands. Brackish lagoons exist where the rivers enter the lake and, in addition to these permanent rivers, there are many brackish or salt springs on the western and eastern shores of the lake. When the lake waters reach a high salinity, proliferation of brine shrimps result in the surface layer having a bright-red coloration.

Much of the lake surface is covered by a surface layer of trona, thermonatrite and halite, similar layers below the surface being interbedded with black organic clays to depths up to 6 m (Guest & Stevens 1951). The clays result from input of volcanic dust into the lake which, combined with organic material from detritus and algal blooms, forms the organic-rich bottom sediments, which also contain gaylussite and pirssonite and up to 10% authigenic analcime (Hay 1966). Although most of the carbonate salts have been leached out of the sodium-rich volcanic rocks surrounding the lake, there has been some direct volcanic input into the lake during natrocarbonatite eruptions of Oldoinyo Lengai. Magnesite has been deposited around one of the springs on the east side of the lake below Gelai. Guest (1951) suggests it was deposited when spring water rich in leached $Mg_2SO_4$ reacted with saline lake waters, an origin supported by the intergrowth of the magnesite with hectorite $(Na_{0.33}[Mg,Ca]_3Si_4O_{10}[OH]_2)$ (McKie 1956). A point of non-geological interest is that the salt-flats on the lake are now the only breeding place for Lesser Flamingo in the rift valley (Brown 1959).

J. B. Dawson, University of Edinburgh (*jbdawson@glg.ed.ac.uk*)
*From*: DAWSON, J. B. 2008. *The Gregory Rift Valley and Neogene–Recent Volcanoes of Northern Tanzania*. Geological Society, London, Memoirs, **33**, 33–38.
0435-4052/08/$15.00 © The Geological Society of London 2008. DOI: 10.1144/M33.6

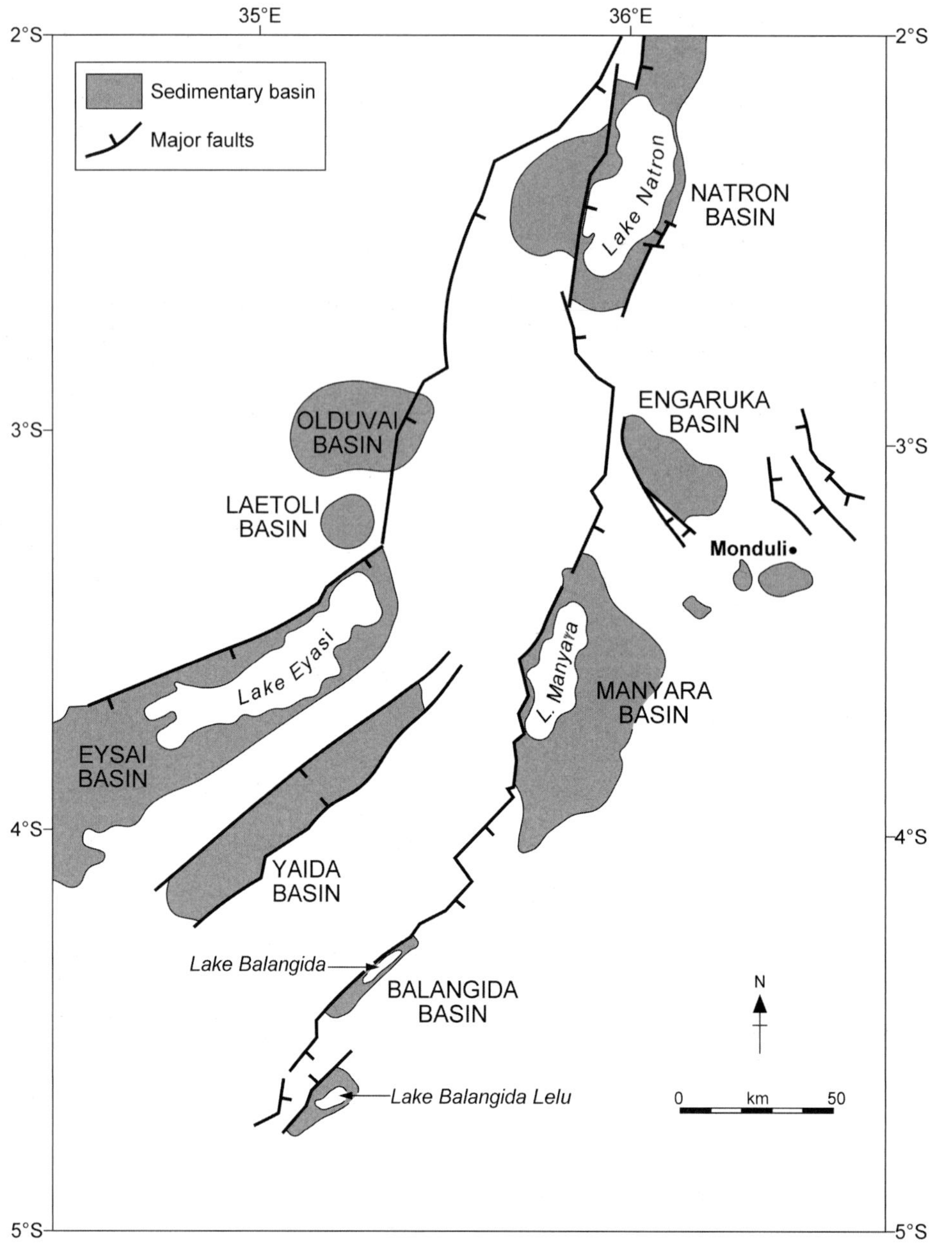


**Fig. 6.1.** Location of the rift-associated sedimentary basins relative to the major faults.

## The Engaruka Basin

The Engaruka Basin is separated from the Natron Basin by the topographic highs of the Oldoinyo Lengai and Kerimasi volcanoes, and from the Manyara Basin by the Engaruka Block (Fig. 5.8). It contains a shallow, seasonal lake fed by the perennial Engaruka River, draining from the Crater Highlands to the west, and smaller seasonal rivers rising on the volcanoes to the east, north and south of the basin. The sediments are poorly exposed, the main exposures being in the gorge of the Emugur Oretati (the river draining westwards between the Ketumbeine and Tarosero volcanoes), and in the gorge of the Engaibataat river that drains into the basin from the SW. These rivers expose tuffs and water-reworked tuffs, the age of which is not precisely known. However, their intercalation with flows of olivine basalt that have baked and reddened the sediments (Dawson 1964*a*) indicates some deposition during at least the later stages of the activity of Ketumbine and/or Tarosero. On the northeastern margin of the basin are exposures of calcite-rich crystal tuffs from Kerimasi.

## The Manyara Basin

The only evidence for the presence of pre-volcanic sedimentary basins in northern Tanzania derives from an exposure at the base of the rift escarpment at the north end of Lake Manyara, where the Manyara Group comprises a boulder conglomerate unconformably overlying metamorphic basement rocks; the conglomerate is overlain by siltstones, sandstones, waterlain tuffs and ashes with an intercalated basalt that has been dated at $4.86 \pm 0.24$ Ma. This succession has been interpreted as indicating some basin subsidence (possibly related to faulting) by around 4.9 Ma (Foster 1997; Foster *et al.* 1997). However, before the formation of the present-day basin at around 1.0 to 1.2 Ma, there was the intervening period of Older Extrusive magmatism.

The modern Manyara Basin, the maximum depth of which has been calculated from Euler deconvolution as being *c.* 2.9 km (Ebinger *et al.* 1997), is one of contrasts. Beneath the veneer of Pleistocene–Recent sediments, the basin at its northern end is underlain at shallow depths by volcanic rocks from the Crater

Highlands volcanoes and from Essimingor, whereas the southern part is underlain solely by Proterozoic metamorphic rocks. Also, the southern part of the basin is characterized by much minor seismic activity, whereas the northern part is relatively quiet seismically.

The present-day Lake Manyara occupies only the western part of a formerly more extensive Pleistocene basin that extended to the east of present-day Lake Burungi (also known as Sereri) that, like Manyara, is another remnant of the former Pleistocene lake (Dawson 2008, fig. 3.1). The eastern margin of the palaeo-lake is at least 40 km east of the present shoreline, suggesting an areal extent of *c.* 1800 $km^2$ compared with the current 480 $km^2$. It is debatable whether the westerly migration of the lake to its present position below the rift escarpment is connected with the general onset of aridity in Africa during the Pleistocene or due to a deepening of the western part of the basin following a series of minor movements on the main boundary fault, or both. The highest level of the former lake, as deduced from guano deposit around a former island of Precambrian quartzite at Minjingu, is around 100 m above the present-day lake (Orridge 1965). The eastern shore of the ancient lake is delineated by a series of at least five terrace-like shoreline features, one of which is characterized by the presence of chert nodules, pisolitic limestones and stromatolites. Some stromatolites show several stages of spongious and oncolitic development (Loftus 1985). However, Casanova (1986) has pointed out that there are no examples of stromatolites currently forming in the sodium carbonate/bicarbonate lakes but, in common with other rift valley lakes where there are fossil stromatolites (Bogoria, Magadi, Natron), their presence is evidence of freshwater palaeolakes containing significant dissolved calcium carbonate.

The fossil lake beds, best exposed around Makuyuni, were first identified as such by Kent (1942) who, from fossil fish and mammals, suggested correlation with Bed II of the Olduvai Gorge succession i.e. Early or Middle Pleistocene. Ring *et al.* (2005) divide the Manyara Beds into lower and upper members, both of which contain much tuffaceous material. The lower member is a lacustrine sequence comprising basal conglomerate overlain by sandstones, marls and thin tuffs. These have yielded fossil mammals, including a hominin skull fragment and a tooth that, together with suid teeth, give a rough age of $<1.7$ to 0.7 Ma, corresponding to the Middle and Upper Bed II and Lower Bed III of Olduvai. The upper member represents an abrupt change from lacustrine to terrestrial sedimentation due to westward migration of the eastern shoreline. It is a dominantly terrestrial facies and consists of mudstones (some showing salt pseudomorphs and desiccation cracks), siltstones and tuffaceous conglomerates and breccias that represent flood-plain, channel-fill and debris flow facies. Mammal fossils suggest correlation with Olduvai Bed III (*c.* 0.7–0.4 Ma, late Pleistocene).

The average depth of the present lake is only around 3 m. In the dry season, the surface of the lake is covered with trona deposits and in times of drought the lake may almost dry up. There are a few freshwater lagoons along the western edge of the lake, where perennial rivers descend from the escarpment, and where water from springs along the foot of the escarpment enters the lake.

Phosphate-bearing horizons in sediments are mined around a former island of basement quartzites at Minjingu. Cormorant skeletons in the beds give grounds for suggesting that the phosphate originated as guano.

### *The Monduli area basins*

In the area to the west of Monduli are three small sedimentary basins, the Yayai, Elunata and Ardai basins (Dawson 2008, fig. 5.8), formed by minor fault-scarps damming streams draining southwards off the Essimingor, Burko, Tarosero and Monduli volcanoes (Dawson 1964*a*). The basins are drained by minor rivers that have cut overflow channels through the escarpments to drain via the Engorika and Loilera rivers into Lake Manyara. Exposure in the basins is poor; olivine-rich tuffs from the Lashaine cone are exposed in a minor river draining southwards across the Ardai Basin.

### *The Balangida and Balangida Lelu basins*

These basins lie at the foot of the same boundary-fault escarpment as the Manyara Basin, but are separated from it by a topographic high of metamorphic rocks. Both basins are underlain by Archaean rocks of the Tanzania Craton. They contain lakes that are seasonally saline, and the surrounding areas are covered with superficial alluvial and aeolian sands and silts. The eastern part of the the Balangida Basin is occupied by the Hanang volcano which, on its northwestern side, sheds fanglomerates into the basin. Both basins are believed to be shallow, with the maximum depth of the Balangida Basin, estimated from Euler deconvolution, being $<1$ km (Ebinger *et al.* 1997).

## The Eyasi Basin

The Eyasi Basin lies within the Eyasi half graben and is bounded to the NW by the Eyasi escarpment; very minor NW-throwing faults form the southern edge of the basin. The basin is occupied by Lake Eyasi, a shallow seasonal lake which, on drying out, leaves mud, trona and halite. The superficial sediments around the lake are of sand–clay size, and are frequently wind-eroded and covered by small dunes. Deeper material is not exposed. The regional Bouguer anomaly map (Dawson 2008, fig. 4.1) shows a low anomaly beneath the east end of the basin; this is separated from a larger anomaly to the west (beneath the Wembere steppe) by a shallow ridge lying beneath the west end of Lake Eyasi. Euler deconvolution indicates the depth to the magnetic basement beneath the East Eyasi basin to be $<1$ km, compared with 3.5 km beneath the Wembere (W. Eyasi) Basin (Ebinger *et al.* 1997).

## The Yaida Basin

The Yaida Basin occupies what is possibly the best example of a graben in East Africa. Earlier called the Hohenlohe graben (Jaeger 1913), it lies between two opposing NE-trending, 100 km long fault scarps that are almost parallel. The northern escarpment has a maximum height of 300 m, the southern of 400 m. The width between the escarpments varies from 10 to 16 km, widening from NE–SW. Nothing is recorded of the sediments within the graben.

## The Olduvai Basin and Laetoli

The Olduvai Basin is a shallow, rift-platform basin situated on the Serengeti Plain to the west of the Crater Highlands. Its sedimentary rocks, exposed in the Olduvai Gorge and some minor tributaries, are justifiably famous for their hominid and Plio-Pleistocene mammalian fossils. The dating of individual formations in connection with anthropological studies at both Olduvai and Laetoli, and the correlation of some of these dated formations with volcanic activity to the east, has been of considerable benefit to volcanic chronology in the Crater Highlands.

The first fossil finds at Olduvai were made by a German entomologist, a Professor Kattwinkel, whilst collecting butterflies in 1911. His small collection aroused interest when he returned to Berlin, and resulted in an expedition to Olduvai in 1913 led by Hans Reck, who made the first systematic ordering of the lithological units from Beds I to V (Reck 1914*a*, *b*). In 1928, L. S. B. Leakey visited Berlin and recognized artefacts amongst the material gathered in 1913, and led to a series of expeditions to collect new material. The history of investigations into the fossils of Olduvai from 1911 to 1975 is reviewed by M. Leakey (1978).

The first attempt to set the fossil finds in their geological context at Olduvai was made by Reck (1914*a*) and various stratigraphic refinements were made up to the 1960s. The most systematic sedimentological work began in 1962 and culminated in the monograph of Hay (1976), from which the following account is mainly drawn. Ashley & Hay (2002) later analysed lateral variations across the basin that elucidate the palaeoenvironent during an early stage of deposition and, more recently, Ashley (2007) has identified five episodes of lake expansion and contraction that may be linked to variations in insolation which, in turn, result from variations in global precession; the lake episodes are thus suggested as being connected to astronomical forcing.

The Olduvai Beds were deposited in a basin that was *c.* 50 km in diameter through much of the Pleistocene. A shallow saline lake some 15–20 km in diameter occupied the axial area of the basin during the early history of the basin. The maximum thickness of the beds is *c.* 140 m, with the greatest thickness in the eastern part of the basin. Above underlying basalt (possibly erupted from Olmoti), the beds are subdivided into seven formations: Beds I–IV, the Masek Beds, the Ndutu Beds and the Naisiusiu Beds; Beds I to IV cover the interval between *c.* 2.1 and 0.6 Ma (Curtis & Hay 1972; Walter *et al.* 1991; and see Appendix 1). In addition, Holocene sediments are present in the gorge and on the surrounding Serengeti Plains.

The earliest deposits (Beds I and II) were derived by airfall, *nueé ardente* and fluviatile input from the east or south. Bed I (the thickest formation, up to 62 m thick, 2.0–1.7 Ma) comprises airfall tuffs, reworked tuffs, ignimbrites and olivine basalt flows into a saline lake, the flows being of local derivation compared with the tuffs and ignimbrites that were derived from the Crater Highlands. The reworked pyroclastic rocks originated in a pyroclastic fan complex on the eastern and southern shores of the lake. Bed II (20–30 m thick) spans the period 1.7 to *c.* 1.15 Ma. The lower part lithologically resembles the tuffs and saline lake deposits of Bed I but, above a disconformity (which marks the onset of widespread faulting and folding in the central part of the basin) the upper part of Bed II records a decrease in the size of the saline lake which disappeared just before the end of Bed II deposition. The disconformity marks a faunal change from swamp- and water-dwelling forms to a predominance of forms favouring open savannah and riverine conditions. Further, Bed II is important from the viewpoint of Palaeolithic industry sites, with Oldowan sites in the lower part, and developed Oldowan and Acheulian sites in the middle and upper sections of the bed. Similarly there is a difference in hominid remains, with remains of *Homo habilis* below the disconformity and *Homo erectus* above the break; remains of *Australopithecus* (*Zinjanthropus*) *boisie* occur both high and low in Bed II (Fig. 6.2). Chert nodules in Bed II have $\delta^{18}O$ values of 31.1 to 38.7‰ (O'Neil & Hay 1973).

A widespread episode of faulting at around 1.15 Ma caused erosion of Bed II and partly changed the direction of sediment provenance. Beds III (7–9 m thick) and IV (*c.* 2 m thick) were deposited on alluvial plains that received metamorphic debris from the north and west, in addition to volcanic debris from the south and east. The sediments are mainly claystones and sandstones containing smectite and illite (Hay & Kyser 2001). The Masek Beds (12 m thick) disconformably overlie Beds III and IV and were the last to be deposited before the the erosion that cut the Olduvai Gorge. They comprise wind-worked tephra erupted from Kerimasi to the NE and detrital sediments that occasionally infill channels in the underlying sediments.

**Fig. 6.2.** Interbedded tuffaceous sandstones and siltstones of Bed II, Olduvai Gorge. The plinth by the standing figure marks the site of the first discovery of *Australopithecus cf. boisei* ('Nutcracker Man').

The Masek Beds were followed by deposition of the Ndutu Beds (up to 18 m thick) during a period of faulting, erosion and partial infilling of the gorge. The lower Ndutu Beds are small patches of sand, conglomerate and tuffs deposited at various levels on the eroded side of the gorge. Deposition of the lower clastic sediments spanned 400 000 to 60 000 year BP, whereas the upper parts of the Ndutu Beds were deposited between 60 000 and 32 000 year BP from early eruptions of Oldoinyo Lengai. The Ndutu Beds also form a thin layer of tuff on the plains around the gorge. The Naisiusiu Beds (*c.* 8 m thick) were deposited after erosion of the Ndutu Beds, and consist of aeolian tuffs that cover the sides of the gorge and the surrounding plains. The beds consist of tephra derived from Oldoinyo Lengai, and are dated at 22 000 to 15 000 year BP.

Calcretes occur in Bed IV and in the Masek, Ndutu and Naisiusiu Beds. Formerly interpreted as surface limestones, they are now recognized as derived from carbonate-rich tephra from Kerimasi and Oldoinyo Lengai (Dawson 1964*b*). The calcretes, often overlie zeolite-rich horizons (Hay 1963); the zeolites arising by breakdown of nepheline and glass within the tephra due to interaction with natrocarbonatite ash; the zeolites are analcime, natrolite, phillipsite and chabazite, and they occur together with calcite, rare dolomite and locally-common dawsonite ($NaAl[OH]_2CO_3$) (Hay & Reeder 1978).

During the lifetime of the basin, the location of the drainage sump migrated eastwards with the Ndutu Beds now confined to the Olbalbal depression following faulting along the most easterly of the faults affecting the basin. The basin was deformed almost continuously from Bed I to the present, with most faults trending NE–SW, mainly parallel to the Olbalbal fault. The eastern margin of the Bed I basin was downwarped and locally faulted, and this continued during the deposition of Bed II. Beds III and IV record movements along new faults that may be correlated with major movements on the Natron–Manyara boundary fault to the east at *c.* 1.0–1.2 Ma; the reversal of drainage and changes in the type of sediment input caused by this episode of faulting is noted above. Several new faults occurred during the deposition of the Masek Beds. The Ndutu Beds record an intense phase of deformation at around 0.4–0.03 Ma when the Olbalbal depression subsided, with southeastwards tilting of the basin. Numerous faults cutting the slopes of the Ngorongoro and Olmoti volcanoes to the east were also active at this time.

**Laetoli** lies to the south of Olduvai and its stratigraphy, hominid and other mammalian fossils have been studied by the same group of investigators as at Olduvai. The Laetoli Beds are Pliocene air-fall and wind-reworked tuffs that are interbedded with lavas on the lower western and southern slopes of the Lemagrut volcano (Dawson 2008, fig. 3.1). The beds, originally called the Vogel River Series, were first mapped by Kent (1941) and divided into the lower Laetoli Beds and the overlying Ngaloba Beds. Importantly, he recognized the volcanoclastic nature of the beds and that the fossils (e.g. elephants, giraffes) represent only terrestrial fauna, contrasting with the swamp- and lake-dwelling fauna at Olduvai. Fossil eggs of francolin, guinea fowl and ostriches (Harrison 2005; Harrison & Msuya 2005) confirm Kent's (1941) interpretation of the environment as savannah or open grassland.

The discovery of fossil hominids (Leakey *et al.* 1978) stimulated more research, including refinement of the stratigraphy and dating. The beds are now known to dip under (and are therefore older than) the Olduvai Beds but overlie lavas from Lemagrut (Hay in Howell 1972). Hay (1987) recognized a lower unit (*c.* 64 m thick) consisting mainly of wind-worked nephelinite tephra, with thin conglomerates and breccias; it is poorly fossiliferous. An upper unit 44–59 m thick consists mainly of wind-worked tuffs, some containing fragments of ijolite and nephelinite, and also minor air-fall tuffs. These tuffs are of phonolitic composition, and several horizons contain calcite, devitrified glass, pyroxene, garnet, melilite and biotite mainly derived from the Sadiman volcano 15 km to the east (Hay 1978). From carbon and oxygen isotope data the calcite was formerly believed to be secondary after original carbonatite alkali-carbonates (Hay & O'Neil 1983) but, on both textural and calcite composition grounds, Barker & Milliken (2008) propose that the cementing calcite is diagenetic, resulting from interaction of glass with groundwater.

Apart from other fossils, these calcite-rich horizons preserve the footprints of a group of hominids (Hay & Leakey 1982), and also delicate bird footprints and insects (Hay 1986). Very rapid cementation by diagenetic calcite, Fe-nontronite and phillipsite is suggested as the cause of the preservation of these delicate prints (Barker & Milliken 2008).

Dates of 3.7 and 3.49 Ma have been obtained on biotites from the lower and upper tuffs, and a vogesite lava from one of a small group of vents to the south, SW and west of Lemagrut (the Ogol lavas), overlying the beds, is dated at 2.41 Ma (Drake & Curtis 1987). Dating of Lemagrut lavas from other localities has given ages ranging from 4.3 to 2.7 Ma, (Appendix 1, Dawson 2008) so both the superposition of the Sadiman-derived tuffs on the lower slopes of Lemagrut lavas at Laetoli, and the bracketing of the radiometric ages of Sadiman by those from Lemagrut, provide evidence for contemporaneous volcanic activity at the two volcanoes.

## Amboseli

Although not strictly comparable with the other fault-bounded rift valley basins nor with Olduvai and Laetoli, the Amboseli Basin owes its origin to the same epoch of crustal deformation and subsequent volcanicity. The Amboseli Basin was created when the eruption of Kilimanjaro interrupted the regional eastwards slope of the deformed sub-Miocene land surface to form a dam, allowing the Amboseli Beds to accumulate (Pickford 1986). The lake itself is *c.* 250 km$^2$ in extent at maximum and considerably less in the dry season. Most of the lake and its larger basin lie in Kenya where Williams (1967) recorded marls, clay limestones, clays and siltstones (the Oltukai Beds, Amboseli Clays and Sinya Beds) overlying basal conglomerates, gravels and sands. In the southern extension of the Amboseli Clays into Tanzania, meerschaum (sepiolite–hydrous magnesium silicate, used for the manufacture of tobacco pipes and ornamental carvings) occurs as banded cavity fillings in dolomitic limestone cropping out near the southern end of the lake (Harris 1961).

## Summary of the rift valley basins

The intitiation of the basins in the rift valley took place during the late Pliocene and early Pleistocene by faulting of the sub-Miocene land surface. During later faulting, a series of lakes formed in the present-day basins at the base of the Natron–Balangida boundary fault when mid-Pleistocene faulting resulted in the disruption of earlier drainage systems. All are now inland drainage systems and have seasonal evaporite deposition. The basins show a consistent pattern in that both the age and depth of the basins decrease from north to south, i.e. from the Natron Basin to the Balangida Lelu Basin. Such a pattern is consistent with a north–south propagation of the main boundary fault responsible for the half graben holding the basins. Similarly, the major fault on the northwestern side of the Eyasi Basin decreases in height to the SW, but the depth of the basin does not show a consistent concomitant decrease.

The lakes in the Natron and Manyara basins are less extensive now than they were in the Pleistocene. Whereas the reason for this in the Manyara Basin is ambiguous (?onset of Pleistocene desiccation or due to tectonic activity), the shrinkage of Lake Natron can be linked to similar decreases in level of some of the lakes in southern Kenya, for which there is evidence of its having been a response to climate change.

## References


Ashley, G. M. 2007. Orbital rhythms, monsoons and playa lake response, Olduvai Basin, equatorial East Africa (ca 1.85–1.74 Ma). *Geology*, **35**, 1091–1094.

Ashley, G. M. & Hay, R. L. 2002. Sedimentation patterns in a Plio-Pleistocene volcaniclastic rift-platform basin, Olduvai Gorge, Tanzania. *In*: Renaut, R. W. & Ashley, G. M. (eds) *Sedimentation in Continental Rifts*. Society for Sedimentary Geology, Special Publications, **73**, 197–122.

Barker, D. & Milliken, K. L. 2008. Cementation of the Footprint Tuff, Laetoli, Tanzania. *Canadian Mineralogist* (in press).

Birt, C. S., Maguire, P. K. H., Khan, M. A., Thybo, H., Keller, G. R. & Patel, J. 1997. The influence on pre-existing structures on the evolution of the southern Kenya Rift Valley – evidence from seismic and gravity studies. *Tectonophysics*, **278**, 211–242.

Brown, L. 1959. *The Mystery of the Flamingos*. Country Life, London.

Butzer, K. W., Isaac, G. L., Richardson, J. L. & Washbourn-Kamau, C. 1972. Radiocarbon dating of East African lake levels. *Science*, **175**, 1069–1076.

Casanova, J. 1986. East African stromatolites. *In*: Frostick, L. E., Renaut, R. W., Reid, I. & Tiercelin, J. J. (eds) *Sedimentation in the African Rifts*. Geological Society, London, Special Publications, **25**, 201–210.

Curtis, G. H. & Hay, R. L. 1972. Further geological studies and K–Ar dating at Olduvai Gorge and Ngorongoro crater. *In*: Bishop, W. W. & Miller, J. W. (eds) *Calibration of Hominoid Evolution*. Scottish Academic Press, Edinburgh, 107–134.

Dawson, J. B. 1964*a*. *Monduli*. Tanzania Geological Survey Quarter Degree Sheet **54**.

Dawson, J. B. 1964*b*. Carbonatitic volcanic ashes in northern Tanganyika. *Bulletin Volcanologique*, **27**, 81–92.

Dawson, J. B. 2008. *The Gregory Rift Valley and Neogone–Recent Volcanoes of Northern Tanzania*. Geological Society, London, Memoirs, **33**.

Drake, P. & Curtis, G. H. 1987. K–Ar geochronology of the Laetoli fossil localities. *In*: Leakey, M. D. & Harris, J. M. (eds) *Laetoli, a Pliocene site in Northern Tanzania*. Clarendon Press, Oxford, 48–52.

Dundas, D. L. & Awadalla, G. W. (unpublished) *Loliondo*. Unpublished report and map of Quarter Degree Sheet 16 and 27, Tanzania Geological Survey, Mapped 1966.

Ebinger, C., Poudjom Djomani, Y., Mbede, E., Foster, A. & Dawson, J. B. 1997. Rifting Archaean lithosphere: the Eyasi–Manyara–Natron rifts, East Africa. *Journal of the Geological Sociey, London*, **154**, 947–60.

FOSTER, A. 1997. *The Seismicity and Tectonics of Africa*. PhD thesis, University of Cambridge.

FOSTER, A., EBINGER, C., MBEDE, E. & REX, D. 1997. Tectonic development of the northern sector of the East African Rift System. *Journal of the Geological Society, London*, **154**, 689–700.

GUEST, N. J. 1951. *Gelai magnesite*. Tanganyika Geological Survey, Mineral Resources Pamphlet **52**.

GUEST, N. J. & STEVENS, J. A. 1951. *Lake Natron, its Springs, Rivers, Brines and Visible Saline Reserves*. Tanganyika Geological Survey, Mineral Resources Pamphlet **58**.

HARRIS, J. F. 1961. *Summary of the Geology of Tanganyika. Memoir 1, Part IV: Economic geology*. Government Printer, Dar es Salaam, 102–103.

HARRISON, T. 2005. Fossil bird eggs from the Pliocene of Laetoli, Tanzania: Their taxonomic and palaeoecological relationships. *Journal of African Earth Sciences*, **41**, 289–302.

HARRISON, T. & MSUYA, C. P. 2005. Fossil struthionid shells from Laetoli, Tanzania: Taxonomic and biostratigraphic significance. *Journal of African Earth Sciences*, **41**, 303–315.

HAY, R. L. 1963. Zeolitic weathering in Olduvai Gorge, Tanganyika. *Bulletin of the Geological Society of America*, **74**, 1281–1286.

HAY, R. L. 1966. *Zeolites and Zeolitic Reactions in Sedimentary Rocks*. Geological Society of America, Special Publications, **85**, 39–44.

HAY, R. L. 1976. *Geology of the Olduvai Gorge*. University of California Press, Berkeley.

HAY, R. L. 1978. Melilitite–carbonatite tuffs in the Laetolil Beds of Tanzania. *Contributions to Mineralogy and Petrology*, **67**, 357–367.

HAY, R. L. 1986. Role of tephra in the preservation of fossils in Cenozoic deposits in East Africa. *In*: FROSTICK, L. E., RENAUT, R. W., REID, I. & TIERCELIN, J. J. (eds) *Sedimentation in the African Rifts*. Geological Society, London, Special Publications, **25**, 339–344.

HAY, R. L. 1987. Geology of the Laetoli area: *In*: LEAKEY, M. D. & HARRIS, J. M. (eds) *Laetoli: a Pliocene Site in Northern Tanzania*. Clarendon Press, Oxford, 23–47.

HAY, R. L. & LEAKEY, M. D. 1982. The fossil footprints of Laetoli. *Scientific American*, **246**, 50–57.

HAY, R. L. & O'NEILL, J. R. 1983. Carbonatite tuffs in the Laetoli Beds of Tanzania and the Kaiserstuhl in Germany. *Contributions to Mineralogy and Petrology*, **82**, 403–406.

HAY, R. L. & KYSER, T. K. 2001. Chemical sedimentology and palaeoenvironmental history of Lake Olduvai, a Pliocene lake in northern Tanzania. *Bulletin of the Geological Society of America*, **113**, 1505–1521.

HAY, R. L. & REEDER, R. J. 1978. Calcretes of the Oldovai Gorge and Ndolanya Beds of northern Tanzania. *Sedimentology*, **25**, 649–673.

HILAIRE-MARCEL, C., CARRO, O. & CASANOVA, J. 1986. $^{14}C$ and Th/U dating of Pleistocene and Holocene stromatolites from East African palaeo-lakes. *Quaternary Research*, **25**, 312–329.

HOWELL, F. C. 1972. Pliocene/Pleistocene hominidae in eastern Africa: absolute and relative ages. *In*: BISHOP, W. W. & MILLER, J. A. (eds) *Calibration of Hominoid Evolution*. Scottish Academic Press, Edinburgh, 331–368.

ISAAC, G. L. 1965. The stratigraphy of the Peninj Beds and the provenance of the Natron Australopithecine mandible. *Quaternaria*, **7**, 101–130.

ISAAC, G. L. & CURTIS, G. H. 1974. Age of the Acheulian industries from the Peninj Group, Tanzania. *Nature*, **249**, 624–627.

JAEGER, F. 1913. Das Hochland der Riesenkrater and die umliegender Hochländer Deutsch Ostafrika, Part 2, *Mitteilungen aus den Deutsches Schutzgebieten, Ergänzungsheft*, **8**, 1–213.

KENT, P. E. 1941. The Recent history and Pleistocene deposits of the plateau north of Lake Eyasi, Tanganyika. *Geological Magazine*, **78**, 173–184.

KENT, P. E. 1942. A note on Pleistocene deposits near Lake Manyara, Tanganyika. *Geological Magazine*, **79**, 72–77.

LEAKEY, L. S. B. & LEAKEY, M. D. 1964. Recent discoveries of fossil hominids at Olduvai and near Lake Natron. *Nature*, **209**, 3–9.

LEAKEY, M. D. 1978. Olduvai Gorge 1911–75: a history of the investigations. *In*: BISHOP, W. W. (ed.) *Geological Background to Fossil Man*. Scottish Academic Press, Edinburgh, 151–155.

LEAKEY, M. D., HAY, R. L., CURTIS, G. H., DRAKE, R. E., JACKES, M. K. & WHITE, T. D. 1978. Fossil hominids from the Laetoli Beds, Tanzania. *In*: BISHOP, W. W. (ed.) *Geological Background to Fossil Man*. Scottish Academic Press, Edinburgh, 157–170.

LOFTUS, G. W. F. 1985. *Petrology and depositional environment of the Dinantian Burdiehouse Limestone formation of Scotland: their relevance to the accumulation of the Oil-Shale Group*. PhD thesis, University College, London.

MCKIE, D. 1956. Mineralogical notes. *Records of the Geological Survey of Tanganyika*, **VI**, 87–93.

O'NEIL, J. R. & HAY, R. L. 1973. $^{18}O/^{16}O$ ratios in cherts associated with the saline lake deposits of East Africa. *Earth and Planetary Science Letters*, **19**, 257–266.

ORRIDGE, G. R. 1965. *Mbulu*. Tanzania Mineral Resources Division Quarter Degree Sheet **69**.

PICKFORD, M. 1986. Sedimentation and fossil preservation in the Nyanza Rift System, Kenya. *In*: FROSTICK, L. E., RENAUT, R. W., REID, I. & TIERCELIN, J. J. (eds) *Sedimentation in the African Rifts*. Geological Society, London, Special Publications, **25**, 345–362.

RECK, H. 1914*a*. Erste vorläufige Mitteilungen über den Fund eines fossil Menschskelets aus Zentralafrika. *Sitzungsberichten der Gesellschaft Naturforschender Freunde*, **3**, 81–95.

RECK, H. 1914*b*. Zweite vorläufige Mitteilungen über fossile Tier- und Menschenfunde aus Oldoway in Zentralafrika. *Sitzungsberichten der Gesellschaft Naturforschender Freunde*, **3**, 305–318.

RING, U., SCHWARTZ, H. L., BROMAGE, T. G. & SANAANE, C. 2005. Kinematic and sedimentological evolution of the Manyara Rift in northern Tanzania, East Africa. *Geological Magazine*, **142**, 355–68.

ROSENDAHL, B. R. 1987. Architecture of continental rifts with special reference to East Africa. *Annual Reviews of Earth and Planetary Sciences*, **15**, 445–503.

THOUVENY, N. & TAIEB, M. 1986. Preliminary magnetostratigraphic record of Pleistocene deposits, Lake Natron basin, Tanzania. *In*: FROSTICK, L. E., RENAUT, R. W., REID, I. & TIERCELIN, J. J. (eds) *Sedimentation in the African Rifts*. Geological Society, London, Special Publications, **25**, 331–336.

WALTER, R. C., MANEGA, P. C., HAY, R. L., DRAKE, R. E. & CURTIS, G. H. 1991. Laser fusion $^{40}Ar/^{39}Ar$ dating of Bed I, Olduvai Gorge, Tanzania. *Nature*, **354**, 145–149.

WILLIAMS, L. A. J. 1967. *Geology of the Amboseli area*. PhD thesis, University of East Africa.

# Chapter 7

# The Neogene–Recent volcanic rocks

The late Tertiary–Recent volcanoes are largely confined to the area of the late Tertiary depression, the exceptions being the most southerly volcanoes of Hanang and Kwaraha, and the Basotu explosion craters of the Mbulu Plateau (Dawson 2008, fig. 3.1).

The first systematic attempt to classify the volcanoes was that of Guest (1953) who, mainly on the basis of rock type, divided them into the Older Extrusives (volcanoes characterized by extrusion of basalt) and the Younger Extrusives (formed by more alkaline rock types). Guest's classification is broadly correct, but modifications to the scheme have resulted from later regional mapping by the Tanzania Geological Survey, together with radiometric dating of rocks from individual centres. This has resulted in a more refined stratigraphy, particularly with respect to the timing of the volcanicity relative to the regional tectonics. For example, some peralkaline volcanoes, formerly included with the Younger Extrusives, are older than, or contemporaneous with, some Older Extrusive basaltic centres. The term 'peralkaline volcano' is used here to indicate that the structure is formed dominantly of peralkaline rocks i.e. those for which whole-rock chemical analyses show $(Na + K)/Al > 1$.

Following the faulting that formed the late Tertiary tectonic depression, a group of major volcanoes erupted within the depression, the lavas largely infilling the depression and eventually overstepping the boundary fault-scarps at several points. In the SW, volcanoes erupting in the Crater Highlands infilled the northeastern end of the Eyasi half graben, and in the east, lavas from the Kilimanjaro massif filled the depression to the north of the Pangani graben.

The earliest date of eruption is poorly constrained since, even in the major fault escarpments, the first extrusive rocks of most of the volcanoes are not exposed. The only exposure of volcanic rocks in contact with the underlying basement rocks within the depression is in the extreme north, immediately south of the Kenya border, where basaltic lavas from Oldoinyo Sambu overlie Usagaran basement quartzites in the escarpment on the west shores of Lake Natron (Dundas & Awadalla, unpublished). One of these basalts has a K–Ar age of $7.7 \pm 0.8$ Ma (Appendix 1). Although these are the oldest extrusions exposed in the escarpment on the eastern side of Oldoinyo Sambu, Dundas & Awadalla (unpublished) state that the oldest lavas from this volcano are melanephelinites and melilitites of the Bast Hills and Birera, to the west of the main volcano. These, as with early lavas from some other centres, are high-alkali, silica-undersaturated types. Like Oldoinyo Sambu, activity at Lemagrut also started with extrusions of nephelinite and subsequently produced abundant alkali basalt lavas.

Major extrusions of the alkali basalt–trachyte–phonolite association typify the activity at many of the large shield volcanoes (Elanairobi, Olmoti, Loolmalasin, Ngorongoro, Oldeani, Gelai, Ketumbeine, Tarosero, Monduli and the Shira and Mawenzi centres of Kilimanjaro). Some of these basaltic volcanoes, Ngorongoro, Olmoti, Elanairobi (collectively termed the Crater Highlands) and Ketumbeine, developed calderas in connection with trachytic eruptions during their later stages.

However, there are exceptions to this dominantly basaltic magmatism. The oldest dated rock within the province is an 8.1 Ma nephelinite from Essimingor (Appendix 1, Dawson 2008), but here undersaturated magmatism persisted with no later basaltic extrusions. In addition to Essimingor, Shombole, Mosonik and Sadiman are nephelinite–phonolite pyroclastic stratovolcanoes, thus contrasting with the basaltic lava-dominated shield volcanoes. Hence, during the Older Extrusive episode, there was an overlap between peralkaline undersaturated magmatism and the more voluminous basaltic magmatism. For example, at Essimingor peralkaline magmatism was the sole type of activity and preceded the dominantly basaltic regional magmatism; it preceded but was then followed by basaltic activity at some centres such as Oldoinyo Sambu and Lemagrut; or it was contemporaneous with some of the basaltic volcanoes (Shombole, Mosonik and Sadiman).

The overall age range of this older phase of magmatism is from 8.1 Ma to around 1.2 Ma. The ages of these older volcanics overlap those of the Kirikiti and Olorgesaile basalts immediately north in Kenya, extruded at 5.0 to 1.0 Ma (Baker *et al.* 1972; Fairhead *et al.* 1972) but there is no equivalent in northern Tanzania of the earlier Kenyan flood phonolites, dated at *c.* 13 Ma (Bishop *et al.* 1969).

The exact timing of the end of activity, and the length of the activity, at most of these older centres is not known. However, lava extrusions at several centres, such as at the Kibo cone on Kilimanjaro, and at Tarosero, continued into the Pleistocene, and lavas from the Crater Highlands and, possibly, Oldoinyo Sambu, are known in the Pleistocene sedimentary successions at Olduvai and Peninj.

## Rock nomenclature

Most of the rock names given hereafter are made on the basis of thin-section examination by the referenced authors. For example, basaltic rocks containing plagioclase more sodic than labradorite are often referred to as 'trachybasalt'; only when chemical analyses are available can they be more correctly identified as mugearite or benmoreite. In naming rocks on the basis of their chemistry, the TAS (total alkalis–silica) system recommended by Le Maitre (1989) has been followed, except in the case of the 'foidite' grouping where, if possible, petrographic observation has allowed a more specific name to be given e.g. nephelinite, melilitite. In the case of olivine- and augite- phyric feldspathoidal lavas, especially from the Essimingor volcano, the broad thin-section criteria in Table 7.1 have been applied.

Some of the most modern geochemical work on volcanic rocks from the area is that of Paslick *et al.* (1995, 1996) who provide major, minor, trace element and Sr, Nd and Pb isotope data. To avoid unnecessary repetition, these analyses are referred to below as 'complete analyses'. In the case of some rock names given by Paslick *et al.* (1995, 1996), these were based on either earlier identifications by the petrographic laboratory of the Tanzania Geological Survey, or on whole-rock chemistry e.g. 'foidite' (as defined on a TAS diagram). Petrographic re-examination by the author in connection with this account has resulted in the renaming of some rocks.

Lava collections from some of the volcanoes have formed the basis for MSc and PhD theses. Space constraints make it impossible to give all the results presented in these theses, but data are summarized to give the interested reader some concept of the content of these valuable data sources.

## Descriptions of individual volcanoes or volcanic areas

When not otherwise referenced, statements pertaining to the occurrence and identification of specific rock-types are based on

J. B. Dawson, University of Edinburgh (*jbdawson@glg.ed.ac.uk*)
*From*: DAWSON, J. B. 2008. *The Gregory Rift Valley and Neogene–Recent Volcanoes of Northern Tanzania*. Geological Society, London, Memoirs, **33**, 39–77.
0435-4052/08/$15.00 © The Geological Society of London 2008. DOI: 10.1144/M33.7

**Table 7.1.** *Rock names based on thin section criteria*

| Rock name | Phenocrysts | Feldspathoid in groundmass |
|---|---|---|
| Limburgite | Mainly olivine + minor Ti-augite | Nepheline, analcime or sodalite |
| Ankaramite | 50% olivine + 50% Ti-augite | Nepheline, analcime or sodalite |
| Augitite | < 10% olivine, mainly Ti-augite | Nepheline, analcime or sodalite |

the author's fieldwork and thin section examination of rocks in his personal collection, now housed in the Natural History Museum, London (Oldoinyo Lengai specimens) and in the Harker Collection at the Sedgwick Museum, University of Cambridge (other Tanzanian samples). The colour plates, illustrate some of the features unique to the volcanoes of northern Tanzania.

Note: In the following accounts of the individual volcanic centres, the abbreviation QDS stands for 'Quarter Degree Sheet'.

## The Older Extrusives

### *Shombole*

*Type*: stratovolcano; *Lat/Long*: 2°09′S 36°06′E; *Elevation*: 1570 m; *Relief*: *c.* 1000 m; *Diameter*: 9.5 km; *Eruptive history*: Pliocene; *Crater diameter*: 2 × 2.5 km; *Composition*: nephelinite/carbonatite; *QDS* (Quarter Degree Sheet): 28 Kibangaini (Mineral Resources Division, Tanzania) and Report 61 (Geological Survey of Kenya).

Straddling the border between Kenya and Tanzania, Shombole lies mainly within Kenya. It is situated on the floor of the rift valley between the Magadi and Natron basins. The volcano is deeply eroded (Fig. 7.1), with exposure of its dominantly pyroclastic rocks in radial ravines and in north–south-trending fault escarpments that cut across the lower eastern slopes (Baker 1963). Fairhead *et al.* (1972) gave two K–Ar dates of 1.96 ± 0.07 and 2.00 ± 0.05 Ma on 'Shombole volcanics'. Kjaarsgaard & Peterson (1991) concluded a 2 Ma age from geological data, namely Quaternary trachytic lavas overlapping the eastern edge of the Shombole volcanics, and dating of trachytic lavas in the area (Crossley 1979).

The most recent and complete description of the volcano is by Peterson (1989*a*). A central vent is the main source of tuffs and agglomerates that are intercalated with the dominantly nephelinite lavas; ijolite blocks occur in the agglomerates (L. A. J. Williams pers. comm.). A caldera-forming event filled the central vent with flat-lying pyroclastics and minor flows, which on the outer flanks dip radially at 10–15°. Carbonatite tuffs are exposed low on the eastern slopes, but the main early lava type is perovskite nephelinite, succeeded by titanite nephelinite and later phonolitic nephelinite. Kjaarsgaard & Peterson (1991) suggested that the carbonatites and nephelinites are linked by liquid immiscibility. Carbonatite-cemented breccia occurs on the lower SW slopes and carbonate replaces wollastonite in some caldera nephelinites. Phonolites, of which three petrographically different types are recognized, occur as minor extrusions and are interpreted, on isotopic grounds, as resulting from the interaction of nephelinite magma with lower crustal granulites (Bell & Peterson 1991).

### *Oldoinyo Sambu*

*Type*: shield volcano; *Lat/Long*: 2.15°S 35.93°E *Elevation*: 2043 m; *Relief*: 1400 m (east side), 700 m (west side); *Diameter*: 42 km; *Eruptive history*: Upper Pliocene to Middle Pleistocene; *Composition*: mainly olivine basalt; *QDS*: 16 and 17 Loliondo.

Oldoinyo Sambu is one of the most northerly of the Tanzanian volcanoes, with its northern slopes descending into Kenya. From the west and east it appears a broad dome. On its western side it rises gradually from the Salei Plain, but its eastern slopes are bisected, and exposed, by the main north–SE-facing 1200 m escarpment of the Natron Basin boundary fault (Fig. 7.2). There is no summit crater and most of the eastern half of the original volcano now lies beneath Lake Natron, though a small hill of basalt, poking up through recent sediments on the western shores of the lake, may have formed part of the summit prior to the faulting.

The volcano is composed mainly of flows of olivine basalt, with phenocrysts of augite, olivine and labradorite, and subordinate trachybasalt and phonolite. Just to the south of the Kenya border, the lowest lavas rest upon basement quartzite (Dundas & Awadalla unpublished); these basal lavas are dated as 7.7 ± 0.8 Ma. Lenses of tuff and agglomerate are locally interbedded with

**Fig. 7.1.** Aerial view of Shombole from the SW. The dominantly pyroclastic deposits are reflected in the deeply-dissected topography. Trona deposits in Lake Natron appear in the foreground.

**Fig. 7.2.** Aerial view of Oldoinyo Sambu from the south. The volcano is bisected by the Natron basin boundary fault. Lake Natron (foreground) is covered with trona deposits and part of the Peninj delta appears in the left foreground.

basalts in the rift escarpment 4–6 km north of the Peninj delta, and deltaic and lacustrine sediments are intercalated with the basalts about 200 m below the top of the lava sequence in the gorge of the Peninj river on the southern lower slopes. A basalt just above these sediments gave a K–Ar age of 2.02 Ma. The lavas at the top of the succession were extruded before the end of the last reversed-polarity episode at 0.7 Ma. Thus, it appears that lavas were extruded for a period of around 5.5 million years.

Small outcrops of nephelinite and olivine nephelinite occur in the Bast Hills and in the Birera and Losanya areas to the west and SW of Oldoinyo Sambu. Their extrusion age is not known precisely, though those in the Bast Hills are overlain by Oldoinyo Sambu basalts. On the lower northern slopes of Oldoinyo Sambu, just over the border in Kenya, are flows of olivine basalt and small outcrops of biotite–olivine nephelinite (ankaratrite) (Saggerson 1966).

Krenkel (1922), quoting Finckh (1903 p.501), refers to, and gives an analysis of, a trachydolerite from Oldoinyo Sambu, which he compares with similar rocks from the Mawenzi peak on Kilimanjaro; compositionally (Appendix 2, Dawson 2008) the rock is a mugearite. Paslick *et al.* (1995) give complete rock analyses for basanites, nephelinites and picro-basalts in the rift escarpment at the south end of Lake Natron; however, it is not clear whether these flows are from Oldoinyo Sambu or from Elanairobi.

### *Mosonik*

*Type*: stratovolcano; *Lat/Long*: 2.61°S 35.80°E; *Elevation*: 1702 m; *Relief*: 480 m; *Volcano diameter*: 7.5 km; *Eruptive history*: Pliocene; *Composition*: nephelinite, phonolite, carbonatite; *QDS*: 39 Angata Salei.

Mosonik stands on the upthrown side of the Natron Basin boundary escarpment near the southern end of Lake Natron, and the cone is faulted on its northern side by the north–south-trending Sanjan Fault. Like Shombole and Essimingor, the volcano is unusual amongst the older volcanoes in being a nephelinitic stratovolcano, rather than a basaltic shield. Reflecting its dominantly pyroclastic make-up, it is strongly dissected, with deep radial valleys and steep intervening ridges (Fig. 7.3). There is now no evidence for a summit crater. Biotite from a nephelinite lava has yielded a date of 3.12 Ma (Isaac & Curtis 1974).

The Quarter Degree Sheet No. 39 (Angata Salei; Guest *et al.* 1961) shows that the cone consists mainly of nephelinitic–phonolitic tuffs and agglomerates, together with a lava lobe that extends east from the cone; the lava lobe may have issued from a breached crater. The flows comprising the lobe are of nephelinite and nephelinitic phonolite, and are cut by a minor fault that parallels the main escarpment. Paslick *et al.* (1995, 1996) give complete rock analyses of a basanite and three nephelinites from the lava lobe, together with analyses of nepheline, clinopyroxene, perovskite, Ti-magnetite and titanite. James (1966) cited the presence of sodalite nephelinite (Rosenbusch 1907) and aegirine sövite (Brögger 1921). The Geological Survey map also shows a small plug of carbonatite intruding the volcanics in a valley on the north side of the mountain. M. S. Garson (pers. comm.) reports that a United Nations geological survey was unable to find this plug but found small arcuate dykes and areas of carbonatite breccia in an area south of the plug locality.

**Fig. 7.3.** Aerial view of Mosonik from the SW. The well-dissected nature of the pyroclastic volcano is apparent. The Sanjan fault (left) and the Natron Basin boundary fault (see Fig. 5.7) appear to the north of the volcano.

J. B. Dawson, A. N. Mariano and R. H. Mitchell visited the volcano in September 2005 and provide the following additional information: (i) carbonate-cemented tuffs and lapilli tuffs blanket the lower northern and western slopes of the mountain, and form an extensive cover on the plains to the west; (ii) within the pyroclastic rocks, fragmented nepheline- and pyroxene-phyric lava is the commonest material but true nephelinite is rare, as most samples have some feldspar in their groundmass and are hence transitional to phonolite. Feldspar-phyric phonolite is present but in relatively small amounts. Phlogopite, titanite and melanite are accessory phases in some lavas; (iii) blocks of mica pyroxenite, ijolite-mantled pyroxenite, ijolite and syenite and xenoliths of biotite quartzite and biotite schist occur in the agglomerates; and (iv) traverses into the valleys on the northern and western sides of the volcano, including the valley where the carbonatite is shown to be located on the QDS map, found no samples of carbonatite, either *in situ* or as detritus. However, in September 2007, Dawson & Mitchell found rare blocks of sövite in the gorge of the Leshuta River that drains eastwards from the volcano into Lake Natron, together with abundant consolidated, Mosonik-derived debris in the walls of the gorge.

### *Gelai*

*Type*: shield volcano; *Lat/Long*: 2.60°S 36.11°E; *Elevation*: 2942 m; *Relief*: 1720 m; *Volcano diameter*: 20 × 36 km *Crater diameter*: none; *Eruptive history*: Pliocene *Composition*: alkali basalt–trachyte; *QDS*: 28 Kibangaini and 40 Gelai and Ketumbeine.

Gelai is a broad, elongate NNE–SSW-trending shield without a summit crater. The lower slopes consist mainly of alkali olivine basalt or basanite, and Guest (1953) noted some that contain 'enclaves of granular green olivine' (? peridotite xenoliths). On the upper slopes and in the well-forested summit area are scattered outcrops of peralkaline trachyte. Numerous minor scoria- and tuff-cones occur on the lower northeastern and southwestern slopes, often following the trend of numerous, small faults trending NNE–SSW (Guest & Pickering 1966*a*, *b*); Those on the southwestern lower slopes are later than the main cone flows, and include peridotite xenolith-bearing tuff cones at Pello Hill and Eledoi (Dawson & Powell 1969; and see 'Natron–Engaruka tuff-cone area' below). The apparent absence of small faults and cones on the upper slopes of the volcano may be due to poor exposure, but could also be due to later extrusion of lavas in the summit area (as on Tarosero). An age for the basalts on the lower northern and eastern slopes is given by their being cut by small faults, which are a continuation of a fault swarm in southern Kenya that has been dated at around 1–1.5 Ma. Hence, at least the lower basalts of Gelai predate the faulting event.

Guest (1953) gave analyses of a barkevikite- and anorthoclase-phyric nepheline benmoreite from the lower SW flank of the mountain and Paslick *et al.* (1996) gave complete analyses of two basanites from the volcano, together with olivine and Ti-magnetite in one of the basanites.

To the north of Gelai is a low-lying, grid-faulted area, Kibangaini, that contains abundant trachytes similar to those in the

**Fig. 7.4.** Ketumbeine from the west. The volcano stands 1800 m above the Engaruka Basin (foreground). Note the break in slope due to higher viscosity lavas on the higher slopes and the flat summit area due to caldera collapse.

Magadi area immediately to the north. Some Kibangaini rocks contain riebeckite and kataphorite; analyses of two trachytes were given by Guest (1953) (Appendix 2, Dawson 2008).

*Ketumbeine*

*Type*: shield volcano; *Lat/Long*: 2.89°S 36.22°E; *Elevation*: 2658 m; *Relief*: 1800 m; *Volcano diameter*: 26 km; *Crater diameter*: 3 × 2 km; *Eruptive history*: Pliocene; *Composition*: alkali olivine basalt and trachyte; *QDS*: 40 Gelai and Ketumbeine.

Ketumbeine is a broad, almost circular, shield volcano. The lower slopes consist mainly of low-angle basalt flows but there is a distinct steepening of the slopes in the summit area due to the extrusion of more viscous lavas (Fig. 7.4). Unlike Gelai, there is a poorly exposed, elongate (north–south-trending) crater on the summit (Guest & Pickering 1966*b*). There are small cones and faults on the lower slopes and Guest (1953) reported that most rocks on the lower flanks are 'basic andesite' and on the upper slopes are trachyandesite and trachyte. A basalt from the lower southeastern slopes contains phlogopite as an intergranular and vesicle-filling phase. (Dawson personal observation). Guest (1953) recorded nephelinite from Lolgurdolgonja, a small late parasitic cone on the lower western slopes of the main volcano.

Paslick *et al.* (1995, 1996) gave complete analyses of four olivine basalts and basanites, with mineral analyses of olivine and clinopyroxene in one specimen and of feldspar and Ti-magnetite in another.

## The Crater Highlands

This major group of dominantly basaltic shield volcanoes runs NE–SW from the Eyasi half graben for a distance of around 90 km and is bounded to the east by the steep, east-facing Natron Basin boundary fault escarpment (Fig. 7.5). The massif slopes down to the Salei Plain and the Olbalbal depression to the west and infills the Eyasi half graben at the southwestern end of the chain. As indicated by the name, some of the volcanoes (Ngorongoro, Olmoti and Elanairobi) have large craters, but several others, including the highest of this group of volcanoes, Loolmalasin, have no summit crater. Because of the thick soil and, particularly, the vegetation cover on the wetter, eastern and southern sides of the Highlands, exposure is not sufficient to establish the relative stratigraphy and age of the individual eruptive centres. It is possible that lavas from some centres are contemporaneous and intercalated.

Thick sequences of lavas presumed to originate from the more easterly centres of Elanairobi, Loolmalasin and Ngorongoro are exposed in the 800–1000 m escarpment between Lake Manyara and Lake Natron. The flows exposed in the escarpment between Oldoinyo Lengai and Lake Natron are distal flows of basanite, olivine basalt, mugearite, benmoreite and nephelinite, possibly from Elanairobi, and phonolitic nephelinites from Mosonik. Some are possibly interdigitated with distal, low-angle flows from Oldoinyo Sambu to the north. The faulted lava plateau on the floor of the rift valley between lakes Natron and Manyara is made up of flows from the Crater Highlands and distal flows from the Ketumbeine, Gelai and ?Tarosero volcanoes to the east.

*Elanairobi/Embagai (Empakai)*

*Type*: shield volcano; *Lat/Long*: 2.92°S 35.83°E; *Elevation*: 3235 m *Relief*: 2000 m above the Salei Plains to the west; *Diameter*: 30 km; *Eruptive history*: Pliocene; *Crater diameter*: 6.3 × 7.8 km; *Composition*: limburgite, trachybasalt, nephelinite, peralkaline phonolite; QDS: 39 Angata Salei.

Elanairobi is the most northerly of the Crater Highlands volcanoes. It is a broad shield, rising to 3222 m at its highest point, the Jaeger summit, and lies on a long NE–SW ridge that is bounded on its eastern side by faults that trend in the same direction. On the southern and south-eastern side of the mountain is the spectacular elliptical caldera of Embagai (Plate 1); this measures 6.3 × 7.8 km (the longer axis being east–west) and it has a maximum depth of 980 m. The eastern part of the crater is occupied by a circular lake 3.5 km in diameter, the level of which stands at 2209 m a.s.l. Recent lake-bottom sediments obtained in two boreholes have been deposited at *c*. 30 cm/ka during the late Pleistocene/early Holocene (Muzuka *et al.* 2004).

Trachybasalt flows form the lower parts, and nephelinite and aegirine phonolite the upper parts, of the Embagai crater walls, and limburgite flows occur at the base of the crater walls. Small parasitic cones on the crater floor have covered the crater floor and parts of the crater wall with trachybasalt tuffs, and tuffs crop out on the western and northeastern outer slopes and across the Jaeger summit. On the lower northern slopes, trachybasaltic tuffs overlie flows of basanite, nephelinite and olivine basalt in the gorge of the Engare Sero (Guest *et al.* 1961).

*Un-named mountain*

*Type*: unknown; *Lat/Long*: 2.98°S 35.89°E; *Elevation*: 2624 m; *Relief*: 600 m above the land to the west, 300 m above the top of the rift escarpment to the east; *Diameter*: *c*.8 km; *QDS*: 39 Angata Salei.

This relatively small, un-named, remote mountain lies between Loolmalasin and the village of Kapenjiro. It slopes down gradually to the Embulbul Depression to the west, and more steeply to the top of the rift escarpment immediately west of Engaruka. It is deeply dissected and has the form of steep, heavily forested ridges incised by ravines that drain eastwards into the Engaruka Basin. Its topography suggests that it is a dissected stratovolcano, similar to Mosonik. The mountain is cut off from the nearest driveable track (that from Nainokanoka to Kapenjiro) by deep ravines.

The Lorngabolo River, drains the eastern part of the mountain into the Engaruka Basin some 3 km north of Engaruka. In addition to basalt boulders derived from the escarpment, there are boulders of agglomerate (containing basalt and abundant pyroxenite clasts and mica), nephelinite and nepheline-phyric phonolite; the river-bed sand contains abundant grains of mica, black pyroxene and an opaque phase (Dawson & Mitchell personal observation, September 2007). These lithologies are similar to those occurring at Mosonik.

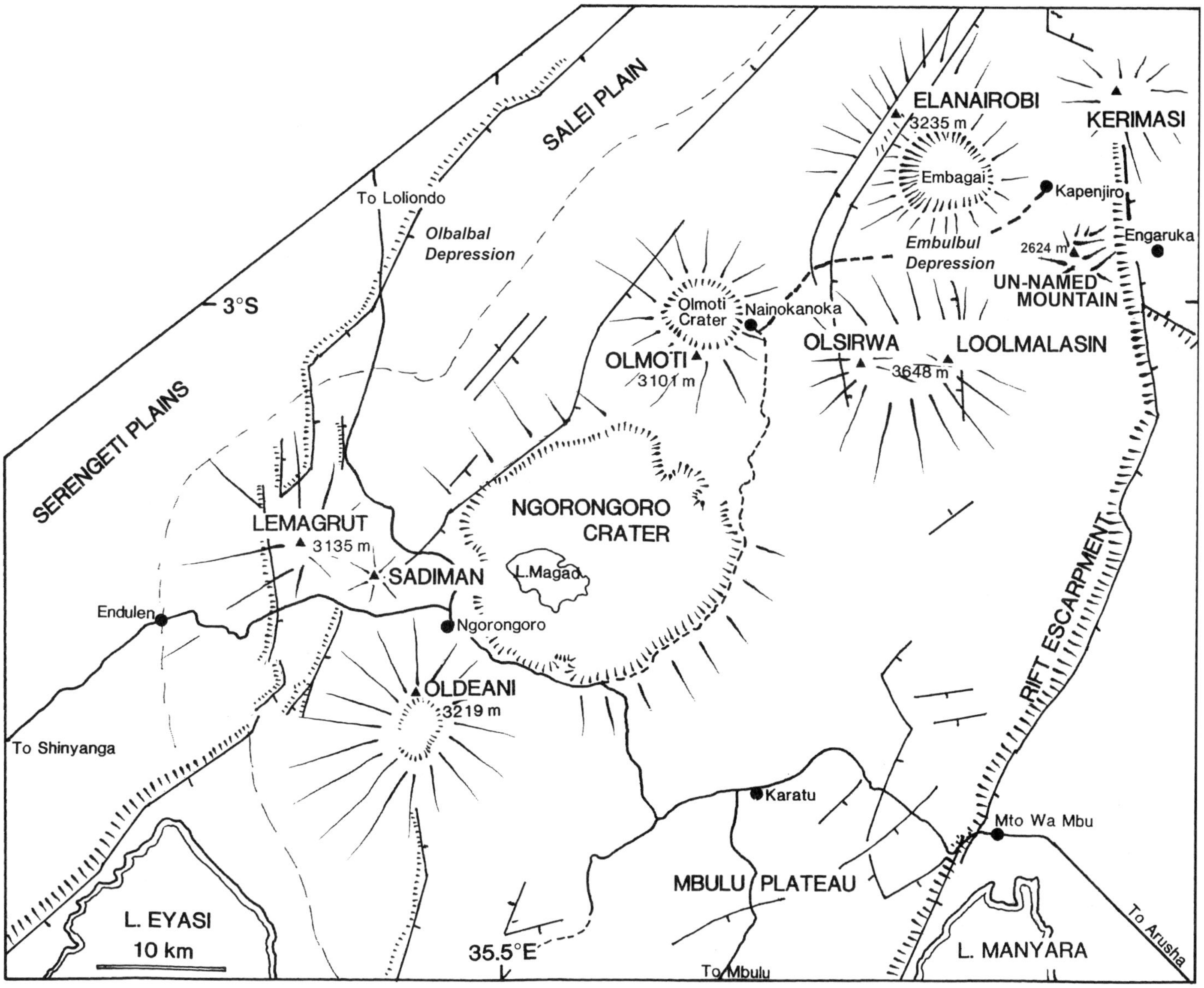


**Fig. 7.5.** Map of the Crater Highlands. Note the apparent absence of faults (solid lines with tick on downthrow side) on the well-forested southern and eastern slopes of Loolmalasin, Ngorongoro and Oldeani.

*Olmoti*

*Type*: shield volcano; *Lat/Long*: 3.02°S 35.65°E; *Elevation*: 3101 m; *Relief*: 1580 m above the Salei Plains to the west; *Diameter*: 30 km; *Eruptive history*: Pliocene; *Crater diameter*: 6.5 km; *Composition*: trachybasalt, trachyte, trachyphonolite; *QDS*: 53 Ngorongoro.

Olmoti is a broad dome that rises gently from the northern rim of the Ngorongoro crater. The southern and eastern outer slopes are heavily forested. The top of the volcano has a shallow circular caldera (Plate 2) on the southern rim of which is the highest point of the mountain. Small streams draining the crater converge at the southeastern corner to form the Munge river which, breaching the southeastern crater wall, plunges down a series of high waterfalls into a deep, highly-forested ravine, before flowing south into the Ngorongoro crater.

The crater, the steep inner walls varying in height from 100 to 200 m, has a small central plug 130 m high, the summit of which is of trachyte, and the lower flanks are of silicified breccia. The rest of the crater floor is reasonably flat and covered by lake sediments (volcanic arenites, some showing slump bedding, and thin limestones) up to 50 m thick (Pickering 1965).

Flows of trachybasalt, trachyte and trachyphonolite, and trachytic tuffs and agglomerates, are exposed on the crater rim and in the inner walls of the crater. James (1966) gave an analysis of an anorthoclase-phyric trachyte (Appendix 2, Dawson 2008). From stream channels at the foot of the northwestern slopes of the mountain, Hay (1976) recorded clasts of sodic trachyandesite and nepheline–aegirine trachyte, a flow of olivine-phyric basalt (containing groundmass biotite), and trachyte tuff in which the glass is altered to zeolite (mainly chabazite).

Trachyte, phonolitic tephrite and ignimbrite from the volcano have given dates of 1.07 to 1.65 Ma (Appendix 1, Dawson 2008), and a trachyandesite in the Olduvai succession, believed to derive from Olmoti, is dated at 1.09 Ma (Manega 1993).

*Loolmalasin*

*Type*: shield volcano; *Lat/Long*: 3.06°S 35.82°E; *Elevation*: 3648 m; *Relief*: 2600 m above the floor of the rift valley to the east; *Volcano diameter*: 25 km; *Crater diameter*: none; *Eruptive history*: Pliocene; *Composition*: basanite, basalt, trachyte; *QDS*: 53 Ngorongoro.

Loolmalasin is the highest of the Crater Highlands volcanoes; 6 km to the west is the subsidiary peak of Olsirwa (3150 m), separated from Loolmalasin by a low col. The mountain is a broad shield and the slopes rise gently from Ngorongoro to the SW and the col between it and Olmoti to the west. The southeastern and eastern slopes, descending to the rift escarpment, are steep, forest-covered, and incised by deep ravines. The summit area, which has no crater, and that of Olsirwa are broken by small north–south fault escarpments.

Exposure is good only in the summit areas of both Loolmalasin and Olsirwa. Basalt and trachyte are the dominant rock-types, with agglomerate in both summit areas. Basalt, trachybasalt and ankaramite of the eastern slopes are exposed in the rift escarpment above Engaruka (Pickering 1965) and, together with mugearite (Paslick *et al.* 1995), are also found as boulders in steams flowing eastwards off the Crater Highlands. In addition, picrobasalt (Paslick *et al.* 1995), basanite and nephelinite (Dawson personal observation), are exposed in small fault-scarps in the area between Engaruka and Mto Wa Mbu.

### *Ngorongoro*

*Type*: shield volcano with caldera; *Lat/Long*: 3.40°S 36.00°E; *Elevation*: 2380 m; *Relief*: 1400 m above the rift valley; *Diameter*: 35 km; *Eruptive history*: Pliocene; *Crater diameter*: 19 × 22 km; *Composition*: olivine basalt, trachybasalt, trachyte; *QDS*: 53 Ngorongoro.

Ngorongoro is a broad, deeply eroded volcano that has a thick cover of rain forest on its southern and eastern slopes. It has a major summit caldera that is justifiably famous as one of Africa's finest wildlife sanctuaries. The crater (Plate 3) measures 19 × 22 km, the longer axis being east–west, and its area of *c.* 370 km$^2$ ranks it amongst the world's largest calderas. The crater floor is at an elevation of around 1700 m a.s.l., and the rim varies between 2100–2400 m. Prior to the collapse that formed the caldera, the volcano may have had an elevation of around 4500–5000 m. A series of small NE-trending faults cut across the northwestern slopes of the volcano, with one on the lower slopes forming the eastern side of the Olbalbal graben (Fig. 7.5).

On the crater floor and at the base of the walls are freshwater springs that are the water supplies for the numerous animals that inhabit the crater. The caldera has no stream outlet and an alkaline seasonal lake (Lake Magad) occurs on the crater floor; it is partly fed by the Munge River that descends from the Olmoti volcano to the north. On the southern and eastern shores of the lake are low eroded bluffs of lacustrine clay, ostracod-bearing clayey limestone and tuffs. These give $^{14}$C ages of 27 990 ± 500 BP to 24 400 ± 690 BP (Hay 1976), providing evidence for a lake highstand at around 24 000 BP, but there is no evidence for a highstand at around 9000 BP when many other East African lakes were much more extensive than at present. Flows of basalt and trachyte have welled out at the base of the crater walls, probably from ring fractures, though much of the crater floor appears to be underlain by tuffs. A gravity survey across the caldera (Searle 1971) showed a negative Bouguer anomaly, attributed to both low-density rocks forming the volcano itself and to sediments on the crater floor. There are several small scoria cones on the floor of the crater, one of which, Engitati Hill, is of basaltic composition, and has a vitroclastic texture suggesting that it was formed during a subaqueous eruption (Pickering 1965).

The volcano comprises mainly lavas of basaltic and trachytic composition, but ignimbrites predominate among the later eruptives. Exposures in the crater walls can be seen on the descent track from the western rim (the Windy Gap track) and on the ascent track to the southern rim (the Lerai track). The following account is abbreviated from Hay (1976). 'On the western track are exposures of trachytic welded tuffs, interbedded with grey trachytic lava. Exposed on the southern track are 340 m of interbedded lavas and pyroclastic rocks; coarsely porphyritic plagioclase–augite–olivine-phyric trachyandesite lavas predominate in the lower part of the section, with slightly porphyritic trachyandesite and olivine basalt in the upper part. Ashes in the lower part of the succession represent voluminous ash eruptions that were erupted before the caldera collapse.' Twelve K–Ar dates range from 2.1 to 2.8 Ma, with a mean of seven dates of 2.46 Ma for a tuff in the lower part of the section, and a mean of five dates of 2.41 Ma for a trachyte 170 m higher in the section (Curtis & Hay 1972). The lower and upper lavas in this section are normally- and reverse-magnetized respectively, with the reversal being dated at 2.45 Ma, corresponding to the boundary between the Gauss normal epoch and the Matuyama reversed epoch (Grommé *et al.* 1970).

On the northwestern slopes of the volcano, stream sections expose trachyte and trachyandesites, together with thin flows of olivine-phyric basalt that overlie red trachytic ignimbrite. A succession of olivine basalt overlying green trachytic ignimbrite, derived from Ngorongoro, is exposed on the fault scarp on the west side of the Olbalbal graben north of Lemagrut (Hay 1976). Anorthoclase from the ignimbrite yielded an age of 2.1 ± 0.1 Ma (Curtis & Hay 1972).

Considering the size of the volcano, there are relatively few chemical data for the lavas. Paslick *et al.* (1995, 1996) gave complete analyses of a basanite and a trachyte, together with analyses of feldspar and Ti-magnetite in the basanite and olivine, clinopyroxene and anorthoclase in the trachyte. Wood (1968) gave analyses of basanite, tephrite, mugearite and phono-tephrite, mainly from the southern rim of the caldera (Appendix 2, Dawson 2008).

Trachytic ignimbrites and tuffs and olivine basalt flows, some interpreted as originating from Ngorongoro and ranging in age from 2.01 to 1.71 Ma, are exposed in the Olduvai Gorge 25 km to the NW. The basal lava in the succession is dated at 2.0 ± 0.1 Ma. (Curtis & Hay 1972). The most distal flow from the volcano is an ignimbrite at Naabi Hill on the Serengeti Plain some 80 km to the NW of Ngorongoro (Hay 1976). Basaltic lavas exposed in the rift escarpment 25 km east of Ngorongoro range in age from 3.8 Ma at the base of the escarpment to 1.4 Ma at the top (Bagdasaryan *et al.* 1973) (Appendix 1, Dawson 2008). However, it is not clear whether the lavas derive from Ngorongoro, the nearby Loolmalasin volcano, or both.

### *Lemagrut*

*Type*: shield volcano; *Lat/Long*: 3.17°S 35.39°E; *Elevation*: 3135 m; *Relief*: 1300 m above the Serengeti Plains to the west, and about 1000 m on the south side above the pass between Lemagrut and the Oldeani volcano; *Diameter*: 20 km; *Eruptive history*: ?Pliocene; *Crater diameter*: no crater; *Composition*: olivine basalt, nephelinite; *QDS*: 52 Endulen.

Lemagrut is a broad craterless dome standing adjacent to, and west of, Ngorongoro. Radial river valleys incise the northern and northwestern slopes in particular. The northern and western lower slopes are cut by small, east-facing escarpments that result from minor faulting associated with reactivation of the major Sonjo–Eyasi Fault.

The first stage in the formation of the volcano was the eruption of tuffs and agglomerates that occur mainly on the southwestern lower slopes and on the plains to the SW. The tuffs are lithic or crystal tuffs of olivine nephelinite composition; these are interbedded with agglomerates containing pebbles of nephelinite, basalt and jacupirangite, and with rare flows of nephelinite (Pickering 1964). The second stage of activity that formed the main part of the volcano was dominated by the extrusion of olivine basalt and trachybasalt.

**Plate 1.** Aerial view (from 11 000 m) from the east of the Embagai caldera and its lake. The caldera is approximately 7.5 km in diameter and has a maximum depth (to lake level) of 950 m.

**Plate 2.** Aerial view from the east of the circular, 6.5 km wide, Olmoti caldera. Objects in the foreground are Maasai cattle pens. Photograph by M. Punkari.

**Plate 3.** Aerial view (from the east at 11 000 m) of the Ngorongoro caldera; the crater measures 19 × 22 km in diameter. Lake Magad is visible on the caldera floor and the Lemagrut volcano appears in the far distance. The dark area in the foreground, going up to the caldera rim, is forest.

**Plate 4.** The Kibo cone of Kilimanjaro from the south. The summit, the highest point in Africa, stands 5000 m above the surrounding plains. The snowfields are less extensive now than in 1960 when the photograph was taken.

**Plate 5.** Mawenzi from the SW, rising 900 m above The Saddle, the col separating it from Kibo. The serrated appearance is due to weathering-out of a radial dyke swarm.

**Plate 6.** Photograph from the air of the Kibo caldera, from the NE. Uhuru Peak (5895 m), the highest point on the African continent, stands on the far ridge.

**Plate 7.** Aerial view of Oldoinyo Lengai, from the NE, showing the Eastern Chasm debris-flow scar. The white material is altered natrocarbonatite, possibly the residue of lava overspill in the early part of the twentieth century. The main pyroclastic deposits are piled against the Natron Basin boundary escarpment that appears in the background. Photograph by M. Punkari.

**Plate 8.** The northern crater of Oldoinyo Lengai (October 1988) looking north from the summit towards Lake Natron in the distance. The crater is dominated by the cone surrounding the black active lava pool. Lava is flowing from the pool into the depression in the foreground. The scale is given by the tents (bottom right).

**Plate 9.** Photograph from the air of the northern crater of Oldoinyo Lengai from the NE (April 2006) showing hornito and collapsed lava cone within the crater, and whitened natrocarbonatite lava flows spilling over from the crater down the eastern and northwestern outer slopes. Compare with Figures 7.9 and 7.10. Photograph by Dean Polley.

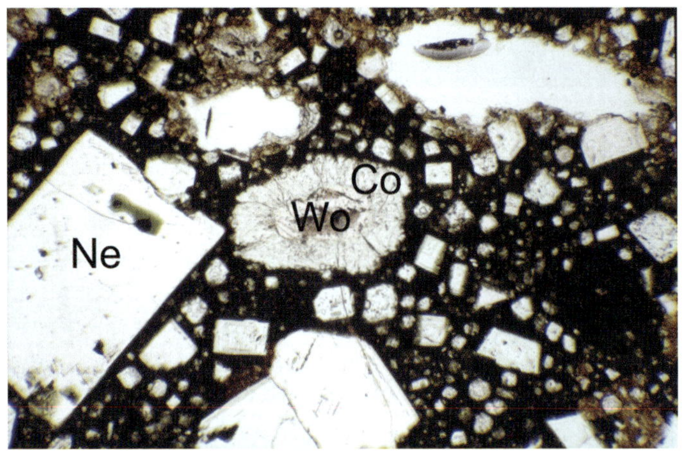


**Plate 10.** Photomicrograph of wollastonite nephelinite BD3955 (Oldoinyo Lengai), in which unstable wollastonite (Wo) is surrounded by a fringe of replacing, metasomatic combeite (Co). Horizontal field width 2.2 mm; plane-polarized light. Ne, nepheline.

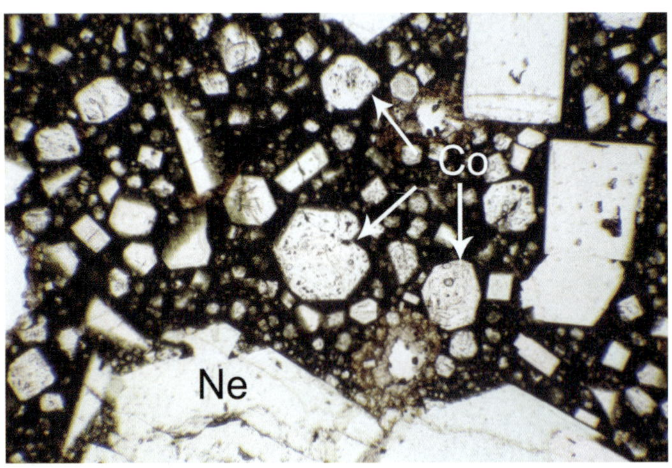


**Plate 11.** Photomicrograph of nephelinite BD3955 (from Oldoinyo Lengai), showing microphenocrysts of magmatic combeite (Co). Horizontal field width 2.2 mm; plane-polarized light. Ne, nepheline.

**Plate 12.** Photomicrograph of nepheline syenite BD59 (Oldoinyo Lengai). Phenocrysts of nepheline (Ne), clinopyroxene (Cpx) and titanite (Ti) are surrounded by poikilitic sanidine (Sa). Horizontal field width 5.4 mm; crossed polarizers.

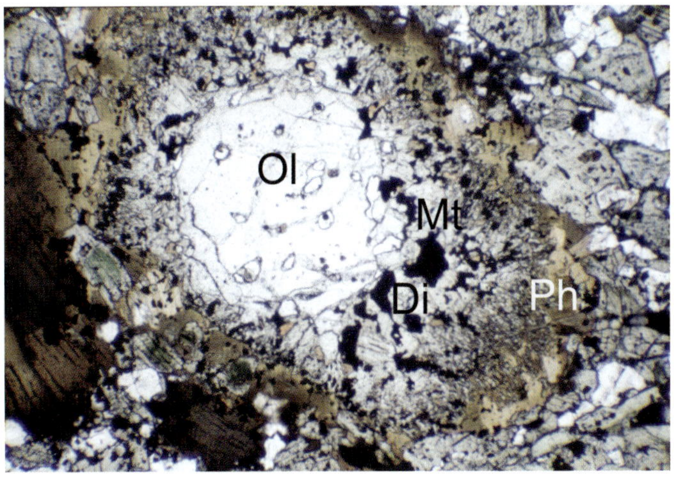


**Plate 13.** Photomicrograph of xenocrystic olivine (Ol) surrounded by a reaction rim of diopside (Di), magnetite (Mt) and phlogopite (Ph) in ijolite BD122 (Oldoinyo Lengai). Horizontal field width 5.4 mm; plane-polarized light.

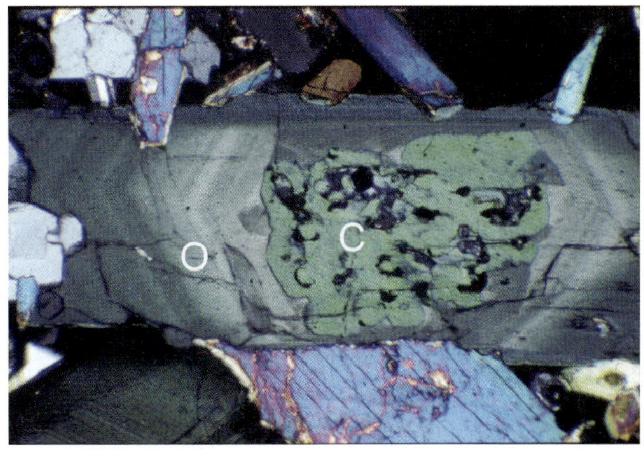


**Plate 14.** Photomicrograph of clinopyroxene in ijolite BD343 (Oldoinyo Lengai). A corroded core (C) is surrounded by complexly-zoned overgrowth (O). Horizontal field width 2.2 mm; crossed polarizers.

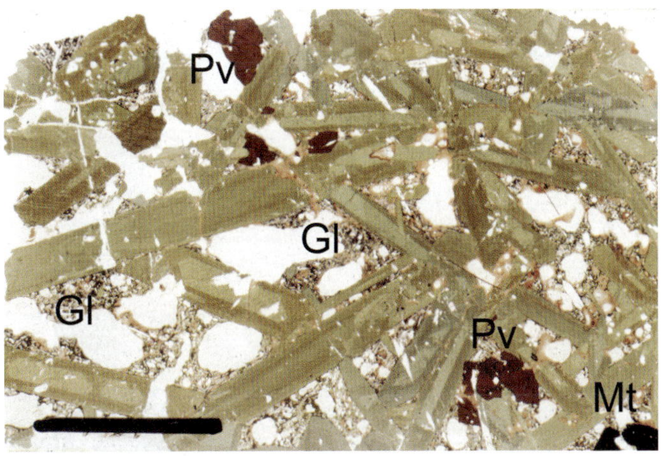


**Plate 15.** Photomicrograph of zoned green clinopyroxene, perovskite (Pv) and magnetite (Mt) cemented by vesiculated glass Gl, in pyroxenite BD878 (Oldoinyo Lengai); white areas are voids. Scale bar is 10 mm; plane-polarized light.

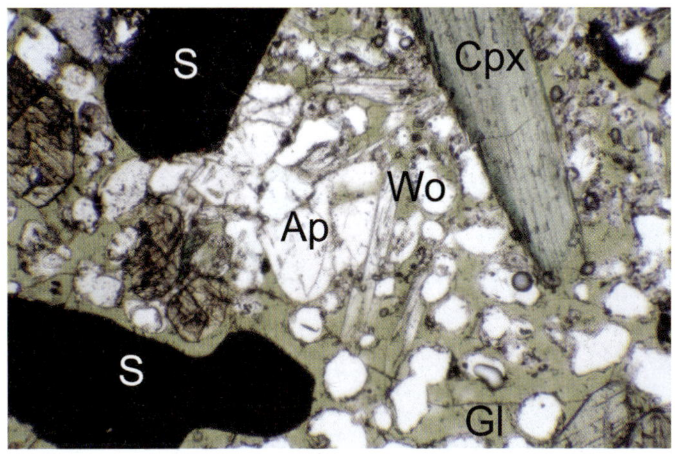


**Plate 16.** Photomicrograph of green vesiculated glass cementing phenocryts of zoned clinopyroxene (Cpx) and sulphide (S) in ijolite BD127(Oldoinyo Lengai). Acicular wollastonite (Wo) and apatite (Ap) have precipitated from the glass (Gl). Horizontal field width 2.2 mm; plane-polarized light.

**Plate 17.** Block of acicular wollastonite, with minor nepheline from Oldoinyo Lengai. The crystals are cemented by vesiculated glass. The texture was formed by rapid crystallization of a compositionally unusual magma that was supersaturated in wollastonite. Wollastonite occurs only rarely as an igneous mineral.

**Plate 18.** Photomicrograph of an olivine–mica pyroxenite BD93 (Oldoinyo Lengai), interpreted as an upper mantle metasomite. Strained and kinked olivine (Ol) is partly replaced by clinopyroxene (Cpx) and phlogopite (Ph). This is a more extreme example of metasomatic replacement of olivine than that shown in Plate 12. The radiogenic strontium and neodymium values for this sample plot to the 'north' of EMI on Figure 7.12. Horizontal field width 2.2 mm; crossed polarizers.

**Plate 19.** Photomicrograph of natrocarbonatite from Oldoinyo Lengai, showing phenocrysts of the complex alkali carbonates nyerereite (Ny, euhedral elongate grains) and gregoryite (Gr, rounded grains, with one basal section showing low-order colour polysynthetic twinning). V, vesicle. Horizontal field width is 5.4 mm; crossed polarizers.

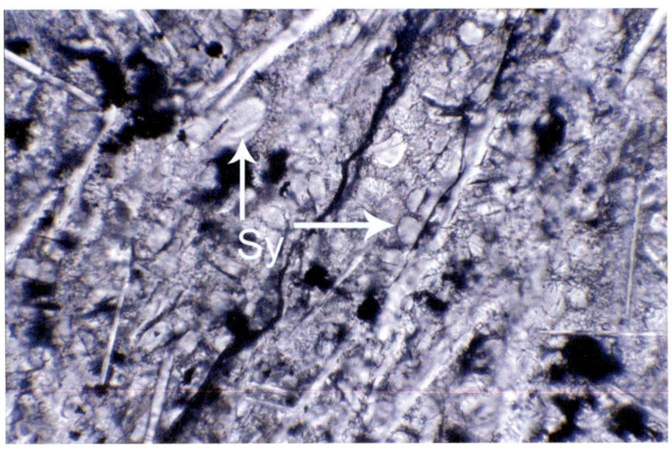


**Plate 20.** Photomicrograph of the groundmass of natrocarbonatite lava, showing rounded microphenocrysts of sylvite (Sy) and slender, elongate microphenocrysts of nyerereite. The rest of the groundmass comprises fine-grained crystals of a complex CaMgSrNaKLa carbonate, apatite, fluorite and opaque sulphides. Horizontal field width 0.5 mm; plane-polarized light.

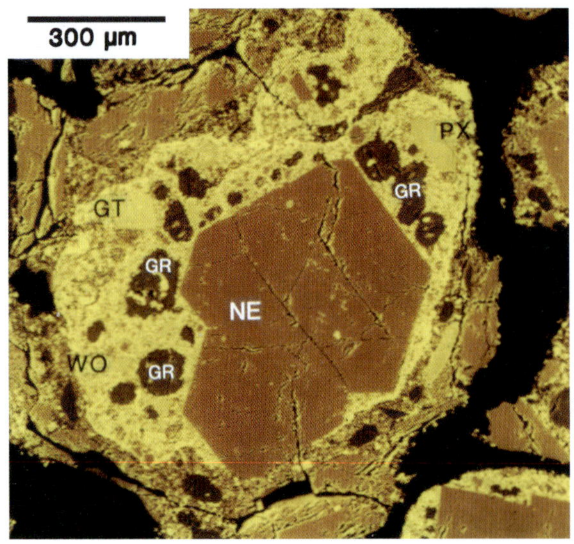


**Plate 21.** Back-scattered, false-colour, image of a lapillus from the 1993 eruption of Oldoinyo Lengai. The silicate lapillus (bright yellow) is encased in less bright natrocarbonatite. The silicate matrix contains a phenocryst of euhedral nepheline (NE) and microphenocrystic wollastonite (WO), garnet (GT) and pyroxene (PX). Set in the silicate matrix are rounded, exsolving globules of natrocarbonatite (GR); see Plate 22.

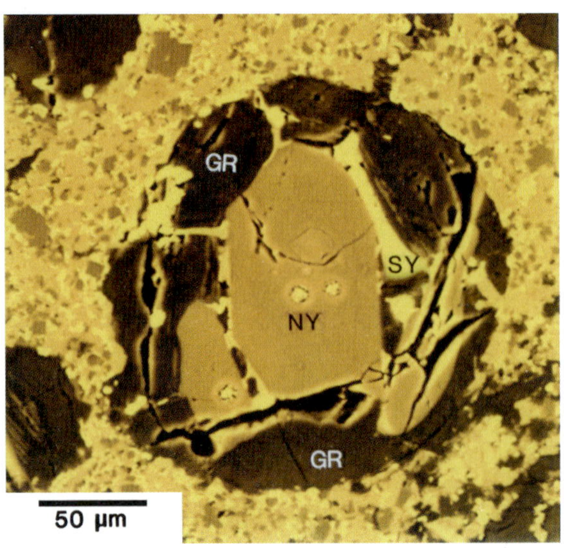


**Plate 22.** Back-scattered, false-colour image of a natrocarbonatite globule in the silicate groundmass of a lapillus from the 1993 eruption (see Plate 21). The globule contains the typical natrocarbonatite minerals gregoryite (GR), nyerereite (NY) and sylvite (SY). The globule is exsolving within the silicate lapillus and is unrelated to the natrocarbonatite surrounding the lapillus.

**Plate 23.** An upper mantle xenolith from the Pello Hill tuff cone (for location see Fig. 7. 18). The peridotite wall rock (light green) is cut by a brown vein consisting of mica, amphibole and ilmenite. The vein contains a median slice of altered peridotite that, due to Fe-metasomatism, is now browner than the original. The xenolith bears testimony to brittle fracture in the upper mantle beneath the rift valley.

**Plate 24.** Meru from the west, showing classic stratovolcano cone shape and deep radial ravines on the upper slopes that may have been initiated by former glaciation. The summit of the volcano stands some 3000 m above the plains in the foreground.

**Plate 25.** Meru from the east, looking into the crater which has been breached with a resulting immense debris avalanche. The recent, central ash cone is visible within the breached crater.

**Plate 26.** Photomicrograph of the kimberlite lava flow at the Igwisi Hills (see Fig. 7.25). A round microxenolith of strained and partly-recrystallized olivine (Xe), and a euhedral microphenocryst of magmatic olivine (Mp), are set in a matrix of dark serpentine and flow-aligned calcite laths. Horizontal field width 5.4 mm; crossed polarizers.

From stream channels on the north side of the volcano, Hay (1976) recorded abundant clasts of trachyandesite and plagioclase–olivine–augite-phyric basalt, and rare trachyte. In the Olduvai Side Gorge are similar clasts, plus clasts of nepheline–pyroxene–titanite-phyric nephelinite, and melilitite tuff containing melilite, nepheline, pyroxene, perovskite, biotite, melanite and titanite.

On the lower southeastern slopes [3°11′S 35°22.5′E], nepheline-phyric phonolite, containing groundmass sodalite, underlies a flow of olivine-barkevikite mugearite (Dawson personal observation). Paslick *et al.* (1995, 1996) provide complete analyses of this phonolite (and its nepheline and clinopyroxene), and of the mugearite (and of olivine, clinopyroxene, feldspar and Ti-magnetite therein).

On the western slopes of the volcano the fossiliferous Laetoli beds (see Dawson 2008, Chapter 5) are overlain by the Ogol lavas that erupted from small, local cones. The lavas, dated at 2.41 Ma, have been called vogesite (Drake & Curtis 1987) but Wirth & Adelsberger (2002) recorded them as being of primitive ($mg\# > 0.6$) alkali olivine basalt (where $mg\# = \mathrm{Mg/Mg + Fe}$).

Stratigraphic relationships at Laetoli indicate that Lemagrut and Sadiman (below) were active at the same time.

### *Sadiman*

*Type*: stratovolcano; *Lat/Long*: 3.20°S 35.42°E; *Elevation*: 2870 m; *Relief*: 500 m; *Diameter*: 6 km; *Eruptive history*: Pliocene; *Crater diameter*: none; *Composition*: nephelinite, phonolite, melilitite, carbonatite; *QDS*: 52 Endulen.

Sadiman is a small, well-forested cone situated between Lemagrut and the Ngorongoro caldera. It is formed mainly of tuffs and agglomerates, containing clasts of nephelinite and melilitite, with subsidiary flows of nephelinite and phonolite (Pickering 1964). From stream channels on the north side of the volcano, Hay (1976) reported rare clasts of ijolite together with abundant clasts of nepheline–anorthoclase–pyroxene-phyric nephelinite and phonolite, in which microphenocrysts of titanite are common.

Nepheline-phyric phonolitic nephelinites containing abundant titanite (some strongly zoned) form the summit ridge (Dawson personal observation). A complete rock analysis, and analyses of nepheline and titanite from one of these nephelinites are given by Paslick *et al.* (1995, 1996). James (1966) described a 'nephelinite boss', which he refers to as 'Malanja'. As Sadiman stands above the Malanja depression and is surrounded by the lavas of Lemagrut, as described by James (1966), it seems probable that they are the same. He gave a whole-rock analysis of a nephelinite (Appendix 2, Dawson 2008) that, in its lower $SiO_2$ content, differs from the phonolitic nephelinite of Paslick *et al.* (1995).

A lava sample from the volcano has given a K–Ar age of 3.7 Ma, which is indirectly confirmed by the presence of clasts of Sadiman lava in the lowest part of Bed I of the Olduvai succession (Hay 1976), indicating that the volcano is older than 2.1 Ma. The volcano is also believed to be the source of the phonolitic and melilititic ash of the fossiliferous Laetoli Beds 15 km to the west (Dawson 2008, Chapter 6).

### *Oldeani*

*Type*: shield volcano; *Lat/Long*: 3.29°S 35.44°E; *Elevation*: 3219 m; *Relief*: 2000 m above Lake Eyasi to the SW, 780 m above the rim of the Ngorongoro caldera to the east; *Diameter*: 20 km; *Eruptive history*: Pliocene; *Crater diameter*: 5 km; *Composition*: olivine basalt, basalt, ankaramite, trachybasalt, trachyte; *QDS*: 52 Endulen.

Oldeani is a broad dome that stands astride and infills the Eyasi half graben. The summit crater is breached to the west, and the lower northwestern slopes are cut by small east-facing escarpments that result from minor faulting associated with reactivation of the major Sonjo–Eyasi fault.

Outcrop is poor, the eastern and southern slopes being covered with dense forest, and there is thick bamboo forest in the summit area. Flows of basalt occur on the western slopes, and of trachybasalt and trachyte in the summit area. Small basalt scoria cones are associated with minor faulting on the lower southeastern slopes (Pickering 1964).

Trachytic ignimbrite, and red soils derived from weathered ignimbrite, interpreted as deriving from Oldeani, are widespread on the Mbulu Plateau for distances up to 30 km to the south of Oldeani (Orridge 1965).

## Angata Salei–Serengeti plains

A large area to the west of the Crater Highlands, including much of the Salei and Serengeti plains, is covered with a brown calcareous tuff that, in places, drapes basement hills (Pickering 1958). This tuff thins and becomes more calcitic westwards towards Lake Nyanza, and in the Olduvai Gorge area it commonly overlies nephelinitic or trachytic tuffs from which it is separated by zones of zeolite. These calcareous tuffs were originally interpreted as calcrete (Pickering 1958; Hay 1963), but Dawson (1964*a*) suggested that the tuffs and, possibly, earlier 'calcretes' in the aeolian horizons in the upper Olduvai Beds might be the result of deposition and recrystallization of carbonatite tephra from the Kerimasi and Oldoinyo Lengai volcanoes that lie upwind of the Serengeti and thus ideally situated as tephra sources, a suggestion endorsed by Hay (1964). Subsequent detailed work (Hay 1976) correlates the Masek Beds exposed in the Olduvai Gorge (age 0.6 and *c.* 0.4 Ma) with eruptions of Kerimasi, and the more extensive Ndutu (60 000 to 400 000 BP) and Naisiusiu Beds (age 15 000 to 22 000 BP) with the earliest material erupted from Oldoinyo Lengai. Later ashes and tephra from Oldoinyo Lengai (the Namorod ashes of Olduvai) are deposited above calcretes at the top of the Naisiusiu Beds.

Holocene black ash derived from Oldoinyo Lengai forms fixed dunes on the Salei Plain (Guest *et al.* 1961) and also occurs as migrating dunes on the Serengeti Plain (Hay 1976); the ash is abundant on the more northerly Crater Highlands but does not appear to have been deposited further south than Olmoti (Dawson personal observation).

An outlier of the Crater Highlands volcanism is Engelosin, an isolated small plug, 500 m in diameter, and 145 m high, on the Serengeti Plain north of Olduvai [2°54.5′S 35°23′E]. It is an anorthoclase–augite-phyric phonolite containing a groundmass of nepheline, anorthoclase, aegirine, a REE silicate and titanite. The age is unknown but its highly eroded nature suggests that it may be Pliocene or older (Hay 1976).

## The Engaruka–Manyara area

*QDS*: 53 Ngorongoro and 54 Monduli.

The main rift escarpment and the escarpments of many minor faults on the Engaruka Block (Fig. 5.8) are notable for the exposure of lava flows forming the lower parts of some of the more easterly Crater Highland volcanoes (Loolmalasin and Ngorongoro). The most common flows are of olivine- and olivine–pyroxene-phyric basalt, but also present are barkevikite trachybasalt and nepheline-phyric olivine nephelinite. In the lower parts of the rift escarpment at Engaruka are flows of limburgite and olivine-phyric basalt (Dawson personal observation). Boulders of the same rock-types are present in outwash in the Engaruka and Engaibataat rivers, and also occur in agglomerate mounds (possibly resulting from erosion of distal outwash sheets from the main escarpment) on the plain NW and north of Essimingor. Wood (1968) gave an analysis of a high-K basanite boulder from Engaruka (Appendix 2, Dawson 2008).

In the gorge of the Loburu River, cutting eastwards across the northern end of the Engaruka Block, at least one boulder bed is interbedded with the basalts, indicating an erosional break with clastic sedimentation within the lava-dominated sequence (Dawson personal observation).

In the area, are many minor flows or spatter cones that were extruded contemporaneously with, or immediately following, the faulting (Fig. 5.8). These include olivine trachybasalt from Engaruka and Narabala (Engaruka also contains barkevikite). At Kitete [3°11′S 35°56.5′E] a number of small flows, ranging from basanite to phonotephrite, were extruded from the main rift fault; these flows have ages of 0.95 to 1.23 Ma (MacIntyre *et al.* 1974). Paslick *et al.* (1995) gave a complete analysis of one of the flows (BD180, an olivine–barkevikite phonotephrite, dated at *c.* 1.0 Ma). In addition, Orkandira hill [3°11′S 36°01′E] is an extrusion of analcitite in which phlogopite occurs as a poikilitic groundmass phase sometimes protruding into calcite-filled amygdales.

### *Essimingor (also known as Losimingori)*

*Type*: stratovolcano; *Lat/Long*: 3.40°S 36.10°E; *Elevation*: 2165 m; *Relief*: 1000 m; *Volcano diameter*: 16 km; *Eruptive history*: Miocene–Pliocene; *Composition*: nephelinite–phonolite; *QDS*: 54 Monduli.

Essimingor is the oldest volcano in northern Tanzania, with K–Ar ages on poorly-documented lavas of 8.1 Ma on a melanephelinite, and of 4.68 and 3.22 Ma on two nephelinites (Appendix 1).

It is a dominantly pyroclastic volcano and, due to deep erosion and thick forest on most of the upper slopes on the eastern, southern and western sides of the mountain, there is no sign of an original crater, the summit area being a series of steep interconnected ridges. A lobe of lavas, possibly flowing from a former, breached crater, extends southwestwards from the main pyroclastic part of the cone (Dawson 1964*b*). The lowest western and southern slopes are blanketed by limestones that were deposited in former extensions of Lake Manyara, and the lower northwestern and northern slopes are partly covered by mounds of poorly-consolidated agglomerate resulting from outwash from the main rift escarpment to the west.

Nephelinitic and phonolitic tuffs and agglomerates, some containing blocks of sodalite-bearing ijolite and nepheline syenite, form the bulk of the volcano. Lavas, both intercalated with the pyroclastics and forming the SW lava lobe, are of three main types: (i) sodalite–anorthoclase–aegirine–augite-phyric phonolites, in which titanite, aenigmatite and magnetite are groundmass phases; (ii) olivine-free nephelinites, in which nepheline, aegirine–augite, melanite and magnetite are the main phenocrysts, and perovskite is a common groundmass mineral; altered melilite is a rare phenocryst phase; and (iii) augitite, in which abundant phenocrysts of titanaugite and rarer perovskite and sodalite, are set in a matrix of nepheline or analcime, perovskite and magnetite; rare olivine grains (? xenocrysts) are surrounded by rims of phlogopite and magnetite. Where analcite is the only groundmass feldspathoid, these rocks would be more correctly described as analcitites. Rarer flows are of ankaramite and trachyte. Dykes of monchiquite, ankaramite, analcitite, nephelinite and phonolite occur on the southeastern ridge of the mountain and in the summit area. Wood (1968) gave analyses of four nephelinites and Paslick *et al.* (1995) gave complete analyses of olivine augitite, melilite nephelinite and augitite.

**Fig. 7.6.** The Older Extrusive cone of Tarosero, from the south. The cone shape results from the volcano being composed of viscous trachyte and phonolite lavas.

### *Tarosero*

*Type*: lava cone; *Lat/Long*: 3.20°S 36.36°E; *Elevation*: 2242 m; *Relief*: 700 m on north slopes, 500 m on south slopes; *Volcano diameter*: 7 km; *Eruptive history*: Pliocene; *Composition*: alkali olivine basalt, hawaiite, peralkaline trachyte, phonolite and comendite; *QDS*: 54 Monduli.

The gentle cone of Tarosero (Fig. 7.6) stands above a northerly-dipping lava plateau that consists of alkali basalt and benmoreite, the eruptive centre(s) for which is (are) not certain. Early trachyte flows extruded from the volcano, dated at 2.2–2.4 Ma (MacIntyre *et al.* 1974), are faulted like the underlying lava plateau and some are well exposed in the escarpment that forms the western side of the Lembolos graben.

The cone itself consists mainly of younger (2.02 Ma), unfaulted, peralkaline trachytes. There are subsidiary trachyte cones on the lower, northern slopes of the volcano, and a crater (2.5 km in diameter), surrounded by trachytic tuffs, perforates its lower, southern flanks. Peralkaline rhyolites (comendites) occur at Oldoinyo Sambu on the lower northwestern slopes (Dawson 1964*b*).

Principal component analysis of major- and trace-elements in 41 lava samples (Cooper 1972) has established two main lava groups, dominated by trachytes, which are alkaline to peralkaline, characteristically containing Na-rich amphiboles, Na-clinopyroxene and aenigmatite; most are acmite-normative and some are nepheline-normative. Group I ranges from benmoreite through trachytes and quartz trachytes to comendites that can be linked by pyroxene and feldspar fractionation. The more siliceous are peralkaline, and chevkinite [$(Ce,La,Ca)_4Fe^2(Fe^2Ti)_2 Ti_2Si_4O_{22}$] occurs in one quartz trachyte (Macdonald *et al.* 2002). Within this group are less common phonolites and phonotephrites. Group II comprises quartz trachytes, trachytes and nepheline-normative trachytes containing sodalite, eudyalite and aenigmatite (Dawson 1997). All are peralkaline and, relative to Group I rocks with the same major element concentrations, are enriched in Cl, Ba, Sr, Rb, REE, Nb, Zr and Pb; some have >3000 ppm Zr (Tables 7.2, 7.3).

The trachytes of both groups comprise phenocrysts of zoned anorthoclase (rims higher in $K_2O$ but lower in BaO than cores), richterite, alkali pyroxene and rare biotite (Table 7.4), the mafic phases being high Na and Fe varieties The groundmass contains anorthoclase (compositions overlap those of the phenocryst rims), amphibole (containing less $Na_2O$ and FeO, but higher CaO and MgO than phenocrysts) and aenigmatite, the presence of which coincides with an absence of FeTi oxides in the groundmass. Sodalite is a common phase in the Group II rocks.

**Table 7.2.** *Tarosero lavas-selected major element analyses (from Cooper 1972)*

| Sample | Group I lavas | | | | | | | | Group II lavas | | | | | |
|---|---|---|---|---|---|---|---|---|---|---|---|---|---|---|
| | 1 | 2 | 3 | 4 | 5 | 6 | 7 | 8 | 9 | 10 | 11 | 12 | 13 | 14 |
| $SiO_2$ | 56.87 | 58.27 | 60.23 | 61.49 | 62.25 | 63.45 | 65.7 | 71.81 | 59.41 | 60.72 | 60.94 | 61.15 | 62.12 | 62.13 |
| $TiO_2$ | 1.65 | 0.83 | 0.89 | 0.52 | 0.62 | 0.44 | 0.41 | 0.20 | 0.28 | 0.48 | 0.45 | 0.50 | 0.30 | 0.29 |
| $Al_2O_3$ | 16.29 | 19.58 | 20.18 | 17.60 | 16.20 | 17.17 | 14.16 | 12.04 | 15.85 | 15.09 | 15.24 | 15.37 | 16.15 | 16.80 |
| $Fe_2O_3$ | 4.11 | 1.94 | 3.69 | 3.45 | 3.72 | 2.70 | 2.79 | 4.51 | 4.99 | 4.87 | 4.76 | 4.89 | 2.36 | 3.41 |
| FeO | 4.63 | 2.74 | 1.07 | 3.76 | 2.76 | 1.18 | 2.87 | 0.06 | 1.27 | 4.04 | 4.04 | 3.85 | 3.91 | 2.41 |
| MnO | 0.16 | 0.16 | 0.10 | 0.14 | 0.13 | 2.87 | 0.22 | 0.09 | 0.25 | 0.25 | 0.28 | 0.27 | 0.26 | 0.24 |
| MgO | 1.82 | 0.61 | 0.49 | 0.07 | 0.06 | 0.19 | 0.27 | 0.00 | 0.38 | 0.12 | 0.20 | 0.18 | 0.11 | 0.22 |
| CaO | 4.62 | 2.81 | 2.24 | 0.48 | 0.58 | 1.08 | 0.70 | 0.28 | 4.22 | 1.24 | 1.15 | 0.69 | 0.91 | 1.12 |
| $Na_2O$ | 5.42 | 7.52 | 5.56 | 6.33 | 6.87 | 7.55 | 6.46 | 5.76 | 6.38 | 7.06 | 6.81 | 6.92 | 7.27 | 6.54 |
| $K_2O$ | 2.85 | 4.64 | 4.78 | 5.31 | 5.97 | 5.37 | 5.66 | 5.24 | 6.35 | 5.30 | 5.30 | 5.38 | 5.72 | 5.11 |
| $P_2O_5$ | 0.93 | 0.23 | 0.24 | 0.00 | 0.06 | 0.13 | 0.00 | 0.00 | 0.00 | 0.00 | 0.00 | 0.00 | 0.00 | 0.00 |
| Sum | 99.35 | 99.33 | 99.47 | 99.15 | 99.22 | 99.34 | 99.24 | 99.99 | 99.38 | 99.17 | 99.17 | 99.20 | 99.11 | 99.31 |
| (Na + K)/Al | 0.74 | 0.88 | 0.76 | 1.00 | 1.17 | 1.11 | 1.27 | 1.35 | 1.22 | 1.21 | 1.17 | 1.18 | 1.19 | 1.16 |

1. BD329 Benmoreite, N end of Tarosero scarp 3°07.9′S 36°24.75′E; 2. BD509. Phonolite, gorge on lower SW slopes 3°13.5′S 36°18.7′E; 3. BD 513 Qtz-normative trachyte, gorge on lower SW slopes, 3°13.5′S 36°18.7′E; 4. BD488 Qtz-hyp-normative trachyte, east-summit ridge, 3°12′S 36°22.8′E, age 1.94 $\pm$ 0.06 Ma.; 5. BD261 Nepheline-normative trachyte, western slopes 3°11.7′S 36°15,4′E; 6. BD255 Nepheline-normative trachyte, Oldoinyo Sambu (small hill on western slopes) 3°08.5′S 36°15′E; 7. BD290 Comenditic trachyte, Oldoinyo Sambu 3°09.5′S 36°17′E; 8. BD289 Comendite, Oldoinyo Sambu 3°09.2′S 36°16.6′E; 9. BD305 Sodalite phonolite, lower north slopes, 3°06′S 36°20.5′E; faulted; age 2.25 $\pm$ 0.06 Ma; 10. BD295 Nepheline normative trachyte, gorge on lower SW slopes, 3°13.5′S 36°18.7′E; 11. BD510 Sodalite trachyte, gorge, lower SW slopes, 3°13.5′S 36°18.7′E, age 2.21 $\pm$ 0.10 Ma; 12. BD490 Sodalite trachyte, summit of Tarosero, 3°11.1′S 36°21.6′E; dated at 2.10 $\pm$ 0.06 Ma; 13. BD306 Sodalite, nepheline-normative trachyte, lower north slopes, 3°06′S 36°20.5′E faulted; age 2.20 $\pm$ 0.06 Ma. Major element chemistry is similar to group I sample BD255; 14. BD512 Quartz-normative sodalite trachyte, gorge on lower SW slopes, 3°13.5′S 36°18.7′E. Major element chemistry is similar to Group I sample BD 261.

The range in lava types on this single volcano covers the same compositional spectrum as the regional trachytes in Kenya. The absence of an overall trend in the extrusion sequence (the latest extrusions are of quartz trachyte), together with interbedding of (a) nepheline-normative and quartz-normative lavas, and (b) Group I and Group II lavas, indicates complex evolution and emptying of the sub-volcanic magma chamber or chambers.

In addition to the analyses of Cooper (1972), Paslick *et al.* (1996) gave complete rock analyses of one phonotephrite and one nephelinite. However, it is possible that the nephelinite sample, from Njanja to the south of Tarosero, may derive from the nearby Burko volcano (Dawson personal observation).

### *Monduli*

*Type*: ?stratovolcano; *Lat/Long*: 3.25°S 36.48°E; *Elevation*: 2660 m; *Relief*: 930 m on south and east sides, 600 m on north side; *Volcano diameter*: 6 × 10 km; *Eruptive history*: not known; *Composition*: olivine basalt and trachyte; *QDS*: 54 Monduli.

**Table 7.3.** *Tarosero lavas-trace element differences between Group I and Group II rocks of similar silica contents and similar alkalinity (from Cooper 1972)*

| Sample | Group I BD255 | Group II BD306 | Group I BD261 | Group II BD512 | Group II maximum concentrations |
|---|---|---|---|---|---|
| $SiO_2$(wt%) | 63.45 | 62.12 | 62.25 | 62.3 | |
| (Na + K)/Al | 1.11 | 1.19 | 1.17 | 1.16 | |
| Ba | 640 | 838 | 517 | 1029 | 1446 |
| Ce | 443 | 992 | 378 | 1205 | 1322 |
| Cl | 173 | 263 | 172 | 178 | 725 |
| Cr | 28 | 7 | 41 | 0 | 13 |
| La | 179 | 476 | 213 | 669 | 868 |
| Nb | 131 | 320 | 118 | 454 | 454 |
| Nd | 82 | 177 | 104 | 238 | 279 |
| Ni | 3 | 4 | 8 | 9 | 9 |
| Pb | 21 | 43 | 5 | 49 | 62 |
| Pr | 38 | 100 | 46 | 139 | 181 |
| Rb | 191 | 360 | 135 | 242 | 391 |
| Sm | 32 | 60 | 26 | 78 | 85 |
| Sr | 156 | 806 | 22 | 199 | 806 |
| Th | 62 | 148 | 42 | 192 | 196 |
| V | 53 | 23 | 19 | 9 | 30 |
| Y | 70 | 173 | 67 | 164 | 295 |
| Zn | 87 | 336 | 201 | 340 | 520 |
| Zr | 1145 | 2611 | 744 | 3355 | 3355 |

Values in ppm; XRF analyses.

**Table 7.4.** *Electron-microprobe analyses of phases in Tarosero trachytes*

| | In Group I trachyte BD255 | | | | | | | | | | | In Group II sodalite trachyte BD295 | | | | | In Group II sodalite trachyte BD512 | | | | |
|---|---|---|---|---|---|---|---|---|---|---|---|---|---|---|---|---|---|---|---|---|---|
| | **1** | **2** | **3** | **4** | **5** | **6** | **7** | **8** | **9** | **10** | **11** | **12** | **13** | **14** | **15** | **16** | **17** | **18** | **19** | **20** | **21** |
| $SiO_2$ | 49.40 | 50.30 | 49.60 | 50.60 | 49.70 | 39.60 | 38.50 | 65.90 | 66.50 | 66.60 | 66.60 | 38.70 | 38.80 | 38.80 | 39.90 | 50.1 | 45.20 | 47.60 | 50.10 | 49.30 | 38.50 |
| $ZrO_2$ | 0.16 | 0.01 | 0.11 | 0.06 | 0.85 | 0.05 | 0.07 | na | na | na | na | na | na | na | na | 11.3 | 0.07 | 1.64 | 0.10 | 0.49 | 0.19 |
| $TiO_2$ | 2.03 | 1.24 | 1.17 | 1.23 | 0.60 | 2.48 | 3.07 | na | na | na | na | na | na | na | na | 0.18 | 1.85 | 1.79 | 0.20 | 0.69 | 6.08 |
| $Al_2O_3$ | 0.57 | 1.16 | 2.15 | 1.43 | 0.26 | 11.20 | 11.60 | 20.40 | 18.20 | 19.00 | 18.20 | 26.30 | 27.80 | 26.40 | 26.00 | 0.20 | 3.91 | 0.65 | 0.46 | 0.60 | 1.19 |
| $Fe_2O_3$ | | | | | | | | 0.25 | 1.40 | 0.71 | 1.26 | 2.99 | 2.45 | 3.15 | 3.97 | | | | | | |
| FeO | 24.90 | 19.10 | 19.50 | 17.70 | 22.20 | 12.80 | 13.30 | | | | | | | | | | 23.40 | 23.80 | 30.70 | 22.70 | 41.90 |
| MnO | 1.4 | 1.14 | 1.15 | 1.09 | 1.24 | 0.42 | 0.44 | na | na | na | na | na | na | na | na | 1.24 | 1.02 | 1.00 | 1.41 | 0.90 | 1.55 |
| MgO | 5.81 | 10.20 | 9.69 | 11.50 | 4.19 | 16.90 | 16.00 | na | na | na | na | 0.09 | 0.05 | 0.20 | 0.12 | na | 7.19 | 1.66 | 2.72 | 3.68 | 0.80 |
| CaO | 4.07 | 6.17 | 6.03 | 6.54 | 15.9 | 0.02 | 0.00 | 1.44 | 0.15 | 0.47 | 0.17 | 0.05 | 0.04 | 0.01 | 0.02 | 8.70 | 7.36 | 10.70 | 1.86 | 16.00 | 0.62 |
| $Na_2O$ | 7.07 | 6.26 | 5.79 | 6.07 | 4.01 | 0.95 | 0.85 | 9.21 | 7.52 | 8.61 | 7.81 | 25.30 | 24.30 | 25.30 | 23.90 | 12.90 | 5.00 | 6.94 | 8.34 | 4.92 | 7.72 |
| $K_2O$ | 1.22 | 1.24 | 1.21 | 1.22 | 0.00 | 8.93 | 8.86 | 3.07 | 6.33 | 4.72 | 5.57 | 0.08 | 0.14 | 0.09 | 0.17 | 0.33 | 1.30 | 0.13 | 1.52 | 0.00 | 0.04 |
| $Nb_2O_5$ | 0.04 | 0.09 | 0.06 | 0.00 | 0.00 | 0.06 | 0.00 | na | na | na | na | na | na | na | na | 0.55 | 0.00 | 0.22 | 0.02 | 0.00 | 0.43 |
| $SO_3$ | na | na | na | na | na | na | na | na | na | na | na | 0.05 | 0.07 | 0.04 | 0.05 | na | na | na | na | na | na |
| Cl | 0.02 | 0.03 | 0.02 | 0.03 | na | 0.02 | 0.02 | na | na | na | na | 7.09 | 6.98 | 7.15 | 7.11 | 2.13 | 0.00 | 0.00 | 0.02 | na | na |
| Sum | | | | | | | | | | | | 100.65 | 100.63 | 101.14 | 101.24 | 98.05 | | | | | |
| Less O≡Cl | | | | | | | | | | | | 1.63 | 1.60 | 1.64 | 1.63 | 0.48 | | | | | |
| Total | 96.69 | 96.94 | 96.48 | 97.47 | 98.95 | 93.43 | 92.71 | 100.27 | 100.10 | 100.11 | 99.61 | 99.02 | 99.03 | 99.30 | 99.61 | 97.57 | 96.30 | 96.13 | 97.45 | 99.28 | 99.10 |

Analyst: J. B. Dawson. na, not analysed.

1, Large dark-blue poikilitic ferri-richterite; 2, 3, Green groundmass ferri-richterite; 4, Brown richterite inclusion in feldspar phenocryst; 5, Aegirine phenocryst; 6, Mica phenocryst; 7, Mica included in feldspar phenocryst; 8, Core of anorthoclase phenocryst; total includes 0.25 wt% BaO; Or 16.8, Ab 76.2, An 6.6, Cs 0.4 mol%; 9, Rim of anorthoclase phenocryst; total includes 0.06 wt% BaO; Or 35.5, Ab 63.8, An 0.7, Cs 0.1 mol%; 10, Core of groundmass anorthoclase; BaO below detection; Or 25.9, Ab 71.9, An 2.2 mol%; 11, Rim of groundmass anorthoclase; total includes 0.18 wt% BaO; Or 31.6, Ab 67.3, An 0.8, Cs 0.3 mol%. This specimen also contains zoned chevkinite (Macdonald *et al.* 2002); 12, 13, Core and rim of sodalite phenocryst; 14, 15, Core and rim of sodalite phenocryst; 16, Groundmass eudialyte; total includes 0.57 wt.% $La_2O_3$, 0.94% $Ce_2O_3$ and 0.58% $Nd_2O_3$; 17, Dark-brown, partly-absorbed kataphorite xenocryst, mantled by corona of light-brown, unidentified phase and green aegirine; 18, Unidentified light-brown phase, in corona round kataphorite xenocryst; 19, Blue groundmass ardvedsonite; 20, Green aegirine in corona around kataphorite xenocryst; 21, Groundmass aenigmatite.

Monduli mountain is an east–west trending structure, with steep southern and eastern slopes, and more gradual northern and western slopes, the northern slopes grading down onto the lava plateau that dips northwards towards Tarosero. The mountain is thickly vegetated, with bamboo forest in the summit area, and also has a thick veneer of tuffs derived from the Meru volcano to the east. Hence much of the structure of Monduli is speculative. The Tarosero escarpment, that forms the eastern boundary of the lava plateau, is obscured by the eastern lower slopes of Monduli, and the steep southern slopes of the mountain may be due in part to faulting. However, the deeply dissected appearance of the eastern and southern slopes suggest that these parts of the structure are formed of poorly consolidated material (? pyroclastics or fault-related volcanic debris) (Dawson 1964*b*). Exposure is poor, and the few patchy outcrops have yielded specimens of olivine basalt and trachyte. A basalt flow on the lower northwestern slopes contains xenoliths of spinel harzburgite that contain primary carbonate (Lee *et al.* 2000).

Paslick *et al.* (1995, 1996) gave complete analyses of one tephrite and two picrobasalts, and analyses of olivine and clinopyroxene in the two picrobasalts. Roberts (2002) gaves analyses of two basanites and two basalts from the volcano and of a tephriphonolite dyke which has affinities with similar rocks at Meru immediately to the east.

### *Kilimanjaro*

*Type*: line of overlapping shield volcanoes; *Lat/Long*: Kibo cone 3.07°S 37.37°E; *Elevation*: 5895 m; *Relief*: 4400 m; *Diameter*: 85 × 45 km; Eruptive history: Pliocene–Recent; *Crater diameter*: Kibo crater 2.5 km (main caldera), 900 m (inner crater); *Composition*: alkali olivine basalt, trachybasalt, perakaline trachyte and peralkaline phonolite; *QDS*: 42, 56 and 57 Kilimanjaro (Downie *et al.* 1965).

Kilimanjaro is a massive mountain formed by the coalescence of three major volcanoes, Shira, Kibo and Mawenzi (Fig. 7.7). The Kibo summit is the highest point on the African continent and, with the exception of some of the highest Hawaiian Islands which rise from the Pacific Ocean floor, in its relative relief is the highest mountain on Earth. An ascent of the mountain goes from the equatorial luxuriance of the foothills to the Arctic-zone glaciers

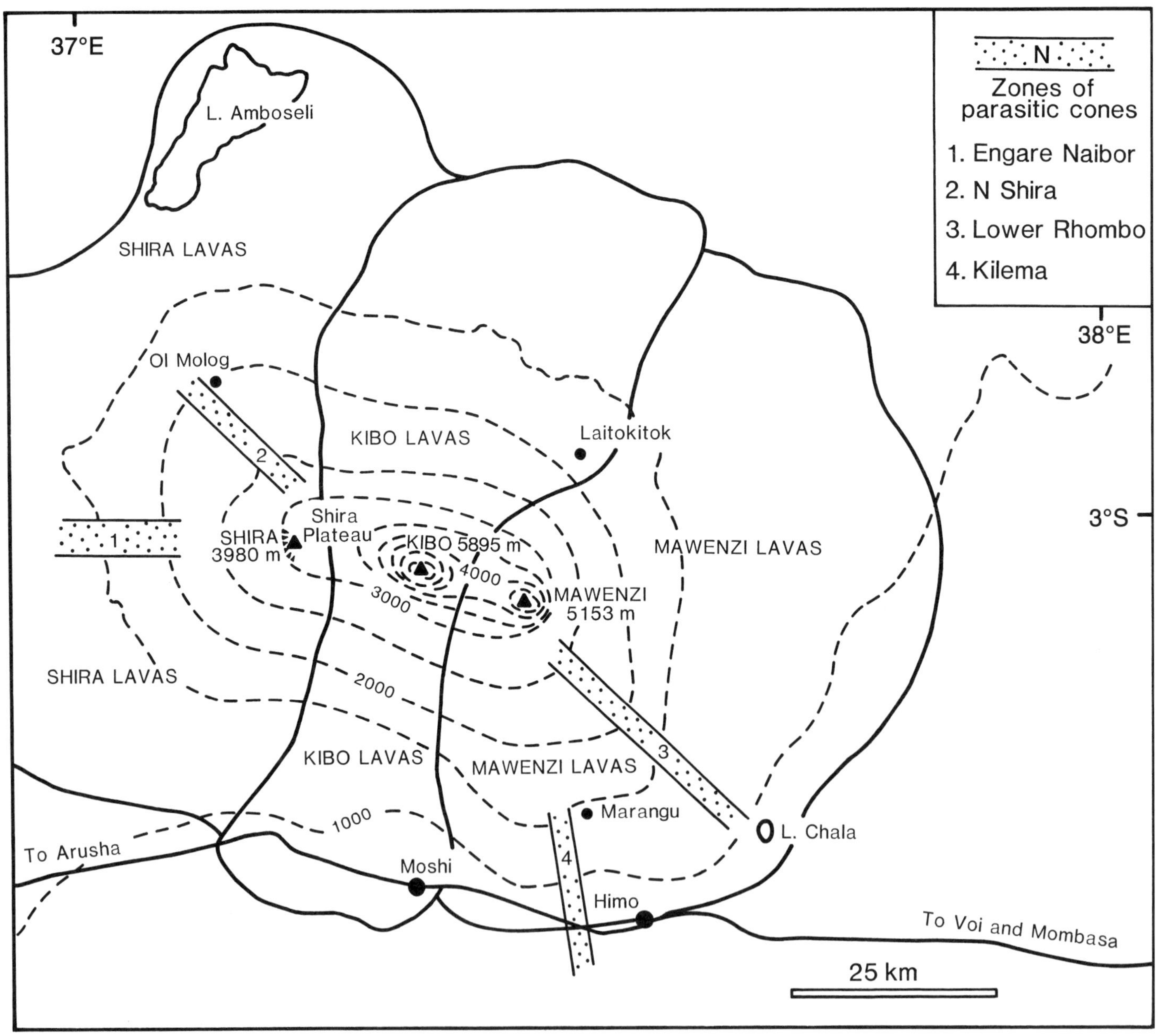


**Fig. 7.7.** Map of the Kilimanjaro massif showing distribution of lavas from the main centres, and the zones of late parasitic cones.

of the Kibo cone (Plate 4), a traverse of the Earth's vegetation zones on a single mountain. Due to the monsoon and counter-monsoon winds, with their associated high rainfall, there is dense forest on the southwestern, southern and eastern slopes of the massif between around 1500 and 3300 m and, due to the poor exposure, the geology of these lower slopes is poorly-known compared with the upper parts of the volcanoes. However, highly mobile extrusions of basanite and olivine basalt have flowed out onto the plains surrounding the massif, and flat-lying, distal lava flows can be found at considerable distances away from the actual topographic mountain, in the Amboseli area on the western, northern and eastern sides of the massif (Williams 1969).

The most comprehensive account of the geology is that of Downie & Wilkinson (1972). The massif overall is elongated NW–SE, but the WNW–ESE alignment of the three, equally spaced, major eruptive centres is oblique to this major trend. There are also numerous small parasitic lava cones, often concentrated along elongate zones on the lower and middle slopes of the massif. Two of the larger zones, on the lower north-western slopes of Shira and the lower southeastern slopes of Mawenzi, follow the main NW–SE trend of the massif (Fig. 7.10).

*Shira.* The oldest and most westerly major centre is a relatively simple cone with a large, eroded summit crater, approximately 4 km in diameter. The crater forms a plateau at around 3700 m, but is bounded on its western and southern sides by ridges formed by the old crater rim, the highest point on which is Klute Peak (3970 m). On its northern and eastern sides, the crater is buried by later extrusions from Kibo. A conical hill, Platzkegel, on the southern part of the Shira Plateau, is a major vent infilled with basaltic breccia and cut by dolerite and essexite intrusions. The lower lavas of Shira, exposed on the northwestern

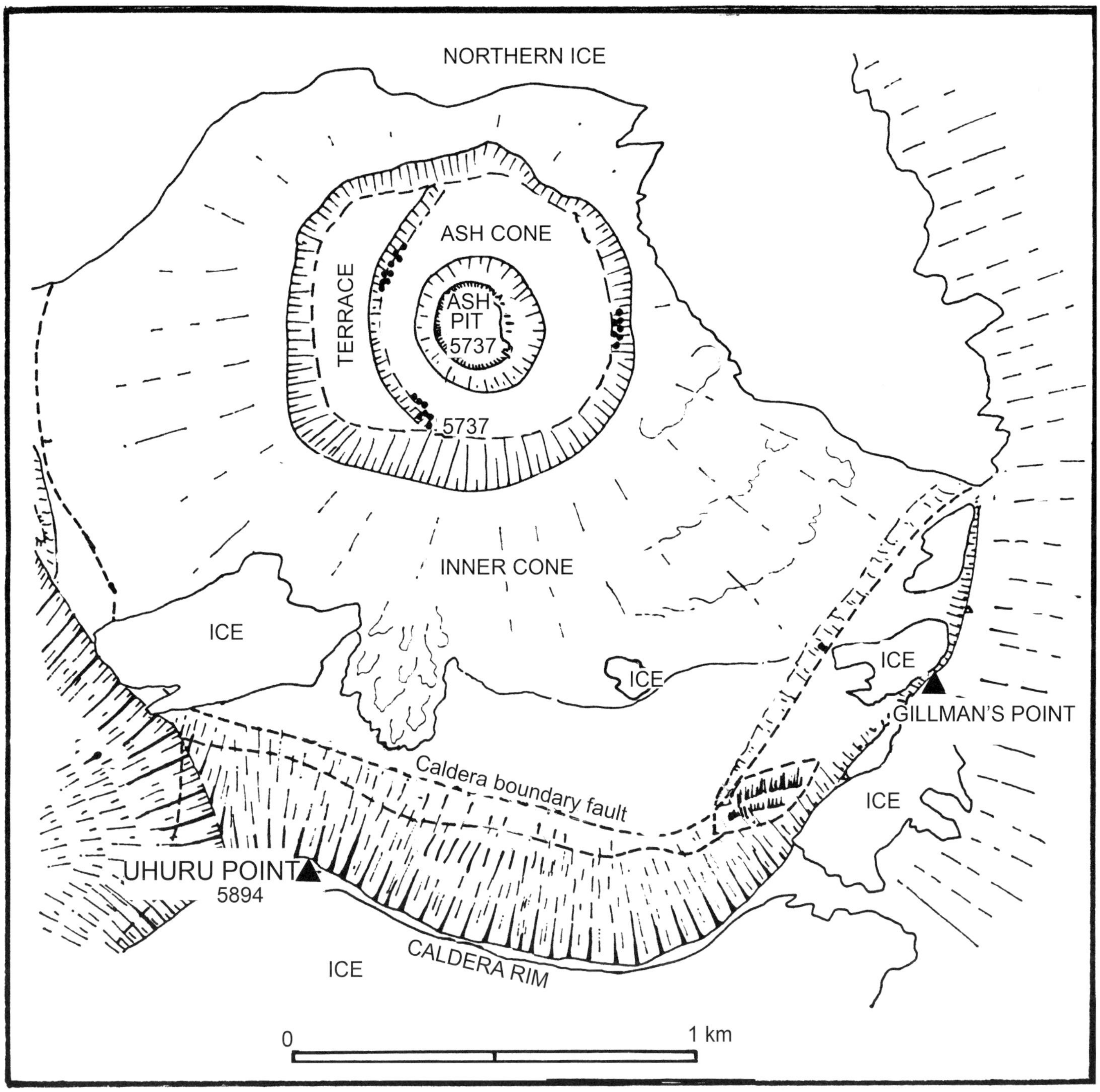


**Fig. 7.8.** Map of the Kibo crater area (after Downie & Wilkinson 1972).

slopes, are of olivine basalt and trachybasalt. Trachybasalts are common on the plateau, with subordinate amounts of basanite and nephelinite. Exposed in the old crater walls is a radial swarm of dykes of similar composition. Volcanic sediments on the east side of the plateau, overlain by lavas from Kibo, include conglomerates containing pebbles of melanephelinite and trachybasalt derived from Shira, and also ijolite, biotite phonolite and basanite, which do not correspond to any known outcrops. The source of this material was to the NE of Shira, and closer to Kibo.

*Mawenzi.* This most easterly of the volcanoes is lower than Kibo, but much more difficult of access, and parts of its summit area are still unexplored. Allowing for burial beneath the later Kibo lavas, the volcano is around 60 km in diameter. Due to erosion and crater wall collapse, the upper part of the cone has been considerably modified, and it is now a highly eroded precipitous group of arêtes (Plate 5), residual between intersecting cirques. They culminate in Hans Meyer Peak (5153 m), the third highest volcano in Africa (after Kibo and Mt Kenya). The formation of the cirques has modified earlier collapse of the NE crater wall of the original cone that has given rise to an extensive lahar deposit on the plains in Kenya to the NE. The largest of the cirques, on the eastern side, lies at the head of the Great Barranco, a deep gorge that provides the most continuous section through the mountain.

There are two main extrusive centres, Mawenzi itself and Neumann Tower (4425 m), a subsidiary peak 2.5 km east of Mawenzi; both centres have been extinct since the early Pleistocene. At Neumann Tower, ankaramite, olivine basalt and trachybasalt lavas dip outwards from an agglomerate-filled vent. Exposures in the wall of the Great Barranco show them to be overlain by lavas from Mawenzi. The extrusives from Mawenzi are of feldspar-phyric basalt and olivine trachybasalt flows and tuff, that dip steeply away from the centre, except on the west side where the youngest flows are flat-lying beneath the Kibo lavas. Plugs of syenogabbro are exposed in the deeply eroded cirques on the eastern side of the centre. The weathering out of numerous dykes of ankaramitic basalt and olivine trachybasalt has given rise to steep pinnacles and ridges (Plate 5). Most belong to an elongate WNW–ESE swarm that intersects the two centres, but radial swarms also emanate from both the Neumann Tower and Mawenzi centres.

*Kibo.* This is the central cone and highest point of Kilimanjaro. Due to a preponderance of viscous trachytic and phonolitic lavas, it has steep slopes. Those on the western, northern and southern sides are partly coated with ice, and a broad flat summit caldera holds snow for much of the year (Plate 6). At its base, the cone merges with, and laps over, the cones of Shira and Mawenzi. Kibo is the only one of the three to have modern glaciers, though Mawenzi shows evidence of earlier glaciation.

The broad, flat-topped summit is due to the collapse of the original cone to form a caldera 2.5 km wide which itself contains the broad shallow Inner Cone (Fig. 7.8). This in turn contains the Inner Crater, 900 m wide, containing a small ash cone punctured by an ash pit 120 m wide and 130 m deep. Fumaroles and associated sulphur deposits occur in the Inner Crater. The caldera rim is highest on its southern edge and culminates in Uhuru Peak (5895 m), the highest point in Africa. One of the two larger glaciers lies on the southern and western outer slopes of the main cone, and another in the northern part of the caldera overtops the caldera rim to flow down the northwestern upper slopes of the cone. Smaller snowfields occur mainly on the western outer slopes, and also along the southern trough of the caldera.

Activity at Kibo began in the Pleistocene and the earliest lavas are possibly intercalated with the youngest flows from Shira and Mawenzi. However, the earliest known lavas from Kibo erupted from a cone that already matched Mawenzi in height and considerably overtopped Shira. The later activity of Kibo produced flows that, due to the constraints of the other two flanking volcanoes, flowed mainly to the north and south, lapping over the older flows from the other centres and obscuring the early flows from Kibo itself. The lava extrusion ranges in age from Lower Pleistocene for the earliest flows to the Recent lava flows from the Inner Crater. Because of the fumarole activity, the volcano is classified as still active.

Ten distinct lava groups have been recognized but, in brief, much of the earliest activity was extrusion of mugearite and benmoreite, interspersed with trachyte. Later extrusions are dominantly of trachyte containing large phenocrysts of alkali feldspar (the so-called 'rhomb porphyrys'), whereas the most recent lavas forming the caldera rim and extruded in and from the Inner Crater are of nepheline-phyric aegirine phonolite. Anorthoclase phenocrysts derived from scoriaceous trachyte on the caldera rim yield an age of $0.21 \pm 0.03$ Ma (Wilkinson *et al.* 1986). Activity was not continuous and the various lava groups were separated by periods of erosion when deep valleys were incised and filled with glacial, fluvio-glacial and alluvial deposits. Despite the relatively high viscosity of the lavas, associated pyroclastics are not abundant, though fine-grained, highly indurated tuffs occur on the upper southern slopes at around 4000 m between the Karanga and South-Eastern valleys (Dawson personal observation). A large lahar deposit, correlated with a huge landslip scar on the upper southwestern slopes (the Kibo Barranco), covers a large area on the plains south and SW of Moshi.

Because of the Recent age of Kibo, exposures of intrusive rocks are scarce. Radial dykes of trachybasalt, trachyte and phonolite crop out in erosion gullies on the upper western slopes, where there is also an outcrop of analcime syenite.

*Parasitic minor cones.* These and associated minor lava flows occur on the lower flanks of the main cones, extending in zones down to the plains, but they also occur on the saddle between Kibo and Mawenzi at 4300 m. The cones are generally only a few hundreds of metres in diameter, but may be up to 1.5 km and up to 200 m high. Most are of scoria or ash, but several have produced extensive lava fields. The cones have apparently been erupted at various periods, as judged by their relationships to extensive lava flows from the three main centres. They have not been systematically studied petrologically, but ankaramite, olivine basalt, trachybasalt, trachyandesite, trachyte and phonolite have been recorded. In the zone of cones on the lower southeastern slopes of Mawenzi (the most accessible, being close to the ascent route from Marangu), the cones above the forest zone are entirely of ankaramite, whereas those lower down in the forest and below are a combination of ankaramite and olivine basalt. At the southeastern extremity of this zone, Lake Chala occupies a caldera 3 km in diameter and 100 m deep. The crater walls expose Mawenzi lavas at their base, overlain by up to 80 m of calcite-cemented grit and agglomerate. It has been suggested that the calcite is of magmatic (carbonatitic) origin.

*Petrochemistry.* Downie & Wilkinson (1972) listed 84 analyses for major and minor oxides, and trace elements (V, Cr, Co, Ni, Cu, Zn, Rb, Sr, Zr, Ba). Of these analyses, eight are of distal basalts and basanites from the Amboseli area, 16 of basalts, trachybasalts ankaramitic basalt and syenogabbro from Mawenzi, nine of basalts, basanite and hawaiite from Shira, 38 of mugearites, benmoreites, trachytes and phonolites from Kibo, and 11 of rocks from parasitic cones and other localities. Recently these data have been supplemented by analyses of 1 basanite, 2 basalts, 1 picrobasalt and 3 trachybasalts from Mawenzi, 2 basanites and a picrobasalt from Shira, and 1 basanite, 3 trachyandesites, 2 tephriphonolites and 5 phonolites from Kibo (Roberts 2002); in addition to the trace elements given in Downie & Wilkinson (1972), these later analyses include values for Ga, Hf, Nb, Pb, Sc, Ta, Th, U and REE.

The three centres are chemically distinct, with Mawenzi being the simplest with most rocks being of hawaiite composition. Kibo is the most silicic and alkaline of the three centres.

The Kilimanjaro rocks have a compatible element range (Cr < 10 to 650 ppm, Ni < 10 to 400 ppm) consistent with olivine and clinopyroxene fractionation. The mafic rocks have concentrations of Rb, Ba Th, U, Nb and Zr in the order of × 100 to × 800 bulk Earth values, and values in the more evolved rocks are one order of magnitude greater than those in the mafic rocks. Further, aegirine phonolites from the latest Inner Crater group on Kibo have high REE concentrations and >3000 ppm Zr and >1500 ppm Nb, suggesting extreme fractional crystallization in the absence of apatite and titanite. Although other evolved rocks can be explained by fractional crystallization of a possible mafic parental magma (Downie & Wilkinson 1972), Roberts (2002) suggests these large increases cannot be due solely to olivine and pyroxene fractionation, and proposes that the high Zr and Nb must result from high concentrations in a melt that was distinct from the parent of the other evolved Kilimanjaro rocks (Roberts 2002). However, Dawson *et al.* (*in press*) find that anorthoclase crystallization in a specimen of Kibo rhomb porphyry results in the residual glass being less silicic, more peralkaline and containing higher Zr (1800 ppm) and Nb (438 ppm) than the bulk trachyte, thus developing a trend that may culminate in the Inner Crater group.

*Glaciation.* Since the first visit to the summit area by H. Meyer and E. Purtscheller in 1889 (Meyer 1900), it has been realized that the ice fields on Kibo are less extensive than in the past; with Meyer recording moraines and glacial striae down to 3500 m. Both Kibo and Mawenzi were glaciated and Downie (1964) records 4 glacial epochs on Kibo. Ice cores taken at six localities on Kibo in 2000 (Thompson *et al.* 2002) have provided a *c.* 12 ka record of Holocene climate variability for the area, including three periods of abrupt climate change at *c.* 8300, *c.* 5200 and *c.* 4000 years ago, the last coinciding with the period of the greatest historically-recorded drought in tropical Africa. Since 1912 the areal extent of the Kibo ice fields has decreased from 12.1 square kilometres to 2.5 in 2003 i.e. by around 80% and Thompson *et al.* (2002) predict they are likely to disappear between 2015 and 2020 if present climate conditions continue. Mote & Kaser (2007) suggested that, although glaciers in mid-latitudes are responding directly to global warming, the disappearance of the Kibo ice fields is more likely to be linked to changes in the surface temperatures of the Indian Ocean. The ensuing changes in the atmospheric circulation lead to lower precipitation; combined with ablation by solar radiation, the effect is retreat of the ice fields

## The Younger Extrusives

Following the Upper Pleistocene faulting at 1.2 Ma, there was another major phase of volcanic activity that, in eruption style (relatively small volume of extruded material, and magma type) contrasts with the relatively quiet, massive extrusions from the earlier basaltic shield volcanoes. Most activity is within the area of the Older Extrusives, but some activity occurred in the Babati–Hanang area, well to the south of the main area of the earlier activity (Fig. 3.1).

The Younger Extrusive activity was highly explosive, giving rise to features varying from major stratovolcanoes dominated by pyroclastic materials (Meru, Oldoinyo Lengai, Kerimasi, Burko, Kwaraha and Hanang) to areas of small tuff cones and explosion craters, many of which occur around the major centres. Tuffs from this phase of activity are widespread throughout the volcanic province. The tuff cone areas often coincide with areas of intense minor faulting, for example at Basotu on the Mbulu Plateau (Downie & Wilkinson 1962) and in the Engaruka–Natron area (Dawson & Powell 1969).

The magma types at most of these later centres are volatile-rich, ultrabasic and high in alkalies, giving rise to olivine-free nephelinites and phonolites. Blocks of plutonic ijolites, feldspathoidal syenite and metasomites occur in the pyroclastic rocks at several centres. Carbonatite lavas and pyroclastics accompany the nephelinites and phonolites at Oldoinyo Lengai, Kerimasi, Ngurdoto, Kwaraha and Hanang, and numerous minor carbonatite tuff cones are known on the Mbulu plateau (Downie & Wilkinson 1962) and in the Monduli area where they are associated with extrusion of ankaramite and olivine melilitite (Dawson 1964*c*; Roberts 2002). Alkali carbonatite lava extrusions and ash eruptions have taken place at Oldoinyo Lengai in modern times (Dawson 1989; Dawson *et al.* 1995*a*), with documented lava extrusion occurring as recently as August 2007 and ash eruption in September 2007 (Mitchell & Dawson 2007). Recent minor flows and tuff cones of olivine nephelinite and olivine melilitite occur on the floor of the rift valley in the Oldoinyo Lengai–Gelai area (Dawson *et al.* 1985; Keller *et al.* 2006), in the Monduli–Arusha area (Dawson *et al.* 1970), and south of Hanang (Dawson *et al.* 1997). Some of the minor tuff cones contain a variety of blocks of upper mantle material that testify to a heterogeneous mantle beneath northern Tanzania.

In addition to these Upper Pleistocene–Holocene centres, activity at some of the older centres persisted after the Middle Pleistocene. At Tarosero, peralkaline trachyte and phonolite flows continued into the Upper Pleistocene. On the Kibo peak of Kilimanjaro (classified as still active), phonolite and trachyte extrusions continued until Recent times, and there is still fumarolic activity in the Inner Crater (Downie & Wilkinson 1972).

### *Oldoinyo Lengai*

*Type*: stratovolcano; *Lat/Long*: 2.75°S 35.90°E; *Elevation*: 2890 m; *Relief*: 2090 m; *Diameter*: 12 km; *Eruptive history*: around 22 Ka to present; *Crater diameter*: south inactive crater 300 × 200 m; north active crater 400 × 300 m; *Composition*: nephelinite, phonolite, natrocarbonatite; *QDS*: 39 Angata Salei.

Oldoinyo Lengai is the only currently active volcano in northern Tanzania and is unique in being the world's only active carbonatite volcano. A distinctive white capping of weathered carbonatite (Plate 7) on the northern side of the mountain lead to its being identified as a 'Snow Mountain' (together with Kilimanjaro and Mt Kenya) on the 1855 map of the interior of East Africa. Due to its remoteness, the volcano was rarely visited before the 1980s. However, the recognition of the unique natrocarbonatite lavas (Dawson 1962*a*, *b*), together with the construction of the Engaruka–Loliondo dirt road in the late 1970s and the boom in geotourism, have resulted in the volcano being visited by many more climbers, both scientists and tourists, in recent years. Since 1983, events at the volcano have been compiled and reported in bulletins published by the Natural History Museum, Smithsonian Institution of Washington, DC 20560, USA. From January 1983 to December 1989, this was the Scientific Event Alert Network (SEAN) Bulletin, and subsequently the eBulletin of the Global Volcanism Network. These bulletins have published reports, diagrams and photographs of activity at the volcano. Many of these reports have been collated by Celia Nyamweru who has also given reviews at various stages of the more recent activity (Nyamweru 1988, 1990, 1997).

*History of activity.* The mountain was first reported as an active volcano in early accounts of the interior of East Africa gleaned from Arab slavers (Wakefield 1870; Farler 1882) and the first eye-witness account of minor ash eruption in 1883 was documented by Gustav Fischer (1884–1885). Before 1960, flows of natrocarbonatite in the active northern crater were recorded in 1904, 1914, and 1954; flows occurred in the period 1960 to 1966, and from 1983 to 2007. Major ash eruptions occurred in 1917, 1926, 1940, 1966–1967 and 2007–2008, with a relatively small ash eruption in

1993. The pre-1960 ash eruptions were interspersed with periods of apparent dormancy (Dawson *et al.* 1995*a*). The word apparent is used advisedly because, whereas ash eruptions can be seen from a distance and recorded, prior to 1960 the summit was rarely visited; hence, lava extrusion within the crater may well have been underestimated. The activity observed during 1966–1967 and September 2007 was a series of Vesuvian- and Plinian-type eruptions (Dawson *et al.* 1968; Dawson & Mitchell 2007), and the description of the 1940 activity (Richard 1942) also suggests that the eruption was of Plinian type. However, a small ash eruption in 1993 erupted hollow lapilli, interpreted as resulting from explosion of vesiculated foam (Church & Jones 1994), and ash samples from the same eruption had the grain-size distribution characteristics that are typical of the products of basaltic Strombolian eruptions (Dawson *et al.* 1996*a*).

Following several earth tremors, ranging in magnitude up to 6, in July and August 2007, remote sensing (Hyperion hyperspectral imager on board the NASA Earth-Observing-01 spacecraft) detected two very bright sources in the summit area on August 27, consistent with erupting lava, and another indicating a flow to the NW from the crater (http://science.jpl.nasa.gov), and on 4 September the volcano erupted forming a significant Vesuvian-type ash cloud. During September 2007 sporadic eruptions were of the Vesuvian and Plinian types (Dawson & Mitchell 2007) and these have continued until at least April 2008. The crater is now occupied by a major ash cone that has spilled over the outer edges of the original crater.

*Morphology.* Morphologically the volcano is a classic cone that stands close to the Natron Basin boundary fault escarpment, against which its pyroclastic deposits are piled (Plate 7). It is now covered by sparse grass and scrub though, prior to the 1917 ash eruption, the volcano was highly vegetated and access on the lower slopes was by game trails; the former luxuriant vegetation was mainly killed during the eruption and has not recovered to the present day. In late 2007, much grass was ignited and burned as lavas spilled down the outer western and northern slopes.

The cone is incised by deep radial gullies and, on its northeastern upper slopes, by a major landslide scar (the Eastern Chasm) (Plate 7). Debris flows from this landslip occur on the plains to the east and NE, and others originating on the northern slopes occur on the plains north of the volcano, some having a sufficiently long run-out to form low islands in Lake Natron some 16 km north of the volcano. Klaudius & Keller (2006) report a $^{14}$C date of 2500 years BP on plant remains embedded in one of the debris flows. Parasitic craters occur on the western and eastern lower flanks of the cone and a north–south trending line of three small scoria cones occurs on the lower northern flanks (Dawson 1962*a*).

The summit is occupied by two small craters: the older, inactive southern crater, which is now a shallow, ash-filled bowl, and the active northern crater. The active crater is some 350 m in a north–south direction and 300 m east–west. When first visited in 1904, the active 'crater' was little more than a flat lava platform from which rose a number of hornitos; there were no crater walls on the northern, western and eastern sides, thereby allowing lava spill down the outer slopes in these directions, and only a shallow arcuate southern wall below the summit (Uhlig 1905; Reck 1924–25). The changes before 1960 are not well documented, except that the shallow 'crater' observed in 1904–1913 was deepened during the 1917 eruption (Reck & Schulze 1921), and the crater was still deep in 1954 (Guest 1956). In 1960, the active crater was a deep (*c.* 200 m) bowl bounded on its northern, western and eastern sides by cliffs (Fig. 7.9), but on its southern side by collapse terraces The changes in the profile of the active crater between the early part of the twentieth century and 1983 are given by Dawson *et al.* (1995*a*) and those between 1960 and 2007 in Figure 7.10. Since 1960, the depth and shape of the

**Fig. 7.9.** The northern crater of Oldoinyo Lengai (1960) looking northwards from the summit. The crater diameter is *c.* 300 m and the north wall is *c.* 200 m high. Recent black spatter surrounds several small vents. The crater has subsequently been infilled by ash from the 1966–67 ash eruption and post-1983 lava flows. Photograph by C. Bristow.

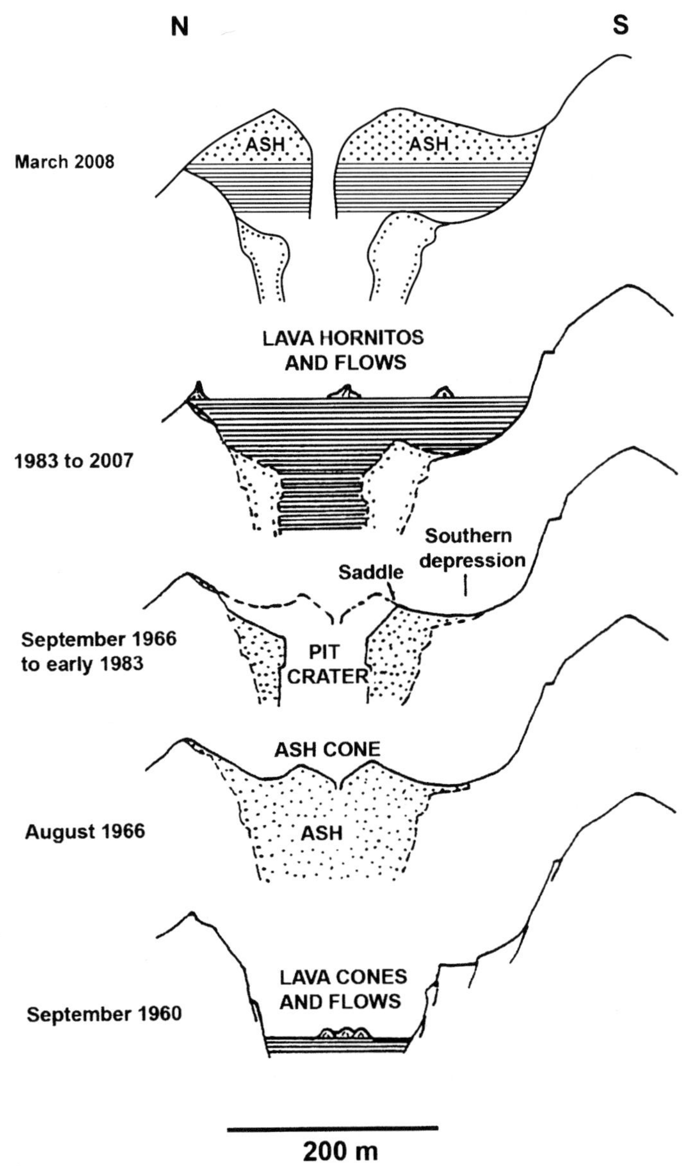


**Fig. 7.10.** Profiles of the northern crater of Oldoinyo Lengai, showing crater changes.

active crater has been modified, first as a result of partial infill by ash during the 1966 ash eruption, followed by partial collapse of this ash cone to form a pit crater. Between 1967 and early 1983, there was no activity but, following the renewal of activity in 1983, the pit crater was gradually filled by lava flows and spatter from a series of small transient cones, hornitos and lava pools (Plate 8). In early 1993, a small ash eruption in which the lapilli were of mixed carbonate–silicate composition accompanied extrusion of exceptionally voluminous and viscous natrocarbonatite lava flows that contained globules of silicate lava (Dawson *et al.* 1994). By November 1998, the crater had become almost infilled by lava, and lava spilled over the lowest northwestern and eastern sectors of the crater rim to flow down the outer slopes (Nyamweru 1999) (Plate 9). Since then, minor lava extrusions and overspill on the northern, northwestern and eastern slopes have continued almost continuously till at least August 2007 including, following a minor ash plume, a major hornito collapse into an underlying lava pool that caused a lava surge that spilled down the western slopes from a low point on the western crater rim on 29 March and 3 April (Bulletin of the Global Volcanism Network 2006; Kervyn *et al.* 2008*a*). Between late 1998 and September 2007, a crater wall was present only on the south side below the summit, and the 'crater' was, as in 1904, a carbonatite lava platform studded with hornitos and lava pools. As a result of the September 2007–April 2008 eruptions, the former crater is (March 2008) occupied by an ash cone with a central ash pit (*Bulletin of Global Volcanism Network*, **32**, 10–16, 2007).

*Geology.* The volcano was formed in two main stages, separated by a period of quiescence, erosion and instability. The bulk of the volcano (Unit I of Dawson 1962*a*) consists of yellow, palagonitized and zeolitized nephelinitic and phonolitic tuffs and agglomerates that were extruded from the now inactive southern crater. A maximum age for the onset of the activity is provided by a date of 0.37 Ma on mica-rich tuffs, surrounding the Loolmurwak crater south of the volcano, that are overlain by Oldonyo Lengai yellow tuffs (MacIntyre *et al.* 1974). Hay (1976) correlated the Unit I tuffs with the Ndutu and Naisiusiu Beds of the Olduvai Gorge to the west of the volcano, which range in age from 0.4 to 0.15 Ma. Phonolitic nephelinite and phonolite flows interbedded with the pyroclastics are exposed in the Eastern Chasm and form a flank flow on the lower eastern slopes, and highly viscous, sanidine-phyric phonolite flows occur on the upper southern and southwestern slopes.

A period of dormancy followed during which there was intense erosion and instability, resulting in numerous debris flows, at least one of which had sufficiently long run-out as to form islands in Lake Natron, 15 km north of the volcano. Klaudius & Keller (2006) proposed that much of the northern face of the original cone was destroyed during a major collapse during this interval; the amount of Unit I material in the lahar field to the north and east of the volcano certainly testify to major collapse(s) (Kervyn *et al.* 2008*b*). The older rock sequence was then penetrated by explosion craters on the western and eastern lower flanks (Unit II pyroclastics of Dawson 1962*a*), and recently Keller *et al.* (2006) showed these rocks to be olivine melilitite tuffs, mineralogically and chemically similar to tuffs extruded from minor tuff cones and lava flows found to the east and north of the volcano (Dawson *et al.* 1985). Thus, the two major stages in the formation of the volcano are separated by extrusion of a primitive magma type.

The period of intense erosion was followed by a shift of the main extrusion site to the modern, active, northern crater (Dawson 1962*a*). Klaudius & Keller (2006) suggested that the following activity filled in the breached crater and then built up the northern side of the cone in two separate but related stages. During this period black nephelinitic tuffs and agglomerates, and wollastonite–combeite nephelinite flows (Unit III of Dawson 1962*a*) were extruded some intercalated with natrocarbonatite ashes and flows in the summit area. The Namorod Ash of the Olduvai succession, dated at around 1250 BP, has been correlated with the oldest of the Unit III pyroclastic units (Hay 1976). Wollastonite–combeite nephelinite, containing cognate xenoliths of ijolite, was also extruded from the small cones on the northern, lower flanks of the main cone (Dawson 1962*a*, 1998). Whereas the equivalents of the earlier phonolites and phonolitic nephelinites, and the associated olivine melilitites, occur at other volcanoes (e.g. in the Homa Bay area of Kenya; Le Bas 1977), the wollastonite–combeite nephelinites were thought to be unique to Oldoinyo Lengai. However, a further example has been found recently at the Meru volcano (see below).

Nephelinite and carbonatite activity alternated in this succession, but observations since the beginning of the twentieth century give the impression that extrusion of carbonatite has predominated, with the exception of: (a) extrusion of a combeite nephelinite flow on the western rim possibly just prior to the 1917 ash eruption; (b) the 1966–67 ash eruption when the ashes were a mixture of silicate and natrocarbonatite material (Dawson *et al.* 1968, 1992); (c) the activity in 1993 when highly viscous natrocarbonatite lavas contained globules of silicate material (Dawson *et al.* 1994), and the erupted ash contained spheroids of silicate material exsolving natrocarbonatite (Dawson *et al.* 1996*a*); and (d) eruption in September 2007 of silicate lapilli rich in combeite, melilite and phosphate (Mitchell & Dawson 2007). There are no records of the nature of the material ejected during the explosive eruptions in 1917, 1926 and 1940.

Extrusions of carbonatite lava have mainly taken place within, and from, the northern crater; exceptions are a small flow that took place from a fracture high on the western slopes in June 1993, and a small flow on the upper eastern slopes, extrusion date unknown (Dawson & Mitchell 2007). Keller & Zaitsev (2006) also record two older, altered natrocarbonatite dykes cutting phonolite lavas high on the upper southern slopes.

*Silicate lavas and plutonic xenoliths.* The phonolites and phonolitic nephelinites of Unit I consist of combinations of nepheline, alkali pyroxene, sanidine, sodalite, vishnevite, melanite garnet, perovskite, apatite, glass and zeolites (Dawson 1962*a*; Donaldson *et al.* 1987). Wollastonite and combeite (ideal formula $Na_2Ca_2Si_3O_9$) are common additional phases in the Unit III nephelinites (Dawson 1998; Peterson 1989*a*, *b*; Klaudius & Keller 2006) which lack sanidine, and one specimen contains exceptionally Fe-rich nepheline, clinopyroxene and sodalite, and also the rarer phases delhayelite ($K_4NaCa_2[Al_2Si_6O_{19}]$F,Cl,H), barian lamprophyllite and Sr–Ba phosphate (Dawson & Hill 1998). However, wollastonite is often replaced by combeite, so there are two types of combeite: that replacing wollastonite (Plate 10) and magmatic phenocrystic (Plate 11), with the magmatic variety having a higher $Na/(Na + Ca)$ composition (Dawson *et al.* 1989).

These mineralogical differences, together with distinct differences in their Sr and Nd isotope composition (Bell & Dawson 1995*a*) have lead to the nephelinites being referred to as Group I (those from Dawson's Unit I) and Group II (from Unit III). Compositionally, the Group I nephelinites and associated phonolites are alkaline whereas the Unit II nephelinites are peralkaline, with peralkalinity indices ($[Na + K]/Al$) of up to 2.36. The groundmass glass in one Group II lava is exceptionally peralkaline, with low $Al_2O_3$ (only $2 - 3$ wt%) leading to $(Na + K)/Al$ up to 13, and contains very high concentrations of iron (FeO up to 18.5 wt%), $SO_3$ (3–5%), Cl up to 0.4%, and F up to 1%; it is relatively potassic with $Na/(Na + K)$ as low as 0.7 (Dawson & Hill 1998).

All the lavas contain low MgO (<2 wt%), Cr and Ni (<10 ppm), but exceptionally high concentrations of the incompatible elements Rb, Ba, Th, U, Nb, Mo, Sr, Zr, and the REE (Fig. 7.11). Zr/Hf, Zr/Nb, Nb/Ta and Hf/Ta ratios all increase from the Unit I phonolites to the Group II wollastonite–combeite nephelinites (Klaudius & Keller 2006). A Sr v. Nd isotope plot

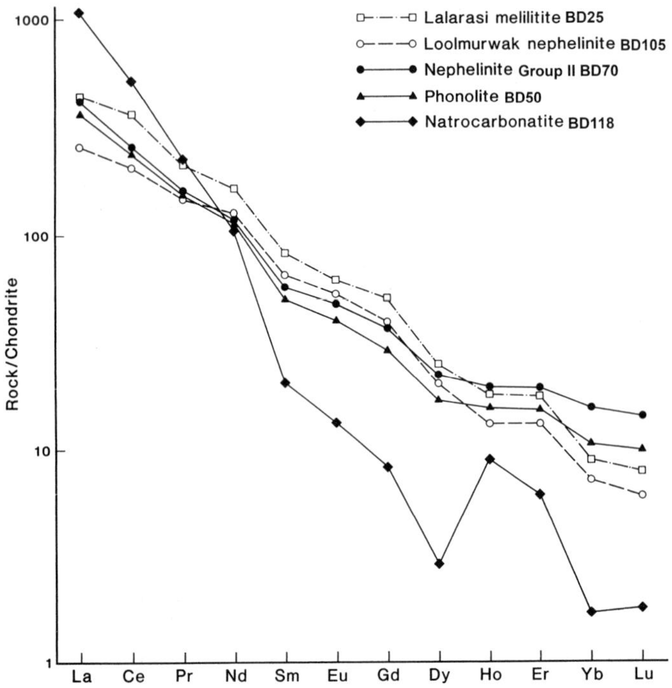


**Fig. 7.11.** Chondrite-normalized rare earth element plot for selected Oldoinyo Lengai lavas, and primary olivine melilitite and olivine melilite nephelinite in the Oldoinyo Lengai area. For the location of Oldoinyo Loolmurwak and Lalarasi, see Figure 7.18. Data from Dawson *et al.* (1985) and Donaldson *et al.* (1987).

(Fig. 7.12) shows that the data for the Group II wollastonite–combeite nephelinites overlap those for the natrocarbonatites, and form a group relatively depleted in radiogenic Sr. The Group I nephelinites form a group more enriched in radiogenic Sr and the data for phonolites lie mainly between the two nephelinite groups (Bell & Dawson 1995*a*).

Plutonic blocks of jacupirangite, ijolite, pyroxenite and nepheline syenite occur in the agglomerates (Plate 12). Mainly cumulates, they comprise combinations of nepheline, haüyne, alkali pyroxene, melanite and schorlomite garnet, apatite, perovskite, titanite, low-Cr–high-Ti spinel, glass, and (in the syenites) sanidine (Dawson *et al.* 1995*b*); eucolite is a rare phase in nepheline syenite (Dawson & Frisch 1971). Blocks containing wollastonite have been found only in the Unit III black agglomerates and tuffs. Ijolite and pyroxenite blocks ejected during the 1966 ash eruption contain the same phases as those in the agglomerates, together with mica-mantled xenocrystal olivine (Dawson *et al.* 1995*b*) (Plate 13). Complex zoning of the phases (Plates 14, 15) and mineral disequilibrium is attributed to convective percolation of fluids through permeable cumulates, possibly complicated by magma replenishment during crystallization of individual magma batches. A common feature of the jacupirangites, pyroxenites and ijolites, including those ejected during the 1966 ash eruption, is that the phases are cemented by vesiculated glass from which minor phases such as acicular apatite and skeletal nepheline and vishnevite have precipitated (Donaldson & Dawson 1978) (Plate 16); low analytical totals for the glasses point to significant concentrations of dissolved volatiles in the parent magmas which, in the absence of hydrous phases, is assumed to be largely $CO_2$

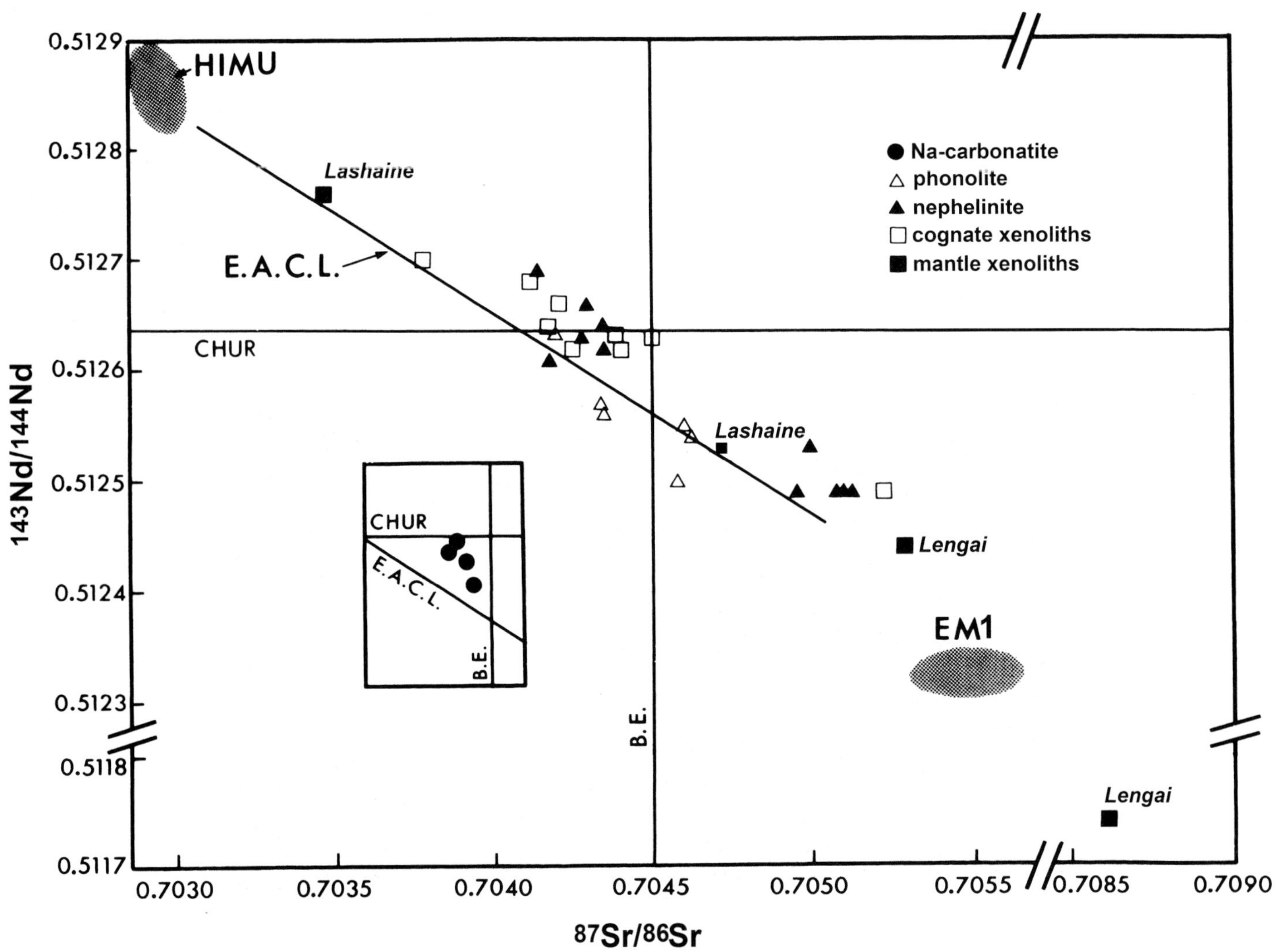


**Fig. 7.12.** Nd v. Sr isotope plot for lavas, cognate xenoliths (ijolites, jacupirangite and nepheline syenite), and two mantle xenoliths from Oldoinyo Lengai (after Bell & Dawson 1995*a*). Also shown are two mantle xenoliths from Lashaine (data from Cohen *et al.* 1984). EACL is the East African Carbonatite Line from Bell & Blenkinsop (1987). HIMU and EM1 values in this, and other Nd v. Sr isotope plots (Dawson 2008, figs 8.1 and 8.2) are from Hart (1988).

(Dawson *et al.* 1995*b*). Blocks of cellular, spinifex-textured wollastonite (Dawson & Sahama 1963) (Plate 17) and titanite pyroxenite possibly derive from dykes within the volcanic pile.

Other deep-seated blocks are of crustal and upper-mantle metasomites, fenites and olivine-mica-pyroxenites, respectively (Morogan & Martin 1985; Dawson & Smith 1992*a*, *b*; Kramm & Sindern 1998). In contrast to the other blocks, the metasomatic olivine-mica-pyroxenites (Plate 18): (a) are often mantled by rinds of ijolite, with which they have reacted; and (b) contrasting with the rounded shape of other blocks, are invariably angular, indicating derivation from a well-jointed mantle protolith.

The accretionary lapilli of the 1966 eruption contained euhedral melanite, nepheline and combeite, corroded clinopyroxene and wollastonite, together with larnite, rankinite and cementing natrite (Dawson *et al.* 1992). The lapilli of the September 2007 eruption contain the same phases but are cemented by a NaCa carbonate–phosphate (Mitchell & Dawson 2007). The lapilli from both these eruptions provide evidence for interaction between Ca-silicates (clinopyroxene and wollastonite) and natrocarbonatite to create metasomatic combeite and melilite. The igneous combeite has higher Na / (Na + Ca) ratios than that replacing wollastonite and clinopyroxene, and nepheline embedded in the igneous combeite is richer in $Fe_2O_3$ (5.0 wt%) and $K_2O$ (8.5%) than nepheline in most silicate lavas (0.8–2.1; 4.4–8.1 wt%) (Dawson *et al.* 1989, 1992). An absence of clinopyroxene in the mantling ash indicates they are neither nephelinite nor melilitite, but result from exteme interaction between nephelinite and natrocarbonatite (Mitchell & Dawson 2007). The lapilli from the September 2007 eruption, in contrast to those in the 1966 eruption, are not accretionary, suggesting the ash cloud was not so buoyant.

Ash erupted in 1993 contained spheroids of nepheline- and garnet- phyric nephelinite intermixed with natrocarbonatite. Importantly, rounded globules of natrocarbonatite occurred both within the silicate matrix of the lava and within silicate glass entrapped in the silicate phenocrysts (Dawson *et al.* 1996*a*).

*The natrocarbonatite lavas.* Natrocarbonatite has not been recognized in the Unit I rocks. The earliest evidence comes from: (i) partly-altered natrocarbonatite lava interbedded with Unit III nephelinitic tuffs and agglomerates in the summit area (Dawson *et al.* 1987); (ii) a natrocarbonatite-nephelinite ash deposit on the lower northern slopes, termed the Footprint Tuff, that was erupted about 600 years ago (Hay 1989); and (iii) an altered block of natrocarbonatite from Unit III agglomerate (Dawson 1993). All historical reports show that natrocarbonatite has been extruded from the northern crater. Pre-1995 studies on the lavas are reviewed in a collection of papers edited by Bell & Keller (1995).

The lavas are jet-black when first extruded, but become lighter in colour after about 24 hours due to reaction with atmospheric moisture. This is most marked by whitening of the surface of pahoehoe flows, whereas aa flows become grey. These colour contrasts have proved useful for determining, from aerial photographs and observations from the crater rim, when individual lava flows were extruded. In addition to other attributes, the small scale and relatively slow advance of the natrocarbonatite flows have made Oldoinyo Lengai a natural laboratory for the measurement of magma temperatures and rheology. Measured extrusion temperatures range from 544 to 592°C (Krafft & Keller 1989; Dawson *et al.* 1990; Pinkerton *et al.* 1995).

Physically, the lavas have the same basic appearances as basaltic lava flows, varying from aa to pahoehoe morphologies (Figs 7.13–7.15), but most are on a relatively small scale; individual pahoehoe flows may be less than 2 m wide and as little as 20 cm thick (Fig. 7.13). In contrast, a 1960 aa flow was *c.* 1 m thick, and an exceptionally viscous and crystal-rich flow in 1993 was 6 m thick. The length of many flows has been constrained by their ponding against the inner walls of the crater, though recent overspill flows have travelled down the outer slopes for distances of up to 3 km before freezing (Kervyn *et al.* 2008*a*). Volumes and extrusion rates are variable, from a volume of 900 $m^3$ and effusion rate of $8.3 \times 10^{-3}\ m^3\ s^{-1}$ for an aa flow in 1960, to a volume of 104 000 $m^3$ and an effusion rate of $86 \times 10^{-3}\ m^3\ s^{-1}$, for a particularly large aa flow in 1993 (Dawson *et al.* 1994). Volumetrically, many of the pahoehoe flows are more comparable to the 1960 flow, though some flows that have spilled down the outer slopes since 1999 must be of considerably greater volume. The least vesicular and crystal-poor flows are the least viscous of any terrestrial magmas, with apparent viscosities as low as 1–5 Pa s (Norton & Pinkerton 1997), whereas the viscosity of the exceptionally-thick and crystal-rich aa flow extruded in 1993 is $3 \times 10^7$ Pa s, falling within the range for rhyolites (Dawson *et al.* 1994). There is the tendency for spatter to build up large hornitos, some of which may be up to 20 m high. These become hollow upon magma withdrawal and most collapse shortly afterwards due to weakening by fumarolic activity and weathering, but a few remain intact sufficiently long to develop stalactites on their interior walls (McFarlane *et al.* 2004).

**Fig. 7.13.** A recent natrocarbonatite aa flow in the northern crater of Oldoinyo Lengai, September 1960. Note the black, pre-alteration colour compared with whiteness of older pahoehoe lava in the foreground.

**Fig. 7.14.** Pahoehoe flows in the northern crater, Oldoinyo Lengai, 1988. Compare the texture and thickness of these flows with those of the aa flow illustrated in Figure 7.13.

The natrocarbonatite lavas contain phenocrysts of the Na–K–Ca carbonates nyerereite ($[Na_{1.64}K_{0.38}]_{2.02}[CaSrBa]_{0.99}(CO_3)_2$) and gregoryite ($[Na_{1.74}K_{0.1}]$ $[CaSrBa]_{0.16}CO_3$) (Plate 19) in which there is considerable substitution of Sr and Ba for Ca, and $(SO_4)^{2-}$, $(PO_4)^{3-}$, Cl and F for $(CO_3)^{2-}$ (McKie & Frankis 1977; Gittins & McKie 1980; Keller & Krafft 1990). The phases are named after Julius Nyerere, first president of Tanzania and J. W. Gregory, of rift valley fame. These two phenocryst phases are set in a groundmass of smaller grains of nyerereite, apatite, Mn–Fe sulphide, Mn–Fe spinel, phase X (a mixed Ca–Mg–Ba–Na carbonate in which the proportions of the individual cations are variable) and Na-sylvite ([KNa]Cl) (Plate 20). They are rare solid solutions that result from high-temperature consolidation of the dominantly carbonate liquid (Dawson *et al.* 1995*c*); phase X may be a variety of khannesite, a member of the burbankite group (Mitchell 1997). The ratio of phenocrysts to groundmass varies considerably, with fewer phenocrysts in the more mobile pahoehoe flows. Other even rarer phases are Fe–Mn-monticellite (Peterson 1990), K-sulphides, Zn-bearing sulphides, spurrite, larnite, and sellaite (Dawson *et al.* 1995*c*), Sr fluorite and sodic tilleyite (Dawson *et al.* 1996*a*), neighborite (Mitchell 1997), halite (Genge *et al.* 2001), Mn–F rasvumite (Jago & Gittins 1999), niocalite-cuspidine and galena (Mitchell & Belton 2004), high-Ba nyerereite (Mitchell 2006*a*) and a sodic phosphate–carbonate (Mitchell 2006*b*). The magnetization of the lavas is dominated by the small amounts of spinel: a jacobsite ($MnFe_2O_4$)–magnetite solid solution (Shive *et al.* 1990).

The unusual chemical composition of the natrocarbonatite lavas is shown in extreme concentrations of Na, Ca, K, P, Cl, F, S, Ba, Sr and $CO_2$ (Table 7.5); Si and Al are very low and, where present, may be due to xenocrystal contamination. Further, they are most unusual in containing more Mn than Fe. The lavas also contain unusually high concentrations of the light REE (Fig. 7.11), Rb, Br, Mo, W, As, Sb, Pb and U. The $La–Yb_{CN}$ (500 to 1000) and U–Th (2 to 3) ratios are the most extreme for terrestrial magmas (Dawson & Gale 1970; papers in Bell & Keller 1995; Simonetti *et al.* 1997; Ohde & Matagario 1999). Li values are also very high (211–294 ppm), though $\delta^7Li$ has a narrow range of $+3.3$ to $+5.1$ (Halama *et al.* 2007). Halogen contents are F 2.5%, Cl 3.6% and I *c.* 1 ppm (Dawson & Fuge 1980). Some compositional variation within individual flows is due to: (a) variation in phenocryst/groundmass ratio (the filter-pressed toes of some

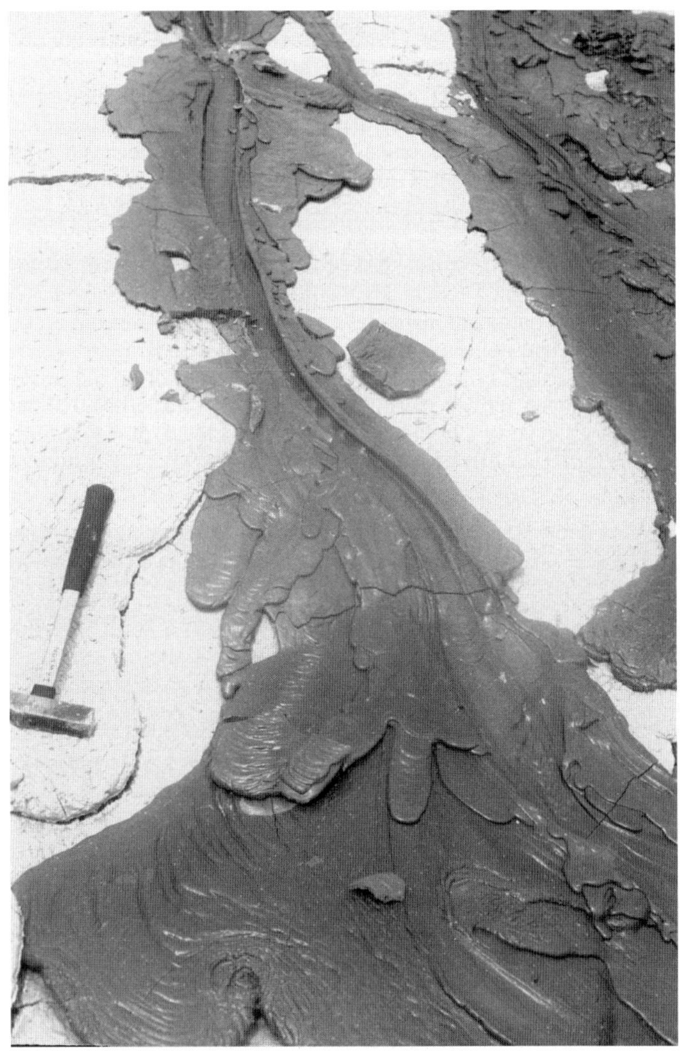

**Fig. 7.15.** Detail of a natrocarbonatite pahoehoe flow, Oldoinyo Lengai, 1988. Despite its very small volume, the low viscosity of the magma has resulted in the flow morphology being similar to that of more voluminous pahoehoe basalt flows.

**Table 7.5.** *Representative analyses of natrocarbonatite lavas*

| | 1 | 2 | 3 | 5 | 6 |
|---|---|---|---|---|---|
| $SiO_2$ | 0.29 | 0.14 | 0.18 | | |
| $TiO_2$ | 0.02 | 0.01 | <0.01 | | |
| $Al_2O_3$ | 0.10 | 0.10 | 0.10 | | |
| $Fe_2O_3$ | 0.50 | 0.00 | 0.21 | | |
| FeO | 0.24 | 0.10 | 0.12 | | |
| MnO | 0.61 | 0.24 | 0.45 | | |
| MgO | 0.42 | 0.46 | 0.22 | | |
| CaO | 13.36 | 13.04 | 14.63 | | |
| BaO | 1.68 | 1.92 | 1.05 | | |
| SrO | 1.74 | 1.79 | 1.63 | | |
| $Na_2O$ | 31.06 | 32.58 | 32.72 | | |
| $K_2O$ | 8.44 | 8.64 | 7.27 | | |
| $H_2O^+$ | 0.05 | 0.23 | 0.22 | | |
| $P_2O_5$ | 0.73 | 0.63 | 1.02 | | |
| $SO_3$ | 3.12 | 3.07 | 3.18 | | |
| $CO_2$ | 32.70 | 29.20 | 34.30 | | |
| Cl | 3.85 | 4.40 | 1.80 | | |
| F | 3.55 | 5.25 | 1.35 | | |
| Sum | 102.47 | 101.8 | 100.45 | | |
| less O=F + Cl | 2.35 | 2.03 | 0.97 | | |
| Total | 100.12 | 99.77 | 99.48 | | |
| V | 232 | 251 | 134 | | |
| Cu | 11 | 11 | 7 | | |
| Zn | 303 | 43 | 136 | | |
| Rb | 259 | 284 | 193 | | |
| Y | 15 | <5 | <5 | | |
| Zr | <20 | 20 | <20 | | |
| Nb | 74 | 13 | 14 | | |
| Pb | 212 | 47 | 108 | | |
| Σ REE | 1878 | 1521 | 1103 | | |
| K/Rb | 270 | 251 | 312 | | |
| Ca/Sr | 6.5 | 6.18 | 7.59 | | |
| La/Yb | 1127 | 1911 | 1657 | | |
| $Na_2O/K_2O$ | 3.68 | 3.77 | 4.5 | | |
| U | | | | 6.73 | 7.31 |
| Th | | | | 2.90 | 2.8 |
| U/Th | | | | 2.32 | 2.61 |

1. Phenocryst-poor pahoehoe flow, near extrusion centre
2. Same phenocryst-poor flow as analysis 1, but distant from extrusion centre (note lower Fe, Mn, Zn, Nb and Pb contents due to sulphide/spinel sedimentation near extrusion centre)
3. Phenocryst-rich aa-type lava. Compared with analysis 1, note differences in Fe, Mn, Mg, Ca, Sr, K, Cl, F, Zn, Rb, Y, Nb, Pb due to higher phenocryst content.

Analyses 1–3, lavas extruded in 1988 (Dawson *et al.* 1995*c*)

5. BD114 Pahoehoe lava
6. BD118 Aa lava

Analyses 5 and 6, lavas extruded in 1960 (Dawson & Gale 1970)

flows are almost phenocryst-free resulting in higher Ba, K, Cl, F, V, Cu and Rb than in phenocryst-rich samples); and (b) flowage differentiation, with the distal parts of individual flows being relatively low in Fe, Mn, Zn, Nb, and Pb due to settling out of denser sulphide and spinel grains close to the extrusion site (Dawson *et al.* 1995*c*). The isotopic compositions of Li, O, C, S, Sr, Nd and Pb (Table 7.6 and considered at greater length in Chapter 8, Dawson 2008) are consistent with a mantle origin for the natrocarbonatites and associated silicate lavas.

Natrocarbonatite lavas are enriched in radon, and have the most extreme disequilibria between U and Th series nuclides yet measured in volcanic rocks. The disequilibria for lavas erupted in 1960 and 1963 suggest gradual formation of the natrocarbonatites (? by separation from a Ra-enriched nephelinite in the

**Table 7.6.** *Stable and radiogenic isotope data for natrocarbonatite and trona*

| Isotope | Range in natrocarbonatite | Trona* | Reference |
|---|---|---|---|
| $\delta^7Li$ | +3.3 to +4.8 | | 1 |
| $\delta^{13}C$(PDB) | −5.9 to −6.49 | +3.8 to +4.7* | 2,3 |
| $\delta^{18}O$ (SMOW) | +6.1 to +7.03 | +38.3 to +40.5* | 2,3,4 |
| $\delta^{34}S$ (CDT) | +2.8 (sulphide), +9.0 (sulphate) | | 2 |
| $^{87}Sr/^{86}Sr$ | 0.7037 to 0.70446 | 0.7056* | 5 |
| $^{143}Nd/^{144}Nd$ | 0.51261 to 0.51268 | 0.51269* | 5 |
| $^{206}Pb/^{204}Pb$ | 19.24 to 19.26 | 18.41 | 5 |
| $^{207}Pb/^{204}Pb$ | 15.60 to 15.63 | 15.62 | 5 |
| $^{208}Pb/^{204}Pb$ | 39.30 to 39.39 | 38.25 | 5 |

References: 1, Halama *et al.* 2007; 2, Dawson *et al.* 1995*c*; 3, Keller & Hoefs 1995; 4, Hay 1989; 5, Bell & Simonetti 1996.*Trona from Lake Magadi: Bell *et al.* 1973; O'Neil & Hay 1973.

sub-volcanic magma chamber) *c.* 15–18 years previously, i.e. shortly after the 1940–41 ash eruption (Williams *et al.* 1986). Similarly, lavas extruded in 1988 formed immediately after the emptying of the magma chamber by the 1966–67 ash eruption (Pyle *et al.* 1991; Pyle 1995). The 1960 Ra–Th disequilibria systematics, together with the other isotopic data (considered at greater length in Dawson 2008, Chapter 8) and major element chemistry, are not consistent with an origin for the natrocarbonatite by mobilization of high-Na evaporites such as those in Lake Natron, as advocated by C. Milton (Bell & Dawson 1995*b*).

The high Na content of the natrocarbonatites lavas in particular has tended to overshadow other significant geochemical features of the overall Oldoinyo Lengai magmatism. The products of the magmatism are also highly potassic e.g. the natrocarbonatite lavas contain 7–8 wt% $K_2O$; and in the silicate lavas and plutonic rocks, sodalites contain 3–4 wt% $K_2O$, nepheline contains 20–25% kalsilite molecule, and feldspar is sanidine rather than anorthoclase. The Mn > Fe nature of the natrocarbonatite whole-rock analyses is mentioned above, and some individual phases in both the silicate rocks and the carbonatite have a relatively-high Mn content e.g. wollastonite contains 0.53 wt% MnO, spinels often contain >1 wt%, and one type of carbonatite sulphide contains up to 74 wt% alabandite (MnS) molecule.

*Natrocarbonatite alteration.* Alteration of the pristine, anhydrous natrocarbonatite lavas has given rise to various mineral parageneses depending on the length of exposure to downgrading influences. The whitening of flows, most readily seen on pahoehoe flows, begins shortly after extrusion and is mainly due to interaction with atmospheric moisture, leading to incipient breakdown of the phases in the matrix to form a thin layer of nahcolite ($NaHCO_3$) and sylvite (KCl) (Dawson 1962*a*; Keller & Krafft 1990). Cracks on lava flows can be fringed with thermonatrite ($Na_2CO_3 \cdot H_2O$), aphthitalite ($NaK_2SO_4$), halite and sylvite as a result of more intense interaction (Genge *et al.* 2001). More prolonged reaction results in the replacement of nyerereite by pirssonite ($Na_2Ca(CO_3)_2 \cdot 2H_2O$) and the formation of secondary calcite and kogarkoite ($Na_3SO_4F$), and whole-rock analyses of this type of altered material show loss of K, Rb, P and Cl compared with fresh natrocarbonatite (Dawson *et al.* 1987; Keller & Zaitsev 2006; Zaitsev & Keller 2006). Even more extensive alteration results in a porous texture, replacement of nyerereite and gregoryite by pseudomorphs consisting of aggregates of tiny calcite grains, precipitation of apatite lower in REE but richer in F compared with that in unaltered natrocarbonatite, and formation of an opaque phase akin to romanèchite ($BaMn_9O_{16}[OH_4]$). Bulk-rock analyses show considerable loss of Na, K, S, Cl, F, Rb and Sr; REE concentrations are higher, Ba is unaltered, (being fixed in the romanèchite), and $\delta^{18}O$ is 24.5‰, compared with 5 to 8‰

for unaltered natrocarbonatite (Dawson 1993; Keller & Zaitsev 2006). A study of hornito stalactites formed by weathering of natrocarbonatite show them to be composed mainly of trona, with lesser amounts of nahcolite, thermonatrite, aphthitalite, kogarkoite, schairerite ($Na_{21}[SO_4]_7F_6Cl$), halite and sylvite; the products of subaerial weathering (pirssonite, gaylussite, shortite and calcite) are not found in the stalactites, though retained in the altered rocks forming the roofs of the hornitos, and the stalactites are interpreted as having formed by evaporation of Ca-free brines seeping from the altered roof lavas (Mitchell 2006*c*).

*Gas species.* Bulk-rock analyses show that fresh lava is anhydrous, leading to the belief that the dominant gas species in the Oldoinyo Lengai magma is $CO_2$. Moreover, during the 1966 and 2007 ash eruptions, despite much lightning and thunder in the ash cloud, no rainfall took place, and lapilli contained anhydrous sodium carbonate (natrite) and NaCa phosphate–carbonate (Dawson *et al.* 1968, 1992; Mitchell & Dawson 2007). However, analyses of gases from fumaroles showed them to be 64–74% $CO_2$, 24–34% $H_2O$, 0.88–1.0% $H_2$, 0.1–0.4% CO and <0.1% $H_2S$, HCl, HF and $CH_4$, and the total $CO_2$ flux is in the order of $0.06 \times 10^{12}$ mol $a^{-1}$ (Koepenick *et al.* 1996). The gas mixture is similar to that from Nyiragongo, and the $CO_2$ flux is second only to that from Mt Etna, although the Etna flux may be enhanced by lava interaction with sedimentary carbonate rocks below the volcano. Another study, which included observations on emissions from a lava pool, found the $CO_2$ level to be even lower, and that the $CO_2/CO$ ratio varied rapidly, interpreted as perhaps reflecting the mixing of gases exsolving from shallow and deeper magma chambers (Oppenheimer *et al.* 2002). These gas analysis studies concluded that $H_2O$ may be more important than previously appreciated. However, the natrocarbonatite lava flows readily hydrate on contact with meteoric moisture, so it is not clear whether the measured $H_2O$ content in the gas analyses is truly magmatic or whether it could have resulted from dehydration of the deep pile of hydrated lava within the crater, due to magma heating.

*Natrocarbonatite formation.* The actual formation of natrocarbonatite is still a matter of debate; hypotheses range from its being: (i) a primitive mantle-derived melt; (ii) fractionation of a parental alkali-bearing sövite magma; (iii) fractionation of silicate-bearing natrocarbonatite; or (iv) an immiscible fraction of the associated peralkaline nephelinites. The views of most authors, in a series of papers edited by Bell *et al.* (1998), incline towards liquid immiscibility. Simultanous ejection of silicate and carbonate material during the 1966, 1993 and 2007 eruptions (Dawson *et al.* 1968, 1992; Church & Jones 1994; Mitchell & Dawson 2007) and the eruption in 1993 of carbonate lavas and tephra containing immiscible globules of silicate magma (within which were immiscible blebs of natrocarbonatite; Plates 21, 22) (Dawson *et al.* 1996*a*) provides evidence for the coexistence of silicate and carbonate magmas at the volcano (possibly in a compositionally- and density-stratified magma chamber). In the specific context of the well-established relationship between natrocarbonatite and the peralkaline nephelinites, Dawson (1998) suggested that $CO_2$ saturation in the sub-volcanic magma chamber resulted in the formation of wollastonite nephelinite plus immiscible natrocarbonatite; interaction of wollastonite nephelinite and natrocarbonatite led to formation of peralkaline combeite nephelinite, with corrosion of wollastonite and clinopyroxene, and the ensuing $CO_2$ release by the decarbonation reaction might lead to explosive eruption. An extreme example of this magma hybridization concept is provided by the melilite- and combeite-rich, but pyroxene-free, lapilli from the 20 September 2007 eruption, the origin of which is envisaged as due to mixing of wollastonite nephelinite and extremely large amounts of natrocarbonatite magmas; the result is precipitation of combeite and melilite rather than wollastonite and clinopyroxene (Mitchell & Dawson 2007).

*Kerimasi*

*Type*: stratovolcano; *Lat/Long*: 2°51′S 35°5′E; *Elevation*: 2607 m; *Relief*: 1450 m; *Diameter*: 14 km; *Crater diameter*: 1.5 km (outer), 0.75 km (inner); *Eruptive history*: Pleistocene; *Composition*: nephelinite, phonolite, carbonatite; *QDS*: 39 Angata Salei.

Kerimasi post-dates the formation of the the Natron Basin boundary fault at 1.2 Ma, and its lower eastern slopes are penetrated by tuff cones that are dated at 0.37 Ma (MacIntyre *et al.* 1974). It initially erupted up, or very close to the Natron Basin boundary fault, then built up alongside the fault escarpment, and eventually overstepped onto the terrain to the west of the fault, a sector of which is now buried beneath the volcano (Fig. 7.16). Hence, it is steeper and has greater relief on its eastern and northern slopes, but has relatively shallow slopes on its western and southwestern flanks. The summit is occupied by a 100 m deep crater, at the eastern end of which is a smaller, 200 m deep crater (Guest *et al.* 1961). A large debris flow (*c.* 5 km wide and 9 km long) originated from a shallow arcuate feature on the upper eastern slopes of the volcano; this deposit is thickest (8 m) at its toe (Hay 1983; Kervyn *et al.* 2008*b*). The debris flow is not recent as it is cut by small NW–SE-trending faults and penetrated by tuff cones and explosion craters on the lower northeastern and eastern slopes (Dawson personal observation).

The core of the volcano is of nephelinitic tuffs and agglomerates that are best exposed in deep, radial gullies on the eastern and northern slopes (Fig. 7.17). Guest (1953) listed blocks of garnet gneiss and various volcanic types (olivine basalt, basanite and dolerite) derived from nearby country rocks, but it is the more exotic rock types in the agglomerates that have attracted the attention of numerous petrographers. Rosenbusch (1907) and Erdmannsdorfer (1935) describe ijolite and syenite collected by Uhlig during the German Central Africa Expedition in 1904–5, and Brögger (1921) described and provided an analysis of a melteigite provided by Finckh from the same collection. Guest (1953) described blocks of melanite urtite, melanite ijolite, amphibole melteigite and biotite pyroxenite, providing a whole-rock analysis of the last. James (1966) gave analyses of an ijolite and a nephelinite, and Dawson (personal observation) has found blocks of ijolite, ijolite-mantled mica pyroxenite, perovskite pyroxenite, mica pyroxenite, calcite-cemented pyroxenite and melanite melteigite. Church (1995) provided: (i) descriptions of melilite nephelinite, phonolitic nephelinite and phonolite, afrikandite, uncompaghrite, pyroxenite and ijolite blocks from agglomerates exposed in gullies on the northeastern lower slopes; (ii) microprobe analyses of clinopyroxene, nepheline, melilite, rutile, perovskite, melanite, Ti-magnetite, apatite, K-feldspar and phlogopite in some rocks; and (iii) major, trace and Nd–Sr isotope data on some rocks.

The silicate core is overlain by a carapace of carbonatite tuffs and agglomerates. Carbonatite tuffs containing variable amounts of phases derived from the silicate core (e.g. melanite, pyroxene, titanite, magnetite, phlogopite, nepheline, perovskite) blanket much of the volcano beneath a layer of brown soil. Carbonatite lapilli tuffs, the lapilli cored with either silicate minerals or rock fragments, occur on the lower eastern and upper southwestern slopes (Guest 1953; Mariano & Roeder 1983) and at Miamoja Hill, a small cone on the plain SE of Kerimasi (Dawson personal observation). Blocks of sövite, some with flow-banding and containing apatite, humite, high-Mn forsterite, high-Mn (up to 6 wt%) magnetite, baddeleyite and melanite are present in the carbonate-tuff-cemented agglomerates and as loose blocks in rivers draining the eastern and north-eastern upper slopes (Dawson *et al.* 1996*b*); some blocks contain zoned calcite phenocrysts (Mariano & Roeder 1983).

Two small plugs of intrusive calcite carbonatite crop out in the northern and eastern walls of the summit crater (James 1966). Abundant blocks of two different types of calcite carbonatite, one consisting of rounded calcite phenocrysts set in a vesicular

**Fig. 7.16.** Kerimasi from the south. The Natron Basin boundary fault escarpment (left) becomes progressively buried beneath the Kerimasi pyroclastics.

matrix characterized by skeletal parallel-growth calcite, the other of spinifex-textured calcite, occur in debris avalanche deposits on a small fault-bounded ridge [2°55′S 36°03′S] on the lower southeastern slopes, and on the eastern slopes where the deposits underlie lapilli and blocks from the Deeti cone (Fig. 7.18) (Dawson personal observation). Church (1995) described carbonatite tuffs, together with blocks of sövite and beforsite in gullies on the northeastern slopes of the mountain, and gave whole-rock major, trace and Nd–Sr isotope data on some carbonatites; one distinctive type contains so-called 'corduroy' calcite with closely-spaced, parallel calcite lamellae. She also provided micro-probe analyses of apatites, Fe–Ti magnetites (containing up to 4% $TiO_2$, 8.1% MnO and 11.4% MgO), pyroclore, zirconolite, calzirtite, perovskite, phlogopite, forsterite, periclase, dolomite and monticellite in the sövites.

Additional chemical data (complete analyses for two sövites) were provided by Paslick *et al.* (1995), whole-rock isotope analyses for three sövite specimens by Kalt *et al.* (1997), and by Keller & Spettel (1995) who gave major, minor and trace element analyses for two carbonatites, one from a carbonatite breccia exposed in the walls of Loolmurwak crater, the other a lava from the summit area of the volcano. The lava contains zoned calcite similar that described by Mariano & Roeder (1983).

Church (1995) proposed that the original magma was a primitive melilite nephelinite, (the volcanic equivalent of uncompaghrite) similar to that at Armykon and Oldoinyo Loolmurwak (Dawson *et al.* 1985). This parent magma is proposed to have fractionated melilite and perovskite in a shallow-level magma chamber to give cumulate afrikandite and fractionated ijolite, and the carbonatites formed by immiscible separation from the primary olivine melilitite at low pressures. On isotopic evidence, Church (1995) suggested that the primary Kerimasi magma had exchanged isotopically with metasomatized mantle, as suggested for Oldoinyo Lengai by Bell & Dawson (1995*a*).

A particular aspect of Kerimasi geology is that Hay (1983) recorded blocks of rock that he interprets as altered natrocarbonatite in the avalanche deposit on the eastern lower slopes. The altered 'natrocarbonatite' contains phenocrysts of elongate primary calcite, and larger porous grains consisting of parallel sheets of calcite, interpreted as pseudomorphs after nyrereite. These are texturally the same as the 'corduroy' calcites described by Church (1995) who disagreed with Hay's interpretation, proposing that the 'corduroy' calcite is an exsolution product of former dolomite. However, Church's disagreement does not take into account oxygen and carbon isotope evidence (Hay & O'Neill 1983). The data are: (i) in the primary calcite $\delta^{18}O$ is

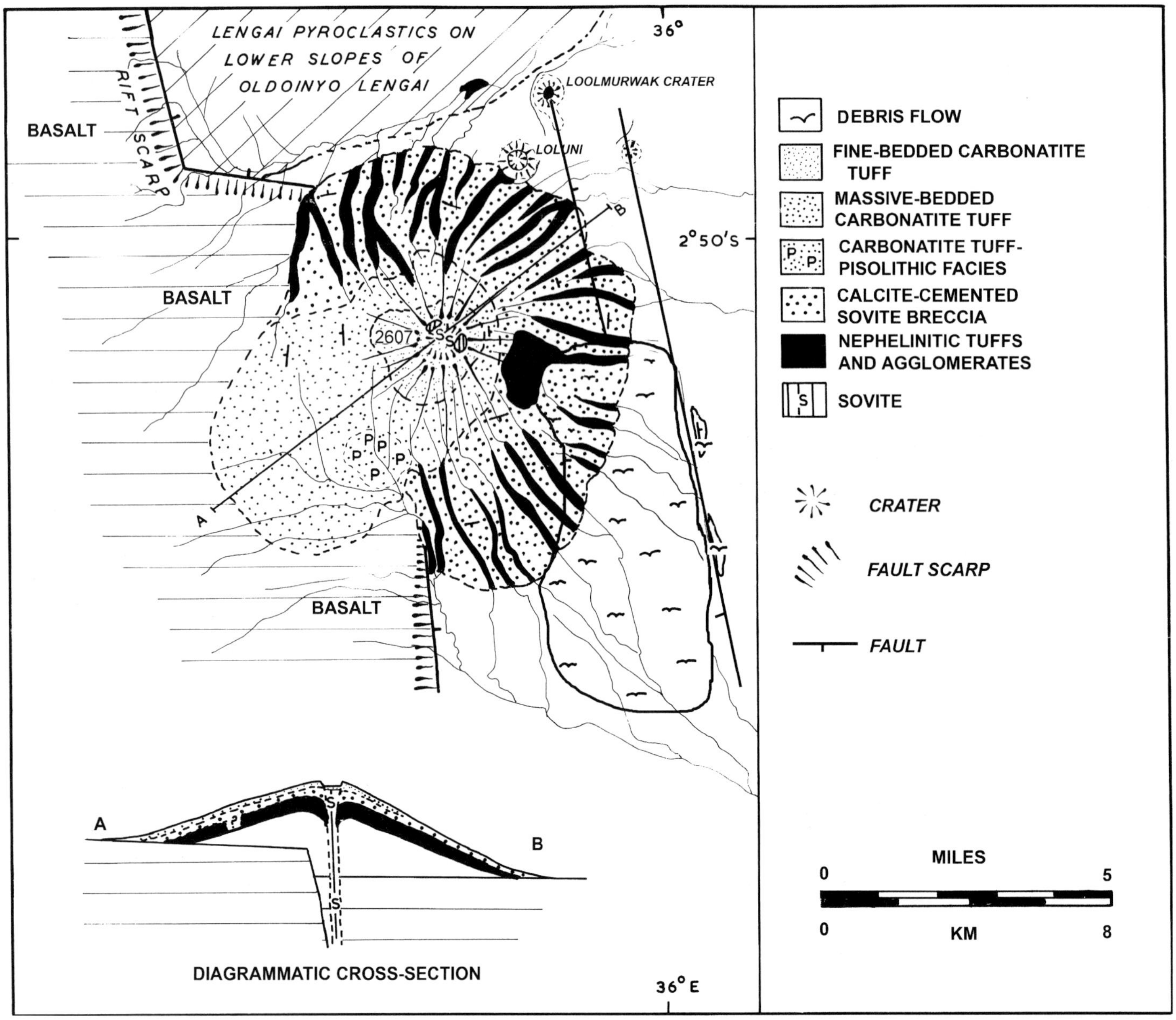


**Fig. 7.17.** Map and diagrammatic cross-section of Kerimasi. Modified from Dawson (1964*b*).

+7.2 to 7.3‰, and $\delta^{13}C$ −4.2 to −4.3‰ (falling within the parameters for primary igneous carbonatite), compared with (ii) $\delta^{18}O$ varying from +21.2 to +25.2‰, and $\delta^{13}C$ from −1.4 to −3.3 in the calcite interpreted as pseudomorphing nyerereite. The main significance arising from Hay's interpretation, if correct, is the occurrence of natrocarbonatite at a locality other than Oldoinyo Lengai.

### *Natron–Engaruka tuff-cone area*

*Type*: monogenetic tuff-cone/crater volcanic field; *Lat/Long*: centred on 3.40°S 36.00°E; *Elevation*: around 700–800 m; *Relief*: individual features up to 50 m; *Eruptive history*: Pleistocene–Holocene; *Composition*: olivine nephelinite, olivine melilitite; *QDS*: 39 Angata Salei, 40 Gelai and Ketumbeine.

The floor of the rift valley between Lake Natron and Engaruka, flanked on the west by Oldoinyo Lengai and Kerimasi, and on the east by Gelai and Ketumbeine, is covered and punctured by numerous small tuff cones, cratered tuff cones, tuff rings and explosion craters (Fig. 7.18). Some are aligned along minor faults cutting across the lower slopes of Kerimasi and Gelai. Many are blanketed by tuff from Oldoinyo Lengai, and most between Oldoinyo Lengai and Lake Natron are partially covered or infilled with lahar deposits and outwash sediment from Oldoinyo Lengai. Most features are asymmetric, with maximum ejecta deposition on their northern or northwestern sides. Major craters, up to 1 km wide and 100 m deep, are Kisetey, Loolmurwak (also known as Swallow Crater), Eledoi and Embulu Sabuk. Cross-lamination and dune-bedding in the tuffs around some major craters (Dawson personal observation) are possibly the result of base surges. Megacrysts of kaersutite, mica and augite are common in the ejecta around these craters, and blocks of perovskite-rich pyroxenite, mica-kaersutite pyroxenite and ijolite occur in beds exposed in the walls of Loolmurwak crater (Dawson & Powell 1969).

Within this area are several small flows and scoria cones of olivine melilitite and olivine nephelinite. Although some are partially buried or obscured by debris flows consisting of Unit I pyroclastic materials from Oldoinyo Lengai, others are not, indicating that their extrusion is a very recent phenomenon. These include Armykon Hill on the southern shores of Lake

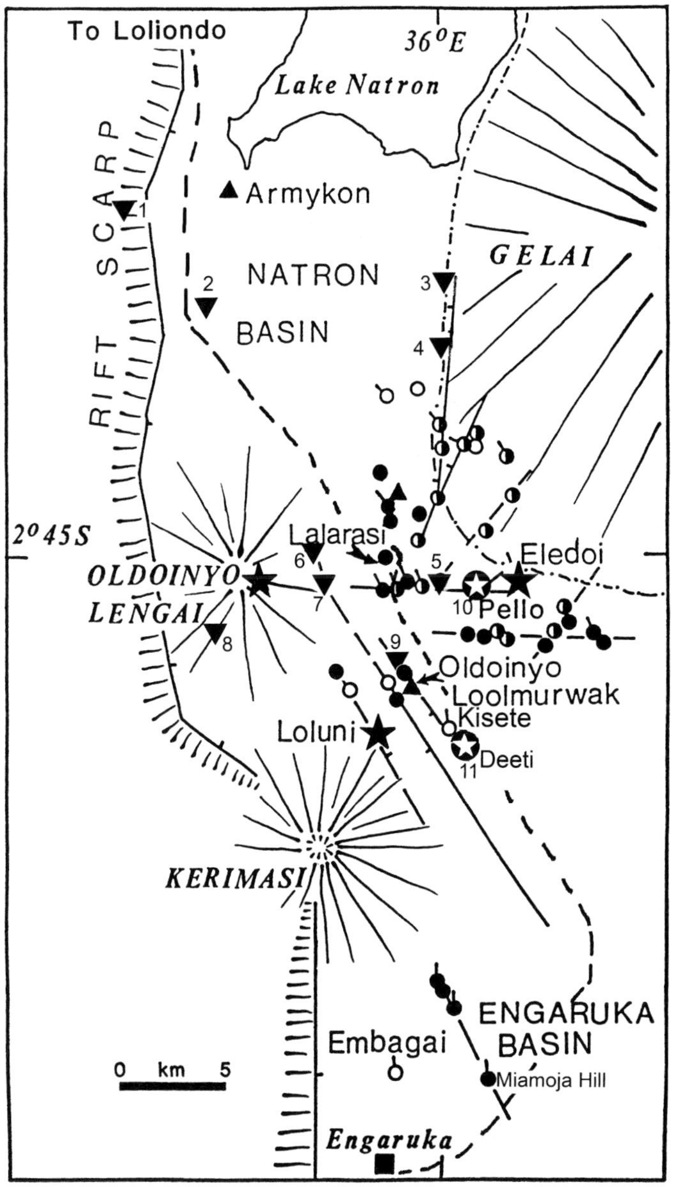


**Fig. 7.18.** The Engaruka–Lake Natron area. Based on Dawson & Powell (1969) with additional data from Keller *et al.* (2006). Explanation of ornaments: solid triangle: melilitite or nephelinite lava flow; inverted solid triangle: melilitite tuff cone; solid star: mantle xenolith locality; open star: mantle xenoliths in melilitite tuff; open circle: crater; half-filled circle: cratered tuff cone; filled circle: tuff cone; tick denotes direction of maximum ejecta deposition. Numbers refer to individual melilitite craters and tuff cones: 1, Sekenge crater; 2, Black Belly crater; 3, Oldoinyo Lolodulal cone; 4, Embululu Sabuk crater; 5, Lalarasi cone; 6, Dorobo cone; 7, Oltawata crater; 8, Kirurum crater; 9, Un-named cone; 10, Pello Hill cratered tuff-cone; 11, Deeti cone.

Natron (a small tholoid with flows of olivine melilitite), a minor flow of olivine melilitite 2 km north of Lalarasi, and an olivine-melilite nephelinite flow at Oldoinyo Loolmurwak (Dawson *et al.* 1985) Keller *et al.* (2006) recorded other occurrences of olivine melilitite tuffs. Importantly some penetrate the older Unit I agglomerates of Oldoinyo Lengai, but are buried by Unit III combeite-nephelinite ash. Thus olivine melilite activity occurs both before and after the debris flows that followed the cessation of the Oldoinyo Lengai Unit I activity, and in the interval between the Unit I and Unit III activity. Klaudius & Keller (2006) gave a $^{14}C$ date of 2500 years BP on plant remains embedded in one of the debris flows. Agglutinated lapillae of carbonated melilitite occur at Miamoja Hill and at a small hill 2 km north of Loolmurwak crater. In the area immediately south and SE of Pello Hill, are numerous, small, asymmetric, cratered tuff-cones consisting of melilitite lapilli; exceptionally the crater of one of these cones {2°47.35′S 36°04′E} is partially infilled by a late extrusion of melilitite (Dawson personal observation, September 2007).

The equivalence of olivine melilitite with biotite- and amphibole-pyroxenites, depending upon depth/pressure of crystallization, has been recognized for some time (Tilley & Yoder 1968). Megacrysts of phlogopite, augite and amphibole (together with cumulate rocks of these phases), occur at several of the olivine melilitite localities in the Engaruka–Natron area and can be interpreted as additional pulses of melilitite magma that crystallized at depth/high pressure.

Major and trace element bulk-rock analyses and mineral analyses for the Oldoinyo Loolmurwak nephelinite lava, and melilitites from Lalarasi and Armykon Hill, are given by Dawson *et al.* (1985), James (1995), Keller *et al.* (2006) and Halama *et al.* (2007). Although of ultramafic composition, some of these lavas are unusually alkaline and have high concentrations of the REE (Fig. 7.11).

Xenoliths of peridotite occur at Eledoi crater (Dawson & Powell 1969) and clinopyroxene formation in a vein in one of these peridotites is of very recent age (Cohen *et al.* 1984). Others occur in katungitic scoria at Pello Hill, where there are also xenoliths of basement granulite. Many of the Pello peridotites are metasomatized, in some cases adjacent to olivine–mica–kaersutite–ilmenite veins (Dawson & Smith 1988) (Plate 23). At Deeti Hill, blocks of both olivine–mica pyroxenite (metasomatized upper mantle peridotite) and cumulate-textured kaersutite pyroxenite form the cores of melilitite autoliths (Johnson *et al.* 1997). Also at Deeti are xenoliths of cumulate-textured phlogopite pyroxenite, in which the main phases are cemented by calcite and glass (Dawson personal observation). Olivine–mica pyroxenite xenoliths (metasomatized peridotite) occur at Loluni crater, on the north-eastern slopes of Kerimasi, and also in the pyroclastics of Oldoinyo Lengai (Dawson & Smith 1992*a*). The metasomatized peridotite xenoliths in this area, some with densities as low as $3140\ kg\ m^{-3}$, provide evidence for anomalously-light upper mantle beneath this sector of the rift (Dawson & Smith 1988).

### *Burko*

*Type*: stratovolcano; *Lat/Long*: 3.31°S 36.23°E; *Elevation*: 2136 m; *Relief*: 900 m; *Volcano diameter*: 7.5 × 6.0 km *Eruptive history*: Pleistocene; *Composition*: olivine-free nephelinite; *QDS*: 54 Monduli.

This small, cone sits above the older faulted lava plateau (Fig. 7.19). It is slightly elongate NE–SW, and has its maximum relief on its NW side, where it falls away to the Engaruka Basin. The cone consists of a core of nephelinitic tuffs and agglomerates containing plutonic xenoliths of barkevikite–phlogopite syenite, nepheline syenite and ijolite (Dawson 1964*b*). The nepheline syenites comprise albite, aegirine-augite and nepheline, and titanite is a constant accessory phase; other minor phases, not present in all specimens, are nosean (Table 7.7), apatite, biotite and an opaque phase. The more abundant ijolites have cumulate textures and comprise mainly zoned nepheline and zoned aegirine-augite; titanite is a common accessory, and other less common phases are nosean, apatite, magnetite and perovskite (in specimens lacking titanite). Astrophyllite occurs in one specimen and another contains primary intergranular calcite.

Nephelinites form blocks in the agglomerates, and also occur as later lava flows. Most are porphritic with varying amounts of nepheline, aegerine-augite, nosean, melanite and titanite. Common groundmass phases are aegirine, nepheline, perovskite and magnetite in a glassy or devitrified matrix. The nosean and

**Fig. 7.19.** The Younger Extrusives volcano Burko from the south. A sector of the Elunata fault (foreground) represents the older faulted basalt terrain overlain by the volcano.

nepheline phenocrysts are commonly zeolitized. Biotite occurs in a few specimens, but it is corroded and surrounded by an opaque phase reaction rim, suggesting that it is xenocrystal. A nephelinite has given a K–Ar age of 0.97 Ma (Evans *et al.* 1971).

Paslick *et al.* (1995, 1996) give complete rock analyses of four nephelinites and of nepheline, titanite and zoned melanite in one of these specimens; analyses of a zoned nosean are given in Table 7.7.

*Meru*

*Type*: stratovolcano; *Lat/Long*: 3°15′S 36°45′E; *Elevation*: 4568 m; *Relief*: *c*. 3100 m; *Volcano diameter*: *c*. 25 km; *Eruptive history*: Pleistocene–Holocene; *Composition*: olivine melilitite, nephelinite, phonolite, carbonatite; *QDS*: 55 Arusha.

Meru (Fig. 7.20) is a major stratovolcano rising to over 3000 m above the surrounding plains. The volcano was first climbed by Jaeger in 1904, and Uhlig (1907) described the recent appearance of the ash cone during a visit to the crater at that time. The volcano is classified as still active, and last erupted ash within the crater in 1910. Fumarolic activity within the crater was recorded as recently as 1953 (Guest & Leedal 1956). Very recently, Meru has assumed new importance in that it, and its subsidiary centre Ngurdoto, have been shown to have previously unrecognized carbonatitic affinities (Roberts 2002).

From the west, Meru appears a classic cone (Plate 24) but its eastern slopes are dominated by a major horseshoe-shaped collapse crater. Many smaller parasitic cones and craters occur in linear zones on the lower slopes and on the surrounding plains particularly to the west and SW. The volcano overlies a faulted basaltic terrain; on its NW side it buries the NW–SE-trending Lembolos graben, and on the southwestern side blankets the north–south-trending Oljoro graben, in the escarpment of which are basalt and mugearite aged 2.49 to 2.3 Ma (Wilkinson *et al.* 1986).

The earliest extrusives are phonolite and phonolitic nephelinite lavas on the lower western slopes (Meru West), aged around 1.5 Ma (Appendix 1, Dawson 2008). Some minor tholoids of phonolite and trachytic phonolite on the northern and lower western slopes, interbedded with pyroclastic rocks, are between 0.4 and 0.3 Ma. (Wilkinson *et al.*1983). Little Meru (*c*. 4000 m), a smaller subsidiary monogenetic cone to the NE of the main cone, comprises nephelinite breccia with late nephelinite flows. The bulk of the main cone is dominated by pumice and tephra, interpreted as the products of Plinian eruptions, interbedded with nephelinite and phonolite flows ranging in age from 0.163 to 0.059 Ma, the youngest being in the summit area (Wilkinson *et al.* 1983).

Erosion of the pyroclastic rocks has given rise to deeply incised radial valleys, those on the upper part of the cone perhaps having been glacially initiated, and the cone has been extremely unstable with major lahar deposits, particularly from the northern and eastern slopes, flowing onto the surrounding plains well beyond the main cone. An especially large debris avalanche deposit on the eastern side resulted from a catastrophic collapse of the east side of the cone. The collapse produced a crater, breached on its eastern side, 5 km wide, 8 km long with almost-parallel sidewalls and a 1300 m high western wall (Plate 25). The debris avalanche has a run-out length of 33 km to the east where it eventually came to rest on the lower slopes of Kilimanjaro. Its areal coverage of *c*. 390 $km^2$ and estimated volume of 28 $km^3$ (Roberts 2002) make it one of the largest sub-aerial debris avalanches yet recorded, comparable with the collapses at Socompa, Chile (36 $km^3$; de Vries *et al.* 2001) and Mt Shasta (26 $km^3$; Crandell *et al.* 1984). For comparison, the 1980 avalache at Mt St Helens had a volume of only 2.8 $km^3$ (Voight *et al.* 1981). Wilkinson *et al.* (1986) correlated the collapse with the vast ash eruption that spread a pumice mantle over the terrain to the west and NW of the cone, but Roberts (2002) suggested that the collapse was tectonically generated. Prior to this event, the mountain must have been considerably higher than its present-day height and by extrapolating the present slopes to a 1 km wide summit, Roberts (2002) estimated the pre-collapse height to be *c*. 5200 m. Part of the avalanche terrain is now occupied by the Momella Lakes; a study on the limnology of the lakes includes radiocarbon dates on bottom sediments of *c*. 8600 BP (Hecky 1971).

Following this collapse event, a lava dome with small flows of phonolitic nephelinite grew up on the crater floor and this, in turn, was followed by the formation of an ash cone (Plate 25). Xenoliths of fenite and kaersutite-bearing pyroxenite occur in one of the phonolitic-nephelinites flowing eastwards through the breach (Rock 1976).

*Petrography and mineralogy.* The most recent and extensive study of the rocks from Meru (Roberts 2002) shows that most lavas range from nephelinites to phonotephrites and tephriphonolites, whereas the Meru pumice is mainly of a more-evolved phonolite

**Table 7.7.** *Burko nosean*

| | 1 | 2 |
|---|---|---|
| $SiO_2$ | 33.60 | 34.90 |
| $TiO_2$ | 0.04 | 0.01 |
| $Al_2O_3$ | 31.10 | 29.10 |
| $Fe_2O_3$* | 0.51 | 0.66 |
| MnO | 0.11 | 0.00 |
| MgO | 0.01 | 0.01 |
| CaO | 2.31 | 1.38 |
| $Na_2O$ | 17.90 | 17.80 |
| $K_2O$ | 3.53 | 3.91 |
| $P_2O_5$ | 0.08 | 0.02 |
| $SO_3$ | 10.60 | 10.00 |
| Cl | 0.81 | 0.81 |
| Sum | 100.60 | 98.60 |
| Less O≡Cl | 0.18 | 0.18 |
| Total | 100.42 | 98.42 |

All iron as $Fe_2O_3$

1, 2. Core and rim of zoned nosean in nephelinite BD266 Burko. Analyst: D.Steele

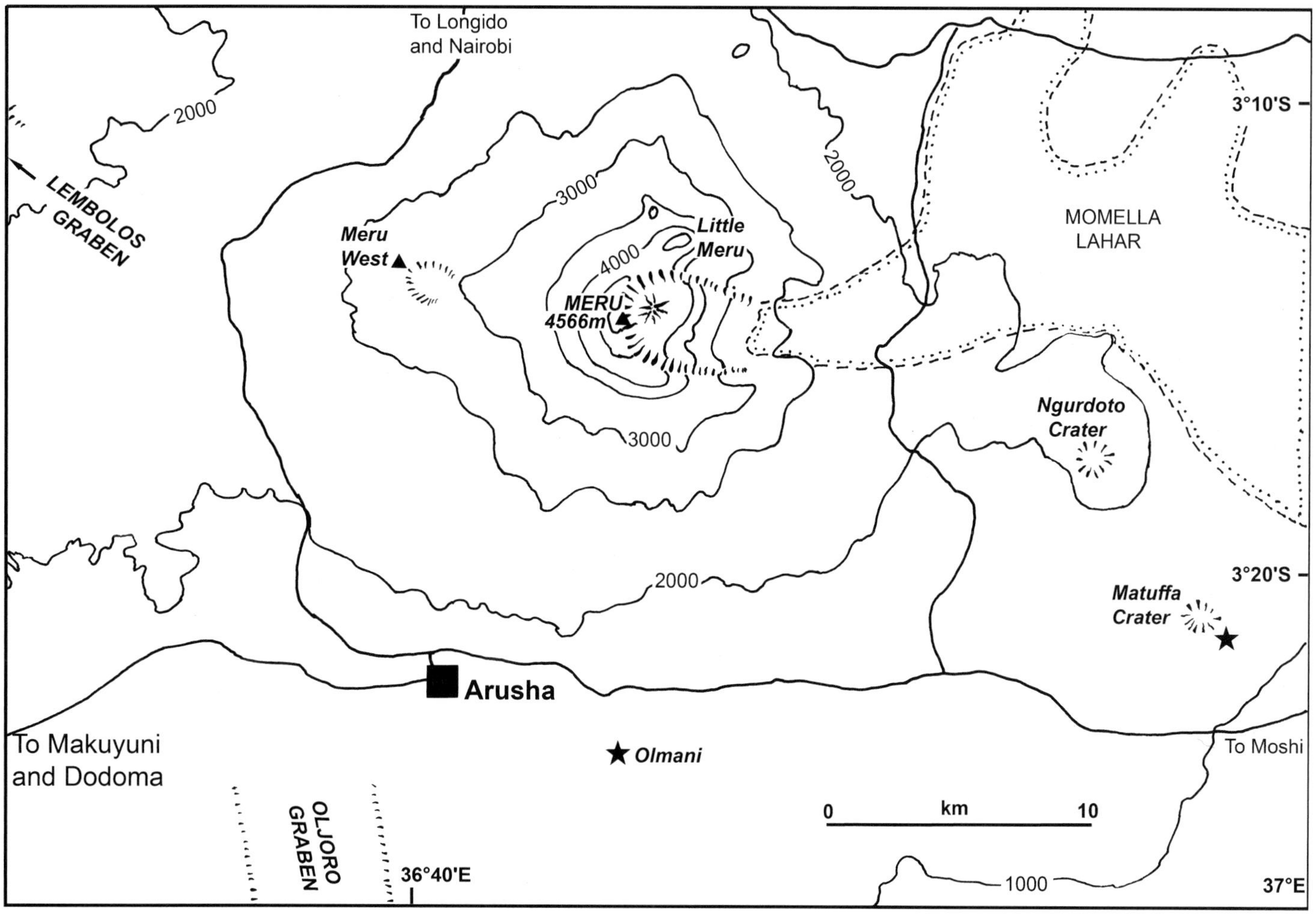


**Fig. 7.20.** Sketch map of Meru showing localities mentioned in the text. Stars indicate mantle xenolith localities.

composition. Nephelinites, including those in the summit area, the late flows within the crater, and from Little Meru, contain alkali clinopyroxene, nepheline, Ti-magnetite, titanite and kaersutite, often in a glassy matrix; rocks transitional to phonolite contain microphenocrystal sanidine. A distinctive type from Meru West contains schorlomite garnet, sodalite, melilite, a lath-shaped phase resembling delhayelite, and wollastonite, the last being replaced by combeite and apophyllite; this paragenesis is highly unusual, being previously identified only in the late combeite–wollastonite nephelinites from Oldoinyo Lengai.

The tephriphonolites and phonotephrites contain phenocrysts of Ti-augite, kaersutite, and Ti-magnetite in a matrix of plagioclase, Ti-magnetite and interstitial alkali feldspar and nepheline. With a gradation into phonolite, nepheline and anorthoclase become phenocryst phases in addition to clinopyroxene; the glassy matrix contains abundant sanidine, together with apatite and nepheline.

Cumulate amphibole–perovskite pyroxenite blocks, similar to those described in one of the later flows from the crater lava dome by Rock (1976), occur as blocks on the col between Little Meru and the main cone (Roberts 2002).

Roberts (2002) gave many analyses of olivine, clinopyroxene, amphibole and mica and a more limited number of analyses of magnetite, anorthoclase, nepheline, melilite, schorlomite, titanite, apatite, perovskite, sodalite, calcite, combeite, apophyllite and delhayelite. Roberts (2002) also gave major, minor and trace element analyses of 52 rocks, ranging from nephelinite to phonolite. Additionally, Paslick *et al.* (1995, 1996) gave complete rock analyses of one tephrite and two phonolites, and of clinopyroxene and amphibole in one of the phonolites. Wood (1968) provided analyses of phonolites from the western slopes, and of phonolites described as being from 'the Kilimanjaro Lakes, 20 miles west of Kilimanjaro' (Appendix 2, Dawson 2008); this locality appears to be the Momella Lakes where the avalanche deposit contains many lava blocks.

Like the other Younger Extrusive volcanoes, the lower flanks of Meru and the surrounding plains are pockmarked by explosion craters and tuff-cones. Of these, three are noteworthy:

(i) *Ngurdoto* is small stratovolcano [3.28°S 36.92°E] lying to the SE of Meru; it has a summit crater 360 m deep and measuring 4 × 3 km. Due to material from the Meru debris avalanche on the lower slopes, and thick forest on the upper slopes, exposure is poor. The very few exposures are of tuff containing numerous mica crystals and blocks of pyroxene–phlogopite nephelinite, nephelinite, phonolite, phlogopite–calcite ijolite, melilitite and carbonatite. Roberts (2002) provided analyses of a nephelinite, carbonate-rich ijolite and sövite, and of pyroxenes, mica (which in its low $TiO_2$ contents is unlike others from the Meru area), kaersutite, schorlomite (in one of the carbonate-rich ijolites), titanite, apatite, perovskite, sodalite, and Sr-calcite (in sövite). Ngurdoto is important in that the overall assemblage indicates that it is a carbonatite-related volcano. Moreover, carbonate–silicate reaction relationships in some of the carbonate-rich blocks are taken by Roberts (2002) to indicate that the original carbonate was natrocarbonatite similar to that of Oldoinyo Lengai. Further, unlike Meru, the rocks at Ngurdoto are potassic, rather than sodic.

(ii) *Matuffa* [3°22′S 37°00′E] is a heavily forested, small crater to the SE of Ngurdoto. On its lower, southern slopes are several small melilitite scoria cones, one of which contains xenoliths of harzburgite (Roberts 2002).

'mani, a cinder cone some 8 km east of Arusha, contains upper mantle dunite, harzburgite and wehrlite, some of ve been partially melted (Jones *et al.* 1983*a*). The pre- low-Al clinopyroxene, monazite and apatite in some of these otherwise highly-refractory rocks indicates that they have been subject to carbonatite metasomatism (Rudnick *et al.* 1993). The Sr isotope values of the clinopyroxenes ($^{87}Sr/^{86}Sr$ 0.70304 to 0.7035) suggest that the metasomatism occurred recently. Originally termed ankaramite, the scoria has more recently been re-classified as olivine melanephelinite (Roberts 2002).

*Monduli–Arusha–Oljoro volcanic field*

*Type*: monogenetic tuff cone field; *Lat/Long*: centred on 3.40°S 36.55°E; *Elevation*: *c.* 1300 m; *Relief*: individual features up to 100 m; *Eruptive history*: ?Late Pleistocene; *Composition*: olivine basalt, olivine melilitite, ankaramite, carbonatite; *QDS*: 54 Monduli and 55 Meru.

The Ardai Plains SW and south of Monduli, the area between Monduli and the town of Arusha, and the area around the Oljoro valley, comprise a lava plateau. The derivation of the basalts, basanites and mugearites forming the plateau is uncertain, though a possible source is the Monduli volcano. Many minor volcanic features stand up above or penetrate the lava plateau (Fig. 5.8; Dawson 1964*a*); they consist of lava and scoria cones, asymmetric tuff rings and cratered tuff cones and explosion craters (always with maximum ejecta deposition on the western side of the features).

The morphology of these minor features suggest that they are of Late Pleistocene age, though in the eastern part of the field around Arusha they are partly covered by yellow tuff originating from Meru. Around Monduli the country is covered by a layer of what was formerly believed to be calcrete but which has been re-interpreted as recrystallized carbonatite ash (Dawson 1964*c*). The lava cones comprise a variety of petrographic types, including phlogopite-bearing basalts, olivine melilitites, and analcitites. Roberts (2002) gave analyses of three olivine melilitites and two basanites from the area, and Paslick *et al.* (1995) gave a whole rock analysis of an olivine analcitite from a lava cone on the Ardai Plain.

Of particular interest in the area are:

(a) the Lashaine asymmetric tuff ring [3°22′S 36°26′E], which lies 8 km south of Monduli township. It comprises a lower unit of olivine melililite scoria overlain by a mantle of carbonatite tuff; both units contain blocks of upper-mantle peridotite and deep crustal granulite. The peridotites, which comprise garnetiferous- and garnet-free dunite, harzburgite, lherzolite, wehrlite and pyroxenite, are very fresh and retain valuable intra-grain and grain-boundary information that is largely lost in similar peridotite blocks in kimberlite. The numerous studies on the peridotites are summarized by Dawson (2002), and a resumé is given in Chapter 8 (Dawson 2008). The granulite blocks are of basic composition, and contain scapolite. Like some of the peridotites, some retain evidence of a metasomatic event at *c.* 2 Ga, and others are partly-melted (Reid *et al.* 1975; Dawson 1977, 1987, 2002; Cohen *et al.* 1984; Jones *et al.* 1983*b*).

(b) Cameron's Crater, also known as Kisima ya Mungu (the Cauldron of God), on the west side of the Oljoro valley[3.53°E 36.54°S]. The explosion crater is 1.5 km in diameter and 200 m deep. Its thin, surrounding tuff-layer is highly calcareous and contains blocks of crustal granulites and granite, and upper-mantle peridotite.

*Basotu maars*

*Type*: maar field; *Lat/Long*: 4.36°S 36.08°E; *Elevation*: around 2000 m; *Relief*: up to 30 m above the surrounding countryside. *Eruptive history*: Pleistocene; *Crater diameter*: up to 1 km; *Composition*: nephelinite/carbonatite; *QDS*: 84 Hanang.

30 craters and maars, and their tuff rings, form two NNE–SSW-trending semi-parallel belts around 15 km long on the southern Mbulu Plateau (Fig. 7.21). The elongation of these volcanic zones is controlled by dyke and fault alignment in the Archaean rocks through which the volcanics have erupted (Downie & Wilkinson 1972). The northwestern belt is the more continuous and contains some overlapping features; the other belt, lying some 8 km to the SE, consists of isolated craters and maars. Both fresh- and salt-water lakes occupy maars in both belts. Crater/maar diameters are up to 1 km (many only up to 300 m) and have depths of up to 100 m.

The fragmental material forming the tuff rings consists mainly of fragments of the Precambrian metasediments and volcanics through which the craters erupted, cemented by carbonate suggested by Downie & Wilkinson (1972) to be of igneous (carbonatite) origin.

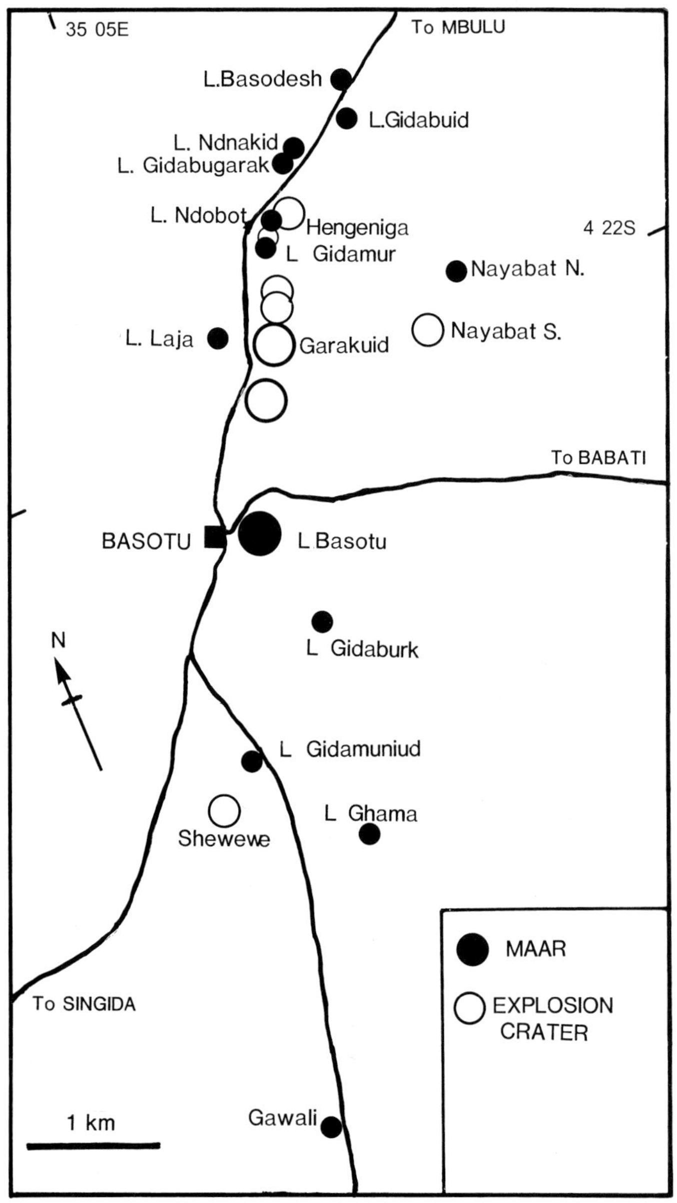


**Fig. 7.21.** Map of the Basotu maars and craters (after Downie & Wilkinson 1972). The overall NE–SW distribution of the features results from their control by shears and dyke intrusions in the underlying basement.

*Kwaraha (sometimes referred to as Ufiome)*

*Type*: stratovolcano; *Lat/Long*: 4.21°S 35.81°E; *Elevation*: 2225 m; *Relief*: 890 m; *Diameter*: 8 km *Eruptive history*: Pleistocene; *Crater diameter*: ?2 km; *Composition*: nephelinite/carbonatite; *QDS*: 85 Babati.

Kwaraha is one of the smaller Younger Extrusive centres. The southern slopes of the mountain are cut by precipitous, forested ravines, but the relatively gentle western and eastern slopes lead to a flat summit area, where the deep soil cover conceals a shallow crater. The mountain is not entirely volcanic, as north–south-trending ridges of Precambrian metasediments occur on the lower western and eastern slopes (Fig. 7.22).

The volcano consists mainly of nephelinite agglomerates and tuffs that contain a high proportion of fenitized metamorphic xenoliths (gneisses and metasediments) Nephelinite lava and crystal tuff occurs in the summit area. James (1966) provided analyses of melilitite, nephelinite and an augite-magnetite rock (Appendix 2, Dawson 2008). A plug of calcite carbonatite occurs in the breached crater and downslope on the lower south-eastern slopes of the mountain. The flow extends to beyond Galappo mission (M. S. Garson pers. comm. 1992) but is best exposed on the road surface at Galappo school (Dawson personal observation). The flow overlies lahar deposits and itself is overlain by nephelinite lava. The carbonatite contains apatite, Ti-magnetites of unusual stratiform shape (Deans & Seager 1978), and tiny (40–100 microns), yellow octahedra of pyrochlore.

Numerous tuff cones and craters occur within a radius of around 10–12 km on the plains NW, east and west of Kwaraha, and on a ridge of Precambrian metamorphic rocks to the south of the volcano. These are composed mainly of lithic and crystal tuffs consisting of pyroxene, olivine, amphibole, mica and plagioclase with a high amount of carbonatitic cement. Also within the explosion crater areas are numerous small hills of 'olivine basalt' (Mudd & Orridge 1966), two of which have proved to be of melilitite. Sora hill [4.17°S 35.88°E] consists of olivine melilitite containing numerous amphibole megacrysts and aggregates, and the prominent hill of Haindadonga [4.15°S 35.9°E], lying within the Tarangire National Park, consists of welded lapilli of phlogopite-phyric melilitite (Dawson *et al.* 1997).

*Hanang*

*Type*: stratovolcano; *Lat/Long*: 4.42°S 35.81°E; *Elevation*: 3420 m; *Diameter*: 10–12 km; *Relief*: 1700 m; *Eruptive history*: Pleistocene; *Crater diameter*: ?2 km; *Composition*: nephelinite/carbonatite; *QDS*: 84 Hanang and 103 Balangida Lelu.

Hanang (Fig. 7.23) is the most southerly of the major volcanoes in northern Tanzania, lying well to the south of the main volcanic area. The adjacent rift escarpment is lower than further north, and Hanang's relative height makes it an impressive mountain (Fig. 7.24). Overlying updomed basement rocks, it is a steep cone with deep, radial ravines, the largest of which, occupied by

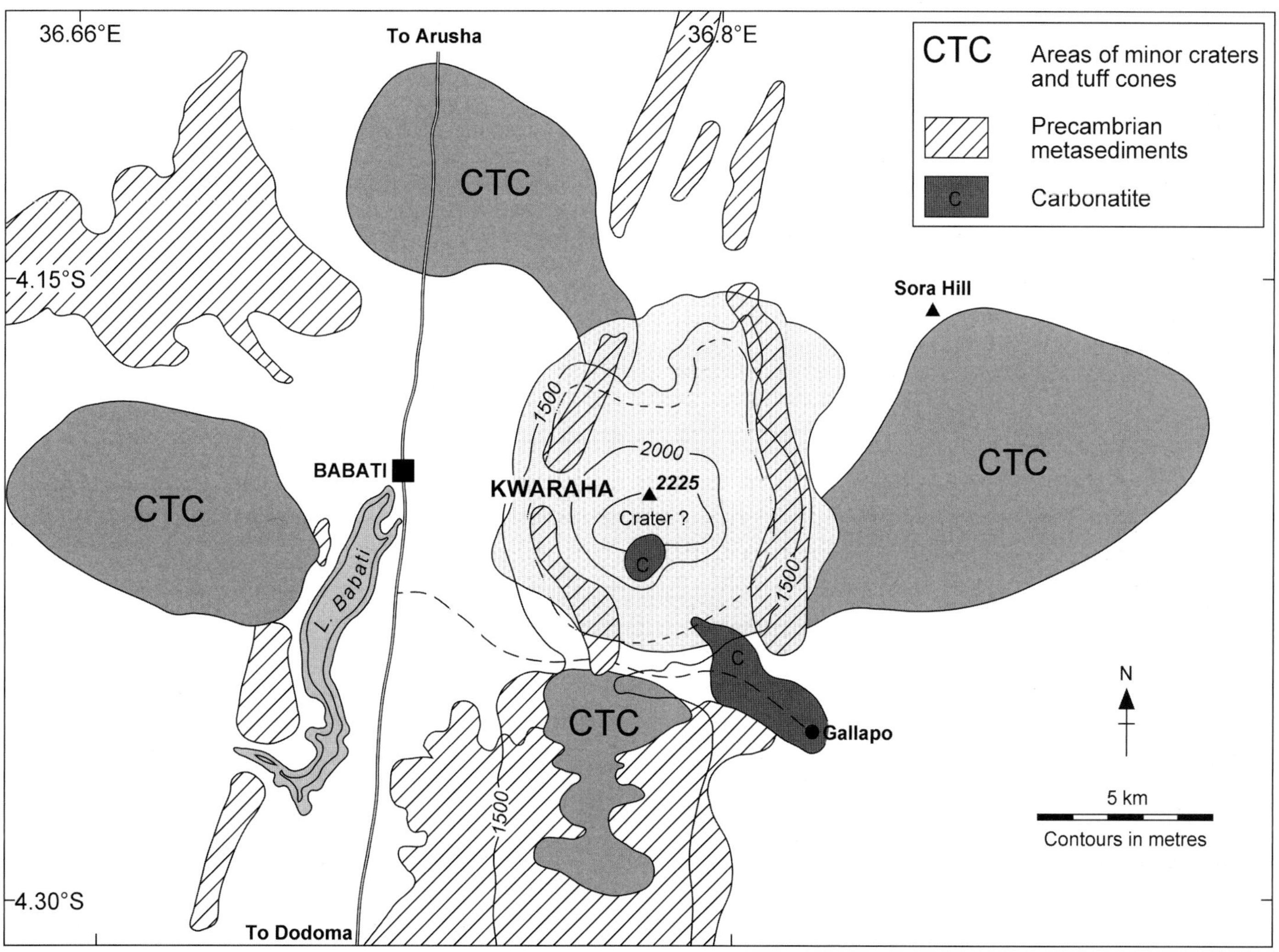


**Fig. 7.22.** Map of the Kwaraha volcano; based on Quarter Degree Sheet 85, information from M. S. Garson, and Dawson (personal observation). Sora Hill is a melilitite cone and is referred to in the text.

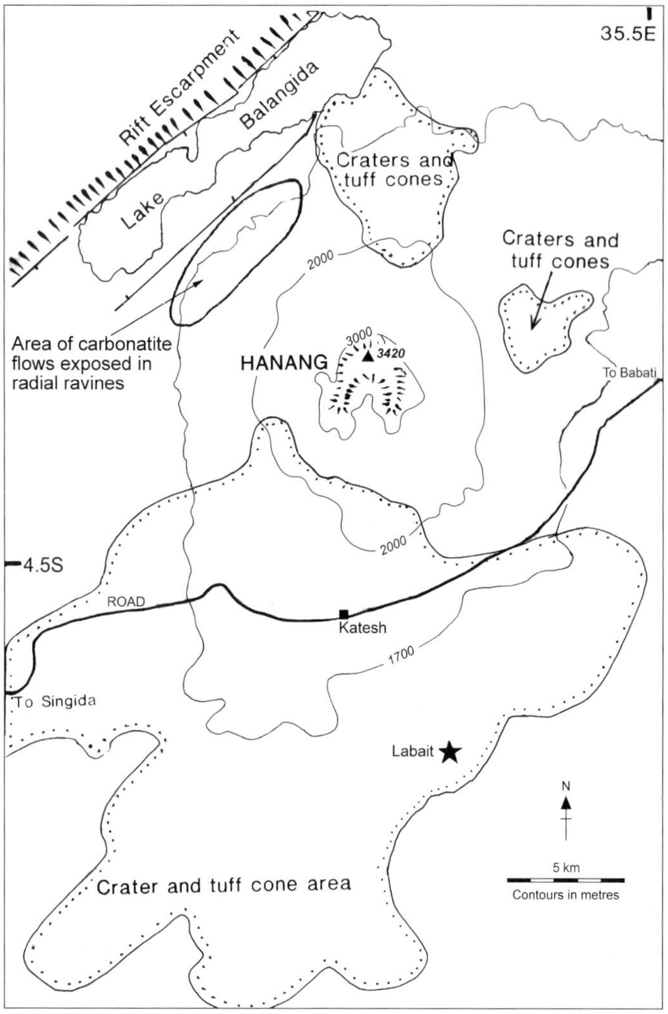


**Fig. 7.23.** Map of Hanang, based on Quarter Degree Sheets 64 and 103, information from M. S. Garson and Dawson (personal observation). Labait is a mantle xenolith locality.

the Hami river, is a major feature on the south side of the volcano. Steep ridges above the Hami gorge, and running SW and SE from the summit, may be the remnants of a breached crater.

Exposure is good only in the summit area and in the radial gorges. The volcano has a core of nephelinitic tuffs and agglomerates, and a partial cover of nephelinite lavas and carbonatitic tuff. Agglomerates in the summit area contain mainly blocks of nephelinite, and rarer melteigite and ijolite, in a calcareous matrix, together with basement-derived xenoliths of granitic and gabbroic rocks, many of which are fenitized. On the summit ridge are nephelinite dykes. The tuffs and lapilli tuffs forming most of the volcano are carbonate-rich and have outward dips of 40° in the summit area, shallowing off downslope (Thomas 1966). Two nephelinites have given ages of 1.5 and 0.9 Ma (Appendix 1, Dawson 2008).

**Fig. 7.24.** Hanang, looking SE across the dried mud-flats of the Balangida basin.

On the lower northwestern slopes, below the 1700 m contour, three calcite carbonatite flows up to 3 m thick are interbedded with a sequence of nephelinitic flows, tuffs and agglomerates. Both the carbonatite flows and the agglomerates are crowded with boulders of fentized basement gneiss and contain sporadically-distributed pyrochlore (M. S. Garson pers. comm. 1992). Many of the tuffs and agglomerates are cemented by calcite, probably of carbonatite origin.

Wood (1968) gave major and minor element analyses of seven nephelinites (Appendix 2, Dawson 2008) and Paslick *et al.* (1995, 1996) gave a complete analysis of one nephelinite from the volcano, and an analysis of its clinopyroxene.

Outcrops of nephelinitic tuffs and agglomerates believed to derive from Hanang occur, together with an interbedded sheet of carbonatite, above the escarpment to the west of Lake Balangida. A small dyke of carbonatite crops out in the escarpment. Numerous, small, parasitic tuff cones and craters occur on the lower northern and southern slopes of Hanang, and on the plains up to 25 km around the volcano. The parasitic cones and craters are mainly of calcite-cemented nephelinitic tuffs, but there are some minor scoria cones and flows of olivine nephelinite, olivine melilitite and monchiquite (Selby & Thomas 1966). Olivine melilitite scoria and lava at Labait hill, 7.5 km SE of Katesh, contain xenoliths of upper-mantle garnet lherzolite, chromite harzburgite and dunite (Dawson *et al.* 1997; Dawson 1999; Lee & Rudnick 1999).

### *Igwisi Hills*

*Type*: monogenetic tuff-cone field; *Lat/Long*: 4.85°S 31.92°E; *Elevation*: *c.* 1300 m; *Relief*: up to 50 m; *Diameter*: single tuff cones up to 0.9 km; *Eruptive history*: ?Pleistocene; *Crater diameter*: up to 0.4 km; *Composition*: carbonatite-kimberlite.

The volcanism discussed hitherto occurs astride and within the rift valley and on the eastern side of the Tanzania Craton. Tertiary–Recent volcanicity is absent on the central part of the craton (although the site of Mesozoic kimberlite intrusion) but towards the western margin of the craton, three small, cratered tuff-cones stand above the peneplain to form the Igwisi Hills (Fig. 7.25). The youthful morphology of the cones indicates they are young, an inference supported by a poorly-constrained radiometric date (below). The volcanicity is highly unusual, in that the extruded material contains a high proportion of micro-xenoliths of upper mantle garnet peridotite in a carbonatitic matrix (Dawson 1994).

The three asymmetric, cratered tuff cones lie along a 2 km long, NE–SW-trending line. They stand some 70 m high above the surrounding countryside, have maximum ejectementa deposition on their northwestern sides, and have steep inner walls and flat grassy crater floors. The outer slopes have karst-like solution topography. The SW hill is relatively simple, being an asymmetric, cratered cone highest on its western side (30 m above the crater floor), with the wall of the 0.23 km wide crater being breached on its eastern side. The northeastern hill is elongated NE–SW, and has two craters, a smaller 0.1 km wide SW crater, and a larger 0.35 km wide northeastern crater. The northeastern crater, the rim of which has a maximum height of 40 m, is breached on its northeastern side, and

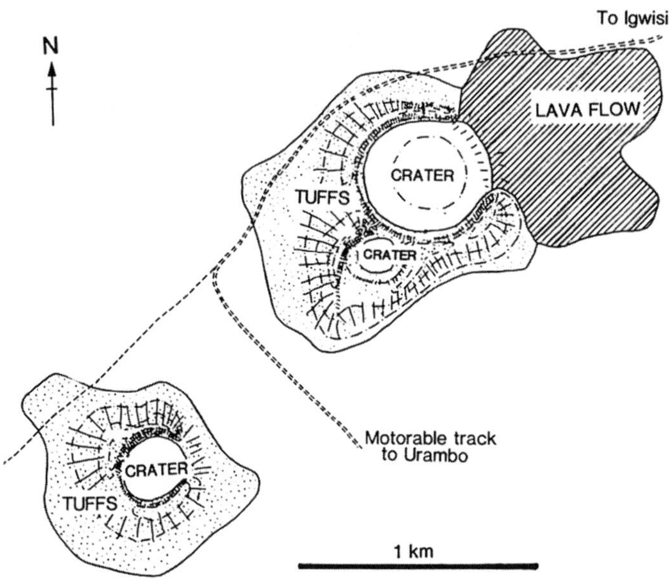


**Fig. 7.25.** Map of the Igwisi Hills (after Dawson 1994).

through this breach has flowed a viscous lava flow measuring around 0.8 km long and 0.7 km wide.

The tuffs are highly calcitic and contain both rounded nodules and fractured clasts of weathered and fresh olivine, together with blocks of biotite- and hornblende-gneiss. The lava is both flow- and gravity-differentiated and, particularly in its lower part, contains numerous rounded microxenoliths (up to 1 cm) (Plate 26) comprising combinations of strained and recrystallized forsterite, together with enstatite, chrome-diopside, pyrope garnet, phlogopite and calcite which, collectively, make up phlogopite-calcite garnet lherzolite. The matrix consists of smaller forsterite grains (unstrained and containing higher CaO than the nodule olivines), calcite, chlorite, spinel, perovskite and apatite. A U–Pb age on perovskite (poorly constrained due to the fine grain-size of the perovskite) is $0 \pm 29$ Ma, consistent with a Quaternary age (Batumike *et al.* 2008). This unusual combination of high-temperature, refractory lherzolite micro-xenoliths with a lower-temperature volatile-rich matrix assemblage, is reminiscent of calcite-rich kimberlites (Dawson 1994), though the highly calcitic matrix and high contents of the rare-earth elements, strontium and barium, indicate some affinities with carbonatites. The conditions of equilibration of the phases in the peridotite xenoliths are 850°C and 35.9 kbar that, together with their mineralogy and the Sr and Nd isotope values of the flow, indicates that the Igwisi Hills magma originated by small degrees of melting of a carbonated peridotite protolith that had isotopic similarities to the source regions for Group I kimberlites and ocean island basalts, at depths of >100 km.

### *Hot springs*

Hot springs (temperatures 18–65°C) occur sporadically along many of the Pleistocene faults, e.g. on either sides of Lakes Natron, Eyasi and Balangdia, at Lake Manyara and in the relatively unrifted area towards the southern end of the Mponde graben and near Kondoa (James 1967). Most contain high concentrations of Na and K, whilst the brines at Mponde and Kondoa contain gases with high concentrations of nitrogen and helium, typical values (percentage by volume) being in the order of He: 5–7; $N_2$: 89–93; Ar: 1–2; $CO_2$: 1–2; $H_2S$, $H_2$ and $CH_4$: <1% (Harris 1958).

## References


BAGDASARYAN, G. P., GERASIMOVSKIY, V. I., POLYAKOV, A. I., GUKASYAN, R. K. & VERNADSKIY, V. I. 1973. Age of volcanic rocks in the rift zones of East Africa. *Geochemistry International*, **10**, 66–71.

BAKER, B. H. 1963. *Geology of the Area South of Magadi*. Geological Survey of Kenya, Report 61.

BAKER, B. H., MOHR, P. A. & WILLIAMS, L. A. J. 1972. *Geology of the Eastern Rift System of Africa*. Geological Society of America, Special Publications, **36**.

BATUMIKE, J. M., GRIFFIN, W. L., BELOUSOVA, E. A., PEARSON, N. J., O'REILLY, S. Y. & SHEE, S. R. 2008. Africa's youngest kimberlite? LAM-ICPMS U–Pb dating of kimberlite perovskite from the Kundelungu Plateau, D.R.Congo. *Earth and Planetary Science Letters*, **267**, 609–619.

BELL, K. & BLENKINSOP, J. 1987. Nd and Sr isotopic compositions of East African carbonatites: implications for mantle heterogeneity. *Geology*, **15**, 99–102.

BELL, K. & DAWSON, J. B. 1995*a*. Nd and Sr isotope systematics of the active carbonatite volcano, Oldoinyo Lengai. *In*: BELL, K. & KELLER, J. (eds) *Carbonatite Volcanism of Oldoinyo Lengai and Petrogenesis of Natrocarbonatite*. Springer Verlag, Berlin, 100–112.

BELL, K. & DAWSON, J. B. 1995*b*. An assessment of the alleged role of evaporites and saline brines in the origin of natrocarbonatite. *In*: BELL, K. & KELLER, J. (eds) *Carbonatite Volcanism of Oldoinyo Lengai and Petrogenesis of Natrocarbonatite*. Springer, Berlin, 137–147.

BELL, K. & KELLER, J. (eds) 1995. *Carbonatite Volcanism of Oldoinyo Lengai and Petrogenesis of Natrocarbonatite*. Springer, Berlin.

BELL, K. & PETERSON, T. 1991. Nd and Sr isotope systematics of Shombole volcano, East Africa, and the links between nephelinites, phonolites and carbonatites. *Geology*, **19**, 582–585.

BELL, K. & SIMONETTI, A. 1996. Carbonatite magmatism and plume activity: implications from the Nd, Pb and Sr isotope systematics of Oldoinyo Lengai. *Journal of Petrology*, **37**, 1321–1339.

BELL, K., FARQUHAR, R. M. & DAWSON, J. B. 1973. Strontium isotope studies of alkalic rocks: the active carbonatite volcano Oldoinyo Lengai, Tanzania. *Bulletin of the Geological Society of America*, **84**, 1019–1030.

BELL, K., KJAARSGAARD, B. A. & SIMONETTI, A. (eds) 1998. *Carbonatites – into the Twenty-First Century. A Volume in Honour of John Gittins*. Special Number of *Journal of Petrology*, **39**, 1839–2151.

BISHOP, W. W., MILLER, J. A. & FITCH, F. H. 1969. New potassium–argon age determinations relevant to the Miocene fossil mammal occurrences in East Africa, *American Journal of Science*, **267**, 669–699.

BULLETIN OF THE GLOBAL VOLCANISM NETWORK 2006. Oldoinyo Lengai, **31**, Number 3, 13–15, March 2006.

BRÖGGER, W. C. 1921. Die Eruptivgesteine des Kristianagebietes. IV Das Fengebiet in Telemark, Norwegen. *Videnskapssels kapets Skrifter, I. Matematisk-natur vitenskapelig Klasse No. 9*, Kristiana.

CHURCH, A. A. 1995. *The Petrology of the Kerimasi Volcano and the Carbonatites of Oldoinyo Lengai with a Review of other Occurrences of Extrusive Carbonatites*. PhD thesis, University College, London.

CHURCH, A. A. & JONES, A. P. 1994. Hollow natrocarbonatite lapilli from the 1992 eruption of Oldoinyo Lengai, Tanzania. *Journal of the Geological Society, London*, **151**, 59–63.

COHEN, R. S., O'NIONS, R. K. & DAWSON, J. B. 1984. Isotope geochemistry of xenoliths from East Africa: implications for the development of mantle reservoirs and their interaction: *Earth and Planetary Science Letters*, **68**, 209–220.

COOPER, J. P. 1972. *Geochemistry of Tarosero volcano, northern Tanzania*. MSc thesis, University of Manchester.

CRANDELL, D. R., MILLER, C. D., GLICKEN, H. X., CHRISTIANSEN, R. L. & NEWHALL, C. G. 1984. Catastrophic debris avalanche from ancestral Mount Shasta volcano, California. *Geology*, **12**, 143–146.

CROSSLEY, R. 1979. The Cenozoic stratigraphy and structure of the western part of the rift valley in southern Kenya. *Journal of the Geological Society of London*, **136**, 393–405.

CURTIS, G. H. & HAY, R. L. 1972. Further geological studies and K–Ar dating at Olduvai Gorge and Ngorongoro crater. *In*: BISHOP, W. W.

& MILLER, J. W. (eds). *Calibration of Hominoid Evolution*. Scottish Academic Press, Edinburgh, 107–134.

DAWSON, J. B. 1962*a*. The geology of Oldoinyo Lengai. *Bulletin Volcanologique*, **24**, 349–387.

DAWSON, J. B. 1962*b*. Sodium carbonate lavas from Oldoinyo Lengai, northern Tanganyika. *Nature*, **196**, 1065–1066.

DAWSON, J. B. 1964*a*. Carbonate tuff cones in northern Tanganyika. *Geological Magazine*, **101**, 129–37.

DAWSON, J. B. 1964*b*. *Monduli*. Tanzania Geological Survey Quarter Degree Sheet **54**.

DAWSON, J. B. 1964*c*. Carbonatitic volcanic ashes in northern Tanganyika. *Bulletin Volcanologique*, **27**, 81–92.

DAWSON, J. B. 1977. Sub-cratonic and upper mantle models based on xenolith suites in kimberlite and nephelinitic diatremes. *Journal of the Geological Society, London*, **134**, 173–184.

DAWSON, J. B. 1987. Metasomatized harzburgites in kimberlite and alkaline magmas: enriched restites and 'flushed' lherzolites. *In*: MENZIES, M. A. & HAWKESWORTH, C. J. (eds) *Mantle Metasomatism*. Academic Press, London, 125–144.

DAWSON, J. B. 1989. Sodium carbonatite extrusions from Oldoinyo Lengai, Tanzania: implications for carbonatite complex genesis: *In*: BELL, K. (ed). *Carbonatites: genesis and evolution*. Unwin Hyman, London, 255–277.

DAWSON, J. B. 1993. A supposed sövite from Oldoinyo Lengai, Tanzania: result of extreme alteration of alkali carbonatite lava. *Mineralogical Magazine*, **57**, 93–101.

DAWSON, J. B. 1994. Quaternary kimberlitic volcanism on the Tanzania craton. *Contributions to Mineralogy and Petrology*, **116**, 473–485.

DAWSON, J. B. 1997. Neogene–Recent rifting and volcanism in northern Tanzania: relevance for comparisons between the Gardar province and the East African Rift valley. *Mineralogical Magazine*, **61**, 543–548.

DAWSON, J. B. 1998. Peralkaline nephelinite–natrocarbonatite relationships at Oldoinyo Lengai, Tanzania. *Journal of Petrology*, **39**, 2077–2094.

DAWSON, J. B. 1999. Melting and metasomatism in spinel peridotite xenoliths from Labait, Tanzania. *In*: GURNEY, J. J., GURNEY, J. L., PASCOE, M. D. & RICHARDSON, S. H. (eds) *Proceedings of the 7th International Kimberlite Conference, Cape Town, 1998*, **1**, Red Roof Design, Cape Town, 164–173.

DAWSON, J. B. 2002. Metasomatism and partial melting in upper-mantle peridotite xenoliths from the Lashaine volcano, northern Tanzania. *Journal of Petrology*, **43**, 1749–1777.

DAWSON, J. B. 2008. *The Gregory Rift Valley and Neogene–Recent Volcanoes of Northern Tanzania*. Geological Society, London, Memoirs, **33**.

DAWSON, J. B. & FRISCH, T. 1971. Eucolite from Oldoinyo L'engai, Tanzania. *Lithos*, **4**, 297–303.

DAWSON, J. B. & FUGE, R. 1980. Halogen content of some African carbonatites. *Lithos*, **13**, 139–143.

DAWSON, J. B. & GALE, N. 1970. Uranium and thorium in alkalic lavas from the active carbonatite volcano Oldoinyo Lengai, northern Tanzania. *Chemical Geology*, **6**, 221–231.

DAWSON, J. B. & HILL, P. G. 1998. Mineral chemistry of a peralkaline combeite–lamprophyllite nephelinite form Oldoinyo Lengai, Tanzania. *Mineralogical Magazine*, **62**, 179–196.

DAWSON, J. B. & MITCHELL, R. H. 2007. Oldoinyo Lengai. Contribution to *Bulletin of the Global Volcanism Network*, 32, November 2007, 14–15.

DAWSON, J. B. & POWELL, D. G. 1969. The Engaruka–Natron explosion crater area, northern Tanzania. *Bulletin Volcanologique*, **33**, 791–817.

DAWSON, J. B. & SAHAMA, T. G. 1963. A note on parawollastonite from Oldoinyo Lengai, Tanganyika. *Schweizer Mineralogisches und Petrographisches Mitteilungen*, **43**, 131–133.

DAWSON, J. B. & SMITH, J. V. 1988. Veined and metasomatized peridotites from Pello Hill, Tanzania: evidence for anomalously-light mantle beneath the Tanzania sector of the eastern Rift Valley. *Contributions to Mineralogy and Petrology*, **100**, 510–527.

DAWSON, J. B. & SMITH, J. V. 1992*a*. Olivine–mica–pyroxenite xenoliths from Oldoinyo Lengai, Tanzania: metasomatic products of upper mantle peridotites. *Journal of Volcanology and Geothermal Research*, **50**, 131–142.

DAWSON, J. B. & SMITH, J. V. 1992*b*. Potassium loss during metasomatic alteration of mica pyroxenite from Oldoinyo Lengai, northern Tanzania; contrasts with fentization. *Contributions to Mineralogy and Petrology*, **112**, 254–260.

DAWSON, J. B., BOWDEN, P. & CLARK, G. C. 1968. Activity of the carbonatite volcano Oldoinyo Lengai, 1966. *Geologisches Rundschau*, **57**, 865–879.

DAWSON, J. B., POWELL, D. G. & REID, A. M. 1970. Ultrabasic xenoliths and lava from the Lashaine volcano, northern Tanzania. *Journal of Petrology*, **11**, 519–548.

DAWSON, J. B., SMITH, J. V. & JONES, A. P. 1985. A comparative study of bulk rock and mineral chemistry of olivine melilitites and associated rocks from East and South Africa. *Neues Jahrbuch für Mineralogie, Abteilung*, **152**, 143–175.

DAWSON, J. B., GARSON, M. S. & ROBERTS, B. 1987. Altered former alkalic carbonatite lava from Oldoinyo Lengai, Tanzania: inferences for calcite carbonatite lavas. *Geology*, **15**, 765–768.

DAWSON, J. B., SMITH, J. V. & STEELE, I. M. 1989. Combeite ($Na_{2.33}Ca_{1.74}others_{0.12}$)$Si_3O_9$ from Oldoinyo Lengai, Tanzania. *Journal of Geology*, **97**, 365–372.

DAWSON, J. B., PINKERTON, H., NORTON, G. E. & PYLE, D. M. 1990. Physico-chemical properties of alkali carbonatite lavas: data from the 1988 eruption of Oldoinyo Lengai, Tanzania. *Geology*, **18**, 260–263.

DAWSON, J. B., SMITH, J. V. & STEELE, I. M. 1992. 1966 ash eruption of the carbonatite volcano, Oldoinyo Lengai; mineralogy of lapilli, and mixing of silicate and carbonate lava. *Mineralogical Magazine*, **56**, 1–16.

DAWSON, J. B., PINKERTON, H., PYLE, D. M. & NYAMWERU, C. 1994. June 1993 eruption of Oldoinyo Lengai, Tanzania: Exceptionally viscous and large carbonatite lava flows and evidence for coexisting silicate and carbonate magmas. *Geology*, **22**, 799–802.

DAWSON, J. B., KELLER, J. & NYAMWERU, C. 1995*a*. Historic and recent eruptive activity of Oldoinyo Lengai. *In*: BELL, K. & KELLER, J. (eds) *Carbonatite Volcanism*. Springer, Berlin, 4–22.

DAWSON, J. B., SMITH, J. V. & STEELE, I. M. 1995*b*. Petrology and mineral chemistry of plutonic igneous xenoliths from the carbonatite volcano Oldoinyo Lengai, Tanzania. *Journal of Petrology*, **36**, 797–826.

DAWSON, J. B., PINKERTON, H., NORTON, G. E., PYLE, D. M., BROWNING, P., JACKSON, D. & FALLICK, A. E. 1995*c*. Petrology and geochemistry of Oldoinyo Lengai lavas extruded in November 1988: magma source, ascent and crystallization. *In*: BELL, K. & KELLER, J. (eds) *Carbonatite Volcanism*. Springer, Berlin, 47–69.

DAWSON, J. B., PYLE, D. M. & PINKERTON, H. 1996*a*. Evolution of natrocarbonatite from a wollastonite nephelinite parent: evidence from the June 1993 eruption of Oldoinyo Lengai, Tanzania. *Journal of Geology*, **104**, 41–54.

DAWSON, J. B., STEELE, I. M., SMITH, J. V. & RIVERS, M. L. 1996*b*. Minor and trace element chemistry of carbonates, apatites and magnetites in some African carbonatites. *Mineralogical Magazine*, **60**, 415–425.

DAWSON, J. B., JAMES, D., PASLICK, C. & HALLIDAY, A. M. 1997. Ultrabasic potassic low-volume magmatism and continental rifting in north-central Tanzania: association with enhanced heat flow. *Russian Geology and Geophysics*, **38**, 69–81.

DAWSON, J. B., HINTON, R. W. & STEELE, I. M. (*in press*). The composition of anorthoclase and nepheline in Mount Kenya phonolite and Kilimanjaro trachyte, and crystal-glass partitioning of elements. *Canadian Mineralogist*.

DE VRIES, B. V., SELF, S., FRANCIS, P. W. & KESZTHELYI, L. 2001. A gravitational spreading origin for the Socompa debris avalanche. *Journal of Volcanological and Geothermal Research*, **105**, 225–247.

DEANS, T. & SEAGER, A. F. 1978. Stratiform magnetite crystals of abnormal morphology from volcanic carbonatites in Tanzania, Kenya, Greenland and India. *Mineralogical Magazine*, **42**, 463–475.

DONALDSON, C. H. & DAWSON, J. B. 1978. Skeletal crystallization and residual glass compositions in a cellular alkalic pyroxenite from

Oldoinyo Lengai: implications for the evolution of the alkalic carbonatite lavas. *Contributions to Petrology and Mineralogy*, **67**, 139–149.

DONALDSON, C. H., DAWSON, J. B., KANARIS-SOTIRIOU, R., BATCHELOR, R. A. & WALSH, J. N. 1987. The silicate lavas of Oldoinyo Lengai. *Neues Jahrbuch für Mineralogie Abhandlungen*, **156**, 247–279.

DOWNIE, C. 1964. Glaciations of Mount Kilimanjaro, northeast Tanganyika. *Bulletin of the Geological Society of America*, **75**, 1–16.

DOWNIE, C. & WILKINSON, P. 1962. The explosion craters of Basotu, Tanganyika Territory. *Bulletin Volcanologique*, **24**, 389–420.

DOWNIE, C. & WILKINSON, P. 1972. *The Geology of Kilimanjaro*. University of Sheffield, Sheffield.

DOWNIE, C., HUMPHRIES, D. W., *ET AL.* 1965. *Kilimanjaro-Moshi (Geological Map of Kilimanjaro)*. Special Sheet covering Tanganyika Geological Survey Quarter Degree Sheets **42, 56** and **57**.

DRAKE, P. & CURTIS, G. H. 1987. K–Ar geochronology of the Laetoli fossil localities. *In*: LEAKEY, M. D. & HARRIS, J. M. (eds) *Laetoli, a Pliocene Site in Northern Tanzania*. Clarendon Press, Oxford, 48–52.

DUNDAS, D. L. & AWADALLAH, G. W. (unpublished ). *Loliondo*. Unpublished report and map of Quarter Degree Sheet 16 and 27, Tanzania Geological Survey. Mapped 1966.

ERDMANNSDORFER, O. 1935. Über wollastoniturtit und die Entstehungsweisen von alkaligesteiner. *Sitzungberichte der Heidelberg Akademie für Wissenschaten, Mathematike und Naturgeschichte*, K 1.2, 1–23.

EVANS, A. L., FAIRHEAD, J. D. & MITCHELL, J. G. 1971. Potassium–argon ages from the volcanic province of northern Tanzania. *Nature*, **229**, 19–20.

FAIRHEAD, J. D., MITCHELL, J. G. & WILLIAMS, L. A. J. 1972. New K/Ar determinations on Rift volcanics of S. Kenya and their bearing on the age of Rift faulting. *Nature*, **238**, 66–9.

FARLER, J. P. 1882. Native routes in East Africa from Pangani to the Masai country. *Proceedings of the Royal Geographical Society, New Series*, **4**, 730–753.

FINCKH, L. 1903. Die trachydolerite des Kibo (Kilimandscharo) und die kenyte des Kenya. *Zeitschrift der Deutsches Geologische Gesellschaft*, **55**, Monatsberich 1–14.

FISCHER, G. A. 1884–1885. Reise in das Masailand. *Mitteilungen der Geographisches Gesellschaft Hamburg*, Pt 1 (1884), 36–99, Pts 2 and 3 (1885) 189–237 and 238–279.

GENGE, M. J., BALME, M. & JONES, A. P. 2001. Salt-bearing fumarole deposits in the summit crater of Oldoinyo Lenga, Northern Tanzania: interactions between natrocarbonatite and meteoric water. *Journal of Volcanological and Geothermal Research*, **106**, 111–122.

GITTINS, J. & MCKIE, D. 1980. Alkalic carbonate magmas: Oldoinyo Lengai and its wider applicability. *Lithos*, **13**, 213–215.

GROMMÉ, C. S., REILLY, T. A., MUSSETT, A. E. & HAY, R. L. 1970. Palaeomagnetism and potassium–argon ages of volcanic rocks of Ngorongoro caldera, Tanzania: *Geophysical Journal of the Royal Astronomical Society*, **22**, 101–115.

GUEST, N. J. 1953. *The Geology of Part of Northern Tanganyika Territory*. PhD thesis, University of Sheffield.

GUEST, N. J. 1956. The volcanic activity of Oldoinyo Lengai 1954. *Records of the Tanganyika Geological Survey*, **4**, 56–59.

GUEST, N. J. & LEEDAL, G. P. 1956. The volcanic activity of Mount Meru. *Records of the Geological Survey of Tanganyika*, **3**, 40–46.

GUEST, N. J. & PICKERING, R. 1966*a*. *Kibangaini*. Tanzania Mineral Resources Division, Quarter Degree Sheet **28**.

GUEST, N. J. & PICKERING, R. 1966*b*. *Gelai and Ketumbeine*. Tanzania Mineral Resources Division, Quarter Degree Sheet **40**.

GUEST, N. J., JAMES, T. C., PICKERING, R. & DAWSON, J. B. 1961. *Angata Salei*. Tanzania Geological Survey Quarter Degree Sheet **39**.

HALAMA, R., MCDONOUGH, W. F., RUDNICK, R. L., KELLER, J. & KLAUDIUS, J. 2007. The Li isotopic composition of Oldoinyo Lengai: Nature of the mantle sources and lack of isotopic fractionation during carbonatite petrogenesis. *Earth and Planetary Science Letters*, **254**, 77–89.

HARRIS, J. F. 1958. Geological investigations, sampling and diamond drilling at Manyeghi helium-bearing hot springs, Singida District. *Records of the Geological Survey of Tanganyika*, **VIII**, 86–98.

HART, S. R. 1988. Heterogeneous mantle domains: signatures, genesis and mixing chronologies. *Earth and Planetary Science Letters*, **90**, 273–296.

HAY, R. L. 1963. Zeolitic weathering in Olduvai Gorge, Tanganyika. *Bulletin of the Geological Society of America*, **74**, 1281–1286.

HAY, R. L. 1964. Discussion in Dawson (1964*b*).

HAY, R. L. 1976. *Geology of the Olduvai Gorge*. University of California Press, Berkeley.

HAY, R. L. 1983. Natrocarbonatite tephra of Kerimasi volcano. *Geology*, **11**, 599–602.

HAY, R. L. 1989. Holocene carbonatite–nephelinite tephra deposits of Oldoinyo Lengai, Tanzania. *Journal of Volcanological and Geothermal Research*, **37**, 77–91.

HAY, R. L. & O'NEILL, J. R. 1983. Carbonatite tuffs in the Laetoli Beds of Tanzania and the Kaiserstuhl in Germany. *Contributions to Mineralogy and Petrology*, **82**, 403–406.

HECKY, R. E. 1971. *The Palaeolimnology of the Alkaline, Saline Lakes on the Mt Meru Lahar*. PhD thesis, Duke University.

ISAAC, G. L. & CURTIS, G. H. 1974. Age of the Acheulian industries from the Peninj Group, Tanzania. *Nature*, **249**, 624–627.

JAGO, B. C. & GITTINS, J. 1999. Mn- and F-bearing rasvumite in natrocarbonatite at Oldoinyo Lengai volcano, Tanzania. *Mineralogical Magazine*, **63**, 53–55.

JAMES, D. E. 1995. *The Geochemistry of Feldspar-free Volcanic Rocks*. PhD thesis, Open University.

JAMES, T. C. 1966. *The Carbonatites of Tanganyika: a Phase of Continental-type Volcanism*. DIC thesis, Imperial College, London.

JAMES, T. C. 1967. Thermal springs in Tanzania. *Transactions of the Institute of Mining and Metallurgy B*, **76**, 1–18.

JOHNSON, L. H., JONES, A. P., CHURCH, A. A. & TAYLOR, W. R. 1997. Ultramafic xenoliths and megacrysts from a melilitite tuff cone, Deeti, northern Tanzania. *Journal of African Earth Sciences*, **25**, 29–42.

JONES, A. P., SMITH, J. V. & DAWSON, J. B. 1983*a*. Glasses in mantle xenoliths from Olmani, Tanzania. *Journal of Geology*, **91**, 167–178.

JONES, A. P., SMITH, J. V., DAWSON, J. B. & HANSEN, E. C. 1983*b*. Metamorphism, partial melting and K-metasomatism of garnet-scapolite-kyanite granulite xenoliths from Lashaine, Tanzania. *Journal of Geology*, **91**, 143–165.

KALT, A., HEGNER, E. & SATIR, M. 1997. Nd, Sr, and Pb isotope evidence for diverse lithospheric mantle sources of East African Rift carbonatites. *Tectonophysics*, **278**, 3145.

KELLER, J. & KRAFFT, M. 1990. Effusive natrocarbonatite activity at Oldoinyo Lengai, June 1988. *Bulletin Volcanologique*, **52**, 629–645.

KELLER, J. & SPETTEL, B. 1995. The trace element composition and petrogenesis of natrocarbonatites *In*: BELL, K. & KELLER, J. (eds) *Carbonatite Volcanism*. Springer, Berlin, 70–86.

KELLER, J. & HOEFS, J. 1995. Stable isotope characteristics of recent natrocarbonatites. *In*: BELL, K. & KELLER, J. (eds) *Carbonatite Volcanism*. Springer, Berlin, 113–123.

KELLER, J. & ZAITSEV, A. N. 2006. Calciocarbonatite dykes at Oldoinyo Lengai, Tanzania: the fate of natrocarbonatite. *Canadian Mineralogist*, **44**, 857–876.

KELLER, J., ZAITSEV, A. & WIEDENMANN, D. 2006. Primary magmas at Oldoinyo Lengai: the role of olivine melilitite. *Lithos*, **91**, 150–172.

KERVYN, M., KLAUDIUS, J., KELLER, J., KERVYN, F., MATTSON, H. B., BELTON, F., MBEDE, E., JACOBS, P. & ERNST, G. J. 2008*a*. Voluminous lava flows at Oldoinyo Lengai in 2006: chronology of events and insights into the shallow magmatic system. *Bulletin of Volcanology*, **70**, 1069–1086.

KERVYN, M., KLAUDIUS, J., KELLER, J., MBEDE, E., JACOBS, P. & ERNST, G. J. 2008*b*. Remote sensing of sector collapse and debris avalanche deposits at Oldoinyo Lengai and Kermasi Volcanoes, Tanzania. *International Journal of Remote Sensing* (in press).

KJAARSGAARD, B. & PETERSON, T. 1991. Nephelinite-carbonate liquid immiscibility at Shombole volcano, East Africa: petrographic and experimental evidence. *Mineralogy and Petrology*, **43**, 293–314.

KLAUDIUS, J. & KELLER, J. 2006. Peralkaline silicate lavas at Oldoinyo Lengai, Tanzania. *Lithos*, **91**, 173–190.

KOEPENICK, K. W., BRANTLEY, S. L., THOMPSON, J. M., ROWE, G. L., NYBLADE, A. A. & MOSHY, C. 1996. Volatile emissions from the crater and flank of Oldoinyo Lengai volcano, Tanzania. *Journal of Geophysical Research*, **101**, 13819–13830.

KRAFFT, M. & KELLER, J. 1989. Temperature measurements in carbonatite lava lakes and flows from Oldoinyo Lengai, Tanzania. *Science* **245**, 168–170.

KRAMM, U. & SINDERN, S. 1998. Nd and Sr isotope signature of fenites from Oldoinyo Lengai, Tanzania, and the genetic relationships between nephelinites, phonolites and carbonatites. *Journal of Petrology*, **39**, 1997–2004.

KRENKEL, E. 1922. *Die Bruchzonen Ostafrikas: Tektonik, Vulkanismus, Erdbeben and Schwereanomalien.* Bornträger, Berlin.

LE BAS, M. J. L. 1977. *Carbonatite-nephelinite Volcanism. An African Case History.* Wiley, London.

LE MAITRE, R. W. (ed.) 1989. *A Classification of Igneous Rocks and Glossary of Terms.* Blackwell Scientific Publications, Oxford.

LEE, C.-T. & RUDNICK, R. L. 1999. Compositionally-stratified cratonic lithosphere: petrology and geochemistry of peridotite xenoliths from the Labait volcano, Tanzania. *In*: GURNEY, J. J., GURNEY, J. L., PASCOE, M. D. & RICHARDSON, S. H. (eds) *Proceedings of the 7th International Kimberlite Conference, Cape Town, 1998*, Vol. 2, Red Roof Design, Cape Town, 503–521.

LEE, C.-T., RUDNICK, R. L., MCDONOUGH, W. F. & HORN, I. 2000. Petrologic and geochemical investigation of carbonates in peridotite xenoliths from northeastern Tanzania. *Contributions to Mineralogy and Petrology*, **139**, 470–484.

MACDONALD, R., MARSHALL, A. S., DAWSON, J. B., HINTON, R. W. & HILL, P. G. 2002. Chevkinite-group minerals from salic volcanic rocks of the East African Rift. *Mineralogical Magazine*, **66**, 287–299.

MCFARLANE, D. A., LUNDBERG, J. & BELTON, F. 2004. An unusual cave from Oldoinyo Lengai, Tanzania. *Journal of Cave and Karst Studies*, **66**, 98–101.

MACINTYRE, R. M., MITCHELL, J. G. & DAWSON, J. B. 1974. Age of the fault movements in the Tanzanian sector of the East African rift system. *Nature*, **247**, 354–356.

MCKIE, D. & FRANKIS, E. J. 1977. Nyerereite: a new volcanic carbonate mineral from Oldoinyo Lengai, Tanzania. *Zeitschrift für Krystallographie*, **145**, 73–95.

MANEGA, P. C. 1993. *Geochronology, Geochemistry and Isotope Study of the Plio-Pleistocene Hominid Sites and the Ngorongoro Volcanic Highlands in Northern Tanzania.* PhD thesis, University of Colorado.

MARIANO, A. N. & ROEDER, P. L. 1983. Kerimasi: a neglected carbonatite volcano. *Journal of Geology*, **91**, 449–455.

MEYER, H. 1900. *Der Kilimandjaro.* Reimer, Berlin.

MITCHELL, R. H. 1997. Carbonate-carbonate immiscibility, neighborite and potassium iron sulphide in Oldoinyo Lengai natrocarbonatite. *Mineralogical Magazine*, **61**, 779–789.

MITCHELL, R. H. 2006*a*. Sylvite and fluorite microcrysts, and fluorite-nyerereite intergrowths from natrocarbonatite, Oldoinyo Lengai, Tanzania. *Mineralogical Magazine*, **70**, 103–114.

MITCHELL, R. H. 2006*b*. An ephemeral pentasodium phosphate carbonate from natrocarbonatite lapilli, Oldoinyo Lengai, Tanzania. *Mineralogical Magazine*, **70**, 211–218.

MITCHELL, R. H. 2006*c*. Mineralogy of stalactites formed by subaerial weathering of natrocarbonatite hornitos at Oldoinyo Lengai, Tanzania. *Mineralogical Magazine*, **70**, 437–444.

MITCHELL, R. H. & BELTON, F. 2004. Niocalite-cuspidine solid solution and manganoan monticellite from natrocarbonatite, Oldoinyo Lengai, Tanzania. *Mineralogical Magazine*, **68**, 787–799.

MITCHELL, R. H. & DAWSON, J. B. 2007. The September 24th, 2007, ash eruption of the carbonatite volcano, Oldoinyo Lengai, Tanzania: mineralogy of the ash and implications for formation of new hybrid magma type. *Mineralogical Magazine*, **71**, 483–492.

MOROGAN, V. & MARTIN, R. F. 1985. Mineralogy and partial melting of fenitzed crustal xenoliths in the Oldoinyo Lengai carbonatitic volcano, Tanzania. *American Mineralogist*, **70**, 1114–1126.

MOTE, P. W. & KASER, G. 2007. The shrinking glaciers of Kilimanjaro: can global warming be blamed? *American Scientist*, **95**, 318–325.

MUDD, G. C. & ORRIDGE, G. R. 1966. *Babati.* Tanzania Mineral Resources Division Quarter Degree Sheet **85**.

MUZUKA, N. N., RYNER, M. & HOLMGREN, K. 2004. 12,000-year preliminary results of the stable nitrogen and carbon isotope record from the Empakai crater lake sediments, northern Tanzania. *Journal of African Earth Sciences*, **40**, 293–303.

NORTON, G. & PINKERTON, H. 1997. Rheological properties of natrocarbonatite lavas from Oldoinyo Lengai, Tanzania. *European Journal of Mineralogy*, **9**, 351–364.

NYAMWERU, C. 1988. Continuing activity of Oldoinyo Lengai volcano, Tanzania: 1983–1985. *Journal of African Earth Sciences*, **7**, 603–610.

NYAMWERU, C. 1990. Observations on changes in the active crater, Oldoinyo Lengai, from 1960–1988. *Journal of African Earth Sciences*, **11**, 385–390.

NYAMWERU, C. 1997. Changes in the crater of Oldoinyo Lengai: June 1993–February 1997. *Journal of African Earth Sciences*, **25**, 43–53.

NYAMWERU, C. 1999. Oldoinyo Lengai. *Bulletin of the Global Volcanism Network*, **24–2**, 12–14.

OHDE, S. & MATAGARIO, J. P. 1999. Instrumental neutron activation analysis of carbonatites from Panda Hill and Oldoinyo Lengai, Tanzania. *Journal of Radioanalytical and Nuclear Chemistry*, **240**, 325–328.

O'NEIL, J. R. & HAY, R. L. 1973. $^{18}O/^{16}O$ ratios in cherts associated with the saline lake deposits of East Africa. *Earth and Planetary Science Letters*, **19**, 257–266.

OPPENHEIMER, C., BURTON, M. R., DURIEUX, J. & PYLE, D. M. 2002. Open-path Fourier transform spectroscopy of gas emissions from Oldoinyo Lengai volcano, Tanzania. *Optics and Lasers in Engineering*, **37**, 203–214.

ORRIDGE, G. R. 1965. *Mbulu.* Tanzania Mineral Resources Division Quarter Degree Sheet **69**.

PASLICK, C., HALLIDAY, A., JAMES, D. & DAWSON, J. B. 1995. Enrichment of the continental lithosphere by OIB melts: Isotopic evidence from the volcanic province of northern Tanzania. *Earth and Planetary Science Letters*, **130**, 109–126.

PASLICK, C., HALLIDAY, A. N., LANGE, R. A., JAMES, D. & DAWSON, J. B. 1996. Indirect crustal contamination: evidence from isotopic and chemical disequilibria in minerals from alkali basalts and nephelinites from northern Tanzania. *Contributions to Mineralogy and Petrology*, **125**, 277–292.

PETERSON, T. D. 1989*a*. Peralkaline nephelinites. I Comparative petrology of Shombole and Oldoinyo Lengai. *Contributions to Mineralogy and Petrology*, **101**, 458–478.

PETERSON, T. D. 1989*b*. Peralkaline nephelinites. I I Low pressure fractionation and the hypersodic lavas of Oldoinyo Lengai. *Contributions to Mineralogy and Petrology*, **102**, 336–346.

PETERSON, T. D. 1990. Petrology and genesis of natrocarbonatite. *Contributions to Mineralogy and Petrology*, **105**, 143–155.

PICKERING, R. 1958. *Oldoinyo Ogol.* Tanganyika Geological Quarter Degree Sheet **38**.

PICKERING, R. 1964. *Endulen.* Tanzania Geological Survey Quarter Degree Sheet **52**.

PICKERING, R. 1965. *Ngorongoro.* Tanzania Geological Survey Quarter Degree Sheet **53**.

PINKERTON, H., NORTON, G. E., DAWSON, J. B. & PYLE, D. M. 1995. Field observations and measurement of the physical properties of Oldoinyo Lengai alkali carbonate lavas, November 1988. *In*: BELL, K. & KELLER, J. (eds) *Carbonatite Volcanism.* Springer, Berlin, 23–46.

PYLE, D. M. 1995. Decay series evidence for transfer and storage times of natrocarbonatite magma. *In*: BELL, K. & KELLER, J. (eds) *Carbonatite Volcanism.* Springer, Berlin, 124–136.

PYLE, D. M., DAWSON, J. B. & IVANOVICH, M. 1991. Short-lived decay series diequilibria in the natrocarbonatite lavas: constraints on the timing of magma genesis. *Earth and Planetary Science Letters*, **105**, 378–396.

RECK, H. 1924–25. L'Engai Bilder. *Zeitschrift für Vulkanologie*, **8**, 172–174.

RECK, H. & SCHULZE, G. 1921. Ein Beitrag zur Kenntnis des Baues an der jüngsten Veränderung des l'Engai-Vulkanes im nordlichen Deutsch-Ostafrika. *Zeitschrift für Vulkanologie*, **6**, 47–71.

REID, A. M., DONALDSON, C. H., BROWN, R. W., RIDLEY, W. I. & DAWSON, J. B. 1975. Mineral chemistry of peridotite xenoliths from the Lashaine volcano, Tanzania. *Physics and Chemistry of the Earth*, **9**, 525–558.

RICHARD, J. J. 1942. Oldoinyo Lengai – the 1940–41 eruption. Volcanological observations in East Africa. *Journal of the East Africa and Uganda Natural History Society*, **XVI**, 89–108.

ROBERTS, M. A. 2002. *The Geochemical and Volcanological Evolution of the Mt. Meru Region, Northern Tanzania*. PhD thesis, University of Cambridge.

ROCK, N. M. S. 1976. Petrogenetic significance of some new xenolithic alkaline rocks from East Africa. *Mineralogical Magazine*, **40**, 611–625.

ROSENBUSCH, H. 1907. *Mikroskopische Physiographie der Mineralien und Gesteine*, 4th edn. E. Schweizerbart'sche, Stuttgart.

RUDNICK, R. L., MCDONOUGH, W. F. & CHAPPELL, B. W. 1993. Carbonatite metasomatism in the northern Tanzanian mantle: petrographic and geochemical characteristics. *Earth and Planetary Science Letters*, **114**, 463–475.

SAGGERSON, E. P. 1966. *Geology of the Loita Hills*. Geological Survey of Kenya, Report **71**.

SEARLE, R. C. 1971. A gravity survey of Ngorongoro caldera. Tanzania. *Bulletin Volcanologique*, **35**, 350–357.

SELBY, J. & THOMAS, C. M. 1966. *Balangida Lelu*. Tanzania Mineral Resources Division, Quarter Degree Sheet **103**.

SHIVE, P. N., NYBLADE, A. A. & WITTKE, J. H. 1990. Magnetic properties of some carbonatites from Tanzania, East Africa. *Geophysical Journal International*, **163**, 103–109.

SIMONETTI, A., BELL, K. & SHRADY, C. 1997. Trace- and rare-earth geochemistry of the June 1993 natrocarbonatite lavas, Oldoinyo Lengai (Tanzania): Implications for the origin of carbonatite magmas. *Journal of Volcanological and Geothermal Research*, **75**, 89–106.

THOMAS, C. M. 1966. *Hanang*. Tanzania Mineral Resources Division, Quarter Degree Sheet **84**.

THOMPSON, L. G., MOSELY-THOMPSON, E., *ET AL*. 2002. Kilimanjaro ice core record: evidence of Holocene change in tropical Africa. *Science*, **298**, 589–593.

TILLEY, C. E. & YODER, H. S. 1968. The pyroxenite facies conversion of volcanic and subvolcanic melilite-bearing and other alkali ultramafic assemblages. *Carnegie Institution of Washington Yearbook*, **66**, 457–460.

UHLIG, C. 1905. Bericht über die Expedition der Otto-Winter Stiftung nach dem Umgebungen des Meru. *Zeitschrift der Gesellschaft für Erdkund*, **1905**, 120–123.

UHLIG, C. 1907. Der sogennante grosse Ostafrikanische Graben zwischen Magad (Natron See) und Lawa ya Mweri (Manyara See). *Geographisches Zeitung*, **15**, 478–505.

VOIGHT, B., GLICKEN, H., JANDA, R. J. & DOUGLASS, P. M. 1981. Catastrophic rockslide avalanche of May 18. *In*: LIPMAN, P. W. & MULLINEAUX, D. R. (eds) The 1980 eruption of Mount St Helens, Washington. *United States Geological Survey Professional Paper*, **1250**, 347–377.

WAKEFIELD, T. 1870. Routes of native caravans from the coast to the interior of East Africa, chiefly from information given by Sadi Bin Ahedi, a native of a district near Gazi, in Udigo, a little north of Zanzaibar. *Journal of the Royal Geographical Society*, **40**, 303–338.

WILKINSON, P., DOWNIE, C., CATTERMOLE, P. J. & MITCHELL, J. G. 1983. *Arusha*. Geological Survey of Tanzania, Quarter Degree Sheet **55**.

WILKINSON, P., MITCHELL, J. G., CATTERMOLE, P. J. & DOWNIE, C. 1986. Volcanic chronology of the Meru-Kilimanjaro region, Northern Tanzania. *Journal of the Geological Society of London*, **143**, 601–605.

WILLIAMS, L. A. J. 1969. Geochemistry and petrogenesis of the Kilimanjaro volcanic rocks of the Amboseli area, Kenya. *Bulletin Volcanologique*, **33**, 862–888.

WILLIAMS, R. W., GILL, J. B. & BRULAND, K. W. 1986. Ra–Th disequilibria systematics: Timescale of carbonatite magma formation at Oldoinyo Lengai volcano, Tanzania. *Geochimica et Cosmochimica Acta*, **59**, 1249–1259.

WIRTH, K. R. & ADELSBERGER, K. A. 2002. Mantle source and magmatic evolution of the late Pliocene Ogol lavas of Laetoli, Tanzania. *Geological Society of America, Abstracts with Programs*, **34**, 363.

WOOD, C. P. 1968. *A Geochemical Study of East African Alkaline Lavas and its Relevance to the Petrogenesis of Nephelinites*. PhD thesis, University of Leeds.

ZAITSEV, A. N. & KELLER, J. 2006. Mineralogical and chemical transformation of Oldoinyo Lengai natrocarbonatites, Tanzania. *Lithos*, **91**, 191–207.

# Chapter 8

# Regional comparisons, petrochemistry and petrogenesis

In a short review of the structure and volcanism of the southern part of the Gregory Rift, Dawson (1992) drew attention to differences in the contemporaneous volcanism in southern Kenya and northern Tanzania, the differences taking place abruptly at a latitude of *c.* 3°S. In Chapter 3 (Dawson 2008), differences were noted in the relationship of the volcanism to their setting relative to the buried cratonic margin and to the thermal structure of the crust and the upper mantle in the two areas. The composition of the mantle and its effect on the nature of the volcanicity are discussed in greater detail here.

## Comparative contemporary volcanism in south Kenya and north Tanzania

Table 8.1 summarizes the volcanic stratigraphy of southern Kenya and northern Tanzania, mainly from *c.* 8 Ma, i.e. the age of the oldest known activity in northern Tanzania at Essimingor. The period 8–*c.* 1.3 Ma (the time of the Older Extrusives) is dominated by central-type volcanoes characterized by effusion of basanites, basalts and minor trachytes, the last being reasonably interpreted as fractionates of the more voluminous basic rocks. There were some peralkaline volcanoes such as Essimingor, Sadiman and Mosonik, which, particularly in the case of Essimingor, the oldest of the northern Tanzania volcanoes, are similar to the peralkaline volcanoes (Mt Elgon, Napak, Yelele) that erupted prior to the onset of basaltic activity in Kenya (Baker *et al.* 1972).

In southern Kenya during this period, volcanic rocks were erupted from both major central volcanoes or from fissures and very minor volcanoes, the latter including the rift basalts and the Magadi flood trachytes. The Lengitoto trachytes (6.9–5.0 Ma) and the Kirikiti basalts (3.1–2.5 Ma) are exposed along the western flanks of the rift (Crossley 1979), whereas the faults along the eastern flank expose the Singaraini, Ol Keja Nero and Ol Tepesi basalts and benmoreites (Baker & Mitchell 1976). Within the rift valley itself, these earlier formations are overlain by extensively block-faulted trachytes. The central volcanoes active during this period are Ngong and Ol Esayeti (basalt and basanite: 6.7–3.6 Ma), Olorgesaile (trachyte 2.7–2.2 Ma), Lenderut (basanite and tephrite *c.* 2.5–2.2 Ma) and Shombole (peralkaline nephelinite, phonolite carbonatite *c.* 2.0 Ma) (Baker 1958, 1963; Fairhead *et al.* 1972; Baker & Mitchell 1976). The basaltic rocks show some resemblances to those in Tanzania (Appendix 2, Dawson 2008) in terms of the dominance of mildly nepheline-normative basalts with minor quantities of hypersthene-normative basalt (Le Roex *et al.* 2001), but the relatively large amounts of trachyte in Kenya find their equivalents in only minor amounts of trachyte in the Kibangaini trachytes in the very north of Tanzania. Shombole, the most southerly of the Kenya central volcanoes, has more petrological affinities with the almost contemporaneous Mosonik and later carbonatite volcanoes of Tanzania.

In the period *c.* 1.2 Ma to the Holocene, following major faulting, the main activity in the southern part of the Kenya rift valley was fissure eruptions of trachyte and, further north, the formation of the major trachytic caldera volcanoes of Suswa, Longonot and Menengai (Williams *et al.* 1984). In stark petrological contrast, in northern Tanzania, the contemporaneous activity was the formation of peralkaline nephelinite-phonolite-carbonatite volcanoes and the Holocene eruption of small volumes of olivine nephelinite and melilitite, some of katungitic composition.

To the east of the rift valley in southern Kenya, during the late Pleistocene–Holocene a linear NW–SE array of small volcanoes erupted to form the Chyulu Range. The activity started at its north-western end with eruption of nephelinite but then extruded rocks that become less silica-undersaturated as the activity progressed southeastwards, eventually producing hypersthere-normative basalts (Späth *et al.* 2000, 2001). The nearest contemporary volcanism in northern Tanzania is the phonolites and nephelinites of Meru, the olivine melilitite activity in the Meru–Monduli area and the highly evolved trachytes and phonolites of the Kibo centre on Kilimanjaro. As in the rift valley further west, there are again significant differences between the contemporary rocks of the Chyulu Volcanic Province and those only a short distance away in northern Tanzania.

Many of the southern Kenya volcanics are believed to have erupted through the buried margin of the Tanzania craton (Smith & Mosley 1993), and those in northern Tanzania through the craton and/or the craton margin. Hence the differences noted above cannot be attributed solely to differences in tectonic setting.

## Petrochemistry

Most volcanoes of the northern Tanzania province have been mapped only on a regional scale, and the sampling of most, and the ensuing chemical data (available for only a few volcanoes), are neither extensive nor systematic; further, most lava suites lack isotopic data. The following commentary should be viewed within these constraints.

The composition of igneous rocks is determined by the composition of the source, the degree and depth of melting, any possible effects of fractionation and wall-rock contamination during ascent from the source area, and any magma mixing. The intraplate tectonic setting of the Tanzania volcanic province and the preponderance of mafic lavas strongly suggest that the Tanzanian volcanics are mainly mantle derived.

## The Northern Tanzania upper mantle

The composition of the upper mantle beneath northern Tanzania has been investigated by: (i) direct observation of the composition of phases in mantle xenoliths brought to the surface, mainly during Younger Extrusive olivine melilitite and olivine nephelinite activity; and (ii) by interpretation of both Older Extrusive and Younger Extrusive lavas.

### *Evidence from mantle xenoliths for mantle heterogeneity*

Xenoliths from the Oldoinyo Lengai, Lashaine, Olmani, Monduli, Labait, Pello Hill, Eledoi, Deeti and Igwisi volcanoes provide direct evidence for the make-up of the upper mantle beneath northern Tanzania. However, most have been brought to the surface during the later stages of the Younger Extrusive activity, so caution must be applied to acceptance of the xenoliths as being representative of the upper mantle at the onset of the Older Extrusive magmatism some 8 Ma earlier.

J. B. Dawson, University of Edinburgh (*jbdawson@glg.ed.ac.uk*)
*From*: DAWSON, J. B. 2008. *The Gregory Rift Valley and Neogene–Recent Volcanoes of Northern Tanzania*. Geological Society, London, Memoirs, **33**, 79–89.
0435-4052/08/$15.00 © The Geological Society of London 2008. DOI: 10.1144/M33.8

**Table 8.1.** *Volcanic stratigraphy of S. Kenya and N. Tanzania*

| | South Kenya | | | | | North Tanzania |
|---|---|---|---|---|---|---|
| | **West flank** | **West rift floor** | **East rift floor** | **East flank** | **Outside rift** | |
| 1 Ma | | Nyokie ignimbrites (0.65) | Suswa volcanics (0–0.4)<br>Barajai trachytes (0.4) | | Chiyulu Hills Nephelinites basanites, basalts | **Younger Extrusives** Nephelinites, olivine melilitites, phonolites, carbonatites<br>*MAJOR FAULTING* |
| | | Magadi trachytes (1.4–0.7) | Plateau trachytes (1.9–0.9) Ol Tepesi basalts & trachytes (1.65–1.4) Ol Kejo Nero basalts | | | **Older Extrusives** Mainly alkali basalts, trachytes, phonolites Some alkaline ultrabasics |
| | *FAULTING* | | | *FAULTING* | | |
| 2 Ma | Sambu Volcanics | Mosiro trachytes 2,1 Shombole (1.9–2.0) Kordiya basalts | Limuru trachytes (2.0–1.9) Singaraini basalts | Limuru trachytes | | Tarosero (2.0–2.2) Shombole (2.0–1.9) |
| 5 Ma | Kirikiti basalts (2.5–3) | | Olorgesaile volcanics (2.2–2.7) | | | Sadiman + Lemagrut (~3–4) |
| | Lengitoto trachytes (5.1–6.9) | | Ol Esayeiti volcanics (6.7–3.2) | Middle trachytes | | |
| | | | | Lower trachytes 6.0 | | |
| | Olivine nephelinites (12–15.2) | | | Kapiti phonolite (13.4) | | Essimingor 3.2–8.1 |

The sub-Tanzania upper mantle, as exemplified by the xenoliths, has had a long and complex history. The Lashaine xenoliths were subject to strong depletion at around 3.2 Ga (Burton *et al.* 2000), whereas at Labait chromites in the xenoliths give Re depletion ages of 2.8 and 2.0 Ga (Chesley *et al.* 1999). At Olmani, although not dated, the main peridotite mineralogy is highly refractory. The depletion event, which gave rise to chromite dunite and harzburgite restites in all of the xenolith suites, was followed by at least one episode of metasomatism, with varying degrees of enrichment of the restite protoliths. At Lashaine (for location see Dawson 2008, fig. 5.8) the first enrichment event was at *c.* 2.0 Ga, when K, Fe, Ca, Ti and Rb metasomatism resulted in the formation of diopside and phlogopite (Dawson *et al.* 1970; Ridley & Dawson 1975). Si metasomatism may have caused the formation of enstatite (Rudnick *et al.* 1994), and it is possible that garnet formed at the same time by exsolution from enstatite combined with REE metasomatism, as has been envisaged for garnet formation in Kaapvaal mantle peridotites (Simon *et al.* 2003). A later metasomatic event added K, Fe, Ca, Ti, Nb, Ta, Sr, Ba, the REE and Zr (Dawson 2002); this second pulse of metasomatism, manifest in mantled and unequilibrated minerals, must have been relatively recent in geological terms. Similar two-stage metasomatism events also took place at Labait (Dawson 1999). At other localities there are distinctive trace-element and volatile-component metasomatic signatures to the xenolith suite, particularly with regard to the relative additions of Ba, Ti and K (Dawson 2002). The presence of primary calcite in xenoliths from Lashaine, Monduli and Igwisi, together with the formation of diopside and monazite in xenoliths at Labait, Olmani and Eledoi, give grounds for believing that the metasomatic fluids at these localities were carbonatitic (Dawson 1987; Rudnick *et al.* 1993; Lee *et al.* 2000). In contrast, at Pello Hill the metasomatic fluids also contained water (forming amphibole and mica) and the $XCO_2/H_2O$ changed during the metasomatism of the wall rock by veins of katungitic composition (Dawson & Smith 1988).

In summary, the sub-Tanzanian upper mantle lithosphere was depleted in the Archaean but has subsequently been subject, at least locally, to variable types and degrees of incompatible-element and volatile enrichment, at different times: one in the Early Proterozoic, the other very recent.

*Xenolith isotope variations*

Diopsides in veined and metasomatized xenoliths from Eledoi and Pello Hill have $^{87}Sr/^{86}Sr$ ratios that range mainly from 0.70344 to 0.70473 (Fig. 8.1), indicating recent metasomatism, as do diopsides from the Olmani peridotites (Rudnick *et al.* 1993). However, other xenoliths from the Oldoinyo Lengai area show considerable isotopic variation. Contrasting with the Pello Hill and Eledoi samples, two blocks of mica pyroxenite (interpreted as metasomatized peridotite, Dawson & Smith 1992) in the agglomerates at nearby Oldoinyo Lengai have $^{87}Sr/^{86}Sr$ ratios of 0.7086 and 0.70529 and $^{143}Nd/^{144}Nd$ 0.5117 and 0.5124 respectively (Bell & Dawson 1995), indicating an older metasomatism event. On a Sr v. Nd isotope diagram (Fig. 7.12, 8.1), one of these samples plots in a more enriched sector than the EM1 values of Hart (1988). Most peridotite samples from Lashaine mainly have Sr and Nd isotope value close to the Pello and Eledoi samples, but one garnet lherzolite (BD738) contains high amounts of radiogenic Sr, Nd and Pb. Its ratios of $^{87}Sr/^{86}Sr$ 0.83604, $^{206}Pb/^{204}Pb$ 15.39 and $^{143}Nd/^{144}Nd$ 0.51127 are the most extreme for any sample of mantle derivation, indicating old enriched lithosphere (Cohen *et al.* 1984) and, again, emphazing the variable metasomatic enrichment of the sub-Tanzanian mantle. The important fact is that the timing of the metasomatism was bimodal. Further, whereas the Pello, Eledoi, Olmani and some Lashaine xenoliths indicate young enrichment, the Oldoinyo Lengai and the extreme Lashaine xenoliths were subject to relatively old metasomatism. It is also apparent from Figure 8.1 that the Sr and Nd isotope values of the olivine melilitites in the Oldoinyo Lengai area overlap those of the recently enriched mantle xenoliths from Pello Hill and Eledoi.

Trace element contents in veins, which do not represent primary melts, cutting some of the Pello Hill peridotite xenoliths show high concentrations of Rb, Pb and Ba, but lower Zr and REE relative to ocean island basalt (Table 8.2).

Another example of mantle heterogeneity is shown by the isotope chemistry of carbonatites from Oldoinyo Lengai and Kerimasi. These two volcanoes are only 10 km apart and both are $<1$ Ma old, yet the carbonatites at these two centres are isotopically different (Fig. 8.1). The $^{87}Sr/^{86}Sr$ and $^{143}Nd/^{144}Nd$ ratios for Oldoinyo Lengai carbonatite are 0.70443 to 0.70437 and

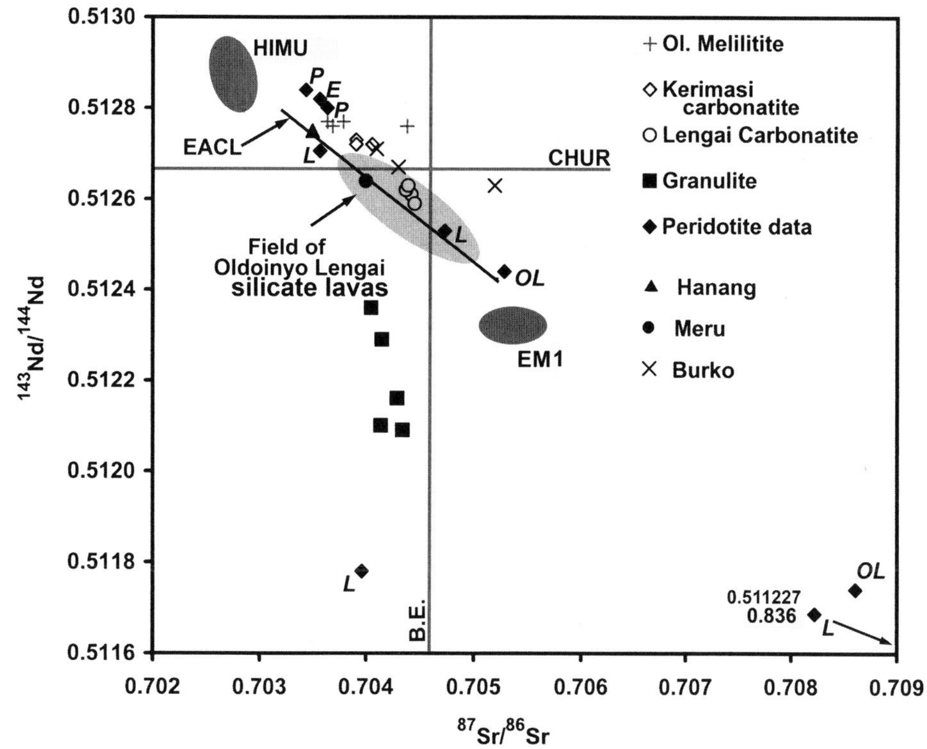


**Fig. 8.1.** Nd v. Sr isotope plot for Younger Extrusive lavas including Oldoinyo Lengai lavas and mantle xenoliths (see Dawson 2008, fig. 7.12), other North Tanzania mantle xenoliths (most data are for peridotite clinopyroxenes) and basic granulites (Cohen *et al.* 1984). Other Younger Extrusive rocks are olivine melilitites (Keller *et al.* 2006, Cohen *et al.* 1984), Kerimasi calcic carbonatites (Paslick *et al.* 1995, Kalt *et al.* 1997) and nephelinites from Hanang, Meru and Burko (Paslick *et al.* 1995). Abbreviations as in Figure 7.12; others P, Pello Hill; E, Eledoi; L, Lashaine; OL, Oldoinyo Lengai. Clinopyroxenes from Olmani peridotite xenoliths (Rudnick *et al.* 1993) plot in the same area as the Pello Hill and Eledoi samples.

0.51263 to 0.51270, respectively (Bell & Dawson 1995), compared with an average of 0.70396 and 0.51273 for five samples from Kerimasi (Paslick *et al.* 1995; Kalt *et al.* 1997). Hence, the underlying lithospheric mantle is isotopically heterogeneous over very small lateral distances.

Some Tanzanian xenoliths analysed for radiogenic isotopes have also been analysed for helium. Apparently contrary to the variability of both the metasomatism and the radiogenic isotopes, the few analysed xenoliths from Lashaine, Eledoi and Pello Hill contain mainly helium with quite constant $^{3}He/^{4}He$ ratios, with $R/R_A$ (the $^{3}He/^{4}He$ ratio normalized to that in atmospheric helium) in the range 6.2 to 7.3; separated olivines give slightly higher values than their host bulk rocks (Porcelli *et al.* 1986). In olivines in one sample (BD3847 from Pello Hill), Stuart & Turner (1992) found abundant $CO_2$-rich inclusions that yielded $R/R_A$ of 8, and also a $^{40}Ar-^{4}He$ value of 0.7, which is close to the value of 0.55 for these isotopes in the upper mantle. The values for helium $R/R_A$ are for undoubtedly metasomatized lithosphere, and most are similar to those in some ocean island basalts that are interpreted as being generated from metasomatized oceanic lithosphere e.g. those from La Grille, Grand Comore ($R/R_A$ 6.75–7.08) but higher than in the plume-derived Comore Karthala volcanics with $R/R_A$ 5–5.41 (Class *et al.* 2005). Most Tanzania xenolith data also overlap $R/R_A$ values of 6.15–6.59 in primitive picrites from Turkana, which are suggested as representing the values of an asthenospheric plume (Furman *et al.* 2006*a*). The conflicting interpretations of the source regions of the mafic rocks from Comore and Turkana, despite the similarities of their helium isotope values, indicates a lack of agreement on the use of helium isotope values for interpreting plume versus lithosphere derivation of basalts. A departure from the generally relatively narrow range for the Tanzanian xenoliths are $R/R_A$ ratios of 13.6 to 15.3 found in the exceptional anciently-enriched Lashaine garnet lherzolite BD738 (see above) (Porcelli *et al.* 1986).

### *The Labait Fe-rich peridotites*

In addition to metasomatized Mg-rich peridotites, the xenolith suite at Labait contains dunite and wehrlite in which the whole-rock and phase compositions are more iron-rich. On the basis of mineral chemistry, and particularly of spinel, Dawson (1999) suggested that these are cumulates from a ponded picritic melt that crystallized at a depth of *c.* 60 km, whereas Lee & Rudnick (1999) proposed that they are due to Fe-metasomatism of Mg-peridotites by asthenospheric melts.

### *Lava compositions*

The composition of the mantle source can also be assessed by study of primary mantle melts, i.e. those in chemical equilibrium with mantle olivine of around $Fo_{90}$. A $Mg/(Mg+Fe)$ ratio of $>0.70$ (*mg*# 70) is taken to be an indication of the required composition (Roeder & Emslie 1970), together with high Ni contents ($>250$ ppm). Table 8.2 lists lavas from northern Tanzania, some of which fulfil these criteria, and other primitive lavas with *mg*# $(Mg/Mg+Fe)>65$ for comparison with lavas of similar composition from Kenya. The table gives selected major element and trace element data for these lavas, and comparative data for reference samples. Although most of the lavas contain U and Zr values similar to those in the reference ocean island basalt (OIB), the Ba, Nb, La, Ce and Th contents of the Tanzanian lavas are higher than in OIB, particularly those in the Younger Extrusives. Further, although some trace element ratios (Ce/Pb and Nb/U) bracket those in OIB, many Ba/Nb, La/Nb and Rb/Ba ratios are higher, supporting the observation of Paslick *et al.* (1995) that the trace element distribution in the northern Tanzanian lavas requires a mantle source that has a partial, but not exclusively, OIB trace element pattern. Similarly, REE inversion modelling for regional alkali basalts in the Meru–Kilimanjaro

**Table 8.2.** *Chemical parameters of Tanzania primary* (mg# > 70) *and primitive* (mg# > 65) *lavas, a Tanzania mantle vein, and some standard reference lithologies*[‡]

| Sample | Rock type | Location | CaO | $Al_2O_3$ | $CaO/Al_2O_3$ | *mg#* | Zr | Th | U | Rb | Ba | Nb | La | Ce | Pb | Ba/Nb | La/Nb | Rb/Ba | Ce/Pb | Nb/U | Reference |
|---|---|---|---|---|---|---|---|---|---|---|---|---|---|---|---|---|---|---|---|---|---|
| BD333* | Ol basalt | Ketumbeine | 8.55 | 11.19 | 0.8 | 68.9 | 198 | 4.0 | na | 23 | 477 | 45 | 45 | 127 | 5.0 | 11 | | 0.05 | 25.4 | | 1 |
| MD93–4* | Picrobasalt | Monduli | 10.04 | 9.59 | 1.1 | 74.4 | 242 | 11.0 | na | 34 | 677 | 88 | 68 | 130 | 3.0 | 7.7 | 0.8 | 0.05 | 43.3 | | 1 |
| MR65* | Basanite | Loljoro | 11.06 | 9.78 | 1.1 | 67.4 | 155 | 10.9 | 1.8 | 18.2 | 475 | 41 | na | na | 14.7 | 11.6 | | 0.04 | | 22.8 | 2 |
| MR105* | Basalt | Lembolos | 10.78 | 10.64 | 1.0 | 65.4 | 122 | 12.2 | 0.7 | 14.7 | 375 | 33 | 34 | 67.1 | 2.7 | 11.4 | 1.03 | 0.04 | 24.9 | 47.1 | 2 |
| MR110* | Basalt | Oldoinyo Liyu | 9.58 | 11.13 | 0.9 | 65.3 | 175 | 14.1 | 1.1 | 19.8 | 567 | 57 | 53 | 108.7 | 3.7 | 9.9 | 0.9 | 0.03 | 29.4 | 51.8 | 2 |
| BD780 | Ol melilitite | Lashaine | 12.24 | 5.84 | 2.1 | 72.8 | na | na | na | na | na | na | na | na | na | | | | | | 3 |
| BD105 | Ol melilitite nephelinite | Loolmurwak | 12.97 | 6.93 | 1.9 | 70.9 | 326 | na | na | 48 | 617 | 97 | na | na | na | 6.4 | | 0.08 | | | 4 |
| BD4200 | Ol melilitite | Labait | 8.87 | 5.43 | 1.6 | 80.4 | 188 | na | na | 65.6 | 695 | 97 | 47 | 82.0 | 5.5 | 7.2 | 0.5 | 0.09 | 14.9 | | 5 |
| BD4225 | Katungite | Sora Hill | 13.70 | 7.67 | 1.8 | 69.3 | 207 | na | na | 80.2 | 1960 | 107 | 117 | 150 | 6.0 | 18.3 | 1.1 | 0.04 | 25.0 | | 5 |
| MR281 | Ol melilitite | Meru | 13.27 | 7.83 | 1.7 | 69.2 | 402 | 14.0 | 2.3 | 57.1 | 855 | 147 | 104 | 203 | 5.2 | 5.8 | 0.7 | 0.07 | 39.3 | 63.9 | 2 |
| | Ankaramite | Olmani | 11.30 | 7.72 | 1.5 | 75.0 | na | na | na | na | na | na | na | na | na | | | | | | 6 |
| OL198 | Ol melilitite | Oldoinyo Lengai | 14.26 | 7.52 | 1.9 | 68.7 | 370 | 11.3 | 3.0 | 59.6 | 956 | 123 | na | na | na | 7.8 | | 0.06 | | 41.0 | 7 |
| Mica-pyroxene vein in mantle xenolith[†] | | Pello Hill | 10.89 | 6.69 | 1.6 | 50.1 | 164 | na | na | 81 | 679 | 53 | 5.5 | 15.3 | 5 | 12.8 | 0.10 | 0.12 | | | 8 |
| Ocean island basalt | | | | | | | 280 | 4.0 | 1.0 | 2.31 | 350 | 48 | 37 | 80.0 | 3.2 | 7.3 | 0.77 | 0.006 | 28.1 | 48.0 | 9 |
| Lower crust | | | | | | | 68 | 1.2 | 0.2 | 11 | 259 | 5 | 8 | 20 | 4 | 51.8 | 1.6 | 0.04 | 5.0 | 25.0 | 10 |
| Upper crust | | | | | | | 193 | 10.5 | 2.7 | 8.2 | 628 | 12 | 31 | 63 | 17 | 52.3 | 2.6 | 0.01 | 3.7 | 4.4 | 10 |

na, not analysed.

*Older Extrusive Series; the rest are Younger Extrusives.

†BD3834B Mica pyroxenite vein in upper mantle peridotite.

References: 1, Paslick *et al.* 1995; 2, Roberts 2002; 3, Dawson *et al.* 1970; 4, Dawson *et al.* 1985; 5, Dawson *et al.* 1997; 6, Jones *et al.* 1983; 7, Keller *et al.* 2006; 8, Dawson & Smith 1988; 9, Sun & McDonough 1989; 10, Rudnick & Gao 2003.

‡*mg#*: (Mg/Mg + Fe).

region shows that whereas their HREE patterns can be produced from primitive mantle by *c.* 5% melting over a depth range 85–105 km at a $T_p$ of *c.* 1500 °C, this type of mantle composition cannot produce the enriched LREE and MREE concentrations in the basalts; hence, an enriched lithospheric source is required for at least the LREE and MREE (Roberts 2002).

A Sr v. Nd isotope plot (Fig. 8.2) for the Tanzanian Older Extrusives (data from Paslick *et al.* 1995) shows that: (i) samples from two of the peralkaline, undersaturated volcanoes (Sadiman and Essimingor) have the most radiogenic Sr and Nd and plot close to EMI. In this they are similar to the sample from Lemagrut which initially erupted peralkaline lavas and also erupted coevally with Sadiman; (ii) other Essimingor samples are quite different, one having similar Sr but lower radiogenic Nd, whilst a third has lower Sr and Nd, resulting in a very wide spread for Essimingor in Sr–Nd isotope space; (iii) basalt and picrobasalt samples from the rift wall near Lake Natron fall into two distinct groups, having similar $^{143}Nd/^{144}Nd$, but distinctly different $^{87}Sr/^{86}Sr$; (iv) the rift wall group with the least radiogenic Sr plots, together with most specimens from Ketumbeine and Monduli intermediate along a broad zone extending from EMI to HIMU values though closer to HIMU; and (v) two primitive lavas (*mg#* >65) from Monduli and Ketumbeine are amongst those having the least radiogenic Sr.

In total, the Nd–Sr isotope data for northern Tanzanian lavas, together those for Pb (Paslick *et al.* 1995), converge on a mantle source that, compared with HIMU, has relatively unradiogenic Nd, radiogenic Sr, and very radiogenic Pb. Unlike HIMU sources which need an enrichment in U and Th (or depletion in Pb) without an accompanying enrichment in Rb, the Tanzania mantle source must have been enriched in Rb relative to Sr, U relative to Pb, and also enriched in LREEs.

Paslick *et al.* (1995) explain these parameters as being due to underplating at around 2 Ga of the continental lithosphere by an OIB-like melt that could develop its radiogenic Pb signature before the underplated material was melted to give the Tanzanian lavas. However, this cannot be the sole source as, compared with values for OIB, the concentrations of incompatible elements such as Ba, Th, U, Nb and Zr are enriched in most Tanzanian Older Extrusive primary and primitive lavas, and especially so in the Younger Extrusive primary lavas (Table 8.2). These data, corroborated by the xenolith evidence, indicate an anciently enriched lithospheric component in the source region of the Tanzanian primary lavas. The even higher concentrations in other lavas can be explained as due to fractionation.

## Effects of crustal contamination and fractional crystallization

In Chapter 3 (Dawson 2008), it was suggested that most of the Tanzanian volcanoes erupted through old cratonic lithosphere overlain by only a thin veneer of mobile-belt rocks. It was further suggested that the potential for crustal contamination of the magmas during ascent must be considered, together with possible compositional variations due to fractional crystallization.

The silica-undersaturated character of the mafic magmas argues against significant assimilation of high-silica crustal material although, from both seismic (see Chapter 4) and xenolith evidence, the Tanzanian lower crust is believed to be of basic composition, the assimilation of which would not show up as high $SiO_2$. To date two studies have been made on Tanzanian volcanic rocks to assess the possibility of contamination, using the AFC (assimilation, fractional crystallization) model of De Paolo (1981). On a range of basalts, Paslick *et al.* (1995) showed that variations in MgO point to some olivine and/or pyroxene fractionation although there is no correlation between $^{87}Sr/^{86}Sr$ and Mg and Sr concentrations. Further, Paslick *et al.* (1995) modelled assimilation and fractional crystallization using Nd and Pb isotope data for a mixture of an uncontaminated magma sample and lower crustal granulite from Lashaine (Cohen *et al.* 1984); the conclusion was that some lavas could, theoretically, have been contaminated with lower crustal material. This was also indicated by disequilibrium between phenocryst and whole-rock Sr, Nd and Pb isotope compositions in a variety of basalts and nephelinites in which some (but not all) 'phenocrysts' of olivine, pyroxene, nepheline and plagioclase appear to be xenocrysts. As these xenocrysts might be expected to be of crustal derivation, and the disequilibria cannot be due to mantle variations, it is possible that the xenocrysts were derived from earlier magma that stagnated or crystallized at depth, where they were contaminated by deep crustal material, before later entrainment into batches of magma that erupted at the surface. Although such disequilbrium can have potentially serious implications when using whole-rock isotopic data to identify source reservoirs for volcanic rocks,

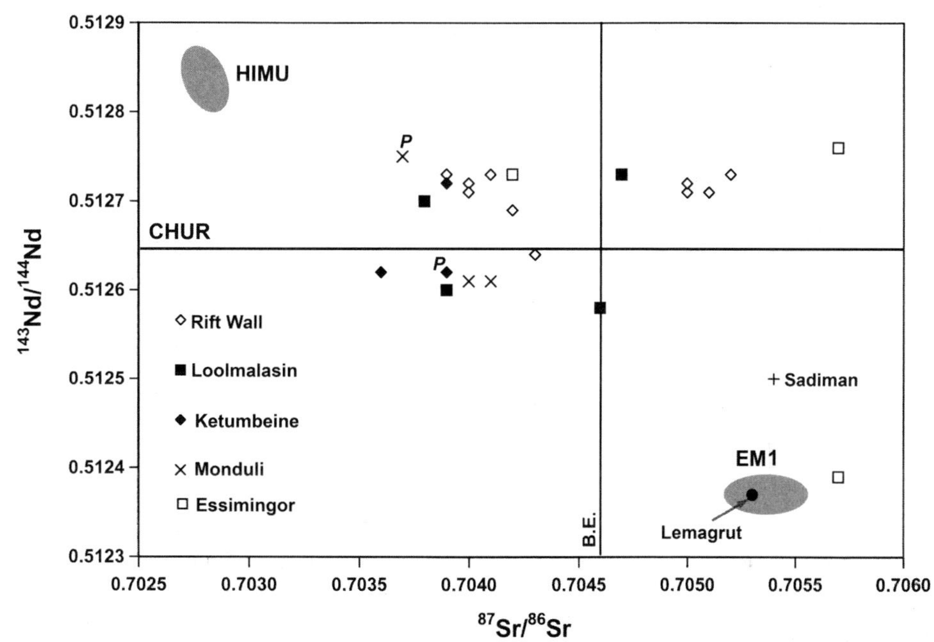


**Fig. 8.2.** Sr–Nd isotope plot for Older Extrusive lavas (data from Paslick *et al.* 1995). 'P' indicates primary or primitive lava (from Table 8.2).

none the less mass balance calculations show that $^{206}Pb/^{204}Pb$ isotope compositions of the Tanzanian lavas would have been changed by only <0.25% by this indirect lower crustal contamination (Paslick *et al.* 1996). Some samples, particularly highly silica-undersaturated lavas, contain Pb that is more highly radiogenic than in any known upper crustal rocks in Tanzania which suggests that the Pb is a feature of the mantle source rather than of crustal contamination. In the lavas listed in Table 8.2, Ba, Th, La and U concentrations are similar to, or bracket, those for the upper crust, but values for Nb, Ce, Rb are much higher, giving rise to distinctively high Rb/Ba, Ce/Pb and Nb/U ratios which are more similar to those for OIB; on these parameters, the cited lavas show no evidence of crustal contamination.

In a study using the Sr and Nd isotopic systematics of three lava compositions, one a nephelinite from Oldoinyo Lengai, and various crustal rocks (including Archaean rocks from the Tanzania craton and an average Tanzanian granulite), Bell & Dawson (1995) found that the spread of isotope date for the lavas at Oldoinyo Lengai could theoretically be modelled using a combination of granulite and nephelinite. However, the mafic melt would need to be of extreme perakalinity and the model was rejected in favour of one involving mixing between a young asthenospheric melt and old enriched lithosphere. This model was reinforced by data from later Pb isotope investigations (Bell & Simonetti 1996; Bell & Tilton 2001).

In summary, the Tanzanian basic lavas show no evidence of upper crustal contamination, but some show theoretical evidence for a very small amount of lower crustal contamination, on the assumption that the contaminating material has the same isotopic characteristics as the Lashaine granulites used in the modelling.

## Petrogenesis

### *Basalt-trachyte volcanoes*

In northern Tanzania, basalts of the Older Extrusives volcanoes form the greatest volume of extruded material, and the associated trachytes and phonolites are volumetrically small. Older Extrusive basalts and basanites from Paslick *et al.* (1995, 1996), together with the Kilimanjaro data of Downie & Wilkinson (1972) and other analyses shown in Appendix 2 (Dawson 2008), show a range through nephelinites, nepheline-normative picrobasalts, basanites and basalts. Fractionation can give rise to minor *hy*-normative basalt, hawaiite and benmoreite and, eventually, to both *ne*- and *qz*-normative trachytes (Fig. 8.3).

Although differing in the relative proportions (see above), the erupted basic rock types in Tanzania and southern Kenya show a similar petrological spectrum and the Tanzanian basalts may be viewed within the context of the basaltic rocks erupted within the overall East African rift system. These have been the subject of considerable research and various factors have been considered as relevant to their origin, including the relative input from convecting asthenospheric mantle, lithospheric mantle and continental crust, and possible connections with plume activity (e.g. Rogers *et al.* 2000; Macdonald *et al.* 2001; Macdonald 2003; Rogers 2006; Furman *et al.* 2006*a*, *b*).

From trace element evidence, Latin *et al.* (1993) and Macdonald *et al.* (2001) concluded that most Kenyan mafic magmas originated in a mantle melt zone extending across the spinel- to garnet-peridotite boundary. *Ne*-normative lavas with higher Ce/Y and lower Zr/Nb ratios originated at greater depths than *hy*- and *qz*-normative mafic rocks that have lower Ce/Y, higher Zr/Nb ratios and different Sr, Nd and lead isotope values. Moreover, Macdonald *et al.* (2001) recognize a correlation between the silica saturation of the magmas, their depth of origin and their tectonic setting. The least silica-saturated were generated at greater depth and erupted in the section of the rift underlain by the Tanzania Craton and its reworked margin, whereas *hy*-normative basalts formed at shallower depths and erupted in the parts of the rift underlain by the Mozambique Belt. The sequence nephelinite → basanite → alkali basalt → *hy*-normative basalt represents progressively shallower melting (i.e. increased spinel-peridotite component) rather than increased degrees of partial melting. This depth/silica-saturation relationship is shown by major element data, namely the $Al_2O_3$ content and the $CaO/Al_2O_3$ ratio, both of which are strongly pressure-dependent;

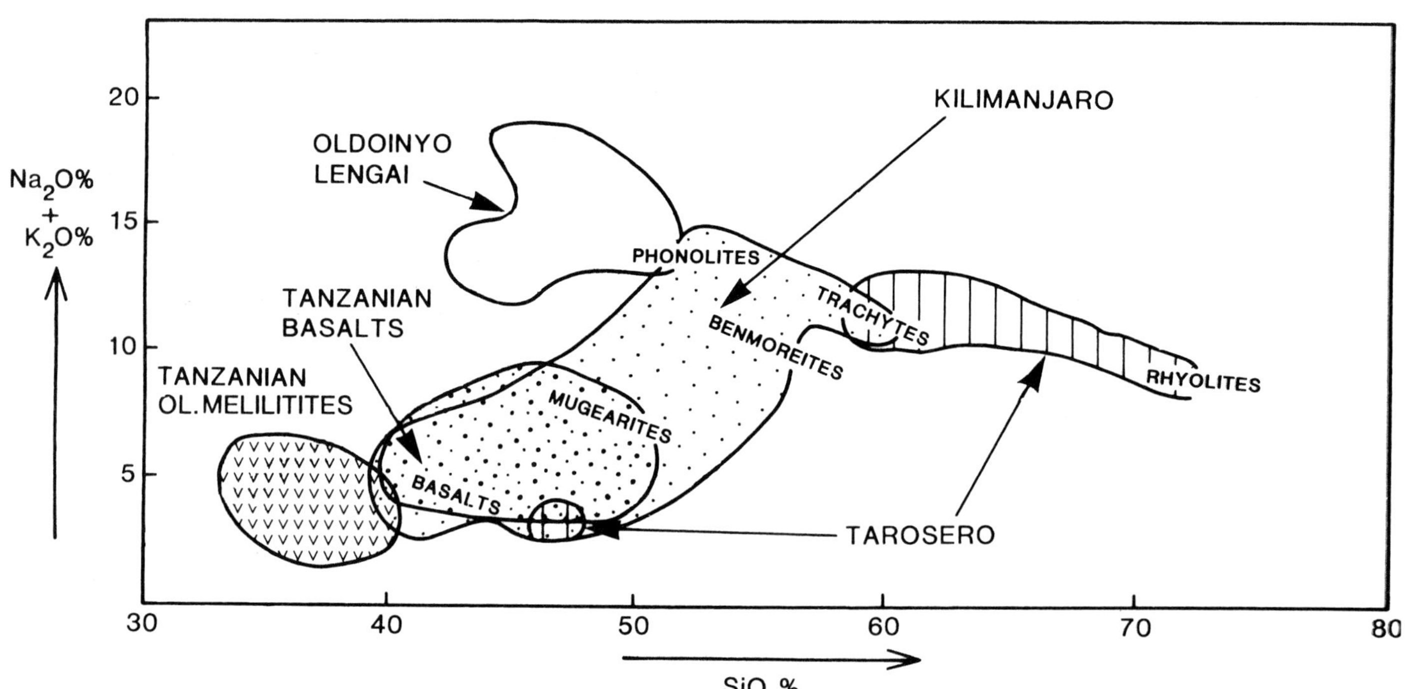


**Fig. 8.3.** Alkali-silica diagram for lavas from northern Tanzania. Data sources: Downie & Wilkinson (1972), Cooper (1972), Dawson *et al.* (1985), Donaldson *et al.* (1987), Paslick *et al.* (1995) and Appendix 2 (Dawson 2008).

the ratio inceases with increasing melting/equilibration pressure, whilst $Al_2O_3$ decreases (Herzberg 1995). Figure 8.4 is an $Al_2O_3$ v. $CaO/Al_2O_3$ plot, modified after Macdonald *et al.* (2001) for mafic rocks from the axial part of the Kenya rift valley with *mg#* (Mg/Mg + Fe) > 65 and therefore potential primary melts. The data set includes some analyses from northern Tanzania (Paslick *et al.* 1995), and additional data for Tanzanian lavas from Table 8.2 are also plotted. As in the case of the Kenyan lavas, the most silica-undersaturated Tanzanian lavas (olivine melilitites and nephelinites) have the highest ratios and lowest $Al_2O_3$ values, and derive from the greatest depth/pressure. The basalts have lower ratios and higher $Al_2O_3$ and are interpreted as being formed at lower pressure/higher levels. The overall trend of the lava types from high to low pressures is immediately apparent, but most data points for the natural rocks do not plot on the line derived for the experimentally derived solidus liquids; most have higher CaO for a given $Al_2O_3$ value, a possible explanation for which is that the natural mantle source rocks are more calcic than the fertile peridotite composition used in Herzberg's experiments.

Overall, the Tanzanian rocks, and the olivine melilitites in particular, derive from greater depths than their Kenyan equivalents. This could be equated with their generation beneath the Tanzanian craton, rather than under the thinned craton margin, and hence deeper in the lithosphere.

A model for the Chyulu Hills Volcanic Province (CHVP), where there is a progression from early minor nephelinites through more extensive basalts to the most recent *hy*-normative minor basalts, is that the lavas result from conductive melting of metasomatized lithospheric mantle. In this model, the early nephelinites result from deep, relatively small amounts of melting, whilst a rise of the thermal source higher into the lithosphere results in greater degrees of melting at shallower depths to produce the more extensive, less silica-undersaturated basalts (Späth *et al.* 2001). Such a model might be applicable to certain Tanzanian volcanoes such as Oldoinyo Sambu and Lemagrut, where the dominantly basaltic activity is preceded by minor nephelinitic magmatism, and hence has some analogies with CHVP.

Although this model, at first sight, might not appear to be applicable to the main basaltic-trachytic volcanoes, such as those of the Crater Highlands, Ketumbeine and the Kilimanjaro massif, where there are rocks intermediate in composition between the basic rocks and trachytes, and also trachytes which have no equivalents in the CHVP, other factors may pertain. First, in the case of most of these central volcanoes, the earliest flows are not exposed so whether the first extrusives were nephelinites, or not, is unknown; none the less, they are known as the earliest extrusives at Oldoinyo Sambu and Lemagrut. Second, the CHVP results from fissure eruptions so lacks the large, central sub-volcanic magma chambers that might enable fractionation towards more silicic

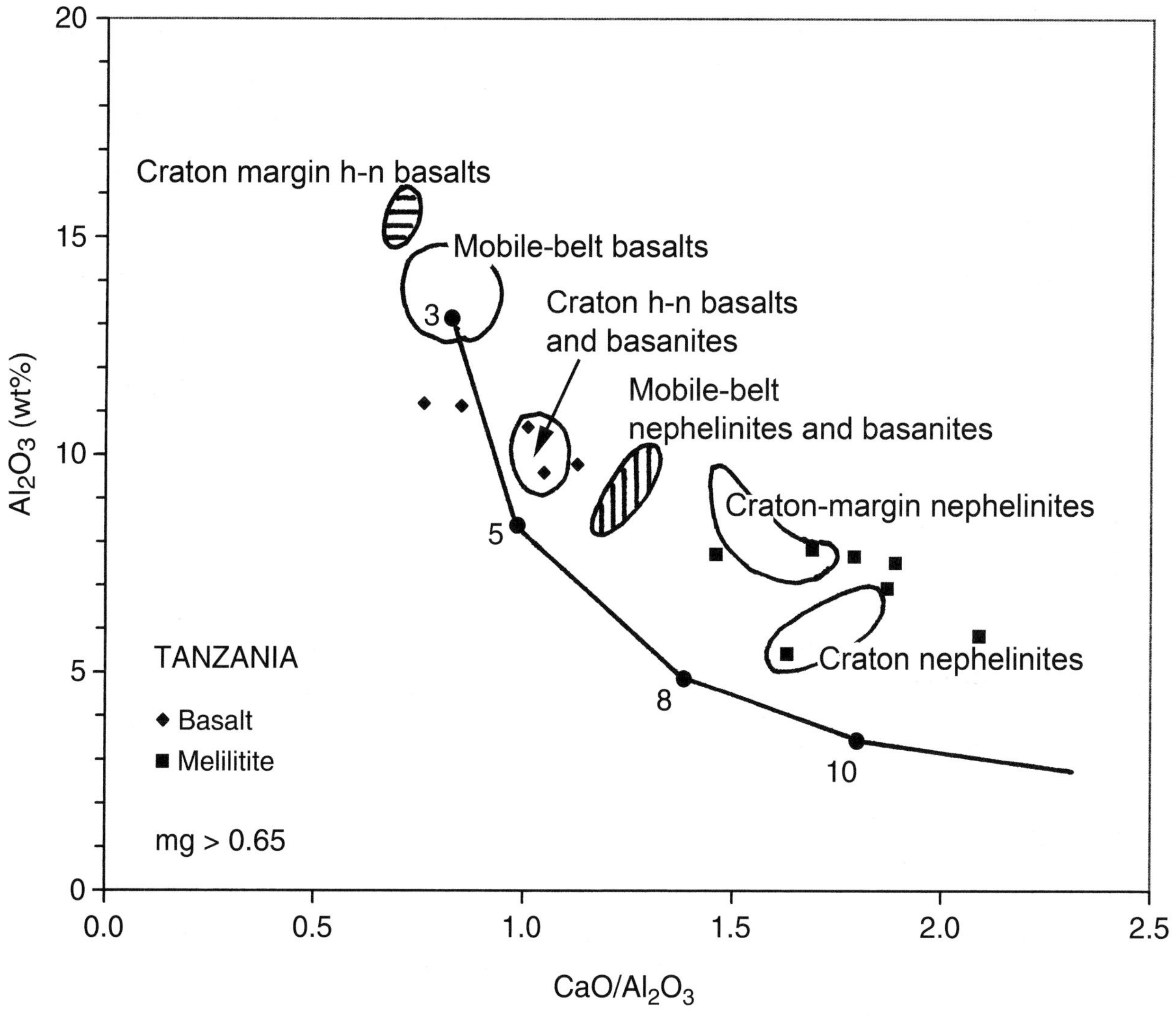


**Fig. 8.4.** $Al_2O_3$ v. $CaO/Al_2O_3$ plot for northern Tanzania lavas inferred to primary or primitive mantle melts ($mg > 0.65$) (data from Table 8.2), compared with similar lavas from different tectonic regimes in Kenya (Macdonald 2003). The solid line with filled circles represents the composition of liquids formed by partial melting of fertile peridotite, the numbers referring to pressure in GPa along the solidus (Herzberg 1995). h-n basalt, hypersthene-normative basalt.

magmas. Third, the CHVP is very young and whether it will develop trachytic magmas is a question for the future. In short, application of the CHVP model to the main basalt-trachyte volcanoes in Tanzania is constrained by lack of knowledge of the detailed eruption sequence at individual volcanoes. However, if a composite or overlapping eruption sequence such as Oldoinyo Sambu plus Ngorongoro were accepted (nephelinite/melilitite → *ne*-normative basalt → *hy*-normative basalt → mugearite → trachyte), the CHVP model is a plausible starting point, provided that the model is extended to allow fractionation in large magma chambers beneath the central-type volcanoes. Essimingor, Sadiman and Mosonik, among the Older Extrusive centres, can be accommodated within this model if it can be accepted that their highly alkaline rocks are the first products of the initial thermal pulse coming into contact with earlier-enriched lithospheric mantle.

*Tarosero trachytes and comendites*

Amongst the Older Extrusives, the alkaline and peralkaline trachytes and comendites of Tarosero are unusual. They erupted from a small isolated cone and, unlike the trachytes at volcanoes such as Ngorongoro, are apparently unconnected with a major basalt centre. To date, Tarosero is the only centre in northern Tanzania known to have erupted such highly silicic lava as comendite and, as noted earlier, the peralkaline members contain unusually high concentrations of incompatible elements (Dawson 2008, table 7.1). Within East Africa, such centres are not common, but Tarosero has some similarities to the Olkaria rhyolitic centre near Lake Naivasha, Kenya. It has been suggested that the Olkaria rhyolites were formed by volatile-induced melting of crustal rocks (Macdonald & Scaillet 2006). Possibly relevant for a similar origin is the very high Cl content in the Tarosero Group II rocks, and the Nd and Pb isotopic evidence for the possible presence of some crustal (granulite) component in other Older Extrusives (Paslick *et al.* 1995). However, to invoke such an origin for the Tarosero comendites, it would be necessary to appeal to some highly selective, and regionally unusual, melting scheme.

*The Younger Extrusives: Oldoinyo Lengai as a model*

The Younger Extrusives magmas are silica-undersaturated, alkaline to peralkaline with high concentrations of incompatible elements, and their dominantly pyroclastic eruption style indicates that their magmas were rich in volatiles, a feature emphasized by the presence of carbonatite at most centres.

Sr and Nd isotope data for samples from Younger Extrusive volcanoes are shown on Figure 8.1 which includes data from Oldoinyo Lengai, the best studied of the Younger Extrusive centres. The data show that: (i) whereas lavas from Oldoinyo Lengai form an elongate zone between EM1 and HIMU, data for olivine melilitite, Kerimasi carbonatite, Hanang nephelinite and two Burko nephelinite samples cluster beyond the HIMU-end of the Oldoinyo Lengai grouping, and very close to recently-enriched peridotites; and (ii) a single sample from Meru falls within the Oldoinyo Lengai field and close to the younger Group II nephelinites, reinforcing the Oldoinyo Lengai–Meru mineralogical link mentioned earlier (Dawson 2008, Chapter 7).

In the specific case of Oldoinyo Lengai, although considerable interest has focused on the natrocarbonatite lavas, the principal problem is the genesis of the alkaline nephelinites and phonolites. These lavas have their counterparts at other Younger Extrusive centres, with the exception of the late peralkaline wollastonite–combeite nephelinites that, to date, have been also found only at Meru. Hence any model for Oldoinyo Lengai is probably broadly applicable to other Younger Extrusive volcanoes.

The Oldoinyo Lengai silicate lavas are silica-undersaturated, and contain high concentrations of Na and K, and other incompatible elements. A REE plot (Dawson 2008, fig. 7.11) illustrates high concentrations particularly of the LREE. Although this pattern is generally believed to be typical of highly-fractionated lavas (and could well be correct in the case of the natrocarbonatite relative to the silicate lavas), this cannot be the sole reason, as similar concentrations exist in the unfractionated primary olivine nephelinite and olivine melilitite from Loolmurwak and Lalarasi (for location see Dawson 2008, fig. 7.18). In terms of their Sr, Nd and Pb isotopes, the Oldoinyo Lengai lavas lie close to a line between HIMU and EM1 mantle sources, which has been interpreted as being due to magma with a primitive mantle signature mixing with enriched lithosphere (Bell & Dawson 1995; Bell & Simonetti 1996). This interpretation finds some support from the presence of enriched peridotite blocks within the ejectementa from Oldoinyo Lengai itself (Dawson 2008, fig. 7.12). Bearing in mind that the Group II nephelinites at Oldoinyo Lengai are (together with natrocarbonatite) the latest eruptives, a temporal connotation can be placed on the distribution of the Oldoinyo Lengai lavas along the EM1-HIMU line. The earlier Group I nephelinites and phonolites can be interpreted as resulting from greater involvement of the EM1 component during the early stage of the volcano, whereas declining EM1 component and/or a greater HIMU (primitive magma) component is apparent in the latest eruptives. Whereas Bell & Dawson (1995) take the EM1 component to be metasomatized peridotite, Kramm & Sindern (1998) show that Nd and Sr data for Oldoinyo Lengai fenites plot in a more enriched part of the Sr–Nd isotope diagram, and suggested that the interaction between a primitive magma and the $^{87}$Sr-enriched fenites would have had the same effect. However, interaction with fenites would also have the effect of increasing the amount of $SiO_2$ in the resulting lavas.

The spread in Nd–Sr space for the Oldoinyo Lengai lavas is mirrored in their $\delta^7$ Li isotope ratios which, although ranging from −0.6 to −1.8 in phonolite, through +2.8 to +3.1 for nephelinite, to +3.3 to +5.1 for natrocarbonatite, all fall within mantle values as represented by data for OIB and MORB (Halama *et al.* 2007).

Turning to the carbonatite–silicate rock relationship, although the stable and radiogenic isotopes indicate that, like the silicate rocks, the carbonatite is of mantle origin (Dawson *et al.* 1995), the actual formation of natrocarbonatite is still a matter of debate. Hypotheses range from its being: (i) a primitive mantle-derived melt; (ii) a product of fractionation of a parental alkali-bearing sövite magma (Gittins 1989); (iii) due to fractionation of silicate-bearing natrocarbonatite; and (iv) an immiscible fraction of the associated peralkaline nephelinites (papers in Bell *et al.* 1998). Points of fact are that there is no evidence for sövite at Oldoinyo Lengai (Dawson 1993) whereas globules of natrocarbonatite are present in silicate glasses in Group II lavas (Dawson *et al.*1996).

The controversy at Oldoinyo Lengai reflects the views regarding the origin of carbonatites in general, with some investigators arguing for a direct origin for carbonatite in the mantle (e.g. Harmer & Gittins 1998). The existence of carbonate-rich mantle magmas, such as those of the Benfontein kimberlite sills, South Africa (Dawson & Hawthorne 1973) or the Igwisi Hills (Dawson 1994), is beyond dispute. Others argue for such an origin based on experiments on alkali-enhanced experimental compositions (Wallace & Green 1988; Sweeney *et al.* 1995). In the case of the carbonate-rich rocks that occur as small-volume components in complexes of evolved alkali-rich rocks, other investigators provide good experimental evidence for low-pressure exsolution of the carbonatites from (formerly) carbonated silicate parents (e.g. Wyllie & Lee 1998; Lee & Wyllie 1998).

*Parent magma(s) of the Younger Extrusives*

A common feature of most of the Younger Extrusive volcanoes is the presence in their vicinities of tuff cones or lava flows of olivine nephelinite or melilitite; primary and, hence, potential parental magma types. Dawson *et al.* (1985) noted that olivine melilitite and olivine melilite nephelinite in the Oldoinyo Lengai area are exceptionally alkaline ([Na + K]/Al 1.25 v. 0.75 for South African olivine melilitites) and suggested that the most primitive of these, an olivine melilite nephelinite (*mg#* 0.71, Ni 350 ppm) from Oldoinyo Loolmurwak, was a potential parent magma for the Oldoinyo Lengai lavas. Dawson (1998) further suggested a model whereby such a carbonated magma could evolve to the Oldoinyo Lengai nephelinites by olivine, pyroxene and phlogopite fractionation. This model is similar to that of Peterson & Kjaarsgaard (1995) who preferred olivine melilitite to be the parent magma, with eventual peralkaline nephelinite giving rise to natrocarbonatite by immiscible separation, the 'Lengai Trend'. An extension of this model forms calcic carbonatite by immiscibilty from nephelinites less alkaline than those at Oldoinyo Lengai, the 'Shombole Trend' (Kjaarsgaard *et al.* 1995).

However, other work has shown that models invoking an olivine melilitite or nephelinite parent magma have certain difficulties, discussed below. But, first, turning to the origin of olivine melilitites, from a series of experiments on a specimen from Tasmania, Brey (1978) concluded that, initially, olivine melilitite could be derived from a 5% melt from a pyrolite composition at *c.* 27 kbar and 1160 °C, with 7–8% $H_2O$ and 6–7% $CO_2$ dissolved in the melt; increased pressure and increased $XCO_2/H_2O$ resulted in increased silica-undersaturation. Two significant results of the experiments were to demonstrate the importance of the presence of volatiles (consistent with the melilitite/carbonatite association) but also the low degree of melting and consequent low volumes of melt. Of the ultrabasic lavas in the Natron area, in terms of their Si and Al contents the Oldoinyo Loolmurwak olivine melilite nephelinite (Dawson *et al.* 1985) and the Dorobo (Oldoinyo Lengai) olivine melilitite (Keller *et al.* 2006) are most similar to melilitite from Sutherland, South Africa, that Brey inferred to originate at *c.* 5 kbar by melting of a pyrolite mantle containing magnesite. This is consistent with the considerable depth of origin shown earlier for the northern Tanzanian olivine melilitites (Fig. 8.4). The Sr and Nd isotope values of the olivine melilitites in the Oldoinyo Lengai area overlap those of the recently-enriched mantle xenoliths from adjacent Pello Hill and Eledoi (Fig. 8.1).

### *The Younger Extrusives paradox, and replenishment of the Tanzanian mantle*

That the lavas of Oldoinyo Lengai, an example of Younger Extrusive volcanic centres, might have an olivine melilitite parent (reviewed above) has been proposed earlier on the basis that no other magmas in the immediate vicinity may be regarded as primary mantle-derived magma.

There is an undoubtedly a close field-relationship between carbonatites and olivine melilitite in northern Tanzania (e.g. at Lashaine and in the Oldoinyo Lengai area) but there are problems linking the Oldoinyo Lengai nephelinites directly with a parental olivine melilitite:

1. There is a compositional gap between the rock-types, with no intermediate types that might indicate a fractional crystallization spectrum (Keller *et al.* 2006).
2. Specifically in the case of Oldoinyo Lengai lavas, the nearby olivine melilitites, although having similar $\delta^7$ Li values (Halama *et al.* 2007), do not have the same Sr and Nd isotopic characteristics (Dawson 2008, fig. 7.1). The olivine melilitites lie closer to a HIMU source than any of the Oldoinyo Lengai lavas and hence, if olivine melilitite was indeed parental, some mixing with more enriched material is required. Fractional crystallization does not explain these isotope data.
3. For an olivine melilitite parent magma to be parental to other Younger Extrusive volcanoes, the prerequisite would be a very large volume of magma (particularly in the case of Meru) at the onset of the Younger Extrusive activity. The olivine melilitites appear late in the overall Younger Extrusive activity and, at present, there is field evidence for only limited volumes, consistent with the experimental evidence that olivine melilitite is generated by very small amounts of mantle melting.
4. For olivine melilite to be parental, repeated or cyclical generation of olivine melilitite magma would be required, both at the beginning and towards the end of the Younger Extrusive activity. Megacrysts of phlogopite, augite and amphibole certainly occur at several of the olivine melilitite localities in the Engaruka–Natron area and can be interpreted as additional pulses of melilitite magma crystallizing at depth but, even so, the volume of material at these occurrences is still very limited compared with the larger Younger Extrusive volcanoes.
5. Experiments and major element composition (Fig. 8.4) indicate that olivine melilitites are formed at considerable depth and the presence of mantle xenoliths in them suggests rapid ascent. This constraint appears at odds with the time necessary for high-level fractionation to form the Younger Extrusives.

If olivine melilitite is not a feasible source material, even allowing for mixing with enriched lithosphere, is there an alternative source for the Younger Extrusives? In many respects, the Younger Extrusives are a petrological repeat of the Older Extrusive volcanoes Essimingor, Sadiman, Shombole and Mosonik. Earlier, it was suggested that the highly alkaline rocks at these volcanoes could be accommodated within a model in which they are the first melting products of enriched lithospheric mantle that was heated by a rising asthenospheric thermal pulse. From the volume constraint, such a model might appear to be more feasible than fractionation of a low-volume olivine melilitite parental magma.

However, like that proposal, to apply such a model to the Youger Extrusives raises a very important question, the necessity for a continued, or re-fertilized, volatile- and incompatibles-enriched mantle source. The earlier Older Extrusive major episode of dominantly basaltic magmatism might be expected to have given rise to depleted mantle beneath northern Tanzania. Yet the Younger Extrusive volcanics are volatile- and incompatibles-enriched and must derive from mantle material that has these very properties. Although, on the timescale available, it might appear unlikely that the largely depleted Tanzanian lithospheric mantle could have been replenished, there is direct evidence from the mantle xenoliths at Pello Hill, Lashaine and Labait for a very young metasomatic event that has resulted in mantle replenishment. Moreover, at Pello Hill, the veined xenoliths provide evidence for an intrusion event. Further, there is seismic evidence for the existence of a mantle plume beneath the Tanzania Craton; a new pulse of upwelling asthenosphere could be the source of the replenishment.

To explain these observations, the following is offered as a tentative, integrated model. First, the Older Extrusives resulted from conductive melting of anciently-enriched lithospheric mantle, with the earliest melts from the deeper parts being peralkaline (e.g. Essimingor) or of limited volume (e.g. earlier nephelinites of Lemagrut and Oldoinyo Sambu). Continuing conductive heating to shallower levels produces larger, less enriched magma batches to produce the Older Extrusive basalt-dominated volcanoes, with high-level fractionation producing derivative benmoreites, mugearites, trachytes and phonolites. Original, enrichment-hetereogeneity of the lithospheric mantle can explain contemporaneous alkaline/carbonatitic (Sadiman) and basaltic (Lemagrut) activity during the later part of the conductive heating event. Second, low-volume, enriched asthenospheric melts (the source of the heat for the Older Extrusive magmatism) eventually rise towards the surface, along channels generated during the mid-Pleistocene extension, to produce the Younger Extrusives. There is evidence from xenoliths for the fracturing, veining, injection and metasomatism of the lithospheric mantle and, from seismicity, for deep, brittle lithosphere fracturing. The asthenosphere melts, although themselves enriched in volatiles and other incompatible components (e.g. $CO_2$, Pb, REE), need to interact with residual, anciently-enriched lithosphere to generate the recent metasomatic effects seen in mantle xenoliths and to produce the Younger Extrusives isotope systematics.

An alternative to the above-preferred model is that enriched magma, generated during the Older Extrusive activity, was ponded in the lithosphere and later enabled to migrate upwards following the mid-Pleistocene faulting to form the Younger Extrusives.

## References


BAKER, B. H. 1958. *Geology of the Magadi Area*. Geological Survey of Kenya, Report **42**.

BAKER, B. H. 1963. *Geology of the Area South of Magadi*. Geological Survey of Kenya, Report **61**.

BAKER, B. H. & MITCHELL, J. G. 1976. Volcanic stratigraphy and geochronology of the Kedong–Olorgesaile area and the evolution of the South Kenya Rift valley. *Journal of the Geological Society, London*, **132**, 467–84.

BAKER, B. H., MOHR, P. A. & WILLIAMS, L. A. J. 1972.*Geology of the Eastern Rift System of Africa*. Geological Society of America, Special Publications, **36**.

BELL, K. & DAWSON, J. B. 1995. Nd and Sr isotope systematics of the active carbonatite volcano, Oldoinyo Lengai. *In*: BELL, K. & KELLER, J. (eds) *Carbonatite Volcanism of Oldoinyo Lengai and Petrogenesis of Natrocarbonatite*. Springer Verlag, Berlin, 100–112.

BELL, K. & SIMONETTI, A. 1996. Carbonatite magmatism and plume activity: implications from the Nd, Pb and Sr isotope systematics of Oldoinyo Lengai. *Journal of Petrology*, **37**, 1321–1339.

BELL, K. & TILTON, G. R. 2001. Nd, Pb and Sr isotopic compositions of East African carbonatites: evidence for mantle mixing and plume inhomogeneity. *Journal of Petrology*, **42**, 1927–1945.

BELL, K., KJAARSGAARD, B. A. & SIMONETTI, A. (eds) 1998. *Carbonatites – into the Twenty-first Century. A Volume in Honour of John Gittins*. Special Number of Journal of Petrology, **39**, 1839–2151.

BREY, G. 1978. Origin of olivine melilitites – chemical and experimental constraints. *Journal of Volcanological and Geothermal Research*, **3**, 61–88.

BURTON, K. W., SCHIANO, P., BIRK, J.-L., ALLÉGRE, C. J., REHKÄMPER, M., HALLIDAY, A. N. & DAWSON, J. B. 2000. The distribution and behaviour of rhenium and osmium amongst mantle minerals and the age of the lithospheric mantle beneath Tanzania. *Earth and Planetary Science Letters*, **183**, 93–106.

CHESLEY, J. T., RUDNICK, R. L. & LEE, C.-T. 1999. Re–Os systematics of mantle xenoliths from the East African Rift: age, structure and history of the Tanzanian craton. *Geochimica et Cosmochimica Acta*, **63**, 1203–1217.

CLASS, C., GOLDSTEIN, STUTE, M., KURZ, M. D. & SCHLOSSER, P. 2005. Grand Comore Island: A well-constrained "low $^3He/^4He$" mantle plume. *Earth and Planetary Science Letters*, **231**, 391–409.

COHEN, R. S., O'NIONS, R. K. & DAWSON, J. B. 1984. Isotope geochemistry of xenoliths from East Africa: implications for the development of mantle reservoirs and their interaction. *Earth and Planetary Science Letters*, **68**, 209–220.

CROSSLEY, R. 1979. The Cenozoic stratigraphy and structure of the western part of the rift valley in southern Kenya. *Journal of the Geological Society, London*, **136**, 393–405.

DAWSON, J. B. 1987. Metasomatized harzburgites in kimberlite and alkaline magmas: enriched restites and "flushed" lherzolites. *In*: MENZIES, M. A. & HAWKESWORTH, C. J. (eds) *Mantle Metasomatism*. Academic Press, London, 125–144.

DAWSON, J. B. 1992. Neogene tectonics and volcanicity in the North Tanzania sector of the Gregory Rift Valley: contrasts with the Kenya sector. *Tectonophysics*, **204**, 81–92.

DAWSON, J. B. 1993. A supposed sövite from Oldoinyo Lengai, Tanzania: result of extreme alteration of alkali carbonatite lava. *Mineralogical Magazine*, **57**, 93–101.

DAWSON, J. B. 1994. Quaternary kimberlitic volcanism on the Tanzania craton. *Contributions to Mineralogy and Petrology*, **116**, 473–485.

DAWSON, J. B. 1998. Peralkaline nephelinite–natrocarbonatite relationships at Oldoinyo Lengai, Tanzania. *Journal of Petrology*, **39**, 2077–2094.

DAWSON, J. B. 1999. Melting and metasomatism in spinel peridotite xenoliths from Labait, Tanzania. *In*: GURNEY, J. J., GURNEY, J. L., PASCOE, M. D. & RICHARDSON, S. H. (eds) *Proceedings of the 7th International Kimberlite Conference, Cape Town, 1998*, Vol. 1, Red Roof Design, Cape Town, 164–173.

DAWSON, J. B. 2002. Metasomatism and partial melting in upper-mantle peridotite xenoliths from the Lashaine volcano, northern Tanzania. *Journal of Petrology*, **43**, 1749–1777.

DAWSON, J. B. 2008. *The Gregory Rift Valley and Neogene–Recent Volcanoes of Northern Tanzania*. Geological Society, London, Memoirs, **33**, 9–12.

DAWSON, J. B. & HAWTHORNE, J. B. 1973. Magmatic sedimentation and carbonatitic differentiation in kimberlite sills at Benfontein, South Africa. *Journal of the Geological Society, London*, **129**, 61–85.

DAWSON, J. B. & SMITH, J. V. 1988. Veined and metasomatized peridotites from Pello Hill, Tanzania: evidence for anomalously-light mantle beneath the Tanzania sector of the eastern Rift Valley. *Contributions to Mineralogy and Petrology*, **100**, 510–527.

DAWSON, J. B. & SMITH, J. V. 1992. Olivine–mica–pyroxenite xenoliths from Oldoinyo Lengai, Tanzania. metasomatic products of upper mantle peridotites. *Journal of Volcanology and Geothermal Research*, **50**, 131–142.

DAWSON, J. B., POWELL, D. G. & REID, A. M. 1970. Ultrabasic xenoliths and lava from the Lashaine volcano, northern Tanzania. *Journal of Petrology* **11**, 519–548.

DAWSON, J. B., SMITH, J. V. & JONES, A. P. 1985. A comparative study of bulk rock and mineral chemistry of olivine melilitites and associated rocks from East and South Africa. *Neues Jahrbuch für Mineralogie, Abhandlungen*, **152**, 143–175.

DAWSON, J. B., PINKERTON, H., NORTON, G. E., PYLE, D. M., BROWNING, P., JACKSON, D. & FALLICK, A. E. 1995. Petrology and geochemistry of Oldoinyo Lengai lavas extruded in November 1988: magma source, ascent and crystallization. *In*: BELL, K. & KELLER, J. (eds) *Carbonatite Volcanism of Oldoinyo Lengai and Petrogenesis of Natrocarbonatite*. Springer, Berlin, 47–69.

DAWSON, J. B., PYLE, D. M. & PINKERTON, H. 1996. Evolution of natrocarbonatite from a wollastonite nephelinite parent: evidence from the June 1993 eruption of Oldoinyo Lengai, Tanzania. *Journal of Geology*, **104**, 41–54.

DAWSON, J. B., JAMES, D., PASLICK, C. & HALLIDAY, A. M. 1997. Ultrabasic potassic low-volume magmatism and continental rifting in north-central Tanzania: association with enhanced heat flow. *Russian Geology and Geophysics*, **38**, 69–81.

DE PAOLO, D. J. 1981. Trace element and isotopic effects of combined wall-rock assimilation and fractional crystallization. *Earth and Planetary Science Letters*, **53**, 189–202.

DONALDSON, C. H., DAWSON, J. B., KANARIS-SOTIRIOU, R., BATCHELOR, R. A. & WALSH, J. N. 1987. The silicate lavas of Oldoinyo Lengai. *Neues Jahrbuch für Mineralogie Abhandlungen*, **156**, 247–279.

DOWNIE, C. & WILKINSON, P. 1972. *The Geology of Kilimanjaro*. University of Sheffield, Sheffield.

FAIRHEAD, J. D., MITCHELL, J. G. & WILLIAMS, L. A. J. 1972. New K/Ar determinations on Rift volcanics of S. Kenya and their bearing on the age of Rift faulting. *Nature*, **238**, 66–69.

FURMAN, T., KALETA, K. M., BRYCE, J. G. & HANAN, B. B. 2006*a*. Tertiary mafic lavas of Turkana, Kenya: constraints on East African plume structure and the occurrence of high-$\mu$ volcanism in Africa. *Journal of Petrology*, **46**, 1221–1244.

FURMAN, T., BRYCE, J., ROONEY, T., HANAN, B., YIRGU, G. & AYELEW, D. 2006*b*. Heads and tails: 30 million years of the Afar plume. *In*: YIRGU, G., EBINGER, C. Y. & MAGUIRE, K. H. (eds) *The Afar Volcanic Province within the East African Rift System*. Geological Society, London, Special Publications, **259**, 95–11.

GITTINS, J. 1989. The origin and evolution of carbonatite magmas. *In*: BELL, K. (ed.) *Carbonatites: Genesis and Evolution*. Unwin Hyman, London, 580–600.

HALAMA, R., MCDONOUGH, W. F., RUDNICK, R. L., KELLER, J. & KLAUDIUS, J. 2007. The Li isotopic composition of Oldoinyo Lengai: Nature of the mantle sources and lack of isotopic fractionation during carbonatite petrogenesis. *Earth and Planetary Science Letters*, **254**, 77–89.

HARMER, R. E. & GITTINS, J. 1998. The case for primary, mantle-derived carbonatite magma. *Journal of Petrology*, **39**, 1895–1903.

HART, S. R. 1988. Heterogeneous mantle domains: signatures, genesis and mixing chronologies. *Earth and Planetary Science Letters*, **90**, 273–296.

HERZBERG, C. 1995. Generation of plume magmas through time: an experimental perspective. *Chemical Geology*, **126**, 1–16.

JONES, A. P., SMITH, J. V. & DAWSON, J. B. 1983. Glasses in mantle xenoliths from Olmani, Tanzania. *Journal of Geology*, **91**, 167–178.

KALT, A., HEGNER, E. & SATIR, M. 1997. Nd, Sr, and Pb isotope evidence for diverse lithospheric mantle sources of East African Rift carbonatites. *Tectonophysics*, **278**, 31–45.

KELLER, J., ZAITSEV, A. & WIEDENMANN, D. 2006. Primary magmas at Oldoinyo Lengai: the role of olivine melilitite. *Lithos*, **91**, 150–172.

KJAARSGAARD, B., HAMILTON, D. L. & PETERSON, T. 1995. Peralkaline nephelinite/carbonatite liquid immiscibility: comparison of phase compositions in experiments and natural lavasa from Oldoinyo Lengai. *In*: BELL, K. & KELLER, J. (eds) *Carbonatite Volcanism of Oldoinyo Lengai and Petrogenesis of Natrocarbonatite*. I.A.V.C.E.I. Proceedings in Volcanology, Vol. 4. Springer, Berlin, 165–190.

KRAMM, U. & SINDERN, S. 1998. Nd and Sr isotope signature of fenites from Oldoinyo Lengai, Tanzania, and the genetic relationships between nephelinites, phonolites and carbonatites. *Journal of Petrology*, **39**, 1997–2004.

LATIN, D., NORRY, M. J. & TARZEY, R. J. E. 1993. Magmatism in the Gregory Rift, East Africa: evidence for melt generation by a plume. *Journal of Petrology*, **34**, 1007–1027.

LE ROEX, A. P., SPÄTH, A. & ZARTMAN, R. E. 2001. Lithospheric thickness beneath the southern Kenya Rift: implications from basalt geochemistry. *Contributions to Mineralogy and Petrology*, **142**, 89–106.

LEE, C.-T. & RUDNICK, R. L. 1999. Compositionally-stratified cratonic lithosphere: petrology and geochemistry of peridotite xenoliths from the Labait volcano, Tanzania. *In*: GURNEY, J. J., GURNEY, J. L., PASCOE, M. D. & RICHARDSON, S. H. (eds) *Proceedings of the 7th International Kimberlite Conference, Cape Town, 1998*, Vol. 2, Red Roof Design, Cape Town, 503–521.

LEE, C.-T., RUDNICK, R. L., MCDONOUGH, W. F. & HORN, I. 2000. Petrologic and geochemical investigation of carbonates in peridotite xenoliths from northeastern Tanzania. *Contributions to Mineralogy and Petrology*, **139**, 470–484.

LEE, W.-J. & WYLLIE, P. J. 1998. Processes of crustal carbonatite formation by liquid immiscibility and differentiation elucidated by model systems. *Journal of Petrology*, **39**, 2005–2014.

MACDONALD, R. 2003. Magmatism of the Kenya Rift Valley: a review. *Transactions of the Royal Society of Edinburgh: Earth Sciences*, **95**, 239–253.

MACDONALD, R. & SCAILLET, B. 2006. The central Kenya peralkaline province: Insights into the evolution of peralkaline salic magmas. *Lithos*, **91**, 59–73.

MACDONALD, R., ROGERS, N. W., FITTON, J. G., BLACK, S. & SMITH, M. 2001. Plume–lithosphere interactions in the generation of basalts of the Kenya Rift, East Africa. *Journal of Petrology*, **42**, 877–900.

PASLICK, C., HALLIDAY, A., JAMES, D. & DAWSON, J. B. 1995. Enrichment of the continental lithosphere by OIB melts: Isotopic evidence from the volcanic province of northern Tanzania. *Earth and Planetary Science Letters*, **130**, 109–126.

PASLICK, C., HALLIDAY, A. N., LANGE, R. A., JAMES, D. & DAWSON, J. B. 1996. Indirect crustal contamination: evidence from isotopic and chemical disequilibria in minerals from alkali basalts and nephelinites from northern Tanzania. *Contributions to Mineralogy and Petrology*, **125**, 277–292.

PETERSON, T. D. & KJAARSGAARD, B. A. 1995. What are the parental magmas at Oldoinyo Lengai? *In*: BELL, K. & KELLER, J. (eds) *Carbonatite Volcanism of Oldoinyo Lengai and Petrogenesis of Natrocarbonatite*. I.A.V.C.E.I. Proceedings in Volcanology, Vol. 4. Springer, Berlin, 148–162.

PORCELLI, D. R., O'NIONS, R. K. & O'REILLY, S. Y. 1986. Helium and strontium isotopes in ultramafic xenoliths. *Chemical Geology*, **54**, 237–249.

RIDLEY, W. I. & DAWSON, J. B. 1975. Lithophile trace element data bearing on the origin of peridotite, ankaramite and carbonatite from the Lashaine volcano, Tanzania. *Physics and Chemistry of the Earth*, **9**, 559–570.

ROBERTS, M. A. 2002. *The Geochemical and Volcanological Evolution of the Mt. Meru Region, Northern Tanzania*. PhD thesis, University of Cambridge.

ROEDER, P. L. & EMSLIE, R. F. 1970. Olivine-liquid equilibrium. *Contributions to Mineralogy and Petrology*, **29**, 275–289.

ROGERS, N. W. 2006. Basalt magmatism and geodynamics of the East African Rift System. *In*: YIRGU, G., EBINGER, C. Y. & MAGUIRE, K. H. (eds) *The Afar Volcanic Province within the East African Rift System*. Geological Society, London, Special Publications, **259**, 77–93.

ROGERS, N. W., MACDONALD, R., FITTON, J. G., GEORGE, R., SMITH, M. & BARREIRO, B. 2000. Two mantle plumes beneath the East African rift system: Sr, Nd and Pb isotope evidence from Kenya Rift basalts. *Earth and Planetary Science Letters*, **176**, 387–400.

RUDNICK, R. L. & GAO, S. 2003. Composition of the continental crust. *In*: HOLLAND, H. D. & TUREKIAN, K. K. (eds) *Treatise on Geochemistry: Volume 3. The Crust*. Elsevier, Amsterdam, 1–64.

RUDNICK, R. L., MCDONOUGH, W. F. & CHAPPELL, B. W. 1993. Carbonatite metasomatism in the northern Tanzanian mantle: petrographic and geochemical characteristics. *Earth and Planetary Science Letters*, **114**, 463–475.

RUDNICK, R. L., MCDONOUGH, W. F. & ORPIN, A. 1994. Northern Tanzanian peridotite xenoliths: a comparison with Kaapvaal peridotites and inferences on metasomatic interactions. *In*: MEYER, H. O. A. & LEONARDOS, O. H. (eds) *Proceedings of the 5th International Kimberlite Conference, Araxa 1991*, Companhia de Pesquisa de Reursos Minerais, Brasilia Vol. 1, 336–353.

SIMON, N. S. C., IRVINE, G. J., DAVIES, G. R., PEARSON, D. G. & CARLSON, R. W. 2003. The origin of garnet and clinopyroxene in "depleted" Kaapvaal peridotites. *Lithos*, **71**, 289–322.

SMITH, M. & MOSLEY, P. 1993. Crustal heterogeneities and basement influence on the development of the Kenya Rift, East Africa. *Tectonics*, **12**, 591–606.

SPÄTH, A., LE ROEX, A. P. & OPIYO-AKECK, N. 2000. The petrology of the Chiyulu Hills Volcanic Province, southern Kenya. *Journal of African Earth Sciences*, **31**, 337–358.

SPÄTH, A., LE ROEX, A. P. & OPIYO-AKECK, N. 2001. Plume–lithosphere interaction and the origin of continental rift-related alkaline volcanism - the Chiyulu Hills Volcanic Province, southern Kenya. *Journal of Petrology*, **42**, 765–787.

STUART, F. M. & TURNER, G. 1992. The abundance and isotopic composition of noble gases in ancient fluids. *Chemical Geology*, **101**, 98–109.

SUN, S.-S. & MCDONOUGH, W. F. 1989. Chemical and isotopic systematics of oceanic basalts: implications for mantle compositions and processes. *In*: SAUNDERS, A. D. & NORRY, M. J. (eds) *Magmatism in the Ocean Basins*. Geological Society, London, Special Publications, **42**, 313–345.

SWEENEY, R. J., FALLOON, T. J. & GREEN, D. H. 1995. Experimental constraints on the possible origin of natrocarbonatite. *In*: BELL, K. & KELLER, J. (eds) *Carbonatite Volcanism of Oldoinyo Lengai and Petrogenesis of Natrocarbonatite*. Springer, Berlin, 191–207.

WALLACE, M. E. & GREEN, D. H. 1988. An experimental determination of primary carbonatite magma composition. *Nature*, **335**, 343–346.

WILLIAMS, L. A. J., MACDONALD, R. & CHAPMAN, G. R. 1984. Late Quaternary volcanoes of the Kenya Rift Valley. *Journal of Geophysical Research*, **89**, B10, 8553–8570.

WYLLIE, P. J. & LEE, W. J. 1998. Model system controls on conditions for formation of magnesiocarbonatite and calciocarbonatite magmas from the mantle. *Journal of Petrology*, **39**, 1885–1893.

# Chapter 9

# Future work

This memoir has attempted to give an account of the present knowledge of this part of the Gregory Rift, but should be regarded as an interim report. Like most reviews, it has served to identify areas that would benefit from further research. Some are listed below.

1. The eastern margin of the Tanzania craton is buried to the east of the mapped surface contact between it and the Mozambique fold belt. As noted in Chapter 3, there is now evidence that Archaean and Neoproterozoic rocks exist intermixed with the rocks of the Mozambique Belt, well to the east of the mapped contact. Seismic traverses across the Maasai Block would be highly desirable to determine whether the eastern extent of the craton can be ascertained, and whether there might be any coincidence between its extent and the identified geophysical properties of the Maasai Block suggesting that it might be a discrete terrain.
2. Investigations into the structure and history of the Tanzanian mantle by means of xenolith studies are still at an early stage. Some xenoliths, such as those at Lashaine and Pello Hill, contain micas, whereas others (e.g. at Olmani) contain monazite. Dating these phases could shed new light on the timing of the upper mantle metasomatism. The age and isotopic characteristics of the host rocks of the xenoliths (mainly olivine melilitites and nephelinites) need investigation to find out whether the metasomatism and magmatism are linked. For example, are the helium isotope characteristics of xenoliths imposed by the host magmas during transportation?
3. There is still a dearth of radiogenic isotope data on the Tanzanian volcanic rocks, particularly on systematic collections from individual volcanoes. One such volcano is Tarosero which, within the Tanzanian province, is a unique and atypical locality for peralkaline rhyolites.
4. U-series and diffusion profiling in phenocrysts can determine the timescales of magma replenishment and crystal fractionation in magmatic systems. Currently, there are virtually no data on the strongly silica-undersaturated systems of which the northern Tanzanian volcanoes offer an almost unparalleled selection e.g. Oldoinyo Lengai, Kerimasi and Meru.
5. Although touched upon in the case of plutonic cognate xenoliths from Oldoinyo Lengai (Dawson 2008, Chapter 7), there is considerable scope for studies of melt and fluid inclusions in phenocrysts and xenoliths with the aim of interpreting melt-fluid evolution in strongly alkaline systems.
6. There are uninvestigated sedimentary sequences exposed in gorges on both the western and eastern sides of the Engaruka Basin, which lies geographically between the Natron and Manyara Basins (where mammalian and hominid fossils have been found). Hence, the Engaruka Basin sediments are worthy of a reconnaissance to assess their potential for fossil studies.

J. B. Dawson, University of Edinburgh (*jbdawson@glg.ed.ac.uk*)
*From*: DAWSON, J. B. 2008. *The Gregory Rift Valley and Neogene–Recent Volcanoes of Northern Tanzania*. Geological Society, London, Memoirs, **33**, 91.
0435-4052/08/$15.00 © The Geological Society of London 2008. DOI: 10.1144/M33.9

# Appendix 1

**Appendix 1A.** *Radiometric dates from Northern Tanzania*

| Sample | Lithology and location | Age (Ma) | Refs |
|---|---|---|---|
| ***Lake Natron Area (see also unpublished dates below)*** | | | |
| 572 | Phonolitic nephelinite, Shombole | 1.96 ± 0.07 | 3 |
| 580 | Phonolitic nephelinite, Shombole | 2.00 ± 0.05 | 3 |
| 138 | Nephelinite, south slopes of Gelai | 0.96 ± 0.03 | 6 |
| | Nephelinite, south slopes of Gelai | 0.99 ± 0.03 | 6 |
| | Biotite lava, Mosonik | 3.18 ± 0.08 | 1 |
| KA 1757 | | | |
| NATM89-07 | Nephelinite, Mosonik | 3.53 ± 0.06 | 2 |
| S11/4 | Nephelinite, flow from Mosonik | 1.28 ± 0.05 | 11 |
| KA 1755 | Basalt, Peninj Gorge | 1.90 ± 0.07 | 1 |
| KA 1755R | Basalt, Peninj Gorge | 1.77 ± 0.04 | 1 |
| KA 2570 | Basalt, Malambo Gorge | 3.5 ± 0.46 | 1 |
| KA 1185 | Basalt, underlying Peninj Group sediments | 2.99 ± 0.09 | 2 |
| KA 1754 | Basalt, Humbu Formation (Lower Peninj Group) | 1.55 ± 0.03 | 1 |
| KA 1754R | Basalt, Humbu Formation | 2.27 ± 0.06 | 1 |
| KA 2382 | Basalt, Humbu Formation | 0.96 ± 0.1 | 1 |
| KA 2382R | Basalt, Humbu Formation | 1.21 ± 0.36 | 1 |
| KA 2578 | Basalt, Humbu Formation | 0.97 ± 0.25 | 1 |
| KA 2646 | Basalt, Humbu Formation | 1.21 ± 0.1 | 1 |
| | Basal tuff, Humbu Formation | 1.70 ± 0.02* | 2 |
| KA 2410 | Intra-Moinik basalt, Moinik Formation (U. Peninj Group) | 1.33 ± 0.05 | 1 |
| KA 2589 | Intra-Moinik basalt, Moinik Formation | 1.38 ± 0.09 | 1 |
| KA 1186 | Intra-Moinik basalt, Moinik Formation | 1.41 ± 0.04 | 2 |
| KA 1187 | Intra-Moinik basalt, Moinik Formation | 1.13 ± 0.06 | 2 |
| S1 1/6 | Nephelinite, interbedded with Upper Moinik Formation. | 1.31 ± 0.26 | 11 |
| | Bird Print Tuff, Moinik Formation | 1.26 ± 0.04* | 2 |
| 3 | Phlogopite, Kisetey crater tuff | 0.57 ± 0.15 | 5 |
| | Phlogopite, Kisetey crater tuff | 0.14 ± 0.15 | 5 |
| 891 | Phlogopite, Loolmurwak crater tuff | 0.53 ± 0.1 | 5 |
| | Phlogopite, Loolmurwak crater tuff | 0.41 ± 0.1 | 5 |
| 105 | Phlogopite, Loolmurwak crater tuff | 0.19 ± 0.08 | 5 |
| | Phlogopite, Loolmurwak crater tuff | 0.14 ± 0.12 | 5 |
| | Calcrete, on reworked tephra, 3–5 km south of Lake Natron | 0.13 Ma | 7 |
| | Calcrete, overlying black tuffs, Loolmurwak crater | 0.35 ± 0.01 | 7 |
| 11031 | Phonolite, Oldonyo Lengai | 0.15 ± 0.02 | 4 |

(*Continued*)

| Sample | Lithology and location | Age (Ma) | Refs |
|---|---|---|---|
| ***Engaruka-Manyara Area*** | | | |
| 155 | Olivine basalt, south slopes of Ketumbeine | 1.54 ± 0.1 | 6 |
| | Olivine basalt, south slopes of Ketumbeine | 1.66 ± 0.07 | 6 |
| 156 | Olivine basalt, south slopes of Ketumbeine | 1.73 ± 0.08 | 6 |
| | Olivine basalt, south slopes of Ketumbeine | 1.87 ± 0.08 | 6 |
| 11255 | Olivine melanephelinite, escarpment base, Engaruka | 3.2 ± 0.5 | 4 |
| 11245 | Olivine basalt, escarpment top, Engaruka | 1.4 ± 0.5 | 4 |
| 11247 | Trachyandesite, escarpment top, Engaruka | 1.8 ± 0.6 | 4 |
| 11 | Olivine trachybasalt, flow, small crater, 1 mile south of Engaruka | 1.15 ± 0.04 | 5 |
| | Olivine trachybasalt, flow, small crater, 1 mile south of Engaruka | 1.11 ± 0.06 | 5 |
| 139 | Olivine trachybasalt, Narabala top of scarp | 1.2 ± 0.04 | 5 |
| | Olivine trachybasalt, Narabala top of scarp | 1.18 ± 0.05 | 5 |
| 137a | Olivine trachybasalt, Narabala, flow from fault | 1.21 ± 0.05 | 5 |
| | Olivine trachybasalt, Narabala, flow from fault | 1.15 ± 0.04 | 5 |
| | Olivine trachybasalt, Narabala, flow from fault | 1.09 ± 0.05 | 5 |
| | Olivine trachybasalt, Narabala, flow from fault | 1.17 ± 0.04 | 5 |
| S13/2 | Olivine basalt, Engaruka Basin escarpment | 1.26 ± 0.07 | 11 |
| BD181 | Olivine basalt, extruded from main rift fault, Kitete | 1.13 ± 0.16 | 5 |
| | Olivine basalt, extruded from main rift fault, Kitete | 1.23 ± 0.05 | 5 |
| BD180 | Olivine basalt, extruded from main rift fault, Kitete | 0.95 ± 0.08 | 5 |
| | Olivine basalt, extruded from main rift fault, Kitete | 1.08 ± 0.05 | 5 |
| S13/3 | Basalt, extruded from foot of Kitete escarpment | 1.0 ± 0.1 | 11 |
| S14/4 | Basalt, top of Kitete escarpment | 0.8 ± 0.4 | 11 |
| S14/5 | Basalt, foot of Kitete escarpment | 1.2 ± 0.06 | 11 |
| 11163 | Picritic basalt, escarpment base, north of Lake Manyara | 3.8 ± 0.2 | 4 |
| 11170 | Basaltic andesite, escarpment top, north of Lake Manyara | 2.25 ± 0.1 | 4 |

(*Continued*)

J. B. Dawson, University of Edinburgh (*jbdawson@glg.ed.ac.uk*)
*From*: Dawson, J. B. 2008. *The Gregory Rift Valley and Neogene–Recent Volcanoes of Northern Tanzania.* Geological Society, London, Memoirs, **33**, 93–96.
0435-4052/08/$15.00 © The Geological Society of London 2008. DOI: 10.1144/M33.App01

| Sample | Lithology and location | Age (Ma) | Refs |
|---|---|---|---|
| 2a | Basalt, interbedded with lacustrine sediments, base of main rift escarpment, Lake Manyara | 4.86 ± 0.24 | 11 |
| 7b | Basalt, top of rift escarpment, Lake Manyara | 1.5 ± 0.1 | 11 |
| ***Essimingor-Burko-Tarosero-Monduli*** | | | |
| | Melanephelinite, Essimingor | 8.1 ± 1.0 | 4 |
| | Melanephelinite, Essimingor | 7.35 ± 0.65 | 4 |
| 142 | Nephelinite, gully on SE slopes, Essimingor | 4.68 ± 0.09 | 6 |
| | Nephelinite, gully on SE slopes, Essimingor | 4.89 ± 0.09 | 6 |
| 143 | Nephelinite, gully on SE slopes, Essimingor | 3.23 ± 0.07 | 6 |
| | Nephelinite, gully on SE slopes, Essimingor | 3.20 ± 0.06 | 6 |
| 137b | Nephelinite, gully on SE slopes, Burko | 1.03 ± 0.04 | 6 |
| | Nephelinite, gully on SE slopes, Burko | 0.91 ± 0.03 | 6 |
| 309 | Olivine basalt, lower north slopes, Tarosero | 2.4 ± 0.62 | 5 |
| 305 | Sodalite phonolite, faulted north slopes of Tarosero | 2.25 ± 0.06 | 5 |
| 306 | Sodalite phonolite, faulted north slopes of Tarosero | 2.2 ± 0.06 | 5 |
| 510 | Phonolite, lower NW slopes of Tarosero | 2.21 ± 0.1 | 5 |
| 490 | Unfaulted trachyte, summit of Tarosero | 2.16 ± 0.06 | 5 |
| | Unfaulted trachyte, summit of Tarosero | 2.04 ± 0.04 | 5 |
| 488 | Unfaulted trachyte, summit ridge, Tarosero | 1.98 ± 0.03 | 5 |
| | Unfaulted trachyte, summit ridge, Tarosero | 1.90 ± 0.04 | 5 |
| 304 | Olivine basalt, highest flow on Matunginini fault escarpment | 2.00 ± 0.05 | 5 |
| 136 | Olivine basalt, north slope of Monduli Mtn, near Tarosero escarpment | 2.09 ± 0.06 | 6 |
| | Olivine basalt, north slope of Monduli Mtn, near Tarosero escarpment | 2.15 ± 0.05 | 6 |
| ***Kwaraha-Hanang*** | | | |
| 11068 | Olivine melanephelinite, Kwaraha | 1.5 ± 1.0 | 4 |
| 11073 | Olivine melilite, Kwaraha | 1.05 ± 0.15 | 4 |
| 11085 | Olivine melanephelinite, Kwaraha | 1.08 ± 0.18 | 4 |
| 11088 | Nephelinite, Kwaraha | 0.7 ± 0.0 | 4 |
| 11001 | Nephelinite, Hanang | 1.5 ± 0.3 | 4 |
| 11030 | Nephelinite, Hanang | 0.9 ± 0.2 | 4 |
| ***Crater Highlands*** | | | |
| | Nephelinite, Sadiman | 3.32 ± 0.06 | 2 |
| 2238 | Nephelinite, Sadiman | 3.7 | 7 |
| Tuff 8 | Tephras, Upper Laetolil Beds, ?from Sadiman | 3.46 ± 0.12 | 8 |
| 11113 | Nephelinite, Sadiman | 4.5 ± 0.4 | 4 |
| NAT90-28 | Ignimbrite, Ngorongoro | 1.83 ± 10.03 | 2 |
| NAT90-30 | Ignimbrite, Ngorongoro | 1.98 ± 0.032 | 2 |
| | Ignimbrite, Ngorongoro | 2.02 ± 0.02* | 2 |
| 20 | Trachyte, Ngorongoro | 2.45 ± 0.15 | 9 |
| | Plagioclase (tuff), Ngorongoro | 2.23 ± 0.1 | 9 |
| | Plagioclase (tuff), Ngorongoro | 2.28 ± 0.14 | 9 |
| | Plagioclase (tuff), Ngorongoro | 2.45 ± 0.15 | 9 |
| | Plagioclase (tuff), Ngorongoro | 2.25 ± 0.15 | 9 |
| | Plagioclase (tuff), Ngorongoro | 2.81 ± 0.11 | 9 |
| 11147 | Olivine basalt, Ngorongoro | 3.7 ± 0.8 | 4 |
| 11124 | Phonolite, north of Lemagrut | 5.5 ± 0.1 | 4 |
| 11130 | Basaltic andesite, Lemagrut | 4.3 ± 1.05 | 4 |
| 11115 | Silicic andesite, Lemagrut | 2.7 ± 0.2 | 4 |
| 7c | Ankaramite, Lemagrut | 3.1 ± 0.3 | 11 |
| | Naibadad beds, Laetoli | 2.26 ± 0.06 | 8 |
| | Vogesite lava flow remnant, Ogol lavas, Laetoli area | 2.41 ± 0.12 | 8 |
| | Biotite in tuff, upper Laetoli Beds | 3.49 ± 0.12 | 8 |
| | Biotite in tuff, lower Laetoli Beds (age mean of 2) | 3.76 ± 0.03 | 8 |
| LMGT89-02 | Ignimbrite, Lemagrut | 1.92 ± 0.04 | 2 |
| LMGT89-03 | Ignimbrite, Lemagrut | 1.82 ± 0.03 | 2 |
| | Ignimbrite, Lemagrut | 1.97 ± 0.01* | 2 |
| NATM89-13 | Trachytic lava, Olmoti | 1.65 ± 0.03 | 2 |
| NATM89-13A | Trachytic lava, Olmoti | 1.11 ± 0.09 | 2 |
| NATM89-11A | Ignimbrite, Olmoti | 1.23 ± 0.04 | 2 |
| NATM89-08 | Phonolitic tephrite, Olmoti | 1.07 ± 0.05 | 2 |
| OGMW89-54 | Trachyandesite flow, Olduvai, from Olmoti? | 1.09 ± 0.03 | 2 |
| MW89-17 | Nephelinite, Embagai | 0.5 ± 0.4 | 2 |
| MW89-15A | Nephelinite, Embagai | 1.05 ± 0.02 | 2 |
| MW89-14A | Nephelinite, Embagai | 1.52 ± 0.03 | 2 |
| | Tephras, Embagai | *c.* 0.6* | 2 |
| ***Olduvai Gorge*** | | | |
| | Naabi ignimbrite below base of Bed I | 2.029 ± 0.005* | 13 |
| | Coarse feldspar crystal tuff (CFCT) above Naabi ignimbrite | 2.018 ± 0.019*14 | 13 |
| | Basalt immediately above CFCT, Bed I (?derived from Olmoti (Hay, 1976) | 1.96 | 12 |
| | Basalt immediately above CFCT, Bed I | 1.865 ± 0.02 | 13 |
| | Tuff IB, above basalt, Bed I | 1.798 ± 0.004* | 13 |
| | Tuff IB, above basalt, Bed I | 1.786 | 12 |
| | Tuff IF top of Bed 1 | 1.749 ± 0.009* | 13 |
| | Tuff IIA base of Bed II | 1.71 | 12 |
| | Ndutu Beds (calcretes and Limicolaria shells) | 30-22 Ka BP | 7 |
| | Naisiusiu Beds (calcretes, collagen and egg-shell) | 20-9 Ka BP | 7 |
| | Calcretes overlying Naisiusiu Beds) | 6.6-4 Ka BP | 7 |
| ***Meru Area*** | | | |
| 135 | Alkali basalt block, collapsed east wall of caldera, Meru | 0.09 ± 0.02 | 6 |
| | Alkali basalt block, collapsed east wall of caldera, Meru | 0.115 ± 0.007 | 6 |
| 141 | Nephelinite block, collapsed east wall of caldera, Meru | 0.099 ± 0.005 | 6 |
| | Nephelinite block, collapsed east wall of caldera, Meru | 0.097 ± 0.004 | 6 |
| 3221 | Pumice from mantling ash, Oldonyo Sambu scoria cone | <0.55 | 10 |

*(Continued)*

| Sample | Lithology and location | Age (Ma) | Refs |
|---|---|---|---|
| | Pumice from mantling ash, Oldonyo Sambu scoria cone | <0.10 | 10 |
| 3403 | Nephelinite, Meru summit | 0.059 ± 0.010 | 10 |
| | Nephelinite, Meru summit | 0.067 ± 0.008 | 10 |
| 3655 | Nephelinite, SW of Meru West (3885 m) | 0.08 ± 0.005 | 10 |
| | Nephelinite, SW of Meru West (3885 m) | 0.08 ± 0.005 | 10 |
| 3930 | Phonolitic nephelinite, summit group, Meru West (2395 m) | 0.079 ± 0.004 | 10 |
| | Phonolitic nephelinite, summit group, Meru West (2395 m) | 0.09 ± 0.004 | 10 |
| 3114 | Phonolitic tephrite, Main Cone group, behind Ash Cone | 0.107 ± 0.006 | 10 |
| | Phonolitic tephrite, Main Cone group, behind Ash Cone | 0.102 ± 0.005 | 10 |
| 4231 | Phonolitic tephrite, Main Cone group, behind Ash Cone | 0.107 ± 0.007 | 10 |
| | Phonolitic tephrite, Main Cone group, behind Ash Cone | 0.111 ± 0.007 | 10 |
| 4191 | Nephelinite, Main Cone group, NW flanks of Meru | 0.163 ± 0.013 | 10 |
| | Nephelinite, Main Cone group, NW flanks of Meru | 0.148 ± 0.010 | 10 |
| 3203 | Sanidine phonolite, Naigonesoit tholoid, north Meru | 0.168 ± 0.009 | 10 |
| | Sanidine phonolite, Naigonesoit tholoid, north Meru | 0.158 ± 0.010 | 10 |
| | Sanidine phonolite, Naigonesoit tholoid, north Meru | 0.179 ± 0.010 | 10 |
| 3193 | Trachytoid phonolite, north of Naigonesoit, north Meru | 0.384 ± 0.009 | 10 |
| | Trachytoid phonolite, north of Naigonesoit, north Meru | 0.377 ± 0.009 | 10 |
| 3127 | Phonolitic nephelinite clast from breccia, Little Meru | 0.273 ± 0.006 | 10 |
| | Phonolitic nephelinite clast from breccia, Little Meru | 0.281 ± 0.005 | 10 |
| 3912 | Phonolite, Oldonyo Sambu (parasitic cone) | 0.311 ± 0.011 | 10 |
| | Phonolite, Oldonyo Sambu (parasitic cone) | 0.3 ± 0.011 | 10 |
| 3303 | Phonolitic nephelinite, Oldonyo Sambu | 0.313 ± 0.009 | 10 |
| | Phonolitic nephelinite, Oldonyo Sambu | 0.299 ± 0.010 | 10 |
| | Phonolitic nephelinite, Oldonyo Sambu | 0.303 ± 0.010 | 10 |
| 3436 | Phonolitic nephelinite, Meru West scarp | 1.53 ± 0.03 | 10 |
| | Phonolitic nephelinite, Meru West scarp | 1.49 ± 0.03 | 10 |
| 4056 | Ankaramitic basalt from Cinderella Cone, Oljoro valley | 1.56 ± 0.05 | 10 |
| | Ankaramitic basalt from Cinderella Cone, Oljoro valley | 1.62 ± 0.05 | 10 |

*(Continued)*

| Sample | Lithology and location | Age (Ma) | Refs |
|---|---|---|---|
| 4054 | Amphibole-rich ankaramitic basalt from Cinderella Cone, Loljoro valley | 1.77 ± 0.04 | 10 |
| | Amphibole-rich ankaramitic basalt from Cinderella Cone, Loljoro valley | 1.67 ± 0.04 | 10 |
| | Amphibole-rich ankaramitic basalt from Cinderella Cone, Loljoro valley | 1.93 ± 0.04 | 10 |
| 3435 | Sanidine phonolite clast from breccia, Meru West scarp | 2.03 ± 0.04 | 10 |
| | Sanidine phonolite clast from breccia, Meru West scarp | 1.99 ± 0.04 | 10 |
| 3706 | Mugearite, Flood lava group, Oljoro valley | 2.3 ± 0.06 | 10 |
| | Mugearite, Flood lava group, Oljoro valley | 2.25 ± 0.04 | 10 |
| 3705 | Alkali basalt, Flood lava group, Oljoro valley | 2.44 ± 0.06 | 10 |
| | Alkali basalt, Flood lava group, Oljoro valley | 2.49 ± 0.06 | 10 |
| ***Kilimanjaro*** | | | |
| | Lower olivine basalts, Naiperra, Amboseli | 0.85 ± 20% | 14 |
| | Lower olivine basalts, Naiperra, Amboseli | 1.10 ± 15% | 14 |
| | Lower olivine basalts, Ngong Naro, Amboseli | 0.420 ± 15% | 14 |
| KA940 | Trachybasalt, Mawenzi | 0.514 | 15,16 |
| KA947 | Basalt, Lava Tower series, Shira | 0.463 | 15,16 |
| KA937 | Obsidian, Rhomb Porphyry Series | 0.365 | 15,16 |
| 134 | Trachyte, Lent Group, Shira Plateau | 0.27 ± 0.02 | 6 |
| | Trachyte, Lent Group, Shira Plateau | 0.23 ± 0.03 | 6 |
| | Trachyte, Lent Group, Shira Plateau | 0.23 ± 0.02 | 6 |
| 200 | Anorthoclase phenocryst (blocky habit), caldera rim, Kibo | 0.22 ± 0.01 | 10 |
| 246 | Anorthoclase phenocryst (blocky habit), caldera rim, Kibo | 0.19 ± 0.01 | 10 |
| | Anorthoclase phenocryst (blocky habit), caldera rim, Kibo | 0.17 ± 0.01 | 10 |
| 223 | Anorthoclase phenocryst (rhombic habit), caldera rim, Kibo | 0.23 ± 0.03 | 10 |
| | Anorthoclase phenocryst (rhombic habit), caldera rim, Kibo | 0.26 ± 0.03 | 10 |

Dates are by the standard K–Ar, except those asterisked for which the method was single crystal laser fusion on feldspars, and some radiocarbon dates (Ref. 7)

**References**: 1, Isaac & Curtis 1974; 2, Manega 1993; 3, Fairhead *et al.* 1972; 4, Bagdasaryan *et al.* 1973; 5, MacIntyre *et al.* 1974; 6, Evans *et al.* 1971; 7, Hay 1976; 8, Drake & Curtis 1987; 9, Grommé *et al.* 1970; 10, Wilkinson *et al.* 1986; 11, Foster *et al.* 1997; 12, Curtis & Hay 1972; 13, Walter *et al.* 1991; 14, Williams 1969; 15, Evernden & Curtis 1965; 16, Downie *et al.* 1956.

**Appendix 1B**. *Unpublished K–Ar dates on olivine basalts in Lake Natron area*

| Sample | Location | Age |
|---|---|---|
| DD1248 | Foot of main escarpment, north end of L. Natron, 2°4′37′S 35°57′37″E. Early Oldoinyo Sambu activity; overlies Usagaran quartzite. | 7.7 ± 0.7 |
| DD1258 | Top of main escarpment, 3 km south of summit of Old. Sambu 2°9′47″S 35°56′32″E. Latest Old. Sambu lava: maximum age of the Natron boundary fault | 2.5 ± 0.15 |
| DD1298 | Top of main escarpment, north of Peninj delta. 2°16′01″S 35°56′21″E. Minimum age of Humbu formation | 2.1 ± 1.2 |
| DD1306 | West slopes of Peninj valley. 2°16′48″S 35°54′57″E. Maximum age of the Humbu Formation | 1.93 ± 0.8 |
| DD1385 | East side of Moinik valley. 2°28′33″S 35°52′23″E. Maximum age of the Moinik Formation | 0.73 ± 0.3 |

Samples collected by D.Dundas; analysed by N.J.Snelling. These data, together with the unpublished map (mapped 1966) and report by D. Dundas of Quarter Degree sheets 16 and 27, Loliondo, were made available by courtesy of the Tanzania Department of Mineral Resources in 1966.

## References


BAGDASARYAN, G. P., GERASIMOVSKIY, V. I., POLYAKOV, A. I., GUKASYAN, R. K. & VERNADSKIY, V. I. 1973. Age of volcanic rocks in the rift zones of East Africa. *Geochemistry International*, **10**, 66–71.

BISHOP, W. W., MILLER, J. A. & FITCH, F.H. 1969. New potassium-argon age determinations relevant to the Miocene fossil mammal occurrences in East Africa. *American Journal of Science*, **267**, 669–699.

CURTIS, G. H. & HAY, R. L. 1972. Further geological studies and K-Ar dating at Olduvai Gorge and Ngorongoro crater. *In*: BISHOP, W. W. & MILLER, J. W. (eds). *Calibration of Hominoid Evolution*. Scottish Academic Press, Edinburgh, 107–134.

DRAKE, P. & CURTIS, G. H. 1987. K-Ar geochronology of the Laetoli fossil localities. *In*: LEAKEY, M. D. & HARRIS, J. M. (eds) *Laetoli, a Pliocene site in northern Tanzania*. Clarendon Press, Oxford, 48–52.

DOWNIE, C., HUMPHRIES, D. W., WILCOCKSON, W. H. & WILKINSON, P. 1956. Geology of Kilimanjaro. *Nature*, **178**, 828–830.

EVANS, A. L., FAIRHEAD, J. D. & MITCHELL, J. G. 1971. Potassium-argon ages from the volcanic province of northern Tanzania. *Nature*, **229**, 19–20.

EVERNDEN, J. F. & CURTIS, G. H. 1965. The potassium-argon dating of Late Cenozoic rocks in East Africa and Italy. *Current Anthropology*, **6**, 343–384.

FAIRHEAD, J. D., MITCHELL, J. G. & WILLIAMS, L. A. J. 1972. New K/Ar determinations on Rift volcanics of S. Kenya and their bearing on the age of Rift faulting. *Nature*, **238**, 66–69.

FOSTER, A., EBINGER, C., MBEDE, E. & REX, D. 1997. Tectonic development of the northern sector of the East African Rift System. *Journal of the Geological Society, London*, **154**, 689–700.

GROMMÉ, C. S., REILLY, T. A., MUSSETT, A. E. & HAY, R. L. 1970. Palaeomagnetism and potassium-argon ages of volcanic rocks of Ngorongoro caldera, Tanzania. *Geophysical Journal of the Royal Astronomical Society*, **22**, 101–115.

HAY, R. L. 1976. *Geology of the Olduvai Gorge*. University of California Press, Berkeley.

ISAAC, G. L. & CURTIS, G.H. 1974. Age of the Acheulian industries from the Peninj Group, Tanzania. *Nature*, **249**, 624–627.

MACINTYRE, R. M., MITCHELL, J. G. & DAWSON, J. B. 1974. Age of the fault movements in the Tanzanian sector of the East African rift system. *Nature*, **247**, 354–356.

MANEGA, P. C. 1993. *Geochronology, geochemistry and isotope study of the Plio-Pleistocene hominid sites and the Ngorongoro Volcanic Highlands in northern Tanzania*. PhD thesis, University of Colorado.

Walter, R. C., Manega, P. C., Hay, R. L., Drake, R. E. & Curtis, G. H. 1991. Laser fusion $^{40}Ar/^{39}Ar$ dating of Bed I, Olduvai Gorge, Tanzania. *Nature*, **354**, 145–149.

WILKINSON, P., MITCHELL, J.G., CATTERMOLE, P.J. & DOWNIE, C. 1986. Volcanic chronology of the Meru-Kilimanjaro region, Northern Tanzania. *Journal of the Geological Society of London*, **143**, 601–605.

WILLIAMS, L. A. J. 1969. Geochemistry and petrogenesis of the Kilimanjaro volcanic rocks of the Amboseli area, Kenya. *Bulletin Volcanologique*, **33**, 862–888.

# Appendix 2

**Appendix 2.** *Unpublished analyses of Northern Tanzania Volcanic rocks*

| | 1 | 2 | 3 | 4 | 5 | 6 | 7 | 8 | 9 | 10 | 11 | 12 | 13 | 14 | 15 | 16 | 17 |
|---|---|---|---|---|---|---|---|---|---|---|---|---|---|---|---|---|---|
| $SiO_2$ | 57.34 | 62.39 | 39.26 | 5.83 | 1.52 | 41.35 | 39.72 | 48.47 | 42.2 | 45.64 | 44.62 | 58.65 | 54.48 | 51.64 | 42.06 | 42.3 | 43.15 |
| $TiO_2$ | 0.31 | 0.75 | 1.66 | 0.30 | 0.15 | 1.26 | 7.03 | 2.23 | 2.61 | 1.34 | 0.68 | 0.64 | 1.36 | 3.02 | 3.28 | 4.24 | 1.28 |
| $Al_2O_3$ | 20.55 | 14.00 | 15.75 | 0.78 | 0.35 | 9.31 | 8.58 | 15.31 | 11.32 | 16.42 | 20.53 | 19.17 | 18.54 | 15.57 | 6.93 | 5.28 | 15.81 |
| $Fe_2O_3$ | 1.92 | 6.26 | 4.88 | 11.36 | 3.24 | 6.90 | 8.05 | 6.69 | 5.82 | 4.74 | 5.7 | 1.8 | 2.38 | 4.83 | 8.15 | 4.72 | 6.87 |
| FeO | 3.06 | 1.77 | 3.14 | 1.08 | 1.13 | 6.37 | 7.42 | 6.36 | 6.01 | 2.43 | 2.22 | 3.78 | 6.1 | 6.87 | 4.14 | 12.6 | 3.9 |
| MnO | 0.46 | 0.15 | | 0.82 | 0.62 | 0.41 | 0.17 | tr | 0.14 | 0.14 | 0.22 | 0.13 | 0.12 | 0.11 | 0.13 | 0.13 | 0.12 |
| MgO | 0.78 | 0.51 | 3.62 | 11.70 | 1.15 | 12.48 | 1.44 | 3.60 | 7.68 | 2.13 | 0.88 | 0.89 | 1.96 | 3.32 | 14.41 | 11.62 | 2.63 |
| CaO | 2.23 | 1.98 | 16.60 | 38.01 | 51.66 | 18.62 | 12.92 | 8.14 | 16.02 | 7.92 | 5.39 | 1.88 | 5.18 | 7.02 | 12.26 | 17.86 | 9.52 |
| $Na_2O$ | 7.39 | 6.23 | 7.50 | 0.12 | tr | 0.93 | 2.25 | 4.49 | 4.22 | 6.94 | 9.04 | 7.02 | 5.44 | 4.13 | 1.96 | 0.91 | 6.98 |
| $K_2O$ | 3.58 | 4.93 | 3.03 | 0.00 | tr | 1.12 | 0.98 | 1.67 | 2.13 | 3.94 | 3.62 | 3.62 | 2.64 | 1.28 | 1.33 | 0.16 | 3.2 |
| $H_2O^+$ | 1.56 | 0.38 | 0.94 | 0.46 | 1.12 | 0.58 | 1.09 | 0.00 | 1.01 | 3.03 | 4.24 | 1.28 | 0.62 | 0.92 | 3.42 | 0.12 | 2.4 |
| $H_2O^-$ | 0.54 | 0.17 | | 0.02 | 0.27 | 0.32 | 0.46 | 1.65 | 0.45 | 2.48 | 2.16 | 0.7 | 0.29 | 0.46 | 1.38 | 0.08 | 1.68 |
| $CO_2$ | Tr | na | | | 35.26 | nil | nil | 0.26 | 0.59 | 2.02 | 0.43 | 0.07 | 0.09 | | | 0.07 | 2.02 |
| $P_2O_5$ | tr | 0.67 | 3.28 | 28.30 | 3.34 | 0.43 | 0.04 | 0.52 | 0.40 | 0.58 | 0.31 | 0.15 | 0.58 | 0.54 | 0.46 | | 0.64 |
| Sum | 99.72 | 100.19 | 99.66 | 99.92 | 99.81 | 100.08 | 100.15 | 99.53 | 100.15 | 99.77 | 100.04 | 99.78 | 99.78 | 99.71 | 99.91 | 100.09 | 100.2 |

| | 18 | 19 | 20 | 21 | 22 | 23 | 24 | 25 | 26 | 27 | 28 | 29 | 30 | 31 | 32 | 33 | 34 |
|---|---|---|---|---|---|---|---|---|---|---|---|---|---|---|---|---|---|
| $SiO_2$ | 44.60 | 44.66 | 44.91 | 54.00 | 53.30 | 51.50 | 40.21 | 40.42 | 42.55 | 40.21 | 30.07 | 48.93 | 43.18 | 40.52 | 43.50 | 48.00 | 37.30 |
| $TiO_2$ | 2.95 | 2.79 | 2.79 | 2.31 | 2.31 | 1.97 | 3.17 | 3.98 | 4.11 | 3.17 | 2.15 | 1.27 | 3.43 | 3.27 | 2.10 | 3.27 | 4.06 |
| $Al_2O_3$ | 11.36 | 14.20 | 14.97 | 14.77 | 15.08 | 16.37 | 12.57 | 12.42 | 12.07 | 12.57 | 15.40 | 18.00 | 10.73 | 12.51 | 15.95 | 14.60 | 9.30 |
| $Fe_2O_3$ | 3.93 | 6.81 | 2.27 | 3.28 | 6.05 | 4.81 | 7.63 | 7.14 | 6.17 | 7.83 | 5.47 | 3.14 | 4.92 | 8.40 | 5.21 | 8.21 | 6.24 |
| FeO | 11.13 | 6.89 | 10.92 | 7.87 | 5.59 | 6.75 | 5.13 | 6.19 | 7.07 | 5.13 | 5.44 | 5.94 | 8.93 | 4.32 | 5.72 | 3.00 | 7.88 |
| MnO | 0.15 | 0.17 | 0.16 | 0.13 | 0.12 | 0.17 | 0.20 | 0.18 | 0.19 | 0.20 | 0.29 | 0.23 | 0.17 | 0.20 | 0.25 | 0.19 | 0.18 |
| MgO | 10.35 | 6.19 | 5.12 | 3.53 | 3.01 | 2.63 | 6.55 | 7.00 | 6.70 | 6.55 | 5.49 | 2.08 | 6.80 | 6.00 | 2.77 | 3.44 | 10.20 |
| CaO | 10.98 | 7.44 | 7.66 | 6.75 | 6.26 | 5.80 | 11.41 | 12.39 | 11.14 | 11.41 | 11.15 | 3.63 | 13.79 | 12.71 | 7.65 | 7.02 | 17.13 |
| $Na_2O$ | 2.73 | 5.17 | 5.81 | 4.59 | 4.90 | 5.90 | 1.26 | 3.28 | 3.60 | 1.26 | 5.24 | 7.70 | 2.72 | 2.40 | 6.96 | 4.28 | 1.48 |
| $K_2O$ | 0.90 | 2.82 | 2.86 | 1.84 | 2.00 | 2.35 | 2.74 | 1.38 | 2.77 | 2.74 | 3.63 | 5.49 | 0.48 | 2.29 | 3.06 | 3.65 | 0.62 |
| $H_2O^+$ | 0.52* | 1.23* | 1.23* | 0.27* | 1.08* | 0.89* | 7.6* | 4.15* | 3.03* | 7.60* | 6.19* | 2.98* | 4.33* | 6.02 | 5.77 | 3.04 | 4.45 |
| $H_2O^-$ | | | | | | | | | | | | | | | | | |
| $CO_2$ | na | na | na | na | na | na | na | na | na | na | na | na | na | na | na | na | na |
| $P_2O_5$ | 0.54 | 0.84 | 0.78 | 0.51 | 0.48 | 0.70 | 1.01 | 1.57 | 1.50 | 1.01 | 0.66 | 0.18 | 1.09 | 1.1 | 0.53 | 0.83 | na |
| Sum | 100.14 | 99.15 | 99.52 | 99.86 | 100.18 | 98.95 | 99.68 | 100.10 | 100.90 | 99.68 | 100.18 | 99.57 | 100.57 | 99.74 | 99.47 | 99.53 | 98.84 |

$H_2O$* is total $H_2O$.

J. B. Dawson, University of Edinburgh (*jbdawson@glg.ed.ac.uk*)
*From*: DAWSON, J. B. 2008. *The Gregory Rift Valley and Neogene–Recent Volcanoes of Northern Tanzania*. Geological Society, London, Memoirs, **33**, 97–98.
0435-4052/08/$15.00 © The Geological Society of London 2008. DOI: 10.1144/M33.App02

| | 35 | 36 | 37 | 38 | 39 | 40 | 41 | 42 | 43 | 44 | 45 | 46 | 47 | 48 |
|---|---|---|---|---|---|---|---|---|---|---|---|---|---|---|
| $SiO_2$ | 48.60 | 46.08 | 45.42 | 44.20 | 46.00 | 46.33 | 47.15 | 53.50 | 55.10 | 55.32 | 54.50 | 50.04 | 53.50 | 57.78 |
| $TiO_2$ | 1.20 | 1.23 | 1.31 | 1.32 | 1.24 | 1.32 | 1.22 | 1.57 | 1.50 | 1.58 | 2.30 | 2.45 | 2.11 | 1.13 |
| $Al_2O_3$ | 12.66 | 11.38 | 10.68 | 12.38 | 11.40 | 12.20 | 15.64 | 18.91 | 17.82 | 17.03 | 14.89 | 16.60 | 16.98 | 17.04 |
| $Fe_2O_3$ | 4.64 | 5.02 | 4.27 | 5.39 | 4.95 | 6.29 | 5.78 | 2.73 | 2.07 | 2.42 | 3.37 | 3.62 | 3.41 | 2.24 |
| FeO | 4.21 | 4.62 | 5.14 | 4.37 | 4.60 | 4.38 | 4.20 | 4.90 | 5.19 | 4.51 | 7.81 | 5.66 | 4.93 | 3.25 |
| MnO | 0.22 | 0.27 | 0.26 | 0.28 | 0.26 | 0.28 | 0.27 | 0.16 | 0.14 | 0.16 | 0.13 | 0.18 | 0.17 | 0.14 |
| MgO | 3.55 | 2.38 | 2.04 | 1.97 | 1.36 | 1.09 | 0.87 | 1.83 | 1.96 | 1.65 | 3.50 | 3.00 | 2.31 | 0.73 |
| CaO | 6.90 | 10.91 | 15.40 | 10.68 | 14.44 | 10.38 | 7.20 | 4.36 | 4.40 | 4.09 | 6.94 | 6.20 | 5.14 | 3.01 |
| $Na_2O$ | 5.28 | 4.75 | 4.73 | 6.51 | 2.90 | 7.76 | 7.86 | 6.92 | 6.70 | 7.05 | 4.45 | 7.50 | 7.10 | 8.13 |
| $K_2O$ | 4.93 | 3.73 | 3.03 | 3.63 | 3.23 | 2.92 | 4.09 | 3.58 | 3.67 | 4.05 | 1.83 | 3.75 | 3.95 | 4.75 |
| $H_2O^+$ | 6.81 | 7.80 | 6.22 | 4.36 | 7.98 | 6.63 | 4.91 | 0.60 | 0.78 | 1.57 | 0.33 | 0.23 | 0.35 | 2.05 |
| $H_2O^-$ | | | | | | | | | | | | | | |
| $CO_2$ | 0.23 | 0.78 | na | 3.95 | na | na | 0.10 | na | na | na | na | na | na | na |
| $P_2O_5$ | 0.34 | 0.51 | 0.95 | 0.54 | 0.87 | 0.44 | 0.20 | 0.37 | 0.42 | 0.32 | 0.46 | 0.59 | 0.50 | 0.26 |
| Sum | 99.57 | 99.46 | 99.45 | 99.58 | 99.23 | 100.02 | 99.49 | 99.53 | 99.75 | 99.75 | 100.81 | 99.82 | 100.45 | 100.51 |

1. Barkevikite phonolite trachyte, Gelai JG923 (Guest 1953).
2. Riebeckite trachyte, Kibangaini JG1145 (Guest,1953).
3. Melanite melteigite, Kerimasi (Finckh 1911). Total includes 0.29% F.
4. Forsterite–magnetite–apatite carbonatite, Kerimasi JG810 (Guest 1953).
5 Apatite carbonatite, Kerimasi JG828 (Guest 1953).
6. Biotite pyroxenite, Kerimasi JG1254 (Guest 1953).
7. Pyroxene amphibolite, Kisetey Hill JG 818 (Guest 1953).
8. Mugearite, Oldoinyo Sambu (Finck 1903). Total includes 0.14% S.
9. Ijolite, Kerimasi (James 1966).
10. Nephelinite, Kerimasi (James 1966).
11. Nephelinite, "Mlanja" (probably Sadiman) (James 1966).
12. Anorthoclase trachyte, Olmoti (James 1966).
13. Mugearite, Lemagrut (James 1966).
14. Hawaiite, Ngorongoro (James 1966).
15. Augite melilitite, Kwaraha (James 1966).
16. Augite magnetite rock, Kwaraha (James 1966).
17. Nephelinite, Kwaraha (James 1966).
18. Basanite, spur on south side of Ngorongo crater (Wood 1968).
19. Tephrite, N.E. side of crater floor, Ngorongoro (Wood 1968).
20. Tephrite, spur on south side of Ngorongoro crater (Wood 1968).
21. Mugearite spur on south side of Ngorongoro crater (Wood 1968).
22. Mugearite spur on south side of Ngorongoro crater (Wood 1968).
23. Phono-tephrite spur on south side of Ngorongoro crater (Wood 1968).
24. High-K basanite BD133, boulder in river, Engaruka (Wood 1968).
25. Nephelinite, L. Manyara (Wood 1968).
26. Nephelinite, L. Manyara (Wood 1968).
27. Zeolitised sodalite nephelinite BD441, flow on SE ridge of Burko (Wood 1968).
28. Nephelinite BD371, Elunata escarpment (Essimingor-derived). (Wood 1968).
29. Clinopyroxene-phyric augitite BD401, flow, NE slopes of Essimingor (Wood 1968).
30. Melanphelinite BD 364, flow, S.slopes Essimingor (Wood 1968).
31. Augitite BD375, flow, SE ridge, Essimingor (Wood 1968).
32. Nephelinite BD400, flow, lower NE slopes, Essimingor (Wood 1968).
33 Nephelinite BD376. N-S dyke, SE ridge, Essimingor (Wood 1968).
34. Analcime ankaramite BD346, Maserani (3°31'S 36°27.7'E) (Wood 1968).
35-41. Nephelinites, Hanang (Wood 1968).
42. Tephriphonolite, western slopes of Meru (Wood 1968).
43. Tephriphonolite, western slopes of Meru (Wood 1968).
44. Tephriphonolite, western slopes of Meru (Wood 1968).
45. Mugearite block in lahar, Momella (Meru-derived) (Wood 1968).
46. Tephri-phonolite block in lahar, Momella (Meru-derived) (Wood 1968).
47. Tephri-phonolite block in lahar, Momella (Meru-derived) (Wood 1968).
48. Phonolite block in lahar, Momella (Meru-derived) (Wood 1968).

## References

FINCKH, L. 1903. Die trachydolerite des Kibo (Kilimandscharo) und die kenyte des Kenya. *Zeitschrift der Deutsches Geologische Gesellschaft*, **55**., Monatsberich 1–14.

FINCKH, L. 1911. Die von F. Jaeger in Deutsch-Ostafrika gesammelten Gesteine. *In*: Jaeger, 1911, 71–85.

GUEST, N. J. 1953. *The geology of part of northern Tanganyika Territory*. PhD thesis, University of Sheffield.

JAEGER, F. 1911. Das Hochland der Riesenkrater and die umliegender Hochländer Deutsch Ostafrika, Part 1, *Mitteilungen aus den Deutsches Schutzgebieten, Ergänzungsheft*, **4**, 1–133.

JAMES, D. E. 1995. *The geochemistry of feldspar-free volcanic rocks.* PhD thesis, Open University.

JAMES, T. C. 1966. *The carbonatites of Tanganyika: a phase of continental-type volcanism.* DIC thesis, Imperial College, London

WOOD, C. P. 1968. *A geochemical study of East African alkaline lavas and its relevance to the petrogenesis of nephelinites.* PhD thesis, University of Leeds.

# Index

Page numbers in *italic* denote figures. Page numbers in **bold** denote tables.